Principles of
Physical Chemistry

L I O N E L   M .   R A F F

OKLAHOMA STATE UNIVERSITY

Principles of
Physical Chemistry

PRENTICE HALL
Upper Saddle River, NJ 07458

Library of Congress Cataloging-in-Publication Data

Raff, Lionel M.
  Principles of physical chemistry / Lionel M. Raff.— 1st ed.
    p.   cm.
  Includes bibliographical references and index.
  ISBN 0-13-027805-X
    1. Chemistry, Physical and theoretical.   I. Title.

QD453.2 .R34 2001
541—dc21   00-041687
                                              CIP

Executive Editor: *John Challice*
Associate Editor: *Kristen Kaiser*
Editorial Assistants: *Gillian Buonanno and Eliana Ortiz*
Senior Marketing Manager: *Steve Sartori*
Assistant Managing Editor: *Beth Sturla*
Copy Editor: *Brian Baker, Writewith, Inc.*
Text Composition and Electronic Page Makeup: *WestWords, Inc.*
Art Studio: *Wellington Studios*
Interior Design: *Joe Sengotta*
Cover Design: *Joe Sengotta*
Manufacturing Buyer: *Michael Bell*
Media Editor: *Paul Draper*
Assistant Managing Editor, Media: *Alison Lorber*

© 2001 by Prentice-Hall, Inc.
Upper Saddle River, New Jersey 07458

Printed in the United States of America
10 9 8 7 6 5 4 3 2 1

ISBN 0-13-027805-X

Prentice-Hall International (UK) Limited, *London*
Prentice-Hall of Australia Pty. Limited, *Sydney*
Prentice-Hall Canada Inc., *Toronto*
Prentice-Hall Hispanoamericana, S.A., *Mexico*
Prentice-Hall of India Private Limited, *New Delhi*
Prentice-Hall of Japan, Inc., *Tokyo*
Pearson Education Asia Pte. Ltd.
Editora Prentice-Hall do Brasil, Ltda., *Rio de Janeiro*

For my parents, Joseph Samuel and Stella Leona Raff, who taught me to love people and to revere knowledge, and for my beloved family—Murna; Aaron and Lyne; and Debra, Mike, and Mara—who have made my life a joy and a delight.

BRIEF CONTENTS

CONTENTS

SECTION 1   Classical Thermodynamics 1

CHAPTER 1

Properties of Gases

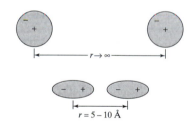

CHAPTER 2

The First Law of Thermodynamics

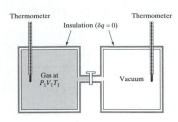

CHAPTER 3

Thermochemistry

CHAPTER 4

The Second Law of Thermodynamics

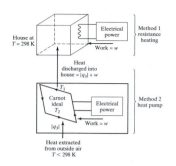

CHAPTER 5

Chemical Equilibrium

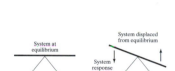

(A)

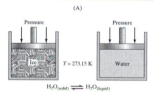

(B)

CHAPTER 6

Phase Equilibrium

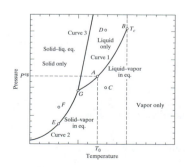

CHAPTER 7

The Thermodynamics of Solids

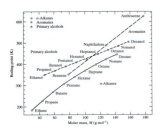

CHAPTER 8

Thermodynamics of Nonelectrolytic Solutions

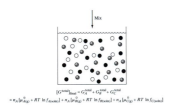

CHAPTER 9

Thermodynamics of Electrolytic Solutions

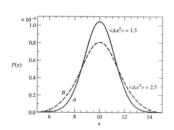

SECTION 2   Quantum Mechanics and Bonding 505

CHAPTER 10

The Mathematics of Chance

CHAPTER 11

Introduction to Quantum Mechanics

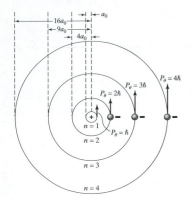

CHAPTER 12

Translational, Rotational, and Vibrational Energies of Molecular Systems

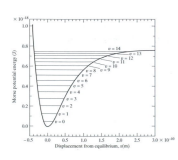

CHAPTER 13

The Electronic Structure of Atoms

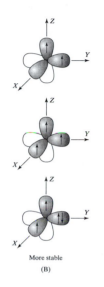

More stable
(B)

CHAPTER 14

Molecular Structure and Bonding

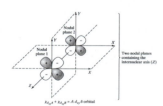

SECTION 3   Spectroscopy 819

CHAPTER 15

Rotational, Vibrational, and Electronic Spectra

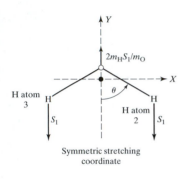

Symmetric stretching coordinate

CHAPTER 16

Magnetic and Diffraction Spectroscopy

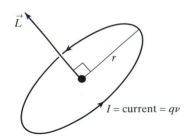

S E C T I O N   4   Statistical Mechanics   965

C H A P T E R 17

Molecular-Energy Distributions: Kinetic Theory of Gases

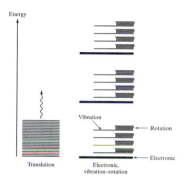

C H A P T E R 18

Statistical Thermodynamics

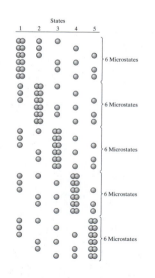

SECTION 5   Kinetics and Dynamics 1083

CHAPTER 19

Phenomenological Kinetics

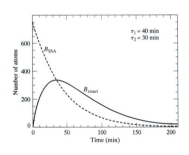

CHAPTER 20

Theoretical Kinetics and Reaction Dynamics

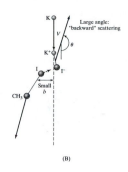

INSTRUCTOR'S PREFACE

This textbook is not written for you. It is written for your students. Its purpose is to teach physical chemistry, as opposed to covering the subject. To cover is not to teach. Pursuant to this objective, the narrative is often informal and relaxed. The material is presented using the same language I might employ if a student were to come to my office asking for help. Indeed, it is my intent that, as students read and study *Principles of Physical Chemistry*, they will feel that I am sitting across the table, providing one-on-one tutorial instruction. The discussion of all of the topics appearing in the text is sufficiently detailed to give the students a reasonable chance of becoming quantitatively competent in the area.

The underlying philosophy of the book is that teaching is a joint enterprise between instructor and author, with a common objective: to bring the students to a functional level of literacy in the use, practice, appreciation, and execution of physical chemistry principles and methods. It is my opinion that this task is extremely difficult—so much so that it cannot be achieved if we bring only half of our weapons to bear. A brilliant textbook coupled with an instructor who devotes little effort to the task of teaching will, at best, produce only poor results. A superb instructor who gives outstanding lectures will also fail to reach many of the students if the textbook he or she uses comprises only pictures, final equations with hand-waving explanations, simplistic examples, and "plug-and-chug" problems that are little more than practice on a calculator. This same superb instructor will also fail to achieve the best possible result if the students cannot or will not read the text. The problem is that once the brilliant lecture is completed, it is gone forever. When the students sit down to study, they have only their incomplete notes that are necessarily flawed because they couldn't listen, understand, and write fast enough to produce accurate notes, and the instructor couldn't speak or write fast enough to cover all the critical points. They need a textbook that makes the same determined effort to teach as the superb instructor does during a lecture and in his or her office in tutorial sessions with the students.

I have done everything my ability permits to produce such a textbook. The derivations of virtually all equations are given in complete detail. All algebraic steps are shown and explanations inserted to help the student understand and learn the derivations, and I take pains to point out where key assumptions or simplifications are made. Figures, diagrams, and drawings are employed when they facilitate learning. They are, however, never used in place of a rigorous presentation of the material. Qualitative explanations and analogies to events that are familiar to the students are frequently used, but both are always backed by a quantitative treatment of the subject. The text assumes that the students have had a one-year university-level course in differential and integral calculus, but the critical mathematical methods are always developed and explained in detail before their use. These presentations are incorporated into the body of the text itself. The text contains 243 fully solved examples that are generally at the same level as the problems at the end of the chapter. The solutions to all 815 end-of-chapter problems are given in the Instructor's Guide (ISBN 0-13-026671-X). These solutions are as detailed as the textbook examples, with all steps shown (A student's solutions manual, with answers to only half the problems, is also available; ISBN 0-13-040664-3). An explanation for each step is inserted, and appropriate comments about the importance of the problem are presented. In addition, the Instructor's Guide contains 351 suggested examination questions that can, if desired, be used as additional homework exercises. All chapters conclude with a summary of the key points and equations. Humor is used without apology.

The use of this textbook gives the instructor great latitude in deciding what to present during the time available in lecture. Since all derivations are given in complete detail, lecture time can be used to work additional examples or to emphasize the key points in the development of the equations. Alternative derivations or additional material can be presented, or question-and-answer sessions with the students can emphasize important concepts. Because the instructor need not "write a textbook on the blackboard," there is considerable flexibility in organizing the lecture material.

My own experience in using prepublication versions of *Principles of Physical Chemistry* suggests that the best option is to present the lecture material as if the textbook did not exist. This means that during the lecture I go over material that is already in the text. I have found that this repetition helps the students a great deal. It is much easier for students to understand the spoken word when combined with body language, voice cadence and inflection, and blackboard derivations than it is to glean the material from a written textbook. When this is done, the material in the book becomes much easier to read and understand. In addition, the text now provides excellent "notes" for study that are accurate and complete and that present good examples at the same level as the problems assigned by the instructor.

It is my sincere hope that you find *Principles of Physical Chemistry* helpful in your efforts to bring your students to a truly functional level of knowledge of, and ability to do, physical chemistry. I extend my apologies for any remaining errors or inaccuracies in the text and would appreciate being informed about them. I invite each of you to submit your favorite problems for publication in the second edition. Full credit will be given at the point the problems appear in the text. Details for submission are given in the Instructor's Guide. Suggestions on any and all points are welcome.

Acknowledgments

It is a pleasure to conclude the writing of this textbook by thanking those individuals who made their expertise available to me in the form of incisive critical commentary and review. Their suggestions and observations played an essential part in the preparation of the book. I would particularly like to thank my friends, colleagues, and family, who constantly provided support and encouragement. When my spirits lagged, they were always there to lift them.

Dr. Fredrick L. Minn, director of pharmaceutical research, McNeil Laboratories (retired), read and commented on the entire manuscript and provided several end-of-chapter problems that appear in the text. His constant good humor, critical comments, and encouragement played a large role in the preparation of the book. The advice and analysis of Professor Richard N. Porter, State University of New York at Stony Brook, on numerous points dealing with quantum theory, electronic spectroscopy, and statistical mechanics were vital. His ability to make complex material very clear is a resource I used frequently. Professor Keith A. Jameson, Loyola University, and Professor Cynthia Jameson, University of Illinois at Chicago, provided critical assistance with the chapter on NMR spectroscopy. Keith also read some of the initial chapters on gas laws and thermodynamics and provided me with annotated corrections and suggestions on those chapters. Professor Nicholas A. Kotov, Oklahoma State University, twice used portions of the first nine chapters in his physical chemistry course. His in-depth review and well-taken comments on this material produced numerous changes in the text. Professor Luis Liz-Marzán, Vigo University, Spain, also used the text in manuscript form in his physical chemistry classes and provided important critical commentary. Professor George Gorin, Oklahoma State University, reviewed the first two chapters and made numerous suggestions for revision that eventually found their way into the final manuscript. In separate discussions, he also suggested the approach that should be used in the description of phenomenological kinetics. His suggested approach is the one employed in Chapter 19. Professor Mark Rockley, Oklahoma State University, reviewed Chapter 15 on microwave, vibrational, and electronic spectra, and Professor Mario Rivera, Oklahoma State University, twice reviewed the section of Chapter 16 dealing with NMR spectroscopy. Their experimental expertise in these areas was essential, and virtually every comment they made was eventually translated into appropriate changes and revisions in the text. Professor Ron Kay, Gordon College, Wenham, Massachusetts, reviewed the sections dealing with thermochemical measurements. His incisive comments led to numerous changes in the text. I am indebted to Professor Isabelle Okuda, Texas Woman's University, for providing source material related to the development of quantum theory and statistical mechanics, as well as general comments on physical chemistry textbooks.

The external reviewers who examined portions of the first draft of the manuscript and provided in-depth criticism and suggestions were particularly helpful. The quality of their reviews, which often comprised over a dozen pages of incisive comment, cannot be overstated. More than 90 percent of their suggestions were incorporated into the final draft. These individuals are

Professor Shawn B. Allin,   *Lamar University*
Professor Mike Barfield,   *University of Arizona*
Professor Ilan Benjamin,   *University of Californiz at Santa Cruz*
Professor Steven L. Bernasek,   *Princeton University*
Professor Eric Bittner,   *University of Houston*
Professor Gary Buckley,   *Cameron University*
Professor David Budil,   *Northeastern University*
Professor D. Allan Cadenhead,   *State University of New York at Buffalo*
Professor Susan Crawford,   *California State University, Sacramento*
Professor Stefan Franzen,   *North Carolina State University*
Professor Brian D. Gilbert,   *Coastal Carolina University*
Professor Donald Harriss,   *University of Minnesota at Duluth*
Professor W. Vernon Hicks,   *Northern Kentucky University*
Professor Gary G. Hoffman,   *Florida International University*
Professor Charles Jaffé,   *West Virginia University*

Professor Frank W. Kutzler,	*Tennessee Technological University*
Professor Howard Mette,	*Youngtown State University*
Professor Mark Ondrias,	*University of New Mexico*
Professor David Ritter,	*Southeastern Missouri State University*
Professor Steven J. Stuart,	*Clemson University*
Professor Joel Tellinghuisen,	*Vanderbilt University*
Professor Michael Trenary,	*University of Illinois at Chicago*
Professor Dean Waldow,	*Pacific Lutheran University*
Professor Rand L. Watson,	*Texas A & M University*
Professor Gary H. Weddle,	*Fairfield University*
Professor Danny Yeager,	*Texas A&M University*
Professor Sidney Young,	*University of Southern Alabama*
Professor Jin Zhang,	*University of California at Santa Cruz*

The students in my physical chemistry classes are due recognition for their assistance in locating errors in the manuscript and for numerous suggestions to improve the clarity of the presentation. These individuals are Mr. Syed S. Alam, Mr. Felix Anyomi, Mr. Rubindra Bariya, Mr. Samuel H. Beuke, Mr. John W. Biava, Ms. Marie S. Coutant, Mr. Thomas J. Crowell, Ms. Melissa E. Hays, Mr. Michael A. Hill, Mr. Travis W. Hill, Ms. Kara A. Karns, Ms. Tamika D. Killian, Ms. Cynthia K. Kuhns, Mr. Tuan A. Le, Ms. Susannah Marley, Mr. Blake Masters, Ms. Debra K. Meyer, Mr. Zachary S. Moore, Mr. Jeffery Mornhinweg, Mr. Sam C. Ngo, Mr. Paul D. O'Reilly, Mr. Michael D. Pararett, Ms. Jill R. Peterson, Mr. Dipesh Pokharel, Mr. Joshua D. Ramsey, Mr. Perry L. Reed, Mr. Chad W. Smith, Ms. Angela C. Taylor, Mr. Tweodrose G. Tirfe, and Mr. Wesley D. Turybury.

My editors at Prentice Hall, Mr. Matthew Hart and Mr. John Challice, and their executive assistants, Ms. Betsy A. Williams, Ms. Gillian Buonanno, and Mr. Sean Hale, played major roles in bringing this textbook to final fruition. In addition to lending their skill and that of the Prentice Hall staff in the technical preparation of textbooks, they provided critical advice on the writing of the book itself. Their advice covered a wide spectrum that included the nature and number of figures and illustrations, the type of ancillary material needed, the ordering of the material, the highlighting of equations, and many other points too numerous to list. They were also responsible for the selection of most of the external reviewers whose excellent advice I have already mentioned. Dr. Paul Draper, media editor, Prentice Hall, played a major role in the development of the ancillary material. The assistance of my copyeditor, Mr. Brian Baker, Writewith, Inc., Damascus, Maryland, 20872, was invaluable. His efforts not only improved the clarity and consistency of the writing, his comments on content also led to several additions and improvements to the text. Mrs. Jami Darby, WestWords, Inc., who was in charge of text composition and page makeup, handled the task with skill and patience. Mr. Dave Wood, Wellingten Studios, reproduced all the beautiful artwork, while carefully retaining the quantitative accuracy of the graphs and illustrations.

Finally, I want to mention my wife, Murna Jean Raff; my son and his wife, Aaron Michael and Lyne Raff; my daughter and her husband, Debra and Michael Katcher; and their daughter, Mara Katcher, to all of whom this volume is dedicated. For four years, they endured my near-total preoccupation with the task of writing the book, the associated Instructor's Guide, and the ancillary material. Without fail, they accepted this with good grace and humor, with their best wishes and their encouragement. Without their support and Mara's marvelous smile, it is doubtful that I could have completed the book.

I welcome your comments and observations.

Lionel M. Raff
lionelraff@hotmail.com

INTRODUCTION TO THE STUDENT

This textbook was written for you. You will find the narrative style informal and relaxed, with frequent use of the first person. Humor is used without apology. It is my intent that as you read and study the text, you will feel that I am sitting across the table, providing one-on-one tutorial instruction.

The text assumes that you have had the equivalent of a one-year university-level course in differential and integral calculus. However, the critical mathematical methods are always developed and explained in detail before their use. These presentations are incorporated into the body of the text, rather then being relegated to an appendix. Two hundred forty-three detailed examples are given, and all steps in their solution are presented and explained. Furthermore, they are generally at the same level as the end-of-chapter problems and thus will be very helpful in your efforts to learn how to work the problems and gain an understanding of physical chemistry in the process.

Let us now turn to the question that is probably uppermost in your mind: "Can I pass this course with a decent grade and learn something in the process?" The answer is "Yes, if you work and study regularly, diligently, and *in the right manner.*" Since the vast majority of you are juniors, seniors, or graduate students, you are already aware that science and engineering courses require a great deal of effort. This comes as no surprise; the real key is in the last prepositional phrase in the preceding sentence: "*in the right manner.*" Each time I teach physical chemistry, a minimum of a half-dozen students come to my office in academic difficulty and say, "I don't know what's wrong. I understand the material very well when you present it in class or when you work examples on the board, but I don't seem to be able to do the problems when I'm taking an examination." The number of times I've heard this statement is well into the hundreds—perhaps as high as a thousand. These are students who are working hard. They want to learn. They want to make a good grade, but things are not working well for them, and they do not understand why that is happening. This is the situation you want to avoid. You know that if you don't work, you will not do well in physical chemistry or any other of life's difficult endeavors. If you do invest the time and effort, however, you have a right to expect reasonably good results, and I would like to try to help you obtain them.

Let me put the problem these students are having in a different perspective with a simple analogy. For years, I have been a fan of the NBA Chicago Bulls, which is to say Michael Jordan and Company. I have watched them play many, many times on television. I have Jordan's moves memorized: He stands at the top of the key and takes the pass from Pippen. One of the most superb players in the league is assigned to guard this incomparable master of the game. Jordan fakes to his right, then dribbles to his left. The player guarding him is not fooled. He has seen the move many times. He follows Jordan step for step, cutting off his move to the basket. With amazing speed, Jordan executes a behind-the-back, crossover dribble and is suddenly moving to his right, driving down the right side of the key toward the basket. His man is left a half-step behind, which is all Jordan needs. The opposing all-pro NBA center, seeing that Jordan has eluded his man, moves out to intercept him. Jordan, realizing that his move to the basket is blocked by the center, pulls up six feet from the basket. He arches himself in an incredibly high fallaway jump that takes him up and back from the opposing center. The center leaves his feet, arms outstretched to block the imminent shot. In midair, Jordan squares his body to the basket and releases the ball in a high, gentle arc that clears the center's fingertips by several inches because of the space created by the backward trajectory of Jordan's jump. The ball approaches the basket with the optimum arc… SWISH… nothing but net, and two more points for "da Bulls."

I understand completely every move that Jordan made. *But I cannot execute a single one of those moves myself!* So it is with physical chemistry. It is not enough to qualitatively understand how your professor works a problem. It is not enough to understand the steps involved in the derivation of an equation. A qualitative understanding using diagrams and pictures that is based on hand-waving, qualitative-type arguments is not close to being enough to permit you to actually execute physical chemistry. If you wish to execute Jordan-like moves in basketball, you must practice those moves. If you wish to execute derivations and solutions to physical chemistry problems, you must practice the execution of such derivations and solutions. With this point in mind, let us now turn our attention to the methods that you should employ in studying physical chemistry.

First, do not cut class. Your professor is your best resource in learning the subject. Listen carefully to the lecture and take the best notes you can. Do not decline to take notes because "the material is covered in the textbook." The process of listening and taking notes focuses your attention on the subject, and writing the equations and the reasons behind them begins the process of learning to execute physical chemistry. Second, purchase a supply of legal notepads. When you begin independent study, use the following techniques: For derivations, read and study the methods presented by your professor or the text to execute the derivation. When you think you know and understand them, close your notes or the text, take out your notepad, and execute every step in the derivation yourself. If you falter, open the text or your notes and study the derivation again. When you see where and why you faltered, close the book or notes and continue the derivation, starting from the point at which you faltered. Do not start the entire derivation over again! Continue this process until you have completed the derivation. Now comes the important part: Discard all the legal sheets on which you have been working, take a fresh sheet, and repeat the derivation without reference to the text or to your notes. If you again falter, you don't yet know the material well enough. Open your notes or the text and study the material once more. Close the book and continue the derivation from the point at which you faltered. When you finish, discard your work and begin again. Continue this process until you can execute the entire derivation without reference to the book or your notes. *Now* you are learning to actually execute physical chemistry.

Use the identical technique to study the examples your professor works in class and those that are contained in the text. When you think you understand how to work a sample problem, close the book, close your notes, take out the legal pads, and work the problem through. If you falter or make an error, use the procedure described for derivations. Do this until you have completed the solution to the example without reference to your notes or the text. Treat the homework problems in the same manner. When your professor makes the solutions available to you in some format, study each one until you can execute the entire problem without referring to the solution sheet. A word of warning at this point is appropriate: *Do not fall into the trap of telling yourself that you know how to do a problem, example, or derivation without actually doing it in the manner just described!* Until you actually do it, you can't. Until the ball hits "nothing but net," you haven't really learned the game.

If you have the will and patience to study physical chemistry in the preceding manner, I guarantee that you will never be one of the students who comes into his or her professor's office and says, "I understand the subject, but I just can't do it." Instead, you will be elated with your results and proud of the fact that you are learning to do something extraordinarily difficult. The method works! It is, in my opinion, the *only* method of study that will work.

There is a cliché that is frequently used by professors that causes an incredible amount of damage to the process of learning physical chemistry: "Do not memorize equations." In this form, the statement is harmful, because it is incomplete and, therefore, very misleading. It leads students to believe that all they really need is a qualitative understanding, since the instructor will provide the equations on examinations. Here is the complete statement: *"Do not memorize equations. Instead, work with them to such an extent that the question of memorization becomes moot."* In this form, the statement is absolutely true. If you follow the study procedure outlined, you will already know the equations. The question of memorization will never arise. You won't need a crib sheet. You won't need to have equations programmed into your graphing calculator. More importantly, you *will* be able to use the equations and theories of physical chemistry to work problems and even extend your knowledge into areas you thought you didn't know. On examinations, you will "hit nothing but net."

I wish each of you the very best of success in physical chemistry and in life.

Lionel M. Raff
lionelraff@hotmail.com

ABOUT THE AUTHOR

Lionel M. Raff is Regents Professor of Chemistry at Oklahoma State University. He did his undergraduate studies at the University of Oklahoma, majoring in chemistry, physics and mathematics, and his graduate studies at the University of Oklahoma and the University of Illinois, Urbana-Champagne, working with Professor Aron Kuppermann on electron impact spectroscopy. Professor Raff completed postdoctoral studies at the University of Illinois and at Columbia University as an NSF Postdoctoral Fellow with Professor Martin Karplus studying theoretical methods for the investigation of reaction dynamics. He then joined the chemistry faculty at Oklahoma State University.

Professor Raff is the author of 165 scientific papers and three book chapters primarily in the area of theoretical studies of reaction dynamics in addition to the text and Instructor's Guide for *Principles of Physical Chemistry*. Dr. Raff was elected to Outstanding Educators in America in 1975. In addition to four teaching awards from Oklahoma State University, he received the Oklahoma Medal for Excellence in College/University teaching from the Oklahoma Foundation for Excellence in 1993.

Professor Raff and his wife Murna have two children, Debra and Aaron.

Principles of
Physical Chemistry

Classical Thermodynamics

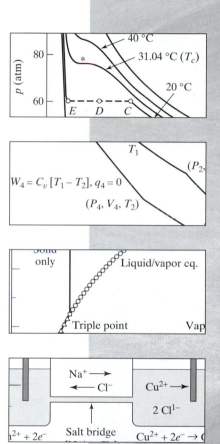

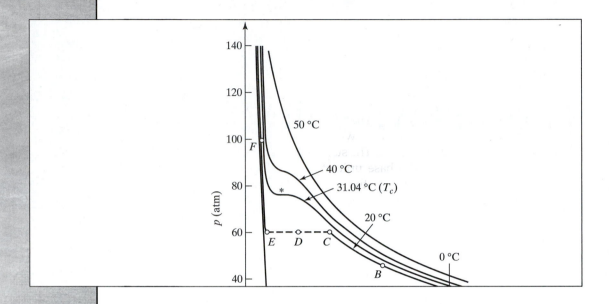

CHAPTER 1

Properties of Gases

The behavior of gases is of central importance to our study of the thermodynamics not only of gases, but of condensed solids, liquids, and solutions as well. Although not apparent at this point, the truth of this statement will become increasingly clear as we proceed. Much of the material in this chapter will be familiar to you from your previous courses in engineering and chemistry. However, you will find some of the concepts to be new and interesting. The mathematical methods we employ for these simple systems will recur repeatedly throughout the text. Therefore, they merit your careful attention as we begin our study of physical chemistry.

1.1 Equations of State: General Considerations

The state of a gas is described by four variables: pressure, volume, temperature, and quantity of material. These variables are generally denoted by P, V, T, and n, respectively, where n measures the amount of gas in terms of the number of moles present. We first consider the definition and measurement of these quantities.

1.1.1 Mass and Chemical Amount

The important factor in chemical reactions is the relative number of atoms and molecules involved. This number can be obtained by dividing the mass of the reactants by their atomic or molecular "weight," which is a number that measures the mass relative to a standard. The standard chosen in the SI system (see Section 1.1.3) is carbon-12. The base unit of chemical amount, the mole, is defined as the amount of a substance that contains the same number of chemically defined particles as exactly 12 g of carbon-12. This number, called Avogadro's constant, is 6.02214×10^{23}. The mass of a substance containing Avogadro's number of particles is termed the molar mass. It follows from the definition of a mole that the molar mass of carbon-12 is 12 g mol^{-1}. The molar masses of the other elements are in proportion to their atomic weights (e.g., 1.00794 g for H and 15.9994 g for O). The molar mass of a compound is the sum of the molar masses of its constituent elements.

1.1.2 Pressure

By definition, pressure is the force exerted per unit area. Mathematically, this can be expressed as

$$P = \frac{F}{A},$$ (1.1)

where F is the force and A is the area against which the force is exerted. A gas exerts a force because of the change in momentum undergone by individual molecules when they collide with the walls of the container. We may see how such collisions result in force by considering the simple example of a single atom moving in the x direction with velocity v inside a cubic container. The situation is illustrated in Figure 1.1.

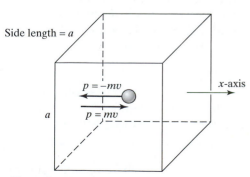

Side length = a

$p = -mv$

$p = mv$

a

x-axis

Change in momentum = $\Delta p = mv - (-mv) = 2\,mv$

◀ FIGURE 1.1
A gas exerts a pressure on the walls of its container by virtue of the change in momentum that occurs when a molecule collides with the wall. The figure shows one such collision of a single molecule moving in the x direction with initial velocity $-v$ and velocity $+v$ after collision with the left-hand wall of a cubic container.

In the figure, a single molecule of mass m is moving toward the left-hand wall parallel to the x-axis with a velocity equal to $-v$. It strikes the wall and rebounds elastically with a new velocity $+v$. The momentum of the molecule prior to collision is $-mv$. After collision, the molecule's momentum is $+mv$. Consequently, the change in momentum, Δp, upon collision is $(+mv) - (-mv) = 2mv$. Newton's first law of classical mechanics defines force as the rate of change of momentum with time. That is,

$$\mathbf{F} = \frac{d\mathbf{p}}{dt}, \tag{1.2}$$

where the boldface type denotes a vector quantity. In our simple example, the molecule is moving with constant speed v between the left- and right-hand walls. It will, therefore, strike the left wall every time it has moved across the container and back, a distance of $2a$. The time between collisions, Δt, is $2a/v$. Using Eq. 1.2, we can determine the force on the wall to be

$$\mathbf{F} = F_x = \frac{d\mathbf{p}}{dt} = \frac{dp_x}{dt} = \frac{\Delta p_x}{\Delta t} = \frac{2mv}{2a/v} = \frac{mv^2}{a}. \tag{1.3}$$

In Eq. 1.3, we have replaced the vector force with the force in the x direction, F_x, since the motion of our molecule is solely along the x-axis.

Because collision with the wall produces a force, we expect the gas to exert a pressure. The area of the wall against which the pressure is exerted is a^2. This fact, along with Eqs. 1.1 and 1.3, show the pressure to be

$$P = \frac{F}{A} = \frac{(mv^2/a)}{a^2} = \frac{mv^2}{a^3} = \frac{mv^2}{V}, \tag{1.4}$$

where $V = a^3$ is the volume of the container. Since the classical kinetic energy of the molecule is $mv^2/2$, Eq. 1.4 demonstrates that the product of pressure and volume has units of energy and that PV for a single particle is twice the kinetic energy of the molecule moving in one dimension.

The units of force may be determined by elaborating Eq. 1.2:

$$\mathbf{F} = \frac{d\mathbf{p}}{dt} = \frac{d(m\mathbf{v})}{dt} = m\left(\frac{d\mathbf{v}}{dt}\right). \tag{1.5}$$

Equation 1.5 shows the unit of force to be mass $\times$ velocity/time. Since velocity has units of distance per unit time, the force unit has dimensions of mass $\times$ distance/time^2. The magnitude of force that moves a 1-kilogram mass with an acceleration of one meter per second squared is called 1 newton (N).

Equation 1.1 makes it clear that pressure has units of force divided by area. The unit of pressure in the International System of units (SI, for the French *Système International*) is the pascal (Pa), which is defined to be a force of 1 newton exerted on an area of 1 square meter. That is,

$$1 \, \text{Pa} = 1 \, \text{N m}^{-2}.$$

The pascal is an inconveniently small unit in which to express pressures on the order of those found at sea level. Therefore, it is common to employ kilopascals (kPa) or bars. The conversion factors between these units are

$$1 \, \text{bar} = 10^5 \, \text{Pa} = 100 \, \text{kPa}.$$

Since many measurements are made near sea level, where the atmospheric pressure is 101.325 kPa, it is common to use this value as a separate unit called one atmosphere (atm). Because of the historical development of chemistry, other units of pressure, such as the torr, are also employed. The conversion factors between these units are

$$1 \text{ atm} = 101.325 \text{ kPa} \quad \text{and} \quad 760 \text{ torr} = 1 \text{ atm},$$

so that 1 atm is about the same as 1 bar. One torr is also the pressure exerted by a column of mercury 1 millimeter in height. Consequently, we may also write

$$1 \text{ atm} = 760 \text{ mm Hg}.$$

EXAMPLE 1.1

Consider the hypothetical situation in which 1 mole of argon atoms is moving in the x direction in a cubic container of volume 10 liters (L). The pressure inside the container is found to be 1 atm. Determine the total kinetic energy of the argon atoms. What is the average kinetic energy of each argon atom? What are the units on the answer?

Solution

Equation 1.4 shows that the product PV is twice the kinetic energy of the system. In order to utilize this equation, we must first convert all quantities to compatible units. Let us use the SI system:

$$P = 1 \text{ atm} \times [1.01325 \times 10^5 \text{ Pa atm}^{-1}] \times [1 \text{ N m}^{-2}/1 \text{ Pa}] = 1.01325 \times 10^5 \text{ N m}^{-2};$$

$$V = 10 \text{ L} \times [1,000 \text{ cm}^3 \text{ L}^{-1}] \times [1 \text{ m}/100 \text{ cm}]^3 = 10^{-2} \text{ m}^3.$$

We may now compute

$$PV = 1.01325 \times 10^5 \text{ N m}^{-2} \times 10^{-2} \text{ m}^3 = 1.01325 \times 10^3 \text{ N-m} = 1,013.25 \text{ N-m.} \quad \textbf{(1)}$$

Consequently, Eq. 1.4 shows that the total kinetic energy of 1 mole of argon atoms is

$$\frac{mv^2}{2} = \frac{PV}{2} = 506.62 \text{ N-m mol}^{-1}. \quad \textbf{(2)}$$

Since the number of molecules or atoms in a mole is equal to Avogadro's number, the average kinetic energy per argon atom is

$$\langle \text{kinetic energy} \rangle = 506.62 \text{ N-m mol}^{-1} \times [1/6.022 \times 10^{23} \text{ mol}^{-1}]$$

$$= 8.413 \times 10^{-22} \text{ N-m atom}^{-1}. \quad \textbf{(3)}$$

We have used the pointed braces, $\langle \rangle$, to denote the average value of a quantity. This notation will be employed throughout the text.

As noted, the unit of energy is newton $\times$ meter. This unit is called a joule (J).

EXAMPLE 1.2

Suppose we have a large right circular cylindrical container of height h and cross-sectional area A filled with a liquid whose density is d. Figure 1.2 illustrates this situation. Determine the pressure exerted on the bottom of the cylinder by the liquid as a function of h, d, and A.

Solution

The volume of the container is

$$V = hA. \quad \textbf{(1)}$$

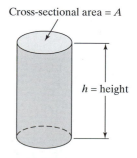

Cross-sectional area = A

h = height

▲ FIGURE 1.2
A column of liquid exerts a pressure on the base of its container that is dependent on the density of the liquid and the height of the column. It is, however, independent of the cross-sectional area of the base. (See Example 1.2.)

Since density is mass per unit volume, the total mass of the liquid in the cylinder is

$$M = dV = hAd. \tag{2}$$

The gravitation force exerted by a mass M close to the earth's surface is

$$F = Mg = hAgd, \tag{3}$$

where g is the gravitational acceleration constant, which is 9.807 m s^{-2}. The pressure is, therefore,

$$P = \frac{F}{A} = hgd. \tag{4}$$

It is important to note that the pressure is independent of the cross-sectional area of the container; it depends only upon the height and density of the material.
For related exercises, see Problems 1.1 and 1.2.

EXAMPLE 1.3

The deepest part of the ocean is approximately 35,000 ft deep. Calculate the pressure at this depth. Assume the water density to be constant at 1 kg L^{-1}. This assumption ignores the effect of NaCl and other solutes in the water. You may also assume water to be incompressible.

Solution

We must first have a compatible set of units. Let us convert all quantities to SI units:

$$h = 35,000 \text{ ft} \times [0.3048 \text{ m ft}^{-1}] = 10,668 \text{ m};$$

$$d = (1 \text{ kg L}^{-1}) \frac{1 \text{ L}}{1,000 \text{ cm}^3} (100 \text{ cm m}^{-1})^3 = 10^3 \text{ kg m}^{-3}.$$

The pressure at this depth can now be obtained from the result of Example 1.2:

$$P = hgd = (10,668 \text{ m})(9.807 \text{ m s}^{-2})(10^3 \text{ kg m}^{-3}) = 1.046 \times 10^8 \text{ (kg m s}^{-2})/\text{m}^2$$

$$= 1.046 \times 10^8 \text{ N m}^{-2} = 1.046 \times 10^8 \text{ Pa} = 1.046 \times 10^5 \text{ kPa}. \tag{1}$$

This pressure may be converted to other units using the conversion factors given earlier:

$$P = 1.046 \times 10^8 \text{ Pa} [1 \text{ bar}/10^5 \text{ Pa}] = 1,046 \text{ bar}; \tag{2}$$

$$P = 1.046 \times 10^5 \text{ kPa} [1 \text{ atm}/101.325 \text{ kPa}] = 1,033 \text{ atm}. \tag{3}$$

In terms of the pressure units often used to report barometric pressures in the United States,

$$P = 1,033 \text{ atm} [14.70 \text{ (lb/in}^2) \text{ atm}^{-1}] = 15,180 \text{ lb/in}^2 = 15,180 \text{ psi}. \tag{4}$$

We may, therefore, deduce that a pressure of 1 atmosphere is produced by a column of water that is 35,000 ft/1033 atm = 33.88 ft high.

EXAMPLE 1.4

What height must a column of mercury have to produce a pressure of 1 atm? The density of mercury is 13.596 kg L^{-1}.

Solution

Example 1.2 shows that $P = hdg$. Consider two cylinders, one filled with water, the second with mercury. Let (h_w, d_w) and (h_m, d_m) be the height and density variables for the water and mercury cylinders, respectively. If both liquid columns produce a pressure of 1 atm, we have

$$h_w d_w g = h_m d_m g, \tag{1}$$

or

$$h_m = \frac{h_w d_w}{d_m}. \tag{2}$$

Example 1.3 shows that a column of water 33.88 ft high will produce a pressure of 1 atm. Thus,

$$h_m = \frac{(33.88 \text{ ft})(1 \text{ kg L}^{-1})}{(13.596 \text{ kg L}^{-1})} = 2.492 \text{ ft.} \tag{3}$$

Conversion to metric units gives

$$h_m = 2.492 \text{ ft} \frac{1{,}000 \text{ mm}}{3.281 \text{ ft}} = 759.5 \text{ mm.} \tag{4}$$

Since 1 atm = 760 torr, 1 torr and 1 mm Hg are very nearly the same.

Alternatively, we could compute h_m directly using the result in Example 1.2. We then have

$$h_m = \frac{P}{dg}, \tag{5}$$

where

$$P = (1 \text{ atm}) \frac{1.01325 \times 10^5 \text{ Pa}}{1 \text{ atm}} \times \frac{1 \text{ N m}^{-2}}{\text{Pa}} = 1.01325 \times 10^5 \text{ N m}^{-2} \tag{6}$$

and

$$d = 13.596 \text{ kg L}^{-1} \times \frac{1\text{L}}{1{,}000 \text{ cm}^3} \times \left[\frac{100 \text{ cm}}{1 \text{ m}}\right]^3 = 1.3596 \times 10^4 \text{ kg m}^{-3}. \tag{7}$$

Substitution into the expression for h_m gives

$$h_m = \frac{1.01325 \times 10^5 \text{ N m}^{-2}}{(1.3596 \times 10^4 \text{ kg m}^{-3})(9.807 \text{ m s}^{-2})} = 0.7599 \text{ m} \approx 760 \text{ mm.} \tag{8}$$

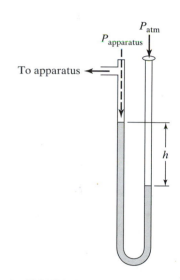

▲ FIGURE 1.3
A simple manometer for measuring pressure. In this version, the right-hand tube is open to the atmosphere so that the gas pressure on the right-hand column is atmospheric pressure, P_{atm}. The left-hand column is connected to an apparatus containing a gas whose pressure is to be measured. In this example, the pressure in the apparatus is less than atmospheric pressure, so the fluid in the manometer, which is often mercury, rises in the left-hand tube. The difference in pressure between the atmosphere and the gas in the apparatus is given by $\Delta P = hdg$, the result derived in Example 1.2. Thus, measurements of h and P_{atm} give the gas pressure inside the apparatus.

The results obtained in Examples 1.2, 1.3 and 1.4 are the basis of a simple pressure-measuring device called a manometer. This device consists of two connected tubes containing some fluid, which is often mercury. Either one tube is open to the atmosphere, so that the pressure on the surface of the fluid is atmospheric pressure, or the tube is closed and evacuated, so that the pressure on the surface of the fluid is near zero. The second tube is connected to the apparatus containing the gas whose pressure is to be measured. In the first case, a measurement of the difference in height of the fluid columns, as shown in Figure 1.3, gives us the pressure difference between the gas in the apparatus and atmospheric pressure, since $\Delta P = hdg$. Measurement of the current atmospheric pressure then yields the pressure inside the apparatus. If the right-hand column is evacuated, the absolute pressure in the apparatus is given by $P_{apparatus} = hdg$. Since h can be measured to an accuracy of about ±1 mm, manometers have an uncertainty that corresponds to the pressure of a 1-mm column of the fluid that is to be measured. A mercury manometer can, therefore, measure pressures to an accuracy of about ±1 mm Hg, or ±1 torr.

1.1.3   Units

The preceding discussion raises the general question of units in scientific calculations. SI units are generally accepted as the units of choice. These are metric units that employ the kilogram (kg), the meter (m), and the second (s) as the base units of mass, distance and time, respectively. In the majority of cases in this text, we will employ this system. Unit conversion tables can be found in Appendix C.

The general acceptance of SI units tends to obscure a very important principle related to the choice of units for a given problem. This principle is the same as that used in the selection of a coordinate system; that is, we always try to choose the coordinate system that makes the problem at hand as simple as possible. We should also choose our units on the basis of the same consideration. This is particularly important in the case of complex computer calculations. All computers are limited with regard to the magnitude of the numbers that can be represented. If a number becomes too large, an overflow error is produced. If a number becomes too small, we obtain an underflow error. Both of these errors can completely destroy the accuracy of the computation or bring it to a halt.

The foregoing considerations suggest that it is particularly important in computer calculations to employ units that make the quantities involved as close to unity as possible. This procedure will minimize the risk of both overflow and underflow errors. Consider a calculation that involves a quantity we shall encounter later in this text, Planck's constant h. This constant has units of energy $\times$ time. If we employ the usual SI units of joules and seconds, its value is approximately 6.626×10^{-34} J s. The use of such a number in computationally intensive calculations will almost certainly produce overflow or underflow errors. Consequently, we need to use a different set of units in this case.

It is important to remember that you are free to use any set of units that may be convenient. You can even invent the units if you wish. At the end of the calculation, the units on the computed quantities can always be converted. Let us consider an illustrative example. Suppose we need an approximate value for the volume occupied by an H_2 molecule. In SI units, the bond length of this molecule is 0.740×10^{-10} m. The electron cloud about the bond axis extends about 5.0×10^{-11} m perpendicular to the bond axis. Since the volume of a right circular cylinder is $\pi r^2 h$, the approximate volume of an H_2 molecule in SI units is $(3.1416)(5.0 \times 10^{-11})^2 (0.740 \times 10^{-10})$ m^3 = 5.8×10^{-31} m^3. Not wishing to use this extremely small number in a computer calculation, we invent a different unit. The barn (b) is a unit of cross section equal to 10^{-28} m^2. The yard (y) is a unit of distance equal to 0.914403 m. We define the barnyard (by) as a unit of volume equal to a cross-sectional area of one barn times a length of one yard.* Clearly, 1 by = 9.14403×10^{-29} m^3. In these units, the volume of an H_2 molecule is 0.0064 barnyard. We could make the unit even better for computer use by using millibarnyards (mby). Accordingly, the H_2 volume is 6.4 mby. We could also address the problem by using derivatives of the base SI units, such as cubic nanometers (nm^3) or cubic picometers (pm^3). In these units, the H_2 volume is 0.00058 nm^3, or 5.8×10^5 pm^3. These procedures are far superior to using the base SI units.

One reviewer of this text pointed out that the barn was originally invented during World War II as a unit in which to express cross sections for

---

* I am indebted to Dr. Oakley H. Crawford for first suggesting this interesting and amusing unit to me.

nuclear reactions so that the enemy would not learn the actual magnitude of these quantities if they happened to gain access to the data. New units can be developed whenever the need arises.

1.1.4   Temperature

Temperature is an intensive variable in that it is independent of the quantity of material present. We shall see that it is, in effect, a measure of the quantity of thermal energy contained in a material. As such, it is a variable that is of central importance in the study of thermodynamics in particular and chemistry in general. It is, therefore, essential that we develop a quantitative definition of temperature.

It is easily observed that objects change their shapes when their temperature is altered. Gases expand when heated. A silver wire elongates as its temperature rises. A well-known example is the column of mercury in a thermometer, which expands at higher temperatures. We can utilize that expansion to develop a phenomenological measure of the temperature. To accomplish this, we need only fix one reference point—the size of the temperature unit—and establish some functional relationship between the object's shape and the change in temperature.

As a first example, let us take the length of a metal rod, L, as our temperature-measuring device. The most commonly used reference point is the melting point of ice under a pressure of 1 atm. The Celsius scale takes this point to be 0°C. The Fahrenheit scale chooses a value of 32°F for the same point. Since scientific work never utilizes the Fahrenheit scale, let us take this reference point to be $T = 0$°C. We denote the length of our rod at this point as L_o. We now need to fix the size of the temperature unit. This is conveniently done by choosing the number of degrees between our base reference point and some second point—for example, the boiling point of pure water under a pressure of 1 atm. The Celsius scale assigns the value 100°C to this point. Finally, we need to establish some functional relationship between L and T. Generally, a linear relationship is chosen, since there is no particular advantage to choosing a more complex, nonlinear dependence. That is, we take T to be

$$T = aL + b. \tag{1.6}$$

Then, since $T = 0$ at $L = L_o$, we must have $b = -aL_o$. With this substitution, Eq. 1.6 becomes

$$T = a(L - L_o). \tag{1.7}$$

The choice of degree size determines a. At $T = 100$, $L = L_{100}$, the length of the rod when immersed in boiling water under a pressure of 1 atm. Insertion of these values into Eq. 1.7 yields

$$a = \frac{100}{(L_{100} - L_o)}$$

and

$$T = 100\,\frac{(L - L_o)}{(L_{100} - L_o)}. \tag{1.8}$$

Equation 1.8 permits a temperature to be determined simply by measuring the length of the rod.

The problem with the preceding procedure is that the numerical value we obtain for the temperature of a given system will vary slightly with our choice of material for the rod. For example, copper would give one value, silver another. This difficulty could be averted by agreeing on a standard choice for the composition of the rod. However, there is a much better method available.

Instead of using the length of a metal rod, we could equally well employ the volume of a rare gas such as helium or neon. With the same choices for reference point and degree size, this alternative would give

$$T = \frac{100 (V - V_o)}{(V_{100} - V_o)}, \tag{1.9}$$

provided that the pressure is held constant. With this procedure, it is found that temperatures computed using Eq. 1.9 are essentially independent of the rare gas used. Further, the relative coefficients of thermal expansion of the gases with respect to the scale established by Eq. 1.9 are all nearly identical. That is, it is found that

$$\frac{(V_{100} - V_o)}{V_o} = \frac{1}{\alpha} \tag{1.10}$$

is nearly the same for all rare gases. Measurement shows α to be about 2.73. As the gas pressure approaches zero, α is found to approach 2.7315 for every gas. This fact may be used to write Eq. 1.9 in a different form. Dividing numerator and denominator by V_o, we obtain

$$T = \frac{100[V/V_o - 1]}{[V_{100} - V_o]/V_o} = 100\, \alpha \left[\frac{V}{V_o} - 1 \right]. \tag{1.11}$$

Equation 1.11 allows us to establish an absolute temperature scale. The lowest temperature will be reached when the volume of the gas approaches zero. At $V = 0$, Eq. 1.11 yields a temperature of $T = -100\alpha$, or $-273.15°C$. We now change our reference point from the normal freezing point of water to this lowest temperature, which we take to be zero degrees. The new scale thereby obtained is termed the absolute, or Kelvin, scale, for which the symbol K is commonly employed. The relationship between the Celsius scale and the absolute scale is

$$T\,(\text{K}) = T\,(°\text{C}) + 273.15. \tag{1.12}$$

EXAMPLE 1.5

Develop an equation analogous to Eq. 1.11 for a rare-gas thermometer using the Fahrenheit scale for the reference point and the size of the degree. What Fahrenheit temperature corresponds to absolute zero?

Solution

Using Eq. 1.6 with the volume of the rare gas substituted for the length of the rod, we have

$$T = aV + b. \tag{1}$$

The Fahrenheit reference point for the melting temperature of ice is $T = 32°F$. Let the volume of the gas at this point be V_{32}. We then have

$$32 = aV_{32} + b, \tag{2}$$

so that

$$b = 32 - aV_{32}. \tag{3}$$

Substitution of b into the equation for T then gives

$$T\,(°\text{F}) = a[V - V_{32}] + 32. \tag{4}$$

The degree size is fixed by setting the normal boiling point of water to 212°F. If the rare-gas volume at this temperature is V_{212}, we have

$$a = \frac{(212 - 32)}{(V_{212} - V_{32})} = \frac{180}{(V_{212} - V_{32})}, \tag{5}$$

and

$$T(°F) = \frac{180(V - V_{32})}{(V_{212} - V_{32})} + 32. \tag{6}$$

Since V_{212} and V_{32} are the same volumes as V_{100} and V_o in Eq. 1.10, we have

$$\frac{(V_{212} - V_{32})}{V_{32}} = \frac{1}{\alpha} = \frac{1}{2.7315}. \tag{7}$$

Substitution into the equation for the Fahrenheit temperature gives

$$T(°F) = \frac{180[V/V_{32} - 1]}{[V_{212} - V_{32}]/V_{32}} + 32 = 180\,\alpha\left[\frac{V}{V_{32}} - 1\right] + 32, \tag{8}$$

which is the desired equation. At absolute zero, the volume V, of the gas goes to zero, and we have $T(°F) = -180\alpha + 32 = -459.67°F$. The absolute temperature based on the Fahrenheit scale is, therefore, $T = T(°F) + 459.67°F$. This scale is termed the Rankine temperature, for which the symbol R is used.

For related exercises, see Problems 1.3, 1.4, and 1.5.

1.1.5   Pressure, Volume, Temperature, and Mass Relationships

Although a given gaseous state is associated with the values of pressure, volume, temperature, and the gaseous mass, experiment demonstrates that only three of these variables are independent. That is, if three of the four variables are known, it is always possible to determine the value of the fourth variable.

A relationship that connects P, V, T, and n is known as an equation of state. In principle, an equation of state can have any mathematical form, such as

$$h(P, V, T, n) = 0. \tag{1.13}$$

However, in the majority of cases, equations of state are written so that either P or V is expressed as an explicit function of the remaining three variables—for example,

$$P = f(T, V, n) \tag{1.14}$$

or

$$V = g(T, P, n). \tag{1.15}$$

Equation 1.14 is explicit for the pressure, while Eq. 1.15 is explicit for the volume.

Some equations of state may be written either in the form of Eq. 1.14 or in the form of Eq. 1.15. The ideal-gas equation considered in the next section is an example. More complex equations of state, however, may be easy to write in one form, but difficult or impossible to write in the other. The van der Waals equation of state is a good example of this duality. It is usually written in the form

$$\left(P + \frac{n^2 a}{V^2}\right)(V - nb) = nRT, \tag{1.16}$$

where a, b, and R are constants. This equation can easily be solved for the pressure, viz.,

$$P = \frac{nRT}{[V - nb]} - \frac{n^2a}{V^2} \tag{1.17}$$

but is difficult to solve for the volume, in which variable the equation is cubic.

1.2   Ideal-gas Equation of State

1.2.1   Boyle's Law

In 1661, Robert Boyle demonstrated experimentally that the pressure exerted by a fixed quantity of gas at a constant temperature varies inversely with the volume occupied by the gas. The mathematical statement for this experimental result is

$$P \propto V^{-1} \qquad \text{at constant } T. \tag{1.18}$$

In view of Eq. 1.4, this result is not surprising. Physically, Eq. 1.18 means that if a gas were to be confined within a piston-and-cylinder arrangement immersed in a constant-temperature bath, the gas pressure would double if the volume were decreased by a factor of two. In the study of thermodynamics, it is very important to think of this condition of constant temperature as one for which the differential change in T, dT, is zero.

In the case of a simple proportionality in which a variable y is directly proportional to x, the relationship may be converted to an equality by multiplying x by a constant. Thus, we express such a proportionality as $y = ax$, where a is a constant. However, when the proportionality is conditioned on some third variable being held constant, we multiply by an arbitrary function of that third variable. That is, if we state that y is proportional to x, provided that z is held constant, then we write $y = a(z)x$, where $a(z)$ is an arbitrary function of z. Since z is being held constant, this condition makes the function $a(z)$ constant, so that the right-hand side is, in effect, being multiplied by a constant.

With these points in mind, we may write Eq. 1.18 in the form

$$P = \frac{k(T)}{V} \qquad \text{at } dT = 0. \tag{1.19}$$

Figure 1.4A shows the hyperbolic dependence of P upon V at $dT = 0$ for temperatures in the range $300 \text{ K} \leq T \leq 1{,}500 \text{ K}$. Equation 1.19 may be graphically represented in a different fashion by multiplying both sides by V. This yields

$$PV = k(T) \qquad \text{at } dT = 0. \tag{1.20}$$

Figure 1.4B shows a plot of PV versus V at 298.15 K. Since $k(T)$ has a fixed value when T is held constant, it is clear that the product PV is a constant and the plot is simply a horizontal line. Consequently, if the product PV is evaluated at two different points, say, (P_1, V_1) and (P_2, V_2), we will have

$$P_1V_1 = k(T) = P_2V_2, \tag{1.21}$$

which is simply an alternative way of stating Boyle's law. Equation 1.21 is important not only because it provides a convenient means of computing pressure–volume changes, but also because it shows that the statements

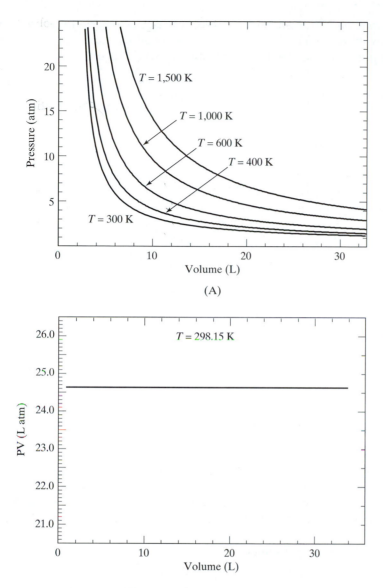

PV = constant and $P_1V_1 = P_2V_2$ are mathematically equivalent. This point will be important in the development of the ideal-gas equation of state.

1.2.2   Charles' Law

The second ideal-gas law was developed by Jacques Charles, who found in 1787 that the volume of a fixed amount of gas is proportional to the absolute temperature, provided that the pressure is held constant. This observation can be expressed as

$$V \propto T \qquad \text{if } dP = 0, \tag{1.22}$$

or

$$V = C(P)T, \tag{1.23}$$

where $C(P)$ is a function of pressure. Equation 1.23 thus shows that if such a gas were to be heated inside a cylinder–piston arrangement such that the pressure remained fixed at some preset value, the experimental data should fall on a straight line whose slope $C(P)$ is dependent upon the value chosen for the pressure. This relationship is illustrated by Figure 1.5.

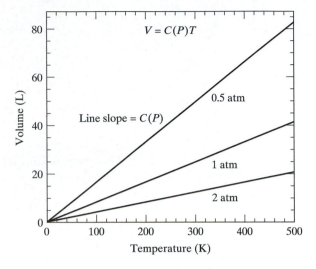

▶ FIGURE 1.5
Charles' law.

Equation 1.23 may be written in the form

$$\frac{V}{T} = C(P). \tag{1.24}$$

Since $C(P)$ is constant if $dP = 0$, the volume–temperature ratio will have the same value at any temperature and volume. That is, for two data points (V_1, T_1) and (V_2, T_2), we will observe that

$$\frac{V_1}{T_1} = C(P) = \frac{V_2}{T_2} \tag{1.25}$$

and once again we note that the mathematical statements $V/T = $ constant and $V_1/T_1 = V_2/T_2$ are equivalent.

1.2.3   Combined Laws

We now wish to combine Boyle's and Charles' laws into one mathematical expression that can be used to compute changes in the variables P, V, and T when the mass of gas is fixed. Several methods can be employed to achieve this combination. The method given next illustrates an extremely important mathematical point that will recur many times throughout your study of physical chemistry and science in general.

We can combine Eqs. 1.19 and 1.24 directly to obtain

$$P = \frac{k(T)}{V} = \frac{k(T)}{C(P)\,T}. \tag{1.26}$$

The problem is to determine the unknown functions $k(T)$ and $C(P)$. Let us rearrange Eq. 1.26 so that all of the pressure-dependent terms are on the left side of the equation while the terms that depend upon temperature are on the right:

$$P\,C(P) = \frac{k(T)}{T}. \tag{1.27}$$

Equations in this form are said to be separable, in that we have separated the variables into different terms.

Separable equations have a very simple solution, the nature of which may be seen by considering an example. Let $f(x) = x + 5$ and $g(y) = 3y - 6$. If we write

$$f(x) = x + 5 = g(y) = 3y - 6, \tag{1.28}$$

then for each x, there exists one unique value of y that satisfies Eq. 1.28, namely, $y = (x + 11)/3$. It is, therefore, clear that with $f(x)$ and $g(y)$ defined in the foregoing manner, the equation $f(x) = g(y)$ cannot hold for *all* values of x and y. For example, Eq. 1.28 is not satisfied if $x = 1$ and $y = 2$. If we require that $f(x) = g(y)$ for *all* values of x and y, then there exists only one class of functions for $f(x)$ and $g(y)$ that satisfies this requirement, namely,

$$f(x) = g(y) = \text{a constant}. \tag{1.29}$$

No other solution is possible. Equation 1.29 is always the solution to any problem in which the variables have been separated.

The method of solution of Eq. 1.27 is now straightforward. Since we require that the equation holds for all values of P and V, we must have

$$P\, C(P) = \frac{k(T)}{T} = \text{a constant} = K. \tag{1.30}$$

Thus, it follows from Eq. 1.30 that the unknown function $C(P)$ must have the form

$$C(P) = \frac{K}{P}. \tag{1.31}$$

Substitution of Eq. 1.31 into Eq. 1.24 gives

$$\frac{PV}{T} = K = \text{a constant}. \tag{1.32}$$

We have already seen that equations of the form of Eq. 1.32 are equivalent to the statement

$$\frac{P_1 V_1}{T_1} = \frac{P_2 V_2}{T_2} \tag{1.33}$$

for two arbitrary sets of data points (P_1, V_1, T_1) and (P_2, V_2, T_2). Equations 1.32 and 1.33 are thus a combined statement of all of the information contained in Boyle's and Charles' laws.

1.2.4 Avogadro's Hypothesis

All that remains to obtain the ideal-gas equation of state is to deduce the value of the constant on the right-hand side of Eq. 1.32. This can be accomplished using a hypothesis formulated by Amedeo Avogadro in 1811, which states that at fixed pressure and temperature, the volume of a gas is proportional to the number of moles of gas present; that is,

$$V \propto n \quad \text{at } dT = dP = 0. \tag{1.34}$$

Since both temperature and pressure must be held constant for Eq. 1.34 to hold, we must write

$$V = f(T, P)\, n. \tag{1.35}$$

Substitution of Eq. 1.35 into Eq. 1.32 yields

$$\frac{P\, n\, f(T, P)}{T} = K. \tag{1.36}$$

For the left side of Eq. 1.36 to be a constant, the proportionality function $f(T, P)$ must have the form $f(T, P) = RT/P$ with R a constant. Substitution of this result into Eq. 1.36 gives $K = nR$, and hence, we have

$$\boxed{PV = nRT},\qquad\qquad (1.37)$$

which is the ideal-gas equation of state.

Since R is a constant, we could, in principle, obtain its value by measuring P, V, T, and n at any convenient point. However, we must work with real gases, which exhibit deviations from Boyle's, Charles', and Avogadro's laws if the pressure is high. For this reason, measurements are made at low pressures, where the deviations from ideality become small. In practice, R is measured at successively lower and lower pressures. The results are then extrapolated to zero pressure to obtain R. The method is illustrated by Problem 1.15 at the end of the chapter. When this is done, we obtain

$$R = 8.31451 \text{ J mol}^{-1}\text{K}^{-1} = 0.0820578 \text{ L atm mol}^{-1}\text{K}^{-1}$$
$$= 0.0831451 \, L \text{ bar mol}^{-1}\text{K}^{-1}.$$

R is generally termed the gas constant. It is always expressed in per mole (mol^{-1}) units. If this is changed to per molecule units by division by Avogadro's constant, viz.,

$$\frac{R}{N} = 8.31451 \text{ J mol}^{-1}\text{K}^{-1} \times \frac{1 \text{ mol}}{6.02214 \times 10^{23}} = 1.38066 \times 10^{-23} \text{ J K}^{-1} = k_b,$$

the resulting constant k_b is called the Boltzmann constant.

The ideal-gas equation of state can be cast in different forms that are often useful in the solution of problems. The number of moles is related to the mass of gas and the molar mass by $n = m/M$, where m is the mass in grams and M is the molar mass, which is numerically equal to the molecular "weight." Substitution of this relationship into Eq. 1.37 produces

$$PV = \frac{mRT}{M}.\qquad\qquad (1.38)$$

Equation 1.38 contains five variables: P, V, T, m, and M. Virtually all problems involving ideal gases give data that permit the computation of four of these quantities and then ask for the fifth quantity. The solution is readily obtained using Eq. 1.38.

The ideal-gas equation may also be expressed in terms of the gas density d. Since density is mass per unit volume, we simply write Eq. 1.38 in the form

$$\frac{m}{V} = d = \frac{MP}{RT}.\qquad\qquad (1.39)$$

1.2.5   Dalton's Law

Let us assume that we have a mixture of s different gases in a container whose total volume is V. If all the gases are ideal, then the total pressure of the mixture will be given by

$$P_{\text{total}} = \frac{n_{\text{total}} RT}{V},\qquad\qquad (1.40)$$

where n_{total} is the total number of moles of gas present. Mass balance requires that

$$n_{\text{total}} = \sum_{i=1}^{s} n_i\qquad\qquad (1.41)$$

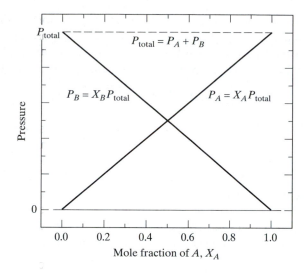

◀ FIGURE 1.6
The partial pressures of gases A and B in a binary mixture of ideal gases as a function of mole fraction. The figure also shows the relationship between the partial and total pressures.

where the notation $\sum_{i=1}^{s} n_i$ represents the sum $n_1 + n_2 + n_3 + \cdots + n_s$. Substitution of Eq. 1.41 into Eq. 1.40 gives

$$P_{\text{total}} = \sum_{i=1}^{s} \frac{n_i RT}{V}. \tag{1.42}$$

The quantity appearing underneath the summation in Eq. 1.42 is labeled P_i and termed the partial pressure of gas i. Physically, it represents the pressure gas i would exert if it were the only gas present in the mixture at temperature T and volume V. In this notation, Eq. 1.42 can be written

$$P_{\text{total}} = \sum_{i=1}^{s} P_i, \tag{1.43}$$

which states that the total pressure is the sum of the individual partial pressures in the gaseous mixture. This statement is Dalton's law, which holds rigorously only for gases that obey the ideal-gas equation of state.

When Dalton's law does hold with sufficient accuracy for a gaseous mixture, we may obtain a very useful relationship between the partial pressure of a gas, the total pressure, and the composition of the mixture. If we define the mole fraction of component k as $X_k = n_k / n_{\text{total}}$, then, using Eq. 1.40 and the definition of the partial pressure, we can write the mole fraction in the form

$$X_k = \frac{n_k}{n_{\text{total}}} = \frac{P_k V / RT}{P_{\text{total}} V / RT} = \frac{P_k}{P_{\text{total}}}. \tag{1.44}$$

Equation 1.44 permits the composition of a gaseous mixture to be determined by measuring the individual partial pressures. This is illustrated in Figure 1.6 for a binary mixture in which we have $X_A + X_B = 1$. Equation 1.44 will prove very useful when we examine the thermodynamic behavior of solutions.

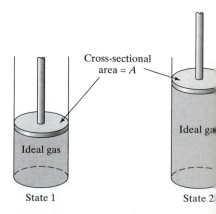

▲ FIGURE 1.7
Piston–cylinder arrangement in which n moles of gas may undergo pressure, volume, and temperature changes as the piston is withdrawn and the gas is simultaneously heated or cooled.

EXAMPLE 1.6

An ideal gas is held in a cylinder–piston arrangement such as that shown in Figure 1.7. In State 1, the temperature is 300 K, the pressure is 3 bar, and the volume of the gas is 10 liters. The piston is moved so as to increase the volume in State 2 to 17.5 liters. The temperature in State 2 is increased to 475 K. What is the pressure inside the cylinder in State 2?

Solution

Problems of this type involving only the variables P, V, and T can be solved using Eq. 1.33. Let $P_1 = 3$ bar, $V_1 = 10$ liters, and $T_1 = 300$ K. In State 2, we have $T_2 = 475$ K and $V_2 = 17.5$ liters. We wish to determine P_2. Using Eq. 1.33, we obtain

$$P_2 = \frac{P_1 V_1 T_2}{V_2 T_1} = \frac{(3 \text{ bar})(10 \text{ L})(475 \text{ K})}{(17.5 \text{ L})(300 \text{ K})} = 2.71 \text{ bar.} \tag{1}$$

(*Note:* If the temperatures had been given in units other than kelvin, it would have been necessary to first convert to the Kelvin scale before using Eq. 1.33, since Charles' law, as expressed by Eq. 1.23, holds only if T is in absolute degrees.)

EXAMPLE 1.7

An ideal gas is initially contained inside a right circular piston–cylinder arrangement such as that shown in Figure 1.7 with a volume of 20,000 cm^3 at $T = 300$ K and a pressure of 1.231 atm. The radius of the piston face is 10 cm. The device is constructed so that the piston is withdrawn (thus increasing the volume of the gas) at a constant rate of A cm per minute. At the same time, heating coils begin to increase the system temperature at the constant rate of B degrees K per minute. **(A)** Obtain an expression in terms of A and B giving the pressure inside the cylinder as a function of time. **(B)** If $A = 2$ cm min^{-1} and $B = 0.5$ K min^{-1}, calculate the pressure at the end of 30 minutes. **(C)** What value would the pressure inside the cylinder approach at extremely long times for the values of A and B given in **(B)**?

Solution

(A) Since this problem involves the variables P, V, and T, Eq. 1.32 should suffice to provide the solution. We have

$$\frac{PV}{T} = K. \tag{1}$$

Substitution of values permits the value of K to be computed:

$$\frac{PV}{T} = \frac{(1.231 \text{ atm})(20,000 \text{ cm}^3)}{300 \text{ K}} = 82.067 \text{ atm cm}^3 \text{ K}^{-1}. \tag{2}$$

The height of the cylinder at the start, h_o, can be obtained from the equation for the volume of a right circular cylinder—that is,

$$V_o = \pi r^2 h_o = 20,000 \text{ cm}^3, \tag{3}$$

where r is the radius of the face. Hence,

$$h_o = \frac{20,000 \text{ cm}^3}{(\pi \, 10^2 \text{ cm}^2)} = 63.662 \text{ cm.} \tag{4}$$

Since we are told that the height increases at a constant rate, we may write

$$h(t) = h_o + At. \tag{5}$$

The cylinder volume is, therefore,

$$V(t) = \pi r^2 h(t) = 10^2 \pi (h_o + At) = 100 \pi \left[\frac{20,000 \text{ cm}^3}{(\pi \, 10^2 \text{ cm}^2)} + At \right]$$

$$= 20,000 + 100 \pi At. \tag{6}$$

The temperature at time t is

$$T(t) = T_o + Bt = (300 + Bt) \text{ K.} \tag{7}$$

Substitution of these results into Eq. 1 gives

$$\frac{P(20,000 + 100\pi At)}{300 + Bt} = K = 82.067. \tag{8}$$

Solving for P, we obtain

$$P = \frac{82.067\,(300 + Bt)}{(20{,}000 + 100\pi At)}, \tag{9}$$

which is the desired function.

(B) With $A = 2$ cm min^{-1} and $B = 0.5$ K min^{-1}, substitution into the preceding expression, followed by evaluation at $t = 30$ minutes, gives

$$P(t = 30 \text{ minutes}) = \frac{82.067\,[300 + 0.5(30)]\text{ atm cm}^3}{[20{,}000 + 100\,(3.14159)(2)(30)]\text{cm}^3} = 0.665 \text{ atm.} \tag{10}$$

(C) We seek the limiting value of the pressure at very long times as $t \longrightarrow \infty$:

$$\lim_{t \to \infty} P = \lim_{t \to \infty} \frac{82.067\,(300 + Bt)}{(20{,}000 + 100\pi At)} = \frac{82.067 B}{100\pi A} = \frac{82.067(0.5)}{100(3.14159)(2)} = 0.0653 \text{ atm.} \tag{11}$$

In the evaluation of the limiting expression, we have used the fact that $(300 + Bt) \approx Bt$ and $(20{,}000 + 100\pi At) \approx 100\pi At$ when t becomes large.

EXAMPLE 1.8

Gas A is contained in one piston–cylinder arrangement, while Gas B is contained in another, similar setup. Both gases may be assumed to obey the ideal-gas equation of state. **(A)** It is experimentally determined that the density of gas A at 0°C and 1 atm pressure is 1.784 g L^{-1}. What is the molar mass of Gas A? **(B)** The temperature and pressure in both cylinders are now changed to identical new values. Under these new conditions, it is determined that the ratio of the gas density of A to that of B is 1.248. That is, $d_A/d_B = 1.248$. Can the molar mass of gas B be obtained from this information? If so, compute it. If not, state what additional information would be needed to compute the molar mass of gas B.

Solution

(A) Using Eq. 1.39, we can compute the molar mass of gas A directly:

$$M = \frac{dRT}{P}. \tag{1}$$

The only danger is failing to put the quantities on the right-hand side into a compatible set of units. The temperature must be in absolute degrees in order to match the gas constant. Since the density is given in units of g L^{-1}, R must be in units of L atm mol^{-1} K^{-1}, and the pressure must be in atmospheres. Hence, we need to convert 0°C to absolute degrees using Eq. 1.12. We have

$$T \text{ (K)} = 0°C + 273.15 = 273.15 \text{ K.} \tag{2}$$

With the proper precautions taken with respect to the units, we may now substitute directly:

$$M = (1.784 \text{ g L}^{-1})(0.08206 \text{ L atm mol}^{-1}\text{ K}^{-1})(273.15 \text{ K})/(1 \text{ atm})$$

$$= 39.99 \text{ g mol}^{-1}. \tag{3}$$

(B) We may write Eq. 1.39 for both gases A and B using subscripts to differentiate between the two:

$$M_A = \frac{d_A R T_A}{P_A}; \tag{4}$$

$$M_B = \frac{d_B R T_B}{P_B}. \tag{5}$$

Division of Eq. 4 by Eq. 5 yields

$$M_A/M_B = d_A/d_B, \text{ since } T_A = T_B \text{ and } P_A = P_B. \tag{6}$$

Solving for M_B, we obtain

$$M_B = M_A[d_B/d_A] = \frac{39.99}{1.248} = 32.04 \text{ g mol}^{-1}. \tag{7}$$

Thus, the problem contains sufficient data to determine the molar mass of gas B. For related exercises, see Problems 1.8, 1.9, 1.10, 1.11, and 1.14.

1.3   Nonideal Gases

1.3.1 P–V–T Behavior of Real Gases

The ideal-gas equation of state accurately represents the pressure, volume, and temperature behavior of real gases only in the limit of low pressure. This is so because Boyle's, Charles', and Avogadro's laws implicitly assume that there are no forces acting between the gaseous molecules and that the molecules themselves occupy no volume. The presence of the first assumption in the ideal-gas equation is obvious from the P–V behavior shown in Figure 1.4A. As the pressure increases, the equation continues to predict a hyperbolic relationship between P and V at any given temperature. At no point does the gas condense to a liquid that is nearly incompressible, so that the relationship between volume and pressure changes drastically from that observed in the gas phase. In reality, because of the existence of attractive forces between the molecules, all real gases can be condensed at a sufficiently low temperature. Such condensation is not possible for an ideal gas, since there are no forces between the molecules.

Figure 1.5 demonstrates the fact that the ideal-gas laws implicitly assume that the gas molecules themselves occupy no volume. As shown in the figure, as $T \rightarrow 0$ K, $V \rightarrow 0$. The gas density is, therefore, predicted to approach infinity at low temperature. Clearly, we cannot produce an infinitely dense system simply by cooling the gas. This anomalous behavior occurs because we are implicitly assuming that the gaseous molecules occupy no volume.

As the pressure approaches zero, the assumptions of no intermolecular forces and a zero volume for the gas molecules become increasingly good approximations. The reason is that the molecular volume becomes insignificant as the total volume of the gas becomes very large. As V increases and P decreases, the average distance between molecules, $\langle r \rangle$, increases. Since intermolecular forces vary approximately with $\langle r^{-7} \rangle$, the forces of attraction become vanishingly small at low pressures. For these reasons, the regime of low pressure and large volume is termed the ideal-gas regime.

The actual P–V behavior at various temperature for a real gas is seen in Figure 1.8, which shows experimental isotherms at several temperatures for CO_2. The prefix "iso" means "the same." Thus, an isotherm is a curve obtained at constant temperature. A comparison of Figures 1.4 and 1.8 illustrates the expected deviations of the behavior of real gases from that predicted by the ideal-gas law. At low P and large V, the two figures are very similar. The point denoted by A in Figure 1.8 is, therefore, in the ideal-gas regime for CO_2. At point B, the behavior is less ideal. At point C on the 20°C isotherm, the P–V behavior deviates dramatically from that predicted

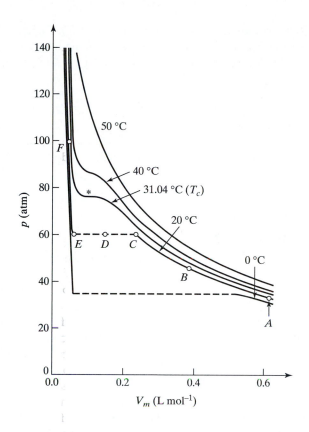

◀ FIGURE 1.8
Experimental isotherms for CO_2 at several temperatures. The critical isotherm is at 31.04°C. The critical point, which is an inflection point in the isotherm, is marked with an asterisk.

by the ideal-gas laws. At this point, CO_2 condenses. As a result, there is a discontinuity in the isotherm. The volume drops significantly, while the pressure remains unchanged. At points E and F, CO_2 is in the liquid state. Since the liquid is nearly incompressible, a huge pressure increase is required to decrease the volume by a slight amount. Consequently, the P–V curve is nearly vertical from point E to point F.

As the temperature is increased, a higher pressure is required to condense CO_2. As a result, the decrease in volume upon condensation becomes smaller at higher temperatures. Finally, a temperature is reached above which the gas cannot be liquefied regardless of the pressure applied. The highest temperature at which liquefaction is still possible is called the critical temperature T_c. For CO_2, T_c is 31.04°C. At this temperature, the discontinuity in the isotherm at the point of condensation will have been reduced to a single point, which will be an inflection point in the curve. The 31.04°C isotherm seen in Figure 1.8 shows this behavior. The asterisk marks the inflection point. The pressure and volume at the inflection point of the critical isotherm are termed the critical pressure P_c and critical volume V_c, respectively. In Chapter 6, we shall discuss the system behavior that gives rise to a critical temperature in more detail.

The P_c, V_c, and T_c values of a real gas depend primarily upon the magnitude of the intermolecular forces. Gases with large intermolecular forces exhibit high critical temperatures; gases with near-zero intermolecular forces have very low values of T_c. For example, helium, which exhibits nearly ideal-gas-type behavior, has a critical temperature of −267.9°C. The normal melting point of materials, T_m^o, also depends upon the magnitude of the intermolecular forces holding the molecules in the solid state. Consequently, it is reasonable to expect a correlation between T_c and T_m°. Figure 1.9 shows that such a correlation does indeed exist. There is a near-linear relationship between the two. The slope of the line indicates that $T_c \approx 2T_m^\circ$, expressed in the Kelvin scale.

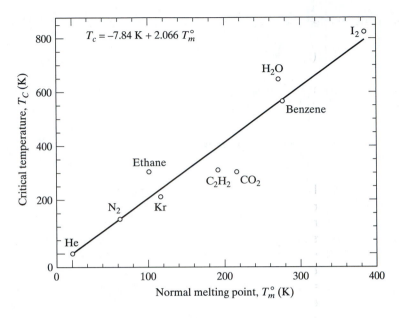

▶ **FIGURE 1.9**
Critical temperature vs. melting point.

1.3.2 Intermolecular Forces and Potentials

The isotherms of Figure 1.8 for CO_2 show that real molecules have intermolecular forces acting between them. In this section, we consider some simple intermolecular forces and introduce the concept of a potential-energy function.

The mathematically simplest intermolecular force occurs between two charged particles such as ions or electrons. The interaction force F between two such particles with charges q_1 and q_2 separated by a distance r in a vacuum is given by Coulomb's law of electrostatic interaction,

$$F = \frac{q_1 q_2}{4\pi\varepsilon_o r^2},$$

(1.45)

where ε_o, the permittivity of the vacuum, has a value of $8.85419 \times 10^{-12}\,J^{-1}\,C^2\,m^{-1}$. In Eq. 1.45, the charges are in coulombs (C), the distance is in meters, and the force is in newtons. Thus, the force between two charged particles decreases as the inverse square of the distance between them. Note that if q_1 and q_2 have the same sign, F will be positive. Repulsive forces are associated with positive values of the interaction force, while attractive forces have negative values of F.

It is now convenient to introduce the concept of potential energy, which is energy a system possesses by virtue of its position in a force field. By definition, the potential-energy function V is given by

$$F_z = -\frac{\partial V(z)}{\partial z},$$

(1.46)

where F_z is the force in the z direction and $V(z)$ is the potential energy of the system. The force is, therefore, the negative of the slope of the potential-energy function. Physically, this means that the force always acts in a direction so as to reduce the potential of the system. Applying Eq. 1.46 to the coulombic force, we obtain the corresponding coulombic potential $V(r)$, given by

$$F_r = \frac{q_1 q_2}{4\pi\varepsilon_o r^2} = -\frac{\partial V(r)}{\partial r}, \tag{1.47}$$

where the variable z has been replaced with r. Integration of Eq. 1.47 gives

$$V(r) = -\int F_r\, dr = -\frac{q_1 q_2}{4\pi\varepsilon_o}\int \frac{dr}{r^2} = \frac{q_1 q_2}{4\pi\varepsilon_o r} + c, \tag{1.48}$$

where c is a constant of integration. If we take $V(r = \infty) = 0$ as our reference point, then $c = 0$, and Eq. 1.48 becomes

$$\boxed{V(r) = \frac{q_1 q_2}{4\pi\varepsilon_o r}}. \tag{1.49}$$

Thus, the potential energy between two charged particles decreases with the inverse first power of the distance between them.

Let us now consider the interaction force between two dipoles. By definition, a dipole occurs in a system in which there exists a separation of charge by a distance d, with the total charge being zero. Figure 1.10 illustrates such a system in which the charges are $+q$ and $-q$. Systems of this type are said to possess a dipole moment $\mu = qd$. The lower half of Figure 1.10 shows two dipoles whose centers are separated by a distance r. We can use Coulomb's law to obtain the potential between two such dipoles. Summing the four interactions, we have

$$V(r) = \frac{q^2}{4\pi\varepsilon_o}\left[-\frac{1}{(r-d)} - \frac{1}{(r+d)} + \frac{2}{r}\right], \tag{1.50}$$

where the first terms are the attractive interactions between the two pairs of unlike charges and the third term is the sum of the two repulsive interactions between like charges. Taking a common denominator between the first two terms, we obtain

$$V(r) = \frac{q^2}{4\pi\varepsilon_o}\left[\frac{2}{r} - \frac{2r}{(r^2 - d^2)}\right]. \tag{1.51}$$

Combining the two fractions with a common denominator produces

$$V(r) = -\frac{2q^2}{4\pi\varepsilon_o r}\left[\frac{d^2}{(r^2 - d^2)}\right]. \tag{1.52}$$

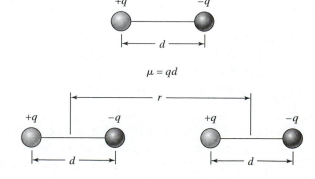

$+q \qquad -q$

$\mu = qd$

$+q \qquad -q \qquad +q \qquad -q$

◀ FIGURE 1.10
The interaction of two dipoles separated by a distance r. Each dipole has a dipole moment $\mu = dq$.

The distance r between the dipoles in the gas phase will generally be much larger than the molecular interatomic distance d. Thus, the quantity $(r^2 - d^2)$ will approach r^2, and Eq. 1.52 may be accurately written as

$$V(r) = -2\frac{q^2}{4\pi\varepsilon_o}\left[\frac{d^2}{r^3}\right] = -\frac{\mu^2}{2\pi\varepsilon_o r^3}. \tag{1.53}$$

Equation 1.53 shows that the dipole–dipole potential is attractive and that it decreases with the cube of the distance between the centers of the dipoles. If the dipoles are not collinear as shown in Figure 1.10, Eq. 1.53 will be modified by the presence of terms that depend upon the angles describing the relative orientation of the two dipoles.

When two molecules, such as CO_2 and benzene, do not possess a permanent dipole moment, there will still be a net attractive force between them at a large separation. This force is called a London dispersion force and arises because the interaction of the electron clouds surrounding the molecules distorts the charge density and induces a temporary dipole. The effect is illustrated in Figure 1.11, which shows two rare-gas atoms at a large separation at the top and in close proximity at the bottom. When the atoms are near each other, the mutual repulsion experienced by each charge cloud distorts the spherical symmetry of the cloud, with the result that a temporary dipole is produced. Quantitative treatment of this effect shows that the resulting potential energy is approximately proportional to the inverse sixth power of the separation; that is, $V(r) = -C/r^6$, where C is a positive constant.

The preceding interactions all produce attractive forces. However, when molecules approach close enough, their electronic charge clouds begin to overlap, and the resulting repulsive forces dominate the interaction. These repulsive interactions usually vary with inverse powers of the separation between 10 and 13. That is, we usually observe that $V_{rep}(r) = K/r^n$, with $n = 10$ to 13 and K a positive constant.

The total potential can be written as the sum of the attractive and repulsive interactions. For a molecule exhibiting only London dispersion interactions, we would expect to have a total potential of the form

$$V(r) = V_{rep}(r) + V_{attractive}(r) = \left[\frac{K}{r^n} - \frac{C}{r^6}\right]. \tag{1.54}$$

The Lennard–Jones (12, 6) potential, usually abbreviated as LJ (12, 6), is a typical example; it has the form

$$V_{LJ}(r) = \varepsilon[(\sigma/r)^{12} - 2\,(\sigma/r)^6], \tag{1.55}$$

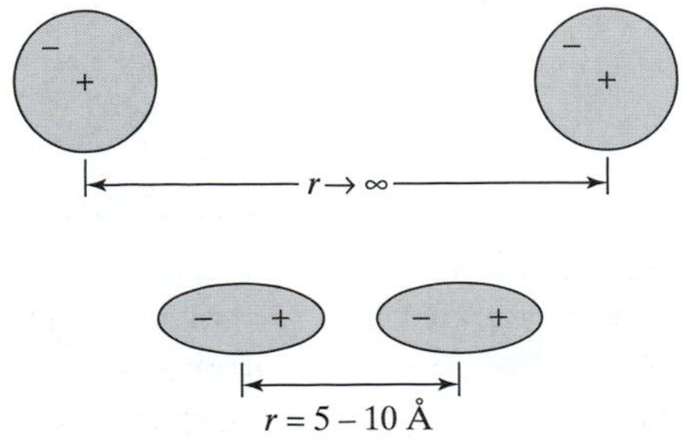

▶ FIGURE 1.11
Schematic illustration of induced dipoles. At the top, the atoms are separated by a large distance. There is no interaction, and the electronic charge density about each atom is spherically symmetric, so that there is no dipole moment. At the bottom, the two atoms have approached each other to a distance of about 5 to 10 Å. The negative electron density on one atom repels that on the other, with the result that two temporary dipoles are produced.

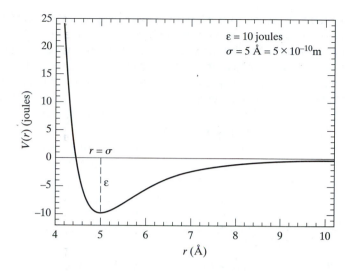

where ε and σ are parameters that determine the attractive well depth and the position of the potential minimum, respectively. This is shown in Figure 1.12 for an LJ(12, 6) potential, with ε and σ having the values of 10 J and 5 Å, respectively, where $1\text{Å} = 1 \times 10^{-10}\,m$. Ideal gases are represented by LJ(12,6) potentials with a zero value for ε. In such a case, σ is physically meaningless.

1.3.3   Virial Equation of State

The pressure, volume, and temperature behavior of real gases can be accurately represented by writing the pressure as an infinite expansion in inverse powers of the volume. That is,

$$P = RT\left[\frac{1}{V_m} + \frac{C_2(T)}{V_m^2} + \frac{C_3(T)}{V_m^3} + \cdots + \frac{C_n(T)}{V_m^n} + \cdots\right], \qquad (1.56)$$

where V_m is the molar volume of the gas. Equation 1.56 is known as the virial equation of state. The expansion coefficients $C_2(T), C_3(T), \ldots, C_n(T)$ are termed the second virial coefficient, the third virial coefficient and so on to the nth virial coefficient. These coefficients are adjusted by least-squares fitting techniques that will be discussed later in the chapter so as to fit the experimental pressure, volume, and temperature data for the real gas as closely as possible. Let us first qualitatively examine the important characteristics of the virial equation of state.

We know that as $V_m \rightarrow \infty$, all real gases must approach ideal behavior, since the interaction forces approach zero and the volume occupied by the molecules becomes negligible relative to the volume of the gas. This means that any equation of state for a real gas must have the property that in the limit of large volume, it approaches the ideal-gas equation of state. That is the case for Eq. 1.56, since V_m^{-2}, V_m^{-3}, and V_m^{-n} all become negligibly small relative to V_m^{-1} as V_m becomes large. Hence, only the first term in the expansion remains in the limit as $V_m \rightarrow \infty$.

It is important to note that each of the virial coefficients is a function of the temperature. Consequently, the value of C_n is different at each temperature, and there are effectively an infinite number of parameters available for adjustment to fit the measured pressure, volume, and temperature behavior of the gas. This is true even if the virial expansion is truncated after the $C_2(T)$

term. As a result, a virial expansion can, in principle, fit the experimental data for any gas to any degree of accuracy desired.

Equation 1.54 indicates that the attractive interactions for a real gas vary as the inverse sixth power of the interparticle distance. The second term in Eq. 1.56, containing V_m^{-2}, also varies as the inverse sixth power of the average distance between the molecules, since the units on V_m^{-2} are m^{-6}. This suggests that it is the second virial coefficient that contains most of the information concerning the attractive interactions present in the real gas. The repulsive interactions are associated with higher powers of the distance. We might, therefore, expect information on these forces to be contained in the values of $C_3(T)$ and $C_4(T)$. If there are no intermolecular forces, we will observe $C_n(T) = 0$ for all n. Because of this, most of the effort in experimental investigations of the pressure, volume, and temperature behavior of real gases is directed toward a determination of the second virial coefficient and its temperature dependence.

1.3.4   van der Waals Equation of State

Although a virial expansion permits nearly exact fitting of the pressure, volume, and temperature data for any gas, it requires more effort to use, since the virial coefficients must be determined at each temperature. If high accuracy is not required, it is often more convenient to employ other equations of state. One that is commonly used is the van der Waals equation, whose functional form is given in Eq. 1.16. In terms of molar volume, this equation is

$$\boxed{(P + a/V_m^2)(V_m - b) = RT}. \tag{1.57}$$

The van der Waals equation is a two-parameter equation of state. The two parameters, a and b, differ from gas to gas, but they are the same at all values of P, V, and T for a given gas. The effort required to determine these constants is considerably less than that needed for a virial expansion. By rearrangement of Eq. 1.57, it is simple to deduce the physical significance of a and b. Division by the first factor in Eq. 1.57, followed by the addition of b to both sides, produces

$$V_m = \frac{RT}{P + a/V_m^2} + b. \tag{1.58}$$

If we now take the limit as $P \to \infty$, the first term on the right-hand side of Eq. 1.58 becomes negligible relative to b, and we have

$$\lim_{P \to \infty} V_m = b. \tag{1.59}$$

Equation 1.59 shows that when the gas is compressed with a very large pressure, the molar volume cannot decrease below the value of b. Thus, b must represent the volume occupied by the molecules themselves. As a result, the van der Waals equation of state compensates to some degree for inaccuracies produced by the ideal-gas-law assumption of zero volume for the molecules of the gas.

We may also rearrange Eq. 1.57 to the form

$$P = \frac{RT}{(V_m - b)} - \frac{a}{V_m^2}. \tag{1.60}$$

At low pressure, when V_m becomes large, $V_m - b$ approaches V_m, since b becomes negligibly small relative to V_m. In this limit, we have

$$\lim_{V_m \to \infty} P = \left[\frac{RT}{V_m} - \frac{a}{V_m^2} \right] = P_{\text{ideal}} - \frac{a}{V_m^2}. \tag{1.61}$$

Table 1.1	van der Waals parameters	
Gas	b **(L mol^{-1})**	a **(L^2 bar mol^{-2})**
He	0.0238	0.0346
Ne	0.01672	0.208
Ar	0.03201	1.355
Kr	0.0396	2.325
Acetylene	0.0522	4.516
N_2	0.0387	1.37
H_2O	0.03049	5.537
CO_2	0.04286	3.658

Source: Handbook of Chemistry and Physics, 78th edition, CRC Press, Boca Raton, Fl, 1997–98

That is, the real-gas pressure is the pressure the gas would exert if it were ideal, minus a quantity that depends upon the inverse sixth power of the distance variable. This observation suggests that the parameter a is a measure of the attractive forces acting between the molecules. Such attractive forces would serve to reduce the pressure exerted on the walls of the container, since they would have the same effect as that expected if there were a spring between pairs of molecules. The existence of such springs would clearly decrease the molecular velocities of impact with the container walls and thereby reduce the pressure. Table 1.1 lists some values of the van der Waals parameters for several gases.

Note in the table that the smaller molecules or atoms, such as He and Ne, have the smallest b values, while larger molecules, such as CO_2 and acetylene, are associated with larger values of b. Figure 1.13 illustrates the correlation between the a values determined from fitting pressure, volume, and temperature data and the measured melting points of the compounds. The correlation is obvious and suggests that our conclusion that a is associated with the magnitude of the attractive interactions between molecules is reasonable.

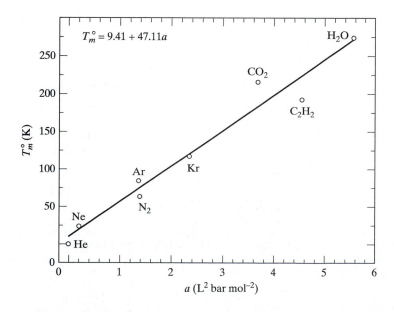

◀ FIGURE 1.13
Melting points vs. a.

The parameters a and b of the van der Waals equation of state are related to the coefficients that would be obtained in a virial expansion. The relationship may be deduced by executing a Taylor series expansion of the van der Waals equation. It will, therefore, be helpful to first review Taylor series expansions.

Let an arbitrary function of x, $F(x)$, be written as a power series expansion about the point x_o. That is, we write

$$F(x) = a_o + a_1(x - x_o) + a_2(x - x_o)^2 + a_3(x - x_o)^3 + a_4(x - x_o)^4 + \cdots$$
$$+ a_n(x - x_o)^n + \cdots .$$

If we evaluate $F(x)$ at the point $x = x_o$, we obtain

$$F(x_o) = a_o.$$

The first derivative of $F(x)$ with respect to x is

$$\frac{dF(x)}{dx} = F^1(x) = a_1 + 2a_2(x - x_o) + 3a_3(x - x_o)^2 + 4a_4(x - x_o)^3 + \cdots$$
$$+ na_n(x - x_o)^{n-1} + \cdots ,$$

so that

$$F^1(x_o) = a_1.$$

The second derivative of $F(x)$ is the derivative of $F^1(x)$:

$$\frac{d^2F(x)}{dx^2} = F^2(x) = 2a_2 + 6a_3(x - x_o) + 12a_4(x - x_o)^2$$
$$+ 20a_5(x - x_o)^3 + \cdots + n(n - 1)a_n(x - x_o)^{n-2} + \cdots .$$

Therefore, we have

$$F^2(x_o) = 2a_2.$$

Continuing in the same manner, we find that the third derivative is

$$F^3(x) = 6a_3 + 24a_4(x - x_o) + 60a_5(x - x_o)^2 + 120a_6(x - x_o)^3 + \cdots$$
$$+ n(n - 1)(n - 2)a_n(x - x_o)^{n-3} + \cdots ,$$

which yields

$$F^3(x_o) = 6a_3 = (3 \times 2)a_3 = 3! \, a_3.$$

For the nth derivative, we will clearly obtain

$$F^n(x_o) = [n \, (n - 1)(n - 2)(n - 3) \, \ldots \, 1] \, a_n = n! \, a_n.$$

Therefore, we can write the power series expansion coefficients about the point x_o in the form

$$a_n = (n!)^{-1} F^n(x_o) \tag{1.62}$$

and the overall power series as

$$F(x) = \sum_{n=0}^{\infty} (n!)^{-1} F^n(x_o) \, [x - x_o]^n. \tag{1.63}$$

Equations 1.62 and 1.63 are the Taylor series expansion formulae. In many cases, the expansion point will be $x_o = 0$, in which case the series is called a Maclaurin expansion.

We now wish to expand the van der Waals equation of state in a series of inverse powers of the molar volume so that it will have virial form. Let us recast Eq. 1.60 by defining $y = V_m^{-1}$. With this transformation, Eq. 1.60 becomes

$$P = \frac{RT}{[y^{-1} - b]} - ay^2,$$

which we can rearrange to

$$P = \frac{RTy}{[1 - by]} - ay^2 = F(y).$$

Since y is inversely related to V_m, the desired virial expansion is simply a power series expansion of $F(y)$ around $y = 0$:

$$P = F(y) = \sum_{n=0}^{\infty} a_n y^n. \tag{1.64}$$

The leading coefficient a_o is $F(0)$, which is zero. The first three nonzero coefficients can easily be shown (see Problem 1.24) to be

$$a_1 = F^1(0) = RT,$$

$$a_2 = \frac{F^2(0)}{2!} = RTb - a,$$

and

$$a_3 = \frac{F^3(0)}{3!} = RTb^2.$$

The van der Waals equation of state, written in virial form, is, therefore,

$$P = RT\left[\frac{1}{V_m} + \frac{(b - a/RT)}{V_m^2} + \frac{b^2}{V_m^3} + \cdots\right]. \tag{1.65}$$

The value of the second virial coefficient predicted by this equation is

$$C_2(T) = b - \frac{a}{RT}. \tag{1.66}$$

It is shown in Problem 1.24 that the van der Waals parameter a appears only in the second virial coefficient. This again suggests that most of the information concerning the attractive intermolecular forces is contained in that coefficient.

1.3.5 Other Equations of State

Many different equations of state have been proposed, with different functional forms and different numbers of parameters available for fitting measured pressure, volume, and temperature data. However, they all have the property that in the limit of zero pressure, the ratio PV/RT approaches unity. Table 1.2 lists some of the more common forms that have been suggested. The parameters a and b in all the two-parameter equations of state have the same physical significance: a is related to the attractive forces between the particles, and b is related to repulsions and the volume of the molecules. However, the units and numerical values of these parameters depend upon which equation of state is being employed.

Table 1.2   Some commonly used equations of state

Equation of State	Functional Form	Number of Parameters
Ideal gas	$PV_m = RT$	0
van der Waals	$(V_m - b)(P + a/V_m^2) = RT$	2
Dieterici	$P(V_m - b)\exp[a/RTV_m] = RT$	2
Berthelot	$(V_m - b)(P + a/TV_m^2) = RT$	2
Virial	$P = RT\left[V_m^{-1} + \sum_{n=2}^{\infty} C_n(T)V_m^{-n}\right]$	∞
Beattie–Bridgman	$PV_m^2 = (1 - \gamma)RT(V_m + \beta) - \alpha,$ with $\gamma = c_o/T^3 V_m,$ $\beta = b_o[1 - b/V_m]$, and $\alpha = a_o[1 + a/V_m]$	5
Redlich–Kwong	$P = \dfrac{RT}{(V_m - b)} - \dfrac{a}{T^{1/2}V_m(V_m + b)}$	2
Reichsanstalt	$PV = RT + AP + BP^2 + CP^3 + \cdots.$	∞

EXAMPLE 1.9

Use the data given in Table 1.1 to compute the pressure exerted by one mole of Kr at a temperature of 400 K and a volume of 20 liters. Assume that the pressure, volume, and temperature behavior of Kr is described by the van der Waals equation of state. What would be the percent error in the pressure if the ideal-gas equation of state were used instead of the van der Waals equation? If the volume were 4 liters instead of 20, what error would result from the use of the ideal-gas equation? Is the ideal-gas equation more accurate at low pressure or high pressure?

Solution

This problem involves no more than substitution of values into the van der Waals equation of state. However, care must be exercised to use a consistent set of units. Using Eq. 1.60, we have

$$P = \frac{RT}{[V_m - b]} - \frac{a}{V_m^2}. \tag{1}$$

Substitution of values from Table 1.1 gives

$$P = \frac{(0.083145 \text{ L bar mol}^{-1}\text{ K}^{-1})(400 \text{ K})}{(20 - 0.0396)\text{L mol}^{-1}} - \frac{2.325 \text{ L}^2\text{ bar mol}^{-1}}{20^2 \text{ L}^2\text{ mol}^{-2}} = 1.660 \text{ bar}. \tag{2}$$

If we were to use the ideal-gas equation of state, the result would be

$$P_{\text{ideal}} = \frac{RT}{V_m} = \frac{(0.083145 \text{ L bar mol}^{-1}\text{ K}^{-1})(400 \text{ K})}{(20) \text{ L mol}^{-1}} = 1.663 \text{ bar}. \tag{3}$$

The percent difference (error) between these results is

$$\% \text{ error} = \frac{100(P_{\text{ideal}} - P)}{P} = \frac{100(1.663 - 1.660)}{1.660} = 0.18 \text{ \%}. \tag{4}$$

At a molar volume of 4 liters, the van der Waals pressure would be

$$P_{\text{vdw}} = \frac{(0.083145)(400)}{(4 - 0.0396)} - \frac{2.325}{4^2} = 8.25 \text{ bar}. \tag{5}$$

The ideal-gas pressure at this volume is

$$P_{ideal} = \frac{RT}{V_m} = (0.083145)(400)/(4) = 8.314 \text{ bar}, \tag{6}$$

which corresponds to a percent difference (error) of

$$\% \text{ error} = \frac{100(8.314 - 8.25)}{8.25} = 0.78 \%. \tag{7}$$

Thus, the error is 4.3 times greater at a pressure of 8.314 bar than at a pressure of 1.660 bar. As expected, the ideal-gas equation of state is more accurate at low pressures and large volumes. For a related exercise, see Problem 1.23.

EXAMPLE 1.10

Some equations of state are not easy to invert. That is, it may be easy to cast the equation in the form of Eq. 1.14, but not Eq. 1.15, or vice versa. The van der Waals equation of state is an example. In such cases, we usually must use numerical means to solve the equation. Let Kr be described by the van der Waals equation of state. A container of one mole of Kr is found to have a pressure of 15 bar at a temperature of 350 K. What is the molar volume of Kr under these conditions?

Solution

We cannot solve the van der Waals equation of state for V_m in any simple manner, since it is cubic in V_m. However, we can easily obtain the solution by using numerical procedures. First, we write the equation in the form

$$F(V_m) = (V_m - b)\left(P + \frac{a}{V_m^2}\right) - RT = 0, \tag{1}$$

so that we seek the value of V_m that makes $F(V_m)$ exactly zero. Next, we substitute the data given in Table 1.1 and in the problem:

$$F(V_m) = (V_m - 0.0396)\left(15 + \frac{2.325}{V_m^2}\right) - (0.083145)(350) = 0. \tag{2}$$

We could now proceed to evaluate $F(V_m)$ at successively larger and larger values of V_m, starting at some low value such as 0.01 L mol^{-1}, until $F(V_m)$ is observed to change sign. We then know the correct value lies between the last two tested. Further calculation will refine the result to any desired accuracy. A more efficient procedure is to compute V_m using the ideal-gas equation and test values of V_m in the region predicted by the perfect-gas laws. We shall use this method:

$$V_{ideal} = \frac{RT}{P} = \frac{(0.083145)(350)}{15} \text{ L mol}^{-1} = 1.940 \text{ L mol}^{-1}. \tag{3}$$

Thus, we will evaluate $F(V_m)$ for V_m values in the range $1.8007 \le V_m \le 2.0007$ L mol^{-1} at intervals of 0.002 L mol^{-1}. The results show that $F(V_m = 1.8987$ L mol$^{-1}) = -0.0440$ and $F(V_m = 1.9007$ L mol$^{-1}) = 0.01350$. Hence, the correct value of V_m lies between 1.8987 and 1.9007 L mol^{-1}. We already have about four significant digits of accuracy. If more accuracy is needed, we repeat the search in the aforementioned range. A second search shows that the sign on $F(V_m)$ changes between $V_m = 1.89976$ and 1.89978 L mol^{-1}. We have, therefore, determined the molar volume to five significant digits (1.8998 L mol^{-1}), which is probably more than the data warrant.

EXAMPLE 1.11

The force in the z direction on an object is found to vary with z according to the equation

$$F_z = az^{-3} - bz^{-2} + ce^{-dz} \text{ newtons,}$$

where a, b, c, and d are constants. What are the units on these constants? If we take the reference point for potential energy to be $V(z) = 0$ at $z = \infty$, determine the functional form of $V(z)$. What are the units on $V(z)$ if z is expressed in meters?

Solution

Since F_z is in newtons, all terms in the equation must have units of newtons. If z is in meters, then the constants must have the following units:

a is in newton m^3, b is in newton m^2, and c is in newtons.

The exponent in the third term must be unitless. Thus, d must have the unit m^{-1}.

Using Eq. 1.46, we have

$$F_z = -\left(\frac{\partial V(z)}{\partial z}\right) = az^{-3} - bz^{-2} + ce^{-dz}. \tag{1}$$

Multiplication by ∂z then gives

$$\int \partial V(z) = -\int [az^{-3} - bz^{-2} + ce^{-dz}]\partial z. \tag{2}$$

Integration yields

$$V(z) = \frac{az^{-2}}{2} - bz^{-1} + \left(\frac{c}{d}\right)e^{-dz} + \text{constant.} \tag{3}$$

If we take $V(\infty) = 0$, the constant must be zero, and

$$V(z) = \frac{az^{-2}}{2} - bz^{-1} + \left(\frac{c}{d}\right)e^{-dz}. \tag{4}$$

The units on $V(z)$ are the same as those of each term on the right side of the equation; that is,

$$az^{-2}/2 - \longrightarrow \text{newton m}^3/\text{m}^2 = \text{newton m} = \text{joules,}$$

and

$$bz^{-1} - \longrightarrow \text{newton m}^2/\text{m} = \text{newton m} = \text{joules,}$$

$$\left(\frac{c}{d}\right)e^{-dz} \longrightarrow \text{newton/m}^{-1} = \text{newton m} = \text{joules.}$$

Hence, $V(z)$ is in joules, and all terms have the correct units.

EXAMPLE 1.12

(A) Use the data in Table 1.1 to estimate the value of the second virial coefficient for Ar gas at 300 K. **(B)** Plot $C_2(T)$ as a function of T, assuming that Ar obeys the van der Waals equation of state. **(C)** At what temperature do the intermolecular forces become predominately repulsive? This will occur whenever $C_2(T)$ becomes positive.

Solution

(A) Equation 1.66 shows that a virial expansion of the van der Waals equation of state gives

$$C_2(T) = b - \frac{a}{RT}. \tag{1}$$

Using the van der Waals parameters for Ar from Table 1.1, we compute

$$C_2(T) = 0.03201 \text{ L mol}^{-1} - \frac{1.355 \text{ L}^2 \text{ bar mol}^{-2}}{(0.083145 \text{ L bar mol}^{-1} \text{ K}^{-1})(300 \text{ K})} \qquad (2)$$

$$= -0.02231 \text{ L mol}^{-1}.$$

(B) The plot is shown in Figure 1.14.

(C) The temperature at which $C_2(T)$ becomes zero can be obtained from Eq. 1:

$$C_2(T_b) = b - \frac{a}{RT_b} = 0, \qquad (3)$$

so that

$$T_b = \frac{a}{bR} = \frac{1.355 \text{ L}^2 \text{ bar mol}^{-2}}{(0.03201 \text{ L mol}^{-1})(0.083145 \text{ L bar mol}^{-1} \text{ K}^{-1})} = 509.1 \text{ K.} \qquad (4)$$

This temperature is called the Boyle temperature. At temperatures above T_b, the second virial coefficient becomes positive and the attractive interactions become less important than the repulsive forces.

For related exercises, see Problems 1.24, 1.25, 1.26, and 1.27.

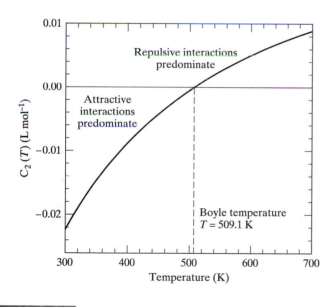

◀ FIGURE 1.14
Second virial coefficient for a van der Waals gas.

1.3.6   The Law of Corresponding States

The ideal-gas equation of state is the most frequently used equation of state in spite of being the least accurate. There are three reasons for this situation: (1) The equation is reasonably accurate if $P < 10$ bar and $T > 100$ K, (2) the simple mathematical form of the ideal-gas equation is very appealing, and (3) the equation is identical for all gases. Consequently, there is no need to investigate the pressure, volume, and temperature behavior of the system if the ideal-gas equation is to be employed. The law of corresponding states permits us to enjoy the third of these advantages while retaining a more accurate expression for the pressure, volume, and temperature behavior of the gas.

We first need to define the concept of a reduced variable. Let X represent a variable such as pressure, temperature, volume, voltage, or distance.

Clearly, X will have units appropriate to the quantity it represents. A reduced variable X_R always has the form

$$X_R = \frac{X}{X^*},$$ (1.67)

where X^* is some specific value of X that usually has a special significance. Since the numerator and denominator of Eq. 1.67 have identical units, the reduced variable will always be unitless. In the case of pressure, volume, and temperature, we define their reduced form by

$$P_R = \frac{P}{P_c}, \qquad V_R = \frac{V}{V_c}, \qquad \text{and} \qquad T_R = \frac{T}{T_c},$$ (1.68)

where P_c, V_c, and T_c are the critical pressure, volume, and temperature, respectively.

Since there exists a linear relationship between pressure, volume, and temperature and the reduced variables, it is clear that any equation of state can be expressed in terms of the reduced variables rather than P, V, and T. When this transformation is made, it is often observed that the equation of state becomes identical for every gas. This is the law of corresponding states: *When expressed in terms of reduced variables, the equation of state becomes identical for all gases.* Let us see how this law works in principle.

Consider a real gas at the inflection point of its critical isotherm, such as that illustrated in Figure 1.8. At this point, where $P = P_c$, $V = V_c$, and $T = T_c$, the equation of state must satisfy three constraints. First, the equation itself must be satisfied by the critical variables. Then, at an inflection point, we must also have

$$\frac{\partial P}{\partial V} = 0 \text{ and } \frac{\partial^2 P}{\partial V^2} = 0 \text{ for } P = P_c, V = V_c, \text{ and } T = T_c.$$ (1.69)

Now consider a two-parameter equation of state. If we adjust the parameters and the ideal-gas constant so as to satisfy the three constraints, the equation of state will no longer contain any parameters. That is, it will become identical for all gases, and the law of corresponding states will hold exactly.

As an example, consider a set of gases described by van der Waals equations of state. From Eq. 1.60, we have

$$P = \frac{RT}{V - b} - \frac{a}{V^2},$$ (1.70)

where, for simplicity, we have written V for the molar volume V_m. The first and second partial derivatives of the pressure with respect to V are, respectively,

$$\frac{\partial P}{\partial V} = -\frac{RT}{(V - b)^2} + \frac{2a}{V^3}$$ (1.71)

and

$$\frac{\partial^2 P}{\partial V^2} = \frac{2RT}{(V - b)^3} - \frac{6a}{V^4}.$$ (1.72)

At the critical point, we must have

$$P_c = \frac{RT_c}{V_c - b} - \frac{a}{V_c^2},$$ (1.73)

$$-\frac{RT_c}{(V_c - b)^2} + \frac{2a}{V_c^3} = 0,$$ (1.74)

and

$$\frac{2RT_c}{(V_c - b)^3} - \frac{6a}{V_c^4} = 0.$$ (1.75)

From Eq. 1.74, we may write

$$\frac{RT_c}{(V_c - b)^2} = \frac{2a}{V_c^3}.$$ (1.76)

Rearrangement of Eq. 1.75 gives

$$\frac{2RT_c}{(V_c - b)^3} = \left[\frac{RT_c}{(V_c - b)^2}\right]\left[\frac{2}{(V_c - b)}\right] = \frac{6a}{V_c^4}.$$ (1.77)

Substitution of Eq. 1.76 into Eq. 1.77 produces

$$\left[\frac{2}{(V_c - b)}\right]\left[\frac{2a}{V_c^3}\right] = \frac{6a}{V_c^4}.$$ (1.78)

Simplification of Eq. 1.78 yields

$$\frac{4}{(V_c - b)} = \frac{6}{V_c}.$$ (1.79)

Solving Eq. 1.79, we obtain $V_c = 3b$. That is, the van der Waals parameter b is one-third the critical molar volume. We may now substitute this result into Eq. 1.76 to obtain

$$\frac{RT_c}{4b^2} = \frac{2a}{27b^3},$$

which we may write as

$$RT_c = \frac{8a}{27b}.$$ (1.80)

Using Eq. 1.80 and $V_c = 3b$ in Eq. 1.73, we obtain

$$P_c = \frac{4a}{27b^2} - \frac{a}{9b^2} = \frac{a}{27b^2}.$$ (1.81)

Equation 1.81 may be written in the form

$$a = 27P_c b^2 = 27P_c \left[\frac{V_c}{3}\right]^2 = 3P_c V_c^2.$$ (1.82)

Thus, from Eq. 1.80, we have

$$RT_c = \frac{24P_c V_c^2}{9V_c} = \frac{8P_c V_c}{3}.$$ (1.83)

Equation 1.83 gives us the value of R in terms of P_c, V_c, and T_c:

$$R = \frac{8P_cV_c}{3T_c}. \tag{1.84}$$

Substitution of the values of b, a, and R in terms of the critical variables into the van der Waals equation of state yields

$$\left[P + \frac{3P_cV_c^2}{V^2}\right]\left[V - \frac{V_c}{3}\right] = RT = \frac{8P_cV_cT}{3T_c}. \tag{1.85}$$

Division of Eq. 1.85 by P_cV_c and substitution of the reduced variables gives

$$\left[P_R + \frac{3}{V_R^2}\right]\left[V_R - \frac{1}{3}\right] = \frac{8T_R}{3}. \tag{1.86}$$

Equation 1.86 contains no parameters. It is identical for all gases that obey a van der Waals equation of state. Thus, if we are permitted to treat R as a parameter, the law of corresponding states holds exactly for all van der Waals gases.

Figure 1.15 shows isotherms for a van der Waals gas in terms of the reduced variables. The curves are obtained directly from Eq. 1.86. The critical isotherm with $T_R = 1.0$ exhibits the expected inflection point, which is marked with an open circle in the figure. At this point, we have $P_R = V_R = T_R = 1.0$. The isotherms with $T_R \geq 1.0$ closely resemble those seen in Figure 1.8 for CO_2. Those with $T_R < 1.0$ qualitatively reflect the measured data above the condensation point and after the gas has condensed. However, the discontinuity in the experimental data produced by the phase transition is replaced by a spurious oscillation. If we wish to use a van der Waals equation of state to describe both the liquid and gas phases, the oscillation must be replaced with a horizontal line. This is generally done by requiring the horizontal line to lie in a position that makes the total area above and below the oscillations equal, as shown in Figure 1.16A. The resulting isotherm, shown in Figure 1.16B, is called a Maxwell construction.

As a test of the accuracy of Eq. 1.86, we may compute the value of R predicted by Eq. 1.84 for several gases. If all gases are indeed van der Waals gases, the result will be the ideal-gas constant in each case. Table 1.3 shows typical results.

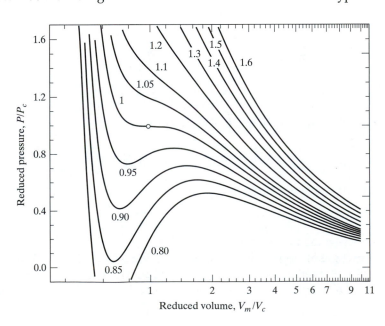

▶ FIGURE 1.15
Van der Waals isotherms in terms of reduced pressure and volume for several reduced temperatures. The open circle marks the position of the inflection point in the critical isotherm with $T_c = 1$. A comparison of the general shape of the isotherms with those of Figure 1.8 shows that the van der Waals equation of state qualitatively resembles the experimental data, except in the region of the phase transition from the gaseous to the liquid state.

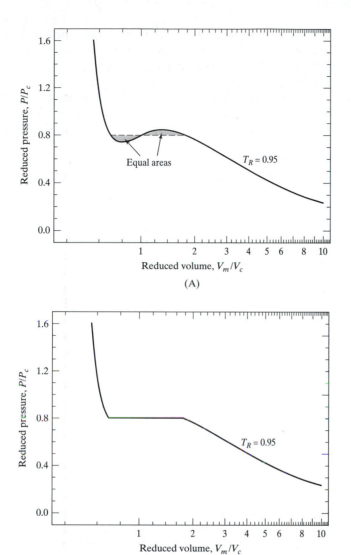

◀ **FIGURE 1.16A**
A Maxwell construction replacing the oscillations of the van der Waals equation of state in the phase transition region with a horizontal straight line drawn so as to make the total area above and below the line equal. The position of this line must be determined by numerical calculation.

◀ **FIGURE 1.16B**
A van der Waals representation of the $T_R = 0.95$ isotherm after the execution of a Maxwell construction.

Table 1.3 Critical constants and R values predicted by Eq. 1.84

Gas	T_c (K)	P_c (atm)	V_c (L mol^{-1})	$R = \dfrac{8 P_c V_c}{T_c}$ (L atm mol^{-1} K^{-1})
Ar	151.15	47.7	0.0752	0.0633
CO_2	304.3	73.0	0.0957	0.0612
C_2H_2	309.15	62	0.1127	0.0603
Ne	44.45	25.9	0.0417	0.0648

The average absolute percent error in R is about 24%. This reflects the fact that real gases do not obey the van der Waals equation of state at the critical point. Nevertheless, the law of corresponding states is a useful concept that represents a significant improvement over the ideal-gas equation of state.

In principle, any two-parameter equation of state will obey the law of corresponding states if R is treated as a parameter and expressed in terms of the critical constants. The problems at the end of the chapter provide several examples. If the equation of state contains three parameters, these parameters may be utilized to satisfy the three constraints at the critical point and thereby eliminate them. However, the resulting equation will usually contain the critical variables themselves and, therefore, not be identical for all gases. Problem 1.30 provides an illustrative example.

Finally, it should be noted that while Eq. 1.86 serves as an illustration of the law of corresponding states, it is usually not as accurate as the van der Waals equation itself [Eq. 1.57]. The reason for this is that Eq. 1.86 is obtained by requiring that the van der Waals constants, a and b, be adjusted to fit the critical point exactly. As can be seen from the data in Table 1.3, this point generally corresponds to one of high pressure and low temperature and volume. Consequently, it is far removed from the ideal-gas regime. Therefore, we expect that Eq. 1.57 will represent the pressure, volume, and temperature behavior more accurately when the system is at lower pressures, provided that a and b have been obtained by fitting pressure, volume, and temperature data in this region.

EXAMPLE 1.13

Measurement shows that at 300 K, Ar has a volume of 17.0 liters at a pressure of 1.115 atm. When Ne is at 88.23 K and a pressure of 0.6055 atm, what volume will Ne have, according to the law of corresponding states?

Solution

Using the data in Table 1.3, we may compute the reduced variables for Ar as

$$(T_R)_{Ar} = \frac{T}{T_c} = \frac{300 \text{ K}}{151.15 \text{ K}} = 1.985, \tag{1}$$

$$(P_R)_{Ar} = \frac{P}{P_c} = \frac{1.115 \text{ atm}}{47.7 \text{ atm}} = 0.0234, \tag{2}$$

and

$$(V_R)_{Ar} = \frac{17.0 \text{ L}}{0.0752 \text{ L}} = 226. \tag{3}$$

For Ne, the corresponding reduced temperature and pressure are

$$(T_R)_{Ne} = \frac{88.23 \text{ K}}{44.45 \text{ K}} = 1.985 \tag{4}$$

and

$$(P_R)_{Ne} = \frac{0.6055 \text{ atm}}{25.9 \text{ atm}} = 0.0234. \tag{5}$$

Since the reduced pressure and temperatures for Ar and Ne are identical, the law of corresponding states requires that their reduced volumes also be identical:

$$(V_R)_{Ar} = (V_R)_{Ne} = 226 = \frac{V_{Ne}}{(V_c)_{Ne}} = \frac{V_{Ne}}{0.0417 \text{ L}}. \tag{6}$$

This gives $V_{Ne} = (226)(0.0417 \text{ L}) = 9.42 \text{ L}$.

1.3.7   The Residual Volume

In the ideal-gas regime at low pressure, where the volume occupied by the molecules becomes negligible and the intermolecular forces approach zero, we expect the ideal-gas equation of state to hold. That is, we expect to observe that

$$\lim_{P \to 0} \left[\frac{RT}{PV_m} \right] = 1. \tag{1.87}$$

For an ideal-gas, we also expect to have

$$V_m - \frac{RT}{P} = 0 \tag{1.88}$$

for all values of P, V, and T. However, we do not expect to see Eq. 1.88 hold for a real gas *even when it is in the ideal-gas region where Eq. 1.87 is satisfied.*

   Although the previous statement may seem to be a contradiction, it is not: It is quite common for the ratio of two functions to approach a limiting value of unity, but their difference not to approach a limiting value of zero. As an example, consider two simple functions

$$f(x) = x + 5$$

and

$$g(x) = x + 3.$$

Then, in the limit as $x \to \infty$, we clearly have

$$\lim_{x \to \infty} \left[\frac{f(x)}{g(x)} \right] = \lim_{x \to \infty} \left[\frac{x + 5}{x + 3} \right] = 1.$$

However, we do not have

$$\lim_{x \to \infty} [f(x) - g(x)] = 0,$$

since $f(x) - g(x) = (x + 5) - (x + 3) = 2$. Thus, a limiting ratio may be unity, while the limiting difference is nonzero. This is exactly the situation that exists for real gases: The ratio of V_m to RT/P is unity as the pressure approaches zero, but the difference between these quantities is not zero.

   The nonzero difference between V_m and RT/P in the limit of zero pressure or infinite volume is defined to be the residual volume

$$\boxed{V_{\text{res}} = \lim_{P \to 0} \left[V_m - \frac{RT}{P} \right].} \tag{1.89}$$

In Chapter 5, it will be shown that the residual volume is required to compute accurate thermodynamic quantities for real gases.

   As an example, let us consider a van der Waals gas whose equation of state is given by Eq. 1.57. To determine the residual volume of this gas, we need an analytic expression for the difference $[V_m - RT/P]$. This can be easily obtained by expanding the left side of Eq. 1.57. Such an expansion produces

$$PV_m - Pb + \frac{a}{V_m} - \frac{ab}{V_m^2} = RT. \tag{1.90}$$

Dividing Eq. 1.90 by P and rearranging terms gives

$$V_m - \frac{RT}{P} = b - \frac{a}{PV_m} + \frac{ab}{PV_m^2}. \tag{1.91}$$

Taking limits of both sides as $P \to 0$, we obtain

$$V_{res} = \lim_{P \to 0} \left[V_m - \frac{RT}{P} \right] = \lim_{P \to 0} \left[b - \frac{a}{PV_m} + \frac{ab}{PV_m^2} \right]. \tag{1.92}$$

From Eq. 1.87, we know that PV_m approaches RT in the limit of low pressure so that Eq. 1.91 may be rewritten as

$$V_{res} = \lim_{P \to 0} \left[b - \frac{a}{RT} + \frac{ab}{RTV_m} \right]. \tag{1.93}$$

Since V_m becomes infinite as the pressure approaches zero, the third term on the right vanishes in the limit, and we obtain

$$V_{res} = b - \frac{a}{RT} \tag{1.94}$$

for a van der Waals gas. It is interesting to note that V_{res} is identical to the second virial coefficient for a van der Waals gas, as shown by Eq. 1.66.

It is always advantageous in science to determine the physical significance of the terms in any equation. An appreciation of the physical significance of the residual volume may be obtained by rewriting Eq. 1.89 in the form

$$V_{res} = \lim_{P \to 0} \left[V_m - \frac{RT}{P} \right] = \lim_{P \to 0} [(V_m)_{real} - (V_m)_{ideal}]. \tag{1.95}$$

Equation 1.95 shows that the residual volume is the difference between the actual volume occupied by the gas and that which the gas would have occupied if it had been ideal. As Eq. 1.94 demonstrates, there are two contributions to this difference: b and $-a/RT$. The first of these is due to the volume occupied by the molecules of the gas themselves. This is zero for an ideal-gas, but is given by the parameter b for a van der Waals gas. Van der Waals molecules occupy a volume of b L mol^{-1}. Consequently, this effect makes a positive contribution to V_{res}. The second contribution to V_{res} arises because of the intermolecular attractive forces that are present for a van der Waals gas, but not for an ideal gas. These attractive forces, whose magnitude is determined by the parameter a, cause the molecules to be pulled closer together. The resulting volume is, therefore, less than would be observed for an ideal-gas, and the contribution to V_{res} is negative, as is seen in Eq. 1.94, which shows that the effect of the intermolecular forces decreases with increasing temperature. This diminution is physically reasonable, since the attractive effect of the intermolecular forces would be less at higher temperatures, where the kinetic energy is large enough to overcome small attractions.

With two conflicting effects in operation, it is possible for the residual volume to be either positive or negative. At low pressures, it is generally observed that the second term is larger in magnitude than the first. As a result, V_{res} is usually negative. Figure 1.14 demonstrates the temperature dependence of V_{res} for a van der Waals gas.

EXAMPLE 1.14

Using the data in Table 1.1, compute the residual volume for Kr at 300 K. At what temperature will the residual volume be zero for this gas?

Solution

Using Eq. 1.94, we may compute

$$V_{res} = b - \frac{a}{RT} = 0.0396 \text{ L mol}^{-1} - \frac{2.325 \text{ L}^2 \text{ bar mol}^{-2}}{(0.083145)(300) \text{ L bar mol}^{-1}} \tag{1}$$

$$= -0.0536 \text{ L mol}^{-1}.$$

Note that the result is negative, as anticipated.

Equation 1 shows that when V_{res} is zero, we must have

$$b = \frac{a}{RT}, \tag{2}$$

or

$$T = \frac{a}{Rb} = \frac{2.325 \text{ L}^2 \text{ bar mol}^{-2}}{(0.083145)(0.0396) \text{ L}^2 \text{ bar mol}^{-2} \text{ K}^{-1}} = 706 \text{ K}. \tag{3}$$

Since V_{res} and the second virial coefficient are identical for a van der Waals gas, this temperature is the Boyle temperature for krypton if we treat it as a van der Waals gas.

For related exercises, see Problems 1.31, 1.32, and 1.33.

1.3.8   The Compression Factor

The residual volume is an experimental measure of the deviation of real gases from ideal behavior in the limit of low pressure. In this limit, the ratio $(PV_m)/(RT)$ approaches unity even though the difference $[V_m - RT/P]$ does not. At higher pressures, $(PV_m)/(RT)$ can deviate significantly from unity. This ratio, which is called the compression factor or compressibility, is frequently used as an experimental measure of the nonideality of real gases. In most texts, the compression factor is given the symbol $Z(T, P)$, where the arguments emphasize the dependence of Z upon temperature and pressure. With this notation, we have

$$\boxed{Z(T, P) = \frac{PV_m}{RT}}. \tag{1.96}$$

Since P/RT is the reciprocal of the molar ideal-gas volume and $V_m/(RT)$ is the reciprocal of the ideal-gas pressure, the compression factor can also be written in the form

$$\boxed{Z(T, P) = \frac{V_m}{V_m^{ideal}} = \frac{P}{P_{ideal}}}, \tag{1.97}$$

which emphasizes the fact that the compression factor is a direct measure of the deviation of a real gas from ideality.

For the same reasons that the residual volume can be either positive or negative, compression factors can be either greater or less than unity. Equation 1.97 shows that when $Z(T, P) < 1$, the actual molar volume of the real gas is less than that of an ideal gas under similar conditions. Compression factors less than unity are usually observed in gas molecules with relatively large intermolecular forces that contract the molecules, thereby producing a volume less than that which would be observed for an ideal gas. Compression factors greater than unity are the result of the nonzero volume of the molecules. When this factor becomes important, compression factors larger than unity are observed.

Figure 1.17 shows measured compression factors for H_2, N_2, and CH_4 as a function of pressure at 273.15 K. The points represent the measured data. The curves are least-squares fits. For H_2, the measured compression factors exceed unity for pressures above 50 atm. They are also greater than unity at much lower pressures. Qualitatively, we should not be surprised by this result, since the very low normal boiling point of H_2 (20.28 K) indicates that the intermolecular forces are very small. Consequently, the nonzero volume of H_2 molecules is the primary reason for deviations from ideality. As a result, we observe that $Z_{H_2}(273.15 \text{ K}, P) > 1$. As the pressure increases, the gaseous volume decreases, and the volume of the molecules becomes an even more important factor. Hence, the deviations from ideality and the compression factor increase with increasing pressure. In contrast, the compression factors for N_2 are very slightly less than unity at lower pressures. This behavior reflects the stronger intermolecular forces present in N_2, which make its normal boiling point (77.35 K) significantly higher than that of H_2. Again, as the pressure increases, the importance of the nonzero molecular volume begins to dominate, and $Z_{N_2}(273.15, P)$ becomes greater than unity around a pressure of 147 atm. With a normal boiling point of 111.668 K, CH_4 exhibits much stronger intermolecular forces than either H_2 or N_2. This leads to compression factors less than unity over the entire pressure range shown in Figure 1.17. At higher pressures, the importance of the nonzero CH_4 molecular volume increases, causing the compression factor to exhibit a minimum around 160 atm. At pressures around 450 atm, $Z_{CH_4}(273.15, P)$ becomes larger than unity.

Figure 1.18 shows the measured compression factors for CO_2 as a function of pressure at temperatures of 304.1 K and 330.7 K. The measured data are shown as points that are connected by straight lines to enhance the visual clarity of the plot. Since CO_2 can be solidified at temperatures around 220 K, we know that the magnitude of its intermolecular forces exceeds those for H_2, N_2, or CH_4. Therefore, we observe compression factors for CO_2 less than unity and significantly less than those for H_2, N_2, or CH_4 under similar conditions.

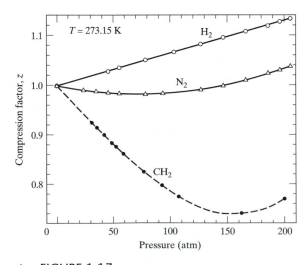

▲ FIGURE 1.17

Measured compression factors for H_2, N_2, and CH_4 at 273.15 K. The points are the experimental data. The solid curves are least-squares fits to the data, which are taken from Landolt-Bornstein, *Physikalisch Chemische Tabellen* (Julius Springer, Berlin, 1931), pp. 43–68.

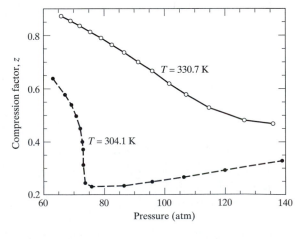

▲ FIGURE 1.18

Measured compression factors for CO_2 at 304.1 K and 330.7 K. The points are the experimental data. The solid curves are straight-line segments drawn between the points to enhance the visual clarity of the graph. Data are taken from Landolt-Bornstein, *Physikalisch Chemische Tabellen* (Julius Springer, Berlin, 1931), pp. 43–68.

Again, the compression factor exhibits a minimum with respect to pressure because of the competing effects of the intermolecular force and the molecular volume. We also see that over the range of pressures shown in Figure 1.18, Z increases with temperature at a given pressure. This type of behavior is similar to that observed for the residual volume, which we found to increase with temperature. At higher temperatures, the molecules possess a larger kinetic energy, which tends to make the intermolecular attractive forces less important. However, at even higher pressures than those shown in Figure 1.18, the opposite trend is often observed. This is because, at very high pressures, the molecules are sufficiently close that, on average, the intermolecular forces are repulsive rather than attractive. This reversal produces the reversal in the qualitative variation of the compression factor with temperature.

It is instructive to examine the compression factor for a van der Waals gas. Multiplying Eq. 1.60 by V_m gives

$$PV_m = \frac{RTV_m}{V_m - b} - \frac{a}{V_m}.$$

Dividing both sides by RT produces the compression factor:

$$Z(T, P) = \frac{PV_m}{RT} = \frac{V_m}{V_m - b} - \frac{a}{RTV_m}. \tag{1.98}$$

Equation 1.98 quantitatively expresses the concepts just discussed. As the intermolecular forces increase, a increases, and the compression factor becomes smaller. If the intermolecular forces are attractive, as is assumed by the van der Waals equation of state, increasing the temperature will increase the compression factor, since the negative second term becomes smaller in magnitude at higher temperatures. Equation 1.98 shows that we always expect compression factors greater than unity when a is negligibly small, since the first term is always larger than unity.

Figure 1.19 illustrates the accuracy of the van der Waals equation of state for CO_2 when the equation is assessed in terms of the compression

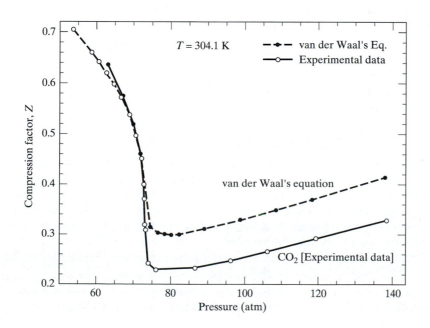

◄ FIGURE 1.19
Comparison of measured compression factors for CO_2 at 304.1 K with values calculated using a van der Waals equation of state with the data given in Table 1.1. The open circles are the experimental data. The solid circles are the computed values. Straight-line segments connect the points to enhance the visual clarity of the graph. Data are taken from Landolt-Bornstein, *Physikalisch Chemische Tabellen* (Julius Springer, Berlin, 1931), pp. 43–68.

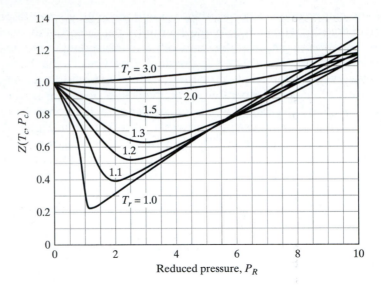

▶ FIGURE 1.20
Average compression factors as a function of reduced pressure at several different reduced temperatures. When plotted in this fashion, the measured compression factors for many gases fall very close to the curves shown in the plot.

factor. The open circles are the experimental data at 304.1 K. These points are connected by solid straight-line segments for clarity. The solid circles are compression factors computed using Eq. 1.98 and the data given in Table 1.1. These points are connected by dashed straight-line segments. To calculate the van der Waals compression factor at a given pressure, we must solve Eq. 1.60 numerically for the molar volume at a fixed pressure and temperature. This result can then be substituted into Eq. 1.98 to obtain $Z(T, P)$. Comparison of the van der Waals result with the data shows that the general character of the compression factor is correctly represented by a van der Waals equation of state, but the computed results lie slightly above the measured data when the pressure exceeds 75 atm.

A better result can be obtained with a virial equation of state or by using the law of corresponding states. When experimental data for the compression factors of various gases are plotted as a function of the reduced pressure rather than the pressure itself, it is found that the results for a wide variety of gases fall on nearly the same curve. Figure 1.20 shows the average compression factors plotted against the reduced pressure at several different reduced temperatures. The measured values for many gases lie very close to these average curves.

For further exercises dealing with the compression factor, see Problems 1.35 and 1.36.

1.4   Linear Least-Squares Analysis

All of the equations of state appearing in Table 1.2 except the ideal-gas equation contain adjustable parameters. These parameters must be determined by fitting the equation to measured pressure, volume, and temperature data on the gas. The question arises as to how this fitting should be done. While there exists no unique method for executing the fitting procedure, most such work utilizes a least-squares method.

Let us assume that we have measured the values of a dependent variable y for N different values of the independent variable x. That is, for $x = x_i$ $(i = 1, 2, 3, \ldots, N)$, we know the corresponding values of y, which we denote as $(y_m)_i$ $(i = 1, 2, 3, \ldots, N)$. In the case of a gas, y might be the volume of the gas and x the pressure at some fixed temperature. Let us further assume that we wish to fit these data to a parametrized equation of the form

$$y = f(x; a_o, a_1, a_2, \ldots, a_K), \qquad \textbf{(1.99)}$$

where the a_i ($i = 0, 1, 2, \ldots, K$) are the parameters that are to be adjusted to fit the data. In a least-squares procedure, this fitting is done by requiring that the a_i be adjusted such that the quantity

$$\sigma^2 = [N(N-1)]^{-1} \sum_{i=1}^{N} [y - y_m]_i^2 \qquad \textbf{(1.100)}$$

be a minimum. The summation in Eq. 1.100 runs over all the data points from 1 to N. The values of y_i and y_{mi} are the calculated and measured values, respectively, of y at the point $x = x_i$, where the calculated value is that given by Eq. 1.99.

The square root of σ^2 is termed the root-mean-square deviation. The fact that the a_i are adjusted so as to minimize the summation of the squared deviations of the values of y computed from Eq. 1.99 from the measured values gives rise to the name "least-squares procedure." The squared deviations are employed rather than the first power of $[y - y_m]$ to prevent fortuitous cancellations of errors. That is, the difference $[y - y_m]$ might be very large in magnitude for many values of x_i, but the summation in Eq. 1.100 fortuitously small due to the cancellation of large negative and positive values. When the squared deviations are employed, such a cancellation cannot occur, since all terms in the sum are positive.

The value of σ^2 is clearly a function of the a_i ($i = 0, 1, 2, \ldots, K$). The general condition for an absolute minimum for σ^2 is that its first partial derivative with respect to each adjustable parameter be zero. That is, we require that

$$\frac{\partial \sigma^2}{\partial a_i} = 0 \text{ for } i = 0, 1, 2, 3, \ldots, K. \qquad \textbf{(1.101)}$$

Equation 1.101 represents a set of $(K + 1)$ equations with $K + 1$ unknown coefficients. Consequently, we can, in principle, solve this set of equations to find the values of the a_i that minimize σ^2 and thereby obtain the "best" fit of Eq. 1.99 to the data. If the fitting function $f(x; a_o, a_1, a_2, \ldots, a_K)$ is complicated, the solution of Eq. 1.101 can be very difficult to determine. Consequently, in many applications, $f(x; a_o, a_1, a_2, \ldots, a_K)$ is chosen to be a power series expansion in x with linear coefficients. That is, we choose

$$y = a_o + a_1 x + a_2 x^2 + a_3 x^3 + \cdots + a_K x^K = \sum_{j=0}^{K} a_j x^j. \qquad \textbf{(1.102)}$$

If this choice is made, the method is termed a linear least-squares fit, where the adjective "linear" refers to the fact that the fitting parameters appear only as multiplicative constants to the first power; the term does not imply that the fit will be a straight line.

When the fitting function has a linear form, it is very easy to solve the set of equations represented by Eq. 1.101. Let us first determine their general form. Direct substitution of Eq. 1.102 into Eq. 1.100 yields

$$\sigma^2 = [N(N-1)]^{-1} \sum_{i=1}^{N} \left[\sum_{j=0}^{K} \{a_j x^j\} - y_m \right]_i^2 \qquad \textbf{(1.103)}$$

where the inner summation runs over the terms in the power series expansion and the outer sum runs over all of the measured data points. To apply Eq. 1.101, we need to obtain the partial derivative of σ^2 with respect to a_n. Because the derivative of a sum is the sum of the derivatives, the derivative and summation operators commute. We may, therefore, differentiate with respect to a_n directly inside the summations. Doing so gives

$$\frac{\partial \sigma^2}{\partial a_n} = 2\left[N(N-1)\right]^{-1} \sum_{i=1}^{N} \left\{ \left[\sum_{j=0}^{K} \{a_j x^j\} - y_m \right]_i x_i^n \right\} = 0, \qquad \textbf{(1.104)}$$

since the only term containing a_n in the second sum has the linear form $a_n x^n$. Although Eq. 1.104 appears to be complicated, it is actually a simple linear algebraic equation. This can be seen by canceling out the multiplicative constants and rearranging the equation into the form

$$\sum_{j=0}^{K} a_j \left[\sum_{i=1}^{N} x_i^{j+n} \right] = \sum_{i=1}^{N} [x^n y_m]_i. \qquad \textbf{(1.105)}$$

Equation 1.105 must hold for all values of n from 0 to K. Consequently, it represents a set of $K + 1$ linear equations whose unknowns are the $K + 1$ fitting parameters $a_0, a_1, a_2, \ldots, a_K$. For example, let us assume that the fitting equation is a quadratic function for which $K = 2$. Then the three equations contained in Eq. 1.105 for $n = 0, 1,$ and 2 are

$$a_o N + a_1 \sum_i x_i + a_2 \sum_i x_i^2 = \sum_i y_{mi} \qquad \text{for } n = 0,$$

$$a_o \sum_i x_i + a_1 \sum_i x_i^2 + a_2 \sum_i x_i^3 = \sum_i x_i y_{mi} \qquad \text{for } n = 1,$$

and

$$a_o \sum_i x_i^2 + a_1 \sum_i x_i^3 + a_2 \sum_i x_i^4 = \sum_i x_i^2 y_{mi} \qquad \text{for } n = 2. \qquad \textbf{(1.106)}$$

In Eq. 1.106, the summations run over all N data points, and we have made use of the fact that $\sum x_i^o$ is equal to N. That is, adding unity to itself N times gives N.

Equation 1.106 may be solved by a variety of methods that include simple substitution techniques, matrix inversion, and the use of determinants. Consider the following two linear equations in two unknowns:

$$5x + 2y = 19;$$

$$4x - y = 10.$$

The value of x that satisfies these equations is the ratio of two determinants. The denominator is the determinant of the coefficients of the variables on the left side of the equation, $\begin{vmatrix} 5 & 2 \\ 4 & -1 \end{vmatrix}$. The numerator is the same determinant, except that the column containing the coefficients of the variable x is replaced with the constants on the right-hand side of the equation. In the example given, this would be $\begin{vmatrix} 19 & 2 \\ 10 & -1 \end{vmatrix}$. Thus, the solution for x is

$$x = \frac{\begin{vmatrix} 19 & 2 \\ 10 & -1 \end{vmatrix}}{\begin{vmatrix} 5 & 2 \\ 4 & -1 \end{vmatrix}} = \frac{-39}{-13} = 3.$$

The solution for y is similarly

$$y = \frac{\begin{vmatrix} 5 & 19 \\ 4 & 10 \end{vmatrix}}{\begin{vmatrix} 5 & 2 \\ 4 & -1 \end{vmatrix}} = \frac{-26}{-13} = 2.$$

This simple method, known as Cramer's rule, not only is useful, but will prove to be important in the development of approximate variational methods to solve quantum mechanical problems that you will encounter later in the text.

Using the preceding method, we may easily write the solution for a_o, a_1, and a_2 in determinant form. For example,

$$a_o = \frac{\begin{vmatrix} \sum y_{mi} & \sum x_i & \sum x_i^2 \\ \sum x_i y_{mi} & \sum x_i^2 & \sum x_i^3 \\ \sum x_i^2 y_{mi} & \sum x_i^3 & \sum x_i^4 \end{vmatrix}}{\begin{vmatrix} N & \sum x_i & \sum x_i^2 \\ \sum x_i & \sum x_i^2 & \sum x_i^3 \\ \sum x_i^2 & \sum x_i^3 & \sum x_i^4 \end{vmatrix}},$$

with similar expressions for a_1 and a_2. Note that in this expression, the sums run over all data points from 1 to N. Inspection of Eq. 1.106 shows that we need to compute seven different sums and the values of four determinants to obtain the fit.

Frequently, the equation to be fitted is a straight line with $K = 1$. In this case, Eq. 1.106 becomes

$$a_o N + a_1 \sum_i x_i = \sum_i y_{mi} \qquad \text{for } n = 0$$

$$a_o \sum_i x_i + a_1 \sum_i x_i^2 = \sum_i x_i y_{mi} \qquad \text{for } n = 1. \qquad \textbf{(1.107)}$$

The intercept of the best straight line is given by

$$a_o = \frac{\sum y_{mi} \sum x_i^2 - \sum x_i \sum x_i y_{mi}}{N \sum x_i^2 - \left\{ \sum x_i \right\}^2}. \qquad \textbf{(1.108)}$$

The slope of the line is

$$a_1 = \frac{N \sum x_i y_{mi} - \sum x_i \sum y_{mi}}{N \sum x_i^2 - \left\{ \sum x_i \right\}^2}. \qquad \textbf{(1.109)}$$

In this case, only four summations must be computed.

Most modern calculators will execute the four summations in Eqs. 1.108 and 1.109 and do the least-squares calculation for you. However, it is important to understand exactly what is being done in the analysis. For example, the minimization of σ^2 weights large values of y much more heavily than smaller values. The reason is that a 1% deviation when y is 1,000 contributes a great deal more to σ^2 than does a similar 1% deviation when $y = 1$. Consequently, the least-squares fitting will naturally tend to give the point where $y = 1,000$ much greater weight than the point where $y = 1$. In order to avoid this and to permit the investigator to assign different weights to points, depending upon the accuracy of measurement at each point, a weighting factor w_i may be assigned to each point. To do this, one simply redefines

$$\sigma^2 = [N(N-1)]^{-1} \sum_{i=1}^{N} w_i \left[\sum_{j=0}^{K} \{a_j x^j\} - y_m \right]_i^2.$$ (1.110)

Minimization of σ^2 with respect to a_n now gives

$$\sum_{j=0}^{K} a_j \left[\sum_{i=1}^{N} w_i x_i^{j+n} \right] = \sum_{i=1}^{N} [x^n w y_m]_i.$$ (1.111)

Thus, the general form for all the foregoing equations remains unaltered. We simply replace summations of the form $\sum x_i^n y_{mi}$ with $\sum x_i^n w_i y_{mi}$ and those of the form $\sum x_i^{n+j}$ with $\sum w_i x_i^{n+j}$ in the equations. A commonly used weighting factor in many applications is $w_i = \varepsilon_i^{-2}$, where ε_i is the estimated error in the measurement of $[y_m]_i$.

In some applications, it is desired to fix one or more of the linear fitting coefficients at some value, rather than adjusting them so as to minimize σ^2. For example, an investigator might wish to set the value of a_2 in Eq. 1.102 to A_o. In that case, Eq. 1.101 will contain only K equations, since we will not set $\partial\sigma^2/\partial a_2$ to zero. In addition, in Eq. 1.105, a_2 will be replaced with A_o wherever a_2 appears. As a result, K linear equations with K unknown coefficients will need to be solved, rather than $K+1$ linear equations with $K+1$ coefficients.

A frequently encountered case involves the fitting of a straight line to a set of data where the investigator wishes to set the intercept a_o to a predetermined value A_o. [This case arises frequently because physical chemists love straight lines, and if the line should go through the origin ($A_o = 0$), it becomes a euphoric experience.] If this is done, the first equation in Eq. 1.107 will be eliminated, and the second equation will take the form

$$A_o \sum_i x_i + a_1 \sum_i x_i^2 = \sum_i x_i y_{mi} \qquad \text{for } n = 1.$$ (1.112)

Under these conditions, the slope of the best line is

$$a_1 = \frac{\sum_{i=1}^{N} x_i y_{mi} - A_o \sum_{i=1}^{N} x_i}{\sum_{i=1}^{N} x_i^2}.$$ (1.113)

EXAMPLE 1.15

An investigator measures the distance of an object from a fixed point as a function of time. She obtains the following data:

time, t (min)	distance, d (m)	time, t (min)	distance, d (m)
0.0	5.40	6.0	6.93
1.0	5.97	7.0	7.15
2.0	6.08	8.0	7.46
3.0	6.19	9.0	8.20
4.0	6.61	10.0	8.51
5.0	6.51	11.0	8.82

The investigator wishes to represent these data with a linear function of the form

$$d = a_o + a_1 t.$$

What two linear equations must be solved to obtain the best values of a_o and a_1 using a linear least-squares procedure? What are the values of the summations appearing in these equations? Solve the equations to obtain a_o and a_1. Plot the results of the least-squares fit and the data on the same graph. Compute the value of σ^2 for the fit. What does the least-squares fit predict the speed of the object to be?

Solution

The two equations are given by Eq. 1.104. In this case, we have $N = 12$ data points. The required summations are

$$\sum_{i=1}^{12} t_i = 66.0, \quad \sum_{i=1}^{12} t_i^2 = 506.0, \quad \sum_{i=1}^{12} d_i = 83.83, \quad \text{and} \quad \sum_{i=1}^{12} t_i d_i = 502.92.$$

The two linear equations that must be solved are, therefore,

$$12a_o + 66.0a_1 = 83.83 \tag{1}$$

and

$$66.0a_o + 506.0a_1 = 502.92. \tag{2}$$

The solution may be obtained from Eqs. 1.105 and 1.106:

$$a_o = \frac{[(83.83)(506.0) - (66.0)(502.92)]}{[(12)(506.0) - (66.0)^2]} = \frac{9225.26}{1716} = 5.38; \tag{3}$$

$$a_1 = \frac{[(12)(502.92) - (83.83)(66.0)]}{[(12)(506.0) - (66.0)^2]} = \frac{502.26}{1716} = 0.293. \tag{4}$$

The best straight line is, therefore,

$$d = 5.38 + 0.293t.$$

Figure 1.21 shows a plot of this result compared to the data itself.

The value of σ^2 is obtained from Eqs. 1.100 and 1.102 with $K = 1$:

$$\sigma^2 = [N(N-1)]^{-1} \sum_{i=1}^{12} [a_o + a_1 t_i - d_i]^2. \tag{5}$$

Setting $N = 12$ and using the values of a_o and a_1 just obtained, we compute

$$\sigma^2 = \frac{0.5358}{(12)(11)} = 0.00406. \tag{6}$$

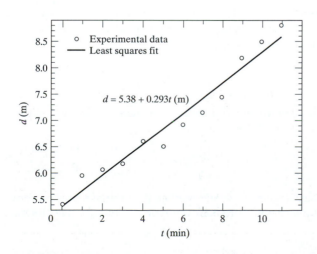

◀ FIGURE 1.21
An example of a least-squares fit.

The speed is given by the derivative of the distance, dd/dt. Using the straight-line fit, we obtain

$$\text{speed} = \frac{dd}{dt} = a_1 = 0.293 \text{ m min}^{-1}. \tag{7}$$

For related exercises, see Problems 1.15 and 1.33.

Summary: Key Concepts and Equations

1. Pressure is the result of molecular collisions with the walls of the container. The product

$$PV = mv^2$$

   is twice the kinetic energy of molecules moving in one dimension and is two-thirds the kinetic energy of molecules moving in three-dimensional space.

2. Boyle's law, $P = k(T)V^{-1}$, can be combined with Charles' law, $V = C(P)T$, by separating the variables to obtain

$$\frac{PV}{T} = \text{a constant,}$$

   or

$$\frac{P_1V_1}{T_1} = \frac{P_2V_2}{T_2}$$

   if the quantity of gas remains constant. When we combine Boyle's law with Avogadro's law, $V = f(T, P)n$, we obtain the ideal-gas equation of state

$$PV = nRT.$$

   This equation implicitly assumes that there are no interaction forces between the molecules and that the molecular volume is zero.

3. The coulombic force acting between particles with charges q_1 and q_2 is

$$F_r = \frac{q_1q_2}{4\pi\varepsilon_o r^2}.$$

   The force is the negative of the first derivative of the potential energy with respect to the distance, $F_r = -\partial V(r)/\partial r$, so that the force always tends to move the system in the direction of lower potential energy. The coulombic potential between point charges is

$$V(r) = \frac{q_1q_2}{4\pi\varepsilon_o r}.$$

   The potential between two identical collinear dipoles with dipole moment μ is approximately

$$V(r) = -\frac{\mu^2}{2\pi\varepsilon_o r^3}.$$

   When molecules approach one another, the mutual repulsion between their electronic charge clouds produces temporary dipoles. The interaction of these dipoles results in a net attractive force that is called a London dispersion force. Such forces generally vary with the inverse sixth power of the dipole separation:

$$V(r) = -\frac{C}{r^6}.$$

4. The most accurate equation of state is the virial equation

$$P = RT\,[V_m^{-1} + C_2(T)\,V_m^{-2} + C_3(T)\,V_m^{-3} + \cdots\,].$$

The virial coefficients $C_n(T)$ are functions of temperature. Consequently, the virial equation of state contains an infinite number of parameters. Most of the information concerning attractive intermolecular forces is contained in the second virial coefficient, $C_2(T)$.

5. The van der Waals equation of state

$$\left(P + \frac{n^2 a}{V^2}\right)(V - nb) = nRT$$

is a two-parameter equation of state in which the parameters, a and b, account, in an approximate manner, for the presence of attractive interactive forces between the molecules and the molecular volume, respectively. A van der Waals gas has a second virial coefficient equal to $b - a/RT$.

6. Any variable X may be put in reduced form by division by a specific value of X, X^*, which usually has some special significance. The reduced variable,

$$X_R = \frac{X}{X^*},$$

is always unitless.

7. The law of corresponding states asserts that, when expressed in terms of reduced pressure, volume, and temperature variables, an equation of state becomes identical for all gases. If R is expressed in terms of the critical constants, any two-parameter equation of state may be expressed in a reduced form that is the same for all gases. However, the accuracy of the reduced form is generally less than that for the equation expressed in terms of the pressure, volume, and temperature variables.

8. The ratio $(RT)/(PV_m)$ approaches unity for all real gases as $P \to 0$. However, the difference $[V_m - RT/P]$ does not approach zero in this limit. This residual volume is defined as

$$V_{res} = \lim_{P \to 0}\left[V_m - \frac{RT}{P}\right].$$

Physically, V_{res} represents the difference between the volume occupied by the real gas and the volume it would occupy if it were ideal.

9. The compression factor $Z(T, P)$ is defined to be the ratio $(PV_m)/(RT)$. Since $P/(RT)$ and $V_m/(RT)$ are the reciprocals of the ideal molar volume and the ideal pressure, respectively, we can express the compression factor in the form

$$Z(T, P) = \frac{V_m}{V_m^{ideal}} = \frac{P}{P_{ideal}}.$$

Therefore, $Z(T, P)$ is a direct measure of the deviation from ideal behavior. Ideal gases have $Z(T, P) = 1$ at all T and P. Nonideal-gases with strong attractive forces usually have compression factors less than unity, except at high pressure, where they become greater than unity due to effect of the molecular volume and repulsive intermolecular forces.

10. A linear least-squares fit is accomplished by minimizing

$$\sigma^2 = [N(N-1)]^{-1} \sum_{i=1}^{N} [y - y_m]_i^2$$

with respect to the linear coefficients in a series expansion of y in terms of the independent variable x:

$$y = \sum_{j=0}^{K} a_j x^j.$$

The equations that must be solved to obtain the fitting coefficients a_j are linear algebraic equations.

Problems

Problems that require the use of some type of computational device are marked with an asterisk (*). Problems that require some type of plotting routine are indicated with a pound sign (#). Unless otherwise stated, all gases may be assumed to behave ideally.

1.1 One mole of Ar atoms are confined in a vessel whose volume is 1,000 cm³. If all of the atoms are moving with identical speeds of 230 m s^{-1} in the x, y, and z directions, compute the expected pressure, in kPa, inside the vessel. What is the pressure in atm?

1.2 A right circular cylinder whose radius is 1 cm contains a mixture of Hg ($d = 13.596$ kg L^{-1}) and H_2O ($d = 1.000$ kg L^{-1}), with the mercury on the bottom of the cylinder and the water on the top. It is found that the pressure at the bottom of the cylinder is 1.200 atm. It is also determined that the mass of Hg in the cylinder is eight times the mass of water present. Determine the length of the Hg column, h_m, and the length of the water column, h_w, in the cylinder. (Be careful with the units.)

1.3 It is found that the lengths of two metal rods, denoted A and B, are linear functions of the absolute temperature T. That is,

$$L_A = a_o + a_1 T,$$

where a_o is the length of rod A at $T = 0$ K and a_1 is a positive constant, and

$$L_B = b_o + b_1 T,$$

with similar definitions for b_o and b_1. An investigator now defines two temperature scales based on the lengths of rods A and B, respectively. The temperatures t_A and t_B are defined by

$$t_A = \frac{100 (L_A - L_o)}{(L_{100} - L_o)}$$

and

$$t_B = \frac{100 (L_B - L_o)}{(L_{100} - L_o)},$$

where L_o and L_{100} are the lengths of the rod at the normal freezing and boiling points of water, respectively.

(A) Determine the relationship between t_A and T.

(B) Show that $t_A = t_B$ at all values of T, even if $a_o \neq b_o$ and $a_1 \neq b_1$.

1.4 If it is found that the lengths of the rods in Problem 1.3 are quadratic functions of T; that is,

$$L_A = a_o + a_1 T + a_2 T^2$$

and

$$L_B = b_o + b_1 T + b_2 T^2.$$

Determine the relationship that will exist between t_A and T, and show that, in general, $t_A \neq t_B$.

1.5 Here is a problem for hot-air balloon enthusiasts. An investigator decides to construct a thermometer using the volume of a balloon filled with Ar as a measuring device. She defines her temperature scale by $t = 75(V - V_{25})/(V_{100} - V_{25})$, where V_{25} and V_{100} are the volumes of the balloon at 25°C and 100°C, respectively, when the pressure is 1 atm. The thermometer is placed outside on the ground and slowly heated with a laser beam. The air temperature is 25°C, and the outside pressure is 1 atm. Heating is continued until the investigator notes that her thermometer has risen off the ground and is floating in the air. The balloon itself weighs 4 grams and contains 15 grams of Ar gas. What is the temperature of the thermometer, t, at the point it rises off the ground? Assume that all gases are ideal and that the air is 20% O_2 and 80% N_2 by mass. Ignore the effect of the balloon's elasticity on the pressure and volume of the balloon.

1.6 Suppose the ideal gas laws in a different galaxy far removed from ours are as follows:

1. At constant temperature, pressure is inversely proportional to the square of the volume.
2. At constant pressure, the volume varies directly with the 2/3 power of the temperature.
3. At 273.15 K and 1 atm pressure, one mole of an ideal gas is found to occupy 22.414 liters.

(A) Under these conditions, show that $V^6 P^3/T^4 = a$ constant, and obtain the form of the ideal-gas equation of state in this galaxy. Be certain to give the value of the ideal-gas constant in the galaxy.

(B) The coefficient of thermal expansion is defined to be $\alpha = V^{-1}(\partial V/\partial T)_P$. Obtain α for the ideal-gas equation of state you found in (A) in terms of the temperature alone.

1.7 The coefficient of thermal expansion is defined to be $\alpha = V^{-1}(\partial V/\partial T)_P$.

(A) Obtain an expression for α for an ideal gas.

(B) Show that, for a Dieterici gas, α may be written in the form

$$\alpha = \frac{RV + a/T}{PV^2 \exp\{a/VRT\} - a}$$

where the notation $\exp\{x\}$ means e^x and a is a parameter in the Dieterici equation of state, which is given in Table 1.2.

1.8 25 grams of Ar and 15 grams of HCl gas are mixed in a 10-liter container at 300 K. Determine the partial pressure of each gas and the total pressure inside the container. You may assume the gases to be ideal.

1.9 A container is divided into two compartments. Compartment A holds ideal gas A at 400 K and 5 atm of pressure. Compartment B is filled with ideal gas B at 400 K and 8 atm. The partition between the compartments is removed and the gases are allowed to mix. (It will be shown in later chapters that this mixing produces no change in temperature if the gases are ideal.) The mole fraction of A in the mixture is found to be $25/43 = 0.581395 \ldots$. The total volume of both compartments is 29 liters. Determine the original volumes of compartments A and B.

1.10 The density of an ideal gas is found to be 1.8813 g L^{-1} at 298 K and 1 atm pressure. What is the molar mass of the gas?

1.11 A container is known to hold a pure rare gas. A 1-liter sample of the gas at 298 K and 1 atm pressure is found to weigh 3.427 grams. What gas is inside the container?

1.12 The ponderosity in whams of a sample of umbo is directly proportional to the number of frenks in the sample and to its muckle, measured in fluggas. The constant of proportionality is 8.31 whams frenk^{-1} fluggas^{-1}. How many whams are there in 4 frenks of umbo whose muckle is 120 fluggas? [The author is indebted to Fredrick L. Minn, M.D., Ph.D, for providing this problem and the associated solution.]

1.13 Consider the equation $Z = A[y^2 + x^2]\exp[-xy/a]$, where Z is a function of x and y, while a and A are constants. The notation $\exp[w]$ represents e^w. Z has units of joules, while x and y each have units of meters.

(A) What are the units on the constant A?

(B) What are the units on the constant a?

(C) Is it possible for Z to be given by the function

$$Z = B[y + x^2]\exp[-xy/a],$$

where B is another constant? Explain.

1.14 An investigator is told that a closed vessel holds a pure rare gas. She measures the density of the gas in the vessel at 298 K and 1 atm pressure and finds it to be 0.8252 g liter^{-1}.

(A) What gas does the investigator think is in the vessel?

(B) After turning in her report, she is told that the vessel actually contains a mixture of He and Ar. What percentage of the gas in the container is He?

1.15* The ideal-gas constant is obtained by measuring P–V data for a real gas at a fixed temperature. The ratio PV/T is then computed at each measured pressure. The result is fitted by an appropriate least-squares procedure and extrapolated to zero pressure, at which point the gas will behave ideally. This problem illustrates the procedure. An investigator measures the pressure of 1 mole of a real gas at various volumes at a fixed temperature of 300 K. Her data are as follows:

Volume (liters)	Pressure (atm)
20	1.223046
21	1.165159
22	1.112504
23	1.064403
24	1.020288
25	0.979685
26	0.942189
27	0.907458
28	0.875196
29	0.845150

30	0.817097
35	0.700794
40	0.613473

(A) Compute the apparent value of R at each of the data points.

(B) Use the last six data points (at volumes $V = 27$ liters to $V = 40$ liters) to execute a least-squares fit of the computed value of R to a linear function of the pressure. That is, fit the function

$$R = a_o + a_1 P$$

to the computed values of R at the six lowest pressures.

(C) Using the fitted function, obtain the limit of R as $P \rightarrow 0$.

(D) Plot the fitted function and compare your curve with the measured data points.

1.16 A container of fixed volume holds two ideal gases, denoted as A and B, such that the mole fraction of gas A is $X_A = \frac{1}{3}$. At a given temperature, the pressure in the container, P_1, is measured. Two additional moles of one of the gases are now added to the container at the same temperature. The new pressure, P_2, is measured. It is found that the ratio $P_2/P_1 = 11/9$. How many moles of A and B were originally present in the container?

1.17 All those concerned about the temperature of hell will find this problem interesting. Because the earth's population is increasing in a near-exponential fashion, it is reasonable to assume that the number of souls in hell is also increasing exponentially with time. That is,

$$n = \text{number of moles of souls in hell} = A \exp\{at\},$$

where A and a are positive constants. Since it is likely that at $t = 0$ we had only one soul (the Devil) in hell, we know that $A = 1/6.022 \times 10^{23} \, \text{mol}^{-1} = 1.66 \times 10^{-24}$ mol. It is also reasonable to assume that souls entering hell do not leave. Let us further assume that we may treat a collection of souls as an ideal gas. Under these conditions, the temperature of hell will be given by

$$T = \frac{PV}{nR} = \frac{PV}{AR \exp\{at\}}.$$

If the pressure of hell is constant at 1 atm, as it is on earth (this assumption is reasonable, since there are no suggestions that hell is a place of very high or very low pressure), the temperature will be dependent on how fast hell is expanding as souls enter it.

(A) There have been approximately 10^{10} people on earth since the Devil entered hell. If we assume that 10% of these people have entered hell and that the Devil was thrown into hell 5,000 years ago, determine the value of the constant a. What are the units on a?

(B) Because at $t = 0$ there was only one occupant of hell, the initial volume must have been rather small. The volume of an average house is probably a good estimate. If this house had 2,000 ft^2 of floor space and 8-ft ceilings, we would have $V_o = 16,000 \, \text{ft}^3 = 4.5307 \times 10^5$ liters. Given that the number of souls in hell is rising exponentially, the volume must also be increasing in exponential fashion; that is

$$V = V_o \exp\{bt\}.$$

Use the foregoing assumptions to obtain the temperature of hell as a function of the parameter b and time.

(C) Compute the initial temperature of hell at $t = 0$. This result shows the origin and the inaccuracy of the popular slang phrase "hot as hell."

(D) Note that if we have $b < a$, hell will eventually freeze over. Let $b = a/2$, and compute how many years must elapse before hell freezes over (reaches the freezing point of water).

(E) Using the same value of b as in (D), determine how long it will take hell to reach a temperature of 1 K. How long will it take to reach a temperature of 0.01 K? How long will it take to reach a temperature of 10^{-10} K? How long will it take to reach a temperature of 0 K? The answer to this last question shows that $T = 0$ K is not attainable. We will later show that the second law of thermodynamics prohibits reaching absolute zero in hell or anywhere else.

1.18 It is clear that we have no clue as to the fraction of people whose souls will enter hell. Consequently, let us generalize the results obtained in Problem 1.17 to include this fact. Let the fraction of people whose souls enter hell be denoted by f. This fraction does not include the Devil, whom we know went to hell.

(A) Use the assumptions contained in Problem 1.17 to obtain the temperature of hell as a function of $b, f,$ and time.

(B) Show that if the Devil is unsuccessful in capturing any souls $(f = 0)$, the temperature of hell must remain constant or increase. Perhaps that is why the Devil seeks souls; he has no other way to cool the place.

1.19 The intermolecular forces between two gas molecules are described by the LJ(12, 6) potential given in Eq. 1.55. Show that the potential minimum occurs at $r = \sigma$ and that the well depth, defined as $[V_{LJ}(r = \infty) - V_{LJ}(r = \sigma)]$, is equal to ε.

1.20 If an intermolecular potential of the form given in Eq. 1.54 is employed, obtain an expression for the value of r at the potential minimum as a function of K, C, and n.

1.21 A helium nucleus is separated from an electron by a distance of 5 Å (5×10^{-10} m). If both particles are treated as point charges, compute the potential energy between them in units of joules and kcal mol^{-1}.

1.22[#] Another functional form often used to represent intermolecular forces is a Morse function, given by

$$V_m = D[1 - \exp\{-\alpha(r - r_e)\}]^2,$$

where D, α, and r_e are constant parameters that depend upon the nature of the particles interacting.

(A) Let $D = 100$ kJ mol^{-1}, $\alpha = 1.5$ Å^{-1} (1 Å = 10^{-10} m) and $r_e = 2.0$ Å. Make a careful plot of V_m versus r over the range 1.4 Å $\leq r \leq$ 6.0 Å.

(B) Show that the minimum for a Morse potential occurs at $r = r_e$.

(C) Show that the well depth, as defined in Problem 1.19, is equal to D.

(D) It will later be shown that the vibrational frequency of an oscillator is dependent upon the second derivative of the intermolecular potential, evaluated at the minimum of the potential. Show that $\partial^2 V_m/\partial r^2|_{r=re} = 2\alpha^2 D$.

1.23 What mass of N_2 gas is present in a 50-liter container at 400 K under 20 atm of N_2 pressure if

(A) the gas is ideal and

(B) the gas obeys the van der Waals equation of state?

1.24 Verify that expansion of the van der Waals equation of state in virial form, Eq. 1.64, gives $a_1 = RT$, $a_2 = RTb - a$, and $a_3 = RTb^2$. Prove that the van der Waals parameter a appears only in the second virial coefficient a_2.

1.25 Using the data in Table 1.1, estimate the second and third virial coefficients for CO_2 at 300 K. At what temperature would we expect the second virial coefficient for CO_2 to be zero?

1.26 A nonideal gas obeys the equation of state $PV_m = RT + \alpha P$, where α is a function of T only. An investigator obtains pressure and volume data for this gas at a fixed temperature. Not knowing the actual equation of state, she fits her data to a virial equation of state. In terms of the quantities appearing in the actual equation of state, what will she obtain for the second virial coefficient?

1.27 Consider a gas that obeys the equation of state

$$P = \frac{RT}{V}\exp\left[-\frac{a}{VRT}\right],$$

where a is a constant and the notation $\exp[x]$ means e^x.

(A) Determine the second and third virial coefficients for this gas as a function of a, R, and T.

(B) Determine the residual volume of the gas as a function of a, R, and T.

1.28 A gas obeys the van der Waals equation of state. The critical parameters for this gas are $P_c = 47.7$ atm, $T_c = 151.15$ K, and $V_c = 0.0752$ liter mol^{-1}. Compute the reduced pressure of the gas at 300 K and a volume of 15 liters.

1.29 A gas is represented by the equation of state

$$P = RT\left[\frac{1}{V} + \frac{a}{V^2} + \frac{b}{V^3}\right],$$

where a and b are constants. By requiring that this equation of state satisfy the three constraints at the critical point, express R, a, and b in terms of the critical variables, and put the equation of state in reduced form.

1.30 A nonideal-gas is represented by the equation of state

$$P = RT\left[\frac{1}{V} + \frac{a}{V^2} + \frac{b}{V^3} + \frac{c}{V^4}\right],$$

where a, b, and c are constants. By requiring that this equation of state satisfy the three constraints at the critical point, express the parameters a, b, and c in terms of the critical variables. Show that substitution of these results into the equation of state does not result in a reduced form that is identical for all gases.

1.31 For a gas represented by a virial equation of state, determine the residual volume in terms of the virial coefficients.

1.32 The equation of state of a nonideal gas is $PV_m(1 - \alpha P) = RT$, where α is a function of temperature only. Obtain an expression for the residual volume of this gas as a function of R, T, and α.

1.33[*] The following set of pressure and volume data is obtained for 1 mole of a nonideal gas at 300 K:

V (liters)	P (atm)	V (liters)	P (atm)
2.000	11.297	5.538	4.313
2.590	8.903	6.128	3.909
3.179	7.342	6.718	3.575
3.769	6.246	7.308	3.293
4.359	5.435	7.897	3.052
4.949	4.809	8.487	2.844

9.077	2.663	17.333	1.407
9.667	2.503	17.923	1.361
10.256	2.362	18.513	1.318
10.846	2.235	19.103	1.278
11.436	2.122	19.692	1.240
12.026	2.019	20.282	1.204
12.615	1.926	20.872	1.170
13.205	1.841	21.462	1.138
13.795	1.763	22.051	1.108
14.385	1.692	22.641	1.079
14.974	1.626	23.231	1.052
15.564	1.565	23.821	1.026
16.154	1.508	24.410	1.002
16.744	1.456		

(A) Using a least-squares method, fit the data to a truncated virial equation of state of the form $P = a_o + a_1 y + a_2 y^2$, where $y = V^{-1}$, and obtain the best values of a_o, a_1, and a_2. Be certain to show the equations that are being solved, and give the values of all required sums.

(B) Plot the data and the fit obtained in (A) on the same graph.

(C) Is the expression used in (A) an appropriate equation of state at low pressures and large volumes? Explain.

(D) Set $a_o = 0$ and $a_1 = RT$ in the equation in (A), and use a least-squares procedure to obtain a_2.

(E) Estimate the residual volume for this nonideal gas.

1.34 The compressibility of a gas is defined to be

$$\beta = -V^{-1}\left(\frac{\partial V}{\partial P}\right)_T$$

where the subscript T means that the partial derivative is taken holding T constant.

(A) Show that the compressibility of an ideal gas is P^{-1}.

(B) Derive an expression for β for a van der Waals gas in terms of $P, V, a,$ and b.

(C) Show that in the limit of infinite volume, β for a van der Waals gas reduces to that for an ideal gas.

1.35 (A) Develop an expression for the compression factor for a gas described by a virial equation of state.

(B) Show that $Z(T, P) \to 1$ as $P \to 0$.

1.36 Use a van der Waals equation of state along with the data given in Table 1.1 to compute the compression factor for N_2 at 273.15 K and 128.09 atm. The measured value is 0.9829. Calculate the percent error in your computed value.

1.37 A young scientist has recently broken up with her boyfriend. The former boyfriend is angry and decides to sabotage her research for revenge. It seems that the young lady is doing pressure, volume, and temperature measurements on gases. Late at night, her former boyfriend enters her laboratory and places a spring inside the cylinder she is using in the experiments. (See accompanying figure.)

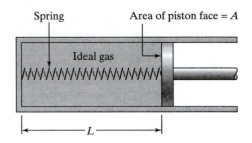

The force on the piston face produced by the spring is

$$F = -k\left[L^{-1} - L_o^{-1}\right] + C\left[L^{-2} - L_o^{-2}\right],$$

where $k, C,$ and L_o are constants. At $L = L_o$, the spring is at equilibrium. The former boyfriend chooses the spring such that when 1 mole of ideal gas is placed in the cylinder at 298 K and 1 atm pressure, the cylinder length is precisely L_o.

The next day, the young scientist enters her laboratory and places 1 mole of an ideal gas in the apparatus at 298 K and 1 atm pressure. She then moves the piston to the right, so as to increase the volume of the gas at constant temperature, and measures the resulting pressure. This procedure is repeated until she has an extensive set of pressure and volume data at 298 K. Being unaware of the presence of the spring, she believes that her gas is behaving in a very nonideal fashion. Therefore, she fits her pressure, volume, and temperature data to a virial equation of state.

(A) Obtain an expression for the apparent "pressure" measured by the investigator as a function of the cylinder volume, the temperature, $k, C, V_o = AL_o$, and A.

(B) When the pressure and volume data are fitted to a virial equation of state, what values, in terms of $T, k, C,$ and V_o, will be obtained for the virial coefficients?

(C) What features of the results will tell the investigator that the data cannot possibly represent the behavior of a nonideal gas?

1.38 The investigator who took the data given in Example 1.15 wishes to set $a_o = 5.500$ m rather than use it as a fitting parameter. Determine the least-squares best value of a_1 under this condition. Compute the value of σ^2 that results from the fit. Is σ^2 larger or smaller than the corresponding value computed in Example 1.15? Explain.

1.39 Don't compute or derive anything. Prepare a brief outline of important points you would discuss if you were presenting a lecture on the van der Waals equation of state. When you're done, relax, have a snack, and get ready to begin your study of the first law of thermodynamics.

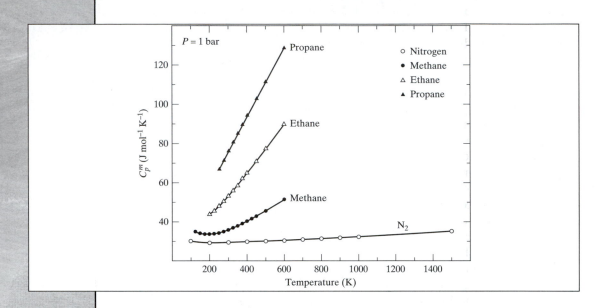

CHAPTER 2

The First Law of Thermodynamics

The first law of thermodynamics is pervasive. There is hardly an area of human existence that is not touched by it. From the electric bills you pay each month to global concerns over the depletion of fossil fuel reserves, it impacts your life. Indeed, it plays a central role in determining the course of history. Together with the second law of thermodynamics, the four postulates of quantum mechanics, the Newtonian laws of classical mechanics, relativity theory and Maxwell's laws of electrodynamics, it represents one of the finest achievements of the human mind.

2.1   The Mathematics of Change

2.1.1   Reference Points and Derivatives

*A*ll *measurements in science are difference measurements.* For example, the measurement of body temperature with a clinical thermometer consists of measuring the difference in length of a column of mercury at two different temperatures. The determination of the height of a table with a meter stick is actually a measurement of the difference between the height of the floor and that of the tabletop. A pressure measurement with a barometer is based upon the difference in force exerted under two different pressures, one of which is often atmospheric pressure. When mass is measured with a two-pan balance, the measurement involves the difference in the amount of weight needed to maintain balance with the object on the right-hand pan and that required when the object is not on the right-hand pan. Since we are making difference measurements, we are concerned primarily with the change in some variable Y, which we write as ΔY.

The consequences of being concerned with ΔY rather than Y itself are profound. One consequence is that we are free to choose our reference point in any manner we desire without fear of biasing the measurement or calculation. For example, consider the measurement of the height of a table, as illustrated in Figure 2.1. The quantity of interest is the height of the tabletop above the floor; we denote that height as d. We could choose our reference point to be the ground. With that choice, we would first need to measure the height of the floor above the ground. Let us say this distance is d_o. Next, we measure the height of the tabletop above the ground. Let this quantity be d_t. The quantity of interest is the difference between the two measured values: $d_t - d_o = d$. The choice of ground level as our reference point is not unique: We could equally well choose the center of the earth as our reference point. The determination of the table height would then require measurements of the distance of the floor and the tabletop from the center of the earth. If these measurements were executed accurately, the difference would again be equal to d. A third choice is to use the floor as the reference point and measure the difference in height directly. This choice also gives the same result. Note, however, that although we get the same answer regardless of our choice of reference point, the measurements are not

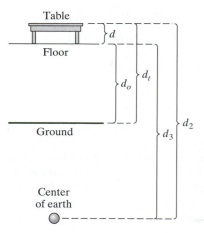

◀ FIGURE 2.1
The invariance of difference measurements with respect to a reference point. The measured height of the table, d, is the same whether we take the floor, the ground, or the center of the earth as our reference point of zero distance. That is, we have $d = d_t - d_o = d_2 - d_3$.

equally easy to make. In fact, if you were to choose the center of the earth as your reference point for the measurement, it would be considered *prima facie* evidence of insanity. The basic principle is to choose the reference point so as to make the measurement or calculation as simple as possible.

The preceding point arises frequently in physical chemistry. In fact, we have already encountered it in the selection of reference points for temperature and the coulombic potential in Chapter 1. Equation 1.48 shows that the coulombic potential between two point charges q_1 and q_2, separated by a distance r is

$$V(r) = -\left[\frac{q_1 q_2}{4\pi\varepsilon_o r}\right] + c, \tag{2.1}$$

where c is a constant. If we choose the reference point to be $V(r) = 10$ joules when $r = 1$ meter, the value of c becomes $c = 10 + [(q_1 q_2)/(4\pi\varepsilon_o)]$ joules. Such a choice would mean we would have to include this cumbersome expression for c in all equations involving $V(r)$. Since c will vanish when differences are taken, it is foolish to use a nonzero value as a reference point. A much more convenient reference is to take $V(r) = 0$ when $r = \infty$. This choice makes $c = 0$.

The foregoing example makes it clear that the value of a quantity Y depends upon the reference point chosen. However, the change in Y, ΔY, produced by some change in the system will always be independent of the choice of reference point.

Let us now determine what information is required to compute ΔY. Consider an automobile traveling down a straight highway, with the distance of the car from its starting point denoted by S. Let us say that at time t_1, $S = 100$ km. If we wish to compute S at some later time t_2, we must know the speed of the automobile as a function of time. That is, we need dS/dt at each point in time. If we know that $dS/dt = f(t)$, then we may compute ΔS simply by integrating both sides of the equation from $t = t_1$ to $t = t_2$:

$$\Delta S = \int_{t_1}^{t_2} \frac{dS}{dt}\, dt = \int_{t_1}^{t_2} f(t)\, dt = S(t_2) - S(t_1) = S(t_2) - 100 \text{ km.}$$

This example demonstrates that the computation of the change produced in a variable Y by a change in the variable x requires that we know the value of the derivative dY/dx. If we are interested in the change in Y produced by a change in x under the condition that some third variable z is held constant, we must know the value of the partial derivative $(\partial Y/\partial x)_z$. Problems in thermodynamics are frequently stated in exactly these terms. For example, we might be asked to determine the change in the internal energy U when the volume of a gas is increased at constant temperature. To solve such a problem, we must obtain the partial derivative $(\partial U/\partial V)_T$. For this reason, the focus of attention in thermodynamics is on derivatives.

EXAMPLE 2.1

A car is 200 km from its starting point 4 hours after beginning its journey. The speed of the car at $t = 4$ hours is 120 km hour^{-1}. At this point, the driver begins to decrease his speed exponentially with time measured from $t = 4$ hours such that at $t = 5$ hours the car's speed is 16.240 km hour^{-1}. Figure 2.2 illustrates the situation. **(A)** If S is the distance the car has traveled, obtain dS/dt as a function of time. **(B)** Compute the distance of the car from its starting point at $t = 6$ hours.

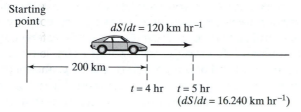

▲ FIGURE 2.2
Illustration of the use of derivatives in the computation of the change in a variable.
In this case, it is the speed, dS/dt, that is the important derivative.

Solution

(A) The critical part of the problem is to obtain the car's speed, which is the derivative dS/dt. By the conditions of the problem, we know that

$$\frac{dS}{dt} = v_o \exp[-a(t - 4)], \tag{1}$$

since we are told that the driver begins to reduce the speed exponentially with time, measured from $t = 4$ hours. At $t = 4$ hours, Eq. 1 gives

$$\left.\frac{dS}{dt}\right|_{t=4} = v_o = 120 \text{ km hour}^{-1}. \tag{2}$$

We also know that at $t = 5$ hours, the derivative is 16.240 km hour^{-1}. Therefore,

$$\left.\frac{dS}{dt}\right|_{t=5} = v_o \exp[-a(5 - 4)] = 120 \exp[-a] = 16.240. \tag{3}$$

Taking logarithms of both sides of Eq. 3 gives

$$\ln(120) - a = \ln(16.240). \tag{4}$$

Therefore,

$$a = \ln\left[\frac{120}{16.240}\right] \text{ hours}^{-1} = 2.0000 \text{ hours}^{-1}. \tag{5}$$

Hence, the required derivative is

$$\frac{dS}{dt} = 120 \exp[-2.0000(t - 4)] \text{ km hour}^{-1}. \tag{6}$$

(B) The total differential of the distance is

$$dS = \frac{dS}{dt} dt = [120 \exp\{-2.0000 (t - 4)\}] \, dt. \tag{7}$$

Integrating both sides between corresponding limits gives

$$\int_{S_1}^{S_2} dS = S_2 - S_1 = \Delta S = \int_4^6 [120 \exp\{-2.0000(t - 4)\}] \, dt$$

$$= \frac{120}{-2.0000} [\exp\{-2.0000(t - 4)\}]_4^6 \tag{8}$$

$$= -60.000[e^{-4} - e^0] = 58.901 \text{ km}.$$

Therefore, at $t = 6$ hours, the total distance of the car from the starting point is

$$S = S(t = 4 \text{ hours}) + \Delta S = 200 + 58.901 = 258.901 \text{ km}. \tag{9}$$

See Problem 2.3 for a related exercise.

Note that, in executing extensive numerical calculations, it is advisable not to round any numbers until all calculations are completed. If rounding to the number of significant digits is done after every operation, the round-off error will accumulate. Accordingly, one should carry extra digits until all computations are completed and then round to the number of digits justified by the data. In this connection, it will be assumed throughout the text that integer data are more accurate than data expressed as a decimal fraction. Therefore, the number of significant digits in the final answer is determined by the data that are expressed as decimal fractions. In the present case, this is the car's speed at $t = 5$ hours, which is 16.240 km hour^{-1}. Consequently, ΔS contains five significant digits.

2.1.2   Total Differentials

We begin by considering a one-variable function $y = f(x)$. If we change x from x_o to $x_o + \Delta x$, we will produce a change in y from y_1 to y_2 as illustrated in Figure 2.3.

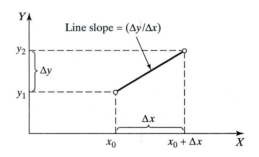

▶ FIGURE 2.3
Change in a variable y produced by a change in the variable x.

If Δx is small, the ratio $\Delta y/\Delta x$ will be a good approximation to the derivative required to compute Δy. In the limit as $\Delta x \rightarrow 0$, the ratio will approach the derivative required for the calculation:

$$\frac{dy}{dx} = \underset{\Delta x \to 0}{\text{Lim}} \frac{\Delta y}{\Delta x}. \tag{2.2}$$

The total differential of y is written as $dy = (dy/dx)\, dx$. Integrating both sides of this equation gives the change in the variable y:

$$\Delta y = \int_{y_1}^{y_2} dy = \int_{x_o}^{x_o + \Delta x} \frac{dy}{dx}\, dx. \tag{2.3}$$

Again, we see that the calculation of Δy requires a knowledge of dy/dx.

The situation becomes considerably more complex when we are dealing with a multivariable function. Let us assume that we have a variable z that is a simultaneous function of x and y. We now ask what change is produced in z if we change x from x_1 to $x_1 + dx$ and y from y_1 to $y_1 + dy$. This process is illustrated in Figure 2.4. Let us assume that we execute the changes in x and y along Paths 1 and 2 as shown in the figure. That is, we first change y from y_1 to $y_1 + dy$ with x held constant at the value x_1 (Path 1). This is followed by changing x from x_1 to $x_1 + dx$ while y is held constant at the value $y_1 + dy$ (Path 2). Under these conditions, we can write

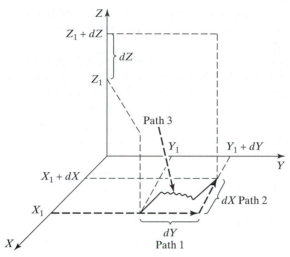

▲ FIGURE 2.4
The change in a variable z produced by simultaneous changes in x from x_1 to $x_1 + dx$ and y from y_1 to $y_1 + dy$.

Change in z along Path 1 = dz_1
  = [rate of change of z with respect to y at constant x] dy
  = $(\partial z / \partial y)_x \, dy$.

For Path 2, we have

Change in z along Path 2 = dz_2
  = [rate of change of z with respect to x at constant y] dx
  = $(\partial z / \partial x)_y \, dx$.

The total change in z is, therefore,

$$dz = dz_1 + dz_2 = \left(\frac{\partial z}{\partial y} \right)_x dy + \left(\frac{\partial z}{\partial x} \right)_y dx. \tag{2.4}$$

The expression for dz in Eq. 2.4 is called the total differential of z. This is the quantity required if we are to compute the change in z. Consequently, for a two-variable function, we need two partial derivatives: $(\partial z / \partial y)_x$ and $(\partial z / \partial x)_y$. If z were dependent upon three variables, we would need three partial derivatives. In general, if z is a function of n variables, $q_1, q_2, q_3, \ldots, q_n$, we have

$$dz = \sum_{i=1}^{n} \left(\frac{\partial z}{\partial q_i} \right) dq_i \, , \tag{2.5}$$

where the partial derivatives are taken with all other variables held constant. Equation 2.5 is the general expression for the total differential of an n-variable function.

Once the total differential is obtained, we can easily compute the desired change in z, Δz, by integrating both sides of Eq. 2.4:

$$\Delta z = \int_{z_1}^{z_1 + dz} dz = \int_{x_1, y_1}^{x_1 + dx, \, y_1 + dy} \left(\frac{\partial z}{\partial y} \right)_x dy + \left(\frac{\partial z}{\partial x} \right)_y dx$$

$$= \int_{y_1}^{y_1 + dy} \left(\frac{\partial z}{\partial y} \right)_x dy + \int_{x_1}^{x_1 + dx} \left(\frac{\partial z}{\partial x} \right)_y dx. \tag{2.6}$$

In using Eq. 2.6, it must be remembered that the integral over y is taken with x constant at the value x_1 and the integral over x is taken with y constant at the value $y_1 + dy$.

If we can obtain the partial derivatives, $(\partial z/\partial y)_x$ and $(\partial z/\partial x)_y$, Eq. 2.6 permits us to calculate the change in z produced by changes in x and y from x_1 to $x_1 + dx$ and y_1 to $y_1 + dy$, respectively, *provided that the changes in x and y are executed along Paths 1 and 2 as shown in Figure 2.4.* But what if we use a different path? Suppose we first change x from x_1 to $x_1 + dx$ with y held constant at y_1 and then execute the variation in y with x held constant at $x_1 + dx$. Will Δz be the same as that computed using Path 1 followed by Path 2? And what if we use a much more complex path, such as Path 3 in Figure 2.4? Will Δz be the same? The answer to this question is of paramount importance in thermodynamics.

2.1.3   Exact and Inexact Differentials

If we are to be successful in the computation of changes in a variable z, we must determine whether Δz depends upon the path describing the changes in the independent variables or whether it is independent of the path. If Δz is independent of the path, we need know only the initial and final states of the independent variables. If, however, Δz is path dependent, we must have detailed knowledge of the precise pathway along which the independent variables are changed.

Let us first consider the differential $dz = y\,dx + x\,dy$. Suppose we wish to compute the change in z that is produced by changing x and y from the point $x = y = 1$ to the point $x = y = 2$. This change can be effected along an infinite number of pathways. One of these is to first change y from 1 to 2 with $x = 1$ and then change x from 1 to 2 with $y = 2$. A second path is to change both variables simultaneously along the path $y = x$. These pathways are shown in Figure 2.5.

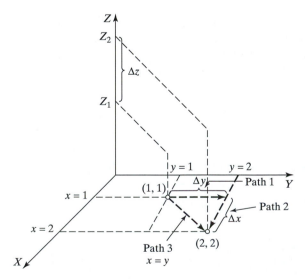

▲ **FIGURE 2.5**
Two possible pathways for changing x and y from the point (1,1) to the point (2,2).

Let us determine Δz along each of these paths. The integral to be evaluated is

$$\Delta z = \int_{x=y=1}^{x=y=2} [y\,dx + x\,dy]. \tag{2.7}$$

If we execute the change in x and y along the sum of Paths 1 and 2 in Figure 2.5, we can write Eq. 2.7 as the sum of two integrals, one along Path 1, the second along Path 2. This yields

$$\Delta z(\text{Path } 1) = \int_{y=1}^{y=2} [y\, dx + x\, dy] = \int_{y=1}^{y=2} x\, dy, \tag{2.8}$$

since the integral over x vanishes along Path 1 with $dx = 0$. (x is constant along this path.) Because x is constant, we factor it out of the integral to obtain

$$\Delta z(\text{Path } 1) = x \int_{y=1}^{y=2} dy = (1)\,[2-1] = 1. \tag{2.9}$$

Along the second segment of the path, we have

$$\Delta z(\text{Path } 2) = \int_{x=1}^{x=2} [y\, dx + x\, dy] = \int_{x=1}^{x=2} y\, dx$$

$$= y \int_{x=1}^{x=2} dx = (2)\,(2-1) = 2, \tag{2.10}$$

since $dy = 0$ along Path 2 and y is constant at the value $y = 2$. The total change in z is, therefore,

$$\Delta z = \Delta z(\text{Path } 1) + \Delta z(\text{Path } 2) = 1 + 2 = 3. \tag{2.11}$$

If we execute the x, y variable change along Path 3 in Figure 2.5 with $x = y$ at all points along the path, we may replace y with x and dy with dx in Eq. 2.7. This gives

$$\Delta z = \int_{x=y=1}^{x=y=2} [x\, dx + x\, dx] = \int_{x=1}^{x=2} 2x\, dx = x^2 \Big|_1^2 = 4 - 1 = 3. \tag{2.12}$$

Consequently, Δz is identical for both paths, and the integral appears to be independent of the pathway used to change x and y.

The underlying reason for the foregoing result can be seen by considering the total differential of the function $z(x, y) = xy + c$, where c is a constant. For this function, we have

$$dz = \left(\frac{\partial z}{\partial x}\right)_y dx + \left(\frac{\partial z}{\partial y}\right)_x dy = y\, dx + x\, dy, \tag{2.13}$$

which is exactly the differential considered before. We can, therefore, write Eq. 2.7 in the form

$$\Delta z = \int_{x=y=1}^{x=y=2} [y\, dx + x\, dy] = \int_{x=y=1}^{x=y=2} dz = z \Big|_{x=y=1}^{x=y=2}$$

$$= [(2)(2) + c] - [(1)(1) + c] = 3, \tag{2.14}$$

which shows that the integral is independent of the path. When this is the case, we say that the differential is *exact*. Consequently, the differential $dz = y\, dx + x\, dy$ is exact in that it is exactly the differential of the function $z(x, y) = xy + c$.

Let us now consider the case $dz = y\, dx$. Suppose we wish to determine the change in z produced by the same changes as those just considered. Along the sum of Paths 1 and 2, we have

$$\Delta z = \Delta z(\text{Path } 1) + \Delta z(\text{Path } 2) = \int_{y=1}^{y=2} y\, dx + \int_{x=1}^{x=2} y\, dx, \tag{2.15}$$

where x is constant in the first integral and y is constant in the second. Since $dx = 0$ in the first integral, the value of that integral is zero. Therefore,

$$\Delta z = y \int_{x=1}^{x=2} dx = 2\,[2 - 1] = 2. \tag{2.16}$$

Along Path 3 with $x = y$, the change in z is

$$\Delta z = \int_{x=1}^{x=2} y\,dx = \int_{x=1}^{x=2} x\,dx = \frac{x^2}{2}\bigg|_{1}^{2} = 2 - \frac{1}{2} = \frac{3}{2}. \tag{2.17}$$

These results show that the change in z now depends upon the pathway along which we change x and y. We term differentials that have this property *inexact* differentials. This nomenclature implies that there exists no function $z(x, y)$ whose differential is exactly $y\,dx$. We will denote inexact differentials with the notation δz; exact differentials will be written dz.

2.1.4   Euler Criterion for Exactness

In computing the change in some variable, it is critically important to know whether its differential is exact or inexact. If it is the former, we need not be concerned with the path along which we do the integration. In fact, we may use an integration path that cannot even physically exist. It doesn't matter: The value of the integral is independent of path; it depends only upon the initial and final states of the independent variables. On the other hand, if the differential is inexact, we must be very careful to execute the integration along the path actually used in the experiment. It is, therefore, important that we have a means to determine the exactness or inexactness of a differential. The Euler criterion provides this.

If the differential $dz = f(x, y)\,dx + g(x, y)\,dy$ is exact, there exists a function $z(x, y)$ whose total differential is exactly $[f(x, y)\,dx + g(x, y)\,dy]$. That is, we must have

$$dz = \left(\frac{\partial z}{\partial x}\right)_y dx + \left(\frac{\partial z}{\partial y}\right)_x dy = f(x, y)\,dx + g(x, y)\,dy. \tag{2.18}$$

For Eq. 2.18 to hold, we must have

$$f(x, y) = \left(\frac{\partial z}{\partial x}\right)_y \tag{2.19}$$

and

$$g(x, y) = \left(\frac{\partial z}{\partial y}\right)_x. \tag{2.20}$$

If we differentiate both sides of Eq. 2.19 with respect to y, holding x constant, we obtain

$$\left(\frac{\partial f(x, y)}{\partial y}\right)_x = \frac{\partial}{\partial y}\left\{\left(\frac{\partial z}{\partial x}\right)_y\right\}_x = \frac{\partial^2 z}{\partial y \partial x}. \tag{2.21}$$

In a similar manner, differentiating both sides of Eq. 2.20 with respect to x, holding y constant, yields

$$\left(\frac{\partial g(x, y)}{\partial x}\right)_y = \frac{\partial}{\partial x}\left\{\left(\frac{\partial z}{\partial y}\right)_x\right\}_y = \frac{\partial^2 z}{\partial x \partial y}. \tag{2.22}$$

Since the order of differentiation does not matter, the right-hand sides of Eqs. 2.21 and 2.22 are equal. Therefore, for dz to be an exact differential, we must have

$$\boxed{\left(\frac{\partial f(x, y)}{\partial y}\right)_x = \left(\frac{\partial g(x, y)}{\partial x}\right)_y.} \qquad (2.23)$$

Equation 2.23 is the *Euler criterion* for exactness. The preceding analysis shows that criterion to be a necessary condition for a differential dz to be exact. It can also be shown that the Euler criterion is a sufficient condition for exactness.

Using the Euler criterion, we may easily show that $[y\, dx + x\, dy]$ is exact, whereas $y\, dx$ is inexact. For the first differential, we have $f(x, y) = y$ and $g(x, y) = x$. Clearly,

$$\left(\frac{\partial f(x, y)}{\partial y}\right)_x = \frac{\partial y}{\partial y} = 1 = \left(\frac{\partial g(x, y)}{\partial x}\right)_y = \frac{\partial x}{\partial x} = 1,$$

and the differential $[y\, dx + x\, dy]$ is exact. On the other hand, for the differential $[y\, dx + 0\, dy]$, we have $f(x, y) = y$ and $g(x, y) = 0$. Now the Euler criterion gives

$$\left(\frac{\partial f(x, y)}{\partial y}\right)_x = \frac{\partial y}{\partial y} = 1 \neq \left(\frac{\partial g(x, y)}{\partial x}\right)_y = \frac{\partial 0}{\partial x} = 0.$$

Therefore, $y\, dx$ is inexact, and we would write the differential in the form $\delta z = y\, dx$.

The points presented in this section are the foundation on which thermodynamics rests; consequently, they should be thoroughly studied and understood before proceeding.

EXAMPLE 2.2

(A) Is the differential $[(3x^2y - 6x)\, dx + (x^3 + 2y)\, dy]$ exact or inexact? **(B)** If it is exact, find the function $z(x, y)$ of which it is the exact differential.

Solution

(A) We may use the Euler criterion to answer this part of the question. We have

$$f(x, y) = 3x^2y - 6x \qquad (1)$$

and

$$g(x, y) = x^3 + 2y. \qquad (2)$$

The two partial derivatives that are required are

$$\left(\frac{\partial f(x, y)}{\partial y}\right)_x = 3x^2 \qquad (3)$$

and

$$\left(\frac{\partial g(x, y)}{\partial x}\right)_y = 3x^2. \qquad (4)$$

Consequently, the Euler criterion is satisfied, and the differential is exact. This means that it may be written in the form

$$dz = \left(\frac{\partial z}{\partial x}\right)_y dx + \left(\frac{\partial z}{\partial y}\right)_x dy = [(3x^2y - 6x)\, dx + (x^3 + 2y)\, dy]. \qquad (5)$$

(B) Equation 5 shows that

$$\left(\frac{\partial z}{\partial x}\right)_y = 3x^2y - 6x. \tag{6}$$

Integration of both sides with respect to x, holding y constant, gives

$$\int \left(\frac{\partial z}{\partial x}\right)_y dx = \int (3x^2y - 6x)\, dx = x^3y - 3x^2 + F(y) = z(x, y), \tag{7}$$

where the "constant" of integration becomes a constant *function* of integration because y is being held constant during the integration. This may be easily verified by taking the partial derivative of the right-hand side of Eq. 7 with respect to x, holding y constant. The result is $3x^2y - 6x$, as required, regardless of the form of $F(y)$. Equation 5 also shows that

$$\left(\frac{\partial z}{\partial y}\right)_x = x^3 + 2y. \tag{8}$$

Integration of both sides with respect to y, holding x constant, gives

$$\int \left(\frac{\partial z}{\partial y}\right)_x dy = \int (x^3 + 2y)\, dy = x^3y + y^2 + G(x) = z(x, y). \tag{9}$$

Since the right-hand sides of Eqs. 7 and 9 are both equal to $z(x, y)$, we must have

$$x^3y - 3x^2 + F(y) = x^3y + y^2 + G(x). \tag{10}$$

This can be true only if $F(y) = y^2 + C$ and $G(x) = -3x^2 + C$, where C is a constant. The required function is, therefore, $z(x, y) = x^3y + y^2 - 3x^2 + C$.
See Problems 2.1 and 2.2 for additional exercises.

2.2   Work and Heat

Thermodynamics is concerned with the determination of the extent to which a process occurs. It also involves the study of energy flow and work. Energy comes in two forms. The first is energy a system possesses by virtue of the motion of its constituent molecules. We term this form *kinetic energy*. The second is energy due to position or chemical composition, which is called *potential energy*. Both forms may be used to produce thermal energy and thereby warm a house. With the right engineering design, it is always possible to generate useful work from the potential energy possessed by a system. Potential energy could, therefore, be used to propel a car down the highway. Kinetic energy, however, can produce work only under certain conditions that we shall examine in detail in Chapter 4.

2.2.1   Work

To produce work, we must displace a system through a distance $d\mathbf{s}$ against a restraining force, $\mathbf{F}_{ext}$, where the boldface indicates a vector quantity. The differential work produced by a differential displacement $d\mathbf{s}$ is defined to be the dot product of the external restraining force and displacement vectors:

$$\delta w = \mathbf{F}_{ext} \cdot d\mathbf{s}. \tag{2.24}$$

Equation 2.24 makes it clear that if we do not displace the system, no work is done, regardless of the force exerted. (Consequently, if you see a muscular giant holding a 200-kg mass over his head, feel free to inform him that he is not doing any work. Of course, you run the risk that the mass will then be on top of *your* head!) We have written the work differential as δw rather than dw because the total work done depends upon the displacement path, which makes the differential inexact. The proof of this statement is provided later in the section.

The dot product of two vectors **A** and **B** is a scalar quantity defined as follows: If θ is the angle between **A** and **B**, then the dot product $\mathbf{A} \cdot \mathbf{B} = A B \cos \theta$. This is illustrated in Figure 2.6.

The preceding definition permits us to write Eq. 2.24 in the form

$$\delta w = F_{ext} \, ds \cos \theta. \tag{2.25}$$

The quantity $F_{ext} \cos \theta$ is the component of the external restraining force in the direction of the displacement, which we write as $\text{comp}_{ds} \mathbf{F}_{ext}$. Consequently, the differential work is given by

$$\delta w = [\text{comp}_{ds} \mathbf{F}_{ext}] \, ds, \tag{2.26}$$

and only the component of force in the direction of the displacement contributes to the work being done. This point is illustrated in Figure 2.7, which shows that the work done when one flies a plane into a crosswind depends only upon the component of external force provided by the crosswind parallel to the direction of flight.

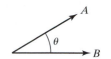

▲ FIGURE 2.6
The dot product of two vectors **A** and **B** is defined to be $\mathbf{A} \cdot \mathbf{B} = AB \cos \theta$, where vectors are indicated by boldface type.

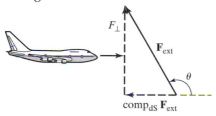

▲ FIGURE 2.7
Work done by a plane flying into a crosswind that exerts an external force equal to $\mathbf{F}_{ext}$ at an angle θ relative to the flight direction. The external force can be broken into two components, one parallel to the flight direction that we label $[\text{comp}_{ds} \mathbf{F}_{ext}]$ and a second component, $F_{\perp}$, that is perpendicular to the flight direction. The differential work is given by $\delta w = \mathbf{F}_{ext} \cdot \mathbf{dS} = [\text{comp}_{ds} \mathbf{F}_{ext}] \, dS = F_{ext} \cos\theta \, dS$. Thus, only the component of the external force parallel to the displacement direction contributes to the work.

In chemical systems, we are generally concerned about two types of work. The first of these is work done when a system expands or contracts under a force produced by external pressure. This type is generally termed *pressure–volume*, or *P–V*, work. When the system contains a source of electrical energy, we shall also be concerned with electrical work. Such systems will be treated in Chapter 9. Here, we shall assume that only *P–V* work is involved. Therefore, we need an expression that gives the work done upon contraction or expansion of a system.

Consider a gas contained in the piston–cylinder apparatus shown in Figure 2.8.

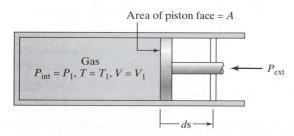

▲ FIGURE 2.8
A gas at an original pressure P_1, volume V_1, and temperature T_1 is expanded against a force produced by an external pressure P_{ext} to a new volume V_2.

The initial internal pressure of the gas in the cylinder is denoted as P_{int}. The initial volume and temperature of the gas are V_1 and T_1, respectively. The piston is displaced to the right by a distance $d\mathbf{s}$ as shown against a restraining force produced by an external pressure P_{ext}. As noted in Eq. 1.1, the force produced by the external pressure is $F_{ext} = AP_{ext}$, where A is the area of the piston face. Using Eq. 2.24, we have

$$\delta w = \mathbf{F}_{ext} \cdot d\mathbf{s} = A\mathbf{P}_{ext} \cdot d\mathbf{s} = AP_{ext}\, ds \cos \pi = -AP_{ext}\, ds, \qquad (2.27)$$

since the angle between the force and displacement vectors is π radians. The quantity $A\, ds$ in Eq. 2.27 is the differential increase in volume produced by the displacement ds, which we write as dV. Therefore, the P–V work may be written in the form.

$$\boxed{\delta w = -P_{ext}\, dV}. \qquad (2.28)$$

In Eq. 2.28, it is important to note that the work is determined by the *external* pressure, not the internal pressure of gas. If the gas were to expand into a vacuum against an external pressure of zero, no work would be done, even though the internal pressure of the gas might be very large.

If the gas expands against the external force, then $dV > 0$ and $\delta w < 0$. That is, if the system expands and does work, such as turning the crankshaft of an automobile, the change in work with respect to the system is negative. It is as if the system lost work to the surroundings. If, however, the external pressure causes a compression of the piston, we have $dV < 0$ and $\delta w > 0$. In this case, the surroundings are doing work on the system, and δw is positive, having "gained" work from the surroundings. This situation is illustrated in Figure 2.9. We conclude that the internal pressure exerts an influence on the process in that its magnitude relative to P_{ext} determines whether we will have expansion ($\delta w < 0$) or compression ($\delta w > 0$).

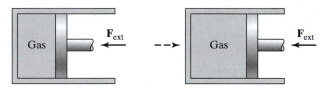

(A) Expansion against an external force:
work done by system on surroundings, $\delta w < 0$

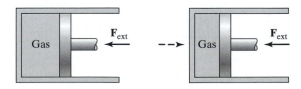

(B) Compression by an external force:
work done by surroundings on system, $\delta w > 0$

▲ FIGURE 2.9
Sign conventions for work.

2.2.2 Irreversible and Reversible Processes

If $P_{int} > P_{ext}$, the piston in Figure 2.8 will expand until the pressures are equalized, and we will observe that $\delta w < 0$. With the piston having expanded and reached the point where $P_{int} = P_{ext}$, nothing further will occur: The piston will not suddenly reverse direction and recompress the gas. Thermodynamically, we say that the expansion is *irreversible*. Furthermore, it is obvious that if $P_{int} > P_{ext}$, expansion of the piston will be spontaneous. Irreversible processes are always spontaneous. If $P_{int} < P_{ext}$, irreversble, spontaneous compression will occur until P_{int} becomes equal to P_{ext} and δw will be positive. Finally, if $P_{ext} = P_{int}$, the system will be in a state of equilibrium, and the piston will not move.

The total work done in the expansion of a gas from volume V_1 to volume V_2 may be obtained by summing all of the differential changes along the expansion path:

$$\int \delta w = -\int_{V_1}^{V_2} P_{ext}\, dV = w. \tag{2.29}$$

As noted, if P_{ext} is zero, $|w|$ will be zero, since the integrand of Eq. 2.29 is zero at all points. As P_{ext} increases, the magnitude of the work done by the system on the surroundings increases. If we wish to maximize $|w|$, we clearly need to maximize P_{ext} at each point along the expansion path. Since the piston will not expand if $P_{ext} > P_{int}$, the maximum value P_{ext} can have at any point during the expansion is $P_{int} - \varepsilon$, where ε is a vanishingly small positive number. Thus, an upper limit to the magnitude of the work that can be obtained by the expansion of a gas, δw_{max}, can be computed by replacing P_{ext} with P_{int} in Eq. 2.28:

$$\boxed{\delta w_{max} = -P_{int}\, dV = -P\, dV}. \tag{2.30}$$

In Eq. 2.30, the notation P_{int} has been replaced with P. By convention, P will always refer to the internal pressure of the gas. If the pressure is to be the external pressure, the subscript "ext" will be specified.

Equation 2.30 gives us the maximum differential work. However, we can never obtain $|w|_{max}$ from any real expansion, since the piston would not expand if $P_{ext} = P_{int} = P$. Instead, the piston would be in equilibrium. Such a hypothetical expansion process that occurs under equilibrium conditions at all points is called a *reversible* process, because the piston is equally likely to move in one direction as the other. The thermodynamic prediction that a process is reversible is equivalent to predicting that the system is in a state of equilibrium.

Since real processes never occur under equilibrium conditions, reversible processes do not take place; they are purely hypothetical concepts. We might, therefore, reasonably ask, Why be concerned with reversible processes if they cannot occur? One plausible answer is that calculation of the reversible work provides us with a useful upper limit to the magnitude of work that can be obtained in the expansion of a gas. Although this is certainly true, it is not the major reason we consider reversible processes. As we shall see, it is much easier to compute changes in a variable along a hypothetical reversible path than to compute similar changes along irreversible, spontaneous paths. If the quantity we are attempting to compute has an exact differential, we can always replace the actual irreversible path with a reversible one without changing the value of the integral. The importance of being able to do this will become apparent as we proceed.

Equation 2.24 implies that the work differential is inexact. It is easy to prove that this is true by writing the reversible work in the form

$$\delta w = -P\, dV + 0\, dP = -P\, dV. \tag{2.31}$$

The Euler criterion for exactness requires that, for the work differential to be exact, we must have

$$-\left(\frac{\partial P}{\partial P}\right)_V = \left(\frac{\partial 0}{\partial V}\right)_P.$$

The left-hand side equals -1, whereas the right-hand side is zero. Consequently, the Euler criterion for exactness is not satisfied, and δw is inexact, as stated. The total work done is, therefore, path dependent.

Let us next consider the example of n moles of an ideal gas being expanded from an initial state (P_1, V_1, T_1) to an arbitrary final state (P_2, V_2, T_1), where $P_2 < P_1$ and $V_2 > V_1$. If δw were an exact differential, the stipulation of the values of $P_1, V_1, T_1, P_2,$ and V_2 would be sufficient to uniquely determine the work, since w would depend only upon the initial and final states and not upon the pathway between them. However, δw is inexact, so we need to specify the path before the calculation can be done. We will examine two possible paths. In the first, Process 1 in Figure 2.10, we stipulate that the path is a reversible isothermal $(dT = 0)$ path. For such a path, we expect P to vary inversely with V, as predicted by the ideal-gas law. This behavior is shown in the figure. The reversible work is given by Eq. 2.31, where

$$P = \frac{nRT_1}{V}. \tag{2.32}$$

Substituting Eq. 2.32 into Eq. 2.31 yields

$$w = -\int_{V_1}^{V_2} \frac{nRT_1}{V}\, dV = -nRT_1 \int_{V_1}^{V_2} \frac{dV}{V} = -nRT_1 \ln\left[\frac{V_2}{V_1}\right]. \tag{2.33}$$

Equation 2.33 gives the reversible isothermal work of expansion of an ideal gas from volume V_1 to volume V_2.

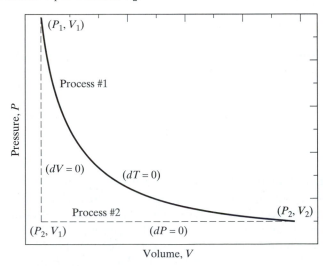

▲ FIGURE 2.10
Gaseous expansion along two different paths.

If the expansion path had not been isothermal and reversible, we would have obtained a different amount of work for the expansion. For example, suppose we first reduce the pressure from P_1 to P_2 by lowering the temperature while keeping the volume constant. Then, once the pressure reaches P_2, we expand the gas reversibly to a final volume V_2 by raising the temperature back to T_1 while holding the pressure constant at the value P_2. This pathway is illustrated by the dashed line in Figure 2.10. For this path, the total work is the sum of the work done in each step:

$$w_{total} = -\int_{V_1}^{V_1} P_{ext}\, dV - \int_{V_1}^{V_2} P_{ext}\, dV. \qquad (2.34)$$

[Step 1]   [Step 2]

In the first step, there was no change in volume ($dV = 0$); consequently, no work is done by the constant-volume cooling of the gas. In the second step, the expansion was specified to be reversible. We may, therefore, replace P_{ext} with P in Eq. 2.34. This gives

$$w_{total} = -\int_{V_1}^{V_2} P\, dV. \qquad (2.35)$$

On the surface, Eq. 2.35 looks identical to Eq. 2.33. For this reason, students of physical chemistry frequently solve Eq. 2.35 incorrectly by proceeding in the manner shown in Eq. 2.33. However, the expansion pathways are different for the two processes. In the first case, which led to Eq. 2.33, the path was isothermal and reversible. In the present case, the path is a reversible constant-pressure path with $P = P_2$. Under these conditions, the solution to Eq. 2.35 is

$$w_{total} = -P_2\int_{V_1}^{V_2} dV = -P_2(V_2 - V_1). \qquad (2.36)$$

Since $-P_2(V_2 - V_1) \neq -nRT_1 \ln[V_2/V_1]$, the total work is dependent upon the path employed to transform the gas from (P_1, V_1, T_1) to (P_2, V_2, T_1).

EXAMPLE 2.3

Five moles of an ideal gas at $P = 10$ atm and $T = 300$ K are expanded reversibly and isothermally to a final volume of 61.545 liters. **(A)** Compute the work done in liter-atm and in joules. **(B)** In a second process, 5 moles of an ideal gas at $P = 10$ atm and 300 K are expanded irreversibly and isothermally against a constant external pressure of 2 atm until equilibrium is reached. Compute the work done in this expansion. **(C)** Is the magnitude of the work done in (A) greater than, less than, or equal to the work done in (B)? **(D)** If δw were an exact differential, what would the answer to (C) have been? Both processes are illustrated in Figure 2.11.

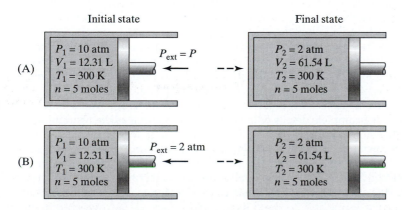

▲ FIGURE 2.11

(A) A reversible isothermal expansion in which $P_{ext} = P$. (B) An irreversible isothermal expansion in which P_{ext} is constant at 2 atm. The expansions are not shown to scale.

Solution

The first step in any problem in thermodynamics is to determine the initial and final states if they are not given. In both (A) and (B), the initial state is $P_1 = 10$ atm, $T_1 = 300$ K, $n = 5$ moles, and

$$V_1 = \frac{nRT_1}{P_1} = \frac{(5\text{ mol})(0.08206\text{ L atm mol}^{-1}\text{ K}^{-1})(300\text{ K})}{(10\text{ atm})} = 12.31\text{ L}. \qquad \textbf{(1)}$$

The final state in both parts is $P_2 = 2$ atm, $T_2 = T_1 = 300$ K (since the expansions are isothermal), $n = 5$ moles, and

$$V_2 = \frac{nRT_1}{P_2} = \frac{(5\text{ mol})(0.08206\text{ L atm mol}^{-1}\text{ K}^{-1})(300\text{ K})}{(2\text{ atm})} = 61.54\text{ L}. \qquad \textbf{(2)}$$

(A) Since the gas is ideal and the expansion is isothermal and reversible, Eq. 2.33 gives the work done:

$$w_{\text{total}} = -nRT_1\ln\left[\frac{V_2}{V_1}\right] = -(5\text{ mol})(0.08206\text{ L atm mol}^{-1}\text{ K}^{-1})(300\text{ K})\ln\left[\frac{61.54}{12.31}\right] \qquad \textbf{(3)}$$

$$= -198.1\text{ L atm}.$$

The conversion factor to joules is the ratio of the gas constants in these units:

$$w_{\text{total}} = -1.98.1\text{ L atm} \times \frac{8.314\text{ J mol}^{-1}\text{ K}^{-1}}{0.08206\text{ L atm mol}^{-1}\text{ K}^{-1}} = -20{,}070\text{ J} = -20.07\text{ kJ}. \qquad \textbf{(4)}$$

(B) The expansion is now irreversible. Consequently, we cannot replace P_{ext} with P. The work must be written as

$$w_{\text{total}} = -\int_{V_1}^{V_2} P_{\text{ext}}\,dV, \qquad \textbf{(5)}$$

and we must know P_{ext} as a function of V in order to execute the integration. In the present case, we are told that P_{ext} is a constant throughout the expansion and that it is equal to 2 atm. This makes the integration easy to do:

$$w_{\text{total}} = -P_{\text{ext}}\int_{V_1}^{V_2} dV = -P_2(V_2 - V_1) = -(2\text{ atm})(61.54 - 12.31)\text{ L}$$

$$= -98.46\text{ L atm} = -9{,}976\text{ J}. \qquad \textbf{(6)}$$

The type of process represented by this part of the problem serves as a useful illustration of the difference between reversible and irreversible processes. However, we usually cannot accurately compute the work associated with such processes by using only thermodynamic methods. Since $P > P_{\text{ext}}$ in (B), the force exerted on the piston by the internal pressure of the gas will exceed the restraining force. There will, therefore, be a net force on the piston, causing it to accelerate. This acceleration will initially move the piston past the equilibrium point at which $P = P_{\text{ext}}$. Once past this point, the net force will be in the opposite direction, and the motion of the piston will reverse itself, setting up an oscillatory motion that will eventually be damped out, bringing the piston to rest at the point $P = P_{\text{ext}}$. Since δw is inexact, an accurate calculation of the work would require that we know the complete details of the acceleration, as well as the restraining force due to the internal pressure, during the oscillatory motion of the piston. This information cannot be obtained from thermodynamics alone. Nevertheless, such examples and problems provide a useful means for gaining a mastery of thermodynamics. Therefore, in all problems involving constant pressure or irreversible expansions or compressions, we shall ignore acceleration effects.

(C) We have $|w_{\text{total}}|$ greater in (A) than in (B). We knew that this would be the result, since the initial and final states of both processes are the same, but the process in (A) is reversible, whereas the process in (B) is not. The maximum magnitude for the work is always obtained for the reversible process.

(D) Although this part is the easiest in terms of amount of calculation that must be done, it is the most important part of the example. If δw were an exact differential, the answers to (A) and (B) would have been identical, since the initial and final states are identical. The pathways, reversible or not, would have made no difference.

For related exercises, see Problems 2.4, 2.5, 2.8, 2.9, and 2.10.

2.2.3   Heat

Thermal energy is a form of kinetic energy in which the center of mass of the system is stationary. This situation is illustrated in Figure 2.12. Both systems shown in the figure possess kinetic energy. However, for System B, the kinetic energy is distributed in such a manner that the center of mass of the entire system is moving. Kinetic energy in this form is termed *translational energy*. In contrast, the kinetic energy in System A is distributed such that the system's center of mass remains stationary. When kinetic energy is distributed in this manner, we call the energy *thermal energy*. When thermal energy flows from one system to another, it is called *heat*, as shown in Figure 2.13.

An important distinction between the systems shown in Figures 2.12 and 2.13 relates to the work that can be obtained from the energy present in the system. With the correct engineering design (for example, placing the gas inside a piston–cylinder arrangement), it is always possible to convert translational energy into work. In sharp contrast, heat can be converted into work only under certain very demanding circumstances that will be examined in detail in Chapter 4.

By convention, we represent heat with the symbol q. The sign convention is chosen to match that for work. When a system "loses" work to the surroundings, δw is negative. We adopt the same convention for heat: If heat escapes to the surroundings, δq will be negative. If heat is added to the system, δq will be positive. With this sign convention, the differential change in heat for System A in Figure 2.13 is negative, while that for System B is positive.

It is important to understand that sign conventions differ from one scientific discipline to another. They also change with time in the same discipline. For example, physicists define the direction of electrical current to be the direction of flow of positive "holes", whereas chemists generally define it to be the direction of negative electron charge flow. Prior to 1970, most physical chemistry textbooks defined the P–V work without the minus sign in Eq. 2.28. Therefore, work done by the system upon expansion was

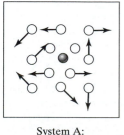

System A:
Thermal energy

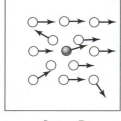

System B:
Transitional energy

▲ FIGURE 2.12
Kinetic energy in the form of thermal energy and translational energy. In both cases, the center of mass is denoted by ●.

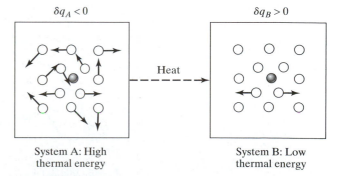

$\delta q_A < 0$

System A: High
thermal energy

Heat →

$\delta q_B > 0$

System B: Low
thermal energy

▲ FIGURE 2.13
The flow of thermal energy from one point to another is called heat.

positive rather than negative. The bottom line is that sign conventions must be examined carefully when one consults different reference sources.

The most important question concerns the differential heat change: Should we write it as dq or as δq? That is, is the heat differential exact? Does the total heat change in a process depend only upon the initial and final states of the system, or does it depend upon the path by which the heat transfer is made? At this point, we give the answer without proof: The heat differential is inexact. Therefore, we write it as δq. The proof of this statement will be deferred until later in the chapter.

2.3   The First Law of Thermodynamics

2.3.1   The Internal Energy U

Although the heat and work differentials are both inexact, their sum is an exact differential. This simple statement is the first law of thermodynamics. As we shall see, the consequences of this law are profound.

Since we know from the first law that $\delta w + \delta q$ forms an exact differential, we may safely infer that there exists a function whose total differential is exactly $\delta w + \delta q$. We are free to denote this function with any symbol and name we choose. Because both heat and work have the units of energy, we name the function the *internal energy* and choose the symbol U. (The reader should be aware that this choice is not uniform across all texts. Occasionally, the symbol E is employed.) With this choice, the first law may be written as

$$dU \equiv \delta w + \delta q ,\tag{2.37}$$

where the exactness of the U differential is denoted by the use of dU rather than δU. Let us now examine some of the important properties of the internal energy.

Consider changing a system from State A to State B along two different paths as shown in Figure 2.14. The change in internal energy is given by $\Delta U = \int dU = \int_A^B (\delta w + \delta q)$ regardless of whether the integration is carried out along Path 1 or Path 2, since dU is exact. This is true even if Path 2 is a purely hypothetical path that cannot actually exist. So long as States A and B are correct, the integral value will be the same for all paths.

Now consider a change in which the initial and final states are the same, as illustrated in Figure 2.15. Here, some complex series of changes is execut-

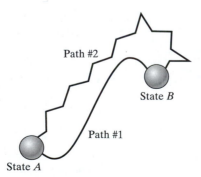

Path #2

State B

Path #1

State A

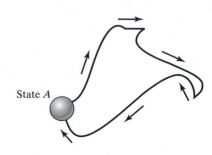

State A

▲ FIGURE 2.14
Two pathways connecting two states of a system denoted A and B.

▲ FIGURE 2.15
Energy changes around a closed path with identical initial and final states.

ed on the system such that the final and initial states are identical. The total change in internal energy for the process is

$$\Delta U = \int_A^A \delta w + \delta q = U(A) - U(A) = 0.$$

That is, the integral of any exact differential about a closed path with identical initial and final states is always zero. This result leads to the statement that *energy is neither created nor destroyed around a closed path*. In fact, this statement is sometimes presented as the first law itself rather than as a *result* of the first law.

Let us now assume that Path 1 in Figure 2.14 is a path of constant volume. In such a case, we obtain a very useful and important result. For a constant-volume path, we have

$$\boxed{\Delta U = \int_A^B \delta w + \delta q = \int_A^B [-P_{ext} dV + \delta q] = \int_A^B \delta q = q_v},\qquad \textbf{(2.38)}$$

where q_v denotes the heat change under conditions of constant volume. Equation 2.38 demonstrates that the total heat associated with any constant-volume process will always be equal to ΔU. This can be important, since it is often much easier to compute ΔU than q_v, due to the fact that dU is exact whereas δq is inexact.

2.3.2   The Enthalpy *H*

If Path 1 in Figure 2.14 is a path of constant pressure, the result is not as convenient, since the $-P_{ext} dV$ term in Eq. 2.38 no longer vanishes. This result suggests that it will be useful to define another thermodynamic function, namely,

$$\boxed{H \equiv U + PV},\qquad \textbf{(2.39)}$$

where H is called the *enthalpy*. The total differential of H is

$$dH = dU + d(PV) = -P_{ext} dV + \delta q + P\, dV + V\, dP$$
$$= (P - P_{ext})\, dV + V\, dP + \delta q.\qquad \textbf{(2.40)}$$

The Euler criterion shows the differential $d(PV) = P\, dV + V\, dP$ to be exact, since we have $(\partial P/\partial P)_V = (\partial V/\partial V)_P = 1$. The first law tells us that dU is exact. Therefore, dH must be exact, since the sum of any two exact differentials is exact. If the calculation is done using a reversible path, we have $P = P_{ext}$, and

$$dH = \delta q + V\, dP.\qquad \textbf{(2.41)}$$

If the path is a constant-pressure path with $dP = 0$, integration of Eq. 2.41 yields

$$\boxed{\Delta H = q_p},\qquad \textbf{(2.42)}$$

since the $V\, dP$ term vanishes. Equation 2.42 shows that the heat associated with a constant-pressure process is always equal to the change in the enthalpy function. The results given in Eqs. 2.38 and 2.42 suggest that it is usually best to work with U when one is considering constant-volume processes and best to work with H if the pressure is being held constant.

Since dU and dH are exact differentials, ΔU and ΔH depend only upon the initial and final states of a process. For this reason, U and H are often termed *state* functions. Such nomenclature implies the exactness of dU and dH.

2.3.3   Total Differentials for *U* and *H*

If we wish to compute changes in U and H, we must have the rates of change of these quantities with respect to the independent variables characterizing the system of interest. In effect, this means that we must have appropriate expressions for the total differentials of U and H.

For a system containing K different substances, there are $K + 3$ possible variables: T, P, V, and the number of moles of each substance present, which we denote as $n_1, n_2, \ldots, n_k$. Since T, P, and V are interrelated through the equation of state of the system, we have $K + 2$ independent variables that we are free to choose as we wish. As usual, our choice is dictated by considerations of convenience and ease of computation. In virtually every chemical system, temperature is a critical variable that is directly measured. Consequently, it would be foolish to choose P, V, and the n_i ($i = 1, 2, 3, \ldots, K$) as our independent variables. Equations 2.38 and 2.42 suggest that we will wish to compute ΔU whenever the volume is constant and ΔH when $dP = 0$. If we express dU in terms of T and V, the term involving dV will vanish for constant-volume processes and thereby simplify the calculations. Similar reasoning suggests that we should choose T and P for the differential dH. Following this reasoning, we write

$$U = f(T, V, n_1, n_2, n_3, \ldots, n_K)$$

and

$$H = g(T, P, n_1, n_2, n_3, \ldots, n_K). \tag{2.43}$$

With these choices, the total differentials needed for the computation of ΔU and ΔH are, respectively,

$$dU = \left(\frac{\partial U}{\partial T}\right)_{V,n} dT + \left(\frac{\partial U}{\partial V}\right)_{T,n} dV + \left(\frac{\partial U}{\partial n_1}\right)_{T,V,nj} dn_1 +$$

$$\left(\frac{\partial U}{\partial n_2}\right)_{T,V,nj} dn_2 + \ldots + \left(\frac{\partial U}{\partial n_K}\right)_{T,V,nj} dn_K$$

$$= \left(\frac{\partial U}{\partial T}\right)_{V,n} dT + \left(\frac{\partial U}{\partial V}\right)_{T,n} dV + \sum_{i=1}^{K} \left(\frac{\partial U}{\partial n_i}\right)_{T,V,nj} dn_i \tag{2.44}$$

and

$$dH = \left(\frac{\partial H}{\partial T}\right)_{P,n} dT + \left(\frac{\partial H}{\partial P}\right)_{T,n} dP + \left(\frac{\partial H}{\partial n_1}\right)_{T,P,nj} dn_1 +$$

$$\left(\frac{\partial H}{\partial n_2}\right)_{T,P,nj} dn_2 + \ldots + \left(\frac{\partial H}{\partial n_K}\right)_{T,p,nj} dn_K$$

$$= \left(\frac{\partial H}{\partial T}\right)_{P,n} dT + \left(\frac{\partial H}{\partial P}\right)_{T,n} dP + \sum_{i=1}^{K} \left(\frac{\partial H}{\partial n_i}\right)_{T,P,nj} dn_i. \tag{2.45}$$

The rates of change of U and H with respect to the number of moles of component i present are given by $(\partial U/\partial n_i)_{T, V, nj}$ and $(\partial H/\partial n_i)_{T, P, nj}$, respectively. The subscript n_j indicates that the number of moles of all substances other than substance i are held constant when the partial derivative is evaluated. When T and P are held constant, such derivatives are called *partial molar quantities*. Therefore, $(\partial H/\partial n_i)_{T, P, nj}$ is the partial molar enthalpy with respect to component i. To simplify the notation, we shall denote partial molar quantities with a bar above the symbol. Thus, $\overline{H}_i = (\partial H/\partial n_i)_{T, P, nj}$. The derivative $(\partial U/\partial n_i)_{T, V, nj}$ will be represented by the symbol $\hat{U}_i$. In this notation, Eqs. 2.44 and 2.45 respectively become

$$dU = \left(\frac{\partial U}{\partial T}\right)_{V, n} dT + \left(\frac{\partial U}{\partial V}\right)_{T, n} dV + \sum_{i=1}^{K} \hat{U}_i \, dn_i \qquad (2.46)$$

and

$$dH = \left(\frac{\partial H}{\partial T}\right)_{P, n} dT + \left(\frac{\partial H}{\partial P}\right)_{T, n} dP + \sum_{i=1}^{K} \overline{H}_i \, dn_i . \qquad (2.47)$$

Equations 2.46 and 2.47 show that we will need $K + 2$ derivatives to compute ΔU and a similar number to obtain ΔH. The need for so many rates of change makes thermodynamic calculations difficult for general, complex systems.

Fortunately, many systems of interest do not contain a large number of different components. This significantly reduces the magnitude of the problem. And even if there are a large number of components, the composition of the system may not change during the process of interest. In that case, $dn_i = 0$ for all i, and all the terms beneath the summations in Eqs. 2.46 and 2.47 vanish. Such systems are termed *closed* systems.

If we are dealing with a closed system, we need find only four derivatives to compute ΔU and ΔH: $(\partial U/\partial T)_{V, n}$ $(\partial U/\partial V)_{T, n}$ $(\partial H/\partial T)_{P, n}$ and $(\partial H/\partial P)_{T, n}$. When the system is closed, it is customary to omit the subscript n on the partial derivatives. With this understanding, most texts, including this one, will write $(\partial U/\partial T)_V$ instead of $(\partial U/\partial T)_{V, n}$ and similarly for the remaining three partial derivatives. In the next section, we shall determine what must be either measured or calculated to obtain these derivatives.

2.4   Heat Capacity

By definition, heat capacity C is the rate of change of a system's heat with temperature:

$$C \equiv \frac{\delta q}{dT} . \qquad (2.48)$$

Examination of Eq. 2.48 immediately indicates that heat capacity is going to be a path-dependent variable, since it depends upon an inexact differential δq. Thus, the question "What is the heat capacity of lead?" has no meaning unless the path of heating or cooling is specified. To illustrate this point, consider adding heat to ice at the normal melting point, 273.15 K. As a quantity of heat $|\delta q|$ is added, some of the ice melts to water at 273.15 K, but the change in temperature, dT, is zero. Thus, for this heating path, we have

$C = |\delta q|/0 = \infty$. If we were freezing water at 273.15 K, it would be necessary to remove heat from system, and the heat change would be $-|\delta q|$. Again, the temperature would not change during the phase transition, so that we would obtain $C = -\infty$.

We can use the first law to express the heat capacity in terms of U or H. From Eqs. 2.37 and 2.46, we have

$$\delta q = dU - \delta w = \left(\frac{\partial U}{\partial T}\right)_V dT + \left(\frac{\partial U}{\partial V}\right)_T dV + P_{ext}\, dV, \qquad \textbf{(2.49)}$$

provided that we are dealing with a closed system for which the $dn_i = 0$ and only pressure–volume work is being done. Substituting Eq. 2.49 into Eq. 2.48 gives

$$C = \left(\frac{\partial U}{\partial T}\right)_V + \left\{\left(\frac{\partial U}{\partial V}\right)_T + P_{ext}\right\}\frac{dV}{dT}. \qquad \textbf{(2.50)}$$

Since the total derivative of V with respect to T appears in Eq. 2.50, we must know the heating or cooling path before C can be evaluated. To illustrate this point, consider the computation of dV/dT for an ideal gas. In such a case, we have $dV/dT = d/dT\,[nRT/P] = nR/P - (nR/P^2)\,(dP/dT)$, and unless we know the path, the derivative dP/dT cannot be obtained.

We may also utilize Eq. 2.41 to express C in terms of H. Solving that equation for δq, we obtain $\delta q = dH - V\,dP$. Substitution of this result and Eq. 2.47 into Eq. 2.48 gives

$$C = \left(\frac{\partial H}{\partial T}\right)_P + \left\{\left(\frac{\partial H}{\partial P}\right)_T - V\right\}\frac{dP}{dT} \qquad \textbf{(2.51)}$$

for a closed system with only P–V work.

We may now obtain two of the four derivatives needed to compute ΔU and ΔH. If we heat or cool a system at constant volume, the heat capacity will have a specific value, since the path has been specified. We denote this value by C_v. An examination of Eq. 2.50 shows that for such a path, the second term vanishes, because $dV = 0$, and we obtain

$$\boxed{C = C_v = \left(\frac{\partial U}{\partial T}\right)_V.} \qquad \textbf{(2.52)}$$

Thus, one of the four derivatives has been determined. Equation 2.52 tells us that we must go into the laboratory and measure the heat capacity of the system for a constant-volume heating or cooling path in order to determine the value of $(\partial U/\partial T)_V$.

Equation 2.51 also shows that if we heat or cool the system along a constant-pressure path, the heat capacity under these conditions is

$$\boxed{C = C_p = \left(\frac{\partial H}{\partial T}\right)_P.} \qquad \textbf{(2.53)}$$

Table 2.1 lists measured molar heat capacities C_p^m for constant-pressure heating for some selected gases at one bar pressure and temperatures close

Table 2.1   Average molar heat capacities C_p^m for selected gases around 298 K and 1 bar pressure

Gas	C_p^m (J mol^{-1} K^{-1})	Gas	C_p^m (J mol^{-1} K^{-1})
CO_2	37.11	HF	29.13
CO	29.14	He	20.786
CH_4	35.31	H_2	28.824
C_2H_6 (ethane)	52.63	H_2O	33.58
C_3H_8 (propane)	73.50	Kr	20.786
C_6H_6 (benzene)	81.67	Ne	20.786
CH_3OH	43.89	N_2	29.125
Ar	20.786	NO	29.844
Br_2	36.02	NO_2	37.20
Cl_2	33.91	NH_3	35.06
HCl	29.12	O_2	29.355
F_2	31.30	Xe	20.786

Table 2.2   Molar heat capacities C_p^m for selected gases as a function of temperature at a pressure of 1 bar[1].

Substance	C_p^m (J mol^{-1} K^{-1})			
	a_o (J mol^{-1} K^{-1})	a_1 (J mol^{-1} K^{-2})	a_2 (J mol^{-1})	Fitting Range (K)
N_2	26.45	0.005529	301.96	100–1,500
O_2	25.35	0.008990	417.34	100–380
He	20.186	0.000646	30.08	5–1,500
CH_4	10.295	0.06281	2,112.71	125–600
C_2H_6	8.489	0.1313	1,691.53	200–600
C_3H_8	30.134	0.1701	−1,652.15	250–600

[1]Experimental data fitted to the function $C_p^m = a_o + a_1T + a_2/T$ by linear least-squares methods. Data taken from the *Handbook of Chemistry and Physics,* 78th edition, CRC Press, Boca Raton, FL, 1997–1998.

to 298 K. Measurement demonstrates that C_p^m is temperature dependent. This is shown by the data in Table 2.2, by Figure 2.16, and by Problem 2.18.

Figure 2.16 shows a plot of the measured constant-pressure molar heat capacities for propane, ethane, methane, and nitrogen as a function of temperature at 1 bar pressure. The points are the measured data. The curves are linear least-squares fits over the ranges given in Table 2.2 of the function $C_p^m = a_o + a_1T + a_2/T$. At higher temperatures, the heat capacities show a near linear variation with temperature. At lower temperatures, the heat capacity often exhibits a minimum and then increases. These data emphasize the fact that heat capacity is a quantity that usually must be

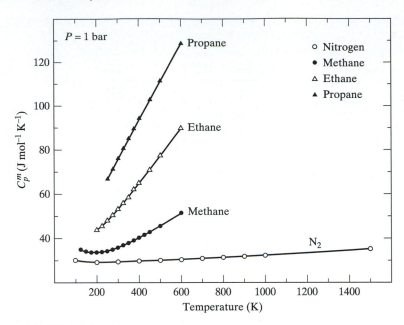

▲ **FIGURE 2.16**

C_p^m for propane, ethane, methane, and nitrogen as a function of temperature at a pressure of one bar. The points are the experimental data. The solid curves are linear least-squares fits of the function $C_p^m = a_o + a_1T + a_2/T$, where the fitting coefficients are those listed in Table 2.2. Data are taken from the *Handbook of Chemistry and Physics,* 78th edition, CRC Press, Boca Raton, FL, 1997–1998.

experimentally measured. We shall find in later chapters that C_p^m and C_v^m can be computed for some systems, but in most cases, heat capacities are measured quantities.

If C_p^m is given by an expression of the form $C_p^m = a_o + a_1T + a_2/T$ at 1 bar pressure, integration of Eq. 2.53 between corresponding limits shows that the enthalpy is given by

$$\int_{\overline{H}(298.15)}^{\overline{H}(T)} d\overline{H} = \overline{H}(T) - \overline{H}(298.15) = \int_{298.15}^{T}\left[a_o + a_1T + \frac{a_2}{T}\right]\partial T$$

$$= a_o(T - 298.15) + \frac{a_1}{2}[T^2 - (298.15)^2] + a_2 \ln\left[\frac{T}{298.15}\right].$$

We shall find in Chapter 3 that the molar enthalpy for $CH_4(g)$ at 298.15 K and 1 bar pressure is $-74,810$ J mol^{-1}. Therefore, the methane enthalpy at temperature T is given by

$$\overline{H}_{CH_4}(T) = -74,810 + a_o(T - 298.15) + \frac{a_1}{2}[T^2 - (298.15)^2]$$

$$+ a_2 \ln\left[\frac{T}{298.15}\right],$$

where a_o, a_1, and a_2 are given in Table 2.2. Figure 2.17 shows a plot of the methane enthalpy as a function of temperature. The slope of this plot at any temperature is C_p^m at that temperature, as illustrated in the figure.

Summarizing our progress to this point, we have

$$dU = C_v\,dT + \left(\frac{\partial U}{\partial V}\right)_T dV$$

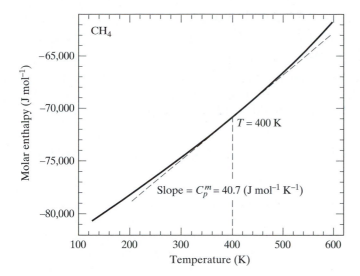

The graph shows: "CH₄" labeled in upper left. Y-axis: Molar enthalpy (J mol⁻¹) with values −65,000, −70,000, −75,000, −80,000. X-axis: Temperature (K) with values 100, 200, 300, 400, 500, 600. Labels on plot: $T = 400$ K, Slope $= C_p^m = 40.7$ (J mol⁻¹ K⁻¹).

▲ **FIGURE 2.17**
Molar enthalpy of methane (solid line) as a function of temperature at 1 bar of pressure. The calculation of $\overline{H}_{CH_4}(T)$ is done in the manner described in the text, using the CH₄ heat capacity data given in Table 2.3 and the measured standard partial molar enthalpy for methane. (See Chapter 3.) C_p^m for methane at temperature T is the slope of $\overline{H}_{CH_4}(T)$ at that temperature, as illustrated in the figure for $T = 400$ K.

and

$$dH = C_p \, dT + \left(\frac{\partial H}{\partial P} \right)_T dP, \qquad (2.54)$$

for closed systems in which only pressure–volume work is being done. (For simplicity, we have dropped the subscript n on the derivatives, as it is unnecessary for closed systems.) We now must obtain $(\partial U / \partial V)_T$ and $(\partial H / \partial P)_T$.

To compute ΔU and ΔH for a closed system, we need to know how U changes with volume and how H varies with pressure when the temperature is held constant. It would, therefore, appear that we must make at least two additional measurements to obtain the required data. However, we may reduce the effort needed by noting that, by combining Eqs. 2.50 and 2.51, we can obtain a relationship between the two remaining derivatives, so that only one of them needs to be measured or calculated. The left sides of Eqs. 2.50 and 2.51 are both equal to C; therefore, the right sides of these equations must be equal, too. This gives

$$C_v + \left\{ \left(\frac{\partial U}{\partial V} \right)_{T,n} + P_{ext} \right\} \frac{dV}{dT} = C_p + \left\{ \left(\frac{\partial H}{\partial P} \right)_{T,n} - V \right\} \frac{dP}{dT}. \qquad (2.55)$$

If we heat the system along a constant-pressure, reversible path, Eq. 2.55 becomes

$$C_p - C_v = \left\{ \left(\frac{\partial U}{\partial V} \right)_{T,n} + P \right\} \left(\frac{\partial V}{\partial T} \right)_P \qquad (2.56)$$

since the total derivative becomes a partial derivative taken for a constant-pressure path. On the other hand, if we heat the system along a constant-volume ($dV = 0$) path, Eq. 2.55 yields

$$C_p - C_v = -\left\{ \left(\frac{\partial H}{\partial P} \right)_{T,n} - V \right\} \left(\frac{\partial P}{\partial T} \right)_v. \qquad (2.57)$$

Combining Eqs. 2.56 and 2.57 produces

$$\left\{ \left(\frac{\partial U}{\partial V} \right)_{T,n} + P \right\} \left(\frac{\partial V}{\partial T} \right)_P = - \left\{ \left(\frac{\partial H}{\partial P} \right)_{T,n} - V \right\} \left(\frac{\partial P}{\partial T} \right)_v. \qquad \textbf{(2.58)}$$

Clearly, we can measure either $(\partial U/\partial V)_{T,n}$ or $(\partial H/\partial P)_{T,n}$ and then use Eq. 2.58 to compute the other partial derivative, provided that we have the equation of state, so that the variations of V and P with T can be obtained.

2.5 The Joule–Thomson Experiment: Measuring $\left(\dfrac{\partial H}{\partial P} \right)_T$

The objective of the Joule–Thomson experiment is to measure the value of $(\partial H/\partial P)_T$. We first describe the nature of the experiment and then present the analysis of the data to demonstrate how $(\partial H/\partial P)_T$ is obtained.

Figure 2.18 illustrates the principle underlying the Joule–Thomson experiment. An insulated apparatus prevents heat from being transferred between the system and the surroundings ($\delta q = 0$). Processes conducted under this condition are called *adiabatic*. Since δq is zero for any adiabatic process, the heat capacity for the system is also zero. The apparatus is divided into two compartments by a porous plug whose pore size permits the slow seepage of gas from Compartment 1 to Compartment 2 when there is a negative pressure difference between the compartments. Gas is contained by pistons in each compartment as shown.

Initially, a gas is placed in Compartment 1 at $T = T_1$, $P = P_1$, and $V = V_1$. Compartment 2 is empty, with the piston flush against the porous plug. Consequently, the initial volume in Compartment 2 is zero. The pistons are electrically controlled such that as the left-hand piston in Compartment 1 is depressed, the piston in Compartment 2 is withdrawn. The speeds of the pistons are adjusted such that the pressure in Compartment 1 remains constant at P_1 as the gas moves through the porous plug into Compartment 2, in which the pressure is maintained at a constant value of P_2. Clearly, we must have $P_2 < P_1$ to induce gas flow through the plug. This process is continued until all the gas has entered Compartment 2. The final state is as shown in Figure 2.18. All the gas is now in Compartment 2 at a temperature T_2 and pressure P_2. The volume of gas in Compartment 1 is zero. The objective of the experiment is to measure the temperature change $\Delta T = T_2 - T_1$ as a

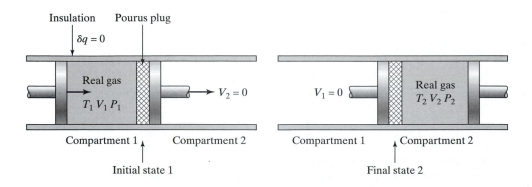

▲ FIGURE 2.18

The Joule–Thomson experiment: Measurement of $(\partial H/\partial P)_T$. All processes are carried out at constant pressure, so that $dP_1 = 0$ and $dP_2 = 0$.

function of the pressure difference $\Delta P = P_2 - P_1$ between the compartments. Since ΔP can be controlled by adjusting the rates of movement of the pistons, the experiment can be repeated using different values of ΔP to obtain ΔT as a function thereof. The data consist of a table of ΔT values for different choices of ΔP.

The first step in the data analysis is to demonstrate that the Joule–Thomson experiment is conducted under conditions of constant enthalpy ($dH = 0$). The change in internal energy during the process is given by

$$\Delta U = \int_{\text{State 1}}^{\text{State 2}} dU = U_2 - U_1 = \int_{\text{State 1}}^{\text{State 2}} [\delta q + \delta w] = \int_{\text{State 1}}^{\text{State 2}} \delta w = w_{\text{total}}, \quad \text{(2.59)}$$

where we have set δq to zero because the process is adiabatic. Since dU is exact, we may compute its value along any path we choose. For computational convenience, we choose a reversible path. The friction of the gas with the porous plug provides the reversible restraining force so that the work done in Compartment 1 is

$$w_1 = -\int_{V_1}^{0} P \, dV = -P_1 \int_{V_1}^{0} dV = P_1 V_1. \quad \text{(2.60)}$$

Because the experiment is conducted under conditions of constant pressure with $P = P_1$, we can factor P out of the integral in Eq. 2.60 and set its value to P_1. In Compartment 2, the right-hand piston provides a reversible restraining force so that the work is

$$w_2 = -\int_{0}^{V_2} P \, dV = -P_2 \int_{0}^{V_2} dV = -P_2 V_2. \quad \text{(2.61)}$$

Combining Eqs. 2.59, 2.60, and 2.61 we obtain

$$U_2 - U_1 = w_{\text{total}} = w_1 + w_2 = P_1 V_1 - P_2 V_2. \quad \text{(2.62)}$$

Rearranging Eq. 2.62 gives

$$U_1 + P_1 V_1 = H_1 = U_2 + P_2 V_2 = H_2. \quad \text{(2.63)}$$

Thus, $dH = 0$, since $H_1 = H_2$. Hence, the Joule–Thomson experiment is conducted under conditions of constant enthalpy.

With $dH = 0$, Eq. 2.54 becomes

$$C_p \, dT + \left(\frac{\partial H}{\partial P}\right)_T dP = dH = 0.$$

We may now simply solve this equation for $(\partial H / \partial P)_T$ to obtain

$$\left(\frac{\partial H}{\partial P}\right)_T = -C_p \left(\frac{\partial T}{\partial P}\right)_H \quad \text{(2.64)}$$

where the derivative is now a partial derivative, because Eq. 2.64 holds only for an experiment conducted under conditions of constant enthalpy. The derivative $(\partial T / \partial P)_H$ is called the *Joule–Thomson coefficient* and is denoted by the symbol μ. In this notation, we have

$$\boxed{\left(\frac{\partial H}{\partial P}\right)_T = -C_p \, \mu}. \quad \text{(2.65)}$$

All that remains is to extract μ from the experimental data. At each value of ΔP that is examined, we may easily compute the ratio $(\Delta T / \Delta P)_H$. The required partial derivative is defined to be

$$\left(\frac{\partial T}{\partial P}\right)_H = \lim_{\Delta P \to 0} \left(\frac{\Delta T}{\Delta P}\right)_H. \tag{2.66}$$

To utilize Eq. 2.66, we plot the measured values of $(\Delta T / \Delta P)_H$ against ΔP. The results are then fitted by a least-squares technique and extrapolated to $\Delta P = 0$, as shown in Figure 2.19A. The intercept gives the value of μ at pressure P_1 and temperature T_1.

The Joule–Thomson coefficient may also be obtained by direct computation of the slope of an isenthalpic curve. In this procedure, the initial and final states, (T_1, P_1) and (T_2, P_2), are plotted on a temperature–pressure diagram. The experiment is then repeated using the same initial temperature and pressure, but a different final pressure, P_3. The new final state, (T_3, P_3), is plotted on the same diagram. After a sufficient number of data points are plotted, a smooth curve is drawn through the data, as illustrated in Figure 2.19B. All of the points have the same enthalpy; therefore, the plot is called an isenthalpic curve. The slope of this curve, $(\partial T / \partial P)_H$, at point (T_0, P_0), is the Joule–Thomson coefficient at that temperature and pressure. The method is straightforward, but the reader should be aware that it is often very difficult to extract accurate slopes. Small errors in the data are frequently greatly magnified in the derivative of the function representing the data. Section 9.6.6 and Problem 9.52 discuss this problem in detail; the section also provides two excellent procedures for obtaining numerical derivatives.

Table 2.3 gives the measured values of μ for some selected gases.

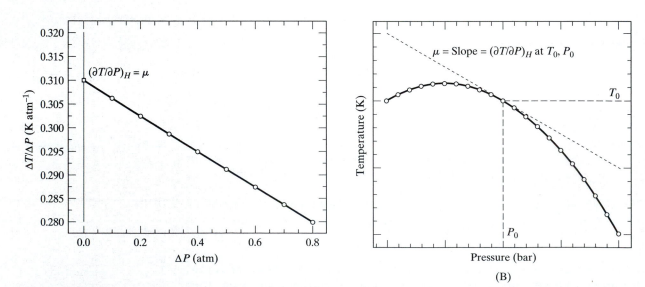

▲ **FIGURE 2.19**
Two methods for obtaining the Joule–Thomson coefficient from the measured changes in temperature, ΔT, for given changes in pressure, ΔP, in a Joule–Thomson experiment. (A) In this method, the ratio $\Delta T / \Delta P$ is plotted against ΔP. The data are fitted to a straight line by a least-squares method and extrapolated to $\Delta P = 0$. The intercept yields the Joule–Thomson coefficient at the temperature and pressure in Compartment 1 in Figure 2.18. (B) An isenthalpic plot in which the final temperature in Compartment 2 is plotted as a function of the final pressure. The slope of the fitted curve at a given temperature and pressure is the Joule–Thomson coefficient at that temperature and pressure.

Table 2.3	Joule–Thomson coefficients for selected gases at 1 atm pressure	
Gas	**T (K)**	**μ (K atm^{-1})**
O_2	298	0.31
N_2	298	0.27
H_2	298	−0.03
He	298	−0.062
CO_2	300	1.11

Once the value of μ is determined, we may obtain the last required derivative using Eq. 2.58 with Eq. 2.65. Combining these equations gives

$$[\mu C_p + V]\left(\frac{\partial P}{\partial T}\right)_V = \left[\left(\frac{\partial U}{\partial V}\right)_T + P\right]\left(\frac{\partial V}{\partial T}\right)_P. \qquad (2.67)$$

Solving for $(\partial U/\partial V)_T$, we obtain

$$\left(\frac{\partial U}{\partial V}\right)_T = \frac{[\mu C_p + V]\left(\dfrac{\partial P}{\partial T}\right)_V}{\left(\dfrac{\partial V}{\partial T}\right)_P} - P. \qquad (2.68A)$$

Equation 2.68A can be put in more compact form by using the cyclic rule for partial derivatives. Consider a function $z = f(x, y)$. The total differential of z is

$$dz = \left(\frac{\partial z}{\partial x}\right)_y dx + \left(\frac{\partial z}{\partial y}\right)_x dy.$$

Division by dy produces

$$\frac{dz}{dy} = \left(\frac{\partial z}{\partial x}\right)_y \frac{dx}{dy} + \left(\frac{\partial z}{\partial y}\right)_x.$$

If we now let z be held constant, so that $dz = 0$, the equation becomes

$$0 = \left(\frac{\partial z}{\partial x}\right)_y \left(\frac{\partial x}{\partial y}\right)_z + \left(\frac{\partial z}{\partial y}\right)_x.$$

Solving for $(\partial x/\partial y)_z$, we obtain

$$\left(\frac{\partial x}{\partial y}\right)_z = -\frac{\left(\dfrac{\partial z}{\partial y}\right)_x}{\left(\dfrac{\partial z}{\partial x}\right)_y}.$$

Since we also have $[(\partial y/\partial z)_x]^{-1} = (\partial z/\partial y)_x$ and $[(\partial x/\partial z)_y]^{-1} = (\partial z/\partial x)_y$, we can write the last equation in the form

$$\left(\frac{\partial x}{\partial y}\right)_z = -\frac{\left(\dfrac{\partial x}{\partial z}\right)_y}{\left(\dfrac{\partial y}{\partial z}\right)_x}.$$

If we now let $z = T$, $x = P$, and $y = V$, we have

$$\left(\frac{\partial P}{\partial V}\right)_T = -\frac{\left(\frac{\partial P}{\partial T}\right)_V}{\left(\frac{\partial V}{\partial T}\right)_P}.$$

Substituting this identity into Eq. 2.68A produces

$$\left(\frac{\partial U}{\partial V}\right)_T = -[\mu C_p + V]\left(\frac{\partial P}{\partial V}\right)_T - P \qquad \textbf{(2.68B)}$$

We now have all derivatives required to compute ΔU and ΔH for closed systems. The results are

$$\boxed{dU = C_v\, dT - \left\{[\mu C_p + V]\left(\frac{\partial P}{\partial V}\right)_T + P\right\} dV} \qquad \textbf{(2.69)}$$

and

$$\boxed{dH = C_p\, dT - \mu C_p\, dP}. \qquad \textbf{(2.70)}$$

Therefore, we require measurement of two heat capacities, C_v and C_p, and the Joule–Thomson coefficient μ. We shall later see that μ can be computed directly from the equation of state using the second law of thermodynamics. At this point, however, its value must be measured.

As a last item of interest, let us consider the physical significance of the Joule–Thomson coefficient. If μ is positive, we must have $(\partial T/\partial P)_H > 0$. The adiabatic expansion occurring in the Joule–Thomson experiment gives $\Delta P < 0$, Therefore, if $\mu > 0$, we must have $\Delta T < 0$ when the gas expands adiabatically. Accordingly, under conditions for which gases have positive Joule–Thomson coefficients, they must cool when expanded adiabatically. If $\mu < 0$, adiabatic expansion will cause the gas to heat. As a practical example, consider the hot air expelled during a political speech. The principal components of air, nitrogen and oxygen, have Joule–Thomson coefficients of 0.27 and 0.31 K atm^{-1}, respectively. Since these values are positive, the political hot air is expected to cool during the near-adiabatic expansion from the lungs. (Perhaps this effect is related to the fact that political campaign promises tend to cool off after the election.)

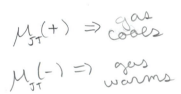

2.6 Joule's Experiment

In 1843, Joule measured the temperature change of a surrounding water bath when a gas expands into a vacuum under adiabatic conditions in which $\delta q = 0$. Since there is no opposing force when a gas expands into a vacuum, δw is also zero. This makes $dU = \delta q + \delta q = 0$ for the experiment. A schematic diagram of the Joule experiment is shown in Figure 2.20.

From the first part of Eq. 2.54, the experimental situation ensures that we will have

$$dU = C_v\, dT + \left(\frac{\partial U}{\partial V}\right)_T dV = 0,$$

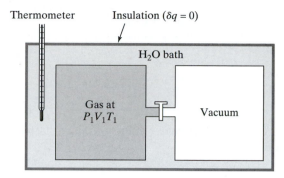

▲ FIGURE 2.20

Joule experiment. At the start of the experiment, the gas at pressure P_1, temperature T_1, and volume V_1 is contained in the left-hand compartment. There is a vacuum in the right-hand compartment. The gas is then allowed to expand adiabatically through the valve into the right-hand compartment. The experiment consists of measuring the change in temperature of the water bath produced by the expansion of the gas.

so that

$$\left(\frac{\partial U}{\partial V}\right)_T = -C_v \left(\frac{\partial T}{\partial V}\right)_U = -C_v \mu_J, \tag{2.71}$$

where the derivative $(\partial T/\partial V)_U$ is written as a partial derivative, since we are holding U constant during the experiment. μ_J is called the *Joule coefficient*. Joule's original experiment showed no temperature change in the surrounding water bath when the gas was expanded under constant U conditions. As a result, he obtained $\mu_J = (\partial T/\partial V)_U = 0$. More accurate measurements in 1924 by Keyes and Sears demonstrated that μ_J is very small, but not zero. Its correct value for a given gas can be obtained from Eq. 2.68B once the Joule–Thomson coefficient of the gas is known.

Although Joule's result for real gases was in error, it is correct for an ideal gas. Since an ideal gas has no forces between the molecules, the internal energy is independent of the distance between them. Because this is true, U will be independent of volume, and we will have $(\partial U/\partial V)_T = 0$. Setting $(\partial U/\partial V)_T$ equal to zero in Eq. 2.68B gives

$$0 = -[\mu C_p + V]\left(\frac{\partial P}{\partial V}\right)_T - P.$$

For an ideal gas, $P = nRT/V$, so that $(\partial P/\partial V)_T = -nRT/V^2 = -P/V$. Substituting this result into the previous equation produces

$$0 = [\mu C_p + V]\frac{P}{V} - P = \frac{\mu C_p P}{V} + P - P = \frac{\mu C_p P}{V}, \tag{2.72}$$

which requires that $\mu = 0$ for an ideal gas. We shall call this statement *Joule's law*. After we have introduced the second law of thermodynamics, we will rigorously derive Joule's law. For the time being, however, we shall have to content ourselves with the physical argument just presented that $(\partial U/\partial V)_T = 0$ for an ideal gas.

Joule's law makes the differentials dU and dH extremely simple for an ideal gas. If $\mu = 0$ in Eqs. 2.69 and 2.70, we have

$$dU = C_v \, dT \tag{2.73}$$

and

$$dH = C_p \, dT \tag{2.74}$$

for an ideal gas. That is, the internal energy and enthalpy of an ideal gas depend only upon T.

We may now use Eq. 2.57 to obtain a very useful result for ideal gases. That equation shows that

$$C_p - C_v = -\left\{ \left(\frac{\partial H}{\partial P} \right)_{T,n} - V \right\} \left(\frac{\partial P}{\partial T} \right)_V = V \left(\frac{\partial P}{\partial T} \right)_V = V \frac{nR}{V} = nR. \tag{2.75}$$

For 1 mole of gas, the difference between the molar heat capacities C_p^m and C_v^m is equal to R. Physically, C_p exceeds C_v because work is done when the heating is executed along a constant-pressure path, whereas there is no work in a constant-volume process.

2.7   Proof of the Inexactness of δq

We are now in a position to prove that the heat differential is inexact and that the heat exchange in a process is path dependent. We first note that for a differential to be exact, it must be exact for any system undergoing any process. If we can find a single process for any system for which a differential is inexact, then that differential is, in general, inexact.

From the first law, we have $dq = dU - \delta w$, where, for the moment, we have written dq as exact until proven otherwise. Let us now consider an ideal gas undergoing a reversible process. For this case, we have, from Eq. 2.73,

$$dq = dU - \delta w = C_v \, dT + P \, dV. \tag{2.76}$$

If dq is to be exact, the Euler criterion must be satisfied. That is, we must have

$$\left(\frac{\partial P}{\partial T} \right)_V = \left(\frac{\partial C_v}{\partial V} \right)_T. \tag{2.77}$$

For an ideal gas, $(\partial P/\partial T)_V = nR/V$. We now need to evaluate the right-hand side of Eq. 2.77. Substituting $(\partial U/\partial T)_V$ for C_v permits us to evaluate

$$\left(\frac{\partial C_v}{\partial V} \right)_T = \frac{\partial}{\partial V} \left\{ \left(\frac{\partial U}{\partial T} \right)_V \right\}_T = \frac{\partial}{\partial T} \left\{ \left(\frac{\partial U}{\partial V} \right)_T \right\}_V, \tag{2.78}$$

since du is exact so that the order of differentiation does not matter. But $(\partial U/\partial V)_T = 0$ for an ideal gas. Therefore, Eq. 2.78 becomes

$$\left(\frac{\partial C_v}{\partial V} \right)_T = \frac{\partial}{\partial T} \left\{ \left(\frac{\partial U}{\partial V} \right)_T \right\}_V = \frac{\partial}{\partial T}[0] = 0.$$

Thus, the Euler condition for exactness is not satisfied, and δq is inexact.

Before considering some example calculations, let us note an important feature of Eq. 2.76. The foregoing analysis shows the differential δq to be inexact. However, if we divide Eq. 2.76 by T, a curious thing occurs. Such division produces

$$\frac{\delta q_{\text{rev}}}{T} = \frac{C_v}{T} \, dT + \frac{P}{T} \, dV, \tag{2.79}$$

where the subscript on δq reminds us that Eq. 2.79 is valid only for an ideal gas undergoing a reversible process. The Euler criterion for exactness now requires that

$$\left(\frac{\partial C_v/T}{\partial V}\right)_T = T^{-1}\left(\frac{\partial C_v}{\partial V}\right)_T = 0 = \left(\frac{\partial P/T}{\partial T}\right)_V = \frac{\partial}{\partial T}\left[\frac{nR}{V}\right]_V = 0,$$

and the Euler criterion is satisfied. Thus, the differential $\delta q_{rev}/T$ is exact for an ideal gas undergoing a reversible process. A mathematician would say that the factor T^{-1} acts as an *integrating factor*, turning an inexact differential δq into an exact differential.

Since the preceding analysis applies only to reversible processes for an ideal gas, the result that $\delta q_{rev}/T$ is an exact differential is not particularly useful at this point. The second law of thermodynamics, however, is going to make this result one of the most important ones in all of science!

EXAMPLE 2.4

One mole of an ideal gas is reversibly heated at $dP = 0$ from $P_1 = 3$ atm and $T_1 = 300$ K to $P_2 = 3$ atm and $T_2 = 500$ K. The molar heat capacity of the system has been measured and found to be a function of temperature given by $C_p^m = a + bT + cT^2$, where a, b, and c are constants. (A) Determine ΔH, ΔU, q, and w for this process as functions of a, b, and c. (B) If the heating were done with $dV = 0$, what heat would be associated with the process?

Solution

Although there are usually several ways to approach a problem in thermodynamics, some guiding principles are useful. It is usually best to first address the calculation of the quantities whose differentials are exact. If pressure is constant, compute ΔH first; if volume is constant, compute ΔU first.

(A) Since $dP = 0$, we first calculate ΔH. The system is an ideal gas. Therefore, we have, from Eq. 2.74,

$$dH = C_p\, dT = nC_p^m\, dT = (a + bT + cT^2)\, dT. \tag{1}$$

Integration of both sides of Eq. 1 gives

$$\Delta H = \int dH = \int_{300}^{500} [a + bT + cT^2]dT = \left[aT + \frac{bT^2}{2} + \frac{cT^3}{3}\right]_{300}^{500} \tag{2}$$

$$= 200a + 80{,}000b + 3.267 \times 10^7 c.$$

Since the pressure is constant, Eq. 2.42 shows that $\Delta H = q_p = 200a + 80{,}000b + 3.267 \times 10^7 c$. It is usually easier to compute ΔU than w. Therefore, from Eq. 2.73,

$$\Delta U = C_v\, dT. \tag{3}$$

Using Eq. 2.75, we have $C_v = C_p - nR = C_p^m - R$, because $n = 1$. Thus,

$$\Delta U = \int_{300}^{500} [C_p^m - R]\, dT = \int_{300}^{500} [\{a - R\} + bT + cT^2]\, dT$$

$$= \left[(a - R)T + \frac{bT^2}{2} + \frac{cT^3}{3}\right]_{300}^{500} \tag{4}$$

$$= 200(a - R) + 80{,}000b + 3.267 \times 10^7 c.$$

The work may now be obtained by difference.

$$\Delta U = q + w, \tag{5}$$

so that

$$w = \Delta U - q = 200(a - R) + 80{,}000b + 3.267 \times 10^7 c$$
$$- [200a + 80{,}000b + 3.267 \times 10^7 c] = -200R = -1{,}663J. \qquad (6)$$

(B) For $dV = 0$, we have $\Delta U = q_v = 200(a - R) + 80{,}000b + 3.267 \times 10^7 c$.

For related problems, see 2.13 and 2.18.

EXAMPLE 2.5

Two moles of a nonideal gas with $C_p^m = 2.5R$ and $\mu = 0.27$ K atm^{-1} are heated from 300 K to 500 K. Simultaneously, the gas is compressed from 3 atm to 4 atm. **(A)** Calculate, if possible, ΔH, q, and w. **(B)** Calculate ΔU if possible.

Solution

We are not told anything about the pathway along which the gas is compressed or heated. Therefore, it is not possible to compute q or w. ΔH, however, can be obtained, since we do not need the pathway. Any pathway can be used. For convenience and ease of computation, we shall use a reversible path in which the gas is first heated from 300 K to 500 K at 3 atm. After heating, we execute a reversible compression from 3 to 4 atm at 500 K. This path is shown in Figure 2.21. dH is given by Eq. 2.70:

$$dH = C_p \, dT - \mu C_p \, dP = nC_p^m \, dT - \mu nC_p^m \, dP. \qquad (1)$$

For the heating step, $dP = 0$, and we have

$$\Delta H_{\text{heating}} = \int_{300}^{500} nC_p^m \, dT = (2)(2.5\,R) \int_{300}^{500} dT = 5(500 - 300)R$$
$$= 1{,}000R = 8{,}314 \text{ J}. \qquad (2)$$

For the compression step, $dT = 0$, and Eq. 1 gives

$$\Delta H_{\text{compression}} = -\int_{3}^{4} \mu nC_p^m dP = -\mu nC_p^m \int_{3}^{4} dP$$
$$= -(0.27 \text{ K atm}^{-1})(2 \text{ moles})(2.5R \text{ J mol}^{-1} \text{ K}^{-1})(4 - 3) \text{ atm} = -11. \text{ J.} \qquad (3)$$

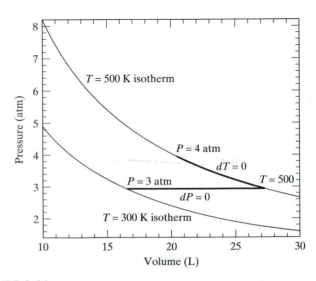

▲ FIGURE 2.21
A possible reversible path for Example 2.5.

The total change in ΔH is

$$\Delta H = \Delta H_{\text{heating}} + \Delta H_{\text{compression}} = 8{,}314 \text{ J} - 11 \text{ J} \approx 8{,}303 \text{ J}. \qquad (4)$$

Note that the heating process makes a much greater contribution to ΔH than compression does.

(B) From Eq. 2.39, we have

$$\Delta H = \Delta U + \Delta(PV). \qquad (5)$$

Therefore,

$$\Delta U = \Delta H - \Delta(PV) = 8{,}303 \text{ J} - \Delta(PV) = 8{,}303 \text{ J} - [(4 \text{ atm})V_2 - (3 \text{ atm})V_1]. \qquad (6)$$

We are given n, P, and T in both the initial and final states. Therefore, if we had the equation of state, we could easily compute the initial and final volumes, V_1 and V_2. However, we know only that the gas is nonideal. Accordingly, without the equation of state, we cannot compute ΔU.

EXAMPLE 2.6

One mole of an ideal gas with $C_v^m = (3/2)R$ is heated reversibly along a path such that $V = A \exp[bT]$, where $A = 1.2257$ liters and $b = 0.01000 \text{ K}^{-1}$. **(A)** How does the pressure vary with T for this system? **(B)** Obtain the molar heat capacity of the gas as a function of T for heating along the given path? **(C)** If 2 moles of the gas are heated reversibly along this path from 300 K to 400 K, compute ΔU, w, and q.

Solution

(A) The gas is ideal; therefore,

$$P = \frac{nRT}{V} = \frac{nRT}{A \exp[bT]} = \frac{nRT}{A} e^{-bT} = (0.06695 \text{ atm K}^{-1}) \, T e^{-bT}. \qquad (1)$$

We see that the pressure tends to vary linearly with T at low temperatures and exponentially with T at higher temperatures.

(B) Clearly, neither the volume nor the pressure is constant during the heating process. Consequently, the molar heat capacity will not equal either C_v^m or C_p^m. Equation 2.50 shows that the molar heat capacity for any path is given by

$$C^m = C_v^m + \left\{ \left(\frac{\partial U}{\partial V} \right)_{T,n} + P_{\text{ext}} \right\} \frac{dV}{dT}. \qquad (2)$$

For an ideal gas, $(\partial U/\partial V)_v$ is zero; therefore,

$$C^m = C_v^m + P_{\text{ext}} \frac{dV}{dT}. \qquad (3)$$

The problem states that the path is reversible. This allows us to replace P_{ext} with P to obtain

$$C^m = C_v^m + P \frac{dV}{dT}. \qquad (4)$$

However, we are told that along this path we have $V = A \exp[bT]$. Thus,

$$\frac{dV}{dT} = Ab \exp[bT] = bV. \qquad (5)$$

Substituting into Eq. 4 gives

$$C^m = C_v^m + bPV = C_v^m + bRT, \qquad (6)$$

since the gas is ideal. Therefore,

$$C^m = 1.5R + (0.01000 \text{ K}^{-1})RT = 12.47 + 0.08314\, T \text{ J K}^{-1} \text{ mol}^{-1}. \tag{7}$$

(C) Following the general principle stated in Example 2.4, we first calculate ΔU. We have

$$dU = C_v\, dT = nC_v^m\, dT = 1.5\, nR\, dT. \tag{8}$$

Integration of both sides then gives

$$\Delta U = \int dU = n\int_{300}^{400} 1.5\, R\, dT = 1.5\, R\, n\,(400 - 300) = 150\,(2)\, R = 2{,}494 \text{ J}. \tag{9}$$

A frequently asked question by students is "How come you use the constant-volume heat capacity to compute ΔU when the volume is not being held constant in this process?" This is a very important question. The answer is that dU is an exact differential, so it doesn't matter what the path is. We are free to use any path we choose, so long as the initial and final states are the same. Since we have an ideal gas, ΔU depends only upon T, so we need only ensure that the integral in Eq. 9 is evaluated between the initial temperature of 300 K and the final temperature of 400 K. However, q and w are path dependent, so in the calculation of these quantities, we must use the actual path.

Equation 2.38 shows that $\Delta U = q_v$. Thus, if the path were one of constant volume, q would be equal to the value of ΔU computed in Eq. 9. However, the process does not have $dV = 0$. Therefore, $q \neq q_v$. Still, q can be computed from the heat capacity for the path computed in (B). We have

$$C = nC^m = \frac{\delta q}{dT}. \tag{10}$$

Hence,

$$\delta q = n\, C^m\, dT = n\,(12.471 + 0.08314\, T)\, dT. \tag{11}$$

Integration of both sides gives

$$\begin{aligned}
q = \int \delta q &= n\int_{300}^{400} [12.47 + 0.08314T]\, dT \\
&= (2 \text{ mol})\left[12.47T + \frac{0.08314T^2}{2} \right]_{300}^{400} \\
&= (2)[12.47(400 - 300) + 0.04157\,(400^2 - 300^2)] = 8{,}314 \text{ J}.
\end{aligned} \tag{12}$$

The total work may now be obtained from the first law:

$$w = \Delta U - q = 2{,}494 - 8{,}314 = -5{,}820 \text{ J}. \tag{13}$$

We could, of course, compute the work directly from

$$w = -\int_{V_1}^{V_2} P\, dV, \tag{14}$$

since the path is reversible and we can replace P_{ext} with P. The pressure is given by Eq. 1 in terms of T. We also have

$$dV = Ab\, e^{bT}\, dT. \tag{15}$$

Converting variables in Eq. 14 to T, we obtain

$$\begin{aligned}
w &= -\int_{T_1}^{T_2} \frac{nRT}{A\exp[bT]} Ab\exp[bT]\, dT = -nbR\int_{300}^{400} T\, dT \\
&= -(2)(0.0100 \text{ K}^{-1})(8.314 \text{ J mol}^{-1}\text{ K}^{-1})(0.5)[400^2 - 300^2] \text{ K}^2 \\
&= -5{,}820 \text{ J mol}^{-1},
\end{aligned} \tag{16}$$

which is the same result as that obtained in Eq. 13.

For related exercises, see Problems 2.27 and 2.28.

2.8 Adiabatic Processes

The Joule–Thomson experiment is an example of an adiabatic process with $\delta q = 0$. In this section, we will obtain the general equation that describes such a process.

The first law gives us $dU = \delta q + \delta w = \delta w$ for any adiabatic process. This equation may be written in the form

$$dU = \left(\frac{\partial U}{\partial T}\right)_{V,n} dT + \left(\frac{\partial U}{\partial V}\right)_{T,n} dV = -P_{ext}\, dV. \tag{2.80}$$

Rearranging Eq. 2.80 and replacing $(\partial U/\partial T)_{V,n}$ with C_v yields

$$C_v\, dT + \left\{\left(\frac{\partial U}{\partial V}\right)_{T,n} + P_{ext}\right\} dV = 0. \tag{2.81}$$

Equation 2.81 is the general equation for adiabatic processes in a closed system.

EXAMPLE 2.7

n moles of an ideal gas are expanded adiabatically and reversibly from $P = P_1$, $V = V_1$, and $T = T_1$ to a final state in which $V = V_2$. **(A)** If $C_v^m = 1.5R$, obtain P_2 and T_2 in terms of n, V_1, T_1, and V_2. **(B)** Obtain an expression for the total work done in the process as a function of n, V_1, T_1, and V_2.

Solution

(A) Since we have an ideal gas, $(\partial U/\partial V)_{T,n} = 0$. The process is reversible, so we may replace P_{ext} with P. Under these conditions, Eq. 2.81 becomes

$$n\, C_v^m\, dT + P\, dV = 0. \tag{1}$$

Replacing P in Eq. 1 with nRT/V then gives

$$n\, C_v^m\, dT = -\frac{nRT}{V}\, dV. \tag{2}$$

Cancellation of n and dividing by T produces

$$\frac{C_v^m}{T}\, dT = -R\frac{dV}{V}. \tag{3}$$

Since the variables are now separated, we may integrate both sides of Eq. 3 between corresponding limits to obtain

$$\int_{T_1}^{T_2} \frac{C_v^m}{T}\, dT = -R\int_{V_1}^{V_2} \frac{dV}{V}. \tag{4}$$

Because C_v^m is a constant, the integration gives

$$C_v^m \ln\left[\frac{T_2}{T_1}\right] = \ln\left[\frac{T_2}{T_1}\right]^{C_v^m} = -R \ln\left[\frac{V_2}{V_1}\right] = \ln\left[\frac{V_2}{V_1}\right]^{-R} = \ln\left[\frac{V_1}{V_2}\right]^{R}. \tag{5}$$

Exponentiation of both sides of Eq. 5 produces

$$\left[\frac{T_2}{T_1}\right]^{C_v^m} = \left[\frac{V_1}{V_2}\right]^{R}. \tag{6}$$

Raising each side of Eq. 6 to the $[C_v^m]^{-1}$ power, we obtain

$$T_2 = T_1 \left[\frac{V_1}{V_2}\right]^{R/C_v^m} = T_1 \left[\frac{V_1}{V_2}\right]^{R/1.5R} = T_1 \left[\frac{V_1}{V_2}\right]^{2/3}, \tag{7}$$

which is the desired expression for T_2. P_2 can now be found from the ideal-gas equation of state. We have

$$P_2 = \frac{nRT_2}{V_2} = \frac{nR}{V_2} \left\{ T_1 \left[\frac{V_1}{V_2}\right]^{2/3} \right\} = \frac{nRT_1 V_1^{2/3}}{V_2^{5/3}}, \tag{8}$$

which is the required expression.

Equation 7 demonstrates that an adiabatic expansion with $V_2 > V_1$ results in a final temperature $T_2 < T_1$. Physically, we observe this result because the energy needed to do the work of expansion must come from the internal energy of the gas. The temperature of the gas must, therefore, decrease.

(B) The work can be obtained directly from the fact that, for an adiabatic process, $dU = \delta w$. Therefore,

$$w = \int dU = \int_{T_1}^{T_2} C_v\, dT = n\, C_v^m\, [T_2 - T_1] = 1.5\, n\, R \left\{ T_1 \left[\frac{V_1}{V_2}\right]^{2/3} - T_1 \right\}$$

$$= 1.5\, nRT_1 \left\{ \left[\frac{V_1}{V_2}\right]^{2/3} - 1 \right\}, \tag{9}$$

which is the required expression.

In an expansion, $T_2 < T_1$, as already noted. Examining Eq. 9 in the light of this fact shows that we have $w < 0$, and the system does work of expansion on the surroundings.

For related exercises, see Problems 2.14, 2.25 and 2.26.

Summary: Key Concepts and Equations

1. In physical science, all measurements are difference measurements. We are, therefore, interested in computing the change in dependent variables for given changes in the independent variables. Since we are always concerned with Δz rather than z itself, we can choose the reference point for the measurement or computation of z in any manner that is convenient.

2. The change in a variable z that depends upon w, x, and y is computed using the total differential

$$dz = \left(\frac{\partial z}{\partial w}\right)_{x,y} dw + \left(\frac{\partial z}{\partial x}\right)_{w,z} dx + \left(\frac{\partial z}{\partial y}\right)_{w,x} dy.$$

The critical quantities that are needed to obtain Δz are the partial derivatives of z with respect to each of the independent variables.

3. If a differential such as $[f(x,y)\, dx + g(x,y)\, dy]$ is *exact*, its definite integral will be independent of the pathway employed to change the independent variables. In addition, there will always exist a function $z(x,y)$ whose total differential is exactly equal to $[f(x,y)\, dx + g(x,y)\, dy]$.

4. If the differential $[f(x,y)\, dx + g(x,y)\, dy]$ is exact, the Euler criterion for exactness requires that

$$\left(\frac{\partial f(x,y)}{\partial y}\right)_x = \left(\frac{\partial g(x,y)}{\partial x}\right)_y.$$

5. The differential work is defined to be the dot product between the external force vector and the vector displacement of the system; that is,

$$\delta w = \mathbf{F}_{ext} \cdot d\mathbf{s},$$

where the subscript "ext" denotes the external restraining force against which the displacement is executed. As noted, the work differential is inexact, so that the work done in a process depends upon the path connecting the initial and final states of the system. If the only work done is expansion or contraction of the system against an external force, the pressure–volume work is given by

$$\delta w = -P_{ext}\, dV.$$

6. In thermodynamics, spontaneous processes are termed *irreversible,* since they will not spontaneously turn around and proceed in the opposite direction. A hypothetical process that occurs under equilibrium conditions is called a *reversible* process, in that it is as likely to proceed in one direction as the other.

7. The maximum magnitude of the work that can be obtained in the expansion of a gas is the *reversible* work that would be obtained if expansion could occur under conditions in which the internal and external pressures were equal at all points during the expansion. Thus, the reversible pressure–volume work is given by

$$\delta w = -P\, dV.$$

8. Heat is denoted by the symbol q. The sign is defined such that δq is negative when the system loses heat. The heat differential is inexact; consequently, the heat associated with a process depends upon the path by which the process is executed.

9. The first law of thermodynamics states that the sum of the work and heat differentials is exact. Therefore, a function exists whose differential is exactly this sum. We call this function the *internal energy* and give it the label U. The first law is

$$dU = \delta q + \delta w.$$

Since dU is exact, the change in internal energy, ΔU, for any process is independent of the path. Furthermore, around any closed path with identical initial and final states, we always have $\Delta U = 0$.

10. When processes occur under conditions of constant volume $(dV = 0)$, the heat associated with the process is always equal to ΔU. That is, we always have $\Delta U = q_v$.

11. The *enthalpy* is defined to be

$$H \equiv U + PV.$$

Consequently, dH is exact. Using a reversible path, we can write the enthalpy differential in the form

$$dH = \delta q + V\, dP$$

so that all processes conducted under conditions of constant pressure have $\Delta H = q = q_p$.

12. Heat capacity is defined as the rate of change of heat with respect to temperature:

$$C \equiv \frac{\delta q}{dT}.$$

The inexactness of δq makes the heat capacity path dependent. The rate of change of the internal energy with temperature at constant volume is equal to the constant-volume heat capacity:

$$\left(\frac{\partial U}{\partial T}\right)_V = C_v.$$

The rate of change of enthalpy with temperature at constant pressure is the heat capacity when the heating is done at constant pressure:

$$\left(\frac{\partial H}{\partial T}\right)_P = C_p.$$

13. The rate of change of enthalpy with pressure at constant temperature can be measured in a Joule–Thomson experiment. The result is

$$\left(\frac{\partial H}{\partial P}\right)_T = -\mu\, C_p,$$

where μ is the Joule–Thomson coefficient, equal to $(\partial T/\partial P)_H$, which is the quantity measured in the experiment.

14. Because there are no interactions between molecules for an ideal gas, we expect to find that

$$\left(\frac{\partial U}{\partial V}\right)_T = 0.$$

This is Joule's law. It leads to the result that μ is zero for an ideal gas. Consequently, the internal energy and enthalpy differentials become $dU = C_v\, dT$ and $dH = C_p\, dT$, respectively, for an ideal gas. Thus, U and H depend only upon T if the gas is ideal. Joule's law is actually a consequence of the second law of thermodynamics.

15. Thermodynamic adiabatic processes are those for which $\delta q = 0$. For any adiabatic process, we have

$$C_v\, dT + \left\{\left(\frac{\partial U}{\partial V}\right)_{T,\,n} + P_{\text{ext}}\right\} dV = 0.$$

If the path of the process is reversible, P_{ext} may be replaced with P.

Problems

Problems that require the use of some type of computational device are marked with an asterisk (*). Problems that require some type of plotting routine are indicated with a pound sign (#). Unless otherwise stated, all gases may be assumed to behave ideally.

2.1 Determine whether the expressions that follow are exact differentials. If so, determine the function F whose total differential is equal to the expression given.

(A) $[\cos(3y) + x\exp(-y^2)]\, dx +$
$[2xy\exp(-y^2) - 3x\sin(3y)]\, dy$

(B) $[2x^3 - xy^2 - 2y + 3]\, dx - [x^2y + 2x]\, dy$

2.2 Consider the differential

$$dZ = [2xy + \sin(y)]\, dx + [x^2 + x\cos(y)]\, dy.$$

(A) The state of the system is changed from $(x = 0, y = 0)$ to $(x = 1, y = 1)$ along the path $x = y$. Calculate the total change in Z along this path.

(B) The change in (A) is now made along the path $x = y^2$. Compute the total change in Z along this path.

(C) Show that dZ is an exact differential.

2.3 An automobile is moving down a straight highway, which we denote as the X-axis. A reasonable person suggests that the distance the car has moved should be measured along that axis, using the starting point as the reference point $X = 0$. Thus, if the car were at the point $X = X_1$, we would simply measure the distance of the car from the starting point, set that distance equal to X_1, and state the location of the car as $X = X_1$. However, a less intelligent person suggests that the position of the car be measured relative to an observer off to one side at point P. He suggests that instead of measuring the value of X_1 directly, we measure the distance d of the car from point P, as shown in the following diagram:

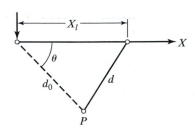

The distance between the observer at point P and the starting point is d_o, and the angle between a vector drawn between point P and the X-axis is given by θ.

(A) Obtain an expression giving the value of X_1, the car's position along the X-axis, in terms of d, d_o, and θ.

(B) Does a measurement of d uniquely determine the position of the car? Under what conditions will it do so?

(C) Comment on the choice of this reference point for the measurement.

2.4 Two moles of an ideal gas at 300 K and 10 atm pressure are expanded isothermally against a constant external pressure of 5 atm until the internal pressure reaches a value of 7 atm. At this point, the expansion is halted. Compute ΔU, ΔH, q, and w for the process if possible. You may ignore acceleration effects. (See explanation in Example 2.3.)

2.5 Two moles of an ideal gas at 300 K and 10 atm pressure are expanded isothermally against a constant external pressure of 5 atm until the internal pressure reaches a value of 7 atm. At this point, the external pressure is reduced to zero and the gas is further expanded into a vacuum until a final state with $P = 1$ atm and $T = 300$ K is reached. Compute, if possible, ΔU, ΔH, q, and w for the process. You may ignore acceleration effects. (See explanatory note in Example 2.3.)

2.6 Two moles of an ideal gas at 300 K and 10 atm pressure are expanded isothermally against a constant external pressure of 5 atm until the internal pressure reaches a value of 7 atm. At this point, the external pressure is reduced to zero and the gas is further expanded into a vacuum until an internal pressure of 1 atm is reached with $T = 300$ K. The gas is then compressed reversibly and isothermally back to its initial state of $T = 300$ K and $P = 10$ atm. Compute, if possible, w, q, ΔU, and ΔH for the total process. After the entire process is finished, has heat been gained or lost by the system? You may ignore acceleration effects. (See explanatory note in Example 2.3.)

2.7 Two moles of an ideal gas at 300 K and 10 atm are expanded isothermally to a final pressure of 2 atm with $T = 300$ K.

(A) What is the minimum magnitude of the work for this expansion?

(B) What is the maximum magnitude of the work for the expansion?

2.8 Four moles of O_2 gas at 400 K and a pressure of 10 atm are expanded isothermally and irreversibly against a constant external pressure of 1 atm. Compute the values of w, q, ΔU, and ΔH after equilibrium is reached. You may ignore acceleration effects. (See explanatory note in Example 2.3.)

2.9 (A) Four moles of O_2 gas at 400 K and an internal pressure of 10 atm are expanded isothermally and irreversibly to a pressure of 4 atm against an external pressure that is equal to $0.5P_{int}$ at all points during the expansion. When P_{ext} reaches a pressure of 2 atm, it is held constant at that pressure, and isothermal expansion continues until the internal pressure reaches 2 atm. Compute q, w, ΔU, and ΔH for the process if O_2 is an ideal gas. You may ignore acceleration effects. (See explanatory note in Example 2.3.)

(B) Does the work computed in (A) have the maximum magnitude possible for the given expansion? If so, prove the result to be a maximum. If not, compute the maximum magnitude the work can have for the expansion.

2.10 Two moles of an ideal gas are expanded isothermally and irreversibly from 20 liters to 30 liters against a constant external pressure. The total work for the process is found to be $-5{,}065.8$ joules.

(A) Determine the external pressure present during the expansion.

(B) Compute ΔU, ΔH, and q for the process.

(C) If the system is at equilibrium at the end of the expansion, what is the temperature of the system? You may ignore acceleration effects. (See explanatory note in Example 2.3.)

2.11 (A) 100 g of N_2 are heated from 300 to 500 K at a constant pressure of 1 atm. Using the data in Table 2.1, compute q, ΔH, w, and ΔU.

(B) The system in (A) is heated from 300 K to 500 K at constant volume. Compute q, w, ΔH, and ΔU.

(C) Why is $|q|_{Part A} > |q|_{Part B}$?

2.12 A gas obeys the equation of state $PV = RT + \alpha P$, where α is a function of T only.

(A) Show that the reversible work done whenever the gas is heated at constant pressure from T_1 to T_2 is given by

$$w = R(T_1 - T_2) + (\alpha_1 - \alpha_2)P,$$

where α_1 and α_2 are the values of α at temperatures T_1 and T_2, respectively.

(B) Show that if the gas is expanded isothermally from V_1 to V_2, the reversible work will be

$$w = RT \ln\left[\frac{V_1 - \alpha}{V_2 - \alpha}\right].$$

2.13 The oxygen in a high-pressure cylinder is at 300 K and 100 atm pressure. The valve of the cylinder is opened, and the gas is allowed to expand to an atmospheric pressure of 1 atm. Assuming that the expansion is adiabatic and conducted under conditions of constant enthalpy, estimate the final temperature of O_2 after expansion. Do not assume O_2 to be an ideal gas.

2.14[#] One mole of an ideal gas at a temperature of 500 K and a pressure of 6 atm is subjected to the following changes:

Step 1. The gas is expanded isothermally and reversibly to a final pressure of 5 atm.

Step 2. After completion of Step 1, the gas is expanded adiabatically and reversibly until the pressure reaches 4 atm.

Step 3. After Step 2 is completed, the gas is compressed isothermally and reversibly to a pressure of 4.800 atm.

Step 4. After Step 3, the gas is compressed adiabatically and reversibly to a pressure of 6 atm, at which point the temperature is found to be 500 K.

The molar heat capacity of the gas at constant volume is $C_v^m = 1.5R$.

(A) Compute w, q, and ΔU for Step 1.

(B) At the completion of Step 2, what are the temperature and volume of the gas? Compute the amount of work done in Step 2.

(C) Compute w, q, and ΔU for Step 3. What is the volume of the gas at the completion of Step 3?

(D) Compute the amount of work done in Step 4.

(E) For the entire process, compute ΔU, w, and q.

(F) How does P vary with V in Steps 1 and 3? How does P vary with V in Steps 2 and 4? Make a careful, quantitative plot of P versus V for all four steps on the same graph. How is the total work done in the overall process related to your graph? This cyclic process is called a *Carnot* cycle. We shall employ it again to prepare for the introduction of the second law of thermodynamics.

2.15 One mole of an ideal gas with $C_p^m = (5/2)R$ is heated reversibly at a constant pressure of 1 atm from 273.15 K to 373.15 K.

(A) Compute the work involved in the process.

(B) If the gas were expanded isothermally and reversibly at 273.15 K from an initial pressure of 1 atm, what would the final pressure need to be in order for the work to be equal to that calculated in (A)?

2.16 Show that $\Delta H = \Delta U$ for any isothermal change of state for an ideal gas.

2.17 Suppose that N_2 gas may be described by a van der Waals equation of state. One mole of N_2 is isothermally and reversibly expanded from a volume of 1 liter to 10 liters at 300 K. Compute the work done in the process. Is the magnitude of the result larger or smaller than that for an ideal gas undergoing the same process? Explain.

2.18[*] For reasons that will be explored in more detail later in the text, the heat capacity of a gas exhibits a temperature dependence. The measured constant-pressure molar heat capacities for N_2 at a pressure of 1 atm are given in the following table:

T (K)	C_p^m (J mol^{-1} K^{-1})
200.000	29.200
300.000	29.200
400.000	29.200
500.000	29.600
600.000	30.100
700.000	30.700
800.000	31.400
900.000	32.000
1,000.000	32.600
1,100.000	33.200
1,200.000	33.600
1,300.000	34.100
1,400.000	34.400
1,500.000	34.700

(A) Using a linear least-squares method, obtain the best fit of C_p^m to the function

$$C_p^m = a_o + a_1T + a_2T^2.$$

(B) 50 grams of N_2 are heated from 300 K to 1,000 K at a constant pressure of 1 atm. Compute q and ΔH for the process.

(C) At 300 K, 50 grams of N_2 are compressed isothermally from 1 atm to 20 atm. Compute ΔH for the process. Do not assume N_2 to be an ideal gas. (*Hint:* You will need the Joule–Thomson coefficient in this part of the problem.)

2.19 Two moles of an ideal gas at 500 K and $P = 10$ atm are contained in a cylinder–piston arrangement. As this gas is expanded isothermally and reversibly, more gas at 500 K is continuously added to the cylinder in a quantity sufficient to maintain the gas pressure constant at 10 atm. The process is continued until the total work done equals -500 L atm. Calculate the number of moles of gas contained in the cylinder at this point.

2.20 In Chapter 1, it was shown that the second virial coefficient for a gas described by a van der Waals equation of state is $C_2(T) = b - a/(RT)$, where a and b are the van der Waals parameters.

(A) Obtain an expression giving the work done in an isothermal, reversible expansion from volume V_1 to volume V_2 for a gas whose equation of state is a virial expansion.

(B) If the virial expansion is truncated after the second virial term, compute the work done for the isothermal expansion of 1 mole of CO_2 at 300 K from 10 liters to 50 liters. The van der Waals parameters for CO_2 are given in Table 1.1.

(C) Using the van der Waals equation of state, determine the work done for the process in (B). What is the percent difference between the answers in (B) and (C)?

2.21 More about the thermodynamics of hell. (*Note*: This problem is dependent upon the analysis and assumptions contained in Problems 1.17 and 1.18). In Problems 1.17 and 1.18, we found that if hell is expanding at a slower rate than the rate at which additional souls enter the place, the temperature of hell will decrease with time. With certain assumptions we made in those problems, it could be determined that the temperature of hell is given by

$$T = 3.326 \times 10^{30} \exp\left[\left(b - \frac{\ln[10^{10}f + 1]}{5,000}\right)t\right]$$

in degrees K,

where b is a constant that determines the rate of expansion of the volume of hell and f is the fraction of people whose souls will enter hell. In Chapter 1, we employed the value $b = 0.0020723$ years^{-1}. We also found that the number of moles of souls in hell is given by

$$n = A \exp[at],$$

where $A = 1.66 \times 10^{-24}$ mol and $a = \ln(10^{10}f + 1)/5,000$ years^{-1}.

(A) If the average molar heat capacity of hell at constant pressure is $3.5R$, determine the total heat change q of hell since the Devil entered it 5,000 years ago (by assumption).

(B) Given that hell is expanding at constant pressure, compute the total amount of pressure–volume work done by the expansion of hell since the Devil entered it.

(C) Compute ΔH and ΔU for hell over the last 5,000 years.

2.22 Two moles of an ideal gas at 500 K and 10 atm pressure are contained in an insulated piston–cylinder arrangement such as the following:

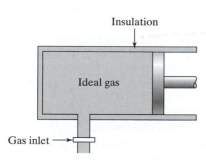

The gas is allowed to expand reversibly and adiabatically. During the expansion, gas at a temperature equal to the current instantaneous temperature of the gas in the cylinder is added through an inlet valve in sufficient quantity to maintain the internal pressure constant at 10 atm throughout the expansion. When the total work done equals -50 L atm, the expansion is halted. Calculate the final temperature of the gas inside the piston and determine the number of moles of gas present after the expansion is complete. You may assume that the molar heat capacity of the gas, C_v^m, is constant at the value $1.5R$.

2.23 Two moles of an ideal gas at 500 K and 10 atm pressure are contained in the insulated piston–cylinder arrangement shown in Problem 2.22. The gas inlet valve is closed and the gas is heated at constant pressure to 800 K. During the heating, the piston expands to allow reversible work to be done on the surroundings. If $C_v^m = 1.5R$, calculate q, w, and ΔU for the process.

2.24 (A) An ideal gas at 300 K and 1 atm pressure fills a 100-liter insulated cylinder fitted with a small escape value. (See accompanying figure.) The external pressure is 1 atm.

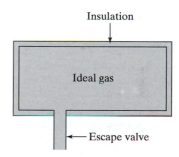

The gas inside the piston is heated from 300 to 400 K. During the heating, sufficient gas escapes through the valve to maintain the gas at a pressure of 1 atm. Assuming that the container does not expand and that C_v^m for the gas is a constant 1.5R, compute the amount of energy that must be added to the gas to effect the stated temperature change.

(B) Compute the fraction of the heat added in (A) that escapes to the surroundings by virtue of the gas that is expelled through the valve.

2.25 One mole of an ideal gas is contained in an insulated piston–cylinder arrangement $(dT = 0)$ in an initial state (T_1, P_1, V_1). The gas is allowed to expand adiabatically and irreversibly against a constant external pressure P_o until a point is reached where the internal pressure becomes equal to P_o. If C_v^m for the gas is constant and equal to 1.5R, derive an expression giving the final temperature of the gas in terms of P_o, V_1, T_1, and R. Ignore acceleration effects. (See explanatory note in Example 2.3.)

2.26 A gas is described by a van der Waals equation of state. However, the attractive forces between the molecules are zero. Therefore, the parameter a is set to zero. Under these conditions, the equation of state becomes

$$P(V_m - b) = RT.$$

One mole of this gas at a temperature T_1 is expanded adiabatically and reversibly from an initial pressure P_1 to a final pressure P_2. The molar heat capacity is $C_v^m = 1.5R$.

(A) Obtain the final temperature of the gas as a function of T_1, P_1, and P_2.

(B) If $P_1 = 15$ atm, $P_2 = 2$ atm, and $T_1 = 400$ K, compute the final temperature.

2.27 One mole of a monatomic ideal gas is heated along a reversible path such that $C = R$. What is the functional relationship between T and V along this path? (For a monatomic ideal gas, $C_v^m = 1.5R$.)

2.28 One mole of an ideal gas with $C_v^m = 1.5R$ is heated along a reversible path such that

$$P = a\,T^{1/3},$$

where a is constant at all points along the path. Obtain an expression for the heat capacity of the gas when heated along this path.

2.29 Consider a gas whose equation of state over a certain temperature range can be represented by

$$PV = RT + aT^2$$

for 1 mole of gas, where a is a constant. Using the second law, we can show that, for such a gas,

$$\left(\frac{\partial U}{\partial V}\right)_T = \frac{aT^2}{V}.$$

(A) The gas is expanded isothermally and reversibly from V_1 to V_2. Derive an expression for the work associated with this process.

(B) Obtain an expression giving the change in the internal energy for the process in (A).

(C) Obtain an expression giving q for the process in (A).

2.30 Everyone who pays a heating bill at the end of the month is concerned about the amount of energy it takes to warm a house. In this problem, you will derive the appropriate equation required to compute that energy. Consider a house of volume V. We wish to heat the house from $T = T_1$ to $T = T_2$ at a constant pressure of P atm. Let us assume that the air in the house may be treated as an ideal gas whose average molar heat capacities, C_v^m and C_p^m, are independent of the temperature.

(A) Derive an expression in terms of P, V, T_1, T_2, and appropriate heat capacities for the amount of energy required to execute the process. [*Hint:* Remember that the amount of air in the house is not constant: As the air is heated, it expands and escapes through the cracks in the windows, doors, etc. Be certain that you take this continuous ejection of air into account in your calculations.]

(B) If $P = 1$ atm, $V = 2{,}000$ m³, $T_1 = 15°C$, $T_1 = 25°C$, and $C_p^m = 2.5R$, compute the heat needed in joules.

2.31 Suppose the house of Problem 2.30 were cooled from temperature T_1 to temperature T_2 with the temperature of the outside air at T_o.

(A) How much heat must be extracted from the air within the house to execute this cooling if you take into account the entry of additional air from the outside upon cooling?

(B) If $T_1 = 25°C$, $T_2 = 15°C$, $T_o = 35°C$, $C_p^m = 2.5R$, $P = 1$ atm, and $V = 2{,}000$ m³, how many joules must be extracted? The amount of thermal energy required to heat the house over a similar temperature range obtained in Problem 2.30 is 1.728×10^7 J. Is it more difficult to cool or to heat a house?

2.32 An insulated cylinder of uniform cross section is equipped with a porous plug and provided with a frictionless piston on one side and a fixed end on the other. (See accompanying figure.)

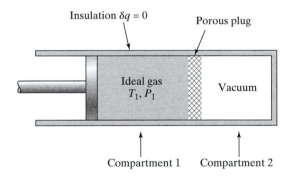

The space between the plug and the fixed end (Compartment 2) is evacuated. The space between the plug and the piston (Compartment 1) contains a large quantity of an ideal gas at temperature T_1 and pressure P_1. The gas on the left is now permitted to effuse through the plug while keeping the pressure on the left-hand side constant at P_1 by depressing the piston. The process continues until the pressure in the right-hand compartment becomes equal to P_1. If the process is adiabatic and there is no heat conduction through the plug, and if we have $P_1 = 1$ atm, $T_1 = 300$ K, $C_v^m = 1.5R$, and $C_p^m = 2.5R$, what is the final temperature of the gas in Compartment 2 at the end of the experiment?

2.33 One mole of an ideal gas is subjected to the following sequence of steps:

Step 1. The gas is heated from 25°C to 100°C at constant volume.

Step 2. The gas is then expanded freely into a vacuum to double its volume.

Step 3. The gas is cooled reversibly to 25°C at constant pressure.

Calculate, if possible, ΔU, ΔH, q, and w for the overall process (Step 1 + Step 2 + Step 3). Do you need to know the heat capacities of the gas?

2.34 Sam is having a great deal of difficulty with thermodynamics, particularly when the systems are not ideal. He complains to his girlfriend that his professor gives homework and exam problems in which the systems are not ideal. "I wish all systems were ideal gases, " he tells his girlfriend. His girlfriend, who is making an A in physical chemistry, attempts to explain that this wouldn't be such a wonderful idea.

As you relax with a pizza and your favorite beverage, consider what would happen to your pizza and to you if Sam gets his wish. Having done this, go to a movie. Thermochemistry is coming up next.

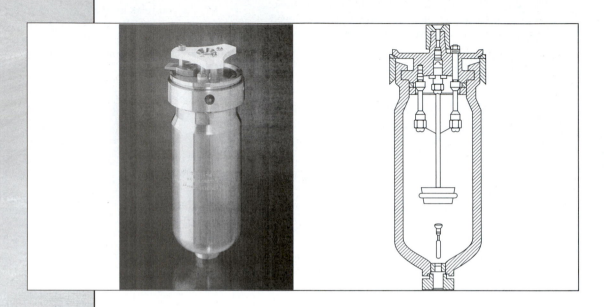

Thermochemistry

In your introductory chemistry courses, you learned how the heat of a chemical reaction occurring at 298 K can be computed using standard molar heats of formation. You also learned how to use Hess' law to sum thermochemical equations to obtain additional information concerning heats of reaction. In this chapter, we shall utilize the first law of thermodynamics to extend your knowledge so that you will be able to compute heats of reactions at any temperature or pressure. In the process, you will encounter the first examples of many to come of the methods used by the physical chemist to choose reference points so as to simplify the calculations and reduce the experimental difficulties.

3.1 Heats of Reaction

When a chemical reaction occurs, heat is almost always either released to or absorbed from the surroundings. In the former case, we term the reaction *exothermic*; in the latter case, the reaction is said to be *endothermic*. In thermodynamic terms, exothermic reactions are associated with processes for which $\delta q < 0$, whereas endothermic reactions have $\delta q > 0$.

Since δq is an inexact differential, it is clear that the heat of reaction will depend upon the pathway by which the reaction occurs. If this path is known, the first law provides the means by which we may compute q for any reaction, provided that the required heat capacities, equations of state, and Joule–Thomson coefficients have been determined.

The ability to execute such calculations is important because we are concerned with the amount of energy that can be produced from reacting a given quantity of fuel. As we shall see in later chapters, the heat of reaction is also directly connected to other critical chemical properties, such as the equilibrium constant and the reaction rate. In this chapter, we shall examine some of the tools available to determine heats of reaction.

3.1.1 Standard Partial Molar Enthalpies

Let us begin by considering a pure substance that we will denote as A. The total differential for the change in enthalpy associated with this compound is given by combining Eqs. 2.47 and 2.70 to obtain

$$dH_A = [C_p]_A \, dT - [\mu C_p]_A \, dP + \overline{H}_A \, dn_A , \tag{3.1}$$

where

$$\overline{H}_A = \left(\frac{\partial H_A}{\partial n_A} \right)_{T \, P \, nj} \tag{3.2}$$

is the partial molar enthalpy of compound A. Physically, $\overline{H}_A$ is the enthalpy per mole of compound A, which will depend upon its environment. One mole of pure ethanol at a given temperature and pressure does not have the same enthalpy as 1 mole of ethanol when mixed with water at the same temperature and pressure, because of the interactions between the ethanol and water molecules. The importance of $\overline{H}_A$ is illustrated by a simple example.

Suppose we have a container holding 10 moles of compound A at temperature T_1 and pressure P_1, as shown in Figure 3.1. We now add 3 more moles of compound A at this same temperature and pressure and ask, What is ΔH for the process? Equation 3.1 provides the answer: Since $dT = dP = 0$ for the process, the first two terms vanish, and we have

$$dH_A = \overline{H}_A \, dn_A. \tag{3.3}$$

Integrating both sides, we obtain

$$\Delta H = \int dH_A = \int_{n_A=10}^{n_A=13} \overline{H}_A \, dn_A = \overline{H}_A \int_{n_A=10}^{n_A=13} dn_A = \overline{H}_A(13 - 10) = 3\overline{H}_A. \tag{3.4}$$

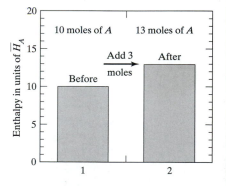

▲ FIGURE 3.1
Change in enthalpy produced by a change in the number of moles of a pure substance. In the example shown, three moles of pure compound A are added at constant temperature and pressure to 10 moles already present. The change in enthalpy produced by the addition is $3\overline{H}_A$, where $\overline{H}_A$ is the partial molar enthalpy of compound A. For a pure substance, $\overline{H}_A$ is effectively the enthalpy per mole of A.

105

(Since $\overline{H}_A$ is independent of the quantity of compound A present, it may be factored out of the integral.) We see, then, that ΔH depends upon the change in the number of moles of compound A and upon the value of $\overline{H}_A$. Consequently, we must find a means of measuring the partial molar enthalpy.

The measurement of the partial molar enthalpy presents a problem: We seem to be out of "partial molar enthalpy meters." To compound the problem, we can never measure the absolute value of anything—only differences can be determined.

To see how these difficulties can be circumvented, let us assume that compound A in this example is $CO_2(g)$ and that all gases are ideal. (The notation denotes the physical state of the compound: Solids, liquids, and gases are indicated by (s), (l) and (g), respectively.) We cannot measure $\overline{H}_{CO_2(g)}$ directly, but we can measure the heat of reaction when $C(s)$ burns in the presence of $O_2(g)$ to form $CO_2(g)$. (Some experimental techniques for executing such measurements will be discussed later in this chapter.) When such a measurement is made, we find that 393.51 kJ of heat are released whenever 1 mole of $CO_2(g)$ is formed by the reaction of 1 mole of $C(g)$ with 1 mole of $O_2(g)$ if the reaction path is one of constant pressure and temperature with $P = 1$ bar and $T = 298.15$ K. We indicate this thermochemical result by writing

$$C(s) + O_2(g) \longrightarrow CO_2(g) + 393.51 \text{ kJ} \qquad [T = 298.15 \text{ K}, P = 1 \text{ bar}].$$

Equation 2.42 shows that, since we have $dP = 0$ for the reaction, $\Delta H = q_p = -393.51 \text{ kJ mol}^{-1}$.

When the pressure is 1 bar for a condensed phase or an ideal gas, we say the material is in its *standard state* at temperature T. The preceding reaction, therefore, results in the formation of 1 mole of $CO_2(g)$ from the elements with all substances in their standard states at 298.15 K. For this reason, ΔH for the reaction is labeled the *standard molar enthalpy of formation* of $CO_2(g)$ and is given the symbol $\Delta H_f^o(298.15)_{CO_2(g)}$, where the superscript o indicates the standard thermochemical state of 1 bar pressure and ideal gases. The subscript f indicates formation from the elements.

We may now utilize Eq. 3.1 to obtain an expression giving ΔH for the foregoing reaction. Since three compounds are present, we have

$$dH(298.15 \text{ K}) = dH_{C(s)}(298.15) + dH_{O_2(g)}(298.15) + dH_{CO_2(g)}(298.15)$$

$$= \overline{H}_{C(s)}^o(298.15) \, dn_{C(s)} + \overline{H}_{O_2(g)}^o(298.15) \, dn_{O_2(g)} + \overline{H}_{CO_2(g)}^o(298.15) dn_{CO_2(g)}. \qquad \textbf{(3.5)}$$

Integrating both sides of Eq. 3.5 gives

$$\Delta H = \Delta H_f^o(298.15)_{CO_2(g)} = \int_1^0 \overline{H}_{C(s)}^o(298.15) dn_{C(s)}$$

$$+ \int_1^0 \overline{H}_{O_2(g)}^o(298.15) dn_{O_2(g)} + \int_0^1 \overline{H}_{CO_2(g)}^o(298.15) dn_{CO_2(g)}. \qquad \textbf{(3.6)}$$

The limits on the integrals in Eq. 3.6 are obtained from the fact that the initial state has 1 mole of $C(s)$ and $O_2(g)$ and no $CO_2(g)$, whereas the final state has no $C(s)$ or $O_2(g)$ and 1 mole of $CO_2(g)$. Since the reactants and products are pure compounds, the $\overline{H}_i^o(298.15)$ are independent of the amount of material and can factored from beneath the integrals. Therefore, we obtain

$$\Delta H = \Delta H_f^o(298.15)_{CO_2(g)} = \overline{H}_{CO_2(g)}^o(298.15)$$

$$- [\overline{H}_{C(s)}^o(298.15) + \overline{H}_{O_2(g)}^o(298.15)] = q_p = -393.51 \text{ kJ}. \qquad \textbf{(3.7)}$$

As expected, we are able to measure the difference between the standard partial molar enthalpies of $CO_2(g)$ and those for $C(s)$ and $O_2(g)$, but we cannot measure $\overline{H}^o_{CO_2(g)}$ itself.

The solution to our dilemma is found in the opening section of Chapter 2. We are free to choose our reference point for measurement in any way we like. Since we wish to simplify the problem as much as possible, we choose the reference point for the measurement of partial molar enthalpies to be the partial molar enthalpies of the elements in their standard states. We assign these standard partial molar enthalpies a value of zero. Therefore, we have $\overline{H}^o_{C(s)}(T) = \overline{H}^o_{O_2(g)}(T) = 0$. With this choice of reference point, Eq. 3.7 becomes

$$\overline{H}^o_{CO_2(g)}(298.15) = \Delta H^o_f(298.15)_{CO_2(g)} = -393.51 \text{ kJ.} \qquad (3.8)$$

Having measured the standard partial molar enthalpy of $CO_2(g)$ at 298.15 K, we may record the value in an appropriate table and use it when needed. The standard partial molar enthalpies of a large number of compounds at 298.15 K have been determined by these methods. Appendix A gives a listing of such results for some common compounds. The 78th edition of the CRC *Handbook of Chemistry and Physics* provides more extensive data. Equation 3.8 shows that the standard partial molar enthalpy of a compound is numerically equal to its standard molar enthalpy of formation at the same temperature, provided that we adopt the reference state defined in the previous paragraph. For this reason, tables giving the values of standard partial molar enthalpies frequently label the data with the symbol $\Delta H^o_f(T)$ and refer to them as standard molar heats of formation. As long as the relationship given in Eq. 3.8 is understood, the nomenclature creates no problem.

Figure 3.2 shows the experimentally determined standard partial molar enthalpies of a series of 12 *n*-alkanes at 298.15 K. The near linear dependence of $\overline{H}^o_i$ upon the number of carbon atoms in the molecule is apparent from the figure. In Section C, on bond enthalpies, we will discuss the qualitative reason for this behavior. Problem 3.36 explores this linear dependence in more quantitative detail.

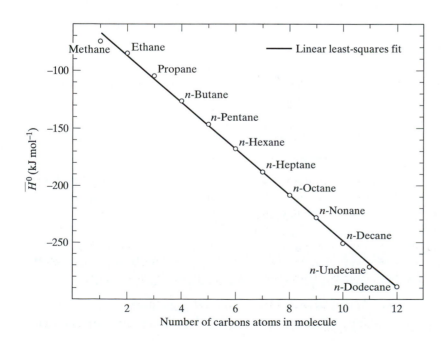

◀ **FIGURE 3.2**
Standard partial molar enthalpies of gas-phase *n*-alkanes at 298.15 K obtained from measured standard enthalpies of formation as described in the text. The near linear dependence of $\overline{H}^o_i$ on the number of carbon atoms in the alkane is apparent. The underlying reason for this behavior is discussed later in the chapter.

3.1.2   Standard Enthalpies of Reaction

Consider a general reaction with all compounds at a pressure of 1 bar. We can write such a reaction in the form

$$\nu_A A + \nu_B B \longrightarrow \nu_C C + \nu_D D \qquad [P = 1 \text{ bar}], \tag{3.9}$$

where the ν_i are the stoichiometric coefficients in the balanced chemical reaction. If the reaction goes to completion, ν_A moles of reactant A will be consumed in the formation of products. We therefore have

$$\Delta n_A = \int_{n_A=\nu_A}^{n_A=0} dn_A = -\nu_A. \tag{3.10}$$

At the same time, ν_C moles of product C will be formed. If there were no C molecules present at the beginning of the reaction, we have

$$\Delta n_C = \int_{n_C=0}^{n_C=\nu_C} dn_C = \nu_C. \tag{3.11}$$

The differential change in enthalpy upon reaction under these standard conditions is given by writing the equivalent of Eq. 3.5 for a general bimolecular reaction. That is,

$$dH^\circ = \overline{H}_C^\circ \, dn_C + \overline{H}_D^\circ \, dn_D + \overline{H}_A^\circ \, dn_A + \overline{H}_B^\circ dn_B, \tag{3.12}$$

where the standard partial molar enthalpies are evaluated at the temperature of the reaction. The standard enthalpy change upon reaction is then obtained by integrating both sides:

$$\Delta H^\circ(T) = \int dH^\circ = \overline{H}_C^\circ \int_{n_C=0}^{n_C=\nu_C} dn_C$$

$$+ \overline{H}_D^\circ \int_{n_D=0}^{n_D=\nu_D} dn_D + \overline{H}_A^\circ \int_{n_A=\nu_A}^{n_A=0} dn_A + \overline{H}_B^\circ \int_{n_B=\nu_B}^{n_B=0} dn_B. \tag{3.13}$$

Using Eqs. 3.10 and 3.11, we find that Eq. 3.13 becomes

$$\Delta H^\circ(T) = \nu_C \, \overline{H}_C^\circ(T) + \nu_D \, \overline{H}_D^\circ(T) - \nu_A \, \overline{H}_A^\circ(T) - \nu_B \, \overline{H}_B^\circ(T). \tag{3.14}$$

The foregoing analysis is easily repeated for reactions that involve an arbitrary number of reactants and products. In general, for J products and K reactants, the standard enthalpy change for a reaction at temperature T is given by

$$\boxed{\Delta H^\circ(T) = \sum_{p=1}^{p=J} \overline{H}_p^\circ(T) \, \nu_p - \sum_{r=1}^{r=K} \overline{H}_r^\circ(T) \, \nu_r}. \tag{3.15A}$$

Since, with our choice of reference state, the standard partial molar enthalpy and the standard molar enthalpy of formation are numerically equal to each other, Eq. 3.15A can also be written in the form

$$\boxed{\Delta H^\circ(T) = \sum_{p=1}^{p=J} [\Delta H_f^\circ(T)]_p \, \nu_p - \sum_{r=1}^{r=K} [\Delta H_f^\circ(T)]_r \, \nu_r}. \tag{3.15B}$$

Equations 3.15A and 3.15B give an extremely simple method for computing the reaction enthalpy with the reactants and products in their standard states at 298.15 K. All that need be done is to balance the chemical equation to obtain the stoichiometric coefficients and then look up the standard partial molar enthalpies for all reactants and products at 298.15 K. Since the reaction occurs under conditions of constant pressure, the heat of reaction, q_p, is equal to $\Delta H^{\circ}(298.15)$.

EXAMPLE 3.1

(A) Compute the standard enthalpy change for the reaction

$$C_2H_4(g) + O_2(g) \longrightarrow CO_2(g) + H_2O(g)$$

at 298.15 K **(B)** What heat will be associated with burning 75 grams of $C_2H_4(g)$ in excess oxygen?

Solution

(A) The first step is to balance the equation to obtain the stoichiometric coefficients. This gives $1\,C_2H_4(g) + 3\,O_2(g) \longrightarrow 2\,CO_2(g) + 2\,H_2O(g)$. We now need the standard partial molar enthalpies for $C_2H_4(g)$, $O_2(g)$, $CO_2(g)$, and $H_2O(g)$ at 298.15 K. These values, given in Appendix A, are as follows:

$$\overline{H}^{o}_{C_2H_4(g)} = \Delta H^{o}_f(C_2H_4(g)) = +52.26 \text{ kJ mol}^{-1};$$

$$\overline{H}^{o}_{CO_2(g)} = \Delta H^{o}_f(CO_2(g)) = -393.51 \text{ kJ mol}^{-1};$$

$$\overline{H}^{o}_{H_2O(g)} = \Delta H^{o}_f(H_2O(g)) = -241.82 \text{ kJ mol}^{-1}.$$

The standard partial molar enthalpy of $O_2(g)$ is zero, since we have chosen the elements in their standard state as our reference point with a value of zero. Using Eq. 3.15A, we obtain

$$\Delta H^{\circ}(298.15) = 2\,(-393.51 \text{ kJ mol}^{-1}) + (2)(-241.82 \text{ kJ mol}^{-1})$$
$$- (1)(52.26 \text{ kJ mol}^{-1}) - (3)(0 \text{ kJ mol}^{-1}) = -1{,}322.9 \text{ kJ mol}^{-1}. \quad \textbf{(1)}$$

Thus, if we burn 1 mole of $C_2H_4(g)$ in excess oxygen with $P = 1$ bar and $T = 298.15$ K, we will have a change in enthalpy of -1322.9 kJ. Because $q = \Delta H$ for constant-pressure processes, we will release 1,322.9 kJ of heat in the process.

(B) If we burn 75 grams of $C_2H_4(g)$ in excess oxygen, the number of moles of $C_2H_4(g)$ burned will be

$$n = 75 \text{ g} \times \frac{1 \text{ mol}}{28.05 \text{ g}} = 2.674 \text{ mol of } C_2H_4(g). \quad \textbf{(2)}$$

For each mole burned, we release 1,322.9 kJ of heat. Therefore, the total heat released will be

$$\text{heat released} = (1{,}322.9 \text{ kJ mol}^{-1}\, C_2H_4(g))\,(2.674 \text{ mol } C_2H_4(g)) = 3537.4 \text{ kJ.} \textbf{(3)}$$

For related exercises, see Problems 3.2, 3.3, and 3.4.

EXAMPLE 3.2

Seventy-five grams of $C_2H_4(g)$ are mixed with 246 grams of $O_2(g)$. Combustion is carried out as far as possible at 298.15 K and $P = 1$ bar. Compute the quantity of heat released in the process.

Solution

The first stage of the calculation is to obtain the solution to Example 3.1 if that has not already been done. Once $\Delta H^{\circ}(298.15)$ is obtained, we must determine which reagent, $O_2(g)$ or $C_2H_4(g)$, is present in a limited amount and which is present in excess. This requires that we obtain the number of moles of each reactant present at the start of the process:

$$\text{no. moles of } C_2H_4(g) = n(C_2H_4(g)) = 75 \text{ g} \times \frac{1 \text{ mol}}{28.05 \text{ g}} = 2.674 \text{ mol of } C_2H_4(g); \quad \textbf{(1)}$$

$$\text{no. moles of } O_2(g) = n(O_2(g)) = 246 \text{ g} \times \frac{1 \text{ mol}}{32.00 \text{ g}} = 7.688 \text{ mol of } O_2(g). \quad \textbf{(2)}$$

The stoichiometry shows that we need 3 moles of $O_2(g)$ for each mole of $C_2H_4(g)$. Consequently, to completely burn the 2.674 moles of $C_2H_4(g)$ present, we require 8.022 moles of $O_2(g)$. Since we have only 7.688 mol of $O_2(g)$ available, oxygen is the limiting reagent that determines the extent of the reaction. Example 3.1 shows that 1322.9 kJ of heat are produced for every 3 moles of $O_2(g)$ consumed. Therefore, if we consume 7.688 mol of $O_2(g)$, the total heat production will be

$$\frac{1322.9 \text{ kJ}}{3 \text{ mol } O_2(g)} \times (7.688 \text{ mol } O_2(g)) = 3{,}390 \text{ kJ of heat produced.} \quad \textbf{(3)}$$

3.1.3   Hess' Law

The treatment in the previous section shows that a measurement of the standard molar heat of formation of a compound yields the standard partial molar enthalpy of that compound at the same temperature, provided that we take the partial molar enthalpies of the elements in their standard states to be zero. Therefore, if we wish to determine the standard partial molar enthalpy of liquid benzene at 298.15 K, we need only measure the change in enthalpy for the reaction

$$6 \text{ C(s)} + 3 \text{ H}_2(g) \longrightarrow C_6H_6(l) \quad \textbf{(A)}$$

at $T = 298.15$ K and $P = 1$ bar. There is, however, a slight problem in executing the desired measurement: The rate at which C(s) reacts with $H_2(g)$ to form liquid benzene is essentially zero, so that, in effect, the reaction does not take place. We must, therefore, seek an indirect means of obtaining the information. Hess' law provides this means.

Although we cannot react C(s) and $H_2(g)$ directly, we can easily burn C(s), $H_2(g)$, and benzene in oxygen. That is, using methods described later in the chapter, we can measure the heats of reaction for the following processes:

$$C(s) + O_2(g) \longrightarrow CO_2(g) + 393.15 \text{ kJ;} \quad \textbf{(B)}$$

$$H_2(g) + 0.5 \, O_2(g) \longrightarrow H_2O(g) + 241.82 \text{ kJ;} \quad \textbf{(C)}$$

$$C_6H_6(l) + 7.5 \, O_2(g) \longrightarrow 6 \, CO_2(g) + 3 \, H_2O(g) + 3133.36 \text{ kJ.} \quad \textbf{(D)}$$

As indicated, the heat released in reactions (B), (C), and (D) are 393.15 kJ, 241.82 kJ, and 3133.36 kJ, respectively, provided that the reactions are run at 298.15 K and a pressure of 1 bar. Note that, when written in the form of a thermochemical equation, exothermic reactions have a positive sign for the heat, which is written on the product side of the reaction. However, when expressed as an enthalpy change ΔH, the exothermic heat loss is expressed with a negative sign.

The preceding data can be combined so as to yield the change in enthalpy for reaction (A) by using *Hess' law*. This law states that thermochemical equations can be treated as if they are algebraic equations, with the arrow serving the function of mathematical equality, provided that all reactions are run at the same temperature and pressure. This means we may add, subtract, divide, or multiply reactions (B), (C), and (D) in any manner permitted by the usual rules of algebra. Bearing in mind that we would like to produce an expression identical to reaction (A), it is clear that we need to have 6 moles of C(s) and 3 moles of $H_2(g)$ on the left side of the equation. This can be achieved by multiplying reaction (B) by a factor of 6 and adding the result to reaction (C) after it has been multiplied by a factor of 3, producing

$$6\,C(s) + 6\,O_2(g) + 3\,H_2(g) + 1.5\,O_2(g) \longrightarrow 6\,CO_2(g) + 6 \times 393.15\,\text{kJ}$$
$$+ 3\,H_2O(g) + 3 \times 241.82\,\text{kJ.}$$

Collecting terms as if we were working with an algebraic equation, we obtain

$$6\,C(s) + 7.5\,O_2(g) + 3\,H_2(g) \longrightarrow 6\,CO_2(g) + 3\,H_2O(g) + 3084.36\,\text{kJ.} \quad \textbf{(E)}$$

Reaction (E) still does not have liquid benzene listed on the right side. We can obtain that result by subtracting reaction (D) from reaction (E):

$$6\,C(s) + 7.5\,O_2(g) + 3\,H_2(g) - C_6H_6(l) - 7.5\,O_2(g) \longrightarrow 6\,CO_2(g) + 3\,H_2O(g)$$
$$+ 3{,}084.36\,\text{kJ} - 6\,CO_2(g) - 3\,H_2O(g) - 3{,}133.36\,\text{kJ.}$$

Collecting terms produces

$$6\,C(s) + 3\,H_2(g) - C_6H_6(l) \longrightarrow -49.00\,\text{kJ.} \quad \textbf{(F)}$$

Rearranging reaction (F) gives $49.00\,\text{kJ} + 6\,C(s) + 3\,H_2(g) \longrightarrow C_6H_6(l)$, which is identical to reaction (A). It is, therefore, clear that when reaction (A) is carried out at 298.15 K and 1 bar pressure, it is an endothermic process in which 49.00 kJ of energy are absorbed. Consequently, $\Delta H = q_p = \Delta H_f^o(298.15)_{C_6H_6(l)} = \overline{H}_{C_6H_6(l)}^o(298.15) = 49.00\,\text{kJ.}$

Hess' law is a consequence of the exactness of the dH differential. Consider the reaction

$$CH_4(g) + 2\,O_2(g) \longrightarrow CO_2(g) + 2\,H_2O(l). \quad \textbf{(Z)}$$

Hess' law tells us that ΔH_Z for reaction (Z) is the sum $\Delta H_X + \Delta H_Y$, where reactions (X) and (Y) are, respectively,

$$CH_4(g) + 1.5\,O_2(g) \longrightarrow CO(g) + 2H_2O(l) \quad \textbf{(X)}$$

and

$$CO(g) + 0.5\,O_2(g) \longrightarrow CO_2(g), \quad \textbf{(Y)}$$

provided that all reactions are carried out at the same temperature and pressure. Suppose, then, that we carry out reaction (Z) along Path (A) as shown in Figure 3.3. Since ΔH_Z is independent of the path, we must have $\Delta H_Z = \Delta H_1 + \Delta H_2$. For the same reason, ΔH_2 must be same as $\Delta H_3 + \Delta H_4$. Therefore, we have $\Delta H_Z = \Delta H_1 + \Delta H_3 + \Delta H_4$. In terms of partial molar enthalpies, $\Delta H_1 = -\overline{H}_{CH_4(g)}, \Delta H_3 = \overline{H}_{CO(g)},$ and $\Delta H_4 = \Delta H_Y + 2\,\overline{H}_{H_2O(l)}.$

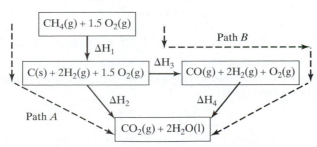

◀ FIGURE 3.3
Chemical transformations along two different paths that illustrate Hess' law.

Combining these expressions gives $\Delta H_Z = [\overline{H}_{CO(g)} + 2\,\overline{H}_{H_2O(l)} - \overline{H}_{CH_4(g)}] + \Delta H_Y$. But the sum of the terms in brackets is just ΔH_X. Therefore, $\Delta H_Z = \Delta H_X + \Delta H_Y$, which is just what Hess' law tells us. In other words, that law is simply a direct consequence of the first law of thermodynamics.

EXAMPLE 3.3

The heats of reaction with all compounds in their standard states at 298.15 K are obtained for the following processes:

$$2\,C_6H_5NH_2(l) + 16.5\,O_2(g) \longrightarrow 12\,CO_2(g) + 7\,H_2O(g) + 2\,NO(g) + 6296.56\ \text{kJ};$$
(A)

$$C(s) + 0.5\,O_2(g) \longrightarrow CO(g) + 110.53\ \text{kJ};$$
(B)

$$CO(g) + 0.5\,O_2(g) \longrightarrow CO_2(g) + 282.98\ \text{kJ};$$
(C)

$$H_2(g) + 0.5\,O_2(g) \longrightarrow H_2O(g) + 241.82\ \text{kJ};$$
(D)

$$NO(g) + 0.5\,O_2(g) \longrightarrow NO_2(g) + 57.07\ \text{kJ}.$$
(E)

In addition, it is known that $\Delta H_f^\circ(298.15)$ for $NO_2(g) = 33.18\ \text{kJ mol}^{-1}$. Use these data to determine $\overline{H}_{C_6H_5NH_2(l)}$ at 298.15 K.

Solution

Knowledge of the standard molar heat of formation of $NO_2(g)$ permits us to write one additional reaction:

$$0.5\,N_2(g) + O_2(g) \longrightarrow NO_2(g) - 33.18\ \text{kJ}.$$
(F)

We wish to obtain ΔH for the following reaction when $T = 298.15$ K and $P = 1$ bar:

$$6\,C(s) + 3.5\,H_2(g) + 0.5\,N_2(g) \longrightarrow C_6H_5NH_2(l).$$

Since all the foregoing reactions are at the same temperature and pressure, Hess' law applies. We seek to combine reactions (A) to (F) so as to form the preceding reaction, whose ΔH is the standard molar heat of formation of $C_6H_5NH_2(l)$. We can produce a reaction with 6 moles of $C(s)$, 3.5 moles of $H_2(g)$, 0.5 moles of $N_2(g)$, and 1 mole of $C_6H_5NH_2(l)$ by forming the combination $6\,(B) + 3.5\,(D) + (F) - 0.5\,(A)$. This linear combination produces

$$6\,C(s) + 3.0\,O_2(g) + 3.5\,H_2(g) + 1.75\,O_2(g)$$

$$+\ 0.5\,N_2(g) + O_2(g) - C_6H_5NH_2(l) - 8.25\,O_2(g)$$

$$\longrightarrow 6\,CO(g) + 6 \times 110.53\ \text{kJ} + 3.5\,H_2O(g)$$

$$+\ 3.5 \times 241.82\ \text{kJ} + NO_2(g) - 33.18\ \text{kJ} - 6\,CO_2(g)$$

$$-\ 3.5\,H_2O(g) - NO(g) - 0.5 \times 6{,}296.56\ \text{kJ}$$
(1)

Collecting terms, we obtain

$$6\,C(s) - 2.5\,O_2(g) + 3.5\,H_2(g) + 0.5\,N_2(g) - C_6H_5NH_2(l)$$

$$\longrightarrow 6\,CO(g) + NO_2(g) - 6\,CO_2(g) - NO(g) - 1{,}671.91\ \text{kJ}.$$
(2)

To produce the desired reaction, we need to remove $CO(g)$, $CO_2(g)$, $NO_2(g)$, and $NO(g)$ on the right-hand side and $O_2(g)$ on the left. Therefore, we add $6(C) - (E)$ to Eq. 2. This operation gives

$$6\,C(s) - 2.5\,O_2(g) + 3.5\,H_2(g) + 0.5\,N_2(g) - C_6H_5NH_2(l)$$

$$+\ 6\,CO(g) + 3.0\,O_2(g) - NO(g) - 0.5\,O_2(g)$$

$$\longrightarrow 6\,CO(g) + NO_2(g) - 6\,CO_2(g) - NO(g) - 1{,}671.91\ kJ$$

$$+ 6\,CO_2(g) + 6 \times 282.98\ kJ - NO_2(g) - 57.07\ kJ.$$

Combining similar terms on each side, we obtain

$$6\,C(s) + 3.5\,H_2(g) + 0.5\,N_2(g) - C_6H_5NH_2(l) + 6\,CO(g)$$

$$- NO(g) \longrightarrow 6\,CO(g) - NO(g) - 31.1\ kJ.$$

$CO(g)$ and $NO(g)$ cancel, and the result, after rearrangement, is

$$6\,C(s) + 3.5\,H_2(g) + 0.5\,N_2(g) \longrightarrow C_6H_5NH_2(l) - 31.1\ kJ. \tag{3}$$

Equation 3 is the reaction we seek. The result shows that we have

$$\Delta H_f^o(298.15)_{C_6H_5NH_2(l)} = \overline{H}_{C_6H_5NH_2(l)}^o(298.15) = 31.1\ kJ.$$

For a related exercise, see Problem 3.10.

3.1.4   Variation of ΔH with Temperature

If all reactions were carried out at 298.15 and 1 bar pressure, Eq. 3.15, combined with tables giving standard partial molar enthalpies at 298.15 K, would permit the calculation of the heat of reaction for any process. However, many reactions are run at temperatures other than 298.15 K. Since it is not realistic to compile extensive tables of standard partial molar enthalpies at every possible temperature, we need to be able to compute the change in ΔH^o produced by a change in the temperature from 298.15 K to some other temperature T. As usual, this means we must find the derivative $(\partial \Delta H^o / \partial T)_P$.

We can obtain the required derivative by straightforward differentiation of Eq. 3.15A. This operation gives

$$\left(\frac{\partial \Delta H^o}{\partial T}\right)_P = \frac{\partial}{\partial T}\left[\sum_{s=1}^{s=J} \overline{H}_s^o \nu_s - \sum_{r=1}^{r=K} \overline{H}_r^o \nu_r\right]_P, \tag{3.16}$$

where the index for the first summation has been changed from p to s to avoid confusion with the subscripts denoting constant pressure. Since the derivative and summation operators commute, we can differentiate inside the summations in Eq. 3.16. This is equivalent to the statement that the derivative of a sum is the sum of the derivatives of each individual term. The interchange produces

$$\left(\frac{\partial \Delta H^o}{\partial T}\right)_P = \sum_{s=1}^{s=J} \nu_s \left(\frac{\partial \overline{H}_s^o}{\partial T}\right)_P - \sum_{r=1}^{r=K} \nu_r \left(\frac{\partial \overline{H}_r^o}{\partial T}\right)_P. \tag{3.17}$$

Let us now consider one of the derivatives under one of the sums in Eq. 3.17. The rate of change of $\overline{H}_s^o$ with temperature at constant pressure may be written in the form

$$\left(\frac{\partial \overline{H}_s^o}{\partial T}\right)_P = \frac{\partial}{\partial T}\left\{\left(\frac{\partial H_s^o}{\partial n_s}\right)_{T,P,nj}\right\}_P. \tag{3.18}$$

The result on the right-hand side of Eq. 3.18 is independent of the order of differentiation. This allows us to reverse the order to obtain

$$\left(\frac{\partial \overline{H}_s^o}{\partial T}\right)_P = \frac{\partial}{\partial n_s}\left\{\left(\frac{\partial H_s^o}{\partial T}\right)_P\right\}_{T,P,nj}. \tag{3.19}$$

Equation 2.53 in Chapter 2 shows that

$$\left(\frac{\partial H_s^o}{\partial T}\right)_P = [C_p]_s = n_s\,[C_p^m]_s. \tag{3.20}$$

That is, the rate of change of enthalpy of compound s with temperature under conditions of constant pressure is the heat capacity, which can be written as the number of moles of compound s, n_s, times the molar heat capacity of compound s at constant pressure. Substituting Eq. 3.20 into Eq. 3.19 yields

$$\left(\frac{\partial \overline{H}_s^o}{\partial T}\right)_P = \frac{\partial}{\partial n_s}\{n_s [C_p^m]_s\}_{T,P,nj} = [C_p^m]_s. \tag{3.21}$$

Equation 3.21 shows that the derivatives needed to compute the change in the heat of reaction with temperature are just the molar heat capacities of the compounds present in the reaction. Combining Eqs. 3.21 and 3.17 gives

$$\left(\frac{\partial \Delta H^o}{\partial T}\right)_P = \sum_{s=1}^{s=J} \nu_s [C_p^m]_s - \sum_{r=1}^{r=K} \nu_r [C_p^m]_r. \tag{3.22}$$

The quantity on the right-hand side of Eq. 3.22 is the sum of the total heat capacities of the products minus that of the reactants. Accordingly, it is the change in total heat capacity that occurs upon reaction. For this reason, it is given the symbol ΔC_p. With this notation, Eq. 3.22 becomes

$$\left(\frac{\partial \Delta H^o}{\partial T}\right)_P = \Delta C_p. \tag{3.23}$$

The form of Eq. 3.23 is so similar to the defining equation for constant pressure heat capacity, $C_p = (\partial H/\partial T)_P$, that it appears that we might derive Eq. 3.23 from the heat capacity equation simply by multiplying both sides by Δ. Such an operation, however, is a dangerous procedure. It would not, for example, tell us what is meant by the notation ΔC_p. Nevertheless, that kind of thinking can serve as a useful mnemonic device once the correct derivation is fully understood.

The variation of ΔH^o with temperature is now easily obtained from Eq. 3.23. After separating the variables, we may integrate both sides between 298.15 K and an arbitrary temperature T:

$$\int_{\Delta H^o(298.15)}^{\Delta H^o(T)} d\Delta H^o = \Delta H^o(T) - \Delta H^o(298.15) = \int_{298.15\ K}^{T} \Delta C_p\, dT. \tag{3.24}$$

Therefore, the change in enthalpy at temperature T is given by

$$\Delta H^o(T) = \Delta H^o(298.15) + \int_{298.15\ K}^{T} \Delta C_p\, dT. \tag{3.25}$$

$\Delta H^o(298.15)$ can be obtained from standard partial molar enthalpies by using Eq. 3.15A. The calculation of $\Delta H^o(T)$ requires a knowledge of the heat capacities of all products and reactants.

It is sometimes convenient to assume that ΔH is independent of temperature for a given process. Equation 3.25 shows that such an assumption is equivalent to assuming that the difference in heat capacities between products and reactants, ΔC_p, is zero at all temperatures or that the temperature range is so small that the integral in Eq. 3.25 is negligible.

We may integrate Eq. 3.24 between any two temperatures; we need not use 298.15 K as the lower limit. If the integration is performed between T_1 and T_2, Eq. 3.25 becomes

$$\boxed{\Delta H^o(T_2) = \Delta H^o(T_1) + \int_{T_1}^{T_2} \Delta C_p\, dT}. \tag{3.26}$$

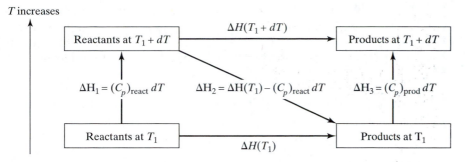

T increases

Reactants at $T_1 + dT$ ——— $\Delta H(T_1 + dT)$ ———→ Products at $T_1 + dT$

$\Delta H_1 = (C_p)_{react}\, dT$ $\Delta H_2 = \Delta H(T_1) - (C_p)_{react}\, dT$ $\Delta H_3 = (C_p)_{prod}\, dT$

Reactants at T_1 ——— $\Delta H(T_1)$ ———→ Products at T_1

◀ **FIGURE 3.4**
Since *dH* is path independent, we must have $\Delta H(T_1 + dT) = \Delta H_2 + \Delta H_3 = \Delta H(T_1) + [(C_p)_{prod} - (C_p)_{react}]dT$. Therefore, if $(C_p)_{prod} > (C_p)_{react}$, the reaction will be less exothermic.

In most cases, we will choose $T_1 = 298.15$ K, since $\Delta H°$ at this temperature is easily computed from data in tables of standard partial molar enthalpies. Equation 3.26 is known as *Kirchhoff's Law.*

It is instructive to consider the physical significance of Eq. 3.23. Suppose we have an exothermic reaction with $\Delta H° < 0$. If the heat capacities of the products are greater than those of the reactants, we will obtain $\Delta C_p > 0$. Equation 3.23 shows that if ΔC_p is positive, $\Delta H°$ will increase as T increases. That is, the reaction will become less exothermic. The pertinent question is, Why does this occur? We may physically understand why by considering what happens when a reaction is run at a higher temperature. First, some energy must be used to heat the reactants to that temperature. Now imagine that the reaction proceeds to produce the products at the lower temperature. The total energy available is the chemical potential energy that is converted into thermal energy and work, plus the thermal energy used to heat the reactants to the higher temperature. Some of this energy must now be used to heat the products to the higher temperature. If $[C_p]_{products} > [C_p]_{reactants}$, the energy needed to heat the products will be greater than that made available by heating the reactants. Consequently, some of the chemical potential energy will have to be used for this heating process, which means that less energy will be available to be released as heat. Therefore, the reaction becomes less exothermic. Figure 3.4 illustrates these ideas; Problem 3.16 puts them in quantitative form.

EXAMPLE 3.4

The esterification reaction of acetic acid with ethanol to give ethylacetate is

$$CH_3COOH(l) + C_2H_5OH(l) \longrightarrow H_2O(l) + CH_3COOC_2H_5(l).$$

(A) Compute $\Delta H°(298.15)$ for this reaction. **(B)** Determine the change in enthalpy for the reaction at 750 K under a pressure of 1 bar.

Solution

(A) Using Eq. 3.15A with standard partial molar enthalpies obtained from Appendix A, we may obtain $\Delta H°(298.15)$. The data at 298.15 K are

$$\overline{H}°_{CH_3COOH(l)} = -484.5 \text{ kJ mol}^{-1},$$
$$\overline{H}°_{C_2H_5OH(l)} = -277.69 \text{ kJ mol}^{-1},$$
$$\overline{H}°_{H_2O(l)} = -285.83 \text{ kJ mol}^{-1},$$

and

$$\overline{H}°_{CH_3COOC_2H_5(l)} = -479.0 \text{ kJ mol}^{-1}.$$

Using Eq. 3.15A, we obtain

$$\Delta H°(298.15) = [(1)(-479.0 \text{ kJ mol}^{-1}) + (1)(-285.83 \text{ kJ mol}^{-1})]$$

$$- [(1)(-484.5 \text{ kJ mol}^{-1}) + (1)(-277.69 \text{ kJ mol}^{-1})] = -2.64 \text{ kJ mol}^{-1}. \quad \textbf{(1)}$$

Thus, the reaction is slightly exothermic.

(B) To compute the temperature dependence of ΔH, we need heat capacities for all reactants and products. These can also be obtained from Appendix A:

	$CH_3COOH(l)$	$C_2H_5OH(l)$	$H_2O(l)$	$CH_3COOC_2H_5(l)$
$[C_p^m] \text{ J mol}^{-1} \text{ K}^{-1}$	124.3	111.46	75.291	170.1

We may compute ΔC_p from Eqs. 3.22 and 3.23:

$$\Delta C_p = [(1)(75.291 \text{ J mol}^{-1} \text{ K}^{-1}) + (1)(170.1 \text{ J mol}^{-1} \text{ K}^{-1})]$$

$$- [(1)(124.3 \text{ J mol}^{-1} \text{ K}^{-1}) + (1)(111.46 \text{ J mol}^{-1} \text{ K}^{-1})] = 9.6 \text{ J mol}^{-1} \text{K}^{-1}. \quad \textbf{(2)}$$

The data assume that C_p^m is a constant. Therefore, ΔC_p is also a constant. Under these conditions, Eq. 3.25 becomes

$$\Delta H°(750) = \Delta H°(298.15) + \int_{298.15 \text{ K}}^{750 \text{ K}} \Delta C_p \, dT = \Delta H°(298.15) + \Delta C_p (750 - 298.15)$$

$$= -2.64 \times 10^3 \text{ J mol}^{-1} + 9.6 \text{ J mol}^{-1} \text{K}^{-1}(451.85 \text{ K})$$

$$= -2.64 \times 10^3 \text{ J mol}^{-1} + 4.3 \times 10^3 \text{ J mol}^{-1}$$

$$= 1.7 \times 10^3 \text{ J mol}^{-1}. \quad \textbf{(3)}$$

Equation 3 shows that the reaction is endothermic at 750 K.

For related exercises, see Problems 3.6, 3.12, 3.13, 3.14, 3.15 and 3.17.

3.1.5 Variation of ΔH with Pressure

If a reaction is carried at a pressure other than 1 bar, the change in enthalpy will not be $\Delta H°$. To determine the new change, we must obtain the derivative $(\partial \Delta H / \partial P)_T$. This can be done using the same procedure employed in the previous section to obtain the temperature dependence of $\Delta H°$.

We first write Eq. 3.15A for an arbitrary pressure by dropping the superscript o. We then differentiate both sides with respect to pressure under conditions of constant temperature. This yields

$$\left(\frac{\partial \Delta H}{\partial P} \right)_T = \frac{\partial}{\partial P} \left[\sum_{s=1}^{s=J} \overline{H}_s \nu_s - \sum_{r=1}^{r=K} \overline{H}_r \nu_r \right]_T, \quad \textbf{(3.27)}$$

where we have again changed the index on the first summation to avoid confusion with the pressure. The derivative and summation operators commute; therefore,

$$\left(\frac{\partial \Delta H}{\partial P} \right)_T = \sum_{s=1}^{s=J} \nu_s \left(\frac{\partial \overline{H}_s}{\partial P} \right)_T - \sum_{r=1}^{r=K} \nu_r \left(\frac{\partial \overline{H}_r}{\partial P} \right)_T. \quad \textbf{(3.28)}$$

The derivative may be written as

$$\left(\frac{\partial \overline{H}_s}{\partial P} \right)_T = \frac{\partial}{\partial P} \left\{ \left(\frac{\partial H_s}{\partial n_s} \right)_{T,P,nj} \right\}_T = \frac{\partial}{\partial n_s} \left\{ \left(\frac{\partial H_s}{\partial P} \right)_T \right\}_{T,P,nj} \quad \textbf{(3.29)}$$

since the order of differentiation does not matter. In Chapter 2, we learned from the Joule–Thomson experiment that

$$\left(\frac{\partial H}{\partial P}\right)_T = -\mu\, C_p = -\mu\, n\, C_p^m. \tag{3.30}$$

Combining Eqs. 3.29 and 3.30, we obtain

$$\left(\frac{\partial H_s}{\partial P}\right)_T = \frac{\partial}{\partial n_s}\{-\mu_s\, n_s\, [C_p^m]_s\}_{T,P,nj} = -\mu_s\, [C_p^m]_s. \tag{3.31}$$

Inserting Eq. 3.31 into Eq. 3.28 gives the result we need:

$$\boxed{\left(\frac{\partial \Delta H}{\partial P}\right)_T = \sum_{s=J}^{s=J} \nu_s[-\mu_s\,[C_p^m]_s] - \sum_{r=1}^{r=K} \nu_r\,[-\mu_r\,[C_p^m]_r] \equiv \Delta[-\mu C_p^m]}. \tag{3.32}$$

The quantity $\Delta[-\mu C_p^m]$ gives the rate of change of the reaction enthalpy with pressure when the temperature is held fixed. It is clear that computing changes in ΔH due to pressure variations requires knowledge of not only the reactant and product heat capacities, but also their corresponding Joule–Thomson coefficients. In many cases, however, the Joule–Thomson coefficients are small. Consequently, ΔH tends to be highly insensitive to changes in the pressure. Indeed, unless the pressure variation is very large, the change in ΔH can often be ignored.

If both the temperature and the pressure are changed from an initial state (T_1, P_1) to (T_f, P_f) the total differential change of ΔH will be given by

$$\boxed{d\Delta H = \Delta C_p\, dT + \Delta(-\mu C_p)\, dP}. \tag{3.33}$$

Since dH is an exact differential, we may compute the change in ΔH along any convenient path. Thus, we could first compute the change as T is varied from T_1 to the final temperature T_f with the pressure at P_1. This calculation will give us ΔH at $T = T_f$ and $P = P_1$. We could then calculate the change in $\Delta H(T, P)$ when the pressure is changed from P_1 to $P = P_f$. In the second computation, the temperature will be fixed at T_f. The final result will be the value of $\Delta H(T_f, P_f)$. We could also reverse the order of these calculations or use any other path. This is illustrated in Figure 3.5. Since dH is exact, we have great latitude in choosing the computational procedure. In many calculations, we will have $T_1 = 298.15$ K and $P_1 = 1$ bar, since standard partial molar enthalpies are available for these values, which makes the computation of $\Delta H^o(298.15)$ relatively easy.

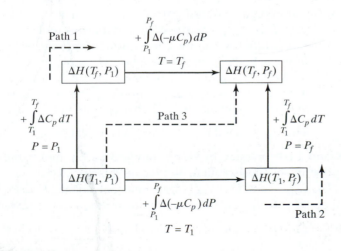

◀ FIGURE 3.5
Calculation of the enthalpy of reaction at temperature and pressure T_f and P_f. As the figure shows, either we can compute $\Delta H(T_f, P_1)$ from the integral of $\Delta C_p\, dT$ over the range T_1 to T_f with $P = P_1$, followed by computing $\Delta H(T_f, P_1)$ from $\Delta H(T_f, P_1)$ by adding the value of the integral of $\Delta(-\mu C_p)\, dP$ over the pressure range P_1 to P_f with $T = T_f$, or we can reverse the procedure and follow Path 2. We could even use Path 3, since dH is an exact differential.

3.1.6 Heats of Reaction in Solution

In many cases, chemical reactions are conducted in solution rather than between pure gases, liquids, or solids. In computing enthalpy changes for such processes, the presence of the solvent must, accordingly, be taken into account. Consider the case of a general bimolecular reaction occurring in a solution at 298.15 K and 1 bar pressure:

$$\nu_A A_{(sol)} + \nu_B B_{(sol)} \longrightarrow \nu_C C_{(sol)} + \nu_D D_{(sol)} \quad [T = 298.15\ \text{K},\ P = 1\ \text{bar}].$$

Here, the subscript (sol) denotes the fact that A, B, C, and D are dissolved in a common solvent. A direct application of Eq. 3.15A gives

$$\Delta H^{\circ}(298.15) = \nu_C \overline{H}^{\circ}_{C(sol)} + \nu_D \overline{H}^{\circ}_{D(sol)} - \nu_A \overline{H}^{\circ}_{A(sol)} - \nu_B \overline{H}^{\circ}_{B(sol)}.$$

The calculation of $\Delta H^{\circ}(298.15)$ requires that we know the standard partial molar enthalpies of A, B, C, and D in solution. Since there are a very large number of possible solvents and an infinite number of possible concentrations in each solvent, it is clear that we can never obtain a sufficiently large tabulation of data to completely solve the problem. Therefore, we must be content with a partial solution.

Let us take a specific example. If we wish to obtain the standard partial molar enthalpy of ethanol in a water solvent at a concentration of 0.75 M with $T = 298.15$ K, we must calculate or measure the enthalpy changes for the processes

$$2\,C(s) + 3\,H_2(g) + 0.5\,O_2(g) \longrightarrow C_2H_5OH(l)\,[T = 298.15\ \text{K},\ P = 1\ \text{bar}] \quad \textbf{(A)}$$

and

$$C_2H_5OH(l) + x\,H_2O(l) \longrightarrow C_2H_5OH_{(sol,0.75M)}[T = 298.15\ \text{K},\ P = 1\ \text{bar}], \quad \textbf{(B)}$$

where x denotes the amount of water required to produce 1.3333 … liters of solution, so that the final concentration will be 0.75 molar. The change in enthalpy for process (A) is the standard molar heat of formation of $C_2H_5OH(l)$. The standard partial molar enthalpy at 298.15 K for a 0.75-M aqueous solution of ethanol is the sum of the enthalpy changes for processes (A) and (B); that is,

$$\overline{H}^{\circ}_{ethanol(sol,0.75M)} = \Delta H^{\circ}_f(298.15)_{C_2H_5OH(l)} + \Delta H_B = -277.69\ \text{kJ mol}^{-1} + \Delta H^{\circ}_B,$$

where ΔH°_B is the enthalpy change for process (B) which is termed the standard enthalpy of solution of ethanol. While there is no fundamental problem associated with measuring this heat of solution, we cannot make such a measurement for every possible concentration in every solvent. Therefore, we compromise by defining a standard partial molar enthalpy at infinite dilution. This quantity, $\overline{H}^{\circ}_{ethanol(sol,\infty)}$, is the standard partial molar enthalpy of ethanol in aqueous solution when the concentration approaches zero. If the concentrations in the actual experiment are not too large, we can use these values to estimate ΔH° for solution reactions with reasonable accuracy.

The situation for ionic reactions in solution requires some special attention. Suppose we wish to estimate the enthalpy change in the reaction

$$HCl_{(sol)} + NaOH_{(sol)} \longrightarrow H_2O(l) + NaCl_{(sol)}$$

at 298.15 K and 1 bar pressure by using standard partial molar enthalpies at infinite dilution in aqueous solution. To execute this calculation, we will need $\overline{H}^{\circ}_{HCl(sol,\infty)}$, $\overline{H}^{\circ}_{NaOH(sol,\infty)}$, and $\overline{H}^{\circ}_{NaCl(sol,\infty)}$ for a water solvent at 298.15 K. The standard partial molar enthalpy of $HCl_{(sol)}$ at infinite dilution can be obtained from

$$0.5\,H_2(g) + 0.5\,Cl_2(g) \longrightarrow HCl(g) + 92.31\ \text{kJ mol}^{-1}[T = 298.15\ \text{K},\ P = 1\ \text{bar}]$$

and

$$HCl(g) + \infty H_2O(l) \longrightarrow HCl_{(sol,\infty)} + 74.85 \text{ kJ mol}^{-1} \ [T = 298.15 \text{ K}, P = 1 \text{ bar}].$$

Adding these thermochemical equations, we obtain

$$0.5 H_2(g) + 0.5 Cl_2(g) + \infty H_2O(l) \longrightarrow HCl_{(sol,\infty)} + 167.16 \text{ kJ mol}^{-1},$$

so that $\overline{H}^o_{HCl(sol,\infty)} = -167.16 \text{ kJ mol}^{-1}$ at 298.15 K.

The preceding analysis ignores the fact that ionic compounds ionize in aqueous solution at infinite dilution. That is, in solution, HCl exists in the form of $H^+_{(sol,\infty)}$ and $Cl^-_{(sol,\infty)}$. This means that $\overline{H}^o_{HCl(sol,\infty)}$ is the sum of $\overline{H}^o_{H^+(sol,\infty)}$ and $\overline{H}^o_{Cl^-(sol,\infty)}$. Since anions and cations always appear together in solution, we are free to partition $\overline{H}^o_{HCl(sol,\infty)}$ between $\overline{H}^+_{(sol,\infty)}$ and $\overline{Cl}^-_{(sol,\infty)}$ in any manner we wish, so long as we maintain consistency. We define $\overline{H}^o_{H^+(sol,\infty)}$ to be zero. This choice of reference points for ions makes $\overline{H}^o_{Cl^-(sol,\infty)} = -167.16 \text{ kJ mol}^{-1}$. With such a start, we can proceed to obtain standard partial molar enthalpies of other ions at infinite dilution. For example, we can infer the value of $\overline{H}^o_{Na^+(sol,\infty)}$ from the following measured heats of reaction:

$$Na(s) + 0.5 Cl_2(g) \longrightarrow NaCl(s) + 411.15 \text{ kJ mol}^{-1}$$
$$[T = 298.15 \text{ K}, P = 1 \text{ bar}];$$

$$NaCl(s) + \infty H_2O(l) \longrightarrow Na^+_{(sol,\infty)} + Cl^-_{(sol,\infty)} - 4.15 \text{ kJ mol}^{-1}$$
$$[T = 298.15 \text{ K}, P = 1 \text{ bar}].$$

Summation of the two reactions gives

$$Na(s) + 0.5 Cl_2(g) + \infty H_2O(l) \longrightarrow Na^+_{(sol,\infty)} + Cl^-_{(sol,\infty)} + 407.3 \text{ kJ mol}^{-1}.$$

Consequently, we have $\overline{H}^o_{Na^+(sol,\infty)} + \overline{H}^o_{Cl^-(sol,\infty)} = -407.3 \text{ kJ mol}^{-1}$. The previous analysis of $HCl_{(sol,\infty)}$ gave $\overline{H}^o_{Cl^-(sol,\infty)} = -167.16 \text{ kJ mol}^{-1}$. Therefore, $\overline{H}^o_{Na^+(sol,\infty)} = -240.14 \text{ kJ mol}^{-1}$. Table 3.1 gives the limiting standard partial molar enthalpies for some common ions at infinite dilution in aqueous solution at 298.15 K.

Table 3.1   Limiting standard partial molar enthalpies of some common ions at infinite dilution in aqueous solution at 298.15 K

Ion	$\overline{H}^o_{(sol,\infty)}$ (kJ mol^{-1})	Ion	$\overline{H}^o_{(sol,\infty)}$ (kJ mol^{-1})
H^+	0.00	Ag^+	105.6
Na^+	-240.1	Li^+	-278.5
Al^{3+}	-531.0	Br^-	-121.6
Ba^{2+}	-537.6	Cl^-	-167.2
Ca^{2+}	-542.8	F^-	-332.6
Cu^{2+}	64.8	I^-	-55.2
Fe^{3+}	-48.5	OH^-	-230.0
Fe^{2+}	-89.1	PO_4^{3-}	-1277.4
K^+	-252.4	SO_4^{2-}	-909.3
Zn^{2+}	-153.9	NO_3^-	-207.4
NH_4^+	-132.5		

Source of data: *Handbook of Chemistry and Physics*, 78th edition, CRC Press, Boca Raton, FL. 1997–98.

Returning to the reaction of HCl with NaOH in aqueous solution, we may write the process as $H^+_{(sol,\infty)} + OH^-_{(sol,\infty)} \longrightarrow H_2O(l)$, since the Na^+ and Cl^- ions play no role in the process at infinite dilution. The standard enthalpy change for the reaction at 298.15 K is

$$\Delta H^o = \overline{H}^o_{H_2O(l)} - \overline{H}^o_{H^+(sol,\infty)} - \overline{H}^o_{OH^-(sol,\infty)}$$

$$= -285.83 \text{ kJ mol}^{-1} - [0 - 230.0 \text{ kJ mol}^{-1}]$$

$$= -55.8 \text{ kJ mol}^{-1}.$$

3.1.7 Heats of Reaction for Constant-Volume Processes

The analysis of the previous section assumes that the reactions are conducted under conditions of constant pressure. Equation 2.42 shows that if this is the case, the heat of reaction will be equal to ΔH. However, for reactions carried out at constant volume, ΔU gives the heat of reaction. The change in the internal energy can always be computed from ΔH, provided that we have the equations of state available. The definition of enthalpy gives

$$\Delta U = \Delta H - \Delta(PV). \tag{3.34}$$

If the process is conducted at constant volume, Eq. 3.34 becomes

$$\Delta U = \Delta H - V\,\Delta P. \tag{3.35}$$

For a reaction involving gases, we will need the equations of state for each gas in order to compute ΔP. If all gases obey the ideal-gas laws, Eq. 3.35 assumes the very simple form

$$\boxed{\Delta U = \Delta H - V\,\Delta P = \Delta H - \Delta(nRT) = \Delta H - RT\,\Delta n}, \tag{3.36}$$

since T is constant. Under these conditions, ΔU will differ from ΔH only if the number of moles of gas for the products differs from that for the reactants.

Physically, ΔH may differ from ΔU because work may be done when the reaction is carried out under conditions of constant pressure, but not under constant volume conditions. Figure 3.6 illustrates this effect for a reaction

▶ FIGURE 3.6
Work done in a reaction conducted at 298.15 K and a constant pressure of 1 bar. Since the pressure is constant, the heat is equal to ΔH^o, and we have $q = q_p = \Delta H^o = \Delta U^o + \Delta(PV) = \Delta U^o + P\,\Delta V = \Delta U^o - w$.

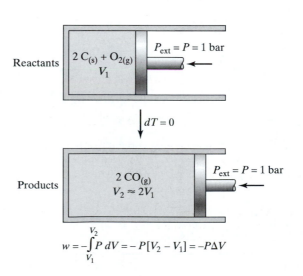

involving an increase in the number of moles of gas present, such as $2\,C(s) + O_2(g) \longrightarrow 2\,CO(g)$.

When the reaction involves only solids and liquids, or when it is a solution reaction, the change in pressure will be so small that the $V\Delta P$ term in Eq. 3.35 may be neglected. If both gases and condensed phases are involved, we need consider only the change in pressure produced by the gases, since the pressure changes due to the solids and liquids will be too small to matter.

EXAMPLE 3.5

Ethene undergoes a hydration reaction with water to yield ethanol via the reaction

$$C_2H_4(g) + H_2O(g) \longrightarrow CH_3\text{–}CH_2OH(l).$$

(A) If the reaction is conducted at 298.15 K and 1 bar pressure, determine the heat of reaction. **(B)** If the reaction temperature is raised to 400 K, compute the heat of reaction if $dT = dP = 0$ and $P = 1$ bar. **(C)** If the reaction is carried out at 400 K in a closed vessel, so that $dV = 0$, determine the heat released.

Solution

(A) $\Delta H^o(298.15)$ may be computed directly from Eq. 3.15A or Eq. 3.15B. The required standard partial molar enthalpies are listed in Appendix A. Since all of the stoichiometric coefficients are unity, the result is

$$\Delta H^o(298.15) = \overline{H}^o_{C_2H_5OH(l)} - \overline{H}^o_{H_2O(g)} - \overline{H}^o_{C_2H_4(g)}$$

$$= -277.69 - [-242.82 + 52.26]\ \text{kJ mol}^{-1} = -87.13\ \text{kJ mol}^{-1}. \quad \textbf{(1)}$$

Since $dP = 0$, we know that $q = q_p = \Delta H^o = -87.13\ \text{kJ mol}^{-1}$.

(B) The enthalpy change at 400 K is given by Eq. 3.25:

$$\Delta H^o(400) = \Delta H^o(298.15) + \int_{298.15\ K}^{400\ K} \Delta C_p\, dT. \quad \textbf{(2)}$$

The required values of C_p^m for reactants and products can be obtained from Appendix A. ΔC_p is given by Eq. 3.22:

$$\Delta C_p = (1)[C_p^m]_{ethanol(l)} - (1)[C_p^m]_{water(g)} - (1)[C_p^m]_{ethene(g)}$$

$$= 111.46 - 33.58 - 43.56\ \text{J mol}^{-1}\,\text{K}^{-1} = 34.32\ \text{J mol}^{-1}\,\text{K}^{-1}. \quad \textbf{(3)}$$

Substituting Eq. 3 into Eq. 2 gives

$$\Delta H_{400} = -8.713 \times 10^4\ \text{J mol}^{-1} + 34.32\ \text{J mol}^{-1}\,\text{K}^{-1}\,(400 - 298.15)\ \text{K}$$

$$= -8.363 \times 10^4\ \text{J mol}^{-1} = -83.63\ \text{kJ mol}^{-1}. \quad \textbf{(4)}$$

Observe that $\Delta H^o(298.15)$ and the heat capacity terms have different units. We must, therefore, convert $\Delta H^o(298.15)$ in Eq. 1 to joules or the heat capacities to kJ before doing the arithmetic. Since $dP = 0$, we have $q = q_p = \Delta H^o(400) = -83.63\ \text{kJ mol}^{-1}$.

(C) If the reaction is carried out at 400 K with $dV = 0$, we must obtain $\Delta U^o(400)$ to compute the heat of reaction. From Eq. 3.36, we have

$$\Delta U^o(400) = \Delta H^o(400) - RT\Delta n. \quad \textbf{(5)}$$

In this reaction, the change in the number of moles of gas is -2, since there are 2 moles of gas among the reactants and none among the products. The volume of the liquid ethanol is very small and may be neglected. Therefore,

$$\Delta U^o(400) = -8.363 \times 10^4 \, \text{J mol}^{-1} - (8.314 \, \text{J mol}^{-1} \, \text{K}^{-1})(400 \, \text{K})(-2)$$

$$= -7.698 \times 10^4 \, \text{J mol}^{-1} = -76.98 \, \text{kJ mol}^{-1}. \qquad (6)$$

The heat of reaction is $q = q_v = \Delta U^o(400) = -76.98 \, \text{kJ mol}^{-1}$.

3.2   Measurements of Heats of Reaction

There are a variety of techniques for obtaining heats of reaction. We shall discuss three methods in this section: direct calorimetric measurement, differential thermal analysis, and differential scanning calorimetry.

3.2.1   Direct Calorimetry

Direct adiabatic calorimetric measurements involving gases are usually conducted under conditions of constant volume. A reaction chamber with a fixed volume V and its surrounding heat bath are insulated from the surroundings such that the heat transfer between the two is near zero. This situation is illustrated in Figure 3.7. The chamber is filled with the reactants—in the figure, $O_2(g)$ and some material whose heat of combustion is to be measured. Reaction is initiated by a pair of ignition wires that pass a current through the sample pellet. The heat generated in the reaction is absorbed by a surrounding water bath that is stirred to produce uniform temperatures. The experiment consists of measuring the temperature change inside the reaction chamber at the completion of the process. The heat of reaction can be obtained directly from this information, or the apparatus can be cooled to its original temperature and a resistance heater used to measure the amount of electrical energy required to produce a similar rise in temperature. Both methods are used. Alternatively, an isoperibol bomb calorimeter can be used. In such an apparatus, the surrounding jacket is maintained at a constant temperature while the temperature of the bomb and bucket rise as heat is released. Here, the measurement consists of determining the temperature rise as a function of time. Illustrations 3.1, 3.2, and 3.3 show photographs of a commercial bomb calorimeter.

▶ FIGURE 3.7
An adiabatic, constant-volume "bomb" calorimeter. The sample and oxidant are placed in the inner chamber. Ignition wires initiate the combustion reaction. The heat released in the reaction is absorbed by the surrounding water bath, and the temperature rise is measured with a high-precision thermometer. Adiabaticity is assured by the surrounding insulation and vacuum space. The heat of reaction can be obtained directly from the temperature increase or by measuring the electrical energy required to produce a similar increase in temperature, as described in the text.

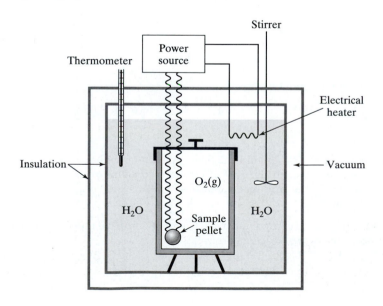

 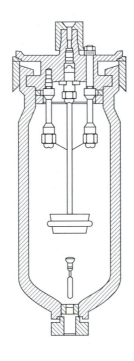

◀ ILLUSTRATION 3.1
Photograph and schematic illustration of the Parr 1136 Oxygen Bomb used in the Parr automatic isoperibol calorimeter, Model 1281. In an isoperibol calorimeter, the surrounding jacket is maintained at a constant temperature while the temperature of the bomb and bucket rise as heat is released during the combustion reaction. A microprocessor-based controller monitors both the temperature of the bucket and that of the jacket and performs the heat-leak corrections required because of the difference in these two temperatures. The corrections are applied continuously throughout the experiment. [Reproduced with permission from Parr Instrument Company.]

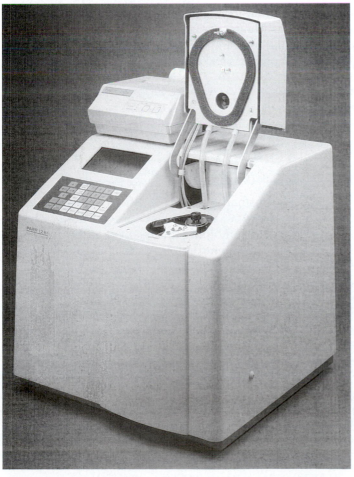

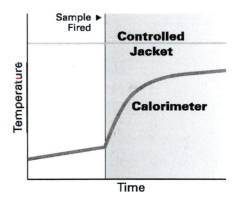

▲ ILLUSTRATION 3.2
Typical curve of temperature versus time obtained in an isoperibol calorimeter. [Reproduced with permission from Parr Instrument Company.]

◀ ILLUSTRATION 3.3
Photograph of the Parr Model 1281 isoperibol calorimeter. The model is shown with printer and built-in calorimeter controller. [Reproduced with permission from Parr Instrument Company.]

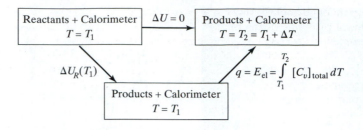

The overall process occurring with the calorimeter is [reactants + calorimeter at T_1] ⟶ [products + calorimeter at T_2], where $T_2 = T_1 + \Delta T$. Since the process is conducted under adiabatic and constant-volume conditions, we have $\delta q = \delta w = 0$ and $dU = 0$. We can use the fact that dU is an exact differential to consider an alternative path for the process, as shown in Figure 3.8.

The exactness of dU ensures that we must have the internal energy change for the reaction ΔU_R at temperature T_1, plus the heat associated with raising the calorimeter and products from T_1 to T_2 equal to zero. This may be expressed mathematically by writing

$$\Delta U_R(T_1) + E_{el} = \Delta U_R(T_1) + \int_{T_1}^{T_2} [C_v]_{\text{total}} \, dT = 0. \tag{3.37}$$

The electrical energy inserted into the calorimeter during the heating is given by $E_{el} = VIt$, where V is the voltage in volts, I is the current in amperes, and t is the time required for the heating to produce a similar temperature change expressed in seconds. Therefore, $\Delta U_R(T_1) = -VIt$.

We can also obtain $\Delta U_R(T_1)$ by using the second portion of Eq. 3.37. The total constant-volume heat capacity is given by

$$[C_v]_{\text{total}} = \{[C_v]_{\text{calorimeter}} + [C_v]_{\text{products}} + [C_v]_{\text{excess reactants}}\}. \tag{3.38}$$

Combining Eqs. 3.37 and 3.38 yields

$$\Delta U_R(T_1) = -\int_{T_1}^{T_2} [C_v]_{\text{total}} \, dT$$

$$= -\{[C_v]_{\text{calorimeter}} + [C_v]_{\text{products}} + [C_v]_{\text{excess reactants}}\} \Delta T, \tag{3.39}$$

provided that ΔT is sufficiently small that the total heat capacity is effectively a constant. If $[C_v]_{\text{total}}$ is known, $\Delta U_R(T_1)$ may be computed from the measured change in temperature. Combining this result with the number of moles of reactant used in the experiment permits us to compute the molar heat of the process, q_v. For a constant-volume process, $q_v = \Delta U$. ΔH may be calculated using Eq. 3.36.

$[C_v]_{\text{total}}$ is the sum of the constant-volume heat capacities of the calorimeter itself, the products of the reaction, and any excess reactants left at the completion of the reaction. The first of these is obtained by running a reaction for which $\Delta U_R(T_1)$ is known and measuring ΔT. If the heat capacities of reactants and products are known, this measurement permits the heat capacity of the calorimeter to be determined. Example 3.6 illustrates the procedure.

EXAMPLE 3.6

An adiabatic constant-volume calorimeter such as that shown in Figure 3.7 is used to measure q_v for the reaction

$$C_2H_5OH(l) + 3\,O_2(g) \longrightarrow 2\,CO_2(g) + 3\,H_2O(l).$$

One gram of ethanol $C_2H_5OH(l)$ is mixed with 10 grams of $O_2(g)$ at 298.15 K. The mixture is placed in the calorimeter. The reaction is initiated and allowed to proceed to completion. At this point, the temperature change is measured and found to be 16.86 K. The calorimeter's heat capacity is determined by mixing 0.05000 mole of $CH_4(g)$ with 0.1000 mole of $O_2(g)$ in the calorimeter and measuring the temperature change upon reaction. The result is $\Delta T = 25.24$ K. **(A)** Use the following data to determine the heat capacity of the calorimeter:

$$\overline{H}^o_{CH_4(g)} = -74.81 \text{ kJ mol}^{-1}; \qquad \overline{H}^o_{H_2O(l)} = -285.83 \text{ kJ mol}^{-1};$$

$$\overline{H}^o_{CO_2(g)} = -393.51 \text{ kJ mol}^{-1}; \qquad [C_v]_{H_2O(l)} = 75.29 \text{ J mol}^{-1}\text{K}^{-1};$$

$$[C_v]_{CO_2(g)} = 28.8 \text{ J mol}^{-1}\text{K}^{-1}; \qquad [C_v]_{O_2(g)} = 21.04 \text{ J mol}^{-1}\text{K}^{-1}.$$

(B) Compute the heat of reaction per mole of ethanol for the reaction of ethanol with oxygen. **(C)** Determine ΔU and ΔH for the reaction of ethanol with oxygen.

Solution

(A) The combustion reaction for $CH_4(g)$ is $CH_4(g) + 2\,O_2(g) \rightarrow CO_2(g) + 2\,H_2O(l)$. The constant-pressure standard heat of reaction for the combustion per mole of methane at 298.15 K is given by

$$q_p = \Delta H^o = \sum_{p=1}^{p=J} \overline{H}^o_p \nu_p - \sum_{r=1}^{r=K} \overline{H}^o_p \nu_r = (1)(-393.51 \text{ kJ mol}^{-1})$$

$$+ (2)(-285.83 \text{ kJ mol}^{-1}) - (1)(-74.81 \text{ kJ mol}^{-1}) = -890.36 \text{ kJ mol}^{-1}. \quad \textbf{(1)}$$

Since the change in the number of moles of gas for the methane reaction is $\Delta n = -2$, we have

$$q_v = \Delta U^o$$

$$= \Delta H^o - RT\Delta n$$

$$= -8.9036 \times 10^5 \text{ J mol}^{-1} - (8.314 \text{ J mol}^{-1}\text{K}^{-1})(298.15 \text{ K})(-2)$$

$$= -8.8540 \times 10^5 \text{ J mol}^{-1} = -885.40 \text{ kJ mol}^{-1}. \quad \textbf{(2)}$$

Since only 0.05 mole of $CH_4(g)$ were used, $q_v = (0.05000 \text{ mol})(-885.40 \text{ kJ mol}^{-1}) = -44.27$ kJ. After the reaction is complete, only 0.05000 mole of $CO_2(g)$ and 0.1000 mole of $H_2O(l)$ are present, since stoichiometric amounts of $CH_4(g)$ and $O_2(g)$ were employed. The total constant-volume heat capacity is, therefore,

$$[C_v]_{\text{total}} = [C_v]_{\text{calorimeter}} + 0.05000 \text{ mol }(28.8 \text{ J mol}^{-1}\text{K}^{-1})$$

$$+ 0.1000 \text{ mol }(75.29 \text{ J mol}^{-1}\text{K}^{-1}) = [C_v]_{\text{calorimeter}} + 8.97 \text{ J K}^{-1}. \quad \textbf{(3)}$$

The constant-volume heat is given by

$$q_v = -\{[C_v]_{\text{calorimeter}} + 8.97 \text{ J K}^{-1}\}\Delta T. \text{ Solving for } [C_v]_{\text{calorimeter}}, \text{ we obtain} \quad \textbf{(4)}$$

$$[C_v]_{\text{calorimeter}} = \frac{-q_v}{\Delta T} - 8.97 \text{ J K}^{-1} = \frac{44{,}270 \text{ J}}{25.24 \text{ K}} - 8.97 \text{ J K}^{-1} = 1{,}745 \text{ J K}^{-1}. \quad \textbf{(5)}$$

(B) One gram of ethanol corresponds to 0.02171 mol. The balanced equation shows that this amount will require 0.06513 mol of $O_2(g)$ for a complete reaction. That amount corresponds to 2.084 g of $O_2(g)$. Since we started with 10 g of oxygen, we will have 7.916 g remaining after the reaction ceases. This is 0.2474 mol of $O_2(g)$. Thus, the reaction will produce 0.04342 mol of $CO_2(g)$ and 0.06513 mol of $H_2O(l)$. Consequently, the total heat capacity of reactants and products at the completion of the reaction is

$$[C_v]_{\text{reactants}} + [C_v]_{\text{products}} = (0.2474 \text{ mol})(21.04 \text{ J mol}^{-1}\text{K}^{-1})$$

$$+ (0.04342 \text{ mol})(28.8 \text{ J mol}^{-1}\text{K}^{-1})$$

$$+ (0.06513 \text{ mol})(75.29 \text{ J mol}^{-1}\text{K}^{-1}) = 11.36 \text{ J K}^{-1}. \quad \textbf{(6)}$$

The total heat produced in the process is given by Eq. (3.39):

$$q_v = \Delta U_R(T_1) = -[C_v]_{\text{total}}\Delta T$$

$$= \{[C_v]_{\text{calorimeter}} + [C_v]_{\text{products}} + [C_v]_{\text{excess reactants}}\}\Delta T$$

$$- [1{,}745 + 11.36]\, \text{J K}^{-1}\, (16.86\, \text{K})$$

$$= -2.961 \times 10^4\, \text{J} = -29.61\, \text{kJ}. \tag{7}$$

Thus, 29.61 kJ are released when 0.02171 mole of ethanol react. If 1 mole of ethanol had reacted, we would have observed that

$$q_v = \frac{-29.61\, \text{kJ}}{0.02171\, \text{mol}} = -1{,}364\, \text{kJ mol}^{-1}. \tag{8}$$

(C) Since the volume is constant in the reaction, we have

$$\Delta U = q_v = -1{,}364\, \text{kJ mol}^{-1}. \tag{9}$$

The enthalpy change is given by

$$\Delta H = \Delta U + \Delta(PV) = \Delta U + RT\, \Delta n \tag{10}$$

if the gases may be treated as ideal. Hence,

$$\Delta H = -1.364 \times 10^6\, \text{J mol}^{-1} + (8.314\, \text{J mol}^{-1}\, \text{K}^{-1})(298.15\, \text{K})(-1)$$

$$= -1.366 \times 10^6\, \text{J mol}^{-1}$$

$$= 1{,}366\, \text{kJ mol}^{-1}, \tag{11}$$

since we have 2 moles of gaseous product and 3 moles of gaseous reactants. For a related exercise, see Problem 3.20.

When the foregoing methods are employed to measure standard molar enthalpies of formation, and thereby standard partial molar enthalpies, for gases, we will, in principle, need to make a very small correction for the deviation of the gaseous reactants and products from ideality. By definition, the standard molar enthalpies of formation require that the gaseous reactants and products be ideal gases at 1 bar pressure. Since ideal gases are hard to come by in the real world, we are forced to run the experiments using real gases at some pressure P at which the gases, of course, do not behave ideally. The standard molar enthalpy of formation is the total enthalpy for the processes shown in Figure 3.9. We need to use ideal gaseous reactants at 1 bar of pressure and end up with ideal gaseous products also at 1 bar of pressure. Since dH is an exact differential, we can imagine conducting this transformation by first expanding the ideal gaseous reactants to zero pressure. Because dH is independent of pressure for ideal gases, ΔH for this first step is zero. We now convert the ideal gases to real gases at zero pressure. At that pressure, the gaseous interactions will be zero, so that ΔH for this conversion is also zero. We now raise the real-gas reactants from zero pressure to the pressure in the calorimeter, P, holding the temperature fixed. Equation 3.33 shows that the change in enthalpy for this process is $\Delta H = \int_0^P (-\mu C_p)_R\, dP$, where the Joule–Thomson coefficients and heat capacities are those for reactants. We now carry out the measurements in the calorimeter to obtain $\Delta H_r(T, P)$ for real-gas reactants forming real-gas products at pressure P. The real-gas products at that pressure are now lowered to zero pressure. The enthalpy change for this process is $\Delta H = \int_P^0 (-\mu C_p)_p\, dP$, where $(-\mu C_p)_p$ is evaluated for the products. Finally, we

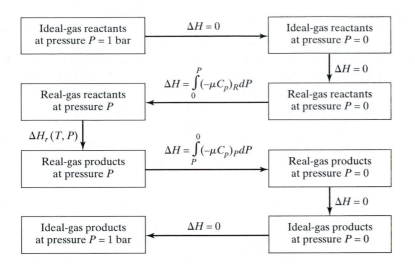

◀ FIGURE 3.9
Analysis of corrections required to obtain standard molar enthalpies of formation and standard partial molar enthalpies of gases from calorimetric measurements.

convert the real-gas products at zero pressure to ideal gases and then raise the pressure to 1 bar. ΔH for both of these processes is zero. Consequently, the correction required to obtain the standard molar enthalpy of formation is

$$\Delta H_f^\circ(T) = \Delta H_r(P, T) + \int_0^P \left[(\mu C_p)_p - (\mu C_p)_R\right] dP. \tag{3.40}$$

In most cases, the correction term in Eq. 3.40 is extremely small, even when the pressure is as large as 30 bar. For $O_2(g)$, $\int_0^{30\,\text{bar}}(\mu C_p)\,dP$ is approximately 0.27 kJ mol^{-1}. This is about 0.02% of the result obtained in Example 3.6. Except in the most precise and accurate work, such corrections can be ignored.

3.2.2   Differential Thermal Analysis/Differential Scanning Calorimetry

Differential thermal analysis (DTA) is well suited to the measurement of heats of reaction in solution. The method depends upon measuring the difference in temperature of two cells. One cell contains the solvent and reactants; the second contains the solvent and perhaps all the reactants save one. Figure 3.10 shows a schematic diagram of a differential thermal analysis

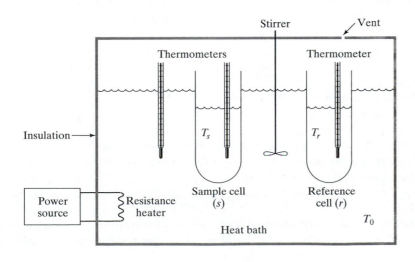

◀ FIGURE 3.10
Schematic diagram of a differential thermal analysis calorimeter.

calorimeter. The cell on the left labeled "Sample Cell" in the figure contains the solvent plus reactants. The chemical reaction of interest takes place in this cell. The reference cell on the right contains only the solvent and perhaps all the reactants, save one so that no reaction will occur. Both cells are equipped with temperature-measuring devices. They are immersed in a large heat bath at temperature T_o. During the course of the experiment, T_o is increased, usually at a constant rate, until the reaction is complete. The increase in T_o serves both to initiate a reaction in the sample cell and to propel it rapidly to completion, thereby expediting the acquisition of data. The cells are vented to the atmosphere so that the reaction occurs under constant-pressure conditions.

At the beginning of the experiment, at time $t = 0$, both the sample and reference cells are in thermal equilibrium with the heat bath, so that $T_s = T_r = T_o$. Let us now consider what occurs when an exothermic reaction proceeds for a length of time dt in the sample cell. The heat generated will produce a temperature increase dT_s whose magnitude will depend on the quantity of heat released minus the heat lost to the heat bath during the time interval dt. That is, we will observe that

$$[C_p]_s \, dT_s = \text{(heat generated)} - \text{(heat lost to heat bath)}$$

$$= -dH - (T_s - T_o)K_s \, dt, \tag{3.41}$$

where K_s is the coefficient of thermal conductivity of the sample cell. The heat generated at constant pressure is $-dH$. If the temperature of the heat bath is less than that of the sample cell, [i.e., if $(T_s - T_o) > 0$], the second term will be negative, denoting heat transfer from the cell to the bath.

Now let us examine the reference cell during this same period. Since no reaction occurs in that cell, we observe only heat transfer to or from the heat bath. Therefore, in the reference cell, we have

$$[C_p]_r \, dT_r = -(T_r - T_o)K_r \, dt, \tag{3.42}$$

where the subscript r refers to the quantities related to the reference cell.

The critical point of a differential thermal analysis measurement is to make the two cells as nearly identical as possible. If the material and construction of both cells are identical, and if we employ very dilute solutions with the same solvent, we will have $K_r = K_s = K$ and $[C_p]_r = [C_p]_s = C_p$. Under these conditions, Eqs. 3.41 and 3.42 become, respectively,

$$C_p \, dT_s = -dH - (T_s - T_o)K \, dt \tag{3.43}$$

and

$$C_p \, dT_r = -(T_r - T_o)K \, dt. \tag{3.44}$$

Subtracting Eq. 3.44 from 3.43 gives

$$C_p \, [dT_s - dT_r] = C_p \, d[T_s - T_r] = -dH - [T_s - T_r] \, K \, dt. \tag{3.45}$$

The measured data in the experiment are the differences $(T_s - T_r) = \Delta T$ at various times. Substituting this notation into Eq. 3.45 and rearranging the equation produces

$$-dH = C_p \, d\Delta T + \Delta T \, K \, dt. \tag{3.46}$$

At the start of the experiment, both cells are in equilibrium with the heat bath, and we have $T_s = T_r = T_o$, so that $\Delta T = 0$ at $t = 0$. After the reaction has been completed in the sample cell and a long time has elapsed, heat transfer will again bring the two cells into equilibrium. Consequently, at $t = \infty$, we also have $\Delta T = 0$. Let us integrate Eq. 3.46 between the corresponding limits ($t = 0$, $\Delta T = 0$) and ($t = \infty$, $\Delta T = 0$):

$$\int_{t=0, \Delta T=0}^{t=\infty, \Delta T=0} -dH = -\Delta H_{total} = C_p \int_{\Delta T=0}^{\Delta T=0} d\Delta T + K \int_{t=0}^{t=\infty} \Delta T\, dt, \qquad \textbf{(3.47)}$$

Here, ΔH_{total} represents the total change in enthalpy for the amount of reactant used in the experiment. The first integral on the right is clearly zero, since it has identical upper and lower limits. Consequently, we have

$$-\Delta H_{total} = K \int_{t=0}^{t=\infty} \Delta T\, dt. \qquad \textbf{(3.48)}$$

To obtain Eq. 3.48, we have assumed that the heat capacities and coefficients of thermal conductivity of the sample and reference cells are identical. We have also assumed that C_p and K are constant over the small temperature interval that will be observed for the dilute solutions employed in a DTA experiment. All these approximations are reasonably accurate, so Eq. 3.48 can be expected to give reliable data for ΔH_{total}.

In most experiments, the measured data ΔT are digitized and fed to a computer that produces a plot of ΔT as a function of time. This plot is termed a DTA curve. A typical example is shown in Figure 3.11 for the unimolecular decomposition of (3a,4,5,6,7,7a-hexahydro-4,7-methano-1H-benzotriazol-1-yl)-phosphonate (see Problem 3.18) in a diglyme solvent. The area under the DTA curve is the integral on the right-hand side of Eq. 3.48. The instrument's computer numerically integrates that equation to produce the area in question. If this area is denoted by A, the result of a DTA measurement is

$$-\Delta H_{total} = K\,A. \qquad \textbf{(3.49)}$$

The coefficient of thermal conductivity can be obtained by executing a DTA measurement on a sample for which ΔH is known. Example 3.7 illustrates this method.

The analysis for endothermic reactions is identical to that for exothermic processes. In the endothermic case, we have $\Delta T = T_s - T_r < 0$, since the process is absorbing heat in the sample cell, but not in the reference cell. Therefore, the area beneath the DTA curve will be negative, and ΔH_{total} will be positive, as is expected for an endothermic reaction.

In most cases, it is much easier to maintain two temperatures the same than it is to accurately measure a small difference between them. Differential

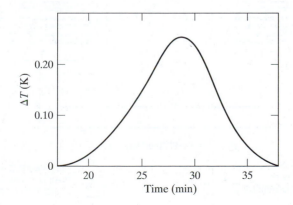

◀ FIGURE 3.11
DTA curve for the thermal decomposition of the triazoline (3a,4,5,6,7,7a-hexahydro-4,7-methano-1H-benzotriazol-1-y1)-phosphonate. The product of the area beneath the curve and the coefficient thermal conductivity gives the negative of the enthalpy change in the process. [Data taken from K. D. Berlin *et al.*, *Tetrahedron*, 23, 965 (1967).]

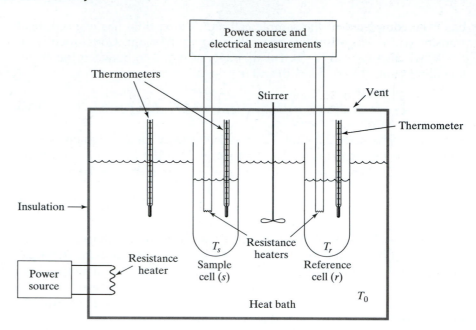

▶ FIGURE 3.12
Schematic diagram of a differential thermal analysis calorimeter modified for use as a differential scanning calorimeter.

scanning calorimetry (DSC) makes use of this fact to improve the accuracy of DTA measurements. A DSC instrument is similar in design to the DTA apparatus diagrammed in Figure 3.10. The major difference is that both cells are equipped with separate electrical heating elements, as shown in Figure 3.12. The rate of heating by these elements is electronically controlled to maintain $\Delta T = 0$ throughout the measurement. The data consist of the measured electrical energy transmitted to the heating elements as a function of time.

It is easy to modify Eqs. 3.41 through 3.49 so that they apply to a DSC experiment. Let us assume that we are dealing with an exothermic reaction, so that in a DTA experiment we would have $T_s > T_r$. To maintain a zero temper-

Power Compensation DSC

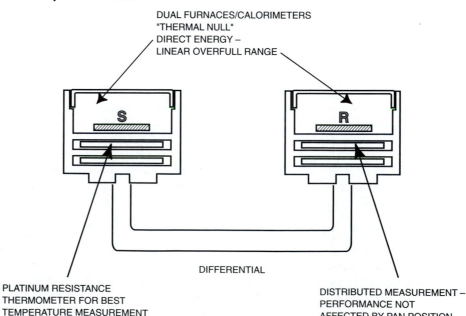

▶ ILLUSTRATION 3.4
Schematic of the power compensation in the reference (*R*) and sample (*S*) cells of a commercial differential scanning calorimeter manufactured by the Perkin-Elmer Corporation. The unit uses platinum resistance thermometers to monitor temperature differences between the cells.
[Reproduced with permission from the Perkin-Elmer Corporation.]

Furnace/Calorimeter Cross Section

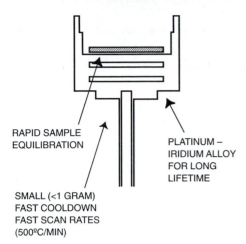

RAPID SAMPLE EQUILIBRATION

PLATINUM – IRIDIUM ALLOY FOR LONG LIFETIME

SMALL (<1 GRAM) FAST COOLDOWN FAST SCAN RATES (500°C/MIN)

◀ ILLUSTRATION 3.5
Schematic of the furnace/calorimeter cross section for a commercial differential scanning calorimeter manufactured by the Perkin-Elmer Corporation. The design produces rapid thermal equilibration so that temperature may be scanned at the rate of 500°C per minute. [Reproduced with permission from the Perkin-Elmer Corporation.]

ature difference between the cells, we must add electrical energy to the reference cell. The differential electrical energy delivered in a time dt is given by

$$dE = VI\,dt = \omega(t)\,dt, \tag{3.50}$$

where V is the voltage, I is the current, and $\omega(t)$ is the electrical power whose time dependence is to be measured in the DSC experiment. If V and I are expressed in volts and amperes, respectively, $\omega(t)$ will be in watts. If t is in seconds, E will be in joules. When electrical energy is delivered to the reference cell, Eq. 3.42 becomes

$$[C_p]_r\,dT_r = -(T_r - T_o)\,K_r\,dt + dE = -(T_r - T_o)\,K_r\,dt + \omega(t)\,dt. \tag{3.51}$$

As in a DTA apparatus, the two cells are identical, and the solutions are dilute, so that we have $K_r = K_s = K$ and $[C_p]_r = [C_p]_s = C_p$. Using this notation, we now subtract Eq. 3.51 from Eq. 3.43 to obtain

$$C_p\,d[T_s - T_r] = -dH - [T_s - T_r]\,K\,dt - \omega(t)\,dt. \tag{3.52}$$

Since $\omega(t)$ will be varied so as to maintain $[T_s - T_r] = 0$, Eq. 3.52 becomes

$$-dH - \omega(t)\,dt = 0. \tag{3.53}$$

Integration of both sides produces

$$\Delta H_{\text{total}} = \int dH = -\int_{t=0}^{t=\infty} \omega(t)\,dt. \tag{3.54}$$

Equation 3.54 shows that in a DSC measurement we plot $[-\omega(t)]$ versus time and compute the area under this curve to obtain ΔH_{total}. Such plots are called *thermograms*. The DSC measurement is more accurate, for the reasons given. It also has the advantage that we do not need to run a calibration experiment to determine the coefficient of thermal conductivity.

Illustration 3.4 shows a schematic of the actual DSC sample and reference cells in a commercial differential scanning calorimeter manufactured by the Perkin-Elmer Corporation. Platinum resistance thermometers are used for sensing the temperature difference between the cells. Illustration 3.5 is a schematic of the furnace/calorimeter cross section. The design is engineered to produce rapid thermal equilibration, so that the temperature of the bath

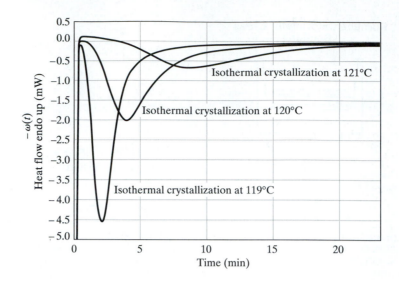

► ILLUSTRATION 3.6
DSC thermograms obtained for the crystallization of 3.733 mg of polyethylene at 121°C, 120°C and 119°C. The negative of the electrical power, in milliwatts, required to maintain temperature equilibrium between the reference and sample cells is given on the ordinate as a function of time, in minutes. The integrated area under these curves yields ΔH_{total} values for the three processes. [Reproduced with permission from the Perkin-Elmer Corporation.]

may be scanned at 500°C per minute. Illustration 3.6 shows three actual DSC thermograms obtained from the crystallization of 3.733 mg of polyethylene at 121°C, 120°C, and 119°C. The ordinate gives the negative of the electrical power in milliwatts required to maintain an equal temperature between the sample and reference cells. The integrated areas under these curves yield ΔH_{total} values for the three processes.

EXAMPLE 3.7

A bimolecular solution reaction between A and B for which $\Delta H = -100$ kJ mol^{-1} is carried out in a differential thermal analysis calorimeter. One hundred ml of a solution that is 1.00 M in both A and B is added to the sample cell, while 100 ml of pure solvent is placed in the reference cell. T_o is gradually increased to initiate a reaction, and a DTA curve is obtained. The area under the curve is found to be 478.0 s K. In a second experiment, 100 ml of a solution that is 1.000 M in both ethanol and acetic acid is placed in the sample cell, while 100 ml of pure solvent is added to the reference cell. T_o is gradually increased from 300 K to initiate the esterification reaction

$$C_2H_5OH_{(sol)} + CH_3COOH_{(sol)} \longrightarrow H_2O(l) + CH_3COOC_2H_{5(sol)}.$$

The area under the resulting DTA curve is found to be 12.60 s K. **(A)** Use the preceding data to determine the coefficient of thermal conductivity for the apparatus. **(B)** Compute $\Delta H(300 \text{ K})$ for the esterification reaction.

Solution

(A) The amount of A and B present in the sample cell is given by

$$\text{moles of } A = \text{moles of } B = V\,M, \tag{1}$$

where V is the volume of the solution, in liters, and M is the molarity of the solution. Therefore, we have

$$n_A = n_B = V\,M = (0.100 \text{ L})(1.00 \text{ mol L}^{-1}) = 0.100 \text{ mol}. \tag{2}$$

Equation 3.49 relates the area under the DTA curve to K and ΔH_{total}:

$$K = -\frac{\Delta H_{total}}{A}. \tag{3}$$

Since we have only 0.100 mol of A and B,

$$\Delta H_{total} = n_A \, \Delta H = (0.100 \text{ mol})(-100{,}000 \text{ J mol}^{-1}) = -1.00 \times 10^4 \text{ J}. \qquad (4)$$

Substituting into Eq. 3 gives

$$K = -\frac{-1.00 \times 10^4 \text{ J}}{478.0 \text{ s K}} = 20.9 \text{ J s}^{-1} \text{K}^{-1}. \qquad (5)$$

(B) Using Eq. 3.49, we may compute

$$-\Delta H_{total} = K \, A = (20.9 \text{ J s}^{-1} \text{K}^{-1})(12.6 \text{ s K}) = 264 \text{ J}. \qquad (6)$$

In this case, the number of moles of reactants are

$$\text{moles } C_2H_5OH = \text{moles } CH_3COOH = VM$$

$$= (0.100 \text{ L})(1.00 \text{ mol L}^{-1}) = 0.100 \text{ mol}. \qquad (7)$$

Using Eq. 4, we obtain

$$\Delta H = \frac{\Delta H_{total}}{n} = \frac{-264 \text{ J}}{0.100 \text{ mol}} = -2.64 \times 10^3 \text{ J mol}^{-1}. \qquad (8)$$

For related exercises, see Problems 3.18 and 3.19.

3.3   Bond Enthalpies

In order to break a chemical bond, energy must be inserted into the molecule. Conversely, when a chemical bond forms, energy is released. It is, therefore, reasonable to associate the total change in enthalpy that occurs in a chemical reaction with the making and breaking of chemical bonds. This leads naturally to the concept of a *bond enthalpy*. For a given molecule, it is always possible to determine a unique set of bond enthalpies that are in accord with the measured standard molar enthalpies of formation.

As an example, let us consider how we might determine the C—H bond enthalpy in methane, H_{C-H}, which we define to be the change in enthalpy when a C—H bond is ruptured. The standard molar enthalpy of formation of methane at 298.15 K is ΔH for the reaction $C(s) + 2 H_2(g) \longrightarrow CH_4(g)$ at 1 bar of pressure. This enthalpy change is -74.81 kJ mol^{-1}. The formation reaction, however, may be carried out along two hypothetical paths, as shown in Figure 3.13. Along Path 2, we first form $C(g)$ and 4 moles of $H(g)$. The change in enthalpy for this process, ΔH_1, can be obtained directly from the standard partial molar enthalpies for $C(s)$ and $H(g)$. Next, the atoms of carbon gas and hydrogen gas are used to form four identical C—H bonds for which the total enthalpy change, ΔH_2, is $-4 \, H_{C-H}$. Since dH is an exact differential whose value is path independent, we must have

$$[\Delta H_f^{o}]_{CH_4(g)} = \overline{H}^{o}_{CH_4(g)} = \Delta H_1 + \Delta H_2 = \overline{H}^{o}_{C(g)} + 4\,\overline{H}^{o}_{H(g)} - 4\,H_{C-H}. \qquad (3.55)$$

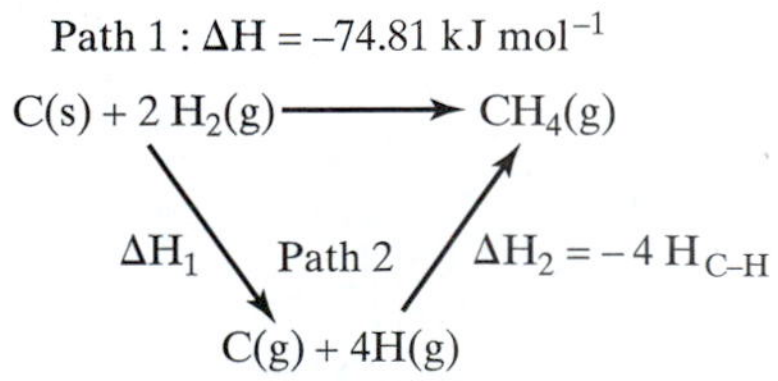

We may now solve Eq. 3.55 for H_{C-H}. The required standard partial molar enthalpies are equal to the standard molar enthalpies of formation given in Appendix A. We have

$$H_{C-H} = \left(\frac{1}{4}\right)\{-\overline{H}^{o}_{CH_4(g)} + \overline{H}^{o}_{C(g)} + 4\,\overline{H}^{o}_{H(g)}\}$$

$$= \left(\frac{1}{4}\right)[-(-74.81) + 716.68 + 4\,(217.97)] \text{ kJ mol}^{-1} = 415.8 \text{ kJ mol}^{-1}.$$

Path 1 : $\Delta H = -74.81$ kJ mol^{-1}

$$C(s) + 2 H_2(g) \xrightarrow{\hspace{2cm}} CH_4(g)$$
$$\Delta H_1 \searrow \quad \text{Path 2} \quad \nearrow \Delta H_2 = -4\,H_{C-H}$$
$$C(g) + 4H(g)$$

▲ FIGURE 3.13
Formation of $CH_4(g)$ along two different reaction paths.

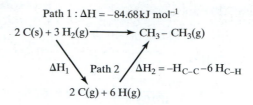

Path 1 : $\Delta H = -84.68\,kJ\,mol^{-1}$

$2\,C(s) + 3\,H_2(g) \longrightarrow CH_3-CH_3(g)$

ΔH_1 Path 2 $\Delta H_2 = -H_{C-C} - 6\,H_{C-H}$

$2\,C(g) + 6\,H(g)$

▶ FIGURE 3.14
Two reaction paths for the formation of ethane.

The analysis suggests that each time 1 mole of C—H bonds forms, the total enthalpy change will be -415.8 kJ.

Now that we have H_{C-H}, it is a simple matter to build up an entire table of bond enthalpies. For example, we could proceed to compute H_{C-C} by using the standard molar enthalpy of formation of ethane, as shown in Figure 3.14. Again, we must have $\overline{H}^{\circ}_{C_2H_6(g)}$ equal to $\Delta H_1 + \Delta H_2$. This permits us to solve for the C—C bond enthalpy. Using our result for H_{C-H} and the data in Appendix A, we obtain

$$H_{C-C} = -\overline{H}^{\circ}_{C_2H_6(g)} - 6\,H_{C-H} + 2\,\overline{H}^{\circ}_{C(g)} + 6\,\overline{H}^{\circ}_{H(g)}$$
$$= [-(-84.68) - 6\,(415.8) + 2\,(716.68) + 6\,(217.97)]\,kJ\,mol^{-1}$$
$$= 331.1\,kJ\,mol^{-1}.$$

The problem with the foregoing procedure is that we would have obtained different results for H_{C-H} and H_{C-C} if we had used different compounds. For example, we could have obtained H_{C-H} using the standard molar enthalpy of formation for the methyl radical, $CH_3(g)$, rather than methane. This situation is a reflection of the fact that the strength of chemical bonds is dependent upon the environment in the molecule. Thus, a C=O bond in CO_2 is not in the same electronic environment as a C=O bond in a ketone such as acetone. As a result, bond enthalpies differ from molecule to molecule.

Although the concept of a bond enthalpy is clearly inexact, it is often found that differences in computed bond enthalpies from molecule to molecule are on the order of about 10%. We might, therefore, adopt an average value for H_{C-H} and other bond enthalpies. Table 3.2 gives a table of such average bond enthalpies. Since there is no general agreement about how the averages should be obtained, tables of bond enthalpies usually exhibit some differences.

Enthalpy changes in chemical reactions may be estimated using tables of bond enthalpies. The procedure is to determine which bonds appear in the products that do not appear in the reactants. These bonds are formed in the reaction. At the same time, we determine which bonds are present in the reactants, but not the products. These bonds have been broken during the reaction. The enthalpy change is then estimated using

$$\Delta H \approx \sum_{\substack{bonds \\ broken}} H_i - \sum_{\substack{bonds \\ formed}} H_j. \tag{3.56}$$

This procedure can be very useful, particularly in qualitative reasoning about effects that concern the relative strengths of chemical bonds. However, it is important to remember that Eq. 3.56 is an approximation that is based on the premise that an average bond enthalpy can be used for all bonds of that type. If an unusual bond is involved, the equation should be used with care. For example, the two C=O bonds in CO_2 are unusual in that CO_2 is the only molecule with two such bonds on the same carbon atom. Consequently, the electronic environment about each C=O bond in CO_2 is very different from

Table 3.2   Average bond enthalpies in kJ mol^{-1}

	H	C	N	O	F	Cl	Br	I	S	P
H	436									
C	415	348(s)								
		612(d)								
		838(t)								
		518(ar)								
N	391	292(s)	161(s)							
		614(d)	413(d)							
		890(t)	946(t)							
O	463	355(s)	166	145(s)						
		734(d)		497(d)						
F	565	484	270	185	156					
Cl	431	333	200	203	254	243				
Br	366	276				219	193			
I	299	238				210	178	151		
S	353	259			496	250	212		265	
P	322									201

(s) single bond, (d) double bond, (t) triple bond, (ar) aromatic bond
Data: Pauling, L. *General Chemistry,* 3d, ed., Freeman, San Francisco, 1970; Atkins, P. *Physical Chemistry,* 5th, ed., Freeman, New York, 1994. Where the data differ, averages are taken.

that for the "average" C=O bond. Likewise, the C—C bond in the highly strained cyclopropane molecule is very different from that present in ethane.

EXAMPLE 3.8

Using bond enthalpies, estimate the enthalpy change at 298.15 K for the hydrogenation reaction of ethene to produce ethane: $H_2C=CH_2(g) + H_2(g) \longrightarrow H_3C-CH_3(g)$. Then, using standard partial molar enthalpies to compute $\Delta H^\circ(298.15)$, determine the percent error in the result.

Solution

We first determine the bonds present in the reactants that are missing in the products. The procedure is illustrated in Figure 3.15. The bonds that are broken are the C=C bond and the H—H bond. The bonds present in ethane that are not pre-

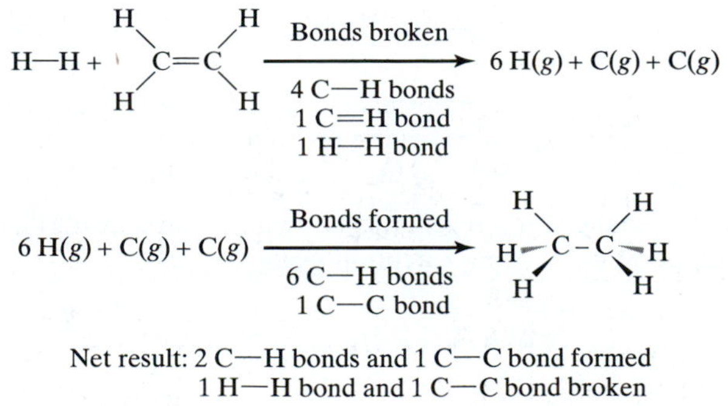

Net result: 2 C—H bonds and 1 C—C bond formed
1 H—H bond and 1 C—C bond broken

◀ FIGURE 3.15
Illustration of bonds broken and bonds formed in the hydrogenation of ethene.

sent in the reactants are two C—H bonds, since ethane has six such bonds and ethene has only four and a C—C single bond. Therefore,

$$\sum_{\substack{\text{bonds} \\ \text{broken}}} H_i = H_{C=C} + H_{H-H} = [612 + 436]\ kJ\ mol^{-1} = 1{,}048\ kJ\ mol^{-1} \qquad (1)$$

and

$$\sum_{\substack{\text{bonds} \\ \text{formed}}} H_i = 2\,H_{C-H} + H_{C-C} = [2(415) + 348]\ kJ\ mol^{-1} = 1{,}178\ kJ\ mol^{-1}. \qquad (2)$$

Using Eq. 3.53, we obtain

$$\Delta H^{\circ}(298.15) \approx (1{,}048 - 1{,}178)\ kJ\ mol^{-1} = -130\ kJ\ mol^{-1}. \qquad (3)$$

Using standard partial molar enthalpies from Appendix A, we compute

$$\Delta H^{\circ}(298.15) = -84.68\ kJ\ mol^{-1} - (52.26\ kJ\ mol^{-1}) = -136.94\ kJ\ mol^{-1}. \qquad (4)$$

The percent error is, therefore,

$$\% \text{ error} = \frac{100 \times (-136.94 - (-130))}{-136.94} = 5.07\ \%. \qquad (5)$$

For related exercises, see Problems 3.7, 3.8, 3.9, 3.30, 3.31 and 3.36.

Summary: Key Concepts and Equations

1. In Chapter 2, it was shown that the differential change in enthalpy of a compound A is given by $dH_A = [C_p]_A\,dT - \mu_A[C_p]_A\,dP + \overline{H}_A\,dn_A$. If a chemical reaction occurs under conditions of constant temperature and pressure, each component of the reaction will exhibit a change given by the last term in this equation. Consequently, the partial molar enthalpy $\overline{H}$ for each substance is the critical factor that determines the total enthalpy change that occurs in the reaction. If J products are formed from K different reactants, then

$$\Delta H = \sum_{p=1}^{p=J} \overline{H}_p \nu_p - \sum_{r=1}^{r=K} \overline{H}_r \nu_r.$$

2. Standard state conditions are defined as 1 bar of pressure for condensed phases and ideal gases. Partial molar enthalpies measured under these conditions are termed standard partial molar enthalpies and are denoted by the superscript o.

3. If we choose our reference point for the measurement of partial molar enthalpies to be the elements in their most stable state under standard state conditions and assign their standard partial molar enthalpies to be zero in this state, we have $\overline{H}_A^{\circ}(T) = [\Delta H_f^{\circ}(T)]_A$, where $[\Delta H_f^{\circ}(T)]_A$ is the standard molar enthalpy of formation of compound A.

4. Hess' law states that thermochemical equations can be treated as if they are algebraic equations, with the arrow serving the function of the mathematical equality, provided that all reactions refer to the same temperature and pressure.

5. If the enthalpy change for a reaction is known at temperature T_1, it may be computed at temperature T_2 using Kirchhoff's law,

$$\Delta H(T_2) = \Delta H(T_1) + \int_{T_1}^{T_2} \Delta C_p\,dT,$$

where

$$\Delta C_p = \sum_{s=1}^{s=J} \nu_s \left[C_p^m \right]_s - \sum_{r=1}^{r=K} \nu_r \left[C_p^m \right]_r.$$

Thus, ΔC_p is the difference in the molar heat capacities of the products and the reactants. In simple cases, ΔC_p will be a constant and can be factored from the integral. In more complex cases, ΔC_p will be a function of temperature. If ΔC_p is zero at all temperatures, ΔH will be independent of temperature.

6. The enthalpy change in a chemical reaction is highly insensitive to pressure changes because of the small magnitude of the Joule–Thomson coefficients. If the pressure changes are large, however, there may be an appreciable variation in ΔH. If so, ΔH at pressure P_2 may be determined from its value at P_1 using

$$\Delta H(P_2) = \Delta H(P_1) + \int_{P_1}^{P_2} \Delta \left[-\mu C_p^m \right] dP,$$

where

$$\Delta \left[-\mu C_p^m \right] = \sum_{s=1}^{s=J} \nu_s \left[-\mu_s \left[C_p^m \right]_s \right] - \sum_{r=1}^{r=K} \nu_r \left[-\mu_r \left[C_p^m \right]_r \right].$$

Consequently, we need both the Joule–Thomson coefficients and the heat capacities for all reactants and products.

7. In principle, enthalpy changes for solution reactions can be determined in the same manner as those involving pure gases, liquids, or solids. However, since we cannot determine standard partial molar enthalpies for all compounds at all possible concentrations in all possible solvents, these quantities must be measured for each individual system of interest. In some cases, we can obtain reasonably accurate values for ΔH by using standard partial molar enthalpies at infinite dilution.

8. For ionic solutions, standard partial molar enthalpies at infinite dilution for individual ions are obtained by choosing the reference point to be the hydrogen ion at infinite dilution under standard state conditions and setting $\overline{H}^o_{H^+(sol,\infty)} = 0$.

9. When chemical reactions are conducted under conditions of constant volume, the heat of reaction is given by ΔU rather than ΔH. We may compute ΔU from measured values for ΔH using $\Delta U = \Delta H - \Delta(PV)$. For reactions between condensed phases, the $\Delta(PV)$ term is negligibly small and may be ignored. For gas-phase reactions, we need the equation of state to obtain $\Delta(PV)$. If all gases behave ideally, we obtain the simple result

$$\Delta U = \Delta H - RT \, \Delta n,$$

where Δn is the change in the number of moles of gas when reactants become products.

10. The bond enthalpy is an approximate concept that associates a given amount of enthalpy change with the making or breaking of a particular type of chemical bond. Since the enthalpy associated with a bond depends not only upon the type of bond, but also upon the overall chemical environment, bond enthalpies differ from molecule to molecule. The magnitude of this variation is on the order of 10%. In most calculations, average bond enthalpies are employed. Using such averages, we find that the enthalpy change for a reaction is approximately

$$\Delta H \approx \sum_{\substack{\text{bonds} \\ \text{broken}}} H_i - \sum_{\substack{\text{bonds} \\ \text{formed}}} H_j.$$

Problems

Problems that require the use of some type of computational device are marked with an asterisk (*). Problems that require some type of plotting routine are indicated with a pound sign (#). Unless otherwise stated, all gases may be assumed to behave ideally.

3.1 At 298.15 K, the molar heat of combustion in oxygen of dipropyl ketone, H_7C_3–CO–C_3H_7, is $-4,395.3$ kJ when the products are $CO_2(g)$ and $H_2O(l)$. The standard partial molar enthalpies of $CO_2(g)$ and $H_2O(l)$ are -393.51 kJ and -285.83 kJ, respectively.

(A) Compute the standard partial molar enthalpy of dipropyl ketone.

(B) What is the standard molar heat of formation of dipropyl ketone?

(C) How much heat is liberated if 10 grams of dipropyl ketone are burned in excess oxygen in a constant-volume calorimeter?

3.2 The standard molar heat of formation of hexadecane, $C_{16}H_{34}$, is -447.97 kJ mol^{-1}. Use this datum and that given in Problem 3.1 to compute the molar heat of combustion of hexadecane at 298.15 K and 1 bar of pressure when the products are $H_2O(l)$ and $CO_2(g)$.

3.3 Using the data in Problems 3.1 and 3.2 and in Appendix A, compute the heat of combustion of hexadecane $[C_{16}H_{34}]$ at 298.15 K and 1 bar of pressure when the products are $CO(g)$ and $H_2O(l)$.

3.4 Fifty grams of hexadecane, $C_{16}H_{34}$, are burned with insufficient oxygen for complete combustion to $CO_2(g)$. As a result, some of the reactant forms $CO(g)$ and $H_2O(l)$ via the reaction

$$C_{16}H_{34}(s) + 16.5\,O_2(g) \longrightarrow 16\,CO(g) + 17\,H_2O(l),$$

and some of the reactant forms $CO_2(g)$ and $H_2O(l)$ in the reaction

$$C_{16}H_{34}(s) + 24.5\,O_2(g) \longrightarrow 16\,CO_2(g) + 17\,H_2O(l),$$

At a constant pressure of 1 bar and 298.15 K, the heat change in the process is found to be $-1,945.2$ kJ.

(A) Compute the percent of hexadecane that reacted to form $CO(g)$ in the process.

(B) If the process had been carried out in a constant-volume calorimeter, how much heat would have been released? Assume that the percent of reactant forming $CO(g)$ would be the same as that computed in part (A) of the problem.

3.5 At 298.15 K and 1 bar of pressure, the molar heat of combustion of benzoic acid, $C_6H_5COOH(s)$, to $CO_2(g)$ and $H_2O(l)$ is $-3,227.5$ kJ mol^{-1}.

(A) Use this fact and the data in Appendix A to determine the standard molar heat of formation of benzoic acid.

(B) If the heat capacity of a constant-volume calorimeter plus the products of combustion is 2,150 J K^{-1}, determine the temperature change that would be observed for the calorimeter if 1.00 gram of benzoic acid were reacted to form $CO_2(g)$ and $H_2O(l)$ at 298.15 K and 1 bar of pressure.

3.6 Molecule A is observed to undergo unimolecular reaction to form molecule B via the reaction $A(g) \longrightarrow B(g)$. The heat capacity at constant pressure for molecule B is found to be

$$[C_p]B = 6.5 + 1.4 \times 10^{-3}\,T$$

over the range 300 K to 900 K. Over this same temperature range, experimental measurement shows that the enthalpy change for the reaction is given by

$$\Delta H(T) = C + 0.1\,T - 0.00005\,T^2,$$

where C is a constant and $\Delta H(T)$ is the enthalpy change at temperature T. Determine the heat capacity at constant pressure of molecule A as a function of temperature.

3.7 Xenon forms a hexafluoride, XeF_6. The standard molar enthalpy of formation of solid $XeF_6(s)$ is -368.2 kJ mol^{-1}. The enthalpy of sublimation of $XeF_6(s)$ at 298.15 K is 62.34 kJ mol^{-1}. [The enthalpy of sublimation is ΔH for the process $XeF_6(s) \longrightarrow XeF_6(g)$.] The standard molar enthalpy of formation of $F(g)$ is 78.99 kJ mol^{-1}. Estimate the Xe–F bond enthalpy.

3.8 Use the average bond enthalpies given in Table 3.2 to estimate ΔH for the reaction

$$C_2H_2(g) + H_2(g) \longrightarrow C_2H_4(g)$$

at 298.15 K. Use standard partial molar enthalpies for these compounds to determine the percent error present in the bond enthalpy calculation.

3.9 Use the average bond enthalpies given in Table 3.2 to estimate ΔH for the reaction

$$C_2H_2(g) + 2.5\,O_2(g) \longrightarrow 2\,CO_2(g) + H_2O(g)$$

at 298.15 K. Use standard partial molar enthalpies for these compounds to determine the percent error present in the bond enthalpy calculation. The percent error in Problem 3.8 is 3.71%. Why is the error in this calculation so much worse?

3.10 One mole of sulfur dioxide, $SO_2(g)$, is dissolved in an infinitely dilute aqueous solution containing two moles of sodium hydroxide at 298.15 K. The dissolved $SO_2(g)$ establishes an equilibrium between $H^+_{(sol,\infty)}$ and $SO_3^{2-}_{(sol,\infty)}$. The $H^+_{(sol,\infty)}$ then reacts completely with $OH^-_{(sol,\infty)}$. Using Table 3.1, Appendix A, and the fact that $H^a_{SO_3^{2-}(sol,\infty)} = -624.3$ kJ mol^{-1}, determine ΔH for the total process.

3.11 N moles of liquid water are supercooled to 270.15 K in a container insulated from the surroundings at a pressure of 1 bar. Some of the water is then allowed to freeze, and the entire system is brought to a final equilibrium temperature of 273.15 K. The process is conducted under constant-pressure, adiabatic conditions. At this point, what fraction of the water is in the form of liquid H_2O? The enthalpy change upon melting is termed the molar enthalpy of fusion, ΔH_{fusion}. For water, $\Delta H^o_{fusion} = 6,004.4 \text{ J mol}^{-1}$ at 273.15 K and 1 bar of pressure. The constant-pressure heat capacities of ice and water are 38.73 J mol^{-1} K^{-1} and 75.94 J mol^{-1} K^{-1}, respectively. Assume that these values are constant from 270 to 273.15 K.

3.12 The heat capacities for HCl(g), $O_2(g)$, $H_2O(g)$, and $Cl_2(g)$ are

$$[C_p]_{HCl} = 6.7319 + 0.4325 \times 10^{-3} T$$
$$+ 3.697 \times 10^{-7} T^2 \text{ cal mol}^{-1} \text{K}^{-1},$$

$$[C_p]_{O_2(g)} = 6.148 + 3.102 \times 10^{-3} T$$
$$- 9.23 \times 10^{-7} T^2 \text{ cal mol}^{-1} \text{K}^{-1},$$

$$[C_p]_{H_2O(g)} = 7.256 + 2.298 \times 10^{-3} T$$
$$+ 2.83 \times 10^{-7} T^2 \text{ cal mol}^{-1} \text{K}^{-1},$$

and

$$[C_p]_{Cl_2(g)} = 7.5755 + 2.4244 \times 10^{-3} T$$
$$- 9.650 \times 10^{-7} T^2 \text{ cal mol}^{-1} \text{K}^{-1}.$$

Using these data and standard partial molar enthalpies from Appendix A, derive an expression for the standard enthalpy change in the reaction

$$2 \text{ HCl(g)} + 0.5 \text{ O}_2(g) \longrightarrow H_2O(g) + Cl_2(g)$$

as a function of temperature.

3.13 Consider a gas-phase dimerization reaction, $2A(g) \longrightarrow B(g)$. An experimental thermodynamicist determines that the enthalpy change in the reaction can be accurately represented by a quadratic function of temperature; that is,

$$\Delta H = aT^2 + bT + c,$$

where a, b, and c are constants.
(A) Obtain an expression for ΔC_p for this reaction as a function of a, b, c, and T.
(B) If C_p for molecule $B(g)$ is a constant equal to 8.368 J mol^{-1} K^{-1}, determine C_p for molecule $A(g)$ as a function of a, b, c, and T.

3.14 Gaseous compounds A and B have the following characteristics:

Quantity	Compound A	Compound B
C_p	25 cal mol^{-1} K^{-1}	35 cal mol^{-1} K^{-1}
μ	1.1 K bar^{-1}	1.5 K bar^{-1}

(μ is the Joule–Thomson coefficient.) The reaction $A \longrightarrow B$ is carried out at 298.15 K, $P = 1$ bar, and the enthalpy change for the process is measured. When the pressure is increased to 10 bar and the temperature is changed to some new value T_2, it is found that the enthalpy change at the new temperature is unchanged from its value at 298.15 K, $P = 1$ bar. Compute the temperature T_2.

3.15* Gaseous compounds A and B react to give gaseous compound C. The balanced chemical equation is

$$3 A(g) + B(g) \longrightarrow 2 C(g).$$

The following standard partial molar enthalpies are found:

$$\overline{H}^o_{A(g)} = 0.0 \text{ kJ mol}^{-1};$$
$$\overline{H}^o_{B(g)} = 0.0 \text{ kJ mol}^{-1};$$
$$\overline{H}^o_{C(g)} = -46.11 \text{ kJ mol}^{-1}.$$

Heat capacities at constant pressure, C_p, have also been measured over the temperature range 300 K $\leq T \leq$ 540 K for $A(g)$, $B(g)$, and $C(g)$. The results of these measurements are given in the table below.

	C_p (J mol^{-1} K^{-1})		
Temp. (K)	A(g)	B (g)	C(g)
300.0	28.7080	29.2610	35.8850
310.0	28.7711	29.2682	36.0414
320.0	28.8352	29.2744	36.1948
330.0	28.9003	29.2796	36.3451
340.0	28.9664	29.2838	36.4922
350.0	29.0335	29.2870	36.6363
360.0	29.1016	29.2892	36.7772
370.0	29.1708	29.2904	36.9150
380.0	29.2408	29.2906	37.0498
390.0	29.3119	29.2898	37.1814
400.0	29.3840	29.2880	37.3100
410.0	29.4751	29.2852	37.4353
420.0	29.5312	29.2814	37.5578
430.0	29.6063	29.2766	37.6771
440.0	29.6824	29.2708	37.7932
450.0	29.7595	29.2640	37.9063
460.0	29.8376	29.2562	38.0162
470.0	29.9167	29.2474	38.1231
480.0	29.9968	29.2376	38.2258
490.0	30.0779	29.2268	38.3275
500.0	30.1600	29.2150	38.4250
510.0	30.2431	29.2022	38.5194

520.0	30.3272	29.1884	38.6108
530.0	30.4123	29.1736	38.6990
540.0	30.4984	29.1578	38.7842

(A) Compute ΔH^o for the given reaction at 298.15 K.

(B) By fitting the heat capacity data with a power series expansion in T of the form

$$C_p(T) = \sum_{n=0}^{n=2} A_n T^n$$

in which the expansion coefficients A_o, A_1, and A_2 are determined using a least-squares procedure, obtain ΔH as an analytic function of temperature. On the same graph, plot the data and the analytic curves obtained from the least-squares fits.

(C) Compute the value of the heat of reaction at 500 K. How much error is made in this value by assuming that ΔH^o is independent of temperature?

3.16 Reactants R at temperature T_1 and pressure P are denoted as $R(T_1,P)$. A chemical reaction converts these reactants into products at temperature T_2 and pressure P, which are represented by the notation $p(T_2,P)$. The following diagram shows this conversion by two pathways:

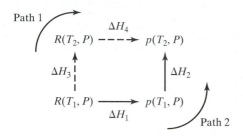

By considering the overall process along Paths 1 and 2 and using the fact that dH is an exact differential, derive Eq. 3.26.

3.17 Consider the reaction $CO(g) + 0.5\,O_2(g) \longrightarrow CO_2(g)$. The heat capacities and Joule–Thomson coefficients are as follows:

Compound	C_p (cal mol^{-1} K^{-1})	μ (K bar^{-1})
$CO(g)$	6. 3423 + 0.0018363 T	1.20
$O_2(g)$	6.148 + 0.003102 T	1.15
$CO_2(g)$	6.369 + 0.0101 T	1.10

Using these data and the data in Appendix A, compute ΔH for the given reaction at 298.15 K with all gases at a pressure of 30 bar. Assume that μ and C_p are independent of pressure.

3.18 The unimolecular decomposition of the triazoline (3a,4,5,6,7,7a-hexahydro-4,7-methano-1H-benzotriazol-1-yl)-phosphonate, has been studied by Berlin *et al.* [Tetrahedron, **1967**, 23, 965] using differential thermal analysis. A 0.1094-gram sample of the triazoline in 0.2885 gram of diglyme solvent was placed in the sample cell of a DTA apparatus. The reference cell contained only the diglyme. The temperature of the heat bath was raised at the rate of 3 K min^{-1}. The area beneath the DTA curve was found to be 238.7 cm^2. In a subsequent calibration experiment, the enthalpy change for melting a 0.1527-gram sample of o-toluic acid ($CH_3C_6H_4COOH$) was measured with the apparatus. In this case, the area beneath the DTA curve was -205.7 cm^2. (The negative sign denotes the fact that ΔT is negative in this endothermic reaction.) The standard molar heat of fusion of o-toluic acid is 20.16 kJ mol^{-1}. If the heat of fusion is assumed to be independent of temperature, compute ΔH for the decomposition of the triazoline.

$$P(O)(OEt)_2$$

3.19* A differential scanning calorimetric measurement is made to determine ΔH for the unimolecular decomposition of a compound. In the experiment, the voltage is maintained at a constant value V_o, while the current is varied so as to maintain a zero temperature differential between the reference and sample cells. The data obtained are as follows:

Time (minutes)	$I(t)$ (amperes)
0	0.00
20	0.0050
22.5	0.0140
25.0	0.0280
27.5	0.0450
28.75	0.0490
30.0	0.0460
32.5	0.0230
35.0	0.0070
40.0	0.0000

A nonlinear least squares fit of the data is made, and it is found that $I(t)$ is accurately represented by the function

$$I(t) = 0.024457 \exp[-0.0809327\,(t - 28.75)^2]$$
$$+ 0.0247461 \exp[-0.0240641(t - 28.75)^2],$$

where t is expressed in minutes. Also, 0.0007929 mole of the compound were used in the experiment, and it is known that $\Delta H = -65.56$ kJ mol^{-1}.

(A) Plot the fitted function $I(t)$ versus t between 0 and 40 minutes. On the same graph, show the measured data.

(B) Using a suitable method, determine the area beneath the curve $I(t)$ versus t.

(C) Determine the constant voltage that was used in the DSC experiment.

3.20 Seven grams of propene ($C_3H_6(g)$) are burned to form $CO_2(g)$ and $H_2O(l)$ at 298.15 K in a constant-volume calorimeter. The temperature of the calorimeter is observed to rise 13.45 K during the process. Compute the heat capacity of the calorimeter and its contents. You may assume that the gases behave ideally.

3.21 N moles of an alkane, C_nH_{2n+2}, are burned in excess oxygen to give $CO_2(g)$ and $H_2O(l)$ at 298.15 K. Obtain a general equation for the total enthalpy change for the process as a function of N, n, and $H^o_{C_nH_{2n+2}}$.

3.22 One mole of an alkane, C_nH_{2n+2}, is burned in excess oxygen to form $CO_2(g)$ and $H_2O(l)$ at 298.15 K and 1 bar of pressure. At constant pressure, the heat released in the process is found to be 2877.04 kJ. It is also found that 20 g of the alkane occupies a volume of 4.210 liters at 298.15 K and 2 atm pressure. Determine the standard partial molar enthalpy of the alkane. Assume that the gaseous alkane behaves ideally.

3.23 One mole of an alkane, C_nH_{2n+2}, is burned in excess oxygen to form $CO_2(g)$ and $H_2O(l)$ at 298.15 K and 1 bar of pressure. At constant pressure, the heat released in the process is found to be 4816.8 kJ. If the products of the combustion at 298.15 K and 1 bar of pressure are $CO_2(g)$ and $H_2O(g)$, the heat released is 4464.7 kJ. Determine the molecular formula of the alkane and its standard partial molar enthalpy.

3.24 N moles of an unknown alkane, C_nH_{2n+2}, are burned in excess oxygen to form $CO_2(g)$ and $H_2O(l)$ at 298.15 K and 1 bar pressure. At constant pressure, the heat released in the process is found to be 8753.0 kJ. If the products of the combustion at 298.15 K and 1 bar of pressure are $CO_2(g)$ and $H_2O(g)$, the heat released is 8119.3 kJ. When N moles of the alkane are burned in limited oxygen at 298.15 K, so that the products are $CO(g)$ and $H_2O(l)$, the heat released is 5130.9 kJ. Determine the molecular formula of the alkane, the number of moles burned in the experiments, and the standard partial molar enthalpy of the unknown alkane.

3.25 Derive a general equation giving the rate of change of the reaction enthalpy with temperature at constant pressure for the combustion of an alkane, C_nH_{2n+2}, to form $CO_2(g)$ and $H_2O(g)$ as a function of

n and C_p^m for the alkane. Assume that the heat capacities are independent of temperature, and use the data given in Table 2.1 as needed.

3.26 One mole of an alkane, C_nH_{2n+2}, whose constant-pressure heat capacity is 120.2 J mol^{-1} K^{-1}, is burned in excess oxygen at 298.15 K to form $CO_2(g)$ and $H_2O(g)$ at 298.15 K. At constant pressure, the heat released is 3,536.09 kJ mol^{-1}. If the same reaction is carried out at 400 K, the heat released is found to be 3,532.83 kJ mol^{-1}. Determine the molecular formula of the alkane.

3.27 Let us assume that a car completely oxidizes its fuel to $CO_2(g)$ and $H_2O(g)$. Let us further assume that gasoline can be accurately represented as n-octane. (Actually, it is a complex mixture of branched-chain octanes and other hydrocarbons, as well as a variety of additives.)

(A) What is the total enthalpy content of a tank of gasoline if the tank holds 20 U.S. gallons?

(B) Ethanol produced by fermentation of grain has been proposed as an alternative automotive fuel. If the ethanol is also completely oxidized to $CO_2(g)$ and $H_2O(g)$ in the car's engine, what is the total enthalpy content of a tank of ethanol?

(C) In order to be cost competitive in terms of enthalpy content per dollar, what is the maximum price that can be charged for ethanol if gasoline is $1.20 per U.S. gallon? In view of the approximation made concerning the composition of gasoline, it is appropriate to ignore the temperature and pressure dependence of the reaction enthalpy. Useful data are as follows: Density of $C_2H_5OH(l) = 0.7893$ g ml^{-1}; density of n-$C_8H_{18}(l) = 0.7036$ g ml^{-1}; 1 U.S. gallon = 3,785.20 ml.

3.28 The first part of this problem is identical to Problem 3.27. Let us assume that a car completely oxidizes its fuel to $CO_2(g)$ and $H_2O(g)$. Let us further assume that gasoline can be accurately represented as n-octane. (Actually, it is a complex mixture of branched-chain octanes and other hydrocarbons, as well as a variety of additives.)

(A) What is the total energy content of a tank of gasoline if the tank holds 20 U.S. gallons?

(B) In order to solve the problem of the limited supply of fossil fuels and to reduce pollution, $H_2(g)$ has been suggested as an alternative automotive fuel. Compute the mass of $H_2(g)$ required to give an energy output equivalent to a 20-U.S.-gallon tank of n-octane.

(C) $H_2(g)$ is normally stored in cylinders of compressed gas at 2,100 lb/in^2 pressure with a volume of 1.75 ft^3. How many such cylinders at 298 K would be required to transport the amount of hydrogen gas computed in Part (B)? You may assume $H_2(g)$ to behave ideally. Useful data are as follows: Density of n-$C_8H_{18}(l) = 0.7036$ g ml^{-1}; 1 U.S. gallon = 3,785.20 ml.

3.29 Let us examine a little of the economics of running hell. Natural gas, which is primarily methane [$CH_4(g)$], is sold by the hundred cubic feet (Cef). The retail price is generally around \$0.30 per Cef.

(A) Compute the energy available from the complete combustion of 1 Cef of $CH_4(g)$ to $CO_2(g)$ and $H_2O(l)$ at 298.15 K and 1 bar of pressure, assuming that the volume of the methane is measured at 1 atm pressure and 298.15 K. (1 ft^3 = 28.3171 liters).

(B) In Problem 1.17, we estimated the volume of hell to be $V = 4.5307 \times 10^5 \exp[0.0020723t]$ liters, where t is in years. The factor 4.5307×10^5 liters is the volume of a 2,000 ft^2 house with 8-ft ceilings. The Devil feels that, to keep up appearances and protect his image, he must keep one roaring fire going in each section of hell whose volume is equal to that of a 2,000 ft^2 house with 8-ft ceilings. A typical home in the winter might use 400 Cef per month of natural gas to provide heat. The Devil, however, wants more intense fires than that found in a normal household. Therefore, he opts to use furnaces that consume 2,000 Cef each month. Compute the total Cef consumption per month needed to keep the fires of hell burning. (*Note:* In Problem 1.17, it was assumed that the Devil entered hell 5,000 years ago; consequently, $t = 5,000$ years at the present time.)

(C) Compute the Devil's gas bill per month at the aforementioned retail price of natural gas. (*Note:* The Devil is far behind in his payments. The gas company has threatened to shut off his gas and thus quench the fires of hell. Congress, however, has decided that the Devil is an endangered species and, therefore, that his gas cannot be turned off. The gas company has appealed to the Supreme Court, but the Devil has an army of lawyers at his disposal, and the hopes of the gas company appear to be dim at the present time.)

3.30 The standard partial molar enthalpy of cyclopropane ($C_3H_6(g)$) is $+53.30$ kJ mol^{-1}.

(A) Using this fact and the data in Appendix A, obtain the C—C bond enthalpy, assuming that the C—H bond enthalpy is that given in Table 3.2.

(B) Compute the percent difference between the value you found in (A) and that given in Table 3.2 for a single C—C bond.

(C) Using the bond enthalpies given in Table 3.2, compute the enthalpy change for the reaction $3 C(g) + 6 H(g) \longrightarrow C_3H_6(g)$. Calculate the correct value for this enthalpy change, and compute the difference between the two calculations. This difference is called the *strain energy* of cyclopropane.

3.31 Benzene [$C_6H_6(g)$], is a six-membered ring of carbon atoms with alternating double and single bonds. Such structures are more stable than might be expected, due to a delocalization of electrons around the ring. The enhanced stability is termed the *resonance energy*

of the molecule. The reasons underlying this effect will be treated in greater detail in Chapters 12 and 14. Here, we will obtain an estimate of the resonance energy in gaseous benzene. Consider the reaction

$$6 C(g) + 6 H(g) \longrightarrow C_6H_6(g).$$

(A) Using the bond enthalpies given in Table 3.2, obtain an estimate of ΔH for this reaction at 298.15 K.

(B) Using standard partial molar enthalpies from Appendix A, compute the correct ΔH for this reaction at 298.15 K.

(C) Compute the difference in the answers obtained in (A) and (B). This difference is the resonance energy of benzene.

3.32 It is well known that diluting sulfuric acid by pouring it into water results in an exothermic reaction. Fifty grams of $H_2SO_4(l)$ are added to a large quantity of water at 298.15 K. Estimate the enthalpy change in the reaction.

3.33 One mole of silver nitrate and 1 mole of NaCl are added to a large quantity of water at 298.15 K. The result is the precipitation of 1 mole of AgCl(s).

(A) Write ionic reactions, showing all processes that occur.

(B) Using Table 3.2 and the data in Appendix A, compute the change in enthalpy for the entire process. Assume that the solutions are sufficiently dilute that standard partial molar enthalpies at infinite dilution may be used for the ions present in the solution.

3.34 One mole of an alkane, C_nH_{2n+2}, is burned in excess $O_2(g)$ to form $CO_2(g)$ and $H_2O(l)$ at 298.15 K. The standard enthalpy change for the reaction is found to be $-2,220.0$ kJ mol^{-1}. The standard partial molar enthalpy of the alkane is known to be in the range

$$-600.0 \text{ kJ mol}^{-1} \leq \overline{H}^o_{C_nH_{n+2}} \leq 500.0 \text{ kJ mol}^{-1}.$$

Using these data and the facts that $\overline{H}^o_{CO_2(g)} = -393.51$ kJ mol^{-1} and $\overline{H}^o_{H_2O(l)} = -285.83$ kJ mol^{-1} *alone*, is it possible to determine the molecular formula of the unknown alkane? If not, state what additional data are required to make the determination. If the data are sufficient, determine the molecular formula of the alkane.

3.35 A hydrocarbon is known to be either an alkene with one C=C double bond or a saturated alkane. When 1 mole of the hydrocarbon is burned in excess $O_2(g)$ at 298.15 K, the enthalpy change in the reaction is found to be 176.04 kJ mol^{-1} larger if the products of the reaction are $CO_2(g)$ and $H_2O(g)$ than is the case if the products are $CO_2(g)$ and $H_2O(l)$. If the amount of $O_2(g)$ is limited so that the products of combustion at 298.15 K are CO(g) and $H_2O(l)$, ΔH^o is 1,132.92 kJ mol^{-1}

larger than the enthalpy change when the products are $CO_2(g)$ and $H_2O(l)$. Using these data and the facts that $\overline{H}^\circ_{CO_2(g)} = -393.51$ kJ mol^{-1}, $\overline{H}^\circ_{CO_2(g)} = -110.53$ kJ mol^{-1}, $\overline{H}^\circ_{H_2O(l)} = -285.83$ kJ mol^{-1}, and $\overline{H}^\circ_{H_2O(g)} = -241.82$ kJ mol^{-1} *alone*, determine the molecular formula of the hydrocarbon.

3.36*(A) Using the data in Appendix A, compute the standard enthalpy change ΔH° for the combustion of 1 mole of alkane, C_nH_{2n+2}, in $O_2(g)$ to form $CO_2(g)$ and $H_2O(l)$ at 298.15 K for all values of n from 1 to 8.

(B) Plot ΔH° as a function of n, and obtain the best linear fit to the data using a least-squares method.

(C) Prove analytically that a plot of ΔH° versus n will be exactly linear if the C—H and C—C bond enthalpies in the alkanes are each the same. Derive analytically the predicted slope and intercept of the line, and compare them with the experimental results obtained in (B) by computing the percent error in each.

(D) Use the analytical result obtained in (C) to compute the heat of combustion of *n*-octane, $C_8H_{18}(l)$, and compare your result with the experimental one obtained in (A) by using standard partial molar enthalpies.

3.37 Which do you consider to be the most energy efficient way to use fodder to heat water?

(A) Burn the fodder under a kettle of water.

(B) Feed the fodder to a horse attached to a friction machine immersed in a kettle of water.

(C) Feed the fodder to a horse immersed in a kettle of water. Explain your answer [This problem is taken from Henry Bent's *The Second Law* (Oxford University Press, New York, 1965), Library of Congress Number 65-15608.]

3.38 *The ice weight-reduction diet.* Ice can be used to counteract the thermal consequences (e.g. storing fat) of eating too much food. The average food intake for adult humans is 2,500 Cal/day. One Cal (large c) is equal to one kcal (small c). How much ice at 273.15 K would one have to consume in order to cancel the effect of eating 2,500 Cal? Assume that body temperature is 310 K (37°C) and that the heat of fusion of ice is 330 kJ kg^{-1}. [The author thanks Fredrick L. Minn, M.D., Ph.D. for contributing this problem.]

3.39 Sam is having great difficulty with physical chemistry. However, he feels that his creativity compensates for his poor performance on examinations. As an example of his talent, he has devised a means of driving ships back and forth across the ocean without using any fuel. His invention is simple: Sam's ship is equipped with an intake port below the water level that allows water to enter the ship. After the water fills a receiving tank, the thermal energy is extracted from the liquid, leaving ice behind. The ice is ejected back into the ocean, where solar energy melts it back into liquid water. In the meantime, the energy extracted from the water is shunted to a boiler, where it is used to provide steam which drives the pistons that turn the crankshaft to the screws, thereby propelling the ship across the ocean with no need to use fuel. Since the sun is, for all practical purposes, an inexhaustible energy source, Sam feels he has "solved" the energy problem for ships. On the basis of this concept, he petitions his professor for a passing grade in physical chemistry.

Put a copy of *Amadeus* in the VCR, sit back with your favorite beverage, and relax as the incomparable music of Mozart fills the room. As you do so, contemplate Sam's suggestion. Will it work? Is he a genius? Should he pass physical chemistry on the basis of his concept? Sam's invention is called a *perpetual-motion machine of the second type*. In the next chapter, we will discuss it in more detail. In the meantime, enjoy *Amadeus* before we begin to discuss the second law of thermodynamics.

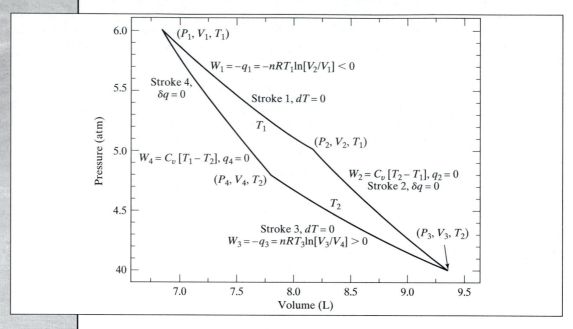

CHAPTER 4

The Second Law of Thermodynamics

The second law of thermodynamics is unique. It is the only scientific law that is stated in negative terms. It tells us what we cannot not do. Together, the first and second laws permit us to determine the conditions under which a process will be spontaneous, when it will be at equilibrium and when work will need to be done to force its occurrence. Because of its unique nature, the second law is more difficult to comprehend and utilize than the first law. In a very real sense, it is the critical junction in your study of thermodynamics. Those who take the right path at this point will find the remainder of this section relatively easy. The wrong path leads to regions best left unexplored.

4.1 General Remarks on Spontaneity

The first law of thermodynamics, in principle, permits us to determine the energy changes associated with any process of interest. If we know the details of the path by which the process occurs, we can determine how much of the computed energy change appears as heat and how much as work. In most cases, however, the first law alone does not permit us to predict the spontaneity or nonspontaneity of the process. That is, with the first law alone, we generally cannot determine whether a particular chemical reaction will occur under a given set of conditions. We can calculate the expected energy changes that will accompany the process if it occurs, but we cannot predict whether it will actually take place.

In some trivial cases, the first law alone allows us to determine that a process will not occur. These cases involve proposed processes that produce work without any expenditure of energy. Since that would violate the first law, which requires energy conservation, we may confidently predict that such processes will not take place. For example, a process in which water runs over a dam, turns the armature of a generator, thereby producing electrical energy, and then turns around and runs back up the dam will not occur spontaneously. Machines that operate in cycles, with each cycle producing work without the expenditure of an equivalent amount of energy in some form, are termed *perpetual-motion machines of the first type.* The nomenclature derives from the fact that the proposed machine violates the first law. The U.S. Patent Office will no longer consider patent applications for such machines.

It is extremely important to have a means of predicting the spontaneity or nonspontaneity of proposed processes that do not violate the first law. For example, consider the process outlined in Problem 3.39. This process suggests that we utilize the thermal energy contained in ocean water to drive a ship. We simply construct an intake port below the waterline, to admit seawater into the ship. The *Titanic* disaster demonstrates that this is a spontaneous process. Once the water is inside a receiving tank, the thermal energy is extracted and sent to the ship's boilers to create steam which will drive the pistons that turn the crankshaft to the ship's screws. The ice formed after extracting the thermal energy from the seawater is ejected back into the ocean, where it is melted by thermal energy from the sun. Therefore, we drive the ship without burning fuel. The overall process is illustrated in Figure 4.1. This "thermal engine"

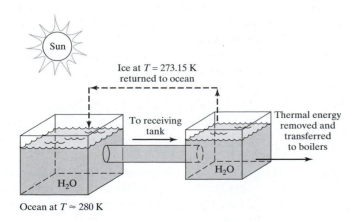

Ice at $T = 273.15$ K returned to ocean

To receiving tank

Thermal energy removed and transferred to boilers

H_2O

H_2O

Ocean at $T \approx 280$ K

◀ FIGURE 4.1

Schematic of a perpetual-motion machine of the second type that removes the thermal energy contained in ocean water and transfers it to a ship's boilers to produce steam to drive the ship's screws while returning ice to the ocean to be melted by the thermal energy of the sun. This apparatus does not violate the first law of thermodynamics, because no net energy is created, but it does violate the second law.

operates in fill–eject cycles and produces useful work in each cycle. It does not, however, violate the first law, since no energy is being created. It simply utilizes the thermal energy from the sun that is present in seawater by virtue of the fact that the water temperature is around 280 K rather than 0 K. Consequently, the proposed engine is not a perpetual-motion machine of the first type. It is, however, a *perpetual motion machine of the second type,* in that it violates the second law of thermodynamics. In this chapter, we shall learn why this is the case.

The second law is concerned with the interconversion of heat and work and the conditions under which it is possible. Together, the first and second laws permit us to predict whether a particular process will be spontaneous under a given set of conditions.

4.2 Carnot Heat Engines

Since the second law deals with the conversion of heat into work, we first need to consider how an engine achieves this conversion. All heat engines operate in cycles. If this were not the case, the engine would be useless. It would be like an automobile driven by a tightly coiled spring. The spring unwinds and drives the car's wheels until its energy is expended. Afterwards, the car coasts to a stop and sits uselessly on the shoulder of the road. To be a useful engine, the spring would have to be able to be rewound. That is, the engine must operate in unwind–rewind cycles. Heat engines differ in the number of distinct steps, or *strokes,* utilized to achieve the required cycle. Some employ two strokes; others use three or four. Engines also differ in the material used to produce the work. A steam engine utilizes liquid and vapor H_2O. An automobile uses the hot gases produced by combustion of the gasoline within the cylinder to drive the engine's pistons. In the past, many air conditioners and heat pumps employed Freon™ as the working material. At present, other materials are being used to eliminate the environmental hazards of Freon™.

Let us consider a hypothetical four-stroke engine that employs n moles of an ideal gas as its working material. We will assume that the engineering design of the engine is perfection itself: The rings and valves do not leak, and there is zero degradation of work to heat because the lubricant totally eliminates friction. Suppose the gas in the cylinders is initially at pressure P_1, volume V_1, and temperature T_1. The four steps, or strokes, used by the engine to achieve the required cyclic operation are as follows:

Stroke 1:   The gas is expanded isothermally and reversibly within the cylinder from (P_1, V_1, T_1) to (P_2, V_2, T_1), thus driving the pistons to provide useful work.

Stroke 2:   The gas is further expanded adiabatically and reversibly from (P_2, V_2, T_1) to (P_3, V_3, T_2) to produce additional work.

Stroke 3:   The gas is now compressed isothermally and reversibly within the cylinder from (P_3, V_3, T_2) to (P_4, V_4, T_2). This compression step requires that work be done on the gas.

Stroke 4:   The gas is further compressed adiabatically and reversibly from (P_4, V_4, T_2) back to its original state $(P_1, V_1, T_1.)$ Like Stroke 3, this operation also requires that we do work on the gas.

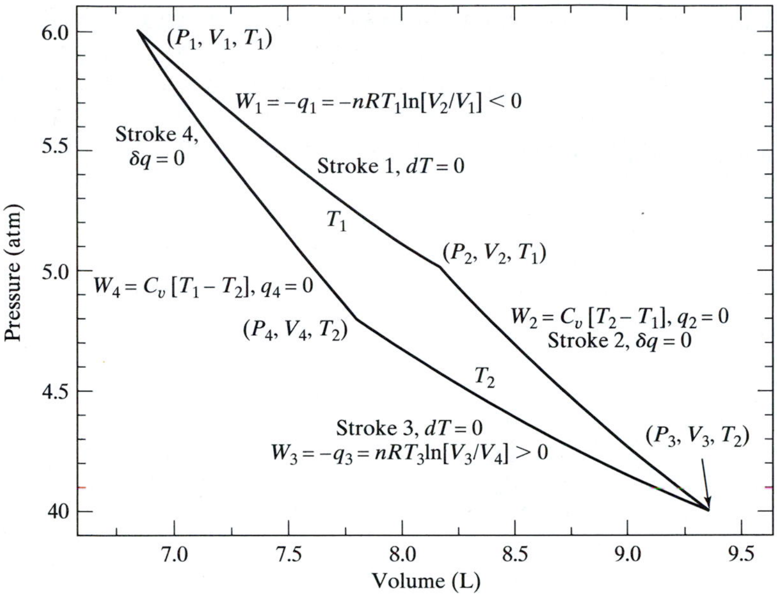

FIGURE 4.2 Carnot cycle.

The (P, V, T) behavior of the gas during these strokes is shown in Figure 4.2. Such an engine is called a *Carnot engine*, and the cyclic process represented by Strokes 1–4 is termed a *Carnot cycle*. Problem 2.14 is a specific example of a Carnot cycle.

It is useful at this point to make a few qualitative observations about the Carnot engine. Since the engineering design is perfection and all processes are reversible, the work obtained from the engine will be the maximum possible. The net work per cycle is

$$\text{Total Work} = w_T = -\int_{\text{cycle}} P_{\text{ext}} \, dV = -\int_{\text{cycle}} P \, dV$$

$$= w_1 + w_2 + w_3 + w_4 \tag{4.1}$$

since the processes are reversible, allowing P_{ext} to be replaced with P. The integrals in Eq. 4.1 are evaluated around the cycle shown in Figure 4.2. w_i ($i = 1, 2, 3, 4$) is the work done in Stroke i. Graphically, w_T is the negative of the area within the cycle's closed loop seen in Figure 4.2. In its operation, the Carnot engine repeats Strokes 1–4 over and over to deliver work in the amount w_T per cycle. It is clear that the Carnot engine is a hypothetical concept: Ideal gases, frictionless parts, perfect valves and rings, and reversible processes do not exist. Nevertheless, the concept is very useful, in that it permits us to determine the maximum possible work per cycle for a perfect engine using an ideal gas as a working material. In doing so, it lays the foundation for the introduction of the second law of thermodynamics.

We now need to obtain an analytic expression for w_T. The first step is to note that the four volumes V_1, V_2, V_3, and V_4 are not all independent. V_2 and V_3 are related because they are connected by the reversible adiabatic expansion occurring in Stroke 2. The same is true of V_4 and V_1. Such a reversible adiabatic expansion of n moles of an ideal gas was treated in detail in Example 2.7. For convenience, we repeat that treatment here.

For an ideal gas undergoing a reversible adiabatic process with $\delta q = 0$, we have

$$dU = C_v \, dT = \delta q + \delta w = \delta w = -P \, dV. \tag{4.2}$$

Replacing P in Eq. (4.2) with nRT/V and C_v with $n\,C_v^m$ gives

$$n\,C_v^m\,dT = -\frac{nRT}{V}\,dV. \tag{4.3}$$

Canceling n and dividing by T produces

$$\frac{C_v^m}{T}\,dT = -R\,\frac{dV}{V}. \tag{4.4}$$

Since the variables are now separated, we may integrate both sides of Eq. 4.4 between corresponding limits for Stroke 2 to obtain

$$\int_{T_1}^{T_2}\frac{C_v^m}{T}\,dT = -R\int_{V_2}^{V_3}\frac{dV}{V}. \tag{4.5}$$

C_v^m is a constant for an ideal gas; therefore, the integration gives

$$C_v^m\ln\left[\frac{T_2}{T_1}\right] = \ln\left[\frac{T_2}{T_1}\right]^{C_v^m} = -R\ln\left[\frac{V_3}{V_2}\right] = \ln\left[\frac{V_3}{V_2}\right]^{-R} = \ln\left[\frac{V_2}{V_3}\right]^{R}. \tag{4.6}$$

Exponentiation of both sides of Eq. 4.6 produces

$$\left[\frac{T_2}{T_1}\right]^{C_v^m} = \left[\frac{V_2}{V_3}\right]^{R}. \tag{4.7}$$

Raising each side of Eq. 4.7 to the R^{-1} power, we obtain

$$\frac{V_2}{V_3} = \left[\frac{T_2}{T_1}\right]^{\frac{C_v^m}{R}}. \tag{4.8}$$

The treatment of the reversible adiabatic compression in Stroke 4 gives analogous results, with V_1 replacing V_2 and V_4 replacing V_3 to yield

$$\frac{V_1}{V_4} = \left[\frac{T_2}{T_1}\right]^{\frac{C_v^m}{R}}. \tag{4.9}$$

Comparison of Eqs. 4.8 and 4.9 shows that the volumes are related by

$$\frac{V_2}{V_3} = \frac{V_1}{V_4}. \tag{4.10}$$

Equation 4.8 demonstrates that an adiabatic expansion with $V_3 > V_2$ results in a final temperature $T_2 < T_1$. Physically, we observe this result because the energy needed to do the work of expansion must come from the internal energy of the gas. The temperature of the gas must, therefore, decrease.

We may now obtain the total work done per cycle by the engine by considering each stroke separately. Stroke 1 is an isothermal expansion between volumes V_1 and V_2 at $T = T_1$. In Eq. 2.33, it was shown that the work done in such a process is

$$w_1 = -nRT_1\ln\left[\frac{V_2}{V_1}\right]. \tag{4.11}$$

Since $dU = 0$ for an isothermal process for an ideal gas, we also have

$$q_1 = -w_1 = nRT_1\ln\left[\frac{V_2}{V_1}\right]. \tag{4.12}$$

Stroke 1 is an expansion, so $V_2 > V_1$, which means that $q_1 > 0$ and heat is added to the engine during that stroke. This is the fuel for the engine. In an internal combustion engine, q_1 is produced by burning gasoline in the cylinders. If we were to postulate that $q_1 = 0$, we would be violating the first law and suggesting that the Carnot engine is a perpetual-motion machine of the first type.

The work produced in the reversible adiabatic expansion occurring in Stroke 2 can be obtained directly from the first law. Since $\delta q = 0$,

$$dU_2 = \delta q_2 + \delta w_2 = \delta w_2 = C_v \, dT = n \, C_v^m \, dT. \tag{4.13}$$

Integrating both sides of Eq. 4.13 gives the total work done in Stroke 2:

$$w_2 = \int_{T_1}^{T_2} nC_v^m dT = nC_v^m \int_{T_1}^{T_2} dT = n \, C_v^m \, [T_2 - T_1]. \tag{4.14}$$

For an adiabatic process, $q_2 = 0$. Thus, no additional thermal energy is added in Stroke 2.

Stroke 3 is identical to Stroke 1, except for the initial and final conditions. In this case, the reversible isothermal compression gives

$$w_3 = -nRT_2 \ln\left[\frac{V_4}{V_3}\right] = nRT_2 \ln\left[\frac{V_3}{V_4}\right]. \tag{4.15}$$

Again, for an isothermal process for an ideal gas, $dU = 0$. Therefore, we obtain

$$q_3 = -w_3 = nRT_2 \ln\left[\frac{V_4}{V_3}\right]. \tag{4.16}$$

It is important to note that in the third stroke we have $w_3 > 0$ and $q_3 < 0$. Work is being done on the gas and heat is being expelled by the engine. All engines have this qualitative characteristic—that heat must be expelled at some point. Anyone who has ever felt the intense heat of the exhaust from an internal combustion engine is painfully aware of that fact.

Equation 4.14 gives the work done in the reversible adiabatic compression of Stroke 4:

$$w_4 = \int_{T_2}^{T_1} nC_v^m \, dT = nC_v^m \int_{T_2}^{T_1} dT = n \, C_v^m \, [T_1 - T_2]. \tag{4.17}$$

Since the process is adiabatic, $q_4 = 0$. The heat and work for each stroke are shown in Figure 4.2.

The total work per cycle is given by Eq. 4.1:

$$w_T = w_1 + w_2 + w_3 + w_4$$

$$= -nRT_1 \ln\left[\frac{V_2}{V_1}\right] + n \, C_v^m \, [T_2 - T_1] - nRT_2 \ln\left[\frac{V_4}{V_3}\right] + n \, C_v^m \, [T_1 - T_2]$$

$$= -nRT_1 \ln\left[\frac{V_2}{V_1}\right] - nRT_2 \ln\left[\frac{V_4}{V_3}\right]. \tag{4.18}$$

We may write Eq. 4.18 in a more compact form using the relationship between the volumes given in Eq. 4.10. From the latter equation, we have $V_4/V_3 = V_1/V_2$. Substituting this result into Eq. 4.18 produces

$$w_T = -nRT_1 \ln\left[\frac{V_2}{V_1}\right] - nRT_2 \ln\left[\frac{V_1}{V_2}\right] = -nR \ln\left[\frac{V_2}{V_1}\right][T_1 - T_2]. \tag{4.19}$$

Since we have $T_1 > T_2$ and $V_2 > V_1$, it follows that $w_T < 0$, and the Carnot engine does work on the surroundings in each cycle, with the magnitude of that work equal to the absolute value of Eq. 4.19. The magnitude is also equal to the area enclosed by the closed loop in Figure 4.2.

A matter of utmost concern is the efficiency of the engine. Ideally, we would like to convert all of the heat added in Stroke 1 to useful work. The engine efficiency is defined to be

$$\mathcal{E} = \frac{\text{magnitude of work done}}{\text{heat added}}. \tag{4.20}$$

For the Carnot engine, the efficiency is

$$\mathcal{E} = \frac{|w_T|}{q_1}. \tag{4.21}$$

Substitution of Eqs. 4.12 and 4.19 into Eq. 4.21 gives

$$\boxed{\mathcal{E} = \frac{nR \ln\left[\dfrac{V_2}{V_1}\right][T_1 - T_2]}{nRT_1 \ln\left[\dfrac{V_2}{V_1}\right]} = \frac{T_1 - T_2}{T_1} = \frac{T_H - T_C}{T_H},} \tag{4.22}$$

where we have written the temperature of the "hot" reservoir as T_H and that of the "cold" reservoir as T_C to emphasize the point that $T_1 > T_2$. Equation 4.22 is a very important result. It demonstrates that unless $T_2 = 0$ K, the efficiency of the Carnot engine is always less than unity. Thus, we can never convert all of the heat added in Stroke 1 to useful work. This is true in spite of the fact that the engine design is perfect and there are no frictional losses of energy. Some of the energy added is discarded in Stroke 3. In a similar manner, some of the energy added to a car's engine is discarded as heat in the exhaust fumes. Complete conversion to work is not obtainable.

Figure 4.3 shows a series of constant efficiency contours for a Carnot engine using an ideal gas as the working fluid. The $\mathcal{E} = 0$ contour has unit slope with $T_1 = T_2$, signifying that we can obtain work from an engine only if there exists a

▶ FIGURE 4.3
Contour lines of constant efficiency for a Carnot engine using an ideal gas as the working fluid. The value of the efficiency for each contour line is shown in the plot. Equation 4.22 indicates that the slope of the contour lines is given by $[1 - \mathcal{E}]^{-1}$.

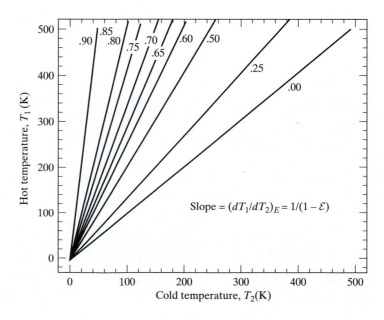

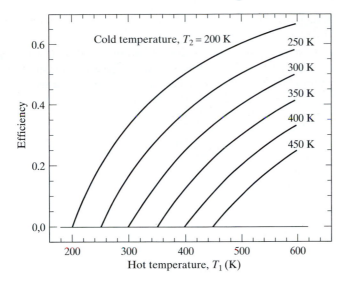

◀ FIGURE 4.4
Variation of the efficiency of a Carnot engine using an ideal gas as the working fluid as a function of the temperature of the high-temperature reservoir T_1 for different temperatures of the low-temperature reservoir T_2.

temperature difference between the hot and cold reservoirs of thermal energy. The $\mathcal{E} = 1$ contour is a vertical line at $T_2 = 0$ K, signifying that the engine can have unit efficiency only if the temperature of the cold reservoir is 0 K. Figure 4.4 shows the Carnot engine efficiency as a function of T_1 for various values of T_2.

EXAMPLE 4.1

(A) A Carnot engine is observed to produce 3.5 kJ of net useful work for every 11 kJ of heat added during the isothermal expansion stroke. If the compression stroke of the engine is executed at 300 K, determine the temperature during the isothermal expansion stroke. **(B)** How much energy is lost to the engine during the compression stroke, Stroke 3?

Solution

(A) The engine efficiency is

$$\mathcal{E} = \frac{\text{magnitude of net work}}{\text{heat added}} = \frac{3.5 \text{ kJ}}{11 \text{ kJ}} = 0.3181818 \ldots . \tag{1}$$

Equation 4.22 shows that

$$\mathcal{E} = \frac{T_1 - T_2}{T_1}. \tag{2}$$

Solving Eq. 2 for T_1, we obtain

$$T_1 = \frac{T_2}{1 - \mathcal{E}} = \frac{300 \text{ K}}{1 - 0.3181818 \ldots} = 440 \text{K}. \tag{3}$$

(B) By energy conservation, we must have

$$|\text{Total energy input}| = |\text{work done}| + |\text{heat expelled}|. \tag{4}$$

Therefore,

$$|\text{Heat expelled}| = |\text{total energy input}| - |\text{work done}|$$
$$= 11 \text{ kJ} - 3.5 \text{ kJ} = 7.5 \text{kJ}. \tag{5}$$

For related exercises, see Problems 4.1, 4.2, and 4.4.

EXAMPLE 4.2

The efficiency of a Carnot engine is dependent upon the two variables T_1 and T_2. **(A)** Obtain the total differential of $\mathcal{E}$, which shows the rate at which the engine efficiency changes as we vary T_1 and T_2. **(B)** $\mathcal{E}$ increases if T_1 is raised or if T_2 is lowered. From the form of the total differential obtained in (A), predict whether $\mathcal{E}$ increases faster as T_1 is raised or as T_2 is lowered. **(C)** A Carnot engine operates between a high-temperature site at 600 K and a low-temperature site at 400 K. Compare the change in efficiency produced by raising the high-temperature site to 650 K while keeping the low-temperature site at 400 K with the change produced by lowering the low-temperature site to 350 K while keeping the high-temperature site at 600 K. Is the result in accord with the prediction made in (B)?

Solution

(A) Equation 4.22 shows that

$$\mathcal{E} = \frac{T_1 - T_2}{T_1}. \tag{1}$$

The total differential of $\mathcal{E}$ is, therefore,

$$d\mathcal{E} = \frac{\partial \mathcal{E}}{\partial T_1}\, dT_1 + \frac{\partial \mathcal{E}}{\partial T_2}\, dT_2. \tag{2}$$

The required partial derivatives can be obtained directly from Eq. 1. The results are

$$\frac{\partial \mathcal{E}}{\partial T_1} = \frac{T_1 - (T_1 - T_2)}{T_1^2} = \frac{T_2}{T_1^2} \tag{3}$$

and

$$\frac{\partial \mathcal{E}}{\partial T_2} = -\frac{1}{T_1}. \tag{4}$$

The total differential of $\mathcal{E}$ is, therefore,

$$d\mathcal{E} = \left[\frac{T_2}{T_1^2}\right] dT_1 - \left[\frac{1}{T_1}\right] dT_2, \tag{5}$$

which is the required result.

(B) The magnitude of the rate of change of $\mathcal{E}$ with T_1 is given by Eq. 3. We may write this equation in the form

$$\left|\frac{\partial \mathcal{E}}{\partial T_1}\right| = \frac{T_2}{T_1^2} = \left[\frac{T_2}{T_1}\right]\left[\frac{1}{T_1}\right]. \tag{6}$$

The magnitude of the rate of change of $\mathcal{E}$ with T_2 is given by Eq. 4:

$$\left|\frac{\partial \mathcal{E}}{\partial T_2}\right| = \left|-\frac{1}{T_1}\right| = \left[\frac{1}{T_1}\right]. \tag{7}$$

Since $T_2 < T_1$, the fraction appearing in Eq. 6, T_2/T_1, is less than unity. Therefore, we have

$$\left|\frac{\partial \mathcal{E}}{\partial T_2}\right| > \left|\frac{\partial \mathcal{E}}{\partial T_1}\right|, \tag{8}$$

and the engine efficiency rises more rapidly when T_2 is decreased than when T_1 is increased by a like amount.

(C) The efficiency at $T_1 = 600$ K and $T_2 = 400$ K is

$$\mathcal{E} = \frac{600 - 400}{600} = 0.3333333. \qquad (9)$$

If we raise T_1 to 650 K, the new efficiency is

$$\mathcal{E}_1 = \frac{650 - 400}{650} = 0.384615. \dots \qquad (10)$$

Lowering T_2 to 350 K while keeping T_1 at 600 K gives an efficiency of

$$\mathcal{E}_2 = \frac{600 - 350}{600} = 0.4166666. \dots \qquad (11)$$

Lowering the low-temperature site by 50 K increases $\mathcal{E}$ more than raising the high-temperature site by 50 K. This is exactly the prediction made in (B). Such behavior is also observed in Figures 4.3 and 4.4. In Figure 4.3, we see that the change in T_1, dT_1, that must be made for a given change in T_2, dT_2, to maintain a constant efficiency is always greater than unity. The engine is thus more sensitive to variations in the temperature of the low-temperature reservoir than to similar variations in the high-temperature reservoir.

4.3   Refrigerators and Heat Pumps

An engine operates by moving thermal energy from a reservoir at a high temperature T_1 through a series of steps (strokes) to produce useful work. In the process, some of the energy is ejected into a low-temperature reservoir at T_2. In the case of a refrigerator or air conditioner, we wish to reverse this process. That is, we want to move thermal energy from a low-temperature reservoir to one at a higher temperature, so that we cool the low-temperature site to an even lower temperature. In effect, we wish to operate the engine in reverse. For a Carnot engine, this means that Strokes 3 and 4 will become the expansion strokes while Strokes 1 and 2 will execute the compression. If the engine is operated in reverse, the first law requires the reversal of all energy terms. Heat will enter the engine in Stroke 3 and be ejected in Stroke 1. In addition, we will now have to insert work into the engine, probably in the form of electrical energy. This situation is illustrated for a Carnot engine in Figure 4.5.

During each cycle of operation, the refrigerator removes heat in the amount $|q_3|$ from the low-temperature site during Stroke 3. During Stroke 1, it ejects heat $|q_1|$ at the high-temperature site. In the process, we must insert work in the amount $|w_T|$ into the engine. Note that $-q_3$ is positive, as it should be, since that amount of heat is being added to the refrigerator. On the other hand, $-q_1$ is negative, since that amount of heat is lost by the refrigerator. The work done, $-w_T$, is positive, because work is being added to the refrigerator.

If we are attempting to cool the low-temperature site, the quantity of interest in judging the efficiency of the refrigerator is the ratio of the amount of heat removed to the work that must be done to remove it. This ratio is called the *coefficient of performance* of the refrigerator and is denoted

$$C = \frac{|q_3|}{|w_T|} = \frac{-q_3}{-w_T}. \qquad (4.23)$$

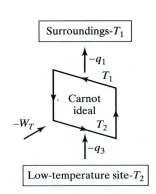

▲ **FIGURE 4.5**
Refrigerator operation to cool a low-temperature site at $T = T_2$ by the movement of heat to a high-temperature site at $T = T_1$.

For a Carnot refrigerator, q_3 and w_T are given by Eqs. 4.16 and 4.19, respectively. Substituting these results into Eq. 4.23 gives

$$C = \frac{-nRT_2 \ln[V_4/V_3]}{nR \ln[V_2/V_1][T_1 - T_2]} = \frac{-T_2 \ln[V_4/V_3]}{[T_1 - T_2]\ln[V_2/V_1]}. \tag{4.24}$$

Using Eq. 4.10, we have $\ln[V_4/V_3] = \ln[V_1/V_2] = -\ln[V_2/V_1]$. Inserting this result into Eq. 4.24 yields

$$C = \frac{T_2}{T_1 - T_2} = \frac{T_C}{T_H - T_C} \tag{4.25}$$

for the coefficient of performance of a Carnot refrigerator. Since the Carnot refrigerator is assumed to have perfect engineering construction with no friction and no leaks, real refrigerators can be expected to exhibit coefficients of performance below that given by Eq. 4.25.

A heat pump operates exactly like a refrigerator, except that the focus of interest is now adding heat to the high-temperature site at temperature T_1 rather than cooling the low-temperature site. The typical example is heating a house during a cold winter day. The house is the high-temperature site that we wish to heat. If we take a quantity of electrical energy w_T and simply discharge it through a resistance heater, the first law tells us that we will add a quantity of thermal energy to the house exactly equal to w_T. If, on the other hand, we use the electrical energy to drive a heat pump, the quantity of heat discharged into the house will be $|q_1|$, as shown in Figure 4.5. Conservation of energy requires that

$$|q_1| = |q_3| + |w_T|. \tag{4.26}$$

Using Eqs. 4.23 and 4.25, we may write Eq. 4.26 in the form

$$|q_1| = C|w_T| + |w_T| = [C + 1]|w_T| = \left[\frac{T_C}{T_H - T_C} + 1\right]|w_T|$$

$$= \left[\frac{T_H}{T_H - T_C}\right]|w_T|. \tag{4.27}$$

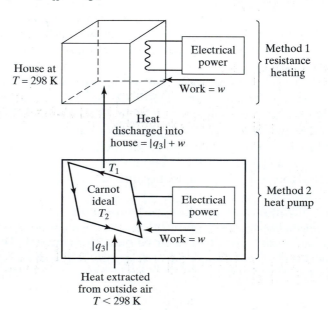

▶ FIGURE 4.6
Two methods for heating a house using electrical power. In Method 1, the electrical power is discharged across a resistance heater inside the house. If the electrical energy expended is w, then an amount of thermal energy equal to w is injected into the house. In Method 2, the electrical power is employed to drive a Carnot engine in reverse. The electrical work is now being used to actively transfer heat from the cold outdoors to the warm house. In this case, the same quantity of electrical energy results in inserting an amount of thermal energy into the house equal to w plus the heat transferred by the heat pump.

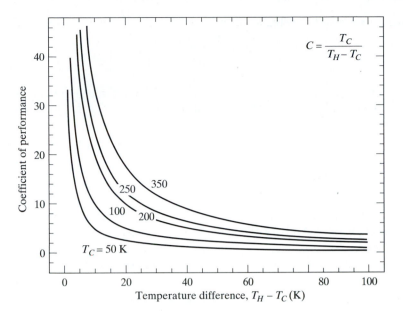

$$C = \frac{T_C}{T_H - T_C}$$

◀ **FIGURE 4.7**
The coefficient of performance of a Carnot refrigerator or heat pump as a function of the temperature difference between hot and cold sites for a series of different cold-site temperatures. The cold-site temperatures for each curve are given in the figure. The results show that it is much easier to transfer thermal energy between sites of nearly the same temperature than between sites with significantly different temperatures. As the temperature difference increases, it becomes much more difficult to transfer heat from the cold site to the hot site.

Equation 4.25 shows that if the house temperature is 300 K and the outside temperature on the winter day is the freezing point of water, 273 K, then $C = 10.111\ldots$. Thus, we gain more than a factor of 11 in heat added to the house by using a heat pump rather than an electrical-resistance heater. Figure 4.6 illustrates this point and the reasons it is true. The dependence of C upon the temperature difference $T_H - T_C$ for various values of T_C is shown in Figure 4.7. Each of the curves rises sharply as $T_H - T_C$ becomes small. It is much easier to transfer thermal energy from a cold site to a hot site when the difference is small than when it is large. In practice, commercial heat pumps are not as efficient as the Carnot heat pump, so the gain will be less.

EXAMPLE 4.3

A closed container holding 100 moles of helium is in equilibrium with its surroundings at a temperature of 300 K. An investigator wishes to cool the helium to 200 K. To effect this change, she connects a Carnot refrigerator to the container. C_v^m for helium is 1.5 R, and the total heat capacity at constant volume for the helium container is 745 J K^{-1}. If these heat capacities are assumed to be independent of temperature, and if the high-temperature site is maintained at a constant temperature of 300 K, compute the amount of work that must be done to cool the helium to 200 K.

Solution

Let the temperature of the high-temperature site be T_H, which is constant at 300 K. The low-temperature site is the helium container. Its temperature, T_C, is continuously decreasing during the process. We must, therefore, consider the situation in which an infinitesimal amount of work $|dw_T|$ is added to the refrigerator to remove a small quantity of heat $|dq_3|$ from the helium container. The coefficient of performance is given by Eqs. 4.23 and 4.25:

$$C = \frac{|dq_3|}{|dw_T|} = \frac{T_C}{T_H - T_C}. \tag{1}$$

Rearranging Eq. 1, we obtain

$$|dw_T| = \left[\frac{T_H - T_C}{T_C}\right]|dq_3| = \left[\frac{T_H}{T_C} - 1\right]|dq_3|. \tag{2}$$

The total heat change of the container plus the helium can be written in terms of the total heat capacity and the differential temperature change of the container and contents:

$$dq_3 = [C_v^{\text{container}} + n\, C_v^m]\, dT_C. \tag{3}$$

Since $dT_C < 0$ (the helium is being cooled), we have

$$|dq_3| = -[C_v^{\text{container}} + n\, C_v^m]\, dT_C. \tag{4}$$

Substituting Eq. 4 into Eq. 2 gives

$$|dw_T| = -\left[\frac{T_H}{T_C} - 1\right][C_v^{\text{container}} + n\, C_v^m]\, dT_C. \tag{5}$$

To obtain the total work done in cooling the container and the helium from 300 K to 200 K, we need to sum up the work for each differential increment of cooling. That is, we must integrate Eq. 5 from 300 K to 200 K. This gives

$$|w_T| = \int |dw_T| = -\int_{300\,K}^{200\,K}\left[\frac{T_H}{T_C} - 1\right][C_v^{\text{container}} + n\, C_v^m]\, dT_C$$

$$= -[C_v^{\text{container}} + n\, C_v^m]\left[300\ln\left[\frac{200}{300}\right] - (200 - 300)\right]. \tag{6}$$

The total work needed may now be obtained by direct substitution into Eq. 6:

$$|w_T| = -[745\,\text{J K}^{-1} + (100\,\text{mol})(1.5)(8.314)\,\text{J mol}^{-1}\,\text{K}^{-1})] \times$$

$$\left[300\ln\left[\frac{200}{300}\right] - (200 - 300)\right]\text{K} = 4.311 \times 10^4\,\text{J of work}. \tag{7}$$

For related exercises, see Problems 4.5, 4.6, 4.7, 4.8, and 4.9.

At this point, it is important to note that the Carnot engine converts heat into work and transfers thermal energy from a hot site to a cold site with high efficiency when $T_1 - T_2$ or $T_H - T_C$ is large. This behavior is observed because heat flows spontaneously from hot sites to cold sites and the driving force for such flow increases as $T_H - T_C$ increases. For the same reason, it becomes increasingly difficult for a heat pump to move thermal energy from a cold site to a hot site as $T_H - T_C$ increases. Figure 4.7 shows that the coefficient of performance steadily decreases as $T_H - T_C$ increases. Therefore, the efficiency $\mathcal{E}$ of a Carnot engine and the coefficient of performance, $\mathcal{C}$, of a Carnot heat pump can be regarded as measures of the spontaneity of the process taking place.

It is useful to state this concept in slightly different terms. If we scatter 10^5 marbles about a room, we might say, "It's total, random chaos in the room." If we gather most of the marbles and place them in a large container and the remainder in a smaller container, we might remark, "The room is now well ordered." If we now think of joules of thermal energy per unit mass as being analogous to one of the marbles, the chaotic distribution cor-

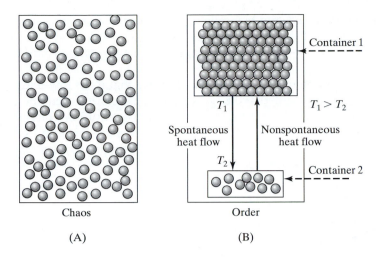

Chaos

(A)

Order

(B)

◀ **FIGURE 4.8**
Order–chaos changes, and the efficiency of a Carnot engine. In (A),
marbles are chaotically distributed
throughout a room. In (B), the
same marbles are neatly ordered
into two containers. If we now
think of the marbles as joules of
thermal energy per unit mass, we
see that the chaotic situation (A)
involves thermal equilibrium in
which all parts of the room are at
the same temperature T. In contrast, the well-ordered situation in
(B) is one in which Container 1 is
at a temperature T_1 that is greater
than that for Container 2, which
has a temperature T_2. In this case,
heat flows spontaneously from
Container 1 to 2.

responds to one of thermal equilibrium, with all parts of the room at the same temperature T, whereas the well-ordered room corresponds to a situation in which part of the room (the large container) is at a high temperature and the rest of the room (the smaller container) is at a low temperature. This analogy is illustrated in Figure 4.8. The efficiency of a Carnot engine operating between these two temperatures will be a measure of the spontaneity of the heat flow process, which is turning order into disorder, since the flow moves the system toward equilibrium at which all temperatures are the same. (See Problem 4.11).

The problem with using the efficiency of a Carnot engine as a measure of spontaneity is that such an engine is a hypothetical concept. Real engines have friction and faulty valves, rings, and pistons that affect the efficiency. Fortunately, the second law of thermodynamics provides us with a much better means of determining spontaneity and the extent to which order is converted into chaos.

4.4   The Second Law of Thermodynamics

The second law of thermodynamics is one of the most unusual and profound laws of science. It states,

> *It is impossible to construct a device that operates in cycles and that converts heat into work without producing some other change in the surroundings.*

The second law is unusual in that it is the only fundamental law of science that is stated in negative terms. It tells us what we cannot do. As we shall see, the ramifications of this law assume awesome proportions.

Let us first examine the Carnot engine in the light of the second law. The engine is a device that operates in cycles. It also converts heat into work with an efficiency equal to $(T_1 - T_2)/T_1 = (T_H - T_C)/T_H$. The second law, therefore, requires that the operation of the engine produce some other change in the surroundings. We can see that such a change does occur. In Stroke 3, a quantity of heat equal to $nRT_2 \ln[V_4/V_3]$ is ejected back into the surroundings. It is precisely such a change in the surroundings that is responsible for the smog problem in many cities. The ejection of exhaust fumes and heat during the compression strokes of the automobile's engine pollute the atmosphere with $CO_2(g)$ and nitrogen oxides, causing smog. The second law tells us that some change in the surroundings is inevitable if the engine converts heat into work in cycles. The best we can hope to do is reduce the damage produced by these changes to the environment.

An examination of Eqs. 4.16 and 4.22 shows that if the low-temperature compressions are executed at 0 K, q_3 will go to zero and the engine efficiency will become unity. Under these conditions, we would be converting the heat inserted in Stroke 1, q_1, quantitatively into work with unit efficiency without producing any change in the surroundings, since q_3 is zero if $T_2 = 0$ K. Such a result violates the second law of thermodynamics. *Therefore, it follows that it must be impossible to attain a temperature of 0 K.* Problem 1.17 alludes to this fact.

EXAMPLE 4.4

A heat reservoir at temperature T_1 is connected to a cylinder–piston arrangement as shown in the following diagram:

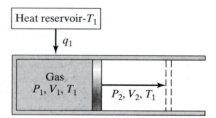

n moles of an ideal gas inside the cylinder are originally at pressure P_1, volume V_1, and temperature T_1. A quantity of heat, q_1, is added reversibly to the cylinder, and the gas is expanded reversibly and isothermally to a new volume V_2 and pressure P_2. **(A)** If P_1, V_1, and T_1 are 10 atm, 12.309 L, and 300 K, respectively, compute the total work and q_1 when $V_2 = 30.0$ L. **(B)** Has any heat been dissipated to the surroundings during the process? **(C)** The process described converts heat q_1 into an equal quantity of work, and no other change occurs in the surroundings. Does the device violate the second law of thermodynamics? Explain.

Solution

(A) The number of moles of gas in the container is

$$n = \frac{PV}{RT} = \frac{(10 \text{ atm})(12.309 \text{ L})}{(0.08206 \text{ L atm mol}^{-1}\text{ K}^{-1})(300 \text{ K})} = 5 \text{ moles.} \tag{1}$$

The reversible work is given by

$$w = -\int_{V_1}^{V_2} P\, dV = -nRT_1 \int_{12.309\text{ L}}^{30.0\text{ L}} \frac{dV}{V} = -nRT_1 \ln\left[\frac{30.0}{12.309}\right]$$

$$= -(5 \text{ mol})(8.314 \text{ J mol}^{-1}\text{ K}^{-1})(300 \text{ K})(0.89087) = -1.111 \times 10^4 \text{ J.} \tag{2}$$

Since $dT = 0$ and the gas is ideal, we have

$$\Delta U = 0 = q_1 + w, \tag{3}$$

so that

$$q_1 = 1.111 \times 10^4 \text{ J.} \tag{4}$$

(B) No heat is added to the surroundings.

(C) The device does not violate the second law, because it does not operate in cycles. Once the gas expands, the process is over. Under these conditions, the second law does not require that a change in the surroundings be produced when heat is converted into work.

4.5   Entropy

The most important consequence of the second law is that it requires that the differential $\delta q_{rev}/T$ be exact for any substance undergoing any reversible process. In this section, we shall derive this result using Carnot and Carnot-like engines.

The proof that $\delta q_{rev}/T$ must be an exact differential consists of four parts. In Part 1, we need to show that $\delta q_{rev}/T$ is exact for any reversible process involving an ideal gas. The proof of this statement has already been given in Section 2.7. It is advisable to review that section. In Part 2, we must show that there can exist no reversible Carnot engine using a different working substance that is either more efficient or less efficient than a Carnot engine using an ideal gas. Part 3 of the proof is to show that unless $\delta q_{rev}/T$ is exact for any Carnot cycle involving any working material, Part 2 of the proof is violated. Finally, we must demonstrate that any arbitrary cycle can be constructed of an infinite set of infinitesimally small Carnot cycles. Since $\int_{cycle} \delta q_{rev}/T$ is zero for all of the Carnot cycles, it must be zero for the overall arbitrary cycle, and hence, the differential $\delta q_{rev}/T$ must be exact.

Part 1

It is shown in Section 2.7 that $\delta q_{rev}/T$ is exact for an ideal gas. Since the integral of any exact differential about a closed path is zero, for the Carnot cycle using an ideal gas as the working fluid, we must have

$$\int_{\text{initial state}}^{\text{final state}} \frac{\delta q_{rev}}{T} = 0 = \int_{\text{Stroke 1}} \frac{\delta q_1}{T} + \int_{\text{Stroke 2}} \frac{\delta q_2}{T} + \int_{\text{Stroke 3}} \frac{\delta q_3}{T} + \int_{\text{Stroke 4}} \frac{\delta q_4}{T}. \quad \textbf{(4.28)}$$

The integrals over Strokes 2 and 4 are zero, because $\delta q_2 = \delta q_4 = 0$ for adiabatic processes. In Stroke 1, T is constant with the value T_1. In Stroke 3, T is constant with the value T_2. Consequently, we may factor T from the integral in these strokes. Equation 4.28, therefore, becomes

$$\int_{\text{Stroke 1}} \frac{\delta q_1}{T} + \int_{\text{Stroke 3}} \frac{\delta q_3}{T} = T_1^{-1} \int_{\text{Stroke 1}} dq_1 + T_2^{-1} \int_{\text{Stroke 3}} dq_3$$

$$= \frac{q_1}{T_1} + \frac{q_3}{T_2} = 0, \quad \textbf{(4.29)}$$

where q_1 and q_3 are the total heat per cycle in Strokes 1 and 3, respectively. q_1 and q_3 are given in Eqs. 4.12 and 4.16. Insertion of these results into Eq. 4.29 yields

$$\frac{nRT_1 \ln[V_2/V_1]}{T_1} + \frac{nRT_2 \ln[V_4/V_3]}{T_2} = nR\left\{ \ln\left[\frac{V_2}{V_1}\right] + \ln\left[\frac{V_4}{V_3}\right] \right\} = 0. \quad \textbf{(4.30)}$$

From Eq. 4.10, we have $[V_2/V_1] = [V_3/V_4]$. Thus, the sum of the logarithms in Eq. 4.30 vanishes, and the integral of $\delta q_{rev}/T$ around the closed Carnot cycle is zero, as it must be.

Equation 4.29 allows us to recast the Carnot engine efficiency in terms of q_1 and q_3. First, we have

$$-\frac{T_2}{T_1} = \frac{q_3}{q_1}.$$ (4.31)

The efficiency of the engine can, therefore, be written in the form

$$\mathcal{E} = \frac{T_1 - T_2}{T_1} = 1 - \frac{T_2}{T_1} = 1 + \frac{q_3}{q_1} = \frac{q_1 + q_3}{q_1} = \frac{q_H + q_C}{q_H},$$ (4.32)

where q_H is the heat removed from the hot reservoir and q_C is that ejected from the engine into the cold reservoir.

The numerator of Eq. 4.32 is just the sum of the heat inserted in Stroke 1 ($q_1 = q_H$) and that ejected in Stroke 3 ($q_3 = q_C$). Conservation of energy according to the first law requires that this be equal to the magnitude of the work done. The denominator is the total heat injected in Stroke 1.

Part 2

We now wish to demonstrate that there can be no reversible Carnot engine using *any* working material that has an efficiency $\mathcal{E}'$ greater or less than the efficiency of a Carnot engine using an ideal gas, which we shall write as $\mathcal{E}_c$. That is, we wish to show that neither $\mathcal{E}' > \mathcal{E}_c$ nor $\mathcal{E}' < \mathcal{E}_c$ can occur.

Let us assume that there exits a Carnot engine using some unspecified working material that has an efficiency $\mathcal{E}_A$ greater than $\mathcal{E}_c$. We will denote this machine as Engine A. Such a machine is shown in Figure 4.9. In the figure, both engines are operating in a clockwise fashion such that work is being produced. Each engine extracts a quantity of heat Q from the energy reservoir, which is at temperature T_1, the same temperature as the engine, to ensure a reversible heat transfer. In Stroke 3, the two engines eject a different quantity of heat into the surroundings, since they are assumed to have different efficiencies. Again, the surroundings are at the same temperature as the engine in Stroke 3, to ensure reversible heat transfer. Engine A produces a quantity of work whose magnitude is W_A, whereas the magnitude of the work produced by the Carnot engine using an ideal gas is W_c.

By assumption, we have $\mathcal{E}_A > \mathcal{E}_c$. Therefore,

$$\mathcal{E}_A = \frac{Q + Q_A}{Q} = \frac{-W_A}{Q} > \mathcal{E}_c = \frac{Q + Q_c}{Q} = \frac{-W_c}{Q}.$$ (4.33)

For the inequality to hold, we must have $-W_A > -W_c$. Because the work done on the surroundings is negative, $-W_A$ and $-W_c$ are positive numbers,

▶ **FIGURE 4.9**
Two reversible Carnot engines. Engine A uses some working material other than an ideal gas. It is assumed that Engine A has an efficiency greater than that for the Carnot engine employing an ideal gas as the working material.

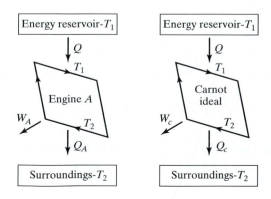

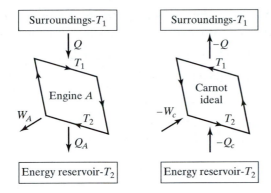

◀ FIGURE 4.10
Two Carnot engines, coupled in tandem, one using an ideal-gas working material, the second any other arbitrary material. The Carnot engine using the ideal gas is run in reverse so that an amount of work W_c is done on the engine. The second engine is run forward so that it produces an amount of work W_A on the surroundings.

and we have $|W_A| > |W_c|$. We also have $Q_A > Q_c$. Since both of these are *negative* numbers (heat leaving the engine is negative), it follows that $|Q_A| < |Q_c|$. That is, more heat is ejected from the Carnot engine using the ideal gas than from Engine A. Thus, the more efficient Engine A is predicted to produce more work than the Carnot engine using an ideal gas as the working material.

Now consider an arrangement in which the two engines are set up to run in tandem, with Engine A running in the forward mode to produce work, but the Carnot engine with the ideal gas running in reverse. (Remember, all strokes in the Carnot engine are reversible.) Because the Carnot engine is running in reverse, the first law requires that the signs on all energy terms change (i.e., Q becomes $-Q$, Q_c changes to $-Q_c$, W_c becomes $-W_c$, and the direction of energy flow changes in each case. This arrangement is shown in Figure 4.10.

The tandem device illustrated in the figure operates as follows: The Carnot engine using ideal gas as the working material is running in reverse. Consequently, Strokes 3 and 4 are the expansion strokes that produce work, whereas Strokes 1 and 2 are the compression strokes that eject heat into the surroundings. The energy input to the engine occurs in Stroke 3, whereupon an amount of heat $-Q_c$ is extracted from the energy reservoir at temperature T_2, which is the same as that of the engine during Stroke 3, thereby permitting reversible heat transfer to occur. During the compression in Stroke 1, an amount of heat $-Q$ is ejected into the surroundings. Since this engine produced a quantity of work W_c when running in the forward direction, the first law requires that we insert an equal amount of work per cycle when the engine runs in reverse. That is, the work when the engine is running in reverse will be $-W_c$. At the same time that the ideal-gas Carnot engine ejects heat $-Q$ into the surroundings, Engine A takes in this heat, runs in the forward mode, and produces a quantity of work W_A while ejecting an amount of heat Q_A back into the energy reservoir at temperature T_2.

Let us now analyze one cycle of this tandem arrangement. The net work done per cycle by the coupled engines is

$$w_{net} = W_A - W_c. \tag{4.34}$$

Our assumption that Engine A is more efficient than the ideal-gas Carnot engine led to the conclusion that $|W_A| > |W_c|$, with both W_A and W_c negative. Therefore, we have $w_{net} = (W_A - W_c) < 0$. That is, the tandem device is operating in cycles and converting heat into net work done on the surroundings. We reach the same conclusion by considering the heat flow for the two engines. Around the closed cycle, we have $dU = 0$. Therefore, the first law requires that

$$\Delta U = 0 = w_{net} + q_{net} = w_{net} + Q_A - Q_c, \qquad (4.35)$$

so that the net work is given by

$$w_{net} = Q_c - Q_A. \qquad (4.36)$$

Since both Q_c and Q_A are negative, but $|Q_A| < |Q_c|$, w_{net} is again predicted to be negative, indicating that heat is being converted into net work done on the surroundings.

The net change in the surroundings is that produced by the Carnot engine plus Engine A:

$$\text{Heat change of the surroundings} = -Q + Q = 0. \qquad (4.37)$$

We are, therefore, converting heat into net work in cycles without producing any change in the surroundings. But the second law prohibits this. Since the original assumption that $\mathcal{E}_A > \mathcal{E}_c$ leads directly to that result, that assumption must be incorrect. Consequently, no Carnot engine using any working material can have an efficiency greater than that of a Carnot engine using an ideal gas.

We may show that the situation in which $\mathcal{E}_A < \mathcal{E}_c$ also leads to a violation of the second law by simply reversing the roles played by Engine A and the Carnot engine using an ideal gas as the working material. That is, we run the two engines in tandem, with Engine A running in reverse and the ideal Carnot engine running in the forward mode. An analysis identical to the one just given then leads to the result that the tandem arrangement is converting heat into work in cycles, without producing any other change in the surroundings. Consequently, we conclude that all Carnot engines have the same efficiency regardless of the working material.

Part 3

In this step, we shall demonstrate that unless the differential $\delta q_{rev}/T$ is exact for a Carnot engine using any working material, we obtain either $\mathcal{E}_A > \mathcal{E}_c$ or $\mathcal{E}_A < \mathcal{E}_c$, both of which are impossible. Accordingly, let us assume that there exists a reversible Carnot engine A with some working material for which $\delta q_{rev}/T$ is inexact, so that

$$\int_{cycle} \frac{\delta q_{rev}}{T} = \frac{q_1}{T_1} + \frac{q_3}{T_2} \neq 0. \qquad (4.38)$$

If the integral in Eq. 4.38 is not zero, it must be either positive or negative. First consider the case in which it is positive. Then

$$\frac{q_1}{T_1} + \frac{q_3}{T_2} > 0. \qquad (4.39)$$

Rearranging terms gives

$$\frac{q_1}{T_1} > -\frac{q_3}{T_2}. \qquad (4.40)$$

We now multiply both sides of the inequality in Eq. 4.40 by the ratio T_2/q_1. Since both T_2 and q_1 are positive, this operation does not reverse the direction of the inequality, and we obtain

$$\frac{T_2}{T_1} > -\frac{q_3}{q_1}. \qquad (4.41)$$

Multiplying both sides of the inequality in Eq. 4.41 by -1 reverses the direction of the inequality and gives

$$\frac{q_3}{q_1} > -\frac{T_2}{T_1}. \tag{4.42}$$

The efficiency of the engine is given by

$$\mathcal{E}_A = \frac{\text{Total work done}}{\text{Total heat added}} = \frac{q_1 + q_3}{q_1} = 1 + \frac{q_3}{q_1} > 1 - \frac{T_2}{T_1}, \tag{4.43}$$

since the inequality in Eq. 4.42 shows that q_3/q_1 is greater than $-T_2/T_1$. However, Eq. 4.22 shows that the efficiency of the Carnot engine using an ideal gas is $1 - T_2/T_1$. Therefore, the inequality in Eq. 4.43 indicates that $\mathcal{E}_A > \mathcal{E}_c$, which we have shown to be impossible. Consequently, the original assumption that $(q_1/T_1) + (q_3/T_2) > 0$ must be incorrect.

If the integral in Eq. 4.38 is assumed to be negative, the same analysis leads to the result that

$$\frac{q_3}{q_1} < -\frac{T_2}{T_1} \tag{4.44}$$

and

$$\mathcal{E}_A = \frac{\text{Total work done}}{\text{Total heat added}} = \frac{q_1 + q_3}{q_1} = 1 + \frac{q_3}{q_1} < 1 - \frac{T_2}{T_1}. \tag{4.45}$$

Equation 4.45 indicates that $\mathcal{E}_A < \mathcal{E}_c$, which is impossible. Consequently, we cannot have $(q_1/T_1) + (q_3/T_2) < 0$. But then, since $(q_1/T_1) + (q_3/T_2)$ cannot be either positive or negative, it must be zero. Therefore, the differential $\delta q_{\text{rev}}/T$ must be exact for any substance undergoing a Carnot cycle.

Part 4

As the final step in the proof, we note that any cycle can be represented by an infinite number of infinitesimally small Carnot cycles. This situation is shown in Figure 4.11. Each of the small rhombuses represents a Carnot cycle. The circle denoted as A represents any arbitrary closed cycle. It is clear that as we make the Carnot cycles infinitesimally small, the fit of the rhombuses to the circle will become exact. We now must show that the sum of the inte-

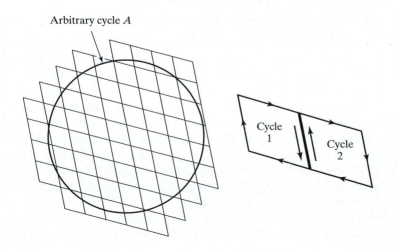

Arbitrary cycle A

Cycle 1

Cycle 2

◀ **FIGURE 4.11**
An arbitrary cycle denoted as A made up of a large number of Carnot cycles represented by the small rhombuses. In the limit of infinitesimally small Carnot cycles, the fit will be exact. The small figure to the right shows how the integrations over sides held in common by two rhombuses cancel out.

grals of $\delta q_{rev}/T$ about all Carnot cycles is equal to the integral of $\delta q_{rev}/T$ about cycle A. In the integration over all Carnot cycles, the integrals over sides held in common will cancel out. Consider the small inset to the right of the figure. When we execute the integration over Carnot cycle 1, the integral over the adiabatic process denoted by the boldface line will be taken from top to bottom. In contrast, when the integration is done over cycle 2, the integral over this path will be taken from bottom to top. Consequently, the limits will be reversed, and the sum of the two integrations will add to zero. This will obtain for all processes held in common between two Carnot cycles. The only cycles that do not have a common side are the ones bordering arbitrary cycle A. Consequently, we will have

$$\sum_{n=1}^{n=\infty} \int_{\text{cycle } n} \frac{\delta q_{rev}}{T} = \int_{\text{cycle } A} \frac{\delta q_{rev}}{T}. \tag{4.46}$$

Since all of the integrals over the Carnot cycles are zero, we have

$$\int_{\text{cycle } A} \frac{\delta q_{rev}}{T} = 0 \tag{4.47}$$

for any arbitrary cycle. But this can be true only if $\delta q_{rev}/T$ is exact, thereby completing the proof.

Since the differential $\delta q_{rev}/T$ is exact, there must exist a function whose total differential is exactly $\delta q_{rev}/T$. We label this function S and give it the name *entropy*. Therefore, the most important consequence of the second law can be said to be that the differential

$$\boxed{dS = \frac{\delta q_{rev}}{T}} \tag{4.48}$$

is exact.

The subscript "rev" in Eq. 4.48 is extremely important. Since dS is exact, the change in entropy for any process depends only upon the initial and final states of the process, not upon the pathway by which the change is effected. However, to determine the value of ΔS, the calculation *must* be done along a reversible path.

4.6   Reversible, Spontaneous, and Nonspontaneous Processes

Our previous discussion suggested that the efficiency of a Carnot engine might be used as a measure of the spontaneity and the extent to which a well-ordered system is converted into a system that is more chaotic or disordered. We also noted that there are problems with such an approach, since ideal Carnot engines do not actually exist. Now that we have the second law, there is a much better solution to the problem of predicting spontaneity and the conversion of ordered systems to disordered ones: Entropy calculations provide us with a means by which to determine whether a process is reversible (at equilibrium), irreversible (spontaneous), or nonspontaneous. They also provide us with a quantitative measure of the extent to which order is convered into chaos in a process.

Consider an infinitesimal change from State A to State B along two paths, one reversible, the other irreversible. (See Figure 4.12.) Along both

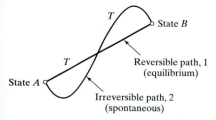

▲ FIGURE 4.12
An infinitesimal change from State A to State B along irreversible and reversible paths. In both cases, the change is sufficiently small that T is unchanged along the paths.

paths, the change is sufficiently small that the temperature may be regarded as being constant. Since the initial and final states are identical for both paths, the first law requires that

$$dU_1 = \delta q_1^{rev} + \delta w_1^{rev} = dU_2 = \delta q_2^{irrev} + \delta w_2^{irrev}. \qquad (4.49)$$

Equation 4.49 may be written in the form

$$\delta q_1^{rev} - P\,dV = \delta q_2^{irrev} - P_{ext}\,dV, \qquad (4.50)$$

or

$$\delta q_1^{rev} = \delta q_2^{irrev} + (P - P_{ext})\,dV. \qquad (4.51)$$

If the process is an expansion, for which $dV > 0$, we must have $P_{ext} < P$, so that the expansion will occur spontaneously along Path 2. If the process is a compression, for which $dV < 0$, we must have $P_{ext} > P$ to give a spontaneous process along Path 2. Therefore, in either case, the second term in Eq. 4.51 is positive, and we have

$$\delta q_1^{rev} > \delta q_2^{irrev}. \qquad (4.52)$$

Since T is positive, we may divide both sides of this inequality by T without reversing the direction of the inequality. This gives

$$\frac{\delta q_{rev}}{T} > \frac{\delta q_{irrev}}{T} \qquad (4.53)$$

for any process. Integrating both sides of the inequality in Eq. 4.53 then yields

$$\boxed{\Delta S = \int \frac{\delta q_{rev}}{T} > \int \frac{\delta q_{irrev}}{T}.} \qquad (4.54)$$

Equation 4.54 is called the *Clausius inequality*. It gives us a direct means of determining whether a proposed process will be spontaneous. If the entropy change for the process is greater than the integral of $\delta q/T$ along the path of the process, the process will be irreversible and spontaneous along that path. If ΔS is equal to $\int \delta q/T$ along the path, the process is reversible and the system is at equilibrium. If $\Delta S < \int \delta q/T$, the process is nonspontaneous and will not occur without work being done on the system. In the absence of such work, a nonspontaneous process is impossible.

EXAMPLE 4.5

One mole of an ideal gas at $T_1 = 300$ K, $P_1 = 2.4618$ atm, and $V_1 = 10$ L is isothermally and reversibly expanded to $V_2 = 20$ L. Compute ΔU, w, q, and ΔS for the process.

Solution

For an ideal gas, we have

$$dU = C_v\,dT = 0, \qquad (1)$$

since $dT = 0$ for an isothermal process. Therefore,

$$dU = \delta q + \delta w = 0, \qquad (2)$$

which gives

$$\delta q = -\delta w = P_{ext}\, dV = P\, dV, \tag{3}$$

because the process is reversible and $P_{ext} = P$. Substituting $P = nRT/V$ and integrating both sides of Eq. 3 gives

$$q_{rev} = -w_{rev} = nRT \int_{10\,L}^{20\,L} \frac{dV}{V} = nRT \ln(2)$$

$$= (1\,mol)(8.314\,J\,mol^{-1}\,K^{-1})(300\,K)\ln(2) = 1{,}729\,J. \tag{4}$$

The change in entropy is given by

$$\Delta S = \int \frac{\delta q_{rev}}{T} = \frac{1}{T}\int \delta q_{rev} = \frac{q_{rev}}{T} = \frac{nRT \ln(2)}{T}$$

$$= (1\,mol)(8.314\,J\,mol^{-1}\,K^{-1})\ln(2) = 5.763\,J\,K^{-1}. \tag{5}$$

It is very important to note that we obtain the result $\Delta S = q_{rev}/T$ because (1) the process is isothermal and T can be factored out of the integral in Eq. 5 and (2) the process is reversible and the calculated q represents the heat change along a reversible path. In this connection, see Example 4.6; for related exercises, see Problems 4.16 and 4.17.

EXAMPLE 4.6

One mole of an ideal gas at $T_1 = 300$ K, $P_1 = 2.4618$ atm, and $V_1 = 10$ L is isothermally expanded to $V_2 = 20$ L against a constant external pressure of 1.2309 atm. Compute ΔU, w, q, and ΔS for the process, and show that Eq. 4.54 is satisfied. Ignore the effects of acceleration of the piston. (In this connection, see the explanatory note in Example 2.3.)

Solution

Since $dU = C_v\, dT$ for an ideal gas and the process is isothermal, we must have $\Delta U = 0$. Consequently,

$$\Delta U = 0 = q + w. \tag{1}$$

Thus,

$$q = -w = P_{ext}\, dV. \tag{2}$$

The work is, therefore,

$$w = -\int_{10\,L}^{20\,L} P_{ext}\, dV = -P_{ext} \int_{10\,L}^{20\,L} dV = -P_{ext}(20 - 10)\,L = -1.2309\,atm\,(10\,L)$$

$$= -12.309\,L\,atm \times \frac{8.314\,J}{0.08206\,L\,atm} = -1{,}247\,J. \tag{3}$$

The heat associated with the process is

$$q = -w = 1{,}247\,J. \tag{4}$$

The entropy calculation must be done along a reversible path. Consequently, we must compute the value of

$$\Delta S = \int \frac{\delta q_{rev}}{T} = \frac{1}{T}\int \delta q_{rev} = \frac{q_{rev}}{T}. \tag{5}$$

This is precisely the calculation done in Example 4.5. Hence, the entropy change is exactly what it was in that example:

$$\Delta S = \frac{q_{rev}}{T} = \frac{nRT \ln(2)}{T} = (1 \text{ mol})(8.314 \text{ J mol}^{-1} \text{K}^{-1}) \ln(2) = 5.763 \text{ J K}^{-1}. \quad \textbf{(6)}$$

Actually, we knew this had to be the case, since the initial and final states in Examples 4.5 and 4.6 are the same. Let us now compute $\int \delta q / T$:

$$\int \frac{\delta q}{T} = T^{-1} \int \delta q = \frac{q}{T} = \frac{q_{irrev}}{T} = \frac{1{,}247 \text{ J}}{300 \text{ K}} = 4.157 \text{ J K}^{-1}. \quad \textbf{(7)}$$

It follows that

$$\Delta S = 5.763 \text{ J K}^{-1} > \int \frac{\delta q}{T} = 4.157 \text{ J K}^{-1}, \quad \textbf{(8)}$$

as is required by the Clausius inequality, Eq. 4.54.

One of the most commonly made mistakes by students taking their first course in thermodynamics is to compute the entropy change in this problem by writing

$$\Delta S = \frac{q}{T} = \frac{1{,}247 \text{ J}}{300 \text{ K}} = 4.157 \text{ J K}^{-1}.$$

This calculation is incorrect because q is not the heat change along a *reversible* path. ΔS ***must*** be computed along a reversible path.

Equation 4.54 gives us one method for determining whether a process is spontaneous: We compute ΔS for the system and compare its value with $\int \delta q / T$ along the path of the process. If $\Delta S > \int \delta q / T$, the process is spontaneous. If the two are equal, the process is reversible and, therefore, at equilibrium. If $\Delta S < \int \delta q / T$, the process is nonspontaneous and will not take place unless work is done on the system.

We may obtain a second method for determining spontaneity by taking advantage of the fact that all spontaneous processes tend to turn the overall order of the universe into a more disordered state. The simple analogy presented in Figure 4.8 makes this point. In Chapter 18, we shall prove that the change in entropy is a quantitative measure of the extent to which order is transformed into chaos. Large positive values for ΔS indicate a substantial increase in the disorder of a system, while large negative values for ΔS signal an increase in order. This means that if we compute the total entropy change for both a system and its surroundings and find it to be positive, order has been transformed into disorder and the process is spontaneous. If, however, we find that the total entropy change for the system plus the surroundings is negative, the overall order has increased. This is always brought about by a nonspontaneous process that cannot occur unless work is done. Example 4.7 illustrates the second method.

EXAMPLE 4.7

(A) One mole of an ideal gas at $T_1 = 300$ K, $P_1 = 2.4618$ atm, and $V_1 = 10$ L is isothermally and reversibly expanded to $V_2 = 20$ L with the surroundings at 300 K. Compute ΔS for the gas and for the surroundings. What is the total entropy change for the universe due to the process? **(B)** One mole of an ideal gas at $T_1 = 300$ K, $P_1 = 2.4618$ atm, and $V_1 = 10$ L is isothermally expanded to

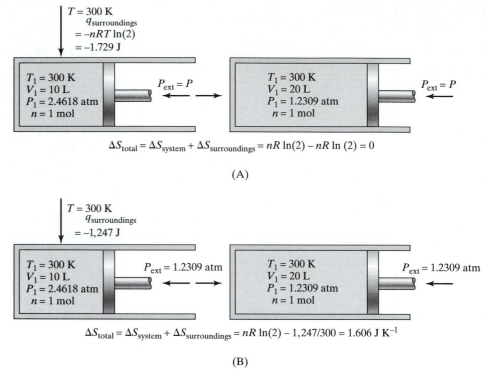

$$\Delta S_{total} = \Delta S_{system} + \Delta S_{surroundings} = nR \ln(2) - nR \ln(2) = 0$$

(A)

$$\Delta S_{total} = \Delta S_{system} + \Delta S_{surroundings} = nR \ln(2) - 1{,}247/300 = 1.606 \text{ J K}^{-1}$$

(B)

▲ **FIGURE 4.13**
Total entropy changes for system plus surroundings for two processes. (A) A reversible expansion has $\Delta S_{total} = 0$, denoting an equilibrium situation. (B) An irreversible, spontaneous expansion has $\Delta S_{total} > 0$.

$V_2 = 20$ L against a constant external pressure of 1.2309 atm. Compute ΔS for the gas and for the surroundings. What is the total entropy change for the universe due to the process? Ignore the effects of acceleration of the piston. (In this connection, see the explanatory note in Example 2.3. Figure 4.13 illustrates the two processes.

Solution

(A) This part of the example is identical to Example 4.5. Thus, for the ideal gas, we have

$$\Delta S_{gas} = nR \ln(2) = (1 \text{ mol}) (8.314 \text{ J mol}^{-1} \text{ K}^{-1}) \ln(2) = 5.763 \text{ J K}^{-1}. \quad \textbf{(1)}$$

For the surroundings, a total heat equal to $RT_1 \ln(2)$ was removed and reversibly added to the gas. Therefore, $q_{surroundings} = -nRT_1 \ln(2)$. The associated entropy change for the surroundings is, therefore,

$$\Delta S_{surroundings} = \int \frac{\delta q_{rev}}{T} = T_1^{-1} \int \delta q_{rev} = \frac{q_{rev}}{T_1} = \frac{q_{surroundings}}{T_1} = -nR \ln(2). \quad \textbf{(2)}$$

The total entropy change for the universe due to this expansion is

$$\Delta S_{total} = \Delta S_{gas} + \Delta S_{surroundings} = nR \ln(2) - nR \ln(2) = 0. \quad \textbf{(3)}$$

This result means that the system is at equilibrium, which is precisely the situation for a reversible process.

(B) The calculation of ΔS_{gas} was done in Example 4.6, where we had to exercise care to use a reversible path. The result was

$$\Delta S_{gas} = nR \ln(2) = 5.763 \text{ J K}^{-1}, \tag{4}$$

the same as in Example 4.5, since the initial and final states are the same. For the surroundings, we again have

$$\Delta S_{surroundings} = \int \frac{\delta q_{rev}}{T} = T_1^{-1} \int \delta q_{rev} = \frac{q_{rev}}{T_1} = \frac{q_{surroundings}}{T_1}. \tag{5}$$

In this case, however, the heat extracted at 300 K from the surroundings was computed in Example 4.6 to be 1,247 J. Since the surroundings and the system must be at the same temperature to ensure an isothermal process, the heat transfer from the surroundings is reversible. Therefore, we have $q_{surroundings} = q_{rev} = -1,247$ J. Substituting into Eq. 5 then gives

$$\Delta S_{surroundings} = \frac{-1,247 \text{ J}}{300 \text{ K}} = -4.157 \text{ J K}^{-1}. \tag{6}$$

The total entropy change for the universe due to this expansion is

$$\Delta S_{total} = \Delta S_{gas} + \Delta S_{surroundings} = nR \ln(2) - 4.157 \text{ J K}^{-1} = 5.763 - 4.157 \text{ J K}^{-1}$$
$$= 1.606 \text{ J K}^{-1}. \tag{3}$$

The total entropy change is positive. This tells us that overall order has been transformed into disorder and that the process is spontaneous.

The results obtained in Example 4.7 are general: *All equilibrium reversible processes have* $\Delta S_{total} = \Delta S_{system} + \Delta S_{surroundings} = 0$. *In contrast, spontaneous, irreversible processes are always associated with positive total entropy changes. Nonspontaneous processes that require work to force them to take place have* $\Delta S_{total} = \Delta S_{system} + \Delta S_{surroundings} < 0$. This general principle rests on two facts: (1) Entropy is a quantitative measure of the extent to which order is converted to disorder and (2) all spontaneous processes cause overall order to be transformed into disorder.

Let us consider the simple example of heat transfer from Site A to Site B, where the site temperatures are T_A and T_B, respectively. We will effect this transfer through Site C in three steps, as shown in Figure 4.14. In Step 1, we extract a small quantity of heat, $-\delta q$, from Site A and transfer it reversibly to Site C, which is at the same temperature. The temperature of Site C is then changed to T_B by a constant-volume process. In Step 2, a quantity of heat, δq, is reversibly added to Site B. Finally, the temperature of Site C is returned to T_A by a constant-volume process. If the quantity of heat trans-

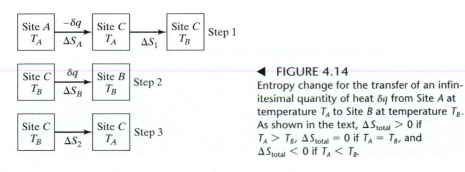

◀ FIGURE 4.14
Entropy change for the transfer of an infinitesimal quantity of heat δq from Site A at temperature T_A to Site B at temperature T_B. As shown in the text, $\Delta S_{total} > 0$ if $T_A > T_B$, $\Delta S_{total} = 0$ if $T_A = T_B$, and $\Delta S_{total} < 0$ if $T_A < T_B$.

ferred, $|\delta q|$, is small relative to the total heat content of the sites, the temperatures of the sites will remain constant. Under these conditions, the entropy changes are

$$\Delta S_A = \frac{\delta q_{rev}}{T_A} = -\frac{\delta q}{T_A}$$

and

$$\Delta S_B = \frac{\delta q_{rev}}{T_B} = \frac{\delta q}{T_B}.$$

In a later section [see Eq. 4.70], we shall find that the entropy changes associated with altering the temperature of Site C are

$$\Delta S_1 = \int_{T_A}^{T_B} \frac{[C_v]_{\text{site } C} \, dT}{T}$$

and

$$\Delta S_2 = \int_{T_B}^{T_A} \frac{[C_v]_{\text{site } C} \, dT}{T}.$$

The total entropy change for the overall process is $\Delta S_{total} = \Delta S_A + \Delta S_B + \Delta S_1 + \Delta S_2$. Since $\Delta S_1 = -\Delta S_2$, the result is

$$\Delta S_{total} = \Delta S_A + \Delta S_B = \delta q \left[\frac{1}{T_B} - \frac{1}{T_A} \right].$$

If $T_A = T_B$, $\Delta S_{total} = 0$, and the system is clearly at thermal equilibrium. If $T_A > T_B$, $\Delta S_{total} > 0$, and the heat transfer is spontaneous. Heat flows spontaneously from a hot site to a cold site. However, if $T_A < T_B$, $\Delta S_{total} < 0$, and the process is nonspontaneous. We can force heat to flow from a cold site to a hot site only by doing work with something like a heat pump. Example 4.3 illustrates this principle.

The preceding analysis demonstrates why the device discussed at the beginning of this chapter will not work. That device proposed to extract thermal energy from seawater and transfer it to a ship's boilers. The temperature of seawater is in the neighborhood of 280 K. The boilers of the ship are much hotter. The device, therefore, requires that we transfer heat from a cold source to a hot source. Such a process is nonspontaneous. It transforms disorder into order, as shown in Figure 4.8. Such a transformation can only be accomplished by doing work on the system. If, however, we are going to do work on the system, then we will need fuel. Hence, the device is a perpetual-motion machine of the second type. It violates the second law.

4.7 Molecular Interpretation of Entropy

The previous section demonstrates how the spontaneity or nonspontaneity of a process under given conditions may be predicted by investigation of entropy changes. In Chapters 17 and 18, we shall prove that entropy changes are quantitatively connected to the degree of disorder exhibited by a system. At this point, however, we lack the tools required to rigorously demonstrate the connection between the entropy, as defined by Eq. 4.48, and disorder. We

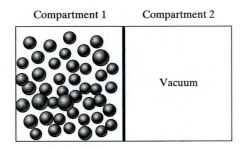

Compartment 1   Compartment 2

Vacuum

◄ FIGURE 4.15
A highly ordered gaseous system. All of the gas molecules are in Compartment 1. None are in Compartment 2.

have, therefore, relied on an appeal to reason and experience to suggest that such a connection exists. Figures 4.8 and 4.14 present two examples. Several more will be given.

Consider a closed container of gas molecules partitioned into two compartments as shown in Figure 4.15. The system is highly ordered in that all of the gas molecules are in Compartment 1. Logic, based on experience, suggests that if the partition separating the two compartments is removed, the gas molecules will spontaneously distribute themselves between the compartments to produce a more disordered system. This result gives rise to the old adage, "Nature abhors a vacuum." Our qualitative arguments to this point suggest that the total entropy increases as a result of the redistribution of molecules between the two compartments. Consequently, the process is indeed spontaneous.

Our everyday experience is replete with other examples that suggest a direct connection between an increase in disorder and the spontaneity of a process. A disordered pile of bricks will not spontaneously assemble itself into a building, but a building left to itself will, in time, spontaneously decay into a jumbled pile of bricks. A cube of sugar placed in a glass of water will spontaneously redistribute the sugar molecules among the water molecules to form a solution, which is a disordered mixture of the two types of molecules. Once formed, the solution will not spontaneously turn back into a cube of sugar and a glass of pure water. The spontaneous process always moves in a direction that turns order into disorder. To reverse this process, work must be done. (To clean your disordered apartment, you must do some work.)

Since spontaneous processes are always associated with an increase in total entropy, it would appear that such an increase is a measure of the increase in disorder that must accompany the spontaneous process. The explicit relationship between the two is given by

$$S = k_b \ln W, \qquad (4.55)$$

where

$$k_b = \frac{R}{N} = \frac{8.314 \text{ J mol}^{-1} \text{ K}^{-1}}{6.02214 \times 10^{23} \text{ mol}^{-1}} = 1.381 \times 10^{-23} \text{ J K}^{-1}$$

is the Boltzmann constant (the gas constant divided by Avogadro's number) and W is a quantity called the *thermodynamic probability*. In Chapters 17 and 18, we shall derive Eq. 4.55 and show that W counts the number of different ways the energy and configuration of a system can be achieved. Highly ordered systems have small values of W, and disordered systems are associ-

▶ FIGURE 4.16

A quantitative evaluation of order–disorder changes as a way to compute the entropy change for a process. In State 1, Compartment 1 contains N molecules of gas while Compartment 2 contains none. There is only one way to achieve this configuration. Therefore, $W_1 = 1$. In State 2, the N molecules are evenly distributed between the two compartments. This configuration can be achieved in $N!/[(N/2)!]^2$ different ways. Therefore, we have $W_2 = N!/[(N/2)!]^2$, which leads to $S_2 = R \ln(2)$. The total entropy change is $\Delta S_{\text{total}} = R \ln(2) - 0 = R \ln 2 > 0$, and the process is spontaneous. In the figure, the different sizes of the circles denote relative positions within the three-dimensional box, not different molecules.

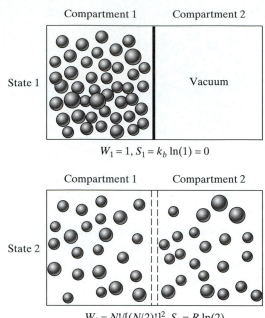

Compartment 1   Compartment 2

State 1

Vacuum

$W_1 = 1, S_1 = k_b \ln(1) = 0$

Compartment 1   Compartment 2

State 2

$W_2 = N!/[(N/2)!]^2, S_2 = R \ln(2)$

ated with larger values of W. Consequently, S increases as we go from order to disorder in a spontaneous process.

Let us use Eq. 4.55 to compute the change in entropy expected for a constant-temperature, constant-volume change in an ideal gas, as illustrated in Figure 4.15, when both compartments have equal volume. Suppose we have 1 mole of molecules of an ideal gas confined in Compartment 1 of the figure. This state is labeled State 1 in Figure 4.16. Since the only way to achieve such a configuration is to place all 6.022×10^{23} molecules in Compartment 1, the thermodynamic probability for this state is $W_1 = 1$. The corresponding entropy of the state is given by Eq. 4.55:

$$S_{\text{State 1}} = S_1 = k_b \ln(1) = 0.$$

In State 2, we have replaced the solid partition separating the two compartments with a porous partition so that molecules can diffuse freely back and forth. On the average, we will now have half of the molecules, $N/2$, in each compartment, since the volumes are equal. The number of ways this configuration can be produced is the number of combinations of N objects (molecules in this case) taken $N/2$ at a time. In general, the number of combinations of N objects taken m at a time is given by $C(N, m) = N!/[(N - m)! \, m!]$. Figure 4.17 illustrates this point for the simple case in which $N = 4$. When all four objects are in Compartment 1, the number of ways to acieve this result is $C(4,4) = 4!/[(4 - 4)! \, 4! = 1]$, as shown in the figure. If we distribute the four objects evenly between the compartments, we will have two objects in each compartment. The number of ways to obtain that result is $C(4, 2) = 4!/[(4 - 2)! \, 2! = 6]$. The six possible ways are illustrated in the figure. Notice that as the disorder in the system increases, so does W. When N is Avogadro's constant, the number of ways to obtain the configuration shown as State 2 in the figure is $C(N, N/2) = N!/[(N - N/2)! \, (N/2)!] = N!/[(N/2)!]^2$. With $N = 6.022 \times 10^{23}$, $C(N, N/2)$ is extremely large, which is indicative of

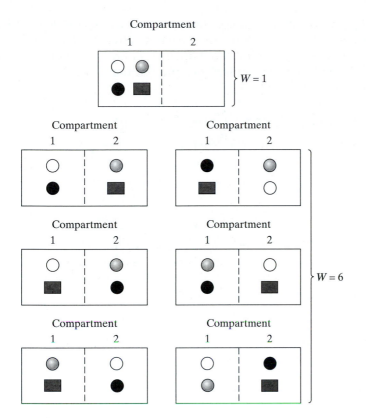

◀ FIGURE 4.17
Thermodynamic probabilities for ordering four objects between two compartments. There is only one way to place all four objects in Compartment 1, but there are six ways to distribute two objects into each compartment as shown.

the chaos, or disorder, present in State 2. Using Eq. 4.55, we can compute the entropy in that state:

$$S_2 = k_b \ln\left[\frac{N!}{[(N/2)!]^2}\right] = k_b \ln N! - k_b \ln[(N/2)!]^2$$

$$= k_b \ln N! - 2k_b \ln(N/2)!. \tag{4.56}$$

To evaluate the expressions $\ln N!$ and $\ln(N/2)!$, we need to introduce a numerical approximation known as Stirling's approximation. For the logarithm of factorials, this approximation is

$$\ln(N!) \approx N \ln N - N.$$

When N is large, the approximation is highly accurate. This will be explicitly shown in Chapter 17. For the moment, we shall be content to use Stirling's approximation to evaluate S_2. Inserting this approximation for the logarithm of the factorials in Eq. 4.56 gives

$$S_2 = k_b\left[N \ln N - N - 2\left\{\frac{N}{2}\ln\frac{N}{2} - \frac{N}{2}\right\}\right]$$

$$= k_b\left[N \ln N - N - 2\left\{\frac{N}{2}(\ln N - \ln 2) - \frac{N}{2}\right\}\right]$$

$$= k_b\left[N \ln N - N - N \ln N + N \ln 2 + N\right] = k_b N \ln(2).$$

However, $Nk_b = R$, so that $S_2 = R \ln(2)$. The overall entropy change for the process of converting the system from State 1 to State 2 is, therefore,

$$\Delta S_{total} = S_2 - S_1 = R \ln(2) - 0 = R \ln(2),$$

and ΔS_{total} is positive. Hence, the process is spontaneous, and overall order is being converted to disorder. We have already found in Example 4.5 that the result $\Delta S_{\text{total}} = R \ln(2)$ obtained from Eq. 4.55 is precisely the correct result for 1 mole of an ideal gas undergoing an isothermal expansion to double its volume.

4.8 Internal Energy and Enthalpy Differentials in Terms of Entropy

It is to our great advantage to develop equations that combine both the first and second laws. Such a combination permits us to bring the power of both laws to bear on thermodynamics problems. From the first law, we have

$$dU = \delta q + \delta w = \delta q - P_{\text{ext}} \, dV \tag{4.57}$$

for a closed system in which only pressure–volume work is being done. Since dU is an exact differential, we may evaluate ΔU along any path we choose. Let us assume that we will choose a reversible path for the calculation. For this choice, Eq. (4.57) becomes

$$dU = \delta q_{\text{rev}} - P \, dV. \tag{4.58}$$

Since the path is reversible, we may express δq_{rev} in terms of dS using Eq. 4.48. This gives

$$\boxed{dU = T \, dS - P \, dV} \,. \tag{4.59}$$

Equation 4.59 contains both the first and second laws. It is one of the most useful relationships in thermodynamics. It holds for any reversible process in a closed system, provided that the only work being done is due to expansion of the volume of a gas.

It is equally simple to obtain dH in terms of dS. The enthalpy is defined as $H = U + PV$. If we take differentials of both sides of this equation, we obtain

$$dH = dU + d(PV) = dU + P \, dV + V \, dP. \tag{4.60}$$

For a reversible path, dU is given by Eq. 4.59. Combining this equation with Eq. 4.60, we have

$$\boxed{dH = T \, dS - P \, dV + P \, dV + V \, dP = T \, dS + V \, dP} \,. \tag{4.61}$$

Equation 4.61 is another extremely useful equation in thermodynamics.

Equations 4.59 and 4.61 hold for closed systems undergoing changes along a reversible path involving only pressure–volume work. We may easily generalize these equations to open systems by adding the terms that give the changes to dU and dH when the number of moles of each component changes by an amount dn_i. The required modifications are given in Eqs. 2.46 and 2.47. For open systems, we have

$$dU = T \, dS - P \, dV + \sum_{i=1}^{K} \widehat{U}_i \, dn_i \tag{4.62}$$

and

$$dH = T \, dS + V \, dP + \sum_{i=1}^{K} \overline{H}_i \, dn_i \tag{4.63}$$

where $\overline{H}_i$ is the partial molar enthalpy for component i and $\widehat{U}_i$ is the derivative $(\partial U/\partial n_i)_{T\,V\,n_j}$ in a system with K different components.

4.9   Temperature, Volume, and Pressure Dependence of the Entropy

Chemical processes often involve changes in temperature, volume, or pressure. Therefore, we need a suitable set of equations that permit changes in the entropy to be computed when T, V, or P is changed. As usual, this means that we require derivatives of S with respect to T, V, and P. Since an equation of state connects T, V, and P, we have only two independent variables for a fixed quantity of material. Consequently, we can write the entropy as a function of either T and V or T and P; that is,

$$S = f(T, V) \tag{4.64A}$$

or

$$S = g(T, P). \tag{4.64B}$$

We could, of course, write S as a function of P and V, but this would be a poor choice, since temperature is almost always one of the variables measured in a chemical experiment. From Eq. 4.64A, the total differential giving the change in entropy is

$$dS = \left(\frac{\partial S}{\partial T}\right)_v dT + \left(\frac{\partial S}{\partial V}\right)_T dV. \tag{4.65}$$

If we express entropy as a function of T and P using Eq. 4.64B, the total differential of S is

$$dS = \left(\frac{\partial S}{\partial T}\right)_P dT + \left(\frac{\partial S}{\partial P}\right)_T dP. \tag{4.66}$$

To compute changes in S when T, V, or P is changed, we need the values of the four derivatives appearing in Eqs. 4.65 and 4.66.

4.9.1   Temperature Derivatives of S

The partial derivatives required by Eqs. 4.65 and 4.66 may be obtained directly from Eqs. 4.59 and 4.61. Dividing Eq. 4.59 by dT, we obtain

$$\frac{dU}{dT} = T\frac{dS}{dT} - P\frac{dV}{dT}. \tag{4.67}$$

If we now invoke the condition that the heating or cooling is done under conditions of constant volume, then $dV = 0$ and the last term in Eq. 4.67 vanishes, while the total derivatives become partial derivatives taken at constant volume:

$$\left(\frac{\partial U}{\partial T}\right)_V = T\left(\frac{\partial S}{\partial T}\right)_V. \tag{4.68}$$

In Chapter 2, we found that $(\partial U/\partial T)_V = C_v = n\,C_v^m$. Therefore, we have

$$\boxed{\left(\frac{\partial S}{\partial T}\right)_V = \frac{C_v}{T} = \frac{nC_v^m}{T}.} \tag{4.69}$$

Equation 4.69 permits us to compute changes in the entropy for constant-volume heating or cooling processes. Separating variables and integrating both sides between the initial temperature T_1 and the final temperature T_2 then gives

$$\int_{S(T_1)}^{S(T_2)} \partial S = S(T_2) - S(T_1) = \Delta S = \int_{T_1}^{T_2} \frac{C_v}{T} \partial T. \tag{4.70}$$

In the special case where C_v can be regarded as being independent of temperature, Eq. 4.70 is easily integrated to give

$$\Delta S = C_v \ln\left[\frac{T_2}{T_1}\right]. \tag{4.71}$$

The rate of change of entropy with temperature under conditions of constant pressure can be obtained directly from Eq. 4.61 by using the same procedure. Dividing that equation by dT, followed by invoking the condition that the heating is done at constant pressure, yields

$$\left(\frac{\partial H}{\partial T}\right)_P = T\left(\frac{\partial S}{\partial T}\right)_P. \tag{4.72}$$

We have shown in Chapter 2 that $(\partial H/\partial T)_P = C_p = n\, C_p^m$. With this substitution, Eq. 4.72 becomes

$$\boxed{\left(\frac{\partial S}{\partial T}\right)_P = \frac{C_p}{T} = \frac{nC_p^m}{T}.} \tag{4.73}$$

The similarity between $(\partial S/\partial T)_V$ and $(\partial S/\partial T)_P$ should be noted. In the former case, the derivative is equal to C_v/T; in the latter case, we simply substitute C_p for C_v.

4.9.2   Volume Derivative of S

Equation 4.65 shows that we need an expression for the rate of change of entropy with volume at constant temperature. Again, this is easily obtained from Eq. 4.59 by dividing both sides by dV. This produces

$$\frac{dU}{dV} = T\frac{dS}{dV} - P. \tag{4.74}$$

If we now invoke the condition that the temperature is to be held constant, we obtain

$$\left(\frac{\partial U}{\partial V}\right)_T = T\left(\frac{\partial S}{\partial V}\right)_T - P \tag{4.75}$$

which gives

$$\boxed{\left(\frac{\partial S}{\partial V}\right)_T = \left(\frac{1}{T}\right)\left[\left(\frac{\partial U}{\partial V}\right)_T + P\right].} \tag{4.76}$$

Equation 4.76 gives the rate of change of entropy with volume under conditions of constant temperature. For an ideal gas, the equation assumes the simple form

$$\left(\frac{\partial S}{\partial V}\right)_T = \frac{P}{T} = \frac{nR}{V} \tag{4.77}$$

since $(\partial U/\partial V)_T = 0$ from Joule's law. Equation 4.77 may be easily integrated to give the entropy change for a constant-temperature volume change for an ideal gas:

$$\Delta S = \int \partial S = \int_{V_1}^{V_2} \frac{nR}{V} \, dV = nR \ln\left[\frac{V_2}{V_1}\right]. \qquad (4.78)$$

Equation 4.78 provides a second route to the entropy change for the process shown in Figure 4.16. There, 1 mole of an ideal gas doubles its volume at constant temperature. Using Eq. 4.78, we find that the entropy change for such a process is $\Delta S = (1)R \ln[2V_1/V_1] = R \ln(2)$, which is precisely the result we obtained using Eq. 4.55. Although this agreement does not establish the validity of the latter equation, it does give us confidence that the change in entropy does indeed provide a quantitative measure of the extent to which a process converts order to disorder.

Direct substitution of Eqs. 4.69 and 4.76 into Eq. 4.65 gives the total differential of S needed to compute changes in the entropy when the temperature and volume change. The result is

$$dS = \left[\frac{nC_v^m}{T}\right] dT + \left\{\left(\frac{1}{T}\right)\left[\left(\frac{\partial U}{\partial V}\right)_T + P\right]\right\} dV. \qquad (4.79)$$

4.9.3   Pressure Derivative of S

Equation 4.66 shows that we need $(\partial S/\partial P)_T$ to obtain the total differential of S expressed as a function of T and P. This derivative can be obtained directly from Eq. 4.61 by dividing both sides by dP and holding the temperature constant. These operations produce

$$\left(\frac{\partial H}{\partial P}\right)_T = T \left(\frac{\partial S}{\partial P}\right)_T + V, \qquad (4.80)$$

which, upon rearrangement, gives

$$\boxed{\left(\frac{\partial S}{\partial P}\right)_T = \left(\frac{1}{T}\right)\left[\left(\frac{\partial H}{\partial P}\right)_T - V\right].} \qquad (4.81)$$

Substituting Eqs. 4.73 and 4.81 into Eq. 4.66 yields the total differential of the entropy expressed as a function of T and P:

$$dS = \left[\frac{nC_p^m}{T}\right] dT + \left\{\left(\frac{1}{T}\right)\left[\left(\frac{\partial H}{\partial P}\right)_T - V\right]\right\} dP. \qquad (4.82)$$

For an ideal gas, $(\partial H/\partial P)_T = -\mu C_p = 0$, since the Joule–Thomson coefficient is zero. Consequently, Eq. 4.81 becomes

$$\left(\frac{\partial S}{\partial P}\right)_T = -\frac{V}{T} = -\frac{nR}{P}. \qquad (4.83)$$

This may be easily integrated between the initial and final pressures, P_1 and P_2, to give the entropy change:

$$\Delta S = \int \partial S = \int_{P_1}^{P_2} -\frac{nR}{P} \, \partial P = -nR \ln\left[\frac{P_2}{P_1}\right] = nR \ln\left[\frac{P_1}{P_2}\right]. \qquad (4.84)$$

EXAMPLE 4.8

Five moles of gas are heated at constant pressure from 300 K to 400 K. The molar heat capacity of the gas is $C_p^m = a + bT + cT^{-2}$, where a, b, and c are constants. Obtain an expression for the entropy change in the process as a function of a, b, and c.

Solution

Equation 4.73 gives the rate of entropy change when the gas is heated at constant pressure:

$$\left(\frac{\partial S}{\partial T}\right)_P = \frac{C_p}{T} = \frac{nC_p^m}{T} = \frac{n[a + bT + cT^{-2}]}{T} = \frac{na}{T} + nb + \frac{nc}{T^3} \tag{1}$$

Separating the variables and integrating between 300 K and 400 K gives

$$\Delta S = \int dS = \int_{300\ K}^{400\ K} \left[\frac{na}{T} + nb + \frac{nc}{T^3}\right] dT$$

$$= na \ln\left[\frac{400}{300}\right] + nb(400 - 300) - \frac{nc}{2}\left[\frac{1}{400^2} - \frac{1}{300^2}\right]. \tag{2}$$

With $n = 5$, Eq. 2 becomes

$$\Delta S = 1.438a + 500b + 0.00001215c. \tag{3}$$

For related exercises, see Problems 4.17, 4.18, 4.25, 4.27, and 4.28.

EXAMPLE 4.9

One mole of an ideal gas with $C_v^m = 1.5R$ is heated from 300 K to 400 K along a reversible path such that $V = AT \exp[0.001T]$, where A is a constant expressed in L K^{-1}. Compute ΔS for the process.

Solution: Method 1

The initial state of the gas is $T_1 = 300$ K, and $V_1 = 300\ A \exp[0.3]$ L. Therefore, the ideal-gas law gives

$$P_1 = \frac{RT_1}{V_1} = \frac{300R}{300e^{0.3}A} = \frac{R}{e^{0.3}A} \text{ atm.} \tag{1}$$

The final state of the gas is $T_2 = 400$ K and $V_2 = 400\ A\ e^{0.4}$ L. Consequently, the final pressure is

$$P_2 = \frac{RT_2}{V_2} = \frac{400R}{400Ae^{0.4}} = \frac{R}{Ae^{0.4}} \text{ atm.} \tag{2}$$

Since dS is an exact differential, ΔS is independent of the path. Let us, therefore, effect the change from the initial to the final state along two different segments, the first being a constant-volume heating from 300 K to 400 K and the second being a constant-temperature expansion from V_1 to V_2 at 400 K. That is, the change will be effected along the path

$$[T = 300\ K, V = V_1] \longrightarrow [T = 400\ K, V = V_1] \longrightarrow [T = 400\ K, V = V_2].$$

This path is illustrated in Figure 4.18. The total entropy change will be the sum of that for the two processes. For the constant-volume heating, Eq. 4.69 gives

$$\left(\frac{\partial S}{\partial T}\right)_V = \frac{C_v}{T} = \frac{nC_v^m}{T}. \tag{3}$$

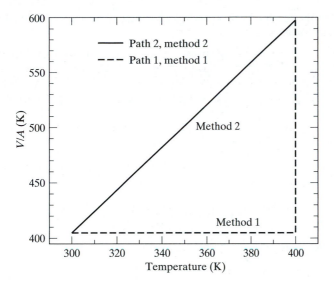

◀ FIGURE 4.18
Two reversible pathways to convert
one mole of an ideal gas from an
initial state given by $T_1 = 300$ K,
$V_1 = ATe^{0.3}$, and $P_1 = R/Ae^{0.3}$ to a
final state in which the variables
are $T_2 = 400$ K, $V_2 = ATe^{0.4}$, and
$P_2 = R/Ae^{0.4}$.

Since C_v^m is constant,

$$\Delta S_1 = \int dS = nC_v^m \int_{300\,K}^{400\,K}\frac{dT}{T} = (1\text{ mol})\,1.5R\ln\left[\frac{400}{300}\right] = 3.588\text{ J K}^{-1}. \quad (4)$$

The entropy change for the constant-temperature expansion at 400 K in the second step can be obtained using Eq. 4.76. We have

$$\left(\frac{\partial S}{\partial V}\right)_T = \left(\frac{1}{T}\right)\left[\left(\frac{\partial U}{\partial V}\right)_T + P\right] = \frac{P}{T} = \frac{nR}{V}, \quad (5)$$

since $(\partial U/\partial V)_T = 0$ for an ideal gas. Integrating both sides with $n = 1$ gives

$$\Delta S_2 = \int dS = R\int_{V_1}^{V_2}\frac{dV}{V} = R\ln\left[\frac{V_2}{V_1}\right] = R\ln\left[\frac{400Ae^{0.4}}{300Ae^{0.3}}\right]$$

$$= R\ln\left[\left(\frac{4}{3}\right)e^{0.1}\right] = R\ln\left(\frac{4}{3}\right) + R\ln(e^{0.1})$$

$$= 0.2877R + 0.1R = 0.3877R = 3.223\text{ J K}^{-1}. \quad (6)$$

The total change in entropy is

$$\Delta S = \Delta S_1 + \Delta S_2 = 3.588\text{ J K}^{-1} + 3.223\text{ J K}^{-1} = 6.811\text{ J K}^{-1}. \quad (7)$$

Method 2

Since the given path is reversible, we can compute ΔS directly along the actual path if we wish. It is not necessary that we use the two-step path employed in Method 1. For any reversible path, we have

$$dS = \frac{\delta q_{rev}}{T}. \quad (8)$$

The definition of heat capacity is

$$C = \frac{\delta q_{rev}}{dT} \quad (9)$$

for temperature changes along a reversible path. Solving for δq_{rev} and inserting the result into Eq. (8), we obtain

$$dS = \frac{C\,dT}{T}. \quad (10)$$

We now must find the heat capacity for heating along this path. In Chapter 2, we saw that the heat capacity is

$$C = \frac{\delta q}{dT} = \frac{dU - \delta w}{dT} = C_v + \left[\left(\frac{\partial U}{\partial V} \right)_T + P \right] \frac{dV}{dT}. \tag{11}$$

For an ideal gas, $(\partial U/\partial V)_T = 0$, so that for 1 mole of gas, Eq. 11 becomes

$$C = C_v^m + P \frac{dV}{dT}. \tag{12}$$

For the given path, dV/dT may be obtained by direct differentiation of $V = AT \exp[0.001T]$:

$$\frac{dV}{dT} = A \exp[0.001T] + 0.001AT \exp[0.001T]. \tag{13}$$

The product $P\,(dV/dT)$ is, therefore given by

$$P \frac{dV}{dT} = \frac{RT}{V} \frac{dV}{dT}$$

$$= \frac{R}{V}[AT \exp(0.001T)] + \frac{0.001RT}{V}[AT \exp(0.001T)]. \tag{14}$$

Since $V = AT \exp(0.001T)$, Eq. 14 becomes

$$P \frac{dV}{dT} = R + 0.001RT. \tag{15}$$

Substituting this result into Eq. 12 then yields

$$C = C_v^m + R + 0.001RT = 1.5R + R + 0.001RT = 2.5R + 0.001RT. \tag{16}$$

Combining Eqs. 10 and 16 produces

$$dS = \frac{[2.5R + 0.001RT]\,dT}{T}. \tag{17}$$

We may now integrate Eq. 17 between 300 K and 400 K to obtain the total entropy change:

$$\Delta S = \int dS = \int_{300\,\text{K}}^{400\,\text{K}} \frac{[2.5R + 0.001RT]\,dT}{T}$$

$$= 2.5R \ln\left[\frac{400}{300}\right] + 0.001R(400 - 300) \text{ J K}^{-1}.$$

$$= 0.8192\,R = 6.811 \text{ J K}^{-1}, \tag{18}$$

which is identical to the result obtained via Method 1 in Eq. 7. The two paths are illustrated in Figure 4.18.

For a related exercise, see Problem 4.18.

EXAMPLE 4.10

One mole of an ideal gas with $C_v^m = 1.5R$ has an initial state defined by $T_1 = 300$ K, $V_1 = 300A \exp[0.3]$ L, and $P_1 = R/(e^{0.3}A)$ atm. The gas is first heated to 400 K at constant volume, after which it is expanded isothermally against a constant external pressure $P_{\text{ext}} = R/Ae^{0.4}$ until the internal pressure and external pressures become equal. Compute w, q, ΔU, and ΔS for the process. Ignore acceleration effects in the expansion. (See the explanatory note in Example 2.3.)

Solution

The change in internal energy depends only upon the temperature, since we have an ideal gas. Therefore,

$$dU = n\, C_v^m\, dT = (1 \text{ mol})\, 1.5R\, dT. \tag{1}$$

Integrating both sides gives

$$\Delta U = n\, 1.5R \int_{300\,\text{K}}^{400\,\text{K}} dT = (1 \text{ mol})1.5R\,(400 - 300)\,\text{K} = 150(8.314)\,\text{J} = 1{,}247\,\text{J}. \tag{2}$$

No work is done in the heating, since $dV = 0$. The work done in the expansion is

$$w = \int \delta w = -\int_{V_1}^{V_2} P_{\text{ext}}\, dV = -P_{\text{ext}}\,(V_2 - V_1), \tag{3}$$

because P_{ext} is constant by the conditions of the process. Using the ideal-gas law, we compute

$$V_2 = \frac{RT_2}{P_2} = \frac{400R}{RA^{-1}e^{-0.4}} = 400Ae^{0.4}. \tag{4}$$

Substituting Eq. 4 into Eq. 3 gives

$$w = -\frac{R}{Ae^{0.4}}\left[400Ae^{0.4} - 300Ae^{0.3}\right] = -R\left[400 - 300e^{-0.1}\right]$$

$$= -(8.314\,\text{J mol}^{-1}\,\text{K}^{-1})(128.55\,\text{K}) = -1{,}069\,\text{J mol}^{-1} = -1{,}069\,\text{J}, \tag{5}$$

since we have 1 mole of gas. The total heat for the process is

$$q = \Delta U - w = 1{,}247 - (-1{,}069)\,\text{J} = 2{,}316\,\text{J}. \tag{6}$$

The initial and final states of the gas in this example are identical to those in Example 4.9. Consequently, the total entropy change is $\Delta S = 6.811\,\text{J K}^{-1}$, as it was in that example. The method of solution for ΔS is identical to that given there as well.

In this connection, it is important to note that the entropy change for the constant-volume heating process from 300 K to 400 K also is identical to that calculated in Example 4.9, viz.,

$$\Delta S_1 = \Delta S_{\text{heating}} = 3.588\,\text{J K}^{-1}. \tag{7}$$

The entropy change in the isothermal expansion is also the same as that computed via Method 1 in the same example:

$$\Delta S_2 = \Delta S_{\text{expansion}} = 3.223\,\text{J K}^{-1}. \tag{8}$$

However,

$$\Delta S_{\text{expansion}} \neq \frac{q_{\text{expansion}}}{T}. \tag{9}$$

The heat associated with the expansion is equal to $-w$, since ΔU for the expansion is zero because $dT = 0$. Therefore,

$$\frac{q_{\text{expansion}}}{T} = \frac{1{,}069\,\text{J}}{400\,\text{K}} = 2.672\,\text{J K}^{-1}. \tag{10}$$

> The result in Eq. 10 is not equal to ΔS_2, which is 3.223 J K^{-1}. The reason is that in this problem, the expansion is irreversible, not reversible, and $dS > \delta q/T$, as expected for an irreversible spontaneous process. The correct equation is $dS = \delta q_{rev}/T$. *The entropy change must be computed along a reversible path whether the actual process is reversible or not.* Failure to take this point into account is the source of many errors on the part of students taking thermodynamics courses.

4.10   Computation of $(\partial U/\partial V)_T$ and $(\partial H/\partial P)_T$ from the Equation of State

In Chapter 2, we demonstrated how the rate of change of the internal energy with volume at a constant temperature and the rate of change of enthalpy with pressure at $dT = 0$ could be obtained from a Joule–Thomson experiment. The results were

$$\left(\frac{\partial H}{\partial P}\right)_T = -C_p\mu \tag{4.85}$$

and

$$\left(\frac{\partial U}{\partial V}\right)_T = \frac{[\mu C_p + V]\left(\dfrac{\partial P}{\partial T}\right)_V}{\left(\dfrac{\partial V}{\partial T}\right)_P} - P = -[\mu C_p + V]\left(\frac{\partial P}{\partial V}\right)_T - P, \tag{4.86}$$

where μ is the Joule–Thomson coefficient. At the time, we noted that it would be much more convenient if μ, $(\partial H/\partial P)_T$, and $(\partial U/\partial V)_T$ could be computed directly from the equation of state. However, since we did not have the second law available at that point, the only option was to measure μ and use its value to obtain $(\partial H/\partial P)_T$ and $(\partial U/\partial V)_T$. Let us now see how the second law permits these quantities to be computed directly from the equation of state.

To obtain $(\partial H/\partial P)_T$, we begin with Eq. 4.82:

$$dS = \left[\frac{C_p}{T}\right]dT + \left\{\left(\frac{1}{T}\right)\left[\left(\frac{\partial H}{\partial P}\right)_T - V\right]\right\}dP.$$

The second law requires that dS be an exact differential. Therefore, the Euler condition for exactness must hold for that equation. That is,

$$\frac{\partial}{\partial P}\left[\frac{C_p}{T}\right]_T = \frac{\partial}{\partial T}\left[\left\{\left(\frac{1}{T}\right)\left[\left(\frac{\partial H}{\partial P}\right)_T - V\right]\right\}\right]_P. \tag{4.87}$$

This requirement will yield the desired equation for $(\partial H/\partial P)_T$. All we need to do is perform the indicated operations carefully. We do so in Problem 4.37, where the various parts of the problem serve as a procedural guide. In Chapter 5, we will introduce two new thermodynamic functions that will greatly simplify this derivation. Here we shall present only the final result.

When Eq. 4.87 is expanded in accordance with the procedure given in Problem 4.37, we obtain

$$\boxed{\left(\frac{\partial H}{\partial P}\right)_T = V - T\left(\frac{\partial V}{\partial T}\right)_P.} \tag{4.88}$$

Since only the variables P, V, and T appear on the right-hand side of Eq. 4.88, it may be evaluated directly from the equation of state. Using Eqs. 4.85 and 4.88, we also have

$$\mu = \frac{T\left(\dfrac{\partial V}{\partial T}\right)_P - V}{C_p}, \qquad (4.89)$$

so that the Joule–Thomson coefficient may be evaluated from the equation of state and the measured heat capacity at constant pressure.

We are now in a position to prove Joule's law, which states that μ for an ideal gas should be zero. Since $V = nRT/P$, we have

$$T\left(\frac{\partial V}{\partial T}\right)_P = T\left[\frac{nR}{P}\right] = V. \qquad (4.90)$$

Consequently, the Joule–Thomson coefficient obtained from Eq. 4.89 is

$$\mu = \frac{V - V}{C_p} = 0, \qquad (4.91)$$

which is Joule's law. We see, then, that Joule's law is merely a consequence of the second law and is not a separate law.

By starting with Eq. 4.79 and requiring that dS be exact, we can obtain $(\partial U/\partial V)_T$ in terms of the variables P, V, and T, which can be evaluated directly from the equation of state. This derivation is Problem 4.38. The result is

$$\boxed{\left(\frac{\partial U}{\partial V}\right)_T = T\left(\frac{\partial P}{\partial T}\right)_V - P.} \qquad (4.92)$$

Equation 4.92 will be formally derived in Chapter 5.

EXAMPLE 4.11

A van der Waals equation of state is being used to describe the behavior of $CO_2(g)$. **(A)** One mole of $CO_2(g)$ is expanded isothermally and reversibly at 300 K from volume V_1 to volume V_2. Obtain w, ΔU, q, and ΔS in terms of the van der Waals parameters a and b and the two volumes V_1 and V_2. **(B)** Using the a and b values given in Table 1.1, compute w, ΔU, q, and ΔS if $V_1 = 10$ L and $V_2 = 20$ L. **(C)** Obtain an expression for the change in internal energy for CO_2 when its volume is changed from 1.0 L to V liters at 300 K if its equation of state is a van der Waals equation whose parameters are given in Table 1.1. Plot $U(V) - U(1\text{ L})$, expressed in joules versus volume, at 300 K over the range $1\text{ L} \le V \le 100$ L. On the same graph, show the result that would be obtained if CO_2 were an ideal gas.

Solution

(A) Following our usual procedure, we begin by solving for the quantities whose differentials are exact. For a closed system,

$$dU = C_v dT + \left(\frac{\partial U}{\partial V}\right)_T dV. \qquad (1)$$

The first term is zero, since $dT = 0$. For an ideal gas, the second term is also zero, because $(\partial U/\partial V)_T = 0$. This, however, is not true for a van der Waals gas. Instead, Eq. 4.92 and Problem 4.38 show that

$$\left(\frac{\partial U}{\partial V}\right)_T = T\left(\frac{\partial P}{\partial T}\right)_V - P. \qquad (2)$$

Solving the van der Waals equation of state for P, we obtain

$$P = \frac{RT}{V - b} - \frac{a}{V^2} \tag{3}$$

for 1 mole of gas. Taking a temperature derivative at $dV = 0$ gives

$$\left(\frac{\partial P}{\partial T}\right)_V = \frac{R}{V - b}. \tag{4}$$

Substituting Eqs. 3 and 4 into Eq. 2 yields

$$\left(\frac{\partial U}{\partial V}\right)_T = \frac{RT}{V - b} - \left[\frac{RT}{V - b} - \frac{a}{V^2}\right] = \frac{a}{V^2}. \tag{5}$$

Therefore,

$$dU = \frac{a}{V^2}\,dV. \tag{6}$$

Integrating both sides between V_1 and V_2 gives

$$\Delta U = \int_{V_1}^{V_2} \frac{a}{V^2}\,dV = a\left[\frac{1}{V_1} - \frac{1}{V_2}\right]. \tag{7}$$

The expansion path is reversible, so the work is given by

$$w = -\int_{V_1}^{V_2} P\,dV = -\int_{V_1}^{V_2}\left[\frac{RT}{V - b} - \frac{a}{V^2}\right]dV$$

$$= -RT\ln\left[\frac{V_2 - b}{V_1 - b}\right] - a\left[\frac{1}{V_2} - \frac{1}{V_1}\right]. \tag{8}$$

The heat for the process can now be obtained from the first law:

$$q = q_{\text{rev}} = \Delta U - w = a\left[\frac{1}{V_1} - \frac{1}{V_2}\right] + RT\ln\left[\frac{V_2 - b}{V_1 - b}\right] + a\left[\frac{1}{V_2} - \frac{1}{V_1}\right]$$

$$= RT\ln\left[\frac{V_2 - b}{V_1 - b}\right]. \tag{9}$$

The change in entropy may be computed by either of two methods:

Method 1

The process is a change in volume at a constant temperature. Therefore, we need $(\partial S/\partial V)_T$. This quantity can be either obtained from Eq. 4.76 or derived quickly using

$$dU = T\,dS - P\,dV. \tag{10}$$

Dividing Eq. 10 by dV, followed by holding T constant, gives

$$\left(\frac{\partial U}{\partial V}\right)_T = T\left(\frac{\partial S}{\partial V}\right)_T - P. \tag{11}$$

Solving for $(\partial S/\partial V)_T$, we obtain

$$\left(\frac{\partial S}{\partial V}\right)_T = \frac{1}{T}\left[\left(\frac{\partial U}{\partial V}\right)_T + P\right]. \tag{12}$$

Substituting Eqs. 3 and 5 into Eq. 12 gives

$$\left(\frac{\partial S}{\partial V}\right)_T = \frac{1}{T}\left[\frac{a}{V^2} + \frac{RT}{V - b} - \frac{a}{V^2}\right] = \frac{R}{V - b}. \tag{13}$$

Separating variables, we may integrate Eq. 13 between V_1 and V_2 to obtain the total entropy change:

$$\Delta S = \int_{V_1}^{V_2} \frac{R}{V - b} \, dV = R \ln\left[\frac{V_2 - b}{V_1 - b}\right]. \tag{14}$$

Method 2

Since T is constant and may be factored from beneath the integral, we may use the basic definition of dS to compute

$$\Delta S = \int \frac{\delta q_{rev}}{T} = \frac{q_{rev}}{T}. \tag{15}$$

Because the process is reversible, the heat computed in Eq. 9 is the heat along a reversible path. Therefore,

$$\Delta S = \frac{1}{T} RT \ln\left[\frac{V_2 - b}{V_1 - b}\right] = R \ln\left[\frac{V_2 - b}{V_1 - b}\right], \tag{16}$$

which is identical to the result obtained in Eq. 14.

It is important to note that if the path had been irreversible, Method 1 would still give the correct answer, because Eq. 10 applies to a reversible path only. However, Method 2 would not work if the heat used is the actual q for the process, since that heat would be an irreversible one, not a reversible heat.

(B) Straightforward calculation using $a = 3.658 \text{ L}^2 \text{ bar mol}^{-2} = 3.610 \text{ L}^2 \text{ atm mol}^{-2}$ and $b = 0.04286 \text{ L mol}^{-1}$ yields

$$\Delta U = a\left[\frac{1}{V_1} - \frac{1}{V_2}\right] = 3.610 \text{ L}^2 \text{ atm mol}^{-2}\left[\frac{1}{10} - \frac{1}{20}\right] \text{L}^{-1} \text{ mol}$$

$$= 0.1805 \text{ L atm mol}^{-1} = 18.29 \text{ J mol}^{-1} \tag{17}$$

and

$$w = -RT \ln\left[\frac{V_2 - b}{V_1 - b}\right] - a\left[\frac{1}{V_2} - \frac{1}{V_1}\right]$$

$$= -(0.08206 \text{ L atm mol}^{-1} \text{ K}^{-1})(300 \text{ K}) \ln\left[\frac{20 - 0.04286}{10 - 0.04286}\right]$$

$$+ 3.610 \text{ L}^2 \text{ atm mol}^{-2}\left[\frac{1}{10} - \frac{1}{20}\right] \text{L}^{-1} \text{ mol}$$

$$= -17.117 \text{ L atm mol}^{-1} + 0.1805 \text{ L atm mol}^{-1}$$

$$= -16.94 \text{ L atm mol}^{-1} = -1{,}716 \text{ J mol}^{-1}. \tag{18}$$

Note that care must be taken here to ensure that both terms are in the same units. Also,

$$q = \Delta U - w = 18.29 \text{ J mol}^{-1} + 1{,}716 \text{ J mol}^{-1} = 1{,}734 \text{ J mol}^{-1} \tag{19}$$

and

$$\Delta S = R \ln\left[\frac{V_2 - b}{V_1 - b}\right] = (8.314 \text{ J mol}^{-1} \text{ K}^{-1}) \ln\left[\frac{20 - 0.04267}{10 - 0.04267}\right]$$

$$= 5.781 \text{ J mol}^{-1} \text{ K}^{-1}. \tag{20}$$

Since we have one mole of $CO_2(g)$,

$$\Delta S = 5.781 \text{ J K}^{-1}, \ \Delta U = 18.29 \text{ J} \ w = -1716 \text{ J, and } q = 1734 \text{ J}. \tag{21}$$

(C) Integrating Eq. 6 between the limits $V = 1$ L and V liters gives

$$\int_{U(1\,L)}^{U(V)} dU = U(V) - U(1\,L) = \int_1^V \frac{a\,dV}{V^2} = a\left[1 - \frac{1}{V}\right]. \tag{22}$$

Substituting the value of a produces

$$U(V) - U(1\,L) = 3.610\left[1 - \frac{1}{V}\right] \text{ L atm mol}^{-1}. \tag{23}$$

In joules, Eq. 23 becomes

$$U(V) - U(1\,L) = 365.8\left[1 - \frac{1}{V}\right] \text{ J mol}^{-1}. \tag{24}$$

If CO_2 were an ideal gas, we would have

$$U(V) - U(1\,L) = 0 \tag{25}$$

for all V, since $(\partial U/\partial V)_T = 0$ for an ideal gas.

The requested plot for 1 mole of CO_2 is shown in Figure 4.19.

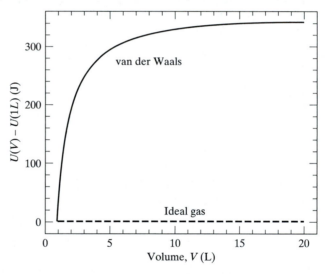

▲ FIGURE 4.19
Variation of the internal energy of one mole of $CO_2(g)$ with volume at 300 K when the equation of state is a van der Waals equation and when it is an ideal gas.

For related exercises, see Problems 4.19, 4.20, 4.22, 4.23, 4.24, and 4.26.

4.11 The Third Law of Thermodynamics

4.11.1 Absolute Entropies

Let us suppose that we wish to compute the entropy change when $O_2(g)$ is heated reversibly at a constant pressure of 1 atm from 0 K to 298.15 K. At first, we will be heating solid oxygen. At the melting point, 54.75 K, the solid undergoes a constant-temperature phase transition to the liquid state. After the solid melts, we will heat liquid oxygen from 54.75 K to the boiling point at 90.15 K. At this temperature, the liquid undergoes another phase transition to gaseous oxygen. The final stage of the process involves heating $O_2(g)$ from 90.15 K to 298.15 K. Figure 4.20 shows a flowchart of the process.

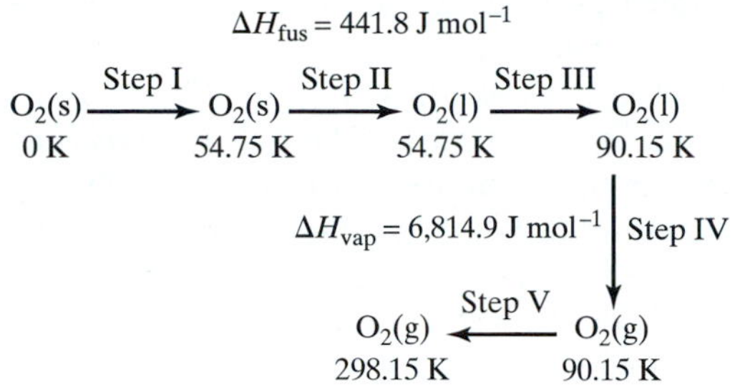

$$\Delta H_{fus} = 441.8 \text{ J mol}^{-1}$$

$$\underset{0 \text{ K}}{O_2(s)} \xrightarrow{\text{Step I}} \underset{54.75 \text{ K}}{O_2(s)} \xrightarrow{\text{Step II}} \underset{54.75 \text{ K}}{O_2(l)} \xrightarrow{\text{Step III}} \underset{90.15 \text{ K}}{O_2(l)}$$

$$\Delta H_{vap} = 6,814.9 \text{ J mol}^{-1} \Big| \text{ Step IV}$$

$$\underset{298.15 \text{ K}}{O_2(g)} \xleftarrow{\text{Step V}} \underset{90.15 \text{ K}}{O_2(g)}$$

◀ **FIGURE 4.20**
The constant-pressure heating of oxygen from 0 K to 298.15 K. Phase transitions to liquid and gas occur at 54.75 K and 90.15 K, respectively.

The total entropy change for the process is the sum of the entropy changes in each of the five steps shown in the figure:

$$\Delta S = \int_{S(0 \text{ K})}^{S(298.15 \text{ K})} dS = S(298.15 \text{ K}) - S(0 \text{ K})$$

$$= \Delta S_I + \Delta S_{II} + \Delta S_{III} + \Delta S_{IV} + \Delta S_V. \qquad (4.93)$$

The entropy changes for the constant-pressure heating, Steps I, III, and V, are given by Eq. 4.73. Applying that equation to the specific processes, we have

$$\Delta S_I = \int_{0 \text{ K}}^{54.75 \text{ K}} \frac{[C_p]_{(s)}}{T} dT,$$

$$\Delta S_{III} = \int_{54.75 \text{ K}}^{90.15 \text{ K}} \frac{[C_p]_{(l)}}{T} dT,$$

and

$$\Delta S_V = \int_{90.15 \text{ K}}^{298.15 \text{ K}} \frac{[C_p]_{(g)}}{T} dT.$$

The entropy changes for the constant-temperature phase transitions can be obtained from the basic definition of the entropy differential:

$$\Delta S_{II} = \int_{fusion} \frac{\delta q_{rev}}{T} = \frac{1}{T_m} \int_{fusion} \delta q_{rev} = \frac{q_{fusion}}{T_m} = \frac{\Delta \overline{H}_{fus}}{T_m}.$$

Here, T_m is the melting point. Since $dP = 0$, $q_{fusion} = \Delta \overline{H}_{fus}$. A similar equation holds for the second phase transition, except that now the temperature will be the boiling point T_b and q_{vap} will be $\Delta \overline{H}_{vap}$, where the subscript denotes the vaporization process:

$$\Delta S_{IV} = \int_{vap} \frac{\delta q_{rev}}{T} = \frac{1}{T_b} \int_{vap} \delta q_{rev} = \frac{q_{vap}}{T_b} = \frac{\Delta \overline{H}_{vap}}{T_b}.$$

Therefore, the desired total entropy change is

$$\Delta S = S(298.15 \text{ K}) - S(0 \text{ K}) = \int_{0 \text{ K}}^{54.75 \text{ K}} \frac{[C_p]_{(s)}}{T} dT + \frac{\Delta \overline{H}_{fus}}{T_m} + \int_{54.75 \text{ K}}^{90.15 \text{ K}} \frac{[C_p]_{(l)}}{T} dT$$

$$+ \frac{\Delta \overline{H}_{vap}}{T_b} + \int_{90.15 \text{ K}}^{298.15 \text{ K}} \frac{[C_p]_{(g)}}{T} dT. \qquad (4.94)$$

If the required heat capacities are available, Eq. 4.94 permits us to compute the entropy difference $S(298.15 \text{ K}) - S(0 \text{ K})$ for oxygen. Similar equations can be easily developed for other materials. As usual, all we can measure is the difference between quantities. In this case, however, the third law of thermodynamics permits us to obtain an absolute value for S(298.15 K) or for any other desired temperature.

The third law of thermodynamics states that the entropy of all perfect crystalline materials approaches the same constant as T approaches 0 K. This happy state of affairs permits us to choose a common reference point for entropy measurements of all substances. Since a perfect crystal has no disorder, Eq. 4.55 and the discussion in Section 4.7 suggest that the appropriate choice for the constant is zero. With this selection, Eq. 4.94 gives the absolute entropy of $O_2(g)$ at 298.15 K, rather than just the difference between the entropy at 298.15 K and 0 K.

The integrations in Eq. 4.94 can be executed by either fitting measured heat capacity data to a suitable function by least-squares methods or by straightforward numerical integration techniques. The integral required to compute ΔS_I in which the lower limit is 0 K generally requires extrapolating the heat capacity data to 0 K, since measurements at very low temperatures cannot be made. The extrapolation is done using the Debye heat capacity theory, which will be discussed in more detail in Chapter 7. This theory predicts that C_p and C_v are proportional to T^3 near $T = 0$ K. The procedure is, therefore, to fit the equation $C_v = aT^3$ to the heat capacity data at the lowest experimental temperature available and then use the equation to execute the required extrapolation to 0 K. At low temperatures, there is no measurable difference between C_v and C_p.

4.11.2   Entropy Changes in Chemical Reactions

The third law makes it possible, in principle, to obtain absolute entropies of any substance at a given temperature and pressure. With the use of such data, it is then straightforward to compute the entropy change for a chemical reaction. For example, consider the reaction

$$CO(g) + 0.5\,O_2(g) \longrightarrow CO_2(g)$$

at 298.15 K and 1 bar of pressure. In the initial state, we have 1 mole of CO(g) and $\frac{1}{2}$ mole of $O_2(g)$, each at 298.15 K and 1 bar of pressure. The total entropy for the system is, therefore, $\Delta S_{initial} = S^o_{CO(g)}(298.15\ \text{K}) + 0.5\,S^o_{O_2(g)}(298.15\ \text{K})$, where $S^o_{CO(g)}(298.15\ \text{K})$ and $S^o_{O_2(g)}(298.15\ \text{K})$ are the absolute entropies of CO(g) and $O_2(g)$, respectively, in their standard states at 298.15 K. After the reaction is complete, the system contains only 1 mole of $CO_2(g)$. Consequently, the entropy of the final state is $S^o_{CO_2(g)}(298.15\ \text{K})$. The entropy change for the reaction is

$$\Delta S = \Delta S_{final} - \Delta S_{initial} = S^o_{CO_2(g)}(298.15\ \text{K})$$

$$- [S^o_{CO(g)}(298.15\ \text{K}) + 0.5\,S^o_{O_2(g)}(298.15\ \text{K})].$$

Appendix A contains a table of selected S^o values for various compounds and elements at 298.15 K.

Let us now consider a general reaction in which there are J different reactants and K different products. We could write such a reaction in the form

$$\nu_1 R_1 + \nu_2 R_2 + \nu_3 R_3 + \cdots + \nu_J R_J \longrightarrow$$

$$\nu_1 P_1 + \nu_2 P_2 + \nu_3 P_3 + \cdots + \nu_K P_K,$$

where the ν_i are the stoichiometric coefficients in the balanced chemical reaction and R_i and P_i denote reactants and products, respectively. The entropy change for this reaction when all compounds are in their standard states is the sum of the absolute, standard-state entropies of the products minus the sum of the absolute, standard-state entropies of the reactants. That is,

$$\Delta S = \sum_{n=1}^{n=K} \nu_n S_n^o - \sum_{r=1}^{r=J} \nu_r S_r^o.$$ **(4.95)**

EXAMPLE 4.12

(A) Compute the entropy change for the following reaction, with all reactants and products in their standard states at 298.15 K:

$$C_2H_6(g) + 3.5\,O_2(g) \longrightarrow 2\,CO_2(g) + 3\,H_2O(l).$$

(B) Discuss the physical significance of the result in terms of order and disorder. **(C)** Does the negative result for ΔS mean that the combustion of ethane is nonspontaneous? Show that the reaction will be spontaneous.

Solution

(A) Using Eq. 4.95, we find that the entropy change for the combustion of ethane gas is

$$\Delta S = 3S_{H_2O(l)}^o + 2S_{CO_2(g)}^o - S_{C_2H_6(g)}^o - 3.5S_{O_2(g)}^o.$$ **(1)**

The standard absolute entropies for these compounds are given in Appendix A. With these values, we obtain

$$\Delta S = 3\,(69.91) + 2\,(213.74) - 229.60 - 3.5\,(205.138)\,\text{J K}^{-1}$$

$$= -310.37\,\text{J K}^{-1}.$$ **(2)**

(B) The decrease in entropy indicates that there is an increase in order in the reaction. This occurs because one of the products, $H_2O(l)$, is a condensed-phase liquid, whereas the reactants are all in the gas phase. Solids are more highly ordered than liquids, which in turn are more ordered than gases (see Appendix A for typical entropies of solids, liquids and gases). Consequently, when 4.5 moles of gaseous reactants are converted to 2 moles of gaseous products and 3 moles of liquid, a decrease in entropy is the expected result.

(C) To address the question of spontaneity, we must compute the total entropy change for the system plus the surroundings, as shown in Figure 4.21. The heat of

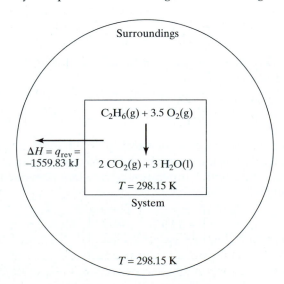

◀ FIGURE 4.21
The oxidation of 1 mole of ethane at 298.15 K with all compounds in their standard states releases 1,559.83 kJ of heat reversibly into the surroundings. To determine whether the process is spontaneous, the total entropy change for the system plus the surroundings must be computed. If $\Delta S_{total} > 0$, the overall process is spontaneous.

the combustion reaction is released into the surroundings. This heat is equal to ΔH for the reaction, since $dP = 0$. The enthalpy change is

$$\Delta H = 3\overline{H}^{\circ}_{H_2O\,(l)} + 2\overline{H}^{\circ}_{CO_2\,(g)} - \overline{H}^{\circ}_{C_2H_6\,(g)}$$

$$= 3\,(-285.83) + 2\,(-393.51) - (-84.68)\;kJ = -1{,}559.83\;kJ. \qquad (3)$$

This heat is added to the surroundings (therefore, relative to the surroundings, the heat change is positive) at $T = 298.15\;K$. Since the system is isothermal at that temperature, the heat transfer is reversible. Consequently, the entropy change for the surroundings is

$$\Delta S_{surroundings} = \frac{q_{rev}}{T} = \frac{1.55983 \times 10^6\;J}{298.15\;K} = 5{,}231.7\;J\,K^{-1}. \qquad (4)$$

The total entropy change for the process is

$$\Delta S_{total} = \Delta S_{system} + \Delta S_{surroundings} = -310.37 + 5{,}231.7\;J\,K^{-1}$$

$$= 4{,}921.3\;J\,K^{-1}. \qquad (5)$$

Since $\Delta S_{total} > 0$, the process is spontaneous, as expected.
For related exercises, see Problems 4.30, 4.31, and 4.32.

If the entropy change for a reaction is desired at some temperature other than 298.15 K, we need to know the rate of change of ΔS with temperature at constant pressure to make the calculation. The appropriate expression for this derivative may be obtained by differentiating Eq. 4.95:

$$\left(\frac{\partial \Delta S}{\partial T}\right)_P = \frac{\partial}{\partial T}\left[\sum_{n=1}^{n=K} \nu_n S^{\circ}_n - \sum_{r=1}^{r=J} \nu_r S^{\circ}_r\right]_P. \qquad (4.96)$$

Since the summation and differentiation operations commute (the derivative of a sum is the sum of the derivatives of the individual terms), Eq. 4.96 may be written in the form

$$\left(\frac{\partial \Delta S}{\partial T}\right)_P = \sum_{n=1}^{n=K} \nu_n \left[\frac{\partial S^{\circ}_n}{\partial T}\right]_P - \sum_{r=1}^{r=J} \nu_r \left[\frac{\partial S^{\circ}_r}{\partial T}\right]_P. \qquad (4.97)$$

The rate of change of entropy with temperature at constant pressure is given by Eq. 4.73. Applying that equation to Eq. 4.97 gives

$$\left(\frac{\partial \Delta S}{\partial T}\right)_P = \sum_{n=1}^{n=K} \nu_n \left[\frac{C^m_p}{T}\right]_n - \sum_{r=1}^{r=J} \nu_r \left[\frac{C^m_p}{T}\right]_r. \qquad (4.98)$$

We may put Eq. 4.98 in a more familiar form by factoring T^{-1} from the summations. This produces

$$\left(\frac{\partial \Delta S}{\partial T}\right)_P = \frac{1}{T}\left\{\sum_{n=1}^{n=K} \nu_n\,[C^m_p]_n - \sum_{r=1}^{r=J} \nu_r\,[C^m_p]_r\right\}. \qquad (4.99)$$

The quantity within the braces is identical to the quantity on the right-hand side of Eq. 3.22; we denote this quantity as ΔC_p for the reaction. Consequently, the rate of change of ΔS with temperature at constant pressure for a chemical reaction is given by

$$\boxed{\left(\frac{\partial \Delta S}{\partial T}\right)_P = \frac{\Delta C_p}{T}.} \qquad (4.100)$$

This is very similar to Eq. 4.73, which gives the rate of change of entropy with temperature at $dP = 0$, viz., $(\partial S/\partial T)_P = C_p/T$. Again, it would appear that Eq. 4.100 can be derived from Eq. 4.73 by multiplying both sides by Δ. But such a "derivation" is improper, although it does serve as a useful mnemonic device.

EXAMPLE 4.13

Use the data given in Appendix A and Example 4.12 to compute the entropy change for the reaction $C_2H_6(g) + 3.5\ O_2(g) \longrightarrow 2\ CO_2(g) + 3\ H_2O(l)$ at 400 K. Assume that the heat capacities are independent of temperature.

Solution

Using Eq. 4.100, we may separate the variables and integrate both sides between corresponding limits to obtain

$$\int_{\Delta S(298.15)}^{\Delta S(400\ \text{K})} d\Delta S = \int_{298.15\ \text{K}}^{400\ \text{K}} \frac{\Delta C_p}{T}\ dT. \tag{1}$$

If the heat capacities are constants, we have

$$\Delta S(400\ \text{K}) - \Delta S(298.15\ \text{K}) = \Delta C_p \ln\left[\frac{400}{298.15}\right] \tag{2}$$

ΔC_p is given by Eq. 4.99:

$$\Delta C_p = 3\ [C_p^m]_{H_2O(l)} + 2\ [C_p^m]_{CO_2(g)} - 3.5\ [C_p^m]_{O_2(g)} - [C_p^m]_{C_2H_6(g)}$$

$$= 3(75.291) + 2(37.11) - 3.5(29.355) - (52.63)\ \text{J K}^{-1} = 144.72\ \text{J K}^{-1}. \tag{3}$$

Inserting this value and the result from Example 4.12 into Eq. 2 gives

$$\Delta S(400\ \text{K}) = -310.37\ \text{J K}^{-1} + 144.72\ \ln\left(\frac{400}{298.15}\right) \text{J K}^{-1} = -267.84\ \text{J K}^{-1}. \tag{4}$$

For a related exercise, see Problem 4.32.

Summary: Key Concepts and Equations

1. Except for processes that violate conservation-of-energy requirements, the first law alone cannot predict whether, under given conditions, a particular process will be irreversible (spontaneous), reversible (at equilibrium), or nonspontaneous. However, the combination of the first and second laws permits such predictions to be made.

2. A Carnot engine is an idealized, perfect four-stroke engine comprising alternating isothermal and adiabatic expansions and compressions. The four strokes make up a cycle called the Carnot cycle. The engine has perfect valves and rings with no frictional energy losses. The efficiency of the engine, $\mathcal{E}$, is defined as the ratio of the magnitude of the work done to the heat added per cycle. For the Carnot engine, $\mathcal{E} = (T_1 - T_2)/T_1$, where T_1 and T_2 are the temperatures of the hot and cold heat reservoirs, respectively. Consequently, even though the engine is "perfect," heat cannot be completely converted into work unless $T_2 = 0$ K.

3. When a Carnot engine is run in reverse, it becomes a refrigerator or a heat pump whose coefficient of performance, defined as the ratio of the heat removed from the cold-temperature site to the magnitude of the work done, is $C = T_2/(T_1 - T_2)$.

4. The second law of thermodynamics states, "*It is impossible to construct a device that operates in cycles and that converts heat into work without producing some other change*

in the surroundings." Since the Carnot engine would violate this law if T_2 were to equal 0 K, it follows that a temperature of absolute zero must be unattainable.

5. The most important consequence of the second law is that the differential $\delta q_{rev}/T$ must be exact. Consequently, there must exist a function whose differential is exactly $\delta q_{rev}/T$. This function is labeled S and called the *entropy*. Therefore, the fundamental equation defining the entropy is $dS = \delta q_{rev}/T$. It is critical to note that the entropy must be calculated along a reversible path whether the actual process is reversible or not.

6. The Clausius inequality requires that $\int dS \geq \int \delta q/T$. If the process is reversible, the equality holds. If the process is irreversible or spontaneous, the inequality holds. If it is found that $\int dS < \int \delta q/T$, the process is nonspontaneous in that it will require work to make it occur.

7. If the changes in entropy of a system and its surroundings are added to obtain ΔS_{total}, we find that

   if $\Delta S_{total} = 0$,   the system is at equilibrium and the process is reversible,

   if $\Delta S_{total} > 0$,   the process is irreversible or spontaneous, and

   if $\Delta S_{total} < 0$,   the process is nonspontaneous.

8. There exists a direct connection between increases in entropy and increases in the disorder of a system. The connecting equation is $S = k_b \ln W$, where k_b is the Boltzmann constant and W is the thermodynamic probability that counts the number of ways the system may be attained.

9. The internal energy and enthalpy differentials may be expressed in terms of the entropy to yield two of the most important equations in thermodynamics. For closed systems and reversible processes, these two equations are

$$dU = T\,dS - P\,dV$$

and

$$dH = T\,dS + V\,dP.$$

10. Using the two equations given in item 9, we may easily develop equations giving the rates of change of entropy with temperature, pressure, and volume. The results are

$$\left(\frac{\partial S}{\partial T}\right)_V = \frac{C_v}{T}, \qquad \left(\frac{\partial S}{\partial T}\right)_P = \frac{C_p}{T},$$

$$\left(\frac{\partial S}{\partial V}\right)_T = \frac{1}{T}\left[\left(\frac{\partial U}{\partial V}\right)_T + P\right],$$

and

$$\left(\frac{\partial S}{\partial P}\right)_T = \frac{1}{T}\left[\left(\frac{\partial H}{\partial P}\right)_T - V\right].$$

11. Using the total differential for dS expressed as a function of T and V and dS expressed as a function of T and P, we may obtain expressions for $(\partial U/\partial V)_T$ and $(\partial H/\partial P)_T$ in terms of the variables P, V, and T that can be evaluated from the equation of state. This is achieved by requiring that the Euler condition for exactness be satisfied for the dS differentials. The two results are

$$\left(\frac{\partial H}{\partial P}\right)_T = V - T\left(\frac{\partial V}{\partial T}\right)_P$$

and

$$\left(\frac{\partial U}{\partial V}\right)_T = T\left(\frac{\partial P}{\partial T}\right)_V - P.$$

12. The third law of thermodynamics states, *"The entropy of all perfect crystalline materials approaches the same constant as T approaches 0 K."* This permits us to choose a common reference point for the entropy at $T = 0\ K$. The connection between disorder and entropy leads us to choose $S(0\ K) = 0$ for all materials.

13. The entropy change for a chemical reaction when all compounds are in their standard states can be obtained by summing the standard-state absolute entropies of the products and subtracting the standard-state absolute entropies of the reactants. That is,

$$\Delta S = \sum_{n=1}^{n=K} \nu_n S_n^o - \sum_{r=1}^{r=J} \nu_r S_r^o,$$

where the ν_i are the stoichiometric coefficients in the balanced chemical equation and S_i^o is the standard-state absolute entropy of substance i.

14. The rate of change of the entropy of a reaction with temperature at a constant pressure is obtained by differentiating the equation in item 13. The result is

$$\left(\frac{\partial \Delta S}{\partial T}\right)_P = \frac{\Delta C_p}{T},$$

where ΔC_p is as defined in Chapter 3.

Problems

Problems that require the use of some type of computational device are marked with an asterisk (*). Problems that require some type of plotting routine are indicated with a pound sign (#).

4.1 A Carnot engine is being used to provide the work needed to compress a large cylinder of gas. It is determined that 2,000 kJ of work will be required to do the job.

(A) If the high-temperature heat reservoir of the engine is at 500 K and heat is discharged at 300 K, how much heat must be added to the engine to obtain the 2,000 kJ of work?

(B) How much methane must be burned in $O_2(g)$ to produce $CO_2(g)$ and $H_2O(g)$ in order to obtain the equivalent quantity of heat? Ignore the variation of the heat of reaction with temperature.

4.2 A Carnot engine operates between $T_1 = 500$ K and $T_2 = 300$ K. The engine contains 100 moles of ideal gas as working material. The initial volume of the gas before the execution of Stroke 1 is 410.3 L. After the first expansion stroke, the volume is 1641.2 L.

(A) How much work per cycle is done by the engine?

(B) If the molar heat capacity of the gas is $C_v^m = 1.5R$, what is the volume of the gas after the adiabatic Stroke 2 of the engine?

(C) What is the volume of the gas after the completion of Stroke 3?

4.3 We have shown that $(\partial H/\partial P)_T = V - T(\partial V/\partial T)_P$ and that $(\partial U/\partial V)_T = T(\partial P/\partial T)_V - P$. Analyze the units on the left and right sides of these two equations, and show that the units are in agreement.

4.4 A Carnot engine is observed to produce 4.5 kJ of net useful work for every 17 kJ of heat added to the engine from the high-temperature reservoir. If the isothermal expansion stroke is executed at 550 K, calculate the temperature of the low-temperature heat reservoir.

4.5 The coefficient of performance of a Carnot refrigerator is $C = T_2/(T_1 - T_2)$.

(A) Derive an expression for the total differential of C that shows the rate at which the coefficient of performance changes as we vary T_1 and T_2.

(B) C increases if T_2 is raised or if T_1 is lowered. Using the result obtained in (A), predict whether C increases faster as T_2 is raised or as T_1 is lowered.

4.6 A family wishes to purchase a heat pump for its home. The Get-It-Cheap Company sells a heat pump that has a coefficient of performance 30% that of an ideal Carnot heat pump. The price is $500. For $950, the Top-of-the-Line Company sells a heat pump whose coefficient of performance is 55% that of an ideal Carnot heat pump. The family wishes to maintain their home at 298 K with an average winter outside temperature of 270 K. The family estimates that it will take an average of 2.2×10^7 kJ of energy per month to maintain the house at this temperature for five cold months. The cost of electrical energy is

$0.0714 per kilowatt-hour (kWh) for the first 600 kWh and $ 0.0410 for each kWh thereafter, figured on a monthly basis.

(A) Compute the coefficients of performance for both heat pumps.

(B) Compute the cost of running each heat pump for the five cold months.

(C) How long will it take for the extra efficiency of the Top-of-the-Line heat pump to make up the cost differential between the pumps? (*Note:* One watt is 1 joule per second. A kilowatt-hour is a power of 10^3 watts used for 1 hour.)

4.7 Refrigeration capacity is usually given in units of tons. One ton of refrigeration capacity is defined to be a rate of energy flow equal to the rate of extraction of the heat of fusion when 1 short ton (2,000 pounds) of ice is produced from water at the same temperature in 24 hours. (Isn't the ton a wonderful unit!) The ton is equal to 200 British thermal units (BTUs) per minute. It is also equal to 3516.85 watts. A family purchases a 5-ton air-conditioning unit for its home.

(A) If the unit has a coefficient of performance of 3.75 under the conditions in which it will be used, how much heat will the family be able to remove from the house each day with this air conditioner?

(B) If the cost of electrical energy is $0.0714 per kilowatt-hour (kWh) for the first 600 kWh and $ 0.0410 for each kWh thereafter, figured on a monthly basis, what will it cost the family to operate the 5-ton unit 12 hours per day for a 31-day month? (See Problem 4.6 for a definition of kWh.).

4.8 A Carnot heat pump with a low-temperature site at 300 K is used to heat a container whose total heat capacity is constant at $1,800 \text{ J K}^{-1}$. If the container is initially at 300 K, how much work must be done to heat the container to 330 K? Assume that the low-temperature heat reservoir is constant at 300 K.

4.9 A Carnot engine's efficiency is found to be numerically equal to its coefficient of performance when the engine is reversed and operated as a refrigerator. What is the engine's efficiency?

4.10 Consider two heat reservoirs, one of finite heat capacity C_1 at an initial temperature T_1 and the other having infinite heat capacity at a lower temperature T_2. A Carnot engine operates between these two reservoirs, with heat flowing into the engine from the hot reservoir at T_1 and being discharged into the colder reservoir at T_2. As the engine runs, the temperature of the hot reservoir decreases because heat is being removed and the reservoir has a finite heat capacity. However, the temperature of the cold reservoir remains constant at T_2 because its heat capacity is infinite.

(A) Obtain an expression giving the total work obtainable from this system in terms of T_1, T_2, and C_1. This situation is similar to that which would exist between a boiler containing a finite amount of thermal energy and the atmosphere, which is so extensive that its heat capacity is essentially infinite.

(B) If $T_1 = 650 \text{ K}$, $T_2 = 298 \text{ K}$, and $C_1 = 10^6 \text{ J K}^{-1}$, compute the total work that the engine will be able to produce. Assume that C_1 is independent of temperature.

4.11 Consider a Carnot heat engine operating between finite energy reservoirs at initial temperatures T_1^o and T_2^o with $T_1^o > T_2^o$. As the engine runs, T_1 decreases and T_2 increases until both reservoirs reach some final common temperature T_f.

(A) If reservoirs 1 and 2 have total heat capacities C_1 and C_2, respectively, which are constant, obtain expressions in terms of C_1, C_2, T_1^o, T_2^o, and T_f giving the total amount of work produced by the engine.

(B) If $C_1 = C_2$, $T_1^o = 650 \text{ K}$, and $T_2^o = 298 \text{ K}$, compute T_f.

4.12 A reversible heat engine uses a three-step cycle consisting of an isothermal expansion at temperature T_1, a constant-volume cooling to temperature T_2, and an adiabatic compression back to the initial state. The process is illustrated in Figure 4.22 on the next page.

(A) In which stroke is the heat for the engine added? Which stroke produces the change in the surroundings required by the second law?

(B) If 1 mole of an ideal gas is used as the working material, obtain an expression for the total work done in one cycle of the engine in terms of volumes, temperatures, and a constant molar heat capacity of the gas, C_v^m.

(C) Show that the efficiency of the engine is given by

$$\mathcal{E} = 1 - \frac{T_1 - T_2}{T_1 \ln[T_1/T_2]}.$$

(D) Show that the efficiency of the engine goes to zero as T_2 approaches T_1.

4.13 A three-stroke, reversible engine like that described in Problem 4.12 contains 10 moles of an ideal gas as a working material. The initial conditions are $T_1 = 500 \text{ K}$ and $V_1 = 20 \text{ L}$. At the end of Stroke 1, the volume, V_2, is 50 L. If C_v^m for the gas is $1.5R$, using the results of Problem 4.12, compute the lower temperature T_2 and the engine's efficiency.

4.14 A reversible, four-stroke engine uses 1 mole of an ideal gas as a working material. The four strokes are as follows:

Stroke 1: Constant-pressure expansion from (P_1, V_1, T_1) to (P_1, V_2, T_2).

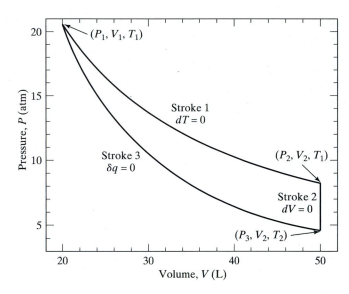

▲ FIGURE 4.22
Three-stroke reversible engine.

Stroke 2: Constant-volume cooling from (P_1, V_2, T_2) to (P_2, V_2, T_3).

Stroke 3: Constant-pressure compression from (P_2, V_2, T_3) to (P_2, V_1, T_4).

Stroke 4: Constant-volume heating from (P_2, V_1, T_4) to (P_1, V_1, T_1).

If C_v^m and C_p^m are $1.5R$ and $2.5R$, respectively,

(A) Sketch a pressure–volume diagram of the operation of the four strokes of the engine.

(B) Obtain an expression for the net work per cycle for the engine in terms of the various temperatures.

(C) Obtain an expression for the total heat added to the engine in one cycle as a function of the various temperatures.

(D) Obtain an expression for the engine efficiency in terms of the temperatures.

(E) If $V_2 = 2V_1$ and $P_2 = 0.5P_1$, compute the engine's efficiency.

4.15 One mole of $CO_2(g)$ is held on the left side of an insulated container similar to the one shown in Figure 4.15. The temperature of the gas is 300 K. The initial volume is 20 L. The dividing partition is removed, and the gas is allowed to expand into the vacuum on the right-hand side such that the final volume is 40 L. If the insulation makes the process adiabatic, compute the final temperature of the gas if

(A) $CO_2(g)$ is an ideal gas and

(B) $CO_2(g)$ obeys a van der Waals equation of state with the parameters being those given in Table 1.1. C_v^m for $CO_2(g)$ is 28.80 J mol^{-1} K^{-1}.

4.16 Sam is still having great difficulty with physical chemistry. However, he has found a problem he can solve:

"An ideal gas is expanded isothermally and adiabatically into a vacuum to double its volume from 20 L to 40 L at 300 K. Compute ΔS for the process." Noting that q for the adiabatic process is zero and that the process is isothermal, Sam computes

$$\Delta S = \int \frac{\delta q}{T} = T^{-1} \int \delta q = \frac{q}{T} = 0.$$

Has Sam finally gotten a problem correct? Explain. Leigh, who is making an A in physical chemistry, obtains a different answer for this problem. What answer does she obtain?

4.17 Two moles of an ideal gas at 300 K and a volume of 35 L expand isothermally and irreversibly against a constantly varying external pressure until the final volume is 90 L. Compute, if possible, ΔU, ΔH, q, w, and ΔS for the process.

4.18 An ideal gas with $C_v^m = 1.5R$ is heated along a path for which the molar heat capacity is $3.5R$. One mole of the gas with an initial volume of 20 L is heated reversibly from 300 K to 500 K along this path.

(A) Compute the final volume of the gas.

(B) Compute ΔS for the process.

4.19 n moles of a gas that obeys a van der Waals equation of state are contained in an insulated piston–cylinder arrangement. The initial state of the gas is $T = T_o$ and $V = V_o$. A reversible adiabatic expansion of the gas is carried out until the volume reaches $V = 2V_o$.

(A) Obtain the final temperature T_f of the gas as a function of n, C_v^m, T_o, V_o, and the van der Waals parameters a and b. Assume that C_v^m is independent of temperature.

(B) If $n = 2 \, \text{mol}$, $C_v^m = 28.80 \, \text{J mol}^{-1} \text{K}^{-1}$, $T_o = 350 \, \text{K}$, and $V_o = 40 \, \text{L}$, compute T_f if the gas is $CO_2(g)$.

4.20 n moles of a gas that obeys a van der Waals' equation of state are contained in an insulated piston–cylinder arrangement. The initial state of the gas is $T = T_o$ and $V = V_o$. An irreversible adiabatic expansion of the gas against a constant external pressure equal to P_{ext} is carried out until the volume of the gas reaches $2V_o$.

(A) Obtain the final temperature T_f of the gas as a function of n, C_v^m, T_o, V_o, P_{ext}, and the van der Waals parameters a and b. Assume that C_v^m is independent of temperature.

4.21 (A) Using the formula $dU = T \, dS - P \, dV$, show that, for a closed system along a reversible path, we must have

$$\left(\frac{\partial T}{\partial V}\right)_S = -\left(\frac{\partial P}{\partial S}\right)_V.$$

(B) Using the formula $dH = T \, dS + V \, dP$, show that, for a closed system along a reversible path, we must have

$$\left(\frac{\partial T}{\partial P}\right)_S = \left(\frac{\partial V}{\partial S}\right)_P.$$

The equations derived in (A) and (B), along with two others, constitute Maxwell's relationships.

4.22 One mole of a gas is described by a virial-type equation

$$P = RT\left[\frac{1}{V} + \frac{C_2}{V^2}\right],$$

where C_2 is the second virial coefficient. The gas is expanded isothermally and reversibly at 300 K from 20 L to 40 L.

(A) Obtain ΔU for the process as a function of C_2 if C_2 is a constant. If the expansion were done irreversibly, how would the result be altered?

(B) If $C_2 = a + bT + cT^2$ is a quadratic function of temperature, obtain ΔU as a function of b and c for the isothermal expansion described in (A).

4.23 A nonideal gas obeys the equation of state $PV = nRT - aP/T$, where a is a positive constant. Obtain an expression for the Joule–Thomson coefficient for this gas in terms of the constant a and the heat capacity of the gas. Does the temperature of the gas increase or decrease in a Joule–Thomson experiment?

4.24 Three moles of a nonideal gas whose equation of state is

$$PV = nRT + nBP,$$

where $B = 30 \, \text{cm}^3 \, \text{mol}^{-1}$, undergoes an irreversible change from ($T = 600 \, \text{K}$, $P = 10 \, \text{atm}$) to ($T = 300 \, \text{K}$, $P = 5 \, \text{atm}$). The constant-pressure molar heat capacity of the gas is given by

$$C_p^m = 54.041 + 9.561 \times 10^{-3}T.$$

Calculate ΔS, ΔH, and ΔU for the process.

4.25 Consider the change in entropy attending the heating of a substance at $dP = 0$ from T_1 to T_2 when $C_p^m = a + bT + cT^2$. What average constant value must the heat capacity have over the same temperature range to give the same entropy change? Express your answer in terms of a, b, c, T_1, and T_2.

4.26 One mole of a nonideal gas obeys the equation of state

$$V = RT\left[\frac{1}{P} + \frac{a}{P^2}\right],$$

where a is a function of temperature only. If the gas is compressed from pressure P_1 to pressure P_2 isothermally, obtain an expression giving the entropy change for the process in terms of P_1, P_2, a, and da/dT.

4.27 One mole of an ideal gas is heated reversibly along a path such that $T = AV^2$, where A is a constant. If the initial temperature is 273 K, what must the final temperature be if the entropy change is equal to 20.785 J K^{-1}? $C_v^m = 1.5R$ J mol^{-1} K^{-1} for the gas.

4.28 The entropy of a particular nonideal gas at 1 bar of pressure is experimentally determined to vary with temperature according to the equation

$$S(T) = S(300 \, \text{K}) + a \ln(T) + bT,$$

where $S(300 \, \text{K})$ is the entropy of the gas at 300 K and a and b are constants. Determine the heat capacity of the gas as a function of T and other constants.

4.29* The heat capacities of Mg between 12 K and 298.15 K are given in the table on the next page.

(A) Plot C_p^m as a function of T. Then, plot C_p^m/T vs. T. Notice that C_p^m approaches $3R$ as T increases. We shall see in a later chapter that this is the limiting value of the heat capacity for an elemental solid.

(B) Assuming the T^3 law to be valid below 12 K, and neglecting the difference between C_p and C_v at these low temperatures, calculate the absolute entropy of Mg at 298.15 K. How is this result related to the plot of C_p^m/T versus T? Compare the result with the data given in Appendix A.

(C) Calculate the entropy of Mg at 550 K. State the approximations you have made. The normal melting point of Mg is about 923 K.

4.30 Compute the entropy change in the following reactions if each occurs at 298.15 K and 1 bar of pressure:

T (K)	C_p^m (cal mol^{-1} K^{-1})	T (K)	C_p^m (cal mol^{-1} K^{-1})
12	0.016	110	4.052
14	0.026	120	4.307
16	0.042	130	4.527
18	0.065	140	4.718
20	0.086	150	4.876
25	0.188	160	5.013
30	0.341	170	5.133
35	0.550	180	5.236
40	0.803	190	5.331
45	1.076	200	5.418
50	1.367	210	5.487
60	1.953	220	5.550
70	2.498	230	5.611
80	2.981	240	5.667
90	3.404	250	5.719
100	3.753	260	5.766
		270	5.811
		280	5.853
		290	5.896
		298.15	5.929

(A) $CaCO_3(s) \longrightarrow CaO(s) + CO_2(g)$
(calcite)

(B) $H_2(g) + Cl_2(g) \longrightarrow 2\,HCl(g)$

(C) $C_4H_{10}(g) + 6.5\,O_2(g) \longrightarrow 4\,CO_2(g) + 5\,H_2O(g)$
(*n*-butane)

Discuss the results in terms of order–disorder changes.

4.31 (A) Using the data in Appendix A, compute ΔS^o for the reaction

$$2\,NO(g) \longrightarrow N_2O_4(g)$$

at 298.15 K and 1 bar of pressure.

(B) Discuss the result in terms of order–disorder concepts.

(C) By computing the total entropy change, determine whether this reaction will be spontaneous at 298.15 K and 1 bar of pressure.

4.32 Consider again the reaction $2\,NO(g) \longrightarrow N_2O_4(g)$, examined in Problem 4.31. In that problem, we found that this reaction is spontaneous under standard thermochemical conditions. In this problem, we will determine the temperature at which the reaction is at equilibrium.

(A) Using the data in Appendix A, obtain ΔS^o for the reaction as a function of temperature. Assume

that the heat capacities are independent of temperature.

(B) Using the data in Appendix A, obtain ΔH^o for the reaction as a function of temperature. Assume the heat capacities to be constants.

(C) Using the result obtained in (B), obtain $\Delta S_{\text{surroundings}}$ for the reaction as a function of T.

(D) Using the results of (A) and (C), obtain ΔS_{total} as a function of T.

(E) Compute the temperature at which the reaction will be at equilibrium.

4.33 A gas obeys the Berthelot equation of state given in Table 1.2. A particular nonideal gas has the Berthelot parameters $a = 1{,}639$ L^2 atm mol^{-2} K and $b = 0.03049$ L mol^{-1}. One mole of this gas is isothermally expanded from 20 L to 50 L at 300 K.

(A) Compute ΔU and ΔS for the process.

(B) If the gas is ideal, compute ΔU and ΔS.

4.34 Why aren't explosions reversible? Why, for example, couldn't the effects produced when the nitrate-laden freighter *Grandcamp* vanished in a thunderclap at 9:13 A.M., April 16, 1947, obliterating its crew and 200 bystanders, flinging a steel barge 50 yards inland, knocking down two light planes circling overhead, breaking windows in every house in the neighboring town of Texas City, and destroying a multimillion dollar chemical plant in an inferno, be reversed in every respect? Why couldn't the atoms and molecules in the bits and pieces of this explosion, in the twisted wreckage, in the dismembered bodies, in the billowing smoke, and in the raging fires spontaneously reassemble themselves into the original objects? [From Henry Bent's, *The Second Law* (Oxford University Press, New York, 1965), Library of Congress Number 65-15608.]

4.35 Knowledge frequently permits you to perform tasks that you might think would be impossible. For example, Clausius stated the first and second laws of thermodynamics succinctly as follows:

Die Energie der Welt ist konstant.

Die Entropie der Welt strebt einem maximum zu.

The German word *Welt* means *world*. Regardless of whether you know any other German words or not, translate Clausius's statements. (Thanks to Fredrick L. Minn, M.D., Ph.D. for this contribution.)

4.36 *Humpty Dumpty sat on a wall.*

Humpty Dumpty had a great fall.

All the king's horses and all the king's men

Couldn't put Humpty together again.

What simple mathematical thermodynamic expression best sums up the situation in this nursery rhyme? (Thanks to Fredrick L. Minn, M.D., Ph.D. for this contribution.)

4.37 In this problem, you will derive Eq. 4.88, starting with Eq. 4.87 and the knowledge that the second law requires that dS be an exact differential.

(A) Rewrite the left-hand side of Eq. 4.87, using the fact that $C_p = (\partial H/\partial T)_P$.

(B) Show that the left-hand side of Eq. 4.87 is given by

$$\frac{\partial}{\partial P}\left[\frac{C_p}{T}\right]_T = \frac{1}{T}\frac{\partial^2 H}{\partial P \partial T}.$$

(C) Show that the right-hand side of Eq. 4.87, when expanded using the standard rules for taking derivatives, is given by

$$\frac{\partial}{\partial T}\left[\left\{\left(\frac{1}{T}\right)\left[\left(\frac{\partial H}{\partial P}\right)_T - V\right]\right\}\right]_P$$

$$= -\frac{1}{T^2}\left[\left(\frac{\partial H}{\partial P}\right)_T - V\right] + \frac{1}{T}\left[\frac{\partial^2 H}{\partial T \partial P} - \left(\frac{\partial V}{\partial T}\right)_P\right].$$

(D) By combining the results obtained in (B) and (C), show that we must have

$$-\frac{1}{T^2}\left[\left(\frac{\partial H}{\partial P}\right)_T - V\right] - \frac{1}{T}\left(\frac{\partial V}{\partial T}\right)_P = 0.$$

(*Hint:* Remember that the order of differentiation in a second derivative does not alter the value of the derivative.)

(E) Use the result obtained in (D) to show that Eq. 4.88 is valid. This completes the derivation.

4.38 In this problem, you will derive Eq. 4.92, starting with Eq. 4.79 and the knowledge that the second law requires that dS be an exact differential.

(A) Use the Euler condition for exactness of dS to obtain an equation analogous to Eq. 4.87.

(B) Rewrite the left-hand side of the equation obtained in (A), using the fact that $C_v = (\partial U/\partial T)_V$.

(C) Show that the left-hand side of the equation obtained in (A) is given by

$$\frac{\partial}{\partial V}\left[\frac{C_v}{T}\right]_T = \frac{1}{T}\frac{\partial^2 V}{\partial V \partial T}.$$

(D) Show that the right-hand side of the equation obtained in (A), when expanded using the standard rules for taking derivatives, is given by

$$\frac{\partial}{\partial T}\left[\left\{\left(\frac{1}{T}\right)\left[\left(\frac{\partial U}{\partial V}\right)_T + P\right]\right\}\right]_V$$

$$= -\frac{1}{T^2}\left[\left(\frac{\partial U}{\partial V}\right)_T + P\right] + \frac{1}{T}\left[\frac{\partial^2 U}{\partial T \partial V} + \left(\frac{\partial P}{\partial T}\right)_V\right].$$

(E) By combining the results obtained in (C) and (D), show that we must have

$$-\frac{1}{T^2}\left[\left(\frac{\partial U}{\partial V}\right)_T + P\right] + \frac{1}{T}\left(\frac{\partial P}{\partial T}\right)_V = 0.$$

(*Hint:* Remember that the order of differentiation in a second derivative does not alter the value of the derivative.)

(F) Use the result obtained in (E) to show that Eq. 4.92 is valid. This completes the derivation.

4.39 Sam is desperate. His grade in physical chemistry has fallen below the limits of detectability. However, he still feels that his creativity should compensate for his inability to take examinations. After thinking deeply about the philosophical implications of the second law of thermodynamics, he presents the following position paper to his professor:

The second law requires the total entropy change for any spontaneous process to be positive. Therefore, since the universe progresses spontaneously from one day to the next, we must have $\Delta S_{universe} = S_{universe}(\text{day } n + 1) - S_{universe}(\text{day } n) > 0$. That is, if we were to compute the total entropy of the universe on a given day, it would necessarily be greater than the total entropy of the universe on the previous day. Consequently, a plot of $S_{universe}$ versus time must be a montonically increasing function. Such a plot might have the form given in Figure 4.23. Since the combination of the third law and the relationship between entropy and disorder requires that $S_{universe} \geq 0$, we know that the plot can never assume a negative value. Hence, if we extrapolate the curve shown in the figure to the point at which $S_{universe} = 0$, we must have the earliest point in time at which the universe could have been created. Of course, the creation point could have been later, because we do not know whether there was perfect order at the point of creation. The total entropy then might have been greater than zero. However, creation cannot have occurred at an earlier date. We have, therefore, determined an upper limit for the age of the universe. Of course, there are technical difficulties in computing the total entropy of the universe, but such difficulties have nothing to do with the fundamental theory.

Comment on Sam's position paper. Does he have a valid point? Are there any flaws in his analysis? Do you think Sam's professor should give his grade special consideration?

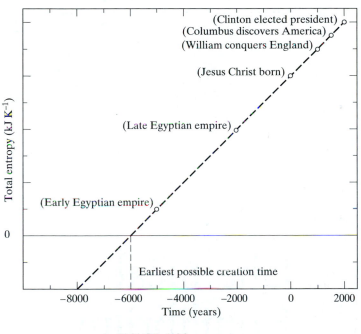

▲ FIGURE 4.23

Entropy of the universe vs. time.

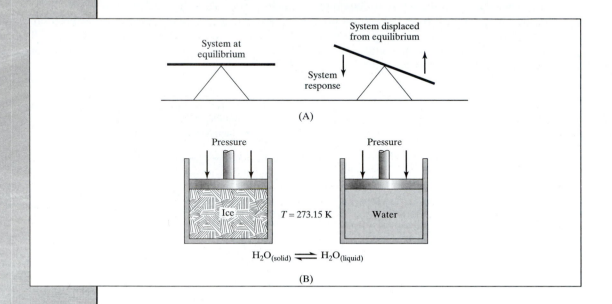

CHAPTER 5

System at equilibrium

System displaced from equilibrium

System response

(A)

Pressure

Pressure

Ice $T = 273.15$ K Water

$H_2O_{(solid)} \rightleftharpoons H_2O_{(liquid)}$

(B)

Chemical Equilibrium

The presentation of the first, second, and third Laws lays the complete foundation of classical thermodynamics. Two major tasks remain: First, we need to recast the combination of the first and second laws into forms that are easier to use; second, we need to apply thermodynamics to a variety of systems that are of importance to chemists and chemical engineers. In this chapter, we shall demonstrate that the introduction of two new thermodynamic functions, the Gibbs free energy and the Helmholtz free energy, makes the application of thermodynamics much easier. We then address the general topic of chemical equilibrium.

5.1 Reversible and Irreversible Processes Revisited

In Chapter 4, we developed two general criteria that permit us to determine whether a process is reversible, irreversible, or nonspontaneous, where "reversible" and "irreversible" are thermodynamic nomenclature for equilibrium and spontaneous processes, respectively. The first of these criteria is the Clausius inequality. If it is found that $\Delta S > \int \delta q/T$ for a process, then the process is irreversible. If the two quantities are equal, then the system is in equilibrium, and we are dealing with a hypothetical, reversible process. If we find that $\Delta S < \int \delta q/T$, then the process is nonspontaneous and will require that work be done in order to force it to occur. While the Clausius inequality is general and exact, it is often difficult to use, because the computation of the integral containing the inexact differential $\delta q/T$ requires that the complete details of the path for the process be known. In most cases, it is easier to use the second criterion: Irreversible processes have $\Delta S_{total} > 0$, reversible processes have $\Delta S_{total} = 0$, and nonspontaneous processes have $\Delta S_{total} < 0$. Since dS is exact, it is usually easier to apply this second method than the Clausius inequality. Both criteria are illustrated in Figure 5.1.

It is instructive to examine carefully how we apply the second criterion to judge the spontaneity or nonspontaneity of a chemical reaction. Example 4.12 and Problems 4.31 and 4.32 are specific illustrations of the method we consider here in more general terms. The entropy change for the system in the reaction

$$\nu_1 R_1 + \nu_2 R_2 + \nu_3 R_3 + \cdots + \nu_J R_J \longrightarrow \nu_1 P_1 + \nu_2 P_2 + \nu_3 P_3 + \cdots + \nu_K P_K,$$

occurring with all compounds in their standard states, is given by Eq. 4.95, viz.,

$$\Delta S_{system} = \sum_{n=1}^{n=K} \nu_n S_n^o - \sum_{r=1}^{r=J} \nu_r S_r^o, \tag{5.1}$$

where the first summation is the total entropy of the products and the second is the total entropy of the reactants in their respective standard states at the

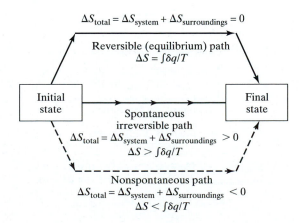

FIGURE 5.1
Thermodynamic criteria involving entropy changes that determine whether a process is irreversible (spontaneous), nonspontaneous, or reversible (at equilibrium).

temperature of the reaction. If the pressure differs from 1 bar, Eq. 4.81 must be used to convert the absolute entropies to the pressure of the reaction.

To determine whether the reaction is spontaneous under the given conditions, we must compute the entropy change of the surroundings. This requires that we first compute the enthalpy change for the reaction, which is

$$\Delta H^o = \sum_{n=1}^{n=K} \nu_n \overline{H}_n^o - \sum_{r=1}^{r=J} \nu_r \overline{H}_r^o. \tag{5.2}$$

If we do not have a pressure of 1 bar, it will, in principle, be necessary to convert ΔH^o to its appropriate value at the pressure of the reaction; for that purpose, we use Eq. 4.85 or 4.88. However, unless highly accurate experimental work is involved, these corrections can usually be ignored. Once ΔH for the reaction has been obtained, the entropy change for the surroundings is calculated along a reversible path using

$$\Delta S_{\text{surroundings}} = \int \frac{\delta q_{\text{rev}}}{T} = \frac{1}{T} \int \delta q_{\text{rev}} = \frac{q_{\text{rev}}}{T} = \frac{-\Delta H}{T}$$

$$\text{(for processes in which } dP = 0). \tag{5.3}$$

Since the heat capacity of the surroundings is effectively infinite, the addition or removal of heat will not affect the temperature. As a result, T will be constant and may be factored out of the integral in Eq. 5.3. The system and the surroundings are at the same temperature; therefore, the heat transfer to or from the surroundings is reversible. Because the reaction is conducted under conditions of constant pressure, we have $q = q_p = \Delta H$. Finally, the heat change relative to the surroundings is the negative of that for the system. That is, heat lost by the system is gained by the surroundings and vice versa. Consequently, q_{rev} for the surroundings is the negative of ΔH. If we are conducting the reaction under constant-volume conditions, wherein $q = q_v = \Delta U$, we will have

$$\Delta S_{\text{surroundings}} = \frac{-\Delta U}{T} \qquad \text{(for reactions in which } dV = 0). \tag{5.4}$$

Figure 5.2 illustrates these points.

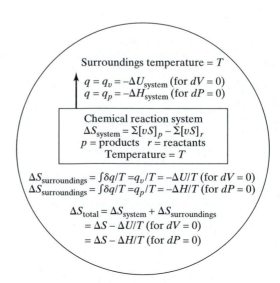

▶ FIGURE 5.2
Diagram illustrating the computation of the total entropy change accompanying a chemical reaction that occurs at a fixed temperature and either constant pressure or constant volume.

The quantity of interest is the total entropy change, $\Delta S_{total} = \Delta S_{system} + \Delta S_{surroundings}$. Using Eqs. 5.3 and 5.4, we may write this in the twin forms

$$\Delta S_{total} = \Delta S_{system} + \frac{-\Delta H_{system}}{T} = \Delta S + \frac{-\Delta H}{T}$$

$$(\text{for reactions in which } dP = 0) \qquad \textbf{(5.5)}$$

and

$$\Delta S_{total} = \Delta S_{system} + \frac{-\Delta U_{system}}{T} = \Delta S + \frac{-\Delta U}{T}$$

$$(\text{for reactions in which } dV = 0), \qquad \textbf{(5.6)}$$

where we have dropped the subscript "system," since all quantities on the right-hand sides of these equations refer to the system. The criteria for equilibrium and spontaneous processes are then

$$\Delta S_{total} = \Delta S - \frac{\Delta H}{T} > 0 \text{ (for irreversible processes with } dP = dT = 0),$$

$$\Delta S_{total} = \Delta S - \frac{\Delta H}{T} = 0 \text{ (for reversible processes with } dP = dT = 0), \qquad \textbf{(5.7)}$$

and

$$\Delta S_{total} = \Delta S - \frac{\Delta H}{T} < 0 \text{ (for nonspontaneous processes with } dP = dT = 0).$$

If the reaction occurs at fixed temperature and volume, the criteria are

$$\Delta S_{total} = \Delta S - \frac{\Delta U}{T} > 0 \text{ (for irreversible processes with } dV = dT = 0),$$

$$\Delta S_{total} = \Delta S - \frac{\Delta U}{T} = 0 \text{ (for reversible processes with } dV = dT = 0), \qquad \textbf{(5.8)}$$

and

$$\Delta S_{total} = \Delta S - \frac{\Delta U}{T} < 0 \text{ (for nonspontaneous processes with } dV = dT = 0).$$

Following standard practice in thermodynamics, we now modify Eqs. 5.7 and 5.8 by multiplying all of the inequalities and equations by $-T$. Since $-T$ is negative, that operation reverses the direction of all inequalities, to give the two sets of equations

$$\Delta H - T\,\Delta S < 0 \quad (\text{for irreversible processes with } dT = dP = 0),$$
$$\Delta H - T\,\Delta S = 0 \quad (\text{for reversible processes with } dT = dP = 0),$$
$$\Delta H - T\,\Delta S > 0 (\text{for nonspontaneous processes with } dT = dP = 0); \qquad \textbf{(5.9)}$$

and

$$\Delta U - T\,\Delta S < 0 \quad (\text{for irreversible processes with } dT = dV = 0),$$
$$\Delta U - T\,\Delta S = 0 \quad (\text{for reversible processes with } dT = dV = 0),$$
$$\Delta U - T\,\Delta S > 0 \text{ (for nonspontaneous processes with } dT = dV = 0). \qquad \textbf{(5.10)}$$

The quantities $\Delta H - T\Delta S$ and $\Delta U - T\Delta S$ are extremely important and useful in that their value determines the spontaneity or nonspontaneity of any process conducted under isothermal conditions with either dP or dV equal to zero. Therefore, we define the two new functions

$$\boxed{G \equiv H - TS},\qquad\qquad\qquad\text{(5.11)}$$

which we call the *Gibbs free energy,* and

$$\boxed{A \equiv U - TS,}\qquad\qquad\qquad\text{(5.12)}$$

which we term the *Helmholtz free energy.* (In some books, F is used for the latter rather than A.) If the temperature is held constant, we have $\Delta G = \Delta H - T\Delta S$ and $\Delta A = \Delta H - T\Delta S$. Therefore, the two sets of conditions for spontaneity given by Eqs. 5.9 and 5.10 can be stated as

$$\boxed{\Delta G < 0 \qquad \text{(for irreversible processes with } dT = dP = 0)},$$

$$\boxed{\Delta G = 0 \qquad \text{(for reversible processes with } dT = dP = 0)},$$

$$\boxed{\Delta G > 0 \text{(for nonspontaneous processes with } dT = dP = 0)}; \text{ (5.13)}$$

and

$$\boxed{\Delta A < 0 \qquad \text{(for irreversible processes with } dT = dV = 0)},$$

$$\boxed{\Delta A = 0 \qquad \text{(for reversible processes with } dT = dV = 0)},$$

$$\boxed{\Delta A > 0 \text{(for nonspontaneous processes with } dT = dV = 0)}. \text{ (5.14)}$$

Figure 5.3 and Table 5.1 summarize the criteria for spontaneity that we have discussed.

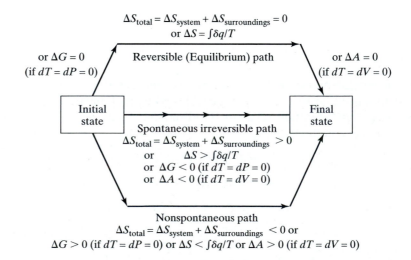

▶ **FIGURE 5.3**
Thermodynamic criteria involving either ΔS, ΔG, or ΔA that determine whether a process is irreversible (spontaneous), nonspontaneous, or reversible (at equilibrium).

Table 5.1	Some thermodynamic criteria of spontaneity and equilibrium	

	Criteria	
Conditions	**Spontaneous Processes**	**Equilibrium Processes**
Constant temperature and constant pressure	$\Delta S_{total} > 0$ $\Delta G < 0$ $\Delta S > \left\lvert \dfrac{\delta q}{T} \right\rvert$	$\Delta S_{total} = 0$ $\Delta G = 0$ $\Delta S = \left\lvert \dfrac{\delta q}{T} \right\rvert$
Constant temperature and constant volume	$\Delta S_{total} > 0$ $\Delta A < 0$ $\Delta S > \left\lvert \dfrac{\delta q}{T} \right\rvert$	$\Delta S_{total} = 0$ $\Delta A = 0$ $\Delta S = \left\lvert \dfrac{\delta q}{T} \right\rvert$

5.2   Total Differentials of G and A

To use the Gibbs and Helmholtz free energies as criteria of spontaneity or equilibrium, we must be able to compute changes in these functions as T, P, V, and n are varied. Such computations require a knowledge of the total differentials for G and A from which the required partial derivatives can be obtained. Let us first address closed systems for which $dn_i = 0$ for all components.

The total differential of the Helmholtz free energy defined by Eq. 5.12 is

$$dA = dU - d(TS) = dU - T\,dS - S\,dT. \qquad (5.15)$$

Since we generally evaluate changes in this exact differential along a reversible path, we impose the condition that the path must be reversible on Eq. 5.15. For such a path, we may use Eq. 4.59 for dU. This gives

$$\boxed{dA = T\,dS - P\,dV - T\,dS - S\,dT = -S\,dT - P\,dV}. \qquad (5.16)$$

Equations 4.59, 4.61, and 5.16 together make up three of the four most important relationships in thermodynamics. These equations hold for any material undergoing a reversible process in a closed system. Equation 5.16 explicitly demonstrates that reversible processes having $dT = dV = 0$ in a closed system are characterized by having $dA = 0$, as indicated by Eq. 5.14, Figure 5.3, and Table 5.1.

The derivatives required to compute changes in the Helmholtz free energy when T and V vary can be obtained directly from Eq. 5.16. For example, division by dT, followed by invoking the condition that the volume is constant, gives

$$\left(\frac{\partial A}{\partial T} \right)_V = -S. \qquad (5.17)$$

A similar operation gives the rate of change of A with volume under constant-temperature conditions:

$$\left(\frac{\partial A}{\partial V} \right)_T = -P.$$

The total differential of the Gibbs free energy defined by Eq. 5.11 is

$$dG = dH - d(TS) = dH - T\,dS - S\,dT. \tag{5.18}$$

For reversible processes, we may substitute Eq. 4.61 for dH to obtain

$$\boxed{dG = T\,dS + V\,dP - T\,dS - S\,dT = -S\,dT + V\,dP} \tag{5.19}$$

Equation 5.19 is the fourth fundamental relationship among the thermodynamic functions. It shows how the Gibbs free energy varies in a closed system as T and P are changed along a reversible path. The rate of change of G with T under conditions of constant pressure is

$$\left(\frac{\partial G}{\partial T}\right)_P = -S. \tag{5.20}$$

The corresponding rate of change of G with pressure at $dT = 0$ is

$$\left(\frac{\partial G}{\partial P}\right)_T = V. \tag{5.21}$$

5.3   Maxwell's Relationships

The four fundamental equations connecting the thermodynamic functions along reversible paths for closed systems involving only pressure–volume work are given by Eqs. 4.59, 4.61, 5.16, and 5.19. To summarize, these equations are

$$\boxed{dU = T\,dS - P\,dV},$$

$$\boxed{dH = T\,dS + V\,dP},$$

$$\boxed{dA = -S\,dT - P\,dV},$$

and

$$\boxed{dG = -S\,dT + V\,dP}. \tag{5.22}$$

Since each of these differentials is exact, the Euler condition for exactness may be applied to each to obtain relationships between various partial derivatives. Application of the Euler criterion to each total differential gives

$$\left(\frac{\partial T}{\partial V}\right)_S = -\left(\frac{\partial P}{\partial S}\right)_V \quad \text{from } dU, \tag{5.23}$$

$$\left(\frac{\partial T}{\partial P}\right)_S = \left(\frac{\partial V}{\partial S}\right)_P \quad \text{from } dH, \tag{5.24}$$

$$\boxed{\left(\frac{\partial S}{\partial V}\right)_T = \left(\frac{\partial P}{\partial T}\right)_V \quad \text{from } dA}. \tag{5.25}$$

and

$$\boxed{\left(\frac{\partial S}{\partial P}\right)_T = -\left(\frac{\partial V}{\partial T}\right)_P \quad \text{from } dG}. \tag{5.26}$$

These relationships between partial derivatives are collectively called *Maxwell's relationships*. Equations 5.25 and 5.26 are the most useful of the four, informing us about how entropy changes with volume and pressure, respectively, at constant temperature in terms of quantities that may be evaluated from the equation of state. Examples 5.1 through 5.3 demonstrate how useful Eqs. 5.22 through 5.26 are.

EXAMPLE 5.1

How does the internal energy vary with volume at constant temperature? Express your answer in terms of the variables P, V, and T only.

Solution

Using the dU differential from Eq. 5.22, we have

$$dU = T\, dS - P\, dV. \tag{1}$$

Dividing both sides by dV gives

$$\frac{dU}{dV} = T\frac{dS}{dV} - P. \tag{2}$$

We now require that the temperature be held constant. This produces

$$\left(\frac{\partial U}{\partial V}\right)_T = T\left(\frac{\partial S}{\partial V}\right)_T - P. \tag{3}$$

Using Eq. 5.25, we find that this expression becomes

$$\left(\frac{\partial U}{\partial V}\right)_T = T\left(\frac{\partial P}{\partial T}\right)_V - P. \tag{4}$$

Equation 4 is the desired result. It is identical to Eq. 4.92. A derivation using the exactness of dS is requested in Problem 4.38. A comparison of the two derivations demonstrates in clear terms how useful the Gibbs and Helmholtz free-energy functions and Maxwell's relationships are.

EXAMPLE 5.2

How does the enthalpy vary with pressure at constant temperature. Express your answer in terms of the variables P, V, and T only. Obtain the Joule–Thomson coefficient in terms of C_p^m and the variables P, V, and T.

Solution

Using the dH differential from Eq. 5.22, we have

$$dH = T\, dS + V\, dP. \tag{1}$$

Dividing by dP gives

$$\frac{dH}{dP} = T\frac{dS}{dP} + V. \tag{2}$$

We now require that $dT = 0$ to obtain

$$\left(\frac{\partial H}{\partial P}\right)_T = T\left(\frac{\partial S}{\partial P}\right)_T + V. \tag{3}$$

Using Eq. 5.26 for the pressure derivative of S, we get

$$\left(\frac{\partial H}{\partial P}\right)_T = V - T\left(\frac{\partial V}{\partial T}\right)_P, \tag{4}$$

which is the desired result. This constitutes the proof of Eq. 4.88 that was promised in Chapter 4. The Joule–Thomson coefficient can now be obtained from Eq. 4:

$$\left(\frac{\partial H}{\partial P}\right)_T = V - T\left(\frac{\partial V}{\partial T}\right)_P = -C_p^m \mu. \tag{5}$$

Solving for μ yields

$$\mu = \frac{T\left(\frac{\partial V}{\partial T}\right)_P - V}{C_p^m}, \tag{6}$$

which is identical to Eq. 4.89. Note how easy the derivation is when we use the four fundamental equations and Maxwell's relationships.

EXAMPLE 5.3

We can combine Eqs. 4.73 and Eq. 5.20 to obtain ΔG as a function of temperature. Consider $O_2(g)$. Use the data given in Appendix A, assuming that C_p^m for $O_2(g)$ is independent of temperature, to obtain an expression for ΔG for oxygen as a function of temperature. Plot the result over the range $298.15\ K \leq T \leq 500\ K$.

Solution

We may compute the entropy of $O_2(g)$ as a function of temperature using Eq. 4.73:

$$\int_{S(298.15)}^{S(T)} dS = S(T) - S(298.15) = \int_{298.15}^{T} \frac{C_p^m}{T}\, dT = C_p^m \ln\left[\frac{T}{298.15}\right]. \tag{1}$$

Therefore,

$$S(T) = S(298.15) + C_p^m \ln\left[\frac{T}{298.15}\right]. \tag{2}$$

Using Eq. 5.20, we find that the Gibbs free energy is

$$\int_{G(298.15)}^{G(T)} dG = G(T) - G(298.15) = \Delta G(T) = -\int_{298.15}^{T} S\, dT$$

$$= -\int_{298.15}^{T} \left[S(298.15) + C_p^m \ln(T) - C_p^m \ln(298.15)\right] dT$$

$$= -[S(298.15) - C_p^m \ln(298.15)][T - 298.15] - C_p^m[T \ln T - T]$$

$$+ C_p^m[298.15 \ln(298.15) - 298.15], \tag{3}$$

because $\int \ell n\, x\, dx = x\, \ell n\, x - x$. Appendix A gives us the following data for oxygen: $S(298.15) = 205.138\ \text{J mol}^{-1}\ \text{K}^{-1}$ and $C_p^m = 29.355\ \text{J mol}^{-1}\ \text{K}^{-1}$. Inserting these data into Eq. 3, followed by some arithmetic, produces

$$G(T) - G(298.15) = 5.241 \times 10^4 - 8.530T - 29.355T \ln T\ (\text{J mol}^{-1}), \tag{4}$$

which is the desired result. Figure 5.4 shows that the Gibbs free energy for oxygen decreases as T increases.

Figure 5.5 qualitatively compares the change in G produced by changes in temperature for gases, liquids, and solids. Since we generally observe that $S_{solid} < S_{liquid} < S_{gas}$, and since Eq. 5.20 shows that the rate of change of G with T is $-S$, the magnitudes of the slopes of the curves decrease as we go from gas to liquid to solid.

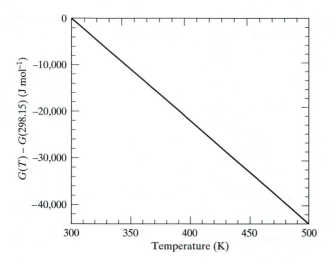

▲ FIGURE 5.4
The change in the Gibbs free energy of $O_2(g)$ from 298.15 K to 500 K. Example 5.3 shows that the variation is given by $\Delta G(T) = 5{,}241 \times 10^4 - 8.530T - 29.355T\ln(T)$ in J mol^{-1}. The linear term in T is the major contributor to $\Delta G(T)$. Therefore, the variation is nearly linear.

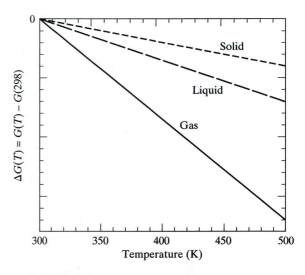

▲ FIGURE 5.5
Qualitative illustration of the variation of ΔG between 298 K and 500 K for a gas, liquid, and solid. Since we generally have $S_{solid} < S_{liquid} < S_{gas}$ and $(\partial G/\partial T)_P = -S$, the magnitude of the slope decreases as we go from a gas to a liquid to a solid.

5.4   Open Systems: The Chemical Potential

The differentials for U and H have already been generalized to open systems undergoing reversible processes involving only pressure–volume work. These results, given in Eqs. 4.62 and 4.63, are

$$dU = T\,dS - P\,dV + \sum_{i=1}^{K} \widehat{U}_i\,dn_i \qquad (5.27)$$

and

$$dH = T\,dS + V\,dP + \sum_{i=1}^{K} \overline{H}_i\,dn_i. \qquad (5.28)$$

The differentials dA and dG may be generalized in a similar fashion. For open systems,

$$dA = -S\,dT - P\,dV + \sum_{i=1}^{K} \widehat{A}_i\,dn_i \qquad (5.29)$$

and

$$dG = -S\,dT + V\,dP + \sum_{i=1}^{K} \overline{G}_i\,dn_i. \qquad (5.30)$$

The quantities appearing inside the summations, which are effectively the internal energy, enthalpy, Helmholtz free energy and Gibbs free energy per mole, respectively, are defined, in order, by

$$\widehat{U}_i = \left(\frac{\partial U}{\partial n_i}\right)_{S,V,nj} = \left(\frac{\partial U}{\partial n_i}\right)_{T,V,nj} \qquad (5.31)$$

$$\overline{H}_i = \left(\frac{\partial H}{\partial n_i}\right)_{S,P,nj} = \left(\frac{\partial H}{\partial n_i}\right)_{T,P,nj} \qquad (5.32)$$

$$\hat{A}_i = \left(\frac{\partial A}{\partial n_i}\right)_{T,V,nj} \tag{5.33}$$

and

$$\overline{G}_i = \left(\frac{\partial G}{\partial n_i}\right)_{T,P,nj}. \tag{5.34}$$

Equation 4.64A shows that the entropy for a closed system can be expressed as a function of T and V. Therefore, holding the entropy and volume fixed also fixes the temperature. Consequently, the two partial derivatives $(\partial U/\partial n_i)_{S,V,nj}$ and $(\partial U/\partial n_i)_{T,V,nj}$ are equal, as indicated in Eq. 5.31. The same logic using Eq. 4.64B shows that $(\partial H/\partial n_i)_{S,P,nj}$ and $(\partial H/\partial n_i)_{T,P,nj}$ are equal, as stated in Eq. 5.32. When T and P are constant, these partial derivatives are called partial molar quantities. We denote such quantities with an overbar as shown earlier.

Examination of Eqs. 5.27 through 5.30 suggests that we need to measure or compute $4K$ different molar derivatives to completely describe the four fundamental differentials for open systems. For example, if we had a system containing three components $(K = 3)$, we would apparently have to measure 12 different molar derivatives. Fortunately, our task is not that difficult, because it is possible to express all the summations appearing in Eqs. 5.27 through 5.30 in terms of the summation appearing in Eq. 5.30. Therefore, we need to measure only the partial molar Gibbs free energies for each component. The analysis that follows shows why this is true.

From the definition of the Gibbs free energy, we may write

$$G = H - TS = U + PV - TS. \tag{5.35}$$

Rearranging Eq. 5.35 produces

$$U = G - PV + TS.$$

Taking total differentials of both sides, we obtain

$$dU = dG - d(PV) + d(TS) = dG - P\,dV - V\,dP + T\,dS + S\,dT.$$

We now may insert the total differentials of dU and dG from Eqs. 5.27 and 5.30. This gives

$$T\,dS - P\,dV + \sum_{i=1}^{K}\hat{U}_i\,dn_i = -S\,dT + V\,dP + \sum_{i=1}^{K}\overline{G}_i\,dn_i - P\,dV - V\,dP + T\,dS + S\,dT.$$

Canceling identical terms on the right-hand side yields

$$T\,dS - P\,dV + \sum_{i=1}^{K}\hat{U}_i\,dn_i = T\,dS - P\,dV + \sum_{i=1}^{K}\overline{G}_i\,dn_i.$$

Finally, canceling the $T\,dS - P\,dV$ terms on both sides, we obtain

$$\sum_{i=1}^{K}\hat{U}_i\,dn_i = \sum_{i=1}^{K}\overline{G}_i\,dn_i. \tag{5.36}$$

Equation 5.36 shows that we do not need to measure the values of all the $\hat{U}_i$; they can be replaced with the summation over the partial molar Gibbs free energies. Example 5.3 shows that this is also true for the molar Helmholtz free energies. In Problem 5.15, you will demonstrate that the same thing holds for the $\overline{H}_i$.

EXAMPLE 5.3

Show that $\sum_{i=1}^{K} \hat{A}_i \, dn_i = \sum_{i=1}^{K} \overline{G}_i \, dn_i$.

Solution

From the definitions of G and A, we have

$$G = H - TS = U + PV - TS = A + PV. \tag{1}$$

Rearranging terms produces

$$A = G - PV. \tag{2}$$

Taking differentials of both sides gives

$$dA = dG - d(PV) = dG - V \, dP - P \, dV. \tag{3}$$

Substituting Eqs. 5.29 and 5.30 yields

$$-S \, dT - P \, dV + \sum_{i=1}^{K} \hat{A}_i \, dn_i = -S \, dT + V \, dP + \sum_{i=1}^{K} \overline{G}_i \, dn_i - V \, dP - P \, dV. \tag{4}$$

Canceling identical terms on the right-hand side gives

$$-S \, dT - P \, dV + \sum_{i=1}^{K} \hat{A}_i \, dn_i = -S \, dT + \sum_{i=1}^{K} \overline{G}_i \, dn_i - P \, dV. \tag{5}$$

Finally, canceling the $[-S \, dT - P \, dV]$ terms on both sides produces

$$\sum_{i=1}^{K} \hat{A}_i \, dn_i = \sum_{i=1}^{K} \overline{G}_i \, dn_i, \tag{6}$$

as desired.

In view of the preceding results, it is clear that only one summation is needed. It is customary to choose the partial molar Gibbs free energy which is given a special symbol and name. By definition, we replace $\overline{G}_i$ with μ_i and call the quantity the *chemical potential*. With this definition, the total differentials of the thermodynamic functions for open systems can be written as

$$\boxed{dU = T \, dS - P \, dV + \sum_{i=1}^{K} \mu_i \, dn_i}, \tag{5.37}$$

$$\boxed{dH = T \, dS + V \, dP + \sum_{i=1}^{K} \mu_i \, dn_i}, \tag{5.38}$$

$$\boxed{dA = -S \, dT - P \, dV + \sum_{i=1}^{K} \mu_i \, dn_i}, \tag{5.39}$$

and

$$\boxed{dG = -S \, dT + V \, dP + \sum_{i=1}^{K} \mu_i \, dn_i}. \tag{5.40}$$

The variation of the chemical potential with temperature and pressure is given by the total differential of that potential:

$$d\mu = \left(\frac{\partial\mu}{\partial T}\right)_P dT + \left(\frac{\partial\mu}{\partial P}\right)_T dP. \tag{5.41}$$

The required partial derivatives in Eq. 5.41 can be obtained from Eqs. 5.20 and 5.21 by differentiating both sides with respect to n. From Eq. 5.20,

$$\left(\frac{\partial\mu}{\partial T}\right)_P = \frac{\partial}{\partial T}\left[\left(\frac{\partial G}{\partial n}\right)_{T,P}\right]_P = \frac{\partial}{\partial n}\left[\left(\frac{\partial G}{\partial T}\right)_P\right]_{T,P} = -\left(\frac{\partial S}{\partial n}\right)_{T,P} = -\overline{S}, \tag{5.42}$$

since the order of differentiation may be reversed without changing the value of the second derivative. We see that the rate of change of the chemical potential with temperature is the negative of the partial molar entropy of the system. A similar result may be obtained from Eq. 5.21 giving the variation of the chemical potential with pressure. From that equation,

$$\left(\frac{\partial\mu}{\partial P}\right)_T = \frac{\partial}{\partial P}\left[\left(\frac{\partial G}{\partial n}\right)_{T,P}\right]_T = \frac{\partial}{\partial n}\left[\left(\frac{\partial G}{\partial P}\right)_T\right]_{T,P} = \left(\frac{\partial V}{\partial n}\right)_{T,P} = \overline{V}, \tag{5.43}$$

where $\overline{V}$ is the partial molar volume. Combining Eqs. 5.41 through 5.43 yields

$$\boxed{d\mu = -\overline{S}\, dT + \overline{V}\, dP}. \tag{5.44}$$

5.5   Standard Chemical Potentials: The Equilibrium Constant

5.5.1   Variation of G with Pressure

Equation 5.13 shows that ΔG is intimately connected with chemical equilibrium when a system is at constant temperature and pressure. Since many chemical reactions are run under these conditions, we are concerned with the calculation of G at a given pressure and temperature. In this section, we shall develop some of the most important equations that describe chemical equilibrium.

Let us first consider 1 mole of a pure substance. Using Eq. 5.22, we have

$$dG = -S\, dT + V\, dP.$$

At constant temperature, the variation in G due to changes in pressure is

$$dG = V\, dP. \tag{5.45}$$

It is now instructive to take indefinite integrals of Eq. 5.45. This produces

$$\int dG = G(T, P) = \int V\, dP. \tag{5.46}$$

Equation 5.46 is as far as we can go if we do not know how V depends upon P. We do know, however, that the molar volumes of solids and liquids are very small and that they are nearly constant with pressure. Therefore, the change in the Gibbs free energy with pressure for condensed phases will usually be small, with G_{liquid} changing faster than G_{solid}, since, in most cases, $V_{liquid} > V_{solid}$. In contrast, the molar volume of a gas is very large. As a result, the Gibbs free energies of gases are strongly dependent upon pressure. The qualitative situation is illustrated in Figure 5.6.

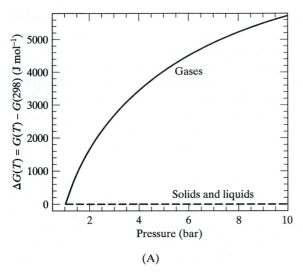

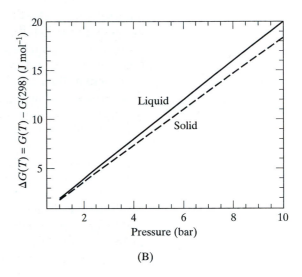

(A)

(B)

▲ **FIGURE 5.6**
The variation of ΔG with pressure over the range 1 bar $\leq P \leq$ 10 bar for gases, liquids, and solids at 298 K. (A) Gases and condensed phases. Due to their small molar volumes, the variations for the condensed phases cannot be seen on the scale of this plot. (B) Typical variations of ΔG for liquids and solids. Compare the ordinate scales in (A) and (B). The illustration assumes that the liquid is less dense than the solid, so that $V_{\text{liquid}} > V_{\text{solid}}$.

If the material under examination is an ideal gas, we may replace V in Eq. 5.46 with RT/P. This gives

$$G(T, P) = \int \frac{RT\, dP}{P} = RT \int \frac{dP}{P} = \Phi(T) + RT \ln(P). \qquad \textbf{(5.47)}$$

For n moles of gas, Eq. 5.47 becomes

$$G(T, P, n) = n\Phi(T) + nRT \ln(P). \qquad \textbf{(5.48)}$$

It is extremely important to note several points concerning Eqs. 5.47 and 5.48. First, the quantity playing the role of a constant of integration, which we have written as $\Phi(T)$, is not actually a constant. Rather, it is a function of the variable that is held constant in order to obtain Eq. 5.45, namely, T. This is easily shown to be true: If we hold T constant and take the differential of both sides of Eq. 5.47, we immediately obtain $dG = (RT\ dP)/P = V\ dP$, which is the correct result. The function $\Phi(T)$ depends only upon the temperature. Physically, it represents the Gibbs free energy per mole of material in the standard state wherein $P = 1$ bar. It is, therefore, customary to represent this function with the notation μ^o and to call it the *standard chemical potential*. It is also important to note that pressure must be expressed in units of bars when one is using Eqs. 5.47 and 5.48 if μ^o is to be interpreted as the Gibbs free energy per mole in the standard state.

The second point that must be explicitly noted concerns the second term on the right-hand side of Eq. 5.47, $RT \ln(P)$. By definition, if $\ln(x) = y$, we have $e^y = x$. Therefore, x cannot have any units. If this were not true, we could ask, What is the natural logarithm of 2 bananas? If the proposed answer were y, then we would have $e^y = 2$ bananas. This would certainly be useful: We would no longer have to purchase bananas or even a Mercedes; we could simply raise e to a power equal to $\ln[\text{Mercedes}]$ and get 1 Mercedes automobile free of charge. Alas, we cannot do that; the argument of the natural logarithm function must be unitless. The question then arises, "What happened to the units on pressure, which appears as the argument of the logarithm function in Eq. 5.47"? The answer is seen in the form of the integral that led to that function. We had the form RT $\int dP/P$, so that the units on dP canceled the units on P. Consequently, the pressure argument of the natural logarithm function is the magnitude of P only. The units vanished inside the integral. As a result, any equation derived from Eqs. 5.47 or 5.48 that contains P will have no units of pressure due to this factor. (Note, however, that the units on P in Eq. 5.45 must still be chosen so as to be consistent with the units on the Gibbs free energy.) In some texts, the notation $\ln(P/\text{bar})$ is used to indicate that the argument is the pressure magnitude only. We shall not do this here as it makes the notation very cumbersome.

The chemical potential for the case of an ideal gas is easily derived from Eq. 5.48. If we differentiate both sides of Eq. 5.48 with respect to n under conditions of constant temperature and pressure, we obtain

$$\left(\frac{\partial G(T, P, n)}{\partial n}\right)_{TP} = \mu_{\text{ideal}} = \Phi(T) + RT \ln(P) = \mu^\circ(T) + RT \ln(P) \,. \qquad \textbf{(5.49)}$$

5.5.2   The Equilibrium Constant: Ideal Systems

Consider a general gas-phase chemical reaction involving four components carried out at fixed temperature and pressure. The equation for this reaction is

$$\nu_A A(g) + \nu_B B(g) \longrightarrow \nu_C C(g) + \nu_D D(g),$$

where the ν_i are the stoichiometric coefficients in the balanced chemical equation and A, B, C, and D denote the compounds in the reaction. We now ask what change in the Gibbs free energy will be produced if an infinitesimal number of moles of $A(g)$ react with $B(g)$ to form more products. The answer is provided by Eq. 5.40:

$$dG = -S \, dT + V \, dP + \sum_{i=1}^{4} \mu_i \, dn_i.$$

Since the reaction is being conducted under conditions of constant temperature and pressure, the first two terms on the right side vanish, leaving

$$dG = \sum_{i=1}^{4} \mu_i \, dn_i. \qquad \textbf{(5.50)}$$

We obtain a similar result if the reaction is carried out at constant temperature and volume. In that case, Eq. 5.39 shows that the differential change in the Helmholtz free energy is

$$dA = \sum_{i=1}^{4} \mu_i \, dn_i, \qquad \textbf{(5.51)}$$

since $dT = dV = 0$.

The changes in the Gibbs and Helmholtz free energies, therefore, depend upon the chemical potentials of the four components of the reaction and the change in the number of moles of each compound. If the number of moles of $A(g)$ that react is denoted by $d\varepsilon$, the stoichiometry of the reaction gives

$$dn_A = -d\varepsilon,$$

$$dn_B = -\frac{\nu_B}{\nu_A} d\varepsilon,$$

$$dn_C = \frac{\nu_C}{\nu_A} d\varepsilon,$$

and

$$dn_D = \frac{\nu_D}{\nu_A} d\varepsilon,$$

where the minus signs appear for dn_A and dn_B because the number of moles of reactants is decreasing. Substituting these results into Eq. 5.50 produces

$$dG = \frac{d\varepsilon}{\nu_A} [\nu_C \mu_C + \nu_D \mu_D - \nu_A \mu_A - \nu_B \mu_B] = \frac{d\varepsilon}{\nu_A} \Delta\mu, \qquad \textbf{(5.52)}$$

where $\Delta\mu$ substitutes for the expression in brackets, which gives the sum of the chemical potentials of the products minus that of the reactants.

The combination of Eqs. 5.13 and 5.14 with Eqs. 5.50, 5.51, and 5.52 allows us to express the criteria for irreversible and reversible processes in terms of changes in the chemical potential. For example, an irreversible process has $dG < 0$ if the temperature and pressure are constant or $dA < 0$ if the temperature and volume are constant. In either case, Eq. 5.52 shows that we must have $\Delta\mu < 0$ for a spontaneous, irreversible process. In general, at constant temperature and either constant volume or pressure, the criteria for irreversible and reversible processes are

$$\boxed{\Delta\mu < 0 \ \ \text{(irreversible, spontaneous process)}},$$

$$\boxed{\Delta\mu = 0 \ \ \ \text{(reversible, equilibrium process)}},$$

and

$$\boxed{\Delta\mu > 0 \qquad \text{(nonspontaneous process)}}. \qquad \textbf{(5.53)}$$

Figure 5.7 summarizes the criteria for spontaneity.

Let us now assume that all gases in the reaction are in the ideal-gas regime, so that they may be accurately described by the ideal-gas equation of state. Under this condition, the chemical potentials appearing in Eq. 5.52 are given by Eq. 5.49. Inserting the latter equation into the former yields

$$dG = \frac{d\varepsilon}{\nu_A} [\nu_C \mu_C^o + \nu_D \mu_D^o - \nu_A \mu_A^o - \nu_B \mu_B^o]$$

$$+ \frac{d\varepsilon RT}{\nu_A} [\nu_C \ln(P_C) + \nu_D \ln(P_D) - \nu_A \ln(P_A) - \nu_B \ln(P_B)]. \qquad \textbf{(5.54)}$$

Since $\ln(x) + \ln(y) - \ln(w) - \ln(u) = \ln\left[\dfrac{xy}{uw}\right]$, we can write Eq. 5.54 in the form

$$dG = \frac{d\varepsilon}{\nu_A} [\nu_C \mu_C^o + \nu_D \mu_D^o - \nu_A \mu_A^o - \nu_B \mu_B^o] + \frac{d\varepsilon RT}{\nu_A} \ln\left[\frac{P_C^{\nu_C} P_D^{\nu_D}}{P_A^{\nu_A} P_B^{\nu_B}}\right]. \qquad \textbf{(5.55)}$$

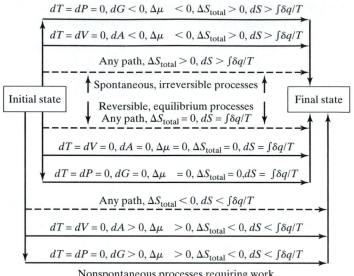

Summary of the criteria for irreversible, reversible, and nonspontaneous processes. Note that the entropy criteria hold for any path. The sign on the chemical potential change, $\Delta\mu$, is decisive for any constant-temperature path at either constant volume or constant pressure. The differentials dG and dA can be used for constant-temperature paths in which the pressure or volume is constant, respectively.

The calculation of dG, therefore, requires that we know the partial pressures of the reactants and products, the standard Gibbs free energies for all components, the temperature, and $d\varepsilon$.

Equation 5.13 shows that if the foregoing reaction is at equilibrium, so that the reaction of $d\varepsilon$ moles of $A(g)$ is reversible, we must have $dG = \Delta\mu = 0$. Under these conditions, we may multiply Eq. 5.55 by $\nu_A/d\varepsilon$ and rearrange the resulting equation to obtain

$$\Delta\mu^o = -RT \ln\left[\frac{P_C^{\nu_C} P_D^{\nu_D}}{P_A^{\nu_A} P_B^{\nu_B}}\right]_{eq}. \tag{5.56}$$

In Eq. 5.56, $\Delta\mu^o$ is the total change in the standard Gibbs free energies when one mole of $A(g)$ reacts to form products; that is,

$$\boxed{\Delta\mu^o = \nu_C \mu_C^o + \nu_D \mu_D^o - \nu_A \mu_A^o - \nu_B \mu_B^o}. \tag{5.57}$$

Note that the partial pressures forming the argument of the natural logarithm function in Eq. 5.56 must be evaluated at equilibrium. This is indicated by the subscript "eq" on the argument.

The most important point related to Eq. 5.56 is that $\Delta\mu^o$ on the left-hand side is a function of temperature only. This means that the right-hand side must also be a function of temperature only. It follows from this requirement that the pressure ratio appearing as the argument of the logarithm function must be independent of the total pressure. It can change only if the temperature is changed. If the temperature is held constant, the pressure ratio must also be constant. This ratio is given the symbol K_p and called the *thermodynamic equilibrium constant*:

$$\boxed{K_p = \left[\frac{P_C^{\nu_C} P_D^{\nu_D}}{P_A^{\nu_A} P_B^{\nu_B}}\right]_{eq}}. \tag{5.58}$$

Equation 5.56 may now be written in terms of K_p to obtain a very simple result:

$$\Delta\mu^o = -RT \ln(K_p).\qquad(5.59)$$

The form of Eq. 5.59 makes it clear that the thermodynamic equilibrium constant has no units. This is not unexpected, since the pressures appearing in Eq. 5.58 had no units. Recall that the pressure units were lost in the integral leading to the $\ln(P)$ term in Eq. 5.47.

Equation 5.51 shows that Eq. 5.59 holds even if the experimental conditions correspond to constant T and V rather than constant T and P. We simply repeat the derivation, this time using dA rather than dG.

Equations 5.57 and 5.59 permit dG for the reaction to be expressed in the more compact form

$$dG = \frac{d\varepsilon}{\nu_A}\left\{-RT \ln(K_p) + RT \ln\left[\frac{P_C^{\nu_C} P_D^{\nu_D}}{P_A^{\nu_A} P_B^{\nu_B}}\right]\right\}.\qquad(5.60)$$

If $d\varepsilon = \nu_A$, Eq. 5.60 becomes

$$\Delta G = -RT \ln(K_p) + RT \ln\left[\frac{P_C^{\nu_C} P_D^{\nu_D}}{P_A^{\nu_A} P_B^{\nu_B}}\right].\qquad(5.61)$$

It is sometimes convenient to express K_p in terms of concentrations in mol L^{-1}. Since the gases have been assumed to be ideal, this can easily be done, provided that we exercise care with respect to the units. The concentration of $A(g)$ in mol L^{-1} is $[A] = n_A/V = P_A/(RT)$, so that $P_A = [A] RT$. The magnitude of P_A is given by the same expression, with the pressure unit eliminated from the gas constant. We may, therefore, replace the partial pressures in Eq. 5.58 with RT times the corresponding concentrations, provided that the pressure unit in R is dropped. This replacement gives

$$K_p = \left[\frac{[C]^{\nu_C} [D]^{\nu_D}}{[A]^{\nu_A} [B]^{\nu_B}}\right]_{eq} [RT]^{(\nu_C+\nu_D-\nu_A-\nu_B)} = \left[\frac{[C]^{\nu_C} [D]^{\nu_D}}{[A]^{\nu_A} [B]^{\nu_B}}\right]_{eq} [RT]^{\Delta n},\quad(5.62)$$

where Δn is defined as the difference between the number of moles of gaseous products and gaseous reactants. The quantity inside the brackets in Eq. 5.62 is given the symbol K_c and called the *concentration equilibrium constant*. In these terms, we have

$$K_p = K_c [RT]^{\Delta n}.\qquad(5.63)$$

EXAMPLE 5.5

Consider the equilibrium between $NO_2(g)$ and $N_2O_4(g)$:

$$2 NO_2(g) = N_2O_4(g).$$

At a total pressure of 1 bar and $T = 298.15$ K, the partial pressures of $NO_2(g)$ and $N_2O_4(g)$ in an equilibrium mixture of the two gases are found to be 0.3181 and 0.6819 bar, respectively. **(A)** Compute the value of K_p assuming that both gases behave ideally. **(B)** Compute the value of K_c for this reaction at 298.15 K. **(C)** Compute $\Delta\mu^o$ for the reaction.

Solution

(A) The thermodynamic equilibrium constant for an ideal system is given by

$$K_p = \left[\frac{P_{N_2O_4(g)}}{P^2_{NO_2(g)}}\right]_{eq} = \frac{0.6819}{0.3181^2} = 6.739. \tag{1}$$

(B) Δn for the reaction is -1, since 1 mole of gaseous product and 2 moles of gaseous reactants are present. Therefore, using Eq. 5.63, we obtain

$$K_c = \frac{[N_2O_4(g)]}{[NO_2(g)]^2} = K_p[RT]^{-\Delta n} = K_p RT = 6.739\,[0.083144 \text{ L mol}^{-1}\,K^{-1}][298.15 \text{ K}]$$

$$= 167.1 \text{ L mol}^{-1}, \tag{2}$$

which is obviously the correct unit for the expression in Eq. 2. Note that, since the pressure in Eq. 5.49 must be expressed in bars, we must use R in units of L bar mol^{-1} K^{-1} with the pressure unit dropped.

(C) Using Eq. 5.59, we obtain

$$\Delta\mu^o = -RT \ln K_p = -(8.314 \text{ J mol}^{-1}\,K^{-1})(298.15 \text{ K}) \ln(6.739) = -4729 \text{ J mol}^{-1}. \tag{3}$$

For a related exercise, see Problem 5.27.

EXAMPLE 5.6

A piston–cylinder arrangement contains a mixture of $NO_2(g)$ and $N_2O_4(g)$, which are in chemical equilibrium at 298.15 K. $K_p = 6.739$ at this temperature, which is the value obtained in Example 5.5. **(A)** If the piston is adjusted so as to make the total pressure inside the cylinder equal to 2 bar, determine the partial pressures of $N_2O_4(g)$ and $NO_2(g)$ inside the cylinder, assuming that both gases behave ideally. **(B)** The piston is now slowly depressed to increase the total pressure inside the cylinder. Plot the equilibrium partial pressure of $N_2O_4(g)$ inside the cylinder as a function of the total pressure over the range $P = 2$ bar to $P = 100$ bar.

Solution

(A) The thermodynamic equilibrium constant for the ideal system is

$$K_p = \left[\frac{P_{N_2O_4(g)}}{P^2_{NO_2(g)}}\right]_{eq} = 6.739. \tag{1}$$

Dalton's law tells us that

$$P_{NO_2(g)} + P_{N_2O_4(g)} = P_T, \tag{2}$$

the total pressure in the cylinder. Direct substitution of Eq. 2 into Eq. 1 gives

$$\frac{P_{N_2O_4(g)}}{[P_T - P_{N_2O_4(g)}]^2} = K_p. \tag{3}$$

Writing Eq. 3 in quadratic form, we obtain

$$K_p P^2_{N_2O_4(g)} - 2K_p P_T P_{N_2O_4(g)} + K_p P^2_T - P_{N_2O_4(g)} = 0. \tag{4}$$

Collecting terms gives

$$K_p P^2_{N_2O_4(g)} - [2K_p P_T + 1] P_{N_2O_4(g)} + K_p P^2_T = 0. \tag{5}$$

Solving for $P_{N_2O_4(g)}$, we obtain

$$P_{N_2O_4(g)} = \frac{[2K_p P_T + 1] - \{[2K_p P_T + 1]^2 - 4K_p^2 P_T^2\}^{1/2}}{2K_p} \tag{6}$$

Substituting the numerical values gives

$$P_{N_2O_4(g)} = \frac{[(2)(6.739)(2) + 1] - \{[(2)(6.739)(2) + 1]^2 - 4(6.739)^2(2)^2\}^{1/2}}{2(6.739)}$$

$$= 1.524 \text{ bar.} \tag{7}$$

Using Eq. 2 with $P_T = 2$ bar, we obtain

$$P_{NO_2(g)} = 2 - 1.524 \text{ bar} = 0.476 \text{ bar.} \tag{8}$$

The results may be checked by substituting back into Eq. 1. This produces

$$K_p = \frac{1.524}{.0476^2} = 6.73, \tag{9}$$

which is correct to three significant digits.

(B) The partial pressure of $N_2O_4(g)$ as a function of P_T is given by Eq. 6. Substituting the numerical data gives

$$P_{N_2O_4(g)} = \frac{[13.478P_T + 1] - \{[13.478P_T + 1]^2 - 181.66\ P_T^2\}^{1/2}}{13.478}. \tag{10}$$

The plot is shown in Figure 5.8.

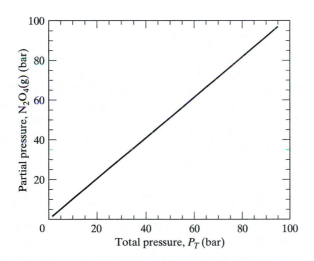

◀ FIGURE 5.8
Variation of the $N_2O_4(g)$ partial pressure as a function of the total pressure in Example 5.6.

For related exercises, see Problems 5.6, 5.7, 5.8, 5.9, 5.10, and 5.11.

EXAMPLE 5.7

Suppose we insert 1 mole of $NO_2(g)$ at 298.15 K into a 1-L container. The $NO_2(g)$ spontaneously dimerizes to form $N_2O_4(g)$ via the reaction $2\ NO_2(g) \longrightarrow N_2O_4(g)$. Calculate the number of moles of $NO_2(g)$ and $N_2O_4(g)$ present in the container after the system reaches equilibrium. All gases are ideal and $K_p = 6.739$.

Solution

Let x moles of $N_2O_4(g)$ be formed when equilibrium is reached. The stoichiometry shows that we will have consumed $2x$ moles of $NO_2(g)$ in the process. Therefore, at equilibrium, we have

$$P_{NO_2(g)} = \frac{nRT}{V} = \frac{(1 - 2x)RT}{1} = (1 - 2x)RT. \tag{1}$$

The partial pressure of $N_2O_4(g)$ is

$$P_{N_2O_4(g)} = \frac{nRT}{V} = \frac{xRT}{1} = xRT. \tag{2}$$

At equilibrium, we must have

$$K_p = \left[\frac{P_{N_2O_4(g)}}{P^2_{NO_2(g)}}\right]_{eq} = 6.739 = \frac{xRT}{(1-2x)^2R^2T^2} = \frac{x}{(1-2x)^2}(RT)^{-1}. \tag{3}$$

Therefore,

$$\frac{x}{(1-2x)^2} = 6.739RT$$

$$= 6.739(0.083144 \text{ L mol}^{-1}\text{ K}^{-1})(298.15 \text{ K}) = 167.06 \text{ L mol}^{-1}. \tag{4}$$

In quadratic form, Eq. 4 becomes

$$668.22x^2 - 669.22x + 167.06 = 0. \tag{5}$$

The root of this equation is

$$x = \frac{669.22 - [669.22^2 - 4(668.22)(167.06)]^{1/2}}{2(668.22)} = 0.4735 \text{ mol}. \tag{6}$$

We also have $1 - 2x$, or 0.0530, mol of $NO_2(g)$ left.

5.5.3   Measurement of Standard Chemical Potentials

We face the same problem in measuring standard chemical potentials as we did in measuring standard partial molar enthalpies in Chapter 3. Therefore, we proceed in the same manner. If we wish to determine the standard chemical potential for $CO_2(g)$, for example, we could carry out the reaction $C(s) + O_2(g) \longrightarrow CO_2(g)$ at constant temperature and pressure with $T = 298.15$ K and $P = 1$ bar. Under these conditions, Eq. 5.40 shows that

$$dG^o = -S\,dT + V\,dP + \sum_{i=1}^{3} \mu_i^o\,dn_i.$$

Since $dT = dP = 0$, the first two terms vanish, and we have $dG^o = \sum_{i=1}^{3} \mu_i^o\,dn_i$.

The superscript zeroes denote standard thermochemical conditions. For the foregoing reaction, the change in the number of moles of each component is $+1$ for $CO_2(g)$ and -1 for $C(s)$ and $O_2(g)$. Therefore,

$$\Delta G^o = \mu^o_{CO_2(g)} - \mu^o_{C(s)} - \mu^o_{O_2(g)} = \Delta H^o - T\Delta S^o. \tag{5.64}$$

Using the third law, we can obtain ΔH^o from standard partial molar enthalpies and ΔS^o from absolute entropies calculated from heat capacity measurements. Since the reaction forms 1 mole of $CO_2(g)$ in its most stable state under standard-state conditions from the elements in their most stable states at standard conditions, the change in Gibbs free energy is called the *standard molar Gibbs free energy of formation*. It is often labeled ΔG_f^o. Our problem is that ΔG_f^o gives us the difference in the standard chemical potentials between $CO_2(g)$ and the elements, not $\mu^o_{CO_2}(g)$ itself. We solve this problem by using the same procedure employed to obtain standard partial molar enthalpies. We choose the reference point for the measurement of standard chemical potentials to be the elements in their most stable state under standard-state conditions. We then assign the standard chemical potentials of the elements a value of zero. With this choice, we have

$$\mu^o_{CO_2(g)} = \Delta G_f^o. \tag{5.65}$$

We may also obtain values of $\Delta\mu^o$ from equilibrium constant measurements by using Eq. 5.59. If all of the standard chemical potentials save one are known from other measurements, the remaining value can be determined from a knowledge of $\Delta\mu^o$. As we shall see in subsequent chapters, standard chemical potentials may also be obtained from electrochemical measurements and computed from spectroscopic data. Appendix A gives $\mu^o = \Delta G_f^o$ values for some selected compounds.

EXAMPLE 5.8

Using the standard partial molar enthalpy of $CH_4(g)$ and the standard absolute entropies of $CH_4(g)$, $H_2(g)$, and $C(s)$ from Appendix A, compute μ^o for $CH_4(g)$ at 298.15 K.

Solution

The data given in Appendix A are $\overline{H}^o_{CH_4(g)} = -74.81 \text{ kJ mol}^{-1}$, $S^o_{CH_4(g)} = 186.26 \text{ J}$ $\text{mol}^{-1}\text{K}^{-1}$, $S^o_C(s) = 5.740 \text{ J mol}^{-1}\text{ K}^{-1}$, and $S^o_{H_2(g)} = 130.684 \text{ J mol}^{-1}\text{K}^{-1}$. The formation reaction for $CH_4(g)$ is

$$C(s) + 2\,H_2(g) \longrightarrow CH_4(g).$$

Therefore, we have

$$\Delta S^o = S^o_{CH_4(g)} - 2\,S^o_{H_2(g)} - S^o_{C(s)} = 186.26 - (2)(130.684) - 5.740 \text{ J mol}^{-1}\text{K}^{-1}$$

$$= -80.848 \text{ J mol}^{-1}\text{ K}^{-1}. \tag{1}$$

ΔH^o for the reaction is

$$\Delta H^o = \overline{H}^o_{CH_4(g)} - 2\overline{H}^o_{H_2(g)} - \overline{H}^o_{C(s)} = -7.481 \times 10^4 \text{ J mol}^{-1} - 0 - 0$$

$$= -7.481 \times 10^4 \text{ J mol}^{-1}. \tag{2}$$

The standard molar Gibbs free energy of formation is, therefore,

$$\Delta G_f^o = \Delta H^o - T\Delta S^o$$

$$= -7.481 \times 10^4 \text{ J mol}^{-1} - (298.15 \text{ K})(-80.848) \text{ J mol}^{-1}\text{ K}^{-1}$$

$$= -5.070 \times 10^4 \text{ J mol}^{-1} = -50.70 \text{ kJ mol}^{-1}. \tag{3}$$

With the choice of reference points described in the preceding section, we have
$$\mu^o_{CH_4(g)} = -50.70 \text{ kJ mol}^{-1}. \tag{4}$$

For related exercises, see Problems 5.4 and 5.5.

EXAMPLE 5.9

Compute ΔG and ΔA for the process described in Example 5.7. Does your answer indicate whether the reaction is spontaneous?

Solution

As shown in Example 5.7, the initial state had 1 mole of $NO_2(g)$ in a 1-L container at 298.15 K. The final state had 0.4735 mol of $N_2O_4(g)$ and 0.0530 mol of $NO_2(g)$ present at equilibrium. The partial pressures of $NO_2(g)$ and $N_2O_4(g)$ at equilibrium were, therefore,

$$P_{NO_2(g)} = \frac{nRT}{V} = 0.0530 \text{ mol } (0.083144 \text{ L bar mol}^{-1}\text{ K}^{-1})(298.15 \text{ K})/1L$$

$$= 1.314 \text{ bar} \tag{1}$$

and

$$P_{N_2O_4(g)} = \frac{0.4735 \text{ mol}(0.083144 \text{ L bar mol}^{-1}\text{K}^{-1})(298.15 \text{ K})}{1 \text{ L}} = 11.74 \text{ bar}, \quad \textbf{(2)}$$

respectively. The Gibbs free energy in the initial state is

$$G(T, P, n)_{\text{initial}} = n\mu^o_{NO_2(g)} + nRT \ln(P), \quad \textbf{(3)}$$

where $n = 1$ mol.

The initial pressure of $NO_2(g)$ is

$$P^{\text{initial}}_{NO_2(g)} = \frac{(1 \text{ mol})(0.083144 \text{ L atm mol}^{-1}\text{K}^{-1})(298.15 \text{ K})}{1 \text{ L}} = 24.789 \text{ bar}. \quad \textbf{(4)}$$

Using the data in Appendix A, we find that $\mu^o_{NO_2(g)} = 5.131 \times 10^4 \text{ J mol}^{-1}$. Therefore, we obtain

$$G(T, P, n)_{\text{initial}} = 1 \text{ mol } (5.131 \times 10^4 \text{ J mol}^{-1})$$

$$+ 1 \text{ mol } (8.314 \text{ J mol}^{-1}\text{K}^{-1})(298.15 \text{ K}) \ln(24.789) = 5.927 \times 10^4 \text{ J}. \quad \textbf{(4)}$$

In the final equilibrium state, the Gibbs free energy is

$$G(T, P, n)_{\text{final}} = n_{NO_2(g)}[\mu^o_{NO_2(g)} + RT \ln(P_{NO_2(g)})]$$

$$+ n_{N_2O_4(g)}[\mu^o_{N_2O_4(g)} + RT \ln(P_{N_2O_4(g)})]. \quad \textbf{(5)}$$

Appendix A gives $\mu^o_{N_2O_4(g)} = 9.789 \times 10^4 \text{ J mol}^{-1}$. Substituting numerical values into Eq. 5 then results in

$$G(T, P, n)_{\text{final}} = 0.0530 \text{ mol}[5.131 \times 10^4 \text{ J mol}^{-1}$$

$$+ (8.314 \text{ J mol}^{-1}\text{K}^{-1})(298.15 \text{ K})\ln(1.314)]$$

$$+ 0.4735 \text{ mol}[9.789 \times 10^4 \text{ J mol}^{-1} + (8.314 \text{ J mol}^{-1}\text{K}^{-1})(298.15 \text{ K})\ln(11.74)]$$

$$= 2.76 \times 10^3 \text{ J} + 4.924 \times 10^4 \text{ J} = 5.200 \times 10^4 \text{ J}. \quad \textbf{(6)}$$

The change in the Gibbs free energy for the process is

$$\Delta G = G_{\text{final}} - G_{\text{initial}} = 5.200 \times 10^4 \text{ J} - 5.927 \times 10^4 \text{ J} = -7.27 \times 10^3 \text{ J}. \quad \textbf{(7)}$$

The change in the Helmholtz free energy can be computed directly from the definitions of G and A:

$$G = H - TS = U + PV - TS = A + PV, \quad \textbf{(8)}$$

so that

$$\Delta G = \Delta A + \Delta(PV) = \Delta A + V\Delta P, \quad \textbf{(9)}$$

since $dV = 0$ for the reaction in the 1-liter container. Solving for ΔA, we obtain

$$\Delta A = \Delta G - V\Delta P. \quad \textbf{(10)}$$

The pressure change is

$$\Delta P = P_{\text{final}} - P_{\text{initial}} = (11.74 + 1.314) \text{ bar} - 24.789 \text{ bar} = -11.74 \text{ bar}. \quad \textbf{(11)}$$

Substituting this result into Eq. 10 produces

$$\Delta A = -7.27 \times 10^3 \text{ J} - (1 \text{ L})(-11.74) \text{ bar} \times \frac{8.314 \text{ J}}{0.083144 \text{ L bar}} = -6.10 \times 10^3 \text{ J}. \quad \textbf{(12)}$$

The reaction is carried out under conditions of constant temperature and volume; therefore, ΔA is the criterion for spontaneity. Since $\Delta A < 0$, the reaction is irreversible and spontaneous.

For related exercises, see Problems 5.7, 5.8, 5.27, 5.28, and 5.29.

5.5.4   Equilibria Involving Condensed Phases

When one or more of the components of an equilibrium reaction is either a solid or a liquid, the form of Eq. 5.54 changes. We may see why this happens by considering the reaction

$$\nu_A A(g) + \nu_B B(g) \longrightarrow \nu_C C(s) + \nu_D D(g),$$

where compound C is a solid. If the reaction is conducted under conditions of constant temperature and pressure, Eq. 5.50 still gives the differential change in G when $d\varepsilon$ moles of $A(g)$ react:

$$dG = \sum_{i=1}^{4} \mu_i \, dn_i.$$

The dn_i are again given by the stoichometric relationships following Eq. 5.51, so that we have

$$dG = \frac{d\varepsilon}{\nu_A} \left[\nu_C \mu_C + \nu_D \mu_D - \nu_A \mu_A - \nu_B \mu_B \right].$$

When all of the compounds involved were described by the ideal-gas law, we substituted Eq. 5.49 for each of the chemical potentials. This led directly to Eq. 5.55 and to the definition of the thermodynamic equilibrium constant in Eq. 5.58.

Since $C(s)$ is a solid, we cannot substitute the ideal-gas result for its chemical potential. In Chapter 6, we shall learn that the chemical potential of a condensed phase of a compound is equal to the chemical potential of the equilibrium vapor associated with the compound. That is,

$$\mu_{C(s)} = \mu_{C(g)} = \mu_{C(g)}^o + RT \ln(P_{C(g)}), \tag{5.66}$$

where the quantities on the right-hand side of the equation refer to the equilibrium vapor phase. $P_{C(g)}$ is the equilibrium vapor pressure of gaseous C over the solid form, and $\mu_{C(g)}^o$ is the standard chemical potential of the vapor. If the temperature is constant, we shall show that all the quantities on the right-hand side are constant. That is, at a fixed temperature, the equilibrium vapor pressure over the solid is a constant, as is $\mu_{C(g)}^o$. As a result, $\mu_{C(s)}$ is a constant at a given temperature. We may, therefore, substitute Eq. 5.49 for the chemical potentials of the gaseous compounds in the reaction and leave the chemical potentials of all condensed phases unaltered, since they are constants. When this is done, we obtain

$$dG = \frac{d\varepsilon}{\nu_A} \left[\nu_C \mu_{C(s)} + \nu_D \mu_D^o - \nu_A \mu_A^o - \nu_B \mu_B^o \right] + \frac{d\varepsilon RT}{\nu_A} \ln\left[\frac{P_D^{\nu_D}}{P_A^{\nu_A} P_B^{\nu_B}} \right]. \tag{5.67}$$

The rest of the treatment follows the same lines as that given in Section E.2. However, we now have

$$\Delta\mu^o = \left[\nu_C \mu_{C(s)} + \nu_D \mu_D^o - \nu_A \mu_A^o - \nu_B \mu_B^o \right], \tag{5.68}$$

which is dependent only upon the temperature, as was the case for gas-phase reactions, and

$$\Delta\mu^o = -RT \ln\left[\frac{P_D^{\nu_D}}{P_A^{\nu_A} P_B^{\nu_B}}\right] = -RT \ln(K_p). \tag{5.69}$$

In simple terms, the expression for the thermodynamic equilibrium constant does not contain pressures for the condensed phases present in the reaction. It follows from this fact that K_c will not contain concentrations for condensed phases.

EXAMPLE 5.10

Consider the equilibrium reaction

$$CaCO_3(s)_{(calcite)} \longrightarrow CaO(s) + CO_2(g).$$

Write the appropriate expression for the thermodynamic constant, assuming that the gaseous compounds obey the ideal-gas equation of state. Evaluate K_p at 298.15 K. What are the units on K_p? Does K_p vary with pressure? Explain.

Solution

There is only one gaseous compound in the reaction. Therefore,

$$K_p = [P_{CO_2(g)}]_{eq}. \tag{1}$$

The value of the equilibrium constant is given by Eq. 5.59, which we may rewrite as

$$K_p = \exp\left[\frac{-\Delta\mu^o}{RT}\right], \tag{2}$$

where $\Delta\mu^o$ is now given by Eq. 5.68:

$$\Delta\mu^o = \mu^o_{CO_2(g)} + \mu_{CaO(s)} - \mu_{CaCO_3(s)}. \tag{3}$$

Figure 5.6 shows that the Gibbs free energies of condensed phases are nearly independent of pressure. Therefore, for any condensed phase, we have $\mu \approx \mu^o$. With this replacement, Eq. 3 becomes

$$\Delta\mu^o = \mu^o_{CO_2(g)} + \mu^o_{CaO(s)} - \mu^o_{CaCO_3(s)}. \tag{4}$$

The required standard chemical potentials are given in Appendix A. We obtain

$$\Delta\mu^o = -394.36 \text{ kJ mol}^{-1} - 604.03 \text{ kJ mol}^{-1} - (-1{,}128.8) \text{ kJ mol}^{-1}$$

$$= 130.41 \text{ kJ mol}^{-1}. \tag{5}$$

Substituting the result in Eq. 5 into Eq. 2 produces

$$K_p = \exp\left[\frac{-1.3041 \times 10^5 \text{ J mol}^{-1}}{(8.314 \text{ J mol}^{-1}\text{ K}^{-1})(298.15 \text{ K})}\right] = \exp(-52.61) = 1.418 \times 10^{-23}. \tag{6}$$

There are no units on K_p, since $P_{CO_2(g)}$ in Eq. 1 is the magnitude of $P_{CO_2(g)}$ only. K_p does not vary with pressure, because $\Delta\mu^o$ is a function of temperature only. This means that we cannot vary $P_{CO_2(g)}$ in Eq. 1 without changing the temperature. For a related exercise, see Problem 5.12.

5.6   The Gibbs–Helmholtz Equations

The discussion presented in the preceding section, combined with data reporting the results of measurements of standard chemical potentials, permits the calculation of changes in the Gibbs and Helmholtz free energies and equilibrium constants for systems of ideal gases at the temperature at which the data are taken, usually 298.15 K. However, many chemical processes are conducted at temperatures other than 298.15 K. This means that we need a convenient procedure that permits us to compute changes in G, A, and μ when the temperature is varied.

For a closed system, the rate of change of G due to temperature changes at constant pressure is given by Eq. 5.20, $(\partial G/\partial T)_P = -S$. We can use this equation directly to compute changes in G with temperature if we first obtain S as a function of T. Example 5.3 illustrates the method. Alternatively, we may cast Eq. 5.20 in a more convenient form by expressing the entropy in terms of H and G. That is, from the definition of G,

$$-S = \frac{G - H}{T}.$$ (5.70)

Substituting Eq. 5.70 into Eq. 5.20 yields

$$\left(\frac{\partial G}{\partial T}\right)_P = \frac{G - H}{T}.$$ (5.71)

Dividing both sides of Eq. 5.71 by T and then rearranging terms gives

$$\left(\frac{1}{T}\right)\left(\frac{\partial G}{\partial T}\right)_P - \frac{G}{T^2} = -\frac{H}{T^2}.$$ (5.72)

The left side of Eq. 5.72 is the exact differential of the function $[G/T]$. This is easily seen by taking the derivative of the $[G/T]$ ratio with respect to temperature. The result is an expression identical to the left side of Eq. 5.72, which we can now express in the form

$$\frac{\partial}{\partial T}\left(\frac{G}{T}\right)_P = -\frac{H}{T^2}.$$ (5.73)

By starting with Eq. 5.17, $(\partial A/\partial T)_V = -S$, and going through the same derivation, but with $-S$ replaced with $(A - U)/T$ [see Problem 5.14], we may also show that

$$\frac{\partial}{\partial T}\left(\frac{A}{T}\right)_V = -\frac{U}{T^2}.$$ (5.74)

Equations 5.73 and 5.74 are known as the *Gibbs–Helmholtz equations*. Together, they permit us to compute changes in G, A, μ, and the equilibrium constant produced by changes in temperature. As we shall see, Eq. 5.73 also provides a very convenient means of obtaining H or ΔH if G or ΔG is known, while Eq. 5.74 allows us to compute U or ΔU if we know A or ΔA.

Let us apply the Gibbs–Helmholtz equations to a chemical reaction. We again consider the generalized four-component reaction

$$\nu_A \, A(g) + \nu_B \, B(g) \longrightarrow \nu_C \, C(g) + \nu_D \, D(g).$$

The change in the Gibbs free energy for this reaction is

$$\Delta G = G_{products} - G_{reactants}.$$ (5.75)

Dividing both sides of Eq. 5.75 by T produces a form to which we can easily apply the Gibbs–Helmholtz equations. Taking the derivative of both sides with respect to temperature at constant pressure produces

$$\frac{\partial}{\partial T}\left[\frac{\Delta G}{T}\right]_P = \frac{\partial}{\partial T}\left[\frac{G_{\text{products}}}{T}\right]_P - \frac{\partial}{\partial T}\left[\frac{G_{\text{reactants}}}{T}\right]_P.$$

Applying of the Gibbs–Helmholtz equation to the right side gives

$$\frac{\partial}{\partial T}\left[\frac{\Delta G}{T}\right]_P = -\left[\frac{H_{\text{products}}}{T^2} - \frac{H_{\text{reactants}}}{T^2}\right] = -\frac{1}{T^2}\left[H_{\text{products}} - H_{\text{reactants}}\right]. \quad \textbf{(5.76)}$$

The difference $H_{\text{products}} - H_{\text{reactants}}$ is easily identified as ΔH for the reaction. This observation allows us to write Eq. 5.76 in the form

$$\boxed{\frac{\partial}{\partial T}\left[\frac{\Delta G}{T}\right]_P = -\frac{\Delta H}{T^2}.} \quad \textbf{(5.77)}$$

Equation 5.77 is nearly identical in appearance to Eq. 5.73. It tells us that the Gibbs–Helmholtz equations hold not only for G, but also for changes in G. It is as if we simply multiplied Eq. 5.73 by Δ. A similar result can be obtained for changes in A:

$$\boxed{\frac{\partial}{\partial T}\left[\frac{\Delta A}{T}\right]_V = -\frac{\Delta U}{T^2}.} \quad \textbf{(5.78)}$$

Equations 5.77 and 5.78 relate ΔH and ΔU for a process to the slopes of plots of $\Delta G/T$ and $\Delta U/T$, respectively, as a function of T. The relationship is shown in Figure 5.9 for hypothetical exothermic and endothermic reactions in which $\Delta H°(298\text{ K})$ is $-20,000$ J and $+20,000$ J, respectively, and ΔC_p is a constant equal to 40 J K^{-1}.

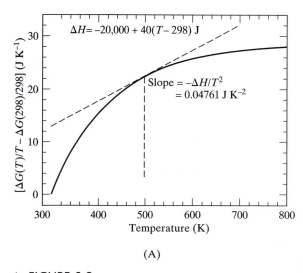
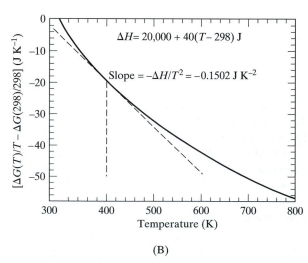

▲ FIGURE 5.9

Variation of $\Delta G(T)/T$ as a function of T. (A) Exothermic reaction with $\Delta H = -20,000 + \Delta C_p(T - 298)$. (B) Endothermic reaction with $\Delta H = 20,000 + \Delta C_p(T - 298)$. In each case, $\Delta C_p = 40$ J K^{-1}. The ordinates of the two plots show the difference between $\Delta G(T)/T$ at temperature T and its value at $T = 298$ K.

EXAMPLE 5.11

Three moles of an ideal gas are isothermally expanded from 50 to 75 L at 300 K. Compute ΔG and ΔH for the process.

Solution

Using the fundamental equation $dG = -S\,dT + V\,dP$, we immediately obtain

$$\left(\frac{\partial G}{\partial P}\right)_T = V = \frac{nRT}{P}, \tag{1}$$

for an ideal gas. Separation of variables and integration then gives the total change in G:

$$\Delta G = \int dG = \int_{P_1}^{P_2} \frac{nRT\,dP}{P} = nRT \int_{P_1}^{P_2} \frac{dP}{P} = nRT \ln\left[\frac{P_2}{P_1}\right]. \tag{2}$$

For an ideal gas undergoing an isothermal change, we have, from Boyle's law,

$$\frac{P_2}{P_1} = \frac{V_1}{V_2}. \tag{3}$$

Substituting Eq. 3 into Eq. 2 gives

$$\Delta G = nRT \ln\left[\frac{V_1}{V_2}\right] = (3\text{ mol})(8.314\text{ J mol}^{-1}\text{ K}^{-1})(300\text{ K}) \ln\left[\frac{50}{75}\right] = -3{,}034\text{ J}. \tag{4}$$

We can obtain ΔH by any of several methods. Let us use the Gibbs–Helmholtz equations. We have, from Eq. 5.77,

$$\frac{\partial}{\partial T}\left[\frac{\Delta G}{T}\right]_P = -\frac{\Delta H}{T^2}. \tag{5}$$

Substituting Eq. 2 into the left side of Eq. 5 gives

$$\frac{\partial}{\partial T}\left[\frac{\Delta G}{T}\right]_P = \frac{\partial}{\partial T}\left[\frac{nRT\ln(P_2/P_1)}{T}\right]_P = \frac{\partial}{\partial T}\left[nR\ln\left(\frac{P_2}{P_1}\right)\right] = 0, \tag{6}$$

since $nR\ln(P_2/P_1)$ is independent of T. Consequently, we obtain

$$-\frac{\Delta H}{T^2} = 0, \tag{7}$$

which requires that $\Delta H = 0$. This is the expected result for an ideal gas, because, for any isothermal process, we have $dH = C_p\,dT = 0$.

EXAMPLE 5.12

Three moles of a nonideal gas are isothermally expanded from P_1 to P_2 at temperature T. Obtain ΔG and ΔH for the process as a function of P_1, P_2, T, α, and derivatives of α if the equation of state of the gas is $PV = nRT + n\alpha P$, where α is a function of temperature only.

Solution

Using the fundamental equation $dG = -S\,dT + V\,dP$, we obtain

$$\left(\frac{\partial G}{\partial P}\right)_T = V. \tag{1}$$

Solving the equation of state for the volume, we get

$$V = \frac{nRT}{P} + n\alpha. \qquad (2)$$

Substituting Eq. 2 into Eq. 1, separating the variables, and integrating between the given limits gives

$$\Delta G = \int dG = \int_{P_2}^{P_1} \left[\frac{nRT}{P} + n\alpha \right] dP = nRT \ln \left[\frac{P_2}{P_1} \right] + n\alpha \left[P_2 - P_1 \right], \qquad (3)$$

which is the required result for ΔG.

We may again obtain ΔH using the Gibbs–Helmholtz equation (Method 1),

$$\frac{\partial}{\partial T} \left[\frac{\Delta G}{T} \right]_P = -\frac{\Delta H}{T^2}. \qquad (4)$$

Substituting Eq. 3 into the left side of Eq. 4 produces

$$\frac{\partial}{\partial T} \left[\frac{\Delta G}{T} \right]_P = \frac{\partial}{\partial T} \left[\frac{nRT \ln[P_2/P_1] + n\alpha \left[P_2 - P_1 \right]}{T} \right]_P = \frac{\partial}{\partial T} \left[\frac{n\alpha[P_2 - P_1]}{T} \right]_P, \qquad (5)$$

since the first term is independent of the temperature. (See Example 5.11.) The derivative of the ratio is

$$\frac{\partial}{\partial T} \left[\frac{n\alpha \left[P_2 - P_1 \right]}{T} \right]_P = -\frac{n\alpha[P_2 - P_1]}{T^2} + \frac{n \left[P_2 - P_1 \right]}{T} \left(\frac{d\alpha}{dT} \right). \qquad (6)$$

Together, the result in Eq. 6 and the Gibbs–Helmholtz equation show that we must have

$$-\frac{\Delta H}{T^2} = -\frac{n\alpha[P_2 - P_1]}{T^2} + \frac{n \left[P_2 - P_1 \right]}{T} \left(\frac{d\alpha}{dT} \right). \qquad (7)$$

Solving for ΔH, we obtain

$$\Delta H = n\alpha \left[P_2 - P_1 \right] - nT[P_2 - P_1] \left(\frac{d\alpha}{dT} \right) = n[P_2 - P_1] \left\{ \alpha - T \left(\frac{d\alpha}{dT} \right) \right\}, \qquad (8)$$

which is the required result for ΔH.

It is not necessary to use the Gibbs–Helmholtz equation to obtain ΔH. We could equally well proceed by using the fundamental equation for dH (Method 2): $dH = T \, dS + V \, dP$. From this, we obtain

$$\left(\frac{\partial H}{\partial P} \right)_T = T \left(\frac{\partial S}{\partial P} \right)_T + V, \qquad (9)$$

which is the derivative needed, since we wish to compute ΔH for an isothermal change of pressure. The rate of change of entropy with pressure at constant temperature is given by the Maxwell relationship obtained from dG, Eq. 5.26:

$$\left(\frac{\partial S}{\partial P} \right)_T = -\left(\frac{\partial V}{\partial T} \right)_P. \qquad (10)$$

Hence,

$$\left(\frac{\partial H}{\partial P} \right)_T = V - T \left(\frac{\partial V}{\partial T} \right)_P. \qquad (11)$$

The equation of state gives

$$V = \frac{nRT}{P} + n\alpha. \qquad (12)$$

Taking the derivative of V with respect to temperature at $dP = 0$ produces

$$\left(\frac{\partial V}{\partial T}\right)_P = \frac{nR}{P} + n\frac{d\alpha}{dT}. \tag{13}$$

Substituting Eqs. 12 and 13 into Eq. 11 yields

$$\left(\frac{\partial H}{\partial P}\right)_T = V - T\left(\frac{\partial V}{\partial T}\right)_P = \frac{nRT}{P} + n\alpha - T\left[\frac{nR}{P} + n\left(\frac{d\alpha}{dT}\right)\right] = n\alpha - nT\left(\frac{d\alpha}{dT}\right). \tag{14}$$

We may now separate the variables in Eq. 14 and integrate both sides to obtain ΔH:

$$\Delta H = \int dH = \int_{P_1}^{P_2}\left[n\alpha - nT\frac{d\alpha}{dT}\right]dP = \left[n\alpha - nT\left(\frac{d\alpha}{dT}\right)\right]\int_{P_1}^{P_2}dP$$

$$= \left[\alpha - T\left(\frac{d\alpha}{dT}\right)\right]n\left[P_2 - P_1\right], \tag{15}$$

which is identical to Eq. 8.

For a related exercise, see Problem 5.13.

5.7   The Temperature Dependence of K_p

The equilibrium constant at 298.15 K can easily be computed from standard chemical potentials measured at that temperature. Example 5.10 illustrates the method. If, however, the temperature differs from 298.15 K, we must have a means of calculating the changes in the standard chemical potentials produced by the temperature variation. This means that we must obtain either $(\partial\mu^o/\partial T)$ or its equivalent. The easiest route to the solution of this problem is via the Gibbs–Helmholtz equations.

Using Eq. 5.48, we have

$$\frac{G(T, P, n)}{T} = \frac{n\,\mu^o(T) + nRT\ln(P)}{T} = \frac{n\,\mu^o}{T} + nR\ln(P). \tag{5.79}$$

Substituting this result into the Gibbs–Helmholtz equation gives

$$\frac{\partial}{\partial T}\left[\frac{G(T, P, n)}{T}\right]_P = \frac{\partial}{\partial T}\left[\frac{n\,\mu^o}{T} + nR\ln(P)\right]_P = n\frac{\partial}{\partial T}\left[\frac{\mu^o}{T}\right] = -\frac{H^o}{T^2}, \tag{5.80}$$

since $nR\ln(P)$ is independent of temperature and μ^o/T is independent of pressure, so that the subscript P may be dropped. To obtain the rate of change of the chemical potential with temperature, we take the derivative of both sides of Eq. 5.80 with respect to n:

$$\frac{\partial}{\partial n}\left\{n\frac{\partial}{\partial T}\left[\frac{\mu^o}{T}\right]\right\}_{T,P} = \frac{\partial}{\partial T}\left[\frac{\mu^o}{T}\right] = -\frac{1}{T^2}\left(\frac{\partial H^o}{\partial n}\right)_{T,P} = -\frac{\overline{H}^o}{T^2}. \tag{5.81}$$

Equation 5.81 tells us that the rate of change of μ^o/T with temperature is the negative of the standard partial molar enthalpy, divided by T^2. If we apply this result to Eq. 5.57, we obtain

$$\frac{\partial}{\partial T}\left[\frac{\Delta\mu^o}{T}\right] = \nu_C\frac{\partial}{\partial T}\left[\frac{\mu_C^o}{T}\right] + \nu_D\frac{\partial}{\partial T}\left[\frac{\mu_D^o}{T}\right] - \nu_A\frac{\partial}{\partial T}\left[\frac{\mu_A^o}{T}\right] - \nu_B\frac{\partial}{\partial T}\left[\frac{\mu_B^o}{T}\right]$$

$$= -\left[\frac{1}{T^2}\right]\left[\nu_C\overline{H}_C^o + \nu_D\overline{H}_D^o - \nu_A\overline{H}_A^o - \nu_B\overline{H}_B^o\right]. \tag{5.82}$$

The quantity in brackets on the right is the standard heat of reaction. Therefore,

$$\frac{\partial}{\partial T}\left[\frac{\Delta\mu^o}{T}\right] = -\frac{\Delta H^o}{T^2}. \tag{5.83}$$

Substituting Eq. 5.59 into Eq. 5.83 gives

$$\frac{\partial}{\partial T}\left[-R\ln(K_p)\right] = -R\frac{\partial\ln(K_p)}{\partial T} = -\frac{\Delta H^o}{T^2}. \tag{5.84}$$

Equation 5.84 provides a simple expression giving the rate at which K_p varies as we change the temperature:

$$\frac{\partial\ln(K_p)}{\partial T} = \frac{\Delta H^o}{RT^2}. \tag{5.85}$$

If we know K_p at temperature T_1, it can be obtained at temperature T_2 by separating the variables in Eq. 5.85 and integrating between corresponding limits:

$$\int_{K_p(T_1)}^{K_p(T_2)} d\ln(K_p) = \ln[K_p(T_2)] - \ln[K_p(T_1)] =$$

$$\ln\left[\frac{K_p(T_2)}{K_p(T_1)}\right] = \int_{T_1}^{T_2}\frac{\Delta H^o\,dT}{RT^2}. \tag{5.86}$$

If we assume that ΔH^o is independent of T, we may factor ΔH^o out of the integral in Eq. 5.86 to obtain a very simple form:

$$\ln\left[\frac{K_p(T_2)}{K_p(T_1)}\right] = -\frac{\Delta H^o}{R}\left[\frac{1}{T_2} - \frac{1}{T_1}\right]. \tag{5.87}$$

The assumption that ΔH^o is independent of temperature is equivalent to assuming that $\Delta C_p = 0$, which requires that the total heat capacity of the products and reactants be the same at all temperatures. If the range from T_1 to T_2 is not too large, Eq. 5.87 will be sufficiently accurate for most calculations. Since K_p at 298.15 K is easily computed from tables of standard chemical potentials, T_1 will usually be 298.15 K.

Equation 5.87 is the first example of a general form that frequently recurs in physical science. Let us write this equation for the case where temperature T_1 changes to some arbitrary temperature T. When this is done, the equation assumes the form

$$\ln\left[\frac{K_p(T)}{K_p(T_1)}\right] = -\Delta H^o\left[\frac{1}{RT} - \frac{1}{RT_1}\right].$$

Performing exponentiation on both sides of this equation, we obtain

$$\frac{K_p(T)}{K_p(T_1)} = \exp\left\{-\Delta H^o\left[\frac{1}{RT} - \frac{1}{RT_1}\right]\right\} = \exp\left\{-\frac{\Delta H^o}{RT_1}\right\}\exp\left\{-\frac{\Delta H^o}{RT}\right\}.$$

This equation may be written as

$$K_p(T) = K_p(T_1) \exp\left\{-\frac{\Delta H^o}{RT_1}\right\} \exp\left\{-\frac{\Delta H^o}{RT}\right\} = C \exp\left\{-\frac{\Delta H^o}{RT}\right\}. \quad (5.88)$$

Equation 5.88 shows that if ΔH^o is independent of temperature, the equilibrium constant varies exponentially with temperature, with the argument of the exponent being $-(\Delta H^o)/(RT)$. As we move through the application of thermodynamics to various experimental systems, we are going to find that if there is an energy requirement for a given process, then any variable X that measures the extent or rate of the process will vary exponentially with the temperature according to the equation

$$X = (\text{constant}) \exp\left[\frac{-\text{Energy Requirement}}{RT}\right], \quad (5.89)$$

as long as the energy requirement is assumed to be independent of temperature. In the current situation, the extent to which the equilibrium process is shifted to the right is measured by the value of K_p. Later in the text, we shall show why we expect a result in the form of Eq. 5.89. If ΔH^o is temperature dependent, the simple form of Eqs. 5.87 and 5.89 is lost.

EXAMPLE 5.13

Consider the equilibrium between $NO_2(g)$ and $N_2O_4(g)$:

$$2\,NO_2(g) = N_2O_4(g).$$

(A) Using the data in Appendix A, compute K_p at 298.15 K. **(B)** Assuming that ΔH^o for the reaction at 298.15 K is independent of temperature, calculate the value of K_p for the preceding reaction at 400 K. Plot $\ln(K_p)$ as a function of temperature from 298.15 K to 500 K. Prepare a second plot of $\ln(K_p)$ versus T^{-1}. What is the slope of this plot?

Solution

(A) The value of K_p at 298.15 K may be obtained directly from the standard chemical potentials of $NO_2(g)$ and $N_2O_4(g)$. We have

$$\Delta\mu^o = \mu^o_{N_2O_4(g)} - 2\mu^o_{NO_2(g)} = 97.89\ \text{kJ mol}^{-1}2(51.31)\ \text{kJ mol}^{-1} = -4.73\ \text{kJ mol}^{-1}. \quad (1)$$

Therefore,

$$K_p = \exp\left[-\frac{\Delta\mu^o}{RT}\right] = \exp\left[\frac{4730\ \text{J mol}^{-1}}{(8.314\ \text{J mol}^{-1}\,\text{K}^{-1})(298.15\ \text{K})}\right]$$

$$= \exp(1.9082) = 6.741. \quad (2)$$

(B) The standard enthalpy of reaction at 298.15 K is given by

$$\Delta H = \Delta H^o = \overline{H}^o_{N_2O_4(g)} - 2\overline{H}^o_{NO_2(g)} = 9.16\ \text{kJ mol}^{-1} - 2(33.18)\ \text{kJ mol}^{-1}$$

$$= -57.20\ \text{kJ mol}^{-1}. \quad (3)$$

If ΔH^o is constant, Eq. 5.87 holds, and we have

$$\ln\left[\frac{K_p(400\ \text{K})}{6.741}\right] = \frac{5.720 \times 10^4\ \text{J mol}^{-1}}{8.314\ \text{J mol}^{-1}\,\text{K}^{-1}}\left[\frac{1}{400} - \frac{1}{298.15}\right]\text{K}^{-1} = -5.876. \quad (4)$$

Performing exponentiation on both sides gives

$$\frac{K_p(400\ \text{K})}{6.741} = \exp(-5.876) = 0.002806. \quad (5)$$

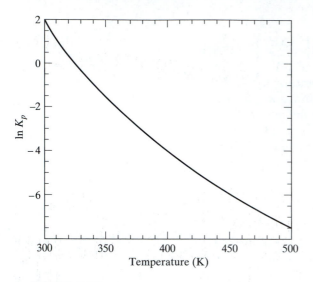

▲ **FIGURE 5.10**
The variation of K_p for the reaction $2\,NO_{2(g)} = N_2O_{4(g)}$ with temperature. Since the reaction is exothermic, K_p decreases with T, as predicted by Le Chatelier's principle. (See Section 5.8)

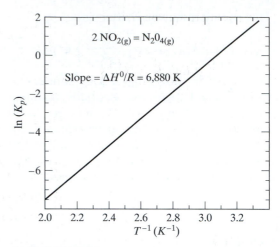

▲ **FIGURE 5.11**
The variation of K_p for the reaction $2\,NO_{2(g)} = N_2O_{4(g)}$ with T^{-1}. When ΔH^o is independent of temperature, this plot should be linear with a slope equal to $-\Delta H^o/R$. As can be seen, this expectation is reflected in the linearity of the plot. The positive slope tells us that ΔH^o must be negative and the reaction exothermic.

Therefore,

$$K_p(400\text{ K}) = 6.741\,(0.002806) = 0.01892. \tag{6}$$

At an arbitrary temperature T, Eq. 4 becomes

$$\ln K_p(T) = \ln(6.741) + 6880\left[\frac{1}{T} - \frac{1}{298.15}\right]. \tag{7}$$

Figure 5.10 shows the requested plot of $\ln(K_p(T))$ versus T. Figure 5.11 shows a plot of $\ln(K_p)$ versus T^{-1}. Equation 7 demonstrates that this plot should be linear, with a slope equal to $\Delta H^o/R$ if ΔH^o is constant.

For related exercises, see Problems 5.18 and 5.19.

EXAMPLE 5.14

For a particular equilibrium process, it is found that $\Delta C_p = a + bT + cT^{-2}$, where a, b, and c are constants. Obtain an expression giving K_p at temperature T in terms of $K_p(298.15\text{ K})$, (the standard enthalpy of reaction at 298.15 K) ΔH^o, a, b, and c. Do we obtain a result in the form of Eq. 5.89? Explain.

Solution

The rate of change of the heat of reaction at constant pressure is given by

$$\left(\frac{\partial \Delta H}{\partial T}\right)_P = \Delta C_p = a + bT + cT^{-2}. \tag{1}$$

Separating variables and integrating between 298.15 and temperature T, we obtain

$$\int_{\Delta H^o(298.15\text{ K})}^{\Delta H^o(T)} d\Delta H = \Delta H^o(T) - \Delta H^o(298.15\text{ K}) = \int_{298.15\text{ K}}^{T} (a + bT + cT^{-2})dT$$

$$= a(T - 298.15) + \left(\frac{b}{2}\right)[T^2 - 298.15^2] - c\left[\frac{1}{T} - \frac{1}{298.15}\right]. \tag{2}$$

The standard enthalpy of reaction at temperature T is, therefore,

$$\Delta H^\circ(T) = \Delta H^\circ(298.15) + a(T - 298.15) + \left(\frac{b}{2}\right)[T^2 - 298.15^2] - c\left[\frac{1}{T} - \frac{1}{298.15}\right]$$

$$= \left[\Delta H^\circ(298.15) - 298.15a - \frac{298.15^2 b}{2} + \frac{c}{298.15}\right] + aT + \frac{bT^2}{2} - \frac{c}{T}$$

$$= C + aT + \frac{bT^2}{2} - \frac{c}{T}, \tag{3}$$

where

$$C = \text{constant} = \left[\Delta H^\circ(298.15) - 298.15a - \frac{298.15^2 b}{2} + \frac{c}{298.15}\right].$$

Substituting $\Delta H^\circ(T)$ into Eq. 5.86, we obtain

$$\ln\left[\frac{K_p(T)}{K_p(298.15\text{ K})}\right] = \frac{1}{R}\int_{298.15\text{ K}}^{T} \frac{\Delta H^\circ\, \partial T}{T^2} = \frac{1}{R}\int_{298.15\text{ K}}^{T} \frac{\left[C + aT + \frac{bT^2}{2} - \frac{c}{T}\right]\partial T}{T^2}$$

$$= -\frac{C}{R}\left[\frac{1}{T} - \frac{1}{298.15}\right] + \frac{a}{R}\ln\left[\frac{T}{298.15}\right] + \frac{b}{2R}[T - 298.15] + \frac{c}{2R}\left[\frac{1}{T^2} - \frac{1}{298.15^2}\right], \tag{4}$$

which is the required expression for $K_p(T)$ in terms of ΔH°, $K_p(298.15\text{ K})$, a, b, and c.

We do not obtain an equation in the form of Eq. 5.89, because the energy requirement for the process, $\Delta H^\circ(T)$, is not independent of temperature.

5.8   Le Chatelier's Principle

In 1884, the French chemist Henri Le Chatelier put forward a useful qualitative concept that has wide application in science. In effect, Le Chatelier's principle states that *when a system in equilibrium is subjected to a stress, the system responds in a manner that will partially relieve that stress.* This principle often permits us to predict qualitatively, without executing any actual calculations, what an equilibrium system will do when the experimental conditions are altered. Let us consider a few examples.

Suppose a tightrope walker is traversing a cable between the twin towers of the World Trade Center 1,400 ft above the ground. The walker and the cable constitute an equilibrium system (at least we hope so; otherwise he will be falling to his death). It is clear that if a wind begins to blow against the walker from the east, he feels a stress. Le Chatelier's principle predicts that the walker will lean slightly toward the east into the wind in order to relieve the stress. And of course, that is precisely what will occur. Note that, using Le Chatelier's principle, we cannot predict the angle at which the walker leans; such quantitative detail requires more information and mathematical calculations. But we can predict that the angle will deviate from the vertical and will be toward the east. Figure 5.12 presents two additional simple illustrations of Le Chatelier's principle.

Now consider a hypothetical chemical reaction at equilibrium:

$$A(g) + B(g) \longrightarrow C(g) + 25\text{ kJ}.$$

Le Chatelier's principle predicts that if we add heat and thereby raise the temperature, the equilibrium will shift to the left ($C(g)$ reacts to form more $A(g)$ and $B(g)$) so as to absorb some of the added heat and thereby relieve the stress). This is exactly what occurs. If, in contrast, we lower the temperature

▶ FIGURE 5.12
Two illustrations of Le Chatelier's principle. (A) Subjected to a stress that displaces it from equilibrium, a balanced beam residing on a fulcrum responds in a manner that attempts to restore the beam to its equilibrium position. (B) Ice and water are at equilibrium at the normal melting point of 273.15 K. The system is subjected to a stress by a piston that applies pressure to the ice. Since ice is less dense than water, the stress can be relieved if the equilibrium shifts toward the water, which occupies less volume. Therefore, Le Chatelier's principle tells us that the ice will melt at 273.15 K when subjected to a large pressure. This is exactly what occurs. The process will be examined in more quantitative detail in Chapter 6.

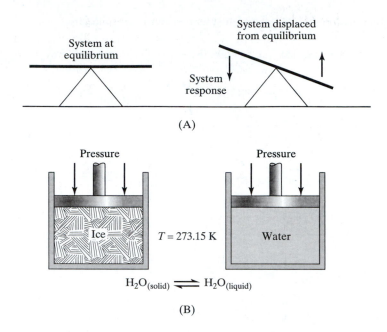

by removing heat, the equilibrium shifts towards the product side. If the total pressure is suddenly increased, producing stress, the equilibrium will shift toward the products, thereby reducing the number of moles of gas present, which will in turn reduce the pressure and thereby relieve the stress. If we add more $A(g)$ to the system, the equilibrium will again shift toward the products to remove some of the added reactant and thereby relieve the stress.

Le Chatelier's principle is nothing new to students of physical chemistry. The principle is usually introduced in beginning chemistry courses. However, what is often not well understood is the underlying thermodynamic reason for the equilibrium shift. There are two possibilities. The first of these arises because the change in conditions threatens to change the equilibrium constant, whereas thermodynamics requires its value to remain fixed. Changing the total pressure in the reaction or adding more reactant $A(g)$ is a change in condition that falls into this category. This can easily be seen by examining the form of K_p. For the foregoing reaction, we have

$$K_p = \left[\frac{P_{C(g)}}{P_{A(g)} P_{B(g)}} \right]_{eq} = \left[\frac{X_{C(g)} P_T}{X_{A(g)} P_T X_{B(g)} P_T} \right]_{eq} = \left[\frac{X_{C(g)}}{X_{A(g)} X_{B(g)}} \right]_{eq} \left[\frac{1}{P_T} \right],$$

where X_i is the mole fraction of gas i and P_T is the total pressure. Since K_p depends only upon temperature, its value cannot be altered by increasing P_T. However, the form of the right-hand side of the equation shows that K_p will decrease if we increase P_T, unless simultaneously $X_{C(g)}$ increases at the expense of $X_{A(g)}$ and $X_{B(g)}$—that is, unless the equilibrium shifts toward the products. The same logic holds if we add more $A(g)$: This will increase $P_{A(g)}$ and, therefore, decrease K_p. But thermodynamics says this cannot happen, since we are not changing the temperature. Consequently, the equilibrium must shift toward the product side to maintain K_p at its original value.

The underlying reason for the equilibrium shift that accompanies a change in temperature is different. In this case, we are changing the value of K_p. The Gibbs–Helmholtz equations lead to Eq. 5.88, $K_p = C \exp[-\Delta H^o / RT]$. If the reaction is exothermic, such as the one we are considering, then $\Delta H^o < 0$, and K_p must decrease as T increases. If K_p decreases, it is clear that $P_{C(g)}$ must decrease, while $P_{A(g)}$ and $P_{B(g)}$ increase. In other words, the equilibrium must shift toward the reactants. Le Chatelier's principle holds in either case, but it is important to distinguish equilibrium shifts caused by an actual change of the equilibrium constant from those shifts that result because of the need to maintain a constant value for K_p.

5.9   Nonideal Gases

5.9.1   General Considerations

In principle, the treatment of equilibria among nonideal gases starts with Eq. 5.46, $G(T,P) = \int V\,dP$. Instead of substituting the ideal-gas expression for V, we simply solve the nonideal equation of state for the volume and substitute the result into Eq. 5.46. After integrating, we obtain the Gibbs free energy in the form

$$G(T,P) = \int V\,dP = n\,\mu^o(T) + f(T, P, n),$$

where $\mu^o(T)$ is the standard Gibbs free energy. The functional form of $f(T, P, n)$ depends upon the equation of state for the gas. Note that while this procedure is straightforward, there may be difficulties associated with its execution. In addition, there are often some unpleasant surprises awaiting in the wings. The nature of these surprises is best seen by considering a simple example.

Suppose all the gases in the equilibrium reaction

$$\nu_A\,A(g) + \nu_B\,B(g) \longrightarrow \nu_C\,C(g) + \nu_D\,D(g)$$

obey an equation of state of the form $PV = nRT + n\alpha_i P$, where α_i is the parameter associated with gas i. Let us assume that α_i is a function of temperature, but not of pressure. If we proceed along the lines used in Sections 5.1 and 5.2, the first step is to obtain an appropriate expression for the Gibbs free energy by using Eq. 5.46. Since we have $V = RT/P + \alpha_i$ for 1 mole of gas, direct substitution into Eq. 5.46 yields

$$G_i(T,P) = \int V\,dP = \int \left[\frac{RT}{P} + \alpha_i\right] dP = \Phi(T) + RT\ln(P) + \alpha_i\,P, \qquad \textbf{(5.90)}$$

where $\Phi(T)$ is the function of integration that depends only upon temperature. If we denote this function by $\mu^o(T)$ and write Eq. 5.90 for n moles of gas, we obtain

$$G_i(T, P, n) = n\mu^o(T) + nRT\ln(P) + n\alpha_i P. \qquad \textbf{(5.91)}$$

Differentiating both sides of Eq. 5.91 with respect to n at constant temperature and pressure yields the following expression for the chemical potential for the nonideal gas:

$$\left(\frac{\partial G_i(T, P, n)}{\partial n}\right)_{T,P} = \mu_i = \mu_i^o(T) + RT\ln(P) + \alpha_i\,P. \qquad \textbf{(5.92)}$$

Because we were able to solve the equation of state for V without difficulty to obtain very simple integrals, the development of an expression for the chemical potential for this type of nonideal gas was easy. Even so, the results are not completely pleasing. Equation 5.52 gives the change in the Gibbs free energy when $d\varepsilon$ moles of $A(g)$ react to form products. The result is

$$dG = \frac{d\varepsilon}{\nu_A}\left[\nu_C\,\mu_C + \nu_D\,\mu_D - \nu_A\,\mu_A - \nu_B\,\mu_B\right].$$

Following our development of the expression for the equilibrium constant for an ideal system, we substitute the chemical potentials given by Eq. 5.92 into this expression for dG. After collecting terms, we obtain

$$dG = \frac{d\varepsilon}{\nu_A} [\nu_C \mu_C^o + \nu_D \mu_D^o - \nu_A \mu_A^o - \nu_B \mu_B^o] + \frac{d\varepsilon RT}{\nu_A} \ln \left[\frac{P_C^{\nu_C} P_D^{\nu_D}}{P_A^{\nu_A} P_B^{\nu_B}} \right]$$

$$+ \frac{d\varepsilon}{\nu_A} [\alpha_C P_C + \alpha_D P_D - \alpha_A P_A - \alpha_B P_B]. \tag{5.93}$$

When the reaction is at equilibrium under conditions of constant temperature and pressure, we must have $dG = 0$. If we write $\Delta\mu^o = [\nu_C \mu_C^o + \nu_D \mu_D^o - \nu_A \mu_A^o - \nu_B \mu_B^o]$ and multiply by $\nu_A/d\varepsilon$, Eq. 5.93 becomes

$$\Delta\mu^o = -RT \ln \left[\frac{P_C^{\nu_C} P_D^{\nu_D}}{P_A^{\nu_A} P_B^{\nu_B}} \right]_{(eq)} - [\alpha_C P_C + \alpha_D P_D - \alpha_A P_A - \alpha_B P_B]_{(eq)}. \tag{5.94}$$

The left-hand side of Eq. 5.94 is dependent only upon temperature. Therefore, the same must be true of the right-hand side. When we had ideal gases, only the first term on the right-hand side was present. This permitted us to make the argument that the pressure ratio had to be a constant at a given temperature, so that we could write an equilibrium constant in the form

$$K_p = \left[\frac{P_C^{\nu_C} P_D^{\nu_D}}{P_A^{\nu_A} P_B^{\nu_B}} \right]_{eq}.$$

However, we can no longer do this, because, while the right-hand side itself must be a constant at a fixed temperature, it is not true that *each term* must be a constant. Only the *sum* of the two terms must be a constant. We have, therefore, lost the concept of an equilibrium constant.

It is indeed an unpleasant surprise to lose the concept of an equilibrium constant. Such a loss makes the treatment of equilibrium systems difficult. Now, every time a gas is described by a different equation of state, we must develop new equations for the chemical potential. In addition, without equilibrium constants, we lose 50% of all of the problems normally assigned in freshman chemistry courses! To avoid these problems, we adopt a different procedure to handle nonideal systems.

5.9.2   Activities and Fugacities

We encountered difficulties in the previous example primarily because the Gibbs free energy and the chemical potential each contained two terms in addition to $\mu^o(T)$. Even though it was simple to obtain the mathematical form of these terms, the fact that there are two terms instead of one leads to the loss of the equilibrium constant concept. To avoid this, and remove the need to develop new equations for every different equation of state, we require that

$$\boxed{G(T, P, n) = n\mu^o(T) + nRT \ln(a)}, \tag{5.95}$$

so that

$$\left(\frac{\partial G(T, P, n)}{\partial n} \right)_{T, P} = \mu = \mu^o(T) + RT \ln(a), \tag{5.96}$$

where the quantity *a* appearing as the argument of the natural logarithm function in both equations is called the *activity*. The activity is whatever it needs to be in order to make Eqs. 5.95 and 5.96 correct for any system of interest. If the system is an ideal gas, the activity will be identical to the magnitude of the pressure, expressed in bars, but without the unit. For nonideal systems, the activity will need to take a different form.

Before we address the critical question of how we determine the activity so that Eqs. 5.95 and 5.96 will hold, let us examine some of the benefits of defining the activity in this fashion. With a single term on the right-hand side, we no longer lose the concept of an equilibrium constant. If Eq. 5.96 is used for the chemical potential, instead of obtaining Eqs. 5.93 and 5.94 for the nonideal system treated in the previous section, we would have

$$\Delta\mu^o = -RT \ln\left[\frac{a_C^{\nu_C} a_D^{\nu_D}}{a_A^{\nu_A} a_B^{\nu_B}}\right]_{eq}, \tag{5.97}$$

and again, we could make the argument that the activity ratio would have to be a constant at a fixed temperature, since the left-hand side of the equation depends only upon temperature. Thus, we retain the concept of a thermodynamic equilibrium constant, which now takes the form

$$K = \left[\frac{a_C^{\nu_C} a_D^{\nu_D}}{a_A^{\nu_A} a_B^{\nu_B}}\right]_{eq}, \tag{5.98}$$

where the activities are evaluated at the equilibrium point. In addition, we no longer need to develop new expressions every time the equation of state changes. Instead, the equations developed for the ideal-gas system can be employed simply by replacing the partial pressures with the corresponding activities.

The use of Eqs. 5.95 and 5.96 makes the task of treating nonideal systems much easier if we know the activities. But alas, therein lies the problem; Mother Nature rarely, if ever, gives you something for nothing. If we are going to avoid the complications of developing new equations for every nonideal system, Mother Nature demands that we pay a price. That price comes in the form of having to find a means to either calculate or measure appropriate values for the activities.

We can simplify the job of obtaining activities by again noting that we are always measuring *differences* in *G* and *μ*. Consequently, we are free to choose our reference point for the measurement or calculation of the activity in any manner that we desire. To do this, we first introduce the terms *reference function* and *reference state*. We shall denote the reference function as *g* and the reference state of this function as *g**. The reference function is the property that we are going to measure to determine the activity. For example, the reference function in the measurement of temperature using a clinical thermometer is the length of a column of mercury. The reference state assigns the value of the reference function at a certain point. For the clinical thermometer, the reference state is the length of the mercury column at the freezing point of water. This length is assigned the value 32°F or 0°C. As will be seen in the discussion that follows, the most convenient reference point for the measurement of activities is to take

$$\lim_{g \to g^*}\left[\frac{a}{g}\right] = 1. \tag{5.99}$$

That is, we take the activity to be equal to the reference function when we approach the reference state. For convenience, we also define an activity coefficient as

$$\gamma = \frac{a}{g},$$

(5.100)

so that our reference point can also be stated as being

$$\underset{g \longrightarrow g^*}{\text{Lim}} (\gamma) = 1.$$

(5.101)

The next decision involves the choice of a specific reference function and reference state. Clever choices will make our life much easier; foolish choices will make it far more difficult. We wish to choose a reference function that is easily measured, and of course, we need to pick a function that is related to the value of the chemical potential in a given state. For gases, the obvious choice is pressure. Pressure is easily measured, and we know that, for an ideal gas, $a = P$. For solutions, other concentration measures, such as the mole fraction (X), molarity (M), or molality (m) are more convenient. The choice for the reference state for gases is suggested by the behavior of gases as the pressure decreases. All gases approach ideal behavior as the pressure approaches zero. We shall, therefore, take that as the reference state. For solutions, we might take infinite dilution ($X \rightarrow 0$, $M \rightarrow 0$, $m \rightarrow 0$) as the reference state. Alternatively, we might choose a solution of unit molarity or molality as the reference state. When the mole fraction is the reference function, $X \rightarrow 1$ is a particularly convenient reference state.

For gases, we choose pressure as our reference function and zero pressure as the reference state. With these choices, our reference point is given by

$$\underset{P \longrightarrow 0}{\text{Lim}} \left[\frac{a}{P} \right] = 1.$$

(5.102)

This choice is in accord with the result that all gases approach ideal behavior as $P \rightarrow 0$. When we approach this limit, the choice made in Eq. 5.102 means that the activity will become equal to the pressure and the chemical potential will become $\mu = \mu^o(T) + RT \ln(P)$, which we know to be the correct result for an ideal gas. Whenever we make the choice $g \equiv P$, and whenever g^* is taken to be a state of zero pressure, for historical reasons, it is customary to replace the activity with the notation f and to call the quantity the *fugacity* of the gas. Then, γ is also termed the *fugacity coefficient* and is written as $\gamma = f/P$. In this notation, Eq. 5.102 becomes

$$\underset{P \longrightarrow 0}{\text{Lim}} \left[\frac{f}{P} \right] = \underset{P \longrightarrow 0}{\text{Lim}} (\gamma) = 1.$$

(5.103)

Having chosen the reference function and reference state for the measurement or calculation of fugacities, we now need an equation that connects f and μ or f and G. From the fundamental equation for the differential of G, we have $dG = -S\,dT + V\,dP$. At constant temperature, this gives Eq. 5.45, $dG = V\,dP$. If we combine that equation with Eq. 5.95, we have

$$dG = nRT\,d\ln(a) = nRT\,d\ln(f) = V\,dP,$$

(5.104)

since μ^o is constant at $dT = 0$. We now integrate both sides between the reference state and some arbitrary pressure and fugacity:

$$nRT \int_{f=f*}^{f} d\ln(f) = \int_{P=P*}^{P} V\, dP. \tag{5.105}$$

The left side may be easily integrated to yield

$$nRT \ln(f) - nRT \ln(f*) = \int_{P*}^{P} V\, dP. \tag{5.106}$$

Our choice of reference point means that we know that as $P* \to 0$, $f* \to P*$, and the ratio $f*/P* \to 1$. In order to form this ratio, let us subtract the quantity $nRT \ln(P/P*)$ from both sides. This operation gives

$$nRT \ln(f) - nRT \ln(P) - nRT \ln(f*) + nRT \ln(P*)$$

$$= \int_{P*}^{P} V\, dP - nRT \ln(P/P*). \tag{5.107}$$

Since we have

$$\int_{P*}^{P} \frac{nRT\, dP}{P} = nRT \ln\left(\frac{P}{P*}\right),$$

the right-hand side of this equation can be written in the form

$$\int_{P*}^{P} V\, dP - nRT \ln\left(\frac{P}{P*}\right) = \int_{P*}^{P}\left[nV_m - \frac{nRT}{P}\right]dP,$$

where V_m is the molar volume of the gas. Combining this expression with Eq. 5.107 and writing the differences of logarithms as the logarithm of a ratio, we obtain

$$nRT \ln(f) - nRT \ln(P) - nRT \ln\left[\frac{f*}{P*}\right] = \int_{P*}^{P}\left[nV_m - \frac{nRT}{P}\right]dP. \tag{5.108}$$

Canceling out n on both sides of Eq. 5.108, dividing by RT, and taking the limit of both sides as $P* \to 0$, we obtain

$$\ln(f) - \ln(P) - \lim_{P* \to 0}\ln\left[\frac{f*}{P*}\right] = (RT)^{-1} \lim_{P* \to 0}\int_{P*}^{P}\left[V_m - \frac{RT}{P}\right]dP. \tag{5.109}$$

Our clever choice of reference state gives $\lim_{P* \to 0} \ln[f*/P*] = \ln(1) = 0$, so that we obtain

$$\boxed{\ln(f) = \ln(P) + \frac{1}{RT}\int_{0}^{P}\left[V_m - \frac{RT}{P}\right]dP}, \tag{5.110}$$

which is the equation connecting the fugacity of a nonideal gas and our reference function P.

   Let us examine Eq. 5.110 qualitatively to verify that it has the correct limiting form at low pressure. First, we note that the quantity appearing inside the integral, $[V_m - RT/P]$ is zero for an ideal gas. Therefore, for such a gas, we have $f = P$ and $\gamma = 1$ at all pressures, and the chemical potential becomes $\mu = \mu^o(T) + RT \ln(P)$. When the gas is not ideal, we must either execute the integral in Eq. 5.110 using the equation of state for the gas or evaluate the integral numerically from measured (P, V, T) data. If we evaluate the integration numerically, we will need the value of the integrand at the lower limit. This value, $\underset{P \longrightarrow 0}{\text{Lim}} [V_m - RT/P]$, is just the residual volume of the nonideal gas. Since the residual volumes of most gases are negative, we expect the integral to have a negative value for moderate pressures. As a result, we will have $\ln(f) < \ln(P)$ and, hence, $f < P$.

EXAMPLE 5.15

One mole of a nonideal gas has the equation of state

$$PV(1 - \beta P) = RT,$$

where β is a function of temperature only. Obtain analytic functions giving the fugacity and the fugacity coefficient for this gas as a function of β and P.

Solution

We need an expression for $V - RT/P$. Solving the equation of state for V, we obtain

$$V = \frac{RT}{P(1 - \beta P)}. \tag{1}$$

Consequently,

$$V - \frac{RT}{P} = \frac{RT}{P(1 - \beta P)} - \frac{RT}{P} = RT \left[\frac{\beta P}{P(1 - \beta P)} \right] = \frac{RT\beta}{1 - \beta P}. \tag{2}$$

Substituting into Eq. 5.110, we obtain

$$\ln(f) = \ln(P) + \beta \int_0^P \frac{dP}{1 - \beta P} = \ln(P) - \ln[1 - \beta P] \Big|_0^P$$

$$= \ln(P) - \ln(1 - \beta P) + \ln(1) = \ln \left[\frac{P}{1 - \beta P} \right]. \tag{3}$$

Performing exponentiation on both sides of Eq. 3 yields

$$f = \frac{P}{1 - \beta P}. \tag{4}$$

The fugacity coefficient is

$$\gamma = \frac{f}{P} = \frac{1}{1 - \beta P}. \tag{5}$$

For a related exercise, see Problem 5.16.

EXAMPLE 5.16

One mole of CO_2 gas is described by the truncated virial equation of state

$$P = RT \left[\frac{1}{V} + \frac{C_2}{V_2} \right],$$

where

$$C_2 = b - \frac{a}{RT}$$

with a and b being the van der Waals parameters for $CO_2(g)$. **(A)** Show that the fugacity of this gas at pressure P and temperature T is given by the expression

$$\ln(f) = \ln(P) + 0.5 \int_0^P \left\{ \left[\frac{1}{P^2} + \frac{4C_2}{RTP} \right]^{1/2} - \frac{1}{P} \right\} dP.$$

(B) Using the van der Waals parameters for $CO_2(g)$ given in Table 1.1, plot the integrand of the preceding integral as a function of pressure from $P = 0$ to $P = 50$ bar. **(C)** Using numerical integration, compute the value of the fugacity coefficient of $CO_2(g)$ at 300 K over the range from 0 to 50 bar of pressure. Plot your results against pressure. What is the fugacity coefficient at 50 bar? Compute the fugacity of $CO_2(g)$ at a pressure of 50 bar at 300 K.

Solution

(A) To utilize Eq. 5.110 for the fugacity, we must obtain an expression for $V - RT/P$. Multiplying the virial equation by V^2, we obtain

$$PV^2 = RTV + RTC_2. \tag{1}$$

Dividing by P and then rearranging terms gives

$$V^2 - \frac{RTV}{P} - \frac{RTC_2}{P} = 0. \tag{2}$$

We may now solve Eq. 2 for V using the quadratic formula:

$$V = \frac{\dfrac{RT}{P} + \left[\dfrac{R^2T^2}{P^2} + \dfrac{4RTC_2}{P} \right]^{1/2}}{2}. \tag{3}$$

The correct root is the one with the plus sign, since, if $C_2 = 0$, the equation must reduce to $V = RT/P$. Using the positive root, we get just that result. Using the negative root, we do not. Subtracting RT/P from both sides of Eq. 3 produces

$$V - \frac{RT}{P} = 0.5 \left[\frac{R^2T^2}{P^2} + \frac{4RTC_2}{P} \right]^{1/2} - \frac{RT}{2P}. \tag{4}$$

Inserting Eq. 4 into Eq. 5.110 gives

$$\ln(f) = \ln(P) + \frac{1}{RT} \int_0^P \left[0.5 \left[\frac{R^2T^2}{P^2} + \frac{4RTC_2}{P} \right]^{1/2} - \frac{RT}{2P} \right] dP. \tag{5}$$

Factoring RT out of the integral and canceling the $(RT)^{-1}$ multiplicative factor, we obtain

$$\ln(f) = \ln(P) + 0.5 \int_0^P \left\{ \left[\frac{1}{P^2} + \frac{4C_2}{RTP} \right]^{1/2} - \frac{1}{P} \right\} dP, \tag{6}$$

which is the desired result.

(B) To perform the integration in Eq. 6 numerically, we need the intercept of the integrand when $P = 0$. As noted in the text, this is the residual volume of the gas. Multiplying the equation of state by V gives

$$PV = RT + \frac{RTC_2}{V}. \tag{7}$$

Dividing Eq. 7 by P, and rearranging terms, we obtain

$$V - \frac{RT}{P} = \frac{RTC_2}{PV}. \tag{8}$$

If we take the limit of both sides as $P \rightarrow 0$, we obtain

$$V_{res} = \lim_{P \rightarrow 0}\left[V - \frac{RT}{P}\right] = \lim_{P \rightarrow 0}\left[\frac{RTC_2}{PV}\right] = C_2 = b - \frac{a}{RT}, \tag{9}$$

since we know that $PV \rightarrow RT$ as $P \rightarrow 0$. Using the van der Waals' parameters given in Table 1.1, we get

$$V_{res} = C_2 = b - \frac{a}{RT} = 0.04286\ \text{L} - \frac{3.659\ \text{L}^2\ \text{bar mol}^{-1}}{(0.083144\ \text{L bar mol}^{-1}\ \text{K}^{-1})(300\ \text{K})}$$

$$= -0.1038\ \text{L}. \tag{10}$$

Since the multiplicative factor of $1/(RT)$ has been divided into the integrand, we expect that the intercept at $P = 0$ will be $C_2/(RT) = -0.004163\ \text{bar}^{-1}$. With this intercept, we may now proceed to compute the integrand at equally spaced pressures from 0 to 50 bar. The results are plotted in Figure 5.13.

(C) Rearranging terms in Eq. 6 gives

$$\ln\left[\frac{f}{P}\right] = \ln(\gamma) = 0.5\int_0^P \left[\frac{1}{P^2} + \frac{4C_2}{RTP}\right]^{1/2} - \frac{1}{P}\,dP. \tag{11}$$

Numerically integrating

$$0.5\int_0^P \left\{\left[\frac{1}{P^2} + \frac{4C_2}{RTP}\right]^{1/2} - \frac{1}{P}\right\}dP$$

with the preceding value of C_2 at 300 K for pressures in the range $P = 0.1$ bar to $P = 50$ bar yields the value of $\ln(\gamma)$ at those pressures. The results are shown in Figure 5.14. When $P = 50$ bar, the value of the integral is -0.2406. The fugacity coefficient is, therefore,

$$\gamma = \exp(-0.2406) = 0.786. \tag{12}$$

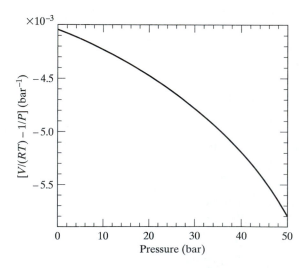

▲ FIGURE 5.13
$[V/(RT) - 1/P]$, in bar^{-1}, as a function of pressure for $CO_2(g)$ when described by a truncated virial expansion with the van der Waals parameters used to estimate the second virial coefficient.

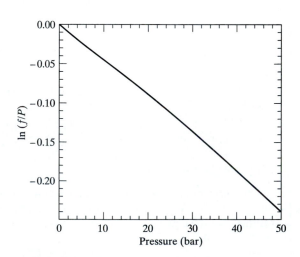

▲ FIGURE 5.14
$\ln(f/P)$, which is the same as $\ln(\gamma)$, as a function of pressure for $CO_2(g)$ when described by a truncated virial expansion with the van der Waals parameters used to estimate the second virial coefficient.

This gives a fugacity of

$$f = \gamma P = (0.786)(50) = 39.3 \qquad \textbf{(13)}$$

for $CO_2(g)$.

For related exercises, see Problems 5.19 and 5.20.

Summary: Key Concepts and Equations

1. The reversibility, spontaneity, or nonspontaneity of a process may be examined by computing ΔS and $\int \delta q/T$ for the process. If we find that $\Delta S > \int \delta q/T$, the process is spontaneous. If the two quantities are equal, the process is reversible. If $\Delta S < \int \delta q/T$, the process is nonspontaneous and will require work to be done to force its occurrence. The relationship between ΔS and $\int \delta q/T$ called the Clausius inequality. Alternatively, we may compute the total change in entropy for the system plus the surroundings. The process will be spontaneous, reversible, or nonspontaneous according to whether we obtain $\Delta S_{total} > 0$, $\Delta S_{total} = 0$, or $\Delta S_{total} < 0$, respectively.

2. The Gibbs and Helmholtz free energies are defined to be $G \equiv H - TS$ and $A \equiv U - TS$, respectively. If a process conducted under conditions of constant pressure and temperature has $\Delta G < 0$, it will occur spontaneously. If $\Delta G = 0$, the system is at equilibrium, and if $\Delta G > 0$, the process is nonspontaneous. If the conditions for the process are constant volume and temperature, the same criteria for spontaneity hold for ΔA.

3. For a closed system, the total differentials for G and A along reversible paths are

$$dG = -S\,dT + V\,dP$$

and

$$dA = -S\,dT - P\,dV.$$

These two equations, along with the differentials for U and H, viz.,

$$dU = T\,dS - P\,dV$$

and

$$dH = T\,dS + V\,dP,$$

constitute the four fundamental equations of classical thermodynamics.

4. Since the differentials of U, H, A, and G are exact, the Euler condition for exactness must hold for each of the total differentials given under item 3. This leads to four equations connecting various partial derivatives. These equations, called Maxwell's relationships, are

$$\left(\frac{\partial T}{\partial V}\right)_S = -\left(\frac{\partial P}{\partial S}\right)_V \quad \text{from } dU,$$

$$\left(\frac{\partial T}{\partial P}\right)_S = \left(\frac{\partial V}{\partial S}\right)_P \quad \text{from } dH,$$

$$\left(\frac{\partial S}{\partial V}\right)_T = \left(\frac{\partial P}{\partial T}\right)_V \quad \text{from } dA,$$

and

$$\left(\frac{\partial S}{\partial P}\right)_T = -\left(\frac{\partial V}{\partial T}\right)_P \quad \text{from } dG.$$

5. For open systems, the differentials given under item 3 must be augmented by including the terms involving $\hat{U}_i$, H_i, $\hat{A}_i$, G_i, and the change in the number of moles

of each component in the system. If we have K different components, the differentials are

$$dU = T\,dS - P\,dV + \sum_{i=1}^{K} \hat{U}_i\,dn_i,$$

$$dH = T\,dS + V\,dP + \sum_{i=1}^{K} \overline{H}_i\,dn_i,$$

$$dA = -S\,dT - P\,dV + \sum_{i=1}^{K} \hat{A}_i\,dn_i,$$

and

$$dG = -S\,dT + V\,dP + \sum_{i=1}^{K} \overline{G}_i\,dn_i,$$

where $\hat{U}_i$, $\overline{H}_i$, $\hat{A}_i$, and $\overline{G}_i$ are the rates of change of the internal energies, enthalpies, Helmholtz free energies, and Gibbs free energies, respectively, with the composition of the system. A particularly important property of the partial molar Gibbs free energy is that $\sum_{i=1}^{K}\overline{G}_i\,dn_i$ is equal to the summations appearing in each of the preceding equations. Therefore, we need to know only the $\overline{G}_i$ to treat open systems. For this reason, $\overline{G}_i$ is a useful quantity. It is given the symbol μ_i and called the chemical potential.

6. The total differential for the chemical potential is $d\mu = -\overline{S}\,dT + \overline{V}\,dP$, where $\overline{S}$ and $\overline{V}$ are the partial molar entropy and partial molar volume of the system, respectively. For an ideal gas, the chemical potential has the form $\mu_{ideal} = \mu^o(T) + RT\ln(P)$, where the standard chemical potential, $\mu^o(T)$, is a function only of the temperature, and P is the magnitude of the pressure when expressed in units of bars.

7. The change in the chemical potential for the chemical reaction

$$\nu_A\,A + \nu_B\,B \longrightarrow \nu_C\,C + \nu_D\,D$$

is $\Delta\mu = [\nu_C\,\mu_C + \nu_D\,\mu_D - \nu_A\,\mu_A - \nu_B\,\mu_B]$. If the reaction is conducted at constant temperature and either constant pressure or constant volume, the spontaneity and equilibrium criteria are

$$\Delta\mu < 0 \qquad \text{(irreversible, spontaneous processes)}$$

$$\Delta\mu = 0 \qquad \text{(reversible, equilibrium processes)}$$

and

$$\Delta\mu > 0 \text{(nonspontaneous processes requiring work)}.$$

8. The equilibrium condition $\Delta\mu = 0$ leads directly to the result that

$$\Delta\mu^o = -RT\ln\left[\frac{P_C^{\nu_C} P_D^{\nu_D}}{P_A^{\nu_A} P_B^{\nu_B}}\right]_{eq} = -RT\ln(K_p).$$

Since $\Delta\mu^o$ is dependent only upon the temperature, the pressure ratio must be a constant for a given temperature. This constant is called the thermodynamic equilibrium constant for an ideal system and is given the symbol K_p.

9. A ratio similar to K_p, but involving the concentrations in mol L^{-1} is defined to be K_c. The relationship connecting K_p and K_c is $K_p = K_c[RT]^{\Delta n}$, where Δn is the change in the number of moles of gas during the reaction.

10. By choosing the reference point for the measurement of standard chemical potentials to be the elements in their most stable state under a pressure of 1 bar and assigning their standard chemical potentials in this state to be zero, we obtain the

result that μ^o for a substance is numerically equal to the standard molar Gibbs free energy of formation. That is,

$$\mu^o = \Delta G_f^o.$$

11. It will be shown in the next chapter that the chemical potential for a condensed phase is a constant at a given temperature. This leads to the result that the equilibrium constant does not contain factors for the solids or liquids involved in the equilibrium reaction.

12. The variations of the Gibbs and Helmholtz free energies with temperature are given by

$$\frac{\partial}{\partial T}\left(\frac{G}{T}\right)_P = -\frac{H}{T^2}$$

and

$$\frac{\partial}{\partial T}\left(\frac{A}{T}\right)_V = -\frac{U}{T^2}, \text{ respectively.}$$

The equations also hold for changes in G and A:

$$\frac{\partial}{\partial T}\left(\frac{\Delta G}{T}\right)_P = -\frac{\Delta H}{T^2};$$

$$\frac{\partial}{\partial T}\left(\frac{\Delta A}{T}\right)_V = -\frac{\Delta U}{T^2}.$$

These relationships are called the Gibbs–Helmholtz equations.

13. The Gibbs–Helmholtz equations lead directly to the temperature dependence of the change in the standard chemical potential and equilibrium constant. These equations are

$$\frac{\partial}{\partial T}\left(\frac{\Delta \mu^o}{T}\right)_P = -\frac{\Delta H^o}{T^2}$$

and

$$\frac{\partial \ln(K_p)}{\partial T} = \frac{\Delta H^o}{RT^2}.$$

14. It is usually true that any variable X that measures the extent or rate of a process for which there is an energy requirement will be found to vary exponentially with temperature via the equation

$$X = (\text{constant}) \exp\left[\frac{-(\text{Energy Requirement})}{RT}\right],$$

provided that the energy requirement is assumed to be independent of temperature.

15. The activity, a, is defined by the equation $G(T, P, n) = n\mu^o(T) + nRT \ln(a)$. With this definition, the equations for nonideal systems can be obtained directly from those for the ideal systems by replacing partial pressures with the corresponding activities. The reference point for the measurement of activities is chosen to be

$$\underset{g \rightarrow g^*}{\text{Lim}}\left[\frac{a}{g}\right] = 1,$$

where g is the reference function and g^* is the reference state. If we choose pressure as the reference function and $P \rightarrow 0$ as the reference state, it is customary to replace a with f and call the quantity the *fugacity*. The activity coefficient is defined as the ratio a/g. The fugacity coefficient is similarly defined as f/P.

16. The equation connecting the fugacity of a nonideal gas to the pressure is

$$\ln(f) = \ln(P) + \frac{1}{RT} \int_0^P \left[V_m - \frac{RT}{P} \right] dP.$$

If the gas is ideal, the integrand vanishes identically, and we have $f = P$.

Problems

Problems that require the use of some type of computational device are marked with an asterisk (*). Problems that require some type of plotting routine are indicated with a pound sign (#). Unless otherwise stated, all gases may be assumed to behave ideally.

5.1 Two moles of an ideal gas are expanded isothermally from 25 L to 125 L at 320 K. Calculate ΔS, ΔG, and ΔA for the process.

5.2 (A) One mole of a van der Waals gas is expanded isothermally from volume V_1 to volume V_2 at temperature T. Obtain an expression giving ΔA for the process in terms of V_1, V_2, T, and the van der Waals parameters a and b.

(B) Obtain ΔS for the process in terms of these same variables.

5.3 One mole of a nonideal gas is described by the virial expansion

$$P = RT \left[\frac{1}{V} + \frac{C_2(T)}{V^2} + \frac{C_3(T)}{V^3} + \cdots \right].$$

If the gas is isothermally expanded from volume V_1 to volume V_2, obtain an expression for ΔA for the process in terms of V_1, V_2, T, and the virial coefficients.

5.4 (A) Using the data in Appendix A, compute the chemical potential of $CO(g)$ at 298.15 K and 10 bar of pressure if the gas is assumed to be ideal.

(B) Compute the chemical potential of $CO(g)$ at 298.15 K and 5 atm of pressure.

5.5 (A) Using the data in Appendix A, compute the chemical potential of $CO(g)$ at 400 K and 10 bar of pressure. Assume that the heat capacity of $CO(g)$ is a constant.

(B) How big is the error if we assume that $\overline{H}^\circ_{CO}(g)$ is a constant equal to its value at 298.15 K?

5.6 In the presence of a metallic catalyst, nitrogen and hydrogen react to form ammonia in a process called the *Haber process*. The Haber reaction is

$$N_2(g) + 3\,H_2(g) \longrightarrow 2\,NH_3(g).$$

(A) Using the data in Appendix A, compute K_p for the Haber reaction at 298.15 K.

(B) Compute the equilibrium constant for the reaction at 500 K. In this calculation, you may assume that ΔH° is a constant equal to its value at 298.15 K.

(C) At what temperature would we have $K_p = 1$? You may again assume that ΔH° is a constant.

5.7 The equilibrium constant K_p for the Haber reaction (see Problem 5.6) at 500 K is 0.1744. Suppose that 0.1 mole of $H_2(g)$ is mixed with 0.15 mole of $N_2(g)$ and 1.5 moles of $NH_3(g)$ in a 2-L fixed-volume reactor at 500 K.

(A) Compute ΔG for the reaction of 0.001 mole of $N_2(g)$ with $H_2(g)$ to form $NH_3(g)$ under these conditions. Ignore the change in the partial pressures produced by this amount of reaction.

(B) Is the process spontaneous under these conditions? (Assume that all gases are ideal.)

5.8 The equilibrium constant K_p for the Haber reaction (see Problem 5.6) at 500 K is 0.1744. Suppose that 0.1 mole of $H_2(g)$ is mixed with 0.15 mole of $N_2(g)$ and 0.11 mole of $NH_3(g)$ in a 2-L fixed-volume reactor at 500 K.

(A) Compute ΔG for the reaction of 0.001 mole of $N_2(g)$ with $H_2(g)$ to form $NH_3(g)$ under these conditions. Ignore the change in the partial pressures produced by this amount of reaction.

(B) Is the process spontaneous under these conditions? (Assume that all gases are ideal.)

5.9* The equilibrium constant K_p for the Haber reaction (see Problem 5.6) at 500 K is 0.1744. Suppose that 3 moles of $H_2(g)$ and 1 mole of $N_2(g)$ are mixed at 500 K in a closed, fixed-volume 2-L container. When equilibrium is reached, what are the partial pressures of $H_2(g)$, $N_2(g)$, and $NH_3(g)$? What is the percent conversion to $NH_3(g)$? (Assume that all gases are ideal.)

5.10* (A) Obtain K_p for the Haber reaction (see Problem 5.6) as a function of temperature. Assume that ΔH° is a constant equal to its value at 298.15 K.

(B) Compute the percent conversion of reactants to $NH_3(g)$ at 50-K intervals between 300 K and 800 K when 3 moles of $H_2(g)$ and 1 mole of $N_2(g)$ are mixed in a closed, fixed-volume 2-L container.

(C) Plot the percent conversion at equilibrium versus T. The Haber process is an extremely important commercial process in that is the principal means by which nitrate fertilizers and nitrogen-based explosives are obtained. In practice, the reaction is run at temperatures above 500 K. In view of your results, why isn't the reaction run at lower temperatures, where the percent conversion is higher? Assume that all gases are ideal.

5.11 Consider the hypothetical reaction

$$A(g) + B(g) \longrightarrow C(g) + 10.00 \text{ kJ}.$$

Suppose the reaction is conducted at constant temperature and pressure between gases that may be accurately described by the ideal-gas law. The standard chemical potentials are $\mu^o = -12.00 \text{ kJ mol}^{-1}$, $-3.00 \text{ kJ mol}^{-1}$, and $-4.00 \text{ kJ mol}^{-1}$ for $C(g)$, $A(g)$, and $B(g)$, respectively, at 298.15 K.

(A) Compute K_p for this reaction at 298.15 K.

(B) Express K_p in terms of the numbers of moles of $A(g)$, $B(g)$, and $C(g)$ present, R, T, and the volume V of the container.

(C) 2 moles of $A(g)$ and 2 moles of $B(g)$ are placed in a 100-L container at 298.15 K. When equilibrium is reached, how many moles of $C(g)$ will be present in the container?

(D) After equilibrium is reached, 1 mole of $A(g)$ and 1 mole of $B(g)$ are added to the container. The temperature of the system is then raised to a point such that when equilibrium is again established, the number of moles of $C(g)$ in the container is exactly the same as it was before the addition of more $A(g)$ and $B(g)$ and the elevation of the temperature. Determine the new temperature of the gas in the container. Assume that ΔH^o is a constant.

5.12 Sam continues to struggle with physical chemistry. His latest problem concerns the equilibrium expression for the reaction

$$CaCO_3(s) \longrightarrow CaO(s) + CO_2(g).$$

His professor has pointed out that, since the chemical potentials of solids are fixed at a given temperature, the equilibrium constant for this reaction is given by

$$K_p = [P_{CO_2(g)}]_{eq}.$$

Sam's professor has also noted in class that, because $\Delta\mu^o$ is a function only of temperature, the equilibrium constant is independent of pressure. That is, $(\partial K_p/\partial P)_T = 0$. This is simply too much for Sam to stomach. He thinks, "How can K_p be equal to the pressure of $CO_2(g)$ and yet not be a function of pressure? After all, if we wrote $y = 2x^2 - 3x + 4$ and then claimed that y was not a function of x, we would fail beginning algebra." Believing that some

professors often wear earplugs when talking to students, Sam decides to prove his point by actual experiment rather than by reasoned argument with his professor. He constructs a piston–cylinder apparatus whose temperature can be controlled. He then places some $CaCO_3(s)$ into the cylinder, adjusts the piston to give a total cylinder volume of 5 L, and raises the temperature sufficiently high to get an easily measured pressure of $CO_2(g)$ inside the cylinder. He now plans to depress the piston to decrease the volume to 2 L, while keeping the temperature constant, thereby producing an increase in the $CO_2(g)$ pressure and, hence, an increase in the value of K_p. Sam feels that this experiment will surely convince his professor that K_p must be a function of pressure if we have $K_p = [P_{CO_2}(g)]_{eq}$. Figure 5.15 at the top of the next page shows a schematic diagram of Sam's apparatus. Is he right? Will his experiment demonstrate the proper pressure dependence of K_p? Why?

5.13 The equation of state for 1 mole of a van der Waals gas is

$$\left(p + \frac{a}{V^2}\right)(V - b) = RT.$$

The gas is expanded isothermally from volume V_1 to volume V_2 at temperature T. Determine ΔA for the process, and then use the Gibbs–Helmholtz equation to obtain ΔU for the process.

5.14 Starting with Eq. 5.17, $\left(\dfrac{\partial A}{\partial T}\right)_V = -S$, use the fact that

$$-S = \frac{A - U}{T} \text{ to show that}$$

$$\frac{\partial}{\partial T}\left(\frac{A}{T}\right)_V = -\frac{U}{T^2}.$$

5.15 Show that $\sum_{i=1}^{K} \overline{H}_i \, dn_i = \sum_{i=1}^{K} \overline{G}_i \, dn_i$.

5.16 A nonideal gas has the equation of state

$$PV = nRT + n\alpha P,$$

where α is a function of temperature only.

(A) Obtain an expression for the fugacity and the fugacity coefficient for this gas as a function of T, P, and α.

(B) If the gases in the Haber reaction

$$N_2(g) + 3\,H_2(g) \longrightarrow 2\,NH_3(g)$$

were all described by equations of state like the preceding one, but with differing values of α for each gas, what would be the appropriate expression for the thermodynamic equilibrium constant?

5.17 The fugacity of a gas is found to obey the equation

$$f = P\exp\left[\frac{AP + BP^2/2}{RT}\right],$$

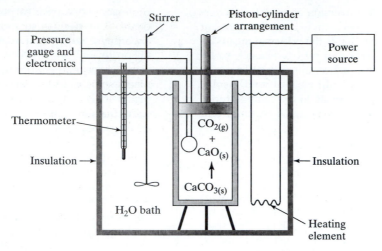

▲ FIGURE 5.15
Sam's apparatus designed to determine the pressure dependence of the equilibrium constant for the reaction $CaCO_3(s) \longrightarrow CaO(s) + CO_2(g)$ at a fixed temperature.

where A and B are constants. Determine the equation of state for this gas.

The next two problems are designed to illustrate the use of fugacities in the treatment of chemical equilibria. The solutions to the problems will be facilitated by the use of a PC or workstation.

5.18* In this and the next problem, we shall examine the equilibrium reaction between $NO_2(g)$ and $N_2O_4(g)$: $2 NO_2(g) \longrightarrow N_2O_4(g)$. In this problem, we assume that both gases behave ideally. The data required in the various parts of the problem either are given directly or can be found in Appendix A.

(A) Determine ΔH° for the reaction at 298.15 K and 1 bar of pressure.

(B) Determine the value of $\Delta \mu^\circ$ for the reaction.

(C) Compute K_p and K_c for the reaction at 298.15 K.

(D) The heat capacities, C_p^m, of $NO_2(g)$ and $N_2O_4(g)$ are 37.20 J mol^{-1} K^{-1} and 77.28 J mol^{-1} K^{-1}, respectively. Obtain ΔH° for the reaction as a function of temperature. Assume that C_p^m is constant for both gases.

(E) Obtain closed-form analytic expressions for K_p and K_c for the reaction as a function of temperature. Carefully plot K_p and K_c versus T over the range 298 K $\leq T \leq$ 1,000 K. Carefully plot ln K_p and ln K_c versus $1/T$ over the range 298 K $\leq T \leq$ 1,000 K. Why are the plots of ln K_p and ln K_C versus $1/T$ nearly linear?

(F) One mole of $NO_2(g)$ is initially placed in a 2-L sealed container whose volume is constant. Obtain an analytic expression for the equilibrium concentration of $N_2O_4(g)$ as a function of temperature. Plot $[N_2O_4(g)]_{eq}$ versus T over the range 298 K $\leq T \leq$ 1,000 K. Use a logarithmic scale for $[N_2O_4(g)]_{eq}$.

(G) Let us assume that you are operating a plant making $N_2O_4(g)$ for sale, using $NO_2(g)$ as your starting material. At high temperatures, the rate at which $N_2O_4(g)$ is formed will be very fast. At low temperatures, it will be very slow. In deciding upon the best operating conditions for the manufacture of $N_2O_4(g)$, should you use very high temperatures, very low temperatures, or an intermediate temperature? Explain.

5.19* In this problem, we shall treat both $NO_2(g)$ and $N_2O_4(g)$ as nonideal. The equilibrium reaction of concern is the same as in Problem 5.18: $2 NO_2(g) \longrightarrow N_2O_4(g)$. The data required in the various parts of the problem either are given directly or can be found in Appendix A. To simplify the calculations, we shall make the following approximations:

1. Both $NO_2(g)$ and $N_2O_4(g)$ obey a modified truncated virial equation of state as described in (C) to follow.

2. Dalton's law holds; that is,

$$P_i = X_i P_T,$$

where

P_i = partial pressure of gas i

P_T = total pressure = $P_{NO_2(g)} + P_{N_2O_4(g)}$

and

X_i = mole fraction of gas i.

3. The fugacity of gas i depends only upon the temperature and the partial pressure of gas i; it is independent of the partial pressures of any other gases present.

(A) If

$$C_p^m \text{ for } NO_2(g) = 6.37 + 0.0101T$$

$$- 34.05 \times 10^{-7}T^2 \text{ cal mol}^{-1} \text{ K}^{-1}$$

and

$$C_p^m \text{ for } N_2O_4(g) = 10.719 + 0.0286T$$

$$- 87.26 \times 10^{-7}T^2 \text{ cal mol}^{-1} \text{ K}^{-1},$$

obtain $\Delta H°$ for the reaction $2 NO_2(g) \longrightarrow N_2O_4(g)$ as a function of T. Plot ΔH as a function of T over the range 298 K $\leq T \leq$ 400 K.

(B) Obtain K_p for the reaction $2 NO_2(g) \longrightarrow N_2O_4(g)$ as a function of T. Plot K_p as a function of T over the range 298 K $\leq T \leq$ 400 K.

(C) We shall use a modified form of the virial equation for the equation of state for both gases. If we truncate the virial expansion after the third term, we have

$$PV = RT \left[1 + \frac{C_2(T)}{V} + \frac{C_3(T)}{V^2} \right]$$

for 1 mole of gas. Equation 1.66 shows that if we employed a van der Waals equation of state, $C_2(T) = b - a/RT$ and $C_3(T) = b^2$. To simplify the calculations, let us replace $1/V$ in the second and third terms with its ideal-gas form, P/RT. Then the truncated virial expression becomes $PV = RT + C_2(T)P + C_3(T)P^2/RT$. The latter expression will represent the equation of state for both gases. This type of expansion, in which the PV product is expressed as a power series in the pressure, is called the *Reichsanstalt equation*. $C_2(T)$ and $C_3(T)$ will be estimated from van der Waals parameters. For $NO_2(g)$, we have $a = 5.354$ L^2 bar mol^{-2} and $b = 0.04424$ L mol^{-1}. For $N_2O_4(g)$, we shall use $a = 6.550$ L^2 bar mol^{-2} and $b = 0.05636$ L mol^{-1}. Obtain an analytic function giving the fugacities of both gases in terms of $C_2(T)$, $C_3(T)$, and the pressure. Prepare plots of the fugacities of both gases at 298.15 K over the pressure range 1 bar $\leq P \leq$ 400 bar.

(D) Compute the fugacity coefficients for both gases at 298.15 K over the pressure range 1 bar $\leq P \leq$ 400 bar. Plot the results.

(E) Assuming both gases to be ideal, obtain the mole fraction of $N_2O_4(g)$ present in the equilibrium system at 298.15 K as a function of the total pressure P_T over the range 1 bar $\leq P_T \leq$ 400 bar. Plot the results, with one graph showing $X_{N_2O_4(g)}$ from 1 to 400 bar of pressure and a second showing $X_{N_2O_4(g)}$ from 1 to 50 bar of pressure.

(F) Assuming $NO_2(g)$ and $N_2O_4(g)$ to be nonideal gases described by the truncated virial equation developed in (C), compute the mole fraction of $N_2O_4(g)$ present at equilibrium as a function of total pressure over the range 1 bar $\leq P_T \leq$ 400 bar.

(G) Prepare a plot of the mole fraction of $N_2O_4(g)$ present at equilibrium versus the total pressure over the range 1 bar $\leq P_T \leq$ 400 bar for the case of ideal and nonideal gases. Compute the percent error made by the ideal-gas assumption as a function of total pressure.

5.20* The following data are obtained for 1 mole of $CO_2(g)$ at $T = 300$ K:

P (atm)	V (L)	$\left[V - \dfrac{RT}{P} \right]$ (L)
0*		−0.1032
1.2239	20.01	−0.10438
1.9993	12.2091	−0.10421
5.732	4.1889	−0.10586
8.0041	2.9688	−0.10687
20.1596	1.1088	−0.11236
50.8051	0.3488	−0.13576
73.2058	0.0940	−0.2423
79.9941	0.0878	−0.2199
90.1682	0.0829	−0.1901
100.2734	0.0798	−0.1657
130.8572	0.0742	−0.1139
215.5771	0.0671	−0.0471
299.9456	0.0635	−0.0185
390.2092	0.0610	−0.0021
498.4209	0.0589	0.0095
610.5791	0.0573	0.0170

*Computed from the van der Waals parameters for $CO_2(g)$.

Use a numerical method to compute the fugacity coefficient for $CO_2(g)$ as a function of pressure at 300 K from 1.2239 atm to 610.5791 atm. Present your results in the form of a graph of γ versus pressure.

5.21 The reference point chosen for the fugacity is

$$\lim_{P \to 0} \left[\frac{f}{P} \right] = 1.$$

Let us assume that there exists a nonideal gas for which $f = P/(1 + bP)$, with b equal to a constant for this choice of reference point.

(A) If we had instead chosen

$$\lim_{P \to 0} \left[\frac{f}{P} \right] = 2$$

as our reference point, determine the form the fugacity would have for this gas.

(B) Show that ΔG for the isothermal expansion of the gas from pressure P_1 to pressure P_2 is the same for either choice of reference point.

5.22 Consider a mixture of two ideal gaseous substances A and B that are in equilibrium:

$$A(g) = B(g).$$

A tube containing a total of n moles of $A(g)$ and $B(g)$ is slowly heated at constant pressure in the presence of a catalyst so that the system is always in equilibrium. If heating the system by an amount dT results in a change in the number of moles of $B(g)$ at equilibrium by an amount dn_B, show that the total heat associated with this change due to the enthalpy of reaction is given by

$$\Delta H \, dn_B = \frac{\Delta H^2 \, nK}{RT^2(K+1)^2} \, dT \, ,$$

where K is the thermodynamic equilibrium constant and ΔH is the molar enthalpy of reaction.

5.23* Consider the equilibrium system

$$A(g) + 2B(g) = 2C(g) + D(g),$$

where all gases are ideal. If the equilibrium constant for this system is 10 at 300 K, and if 0.5 mol L^{-1} of $A(g)$, 5 mol L^{-1} of $B(g)$, 2 mol L^{-1} of $C(g)$, and 0.2 mol L^{-1} of $D(g)$ are simultaneously mixed together in a 1-L container at 300 K,

(A) will $A(g) + 2B(g)$ spontaneously react to form $2C(g) + D(g)$, or will $2C(g) + D(g)$ spontaneously react to form $A(g) + 2B(g)$? Show that your answer is correct.

(B) When equilibrium is reached, how many moles of each gas will be present?

5.24 A chemist measures the equilibrium constant for a particular reaction at several different temperatures. Her results are as follows:

T (K)	K_p	T (K)	K_p
300	5.84	500	6.33×10^{-4}
350	0.227	550	1.89×10^{-4}
400	1.93×10^{-2}	600	6.68×10^{-5}
450	2.81×10^{-3}		

(A) Plot $\ln (K_p)$ versus T^{-1}, and use a least-squares procedure to obtain the best straight-line fit to the data.

(B) Determine the enthalpy change for the reaction from the results of the least-squares fit to the data.

5.25 State the direction of the equilibrium shift (if any) in each of the following cases, and stipulate whether the shift is taking place because the equilibrium constant is increasing or decreasing or because of the

necessity to maintain the equilibrium constant's value unchanged:

(A) In the reaction $2 NO_2(g) \longrightarrow N_2O_4(g) + 57.2$ kJ, the total pressure is decreased.

(B) In the reaction $2 NO_2(g) \longrightarrow N_2O_4(g) + 57.2$ kJ, the temperature is decreased.

(C) In the reaction $2 NO_2(g) \longrightarrow N_2O_4(g) + 57.2$ kJ, some of the $N_2O_4(g)$ is removed.

5.26 Initial concentrations of 0.100 mol L^{-1} of $H_2S(g)$ and $H_2(g)$ are mixed in a 1.00-L container with excess solid sulfur, and the mixture is allowed to come to equilibrium at 380 K. For the reaction

$$H_2S(g) \longrightarrow H_2(g) + S(g),$$

the equilibrium constant is $K = 0.0700$. As written, the reaction is endothermic.

(A) What is the concentration of $H_2(g)$ at equilibrium?

(B) Would the amount of hydrogen sulfide increase, decrease, or remain unchanged if (i) more sulfur were added, (ii) the volume of the container were decreased, (iii) the temperature were decreased? [Thanks to Fredrick L. Minn, M.D., Ph.D. for this contribution.]

5.27 Using the data in Appendix A, Sam computes $\Delta \mu^o$ for the reaction

$$2 HI(g) \longrightarrow H_2(g) + I_2(g)$$

at 298.15 K. He obtains the result $\Delta \mu^o = -3.40$ kJ mol^{-1}. Since $\Delta \mu^o < 0$, Sam concludes that the reaction at 298.15 K will be spontaneous and irreversible if carried out at either constant volume or constant pressure.

(A) Has Sam done the calculation of $\Delta \mu^o$ correctly?

(B) Are his conclusions concerning the spontaneity of the reaction correct? Explain.

5.28 If the reaction in Problem 5.27 is conducted at constant volume with $T = 298.15$ K and initial partial pressures of $H_2(g)$ and $I_2(g)$ of 0.100 bar, under what conditions of partial pressure for HI(g) will the reaction to form $H_2(g) + I_2(g)$ be spontaneous? Assume that all gases are ideal.

5.29 If the reaction in Problem 5.27 is conducted at constant volume with $T = 298.15$ K and initial partial pressures of $H_2(g)$ and $I_2(g)$ of 0.100 bar, under what conditions of partial pressure for HI(g) will the reaction to form $H_2(g) + I_2(g)$ be spontaneous? Assume that $H_2(g)$ and $I_2(g)$ are ideal, but that HI(g) is described by the equation of state given in Example 5.15, where the pressure refers to the partial pressure of HI(g) and $\beta = -0.0300$ at 298.15 K.

5.30 Consider a physical chemistry student whose grade expectations and performance are in equilibrium. Everything is proceding according to plan. Then the third hour examination arrives, and the student makes a grade of F. Obviously, a stress has been applied to an equilibrium system, and Le Chatelier's principle predicts that the equilibrium will shift so as to relieve that stress. However, when there is more than one way to relieve the stress, Le Chatelier's principle does not tell us *how* the equilibrium will

shift; it simply assures us that a shift will occur. Perhaps the student will go out and get drunk. Perhaps he or she will resolve to study harder. As you relax with a good book, contemplate how your equilibrium shifts to accommodate stress. Is the shift productive? Now enjoy the book. Phase equilibrium is right around the corner.

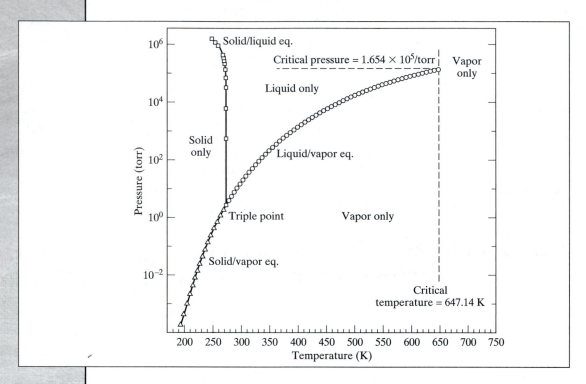

CHAPTER 6

Phase Equilibrium

In Chapter 5, we developed the concept of chemical equilibrium. In the current chapter, we apply these concepts to equilibria between different phases of the same compound. In the process, we shall learn how the thermodynamics of liquids and solids may be expressed in terms of their equilibrium vapors. Phase diagrams will be introduced as a means of graphically illustrating the thermodynamic results.

6.1 Qualitative Description of Phase Transitions and Definition of Terms

At sufficiently low temperatures, all compounds and elements condense to solids. Condensation occurs when the internal kinetic energy becomes so small that the molecules can no longer overcome the forces of mutual attraction and are consequently frozen into fixed positions in a solid lattice structure. The solid is called a *phase,* in that it is uniform in chemical composition and physical structure throughout the material. Equations 5.13 and 5.53 showed that spontaneous processes at a given temperature and pressure are characterized by $\Delta G < 0$ and $\Delta \mu < 0$. Therefore, we expect that the lattice will spontaneously assume the form that makes the Gibbs free energy and the chemical potential as low as possible. Thus, if we have two solid phases A and B that have different lattice structures with $\mu_A < \mu_B$, phase A will be the more thermodynamically stable and will tend to form spontaneously.

As we change the temperature and pressure, the chemical potentials of solid phases A and B will vary in the manner described by Eq. 5.44 and illustrated in Figures 5.5 and 5.6; that is,

$$ d\mu = -\overline{S}\, dT + \overline{V}\, dP. $$

Since $-\overline{S}$ is always negative, both the chemical potential and the Gibbs free energy decrease as the temperature rises. The molar volume $\overline{V}$, however, is positive, so that G and μ increase with pressure. For condensed phases, $\overline{V}$ is small relative to the gas phase; therefore, G and μ for solids and liquids increase very slowly with pressure. As long as μ_A remains lower than μ_B, A will continue to be the stable form. If we should reach a temperature and pressure at which we have $\mu_A = \mu_B$, a phase transition from A to B would have $\Delta\mu = \Delta G = 0$, which is the condition for equilibrium. Consequently, the two phases are in equilibrium at that "transition" temperature and pressure. Because $\Delta G = \Delta H - T_o \Delta S = 0$ at this point, we must have $\Delta S = \Delta H / T_o$, where T_o is the transition temperature. Endothermic transitions with $\Delta H > 0$ always have $\Delta S > 0$, which means that the system becomes more disordered as a result of the transition. Exothermic phase transitions, on the other hand, always cause an increase in the order of the system. Figures 6.1

▶ FIGURE 6.1
Qualitative illustration of the variation of chemical potential with temperature for two solid phases. In this illustration, the partial molar entropy of Phase B is greater than that of Phase A. Consequently, μ_B decreases faster than μ_A as the temperature rises. As a result, Phase A, which is the thermodynamically stable phase at lower temperature, becomes a metastable phase at higher temperatures, while Phase B becomes the stable phase at higher temperatures. At $T = T_o$, when their chemical potentials are equal, the two phases are in equilibrium.

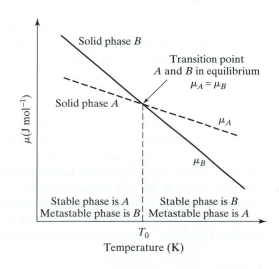

253

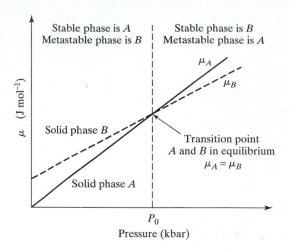

▲ FIGURE 6.2

Qualitative illustration of the variation of chemical potential with pressure for two solid phases. In this illustration, the partial molar volume of Phase A is greater than that for Phase B. Consequently, μ_A increases faster than μ_B as the pressure rises. As a result, Phase A, which is the thermodynamically stable phase at lower pressures, becomes a metastable phase at higher pressures, where Phase B has a lower chemical potential and is therefore the stable phase. At pressure P_0 where $\mu_A = \mu_B$, the two phases are in equilibrium. The diamond–graphite system exhibits this behavior. At low pressures, graphite is the stable phase. However, the partial molar volume of graphite is larger than that of diamond, since the density of the graphite is lower. Therefore, $\mu_{graphite}$ increases faster than $\mu_{diamond}$, which produces a crossing of the two chemical potentials at high pressures, thus making diamond the stable phase. (See Example 6.1.)

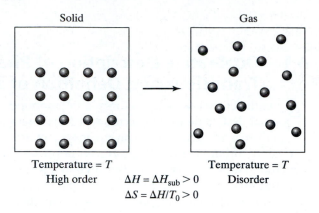

▲ FIGURE 6.3

Order–disorder variations in equilibrium phase transitions. For exothermic phase transitions at the equilibrium temperature T_0, ΔH is negative. Therefore, the entropy change, which is $\Delta H/T_0$, is also negative, indicating that the transition produces an increase in order. When the transition is endothermic, the reverse is true, and there is a decrease in the order of the system, since ΔS will be positive. The figure illustrates this principle with the sublimation of a solid into the gas phase. Such a transition is endothermic. Therefore, we expect a decrease in the order of the phase. This is obviously true for the solid $\longrightarrow$ gas transition.

and 6.2 illustrate these principles for variations in temperature and pressure, respectively. Figure 6.3 shows the relationship between ΔH and ΔS for phase transitions occurring at equilibrium.

   It sometimes happens that a solid phase B will be formed under conditions in which $\mu_B > \mu_A$. In such a case, the transition $B \longrightarrow A$ will be spontaneous. However, the rate at which the transition occurs may be so slow as to be unobservable. This situation can occur between solid phases because the intermolecular forces are so strong that the atoms do not have sufficient mobility to transform the structure to the more stable phase A. In such a situation, we term phase B a *metastable* phase or a *kinetically stable* phase. Since the molecular mobilities in liquid and gas phases are much greater than in solids, we usually do not observe metastable states for these phases. An exception is crystalline phases, which can be metastable. One example of a metastable state is frequently observed in the graphite and diamond system. Graphite is a crystal structure of carbon in which the atoms are arranged in planar hexagons. The diamond structure, by contrast, is a three-dimensional tetrahedral configuration similar to that of CH_4. At room temperature and 1 atm of pressure, $\mu_{graphite} < \mu_{diamond}$. Graphite is, therefore, the stable solid phase for carbon under these conditions. However, the rate at which diamond transforms itself into graphite at that temperature and pressure is essentially zero; diamond is a metastable phase. Individuals who own expensive diamond jewelry are particularly happy about this situation. Another example of a metastable state is a mountain climber who is suspended 2,000 ft above the ground by a rope attached to a piton driven into the face of the mountain. The force produced by the rope restrains the climber's motion sufficiently that he cannot undergo the transition to the more stable thermodynamic state in which he would be splattered over the rocks 2,000 ft below. The mountain climber is also happy about this situation.

At low temperatures, the solid is generally the thermodynamically stable phase of a compound or element. Therefore, in this temperature range, $\mu_{solid} < \mu_{liquid}$. As the temperature increases, the chemical potential of both solid and liquid decrease. The rate of decrease is given by Eq. 5.44, which shows that $(\partial\mu/\partial T)_P = -\bar{S}$. Since the partial molar entropy of the more disordered liquid phase always exceeds that of the solid, μ_{liquid} decreases more rapidly as the temperature increases than does μ_{solid}. Because of this, we will eventually reach a temperature at which $\mu_{solid} = \mu_{liquid}$, so that the two phases come into equilibrium. The point at which this occurs is termed the *melting* or *freezing* point. If the pressure is 1 atm, it is called the *normal* melting or freezing point. If the pressure is 1 bar, we refer to the temperature as the *standard* melting or freezing point.

Figure 6.4 shows an approximate calculation of the standard melting point for aluminum under an applied pressure of 1 bar. At room temperature, the chemical potential for solid aluminum lies below that for the liquid. Therefore, the stable phase is the solid. As the temperature rises, the chemical potentials for both phases decrease, as expected. The partial molar entropy for the solid is less than that for liquid due to the greater order present in the solid phase. Consequently, μ_{solid} decreases less rapidly than μ_{liquid}. As a result, a temperature is finally reached at which we have $\mu_{solid} = \mu_{liquid}$. This temperature is the standard melting point for aluminum. In the figure, we have assumed that $\bar{S}$ is independent of temperature. This assumption leads to a predicted melting point of 940 K, which is too high by 6.53 K. Nevertheless, the calculation is sufficiently accurate to make the point that melting occurs when the solid and liquid chemical potentials become equal.

Sometimes we can bring the solid and liquid phases into equilibrium at temperatures below the normal or standard melting points by increasing the applied pressure. Equation 5.44 shows us that $(\partial\mu/\partial P)_T = \bar{V}$, so that the chemical potential increases with increasing pressure at a rate equal to the partial molar volume. At a temperature below the melting point, $\mu_{solid} < \mu_{liquid}$; therefore, μ_{solid} will need to increase more rapidly with pres-

◀ **FIGURE 6.4**

Computed chemical potentials for solid and liquid aluminum under 1 bar of pressure as a function of temperature. The computations are done using the fact that $(\partial\mu/\partial T)_P = -\bar{S}$, with the assumption that $\bar{S}$ is independent of temperature (see Problem 6.1). Below a temperature of 940 K, the chemical potential of the solid lies below that for the liquid. It is, therefore, the stable phase. At 940 K, the two phases have equal chemical potentials so that they are in equilibrium and the standard melting point is predicted to be 940 K. This result is too high by 6.53 K. The error is due primarily to our assumption of a constant value for $\bar{S}$.

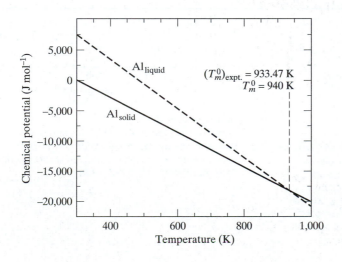

$(T_m^0)_{expt.} = 933.47$ K
$T_m^0 = 940$ K

▶ FIGURE 6.5
Computed chemical potentials for solid and liquid aluminum at 298.15 K as a function of applied pressure. The calculations are carried out assuming that the partial molar volumes are independent of pressure. (See Problem 6.2.) Since $\overline{V}_{liquid} > \overline{V}_{solid}$, μ_{liquid} increases more rapidly than μ_{solid}, and the two quantities never become equal at any pressure. Consequently, aluminum cannot be forced to melt at a temperature below the standard melting point by increasing the pressure.

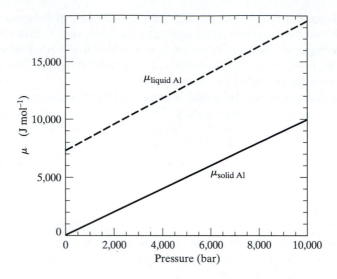

sure than μ_{liquid} if we are to bring the two phases into equilibrium. This means we must have $\overline{V}_{solid} > \overline{V}_{liquid}$. Figure 6.5 shows the variation of μ_{solid} and μ_{liquid} for aluminum at 298.15 K. In this case, the density of the solid exceeds that of the liquid, and we have $\overline{V}_{solid} < \overline{V}_{liquid}$. As a result, μ_{liquid} increases with pressure faster than μ_{solid} and the two chemical potentials never become equal. Accordingly, we cannot melt aluminum at temperatures below the standard or normal melting points by raising the pressure. In contrast, ice is less dense than water, so that $\overline{V}_{solid} > \overline{V}_{liquid}$. This means it is possible to melt ice by applying a sufficiently high pressure. Figure 6.6 shows the computed chemi-

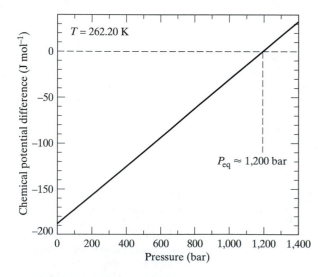

▲ FIGURE 6.6
Computed chemical potential difference between ice and water at 262.20 K as a function of applied pressure. The calculations are carried out assuming that the partial molar volumes are independent of pressure. (See Problem 6.3.) Since we now have $\overline{V}_{liquid} < \overline{V}_{solid}$, μ_{liquid} increases less rapidly that μ_{solid}, so that the two chemical potentials become equal at an applied pressure of 1,200 bar. Consequently, we would observe ice to melt at 262.20 K if the applied pressure is raised to 1,200 bar.

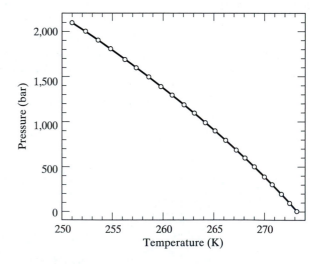

▲ FIGURE 6.7
Experimental pressures at which ice and water attain equilibrium at the temperature shown on the abscissa. Source of data: *Handbook of Chemistry and Physics,* 78th edition, CRC Press, Boca Raton, FL. 1997–1998.

cal potential difference between $H_2O(s)$ and $H_2O(l)$ as a function of pressure at 262.20 K if we assume that the partial molar volumes of the condensed phases are independent of pressure. We see that the chemical potentials become equal when the applied pressure reaches 1,200 bar. Therefore, ice can be forced to melt at temperatures below the standard freezing point if the applied pressure is sufficiently large. Figure 6.7 shows the experimentally measured pressures required to bring liquid and solid H_2O into equilibrium as a function of temperature.

Solids and liquids below the critical temperature have a characteristic equilibrium vapor pressure, as illustrated in Figure 6.8. As the figure shows, the molecules possess kinetic energy in an amount that occasionally permits their escape from the attractive forces that produce the condensed phase. Once the material is in the gas phase, the random motion of the molecules will sometimes result in their reentry into the condensed phase. When the rate of escape and the rate of reentry become equal, equilibrium will be established. The gas-phase pressure that exists at that point, P^{eq}, is termed the *equilibrium vapor pressure* of the solid or liquid. Since the average intermolecular force is greater for the solid than for the liquid, the rate of escape of molecules from the solid will be less, and the equilibrium vapor pressure will be correspondingly lower. As the temperature is raised, the kinetic energy increases and the rate of escape rises. Therefore, we expect the equilibrium vapor pressure to increase as T increases.

It is very important to understand that when a condensed phase in an enclosed container is in equilibrium with its vapor phase at pressure P^{eq} and temperature T, it is impossible to change the pressure if the temperature is held constant. It is also impossible to vary the temperature if the vapor pressure is held constant. So long as both phases are present, these generalizations will hold.

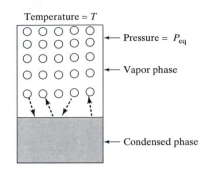

▲ FIGURE 6.8
Because the kinetic energy of the molecules occasionally permits their escape into the vapor phase, all condensed phases will exhibit a characteristic equilibrium vapor pressure P^{eq} at a given temperature. Equilibrium will be reached when the number of molecules escaping per unit time equals that reentering the condensed phase.

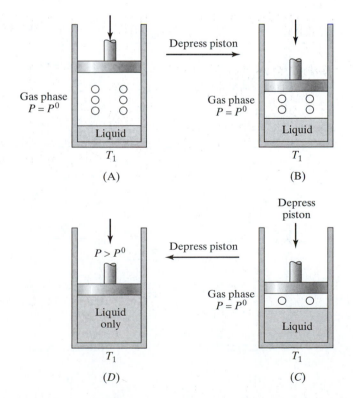

◀ FIGURE 6.9
(A) Schematic diagram of a liquid and its equilibrium vapor contained in a piston–cylinder arrangement at a fixed temperature T_1. (B and C) Depression of the piston always converts sufficient vapor to liquid to maintain the pressure at its equilibrium value P^o. Only after all vapor has been condensed to a liquid, thereby converting the system into one containing a single phase, can the pressure be increased by the application of additional force on the piston as shown in D.

Figure 6.9 illustrates what occurs if we attempt to vary the vapor pressure at a fixed temperature. Figure 6.9A shows a liquid and vapor in equilibrium in a piston–cylinder apparatus at temperature T_1 and pressure P^o. Suppose we depress the piston in an effort to raise the vapor pressure while keeping the temperature at T_1. Initially, there will be an instantaneous increase in pressure. However, this transitory increase is soon lost as some of the vapor condenses to a liquid to restore the vapor pressure to original value, P^o, as shown in Figure 6.9B. Since heat will be released upon condensation of the vapor, arrangements will need to be made to conduct the heat away from the system in order to maintain a constant temperature. If we continue to depress the piston, more and more vapor will condense, as illustrated in Figure 6.9C. Finally, a point will be reached where the piston is flush against the surface of the liquid and all vapor has been condensed (Figure 6.9D). With only a single phase present, it is now possible to raise the pressure by applying additional force to the piston. This cannot be accomplished with two phases present.

Figure 6.10 demonstrates that we cannot change the temperature of a liquid–vapor system that is at equilibrium if we maintain a constant pressure. As we add heat to the system shown in Figure 6.10A, the heat is used to provide the thermal energy required to vaporize sufficient liquid to maintain the temperature fixed at T_1. To keep the vapor pressure constant at P^o, the piston is withdrawn as heat is added. (See Figures 6.10B and 6.10C.) This process continues until all the liquid has been converted to vapor at temperature T_1 (Figure 6.10D). Only then, with a single phase present, can we effect an increase in temperature by the addition of more heat.

As we continue to heat a liquid, the equilibrium vapor pressure P^{eq} and the gas-phase density increase. At the same time, the volume of the liquid expands and its density decreases. Eventually, we must reach a temperature at

▶ FIGURE 6.10
(A) Schematic diagram of a liquid and its equilibrium vapor contained in a piston–cylinder arrangement at a fixed pressure P^o and temperature T_1. (B and C)The addition of heat to the system in an effort to raise the temperature above T_1 results in vaporization of some of the liquid, with the added heat being absorbed as heat of vaporization so that the temperature remains at T_1 as long as the piston is moved to keep the vapor pressure fixed at P^o. Only after all liquid is converted to vapor, thereby converting the system to one containing a single phase, can the temperature be increased by the addition of more heat as shown in D.

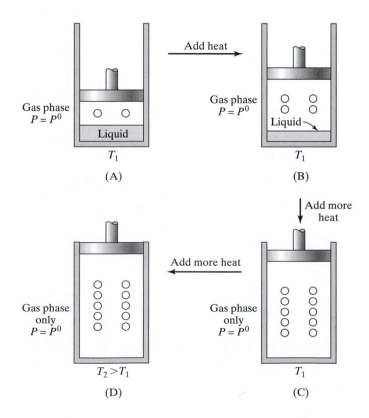

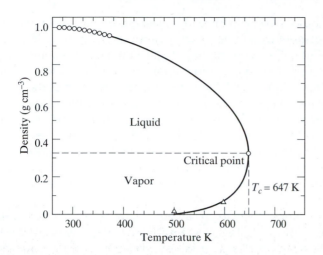

◀ FIGURE 6.11
Liquid and vapor densities of H_2O as a function of temperature. At the critical temperature, 647 K, the densities of liquid and vapor become equal, so that the boundary between the two phases vanishes. The plotted points are experimental data from the *Handbook of Chemistry and Physics,* 78th edition, CRC Press, Boca Raton, FL, 1997–1998. The solid curve qualitatively illustrates the approximate variation in the density with temperature.

which the liquid and gas densities are equal and the boundary between the two disordered phases consequently vanishes. The temperature at which this occurs is called the *critical temperature* T_c. The equilibrium vapor pressure at $T = T_c$ is termed the *critical pressure* P_c. Figure 6.11 shows the variation of the liquid- and gas-phase density of water as a function of temperature. When the temperature reaches the critical temperature of 647.15 K for H_2O, the densities become equal, and the demarcation line between the liquid and vapor phases disappears. When this occurs, the change in enthalpy, ΔH_{vap}, associated with conversion of the liquid phase to the vapor goes to zero, as shown in Figure 6.12. If either of the phases were highly ordered, the boundary line between the phases would not vanish. For example, the grain boundary between two ordered solid phases is observable even when the densities of the phases are equal. A critical point is observable only between *disordered* phases.

If the container illustrated in Figure 6.8 is opened to the atmosphere, the situation changes. Under this condition, the vapor can diffuse into the surroundings so that the gaseous molecules do not have the opportunity to reenter the condensed phase. However, molecules continue to escape the surface of the condensed phase at a rate that depends upon the temperature. As a result, the condensed phase is observed to slowly disappear. If the condensed

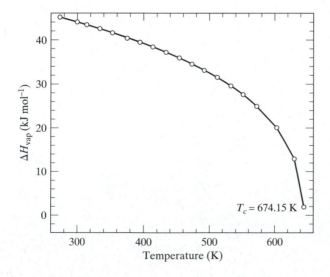

◀ FIGURE 6.12
Partial molar enthalpy of vaporization of H_2O as a function of temperature. At the critical temperature, 647.15 K, $\Delta \overline{H}_{vap}$ goes to zero as the boundary between the liquid and the vapor phases vanishes. Source of data: *Handbook of Chemistry and Physics,* 78th edition, CRC Press, Boca Raton, FL, 1997–1998.

phase is a liquid, we say the liquid *evaporates*. If it is a solid, we say the solid *sublimes*. A glass of water will slowly evaporate if left to itself. A chunk of dry ice (solid CO_2) will slowly sublime if placed in a dish. As the temperature of a liquid is raised, the equilibrium vapor pressure increases, causing the rate of evaporation to increase. When the temperature reaches the point at which the equilibrium vapor pressure equals the externally applied pressure, vaporization can occur throughout the bulk of the liquid, resulting in a roiling motion as the gas expands freely into the surroundings. This phenomenon is called *boiling*, and the temperature at which it occurs is referred to as the *boiling point*. If the externally applied pressure is 1 atm, we call the temperature the *normal* boiling point. If it is 1 bar, the term *standard* boiling point is used.

We can qualitatively express the preceding concepts using a sketch showing the pressure of a single-component system as a function of temperature. Figure 6.13 shows such a sketch, which is called a *phase diagram*. The foregoing discussion shows that when both the liquid and vapor are present at temperature T_o, the pressure must be the equilibrium vapor pressure P^{eq}. This corresponds to Point A in the figure. As the temperature increases, P^{eq} increases. Curve 1 shows the qualitative form of this variation. Liquid and vapor are in equilibrium at every point on this curve, which continues to rise until the critical temperature, T_c, is reached at point B. When only the vapor phase is present, it is possible to raise the temperature at a fixed pressure. This means that the system can be at Point C or any point below Curve 1 only if vapor is the only phase present. If only the liquid phase is present, the pressure can exceed the equilibrium vapor pressure. Therefore, if the system is at point D or any point above Curve 1, liquid will be the only phase present.

The same considerations hold for the solid–vapor equilibrium. Curve 2 shows the qualitative variation of the equilibrium pressure over the solid phase as a function of temperature. At any point on this curve, both solid and vapor are present in equilibrium. Above Curve 2, only the solid phase can be present; below Curve 2, we have only the vapor phase.

Point G lies on both Curves 1 and 2. Therefore, at this single point, which we call the *triple point*, we have solid, liquid, and vapor all present and in equi-

▶ **FIGURE 6.13**
Qualitative sketch of a single-component phase diagram showing the loci of points at which liquid–vapor and solid–vapor are in equilibrium at Curves 1 and 2, respectively. The critical point at $T = T_c$ is shown at Point B. Points C, D, and F are in single-component regions where only vapor, liquid, and solid, respectively, are present. The triple point at which solid, liquid, and vapor are all present and in equilibrium occurs at the intersection of Curves 1 and 2. It is labeled as Point G in the diagram. When $T > T_c$, only supracritical gas can be present.

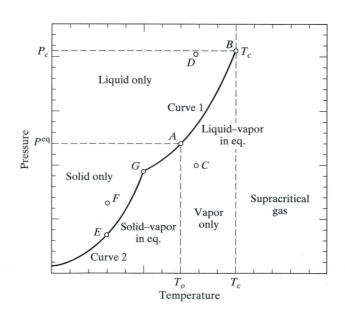

librium. At that point, the chemical potentials of solid, liquid, and vapor are all equal. Since $\mu_{\text{solid}} = \mu_{\text{liquid}}$ at the triple point, Point G must lie on the loci of points at which solid and liquid are in equilibrium. Our previous discussion demonstrated that we can use applied pressure to melt a solid if we have $\overline{V}_{\text{solid}} > \overline{V}_{\text{liquid}}$, but not otherwise. This tells us that if that inequality holds, as it does for the ice–water system, the points at which liquid and solid are in equilibrium must form a curve that intersects the triple point with a negative slope, as shown in Figure 6.14. If this is the case, the solid phase present at Point H could be converted to Point I, where only liquid is present, by increasing the pressure. If we have a system for which $\overline{V}_{\text{solid}} < \overline{V}_{\text{liquid}}$, the solid–liquid equilibrium Curve 3 must intersect the triple point with a positive slope, as shown in Figure 6.15. Inspection of this figure shows that there is no way to convert solid into liquid by applying increased pressure at a fixed temperature.

This completes the qualitative description of single-component phase equilibrium. What is missing are quantitative methods whereby we can accurately compute Curves 1, 2, and 3 in Figures 6.13, 6.14, and 6.15. We address this question in the sections that follow.

6.2   General Condition for Phase Equilibrium

The discussion of the previous section tells us qualitatively what to expect as we heat a pure substance. At low temperatures, we will have a solid phase that may undergo a phase transition to a different solid phase as we change the temperature and pressure. Eventually, we will observe a transition to a liquid. These condensed phases will all have characteristic equilibrium vapor

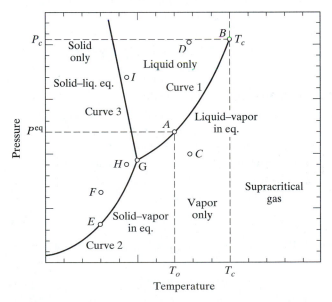

▲ FIGURE 6.14
Qualitative sketch of a single-component phase diagram showing the loci of points at which liquid–vapor, solid–vapor, and solid–liquid are in equilibrium as Curves 1, 2, and 3, respectively, for the case where the solid phase is less dense than the liquid phase. In this case, a solid phase at Point H can be converted to a liquid at Point I by increasing the pressure without changing the temperature. The other points are as described in the text and in the caption to Figure 6.13.

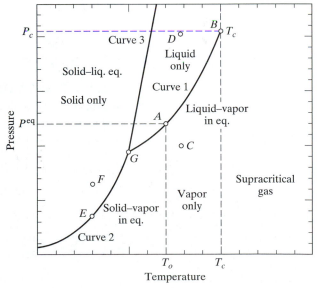

▲ FIGURE 6.15
Qualitative sketch of a single-component phase diagram showing the loci of points at which liquid–vapor, solid–vapor, and solid–liquid are in equilibrium as Curves 1, 2, and 3, respectively, for the case where the solid phase is more dense than the liquid phase. In this case, a solid phase cannot be converted to a liquid by increasing the pressure without changing the temperature. The other points are as described in the text and in the caption to Figure 6.13.

pressures. If the container holding the liquid phase is open, boiling will occur when the equilibrium vapor pressure reaches the applied external pressure. At the moment, we have no mechanism for computing the temperatures and pressures at which these events will occur. Nor do we have a means of calculating the equilibrium vapor pressure. In this section, we shall develop the thermodynamic equations required to execute such calculations.

Consider two phases A and B, each at its own temperature and pressure, as shown in Figure 6.16. In Chapter 5, it was explicitly shown that for two states to be in equilibrium at constant temperature and either constant volume or pressure, the chemical potentials of the states must be equal, so that $\Delta \mu = 0$. Applying this general principle to the two phases in Figure 6.16 means that they will be at equilibrium at a fixed temperature and pressure or a fixed temperature and volume when $\mu_A = \mu_B$. Taking differentials of both sides shows that we may state this requirement in the form $d\mu_A = d\mu_B$. Equation 5.44 relates the differential change in the chemical potential to changes in the temperature and pressure. Applying this equation to the aforesaid phase equilibrium gives

$$\boxed{d\mu_A = -\overline{S}_A \, dT_A^{eq} + \overline{V}_A \, dP_A^{eq} = d\mu_B = -\overline{S}_B \, dT_B^{eq} + \overline{V}_B \, dP_B^{eq}}, \qquad (6.1)$$

where $\overline{S}_i$ and $\overline{V}_i$ are the partial molar entropy and volume, respectively, and dT_i^{eq} and dP_i^{eq} are the changes in the equilibrium temperature and pressure, respectively, for component i. Equation 6.1 is the general equation governing changes in the equilibrium temperatures and pressures between phases. It holds regardless of the nature of phases A and B.

Under most conditions, the two phases will be in thermal and pressure equilibrium. That is, we will have $dT_A^{eq} = dT_B^{eq}$ and $dP_A^{eq} = dP_B^{eq}$. However, this need not be the case, as we shall see in Section 6.5.

6.3 The Clapeyron and Clausius–Clapeyron Equations

6.3.1 The Clapeyron Equation

In this section, we will examine the general condition for phase equilibrium when the two phases are at the same temperature and pressure. When this condition is present, we have $dT_A^{eq} = dT_B^{eq} = dT^{eq}$ and $dP_A^{eq} = dP_B^{eq} = dP^{eq}$. These equations describe the experimental situation in most cases. Under such conditions, we may drop the subscripts A and B on the temperature and pressure differentials. Equation 6.1 then becomes

$$-\overline{S}_A \, dT^{eq} + \overline{V}_A \, dP^{eq} = -\overline{S}_B \, dT^{eq} + \overline{V}_B \, dP^{eq}. \qquad (6.2)$$

Rearranging Eq. 6.2 so that the temperature and pressure differentials are on different sides gives

$$[\overline{S}_B - \overline{S}_A] \, dT^{eq} = [\overline{V}_B - \overline{V}_A] \, dP^{eq}. \qquad (6.3)$$

We may identify $\overline{S}_B - \overline{S}_A$ as the change in entropy when 1 mole of phase A converts to 1 mole of phase B. We denote this quantity as $\Delta\overline{S}$. The factor $\overline{V}_B - \overline{V}_A$ is the change in volume for the same transition, which we write as $\Delta\overline{V}$. In this notation, Eq. 6.3 becomes

$$\boxed{dP^{eq} = \frac{\Delta\overline{S}}{\Delta\overline{V}} \, dT^{eq}}. \qquad (6.4)$$

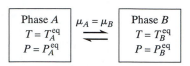

▲ FIGURE 6.16
Thermodynamic condition for equilibrium between two arbitrary phases, each at its own equilibrium temperature and pressure.

Equation 6.4 is the *Clapeyron equation*. Physically, it tells us by how much the equilibrium pressure will change for a given change in the equilibrium temperature.

In our qualitative treatment of phase equilibrium in Section 6.1, we found that a phase transition occurs at a fixed transition temperature, T^{eq}. The Clapeyron equation may be viewed as telling us how this temperature changes if we vary the equilibrium pressure. Since T^{eq} does not vary during a phase transition occurring at a given pressure P^{eq}, $\Delta \overline{S}$ may be expressed in terms of the transition enthalpy change as

$$\Delta \overline{S} = \int \frac{\delta q_{rev}}{T} = \frac{1}{T^{eq}} \int \delta q_{rev} = \frac{q_{rev}}{T^{eq}} = = \frac{\Delta \overline{H}}{T^{eq}}, \qquad (6.5)$$

where $\Delta \overline{H}$, the change in enthalpy per mole for the transition, is equal to q_{rev}, since the transition occurs at some fixed pressure. Combining Eqs. 6.4 and 6.5 produces

$$\boxed{dP^{eq} = \frac{\Delta \overline{H}}{T^{eq} \Delta \overline{V}} dT^{eq}}, \qquad (6.6)$$

which is another form of the Clapeyron equation. Equation 6.6 shows that, for an endothermic phase transition ($\Delta \overline{H} > 0$), the equilibrium pressure will increase as the equilibrium temperature rises if the change in volume for the transition is positive. The opposite behavior will be observed if the change in volume is negative.

EXAMPLE 6.1

Consider the graphite $\longrightarrow$ diamond phase transition. **(A)** Use the data given in Appendix A to compute $\Delta \overline{H}^{o}$ for the transition at 298.15 K and 1 bar of pressure. **(B)** The density of graphite and diamond are 2.25 g cm^{-3} and 3.51 g cm^{-3}, respectively. Compute $\Delta \overline{V}$ for the transition. **(C)** Given the fact that graphite and diamond are in equilibrium at 3,000 K and 235 kbar of pressure, compute the rate at which the equilibrium pressure changes with temperature at that temperature and pressure. Assume $\Delta \overline{H}^{o}$ to be a constant.

Solution

(A) The enthalpy change per mole under standard conditions is given by

$$\Delta \overline{H}^{o} = \overline{H}^{o}_{diamond} - \overline{H}^{o}_{graphite} = 1{,}895 \text{ J mol}^{-1} - 0 = 1{,}895 \text{ J mol}^{-1}. \qquad (1)$$

(B) The volume for 1 mole is given by

$$\overline{V} = \frac{\text{mass of one mole}}{\text{density}} = \frac{12.01 \text{ g}}{\text{density}}. \qquad (2)$$

Substituting the given data yields

$$\overline{V}_{graphite} = \frac{12.01 \text{ g mol}^{-1}}{2.25 \text{ g cm}^{-3}} = 5.34 \text{ cm}^3 \text{ mol}^{-1} = 0.00534 \text{ L mol}^{-1} \qquad (3)$$

and

$$\overline{V}_{diamond} = \frac{12.01 \text{ g mol}^{-1}}{3.51 \text{ g cm}^{-3}} = 3.42 \text{ cm}^3 \text{ mol}^{-1} = 0.00342 \text{ L mol}^{-1}. \qquad (4)$$

Therefore,

$$\Delta \overline{V} = 0.00342 - 0.00534 \text{ L mol}^{-1} = -0.00192 \text{ L mol}^{-1}. \qquad \textbf{(5)}$$

(C) Using the Clapeyron equation, we find that the rate of change of the equilibrium pressure with equilibrium temperature at 3,000 K is

$$\frac{dP^{eq}}{dT^{eq}} = \frac{\Delta \overline{H}^{o}}{T^{eq}\Delta \overline{V}} = \frac{1{,}895 \text{ J mol}^{-1}}{(3{,}000 \text{ K})(-0.00192 \text{ L mol}^{-1})} \times \frac{0.083144 \text{ L bar mol}^{-1} \text{ K}^{-1}}{8.314 \text{ J mol}^{-1} \text{ K}^{-1}}$$

$$= -3.29 \text{ bar K}^{-1}. \qquad \textbf{(6)}$$

Consequently, the equilibrium pressure increases as the equilibrium temperature decreases. Experiment confirms this result; note how the units must be converted in Eq. 6.

6.3.2   The Clausius–Clapeyron Equation: Application to Liquid–Vapor Equilibrium

Let us now apply the Clapeyron equation to the situation in which phase B is a vapor in equilibrium with liquid phase A. In that case, the change in enthalpy is the molar heat of vaporization, which we denote by $\Delta \overline{H}_{vap}$. With this substitution, the Clapeyron equation becomes

$$dP^{eq} = \frac{\Delta \overline{H}_{vap}}{T^{eq}\Delta \overline{V}} dT^{eq}. \qquad \textbf{(6.7)}$$

The change in molar volume taking place upon vaporization is $\Delta \overline{V} = \overline{V}_{(g)} - \overline{V}_{(l)}$. If we are at a temperature and pressure well below the critical values, T_c and P_c, the molar volume of the vapor will usually be on the order of 1,000 times greater than that for the liquid. Under such conditions, we may ignore the volume of the liquid and incur an error of no more than 0.1% for $\Delta \overline{V}$. With this approximation, we have

$$dP^{eq} = \frac{\Delta \overline{H}_{vap}}{T^{eq}\overline{V}_{(g)}} dT^{eq}. \qquad \textbf{(6.8)}$$

For temperatures below or only slightly above the standard boiling point, typical equilibrium vapor pressures range between zero and 10 bar. At these pressures, the vapor will obey the ideal-gas equation of state to a high degree of accuracy. We may, therefore, replace $\overline{V}_{(g)}$ in Eq. 6.8 with RT^{eq}/P^{eq}. When this is done, after rearranging terms, we obtain

$$\boxed{\frac{dP^{eq}}{P^{eq}} = d(\ln P^{eq}) = \frac{\Delta \overline{H}_{vap}}{R(T^{eq})^2} dT^{eq}}. \qquad \textbf{(6.9)}$$

Equation 6.9 is called the *Clausius–Clapeyron equation;* it is an approximation whose accuracy depends upon the error incurred by ignoring the volume of the condensed phase relative to that of the vapor. The accuracy is also limited by deviations of the pressure, volume, and temperature behavior of the vapor from the ideal-gas equation of state. Usually, these are very small.

  If we ignore the tiny pressure dependence of $\Delta \overline{H}_{vap}$, the variables are separated in Eq. 6.9, so we may integrate both sides to obtain the equilibrium vapor pressure as a function of temperature. Let us first take indefinite integrals. This gives

$$\int \frac{dP^{eq}}{P^{eq}} = \ln P^{eq} = \frac{1}{R} \int \frac{\Delta \overline{H}_{vap}}{(T^{eq})^2} \, dT^{eq} + \text{constant.} \tag{6.10}$$

The integration on the right side of Eq. 6.10 cannot be executed until we have determined the temperature dependence of $\Delta \overline{H}_{vap}$. This determination requires that we know C_p^m for the vapor and liquid phases. To evaluate the constant of integration, we must know the equilibrium vapor pressure at some temperature.

A superficial examination of Eq. 6.10 might suggest that we are taking the logarithm of a quantity, P^{eq}, that has units. This is not the case; the units of pressure canceled in the ratio dP^{eq}/P^{eq} before we took the integral. Therefore, the quantity P^{eq} that appears in the equation is the magnitude of the pressure without the unit. We encountered an identical situation in Chapter 5, in Eqs. 5.47 and 5.48, where the pressure units canceled before the integration. This led us to a unitless equilibrium constant. Since the magnitude of P^{eq} depends upon the unit used to express the equilibrium pressure, the constant will depend upon our choice of units, but the argument of the natural logarithm function in Eq. 6.10 is always unitless. In many texts, the superscripts denoting equilibrium pressures and temperatures are omitted for convenience. We have retained them to continue to emphasize the fact that equations such as this hold only when we have $d\mu_A = d\mu_B$, so that a state of equilibrium between the phases exists.

The constant of integration in Eq. 6.10 can be incorporated directly by taking a definite integral of Eq. 6.9 in which the lower limit is chosen to reflect our knowledge of the equilibrium vapor pressure P_1^{eq} at some temperature T_1^{eq}. The procedure produces

$$\int_{P_1^{eq}}^{P_2^{eq}} \frac{dP^{eq}}{P^{eq}} = \ln\left[\frac{P_2^{eq}}{P_1^{eq}} \right] = \frac{1}{R} \int_{T_1^{eq}}^{T_2^{eq}} \frac{\Delta \overline{H}_{vap}}{(T^{eq})^2} \, dT^{eq}. \tag{6.11}$$

As in Eq. 6.10, the pressure unit cancels in the ratio dP^{eq}/P^{eq}, so that the equilibrium pressures in Eq. 6.11 are the magnitudes of the pressure only, with no units.

If we make the approximation $\Delta C_p \approx 0$, $\Delta \overline{H}_{vap}$ will be independent of temperature, so that it may be factored from beneath the integral in Eqs. 6.10 and 6.11. We may do this without appreciable error if the temperature range involved in the integral is small. The approximation leads to a very simple form for the temperature dependence of the equilibrium vapor pressure:

$$\ln\left[\frac{P_2}{P_1} \right]^{eq} = \frac{\Delta \overline{H}_{vap}}{R} \int_{T_1^{eq}}^{T_2^{eq}} \frac{dT^{eq}}{(T^{eq})^2} = -\frac{\Delta \overline{H}_{vap}}{R} \left[\frac{1}{T_2^{eq}} - \frac{1}{T_1^{eq}} \right]. \tag{6.12}$$

Equation 6.12 permits the equilibrium vapor pressure to be computed at temperature T_2 if the equilibrium pressure is known at some other temperature T_1.

Most tabulations of boiling points list the normal ones. Consequently, at that temperature, we know that P^{eq} is 1 atm, or 760 torr. Substituting these values for the temperature and pressure into Eq. 6.12 for P_1^{eq} and T_1 gives

$$\ln P^{eq} = \ln(760) - \frac{\Delta \overline{H}_{vap}}{R} \left[\frac{1}{T^{eq}} - \frac{1}{T_b^o} \right], \tag{6.13A}$$

where we have written the normal boiling temperature as T_b^o. Note that both P^{eq} and 760 in Eq. 6.13A are magnitudes only. And since the original unit was torr, the equilibrium vapor pressure computed with the use of that equation is its magnitude when the pressure unit is torr. If we express P_1^{eq} in atm, the result is

$$\ln P^{eq} = \ln(1) - \frac{\Delta \overline{H}_{vap}}{R}\left[\frac{1}{T^{eq}} - \frac{1}{T_b^o}\right] = -\frac{\Delta \overline{H}_{vap}}{R}\left[\frac{1}{T^{eq}} - \frac{1}{T_b^o}\right], \quad \textbf{(6.13B)}$$

where now P^{eq} is the magnitude of the equilibrium vapor pressure at T^{eq} when the unit is atm.

EXAMPLE 6.2

The normal boiling temperature of water is 373.15 K. Compute the equilibrium vapor pressure of water at 50°C or 323.15 K, assuming that $\Delta C_p = 0$.

Solution

We first need to calculate the molar enthalpy of vaporization of water. For the reaction $H_2O(l) \longrightarrow H_2O(g)$,

$$\Delta \overline{H}^o = \overline{H}^o_{H_2O(g)} - \overline{H}^o_{H_2O(l)} = -241.82 - (-285.83)\ kJ\ mol^{-1} = 44.01\ kJ\ mol^{-1}$$

$$= 4.401 \times 10^4\ J\ mol^{-1}, \quad \textbf{(1)}$$

where the standard partial molar enthalpies are obtained from Appendix A. If we assume that $\Delta C_p = 0$, then $\Delta \overline{H}^o$ is a constant, and Eq. 6.13A may be employed. This gives

$$\ln P^{eq} = \ln(760) - \frac{4.401 \times 10^4\ J\ mol^{-1}}{8.314\ J\ mol^{-1}\ K^{-1}} - \left[\frac{1}{323.15} - \frac{1}{373.15}\right]$$

$$= 6.6333 - 2.1949 = 4.4384. \quad \textbf{(2)}$$

Performing exponentiation yields

$$P^{eq}\ at\ 323.15\ K = 84.64\ torr = 0.1114\ atm = 0.1129\ bar. \quad \textbf{(3)}$$

For related exercises, see Problems 6.5 and 6.6.

It is instructive to put Eq. 6.13A in exponential form. Taking exponentials of both sides, we obtain

$$P^{eq} = 760\ exp\left\{-\frac{\Delta \overline{H}_{vap}}{R}\left[\frac{1}{T^{eq}} - \frac{1}{T_b^o}\right]\right\} = 760\ exp\left[\frac{\Delta \overline{H}_{vap}}{RT_b^o}\right]exp\left[-\frac{\Delta \overline{H}_{vap}}{RT^{eq}}\right].$$

Since $\Delta \overline{H}_{vap}$ is a constant if $\Delta C_p \approx 0$, and since T_b^o is fixed for a given liquid, the first exponential is a constant, and we have

$$P^{eq} = C\ exp\left[-\frac{\Delta \overline{H}_{vap}}{RT^{eq}}\right].$$

It is important to note that this equation has the form of Eq. 5.89. This is not surprising, because the process $A(l) + \Delta \overline{H}_{vap} \longrightarrow A(g)$ has an energy requirement equal to the molar heat of vaporization and the equilibrium vapor pressure measures the extent to which the process proceeds to the right. Consequently, when we assume that the energy requirement is independent of temperature, we get the general result given by Eq. 5.89. This is the second example we have seen of this type. We shall encounter numerous others as we proceed.

Equation 6.13A shows that, within the framework of the approximations made to obtain the Clausius–Clapeyron equation, we expect the logarithm of the equilibrium vapor pressure to vary linearly when plotted as a function of $(T^{eq})^{-1}$. Furthermore, that equation demonstrates that the slope of the line will be $-[\Delta \overline{H}_{vap}/R]$ and the intercept will be $\ln(760) - \Delta \overline{H}_{vap}/(RT_b^o)$. We may, therefore, obtain the molar heat of vaporization from the slope of an equilibrium vapor pressure plot against $(T^{eq})^{-1}$. Once $\Delta \overline{H}_{vap}$ has been determined, the normal boiling point can be computed from the intercept. In the problems at the end of the chapter, you will assess the magnitude of the error we expect from our assumptions of $\Delta C_p \approx 0$ and ideal behavior for the vapor.

EXAMPLE 6.3

The vapor pressure of a liquid is measured as a function of temperature. The investigator obtains the data at right.

(A) Plot $\ln(P^{eq})$ versus $(T^{eq})^{-1}$. Using a least-squares method, obtain the best straight-line fit to the data. **(B)** Use the Clausius–Clapeyron equation to compute $\Delta \overline{H}_{vap}$ and the normal boiling point of the liquid.

Solution

(A) Figure 6.17 shows a plot of the data, along with the best straight-line fit to the data. The best line is given by

$$\ln(P^{eq}) = 21.120 - \frac{5{,}097.4 \text{ K}}{T}.$$

(B) The slope of the line is $-5{,}097.4$ K. Therefore, we obtain

$$\text{slope} = \frac{-\Delta \overline{H}_{vap}}{R} = -5{,}097.4 \text{ K}. \qquad (1)$$

Solving for $\Delta \overline{H}_{vap}$ yields

$$\Delta \overline{H}_{vap} = 5{,}097.4 \text{ K} (8.314 \text{ J mol}^{-1} \text{ K}^{-1})$$

$$= 4.238 \times 10^4 \text{ J mol}^{-1} = 42.38 \text{ kJ mol}^{-1}. \qquad (2)$$

T^{eq} (K)	P^{eq} (torr)
275	13.8
280	17.9
285	25.2
290	34.7
295	46.1
300	61.2
305	82.1
310	107.4
315	139.1
320	180.3
325	231.0
330	292.4

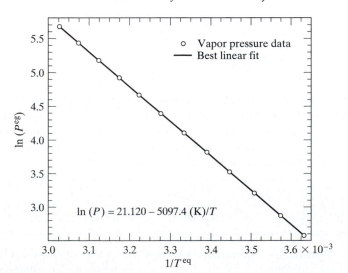

$$\ln(P) = 21.120 - 5097.4 \text{ (K)}/T$$

◄ FIGURE 6.17
Variation of $\ln(P^{eq})$ with $1/T^{eq}$ The Clausius–Clapeyron equation predicts that the result should be linear if $\Delta \overline{H}_{vap}$ is independent of temperature and the vapor is described accurately by the ideal-gas equation of state.

The boiling point can be computed from the intercept:

$$\ln(760) + \frac{\Delta \overline{H}_{vap}}{RT_b^o} = 21.120. \tag{3}$$

Solving for T_b^o, we obtain

$$T_b^o = \frac{\Delta \overline{H}_{vap}}{R[21.120 - \ln(760)]} = \frac{4.238 \times 10^4 \text{ J mol}^{-1}}{8.314 \text{ J mol}^{-1} \text{ K}^{-1}[21.120 - 6.6333]}$$

$$= 351.9 \text{ K} = 78.72°\text{C} \tag{4}$$

We could also compute the normal boiling point by calculating the temperature at which $P^{eq} = 760$ torr. Substituting into the linear fit for the vapor pressure produces

$$\ln(760) = 21.120 - \frac{5{,}097.4 \text{ K}}{T_b^o}. \tag{5}$$

Solving for T_b^o, we obtain

$$T_b^o = \frac{5{,}097.4}{[21.120 - \ln(760)]} = 351.9 \text{ K}, \tag{6}$$

which is the same result as obtained in Eq. 4.

For related exercises, see Problems 6.7, 6.9, and 6.10.

6.3.3   Application to Solid–Vapor Equilibrium

If a solid and a vapor are in equilibrium under conditions in which both phases have the same temperature and pressure, the Clapeyron equation may be applied to give

$$dP^{eq} = \frac{\Delta \overline{H}_{sub}}{T^{eq} \Delta \overline{V}} dT^{eq}. \tag{6.14}$$

In Eq. 6.14, the enthalpy change for the transition is now termed the enthalpy of sublimation, which we denote by $\Delta \overline{H}_{sub}$, and the change in molar volume taking place upon sublimation is $\Delta \overline{V} = \overline{V}_{(g)} - \overline{V}_{(s)}$. As in the case of liquid–vapor equilibrium, the molar volume of the solid is usually much smaller than that of the vapor. Consequently, we may use the approximation $\Delta \overline{V} = \overline{V}_{(g)} - \overline{V}_{(s)} \approx \overline{V}_{(g)}$ with confidence and assume the vapor to behave ideally. This again leads to the Clausius–Clapeyron equation:

$$\frac{dP^{eq}}{P^{eq}} = d(\ln P^{eq}) = \frac{\Delta \overline{H}_{sub}}{R(T^{eq})^2} dT^{eq}. \tag{6.15}$$

If we assume that the difference between the heat capacities of solid and vapor times dT^{eq} is small relative to $\Delta \overline{H}_{sub}$ ($\Delta C_p\, dT^{eq} \ll \Delta \overline{H}_{sub}$), $\Delta \overline{H}_{sub}$ will be nearly constant, and we obtain formulas identical to equations 6.12 and 6.13, but with $\Delta \overline{H}_{vap}$ replaced by $\Delta \overline{H}_{sub}$. Under these conditions, the equilibrium vapor pressure above the solid phase is given by

$$\ln\left[\frac{P_2^{eq}}{P_1^{eq}} \right]_{eq} = \frac{\Delta \overline{H}_{sub}}{R} \int_{T_1^{eq}}^{T_2^{eq}} \frac{dT^{eq}}{(T^{eq})^2} = -\frac{\Delta \overline{H}_{sub}}{R} \left[\frac{1}{T_2^{eq}} - \frac{1}{T_1^{eq}} \right]. \tag{6.16}$$

The only significant difference between the case of liquid–vapor and solid–vapor equilibrium is that we have a very convenient means of obtaining the equilibrium vapor pressure above the liquid at the normal boiling point, where we know that $P^{eq} = 1$ atm $= 760$ torr. This provides the values for the lower limit in Eq. 6.12, to give Eq. 6.13. We have no such convenient means of obtaining the lower limits in Eq. 6.16. The equilibrium vapor pressure above the solid must be measured at some convenient temperature. Once this single data point is available, Eq. 6.16 permits us to compute the equilibrium vapor pressure at any other temperature desired.

In Chapter 5, we noted that we do not include factors in the expression for the equilibrium constant for pure solids and liquids that appear in the equilibrium reaction, because their chemical potentials are constants at a given temperature. Equations 6.7 and 6.14 provide the proof of this statement promised in Chapter 5. For example, consider the case of a solid phase A. The chemical potential of solid A, $\mu_{(s)}^A$, must be equal to the chemical potential of its equilibrium vapor, $\mu_{(g)}^A$. That is, we must have

$$\mu_{(s)}^A = \mu_{(g)}^A = \mu^o + RT \ln(P_A^{eq}), \qquad (6.17)$$

where the assumption of ideal vapors is essentially exact at the low vapor pressures observed over solids. Equation 6.14 shows that P_A^{eq} can be changed only if we alter the temperature. Therefore, the second term in Eq. 6.17 is a constant at a given temperature. The standard chemical potential, μ^o, is also a function of temperature only. Consequently, $\mu_{(s)}^A$ is a constant at a given temperature. The same analysis holds for a liquid.

Solid–vapor and liquid–vapor equilibria provide examples of the general principle concerning the relationship between ΔH and ΔS for phase transitions that was discussed in Section 6.1 of this chapter. Both vaporization and sublimation are endothermic process that have $\Delta H > 0$. Consequently, we expect that the phase transition will be one for which ΔS increases, indicating an increase in the degree of disorder in the system. This is obviously the case for both solid $\longrightarrow$ vapor and liquid $\longrightarrow$ vapor transitions.

6.3.4   Application to Solid–Liquid Equilibrium

In our treatment of solid–vapor and liquid–vapor equilibria, the pressure exerted on the condensed phase is exactly equal to the equilibrium vapor pressure, thereby producing pressure equilibrium between the phases, so that $dP_A^{eq} = dP_B^{eq} = dP^{eq}$. In solid–liquid equilibrium, the pressure on the condensed phases is produced by the atmosphere or some other external force. This point is illustrated in Figure 6.18. By applying downward force to the piston shown in the figure, we may alter the equilibrium temperature between the solid and the liquid. Since we call the equilibrium temperature the melting or freezing point, the previous statement is equivalent to saying that the melting point of a substance depends upon the pressure applied to the condensed phases of the substance. The nature of the dependence is described by the Clapeyron equation.

The Clapeyron equation describing the equilibrium between solid and liquid is

$$dP^{eq} = \frac{\Delta \overline{H}_{fus}}{T^{eq} \Delta \overline{V}} dT^{eq}, \qquad (6.18)$$

where the enthalpy change is now called the heat of fusion, which we denote by $\Delta \overline{H}_{fus}$, and the change in volume is $\Delta \overline{V} = \overline{V}_{(l)} - \overline{V}_{(s)}$. In this case, we cannot put Eq. 6.18 into the form of the Clausius–Clapeyron equation, since neither

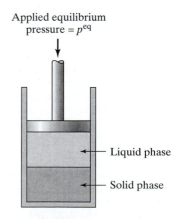

Applied equilibrium pressure $= p^{eq}$

Liquid phase

Solid phase

▲ FIGURE 6.18
The application of an applied external pressure P^{eq} affects the solid–liquid equilibrium temperature in a manner predicted by the Clapeyron equation.

phase is a gas. Nor can we ignore either $\overline{V}_{(s)}$ or $\overline{V}_{(l)}$, as these volumes are about the same size. Because fusion is always an endothermic process with $\Delta\overline{H}_{fus} > 0$, Eq. 6.18 shows that the equilibrium temperature (melting point) will increase as the pressure increases if $\Delta\overline{V} > 0$. The opposite behavior will be observed if $\Delta\overline{V} < 0$. A particularly important example of the latter behavior is seen in Figure 6.6, which shows the variation of the liquid and solid chemical potentials as a function of pressure for the water–ice system. Since the density of water is greater than that of ice (ice floats), we have $\overline{V}_{(s)} > \overline{V}_{(l)}$, and hence, $\Delta\overline{V} < 0$. An increase in pressure, therefore, produces a decrease in the equilibrium temperature at which melting occurs. That is, the melting point of ice is lower than 0°C if the pressure is greater than one atm. Figure 6.7 shows the variation of the ice–water melting point as a function of applied pressure.

We may obtain a very useful expression relating the equilibrium temperature to pressure by using Eq. 6.18. Integration of both sides gives

$$\int_{P_1^{eq}}^{P_2^{eq}} dP^{eq} = [P_2^{eq} - P_1^{eq}] = \int_{T_1^{eq}}^{T_2^{eq}} \frac{\Delta\overline{H}_{fus}}{T^{eq}\Delta\overline{V}} dT^{eq}. \qquad (6.19)$$

Since the melting point will vary only over a very small range for most applied pressures, we may regard $\Delta\overline{H}_{fus}$ as being constant over this range with virtually no error. If we also assume that the partial molar volumes of the solid and liquid phases are independent of pressure, we may factor the ratio $\Delta\overline{H}_{fus}/\Delta\overline{V}$ out of the integral in Eq. 6.19 to obtain

$$[P_2^{eq} - P_1^{eq}] = \frac{\Delta\overline{H}_{fus}}{\Delta\overline{V}} \int_{T_1^{eq}}^{T_2^{eq}} \frac{dT^{eq}}{T^{eq}} = \frac{\Delta\overline{H}_{fus}}{\Delta\overline{V}} \ln\left[\frac{T_2^{eq}}{T_1^{eq}}\right]. \qquad (6.20)$$

If pressure $P_1^{eq} = 1$ atm, T_1^{eq} will be the normal melting point, T_m^o. Insertion of this point for the lower limits in Eq. 6.20 gives

$$P^{eq} = 1 \text{ atm} + \frac{\Delta\overline{H}_{fus}}{\Delta\overline{V}} \ln\left[\frac{T^{eq}}{T_m^o}\right]. \qquad (6.21)$$

▶ FIGURE 6.19
The variation of the melting point as a function of pressure for the ice–water system. The points correspond to the experimental data (Source: *Handbook of Chemistry and Physics*, 78th edition, CRC Press, Boca Raton, FL, 1997–1998). The line is a least-squares fit of a straight line to the data. Over this pressure range, the linearity is excellent and in agreement with the predictions of Eq. 6.22. The slope of the line predicts a molar heat of fusion for ice of 5,998 J mol^{-1}, which is 0.20% below the measured value of 6,010 J mol^{-1}.

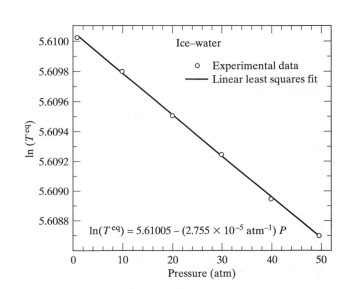

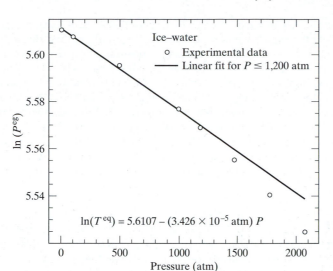

◀ FIGURE 6.20
The variation of melting point as a function of pressure over an extended pressure range for the ice–water system. The points correspond to the experimental data (Source: *Handbook of Chemistry and Physics*, 78th edition, CRC Press, Boca Raton, FL, 1997–1998). The line is a least-squares fit of a straight line to the data at pressures below 1,200 atm. Over this pressure range, significant deviations from linearity can be seen. These deviations are the result of the fact that the partial molar volumes of ice and water are not constant, as is assumed in the derivation of Eq. 6.22.

By writing the logarithm of the temperature ratio as the difference of two logarithms, we may solve Eq. 6.21 for $\ln(T^{eq})$ to obtain

$$\ln(T^{eq}) = \left[\ln(T_m^o) - \frac{\Delta\overline{V}}{\Delta\overline{H}_{fus}}\right] + \left[\frac{\Delta\overline{V}}{\Delta\overline{H}_{fus}}\right] P^{eq}, \qquad (6.22)$$

which predicts that a plot of $\ln(T^{eq})$ versus P^{eq} will be linear with a slope equal to $\Delta\overline{V}/\Delta\overline{H}_{fus}$ and an intercept $\ln(T_m^o) - (\Delta\overline{V}/\Delta\overline{H}_{fus})$. Again, we note that the units on T^{eq} and T_m^o canceled in the temperature ratio that appears in Eqs. 6.19, 6.20, and 6.21. Therefore, T^{eq} and T_m^o in Eq. 6.22 are magnitudes only. The accuracy of Eq. 6.22 is illustrated in Figure 6.19, which plots $\ln(T^{eq})$ versus pressure for the ice–water system. The data are represented with good accuracy by a straight line. From the slope of the line, we obtain

$$\left[\frac{\Delta\overline{V}}{\Delta\overline{H}_{fus}}\right] = -2.755 \times 10^{-5}\ atm^{-1}.$$

Using the measured densities for water and ice, we compute $\Delta\overline{V} = -0.001631\ L\ mol^{-1}$. This result means that the partial molar heat of fusion of ice is

$$\Delta\overline{H}_{fus} = \frac{-0.001631\ L\ mol^{-1}}{-2.755 \times 10^{-5}\ atm^{-1}} = 59.20\ L\ atm\ mol^{-1} = 5{,}998\ J\ mol^{-1}.$$

The measured value is 6,010 J mol^{-1}, so our result is in error by -0.20%. At higher pressures, Eq. 6.22 becomes less accurate, because the partial molar volumes vary with pressure and, therefore, cannot be factored out of the integral in Eq. 6.19. Figure 6.20 shows that as a result, a plot of $\ln(T^{eq})$ is no longer linear.

EXAMPLE 6.4

When I was taking general science in junior high school, my teacher performed the following demonstration for the class: When we arrived for class, there was a block of ice sitting in a tray resting on a small stand. A thin wire was stretched across the top of the ice block, with large weights attached to each end, as shown

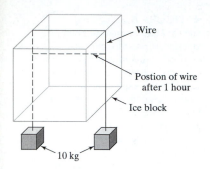

▲ FIGURE 6.21
A simple experiment showing the
Clapeyron equation in action.

in Figure 6.21. The teacher told us to observe the block of ice from time to time. As the class progressed, we noticed the wire beginning to "cut through" the ice. By the end of the period, it had reached a position about 20% through the block. However, the ice was still solid above the wire, which was now embedded within the block. By the end of the morning, I returned to class to find that the wire had sliced all the way through the ice, allowing the weights to fall to the floor, but the block of ice was still intact. I was very puzzled. **(A)** Explain what is happening in this experiment. **(B)** If the area of the wire in contact with the ice is 10^{-5} m^2 and the weights each have a mass of 10 kg, compute the melting point of the ice beneath the wire. Assume that $\Delta \overline{H}_{\text{fus}}$ for ice is 6,010 J mol^{-1}. Assume also that the densities of water and ice are 0.9998 g cm^3 and 0.9168 g cm^3, respectively.

Solution

(A) Equation 6.18 and the discussion that follows show that the melting point of ice drops as the pressure is increased, since $\Delta \overline{V} < 0$ for ice–water equilibrium. The temperature of the ice is at or slightly below 273.15 K, the normal melting point of ice. Directly beneath the wire, however, the melting point is below this value, because of the pressure exerted by the wire. Consequently, the temperature of the ice is above the melting temperature, and the ice melts, allowing the wire to descend into the block. Once the pressure is relieved, the melting point of the water above the wire returns to 273.15 K, and the water refreezes. This process continues until the wire completes its journey through the ice block. Science is simple with the right equations!

(B) Equation 1.1 defines pressure as force per unit area. The wire exerts a pressure against the ice given by

$$P = \frac{F}{A} = \frac{mg}{A} = \frac{(20 \text{ kg})(9.806 \text{ m s}^{-2})}{10^{-5} \text{ m}^2} = 1.961 \times 10^7 \text{ N m}^{-2} = 1.961 \times 10^7 \text{ Pa}$$

$$= 196.1 \text{ bar} = 193.5 \text{ atm}. \tag{1}$$

From the data given in the example, we may compute

$$\overline{V}_{(l)} = \frac{18.016 \text{ g mol}^{-1}}{0.9998 \text{ g cm}^{-3}} = 18.02 \text{ cm}^3 \text{ mol}^{-1} = 0.01802 \text{ L mol}^{-1}. \tag{2}$$

For ice, we have

$$\overline{V}_{(s)} = \frac{18.016 \text{ g mol}^{-1}}{0.9168 \text{ g cm}^{-3}} = 19.65 \text{ cm}^3 \text{ mol}^{-1} = 0.01965 \text{ L mol}^{-1}, \tag{3}$$

which gives

$$\Delta \overline{V} = \overline{V}_{(l)} - \overline{V}_{(s)} = -0.001630 \text{ L mol}^{-1}. \tag{4}$$

Solving Eq. 6.21 for T^{eq}, we obtain

$$T^{\text{eq}} = T_m^o \exp\left[\frac{\Delta \overline{V}(P^{\text{eq}} - 1)}{\Delta \overline{H}_{\text{fus}}}\right]. \tag{5}$$

Converting $\Delta \overline{H}_{\text{fus}}$ to L atm mol^{-1} yields

$$\Delta \overline{H}_{\text{fus}} = 6{,}010 \text{ J mol}^{-1} \times \frac{0.08206 \text{ L atm mol}^{-1} \text{ K}^{-1}}{8.314 \text{ J mol}^{-1} \text{ K}^{-1}} = 59.32 \text{ L atm mol}^{-1}. \tag{6}$$

This gives

$$T^{\text{eq}} = 273.15 \text{ K} \exp\left[\frac{(-0.001630 \text{ L mol}^{-1})(193.5 - 1)\text{atm}}{59.32 \text{ L atm mol}^{-1}}\right] = 271.7 \text{ K}. \tag{7}$$

The melting point of the ice beneath the wire is 1.4 K below the temperature of block. As a result, the ice melts as described.

6.4   Triple Point: Phase Diagrams
6.4.1   The Triple Point

The equilibrium vapor pressure above a pure liquid is given by Eq. 6.11, viz.,

$$\ln\left[\frac{P_2}{P_1}\right]^{eq} = \frac{1}{R} \int_{T_1^{eq}}^{T_2^{eq}} \frac{\Delta \overline{H}_{vap}\, dT^{eq}}{(T^{eq})^2},$$

provided that the Clausius–Clapeyron equation may be used. If we take $\Delta \overline{H}_{vap}$ to be independent of temperature, we obtain Eq. 6.12:

$$\ln\left[\frac{P_2^{eq}}{P_1^{eq}}\right] = -\frac{\Delta \overline{H}_{vap}}{R}\left[\frac{1}{T_2^{eq}} - \frac{1}{T_1^{eq}}\right].$$

Therefore, at every point along a curve whose equation is

$$P_2^{eq} = P_1^{eq} \exp\left\{-\frac{\Delta \overline{H}_{vap}}{R}\left[\frac{1}{T_2^{eq}} - \frac{1}{T_1^{eq}}\right]\right\}, \tag{6.23}$$

the liquid and vapor are present in equilibrium. If the approximations we have made to obtain the Clausius–Clapeyron equation are accurate, and if $\Delta \overline{H}_{vap}$ is nearly independent of temperature, Eq. 6.23 provides a means by which Curve 1 in Figures 6.13, 6.14, and 6.15 may be computed.

A similar situation holds for the equilibrium vapor pressure above the pure solid. In this case, the equilibrium vapor pressure is given by

$$P_{(s)}^{eq} = P_o \exp\left\{-\frac{\Delta \overline{H}_{sub}}{R}\left[\frac{1}{T^{eq}} - \frac{1}{T^o}\right]\right\}, \tag{6.24}$$

where P_o is the equilibrium vapor pressure over the solid at temperature T^o. At all points $(T^{eq}, P_{(s)}^{eq})$ the solid and vapor are both present in equilibrium. When our approximations are accurate, the loci of points that satisfy Eq. 6.24 form Curve 2 in Figures 6.13, 6.14, and 6.15.

At the point the two curves described by Eqs. 6.23 and 6.24 intersect, the equilibrium vapor pressures over the liquid and solid will be identical, and we will have all three phases present in simultaneous equilibrium. The unique temperature at which this occurs is the triple point, first introduced in Section 6.1. It is represented by Point G in the aforementioned figures. We denote the temperature at the triple point by T_t. The pressure at the triple point, P_t, may be obtained using either Eq. 6.23 or Eq. 6.24:

$$P_t = P_1^{eq} \exp\left\{-\frac{\Delta \overline{H}_{vap}}{R}\left[\frac{1}{T_t} - \frac{1}{T_1^{eq}}\right]\right\} = P_o \exp\left\{-\frac{\Delta \overline{H}_{sub}}{R}\left[\frac{1}{T_t} - \frac{1}{T^o}\right]\right\}. \tag{6.25}$$

Equation 6.25 provides a direct means for locating the triple point. Rearranging that equation yields

$$\frac{P_1^{eq}}{P_o} = \exp\left[\left\{-\frac{\Delta \overline{H}_{sub}}{R}\left[\frac{1}{T_t} - \frac{1}{T^o}\right]\right\} + \frac{\Delta \overline{H}_{vap}}{R}\left[\frac{1}{T_t} - \frac{1}{T_1^{eq}}\right]\right].$$

Collecting terms within the argument of the exponential produces

$$\frac{P_1^{eq}}{P_o} = \exp\left[\frac{1}{T_t}\left[\frac{\Delta \overline{H}_{vap} - \Delta \overline{H}_{sub}}{R}\right] + \frac{\Delta \overline{H}_{sub}}{RT^o} - \frac{\Delta \overline{H}_{vap}}{RT_1^{eq}}\right].$$

Table 6.1 Measured equilibrium vapor pressures for ice and water

Temp. (K)	ice	water
	(Eq.) vapor pressure (torr)	
263.15	1.950	
264.15	2.131	
265.15	2.326	
266.15	2.537	
267.15	2.765	
278.15		6.543
279.15		7.013
280.15		7.513
281.15		8.045
282.15		8.609

Taking logarithms of both sides and solving the resulting expression for T_t, we obtain

$$T_t = \frac{\Delta \overline{H}_{vap} - \Delta \overline{H}_{sub}}{R \ln\left[\dfrac{P_1^{eq}}{P_o}\right] - \dfrac{\Delta \overline{H}_{sub}}{T^o} + \dfrac{\Delta \overline{H}_{vap}}{T_1^{eq}}}. \tag{6.26}$$

As an example, let us use Eq. 6.26 to determine the triple point for the specific case of the ice–water-vapor equilibrium. We must be careful because of our assumption that $\Delta \overline{H}_{vap}$ and $\Delta \overline{H}_{sub}$ are constants. This assumption has limited accuracy, and the exponential functions contained in Eqs. 6.23 and 6.24 are highly sensitive to the values of $\Delta \overline{H}_{sub}$ and $\Delta \overline{H}_{vap}$. Therefore, we need to evaluate these quantities in a region near the triple point, so that the errors will be minimized. The measured equilibrium vapor pressures over ice and water at temperatures near the triple point are given in Table 6.1.

Equations 6.13A and 6.16 show that $\Delta \overline{H}_{vap}$ and $\Delta \overline{H}_{sub}$ may be obtained from the slope of plots of the logarithm of the equilibrium vapor pressures versus T^{-1}. Example 6.3 illustrates the method. Figure 6.22 shows such plots for ice and water using the data in Table 6.1. Equations 6.13A and 6.16 show that the slopes of the least-squares fits to the data equal $-\Delta \overline{H}_{vap}/R$ and $-\Delta \overline{H}_{sub}/R$, respectively. Direct calculation gives

$$\Delta \overline{H}_{sub} = -(8.314 \text{ J mol}^{-1} \text{ K}^{-1})(-6{,}135.2 \text{ K}) = 5.101 \times 10^4 \text{ J mol}^{-1}$$

and

$$\Delta \overline{H}_{vap} = -(8.314 \text{ J mol}^{-1} \text{ K}^{-1})(-5{,}381.2 \text{ K}) = 4.474 \times 10^4 \text{ J mol}^{-1}.$$

The data in Table 6.1 show that if we take T^o and T_1^{eq} to be 263.15 K and 282.15 K, respectively, we have $P_o = 1.950$ torr and $P_1^{eq} = 8.609$ torr. Substituting these values into Eq. 6.26 gives

$$T_t = \frac{4.474 \times 10^4 - 5.101 \times 10^4 \text{ J mol}^{-1}}{\left[8.314 \ln\left[\dfrac{8.609}{1.950}\right] - \dfrac{5.101 \times 10^4}{263.15} + \dfrac{4.474 \times 10^4}{282.15}\right] \text{J mol}^{-1} \text{ K}^{-1}} = 273.4 \text{ K}.$$

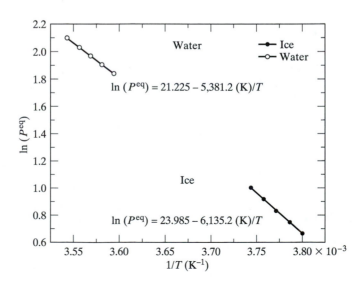

▶ FIGURE 6.22
Plots of the logarithm of the equilibrium vapor pressure as a function of $1/T$ for ice and water. Data are taken from the *Handbook of Chemistry and Physics,* 78th edition, CRC Press, Boca Raton, FL, 1997–1998. The slopes of the least-squares linear fits to the data give $-(\Delta \overline{H}/R)$.

Water
$\ln(P^{eq}) = 21.225 - 5{,}381.2 \text{ (K)}/T$

Ice
$\ln(P^{eq}) = 23.985 - 6{,}135.2 \text{ (K)}/T$

The measured value of the triple point of H_2O is 273.16 K. The 0.3 K error is due primarily to the temperature variation of $\Delta \overline{H}_{sub}$ and $\Delta \overline{H}_{vap}$, which we have not taken into account.

6.4.2   Quantitative Construction of Phase Diagrams

Within the constraints imposed by our approximations, the preceding equations completely describe the liquid–vapor, solid–vapor, and liquid–solid equilibria for a pure compound or element. The variation of equilibrium pressures and temperatures for these equilibria are conveniently illustrated by a plot of P^{eq} versus T^{eq}, which can record either measured data or data calculated using the Clapeyron or the Clausius–Clapeyron equation. In Section 6.1, we called such a plot a *phase diagram*, in that it shows the conditions under which various phases are in thermodynamic equilibrium. As an illustration, let us quantitatively construct the phase diagram for the H_2O system.

We first plot the curve along which the liquid and vapor are in equilibrium. This curve is given by Eq. 6.11:

$$\ln \left[\frac{P_2^{eq}}{P_1^{eq}} \right] = \frac{1}{R} \int_{T_1^{eq}}^{T_2^{eq}} \frac{\Delta \overline{H}_{vap} \, dT^{eq}}{(T^{eq})^2}.$$

Rather than obtain an approximate result by assuming $\Delta \overline{H}_{vap}$ to be a constant, we shall use the measured heat capacities of $H_2O(l)$ and $H_2O(g)$ to obtain $\Delta \overline{H}_{vap}$ as a function of temperature. The variation of $\Delta \overline{H}_{vap}$ with temperature is given by

$$\left(\frac{\partial \overline{H}_{vap}}{\partial T} \right)_P = \Delta C_p = [C_p^m]_{H_2O(g)} - [C_p^m]_{H_2O(l)}. \qquad (6.27)$$

In the most precise work, the liquid and vapor heat capacities are obtained as functions of temperature over the range of interest. Here, we will be content to assume that ΔC_p is a constant. This assumption gives

$$\int_{\Delta \overline{H}_{vap}(282.15)}^{\Delta \overline{H}_{vap}(T)} d\overline{H}_{vap} = \Delta \overline{H}_{vap}(T) - \Delta \overline{H}_{vap}(282.15)$$

$$= \int_{282.15}^{T} \Delta C_p \, dT = \Delta C_p [T - 282.15] \qquad (6.28)$$

Rearranging terms yields $\Delta \overline{H}_{vap}(T)$ as a function of temperature:

$$\Delta \overline{H}_{vap}(T) = \Delta \overline{H}_{vap}(282.15) + \Delta C_p [T - 282.15]. \qquad (6.29)$$

Using the data given in Figure 6.21, Table 6.1, and Appendix A, we obtain

$$\Delta \overline{H}_{vap}(T) = 4.474 \times 10^4 - 41.71[T - 282.15] \text{ J mol}^{-1}$$

$$= 5.651 \times 10^4 - 41.71T \text{ J mol}^{-1}. \qquad (6.30)$$

Inserting this result into Eq. 6.11 gives us the equilibrium vapor pressure over the liquid with reasonable accuracy. The result is

$$\ln \left[\frac{P_2^{eq}}{8.609} \right]_{eq} = \frac{1}{R} \int_{282.15}^{T_2^{eq}} \frac{[5.651 \times 10^4 - 41.71T] \, dT^{eq}}{(T^{eq})^2}$$

$$= -\frac{5.651 \times 10^4}{R} \left[\frac{1}{T_2} - \frac{1}{282.15} \right] - \frac{41.71}{R} \ln \left[\frac{T_2}{282.15} \right]. \qquad (6.31)$$

Dropping the subscript "2," we find that the equilibrium vapor pressure over the liquid is

$$P_{(l)}^{eq} = 8.609 \exp\left\{ -\frac{5.651 \times 10^4}{R}\left[\frac{1}{T^{eq}} - \frac{1}{282.15}\right] - \frac{41.71}{R}\ln\left[\frac{T^{eq}}{282.15}\right] \right\} \text{ torr.}$$

(6.32)

If we restrict the range of temperatures over which we intend to illustrate the equilibrium behavior between solid and vapor, we may utilize Eq. 6.24 without concern for the temperature variation of ΔH_{sub}. This result, with the data from Figure 6.21 and Table 6.1 inserted, is

$$P_{(s)}^{eq} = 1.950 \exp\left\{ -\frac{5.101 \times 10^4}{R}\left[\frac{1}{T^{eq}} - \frac{1}{263.15}\right] \right\} \text{ torr.}$$ (6.33)

The first portion of the phase diagram is obtained by plotting the two curves given by Eqs. 6.32 and 6.33. The result is shown in Figure 6.23A. The curve from Point A to Point B is a plot of Eq. 6.33, while the curve from Point B to Point C is a plot of Eq. 6.32. These results are represented in a qualitative fashion by Curves 2 and 1 in Figures 6.13, 6.14, and 6.15. The two curves intersect at the triple point denoted by Point B in Figure 6.23A.

All that remains is to illustrate how the equilibrium temperature for the solid–liquid equilibrium varies with pressure. This is given by Eq. 6.20:

$$[P_2^{eq} - P_1^{eq}] = \frac{\Delta \overline{H}_{fus}}{\Delta \overline{V}}\ln\left[\frac{T_2^{eq}}{T_1^{eq}}\right].$$

We know that the solid and liquid are in equilibrium at the triple point, which occurs at 273.16 K and a pressure of 4.588 torr. Insertion of these values along with $\Delta \overline{H}_{fus}$ and $\Delta \overline{V}$ given in Example 6.4 produces

$$P^{eq} = 4.588 + \frac{59.32 \text{ L atm mol}^{-1}}{-0.001630 \text{ L mol}^{-1}} \times \left[\frac{760 \text{ torr}}{1 \text{ atm}}\right]\ln\left[\frac{T^{eq}}{273.16}\right]$$

$$= 4.588 - 2.766 \times 10^7 \ln\left[\frac{T^{eq}}{273.16}\right] \text{ torr.}$$ (6.34)

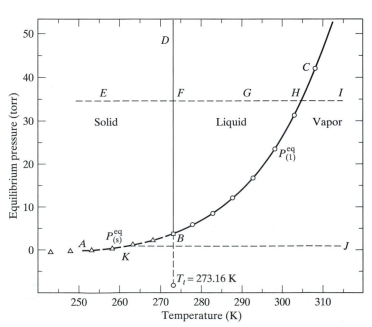

► FIGURE 6.23A
Quantitatively accurate phase diagram for the H_2O system over the temperature range from 250 K to 310 K. The liquid–vapor equilibrium curve is shown as Curve B–C. The curve from A to B shows the loci of points at which solid and vapor are in equilibrium. The solid–liquid equilibrium line is shown as the curve from B to D. The triple point is at B. The plotted points are measured equilibrium vapor pressures.

Equation 6.34 describes a curve that rises almost vertically from the triple point with a negative slope, as illustrated qualitatively in Figure 6.14. The line from Point B to Point D in Figure 6.23A is a plot of Eq. 6.34. The steep slope of the line tells us that applied pressure has very little effect upon the melting point, a fact that was clear from the results of Example 6.4.

Figure 6.23A shows the completed phase diagram for the H_2O system over the narrow temperature range for which our approximations for $\Delta \overline{H}_{vap}$ and $\Delta \overline{H}_{sub}$ are accurate. There are some additional phase transitions between different crystalline forms of ice at pressures above 1,900 bar and temperatures below 273 K. These transitions are not shown. The curve from Point B to Point C continues up to the critical temperature at 647.1 K. However, to extend this curve to temperatures somewhat above the normal boiling point, 373.15 K, we need a more accurate description of the temperature dependence of $\Delta \overline{H}_{vap}$. When the temperature approaches the critical temperature, the Clausius–Clapeyron equation breaks down completely, since the partial molar volumes of liquid and vapor are about the same, so that we can no longer assume that $\Delta \overline{V}$ is equal to $\overline{V}_{(g)}$. In general, the form of the phase diagram in these regions must be determined by experimental measurements. A plot of such experimental data is shown in Figure 6.23B for the H_2O system.

We may see how the phase diagram is used by considering some water vapor contained in a piston–cylinder arrangement (see Figure 6.10) at a temperature of 315 K and a pressure of 35 torr. This point is Point I on the horizontal dashed line in Figure 6.23A. Since Point I does not lie on the liquid–vapor equilibrium curve, the two phases are not in equilibrium at this point. Hence, only vapor will be in the cylinder. Let us now ask what happens as we begin to remove heat from the system while manipulating the piston so as to maintain the pressure inside the cylinder at 35 torr. Such a process corresponds to moving to the left along the dashed line from Point I to Point E in the figure. Initially, as the temperature is lowered in the cylinder, only vapor

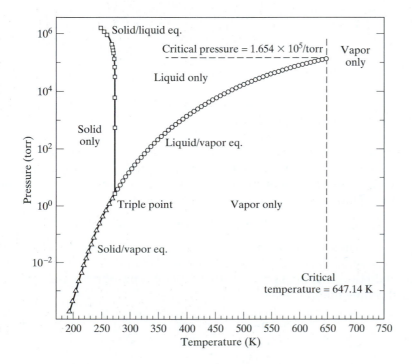

◀ **FIGURE 6.23B**
Experimentally measured H_2O phase diagram over the temperature range from 190 K to the critical temperature, 647.14 K. The plotted points are the measured data. The points are connected by straight-line segments to enhance the visual clarity of the graph. (Source of data: *The Handbook of Chemistry and Physics*, CRC Press, Boca Raton, FL, 1997–1998.)

is present, and the piston must move inward to decrease the volume so as to maintain the pressure constant at 35 torr. When Point H is reached, around 305 K, liquid and vapor come into equilibrium. As we now continue to remove heat from the cylinder, vapor condenses to form liquid, but the temperature remains unchanged. Since the amount of vapor is decreasing, the piston will continue to diminish the system volume to keep the pressure at 35 torr. This process continues until we have condensed all of the vapor. The piston now rests on the surface of the liquid, and only that phase is present. As heat continues to be removed, the system moves to Point G on the phase diagram, where only liquid is present. The liquid phase continues to cool until we reach Point F, at 273 K. At this temperature, liquid begins to freeze to solid, and the two phases are in equilibrium. The temperature remains at 273 K until all the liquid has frozen to solid. During the process, the piston must rise to accommodate the increased volume of the solid ice, since $(\overline{V}_{(s)} - \overline{V}_{(l)}) > 0$. Only then will the solid begin to cool and the system move to Point E, where just the solid phase is present.

If the same experiment is performed starting at Point J with $T = 315$ K and $P = 1.462$ torr, the system will remain in the vapor state until a temperature of 260 K is reached. At Point K, the vapor will condense to a solid, and equilibrium between these two phases will be established. When all of the vapor has condensed to a solid, the system will again begin to cool, so that Point A can be reached. No liquid water will form during the process.

6.5   Effect of Total Pressure on the Equilibrium Vapor Pressure

In all the systems considered to this point, the phases that are in equilibrium possess a common pressure and temperature. Indeed, that condition permitted us to drop the subscripts on the general equation for phase equilibrium, Eq. 6.1. In some cases, however, the condition does not exist. An example is a glass of water sitting on a table. The equilibrium pressure of H_2O above the liquid and the pressure on the liquid phase are not the same. The pressure on the liquid is that exerted by the equilibrium vapor plus the pressure exerted by the atmosphere. We might, therefore, ask what effect the extra pressure on the liquid or solid has on the equilibrium vapor pressure. The answer is contained in Eq. 6.1.

Consider the system shown in Figure 6.24: A condensed phase A is in equilibrium with its vapor. However, another gas, B, perhaps argon, is also present, with a partial pressure of P_B. The entire system is at temperature T. Under those conditions, we may drop the subscripts on the temperature differentials in Eq. 6.1, since we have thermal equilibrium. But we cannot do this for the pressure differentials, because the two phases are at different pressures. Under these conditions, Eq. 6.1 becomes

$$d\mu_C = -\overline{S}_c \, dT^{eq} + \overline{V}_c \, dP_c^{eq} = d\mu_v = -\overline{S}_v \, dT^{eq} + \overline{V}_v \, P_v^{eq}, \qquad \textbf{(6.35)}$$

where the subscript c denotes the condensed phase, and v indicates the vapor phase, of the gas A. If we hold the temperature fixed so that $dT^{eq} = 0$, we obtain the governing equation that describes the effect of gas B upon the equilibrium—that is,

$$\overline{V}_c \, dP_c^{eq} = \overline{V}_v \, dP_v^{eq} = RT \frac{dP_v^{eq}}{P_v^{eq}}, \qquad \textbf{(6.36)}$$

▶ **FIGURE 6.24**
Equilibrium between a condensed phase A and its vapor (open circles) in the presence of a second gas B (filled circles). The total pressure on the condensed phase is the sum of the partial pressures of A and B.

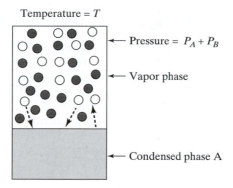

Temperature = T

← Pressure = $P_A + P_B$

← Vapor phase

← Condensed phase A

where we have assumed that the vapor may be treated as an ideal gas. At the low pressures characteristic of liquid–vapor equilibrium, this is usually an excellent approximation.

When there is no gas B present and the pressure on the condensed phase is $P_c^{eq} = P_v^{eq}$, the equilibrium vapor pressure is that described by Eq. 6.11. Let us denote this pressure as P_o. We may now obtain the dependence of P_v^{eq} upon P_c^{eq} by integrating Eq. 6.36 between corresponding limits:

$$\int_{P_o}^{P_c^{eq}} \overline{V}_c \, dP_c^{eq} = RT \int_{P_o}^{P_v^{eq}} \frac{dP_v^{eq}}{P_v^{eq}} = RT \ln\left[\frac{P_v^{eq}}{P_o}\right]. \tag{6.37}$$

If the liquid is incompressible, so that its molar volume is independent of the applied pressure, we may factor $\overline{V}_c$ out of the integral on the left side of Eq. 6.37 to obtain

$$P_c^{eq} = P_o + \frac{RT}{\overline{V}c} \ln\left[\frac{P_v^{eq}}{P_o}\right]. \tag{6.38}$$

Equation 6.38 gives the relationship between the total applied equilibrium pressure on the condensed phase, P_c^{eq}, and the equilibrium vapor pressure, P_v^{eq}, which will be produced with that applied pressure. Example 6.5 illustrates the use of that equation and provides an example of the magnitude of this effect.

EXAMPLE 6.5

A 2-L closed container at 300 K holds 100 cm³ of H_2O(l). The equilibrium vapor pressure is found to be 0.0328 atm. What pressure of Ar gas must be placed in the container to double the equilibrium water vapor pressure to 0.0656 atm?

Solution

The molar volume of water is 0.018 L mol⁻¹. Direct substitution into Eq. 6.38 gives

$$P_c^{eq} = 0.0328 \text{ atm} + \frac{(0.08206 \text{ L atm mol}^{-1} \text{ K}^{-1})(300 \text{ K})}{0.018 \text{ L mol}^{-1}} \ln\left[\frac{0.0656}{0.0328}\right]$$

$$= 0.0328 \text{ atm} + 948.0 \text{ atm} = 948.0 \text{ atm}. \tag{1}$$

Consequently, we must have an argon pressure of 948.0 atm inside the container to double the equilibrium vapor pressure of water. Obviously, the effect is very small. The near-zero effect of atmospheric pressure upon the equilibrium vapor pressure of a substance explains why most texts totally ignore this effect.
    For a related exercise, see Problem 6.29.

Summary: Key Concepts and Equations

1. Since spontaneous processes at a given temperature and pressure are characterized by $\Delta G < 0$ and $\Delta\mu < 0$, a pure substance will spontaneously assume the form that makes the Gibbs free energy and the chemical potential as low as possible. The phase with the lowest value of the chemical potential under given conditions is termed the *stable* phase. Solid-state phases that exist in spite of having a chemical potential greater than that of the stable phase are termed *metastable* or *kinetically stable*. Such phases exist because the rate of the transition to the stable phase is often extremely slow.

2. When the chemical potentials of two phases become equal, equilibrium between them exists, and the temperature and pressure that characterize that state are called the transition temperature and pressure. For solid–liquid equilibrium, the transition temperatures under pressures of 1 atm or 1 bar are defined as the *normal* and *standard* melting points, respectively. Since $\Delta\mu_{transition} = 0$ at equilibrium, we have $\Delta\overline{S}_{transition} = \Delta\overline{H}_{transition}/T_o$, where T_o is the transition temperature. Therefore, exothermic transitions with $\Delta\overline{H}_{transition} < 0$ have $\Delta\overline{S}_{transition} < 0$. The decrease in entropy tells us that the phase transition is associated with a decrease in the disorder of the system. By contrast, endothermic transitions are associated with an increase in the disorder of the system.

3. Liquids and solids have characteristic equilibrium vapor pressures that increase with temperature because the kinetic energy of the molecules increases, thereby raising the probability of escape from the condensed phase. As the equilibrium temperature rises, the density of the equilibrium vapor increases, while the densities of the liquid decreases. When a temperature is reached at which the density of both phases become equal, the phase boundary vanishes. The temperature and pressure at which this occurs are termed the *critical* values. Boiling takes place whenever the equilibrium vapor pressure equals the applied pressure. If this pressure is 1 atm, the temperature is termed the *normal* boiling point; if it is 1 bar, the term *standard* boiling point is used.

4. The general equation for phase equilibrium is

$$d\mu_A = -\overline{S}_A\, dT_A^{eq} + \overline{V}_A\, dP_A^{eq} = d\mu_B = -\overline{S}_B\, dT_B^{eq} + \overline{V}_B\, dP_B^{eq},$$

where $\overline{S}_i$ and $\overline{V}_i$ are the partial molar entropy and volume, respectively, and dT_i^{eq} and dP_i^{eq} are the changes in the equilibrium temperature and pressure, respectively, for component i. Whenever both phases possess the same temperature and pressure, we may drop the subscripts on dT and dP in this equation. The result is the Clapeyron equation:

$$-\overline{S}_A\, dT^{eq} + \overline{V}_A\, dP^{eq} = -\overline{S}_B\, dT^{eq} + \overline{V}_B\, dP^{eq}.$$

Combining the dP^{eq} and dT^{eq} terms, we obtain

$$dP^{eq} = \frac{\Delta\overline{S}}{\Delta\overline{V}}\, dT^{eq}.$$

Since the phase transition occurs at a fixed transition temperature T^{eq}, $\Delta\overline{S}$ may be expressed in terms of the transition enthalpy. This gives

$$dP^{eq} = \frac{\Delta\overline{H}}{T^{eq}\Delta\overline{V}}\, dT^{eq},$$

which is the usual form of the Clapeyron equation.

5. If we assume that the volume of condensed phases may be neglected relative to the volume of the vapor phase and substitute the ideal-gas equation of state for the vapor volume, the Clapeyron equation becomes

$$\frac{dP^{eq}}{P^{eq}} = d(\ln P^{eq}) = \frac{\Delta \overline{H}}{R(T^{eq})^2} dT^{eq},$$

which is called the Clausius–Clapeyron equation. It is important to note that the pressure units in the expression dP^{eq}/P^{eq} have canceled. If it is now assumed that $\Delta \overline{H}$ is constant, this equation may be easily integrated to yield

$$\ln\left[\frac{P_2^{eq}}{P_1^{eq}}\right] = -\frac{\Delta \overline{H}}{R}\left[\frac{1}{T_2^{eq}} - \frac{1}{T_1^{eq}}\right].$$

If the condensed phase is a liquid and T_1 is the normal boiling point T_b^o, this equation becomes

$$\ln P^{eq} = \ln(760) - \frac{\Delta \overline{H}_{vap}}{R}\left[\frac{1}{T^{eq}} - \frac{1}{T_b^o}\right].$$

As noted, P^{eq} is the pressure magnitude without units. In this form, the magnitude of P^{eq} is that when pressure is expressed in torr.

6. The Clausius–Clapeyron equation may be cast in the form

$$P^{eq} = C \exp\left[-\frac{\Delta \overline{H}}{RT^{eq}}\right],$$

which is the general form that usually occurs for the temperature dependence of a variable that measures the extent or rate of a process with a temperature-independent energy requirement $\Delta \overline{H}$. In this case, the vaporization process has an energy requirement, either $\Delta \overline{H}_{vap}$ or $\Delta \overline{H}_{sub}$, and the equilibrium vapor pressure is a variable that measures the extent of vaporization.

7. The chemical potential of a condensed phase A is equal to that of its equilibrium vapor. Therefore,

$$\mu_{(s)}^A = \mu_{(g)}^A = \mu^o + RT \ln(P_A^{eq}),$$

since vapor pressures are usually sufficiently low that the ideal-gas equation of state is essentially exact. For all practical purposes, the equilibrium vapor pressure, P_A^{eq}, depends only upon the temperature, as does the standard chemical potential μ^o. Consequently, the chemical potential of a condensed phase is a constant at a given temperature.

8. Application of the Clapeyron equation to solid–liquid equilibrium shows that the equilibrium temperature (melting point) varies with the applied pressure. The relationship is

$$[P_2^{eq} - P_1^{eq}] = \frac{\Delta \overline{H}_{fus}}{\Delta \overline{V}} \ln\left[\frac{T_2^{eq}}{T_1^{eq}}\right],$$

assuming that $\Delta \overline{H}_{fus}$ is independent of temperature and that the condensed phases are incompressible. This equation shows that the qualitative effect of pressure depends upon the sign of $\Delta \overline{V}$. If the solid is less dense than the liquid, the melting point will decrease with increasing pressure. Water–ice is an example. If the opposite is true, the melting point increases as the pressure rises. However, in both cases, the magnitude of the effect is very small.

9. The Clapeyron and Clausius–Clapeyron equations may be used to plot the loci of (P, T) points at which liquid–vapor, solid–vapor, and liquid–solid equilibria exist. Such a plot is called a phase diagram. At all points on one of the lines, two phases are simultaneously present in equilibrium. Off the lines, only one phase can be present. When the liquid–vapor, solid–vapor, and liquid–solid lines intersect, liquid, solid, and vapor are all present in simultaneous equilibria. This intersection point is called the *triple point*. By equating the Clapeyron equation for

equilibrium between phases and solving for the intersection point, we may predict the temperature and pressure at the triple point.

10. The equilibrium vapor pressure is weakly dependent upon the total applied pressure. The Clapeyron equation shows this dependence to be

$$P_c^{eq} = P_o + \frac{RT}{\overline{V}_c} \ln\left[\frac{P_v^{eq}}{P_o}\right],$$

where P_v^{eq} is the equilibrium vapor pressure above the condensed phase, P_c^{eq} is the total pressure applied to the condensed phase, and P_o is the equilibrium vapor pressure if that is the only pressure being exerted on the condensed phase. This equation holds if the liquid or solid is incompressible, so that its molar volume $\overline{V}_c$ is a constant. The logarithmic dependence makes the effect very small. For example, to double the equilibrium vapor pressure over liquid water at 300 K requires an externally exerted pressure of about 948 atm on the water.

Problems

Problems that require the use of some type of computational device are marked with an asterisk (*). Problems that require some type of plotting routine are indicated with a pound sign (#). Unless otherwise stated, all gases may be assumed to behave ideally.

6.1 Using the data given in Appendix A, along with the assumption that the partial molar entropies of solid and liquid aluminum are constants with a value equal to their values at 298.15 K and 1 bar of pressure, obtain the chemical potentials of Al(s) and Al(l) as a function of temperature, and compute the temperature at which these two quantities are equal. Compare your results with those shown in Figure 6.1.

6.2 Using the data in Appendix A, along with the assumption that the partial molar volumes of liquid and solid aluminum are independent of pressure, obtain the chemical potentials of Al(s) and Al(l) as a function of pressure in bars at 298.15 K. Show that aluminum cannot be melted at 298.15 K by applying pressure if our assumptions are accurate. Assume that the density of solid aluminum is 2.70 g cm^{-3} and the density of liquid aluminum is 2.357 g cm^{-3}.

6.3 Using the data in Appendix A, along with the assumption that the partial molar volumes of liquid and solid H_2O are independent of pressure, obtain the chemical potentials of H_2O(s) and H_2O(l) as a function of pressure in bars at 262.65 K. Determine the pressure required to melt ice at 262.65 K. Compare your result with the data given in Figure 6.6. Assume that the density of solid H_2O is 0.9168 g cm^{-3}, the density of liquid H_2O is 0.9998 g cm^{-3}, μ^o(262.65 K) for liquid H_2O is 195.6 J mol^{-1}, and μ^o(262.65 K) for solid H_2O is 0.000 J mol^{-1}.

6.4 Consider a single-component system whose phase diagram is represented by that shown in Figure 6.15. Suppose the state of the system is that represented by Point C on the diagram.

(A) Describe the phase(s) present.

(B) If the system in (A) is now converted to the state represented by Point D on the phase diagram along the dashed-line path shown in the following diagram, describe the phases present in the system at each point during the transformation:

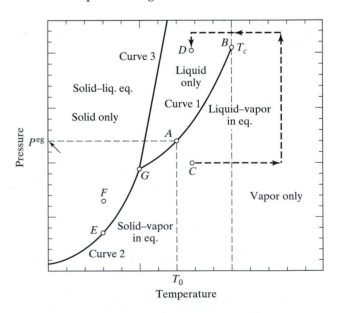

6.5 The normal boiling point of ethanol is 352.6 K. Using the data in Appendix A, along with the assumption that $\Delta C_p = 0$ for the vaporization process, compute the equilibrium vapor pressure of ethanol at 298 K.

6.6 The equilibrium vapor pressure over a solid at 300 K is known to be 20 torr, and the enthalpy of sublimation is 8,000 cal mol^{-1}. Determine the equilibrium vapor pressure over the solid at 350 K. Assume that ΔH_{sub} is a constant.

6.7 Carbon disulfide (CS_2) has vapor pressures of 40 and 100 torr at 250.65 K and 268.05 K, respectively.

(A) Compute the enthalpy vaporization and the normal boiling point of CS_2, assuming that $\Delta C_p = 0$.

(B) The standard partial molar enthalpy of $CS_2(l)$ is 89.70 kJ mol^{-1}. What does the calculation in (A) predict the standard partial molar enthalpy of $CS_2(g)$ to be?

6.8 Problem 6.7 predicts the partial molar enthalpy of vaporization for CS_2 to be 29.42 kJ mol^{-1}, and the normal boiling point is 319.7 K. Using these values for ΔH_{vap} and the normal boiling point, along with the assumption that $\Delta C_p = 0$, compute the percent difference between the normal and standard boiling points for CS_2.

6.9 Some measured vapor pressures of $CCl_4(l)$ are given in the following table:

T (K)	Vapor pressure (torr)
253.55	10
277.45	40
296.15	100
330.95	400

(A) Determine ΔH_{vap} for $CCl_4(l)$ from a plot of ln P^{eq} vs. $1/T$. Use a least-squares fitting procedure.

(B) The measured heat of fusion of $CCl_4(s)$ at 249.15 K is 2677 J mol^{-1}. If we assume that the heat capacities of solid, liquid, and vapor CCl_4 are all equal, we will have $\Delta H_{sub} = \Delta H_{vap} + \Delta H_{fus}$. The measured vapor pressure over $CCl_4(s)$ at 223.15 K is 1 torr. With these data, the results obtained in (A), and the assumption that the heat capacities of all phases are equal, compute the melting point of CCl_4. The measured result is 250.55 K. Compute the percent error introduced by the heat capacity assumption.

6.10 In this problem, you will examine the magnitude of the error that can be introduced by the assumption that ΔC_p for the vaporization process is zero, which leads directly to a constant value for ΔH_{vap}. We will take water as our test case.

(A) Evaluate ΔH_{vap} at 298.15 K for the process $H_2O(l) \longrightarrow H_2O(g)$, using standard partial molar enthalpies from Appendix A.

(B) The measured vapor pressure of $H_2O(l)$ at 298.15 K is 23.756 torr. Use these data and the

assumption that ΔH_{vap} is a constant to compute the normal boiling point of water. What is the percent error in the calculation?

(C) Use the same assumptions as in (B) to compute the equilibrium vapor pressure at 373.15 K, the measured normal boiling point of water. What percent error is present in this calculation? Why is the percent error in (C) so much larger than that in (B)?

6.11 Consider a vapor whose equation of state for 1 mole is $PV(1 - bP) = RT$, where b is a constant. For this vapor, derive an expression for the equilibrium pressure as a function of temperature, assuming that the enthalpy of vaporization is independent of temperature and that the partial molar volume of the liquid is negligible relative to that of the vapor.

6.12 Liquid A is in equilibrium with its vapor at 300 K, at which temperature the equilibrium vapor pressure is 40 torr. The enthalpy of vaporization at 300 K is 8,000 J mol^{-1}. The heat capacities of $A(l)$ and $A(g)$ are 67.0 J mol^{-1} K^{-1} and 35 J mol^{-1} K^{-1}, respectively. Compute the equilibrium vapor pressure over liquid A at 350 K. Do not assume ΔH_{vap} to be independent of temperature.

6.13 Some liquid ethanol (C_2H_5OH) is placed inside a sealed 10-L container. At 292.15 K, the vapor density above the liquid is found to be 0.1011 g L^{-1}. At 308.05 K, the vapor density above the liquid is 0.2398 g L^{-1}. Compute ΔH_{vap} for ethanol, assuming it to be independent of temperature.

6.14* 10 grams of ethanol are placed in a 10-L sealed container at 292.15 K. At this temperature, the equilibrium vapor pressure in the container is 10 torr.

(A) If we assume that the enthalpy of vaporization of ethanol is 43 kJ mol^{-1} and constant, compute the equilibrium vapor pressure of ethanol inside the container as a function of temperature from 300 K up to the temperature at which all of the ethanol is in the vapor phase.

(B) Plot the pressure inside the container as a function of temperature over the range 300 K $\leq T \leq$ 450 K. (Hint: You first need to compute the temperature at which all 10 grams of ethanol are in the vapor phase.)

6.15 If y is a function of x and we wish to measure y, it may be advantageous to measure x and then compute y using the relationship $y = f(x)$. This will be the case if it is much easier to measure x than y or if small changes in y produce large changes in x that are more easily measured. A chemist wishes to use this fact to determine the temperature of a substance by measuring its vapor pressure. To execute this measurement, she constructs a small closed sphere containing pure water and its equilibrium vapor. The device is equipped with a pressure-measuring device that can easily detect pressure changes of 10^{-3} torr. (Actually, it is possible to do much better than this.) She places the device in a gas-filled chamber whose temperature is to be monitored.

(A) Assuming that the enthalpy of vaporization of water is a constant equal to 44010 J mol^{-1}, obtain the equation relating the measured vapor pressure inside the sphere to the chamber's temperature. The equilibrium vapor pressure of water at 298.15 K is 23.756 torr.

(B) Use the result obtained in (A) to compute the temperature inside the chamber if the pressure is 475 torr.

(C) Use the Clausius–Clapeyron equation to show that if the percent error in the measurement of pressure is $X\%$, the error in the temperature will be only $0.057X\%$ at 300 K and $0.076X\%$ at 400 K. Consequently, the pressure measurement affords a very accurate means of measuring the temperature. (*Hint:* The percent error in x is $100 \, dx/x$, where x is the measured value and dx is the uncertainty in the measurement.)

6.16 A closed piston–cylinder arrangement holds pure H_2O under a pressure of 50 torr at 290 K.

(A) What phases are present inside the cylinder?

(B) The pressure produced by the piston on the H_2O is decreased at a constant rate of 1 torr per minute while maintaining the temperature at 290 K. How long will it take before water vapor appears inside the cylinder?

6.17* A closed piston–cylinder arrangement containing pure H_2O is equipped with temperature and pressure controls. By careful manipulation of the temperature and pressure controls, the system is cooled from 300 K to 257 K along a line whose equation is

$$P = [0.46345 \text{ torr K}^{-1}] \, T - 119.035 \text{ torr}.$$

Describe all the phase changes that occur as the system is cooled along this path. Give the temperatures and pressures at which these changes take place.

6.18 Three identical closed containers are labeled A, B, and C, respectively. Each container holds a pure liquid in equilibrium with its vapor. One container holds acetone, the second holds ethanol, and the third contains water. Each container is equipped with a pressure-measuring device that records the equilibrium vapor pressure. It is known that all containers are at the same temperature, but that common temperature is unknown. The partial molar enthalpies of vaporization are $44,010 \text{ J mol}^{-1}\text{K}^{-1}$ for water, $30,250 \text{ J mol}^{-1}\text{K}^{-1}$ for acetone, and $39,322 \text{ J mol}^{-1}\text{K}^{-1}$ for ethanol. The normal boiling points are 373.15 K for water, 329.65 K for acetone, and 351.65 K for ethanol. Using these data and measured vapor pressures <u>alone</u>, is it possible to determine which container holds

which liquid? Discuss your analysis in detail and show that your conclusions are correct. You may assume that the partial molar enthalpies of vaporization are independent of temperature.

6.19 In the dead of winter, Sam finds his driveway covered with ice. The outside temperature is 263 K. Being a student of physical chemistry, Sam knows that the melting point of ice decreases when pressure is exerted on the solid. Therefore, he decides to clear his driveway by exerting sufficient pressure, using a sledgehammer to cause the ice to melt. Later in the day, his physical chemistry classmates find Sam collapsed from exhaustion on his ice-covered, shattered driveway.

(A) Was Sam incorrect in his reasoning concerning the effect of pressure on the melting point of ice?

(B) Why has Sam collapsed from exhaustion, and why is his driveway still covered with ice?

6.20 A high-pressure cylinder holds $H_2(g)$ gas at a pressure of 150 atm. Some water has condensed to a liquid inside the cylinder. What is the equilibrium vapor pressure of H_2O inside the cylinder at 298 K if its equilibrium vapor pressure at that temperature would be 23.756 torr in the absence of $H_2(g)$?

6.21* It is possible to use vapor pressure data to determine the temperature dependence of the difference in heat capacity between a vapor and a liquid or a vapor and a solid. If the temperature dependence of either phase is known, such a determination provides the temperature dependence of the heat capacity for the other phase. This problem serves as an example of one possible method for making such a determination.

Some measured equilibrium vapor pressures for $H_2O(l)$ are given in the following table:

T (K)	P^{eq} (torr)	T (K)	P^{eq} (torr)
303.15	31.824	343.15	233.70
313.15	55.324	353.15	355.10
323.15	92.510	363.15	525.76
333.15	149.38	373.15	760.00

(A) Assume that ΔC_p for the vaporization of water can be regarded as being constant over a small range of temperatures. Using standard partial molar enthalpies in Appendix A, obtain $\Delta \overline{H}_{vap}$ as a function of ΔC_p and the temperature.

(B) By substituting the result of (A) into Eq. 6.11 for $\Delta \overline{H}_{vap}$, obtain an expression for $\ln[P_2/P_1]^{eq}$ in terms of ΔC_p and the temperatures T_1 and T_2 at which the vapor pressures are P_1 and P_2, respectively.

(C) If we take T_1 and T_2 to be adjacent temperatures in the preceding table of temperatures and pressures, we can easily compute ΔC_p by using the result obtained in (B). We can now repeat this calculation for each pair of adjacent temperatures in the table. This will give ΔC_p over each temperature interval. Execute these calculations for all temperature intervals in the table, and plot the resulting values of ΔC_p versus the temperature at the center point of each interval considered. Obtain the best straight-line fit to the data, and thereby obtain ΔC_p as a linear function of temperature.

(D) Using the results obtained in (C) that give $\Delta C_p = A + BT$, calculate the equilibrium vapor pressure of water at 393.15 K. The experimental result at this temperature is 1489.14 torr. Calculate the percent error in your result. How much error would be present if you assumed $\Delta \overline{H}_{vap}$ to be constant at its value at 298.15 K?

The next five problems are related to phase equilibrium for the CO_2 system. If the entire set is assigned, the students will be able to use the results to construct a reasonably accurate phase diagram for CO_2. Alternatively, the problems can be used individually.

6.22 The 71st edition of the *Handbook of Chemistry and Physics* gives the following values for the vapor pressure over solid CO_2:

T (K)	P^{eq} (kPa)
190	68.4
195	104
200	155
205	227
210	327
215	465
220	600

Plot $\ln(P^{eq})$ versus T^{-1} from $T = 190\ K$ to $T = 220\ K$. Use a least-squares linear fit to the data to determine $\Delta \overline{H}_{sub}$ for $CO_2(s)$. Use your results to obtain an analytic expression for the equilibrium vapor pressure over solid CO_2. Assume that $\Delta \overline{H}_{sub}$ is constant. What is the normal boiling point for $CO_2(s)$?

6.23 The partial molar enthalpy of fusion of $CO_2(s)$ is 7,950 J mol^{-1} at the melting point. At this same temperature, the partial molar enthalpy of sublimation is about 25,505 J mol^{-1}. At 220 K, the equilibrium vapor pressure over liquid CO_2 is 590.6 kPa. Use the Clausius–Clapeyron equation to obtain an analytic expression for the vapor pressure over liquid CO_2, assuming the enthalpy of vaporization to be independent of temperature.

6.24 Using the results obtained in Problems 6.22 and 6.23, determine the temperature at which the equilibrium vapor pressure over $CO_2(s)$ and $CO_2(l)$ are equal. What is the value of the equilibrium vapor pressure at this temperature? What are the predicted temperature and pressure at the CO_2 triple point?

6.25 The densities of solid and liquid CO_2 are 1.56 kg L^{-1} and 1.101 kg L^{-1}, respectively. The triple point was predicted in Problem 6.24 to occur at $T = 216.8\ K$, with a pressure of 5.12 bar. The partial molar enthalpy of fusion of $CO_2(s)$ is 7,950 J mol^{-1} = 79.504 L bar mol^{-1}. Use these data to obtain an analytical equation showing the dependence of the equilibrium temperature (melting point) of solid $CO_2(s)$ on pressure. At what pressure will the $CO_2(s)$ melting point be 222 K?

6.26 Use the results of Problems 6.22 through 6.25 to make a careful plot of the CO_2 phase diagram.

6.27 10 moles of $H_2O(l)$ at 330 K are inserted into an evacuated container that is insulated so as to prevent heat transfer to or from the surroundings. The volume of vacuum above the water is 20 L. As the $H_2O(l)$ vaporizes to establish equilibrium with $H_2O(g)$, the temperature drops because the heat of vaporization must be provided by the internal energy present in the $H_2O(l)$, since there is no heat transfer from the surroundings. If we ignore the change in the volume, mass, and heat capacity of the liquid, what is the final temperature of the system when equilibrium is established between $H_2O(l)$ and $H_2O(g)$? Assume that $\Delta \overline{H}_{vap}$ for $H_2O(l)$ is constant and equal to 44,010 J mol^{-1} and that the total heat capacity of the $H_2O(l)$ is 752.9 J K^{-1}.

6.28 Two crystalline forms of a compound, A and B, are in equilibrium. The density of A is greater than the density of B. The conversion of A to B is found to be an exothermic process.

(A) If we wish to shift the equilibrium toward crystal B, should we use high or low temperature and high or low pressure? Explain.

(B) Which crystal is more highly ordered, A or B? How do you know?

6.29# A scientist has water in an enclosed container at 300 K. (See accompanying figure.) At this temperature, the

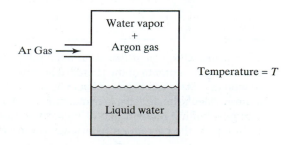

equilibrium vapor pressure is 0.0328 atm. The scientist wishes to conduct an experiment in which the temperature of the system will be gradually lowered, but he desires to keep the equilibrium vapor pressure fixed at 0.0328 atm. He intends to accomplish this by introducing argon gas into the chamber in sufficient quantity to hold the equilibrium vapor pressure constant as the temperature is lowered.

(A) Obtain an expression for the argon pressure that must be present in the container as a function of temperature. Assume that the partial molar enthalpy of vaporization of water is constant and that the liquid is incompressible.

(B) Plot the pressure of argon needed versus T over the range $290 \text{ K} \le T \le 300 \text{ K}$. Prepare a second plot over the temperature range $50 \text{ K} \le T \le 300 \text{ K}$.

6.30 Sam is having a problem understanding the Clapeyron equation. He notes that the equation predicts that the rate of change of the equilibrium vapor pressure with temperature is

$$\frac{dP^{\text{eq}}}{dT^{\text{eq}}} = \frac{\Delta \overline{H}_{\text{vap}}}{T^{\text{eq}} \Delta \overline{V}},$$

where $\Delta \overline{V} = \overline{V}_{(g)} - \overline{V}_{(l)}$. Sam is concerned about the system's behavior as T^{eq} approaches the critical temperature. He points out that Figure 6.11 shows that as $T^{\text{eq}} \to T_c$ the partial molar volumes of liquid and vapor approach a common value. Therefore, $\Delta \overline{V}$ approaches zero as $T^{\text{eq}} \to T_c$. Under these conditions, $dP^{\text{eq}}/dT^{\text{eq}}$ approaches an infinite value. This means that the equilibrium vapor pressure P^{eq} increases a near infinite amount for a small change in the equilibrium temperature T^{eq}. Sam observes that although P^{eq} is large near the critical temperature, it is neither infinite nor even close to being infinite. He suggests that the Clapeyron equation must be incorrect. Is Sam's analysis correct? If so, why does the Clapeyron equation break down? If he is wrong, what is his mistake?

6.31 The following thermodynamic data for liquid and gaseous mercury are available at 298.15 K and 1 bar

Substance	μ^o (J mol^{-1})	$\overline{S}$ (J mol^{-1} K^{-1})
Hg(l)	0	76.02
Hg(g)	31,820	174.96

of pressure: *Using only these data*, obtain a reasonably accurate estimate of the standard boiling point of mercury. State any assumptions you are making in your calculations. The experimentally measured standard boiling temperature of mercury is 629.88 K. Compute the percent error in your result. What

is the major source of the error?

6.32* The following thermodynamic data for liquid and

Substance	Hg(l)	Hg(g)
μ^o (J mol^{-1})	0	31,820
$\overline{S}$ (J mol^{-1} K^{-1})	76.02	174.96
C_p^m (J mol^{-1} K^{-1})	27.983	20.786

gaseous mercury are available at 298.15 K and 1 bar of pressure. *Using only these data*, obtain an accurate value for the standard boiling point of mercury. State any assumptions you are making in your calculations. The experimentally measured standard boiling temperature of mercury is 629.88 K. Compute the percent error in your result. What is the major source of the error?

6.33 The following thermodynamic data for liquid and

Substance	μ^o (J mol^{-1})	$\overline{H}^o$ (J mol^{-1})
Hg$_{(l)}$	0	0
Hg$_{(g)}$	31,820	61,320

gaseous mercury are available at 298.15 K and 1 bar of pressure. *Using only these data*, obtain a reasonably accurate estimate for the equilibrium vapor pressure of mercury at 400 K. The experimentally measured equilibrium vapor pressure at 400 K is 0.00138 bar. Compute the percent error in your estimate. What is the major source of the error in your calculations?

6.34 Sam suggests that a simple dunking bird found in many novelty stores is a perpetual-motion machine of the second type. He prepares the schematic of this dunking bird below.

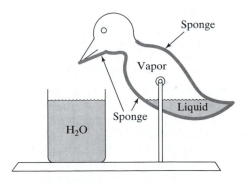

The plastic bird is mounted on a stand as shown. Inside the plastic exterior is a sealed cylinder containing a volatile liquid. When the liquid is in the bottom of the container, as shown in the sketch, the balance is such that the bird's head is elevated above the water container. As the thermal energy of the air heats the liquid, it vaporizes into the upper portion of the chamber, thereby shifting the center of gravity of the assembly and causing it to tilt so as to bring the bird's head down into the water. When the bird's head is immersed, water is absorbed into the sponge coating the cylinder. As the water evaporates from the sponge, it absorbs the enthalpy of vaporization and thus cools the liquid inside the cylinder, causing it to condense and return to bottom of the cylinder,

thereby again shifting the center of gravity, this time lifting the bird's head from the water. This cycle is repeated over and over.

Sam notes that we could, in principle, use the up-and-down motion of the bird to produce useful work from the thermal energy of the air. Thus, we have a machine that operates in cycles that converts heat into work. He poses the question to the class, "Why doesn't this machine violate the second law of thermodynamics?" How do you respond?

Now that you've analyzed the thermodynamics of the dunking bird, go to a movie or a sporting event. The thermodynamics of solids is next.

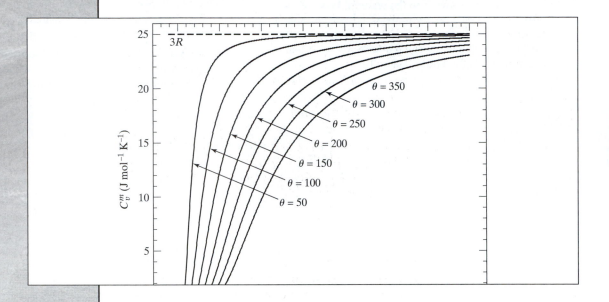

CHAPTER 7

Thermodynamics of Solids

Before we turn our attention to solutions, which are often mixtures of liquids and solids, we shall consider some of the thermodynamic properties of pure solids. Sublimation has already been described in Chapter 6. In the current chapter, we will study the binding energies of ionic crystals in detail. The Born–Haber cycle and the concepts of ionization potentials and electron affinities will be introduced. We shall also examine the variation of the heat capacity of solids with temperature—the first example we will encounter of the influence of quantum mechanics on the quantities of interest in thermodynamics. The Einstein and Debye theories of solid-state heat capacities will be presented.

7.1   Types of Crystal Forces

Since a solid maintains its shape, generally has a lower equilibrium vapor pressure than a liquid, and requires a significant amount of energy to vaporize, the binding forces operating in the solid are large relative to those present in most liquids and are vastly larger than the intermolecular forces in the gas phase. Solid-state forces vary from extremely strong to relatively weak. Ionic crystals and covalently bonded materials, such as sodium chloride, diamond, and silicon crystals, are examples of the former; organic solids, such as naphtalene, usually fall into the second class of weak forces. This difference is reflected in the high boiling and melting points exhibited by ionic crystals, diamond, and silicon compared with the much lower temperatures required to melt and boil organic substances.

The coulombic forces existing between charged particles are responsible for producing the binding in ionic crystals. These forces were discussed in detail in Chapter 1. The force acting between two particles with charges q_1 and q_2 separated by a distance r is given by Eq. 1.45,

$$F = \frac{q_1 q_2}{4\pi \varepsilon_o r^2},$$

where ε_o, the permittivity of the vacuum, has a value of $8.854 \times 10^{-12}\,\mathrm{J^{-1}\,C^2\,m^{-1}}$ if the charges are expressed in coulombs and the distance in meters. With a potential defined by Eq. 1.46, this force leads to a coulombic potential energy

$$V(r) = \frac{q_1 q_2}{4\pi \varepsilon_o r}.$$

A simple calculation demonstrates that ionic forces are very strong. Consider a proton separated by one angstrom (10^{-10} m) from an electron. The protonic and electronic charges are 1.602×10^{-19} C and -1.602×10^{-19} C, respectively, or, in atomic charge units (acu; 1 acu equals 1.602×10^{-19} C), $+1$ and -1, respectively. Substitution into the coulombic potential gives

$$V(r = 10^{-10}\,\mathrm{m}) = -\frac{(1.602 \times 10^{19})^2\,\mathrm{C^2}}{4(3.14159)(8.854 \times 10^{-12}\,\mathrm{J^{-1}\,C^2\,m^{-1}})(10^{-10}\,\mathrm{m})}$$

$$= -0.0231 \times 10^{-16}\,\mathrm{J}.$$

If we have 1 mole of such pairs, the total potential energy per mole will be $-0.0231 \times 10^{-16}\,\mathrm{J\,pair^{-1}} \times 6.022 \times 10^{23}\,\mathrm{pairs\,mol^{-1}} = -1.391 \times 10^6\,\mathrm{J\,mol^{-1}} = -1{,}391\,\mathrm{kJ\,mol^{-1}}$. Thus, it would require 1,391 kJ of energy to overcome the binding forces holding the pairs together. This is approximately the energy needed to dissociate 4 moles of $C-C$ single bonds.

Table 7.1 lists the measured melting points of several ionic crystals. The high temperatures required to melt these crystals are obvious from the data. The existence of several qualitative trends is also clear. First, the crystals with larger $q_1\,q_2$ products have higher melting points. For example, $BaSO_4$, with $+2$ and -2 charges on the cation and anion, respectively, melts at about 800 K higher than any of the crystals with $+1$ and -1 charges. Second, the crystals with a $+2$ charge on the cation and a -1 charge on the anion have melting

Table 7.1	Melting points of some ionic and covalently bonded crystals		
Crystal	**Cation Charge [1]**	**Anion Charge[1]**	**Melting Point (K)**
NaCl	+1	−1	1,074
KCl	+1	−1	1,049
RbCl	+1	−1	988
NaBr	+1	−1	1,028
KBr	+1	−1	1,003
RbBr	+1	−1	955
$CaCl_2$	+2	−1	1,045
$SrCl_2$	+2	−1	1,146
$BaCl_2$	+2	−1	1,235
$BaSO_4$	+2	−2	1,853
Diamond	0	0	> 3,773
Silicon	0	0	1,693

[1] The charges are given in atomic charge units (acu), where
      1 acu = 1.602×10^{-19}C.
Data are from the *Handbook of Chemistry and Physics*, 78th ed., CRC Press, Boca Raton, FL, 1997-1998.

temperatures that are intermediate between the other two classes. The dependence of the coulombic potential on the $q_1 q_2$ product provides the explanation of this observation.

Third, note that melting points decrease as we descend down Groups IA or VIIA in the periodic table. For instance, NaCl has a higher melting point than KCl, which melts above RbCl, and the bromides melt at lower temperatures than the chlorides. In contrast, the opposite trend is observed in the chlorides of Group IIA metals: The melting points increase in the order $CaCl_2 < SrCl_2 < BaCl_2$. This behavior is related to two competing effects: ionic size and the ionization potential of the metal. Table 7.2 gives the values of these quantities for elements in the A groups that commonly form ionic crystals. The nth ionization potential measures the energy required to produce the cation with charge $+n$ from the particle whose charge is $+n - 1$. Thus, the nth ionization potential, I_n, is the endothermicity of the process $A^{+n-1} \longrightarrow e^- + A^{n+}$. I_1 is the energy required to produce the +1 ion from the neutral particle.

As the ionization potentials increase, more energy is required to completely remove the electron. As a result, the extent of negative charge transfer from the cation to the anion may correspond to less than a full electronic charge, thus making the final cation charge somewhat less than +1 or +2. If this should occur, the coulombic potential becomes less attractive and the binding energy of the crystal decreases. As the ionic radii increase, the separation of the charges increases, which reduces the coulombic attractive forces and the binding energies. The difference in trends noted in the previous two paragraphs reflects a switch in the magnitude of these competing effects. For crystals involving Group IA metals, the ionic radius of the cation increases 0.47 Å in going from Na^{1+} to Rb^{1+}. The first ionization potential of Na is 0.961 eV (92.72 kJ mol^{-1}) larger than that for Rb. In Group IIA, the size increase from

Table 7.2 First and second ionization potentials and ionic radii for some group A elements

Element	Group	Ionization Potential (eV)[1] First	Second	Ionic Radii (Å)
H	IA	13.60	—	—
Li		5.363	75.26	0.59 (+1)[2]
Na		5.12	47.06	1.02
K		4.318	31.66	1.38
Rb		4.159	27.36	1.49
Cs		3.87	23.4	1.67
Be	IIA	9.28	18.12	0.12 (+2)
Mg		7.61	14.96	0.72
Ca		6.09	11.82	1.00
Sr		5.667	10.98	1.16
Ba		5.19	9.95	1.36
B	IIIA	8.257	25.00	0.12 (+3)
Al		5.96	18.74	0.53
Ga		5.97	20.43	0.62
In		5.76	18.79	0.79
Tl		6.07	20.32	0.88
N	VA	14.48	29.47	1.71 (−3)
P		10.9	19.56	2.12
As		10.5	20.1	2.22
O	VIA	13.550	34.93	1.40 (−2)
S		10.30	23.3	1.84
Se		9.7	21.3	1.98
Te		8.96	—	2.21
F	VIIA	17.34	34.81	1.33 (−1)
Cl		12.952	23.67	1.81
Br		11.80	19.1	1.96
I		10.6	19.4	2.20

[1] 1 eV = 1 electron-volt atom^{-1} = 23.06 kcal mol^{-1} = 96.48 kJ mol^{-1}.
[2] Formal charge on the ion. The actual magnitude may be less.
Note: Dash indicates datum not available.
Source of data: *Handbook of Chemistry and Physics*, 78th ed., CRC Press, Boca Raton, FL, 1997-1998; R. D. Shannon and C. T. Prewitt, *Acta. Cryst.* **B25**, 925 (1969).

Ca^{2+} to Ba^{2+} is 0.36 Å The decrease in the second ionization potential is 1.87 eV (180.42 kJ mol^{-1}). It is, therefore, not surprising to find that the variation in ionic size dominates the binding energy trend in Group IA, whereas the decrease in ionization potential is the major factor for Group IIA.

These points are illustrated by Figures 7.1 and 7.2, which show the variations of ionization potentials and ionic radii for Group IA and Group IIA cations as a function of row number in the periodic table. The first ionization

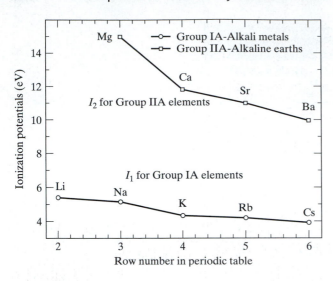

▲ **FIGURE 7.1**
Variation of the first and second ionization potentials for Group IA and Group IIA elements, respectively, as a function of row number in the periodic table.

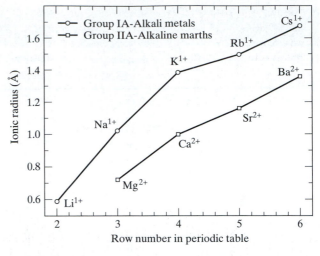

▲ **FIGURE 7.2**
Variation of ionic radii for Group IA and Group IIA cations with formal charges of 1+ and 2+, respectively, as a function of row number in the periodic table.

potential for Group IA cations shows very little variation, but the change in ionic radius is sizable. For Group IIA cations, the change in ionic radii is about the same as that for Group IA, but the variation of the second ionization potential is much larger for the alkaline earth metals in Group IIA than for those in Group IA.

EXAMPLE 7.1

Using the patterns shown in Table 7.1, predict the melting points of CsCl and CsBr. Explain your reasoning and procedure. Determine the error in your predictions. Do you expect $RaCl_2$ to melt above or below 1,235 K?

Solution

Qualitatively, the predictions are not hard to make. The melting points decrease with atomic number for the alkali metal halides. Therefore, we expect CsCl and CsBr to melt below 988 K and 955 K, respectively. We also expect CsCl to have a higher melting point than CsBr.

There is no unique answer to the question of how we should quantitatively predict these melting points. The simplest procedure is a linear extrapolation based on ionic radii for the alkali metal halides.

Since the coulombic potential varies with $1/r$, we might expect the melting points to decrease as $1/r$ decreases. The following table lists r, $1/r$, and the melting points for KCl, RbCl, and CsCl:

Ion	Ionic Radius, r (Å)	r^{-1} (Å^{-1})	Melting Temperature of the Chloride
K^{1+}	1.38	0.725	1049
Rb^{1+}	1.49	0.671	988
Cs^{1+}	1.67	0.600	?

Figure 7.3 shows the melting points of KCl and RbCl plotted against the reciprocals of the cationic radii. Assuming a linear variation of the melting point with r^{-1}, we can extrapolate the line as shown in the figure, to r^{-1} for Cs^{1+}. This extrapolation predicts that CsCl should melt at 907.8 K. The measured result is 919 K. The prediction is, therefore, low by 1.2%. A similar calculation for the bromides, using the data in Table 7.1 and 7.2, suggests a melting point for CsBr of 892 K. The measured result is 909 K. In this case, our simple procedure yields a result that is low by 1.9%. We could do better if we attempted a double correlation with both ionic size and ionization potential. However, even the simple procedure is reasonably accurate and the melting point of CsBr is properly predicted to lie below that of CsCl.

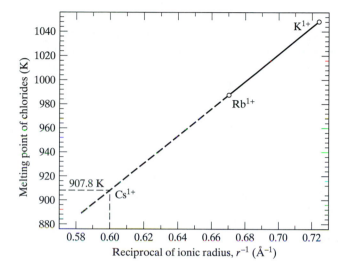

◀ **FIGURE 7.3**
Linear extrapolation based on the reciprocal of the cationic radii to predict the melting point of CsCl from the measured values for KCl and RbCl. Extrapolation of the line to $1/r = 0.60$, the reciprocal of the ionic radius of Cs^{1+}, yields a predicted value of 907.8 K for the melting point of CsCl. The measured value is 919 K.

The melting points for the alkaline earth chlorides increase with atomic number. Consequently, $RaCl_2$ should melt at a higher temperature than $BaCl_2$. Hence, we expect the melting point of $RaCl_2$ to be above 1,235 K. This is indeed the case; the measured melting point of $RaCl_2$ is 1273 K.

For a related exercise, see Problem 7.1.

The melting points of diamond and silicon listed in Table 7.1 are much higher than those of ionic crystals. Obviously, the attractive forces produced by the C–C and Si–Si covalent chemical bonds are even stronger than coulombic forces between ions. We shall defer discussion of covalent bonding forces to Chapter 14.

The binding forces operating in organic crystals are a combination of dipole–dipole interactions and London dispersion forces. Both of these were discussed in Chapter 1. Equation 1.53 indicates that the dipole–dipole potential is attractive and varies with the square of the dipole moment μ and the inverse third power of the dipole separation for collinearly aligned dipoles. For such an alignment, the dipole–dipole potential is

$$V(r) = -\frac{\mu^2}{2\pi\varepsilon_o r^3}.$$

Consequently, we expect molecules possessing large dipole moments to have greater binding energies and higher melting and boiling points.

We noted in Chapter 1 that London dispersion forces are the result of small induced dipole interactions. These interactions lead to attractive potentials of the form

$$V(r) = -\frac{C}{r^6},$$

where C is a positive constant whose value is dependent upon the ease with which induced dipoles are formed, which is in turn measured by the molecule's polarizability. In the presence of an electric field $\mathcal{E}$, either applied or produced by a molecular charge cloud, the induced dipole moment is given by

$$\mu^* = \alpha\mathcal{E}. \tag{7.1}$$

The proportionality constant α is called the *polarizability* of the molecule. If the electric field is expressed in units of volt m^{-1} and the dipole moment in coulomb-meters (C m), the unit on the polarizability will be C m^2 volt^{-1}. Since a joule is a volt-coulomb, this unit may be written in the form $C^2 m^2 J^{-1}$. Values of α for some atoms and molecules are given in Table 7.3.

Qualitatively, the electrons in small molecules are tightly held by the positive charges on the nuclei. As a result, they are distorted relatively little by electric fields, and their polarizabilities are small. This leads to small values of

Table 7.3   Molar masses and polarizabilities of some selected molecules and atoms

Molecule/Atom	Molar Mass (g mol^{-1})	Polarizability ($C^2m^2J^{-1}$)
He	4.003	0.22
H_2	2.016	0.911
CH_4	16.042	2.89
NH_3	17.034	2.47
H_2O	18.016	1.65
HF	20.008	0.57
N_2	28.02	1.97
CO	30.01	2.20
CH_3OH	32.042	3.59
HCl	36.458	2.93
Ar	39.95	1.85
CO_2	44.01	2.93
CH_3Cl	50.484	5.04
C_6H_6	78.11	11.6
HBr	80.91	4.01
CH_2Cl_2	84.93	7.57
$CHCl_3$	119.37	9.46
HI	127.9	6.06
CCl_4	153.81	11.7

Data from the *Handbook of Chemistry and Physics,* 78th ed., CRC Press, Boca Raton, FL, 1997-1998 and C. J. F. Böttcher and P. Bordewijk, *Theory of Electric Polarization,* Elsevier, Amsterdam (1978).

α, weak London dispersion forces, and low boiling and melting points. As the molecule becomes larger, the number of electrons increases, and there are more opportunities to form induced dipoles. This leads naturally to larger polarizabilities. Atoms with larger atomic numbers have much of their negative charge at larger distances from the nucleus. The electronic charge is, therefore, easier to distort, yielding larger polarizabilities, greater London dispersion forces, and higher melting and boiling points.

Both of the effects giving rise to increased values of the polarizability are related to molecular mass. Atoms with larger atomic number are usually associated with larger atomic masses. As the number of atoms and electrons in a molecule increases, we expect a corresponding increase in the mass. Consequently, it is reasonable to expect a correlation between α and the molar mass. Figure 7.4 shows a plot of α versus molar mass for the molecules and atoms in Table 7.3. The line is a least-squares fit to the data. A correlation appears to be present, but the data are considerably scattered, because some of the molecules have permanent dipoles and some do not. Others have heavy elements, while some contain only second-row elements. If we restrict the data to compounds in a homologous series, such as primary alcohols, alkyl chlorides, etc., the correlation becomes much stronger.

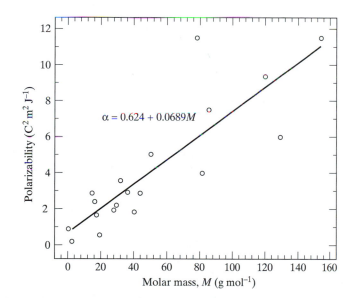

◀ FIGURE 7.4
Polarizability versus molar mass for the molecules and atoms listed in Table 7.3. The line is a least-squares fit to the data.

Figure 7.5 illustrates this point by plotting the measured boiling points of a homologous series of n-alkanes, primary alcohols, and aromatics against molar mass. Clearly, the correlation is strong, with very little scatter in the data. The lines are least-squares fits to the data with equations of

$$T_b^o(n - \text{alkanes}) = 81.518 + 3.741\,M - 0.00831\,M^2\ \text{K},$$

$$T_b^o(\text{primary alcohols}) = 287.68 + 1.377\,M\,\text{K},$$

and $$T_b^o(\text{aromatics}) = 139.83 + 2.73\,M\,\text{K},$$

where M is the molar mass in units of g mol^{-1} The normal boiling points are given in degrees kelvin. As expected, the n-alkanes with relatively small dipole moments have lower boiling points than primary alcohols, whose dipole–dipole interactions are generally larger. The planar aromatics are nonpolar, with only London dispersion forces. However, the π electrons in the aromatic ring are less tightly bound than the electrons in either the n-alkanes or the primary alcohols. Consequently, the polarizabilities of aromatics are larger and the boiling points are uniformly higher than those for the alkanes. For low molar masses, the dipole–dipole forces present in the alcohols are more important than the higher polarizabilities of the aromatics. Therefore, the boiling points are higher for the alcohols. By the time we reach naphthalene, however, London forces have become dominant, and we observe higher boiling points for the aromatics.

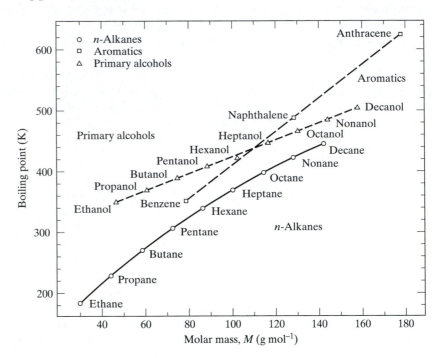

▶ FIGURE 7.5
Boiling point versus molar mass in a homologous series of n-alkanes, primary alcohols, and aromatics.

EXAMPLE 7.2

Use the correlation shown in Figure 7.5 to predict the normal boiling point of n-dodecane ($C_{12}H_{26}$). Evaluate the percent error in your prediction.

Solution

Using the least-squares fit to the n-alkane curve in Figure 7.5, we have

$$T_b^\circ = 81.518 + 3.741\,M - 0.00831\,M^2. \tag{1}$$

The molar mass of n-dodecane is 170.328 g mol^{-1}. Substituting into Eq. 1 we obtain

$$T_b^\circ = 81.518 + 3.741\,(170.328) - 0.00831\,(170.328)^2 = 477.63\ \text{K} = 204.48°\text{C} \tag{2}$$

The measured normal boiling point is 214.5°C. The error in our computed result is therefore

$$\% \text{ error} = \frac{100(204.48 - 214.5)}{214.5}\ \% = -4.7\% \tag{3}$$

A word of caution at this point is appropriate. The least-squares fit used in this example was obtained for data with molar masses in the range from 46 to 158 g mol^{-1}. When such a fit is used outside that range, the results can often be very poor. With a molar mass of 170.33 g mol^{-1}, n-dodecane is sufficiently close to the fitting range that the results are reasonably good. However, it would be poor practice to utilize this fit for $C_{20}H_{42}$, with a molar mass of 282.54 g mol^{-1}.

For related exercises, see Problems 7.2 and 7.3.

7.2   Determination of Binding Energies in Ionic Crystals

7.2.1   The Born–Haber Cycle

A quantitative measure of the forces operating within an ionic crystal may be obtained by measuring the binding enthalpy of the crystal, which we denote by ΔH_c. This quantity corresponds to the enthalpy change for the process

$$\text{Ionic crystal} \longrightarrow \text{cations} + \text{anions.}$$

Let us take NaCl(c) as an example. The notation (c) denotes a crytalline phase. We wish to determine the enthalpy change for the process

$$\text{NaCl(c)} \longrightarrow \text{Na}^{1+}(g) + \text{Cl}^{1-}(g),$$

where the ions are separated by an infinite distance, so that their interaction potential is zero. There is no simple way to run this process directly in a calorimetric experiment. Even if we were able to insert sufficient energy to dissociate NaCl(c) into its ions, we could not prevent the formation of either ion pairs in the gas phase or gaseous NaCl. Such reactions would significantly alter the measured endothermicity of the reaction. If we could obtain Na$^+$(g) and Cl$^-$(g) from the stockroom, we could mix the two and easily measure the energy release for the reverse reaction. Unfortunately, all stockrooms are out of bottles of Na$^+$(g) and Cl$^-$(g); the ions generally come in pairs of one cation and one anion. At present, we have no practical means of separating the ions from 1 mole of NaCl(c). Consequently, we must utilize the first law to measure ΔH_c.

Let us imagine that we dissociate NaCl(c) into its ions by first dissociating it into its elements in their standard thermochemical states at 298.15 K and 1 bar of pressure. This gives Na(s) and 0.5 mole of Cl$_2$(g). Sodium is then vaporized to yield Na(g), while Cl$_2$(g) is dissociated into atomic chlorine. Na(g) is now ionized to yield Na$^+$(g), and an additional electron is added to Cl(g) to produce Cl$^-$(g). The entire process is illustrated in Figure 7.6. This cyclic process is called the *Born–Haber cycle*. Although we cannot measure the enthalpy change for the direct dissociation of NaCl(c) into its constituent ions, we can determine the enthalpy changes for all of the remaining steps in the cycle. Since the first law of thermodynamics requires that we have

$$\Delta H_c = \Delta H_1 + \Delta H_2 + \Delta H_3 + \Delta H_4 + \Delta H_5, \tag{7.2}$$

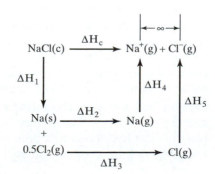

▶ **FIGURE 7.6**
The Born–Haber cycle for NaCl. The first law requires that the sum $\Delta H_1 + \Delta H_2 + \Delta H_3 + \Delta H_4 + \Delta H_5$ be equal to ΔH_c.

Table 7.4   Electron affinities of group A elements

Element	Group	A (kJ mol^{-1})	Element	Group	A (kJ mol^{-1})
H	I	72.8	Be	II	−18
Li		59.8	Mg		−21
Na		52.9	Ca		−186
K		48.3	Sr		−146
Rb		46.9	Ba		−46
Cs		45.5			
B	III	23	C	IV	122.5
Al		44	Si		133.6
Ga		36	Ge		116
In		34	Sn		121
Tl		30	Pb		35.2
N	V	−7	O	VI	141
P		71.7			−844[1]
As		77	S		200.4
Sb		101			−532[1]
Bi		101	Se		195.0
			Te		190.2
			Po		186
F	VII	322	He	VIII	−21
Cl		348.7	Ne		−29
Br		324.5	Ar		−35
I		295.3	Kr		−39
At		270	Xe		−41
			Rn		−41

[1] The second number refers to the energy associated with removal of the second electron to change the doubly charged ion to an ion that has a single charge.
Data from J. Emsley, *The Elements*, Clarendon Press, Oxford (1991).

we may compute the value of the crystal enthalpy from the measured values of the $\Delta H_i(i = 1, 2, \ldots, 5)$. Equation 7.2 is an example of Hess' law in use.

The enthalpy change for the reverse of the first step in the cycle is the standard molar heat of formation of NaCl(c). With the choice of reference state made in Chapter 3, that quantity is numerically equal to the standard partial molar enthalpy of NaCl(c). Therefore, we have $\Delta H_1 = -\overline{H}^o_{\text{NaCl(c)}}$. The enthalpy change for the vaporization of solid sodium is the standard partial molar enthalpy of Na(g), and that for the dissociation of 0.5 mole of Cl$_2$(g) is the standard partial molar enthalpy of Cl(g). That is, we have $\Delta H_2 = -\overline{H}^o_{\text{Na(g)}}$ and $\Delta H_3 = -\overline{H}^o_{\text{Cl(g)}}$. The energy required to remove the first electron from Na(g) is the first ionization potential of the element. Therefore, $\Delta U_4 = I_1^{\text{Na}}$. Using the definition of the enthalpy function, we may relate ΔU_4 to ΔH_4:

$$\Delta H_4 = \Delta U_4 + \Delta(PV).$$

Since the ions are separated by an infinite distance, we expect them to behave ideally, so that

$$\Delta H_4 = \Delta U_4 + \Delta(nRT) = \Delta U_4 + RT\Delta n = I_1^{Na} + RT,$$

because $\Delta n = +1$ for the process $Na(g) \rightarrow Na^{1+}(g) + e^-(g)$.

The final step in the Born–Haber cycle involves the addition of an electron to $Cl(g)$ to form the anion:

$$Cl(g) + e^-(g) \longrightarrow Cl^{1-}(g).$$

The internal energy change for the reverse of this process is called the *electron affinity* of $Cl(g)$ and is usually given the symbol A_{Cl^-}. In these terms, we have $\Delta U_5 = -A_{Cl^-}$. As in the case of the ionization of sodium, the enthalpy change is related to ΔU_5 by

$$\Delta H_5 = \Delta U_5 + \Delta(PV) = \Delta U_5 + RT\Delta n = \Delta U_5 - RT = -A_{Cl^-} - RT.$$

Table 7.4 gives the measured electron affinities of the group A elements.

Using Eq. 7.2, we may now obtain the crystal enthalpy for $NaCl(c)$ as

$$\Delta H_c(NaCl) = -\overline{H}^o_{NaCl(c)} + \overline{H}^o_{Na(g)} + \overline{H}^o_{Cl(g)} + I_1^{Na} + RT - A_{Cl^-} - RT.$$

Substituting values from Appendix A and Tables 7.2 and 7.4 gives

$$\Delta H_c(NaCl) = 411.15 + 107.32 + 121.68 + 5.12 \text{ eV}(96.48 \text{ kJ mol}^{-1}\text{eV}^{-1})$$

$$-348.7 \text{ kJ mol}^{-1}$$

$$= 785.43 \text{ kJ mol}^{-1}.$$

EXAMPLE 7.3

Compute the crystal enthalpy for $MgCl_2$, using the Born–Haber cycle.

Solution

The steps in the cycle are as follows:

Step 1:	$MgCl_2(c) \longrightarrow Mg(s) + Cl_2(g)$	$\Delta H_1 = -\overline{H}^o_{MgCl_2(c)}$
Step 2:	$Mg(s) \longrightarrow Mg(g)$	$\Delta H_2 = \overline{H}^o_{Mg(g)}$
Step 3:	$Cl_2(g) \longrightarrow 2Cl(g)$	$\Delta H_3 = 2\,\overline{H}^o_{Cl(g)}$
Step 4:	$Mg(g) \longrightarrow Mg^{2+}(g) + 2e^-(g)$	$\Delta H_4 = I_1^{Mg} + I_2^{Mg} + 2RT$
Step 5:	$2Cl(g) + 2e^-(g) \longrightarrow 2Cl^-(g)$	$\Delta H_5 = -2\,A_{Cl^-} - 2RT.$

In Step 4, we are removing two electrons from $Mg(g)$. The energy required to remove the first electron is the first ionization potential, I_1^{Mg}. The energy required to remove the second electron is the second ionization potential, I_2^{Mg}. The sum of the enthalpy changes for these processes yields the crystal enthalpy of $MgCl_2(c)$. Using Appendix A and Tables 7.2 and 7.4, we may compute

$$\Delta H_c(MgCl_2) = -\overline{H}^o_{MgCl_2(c)} + \overline{H}^o_{Mg(g)} + 2\,\overline{H}^o_{Cl(g)} + I_1^{Mg} + I_2^{Mg} + 2RT - 2A_{Cl^-}$$

$$-2RT = 641.32 + 147.70 + 2(121.68) + 7.61(96.48)$$

$$+ 14.96(96.48) - 2(348.7) \text{ kJ mol}^{-1}$$

$$= 2,512.53 \text{ kJ mol}^{-1}.$$

It is important to note that it is the ionization potential terms that make the largest contribution to this endothermicity. The sum of the ionization potentials, $I_1^{Mg} + I_2^{Mg}$, has a value of 2,177.55 kJ mol^{-1}. This makes it clear why the decrease in ionization potential dominates the trends in melting points for the alkaline earth halides.

For related exercises, see Problems 7.4, 7.5, and 7.6.

7.2.2   Qualitative Description of Crystal Forms

Crystals are constructed from a regular, three-dimensional arrangement of their basic unit, which can be a single atom, a molecule, or an even larger entity, such as a cluster, a single biological cell, or a virus. In a perfectly formed crystal, these units are positioned at equally spaced intervals along parallel lines in three different directions. We may, therefore, choose an arbitrary point in the interior of the crystal as the origin and draw three vectors that specify the directions of the parallel lines along which the basic crystal units are arranged. These vectors are called the *primitive translations* of the crystal. The angles between the primitive translations are generally labeled α, β, and γ, as shown in Figure 7.7.

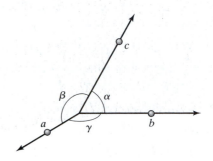

▶ FIGURE 7.7
The three primitive translations (⟶) of a crystal. The angles between the translations are denoted by α, β, and γ as shown. The unit cell spacings along the three translational vectors are represented by a, b, and c.

All perfect crystals have the property of translational symmetry, according to which there exists a translationally repeating pattern of a small, three-dimensional group of basic units. The situation is similar to the two-dimensional repeating patterns present in many wallpapers. The smallest three-dimensional, repeating group of basic units from which the entire crystal may be constructed by simple translational displacements along the primitive vectors is termed the *unit cell*. The lengths of the unit cell along each of the primitive translations are denoted by a, b, and c, as shown in Figure 7.7. In Chapter 16, we shall learn how these dimensions and the angles connecting the primitive translations may be determined. At this point, however, we shall simply describe the different crystal forms, as that is all that is required to permit us to compute the ionic crystal enthalpies.

There are seven basic crystal systems. The most symmetric is called the *cubic* system and is characterized by having mutually perpendicular primitive translations with $\alpha = \beta = \gamma = 90°$ and all unit cell dimensions equal. Thus, the unit cell resembles a cube. The remaining systems are progressively less symmetric. Crystals having $a \neq b \neq c$ and $\alpha \neq \beta \neq \gamma \neq 90°$ are called *triclinic* crystals. The seven systems are summarized in Table 7.5.

Table 7.5   The seven crystal systems

Dimensions	Angles	System
$a = b = c$	$\alpha = \beta = \gamma = 90°$	Cubic
$a = b = c$	$\alpha = \beta = \gamma$	Rhombohedral (trigonal)
$a = b \neq c$	$\alpha = \beta = \gamma = 90°$	Tetragonal
$a = b \neq c$	$\alpha = \beta = 90°, \gamma = 120°$	Hexagonal
$a \neq b \neq c$	$\alpha = \beta = \gamma = 90°$	Orthorhombic
$a \neq b \neq c$	$\alpha = \beta = 90° \neq \gamma$	Monoclinic
$a \neq b \neq c$	$\alpha \neq \beta \neq \gamma \neq 90°$	Triclinic

Within each crystal system, there are often several subclasses. The cubic system has three such subclasses. The first of these is a primitive cubic lattice with lattice sites at each vertex of the cube. The second has lattice sites at each vertex, plus an addition site in the center of the cube. This is called a *body-centered* cubic lattice. The third class has lattice sites at each vertex and in the center of each of the six faces of the cube. This type is termed a *face-centered* cubic lattice. Each of these subclasses is illustrated in Figure 7.8. NaCl(c), an example of a face-centered cubic crystal, is shown in Figure 7.9.

There are two subclasses of the tetragonal system, four of the orthorhombic system, two of monoclinic crystals, and one each of triclinic, hexagonal, and trigonal systems. There are, therefore, seven crystal systems, containing a total of 14 subclasses, which are called the *Bravais lattices*. The structures of the Bravais lattices are illustrated in Figures 7.8 and 7.10 through 7.15.

7.2.3   Calculation of Binding Energies in Ionic Crystals

Originally, the Born–Haber cycle was used to obtain electron affinities, since there was, at the time, no good experimental method for measuring them directly. All the quantities in the cycle could be easily measured, except the crystal enthalpy, ΔH_c. This quantity was computed theoretically, by summing the total coulombic interactions present in the lattice and by making appropriate corrections for the repulsive forces that become important when the ions approach to close distances. Today, electron affinities can be obtained by more direct measurement, so that it is no longer necessary to extract their values using theory. Nevertheless, the method still serves as an excellent example of how theoretical models can be combined with experimental data to study a system. We shall, therefore, describe the process in some detail.

Consider 1 mole of a 1:1 ionic crystal in which the cation has a charge $+q$ while the anionic charge is $-q$. Examples are NaCl(c), KBr(c), and MgO(c). One mole of the crystal contains $2N$ ions, where N is Avogadro's constant. The coulombic potential between ion i and ion j is given by

$$V_{ij}(r_{ij}) = \frac{q_i q_j}{4\pi\varepsilon_o r_{ij}} \tag{7.3}$$

where r_{ij} is the distance between the two ions. Ion i interacts with all of the other $2N - 1$ ions in the crystal. The total coulombic potential acting on ion i can, therefore, be obtained by summing interactions of the form given in Eq. 7.3 over all other ions. That is,

$$V_i^T = \text{total coulombic potential for ion } i = \sum_{j=1}^{2N}{}' \left[\frac{q_i q_j}{4\pi\varepsilon_o r_{ij}}\right] \tag{7.4}$$

where the prime on the summation means that the term $j = i$ is omitted from the sum. Since $|q_i| = |q_j|$ and $4\pi\varepsilon_0$ are constants, we may write Eq. 7.4 in the form

$$V_i^T = -\frac{q^2}{4\pi\varepsilon_0} \sum_{j=1}^{2N}{}' \pm r_{ij}^{-1}. \tag{7.5}$$

In executing the summation in Eq. 7.5, care must be taken to use the correct sign on each term. If the ions involved in the interaction for a particular term have the same sign, we wish the contribution of that term to V_i^T to be positive. With the minus sign we have elected to write in front of the constant term, this means that we need to use the minus sign inside the summation. If the ions have opposite sign such that the final result should be negative, we will employ the plus sign inside the summation.

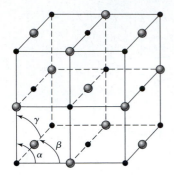

▲ FIGURE 7.9
The face-centered cubic structure of a sodium chloride (NaCl) crystal. The black dots represent Na$^+$ ions. The shaded dots are Cl$^-$ ions. Examination shows both the Na$^+$ and Cl$^-$ ions to be arranged in a face-centered cubic structure with $\alpha = \beta = \gamma = 90°$.

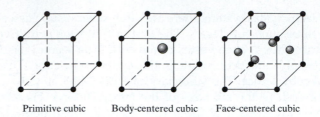

Primitive cubic   Body-centered cubic   Face-centered cubic

▲ FIGURE 7.8
The three subclasses of the cubic crystal system. The black dots represent lattice sites at the vertices of the cube. The shaded dots denote lattice sites either in the center or on the faces of the cube.

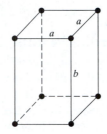
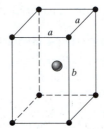

▲ FIGURE 7.10
The Bravais lattices for the tetragonal crystal system. The primitive lattice has an atom, a molecule, or some other unit at each vertex of the rectangular parallelepiped. The body-centered lattice has an additional unit in the center of the unit cell.

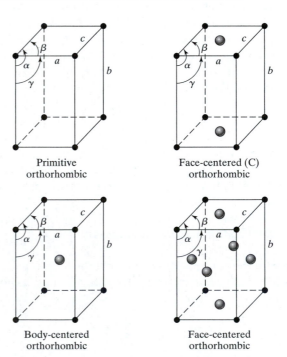

Primitive
orthorhombic

Face-centered (C)
orthorhombic

Body-centered
orthorhombic

Face-centered
orthorhombic

▲ FIGURE 7.11
The four Bravais lattices of the orthorhombic system. The primitive lattice has an atom, a molecule, or some other unit at each vertex. The face-centered (C) lattice has units at each vertex and also on two opposite faces. The body-centered lattice has units at each vertex and an additional unit in the center of the unit cell. The face-centered lattice has units at each vertex and on each face. In all lattices, we have $a \neq b \neq c$ and $\alpha = \beta = \gamma = 90°$.

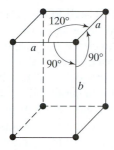

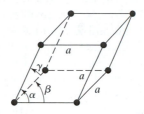

▲ FIGURE 7.12
A hexagonal lattice system with two unit cell distances equal, two angles between the primitive translational vectors equal to 90°, and the third angle equal to 120°.

▲ FIGURE 7.13
A rhombohedral (trigonal) lattice in which $a = b = c$ and $\alpha = \beta = \gamma \neq 90°$. Each face of the lattice is a rhombus.

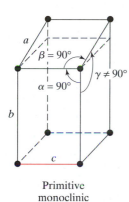

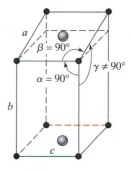

Primitive
monoclinic

Face-centered
monoclinic (C)

▲ FIGURE 7.14
The monoclinic crystal system with $a \neq b \neq c$ and $\alpha = \beta = 90° \neq \gamma$. The two Bravais lattices are a primitive lattice with atoms, molecules, or some other units at each vertex and a face-centered lattice with atoms, molecules, or some other units at each vertex and on two opposite faces. This type of lattice is generally denoted by C.

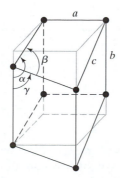

▲ FIGURE 7.15
A triclinic crystal system for which $a \neq b \neq c$ and $\alpha \neq \beta \neq \gamma \neq 90°$. The solid lines show the sides of the triclinic unit cell. The dashed lines show a superimposed orthorhombic crystal, to enhance the visual clarity of the three-dimensional shape of the triclinic crystal.

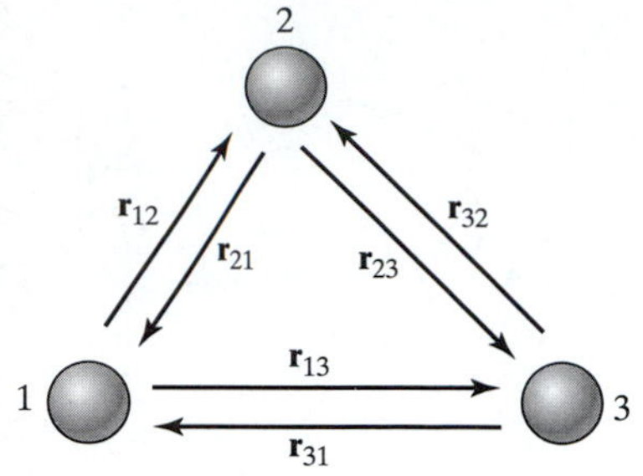

▲ FIGURE 7.16

The pairwise interactions among three particles. Note that a simple summation of all interactions counts each interaction twice.

Equation 7.5 gives the total coulombic interaction for an arbitrary ion in the crystal lattice. Because there are $2N$ total ions, it would appear that the total coulombic interaction for the entire lattice can be obtained by multiplying Eq. 7.5 by $2N$. The problem with this procedure is that it counts every pairwise coulombic interaction twice. This is easily seen to be the case by considering Figure 7.16, which illustrates the interaction of three particles. Since $r_{12} = r_{21}$, $r_{13} = r_{31}$, and $r_{23} = r_{32}$, a pairwise summation of all interactions for each of the three particles will count every interaction twice. The correct result can, therefore, be obtained by dividing the sum by a factor of two. If we apply this logic to the summation of all the coulombic interactions among the $2N$ ions in our 1:1 crystal, the total interaction is

$$V_c^T = \frac{2N V_i^T}{2} = -\frac{2Nq^2}{2 \times 4\pi\varepsilon_0} \sum_{j=1}^{2N}{}' \pm r_{ij}^{-1} = -\frac{Nq^2}{4\pi\varepsilon_0} \sum_{j=1}^{2N}{}' \pm r_{ij}^{-1}. \qquad (7.6)$$

where ion i can be any arbitrary ion in the crystal.

Let us now denote the shortest distance between ions in the crystal by d. Multiplying by d inside the summation and dividing by d outside yields

$$V_c^T = -\frac{Nq^2}{4\pi d\varepsilon_o} \sum_{j=1}^{2N}{}' \pm dr_{ij}^{-1}. \qquad (7.7)$$

It would appear that the computation of V_c^T requires the evaluation of the summation in Eq. 7.7 for each crystal of interest. However, the problem is not nearly that difficult. The reason is that the value of the summation is dependent only upon the type of crystal and the details of the lattice parameters, not upon the specific crystal. That is, body-centered cubic crystals usually have the same value for the sum. Consequently, the summation does not need to be evaluated for every crystal. The value of the sum is called the *Madelung* constant, $\mathcal{M}$. With this notation, Eq. 7.7 may be written in the form

$$V_c^T = -\frac{N\mathcal{M}q^2}{4\pi d\varepsilon_o}. \qquad (7.8)$$

Let us take a simple example to see why the value of the summation is dependent only upon the type of crystal and the details of the lattice parameters. Suppose we are in a different universe, where NaCl(c) is a one-dimensional, linear crystal with alternating plus and minus charges separated by distance d, as shown in Figure 7.17.

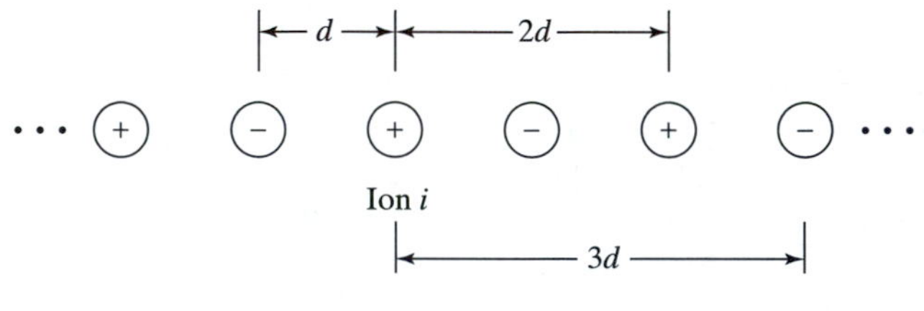

▶ FIGURE 7.17

Hypothetical infinite one-dimensional crystal with a characteristic spacing of d between alternating plus and minus charges.

We may obtain the Madelung constant for this hypothetical linear crystal very easily. We wish to evaluate the sum of the interactions undergone by ion i. The required summation is

$$\sum_{j=1}^{2N}{}' \pm dr_{ij}^{-1} = 2\left[\frac{d}{d} - \frac{d}{2d} + \frac{d}{3d} - \frac{d}{4d} + \cdots\right],$$

where we have summed the interactions for ions located to the right of ion i in the figure and multiplied the result by two to account for the interaction of ions to the left of ion i. We have also been careful to use the plus sign when the charges are different and the minus sign when the charges are the same. The important thing to note is that the lattice spacing, d, cancels out in every term. The summation will, therefore, have the same value for *all* linear crystals, regardless of the spacing d. After cancellation of d, the resulting sum is

$$\sum_{j=1}^{2N}{}' \pm d r_{ij}^{-1} = 2\left[1 - \frac{1}{2} + \frac{1}{3} - \frac{1}{5} - \cdots\right] = 2\sum_{n=1}^{n=\infty}(-1)^{n+1}n^{-1}.$$

Since this series alternates, and since

$$\underset{n\to\infty}{\text{Lim}}[(-1)^{n+1}n^{-1}] = 0,$$

the summation is known to converge. A FORTRAN or C code for a small PC or workstation very quickly sums the first 10,000 terms of the series to find that, correct to five significant digits, the sum is 1.3862.... In this simple case, we can also obtain an analytic expression for the summation. If the function $F(x) = \ln(1 + x)$ is expanded in a Taylor series about the point $x = 0$, the result is

$$\ln(1 + x) = x - \frac{x^2}{2} + \frac{x^3}{3} - \frac{x^4}{4} + \cdots + \frac{(-1)^{n+1}x^n}{n} + \cdots$$

Setting $x = 1$, we obtain

$$\ln(1 + 1) = \ln(2) = 1 - \frac{1}{2} + \frac{1}{3} - \frac{1}{4} + \frac{1}{5} - \cdots,$$

which we recognize as the required summation. Therefore, the exact value of the Madelung constant for this hypothetical one-dimensional crystal is $\mathcal{M} = 2\ln(2) = 1.38629436\ldots$ Table 7.6 lists the Madelung constants for various 1:1 types of crystal.

Equation 7.8 gives the sum of all the coulombic interactions in a 1:1 ionic crystal. The form of this equation indicates that the potential becomes more and more negative as the distance between the ions decreases toward zero. Experiment, however, shows that this does not happen. If it did, the ions would collapse to a single point in an infinitely deep potential well and thereby form a black hole. This simple analysis indicates that repulsive forces other than coulombic forces must be present and become important when the ions approach to small values of d. Such repulsive forces that make positive contributions to the potential are indeed present. They arise for several reasons. First, when the ions are in close proximity, they no longer appear to be point charges, as is assumed by our use of the coulombic potential form. The individual electrons interact and create a repulsive potential.

Table 7.6 Madelung constants		
Type of Crystal	**Example**	**Madelung Constant, $\mathcal{M}$**
Face-centered Cubic	NaCl	1.74756
Body-centered Cubic	CsCl	1.76267
Hexagonal	ZnO	1.4985
Hexagonal	ZnS	1.64132

Source of data: *Handbook of Chemistry and Physics*, CRC Press, 60th ed. (1980), Boca Raton, FL.

Second, when we turn our attention to quantum mechanics and bonding, we shall find that confining particles to small volumes causes a rapid increase in the kinetic energy of the system, which is another positive contribution to the energy. These considerations indicate that another positive term must be added to Eq. 7.8 if crystal enthalpies are to be successfully computed.

As noted in Chapter 1 in Eqs. 1.54 and 1.55, the repulsive contributions to the potential at small distances are often represented by terms of the form $V_{rep}(d) = K/d^n$, where K and n are parameters. There is nothing unique about such a choice: We could equally well assume that the repulsive contributions to the potential have the exponential form $V_{rep}(d) = K \exp[-d/\rho]$, where K and ρ and now the potential parameters. In the analysis set forth here, we shall take the first form to represent the repulsive contributions to the potential. The results obtained using an exponential form will be explored in the problems at the end of the chapter.

If we represent the repulsive contributions to the potential with the form used in Eqs. 1.54 and 1.55, the total potential for the crystal becomes

$$V(d) = \frac{K}{d^n} - \frac{N\mathcal{M}q^2}{4\pi d\varepsilon_o}. \tag{7.9}$$

With the total potential in this form, there is no longer any problem with the crystal forming a black hole: The repulsive forces prevent the approach of the ions beyond a certain distance. The minimum potential attainable by the ions occurs at a distance d_o, as shown in Figure 7.18. If d decreases below d_o, the crystal potential is seen to rise because of the rapid increase in the magnitude of the repulsive interactions. As d increases beyond d_o, the potential rises because of a decrease in the total attractive coulombic interactions.

At the potential minimum, $d = d_o$, the slope of the potential with respect to d is exactly zero. That is,

$$\frac{dV(d)}{dd} = -\frac{nK}{d^{n+1}} + \frac{N\mathcal{M}q^2}{4\pi d^2\varepsilon_o}\Bigg|_{d=d_o} = -\frac{nK}{d_o^{n+1}} + \frac{N\mathcal{M}q^2}{4\pi d_o^2\varepsilon_o} = 0. \tag{7.10}$$

Equation 7.10 permits us to obtain one of the potential parameters, K, in terms of the remaining parameter, n. Solving Eq. 7.10 for K, we obtain

$$K = \frac{N\mathcal{M}q^2 d_o^{n-1}}{4\pi n\varepsilon_o}. \tag{7.11}$$

▶ FIGURE 7.18
Typical crystal potential for an assumed repulsive interaction of the form $V_{rep}(d) = K/d^n$. The total attractive interaction is given by Eq. 7.8. These are respectively shown as the solid and short dashed curves in the figure. The sum of these interactions yields the total potential V(d), which is shown as the long dashed curve. In this hypothetical example, the equilibrium ion–ion separation occurs at 2.81 Å.

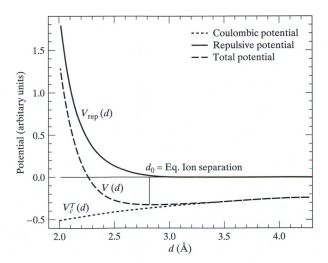

Substituting Eq. 7.11 into Eq. 7.9 gives

$$V(d) = \frac{N\mathcal{M}q^2 d_o^{n-1}}{4\pi n \varepsilon_o d^n} - \frac{N\mathcal{M}q^2}{4\pi d \varepsilon_o} = \frac{N\mathcal{M}q^2}{4\pi\varepsilon_o}\left[\frac{d_o^{n-1}}{nd^n} - \frac{1}{d}\right]. \tag{7.12}$$

The change in the internal energy for the process described by Figure 7.6 is the difference in potential when the ions are separated by an infinite distance in the product state and when they are at their equilibrium separation d_o. That is,

$$\Delta U_c = V(d = \infty) - V(d = d_o) = 0 - \frac{N\mathcal{M}q^2}{4\pi\varepsilon_o}\left[\frac{1}{nd_o} - \frac{1}{d_o}\right]$$

$$= \frac{N\mathcal{M}q^2}{4\pi\varepsilon_o d_o}\left[1 - \frac{1}{n}\right]. \tag{7.13}$$

Since we have $\Delta H_c = \Delta U_c + \Delta(PV) = \Delta U_c + RT\Delta n = \Delta U_c + 2RT$, the crystal enthalpy is given by

$$\boxed{\Delta H_c = \frac{N\mathcal{M}q^2}{4\pi\varepsilon_o d_o}\left[1 - \frac{1}{n}\right] + 2RT}. \tag{7.14}$$

Equation 7.14 provides the desired expression for the crystal enthalpy when we use $V_{rep}(d) = K/d^n$ as the repulsive interaction. To obtain numerical results, we need only determine the value of the exponent n. The method of choice is generally to select a subset of crystals of a given type and use a least-squares procedure to determine the value of n that best fits their measured crystal enthalpies. As an example, let us take NaCl, KCl, and NaBr, all of which are face-centered cubic crystals. The measured cation–anion equilibrium distances for these crystals are 2.184×10^{-10} m, 3.138×10^{-10} m, and 2.981×10^{-10}, respectively. The charge is 1.602×10^{-19} C, and the Madelung constant from Table 7.6 is 1.74756. The crystal enthalpy for NaCl at 298.15 K computed from Eq. 7.14 is thus

$$\Delta H_c(\text{NaCl}) = \frac{(6.022 \times 10^{23})(1.74756)(1.602 \times 10^{-19})^2}{4(3.14159)(8.85419 \times 10^{-12})(2.814 \times 10^{-10})}\left[1 - \frac{1}{n}\right]$$

$$+ 2(8.314)(298.15) \text{ J mol}^{-1}$$

$$= 8.626 \times 10^5\left[1 - \frac{1}{n}\right] + 4958 \text{ J mol}^{-1}$$

$$= 862.6\left[1 - \frac{1}{n}\right] + 4.96 \text{ kJ mol}^{-1}.$$

Similar calculations for KCl and NaBr yield

$$\Delta H_c(\text{KCl}) = 773.5\left[1 - \frac{1}{n}\right] + 4.96 \text{ kJ mol}^{-1}$$

and

$$\Delta H_c(\text{NaBr}) = 814.3\left[1 - \frac{1}{n}\right] + 4.96 \text{ kJ mol}^{-1}.$$

The measured enthalpies for these three crystals are $\Delta H_c(\text{NaCl}) = 787$ kJ mol^{-1}, $\Delta H_c(\text{KCl}) = 717$ kJ mol^{-1}, and $\Delta H_c(\text{NaBr}) = 752$ kJ mol^{-1}. To obtain the best least-squares fit of n to these data, we need the sum of the squares of the

▶ FIGURE 7.19
Least-squares computation of the best value of n for the repulsive interaction in a 1:1 ionic crystal using the experimentally measured crystal enthalpies of NaCl, KCl, and NaBr as data to which the theoretical expression of the crystal enthalpy is fitted. The best value of n is the value that minimizes the summation of the squared deviations of the computed values from the experimental results. With our choice of data, the best value of n is 11.67.

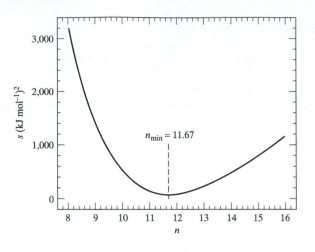

▶ FIGURE 7.20
Comparison of experimental and computed crystal enthalpies at 298.15 K for the chlorides and bromides of the alkali metals using Eq. 7.14 with $n = 11.67$. The 45° line represents the loci of points if the agreement between theory and experiment were perfect.

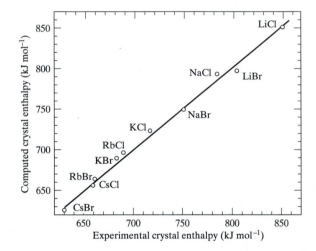

deviations of the predicted crystal enthalpies from the measured values to be a minimum. That is, we require that the sum

$$S = \left[787 - 862.6\left[1 - \frac{1}{n}\right] - 4.96\right]^2 + \left[717 - 773.5\left[1 - \frac{1}{n}\right] - 4.96\right]^2$$

$$+ \left[752 - 814.3\left[1 - \frac{1}{n}\right] - 4.96\right]^2$$

be a minimum. Figure 7.19 shows the variation of S with n. A minimum occurs at $n = 11.67$, for which the value of S is 73.72 kJ^2 mol^{-2}. This corresponds to a root-mean-square deviation of the computed values from the measured data of ± 6.1 kJ mol^{-1} The computed values of crystal enthalpies obtained using this value of n are 794.6 kJ mol^{-1}, 712 kJ mol^{-1}, and 749 kJ mol^{-1} for NaCl, KCl, and NaBr, respectively.

Figure 7.20 compares the crystal enthalpies at 298.15 K computed using Eq. 7.14 with $n = 11.67$ with those measured experimentally using the Born–Haber cycle. The 45° line in the figure represents the loci of points if there were perfect agreement between experiment and theory. Considering the approximate character of the repulsive potential, the results are reasonably accurate. Example 7.4 provides an additional illustration of such calculations.

EXAMPLE 7.4

The shortest distance between K^{1+} and Br^{1-} ions in a face-centered cubic KBr crystal is 3.293×10^{-10}m. Compute the crystal enthalpy of KBr(c) at 298.15 K, using the value of n obtained by the fit to NaCl(c), KCl(c), and NaBr(c) and the Madelung constant given in Table 7.6. Use the data in Appendix B to determine the percent error in your result.

Solution

Using Eq. 7.14 with $n = 11.67$, we have

$$\Delta H_c(KBr) = \frac{N\mathcal{M}q^2}{4\pi\varepsilon_o d_o}\left[1 - \frac{1}{n}\right] + 2RT$$

$$= \frac{(6.022 \times 10^{23})(1.74756)(1.602 \times 10^{-19})^2}{4(3.14159)(8.85419 \times 10^{-12})(3.293 \times 10^{-10})}\left[1 - \frac{1}{11.67}\right]$$

$$+ 2(8.314)(298.15) \text{ J mol}^{-1}$$

$$= [6.740 \times 10^5 + 4958] \text{ J mol}^{-1} = 6.790 \times 10^5 \text{ J mol}^{-1}$$

$$= 679.0 \text{ kJ mol}^{-1}. \tag{1}$$

The result from the Born–Haber cycle is $\Delta H_c(KBr) = 682$ kJ mol^{-1}. The percent error in our computed result is

$$\% \text{ error} = \frac{100 \times (679.0 - 682)}{682} = -0.44\%. \tag{2}$$

This example illustrates the difference between significant digits and the accuracy of an approximation. Every quantity appearing in Eq. 1 contains at least four significant digits. Consequently, the result, 679.0 kJ mol^{-1}, also contains four significant digits. However, this value is not in fact correct to four significant digits; it differs from the experimentally measured crystal enthalpy in the third digit.

For related exercises, see Problems 7.9 and 7.10.

7.3   Heat Capacity of Solids

7.3.1   Empirical Law of Dulong and Petit

Prior to the advent of quantum mechanics in the 1920s, the world was viewed as being described exactly by the Newtonian laws of physics, which we shall discuss in greater detail in Chapter 11. These laws require that the energy a system possesses be a continuous variable which can have any value subject only to the constraints imposed by the first and second laws of thermodynamics. Equation 1.5 shows that, in rectangular Cartesian coordinates, the classical Newtonian momentum is related to the velocity by

$$\mathbf{P} = m\,\mathbf{v},$$

where m is the mass and $\mathbf{v}$ is the velocity. (Boldface denotes a vector quantity.) The classical kinetic energy of a system is given by

$$\text{Kinetic Energy} = 0.5\,m\,\mathbf{v \cdot v} = 0.5\,m^{-1}\,\mathbf{P \cdot P} = 0.5\,m^{-1}[P_x^2 + P_y^2 + P_z^2], \tag{7.15}$$

where P_x, P_y, and P_z are the x, y, and z components of momentum, respectively.

The classical theory of the equipartition of energy states that, on the average, the total translational energy will be divided equally between the three squared momentum components. Furthermore, the analysis leading to Eq. 1.4 indicates that we expect the energy in each component to be $PV/2$ or $RT/2$. This leads to a total translational energy of $U = 3RT/2$ per mole. The energy per molecule is $3RT/2N = 3k_bT/2$, where N is Avogadro's number and k_b is the Boltzmann constant introduced in Chapter 1.

Since $C_v^m = (\partial U/\partial T)_V$, we expect C_v^m for systems having only translational energy to be $1.5R$, or $12.471 \text{ J mol}^{-1} \text{ K}^{-1}$. If the system behaves ideally, we have $C_p^m = C_v^m + R$, so that we would expect C_p^m for these systems to be $20.785 \text{ J mol}^{-1}\text{K}^{-1}$. This number is nearly exact for the monatomic rare gases, all of which have C_p^m values of $20.786 \text{ J mol}^{-1}\text{K}^{-1}$; it is reasonably accurate for $C(g)$, whose constant-pressure heat capacity is $20.838 \text{ J mol}^{-1}\text{K}^{-1}$. The difference reflects a departure of $C(g)$ gas from ideality.

For diatomic molecules, the situation becomes more complex. Each atom can move in three different directions, so that six types of motion are present. The situation is described by saying that the system possesses six degrees of freedom. In general, the total number of degrees of freedom for three-dimensional systems is $3N$, where N is the number of particles present. Three degrees of freedom are associated with the center-of-mass translational motion for all systems. If the molecule is linear, significant rotational energy is present only when the molecule is rotating about an axis perpendicular to the molecular axis. There are two such axes, as shown in Figure 7.21. Therefore, two degrees of rotational freedom are present for linear molecules. The remaining $3N - 5$ degrees of freedom are associated with vibrational motions of the molecule. If the N-atom molecule is nonlinear, energy of rotation is present about any of three axes through the center of mass. With three rotational degrees of freedom present, we have $3N - 6$ vibrational degrees of freedom.

The six degrees of freedom of a diatomic molecule can, therefore, be divided into three translational motions of the center of mass of the system (in the x, y, and z directions), two rotational motions, and one vibrational motion. These components are illustrated in Figure 7.21, where the z-axis is taken to be the diatomic bond axis with the center of mass at the origin. The x- and y-axes are mutually perpendicular to this axis. As the diagram indicates, the molecular center of mass can move in three different directions with momentum compo-

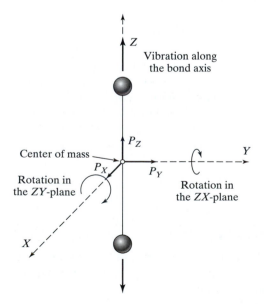

◀ FIGURE 7.21
Translational, rotational, and vibrational components of the molecular energy for a diatomic molecule.

nents P_x, P_y, and P_z. The molecule can rotate around the x-axis in the yz plane, or it can rotate about the y-axis in the xz plane. Finally, the molecule can vibrate along the z-axis.

The classical equipartition theorem requires that each of the components of motion involving only kinetic energy have an average internal energy equal to $k_bT/2$. Since the three translational and two rotational motions involve only kinetic energy, we would expect to have $U = 5k_bT/2$ from those motions. The equipartition theorem further requires that components of motion involving both potential and kinetic energy have an average internal energy equal to k_bT. Our previous discussions of interatomic forces and potentials makes it clear that a bond potential energy is present. Therefore, the vibrational motion should have an average internal energy of k_bT to make $U_{total} = 7k_bT/2$ per molecule, or $7RT/2$ per mole, if the classical equipartition theorem holds.

With $U_{total} = 7RT/2$, we expect to have $C_v^m = (\partial U/\partial T)_V = 7R/2$ and $C_p^m = 9R/2 = 37.413$ J mol^{-1} K^{-1}. The accuracy of this prediction may be easily checked by examining the heat capacities of some typical gas-phase diatomic molecules. Table 7.7 gives some relevant data. All of the systems shown have C_p^m values significantly below that predicted by the equipartition theorem. Clearly, then, something is wrong with this classical theorem. It is interesting and important to note that, although the results deviate from the classical equipartition prediction, if the vibrational mode is ignored, so that we have $U_{total} = 5RT/2$, the predictions are much more accurate. With this internal energy, we should have $C_v^m = 5R/2$ and $C_p^m = 7R/2 = 29.099$ J mol^{-1} K^{-1}. This value is much closer to the measured data given in the table, which suggests that the major problem lies in the classical treatment of the energy in the vibrational mode.

Let us now turn our attention to one mole of a solid containing Avogadro's number of particles (e.g., one mole of Mg(s) or Xe(s)). A macroscopic solid does not rotate or translate through space. Therefore, the three translational and rotational degrees of freedom make no contribution to the internal energy of the solid. The only energy is present as vibrational energy. With N particles, we have a total of $3N - 6$ vibrational degrees of freedom. However, Avogadro's number is so large, that $3N - 6 \approx 3N$. If the classical equipartition theorem holds, each of these vibrational degrees of freedom has an average energy of k_bT. Under these conditions, the total internal energy of the solid is $3Nk_bT = 3RT$ This yields a solid-state heat capacity $C_v^m = (\partial U/\partial T)_V = 3R$. Consequently, the classical equipartition theorem predicts a solid-state heat capacity of $3R$ at all temperatures. This result is the empirical law first introduced by Dulong and Petit.

When solid-state heat capacities are measured in the laboratory at temperatures ranging from a few degrees to 300 K, we find that the law of Dulong and Petit fails badly. The heat capacities vary dramatically as the temperature is lowered. In fact, they approach zero as T approaches 0 K. For temperatures above 300 K, the Dulong and Petit approximation that $C_v^m = 3R = 24.942$ J mol^{-1} K^{-1} is reasonably accurate for many, but not all, solids. Table 7.8 illustrates this point.

Figure 7-22 shows measured constant-volume heat capacity data for Mg(s) from 12 K to 298.15 K. The experimental data show that, contrary to Dulong and Petit's "law," C_v^m is not constant over this range. The collapse of the classical equipartition theorem at low temperatures is the first example of many to come of the failure of classical Newtonian principles to describe the physics and chemistry of molecular systems.

Table 7.7 Constant-pressure heat capacities for some gas-phase diatomic molecules at 298.15 K

Molecule	C_p^m (J mol^{-1} K^{-1})
HI	29.158
O_2	29.355
P_2	32.05
HCl	29.12
HF	29.13
HBr	29.142
H_2	28.824
HD	29.196

Table 7.8 Constant-volume heat capacities for some solids at 298.15 K. Heat capacities are given m units of J mol^{-1}K^{-1}.

Solid	C_v^m	$C_v^m/3R$
Al	24.35	0.976
Be	16.44	0.659
Bi	25.52	1.023
Ca	25.31	1.015
C	8.527	0.342
Cr	23.35	0.936
Cu	23.35	0.936
Fe	25.10	1.006
Mg	24.89	0.998
Ag	25.351	1.016
Zn	25.40	1.018

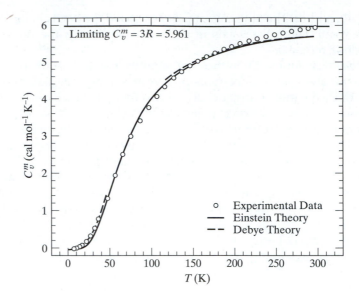

▶ FIGURE 7.22
The constant-volume heat capacity of Mg(s) from 12 K to 298.15 K. The solid points are the measured data. The solid and dashed curves are the predictions of the Einstein and Debye heat capacity theories, respectively.

7.3.2   Einstein Theory

The low-temperature deviations of the measured heat capacity of Mg(s) and other solids from the Dulong–Petit prediction of $3R$ are the result of quantum mechanics, which requires that molecular energies have discrete values rather than a continuum of values. This situation is generally described by the statement that molecular energy states are *quantized*. To obtain a more accurate description of crystal heat capacities, we will, therefore, need to introduce two quantum mechanical results in advance of the detailed discussions and derivations contained in the latter half of this text.

In 1923, de Broglie postulated that matter needs to be described by a wave theory similar to that used for electromagnetic radiation, rather than the particulate theory employed by Newtonian physics. He further suggested that the momentum is related to the wavelength of the wave describing matter by the relationship

$$P = \frac{h}{\lambda},$$ (7.16)

where h is Planck's constant, with a value of 6.62608×10^{-34} J s, and λ is the wavelength. Since a joule has units of kg m²s⁻², Eq. 7.16 can be seen to be unitwise correct in that h/λ has units of kg m s⁻¹, which are the correct units for momentum. The wavelength is related to the wave's frequency ν by $\lambda^{-1} = \nu/c$, where c is the speed of light in a vacuum. Equation 7.16 may, therefore, be written as

$$P = \frac{h\nu}{c}.$$ (7.17)

Equation 7.17 is the first of the two quantum results we require to treat solid-state heat capacity properly.

We shall show in the second half of the text that, once the decision is made to treat matter as a wave, the quantization of energy in a bound molecular system is an inescapable consequence of that decision. We shall also show that a vibrational degree of freedom oscillating with a frequency ν at temperature T has an average internal energy

$$U = \frac{h\nu}{2} + \frac{h\nu}{\exp[h\nu/k_bT] - 1}$$ (7.18)

rather than k_bT as suggested by classical equipartition theory. Equation 7.18 is the second quantum mechanical result needed to treat vibrational heat capacities of solids. We present these two results here without proof or explanation; both will be provided in subsequent chapters.

The second term in Eq. 7.18 approaches zero as the temperature approaches 0 K, and the internal energy becomes $U = h\nu/2$ in this limit. Consequently, the energy present in a vibrational degree of freedom can never be zero, even at absolute zero. For this reason, the first term in Eq. 7.18 is called the *zero-point* vibrational energy.

Einstein used Eq. 7.18 to obtain a simple description of solid-state heat capacities. He assumed that all $3N$ vibrational degrees of freedom in one mole of the crystal are vibrating with the same fundamental frequency ν_o. Under this condition, the total internal energy of the crystal is just that given by Eq. 7.18 multiplied by $3N$, with ν replaced by ν_o. That is,

$$U_{\text{total}} = \frac{3Nh\nu_o}{2} + \frac{3Nh\nu_o}{\exp[h\nu_o/k_bT] - 1}. \tag{7.19}$$

With the internal energy described by Eq. 7.19, the molar heat capacity is

$$C_v^m = \left(\frac{\partial U}{\partial T}\right)_V = \frac{-3Nh\nu_o}{\{\exp[h\nu_o/k_bT] - 1\}^2}\left[-\frac{h\nu_o}{k_bT^2}\right]\exp\left[\frac{h\nu_o}{k_bT}\right]$$

$$= \frac{3Nh^2\nu_o^2}{k_bT^2\{\exp[h\nu_o/k_bT] - 1\}^2}\exp\left[\frac{h\nu_o}{k_bT}\right]. \tag{7.20}$$

Note that the zero-point energy term makes no contribution to the heat capacity of the crystal. Multiplying numerator and denominator of Eq. 7.20 by k_b produces

$$C_v^m = \frac{3Nk_b}{\{\exp[h\nu_o/k_bT] - 1\}^2}\left[\frac{h\nu_o}{k_bT}\right]^2\exp\left[\frac{h\nu_o}{k_bT}\right]. \tag{7.21}$$

We now define the variable $\theta \equiv h\nu_o/k_b$. Since $h\nu_o/(k_bT)$ is unitless, θ has units of temperature. It is therefore called the *characteristic temperature*. Inserting this definition into Eq. 7.21 and replacing Nk_b with R, we obtain

$$\boxed{C_v^m = \frac{3R}{\{\exp[\theta/T] - 1\}^2}\left[\frac{\theta}{T}\right]^2\exp\left[\frac{\theta}{T}\right].} \tag{7.22}$$

Equation 7.22 is the Einstein expression for the molar heat capacity of a crystal.

It is instructive to examine the limiting values of the Einstein heat capacity at high and low temperatures. To examine the limit as $T \rightarrow 0$ K we begin by rewriting Eq. 7.22 by expanding the denominator and dividing numerator and denominator by $\exp[\theta/T]$. These operations yield

$$C_v^m = \frac{3R\theta^2T^{-2}}{\exp[\theta/T] - 2 + \exp[-\theta/T]}. \tag{7.23}$$

As T approaches zero, θ/T becomes infinite, so that $\exp[-\theta/T]$ approaches zero. Hence, the limiting value of C_v^m is

$$\text{Lim}_{T \rightarrow 0} C_v^m = \text{Lim}_{T \rightarrow 0}\left[\frac{3R\theta^2T^{-2}}{\exp[\theta/T] - 2 + \exp[-\theta/T]}\right]$$

$$= 3R \text{ Lim}_{T \rightarrow 0}\left[\frac{(\theta/T)^2}{\exp[\theta/T] - 2}\right] = 3R \text{ Lim}_{T \rightarrow 0}\left[\frac{(\theta/T)^2}{\exp[\theta/T]}\right] = 0, \tag{7.24}$$

since 2 is negligible relative to $\exp[\theta/T]$ as T goes to zero. Both the numerator and denominator of Eq. 7.24 become infinite in the limit as $T \to 0$, but the exponential increases much faster than $(\theta/T)^2$. Consequently, the limiting value is zero, in exact agreement with experiment. Thus, solid-state heat capacities approach zero at low temperatures.

We now wish to examine the limiting value of C_v^m at high temperatures. Writing $x = \theta/T$, we find that Eq. 7.22 becomes

$$C_v^m = \frac{3Rx^2e^x}{\{e^x - 1\}^2}. \tag{7.25}$$

As $T \to \infty$, $x \to 0$. To obtain the limiting value of C_v^m as $T \to \infty$, we need the limit of Eq. 8.25 as x approaches zero. This is most easily obtained by using a Taylor series expansion about $x = 0$ for the exponential. (See Chapter 1 for a review of such expansions.) The result is

$$e^x = 1 + x + \frac{x^2}{2!} + \frac{x^3}{3!} + \cdots + \frac{x^n}{n!} + \ldots . \tag{7.26}$$

Substituting Eq. 7.26 into Eq. 7.27 gives

$$C_v^m = \frac{3Rx^2\left[1 + x + \dfrac{x^2}{2!} + \dfrac{x^3}{3!} + \cdots + \dfrac{x^n}{n!} + \cdots\right]}{\left[x + \dfrac{x^2}{2!} + \dfrac{x^3}{3!} + \cdots + \dfrac{x^n}{n!} + \cdots\right]^2}.$$

In the limit of small x, $x \gg x^2$ and higher powers of x. Therefore, the limiting value of C_v^m is

$$\lim_{T \to \infty} C_v^m = \lim_{x \to 0}\left[\frac{3Rx^2}{(x)^2}\right] = 3R. \tag{7.27}$$

Equation 7.27 shows that the Einstein expression for the solid-state heat capacity approaches the value predicted by Dulong and Petit in the limit of high temperature. As can be seen from the data in Figure 7.22, this is the observed result.

The preceding analysis demonstrates that the Einstein expression has the correct limiting values for C_v^m. The only remaining question is how well it describes the measured heat capacities at intermediate temperatures. This can be determined by executing a least-squares fit of Eq. 7.22 to the experimental data, using θ as the adjustable parameter. The solid curve in Figure 7.22 shows the result of this calculation for Mg(s). The fit yields a characteristic temperature of 230.39 K for Mg(s). As can be seen, the fit is reasonably accurate. The major defects occur at high temperatures, where the heat capacity does not rise to its limiting value rapidly enough, and at very low temperatures, where the Einstein result falls off too rapidly toward zero.

EXAMPLE 7.5

T (K)	C_v^m (J mol^{-1} K^{-1})
15	1.474
30	7.375
70	8.31
130	22.97

The measured heat capacities of Au at four temperatures are as shown in the table to the left.
(A) Obtain an expression for the sum of the squared differences between the data given in this table and the results predicted by the Einstein heat capacity equation.
(B) Plot the expression obtained in (A) as a function of the characteristic temperature. **(C)** Find the value of the characteristic temperature θ that minimizes the value of the expression plotted in (B). **(D)** Using this value of θ, plot the Einstein heat capacity as a function of T, and compare your curve with the data in the table.

Solution

(A) Let $C_v^{\text{expt}}(T)$ represent the experimental data given in the table and $C_v^E(T)$ the Einstein prediction of the heat capacity. The desired sum is

$$S = \sum_{i=1}^{i=4} [C_v^{\text{expt}}(T_i) - C_v^E(T_i)]^2 = \sum_{i=1}^{i=4} \left[C_v^{\text{expt}}(T_i) - \frac{3R(\theta/T_i)^2 \exp[\theta/T_i]}{\{\exp[\theta/T_i] - 1\}^2} \right]^2, \quad \textbf{(1)}$$

where the summations run over the four data points given in the table.

(B) and **(C)** Figure 7.23 shows the variation of S in Eq. 1 with the value of the characteristic temperature for Au(s). As can be seen, the best fit of the Einstein formula to the measured data occurs with $\theta = 124.98$ K. At this point, the value of S is 2.207, which corresponds to the a root-mean-square deviation of ± 0.43 J mol^{-1} K^{-1}.

(D) Figure 7.24 compares the predictions of Einstein's theory (solid line) with the measured data (points) for Au(s), using $\theta = 124.98$ K. The points are the data given in the table.

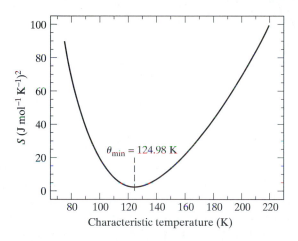

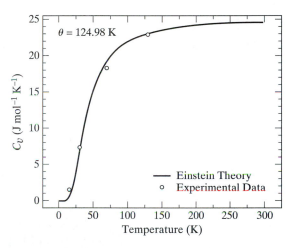

▲ **FIGURE 7.23**

Sum of the squares of the deviations of the computed Einstein heat capacity for Au(s) from the measured data as a function of the characteristic temperature for Au(s). The result shows that the best characteristic temperature for Au(s) is $\theta = 124.98$ K.

For a related example, see Problem 7.14.

▲ **FIGURE 7.24**

Comparison of the predicted Einstein heat capacity (solid curve) for Au(s) with the experimental data (points). The computation are carried out using a characteristic temperature of 124.98 K for Au(s).

7.3.3   Debye Theory

The excellence of the fit between Einstein's theory and the measured heat capacity data clearly demonstrates that the major flaw in the empirical law suggested by Dulong and Petit is the failure to take quantum effects into account. Dulong and Petit could not have done this, since quantum theory was not formulated until the late 1920s. Nevertheless, the fact that the calculated results using Eq. 7.22 decrease too rapidly at low temperatures shows that the Einstein formulation contains some flaws. One of these was pointed out by Debye.

The clue to the problem is contained in the definition of the characteristic temperature and the form of the exponent in the equation. The analysis resulting in Eq. 7.24 shows that C_v^m goes to zero as the temperature approaches

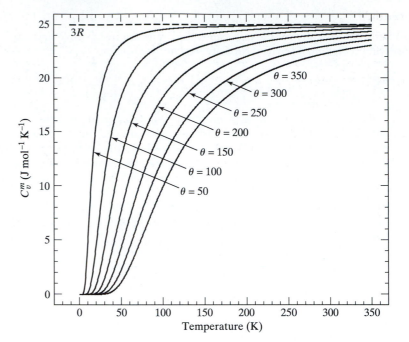

▶ FIGURE 7.25
The predicted Einstein constant-volume heat capacity as a function of temperature for several values of the characteristic temperature in the range $50\ K \le \theta \le 350\ K$. The heat capacity falls off too slowly with temperature when θ is small and approaches zero too rapidly when θ is large. The Debye theory corrects these deficiencies to a large extent.

zero. Since $\theta/T = h\nu/k_bT$, decreasing T is equivalent to increasing ν or θ. Therefore, C_v^m will also approach zero if we make ν large. In the Einstein theory, all $3N$ vibrational degrees of freedom in the crystal are assumed to vibrate with exactly the same frequency. If this common frequency is taken to be large, the calculated heat capacity will approach zero too rapidly. On the other hand, if a small frequency is used, the computed heat capacity will not decrease fast enough at higher temperatures. These problems are illustrated in Figure 7.25, which shows the predicted Einstein heat capacity for various characteristic temperatures. When θ and ν are large, C_v^m approaches zero very rapidly at low temperatures. Small vaues of θ and ν yield heat capacities that decrease very slowly with decreasing temperature. For example, with $\theta = 50$ K, we do not observe a significant decrease in C_v^m until the temperature falls below 100 K.

Debye's solution to this problem was to remove the Einstein assumption that all vibrational degrees of freedom in the crystal had to exhibit the same frequency. Instead, he postulated that so many vibrational degrees of freedom are present, that they exhibit a continuous range of frequencies from zero to a maximum value ν_{max}. This assumption permits both high and low frequencies to be present, which will tend to correct the deviations of Einstein's theory from experiment at both high and low temperatures.

We may develop the Debye expression for the solid-state heat capacity in the same manner as we developed the Einstein equation. The first step is to obtain the expression for the total internal energy of the $3N$ vibrational degrees of freedom. The partial derivative of this quantity with respect to T at constant volume then gives C_v^m. However, we immediately encounter a problem when we attempt to execute this derivation: We can no longer simply multiply Eq. 7.18 by 3N to obtain U_{total}, since not all of the vibrational degrees of freedom have the same frequency. Instead, we must add the contributions of those vibrators with frequency ν' to those with frequency ν'' and so on, up to the maximum frequency ν_{max}. This procedure is analogous to computing the sum of the grades made by a large group of students on an examination: We multiply the exam score by the number of students who made that score and add that product to similar products for all scores from 0 to the maximum score of 100. A similar process gives us the total internal energy for the $3N$ vibrational degrees of freedom in our crystal. That is, we must execute the summation

$$U_{total} = \text{(number of vibrators with frequency } \nu') \quad \times \text{(energy of vibrators with frequency } \nu')$$

$$+ \quad \text{(number of vibrators with frequency } \nu'') \quad \times \text{(energy of vibrators with frequency } \nu'')$$

$$+ \quad \text{(number of vibrators with frequency } \nu''') \quad \times \text{(energy of vibrators with frequency } \nu''')$$

$$+ \cdots + \text{(number of vibrators with frequency } \nu_{max}) \times \text{(energy of vibrators with frequency } \nu_{max}).$$

We can write this summation in the form

$$U_{total} = \sum_{\nu=0}^{\nu=\nu_{max}} g(\nu)U(\nu), \tag{7.28}$$

where $g(\nu)$ is the number of vibrational degrees of freedom that have frequency ν and $U(\nu)$ is the average internal energy of vibrators with frequency ν given by Eq. 7.18. The quantity $g(\nu)$ is called the *degeneracy*; it is a quantity that we shall encounter repeatedly in our study of quantum mechanics and statistical mechanics later in the text.

Equation 7.28 describes what must be done to obtain the total internal energy for the $3N$ vibrational degrees of freedom in the crystal. If, however, it is assumed that the frequencies form a continuous distribution, as opposed to a discrete set, the summation must be replaced with an integral. That is, for the continuous distribution of vibrational frequencies that Debye assumed to be present, Eq. 7.28 becomes

$$U_{total} = \int_{\nu=0}^{\nu=\nu_{max}} g(\nu)U(\nu)d\nu. \tag{7.29}$$

$U(\nu)$ is given by Eq. 7.18. However, we must still determine the degeneracy $g(\nu)$ before we integrate. We shall obtain the degeneracy by using the de Broglie relationship between momentum and frequency.

Consider a single particle moving in three-dimensional space with momentum components P_x, P_y, and P_z. The total momentum is a vector whose magnitude is given by

$$P = [P_x^2 + P_y^2 + P_z^2]^{1/2}. \tag{7.30}$$

The number of ways a vector of length P can exist is illustrated in Figure 7.26. A vector of magnitude P may be drawn at any pair of spherical polar angles θ and ϕ, as shown in the figure. If all such orientations are drawn, the loci of points denoting the end of the vector will form a sphere of radius P. The number of ways we might observe a vector of magnitude P is, therefore, proportional to the surface area of this sphere, which is $4\pi P^2$. We expect that the number of systems that exhibit the momentum P will be proportional to this quantity. That is,

$$g(P) \propto 4\pi P^2. \tag{7.31}$$

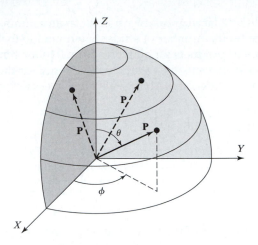

Consequently, the degeneracy of a momentum vector with magnitude P is

$$g(P) = 4\pi C P^2, \tag{7.32}$$

where C is a constant. If we now replace the momentum with the de Broglie expression given in Eq. 7.17, we obtain:

$$g(\nu) = 4\pi C \left[\frac{h\nu}{c} \right]^2 = K\nu^2, \tag{7.33}$$

where K is a constant.

We may obtain the constant K by taking advantage of the fact that the sum of all the degeneracies must be the number of degrees of vibrational freedom in the crystal, $3N$. That is, we must have

$$3N = \int_{\nu=0}^{\nu=\nu_{max}} g(\nu)d\nu = K\int_{\nu=0}^{\nu=\nu_{max}} \nu^2 d\nu = \frac{K}{3}\nu^3 \Big|_0^{\nu_{max}} = \frac{K\nu_{max}^3}{3}. \tag{7.34}$$

This gives $K = 9N/\nu_{max}^3$.

Substituting of Eqs. 7.18, 7.33, and 7.34 into Eq. 7.29 gives the desired integral expression for the total internal energy of the crystal:

$$U_{total} = \frac{9N}{\nu_{max}^3} \int_{\nu=0}^{\nu=\nu_{max}} \left[\frac{h\nu^3}{2} + \frac{h\nu^3}{\exp[h\nu/k_b T] - 1} \right] d\nu$$

$$= \frac{9Nh\nu_{max}}{8} + \frac{9Nh}{\nu_{max}^3} \int_{\nu=0}^{\nu=\nu_{max}} \left[\frac{\nu^3}{\exp[h\nu/k_b T] - 1} \right] d\nu. \tag{7.35}$$

Making the substitution $x = h\nu/(k_b T)$, which gives $dx = (h/k_b T)d\nu$, in Eq. 7.35 produces

$$U_{total} = \frac{9Nh\nu_{max}}{8} + \frac{9Nh}{\nu_{max}^3}\left[\frac{k_b T}{h} \right]^4 \int_{x=0}^{x=\frac{h\nu_{max}}{k_b T}} \left[\frac{x^3}{e^x - 1} \right] dx. \tag{7.36}$$

Substitution of the Debye temperature $\theta = h\nu_{max}/k_b$ into Eq. 7.36 after multiplying the numerator and denominator of the first term by k_b and replacing Nk_b with R yields

$$U_{total} = \frac{9R\theta}{8} + \frac{9RT^4}{\theta^3} \int_{x=0}^{x=\frac{\theta}{T}} \left[\frac{x^3}{e^x - 1} \right] dx. \tag{7.37}$$

The definition of the Debye temperature is similar to that of the characteristic temperature used in the Einstein formulation. However, the Debye temperature is related to the maximum frequency of a continuous range of vibration frequencies, whereas the Einstein characteristic temperature is defined in terms of a common frequency for all oscillators. The two are, therefore, not the same.

Before integrating Eq. 7.37, let us examine the low-temperature limit of the expression. As $T \to 0$ K, the upper limit of the integral approaches infinity. In this limit, we have

$$\lim_{T \to 0 \, K} U_{\text{total}} = \frac{9R\theta}{8} + \frac{9RT^4}{\theta^3} \int_{x=0}^{x=\infty} \left[\frac{x^3}{e^x - 1} \right] dx. \tag{7.38}$$

We now need to expand the denominator of the integrand in a power series in e^{-x}. This can easily be done by defining $y = e^{-x}$, so that $e^x = y^{-1}$. The denominator of the integrand in Eq. 7.38 may now be written as

$$\left[\frac{1}{e^x - 1} \right] = \frac{1}{1/y - 1} = \frac{y}{1 - y}.$$

If we define the function $F(y) = y + y^2 + y^3 + y^4 + \dots$, we can express $F(y)$ in the form

$$F(y) = y[1 + y + y^2 + y^3 + y^4 + \cdots] = y[1 + F(y)].$$

Solving this equation for $F(y)$, we obtain $F(y) = y/(1 - y)$, which is exactly the expression for $1/(e^x - 1)$. This result permits us to write

$$F(y) = y + y^2 + y^3 + \cdots + y^n + \cdots$$

$$= e^{-x} + e^{-2x} + e^{-3x} + e^{-4x} + \cdots = \left[\frac{1}{e^x - 1} \right]. \tag{7.39}$$

Substitution of Eq. 7.39 into Eq. 7.38 produces

$$\lim_{T \to 0} U_{\text{total}} = \frac{9R\theta}{8} + \frac{9RT^4}{\theta^3} \int_{x=0}^{x=\infty} x^3 [e^{-x} + e^{-2x} + e^{-3x} + \cdots] \, dx. \tag{7.40}$$

The integral in Eq. 7.40 may be executed in closed form by using the general result

$$\int_0^\infty x^n e^{-ax} dx = \frac{n!}{a^{n+1}} \text{ for } a \geq 0 \text{ and integer } n. \tag{7.41}$$

Equation (7.41) is an extremely useful result that we shall have occasion to use frequently in later sections of the text. Applying it now to Eq. 7.40 gives

$$\lim_{T \to 0} U_{\text{total}} = \frac{9R\theta}{8} + \frac{9RT^4}{\theta^3} 3![1 + 2^{-4} + 3^{-4} + 4^{-4} + \cdots + n^{-4} + \cdots]. \tag{7.42}$$

The series contained in Eq. 7.42 converges very rapidly. After 10 terms, the sum is $1.08203658\ldots$, after 50 terms, we obtain $1.08232064597\ldots$, and after 100 terms, the result is $1.0823229053\ldots$. By summing certain Fourier series, it can be shown that this sum converges to $\pi^4/90$. Consequently, the limiting value of U_{total} at low temperatures is

$$\lim_{T \to 0} U_{\text{total}} = \frac{9R\theta}{8} + \frac{3\pi^4 RT^4}{5\theta^3}.$$

The limiting low-temperature Debye heat capacity is, therefore,

$$\lim_{T \to 0} C_v^m = \left(\frac{\partial U_{\text{total}}}{\partial T} \right)_V = \frac{12\pi^4 R}{5} \frac{T^3}{\theta^3} = 1{,}943.7 \left[\frac{T}{\theta} \right]^3 \text{ J mol}^{-1} \text{ K}^{-1}. \tag{7.43}$$

Once again, we see that the first term arising from the contribution of the zero-point vibrational energy makes no contribution to the heat capacity. Equation 7.43 shows that we expect the solid-state heat capacity to vary as T^3 at low temperatures. This result was previously introduced in Chapter 4 to permit third-law calculations of the absolute entropy.

To obtain a useful expression for the heat capacity at higher temperatures, we must perform the integration indicated in Eq. 7.37. This integration cannot be done in closed analytic form, but we can obtain a sufficiently accurate result by first expanding the integrand in a power series in x. That is, we write

$$F(x) = \frac{x^3}{e^x - 1} = \sum_{n=0}^{n=\infty} a_n x^n.$$

The expansion coefficients can be obtained in the usual manner. (See Chapter 1.) However, the easiest way to execute this expansion is to first expand e^x in a power series to obtain $e^x = 1 + x + x^2/2! + x^3/3! + \cdots + x^n/n! + \ldots$. Accordingly, we write $F(x)$ in the form

$$F(x) = \frac{x^3}{x + x^2/2 + x^3/3! + x^4/4! + \cdots}$$

and expand by performing long division as follows:

$$
x + x^2/2! + x^3/3! + x^4/4! + x^5/5! + x^6/6! + x^7/7!\overline{)x^3}
$$

$$x^2 - x^3/2 + x^4/12 - x^6/720 + x^8/30{,}240 + \ldots.$$

$$-(x^3 + x^4/2 + x^5/6 + x^6/24 + x^7/120 + x^8/720 + x^9/5{,}040 + \cdots)$$

$$-x^4/2 - x^5/6 - x^6/24 - x^7/120 - x^8/720 - x^9/5{,}040 - \cdots$$

$$-(-x^4/2 - x^5/4 - x^6/12 - x^7/48 - x^8/240 - x^9/1{,}440 - \cdots)$$

$$x^5/12 + x^6/24 - x^7/80 - x^8/360 + x^9/2{,}016 + \cdots$$

$$-x^5/12 + x^6/24 + x^7/72 + x^8/288 + x^9/1{,}440 + \cdots)$$

$$-x^7/720 - x^8/1{,}440 - x^9/5040 - \cdots$$

$$-(-x^7/720 - x^8/1{,}440 - x^9/4{,}320 - \cdots)$$

$$+ x^9/30{,}240 + \cdots$$

The desired expansion is, therefore,

$$F(x) = \frac{x^3}{e^x - 1} = x^2 - \frac{x^3}{2} + \frac{x^4}{12} - \frac{x^6}{720} + \frac{x^8}{30{,}240} + \ldots. \tag{7.44}$$

Substituting the expansion from Eq. 7.44 into Eq. 7.37 yields

$$U_{\text{total}} = \frac{9R\theta}{8} + \frac{9RT^4}{\theta^3} \int_{x=0}^{x=\frac{\theta}{T}} \left[x^2 - \frac{x^3}{2} + \frac{x^4}{12} - \frac{x^6}{720} + \frac{x^8}{30{,}240} + \cdots \right] dx.$$

The integration in this equation may be easily carried out to obtain

$$U_{\text{total}} = \frac{9R\theta}{8} + \frac{9RT^4}{\theta^3} \left[\frac{\theta^3}{3T^3} - \frac{\theta^4}{8T^4} + \frac{\theta^5}{60T^5} - \frac{\theta^7}{5040T^7} + \frac{\theta^9}{272160T^9} + \cdots \right]$$

$$= \frac{9R\theta}{8} + 3R \left[T - \frac{3\theta}{8} + \frac{\theta^2}{20T} - \frac{\theta^4}{1{,}680T^3} + \frac{\theta^6}{92{,}720T^5} + \cdots \right]. \tag{7.45}$$

The heat capacity is now obtained by taking the partial derivative of U_{total} with respect to T. This operation gives

$$C_v^m = 3R\left[1 - \frac{\theta^2}{20T^2} + \frac{\theta^4}{560T^4} - \frac{\theta^6}{18{,}544T^6} + \cdots\right]. \qquad (7.46)$$

Equation 7.46 provides a simple expression for solid-state heat capacities at temperatures in the range $T > \theta/2$. At temperatures below $\theta/2$, the expansion must be extended to include higher order terms. (See Problem 7.19.)

An examination of Eq. 7.46 shows that the Debye heat capacity approaches $3R$ as T becomes large. This again emphasizes the fact that the law of Dulong and Petit is a high-temperature limiting form valid only in the range of temperatures for which quantum mechanical effects become negligible.

Figure 7.22 compares the Debye and Einstein results for the heat capacities for Mg(s). A fit of the low-temperature limiting form given by Eq. 7.43 yields a Debye temperature of 343.7 K. The Debye results for temperatures in excess of 120 K were computed using Eq. 7.46 with a Debye temperature of 294.2 K. The principal advantage of the Debye formulation lies in its greater accuracy at low temperatures. Figure 7.27 shows an expanded plot of the low-temperature results for Mg(s). The superior fit of the Debye theory is obvious.

Although the Einstein and Debye formulations yield C_v^m, the results are essentially unchanged for C_p^m. For gases, C_p^m exceeds C_v^m because, at constant pressure, work is done when the gas expands upon heating. Energy must be added to do this work; therefore, the energy required to raise the temperature of the gas when heating at constant pressure is greater than that needed for constant-volume heating. For condensed phases, however, the coefficients of thermal expansion are very small, so the work done by expanding the volume is negligible. This makes C_v^m and C_p^m nearly identical.

7.3.4   Application to Gas-phase Heat Capacities

The previous discussion shows that the empirical law suggested by Dulong and Petit is reasonably accurate for translational and rotational degrees of freedom, but not for vibrational motions. This observation suggests that

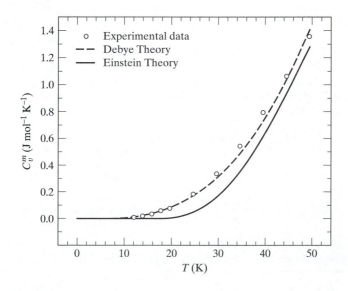

◀ FIGURE 7.27
The constant-volume heat capacity of Mg(s) from 12 K to 50 K. The solid points are the measured data. The solid and dashed curves are the predictions of the Einstein and Debye heat capacity theories, respectively. The Einstein result is obtained using a characteristic temperature of 230.39 K. The Debye temperature used in the calculation is 343.7 K.

more accurate gas-phase heat capacities might be obtained by using the classical equipartition theorem for translations and rotations and the Einstein quantum statistical result for vibrational degrees of freedom.

Let us first consider diatomic molecules with three translational, two rotational, and one vibrational degree of freedom as illustrated in Figure 7.21. If we proceed as suggested in the preceding paragraph, we will find that the total internal energy for 1 mole will be

$$U_{total} = \frac{5RT}{2} + \frac{Nh\nu}{2} + \frac{Nh\nu}{\exp[h\nu/k_bT] - 1}. \tag{7.47}$$

For gas-phase molecules, the Einstein assumption that all vibrational degrees of freedom have the same frequency will be exact. Therefore, we can simply multiply the quantum statistical expression for the vibrational energy of a vibrator with frequency ν by Avogadro's number to obtain the energy for one mole of such vibrators.

With the total internal energy given by Eq. 7.47, the constant-volume heat capacity will be

$$C_v^m = \left(\frac{\partial U_{total}}{\partial T}\right)_V = \frac{5R}{2} + \frac{R}{\{\exp[\theta/T] - 1\}^2}\left[\frac{\theta}{T}\right]^2 \exp[\theta/T], \tag{7.48}$$

where the factor of 3 appearing in Eq. 7.22 has been dropped, since we now have only N vibrational degrees of freedom for one mole of gas-phase diatomic molecules. If we replace θ/T with x, Eq. 7.48 becomes

$$C_v^m = 2.5R + \frac{Rx^2 e^x}{\{e^x - 1\}^2}. \tag{7.49}$$

Since we are now in the gas phase, C_p^m will exceed C_v^m by R if the gas behaves ideally. Therefore,

$$C_p^m = 3.5R + \frac{Rx^2 e^x}{\{e^x - 1\}^2}. \tag{7.50}$$

EXAMPLE 7.6

Let us test the accuracy of Eq. 7.50 by computing C_p^m for $P_2(g)$ at 298.15 K. As the data in Table 7.7 show, the measured result is 32.05 J mol^{-1} K^{-1}. If we use classical equipartition theory for all degrees of freedom, we obtain $C_p^m = 4.5R = 37.413$ J mol^{-1} K^{-1}. If we simply ignore the vibrational degree of freedom, the result is $C_p^m = 3.5R = 29.099$ J mol^{-1} K^{-1}. Neither of these results is very good. Using spectroscopic methods that will be discussed in detail in Chapter 15, we may easily measure the vibrational frequency of $P_2(g)$ molecules. The result is $\nu = 2.339 \times 10^{13}$ s^{-1}. With the use of Eq. 7.50, we obtain

$$x = \frac{h\nu}{k_bT} = \frac{(6.62608 \times 10^{-34} \text{ J s})(2.339 \times 10^{13} \text{ s}^{-1})}{(1.38066 \times 10^{-23} \text{J K}^{-1})(298.15 \text{ K})} = 3.7650, \tag{1}$$

so that

$$C_p^m = 3.5R + \frac{R(3.7650)^2 e^{3.7650}}{(e^{3.7650} - 1)^2} = 3.844R = 31.96 \text{ J mol}^{-1} \text{ K}^{-1}. \tag{2}$$

This result is excellent: The percent error is -0.28%.

For related exercises, see Problems 7.21, 7.22, and 7.23.

A word of caution is needed at this point. Our assumption that rotational degrees of freedom may be treated classically is quite accurate at temperatures at or above 298 K. As T decreases, however, it becomes less accurate because quantum effects manifest themselves in the rotational motions as well as in the vibrational degrees of freedom. These effects will be explored in more detail later in the text.

Let us now consider polyatomic molecules. In the gas phase, all such molecules will have three degrees of translational freedom for center-of-mass motion. Linear molecules have two degrees of rotational freedom, since there are two axes about which rotation can occur, as illustrated in Figure 7.21. Nonlinear polyatomic molecules have three degrees of rotational freedom, because there are three possible axes of rotation. Since the total number of degrees of freedom for a molecule having N atoms is $3N$, linear molecules will have $3N - 5$ vibrational degrees of freedom, while nonlinear polyatomic molecules will possess $3N - 6$ vibrational motions. Each of the vibrational modes has its own characteristic vibrational frequency. To obtain the total internal energy of the molecule, we must sum over all these degrees of freedom.

The same analysis used for diatomics leads to the result

$$U_{\text{total}} = \frac{5k_bT}{2} + \sum_{i=1}^{3N-5} \left[\frac{h\nu_i}{2} + \frac{h\nu_i}{\exp[h\nu_i/k_bT] - 1} \right] \tag{7.51}$$

for linear molecules and

$$U_{\text{total}} = 3k_bT + \sum_{i=1}^{3N-6} \left[\frac{h\nu_i}{2} + \frac{h\nu_i}{\exp[h\nu_i/k_bT] - 1} \right] \tag{7.52}$$

for nonlinear polyatomic molecules. For one mole of molecules, we simply multiply Eqs.7.51 and 7.52 by Avogadro's constant to obtain

$$U_{\text{total}} = \frac{5RT}{2} + \sum_{i=1}^{3N-5} \left[\frac{Nh\,\nu_i}{2} + \frac{Nh\nu_i}{\exp[h\nu_i/k_bT] - 1} \right], \tag{7.53}$$

for linear molecules and

$$U_{\text{total}} = 3RT + \sum_{i=1}^{3N-6} \left[\frac{Nh\nu_i}{2} + \frac{Nh\nu_i}{\exp[h\nu_i/k_bT] - 1} \right], \tag{7.54}$$

for nonlinear molecules.

The heat capacities are obtained by taking the partial derivatives of these expressions with respect to temperature at constant volume. The results are

$$C_v^m = 2.5R + \sum_{i=1}^{3N-5} \left[\frac{Rx^2e^x}{\{e^x - 1\}^2} \right]_i \tag{7.55}$$

for linear molecules and

$$C_v^m = 3R + \sum_{i=1}^{3N-6} \left[\frac{Rx^2e^x}{\{e^x - 1\}^2} \right]_i \tag{7.56}$$

for nonlinear molecules. The corresponding values of C_p^m can be obtained by adding R to the result for C_v^m.

EXAMPLE 7.7

The spectroscopically measured vibration frequencies for H_2O are $\nu_1 = 1.1257 \times 10^{14}$ s^{-1}, $\nu_2 = 1.0945 \times 10^{14}$ s^{-1}, and $\nu_3 = 4.7802 \times 10^{13}$ s^{-1}. Compute C_p^m at 298.15 K for $H_2O(g)$.

Solution

Water is nonlinear, so we expect $3(3) - 6 = 3$ vibrational degrees of freedom. This is exactly what is found spectroscopically. From the vibration frequencies of the three modes just given, we will have three terms in the summation in Eq. 7.56. The values of the x_i are

$$x_1 = \frac{h\nu_1}{k_bT} = \frac{(6.62608 \times 10^{-34}\text{J s})(1.1257 \times 10^{14}\text{s}^{-1})}{(1.38066 \times 10^{-23}\text{J K}^{-1})(298.15\text{ K})} = 18.120, \qquad (1)$$

$$x_2 = \frac{h\nu_2}{k_bT} = \frac{(6.62608 \times 10^{-34}\text{J s})(1.0945 \times 10^{14}\text{ s}^{-1})}{(1.38066 \times 10^{-23}\text{J K}^{-1})(298.15\text{ K})} = 17.62, \qquad (2)$$

and

$$x_3 = \frac{h\nu_3}{k_bT} = \frac{(6.62608 \times 10^{-34}\text{J s})(4.7802 \times 10^{13}\text{ s}^{-1})}{(1.38066 \times 10^{-23}\text{J K}^{-1})(298.15\text{ K})} = 7.695. \qquad (3)$$

Inserting these values into Eq. 7.56 gives

$$C_p^m = 4R + \left[\frac{R(18.120)^2 e^{18.12}}{\{e^{18.12} - 1\}^2}\right] + \left[\frac{R(17.62)^2 e^{17.62}}{\{e^{17.62} - 1\}^2}\right] + \left[\frac{R(7.695)^2 e^{7.695}}{\{e^{7.695} - 1\}^2}\right]$$

$$= 4R + 0.00000444R + 0.00000691R + 0.02697R = 4.027R = 33.48 \text{ J mol}^{-1}\text{K}^{-1}. \quad (4)$$

The measured value is 33.58 J mol^{-1} K^{-1}. Thus, the calculated result is in error by 0.30%.

For related exercises, see Problems 7.21, 7.22, and 7.23.

Summary: Key Concepts and Equations

1. The forces present in a solid vary from extremely strong to relatively weak. Ionic crystals and covalently bonded materials have very strong binding forces that lead to high melting and boiling points. London dispersion forces and dipole–dipole interactions provide the cohesive forces holding organic crystals together. These forces are much weaker than those found in ionic crystals and covalently bonded solids. Consequently, the melting and boiling points of organic solids are usually much lower.

2. The magnitudes of ionic forces are strongly influenced by the ionic radius and the ionization potential. As the ionic radius increases, the distance between the charges increases, thereby reducing the coulombic forces. This effect tends to make ionic crystals involving larger ions have lower boiling and melting points and smaller crystal enthalpies. The ionization potential measures the amount of energy required to form an ion. As the ionization potential increases, the extent of charge transfer tends to decrease, in turn lowering the coulombic force and potential. As a result, ionic crystals involving metals with higher ionization potentials tend to have lower melting and boiling points. Since the ionization potential tends to decrease as the ionic size increases, the two effects are competitive. If the variation in ionic size is the dominant factor, we observe melting and boiling points and crystal enthalpies decreasing as we move down a group in the periodic table. If, however, the change in ionization potential is the most important factor, the opposite trend is seen.

3. London dispersion forces arise because of induced dipoles formed by distortion of the electronic charge clouds under the influence of an electric field. The magnitude of the induced dipole formed by a given electric field is proportional to the polarizability of the system, α. Qualitatively, α increases as the number of electrons in the molecule increases and as the size of the atoms increases. Both of these variables are related to the molar mass of the molecule; consequently, in a homologous series of molecules such as primary alcohols, n-alkanes, etc., we generally observe a strong correlation between molar mass and melting and boiling points.

4. Crystal enthalpies can be conveniently measured by summing the enthalpy changes around the Born–Haber cycle.

5. A crystal system is defined by the relationship between the three unit cell distances and the three angles formed by the primitive translations of the crystal. There are seven crystal systems. The most symmetric is the cubic system, which has all unit cell distances equal and the primitive translational vectors mutually perpendicular. The least symmetric system is the triclinic, in which all unit cell dimensions and the angles between the primitive translations are different, with no angle being equal to 90°. A crystal system may have subclasses, including primitive cubic, body-centered cubic, and face-centered cubic. There are 14 possible subclasses, called *Bravais lattices*.

6. The total coulombic interaction present in a given crystal may be obtained by summing over all possible interaction pairs. When this is done, the resulting potential for a 1:1 system is

$$V_c^T = -\frac{Nq^2}{4\pi d\varepsilon_o} \sum_{j=1}^{2N} \pm dr_{ij}^{-1} = -\frac{N\mathcal{M}q^2}{4\pi d\varepsilon_o}$$

where ion i is any arbitrary ion in the crystal, N is Avogadro's constant, d is the shortest distance between ions in the crystal, ε_o is the permittivity of the vacuum and $\mathcal{M}$ is the Madelung constant, which is equal to the value of the summation. The Madelung constant depends only upon the type of crystal and the details of the lattice parameters.

7. To compute the crystal enthalpy, we must include a repulsive term in the potential in order to account for short-range repulsive forces that arise because of electronic interactions between individual electrons in the charge clouds and quantum effects that cause the kinetic energy to increase when the ions are confined to small spaces. The most commonly used functional forms for these repulsive forces are inverse powers of the distance d and exponential forms. If the short-range repulsive interaction is represented by $V_{\text{rep}}(d) = K/d^n$, where K is a positive constant, the crystal enthalpy for a 1:1 system has the form

$$\Delta H_c = \frac{N\mathcal{M}q^2}{4\pi\varepsilon_o d_o}\left[1 - \frac{1}{n}\right] + 2RT,$$

where d_o is the value of d at equlimibrium.

8. If the classical equipartition theorem is used for vibrational degrees of freedom of a crystal, the solid-state heat capacity is predicted to have a constant value of $3R$. This is the empirical law of Dulong and Petit. Experimental measurements show that the law holds only at higher temperatures.

9. To describe the variation of the solid-state heat capacity with temperature accurately, it is necessary to utilize the average energy of a vibrational degree of freedom obtained by combining a quantum mechanical treatment of vibration with statistical theory. The result is

$$U = \frac{h\nu}{2} + \frac{h\nu}{\exp[h\nu/k_b T] - 1},$$

where ν is the frequency of the vibrational degree of freedom, h is Planck's constant, and k_b is the Boltzmann constant, which is equal to R/N. The first term is called the *zero-point energy* in that the limiting value of the preceding equation as $T \rightarrow 0\ K$ is $h\nu/2$.

10. Einstein assumed that all $3N$ vibrational degrees of freedom in one mole of a crystal have the same vibrational frequency ν_o. The total energy can, therefore, be obtained by multiplying the equation in item 9 by $3N$. Differentiating of the total energy with respect to T then yields the heat capacity. The Einstein result is

$$C_v^m = \frac{3R}{\{\exp[\theta/T] - 1\}^2} \left[\frac{\theta}{T}\right]^2 \exp\left[\frac{\theta}{T}\right],$$

where the value of $\theta = h\nu_o/k_b$, called the *characteristic temperature*, is determined by fitting measured heat capacity data to the foregoing expression. The limiting values of the Einstein heat capacity as the temperature approaches $0\ K$ or a high temperature are zero and $3R$, respectively, in agreement with experimental data for a wide variety of solids.

11. The major problem with the Einstein expression is that it decreases to zero too rapidly at low temperatures. Debye corrected this flaw by assuming that the $3N$ vibrators in the crystal exhibit a continuous distribution of frequencies from 0 to a maximum value, ν_{max}. The total crystal energy is then obtained by summing (i.e., integrating) the energy of a vibrator with frequency ν times the number of vibrators that have this frequency over all possible frequencies, or

$$U_{total} = \int_{\nu=0}^{\nu=\nu_{max}} g(\nu)U(\nu)d\nu,$$

where $g(\nu)$, the number of vibrators that have frequency ν, is called the *degeneracy*.

12. The degeneracy $g(\nu)$ is obtained by using the de Broglie relationship between the momentum P and the wavelength λ of the wave representing the system,

$$P = \frac{h}{\lambda}.$$

By considering the number of ways a momentum vector of length P can exist in a three-dimensional system, the degeneracy is found to be

$$g(\nu) = K\nu^2,$$

where K is a constant.

13. Executing the Debye integral in Item 11 gives a low-temperature limiting form for the heat capacity which shows that C_v^m depends upon T^3 at low temperatures. The result is

$$\underset{T\rightarrow 0}{Lim}\ C_v^m = \frac{12\pi^4R\ T^3}{5\ \theta^3} = 1{,}943.7 \left[\frac{T}{\theta}\right]^3 J\ mol^{-1}K^{-1},$$

where $\theta = h\nu_{max}/k_b$ is the *Debye temperature*. This quantity is similar to, but is not the same as, the Einstein characteristic temperature. The latter is defined in terms of average vibrational frequency, whereas the Debye temperature involves the maximum frequency of a continuous distribution of frequencies. The preceding low-temperature form is considerably more accurate than that predicted by the Einstein formulation.

14. At temperatures greater than half the Debye temperature, the Debye heat capacity is given by

$$C_v^m = 3R\left[1 - \frac{\theta^2}{20T^2} + \frac{\theta}{560T^4} - \frac{\theta^6}{18{,}544\ T^6} + \cdots\right].$$

The fifth term in this expansion is obtained in Problem 7.19. C_v^m is given in Problem 7.20.

15. Accurate values for gas-phase heat capacities above room temperature may be obtained by treating the translational and rotational degrees of freedom via the classical equipartition theorem and by using the quantum statistical theory expression to obtain the contribution of the vibrations. The resulting equation for the gas-phase heat capacity of nonlinear polyatomic molecules is

$$C_v^m = 2.5R + \sum_{i=1}^{3N-6}\left[\frac{Rx^2e^x}{\{e^x-1\}^2}\right]_i.$$

For linear polyatomic molecules, the result is

$$C_v^m = 2.5R + \sum_{i=1}^{3N-5}\left[\frac{Rx^2e^x}{\{e^x-1\}^2}\right]_i$$

where $x = h\nu/k_bT$.

Problems

Problems that require the use of some type of computational device are marked with an asterisk (*). Problems that require some type of plotting routine are indicated with a pound sign (#). Unless otherwise stated, all gases may be assumed to behave ideally.

7.1 A chemist obtains the following data:

Compound	Normal Melting Point (°C)
Na_2O	1275
K_2O	decomposes before melting
CaO	2614
BaO	1918

Using these data alone, answer the following questions and give reasons for your answers:

(A) Is the normal melting point of Li_2O higher or lower than 1,275°C?

(B) Can you provide upper and lower limits for the normal melting point of SrO? If not, why not? If so, provide these limits and state why they are appropriate.

(C) Can we determine whether Li_2O or MgO has the higher normal melting point? Explain.

(D) Can we determine whether Li_2O or BaO has the higher normal melting point? Explain.

7.2* The normal boiling points of various aliphatic acids are given in the following table:

Acid	Normal Boiling Point (K)
formic acid	373.8
acetic acid	391.8
propanoic acid	414.1
pentanoic acid	459.2
hexanoic acid	478.2
heptanoic acid	496.2
nonanoic acid	528.2
decanoic acid	543.2

(A) By plotting these normal boiling points against the molar masses of the acids, determine whether a correlation exists between molar mass and boiling point. Using a least-squares method, find the best straight-line fit to the data.

(B) Use the results obtained in (A) to predict the normal boiling points for butanoic acid and octanoic acid. Compute the percent error in your predictions.

7.3* The normal boiling points of various aliphatic 2-ketones are given in the following table:

Ketone	Normal Boiling Point (K)
acetone	329.4
2-butanone	352.8
2-pentanone	375.2
2-hexanone	401.2

2-heptanone	424.6
2-nonanone	468.4
2-decanone	483.2

(A) By plotting these normal boiling points against the molar masses of the ketones, determine whether a correlation exists between molar mass and boiling point. Using a least-squares method, find the best straight-line fit to the data.

(B) Use the results obtained in (A) to predict the normal boiling point for 2-octanone. Compute the percent error in your prediction.

7.4 Using the data in Appendix A and Table 7.4 and the Born-Haber cycle, compute the crystal enthalpy of AgBr(c). The first ionization potential of Ag(g) is 7.576 eV. The value listed in the 78 edition, of the *Handbook of Chemistry and Physics* is 904 kJ mol^{-1}. Compute the percent difference between this value and your result.

Problems 7.5 and 7.6 explore the reasons we do not observe ionic crystals of HCl.

7.5 The alkali metals all have one electron in a valence *s* orbital. All of them form high-melting ionic crystals with the halogens, whose melting points increase as the atomic number decreases. (See Table 7.1.) Hydrogen also has a single electron in its valence *s* orbital and is listed in Group IA in most periodic tables. However, hydrogen does not form an ionic crystal with any of the halogens. Why not?

7.6 In this problem, you will attempt a more quantitative assessment of why we do not observe ionic crystals of HCl.

(A) Assume that HCl forms a face-centered ionic crystal similar to those of the alkali metal chlorides. The ionic radius of Cl^{1-}, from Table 7.2, is 1.81 Å. H^{+} is a bare proton. Consequently, we will take its ionic radius to be near zero. This suggests that the HCl ionic crystal should have $d \approx 1.81$Å. Use this value to compute the expected crystal enthalpy for a hypothetical ionic HCl crystal from Eq. 7.14.

(B) Using the Born–Haber cycle, the results from (A), the data in Appendix A, and Tables 7.2 and 7.4, compute the standard partial molar enthalpy of an ionic HCl crystal. That is, compute the standard molar heat of formation for such a crystal.

(C) Using absolute entropies from Appendix A, together with the assumption that the absolute entropy at 298.15 K of NaCl(c) is close to that of a hypothetical ionic HCl(c) crystal, compute $\Delta \overline{S}$ for the reaction

$$0.5 \, H_2(g) + 0.5 \, Cl_2(g) \longrightarrow HCl(c).$$

(D) Using the results from (B) and (C), compute $\Delta \mu$ for the preceding reaction. Is $\Delta \mu$ positive or negative? Will we observe the spontaneous formation of HCl(c)?

7.7 (A) Using the method described in the text with a repulsive potential of the form $V_{rep}(d) = K/d^n$ with $n = 11.67$, compute the crystal enthalpy of KCl(c) at 298.15 K.

(B) Use the Born–Haber cycle, along with the result obtained in (A), the data in Appendix A, and Table 7.2, to obtain an estimate of the electron affinity of Cl. Compute the percent error in your estimate. KCl forms a face-centered cubic crystal, with the shortest distance between ions being 3.14×10^{-10} m.

The next three problems explore results obtained using an exponential form for the repulsive interactions when computing crystal enthalpies.

7.8 Equation 7.14 for the crystal enthalpy was obtained by assuming a repulsive term whose form is $V_{rep}(d) = K/d^n$. As noted in the text, we could have equally well used an exponential form, $V_{rep}(d) = B \exp[-d/\rho]$, where ρ and B are parameters. With this assumption, the total potential becomes

$$V(d) = B \exp[-d/\rho] - \frac{N \mathcal{M} q^2}{4 \pi d \varepsilon_0}.$$

Use this form to obtain an analogous equation to eq. 7.14 for the crystal enthalpy of a 1:1 ionic crystal as a function of ρ.

7.9* The crystal enthalpy of a 1:1 ionic crystal is

$$\Delta H_c = \frac{N \mathcal{M} q^2}{4 \pi d_o \varepsilon_o} \left[1 - \frac{\rho}{d_o} \right] + 2RT$$

if an exponential repulsive potential form is assumed. Use a least-squares procedure with the following data at 298.15 K to obtain the best value of the parameter ρ in the foregoing expression:

Crystal	Crystal Enthalpy (kJ mol^{-1}K^{-1})	d_o(m)
NaCl	787	2.814×10^{-10}
KCl	717	3.138×10^{-10}
NaBr	752	2.981×10^{-10}

All of these crystals have face-centered cubic form. Compute ΔH_c for each of these crystals, using the value of ρ determined in the least-squares analysis.

7.10 In Example 7.4, we computed the KBr crystal enthalpy at 298.15 K using a repulsive term of the form $V_{rep}(d) = K/d^n$, with the value of n determined by fitting the data for NaCl, KCl, and NaBr. The result was an error of -0.44%. Using the exponential form for the repulsion given in Problem 7.8, with the value of ρ fitted to the NaCl, KCl, and NaBr data in Problem 7.9, compute the KBr crystal energy. The best value of ρ is 0.246×10^{-10} m. Calculate the percent error in this result. Is the result more or less accurate than that obtained with the inverse–power repulsive form?

7.11* In the text, we computed the value of the Madelung constant for a hypothetical linear crystal. Let us assume that we are in the universe of the Flatlanders, where everything is two dimensional. (See Edwin A. Abbot, *Flatland* (6th edition, Dover, New York, 1952; 1st edition, Seeley & Co., Great Britian, 1884).) In this universe, a common crystal system is the square lattice. One of the subclasses of such a system is the primitive square, which has an ion at each vertex. The following figure illustrates this subclass:

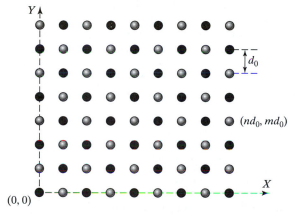

▲ A primitive square lattice in the Flatland universe. The black dots denote positive cations. The lighter dots indicate negatively charged anions in the crystal. The lattice spacing is denoted by d_o.

Calculate the Madelung constant for the square primitive lattice for the case where the anions and the cations have the same magnitude of charge. You will need to use a PC, a workstation, or some type of spreadsheet for this calculation. The various parts of the problem serve as a procedure guide. [*Note to instructor:* If you wish to omit the numerical evaluation of $\mathcal{M}$, just skip Part (E).]

(A) To obtain the Madelung constant, we need to sum all of the interactions with ion i which we shall take to be the ion at the origin of the coordinate system shown in the preceding figure. For interactions involving ions of the same charge, we sum with a negative sign; interactions between ions of different charge we sum with a positive sign. Obtain an equation for the term in the Madelung sum involving ion i and the ion in the figure whose (X, Y) coordinates are (nd_o, md_o). Show that this term is a function of n and m alone.

(B) Show that the sign on the term in the Madelung sum will be negative if $(n + m)$ is even and positive if $(n + m)$ is odd.

(C) Note that the terms in the Madelung sum for the interactions with the ions whose coordinates are (md_o, nd_o) or $(-nd_o, md_o)$ are identical to the term for the ion whose coordinates are (nd_o, md_o). If $n \neq m$ and neither n nor m is zero, how many terms have the same value as the term for the (nd_o, md_o) ion? If $n = m \neq 0$, how many terms have the same value as the term for the (nd_o, md_o) ion? If either n or $m = 0$, how many terms have the same value as the term for the (nd_o, md_o) ion? Let us call these quantities the *degeneracy* for the (nd_o, md_o) lattice point and denote it by the notation $g(n, m)$.

(D) If we multiply the interaction for the (nd_o, md_o) ion by its degeneracy factor $g(n, m)$, we need sum only over ions for which $n \geq 0$ and $m \geq 0$ and for which $m \geq n$. Using this fact and the results of (A), (B), and (C), express the Madelung constant as a double summation over n and m with $m \geq n$.

(E) Use a PC, a workstation, or a spreadsheet to execute the summation obtained in (D) with the upper limits on each sum equal to K. Execute the summation for $K = 500, 750, 1,000, 1,500, 3,000$, and $5,000$. From the results, determine the number of significant digits to which $\mathcal{M}$ has been evaluated for the square primitive lattice.

7.12 In the Flatlander universe, NaCl(c) forms a primitive square crystal. The value of the Madelung constant for this lattice is $\mathcal{M} = 1.6155$. Compute the crystal enthalpy listed in the Flatlander's *Handbook of Chemistry and Physics* for NaCl(c) at 298.15 K. Use an inverse-power repulsive term. You may assume that $n = 11.67$ in the Flatlander's universe and that the shortest distance between ions is 2.814×10^{-10} m, as it is in our universe.

7.13 (A) Use the data in Appendix A and Tables 7.2, 7.4, and 7.6, along with the fact that the first two ionization potentials of Zn are $I_1 = 9.394$ eV and $I_2 = 17.964$ eV, to compute the crystal enthalpy of ZnO at 298.15 K.

(B) Using the results of (A) and Eq. 7.14, determine the shortest distance between the Zn^{2+} and O^{2-} ions in a crystal of ZnO.

7.14* The measured heat capacities of Cu at four temperatures are as follows:

Temperature (K)	C_v^m (J mol^{-1}K^{-1})
30	1.693
50	6.154
70	10.86
150	20.51

330   Chapter 7   Thermodynamics of Solids

(A) Obtain an expression for the sum of the squares of the differences between the data given in the table and the results predicted by the Einstein heat capacity equation.

(B) Plot the expression obtained in (A) as a function of the characteristic temperature.

(C) Find the value of the characteristic temperature that minimizes the value of the expression plotted in (B).

(D) Using the value of θ you obtained in (C), plot the Einstein heat capacity as a function of T, and compare it with the data in the foregoing table. (*Hint:* See Example 7.5.)

7.15 (A) By fitting the low-temperature limiting form of the Debye heat capacity to the measured value at 30 K for Cu given in Problem 7.14, obtain an expression for C_v^m for Cu for temperatures in the range 0 K to 30 K.

(B) Use the expression for C_v^m that you obtained in (A) to compute the absolute entropy of Cu(s) at 30 K.

7.16* (A) The Einstein characteristic temperature for Cu(s) is 224.25 K. Use this value and the Einstein expression for C_v^m to compute $\Delta \bar{S}$ for the constant-volume heating of Cu(s) from 30 K to 298.15 K. (*Hint:* The integration cannot be done analytically; you will need to employ a numerical method.)

(B) The absolute entropy of Cu(s) at 30 K is 0.5643 J mol^{-1}K^{-1}. Use this fact and the result you obtained in (A) to compute the absolute entropy of Cu(s) at 298.15 K. Compare your answer with the value given in Appendix A. Calculate the percent error in the computation.

7.17* The measured heat capacities of Ag at four temperatures are as follows:

Temperature (K)	C_v^m (J mol^{-1}K^{-1})
20	1.647
35	6.612
60	14.27
130	22.07

(A) Obtain an expression for the sum of the squares of the differences between the data given in the table and the results predicted by the Einstein heat capacity equation.

(B) Plot the expression obtained in (A) as a function of the characteristic temperature.

(C) Find the value of the characteristic temperature that minimizes the value of the expression plotted in (B). (*Hint:* See Example 7.5.)

7.18# In later chapters of this text, we shall learn that the vibration frequency ν of a homonuclear diatomic molecule at low vibrational energy is related to the atomic mass M by $\nu = [k/4\pi^2 M]^{1/2}$, where k is called the *vibrational force constant*. This constant measures the magnitude of the force acting between the two atoms. Since the characteristic temperature is directly proportional to the frequency exhibited by the lattice vibrations, and since the melting point of the crystal is related to the forces acting between the atoms, it seems reasonable that there should exist a correlation between the melting point of a crystal and its characteristic Einstein temperature. In the text and the foregoing problems, we found that the characteristic temperatures of Mg, Cu, Ag, and Au are 230.39 K, 224.25 K, 153.0 K, and 124.98 K, respectively. Use these data to explore the possibility of a correlation between the characteristic temperature of a crystal and its melting point. If such a correlation is possible, develop an equation whereby a reasonable value of the characteristic temperature of a crystal may be obtained from a measurement of its melting point. Predict a characteristic temperature for Fe(s). The normal melting point of Fe(s) is 1808.2 K.

The next two problems expand the high-temperature expression for the Debye heat capacity and test its convergence properties.

7.19 Equation 7.46 gives the first four terms of the high-temperature limiting form for the Debye heat capacity. By expanding the derivation given in the text, obtain the fifth term of this expansion.

7.20 The high-temperature expansion of the Debye heat capacity is

$$C_v^m = 3R\left[1 - \frac{\theta^2}{20T^2} + \frac{\theta^4}{560T^4} - \frac{\theta^6}{18,544\,T^6}\right.$$
$$\left. + \frac{\theta^8}{633,600\,T^8} - \cdots\right].$$

In this problem, you will examine the convergence properties of this expansion as a function of temperature to determine the range of temperatures over which the expansion may be used. Let us define T in terms of the Debye temperature by writing $T = f\theta$. In these terms,

$$C_v^m = 3R\left[1 - \frac{1}{20f^2} + \frac{1}{560f^4} - \frac{1}{18,544f^6}\right.$$
$$\left. + \frac{1}{633,600f^8} - \cdots\right].$$

Compute the ratio $C_v^m/3R$ after one, two, three, four, and five terms are added in this expansion, for values of f of 0.30, 0.35, 0.40, 0.45, and 0.50. From your results, determine how many terms are needed to obtain convergence to three significant digits for each value of f.

7.21 By treating translations and rotations by means of the classical equipartition theorem and vibrations with quantum/statistical theory, compute the gas-phase heat capacities of (A) $I_2(g)$ and (B) $Cl_2(g)$ at 298.15 K. The fundamental vibrational frequencies for $I_2(g)$ and $Cl_2(g)$ are $6.431 \times 10^{12}\,s^{-1}$ and $1.6921 \times 10^{13}\,s^{-1}$, respectively. Use the data given in Appendix A to compute the percent error in your results.

7.22 Gas-phase $CO_2(g)$ is a linear molecule with four vibrational degrees of freedom. The measured frequencies for these vibrations are $7.03995 \times 10^{13}\,s^{-1}$, $4.01598 \times 10^{13}\,s^{-1}$, $1.9990 \times 10^{13}\,s^{-1}$, and $1.9990 \times 10^{13}\,s^{-1}$. The last two vibrational degrees of freedom have the same frequency and are, therefore, said to be twofold degenerate. Compute the heat capacity of $CO_2(g)$ at 298.15 K.

7.23* Sam is beginning to get the hang of physical chemistry. His grades are improving rapidly, and his creative ideas are starting to move on target. His latest proposal is for a "heat capacity spectrometer." He correctly notes that Raman and IR spectrometers needed to obtain molecular spectra are very expensive. On the other hand, calorimeters needed to measure heat capacities are much less expensive. Sam suggests that we place the compound whose gas-phase spectrum is to be measured in a calorimeter and measure its heat capacity at some convenient temperature. The data are then fed into a dedicated computer, which computes the vibrational frequencies of the molecule.

(A) Obtain a calibration curve at 298.15 K for the vibrational frequency of gas-phase diatomic molecules, which could be used to convert measured heat capacity data into vibration frequencies. That is, plot ν versus C_p^m at $T = 298.15$ K.

(B) Compute the fundamental vibrational frequencies of $HCl(g)$, $S_2(g)$, and $I_2(g)$, from their measured

heat capacities given in Appendix A. Comment on Sam's proposed heat capacity spectrometer.

7.24 This problem provides important insight into the nature of the second law of thermodynamics. Let us assume that we have one mole of $Mg(s)$ cooled to a temperature of 10 K. An individual who does not understand the limitations imposed by the second law wishes to cool the magnesium to 0 K. To execute this process, he has available a Carnot-type refrigerator whose coefficient of performance is a fraction f that of an ideal Carnot refrigerator.

(A) Use the low-temperature limiting form of the Debye heat capacity for $Mg(s)$ with a Debye temperature of 343.7 K, as computed in the text, to obtain an expression in terms of f for the amount of work that the refrigerator will need to do in order to cool the magnesium to 0 K. Assume that the high-temperature site is at 298 K.

(B) Compute the work required if $f = 0.01$.

(C) The quantity of work computed in (B) is finite. Furthermore, it will be finite no matter what the value of f is, so long as it is not zero. Thus, the work required to cool an object to absolute zero is finite. Yet, the second law tells us that we cannot reach absolute zero. Is the result obtained in (B) incompatible with the second law? Explain.

7.25 One mole of a metal is heated from a temperature $T = \alpha\theta$ to $T = \beta\theta$, where α and β are constants greater than or equal to unity and θ is the Debye temperature of the metal.

(A) Use the high-temperature limiting form of the Debye heat capacity expression to show that $\Delta \bar{S}$ for the process is the same for all metals.

(B) For $\alpha = 1$ and $\beta = 2$, compute $\Delta \bar{S}$ for the process.

Had enough of solids? Me, too. So let's move on to the thermodynamics of solutions. And speaking of solutions, how about a pizza and your favorite soft drink?

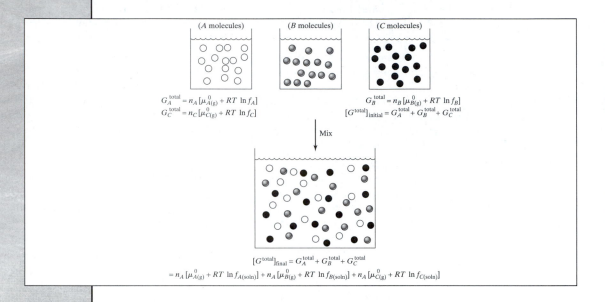

CHAPTER 8

$$G_A^{\text{total}} = n_A\,[\mu_{A(g)}^0 + RT\,\ln f_A]$$
$$G_C^{\text{total}} = n_C\,[\mu_{C(g)}^0 + RT\,\ln f_C]$$

$$G_B^{\text{total}} = n_B\,[\mu_{B(g)}^0 + RT\,\ln f_B]$$
$$[G^{\text{total}}]_{\text{initial}} = G_A^{\text{total}} + G_B^{\text{total}} + G_C^{\text{total}}$$

Mix

$$[G^{\text{total}}]_{\text{final}} = G_A^{\text{total}} + G_B^{\text{total}} + G_C^{\text{total}}$$
$$= n_A\,[\mu_{A(g)}^0 + RT\,\ln f_{A(\text{soln})}] + n_A\,[\mu_{B(g)}^0 + RT\,\ln f_{B(\text{soln})}] + n_A\,[\mu_{C(g)}^0 + RT\,\ln f_{C(\text{soln})}]$$

Thermodynamics of Nonelectrolytic Solutions

In most chemical applications, we are not dealing with pure substances. Indeed, the very nature of chemistry requires that compounds be mixed. In some cases, the mixture is of gases. Occasionally, we may have heterogeneous mixtures of solids or of solids and liquids. Most frequently, however, mixtures are of homogeneous liquid solutions. The thermodynamics of such systems is, therefore, of great importance. We will begin an examination of these systems by introducing the ideal-solution law, the solution counterpart of the ideal-gas law. Although much more restrictive and far less accurate than the ideal-gas law, the ideal-solution law nevertheless provides a very convenient platform from which to launch our study of solution thermodynamics. Using this law, we shall study fractional distillation, the thermodynamics of mixing, solubility, and a variety of colligative properties that include the lowering of vapor pressure, the depression of freezing points, the elevation of boiling points, and osmotic pressure. With these results providing the foundation, we turn our attention to nonideal solutions and the concept of solution activities.

8.1   Ideal Solutions

8.1.1   Raoult's Law: Vapor Pressures

In Chapter 1, we learned that if gaseous molecules occupy no volume and if the interactions between them are all identical and equal to zero, the equation of state assumes the simple form $PV = nRT$, which we call the ideal-gas equation of state. The solution counterpart of the ideal-gas laws makes a similar assumption that the interactions between all molecules in the solution are identical. However, the magnitude of these interactions cannot be zero, or the system could not exist as a condensed phase.

Let us assume that we have a binary solution of two compounds A and B in which the interactions A–A, B–B, and A–B are all identical. Figure 8.1 illustrates this situation. At temperature T, the equilibrium vapor pressure of pure liquid A will be a value P_A^o that we can compute using the Clausius–Clapeyron equation. In Figure 8.1A, this vapor pressure is illustrated by having 25% of the molecules of A in the vapor phase. Now suppose that we have a second pure liquid B whose intermolecular forces are identical to those present in liquid A. Under these conditions, the equilibrium vapor pressure of B, P_B^o, will be equal to P_A^o. If we form an A–B solution by removing half of the A molecules and replacing them with B molecules, a mixture with $X_A = X_B = 0.5$ will result. If the solution is ideal, all interactions will be identical, and the total equilibrium vapor pressure above the solution will be identical to that above pure liquid A. However, only half the gas-phase molecules will be A; the other half will be B. This situation is illustrated in Figure 8.1B.

Under the special conditions just described, the equilibrium vapor pressure of component A above the solution will be $P_A = X_A P_A^o = 0.5 P_A^o$. We will also have $P_B = X_B P_B^o = 0.5 P_B^o$. It is clear that if we were to replace only 25% of the A molecules in Figure 8.1A with B molecules, so that $X_A = 0.75$ and $X_B = 0.25$, the total equilibrium vapor pressure above the solution

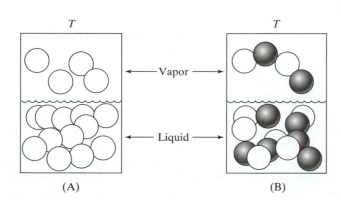

(A)   (B)

◀ FIGURE 8.1

Part A shows a hypothetical situation in which pure liquid A (open circles) has an equilibrium vapor pressure at a temperature T such that 25% of the molecules are in the vapor phase. Part B illustrates the situation that will result if half of the A molecules are replaced by B molecules (shaded circles) that form an ideal solution. If the equilibrium vapor pressure of pure liquid B is the same as that of A, 25% of the molecules will still be in the vapor phase, but half of the gas-phase molecules will now be B molecules.

would remain the same, but we would now observe 75% of the gas-phase molecules to be A and 25% to be B. That is, we would now have

$$P_A = X_A P_A^o = 0.75 P_A^o \tag{8.1}$$

and

$$P_B = X_B P_B^o = 0.25 P_B^o. \tag{8.2}$$

Now consider a situation in which pure liquid A is more volatile than pure liquid B; that is, $P_A^o > P_B^o$. Our discussion of phase equilibrium in Chapter 6 shows that this inequality means that the B–B interactions are more attractive than the A–A interactions, thereby making it more difficult for B molecules to escape into the vapor phase. In this situation, we would not expect an A–B solution to be ideal, since the interactions cannot all be identical. If, however, the A–A and B–B interactions are nearly the same, Eqs. 8.1 and 8.2 might still be reasonably good approximations. The assumption that we still have $P_A = X_A P_A^o$ and $P_B = X_B P_B^o$ is called the *ideal-solution approximation*. The situation is similar to that for a gas which is known to condense to a liquid at sufficiently low temperatures. We know that the interactions between molecules cannot possibly be zero. Nevertheless, if they are sufficiently small, we can still write $PV = nRT$ and obtain reasonably accurate results. In general, for an ideal solution containing K components,

$$\boxed{P_i = X_i P_i^o}. \tag{8.3}$$

The ideal-solution approximation contained in Eq. 8.3 is called *Raoult's law*. In Eq. 8.3 and all subsequent equations, X_i refers to the mole fraction of component i in the solution. The notation Y_i will be used for the mole fraction of this component in the equilibrium vapor phase.

Before proceeding, it is important to note that, although the assumptions leading to Eq. 8.3 are analogous to those contained in the ideal-gas law, the magnitude of the errors inherent in the approximations are very different. In the case of gases, if the pressure is low, the interactions are very close to zero. As a result, deviations of the ideal-gas equation of state from more accurate forms are generally small. The examples and problems in Chapter 1 clearly demonstrate this point. Example 1.9 shows that the ideal-gas and van der Waals equations of state differ in their predictions by only 0.78% when the pressure is 8.14 atm. In contrast, the differences in the intermolecular interactions in acetone, with a normal boiling point of 329.4 K and H_2O, which boils at 373.15 K, are large compared with gas-phase interactions. Consequently, Raoult's law is often far less accurate than the ideal-gas equation of state. The approximation will be accurate only when the components of the solution are very similar, as illustrated in Figure 8.2.

As the pressure approaches zero for a gas, the ideal-gas assumptions are more nearly obeyed. Therefore, the ideal-gas equation of state becomes more accurate. In a similar manner, as X_A approaches unity, all the solution interactions become identical, since nearly all of them are A–A interactions. As a result, Raoult's law is more accurate for any component whose mole fraction approaches unity. We call the region in which this takes place the *ideal-solution regime*. In spite of the lack of accuracy of Raoult's law, it is

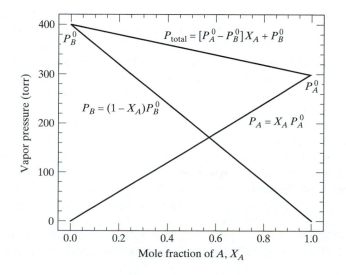

very instructive to utilize that law to study the thermodynamic properties of solutions. Doing so provides important physical insights into the processes involved.

Figure 8.3 shows the variation of equilibrium vapor pressures over an ideal binary solution of components A and B as a function of the mole fraction of A. For illustrative purposes, we have taken the pure component vapor pressures of A and B to be 300 torr and 400 torr, respectively. Using Raoult's law, we have $P_A = X_A P_A^o$ and $P_B = X_B P_B^o = (1 - X_A)P_B^o$. Consequently, both P_A and P_B are linear functions of X_A with slopes P_A^o and

$-P_B^o$, respectively. If we assume the equilibrium vapor to be ideal, the total vapor pressure is given by Dalton's law:

$$P_{\text{total}} = P_A + P_B = X_A P_A^o + (1 - X_A)P_B^o = [P_A^o - P_B^o]X_A + P_B^o. \quad \textbf{(8.4)}$$

Equation 8.4 shows that the total vapor pressure over an ideal binary solution is also a linear function of X_A or X_B, as illustrated in Figure 8.3.

Vapor pressure lowering is an immediate consequence of Raoult's law. The equilibrium vapor of any component over a solution is lower than that over the pure liquid. The magnitude of the lowering can easily be shown to be a function of the amount of the other components added and not of what is added. The change in vapor pressure of component A upon forming a solution is $\Delta P_A = P_A - P_A^o$. Using Raoult's law, we may express ΔP_A as a function of the mole fractions of all of the other solution components. If K components are present in the solution, we have

$$\Delta P_A = P_A - P_A^o = X_A P_A^o - P_A^o = (X_A - 1)P_A^o = -\sum_{i=1}^{i=K}{}' X_i P_A^o, \quad \textbf{(8.5)}$$

where the prime on the summation indicates that the index runs over all values of i from 1 to K, except $i = A$. For the special case of a binary solution, Eq. 8.5 becomes

$$\Delta P_A = -X_B P_A^o. \quad \textbf{(8.6)}$$

Note that ΔP_A will always be negative. Therefore, the equilibrium vapor pressure of any component of an ideal solution is always decreased upon forming a solution. Furthermore, the magnitude of the lowering depends only on the amount of component B added, not upon its chemical nature. Solution properties that have this characteristic are termed *colligative* properties.

Let us now consider the case in which P_B^o is very small. Such solution components are said to be *nonvolatile*. Under this condition, Eq. 8.4 shows that $P_{\text{total}} = P_A$, as expected. Since we know that $P_A < P_A^o$ and that boiling occurs whenever the equilibrium vapor pressure equals the applied external pressure, we may confidently predict that the ideal solution will exhibit a higher boiling temperature than that of pure component A. This may be demonstrated even at home, by conducting a simple experiment. Water is placed in a saucepan and brought to a boil on the stove. At this point, we have $P_{\text{H}_2\text{O}}^o = 1$ atm if we are at atmospheric pressure at sea level. We now pour NaCl into the boiling water, forming a solution. Immediately, the equilibrium vapor pressure of the H_2O drops below 1 atm, and we observe a cessation of boiling. Only after the fire heats the water to a higher temperature at which we again have $P_{\text{H}_2\text{O}} = 1$ atm will boiling recommence. The magnitude of the elevation of the boiling point will be examined quantitatively later in the chapter.

The fact that the magnitude of a colligative property depends only upon how much material is dissolved and not upon its chemical nature must be taken into account when dealing with compounds that ionize in solution. For example, consider an aqueous solution of NaCl. Upon dissolution in the water solvent, the equilibrium reaction

$$\text{NaCl}(c) + x\,\text{H}_2\text{O}(l) = \text{Na}^+(aq) + \text{Cl}^-(aq)$$

takes place, where the subscript (aq) denotes an aqueous solution and $\text{Na}^+(aq)$ and $\text{Cl}^-(aq)$ represent hydrated ions in aqueous solution, as illustrated in Figure 8.4. Therefore, NaCl(aq), Na$^+$(aq), and Cl$^-$(aq) are all present. The quantity of interest is the sum of the mole fractions of NaCl(aq), Na$^+$(aq), and Cl$^-$(aq), as indicated by Eq. 8.5. If we add n_1 moles of NaCl(c) to n_o moles of water, and if f is the fraction of the NaCl(c) that ionizes, we have $n_{\text{NaCl(aq)}} = (1 - f) n_1$ and $n_{\text{Na}^+} = n_{\text{Cl}^-} = f n_1$. The individual mole fractions are, therefore,

$$X_{\text{NaCl(aq)}} = \frac{(1 - f) n_1}{(1 - f) n_1 + 2f n_1 + n_o}$$

and

$$X_{\text{Na}^+\text{(aq)}} = X_{\text{Cl}^-\text{(aq)}} = \frac{f n_1}{(1 - f) n_1 + 2f n_1 + n_o},$$

which gives

$$\sum_{i=1}^{i=3} X_i = \frac{(1 - f) n_1 + 2f n_1}{(1 - f) n_1 + 2f n_1 + n_o} = \frac{n_1 + f n_1}{n_1 + f n_1 + n_o}. \tag{8.7}$$

The associated lowering of the vapor pressure is

$$\Delta P_{\text{H}_2\text{O}} = -\frac{n_1 + f n_1}{n_1 + f n_1 + n_o} P^o_{\text{H}_2\text{O}}, \tag{8.8}$$

which reduces to Eq. 8.6 if $f = 0$.

Since the intermolecular forces in NaCl(c) and H$_2$O(l) are vastly different, Raoult's law will be highly inaccurate, unless we are in the ideal-solution regime, wherein $X_{\text{H}_2\text{O}}$ approaches unity. For this reason, we usually employ Eq. 8.8 only for the solvent in dilute solutions, for which $n_o \gg n_1$. For such solutions, Eq. 8.8 reduces to

$$\Delta P_{\text{H}_2\text{O}} = -\frac{n_1 + f n_1}{n_o} P^o_{\text{H}_2\text{O}} = -\frac{(1 + f) n_1}{n_o} P^o_{\text{H}_2\text{O}}. \tag{8.9}$$

If we now recognize the fact that the ratio n_1/n_o is the dilute-solution mole fraction of NaCl(aq) that would be present if no ionization occurred, we can write Eq. 8.9 in the form

$$\Delta P_{\text{H}_2\text{O}} = -(1 + f) X_{\text{NaCl}} P^o_{\text{H}_2\text{O}} \tag{8.10}$$

where X_{NaCl} is computed as if there were no ionization. If we have complete ionization ($f = 1$), then $\Delta P_{\text{H}_2\text{O}} = -2X_{\text{NaCl}} P^o_{\text{H}_2\text{O}}$ in the limit of dilute solutions.

◀ **FIGURE 8.4**
Schematic illustration of hydrated Na$^+$ and Cl$^-$ ions. The negative end (oxygen) of the water dipole will generally be aligned to point toward the positive sodium ion, whereas the positive end (hydrogen atoms) will be directed toward the negatively charged chloride ion. The smaller Na$^+$ ion has a greater charge density than the larger Cl$^-$ ion. It will, therefore, be hydrated by a larger number of water molecules. This is illustrated qualitatively in the figure.

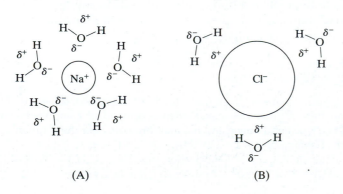

(A)   (B)

▶ FIGURE 8.5

The percent error as a function of the mole fraction of an added 1 : 1 salt such as NaCl or KBr that is produced by using the dilute solution limiting form of Eq. 8.8 to compute vapor pressure lowering. The error is shown for a maximum mole fraction of $X = 0.005$, since the ideal-solution approximation becomes very inaccurate at higher concentrations.

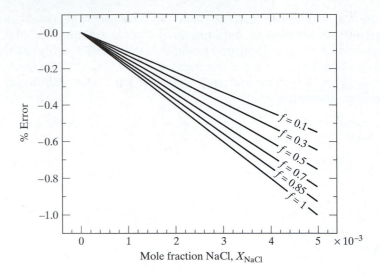

Figure 8.5 shows the percent error produced by using the dilute-solution limit rather than Eq. 8.8 as a function of X_{NaCl} for several values of f. The figure shows that the percent error is always less than 1%, provided that the mole fraction of added salt is 0.005 or less.

If we had used $BaCl_2(c)$ as our example instead of $NaCl(c)$, Eq. 8.9 would have assumed the form

$$\Delta P_{H_2O} = -\frac{n_1 + 2fn_1}{n_o} P^o_{H_2O} = -\frac{(1 + 2f)\,n_1}{n_o}\, P^o_{H_2O}.$$

With $X_{BaCl_2(c)}$ defined to be the dilute-solution mole fraction of $BaCl_2(c)$ that would be present if no ionization occurred, we would have obtained

$$\Delta P_{H_2O} = -(1 + 2f)\, X_{BaCl_2}\, P^o_{H_2O}$$

which, in the limit of complete ionization in a dilute solution, yields

$$\Delta P_{H_2O} = -3X_{BaCl_2} P^o_{H_2O}.$$

In general, if an ionic salt ionizes completely to form i ions, the limiting dilute-solution equation for the vapor pressure lowering will be (see Problem 8.6)

$$\Delta P_A = -i\, X_{salt}\, P^o_A, \tag{8.11}$$

where X_{salt} is the mole fraction of the ionic salt in solution if no ionization had occurred. Equation 8.11 permits the degree of ionization to be obtained from measurements of colligative properties.

EXAMPLE 8.1

Twenty grams of benzene (C_6H_6) are added to 250 grams of toluene (C_7H_8) to form a solution. If the solution may be regarded as being ideal, compute the partial pressures of benzene and toluene and the total pressure over the solution at 298.15 K. The normal boiling points are $T^o_b(C_6H_6) = 353.25$ K and $T^o_b(C_7H_8) = 383.78$ K. The partial molar enthalpies of vaporization are $\Delta \overline{H}_{vap}(C_6H_6) = 30.82$ kJ mol^{-1} and $\Delta \overline{H}_{vap}(C_7H_8) = 39.20$ kJ mol^{-1}. Assume $\Delta \overline{H}_{vap}$ to be constant.

Solution

Using Raoult's law, we have

$$P_{benzene} = X_{benzene} P^o_{benzene} \qquad (1)$$

and

$$P_{toluene} = X_{toluene} P^o_{toluene}. \qquad (2)$$

Therefore, we need to compute the mole fractions of the two solution components and their pure component vapor pressures at 298.15 K. Accordingly,

$$n_{benzene} = \text{number of moles benzene} = (20 \text{ g}) \times \frac{1 \text{ mole}}{78.108 \text{ g}}$$
$$= 0.2561 \text{ mole of benzene,} \qquad (3)$$

and

$$n_{toluene} = \text{number of moles of toluene} = (250 \text{ g}) \times \frac{1 \text{ mole}}{92.134 \text{ g}}$$
$$= 2.7134 \text{ moles of toluene.} \qquad (4)$$

Thus,

$$X_{benzene} = \frac{0.2561}{0.2561 + 2.7134} = 0.08624, \qquad (5)$$

which gives

$$X_{toluene} = 1 - X_{benzene} = 0.9138. \qquad (6)$$

The pure component vapor pressures at 298.15 K may be obtained from the Clausius–Clapeyron equation 6.12:

$$P = 760 \exp\left[\frac{-\Delta \overline{H}_{vap}}{R}\left\{\frac{1}{T} - \frac{1}{T^o_b}\right\}\right] \text{torr.} \qquad (7)$$

Substituting the given data yields

$$P^o_{benzene} = 760 \exp\left[\frac{-30,820}{8.314}\left\{\frac{1}{298.15} - \frac{1}{353.25}\right\}\right] \text{torr} = 109.3 \text{ torr} \qquad (8)$$

and

$$P^o_{toluene} = 760 \exp\left[\frac{-39,200}{8.314}\left\{\frac{1}{298.15} - \frac{1}{383.78}\right\}\right] \text{torr} = 22.30 \text{ torr.} \qquad (9)$$

Therefore, the equilibrium partial pressures over the solution are

$$P_{benzene} = 0.08624 \,(109.3) \text{ torr} = 9.426 \text{ torr} \qquad (10)$$

and

$$P_{toluene} = 0.9138 \,(22.30) \text{ torr} = 20.38 \text{ torr.} \qquad (11)$$

The total vapor pressure is

$$P_{total} = P_{benzene} + P_{toluene} = 29.81 \text{ torr.} \qquad (12)$$

For related exercises, see Problems 8.1, 8.2, 8.3, and 8.4.

EXAMPLE 8.2

Fifteen grams of NaCl(c) are dissolved in one kg of H_2O at 363.15 K. The equilibrium H_2O vapor pressure over the solution is found to be 521.00 torr. Pure H_2O at 363.15 K has an equilibrium vapor pressure of 525.76 torr. Determine the percent ionization of NaCl(c) in this solution, assuming that the dilute-solution limit of the ideal-solution law holds.

Solution

If the dilute-solution limiting form of the ideal-solution law is used, Eq. 8.10 holds, and we have

$$\Delta P_{H_2O} = -(1 + f) X_{NaCl} P^o_{H_2O}, \tag{1}$$

where f is the fraction of NaCl(c) that ionizes. Substituting the given data yields

$$521.00 - 525.76 = -(1 + f) X_{NaCl} (525.76), \tag{2}$$

where

$$X_{NaCl} = \frac{n_{NaCl}}{n_{NaCl} + n_{H_2O}}. \tag{3}$$

The number of moles of each component is

$$n_{NaCl} = (15 \text{ g}) \times \frac{1 \text{ mole}}{58.44 \text{ g}} = 0.2567 \text{ mole of NaCl}, \tag{4}$$

$$n_{H_2O} = 1,000 \text{ g} \times \frac{1 \text{ mole}}{18.016 \text{ g}} = 55.51 \text{ moles of } H_2O. \tag{5}$$

Substituting into Eq. 3 then gives

$$X_{NaCl} = \frac{0.2567}{0.2567 + 55.51} = 0.004603, \tag{6}$$

and substitution into Eq. 2 produces

$$1 + f = -\frac{521.00 - 525.76}{0.004603(525.76)} = 1.967. \tag{7}$$

Therefore, we have $f = 0.967$, and the NaCl is 96.7% ionized in the solution.
      For related exercises, see Problems 8.5 and 8.6.

EXAMPLE 8.3

Compute the fraction of the NaCl(c) ionized in Example 8.2, assuming that the solution to behaves ideally, but without making the dilute-solution approximation. Compute the percent error involved in making the dilute-solution approximation in this problem.

Solution

The general equation for lowering the vapor pressure without incorporating a dilute-solution approximation is Eq. 8.8:

$$\Delta P_{H_2O} = -\frac{n_1 + f n_1}{n_1 + f n_1 + n_o} P^o_{H_2O}. \tag{1}$$

Rearranging Eq. 1 gives

$$[n_1 + f n_1 + n_o] \Delta P_{H_2O} = -n_1 P^o_{H_2O} - n_1 f P^o_{H_2O}. \tag{2}$$

Solving Eq. 2 for f, we obtain

$$f [n_1 \Delta P_{H_2O} + n_1 P^o_{H_2O}] = -n_1 P^o_{H_2O} - n_1 \Delta P_{H_2O} - n_o \Delta P_{H_2O}, \tag{3}$$

from which it follows that

$$f = \frac{-n_1 P_{H_2O}^o - n_1 \Delta P_{H_2O} - n_o \Delta P_{H_2O}}{[n_1 \Delta P_{H_2O} + n_1 P_{H_2O}^o]}$$

$$= \frac{-(0.2567)(525.76) - (0.2567)(-4.76) - (55.51)(-4.76)}{0.2567(-4.76) + (0.2567)(525.76)} = 0.976. \quad \textbf{(4)}$$

Therefore, the percent ionization of NaCl(c) computed without the dilute-solution approximation is 97.6%. The percent error associated with the dilute-solution approximation is

$$\% \text{ error} = \frac{(96.7 - 97.6) \times 100}{97.6} = -0.92\%. \quad \textbf{(5)}$$

Note that deviations from ideality in ionic solutions are very large even when the concentrations are low. This state of affairs is the result of the exceptionally strong coulombic forces that operate between ions. In Chapter 9, we will quantitatively examine the effect of such forces on the ideality of a solution. In the meantime, the student should be aware that the ideal-solution assumption made in Examples 8.2 and 8.3 will be inaccurate even at very low concentrations of NaCl.

8.1.2   Fractional Distillation

A careful examination of Raoult's law shows that it predicts that the vapor in equilibrium with a solution will have a different composition than that present in the solution. This simple statement is the basis of the major production process used by the petroleum industry to produce gasoline, fuel oil, and other petrochemical products.

Let Y_i denote the mole fraction of component i in an equilibrium vapor whose solution mole fraction is X_i. For an ideal solution with K components, Raoult's law gives $P_i = X_i P_i^o$ for every component of the solution. Usually, the equilibrium vapor pressures are so low that we may accurately assume that the vapor behaves ideally. Under this condition, Dalton's law holds, and the mole fraction of component i in the vapor is given by Eq. 1.44, where we now replace the notation of X_i with Y_i:

$$Y_i = \frac{P_i}{P_{\text{total}}} = \frac{X_i P_i^o}{\sum\limits_{n=1}^{n=K} X_n P_n^o}. \quad \textbf{(8.12)}$$

For a binary solution of A and B, Eq. 8.12 becomes

$$\boxed{Y_A = \frac{X_A P_A^o}{X_A P_A^o + X_B P_B^o} = \frac{X_A P_A^o}{X_A(P_A^o - P_B^o) + P_B^o}.} \quad \textbf{(8.13)}$$

Inspection of Eq. 8.13 shows that we will have $Y_A = X_A$ only in the trivial case $X_A = 1$ or in the case where $P_A^o = P_B^o$ and the two components of the solution have equal volatility.

Since the sum of the mole fractions must add to unity, the mole fraction of component B in the vapor phase of a binary solution is

$$Y_B = 1 - Y_A = \frac{X_B P_B^o}{X_A P_A^o + X_B P_B^o}. \quad \textbf{(8.14)}$$

Dividing Eq. 8.13 by Eq. 8.14 gives the ratio

$$\frac{Y_A}{Y_B} = \frac{X_A\,P_A^o}{X_B\,P_B^o} = \frac{P_A}{P_B}.$$ (8.15)

Rearranging Eq. 8.15 produces

$$\frac{Y_A}{X_A\,P_A^o} = \frac{Y_B}{X_B\,P_B^o}.$$ (8.16)

Equation 8.16 shows that the vapor becomes richer in the more volatile component. That is, if $P_A^o > P_B^o$, we must have $Y_A > X_A$ in order to satisfy Eq. 8.16. The next example demonstrates this principle.

EXAMPLE 8.4

In Example 8.1, we found that $X_{\text{benzene}} = 0.08624$. Compute Y_{benzene} at 298.15 K, and show that the mole fraction of benzene in the vapor is greater than that in the solution. Show that the opposite is true for toluene.

Solution

The mole fraction of component A in the vapor is given by Eq. 8.13:

$$Y_A = \frac{X_A\,P_A^o}{X_A\,P_A^o + X_B\,P_B^o}.$$ (1)

Letting A denote benzene and B toluene, the data and calculations in Example 8.1 give

$$Y_{\text{benzene}} = \frac{(0.08624)(109.3)}{(0.08624)(109.3) + (0.9138)(22.89)} = 0.3107$$ (2)

and

$$Y_{\text{toluene}} = 1 - Y_{\text{benzene}} = 0.6893.$$ (3)

The equilibrium vapor is thus enriched in the more volatile component (benzene) and depleted in the less volatile material (toluene).

In principle, a solution may be separated by a succession of vaporization–condensation steps at any temperature. At each step, the condensate will become progressively more and more enriched in the most volatile component. After a sufficient number of such cycles, the condensate will contain essentially only that component. Example 8.4 provides a numerical illustration of one such step at 298.15 K for a benzene–toluene solution. In practice, however, it is inefficient to conduct the separation at temperatures far below the boiling point of the solution, since the total mass of equilibrium vapor produced at each step is unacceptably small. A much more rapid and efficient separation can be achieved by executing each step at the boiling point of the condensate obtained in the previous step. If this is done, the process is called *fractional distillation*. When a liquid mixture is boiled in a fractionating column, the original equilibrium vapor rises up the column, whereupon it cools and condenses on the packing contained within the column. Since this condensate is richer in the more volatile component, its boiling point is lower than that of the original mixture. As hot vapors continue to conduct thermal energy up the column, this first condensate reaches its boiling point, and the new equilibrium vapor rises further up the column, whereupon it, too, condenses. The cycle is repeated many times, until the liquid emerging at the top of the column is nearly pure in the most volatile component. Each vaporization–condensation cycle occurring in the column is called a *theoretical plate*. Figure 8.6 shows a schematic diagram of a simple distillation column.

To examine fractional distillation more quantitatively, we must first obtain an appropriate expression for the boiling point of an ideal solution. Boiling occurs when the total equilibrium vapor pressure equals the applied pressure P_o. For a system with ideal vapors, the boiling condition is, therefore,

$$P_{\text{total}} = \sum_{n=1}^{n=K} P_n = \sum_{n=1}^{n=K} X_n P_n^o = P_o. \qquad (8.17)$$

If we ignore the volume of the liquid phase relative to the vapor and assume the vapors to be ideal, the pure component vapor pressures will be given by the Clausius–Clapeyron equation, Eq 6.12. Inserting this result into Eq. 8.17 gives

$$P_{\text{total}} = \sum_{n=1}^{n=K} 760\, X_n \exp\left[\frac{-\Delta \overline{H}_{\text{vap}}^n}{R}\left\{\frac{1}{T} - \frac{1}{T_{bn}}\right\}\right] = P_o \text{ torr}, \qquad (8.18)$$

where $\Delta \overline{H}_{\text{vap}}^n$ is the partial molar enthalpy of vaporization of component n, which is assumed to be a constant, and T_{bn} is the normal boiling point of

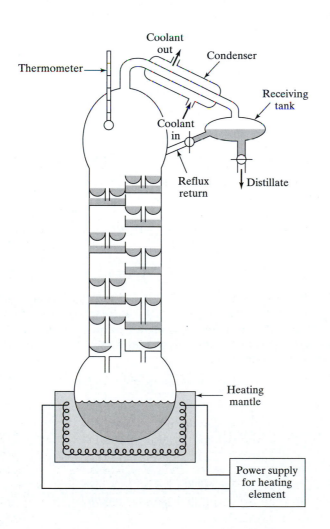

Thermometer

Coolant out

Condenser

Receiving tank

Coolant in

Reflux return

Distillate

Heating mantle

Power supply for heating element

◀ FIGURE 8.6
Schematic diagram of a fractional distillation column. The solution to be fractionated is placed in the lower reservoir, where it is heated to its boiling point by the mantle. Vapors enriched in the more volatile component rise up the column, where they condense on the packing, which is represented by collection dishes in the figure. When vaporization occurs from these solution pockets, the new vapor becomes further enriched in the more volatile component. Vapor reaching the top of the column enters a condenser that converts the hot vapor to the liquid phase, after which it collects in a receiving tank. This liquid is then returned to the column via the reflux line. When the column attains equilibrium, the reflux line is closed and the highly enriched distillate is removed.

component n. Let T^* be the temperature that satisfies Eq. 8.18. Then, at $T = T^*$, the solution will boil. The pure component vapor pressures at T^* are

$$P_n^o = 760 \exp\left[\frac{\Delta H_{vap}^n}{R}\left\{\frac{1}{T^*} - \frac{1}{T_{bn}}\right\}\right] \text{ torr.} \qquad \textbf{(8.19)}$$

The composition of the vapor at this step in the fractional distillation can now be computed by using Eq. 8.12 or Eq. 8.13 for a binary solution.

We may utilize the benzene–toluene solution described in Example 8.1 to provide a numerical illustration of fractional distillation. This is done in Example 8.5.

EXAMPLE 8.5

Twenty grams of benzene (C_6H_6), are added to 250 grams of toluene (C_7H_8) to form an ideal solution. The normal boiling points are $T_b^o(C_6H_6) = 353.25$ K and $T_b^o(C_7H_8) = 383.78$ K. The partial molar enthalpies of vaporization are $\Delta \overline{H}_{vap}(C_6H_6) = 30.82$ kJ mol^{-1} and $\Delta \overline{H}_{vap}(C_7H_8) = 39.20$ kJ mol^{-1}. Assume the $\Delta \overline{H}_{vap}$ are constant. **(A)** Compute the boiling point of this solution if the applied external pressure is 760 torr (1 atm). **(B)** Compute the pure component vapor pressures at the boiling point of the original solution. **(C)** Calculate the composition of the equilibrium vapor at the boiling point of the solution. **(D)** Repeat (A), (B), and (C), using the condensate computed in (C) as the original solution.

Solution

(A) As determined in Example 8.1, the solution mole fraction of benzene is $X_{benzene} = 0.08624$. Therefore, we need the temperature for which

$$760 X_{benzene} \exp\left[\frac{-30,820}{R}\left\{\frac{1}{T} - \frac{1}{353.25}\right\}\right]$$

$$+ 760 X_{toluene} \exp\left[\frac{-39,200}{R}\left\{\frac{1}{T} - \frac{1}{383.78}\right\}\right] = 760. \qquad \textbf{(1)}$$

Canceling the factor of 760 on both sides and using the fact that $X_{toluene} = 1 - X_{benzene}$, we can write Eq. 1 in the form

$$F(T) = X_{benzene}\left[\exp\left[\frac{-30,820}{R}\left\{\frac{1}{T} - \frac{1}{353.25}\right\}\right] - \exp\left[\frac{-39,200}{R}\left\{\frac{1}{T} - \frac{1}{383.78}\right\}\right]\right]$$

$$+ \exp\left[\frac{-39,200}{R}\left\{\frac{1}{T} - \frac{1}{383.78}\right\}\right] - 1 = 0. \qquad \textbf{(2)}$$

With $X_{benzene} = 0.08624$, we may now conduct a one-dimensional grid search to find the root of $F(T) = 0$. The procedure is described in Example 1.10. The result is $T = T^* = 380.36$ K. This temperature is slightly below the normal boiling point of toluene, as expected, since the original solution is only 91.4% toluene and benzene has a lower normal boiling point.

(B) The pure component vapor pressures are given by Eq. 8.19 with $T^* = 380.36$ K. We have

$$P_{benzene}^o = 760 \exp\left[\frac{-30,820}{R}\left\{\frac{1}{380.36} - \frac{1}{353.25}\right\}\right] = 1,605.52 \text{ torr} \qquad \textbf{(3)}$$

and

$$P_{toluene}^o = 760 \exp\left[\frac{-39,200}{R}\left\{\frac{1}{380.36} - \frac{1}{383.78}\right\}\right] = 680.46 \text{ torr.} \qquad \textbf{(4)}$$

(C) The composition of the vapor at 380.36 K is given by Eq. 8.13:

$$Y_{benzene} = \frac{X_{benzene}\, P^o_{benzene}}{X_{benzene}\, P^o_{benzene} + X_{toluene}\, P^o_{toluene}}$$

$$= \frac{(0.08624)(1{,}605.52)}{(0.08624)(1{,}605.52) + (0.9138)(680.46)} = 0.1821. \qquad (5)$$

Consequently,

$$Y_{toluene} - 1 - X_{benzene} = 0.8179. \qquad (6)$$

We see that the vapor is enriched in the more volatile component, benzene.

(D) We now repeat (A), (B), and (C) with an original solution whose composition is $X_{benzene} = 0.1821$. In other words, we condense the equilibrium vapor obtained in (C) to a new solution and execute another vaporization–condensation cycle. Using Eq. 2 with $X_{benzene} = 0.1821$, we conduct a new one-dimensional grid search to obtain the new boiling point. The result is $T^* = 376.72$ K. The pure component vapor pressures at this temperature are given by Eqs. 3 and 4, with 380.36 K replaced with 376.72 K. The results are $P^o_{benzene} = 1461.47$ torr and $P^o_{toluene} = 603.78$ torr Substituting these results into Eq. 5 gives the new composition of the equilibrium vapor, viz.,

$$Y_{benzene} = \frac{(0.1821)(1461.47)}{(0.1821)(1461.47) + (0.8179)(603.78)} = 0.3502 \qquad (7)$$

and

$$Y_{toluene} = 1 - X_{benzene} = 0.6498. \qquad (8)$$

For related exercises, see Problems 8.7, 8.8, and 8.9.

Example 8.5 shows that two successive vaporization–condensation steps at the solution boiling point for a benzene–toluene mixture originally only 8.62% benzene produces a condensate at the end of the second step that is 35.0% benzene. A fractionating column that produces these two vaporization–condensation steps is said to have two *theoretical plates* of separation power.

We may conveniently illustrate the fractional distillation process by using a temperature–composition phase diagram in which we plot the boiling point of the solution as a function of the solution mole fraction of one component and, on the same graph, plot the solution boiling point as a function of the mole fraction of that component in the equilibrium vapor. Example 8.6 illustrates the procedure for the benzene–toluene system.

EXAMPLE 8.6

Twenty grams of benzene (C_6H_6) are added to 250 grams of toluene (C_7H_8) to form an ideal solution. The normal boiling points are $T^o_b(C_6H_6) = 353.25$ K and $T^o_b(C_7H_8) = 383.78$ K. The molar enthalpies of vaporization are $\Delta \overline{H}_{vap}(C_6H_6) = 30.82$ kJ mol^{-1} and $\Delta \overline{H}_{vap}(C_7H_8) = 39.20$ kJ mol^{-1}. Assume the $\Delta \overline{H}_{vap}$ are constant. **(A)** Use Eqs. 1 and 2 of Example 8.5 to compute the solution boiling point T^* as a function of the mole fraction under an applied pressure of 1 atm. Plot T^* vs. $X_{benzene}$. **(B)** Use Eqs. 3, 4, and 5 of Example 8.5 to obtain the mole fraction of benzene in the equilibrium vapor above the solution at the boiling point as a function of the mole fraction of benzene in the solution. On the same graph used in (A) with the same abscissa, plot T^* as a function of $Y_{benzene}$.

Solution

(A) By repeatedly solving Eqs. 1 and 2 of Example 8.5, we obtain the following table of solution boiling points T^* as a function of $X_{benzene}$:

$X_{benzene}$	$T^*(K)$	$Y_{benzene}$	$X_{benzene}$	$T^*(K)$	$Y_{benzene}$
0.00	383.78	0.0000	0.500	366.13	0.7232
0.08624	380.36	0.1821	0.600	363.23	0.8007
0.1821	376.72	0.3501	0.700	360.49	0.8646
0.250	374.29	0.4150	0.800	357.92	0.9175
0.300	372.55	0.5167	0.900	355.52	0.9548
0.350	370.87	0.5762	1.000	353.25	1.0000
0.400	369.24	0.6302			

Figure 8.7 shows the variation of T^* with $X_{benzene}$.

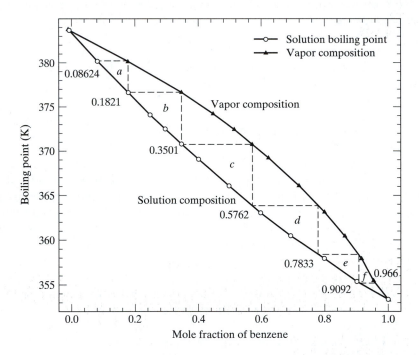

▲ **FIGURE 8.7**
Temperature–composition diagram for a benzene–toluene solution assumed to be ideal. The horizontal tie lines between liquid and equilibrium vapor composition curves give the effect of one theoretical plate of fractionating power. The lines are shown for an initial benzene mole fraction of 0.08624. The figure indicates that six theoretical plates of separation power will yield a distillate whose benzene mole fraction is 0.966.

(B) Using Eqs. 3, 4, and 5 of Example 8.5, along with the boiling points computed in (A), we may calculate the mole fraction of benzene in the equilibrium vapor, $Y_{benzene}$, at each of the boiling points. The results are shown in the right-hand columns of the table and in Figure 8.7, where we use the same abscissa to plot both $X_{benzene}$ and $Y_{benzene}$.

For related exercises, see Problems 8.7, 8.8, and 8.9.

The boiling point–composition diagram developed in Examples 8.5 and 8.6 may be employed to determine the extent to which a fractionating column containing N theoretical plates will separate any given initial mixture of benzene and toluene. To do this, we simply compute the mole fraction of benzene in the initial solution and determine the solution boiling point from the graph. In the examples, this initial value is $X_{benzene} = 0.08624$. The composition of the equilibrium vapor is found by drawing a horizontal tie-line between that initial point and the corresponding point on the vapor composition curve at the same boiling point. This line, denoted by a in the figure, yields a value $Y_{benzene} = 0.1821$. When this vapor is condensed to a liquid, it will form a solution whose boiling point is located by a vertical tie-line from the vapor composition curve to the boiling point curve at the abscissa value $X_{benzene} = 0.1821$. A horizontal tie-line from this point back to the vapor composition curve gives the benzene vapor composition in equilibrium with the new solution. This is shown as tie-line b in the figure. The intercept of b with the vapor composition curve shows the vapor composition to be $Y_{benzene} = 0.3501$. Each horizontal tie-line represents one theoretical plate of fractionating power. The figure shows six such plates. After the sixth one (shown as tie-line f), the distillate will have a composition $X_{benzene} = 0.966$.

8.1.3   Vacuum Distillation

In most cases, fractional distillation is conducted under conditions where the applied pressure is the current barometric pressure. Thus, P_o in Eq. 8.18 is usually about 760 torr. In some situations, however, it is highly advantageous to conduct the distillation with P_o significantly less than atmospheric pressure. If the material to be fractionated decomposes at temperatures below the normal boiling point, the separation process will fail. If, however, P_o is reduced, the boiling temperatures can be decreased below the decomposition point, and the distillation can be successfully conducted. In addition, the separation power of the fractionating column increases when boiling occurs at a lower temperature. This occurs because the equilibrium vapor pressures of the less volatile components decrease with temperature faster than that of the most volatile component. The relative difference in vapor pressure, therefore, increases as T decreases. An examination of the Clausius–Clapeyron equation, Eq. 6.12, shows this to be the case. Example 8.7 provides a numerical illustration of the point. The increased separation power achieved at lower temperatures is purchased at the cost of distillation volume per unit time, since the equilibrium vapor pressures of all components are reduced exponentially with declining temperature. To circumvent this problem and thereby increase the amount of distillate, the condenser shown in Figure 8.6 must be placed close to the surface at which evaporation is occurring. When fractional distillation is conducted with P_o less than barometric pressure, the process is called *vacuum distillation*.

EXAMPLE 8.7

The benzene–toluene solution used in the previous examples is to be vacuum distilled, with P_o equal to 100 torr in a column having two theoretical plates. Compute the composition of the solution produced by the column.

Solution

The original solution composition was shown in Example 8.1 to be $X_{benzene} = 0.08624$. We now need to compute the solution boiling point with $P_o = 100$ torr.

Equation 8.18 shows the boiling condition to be

$$760 X_{\text{benzene}} \exp\left[\frac{-30{,}820}{R}\left\{\frac{1}{T} - \frac{1}{353.25}\right\}\right]$$

$$+ 760 X_{\text{toluene}} \exp\left[\frac{-39{,}200}{R}\left\{\frac{1}{T} - \frac{1}{383.78}\right\}\right] = 100. \tag{1}$$

Using the fact that $X_{\text{toluene}} = 1 - X_{\text{benzene}}$, we can write Eq. 1 in the form

$$F(T) = X_{\text{benzene}}\left[\exp\left[\frac{-30{,}820}{R}\left\{\frac{1}{T} - \frac{1}{353.25}\right\}\right] - \exp\left[\frac{-39{,}200}{R}\left\{\frac{1}{T} - \frac{1}{383.78}\right\}\right]\right]$$

$$+ \exp\left[\frac{-39{,}200}{R}\left\{\frac{1}{T} - \frac{1}{383.78}\right\}\right] - \frac{100}{760} = 0. \tag{2}$$

With $X_{\text{benzene}} = 0.08624$, we may now conduct a one-dimensional grid search to find the root of $F(T) = 0$. This procedure is fully described in Example 1.10. The result is $T = T^* = 324.63$K. The pure component vapor pressures are given by Eq. 8.19 with $T^* = 324.63$ K. We have

$$P^o_{\text{benzene}} = 760 \exp\left[\frac{-30{,}820}{R}\left\{\frac{1}{324.63} - \frac{1}{353.25}\right\}\right] = 301.35 \text{ torr} \tag{3}$$

and

$$P^o_{\text{toluene}} = 760 \exp\left[\frac{-39{,}200}{R}\left\{\frac{1}{324.63} - \frac{1}{383.78}\right\}\right] = 81.05 \text{ torr}. \tag{4}$$

The composition of the vapor at 324.63 K is given by Eq. 8.13:

$$Y_{\text{benzene}} = \frac{X_{\text{benzene}} P^o_{\text{benzene}}}{X_{\text{benzene}} P^o_{\text{benzene}} + X_{\text{toluene}} P^o_{\text{toluene}}} = \frac{(0.08624)(301.35)}{(0.08624)(301.35) + (0.9138)(81.05)}$$

$$= 0.2599. \tag{5}$$

Repeating the process with a new benzene mole fraction $X_{\text{benzene}} = 0.2599$, we find that the boiling point is now $T^* = 316.55$ K. The equilibrium vapor pressures at this temperature are $P^o_{\text{benzene}} = 225.15$ torr and $P^o_{\text{toluene}} = 55.94$ torr. Using these results in Eq. 5 yields a benzene mole fraction of $Y_{\text{benzene}} = 0.5852$ after the second theoretical plate. When the fractionation was conducted with $P_o = 760$ torr in Example 8.6, the second theoretical plate gave $Y_{\text{benzene}} = 0.3501$. The separation power of the column is increased at the lower applied pressure. We pay for this advantage with a lower yield of the product, since the equilibrium vapor pressures are much lower.

8.1.4 Thermodynamics of Mixing

In Chapter 4, we noted that changes in entropy can be associated with variations in the extent of disorder of a system. As disorder increases, we expect positive changes in the entropy. A decrease in the extent of disorder results in a corresponding decrease in the system entropy. Since a system consisting of one beaker of benzene and a second beaker of toluene is more highly ordered than a solution of the contents of the two beakers, we expect the formation of a solution to produce ΔS values greater than zero. A simple everyday example occurs when a parent carefully arranges his or her child's toys, each in its proper place on the shelf. The resulting arrangement is a state of low entropy. Once the child is turned loose in the room, the toys are mixed and jumbled like the contents of a benzene–toluene solution. This is a state of higher entropy. The formation of a solution always has $\Delta S > 0$. Consequently, we may advise the child to tell the parent when he or she is scolded for messing up the playroom that it is not their fault, because *"Die Entropie der Welt strebt einem maximum zu."* (See Problem 4.35.)

Since ΔG and ΔA both depend upon ΔS, we would expect nonzero values for these quantities upon the formation of a solution. If ΔS, ΔG, and ΔA all do exhibit nonzero values, we might also observe changes in the internal energy and the enthalpy of the system. In this section, we shall determine the changes in the various thermodynamic state functions that accompany solution formation.

To compute the change in any thermodynamic quantity, we must be able to evaluate that quantity in both the initial and final state. Let us first consider the initial state of a system consisting of K components, each of which is held in a separate container at a common temperature T, as shown in Figure 8.8. In each case, the general condition for phase equilibrium given by Eq. 6.1 requires that the chemical potential of the pure condensed phase A, μ_A^c, be equal to the chemical potential of its equilibrium vapor, μ_A^v. That is, we must have

$$\mu_A^c = \mu_A^v = \mu_{A(g)}^o + RT \ln f_A, \tag{8.20}$$

where $\mu_{A(g)}^o$ is the standard chemical potential of gaseous A and f_A is the fugacity of the vapor. Since the chemical potential is the partial molar Gibbs free energy, which is the Gibbs free energy per mole, the total Gibbs free energy for n_A moles of component A is

$$G_A = n_A \mu_A^c = n_A \mu_{A(g)}^o + n_A RT \ln f_A. \tag{8.21}$$

The Gibbs free energy for the system is obtained by summing Eq. 8.21 over all K components:

$$G_{\text{initial}}^{\text{total}} = \sum_{i=1}^{i=K} G_i = \sum_{i=1}^{i=K} n_i [\mu_{i(g)}^o + RT \ln f_i]. \tag{8.22}$$

(*A* molecules)   (*B* molecules)   (*C* molecules)

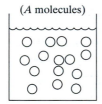 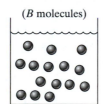 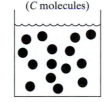

$G_A^{\text{total}} = n_A [\mu_{A(g)}^0 + RT \ln f_A]$

$G_C^{\text{total}} = n_C [\mu_{C(g)}^0 + RT \ln f_C]$

$G_B^{\text{total}} = n_B [\mu_{B(g)}^0 + RT \ln f_B]$

$[G^{\text{total}}]_{\text{initial}} = G_A^{\text{total}} + G_B^{\text{total}} + G_C^{\text{total}}$

Mix

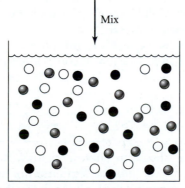

$[G^{\text{total}}]_{\text{final}} = G_A^{\text{total}} + G_B^{\text{total}} + G_C^{\text{total}}$

$= n_A [\mu_{A(g)}^0 + RT \ln f_{A(\text{soln})}] + n_A [\mu_{B(g)}^0 + RT \ln f_{B(\text{soln})}] + n_A [\mu_{C(g)}^0 + RT \ln f_{C(\text{soln})}]$

◀ FIGURE 8.8
The Gibbs free energy of mixing is obtained by summing the Gibbs free energies of the pure components before mixing and subtracting that total from the Gibbs free energy of the mixed system. In each case, the Gibbs free energy of the condensed-phase components is obtained from the Gibbs free energy of the equilibrium vapor phase.

Equation 8.22 provides the required expression for the Gibbs free energy of the K components before they are mixed. We now need a corresponding expression for the Gibbs free energy of the solution. The general condition for phase equilibrium, Eq. 6.1, holds for the components in the solution as well as for the pure compounds. That is, we must have

$$\mu^c_{A(soln)} = \mu^v_{A(soln)} = \mu^o_{A(g)} + RT \ln f_{A(soln)} \tag{8.23}$$

where $\mu^c_{A(soln)}$ is the chemical potential of A in the liquid solution, $\mu^v_{A(soln)}$ is the corresponding chemical potential for the equilibrium vapor A over the solution, and $f_{A(soln)}$ is the fugacity of the equilibrium vapor A. The Gibbs free energy for n_A moles of component A is

$$G_{A(soln)} = n_A \mu^c_{A(soln)} = n_A \mu^o_{A(g)} + n_A RT \ln f_{A(soln)}. \tag{8.24}$$

The total Gibbs free energy for the final state after the solution is formed is given by summing Eq. 8.24 over all of the components of the solution:

$$G^{total}_{final} = \sum_{i=1}^{i=K} G_{i(soln)} = \sum_{i=1}^{i=K} n_i[\mu^o_{i(g)} + RT \ln f_{i(soln)}]. \tag{8.25}$$

Using Eqs. 8.22 and 8.25, we may now easily obtain an expression for the change in the Gibbs free energy upon mixing. This is given by

$$\Delta G_{mix} = G^{total}_{final} - G^{total}_{initial}$$

$$= \sum_{i=1}^{i=K} n_i[\mu^o_{i(g)} + RT \ln f_{i(soln)}] - \sum_{i=1}^{i=K} n_i[\mu^o_{i(g)} + RT \ln f_i] \tag{8.26}$$

$$= RT \sum_{i=1}^{i=K} n_i[\ln f_{i(soln)} - \ln f_i] = RT \sum_{i=1}^{i=K} n_i \ln\left[\frac{f_{i(soln)}}{f_i}\right].$$

Equation 8.26 is a rigorous expression for the change in chemical potential upon mixing. To use that equation, we must compute or measure the fugacities of the vapors over the pure components and over the solutions.

Since the vapor pressures are usually sufficiently low that the vapor is accurately described by the ideal-gas equation of state, we may replace the fugacities in Eq. 8.26 with the corresponding vapor pressures, producing

$$\boxed{\Delta G_{mix} = RT \sum_{i=1}^{i=K} n_i \ln\left[\frac{P_{i(soln)}}{P^o_i}\right].} \tag{8.27}$$

Equation 8.27 accurately describes the change in the Gibbs free energy upon formation of a solution. If the required vapor pressures have been measured, ΔG_{mix} may be computed directly.

For ideal solutions, Eq. 8.27 assumes a very simple form, since the ratio $P_{i(soln)}/P^o_i$ is just the mole fraction of component i in the solution. Therefore, we have

$$\boxed{\Delta G^{ideal}_{mix} = RT \sum_{i=1}^{i=K} n_i \ln X_i.} \tag{8.28}$$

Since RT and n_i are always positive and $\ln X_i < 0$ for all components, we always have $\Delta G^{ideal}_{mix} < 0$, and the formation of an ideal solution is always spontaneous for systems in which all the interactions are identical.

Equation (5.40) relates changes in the Gibbs free energy to changes in temperature and pressure at fixed composition:

$$dG = -S\,dT + V\,dP.$$

Taking partial derivatives of Eq. 8.26 with respect to temperature while holding pressure constant, we obtain

$$\left(\frac{\partial \Delta G_{\text{mix}}}{\partial T}\right)_P = \left(\frac{\partial G_{\text{final}}^{\text{total}}}{\partial T}\right)_P - \left(\frac{\partial G_{\text{initial}}^{\text{total}}}{\partial T}\right)_P$$

$$= \sum_{i=1}^{i=K}\left(\frac{\partial G_{i(\text{soln})}}{\partial T}\right)_P - \sum_{i=1}^{i=K}\left(\frac{\partial G_i}{\partial T}\right)_P. \qquad (8.29)$$

Substituting Eq. 5.40 into the right-hand side of Eq. 8.29 gives

$$\left(\frac{\partial \Delta G_{\text{mix}}}{\partial T}\right)_P = -\sum_{i=1}^{i=K} S_{i(\text{soln})} + \sum_{i=1}^{i=K} S_i = -\Delta S_{\text{mix}}. \qquad (8.30)$$

The derivative of ΔG_{mix} for an ideal solution can be obtained from Eq. 8.28. We have

$$\left(\frac{\partial \Delta G_{\text{mix}}^{\text{ideal}}}{\partial T}\right)_P = -\Delta S_{\text{mix}}^{\text{ideal}} = R\sum_{i=1}^{i=K} n_i \ln X_i,$$

so that

$$\boxed{\Delta S_{\text{mix}}^{\text{ideal}} = -R\sum_{i=1}^{i=K} n_i \ln X_i}. \qquad (8.31)$$

Equation 8.31 shows that the entropy change upon mixing is always positive, since $\ln X_i < 0$ for all components. This equation also demonstrates that if all interactions are identical, then the formation of an ideal solution is always spontaneous.

Equation 5.40 shows that if we take partial derivatives of Eq. 8.26 with respect to pressure at a fixed temperature, then instead of obtaining the entropies in Eq. 8.30, the same equation with volumes results, along with a sign change:

$$\left(\frac{\partial \Delta G_{\text{mix}}}{\partial P}\right)_T = \sum_{i=1}^{i=K} V_{i(\text{soln})} - \sum_{i=1}^{i=K} V_i = \Delta V_{\text{mix}}. \qquad (8.32)$$

Using Eq. 8.28, we find that, for ideal solutions,

$$\left(\frac{\partial \Delta G_{\text{mix}}}{\partial P}\right)_T = \frac{\partial}{\partial P}\left(RT\sum_{i=1}^{i=K} n_i \ln X_i\right)_T = \Delta V_{\text{mix}}^{\text{ideal}} = 0, \qquad (8.33)$$

since the quantity inside the rightmost set of parentheses is independent of pressure. Equation 8.33 shows that the formation of an ideal solution involves no change in volume. Thus, if a nonzero change in volume is observed, as it is for a mixture of ethanol and water, we know that the resulting solution cannot be ideal.

The enthalpy change upon mixing can be obtained by applying the Gibbs–Helmholtz equations to Eqs. 8.27 and 8.28. Equation 5.76 relates the derivative of $\Delta G/T$ with respect to temperature to ΔH by

$$\frac{\partial}{\partial T}\left(\frac{\Delta G}{T}\right)_P = -\frac{\Delta H}{T^2}. \qquad (8.34)$$

Applying Eq. 8.34 to mixing, we obtain

$$\frac{\partial}{\partial T}\left(\frac{\Delta G_{mix}}{T}\right)_P = -\frac{\Delta H_{mix}}{T^2} = \frac{\partial}{\partial T}\left\{R\sum_{i=1}^{i=K} n_i \ln\left[\frac{P_{i(soln)}}{P_i^o}\right]\right\}_P. \qquad (8.35)$$

Since the equilibrium vapor pressures over the solution and over the pure components are both functions of temperature, the enthalpy of mixing will not, in general, be zero. If the solution is ideal, however, ΔG_{mix} is given by Eq. 8.28, and the result is

$$\frac{\partial}{\partial T}\left(\frac{\Delta G_{mix}^{ideal}}{T}\right)_P = -\frac{\Delta H_{mix}^{ideal}}{T^2} = \frac{\partial}{\partial T}\left[R\sum_{i=1}^{i=K} n_i \ln X_i\right]_P = 0, \qquad (8.36)$$

so that $\Delta H_{mix}^{ideal} = 0$. This is the expected result, since if all interactions are identical, there would be no enthalpy change upon mixing.

The changes in internal energy and the Helmholtz free energy can be obtained directly from the foregoing results. For the internal energy, we have

$$\Delta U_{mix} = \Delta H_{mix} - \Delta(PV)_{mix} = \Delta H_{mix} - P\,\Delta V_{mix}, \qquad (8.37)$$

for constant-pressure mixing. For ideal solutions, we have $\Delta H_{mix}^{ideal} = \Delta V_{mix}^{ideal} = 0$. Therefore, $\Delta U_{mix}^{ideal} = 0$. A similar result is obtained for ΔA_{mix}:

$$\Delta A_{mix} = \Delta G_{mix} - \Delta(PV)_{mix} = \Delta G_{mix} - P\,\Delta V_{mix}. \qquad (8.38)$$

For an ideal solution, Eqs. 8.28 and 8.33 show that

$$\boxed{\Delta A_{mix}^{ideal} = \Delta G_{mix}^{ideal} - P\Delta V_{mix}^{ideal} = RT\sum_{i=1}^{i=K} n_i \ln X_i}. \qquad (8.39)$$

EXAMPLE 8.8

For the benzene–toluene solution described in Example 8.1, compute ΔG_{mix}^{ideal}, ΔS_{mix}^{ideal}, ΔA_{mix}^{ideal}, ΔU_{mix}^{ideal}, ΔH_{mix}^{ideal}, and ΔV_{mix}^{ideal} at 298.15 K.

Solution

The mole fractions of benzene and toluene were computed in Example 8.1 and are $X_{benzene} = 0.08624$ and $X_{toluene} = 0.9176$. The number of moles of each component were also calculated in Example 8.1 and found to be $n_{benzene} = 0.2561$ and $n_{toluene} = 2.7134$. Equation 8.28 gives

$$\Delta G_{mix}^{ideal} = RT\sum_{i=1}^{i=K} n_i \ln X_i$$

$$= (8.314\ \text{J mol}^{-1}\ \text{K}^{-1})(298.15\ \text{K})\left[(0.2561)\ln(0.08624) + (2.7134)\ln(0.9176)\right]\text{mol}$$

$$= -2{,}134\ \text{J}. \qquad (1)$$

The entropy change upon mixing is given by Eq. 8.31:

$$\Delta S_{mix}^{ideal} = -R\sum_{i=1}^{i=K} n_i \ln X_i = -\frac{\Delta G_{mix}^{ideal}}{T} = \frac{2{,}134\ \text{J}}{298.15\ \text{K}} = \underline{7.158\ \text{J K}^{-1}}. \qquad (2)$$

Equation 8.39 shows that $\Delta A_{mix}^{ideal} = \Delta G_{mix}^{ideal} = -2{,}134$ J. Finally, for an ideal solution, we have

$$\Delta U_{mix}^{ideal} = \Delta H_{mix}^{ideal} = \Delta V_{mix}^{ideal} = 0. \qquad (3)$$

For related exercises, see Problems 8.13, 8.14, and 8.15.

8.1.5   Solubility

Whenever we work with solutions, the question of solubility eventually arises. We cannot always dissolve any quantity of solute in a fixed amount of a given solvent. At constant temperature and constant pressure or volume, the equilibrium condition $\Delta\mu = 0$ frequently sets limits on the solubility of a solute. If the solution formed upon dissolution of the solute is ideal or nearly ideal, we can predict the limit of solubility or, more precisely, the composition of the saturated solution. We can also predict how that limiting composition will vary with temperature and pressure.

Let us first consider the trivial case of two liquids, A and B, that form an ideal solution. For the solution to be ideal, the A–A, A–B and B–B interactions must be identical or nearly identical. Therefore, A and B must be chemically similar molecules, as illustrated in Figure 8.2. The most favorable case is that of two isotopically substituted molecules. For example, a solution of benzene and benzene-d_1 (benzene with one hydrogen atom replaced with deuterium) will be so nearly ideal that it will be difficult to detect any deviation from Raoult's law. The benzene–toluene solution used in many of the previous examples will be close to an ideal solution, due to the similar chemical structures of the two molecules. A solution of methanol (CH_3OH) and ethanol (CH_3–CH_2OH) will also reflect ideal behavior to a significant extent. Since all interactions are identical, we have $\Delta H_{mix}^{ideal} = 0$, as demonstrated in the previous section. We also always observe that $\Delta S_{mix} > 0$ and $\Delta G_{mix}^{ideal} < 0$. This means that a solution always forms spontaneously in the case of any quantity of two liquids that form an ideal solution. Consequently, the solubility of A in B or B in A is infinite. We generally express this result by saying that the two liquids are *miscible in all proportions*. The corollary is that liquids that are not miscible in all proportions (e.g., oil and water) do not form ideal solutions. We shall examine such cases in more detail later in the chapter.

8.1.5.1   Solids in Liquids

Let us now consider the extent to which we may dissolve solid A in liquid solvent B under the condition that the A–B solution is ideal. When the solution reaches the saturation point at some temperature and pressure, we have established an equilibrium between solid $A(s)$ and A dissolved in solution, which we shall denote by $A(soln)$. Consequently, the concentration of A dissolved in the solution ceases to change with time or with added $A(s)$, thereby defining the limit of solubility, or the *saturation composition*, of the solution.

The condition for equilibrium when $dT = dP = 0$ or $dT = dV = 0$ is that the chemical potentials of the reactants and products be identical. [See Eq. 5.53.] Therefore, at the saturation point, we have

$$\mu_{A(soln)} = \mu_{A(s)}. \tag{8.40}$$

Equation 8.40 defines the composition of the saturated solution. We need only write it in a form that permits computations to be done. Since the chemical potential of a condensed phase must be identical to that of its equilibrium vapor, we may write both $\mu_{A(soln)}$ and $\mu_{A(s)}$ in terms of the chemical potentials of the equilibrium A vapor over the solution and the solid, respectively. This gives

$$\mu_{A(soln)} = \mu_{A(g)}^{o} + RT \ln f_{A(soln)} \tag{8.41}$$

and

$$\mu_{A(s)} = \mu_{A(g)}^o + RT \ln f_{A(s)} \tag{8.42}$$

where $\mu_{A(g)}^o$ is the standard chemical potential of vapor A, and $f_{A(soln)}$ and $f_{A(s)}$ are the fugacities of the equilibrium A vapor over the solution and solid A, respectively. Substituting Eqs. 8.41 and 8.42 into Eq. 8.40 yields

$$\ln f_{A(soln)} = \ln f_{A(s)} \tag{8.43}$$

as the condition for saturation. Equation 8.43 requires that the fugacities of the equilibrium vapors over the solid and the solution be equal. The solution composition that produces this equality is the solubility limit.

Equation 8.43 rigorously defines the solubility condition for any solution, ideal or otherwise. Unfortunately, we do not have a simple method of determining the solution composition which satisfies that equation. We can simplify Eq. 8.43 by noting that the vapor pressures are generally so low that we make effectively no error by assuming the vapors to be ideal. This assumption permits us to replace fugacities with pressures. The condition for saturation then becomes

$$P_{A(soln)} = P_{A(s)}. \tag{8.44}$$

The equilibrium vapor pressures of A over the solution and the pure solid must be identical. The Clausius–Clapeyron equation provides a convenient method by which we may compute $P_{A(s)}$, but we have no simple way to calculate $P_{A(soln)}$ unless the solution is ideal.

If the solution is ideal, we may use Raoult's law to write the equilibrium vapor pressure of A over the solution in terms of the solution composition and the pure component *liquid* vapor pressure of A. That is,

$$P_{A(soln)} = X_A P_{A(l)}^o, \tag{8.45}$$

where $P_{A(l)}^o$ represents the equilibrium vapor pressure over pure liquid A at the temperature of the system. This point requires careful attention. Since we are discussing the solubility of solid A in a liquid solvent, clearly compound A is a solid at the temperature of the experiment. Therefore, $P_{A(l)}^o$ must be the equilibrium vapor pressure over supercooled liquid A.

Combining Eqs. 8.44 and 8.45 yields the condition for saturation

$$X_A^{sat} P_{A(l)}^o = P_{A(s)}. \tag{8.46}$$

Solving Eq. 8.46 for X_A^{sat}, we obtain the composition of the saturated solution:

$$X_A^{sat} = \frac{P_{A(s)}}{P_{A(l)}^o}. \tag{8.47}$$

We may now employ the Clausius–Clapeyron equation to obtain expressions for each of the equilibrium vapor pressures in Eq. 8.47. This procedure gives

$$P_{A(s)} = P_o \exp\left[-\frac{\Delta \overline{H}_{sub}}{R}\left\{\frac{1}{T} - \frac{1}{T_o}\right\}\right] \text{torr}, \tag{8.48}$$

where P_o is the sublimation equilibrium vapor pressure at temperature T_o. For the supercooled liquid, we have

$$P_{A(l)}^o = 760 \exp\left[-\frac{\Delta \overline{H}_{vap}}{R}\left\{\frac{1}{T} - \frac{1}{T_b}\right\}\right] \text{torr}, \tag{8.49}$$

where T_b is the normal boiling point of liquid A. Combining Eqs. 8.48, 8.49, and 8.47 yields

$$X_A^{sat} = \frac{P_o \exp\left[-\dfrac{\Delta \overline{H}_{sub}}{R}\left\{\dfrac{1}{T} - \dfrac{1}{T_o}\right\}\right]}{760 \exp\left[-\dfrac{\Delta \overline{H}_{vap}}{R}\left\{\dfrac{1}{T} - \dfrac{1}{T_b}\right\}\right]}$$

$$= \frac{P_o}{760} \exp\left[\frac{\Delta \overline{H}_{sub}}{RT_o}\right]\exp\left[-\frac{\Delta \overline{H}_{vap}}{RT_b}\right]\exp\left[-\frac{(\Delta \overline{H}_{sub} - \Delta \overline{H}_{vap})}{RT}\right]. \quad (8.50)$$

The first three factors in Eq. 8.50 are constants, and $\Delta \overline{H}_{sub} - \Delta \overline{H}_{vap} = \Delta \overline{H}_{fus}$ if $\Delta \overline{H}$ is assumed to be constant, as we have done in using the Clausius–Clapeyron equation. Therefore, the saturation mole fraction is

$$X_A^{sat} = C \exp\left[-\frac{\Delta \overline{H}_{fus}}{RT}\right], \quad (8.51)$$

where C is a constant. To determine the value of C, we need to know or measure the solubility at one temperature. For ideal solutions, there is one temperature that is particularly convenient. When the temperature of the solution is the normal melting point of solid A, T_m, we will be mixing two liquids that form an ideal solution. In that case, the discussion at the beginning of this section shows that we have infinite solubility, so that X_A^{sat} at $T = T_m$ is unity. This allows the constant in Eq. 8.52 to be evaluated. The result is

$$C = \exp\left[\frac{\Delta \overline{H}_{fus}}{RT_m}\right].$$

Substituting into Eq. 8.51 gives

$$X_A^{sat} = \exp\left[-\frac{\Delta \overline{H}_{fus}}{R}\left\{\frac{1}{T} - \frac{1}{T_m}\right\}\right]. \quad (8.52)$$

Equation 8.52 permits the composition of the saturated solution to be determined from a knowledge of the enthalpy of fusion and melting point of the solute A.

EXAMPLE 8.9

The melting point of naphthalene ($C_{10}H_8$) is 353.05 K. Its molar heat of fusion is 19.09 kJ mol^{-1}. How many grams of naphthalene can be dissolved in 1,000 g of benzene at 298.15 K. Assume that the solution is ideal and $\Delta \overline{H}_{fus}$ is independent of temperature.

Solution

This problem is a direct application of Eq. 8.52. The limiting saturation mole fraction of naphthalene is

$$X_{C_{10}H_8}^{sat} = \exp\left[-\frac{\Delta \overline{H}_{fus}}{R}\left\{\frac{1}{T} - \frac{1}{T_m}\right\}\right] = \exp\left[-\frac{19,090}{8.314}\left\{\frac{1}{298.15} - \frac{1}{353.05}\right\}\right]$$

$$= 0.302. \quad (1)$$

The number of moles of benzene present is

$$n_{C_6H_6} = (1,000 \text{ g}) \times \frac{1 \text{ mole}}{78.11 \text{ g}} = 12.80 \text{ moles of benzene.} \quad (2)$$

If n_o is the number of moles of naphthalene present at saturation, we have

$$X_{C_{10}H_8}^{sat} = \frac{n_o}{n_o + 12.80} = 0.302. \tag{3}$$

Solving Eq. 3 for n_o, we obtain

$$n_o = \frac{(0.302)(12.80)}{0.698} = 5.538 \text{ moles of naphthalene.} \tag{4}$$

Consequently, the limiting solubility at 298.15 K is

number of grams of naphthalene per 1,000 g of benzene

$$= 5.538 \text{ moles} \times \frac{128.16 \text{ g}}{\text{mole}} = 709.8 \text{ g.} \tag{5}$$

For a related exercise, see Problem 8.16.

Equation 8.52 predicts that the limiting composition of a saturated solution varies exponentially with T^{-1}. Therefore, solubility is a highly sensitive function of temperature. In contrast, the solubility of solids is nearly independent of the applied pressure. In physical terms, this means that we cannot appreciably affect the solubility of naphthalene in benzene by placing the solution in a piston–cylinder apparatus and applying pressure.

Quantitatively, we may evaluate the effect of pressure by using Eq. 8.47. Taking logarithms of both sides, we obtain

$$\ln X_A^{sat} = \ln P_{A(s)} - \ln P_{A(l)}^o. \tag{8.53}$$

For ideal vapors, the right-hand side of Eq. 8.53 can be expressed in terms of the corresponding chemical potentials as

$$\ln P_{A(s)} = \frac{\mu_{A(s)}^o - \mu_{A(g)}^o}{RT}$$

and

$$\ln P_{A(l)}^o = \frac{\mu_{A(l)} - \mu_{A(g)}^o}{RT}.$$

We are interested in how X_A^{sat} varies with pressure. To obtain this dependence, we need the partial derivative of the saturation mole fraction with respect to P with the temperature held constant. Taking this derivative in Eq. 8.53 produces

$$\left(\frac{\partial \ln X_A^{sat}}{\partial P} \right)_T = \left(\frac{\partial \ln P_{A(s)}}{\partial P} \right)_T - \left(\frac{\partial \ln P_{A(l)}^o}{\partial P} \right)_T$$

$$= \frac{\partial}{\partial P} \left[\frac{\mu_{A(s)} - \mu_{A(g)}^o}{RT} \right]_T - \frac{\partial}{\partial P} \left[\frac{\mu_{A(l)} - \mu_{A(s)}^o}{RT} \right]_T. \tag{8.54}$$

Since the standard chemical potential is independent of pressure, the derivatives of $\mu_{A(g)}^o$ with respect to pressure vanish. Using $d\mu = -\overline{S}\,dT + \overline{V}\,dP$ [see Eq. 5.44], we have

$$\left(\frac{\partial \ln X_A^{sat}}{\partial P} \right)_T = \frac{\overline{V}_{(s)} - \overline{V}_{(l)}}{RT} = -\frac{\Delta \overline{V}}{RT}, \tag{8.55}$$

where $\overline{V}_{(s)}$ and $\overline{V}_{(l)}$ are the partial molar volumes of solid and liquid A, respectively, and $\Delta \overline{V}$ is the partial change in molar volume upon melting. Since $\Delta \overline{V}$ is very small for condensed phases, the effect of pressure on the solubility of solids is negligible. For example, for water, we have $\overline{V}_{(s)} = 0.019647$ L mol^{-1} and $\overline{V}_{(l)} = 0.018016$ L mol^{-1}. Therefore, $\Delta \overline{V} = -0.00163$ L mol^{-1} and $-\Delta \overline{V}/(RT) = 7.28 \times 10^{-5}$ atm^{-1}. Under normal pressure conditions, we may ignore pressure effects on the solubility of solids.

8.1.5.2   Gases in Liquids

The treatment of gaseous solubility in liquids follows the same lines as that of solids, except that the notation (s) is changed to (g). Thus, the equilibrium condition that the chemical potentials of $A(g)$ and $A(soln)$ be equal at saturation is given by Eq. 8.40 with the given subscript change:

$$\mu_{A(soln)} = \mu_{A(g)}. \tag{8.56}$$

By replacing fugacities with partial pressures and assuming the solution to be ideal, we obtain the analogue of Eq. 8.47 giving the composition of the saturated solution:

$$X_A^{sat} = \frac{P_{A(g)}}{P^o_{A(l)}}. \tag{8.57}$$

$P^o_{A(l)}$ is now the equilibrium vapor pressure over the superheated pure liquid A that is normally a gas at the temperature of the experiment. Using the Clausius–Clapeyron equation, we can compute

$$P^o_{A(l)} = P^* \exp\left[-\frac{\Delta \overline{H}_{vap}}{R} \left\{ \frac{1}{T} - \frac{1}{T^*} \right\} \right], \tag{8.58}$$

where P^* is the equilibrium vapor pressure at temperature T^*. Substituting Eq. 8.58 into Eq. 8.57 gives

$$X_A^{sat} = \frac{P_{A(g)}}{P^*} \exp\left[\frac{\Delta \overline{H}_{vap}}{R} \left\{ \frac{1}{T} - \frac{1}{T^*} \right\} \right]. \tag{8.59}$$

Equation 8.59 shows that the composition of the saturated solution at a given temperature is proportional to the applied pressure of the gas to be dissolved. This is in marked contrast to the situation with solids, where pressure has very little effect. The increased solubility of gases at higher partial pressures is seen in everyday life by examining bottles containing carbonated beverages. When the bottle is sealed, no $CO_2(g)$ bubbles are visible on the walls of the container. All the $CO_2(g)$ is in solution. When the seal is broken, however, the pressure of the CO_2 is released and the $CO_2(g)$ solubility decreases. The result is the formation of $CO_2(g)$ bubbles on the container walls. This same effect is responsible for the condition known as "the bends" in deep-sea diving. Divers rely on compressed air for their oxygen supply. At the higher pressure, both $N_2(g)$ and $O_2(g)$ exhibit higher solubility in the blood. If the pressure is suddenly reduced by exposing the diver to atmospheric pressure, bubbles of $N_2(g)$ and $O_2(g)$ form in the bloodstream, where they function to block the flow of blood, thereby damaging the nervous system. Of the two gases, $N_2(g)$ causes the more serious problem, because it has the higher partial pressure in air and because $O_2(g)$ can be removed through the metabolic process, whereas $N_2(g)$ can be removed only through the respiratory system. To avoid the bends, divers must surface slowly, with frequent stops along the way, to permit the respiratory system time to gradually remove $N_2(g)$ as it forms. The problem can also be reduced by using a mixture of $O_2(g)$ and $He(g)$ instead of air. The solubility of $He(g)$ in biological fluids is much lower than that for $N_2(g)$. Jacques Cousteau's divers used a mixture of 98% $He(g)$ and 2% $O_2(g)$.

It is possible to put Eq. 8.59 in a more compact form by choosing the temperature T^* to be that temperature for which P^* is exactly equal to the applied pressure of $A(g)$. If this is done, we will have $P_{A(g)} = P^*$, and the two pressures will cancel. If the only pressure on the solution is that due to $A(g)$, then the temperature T^* at which we have $P_{A(g)} = P^*$ will be the boiling point of liquid A under an applied pressure equal to $P_{A(g)}$. This permits us to write $T^* = T_b$, so that Eq. 8.59 becomes

$$X_A^{\text{sat}} = \exp\left[\frac{\Delta \overline{H}_{\text{vap}}}{R}\left\{\frac{1}{T} - \frac{1}{T_b}\right\}\right]. \tag{8.60}$$

Although Eq. 8.60 appears to be a more convenient form than Eq. 8.59, it has some major disadvantages. First, the equation conceals the dependence of the gaseous solubility on the applied pressure of $A(g)$, since $P_{A(g)}$ no longer appears. The dependence is still present, of course, in that T_b depends upon that pressure. However, there is a tendency to overlook this fact and begin to think of T_b as the normal boiling point of $A(l)$. This will be the case, however, if and only if the applied pressure of gaseous A on the solution is 760 torr. Such misidentification of symbols can create serious problems.

Equations 8.59 and 8.60 show that the solubility of gases decreases as T increases. Everyone who has ever opened a can of a carbonated beverage has learned this fact from experience. When the can is taken from the refrigerator and opened, most of the $CO_2(g)$ carbonation is in solution, and very little gas escapes as the can is opened. However, if the can is opened while it is hot, the beverage often spews out under the force of escaping $CO_2(g)$ that was under high pressure because the gas had come out of solution as the temperature increased.

It is frequently observed that the solubilities of real gases in liquids appear to obey Eq. 8.60 in that a plot of $\ln X_A^{\text{sat}}$ vs. T^{-1} is linear or nearly linear. However, unless the compounds involved are very similar, so that ideality is approached, the slope of the plot will not be the partial molar enthalpy of vaporization as suggested by Eq. 8.60. The reason for this behavior may be seen by using Eq. 5.89 and considering the solution process to consist of two steps. In the first step, the gas is condensed to a liquid, after which the two liquids are mixed in the second step. The energetics of these steps are illustrated in the following reaction and in Figure 8.9:

$$
\begin{array}{ccc}
\text{Step \#1} & & \text{Step \#2} \\
& & B_{(l)} \\
A_{(g)} \xrightarrow{\hspace{2cm}} A_{(l)} & \xrightarrow{\hspace{2cm}} & (A-B)_{\text{soln}} \\
\Delta H_1 = -\Delta H_{\text{vap}} & \Delta H_2 = \Delta H_{\text{mix}} &
\end{array}
$$

For an ideal solution, $\Delta \overline{H}_{\text{mix}} = 0$, so that the total energy requirement for the process is simply the negative of the enthalpy of vaporization of gaseous A. For a real solution, however, the total energy requirement is $-\Delta \overline{H}_{\text{vap}} + \Delta \overline{H}_{\text{mix}}$, where $\Delta \overline{H}_{\text{mix}}$ refers to the enthalpy change that occurs when 1 mole of liquid A is mixed with sufficient solvent $B(l)$ to form a solution whose concentration is X_A^{sat}. Since the mole fraction at saturation measures the extent to which the pre-

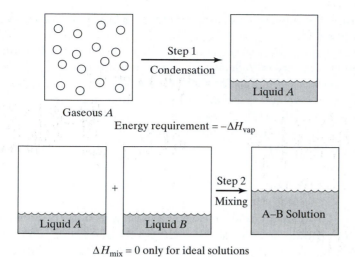

◀ FIGURE 8.9
Steps involved in solution formation of a gas in a liquid solvent. If the solution is near ideal, ΔH_{mix} in the second step will be near zero, so that the total energy requirement for the process will be $-\Delta H_{vap}$. For nonideal solutions, the total energy requirement is the sum $-\Delta H_{vap} + \Delta H_{mix}$. ΔH_{mix} can be very large in magnitude.

ceding solution process occurs, Eq. 5.89 suggests that if the energy requirement for the process is nearly independent of temperature over the temperature range of the experiments, we will observe

$$X_A^{sat} = C \exp\left[-\frac{\text{energy requirement}}{RT}\right],$$

so that a plot of $\ln X_A^{sat}$ versus T^{-1} will be linear with a slope equal to $-(\text{energy requirement})/RT$. If the solution is ideal, this slope will be $\Delta \overline{H}_{vap}/R$. If, however, the solution deviates substantially from ideality, we expect a slope equal to $(\Delta \overline{H}_{vap} - \Delta \overline{H}_{mix})/R$. If $\Delta \overline{H}_{vap}$ is known, then we may simultaneously assess the extent of ideality of the solution and obtain an estimate of the enthalpy of mixing of the liquids, using solubility measurements on gases. Example 8.10 and Problems 8.18 and 8.19 serve as illustrations of this concept. Figure 8.10 shows the measured solubilities of six gases in H_2O. In each case, $\ln X_A^{sat}$ is plotted against T^{-1}. The near linearity of the plot for each gas is apparent,

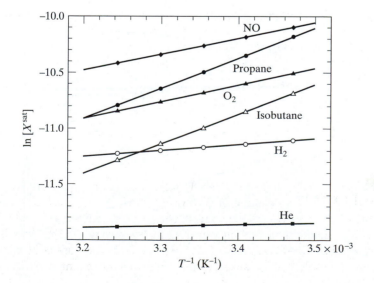

◀ FIGURE 8.10
The solubilities of six gases in H_2O as a function of temperature. In each case, the logarithm of the saturation mole fraction is plotted against T^{-1}. The points are the measured data. The straight lines are least-squares fits. The excellent linearity of each plot is apparent, even though the solutions are not ideal. The reasons underlying this behavior are explained in the text. (Source of data: The *Handbook of Chemistry and Physics*, 78th edition, CRC Press, Boca Raton, FL, 1997–1998.)

even though the solutions are, in some cases, such as propane and isobutane, far from being ideal.

The foregoing analysis provides an explanation of why the solubility of some solids is observed to decrease with temperature whereas Eq. 8.52 indicates that the solubility of solids should increase with temperature and that a plot of $\ln X_{A(s)}^{\text{sat}}$ versus T^{-1} should be linear with a slope equal to $-\Delta\overline{H}_{\text{fus}}^{A}/R$. Replacing $A(g)$ with $A(s)$, in the analysis indicates that, for real solutions, the energy requirement for the dissolution of solid A is $\Delta\overline{H}_{\text{fus}}^{A} + \Delta\overline{H}_{\text{mix}}$. This leads to the result that

$$\ln X_{A_{(s)}}^{\text{sat}} \approx -\frac{\Delta\overline{H}_{\text{fus}} + \Delta\overline{H}_{\text{mix}}}{RT} + \text{constant}.$$

If we have a solid compound for which mixing of its liquid melt with solvent is a highly exothermic process with $|\Delta\overline{H}_{\text{mix}}| > \Delta\overline{H}_{\text{fus}}$, then an inverse temperature dependence of the solubility will be the result. Problem 8.20 illustratives this effect. Figure 8.11 shows the measured solubilities of five ionic solids in H_2O as a function of temperature. In each case, $\ln X_A^{\text{sat}}$ is plotted against T^{-1}. The results are not quite as linear as those in Figure 8.10, which is not surprising, since the solutions formed by these ionic compounds are far from ideal. However, the deviations from linearity are not large. Li_2CO_3 is an example of a solid whose solubility shows a negative temperature dependence. This observation permits us to conclude that the mixing of liquid Li_2CO_3 with H_2O must be a highly exothermic process. The large enthalpy of hydration of the Li^+ ion is the primary reason underlying this behavior.

It is also possible to have a gas whose solubility in a liquid increases as the temperature rises. The previous discussion suggests that the temperature dependence of gaseous solubility has the form

$$X_{A(\text{gas})}^{\text{sat}} = C \exp\left[\frac{\Delta\overline{H}_{\text{vap}} - \Delta\overline{H}_{\text{mix}}}{RT}\right].$$

If the enthalpy of mixing is sufficiently endothermic, such that we have $\Delta\overline{H}_{\text{mix}} > \Delta\overline{H}_{\text{vap}}$, then $X_{A(\text{gas})}^{\text{sat}}$ will increase with increasing temperature. Such behavior is in fact observed: Gas solubilities often decrease with temperature, go through a minimum, and then increase as the temperature continues to rise.

▶ FIGURE 8.11
The solubilities of five ionic solids in H_2O as a function of temperature. In each case, the logarithm of the saturation mole fraction is plotted against T^{-1}. The points are the measured data. The straight lines are least-squares fits. The results for $MgCl_2$, KBr, PbI_2, and Li_2CO_3 exhibit good linearity. Those for $AgClO_2$ are less linear. The negative temperature dependence of the solubility of Li_2CO_3 is due to the highly exothermic enthalpy of hydration of the Li^+ ion. (Source of data: The *Handbook of Chemistry and Physics*, 78th edition, CRC Press, Boca Raton, FL, 1997–1998.)

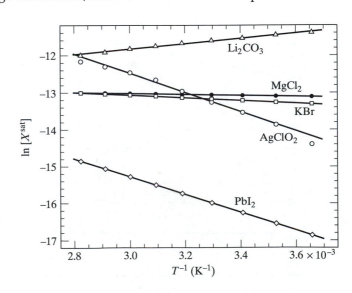

It should be apparent that the preceding analysis is only semiquantitatively correct. The use of Eq. 5.89 requires that the energy requirement for the process be independent of temperature. If the heat capacities of the solid, liquid, and gaseous phases of the solute are not identical, both $\Delta \overline{H}_{vap}$ and $\Delta \overline{H}_{fus}$ will vary with temperature. The errors introduced by these assumptions will usually be small; therefore, this simple analysis is often useful.

Finally, we emphasize that the accuracy of Eqs. 8.52 and 8.60 is very limited because of failure of the ideal-solution approximations. If the solute and solvent are similar, as in the case of $H_2S(g)$ dissolved in $H_2O(l)$ (see Example 8.10), the results are fair. If the equations are applied to dissimilar solutes and solvents, such as N_2O and H_2O (see Problem 8.19) or CO_2 and H_2O (see Problem 8.20), the errors will be quite large. Problem 8.20 illustrates how bad the ideal-solution assumption can be when we are dealing with an ionic solute, wherein the interparticle forces are very different.

EXAMPLE 8.10

The solubility of $H_2S(g)$ in $H_2O(l)$ has been measured at temperatures between 273.15 K and 373.15 K. Some of the data are given in the table to the right, in which the solubility S is reported in units of grams of $H_2S(g)$ dissolved per 100 grams of $H_2O(l)$ when the applied pressure of $H_2S(g)$ is 760 torr:

(A) Prepare a plot of $\ln X_{H_2S}^{sat}$ vs. T^{-1}. Obtain the best straight-line fit to the results. **(B)** Using the slope of the best straight-line fit, compute the molar heat of vaporization of H_2S. The measured result is 18,670 J mol^{-1}. What is the percent error in your result? **(C)** Use the result of (B) to predict the value of S at $T = 343.15\ K$. The measured result is 0.1101 g H_2S per 100 g of H_2O. Calculate the percent error in your predicted value. **(D)** How well does the ideal-solution approximation hold for this system? Why do you think this is the case? **(E)** Use the data to estimate the molar enthalpy of mixing for $H_2S(l)$ and $H_2O(l)$ to form saturated solutions.

T (K)	S
273.15	0.7066
283.15	0.5112
293.15	0.3846
303.15	0.2648
313.15	0.2361
333.15	0.1480

Solution

(A) To execute the required plot, we must convert S to mole fraction. The required equation is

$$X_{H_2S}^{sat} = \frac{S/M}{S/M + 100.0/18.016},$$ (1)

where M is the molar mass of H_2S, 34.086 g mol^{-1}.

The data required for the plot are given in the table below:

T^{-1} (K^{-1})	$X_{H_2S}^{sat}$	$\ln X_{H_2S}^{sat}$
0.0036610	0.0037208	−5.59381
0.0035317	0.0026946	−5.91649
0.0034112	0.0020287	−6.20038
0.0032987	0.0013976	−6.57298
0.0031934	0.0012463	−6.68754
0.0030017	0.0007816	−7.15412

The plot is shown in Figure 8.12. The line is the best linear fit to the data in the table. The equation of the fit is $\ln X_{H_2S}^{sat} = -14.284 + 2367.32\,(K)/T$.

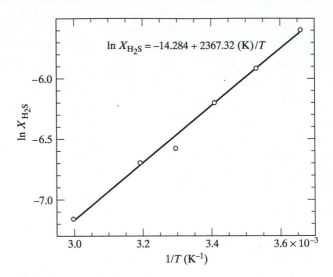

▲ FIGURE 8.12

Measured solubility of $H_2S(g)$ in $H_2O(l)$ as a function of temperature. The logarithm of the saturation mole fraction if plotted against T^{-1}. The points are the measured data. The straight line is a least-squares fit whose equation is given in the figure. The slope of the line provides an estimate of the partial molar enthalpy of vaporization of $H_2S(g)$, as described in Example 8.10.

(B) Equation 8.60 shows that the slope of the line should be $\Delta\overline{H}_{vap}/R$. This gives

$$\Delta\overline{H}_{vap} = (2367.32\ \text{K})(8.314\ \text{J mol}^{-1}\,\text{K}^{-1}) = 19682\ \text{J mol}^{-1}. \tag{2}$$

The percent error in this result is

$$\%\ \text{error} = \frac{100 \times (19682 - 18670)}{18670} = 5.42\%. \tag{3}$$

(C) The solubility at 343.15 K is predicted to be

$$\ln X_{H_2S}^{sat} = -14.282 + \frac{2{,}367.32}{343.15} = -7.3832, \tag{4}$$

which gives

$$X_{H_2S}^{sat} = 0.0006216. \tag{5}$$

Solving Eq. 1 for S, we obtain

$$S = \frac{100\,X_{H_2S}^{sat}\,M}{18.016\,(1 - X_{H_2S}^{sat})} = \frac{(100)(0.0006216)(34.086)}{18.016(1 - 0.0006216)}$$

$$= 0.1177\ \text{g}\ H_2S\ \text{per 100 g}\ H_2O. \tag{6}$$

The percent error in the result is

$$\%\ \text{error} = \frac{100 \times (0.1177 - 0.1101)}{0.1101} = 6.90\%. \tag{7}$$

(D) The ideal-solution approximation holds reasonably well for the H$_2$S–H$_2$O solution. The reason for this is the similarity of the two compounds. Both are bent polar molecules with nearly the same structure; consequently, the interactions are close to being the same.

(E) The discussion in the text indicates that the difference between the molar enthalpy of vaporization computed from the slope of a plot of ln $X_{H_2S}^{sat}$ versus T^{-1} and the measured value reflects the molar enthalpy of mixing of H$_2$S$_{(l)}$ and H$_2$O$_{(l)}$. Using this concept, we have

$$-\Delta\overline{H}_{mix} \approx (\text{slope})\,R - (\Delta\overline{H}_{vap})_{expt} = 19{,}682 - 18{,}670 \text{ J mol}^{-1} = 1{,}012 \text{ J mol}^{-1}. \quad \textbf{(8)}$$

This result suggests that the molar enthalpy change for the mixing of 1 mole of H$_2$S(l) with sufficient H$_2$O(l) to form saturated solutions is about -1 kJ. Such a small value indicates that the solution is reasonably well described by the ideal-solution laws. For contrasting situations, examine the results obtained in Problems 8.18 and 8.19.

8.1.6 Colligative Properties
8.1.6.1 Depression of Freezing Point

Whenever a nonvolatile solute is dissolved in a solvent, the freezing point of the solvent is lowered by an amount that depends upon the quantity of solute dissolved. This occurs because the addition of the solute lowers the equilibrium vapor pressure of the solvent, thereby decreasing the fugacity of the solvent, which in turn leads, as we shall see, to a lowering of its freezing point.

Consider a solution of a nonvolatile solute A in a solvent B. Let the mole fraction of A be X_A. Let T_f be the freezing point of the solution. At this temperature, solid solvent $B(s)$ is in equilibrium with liquid solvent $B(l)$. At either constant pressure or volume, the condition for equilibrium is

$$\mu_{B(soln)} = \mu_{B(s)}. \quad \textbf{(8.61)}$$

Once again, we can express the chemical potential of a condensed phase in terms of the chemical potential of its equilibrium vapor, since the two must be equal. This gives

$$\mu_{B(soln)} = \mu_{B(soln)}^{o} + RT \ln f_{B(soln)} \quad \textbf{(8.62)}$$

and

$$\mu_{B(s)} = \mu_{B(g)}^{o} + RT \ln f_{B(soln)}, \quad \textbf{(8.63)}$$

where $f_{B(soln)}$ and $f_{B(s)}$ are the fugacities of the equilibrium B vapor over the solution and solid, respectively. Combining the preceding three equations shows that at $T = T_f$, we must have

$$f_{B(soln)} = f_{B(s)}. \quad \textbf{(8.64)}$$

Since the equilibrium vapor pressures are usually very low, the fugacities may be replaced with partial pressures without significant loss of accuracy. This produces

$$P_{B(soln)} = P_{B(s)}. \quad \textbf{(8.65)}$$

We can use the Clausius–Clapeyron equation to obtain $P_{B(s)}$, but there is no convenient method for obtaining $P_{B(soln)}$ unless the solution is ideal. Since the nonvolatile solute A usually has intermolecular forces that are significantly different from those of the volatile solvent, we expect the solution to be ideal with respect to the solvent only in the ideal-solution regime, where $X_B \to 1$ and $X_A \to 0$. That is, the solution must be very dilute in solute A. If this is the case, the vapor pressure of B over the solution is given by Raoult's law, and we have

$$X_B P^o_{B(l)} = P_{B(s)},\tag{8.66}$$

where $P^o_{B(l)}$ is the equilibrium vapor pressure over the pure liquid solvent B. This pressure can be obtained from the Clausius–Clapeyron equation. For pure liquid B, we have

$$P^o_{B(l)} = 760 \exp\left[-\frac{\Delta \overline{H}^B_{vap}}{R}\{T_f^{-1} - (T_b^o)^{-1}\}\right] \text{torr},\tag{8.67}$$

where T_b^o is the normal boiling point of $B(l)$ and $\Delta \overline{H}^B_{vap}$ is the partial molar enthalpy of vaporization of $B(l)$. For the solid, applying the Clausius–Clapeyron equation yields

$$P_{B(s)} = P^* \exp\left[-\frac{\Delta \overline{H}^B_{sub}}{R}\{T_f^{-1} - (T^*)^{-1}\}\right] \text{torr},\tag{8.68}$$

where $\Delta \overline{H}^B_{sub}$ is the partial molar enthalpy of sublimation of $B(s)$ and P^* is the equilibrium sublimation pressure at temperature T^*. Substituting Eqs. 8.67 and 8.68 into Eq. 8.66 produces

$$X_B = \frac{P^*}{760} \frac{\exp\left[-\dfrac{\Delta \overline{H}^B_{sub}}{R}\{T_f^{-1} - (T^*)^{-1}\}\right]}{\exp\left[-\dfrac{\Delta \overline{H}^B_{vap}}{R}\{T_f^{-1} - (T_b^o)^{-1}\}\right]}$$

$$= \frac{P^*}{760} \exp\left[\frac{\Delta \overline{H}^B_{sub}}{RT^*}\right]\exp\left[-\frac{\Delta \overline{H}^B_{vap}}{RT_b^o}\right]\exp\left[-\frac{\Delta \overline{H}^B_{sub} - \Delta \overline{H}^B_{vap}}{RT_f}\right].\tag{8.69}$$

If $\Delta \overline{H}^B_{sub}$ and $\Delta \overline{H}^B_{vap}$ are independent of temperature, $\Delta \overline{H}^B_{sub} - \Delta \overline{H}^B_{vap} = \Delta \overline{H}^B_{fus}$. We have already made this assumption in developing the Clausius–Clapeyron equation. Since the first three factors in Eq. 8.69 contain nothing but constants characteristic of the solvent, we may express the result in the form

$$X_B = C \exp\left[-\frac{\Delta \overline{H}^B_{fus}}{RT_f}\right],\tag{8.70}$$

where C is a constant. To evaluate C, we need the freezing point for one value of X_B. A particularly convenient value to use is $X_B = 1$, with $T_f = T_f^o$, the normal freezing point of pure solvent. This gives $C = \exp[\Delta \overline{H}^B_{fus}/RT_f^o]$ and

$$\boxed{X_B = \exp\left[-\frac{\Delta \overline{H}^B_{fus}}{R}\left\{\frac{1}{T_f} - \frac{1}{T_f^o}\right\}\right].}\tag{8.71}$$

Equation 8.71 gives the freezing point of the solution, T_f, in terms of the composition of the solution, the heat of fusion, and the normal freezing point of the pure solvent. The result is expected to be accurate whenever X_B is sufficiently close to unity that the solution is in the ideal-solution regime. With a little algebra, we may easily take logarithms of both sides of Eq. 8.71 and solve the resulting equation for T_f. We obtain

$$T_f = \frac{T_f^o \Delta \overline{H}^B_{fus}}{\Delta \overline{H}^B_{fus} - RT_f^o \ln X_B}.\tag{8.72}$$

If we define the freezing-point depression to be $\Delta T_f = T_f - T_f^o$, Eq. 8.72 gives

$$\Delta T_f = \frac{T_f^o \Delta \overline{H}_{fus}^B}{\Delta \overline{H}_{fus}^B - RT_f^o \ln X_B} - T_f^o = \frac{R(T_f^o)^2 \ln X_B}{\Delta \overline{H}_{fus}^B - RT_f^o \ln X_B}. \qquad (8.73)$$

Since the denominator of Eq. 8.73 is always positive and $\ln X_B$ is always negative, we have $\Delta T_f < 0$ for any solute at any composition. The freezing point for a solution whose solvent is in the ideal-solution regime is always depressed below the normal freezing point. Note that the right-hand side of Eq. 8.73 is dependent upon X_B, which in turn is dependent upon the molar mass of the solute. Therefore, measurements of ΔT_f can be used as a means of determining solute molecular masses. Example 8.11 is illustrative.

If the solute ionizes in solution, the concentration of all species in solution must be taken into account in the computation of X_B in Eq. 8.73. For the example of NaCl(c) used to obtain Eqs. 8.7 through 8.10, we would have

$$X_B = \frac{n_o}{n_1 + fn_1 + n_o},$$

where n_0 and n_1 are the number of moles of solvent and NaCl(c), respectively, and f is the fraction of the NaCl(s) that ionizes.

EXAMPLE 8.11

5.0000 g of a nonvolatile solute are dissolved in 100 g of $H_2O(1)$. The solution is found to freeze at 272.880 K. The normal freezing point of H_2O is 273.150 K, and the molar enthalpy of fusion of $H_2O(s)$ is 6003.1 J mol^{-1}. Compute the molar mass of the solute.

Solution

The freezing-point depression is

$$\Delta T_f = T_f - T_f^o = 272.880 \text{ K} - 273.150 \text{ K} = -0.270 \text{ K}. \qquad (1)$$

Using Eq. 8.71, we may now compute the mole fraction of solvent, X_B, in the solution:

$$X_B = \exp\left[-\frac{\Delta H_{fus}^B}{R}\left\{\frac{1}{T_f} - \frac{1}{T_f^o}\right\}\right] = \exp\left[-\frac{6003.1}{8.3145}\left\{\frac{1}{272.880} - \frac{1}{273.150}\right\}\right]$$

$$= 0.99739. \qquad (2)$$

We may also write X_B in the form

$$X_B = \frac{n_B}{n_A + n_B} = \frac{100/18.016}{100/18.016 + 5/M}, \qquad (3)$$

where M is the molar mass of the solute. Rearranging Eq. 3 produces

$$\frac{100 X_B}{18.016} + \frac{5 X_B}{M} = \frac{100}{18.016}. \qquad (4)$$

$$\frac{5 X_B}{M} = \frac{100(1 - X_B)}{18.016}. \qquad (5)$$

Solving Eq. 5 for M, we obtain

$$M = \frac{(5)(18.016)(X_B)}{100(1 - X_B)} = \frac{(5)(18.016)(0.99739)}{100(1 - 0.99739)} = 344.2 \text{ g mol}^{-1}. \qquad (6)$$

Since the ideal-solution law will be accurate for the solvent only if $X_B \to 1$, Eq. 8.73 holds only for very dilute solutions. In view of this, it is customary to write that equation in its dilute-solution limiting form. When $X_B \to 1$, we generally have $|\Delta H_{fus}^B| \gg |RT_f^\circ \ln X_B|$, so that Eq. 8.73 may be written in the form

$$\Delta T_f \approx \frac{R(T_f^\circ)^2 \ln X_B}{\Delta \overline{H}_{fus}^B}. \tag{8.74}$$

The logarithm term can be simplified by expansion in a Taylor series about the point $X_A = 0$:

$$\ln X_B = \ln[1 - X_A] = -X_A - \frac{X_A^2}{2} - \frac{X_A^3}{3} + \cdots .$$

When X_A approaches zero, we may truncate this expansion after the first term and write

$$\ln X_B \approx -X_A = -\frac{n_A}{n_A + n_B} \approx -\frac{n_A}{n_B}.$$

Substituting these approximations into Eq. 8.74 produces

$$\Delta T_f \approx -\frac{R(T_f^\circ)^2 \, n_A}{n_B \, \Delta \overline{H}_{fus}^B}. \tag{8.75}$$

In most applications of this limiting form, the ratio n_A/n_B is written in terms of the solution molality m, which is defined as the number of moles of solute per kilogram of solvent. For a solution that is m molal, we have $n_A = m$ and $n_B = 1000/M_B$, where M_B is the molar mass of the solvent. In these terms, Eq. 8.75 becomes

$$\boxed{\Delta T_f \approx -\left[\frac{R(T_f^\circ)^2 M_B}{1000 \Delta \overline{H}_{fus}^B}\right] m = -K_f m}, \tag{8.76}$$

where K_f is called the *molal freezing-point depression constant* or the *cryoscopic constant*. Note that its value depends only upon the normal freezing point, the molar heat of fusion, and the molar mass of the solvent. No properties of the solute appear. For a solute that ionizes, n_A must include the contributions of all ions and undissociated solute. The required modification for NaCl(c) is to replace m with $(1 + f)m$, where m is the molality of the NaCl(c) solution if no ionization occurs and f is the fraction of NaCl(c) that ionizes.

EXAMPLE 8.12

Use the dilute-solution limiting form for the freezing-point depression to solve the problem in Example 8.11. Compute the percent error produced by the limiting form.

Solution

We first need to calculate the molal freezing-point depression constant for H_2O:

$$K_f = \left[\frac{R(T_f^\circ)^2 M_B}{1{,}000 \, \Delta \overline{H}_{fus}^B}\right] = \frac{(8.314 \text{ J mol}^{-1}\text{K}^{-1})(273.15 \text{ K})^2 (18.016 \text{ g mol}^{-1})}{(1{,}000 \text{ g})(6003.1 \text{ J mol}^{-1})}$$

$$= 1.862 \text{ K mol}^{-1}. \tag{1}$$

The molality of the solution is

$$m = \frac{-\Delta T_f}{K_f} = -\frac{272.880 \text{ K} - 273.150 \text{ K}}{1.862 \text{ K mol}^{-1}} = 0.145 \text{ molal}. \tag{2}$$

The molality of a solution containing 5 grams of solute with a molar mass M g mol^{-1} in 100 g of H_2O is

$$m = \frac{(5 \text{ g}/M)}{0.1 \text{ kg}} = 0.145 \text{ mol kg}^{-1}. \tag{2}$$

Solving Eq. 2 for M, we obtain

$$M = \frac{(5 \text{ g})}{0.1(0.145 \text{ mol})} = 345 \text{ g mol}^{-1}. \tag{3}$$

The percent error in the result relative to that obtained using the full equation is

$$\% \text{ error} = \frac{100 \times (345 - 344.2)}{344.2} = 0.23\%. \tag{4}$$

The ideal-solution assumption probably produces more error than this.
    For a related exercise, see Problem 8.22.

8.1.6.2   Elevation of Boiling Point

We have already noted that, because the formation of a solution with a non-volatile solute depresses the vapor pressure, we expect the boiling point to rise. Let us again consider a solution of a nonvolatile solute A in solvent B, whose mole fraction is X_A.

If the applied pressure is P_o, then, at the boiling point $T = T_b$, we must have

$$P_{B(\text{soln})} = P_o. \tag{8.77}$$

If the solution is sufficiently dilute that Raoult's law holds for the solvent, we also have

$$X_B P_{B(l)}^o = P_o \tag{8.78}$$

as the boiling condition. The vapor pressure over pure liquid B, $P_{B(l)}^o$, is given by the Clausius–Clapeyron equation. Inserting this result into Eq. 8.78 and solving for X_B gives

$$X_B = \frac{P_o}{P^*} \exp\left[\frac{\Delta \overline{H}_{\text{vap}}^B}{R}\left\{\frac{1}{T_b} - \frac{1}{T^*}\right\}\right], \tag{8.79}$$

where P^* is the equilibrium vapor pressure over pure liquid B at temperature T^*. Taking logarithms of both sides, we may easily solve this equation for T_b:

$$T_b = \frac{T^* \Delta \overline{H}_{\text{vap}}^B}{\Delta \overline{H}_{\text{vap}}^B + RT^* \ln\left[\dfrac{P^* X_B}{P_o}\right]}. \tag{8.80}$$

The elevation of the boiling point is $\Delta T_b = T_b - T_{bo}$, where T_{bo} is the boiling point of the pure solvent under an applied pressure of P_o. Subtracting T_{bo} from both sides of Eq. 8.80 yields

$$\Delta T_b = \frac{\Delta \overline{H}_{\text{vap}}^B (T^* - T_{bo}) - RT^* T_{bo} \ln\left[\dfrac{P^* X_B}{P_o}\right]}{\Delta \overline{H}_{\text{vap}}^B + RT^* \ln\left[\dfrac{P^* X_B}{P_o}\right]} \tag{8.81}$$

We may simplify Eq. 8.81 by taking T^* to be that temperature which produces an equilibrium vapor pressure over the pure solvent of P_o. With this choice, we have $T^* = T_{bo}$ and $P^* = P_o$. Inserting these equalities into Eq. 8.81 gives

$$\Delta T_b = \frac{-RT_{bo}^2 \ln X_B}{\Delta \overline{H}_{vap}^B + RT_{bo} \ln X_B}. \tag{8.82}$$

When one uses Eq. 8.82, it is very important to remember that T_{bo} is not the normal boiling point of the solvent; it is the boiling point of the pure solvent under an applied external pressure P_o. Therefore, T_{bo} will be the normal boiling point only if $P_o = 760$ torr. The same warning about misinterpreting symbols as discussed in our treatment of solubility is appropriate here: By taking T^* to be T_{bo}, so that $P^* = P_o$, we obtain Eq. 8.82, which seems to suggest that the change in boiling point is independent of the applied pressure, since P_o no longer appears. As long as the proper interpretation of T_{bo} is used, it is clear that P_o is implicitly contained in Eq. 8.82 in that T_{bo} must be computed before one uses the equation. If this point is overlooked, disaster ensues. (See Problem 8.41.)

Equation 8.82 is of the same form as Eq. 8.73 for the freezing-point depression. The only differences are a minus sign, the replacement of $\Delta \overline{H}_{fus}^o$ with $\Delta \overline{H}_{vap}^o$, and the replacement of T_f^o with T_{bo}. The limiting dilute-solution form of Eq. 8.82 is, therefore, the same as Eq. 8.76 with the aforesaid replacements:

$$\boxed{\Delta T_b \approx \left[\frac{RT_{bo}^2 M_B}{1,000 \, \Delta \overline{H}_{vap}^B} \right] m = K_b(P_o) \, m}, \tag{8.83}$$

where $K_b(P_o)$ is called the *molal boiling-point elevation constant* or the *ebullioscopic constant*. Since T_{bo} appears in the definition of $K_b(P_o)$, the latter "constant" depends upon the applied pressure. The values listed for $K_b(P_o)$ in many texts are those for $P_o = 760$ torr.

As usual, if the nonvolatile solute ionizes, appropriate care must be taken in the computation of X_B in Eq. 8.82 and of m in Eq. 8.83.

EXAMPLE 8.13

(A) Determine the boiling point of the solution considered in Example 8.11 if the applied pressure is 760 torr. **(B)** Determine the boiling point of the solution if the applied pressure is 500 torr. **(C)** Determine the boiling point of the solution if the applied pressure is 1,520 torr. **(D)** Compute $K_b(P_o)$ for each of these cases. What is the percent change in $K_b(P_o)$ from 500 to 1,520 torr? Is boiling-point elevation a more pronounced colligative property at high or low applied pressure? The molar heat of vaporization of $H_2O(l)$ is 40670.7 J mol^{-1}. Assume this value to be a constant.

Solution

(A), (B), and (C): In all parts of the problem, we need to determine the temperature at which pure $H_2O(l)$ has an equilibrium vapor pressure equal to the given applied pressure. The equilibrium vapor pressure is given by the Clausius–Clapeyron equation:

$$P_{eq} = 760 \exp\left[-\frac{\Delta \overline{H}_{vap}}{R} \left\{ \frac{1}{T} - \frac{1}{373.15} \right\} \right] \text{ torr.} \tag{1}$$

Solving Eq. 1 for T, we obtain

$$T = \frac{373.15 \, \Delta \overline{H}_{vap}}{\Delta \overline{H}_{vap} - 373.15 \, R \ln(P_{eq}/760)} = \frac{(373.15)(40670.7)}{40670.7 - (373.15)(8.314) \ln(P_{eq}/760)} \text{ K.} \tag{2}$$

The temperatures required to produce equilibrium vapor pressures of 760 torr, 500 torr, and 1,520 torr are, respectively,

$$T = \frac{(373.15)(40670.7)}{40670.7 - (373.15)(8.314)\ln(760/760)} \text{ K} = 373.15 \text{ K}, \tag{3}$$

$$T = \frac{(373.15)(40670.7)}{40670.7 - (373.15)(8.314)\ln(500/760)} \text{ K} = 361.60 \text{ K}, \tag{4}$$

and

$$T = \frac{(373.15)(40670.7)}{40670.7 - (373.15)(8.314)\ln(1520/760)} \text{ K} = 393.98 \text{ K}. \tag{5}$$

Using the results of Example 8.11, we find that the mole fraction of the solvent in solution is

$$X_{H_2O} = \frac{100/18.016}{100/18.016 + 5/344.2} = 0.99739. \tag{6}$$

The boiling-point elevation at each of the applied pressures given in the problem can be computed directly from Eq. 8.82:

$$\Delta T_b = \frac{-RT_{bo}^2 \ln X_B}{\Delta \overline{H}_{vap}^B + RT_{bo} \ln X_B} = -\frac{(8.314)(373.15)^2 \ln(0.99739)}{40670.7 + (8.314)(373.15)\ln(0.99739)}$$

$$= 0.0744 \text{ K at 760 torr applied pressure.} \tag{7}$$

The solution therefore boils at $T_b = 373.15 \text{ K} + 0.0744 \text{ K} = 373.22 \text{ K}$. If the applied pressure is 500 torr, the result is

$$\Delta T_b = -\frac{(8.314)(361.60)^2 \ln(0.99739)}{40670.7 + (8.314)(361.60)\ln(0.99739)} = 0.0697 \text{ K}. \tag{8}$$

The solution now boils at $T_b = 361.60 \text{ K} + 0.0697 \text{ K} = 361.67 \text{ K}$. For an applied pressure of 1520 torr, we obtain

$$\Delta T_b = -\frac{(8.314)(393.98)^2 \ln(0.99379)}{40670.7 + (8.314)(393.98)\ln(0.99739)} = 0.0829 \text{ K}, \tag{9}$$

which gives a boiling point of $T_b = 393.98 + 0.0829 \text{ K} = 394.06 \text{ K}$.

(D) The molal boiling-point elevation constants are given by Eq. 8.83:

$$K_b(P_o) = \left[\frac{RT_{bo}^2 M_B}{1,000 \, \Delta \overline{H}_{vap}^B} \right]. \tag{10}$$

Therefore, we have

$$K_b(760 \text{ torr}) = \frac{(8.314)(373.15)^2(18.016)}{1000(40670.7)} = 0.5128 \text{ K mol}^{-1}, \tag{11}$$

$$K_b(500 \text{ torr}) = \frac{(8.314)(361.60)^2(18.016)}{1000(40670.7)} = 0.4816 \text{ K mol}^{-1}. \tag{12}$$

and

$$K_b(1,520 \text{ torr}) = \frac{(8.314)(393.98)^2(18.016)}{1,000(40670.7)} = 0.5717 \text{ K mol}^{-1}. \tag{13}$$

The percent change in $K_b(P_o)$ from $P_o = 500$ torr to $P_o = 1,520$ torr is

$$\% \text{ change} = \frac{100 \times (0.5717 - 0.4816)}{0.4816} = 18.71\%. \tag{14}$$

Obviously, the magnitude of the boiling-point elevation increases with applied pressure. Figure 8.13 shows a more detailed plot of $K_b(P_o)$ versus P_o.

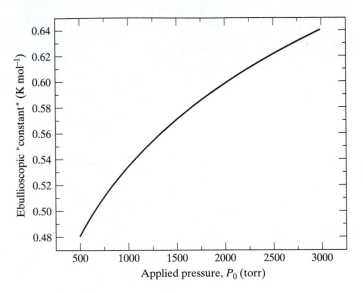

▲ FIGURE 8.13
The variation of the ebullioscopic "constant" for H_2O as a function of applied pressure.

For a related exercise, see Problem 8.41.

8.1.6.3   Osmotic Pressure

The magnitudes of the freezing-point depression and boiling-point elevation for dilute solutions are relatively small. In contrast, osmotic pressure is an extremely large and very important effect.

Consider two compartments, denoted 1 and 2 in Figure 8.14. Compartment 1 contains pure liquid solvent B. Compartment 2 contains a dilute solution of solute A in solvent B with mole fraction $X_B < 1$. The two compartments are separated by a molecular sieve whose pore size permits the transmission of solvent B molecules, but not solute A molecules, between compartments. Let us ask what will happen in such a situation.

The chemical potential of $B(l)$ and $B(\text{soln})$ in each compartment must be equal to the chemical potentials of the equilibrium vapors. Therefore,

$$\mu_{B(l)} = \mu^o_{B(\text{soln})} + RT \ln f_{B(l)} \tag{8.84}$$

▶ FIGURE 8.14
Schematic diagram of an apparatus for measuring osmotic pressure. The molecular sieve permits the transmission of B molecules, but not solute A molecules. Because $\mu_{B(l)} > \mu_{B(\text{soln})}$, solvent flows from Compartment 1 to Compartment 2, building up a hydrostatic pressure head equal to the osmotic pressure π when equilibrium is reached.

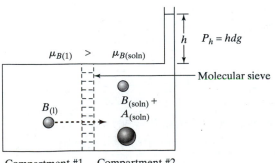

and

$$\mu_{B(soln)} = \mu_{B(g)}^o + RT \ln f_{B(soln)}. \tag{8.85}$$

At the low vapor pressure expected for most systems, we may accurately replace the fugacities with partial pressures. If the solution in Compartment 2 is sufficiently dilute that Raoult's law holds for the solvent, we have

$$\mu_{B(l)} = \mu_{B(g)}^o + RT \ln P_{B(l)}^o \tag{8.86}$$

and

$$\mu_{B(soln)} = \mu_{B(g)}^o + RT \ln P_{B(soln)} = \mu_{B(g)}^o + RT \ln [X_B P_{B(l)}^o]. \tag{8.87}$$

Since $X_B < 1$, we have $\mu_{B(l)} > \mu_{B(soln)}$, so that the system is not in a state of equilibrium. B molecules will spontaneously flow from Compartment 1, where they have a high chemical potential, into Compartment 2, where their chemical potential is lower. Such a transfer process has $\Delta\mu < 0$ and is, therefore, spontaneous. The flow will continue until the two chemical potentials become equal.

Equations 8.86 and 8.87 make it appear that we can have $\mu_{B(l)} > \mu_{B(soln)}$ only at the point $X_B = 1$. As B molecules flow into Compartment 2, X_B will gradually increase, but it can never become equal to unity. This suggests that B molecules will continue to flow into Compartment 2 indefinitely. However, such an analysis is flawed, because it ignores the effect of the hydrostatic pressure head that will build up in Compartment 2. As liquid B flows into Compartment 2, liquid must continuously rise up the tube on the right, thereby building up a hydrostatic pressure P_h, which Pascal's principle tells us is transmitted equally throughout the solution. Example 1.2 shows that this hydrostatic pressure is equal to hdg, where h is the height of the column, d is the density of the solution, and g is the acceleration due to gravity close to the earth's surface. The analysis that follows shows that the thermodynamic effect of this hydrostatic pressure is to increase the chemical potential of B molecules in the solution. When the pressure is sufficiently high, the chemical potential of B molecules in Compartment 2 will become equal to that in Compartment 1. At that point, equilibrium is established, and the flow of B molecules into Compartment 2 ceases. The hydrostatic pressure at this point is called the *osmotic pressure*. Its value is generally denoted by the symbol π.

To quantitatively determine the osmotic pressure, we must find the rate of change of chemical potential with applied pressure under constant-temperature conditions. This is given by Eq. 5.44:

$$d\mu = -\overline{S} \, dT + \overline{V} \, dP.$$

The required rate of change of μ is

$$\left(\frac{\partial\mu}{\partial P}\right)_T = \overline{V} \tag{8.88}$$

where $\overline{V}$ is the partial molar volume. Since $\overline{V}$ is always positive, Eq. 8.88 confirms the statement made in the previous paragraph that the presence of the hydrostatic pressure will increase the chemical potential in Compartment 2. Taking partial derivatives of Eq. 8.87 with respect to pressure at constant temperature, we obtain

$$\left(\frac{\partial\mu_{B(soln)}}{\partial P_h}\right)_T = \left(\frac{\partial\mu_{B(g)}^o}{\partial P_h}\right)_T + RT\left(\frac{\partial \ln P_{B(soln)}}{\partial P_h}\right)_T = \overline{V}_{B(l)}. \tag{8.89}$$

Since $\mu^0_{B(g)}$ is independent of pressure, the first term on the right side of Eq. 8.89 vanishes, and we have

$$\left(\frac{\partial \ln P_{B(soln)}}{\partial P_h}\right)_T = \frac{\overline{V}_{B(l)}}{RT}. \tag{8.90}$$

Separating variables and then integrating both sides between corresponding limits produces

$$\int_{X_B P^o_{B(l)}}^{P^o_{B(l)}} d\ln P_{B(soln)} = \int_0^\pi \frac{\overline{V}_{B(l)}}{RT} dP_h. \tag{8.91}$$

Before proceeding, a word about the limits on the integrals in Eq. 8.91 is appropriate. When the hydrostatic pressure in Compartment 2 is zero, the equilibrium vapor pressure of B in that compartment is given by Raoult's law. That is, we have $P_{B(soln)} = X_B P^o_{B(l)}$ when $P_h = 0$. These corresponding values give us the lower limits on the integrals. When the hydrostatic pressure becomes equal to the osmotic pressure, the chemical potentials of B in both compartments are equal and equilibrium is attained. At that point, we have $P_h = \pi$ and $P_{B(soln)} = P^o_{B(l)}$. This point provides the upper limits on the integrals.

If we assume the liquid to be incompressible, $\overline{V}_{B(l)}/(RT)$ may be factored out of the integral. Under this condition, integrating Eq. 8.91 gives

$$\ln\left[\frac{P^o_{B(l)}}{X_B P^o_{B(l)}}\right] = \ln\left[\frac{1}{X_B}\right] = -\ln X_B = \frac{\overline{V}_{B(l)}\pi}{RT}. \tag{8.92}$$

Solving for the osmotic pressure, we obtain

$$\boxed{\pi = \frac{-RT \ln X_B}{\overline{V}_{B(l)}}}. \tag{8.93}$$

If the novolatile solute ionizes, appropriate care must be taken in the computation of X_B in equation 8.93. The magnitude of the osmotic-pressure effect may be seen in the next, numerical example.

EXAMPLE 8.14

Five grams of a nonvolatile solute whose molar mass is 344.2 g mol^{-1} are dissolved in 100 g of H_2O at 298.15 K. The solution is placed in an osmotic-pressure apparatus with a molecular sieve that permits the flow of H_2O, but not the solute, between compartments. Compute the osmotic pressure of this solution. How high is the liquid column shown in Figure 8.14 when equilibrium is established?

Solution

The mole fraction of B in the solution is given by

$$X_B = \frac{n_B}{n_A + n_B} = \frac{100/18.016}{5/344.2 + 100/18.016} = 0.99739. \tag{1}$$

The molar volume of H_2O is 0.018016 L mol^{-1}. Substituting the given data into Eq. 8.93 yields

$$\pi = \frac{-(0.08206 \text{ L atm mol}^{-1} \text{K}^{-1})(298.15 \text{ K}) \ln(0.99739)}{0.018016 \text{ L mol}^{-1}} = 3.549 \text{ atm}. \tag{2}$$

If the solvent were mercury, the height of the liquid column would be $(760 \text{ mm Hg atm}^{-1})(3.549 \text{ atm}) = 2697.2 \text{ mm Hg}$. Since the density of mercury is 13.55 times that of H_2O, the height of the liquid water column is $13.55 \times 2697.2 \text{ mm} = 36547 \text{ mm} = 36.55 \text{ m}$!

The magnitude of the osmotic pressure in this example is so large that the hydrostatic pressure could rupture the apparatus before equilibrium is attained. A much more dilute solution would have to be employed to make the measurement practical.

For a related exercise, see Problem 8.22.

Example 8.14 makes it clear that osmotic-pressure measurements can be made only on very dilute solutions. For that reason, it is appropriate to use a limiting form of Eq. 8.93. Because X_B approaches unity, we have

$$\ln X_B = \ln(1 - X_A) \approx -X_A = -\frac{n_A}{n_A + n_B} \approx -\frac{n_A}{n_B},$$

as shown previously when we obtained the limiting dilute-solution form of the freezing-point depression equation. Substituting into Eq. 8.93 gives

$$\pi = \frac{n_A RT}{n_B \overline{V}_{B(l)}}. \tag{8.94}$$

Since $n_B \overline{V}_{B(l)}$ is the total volume of solvent in Compartment B, Eq. 8.94 becomes

$$\pi = \frac{n_A RT}{V_B}. \tag{8.95}$$

If Eq. 8.95 is written in the form

$$\boxed{\pi V_B = n_A RT}, \tag{8.96}$$

it is called the *van't Hoff equation*. This equation bears a striking resemblance to the ideal-gas equation of state. However, in using this limiting form, it is important to remember that V_B refers to the total volume of the solvent, whereas n_A is the number of moles of solute present.

The magnitude of the osmotic-pressure effect makes it an ideal method for determining the molar masses of large macromolecules. If we express Eq. 8.95 in terms of the mass of solute dissolved, w, and its molar mass M_A, the result is

$$\pi = \frac{wRT}{M_A V_B},$$

which we can write in the form

$$\boxed{\frac{\pi}{w} = \frac{RT}{M_A V_B}}. \tag{8.97}$$

If we measure the temperature and the osmotic pressure for a given mass of solute, we can compute the molar mass from Eq. 8.97. However, solutions of large molecules frequently show significant deviations from Raoult's law. Therefore, the best practice is to measure the ratio π/w as a function of w and then extrapolate the results to $w \rightarrow 0$, where the solvent will behave ideally, so that Eq. 8.97 will be accurate. Example 8.15 illustrates the procedure.

w g L^{-1}	π (atm)
1.00	0.0002656
2.00	0.0006734
4.00	0.0019064
7.00	0.0048372
9.00	0.0075878

w (g L^{-1})	$\dfrac{\pi}{w}$ (atm L g^{-1})
1.00	0.0002656
2.00	0.0003367
4.00	0.0004766
7.00	0.0006910
9.00	0.0008431

EXAMPLE 8.15

The osmotic pressures of a large polymer in cyclohexanone have been measured at several concentrations at 298 K. Some of the data are shown in the table to the left. **(A)** Plot π/w vs. w and extrapolate the curve to $w \rightarrow 0$. **(B)** Using the result obtained in (A), compute the molar mass of the polymer.

Solution

(A) Let us assume that we have 1 liter of solution. Then the data needed for the plot are shown in the second table to the left. Figure 8.15 shows the variation of π with w. The line is the best linear least-squares fit to the data. The equation of the line is $\pi/w = [0.00019180 + 7.1914 \times 10^{-5}\,(\text{L g}^{-1})\,w]$ atm L g^{-1}. The intercept at $w \rightarrow 0$ is $(\pi/w)_o = 0.00019180$ atm L g^{-1}.

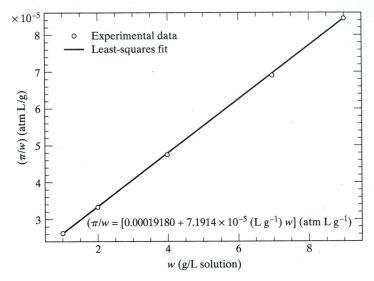

▲ **FIGURE 8.15**
Example of the extrapolation of the measured ratio π/w as a function of w to infinite dilution ($w = 0$), at which point the solution will behave ideally, and Eq. 8.97 will be accurate.

(B) Using Eq. 8.97 with the extrapolated value for π/w, we obtain

$$M_{\text{polymer}} = \frac{RT}{V(\pi/w)_o} = \frac{(0.08206\ \text{L atm mol}^{-1}\,\text{K}^{-1})(298\ \text{K})}{1.0\ \text{L}(0.00019180\ \text{atm L g}^{-1})} = 1.275 \times 10^5\ \text{g mol}^{-1}.$$

Osmotic pressure gradients are vital to many life processes in plants and animals. *In vivo*, the molecular sieve shown in Figure 8.14 is replaced with semipermeable membranes. In some biological processes, active work must be done by the cell to overcome natural osmotic-pressure gradients and thereby force materials to move in nonspontaneous directions.

8.2 Nonideal Solutions

When the concentrations of solutions are sufficiently high that Raoult's law is no longer accurate, we have no reliable method of computing the equilibrium partial vapor pressures and the associated fugacities above the solution. Consequently, vapor pressures and vapor composition, solubilities, ΔU, ΔH, ΔS, ΔA, and ΔG for mixing, freezing-point depression, boiling-point elevation, and osmotic pressures must be determined by experimental measurement. As we shall see, it is not necessary that we measure all these quantities separately, but the basis of their determination must rest on the results of laboratory experiments.

When any of the preceding quantities are measured as a function of temperature, pressure, composition, and perhaps other variables, the results are generally represented by a huge array of numerical data. For the most part, the human mind has great difficulty in grasping the significance of large numerical compilations. For this reason, it is common practice to utilize various types of diagrams to convey the essential details of the experimental results. Vapor pressure–composition and temperature–composition phase diagrams can be used effectively to convey experimentally determined information about nonideal systems. In this section, we shall examine a variety of such diagrams for nonideal solutions and thereby gain an appreciation for the type of behavior that is observed.

8.2.1   Vapor Pressures for Nonideal Solutions

The intermolecular interactions present in many solutions will be sufficiently different that deviations from Raoult's law will be observed. When pairs of unlike molecules attract each other more than pairs of like molecules, negative deviations from Raoult's law will generally be seen. That is, the equilibrium vapor pressures will be less than those for an ideal solution with the same concentration. Moreover, the solution will tend to have a more uniform composition on a microscopic scale than might be expected from purely statistical considerations. Extreme cases might even involve actual A–B association due to hydrogen bonding or other strong dipole–dipole interactions. On the other hand, if the A–A and B–B interactions are much more attractive than A–B interactions, each component will tend to cluster about itself on a microscopic scale, thereby making the solution less uniform and with more pronounced microscopic fluctuations than would be predicted from purely random statistical theory. This type of solution usually exhibits positive deviations from Raoult's law, with the equilibrium vapor pressures being greater than those for an ideal solution. If the A–A and B–B interactions are greatly different from the A–B interactions, the mixture will separate into two phases: an A phase that is saturated with B and a B phase that is saturated with A.

Figure 8.16 shows a typical example of a system that exhibits negative deviations from Raoult's law. For this system and for those solutions exhibiting

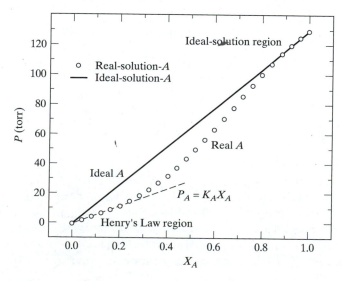

◀ FIGURE 8.16
A hypothetical solution that exhibits negative deviations from Raoult's law. As X_A approaches unity, the measured data are seen to approach the ideal-solution line. The range of concentrations for which $X_A \longrightarrow 1$ is termed the ideal-solution region for component A. We also observe the measured vapor pressures to vary linearly with X_A as $X_A \longrightarrow 0$, but with a different slope. This is called the Henry's law region for component A.

positive deviations, there are certain definite common features. First, we note that as $X_A \to 1$, the partial pressure of the nearly pure liquid component asymptotically approaches the ideal-solution line. That is, the real partial-pressure curve becomes tangent to the line representing Raoult's law. This type of behavior is clearly seen in the figure. As we have previously noted, such behavior occurs because almost all the interactions involving A molecules will be identical as $X_A \to 1$, which is termed the ideal-solution region. On the other hand, at sufficient low concentrations, the equilibrium vapor pressure of the dilute component is found to vary linearly with X_A but with a slope that is different from the line representing Raoult's law because now the interactions are almost all A–B interactions. That is, in the limit as $X_A \to 0$, we observe that

$$\boxed{P_A = K_A X_A}. \tag{8.98}$$

Equation 8.98 is known as *Henry's law*, and K_A is called the *Henry's law constant*. This feature of real solutions is illustrated in Figure 8.16.

In a subsequent section of this chapter, we shall show that when Raoult's law holds for component B in a solution, Henry's law must hold for component A over the same range of concentrations. That is, if component B is in the ideal-solution region when X_B lies in the range from 0.9 to 1.0, the equilibrium vapor pressure of component A must be given by Eq. 8.98 over this same range. For an ideal solution, this statement is obviously true, since $P_A = X_A P_A^o$ for all concentrations, so that the Henry's law constant is simply equal to P_A^o.

A solution of acetone and chloroform is an example of a solution that exhibits negative deviations from Raoult's law. The measured vapor pressures as a function of composition at 308.35 K are shown in Figure 8.17. Although the deviations from the ideal-solution approximation are significant, the measured equilibrium partial pressures are seen to approach the Raoult's law lines when the respective mole fractions approach unity. It is also apparent that Henry's law holds in the regions where the mole fractions approach zero.

Large positive deviations from Raoult's law are seen in solutions of acetone and carbon disulfide. The vapor pressure–composition diagram for this

▶ **FIGURE 8.17**
Measured equilibrium partial vapor pressures at 308.35 K over a chloroform–acetone solution as a function of the composition of the solution. The experimental data and ideal-solution predictions are shown as solid and dashed lines, respectively. Since the measured vapor pressures lie below those predicted by Raoult's law, the solution exhibits negative deviations from the idea-solution approximation. Note that the measured data coincide with the Raoult's law lines as the mole fractions approach unity. The Henry's law slopes are indicated by the arrows.

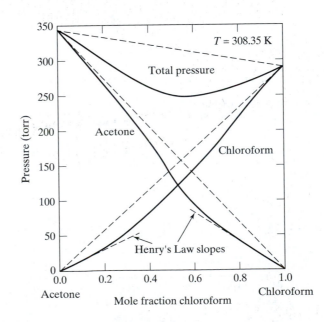

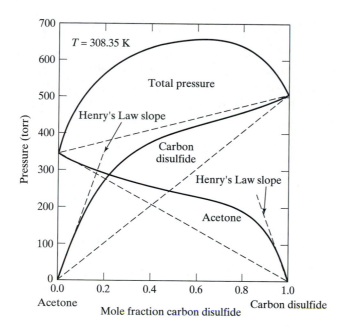

◄ FIGURE 8.18
Measured equilibrium partial vapor pressures at 308.35 K over a carbon disulfide–acetone solution as a function of the composition of the solution. The experimental data and ideal-solution predictions are shown as solid and dashed lines, respectively. Since the measured vapor pressures lie above those predicted by Raoult's law, the solution exhibits positive deviations from the ideal-solution approximation. Note that the measured data coincide with the Raoult's law lines as the mole fractions approach unity. The Henry's law slopes are indicated by the arrows.

system at 308.35 K is shown in Figure 8.18. Again, we note that as the mole fractions approach unity, the measured data coincide with the Raoult's law lines and in the dilute region, Henry's law is obeyed.

Sometimes the deviations from Raoult's law are so large that the total vapor pressure of the solution plotted as a function of mole fraction will exhibit a maximum or a minimum. Such extrema are seen in Figures 8.17 and 8.18. When this occurs, separation of the mixture by fractional distillation will be impossible. Instead, the distillate will form what is known as a *constant-boiling mixture* or a *constant-boiling azeotrope*. Once this composition is reached, no further separation by distillation is possible, because the liquid and vapor will have the same composition. The point is illustrated by the boiling point, solution–vapor composition diagrams for acetone–chloroform and ethanol-benzene shown in Figures 8.19 and 8.20, respectively.

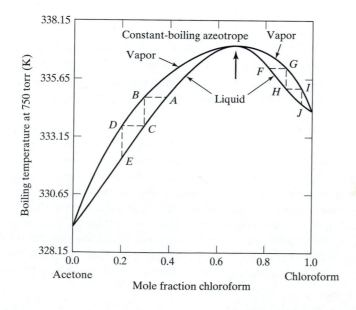

◄ FIGURE 8.19
Measured boiling points of acetone–chloroform solutions at 750 torr applied pressure as a function of the composition of the solution. The curve marked "vapor" gives the composition of the vapor in equilibrium with the solution at the temperature shown on the ordinate. The composition at the maximum (indicated by the heavy arrow) corresponds to the constant-boiling azeotrope.

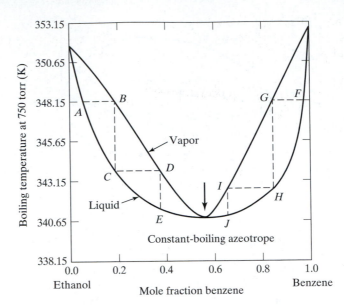

► FIGURE 8.20

Measured boiling points of ethanol–benzene solutions at 750 torr applied pressure as a function of the composition of the solution. The curve marked "vapor" gives the composition of the vapor in equilibrium with the solution at the temperature shown on the ordinate. The composition at the minimum (indicated by the heavy arrow) corresponds to the constant-boiling azeotrope.

The total pressure for an acetone–chloroform solution exhibits a minimum. Figure 8.17 shows such a minimum at $X_{CHCl_3} \approx 0.55$ when the temperature is 308.35 K. The minimum in the total vapor pressure produces a maximum in the boiling point–composition diagram in Figure 8.19. Suppose we have an acetone–chloroform solution whose composition corresponds to the maximum in this figure, $X_{chloroform} = 0.64$. If we attempt to separate this solution by fractional distillation, we will fail, because the equilibrium vapor has the same composition as the solution. Therefore, the fractionation process will produce a distillate of the same composition. An acetone–chloroform solution with $X_{chloroform} = 0.64$ is a constant-boiling azeotrope. If we begin the fractionation process with the solution represented by Point A in the figure, which is rich in acetone, we can obtain pure acetone in the distillate, provided that the column has a sufficient number of theoretical plates. The first step will produce a vapor whose composition is that at Point B. Condensation of this vapor produces the solution at Point C. The next vaporization–condensation cycle produces the solution at point E. Eventually, we will reach a point where the vapor distillate is essentially pure acetone. The solution remaining in the distillation pot will then be the constant-boiling azeotrope, which cannot be separated by fractionation. If we start with a solution rich in chloroform, such as that at Point F, fractionation up the column produces solutions with the compositions at Points H and J. With enough plates, the distillate will be pure chloroform. Again, the distillation pot will contain the constant-boiling azeotrope.

An ethanol–benzene solution has a maximum in its total vapor pressure curve when plotted as a function of solution composition. Therefore, its boiling point–composition diagram has a minimum, as seen in Figure 8.20. In this case, if we fractionate a solution originally rich in ethanol, the vaporization–condensation cycles taking place within the column will produce solutions such as those represented by Points C and E in the figure. If there are enough theoretical plates in the column, the distillate will be the azeotropic mixture containing about 0.58 mole percent benzene, while the solution in the distillation pot will become nearly pure ethanol. If we start with a solution rich in benzene, such as that represented by Point F, fractionation will produce the solutions at points *H* and *J* as we move up the column. With enough theoretical plates, we will again distill the constant-boiling azeotrope. In this case, the distillation pot will contain nearly pure benzene.

8.2.2   Liquid–Liquid Solubility

When two liquids form an ideal solution, the solubility of either liquid in the other is infinite. That is, the liquids are miscible in all proportions. Highly similar liquids, such as benzene–toluene and ethanol–water, exhibit this behavior. In most cases, however, the solubility of one liquid in another is limited. For example, if we mix 0.1 kg of water and 0.1 kg of *n*-butanol at 313 K, we observe the liquids to separate into two phases. One phase is primarily *n*-butanol saturated with water. The second phase is water saturated with *n*-butanol. If we raise the temperature of this mixture to 405 K, the two phases coalesce, forming a single, homogeneous solution because the solubility has increased.

The complete phase diagram for the *n*-butanol–water system can be determined by preparing a mixture of known composition, heating the mixture until only one homogeneous phase is present, and then cooling the system until two liquid phases appear. The temperature at which this occurs is recorded, and the experiment is repeated with a mixture of a different composition. By repeating the procedure numerous times and plotting the temperature at which a solution forms as a function of the composition of the mixture, we obtain the phase diagram for the system. Figure 8.21 shows the result for the *n*-butanol–water system. The measured data points are represented as open circles, with the curve connecting the experimental data.

The phase diagram shows that when *n*-butanol is mixed with pure water at 293.15 K, it initially forms a single-phase solution. As the mass percent of *n*-butanol added is increased, we eventually reach the limit of solubility. The diagram shows that this occurs at around 8% by mass of *n*-butanol. At that point, the chemical potential of pure *n*-butanol and *n*-butanol in the solution are equal, and we are in a state of equilibrium for the saturated solution. The phase diagram also shows that if the temperature is raised to 393.15 K, the saturated water solution now contains about 20% by mass of *n*-butanol. If we start with pure *n*-butanol on the right of the diagram and add water at 293.15 K, the solubility limit is reached when we have 20% by mass of water. The smooth curve through the data points represents the solubility of *n*-butanol in water and water in *n*-butanol as a function of temperature. Above the

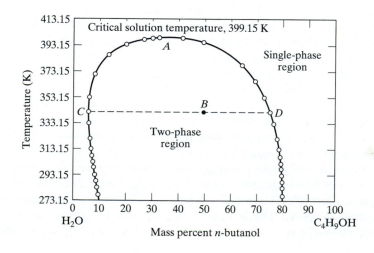

◀ FIGURE 8.21
Solubility phase diagram for the *n*-butanol–water system.

curve, the two components are in solution and we have only a single phase. Below the curve, we have two phases. One phase is a water solution saturated with n-butanol; the second is an n-butanol solution saturated with water.

When we are in the two-phase region of Figure 8.21, the diagram allows us to predict the composition of both phases, and if we know the initial amounts of material mixed, we can also compute the actual amounts of each compound present in each phase. For example, suppose that we mix 100 g of n-butanol with 100 g of water at 343.15 K. This temperature corresponds to point B in the phase diagram, which is below the saturation curve. The mixture will, consequently, separate into two phases. The compositions of these phases must be the saturated solutions at 343.15 K. The compositions of the saturated solutions are found by drawing a tie-line at 343.15 K through Point B and locating the intersections of the line with the saturation curve at Points C and D. Point C gives the composition of the n-butanol-saturated water solution, which we see is about 6% n-butanol and 94% water. Point D gives the composition of the water-saturated n-butanol solution, which the phase diagram indicates is about 76% n-butanol and 24% water. The actual amounts of each phase can be easily determined by mass balance. The phase diagram shows that if the water-saturated n-butanol solution contains w_1 grams of H_2O and w_2 grams of n-butanol, then $w_2/w_1 = 76/24$. Mass balance requires that the mass of water and the mass of n-butanol in the n-butanol-saturated water solution must be $100 - w_1$ and $100 - w_2$, respectively. Therefore, $(100 - w_1)/(100 - w_2) = 94/6$. If we solve these two equations for w_1 and w_2, the result is $w_1 = 30.17$ g and $w_2 = 95.54$ g. This means that 125.71 g of the water-saturated n-butanol solution and 74.29 g of the n-butanol-saturated water solution are present. (See Problem 8.35 for an additional example.)

Figure 8.21 shows that when the temperature exceeds 399.15 K, which is labeled Point A in the diagram, we will always be in the single-phase region regardless of the solution composition. Therefore, above that temperature, n-butanol and water are miscible in all proportions. The temperature for which this is true is called the *critical-solution temperature*.

8.2.3   Solid–Liquid Solubility

If the solution were ideal, we could employ Eq. 8.52 to compute the solubility of a solid in a liquid and Eq. 8.72 to determine the temperature at which the solvent freezes. For nonideal solutions, these quantities must be determined by experimental measurements. Figure 8.11 shows the results of solubility measurements for aqueous solutions of five different ionic salts.

A two-component system of great interest involves two metals that are completely miscible in the liquid state at high temperatures. The interest derives from the many practical applications for metals and their alloys. The discussion that follows illustrates how we may determine the temperature–concentration phase diagram for such systems.

Consider a system in which 400 grams of bismuth are mixed with 100 grams of cadmium at 600 K. Observation shows that at this temperature the system is in the liquid state and that the two metals are miscible in all proportions. We now begin to extract thermal energy from the system at a constant rate. If we ignore the temperature dependence of the heat capacities, we expect this operation to cause a steady, linear decrease in the temperature, as shown in Figure 8.22. At time $t = 0$ minutes, the temperature of the solution is 600 K. The temperature then steadily decreases with time until it reaches 478 K. At this point, the rate of decrease of the temperature

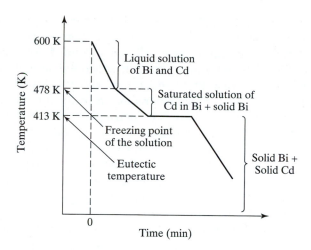

◀ FIGURE 8.22
Schematic diagram of a cooling curve for a Bi + Cd mixture that is originally 20 mass percent Cd.

changes because Bi begins to freeze out of the solution. The heat of fusion released in the process serves to slow the rate at which the mixture cools, causing a change in the slope of the curve. This tells us that the freezing point of the 80–20 mass-percent Bi–Cd solution is 478 K. Experimental data such as that schematically represented in the figure are called *cooling curves*. By repeating the experiment with different mixture compositions, we can determine the freezing point for solutions of all compositions. The points so determined can then be plotted to create a temperature–composition phase diagram such as that shown in Figure 8.23.

Above the lines *AB* and *AD*, the Bi–Cd system forms a single, homogeneous phase. In this region, liquid bismuth and cadmium are miscible in all proportions. When the 80–20 mass-percent Bi–Cd solution discussed in the previous paragraph is cooled, the cooling curve shows that freezing of bismuth occurs at 478 K. This is Point *C* in the phase diagram. At that point, Bi and the saturated Bi–Cd solution have equal chemical potentials and are,

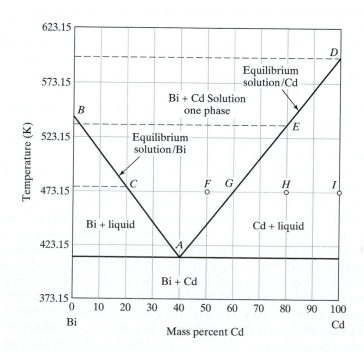

◀ FIGURE 8.23
Temperature–concentration phase diagram for the bismuth–cadmium system. The eutectic corresponds to a 40–60 mass percent Cd–Bi solution that freezes at 413 K.

therefore, in equilibrium. The AB line is the loci of freezing points obtained in the cooling-curve experiments. The corresponding freezing points of Cd from systems rich in this element are shown as line AD. Points below line AB at temperatures above 413 K have compositions in which the amount of Bi present exceeds the solubility of Bi at that temperature. Points below line AD at temperatures above 413 K have compositions in which the amount of cadmium exceeds the solution solubility. Point H is such a point: With 80% Cd present at 473.15 K, not all the cadmium can dissolve. Consequently, this is a two-phase region in which a saturated Cd–Bi solution and solid Cd are present. The composition of the saturated solution is found by using the same technique as that described for the liquid–liquid phase diagram shown in Figure 8.21. A tie-line is drawn at 473.15 K through Point H. Its intersection with line AD at Point G gives the saturated solution composition. The intersection on the right shows that the second phase is pure Cd. In this case, we see that the saturated solution contains 60% Cd and 40% Bi by mass. Beneath line AB at temperatures above 413 K, pure Bi and a saturated Bi–Cd solution are present. If we know the total mass of each component present at any point on the phase diagram, the quantities of each phase at any other point may be computed in a manner analogous to that illustrated for liquid–liquid solutions.

Along line AB, solid bismuth and a bismuth-saturated cadmium solution are in equilibrium at all points; along line AD, solid cadmium and a cadmium-saturated bismuth solution are in equilibrium. When these two lines intersect at Point A, we have equilibrium between three phases: solid Bi, solid Cd, and a saturated solution whose composition is that at Point A. This situation is analogous to the triple point in the phase diagram for a single component, at which solid, vapor, and liquid are in equilibrium. For two-component systems, such points in the phase diagram are called *eutectic points,* and the composition and temperature of the solution at eutectic point are called the *eutectic composition and temperature,* respectively. When a 20%–80% Cd–Bi solution is cooled from 600 K, we observe a break in the cooling curve at the bismuth freezing point of 478 K. This phase transition produces the first break in the cooling curve shown in Figure 8.22. When the temperature reaches 413 K, the eutectic temperature, the entire system will freeze to solid Bi and solid Cd. This is reflected by the very flat region of the cooling curve at 413 K. Only after the entire system has frozen to solid Bi and Cd will the temperature drop below the eutectic temperature. If a system with the eutectic composition is cooled, nothing will freeze out of solution until we reach the eutectic temperature. Further removal of thermal energy will cause the entire system to freeze to a solid mixture. Figure 8.24 illustrates the cooling curve in this case for the Cd–Bi eutectic mixture. Problem 8.36 provides an additional example of the behavior of the Bi–Cd solution system.

The phase diagrams for some two-component systems are considerably more complex than that for the Bi–Cd system. For example, consider the phase diagram for the Mg–Zn system shown in Figure 8.25, where, in this case, the horizontal axis gives the magnesium composition in mole percent rather than mass percent. Viewed correctly, the interpretation of this diagram is no more difficult than that for the Bi–Cd system. First, consider what happens when we combine 1 mole of Mg with two moles of Zn to form a 1:2 molar mixture at 970 K. Since we are above the freezing curves, we have a homogeneous, single solution phase. Mg and Zn are miscible in all proportions at this temperature. The solution is now gradually cooled until a temperature of 863 K is reached. This is Point C in the diagram. Since we are now on a line along which liquid solution and solid are in equilibri-

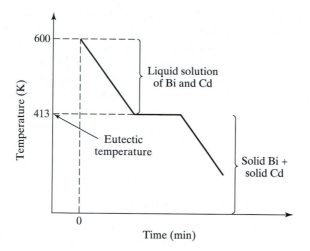

◀ FIGURE 8.24
Schematic diagram of the cooling
curve for the 40%–60% (by mass)
eutectic composition of Cd–Bi.
Only a single break in the curve is
observed at the eutectic temperature when the entire solution
freezes to solid Cd and Solid Bi.

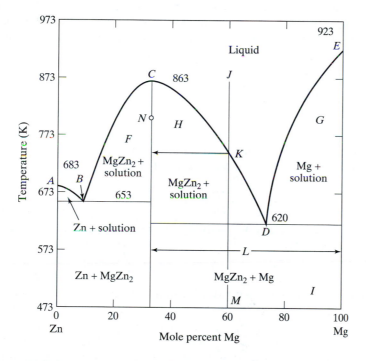

◀ FIGURE 8.25
Temperature–concentration phase
diagram for the Mg–Zn system. The
diagram shows the formation of a
compound corresponding to the composition at the maximum of the figure, Point C, which is $MgZn_2$. The
overall diagram can be understood
best by considering it to be a composite of two diagrams: one for a
Zn–$MgZn_2$ system with a eutectic at
653 K and a second for an
Mg–$MgZn_2$ system that exhibits a
eutectic point at 620 K. The vertical
line through Point C separates the
phase diagrams for these two systems.

um, something must freeze out of solution. What is it? Suppose we lower
the temperature to that at Point N. A horizontal tie-line through N does not
first intersect either the pure Mg or pure Zn lines on the right and left,
respectively. Therefore, neither solid Mg nor solid Zn freezes out of solution. Experimental measurement shows that the solution freezes to solid
$MgZn_2$, which is an intermetallic compound. The composition at a maximum in a temperature–composition phase diagram corresponds to that of a
chemical compound. Since this compound has the precise composition of
the solution at Point C, nothing else can be present at Point N except solid
$MgZn_2$. This situation is identical to that at Points A and E, where we have
only solid Zn and solid Mg, respectively.

Once it is realized that this system has the possibility of forming three pure
substances—Zn, Mg, and $MgZn_2$—we can now view the phase diagram in
Figure 8.25 as two phase diagrams, each similar to that for the Bi–Cd system in
Figure 8.23. We simply extend the vertical line through Point C upwards. To the

left of this vertical line, we have the phase diagram for a simple two-component system (Zn and MgZn$_2$) exhibiting a single eutectic point at 653 K. To the right of the vertical line, we have a second phase diagram for the Mg–MgZn$_2$ system, which also exhibits a single eutectic at 620 K. The interpretation of each of these diagrams follows that described for the Bi–Cd phase diagram. For example, if the liquid solution at Point J is cooled, equilibrium between solid MgZn$_2$ and a MgZn$_2$-saturated magnesium solution will be established at Point K. At temperatures below that at Point K, the solubility of MgZn$_2$ in the Mg solution will decrease, and MgZn$_2$ will precipitate out of the solution. Thus, we will have two phases present: solid MgZn$_2$ and a Mg solution saturated with MgZn$_2$. Problems 8.37 and 8.38 explore various features of the Mg–Zn phase diagram.

8.2.4 The Gibbs–Duhem Equation

The chemical potentials of solution components are not all independent. In general, if a mixture contains K components, only $K - 1$ of the chemical potentials can be arbitrarily varied. This fact permits us to determine the chemical potential of one component by knowing the chemical potentials of the other $K - 1$ components. Consider a mixture of K components whose chemical potentials are $\mu_1, \mu_2, \mu_3, \ldots, \mu_K$. Since the chemical potential is the partial molar Gibbs free energy or Gibbs free energy per mole, the total Gibbs free energy for the mixture is

$$G = \sum_{i=1}^{i=K} n_i \mu_i. \tag{8.99}$$

The total differential change in G when we simultaneously vary the solution composition and the chemical potentials of all components is

$$dG = \sum_{i=1}^{i=K} \mu_i \, dn_i + \sum_{i=1}^{i=K} n_i \, d\mu_i. \tag{8.100}$$

However, it was shown in Eq. 5.40 that the total differential for G is

$$dG = -S \, dT + V \, dP + \sum_{i=1}^{i=K} \mu_i \, dn_i. \tag{8.101}$$

If we equate Eqs. 8.100 and 8.101, we obtain

$$\sum_{i=1}^{i=K} n_i \, d\mu_i = -S \, dT + V \, dP. \tag{8.102}$$

Equation 8.102 shows that when processes are conducted under conditions of constant temperature and pressure, we must have

$$\boxed{\sum_{i=1}^{i=K} n_i \, d\mu_i = 0 \quad \text{at } dT = dP = 0} \; . \tag{8.103}$$

Equation 8.103 is known as the *Gibbs–Duhem equation*. It tells us that we cannot vary all of the chemical potentials in a mixture independently. After arbitrary changes are made in $K - 1$ of the μ_i, Eq. 8.103 fixes the change that must occur in the remaining chemical potential. Only $K - 1$ of the potentials can be regarded as being independent.

Let us now consider a binary solution of A and B. For this special case, we have

$$n_A \, d\mu_A + n_B \, d\mu_B = 0 \qquad \text{for } dT = dP = 0. \tag{8.104}$$

Let dX_A represent some arbitrary change in the mole fraction of component A. Dividing both sides of Eq. 8.104 by dX_A and by the total number of moles, $n_A + n_B$, we obtain

$$X_A \left(\frac{\partial \mu_A}{\partial X_A} \right)_{T,P} + X_B \left(\frac{\partial \mu_B}{\partial X_A} \right)_{T,P} = 0, \tag{8.105}$$

where the derivatives are partial derivatives, since Eq. 8.105 holds only under conditions of constant temperature and pressure. This result shows that any change in μ_A produces a corresponding change in μ_B given by

$$\left(\frac{\partial \mu_B}{\partial X_A} \right)_{T,P} = - \frac{X_A}{X_B} \left(\frac{\partial \mu_A}{\partial X_A} \right)_{T,P}.$$

The two chemical potentials are not independent.

We may now use Eq. 8.105 to prove that if Raoult's law holds for one component, Henry's law must hold for the other component. In general, the chemical potentials of A and B must be equal to those of their equilibrium vapors. Therefore,

$$\mu_A = \mu_{A(g)}^o + RT \ln f_{A(g)} = \mu_{A(g)}^o + RT \ln P_A$$

and

$$\mu_B = \mu_{B(g)}^o + RT \ln f_{B(g)} = \mu_{B(g)}^o + RT \ln P_B.$$

The replacement of fugacities with vapor pressures will be sufficiently accurate for most systems. Differentiating both sides with respect to X_A at constant temperature and pressure produces

$$\left(\frac{\partial \mu_B}{\partial X_A} \right)_{T,P} = RT \left(\frac{\partial \ln P_B}{\partial X_A} \right)_{T,P}$$

and

$$\left(\frac{\partial \mu_A}{\partial X_A} \right)_{T,P} = RT \left(\frac{\partial \ln P_A}{\partial X_A} \right)_{T,P} \tag{8.106}$$

since the standard chemical potentials depend only upon temperature. Combining Eqs. 8.105 and 8.106 gives

$$X_A \left(\frac{\partial \ln P_A}{\partial X_A} \right)_{T,P} + X_B \left(\frac{\partial \ln P_B}{\partial X_A} \right)_{T,P} = 0. \tag{8.107}$$

If Raoult's law holds for component A, we have $P_A = X_A P_A^o$, so that $(\partial \ln P_A / \partial X_A)_{T,P} = 1/X_A$. This gives

$$1 + X_B \left(\frac{\partial \ln P_B}{\partial X_A} \right)_{T,P} = 0.$$

Rearranging terms produces

$$d \ln P_B = - \frac{dX_A}{X_B} = - \frac{dX_A}{1 - X_A}.$$

Taking indefinite integrals of both sides yields

$$\int d \ln P_B = \ln P_B = - \int \frac{dX_A}{1 - X_A} = \ln[1 - X_A] + F(T, P)$$

$$= \ln X_B + F(T, P). \tag{8.108}$$

Since temperature and pressure are being held fixed during the integration, the constant of integration becomes a function that depends upon T and P and is thus denoted $F(T, P)$. Performing exponentiation on both sides of Eq. 8.108 gives

$$P_B = \exp[F(T, P)] X_B = K_B X_B. \tag{8.109}$$

The function $\exp[F(T, P)]$ is the Henry's law "constant, " K_B. This completes the proof. If Raoult's law holds for one component of a binary solution, Henry's law must hold for the second component. The analysis also demonstrates that K_B is dependent upon both T and P.

The dependence of K_B upon pressure is very weak, but it is highly sensitive to variations in temperature. This is easily seen by first taking logarithms of both sides of Eq. 8.109 and then taking the derivative with respect to temperature. With the use of the Clausius–Clapeyron equation, these two operations give

$$\left(\frac{\partial \ln P_B}{\partial T}\right)_P = \left(\frac{\partial \ln K_B}{\partial T}\right)_P = \frac{\Delta \overline{H}_{vap}^B}{RT^2}, \tag{8.110}$$

where $\Delta \overline{H}_{vap}^B$ refers to the molar heat of vaporization of component B from the solution, which is not necessarily the same as that for the pure liquid. Consequently, we expect the Henry's law constant to vary with temperature in the same manner as vapor pressure and exhibit about the same sensitivity.

EXAMPLE 8.16

(A) Estimate the value of the Henry's law constant for component A whose equilibrium vapor pressure is shown as a function of X_A in Figure 8.16. (B) If the data shown in the figure were obtained at 345 K, compute the expected value of the Henry's law constant at 320 K if the molar enthalpy of vaporization of A in solution is a constant equal to 20,000 J mol^{-1}.

Solution

(A) The analysis leading to Equation 8.109 shows that K_A is the slope of the vapor pressure curve in Figure 8.16 as $X_A \to 0$. This slope can be computed approximately from the graph. The result is

$$K_A \approx \frac{27.5 \text{ torr}}{0.5} = 55.0 \text{ torr.} \tag{1}$$

(B) The temperature dependence of K_A is given by Eq. 8.110:

$$\left(\frac{\partial \ln K_A}{\partial T}\right)_P = \frac{\Delta \overline{H}_{vap}}{RT^2}. \tag{2}$$

Separating variables, we have

$$\partial \ln K_A = \frac{\Delta \overline{H}_{vap}}{RT^2} \partial T. \tag{3}$$

Integrating between corresponding limits gives

$$\int_{K_A=55}^{K_A(320)} \partial \ln K_A = \ln K_A(320) - \ln(55) = \frac{\Delta \overline{H}_{vap}}{R} \int_{345 \text{ K}}^{320 \text{ K}} \frac{\partial T}{T^2} = \frac{-20,000}{8.314}\left[\frac{1}{320} - \frac{1}{345}\right]$$

$$= -0.54474. \tag{4}$$

Therefore,

$$\ln K_A(320) = \ln(55) - 0.54474 = 3.4626, \qquad (5)$$

and it follows that

$$K_A(320) = 31.90. \qquad (6)$$

For a related exercise, see Problem 8.24.

8.2.5   Solution Activities

8.2.5.1   Definition

The basic approach to thermodynamics in nonideal solutions involves the use of the "activity," which has been discussed in detail in Chapter 5. (See especially Section 9.2.) An examination of all of the derivations contained in the current chapter shows that the starting point is to note that equilibrium at constant temperature and either constant pressure or constant volume requires equality between the chemical potentials of the equilibrium states. Therefore, we need an expression for the chemical potential of the components of a nonideal solution. Using the activity, a, we may write this as

$$\mu_{B(\text{soln})} = \mu_{B(\text{soln})}^o + RT \ln a_B, \qquad (8.111)$$

where a_B is the activity of component B and $\mu_{B(\text{soln})}^o$ is the standard chemical potential of B *in the solution.* It is important to distinguish $\mu_{B(\text{soln})}^o$ from its gas-phase counterpart, $\mu_{B(g)}^o$. The former quantity is the standard chemical potential of component B in solution, whereas the latter is its standard chemical potential for the vapor. These two quantities are not the same; both depend only upon the temperature, but their magnitudes differ.

We now need to select a reference function for the measurement of the solution activity and a reference state that will set the relative values of the activities. As usual, we are free to make any convenient choice we wish for these quantities, since we are interested only in *changes* in μ and G, not their absolute values. For gases, the best choice for the reference function is pressure, because the fugacity becomes equal to the pressure as $P \rightarrow 0$. This consideration also suggests that we choose $P \rightarrow 0.$ as the reference state. A similar approach leads us to the best choice for solutions.

For ideal solutions, the measured quantity that determines the chemical potential of component B for a given solution is the mole fraction X_B. This fact, coupled with the ease with which the mole fraction can be measured, suggests that we choose the mole fraction as our reference function for the solvent. Since component B of the mixture will obey Raoult's law as $X_B \rightarrow 1$, the latter will be the most convenient choice for the reference state. With these concepts in mind, we define the solvent activity as

$$\operatorname*{Lim}_{X_B \rightarrow 1} \left[\frac{a_B}{X_B} \right] = 1. \qquad (8.112)$$

The corresponding activity coefficient is

$$\gamma_B = \frac{a_B}{X_B}. \qquad (8.113)$$

8.2.5.2. Measurement from Vapor Pressures

The activity of the solvent can be computed directly from the measured equilibrium partial pressures of the solvent. First consider the pure liquid B. For this system, the chemical potential is given by

$$\mu_B = \mu^o_{B(soln)} + RT \ln a_B = \mu^o_{B(soln)} + RT \ln(1) = \mu^o_{B(soln)}, \quad (8.113)$$

since the activity of the pure liquid is unity because of our choice of reference state and reference function. The chemical potential of pure liquid B is also equal to that of its equilibrium vapor, so that we have

$$\mu_B = \mu^o_{B(soln)} = \mu^o_{B(g)} + RT \ln f^o_B = \mu^o_{B(g)} + RT \ln P^o_B \quad (8.114)$$

where we have replaced the fugacity with the equilibrium partial pressure because the pressure is usually so low that we are near the reference state for gases.

Now consider a binary solution with mole fraction X_B. The chemical potential of solvent B in this solution is

$$\mu_{B(soln)} = \mu^o_{B(soln)} + RT \ln a_B = \mu^o_{B(g)} + RT \ln P^o_B + RT \ln a_B, \quad (8.115)$$

where we have used Eq. 8.114 to express $\mu^o_{B(soln)}$ in terms of $\mu^o_{B(g)}$ and P^o_B. Again, the chemical potential of B in the solution must be equal to that of its equilibrium vapor. Therefore,

$$\mu_{B(soln)} = \mu^o_{B(g)} + RT \ln P^o_B + RT \ln a_B = \mu^o_{B(g)} + RT \ln P_B. \quad (8.116)$$

Solving Eq. 8.116 for a_B, we obtain

$$\boxed{a_B = \frac{P_B}{P^o_B}}. \quad (8.117)$$

Equation 8.117 shows that the activity of solvent B may be easily computed if its equilibrium partial pressure above the solution has been measured.

8.2.5.3 Measurement from Freezing-Point Depression

Solvent activities may be determined by measuring any of the colligative properties discussed in the previous section on ideal solutions. We consider each of these, starting with freezing-point depression.

At the freezing point $T = T_f$, pure solid solvent $B(s)$ is in equilibrium with B molecules in solution. Therefore,

$$\mu_{B(s)} = \mu_{B(soln)}. \quad (8.118)$$

The chemical potential of the pure solid is equal to that of its equilibrium vapor, so

$$\mu_{B(s)} = \mu^o_{B(g)} + RT \ln f_{B(s)} = \mu^o_{B(g)} + RT \ln P_{B(s)}, \quad (8.119)$$

where $f_{B(s)}$ is the fugacity of the equilibrium vapor over the solid at the freezing point. In most cases, this may be replaced with the equilibrium vapor pressure, since the vapor will behave ideally in this pressure regime. The chemical potential of B in solution is given by Eq. 8.115. Combining Eqs. 8.115, 8.118, and 8.119 results in

$$\mu^o_{B(g)} + RT \ln P_{B(s)} = \mu^o_{B(g)} + RT \ln P^o_B + RT \ln a_B. \quad (8.120)$$

The desired activity is given by

$$\ln a_B = \ln P_{B(s)} - \ln P^o_B. \quad (8.121)$$

If we assume that the enthalpies of sublimation and vaporization are constants, we may use the Clausius–Clapeyron equation to evaluate the right-hand side of Eq. 8.121. When we express the pressure in torr, the required equilibrium vapor pressures are

$$\ln P_{B(s)} = \ln P^* - \frac{\Delta \overline{H}_{sub}^{B}}{R}\left[\frac{1}{T_f} - \frac{1}{T^*}\right]$$

and

$$\ln P_B^o = \ln 760 - \frac{\Delta \overline{H}_{vap}^{B}}{R}\left[\frac{1}{T_f} - \frac{1}{T_b^o}\right],$$

where P^* is the equilibrium sublimation pressure at temperature T^* and T_b^o is the normal boiling point of pure liquid B. Substitution into Eq. 8.121 yields

$$\ln a_B = \ln\frac{P^*}{760} - \frac{\Delta \overline{H}_{sub}^{B} - \Delta \overline{H}_{vap}^{B}}{RT_f} + \frac{\Delta \overline{H}_{sub}^{B}}{RT^*} - \frac{\Delta \overline{H}_{vap}^{B}}{RT_b^o}. \tag{8.122}$$

Replacing $[\Delta \overline{H}_{sub}^{B} - \Delta \overline{H}_{vap}^{B}]$ with $\Delta \overline{H}_{fus}^{B}$ and performing exponentiation on both sides produces

$$a_B = \frac{P^*}{760}\exp\left[\frac{\Delta \overline{H}_{sub}^{B}}{RT^*} - \frac{\Delta \overline{H}_{vap}^{B}}{RT_b^o}\right]\exp\left[-\frac{\Delta \overline{H}_{fus}^{B}}{RT_f}\right]. \tag{8.123}$$

The first two factors on the right of Eq. 8.123 are constants. This allows us to express the equation in the form

$$a_B = C\exp\left[-\frac{\Delta \overline{H}_{fus}^{B}}{RT_f}\right]. \tag{8.124}$$

If we can evaluate Eq. 8.124 at any point, the value of the constant C can be determined. In the solvent reference state, we have pure B, for which the freezing point is the normal freezing point T_f^o and the activity is unity. Inserting these values in Eq. 8.124 yields

$$C = \exp\left[\frac{\Delta \overline{H}_{fus}^{B}}{RT_f^o}\right],$$

and the required equation for the activity of solvent B is

$$\boxed{a_B = \exp\left[-\frac{\Delta \overline{H}_{fus}^{B}}{R}\left\{\frac{1}{T_f} - \frac{1}{T_f^o}\right\}\right].} \tag{8.125}$$

Equation 8.125 allows the activity of solvent B to be easily calculated from a measurement of the freezing point of B when the composition of the solution is X_B.

The activity computed with the use of Eq. 8.125 is the value of a_B at the freezing point of the solution. We might, therefore, reasonably inquire how we can utilize this information to compute a_B at any other temperature that might be desired. Such a calculation requires that we obtain the rate of change of the activity with temperature. This can be accomplished by utilizing the Gibbs–Helmholtz equation. Rearranging of Eq. 8.111 produces

$$\ln a_B = \frac{\mu_{B(soln)} - \mu_{B(l)}}{RT}$$

Since Eq. 8.113 shows that $\mu_{B(soln)}^o$ is the same as $\mu_{B(l)}$ for pure liquid B.

Taking derivatives of both sides of this expression with respect to temperature gives

$$\left(\frac{\partial \ln a_B}{\partial T}\right)_P = \frac{1}{R}\left[\left(\frac{\partial(\mu_{B(soln)}/T)}{\partial T}\right)_P - \left(\frac{\partial(\mu_{B(l)}/T)}{\partial T}\right)_P\right].$$

The Gibbs-Helmholtz equation allows the right-hand side of the foregoing equation to be evaluated. The result is

$$\left(\frac{\partial \ln a_B}{\partial T}\right)_P = \frac{-\overline{H}_{B(soln)} + \overline{H}_{B(l)}}{RT^2}, \tag{8.126}$$

where $\overline{H}_{B(soln)}$ and $\overline{H}_{B(l)}$ are the partial molar enthalpies of solvent B in the solution and in the pure liquid state. If these are the same, the activity will be independent of temperature.

EXAMPLE 8.17

Consider the nonideal solution illustrated in Figure 8.16. Assume that compound A acts as the solvent, whose activity is defined by Eqs. 8.112 and 8.113. When $X_A = 0.720$, the measured equilibrium partial pressure of A is 85.73 torr. When $X_A = 1$, the measured vapor pressure is 128.57 torr. If the molar enthalpy of fusion of compound A is independent of temperature and equal to 6000 J mol^{-1} and the normal freezing point of compound A is 273 K, compute the freezing point of compound A in this nonideal solution. Assume that a_A is independent of temperature.

Solution

The activity of compound A at the temperature at which the vapor pressure data were taken is given by Eq. 8.117, applied to compound A:

$$a_A = \frac{P_A}{P_A^o} = \frac{85.73 \text{ torr}}{128.57 \text{ torr}} = 0.667. \tag{1}$$

The corresponding activity coefficient is

$$\gamma_A = \frac{a_A}{X_A} = \frac{0.667}{0.720} = 0.9264. \tag{2}$$

If the activity is independent of temperature, we will have $a_A = 0.667$ at all temperatures. Using Eq. 8.125, we have

$$a_A = \exp\left[-\frac{\Delta \overline{H}_{fus}^A}{R}\left\{\frac{1}{T_f} - \frac{1}{T_f^o}\right\}\right]. \tag{3}$$

Substituting the given data yields

$$0.667 = \exp\left[-\frac{6000 \text{ J mol}^{-1}}{8.314 \text{ J mol}^{-1}\text{K}^{-1}}\left\{\frac{1}{T_f} - \frac{1}{273}\right\}\right]. \tag{4}$$

Solving for T_f^{-1}. we obtain

$$\frac{1}{T_f} = \frac{1}{273} - \frac{8.314 \ln(0.667)}{6000} = 0.004224, \tag{5}$$

so that

$$T_f = 236.7 \text{ K.} \tag{6}$$

Note that when we use the activity, we do not have to measure all the properties of a nonideal solution. In this case, measuring the equilibrium vapor pressures permits us to compute the solvent freezing point in the solution.

8.2.5.4   Measurement from Boiling-point Elevation

The activity of the solvent may be related to its measured boiling point by analyzing the system along the same lines as those employed for freezing-point depression. The chemical potential of solvent B in the solution must be equal to that of its equilibrium vapor. That is,

$$\mu_{B(soln)} = \mu_{B(g)}.$$

Using Eq. 8.116, we may write this equation as

$$\mu_{B(g)}^o + RT \ln f_B^o + RT \ln a_B = \mu_{B(g)}^o + RT \ln f_B, \qquad (8.127)$$

where f_B^o and f_B are the fugacities of B vapor over pure liquid B and over the solution, respectively. This leads directly to

$$\ln a_B = \ln f_B - \ln f_B^o. \qquad (8.128)$$

At pressures generally characteristic of equilibrium vapor pressures, we expect that $f = P$, so that

$$\ln a_B = \ln P_B - \ln P_B^o.$$

At the boiling point T_b, P_B will be equal to the applied external pressure P_o, and the vapor pressure over the pure liquid will be given by the Clausius–Clapeyron equation, provided that the enthalpy of vaporization is a constant. Thus,

$$\ln P_B^o = \ln(760) - \frac{\Delta \overline{H}_{vap}^B}{R}\left[\frac{1}{T_b} - \frac{1}{T_b^o}\right],$$

and we obtain

$$\boxed{\ln a_B = \ln\left[\frac{P_o}{760}\right] + \frac{\Delta \overline{H}_{vap}^B}{R}\left[\frac{1}{T_b} - \frac{1}{T_b^o}\right].} \qquad (8.129)$$

Equation (8.129) relates the measured boiling point T_b to the activity of solvent B in solution at any applied pressure P_o. If P_o is 760 torr, the first term on the right vanishes, and we have

$$\boxed{\ln a_B = \frac{\Delta \overline{H}_{vap}^B}{R}\left[\frac{1}{T_b} - \frac{1}{T_b^o}\right] \text{ for } P_o = 760 \text{ torr}.} \qquad (8.130)$$

8.2.5.5   Measurement from Osmotic Pressure

The procedure for relating the osmotic pressure to the solution activity is identical to that given from Eq. 8.88 through 8.91, except that Eq. 8.89 now becomes

$$\left(\frac{\partial \mu_{B(soln)}}{\partial P_h}\right)_T = \left(\frac{\partial \mu_{B(soln)}^o}{\partial P_h}\right)_T + RT\left(\frac{\partial \ln a_B}{\partial P_h}\right)_T = \overline{V}_{B(l)}. \qquad (8.131)$$

The standard chemical potential is independent of pressure, so

$$\left(\frac{\partial \ln a_B}{\partial P_h}\right)_T = \frac{\overline{V}_{B(l)}}{RT}. \qquad (8.132)$$

Separating variables in the usual manner and integrating between zero pressure head, where the activity of solvent B is a_B, and a pressure head of π, where the activity of B is equal to that for the pure component, which is unity by our choice of reference state, we have

$$\int_{a_B}^{a_B=1} d\ln a_B = \int_{P_h=0}^{P_h=\pi} \frac{\overline{V}_{B(l)}}{RT} dP_h. \tag{8.133}$$

If the liquid is incompressible, the resulting equation is

$$\boxed{\ln a_B = -\frac{\pi \overline{V}_{B(l)}}{RT}}. \tag{8.134}$$

8.2.5.6   Comparison of Ideal and Nonideal Colligative-Property Equations

Before continuing our discussion of solvent activities, it is instructive to compare the colligative-property equations for ideal and nonideal solutions. This comparison is given in Table 8.1, in which it is assumed that the enthalpies of sublimation, fusion, and vaporization are all independent of temperature.

A quick examination of the table shows that the equations for the solvent activity in nonideal solutions are identical to those for ideal systems if the mole fraction X_B is replaced with the activity a_B. This happy result occurs because of our choice of reference function and reference state. A different choice would have made things more difficult.

8.2.5.7   Determination of the Activity of the Solute from the Activity of the Solvent

The equations developed in the preceding sections provide several methods for measuring the activity of the solvent of a solution for which we adopt the mole fraction as the reference function and the pure component as the reference state. We now ask, How do we determine the activity of the solute? Must we conduct a new series of measurements to obtain these data? The answer is no: For a two-component solution, a determination of the activity of the solvent permits the activity of the solute to be computed using the Gibbs–Duhem equation.

Table 8.1   Comparison of colligative-property equations for ideal and nonideal solutions

Property	Ideal Solutions	Nonideal Solutions
Vapor Pressure	$X_B = \dfrac{P_B}{P_B^o}$	$a_B = \dfrac{P_B}{P_B^o}$
Freezing-point Depression	$\ln X_B = -\dfrac{\Delta \overline{H}_{fus}^B}{R}\left[\dfrac{1}{T_f} - \dfrac{1}{T_f^o}\right]$	$\ln a_B = -\dfrac{\Delta \overline{H}_{fus}^B}{R}\left[\dfrac{1}{T_f} - \dfrac{1}{T_f^o}\right]$
Boiling-point Elevation*	$\ln X_B = \dfrac{\Delta \overline{H}_{vap}^B}{R}\left[\dfrac{1}{T_b} - \dfrac{1}{T_b^o}\right]$	$\ln a_B = \dfrac{\Delta \overline{H}_{vap}^B}{R}\left[\dfrac{1}{T_b} - \dfrac{1}{T_b^o}\right]$
Osmotic Pressure	$\ln X_B = -\dfrac{\pi \overline{V}_{B(l)}}{RT}$	$\ln a_B = -\dfrac{\pi \overline{V}_{B(l)}}{RT}$

* These equations assume that the applied pressure is 760 torr.

Let us assume that we have measured the activity of component B (the solvent) at a series of different compositions spanning the range from $X_B = 0$ to $X_B = 1$. Equation 8.105 requires that we have $X_A(\partial\mu_A/\partial X_A)_{T,P} + X_B(\partial\mu_B/\partial X_A)_{T,P} = 0$. Since $\mu_i = \mu^o_{i(\text{soln})} + RT \ln a_i$, we may write Eq. 8.105 in the form

$$X_A \left(\frac{\partial \ln a_A}{\partial X_A} \right)_{T,P} + X_B \left(\frac{\partial \ln a_B}{\partial X_A} \right)_{T,P} = 0, \tag{8.135}$$

because the standard chemical potentials are independent of composition of the solution.

We now need to express Equation 8.135 in terms of the activity coefficients a/X. To do this, we note that

$$X_B \left(\frac{\partial \ln X_B}{\partial X_A} \right)_{T,P} = \frac{X_B}{X_B} \left(\frac{\partial X_B}{\partial X_A} \right)_{T,P} = \left(\frac{\partial(1 - X_A)}{\partial X_A} \right)_{T,P} = -\left(\frac{\partial X_A}{\partial X_A} \right)_{T,P} = -1$$

and

$$X_A \left(\frac{\partial \ln X_A}{\partial X_A} \right)_{T,P} = \frac{X_A}{X_A} \left(\frac{\partial X_A}{\partial X_A} \right)_{T,P} = 1.$$

Hence,

$$X_A \left(\frac{\partial \ln X_A}{\partial X_A} \right)_{T,P} + X_B \left(\frac{\partial \ln X_B}{\partial X_A} \right)_{T,P} = 0.$$

We may, therefore, subtract this sum from Eq. 8.135 without changing the latter's value. This procedure gives

$$X_A \left(\frac{\partial \ln a_A}{\partial X_A} \right)_{T,P} - X_A \left(\frac{\partial \ln X_A}{\partial X_A} \right)_{T,P} + X_B \left(\frac{\partial \ln a_B}{\partial X_A} \right)_{T,P} - X_B \left(\frac{\partial \ln X_B}{\partial X_A} \right)_{T,P} = 0,$$

which we may write in the form

$$X_A \left(\frac{\partial \ln(a_A/X_A)}{\partial X_A} \right)_{T,P} + X_B \left(\frac{\partial \ln(a_B/X_B)}{\partial X_A} \right)_{T,P} = 0. \tag{8.136}$$

Rearranging of Eq. 8.136 produces

$$d \ln [a_A/X_A] = -\frac{X_B}{X_A} \left(\frac{\partial \ln(a_B/X_B)}{\partial X_A} \right)_{T,P} dX_A$$

$$= \frac{X_A - 1}{X_A} \left(\frac{\partial \ln(a_B/X_B)}{\partial X_A} \right)_{T,P} dX_A. \tag{8.137}$$

We must now choose a reference function and a reference state for the measurement of the activity of the solute A. Once again, the most convenient reference function is the mole fraction. However, it is now more convenient to take infinite dilution where $X_A \to 0$ as the reference state. This choice corresponds to our choice of $X_B \to 1$ for the reference state of the solvent, since if X_B approaches unity, X_A must approach zero. With this reference state, we have

$$\boxed{\lim_{X_A \to 0} \left[\frac{a_A}{X_A} \right] = 1}. \tag{8.138}$$

We may now integrate Eq. 8.137 from $X_A = 0$ to some value of X_A at which we wish to know the solute activity. This gives

$$\int_{a_A/X_A=1}^{a_A/X_A} d\ln[a_A/X_A] = \ln[a_A/X_A] - \ln(1) = \ln[a_A/X_A]$$

$$= \int_{X_A=0}^{X_A} \frac{X_A - 1}{X_A} \frac{\partial \ln(a_B/X_B)}{\partial X_A} dX_A,$$

so that we have

$$\ln\left[\frac{a_A}{X_A}\right] = \ln\gamma_A = \int_{X_A=0}^{X_A} \frac{X_A - 1}{X_A}\left(\frac{\partial \ln(a_A/X_A)}{\partial X_A}\right). \qquad (8.139)$$

Since we know a_B as a function of X_A, the right-hand side of Eq. 8.139 can always be evaluated by numerical integration.

Before proceeding to an example of such a numerical integration to obtain γ_A, some attention must be paid to what appears to be a singularity in Eq. 8.139 at the point $X_A = 0$. In reality, no singularity exits. The integral we wish to evaluate is

$$\ln\gamma_A = \int_{X_A=0}^{X_A} \frac{X_A - 1}{X_A}\left(\frac{\partial \ln(a_B/X_B)}{\partial X_A}\right) dX_A.$$

To perform the integration, we might plot $\left[\left(\dfrac{X_A - 1}{X_A}\right)\left(\dfrac{\partial \ln(a_B/X_B)}{\partial X_A}\right)\right]$ against X_A and compute the area under the curve from zero to X_A. The point of interest is the behavior of the integrand as X_A approaches zero. The limiting value of

$$\left[\frac{X_A - 1}{X_A}\left(\frac{\partial \ln(a_B/X_B)}{\partial X_A}\right)\right]$$

as $X_A \to 0$ is

$$\lim_{X_A \to 0}\left[\frac{X_A - 1}{X_A}\left(\frac{\partial \ln(a_B/X_B)}{\partial X_A}\right)\right] = -\lim_{X_A \to 0}\left[\frac{1}{X_A}\left(\frac{\partial \ln(a_B/X_B)}{\partial X_A}\right)\right], \qquad (8.140)$$

since $X_A - 1$ approaches -1 as $X_A \to 0$. Using Eq. 8.117, the logarithm term may be written in the form

$$\ln(a_B/X_B) = \ln\left[\frac{P_B}{X_B P_B^o}\right]. \qquad (8.141)$$

In the limit as $X_A \to 0$, Raoult's law becomes exact for the solvent B, and we have

$$\lim_{X_A \to 0} \ln(a_B/X_B) = \ln\left[\frac{P_B}{P_B}\right] = \ln[1] = 0. \qquad (8.142)$$

We also know that a plot of P_B versus X_A must asymptotically approach the Raoult's Law line as X_A approaches zero. Therefore, the slope of a plot of (a_B/X_B) versus X_A must approach zero in this limit. For these reasons, the limit in Eq. 8.140 becomes

$$-\lim_{X_A \to 0}\left[\frac{1}{X_A}\left(\frac{\partial \ln(a_B/X_B)}{\partial X_A}\right)\right] = -\lim_{X_A \to 0}\left[\frac{\left(\dfrac{\partial \ln(1)}{\partial X_A}\right)}{X_A}\right]. \qquad (8.143)$$

Since the slope is 0, this limit has the indeterminate form 0/0. We can still evaluate the limit by using L'Hôpital's rule, which tells us that the limit of an indeterminate form is the same as the limit of the ratio of the derivatives of numerator and denominator taken with respect to the original variable whose limit is being taken. Therefore, we have

$$\underset{X_A \to 0}{\text{Lim}}\left[\frac{X_A - 1}{X_A}\left(\frac{\partial \ln(a_B/X_B)}{\partial X_A}\right)\right] = -\underset{X_A \to 0}{\text{Lim}}\left[\frac{\frac{\partial}{\partial X_A}\left(\frac{\partial \ln(a_B/X_B)}{\partial X_A}\right)}{1}\right]$$

$$= \text{a finite number,} \qquad (8.144)$$

since the second derivative is the curvature, which is certainly not infinite. Consequently, there is no singularity in the integrand at $X_A = 0$.

Although the preceding analysis demonstrates that the integrals in Eq. 8.139 converge properly, there are still some numerical difficulties that must be addressed. These difficulties arise because computers do a terrible job of taking analytical limits. That is, we know that there is no singularity in the integrand of Eq. 8.139 as $X_A \to 0$, but our computers don't "know" that unless we tell them. Basically, we need to ensure that the computer "knows" that the intercept of the quantity $\ln[a_B/X_B]$ is zero and that the slope $\partial \ln(a_B/X_B)/\partial X_A$ also approaches zero as $X_A \to 0$. This can be done by using least-square methods to fit the measured data for the solvent, $\ln[a_B/X_B]$, to an analytic function that has those characteristics. Example 8.18 illustrates the technique.

X_B	P_B (torr)
0.000	0.000
0.040	2.298
0.080	4.565
0.120	6.914
0.160	9.445
0.200	12.245
0.240	15.385
0.280	18.920
0.320	22.894
0.360	27.334
0.400	32.254
0.440	37.653
0.480	43.515
0.520	49.811
0.560	56.495
0.600	63.510
0.640	70.782
0.680	78.223
0.720	85.731
0.760	93.190
0.800	100.470
0.840	107.423
0.880	113.891
0.920	119.700
0.960	124.660
1.000	128.568

EXAMPLE 8.18

The measured data represented by the points shown in Figure 8.16 are given in the accompanying table. Although labeled component A in the figure, let us assume that these data are for the solvent B.

(A) Compute (a_B/X_B) and $\ln[a_B/X_B]$ for each of the values given in the table. Let us fit these data to an analytic function that reflects our knowledge that the limiting values of $\ln[a_B/X_B]$ and $\partial \ln[a_B/X_B]/\partial X_A$ as $X_A \to 0$ are zero. A simple polynomial could be employed. That is, we could write

$$\ln\left[\frac{a_B}{X_B}\right] = c_2 X_A^2 + c_3 X_A^3 + c_4 X_A^4 + c_5 X_A^5,$$

which gives

$$\frac{\partial \ln[a_B/X_B]}{\partial X_A} = 2c_2 X_A + 3c_3 X_A^2 + 4c_4 X_A^3 + 5c_5 X_A^4.$$

This choice satisfies both of the preceding requirements. Use a linear least-squares method to obtain the c_i ($i = 2, 3, 4, 5$), and plot the fitted function versus X_A. Show the actual data on the same graph. **(B)** Using the results of (A), obtain an analytic expression for

$$\frac{X_A - 1}{X_A}\left(\frac{\partial \ln(a_B/X_B)}{\partial X_A}\right),$$

and plot this function versus X_A over the range $0 \le X_A \le 1$. **(C)** Use the results obtained in (B) to obtain γ_A as a function of X_A over the range $0 \le X_A \le 1$. Plot γ_A versus X_A.

Solution

(A) Since the activity of the solvent B is given by $a_B = P_B/P_B^o$, straightforward arithmetic permits calculations of a_B and $\ln[a_B/X_B]$. The results are as follows:

X_B	X_A	a_B	$\ln[a_B / X_B]$	X_B	X_A	a_B	$\ln[a_B / X_B]$
0.04	0.96	0.017874	−0.80554	0.56	0.44	0.439417	−0.24249
0.08	0.92	0.035506	−0.81231	0.60	0.40	0.493980	−0.19443
0.12	0.88	0.053777	−0.80265	0.64	0.36	0.550541	−0.15057
0.16	0.84	0.073463	−0.77839	0.68	0.32	0.608417	−0.11123
0.20	0.80	0.095241	−0.74190	0.72	0.28	0.666814	−0.07674
0.24	0.76	0.119664	−0.69595	0.76	0.24	0.724830	−0.04738
0.28	0.72	0.147159	−0.64327	0.80	0.20	0.781454	−0.02346
0.32	0.68	0.178069	−0.58615	0.84	0.16	0.835534	−0.00533
0.36	0.64	0.212603	−0.52668	0.88	0.12	0.885843	0.00662
0.40	0.60	0.250871	−0.46653	0.92	0.08	0.931025	0.01191
0.44	0.56	0.292864	−0.40706	0.96	0.04	0.969604	0.00995
0.48	0.52	0.338459	−0.34938	1.00	0.00	1.000000	0.00000
0.52	0.48	0.387429	−0.29430				

A linear least-squares fitting of $\ln[a_B/X_B]$ to the polynomial function yields

$$\ln[a_B/X_B] = 0.134977X_A^2 - 5.379975X_A^3 + 5.7272245X_A^4 - 1.2734378\,X_A^5. \quad \textbf{(1)}$$

Figure 8.26 compares the measured data with the analytic fit. If the data used in this example were actual laboratory data, the same maximum seen around $X_A \approx 0.08$ would suggest possible complex formation or the presence of some errors in the measurements.

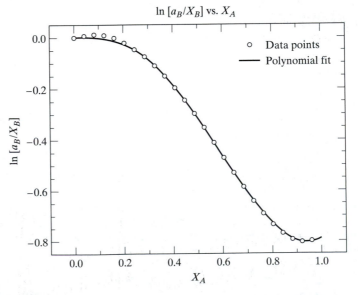

$\ln[a_B/X_B]$ vs. X_A

▲ FIGURE 8.26

Analytical fit to $\ln[a_B/X_B]$ which ensures that $\underset{X_A \to 0}{\text{Lim}}\,[\ln a_B/X_B]$ and $\underset{X_A \to 0}{\text{Lim}}\left[\dfrac{\partial \ln(a_B/X_B)}{\partial X_A}\right]$

are both zero. The points are the hypothetical measured data. The curve is a least-squares fit of the function $\ln[a_B/X_B] = 0.134977\,X_A^2 - 5.379975\,X_A^3 + 5.7272245\,X_A^4 - 1.2734378\,X_A^5$.

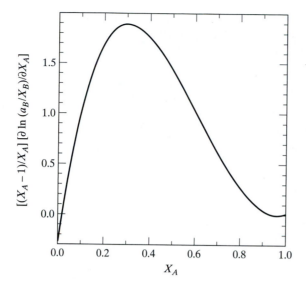

The integrand of Eq. 8.139,

$$\left[\frac{X_A - 1}{X_A} \frac{\partial \ln(a_B/X_B)}{\partial X_A}\right], \text{ plotted}$$

as a function of X_A for the data given in Example 8.18. As can be seen, there is no singularity in this integrand at $X_A = 0$.

(B) The required derivative is

$$\frac{\partial \ln[a_B/X_B]}{\partial X_A} = 2c_2 X_A + 3c_3 X_A^2 + 4c_4 X_A^3 + 5c_5 X_A^4. \tag{2}$$

It should be noted that the computation of derivatives from numerical data using these methods places great demands on the accuracy of the data. See section 9.6.6.3 for an extended discussion of this point.

Therefore, we have

$$\frac{X_A - 1}{X_A}\left(\frac{\partial \ln(a_B/X_B)}{\partial X_A}\right) = (X_A - 1)\left[2c_2 + 3c_3 X_A + 4c_4 X_A^2 + 5c_5 X_A^3\right]$$

$$= (X_A - 1)[0.269954 - 16.139925 X_A + 22.90898 X_A^2 - 6.367189 X_A^3]. \tag{3}$$

Figure 8.27 shows a plot of this result as a function of X_A.

(C) Equation 8.139 indicates that the area under the plot in Figure 8.27 from $X_A = 0$ to any value X_A yields the logarithm of the activity coefficient for component A at a concentration of X_A. The integration producing the area could be carried out using a trapezoidal rule with upper limits equal to each of the values of X_A listed in the table given in (A). However, in this case, the integration may be performed analytically using Eq. 3:

$$\ln \gamma_A = \int_{X_A=0}^{X_A} \frac{X_A - 1}{X_A}\left(\frac{\partial \ln(a_B/X_B)}{\partial X_A}\right)dX_A$$

$$= \int_{X_A=0}^{X_A} (X_A - 1)[0.269954 - 16.139925 X_A + 22.90898 X_A^2 - 6.367189 X_A^3]dX_A. \tag{4}$$

Integrating between the limits gives

$$\ln \gamma_A = -0.269954 X_A + \left(\frac{1}{2}\right)[16.39924 + 0.269954]X_A^2 - \frac{1}{3}[22.90898$$

$$+ 16.139925]X_A^3 + \left(\frac{1}{4}\right)[6.367189 + 22.90898]X_A^4 - \frac{1}{5}(6.3971890) X_A^5$$

$$= -0.269954 X_A + 8.334597 X_A^2 - 13.016302 X_A^3 + 7.319042 X_A^4 - 1.279438 X_A^5.$$

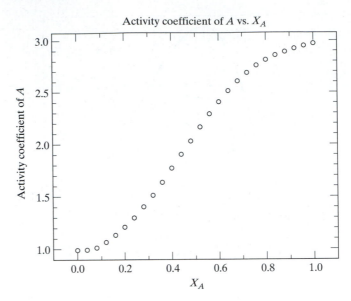

▲ FIGURE 8.28
Activity coefficients for solute A as a function of X_A. Example 8.18 shows how these values can be obtained from measured activities of the solvent B.

Figure 8.28 shows the computed activity coefficients γ_A as a function of X_A. For a related exercise, see Problem 8.28.

8.2.5.8   Conversion of Activity Coefficients from One Reference Function to Another

Although the mole fraction is the most convenient choice of reference function for the measurement of activity coefficients, once their values are obtained it is frequently more convenient to transform those values to another reference function, such as molality or molarity, since these are the measures of concentration usually employed in most laboratory work. The conversion is based on the fact that the difference in chemical potentials between two states is independent of the choice of reference function, provided that the reference states are the same. Therefore, if μ and μ^* are the chemical potentials for two states, we must have the same value of $\mu - \mu^*$ regardless of whether we choose the mole fraction, molality (m) or molarity (c) as the reference function.

Let a_X and a_m be the activities when the mole fraction and molality are chosen as the reference function, with infinite dilution being the reference state in each case. The preceding considerations show that we must have

$$
\begin{aligned}
\mu - \mu^* &= \left[\mu_X^o + RT \ln a_X\right] - \left[\mu_X^o + RT \ln a_X{}^*\right] \\
&= \left[\mu_m^o + RT \ln a_m\right] - \left[\mu_m^o + RT \ln a_m{}^*\right] \\
&= RT \ln\left[\frac{a_X}{a_X{}^*}\right] = RT \ln\left[\frac{a_m}{a_m{}^*}\right].
\end{aligned}
\tag{8.145}
$$

Equation 8.145 shows that the ratio of activities for two states is independent of the reference function chosen for the measurement of the values.

Let us now choose the state denoted by the asterisk in Eq. 8.145 to be one that is approaching the reference state of infinite dilution for the solute. Then the equation shows that

$$\frac{a_X}{a_X{}^*} = \frac{a_m}{a_m{}^*} = \frac{\gamma_X X}{\gamma_X{}^* X^*} = \frac{\gamma_m m}{\gamma_m{}^* m^*}. \tag{8.146}$$

Rearranging Eq. 8.146 gives

$$\gamma_m = \gamma_X \left[\frac{X}{m}\right]\left[\frac{m^*}{X^*}\right]\left[\frac{\gamma_m^*}{\gamma_X^*}\right]. \tag{8.147}$$

The mole fraction is easily related to the molality. For an m molal solution, we have m moles of solute per 1000 g of solvent. If M_o is the molar mass of the solvent, the mole fraction of this solution will be

$$X = \frac{m}{m + 1000/M_o},$$

so that

$$\frac{X}{m} = \frac{1}{m + 1000/M_o}. \tag{8.148}$$

In the limit of infinite dilution as $m \to 0$, we have

$$\frac{X^*}{m^*} = \frac{M_o}{1000}. \tag{8.149}$$

Substituting Eqs. 8.148 and 8.149 into Eq. 8.147 produces

$$\gamma_m = \gamma_X \frac{1}{m + 1000/M_o}\left[\frac{1000}{M_o}\right]\left[\frac{\gamma_m^*}{\gamma_X^*}\right] = \gamma_X \frac{1000}{M_o m + 1000}\left[\frac{\gamma_m^*}{\gamma_X^*}\right].$$

In the reference state, the activity coefficients each approach unity. Equating $[\gamma_m^*/\gamma_X^*]$ to unity, we obtain

$$\gamma_m = \gamma_X \frac{1000}{M_o m + 1000}. \tag{8.150}$$

Dividing both sides by γ_x and then inverting the fractions gives the very convenient result

$$\boxed{\frac{\gamma_x}{\gamma_m} = 1 + 0.001 M_o m} \tag{8.151}$$

as the conversion equation between activity coefficients for which the mole fraction and molality are the reference functions. Similar conversion equations can be obtained for molarity. These will involve the densities of the solution and the pure solvent. (See Problem 8.32.)

Summary: Key Concepts and Equations

1. The ideal-solution approximation assumes that all solution interactions are identical. When this is so, the equilibrium vapor pressure of component B over the solution is related to that for pure liquid B by simple statistical considerations that lead to $P_B = X_B P_B^o$, which is known as Raoult's law. We observe solution ideality

only when the components of the solution are highly similar. When the mole fraction of one component approaches unity, Raoult's law will be obeyed by that component. Consequently, the composition region for which $X_B \rightarrow 1$ is called the ideal-solution regime for component B.

2. When Raoult's law holds for component B, we observe a lowering of the vapor pressure for that component of the solution which depends only upon the mole fractions of all the other components in the solution. If there are K compounds in the solution, the vapor pressure lowering is

$$\Delta P_B = P_B - P_B^o = - \sum_{i=1}^{i=K}{}' X_i P_B^o.$$

If $K = 2$, then $\Delta P_A = -X_B P_A^o$. When the solution contains ionic compounds that ionize into several particles, the mole fractions of all ions must be included in the summation.

3. If the components of the solution have different pure component vapor pressures, the equilibrium vapor composition will differ from that of the solution. The mole fraction of component A in the vapor, Y_A, is related to the composition of the solution by

$$Y_A = \frac{X_A P_A^o}{X_A P_A^o + X_B P_B^o} = \frac{X_A P_A^o}{X_A(P_A^o - P_B^o) + P_B^o}$$

for a binary solution. When $Y_A \neq X_A$, we can usually separate a liquid mixture by using fractional distillation at the boiling point. In effect, a fractionating column executes a series of vaporization–condensation steps, each at the new boiling point of solution. At each step, the new condensate becomes progressively richer in the most volatile component of the solution. If enough such steps are performed, a near complete separation can be achieved. A fractionating column that effectively executes N vaporization–condensation steps is said to have N theoretical plates of separation power.

4. If fractional distillation is conducted under conditions of reduced applied pressure, it is called vacuum distillation. When a solution is fractionated at reduced applied pressures, the separation power of the column increases, but the throughput of vapor per unit time decreases because of the lower vapor pressures at the boiling point.

5. When the K components of an ideal solution are mixed, we observe changes in the entropy, Gibbs free energy, and Helmholtz free energy, but not in the enthalpy, internal energy, or total volume. The changes are

$$\Delta G_{\text{mix}}^{\text{ideal}} = \Delta A_{\text{mix}}^{\text{ideal}} = RT \sum_{i=1}^{i=K} n_i \ln X_i$$

and

$$\Delta S_{\text{mix}}^{\text{ideal}} = -R \sum_{i=1}^{i=K} n_i \ln X_i,$$

but $\Delta V_{\text{mix}} = \Delta H_{\text{mix}}^{\text{ideal}} = \Delta U_{\text{mix}}^{\text{ideal}} = 0$.

6. The solubility of solids and gases in liquids can be easily computed for ideal solutions. For a solid A dissolving in solvent B, the saturation mole fraction is given by

$$X_A^{\text{sat}} = \exp\left[-\frac{\Delta \overline{H}_{\text{fus}}}{R}\left\{\frac{1}{T} - \frac{1}{T_m}\right\}\right],$$

where the partial molar enthalpy of fusion is for solute A and T_m is the melting point of the solute. The solubility of solids is nearly independent of the applied pressure. For gaseous A, the solubility is

$$X_A^{\text{sat}} = \frac{P_{A(g)}}{P^*} \exp\left[\frac{\Delta \overline{H}_{\text{vap}}}{R}\left\{\frac{1}{T} - \frac{1}{T^*}\right\}\right],$$

where P^* is the equilibrium vapor pressure of liquid A at temperature T^*, $\Delta \overline{H}_{vap}$ is the partial molar enthalpy of vaporization of liquid A, and $P_{A(g)}$ is the pressure of gaseous A above the solution. Thus, gaseous solubility in ideal solutions depends linearly upon the pressure of the gas to be dissolved. It also decreases with temperature, whereas the solubility of solids increases with temperature. If T^* is chosen such that $P^* = P_{A(g)}$ the solubility equation becomes

$$X_A^{sat} = \exp\left[\frac{\Delta \overline{H}_{vap}}{R}\left\{\frac{1}{T} - \frac{1}{T_b}\right\}\right].$$

If this form is used, it is essential to remember that T_b is not the normal boiling point of liquid A. Rather, it is the temperature that yields an equilibrium vapor pressure equal to that of gaseous A above the solution. These solubility equations will be accurate only if the components of the solution are highly similar. If they are not, deviations from the ideal-solution predictions will become large.

7. Solutions containing a nonvolatile solute exhibit a depression of their freezing points, an elevation of their boiling points and osmotic-pressure gradients. These effects are called colligative properties. If the solution is sufficiently dilute that the solvent is in the ideal-solution regime, the magnitude of the colligative properties can be related to the composition of the solution. The freezing-point depression is given by

$$\Delta T_f = \frac{R(T_f^o)^2 \ln X_B}{\Delta \overline{H}_{fus}^B - RT_f^o \ln X_B}.$$

The limiting form for this equation for very dilute solutions is

$$\Delta T_f \approx -\left[\frac{R(T_f^o)^2 M_B}{1,000 \, \Delta \overline{H}_{fus}^B}\right] m = -K_f m,$$

where K_f is called the cryoscopic constant and m is the solution molality. The boiling-point elevation has the similar form.

$$\Delta T_b = \frac{-RT_{bo}^2 \ln X_B}{\Delta \overline{H}_{vap}^B + RT_{bo} \ln X_B},$$

where T_{bo} is the temperature at which the solution boils. T_{bo} will be the normal boiling point only if the applied pressure is 760 torr. The dilute-solution limiting form is

$$\Delta T_b \approx \left[\frac{RT_{bo}^2 M_B}{1,000 \, \Delta \overline{H}_{vap}^B}\right] m = K_b(P_o)m,$$

where $K_b(P_o)$ is the ebullioscopic constant, whose value is a function of the applied pressure P_o.

The osmotic pressure π of an ideal solution is given by

$$\pi = \frac{-RT \ln X_B}{\overline{V}_{B(l)}},$$

where $\overline{V}_{B(1)}$ is the molar volume of the pure solvent B. For dilute solutions, this equation can be cast in a form very similar to the ideal-gas equation of state, viz., $\pi V_B = n_A RT$, where V_B is the total volume of solvent and n_A is the number of moles of solute present. Whereas the freezing-point depression and boiling-point elevation have relatively small magnitudes, osmotic pressures can be extremely large.

8. Real solutions exhibit negative deviations from ideality if the A–B interactions are stronger than the interactions between the pure components. If the reverse is true, the deviations are positive. That is, the actual equilibrium vapor pressures are greater than those predicted by Raoult's law when the deviations are

positive and are less that those predicted by Raoult's law when the deviations are negative. The measured equilibrium partial vapor pressure asymptotically approaches the Raoult's law line as its mole fraction in the solution approaches unity. That is, Raoult's law holds in real solutions for the component whose mole fraction approaches unity. In the region of concentration where Raoult's law holds for the concentrated component in a binary solution, Henry's law holds for the dilute component. Henry's law states that the equilibrium vapor pressure of component A is proportional to its mole fraction in the solution, or, mathematically, $P_A = K_A X_A$, where K_A is called the Henry's law constant. In general, K_A is a function of temperature and total applied pressure.

9. The vapor pressures, solubilities, mixing thermodynamics, and colligative properties of nonideal solutions must be experimentally measured. The resulting data are conveniently represented using temperature–composition phase diagrams. In general, it is not necessary to measure all solution properties. If solution activities are obtained from one set of measurements, other solution properties can be computed using those measurements.

10. The chemical potentials for the K components of a solution are not all independent, because they are related by the Gibbs–Duhem equation, which requires that

$$\sum_{i=1}^{i=K} n_i \, d\mu_i = 0 \quad \text{at } dT = dP = 0.$$

For a binary solution, this equation takes the form

$$X_A \left(\frac{\partial \mu_A}{\partial X_A} \right)_{T, P} + X_B \left(\frac{\partial \mu_B}{\partial X_A} \right)_{T, P} = 0.$$

Using the Gibbs–Duhem equation, we proved that if Raoult's law holds for component B, Henry's law must hold for component A. The importance of the equation lies in the fact that it permits us to compute the chemical potential or activity of component A from measured data on component B. Therefore, we do not need to conduct activity measurements on both components.

11. For the solvent of a binary solution, the most convenient reference function for the measurement of activity is the mole fraction and the most useful reference state is the pure liquid, for which $X_B \rightarrow 1$. With these choices, the activity of the solvent is

$$\lim_{x_B \rightarrow 1} \left[\frac{a_B}{X_B} \right] = 1,$$

and the corresponding activity coefficient is

$$\gamma_B = \frac{a_B}{X_B}.$$

Activities can be obtained from vapor pressure measurements or the measurement of any colligative property. The equations relating the activity to these solution properties are summarized in Table 8.1.

12. The most convenient reference state for the measurement of solute activities is infinite dilution wherein $X_A \rightarrow 0$. With this choice, the Gibbs–Duhem equation permits us to calculate γ_A from measured values of γ_B. The connecting equation is

$$\ln \left[\frac{a_A}{X_A} \right] = \ln \gamma_A = \int_{X_A=0}^{X_A} \frac{X_A - 1}{X_A} \left(\frac{\partial \ln(a_B/X_B)}{\partial X_A} \right) dX_A = \int_0^{X_A} \frac{X_A - 1}{X_A} \, d \ln \left(\frac{a_B}{X_B} \right).$$

The integral can be conveniently computed by calculating the area beneath a plot of $\left(\dfrac{X_A - 1}{X_A} \right) \left(\dfrac{\partial \ln(a_B/X_B)}{\partial X_A} \right)$ versus X_A, provided that the measured data for $\ln[a_B/X_B]$ versus X_A are fitted to an analytic function for which $\ln a_B/X_B$ and $(\partial \ln(a_B/X_B)/\partial X_A)$ both approach zero as $X_A \rightarrow 0$, as is required by the fact that $[a_B/X_B]$ approaches unity asymptotically as X_A approaches zero.

13. Once activity coefficients are obtained using one reference function, it is a simple matter to convert them to corresponding values for a different reference function. For example, if γ_x is the activity coefficient obtained using the mole fraction as the reference function, the activity coefficient with molality as the reference function is given by

$$\gamma_m = \gamma_x \left[\frac{1000}{M_o m + 1000} \right],$$

where m is the molality and M_o is the molar mass of the solvent.

Problems

Problems that require the use of some type of computational device are marked with an asterisk (*). Problems that require some type of plotting routine are indicated with a pound sign (#). Unless otherwise stated, all gases may be assumed to behave ideally.

8.1 An ideal binary solution is formed from liquids A and B. The pure component vapor pressures at 300 K are $P_A^o = 36.5$ torr and $P_B^o = 19.5$ torr. Determine the composition of the equilibrium vapor for an equimolar mixture of A and B.

8.2 An ideal binary solution is formed from liquids A and B. The pure component vapor pressures at 340 K are 275 torr and 167 torr for A and B, respectively.

(A) Make a careful plot of the equilibrium partial vapor pressure of A and B as a function of X_A.

(B) Compute the total vapor pressure over the solution at $X_A = 0.60$.

(C) At what value of X_A will we observe $P_A = P_B$?

8.3 Consider an ideal binary solution of A and B.

(A) Derive a general equation in terms of P_A^o and P_B^o that gives the mole fraction of A at which $P_A = P_B$. Note that this point corresponds to the crossing point of the two lines representing Raoult's law on a plot of ideal vapor pressure vs. mole fraction.

(B) Show that the total equilibrium vapor pressure over the solution at the point where $P_A = P_B$ is given by

$$P_T = \frac{2P_A^o P_B^o}{P_A^o + P_B^o}.$$

8.4 An ideal binary solution of A and B is observed to have $P_A = P_B$ when $X_A = 0.57895$. The total equilibrium vapor pressure over the solution at this point is 231.58 torr. Compute the pure component equilibrium vapor pressures of A and B at the temperature of the solution. (*Hint:* See Problem 8.3.)

8.5 Forty grams of $Ba(NO_3)_2$ are dissolved in 1 kilogram of water.

(A) If no ionization of the $Ba(NO_3)_2$ takes place, compute the vapor pressure of the solution at 313.15 K, where the equilibrium vapor pressure of pure water is 55.324 torr. Assume that the solution is ideal.

(B) If the vapor pressure of the solution is found to be 54.909 torr, compute the fraction of the $Ba(NO_3)_2$ that is ionized in the solution.

(C) What would the equilibrium H_2O vapor pressure be if we had 100% ionization of the $Ba(NO_3)_2$?

(D) What percent error is made in computating of the equilibrium vapor pressure using the limiting form given by Eq. 8.11?

8.6 Show that if a nonvolatile ionic compound ionizes completely into i ions, the limiting dilute-solution equation for the vapor pressure lowering is $\Delta P_B = -i\, X_{salt}\, P_B^o$, where X_{salt} is the mole fraction of the ionic salt in solution if no ionization occurs.

The next three problems permit the student to conduct a detailed examination of fractional distillation for an ideal CCl_4–$CHCl_3$ solution. The problems are constructed so that they may be assigned as a single unit or separately, as desired by the instructor.

8.7* Chloroform ($CHCl_3$) and carbon tetrachloride (CCl_4) have similar structures, and their boiling points are within 16 K of one another. We might, therefore, expect that a mixture of the two compounds will be nearly ideal. Chloroform boils at 334.41 K and has a partial molar enthalpy of vaporization of 29,469 J mol^{-1}. Carbon tetrachloride boils at 349.95 K and has a partial molar enthalpy of vaporization of 29,863 J mol^{-1}. Compute the boiling points of mixtures of these two compounds with mole fractions of CCl_4 from $X_{CCl_4} = 0$ to $X_{CCl_4} = 1$ in increments of 0.1. Assume that the applied pressure is 760 torr. Plot your results as a function of X_{CCl_4} and connect the data points by fitting them to a quadratic function of X_{CCl_4} using a least-squares method. Assume that the solution is ideal, that the molar enthalpies of vaporization are constant, and that the volume of the liquid is negligible relative to that of its equilibrium vapor.

8.8* (A) Compute the composition of the equilibrium vapor over solutions of CCl_4 and $CHCl_3$ at each of the boiling points determined in Problem 8.7. That is, determine the mole fraction of CCl_4 in the vapor,

Y_{CCl_4}, at each boiling point. These boiling temperatures are given in the following table:

X_{CCl_4}	T_b (K)	X_{CCl_4}	T_b (K)
0.00	334.41	0.60	342.70
0.10	335.63	0.70	344.36
0.20	336.91	0.80	346.10
0.30	338.24	0.90	347.97
0.40	339.67	1.00	349.95
0.50	341.14		

Assume that the solution is ideal, that the partial molar enthalpies of vaporization given in Problem 8.7 are constant, and that the volume of the liquid is negligible relative to that of its equilibrium vapor. The normal boiling points are given in Problem 8.7.

(B) Plot the boiling point vs. X_{CCl_4} and Y_{CCl_4} on the same graph. Fit the new data to a quadratic function of Y_{CCl_4}.

(C) By drawing appropriate tie-lines on the boiling point-vapor composition diagram obtained in (B), determine the minimum number of theoretical plates a fractionating column must have to produce a distillate whose mole fraction of CCl_4 is equal to or less than 0.2 if the initial mixture has a mole fraction of CCl_4 of 0.90.

8.9 The calculations of the solution boiling point as a function of X_{CCl_4} in Problem 8.7 show that the data are accurately fitted by the equation

$$T_b = 334.453 + 11.298X_{CCl_4} + 4.1562X_{CCl_4}^2(K),$$

where X_{CCl_4} is the mole fraction of CCl_4 in the CCl_4–$CHCl_3$ solution. The computations carried out in Problem 8.8 show that the boiling points as a function of the mole fraction of CCl_4 in the vapor phase, Y_{CCl_4}, are accurately represented by the equation

$$T_b = 334.44 + 18.634Y_{CCl_4} - 3.1590Y_{CCl_4}^2 \text{ (K)}.$$

Using these results, determine the composition of the distillate of a CCl_4–$CHCl_3$ solution that originally had $X_{CCl_4} = X_{CHCl_3} = 0.50$ if the fractionating column has two theoretical plates. At what temperature does this solution boil? (*Note:* Least-squares fitting is extremely sensitive to round-off errors. Consequently, the fits obtained will vary slightly, depending upon the computer software employed).

The next three problems permit the student to conduct a detailed examination of vacuum distillation for an ideal CCl_4–$CHCl_3$ solution. The problems are constructed so that they may be assigned as a single unit or separately, as desired by the instructor.

8.10* Chloroform ($CHCl_3$) and carbon tetrachloride (CCl_4) have similar structures, and their boiling points are within 16 K of one another. We might, therefore, expect that a mixture of the two compounds will be nearly ideal. Chloroform boils at 334.41 K and has a partial molar enthalpy of vaporization of 29,469 J mol^{-1}. Carbon tetrachloride boils at 349.95 K and has a partial molar enthalpy of vaporization of 29,863 J mol^{-1}. Compute the boiling points of mixtures of these two compounds with mole fractions of CCl_4 from $X_{CCl_4} = 0$ to $X_{CCl_4} = 1$ in increments of 0.1 if the applied pressure is 100 torr. Plot your results as a function of X_{CCl_4}, and connect the data points by fitting them to a quadratic function of X_{CCl_4} using a least-squares method. Assume that the solution is ideal, that the molar enthalpies of vaporization are constant, and that the volume of the liquid is negligible relative to that of its equilibrium vapor.

8.11* (A) Compute the composition of the equilibrium vapor over solutions of CCl_4 and $CHCl_3$ at each of the boiling points determined in Problem 8.10. That is, determine the mole fraction of CCl_4 in the vapor, Y_{CCl_4}, at each boiling point. These boiling temperatures are given in the following table:

X_{CCl_4}	T_b (K)	X_{CCl_4}	T_b (K)
0.00	280.71	0.60	286.82
0.10	281.61	0.70	288.04
0.20	282.55	0.80	289.34
0.30	283.53	0.90	290.72
0.40	284.57	1.00	292.20
0.50	285.67		

Assume that the solution is ideal, that the partial molar enthalpies of vaporization given in Problem 8.10 are constant, and that the volume of the liquid is negligible relative to that of its equilibrium vapor. The normal boiling points are given in Problem 8.10.

(B) Plot the boiling point vs. X_{CCl_4} and Y_{CCl_4} on the same graph. Fit the new data to a quadratic function of Y_{CCl_4}.

8.12 The calculations of the solution boiling point as a function of X_{CCl_4} in Problem 8.10 show that the data are accurately fitted by the equation

$$T_b = 280.75 + 8.252X_{CCl_4} + 3.1597X_{CCl_4}^2 \text{ (K)}$$

where X_{CCl_4} is the mole fraction of CCl_4 in the CCl_4–$CHCl_3$ solution. The computations carried out in Problem 8.11 show that the boiling points as a function of the mole fraction of CCl_4 in the vapor phase, Y_{CCl_4}, are accurately represented by the equation

$$T_b = 280.788 + 13.9031 Y_{CCl_4} - 2.5248 Y_{CCl_4}^2 \text{ (K)}.$$

Using these results, determine the composition of the distillate of a CCl_4–$CHCl_3$ solution that originally had $X_{CCl_4} = X_{CHCl_3} = 0.50$ if the fractionating column has two theoretical plates. At what temperature does this solution boil? (*Note:* Least-squares fitting is extremely sensitive to round-off errors. Consequently, the fits obtained will vary slightly, depending upon the computer software employed.)

8.13 Seventy-five grams of CCl_4 are mixed with 10 grams of $CHCl_3$ at 298 K to form a solution. If the solution is ideal, compute ΔG_{mix}, ΔS_{mix}, ΔH_{mix}, ΔU_{mix}, ΔV_{mix}, and ΔA_{mix} for the process.

8.14* Twenty grams of benzene (C_6H_6) are mixed with 30 grams of toluene (C_7H_8) and 40 grams of an unknown compound. The entropy change upon mixing is measured and found to be 8.6559 J K^{-1}. Assuming the solution to be ideal, determine the molar mass of the unknown compound.

8.15 An ideal solution of compounds A and B is to be formed by mixing a total of n_o moles of material, divided such that we have n_A and n_B moles of A and B, respectively. Prove that the entropy of mixing is a maximum if

$$f = \text{fraction of moles of type } A = \frac{n_A}{n_A + n_B} = 0.5.$$

That is, prove that, for a fixed number of total moles, n_o, the entropy of mixing is a maximum if A and B are mixed in equal molar amounts.

8.16 (A) Let us assume that naphthalene$(C_{10}H_8)$, which melts at 353.15 K, forms an ideal solution in benzene. If the enthalpy of fusion of naphthalene is 148.95 J g^{-1}, calculate the solubility of naphthalene in benzene at 298 K.

(B) At what temperature would the solubility of naphthalene be 70 g of naphthalene per mole of benzene?

8.17 Gaseous A forms an ideal solution with liquid solvent B. At 350 K, a saturated solution is found to have a mole fraction of A of 0.3017 when the pressure of A above the solution is 760 torr. At a temperature of 200 K, gaseous A has condensed to a liquid that is found to have an equilibrium vapor pressure of 11.466 torr.

(A) Compute the partial molar enthalpy of vaporization of A. Assume that the vapor is ideal and that the heat capacity of the liquid and gaseous A are the same at all temperatures.

(B) Compute the normal boiling point of compound A. The same assumptions made in (A) may be made here.

8.18 The following table gives solubility data for nitrous oxide, $N_2O(g)$, in water when the applied pressure of $N_2O(g)$ is 760 torr:

T (K)	Solubility, $X_{N_2O}^{sat}$
288.15	5.948×10^{-4}
293.15	5.068×10^{-4}
298.15	4.367×10^{-4}
303.15	3.805×10^{-4}
308.15	3.348×10^{-4}

(A) Plot $\ln X_{N_2O}^{sat}$ vs. T^{-1}. Does the degree of linearity indicate that the solution is ideal?

(B) Assuming the solution to be ideal, what is the predicted partial molar enthalpy of vaporization of N_2O? The experimental result is $16,530 \text{ J mol}^{-1}$. What is the percent error in the result obtained by the assumption of an ideal solution for N_2O?

(C) Use the results of (A) and (B) to obtain an estimate for the enthalpy of mixing of 1 mole of liquid N_2O with sufficient $H_2O(l)$ to form a saturated solution.

8.19 The following table gives solubility data for $CO_2(g)$ in H_2O as a function of temperature when the partial pressure of $CO_2(g)$ over the solution is 760 torr:

T (K)	Solubility (g of CO_2 per 100 g of H_2O)
273.15	0.3346
283.15	0.2318
288.15	0.1970
293.15	0.1688
298.15	0.1449
303.15	0.1257
313.15	0.0973
323.15	0.0761
333.15	0.0576

(A) Plot $\ln X_{CO_2}^{sat}$ vs. T^{-1}. Does the degree of linearity indicate that the solution is ideal?

(B) Assuming the solution to be ideal, what is the predicted partial molar enthalpy of vaporization of CO_2?

(C) *The Handbook of Chemistry and Physics* lists the partial molar enthalpy of fusion of CO_2 as $9,020 \text{ J mol}^{-1}$. It also lists the sublimation pressure of $CO_2 (s)$ as 227.1 kPa and 518.0 kPa at 205 K and 216.58 K, respectively. Use these data to obtain $\Delta \overline{H}_{vap}$ for CO_2, assuming that $\Delta \overline{H}_{vap}$ and $\Delta \overline{H}_{fus}$ are independent of temperature. Using this result, compute the percent error for the result obtained in (B), where we assumed a solution of CO_2 and water to be ideal. Does this result suggest near ideality?

(D) Use the results of (A), (B), and (C) to obtain an estimate for the enthalpy of mixing of 1 mole of liquid CO_2 with sufficient $H_2O(l)$ to form a saturated solution.

8.20 The following table gives data on the solubility of solid Li_2SO_4 in water at temperatures between 273.15 and 373.15 K in terms of the mass percent of Li_2SO_4, which is defined as $100m_1/(m_1 + m_2)$, where m_1 is the mass of Li_2SO_4 and m_2 is the mass of water:

T (K)	mass % Li_2SO_4
273.15	26.3
293.15	25.6
313.15	25.3
333.15	24.8
353.15	24.0
373.15	23.6

(A) Plot $\ln X_{Li_2SO_4}^{sat}$ vs. T^{-1}. Does the degree of linearity indicate that the solution is ideal?

(B) Assuming the solution to be ideal, what is the predicted partial molar enthalpy of fusion of Li_2SO_4? Does this result make any physical sense? What is wrong?

(C) *The Handbook of Chemistry and Physics* lists the partial molar enthalpy of fusion of Li_2SO_4 as 7,500 J mol^{-1}. Using this result, obtain an estimate for the enthalpy of mixing of 1 mole of liquid Li_2SO_4 with sufficient $H_2O(l)$ to form a saturated solution.

8.21 Compound A with a normal melting point of 400 K and a partial molar enthalpy of fusion of 3,500 cal mol^{-1}, forms ideal solutions with solvent B. It is found that, at $T = 300$ K, dissolving 150 g of A in 10 moles of B forms a saturated solution. Compute the molar mass of compound A.

8.22 Ten grams of a nonvolatile solute are added to 5 moles of a solvent whose total volume is 200 cm^3. The partial molar enthalpy of fusion of the solvent is 2,000 cal mol^{-1}, and its normal freezing point is 280 K. The solvent in the solution is found to freeze at 279.894 K. Calculate the osmotic pressure of the solution at 300 K. Assume that the solution is ideal.

The next nine problems explore the thermodynamics of a hypothetical nonideal binary solution. The problems are constructed so that they may be assigned individually. However, the interrelationships between the various properties of solutions are best illustrated by assigning the entire set as one homework exercise.

8.23 Solute A is dissolved in solvent B at 320 K. The following table gives the measured equilibrium partial pressures of B over the solution as a function of its mole fraction in the solution, X_B:

X_B	$P_{B(soln)}$ (torr)	X_B	$P_{B(soln)}$ (torr)
0.000	0.000	0.550	21.800
0.050	1.000	0.600	25.500
0.100	2.200	0.650	28.900
0.150	3.500	0.700	33.300
0.200	5.000	0.750	36.900
0.250	7.000	0.800	40.600
0.300	8.500	0.850	43.300
0.350	10.000	0.900	46.100
0.400	13.000	0.950	49.240
0.450	15.200	1.000	51.858
0.500	18.600		

Is the solution ideal? Plot $P_{B(soln)}$ vs. X_B at 320 K, and on the same graph plot the ideal-solution result. What can we conclude about the nature of the A–B intermolecular forces relative to the pure component intermolecular forces?

8.24* The equilibrium partial vapor pressures of compound B at 320 K above a binary solution of compound A in solvent B are measured and found to be those listed in the table in Problem 8.23. Assume that these measured pressures can be represented by the expression

$$P_{B(soln)} = P_B^o \sum_{i=1}^{i=4} C_i X_B^i,$$

where P_B^o is the pure component equilibrium vapor pressure over liquid B at 320 K. If the solution were ideal, we would have $C_1 = 1$ and $C_2 = C_3 = C_4 = 0$. For the measured data, determine the best values of the C_i ($i = 1, 2, 3, 4$) by using a linear least-squares fitting procedure. Prepare a plot showing the calculated fit and the data points on the same graph. Use the analytical fit to the vapor pressure data to determine the Henry's law constant for component B. (*Note:* Least-squares fitting is extremely sensitive to round-off errors. Consequently, the fits obtained will vary slightly depending upon the computer software employed.)

8.25# The equilibrium partial vapor pressures of compound B at 320 K above a binary solution of compound A in solvent B are measured and found to be accurately described by the analytical function

$$P_{B(soln)} = P_B^o[0.442067X_B - 0.311495X_B^2$$
$$+ 2.60907X_B^3 - 1.74685X_B^4] \text{ torr},$$

where P_B^o is the pure component equilibrium vapor pressure at 320 K, which has been measured and found to be 51.858 torr. If the mole fraction of B is taken as the reference function and $X_B \to 1$ is taken as the reference state, determine the activity and the activity coefficient of B in the A–B solution at 320 K as a function of X_B. Prepare a plot of γ_B vs. X_B for this system. Based on the appearance of the plot, comment on the accuracy of the analytical fit .

8.26 The equilibrium partial vapor pressures of compound B at 320 K above a binary solution of compound A in solvent B are measured and found to be accurately described by the analytical function

$$P_{B(\text{soln})} = P_B^o [0.442067 X_{B_2} - 0.311495 X_B^2$$
$$+ 2.60907 X_B^3 - 1.74685 X_B^4]\ \text{torr},$$

where P_B^o is the pure component equilibrium vapor pressure at 320 K, which has been measured and found to be 51.858 torr. If the partial molar enthalpies of B in the solution and in the pure liquid state are the same, compute the temperature at which solvent B will freeze in a solution whose mole fraction is $X_B = 0.80$. The normal freezing point of B is 250 K, and its partial molar enthalpy of fusion is 18,000 J mol^{-1}.

8.27 The equilibrium partial vapor pressures of compound B at 320 K above a binary solution of compound A in solvent B are measured and found to be accurately described by the analytical function

$$P_{B(\text{soln})} = P_B^o [0.442067 X_B - 0.311495 X_B^2$$
$$+ 2.60907 X_B^3 - 1.74685 X_B^4]\ \text{torr},$$

where P_B^o is the pure component equilibrium vapor pressure at 320 K, which has been measured and found to be 51.858 torr. Compute the osmotic pressure for this solution at 320 K when $X_B = 0.85$. The partial molar volume of B is 0.089 L mol^{-1}. What would the osmotic pressure be if the solution were ideal?

8.28* The activities of the solvent B at 320 K have been determined by vapor pressure measurements. The results for $\ln[a_B/X_B]$ have been fitted to a polynomial function by linear least-square methods. The result is

$$\ln[a_B/X_B] = -0.359427 X_A^2 - 3.428316 X_A^3$$
$$+ 3.156032 X_A^4 - 0.1863966 X_A^5.$$

In this problem, we wish to use the Gibbs–Duhem equation to compute the activity coefficient of the solute A as a function of X_A at 320 K. We will take X_A as the reference function and $X_A \to 0$ as the reference state. The various parts of the problem serve as a guide.

(A) Use the fitted data to develop an analytic expression for $\left(\dfrac{X_A - 1}{X_A}\right)\left(\dfrac{\partial \ln(a_B/X_B)}{\partial X_A}\right)$. Show that there is no discontinuity in this expression as $X_A \to 0$.

(B) Make a careful plot of $\left(\dfrac{X_A - 1}{X_A}\right)\left(\dfrac{\partial \ln(a_B/X_B)}{\partial X_A}\right)$ as a function of X_A, using the result obtained in (B).

(C) Using the Gibbs–Duhem equation, obtain an expression for $\ln \gamma_A$ as a function of X_A. Prepare a plot of γ_A versus X_A.

8.29 The activities of the solvent B at 320 K have been determined by vapor pressure measurements. The results for $\ln[a_B/X_B]$ have been fitted by linear least-square methods to a polynomial function. The result is

$$\ln[a_B/X_B] = -0.359427 X_A^2 - 3.428316 X_A^3$$
$$+ 3.156032 X_A^4 - 0.1863966 X_A^5.$$

From these data, together with the Gibbs–Duhem equation, the activity coefficients of solute A are computed as a function of the mole fraction of A in Problem 8.28. The result is

$$\ln \gamma_A = 0.718854\ X_A + 4.783047 X_A^2 - 7.636359 X_A^3$$
$$+ 3.389028 X_A^4 - 0.186397 X_A^5.$$

Compute ΔG_{mix} when 3 moles of solute A are mixed with 20 moles of solvent B at 320 K. Compare your result with that expected for an ideal solution.

8.30 In Problem 8.24, the Henry's law constant for the nonideal solution considered in Problems 8.23 through 8.30 is calculated to be 22.92 torr at 320 K. If the partial molar enthalpy of vaporization of component B from the solution is found to be 48,000 J mol^{-1}, determine the value the Henry's law constant will have at 340 K.

8.31 For the solution considered in Problem 8.29, the activity coefficient of solute A at a mole fraction $X_A = 0.360$ is 1.783. If we change the reference function for solute A from mole fraction to molality, with the reference state taken to be $m \to 0$, compute the activity for solute A. The molar mass of the solvent is 78.11 g mol^{-1}.

8.32 Let γ_c be the activity coefficient for solute A when the reference function is taken to be the molarity c of the solution and infinite dilution ($c \to 0$) is taken as the reference state.

(A) Obtain an equation permitting γ_c to be computed from γ_m, the activity coefficient when molality (m) is used as the reference function with $m \to 0$ as the reference state. Let the solution and pure solvent densities be d and d_o, respectively. Let the molar masses of solute and solvent be M_1 and M_o, respectively.

(B) Obtain an equation permitting γ_c to be computed from γ_X, the activity coefficient when the mole fraction is used as the reference function with $X \to 0$ as the reference state.

8.33 Osmotic-pressure measurements on solutions containing a high-molecular-weight polymer are taken as a function of the mass of polymer dissolved per liter of solution, w, at 298 K. The data are as follows:

w (g L^{-1})	π (torr)	w (g L^{-1})	π (torr)
1.000	0.02746	3.106	0.1219
1.897	0.06162	4.513	0.2127
2.170	0.07381		

Determine the molar mass of the polymer.

8.34 Figure 8.11 shows that the solubilities of $MgCl_2$ and KBr are almost independent of temperature. What qualitative conclusions can we draw from these experimental data?

8.35 A chemist mixes 20 g of *n*-butanol with 80 g of H_2O at a temperature of 410 K.

(A) How many phases are present? Are any of the solutions saturated? How do you know?

(B) The chemist lowers the temperature of the mixture to 373.15 K. How many phases are now present? If there are two phases, what masses of H_2O and *n*-butanol are present in each phase? If there is only one phase, at what temperature will two phases appear. (See the phase diagram shown in Figure 8.21. The calculations need be done only as accurately as permitted by this diagram.)

8.36 A chemist adds 10 g of Cd to 90 g of Bi at 473.15 K.

(A) How many phases are present? What are the phases?

(B) The chemist now begins to add additional Cd to the mixture while keeping the temperature constant at 473.15 K. How much Cd must be added before the chemist sees a change in the number of phases present? What phase(s) is (are) now present?

(C) After the number of phases has changed in (B), additional Cd is added at 473.15 K. How much additional Cd must be added before the number of phases changes once again? What phase(s) is (are) now present? What mass of Cd is present in each phase? (See the phase diagram shown in Figure 8.23. The calculations need be done only as accurately as permitted by this diagram.)

8.37 A chemist mixes Mg and Zn. Starting at a temperature of 950 K, she obtains the cooling curve shown in Figure 8.29 for her mixture.

(A) If we know that the number of moles of Mg in the mixture is greater than the number of moles of Zn, what is the mole percent of magnesium in the mixture, and at what temperature does the flat portion of the cooling curve in the figure occur?

(B) If know that the number of moles of Mg in the mixture is less than the number of moles of Zn, can we determine the mole percent of Mg present? If so,

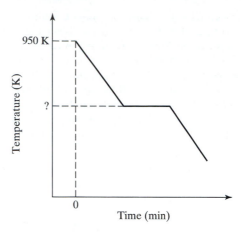

▲ FIGURE 8.29
Cooling curve obtained by a chemist using *Mg–Zn* with an initial temperature of 950 K.

obtain this percentage. If not, state what information we can obtain about the mole percent of Mg present and the temperature at which the flat portion of the cooling occurs. (See the phase diagram in Figure 8.25.)

8.38 A chemist has two mixtures of Mg and Zn. Both contain 0.1 mole of Mg and 0.9 mole of Zn. Mixture 1 is at a temperature of 935 K, while Mixture 2 is at 773 K. The chemist begins to add additional Mg to each mixture while maintaining the temperature of each mixture constant at its initial value. Describe the phase changes that occur in each mixture and determine how much magnesium must be added to produce each phase change. (See the phase diagram in Figure 8.25. The calculations need be done only as accurately as permitted by this diagram.)

8.39 The H_2O–H_2SO_4 phase diagram is shown in Figure 8.30 on the next page. How many compounds of H_2O and H_2SO_4 can be formed? What are their chemical formulae? Label the phases present in each region of the diagram.

8.40 The phase diagram for two metals A and B contains one eutectic and no compound between A and B forms. The freezing points of solutions of A and B are found to vary linearly with the mole percent of B in the solution. The melting points of pure A and B are 650 K and 600 K, respectively. The cooling curve for a mixture containing 40 mole percent B exhibits a single flat region at 400 K. Using this information, sketch the $A-B$ phase diagram and indicate the phases present in each region.

8.41 Sam's grades in physical chemistry have been increasing rapidly. He is now one of the better students in the class. He has noticed that a popular laboratory manual for freshman chemistry contains the following simple boiling-point elevation experiment: Students place 100 grams of an unknown compound, which is actually sucrose ($C_{12}H_{22}O_{11}$), into 1000 grams of H_2O and then measure the elevation of the boiling point.

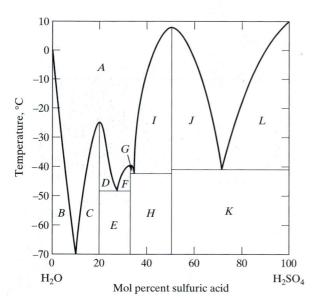

▲ FIGURE 8.30
The H_2O–H_2SO_4 phase diagram. The letters mark the various regions of the diagram. Note that all temperatures are given in degrees Celsius.

Using the limiting form of the boiling-point elevation equation, $\Delta T_b = K_b m$ with $K_b = 0.51 \text{ K mol}^{-1}$, the students are required to determine the molar mass of the unknown solute. Sam points out to the class that when students in New Orleans and Miami perform this experiment, they generally obtain molar masses in the range 336.7 to 343.5, an error of about $\pm 1\%$. However, when students in Lhasa, Tibet, perform this same experiment using the same equipment, they usually obtain results in the range 364.6 to 372.0, an error of about $\pm 7.6\%$. Before beginning the study of the thermodynamics of electrolytic solutions, Sam offers to buy lunch for anyone in the class who can explain why the students in Lhasa are getting such poor results, provided that those who cannot explain this unexpected result buy him lunch. Are there any gamblers in the class? (The partial molar enthalpy of vaporization of H_2O is 40,657 J mol^{-1}.)

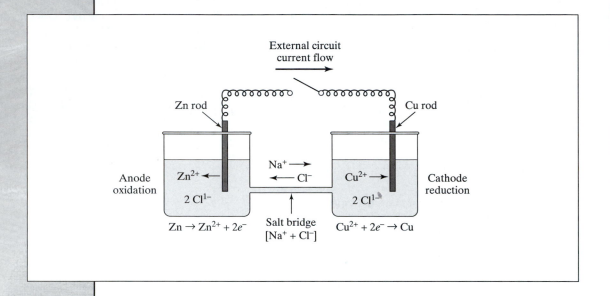

CHAPTER 9

Thermodynamics of Electrolytic Solutions

In many experimental situations, the solute is a material that forms charged particles called ions in solution. Solutions in which this occurs are called *electrolytic solutions*. The presence of ions produces properties not seen in nonelectrolytic solutions that permit a wide variety of electrochemical measurements to be made.

With nonelectrolytic solutions, deviations from ideal behavior occur primarily because of differences in size of the solution components. This size difference produces significantly different polarizabilities, which in turn lead to differences in the intermolecular forces and deviations from Raoult's law. In electrolytic solutions, deviations from ideal behavior are mainly the result of strong coulombic forces between the charged ions—forces that are absent with uncharged solvent molecules. The difference in the magnitude of the London dispersion forces between solvent molecules and the coulombic interaction between the ions is so large that ideal behavior is observed in ionic solutions only in the limit of extreme dilution.

In this chapter, we shall discuss how we may express the chemical potential of an electrolytic solute in terms of activity coefficients and concentrations. We shall then develop the Debye–Hückel theory, which permits us to compute ionic activity coefficients for dilute solutions. This treatment will introduce many of the mathematical concepts and equations that recur in the study of quantum mechanics. Methods for measuring these quantities are described in some detail. The importance of activity coefficients in ionic equilibria is illustrated using acid–base equilibria, hydrolysis, and solubility as examples. The chapter presents a careful treatment of solution conductivity before beginning a discussion of electrochemistry.

9.1   The Chemical Potential of Electrolytic Solutes

If a solute produces ions in solutions, it is called an *electrolyte,* and, as mentioned, its solutions are referred to as electrolytic solutions. Whether ions have formed in a given solution can be easily determined by measuring the conductivity of the solution. If the conductivity is large, the solution is termed a *strong* electrolyte; if the conductivity is small, the term *weak* electrolyte is used. Strong acids and bases and most salts are examples of strong electrolytes, whereas weak acids and weak bases, along with a few salts containing primarily transition metals and Group IIIA and IVA metals, are weak electrolytes. Figure 9.1 qualitatively illustrates these points.

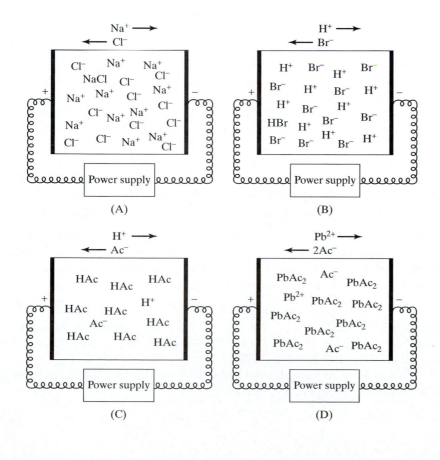

◄ FIGURE 9.1
Weak and strong electrolytes. Most salts, such as NaCl in (A), and strong acids, such as HBr in (B), ionize almost completely in aqueous solution, producing large numbers of ions that act as charge carriers as the positive ions move toward the negative electrode under the influence of an electromotive force while the negatively charged anions move toward the positive electrode. This motion allows current to be conducted easily through the circuit. These solutions are called strong electrolytes. In contrast, weak acids, such as acetic acid (HAc) in (C), and some salts, such as $PbAc_2$ in (D), exhibit a very low degree of ionization and consequently very few ions. Solutions of these materials have a much lower ability to conduct current and hence are called weak electrolytes.

411

The salts classified as strong electrolytes generally exist as ionic compounds in the solid state. Consequently, they exhibit high melting and boiling temperatures. These points were covered in detail in Chapter 7. When such salts dissolve in a polar solvent, they split into their component ions, which then react to form solvated ions in solution. (See Figure 8.4.) The *solvation number* refers to the number of solvent molecules usually associated with a given ion in solution. Such solvation occurs because of the very strong, attractive ion–dipole forces between the bare ion and the solvent. In the usual case, ions with high charge density tend to undergo solvation to a greater extent, as shown in Figure 9.2. Because these forces are attractive, solvation is an exothermic process for which $\Delta \overline{H}$ is negative. This negative value of $\Delta \overline{H}$ in turn tends to make $\Delta \mu$ negative. When $\Delta \mu$ is negative, the formation of a solution at a constant temperature and constant pressure or volume is spontaneous. This provides the qualitative explanation of why ionic solids tend to dissolve in polar solvents. In nonpolar solvents, the attractive ion–dipole forces are absent, and the solvation driving force is, therefore, greatly reduced.

Some solutes are covalently bonded in their pure state, but tend to form ions in solution because of the driving force produced by solvation. Examples are acetic acid ($HC_2H_3O_2$) and hydrochloric acid (HCl). In the case of hydrochloric acid, extensive ionization occurs in water via the reaction $HCl(g) + x\,H_2O(l) = H_3O^+(aq) + Cl^-(aq)$. This equilibrium is shifted far to the right; therefore, HCl is classified as a strong electrolyte. The same type of reaction occurs with acetic acid: $HC_2H_3O_2(l) + x\,H_2O(l) = H_3O^+(aq) + C_2H_3O_2^-(aq)$, but now the equilibrium is usually shifted to the left, and acetic acid is termed a weak electrolyte. Both cases are illustrated in Figure 9.1. Note that, in solvation reactions, the number of solvent molecules involved depends upon the extent of solvation, or the *solvation number,* of the solute ions. In this text, we will write hydrated ions with the notation "aq" and reflect the uncertainty in the hydration numbers by using a coefficient x for H_2O on the left side of the chemical equation.

Interionic coulombic forces are much stronger than the London dispersion forces or dipole–dipole interactions between solvent molecules. Therefore, we expect substantial deviations from ideality, so a consideration of solute activities is critical. Let us begin by considering a strong electrolyte whose molecular formula is $C_{\nu_+} A_{\nu_-}$, where C represents the particle that will form a cation with charge z_+ in solution and A is the particle that will form an anion with charge $-z_-$ in solution, so that z_- is the magnitude of the charge. For a strong electrolyte, we expect near complete ionization in a polar solvent:

$$C_{\nu_+} A_{\nu_-} + x\,H_2O(l) \longrightarrow \nu_+ C^{z_+}(aq) + \nu_- A^{-z_-}(aq).$$

In the general case, we might also observe ion-pair formation; that is, species with the formula $[C - A]^{z_+ - z_-}(aq)$ could be formed. In our initial treatment of the problem, we will assume that such ion-pair formation does not occur. Since the solution must be electrically neutral, we have

$$z_+ \nu_+ - z_- \nu_- = 0. \tag{9.1}$$

Equation 9.1 requires that $z_+ = \nu_-$ and $z_- = \nu_+$. For example, if the salt were $CaBr_2$, we would have $CaBr_2 + x\,H_2O(l) \longrightarrow Ca^{2+}(aq) + 2\,Br^-(aq)$, so that $\nu_+ = 1, \nu_- = 2, z_+ = 2,$ and $z_- = 1$.

We now need to formulate an expression for the total Gibbs free energy of the solution. Using Eq. 8.99, we have

$$G = \sum_{i=1}^{i=K} n_i \mu_i \tag{9.2}$$

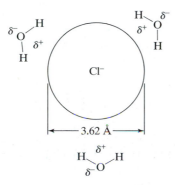

where μ_i is the chemical potential of component i and n_i is the number of moles of that component present in the solution containing K components. For a solution of the preceding strong electrolyte, Eq. 9.2 gives

$$G = n_B[\mu^o_{B(soln)} + RT \ln a_B] + n_e \nu_+[\mu^o_+ + RT \ln a_+] + n_e \nu_-[\mu^o_- + RT \ln a_-], \tag{9.3}$$

where μ^o_+ and μ^o_- are the standard chemical potentials for the cation and anion, respectively, a_+ and a_- are the corresponding activities, and n_e is the number of moles of $C_{\nu_+} A_{\nu_-}$ added. The first term represents the contribution of the solvent B to the Gibbs free energy. Since the cation and anion always appear together, it is customary to combine the last two terms and write Eq. 9.3 in the form

$$G = n_B[\mu^o_{B(soln)} + RT \ln a_B] + n_e[\nu_+ \mu^o_+ + \nu_- \mu^o_-] + n_e RT \ln[a_+^{\nu_+} a_-^{\nu_-}]$$
$$= n_B[\mu^o_{B(soln)} + RT \ln a_B] + n_e \mu^o_e + n_e RT \ln[a_+^{\nu_+} a_-^{\nu_-}], \tag{9.4}$$

where $\mu^o_e = [\nu_+ \mu^o_+ + \nu_- \mu^o_-]$, with the superscript denoting the standard chemical potential of the electrolyte.

The first step in the computation or measurement of an activity is the choice of reference function and reference state. The most convenient reference function for this purpose is the molality (m). For the reference state, we wish to choose a state in which the deviations from ideality will approach zero for all ionic species. Since these deviations are due primarily to the coulombic interactions between ions, this consideration suggests that the reference state be taken as infinite dilution, wherein the mean distance between ions is so large that their coulombic interaction is effectively zero. Consequently, we define the ionic activity scale for a_+ and a_- by taking

$$\underset{m\to 0}{\text{Lim}} \frac{a_+}{g(m)} = 1 \quad \text{and} \quad \underset{m\to 0}{\text{Lim}} \frac{a_-}{g(m)} = 1, \tag{9.5}$$

where $g(m)$ is some appropriate function of the molality. Equation 9.3 shows that we cannot choose $g(m)$ to be m itself. Since both a_+ and a_- appear as arguments of the logarithm function, both must be unitless quantities. Therefore, Eq. 9.5 cannot be satisfied with the choice $g(m) = m$, because the ratios a_+/m and a_-/m can never be equal to the unitless value of unity at any concentration. Accordingly, we must choose $g(m)$ so as to ensure that our limiting expression in Eq. 9.5 is unitless. The simplest choice is

$$\boxed{g(m) = \frac{m}{m^o}},\tag{9.6}$$

where $m^o = 1$ mol kg^{-1}. The corresponding activity coefficients are

$$\gamma_+ = \frac{a_+}{g(m)} = \frac{m^o a_+}{m_+}\tag{9.7A}$$

and

$$\gamma_- = \frac{a_-}{g(m)} = \frac{m^o a_-}{m_-},\tag{9.7B}$$

where m_+ and m_- are the molalities of the cation and anion, respectively.

Using Eqs. 9.7A and 9.7B, we may now express the Gibbs free energy of the solution in terms of concentrations, activity coefficients, and stoichiometric coefficients. Directly substituting those equations into Eq. 9.4 yields

$$G = n_B\left[\mu^o_{B(\text{soln})} + RT \ln a_B\right] + n_e\left[\nu_+ \, \mu^o_+ + \nu_- \, \mu^o_-\right]$$

$$+ n_e \, RT \ln\left\{\gamma_+^{\nu_+} \, \gamma_-^{\nu_-}\left[\frac{m_+}{m^o}\right]^{\nu_+}\left[\frac{m_-}{m^o}\right]^{\nu_-}\right\}.\tag{9.8}$$

We may express the ionic molalities in terms of the molality the compound $C_{\nu_+} A_{\nu_-}$ would have if it did not ionize in solution. If m represents this quantity, we have $m_+ = \nu_+ m$ and $m_- = \nu_- m$, since the ionization of each mole of $C_{\nu_+} A_{\nu_-}$ produces ν_+ moles of the cation and ν_- moles of the anion. Making this substitution into Eq. 9.8, we obtain

$$G = n_B\left[\mu^o_{B(\text{soln})} + RT \ln a_B\right] + n_e \, \mu^o_e$$

$$+ n_e \, RT \ln\left\{\gamma_+^{\nu_+}\gamma_-^{\nu_-} \, \nu_+^{\nu_+}\nu_-^{\nu_-}\left[\frac{m}{m^o}\right]^{\nu_+ +\nu_-}\right\}.\tag{9.9}$$

The quantities $\left[a_+^{\nu_+} a_-^{\nu_-}\right]$ or $\left[\gamma_+^{\nu_+} \gamma_-^{\nu_-}\right]$, and $\left[\nu_+^{\nu_+} \nu_-^{\nu_-}\right]$ always appear in the expression for the Gibbs free energy for any electrolytic solution. Equation 9.9 may, therefore, be simplified by respectively defining the mean activity $a_\pm$, the mean activity coefficient $\gamma_\pm$, and the mean stoichiometric coefficient $\nu_\pm$ as

$$\boxed{a_\pm^\nu = a_+^{\nu_+} a_-^{\nu_-}},$$

$$\boxed{\gamma_\pm^\nu = \gamma_+^{\nu_+} \gamma_-^{\nu_-}},$$

and

$$\boxed{\nu_\pm^\nu = \nu_+^{\nu_+} \nu_-^{\nu_-}},\tag{9.10}$$

where $\nu = \nu_+ + \nu_-$. In these terms, Eq. 9.9 becomes

$$G = n_B[\mu^o_{B(soln)} + RT \ln a_B] + n_e\,\mu^o_e + n_e\nu\,RT\,\ln\left[\nu_\pm\,\gamma_\pm\left(\frac{m}{m^o}\right)\right]. \quad (9.11)$$

The corresponding chemical potential of the solute is given by

$$\mu_e = \left(\frac{\partial G}{\partial n_e}\right)_{T,P,n_B} = \mu^o_e + \nu\,RT\,\ln\left[\nu_\pm\,\gamma_\pm\left(\frac{m}{m^o}\right)\right]. \quad (9.12)$$

EXAMPLE 9.1

Consider a solution of $CaBr_2$ in water containing 0.05 mole of $CaBr_2$ and 500 g of $H_2O(l)$. **(A)** Obtain an expression for the Gibbs free energy of the solute of this solution in terms of the activities, the standard Gibbs free energies of the Ca^{2+} and Br^- ions, and the temperature. **(B)** Express the result obtained in (A) in terms of the mean ionic activity coefficient and the mean stoichiometric coefficient if m/m^o is used as the reference function and infinite dilution as the reference state. **(C)** Obtain an expression for the chemical potential of the solute in terms of the mean ionic activity coefficient, the temperature, and the standard chemical potentials.

Solution

(A) The ionization reaction is $CaBr_2 \longrightarrow Ca^{2+} + 2\,Br^-$. The Gibbs free energy of the solute is given by Eq. 9.4:

$$G_{solute} = n_e[\nu_+\,\mu^o_+ + \nu_-\,\mu^o_-] + n_e\,RT\,\ln[a^{\nu_+}_+\,a^{\nu_-}_-]$$

$$= 0.05[\mu^o_{Ca^{2+}} + 2\mu^o_{Br^-}] + 0.05\,RT\,\ln[a_{Ca^{2+}}\,a^2_{Br^-}]. \quad (1)$$

(B) Using m/m^o as the reference function and $m \to 0$ as the reference state, we have

$$G_{solute} = 0.05[\mu^o_{Ca^{2+}} + 2\mu^o_{Br^-}] + n_e\nu\,RT\,\ln\left[\nu_\pm\,\gamma_\pm\left(\frac{m}{m^o}\right)\right]$$

$$= 0.05[\mu^o_{Ca^{2+}} + 2\mu^o_{Br^-}] + 0.05(3)\,RT\,\ln\left[\nu_\pm\,\gamma_\pm\left(\frac{m}{m^o}\right)\right], \quad (2)$$

where

$$\nu_\pm = [\nu^{\nu_+}_+\,\nu^{\nu_-}_-]^{1/\nu} = [1^1\,2^2]^{1/3} = 4^{1/3} \quad (3)$$

and

$$m = \frac{0.05\,\text{mol}}{0.5\,\text{kg}} = 0.10\,\text{mol kg}^{-1}. \quad (4)$$

Substituting into Eq. 2, we obtain

$$G_{solute} = 0.05[\mu^o_{Ca^{2+}} + 2\mu^o_{Br^-}] + 0.15RT\,\ln[4^{1/3}\,(0.1)\,\gamma_\pm]$$

$$= 0.05[\mu^o_{Ca^{2+}} + 2\mu^o_{Br^-}] + 0.15RT\,\ln[0.1587\,\gamma_\pm]. \quad (5)$$

(C) The chemical potential is given by Eq. 9.12:

$$\mu_e = \mu^o_e + \nu\,RT\,\ln\left[\nu_\pm\,\gamma_\pm\left(\frac{m}{m^o}\right)\right] = [\mu^o_{Ca^{2+}} + 2\mu^o_{Br^-}] + 3\,RT\,\ln[0.1587\gamma_\pm]. \quad (6)$$

For related exercises, see Problems 9.1 and 9.2.

Let us now assume that ions can associate in an equilibrium process to form pairs. With ion-pair formation included, the dissolution of the solute is described by

$$C_{\nu_+} A_{\nu_-} + x\, H_2O(l) \longrightarrow \nu_+\, C^{z_+}(aq) + \nu_-\, A^{z_-}(aq)$$

followed by

$$C^{z_+}(aq) + A^{z_-}(aq) = [C - A]^{z_+ - z_-}(aq),$$

where the last process is at equilibrium in the ionic solution. The differential change in the Gibbs free energy is given by Eq. 5.40:

$$dG = -S\, dT + V\, dP + \sum_{i=1}^{K} \mu_i\, dn_i.$$

Hence,

$$dG = -S\, dT + V\, dP + \mu_B\, dn_B + \mu_+\, dn_+ + \mu_-\, dn_- + \mu_p\, dn_p, \quad \text{(9.13)}$$

where the subscript p stands for the ion pair. We now let n_e be the total number of moles $C_{\nu_+} A_{\nu_-}$ added to the solution. Balancing masses then requires that

$$n_+ = \nu_+ n_e - n_p$$

and

$$n_- = \nu_- n_e - n_p.$$

Taking differentials of both sides, we obtain

$$dn_+ = \nu_+\, dn_e - dn_p$$

and

$$dn_- = \nu_-\, dn_e - dn_p.$$

Substituting these results into Eq. 9.13 gives

$$dG = -S\, dT + V\, dP + \mu_B\, dn_B + \mu_+[\nu_+\, dn_e - dn_p]$$
$$+ \mu_-[\nu_-\, dn_e - dn_p] + \mu_p\, dn_p \quad \text{(9.14)}$$

If the temperature and pressure are held constant, then

$$dG = \mu_B\, dn_B + [\mu_+ \nu_+ + \mu_- \nu_-]dn_e - [\mu_+ + \mu_-]dn_p + \mu_p\, dn_p. \quad \text{(9.15)}$$

If the ion-pair formation reaction is at equilibrium, we must have $\mu_+ + \mu_- = \mu_p$, so that the last two terms in Eq. 9.15 add to zero. Therefore, the chemical potential of the solute is given by the second term; that is,

$$\mu_e = \mu_+ \nu_+ + \mu_- \nu_- = \nu_+[\mu_+^o + RT \ln a_+] + \nu_-[\mu_-^o + RT \ln a_-]. \quad \text{(9.16)}$$

Equation 9.16 is essentially identical to Eq. 9.4, so by choosing $[m/m^o]$ as the reference function and $m \to 0$ as the reference state, we obtain

$$\mu_e = [\nu_+ \mu_+^o + \nu_- \mu_-^o] + RT \ln\left\{ \gamma_+^\nu \gamma_-^\nu \left[\frac{m_+}{m^o}\right]^{\nu+} \left[\frac{m_-}{m^o}\right]^{\nu-} \right\}. \quad \text{(9.17)}$$

When there was no ion-pair formation, we had $n_+ = \nu_+ n_e$ and $n_- = \nu_- n_e$. When ion pairs form, however, both n_+ and n_- will be less than those values. Let f be the fraction of the cations that do not associate into ion pairs. With this definition, we can write $n_+ = f \nu_+ n_e$ and $n_p = (1 - f)\nu_+ n_e$. The number of moles of the anion will be the total number of anions, minus the number of ion pairs formed; that is, $n_- = \nu_- n_e - n_p = \nu_- n_e - (1-f)\nu_+ n_e = n_e[\nu_- - (1-f)\nu_+]$. Dividing these expressions for n_+ and n_- by the number of kilograms of solvent, we obtain the new equations for m_+ and m_- that include the effect of ion-pair formation:

$$m_+ = f\,\nu_+\,m$$

and

$$m_- = [\nu_- - (1-f)\nu_+]\,m. \qquad (9.18)$$

Substituting these results into Eq. 9.17 produces the required expression for the chemical potential of the solute when a fraction $(1-f)$ of the cations form ion pairs:

$$\mu_e = \mu_e^o + RT \ln\left\{\gamma_+^{\nu_+}\,\gamma_-^{\nu_-}\left[f\,\nu_+\frac{m}{m^o}\right]^{\nu_+}\left[(\nu_- - (1-f)\nu_+)\frac{m}{m^o}\right]^{\nu_-}\right\}. \qquad (9.19)$$

Insertion of the mean ionic activity coefficient into Eq. 9.19 gives

$$\mu_e = \mu_e^o + RT \ln\left\{\gamma_\pm^{\nu}\,[f\nu_+]^{\nu_+}\,[\nu_- - (1-f)\nu_+]^{\nu_-}\left(\frac{m}{m^o}\right)^{\nu}\right\} \qquad (9.20)$$

We may force Eq. 9.20 to assume the same form as Eq. 9.12, when there was no ion-pair formation, by defining

$$\gamma_i^{\nu} = \nu_\pm^{-1}[f\nu_+]^{\nu_+}\,[\nu_- - (1-f)\nu_+]^{\nu_-}\,\gamma_\pm^{\nu}. \qquad (9.21)$$

Substituting this definition into Eq. 9.20 produces

$$\mu_e = \mu_e^o + \nu\,RT \ln\left\{\nu_\pm\,\gamma_i\left(\frac{m}{m^o}\right)\right\}, \qquad (9.22)$$

which has the same form as Eq. 9.12, except that γ_i replaces γ_+. When there is no ion-pair formation ($f = 1$), Eq. 9.22 reduces identically to Eq. 9.12. For 1:1 electrolytes such as NaCl, Eq. 9.21 shows that there is a very simple relationship between γ_i and $\gamma_\pm$. In that case, $\nu_+ = \nu_- = 1$ and $\nu = 2$, and under those conditions, we have $\nu_\pm = (1^1\,1^1)^{1/2} = 1$. Substituting these values into Eq. 9.21 then gives

$$\gamma_i^2 = (1)^{-1}\,f\,[1 - (1-f)1]^1\gamma_\pm^2 = f^2\,\gamma_\pm^2,$$

so that $\gamma_i = f\gamma_\pm$.

9.2   The Calculation of Ionic Activity Coefficients: Debye–Hückel Theory

In Chapter 7, we computed the binding energy of ionic crystals by assuming that the only forces we need to consider are the coulombic interactions, with an additional term to represent the short-range repulsions. This procedure resulted in calculated binding energies that were generally within a few percent of those measured using the Born–Haber cycle. A similar approach can be used to treat deviations from ideality in ionic solutions. The electrostatic interactions are so large, that other effects are essentially negligible.

The basic premise of Debye–Hückel theory is that deviations from ideality in electrolytic solutions are due solely to the presence of ionic charges. It is assumed that if NaCl were to dissolve in water by dissociating into neutral Na and Cl atoms, the solution would be ideal, in that Raoult's law would be obeyed by the solvent and Henry's law would describe each type of solute particle. If this is true, then we may determine ionic activity coefficients simply by computing the change in the Gibbs free energy that occurs when hypothetical neutral particles such as Na and Cl atoms are electrically charged so that they exhibit the charges of the actual ions in solution. That is, we write the chemical potential of the solute in the form

$$\mu_e = \mu_{ideal} + \Delta\mu_{elect}, \tag{9.23}$$

where μ_{ideal} is the chemical potential for the ideal solution that would exist if the particles were not charged and $\Delta\mu_{elect}$ is the change in the chemical potential associated with the charging of the neutral particles. If we now substitute Eq. 9.22 for μ_e, we obtain

$$\mu_e^o + \nu RT \ln\left\{\nu_{\pm}\, \gamma_i \left(\frac{m}{m^o}\right)\right\} = \mu_{ideal} + \Delta\mu_{elect}. \tag{9.24}$$

Expanding the logarithm term produces

$$\mu_e^o + \nu RT \ln\left\{\nu_{\pm} \left(\frac{m}{m^o}\right)\right\} + \nu RT \ln[\gamma_i] = \mu_{ideal} + \Delta\mu_{elect}. \tag{9.25}$$

If $\gamma_i = 1$, we have an ideal solution. Therefore, the first two terms in Eq. 9.25 must correspond to μ_{ideal} and the last term to $\Delta\mu_{elect}$. This observation allows us to write

$$\ln[\gamma_i] = \frac{\Delta\mu_{elect}}{\nu RT}. \tag{9.26}$$

Equation 9.26 indicates that we may obtain the activity coefficient if we can successfully calculate the change in the chemical potential attending the hypothetical charging of neutral solute particles up to the charges of the actual ions.

Although the solution problem is analogous to that of computing the total coulombic interaction energy for an ionic crystal, it is much more difficult. This difference in difficulty arises because, in the case of the solid, we know precisely where the charges are located if we know the type of crystal. In the solution, however, the charges have translational freedom, so information on their locations is not readily available. This is the crux of the problem: We must determine how the charges are distributed in the solution before the interaction energy can be calculated. This determination requires that we utilize the Boltzmann distribution and Poisson's equation. We shall have numerous occasions throughout the remainder of the text to utilize the Boltzmann distribution, and the Poisson equation is very similar to equations we shall encounter in our study of quantum mechanics in Chapters 11 through 14. Consequently, although the discussion that follows is mathematically involved, it serves not only to develop to the concepts associated with solution activities, but also as a precursor to much of the formalism in the second half of the text. Therefore, it merits careful attention and study.

In Chapter 17, we shall carefully derive the Boltzmann distribution. For the moment, however, let us be content with a superficial treatment of this important result. In Eq. 5.89, we noted that the temperature dependence of any variable X that measures the extent or rate of a process generally has the form

$$X = (\text{constant}) \exp\left[-\frac{(\text{energy requirement})}{RT}\right],$$

provided that the energy requirement for the process is independent of temperature. Up to now, we have seen numerous examples of this type of result, including the temperature dependence of the equilibrium constant, the variation of the equilibrium vapor pressure and solubility with temperature, the depression of the freezing point of the solvent of a solution, and the dependence of the Henry's law constant on temperature.

The underlying principle leading to such results is the Boltzmann distribution law. In its simplest form, this law states that the probability $P(E)$ of observing a molecular system with energy in the range from E to $E + dE$ at temperature T is proportional to $\exp[-E/RT]\,dE$, or, mathematically,

$$P(E)dE \propto \exp\left[-\frac{E}{RT}\right]dE. \tag{9.27}$$

If we have a process that has an energy requirement E_o, the probability that the process will occur will be proportional to the probability that the molecular system will contain an energy equal to or greater than E_o. Using the Boltzmann distribution law, we find that this cumulative probability is

$$P(E \geq E_o) \propto \int_{E=E_o}^{\infty} P(E)\,dE = \int_{E=E_o}^{\infty} \exp\left[-\frac{E}{RT}\right]dE = -RT\,\exp\left[-\frac{E}{RT}\right]\Big|_{E_o}^{\infty}$$

$$= RT\,\exp\left[-\frac{E_o}{RT}\right]. \tag{9.28}$$

Generally, the exponential temperature dependence is much stronger than that produced by the preexponential factor. If the preexponential temperature dependence is neglected, the probability that the process will occur is described by Eq. 5.89.

Equation 9.28 expresses the energy requirement E_o in terms of energy per mole. If we divide numerator and denominator of the argument of the exponential function by Avogadro's constant, N_A, we obtain

$$\frac{E_o/N_A}{(R/N_A)T} = \frac{\varepsilon_o}{k_bT},$$

where ε_o is now the energy requirement per molecule and k_b is the Boltzmann constant. This permits us to write Eq. 9.28 in the form

$$P(\varepsilon \geq \varepsilon_o) \propto k_bT\,\exp\left[-\frac{\varepsilon_o}{k_bT}\right] \tag{9.29}$$

and the Boltzmann distribution law itself in the form

$$P(\varepsilon) \propto \exp\left[-\frac{\varepsilon}{k_bT}\right]. \tag{9.30}$$

To determine the charge distribution about an ion in an ionic solution, Debye assumed that the distribution would be accurately described by the Boltzmann distribution. The electrolyte $C_{\nu_+}A_{\nu_-}$ dissociates into ν_+ positive ions of charge z_+e and ν_- negative ions of charge $-z_-e$, where e is the electronic charge. Let the electrostatic potential at an ion (measured in volts) be Φ. Since the electrostatic potential energy is the product of the charge and the electric potential (1 J = 1 coulomb × 1 volt), the potential energy for a positive ion will be $z_+e\Phi$, and that for a negative ion will be $-z_-e\Phi$. The Boltzmann distribution law tells us that if n_+ and n_- are the total number of positive and negative ions, respectively, per kilogram of solvent, then the number of positive ions per kilogram of solvent that have an electrostatic potential energy $z_+e\Phi$ will be $n_+\exp[-z_+e\Phi/k_bT]$, and the number of negative ions with potential energy $-z_-e\Phi$ will be $n_-\exp[-(-z_-e\Phi/k_bT)] = n_-\exp[z_-e\Phi/k_bT]$. The total contribution of these positive ions to the local charge density about a specific ion is the charge on each ion multiplied by the total number of such ions. That is,

$$\rho_+ = \text{positive-ion contribution to charge density}$$

$$= (\text{charge per ion}) \times (\text{number of ions})$$

$$= z_+e\,n_+\exp\left[-\frac{z_+e\Phi}{k_bT}\right].$$

Using similar logic, we also have

$$\rho_- = \text{negative-ion contribution to charge density}$$

$$= (\text{charge per ion}) \times (\text{number of ions})$$

$$= -z_-e\,n_-\exp\left[\frac{z_-e\Phi}{k_bT}\right].$$

The total charge density about a specific ion is the sum of ρ_+ and ρ_-:

$$\rho = \rho_+ + \rho_- = z_+e\,n_+\exp\left[-\frac{z_+e\Phi}{k_bT}\right] - z_-e\,n_-\exp\left[\frac{z_-e\Phi}{k_bT}\right]. \qquad \textbf{(9.31)}$$

If we knew the functional form of the electrostatic potential Φ, Eq. 9.31 would tell us the charge distribution about a specific ion in the ionic solution. This is as close as we can come to the situation in a solid ionic crystal where we know the positions of all the charges if we know the crystal form. The next required step is to determine the electrostatic potential for a system whose charge density is given by Eq. 9.31. The relationship between Φ and ρ is expressed in Poisson's equation,

$$\frac{\partial^2\Phi}{\partial x^2} + \frac{\partial^2\Phi}{\partial y^2} + \frac{\partial^2\Phi}{\partial z^2} = -\frac{\rho}{\varepsilon_o D}, \qquad \textbf{(9.32)}$$

where D is the dielectric constant of the solution and ε_o is the permittivity of the vacuum which equals $8.854 \times 10^{-12}\,\text{J}^{-1}\,\text{C}^2\,\text{m}^{-1}$. The dielectric constant is simply the unitless ratio of the actual permittivity of the medium to that for the vacuum. As such, it is sometimes called the *relative permittivity*. Therefore, we have $\varepsilon = \varepsilon_o D$, and the denominator of the right-hand side of Eq. 9.32 is just the permittivity of the medium. The solution of this second-order partial differential equation gives the electrostatic potential Φ as a function of x, y, and z for some arbitrary coordinate system for the solution. This is the first time we have encountered a partial differential equation of this form. We shall, however, see many more of similar form when we begin our study of quantum mechanics. (See Problem 9.7.) By factoring the electrostatic potential to the right of the differential operators, we may write Eq. 9.32 in the form

$$\left[\frac{\partial^2}{\partial x^2} + \frac{\partial^2}{\partial y^2} + \frac{\partial^2}{\partial z^2}\right]\Phi = -\frac{\rho}{\varepsilon_o D}.$$

The differential operator in brackets is called the *Laplacian* operator or sometimes the *del-squared* operator. It is given the symbol ∇^2, so the Poisson equation may alternatively be written as

$$\nabla^2 \Phi = -\frac{\rho}{\varepsilon_o D}. \tag{9.33}$$

If we are to compute activity coefficients for ionic solutions, Eq. 9.33 must be solved to obtain Φ. The student should be aware of the fact that we have not derived Poisson's equation. It can be obtained from Maxwell's laws of electrodynamics. A discussion of these laws is, however, beyond the scope of the text.

The first step in the solution of Eq. 9.33 is to pick a convenient coordinate system. We take the origin to be the specific ion about which Eq. 9.31 gives the charge density. It is immediately clear that the electrostatic potential exhibits spherical symmetry about this ion. That is, the potential is the same in any direction about the ion. This consideration suggests that the differential equation will be easier to solve if we use spherical polar coordinates than if we use rectangular Cartesian coordinates. Figure 9.3 shows the spherical polar coordinate system, with transformations

$$X = r \sin \theta \cos \phi$$

$$Y = r \sin \theta \sin \phi$$

and

$$Z = r \cos \theta. \tag{9.34}$$

We shall find repeated application for this coordinate system throughout the remainder of the text. It is straightforward (although very messy) to transform the ∇^2 operator to spherical polar coordinates. If this transformation is executed, we obtain

$$\nabla^2 = \frac{1}{r^2} \frac{\partial}{\partial r} \left[r^2 \frac{\partial}{\partial r} \right] + \frac{1}{r^2 \sin \theta} \frac{\partial}{\partial \theta} \left[\sin \theta \frac{\partial}{\partial \theta} \right] + \frac{1}{r^2 \sin^2 \theta} \frac{\partial^2}{\partial \phi^2}. \tag{9.35}$$

Since we expect Φ to be spherically symmetric, it will be independent of both spherical polar angles θ and ϕ. As a result, the angular derivatives of Φ are zero, and Eq. 9.33 becomes

$$\frac{1}{r^2} \frac{\partial}{\partial r} \left[r^2 \frac{\partial}{\partial r} \right] \Phi = -\frac{\rho}{\varepsilon_o D}, \tag{9.36}$$

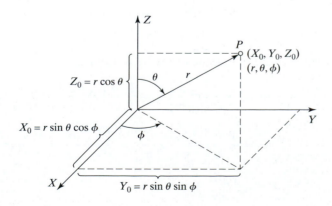

◀ FIGURE 9.3
The spherical polar coordinate system. The relationships between the rectangular coordinates (X_o, Y_o, Z_o) and the spherical polar coordinates (r, θ, ϕ) are illustrated in the figure. The point P may be described by either set of coordinates.

where the charge density is given by Eq. (9.31). Although we are making progress, Eq. 9.36 brings us to a stop, since it has no analytic solution. Debye proceeded by expanding the charge density in a power series. Using the fact that the Taylor series expansion of e^{ax} is $[1 + ax/1! + a^2x^2/2! + a^3x^3/3! + \dots + a^nx^n/n! + \cdots]$, he wrote the charge density in the form

$$\rho = z_+e\, n_+\exp\left[-\frac{z_+e\Phi}{k_bT}\right] - z_-e\, n_-\exp\left[\frac{z_-e\Phi}{k_bT}\right]$$

$$= z_+en_+\left[1 - \frac{z_+e\Phi}{k_bT} + \cdots\right] - z_-e\, n_-\left[1 + \frac{z_-e\Phi}{k_bT} + \cdots\right]$$

$$= e[z_+n_+ - z_-n_-] - \left[\frac{n_+z_+^2e^2 + n_-z_-^2e^2}{k_bT}\right]\Phi + \cdots \qquad (9.37)$$

The first term in this expansion is zero, since electrical neutrality of the solution requires that $z_+n_+ - z_-n_- = 0$. Debye now assumed that the electrostatic potential Φ would be sufficiently small relative to k_bT that the higher terms in the series expansion could be neglected. It is this assumption, among others, that limits the accuracy of Debye–Hückel theory to dilute solutions containing sufficiently few ions that Φ will be small.

Using the truncated expansion for the charge density, we can now obtain a solution of Eq. 9.36. The stoichiometry of the ionization shows that $n_+ = \nu_+ N_e$ and $n_- = \nu_- N_e$, where N_e is the number of solute molecules per kilogram of solvent. Inserting these expressions into Eq. 9.37 gives

$$\rho = -e^2N_e\left[\frac{\nu_+z_+^2 + \nu_-z_-^2}{k_bT}\right]\Phi \qquad (9.38)$$

We now simplify the notation by letting

$$\kappa^2 = e^2N_e\left[\frac{\nu_+z_+^2 + \nu_-z_-^2}{\varepsilon_oDk_bT}\right]. \qquad (9.39)$$

With this definition, Eq. 9.36 becomes

$$\frac{1}{r^2}\frac{\partial}{\partial r}\left[r^2\frac{\partial}{\partial r}\right]\Phi = \kappa^2\Phi. \qquad (9.40)$$

Equation 9.40 is an example of an extremely important class of differential equations that we shall encounter repeatedly throughout much of the remainder of the text. It is, therefore, worthwhile to consider the general form of the solutions at this point. We begin by putting the equation into a simpler form by transforming the variables. Let our unknown electrostatic potential be

$$\Phi = \frac{F}{r}. \qquad (9.41)$$

The left-hand side of Eq. 9.40 may now be written as

$$\frac{1}{r^2}\frac{\partial}{\partial r}\left[r^2\frac{\partial}{\partial r}\right]\Phi = \frac{1}{r^2}\frac{\partial}{\partial r}\left[r^2\frac{\partial}{\partial r}\right]\frac{F}{r} = \frac{1}{r^2}\frac{\partial}{\partial r}\left[r^2\left\{-\frac{F}{r^2} + \frac{1}{r}\frac{\partial F}{\partial r}\right\}\right]$$

$$= \frac{1}{r^2}\frac{\partial}{\partial r}\left[-F + r\frac{\partial F}{\partial r}\right] = \frac{1}{r^2}\left[-\frac{\partial F}{\partial r} + \frac{\partial F}{\partial r} + r\frac{\partial^2 F}{\partial r^2}\right]$$

$$= \frac{1}{r}\frac{\partial^2 F}{\partial r^2}.$$

Substituting this result into Eq. 9.40 gives

$$\frac{\partial^2 F}{\partial r^2} = \kappa^2 \, r\Phi = \kappa^2 \, F. \tag{9.42}$$

The form of the solutions of Eq. 9.42 and hence to Eq. 9.40 depends upon the sign in front of κ^2. If we have $+\kappa^2$ with κ^2 positive, the solutions are

$$F = C_1 \exp(-\kappa r) + C_2 \exp(\kappa r), \tag{9.43}$$

where C_1 and C_2 are constants that are employed to satisfy the boundary conditions of the physical problem. We may easily verify the fact that Eq. 9.43 is a solution of Eq. 9.42 by taking the second derivative of F. The first derivative is

$$\frac{\partial F}{\partial r} = -\kappa C_1 \exp(-\kappa r) + \kappa C_2 \exp(\kappa r).$$

The second derivative is

$$\frac{\partial^2 F}{\partial r^2} = \kappa^2 C_1 \exp(-\kappa r) + \kappa^2 C_2 \exp(\kappa r)$$

$$= \kappa^2 \{ C_1 \exp(-\kappa r) + C_2 \exp(\kappa r) \} = \kappa^2 F,$$

as required. If, however, the right side of Eq. 9.42 is $-\kappa^2$ with κ^2 positive, the solutions have the form

$$F = C_1 \sin(\kappa r) + C_2 \cos(\kappa r). \tag{9.44}$$

(See Problem 9.5.)

In the present case, Eq. 9.39 shows that κ is positive since every term and factor in that equation is positive. Therefore, the solutions to the Poisson equation are

$$F = C_1 \exp(-\kappa r) + C_2 \exp(\kappa r) = \Phi r, \tag{9.45}$$

and the electrostatic potential about the positive ion at the origin is

$$\Phi = \frac{C_1 \exp(-\kappa r) + C_2 \exp(\kappa r)}{r}.$$

Since the physics of the problem require that Φ approach zero when r becomes infinite, we must have the constant C_2 equal to zero. This boundary condition yields

$$\Phi = \frac{C_1 \exp(-\kappa r)}{r}. \tag{9.46}$$

In the limit as $r \to 0$, the environment around our positive ion at the origin will be irrelevant. The dominant factor in determining Φ as we approach the positive ion will be the coulombic potential of the single ion. Coulomb's law tells us that the potential for a bare ion of charge $z_+ e$ is $z_+ e / (4\pi\varepsilon_o D r)$. Therefore, we must have

$$\lim_{r \to 0} \Phi = \frac{C_1}{r} = \frac{z_+ e}{4\pi\varepsilon_o D r}. \tag{9.47}$$

Consequently, the constant $C_1 = z_+ e / (4\pi\varepsilon_o D)$. If we had taken a negative ion as our origin, we would have obtained $C_1 = -z_- e / (4\pi\varepsilon_o D)$. The student should notice that it is the boundary conditions at $r \to \infty$ and $r \to 0$ which determine the constants in the solution of the partial differential equation. This is always the case; we shall see it repeatedly in our study of quantum mechanics.

We have already assumed that κr is very small in truncating the expansion of the charge density in Eq. 9.37. Let us then make use of this assumption again to expand the exponential in Eq. 9.46 to obtain

$$\Phi = \frac{C_1}{r}\left[1 - \kappa r + \frac{(\kappa r)^2}{2!} + \cdots\right] \approx \frac{C_1}{r}[1 - \kappa r] = \frac{C_1}{r} - C_1\kappa$$

$$= \frac{z_+e}{4\pi\varepsilon_o Dr} - \frac{z_+e\kappa}{4\pi\varepsilon_o D}. \tag{9.48}$$

The Debye–Hückel approximations for the electrostatic fields are

$$\Phi = \frac{z_+e}{4\pi\varepsilon_o Dr} - \frac{z_+e\kappa}{4\pi\varepsilon_o D} \qquad \text{about a positive ion}$$

and

$$\Phi = -\frac{z_-e}{4\pi\varepsilon_o Dr} + \frac{z_-e\kappa}{4\pi\varepsilon_o D} \qquad \text{about a negative ion.} \tag{9.49}$$

The first term on the right of Eq. 9.49 is the electrostatic potential that would be produced by a single cation or anion at the origin in the absence of the electrolytic solution. The second term reflects the influence of the ionic environment, which we expect to be a cloud of counterions surrounding the ion at the origin. With the simplifying assumption of a dilute solution, this ionic cloud acts as if it were a spherical shell of charge of radius κ^{-1} with a total charge equal to that on the central ion, but of opposite sign. Since it is the interaction of the ionic charges with their environment that produces the energy needed to charge neutral particles up to the charges actually contained on the ions, it is the second term that we shall consider in the calculation of $\Delta\mu_{\text{elect}}$ that appears in Eq. 9.26.

EXAMPLE 9.2

Consider a 0.10-molal KCl solution in water at 293.15 K. Using the Debye formalism, compute the radius κ^{-1} associated with the hypothetical sphere of counterions surrounding a central K^+ ion. Since the solution must be dilute in order for the Debye–Hückel theory to apply, assume that the dielectric constant is that for water at 293.15 K ($D = 78.54$).

Solution

κ^2 is given by Eq. 9.39, viz.,

$$\kappa^2 = e^2 N_e\left[\frac{\nu_+z_+^2 + \nu_-z_-^2}{\varepsilon_o Dk_bT}\right], \tag{1}$$

where, for the specified problem, we have $N_e = (0.1 \text{ mol kg}^{-1})N_A$, in which N_A is Avogadro's constant, $\nu_+ = \nu_- = z_+^2 = z_-^2 = 1$, $e = 1.6022 \times 10^{-19}$ coulombs, $k_b = 1.3807 \times 10^{-23}$ J K^{-1}, and $D = 78.54$. Substituting these values and then converting mol to molecules and L^{-1} to m^{-3} gives

$$\kappa^2 = (1.6022 \times 10^{-19} \text{ C})^2(0.1 \text{ mol kg}^{-1})(6.022 \times 10^{23} \text{ mol}^{-1})$$

$$\times \left(\frac{1 \text{ kg}}{1 \text{ L}}\right)\left(\frac{1 \text{ L}}{1,000 \text{ cm}^3}\right)\left[\frac{100 \text{ cm}}{\text{m}}\right]^3$$

$$\times \left[\frac{1(1)^2 + (1)(1)^2}{(8.854 \times 10^{-12} \text{ J}^{-1} \text{ C}^2 \text{ m}^{-1})(78.54)(1.3807 \times 10^{-23} \text{ J K}^{-1})(293.15 \text{ K})}\right]$$

$$= 1.098 \times 10^{18} \text{ m}^{-2}. \tag{2}$$

Taking the square roots of both sides of Eq. 2 gives

$$\kappa = 1.048 \times 10^9 \, \text{m}^{-1}. \tag{3}$$

The radius of the environmental charge shell is, therefore,

$$\kappa^{-1} = \frac{1}{1.048 \times 10^9 \, \text{m}^{-1}} = 9.541 \times 10^{-10} \, \text{m} = 9.541 \times 10^{-8} \, \text{cm} = 9.541 \, \text{Å}. \tag{4}$$

For a related exercise, see Problem 9.6.

EXAMPLE 9.3

(A) Calculate and plot the electrostatic potential Φ about a positive ion in an aqueous solution of KCl at 293.15 K as a function of the radial distance from the ion over the range $2.5 \times 10^{-10} \, \text{m} \leq r \leq 1.0 \times 10^{-9} \, \text{m}$ when the KCl molality has the following values: m = 0.005 molal, 0.01 molal, 0.05 molal, 0.10 molal, and 0.5 molal. Put the five requested plots on the same graph and comment on the physical significance of the results. D is 78.54 for H_2O. **(B)** Obtain an expression for the expected electrostatic potential field about the K^+ ion if there were no counterion field about the K^+ ion. Plot the result over the range $2.5 \times 10^{-10} \, \text{m} \leq r \leq 1.0 \times 10^{-9} \, \text{m}$, and on the same graph, plot the results obtained in (A) for the 0.005- and 0.5-molal solutions. Comment on the results.

Solution

(A) Using Eqs. 9.46 and 9.47, we have

$$\Phi = \frac{z_+ e}{4\pi\varepsilon_o D} \frac{\exp(-\kappa r)}{r}, \tag{1}$$

where

$$\kappa = \left\{ e^2 N_e \left[\frac{\nu_+ z_+^2 + \nu_- z_-^2}{\varepsilon_o D k_b T} \right] \right\}^{1/2} = \left\{ e^2 N_A m_s \left[\frac{\nu_+ z_+^2 + \nu_- z_-^2}{\varepsilon_o D k_b T} \right] \right\}^{1/2}. \tag{2}$$

In Eq. 2, N_A is Avogadro's constant and m_s is the molality of the solution. Insertion of the constants gives

$$\kappa^2 = (1.6022 \times 10^{-19} \, \text{C})^2 (6.022 \times 10^{23} \, \text{mol}^{-1}) \, m_s \left(\frac{1 \, \text{kg}}{1 \, \text{L}} \right) \left(\frac{1 \, \text{L}}{1{,}000 \, \text{cm}^3} \right) \left[\frac{100 \, \text{cm}}{\text{m}} \right]^3$$

$$\times \left[\frac{1(1)^2 + (1)(1)^2}{(8.854 \times 10^{-12} \, \text{J}^{-1} \, \text{C}^2 \, \text{m}^{-1})(78.54)(1.3807 \times 10^{-23} \, \text{J K}^{-1})(293.15 \, \text{K})} \right]$$

$$= 1.098 \times 10^{19} \, m_s \, \text{mol}^{-1} \, \text{kg m}^{-2}. \tag{3}$$

Thus,

$$\kappa = 3.314 \times 10^9 \, m_s^{1/2} \, \text{mol}^{-1/2} \, \text{kg}^{1/2} \, \text{m}^{-1}. \tag{4}$$

The quantity

$$\frac{z_+ e}{4\pi\varepsilon_o D} = \frac{(1)(1.6022 \times 10^{-18} \text{C})}{4(3.14159)(8.854 \times 10^{-12} \, \text{J}^{-1} \, \text{C}^2 \, \text{m}^{-1})(78.54)} = 1.833 \times 10^{-11} \, \text{J m C}^{-1},$$

$$= 1.833 \times 10^{-11} \, \text{volt m} \tag{5}$$

since $1 \, \text{J C}^{-1} = 1$ volt. Combining the results in Eqs. 1 through 5, we obtain

$$\Phi = (1.833 \times 10^{-11}) \frac{\exp(-3.314 \times 10^9 \, m_s^{1/2} r)}{r} \, \text{volts}. \tag{6}$$

Notice that the units in the exponential cancel, as expected. Equation 6 shows that the electrostatic potential about an ion decreases exponentially with the square root of the molality of the solution. Figure 9.4 shows Φ as a function of r for five different concentrations. At large distances, Φ approaches zero, as required. At small values of r, Φ becomes proportional to r^{-1}, as predicted by Coulomb's law. As the concentration increases, the number of counterions surrounding the central positive ion increases and effectively reduces the positive electrostatic potential field.

(B) The coulombic potential field about a central K^+ ion is

$$\Phi_{coulombic} = \frac{z_+ e}{4\pi\varepsilon_o D r}. \tag{7}$$

Using the results obtained in (A), we may write this as

$$\Phi_{coulombic} = \frac{1.833 \times 10^{-11} \text{ volt m}}{r}. \tag{8}$$

Figure 9.5 shows a plot of $\Phi_{coulombic}$ versus r over the requested range. The results obtained for the 0.005- and 0.5-molal solutions are also shown on this graph. The effect of the counterion atmosphere about the K^+ ion is dramatic: As the concentration of Cl^- increases with the concentration of the solution, the surrounding negative counterions significantly reduce the electrostatic potential field produced by the central K^+ ion.

We must now compute the energy required to charge the ions from zero up to their actual charges in the electrostatic field produced by the ionic environment. To execute this calculation, let us replace the electronic charge e with βe, where β will be allowed to vary from zero to unity, to reflect the

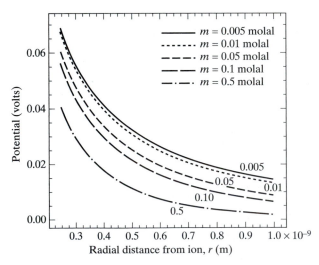

▲ FIGURE 9.4
The electrostatic potential field about a K^+ ion in aqueous solutions of KCl at 293.15 K as a function of the radial distance from the ion and the concentration of the solution. At higher concentrations, the potential field produced by the Cl^- counterions cancels a portion of the K^+ potential.

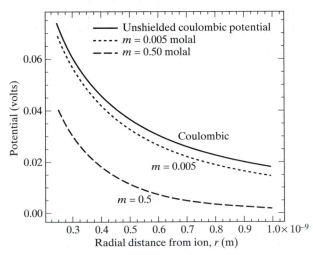

▲ FIGURE 9.5
The electrostatic potential field about a K^+ ion in aqueous solutions of KCl at 293.15 K as a function of the radial distance from the ion and the concentration of the solution. The curve labeled "coulombic" is the electrostatic field for a K^+ ion in the absence of counterions. It therefore corresponds to the m = 0 case. The electrostatic potentials for 0.005-molal and 0.5-molal solutions are also shown for comparison purposes. Notice the large magnitude of the counterion field.

charging of neutral particles up to the actual charges contained on the ions. With this alteration, Eq. 9.39 indicates that κ^2 becomes

$$\beta^2 e^2 N_e \left[\frac{\nu_+ z_+^2 + \nu_- z_-^2}{\varepsilon_o D k_b T} \right] = \beta^2 \kappa^2. \tag{9.50}$$

Using Eq. 9.49, we find that the environmental field against which the charging must be done has the form $-z_+\beta^2 e\kappa/(4\pi\varepsilon_o D)$ for a positive ion and $z_-\beta^2 e\kappa/(4\pi\varepsilon_o D)$ for a negative ion. If we bring an infinitesimal positive charge of magnitude $z_+e\,d\beta$ into a region where the environmental potential is $-z_+\beta^2 e\kappa/(4\pi\varepsilon_o D)$, then the differential amount of work that must be done on the system is the product of the electrostatic field and the charge; that is,

$$\delta w_{\text{positive}} = -\frac{z_+\beta^2 e\kappa}{4\pi\varepsilon_o D} z_+ e\,d\beta = -\frac{z_+^2 e^2 \kappa}{4\pi\varepsilon_o D} \beta^2\,d\beta. \tag{9.51A}$$

For a negative ion, the differential work will be

$$\delta w_{\text{negative}} = \frac{z_-\beta^2 e\kappa}{4\pi\varepsilon_o D} (-z_- e\,d\beta) = -\frac{z_-^2 e^2 \kappa}{4\pi\varepsilon_o D} \beta^2\,d\beta. \tag{9.51B}$$

Equations 9.51A and 9.51B give the differential work for charging one positive and one negative ion by an amount $z_+e\,d\beta$ and $-z_-e\,d\beta$, respectively. The number of positive ions per kilogram of solvent is $\nu_+ N_e$. The concentration of negative ions is $\nu_- N_e$. The total differential work of charging is obtained by multiplying Eqs. 9.51A and 9.51B by the total number of such ions present and summing the two results. This operation gives

$$\delta w_{\text{elect}}^{\text{total}} = \nu_+ N_e M \delta w_{\text{positive}} + \nu_- N_e M \delta w_{\text{negative}}$$

$$= -\frac{N_e e^2 \kappa M}{4\pi\varepsilon_o D} [\nu_+ z_+^2 + \nu_- z_-^2] \beta^2 d\beta, \tag{9.52}$$

where M is the mass of the solution, expressed in kilograms.

Later in the chapter, we shall show that

$$dG = \delta w_{\text{elect}}^{\text{total}} \tag{9.53}$$

for any process in a closed system conducted under conditions of constant temperature and pressure. The total change in the Gibbs free energy due to charging neutral particles can, therefore, be obtained by integrating (summing) Eq. 9.52 from zero charge ($\beta = 0$) to the full charge on the ions ($\beta = 1$). This operation produces

$$\Delta G_{\text{elect}} = \int dG_{\text{elect}} = -\frac{N_e e^2 \kappa M}{4\pi\varepsilon_o D} [\nu_+ z_+^2 + \nu_- z_-^2] \int_{\beta=0}^{\beta=1} \beta^2 d\beta$$

$$= -\frac{N_e e^2 \kappa M}{12\pi\varepsilon_o D} [\nu_+ z_+^2 + \nu_- z_-^2]. \tag{9.54}$$

Equation 9.54 gives the total change in the Gibbs free energy attending the charging of neutral particles. We may now obtain the change in the chemical potential by differentiating ΔG_{elect} with respect to the number of moles of solute present. We have

$$\Delta\mu_{\text{elect}} = \left(\frac{\partial \Delta G_{\text{elect}}}{\partial n_e} \right)_{T,P} = \left(\frac{\partial \Delta G_{\text{elect}}}{\partial N_e} \right) \left(\frac{\partial N_e}{\partial n_e} \right), \tag{9.55}$$

where the number of solute moles is related to the number of solute molecules per kilogram of solvent, N_e, by $n_e = N_e M / N_A$. Therefore, $N_e = n_e N_A / M$, and Eq. 9.55 gives

$$\Delta \mu_{\text{elect}} = \frac{N_A}{M} \left(\frac{\partial \Delta G_{\text{elect}}}{\partial N_e} \right). \tag{9.56}$$

In taking the derivative of ΔG_{elect} with respect to N_e, we must keep in mind that κ depends upon N_e in the manner described by Eq. 9.39. Using Eq. 9.54, we obtain

$$\Delta \mu_{\text{elect}} = -\frac{N_A e^2}{12 \pi \varepsilon_o D} [\nu_+ z_+^2 + \nu_- z_-^2] \frac{\partial}{\partial N_e} \{N_e \kappa\}$$

$$= -\frac{N_A e^2}{12 \pi \varepsilon_o D} [\nu_+ z_+^2 + \nu_- z_-^2] \left[\kappa + N_e \frac{\partial \kappa}{\partial N_e} \right]. \tag{9.57}$$

Now using Eq. 9.39, we may obtain $\dfrac{\partial \kappa}{\partial N_e}$. We have

$$2\kappa \frac{\partial \kappa}{\partial N_e} = e^2 \left[\frac{\nu_+ z_+^2 + \nu_- z_-^2}{\varepsilon_o D k_b T} \right] = \frac{e^2 N_e}{N_e} \left[\frac{\nu_+ z_+^2 + \nu_- z_-^2}{\varepsilon_o D k_b T} \right] = \frac{\kappa^2}{N_e}$$

so that

$$\frac{\partial \kappa}{\partial N_e} = \frac{\kappa}{2N_e}. \tag{9.58}$$

Combining Eqs. 9.57 and 9.58 yields

$$\Delta \mu_{\text{elect}} = -\frac{N_A e^2 \kappa}{8 \pi \varepsilon_o D} [\nu_+ z_+^2 + \nu_- z_-^2]. \tag{9.59}$$

Equation 9.59 gives the chemical potential due to the interaction of the charged ions of the solute. As long as the solution is sufficiently dilute to justify the truncations made in Eqs. 9.37 and 9.48, the use of Eq. 9.59 in Eq. 9.26 yields the activity coefficient for the solute. We may simplify Eq. 9.59 by making use of the electrical neutrality of the solution, which requires that $\nu_+ z_+ = \nu_- z_-$. Multiplying of both sides by z_+ shows that we must also have $\nu_+ z_+^2 = \nu_- z_- z_+$. Similarly, multiplying both sides by z_- gives $\nu_+ z_+ z_- = \nu_- z_-^2$. Adding the latter two equations shows that

$$\nu_+ z_+^2 + \nu_- z_-^2 = \nu_- z_- z_+ + \nu_+ z_+ z_- = (\nu_- + \nu_+) z_+ z_- = \nu z_+ z_-. \tag{9.60}$$

Using this result in Eq. 9.59 gives

$$\Delta \mu_{\text{elect}} = -\frac{N_A e^2 \kappa \nu \, z_+ z_-}{8 \pi \varepsilon_o D}. \tag{9.61}$$

The ionic activity coefficient is now given by Eq. 9.26; that is,

$$\ln (\gamma_i) = \frac{\Delta \mu_{\text{elect}}}{\nu R T} = -\frac{N_A e^2 \kappa \, z_+ z_-}{8 \pi \varepsilon_o D R T} = -\frac{e^2 \kappa \, z_+ z_-}{8 \pi \varepsilon_o D k_b T} \tag{9.62}$$

since $N_A / R = 1/k_b$. Note that the expression on the right of Eq. 9.62 is unitless, as required. The numerator has units of $C^2 \, m^{-1}$. The units of the denominator are $(J^{-1} C^2 \, m^{-1})(J K^{-1})(K) = C^2 \, m^{-1}$, so the units cancel properly. Although single-ion activity coefficients cannot be directly measured, the Debye–Hückel theory permits these quantities to be calculated. From the definition of the mean ionic activity coefficient, we have

$$\ln \gamma_i = \ln [\gamma_+^{\nu_+/\nu} \, \gamma_-^{\nu_-/\nu}] = \frac{\nu_+}{\nu} \ln [\gamma_+] + \frac{\nu_-}{\nu} \ln [\gamma_-]. \tag{9.63A}$$

Using Eqs. 9.60 and 9.62, we also have

$$\ln \gamma_i = -\frac{\nu_+}{\nu}\left[\frac{e^2 \kappa\, z_+^2}{8\pi\varepsilon_o D k_b T}\right] - \frac{\nu_-}{\nu}\left[\frac{e^2 \kappa\, z_-^2}{8\pi\varepsilon_o D k_b T}\right]. \tag{9.63B}$$

Comparing Eqs. 9.63A and 9.63B shows that the single-ion activity coefficients must be given by

$$\ln \gamma_+ = -\frac{e\kappa\, z_+^2}{8\pi\varepsilon_o D k_b T} \tag{9.63C}$$

and

$$\ln \gamma_- = -\frac{e\kappa\, z_-^2}{8\pi\varepsilon_o D k_b T}. \tag{9.63D}$$

Using Eq. 9.39, we may write κ^2 for an aqueous solution in the form

$$\kappa^2 = e^2 N_e \left[\frac{\nu_+ z_+^2 + \nu_- z_-^2}{\varepsilon_o D k_b T}\right]$$

$$= \frac{(1.6022 \times 10^{-19}\text{ C})^2 (6.022 \times 10^{23}\text{ mol}^{-1})\left(\dfrac{1\text{ kg}}{1\text{ L}}\right)\left(\dfrac{1\text{ L}}{1{,}000\text{ cm}^3}\right)\left[\dfrac{100\text{ cm}}{\text{m}}\right]^3}{(8.854 \times 10^{-12}\text{ J}^{-1}\text{ C}^2\text{ m}^{-1})(78.54)(1.3807 \times 10^{-23}\text{ J K}^{-1})}$$

$$\times \frac{c_e}{T}[\nu_+ z_+^2 + \nu_- z_-^2]$$

$$= 1.6101 \times 10^{21}\,\frac{c_e}{T}[\nu_+ z_+^2 + \nu_- z_-^2]\text{ m}^{-2}\text{ mol}^{-1}\text{ kg K},$$

where c_e is the concentration of the solute in mol kg^{-1}. It is useful to express this result in terms of the *ionic strength*, which we define as

$$S = 0.5 c_e \sum_{i=1}^{i=K} \nu_i z_i^2 = 0.5 \sum_{i=1}^{i=K} c_i z_i^2, \tag{9.64}$$

where the summation runs over all the ions present in the solution and c_i is the concentration of ion i in mol kg^{-1}, which is equal to $\nu_i c_e$. Note that the units on c_e and c_i are canceled by the other factors in κ^2, so that the ionic strength is a unitless quantity. In terms of the ionic strength, for an aqueous solution,

$$\kappa = 5.67468 \times 10^{10}\left[\frac{S}{T}\right]^{1/2}\text{ m}^{-1}. \tag{9.65}$$

The final Debye–Hückel expression for the activity coefficient can now be obtained by substituting Eq. 9.65 into Eq. 9.62. We get

$$\ln(\gamma_i) = -\frac{e^2 \kappa\, z_+ z_-}{8\pi\varepsilon_o D k_b T}$$

$$= -\left[\frac{(1.6022 \times 10^{-19}\text{ C})^2}{8(3.14159)(8.854 \times 10^{-12}\text{ J}^{-1}\text{ C}^2\text{ m}^{-1})(78.54)(1.3807 \times 10^{-23})T}\right] 5.67468 \times 10^{10} z_+ z_-\left[\frac{S}{T}\right]^{1/2},$$

which gives

$$\ln(\gamma_i) = -6036.78\, z_+ z_-\, T^{-3/2}\, S^{1/2}. \qquad \textbf{(9.66)}$$

Equation 9.66 is the final Debye–Hückel expression for the ionic activity coefficient in aqueous solutions. For other solvents, the appropriate result is easily obtained by replacing the aqueous dielectric constant of 78.54 in Eqs. 9.63 and 9.66 with the appropriate value for the solvent being used. The more familiar form of the Debye–Hückel equation is obtained by evaluating Eq. 9.66 at 298.15 K and converting to a base-10 logarithm. This gives

$$\log(\gamma_i) = -\frac{6036.78}{(298.15)^{3/2}\,(2.30258)} z_+ z_-\, S^{1/2} = -0.50926\, z_+ z_-\, S^{1/2}, \qquad \textbf{(9.67A)}$$

$$\log\gamma_+ = -0.50926 z_+^2 S^{1/2}, \qquad \textbf{(9.67B)}$$

and

$$\log\gamma_- = -0.50926 z_-^2\, S^{1/2}. \qquad \textbf{(9.67C)}$$

Note that Eq. 9.67 is in accord with our choice of reference state: When the solute molality approaches zero, $S \to 0$ and γ_i and the single-ion activity coefficients are predicted to approach unity, as we expect.

The foregoing derivation assumes that we are dealing with a solution of a single electrolyte. This assumption is not necessary. If we have a mixture of electrolytes, we need only modify the expression for the charge density in Eq. 9.31 by including all positive and negative ions. The series expansion in Eq. 9.37 will then include a separate term having the form $\nu_i z_i^2$ for each ion. These extra terms will simply change the upper limit of the summation in Eq. 9.64 for the ionic strength of the medium. The final expressions for the activity coefficient will remain those given in Eqs. 9.66 and 9.67. Of course, the actual values of the activity coefficients will be altered because of the increased value of the ionic strength.

EXAMPLE 9.4

(A) Compute the activity coefficient predicted by the Debye–Hückel theory at 298.15 K for an aqueous solution of KCl whose concentration is 0.01 mol kg^{-1}. **(B)** If we add 0.015 mol kg^{-1} of BaCl$_2$ to the solution in (A), compute the new KCl activity coefficient, using the Debye–Hückel theory.

Solution

(A) Since the solvent is water and the temperature is 298.15 K, Eq. 9.67 gives the limiting value of γ_i for a dilute solution. The required ionic strength is

$$S = 0.5 \sum_{i=1}^{i=K} c_i z_i^2 = 0.5\,[0.01(1)^2 + 0.01(1)^2] = 0.01. \qquad \textbf{(1)}$$

Therefore,

$$\log(\gamma_{KCl}) = -0.50926 z_+ z_-\, S^{1/2} = -0.50926(1)(1)(0.01)^{1/2} = -0.050926, \qquad \textbf{(2)}$$

so that

$$\gamma_{KCl} = 10^{-0.050926} = 0.8894. \qquad \textbf{(3)}$$

(B) With the addition of $BaCl_2$, we have the reaction $BaCl_2 \longrightarrow Ba^{2+}(aq) + 2Cl^{1-}(aq)$, so the ionic strength is now given by

$$S = 0.5[(0.01(1)^2 + (0.01)(1)^2 + (0.015)(2)^2 + (0.015)(2)(1)^2] = 0.055. \quad (4)$$

Therefore, the KCl activity coefficient is

$$\log(\gamma_{KCl}) = -0.50926(1)(1)(0.055)^{1/2} = -0.11943, \quad (5)$$

and it follows that

$$\gamma_{KCl} = 10^{-0.11943} = 0.760. \quad (6)$$

For a related exercise, see Problem 9.8.

The Debye–Hückel theory is accurate only for very dilute solutions, because of the series truncations used in the derivation, ion-pair formation, and solvation effects. Numerous attempts have been made to extend the range of applicability of the theory. Examination of the deviations of Debye–Hückel predictions from experiment has shown that the addition of a term that is linear in the ionic strength improves the extent of agreement with experiment. Using this idea, C. W. Davies [*Ion Association* (London: Butterworth, 1962)] proposed the expression

$$\log(\gamma_i) = -0.50926z_+z_- \left[\frac{S^{1/2}}{1 + S^{1/2}} - 0.30S \right]. \quad (9.68)$$

Figures 9.6 and 9.7 show comparisons between activity coefficients for NaCl in aqueous solution at 298.15 K computed using the Debye–Hückel theory, Eq. 9.67, the empirical equation 9.68 suggested by Davies, and values measured using methods discussed in the next section. At a concentration of

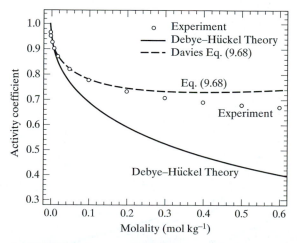

▲ **FIGURE 9.6**
Activity coefficients for aqueous solutions of NaCl at 298.15 K as a function of solution molality over the range $0 \le m \le 0.6$ molal. The points are the experimental data. The solid curve is the Debye–Hückel result obtained from Eq. 9.67A. The dashed curve is obtained from the Davies empirical equation given in the text as Eq. 9.68.

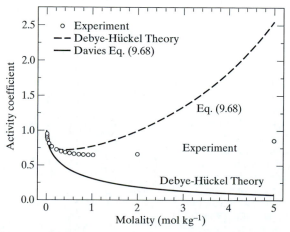

▲ **FIGURE 9.7**
Activity coefficients for aqueous solutions of NaCl at 298.15 K as a function of solution molality over the range $0 \le m \le 5.0$ molal. The points are the experimental data. The solid curve is the Debye–Hückel result obtained from Eq. 9.67A. The dashed curve is obtained from the Davies empirical equation given in the text as Eq. 9.68.

0.02 molal, the Debye–Hückel result is $\gamma_i = 0.847$, whereas experiment yields a result of 0.872. The theoretical result is, therefore, in error by -2.9% at that concentration. When the concentration reaches 0.1 molal, the error is -11.4%. At higher concentrations, the error becomes even larger. The empirical equation suggested by Davies fares much better. At $m = 0.02$ molal, the difference is -0.11%. At a concentration of 0.1 molal, the error is $+0.26\%$. However, at higher concentrations, the deviations become quite large. Figure 9.8 shows a plot of $-\log(\gamma)/(z_+ z_- \, S^{1/2})$ versus molality for seven compounds. If the Debye–Hückel theory were exact, all points would fall on the horizontal line. As can be seen, this is not the case. We also see that compounds containing multiply charged ions show greater deviations from the predictions of Debye–Hückel theory than do those with only singly charged ions. Obviously, the best procedure is to use experimentally measured activity coefficients. The *Handbook of Chemistry and Physics* provides tabulations of these values for many of the common acids, bases, and salts. However, if these values are not available, the Debye–Hückel theory and several empirical modifications will provide reasonably accurate results if the concentrations are not too large.

The experimental data for NaCl seen in Figures 9.6 and 9.7 show that γ_{NaCl} decreases, as expected, as the concentration increases. However, between $m = 1$ mol kg^{-1} and $m = 2$ mol kg^{-1}, the activity coefficient goes through a minimum and increases thereafter. Such behavior is typical for many ionic solutions. A second example is provided by the CaCl$_2$ data given in Problem 9.8. In this case, the minimum occurs around $m = 0.5$ mol kg^{-1}. Figure 9.9 shows the experimentally measured activity coefficients for six additional ionic compounds. The results for HClO$_4$, LiBr, BaI$_2$, NiCl$_2$, and BaBr$_2$ all exhibit minima at concentrations below 2.0 molal. The minimum for Na$_2$SO$_4$ occurs at a higher concentration. Whenever we observe an extremum (minimum or maximum) in the variation of a physical quantity, it is generally a consequence of competing effects. One set of effects tends to make the quantity increase, while a sec-

▶ FIGURE 9.8
The variation of $-\log(\gamma)/z_+ z_- \, S^{1/2}$ with concentration for seven compounds. If the Debye–Hückel theory were exact, all points would fall on the horizontal line. As can be seen, compounds with multiply charged ions show greater deviations from the theory than those with only singly charged ions. (●) HClO$_4$, (✚) BaBr$_2$, (▲) LiBr, (■) NiCl$_2$, (▼) HCl, (◆) CsCl, and (✖) CoCl$_2$. The experimental points are connected by straight-line segments to enhance the visual clarity of the graph. (Source of data: *The Handbook of Chemistry and Physics*, 78th edition, CRC Press, Boca Raton, FL, 1997–1998.)

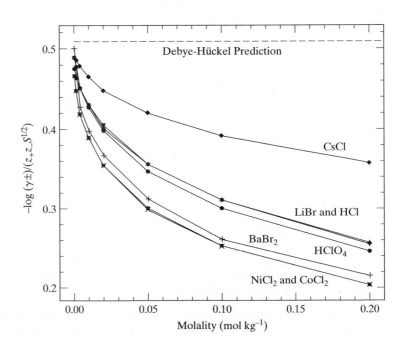

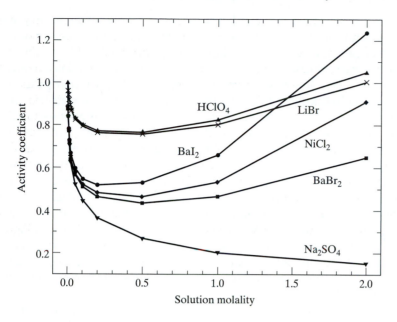

◀ **FIGURE 9.9**
Experimentally measured activity coefficients for six ionic compounds in aqueous solution at 298.15 K as a function of the solution molality. To increase the visual clarity of the plot, the measured values are represented by the points that are connected by straight-line segments. At low concentrations, the results for all compounds become nearly identical, which is in accord with the predictions of Debye–Hückel theory. (Source of data: *The Handbook of Chemistry and Physics,* 78th edition, CRC Press, Boca Raton, FL, 1997–1998.)

ond set acts in the opposite direction. The extremum occurs at the point where the magnitudes of the competing effects become equal.

As an example, suppose we take a very large statistical sample of the age of the male population by asking all males who are old enough to speak to give us their age a. Let us assume that we have some means of determining the actual age of each person we ask, which we denote as A. We now define the function $f(A) = \langle a/A \rangle$, where the braces denote the average value, make a plot of $f(A)$ versus A, and inquire about the shape of the curve thereby produced. If human beings were ideal, we would clearly have $f(A) = 1$ for all A. If ionic solutions were ideal for all m, we would have $\gamma_i = 1$ for all m. However, human beings are not ideal and neither are ionic solutions. When A is small, there is a group of factors that tend to make $a > A$. Among these factors are the desire of males to obtain a driver's license prior to being of legal age, the desire to be allowed to purchase alcoholic beverages, the desire to persuade your father that you really are old enough to be trusted with the car, and the desire to have that gorgeous girl in algebra class think you're more mature than you really are. Moreover, these factors become more important as A increases, since, as $A \to 0$, the human being approaches ideality and $a \to A$. (Newborn babies are totally innocent and do not know how to lie, which argues that we should choose our reference state for human behavior to be $A \to 0$.) Therefore, our plot of $f(A)$ versus A will initially increase as A increases.

There also exists a second group of factors that tend to make $a < A$. These factors include a desire to be young enough to be considered a good candidate for an entry-level job position and a desire to be considered a desirable partner by the young divorcee down the block. These factors become more powerful when A exceeds 30. Consequently, the plot of $f(A)$ will probably exhibit a maximum somewhere between $A = 30$ and $A = 50$. If it were possible to begin drawing Social Security benefits simply by stating that your age exceeds 62, we would also see a minimum in the curve of $f(A)$ versus A somewhere in the neighborhood of $A = 55$. The probability of the existence of such a minimum would be enhanced by the mistaken notion that age always denotes wisdom.

The foregoing analysis shows that the observation of extrema in science or engineering indicates the existence of competing factors. It also shows that if human beings were ideal, we wouldn't have any politicians!

In the present case, the electrostatic interactions substantially reduce ionic mobilities in the solution and thereby decrease the electrical conductance of the solution. The reduced mobilities make it appear as if the concentration of the electrolyte is significantly less than it actually is. That is, the solution activity is less than the solution molality. Therefore, the activity coefficient, which is the ratio of the ionic activity to the molality, decreases below its ideal value of unity at low concentrations. If this were the only effect present, γ_i would be a monotonically decreasing function of concentration, as predicted by the Debye–Hückel theory. However, there is a second important effect present that is ignored by the Debye–Hückel theory: The dipoles of the solvent molecules interact with the charged ions, producing solvated species such as those illustrated in Figure 9.2. In Chapter 8, we noted that such solvation effects can produce an inverse temperature dependence for the solubility. When solvation occurs, it effectively increases the concentration of the solution. For example, consider an aqueous solution of NaCl. If we have a 1-molal solution with no solvation, the concentration is 1 mole of Na^+ and 1 mole of Cl^- ions per $1{,}000/18 = 55.5$ moles of H_2O. If however, solvation were to produce $[Na \cdot 4\,H_2O]^+$ and $[Cl \cdot 2\,H_2O]^-$ ions, we would then have 1 mole of $[Na \cdot 4\,H_2O]^+$ and 1 mole of $[Cl \cdot 2\,H_2O]^-$ in $55.5 - 6 = 49.5$ moles of H_2O. Thus, the effective concentration of charged particles is increased by the solvation effect. Such an effective increase acts in the direction opposite of the electrostatic interaction effect, thereby producing a minimum in the plot of γ_i versus molality.

It is interesting to note that Eq. 9.68 predicts a minimum in a plot of γ_i versus molality. In Problem 9.9, it is shown that this minimum occurs whenever the concentration is $c_e = 0.789959 \ldots \left[\sum_{i=1}^{i=K} \nu_i z_i^2 \right]^{-1}$. In most cases (see Figure 9.7), this concentration is below the value at which the minimum is experimentally observed, so Eq. 9.68 overestimates the importance of solvation effects.

It is instructive to consider the physical significance of the various factors appearing in the Debye–Hückel expression for the activity coefficient. Since the right-hand side of Eq. 9.62 is negative, γ_i increases as $e^2 \kappa\, z_+ z_- /(8\pi\varepsilon_o D k_b T)$ becomes smaller. Therefore, we expect solutions of a given concentration to be closer to ideal ($\gamma_i = 1$) if the ionic charges are ± 1 rather than ± 2 or ± 3. Since deviations from ideality are primarily the result of the electrostatic interactions, and since these are larger for multiply charged ions, this behavior is reasonable. The data in Figure 9.9 illustrate that point: The activity coefficients for compounds with multiply charged ions deviate from unity by more than those with only singly charged ions. Figure 9.10 shows the experimentally measured activity coefficients for 73 compounds at 298.15 K and 0.20 molal concentration. In all cases, the coefficients for compounds with multiply charged ions are below those for compounds containing only singly charged ions. The effect of concentration is contained in κ. As the concentration increases, the number of electrostatic interactions, naturally, increases as well. This increased interaction makes the solution less ideal, so that γ_i decreases. Equation 9.62 also shows that the solution becomes more ideal if we raise the temperature or use a solvent with a higher dielectric constant. A higher temperature produces more random thermal motion of the ions, thereby tending to break up the counterionic atmosphere that builds up around the central ion. Since it is this counterionic environment against which the work of charging the central ion must be done, thermal motion reduces the work and, as a result, increases the activity coefficient toward its ideal value of unity. The effect of the dielectric

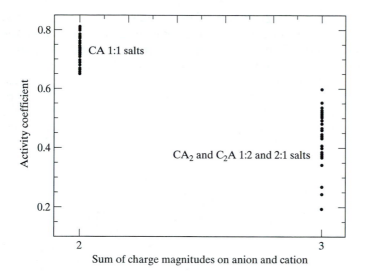

◀ **FIGURE 9.10**
The experimentally measured activity coefficients for 73 different compounds in aqueous solution at 298.15 K and 0.20 molal concentration. In all cases, the coefficients for compounds containing multiply charged ions are less than those for compounds with only singly charged ions. (Source of data: *The Handbook of Chemistry and Physics*, 78th edition, CRC Press, Boca Raton, FL, 1997–1998.)

constant is related to the polarity of the solvent: Highly polar solvents have high dielectric constants. Polar solvent molecules will tend to arrange their dipoles about the ions, producing a shielding effect that reduces the electrostatic interactions between ions. As these interactions are reduced, the solution becomes more ideal and γ_i increases toward unity. (See Problem 9.10.)

9.3   Measurement of Ionic Activity Coefficients

Since ions always come in combinations that are electrically neutral, there is no method whereby we can measure the activity of a single ion. We can, however, measure γ_i or, if there is no ion-pair formation, the mean ionic activity coefficient $\gamma_\pm$. In principle, these quantities can be obtained by measuring the activity of the solvent, using any convenient colligative property, and then employing the Gibbs–Duhem equation to compute the corresponding solute activities. These methods were described in detail in Chapter 8. Since some modifications are required for electrolytes, we shall describe the procedure here using freezing-point depression and osmotic pressure as examples.

The Gibbs–Duhem equation (8.105) requires that we have

$$X_A \frac{\partial \mu_A}{\partial X_A} + X_B \frac{\partial \mu_B}{\partial X_A} = 0,$$

where A and B denote the electrolytic solute and solvent, respectively. We may rewrite this equation in the form

$$\frac{n_A}{n_T} \frac{\partial \mu_A}{\partial n_A} \frac{\partial n_A}{\partial X_A} + \frac{n_B}{n_T} \frac{\partial \mu_B}{\partial n_A} \frac{\partial n_A}{\partial X_A} = 0 \qquad (9.69)$$

if n_T is the total number of moles in the system. Using the fact that $n_A = n_T X_A$, we have $\partial n_A / \partial X_A = n_T$. Inserting this result into Eq. 9.69 gives

$$n_A \frac{\partial \mu_A}{\partial n_A} + n_B \frac{\partial \mu_B}{\partial n_A} = 0,$$

from which we can write

$$\frac{\partial \mu_A}{\partial n_A} = -\frac{n_B}{n_A} \frac{\partial \mu_B}{\partial n_A}. \qquad (9.70)$$

Since our reference function and reference state have been chosen in terms of the molality, let us convert Eq. 9.70 to that measure of concentration. If the solution is m molal, we will have $n_A = m$ and $n_B = 1,000/M_B$, where M_B is the molar mass of the solvent. This gives

$$\frac{\partial \mu_A}{\partial m} = -\frac{1,000}{M_B m} \frac{\partial \mu_B}{\partial m}. \tag{9.71}$$

The chemical potential of the solvent may be expressed in terms of its activity as

$$\mu_B = \mu_B^o + RT \ln a_B,$$

and the chemical potential of the solute is given by Eq. 9.22:

$$\mu_e = \mu_e^o + \nu RT \ln \left\{ \nu_\pm \gamma_i \left(\frac{m}{m^o} \right) \right\}$$

$$= [\mu_e^o + \nu RT \ln(\nu_\pm)] + \nu RT \ln(\gamma_i) + \nu RT \ln \left(\frac{m}{m^o} \right).$$

If we insert these expressions into Eq. 9.71 and remember that the standard chemical potentials and the stoichiometry are independent of concentration, we obtain

$$\nu RT \frac{\partial \ln(\gamma_i)}{\partial m} + \frac{\nu RT m^o}{m} = -\frac{1,000 RT}{M_B m} \frac{\partial \ln a_B}{\partial m}.$$

Rearranging this equation produces

$$\boxed{\frac{\partial \ln(\gamma_i)}{\partial m} = -\left[\frac{1,000}{M_B \nu m} \right] \left(\frac{\partial \ln a_B}{\partial m} \right) - \frac{m^o}{m}.} \tag{9.72}$$

Equation 9.72 is the connection between the solvent activity and the ionic activity coefficient. If there is no ion-pair formation, the left side of the equation becomes $\partial \ln(\gamma_\pm)/\partial m$.

It was shown in Chapter 8 that the solvent activity is given by

$$\ln a_B = -\frac{\Delta \overline{H}_{\text{fus}}^B}{R} \left[\frac{1}{T_f} - \frac{1}{T_f^o} \right] = -\frac{\Delta \overline{H}_{\text{fus}}^B [T_f^o - T_f]}{RT_f T_f^o}, \tag{9.73}$$

provided that the molar enthalpy of fusion of the solvent is assumed to be independent of temperature. If we define the freezing-point depression as $\Delta T_f = T_f^o - T_f$ and make use of the fact that the depression is small, so that we have $T_f \approx T_f^o$, we may write Eq. 9.73 in the form

$$\ln a_B = -\frac{\Delta \overline{H}_{\text{fus}}^B \Delta T_f}{R(T_f^o)^2}. \tag{9.74}$$

We now combine Eqs. 9.72 and 9.74 to relate the solute activity coefficient to the freezing-point depression. The result is

$$\frac{\partial \ln(\gamma_i)}{\partial m} = \left[\frac{1,000 \Delta \overline{H}_{\text{fus}}^B}{M_B R(T_f^o)^2} \right] \left[\frac{1}{\nu m} \right] \frac{\partial \Delta T_f}{\partial m} - \frac{m^o}{m}. \tag{9.75}$$

The quantity in the first set of brackets on the right side of Eq. 9.75 is the inverse of the molal freezing-point depression constant defined in Eq. 8.76, for which we employed the symbol K_f. In these terms, we have

$$\frac{\partial \ln(\gamma_i)}{\partial m} = \left[\frac{1}{K_f \nu m} \right] \frac{\partial \Delta T_f}{\partial m} - \frac{m^o}{m}. \tag{9.76}$$

For purposes of numerical integration, it is useful to write Eq. 9.76 in the form

$$\frac{\partial \ln(\gamma_i)}{\partial m} = -\frac{\partial}{\partial m}\left[1 - \frac{\Delta T_f}{K_f \nu m}\right] - \left[m^o - \frac{\Delta T_f}{K_f \nu m}\right]\frac{1}{m}. \qquad (9.77)$$

By expanding the right-hand side of Eq. 9.77, we may easily show that that equation is identical to Eq. 9.76:

$$-\frac{\partial}{\partial m}\left[1 - \frac{\Delta T_f}{K_f \nu m}\right] - \left[m^o - \frac{\Delta T_f}{K_f \nu m}\right]\frac{1}{m}$$

$$= \left[\frac{1}{K_f \nu m}\right]\frac{\partial \Delta T_f}{\partial m} - \frac{\Delta T_f}{K_f \nu m^2} - \frac{m^o}{m} + \frac{\Delta T_f}{K_f \nu m^2}$$

$$= \left[\frac{1}{K_f \nu m}\right]\frac{\partial \Delta T_f}{\partial m} - \frac{m^o}{m}.$$

The ionic activity coefficients may now be calculated by separating the variables in Eq. 9.77 and integrating both sides between the limits $m = 0$, at which $\gamma_i = 1$, and $m = m$. This gives

$$\int_{\gamma_i=1}^{\gamma_i} d\ln(\gamma_i) = \ln(\gamma_i) = -\int_{m=0}^{m=m} d\left[1 - \frac{\Delta T_f}{K_f \nu m}\right] - \int_{m=0}^{m}\left[m^o - \frac{\Delta T_f}{K_f \nu m}\right]\frac{dm}{m}$$

$$= -\left[1 - \frac{\Delta T_f}{K_f \nu m}\right] - \int_0^m\left[m^o - \frac{\Delta T_f}{K_f \nu m}\right]\frac{dm}{m}, \qquad (9.78)$$

since the limiting value of $1 - \Delta T_f/(K_f \nu m)$ as $m \to 0$ will be zero because, at this point, the solution is ideal, so that we have complete ionization of the solute, no ion-pair formation, and $\Delta T_f = K_f \nu m$.

Equation 9.78 provides us with a means of evaluating the ionic activity coefficients. The first term on the right may be computed directly from the measured freezing-point depression at molality m. The integral must be obtained numerically. If we plot $m^o - \Delta T_f/(K_f \nu m)$ versus $\ln(m)$, the area under the curve between $m = 0$ and $m = m$ will be the value of the integral. This integral will converge because the quantity $m^o - \Delta T_f/(K_f \nu m)$ goes to zero when m goes to zero. An alternative procedure is to define $x = m^{1/2}$ so that $dx = dm/2m^{1/2}$ and then write the integral in Eq. 9.78 in the form

$$\int_0^m\left[m^o - \frac{\Delta T_f}{K_f \nu m}\right]\frac{dm}{m} = 2\int_0^{x^2}\left[m^o - \frac{\Delta T_f}{K_f \nu x^2}\right]x^{-1}\,dx,$$

so that the area under a plot of $[m^o - \Delta T_f/(K_f \nu x^2)]x^{-1} = [m^o - \Delta T_f/K_f \nu m)]m^{-1/2}$ versus $m^{1/2}$ is one-half the value of the integral. A third option to use a least-squares method to fit the experimental data and thereby obtain an expression for ΔT_f that permits the integral to be evaluated analytically. Problems 9.12 through 9.15 explore this option.

Somewhat more accurate results can be obtained by eliminating the assumption that $\Delta \overline{H}_{fus}^B$ is independent of temperature and using the difference in heat capacities of the liquid and solid solvent to obtain that quantity as a function of T. However, in most cases, this is a pointless refinement, since ΔT_f is usually small and the variation of $\Delta \overline{H}_{fus}^B$ over that temperature interval is negligible.

EXAMPLE 9.5

The freezing points of aqueous solutions of a hypothetical 1:2 chloride salt, XCl_2, are measured. The measured freezing-point depressions, $\Delta T_f = T_f^o - T_f$, are reported in the following table:

m (mol kg^{-1})	ΔT_f (K)	m (mol kg^{-1})	ΔT_f (K)
0.000	0.000	0.020	0.09861
0.00001	0.0000557	0.050	0.2384
0.0001	0.000555	0.100	0.4659
0.001	0.00536	0.200	0.9151
0.002	0.01057	0.500	2.2800
0.005	0.02582	1.000	4.7421
0.010	0.05053	2.000	11.4180

Determine the ionic activity coefficient for XCl_2 as a function of m between 0 and 2.00 molal.

Solution

The activity coefficient is given by Eq. 9.78:

$$\ln(\gamma_{XCl_2}) = -\left[1 - \frac{\Delta T_f}{K_f \nu m}\right] - \int_0^m \left[m^o - \frac{\Delta T_f}{K_f \nu m}\right] \frac{\partial m}{m}. \tag{1}$$

The quantities needed for the calculation are $1 - \Delta T_f/(K_f \nu m)$ and $\ln(m)$. We will take the value of K_f to be that of H_2O, 1.86 K mol^{-1}. For XCl_2, $\nu = 3$. These choices and the data give the results shown in the following table.

m (mol kg^{-1})	ΔT_f (K)	$\ln(m)$	$\left[m^o - \dfrac{\Delta T_f}{K_f \nu m}\right]$	Integral	γ_{XCl_2}
0.000	0.000	$-\infty$	0.0000	0.0000	1.000
0.00001	0.0000557	-11.513	0.00179	0.0000	0.998
0.0001	0.000555	-9.2103	0.00538	0.0083	0.986
0.001	0.00536	-6.9077	0.03942	0.0598	0.906
0.002	0.01057	-6.2146	0.05287	0.0918	0.865
0.005	0.02582	-5.2983	0.07455	0.1502	0.799
0.010	0.05053	-4.6052	0.09444	0.2088	0.738
0.020	0.09861	-3.9120	0.1164	0.2818	0.671
0.050	0.2384	-2.9957	0.1455	0.4018	0.578
0.100	0.4659	-2.3026	0.1651	0.5095	0.509
0.200	0.9151	-1.6094	0.1800	0.6291	0.445
0.500	2.2800	-0.6932	0.1828	0.7953	0.376
1.000	4.7421	0.0000	0.1502	0.9107	0.346
2.000	11.4180	0.6932	-0.0231	0.9547	0.394

Figure 9.11 shows a plot of $m^o - \Delta T_f/(K_f \nu m)$ versus $\ln(m)$. The area under this curve between $m = 0$ and $m = m$ is the second term in Eq. 1. The integral

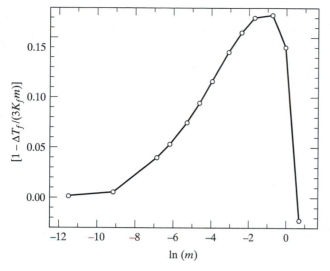

◀ **FIGURE 9.11**
Plot of $1 - \Delta T_f/(K_f\nu m)$ versus $\ln(m)$. The area beneath this curve from $-\infty$ to $\ln(m)$ is the quantity required to compute the ionic activity coefficient using Eq. (9.78). The area obtained from this plot is the same as that under a plot of $[m_o - \Delta T_f/(K_f\nu m)](1/m)$ versus m between 0 and m, since $dm/m = d\ln(m)$. It should be noted that the integrand is well behaved and the required area finite. In practice, the numerical integration can be carried out between $\ln(m)$ and a negative value sufficiently small that $1 - \Delta T_f/(K_f\nu m)$ is essentially zero.

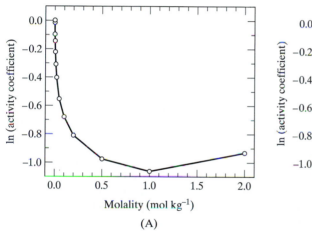

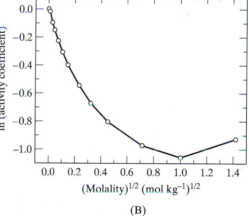

▲ **FIGURE 9.12**
(A) Plot of $\ln(\gamma_{XCl_2})$ versus m for the data given in Example 9.5. (B) Plot of $\ln(\gamma_{XCl_2})$ versus $m^{1/2}$ for the data given in Example 9.5. The Debye–Hückel theory indicates that the variation in $\ln(\gamma_{XCl_2})$ with $m^{1/2}$ should be linear over a wider concentration range than is the case for a plot of $\ln(\gamma_{XCl_2})$ versus m. These results show this characteristic.

was computed via a trapezoidal rule. The results are given in the preceding table. The value of the first term in Eq. 1 is given in column four of the table. The negative of the sum of these two terms is $\ln(\gamma_{XCl_2})$. The calculated activity coefficients are given in the last column of the table. The results for $\ln(\gamma_{XCl_2})$ are shown as a function of m and $m^{1/2}$ in Figures 9.12A and 9.12B, respectively. The Debye–Hückel theory suggests that the variation of $\ln(\gamma_{XCl_2})$ with $m^{1/2}$ should be more nearly linear over a wider range of concentrations than is the case when $\ln(\gamma_{XCl_2})$ is plotted against m. Figures 9.12A and 9.12B show this characteristic.

For related exercises, see Problems 9.12, 9.13, 9.14, and 9.15.

EXAMPLE 9.6

Compare the predictions of the limiting form of the Debye–Hückel theory for the activity coefficient of the hypothetical salt XCl_2 treated in Example 9.5 with the results obtained from the freezing-point data.

Solution

The Debye–Hückel limiting equation for the activity coefficient is

$$\log(\gamma_{XCl_2}) = -0.50926\, z_+ z_-\, S^{1/2}, \tag{1}$$

where

$$S^{1/2} = [0.5m\{(1)(2)^2 + (2)(1)^2\}]^{1/2} = (3m)^{1/2}, \tag{2}$$

since $z_+ = 2$, $z_- = 1$, $\nu_+ = 1$, and $\nu_- = 2$. Combining Eqs. 1 and 2 gives

$$\log(\gamma_{XCl_2}) = -1.764128\, m^{1/2}. \tag{3}$$

The following table shows the comparison:

m (mol kg^{-1})	γ_{XCl_2} (data)	γ_{XCl_2} (Debye–Hückel)	m (mol kg^{-1})	γ_{XCl_2} (data)	γ_{XCl_2} (Debye–Hückel)
0.00001	0.998	0.987	0.050	0.578	0.403
0.0001	0.986	0.960	0.100	0.509	0.277
0.001	0.906	0.879	0.200	0.445	0.163
0.002	0.865	0.834	0.500	0.376	0.057
0.005	0.799	0.750	1.000	0.346	0.017
0.010	0.738	0.666	2.000	0.394	0.003
0.020	0.671	0.563			

Plainly, at concentrations in excess of $m = 0.01$ molal, the results obtained from the Debye–Hückel theory begin to deviate significantly from those computed using the freezing-point depression data. Although the salt in these examples is hypothetical, the nature of the results obtained are typical of real ionic systems.
For a related exercise, see Problem 9.8.

Equation 9.72 makes it obvious that we may determine ionic activity coefficients using any of the methods described in Chapter 8 to obtain the solvent activity. We conclude this section by illustrating how this is done when solvent activities are obtained from osmotic-pressure measurements.

The solvent activity is related to the osmotic pressure by Eq. 8.134, viz.,

$$\ln a_B = -\frac{\overline{V}_{B(l)}\, \pi}{RT},$$

where $\overline{V}_{B(l)}$ is the partial molar volume of the pure liquid solvent. Inserting this expression into Eq. 9.72, we obtain

$$\frac{\partial \ln(\gamma_i)}{\partial m} = -\left[\frac{1,000\overline{V}_{B(l)}}{M_B RT \nu m}\right]\left(\frac{\partial \pi}{\partial m}\right) - \frac{m^o}{m}. \tag{9.79}$$

Using the same technique as that employed for freezing-point depression measurements, we may write Eq. 9.79 in the form

$$\frac{\partial \ln(\gamma_i)}{\partial m} = -\frac{\partial}{\partial m}\left[1 - \frac{C\pi}{\nu m}\right] - \left[m^o - \frac{C\pi}{\nu m}\right]\frac{1}{m}, \tag{9.80}$$

where $C = 1{,}000\overline{V}_{B(l)}/(M_B RT)$. Separating the variables and then integrating between the limits $m = 0$ and $m = m$ gives

$$\ln(\gamma_i) = -\left[1 - \frac{C\pi}{vm}\right] - \int_0^m \left[m^o - \frac{C\pi}{vm}\right]\frac{\partial m}{m}. \qquad (9.81)$$

The first term on the right of Eq. 9.81 may be computed directly from the measured osmotic pressures. The integral is the area under a plot of $m^o - C\pi/(vm)$ versus $\ln(m)$ from $m = 0$ to $m = m$. This integral can be evaluated by a variety of straightforward methods. As before, we note that the integral converges and its lower limit vanishes, since $1 - C\pi/(vm)$ approaches zero as m goes to zero. This can be seen from the van't Hoff equation 8.96. Using that equation, we have

$$\frac{\pi V_B}{n_A RT} = 1.$$

Replacing V_B with $n_B \overline{V}_{B(l)}$ gives

$$\frac{\pi n_B \overline{V}_{B(l)}}{n_A RT} = 1.$$

Since we have 1 kg of solvent, $n_B = 1{,}000/M_B$. Also, the number of moles of dissolved ions near infinite dilution, where we have complete ionization, is vm. Substituting these results into the foregoing equation produces

$$\frac{1{,}000\pi \overline{V}_{B(l)}}{M_B vm RT} = \frac{C\pi}{vm} = 1$$

in the limit of infinite dilution. Therefore, $1 - C\pi/(vm)$ approaches zero in this limit.

Ionic activities can also be conveniently measured using electrochemical techniques that we shall discuss in detail later in the chapter.

9.4   Ionic Equilibria

9.4.1   Standard Chemical Potentials for Ionic Solutions

Up to this point in the chapter, the discussion makes it clear that if we wish to investigate equilibrium processes involving ionic solutions, we will need to include a proper treatment of ionic activity, since a solution cannot be regarded as ideal unless it is so dilute that it is, in reality, little more than "slightly contaminated solvent." (Shortly after of the Debye–Hückel theory was presented, this phrase was often used as a derogatory comment on the theory.)

We will also need the values of the standard chemical potentials. The physical significance of these quantities can be seen by examining Eq. 9.22:

$$\mu_e = \mu_e^o + v\,RT\ln\left\{v_\pm \gamma_i \left(\frac{m}{m^o}\right)\right\}.$$

From a mathematical point of view, we will have $\mu_e^o = \mu_e$ at the point $v_\pm \gamma_i(m/m^o) = 1$. The problem with this definition of the standard state is that it will differ for every solute. In fact, there is no guarantee that such a state even exists for a given solute: It is entirely possible that the standard state for ionic solutes may be purely hypothetical! Therefore, we choose the standard state to be a state with $m = m^o = 1$ molal and $\nu_{\pm} \gamma_i = 1$. In most—indeed, perhaps all—cases, this state will not actually exist; but this creates no problems, since we are free to choose our reference point in any manner we wish, as it is only differences that are important.

Because ions are always formed in combinations that are electrically neutral, the only quantities that we can measure are ones associated with the neutral combination $\nu_+ C^{z+} + \nu_- A^{-z-}$. We can never measure μ^o for individuals ions. To circumvent this problem, we adopt a strategy similar to one we used in Chapter 3, where we defined the standard partial molar enthalpy of $H^+_{(aq,\infty)}$ to be zero. This choice permitted us to obtain values for the standard partial molar enthalpies of all other ions at infinite dilution. In the present case, we will take the standard chemical potential for $H^+_{(aq,1m)}$, $\mu^o_{H^+}$, to be zero. This procedure permits us to obtain values for the standard chemical potentials of all other ions. Measurements of these quantities are best done using electrochemical techniques that we will discuss later in the chapter.

9.4.2   Equilibrium Constants in Ionic Solutions

The general expression for the equilibrium constant was developed in Chapter 5 and given in Eqs. 5.97 and 5.98. In all cases, the value of the thermodynamic equilibrium constant is determined by the difference in standard chemical potentials between products and reactants, which we denote by $\Delta\mu^o$. The general relationship, which we repeat here for convenience, is

$$\Delta\mu^o = -RT \ln \left[\frac{a_C^{\nu_c} a_D^{\nu_d}}{a_A^{\nu_a} a_B^{\nu_b}} \right]_{eq}, \tag{9.82}$$

where C and D represent all products, A and B represent all reactants and the ν_i are the corresponding stoichiometric coefficients. The subscript eq indicates that the ratio of activities must be evaluated at the equilibrium point. The quantity inside the brackets in Eq. 9.82 is called the thermodynamic equilibrium constant:

$$K = \left[\frac{a_C^{\nu_c} a_D^{\nu_d}}{a_A^{\nu_a} a_B^{\nu_b}} \right]_{eq}. \tag{9.83}$$

Since the activities have no units, K is unitless. If one or more of the products or reactants are gases, we replace the activity with the corresponding fugacity for the gas, as discussed in Chapter 5. Whenever the species involved in the equilibrium is an ion, the activity coefficient is defined by Eqs. 9.7A and 9.7B. Because the ions always occur in pairs, the individual ionic activity coefficients, which we cannot measure, can generally be replaced by the mean ionic activity coefficient $\gamma_{\pm}$ or, in the case of ion-pair formation, γ_i.

9.4.3   Acid–Base Equilibria

For purposes of the discussion that follows, we shall adopt the Brønsted definition according to which an acid is a proton donor while a base is a proton acceptor. The most common example of such an acid–base reaction is the self-ionization of water, wherein water acts as both the acid and the base:

$$H_2O + H_2O = H_3O^+ + OH^-.$$

Substances that behave in this fashion are said to be *amphoteric*. The thermodynamic equilibrium constant for the reaction is generally given the symbol K_w; its formal expression is

$$K_w = \left[\frac{a_{H_3O^+}\, a_{OH^-}}{a_{H_2O}^2} \right]_{eq}.$$ (9.84)

Using m/m^o as the reference function for the ions, with $m \to 0$ as the reference state, we have

$$K_w = \left[\frac{m(H_3O^+)m(OH^-)(m^o)^{-2}\gamma_{H_3O^+}\gamma_{OH^-}}{a_{H_2O}^2} \right]_{eq}$$

$$= \left[\frac{m(H_3O^+)m(OH^-)(m^o)^{-2}\gamma_\pm^2}{a_{H_2O}^2} \right]_{eq}$$ (9.85)

where $m(H_3O^+)$ and $m(OH^-)$ denote the molalities of these ions. For the water solvent, the reference function is the mole fraction, with $X_{H_2O} = 1$ the reference state. For pure water, we therefore have $a_{H_2O} = 1$. For dilute solutions, we also expect the activity of the solute to be very close to unity, since it is an uncharged species. With a_{H_2O} set to unity, Eq. 9.85 becomes

$$K_w = [m(H_3O^+)m(OH^-)(m^o)^{-2}\,\gamma_\pm^2]_{eq}.$$ (9.86)

Equations 9.82 and 9.83 show that $K_w = \exp[-\Delta\mu^o/RT]$. Using the data in Appendix A, we find that for the reaction $H_2O(l) \to H^+(aq) + OH^-(aq)$

$$\Delta\mu^o = \mu_{H_3O^+}^o + \mu_{OH^-}^o - \mu_{H_2O}^o = 0 - 157.24 \text{ kJ mol}^{-1} - (-237.13) \text{ kJ mol}^{-1}$$

$$= 79.89 \text{ kJ mol}^{-1}.$$

Thus, at 298.15 K, we have

$$K_w = \exp\left[\frac{-79,890 \text{ J mol}^{-1}}{(8.314 \text{ J mol}^{-1} \text{ K})(298.15 \text{ K})} \right] = 1.01 \times 10^{-14}.$$

If we have only H_2O present, the stoichiometry shows that $m(H_3O^+) = m(OH^{-1}) = m$, so that Eq. 9.86 gives

$$K_w = \left[\left(\frac{m}{m^o} \right)^2 \gamma_\pm^2 \right]_{eq} = 1.01 \times 10^{-14} \qquad \text{at } T = 298.15 \text{ K}.$$

The purpose of $m^o = 1 \text{ mol kg}^{-1}$ is simply to cancel the units on m. If we regard the units as having been canceled, so that m^o can be dropped, the result is

$$K_w = [m^2\, \gamma_\pm^2]_{eq} = 1.01 \times 10^{-14}.$$ (9.87)

We may solve Eq. 9.87 by an iterative procedure to obtain m at the equilibrium point. In this procedure, we first assume that $\gamma_\pm$ is unity and solve for m_{eq}. The result is inserted into an appropriate expression that permits $\gamma_\pm$ to be computed, or else an experimental value at that molality is obtained. Equation 9.87 is then solved again, with $\gamma_\pm$ replaced by its computed or measured value at the concentration obtained in the first step. The procedure is repeated until the new value of m is identical to the previous value, at which point the iteration is said to have converged to the final answer.

Setting $\gamma_{\pm}$ equal to unity in Eq. 9.87, we obtain $m = 1.005 \times 10^{-7}$ molal. At this concentration, the Debye–Hückel theory will yield an accurate value for $\gamma_{\pm}$. Using Eq. 9.67, we compute

$$\log(\gamma_{\pm}) = -0.50926 \, z_+ z_- \, S^{1/2}$$

$$= -0.50926(1)(1)[0.5\{(1)(1)^2 + (1)(1)^2\}1.005 \times 10^{-7}]^{1/2}$$

$$= -0.000161,$$

so that $\gamma_{\pm} = 0.9996$. With this value of $\gamma_{\pm}$ inserted into Eq. 9.87, we compute

$$m = \left[\frac{1.01 \times 10^{-14}}{(0.9996)^2}\right]^{1/2} = 1.005 \times 10^{-7} \text{ molal,}$$

which is the same as that obtained in the first iteration to four significant digits. The process has therefore converged, and we have $m = m(H_3O^+) = m(OH^-) = 1.005 \times 10^{-7}$ molal at equilibrium in water.

Let us now consider the more complex case of a C_o molal aqueous solution of a weak acid and ask what the H_3O^+ molality is in such a solution. We now have two simultaneous, coupled equilibria involved, viz.,

$$H_2O + H_2O = H_3O^+ + OH^-$$

and

$$HX + H_2O = H_3O^+ + X^-,$$

where HX represents the weak acid. The equilibrium constant for the first reaction is

$$K_w = [m(H_3O^+) \, m(OH^-) \, \gamma_{\pm}^2(1)], \tag{9.88}$$

where we have again dropped m^o and assumed that the activity of the solvent is unity. We have also dropped the subscript eq for simplicity and denoted the mean ionic activity coefficient for the H_3O^+–OH^- pair as $\gamma_{\pm}(1)$. For the ionization of the acid, we have

$$K_a = \frac{a_{H_3O^+} \, a_{X^-}}{a_{H_2O} \, a_{HX}},$$

where K_a denotes the ionization constant for HX. If the solution is sufficiently dilute, both a_{H_2O} and the activity coefficient of the uncharged HX acid will approach unity. Under these conditions, the ionization constant may be written as

$$K_a = \frac{\gamma_{H_3O^+} \, m(H_3O^+) \, \gamma_{X^-} \, m(X^-)}{m(HX)} = \frac{\gamma_{\pm}^2(2)m(H_3O^+)m(X^-)}{m(HX)}. \tag{9.89}$$

Equations 9.88 and 9.89 are coupled because of the presence of the common ion H_3O^+ in both equilibria and also because each reaction exerts an influence upon the ionic strength that affects the mean ionic activity coefficients. Consequently, the equations must be solved simultaneously for $m(H_3O^+)$.

We will again use an iterative procedure to obtain the solution of Eqs. 9.88 and 9.89. Let x be the number of moles per kg of H_3O^+ and OH^- that are produced by the self-ionization of H_2O. Let y denote the number of moles per kg of H_3O^+ and X^- that are generated by the ionization of the weak acid. These definitions give $m(H_3O^+) = x + y$, $m(OH^-) = x$, $m(X^-) = y$, and $m(HX) = C_o - y$. In the first iteration, we set the activity coefficients equal to unity, so that the equations we need to solve are

$$(x + y)x = K_w \tag{9.90A}$$

and

$$\frac{(x + y)y}{C_o - y} = K_a. \tag{9.90B}$$

In the first iteration, we also assume that y is small relative to C_o and write Eq. 9.90B in the form

$$\frac{y(x + y)}{C_o} = K_a. \tag{9.90C}$$

Dividing Eq. 9.90A by 9.90C produces

$$\frac{C_o x}{y} = \frac{K_w}{K_a},$$

so that x is given by

$$x = \frac{K_w y}{C_o K_a}. \tag{9.90D}$$

Substituting this result into Eq. 9.90C yields

$$\left[\frac{K_w}{C_o K_a} + 1 \right] y^2 = C_o K_a,$$

which may be easily solved for y to give

$$y = \frac{C_o K_a}{[K_w + C_o K_a]^{1/2}}. \tag{9.90E}$$

Combining Eqs. 9.90D and 9.90E, we obtain

$$x = \frac{K_w}{[K_w + C_o K_a]^{1/2}}. \tag{9.90F}$$

These results are used to obtain the molalities in the first iteration. We now use the Debye–Hückel theory or the Davies modification to calculate the two activity coefficients. These are inserted into Eqs. 9.88 and 9.89, which are then solved again for new molalities. In the second iteration and all iterations thereafter, C_o is replaced with $C_o - y$, where y is the value computed in the previous iterative pass. Such iterations are continued until convergence is achieved.

Before turning our attention to a couple of specific examples, it is instructive to examine Eq. 9.90E. If, in a particular problem, we have $K_w \ll C_o K_a$, we may ignore K_w in that equation and simply write $y = [C_o K_a]^{1/2}$, which is precisely the result we would have obtained if we had ignored the self-ionization of water from the start. The inequality is the key test in the decision as to whether we may ignore the ionization of water without incurring a large error. This test is explored in more quantitative detail in Problem 9.18.

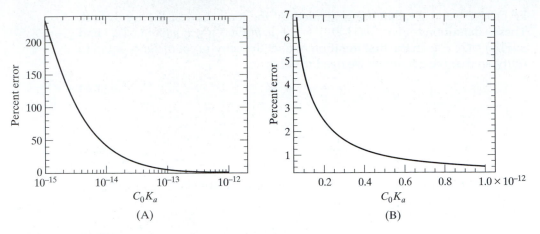

▲ **FIGURE 9.13**
Percent error in the computation of the H_3O^+ concentration for a weak acid in aqueous solution when the contribution from the self-ionization of water is ignored. The figures show the error as a function of the product of the acid concentration and the acid ionization constant, C_oK_a, for a system in which all activity coefficients are unity. In Figure (A), the product C_oK_a is shown on a logarithmic scale, while the scale in Figure (B) is linear.

Figure 9.13 shows the percent error as a function of C_oK_a for a hypothetical system in which the activity coefficients are unity at all concentrations.

EXAMPLE 9.7

What is the H_3O^+ molality in a 0.01-molal aqueous solution of formic acid ($HCHO_2$) at 298.15 K? K_a for formic acid at 298.15 K is 1.80×10^{-4}.

Solution

We first note that $C_oK_a = 0.01(1.80 \times 10^{-4}) = 1.80 \times 10^{-6}$, which is almost eight orders of magnitude (one order of magnitude is 10^1) larger than K_w. We may, therefore, safely ignore the contribution from the ionization of water. The important equilibrium is

$$HCHO_2 + H_2O = H_3O^+ + CHO_2^-.$$

The equilibrium constant for this reaction is given by Eq. (9.89), or

$$\frac{\gamma_\pm^2\, m(H_3O^+)m(X^-)}{m(HX^-)} = K_a, \tag{1}$$

provided that the activity of the water is unity and the activity coefficient of uncharged formic acid may be set to unity. Following the procedure previously outlined, we let $m(H_3O^+) = m(X^-) = y$ and $m(HX) = 0.01 - y$. In the first iteration, we set $\gamma_\pm^2 = 1$ and $m(HX) = 0.01 - y \approx 0.01$. This yields

$$y^2 = 0.01K_a, \tag{2}$$

so that

$$y = [0.01K_a]^{1/2} = [(0.01)(1.80 \times 10^{-4})]^{1/2} = 1.342 \times 10^{-3} \text{ mol kg}^{-1}. \tag{3}$$

For simplicity, let us employ the Debye–Hückel theory to compute $\gamma_\pm$. The only charged particles are H_3O^+ and CHO_2^-. Therefore,

$$\log(\gamma_\pm) = -0.50926z_+z_-\, S^{1/2}$$
$$= -0.50926(1)(1)[0.5\{(1)(1)^2 + (1)(1)^2\}1.342 \times 10^{-3}]^{1/2}$$
$$= -0.018656. \tag{4}$$

This gives $\gamma_\pm = 0.9579$.

In the second iteration, we use $(\gamma_{\pm})^2 = (0.9579)^2 = 0.9176$ and replace C_o with $C_o - y = 0.01 - 1.342 \times 10^{-3} = 0.008658$. The equation to be solved is now

$$\frac{0.9176 \, y^2}{0.008658} = K_a, \tag{5}$$

and the solution is

$$y = \left[\frac{0.008658 \, K_a}{0.9176} \right]^{1/2} = \left[\frac{(0.008658)(1.80 \times 10^{-4})}{0.9176} \right]^{1/2}$$

$$= 0.001303 \text{ mol kg}^{-1}. \tag{6}$$

This value for y is different from the initial result by 4%. If this deviation is larger than desired, we might execute a third iteration, with new activity coefficient given by

$$\log(\gamma_{\pm}) = -0.50926 z_+ z_- \, S^{1/2}$$

$$= -0.50926 \, (1)(1)[0.5\{(1)(1)^2 + (1)(1)^2\} \, 1.303 \times 10^{-3}]^{1/2}$$

$$= -0.018383. \tag{7}$$

The new activity coefficient is, therefore, $\gamma_{\pm} = 0.9586$. Using this value and $y = 1.303 \times 10^{-3} \text{ mol kg}^{-1}$, we now solve

$$\frac{0.9189 y^2}{(0.01 - 0.001303)} = K_a. \tag{8}$$

The solution is $y = 1.305 \times 10^{-3} \text{ mol kg}^{-1}$, which differs by only two in the fourth significant digit, so that it is probably sufficiently accurate. We conclude that at equilibrium $m(H_3O^+) = 1.305 \times 10^{-3} \text{ mol kg}^{-1}$, and the mean ionic activity coefficient is 0.9586.

For related exercises, see Problems 9.16, 9.17, 9.18, 9.19, and 9.20.

EXAMPLE 9.8

What is the H_3O^+ molality in a 0.001-molal aqueous solution of hypoiodous acid (HIO) at 298.15 K? K_a for hypoiodous acid at 298.15 K is 2.30×10^{-11}.

Solution

In one way, this example is harder than the previous one, but in another way, it is easier. First, we check to see if we must include the contribution to H_3O^+ from the self-ionization of water. Here we have $C_o K_a = 0.001(2.3 \times 10^{-11}) = 2.3 \times 10^{-14}$, which is not small relative to $K_w = 1.00 \times 10^{-14}$. We must, therefore, consider both equilibria simultaneously. However, the concentrations of the ions are now so low that we may confidently set $\gamma_{\pm} = 1$.

The equations that we must solve are given by Eq. 9.90A and 9.90B; viz.,

$$(x + y)x = K_w \tag{1}$$

and

$$\frac{(x + y)y}{C_o - y} = K_a.$$

In the first iteration, we take $C_o - y \approx C_o = 0.001 \text{ mol kg}^{-1}$. The solutions for x and y are then given in Eqs. 9.90E and 9.90F:

$$y = \frac{C_o K_a}{[K_w + C_o K_a]^{1/2}} = \frac{(0.001)(2.3 \times 10^{-11})}{[10^{-14} + (0.001)(2.3 \times 10^{-11})]^{1/2}} = 1.266 \times 10^{-7} \text{ mol kg}^{-1};$$

$$\tag{2}$$

$$x = \frac{K_w}{[K_w + C_o K_a]^{1/2}} = \frac{10^{-14}}{[10^{-14} + (0.001)(2.3 \times 10^{-11})]^{1/2}} = 5.505 \times 10^{-8} \text{ mol kg}^{-1}.$$

(3)

The corresponding molality of the H_3O^+ is

$$m(H_3O^+) = x + y = 1.8165 \times 10^{-7} \text{ mol kg}^{-1}. \tag{4}$$

We could now make a second iterative pass by replacing $C_o = 0.001$ with $C_o - y = 0.001 - 1.8165 \times 10^{-7} = 0.0009998 \text{ mol kg}^{-1}$. However, this difference is only 0.02%. A second iteration is, therefore, not needed; Equation 4 provides the desired result.

For related exercises, see Problems 9.16, 9.17, 9.18, 9.19, and 9.20.

In Examples 9.7 and 9.8, the activity coefficients made very little difference because the solutions were so dilute. As a counterexample, consider the case of a common ion effect that is often used in introductory chemistry courses. Suppose we wish to compute the H_3O^+ concentration in a formic acid solution to which we have also added 0.01 mol kg^{-1} of $NaCHO_2$. If deviations from ideality are ignored in this situation, the results will be poor. The next example demonstrates the magnitude of the error that can occur.

EXAMPLE 9.9

What is the H_3O^+ molality in a 0.01-molal aqueous solution of formic acid ($HCHO_2$) at 298.15 K that also contains 0.01 mol kg^{-1} of $NaCHO_2$? K_a for formic acid at 298.15 K is 1.80×10^{-4}.

Solution

As in Example 9.7, the value of C_oK_a is much larger than K_w. Consequently, we have only two reactions to consider:

$$HCHO_2 + H_2O = H_3O^+(aq) + CHO_2^-(aq)$$

and

$$NaCHO_2 + x\,H_2O \longrightarrow Na^+(aq) + CHO_2^-(aq).$$

The 0.01-molal $NaCHO_2$ solution will be almost completely ionized. In solving this problem, it will be assumed that ionization is complete. Under these conditions, the equilibrium concentrations are $m(H_3O^+) = y$, $m(CHO_2^-) = 0.01 + y$, and $m(HCHO_2) = C_o - y = 0.01 - y$, where y is the number of moles of H_3O^+ and CHO_2- formed by the ionization of formic acid. Equation 9.89 must now be solved for y:

$$\frac{\gamma_\pm^2\, m(H_3O^+)m(X^-)}{m(HX)} = \frac{\gamma_\pm^2\, y(0.01 + y)}{C_o - y} = K_a. \tag{1}$$

In the first iteration, we set $\gamma_\pm^2 = 1$ and let $C_o - y \approx C_o = 0.01$ and $0.01 + y \approx 0.01$. This gives

$$y = \frac{0.01K_a}{0.01} = K_a = 1.80 \times 10^{-4} \text{ mol kg}^{-1}. \tag{2}$$

Equation 2 is the usual answer in an introductory chemistry examination. It is, however, not the correct answer. For the second iteration, we compute the activity coefficient using $y = 1.80 \times 10^{-4}$ mol kg^{-1}. For illustrative purposes, let us use the Davies empirical equation for this computation. Equation 9.68 gives

$$\log(\gamma_i) = -0.50926z_+z_-\left[\frac{S^{1/2}}{1 + S^{1/2}} - 0.30S\right]. \tag{3}$$

The ionic strength is given by

$$S = 0.5 \sum_{i=1}^{i=K} c_i z_i^2. \tag{4}$$

The charged species are Na^+, H_3O^+, and CHO_2^-. Therefore,

$$S = 0.50[0.01(1)^2 + (1.8 \times 10^{-4})(1)^2 + (0.01 + 0.00018)(1)^2] = 0.01018. \tag{5}$$

Substituting into Eq. 3 produces

$$\log(\gamma_i) = -0.50926\,(1)(1)\left[\frac{(0.01018)^{1/2}}{1 + (0.01018)^{1/2}} - 0.30\,(0.01018)\right] = -0.045118, \tag{6}$$

which gives $\gamma_i = 0.9013$. If we had used the Debye–Hückel theory, the result would have been $\gamma_i = 0.8884$. In the second iteration, we use $\gamma_\pm^2 = (0.9013)^2 = 0.8123$, $C_o - y = 0.01 - 0.00018 = 0.00982$, and $0.01 + y = 0.01 + 0.00018 = 0.01018$. Substituting into Eq. 1, we obtain

$$y = \frac{(0.00982)K_a}{(0.8123)(0.01018)} = 2.138 \times 10^{-4}\ \text{mol kg}^{-1}, \tag{7}$$

which differs from our previous answer by 18.78%. A third iteration uses $C_o - y = 0.01 - 0.0002138 = 0.009786$, $0.01 + y = 0.01 + 0.0002138 = 0.0102138$, and

$$S = 0.50[0.01(1)^2 + (2.138 \times 10^{-4})(1)^2 + (0.01 + 0.0002138)(1)^2] = 0.0102138. \tag{8}$$

Substituting into Eq. 3 then produces $\gamma_\pm = 0.9012$ and $\gamma_\pm^2 = 0.8122$. When these values are substituted into Eq. 1, we obtain

$$y = \frac{(0.009786)K_a}{(0.8122)(0.0102138)} = 2.123 \times 10^{-4}. \tag{9}$$

This result differs by only 0.7% from the answer obtained in the second iteration. Thus, the problem is essentially solved: The concentration of H_3O^+ is 2.123×10^{-4} mol kg^{-1}, and the mean ionic activity coefficient is 0.9012.

For related exercises, see Problems 9.16, 9.17, 9.18, 9.19, and 9.20.

9.4.4 Hydrolysis of Strong Conjugate Acids and Bases

When a Brønsted acid ionizes in water, it produces H_3O^+ and a solvated anion:

$$HX + H_2O = H_3O^+(aq) + X^-(aq).$$

The solvated anion, $X^-(aq)$, is called the *conjugate base,* in that, if the reaction is viewed from right to left, $X^-(aq)$, acts as a Brønsted base, accepting a proton from $H_3O^+(aq)$. When a Brønsted base ionizes in water, it produces OH^- and a solvated cation:

$$B + H_2O = BH^+(aq) + OH^-(aq).$$

The solvated cation, $BH^{1+}(aq)$, is termed the *conjugate acid.*

In Brønsted terms, we may view a salt as the combination of a conjugate acid and a conjugate base. Ammonium acetate ($NH_4C_2H_3O_2$) is a typical example. The ammonium ion is the conjugate acid formed when ammonia (NH_3) undergoes a basic reaction with water to form $NH_4^+(aq)$ and $OH^-(aq)$, and the acetate ion is the conjugate base formed by the ionization of acetic acid. When such a salt is dissolved in water in a relatively low concentration, complete ionization occurs. The conjugate acid and conjugate base then undergo Brønsted acid–base reactions with water. For example, for $NH_4C_2H_3O_2$, we would observe

$$NH_4^+(aq) + H_2O = H_3O^+(aq) + NH_3$$

and

$$C_2H_3O_2^-(aq) + H_2O = HC_2H_3O_2 + OH^-(aq).$$

Such processes are generally termed *hydrolysis* reactions. Consequently, the $NH_4C_2H_3O_2$ solution can be acidic, basic, or neutral, depending upon which of the preceding hydrolysis reactions has the larger equilibrium constant. When the equilibrium constant for the hydrolysis reaction is relatively large, we say that the conjugate acid or base is *strong*. If the constant is relative small, we term the conjugate acid or base *weak*. The discussion that follows will show that weak acids produce strong conjugate bases, whereas strong acids produce very weak conjugate bases. In the same manner, weak bases produce strong conjugate acids and strong bases give weak conjugate acids.

Let us first consider the case in which the salt contains a weak conjugate acid combined with a strong conjugate base. An example is $NaC_2H_3O_2$. The $Na^+(aq)$ cation functions as a weak conjugate acid, since its corresponding base, NaOH, is strong. On the other hand, $C_2H_3O_2^-(aq)$ functions as a strong conjugate base, because acetic acid is weak. We may, therefore, ignore the hydrolysis of $Na^+(aq)$. To simplify the notation, let us denote the strong conjugate base as X^-. The two reactions we need to consider are

$$H_2O + H_2O = H_3O^+(aq) + OH^-(aq)$$

and

$$X^-(aq) + H_2O = HX + OH^-(aq).$$

The equilibrium constant for the self-ionization of water is K_w. The equilibrium constant for the hydrolysis reaction is generally denoted by K_h. We consider the latter equilibrium first. As usual, we have

$$K_h = \frac{a_{HX}\, a_{OH^-}}{a_{X^-}\, a_{H_2O}} = \frac{a_{HX}\, a_{OH^-}}{a_{X^-}}, \tag{9.91}$$

since the activity of the solvent will be essentially unity in a dilute solution. We now note that Eq. 9.91 may be written as the ratio of K_w to the ionization constant for HX, which we denote by K_a. That is,

$$K_h = \frac{K_w}{K_a} = \frac{a_{H_3O^+}\, a_{OH^-}}{\dfrac{a_{H_3O^+}\, a_{X^-}}{a_{HX}}}. \tag{9.92}$$

Equation 9.92 shows why weak acids produce strong conjugate bases whose hydrolysis equilibrium is shifted toward the right. The weaker the acid HX, the smaller the value of K_a, and consequently, the larger the hydrolysis constant K_h will be. Replacement of the activities in Eq. 9.92 with the corresponding products of concentration and activity coefficients produces

$$K_h = \frac{\gamma_\pm^2(H_2O)\, m(H_3O^+)\, m(OH^-)}{\dfrac{\gamma_\pm^2(HX)\, m(H_3O^+)\, m(X^-)}{\gamma(HX)\, m(HX)}}. \tag{9.93}$$

In Eq. 9.93, $\gamma_{\pm}^2(H_2O)$ and $\gamma_{\pm}^2(HX)$ are the mean ionic activity coefficients for the H_2O self-ionization and the ionization of HX, respectively. $\gamma(HX)$ is the activity coefficient for neutral HX. Since the charges on the ions are identical for both ionization reactions and both experience the same ionic strength, we expect to have $\gamma_{\pm}^2(H_2O) = \gamma_{\pm}^2(HX)$. If the solution is dilute, we also expect the activity coefficient for the uncharged HX species to be near unity. Inserting these expected values into Eq. 9.93 produces

$$K_h = \frac{m(HX)\, m(OH^-)}{m(X^-)}. \qquad (9.94)$$

We may now determine $m(OH^-)$ by solving the two simultaneous equations obtained from the two equilibria. Let x be the number of mol kg^{-1} of H_3O^+ and OH^- obtained from the self-ionization of water and y the number of mol kg^{-1} of HX and OH^- resulting from the hydrolysis reaction of X^-. In this notation, we have

$$m(OH^-) = x + y, \quad m(H_3O^+) = x, \quad m(HX) = y, \text{ and } m(X^-) = C_o - y,$$

where C_o is the number of mol kg^{-1} of the salt added to the water. We seek the values of two unknowns, x and y, and we have two equilibrium equations available from K_w and Eq. 9.94. That is,

$$K_w = \gamma_{\pm}^2(H_2O)\, m(OH^-)\, m(H_3O^+) = \gamma_{\pm}^2(H_2O)\,(x+y)x \qquad (9.95A)$$

and

$$K_h = \frac{y\,(x+y)}{C_o - y}. \qquad (9.95B)$$

The method of solving Eqs. 9.95A and 9.95B is the same at that employed to treat acid–base equilibria. We use an iterative method in which the first iteration is made with $\gamma_{\pm}^2(H_2O) = 1$ and $C_o - y \approx C_o$. The resulting equations are identical to those solved in Example 9.8. Therefore, the solutions are identical, except for the substitution of K_h for K_a. The result is

$$y = \frac{C_o K_h}{[K_w + C_o K_h]^{1/2}}$$

and

$$x = \frac{K_w}{[K_w + C_o K_h]^{1/2}}. \qquad (9.96)$$

The OH^- concentration is given by

$$m(OH^-) = x + y = \frac{K_w}{[K_w + C_o K_h]^{1/2}} + \frac{C_o K_h}{[K_w + C_o K_h]^{1/2}} = [K_w + C_o K_h]^{1/2}. \qquad (9.97)$$

We may now use the concentrations predicted by Eq. 9.96 to compute the activity coefficient $\gamma_{\pm}^2(H_2O)$ and then conduct a second iteration. Example 9.10 shows that if we have $K_w \ll C_o K_h$, the contribution from the self-ionization of H_2O may be ignored.

Note that the same result will be obtained for the case of a salt containing a strong conjugate acid and weak conjugate base, such as NH_4Cl. Equation 9.97 will now yield $m(H_3O^+)$ instead of $m(OH^-)$, provided that we compute $K_h = K_w/K_b$, where K_b is the ionization constant for the weak base that yields the strong conjugate acid.

EXAMPLE 9.10

Compute the OH^- concentration in a 0.3-molal aqueous solution of $NaC_2H_3O_2$ at 298.15 K. K_a for acetic acid at 298.15 K is 1.80×10^{-5}.

Solution

The hydrolysis constant for $C_2H_3O_2^-$ is given by Eq. 9.92:

$$K_h = \frac{K_w}{K_a} = \frac{10^{-14}}{1.8 \times 10^{-5}} = 5.555 \times 10^{-10}. \tag{1}$$

In the first iteration, with $\gamma_\pm^2(H_2O) = 1$ and $C_o - y \approx C_o = 0.3$ molal, Eq. 9.97 gives the OH^- concentration:

$$m(OH^-) = [K_w + C_o K_h]^{1/2} = [10^{-14} + (0.3)(5.555 \times 10^{-10})]^{1/2}$$

$$= 1.29 \times 10^{-5} \text{ molal.} \tag{2}$$

Notice that the self-ionization of water contributes very little. If we set $K_w = 0$ in Eq. 2, the result is $m(OH^-) = 1.29 \times 10^{-5}$ molal. For this reason, further iteration is a needless refinement: It affects only the contribution from the self-ionization of water, which is negligible. In addition, changing C_o to $C_o - y$ merely changes the quantity in the fifth significant digit. The correction is, therefore, of no importance.

For related exercises, see Problems 9.21, 9.22, 9.23, and 9.24.

The aforesaid case of $NaC_2H_3O_2$ is the one usually treated in all introductory textbooks and many advanced texts as well. There is, however, no problem in computing the OH^- or H_3O^+ concentration for the more complex case of a salt containing both a strong conjugate base and a strong conjugate acid, such as $NH_4C_2H_3O_2$. We simply need to treat three simultaneous equilibria rather than two. Let us take $NH_4C_2H_3O_2$ as an illustrative example.

The three equilibrium processes that need to be considered are

$$H_2O + H_2O = H_3O^+(aq) + OH^-(aq),$$

$$NH_4^+(aq) + H_2O = NH_3 + H_3O^+(aq),$$

and

$$C_2H_3O_2^-(aq) + H_2O = HC_2H_3O_2 + OH^-(aq).$$

The corresponding equilibrium relationships are

$$K_w = \gamma_\pm^2(H_2O)\, m(OH^-)\, m(H_3O^+), \tag{9.98A}$$

$$K_h^a = \frac{m(NH_3)\, m(H_3O^+)}{m(NH_4^+)}, \tag{9.98B}$$

and

$$K_h^b = \frac{m(HC_2H_3O_2)\, m(OH^-)}{m(C_2H_3O_2^-)}. \tag{9.98C}$$

In Eq. 9.98, we have denoted the hydrolysis constants for the strong conjugate acid and base as K_h^a and K_h^b, respectively. We now define the three unknowns needed to solve the problem. Let

x = number of mol kg^{-1} of NH_3 and H_3O^+ from the hydrolysis of NH_4^+,

y = number of mol kg^{-1} of $HC_2H_3O_2$ and OH^- from the hydrolysis of $C_2H_3O_2^-$,

and

z = number of mol kg^{-1} of H_3O^+ and OH^- from the self-ionization of water.

In this notation, we have

$$m(H_3O^+) = x + z, \quad m(OH^-) = y + z, \quad m(NH_3) = x,$$

$$m(HC_2H_3O_2) = y, \quad m(NH_4^+) = C_o - x, \quad \text{and} \quad m(C_2H_3O_2^-) = C_o - y.$$

Direct substitution into Eq. 9.98 yields

$$K_w = \gamma_{\pm}^2(H_2O)\,(x + z)(y + z), \tag{9.99A}$$

$$K_h^a = \frac{x(x + z)}{C_o - x}, \tag{9.99B}$$

and

$$K_h^b = \frac{y(y + z)}{C_o - y}, \tag{9.99C}$$

where C_o is the number of mol kg^{-1} of $NH_4C_2H_3O_2$ added to the solution.

In the first iteration to solve Eq. 9.99, we set $\gamma_{\pm}^2(H_2O) = 1$ and assume that y and x are negligible relative to C_o, so that $C_o - y \approx C_o$ and $C_o - x \approx C_o$. The solution is straightforward, but somewhat messy algebraically. We outline it as follows:

We first solve Eq. 9.99B for z in terms of x:

$$z = \frac{C_o K_h^a}{x} - x. \tag{9.100A}$$

We now use Eqs. 9.99A and 9.100A to write y in terms of x:

$$y = \frac{K_w}{x + z} - z = \frac{K_w}{x + \dfrac{C_o K_h^a}{x} - x} - \left[\frac{C_o K_h^a}{x} - x\right]$$

$$= \frac{K_w x}{C_o K_h^a} - \frac{C_o K_h^a}{x} + x. \tag{9.100B}$$

Next, we use Eq. 9.99C and substitute into Eq. 9.100B to obtain

$$K_h^b = \frac{y(y + z)}{C_o} = \left[\frac{K_w x}{C_o K_h^a} - \frac{C_o K_h^a}{x} + x\right]\left[\frac{K_w x}{C_o K_h^a} - \frac{C_o K_h^a}{x} + x + z\right]C_o^{-1}. \tag{9.100C}$$

Substituting Eq. 9.100A for z in Eq. 9.100C produces

$$K_h^b = \left[\frac{K_w x}{C_o K_h^a} - \frac{C_o K_h^a}{x} + x\right]\left[\frac{K_w x}{C_o^2 K_h^a}\right]. \tag{9.100D}$$

Expanding Eq. 9.100D and then multiplying both sides by C_o gives

$$C_o K_h^b = \left[\frac{K_w}{C_o K_h^a}\right]^2 x^2 - K_w + \left[\frac{K_w}{C_o K_h^a}\right]x^2. \tag{9.100E}$$

Collecting terms in x^2, we may easily solve for x^2 and for x. The result is

$$x = \left[\frac{C_o K_h^b + K_w}{\left[\dfrac{K_w}{C_o K_h^a}\right]^2 + \left[\dfrac{K_w}{C_o K_h^a}\right]}\right]^{1/2}. \tag{9.100F}$$

Using Eq. 9.100A, we find that the H_3O^+ concentration is

$$m(H_3O^+) = x + z = x + \frac{C_oK_h^a}{x} - x = \frac{C_oK_h^a}{x}$$

$$= C_oK_h^a \left[\frac{\left[\frac{K_w}{C_oK_h^a}\right]^2 + \left[\frac{K_w}{C_oK_h^a}\right]}{C_oK_h^b + K_w} \right]^{1/2}. \qquad (9.100G)$$

Equation 9.100G is the final result of the first iteration. We may now use that equation to compute x and, with the value obtained, calculate y using Eq. 9.100B and z using Eq. 9.100A. The concentrations of all the ions may then be computed and the activity coefficient $\gamma_\pm^2(H_2O)$ obtained from the Debye–Hückel expression or the Davies empirical equation. The coefficient is then substituted into Eq. 9.98A and the solution repeated. The second iteration requires nothing more than replacing K_w with $K_w/\gamma_\pm^2(H_2O)$ and substituting into Eq. 9.100G. We also replace C_o with $C_o - x$ and $C_o - y$. This iterative procedure is continued until convergence is obtained. In most cases, the first solution will be sufficiently accurate that iteration will be unnecessary.

In the majority of cases, the self-ionization of water will be unimportant. This will certainly be so if we have $K_w \ll C_oK_h^a$ and $K_w \ll C_oK_h^b$. When these inequalities hold, Eq. 9.100G assumes a very simple form. When K_w is very small relative to $C_oK_h^a$ and $C_oK_h^b$, we have $[K_w/(C_oK_h^a)] \ll 1$, so that

$$\left[\frac{K_w}{C_oK_h^a}\right]^2 \ll \left[\frac{K_w}{C_oK_h^a}\right].$$

We also have $C_oK_h^b + K_w \approx C_oK_h^b$. Inserting these results into Eq. 9.100G produces

$$m(H_3O^+) = C_oK_h^a \left[\frac{\left[\frac{K_w}{C_oK_h^a}\right]}{C_oK_h^b} \right]^{1/2} = \left[\frac{C_oK_h^aK_w}{C_oK_h^b} \right]^{1/2} = \left[\frac{K_wK_h^a}{K_h^b} \right]^{1/2}. \qquad (9.100H)$$

Equation 9.100H demonstrates that the acidity of a solution of a salt containing both a strong conjugate acid and a strong conjugate base is independent of the salt concentration C_o, provided that the concentration is not so high that the activity coefficients deviate significantly from unity and provided that K_w is relatively small compared with $C_oK_h^a$ and $C_oK_h^b$.

EXAMPLE 9.11

Compute the molality of H_3O^+ when 0.1 mol kg^{-1} of $NH_4C_2H_3O_2$ is added to water at 298.15 K.

Solution

The first step is to compute the hydrolysis constants for NH_4^+ and $C_2H_3O_2^-$. These are given by

$$K_h^a = \frac{K_w}{K_b}, \qquad (1)$$

where K_b is the ionization constant for NH_3. Using the data in Appendix D, we obtain

$$K_h^a = \frac{K_w}{K_b} = \frac{1.01 \times 10^{-14}}{1.80 \times 10^{-5}} = 5.61 \times 10^{-10}. \qquad (2)$$

For the acetate ion, we have

$$K_h^b = \frac{K_w}{K_a} = \frac{1.01 \times 10^{-14}}{1.80 \times 10^{-5}} = 5.80 \times 10^{-10}. \tag{3}$$

The values of $C_o K_h^a$ and $C_o K_h^b$ are 5.61×10^{-11} and 5.80×10^{-11}, respectively. Both of these are more than three orders of magnitude larger than K_w at 298.15 K. Consequently, we may safely ignore the contribution from the self-ionization of water. As a result, the H_3O^+ concentration is given by Eq. 9.100H:

$$m(H_3O^+) = \left[\frac{K_w K_h^a}{K_h^b} \right]^{1/2}. \tag{4}$$

Substituting the values of the constants yields

$$m(H_3O^+) = \left[\frac{(1.01 \times 10^{-14})(5.61 \times 10^{-10})}{5.80 \times 10^{-10}} \right]^{1/2} = 9.88 \times 10^{-8} \, \text{mol kg}^{-1}. \tag{5}$$

For related exercises, see Problems 9.21, 9.22, 9.23, and 9.24.

9.4.5   Solubility of Ionic Compounds

Many ionic compounds exhibit only slight solubility in solution. When small amounts dissolve in highly polar solvents such as water, ionization is almost complete. That is, the dissolution reaction involves the equilibrium process

$$C_{\nu+}A_{\nu-} + x\,H_2O = \nu_+\,C^{z+}(aq) + \nu_-\,A^{z-}(aq).$$

The equilibrium constant for this reaction is

$$K = \frac{a_{C^{z+}}^{\nu_+}\, a_{A^{z-}}^{\nu_-}}{a_{C_{\nu+}A_{\nu-}}}. \tag{9.101}$$

Since the ionic salt is a pure compound whose activity is a constant at a given temperature, Eq. 9.101 may be written as

$$K = K_{sp} = \gamma_\pm^{(\nu_+ + \nu_-)} \left[\nu_+ \left(\frac{m}{m^o} \right) \right]^{\nu_+} \left[\nu_- \left(\frac{m}{m^o} \right) \right]^{\nu_-}, \tag{9.102}$$

where m is the number of moles per kilogram of $C_{\nu+}A_{\nu-}$ that dissolve. Because the expression on the right-hand side of the equation has the form of a simple product function, the equilibrium constant is customarily denoted as the *solubility product constant* with the symbol K_{sp}.

As usual, the equilibrium constant at 298.15 K can be determined from standard chemical potentials. For example, if the insoluble salt is AgCl, we have

$$AgCl + x\,H_2O = Ag^+(aq) + Cl^-(aq),$$

so that $\Delta\mu^o = \mu_{Cl^-}^o + \mu_{Ag^+}^o - \mu_{AgCl}^o$. Using the data given in Appendix A, we obtain $\Delta\mu^o = -131.23 + 77.11 - (-109.79)$ kJ mol^{-1} = 55.67 kJ mol^{-1}. Therefore, we have

$$K_{sp} = \exp\left[-\frac{55{,}670}{RT} \right] = \exp\left[-\frac{55{,}670}{(8.314)(298.15)} \right]$$

$$= \exp[-22.4583] = 1.764 \times 10^{-10}.$$

Note that, although the activity of the salt does not appear in the expression for the solubility product constant (because its value is a constant), it is still important to include its standard chemical potential in the computation of

K_{sp}. If K_{sp} is needed at a temperature other than 298.15 K, we use the Gibbs–Helmholtz equations to execute the computation. This type of calculation has been discussed in detail in Chapter 5. The data required for the computation include $\Delta \overline{H}$ for the reaction at 298.15 K and the heat capacities of the ions in solution and of the undissolved salt.

EXAMPLE 9.12

Determine the solubility of $Ni_3(PO_4)_2$ in water at 298.15 K. K_{sp} for this salt is 4.74×10^{-32}.

Solution

The ionization reaction is $Ni_3(PO_4)_2 + x\,H_2O = 3\,Ni^{2+}(aq) + 2\,PO_4^{3-}(aq)$. Therefore, the required expression for the solubility product is

$$K_{sp} = \gamma_{\pm}^5 \left[3\left(\frac{m}{m^o}\right)\right]^3 \left[2\left(\frac{m}{m^o}\right)\right]^2 = 108\gamma_{\pm}^5 \left[\left(\frac{m}{m^o}\right)\right]^5. \tag{1}$$

If we assume that we have canceled the units properly on both sides, we may set m^o to unity. In the first iteration to obtain the solution, we take the activity coefficient to be unity. This gives

$$m = \left[\frac{K_{sp}}{108}\right]^{1/5} = \left[\frac{4.74 \times 10^{-32}}{108}\right]^{1/5} = 2.1304 \times 10^{-7}\,\text{mol kg}^{-1}. \tag{2}$$

In the second iteration, we compute the mean ionic activity coefficient, assuming that $m = 2.1304 \times 10^{-7}$ mol kg^{-1}. The ionic strength of the solution is

$$S = 0.5 \sum_{i=1}^{i=K} c_i z_i^2 = 0.5\{(3)(2.1304 \times 10^{-7})(2)^2 + (2)(2.1304 \times 10^{-7})(3)^2\}$$

$$= 3.196 \times 10^{-6}. \tag{3}$$

At this ionic strength, we may use the Debye–Hückel theory to compute the activity coefficient:

$$\log(\gamma_{\pm}) = -0.50926 z_+ z_- \, S^{1/2} = -0.50926(2)(3)(3.196 \times 10^{-6})^{1/2} = -0.005463. \tag{4}$$

Hence, $\gamma_{\pm} = 0.9875$. Inserting this value into Eq. 1 gives

$$K_{sp} = 108\gamma_{\pm}^5 m^5 = 108(0.9875)^5 m^5 = 4.74 \times 10^{-32}. \tag{5}$$

Solving for m, we obtain

$$m = \left[\frac{4.74 \times 10^{-32}}{(108)(0.9875)^5}\right]^{1/5} = 2.16 \times 10^{-7}\,\text{mol kg}^{-1}. \tag{6}$$

The difference between this answer and that obtained in the first iteration is about 1.4%, so we may regard the result as having converged to three significant digits.

For related exercises, see Problems 9.25 and 9.26.

EXAMPLE 9.13

Determine the solubility of $Ni_3(PO_4)_2$ in a 0.1-molal aqueous solution of $BaCl_2$ at 298.15 K. K_{sp} for this salt is 4.74×10^{-32}. Assume that the $BaCl_2$ ionizes completely and ignore the possibility of precipitation of $Ba_3(PO_4)_2$.

Solution

This example is similar to Example 9.12. The difference is the presence of the $BaCl_2$, which is soluble. Consequently, we have

$$BaCl_2 + x\, H_2O \longrightarrow Ba^{2+}(aq) + 2\, Cl^-(aq),$$

and the concentration of $Ba^{2+}(aq)$ and $Cl^{1-}(aq)$ will be 0.1 molal and 0.2 molal, respectively. These concentrations are so large that the ionic strength of the medium is completely determined by the $BaCl_2$. The contributions of $Ni^{2+}(aq)$ and the $PO_4^{3-}(aq)$ are negligible. For this reason, we do not need to solve the problem iteratively. The ionic strength is

$$S = 0.5 \sum_{i=1}^{i=K} c_i z_i^2 \approx 0.5\{(0.1)(2)^2 + (0.2)(1)^2\} = 0.30, \qquad (1)$$

provided that we ignore the contributions of the $Ni^{2+}(aq)$ and $PO_4^{3-}(aq)$ ions. This ionic strength is much too large for us to make use of the Debye–Hückel theory. If we employ the Davies approximation, Eq. 9.68, we obtain

$$\log(\gamma_i) = -0.50926 z_+ z_- \left[\frac{S^{1/2}}{1 + S^{1/2}} - 0.30S \right]$$

$$= -0.50926(2)(1) \left[\frac{0.3^{1/2}}{1 + 0.3^{1/2}} - 0.30\,(0.3) \right] = -0.26878. \qquad (2)$$

This gives $\gamma_{Ni_3(PO_4)_2} = 0.539$. The solubility product for $Ni_3(PO_4)_2$ is

$$K_{sp} = \gamma_\pm^5 \left[3\left(\frac{m}{m^o} \right) \right]^3 \left[2\left(\frac{m}{m^o} \right) \right]^2 = 108\gamma_\pm^5 \left[\left(\frac{m}{m^o} \right) \right]^5 = (108)(0.539)^5\, m^5. \qquad (3)$$

Solving for m, we obtain

$$m = \left[\frac{1}{0.539} \right]\left[\frac{K_{sp}}{108} \right]^{1/5} = 1.86 \left[\frac{4.74 \times 10^{-32}}{108} \right]^{1/5} = 3.95 \times 10^{-7}\, \text{mol kg}^{-1}. \qquad (4)$$

Note that the solubility of $Ni_3(PO_4)_2$ is 1.83 times greater in the $BaCl_2$ solution than in pure H_2O. In this case, the activity coefficient makes a very large difference.

For related exercises, see Problems 9.25 and 9.26.

In most introductory chemistry textbooks, the solubility product is employed to compute the solubilities of only nearly insoluble compounds. This restriction is unnecessary; it is made in beginning texts solely to avoid the problem of dealing with activity coefficients. We may use the solubility product to compute the solubility of *any* ionic compound. If, however, the compound exhibits a high solubility, the mean ionic activity coefficient must be included if the answer is to be accurate. Problems 9.27, 9.28, and 9.29 serve as examples.

9.5   Conductivity of Ionic Solutions

One of the distinguishing features of ionic solutions is their ability to conduct an electrical current. The ions act as charge carriers, with part of the total charge being conducted by the solvated cations and the remainder by the solvated anions. In principle, some of the charge can be conducted by the solvent. However, unless the solution is very dilute, solvent conduction is generally negligible relative to that of the ions.

9.5.1   Electrical Quantities and Units

The *current*, generally denoted by the symbol I, is defined as the rate at which a charge dQ is conducted through a medium. That is, the current is the derivative:

$$I = \frac{dQ}{dt}. \tag{9.103}$$

The SI unit for current is the *ampere*, which is often abbreviated *amp*. The ampere is defined to be the current flowing through two parallel wires of infinite length separated by 1 meter that produces a force per unit length between the wires of exactly $2 \times 10^{-7}\,\text{N m}^{-1}$.

The unit of charge is called the *coulomb*, which is defined to be the total charge passing a point when a constant current of 1 ampere flows for 1 second. Using Eq. 9.103, we may obtain a simple integral expression for the total charge passing a given point in time Δt:

$$Q = \int_{0}^{\Delta t} I\, dt. \tag{9.104}$$

If the current is constant, I can be factored out of the integral, and we have $Q = I\,\Delta t$, which is the form generally used in introductory chemistry. There are other units used for current and charge, but we shall restrict the treatment in this chapter to these SI units.

The current density j is defined to be current per unit cross-sectional area. That is,

$$j = \frac{I}{A} \tag{9.105}$$

where A is the total cross-sectional area of the medium through which the current is passing. In the SI system, the units on j are amperes per square meter, amp m^{-2}.

Since most conducting media produce resistance to the flow of current, there must be an electric field $\mathcal{E}$ present to drive the charge carriers through the medium. In SI units, $\mathcal{E}$ is measured in newtons of force per coulomb (N C^{-1}). An exception to this general condition occurs in *superconductors*, which have zero resistance to charge flow. Once initiated, current can continue to flow indefinitely in a superconductor without the presence of a driving force, so long as no work is done by the system. This fact is very important in magnetic resonance spectroscopy, which we shall discuss in Chapter 16. In general, we expect the current density to double if we double the driving force. That is, we expect j to be proportional to $\mathcal{E}$. This expectation can be expressed in the form

$$\boxed{j = \kappa \mathcal{E}}, \tag{9.106}$$

where the constant of proportionality, κ, is called the *specific conductivity* of the medium. The reciprocal, $\rho = 1/\kappa$, is called the *resistivity*. Since the units on j are amp m^{-2}, κ must have units of amp C $\text{m}^{-2}\,\text{N}^{-1}$. Because an ampere has units of C s^{-1} and 1 N is $1\,\text{kg m s}^{-2}$, the base SI units on κ are $\text{C}^2\,\text{s kg}^{-1}\,\text{m}^{-3}$.

In Chapter 1, we introduced the concept of potential energy by defining the potential to be that function whose negative derivative with respect to coordinate q yields the force in the q direction. This relationship was expressed in Eq. 1.46, $F_z = -\partial V(z)/\partial z$, where $V(z)$ was the potential-energy function and F_z was the corresponding force in the z direction. Using this same general concept, we may define an electric potential Φ such that

$$\mathcal{E}_x = -\frac{\partial \Phi}{\partial x} \tag{9.107}$$

provided that x denotes the direction of the electric field. Separating variables in Eq. 9.107 and then integrating between corresponding limits produces

$$\int_{\Phi_1}^{\Phi_2} d\Phi = -\int_{x_1}^{x_2} \mathcal{E}_x \, dx.$$

If we take x_2 to be sufficiently close to x_1, we may regard the electric field to be constant over this small interval. Under such conditions, $\mathcal{E}_x$ may be factored out of the integral, and we obtain

$$\Phi_2 - \Phi_1 = \Delta \Phi = -\mathcal{E}_x (x_2 - x_1) = -\mathcal{E}_x \Delta x,$$

from which we see that $\mathcal{E}_x = -\Delta \Phi / \Delta x$. This result shows that the unit on the electric potential must be N m C^{-1}, or J C^{-1}, since a newton-meter is a joule. One J C^{-1} is defined to be an electric potential of 1 volt. Combining this result with Eqs. 9.105 and 9.106, we obtain

$$j = \kappa \mathcal{E} = -\kappa \frac{\Delta \Phi}{\Delta x} = \frac{I}{A}. \tag{9.108}$$

The magnitude of the electric potential can, therefore, be written in the form

$$|\Delta \Phi| = \frac{I \Delta x}{\kappa A} = \left[\frac{\rho \, \Delta x}{A} \right] I. \tag{9.109}$$

The quantity in brackets in Eq. 9.109 is called the *resistance R* of the conductor. That is, by definition,

$$R = \frac{\rho \, \Delta x}{A}. \tag{9.110}$$

This result allows us to write Eq. 9.109 in the form $|\Delta \Phi| = IR$. The analysis also shows that if the electric field varies with x, R, too, will be a function of position along the conductor. If, however, we have a conductor with homogeneous composition whose cross-sectional area A is constant, then the current density and the electric field will be constants. For such a system, we may take Δx to be the total length of the conductor, L, and write

$$\boxed{R = \frac{\rho L}{A}.} \tag{9.111}$$

The SI unit for resistance is the *ohm*, for which the symbol Ω is used. Equation 9.111 shows the connection between the ohm and mass–distance–time units: Since $\rho = 1/\kappa$, the units on R are $(\text{kg m}^3 \, \text{C}^{-2} \, \text{s}^{-1})(\text{m/m}^2) = \text{kg m}^2 \, \text{C}^{-2} \, \text{s}^{-1}$. In terms of electrical SI units, one ohm = one volt/one amp. (See Example 9.14.)

In many materials, the resistivity is independent of the electric field. This is true for metals and for electrolytic solutions if the current density is not too high and steady-state conditions are maintained. When the resistivity of the conducting medium is independent of the electric field, *Ohm's law* is said to obtain. Equation 9.106 shows that a plot of the current density versus $\mathcal{E}$ will be a straight line with slope $\kappa = 1/\rho$ if Ohm's law holds.

The reciprocal of the resistance is called the *conductivity*. This quantity obviously has the unit Ω^{-1}, which is often called the *mho*. The correct name for Ω^{-1}, however, is the *siemens*, for which the symbol S is used.

Table 9.1 summarizes electrical quantities, symbols, units, and unit conversions.

Table 9.1   Electrical quantities, symbols and units[1]

Electrical Quantity	Symbol	Unit Names, Abbreviations and Unit Conversions
Current	I	ampere (amp) $1 \text{ amp} = 1 \text{ C s}^{-1}$
Charge	Q	coulomb (C) $1 \text{ C} = 1 \text{ amp s}$
Current density	j	amp m^{-2}
Electric field	$\mathcal{E}$	$1 \text{ N C}^{-1} = 1 \text{ kg m s}^{-2} \text{C}^{-1} = 1 \text{ V m}^{-1}$
Specific conductance	κ	$1 \text{ amp C m}^{-2} \text{N}^{-1} = 1 \text{ C}^2 \text{ s kg}^{-1} \text{m}^{-3}$
Resistivity	ρ	$1 \text{ amp}^{-1} \text{C}^{-1} \text{m}^2 \text{N} = 1 \text{ C}^{-2} \text{s}^{-1} \text{kg m}^3$
Electric potential	Φ	$1 \text{ volt (V)} = 1 \text{ J C}^{-1} = 1 \text{ N m C}^{-1}$
Resistance	R	$1 \text{ ohm } (\Omega) = 1 \text{ C}^{-2} \text{s}^{-1} \text{kg m}^2 = 1 \text{ V amp}^{-1}$
Conductivity	S	$1 \text{ siemens (S)} = 1 \Omega^{-1} = \text{C}^2 \text{ s kg}^{-1} \text{m}^{-2} = 1 \text{ V}^{-1} \text{amp}$

[1]The abbreviations used in this table are 1 newton = 1 N, 1 second = 1 s, and 1 volt = 1 V.

EXAMPLE 9.14

The preceding discussion shows that $1 \Omega = 1 \text{ kg m}^2 \text{ C}^{-2} \text{s}^{-1}$. Use this fact to demonstrate that 1 ohm is the same as 1volt amp^{-1}.

Solution

We know that $1 \text{ amp} = 1 \text{ C s}^{-1}$. Therefore,

$$(1 \text{ amp})^{-1} = 1 \text{ C}^{-1} \text{s}. \tag{1}$$

Substituting this result into the definition for the ohm gives

$$1 \Omega = 1 \text{ kg m}^2 \text{ C}^{-1} \text{s}^{-2} \text{amp}^{-1}. \tag{2}$$

However, $1 \text{ kg m}^2 \text{ s}^{-2}$ is a joule. Hence,

$$1 \Omega = 1 \text{ J C}^{-1} \text{amp}^{-1}. \tag{3}$$

Finally, we know that 1 volt = 1 J C^{-1}. This gives

$$1 \Omega = 1 \text{ volt amp}^{-1}, \tag{4}$$

as required.

9.5.2   Molar Conductivity

The conductivity κ of a solution depends upon the concentration of charge carriers in the solution. To obtain a measure of the intrinsic ability of a particular type of charge carrier to conduct current, we define the molar conductivity as

$$\Lambda_m = \frac{\kappa}{c},$$ **(9.112)**

where c is the electrolyte concentration. If c is expressed in units of mol cm^{-3} and κ in Ω^{-1} cm^{-1}, Λ_m will be in Ω^{-1} cm^2 mol^{-1} or S cm^2 mol^{-1}.

Since the definition of Λ_m tends to divide out the dependence of the conductivity upon concentration, we might anticipate that Λ_m will be independent of the electrolyte concentration and, therefore, be a true measure of the conductivity of a particular combination of charge carriers. That is, if we were to double the concentration of charge carriers, we would expect κ to double, but since c would also double, Λ_m would remain unchanged. This expectation would be borne out in observation if doubling the concentration of electrolyte actually doubled the concentration of charge carriers and if the carriers did not interact in a way that affects their relative motion in the solution. However, such is not the case; instead, as the concentration increases, the propensity of the ions to form ion pairs and thereby *reduce* the effective number of charge carriers increases. The result is a decrease in the molar conductivity of the solution. In addition, the electrostatic interaction between ions increases with concentration. This attractive interaction tends to reduce the ionic mobility and thereby reduce the ability of the ions to conduct charge through the solution. Figure 9.14, which shows the variation of Λ_m with c for an aqueous solution of HCl at 298.15 K, illustrates this behavior.

To obtain a true measure of the intrinsic ability of the ions to conduct current, we need to obtain the molar conductivity of a solution in which there is no ion-pair formation or attractive interactions between the ions that reduce their mobilities. This condition can be achieved only at infinite dilution. Therefore, we need to extrapolate the measured conductivities to $c \rightarrow 0$. We denote this limiting molar conductivity at infinite dilution as Λ_m^∞. There is, however, some difficulty associated with executing the extrapolation: Figure 9.14 shows that Λ_m does not vary linearly with c; consequently, it is not clear how the extrapolation should be carried out.

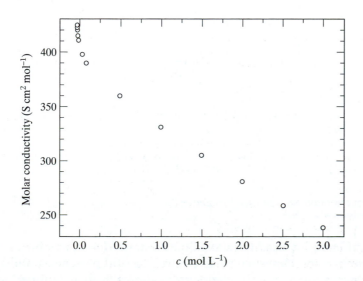

◀ FIGURE 9.14
The molar conductivity of aqueous solutions of HCl at 298.15 K as a function of HCl concentration in mol L^{-1}. (Source of data: *The Handbook of Chemistry and Physics*, 78th edition, CRC Press, Boca Raton, FL, 1997–1998.)

We may obtain a partial solution to the extrapolation problem by noting that as $c \to 0$, ion-pair formation will rapidly approach zero. However, the electrostatic interactions that give rise to mean ionic activity coefficients less than unity persist even at very low concentrations. The Debye–Hückel theory indicates that these interactions exhibit a dependence on $c^{1/2}$ at low concentrations. This suggests that a plot of Λ_m versus $c^{1/2}$ should be linear in the region of low concentrations. If so, accurate extrapolation to infinite dilution can be easily done. The next example serves as an illustration.

EXAMPLE 9.15

The measured molar conductivities of aqueous solutions of HCl at 298.15 K are 424.5, 422.6, 421.2, 415.7, and 411.9 S cm^2 mol^{-1} at concentrations of 0.0001, 0.0005, 0.0010, 0.0050, and 0.0100 mol L^{-1}, respectively. Use these data to obtain Λ_m^∞.

Solution

As expected, a plot of Λ_m against c is nonlinear (see Figure 9.14). However, if the measured values are plotted against $c^{1/2}$, we obtain the table shown to the left.

Figure 9.15 shows a plot of these data. It is clearly seen that Λ_m varies in a near linear manner with $c^{1/2}$. The equation of the fit is

$$\Lambda_m = 425.74 - 139.62 c^{1/2}.$$

Therefore, we obtain $\Lambda_m^\infty = 425.74$ S cm^2 mol^{-1}. The value reported in the *Handbook of Chemistry and Physics* is 426.1 S cm^2 mol^{-1}.

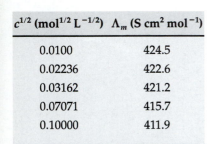

$c^{1/2}$ (mol$^{1/2}$ L$^{-1/2}$)	Λ_m (S cm^2 mol^{-1})
0.0100	424.5
0.02236	422.6
0.03162	421.2
0.07071	415.7
0.10000	411.9

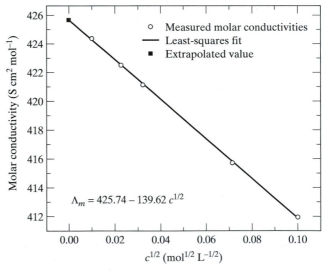

$$\Lambda_m = 425.74 - 139.62\, c^{1/2}$$

▲ FIGURE 9.15
The molar conductivity of aqueous solutions of HCl at 298.15 K as a function of the square root of the HCl concentration expressed in mol L^{-1}. The points are the measured conductivities. The line is a least-squares fit to the measured data. The equation of this line is given in the figure and in the text. (Source of data: *The Handbook of Chemistry and Physics,* 78th edition, CRC Press, Boca Raton, FL, 1997–1998.)

For a related exercise, see Problem 9.31.

The aforesaid procedure allows an accurate determination of Λ_m^∞ for strong electrolytes. However, weak electrolytes still present a problem: The molar conductivities of such compounds approach their limiting value at

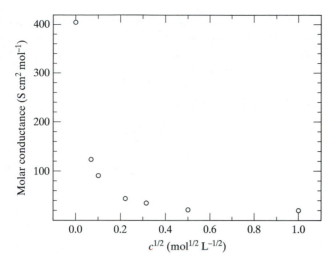

◀ **FIGURE 9.16**
The molar conductivity of aqueous solutions of HF at 298.15 K as a function of the square root of the HF concentration expressed in mol L^{-1}. (Source of data: *The Handbook of Chemistry and Physics*, 78th edition, CRC Press, Boca Raton, FL, 1997–1998.)

infinite dilution asymptotically, which makes accurate extrapolation essentially impossible. Figure 9.16 shows the measured molar conductivity of aqueous solutions of HF at 298.15 K as a function of $c^{1/2}$. Obviously, extrapolating the data to infinite dilution cannot be done accurately. The problem with these compounds is that they exist essentially in un-ionized form except at very low concentrations, where the percent ionization increases rapidly with dilution. In Problem 9.31, it is shown that weak electrolytes ionize completely at infinite dilution.

To obtain the limiting molar conductance for weak electrolytes, we must rely on a different procedure. If we can obtain the molar conductance of individual ions at infinite dilution, these quantities can simply be added in an appropriate manner to obtain Λ_m^∞, since at infinite dilution, the ions act independently. The next section addresses this problem.

9.5.3   Conductance of Individual Ions

The principle of conservation of total current requires that if we measure the molar conductance of a solution of electrolyte $C_{\nu_+} A_{\nu_-}$ that ionizes to yield $\nu_+ C^{z_+}$ cations and $\nu_- A^{z-}$ anions, then we have $j = j_+ + j_-$, where j_+ and j_- denote the current densities of the cations and anions, respectively. These current densities may be related to the terminal velocities of the ions in a straightforward manner. Let us suppose that we place an aqueous solution containing N_e molecules of $C_{\nu_+} A_{\nu_-}$ in the conductance cell shown in Figure 9.17. We then apply a voltage across the cell that generates an electric field $\mathcal{E}$, which in turn produces a force on the individual ions that causes the cations and anions to move toward the negative and positive electrodes, respectively. The motion of the cations and anions then generates a current, the measurement of which permits the molar conductivity of $C_{\nu_+} A_{\nu_-}$ to be determined.

Let v_+ and v_- be the terminal velocities of the cations and anions, respectively, in the solution. During a short time dt, each cation will move a distance

$$d_+ = v_+ \, dt. \tag{9.113}$$

Therefore, if a cation is within a distance d_+ of the negative electrode at the start of the time interval, it will reach the electrode before the end of the interval. If N_+ is the total number of cations in the cell, then the number within a distance d_+ of the negative electrode will be $d_+ N_+/D = v_+ N_+ dt/D$. The total positive charge dQ_+ reaching the negative electrode in time dt is the

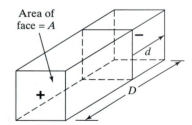

▲ **FIGURE 9.17**
Conductance cell of length D. Positive ions are moving toward the negative electrode, negative ions toward the positive electrode. The area of the electrode face is A.

product of this number and the charge carried by each cation, z_+e, where e is the charge on the electron, 1.602177×10^{-19} coulomb. That is,

$$dQ_+ = \frac{N_+ z_+ e\, v_+ dt}{D}. \tag{9.114}$$

The current produced by this charge transport is

$$I_+ = \frac{\partial Q_+}{dt} = \frac{N_+ z_+ e\, v_+}{D}. \tag{9.115}$$

The contribution to the current density from the positive ion flow is obtained by combining Eqs. 9.105 and 9.115; that is,

$$j_+ = \frac{I_+}{A} = \frac{N_+ z_+ e\, v_+}{DA} = \frac{N_+ z_+ e\, v_+}{V}, \tag{9.116}$$

since the volume of the conductance cell, V, is DA. The ratio of N_+ to V in Eq. 9.116 is the concentration of the positive ions in terms of number of ions per unit volume. If we divide the numerator and denominator by Avogadro's constant, N_A, the concentration units may be converted to moles L^{-1}, which we denote by c_+. These operations yield

$$j_+ = c_+ z_+ N_A e\, v_+. \tag{9.117}$$

A similar analysis for the anion current yields $j_- = c_- z_- N_A e\, v_-$, so that the total current density is

$$j = j_+ + j_- = c_+ z_+ N_A e\, v_+ + c_- z_- N_A e\, v_-. \tag{9.118}$$

If more than two ions are present in the solution, the total current density is obtained by summing expressions similar to Eq. 9.117 for each ion. We obtain

$$j = N_A e \sum_{i=1}^{i=K} c_i z_i v_i \tag{9.119}$$

where K is the number of different ions in the solution.

The quantity $N_A e$ appearing in Eqs. 9.117 through 9.119 occurs frequently in problems related to electrolytic solutions. Since N_A is the number of particles in a mole and e is the magnitude of the charge on the electron, $N_A e$ is the total charge carried by 1 mole of particles with unit charge. The value of this product is

$$N_A e = (6.02214 \times 10^{23}\ mol^{-1})(1.602177 \times 10^{-19}\ coulomb)$$

$$= 96{,}485\ coulombs\ mol^{-1}.$$

This quantity of charge is called the *faraday* and is commonly denoted with the symbol $\mathcal{F}$. In terms of the faraday, the total current density is

$$j = \mathcal{F} \sum_{i=1}^{i=K} c_i z_i v_i. \tag{9.120}$$

The force on the ions, and hence their terminal velocities, depends upon the magnitude of the electric field produced by the potential placed across the conductance cell. Measurements show that electrolytic solutions obey Ohm's law. Therefore, the specific conductivity is independent of the electric field. Combining Eqs. 9.106 and 9.120 gives

$$j = \mathcal{F} \sum_{i=1}^{i=K} c_i z_i v_i = \kappa\, \mathcal{E}.$$

Solving for κ, we obtain

$$\kappa = \mathcal{F} \sum_{i=1}^{i=K} c_i z_i \left(\frac{v_i}{\mathcal{E}} \right).$$

Since κ is independent of $\mathcal{E}$, the ratio $v_i/\mathcal{E}$ must be a constant characteristic of the ion itself; that is,

$$\frac{v_i}{\mathcal{E}} = u_i. \tag{9.121}$$

where the constant u_i is called the *electric mobility* of the ion. Since the units on v_i are cm s^{-1} or m s^{-1} and $\mathcal{E}$ is generally expressed in volts cm^{-1}(V cm^{-1}), the units on the electric mobility are cm^2 V^{-1} s^{-1}. In these terms, the specific conductivity of the solution is

$$\boxed{\kappa = \mathcal{F} \sum_{i=1}^{i=K} c_i z_i u_i}. \tag{9.122}$$

Equations 9.106 and 9.122 show that we can obtain the current density and conductance due to a particular ion in the solution if we know its concentration and electric mobility. If these quantities are known, we can also determine the fraction of the total current carried by each type of ion. This fraction, t, is generally called the *transference number* or *transport number* of the ion. The fraction for ion i is $t_i = j_i/j$. In terms of the electric mobilities given in Eq. 9.122,

$$t_i = \frac{\mathcal{F} c_i z_i u_i}{\mathcal{F} \sum_{i=1}^{i=K} c_i z_i u_i} = \frac{c_i z_i u_i}{\sum_{i=1}^{i=K} c_i z_i u_i}. \tag{9.123}$$

The expression for the transference number assumes a particularly simple form for a solution with only two types of ions. In that case, Eq. 9.123 becomes

$$\boxed{t_+ = \frac{c_+ z_+ u_+}{c_+ z_+ u_+ + c_- z_- u_-} = \frac{u_+}{u_+ + u_-}}, \tag{9.124}$$

since electrical neutrality requires that we have $c_+ z_+ = c_- z_-$. For the anion, we have a similar equation: $t_- = 1 - t_+ = u_-/(u_+ + u_-)$.

The preceding analysis shows that we have two options in measuring ionic conductances: Either we can measure their electric mobilities directly, or we can measure one mobility and one transference number. In practice, both methods are employed. Once the total ionic conductance is obtained by either of these methods, the molar ionic conductance can be computed by using a definition analogous to that for the molar conductance of the electrolyte given in Eq. 9.112. That is, we define the molar ionic conductance of ion i to be

$$\lambda_i = \frac{\kappa_i}{c_i}. \tag{9.125}$$

Using Eq. 9.122, we see that λ_i may be expressed in terms of the electric mobility as

$$\lambda_i = \mathcal{F} z_i u_i.$$

The molar conductance of the electrolyte, given by Eq. 9.112 is, therefore,

$$\boxed{\Lambda_m = \frac{\kappa}{c} = c^{-1} \sum_{i=1}^{K} \kappa_i = c^{-1} \sum_{i=1}^{K} c_i \lambda_i}. \tag{9.126}$$

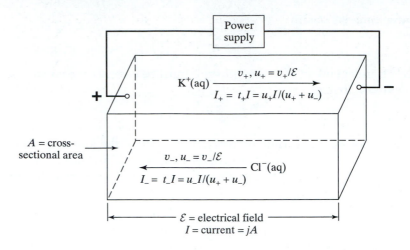

► FIGURE 9.18
Conductivity in an aqueous solution of KCl under the influence of an applied electric field $\mathcal{E}$. The figure shows the relationships between ionic velocity, mobility, transference number, electric field, cross-sectional area, current density, anionic current, I_-, cationic current I_+, and total current.

For a 1:1 electrolyte that is completely ionized in solution with no ion pairs, we will have $c_+ = c_- = c$, and hence, from Eq. 9.126, $\Lambda_m = \lambda_+ + \lambda_-$. For a 1:2 electrolyte, such as $BaCl_2$, for which the concentrations are $c_+ = c$ and $c_- = 2c$, Eq. 9.126 gives $\Lambda_{BaCl_2} = \lambda_{Ba^{2+}} + 2\lambda_{Cl^-}$.

Figure 9.18 summarizes the relationships between the ionic velocity, mobility, transference number, electric field, cross-sectional area, current density, and total current, using an aqueous solution of KCl as an example.

EXAMPLE 9.16

The measured electric mobilities of $K^+(aq)$ and $Cl^-(aq)$ at a KCl concentration of $0.20\ mol\ L^{-1}$ are, respectively, 62.9×10^{-5} and $65.6 \times 10^{-5}\ cm^2\ V^{-1}\ s^{-1}$ at 298.15 K. Compute the transference numbers and molar ionic conductances of $K^+(aq)$ and $Cl^-(aq)$. What is the molar conductivity of KCl at this concentration and temperature? Assume complete ionization of KCl and no ion pairing.

Solution

The transference numbers are given by Eq. 9.124:

$$t_{K^+} = \frac{u_{K^+}}{u_{K^+} + u_{Cl^-}} = \frac{62.9 \times 10^{-5}}{62.9 \times 10^{-5} + 65.6 \times 10^{-5}} = 0.489. \tag{1}$$

Since the sum of the transference numbers must be unity, we have

$$t_{Cl^-} = 1 - t_{K^+} = 1 - 0.489 = 0.511. \tag{2}$$

The ionic conductances are given by

$$\kappa_i = \mathcal{F}\, c_i\, z_i\, u_i. \tag{3}$$

With complete ionization and no ion pairing, we have

$$c_{K^+} = c_{Cl^-} = c = 0.20\ mol\ L^{-1} = 0.00020\ mol\ cm^{-3}. \tag{4}$$

The molar ionic conductances are given by Eq. 9.125:

$$\lambda_i = \frac{\kappa_i}{c_i} = \mathcal{F}\, z_i\, u_i. \tag{5}$$

Therefore,

$$\lambda_{K^+} = (96{,}485\ C\ mol^{-1})(1)(62.9 \times 10^{-5}\ cm^2\ V^{-1}\ s^{-1})$$

$$= (96{,}485\ amp\ s\ mol^{-1})(62.9 \times 10^{-5}\ cm^2\ V^{-1}\ s^{-1}) = 60.7\ amp\ cm^2\ V^{-1}\ mol^{-1}. \tag{6}$$

But an ohm (Ω) is a volt amp^{-1}, so that 1 amp V^{-1} = 1 Ω^{-1} = 1 S. Therefore,

$$\lambda_{K^+} = 60.7 \text{ S cm}^2 \text{ mol}^{-1}. \tag{7}$$

For the Cl$^-$ ion, we obtain

$$\lambda_{Cl^-} = (96{,}485 \text{ C mol}^{-1})(1)(65.6 \times 10^{-5} \text{ cm}^2 \text{ V}^{-1} \text{ s}^{-1}) = 63.3 \text{ S cm}^2 \text{ mol}^{-1}. \tag{8}$$

The molar conductivity is given by Eq. 9.126:

$$\Lambda_{KCl} = c^{-1}[c_{K^+}\lambda_{K^+} + c_{Cl^-}\lambda_{Cl^-}] = \lambda_{K^+} + \lambda_{Cl^-} = 60.7 + 63.3$$

$$= 124.0 \text{ S cm}^2 \text{ mol}^{-1}. \tag{9}$$

In Eq. 9, the concentrations cancel out because of the relationship given in Eq. 4. For related exercises, see Problems 9.37 and 9.38.

EXAMPLE 9.17

The KCl solution at 298.15 K described in Example 9.16 is placed in a conductivity cell such as that shown in Figure 9.17. If $A = 25$ cm^2, $D = 10$ cm, and the potential applied across the cell is 50 volts, compute the expected current density and the current in the cell. Assume complete ionization of KCl and no ion pairing.

Solution

Combining of Eqs. 9.106 and 9.120 gives the current density:

$$j = \mathcal{F} \sum_{i=1}^{i=K} c_i z_i v_i = \kappa \mathcal{E} = \frac{I}{A}. \tag{1}$$

For the KCl solution with complete ionization and no ion pairing, we have $c = c_{K^+} = c_{Cl^-} = 0.00020$ mol cm^{-3}. We also have $z_+ = z_- = 1$ and $v_i/\mathcal{E} = u_i$. Therefore, Eq. 1 becomes

$$j = \mathcal{F} \mathcal{E}(0.00020 \text{ mol cm}^{-3})[u_{K^+} + u_{Cl^-}]$$

$$= (96{,}485 \text{ C mol}^{-1})(50 \text{ V}/10 \text{ cm})(0.00020 \text{ mol cm}^{-3})[62.9 + 65.6] 10^{-5} \text{ cm}^2 \text{ V}^{-1} \text{ s}^{-1}$$

$$= 0.124 \text{ amp cm}^{-2}. \tag{2}$$

The current is the product of the current density and the area of the face of the electrode:

$$I = jA = 0.124 \text{ amp cm}^{-2} (25 \text{ cm}^2) = 3.10 \text{ amp}. \tag{3}$$

For a related exercise, see Problem 9.33.

It is possible to theoretically compute ionic electric mobilities and molar ionic conductivities for dilute solutions. At infinite dilution, the ions act independently, and the molar ionic conductivity approaches a limiting value for a given solvent at a fixed temperature. As the concentration increases, both the electric mobility and the molar ionic conductance decrease, due to coulombic interactions whose magnitude increases as the average distance between ions decreases. The increased coulombic interactions produce a relaxation effect and an electrophoretic effect, both of which serve to diminish the ion's mobility.

The relaxation effect is related to the counterion atmosphere that surrounds an ion. Equation 9.49 indicates that the total counterion charge equals that on the central ion and is a spherical shell of charge whose radius is κ^{-1}. This spherically symmetric counterion charge cloud produces no net attractive force on the central ion. However, when an electric field is applied to the solution, the central ion moves in one direction, the counterion atmosphere in the opposite direction. This motion distorts the spherical symmetry of the counterion atmosphere, producing a net electrical attraction between the asymmetric counterion atmosphere and the central ion that retards the motion of the central ion and thereby reduces its electric mobility.

The electrophoretic effect is due to the fact that the counterions are solvated. Because of this, they carry solvent molecules in a direction opposite to the motion of the central ion. As a result, the velocity of the central ion relative to the solvent is greater than its velocity relative to the stationary electrodes. Since frictional drag through a viscous medium increases with velocity, the net result of the central ion's greater velocity relative to the solvent is increased friction, which decreases the electric mobility of the ions in the solution.

Debye and Hückel applied their theory of electrostatic interactions in ionic solutions to compute the electric mobilities of ions. Onsager improved the treatment in 1927. The result of their efforts is the *Debye–Hückel–Onsager limiting law*, which quantitatively evaluates the retarding effects of both counterion relaxation and electrophoresis, thereby permitting the calculation of electric mobilities and molar ionic conductances for dilute solutions. For the case of a 1:1 electrolyte with $z_+ = z_-$, the limiting law is

$$\lambda_m = \lambda_m^\infty - \left[\left(2.92 \times 10^{-4} \frac{C^2 K^{1/2}}{\text{mol m}} \right) \left(\frac{z^3}{\eta(DT)^{1/2}} \right) + (5.80 \times 10^5 \, K^{3/2}) \left(\frac{z^3 \lambda_m^\infty}{(DT)^{3/2}} \right) \right] \left[\frac{2c}{c^o} \right]^{1/2},$$

$$\text{(9.127)}$$

where D and η are the dielectric constant and viscosity of the solvent, respectively, and c^o is 1 mol dm^{-3}, which is the same as 1 mol L^{-1}.

Some discussion of the units on the quantities in Eq. 9.127 is appropriate at this point. Viscosity is usually expressed in units of Pa s or mPa s. Since 1 Pa $= 1 N/m^2 = 1 \text{ kg m}^{-1} \text{s}^{-2}$, the SI base units on viscosity are kg m^{-1} s^{-1}. The dielectric constant is the ratio of the medium's permittivity to that of vacuum, so it is unitless. Consequently, the units on the first term inside the left most set of brackets in Eq. 9.127 are C^{2}K$^{1/2}$/(kg m^{-1} s^{-1} K$^{1/2}$ mol m), or C^2/(kg s^{-1} mol). We may write this unit in the form m^2 mol^{-1}/(kg m^2 s^{-1} C^{-2}). The discussion following Eq. 9.111 shows that the base SI units for the ohm (Ω) are kg m^2 s^{-1} C^{-2}. Therefore, the units on the first term within the leftmost set of brackets in Eq. 9.127 are Ω^{-1} m^2 mol^{-1}, which are the correct units for the molar ionic conductance. Hence, we need to express the solvent viscosity in Pa s and the concentration in mol L^{-1} in order to use the Debye–Hückel–Onsager limiting law.

If the temperature is 298.15 K and the solvent is water, we have $\eta = 0.000890$ Pa s $= 0.000890$ kg m^{-1} s^{-1} and $D = 78.54$. Substituting into the aforesaid first term in brackets produces

$$\frac{2.92 \times 10^{-4} z^3}{\eta(DT)^{1/2}} \frac{C^2 K^{1/2}}{\text{mol m}} = \frac{2.92 \times 10^{-4} z^3}{(0.000890)[(78.54)(298.15)]^{1/2}} \Omega^{-1} \text{ m}^2 \text{ mol}^{-1}$$

$$= 0.002144 \, z^3 \, \Omega^{-1} \text{ m}^2 \text{ mol}^{-1}.$$

The second term in the same set of brackets is

$$(5.80 \times 10^5 \, K^{3/2}) \left(\frac{z^3 \lambda_m^\infty}{(DT)^{3/2}} \right) = 0.1619 z^3 \, \lambda_m^\infty.$$

Multiplying these results by the factor $2^{1/2}$ contained in the rightmost set of brackets in Eq. 9.127 produces

$$\lambda_m = \lambda_m^\infty - [0.003032z^3 + 0.2290z^3\lambda_m^\infty]\left[\frac{c}{c^o}\right]^{1/2} \Omega^{-1}\,m^2\,mol^{-1} \quad \text{(9.128)}$$

for the limiting law in aqueous solutions at 298.15 K. If the molar ionic conductivity is expressed in units of $\Omega^{-1}\,cm^2\,mol^{-1}$, then the equation becomes

$$\lambda_m = \lambda_m^\infty - [30.32z^3 + 0.2290z^3\,\lambda_m^\infty]\left[\frac{c}{c^o}\right]^{1/2} \Omega^{-1}\,cm^2\,mol^{-1} \quad \text{(9.129)}$$

Experiment shows that the limiting law is accurate for $c \leq 0.001$ mol L^{-1}. At higher concentrations, the error increases.

The dependence of λ_m on $c^{1/2}$ given by the limiting law provides the theoretical justification for the empirical results obtained in Example 9.15, in which it was found that the molar conductance varies linearly with $c^{1/2}$. A full derivation of the limiting law may be found in Erying, Henderson, and Jost (Eds.), *Physical Chemistry: An Advanced Treatise*, Vol. IXA, *Electrochemistry* (New York: Academic Press, 1970).

EXAMPLE 9.18

Using the Debye–Hückel–Onsager limiting law, calculate the molar ionic conductance of $K^+(aq)$ in a 0.01-molar KCl solution at 298.15 K. The extrapolated molar ionic conductivity of $K^+(aq)$ at infinite dilution is 73.48 S cm^2 mol^{-1}. From your result for the molar ionic conductance, obtain the electric mobility of $K^+(aq)$ in the solution. The measured result is 7.18×10^{-4} cm^2 V^{-1} s^{-1}. Compute the percent error in the limiting law at the stated concentration.

Solution

Since we have an aqueous solution of a 1:1 electrolyte at 298.15 K, we may use Eq. 9.129 directly:

$$\lambda_m = \lambda_m^\infty - [30.32z^3 + 0.2290z^3\,\lambda_m^\infty]\left[\frac{c}{c^o}\right]^{1/2}$$

$$= 73.48 - [30.32(1)^3 + 0.2290\,(1)^3\,(73.48)](0.01)^{1/2}\,\Omega^{-1}\,cm^2\,mol^{-1}$$

$$= 68.76\,\Omega^{-1}\,cm^2\,mol^{-1}. \tag{1}$$

The ionic electric mobility is related to the ionic molar conductivity by

$$\lambda_m = \mathcal{F}\,z\,u. \tag{2}$$

Therefore,

$$u = \frac{\lambda_m}{\mathcal{F}z} = \frac{68.76\,\Omega^{-1}\,cm^2\,mol^{-1}}{96,485\,C\,mol^{-1}\,(1)} = 7.126 \times 10^{-4}\,\frac{cm^2}{\Omega\,C}. \tag{3}$$

Since an ohm is a volt per ampere and a coulomb is an ampere-seconds, $\Omega\,C = V\,s$. This gives

$$u = 7.126 \times 10^{-4}\,cm^2\,V^{-1}\,s^{-1}. \tag{4}$$

The percent error in this result is

$$\frac{100(7.126 \times 10^{-4} - 7.18 \times 10^{-4})}{7.18 \times 10^{-4}} = -0.75\%.$$

For related exercises, see Problems 9.34 and 9.35.

9.5.4 Measurement of Conductivity, Electric Mobility, and Transference Numbers

In principle, we may obtain the conductivity of an electrolytic solution by measuring its resistance with the use of a resistance bridge such as that illustrated in Figure 9.19. The solution whose conductivity is to be measured is placed in the conductance cell. The cell, whose resistance is R, forms one arm of the resistance bridge shown in the upper right of the figure. A potential difference, $\Delta\phi$ is applied across Points A–B, and the variable resistance R_2 is adjusted so as to produce a zero current (called a *current null point*) between Points C and D. In this situation, the electric potential at Points C and D must be equal. Let the current flowing through Points A–C–B be I_1 and that flowing through Points A–D–B be I_2. In order for the potential at Point C to be equal to that at Point D, the voltage drop IR from A to C must equal that from A to D. That is, we must have $I_1R_1 = I_2R_2$. For the same reason, the IR drop from B to C must equal that from B to D. Therefore, $I_1R = I_2R_3$. Dividing the first equation by the second yields

$$\frac{R_1}{R} = \frac{R_2}{R_3},\qquad(9.130)$$

so that the total resistance of the electrolytic solution is given by

$$R = \frac{R_1R_3}{R_2}.\qquad(9.131)$$

Substituting Eq. 9.111 into 9.131 gives

$$\frac{\rho d}{A} = \frac{d}{\kappa A} = \frac{R_1R_3}{R_2},\qquad(9.132)$$

where d and A are the length and cross-sectional area of the cell, respectively, as shown in Figure 9.19. The molar conductance of the solution can be obtained by solving Eq. 9.132 for the ratio κ/c:

$$\Lambda_m = \frac{\kappa}{c} = \frac{dR_2}{AcR_1R_3}.\qquad(9.133)$$

Although the foregoing procedure works in principle, it is more accurate to calibrate the cell by means of a solution of known conductivity than to attempt to measure d and A. That is, we solve Eq. 9.133 for the ratio d/A and

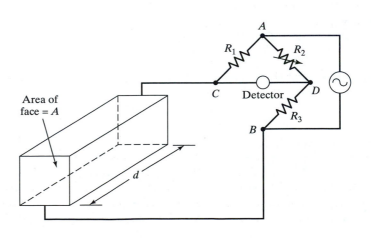

▶ FIGURE 9.19
Direct resistance measurement of an electrolytic solution in a conductance cell of length d and cross-sectional area A. In the resistance bridge, resistance R_2 is variable.

obtain its value from data taken on a solution whose molar conductivity is well known. This ratio is

$$\frac{d}{A} = \frac{\Lambda_m c R_1 R_3}{R_2} = C,$$

(9.134)

where C is called the *cell constant*. Once the value of C is determined, conductivities of other solutions can be obtained from Eq. 9.133 with d/A replaced by C.

EXAMPLE 9.19

The specific conductance of a 0.01-molar aqueous KCl solution at 298.15 K is 0.00141 S cm^{-1}. When this solution is placed in a conductance cell and its resistance is measured using a bridge, it is found that a current null point in the bridge is produced whenever $R_1 = 100\ \Omega$, $R_2 = 52.796\ \Omega$, and $R_3 = 150\ \Omega$. Determine the cell constant.

Solution

Using Eq. 9.132, we have

$$C = \frac{d}{A} = \frac{\kappa R_1 R_3}{R_2} = \frac{(0.00141\ \Omega^{-1}\ \mathrm{cm}^{-1})(100\ \Omega)(150\ \Omega)}{52.796\ \Omega} = 0.4006\ \mathrm{cm}^{-1}.$$

EXAMPLE 9.20

A 0.01000-molar solution of HCl is placed in the cell whose cell constant was determined in Example 9.19. Using a resistance bridge, we find that the resistance of the solution is 97.26 Ω. Determine the molar conductivity of HCl at the given concentration and temperature.

Solution

In terms of the cell constant, the specific conductivity is found by combining Eqs. 9.132 and 9.133:

$$\kappa = \frac{C}{R} = \frac{0.4006\ \mathrm{cm}^{-1}}{97.26\ \Omega} = 0.004119\ \Omega^{-1}\ \mathrm{cm}^{-1}.$$

(1)

The molar conductivity is given by Eq. 9.112:

$$\Lambda_m = \frac{\kappa}{c} = \frac{0.004119\ \Omega^{-1}\ \mathrm{cm}^{-1}}{0.01000\ \mathrm{mol\ L^{-1}}(1\mathrm{L}/1{,}000\ \mathrm{cm}^3)} = 411.9\ \Omega^{-1}\ \mathrm{cm}^2\ \mathrm{mol}^{-1}.$$

(2)

For related exercises, see Problems 9.36 and 9.37.

Electric mobilities of individual ions may be measured directly using the *moving-boundary method*. In this technique, two electrolytic solutions with one ion in common are layered, with the solution containing the ion with the greater electric mobility on top. The layering must be done carefully to prevent mixing of the solutions. The ions are then subjected to an electric field $\mathcal{E}$ that forces the anions and cations to move in opposite directions. The velocity of one of the ions is determined by measuring the movement of the boundary separating the two solutions as a function of time. The boundary can be located by measuring the change in the refractive index that occurs as one passes through the boundary region. If the two solutions have different acidities, a

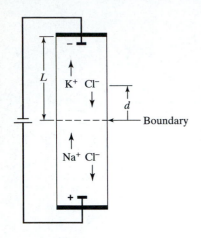

▲ FIGURE 9.20
Moving-boundary apparatus.

colored indicator can be used to mark the boundary position. Figure 9.20 shows a schematic diagram of a moving-boundary apparatus.

The configuration shown in the figure is arranged to permit measurement of $u(K^+)$. The two solutions have the chloride ion in common, and $u(K^+)$ is greater than $u(Na^+)$, so the KCl solution is above the NaCl solution. When an electric field is applied, the cations are forced upward toward the negative electrode, while the Cl^- ions move toward the positive terminal. The speed of the K^+ ions, $v(K^+)$, is determined by measuring the distance d the boundary moves in a time interval Δt. The speed is then given by $v(K^+) = d/\Delta t$. The electric mobility is obtained with the use of Eq. 9.121, $u(K^+) = v(K^+)/\mathcal{E}$. Since the electric field will be uniform throughout the KCl solution, we have $\mathcal{E} = |\Delta\Phi|/L$, where L is the length of the KCl solution and $|\Delta\Phi|$ is the magnitude of the potential difference across the solution. Using Eq. 9.108, we may relate the potential difference to the cross-sectional area A of the apparatus, the specific conductivity κ of the KCl solution, the current I, and the length L:

$$|\Delta\Phi| = \frac{IL}{\kappa A}. \tag{9.135}$$

Substituting this result into the preceding equations gives

$$u(K^+) = \frac{v(K^+)}{\mathcal{E}} = \frac{v(K^+)L}{|\Delta\Phi|} = \frac{\kappa A v(K^+)}{I} = \frac{\kappa A d}{I\Delta t}. \tag{9.136}$$

The product $I\Delta t$ is the total charge Q that flows through the solution. This quantity is conveniently measured with a coulometer. Therefore, the mobility of K^+ is

$$u(K^+) = \frac{\kappa A d}{Q}. \tag{9.137}$$

The specific conductivity of the KCl solution can be measured using a conductance cell, as previously described.

It is instructive to consider the moving-boundary experiment in more detail. The method permits the electric mobility of the ion with the greater mobility to be determined, but not that for the other ion. From Eq. 9.121, we know that

$$v(K^+) = u(K^+)\,\mathcal{E}(KCl) \text{ and } v(Na^+) = u(Na^+)\,\mathcal{E}(NaCl). \tag{9.138}$$

Substituting $|\Delta\Phi|/L$ for $\mathcal{E}$ and using Eq. 9.135 in 9.138, we obtain

$$v(K^+) = u(K^+)\frac{I}{\kappa(KCl)A} \text{ and } v(Na^+) = u(Na^+)\frac{I}{\kappa(NaCl)A}. \tag{9.139}$$

If the NaCl solution is chosen with a sufficiently low concentration to reduce $\kappa(NaCl)$ to a point where we have $v(Na^+) > v(K^+)$, we will also have $\mathcal{E}(NaCl) > \mathcal{E}(KCl)$. Under these conditions, the Na^+ ions will move faster than the K^+ ions. As a result, Na^+ ions will cross the solution boundary and enter the KCl solution. As soon as this occurs, the Na^+ ions will be acted upon by the lower electric field within the KCl solution. Since $u(Na^+) < u(K^+)$, Eq. 9.138 shows that once the boundary is crossed, the speed of the Na^+ ions will be reduced below that of the K^+ ions. The Na^+ ions will, therefore, recross the boundary back into the NaCl solution, where they again experience a higher field. Accordingly, their speed increases and the ions recross the boundary once more, only to be repelled by the lower electric field within the

KCl solution. The result of this continuous crossing–repelling process is a build-up of Na^+ ions at the solution boundary, compared with the concentration in the bulk of the NaCl solution. Consequently, a concentration gradient is established within the solution, and the electric field is, therefore, not uniform. Its value is unknown and we cannot obtain the electric mobility of Na^+ from the results of the experiment.

A second consequence of the crossing–repelling process is that the solution boundary remains sharp, which permits its position to be accurately measured. The motion of the common ion does not distort the boundary, since that ion is contained within both solutions.

If, on the other hand, the initial concentration of the NaCl solution is chosen so as to make κ(NaCl) large, Eq. 9.139 shows that $v(Na^+)$ will be less than $v(K^+)$ and consequently, the Na^+ ions will lag behind the K^+ ions, thus depleting the concentration of Na^+ ions near the boundary. As the concentration near the solution boundary decreases, the electric field within the NaCl solution in the vicinity of the boundary increases, thereby increasing $v(Na^+)$. This effect continues until we have $v(Na^+) = v(K^+)$, and once again, the boundary remains sharp.

If we wish to measure $u(Na^+)$, we must use a second solution whose cation has a lower electric mobility than that of Na^+. For example, we could make the upper solution in Figure 9.20, NaCl and the lower solution LiCl.

EXAMPLE 9.21

A moving-boundary apparatus with a cross-sectional area of 9 cm² contains a 1.000×10^{-2} M $BaCl_2$ solution above a $CaCl_2$ solution. When the $BaCl_2$ solution is placed in a conductance cell whose cell constant is 0.350 cm⁻¹, the resistance of the solution is found to be 141.24 Ω. An electric potential is applied across the moving-boundary apparatus such that a current of 5 A is flowing through the cell. After 200 seconds, the $BaCl_2$–$CaCl_2$ boundary has moved 26.17 cm. Determine the electric mobility of Ba^{2+} from these data.

Solution

The electric mobility of Ba^{2+} is given by Eq. 9.137:

$$u(Ba^{2+}) = \frac{\kappa A d}{Q}. \tag{1}$$

The total charge that has moved through the apparatus is

$$Q = I\,\Delta t = (5\text{ amp})(200\text{ s}) = 1{,}000\text{ C}. \tag{2}$$

The conductivity of the $BaCl_2$ solution can be obtained from

$$\kappa = \frac{C}{R} = \frac{0.350\text{ cm}^{-1}}{141.24\ \Omega} = 0.002478\ \Omega^{-1}\text{cm}^{-1} = 0.002478\text{ amp V}^{-1}\text{cm}^{-1}, \tag{3}$$

since an ohm is a volt per ampere. Substituting the data into Eq. 1 then produces

$$u(Ba^{2+}) = \frac{(0.002478\text{ amp V}^{-1}\text{cm}^{-1})(9\text{ cm}^2)(26.17\text{ cm})}{1{,}000\text{ amp s}} = 0.0005836\text{ cm}^2\text{V}^{-1}\text{s}^{-1}. \tag{4}$$

For a related exercise, see Problem 9.38.

Transference numbers may be obtained from measured electric mobilities as illustrated by Problem 9.37. They may also be measured directly using the

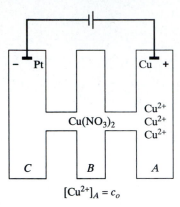

$[Cu^{2+}]_A = c_o$

(A) Initial state

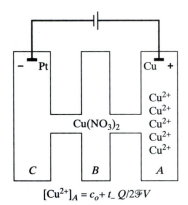

$[Cu^{2+}]_A = c_o + t_- Q/2\mathcal{F}V$

(B) Final state

▲ **FIGURE 9.21**
Hittorf apparatus for the measurement of transference numbers.
(A) Compartments A, B, and C all contain a c_o molar solution of $Cu(NO_3)_2$ at the start of the experiment. (B) After the passage of Q coloumbs of charge, Compartment A contains a $c_o + t_- Q/2\mathcal{F}V$ molar solution of Cu^{2+} ions.

Hittorf method. Figure 9.21 illustrates a typical Hittorf measurement of the transference number of Cu^{2+} in a $Cu(NO_3)_2$ solution. When an electric potential is placed across Compartments A and C, the apparatus acts as an electrolytic cell. Water is electrolyzed in the cathode compartment (C), with the net cell reaction being $2e^- + 2\,H_2O \longrightarrow H_2(g) + 2\,OH^-$. In the anode compartment (A), Cu from the electrode surface is oxidized to Cu(II) ion via the reaction $Cu \longrightarrow 2e^- + Cu^{2+}$. In the experiment, the amount of charge Q passed through the cell is measured with a coulometer. Since one mole of Cu^{2+} will be produced for every two faradays of charge passed through the cell, the number of moles of Cu^{2+} released into compartment A by the process will be $Q/2\mathcal{F}$. The Cu^{2+} and NO_3^- ions conduct the charge within the apparatus. The fraction of the total charge carried by the Cu^{2+} ions is t_+. Therefore, Cu^{2+} carries t_+Q of charge from compartment A toward the negative cathode into compartment B during the process. This requires that $t_+Q/2\mathcal{F}$ moles of Cu^{2+} pass from compartment A to compartment B. If the volume of compartment A is V liters and the initial concentration of Cu^{2+} in compartment A was c_o mol L^{-1}, then the number of moles of Cu^{2+} in the compartment after the passage of charge Q will be

$$n(Cu^{2+}) = \text{number of moles at the start} + \text{number of moles of } Cu^{2+}$$
$$\text{produced} - \text{number of moles of } Cu^{2+} \text{ that leave}$$

$$= Vc_o + \frac{Q}{2\mathcal{F}} - \frac{t_+Q}{2\mathcal{F}} = Vc_o + (1 - t_+)\frac{Q}{2\mathcal{F}}. \qquad (9.140)$$

The final concentration of Cu^{2+} in the compartment, from Eq. 9.140, is

$$c = \frac{n(Cu^{2+})}{V} = c_o + (1 - t_+)\frac{Q}{2\mathcal{F}V} = c_o + \frac{t_-Q}{2\mathcal{F}V} \qquad (9.141)$$

since $1 - t_+ = t_-$. Therefore, we need measure only the total number of coulombs of charge passed, the initial and final concentrations, and the volume of the compartment in order to determine the transference numbers.

During the experiment, care must be taken to avoid electrical heating of the solution by the passage of the current. This means that the current density should be low. Diffusion in liquids is very slow and will create no problem, provided that care is exercised not to mechanically disturb the solutions. There is, however, a more fundamental problem with the Hittorf method. Since the transference number can be expressed as a ratio of electric mobilities [see Eq. 9.124], and since electric mobilities depend upon the concentration of the material in question, it follows that transference numbers are also dependent upon the concentration. During the Hittorf experiment, the concentration of the cation in compartment A changes from c_o to c. Therefore, the resulting transference number is an average of transference numbers over the range of concentrations sampled during the experiment. For this reason, the experiment needs to be conducted in a manner that avoids large changes in concentration. Problem 9.40 explores this point in more quantitative detail.

9.5.5 Applications of Conductivity Measurements

Conductivity measurements have both analytical and thermodynamic applications. They can be utilized to determine the endpoint in titrations. For example, if we are titrating a solution of HCl with NaOH, the initial conductance of the solution is very large, because the H_3O^+ ion has an extraordinarily large

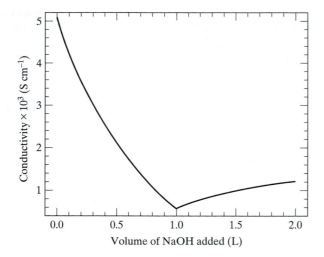

A conductiometric determination of the endpoint in the titration of 1 liter of 0.01 M HCl with 0.01 M NaOH. The point at which the slope of the conductivity as a function of the volume of NaOH added changes is the endpoint, as described in the text and illustrated in Problem 9.41.

ionic molar conductivity. As this ion is neutralized by the OH^- ion to form nonconducting H_2O, the conductivity decreases. At the endpoint, essentially all of the H_3O^+ is gone. The further addition of NaOH changes the concentra-

tion of charge carriers, thereby producing an abrupt change in the slope $d\kappa/dV$ of a plot of κ versus the volume V of NaOH added. The volume at which this change in slope occurs is the endpoint. Figure 9.22 shows the result of such a conductive determination of the endpoint when 1 liter of 0.01 M HCl is titrated with 0.01 M NaOH. Problem 9.41 examines this case in more detail.

Conductivity measurements also can be used to obtain ionization constants and solubility products. The procedure is straightforward. The system is permitted to reach equilibrium after which a conductivity measurement is made. Knowing κ, we solve the Debye–Hückel–Onsager equation to obtain the ionic concentrations, which are then used in the Debye–Hückel equation to compute the ionic activity coefficient. The combination of these data allows K_i or K_{sp} to be computed. Problems 9.42, 9.43, and 9.44 are examples of this application. The only problem is the solution of the Debye–Hückel–Onsager equation, which is cubic in the ionic concentration variable. This problem is minor since the solution may easily be obtained using a one-dimensional grid search on a small computer.

In principle, conductivity measurements can serve as the experimental probe in any situation where concentrations of ionic species are required. Table 9.2 lists selected values for λ^o at 298.15 K.

9.6 Electrochemistry

9.6.1 Qualitative Description of Galvanic and Electrolytic Cells

An oxidation–reduction reaction describes a chemical change in which there is a transfer of electronic charge from one reactant to another. This transfer produces changes in the formal oxidation numbers of some of the reactant atoms. Such a process is often called a *redox* reaction, which is pronounced (rē-däks) to avoid confusion with (red-äks), a reddish-colored beast of burden. A simple example of a redox reaction is

$$Zn + Cu^{2+} \longrightarrow Zn^{2+} + Cu.$$

Table 9.2   Ionic molar conductances at 298.15 K and infinite dilution

Ion	λ^o (S cm^2 mol^{-1})	Ion	λ^o (S cm^2 mol^{-1})
Ag^+	61.9	Br^-	78.1
Al^{3+}	183	CN^-	78
Ca^{2+}	118.94	Cl^-	76.31
Co^{2+}	110	ClO_4^-	67.3
Cr^{3+}	201	F^-	55.4
Cs^+	77.2	I^-	76.8
Cu^{2+}	107.2	NO_3^-	71.42
D^+	249.9	OH^-	198
Fe^{2+}	108	SO_3^{2-}	159.8
Fe^{3+}	204	SO_4^{2-}	160
H^+	349.65	NO_2^-	71.8
K^+	73.48	acetate	40.9
Li^+	38.66	benzoate	32.4
Mg^{2+}	106	formate	54.6
NH_4^+	73.5	oxalate	148.22
Na^+	50.08	picrate	30.37
Ni^{2+}	99.2	tartrate	119.2
Pb^{2+}	142		
Rb^+	77.8		
Zn^{2+}	105.6		

Source of Data: *Handbook of Chemistry and Physics*, 78th Ed., CRC Press, Boca Raton, FL, 1997–1998.

In this process, each zinc atom transfers two electrons to a Cu^{2+} ion, thereby forming neutral Cu while Zn^{2+} is produced as a product. In beginning chemistry courses, you learned that the compound or ion that loses electrons is said to be *oxidized.* The one that gains electrons is said to be *reduced.* The oxidized substance is called the *reducing agent,* in that it causes something else to be reduced, while the substance that is reduced is called the *oxidizing agent.*

Any redox reaction may be viewed as consisting of two *half-reactions:* an oxidation half and a reduction half. For the preceding reaction, the two half-reactions are

$$Zn \longrightarrow Zn^{2+} + 2e^- \quad \text{(oxidation)}$$

and

$$Cu^{2+} + 2e^- \longrightarrow Cu \quad \text{(reduction).}$$

Although we can never carry out a single half-reaction, it is convenient and often useful to view the redox process as consisting of these two parts.

In principle, any redox reaction can be used to produce electrical energy by setting up a galvanic cell or battery. Since there is a transfer of electronic charge, all that need be done is to physically separate the oxidizing and reducing agents so that the charge transfer is forced to occur along a conducting circuit connecting the two reactants. The galvanic cell that utilizes the foregoing Zn–Cu^{2+} reaction is called a *Daniel Cell;* it is illustrated in Figure 9.23. As seen in the figure, the oxidation half-reaction occurs in the left-hand compartment while the reduction process takes place in the right-hand compartment. The difference in chemical potentials for the two half-reactions drives the electrons through the external circuit from the point of their release in the Zn compartment to the Cu^{2+} compartment, thereby producing a current that could be utilized to drive a small motor or light the bulb in a flashlight. By definition, the electrode at which oxidation occurs is called the *anode.* The electrode at which we observe reduction is termed the *cathode.* To complete the circuit, provision must be made for charge transfer within the cell. In the figure, this is accomplished by means of a *salt bridge,* which is simply a tube connecting anode and cathode compartments that is filled with a concentrated conducting electrolytic solution. The figure shows an aqueous solution of NaCl being used. $Cl^-(aq)$ ions flow from the salt bridge into the anode compartment in sufficient number to maintain electrical neutrality with the newly formed Zn^{2+} ions. At the same time, $Na^+(aq)$ ions flow into the cathode compartment to replace the Cu^{2+} that is being removed to form free copper and thereby maintain the electrical neutrality of that compartment.

It is inconvenient to draw an elaborate diagram such as that shown in Figure 9.23 every time we wish to describe a galvanic cell. For that reason, chemists use a compact notation to describe such a cell. This notation for the Daniel cell with the connecting salt bridge is

$$Zn|ZnCl_2\,(m_1)||CuCl_2\,(m_2)|Cu,$$

where the single vertical line indicates materials in direct contact and the double vertical line represents the salt bridge. Sometimes the material contained in the bridge is specified between a pair of double lines. The notation (m_1) and (m_2) gives the molalities of the electrolytic solutions. Unless otherwise indicated, the solvent is assumed to be water. By convention, the contents of the anode compartment are written first.

Suppose we wish to produce a galvanic cell utilizing the redox reaction

$$6\,Fe^{2+}(aq) + Cr_2O_7^{2-}(aq) + 14\,H^+(aq) \longrightarrow 2\,Cr^{3+}(aq) + 6\,Fe^{3+}(aq) + 7\,H_2O.$$

The two half-reactions are

$$6e^- + Cr_2O_7^{2-}(aq) + 14\,H^+(aq) \longrightarrow 2\,Cr^{3+}(aq) + 7\,H_2O \quad \text{(reduction)}$$

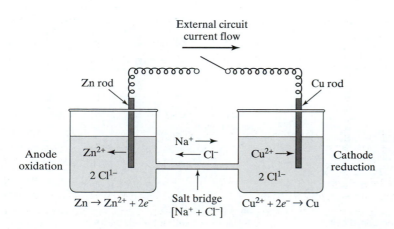

◀ **FIGURE 9.23**
The Daniel cell.

and

$$Fe^{2+}(aq) \longrightarrow Fe^{3+}(aq) + e^- \quad \text{(oxidation)}.$$

All we would need to do is place a solution of a soluble $Fe^{2+}(aq)$ salt in the anode compartment that contains an inert conducting electrode such as Pt. In the cathode, we would provide a $Cr_2O_7^{2-}(aq)$ salt, a source of $H^+(aq)$ ions which does not contain an anion that will react with the dichromate and an inert conducting electrode. The two compartments might then be connected by a concentrated salt bridge containing K_2SO_4. We would avoid NaCl, since $Cl^{1-}(aq)$ will react with the dichromate solution to form $Cl_2(aq)$. The cell diagram might be

$$Pt|FeSO_4\ (m_1),\ Fe_2(SO_4)_3\ (m_2)||K_2Cr_2O_7\ (m_3),\ H_2SO_4\ (m_4)|Pt,$$

or the material in the salt bridge could be specified by writing

$$Pt|FeSO_4\ (m_1),\ Fe_2(SO_4)_3\ (m_2)||K_2SO_4||K_2Cr_2O_7\ (m_3),\ H_2SO_4\ (m_4)|Pt.$$

A galvanic cell utilizes a spontaneous redox reaction to produce electrical energy. If a source of energy is connected across the electrodes of a cell such that energy is expended to force a nonspontaneous redox reaction to occur, the cell is called an *electrolytic* cell. In effect, this process reverses everything in a galvanic cell. The anode becomes the cathode, the cathode becomes the anode, the cell reaction reverses, and the current flows in the opposite direction. Figure 9.24 illustrates a Daniel cell being operated as an electrolytic cell. A comparison of that figure with Figure 9.23 shows that the overall cell reaction has been reversed by the presence of the power source in the external circuit. Zn^{2+} is now being reduced, while Cu is being oxidized. Consequently, the Zn compartment is now the cathode, while the compartment containing the copper solution functions as the anode. Negative charge is forced to flow in a counterclockwise direction.

We can also set up galvanic cells without using a salt bridge. In this case, electric conduction between the solutions in the anode and cathode compartments is achieved by permitting ions from each solution to diffuse into the other one. Figure 9.25 illustrates a Daniel cell without a salt bridge. Such an arrangement is termed a *galvanic cell with transference*, because ions from

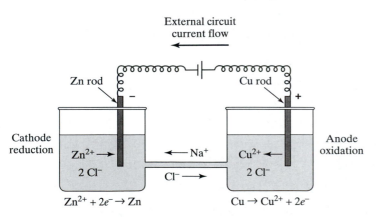

▲ FIGURE 9.24
An electrolytic cell that reverses the operation of the Daniel cell.

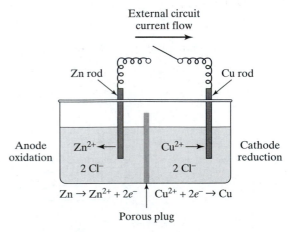

▲ FIGURE 9.25
A Daniel cell without a salt bridge.

one electrolyte solution transfer charge into the other solution. In effect, the cell functions like a Hittorf apparatus. The cell notation is

$$\mathrm{Zn}|\mathrm{ZnCl_2}\,(m_1)|\mathrm{CuCl_2}\,(m_2)|\mathrm{Cu},$$

since the electrolytic solutions are in physical contact. One result of such transference is the establishment of a junction potential at the solution boundary. This will be described in more detail in a subsequent section.

9.6.2   Thermodynamics of Electrochemical Systems

The derivation of the equations for dU, dH, dS, dA, and dG obtained in Chapters 2–5 implicitly assume that the only work being done is pressure–volume work for which $\delta w = -P_{ext}\,dV$. For a reversible path, the work could be written as $\delta w = -P\,dV$. In order to treat electrochemical systems, we must modify these equations so that they include the electrical work.

Whenever we move a charge dQ through an electric potential Φ, electrical work in the amount $\delta w = -\Phi\,dQ$ is done. The minus sign is included in the definition to be consistent with our convention that work done by the system on the surroundings is negative. If the potential is expressed in volts and the charge in coulombs, the work will be in joules. When the electrical work is included, the first law for a closed system becomes

$$dU = \delta q + \delta w = \delta q - P_{ext}\,dV - \Phi\,dQ. \qquad \textbf{(9.142)}$$

If the path is reversible, we have $\delta q = \delta q_{rev} = T\,dS$ and $P_{ext} = P$, so that

$$dU = T\,dS - P\,dV - \Phi\,dQ. \qquad \textbf{(9.143)}$$

If the system is open, we must include the changes in U due to variations in the composition of the system. Previously, we found that

$$\sum_{i=1}^{K} \hat{U}_i\,dn_i = \sum_{i=1}^{K} \mu_i\,dn_i$$

for any K-component system. Hence, the change in internal energy for an open system is

$$\boxed{dU = T\,dS - P\,dV - \Phi\,dQ + \sum_{i=1}^{K} \mu_i\,dn_i}. \qquad \textbf{(9.144)}$$

The differentials for dH, dA, and dG can be obtained directly from Eq. 9.144. Since $H = U + PV$, it follows that

$$\boxed{dH = dU + P\,dV + V\,dP = T\,dS + V\,dP - \Phi\,dQ + \sum_{i=1}^{K} \mu_i\,dn_i}. \qquad \textbf{(9.145)}$$

The Helmholtz free energy is defined to be $A = U - TS$. Therefore,

$$\boxed{dA = dU - T\,dS - S\,dT = S\,dT - P\,dV - \Phi\,dQ + \sum_{i=1}^{K} \mu_i\,dn_i}. \qquad \textbf{(9.146)}$$

Finally, the Gibbs free energy is defined as $G = H - TS$, so we obtain

$$\boxed{dG = dH - T\,dS - S\,dT = -S\,dT + V\,dP - \Phi\,dQ + \sum_{i=1}^{K} \mu_i\,dn_i}. \qquad \textbf{(9.147)}$$

Equations 9.144 through 9.147 permit the computation of changes in all of the thermodynamic potentials for open electrochemical systems.

Let us now apply Eq. 9.147 to a half-reaction occurring at the anode of an electrochemical cell. We might represent the general cell half-reaction as

$$\nu_A A + \nu_B B = \nu_C C + \nu_D D + ne^-,$$

where the ν_i are the stoichiometric coefficients for species A, B, C, and D in a half-reaction in which n moles of electrons are released at the anode for each ν_A moles of species A that react. When the circuit is open, the reaction will rapidly reach equilibrium. As negative electronic charge builds up on the anode, a potential difference Φ_a will be established between the electrolytic solution that is deficient in electronic charge and the surface of the electrode. Any charge transferred through this potential will result in electrical work being done.

At the equilibrium point, let $d\chi$ moles of species A react under conditions of constant temperature and pressure. For such a change, we have $d\nu_A = -d\chi$, $d\nu_B = -\nu_B d\chi/\nu_A$, $d\nu_C = \nu_C d\chi/\nu_A$, $d\nu_D = \nu_D d\chi/\nu_A$, and $dQ = n\mathcal{F} d\chi/\nu_A$, since n moles of electrons contains a total charge of n faradays. Substituting these expressions into Eq. 9.147 produces

$$dG = \frac{d\chi}{\nu_A}[\nu_C\mu_C + \nu_D\mu_D - \nu_A\mu_A - \nu_B\mu_B - \Phi_a n\mathcal{F}], \tag{9.148}$$

where the subscript a denotes the potential at the anode. At equilibrium at constant temperature and pressure, we must have $dG = 0$. Therefore,

$$[\nu_C\mu_C + \nu_D\mu_D - \nu_A\mu_A - \nu_B\mu_B]_{eq} - \Phi_a n\mathcal{F} = 0. \tag{9.149}$$

The chemical potentials appearing in Eq. 9.149 may be expressed in terms of the standard potentials and the corresponding activities; that is, $\mu_i = \mu_i^o + RT \ln a_i$. Substituting this expression into Eq. 9.149 gives

$$\Phi_a n\mathcal{F} = \Delta\mu^o + RT \ln \left[\frac{a_C^{\nu_C} a_D^{\nu_D}}{a_A^{\nu_A} a_B^{\nu_B}}\right]_{eq}. \tag{9.150}$$

The subscript eq emphasizes the fact that the activities must be evaluated at the equilibrium point.

For electrolytic solutions, the standard state corresponds to a hypothetical situation in which all solute concentrations are 1 molal with unit activity coefficients. In such a state, each of the activities appearing in Eq. 9.150 is unity, and we have

$$\Phi_a^o n\mathcal{F} = \Delta\mu^o, \tag{9.151}$$

where Φ_a^o denotes the electric potential present in the standard state at the anode. Substituting Eq. 9.151 into 9.150 and then dividing by $n\mathcal{F}$ yields

$$\boxed{\Phi_a = \Phi_a^o + \frac{RT}{n\mathcal{F}} \ln \left[\frac{a_C^{\nu_C} a_D^{\nu_D}}{a_A^{\nu_A} a_B^{\nu_B}}\right]_{eq}} \tag{9.152}$$

for any half-reaction at the anode.

We may apply the same analysis to a half-reaction at the cathode. However, care must be taken with the sign on the dQ term. Since the n moles of electrons appear on the reactant side of the cathode half-reaction, we will have $dQ = -n\mathcal{F}\,d\chi/\nu_A$. The inclusion of this additional minus sign in Eqs. 9.148 through 9.150 produces $-\Phi_c^o\,n\mathcal{F} = \Delta\mu^o$ and

$$\Phi_c = \Phi_c^o - \frac{RT}{n\text{F}}\ln\left[\frac{a_C^{\nu_C}\,a_D^{\nu_D}}{a_A^{\nu_A}\,a_B^{\nu_B}}\right]_{eq}. \tag{9.153}$$

Although Eqs. 9.148 to 9.153 are formally correct, there is no convenient experimental method for measuring the potentials associated with half-reactions. Fortunately, this presents no real difficulty, since we never observe oxidation without reduction and vice versa. That is, we are always measuring the *difference* in potential between the half-reaction at the anode and that at the cathode. This difference can be very accurately measured.

Let us apply Eqs. 9.152 and 9.153 to the case of the Daniel cell. For this cell, the reaction at the anode is $Zn \longrightarrow Zn^{2+} + 2e^-$. Applying Eq. 9.152 gives

$$\Phi_a = \Phi_a^o + \frac{RT}{2\mathcal{F}}\ln\left[\frac{a_{Zn^{2+}}}{a_{Zn}}\right]_{eq}. \tag{9.154}$$

The reaction at the cathode is $Cu^{2+} + 2e^- \longrightarrow Cu$. Therefore, from Eq. 9.153, we obtain

$$\Phi_c = \Phi_c^o - \frac{RT}{2\mathcal{F}}\ln\left[\frac{a_{Cu}}{a_{Cu^{2+}}}\right]_{eq}. \tag{9.155}$$

If we open the circuit and measure the potential between anode and cathode with no current flowing, we are actually measuring the difference

$$|\Phi_c - \Phi_a| = |\Phi_c^o - \Phi_a^o| - \frac{RT}{2\mathcal{F}}\ln\left[\frac{a_{Zn^{2+}}\,a_{Cu}}{a_{Zn}\,a_{Cu^{2+}}}\right]_{eq}. \tag{9.156}$$

The cell potential, Φ_{cell}, is $|\Phi_c - \Phi_a|$, and the standard cell potential, Φ_{cell}^o, is $|\Phi_c^o - \Phi_a^o|$. Furthermore, because of our choice of reference state, we know that the activities of the pure substances, Zn and Cu, will be unity. Consequently, Eq. 9.156 becomes

$$\Phi_{cell} = \Phi_{cell}^o - \frac{RT}{2\mathcal{F}}\ln\left[\frac{a_{Zn^{2+}}}{a_{Cu^{2+}}}\right]_{eq}. \tag{9.157}$$

Equation 9.157 is called the *Nernst equation* for a Daniel cell. Its more general form is

$$\Phi_{cell} = \Phi_{cell}^o - \frac{RT}{n\mathcal{F}}\ln\left[\frac{\Pi_p a_p^{\nu_p}}{\Pi_r a_r^{\nu_r}}\right], \tag{9.158}$$

where the capital Π denotes the product function. In the numerator, it runs over the activities of the cell reaction products and in the denominator, over the reactants. The factor $RT/\mathcal{F}$ at 298.15 K has the value $(8.314\text{ J mol}^{-1}\text{K}^{-1})(298.15\text{ K})/96,485\text{ C mol}^{-1} = 0.02569$ volts, so that Eq. 9.158 may be written in the form

$$\Phi_{cell} = \Phi_{cell}^o - \frac{0.02569}{n}\ln\left[\frac{\pi_p a_p^{\nu_p}}{\pi_r a_r^{\nu_r}}\right] = \Phi_{cell}^o - \frac{0.05916}{n}\log_{10}\left[\frac{\Pi_p a_p^{\nu_p}}{\Pi_r a_r^{\nu_r}}\right] \tag{9.159}$$

if the temperature is 298.15 K.

Note that Eq. 9.159 gives the cell potential only in the case when the circuit is open and no current is flowing. Under these conditions there is no *IR* drop across the circuit, and the half-reactions are both at equilibrium, so that Eq. 9.149 is valid. When the circuit is closed and current is drawn, the potential will deviate from that predicted by Eq. 9.159. The cell potential will also be affected by junction potentials. We shall defer discussion of such effects to a later section.

9.6.3   Standard Half-Reaction Potentials

As mentioned in the previous section, the half-reaction potential cannot be measured directly. However, since we are always measuring a difference in potential between the anode and cathode, we do not need to know the actual values of Φ_a^o and Φ_c^o. As usual when measuring a difference, we are free to choose our reference point in any manner that is convenient. For this purpose, we take the standard potential for the half-reaction

$$H_2(g)\,(1\text{ atm}) \longrightarrow 2\,H^+(m = 1\text{ molal}) + 2e^-$$

at 298.15 K to be zero. That is, Φ^o for the hydrogen electrode $Pt\,|H_2(g)|H_3O^+$ ($m = 1$ molal) at 298.15 K is assigned a value of zero.

Once the aforesaid reference point is set, it is a simple matter to obtain standard half-reaction potentials for all other possible half-reactions. Example 9.22 illustrates the procedure.

EXAMPLE 9.22

The voltage of the following galvanic cell is measured at 298.15 K using a bridge circuit, so that no current flows during the measurement:

$$Ag(s)|H_2(g)(P = 1\text{ bar}),\ HCl\,(m = 0.01\text{ molal}),\ AgCl(s)|Ag(g).$$

The voltage is found to be 0.4617 volt. Using these data, determine the standard half-reaction potential for $AgCl(s)|Ag(s)$. Ignore junction potentials.

Solution

The anode reaction is $0.5\,H_2(g)\,(P = 1\text{ bar}) \longrightarrow H^+\,(m = 0.01) + 1e^-$. At the cathode, the reaction is $AgCl(s) + 1e^- \longrightarrow Ag(s) + Cl^-(m = 0.01)$. The overall cell reaction is, therefore,

$$0.5\,H_2(g)\,(P = 1\text{ bar}) + AgCl(s) \longrightarrow HCl(m = 0.01\text{ molal}) + Ag(s).$$

From Eq. 9.159, the cell voltage is given by

$$\Phi_{\text{cell}} = \Phi_{\text{cell}}^o - \frac{0.02569}{n}\ln\left[\frac{a_{HCl}\,a_{Ag}}{f_{H_2}^{1/2}\,a_{AgCl}}\right]. \tag{1}$$

For this cell reaction, $n = 1$ and the fugacity of the H_2 gas may be equated to the magnitude of the pressure at this low value of the pressure. If we take the activities of the pure components, $Ag(s)$ and $AgCl(s)$, to be unity, Eq. 1 becomes

$$\Phi_{\text{cell}} = \Phi_{\text{cell}}^o - 0.02569\ln[a_{HCl}] = |\Phi_{AgCl|Ag}^o - \Phi_{H_2|H^+}^o| - 0.02569\ln[\gamma_\pm^2\,m^2]$$

$$= 0.4617\text{ volt}. \tag{2}$$

We may obtain a reasonably good answer by using the Debye–Hückel theory to compute the mean ionic activity coefficient for HCl at a concentration of 0.01 molal:

$$\log\gamma_\pm = -0.50926\,(1)(1)(0.01)^{1/2} = -0.050926. \tag{3}$$

This gives $\gamma_\pm = 0.88935$ and $\gamma_\pm^2 = 0.79095$. Using $\Phi^o_{H_2|H^+} = 0$, we obtain

$$\Phi^o_{AgCl|Ag} = 0.4617 + 0.02569 \ln[(0.79095)(0.01)^2] = 0.219 \text{ volt}. \quad \textbf{(4)}$$

A slightly better answer may be obtained by employing the measured mean ionic activity coefficient for HCl at $m = 0.01$ molal. This value is $\gamma_\pm = 0.905$, the use of which in Eq. 2 gives

$$\Phi^o_{AgCl|Ag} = 0.4617 + 0.02569 \ln[(0.905)^2(0.01)^2] = 0.220 \text{ volt}. \quad \textbf{(5)}$$

For a related exercise, see Problem 9.45.

Table 9.3 lists measured standard half-reaction potentials at 298.15 K.

Note that all the half-reactions in the table are written as if they are reductions occurring at the cathode. If the reverse reaction of the one listed is occurring at the anode, the standard half-reaction potential Φ^o_a will be the negative of the value listed.

With the sign convention adopted in the table, the cell potential may be written as

$$\Phi_{cell} = |\Phi_c - \Phi_a| = \Phi_c - \Phi_a. \quad \textbf{(9.160)}$$

In a similar manner, the difference in standard half-reaction potentials is

$$\Phi^o_{cell} = |\Phi^o_c - \Phi^o_a| = \Phi^o_c - \Phi^o_a. \quad \textbf{(9.161)}$$

A much more extensive listing of standard half-reaction potentials may be found in the *Handbook of Chemistry and Physics*.

9.6.4   Cells Requiring Single-ion Activities

The potential for some galvanic cells depends upon single-ion activities. The Daniel cell is a typical example. Equation 9.157 shows that the potential for this cell is

$$\Phi_{cell} = \Phi^o_{cell} - \frac{RT}{2\mathcal{F}} \ln\left[\frac{a_{Zn^{2+}}}{a_{Cu^{2+}}}\right]_{eq}.$$

The required activities may be expressed in terms of their respective activity coefficients to give

$$\Phi_{cell} = \Phi^o_{cell} - \frac{RT}{2F} \ln\left[\frac{\gamma_+(Zn^{2+})\, m(Zn^{2+})}{\gamma_+(Cu^{2+})\, m(Cu^{2+})}\right]_{eq}. \quad \textbf{(9.162)}$$

Therefore, we need the single-ion activity coefficients for both Zn^{2+} and Cu^{2+} in order to compute the cell voltage. Neither of these quantities can be measured, but if the solutions are sufficiently dilute, we may use Eq. 9.67B to calculate their values. Since both Zn^{2+} and Cu^{2+} have the same charge, the activity coefficient ratio is

$$\frac{\gamma_+(Zn^{2+})}{\gamma_+(Cu^{2+})} = \frac{-0.50926\,(2)^2\,S_a^{1/2}}{-0.50926\,(2)^2\,S_c^{1/2}} = \left[\frac{S_a}{S_c}\right]^{1/2}, \quad \textbf{(9.163)}$$

where S_a and S_b are the ionic strengths in the anode and cathode compartments, respectively. The next example shows how these results may be used to calculate the Daniel cell potential when the electrolyte solutions are sufficiently dilute to permit the Debye–Hückel theory to be used.

Table 9.3 Standard half-reaction potentials at 298.15 K

Half-reaction	Φ_c^o (volts)
$XeF + e^- \longrightarrow Xe + F^-$	3.4
$F_2 + 2H^+ + 2e^- \longrightarrow 2\,HF$	3.053
$F_2 + 2e^- \longrightarrow 2\,F^-$	2.866
$O(g) + 2H^+ + 2e^- \longrightarrow H_2O$	2.421
$O_3(g) + 2H^+ + 2e^- \longrightarrow O_2(g) + H_2O$	2.076
$Co^{3+} + e^- \longrightarrow Co^{2+}$	1.92
$H_2O_2 + 2H^+ + 2e^- \longrightarrow 2\,H_2O$	1.776
$N_2O + 2H^+ + 2e^- \longrightarrow N_2 + H_2O$	1.766
$PbO_2 + SO_4^{2-} + 4H^+ + 2e^- \longrightarrow PbSO_4 + 2\,H_2O$	1.6913
$MnO_4^- + 4H^+ + 3e^- \longrightarrow MnO_2 + 2\,H_2O$	1.679
$HClO_2 + 3\,H^+ + 3e^- \longrightarrow 0.5\,Cl_2 + 2\,H_2O$	1.628
$HClO + H^+ + e^- \longrightarrow 0.5\,Cl_2 + H_2O$	1.611
$MnO_4^- + 8\,H^+ + 5e^- \longrightarrow Mn^{2+} + 4\,H_2O$	1.507
$HClO + H^+ + 2e^- \longrightarrow Cl^- + H_2O$	1.482
$PbO_2 + 4\,H^+ + 2e^- \longrightarrow Pb^{2+} + 2\,H_2O$	1.455
$HCrO_4^- + 7\,H^+ + 3e^- \longrightarrow Cr^{3+} + 4\,H_2O$	1.350
$Cr_2O_7^{2-} + 14H^+ + 6e^- \longrightarrow 2Cr^{3+} + 7\,H_2O$	1.232
$MnO_2 + 4H^+ + 2e^- \longrightarrow Mn^{2+} + 2\,H_2O$	1.224
$Br_2(aq) + 2e^- \longrightarrow 2\,Br^-$	1.0873
$IO_{3^-} + 6H^+ + 6e^- \longrightarrow I^- + 3\,H_2O$	1.085
$Br_2(l) + 2e^- \longrightarrow 2\,Br^-$	1.066
$NO_3^- + 3\,H^+ + 2e^- \longrightarrow HNO_2 + H_2O$	0.934
$Ag^+ + e^- \longrightarrow Ag$	0.7996
$Fe^{3+} + e^- \longrightarrow Fe^{2+}$	0.771
$2NO + H_2O + 2e^- \longrightarrow N_2O + 2\,OH^-$	0.76
$ClO_3^- + 3\,H_2O + 6e^- \longrightarrow Cl^- + 6\,OH^-$	0.62
$MnO_4^- + 2H_2O + 3e^- \longrightarrow MnO_2 + 4\,OH^-$	0.595
$I_2 + 2e^- \longrightarrow 2I^-$	0.5355
$Cu^+ + e^- \longrightarrow Cu$	0.521
$O_2 + 2H_2O + 4e^- \longrightarrow 4\,OH^-$	0.401
$Cu^{2+} + 2e^- \longrightarrow Cu$	0.345
Calomel electrode, 0.1 molar KCl	0.3337
$Bi^{3+} + 3e^- \longrightarrow Bi$	0.308
Calomel electrode, Saturated KCl	0.2360
$AgCl + e^- \longrightarrow Ag + Cl^-$	0.22233
$Cu^{2+} + e^- \longrightarrow Cu^{1+}$	0.153
$Sn^{4+} + 2e^- \longrightarrow Sn^{2+}$	0.151
$AgBr + e^- \longrightarrow Ag + Br^-$	0.07133
$2H^+ + 2e^- \longrightarrow H_2$	0.00000
$Fe^{3+} + 3e^- \longrightarrow Fe$	−0.037

$Hg_2I_2 + 2e^- \longrightarrow 2\,Hg + 2\,I^-$	-0.0405
$Pb^{2+} + 2e^- \longrightarrow Pb$	-0.1262
$CrO_4^{2-} + 4\,H_2O + 3e^- \longrightarrow Cr(OH)_3 + 5\,OH^-$	-0.13
$Sn^{2+} + 2e^- \longrightarrow Sn$	-0.1375
$AgI + e^- \longrightarrow Ag + I^-$	-0.15224
$Ni^{2+} + 2e^- \longrightarrow Ni$	-0.257
$PbCl_2 + 2e^- \longrightarrow Pb + 2Cl^-$	-0.2675
$Co^{2+} + 2e^- \longrightarrow Co$	-0.28
$PbSO_4 + 2e^- \longrightarrow Pb + SO_4^{2-}$	-0.3588
$Cd^{2+} + 2e^- \longrightarrow Cd$	-0.4030
$Fe^{2+} + 2e^- \longrightarrow Fe$	-0.447
$Ga^{3+} + 3e^- \longrightarrow Ga$	-0.549
$Fe(OH)_3 + e^- \longrightarrow Fe(OH)_2 + OH^-$	-0.56
$Cr^{3+} + 3e^- \longrightarrow Cr$	-0.744
$Zn^{2+} + 2e^- \longrightarrow Zn$	-0.7618
$2H_2O + 2e^- \longrightarrow H_2 + 2\,OH^-$	-0.8277
$Se + 2e^- \longrightarrow Se^{2-}$	-0.924
$Mn^{2+} + 2e^- \longrightarrow Mn$	-1.185
$Ti^{3+} + 3e^- \longrightarrow Ti$	-1.37
$Cr(OH)_3 + 3e^- \longrightarrow Cr + 3\,OH^-$	-1.48
$Al^{3+} + 3e^- \longrightarrow Al$	-1.662
$H_2 + 2e^- \longrightarrow 2H^-$	-2.23
$Ce^{3+} + 3e^- \longrightarrow Ce$	-2.336
$Mg^{2+} + 2e^- \longrightarrow Mg$	-2.372
$Na^+ + e^- \longrightarrow Na$	-2.71
$Ra^{2+} + 2e^- \longrightarrow Ra$	-2.8
$Ca^{2+} + 2e^- \longrightarrow Ca$	-2.868
$Sr^{2+} + 2e^- \longrightarrow Sr$	-2.89
$Fr^+ + e^- \longrightarrow Fr$	-2.9
$Ba^{2+} + 2e^- \longrightarrow Ba$	-2.912
$Rb^+ + e^- \longrightarrow Rb$	-2.98
$Cs^+ + e^- \longrightarrow Cs$	-3.026
$Li^+ + e^- \longrightarrow Li$	-3.0401

Source of data: *Handbook of Chemistry and Physics*, 78th ed., CRC Press, Boca Raton, FL, 1997–1998.

EXAMPLE 9.23

Compute the potential for the following Daniel cell at 298.15 K:

$$Zn(s)|ZnCl_2(m = 0.01) \| CuCl_2 \,(m = 0.005)|Cu(s)$$

Ignore the effect of all junction potentials.

Solution

The electrolyte concentrations are sufficiently low in this example that we can expect to obtain reasonable results using the Debye–Hückel theory. Therefore, we have, from Eq. 9.162,

$$\Phi_{cell} = \Phi_{cell}^o - \frac{RT}{2\mathcal{F}} \ln\left[\frac{\gamma_+(Zn^{2+})\, m(Zn^{2+})}{\gamma_+(Cu^{2+})\, m(Cu^{2+})}\right]_{eq}$$

$$= \Phi_{cell}^o - \frac{0.02569}{2} \ln\left[\frac{S_a^{1/2}\, m(Zn^{2+})}{S_c^{1/2}\, m(Cu^{2+})}\right]_{eq}. \tag{1}$$

The ionic strengths are given by Eq. 9.64:

$$S = 0.5\, c_e \sum_{i=1}^{i=K} \nu_i z_i^2 = 0.5 \sum_{i=1}^{i=K} c_i z_i^2. \tag{2}$$

Therefore, in the anode compartment,

$$S_a = 0.5[(0.01)(2)^2 + (0.02)(1)^2] = 0.03, \tag{3}$$

since $m(Zn^{2+}) = 0.01$ and $m(Cl^{-1}) = 0.02$. In the cathode compartment, we have

$$S_c = 0.5[(0.005)(2)^2 + (0.01)(1)^2] = 0.015. \tag{4}$$

The standard cell voltage at 298.15 K is given by Eq. 9.161 and the data in Table 9.3:

$$\Phi_{cell}^o = |\Phi_c^o - \Phi_a^o| = \Phi_c^o - \Phi_a^o = 0.345 - (-0.7618)\text{ volts} = 1.107\text{ volts}. \tag{5}$$

Substituting into Eq. 1 produces

$$\Phi_{cell} = 1.107\text{ volts} - 0.012845 \ln\left[\frac{(0.03)^{1/2}(0.01)}{(0.015)^{1/2}(0.005)}\right] = 1.094\text{ volts}. \tag{6}$$

If the electrolyte concentrations are too large to permit the use of the Debye–Hückel theory to compute the single-ion activity coefficients, we cannot calculate the cell voltage. In addition, for concentrated solutions, we would need information on the extent of ion-pair formation.

9.6.5 Concentration Cells and Junction Potentials

Throughout much of the foregoing discussion, frequent reference has been made to junction potentials that exist whenever two dissimilar materials or solutions are in physical contact. We may see how such junction potentials are produced by first considering the following galvanic cell with transference:

$$Pt(s)|H_2(g)(P = P_1)|HCl\,(m_1)|HCl\,(m_2)|H_2(g)\,(P = P_1)|Pt(s).$$

Figure 9.26 shows a schematic setup for such a cell. As charge passes through this cell, $H_2(g)$ is oxidized in the anode compartment, while $2H^+$ is reduced to $H_2(g)$ at the cathode. In the external circuit, negative charge flows from anode to cathode. Within the cell, an equal amount of charge is conducted from cathode to anode. A fraction t_- of this charge is carried by the Cl^- ions as they move from the cathode to the anode compartment. At the same time, H^+ ions are moving from anode to cathode, conducting a fraction of the total charge equal to t_+. Our previous discussion of conductivity shows that t_- and t_+ are the transference numbers of Cl^- and H^+, respectively. The processes taking place within the cell are, therefore,

$$0.5\, H_2(g)\,(P = P_1) \longrightarrow H^+(m_1) + e^-, \tag{A}$$

$$t_+ H^+(m_1) \longrightarrow t_+ H^+(m_2), \tag{B}$$

$$t_- Cl^-(m_2) \longrightarrow t_- Cl^-(m_1), \tag{C}$$

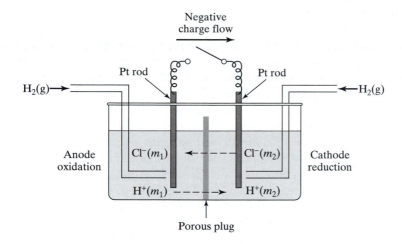

and

$$H^+(m_2) + e^- \longrightarrow 0.5\,H_2(g)(P = P_1), \qquad \textbf{(D)}$$

provided that the variation of transference numbers with concentration may be ignored. The total cell reaction is the sum of processes (A)–(D). Using Hess' law, we find this sum to be

$$t_+\,H^+(m_1) + t_-\,Cl^-(m_2) + H^+(m_2) \longrightarrow H^+(m_1) + t_+\,H^+(m_2) + t_-\,Cl^-(m_1).$$

Rearranging this chemical equation gives

$$H^+(m_2)\,[1 - t_+] + t_-\,Cl^-(m_2) \longrightarrow H^+(m_1)[1 - t_+] + t_-\,Cl^-(m_1).$$

Using the fact that $1 - t_+ = t_-$, we see that the overall cell reaction is

$$t_-\,HCl\,(m_2) \longrightarrow t_-\,HCl\,(m_1).$$

That is, the passage of one faraday of charge through the cell results in the transfer of t_- moles of HCl from the cathode to the anode compartment. Equation 9.159 shows us that the cell potential will be

$$\Phi_{cell} = \Phi^o_{cell} - 0.02569 \ln\left[\frac{a^{t_-}_{HCl(m_1)}}{a^{t_-}_{HCl(m_2)}}\right] = \Phi^o_{cell} - 0.02569 \ln\left[\frac{a_{HCl(m_1)}}{a_{HCl(m_2)}}\right]^{t_-}$$

$$= \Phi^o_{cell} - 0.02569\,t_- \ln\left[\frac{a_{HCl(m_1)}}{a_{HCl(m_2)}}\right]. \qquad \textbf{(9.164)}$$

Since $\Phi^o_{cell} = \Phi^o_c - \Phi^o_a$, we will have $\Phi^o_{cell} = 0$, because the standard half-reaction potentials are the same for both anode and cathode. Substituting this result and writing the activities in Eq. 9.164 in terms of the activity coefficients and molalities produces

$$\Phi_{cell} = -0.02569\,t_- \ln\left[\frac{\gamma^2_{\pm(1)}\,m_1^2}{\gamma^2_{\pm(2)}\,m_2^2}\right]. \qquad \textbf{(9.165)}$$

If Φ_{cell} is to be positive, we must have $a_{HCl(m_1)} < a_{HCl(m_2)}$. That is, the net flow of HCl must be from a region of high activity to one of low activity. Such a galvanic cell is called a *concentration cell*, since the potential is produced by the HCl concentration gradient that exists at the boundary of the anode and cathode compartments.

We can use Eq. 9.165 for a variety of purposes. If the transference number of Cl^- is known, we can make m_1 sufficiently small that $\gamma_{\pm(1)}$ can be computed from the Debye–Hückel theory and then use the measured value of the cell potential to obtain mean ionic activity coefficients for HCl as a function of the concentration m_2. Alternatively, if the activity coefficients are known, we may use the measured potential to obtain an approximate transference number for Cl^-. The result will be approximate because in deriving Eq. 9.165, we have assumed that t_- is independent of the concentration.

It is instructive to compute the approximate magnitude of the potential produced by this concentration cell for reasonable values of m_1 and m_2. At infinite dilution, t_- for Cl^- in HCl is 0.18. Since transference numbers are rather insensitive to concentration (see Problem 9.40), we will make only a small error by using this value at higher concentrations. Suppose $m_1 = 0.20$ mol kg^{-1} and $m_2 = 0.50$ mol kg^{-1}. The measured mean ionic activity coefficients for HCl at these concentrations are 0.768 and 0.759, respectively. Substituting of these values into Eq. 9.165 yields

$$\Phi_{cell} = -0.02569\,(0.18)\,\ln\left[\frac{(0.768)^2(0.20)^2}{(0.759)^2(0.50)^2}\right] = 0.00836 \text{ volt.}$$

Thus, we see that voltages produced by concentration gradients are on the order of millivolts.

The example presented in the previous paragraph demonstrates quantitatively the effect of junction potentials. When two electrolyte solutions are in physical contact so that diffusion between the solutions may occur, a liquid junction potential on the order of millivolts can be produced. Physically, this potential is the result of a difference in the diffusion rates of ions between the solutions that occurs because of differences in concentration and ionic mobility. In the example, H^+ ions in the more dilute solution have a higher mobility than those in the more concentrated solution. This difference enhances the diffusion rate from the solution with molality m_1 to that with molality m_2 (assuming that $m_1 < m_2$, as indicated in the example). On the other hand, the greater concentration of H^+ ions in the more concentrated solution increases the diffusion rate in the opposite direction. In general, these competing effects will produce a situation in which one solution gains a small excess of positive charge while the other acquires a corresponding excess of negative charge. This difference establishes a charge gradient and an electric potential that retards further diffusion of H^+ into the compartment with the excess of positive charge while increasing the diffusion rate of H^+ in the reverse direction. Eventually, a steady state is achieved at which the two diffusion rates are equal. The charge gradient that exists at the steady-state point is responsible for the liquid junction potential.

The same effect occurs when two different conductors are in physical contact. For example, if a copper rod is placed in contact with a zinc rod, a junction potential will be established between them. In this case, the charge gradient results because of the difference in the electronegativities of the two materials. Copper has a higher electronegativity (1.90) than zinc (1.65). This means that copper's attraction for electrons exceeds that of zinc. As a result, there is a net transfer of electronic charge from the zinc to the copper conductor that establishes a charge gradient and an electric potential. This type of junction potential is sometimes called a *contact potential*.

The only way to eliminate contact potentials between electrodes in electrochemical measurements is to employ identical electrode material for both the anode and cathode. Liquid junction potentials, however, can be nearly completely eliminated by the use of a concentrated salt bridge containing ions whose mobilities are nearly the same. If the electrolyte in the bridge is highly concentrated, the magnitude of the liquid junction potential will be determined almost solely by the gradients produced by diffusion of ions from the bridge into the dilute electrolytes in the anode and cathode compartments. If the cation and anion have the same mobilities, no charge gradient will occur, and hence, there will be no junction potential. If there is a small difference between cation and anion mobilities for the ions in the bridge, then a small liquid junction potential will be established between the anode compartment and the bridge. However, we will also have an equal, but opposite, junction potential established between the bridge and the cathode compartment. The two junction potentials can be expected to almost cancel. Galvanic cells with transference typically exhibit liquid junction potentials on the order of tens of millivolts. When a concentrated salt bridge is used, these junction potentials are reduced to 1 or 2 millivolts.

9.6.6   Applications of Electrochemical Measurements

9.6.6.1   Determination of Activity Coefficients

Since the activity coefficients appear in the expression for the potential of an electrochemical cell, we may, in principle, obtain their values by measuring the cell potential for a given concentration of electrolyte. Problems 9.47 through 9.49 are examples.

9.6.6.2   Potentiometric Titrations

Electrochemical measurement of cell potentials as a function of the volume of titrant added can serve as a convenient means of detecting the endpoint of a titration. This method is based on the characteristics of the derivative of the ln function. If we have $y = A \ln(ax)$, then $dy/dx = A/x$. Consequently, the derivative increases without bound as $x \rightarrow 0$, and the rate at which y is changing becomes very large (infinite) in this limit. The Nernst equation shows that Φ_{cell} is a logarithmic function of the concentrations of the reactants and products of the cell reaction. If one of these is being titrated, its concentration will continuously decrease as titrant is added. At the endpoint, its concentration will approach zero. As a result, the argument of the ln function will approach zero, and the magnitude of the rate of change of Φ_{cell} will become very large. Measuring this rate of change, therefore, affords a means of detecting the endpoint.

As an example, consider the titration of a solution of $SnCl_2$ with an acidic solution of $KMnO_4$. The titration reaction is

$$5\,Sn^{2+} + 2\,MnO_4^{1-} + 16\,H^+ \longrightarrow 5\,Sn^{4+} + 2\,Mn^{2+} + 8\,H_2O.$$

The endpoint of this reaction could be determined by measuring the voltage of an electrochemical cell in which the container holding the $SnCl_2$ solution is connected via a salt bridge to an $AgNO_3|Ag(s)$ cathode to form the cell

$$Ag(s)|Sn(NO_3)_2(m_1),Sn(NO_3)_4(m_2) \parallel AgNO_3(m_3)|Ag(s)$$

whose overall cell reaction is $Sn^{2+} + 2\,Ag^{1+} = Sn^{4+} + 2\,Ag$. Since the activity of pure solid Ag is unity, the potential generated by this cell is

$$\Phi_{cell} = \Phi_{cell}^o + \Phi_J - \frac{0.02569}{n}\ln\left[\frac{a(Sn^{4+})}{a^2(Ag^{1+})\,a(Sn^{2+})}\right],$$

Where Φ_J represents the junction potential. Rearranging this equation produces

$$\Phi_{cell} = \Phi_{cell}^o + \Phi_J + \frac{0.02569}{n}\ln[a^2(Ag^{1+})] + \frac{0.02569}{n}\ln\left[\frac{a(Sn^{2+})}{a(Sn^{4+})}\right].$$

As the titration proceeds, the argument of the ln function in the last term will become very small, because the concentration of Sn^{2+} will approach zero. Consequently, at the endpoint, the slope of a plot of Φ_{cell} versus the volume of titrant added will become large. The point at which the slope attains a maximum is the endpoint. Problem 9.50 examines the potentiometric titration of 1 liter of 0.01 M HCl with 0.01 M NaOH in the anode compartment of the concentration cell shown in Figure 9.26. Figure 9.27 shows the variation of the cell voltage as a function of the volume of NaOH added. As can be seen, there is an extremely rapid rise in this voltage as the endpoint point at $V_{NaOH} = 1.00$ L is approached.

9.6.6.3   Determination of Thermodynamic Quantities

The total change in the standard chemical potential for a cell reaction is the sum of that for the reaction at the anode and that for the reaction at the cathode:

$$\Delta\mu_{cell}^o = \Delta\mu_a^o + \Delta\mu_c^o. \tag{9.166}$$

Using Eq. 9.151 and the analogous expression for the cathode reaction, we can put Eq. 9.166 into the form

$$\Delta\mu_{cell}^o = n\mathcal{F}\Phi_a^o - n\mathcal{F}\Phi_c^o = -n\mathcal{F}[\Phi_c^o - \Phi_a^0] = -n\mathcal{F}\Phi_{cell}^o. \tag{9.167}$$

Consequently, once the standard half-reaction potentials have been determined, they can be combined in the manner described by Eq. 9.161 to compute

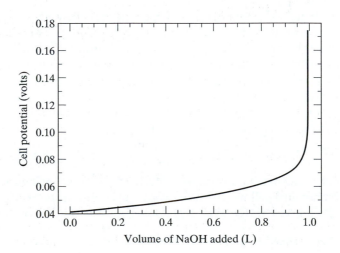

▶ FIGURE 9.27

A potentiometric determination of the endpoint in the titration of 1 liter of 0.01 M HCl with 0.01 M NaOH in the anode compartment of the concentration cell shown in Figure 9.26. Note the steep rise in the cell voltage as the endpoint at $V_{NaOH} = 1.00$ L is approached.

$\Delta\mu^o$ for the reaction. Using Eq. 9.82, we can then calculate the equilibrium constant for the reaction, since

$$K = \exp\left[-\frac{\Delta\mu^o}{RT}\right]. \tag{9.168}$$

We can also obtain the changes in the partial molar entropy and partial molar enthalpy when the cell reaction occurs with all reactants and products in their standard states. The variation of $\Delta\mu^o$ with temperature is

$$\left(\frac{\partial\Delta\mu^o}{\partial T}\right)_P = -\Delta\overline{S}^o. \tag{9.169}$$

Combining Eqs. 9.167 and 9.169 gives

$$\Delta\overline{S}^o = n\mathcal{F}\left(\frac{\partial\Phi^o_{cell}}{\partial T}\right)_P. \tag{9.170}$$

The change in the standard partial molar enthalpy can then be obtained using

$$\Delta\overline{H}^o = \Delta\mu^o + T\Delta\overline{S}^o = -n\mathcal{F}\Phi^o_{cell} + n\mathcal{F}T\left(\frac{\partial\Phi^o_{cell}}{\partial T}\right)_P. \tag{9.171}$$

In principle, the difference in heat capacities between reactants and products in their standard states can also be calculated, using

$$\Delta C^o_p = \left(\frac{\partial\Delta\overline{H}^o}{\partial T}\right)_P.$$

Equation 9.171 gives

$$\Delta C^o_p = -n\mathcal{F}\left(\frac{\partial\Phi^o_{cell}}{\partial T}\right)_P + n\mathcal{F}\left(\frac{\partial\Phi^o_{cell}}{\partial T}\right)_P + n\mathcal{F}T\left(\frac{\partial^2\Phi^o_{cell}}{\partial T^2}\right)_P,$$

so that

$$\Delta C^o_p = n\mathcal{F}T\left(\frac{\partial^2\Phi^o_{cell}}{\partial T^2}\right)_P. \tag{9.172}$$

The calculation of $\Delta\overline{S}^o$ and $\Delta\overline{H}^o$, therefore, requires that the variation of Φ^o_{cell} with temperature be measured. The determination of ΔC^o_p requires the second derivative or the rate of change of the first derivative of $(\partial\Phi^o_{cell}/\partial T)_P$ with temperature. It is generally difficult to obtain accurate derivatives, and extremely difficult to obtain second derivatives, from measured data. This is a consequence of the fact that if a measurement of y as a function of x contains p percent error, the error in the numerically extracted derivative dy/dx will be greatly magnified. The magnification of error continues as we attempt to extract the second derivative.

Since numerical derivatives are needed in numerous areas of physical chemistry, at this point we present two methods that permit such calculations to be made. The methods are similar in concept, but different in practice.

Suppose we have measured a quantity y at N different values of x and wish to obtain the derivative dy/dx at the point $x = x^*$. In the first procedure, we execute a linear least-squares fit of the data to the function

$$y = a_o + a_1(x - x_o) + a_2(x - x_o)^2 + a_3(x - x_o)^3 + \dots,$$

where x_o is any convenient expansion point. The Taylor series expansion can include as many terms as the data warrant. Once the expansion coefficients are determined using the method described in Chapter 1, the desired derivative is

$$\frac{dy}{dx}\bigg|_{x=x^*} = a_1 + 2a_2(x^* - x_o) + 3a_3(x^* - x_o)^2 + \cdots$$

If we choose $x_o = x^*$, then $(dy/dx)|_{x=x^*} = a_1$, and only a_1 need be determined in executing the least-squares fitting. The second derivative is obtained by direct differentiation of dy/dx. As noted, its accuracy will be very low unless the original data are exceptionally accurate.

The second procedure requires that we have values of y at equally spaced intervals for x. That is, we need the value of y at $x_1, x_1 + h, x_1 + 2h, \dots$, where the spacing between x values is h. Let y_n denote the value of y at the point $x = x^* + nh$. It can be shown (see Problem 9.53) that if the function $y = f(x)$ can be accurately represented by a polynomial in x of order 6 or less, then the derivative at the point $x = x^*$ is given exactly by

$$\boxed{\frac{dy}{dx}\bigg|_{x=x^*} = 0.75S_1 - 0.15S_2 + \frac{1}{60}S_3}, \qquad \textbf{(9.173)}$$

where

$$\boxed{S_1 = \frac{y_1 - y_{-1}}{h}}$$

$$\boxed{S_2 = \frac{y_2 - y_{-2}}{h}}$$

and

$$\boxed{S_3 = \frac{y_3 - y_{-3}}{h}}. \qquad \textbf{(9.174)}$$

Equations 9.173 and 9.174 provide a very convenient method for obtaining numerical derivatives, since only a few multiplications and divisions need be done to compute the desired derivative. Second derivatives can be calculated by using the method a second time on the first derivatives. That is, we simply replace y_i with the first derivative at that point. Usually, the method becomes increasingly accurate as h decreases. However, if h is made so small that the differences in Eq. 9.174 cannot be accurately evaluated, the accuracy of the method will *decrease* with decreasing h. Therefore, there is an optimum value of h that must be determined empirically.

It is important to bear in mind that once experimental data have been fitted to an assumed form by a least-squares method or smoothed in some manner prior to the use of Eqs. 9.173 and 9.174 to obtain derivatives, systematic and random errors in the data will be obscured because the data will now appear to be noise free. This can sometimes lead to situations in which too much confidence is placed in the data or the model.

EXAMPLE 9.24

Compute the equilibrium constant for the cell reaction taking place in the Daniel cell at 298.15 K.

Solution

The equilibrium constant is given by Eq. 9.168. $\Delta\mu^o$ can be obtained from Eq. 9.167. Using the data in Table 9.3, we obtain

$$\Delta\mu^o = -n\mathcal{F}\,\Phi^o_{cell} = -n\mathcal{F}\,[(0.345) - (-0.7618)]\text{ volts} = -1.107n\mathcal{F}\text{ volts}$$

$$= (-1.107\text{ volts})\,(2)\,(96{,}485\text{ C mol}^{-1}) = -2.136 \times 10^5\text{ volt C mol}^{-1}$$

$$= -2.136 \times 10^5\text{ J mol}^{-1}. \tag{1}$$

The corresponding equilibrium constant is

$$K = \exp\left[\frac{-\Delta\mu^o}{RT}\right] = \exp\left[\frac{2.136 \times 10^5\text{ J mol}^{-1}}{(8.314\text{ J mol}^{-1}\text{ K}^{-1})(298.15\text{ K})}\right] = 2.65 \times 10^{37}. \tag{2}$$

6.2.2.4   pH Measurements

In beginning courses, it is customary to define

$$\text{pH} = -\log_{10}[c(\text{H}^+)/c^o], \tag{9.175}$$

where $c(\text{H}^+)$ is the molarity of the hydrogen ion or hydronium ion concentration and c^o is 1 mol dm^{-3}, which is inserted to remove the units within the logarithm function. In fact, in many introductory texts, one finds the pH defined as $-\log_{10}[c(\text{H}^+)]$, totally ignoring the fact that the expression is meaningless, since the argument contains units. More properly, the definition given in Eq. 9.175 defines the *free-hydrogen ion scale,* which is often written as $\text{pH}_\text{F} = -\log_{10}[c(\text{H}^+)/c^o]$. In some applications, it may be useful to work with pH_F, but this quantity is not the pH, nor is it the quantity that is measured when one employs a pH meter.

The formal definition of pH is

$$\text{pH} = -\log_{10}[a(\text{H}^+)], \tag{9.176}$$

where $a(\text{H}^+)$ is the single-ion activity of the hydrogen or hydronium ion in solution. If we use m/m^o as our reference function and infinite dilution as the reference state, Eq. 9.176 may be written in the form

$$\text{pH} = -\log_{10}[\gamma_+ \, m(\text{H}^+)/m^o]. \tag{9.177}$$

Having defined the pH in this fashion, we immediately encounter a problem: Single-ion activities cannot be measured. If the solution is sufficiently dilute, they can be computed using the Debye–Hückel theory and Eq. 9.67B, but this does not solve our problem in the general case. Since it is impossible to determine the thermodynamic single-ion activity, we use a nonthermodynamic, operational definition of pH that is believed to yield results within ± 0.02 of Eq. 9.177.

The operational procedure first sets up the electrochemical cell

$$\text{Pt(Pd)(s)}|\text{H}_2(\text{g})\,(P = 1\text{ bar}),\,\text{H}^+(m_s)|\text{KCl (sat)}|\text{AgCl(s)}|\text{Ag(s)},$$

where the H^+ ion is that in a standard solution of molality m_s. In the most careful work, a 0.05-mol kg^{-1} solution of aqueous potassium hydrogen phthalate is used as the standard solution. The electrode in the anode compartment is a palladised-platinum rod that retards reaction of the hydrogen gas with the phthalate ion. The cell reaction is

$$0.5\, H_2(g)\,(P = 1\, bar) + AgCl(s) \longrightarrow Ag(s) + H^+(m_s) + Cl^-(aq).$$

If we include the junction potential Φ_J, the total measured cell potential will be

$$\Phi_{cell}^s = \Phi_{cell}^o + \Phi_J - \frac{RT}{\mathcal{F}} \ln\left[\frac{a(H^+(m_s))\, a(Cl^-)}{f^{1/2}(H_2(g))}\right].$$

provided that the reference states of the pure solids are chosen so as to make their activities equal to unity. We now set up the identical cell, but replace the standard solution with the one whose pH we desire to measure. If the H^+ molality in this solution is m_x, the cell potential will be

$$\Phi_{cell}^x = \Phi_{cell}^o + \Phi_J - \frac{RT}{\mathcal{F}} \ln\left[\frac{a(H^+(m_x))\, a(Cl^-)}{f^{1/2}(H_2(g))}\right].$$

The measurement rests on the difference between Φ_{cell}^x and Φ_{cell}^s, which is

$$\Phi_{cell}^x - \Phi_{cell}^s = -\frac{RT}{\mathcal{F}}\left[\ln\{a(H^+(m_x))\} - \ln\{a(H^+(m_s))\}\right].$$

Solving for $\ln[a(H^+(m_x))]$, we obtain

$$\ln[a(H^+(m_x))] = \ln[a(H^+(m_s))] - \frac{[\Phi_{cell}^x - \Phi_{cell}^s]\mathcal{F}}{RT}.$$

Dividing both sides of this equation by $-\ln 10 = -2.302585$ gives

$$-\log_{10}[a(H^+)_x] = -\log[a(H^+)_s] + \frac{[\Phi_{cell}^x - \Phi_{cell}^s]\mathcal{F}}{\ln(10)\, RT}$$

so that we have

$$\boxed{pH_x = pH_s + \frac{[\Phi_{cell}^x - \Phi_{cell}^s]\mathcal{F}}{\ln(10)\, RT}.} \qquad \textbf{(9.178)}$$

Equation 9.178 is the operational definition for the pH of solution x. For the equation to be useful, the value of the pH of the standard solution must be obtained. Values for this quantity are obtained by using the Bates–Guggenheim expression for single-ion activities. [R. G. Bates, *Measurement of pH: Theory and Practice*, 2d ed., John Wiley & Sons, New York, 1973; and A. K. Covington, R. G. Bates, and R. A. Durst, *Pure Appl. Chem.* **57**, 531, 1985.] At 298.15 K, pH_s is 4.005 when 0.05-molal potassium hydrogen phthalate is used as the standard solution. At that temperature, Φ_{cell}^s is 0.22234 volt. Values at other temperatures are given in the *Handbook of Chemistry and Physics*, 78th ed., CRC Press, Boca Raton, Florida, pages 8–36. Subsequent pages list other reference values.

In most laboratory measurements of pH, a glass electrode is used in place of the $Pt(Pd)|H_2(g)$ electrode for convenience. The fundamental measurement is the difference $\Phi_{cell}^x - \Phi_{cell}^s$, after which Eq. 9.178 can be used to determine the pH.

Summary: Key Concepts and Equations

1. If 1 mole of a strong electrolyte $C_{\nu_+} A_{\nu_-}$ ionizes completely to $\nu_+ C^{z_+}(aq) +$ $\nu_- A^{z_-}(aq)$ moles of ions, the chemical potential of the solute is given by

$$\mu_{solute} = [\nu_+ \mu_+^o + \nu_- \mu_-^o] + RT \ln[a_+^{\nu_+} a_-^{\nu_-}],$$

where μ_+^o and μ_-^o denote the standard chemical potentials of the cation and anion, respectively, and a_+ and a_- are the corresponding activities. If m/m^o with $m^o = 1$ mol kg^{-1} is taken as the reference function and infinite dilution as the reference state, the chemical potential becomes

$$\mu_{solute} = [\nu_+ \mu_+^o + \nu_- \mu_-^o] + RT \ln\left\{ \gamma_+^{\nu_+} \gamma_-^{\nu_-} \nu_+^{\nu_+} \nu_-^{\nu_-} \left[\frac{m}{m^o}\right]^{\nu_+ + \nu_-}\right\},$$

where m is now the molality of the $C_{\nu_+} A_{\nu_-}$ that is added. By defining the mean ionic activity coefficient and the mean stoichiometric coefficient as $\gamma_\pm^\nu = \gamma_+^{\nu_+} \gamma_-^{\nu_-}$ and $\nu_\pm^\nu = \nu_+^{\nu_+} \nu_-^{\nu_-}$, respectively, the solute activity can be written in the form

$$\mu_{solute} = [\nu_+ \mu_+^o + \nu_- \mu_-^o] + \nu RT \ln\left[\nu_\pm \gamma_\pm \left(\frac{m}{m^o}\right)\right].$$

2. By assuming that all deviations from ideality for charged solutes occur because of the electrostatic interactions between the ions, Debye and Hückel were able to show that for very dilute solutions, the ionic activity coefficient in aqueous solutions at 298.15 K is given by

$$\log(\gamma_i) = -0.50926 \, z_+ z_- \, S^{1/2},$$

where

$$S = 0.5 \sum_{i=1}^{i=K} c_i z_i^2$$

is the ionic strength. The summation runs over all the ions present in the solution, and c_i is the concentration of ion i in mol kg^{-1}. The corresponding single-ion activity coefficients are

$$\log \gamma_+ = -0.50926 \, z_+^2 \, S^{1/2}$$

and

$$\log \gamma_- = -0.50926 \, z_-^2 \, S^{1/2}.$$

3. Ionic activity coefficients can be measured using the Gibbs–Duhem equation and any convenient colligative property of the solution. If the freezing-point depression ΔT_f is used, the ionic activity coefficient is given by

$$\ln(\gamma_i) = -\left[1 - \frac{\Delta T_f}{K_f \nu m}\right] - \int_0^m \left[m^o - \frac{\Delta T_f}{K_f \nu m}\right]\frac{dm}{m}.$$

If, instead, we use the osmotic pressure π, the result is

$$\ln(\gamma_i) = -\left[1 - \frac{C\pi}{\nu m}\right] - \int_0^m \left[m^o - \frac{C\pi}{\nu m}\right]\frac{dm}{m}.$$

4. The equilibrium constant for any process is given by

$$K = \left[\frac{a_C^{\nu_C} a_D^{\nu_D}}{a_A^{\nu_A} a_B^{\nu_B}}\right]_{eq}$$

where C and D are products and A and B are the reactants in the equilibrium process. The ν_i are the stoichiometric coefficients. The subscript eq refers to the fact that the activities must be measured at the equilibrium point. The equilibrium constant is related to the change in the standard chemical potential by

$$\Delta\mu^\circ = -RT \ln(K).$$

Acid–base equilibria, hydrolysis, and the solubility of ionic compounds can all be examined quantitatively with the use of these equations, provided that the activities are properly evaluated and all important equilibrium processes are taken into account.

5. The electrical quantities of importance are as follows:

$$\text{Current } I: \qquad I = \frac{dQ}{dt},$$

where Q is the charge in coulombs, t is the time in seconds, and I is measured in amperes.

$$\text{Current density } j: \qquad j = \frac{I}{A} = \kappa\,\mathcal{E},$$

where A is the cross-sectional area through which the current passes, κ is the specific conductivity of the system, and $\mathcal{E}$ is the electric field in $N\,C^{-1}$ or volt m^{-1}.

$$\text{Electric potential } \Phi: \qquad \mathcal{E} = -\frac{\partial\Phi}{\partial x}.$$

If the distance is measured in meters and $\mathcal{E}$ in volt m^{-1}, Φ will be in volts.

$$\text{Resistance } R: \qquad R = \frac{L}{\kappa A},$$

where L is the length of the resistor and A is its cross-sectional area.

6. The molar conductivity Λ is defined in terms of the specific conductivity and the concentration of the solution, c:

$$\Lambda = \frac{\kappa}{c}.$$

If c is expressed in units of mol cm^{-3} and κ is in $\Omega^{-1}\,cm^{-1}$, Λ will be in $\Omega^{-1}\,cm^2\,mol^{-1}$ or S $cm^2\,mol^{-1}$. The specific conductivity of a solution is related to the mobilities μ of the ions present in the solution. The connecting equation is

$$\kappa = \mathcal{F} \sum_{i=1}^{i=K} c_i\,z_i\,u_i,$$

where $\mathcal{F}$ is the faraday. The molar ionic conductivity of ion i, is defined as

$$\lambda_i = \kappa_i/c_i.$$

λ_i may also be expressed in terms of the electric mobility:

$$\lambda_i = \mathcal{F}\,z_i\,u_i.$$

The molar conductance of the electrolyte is therefore

$$\Lambda = \frac{\kappa}{c} = c^{-1} \sum_{i=1}^{K} \kappa_i = c^{-1} \sum_{i=1}^{K} c_i\,\lambda_i.$$

7. For dilute solutions, the molar ionic conductivity may be computed using the Debye–Hückel–Onsager limiting law, viz.,

$$\lambda_m = \lambda_m^\infty - \left[\left(2.92\times10^{-4}\frac{C^2K^{1/2}}{\text{mol m}}\right)\left(\frac{z^3}{\eta(DT)^{1/2}}\right) + (5.80\times10^5\,K^{3/2})\left(\frac{z^3\lambda_m^\infty}{(DT)^{3/2}}\right)\right]\left[\frac{2c}{c^\circ}\right]^{1/2},$$

where D and η are the dielectric constant and viscosity of the solvent, respectively, and c° is 1 mol dm^{-3}, which is the same as 1 mol L^{-1}.

8. The change in the Gibbs free energy for a system in which electrical work is being done is

$$dG = dH - T\,dS - S\,dT = -S\,dT + V\,dP - \Phi\,dQ + \sum_{i=1}^{K} \mu_i\,dn_i.$$

Applying this equation to half-reactions at equilibrium in electrochemical cells shows that the cell potential will be given by the Nernst equation, provided that there are no junction potentials present:

$$\Phi_{cell} = \Phi_{cell}^o - \frac{RT}{n\mathcal{F}} \ln\left[\frac{\Pi_p\,a_p^{\nu_p}}{\Pi_r\,a_r^{\nu_r}}\right].$$

If junction potentials are present, these must be added to Φ_{cell} to obtain the total potential.

9. The change in a variety of thermodynamic quantities for the cell reaction in the standard state can be obtained by measuring the standard cell potential and its derivative with respect to temperature. The results are as follows:

Change in standard chemical potential: $\Delta\mu_{cell}^o = -n\mathcal{F}\Phi_{cell}^o,$

Equilibrium constant: $K = \exp\left[-\dfrac{\Delta\mu^o}{RT}\right],$

Change in partial molar entropy: $\Delta\bar{S}^o = n\mathcal{F}\left(\dfrac{\partial\Phi_{cell}^o}{\partial T}\right)_P,$

Change in partial molar enthalpy: $\Delta\bar{H}^o = -n\mathcal{F}\Phi_{cell}^o + n\mathcal{F}T\left(\dfrac{\partial\Phi_{cell}^o}{\partial T}\right)_P,$

Difference in heat capacity upon reaction: $\Delta C_P^o = n\mathcal{F}T\left(\dfrac{\partial^2\Phi_{cell}^o}{\partial T^2}\right)_P.$

10. The pH is defined as

$$pH = -\log_{10}[a(H^+)] = -\log_{10}\left[\frac{\gamma_+ m(H^+)}{m^o}\right],$$

where $a(H^+)$ is the single-ion activity of the hydrogen or hydronium ion in solution. Since single-ion activities cannot be measured, an operational definition of the pH is used. This definition is based on the measured difference in potential between two electrochemical cells, one of which contains a standard solution of 0.05-molal potassium hydrogen phthalate and the other the solution whose pH is to be measured. The pH of the latter solution is given by

$$pH_x = pH_s + \frac{[\Phi_{cell}^x - \Phi_{cell}^s]\mathcal{F}}{\ln(10)RT},$$

where the subscripts x and s represent the unknown and standard solutions, respectively. At 298.15 K, pH_s is 4.005 and Φ_{cell}^s is 0.22234 volt.

Problems

Problems that require the use of some type of computational device are marked with an asterisk (*). Problems that require some type of plotting routine are indicated with a pound sign (#).

9.1 Show that, for a one-to-one electrolyte such as NaCl with $\nu_+ = \nu_- = 1$, the mean ionic activity, the mean ionic activity coefficient, and the mean stoichiometric coefficient correspond to the geometric mean value of the individual ionic activities, activity coefficients, and ionic stoichiometric coefficients, respectively.

9.2 Twenty-five grams of $AlCl_3$ are dissolved in 2 kg of $H_2O(1)$. Assuming complete ionization and no ion-pair formation, obtain an expression for the chemical potential of the solute in terms of the standard chemical potentials, the temperature, and the mean ionic activity coefficient.

9.3 Suppose it is found that an aqueous solution of $BaCl_2$ completely dissociates into ions and that 10% of the Ba^{2+} ions form ion pairs with Cl^{1-} ions. Show that, under these conditions, we will have $\gamma_i = 1.2696\gamma_{\pm}$.

9.4 Solution concentrations are often measured in terms of their molarity c (moles of solute per liter of solution). Therefore, we need to be able to convert from activity coefficients for which the reference function is m/m° (γ_m) to an activity state in which the reference function is taken to be c/c°, with $c^{\circ} = 1$ mol L^{-1}, and the reference state is $c \to 0$ (γ_c). Obtain an expression for γ_c in terms of γ_m and the density of the solvent, d_o.

9.5 (A) Show that the solution of $\partial^2 F/\partial r^2 = -\kappa^2 F$ has the form $F = C_1 \sin[\kappa r] + C_2 \cos[\kappa r]$, where C_1 and C_2 are constants. (B) Show that we may also express this solution in the complex form $F = c_1 \exp[i\kappa r] + c_2 \exp[-i\kappa r]$, where i is the imaginary quantity $(-1)^{1/2}$. Obtain the relationship between the C_i and the c_i.

9.6 Consider three $MgCl_2$ solutions whose concentrations are 0.1, 0.05, and 0.005 molal in water at 293.15 K. Using the Debye formalism, compute the radius κ^{-1} associated with the hypothetical sphere of counterions surrounding a central Mg^{2+} ion for each of these solutions. Since the solution must be dilute in order for us to use the Debye–Hückel theory, assume that the dielectric constant is that for water at 293.15 K ($D = 78.54$).

9.7 Consider a spherical cavity of radius r_o. It is possible to set up spherically symmetric standing waves inside this cavity, which we shall see are very closely related to the quantum mechanical description of translational motion. The differential equation these standing waves must satisfy is

$$\nabla^2 \Psi = -\beta^2 \Psi.$$

The boundary conditions on the standing waves require that $\Psi = \beta$ when $r = 0$ and that $\Psi = 0$ at $r = r_o$. Using the methods employed to solve the Poisson equation, find the equation for the spherical symmetric standing waves inside the cavity.

9.8[#] The *Handbook of Chemistry and Physics* lists the following data for the mean ionic activity coefficients for aqueous solutions of $CaCl_2$ at 298.15 K:

(A) Compute the value of $\gamma_{\pm}$ predicted by the Debye–Hückel theory at each of these concentrations, and determine the percent error in the theory.

(B) Use Eq. 9.68 to compute $\gamma_{\pm}$ at each of the given concentrations, and determine the percent error in the equation.

(C) Prepare a plot showing all of the data on the same graph over the range of concentrations $0 \le m \le 2.0$ molal.

m (mol kg^{-1})	$\gamma_{\pm}$
0.001	0.888
0.002	0.851
0.005	0.787
0.010	0.727
0.020	0.664
0.050	0.577
0.100	0.517
0.200	0.469
0.500	0.444
1.000	0.495
2.000	0.784
5.000	5.907
10.000	43.1

9.9 The Debye–Hückel theory predicts that the mean ionic activity coefficient is a monotonically decreasing function of the ionic strength (i.e., as S increases, $\gamma_{\pm}$ steadily decreases). In contrast, the empirical relationship given in Eq. 9.68 exhibits a minimum when $\gamma_{\pm}$ is plotted against the molality of the solution. In many cases, the experimentally measured activity coefficients also exhibit such a minimum.

(A) Show that the minimum in $\gamma_{\pm}$ predicted by the equation occurs at the point when the molality of the solute is

$$c_e = \frac{0.789959 \dots}{\sum\limits_{i=1}^{i=K} \nu_i z_i^2}$$

(B) At what concentration would we obtain a minimum value of $\gamma_{\pm}$ for $CaCl_2$?

9.10 The dielectric constant of methanol (CH_3OH) at 298.15 K is 32.63.

(A) Obtain the form of the limiting Debye–Hückel law for methanol at 298.15 K?

(B) Obtain γ_i in methanol as a function of γ_i in H_2O.

9.11 An investigator determines the activity coefficients for a salt and finds the following data:
If the salt is represented by the molecular formula $C_{\nu_+} A_{\nu_-}$, what are the possible values of ν_+ and ν_-? Show that your answer is correct.

m (kg mol^{-1})	γ_i
0.00001	0.917
0.00010	0.761
0.00100	0.422

9.12 An investigator measures the freezing-point depression for various molalities of a salt. She then fits her data by least-squares techniques to a function of the form

$$\Delta T_f = \nu K_f[(m + a_2 m^2 + a_3 m^3 + a_4 m^4],$$

where m is the molality and the a's are the fitting coefficients. Obtain the activity coefficient of the salt as a function of ν, K_f, and the fitting coefficients.

9.13[#] An investigator chooses to employ a least-squares method to fit the freezing-point depression data given in Example 9.5 between $m = 0$ and $m = 0.02$ molal to an analytic function whose form is that given in Problem 9.12. The result using $\nu K_f = 5.58 \text{ K mol}^{-1}$ is

$$\Delta T_f = \nu K_f [\, m - 23.84066 m^2 + 1980.933 m^3$$
$$- 53995.094 m^4].$$

Use this equation to compute the activity coefficient for the solute at molalities between 0 and 0.02. Plot the results, and on the same plot, show those obtained from the numerical integration carried out in Example 9.5.

9.14 The problem with the procedure used in Problems 9.12 and 9.13 is that the method leads to the prediction that $\ln(\gamma_i)$ depends upon integral values of the molality, whereas the Debye–Hückel theory indicates that $\ln(\gamma_i)$ should depend upon $m^{1/2}$ at low values of m.

(A) Show that if we fit ΔT_f to the function

$$\Delta T_f = \nu K_f[m + b_1 m^{3/2} + b_2 \, m^2 + b_3 \, m^{5/2} + b_4 \, m^3],$$

then

$$\ln(\gamma_i) = \left[3b_1 m^{1/2} + 2b_2 m + \frac{5 b_3 m^{3/2}}{3} + \frac{3 b_4 m^2}{2} \right].$$

(B) Obtain an appropriate expression for b_1 such that the preceding expression for γ_i is in accord with the predictions of the Debye–Hückel theory.

9.15[#] The *Handbook of Chemistry and Physics* provides activity coefficients for $Ni(ClO_4)_2$. This problem demonstrates how we may extract freezing-point depression data from measured activity coefficients. In effect, the problem shows how to work the usual activity coefficient problem in reverse. The measured values of γ_i for $Ni(ClO_4)_2$ are as follows:

m (mol kg^{-1})	γ_i
0.000	1.000
0.001	0.891
0.002	0.855
0.005	0.797
0.010	0.745
0.020	0.690
0.050	0.621
0.100	0.582
0.200	0.567
0.500	0.639
1.000	0.946

In Problem 9.14, it was found that if freezing-point depression data are fitted to the function $\Delta T_f = \nu K_f [m + b_1 m^{3/2} + b_2 \, m^2 + b_3 \, m^{5/2} + b_4 \, m^3]$, the activity coefficient is given by

$$\ln(\gamma_i) = \left[3b_1 \, m^{1/2} + 2b_2 m + \frac{5 b_3 m^{3/2}}{3} + \frac{3 b_4 m^2}{2} \right],$$

where

$$3b_1 = -1.1726 z_+ z_- \, [0.5\{\nu_+ z_+^2 + \nu_- z_-^2\}]^{1/2}$$

if the result is to be compatible with the Debye-Hückel theory.

(A) For the $Ni(ClO_4)_4$ system, determine the value of $3b_1$ if the fitted results are to agree with the Debye–Hückel theory.

(B) Using the value of $3b_1$ determined in (A), we now conduct a least-squares fitting of the expression for $\ln(\gamma_i)$ to the data given in the foregoing table. The result of this fitting is $b_2 = 5.4184 \text{ mol}^{-1} \text{ kg}$, $b_3 = -7.50582 \text{ mol}^{-3/2} \text{ kg}^{3/2}$, and $b_4 = 3.78647 \text{ mol}^{-2} \text{ kg}^2$. Plot the fitted results for γ_i as a function of m. On the same graph, show the data obtained from the *Handbook of Chemistry and Physics*.

(C) Using the fitted results, obtain the expected freezing-point depression for solutions of $Ni(ClO_4)_2$ with molalities between zero and 1 molal. Plot ΔT_f versus m.

(D) At what temperature would an aqueous solution of $Ni(ClO_4)_2$ containing 69.55 grams of $Ni(ClO_4)_2$ per 1,000 grams of H_2O freeze under an applied pressure of 1 bar?

9.16 What is the H_3O^+ molality in a 0.15-molal aqueous solution of hydrofluoric acid (HF) at 298.15 K? K_a for HF at 298.15 K is 6.31×10^{-4}. You may use the Debye–Hückel expression to obtain the activity coefficients, if needed.

9.17 0.1 mole of formic acid ($HCHO_2$) and 0.15 mole of lactic acid ($HC_3H_5O_3$) are added to 1 kg of water at 298.15 K in the same container. The ionization constants are $K_a(HCHO_2) = 1.8 \times 10^{-4}$ and $K_a(HC_3H_5O_3) = 1.4 \times 10^{-4}$. Compute the H_3O^+ concentration in the solution. You may use the Debye–Hückel theory to compute mean ionic activity coefficients if needed.

9.18[#] Let us consider the case of a hypothetical weak acid HX whose ionization constant at 298.15 K is 1.0×10^{-10}.

(A) Compute the concentration of H_3O^+ in an aqueous solution of HX at 298.15 K when 0.10 mole of HX is added per kg of water ($C_o = 0.1$ mol kg^{-1}).

(B) Repeat (A) for $C_o = 0.01$ mol kg^{-1}, 0.005 mol kg^{-1}, 0.0001 mol kg^{-1}, 0.0005 mol kg^{-1}, and 0.0001 mol kg^{-1}.

(C) Obtain the answers to (A) and (B) under the condition that the contribution from the self-ionization of water is ignored for all cases. Calculate the percent error that results from ignoring this contribution.

(D) Make a careful plot of the percent error as a function of $K_w/(C_oK_a)$. What is the largest value $K_w/(C_oK_a)$ can have and not have an error greater than 10% when the contribution from the self-ionization of water is ignored? Activity coefficients, if needed, may be computed using the Debye–Hückel theory.

9.19 Compute the H_3O^+ concentration in mol kg^{-1} if 10^{-7} mol kg^{-1} of HCl is added to water at 298.15 K. Assume that the HCl is completely ionized and that the mean ionic activity coefficient is unity at these low concentrations of ionic species.

9.20 Compute the H_3O^+ concentration in mol kg^{-1} if 10^{-7} mol kg^{-1} of HCl is added to a water solution at 298.15 K containing 0.1 mol kg^{-1} of CaBr$_2$. Assume that HCl and CaBr$_2$ are completely ionized. Use the Davies approximation to compute activity coefficients if needed.

9.21 A quantity of 0.2 mol kg^{-1} of trimethylammonium chloride [$(CH_3)_3NHCl$] is added to water at 298.15 K. Compute the H_3O^+ concentration in the solution. Assume complete ionization of the $(CH_3)_3NHCl$ salt. K_b for $(CH_3)_3N$ is 6.25×10^{-10} at 298.15 K.

9.22 A quantity of 5×10^{-4} mol kg^{-1} of trimethylammonium chloride [$(CH_3)_3NHCl$] is added to water at 298.15 K. Compute the H_3O^+ concentration in the solution. Assume complete ionization of the $(CH_3)_3NHCl$ salt. K_b for $(CH_3)_3N$ is 6.25×10^{-10} at 298.15 K.

9.23 A quantity of 5×10^{-4} mol kg^{-1} of trimethylammonium chloride [$(CH_3)_3NHCl$] is added to a water solution at 298.15 K containing 0.2 mol kg^{-1} of NaCl. Compute the H_3O^+ concentration in the solution. Assume complete ionization of NaCl and $(CH_3)_3NHCl$. K_b for $(CH_3)_3N$ is 6.25×10^{-10} at 298.15 K.

9.24 A quantity of 0.2 mol kg^{-1} of trimethylammonium hypochlorite [$(CH_3)_3NHClO$] is added to water at 298.15 K. Compute the H_3O^+ concentration in the solution. Assume complete ionization of the $(CH_3)_3NHClO$ salt. K_b for $(CH_3)_3N$ is 6.25×10^{-10} at 298.15 K, and K_a for hypochlorous acid (HClO) is 3.0×10^{-8} at 298.15 K.

9.25 Using the data in Appendix A, compute the K_{sp} values for the following compounds at 298.15 K:

(A) calcite (CaCO$_3$(s));

(B) NaCl(s);

(C) BaCl$_2$(s).

9.26 The solubility product constant for calcite (CaCO$_3$(s)) is 4.94×10^{-9} at 298.15 K. Compute the solubility of CaCO$_3$(s) in H$_2$O at 298.15 K. You may use the Debye–Hückel theory to compute activity coefficients if needed.

9.27[#] The measured mean ionic activity coefficients for NaCl in aqueous solution at 298.15 K are reported in the *Handbook of Chemistry and Physics*. The following table shows the data at different concentrations.

m (mol kg^{-1})	$\gamma_\pm$
0.000	1.000
0.001	0.965
0.002	0.952
0.005	0.928
0.010	0.903
0.020	0.872
0.050	0.822
0.100	0.779
0.200	0.734
0.500	0.681
1.000	0.657
2.000	0.668
5.000	0.874

In Problem 9.14, it was shown that we may obtain an accurate fit of measured activity coefficients by using the function

$$\gamma_i = \exp\left[3b_1m^{1/2} + 2b_2m + \frac{5b_3m^{3/2}}{3} + \frac{3b_4m^2}{2}\right],$$

where, for compatibility with the Debye-Hückel theory, we must have

$$3b_1 = -1.17261z_+z_-[0.5\{\nu_+z_+^2 + \nu_-z_-^2\}]^{1/2}$$

When the data for NaCl are fitted to the function

$$\gamma_\pm = \exp[-1.17261m^{1/2} + c_2m + c_3m^{3/2} + c_4m^2]$$

using least-squares methods, the result is

$$\gamma_\pm = \exp[-1.17261m^{1/2} + 1.31903m$$
$$- 0.704482m^{3/2} + 0.150754m^2].$$

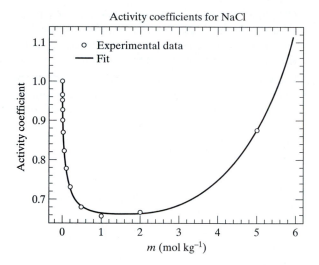

Activity coefficients for NaCl

▲ PROBLEM 9.27

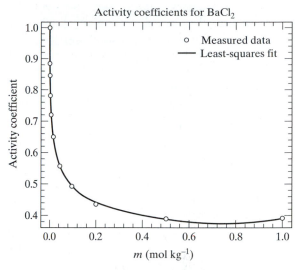

Activity coefficients for BaCl₂

▲ PROBLEM 9.28

The plot shows the quality of this fit. The solubility product constant for NaCl(s) is 37.74 at 298.15 K. Compute the solubility of NaCl(s) in H_2O at 298.15 K in terms of grams per kg of water. Use the fitted experimental data for NaCl to obtain activity coefficients if needed. The datum reported in the *Handbook of Chemistry and Physics* is 359.62 g NaCl per kg of water at 298.15 K. Compute the percent error in your answer.

9.28[#] The measured mean ionic activity coefficients for BaCl₂ in aqueous solution at 298.15 K are reported in the *Handbook of Chemistry and Physics*. The following table shows the data at different concentrations.

m (mol kg⁻¹)	$\gamma_\pm$
0.000	1.000
0.001	0.887
0.002	0.849
0.005	0.782
0.010	0.721
0.020	0.653
0.050	0.559
0.100	0.492
0.200	0.436
0.500	0.391
1.000	0.393

In Problem 9.14, it was shown that we may obtain an accurate fit of measured activity coefficients by using the function

$$\gamma_i = \exp\left[3b_1 m^{1/2} + 2b_2 m + \frac{5b_3 m^{3/2}}{3} + \frac{3b_4 m^2}{2}\right],$$

where, for compatibility with the Debye–Hückel theory, we must have

$$3b_1 = -1.17261 z_+ z_- \left[0.5\{\nu_+ z_+^2 + \nu_- z_-^2\}\right]^{1/2} = -4.06204$$

for BaCl₂.

When the data for BaCl₂ are fitted to the function

$$\gamma_\pm = \exp\left[-4.06204 m^{1/2} + c_2 m + c_3 m^{3/2} + c_4 m^2\right]$$

using least-squares methods, the result is

$$\gamma_\pm = \exp\left[-4.06204 m^{1/2} + 8.14202 m\right.$$
$$\left. - 8.62111 m^{3/2} + 3.60842 m^2\right].$$

The accompanying above plot shows the quality of this fit. The solubility product constant for BaCl₂(s) is 176.94 at 298.15 K. Compute the solubility of BaCl₂(s) in H_2O at 298.15 K in terms of grams per kg of water. Use the fitted experimental datum for BaCl₂ to obtain activity coefficients if needed. The data reported in the *Handbook of Chemistry and Physics* is 370.43 g BaCl₂ per kg of water at 298.15 K. Compute the percent error in your answer. What is the major source of the error?

9.29 The solubility product constant for BaCl₂(s) is 176.94 at 298.15 K. The measured solubility of BaCl₂(s) in water at the temperature is 370.43 g kg⁻¹ water. Determine the mean ionic activity coefficient for BaCl₂ at saturation.

9.30 The current through a conductor varies with time according to the equation

$$I = (5s^{-1})t \exp[-(0.2s^{-1})t] \text{ amps.}$$

Compute the total charge that flows through the conductor in 30 seconds.

9.31 The measured molar conductivities of AgNO₃ at 298.15 K are given in the table on the next page.

Data for Problem 9.31	
c (mol L^{-1})	Λ_m (S cm^2 mol^{-1})
0.0005	131.29
0.0010	130.45
0.0050	127.14
0.0100	124.70
0.0200	121.35

Using an appropriate method, obtain Λ_m^∞ for AgNO$_3$ in aqueous solution at 298.15 K.

9.32 The discussion in the text regarding the conductivity of weak electrolytes indicates that at infinite dilution, they are completely ionized. Let us consider acetic acid as an example. The ionization reaction is HC$_2$H$_3$O$_2$ + H$_2$O = H$_3$O$^+$ + C$_2$H$_3$O$_2^{1-}$. Let K_i be the ionization constant for this acid at the temperature of the solution. If m(HC$_2$H$_3$O$_2$) is the molality of the acetic acid in the absence of dissociation, show that ionization must approach 100% as m(HC$_2$H$_3$O$_2$) approaches zero.

9.33 The electric mobility of Mg^{2+}(aq) and Cl^{1-}(aq) at infinite dilution and 298.15 K are 55.0×10^{-5} cm^2 V^{-1} s^{-1} and 79.1×10^{-5} cm^2 V^{-1} s^{-1}, respectively. Compute the transference numbers and molar ionic conductances of Mg^{2+}(aq) and Cl^{1-}(aq) at infinite dilution. What is the value of $\Lambda_{\text{MgCl}_2}^\infty$ at 298.15 K?

9.34 Using the Debye–Hückel–Onsager limiting law, compute the ionic molar conductivity for Li^{1+}(aq) and Cl^{1-}(aq) ions in a 0.01-molar aqueous solution of LiCl at 298.15 K. The limiting ionic molar conductivities are $\lambda_{\text{Li}^+}^\infty = 38.66$ S cm^2 mol^{-1} and $\lambda_{\text{Cl}^-}^\infty = 76.31$ S cm^2 mol^{-1}. Use your results to compute λ_m(LiCl) at the given concentration. The value listed in the *Handbook of Chemistry and Physics* is 107.27 S cm^2 mol^{-1}. Calculate the percent error in your result.

9.35 Using the Debye–Hückel–Onsager limiting law, compute the ionic molar conductivity for Zn^{2+}(aq) and SO$_4^{2-}$(aq) ions in a 0.01-molar aqueous solution of ZnSO$_4$ at 298.15 K. The limiting ionic molar conductivities are $\lambda_{\text{Zn}^{2+}}^\infty = 105.6$ S cm^2 mol^{-1} and $\lambda_{\text{SO}_4}^\infty = 160.0$ S cm^2 mol^{-1}. Use your results to compute Λ_m(ZnSO$_4$) at the given concentration. The value listed in the *Handbook of Chemistry and Physics* is 169.74 S cm^2 mol^{-1}. Calculate the percent error in your result.

9.36 The molar conductivity of an aqueous 0.10-molar solution of AgNO$_3$ is 109.09 S cm^2 mol^{-1} at 298.15 K. When this solution is placed in a particular conductance cell, the resistance of the solution is found to be 35 Ω.

(A) Compute the specific conductivity of the AgNO$_3$ solution.

(B) Determine the cell constant for the conductivity cell.

(C) When a 0.02-molar aqueous solution of KBr is placed in the cell at 298.15 K, the resistance is found to be 135.96 ohms. Determine the molar conductivity of KBr at this concentration and temperature.

9.37 When a 0.005-molar aqueous solution of BaCl$_2$ at 298.15 K is placed in a conductivity cell whose cell constant is 0.280 cm^{-1}, its resistance is found to be 218.82 ohms. The solution is now placed above LiCl in a moving-boundary apparatus whose cross-sectional area is 25 cm^2. An electric potential is applied across the apparatus such that a current of 2.5 amperes flows. After 135 seconds have elapsed, how far has the BaCl$_2$–LiCl boundary moved if the electric mobility of Ba^{2+} in the solution is 0.0006029 V^{-1} cm^2 s^{-1}?

9.38 (A) Using the data given in Problem 9.37, calculate the transference number of the Ba^{2+} ion in a 0.005-molar solution of BaCl$_2$ at 298.15 K.

(B) Compute the electric mobility of the Cl$^-$ ion in the BaCl$_2$ solution.

9.39 A 0.01-molar aqueous solution of CuSO$_4$ at 298.15 K is placed in the Hittorf apparatus shown in Figure 9.21. A current of 0.1 amp is passed through the cell for 30 minutes, at which point the concentration of Cu^{2+} in the anode compartment is found to be 0.0286 molar. If the volume of the anode compartment is 30 ml, determine the transference numbers of Cu^{2+} and SO$_4^{2-}$.

9.40 The discussion of the Hittorf method noted that large changes in concentration need to be avoided during the experiment, since transference numbers vary with concentration. In this problem, we will assess the magnitude of this variation. The measured electric mobilities of the K^{1+}(aq) and Cl^{1-}(aq) in KCl

c (mol L^{-1})	u(K$^+$)	u(Cl$^-$)
0.00	0.000762	0.000791
0.01	0.000718	0.000746
0.10	0.000654	0.000682
0.20	0.000629	0.000656
1.00	0.000566	0.000593

solutions at 298.15 K are given in the following table in units of cm^2 V^{-1} s^{-1}.

Compute the K$^+$ transference number at each of the given concentrations.

9.41[#] One liter of 0.01-molar HCl is being titrated with 0.01-molar NaOH.

(A) Obtain expressions giving the composition of the solution in terms of the molarity of HCl, NaCl, and NaOH as a function of the volume V of NaOH added. Assume that the volumes are additive, and ignore the contribution of the self-ionization of H_2O.

(B) The molar conductivities of HCl, NaCl, and NaOH are given in the following table in units of $\Omega^{-1}\,cm^2\,mol^{-1}$. The Debye–Hückel–Onsager theory indicates that the molar conductivities should vary linearly at low concentrations with $c^{1/2}$. Plot the data sets given in the table against $c^{1/2}$, and obtain the best linear fit for each compound.

c (mol L^{-1})	Λ_{HCl}	Λ_{NaCl}	Λ_{NaOH}
0.000	425.95	126.39	247.7
0.0005	422.53	124.44	245.5
0.001	421.15	123.68	244.6
0.005	415.59	120.59	240.7
0.010	411.80	118.45	237.9

Data from the *Handbook of Chemistry and Physics*, 78th ed., CRC Press, Boca Raton, FL, 1997–1998.

(C) Using the results of (A) and (B), compute the conductivity of the solution as a function of V, the volume in liters of NaOH added, from $V = 0$ to $V = 2$ L. Assume that the conductivities of HCl, NaCl, and NaOH are additive at all concentrations. Plot the calculated conductivity versus V over the range $0 \leq V \leq 2$ L.

9.42* A 0.01-molar aqueous solution of formic acid (HCO_2H) at 298.15 K is placed in a conductivity cell whose cell constant is 0.460 cm^{-1}. The resistance of the solution is measured with a bridge and is found to be 884.11 Ω.

(A) Determine the specific conductivity of the solution.

(B) The limiting molar ionic conductance of H_3O^+ and the formate anion are 349.65 S cm^2 mol^{-1} and 54.6 S cm^2 mol^{-1}, respectively. Use the Debye–Hückel–Onsager equation to determine the concentration of H_3O^+ and $COOH^-$ in the given solution.

(C) Using the results of (A) and (B), compute the ionization constant for formic acid. The value reported in the *Handbook of Chemistry and Physics* is 1.8×10^{-4} at 298.15 K.

9.43 The ionization constant for HF in aqueous solution at 298.15 K is reported to be 3.5×10^{-4}. Use the Debye–Hückel theory and the Debye–Hückel–Onsager equation to estimate the specific conductivity of a 0.02-molar HF solution at 298.15 K. The molar ionic conductance of H_3O^+ and F^- at infinite dilution are listed in the *Handbook of Chemistry and Physics* as 349.65 S cm^2 mol^{-1} and 55.4 S cm^2 mol^{-1}, respectively.

9.44* AgCl is a slightly soluble salt in aqueous solution. A saturated aqueous solution of AgCl at 298.15 K is placed in a conductivity cell whose cell constant is

0.430 cm^{-1}. The resistance of the solution is found to be $2.33 \times 10^5\,\Omega$. The limiting ionic molar conductance of Ag^+ and Cl^- at 298.15 K and infinite dilution are 61.9 and 76.31 S cm^2 mol^{-1}, respectively. Using these data, compute K_{sp} for AgCl.

9.45 Using the data given in Table 9.3, determine the voltage that would be obtained from the galvanic cell

$$Pt|Zn|ZnSO_4\,(m = 0.005)\|PbSO_4(s)|Pb(s)$$

at 298.15 K if junction potentials are ignored. You may use the Debye–Hückel theory to estimate any activity coefficients needed.

9.46 (A) Show a schematic diagram of a galvanic cell whose cell reaction is

$$5\,Cd + 2\,M_nO_4^- + 16\,H^+ \longrightarrow 5\,Cd^{2+} + 2\,Mn^{2+} + 8\,H_2O$$

if a salt bridge is employed.

(B) Show a schematic diagram of a galvanic cell with transference and with the same cell reaction. Compute the standard electric potential for this cell at 298.15 K.

9.47 Consider a Daniel cell such as that shown in Figure 9.23 with KCl replacing the NaCl in the salt bridge to eliminate the liquid junction potential more completely. The cell potential at 298.15 K will, therefore, be that given in Eq. 9.159, plus the contact potential between the Cu and Zn electrodes, which we denote as Φ_J. The cell potential at 298.15 K, Φ_1, is now measured with a potentiometer under conditions of zero current flow when the concentrations of $ZnCl_2$ and $CuCl_2$ are m_o and m_1, respectively. A second cell potential at 298.15 K, Φ_2, is obtained in the same manner, with m_1 changed to m_2 for the same concentration of $ZnCl_2$.

(A) Obtain an expression giving the difference $\Phi_2 - \Phi_1$ in terms of m_1, m_2, and the corresponding Cu^{2+} activity coefficients.

(B) If m_1 is adjusted to 0.0001 mol kg^{-1} so that the Debye–Hückel theory may be accurately used to compute the single-ion activity coefficient, obtain an expression giving that coefficient when the molality of the $CuCl_2$ electrolyte is m_2.

(C) When $m_2 = 1.0$ mol kg^{-1}, the measured cell voltage is 0.0961 volt. Compute γ_+ for the Cu^{2+} ion.

9.48 The voltage of the galvanic cell

$$Pt|H_2(g)(P = 1\,bar),\,HCl\,(m = 1.000\,molal),$$
$$AgCl(s)|Ag(s)$$

is measured at 298.15 K using a bridge circuit so that no current flows during the measurement. The voltage is found to be 0.2411 volt.

(A) Determine the ionic activity coefficient for HCl with $m = 1.000$ mol kg^{-1} if the contact potential between the Pt and Ag electrodes is ignored.

(B) If the contact potential is -8 millivolts, what percent error does ignoring the contact potential introduce into the value of γ_{HCl}?

9.49 If we know the activity coefficient at one concentration, then the effect of junction potentials may be eliminated by working with the difference in voltage obtained at two concentrations. From previous measurements, the activity coefficient for HCl at $m = 0.500$ mol kg^{-1}, is known to be 0.759. The voltage of the galvanic cell

$$Pt|H_2(g)(P = 1 \text{ bar}), HCl(m), AgCl(s)|Ag(s)$$

is measured at 298.15 K with $m = 0.500$ mol kg^{-1} and again with $m = 2.000$ mol kg^{-1}, using a bridge circuit so that no current flows during the measurement. The cell voltage with $m = 2.000$ mol kg^{-1} is found to be higher by 0.08586 volt than the voltage for the cell with $m = 0.5000$ mol kg^{-1}.

(A) Show that the voltage difference is independent of junction potentials.

(B) Determine the activity coefficient for HCl at a concentration of 2.000 mol kg^{-1} at 298.15 K.

9.50# One liter of 0.01-molar HCl is being titrated with 0.01-molar NaOH.

(A) Obtain expressions giving the composition of the solution in terms of the molarity of HCl and NaCl as a function of the volume V of NaOH added. Assume that the volumes are additive and that, at these low concentrations, the molarity and molality of the solution are identical. Ignore the contribution from the self-ionization of water.

(B) If the titration is being carried out at 298.15 K in the anode compartment of the concentration cell shown in Figure 9.26 with $m_2 = 1$ mol kg^{-1} and $\gamma_{\pm(2)}$ equal to its measured value of 0.811, compute the cell voltage as a function of V. Use the Debye–Hückel theory to compute the required activity coefficients. Assume that the transference numbers are independent of concentration. Plot Φ_{cell} versus V over the range $0 \le V \le 1.0$ L.

9.51 Φ^o_{cell} for the cell

$$Ag(s)|H_2(g)(P = 1 \text{ bar}), HCl(m = 1.00 \text{ molal}),$$
$$AgCl(s)|Ag(s)$$

is measured at different temperatures. The data are fitted to a Taylor series expansion. The result of the fitting is

$$\Phi^o_{cell} = 0.22233 - 0.0006477(T - 298.15)$$
$$- 3.241 \times 10^{-6}(T - 298.15)^2 \text{ volts.}$$

Determine the equilibrium constant, $\Delta \bar{S}^o$, $\Delta \bar{H}^o$, and ΔC^o_p for the cell reaction at 280 K.

9.52 An investigator measures the standard potential Φ^o_{cell} for an electrochemical cell as a function of temperature over the range 280 K $\le T \le$ 310 K and obtains the data shown in the following table.

(A) Using these data, determine $\Delta \bar{S}^o$ for the cell reaction at 290 K, 292 K, 294 K, 298 K, 300 K, and 302 K. Assume that $n = 1$ for the cell reaction.

(B) Using the results obtained in (A), determine the value of ΔC^o_p at 296 K.

T(K)	Φ^o_{cell} volts	T(K)	Φ^o_{cell} volts
280	0.23302	296	0.22371
282	0.23195	298	0.22243
284	0.23085	300	0.22112
286	0.22972	302	0.21979
288	0.22857	304	0.21843
290	0.22739	306	0.21705
292	0.22619	308	0.21564
294	0.22496	310	0.21420

9.53 In this problem, you will prove the theorem embodied in Eqs. 9.173 and 9.174. The various steps in the problem serve as a road map for the proof.

(A) Let the function $y = f(x)$ be represented by a polynomial in x of order 6. That is, let $y = \sum_{i=0}^{6} a_i x^i$, where we will take the origin of the x-coordinate system to be at the point at which we wish to evaluate the derivative dy/dx. Show that dy/dx at the point $x = 0$ is equal to a_1.

(B) Let us assume that we know the value of y at the points $x = -3h, -2h, -h, h, 2h$, and $3h$. Let these values be denoted by $y_{-3}, y_{-2}, y_{-1}, y_1, y_2$, and y_3, respectively. Obtain expressions for each of the y_i in terms of h and the a_i.

(C) Using the result of (B), obtain expressions for S_1, S_2, and S_3 as defined in Eq. 9.174. Be certain to collect terms in the various powers of h.

(D) We now seek to find a linear combination of S_1, S_2, and S_3 that is exactly equal to the desired derivative, which we have shown in (A) to be equal to a_1. That is, we write

$$AS_1 + BS_2 + CS_3 = a_1 = \frac{dy}{dx} \text{ at the point } x = 0$$

and ask what must be the values of A, B, and C to make this equation correct. Using the results obtained in (C), find the values of A, B, and C, and show that Eq. 9.173 is correct as written.

9.54 An acidic solution X is placed in the anode compartment of the electrochemical cell

$$Pt(Pd)(s)|H_2(g)(P = 1 \text{ bar}),$$
$$H^+(m_x)|KCl(sat)|AgCl(s)|Ag(s),$$

and the voltage of the cell is measured at 298.15 K. This potential is found to be 0.16067 volt higher than that obtained using a 0.05-molal potassium hydrogen phthalate solution in place of solution X. What is the pH of solution X?

Quantum Mechanics and Bonding

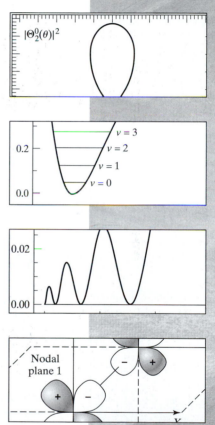

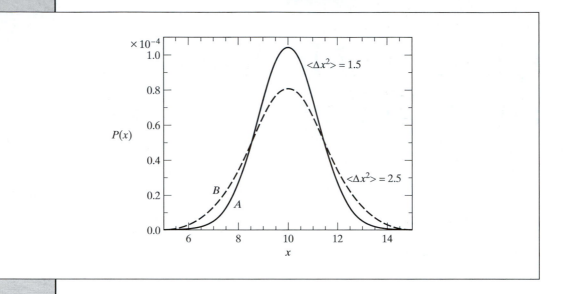

The Mathematics of Chance

Much of the remainder of this text deals with quantum mechanics, statistical mechanics, and the application of these theories to the treatment of chemical bonding, spectroscopy, and kinetics. Quantum theory and statistical mechanics are, in essence, probability theories. If you, the physical chemistry student, understands the mathematics and principles of probability, you will find it relatively easy to learn quantum and statistical mechanics. If, however, the opposite is true, these theories will appear to be mystical machinations that defy understanding.

When quantum theory was introduced by Schrödinger in 1926, Albert Einstein is reputed to have said, "God does not play dice!" The thrust of Einstein's remark was to express doubt that a fundamental description of the behavior of atomic and molecular systems would involve probability theory. Like most physicists of his day, Einstein favored a more definite, deterministic theory. However, while God may not play dice, atoms and molecules apparently do. At the present time, the best description of the laws governing atomic and molecular behavior is provided by quantum theory.

This chapter introduces the concepts of discrete and continuous probability distributions, normalization, the computation of average values, and uncertainty. The effect of measurement upon probability is discussed in detail. In many cases, the examples and problems are taken from topics that will be more fully treated in subsequent chapters. This emphasizes the close connection between probability theory and quantum and statistical mechanics.

10.1   Probability Distribution Functions: Discrete Variables

10.1.1   Normalized Probability Distributions

A variable is said to be *discrete* if it can have only certain predetermined values. A common example is the number that appears on the up side of a normal cubic die. This number can be only 1, 2, 3, 4, 5 or 6. A second example is the infinite set of all positive integers: 1, 2, 3, 4, 5, 6, 7, 8, … A variable is said to be *continuous* if it can take any value within the domain it spans. The set of points on the x-axis with positive values of the x-coordinate is an example of such a set. It can be proven that this set contains more members than the discrete set of positive integers; that is, the number of members of the continuous set is a higher order of infinity than that of the discrete set.

Let us consider a discrete variable X that can have M different values which we denote by $X_1, X_2, X_3, \ldots, X_M$. Suppose that the probability that a measurement of X will result in the value X_i is proportional to a quantity f_i, which we might call a probability factor. That is, if $P(X_i)$ is the probability that X will be equal to X_i, we have

$$P(X_i) \propto f_i. \tag{10.1}$$

The set of $P(X_i)$ for $i = 1, 2, 3, \ldots, M$ is called the *probability distribution*.

Before proceeding, it is worthwhile to consider some methods whereby we might determine the values of f_i for $i = 1, 2, 3, \ldots, M$. Although our discussion of probability does not require that we know how these values are obtained, such knowledge will facilitate an understanding of the concepts. In many cases, we shall find that there exist explicit mathematical formulae that permit us to compute values for all the f_i. This will be the case in particular when we turn our attention to quantum mechanics and statistical mechanics. An example that we have already encountered involves Raoult's law. Suppose we wish to obtain the probability that a randomly selected molecule in a binary A–B solution will be an A molecule. This probability will be proportional to the mole fraction of A molecules in the solution. Raoult's law tells us that $X_A = P_A/P_A^o$. Therefore, if the solution is ideal, we can confidently take f_A to be equal to P_A/P_A^o.

In simple cases, we can determine an appropriate set of probability factors by counting the number of ways X can take on the value X_i. For example, in the case of a cubic die, there is exactly 1 way we can obtain an up face with one dot, 1 way to obtain an up face with two dots, etc. In fact, there is exactly one way to obtain *any* given number of dots on the up face of the die. An appropriate set of f_i values is, therefore, $f_1 = f_2 = f_3 = f_4 = f_5 = f_6 = 1$. A second example of using a simple count to obtain f_i values may be found in card games in which all face cards and 10's have a value of 10, while the remaining cards have a value equal to the number of spots on the face of the card. Therefore, the number of ways to obtain a count of 1 is 4, since there are four aces with a single spot in a normal deck of 52 cards. There are also 4 ways to obtain counts of 2 through 9. However, since there are 12 face cards and four 10's, there are 16 ways to obtain a count of 10. As a result, an appropriate set of probability factors for such games could be $f_1 = f_2 = f_3 = f_4 = f_5 = f_6 = f_7 = f_8 = f_9 = 4$ and $f_{10} = 16$.

Many methods for obtaining values of the f_i are approximations that are based upon information related to the event whose probability is desired. For example, a person at a race track who intends to place a wager on a horse race will often consult a racing form that gives the entire history of every horse entered in each race. The information includes the performance records of all horses in all previous races, the sire and mare (parents) of each horse in the race, and the racing records of the jockeys riding the horses. The basic idea is to utilize this information to obtain an estimated set of f_i factors that provide direction as to which horse to select in each race. A similar situation arises when the local weatherman or weatherwoman utilizes information on pressure fronts, temperatures, wind velocities, and precipitation maps to obtain probability factors for the weather during the coming week.

Statistical sampling is a well-known method for obtaining an approximate set of f_i values. Suppose there are five candidates seeking the presidential nomination from a particular party. Prior to the primary elections and nominating convention, many organizations will ask a small group of potential voters which candidate they prefer. If 137 select Candidate 1, 252 Candidate 2, 75 Candidate 3, 369 Candidate 4, and 167 Candidate 5, a set of f_i values that are approximately proportional to the probability that candidate i will win the nomination are $f_1 = 137$, $f_2 = 252$, $f_3 = 75$, $f_4 = 369$, and $f_5 = 167$. The accuracy of these probability factors depends strongly on how the set of voters to be sampled was selected and how the question was posed to members of the set.

In science, we are usually concerned with methods that permit the exact or near exact computation of f_i values. We shall discuss such methods in detail in the chapters that follow. Nevertheless, approximation methods are frequently employed in many areas of human endeavor. If appropriate caution is used in interpreting the results of such methods, they can be very useful. Figure 10.1 illustrates various methods for the evaluation of probability factors.

Let us now return to Eq. 10.1 and see how we can obtain useful results once an appropriate set of probability factors is determined. That proportionality may be written as an equality by multiplying on the right side by a proportionality constant:

$$P(X_i) = Cf_i. \qquad (10.2)$$

Since the sum of the probabilities of all possible events must be unity, we may write

$$\sum_{i=1}^{M} P(X_i) = 1. \qquad (10.3)$$

Class of methods	Examples
Exact or near exact	Formulae from quantum and statistical mechanics
	Raoult's Law for ideal solutions $f_A = X_A = P_A/P_A^0$
	Complete counting

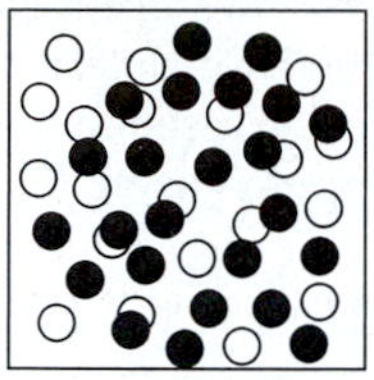

24 black balls
20 white balls

Thus, we can take
$f_{Black} = 24$ and
$f_{White} = 20$.

Approximate	Use of related information
	(a) Baseball statistics to predict team performance
	(b) Temperature profiles, barometric pressure, wind velocities, precipitation maps for weather forecasting
	Statistical sampling
	(a) Gallup Poll
	(b) Harris Poll

▲ FIGURE 10.1
Methods for obtaining probability factors.

A probability distribution that satisfies Eq. 10.3 is said to be *normalized*. The normalization condition represented by Eq. 10.3 places a mathematical constraint on Eq. 10.2 that permits the proportionality constant to be determined. Substituting Eq. 10.2 into Eq. 10.3 gives

$$\sum_{i=1}^{M} C f_i = C \sum_{i=1}^{M} f_i = 1. \tag{10.4}$$

Therefore, we obtain

$$C = \frac{1}{\sum_{i=1}^{M} f_i} = \left[\sum_{i=1}^{M} f_i \right]^{-1}. \tag{10.5}$$

When C is assigned the value given by Eq. 10.5, the probability distribution will satisfy the normalization condition required by Eq. 10.3. C is, therefore, called the *normalization constant*, and the process of computing its value is referred to as *normalization*. The normalized probability distribution is

$$P(X_i) = \left[\sum_{i=1}^{M} f_i \right]^{-1} f_i. \tag{10.6}$$

EXAMPLE 10.1

A normal die can exhibit six different faces when thrown. Each face is characterized by having a different number of dots, from 1 to 6. If the die is "honest," all faces are equally probable to be on top when the die comes to rest after being tossed. Therefore, we expect all the f_i to be equal. Determine the normalized probability distribution for the appearance of the various faces when an honest die is thrown.

Solution

Since all faces are equally probable, we may represent this situation by taking all the f_i to be the same. For example, if we use a count of the number of ways to obtain each outcome as our method for determining the f_i, the result will be $f_1 = f_2 = f_3 = f_4 = f_5 = f_6 = 1$, as discussed in the statement of the problem. The normalization constant can be computed using Eq. 10.5:

$$C = \left[\sum_{i=1}^{6} f_i \right]^{-1} = [1 + 1 + 1 + 1 + 1 + 1]^{-1} = 0.166666\ldots. \tag{1}$$

The normalized probability distribution is obtained using Eq. 10.6:

$$P(X_i) = \frac{1}{6} f_i = \frac{1}{6}. \tag{2}$$

While this result is correct, it is important to realize that *any* value for the probability factors can be used, so long as our choice reflects the basic condition that all faces are equally probable. We could, for example, take each f_i to be equal to a constant K, in which case we would have

$$C = \left[\sum_{i=1}^{6} f_i \right]^{-1} = [K + K + K + K + K + K]^{-1} = (6K)^{-1}. \tag{3}$$

The normalized probability distribution is then

$$P(X_i) = \frac{1}{6K} f_i = \frac{K}{6K} = \frac{1}{6}, \tag{4}$$

which is identical to Eq. 2. Thus, all results are equally probable and all have a probability of exactly 1 chance in 6. That is, one-sixth of the time the face with one dot will be on top, one-sixth of the time the face with two dots will be on top, etc.

For a related exercise, see Problem 10.2.

EXAMPLE 10.2

A cubic die is loaded and the edges are shaved so that it is no longer "honest." The loading is such that the face with two dots is twice as likely to be on top as the face with a single dot. The face with three dots is 1.5 times as likely to be on top as the face with two dots. The remaining faces are all equally likely to be on top after the die comes to rest, and the probability that the face with four dots will be on top is 1.2 times as likely as that of the face with a single dot. Determine the normalized probability distribution for the appearance of the various faces when this die is thrown. Make bar graphs of the normalized and unnormalized probability distributions.

Solution

The solution is easy to obtain, provided that we choose a set of probability factors which reflect the relative probability values given in the problem. Let $f_1 = 10$. Since the two-dot face is twice as likely to be on top as the single-dot face, we must have

$$f_2 = 20. \tag{1}$$

The three-dot face is 1.5 times as likely to be on top as the two-dot face; therefore,

$$f_3 = 1.5 f_2 = 30. \tag{2}$$

By the conditions of the problem, we also have

$$f_4 = f_5 = f_6 = 1.2f_1 = 12. \qquad (3)$$

The normalization constant is given by Eq. 10.5:

$$C = \left[\sum_{i=1}^{6} f_i \right]^{-1} = [10 + 20 + 30 + 12 + 12 + 12]^{-1} = (96)^{-1}. \qquad (4)$$

The normalized probability distribution is

$$P(X_i) = Cf_i = \frac{f_i}{96}. \qquad (5)$$

Thus,

$$P(X_1) = \frac{10}{96}, P(X_2) = \frac{20}{96}, P(X_3) = \frac{30}{96}, P(X_4) = P(X_5) = P(X_6) = \frac{12}{96}. \qquad (6)$$

Notice that we obtain the same result if we set $f_1 = K$, where K is any constant. We then have $f_2 = 2K$ and $f_3 = 1.5f_2 = 3K$. Also from Eq. 3, $f_4 = f_5 = f_6 = 1.2f_1 = 1.2K$, and Eq. 4 now becomes

$$C = \left[\sum_{i=1}^{6} f_i \right]^{-1} = [K + 2K + 3K + 1.2K + 1.2K + 1.2K]^{-1} = (9.6K)^{-1}. \qquad (7)$$

Using Eq. 5, we then have

$$P(X_1) = \frac{K}{9.6K} = \frac{10}{96}, P(X_2) = \frac{2K}{9.6K} = \frac{20}{96}, P(X_3) = \frac{3K}{9.6K} = \frac{30}{96},$$

$$P(X_4) = P(X_5) = P(X_6) = \frac{1.2K}{9.6K} = \frac{12}{96}, \qquad (8)$$

which are the same results obtained in Eq. 6. The unnormalized probability distribution is just a plot of f_i versus i. The normalized distribution is a plot of $P(X_i)$ versus i. These plots are shown in Figures 10.2 and 10.3, respectively.

For related exercises, see Problems 10.18, 10.19, and 10.22.

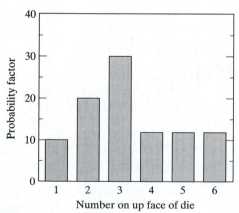

▲ FIGURE 10.2
Unnormalized probability distribution for the dishonest die described in Example 10.2.

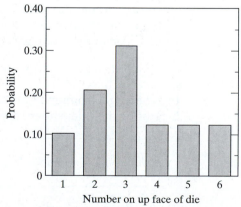

▲ FIGURE 10.3
Normalized probability distribution for the dishonest die described in Example 10.2. This plot appears to be identical to the unnormalized distribution shown in Figure 10.2. The difference lies in the ordinate values. In Figure 10.2, these values are those for the probability factors, which do not sum to unity. In the current figure, the ordinate values are the properly normalized probabilities, which do sum to unity.

It should be obvious from Examples 10.1 and 10.2 that an essential part of any probability problem is to have some means by which an appropriate set of f_i values may be either calculated or measured. In Example 10.2, the determination of the relative probabilities that each face will be on top is the crucial part of the calculation or experiment. As we shall see in subsequent chapters, this is the real power of quantum theory and statistical mechanics: They provide the means by which an accurate set of f_i values can be obtained. For the moment, however, we shall assume that a method exists for determining these quantities, and we shall simply investigate how such values are used to solve problems of interest.

10.1.2 Combined Probabilities

In some cases, we will be concerned with the probability that two or more events will occur or with the probability that one of two or more possible events will take place. The method for handling such situations depends upon whether the probabilities of the events are independent or dependent upon one another. The throw of two dice is an example of the former. The probability of obtaining a single dot on the up face of the first die is in no way dependent upon what happens to the second die, and vice versa. We term such probabilities *independent*. On the other hand, consider a situation in which we are interested in the probability of the next person entering a room being male or female and the probability that the next person entering the room will be carrying a handbag. These two probabilities are not independent. If the next person entering the room is female, the probability that that person will be carrying a handbag is greater than would be the case if the person entering the room were male. Such probabilities are called *conditional* probabilities. In most cases, the probabilities of concern in scientific applications are independent, although conditional probabilities are occasionally of interest. For this reason, we shall be primarily concerned with independent probabilities. Problem 10.22 serves as an example of conditional probabilities.

If the probabilities of occurrence of Events 1 and 2 are $P(X_1)$ and $P(X_2)$, respectively, the combined probability that both Event 1 *and* Event 2 will occur in different, independent experiments is

$$P(X_1 \text{ and } X_2) = P(X_1)P(X_2). \tag{10.7}$$

In general, the combined probability of the occurrence of S specified events in S different, independent experiments is

$$\boxed{P(X_1, X_2, X_3, \dots, \text{ and } X_s) = \prod_{i=1}^{S} P(X_i) = P(X_1)P(X_2)P(X_3) \dots P(X_s)}, \tag{10.8}$$

provided that the probabilities of all the events are independent.

If, on the other hand, we wish to compute the probability that either Event 1 *or* Event 2 will occur in *one* particular experiment, we obtain

$$P(X_1 \text{ or } X_2) = P(X_1) + P(X_2) \tag{10.9}$$

if the probabilities of the two events are independent. For the general case of independent probabilities for which we wish to know the probability that either Event 1, or Event 2, or Event 3, ... , *or* Event S will occur in a single experiment, the desired result is

$$P(X_1, X_2, \ldots, \text{or } X_s) = \sum_{i=1}^{S} P(X_i) = P(X_1) + P(X_2) + \cdots + P(X_s).$$

(10.10)

Notice how the result depends upon the conjunction used. The combined probability that Events 1 *and* 2 will occur in two independent experiments is $P(X_1)P(X_2)$. The combined probability that either Event 1 *or* Event 2 will occur in a single experiment is $P(X_1) + P(X_2)$. This generalization assumes that the probabilities are independent.

These general rules may be utilized to compute the probability of more complex situations in a straightforward manner. For example, the probability that either Events 1 and 2 or Events 3 and 4 will occur in that order in two independent experiments is

$$P(1 \text{ and } 2, \text{ or } 3 \text{ and } 4) = P(X_1)P(X_2) + P(X_3)P(X_4).$$

The first term represents the probability that Events 1 and 2 will occur in that order. The second term is the probability that Events 3 and 4 will occur in that order. The sum represents the combined probability that we will observe either Event 1 followed by Event 2 or Event 3 followed by Event 4 in the two experiments. If we are unconcerned about the order in which the events occur, the result changes. For example, the probability that events 1 and 2 will occur in two independent experiments is

$$P(1 \text{ and } 2) + P(2 \text{ and } 1) = P(X_1)P(X_2) + P(X_2)P(X_1) = 2P(X_1)P(X_2),$$

since we can achieve our objective if Event 1 occurs in the first experiment followed by Event 2 in the second experiment, or vice versa.

EXAMPLE 10.3

A nickel (N), a dime (D), and a quarter (Q) are randomly tossed onto a table. Assume that none of the coins can stand on edge. **(A)** Compute the probability that we will observe heads (H) on the nickel, tails (T) on the dime, and heads (H) on the quarter. **(B)** If the act of throwing the coins onto the table is viewed as a single experiment, determine all possible results of that experiment, and obtain the normalized probability distribution for the single event. **(C)** What is the probability that we will observe heads on the nickel or tails on the dime or heads on the quarter?

Solution

(A) The act of throwing the three coins on the table can be viewed as consisting of three independent experiments. First, the nickel is thrown, then the dime, and finally the quarter. If the coins are honest, the normalized probability distribution for each coin is

$$P_N(H) = P_N(T) = 0.5, \tag{1}$$

where $P_N(H)$ and $P_N(T)$ are the probabilities that the nickel will show a head or tail, respectively, when tossed onto the table. Similarly, we also have

$$P_D(H) = P_D(T) = 0.5 \text{ and } P_Q(H) = P_Q(T) = 0.5. \tag{2}$$

Since we are dealing with independent events, the probability that we will observe a head on the nickel and a tail on the dime and a head on the quarter is given by Eq. 10.8:

$$P(H_N, T_D, H_Q) = P_N(H)P_D(T)P_Q(H) = (0.5)^3 = 0.125. \tag{3}$$

(B) If we view the process as a single experiment, we find that the possible outcomes are given in the following table:

Event	Nickel	Dime	Quarter	Probability Factor f_i
1	H	H	H	1
2	H	H	T	1
3	H	T	H	1
4	H	T	T	1
5	T	H	H	1
6	T	H	T	1
7	T	T	H	1
8	T	T	T	1

If all the coins are honest, the probabilities of these events are the same, so we choose a set of probability factors that are all equal. The normalized probability distribution for this single experiment is obtained using Eqs. 10.5 and 10.6. We get

$$C = \text{normalization constant} = \left[\sum_{i=1}^{8} f_i \right]^{-1} = \frac{1}{8}. \tag{4}$$

Hence, the probability of each independent event is

$$P_i = \frac{1}{8}(1) = \frac{1}{8}. \tag{5}$$

(C) We are now interested in different outcomes of a single experiment (throwing a nickel, a dime, and a quarter onto the table). Equation 10.10 provides the answer. We sum all probabilities that satisfy the condition we are seeking—either heads on the nickel or tails on the dime or heads on the quarter. The only event that does not satisfy one of these conditions is Event 6 in the table. Therefore, our desired probability is

$$P(H_N \text{ or } T_D \text{ or } H_Q) = P_1 + P_2 + P_3 + P_4 + P_5 + P_7 + P_8 = \frac{7}{8} = 0.875. \tag{6}$$

We can also solve this problem by viewing the process as consisting of three independent experiments. (First we toss the nickel, then the dime, and finally the quarter.) However, if we do this, we must work with Eq. 10.8, not with Eq. 10.10. In order *not* to satisfy the requirements of the problem, we must obtain a tail when we toss the nickel in Experiment 1, a head when we toss the dime in Experiment 2, and finally a tail when we toss the quarter in the Experiment 3. The combined probability that we will observe all three events is

$$P(T_N \text{ and } H_D \text{ and } T_Q) = P_N(T)P_D(H)P_Q(T) = (0.5)(0.5)(0.5) = 0.125. \tag{7}$$

The probability that this will *not* occur is $1 - P(T_N \text{ and } H_D \text{ and } T_Q) = 1 - 0.125 = 0.875$, which is the same result we obtained in Eq. 6.

We can glean three important generalizations from this example: (1) Decide if we are dealing with different outcomes of a single experiment or combined probabilities for several independent experiments. (2) Obtain the normalized probability distribution for the relevant case. (3) Use either Eq. 10.8 or Eq. 10.10, depending upon which situation applies.

For a related exercise, see Problem 10.23.

EXAMPLE 10.4

Suppose we have a pair of dice with each die loaded and shaved in the manner described in Example 10.2. **(A)** Compute the probability that if both dice are thrown, the top face of Die 1 will have four dots while the top face of Die 2 has three dots. In this event, the total number of dots on the top faces is seven. **(B)** Compute the probability that if both dice are thrown, a total of seven dots will show on the top faces of the two dice.

Solution

(A) We are dealing with independent probabilities in two experiments. Therefore, Eq. 10.7 is applicable. We wish to have four dots on Die 1 *and* three dots on Die 2. Using Eq. 10.7, we find that the desired probability is

$$P(X_4 \text{ and } X_3) = P(X_4)\,P(X_3) = \frac{12}{96} \times \frac{30}{96} = \frac{(12)(30)}{(96)^2} = \frac{5}{128} = 0.0390625, \quad \textbf{(1)}$$

where the individual probabilities are those determined in Example 10.2.

(B) There are six ways we can observe a total of seven dots on the two faces:

Case	Dots on Die 1	Dots on Die 2	Combined Probability
1	1	6	$P(1)P(6)$
2	6	1	$P(6)P(1)$
3	2	5	$P(2)P(5)$
4	5	2	$P(5)P(2)$
5	3	4	$P(3)P(4)$
6	4	3	$P(4)P(3)$

To see a total of seven dots on the two faces, we must observe either Case 1 *or* Case 2 *or* Case 3 *or* Case 4 *or* Case 5 *or* Case 6. The desired probability is, therefore,

$$P(7 \text{ dots total}) = P(X_1)P(X_6) + P(X_6)P(X_1) + P(X_2)P(X_5) + P(X_5)P(X_2) \quad \textbf{(2)}$$
$$+ P(X_3)P(X_4) + P(X_4)P(X_3)$$
$$= \left[\frac{10}{96} \times \frac{12}{96}\right] + \left[\frac{12}{96} \times \frac{10}{96}\right] + \left[\frac{20}{96} \times \frac{12}{96}\right]$$
$$+ \left[\frac{12}{96} \times \frac{20}{96}\right] + \left[\frac{30}{96} \times \frac{12}{96}\right] + \left[\frac{12}{96} \times \frac{30}{96}\right] = \frac{5}{32} = 0.15625.$$

Notice that we can write Eq. 2 in the form

$$P(7 \text{ dots total}) = 2P(X_1)P(X_6) + 2P(X_2)P(X_5) + 2P(X_3)P(X_4). \quad \textbf{(3)}$$

The factor 2 in front of each term is called the *degeneracy*; it counts the number of ways, each having the same probability, that a given event can occur. For example, one die having a six and the other having a one can occur in the two ways given by Cases 1 and 2 in the table. The degeneracy for that situation is, therefore, 2.

For related exercises, see Problems 10.15, 10.18, 10.19, and 10.22.

Example 10.4 again introduces the concept of degeneracy that we first encountered in our derivation of the Debye heat capacity equations for solids. The degeneracy is, by definition, the number of states or events that have the same value of f_i or $P(X_i)$ in a probability distribution. This quantity is generally denoted by g_i. Thus, if only one state or event has probability $P(X_i)$, the degeneracy for that state is 1. If there are two states or events that have equal probabilities, then $g = 2$ for that state or event. A more restricted definition of degeneracy defines two states to be degenerate if they have the same energy. In Chapters 17 and 18, we shall learn that such states also have the same probability of being observed at a given temperature. Therefore, the two definitions are consistent, but the latter is more restrictive. We shall use the broader definition.

The use of degeneracy gives us two different ways to express normalization constants and probabilities in general. Consider Example 10.2 again. For the "dishonest" die described therein, the normalization constant for the probability distribution is

$$C = \left[\sum_{i=1}^{6} f_i \right]^{-1} = [f_1 + f_2 + f_3 + f_4 + f_5 + f_6]^{-1}$$

$$= [10 + 20 + 30 + 12 + 12 + 12]^{-1} = (96)^{-1}.$$

In this case, we have $f_4 = f_5 = f_6 = 12$ and $P(X_4) = P(X_5) = P(X_6) = 12/96$. The degeneracy for these states is, therefore, 3. Consequently, we could write the equation for the normalization constant in the form

$$C = \left[\sum_{i=1}^{4} g_i f_i \right]^{-1} = [f_1 + f_2 + f_3 + 3f_4]^{-1} = [10 + 20 + 30 + 3(12)]^{-1}$$

$$= (96)^{-1},$$

where the degeneracies are $g_1 = g_2 = g_3 = 1$ and $g_4 = 3$. The two expressions for C give the same result and they look very much the same. The essential difference is that the summation in the first expression without the degeneracy factor runs over all possible events. In the second expression, with the degeneracy included, the *summation runs only over those states whose probabilities differ*. In general, for a system with M different states, the normalization constant can be written in the form of Eq. 10.5, viz.,

$$C = \left[\sum_{i=1}^{M} f_i \right]^{-1},$$

or it can be written as

$$C = \left[\sum_{i=1}^{M'} g_i f_i \right]^{-1}. \tag{10.11}$$

In Eq. 10.11, the upper limit on the summation is written as M' rather than M to emphasize the fact that the sum includes only events with different values of f_i, whereas the summation in Eq. 10.5 includes the f_i values for all the events. If the equation is written with the degeneracy factor included, the summation is always that in Eq. 10.11.

10.1.3   Average Values

The computation of average values is a matter of central importance in both chemistry and physics. Molecular systems usually comprise on the order of 10^{23} molecules. Such a collection of a large number of identical objects or systems is called an *ensemble*. The quantities that are measured in laboratory experiments are the averages of those quantities for the individual molecules in the ensemble. Such averages are often called *ensemble averages*. If the normalized probability distribution is known, the appropriate average values can always be computed.

Let us begin by considering the dishonest, loaded die described in Example 10.2. In that example, we found that the probabilities for obtaining one, two, three, four, five, and six dots on the up face of the die are 10/96, 20/96, 30/96, 12/96, 12/96, and 12/96, respectively. Let us imagine that we throw this die 96,000,000 times and record the result of each throw. On average, we would expect the face with one dot to show 10,000,000 times, the face with two dots 20,000,000 times, the face with three dots 30,000,000 times, and the remaining faces 12,000,000 times each. We now ask, "What is the average number of dots that appear on a throw of the die?" This average is calculated in the same manner one computes the average score on a physical chemistry examination: The scores of all the students are totaled and then divided by the number of students in the class who took the examination. Therefore, the average number of dots is given by

$$\langle \text{number of dots} \rangle = \frac{\text{sum of all dots}}{\text{number of dice thrown}}$$

$$= [(1)(10000000) + (2)(20000000) + (3)(30000000) + (4)(12000000)$$
$$+ (5)(12000000) + (6)(12000000)]/96000000$$

$$= (1)\frac{10}{96} + (2)\frac{20}{96} + (3)\frac{30}{96} + (4)\frac{12}{96} + (5)\frac{12}{96} + (6)\frac{12}{96} = \frac{10}{3} = 3.3333333\ldots,$$

where the angle brackets $\langle \ \rangle$, denote the average value of the quantity within them. Notice that the right-hand side of this equation may be written so that the equation takes the following form:

$$\langle \text{number of dots} \rangle = (1)P(X_1) + (2)P(X_2) + (3)P(X_3)$$
$$+ (4)P(X_4) + (5)P(X_5) + (6)P(X_6)$$
$$= \sum_{i=1}^{6} iP(X_i).$$

The foregoing expression is a specific example of the general definition of the average value. If $P(X_i)$ is the normalized probability of observing a value X_i of the discrete variable X, the average value of any function $h(X)$ is given by

$$\langle h(X) \rangle = \sum_{i=1}^{M} h(X_i)P(X_i) \, . \tag{10.12}$$

Substiting for $P(X_i)$ in terms of the normalization constant and the f_i yields

$$\langle h(X) \rangle = C \sum_{i=1}^{M} h(X_i) f_i = \frac{\sum_{i=1}^{M} h(X_i) f_i}{\left[\sum_{i=1}^{M} f_i \right]}. \tag{10.13}$$

If $h(X) = X$, Eq. 10.13 gives $\langle X \rangle$, which is called the *first moment* of the distribution. If $h(X) = X^2$, the result is the average of the square of X, $\langle X^2 \rangle$, which is termed the *second moment* of the distribution.

EXAMPLE 10.5

Consider the following game of chance: A small ball whose radius is 1.0 cm is randomly thrown onto a large board, on which it bounces and rolls around until it falls into one of many holes on the board. The board contains $5N$ holes evenly divided between those with radii 5 cm, 10 cm, 15 cm, 20 cm, and 25 cm. The player pays $1.00 to play the game. If the small ball falls into a hole with a 25-cm radius, the player receives nothing. If the ball enters a hole whose radius is 20 cm, the player receives $1.00. The payoffs for the ball entering the other holes are $2.00 for a 15-cm hole, $3.00 for a 10-cm hole, and $5.00 for a 5-cm hole. Assume that the game board is perfectly machined with exactly circular holes whose perimeters are sharp. **(A)** Calculate the average return to the player on a single play and an investment of $1.00. **(B)** What would the payoff for the 5-cm hole have to be to make the average return to the player equal to his or her investment? **(C)** What is the probability that a player will have a profit of $4.00 after playing the game twice? Figure 10.4 shows a schematic diagram of the game board for the case $N = 4$.

Solution

(A) The first step in any problem involving probabilities is to obtain the normalized probability distribution. In this case, the probability that the small ball will enter a hole of radius r is proportional to the area of the hole. The area itself is proportional to r^2, so an appropriate set of probability factors is $f_1 = 5^2 = 25$ for the 5-cm hole, $f_2 = 10^2 = 100$ for the 10-cm hole, $f_3 = 15^2 = 225$ for the 15-cm hole, $f_4 = 20^2 = 400$ for the 20-cm hole, and $f_5 = 25^2 = 625$ for the 25-cm hole. The normalization constant can now be obtained using Eq. 10.5:

$$C = \left[\sum_{i}^{5} f_i \right]^{-1} = (25 + 100 + 225 + 400 + 625)^{-1} = (1{,}375)^{-1}. \tag{1}$$

The returns to the player are $X_5 = 0$, $X_4 = \$1.00$, $X_3 = \$2.00$, $X_2 = \$3.00$, and $X_1 = \$5.00$. The average return is given by Eq. 10.13, where $h(X) = X$:

$$\langle X \rangle = \frac{\sum_{i=1}^{5} X_i f_i}{1{,}375}$$

$$= \frac{(\$5.00)(25) + (\$3.00)(100) + (\$2.00)(225) + (\$1.00)(400) + (0)(625)}{1{,}375}$$

$$= \frac{\$1{,}275}{1{,}375} = \$0.9273\ldots. \tag{2}$$

The player, therefore, loses an average of 7.27 cents each time the game is played.

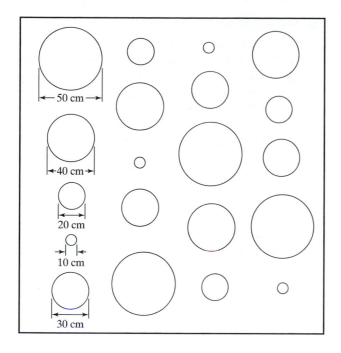

◀ FIGURE 10.4
Gaming board used in Example 10.5. In this case, $N = 4$, since there are four holes of each radius. (The diameters are given in the figure.) The solid lines represent the walls of the gaming board, which are elastic so that the ball will bounce away when it strikes a wall.

(B) We wish to know to what value X_1 must be changed in order to make $\langle X \rangle = \$1.00$. With X_1 unknown, Eq. 2 becomes

$$\langle X \rangle = \frac{(X_1)(25) + (\$3.00)(100) + (\$2.00)(225) + (\$1.00)(400) + (0)(625)}{1,375} = \$1.00. \tag{3}$$

Solving Eq. 3 for X_1, we obtain

$$X_1 = \frac{\$1375 - (\$1.00)(400) - (\$2.00)(225) - (\$3.00)(100)}{25} = \frac{\$225}{25} = \$9.00. \tag{4}$$

(C) To show a profit of \$4.00 in two plays, the player must receive a return of \$6.00, since he or she paid \$2.00 to play the game twice. There are three ways to receive a return of \$6.00 in two plays: (1) Win \$5.00 on the first play *and* \$1.00 on the second; (2) win \$1.00 on the first play *and* \$5.00 on the second; and (3) win \$3.00 on the first play *and* \$3.00 on the second. Since the probabilities are independent and each case involves the occurrence of two events in two sucessive experiments (plays), we have

$$\text{Probability of Case (1)} = P(X_1)P(X_4), \tag{5}$$

$$\text{Probability of Case (2)} = P(X_4)P(X_1), \tag{6}$$

$$\text{Probability of Case (3)} = P(X_2)P(X_2), \tag{7}$$

To show a net profit of \$4.00 on two plays, we must have either Case (1) *or* Case (2) *or* Case (3). Therefore, the total probability of obtaining a \$4.00 profit is

$$P(\$4.00 \text{ profit}) = P(X_1)P(X_4) + P(X_4)P(X_1) + P(X_2)P(X_2)$$

$$= \frac{25}{1,375} \times \frac{400}{1,375} + \frac{400}{1,375} \times \frac{25}{1,375} + \frac{100}{1,375} \times \frac{100}{1,375} = 0.015867\ldots, \tag{8}$$

or about a 1.59% chance.

For related exercises, see Problems 10.1 and 10.17.

10.1.4 The Effect of Measurement

In one sense, the fact that measurement affects the probability distribution is obvious. However, this point often causes great concern and difficulty when one is studying quantum mechanics. Let us begin with a simple example. Someone offers you a shuffled deck of normal playing cards and asks that you choose one card at random and place it face down on a glass-top table. You do so and are then asked what is the probability that the card on the table is the ace of spades. Since all 52 cards in the deck have an equal chance of being selected, you quickly assign $f_1 = 1$ for all 52 cards and with equal speed compute the normalization constant from Eq. 10.5:

$$C = \left[\sum_{i=1}^{52} f_i \right]^{-1} = \frac{1}{52}.$$

You then announce with complete confidence that the chance that the ace of spades lies face down on the table is $Cf_i = (1/52)(1) = 1/52$. Your answer is, of course, correct.

At this point, you are permitted to crawl under the table and examine the card lying on the surface through the glass top. That is, you are allowed to make a measurement in which the experimental apparatus is your eyes. Having made this measurement, you are once again asked to state the probability that the card lying face down on the table is the ace of spades. Obviously, your answer will now not be 1/52! It will either be zero or unity, depending upon the result of your "measurement." Thus, the measurement has drastically altered the values of the f_i and the probability distribution! If you saw the ace of spades when you performed the measurement, all the f_i become zero save that for the ace of spades, which is still unity. Consequently, we must renormalize the distribution, since the f_i have changed. We now obtain

$$C = \left[\sum_{i=1}^{52} f_i \right]^{-1} = 1,$$

and the probability that the ace of spades lies on the table is $Cf_{\text{spade ace}} = (1)(1) = 1$. No one in the world is surprised by this fact.

The same situation occurs in quantum mechanics. In the next chapter, we shall learn that the probability distribution for a quantum mechanical system is proportional to the absolute square of a quantity that we will call the wave function whose symbol will be ψ. That is, we will find that $f \propto |\psi|^2$. If this is true, then it should come as no shock to anyone that if we perform a measurement on a system whose wave function is ψ, the result of the measurement may well be that we have changed the probability distribution and, therefore, the wave function.

In the example of the card lying face down on the table, our measurement provided such complete information that all the f_i save one became zero. In many cases, the measurement provides only partial information. If so, the probability factors are altered so as to include the information obtained in the measurement, the distribution is then renormalized, and appropriate probability calculations are then performed. The next example illustrates the procedure.

EXAMPLE 10.6

A large container holds a very large number of black, red, green, yellow, and white balls. There are twice as many red balls as black. The ratio of the number of green balls to red balls is 1.4. The ratio of yellow balls to green balls is 0.85, and there are the same number of yellow and white balls in the container. With eyes closed, you draw two balls from the container at random and place them inside a box. **(A)** What is the probability that the box holds exactly one green ball and one red ball? **(B)** One of your friends looks inside the box and then closes it. You ask if there is one red and one green ball inside the box. Your friend responds that it wouldn't be fair to tell you. You become irate and tell him that you do all sorts of favors for him and now he won't even do one little favor for you. Your friend compromises by telling you that there are no black balls inside the box. Now what is the probability that the box holds exactly one green ball and one red ball?

Solution

(A) As usual, the initial step is to obtain the normalized probability distribution. To do this, we need the f_i. Let $f_1 = f_{\text{Black}} = 10$. Then we must have $f_2 = f_{\text{Red}} = 20$, since there are twice as many red balls as black balls in the container. We are told that there are 1.4 times as many green balls as red, so $f_3 = f_{\text{Green}} = (1.4)(20) = 28$. Since the ratio of yellow balls to green balls is 0.85, we have $f_4 = f_{\text{Yellow}} = (0.85)(28) = 23.8$, which is also the value $f_5 = f_{\text{White}}$ for the white balls, because there are equal numbers of white and yellow balls. Summarizing, we have the following table:

Color of Ball	f_i	Value of f
Black	f_1	10
Red	f_2	20
Green	f_3	28
Yellow	f_4	23.8
White	f_5	23.8

The normalization constant given by Eq. 10.5 is

$$C = \left[\sum_{i=1}^{5} f_i \right]^{-1} = (105.6)^{-1}. \tag{1}$$

The corresponding probabilities are

$$P_{\text{Black}} = C f_{\text{Black}} = \frac{10}{105.6}, \; P_{\text{Red}} = C f_{\text{Red}} = \frac{20}{105.6}, \; P_{\text{Green}} = C f_{\text{Green}} = \frac{28}{105.6}, \; \text{and}$$

$$P_{\text{Yellow}} = C f_{\text{Yellow}} = P_{\text{White}} = C f_{\text{White}} = \frac{23.8}{105.6}. \tag{2}$$

There are two ways to obtain exactly one green and one red ball in the box: Ball 1 is red *and* Ball 2 is green, *or* Ball 1 is green *and* Ball 2 is red. Therefore, we have

$$P(1 \text{ red and } 1 \text{ green}) = P_{\text{Red}} P_{\text{Green}} + P_{\text{Green}} P_{\text{Red}}$$

$$= \frac{20}{105.6} \times \frac{28}{105.6} + \frac{28}{105.6} \times \frac{20}{105.6} = 0.1004 \ldots, \tag{3}$$

or roughly a 10% chance. The point that the container holds a very large number of balls allows us to assume that choosing a ball does not appreciably alter the ratios of colors in the box for the second pick. If this is not the case, we must adjust the probability distribution after a ball is removed. That is, the probabilities become conditional. (See Problem 10.23.)

(B) Your friend has performed a measurement. As a result of this measurement, you are provided with the information that no black balls are in the box. This changes f_1 to zero. The new set of probability factors is as follows:

Color of Ball	f_i	Value of f
Black	f_1	0
Red	f_2	20
Green	f_3	28
Yellow	f_4	23.8
White	f_5	23.8

The new normalization constant is thus

$$C = \left[\sum_{i=1}^{5} f_i \right]^{-1} = (95.6)^{-1}. \tag{4}$$

The new probabilities are

$$P_{\text{Black}} = \frac{0}{95.6}, \; P_{\text{Red}} = \frac{20}{95.6}, \; P_{\text{Green}} = \frac{28}{95.6}, \; \text{and} \; P_{\text{Yellow}} = P_{\text{White}} = \frac{23.8}{95.6}. \tag{5}$$

Equation 3 now produces

$$P(1 \text{ red and 1 green}) = P_{\text{Red}}P_{\text{Green}} + P_{\text{Green}}P_{\text{Red}}$$

$$= \frac{20}{95.6} \times \frac{28}{95.6} + \frac{28}{95.6} \times \frac{20}{95.6} = 0.1225\ldots. \tag{6}$$

For related exercises, see Problems 10.3, 10.5, 10.6, 10.7, and 10.10.

10.1.5   Distribution Uncertainty

Some probability distributions are such that we are very confident about the result that will be obtained if a measurement were to be made. For example, if purple balls were added to the container in Example 10.6 in an amount that makes the ratio of the number of purple balls to that of any other color equal to 10^6 or greater, we would be very confident that if we pick a ball from the container at random and measure its color, the result will be purple. Other distributions are more uncertain.

Clearly, it would be useful to have a quantitative measure of the uncertainty of a distribution. Numerous quantities might be employed to provide such a measure; there is no obvious unique choice. The quantity most frequently used in scientific applications is the average of the square of the deviation of a measured value from the mean, where "mean" is another term for the first moment, or average, of a distribution. This quantity is also called the *variance* or *square uncertainty*. The square of the deviation of a value of $X = X_i$ from the mean is $(X_i - \langle X \rangle)^2$. The average of this function can be computed by direct application of Eq. 10.12 if we take $h(X) = (X_i - \langle X \rangle)^2$. Inserting this choice into Eq. 10.12 produces

$$\boxed{ \langle \Delta X^2 \rangle = \sum_{i=1}^{M} (X_i - \langle X \rangle)^2 P(X_i) }, \tag{10.14}$$

where we define the symbol $\langle \Delta X^2 \rangle$ to be the variance or the square uncertainty. Expanding the square in Eq. 10.14 gives

$$\langle \Delta X^2 \rangle = \sum_{i=1}^{M} [X_i^2 - 2\langle X \rangle X_i + \langle X \rangle^2] P(X_i). \qquad \textbf{(10.15)}$$

Since the summation of a group of terms may be written as the sum of each individual term, Eq. 10.15 can be expressed in the form

$$\langle \Delta X^2 \rangle = \sum_{i=1}^{M} X_i^2 P(X_i) - 2\langle X \rangle \sum_{i=1}^{M} X_i P(X_i) + \langle X \rangle^2 \sum_{i=1}^{M} P(X_i). \qquad \textbf{(10.16)}$$

To obtain Eq. 10.16, we have utilized the fact that the mean, $\langle X \rangle$, is a constant, which permits us to factor it out of the summations. The first term on the right-hand side of Eq. 10.16 is just the second moment of the distribution, or $\langle X^2 \rangle$. The summation in the second term produces the first moment, $\langle X \rangle$, so that the term is equal to $-2\langle X \rangle^2$. The summation appearing in the third term is unity, since the probability distribution is normalized. Thus, Eq. 10.16 becomes

$$\boxed{\langle \Delta X^2 \rangle = \langle X^2 \rangle - 2\langle X \rangle^2 + \langle X \rangle^2 = \langle X^2 \rangle - \langle X \rangle^2}. \qquad \textbf{(10.17)}$$

The square uncertainty of the distribution is, therefore, measured by the difference of the second moment and square of the first moment.

Qualitatively, it is easy to see that the value of $\langle \Delta X^2 \rangle$ is a measure of the distribution uncertainty. If all values of X are equal to X_o, we will have $\langle X \rangle = X_o$. As a result, $(X_i - \langle X \rangle)^2$ will be zero for all measured values of X, which will produce $\langle \Delta X^2 \rangle = 0$. In this situation, there is no uncertainty about what we will obtain when X is measured. Such a distribution is called a *Kronecker delta function*. As the possible values of X deviate more and more from $\langle X \rangle$, the sum of the squares of the deviations and $\langle \Delta X^2 \rangle$ increase monotonically, reflecting the increasing uncertainty in the result of a measurement of X. Example 10.7 illustrates this relationship in a more quantitative manner.

EXAMPLE 10.7

Consider two introductory chemistry classes, each with 20 students. The professor in Class *A* gives a 100-point examination and obtains the following results:

$$88, 8, 94, 75, 99, 74, 85, 66, 47, 86, 92, 94, 98, 79, 73, 70, 90, 74, 68, 67.$$

The professor in Class *B* also gives a 100-point examination. Her results are as follows:

$$17, 99, 45, 23, 38, 52, 68, 75, 88, 19, 8, 94, 76, 59, 68, 83, 43, 41, 77, 91.$$

Compute the square uncertainty or variance for the distribution of exam scores for the two classes.

Solution

Since a measurement has been made (the examinations have been given and graded), we know that the relative probabilities of the grades,

$$88, 8, 94, 75, 99, 74, 85, 66, 47, 86, 92, 94, 98, 79, 73, 70, 90, 74, 68, 67,$$

in Class A are $f_{88} = f_8 = f_{75} = f_{99} = f_{85} = f_{66} = f_{47} = f_{86} = f_{92} = f_{98} = f_{79} = f_{73} = f_{70} = f_{90} = f_{68} = f_{67} = 1, f_{74} = f_{94} = 2$ (since the latter two grades appear twice), and $f_i = 0$ for all other grades. Using Eq. 10.13 with $h(X)$ equal to the grade, G, we obtain

$$\langle G \rangle_A = (20)^{-1} \sum_{i=1}^{100} G_i f_i$$

$$= [88 + 8 + 2(94) + 75 + 99 + 2(74) + 85 + 66$$

$$+ 47 + 86 + 92 + 98 + 79 + 73 + 70 + 90 + 68 + 67]/20$$

$$= 76.35 \tag{1}$$

and

$$\langle G^2 \rangle_A = (20)^{-1} \sum_{i=1}^{100} G_i^2 f_i$$

$$= [88^2 + 8^2 + 2(94^2) + 75^2 + 99^2 + 2(74^2) + 85^2 + 66^2$$

$$+ 47^2 + 86^2 + 92^2 + 98^2 + 79^2 + 73^2 + 70^2 + 90^2 + 68^2 + 67^2]/20$$

$$= 6{,}239.75. \tag{2}$$

Using Eqs. 1 and 2, we get the square uncertainty of the distribution:

$$\langle \Delta G^2 \rangle = 6{,}239.75 - (76.35)^2 = 410.4. \tag{3}$$

The uncertainty is, then,

$$\langle \Delta G^2 \rangle^{1/2} = 20.26. \tag{4}$$

For Class B, the grades are as follows:

$$17, 99, 45, 23, 38, 52, 68, 75, 88, 19, 8, 94, 76, 59, 68, 83, 43, 41, 77, 91.$$

An appropriate set of f_i for grades over the range 0–100 is thus

$$f_{17} = f_{99} = f_{45} = f_{23} = f_{38} = f_{52} = f_{75} = f_{88} = f_{19} = f_8 = f_{94} = f_{76}$$

$$= f_{59} = f_{83} = f_{43} = f_{41} = f_{77} = f_{91} = 1, f_{68} = 2, \text{ and all other } f_i = 0.$$

The average, or first moment, of the distribution is obtained from Eq. 10.13 with $h(X_i) = G_i$. This gives

$$\langle G \rangle_B = (20)^{-1} \sum_{i=1}^{100} G_i f_i$$

$$= [17 + 99 + 45 + 23 + 38 + 52 + 2(68) + 75 + 88$$

$$+ 19 + 8 + 94 + 76 + 59 + 83 + 43 + 41 + 77 + 91]/20 = 58.20. \tag{5}$$

and

$$\langle G^2 \rangle_B = (20)^{-1} \sum_{i=1}^{100} G_i^2 f_i$$

$$= 17^2 + 99^2 + 45^2 + 23^2 + 38^2 + 52^2 + 2(68)^2 + 75^2 + 88^2$$

$$+ 19^2 + 8^2 + 94^2 + 76^2 + 59^2 + 83^2 + 43^2 + 41^2 + 77^2 + 91^2/20$$

$$= 4{,}127.80. \tag{6}$$

The square uncertainty in the Class B distribution is

$$\langle \Delta G^2 \rangle = 4{,}127.80 - (58.20)^2 = 740.6. \tag{7}$$

The associated uncertainty is

$$\langle \Delta G^2 \rangle^{1/2} = 27.21. \tag{8}$$

Equations 4 and 8 show that the uncertainty in the Class B distribution is 34.3% greater than that Class A. From an operational point of view, this means that if you were to guess the grade of a randomly chosen student in Class A to be the class average, you would, on average, be closer to the correct answer than would be the case for such a student in Class B. In both cases, the grades span the range from 8 to 99. However, most of the grades in Class A are bunched in the range from 45 to 99, whereas in Class B they are more evenly spread over the entire range. In the language of statistics, we say that the grade distribution for Class A exhibits less uncertainty than that for Class B or that the grade distribution for Class B exhibits more *dispersion*.

For related exercises, see Problems 10.1 and 10.14.

10.2   Probability Distribution Functions: Continuous Variables

Let x represent a continuous variable over the range $X_1 \leq x \leq X_2$. The question we wish to address is "What is the probability of observing a value of x in the range x_o to $x_o + dx$ in a measurement of x?" Before proceeding, it is important to note the difference in the way this question is phrased for continuous and discrete variables. If x were a discrete variable with possible values $x_1, x_2, x_3, \ldots, x_m$ spanning the same range, we might reasonably ask, "What is the probability of observing a value $x = x_3$ in a measurement of x?" When x is a continuous variable, such a question is pointless, because the answer is always zero. For example, if x represents the points on the x-axis between -10 and $+10$, the probability that a measurement of x will result in precisely the value $x_o = 3.576982567812456$ is zero, because there are an infinitude of possible values of x that might result from the measurement and there is only one value that yields precisely x_o. The probability $P(x_o)$ is, therefore, $1/\infty = 0$. Thus, if x is continuous, we must discuss the probability that a measurement of x will result in a value in the *range* x_o to $x_o + dx$. Since the number of points in this range is an infinity of the same order as the total number of points on the line segment from -10 to $+10$, the outcome can be nonzero. This concept is illustrated in Figure 10.5.

Let $P(x)dx$ be the probability that a value of x in the range x to $x + dx$ will be observed in a measurement of x. Once again, we assume that we have some method of obtaining a function $f(x)dx$ to which $P(x)dx$ is proportional. The method might be simple logic, experimental measurement, quantum theory, or statistical mechanics. At this point, we simply assume that these functions can be obtained. If that is not the case, the probability problem cannot be solved. The proportionality condition allows us to write

$$P(x)dx = Cf(x)dx, \tag{10.18}$$

where, mathematically, C plays the role of a proportionality constant. As in the case of discrete variables, the sum of all possible probabilities must be unity. However, since x is continuous, the summation of $P(x)dx$ from X_1 to X_2 must be done by integration:

$$\int_{X_1}^{X_2} P(x)dx = \int_{X_1}^{X_2} Cf(x)dx = C\int_{X_1}^{X_2} f(x)dx = 1. \tag{10.19}$$

▶ FIGURE 10.5
▶ FIGURE 10.5

Probability distributions for continuous variables. Part A shows that the probability of randomly selecting a specific value of X equal to X_0 in the range $-10 \leq X \leq 10$ is zero because the total number of points on the line segment is infinite. B illustrates the fact that the probability of randomly selecting a point in the range $X = X_0$ to $X = X_0 + dX$ is nonzero. The analysis assumes that all points on the line segment have an equal probability of being selected. Although the number of points per unit length is infinite, it is the same order of infinity for the total line segment and for the range X_0 to $X_0 + dX$. Therefore, it cancels, and $P(X_0)\, dX$ can be nonzero.

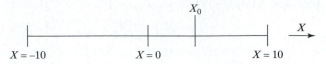

$X = -10$ $X = 0$ $X = 10$

$P(X_0) = 1$ point/total points in range $= 1/\infty = 0$

(A)

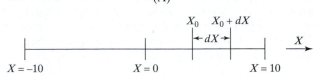

$P(X_0)dX$ = Number of points between X_0 and $X_0 + dX$/total points

in range $= \dfrac{Z\, dX}{20\, Z} = \dfrac{dX}{20}$, where Z = number of points per unit length

(B)

Equation 10.19 is the normalization condition for the probability distribution for a continuous variable. Solving for C, we obtain

$$C = \frac{1}{\displaystyle\int_{X_1}^{X_2} f(x)\,dx} = \left[\int_{X_1}^{X_2} f(x)\,dx\right]^{-1}. \tag{10.20}$$

Once again, we call C the *normalization constant*. Its functional form is identical to Eq. 10.5, except that the summation must be replaced with an integral to sum a continuous variable. Combining Eqs. 10.18 and 10.20, we have

$$P(x)\,dx = \frac{f(x)\,dx}{\displaystyle\int_{X_1}^{X_2} f(x)\,dx}. \tag{10.21}$$

All the principles and derivations that apply to discrete variables may now be transformed to equivalent expressions for a continuous variable by replacing discrete probabilities $P(X_i)$ with $P(x)\,dx$ and substituting integrals for summations where the limits are those appropriate to the problem. For example, the average value of a function $h(x)$ over the range $X_1 \leq x \leq X_2$ is given by Eqs. 10.12 and 10.13 if those replacements are made. The result is

$$\langle h(x)\rangle = \int_{X_1}^{X_2} h(x)P(x)\,dx. \tag{10.22}$$

Substituting Eq. 10.21 for $P(x)\,dx$ gives

$$\langle h(x)\rangle = \frac{\displaystyle\int_{X_1}^{X_2} h(x)f(x)\,dx}{\displaystyle\int_{X_1}^{X_2} f(x)\,dx}. \tag{10.23}$$

As noted previously, we shall learn in the next chapter that the absolute square of a quantum mechanical quantity called the wave function ψ is proportional to the probability distribution for the system. That is, $|\psi|^2$ plays the role of $f(x)$ in Eq. 10.23. We might, therefore, expect average quantum mechanical values to be calculated using an equation that has the form

$$\langle h(x) \rangle = \frac{\int_{X_1}^{X_2} h(x)|\psi|^2 dx}{\int_{X_1}^{X_2} |\psi|^2 dx}.$$

This expectation is correct for many functions $h(x)$, although for some, it must be slightly modified. However, the fact that we can almost anticipate the mathematical form of such quantum mechanical equations using simple probability theory emphasizes one of the essential features of quantum theory.

The uncertainty of a continuous probability distribution is still given by Eq. 10.17. The only difference is that the first and second moments of the distribution must be computed using Eq. 10.23 rather than the discrete form in Eq. 10.13. All the points concerning combined probabilities and the effect of measurement also apply to continuous probability distributions.

EXAMPLE 10.8

By using the results of IQ tests on a large sample of the population in a given area of the country, it is found that the probability $P(q)dq$ that a randomly chosen individual from that area will have an IQ in the range from q to $q + dq$ is proportional to $q^4 \exp[-q/25]dq$, where q can assume values in the range $0 \le q \le \infty$. **(A)** Obtain the normalized probability distribution function, and plot $P(q)$ as a function of q over the range $0 \le q \le 300$. **(B)** Determine the most probable IQ in the given area. **(C)** What is the average IQ of inhabitants in the area? **(D)** Compute the uncertainty in the distribution of IQs in this area of the country. **(E)** If the inhabitants of the area expel all the politicians and administrators, so that no one is left with $q < 50$, what is the new average IQ of inhabitants in that area of the country? Plot the normalized probability distribution after the expulsion of politicians and administrators, and, on the same graph, show the original distribution.

Solution

(A) The conditions of the problem are

$$P(q)dq = Cq^4 \exp\left[-\frac{q}{25}\right]dq. \tag{1}$$

The normalization condition requires that

$$C \int_0^\infty q^4 \exp\left[-\frac{q}{25}\right]dq = 1. \tag{2}$$

To evaluate the normalization constant, we must integrate $\int_0^\infty x^n \exp[-ax]dx$. Integrals of the form $\int_0^{x_o} x^n \exp[-ax]dx$ for integral values of $n \ge 0$ and $a > 0$ appear frequently in the study of atomic structure. The integral can be executed by repeated integration by parts.

The result is:

$$I = -\exp[-ax_o]\left[\frac{x_o^n}{a} + \frac{nx_o^{n-1}}{a^2} + \frac{n(n-1)x_o^{n-2}}{a^3} + \cdots + \frac{n!x_o}{a^n} + \frac{n!}{a^{n+1}}\right] + \frac{n!}{a^{n+1}}. \tag{3}$$

We now note that if $x_o = \infty$, $\exp[-ax_o]$ goes to zero, and we have the very useful result,

$$\int_0^\infty x^n \exp[-ax]\,dx = \frac{n!}{a^{n+1}} \qquad \text{for integral values of } n \geq 0 \text{ and for } a > 0. \quad \textbf{(4)}$$

Applying Eq. 4 to the integral in Eq. 2 where we have $n = 4$ and $a = 1/25$, the result is

$$C\left[\frac{4!}{(1/25)^5}\right] = C(24)(25)^5 = C(234{,}375{,}000) = 1. \quad \textbf{(5)}$$

Solving Eq. 5 for C, we obtain

$$C = \frac{1}{234{,}375{,}000}. \quad \textbf{(6)}$$

The normalized probability distribution function is

$$P(q)\,dq = \frac{q^4 \exp\left[-\dfrac{q}{25}\right]dq}{234{,}375{,}000}. \quad \textbf{(7)}$$

The plot of $P(q)$ versus q is shown in Figure 10.6.

(B) The most probable IQ occurs at the point where $P(q)$ attains a maximum. At that point in the plot, we have

$$\frac{dP(q)}{dq} = (234{,}375{,}000)^{-1}\left\{4q^3 \exp\left[-\frac{q}{25}\right] - \frac{q^4}{25}\exp\left[-\frac{q}{25}\right]\right\} = 0. \quad \textbf{(8)}$$

This requires that at the maximum we have

$$4q_{max}^3 \exp\left[-\frac{q_{max}}{25}\right] = \frac{q_{max}^4}{25}\exp\left[-\frac{q_{max}}{25}\right]. \quad \textbf{(9)}$$

Solving for q_{max}, we obtain $q_{max} = 100$, which is in agreement with the result shown in Figure 10.6.

(C) The average IQ is given by Eq. 10.23:

$$x\langle q\rangle = \int_0^\infty qP(q)\,dq = (234{,}375{,}000)^{-1}\int_0^\infty q^5 \exp\left[-\frac{q}{25}\right]dq \quad \textbf{(10)}$$

$$= (234{,}375{,}000)^{-1}\left[\frac{5!}{(1/25)^6}\right] = \frac{5!(25)^6}{4!(25)^5} = 5(25) = 125. \quad \textbf{(11)}$$

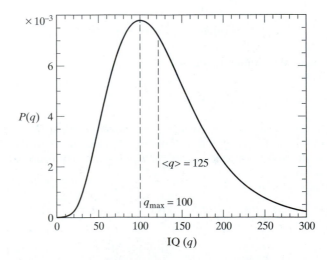

▶ FIGURE 10.6
Normalized probability distribution of IQ for the region described in Example 10.8. The total area under this curve is unity, as required by Eq. 10.19. The positions of the most probable IQ and the average IQ are also indicated in the plot. Because the distribution is skewed towards higher IQ's, the average lies above the most probable result.

(D) The square uncertainty is given by Eq. 10.17. To compute its value, we need both the first and second moments of the distribution. The first moment, or average, was computed in (C). The average square IQ, or the second moment, is

$$\langle q^2 \rangle = \int_0^\infty q^2 P(q)\, dq = (234{,}375{,}000)^{-1} \int_0^\infty q^6 \exp\left[-\frac{q}{25}\right] dq$$

$$= (234{,}375{,}000)^{-1} \left[\frac{6!}{(1/25)^7}\right] = \frac{6!(25)^7}{4!(25)^5} = 5(6)(25)^2 = 18{,}750. \qquad \textbf{(12)}$$

Using the values of $\langle q \rangle$ and $\langle q^2 \rangle$ in Eqs. 11 and 12, we can compute the square uncertainty:

$$\langle \Delta q^2 \rangle = \sigma^2 = \langle q^2 \rangle = \langle q \rangle^2 = 18{,}750 - 125^2 = 3{,}125. \qquad \textbf{(13)}$$

The uncertainty is, therefore,

$$\langle \Delta q^2 \rangle^{1/2} = 55.901 \ldots. \qquad \textbf{(14)}$$

(E) Now that a "measurement" has been made (no one with $q < 50$ lives in the region any longer), we have

$$\begin{array}{ll} P(q)dq = Cq^4 \exp\left[-\dfrac{q}{25}\right] dq & \text{for } q \ge 50 \\ P(q)dq = 0\, dq & \text{for } q < 50 \end{array} \Bigg\} . \qquad \textbf{(15)}$$

Consequently, we must renormalize the distribution. The normalization requirement is now

$$\int_0^\infty P(q)\, dq = \int_0^{50} 0\, dq + \int_{50}^\infty Cq^4 \exp\left[-\frac{q}{25}\right] dq = 1. \qquad \textbf{(16)}$$

Solving Eq. 16 for C, we obtain

$$C = \left[\int_{50}^\infty q^4 \exp\left[-\frac{q}{25}\right] dq\right]^{-1} \qquad \textbf{(17)}$$

The integral in Eq. 17 is given by

$$\int_{50}^\infty q^4 \exp\left[-\frac{q}{25}\right] dq = \int_0^\infty q^4 \exp\left[-\frac{q}{25}\right] dq - \int_0^{50} q^4 \exp\left[-\frac{q}{25}\right] dq. \qquad \textbf{(18)}$$

The values of the two integrals on the right of Eq. 23 are given by Eqs. 3 and 4:

$$\int_{50}^\infty q^4 \exp\left[-\frac{q}{25}\right] dq$$

$$= \frac{4!}{(1/25)^5} + \exp\left[-\frac{50}{25}\right] [(25)(50)^4 + 4(25)^2(50)^3 + 4(3)(25)^3(50)^2$$

$$+ 4!(25)^4(50) + 4!(25)^5] - \frac{4!}{(1/25)^5}$$

$$= \exp\left[-\frac{50}{25}\right] [(25)(50)^4 + 4(25)^2(50)^3 + 4(3)(25)^3(50)^2 + 4!(25)^4(50) + 4!(25)^5]$$

$$= e^{-2}[156{,}250{,}000 + 312{,}500{,}000 + 468{,}750{,}000 + 468{,}750{,}000 + 234{,}375{,}000]$$

$$= e^{-2}[1{,}640{,}625{,}000] \qquad \textbf{(19)}$$

The required new normalization constant is, therefore,

$$C = \frac{e^2}{1{,}640{,}625{,}000}. \qquad \textbf{(20)}$$

The new average IQ after removing undesirables from the area is

$$\langle q \rangle = \int_0^\infty q\, P(q)\, dq = \frac{e^2}{1{,}640{,}625{,}000} \int_{50}^\infty q^5 \exp\left[-\frac{q}{25}\right] dq$$

$$= \frac{e^2}{1{,}640{,}625{,}000} \left[\int_0^\infty q^5 \exp\left[-\frac{q}{25}\right] dq - \int_0^{50} q^5 \exp\left[-\frac{q}{25}\right] dq)\right]$$

$$= \frac{e^2}{1{,}640{,}625{,}000} \exp\left[-\frac{50}{25}\right] [(25)(50)^5 + 5(25)^2(50)^4 + 5(4)(25)^3(50)^3$$

$$+ 5(4)(3)(25)^4(50)^2 + 5!(25)^5(50) + 5!(25)^6]$$

$$= [78{,}125{,}000{,}000 + 19{,}531{,}250{,}000 + 39{,}062{,}500{,}000$$

$$+ 58{,}593{,}750{,}000 + 58{,}593{,}750{,}000 + 29{,}296{,}875{,}000]/1{,}640{,}625{,}000$$

$$= 129.76 \ldots . \tag{21}$$

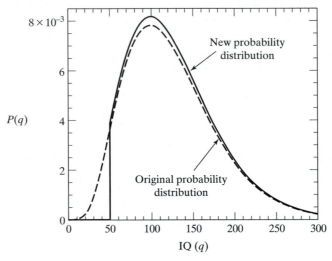

▲ FIGURE 10.7
New normalized probability distribution of IQ for the region described in Example 10.8 after the expulsion of politicians and administrators. The distribution before expulsion is shown as the dashed curve for comparison. The total area under both distributions is unity, as required by the normalization condition.

Figure 10.7 shows a plot of the new normalized distribution, compared with the original distribution before the expulsion of politicians and administrators.
    For related exercises, see Problems 10.4, 10.8, 10.9, 10.12, 10.14, 10.20, and 10.21.

As a final point of interest, the "normal" distribution upon which many teachers and some professors base their grades is a Gaussian probability distribution which assumes that the probability that a value of x will deviate from the average by an amount $x - \langle x \rangle$ is

$$P(x)\, dx = C \exp\left[-\frac{(x - \langle x \rangle)^2}{2\sigma^2}\right] dx \qquad \text{for } -\infty \le x \le \infty, \tag{10.24}$$

where σ^2 is the square uncertainty, $\langle \Delta x^2 \rangle$, defined by Eq. 10.17. The normalization constant can be shown to be $C = [(2\pi)^{1/2}\sigma]^{-1}$. (See Problem 10.20.) This distribution has the characteristic that the probability that a value of x will lie within σ of the mean is approximately 0.683 and the probability that it will lie within 2σ of the mean is about 0.955. (See Problem 10.21.) Figure 10.8 shows a plot of this distribution for the specific case where $\langle x \rangle = 10$ and $\sigma^2 = 1.5$.

The effect of increased uncertainty is to broaden the probability distribution. This can be easily illustrated using the normal distribution. Figure 10.9 shows two such distributions, each with a mean at $\langle x \rangle = 10$. Distribution A has a square uncertainty equal to 1.5, whereas Distribution B has $\langle \Delta x^2 \rangle = 2.5$. A comparison of the results shows that distributions associated with greater uncertainty are broader and more diffuse.

10.3   Continuous Representations of Discrete Variables

It is often much easier to execute computations when the probability distributions are continuous rather than discrete. This is a consequence of the fact that integrals can frequently be performed more easily than the corresponding summations for discrete variables.

Sometimes we can assume a discrete variable to be continuous without significant loss of accuracy. For example, consider a probability distribution for a discrete variable x that assumes only integral values over the range $0 \leq x \leq 100$. This distribution has the form

$$P(x_i) = C \exp[-0.01x_i] \quad \text{for integral } x_i \text{ with } 0 \leq x_i \leq 100.$$

In this case, the normalization constant given by Eq. 10.5 is

$$C = \left[\sum_{n=0}^{100} \exp[-0.01n] \right]^{-1} = \frac{1}{63.896\ldots} = 0.015650\ldots. \quad \textbf{(10.25)}$$

The average value of x given by Eq. 10.13 is

$$\langle x \rangle = (0.015650\ldots) \sum_{n=0}^{100} n \exp[-0.01n]. \quad \textbf{(10.26)}$$

The above sums can be easily evaluated using an Excel spreadsheet or a simple FORTRAN or C computer code. The result for Eq. 10.26 is $\langle x \rangle = 41.641\ldots$.

Let us now assume that we may treat x as a continuous variable over the same range. With this assumption, the normalization constant will be given by Eq. 10.20:

$$C = \left[\int_0^{100} \exp[-0.01x]\,dx \right]^{-1} = \frac{1}{[-100 \exp[-0.01x]]_0^{100}}$$

$$= \frac{1}{-100e^{-1} + 100} = 0.015819\ldots.$$

This result is within 1.09% of the correct result obtained by summing the series in Eq. 10.25. Although it is not as accurate, it is much easier to obtain. If x is continuous, the average value is given by Eq. 10.23:

$$\langle x \rangle = C \int_0^{100} x \exp[-0.01x]\,dx = C \left[\frac{xe^{-0.01x}}{0.01} - \frac{e^{-0.01x}}{(0.01)^2} \right]_0^{100}$$

$$= C[-10,000e^{-1} - 10,000e^{-1} + 10,000]$$

$$= (0.015819\ldots)[10,000 - 20,000e^{-1}]$$

$$= 41.800\ldots.$$

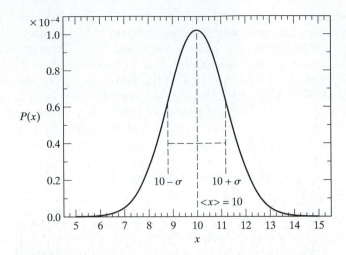

▲ **FIGURE 10.8**
Example of a normal Gaussian probability distribution defined by
Eq. 10.24. In this example, $\langle x \rangle = 10$ and $\langle \Delta x^2 \rangle = \sigma^2 = 1.5$.
Problem 10.21 shows that a random selection of x from this distribu-
tion will result in a value in the range from $\langle x \rangle - \sigma$ to $\langle x \rangle + \sigma$ 68.3% of
the time. This region is shown by the vertical dashed lines in the figure.

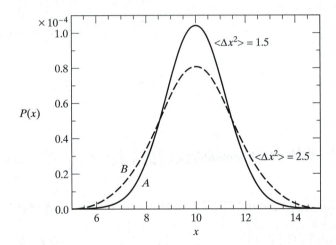

▲ **FIGURE 10.9**
Comparison of two normal Gaussian probability distributions
given by Eq. 10.24. Both distributions exhibit a mean at $x = 10$.
The distribution labeled A has a squared uncertainty of 1.5,
whereas distribution B has $\langle \Delta x^2 \rangle = 2.5$. As can be seen, the
effect of the increased uncertainty is to broaden the distribution.

The percent error in this result, compared with the correct answer obtained
by summing Eq. 10.26, is $0.382\ldots\%$.

If we rewrite the sum in Eq. 10.25 in a somewhat different form, a useful
physical picture of the difference between the continuous and discrete distributions may be obtained. The summation from $n = 0$ to $n = 9$ can be
written as

$$\sum_{n=0}^{9} \exp[-0.01n] = \sum_{n=0}^{9} \exp[-0.01n]\Delta n, \qquad \textbf{(10.27)}$$

provided that we set $\Delta n = 1$ for all values of n. The right-hand side of Eq. 10.27 represents the area beneath the histogram shown in Figure 10.10. The smooth dashed curve in the figure is a plot of $\exp[-0.01x]$ over the same range. The first part of the integral required to normalize the continuous distribution is $\int_0^9 \exp[-0.01x]\,dx$. This integral is the area beneath the smooth dashed curve. Inspection of Figure 10.10 shows that the area beneath the histogram is greater than that beneath the smooth curve. Consequently, the normalization constant for the discrete distribution is smaller than that computed assuming x to be a continuous variable. This conclusion is in accord with the results of the preceding calculations. The error is represented by the difference in the area beneath the histogram and that beneath the smooth curve in the figures. This difference is approximately the sum of 100 nearly triangular areas, each of which has a height $\Delta n = 1$ and a base that is the difference $[P(x_i + 1) - P(x_i)]/C$, plus the area beneath the final histogram block, whose height is that associated with the final term in the discrete sum appearing in Eq. 10.25. Nine of these near-triangular areas and the final histogram block are seen in Figure 10.10. The difference between the normalization sum and the normalization integral for the continuous distribution may be estimated by summing these areas and the final histogram block as follows:

$$\sum_{n=0}^{100} \exp[-0.01n] - \int_0^{100} \exp[-0.01x]\,dx \approx \frac{\Delta n}{2C} \sum_{i=0}^{99} [P(x_i) - P(x_{i+1})]$$

$$+ \frac{P(x_{100})\,\Delta n}{C} = \frac{\Delta n}{2C}\{[P(x_0 - P(x_1)] + [P(x_1) - P(x_2)] + [P(x_2 - P(x_3)] + \cdots$$

$$+ [P(x_{99}) - P(x_{100})]\} \quad + P(x_{100})\frac{\Delta n}{C} = \frac{\Delta n}{2C}[P(x_0) - P(x_{100})] + P(x_{100})\frac{\Delta n}{C}.$$

$$(10.28)$$

For the probability distribution in the example we are considering, $P(x_0)/C = \exp[-0.01(0)] = 1.000$ and $P(x_{100})/C = \exp[-0.01(100)] = e^{-1} = 0.3678\ldots$. With $\Delta n = 1$, Eq. 10.28 states that the difference between the discrete sum and the integral is approximately $(1)(1.00 - 0.3678)/2 + (1)(0.3678) = 0.6839$. The actual result is $0.684\ldots$. The equation shows that if the differences

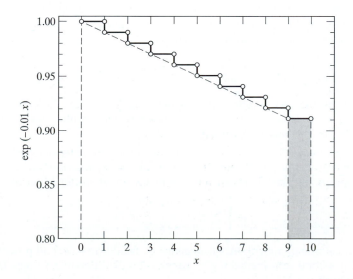

◀ FIGURE 10.10
Continuous versus discrete distributions. The area beneath the histogram represents part of the sum required to normalize the discrete distribution described in the text. The area beneath the dashed curve is the corresponding area if the discrete distribution is assumed to be continuous. The difference between the two is represented by the sum of the near-triangular areas between the histogram and the dashed curve plus the area of the final histogram block, which is shaded in the figure.

between $P(x_i)$ and $P(x_{i+1})$ are very small for all values of i, and if $P(x_{final})/C$ is small relative to the entire sum or integral, the difference between the discrete sum and the integral will be negligible. When this is the case, very little error will be made by assuming that the discrete distribution may be replaced with a distribution that is continuous over the same range. Problem 10.13 is a second case in point. We shall make use of this type of approximation when we treat the statistical thermodynamics of translational and rotational energies in Chapters 17 and 18.

Summary: Key Concepts and Equations

1. Let $P(X_i)$ be the probability that a measurement of a discrete variable X will produce the result X_i. If $P(X_i)$ is known to be proportional to f_i, then

$$P(X_i) = Cf_i$$

where the proportionality constant C is called the normalization constant. The fact that the sum of $P(X_i)$ for all i must be unity requires that

$$C = \frac{1}{\displaystyle\sum_{i=1}^{M} f_i} = \left[\sum_{i=1}^{M} f_i \right]^{-1},$$

where M is the number of possible values for X_i. The probability is, therefore, given by

$$P(X_i) = \left[\sum_{i=1}^{M} f_i \right]^{-1} f_i.$$

This expression is called the normalized probability distribution.

2. In general, the combined probability of the occurrence of S specified events in S different independent experiments is

$$P(X_1, X_2, X_3, \ldots, \text{ and } X_s) = \prod_{i=1}^{S} P(X_i) = P(X_1)P(X_2)P(X_3)\ldots P(X_S),$$

provided that the probabilities of all the events are independent. If we wish to know the probability that either event 1, event 2, event 3, ..., *or* event S will occur in a single experiment, the desired result is

$$P(X_1, X_2, \ldots, \text{ or } X_s) = \sum_{i=1}^{S} P(X_i) = P(X_1) + P(X_2) + \cdots + P(X_S),$$

as long as the probabilities are independent. Therefore, the key points are whether we are dealing with a series of experiments or with a single experiment and whether the logical connective is "and" or "or."

3. The degeneracy of event i, denoted g_i, is the number of states or events that have the same probability factor, or $P(X_i)$, in a probability distribution. We may use this concept to write the normalization constant in two different ways:

$$C = \left[\sum_{i=1}^{M} f_i \right]^{-1} = \left[\sum_{i=1}^{M'} g_i f_i \right]^{-1}.$$

The first summation, without the degeneracy factor, runs over all possible f_i values. The second sum, containing g_i, runs over only different f_i values. Therefore, we always have $M' \leq M$.

4. If x is a continuous variable over the range $X_1 \le x \le X_2$ and $f(x)dx$ is a function that is proportional to the probability of observing x in the range from x to $x + dx$, then the normalized probability distribution is

$$P(x)dx = Cf(x)dx,$$

where the normalization constant is

$$C = \frac{1}{\int_{X_1}^{X_2} f(x)dx} = \left[\int_{X_1}^{X_2} f(x)dx \right]^{-1}.$$

The normalized probability distribution can, therefore, be written in the form

$$P(x)dx = \frac{f(x)dx}{\int_{X_1}^{X_2} f(x)dx}.$$

It is important to keep in mind that, for a continuous variable, the probability distribution function always gives the probability that x will lie in a range dx about some value x_o. The probability that x will be precisely equal to x_o is always zero. This fact, although correct, has no practical significance.

5. The average value of a function $h(x)$ is given by

$$\langle h(x) \rangle = \frac{\sum_{i=1}^{M} h(x_i) f_i}{\left[\sum_{i=1}^{M} f_i \right]},$$

as long as x is a discrete variable with M possible values. The denominator of the equation could be written in the form with the degeneracy factor g_i, as described under Point 3. If x is a continuous variable, the appropriate expression for the average value of $h(x)$ over the interval from X_1 to X_2 is

$$\langle h(x) \rangle = \frac{\int_{X_1}^{X_2} h(x) f(x) dx}{\int_{X_1}^{X_2} f(x) dx}.$$

6. Whenever a measurement is performed or additional information is obtained, the normalized probability distribution is almost always altered in some manner. The appropriate procedure for incorporating additional information is to adjust the values of the f_i so as to reflect the added data, renormalize the new probability distribution, and then proceed in the usual manner to compute new average values, most probable values, uncertainties, etc.

7. The square uncertainty, $\langle \Delta x^2 \rangle$, is defined to be the mean square deviation from the mean, where "mean" denotes the average. In statistics, this quantity is often called the variance and is given by

$$\langle \Delta x^2 \rangle = \langle x^2 \rangle - \langle x \rangle^2.$$

Therefore, we require the average square and the average; the second and first moments of the distribution, respectively, to compute the square uncertainty. These averages are computed by appropriate summations or integrations, depending upon whether x is a discrete or continuous variable.

8. If the magnitude of the differences between successive normalized probabilities, $|P(x_{i+1}) - P(x_i)|$, and $P(x_{\text{final}})$ of a discrete distribution are very small relative to the normalization sum or integral, x can be treated as if it is a continuous variable without introducing appreciable error into the calculations.

Problems

Problems that require the use of some type of computational device are marked with an asterisk (*). Problems that require some type of plotting routine are indicated with a pound sign (#).

10.1 There are 10 students in a class. On a 100-point hour examination, their grades are as follows:

$$58, 26, 88, 95, 71, 40, 15, 52, 85, 64.$$

(A) Compute the average grade.

(B) Compute the square uncertainty in this distribution of grades.

10.2 (A) There are 10 horses numbered from 1 to 10 in a race. If the probability that a given horse will win is proportional to the number the horse wears, obtain the normalized probability distribution for each horse winning the race.

(B) For the horses in (A), what is the probability that an odd-numbered horse will win the race?

(C) Repeat (B) under the condition that the probability that the horse will win is proportional to the square root of the number the horse wears.

10.3 Jim, John, and Bob are reporters for a newspaper. Jim gets the facts correct 88% of the time, John gets them right 80% of the time, and Bob gets them right 55% of the time (Bob has now gone on to a political career). Jim and John each cover a story independently and submit stories that agree as to the facts.

(A) What is the probability that Jim and John have the facts of the story correct?

(B) The editor, wishing confirmation, sends Bob to cover the same story, and Bob's facts also agree with those of Jim and John. What is the probability that the facts, as reported, are correct?

(C) If Bob's facts had disagreed with those of Jim and John, what would be the probability that the facts, as reported by Jim and John, are correct?

10.4 In a given area of the country, the ages of all the inhabitants are found to lie in the range $0 \le A \le 50$, where A is the age of an inhabitant. It is further determined that the probability of an inhabitant having an age in the range from A to $A + dA$ is proportional to the quantity $[50A - A^2]dA$.

(A) Obtain the normalized probability distribution for the ages of inhabitants in the given area of the country.

(B) If an inhabitant is chosen randomly, what is the probability that his or her age will be in the range $10 \le A \le 20$?

(C) Compute the average age of the inhabitants of the area.

(D) If all of the inhabitants over the age of 25 in the area were to move out of the area, obtain an expression giving the new normalized probability distribution for the ages of the remaining inhabitants of the area.

(E) Compute the new average age of the remaining inhabitants of the area under the conditions given in (D).

(F) Plot the normalized probability distributions from (A) and (D) on the same graph.

10.5 Consider three playing cards. One card is white on both sides. A second card is white on one side and black on the other. The third card is black on both sides. The cards are shuffled and one card is chosen randomly. With eyes closed, you place this card on the table. Upon opening your eyes, you find that the card on the table has a black side face up. What is the probability that the down side of the card on the table is also black?

10.6 Mr. Jones is known to have two children. You are told in advance that at least one of Mr. Jones' children is a boy. What is the probability that both children are boys? This is a problem that was reported in Marilyn vos Savant's newspaper column ("Ask Marilyn", *Parade*, October 19, 1997, p. 8). She solved the problem correctly and was besieged with letters informing her that her solution was incorrect. Apparently, a lot of people need to take physical chemistry, or, at least, a course in logic. (The author thanks Ms. vos Savant and Parade Publications for permission to cite this material.)

10.7 Mr. Jones is known to have two children. You are told in advance that Mr. Jones' oldest child is a boy. What is the probability that both children are boys?

A professor wishes to evaluate the probability that a student who has scored X_1 points in first-semester physical chemistry out of a possible score of 667 will score X_2 points during the second semester using the same grading system. The system has 300 points possible on three 100-point hour exams, 167 points on the final examination, and 200 points on the homework.

The professor makes the following assumptions to estimate the desired probability:

(A) The variation in the student's score will come entirely from the examinations. The homework score will be identical to that made during the first semester.

(B) The probability distribution for the exam scores will be a normal Gaussian distribution centered at the student's total exam score during the first semester, X_{T1}. The functional form of this distribution is

$$P(X_e) = C \exp\left[-\frac{(X_e - X_{T1})^2}{2\sigma^2}\right],$$

where X_e is the number of points the student will score on exams during the second semester, C is the normalization constant, and σ^2 is the square of the uncertainty in the distribution of the student's exam scores during the first semester. That is, if F_i is the fraction of total points scored by the student on exam i during the first semester, then

$$\sigma^2 = [\langle F^2 \rangle - \langle F \rangle^2](467)^2,$$

where $\langle \ \rangle$ stands for the average value. As an example, if the student has scores of 75, 84, and 71 on the three hour exams and 101 on the final, then $F_1 = 75/100 = 0.75$, $F_2 = 84/100 = 0.84$, $F_3 = 71/100 = 0.71$, and $F_4 = 101/167 = 0.6048$. We then have $\langle F \rangle = (0.75 + 0.84 + 0.71 + 0.6048)/4 = 0.7262$ and $\langle F^2 \rangle = [(0.75)^2 + (0.84)^2 + (0.71)^2 + (0.6048)^2]/4 = 0.5345$. Therefore, $\sigma^2 = [0.5345 - (.7262)^2](467)^2 = 1,556$.

(C) The distribution $P(X_e)$ can be treated as if X_e is a continuous variable over the range $0 \le X_e \le 467$.

The next six problems are based on the above type of probability distribution.

10.8[#] A student in first-semester physical chemistry scored 511 points with 180 on the homework and exam scores of 75, 84, 71, and 101 on the three hour exams and final, respectively. Determine the normalized probability distribution $P(X_e)$ for this student's scores in the second-semester course. Plot $P(X_e)$ versus X_e.

10.9* What is the probability that the student described in the previous problem will make a grade of A in the second-semester course if the cutoff lines are as follows:

Range	Grade
$500 \le X_2 \le 667$	A
$400 \le X_2 \le 499$	B
$320 \le X_2 \le 399$	C
$265 \le X_2 \le 319$	D
$0 \le X_2 \le 264$	F

10.10* The student described in the previous two problems does not keep track of his grades during the second semester. When the course is over, he has no idea what his grade is. One of his friends asks the professor about the final grade. The professor, of course, refuses to reveal another student's grade to the friend, but as a consolation prize, the professor says, "I won't tell you his final grade, but I will tell you that it is not a B." What is the probability that the student has a grade of A? (The grade cutoff lines are given in the previous problem.)

10.11* Let us now consider a second student in the class, whom we shall call Student 2. The student in the previous three problems will be Student 1. Student 2 made a B during the first semester, having a total homework score of 165 and exam scores of 67, 73, and 69, on the hour exams and 112 on the final.

(A) Obtain the probability distribution for Student 2's exam grades during the second semester.

(B) Which student, 1 or 2, is the more consistent performer on examinations? How do you know?

(C) What is the probability that Student 2 will make a higher grade than Student 1 during the second semester? (Grade cutoff lines are as given in Problem 10.9.)

10.12* The student described in Problem 1 was found to have a normalized distribution of examination grades given by

$$P(X_e) = 0.010117 \exp\left[-\frac{(X_e - 331)^2}{3111.6}\right].$$

With this distribution and his previous homework score of 180, he has a 60.97% chance of making a grade of A during the second semester. Assuming that his distribution of exam grades is unaffected by his total homework score, compute the probability that the student will make a grade of A during the second-semester if his second-semester homework scores are 160, 165, 170, 175, 180, 185, 190, and 195. Plot the probability of making an A as a function of the homework score, and comment on the importance of the homework grade. The grade cutoff lines are as given in Problem 10.9.

10.13* In Problems 10.8 through 10.12, the professor treated the probability distribution as if the total exam score X_e were a continuous variable when, in reality, it is discrete. (That is, exam scores are usually integers.) However, if the difference $|P(X_e + 1) - P(X_e)|$ is very small relative to the total integrated probability (i.e., if the probabilities are closely spaced), the results obtained by treating the distribution as if it involved a continuous variable will be nearly identical to those obtained by treating the variable as discrete. In this problem, we illustrate this point by repeating Problems 10.8 and 10.9 with X_e being a discrete variable. When X_e is continuous over the range $0 \le X_e \le 467$, the normalized probability distribution is

$$P(X_e) = 0.010117 \exp\left[-\frac{(X_e - 331)^2}{3111.6}\right].$$

(A) Compute the nomalization constant for the probability distribution

$$P(X_e) = C \exp\left[-\frac{(X_e - 331)^2}{3111.6}\right],$$

assuming that X_e is a discrete variable which takes the values 0, 1, 2, 3, 4, ..., 467.

(B) Compute the probability that the student will make a grade of A during the second semester. The result using the assumption that X_e may be treated as a continuous variable is 0.6097. How much error was made by the continuous-variable assumption? The grade cutoffs are given in Problem 10.9. In

Problem 10.8, it was stated that the student will make 180 points on the homework.

10.14 A ball is moving back and forth between two points on the x-axis, $x = 0$ and $x = a$. It is found that the probability of the ball lying in the range from x to $x + dx$ is proportional to $\sin^2[\pi x/a]$; that is

$$P(x)\,dx = C \sin^2\left[\frac{\pi x}{a}\right] dx \quad \text{for } 0 \le x \le a.$$

(A) Determine the normalization constant for this probability distribution.

(B) Compute the average value of x.

(C) Compute the average value of x^2.

(D) Compute the uncertainty in the position of the ball. (This problem is taken directly from the quantum mechanical solutions given in Chapter 12 to one-dimensional translational motion inside a container of length a. Its presence in this set of problems on probability calculations emphasizes the fact that quantum theory is closely related to probability theory.)

10.15 In Chapter 17, which deals with statistical mechanics, we shall find that the probability that a system will occupy an energy state with energy E is proportional to $\exp[-E/RT]$ if E is given in units of energy per mole. Consider a simple system having only four possible energy states that we denote as States 1, 2, 3, and 4. The energies of these states are $E_1 = 0$ J mol^{-1}, $E_2 = 1{,}000$ J mol^{-1}, $E_3 = 2{,}000$ J mol^{-1}, and $E_4 = 3{,}000$ J mol^{-1}.

(A) Obtain the normalized probability distribution for the system in state i. Evaluate the normalization constant if $T = 298$ K.

(B) Calculate the average energy per system that a large number of such systems would have at 298 K.

(C) Which energy state makes the greatest contribution to the average energy?

(D) Which energy state is the most heavily populated?

(E) What is the probability that the combined energy of two separate systems described by the given probability distribution will add to 3000 joules?

(F) Compute the average total energy the two systems in (E) will have at 298 K. What relationship does this result have to that obtained in (B)?

10.16 The average energy of two systems described by Problem 10.15 is exactly twice the average energy of a single system. This is a general property of any system in which the probability that the system will occupy a state of energy E is proportional to $\exp[-E/RT]$. Suppose such a system has M possible energy states.

(A) Obtain an expression for the average energy of one such system, $\langle E \rangle_1$

(B) Show that the average energy for N such systems is just $N \langle E \rangle_1$.

10.17 Example 10.5 examined the probabilities involved in a game of chance in which there were an equal number of each of five different-sized holes in a game board. (see Figure 10.4.) Without informing the players, an enterprising gambler changes the ratios of different sizes of holes. The new game board contains 11 25-cm holes and only 10 of each of the other sizes. Nothing else is changed. Compute the average return to the players on this new game board.

10.18 Craps is a popular dice game played in all major casinos in the world. In this game, the player throws two dice. If the first roll produces a 7 or an 11, the player wins. If the first roll produces a 2, 3, or 12, he or she loses. If any other sum of the dots on the up faces of the two dice occurs on the first roll, the player continues to roll the dice either until a 7 comes up, in which case the player loses, or until the sum of the dots on the up faces of the dice equals that which appeared on the first roll, in which case the player wins. When the player wins, he or she is said to have made a pass. Compute the probability that the player will make a pass at craps. (There are many other bets available to the player at the craps table; all save one are inferior to simply betting that you will make a pass.)

10.19 The sole bet available at the craps table which is better for the player than betting that a "pass" will be made is for the player to bet that a pass will not be made. If the player bets in this manner and rolls a 12 on the first roll, there is no action. That is, it is as if the dice had not been thrown at all. The player does not lose or win; instead, he or she rolls the dice again, and this new roll is counted as the first roll. Compute the probability that a player who bets on "Don't Pass" will win the bet at craps. (See Problem 10.18 for the rules of craps.)

10.20 Show that the normalization constant for a normal Gaussian distribution, as given in Eq. 10.24, is $[(2\pi)^{1/2}\sigma]^{-1}$, as stated in the text. (*Hint:* Transform x to the variable z, where $z = (x - \langle x \rangle)/\sigma$, and use a table of definite integrals.)

10.21* (A) Show that the probability that x will lie in the range $\langle x \rangle - \sigma \le x \le \langle x \rangle + \sigma$ for a normal Gaussian probability distribution, as given in Eq. 10.24, is approximately 0.683.

(B) Compute the probability that x will lie in the range $\langle x \rangle - 2\sigma \le x \le \langle x \rangle + 2\sigma$.

(C) Compute the probability that x will lie in the range $\langle x \rangle - 3\sigma \le x \le \langle x \rangle + 3\sigma$ if the probability distribution is that given in Eq. 10.24. (See hint in Problem 10.20; as an additional hint, use a table of error function integrals, or do the integrations numerically.)

10.22 A container holds four red balls, four black balls, four white balls, and four green balls. All the balls are of equal size and mass.

(A) If three balls are drawn from the container at random, what is the probability that they will all be red?

(B) What is the probability that the three balls drawn in (A) will comprise two red balls and one green ball? (*Hint:* This problem is similar to Example 10.6, except for the important fact that the probabilities are now conditional rather than independent. The dependency, which must be taken into account in working the problem, arises because the probability of selecting a particular color of ball on the second draw is dependent upon the color of ball obtained in the first draw.)

10.23* There are N randomly chosen people in a room.

(A) Ignoring the effect of leap years and assuming that all years have 365 days, obtain an expression as a function of N giving the probabiliity that no two birthdays will fall on the same day of the year. Plot the result as a function of N over the range $2 \le N \le 60$.

(B) What value of N makes the probability as close to 0.5 as possible? If $N = 365$, what is the probability that no two persons in the room have their birthdays on the same day of the year. (*Hint:* This problem is similar to Example 10.3, but it is much easier to solve the problem if it is viewed as comprising N separate experiments rather than one experiment in which all N persons state their birthdays. The solution to the problem is rather surprising—so much so, that many persons refuse to believe it is correct!)

10.24 Glrrp settled his intergalactic spacecraft down in a field on the Kent Farm in Smallville, out of fuel. He acertained that the Kents had a can of pea soup, just the fuel he needed to get back home. Mr. Kent offered to play craps for the fuel (see Problems 10.18 and 10.19) in exchange for a chance to win the spacecraft. Mr. Kent proposed using standard earth dice, shown in the accompanying diagram as they would look if they were constructed out of paper and unfolded. Glrrp wished to use his own dice, also shown in the diagram.

(A) What are the chances that Mr. Kent will throw a 7 or an 11 on the first throw of two earth dice?

(B) What are the chances that Glrrp will throw a 7 or an 11 on the first throw of two alien dice?

(C) What are the chances that Mr. Kent will throw a 2, 3, or 12 on the first throw of two earth dice?

(D) What are the chances that Glrrp will throw a 2, 3, or 12 on the first throw of two alien dice?

(E) What interesting hypothesis do the preceding questions suggest to you? Prove it. (You may assume that both the earth and alien dice are "honest," in that each face has an equal probability of being on top when the die is thrown.)

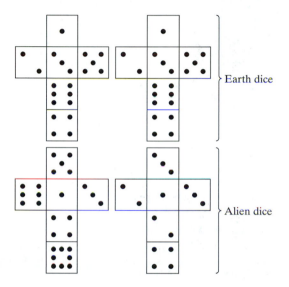

▲

The author is indebted to Fredrick L. Minn, M.D., Ph.D, for providing this problem and the associated solution.

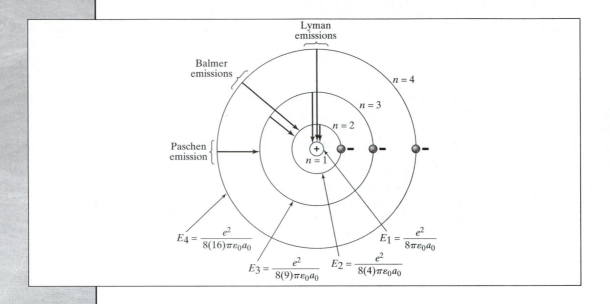

Introduction to Quantum Mechanics

Quantum mechanics is one of the most profound and beautiful creations of the human mind. To the best of our current knowledge, the theory provides the basis for understanding and predicting the behavior of all atomic and molecular systems. When quantum theory is combined with the principles of statistical mechanics, all the thermodynamic properties discussed in Chapters 1 through 9 of this text can, in principle, be quantitatively computed. The theory provides the underlying foundation for every type of spectroscopy. Beauty comes in many forms. The exquisite art and sculpture of da Vinci and Michelangelo, the incomparable music of Mozart, and the intricate complexity of quantum theory are among the greatest of human achievements. Their beauty is transcended only by those things that are beyond the reach of man, such as the smile of a child.

In this chapter, we will first discuss classical (Newtonian) mechanics in sufficient detail to permit its application to systems of interest throughout the remainder of the text. The discussion will also provide you a basis for comparing and contrasting the predictions of quantum theory with those of the more familiar classical world. We then discuss the history of the development of quantum theory. This discussion sets the stage for the introduction of the formal theory and affords insight into the magnitude of human achievement that quantum mechanics represents. Just as it was necessary to present a thorough discussion of the significance of exact differentials before introducing the first law of thermodynamics, likewise, some mathematical points must be understood before quantum theory can be presented. The chapter provides this mathematical review. Finally, the quantum mechanical postulates are introduced, and the fundamental equations of the theory are developed. Later chapters apply these equations to systems of interest to the physical chemist.

11.1   Classical Mechanics

11.1.1   Fundamental Postulates

Classical mechanics is primarily the creation of Sir Isaac Newton (1642–1727), an English mathematician and physicist who was knighted for his contributions to human knowledge. Because of its origin, classical mechanics is sometimes called *Newtonian* mechanics. In classical mechanics, all variables are continuous over the range they span. Position and momentum coordinates, mass, time, and energy can assume any value within the ranges permitted by the experimental situation.

The basis of classical mechanics is the three Newtonian postulates, or laws. Like the first, second, and third laws of thermodynamics, these postulates are not proven. They are simply accepted as fact. The consequences of the postulates are then deduced and compared with experimental observations. If the observations agree with the predictions of the theory, the postulates are advanced to the status of laws; if the observations do not agree, the postulates may be either totally discarded or retained as excellent approximations in some situations. Prior to 1900, all observations appeared to be in agreement with the predictions of classical mechanics to such an extent, that some physicists were reputed to have remarked that "The future of physics lies in the fourth, fifth, and sixth decimal places." The thrust of this remark was to emphasize the "fact" that all fundamental theory was in place and the only remaining challenge was to refine experimental measurements to the point that four-, five-, and six-digit accuracy could be obtained. Obviously, these scientists did not anticipate the cataclysm that was to come between 1900 and 1927.

The first Newtonian postulate states that the rate of change of the momentum of any particle i is equal to the force acting on the particle. The mathematical statement of this postulate is

$$\mathbf{F}_i = \frac{d\mathbf{P}_i}{dt},$$

(11.1)

where the boldface denotes a vector quantity. In nonvector form, Eq. 11.1 becomes the three equations

$$F_{xi} = \frac{dP_{xi}}{dt} \tag{11.2A}$$

$$F_{yi} = \frac{dP_{yi}}{dt} \tag{11.2B}$$

and

$$F_{zi} = \frac{dP_{zi}}{dt}. \tag{11.2C}$$

In Eq. 11.2, F_{xi}, F_{yi}, and F_{zi} are, respectively, the x, y, and z components of force on particle i, with a similar notation for the various momentum components. We will soon find that in rectangular Cartesian coordinates, $P_{xi} = m_i\, dx_i/dt = m_i v_{xi}$, where m_i is the mass of particle i and v_{xi} is the x-component of velocity of the particle. Substituting this equation into Eq. 11.2A, we obtain

$$F_{xi} = \frac{d\left(m_i \dfrac{dx_i}{dt}\right)}{dt} = m_i \frac{d^2 x_i}{dt^2} = m_i a_{xi} \tag{11.3}$$

where a_{xi} is the acceleration of particle i in the x-direction. This equation is the familiar form taught in all high school physics courses.

The second Newtonian postulate is commonly stated in the form "For every action, there is an equal, but opposite, reaction." If $\mathbf{F_{ji}}$ is the force on particle i produced by the presence of particle j, then the mathematical statement of the second postulate is

$$\boxed{\mathbf{F}_{ji} = -\mathbf{F}_{ij}}. \tag{11.4}$$

Standing alone, the second postulate suggests that we might observe the situation shown in Figure 11.1.

The third postulate shows that the situation illustrated in Figure 11.1 cannot exist. It requires that the interparticle forces act along the line of centers of the two particles (the line connecting the particles). The third postulate is, therefore, called the *collinearity law*. The required directions of the forces are shown in Figure 11.2. In the upper diagram, the interparticle force is repulsive along the line of centers; in the lower diagram, the force is attractive.

11.1.2   Newtonian Equations of Motion

Classical mechanics is *deterministic*. This statement means that if we know the position and momentum coordinates of all particles in any system at any moment in time, t_o, we can, in principle, compute all future behavior of the system, as well as its entire previous history. This point was the focus of considerable philosophical debate in the years preceding quantum theory. If Newtonian mechanics truly describes the nature of the universe, it becomes conceptually possible to compute exactly what everyone and everything in the universe will be doing tomorrow from the knowledge of the positions and momenta of all particles today and, as we shall see, a complete knowledge of the potential-energy function for the universe. Freedom of choice and individual responsibility are completely removed. No one has any choice. All

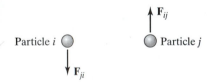

▲ FIGURE 11.1
Equal and opposite forces applied by one particle on another. Standing alone, the second Newtonian postulate might permit this situation.

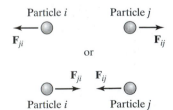

▲ FIGURE 11.2
The two possible directions for the interparticle forces permitted by the collinearity law.

actions are predetermined by the momenta and coordinates of the system today. Moreover, we can also determine what secret things everyone did yesterday. The fact that we can never know all the positions and momenta of all the particles in the universe nor the overall potential-energy function does not alter the philosophical point that if such things were known, the computations could, in principle, be made. Such a situation was disquieting to many people who refused to accept the notion that everything is predetermined. Quantum theory fully resolved this debate. We shall defer discussion of the manner by which quantum theory achieves this resolution until Chapter 12.

The equations required to compute the future and past behavior of a system can be obtained by starting with Eq. 11.3 and using the general definition of the potential-energy function given in Chapter 1. If we employ a rectangular Cartesian coordinate system, Eq. 11.3 gives

$$F_{xi} = \frac{dP_{xi}}{dt} = \frac{d\left(m_i \dfrac{dx_i}{dt}\right)}{dt} = m_i \frac{d^2 x_i}{dt^2}.$$

The left side of this expression may be written in terms of the system potential V by using Eq. 1.46:

$$F_{xi} = -\frac{\partial V}{\partial x_i}. \tag{11.5}$$

In general, for an N-particle system, the potential is a function of all $3N$ coordinates. Combining of Eqs. 11.3 and 11.5 yields

$$\boxed{m_i \frac{d^2 x_i}{dt^2} + \frac{\partial V}{\partial x_i}} = 0. \tag{11.6A}$$

Similar equations can be easily obtained for the y- and z-coordinates. The results are

$$\boxed{m_i \frac{d^2 y_i}{dt^2} + \frac{\partial V}{\partial y_i}} = 0 \tag{11.6B}$$

and

$$\boxed{m_i \frac{d^2 z_i}{dt^2} + \frac{\partial V}{\partial z_i}} = 0. \tag{11.6C}$$

Equations 11.6A through 11.6C are called the *Newtonian equations of motion*. They constitute a set of $3N$, second-order, coupled ordinary differential equations. If we know the positions and velocities of all N particles at time t_o and the functional form of V, we can calculate the entire future and past behavior of the system. The example that follows illustrates this point.

EXAMPLE 11.1

An artillery piece located at coordinates $x = y = z = 0$ fires a projectile with a muzzle velocity v_o. If the muzzle of the artillery piece lies in the x–y plane and makes an angle of θ_o, with the x-axis, determine the maximum height of the projectile, the point of impact of the projectile with the ground, and the equation of

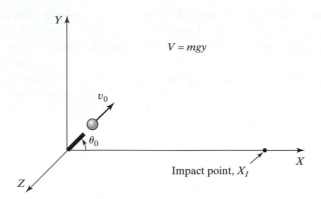

▲ FIGURE 11.3
Trajectory of a projectile fired into the X–Y plane in the earth's gravitational field.

the projectile's trajectory as a function of v_o, θ_o, m (the mass of the projectile), and the gravitational constant g. Assume that the gravitational potential is given by $V = mgy$ and that the effect of friction from the air may be ignored. Figure 11.3 illustrates the problem.

This example illustrates the basic technique used to compute classical trajectories to study chemical reactions. The technique will be discussed in more detail in Chapter 20.

Solution

Since we have one particle (the projectile), there are three Newtonian equations of motion. In this simple example, the equations are uncoupled and can be solved separately. Let us first solve the y-equation. Equation 11.6B gives

$$m\frac{d^2y}{dt^2} + \frac{\partial V}{\partial y} = 0, \tag{1}$$

where we have dropped the subscript, since we have only one particle. The derivative of V with respect to y is

$$\frac{\partial V}{\partial y} = mg. \tag{2}$$

Substituting Eq. 2 into Eq. 1 and then dividing by m gives

$$\frac{d^2y}{dt^2} + g = \frac{d}{dt}\left(\frac{dy}{dt}\right) + g = 0. \tag{3}$$

Equation 3 shows that the motion in the y direction is independent of mass, because m no longer appears in the equation. We now move g to the right side, multiply by dt, and take the indefinite integral of both sides, producing

$$\int d\left(\frac{dy}{dt}\right) = -g \int dt. \tag{4}$$

Since the integral of dx is x, the integral of $d(dy/dt)$ is (dy/dt). Equation 4, therefore, gives

$$\left(\frac{dy}{dt}\right) = y \text{ component of velocity} = v_y = -gt + C, \tag{5}$$

where C is a constant of integration. If we take our zero of time to be the instant the projectile is fired, Eq. 5, evaluated at $t = 0$, yields

$$v_y = v_o \sin(\theta_o) = C. \tag{6}$$

Substituting this result into Eq. 5 produces

$$v_y = \left(\frac{dy}{dt}\right) = v_o \sin(\theta_o) - gt. \tag{7}$$

Again, we separate variables and integrate:

$$\int dy = \int [v_o \sin(\theta_o) - gt]dt. \tag{8}$$

This produces

$$y = v_o \sin(\theta_o)t - \frac{gt^2}{2} + C', \tag{9}$$

where C' is a second constant of integration. Since we know that at $t = 0$, $y = 0$, the value of C' must be zero. In that case, the variation of the y-coordinate with time is given by

$$y = v_o \sin(\theta_o)t - \frac{gt^2}{2}. \tag{10}$$

The equation of motion for the x-coordinate is even easier to solve. Equation 11.6A gives

$$m\frac{d^2x}{dt^2} + \frac{\partial V}{\partial x} = m\frac{d^2x}{dt^2} = 0, \tag{11}$$

because $\partial V/\partial x = 0$. Again, we may divide by m to show that the motion along x does not depend upon mass. We now have

$$\frac{d}{dt}\left(\frac{dx}{dt}\right) = 0. \tag{12}$$

But the only way Eq. 12 can hold is to have $(dx/dt) =$ a constant $= K$. At $t = 0$, the x component of velocity is $v_o \cos(\theta_o)$. Since this velocity is a constant, it follows that

$$\left(\frac{dx}{dt}\right) = v_x = v_o \cos(\theta_o). \tag{13}$$

Separating the variables and integrating gives

$$\int dx = v_o \cos(\theta_o) \int dt, \tag{14}$$

so that we obtain

$$x = v_o \cos(\theta_o)t + K'. \tag{15}$$

Now, at $t = 0$, the projectile is at the origin, $x = 0$; hence, the constant of integration K' must be zero, and we have

$$x = v_o \cos(\theta_o)t. \tag{16}$$

The z equation is identical to the x equation, since we also have $\partial V/\partial z = 0$; therefore, we obtain a result similar to Eq. 12:

$$\frac{d}{dt}\left(\frac{dz}{dt}\right) = 0. \tag{17}$$

Accordingly, (dz/dt) must be a constant. Also, at $t = 0$, we have $v_z = 0$, which means that the constant must be zero, so that at all times,

$$\frac{dz}{dt} = 0. \tag{18}$$

This shows that z itself must be a constant. Further, because $z = 0$ at $t = 0$, the constant is zero. As a result, $z = v_z = 0$, and the motion of the projectile is confined to the x–y plane. Summarizing the results, we have

$$y = v_o \sin(\theta_o)t - \frac{gt^2}{2} \quad \text{and} \quad v_y = v_o \sin(\theta_o) - gt,$$

$$x = v_o \cos(\theta_o)t \quad \text{and} \quad v_x = v_o \cos(\theta_o),$$

and

$$z = 0 \quad \text{and} \quad v_z = 0.$$

To obtain the maximum height of the projectile, we wish to maximize y. This requires that we have

$$\frac{dy}{dt} = v_o \sin(\theta_o) - gt = 0. \tag{19}$$

The time at which the maximum is attained is

$$t_{\max} = \frac{v_o \sin(\theta_o)}{g}. \tag{20}$$

Substituting $t_{\max}$ into Eq. 10 gives

$$y_{\max} = v_o \sin(\theta_o)\frac{v_o \sin(\theta_o)}{g} - \frac{g}{2}\left[\frac{v_o \sin(\theta_o)}{g}\right]^2 = \frac{v_o^2 \sin(\theta_o)}{2g}. \tag{21}$$

We can find the point of impact by solving Eq. 10 for the time at which the projectile strikes the ground when y becomes zero. That is, we seek the roots of the equation

$$v_o \sin(\theta_o)t - \frac{gt^2}{2} = 0 = t\left[v_o \sin(\theta_o) - \frac{gt}{2}\right]. \tag{22}$$

One root is $t = 0$, when the projectile is at the origin. The root of interest, however, is the one obtained by requiring that

$$v_o \sin(\theta_o) - \frac{gt}{2} = 0. \tag{23}$$

Solving for $t = t_I$ (the time of impact), we obtain

$$t_I = \frac{2v_o \sin(\theta_o)}{g}. \tag{24}$$

Substituting Eq. 24 into Eq. 16 provides the value of x at the point of impact, x_I:

$$x_I = v_o \cos(\theta_o)\, t_I = v_o \cos(\theta_o)\frac{2v_o \sin(\theta_o)}{g}$$

$$= \frac{v_o^2 2 \sin(\theta_o) \cos(\theta_o)}{g} = \frac{v_o^2 \sin(2\theta_o)}{g}. \tag{25}$$

The equation describing the trajectory in terms of y and x can be found by using Eq. 16 to obtain t in terms of x and then substituting the result into Eq. 10. First, we have

$$t = \frac{x}{v_o \cos(\theta_o)}. \tag{26}$$

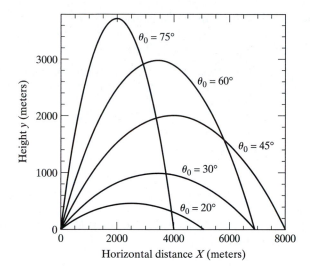

▲ FIGURE 11.4
Trajectories for projectiles moving in the earth's gravitational field with an initial velocity of 280 m s^{-1} for various initial angles θ_0. The curves are obtained by direct solution of the Newtonian equations of motion, as illustrated in Example 11.1. The maximum horizontal distance for the trajectory is achieved when $\theta_0 = 45°$.

Then substituting into Eq. 10 gives

$$y = v_o \sin(\theta_o)\left[\frac{x}{v_o \cos(\theta_o)}\right] - \frac{g}{2}\left[\frac{x}{v_o \cos(\theta_o)}\right]^2$$

$$= x \tan(\theta_o) - \frac{gx^2}{2v_o^2 \cos^2(\theta_o)} = x\left[\tan(\theta_o) - \frac{gx}{2v_o^2 \cos^2(\theta_o)}\right]. \qquad (27)$$

Equation 27 is a parabola that intersects the origin and crosses the x-axis again at $x = v_o^2 \sin(2\theta_o)/g$, as Eq. 25 demonstrates. Figure 11.4 shows several trajectories with varying values of θ_o and with $v_o = 280$ m s^{-1}.

For related exercises, see Problems 11.1, 11.2, 11.3, and 11.6.

11.1.3   Hamilton's Equations of Motion

It is often useful to express the Newtonian equations of motion in a form that involves first-order, rather than second-order, differential equations. It is even more important to obtain a set of equations that is valid for any coordinate system, not just rectangular Cartesian coordinates. In seeking to meet these two aims, we begin by first defining the classical Hamiltonian, **H**, as the sum of the kinetic and potential energies of the system:

$$\boxed{\mathbf{H} = T(\mathbf{v}_1, \mathbf{v}_2, \dots, \mathbf{v}_N) + V(q_1, q_2, \dots, q_{3N}) = \mathbf{E}}. \qquad \textbf{(11.7)}$$

In Eq. 11.7, T is the kinetic energy of the system, which depends upon the velocities of the N particles of the system, and V is the potential energy of the system, which depends upon the $3N$ position coordinates denoted by the q_i. If the system is isolated, with no input or withdrawal of external forces or energy, the sum of the kinetic and potential energies will be equal to the total energy E. Equation 11.7 is not the usual manner in which the classical Hamiltonian is introduced, but it will suffice for the limited treatment of classical mechanics we are presenting here.

If we are in Cartesian coordinates, the kinetic energy takes the form

$$T = \sum_{i=1}^{N}\left(\frac{m_i}{2}\right)[v_{xi}^2 + v_{yi}^2 + v_{zi}^2]. \qquad \textbf{(11.8)}$$

Since the Cartesian linear momentum is given by

$$P_{xi} = m_i v_{xi} \tag{11.9}$$

with similar equations for P_{yi} and P_{zi}, we may write the total kinetic energy in the form

$$T = \sum_{i=1}^{N} \left[\frac{1}{2m_i}\right][P_{xi}^2 + P_{yi}^2 + P_{zi}^2], \tag{11.10}$$

so that the classical Hamiltonian becomes

$$\mathbf{H} = \sum_{i=1}^{N} \left[\frac{1}{2m_i}\right][\,[P_{xi}^2 + P_{yi}^2 + P_{zi}^2] + V(x_1, y_1, z_1, x_2, y_2, z_2, \ldots). \tag{11.11}$$

Using Eq. 11.1, 11.5, and 11.11, we can immediately obtain the following six equations for a particle k:

$$\frac{\partial \mathbf{H}}{\partial P_{xk}} = \frac{P_{xk}}{m_k} = v_{xk} = \frac{dx_k}{dt}; \tag{11.12A}$$

$$\frac{\partial \mathbf{H}}{\partial P_{yk}} = \frac{P_{yk}}{m_k} = v_{yk} = \frac{dy_k}{dt}; \tag{11.12B}$$

$$\frac{\partial \mathbf{H}}{\partial P_{zk}} = \frac{P_{zk}}{m_k} = v_{zk} = \frac{dz_k}{dt}; \tag{11.12C}$$

$$\frac{\partial \mathbf{H}}{\partial x_k} = \frac{\partial V}{\partial x_k} = -F_{xk} = -\frac{dP_{xk}}{dt}; \tag{11.12D}$$

$$\frac{\partial \mathbf{H}}{\partial y_k} = \frac{\partial V}{\partial y_k} = -F_{yk} = -\frac{dP_{yk}}{dt}; \tag{11.12E}$$

$$\frac{\partial \mathbf{H}}{\partial z_k} = \frac{\partial V}{\partial z_k} = -F_{zk} = -\frac{dP_{zk}}{dt}. \tag{11.12F}$$

Equations 11.12A through 11.12F are a set of $6N$ first-order, coupled ordinary differential equations for the N particles in the system. Basically, we have converted $3N$ second-order equations into an equivalent set of $6N$ first-order equations, known as *Hamilton's equations of motion*. There are two advantages to expressing equations of motion in Hamilton's form. First, it is easier to numerically solve first-order equations than second-order equations. In many cases we shall examine, analytic solutions will not be possible and numerical methods will be needed. When this is the case, the first-order form of Eq. 11.12 is ideal.

At this point, we need to introduce the general definition of momentum. Consider a general coordinate q that might be x, y, z, an angle, or some other type of coordinate. Let the rate of change of this coordinate with time, dq/dt, be denoted by v_q. Then, by definition, the momentum conjugate to coordinate q is given by

$$\boxed{P_q = \frac{\partial T}{\partial v_q}}. \tag{11.13}$$

In general, we say that q is the conjugate coordinate to momentum P_q and vice versa. The (q, P_q) pair is said to be a pair of conjugate variables. Equation 11.13

is the reason we say that the momentum conjugate to the x-coordinate of particle k in a Cartesian system, P_{xk}, is given by $m_k v_{xk}$. From Eq. 11.8, we have

$$\frac{\partial T}{\partial v_{xk}} = \frac{m_k}{2} \frac{\partial v_{xk}^2}{\partial v_{xk}} = m_k v_{xk}.$$

Therefore, by definition, $P_{xk} = m_k v_{xk}$ for the Cartesian x-coordinate of particle k. The importance of Eq. 11.13 lies in the fact that it gives us the form of the momentum for *any* coordinate, not just Cartesian coordinates.

The second advantage of the Hamiltonian form of equations of motion is that *they hold regardless of the coordinate system employed.* Let us say we transform from rectangular Cartesian coordinates to some other system of coordinates, $q_1, q_2, \ldots, q_{3N}$, using the $3N$ transformation equations

$$\begin{aligned}
x_1 &= f(q_1, q_2, q_3, \ldots, q_{3N}), \\
y_1 &= h(q_1, q_2, q_3, \ldots, q_{3N}), \\
z_1 &= g(q_1, q_2, q_3, \ldots, q_{3N}), \\
&\vdots \\
z_N &= u(q_1, q_2, q_3, \ldots, q_{3N}).
\end{aligned}$$

By directly substituting these equations into the expression for **H** in rectangular Cartesian coordinates given by combining Eqs. 11.7 and 11.8, we can obtain **H** in terms of the new coordinates and rates of change of those coordinates with respect to time. The resulting expression for **H** can then be used with Eq. 11.13 to obtain the functional form of the conjugate momenta. This knowledge permits **H** to be converted to an expression involving the momenta and coordinates of the new system. If that is done, it will still be true that

$$\boxed{\frac{\partial \mathbf{H}}{\partial P_{qi}} = \frac{dq_i}{dt}} \qquad \text{(11.14A)}$$

and

$$\boxed{\frac{\partial \mathbf{H}}{\partial q_i} = -\frac{dP_{qi}}{dt}} \qquad \text{for } i = 1, 2, 3, \ldots, 3N. \qquad \text{(11.14B)}$$

That is, Hamilton's equations hold for any coordinate system. The proof of this statement is not difficult, but it requires a more extensive treatment of classical mechanics than can be justified in an introductory physical chemistry text. Therefore, we present Eqs. 11.14A and 11.14B without proof. (For a derivation, see A. P. Arya, *Introduction to Classical Mechanics* (Allyn and Bacon, Needham Haights, MA, 1990), Chapter 12, pp. 441-446,)

To gain an appreciation and understanding of the foregoing points, we need to apply them to some simple systems. Let us begin by treating the case of a single particle of mass m moving in two-dimensional space about the origin. This might be, for example, the earth moving about a stationary sun of infinite mass. Figure 11.5 illustrates such a system.

If we are using rectangular Cartesian coordinates, the Hamiltonian for the system is

$$\mathbf{H} = \frac{m}{2} [v_x^2 + v_y^2] + V(X, Y), \qquad \text{(11.15)}$$

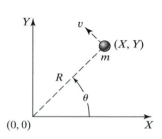

▲ FIGURE 11.5
A single particle of mass m with linear velocity v moving in two-dimensional space. Both rectangular Cartesian and polar coordinate systems are illustrated.

where $V(X, Y)$ is the system potential, which might, for example, be a gravitational potential. If we use this coordinate system, Hamilton's equations will simply give us the Newtonian equations of motion in first-order form. If the system is something like rotation of a planet about a stationary sun, it is probably better to use polar coordinates, which utilizes an angle, θ, that is very convenient for describing rotational motion.

The transformation equations between (X, Y) coordinates and (R, θ) coordinates are

$$X = R \cos \theta$$

and

$$Y = R \sin \theta. \tag{11.16}$$

We now need to convert the Hamiltonian from (X, Y) coordinates and velocities to the polar system. Taking derivatives of both sides of Eq. 11.16 with respect to time, we obtain

$$\frac{dX}{dt} = \frac{dR}{dt} \cos \theta - \frac{d\theta}{dt} R \sin \theta = v_x = v_R \cos \theta - v_\theta R \sin \theta,$$

and

$$\frac{dY}{dt} = \frac{dR}{dt} \sin \theta + \frac{d\theta}{dt} R \cos \theta = v_y = v_R \sin \theta + v_\theta R \cos \theta. \tag{11.17}$$

Next we need an expression for $[v_x^2 + v_y^2]$ that appears in the Hamiltonian in Eq. 11.15. Squaring both sides of each equation in 11.17 and adding, we obtain

$$[v_x^2 + v_y^2] = v_R^2 [\cos^2\theta + \sin^2\theta] + v_\theta^2 R^2 [\cos^2\theta + \sin^2\theta] - 2v_R v_\theta R \sin \theta \cos \theta$$
$$+ 2v_R v_\theta R \sin \theta \cos \theta = v_R^2 [\cos^2\theta + \sin^2\theta] + v_\theta^2 R^2 [\cos^2\theta + \sin^2\theta]$$
$$= v_R^2 + v_\theta^2 R^2, \tag{11.18}$$

since $[\cos^2\theta + \sin^2\theta] = 1$. The Hamiltonian in the polar coordinate system is, therefore,

$$\mathbf{H} = \frac{m}{2} [v_R^2 + v_\theta^2 R^2] + V(R, \theta). \tag{11.19}$$

Using Eq. 11.13, we can easily determine the proper form of the momentum conjugate to the radial coordinate R and the angular coordinate θ. For the radial coordinate, we obtain

$$P_R = \frac{\partial T}{\partial v_R} = mv_R, \tag{11.20}$$

so that the result is similar to that for Cartesian coordinates. However, for the angular coordinate, the result is

$$P_\theta = \frac{\partial T}{\partial v_\theta} = mv_\theta R^2. \tag{11.21}$$

P_θ is called the *angular momentum of the system*. A more proper name is the *momentum conjugate to the angle* θ. With the use of Eqs. 11.20 and 11.21, we can express $\mathbf{H}$ in terms of the conjugate momenta. The result is

$$\mathbf{H} = \frac{P_R^2}{2m} + \frac{P_\theta^2}{2mR^2} + V(R, \theta). \tag{11.22}$$

Hamilton's equations of motion for this system are given by Eq. 11.14. The four first-order equations are

$$\frac{\partial \mathbf{H}}{\partial P_R} = \frac{P_R}{m} = \frac{dR}{dt}, \qquad \text{(11.23A)}$$

$$\frac{\partial \mathbf{H}}{\partial P_\theta} = \frac{P_\theta}{mR^2} = \frac{d\theta}{dt}, \qquad \text{(11.23B)}$$

$$\frac{\partial \mathbf{H}}{\partial R} = -\frac{P_\theta^2}{mR^3} + \frac{\partial V}{\partial R} = -\frac{dP_R}{dt}, \qquad \text{(11.23C)}$$

and

$$\frac{\partial \mathbf{H}}{\partial \theta} = \frac{\partial V}{\partial \theta} = -\frac{dP_\theta}{dt}. \qquad \text{(11.23D)}$$

If we know $V(R, \theta)$, we may be able to solve these equations for the motion of the system as we did in the case of the projectile problem of Example 11.1. If an analytic solution is not possible, the four equations can always be solved by numerical means.

A particularly important result is obtained if the potential is a function of R only, as it would be in the case of a planet rotating about a stationary sun. Then we have $\partial V/\partial \theta = 0 = dP_\theta/dt$. That is, the angular momentum of the system does not change with time. Rather, it is conserved. It is, therefore, called a *constant of the motion.* Quantities that are constants of the motion play a very important role in both classical and quantum mechanics. (We have already seen that in thermodynamics, the internal energy U of an isolated system is such a conserved quantity.)

The force acting along the radial coordinate R can be obtained by combining Eqs. 11.1 and 11.23C. From the first Newtonian postulate, we know that

$$F_R = \text{radial force} = \frac{dP_R}{dt}.$$

Inserting Eq. 11.23C into this equation yields

$$F_R = -\frac{\partial V}{\partial R} + \frac{P_\theta^2}{mR^3}. \qquad \text{(11.24)}$$

The first term on the right-hand side of Eq. 11.24 gives the radial force due to the potential. This force might be a gravitational attraction, an electrostatic interaction between charged particles, or perhaps a van der Waals dispersion force. The second term represents a radial force produced by the presence of the angular momentum. Since this term is positive, it acts to move the particle outward to larger values of R. A force that behaves in this manner is called *centrifugal force.* As we shall see in the next section, Eqs. 11.23A through 11.23D played a critical role in one of the early attempts to move from the world of classical mechanics toward quantum theory.

11.1.4   Angular Momentum

In the previous section, we found that if the potential acting on a particle is independent of the angle of rotation, then the particle's angular momentum is a constant of the motion. Moreover, it is possible to show that the x, y, and z components of the angular momentum about the center of mass for any isolated classical system are always constants of the motion. As such, they are important quantities in both classical and quantum mechanics.

In classical mechanics, the angular momentum is defined to be the cross product of the radial vector locating the particle and the particle's momentum vector:

$$\boxed{\mathbf{M} = \mathbf{r} \times \mathbf{p}}. \qquad (11.25A)$$

The direction of the vector representing the cross product of two vectors **A** and **B** is perpendicular to the plane containing **A** and **B**, as shown in Figure 11.6. Consequently, **M** is perpendicular to the **r**–**p** plane. In Figure 11.5, **r** and **p** lie in the *x*–*y* plane. Therefore, **M** is parallel to the *z*-axis. The cross product of two vectors can be written in terms of the Cartesian components of each vector. In these terms, **M** is given by the determinant

$$\mathbf{M} = \begin{vmatrix} \mathbf{i} & \mathbf{j} & \mathbf{k} \\ x & y & z \\ P_x & P_y & P_z \end{vmatrix}, \qquad (11.25B)$$

where **i**, **j**, and **k** are unit vectors in the *x*, *y*, and *z* directions, respectively. Expanding this determinant by elements of the first row gives

$$\mathbf{M} = [yP_z - zP_y]\mathbf{i} + [zP_x - xP_z]\mathbf{j} + [xP_y - yP_x]\mathbf{k} = M_x\mathbf{i} + M_y\mathbf{j} + M_z\mathbf{k}. \qquad (11.25C)$$

If there are *N* particles in the system, the Cartesian components of **M** are given by

$$M_x = \sum_{i=1}^{N} [y_i P_{zi} - z_i P_{yi}] \qquad (11.26A)$$

$$M_y = \sum_{i=1}^{N} [z_i P_{xi} - x_i P_{zi}] \qquad (11.26B)$$

and

$$M_z = \sum_{i=1}^{N} [x_i P_{yi} - y_i P_{xi}]. \qquad (11.26C)$$

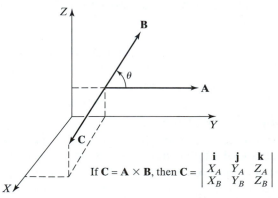

▶ FIGURE 11.6
The cross product of vectors **A** and **B**. The resulting vector, **C**, is directed perpendicular to the plane containing **A** and **B**. In the case shown in the figure, **A** and **B** lie in the *Y–Z* plane. **C** is parallel to the *X*-axis.

If $\mathbf{C} = \mathbf{A} \times \mathbf{B}$, then $\mathbf{C} = \begin{vmatrix} \mathbf{i} & \mathbf{j} & \mathbf{k} \\ X_A & Y_A & Z_A \\ X_B & Y_B & Z_B \end{vmatrix}$

The magnitude of **C** is given by $C = AB \sin \theta$

We may easily verify that Eq. 11.26C is equivalent to Eq. 11.21. In this example, the rotational motion is in the x–y plane, and $\mathbf{M}$ is therefore parallel to the z-axis and equal to M_z. Since we have only one particle in the system, Eq. 11.26C becomes

$$M_z = x\,P_y - y\,P_x.$$

With the use of Eqs. 11.16 and 11.17, this expression becomes

$$M_z = (R\cos\theta)mv_y - (R\sin\theta)mv_x$$

$$= mR[\cos\theta\{v_R\sin\theta + v_\theta R\cos\theta\} - \sin\theta\{v_R\cos\theta - v_\theta R\sin\theta\}]$$

$$= mR[v_R\cos\theta\sin\theta + v_\theta R\cos^2\theta - v_R\cos\theta\sin\theta + v_\theta R\sin^2\theta]$$

$$= mR^2 v_\theta[\cos^2\theta + \sin^2\theta] = mR^2 v_\theta.$$

The last expression is identical to that obtained in Eq. 11.21, so we have $M_z = P_\theta$ for this simple system.

EXAMPLE 11.2

In the planetary model of atomic structure developed by Niels Bohr (see Section 11.2.5), electrons rotate in circular orbits about the nucleus, much like planets moving about the sun (actually the planetary orbits are ellipses, but they are close to circular). We would, therefore, expect the electrons to possess angular momentum. Let us compute the angular momentum of the earth, assuming its orbit to be circular. The mass of the earth is 5.9742×10^{24} kg. The average distance of the earth from the sun is 149,597,870 km, and the time required for one complete rotation of the earth about the sun is approximately 365.25 days. Assuming that the earth rotates in a circular orbit about the sun, compute the angular momentum of the earth.

Solution

Angular momentum is given by Eq. 11.21:

$$P_\theta = mR^2 v_\theta. \tag{1}$$

With regard to the earth, we have all quantities save $v_\theta = d\theta/dt$. Since the gravitational potential producing a circular orbit is independent of the angle θ, the angular momentum is a constant of the motion, and hence v_θ must be a constant. We know that one complete rotation moves the angle θ through 2π radians. The rate of change of angle is, therefore,

$$v_\theta = \frac{d\theta}{dt} = \frac{2\pi}{365.25\text{ days}} \times \frac{1\text{ day}}{24\text{ hours}} \times \frac{1\text{ hour}}{3600\text{ s}} = 1.99102 \times 10^{-7}\text{ s}^{-1}. \tag{2}$$

The total angular momentum is, consequently,

$$P_\theta = mR^2 v_\theta = (5.9742 \times 10^{24}\text{ kg})(1.4959787 \times 10^{11}\text{ m})^2 (1.99102 \times 10^{-7}\text{ s}^{-1})$$

$$= 2.6620 \times 10^{40}\text{ kg m}^2\text{ s}^{-1}. \tag{3}$$

For related exercises, see Problems 11.4 and 11.5.

11.2   The Birth of Quantum Theory

I never saw a moor,
I never saw the sea;
Yet know I how the heather looks,
And what a wave must be.

Emily Dickenson from Time and Eternity CXXVII

Quantum theory was not created in one lightning stroke of genius. The concept that classical mechanics was exact for all systems had become such an ingrained part of physics that it took almost three decades of experimental and theoretical effort to dispel this notion and to create the body of science now known as quantum or wave mechanics. This section describes some of the landmark events which occurred during that period.

11.2.1 Blackbody Radiation

Light is a form of electromagnetic energy having a wave nature characterized by a particular wavelength λ. A wave generally exhibits periodic behavior in that the entire wave is made up of simple repeating units, as illustrated in Figure 11.7. In the figure, the portion of the wave lying between $x = 7.5$ Å (1Å $= 10^{-10}$m) and $x = 12.5$ Å is such a repeating unit. By duplicating this portion of the wave many times, the entire wave can be generated. The length of the repeating unit, 5Å in this example, is λ. The time required for the light wave to propagate a distance λ is λ/c, where c is the speed of light, 2.9979×10^8 m s^{-1} in a vacuum. This time is called the *period* τ. The frequency of the wave, ν, is the reciprocal of the period; that is, $\nu = 1/\tau$.

▶ **FIGURE 11.7**
The properties of a wave. Waves exhibit oscillatory behavior. The distance λ between equivalent portions of the wave is called the *wavelength*. The magnitude of the oscillations is called the *amplitude*. The time required for a complete oscillation is called the *period* τ, equal to λ/c, where c is the speed of the wave. For electromagnetic radiation, c is the speed of light. The wave frequency ν is defined as τ^{-1}.

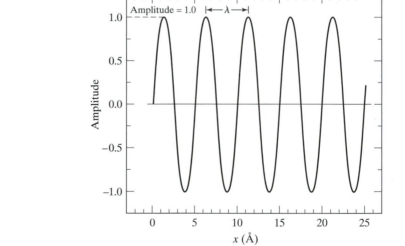

Differences in color perceived by the human eye are due to differences in the wavelength of light. The visible range lies between 4,000 Å and 8,000 Å. The longer wavelengths are seen by the eye as red light; the shorter wavelengths appear to be blue to violet. When the wavelength exceeds 8,000 Å, it is said to be in the *infrared* range. The human eye can no longer sense the presence of radiation in this range. However, the skin can, since infrared light produces heat. Even longer wavelengths are said to lie in the microwave and radio wave regions of the spectrum. When the wavelength is less than 4,000 Å, it is said to be in the *ultraviolet* range of the spectrum. Radiation in this range is primarily responsible for the sunburn a person gets at the lake or beach when the skin is not protected. Such sunburn is now known to be a primary cause of skin cancer. Even shorter wavelengths are termed X rays, and still shorter wavelengths are called gamma rays. Figure 11.8 shows the various regions of the electromagnetic spectrum.

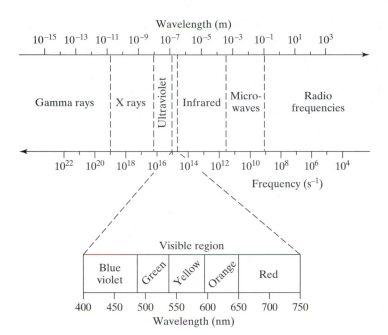

◀ FIGURE 11.8
The electromagnetic spectrum. The dividing lines between various regions are approximate. The colors perceived by the eye in the range between 400 and 750 nm are qualitative. There is no sharp dividing line. Violet-to-blue colors gradually convert to green as the wavelength approaches 500 nm.

When a material is heated, it emits electromagnetic radiation that is sometimes in the visible region of the spectrum. A common example is a red-hot piece of iron. That fact that the iron appears to be red is the result of emission of radiation near the long-wavelength end of the visible range around 700 nm. The intensity of emitted radiation at wavelength λ is a function of temperature and the nature of the material. This intensity is expressed as the *emissive power*, $e(T, \lambda)$, which is defined as the amount of radiation at wavelength λ and temperature T emitted per second per unit area per unit wavelength.

Matter not only emits radiation; it also absorbs a fraction of the radiation incident upon it. A typical example of this effect is the burn you receive when you pick up a piece of metal that has absorbed radiation from lying under a blazing summer sun. The fraction of radiation absorbed depends upon the wavelength, temperature, and nature of the material. Quantitatively, the absorption power, $a(T, \lambda)$, of a material is defined as the ratio Q_λ/Q_λ', where Q_λ' is the incident radiant energy per unit wavelength at wavelength λ and Q_λ is the absorbed radiant energy per unit wavelength at that same wavelength. If $a(T, \lambda) = 1$ for any material at all wavelengths, that material is called a *blackbody*. A blackbody absorbs all radiation incident upon it.

In 1859, Kirchhoff found that the ratio $e(T, \lambda)/a(T, \lambda) = E(T, \lambda)$ is independent of the nature of the material used in the experiments. In the years following Kirchhoff's discovery, the values of $E(T, \lambda)$ were measured as a function of wavelength and temperature for a variety of materials. The results were generally reported in terms of the energy density per unit wavelength at temperature T, defined as $u(T, \lambda) = (4\pi/c)E(T, \lambda)$, so that $u(T, \lambda)$ is proportional to $E(T, \lambda)$. Figure 11.9 shows some typical experimental results.

As the years passed, scientists tried to develop a fundamental theory that would both explain and predict the behavior seen in the experiments. For 41 years, these attempts met with only limited success. In 1896, Wien was able to show that the wavelength at which $u(T, \lambda)$ exhibits a maximum, λ_{max}, was given by $\lambda_{max} = 0.294$ cm K$/T$. This prediction was verified experimentally by

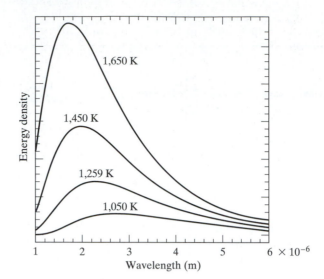

Energy density

1,650 K

1,450 K

1,259 K

1,050 K

1   2   3   4   5 6×10^{-6}

Wavelength (m)

▶ FIGURE 11.9
Energy density of blackbody radiation. In 1900, Planck found that these data are accurately described by Eq. 11.29.

Lummer and Pringsheim in 1899. Predicting the variation of $u(T, \lambda)$ with λ at a given temperature proved to be a more difficult problem. Wien suggested that

$$u(T, \lambda) = \frac{c_1}{\lambda^5} \exp\left[-\frac{c_2}{\lambda T}\right], \tag{11.27}$$

where c_1 and c_2 are constants. Wien's equation works well when λT is small, but not for larger values. Rayleigh and Jeans analyzed the phenomenon using classical theory. Their analysis suggested that the dependence of $u(T, \lambda)$ upon λ and T is given by

$$u(T, \lambda) = 8\pi k_b \frac{T}{\lambda^4}, \tag{11.28}$$

where k_b is the Boltzmann constant. This equation was found to be accurate for large values of λT, but it failed when λT was small.

In 1900, Planck solved the problem by using the analysis advanced by Rayleigh and Jeans, but with a new hypothesis that radiation had to be emitted in integer multiples of $h\nu$, where $\nu = c/\lambda$. That is, Planck assumed that radiation came in *quantized* increments of $h\nu$. A material could emit radiation in the amounts $h\nu, 2h\nu, 3h\nu, \ldots$, but not $1.457h\nu$ or any other nonintegral multiple of $h\nu$. This bold assumption led to the equation

$$u(T, \lambda) = \frac{8\pi hc}{\lambda^5} \frac{1}{\exp\left[\dfrac{hc}{\lambda k_b T}\right] - 1}. \tag{11.29}$$

Equation 11.29 is known as *Planck's law*. The constant h appearing in the equation had to be determined by fitting the equation to measured data such as that shown in Figure 11.9. It later became known as *Planck's constant*. Its measured value is 6.62608×10^{-34} J s. It can easily be shown that Eq. 11.29 reduces to Eqs. 11.27 and 11.28 in the limit of small and large values of λT, respectively. (See Problems 11.7 and 11.8.) Planck presented his findings to the German Physical Society on December 14, 1900. As that was the introduction of the concept of quantization to the scientific community, it may be properly regarded as the birth date of quantum theory. In 1907, Einstein used Planck's concept of quantized energies to explain the heat capacity of solids. (Einstein's heat capacity theory was discussed in detail in Chapter 7.) In 1920, Planck was awarded the Nobel prize for his contributions to physics.

11.2.2   The Photoelectric Effect

When ultraviolet light with wavelengths in the range 2,000 to 4,000 Å ($1 \text{ Å} = 10^{-10}$ m) is incident upon a metallic surface, electrons are emitted with some velocity v. This phenomenon is known as the *photoelectric effect*. (See Figure 11.10.) Experimental measurements demonstrated that the energy of the ejected photoelectrons is dependent upon the frequency of the incident radiation, but not upon its "intensity," or rate at which energy is propagated to the surface by the electromagnetic waves. Higher intensity of monochromatic light only produced more photoelectrons, each with the same energy. This result was contrary to wave theory as it existed before 1900, which held that a high intensity should increase the probability that an ejected electron will acquire a large energy.

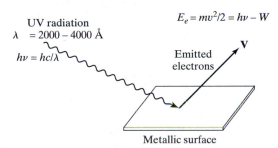

◄ FIG. 11.10
The photoelectric effect. When electrons are ejected from a metallic surface due to interaction with incident electromagnetic radiation, it is found that the energy of the ejected electrons, E_e, is dependent upon the frequency of the incident radiation, not upon the rate at which energy is propagated to the surface. It is as if each electron interacts with a "particle" of light.

To explain the dependence of the kinetic energies of the ejected electrons upon the frequency of the radiation, Einstein assumed that radiant energy could be treated as a thermodynamic system of particles whose total energy can be written in the form $E = nh\nu$, where ν is the frequency of the radiation. When Einstein compared his derived expression for the entropy of the radiation with that of a system of n gaseous particles, he saw that the integer n appeared in the energy expression. Thus, the incident radiation appeared to be quantized in units of $h\nu$ that we now call *photons*.

Using the notion of quantization, Einstein derived the basic equation describing the photoelectric effect:

$$h\nu = W + \frac{mv^2}{2}. \qquad \textbf{(11.30)}$$

In Eq. 11.30, W is the energy required to remove an electron from the metal, called the *work function*, and v is the velocity of the ejected electron. In 1921, Einstein was awarded the Nobel prize for this contribution. Millikan received the Nobel prize in 1923 for his experimental verification of Einstein's equation.

11.2.3   The Rutherford Atom

In 1909, Rutherford and Geiger began a series of studies to probe the structure of the atom. Their concept was to fire a beam of alpha particles (He^{2+} ions) at a very thin gold foil and observe the small deflections of the α particles produced by interactions with the electrons and protons of the gold atoms.

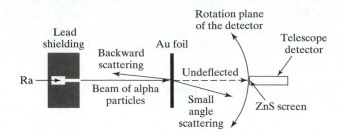

► FIGURE 11.11
Rutherford–Geiger apparatus. The apparatus is mounted in a vacuum chamber. The telescope detector rotates in the circular track shown in the figure.

A schematic of their apparatus is shown in Figure 11.11. When the alpha particles strike the ZnS screen mounted in front of the telescope, a flash of light is produced that can be observed through the telescope. By rotating the telescope in a circular fashion about the point of impact with the gold film, the angular distribution of scattered alpha particles can be determined.

The results of the experiments were that (1) the majority of the α particles penetrated the foil undeflected, (2) a few α particles were scattered through a small angle, (3) a few (about 1 in 20,000) suffered a serious deflection as they passed through the Au foil, and (4) a few underwent direct backward scattering along the original path. The final observation shocked Rutherford, who later commented that the observation was "about as credible as if you had fired a 15-inch shell at a piece of tissue paper and it came back and hit you."

In order to explain the results of the experiments, Rutherford was forced to conclude that most of the mass of the atom and all of the positive charge were centered in a very small region he called the *nucleus*. The positive center was said to be surrounded by an equal number of negatively charged electrons outside the nucleus. This model has since become known as the *nuclear atom*; it is the picture currently used to describe the atom's structure.

The development of the Rutherford model of the atom created even more problems for the proponents of classical mechanics and electrodynamics. How could such a structure possibly exist? If the electrons are stationary, what keeps the attractive forces existing between positively charged protons and negatively charged electrons from spontaneously pulling the electrons into the nucleus? The suggestion that perhaps the electrons are in orbital motion about the nucleus, like the planets about the sun, with velocities sufficient to permit the centrifugal force to exactly balance the electrostatic attraction and thereby keep the electrons outside the nucleus did not help. The problem with this suggestion comes from one of the fundamental laws of classical electrodynamics: When a charged particle accelerates through an electric field, it must radiate energy. A curved electron path requires that the direction of motion of electrons constantly change. This means that the electrons must be continuously accelerated. The electrons are charged particles, and the protons within the nucleus create an electric field through which the electrons would have to accelerate. Thus, energy would have to radiate from the rotating electrons.

The classical laws of electrodynamics require that the frequency of the emitted radiation equal the frequency of rotation, $\nu_{rot} = v_\theta/2\pi$, where v_θ is the angular velocity of an electron. The rate of emission of energy must be proportional to the square of the acceleration. The actual result is $dE/dt = -(e^2 v_\theta^4 r^2)/(6\pi\varepsilon_o c^3)$ for an orbit of radius r. This energy loss would slow the rotational motion and thereby reduce the centrifugal force. As the process continued, the velocity of the electrons would inexorably decrease to the point where the electrons would spiral into the nucleus. The Rutherford atom simply cannot exist if classical mechanics and electrodynamics hold. (See Problem 11.27 for a quantitative treatment of this point.) Through Rutherford's efforts, another nail had been driven into the coffin of the concept that classical physics is an exact description of the behavior of all systems.

11.2.4   Atomic Spectra

Lurking in the background of theoretical efforts to reconcile classical physics with the experimental observations that were being made was the discovery of atomic spectra in the late 1880s. When an atom is excited to higher energy levels by the input of energy, the energy that is absorbed is subsequently emitted in the form of light at a later time when the atom returns to its original state. The measurement of the wavelengths or frequencies of the emitted radiation produces an *atomic spectrum*. The observation of atomic spectra caused enormous concern to scientists, because the nature of the observed spectra was contrary to the laws of classical Newtonian physics.

According to classical theory, the absorption–emission process should have worked in the same fashion as hurling a rock into the air. For example, if 100 J of energy are expended by throwing a 0.1-kg rock upward in a vacuum, the rock would rise to a height d such that $E = mgd$, which gives

$$d = \frac{100 \text{ kg m}^2 \text{ s}^{-2}}{(0.1 \text{ kg})(9.80665 \text{ m s}^{-2})} = 101.9716 \text{ m}.$$

At this point, the rock would begin to fall, picking up speed in the process. At the instant of impact with the ground, we would have

$$E = 100 \text{ J} = \frac{mv^2}{2},$$

so that the final velocity of the rock at impact would be

$$v = \left[\frac{2E}{m}\right]^{1/2} = \left[2\left(\frac{100}{0.1}\right)\right]^{1/2} = 44.721 \text{ m s}^{-1}.$$

If we expended 200 J of energy, d would increase by a factor of 2, and the final velocity would increase to 63.246 m s^{-1}. In other words, the energy emitted in the form of translational energy of the rock on impact can be varied at will by choosing different input energies in throwing the rock upward.

Using this analogy, scientists expected the frequency of the emitted radiation in an atomic spectrum to vary in proportion to the energy used to excite the atoms. This did not occur. For example, in the case of hydrogen atoms, the emission spectrum comprises a discrete set of frequencies whose values are independent of the excitation energy. Empirical fitting of the observed emitted frequencies by Balmer in 1885 demonstrated that the only wavelengths emitted were those given by

$$\lambda^{-1} = 109{,}677 \left[\frac{1}{n^2} - \frac{1}{m^2}\right] \text{ cm}^{-1} \text{ for } n = 1, 2, 3, 4, \ldots$$

and $m = n + 1, n + 2, n + 3, \ldots$,

regardless of the excitation energy employed in the experiment. Clearly, the emission energies were quantized. In other words, not all energies were permitted. The set of emission frequencies computed by setting $n = 1$ is called the *Lyman series*. Those frequencies obtained using $n = 2$ are termed the *Balmer series*. The $n = 3$ frequencies make up the *Paschen series*. The constant 109,677 cm^{-1} is now called the *Rydberg constant* and is usually given the symbol R_H. The concept that classical theory is exact for all systems was in its death throes.

EXAMPLE 11.3

The reciprocal of the wavelength for one of the emissions from an excited hydrogen atom is found to be $5{,}331.52 \text{ cm}^{-1}$. What values of n and m are responsible for this emission?

Solution

Using the experimentally observed result, we have

$$\frac{1}{n^2} - \frac{1}{m^2} = \frac{5{,}331.52}{109{,}677} = 0.0486111. \tag{1}$$

If $n = 1$, the smallest value $(1/n^2) - (1/m^2)$ can have is $1 - (1/2^2) = 0.75$. Consequently, we cannot have $n = 1$.

If $n = 2$, the smallest value possible for $(1/n^2) - (1/m^2)$ is $(1/2^2) - (1/3^2) = (1/4) - (1/9) = 0.13888 \dots$. Therefore, $n \neq 2$.

If $n = 3$, the smallest value possible for $(1/n^2) - (1/m^2)$ is $(1/9) - (1/16) = 0.048611 \dots$, which is exactly the value given by Eq. 1. Therefore, the emission is fitted by using $n = 3$ and $m = 4$. As a result, this wavelength is one of the Paschen emissions.

For a related exercise, see Problems 11.10 and 11.11.

11.2.5 The Bohr Theory of Atomic Structure

In 1913, Niels Bohr (1885–1962) used the concept of quantization introduced by Planck, along with classical physics, in an attempt to explain the appearance of the spectrum of the hydrogen atom. It was a remarkable effort. Faced with the Rutherford atom and the impossibility of keeping the electrons out of the nucleus using classical mechanics and electrodynamics, Bohr combined Planck's concept of quantization with the bold postulate of a radiationless orbit. In doing so, he set aside many of the most fundamental concepts of classical theory. The assumptions of the Bohr theory are as follows:

1. Electrons revolve about the nucleus in circular orbits that do not radiate energy.
2. The laws of classical mechanics hold, except that radiation due to acceleration does not occur.
3. The angular momentum of the electron is quantized in units of $h/(2\pi)$. That is, electrons may only have angular momentum equal to $nh/(2\pi)$, where n takes on only integer values $1, 2, 3, 4, \dots$. Bohr was led to this postulate by assuming that, in the limit of large angular momentum (large n), the classical emission frequency must become equal to the electron's orbital frequency, as required by the laws of classical electrodynamics. This assumption is now called the *Bohr correspondence principle*. (See Problem 11.24.)
4. The potential energy is coulombic.

With these four assumptions, Bohr was able to quantitatively predict the spectrum for any hydrogen-like atom containing only one electron (H, He^+, Li^{2+}, etc.).

The Bohr picture is essentially that described by Figure 11.3, where the rotating particle is an electron and the massive, stationary object at the origin is the nucleus. By the second assumption, Hamilton's equations for this system hold. Therefore, Eqs. 11.23A through 11.23D describe the electron's motion and energy. The fourth assumption tells us that the potential is

$$V(R) = \frac{(Ze)(-e)}{4\pi\varepsilon_o R} = -\frac{Ze^2}{4\pi\varepsilon_o R} \tag{11.31}$$

where Z is the atomic number of the atom or ion. Equation 11.23D tells us that, since the potential does not depend upon θ, the angular momentum of the electron is a constant. Moreover, by the third assumption,

$$P_\theta = \frac{nh}{2\pi} \qquad \text{for } n = 1, 2, 3, 4, \dots. \tag{11.32}$$

If the orbit is to be circular, as the first assumption contends, the net radial force must be zero. Otherwise, the electron would not stay at the same radius as it rotated about the nucleus. Therefore, using Eq. 11.23C, we have

$$-\frac{P_\theta^2}{mR^3} + \frac{\partial V}{\partial R} = -\frac{dP_R}{dt} = -F_R = -(\text{radial force}) = 0. \tag{11.33}$$

Rearranging Eq. 11.33 gives

$$\frac{P_\theta^2}{mR^3} = \frac{\partial V}{\partial R}. \tag{11.34}$$

Substituting of Eqs. 11.31 and 11.32 into 11.34 produces

$$\frac{n^2 h^2}{4\pi^2 mR^3} = \frac{Ze^2}{4\pi\varepsilon_o R^2}. \tag{11.35}$$

Solving for the radius of the circular orbit, we obtain

$$R = \frac{n^2 h^2 \varepsilon_o}{\pi Z e^2 m} = \frac{n^2}{Z}\left[\frac{4\pi\hbar^2\varepsilon_o}{e^2 m}\right] \tag{11.36}$$

where $\hbar$ (pronounced "aitch-bar") $= h/2\pi = (6.62608 \times 10^{-34}\,\text{J s})/2(3.1415927) = 1.05457 \times 10^{-34}\,\text{J s}$. The quantity inside the brackets in Eq. 11.36, which has the units of distance, is given the symbol a_o and is called the *Bohr radius*. Its value is

$$a_o = \left[\frac{4\pi\hbar^2\varepsilon_o}{e^2 m}\right]$$

$$= \frac{4(3.1415927)(1.05457 \times 10^{-34}\,\text{J s})^2(8.854190 \times 10^{-12}\,\text{J}^{-1}\,\text{C}^2\,\text{m}^{-1})}{(1.602177 \times 10^{-19}\,\text{C})^2(9.10930 \times 10^{-31}\,\text{kg})}$$

$$= 5.29175 \times 10^{-11}\,\text{m}. \tag{11.37}$$

Equation 11.36 was a stunning result, showing that not all radii for the circular orbits are permitted. The electrons can rotate only in circular orbits whose radii are $a_o/Z, 4a_o/Z, 9a_o/Z, 16a_o/Z, \dots$, as shown in Figures 11.12 and 11.13. For the hydrogen atom, with $Z = 1$, the smallest possible radius is a_o.

With the possible orbital radii known, the total energy of the system can now be obtained by directly substituting Eq. 11.36 into 11.22:

$$\mathbf{H} = T + V(R, \theta) = E = \frac{P_R^2}{2m} + \frac{P_\theta^2}{2mR^2} + V(R, \theta).$$

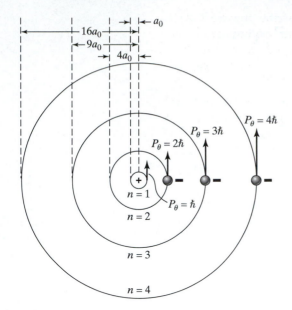

▲ FIGURE 11.12

The Bohr orbits. The first four circular Bohr orbits are shown to scale. The angular momentum possessed by the electron in each of these orbits is given in each case.

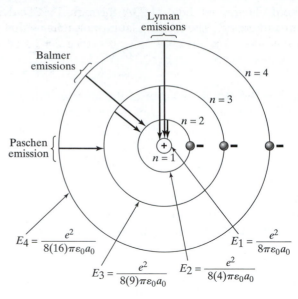

▲ FIGURE 11.13

The Lyman, Balmer, and Paschen emissions in the Bohr model of the hydrogen atom for the first four Bohr orbits. The energies of these circular orbits are also given.

The first term on the right is zero, since the circular orbit ensures that $P_R = 0$. Therefore,

$$E = \frac{P_\theta^2}{2mR^2} + V(R, \theta) = \frac{P_\theta^2}{2mR^2} + V(R), \tag{11.38}$$

because the potential depends only upon R. Substitution of Eqs. 11.31 and 11.32 into Eq. 11.38 gives

$$E = \frac{n^2\hbar^2}{2mR^2} - \frac{Ze^2}{4\pi\varepsilon_o R}. \tag{11.39}$$

Inserting R from Eq. 11.36 produces

$$E = \frac{n^2\hbar^2}{2m}\left[\frac{Ze^2m}{4\pi n^2\hbar^2\varepsilon_o}\right]^2 - \frac{Ze^2}{4\pi\varepsilon_o}\left[\frac{Ze^2m}{4\pi n^2\hbar^2\varepsilon_o}\right]$$

$$= \frac{Z^2e^4m}{32\pi^2 n^2\hbar^2\varepsilon_o^2} - \frac{Z^2e^4m}{16\pi^2 n^2\hbar^2\varepsilon_o^2} = -\frac{Z^2e^4m}{32\pi^2 n^2\hbar^2\varepsilon_o^2}. \tag{11.40}$$

In terms of the Bohr radius, the energy is

$$E_n = -\frac{Z^2e^2}{8\pi n^2\varepsilon_o a_o}, \tag{11.41}$$

where the subscript on E_n emphasizes the fact that the energy depends upon the value of n.

The result of Bohr's theorization was a quantized energy for hydrogen-like atoms! For the hydrogen atom itself, with $Z = 1$, the energies could take on only the values

$$E_1 = -\frac{e^2}{8\pi\varepsilon_o a_o}, E_2 = -\frac{e^2}{8(4)\pi\varepsilon_o a_o}, E_3 = -\frac{e^2}{8(9)\pi\varepsilon_o a_o}, E_4 = -\frac{e^2}{8(16)\pi\varepsilon_o a_o}, \ \dots.$$

Bohr now recognized that the spectral emission lines from hydrogen atoms were due to transitions from higher energy states to lower ones. Figure 11.13 illustrates this point. The energy difference ΔE between the states shown would have to be equal to the energy of the emitted radiation, which, by Planck's theory, is $h\nu = hc/\lambda$. Therefore, for a given transition between an upper hydrogen state with $n = m$ and a lower state with $n = n$, Eq. 11.41 produces

$$\Delta E(m, n) = E_m - E_n = -\frac{e^2}{8\pi\varepsilon_o a_o}\left[\frac{1}{m^2} - \frac{1}{n^2}\right] = h\nu. \qquad \textbf{(11.42)}$$

The predicted spectral frequencies are

$$\nu = -\frac{e^2}{8h\pi\varepsilon_o a_o}\left[\frac{1}{m^2} - \frac{1}{n^2}\right] = \left\{\frac{e^2}{8h\pi\varepsilon_o a_o}\right\}\left[\frac{1}{n^2} - \frac{1}{m^2}\right]. \qquad \textbf{(11.43)}$$

The radiation frequency is related to the wavelength by

$$\nu = \frac{c}{\lambda} = c\lambda^{-1}.$$

Therefore, the emitted wavelengths were predicted to be given by

$$\lambda^{-1} = \left\{\frac{e^2}{8hc\pi\varepsilon_o a_o}\right\}\left[\frac{1}{n^2} - \frac{1}{m^2}\right]. \qquad \textbf{(11.44)}$$

The calculated value of the constants in the curled braces in Eq. 11.44 is

$$\left\{\frac{e^2}{8hc\pi\varepsilon_o a_o}\right\} = \frac{(1.602177 \times 10^{-19}\,\text{C})^2}{(6.62608 \times 10^{-34}\,\text{J s})(2.99792 \times 10^8\,\text{m s}^{-1})} \times$$

$$\frac{1}{8(3.1415927)(8.85419 \times 10^{-12}\,\text{J}^{-1}\,\text{C}^2\,\text{m}^{-1})(5.29175 \times 10^{-11}\,\text{m})}$$

$$= 1.09738 \times 10^7\,\text{m}^{-1} = 109738\,\text{cm}^{-1}.$$

Therefore, the Bohr model predicts that the emitted wavelengths for hydrogen will be

$$\boxed{\lambda^{-1} = 109{,}738\left[\frac{1}{n^2} - \frac{1}{m^2}\right]\text{cm}^{-1}.} \qquad \textbf{(11.45)}$$

If Eq. 11.45 is corrected for reduced mass effects, which we shall discuss in Chapter 13, the predicted Rydberg constant becomes $109{,}677\,\text{cm}^{-1}$, in precise agreement with the experimentally observed results! Actually, Bohr reported a value of $R_H = 103{,}500\,\text{cm}^{-1}$ because the fundamental constants were not accurately known in 1913. The essential results of Bohr's model are presented in Figures 11.12 through 11.14.

After Bohr presented his theory of atomic structure, a symposium was held in Zurich to discuss the ramifications of the theory. At this meeting, von Laue commented, "It is nonsense!" Einstein, however, replied, "Very remarkable! There must be something behind it. I do not believe that the derivation of the absolute value of the Rydberg constant is purely fortuitous." In 1922, Bohr received the Nobel prize for his work.

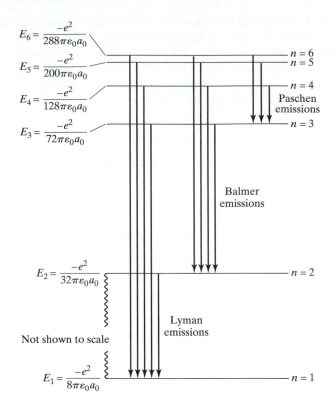

▶ FIGURE 11.14
The hydrogen-atom energy levels predicted by the Bohr model. The transitions from upper energy states to the $n = 1$ state constitute the Lyman bands of the H-atom spectrum. Transitions from upper states to $n = 2$ are the Balmer emission bands, and transitions from higher states to $n = 3$ are called the Paschen emission bands. The spacing between the $n = 1$ and $n = 2$ states is not shown to scale. The spacings between the other bands are drawn to scale.

The Bohr theory demonstrated that Planck's ideas of quantization could be used to explain and predict the complete details of the emission spectra of hydrogen-like atoms. Accordingly, it was a giant step in the right direction. However, when efforts were made to apply the theory to many-electron atoms, problems arose. The answers were no longer exact. Numerous efforts were made to determine the reasons for this failure. At first it was thought that the difficulty resided in the assumption of circular orbits. Replacing this assumption with elliptical orbits improved the results, but they were still not exact. Moreover, the Bohr theory provided no means by which the transition probabilities from one state to another could be calculated. With the benefit of hindsight, it was "obvious" that something was still missing. It remained for Compton, de Broglie, Davisson, Germer, and G. P. Thomson to provide the last piece of the puzzle.

11.2.6   The Wave Nature of Matter

In 1922, Compton performed experiments in which X rays of wavelength λ were scattered from a particle as shown in Figure 11.15. Compton observed that the scattered wavelength λ' is greater than the original wavelength. To explain this observation, he assumed that the incident electromagnetic radiation could be treated as a particle, called a *photon*, whose energy is $h\nu = hc/\lambda$. He then assumed that the increase in wavelength is due to the fact that energy is lost by the photon to an electron with mass m. The electron gains energy, while the X ray loses energy. As a consequence, the wavelength of the scattered X ray is longer than that of the incident radiation.

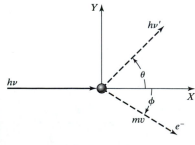

▲ FIGURE 11.15
Compton scattering.

Let us go through the mathematical reasoning behind Compton's explanation. If the velocity of the scattered electron is v, conservation of energy requires that

$$h\nu = h\nu' + \frac{mv^2}{2}. \qquad \text{(11.46A)}$$

In addition, both the x and y components of momentum must be conserved. This requirement produces two more equations:

$$p = p' \cos\theta + mv \cos\phi \quad \text{(conservation of x component of momentum)} \qquad \text{(11.46B)}$$

and

$$0 = p' \sin\theta - mv \sin\phi \quad \text{(conservation of y component of momentum).} \qquad \text{(11.46C)}$$

In Eqs. 11.46B and 11.46C, p and p' are the incident and scattered momenta of the photon. The direction of the incident photon is taken to be the x-axis, as shown in Figure 11.15.

If we isolate the term containing ϕ in Eqs. 11.46B and 11.46C, square both sides of each equation and add the two equations, we obtain

$$m^2v^2[\sin^2\phi + \cos^2\phi] = m^2v^2 = (p')^2 + p^2 - 2pp' \cos\phi.$$

From Eq. 11.46A, we also have

$$m^2v^2 = 2m(h\nu - h\nu').$$

Substituting this result into the previous equation yields

$$2m(h\nu - h\nu') = (p')^2 + p^2 - 2pp' \cos\theta. \qquad \text{(11.47)}$$

Compton now postulated that the momentum of the photon is given by $h\nu/c = h/\lambda$. Substituting this expression into Eq. 11.47 and replacing the frequencies with c/λ produces

$$2mch\left(\frac{1}{\lambda} - \frac{1}{\lambda'}\right) = h^2\left[\frac{1}{\lambda^2} + \frac{1}{(\lambda')^2}\right] - \frac{2h^2}{\lambda\lambda'} \cos\theta. \qquad \text{(11.48)}$$

Taking a common denominator on the left side of Eq. 11.48 and rearranging, we obtain

$$\lambda' - \lambda = \frac{\lambda\lambda'h}{2mc}\left[\frac{1}{\lambda^2} + \frac{1}{(\lambda')^2}\right] - \frac{h}{mc} \cos\theta. \qquad \text{(11.49)}$$

If the change in wavelength is small, we will have $\lambda \approx \lambda'$ and $\lambda\lambda' \approx \lambda^2 \approx (\lambda')^2$. With this assumption, Eq. 11.49 becomes

$$\boxed{\lambda' = \lambda + \frac{h}{mc}[1 - \cos\theta]}. \qquad \text{(11.50)}$$

Compton found that his experimental results were in excellent agreement with Eq. 11.50. The factor $h/(mc) = 2.4263 \times 10^{-12}$ meters $= 2.4263 \times 10^{-10}$ cm $= 0.024263$ Å is now known as the *Compton wavelength*. Equation 11.50 shows that the maximum change in wavelength occurs when $\theta = \pi$, for which we have $\Delta\lambda = 0.048526$ Å. Thus, the approximation that $\lambda \approx \lambda'$ is reasonably good since typical wavelengths for X-rays are 1–3 Å. Furthermore, we note that if the ejected electron is treated relativistically, Eq. 11.50 can be derived without using this approximation. These experiments established the fact that photons have a particle-like nature, with a momentum equal to $h\nu/c$. For this discovery, Compton received the Nobel prize in 1927.

If electromagnetic waves can have momentum and particle character, why shouldn't particles of matter have a wavelength and wave character? This thought was put in concrete form by de Broglie in 1923 in his doctoral dissertation. De Broglie advanced the hypothesis that a wave is associated with every particle. He further proposed that the wavelength of these "matter waves" is given by

$$\boxed{\lambda = \frac{h}{p}} , \tag{11.51}$$

▲ From left to right,

Back row: Auguste Piccard, E. Henriot, Paul Ehrenfest, E. Herzen, Th. de Donder, Erwin Schrödinger, E. Verschaffelt, Wolfgang Pauli, Werner Heisenberg, Ralph H. Fowler, Marcel L. Brillouin

Second row: Peter Debye, Martin Knudsen, William L. Bragg, Hendrik A. Karmers, Paul A. M. Dirac, Arthur H. Compton, Louis V. de Broglie, Max Born, Neils Bohr

Front row: Irving Langmuir, Max Planck, Mme. Marie Curie, H. A. Lorentz, Albert Einstein, Paul Langevin, Charles E. Guye, Charles T. R. Wilson, Owen W. Richarson

The Architects of Modern Physics and Chemistry. This unique photograph shows the delegates to the Fifth International Congress on Physics, which took place October 23–29, 1927, at the Solvay Institute in Brussels, Belgium. It was attended by the most important physicists of the era. The accomplishments of the delegates have, in large part, been described in the foregoing section. The people pictured were awarded eighteen Nobel Prizes, fifteen in physics and three in chemistry between 1902 and 1954.

where p is the linear momentum of the particle. Equation 11.51 is now called the *de Broglie hypothesis.*

During de Broglie's final Ph.D. oral examination, one of the examiners asked how one might verify experimentally that matter did indeed exhibit wave character such as that proposed in his dissertation. De Broglie replied that just as X rays exhibit a diffraction pattern when they are scattered from crystals, a beam of electrons should show the same effect. This suggestion was soon tested and confirmed by Davisson and Germer and also by Thomson. De Broglie received the Nobel prize in 1929 for his hypothesis of matter waves. Davisson and Thomson received the same prize jointly in 1937 for their confirmation of the wave nature of matter.

The stage was now set. The problem with the Bohr theory and all of the modifications introduced in a series of attempts to make the theory applicable to many-electron systems was their insistence on a classical, particle-like description of the electron. Once it became clear that only a wavelike theory would adequately solve the problem, Heisenberg and Schrödinger developed the theoretical foundation of what is now known as quantum or wave mechanics. Schrödinger's first paper, entitled "Quantization as a Problem of Proper Values," was received by the editor of *Annalen der Physik* on January 27, 1926. In 1933, Schrödinger received the Nobel prize for this landmark work. The cataclysm, which spanned 33 years, was complete. The reader may assess the magnitude of this monumental achievement by simply counting the number of Nobel prizes awarded in connection with the effort. There is nothing else remotely comparable in the annals of science.

The accompanying photograph was taken at the Solvay Conference in 1927. It shows many of the individuals whose work contributed to the development of quantum theory. They are, indeed, the "architects of modern physics and chemistry."

EXAMPLE 11.4

An incident X ray with wavelength 1.5400 Å undergoes Compton scattering at an angle of 28° relative to its direction of motion. **(A)** Compute the wavelength of the scattered X ray. **(B)** Compute the momentum of the scattered electron. **(C)** Calculate the wavelength of the scattered electron.

Solution

(A) Using Eq. 11.50, we obtain

$$\lambda' = \lambda + \frac{h}{mc}[1 - \cos\theta] = \{1.5400 \times 10^{-10} + 2.4263 \times 10^{-12}[1 - \cos(28°)]\}\ \text{m}$$

$$= [154.00 \times 10^{-12} + 2.8400 \times 10^{-13}]\ \text{m} = 154.28 \times 10^{-12}\ \text{m}. \qquad (1)$$

(B) Using Eq. 11.46A, we have

$$m^2 v^2 = P^2 = 2m[h\nu - h\nu'] = 2mhc[1/\lambda - 1/\lambda']$$

$$= 2(9.10939 \times 10^{-31}\ \text{kg})(6.62608 \times 10^{-34}\ \text{J s})(2.9979 \times 10^{8}\ \text{m s}^{-1})$$

$$\times \left[\frac{1}{154 \times 10^{-12}} - \frac{1}{154.28 \times 10^{-12}}\right]\text{m}^{-1})$$

$$= 4.2650 \times 10^{-48}\ \text{kg}^2\,\text{m}^2\,\text{s}^{-2}. \qquad (2)$$

Therefore, the momentum is

$$P = 2.065 \times 10^{-24} \text{ kg m s}^{-1}. \tag{3}$$

(C) The de Broglie hypothesis is

$$\lambda = \frac{h}{P} = \frac{6.62608 \times 10^{-34} \text{ J s}}{2.065 \times 10^{-24} \text{ kg m s}^{-1}} = 3.209 \times 10^{-10} \text{ m} = 3.209 \text{ Å}. \tag{4}$$

For related exercises, see Problems 11.15, 11.16, and 11.17.

11.3 The Mathematics of Quantum Mechanics

Most of the mathematical difficulties associated with quantum theory involve solving partial differential equations, either exactly or approximately. These difficulties will be addressed as they arise. There are, however, several mathematical principles that are critical to learning and understanding quantum mechanics. The most important is the development and use of probability distributions. This subject has already been covered in depth in Chapter 10. In this section, we will address several additional mathematical concepts. For students well versed in operator algebra, the section will serve as a useful review.

11.3.1 Complex Conjugates and Absolute Squares

Many of the quantities that we will encounter in quantum mechanics are complex. That is, they consist of a real and an imaginary part. Such a quantity may always be represented in the form $X = a + bi$, where a and b are real numbers and i is the imaginary unit $(-1)^{1/2}$. We may also represent complex numbers in the form $X = Ae^{iB}$. Expanding this exponential in a Maclaurin series gives

$$X = A\left[1 + \frac{iB}{1!} + \frac{(iB)^2}{2!} + \frac{(iB)^3}{3!} + \frac{(iB)^4}{4!} + \cdots + \frac{(iB)^n}{n!} + \cdots \right].$$

Since $i^2 = -1$, the odd powers of i will produce an alternating series in which the terms have the form $i(-1)^k B^{2k+1}/(2k+1)!$, where k is a positive integer. The even powers of i will give an alternating series whose general term is $(-1)^k B^{2k}/(2k)!$, where k is a positive integer. Therefore,

$$X = A\left[1 - \frac{B^2}{2!} + \frac{B^4}{4!} - \cdots + \frac{(-1)^k B^{2k}}{(2k)!} + \cdots \right]$$

$$+ iA\left[\frac{B}{1!} - \frac{B^3}{3!} + \frac{B^5}{5!} - \cdots + \frac{(-1)^k B^{2k+1}}{(2k+1)!} + \cdots \right].$$

The first series expansion is just $\cos(B)$ while the second is $\sin(B)$. As a result,

$$X = Ae^{iB} = A\cos(B) + i A \sin(B),$$

which is in the general form $a + bi$ of a complex number.

The complex conjugate of a complex number $X = a + bi$ is, by definition, $X^* = a - bi$. In effect, the signs of all imaginary units are changed to obtain the complex conjugate.

EXAMPLE 11.5

What is the complex conjugate of $X = Ae^{5i} - 3i + (6 + 3i)^2$?

Solution

The solution can be obtained by simply replacing i with $-i$ and $-i$ with i. This operation gives

$$X^* = Ae^{-5i} + 3i + (6 - 3i)^2.$$

For related exercises, see Problems 11.32 and 11.33.

The absolute square of X, is, by definition,

$$|X|^2 = X^*X. \qquad \textbf{(11.52)}$$

It is important to note that the absolute square of any number is always real, even if the number is complex or imaginary. The proof is straightforward: If $X = a + bi$, then

$$|X|^2 = (a + bi)(a - bi) = a^2 + abi - abi - b^2i^2.$$

The cross terms cancel and $i^2 = -1$, so that

$$|X|^2 = a^2 + b^2,$$

which is real. The positive square root of the absolute square of X is also called the *modulus* of X. If X is real, the absolute square reduces to a simple square, since we have $X = X^*$.

EXAMPLE 11.6

Show that the absolute square of X as defined in Example 11.5, is real.

Solution

As defined in Example 11.5, $X = Ae^{5i} - 3i + (6 + 3i)^2 = Ae^{5i} - 3i + 36 + 36i + 9i^2 = Ae^{5i} + 33i + 27$. We can solve this problem directly or by first converting the complex exponential to sines and cosines. Both methods will be shown.

The complex conjugate of X is $X^* = Ae^{-5i} - 33i + 27$. Direct multiplication gives

$$|X|^2 = X^*X = [Ae^{-5i} - 33i + 27][Ae^{5i} + 33i + 27]. \qquad \textbf{(1)}$$

Expansion of the right-hand side of Eq. 1 produces

$$|X|^2 = A^2e^o + 33Aie^{-5i} + 27Ae^{-5i} - 33Aie^{5i} - (33)^2i^2 - (27)(33)i$$

$$+ 27Ae^{5i} + (27)(33)i + (27)^2$$

$$= A^2 + 33Ai[e^{-5i} - e^{5i}] + 27A[e^{-5i} + e^{5i}] + 1{,}089 + 729. \qquad \textbf{(2)}$$

Using the facts that $[e^{-5i} - e^{5i}] = -2i\sin(5)$ and $[e^{-5i} + e^{5i}] = 2\cos(5)$, we obtain

$$|X|^2 = A^2 + 66A\sin(5) + 54A\cos(5) + 1{,}818, \qquad \textbf{(3)}$$

which is clearly real.

Alternatively, we can write the exponential in sine and cosine terms immediately, collect terms, and then multiply as follows:

$$X = A\cos(5) + Ai\sin(5) - 3i + 36 + 36i + 9i^2$$

$$= A\cos(5) + Ai\sin(5) + 33i + 27$$

$$= [A\cos(5) + 27] + i[A\sin(5) + 33] \qquad \textbf{(4)}$$

(since $i^2 = -1$). Now, the complex conjugate of X is

$$X^* = [A \cos (5) + 27] - i [A \sin(5) + 33]. \tag{5}$$

Therefore, we obtain

$$|X|^2 = X^*X = \{[A \cos (5) + 27] - i [A \sin(5) + 33]\} \times$$

$$\{[A \cos (5) + 27] + i [A \sin(5) + 33]\}$$

$$= [A \cos (5) + 27]^2 - i^2 [A \sin(5) + 33]^2 + i [A \cos (5) + 27] [A \sin(5) + 33]$$

$$-i [A \cos (5) + 27] [A \sin(5) + 33]. \tag{6}$$

Collecting terms, we obtain

$$|X|^2 = X^*X = [A \cos (5) + 27]^2 + [A \sin(5) + 33]^2$$

$$= A^2 [\cos^2(5) + \sin^2(5)] + 54A \cos(5) + 729 + 66A \sin(5) + 1,089$$

$$= A^2 + 54A \cos(5) + 66A \sin(5) + 1,818, \tag{7}$$

which is the same result as that obtained by the first method.
For related exercises, see Problems 11.18, 11.32, and 11.33.

11.3.2 Eigenvalue Equations

Almost all the differential equations that we shall encounter in quantum mechanics have the form

$$\mathcal{G}\phi = g\phi, \tag{11.53}$$

where $\mathcal{G}$ is a differential operator such as $\partial/\partial x$ or $\partial^2/\partial x^2$, ϕ is a function of the variables of the problem, and g is a constant. Equations of this form, in which the result of operating with $\mathcal{G}$ upon ϕ is to produce a constant times ϕ, are called *eigenvalue equations*. The constant g is called the *eigenvalue* of $\mathcal{G}$ associated with ϕ, and the function ϕ is said to be an *eigenfunction* of the operator $\mathcal{G}$.

EXAMPLE 11.7

Show that $\sin(6x)$ is not an eigenfunction of the operator $\mathcal{G} = \partial/\partial x$, but that it is an eigenfunction of the operator $\mathcal{G} = \partial^2/\partial x^2$. What is the eigenvalue when $\mathcal{G} = \partial^2/\partial x^2$ for the eigenfunction $\sin(6x)$?

Solution

If $\mathcal{G} = \partial/\partial x$, we have

$$\mathcal{G} \sin(6x) = \frac{\partial}{\partial x} \sin(6x) = 6 \cos(6x). \tag{1}$$

Since $6 \cos(6x)$ is not equal to a constant times $\sin(6x)$, we have shown that $\sin(6x)$ is not an eigenfunction of the operator $\mathcal{G} = \partial/\partial x$. However, if $\mathcal{G} = \partial^2/\partial x^2$, we obtain

$$\mathcal{G} \sin(6x) = \frac{\partial^2}{\partial x^2} \sin(6x) = 6\frac{\partial}{\partial x} \cos(6x) = -36 \sin(6x). \tag{2}$$

Equation 2 shows that $\sin(6x)$ is an eigenfunction of $\mathcal{G}$, since $\mathcal{G}$ operating on $\sin(6x)$ produces a constant times $\sin(6x)$. The constant on the right-hand side of Eq. 2 is -36; therefore, the eigenvalue of $\mathcal{G}$ associated with $\sin(6x)$ is -36.

For related exercises, see Problems 11.19, 11.23, 11.24, 11.25, 11.29, 11.30, and 11.31.

Many of the operators that occur in quantum mechanics belong to a class called *Hermitian operators*, which have the following property: If $\psi(x)$ and $\phi(x)$ are functions of a variable x and G is a Hermitian operator, then

$$\int_{\text{all space}} \psi^*(x)\, G\phi(x)\, dx = \int_{\text{all space}} \{G\psi(x)\}^*\phi(x)\, dx.$$

If ψ and ϕ are functions of several variables, $q_1, q_2, q_3, \ldots q_n$ and G operates on any or all of these variables, then if G is Hermitian,

$$\int_{\text{all space}} \psi^*(q_1, q_2, \ldots, q_n)\, G\, \phi(q_1, q_2, \ldots, q_n)\, dq_1 dq_2 \ldots dq_n$$

$$= \int_{\text{all space}} \{G\psi(q_1, q_2, \ldots, q_n)\}^*\, \phi(q_1, q_2, \ldots, q_n)\, dq_1 dq_2 \ldots dq_n$$

$$= \int_{q_1}\int_{q_2} \cdots \int_{q_n} \{G\psi(q_1, q_2, \ldots, q_n)\}^*\, \phi(q_1, q_2, \ldots, q_n)\, dq_1 dq_2 \ldots dq_n, \quad \textbf{(11.54)}$$

where the notation $\int_{\text{all space}}$ denotes an n-dimensional integration. Equation 11.54 tells us that in executing the integral, we can first operate with G on ϕ, then multiply the result by ψ^* and integrate, or we can first operate with G on ψ, take the complex conjugate of the result, multiply by ϕ, and integrate. The value of the integral will be the same regardless of which procedure we use.

Equations such as Eq. 11.54 are laborious and take up much space. We will, therefore, adopt some standard notation to denote integrations over all space. The multidimensional volume element $dq_1 dq_2 dq_3 \ldots dq_n$ will be written as $d\tau$ in all coordinate systems. The explicit statement that the functions ψ and ϕ depend upon $q_1, q_2, \ldots, q_n$ will be omitted. With these conventions, the definition of a Hermitian operator, Eq. 11.54, can be written in the more compact notation

$$\int_{\text{all space}} \psi^*\, G\, \phi\, d\tau = \int_{\text{all space}} \{G\psi\}^*\, \phi\, d\tau. \quad \textbf{(11.55)}$$

An even more compact notation for the integrals in Eq. 11.55 is

$$\langle \psi|G|\phi \rangle = \langle G\psi|\phi \rangle, \quad \textbf{(11.56)}$$

in which it is understood that if the angle brackets $\langle \rangle$ (called "bra" and "ket," respectively) contain operators and functions, they represent an integral over all the space spanned by the variables of the problem. It is further understood that the quantities to the left of the first vertical line are complex conjugates. A comparison of Eqs. 11.54 and 11.56 makes the advantage of such condensed notation obvious.

EXAMPLE 11.8

Let $\psi = \sin(\theta)$ and $\phi = \cos(\theta)$. Show that $G = i\dfrac{\partial}{\partial \theta}$ behaves like a Hermitian operator if θ spans the angular range from 0 to 2π radians.

Solution

To show that $\mathcal{G}$ behaves like a Hermitian operator, we must show that Eq. 11.56 holds. We first evaluate the left side of the equation:

$$\langle\psi|\mathcal{G}|\phi\rangle = \left\langle \sin(\theta)\left|i\frac{\partial}{\partial\theta}\right|\cos(\theta)\right\rangle = i\int_0^{2\pi}\sin^*(\theta)\frac{\partial}{\partial\theta}\cos(\theta)\,d\theta$$

$$= -i\int_0^{2\pi}\sin^2(\theta)d\theta = -i\int_0^{2\pi}\frac{1-\cos(2\theta)}{2}\,d\theta = -i\left[\frac{\theta}{2}-\frac{\sin(2\theta)}{4}\right]_0^{2\pi}$$

$$= -\pi i. \tag{1}$$

The right-hand side of Eq. 11.56 is

$$\langle\mathcal{G}\psi|\phi\rangle = \left\langle\left\{i\frac{\partial}{\partial\theta}\sin(\theta)\right\}\Big|\cos(\theta)\right\rangle = -i\int_0^{2\pi}\cos^*(\theta)\cos(\theta)\,d\theta = -i\int_0^{2\pi}\cos^2(\theta)d\theta$$

$$= -i\int_0^{2\pi}\frac{\cos(2\theta)+1}{2}d\theta = -i\left[\frac{\theta}{2}+\frac{\sin(2\theta)}{4}\right]_0^{2\pi} = -\pi i. \tag{2}$$

Thus, $\langle\psi|\mathcal{G}|\phi\rangle = \langle\mathcal{G}\psi|\phi\rangle$, and $\mathcal{G}$ behaves as if it is Hermitian. If we choose arbitrary functions for ψ and ϕ, we can prove that $\mathcal{G}$ is indeed Hermitian. (See Problem 11.20.)

11.3.3   Properties of Hermitian Operators

All Hermitian operators have two very important properties: (1) Their eigenvalues are always real and (2) their eigenfunctions are orthogonal if they are associated with different eigenvalues. By definition, two functions ψ and ϕ are orthogonal if $\langle\psi|\phi\rangle = \langle\phi|\psi\rangle = 0$. In this section, we prove these assertions.

Theorem 1: If ϕ is an eigenfunction of a Hermitian operator $\mathcal{G}$, its eigenvalue must be real.

Proof:
Since ϕ is an eigenfunction of $\mathcal{G}$, we have

$$\mathcal{G}\phi = g\phi. \tag{1}$$

We now multiply both sides of Eq. 1 on the left by ϕ^*. This gives

$$\phi^*\mathcal{G}\phi = \phi^*g\phi. \tag{2}$$

Integrating both sides of Eq. 2 over all space produces

$$\langle\phi|\mathcal{G}|\phi\rangle = \langle\phi|g\phi\rangle = g\langle\phi|\phi\rangle, \tag{3}$$

where we have factored g out of the integral because it is a constant. Since $\mathcal{G}$ is Hermitian, Eq. 3 may be written in the form

$$\langle\mathcal{G}\phi|\phi\rangle = g\langle\phi|\phi\rangle. \tag{4}$$

By Eq. 1, we can write the left side of Eq. 4 as

$$\langle\mathcal{G}\phi|\phi\rangle = \langle g\phi|\phi\rangle = g^*\langle\phi|\phi\rangle. \tag{5}$$

Again, we may factor g out of the integral, but since it is to the left of the first vertical bar, it comes out as the complex conjugate. Combining Eqs. 4 and 5, we obtain

$$g^*\langle\phi|\phi\rangle = g\langle\phi|\phi\rangle. \tag{6}$$

Unless $\phi = 0$ for all values of the variables, so that $\langle \phi | \phi \rangle$ is zero as well, Eq. 6 shows that $g^* = g$. But the only way this can be true is for the eigenvalue g to be real. Hence, Hermitian operators always have real eigenvalues.

Theorem 2: If ψ and ϕ are both eigenfunctions of a Hermitian operator $\mathcal{G}$ associated with different eigenvalues, the integral over all space, $\langle \psi | \phi \rangle$ or $\langle \phi | \psi \rangle$, must be zero. That is, ψ and ϕ must be orthogonal.

Proof:

Since ψ and ϕ are eigenfunctions of $\mathcal{G}$ with different eigenvalues, we must have

$$\mathcal{G}\phi = a\phi \tag{1}$$

and

$$\mathcal{G}\psi = b\psi, \tag{2}$$

where a and b are the eigenvalues associated with ϕ and ψ, respectively. We now multiply Eq. 2 on the left by ϕ^* to obtain

$$\phi^*\mathcal{G}\psi = \phi^*b\psi. \tag{3}$$

Integrating both sides of Eq. 3 over all space gives

$$\langle \phi | \mathcal{G} | \psi \rangle = \langle \phi | b\psi \rangle = b\langle \phi | \psi \rangle, \tag{4}$$

since b is a real constant that may be factored out of the integral. Using the Hermitian character of $\mathcal{G}$, we may write Eq. 4 in the form

$$\langle \phi | \mathcal{G} | \psi \rangle = \langle \mathcal{G}\phi | \psi \rangle = b\langle \phi | \psi \rangle. \tag{5}$$

From Eq. 1, $\mathcal{G}\phi = a\phi$. Therefore,

$$\langle \mathcal{G}\phi | \psi \rangle = \langle a\phi | \psi \rangle = b\langle \phi | \psi \rangle. \tag{6}$$

The constant a may now be factored out of the left-hand integral. Since it is real, the result is

$$a\langle \phi | \psi \rangle = b\langle \phi | \psi \rangle. \tag{7}$$

Rearranging terms produces

$$(a - b)\langle \phi | \psi \rangle = 0. \tag{8}$$

If $a \neq b$, we must have $\langle \phi | \psi \rangle = 0$. Functions that have this property are said to be orthogonal. The student should note that if $a = b$, Eq. 8 does not forbid us from having $\langle \phi | \psi \rangle = 0$; it just does not require that to be true.

In Chapters 12 and 13, we shall see numerous applications of Theorems 1 and 2. The problems at the end of those chapters are especially intended to point up the utility of Theorem 2.

11.3.4 Commutators

Two variables or operators $\mathcal{F}$ and $\mathcal{G}$ are said to commute if they have the property that $\{\mathcal{F}\mathcal{G} - \mathcal{G}\mathcal{F}\}\psi = 0$, where ψ is an arbitrary function not equal to zero. The Cartesian variables x and y are examples of quantities that commute. This can be seen by noting that $\{xy - yx\}\psi(x, y) = 0$, since xy is exactly equal to yx. The commutator of two variables or operators $\mathcal{F}$ and $\mathcal{G}$ is defined to be the difference $\mathcal{F}\mathcal{G} - \mathcal{G}\mathcal{F}$ This difference is written as $[\mathcal{F}, \mathcal{G}]$. Therefore,

$$[\mathcal{F}, \mathcal{G}]\psi = \{\mathcal{F}\mathcal{G} - \mathcal{G}\mathcal{F}\}\psi. \tag{11.57}$$

Let us now consider a few examples of commutators. The preceding discussion shows that $[x, y] = 0$, which means that x and y commute. If, however, we have $\mathcal{F} = x$ and $\mathcal{G} = \partial/\partial x$, then $\mathcal{F}$ and $\mathcal{G}$ do not commute, since

$$[\mathcal{F}, \mathcal{G}]\,\psi(x) = \{\mathcal{F}\mathcal{G} - \mathcal{G}\mathcal{F}\}\psi(x) = \left\{ x\frac{\partial}{\partial x} - \frac{\partial}{\partial x}x \right\}\psi(x) = x\frac{\partial\psi(x)}{\partial x} - \frac{\partial}{\partial x}\{x\psi(x)\}$$

$$= x\frac{\partial\psi(x)}{\partial x} - \psi(x) - x\frac{\partial\psi(x)}{\partial x} = -\psi(x). \tag{11.58}$$

A comparison of the left and right sides of Eq. 11.58 shows that $[\mathcal{F}, \mathcal{G}] = -1$ for this choice of $\mathcal{F}$ and $\mathcal{G}$. Therefore, x and $\partial/\partial x$ do not commute. In contrast, if we take $\mathcal{F} = \partial/\partial x$ and $\mathcal{G} = \partial^2/\partial x^2$, then

$$[\mathcal{F}, \mathcal{G}]\,\psi(x) = \{\mathcal{F}\mathcal{G} - \mathcal{G}\mathcal{F}\}\psi(x) = \left\{ \frac{\partial}{\partial x}\frac{\partial^2}{\partial x^2} - \frac{\partial^2}{\partial x^2}\frac{\partial}{\partial x} \right\}\psi(x)$$

$$= \left\{ \frac{\partial^3}{\partial x^3} - \frac{\partial^3}{\partial x^3} \right\}\psi(x) = 0,$$

so that $\partial/\partial x$ commutes with $\partial^2/\partial x^2$.

11.4   The Fundamental Postulates of Quantum Mechanics

With our discussions of probability theory and operator algebra complete, we are now ready to introduce quantum theory. In Section 11.2, we found that the de Broglie hypothesis and the experiments of Davisson and Germer and of Thomson demonstrated that a wave theory of matter is required if we are to understand and predict the results of experimental measurements. In 1926, Schrödinger provided the basis of such a theory. It is this theory that we shall now describe and use throughout the remainder of the text.

There are four basic postulates of quantum theory. The first states that matter is described by a wave (wave function) that must have certain properties. The second and third postulates tell us how to obtain the wave function. The fourth postulate tells us how to obtain all the experimentally measurable properties of a system from its wave function once we have it.

11.4.1   The First Postulate

Matter is described by a wave $\Psi(q, t)$, *called the wave function, that depends upon the coordinates* q *and time. In general,* $\Psi(q, t)$ *is finite, single valued, and continuous at all points.*

The importance of the first postulate is at least equal to that of the first and second laws of thermodynamics combined. Not only does the first postulate unequivocally state that matter is described by a wave, it also places three critical constraints on the nature of such waves. As we shall see, these constraints give rise to the quantization of energy, momentum, and angular momentum and, consequently, to the science we call spectroscopy. The entire structure of the periodic table of elements rests on the three constraints. Let us examine them.

In order to describe matter properly, the wave function must be finite at all points in coordinate space. That is, $\Psi(q, t)$ cannot be $\pm\infty$ anywhere. The wave function must also be continuous at all points in space. Consequently, the wave

$$\Psi(x) = \frac{\sin[2\pi x]}{|x - 0.4|}$$

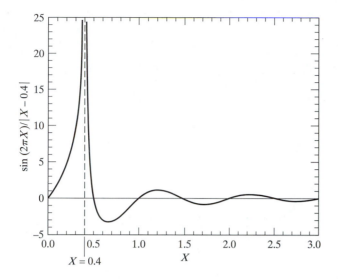

◀ FIGURE 11.16
An example of a function that exhibits both an infinity and a discontinuity, both occurring at the point $x = 0.4$. The function attains an infinite value at this point. Furthermore, the derivative is undefined there. If $x = 0.4$ is approached from below, the derivative approaches $+\infty$. If $x = 0.4$ is approached from above, the derivative approaches $-\infty$. The first postulate forbids such functions from being wave functions for any real system.

can never be the wave function of any system. Figure 11.16 shows a plot of this function over the range $0 \le x \le 3.0$. As can be seen, the wave has an infinite value at the point $x = 0.4$. In addition, $\Psi(x)$ is discontinuous at this point. The discontinuity is shown by considering the derivative of $\Psi(x)$. For $x < 0.4$,

$$\Psi(x) = \frac{\sin[2\pi x]}{(0.4 - x)},$$

and its derivative is given by

$$\frac{\partial \Psi(x)}{\partial x} = \frac{2\pi \cos[2\pi x]}{(0.4 - x)} + \frac{\sin[2\pi x]}{(0.4 - x)^2}.$$

When $x > 0.4$,

$$\Psi(x) = \frac{\sin[2\pi x]}{(x - 0.4)}$$

and

$$\frac{\partial \Psi(x)}{\partial x} = \frac{2\pi \cos[2\pi x]}{(x - 0.4)} - \frac{\sin[2\pi x]}{(x - 0.4)^2}.$$

As we approach the point $x = 0.4$, the second term in the derivative dominates, since it approaches an infinite value much faster than the first term. Since $\sin[2\pi x]$ is $0.58778\ldots$ at $x = 0.4$, $\partial\Psi(x)/\partial x$ is positive if we approach that point from below. If, however, we approach the point from above, $\partial\Psi(x)/\partial x$ is negative. There is, therefore, a discontinuity in $\partial\Psi(x)/\partial x$ at $x = 0.4$. This behavior is evident in Figure 11.16. Figure 11.17 shows a second example of a discontinuous wave that the first postulate forbids from being the wave function for any system. In this case, the discontinuity appears as a cusp at $x = 1.57$. To the left of the cusp, the derivative is positive. At the cusp, the derivative is undefined. To the right, it abruptly becomes negative.

The requirement that a wave function have a continuous first derivative requires some additional discussion. On occasion, scientists employ mathematical models that contain infinities in the potential-energy function. A common example is the coulombic potential, which is infinite when the distance

▶ **FIGURE 11.17**
An example of a function that has a discontinuous derivative at the point $x = 1.57$. If $x = 1.57$ is approached from below, the derivative is positive. If $x = 1.57$ is approached from above, the deriative is negative. There is, therefore, a discontinuity in the derivative at $x = 1.57$. Such points are called *cusps*. The first postulate forbids wave functions from possessing these characteristics.

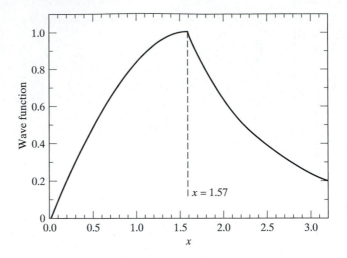

between the charged particles approaches zero. For a tritium ion and a proton, this potential has the form $V(r) = e^2/(4\pi\varepsilon_o r)$. Consequently, $V(r)$ approaches infinity as r approaches zero. (See Figure 11.18A.) We know, however, that this result is aphysical: Infinite potentials do not exist in nature. In fact, when a tritium ion and a proton are given sufficient energy to approach to very close distances, nuclear fusion can occur to form a helium ion. Such energies are commonly obtained in nuclear accelerators. The fact that nuclear fusion can occur, with its subsequent release of huge quantities of energy, means that the real potential-energy curve qualitatively resembles something like that shown in Figure 11.18B. At sufficiently small distances, $V(r)$ must attain a maximum and then decrease rapidly as fusion takes place. Employing a model containing an aphysical infinite potential can produce cusps in the quantum mechanical wave function at the points where the potential becomes infinite. This is the way quantum theory responds to a mathematical model that uses potentials containing infinities. In effect, the

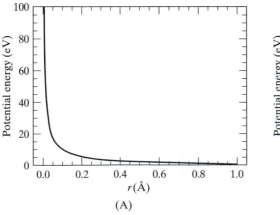

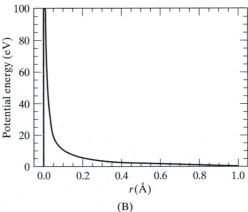

(A)   (B)

▲ **FIG. 11.18**
(A) Coulombic potential between two protons that contains an aphysical infinity at the point $r = 0$. In reality, such an infinite potential does not exist in the real system. At sufficiently close distances at which nuclear fusion can occur, the potential must attain a maximum value and then rapidly decrease as shown in (B). The use of mathematical models containing potentials that assume infinite values can lead to wave functions that exhibit cusps at the point(s) at which the potential becomes infinite.

cusps in the wave function tell us that our model potential does not correctly represent the real physical system.

Figure 11.19 shows a plot of the function $\Psi(\theta) = \sin(1.67\theta)$ versus θ. This wave is finite at all points. It has no discontinuities in either $\Psi(\theta)$ or $\partial\Psi(\theta)/\partial\theta$, but it still does not satisfy the conditions of the first postulate, because it is double valued. The angles θ and $\theta + 2\pi$ are physically identical. Therefore, we must have $\Psi(\theta) = \Psi(\theta + 2\pi)$ if we are to satisfy the requirement of the first postulate that Ψ be single valued. Figure 11.19 shows that at almost all points the function is double valued. Such equations are not acceptable wave functions.

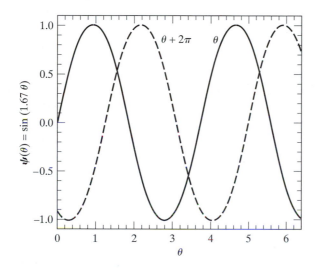

◀ FIG. 11.19
Plot of $\sin(1.67\theta)$ versus θ expressed in radians. This is an example of a double-valued function. Since the angles θ and $\theta + 2\pi$ are physically identical, the first postulate of quantum mechanics requires that we have $\Psi(\theta) = \Psi(\theta + 2\pi)$. The figure shows that the function $\sin(1.67\theta)$ does not have this property. Therefore, it cannot serve as a quantum mechanical wave function.

11.4.2   The Second Postulate

A rigorous statement of the second postulate requires a more detailed treatment of classical mechanics than is justified in a physical chemistry text. We shall, therefore, present a modified version of this postulate that will suffice for our purposes:

> *For every observable physical variable (position, momentum, energy, etc.), there exists a corresponding quantum mechanical operator that is Hermitian. The commutator between the operators representing conjugate variables must equal either $i\hbar$ or $-i\hbar$, with the sign depending upon the variables involved.*

For a more detailed discussion of the second postulate, consult J. C. Davis, Jr., *Advanced Physical Chemistry* (Ronald Press Co., New York, 1965) Chapter 3, pp. 58–65.

Let us take the coordinate x and its conjugate momentum p_x to see how the second postulate defines the operators. For the case of coordinates and conjugate momenta, the commutator must be $i\hbar$. We cannot take both x and p_x to be multiplicative variables as they are in classical mechanics, since we would then have

$$[x, p_x]\psi = [xp_x - p_x x]\psi = 0,$$

because multiplicative variables commute. The example shown in Eq. 11.58 gives us a hint about what is needed. Let us take x to be identical to the classical variable x. Then equation 11.58 suggests that we need p_x to have the form $p_x = b(\partial/\partial x)$ where b is a constant. The second postulate requires that we have

$$[x, p_x]\psi = \left[xb\frac{\partial}{\partial x} - b\frac{\partial}{\partial x}x \right]\psi = i\hbar\psi. \tag{11.59}$$

Factoring out the constant b, we obtain

$$b\left[x\frac{\partial\psi}{\partial x} - \frac{\partial}{\partial x}(x\psi) \right] = bx\,\frac{\partial\psi}{\partial x} - bx\,\frac{\partial\psi}{\partial x} - b\psi = -b\psi = i\hbar\psi. \tag{11.60}$$

Equation 11.60 will be satisfied only if we take $b = -i\hbar = \hbar/i$. The quantum mechanical Hermitian operator corresponding to p_x is, therefore, $(\hbar/i)\,\partial/\partial x$. Table 11.1 summarizes the results for the operators we will need in this text.

11.4.3   The Third Postulate

The third postulate tells us how to use the operators defined by the second postulate to obtain the quantum mechanical differential equations whose solutions give us the possible wave functions of the system:

> *The differential equations of quantum mechanics are obtained by direct substitution of the quantum variables and operators into the classical equations in Cartesian coordinates and then allowing the operators thus formed to operate upon* $\Psi(q, t)$.

In the next section, we illustrate how this postulate is used to obtain the Schrödinger wave equation.

11.4.4   The Fourth Postulate

The fourth postulate tells us how to connect the wave functions obtained from the solutions of the differential equations formulated by the second and third postulates to measured quantities in the laboratory:

> *The product of the absolute square of the wave function,* $|\Psi(q, t)|^2$, *and the volume element* $d\tau$ *is proportional to the probability distribution for*

Table 11.1   Classical variables and quantum mechanical operators (E denotes energy and t is time)

Classical Variable	Quantum Mechanical Operator
x	x
y	y
z	z
t	t
p_x	$\dfrac{\hbar}{i}\dfrac{\partial}{\partial x}$
p_y	$\dfrac{\hbar}{i}\dfrac{\partial}{\partial y}$
p_z	$\dfrac{\hbar}{i}\dfrac{\partial}{\partial z}$
E	$i\hbar\dfrac{\partial}{\partial t}$

the system. That is, $|\Psi(q, t)|^2 d\tau$ is proportional to the probability that the system will be in the volume element $d\tau$ at time t.

Since $|\Psi(q, t)|^2 d\tau$ is proportional to the probability distribution, it plays the role of the probability function $f(x) dx$ discussed in chapter 10. Average values are, therefore, obtained using methods essentially identical to those described in that chapter. For example, suppose $\Psi(x, t)$ is the wave function for a one-dimensional system. Then the average value of some multiplicative function of x, $h(x)$, is given by Eq. 10.23 if we replace $f(x) dx$ with $|\Psi(x, t)|^2 dx$. That is,

$$<h(x)> = \frac{\displaystyle\int_{\text{all } x} h(x)|\Psi(x, t)|^2 dx}{\displaystyle\int_{\text{all } x} |\Psi(x, t)|^2 dx}. \tag{11.61}$$

If we wish to compute the average value of a variable whose quantum mechanical variable is a differential operator, we must permit the operator to operate upon $\Psi(x, t)$, as required by the third postulate. That is, if the operator is $\mathcal{F}_{op}$, we must write $\Psi^*(x, t) \mathcal{F}_{op} \Psi(x, t) dx$ instead of $\mathcal{F}_{op} |\Psi(x, t)|^2 dx$. For example, the average value of p_x is given by

$$<p_x> = \frac{\displaystyle\int_{\text{all } x} \Psi^*(x, t)\frac{\hbar}{i}\frac{\partial}{\partial x}\Psi(x, t) dx}{\displaystyle\int_{\text{all } x} |\Psi(x, t)|^2 dx}, \tag{11.62}$$

since the operator for p_x is $(\hbar/i)\partial/\partial x$.

11.5   The Schrödinger Wave Equations

11.5.1   Time-dependent Wave Equation

The energy E of an isolated classical system is given by the classical Hamiltonian, Eq. 11.11. In rectangular Cartesian coordinates, this is

$$\mathbf{H} = \sum_{i=1}^{N} \left[\frac{1}{2m_i}\right] [P_{xi}^2 + P_{yi}^2 + P_{zi}^2] + V(x_1, y_1, z_1, x_2, y_2, z_2, \ldots, t) = E. \tag{11.63}$$

The third postulate tells us that we may obtain the differential equations whose solutions are the possible wave functions of the system by replacing the classical variables with the corresponding quantum mechanical expressions given in Table 11.1 and then allowing the operators thus formed to operate upon $\Psi(x_1, y_1, z_1, x_2, y_2, z_2, \ldots, t)$. To execute this procedure on Eq. 11.63, we need the operators for P_x^2, P_y^2, and P_z^2. Since $P_x^2 = P_x P_x$, we obtain

$$[P_x^2]_{\text{operator}} = [P_x]_{\text{operator}} [P_x]_{\text{operator}} = \left[\frac{\hbar}{i}\frac{\partial}{\partial x}\right]\left[\frac{\hbar}{i}\frac{\partial}{\partial x}\right] = -\hbar^2 \frac{\partial^2}{\partial x^2}. \tag{11.64A}$$

A similar procedure gives

$$[P_y^2]_{\text{operator}} = -\hbar^2 \frac{\partial^2}{\partial y^2} \tag{11.64B}$$

and

$$[P_z^2]_{\text{operator}} = -\hbar^2 \frac{\partial^2}{\partial z^2}. \tag{11.64C}$$

Following the procedure mandated by the third postulate, we now substitute Eqs. 11.64A through 11.64C into the kinetic-energy term in Eq. 11.63 to obtain

$$T = \sum_{i=1}^{N} -\frac{\hbar^2}{2m_i}\left[\frac{\partial^2}{\partial x_i^2} + \frac{\partial^2}{\partial y_i^2} + \frac{\partial^2}{\partial z_i^2}\right]. \tag{11.65}$$

This is the second time we have encountered the sum of the three second partial derivatives with respect to x, y, and z. The first time was in the development of the Debye–Hückel theory of ionic activities. [See Eqs. 9.32 and 9.33.] This sum is called the *Laplacian*, or *del-squared*, operator. It is commonly assigned the symbol ∇^2. With this notation, Eq. 11.65 becomes

$$T = \sum_{i=1}^{N} -\frac{\hbar^2}{2m_i}\nabla_i^2, \tag{11.66}$$

and the quantum mechanical operator for the kinetic energy of a single particle is $-(\hbar^2/2m)\nabla^2$.

Since the potential-energy function $V(x_1, y_1, z_1, x_2, y_2, z_2, \ldots, t)$ depends only upon the x-, y-, and z-coordinates of the particles and possibly time, its quantum mechanical form is identical to the classical form. The energy, however, is replaced with $i\hbar\,(\partial/\partial t)$. The quantum mechanical Hamiltonian is, therefore,

$$\mathcal{H} = \sum_{i=1}^{N} -\frac{\hbar^2}{2m_i}\nabla_i^2 + V(x_1, y_1, z_1, x_2, y_2, z_2, \ldots, t). \tag{11.67}$$

The differential equation that must be solved to find the wave function is obtained by allowing $\mathcal{H}$ and $i\hbar\,(\partial/\partial t)$ to operate upon the wave function. This gives

$$\mathcal{H}\Psi(x_1, y_1, z_1, x_2, y_2, z_2, \ldots, t) = i\hbar\frac{\partial}{\partial t}\Psi(x_1, y_1, z_1, x_2, y_2, z_2, \ldots, t),$$

or

$$\mathcal{H}\Psi(q, t) = i\hbar\frac{\partial}{\partial t}\Psi(q, t), \tag{11.68}$$

where q represents all the coordinates. Equation 11.68 is called the *time-dependent Schrödinger equation*. When solved, it yields $\Psi(q, t)$.

11.5.2 Stationary-state Schrödinger Equation

Let us assume that we have solved Eq. 11.68 and found $\Psi(q, t)$. The fourth postulate tells us that the probability distribution for the system will be proportional to $\Psi^*(q, t)\Psi(q, t)\,d\tau$. A cursory examination of this expression seems to suggest that it cannot be correct, since time appears in the expression. Let us say that we are interested in how the 22 electrons and three nuclei in CO_2 are distributed in space. We proceed to solve Eq. 11.68 by some appropriate method and thereby obtain $\Psi(q, t)$ for the 25 "particles" forming CO_2. According to the fourth postulate, the probability distribution for these 25 particles is proportional to $|\Psi(q, t)|^2\,d\tau$. The appearance of time in this expression suggests that the electron–nuclei distribution for CO_2 was one thing when King John was signing the Magna Carta in June 1215, another when Columbus stepped ashore on October 12, 1492, and still another on

Independence Day, July 4, 1776. Such a concept is nonsense. Therefore, we are led to conclude either that the fourth postulate and probably all of quantum theory is nonsense or that time does not really appear in the probability expression for a stable molecule such as CO_2.

In Chapter 1, we obtained the combined ideal-gas law from Boyle's and Charles' laws by separating the variables and noting that if $f(x) = g(y)$ for all values of the independent variables x and y, then we must have $f(x) = g(y) = a$ constant. This method of separation of variables can be used to great advantage in solving quantum mechanical problems. For partial differential equations, the general procedure is as follows:

Step 1:   We assume that our unknown function, say, $f(x, y)$, can be written in the separable form $f(x, y) = h(x) g(y)$.

Step 2:   We substitute the separable form into the differential equation and execute the indicated operations to the extent possible.

Step 3:   We divide both sides of the equation by the separable form.

If this three-step procedure produces a separation of variables, a separable solution of the form $h(x)g(y)$ exists. By equating the terms involving x and those involving y to a common constant K, separate equations for $h(x)$ and $g(y)$ can be obtained. The solution of these two equations gives $h(x)$ and $g(y)$, and their product is $f(x, y)$. On the other hand, if the three-step procedure does not separate the variables, then a separable solution does not exist, and other means of solving the partial differential equation must be found.

Let us apply this technique to the time-dependent Schrödinger equation. We seek the wave function $\Psi(q, t)$. Perhaps a separable solution exists. To investigate this possibility, we write

$$\Psi(q, t) = \psi(q) \, \Phi(t). \qquad \textbf{(11.69)}$$

We now substitute the separable form into Eq. 11.68, giving

$$\mathcal{H}[\psi(q) \, \Phi(t)] = i\hbar \frac{\partial}{\partial t}[\psi(q) \, \Phi(t)]. \qquad \textbf{(11.70)}$$

Since $\mathcal{H}$ contains no differential operators involving time, $\Phi(t)$ can be factored to the left of $\mathcal{H}$. On the right-hand side, $\psi(q)$ is independent of time, so it can be factored to the left of the time derivative. These operations produce

$$\Phi(t) \, \mathcal{H}\psi(q) = \psi(q) \, i\hbar \frac{\partial}{\partial t} \Phi(t). \qquad \textbf{(11.71)}$$

Executing Step 3, we divide both sides of Eq. 11.71 by $\psi(q) \, \Phi(t)$ to obtain

$$\frac{\Phi(t) \, \mathcal{H}\psi(q)}{\psi(q) \, \Phi(t)} = \frac{\psi(q) \, i\hbar \frac{\partial}{\partial t}\Phi(t)}{\psi(q) \, \Phi(t)}. \qquad \textbf{(11.72)}$$

On the left, we can cancel $\Phi(t)$, because it is not within the scope of the differential operator. On the right, we can cancel $\psi(q)$ for the same reason. The result is

$$\frac{\mathcal{H}\psi(q)}{\psi(q)} = \frac{i\hbar \frac{\partial}{\partial t}\Phi(t)}{\Phi(t)}. \qquad \textbf{(11.72)}$$

The three-step procedure is now completed, so we ask, "Are the variables separated?" Inspection of Eq. 11.72 shows that the right-hand side contains only time variables. On the left-hand side, $\psi(q)$ contains only coordinate variables. The kinetic-energy operator in $\mathcal{H}$ contains just constants and derivatives with respect to coordinates. The only question has to do with the potential function $V(x_1, y_1, z_1, x_2, y_2, z_2, \ldots, t)$. If this function depends explicitly upon time, the variables are not separated and no separable solution exists. If, however, the potential is independent of time, separation has been achieved.

If V depends only upon coordinates, both sides of Eq. 11.72 must be equal to a constant K. This gives two equations:

$$\frac{\mathcal{H}\psi(q)}{\psi(q)} = K,$$

or

$$\mathcal{H}\psi(q) = K\,\psi(q), \tag{11.73}$$

and

$$\frac{i\hbar\frac{\partial}{\partial t}\Phi(t)}{\Phi(t)} = K,$$

or

$$i\hbar\frac{\partial}{\partial t}\Phi(t) = K\Phi(t). \tag{11.74}$$

Equation 11.74 is easy to solve. Rearranging terms produces

$$\frac{\partial\Phi(t)}{\Phi(t)} = \frac{K}{i\hbar}\partial t.$$

Taking indefinite integrals of both sides, we obtain

$$\int\frac{\partial\Phi(t)}{\Phi(t)} = \int\frac{K}{i\hbar}\partial t,$$

which gives

$$\ln\Phi(t) = \frac{Kt}{i\hbar} + \text{constant.}$$

Exponentiation of both sides yields

$$\Phi(t) = C\exp\left[\frac{Kt}{i\hbar}\right] = C\exp\left[-\frac{iKt}{\hbar}\right], \tag{11.75}$$

where C is a constant. The wave function is, therefore, given by

$$\boxed{\Psi(q, t) = \psi(q)\,\Phi(t) = C\,\psi(q)\exp\left[-\frac{iKt}{\hbar}\right]} \tag{11.76}$$

if the potential-energy function does not depend explicitly upon time.

Let us again consider CO_2. Clearly, the interparticle potential energy operating between the 22 electrons and 3 nuclei is the same on Monday as it is on Tuesday. This observation means that the potential is independent of time and Eq. 11.76 is the wave function. The probability distribution for CO_2 is, therefore,

$$|\Psi(q, t)|^2 \, d\tau = \Psi^*(q, t)\Psi(q, t) \, d\tau = C^* \, \psi^*(q) \exp\left[\frac{iKt}{\hbar}\right] C \, \psi(q) \exp\left[-\frac{iKt}{\hbar}\right] d\tau$$

$$= |C|^2 \, |\psi(q)|^2 \, d\tau, \tag{11.77}$$

which is independent of time, as we would expect. Consequently, all that is needed to obtain all the measurable properties of CO_2 or any other system of interest for which the potential is independent of time is the solution of Eq. 11.73. This equation is called the *stationary-state Schrödinger equation*.

It is important to note what has been said in this section and what has not been said. The quantum mechanical wave function always depends upon time in the manner described by Eq. 11.76. However, if the potential energy contains no *explicit* dependence on time, the probability distribution for the system will be independent of time and we will have $|\Psi(q, t)|^2 \, d\tau = C^2|\psi(q)|^2 \, d\tau$.

We may use the fourth postulate to identify the constant K in Eq. 11.76. From Table 11.1, we see that the quantum mechanical operator for the energy is $i\hbar(\partial/\partial t)$. Using this fact, we find that the average energy of the system is given by

$$\langle E \rangle = \frac{\displaystyle\int_{\text{all space}} C^* \, \psi^*(q) \exp\left[\frac{iKt}{\hbar}\right] i\hbar \frac{\partial}{\partial t} C \, \psi(q) \exp\left[-\frac{iKt}{\hbar}\right] d\tau}{\displaystyle\int_{\text{all space}} C^* \, \psi^*(q) \exp\left[\frac{iKt}{\hbar}\right] C \, \psi(q) \exp\left[-\frac{iKt}{\hbar}\right] d\tau}$$

$$= \frac{|C|^2 K \displaystyle\int_{\text{all space}} \psi^*(q) \, \psi(q) \, d\tau}{|C|^2 \displaystyle\int_{\text{all space}} \psi^*(q) \, \psi(q) \, d\tau} = K.$$

Since K is the average energy of the system, it is customary to write the stationary-state Schrödinger equation in the form

$$\boxed{\mathcal{H}\psi(q) = E \, \psi(q)}. \tag{11.78}$$

Equation 11.78 has the form of an eigenvalue equation in which E is the eigenvalue of the Hamiltonian operator associated with the eigenfunction $\psi(q)$. Because $\mathcal{H}$ is guaranteed to be Hermitian by the second postulate, we know that E is always real and that all eigenfunctions of $\mathcal{H}$ associated with different energies must be orthogonal. In the remaining chapters of this section of the text, we shall use Eq. 11.78 to study the nature of molecular translational, rotational, vibrational, and electronic energies, atomic structure, and chemical bonding.

Summary: Key Concepts and Equations

1. In classical mechanics, all variables are multiplicative and continuous. They can assume any value within the range they span. The equations of classical mechanics are based on the three Newtonian postulates. The first postulate states that the force acting on a particle i is the rate of change of its momentum:

$$\mathbf{F}_i = \frac{d\mathbf{P}_i}{dt}.$$

The second postulate states that the force on particle i due to the presence of particle j, $\mathbf{F}_{ji}$, is the negative of the force on particle j produced by particle i. That is,

$$\mathbf{F}_{ji} = -\mathbf{F}_{ij}.$$

This is equivalent to the statement, "For every action, there is an equal but opposite reaction." The third postulate requires that $\mathbf{F}_{ij}$ act along a line connecting the two particles. This statement is called the collinearity law.

2. The first Newtonian postulate leads directly to the Newtonian equations of motion:

$$m_i \frac{d^2 x_i}{dt^2} + \frac{\partial V}{\partial x_i} = 0,$$

$$m_i \frac{d^2 y_i}{dt^2} + \frac{\partial V}{\partial y_i} = 0,$$

and

$$m_i \frac{d^2 z_i}{dt^2} + \frac{\partial V}{\partial z_i} = 0,$$

for each particle in the system. There are, therefore, $3N$ second-order, coupled ordinary differential equations for an N-particle system. These equations are deterministic, in that a measurement of the position and momenta of all particles at any point in time determines all future and past behavior of the system.

3. The classical Hamiltonian, $\mathbf{H}$, is the sum of the kinetic and potential energies of a system. In rectangular Cartesian coordinates,

$$\mathbf{H} = \sum_{i=1}^{N} \left[\frac{1}{2m_i} \right] [P_{xi}^2 + P_{yi}^2 + P_{zi}^2] + V(q_1, q_2, \dots, q_{3N}).$$

The general definition of the momentum conjugate to any coordinate is

$$P_q = \frac{\partial T}{\partial v_q},$$

where T is the expression for the kinetic energy, $v_q = dq/dt$ is the rate of change of coordinate q with time, and P_q is called the momentum conjugate to coordinate q. When this definition is used to express the classical Hamiltonian in any arbitrary coordinate system, we always have

$$\frac{\partial \mathbf{H}}{\partial P_{qi}} = \frac{dq_i}{dt} \quad \text{and} \quad \frac{\partial \mathbf{H}}{\partial q_i} = -\frac{dP_{qi}}{dt} \quad \text{for } i = 1, 2, 3, \dots, 3N.$$

These equations represent $6N$ first-order, coupled ordinary differential equations that describe the classical motion of the system. The equations are called Hamilton's equations of motion.

4. Variables that remain constant throughout the motion of a system are called constants of the motion. Such quantities play important roles in both classical and quantum mechanics. If a classical system is isolated, the total energy, all components of the linear momentum of the center of mass, and all components of the angular momentum, $\mathbf{M}$, about the center of mass are constants of the motion. In Cartesian coordinates, the components of the angular momentum are

$$M_x = \sum_{i=1}^{N} [y_i P_{zi} - z_i P_{yi}],$$

$$M_y = \sum_{i=1}^{N} [z_i P_{xi} - x_i P_{zi}],$$

and

$$M_z = \sum_{i=1}^{N} [x_i\, P_{yi} - y_i\, P_{xi}].$$

5. The important discoveries and theoretical developments leading up to the development of quantum theory were as follows:

(A) The Planck equation, which accurately described the energy density of blackbody radiation, introduced the concept of quantization to the scientific community. The Planck equation for the energy density is

$$u(T, \lambda) = \frac{8\pi hc}{\lambda^5} \frac{1}{\exp\left[\dfrac{hc}{\lambda k_b T}\right] - 1}.$$

(B) It was observed that when electrons are ejected from a metal by ultraviolet radiation, the kinetic energy of the ejected electrons depends only upon the frequency of the incident radiation and not upon its intensity. To explain this observation, Einstein had to postulate that light can behave as a particle that has energy $h\nu$. Using this concept, Einstein was able to demonstrate that the kinetic energy of the ejected photoelectron is given by

$$T = \frac{mv^2}{2} = h\nu - W,$$

where W is the work function of the metal, which is the amount of energy required to overcome the attractive forces within the metal and eject the electron.

(C) The discovery of the structure of the atom by Rutherford and Geiger presented a picture that could not exist according to the laws of classical mechanics and electrodynamics. Rutherford and Geiger's experiments demonstrated that the positive charge on the atom is contained within a small region of space called the nucleus. The negative electrons had to be outside the nucleus. The classical electrostatic attraction between positive and negative particles would be expected to pull stationary electrons into the nucleus. If it is postulated that the electrons rotate about the nucleus like planets about the sun, such that the centrifugal force is sufficient to balance the electrostatic attraction, the electrons would have to be continuously accelerated in the electric field produced by the positively charged nucleus. In this case, the laws of classical electrodynamics require that the electrons radiate energy at a frequency equal to their rotational frequency. This radiation of energy would slow the electrons and reduce the centrifugal force, so that the electrons would gradually spiral into the nucleus.

(D) The observation of atomic spectra demonstrated that only certain wavelengths of radiation are emitted when an atom is excited. These wavelengths were found to be independent of the quantity of excitation energy used. Such a situation cannot exist if classical mechanics describes nature. The wavelengths for the hydrogen atom were observed to fit the equation

$$\lambda^{-1} = 109{,}677 \left[\frac{1}{n^2} - \frac{1}{m^2}\right] \text{cm}^{-1}$$

for $n = 1, 2, 3, 4, \ldots$ and $m = n + 1, n + 2, n + 3, \ldots$.

The constant 109677 cm^{-1} is now called the Rydberg constant R_H. Clearly, the emission lines were quantized.

(E) Faced with the Rutherford atom and the impossibility of keeping the electrons out of the nucleus if classical mechanics and electrodynamics described the atom, Bohr combined Planck's concept of quantization with the bold postulate of a radiationless orbit. In doing so, he set aside many of the most fundamental concepts of classical theory. The assumptions of the Bohr theory are as follows:

1. Electrons revolve about the nucleus in circular orbits that do not radiate energy.
2. The laws of classical mechanics hold, except that radiation due to acceleration does not occur.
3. The angular momentum of the electron is quantized in units of $\hbar$. That is, electrons may only have angular momentum equal to $n\hbar$, where n takes on only integer values $1, 2, 3, 4, \ldots$.
4. The potential energy is coulombic.

With these assumptions, Bohr was able to compute the value of the Rydberg constant and quantitatively predict the nature of the spectrum of the hydrogen atom. However, when efforts were made to apply the theory to many-electron atoms, problems arose. The answers were no longer exact. Apparently, something was still missing.

(F)   When X rays are scattered from a particle, the wavelengths of the scattered X rays increase. This effect is called Compton scattering. To explain the effect, Compton assumed that the incident electromagnetic radiation could be treated as a particle called a photon whose energy was $h\nu = hc/\lambda$. He then assumed that the increase in wavelength is due to the fact that energy is lost by the photon to an electron whose mass is m. The electron gains energy, while the X ray loses energy. As a consequence, the wavelength of the scattered X ray is longer than that of the incident radiation. When the requirements of momentum and energy conservation are applied, this concept quantitatively predicts the change in wavelength of the X ray, provided that the photon is assigned a momentum equal to h/λ.

De Broglie reasoned that if electromagnetic waves can have momentum and a particle character, matter could have a wavelength and a wave character. He supported his reasoning by postulating that a wave is associated with every particle and that the wavelength of these "matter waves" is given by

$$\lambda = \frac{h}{p}.$$

This equation is now known as the de Broglie hypothesis. The wave character of matter was experimentally demonstrated by Davisson and Germer and by Thomson. These developments led directly to the development of wave or quantum mechanics by Heisenberg and Schrödinger in 1926.

6. A differential equation of the form

$$\mathcal{G}\phi = g\phi,$$

where $\mathcal{G}$ is a differential operator, ϕ is a function of the variables of the problem, and g is a constant, is called an eigenvalue equation. The constant g is called the eigenvalue of $\mathcal{G}$ associated with ϕ, and the function ϕ is said to be an eigenfunction of the operator $\mathcal{G}$.

7. A Hermitian operator $\mathcal{G}$ has the property that

$$\int_{\text{all space}} \psi^*(q_1, q_2, \ldots, q_n)\, \mathcal{G}\phi(q_1, q_2, \ldots, q_n)\, dq_1 dq_2 \ldots dq_n$$

$$= \int_{\text{all space}} \{\mathcal{G}\psi(q_1, q_2, \ldots, q_n)\}^*\, \phi(q_1, q_2, \ldots, q_n)\, dq_1 dq_2 \ldots dq_n,$$

where the complex conjugate $\psi^* = F(q) - iG(q)$ if $\psi = F(q) + iG(q)$, with $F(q)$ and $G(q)$ being real functions. The foregoing integral equation is generally expressed using the compact notation

$$\langle \psi | \mathcal{G} | \phi \rangle = \langle \mathcal{G}\psi | \phi \rangle.$$

The absolute square of any complex function, $|\psi(q)|^2 = \psi^*(q)\, \psi(q)$, is always real.

8. The eigenvalues of all Hermitian operators are always real. Also, if ϕ and ψ are eigenfunctions of a common Hermitian operator $\mathcal{G}$ with different eigenvalues, then

$$\langle \phi | \psi \rangle = \langle \psi | \phi \rangle = 0.$$

Functions that have this property are said to be orthogonal. The commutator of two variables or operators $\mathcal{F}$ and $\mathcal{G}$ is defined to be

$$[\mathcal{F}, \mathcal{G}]\psi = \{\mathcal{F}\mathcal{G} - \mathcal{G}\mathcal{F}\}\psi.$$

If $[\mathcal{F}, \mathcal{G}] = 0$, the operators or variables are said to commute.

9. The foundation of quantum or wave mechanics is contained in four postulates:

(A) Matter is described by a wave $\Psi(q, t)$, called the wave function, that depends upon the coordinates q and time. In general, $\Psi(q, t)$ is finite, single valued, and continuous at all points.

(B) For every observable physical variable (position, momentum, energy, etc.), there exists a corresponding quantum mechanical operator that is Hermitian. The commutator between the operators representing the conjugate variables must equal either $i\hbar$ or $-i\hbar$, with the sign depending upon the variables involved.

The forms of the quantum mechanical operators that result from this postulate are summarized in Table 11.1.

(C) The differential equations of quantum mechanics are obtained by direct substitutuion of the quantum variables and operators into the classical equations in Cartesian coordiantes and then allowing the operators thus formed to operate upon $\Psi(q, t)$.

(D) The product of the absolute square of the wave function, $|\Psi(q, t)|^2$, and the volume element $d\tau$ is proportional to the probability distribution for the system. That is, $|\Psi(q, t)|^2 d\tau$ is proportional to the probability that the system will be in the volume element $d\tau$ at time t.

10. Application of the quantum mechanical postulates shows that the wave function of a system can be obtained by solving the time-dependent Schrödinger equation

$$\mathcal{H}\Psi(q, t) = i\hbar \frac{\partial}{\partial t} \Psi(q, t),$$

where q represents all the coordinates, t is the time, and the quantum mechanical Hamiltonian operator is

$$\mathcal{H} = \sum_{i=1}^{N} -\frac{\hbar^2}{2m_i} \nabla_i^2 + V(x_1, y_1, z_1, x_2, y_2, z_2, \ldots, t),$$

in which the first term is the kinetic energy for all N particles in the system and the second term is the total potential energy of the system. This means that the quantum mechanical operator for the kinetic energy of particle i is $-\hbar^2/(2m_i) \nabla_i^2$.

11. If the potential does not depend explicitly upon time, the variables in the time-dependent Schrödinger equation can be separated and the solution put in the form

$$\Psi(q, t) = \psi(q) \Phi(t) = C \psi(q) \exp\left[-\frac{iEt}{\hbar}\right].$$

The absolute square of this separable wave function is independent of time, as we would expect. To find the part of the wave function that depends upon the coordinates, we solve the stationary-state Schrödinger equation

$$\mathcal{H} \psi(q) = E \psi(q).$$

This equation has the form of an eigenvalue equation in which E is the eigenvalue of the Hamiltonian operator associated with the eigenfunction $\psi(q)$. Since $\mathcal{H}$ is guaranteed to be Hermitian by the second postulate, it follows that E is always real and that all eigenfunctions of $\mathcal{H}$ associated with different energies must be orthogonal.

Problems

Problems that require the use of some type of computational device are marked with an asterisk (*). Problems that require some type of plotting routine are indicated with a pound sign (#).

11.1 This problem is based on the results obtained in Example 11.1. A quarterback is poised to throw a football in the x–y plane with an initial velocity v_o that we will assume is the result of the maximum force he can generate. Let us assume that a receiver is located at the point $x = x_o$, $y = 0$, and $z = -z_o$ at time $t = 0$. The receiver is running at a constant velocity equal to S_o in the $+z$ direction. The quarterback wishes to complete a pass to the receiver when he reaches the point $x = x_o$ and $y = z = 0$. (See the diagram at the bottom of the page.)

(A) Show that if $x_o > v_o^2/g$, the receiver is out of range of the quarterback's ability to throw the football.

(B) Show that if $z_o/S_o < (2v_o/g) \sin[0.5 \sin^{-1}\{gx_o/v_o^2\}]$, the receiver is moving too fast for the quarterback to complete the pass at the designated point.

(C) If $x_o < v_o^2/g$ and $z_o/S_o > (2v_o/g) \sin[0.5 \sin^{-1}\{gx_o/v_o^2\}]$, show that the pass may be completed if the quarterback throws the ball when

$$t = (z_o/S_o) - (2v_o/g) \sin[0.5 \sin^{-1}\{gx_o/v_o^2\}]$$

at an angle given by $\theta_o = 0.5 \sin^{-1}\{gx_o/v_o^2\}$. Figure 11.20 illustrates the situation. You may use the results obtained in Example 11.1. Ignore the effects of air resistance.

11.2 In three-dimensional space, a classical particle moves in a potential field given by

$$V(x, y, z) = ax^3 + by^3 + cz^3 + dxy + exz + fyz,$$

where a, b, c, d, e, and f are constants.

(A) When the particle is at the point $x = 1$ cm, $y = 2$ cm, and $z = 3$ cm, what is the x component of force acting on it in terms of the constants of the problem?

(B) Consider a vector $\mathbf{F}$ as shown in the following diagram:

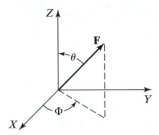

The direction of $\mathbf{F}$ may be specified by its spherical polar angles θ and ϕ, where θ is the angle between $\mathbf{F}$ and the z-axis and ϕ is the angle between the projection of $\mathbf{F}$ into the $(x$–$y)$ plane and the x-axis. If $\mathbf{F}$ is the force vector on a particle at the point given in (A), calculate the direction of $\mathbf{F}$ by computing the spherical polar angles θ and ϕ. Let $a = b = c = 1$ dyne cm^{-2} and $d = e = f = 1$ dyne cm^{-1}.

11.3 Example 11.1 requested that we solve the classical equations of motion for a projectile moving in the x–y plane subject to the gravitational potential $V = mgy$ with $x = y = z = 0$ at $t = 0$ and with

$$\frac{dx}{dt} = v_o \cos \theta_o \quad \text{and} \quad \frac{dy}{dt} = v_o \sin \theta_o \quad \text{at} \quad t = 0.$$

The effect of air friction was ignored.

(A) Obtain the solution to this problem if the potential field is $V(x, y) = mgy + mkx$ instead of the gravitational potential. (Take m, g, and k to be constants.)

(B) For the potential in (A), determine x at the point of impact with the ground in terms of v_o, θ, k, and g.

(C) For the potential in (A), determine the maximum height attained by the projectile in terms of v_o, θ, k, and g.

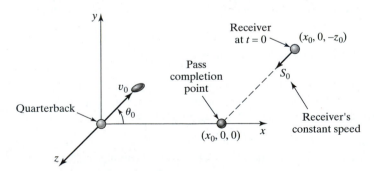

▲ FIGURE 11.20
Diagram of quarterback and receiver in Problem 11.1 who are attempting to complete a pass at the point $x = x_o$.

11.4   Assume that the hydrogen atom behaves like a planetary system, with the electron rotating about an extremely heavy nucleus. The mass of the electron is 9.1094×10^{-31} kg. The kinetic energy of the electron is 2.1792×10^{-18} J. If the nuclear mass is assumed to be infinite and the orbit circular with a radius of 0.519×10^{-10} m, compute the angular momentum of the electron.

11.5   Assume the particle illustrated in Figure 11.5 is moving in three-dimensional space. Then, in rectangular Cartesian coordinates, the kinetic energy of the particle is

$$T = \frac{m}{2}[v_x^2 + v_y^2 + v_z^2].$$

Suppose we wish to transform to a spherical polar coordinate system with coordinates (R, θ, ϕ) instead of (x, y, z). This coordinate system is illustrated in the figure given in Problem 11.2 if we replace F with R. The transformation equations between the spherical polar coordinate system and a rectangular Cartesian system, given in Chapter 9, Figure 9.3, are

$$x = R \sin \theta \cos \phi,$$

$$y = R \sin \theta \sin \phi,$$

and

$$z = R \cos \theta.$$

(A) Obtain the kinetic energy of the particle in terms of the spherical polar coordinates and their rates of change with time.

(B) Obtain expressions for the momenta conjugate to the spherical polar coordinates.

(C) Show that the momentum conjugate to ϕ, P_ϕ, is the same as the Cartesian z-component of angular momentum.

(D) Express the kinetic energy in terms of the spherical polar conjugate momenta and coordinates.

11.6   Consider a single particle of mass m attached to a wall of infinite mass by a spring as shown in the following diagram:

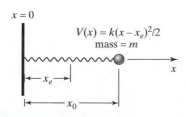

The particle moves in one dimension (x). The potential for the system is $V(x) = k(x - x_e)^2/2$, where x_e is a constant. At time $t = 0$, the particle is at position x_o and its velocity is zero.

(A) Set up and solve the Newtonian equations of motion for this system. Plot the position of the particle as a function of time from $t = 0$ to $t = 4\pi(m/k)^{1/2}$. Make the abscissa of the plot be in units of $4\pi(m/k)^{1/2}$ so that the abscissa will go from 0 to 1. (*Hint*: Refer to Problem 9.5 for assistance in solving the differential equation.)

(B)   As can be seen from the plot made in (A), the classical motion of the spring is oscillatory. In fact, this system is called a harmonic oscillator. The period τ of the oscillator is the time required to execute one complete vibrational cycle. The vibrational frequency ν is the reciprocal of the period. Determine the vibrational period and frequency of this harmonic oscillator in terms of k and m.

11.7   Show that Planck's equation for the energy density of blackbody radiation reduces to the equation suggested by Wien for small values of λT.

11.8   Show that Planck's equation for the energy density of blackbody radiation reduces to the equation suggested by Rayleigh and Jeans for large values of λT. (*Hint*: Expand the exponential in a power series.)

11.9   Show that Planck's equation predicts a maximum in the energy density of blackbody radiation at the point $\lambda \approx 0.290$ cm K$/T$ when $u(T, \lambda)$ is plotted against λ. [*Hint*: You may solve the problem either by a one-dimensional grid search or by iterative methods.]

11.10 (A) Calculate the frequencies and wavelengths of the four Lyman spectral lines of lowest energy for the hydrogen atom.

(B) Compute the frequencies and wavelengths for the four Balmer spectral lines of lowest energy for the hydrogen atom.

11.11 Compute the possible range of emisson energies seen for the Lyman, Balmer, and Paschen series for the hydrogen atom.

11.12 Compute the angular momentum of the electron in the third Bohr orbit $(n = 3)$.

11.13 (A) What is the magnitude of the centrifugal force on an electron in the second Bohr orbit $(n = 2)$ for a hydrogen atom?

(B)   What is this centrifugal force for the He$^+$ ion?

11.14 The ionization potential of an atom is the amount of energy required to completely remove the electron from the lowest lying energy state (ground state). Compute the ionization potential of the hydrogen atom predicted by the Bohr theory. The experimentally measured result is 2.17866×10^{-18} J. Compute the percent error in the Bohr result.

11.15 (A) Compute the energy of an electron in the $n = 1$ Bohr orbit. The linear velocity v of a particle moving in a circular orbit of radius R is related to its angular velocity v_θ by $v = v_\theta R$.

(B) What is the de Broglie wavelength of this $n = 1$ electron?

11.16 Compute the de Broglie wavelength for a 100-gram ball moving at 100 miles per hour. Is there any experimental method by which a wavelength of this size might be measured?

11.17 An incident X ray with wavelength 1.540 Å undergoes Compton scattering at an angle θ relative to the direction of motion of the ray. The wavelength of the scattered electron is 3.000 Å. Compute the wavelength of the scattered X ray and the scattering angle of the X ray.

11.18 Show that the absolute square of $X = 3e^{-6i} + 2i + (-3 + 4i)^2$ is real.

11.19 (A) Show that the function $\phi = A[e^{ax} + e^{-ax}]$, where A and a are constants, is an eigenfunction of the operator $\mathcal{G} = \partial^2/\partial x^2$, but not of the operator $\mathcal{F} = \partial/\partial x$.

(B) What is the eigenvalue of $\partial^2/\partial x^2$ associated with the eigenfunction ϕ?

11.20 Let $\psi(\theta)$ and $\phi(\theta)$ be arbitrary functions of the angle θ which have the property that $\psi(\theta) = \psi(\theta + 2\pi)$ and $\phi(\theta) = \phi(\theta + 2\pi)$, since the angles θ and $\theta + 2\pi$ are physically identical. Show that the operator $\mathcal{G} = i(\partial/\partial\theta)$ is Hermitian if θ spans the angular range from 0 to 2π radians. (*Hint*: Integrate the integral $\langle\psi|\mathcal{G}|\phi\rangle$ by parts to show that it equals $\langle\mathcal{G}\psi|\phi\rangle$.)

11.21 Energy and time are conjugate variables in quantum mechanics. Using the operators given in Table 11.1, show that the commutator of t and $i\hbar\,(\partial/\partial t)$ obeys the requirements of the second postulate in that it is equal to $-i\hbar$.

11.22 Scientist A makes the suggestion that there exists a one-dimensional system for which the wave function is

$$\psi = N\tan(ax)\ for\ 0 \le x \le \infty,$$

where a and N are constants.

Scientist B disagrees with A and states that the true wave function for the system in question has the form

$$\psi = Nx^{1/2}\exp[-ax]\quad for\quad 0 \le x \le \infty,$$

where a and N are again constants.

Scientist C disagrees with both A and B and suggests that

$$\psi = N\sin(ax),\quad where\ a\ and\ N\ are\ constants.$$

Which scientist is most likely to be correct. Justify your answer.

11.23 Let ψ and ϕ be degenerate eigenfunctions of the operator $\mathcal{G}$. That is, the eigenvalues of $\mathcal{G}$ associated with ψ and ϕ are both the same. Show that any linear combination of ψ and ϕ is also an eigenfunction of $\mathcal{G}$ with the same eigenvalue.

11.24 Let ϕ be any arbitrary function that satisfies the conditions of the first postulate. If it found that ϕ is a simultaneous eigenfunction of the operators $\mathcal{G}$ and $\mathcal{F}$, with eigenvalues a and b, respectively, show that $\mathcal{G}$ and $\mathcal{F}$ must be commuting operators. We might also state this problem in the following way: ϕ is any arbitrary function that satisfies the conditions of the first postulate. The function ϕ is found to be an eigenfunction of the operator $\mathcal{G}$ with eigenvalue a. It is also found to be an eigenfunction of the operator $\mathcal{F}$ with eigenvalue b. Show that the operators $\mathcal{G}$ and $\mathcal{F}$ must commute. That is, show that $[\mathcal{G}, \mathcal{F}] = [\mathcal{F}, \mathcal{G}] = 0$.

11.25 Classical electrodynamics requires that the frequency of radiation emitted from a charged particle accelerating through an electric field be equal to the frequency of the particle's rotation. In this problem, you will show that the Bohr correspondence principle does indeed make the orbital frequency of the electron in the Bohr model equal to the radiation frequency as n becomes large. Consequently, we expect the correspondence principle to hold as the energy becomes large.

(A) Compute the angular momentum of the hydrogen-atom electron in the first Bohr orbit $(n = 1)$. Use Eq. 11.21 to obtain the angular velocity v_θ of the electron in this orbit. What is the rotational frequency of the electron about the nucleus in this orbit?

Next, assume that emission occurs from a hydrogen-atom electron excited to the $n = 2$ orbit undergoing a transition to the adjacent $n = 1$ orbit. Compute the frequency of the radiation emitted. How does this frequency compare with the orbital frequency of the electron?

(B) Repeat (A) with the electron in the $n = 100$ orbit and the emission occurring from a transition from the $n = 101$ orbit to the $n = 100$ orbit. How does the emission frequency compare with the electron's orbital frequency? What does this tell us about the Bohr correspondence principle?

11.26 (A) Compute the energy of an electron in the $n = 1$ Bohr orbit for a hydrogen atom. Calculate the linear velocity of the electron in this orbit. (The linear velocity is the product of the angular velocity v_θ and the radius of the orbit.)

(B) Repeat (A) for an electron in the $n = 1$ Bohr orbit of the Li^{2+} ion.

(C) Repeat (A) for the Co^{26+} ion.

(D) When a particle approaches the speed of light, relativistic effects become very important. What is the ratio of the linear velocity of the electron in (C) to the speed of light in a vacuum? What does this tell us about the importance of relativistic effects for electrons in the $n = 1$ orbits of the heavier elements?

11.27 The discussion of the Rutherford atom in the text points out that such a system cannot exist in a classical world. In this problem, we will quantitatively determine what would happen to the electron in a hydrogen-like atom if classical mechanics and electrodynamics were valid. The various parts of the problem serve as a procedure guide.

(A) For a circular orbit, the radial momentum P_R and its derivative dP_R/dt are both zero. Use this fact together with Eqs. 11.22 and 11.23C to show that the classical Hamiltonian, and hence the energy E, of a hydrogen-like electron is given by

$$E = -\frac{Ze^2}{8\pi\varepsilon_o R}.$$

(B) Using Eq. 11.21 and the equations obtained in (A), show that the angular velocity is given by $v_\theta = [Ze^2/(4\pi\varepsilon_o mR^3)]^{1/2}$.

(C) Use the fact that classical electrodynamics requires that the rate of energy loss due to radiation as the electron accelerates through the electric field be

$$\frac{dE}{dt} = -\frac{e^2\,v_\theta^4 R^2}{6\pi\varepsilon_o c^3}$$

to show that we must have

$$\frac{dE}{dt} = -\frac{Z^2 e^6}{96\pi^3\varepsilon_o^3 c^3 m^2 R^4}.$$

(D) Use the expression for E obtained in (A) to derive an expression for dE/dt in terms of R and dR/dt.

(E) Combine the results of (C) and (D) to show that we must have

$$R^2\,dR = -\frac{Ze^4}{12\pi^2\varepsilon_o^2 c^3 m^2}dt.$$

(F) Assuming that the radial position of the electron is at a distance equal to the Bohr radius a_o at time $t = 0$, integrate the result of (E) to obtain R as a function of time.

(G) Compute how long it will take for the electron in a hydrogen atom to collapse into the nucleus.

11.28 An eigenvalue equation that we will encounter in Chapter 12 has the form

$$\frac{d^2}{dx^2}\psi(x) = C\psi(x),$$

where C is a positive constant. Find the general form for the eigenfunction $\psi(x)$. To the extent possible, evaluate the constants that appear in the equation in terms of C. (*Hint*: The only two common functions whose second derivative is a constant times the function itself are exponentials and trigonometric sine or cosine functions.)

11.29 An eigenvalue equation that we will encounter in Chapter 12 has the form

$$\frac{d^2}{dx^2}\psi(x) = -C\,\psi(x),$$

where C is a positive constant. Find the general form for the eigenfunction $\psi(x)$. To the extent possible, evaluate the constants that appear in the equation in terms of C. (*Hint*: The only two common functions whose second derivative is a constant times the function itself are exponentials and trigonometric sine or cosine functions.)

11.30 An eigenvalue equation that we will encounter in Chapter 12 has the form

$$\frac{d}{dx}\psi(x) = -C\,\psi(x),$$

where C is any real constant. Obtain a general form for the eigenfunctions of this equation.

11.31 Table 11.1 shows that the quantum mechanical operator for P_x is $(\hbar/i)\,(\partial/\partial x)$.

(A) Write down an eigenvalue equation with eigenvalue a whose solution will give the eigenfunctions of the operator for P_x.

(B) Solve the eigenvalue equation developed in (A) for the eigenfunctions of the momentum operator P_x.

(C) Separate the eigenfunction obtained in (B) into real and imaginary parts, and plot each on a separate graph as a function of the variable $z = ax/(\pi\hbar)$. Show the plots for the range $0 \leq z \leq 8$. Take the amplitudes of the waves to be unity.

(D) Determine the wavelength for both the real and imaginary parts of the wave in terms of z. What is the wavelength in terms of the distance variable x? Use the de Broglie hypothesis to interpret the physical significance of the eigenvalue of P_x.

11.32 Let $y = \ln(-5)$. Is y real? Is y complex? Is y purely imaginary? Prove that

$$y = 1.609437912.. + 3.1415927.. i.$$

What is the complex conjugate of y? What is the value of $|y|^2$?

11.33 Sam has just noticed an unusual expression in an article he was reading. The expression involved the sine of an imaginary quantity. Specifically, the expression in question was

$$y = A\sin[iax].$$

(A) Is it necessary that this expression be a misprint? Does the sine of an imaginary quantity have any meaning? If so, what is the meaning? Is y real? Is y imaginary? Is y complex? What is the complex conjugate of y? Is $|y|^2$ real? Prove that all your responses are correct. (Hint: See Section 11.3.1.)

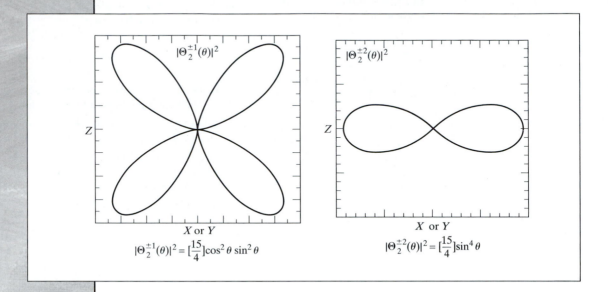

CHAPTER 12

$$|\Theta_2^{\pm 1}(\theta)|^2 = [\frac{15}{4}]\cos^2\theta\sin^2\theta \qquad |\Theta_2^{\pm 2}(\theta)|^2 = [\frac{15}{4}]\sin^4\theta$$

Translational, Rotational, and Vibrational Energies of Molecular Systems

The molecular energies in translational, rotational, and vibrational motion are all quantized. If this quantization is ignored, it becomes impossible to accurately compute changes in the thermodynamic potentials, such as ΔU, ΔH, ΔA, ΔS, and ΔG, from the molecular properties of the molecules. We have already seen in Chapter 7 that heat capacities cannot be accurately predicted if quantization effects are not properly considered. Spectroscopic observations that depend entirely upon the spacings between the quantized energy levels cannot be understood or predicted without a quantum treatment of those energy states.

In this chapter, we shall investigate the quantized nature of translational, rotational, and vibrational energy. In the process, we shall see that the wave treatment of matter automatically introduces an uncertainty with respect to momentum and position that fully resolves the philosophical difficulties associated with the deterministic nature of classical mechanics.

12.1 Translational Energy

12.1.1 The Free Particle

We begin by considering the case of a single isolated "particle" of mass m, which might be a molecule whose position is given by its center of mass. This system is called a *free particle* in that it is not restrained or bound by the presence of a potential. Since we always have more than one "particle" in any laboratory experiment, the free-particle system does not exist, but the solutions to the corresponding stationary-state Schrödinger equation are instructive and sometimes useful. From this point on, we shall speak of "particles" without using the quotes. However, the reader must understand that the systems which comprise these hypothetical entities are better described by the matter waves dictated by the quantum mechanical postulates.

If a particle is truly isolated, it will not be in a potential field, since there is nothing around that could produce such a field. Thus, we have

$$V(x, y, z) = 0 \qquad \text{for all } x, y, \text{ and } z. \qquad (12.1)$$

When the potential is zero at all points, the force on the particle will also be zero, because

$$\mathbf{F} = -\left(\frac{\partial V}{\partial x}\right)\mathbf{i} - \left(\frac{\partial V}{\partial y}\right)\mathbf{j} - \left(\frac{\partial V}{\partial z}\right)\mathbf{k} \qquad (12.2)$$

and all derivatives of $V(x, y, z)$ are zero. Under these conditions, Eq. 11.67 shows that the quantum mechanical Hamiltonian for the free particle is

$$\mathcal{H} = -\frac{\hbar^2}{2m}\nabla^2. \qquad (12.3)$$

The corresponding stationary-state Schrödinger equation 11.78 is

$$-\frac{\hbar^2}{2m}\nabla^2 \psi(x, y, z) = E\,\psi(x, y, z), \qquad (12.4)$$

where the $\psi(x, y, z)$ are the possible eigenfunctions of the Hamiltonian with corresponding energy eigenvalues E. Since there is no potential energy and a single particle cannot exhibit either rotational or vibrational motion, E represents pure translational energy, which is just kinetic energy involving motion of the center of mass. Figure 2.12 illustrates the distinction between translational and thermal energy.

To make our life somewhat easier (every little bit helps), let us first restrict our free particle to move in only one dimension. Under this restrictive condition, Eq. 12.4 becomes

$$-\frac{\hbar^2}{2m}\frac{\partial^2}{\partial x^2}\,\psi(x) = E_x\,\psi(x), \tag{12.5}$$

where we now write E_x instead of E to emphasize the fact that this energy eigenvalue includes only the translational energy for motion in the x direction. Multiplying both sides of Eq. 12.5 by $-2m/\hbar^2$ and then rearranging terms yields

$$\frac{\partial^2}{\partial x^2}\,\psi(x) + k^2\,\psi(x) = 0, \tag{12.6}$$

where we have defined

$$k^2 = \frac{2mE_x}{\hbar^2}. \tag{12.7}$$

The units of k^2 are (length)$^{-2}$, so that k has units of inverse length, such as m^{-1}, cm^{-1}, or $Å^{-1}$. These are the same units as the wave number $\bar{\nu} = \nu/c$ for an electromagnetic wave. Therefore, k is often called the free-particle wave number.

We have previously encountered equations of the form of Eq. 12.6 in our treatment of the Debye-Hückel theory. The discussion surrounding the solution of Eq. 9.42 together with Problem 9.5, shows that the solution of Eq. 12.6 has the form

$$\psi(x) = C_1 \sin(kx) + C_2 \cos(kx) \tag{12.8}$$

if k^2 is positive. Since the translational energy E_x is always positive, this is the case for Eq. 12.6. It is a simple matter to show that Eq. 12.8 may be written in the complex exponential form

$$\psi(x) = Ae^{ikx} + Be^{-ikx}. \tag{12.9}$$

This may be seen by expanding Eq. 12.9 using the fact that $e^{iax} = \cos(ax) + i\sin(ax)$, to produce

$$\psi(x) = A[\cos(kx) + i\sin(kx)] + B[\cos(-kx) + i\sin(-kx)]. \tag{12.10}$$

We also know that $\cos(-kx) = \cos(kx)$ and $\sin(-kx) = -\sin(kx)$. These relationships allow us to write Eq. 12.10 in the form

$$\psi(x) = (A + B)\cos(kx) + i(A - B)\sin(kx). \tag{12.11}$$

If we identify the constant C_2 in Eq. 12.8 with $(A + B)$ and the constant C_1 with $(A - B)i$, Eqs. 12.8 and 12.9 become identical.

If we know the value of either the eigenfunction or its derivative at some point, the constants C_1 and C_2, or A and B, can be determined. As shown in Example 12.1, this determination places no restriction upon the value of k^2. Consequently, k^2 is a continuous variable that can assume any value between zero and infinity. Thus, the translational energy of a free particle is not quantized. This is a specific case of a general quantum mechanical principle: *The energies of unbound systems with energies greater than the potential are continuous variables, whereas bound systems exhibit quantized energy levels.*

EXAMPLE 12.1

(A) The eigenfunction for a free particle is zero at $x = 0$. With this restriction, what are the values of C_1 and C_2 in Eq. 12.8? What are the values of A and B in Eq. 12.9? **(B)** If the eigenfunction for the free particle exhibits a maximum at the point $x = 0$, what are the values of C_1 and C_2 in Eq. 12.8? **(C)** Do the restrictions imposed in either (A) or (B) place any constraints on k^2? What is the energy of the system in terms of m and k?

Solution

(A) At the point $x = 0$, Eq. 12.8 yields

$$\psi(x = 0) = C_1 \sin(k\{0\}) + C_2 \cos(k\{0\}) = C_2 = 0. \tag{1}$$

Therefore, the eigenfunction under these conditions is

$$\psi(x) = C_1 \sin(kx), \tag{2}$$

where C_1 can have any value other than $\pm\infty$. If we represent the eigenfunction by using Eq. 12.9, the result at $x = 0$ is

$$\psi(x = 0) = Ae^{ik(0)} + Be^{-ik(0)} = A + B = 0. \tag{3}$$

Therefore, we must have $B = -A$, and the eigenfunction becomes

$$\psi(x) = Ae^{ikx} - Ae^{-ikx} = A[e^{ikx} - e^{-ikx}]$$

$$= A[\cos(kx) + i\sin(kx) - \cos(-kx) - i\sin(-kx)] = 2Ai\sin(kx), \tag{4}$$

which is identical to the result given in Eq. 2 if we identify C_1 with $2Ai$.

(B) A maximum in $\psi(x)$ occurs at the point for which $\partial\psi(x)/\partial x = 0$. Using Eq. 12.8, we find that the derivative of $\psi(x)$ is

$$\frac{\partial\psi(x)}{\partial x} = kC_1 \cos(kx) - kC_2 \sin(kx). \tag{5}$$

At the point $x = 0$, this equation becomes

$$\left.\frac{\partial\psi(x)}{\partial x}\right|_{x=0} = 0 = kC_1 \cos(0) - kC_2 \sin(0) = kC_1. \tag{6}$$

Thus, $C_1 = 0$, and the eigenfunction takes the form

$$\psi(x) = C_2 \cos(kx), \tag{7}$$

with the value of C_2 arbitrary.

(C) The boundary conditions imposed in (A) and (B) place no restrictions on k. Therefore, k may have any value, and the energy, given by Eq. 12.7, viz.,

$$E_x = \frac{k^2\hbar^2}{2m}, \tag{8}$$

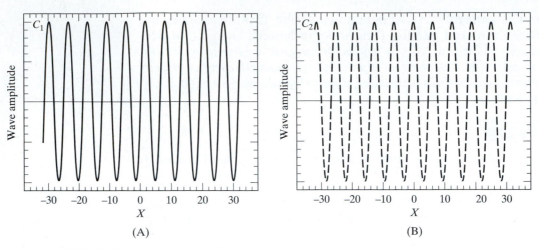

▲ FIGURE 12.1

Free-particle eigenfunctions, each with a wave number $k = 1\,\text{Å}^{-1}$. The eigenfunction shown in (A) is one that satisfies the condition that $\psi(x = 0) = 0$, while the one shown in (B) satisfies the condition that $\psi(x)$ exhibit a maximum at the point $x = 0$. In each case, the eigenfunction extends over all space. The plots show only the range $-32\,\text{Å} \leq x \leq 32\,\text{Å}$. In both cases, the energy is continuous, since the system is unbound.

is continuous. Such continuous energy levels are characteristic of unbound systems. Figure 12.1 illustrates the eigenfunctions for the free particle described in (A) and (B) for the specific case where $k = 1\,\text{Å}^{-1}$.

For related exercises, see Problems 12.1, 12.2, and 12.3.

The preceding analysis demonstrates that A and B in Eq. 12.9 may have any value and we will still have an eigenfunction of the Hamiltonian for the free particle. Let us now consider two important special cases. In the first case, we take $B = 0$, so that

$$\psi(x) = Ae^{ikx}.\tag{12.12}$$

The average momentum for such a free particle can be obtained using Eq. 11.62 with the stationary-state wave function replacing $\Psi(x, t)$:

$$\langle p_x \rangle = \frac{\displaystyle\int_{\text{all }x} \psi^* \frac{\hbar}{i}\frac{\partial}{\partial x}\,\psi(x)\,dx}{\displaystyle\int_{\text{all }x} |\psi(x)|^2\,dx}.\tag{12.13}$$

Substituting Eq. 12.12 into 12.13 gives

$$\langle p_x \rangle = \frac{\displaystyle\int_{\text{all }x} A^* e^{-ikx}\frac{\hbar}{i}\frac{\partial}{\partial x} A\,e^{ikx}\,dx}{\displaystyle\int_{\text{all }x} A^* e^{-ikx} A\,e^{ikx}\,dx} = \frac{\displaystyle\int_{\text{all }x} A^* e^{-ikx}\frac{\hbar}{i} Aike^{ikx}\,dx}{\displaystyle\int_{\text{all }x} A^* A\,dx}$$

$$= \frac{\hbar k|A|^2 \displaystyle\int_{\text{all }x} dx}{|A|^2 \displaystyle\int_{\text{all }x} dx} = \hbar k.\tag{12.14}$$

Equations 12.12 through 12.14 show that the average momentum of a free particle whose wave function is given by Eq. 12.12 is positive and equal to $k\hbar$. Furthermore (see Problem 12.2), the average square momentum for a system with this eigenfunction is $k^2\hbar^2$. Therefore, the square uncertainty in the momentum is

$$\langle \Delta p_x^2 \rangle = \langle p_x^2 \rangle - \langle p_x \rangle^2 = k^2\hbar^2 - (k\hbar)^2 = 0.$$

The only distribution of momentum for which the square uncertainty is zero is a delta function—that is, a distribution in which the only possible momentum is $k\hbar$. Consequently, not only is the average momentum of the free particle $k\hbar$ when the wave function is $\psi(x) = Ae^{ikx}$, but also, this is the only momentum the particle ever has in that eigenstate. Problem 12.3, requests that you show that if we choose $A = 0$ in Eq. 12.9, we obtain an eigenfunction for which p_x is always $-k\hbar$. If this is so, quantum free particles moving in the positive x direction with a fixed momentum $k\hbar$ are represented by the eigenfunction $\psi(x) = Ae^{ikx}$, whereas such particles moving in the negative x direction with a well-defined, fixed momentum $-k\hbar$ have the wave function $\psi(x) = Be^{-ikx}$. (See Figure 12.2.)

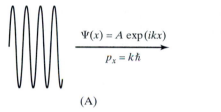

$\Psi(x) = A\exp(ikx)$

$p_x = k\hbar$

(A)

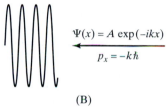

$\Psi(x) = A\exp(-ikx)$

$p_x = -k\hbar$

(B)

◀ FIGURE 12.2
(A) Quantum mechanical description of a free particle moving in the $+x$ direction with momentum $p_x = k\hbar$. (B) Quantum mechanical description of a free particle moving in the $-x$ direction with momentum $p_x = -k\hbar$.

The foregoing treatment can easily be generalized to three-dimensional motion. Without any mathematics, we may deduce certain characteristics of this generalization. Since motion in the x, y, and z directions are equivalent, the wave functions describing the y- and z-components of motion must be functionally identical to Eqs. 12.8 and 12.9. For the same reason, the translational energies for y and z motion must be continuous and given by equations analogous to Eq. 12.7. The total translational energy should be the sum of the translational energies in the x, y, and z directions. Using Eq. 12.4, we can show that these are exactly the results predicted by quantum theory.

Expanding the ∇^2 operator in Eq. 12.4 gives

$$-\frac{\hbar^2}{2m}\left[\frac{\partial^2}{\partial x^2} + \frac{\partial^2}{\partial y^2} + \frac{\partial^2}{\partial z^2}\right]\psi(x, y, z) - E\,\psi(x, y, z) = 0. \qquad \textbf{(12.15)}$$

We now attempt to write the solution in separable form, using the three-step procedure described in Chapter 11 to obtain the stationary-state Schrödinger equation. We first assume that a solution of the form

$$\psi(x, y, z) = F(x)G(y)H(z) \qquad \textbf{(12.16)}$$

exists. Next, we substitute Eq. 12.16 into Eq. 12.15, perform the indicated operations and then divide the equation by the separated form. If these manipulations result in a separation of variables in the differential equation, we know that a separable solution exists.

Substituting Eq. 12.16 into 12.15 produces

$$-\frac{\hbar^2}{2m}\left[\frac{\partial^2}{\partial x^2} + \frac{\partial^2}{\partial y^2} + \frac{\partial^2}{\partial z^2}\right]F(x)G(y)H(z) - E\,F(x)G(y)H(z) = 0.$$

When the second derivative with respect to x operates on $F(x)G(y)H(z)$, $G(y)$ and $H(z)$ may be factored out to the left of the operator, since these functions do not depend upon x. Similar statements apply for the other second derivatives. These operations give

$$G(y)\,H(z)\left[-\frac{\hbar^2}{2m}\right]\frac{\partial^2}{\partial x^2}F(x) + F(x)\,H(z)\left[-\frac{\hbar^2}{2m}\right]\frac{\partial^2}{\partial y^2}G(y)$$

$$+ F(x)G(y)\left[-\frac{\hbar^2}{2m}\right]\frac{\partial^2}{\partial z^2}H(z) - E\,F(x)G(y)H(z) = 0.$$

Dividing both sides by $F(x)G(y)H(z)$ produces

$$\left\{\frac{\left[-\dfrac{\hbar^2}{2m}\right]\dfrac{\partial^2}{\partial x^2}F(x)}{F(x)}\right\} + \left\{\frac{\left[-\dfrac{\hbar^2}{2m}\right]\dfrac{\partial^2}{\partial y^2}G(y)}{G(y)}\right\} + \left\{\frac{\left[-\dfrac{\hbar^2}{2m}\right]\dfrac{\partial^2}{\partial z^2}H(z)}{H(z)}\right\} - E = 0.$$

(12.17)

Inspection of Eq. 12.17 shows that the variables are now separated: The first term depends only upon x, the second only upon y, and the third only upon z. This equation must hold for *all* values of x, y, and z. The only way this can be is for each of the first three terms to be equal to a constant such that the sum of the three constants is exactly E. That is, we must have

$$\left\{\frac{\left[-\dfrac{\hbar^2}{2m}\right]\dfrac{\partial^2}{\partial x^2}F(x)}{F(x)}\right\} = E_x,$$

(12.18A)

$$\left\{\frac{\left[-\dfrac{\hbar^2}{2m}\right]\dfrac{\partial^2}{\partial y^2}G(y)}{G(y)}\right\} = E_y,$$

(12.18B)

$$\left\{\frac{\left[-\dfrac{\hbar^2}{2m}\right]\dfrac{\partial^2}{\partial z^2}H(z)}{H(z)}\right\} = E_z,$$

(12.18C)

and

$$E_x + E_y + E_z = E.$$

(12.18D)

Equations 12.18A through 12.18C may be written in the forms

$$\left[-\frac{\hbar^2}{2m}\right]\frac{\partial^2}{\partial x^2}F(x) = E_x\,F(x),$$

(12.19A)

$$\left[-\frac{\hbar^2}{2m}\right]\frac{\partial^2}{\partial y^2}G(y) = E_y\,G(y),$$

(12.19B)

and

$$\left[-\frac{\hbar^2}{2m}\right]\frac{\partial^2}{\partial z^2}H(z) = E_z\,H(z).$$

(12.19C)

Equations 12.19A, 12.19B, and 12.19C are all identical in form to Eqs. 12.5 and 12.6. Therefore, the solutions must have the same form as Eqs. 12.7 through 12.9. That is,

$$F(x) = A_x e^{ik_x x} + B_x e^{-ik_x x},$$

(12.20A)

$$G(y) = A_y e^{ik_y y} + B_y e^{-ik_y y},$$

(12.20B)

and

$$H(z) = A_z e^{ik_z z} + B_z e^{-ik_z z},$$

(12.20C)

with

$$k_x^2 = \frac{2mE_x}{\hbar^2},$$

(12.21A)

$$k_y^2 = \frac{2mE_y}{\hbar^2},$$

(12.21B)

and

$$k_z^2 = \frac{2mE_z}{\hbar^2},$$

(12.21C)

Using Eqs. 12.18D and 12.21, we see that the total translational energy is

$$E = \frac{k_x^2 \hbar^2}{2m} + \frac{k_y^2 \hbar^2}{2m} + \frac{k_z^2 \hbar^2}{2m}.$$

(12.22)

The preceding results show that quantum mechanics fulfills all our expectations: The total translational energy is simply the sum of that for motion in each of the three directions; furthermore, the wave functions describing the system have equivalent mathematical forms for each direction.

EXAMPLE 12.2

A quantum free particle of mass m is moving in three-dimensional space in the positive $x, y,$ and z directions. **(A)** What is the wave function of the system? **(B)** If the translational energies in the $x, y,$ and z directions are 1×10^{-20} J, 2×10^{-20} J, and 3×10^{-20} J, respectively, obtain the magnitude of the total momentum as a function of m. **(C)** What angle does the momentum vector for the particle make with the z-axis?

Solution

(A) Since the particle is moving in the positive $x, y,$ and z directions, $B_x, B_y,$ and B_z in Eqs. 12.20A through 12.20 C, respectively, must be zero. Therefore, using Eq. 12.16, we know that the wave function is

$$\psi(x, y, z) = F(x)G(y)H(z) = A_x A_y A_z e^{ik_x x} e^{ik_y y} e^{ik_z z}$$

$$= A_x A_y A_z \exp[i(k_x x + k_y y + k_z z)].$$

(1)

(B) The momentum of the free particle moving in the positive x direction is $k_x \hbar$, with similar expressions for y and z. The square of the total momentum is the sum of the squares of the individual Cartesian components:

$$P_{total}^2 = P_x^2 + P_y^2 + P_z^2 = k_x^2 \hbar^2 + k_y^2 \hbar^2 + k_z^2 \hbar^2.$$

(2)

Using Eqs. 12.21A through 12.21C, we also have

$$k_x^2 \hbar^2 = 2mE_x, \tag{3}$$

$$k_y^2 \hbar^2 = 2mE_y, \tag{4}$$

and

$$k_z^2 \hbar^2 = 2mE_z. \tag{5}$$

Combining Eqs. 2–5, we obtain

$$P_{total}^2 = 2m[E_x + E_y + E_z] = 2m \times 10^{-20} (1 + 2 + 3) = 12m \times 10^{-20} \text{ kg}^2 \text{ m}^2 \text{ s}^{-2}. \tag{6}$$

Taking square roots of both sides of Eq. 6, we find that the magnitude of the momentum is

$$P_{total} = (12m)^{1/2} \times 10^{-10} \text{ kg m s}^{-1}. \tag{7}$$

(C) The cosine of the angle that P_{total} makes with the z-axis is

$$\cos \theta = \frac{P_z}{P_{total}} = \frac{k_z \hbar}{P_{total}} = \frac{(2mE_z)^{1/2}}{P_{total}} = \frac{(2m)^{1/2} (3 \times 10^{-20})^{1/2} \text{ kg}^{1/2} \text{ J}^{1/2}}{(12m)^{1/2} \times 10^{-10} \text{ kg m s}^{-1}} = (1/2)^{1/2}. \tag{8}$$

This gives

$$\theta = \cos^{-1}[(0.5)^{1/2}] = 45°. \tag{9}$$

For related exercises, see Problems 12.4 and 12.7.

12.1.2   Particle in an Infinite Potential Well

When experiments are conducted in chemistry laboratories, we generally have the molecules contained within some type of apparatus. They are not unbound and free to roam the universe. If the molecules should happen to be something like HCN, phosgene, some selenium compounds, or some other highly toxic material, it is particularly important that the molecules be bound instead of free. In the laboratory, the molecular system is bound by the walls of the apparatus. Mathematically, we may model such a system by using an extremely high and very thick potential barrier to represent the walls. If this potential barrier is assumed to be infinite, the mathematical difficulties associated with solution of the Schrödinger equation are greatly reduced. As we shall see, neither classical nor quantum particles can penetrate such an infinite potential barrier. The particles will, therefore, be bound. Figure 12.3 illustrates such a model for a one-dimensional system. The point particle of mass m shown in the figure is contained within the region $0 \leq x \leq a$, since it is unable to penetrate the infinite potentials existing for

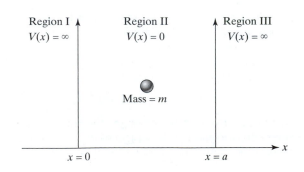

▶ FIGURE 12.3
Particle in an infinite one-dimensional well. The particle is bound by virtue of the fact that it cannot penetrate the infinitely high and infinitely thick potential barriers for $x < 0$ or $x > a$.

$x > a$ and $x < 0$. We know that a single point mass cannot possess rotational or vibrational energy, and the potential in Region II is zero. Therefore, the energy possessed by the particle will be pure translational energy. Unlike the energy states of the free particle, however, its energy states will be quantized, because the system is bound. A system such as this is sometimes called a *particle in a box*.

It is important to realize that this mathematical model represents a limiting case that does not actually exist. No laboratory apparatus has walls that cannot be penetrated. Every child has watched in dismay as his or her helium-filled balloon slowly deflates due to helium diffusion through the balloon wall. The Rutherford experiment described in Chapter 11 clearly shows that alpha particles easily penetrate a thin gold foil. As the walls of the apparatus become thicker and more dense, however, the probability of penetration rapidly approaches zero. Mathematically, this is equivalent to saying that the potential barrier at the boundary becomes very large. Still, is never infinite. Therefore, the model shown in Figure 12.3 represents a limiting case. If the real potential barrier is large, we expect the model to give accurate results, but the use of an aphysical infinite potential may produce cusps in the wave functions, as discussed in Chapter 11.

We first demonstrate that the particle cannot penetrate into Region I or Region III, where the potential is infinite. The one-dimensional stationary-state Schrödinger equation in both Region I and Region III has the form

$$-\frac{\hbar^2}{2m}\frac{\partial^2}{\partial x^2}\psi_I(x) + V(x)\psi_I(x) = -\frac{\hbar^2}{2m}\frac{\partial^2}{\partial x^2}\psi_I(x) + \infty\psi_I(x) = E_x\,\psi_I(x),\ \textbf{(12.23)}$$

where $\psi_I(x)$ is the wave function in Region I of Figure 12.3. The first postulate requires that $\psi_I(x)$ be finite. Also, we know that the translational energy, E_x, cannot be infinite. Therefore, the only way Eq. 12.23 can be satisfied is to have $\psi_I(x) = 0$ for all values of x in Region I. A similar argument shows that we must also have $\psi_{III}(x) = 0$ throughout Region III. The fourth postulate tells us that the probability of finding the particle in the region between x and $x + dx$ is given by $\psi^*(x)\psi(x)\,dx$. Hence, the probability that the particle will penetrate the infinite potential barrier and move into either Region I or Region III is zero, as noted in the preceding paragraph.

In Region II, the potential is zero, so the Schrödinger equation takes the same form as that for the free particle, viz.,

$$-\frac{\hbar^2}{2m}\frac{\partial^2}{\partial x^2}\psi_{II}(x) = E_x\,\psi_{II}(x),\qquad\textbf{(12.24)}$$

where $\psi_{II}(x)$ represents the set of possible eigenfunctions in Region II. Once more, we define $k^2 = 2mE_x/\hbar^2$, so that Eq. 12.24 becomes

$$\frac{\partial^2}{\partial x^2}\psi_{II}x + k^2\,\psi_{II}(x) = 0,\qquad\textbf{(12.25)}$$

which is identical to the Schrödinger equation for the free particle. The solutions are, therefore, given by Eq. 12.8:

$$\psi_{II}(x) = C_1\sin(kx) + C_2\cos(kx).\qquad\textbf{(12.26)}$$

We must now consider the boundary conditions imposed by the first postulate. Since the wave function must be continuous at all points, we must have

$$\psi_I(x = 0) = 0 = \psi_{II}(x = 0)\qquad\textbf{(12.27A)}$$

and

$$\psi_{\text{II}}(x = a) = 0 = \psi_{\text{III}}(x = a). \tag{12.27B}$$

Using Eqs. 12.26 and 12.27A, we find that

$$\psi_{\text{II}}(x = 0) = C_1 \sin(0) + C_2 \cos(0) = C_2 = 0. \tag{12.28}$$

Although the differential equation is satisfied by the general form given in Eq. 12.26, the boundary conditions are not. To satisfy the boundary condition that $\psi_{\text{I}}(x) = \psi_{\text{II}}(x)$ at the point $x = 0$, we must have $C_2 = 0$. Consequently, the wave function in Region II is forced to assume the form

$$\psi_{\text{II}}(x) = C_1 \sin(kx). \tag{12.29}$$

The second boundary condition given by Eq. 12.27B requires that

$$\psi_{\text{II}}(x = a) = C_1 \sin(ka) = 0. \tag{12.30}$$

There are two possible solutions of Eq. 12.30. The most obvious one is to take $C_1 = 0$. This certainly satisfies both the boundary conditions and the Schrödinger equation. However, it yields $\psi(x) = 0$ for all x. Since the stationary-state Schrödinger equation is $\mathcal{H}\psi(x) = E\psi(x)$, taking $\psi(x) = 0$ for all x gives $0 = 0$, which is correct. While such a solution might be of interest to numerologists, it is not the sort of result that excites chemists. In general, the solution $\psi(x) = 0$ is called a trivial solution. Such a solution violates the fourth postulate, which requires $\Psi^*(x)\Psi(x)\,d\tau$ to be the probability distribution for the system. If $\psi(x) = 0$ for all x, the probability distribution will also be zero at all values of x, which cannot be correct. Therefore, the trivial solution is unacceptable.

The solution of chemical interest is the one which requires that $\sin(ka) = 0$, which is the case only if ka is an integral number of π radians. The boundary conditions require that

$$ka = n_x\pi \quad \text{for } n_x = 1, 2, 3, 4, 5, 6, 7, \ldots . \tag{12.31}$$

Note that we cannot permit $n_x = 0$, since that choice generates the trivial solution once again. (See Problem 12.8.) The quantity n_x is called a *quantum number*. Such nomenclature is applied to any discrete variable that is produced by the boundary conditions on the solutions of the Schrödinger equation. In this particular case, we are considering translational energy along the x-axis. Consequently, we refer to n_x as the *x translational quantum number*.

Substituting Eq. 12.7 into Eq. 12.31 gives

$$\frac{(2mE_x)^{1/2}}{\hbar} a = n_x\pi.$$

Squaring both sides produces

$$\frac{2mE_x a^2}{\hbar^2} = n_x^2 \pi^2,$$

so that the energy eigenvalues are

$$\boxed{E_x = \frac{n_x^2 \hbar^2 \pi^2}{2ma^2} = \frac{n_x^2 h^2}{8ma^2}} \quad \text{for } n_x = 1, 2, 3, \ldots . \tag{12.32}$$

Viewed from the perspective of classical mechanics, Eq. 12.32 is a stunning result! It tells us at least two very interesting things. First, the translational energy states of a bound particle in an infinite well are quantized. The particle can only exhibit energies $h^2/(8ma^2)$, $4h^2/(8ma^2)$, $9h^2/(8ma^2)$, $16h^2/(8ma^2)$, ... ; no other values can ever be observed in the laboratory. This result is not unexpected, since we have already learned that the energy levels of bound systems are always quantized. Note in particular, that it is the boundary conditions required by the first postulate that produce the quantization. This is always the case. Second, the quantum particle can never be stationary: The smallest translational energy that it can have is $h^2/(8ma^2)$. It is as if quantum particles are afflicted with the fidgets; they can't sit still. Later in this chapter, we shall learn that this nonzero minimum energy is intimately connected with a fundamental result of quantum theory called the uncertainty principle.

We can now utilize Eq. 12.31 to express the wave number k in terms of the translational quantum number and the width of the infinite potential well. The result is

$$k = \frac{n_x \pi}{a}. \tag{12.33}$$

Substituting Eq. 12.33 into Eq. 12.29 shows that the eigenfunctions for a particle in an infinite potential well are

$$\psi(x) = \begin{cases} C_1 \sin\left[\dfrac{n_x \pi x}{a}\right] & \text{for } 0 \leq x \leq a \\ 0 & \text{otherwise} \end{cases}. \tag{12.34}$$

Equation 12.34 demonstrates why the translational quantum number is restricted to positive values. If we replace n_x in Eq. 12.34 with $-n_x$, the eigenfunction becomes

$$\psi(x) = C_1 \sin\left[\frac{-n_x \pi x}{a}\right] = -C_1 \sin\left[\frac{n_x \pi x}{a}\right].$$

Since C_1 is an arbitrary constant, there is no physically significant difference between it and $-C_1$. Consequently, the eigenfunction for $-n_x$ is the same as that for $+n_x$. This means that only positive values of the translational quantum number are needed to obtain all possible eigenfunctions.

The constant C_1 serves as the normalization constant for the probability distribution. The fourth postulate tells us that $|\psi(x)|^2 dx$ is the probability distribution for the particle. Since the summation of all the probabilities must be unity, we have

$$\langle \psi(x) | \psi(x) \rangle = \int_{-\infty}^{\infty} \psi^*(x)\psi(x)\,dx = 1. \tag{12.35}$$

Because $\psi(x)$ has different forms in different regions of space, the integral in Eq. 12.35 must be broken into three different integrals:

$$\langle \psi(x) | \psi(x) \rangle = \int_{-\infty}^{0} \psi_I^*(x)\psi_I(x)\,dx + \int_{0}^{a} \psi_{II}^*(x)\psi_{II}(x)\,dx$$

$$+ \int_{a}^{\infty} \psi_{III}^*(x)\psi_{III}(x)\,dx = 1.$$

Since $\psi_I(x) = \psi_{III}(x) = 0$, only the second integral makes a nonzero contribution. Therefore,

$$\langle \psi(x)|\psi(x) \rangle = \int_0^a \psi_{II}^*(x)\psi_{II}(x)\,dx = C_1^2 \int_0^a \sin^2\left[\frac{n_x\pi x}{a}\right]dx = 1. \quad \textbf{(12.36)}$$

Usually, textbooks omit without comment regions of space for which $\psi(x) = 0$ and simply write Eq. 12.36 directly. The student, however, needs to be aware that the integrals are always taken over all the available space, which, for x, is $-\infty$ to $+\infty$. The fact that certain regions may make no contribution is a result of the specific problem being considered.

Using the relationship

$$\sin^2\left[\frac{n_x\pi x}{a}\right] = \frac{1}{2}\left\{1 - \cos\left[\frac{2n_x\pi x}{a}\right]\right\},$$

we can easily integrate Eq. 12.36. The result is

$$\langle \psi(x)|\psi(x) \rangle = C_1^2 \int_0^a \frac{1}{2}\left\{1 - \cos\left[\frac{2n_x\pi x}{a}\right]\right\}dx$$

$$= C_1^2 \left[\frac{x}{2} - \frac{a}{2n_x\pi}\sin\left[\frac{2n_x\pi x}{a}\right]\right]_0^a = \frac{a}{2}C_1^2 = 1.$$

Solving for the normalization constant, we obtain

$$C_1 = \left[\frac{2}{a}\right]^{1/2}. \quad \textbf{(12.37)}$$

Equation 12.37 is unusual in that the normalization constant is independent of the quantum number n_x. In most quantum systems, this is not the case.

The problem is now completely done. The possible normalized eigenfunctions for a quantum particle in an infinite well are

$$\boxed{\psi_{n_x}(x) = \left[\frac{2}{a}\right]^{1/2}\sin\left[\frac{n_x\pi x}{a}\right] \quad \text{for} \quad 0 \le x \le a}$$

and

$$\boxed{\psi(x) = 0 \quad \text{otherwise}}, \quad \textbf{(12.38)}$$

and the corresponding energy eigenvalues are

$$\boxed{E_{n_x} = \frac{n_x^2 h^2}{8ma^2}} \quad \text{for } n_x = 1, 2, 3, \ldots. \quad \textbf{(12.39)}$$

The allowed energy states for the system are illustrated in Figure 12.4. Note that they are not equally spaced.

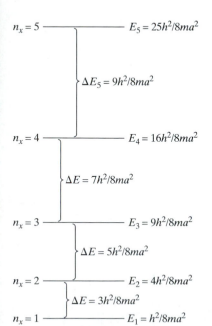

$n_x = 5$ ——— $E_5 = 25h^2/8ma^2$

$\Delta E_5 = 9h^2/8ma^2$

$n_x = 4$ ——— $E_4 = 16h^2/8ma^2$

$\Delta E = 7h^2/8ma^2$

$n_x = 3$ ——— $E_3 = 9h^2/8ma^2$

$\Delta E = 5h^2/8ma^2$

$n_x = 2$ ——— $E_2 = 4h^2/8ma^2$

$\Delta E = 3h^2/8ma^2$

$n_x = 1$ ——— $E_1 = h^2/8ma^2$

▲ FIGURE 12.4

Allowed energy states for a one-dimensional quantum particle of mass m in an infinite potential well of width a. n_x is the translational quantum number for the state.

EXAMPLE 12.3

In Chapter 11, it was shown that eigenfunctions of a Hermitian operator that have different eigenvalues must be orthogonal. Show that the eigenfunction for a quantum particle in an infinite well with $n_x = 1$ is orthogonal to the eigenfunction with $n_x = 2$.

Solution

Since, by the second postulate, $\mathcal{H}$ is guaranteed to be Hermitian, $\psi_1(x)$ must be orthogonal to $\psi_2(x)$. To prove that this is the case, we must show that

$$\langle \psi_1(x) | \psi_2(x) \rangle = 0. \tag{1}$$

Substituting the eigenfunctions, we obtain

$$\int_0^a \left[\frac{2}{a} \right]^{1/2} \sin\left[\frac{\pi x}{a} \right] \left[\frac{2}{a} \right]^{1/2} \sin\left[\frac{2\pi x}{a} \right] dx = \frac{2}{a} \int_0^a \sin\left[\frac{\pi x}{a} \right] \sin\left[\frac{2\pi x}{a} \right] dx, \tag{2}$$

which we wish to show is zero. Using the fact that $\sin(2x) = 2\sin(x)\cos(x)$, we may write Eq. 2 in the form

$$\frac{4}{a} \int_0^a \sin^2\left[\frac{\pi x}{a} \right] \cos\left[\frac{\pi x}{a} \right] dx = \frac{4}{3\pi} \left\{ \sin^3\left[\frac{\pi x}{a} \right] \right\}_0^a = 0, \tag{3}$$

where we have utilized the relationship $\sin^3(0) = \sin^3(\pi) = 0$. Thus, $\psi_1(x)$ is orthogonal to $\psi_2(x)$, as we knew it must be.

For related exercises, see Problems 12.5 and 12.6.

Let us now examine the features of the matter waves that can describe a quantum particle confined in a one-dimensional infinite well. The mathematical forms for the various eigenfunctions are given by Eq. 12.38. Figure 12.5 shows plots of the eigenfunctions for $n_x = 1, 2, 3, 4,$ and 5 when the width of the well is 1 Å. The eigenfunction with the lowest energy is called the *ground state,* which, in this case, corresponds to the eigenfunction with $n_x = 1$. The figure shows that $\psi_1(x)$ has a maximum at $x = a/2 = 0.5$ Å. Therefore, a quantum particle in this state is most likely to be found in the center of the well. We also note that $\psi_1(x)$ is not zero for any value of x except the boundary values at $x = 0$ and $x = a = 1$ Å. Points other than the boundaries for which $\psi(x)$ and $|\psi(x)|^2$ are zero are called *nodes.* Consequently, the ground-state eigenfunction for a particle in a one-dimensional infinite

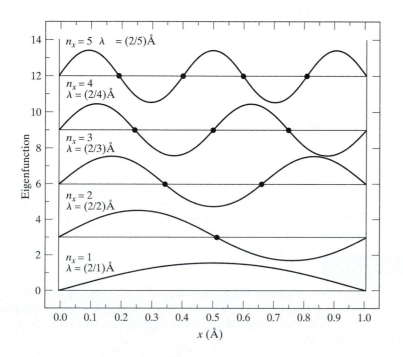

◀ **FIGURE 12.5**
The eigenfunctions for a quantum particle in an infinite potential well of width 1 Å. The plots show the eigenfunctions for the first five quantum states. Each plot is displaced upwards to enhance the visual clarity of the graph. Nodes are indicated by black dots. The de Broglie wave length associated with each wave is also given.

well is nodeless. This is a general result for any quantum mechanical system. *The ground-state eigenfunction is always nodeless.*

The eigenfunction with the lowest energy other than the ground state is called the *first excited state.* The next energy state is termed the *second excited state,* and so on. The first excited state for this system is the $n_x = 2$ eigenfunction with energy $E_2 = 4h^2/(8ma^2)$. Figure 12.5 shows that this eigenfunction has one node, at $x = a/2 = 0.5$ Å. The second excited state, $n_x = 3$, has two nodes, one at $x = a/3 = \frac{1}{3}$ Å and the second at $x = 2a/3 = \frac{2}{3}$ Å. *In a one-dimensional system, the mth excited state will have m nodes.* For the current system, the *m*th excited state corresponds to the $n_x = (m + 1)$ eigenfunction. Therefore, we expect the n_xth eigenfunction to exhibit $n_x - 1$ nodes. Figure 12.5, which indicates the nodes by black dots, shows that that is indeed the case. This characteristic of eigenfunctions permits us to determine the energy state by simply counting nodes if the system is one dimensional or if the energy is dependent upon a single quantum number. However, if the system is multidimensional such that the energy is dependent upon two or more quantum numbers, this is not true. One reason is that the nodes can appear along any of the coordinates, and nodes in one coordinate may make a larger contribution to the energy than nodes in another coordinate. In addition, in a multidimensional system, it is possible for two nodes to occur at the same point, so that they appear to be a single node. When this happens, a simple node count can lead to an erroneous conclusion concerning the energy state. Consequently, we usually cannot determine the relative energy state in a multidimensional system by a simple node count. Problem 12.14 illustrates this point.

In Chapter 11, we learned that de Broglie postulated that matter was associated with a wave with wavelength given by

$$\lambda = \frac{h}{p}$$

where p is the momentum. Let us examine the system we have been discussing to see if quantum mechanics is compatible with the de Broglie hypothesis. The general form for a sine wave with wavelength λ is $\sin[2\pi x/\lambda]$. Since the eigenfunctions have the form $\sin[n_x\pi x/a]$, we must have $2/\lambda = n_x/a$, so that

$$\lambda = \frac{2a}{n_x}. \tag{12.40}$$

An examination of Figure 12.5 where $a = 1$ Å shows that this is indeed the case. Using Eq. 12.40 with the de Broglie hypothesis shows that we must have

$$p = \frac{h}{\lambda} = \frac{n_x h}{2a}. \tag{12.41}$$

Classically, we expect the translational energy to be given by $E = p^2/(2m)$. If we use Eq. 12.41 for the momentum, the result is $E = n_x^2 h^2/(8ma^2)$, which is identical to the quantum mechanical result given in Eq. 12.39. Therefore, quantum mechanics is consistent with the de Broglie hypothesis.

EXAMPLE 12.4

A particle in a one-dimensional infinite potential well is described by the eigenfunction shown in Figure 12.6. **(A)** What is the translational energy of this particle in terms of m, a, and h? **(B)** What is the wavelength of the matter wave in terms of these same variables?

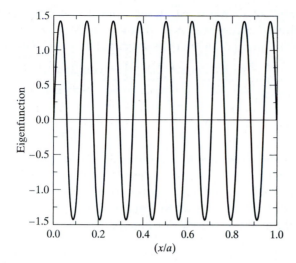

Solution

(A) In this one-dimensional system, all nodes must appear in the x-coordinate. Consequently, we expect that a node count along that coordinate will permit us to determine the quantum state. The number of nodes in the eigenfunction, excluding the boundaries, is 16. Therefore, we must have $n_x = 17$, and it follows that

$$E_{17} = \frac{n_x^2 h^2}{8ma^2} = \frac{17^2 h^2}{8ma^2} = \frac{36.125 h^2}{ma^2}. \tag{1}$$

(B) The wavelength of the matter wave given by Eq. 12.40 is

$$\lambda = \frac{2a}{n_x}. \tag{2}$$

Therefore,

$$\lambda = \frac{2a}{n_x} = \frac{2a}{17} = 0.1176\ldots a. \tag{3}$$

For a related exercise, see Problem 12.9.

The fourth postulate permits us to compute the average properties of the system. As an illustrative example, let us assume we know that a particle in a one-dimensional infinite well is in its ground state. Then the average position, $\langle x \rangle$, and the average square position, $\langle x^2 \rangle$, may be easily computed to obtain a measure of the uncertainty of the spatial distribution given by $|\psi(x)|^2 dx$. According to the fourth postulate, the average position is given by

$$\langle x \rangle = \frac{\langle \psi_1(x)|x|\psi_1(x)\rangle}{\langle \psi_1(x)|\psi_1(x)\rangle} = \frac{\displaystyle\int_0^a \psi_1^*(x)\,x\,\psi_1(x)\,dx}{\displaystyle\int_0^a \psi_1^*(x)\psi_1(x)\,dx}. \tag{12.42}$$

Since we have already normalized the eigenfunction $\psi_1(x)$, the denominator of Eq. 12.42 is known to be unity. Therefore,

$$\langle x \rangle = \int_0^a \psi_1^*(x)\,x\,\psi_1(x)\,dx = \frac{2}{a}\int_0^a x \sin^2\left[\frac{\pi x}{a}\right] dx. \tag{12.43}$$

In Eqs. 12.42 and 12.43, the integrations from $-\infty \leq x \leq 0$ and $a \leq x \leq \infty$ have been omitted because we know that $\psi(x)$ is zero in those regions. The integral appearing in Eq. 12.43 may be executed by parts. The result is

$$\langle x \rangle = \frac{2}{a} \left[\frac{x^2}{4} - \frac{x \sin\{2\pi x/a\}}{(4\pi/a)} - \frac{\cos\{2\pi x/a\}}{(8\pi^2/a^2)} \right]_0^a$$

$$= \frac{2}{a} \left\{ \left[\frac{a^2}{4} - 0 - \frac{a^2}{8\pi^2} \right] - \left[0 - 0 - \frac{a^2}{8\pi^2} \right] \right\} = \frac{a}{2}.$$

On the average, the particle is found in the center of the well. This is hardly surprising, considering the symmetric distributions that will result from the absolute squares of the eigenfunctions shown in Figure 12.5.

A similar analysis permits us to compute the average square position. Since the eigenfunction is already normalized, we have

$$\langle x^2 \rangle = \int_0^a \psi_1^*(x) \, x^2 \, \psi_1(x) \, dx = \frac{2}{a} \int_0^a x^2 \sin^2\left[\frac{\pi x}{a} \right] dx. \tag{12.44}$$

Integrating by parts twice or using a standard table of integrals gives

$$\langle x^2 \rangle = \frac{2}{a} \left[\frac{x^3}{6} - \left\{ \frac{ax^2}{4\pi} - \frac{a^3}{8\pi} \right\} \sin\left\{ \frac{2\pi x}{a} \right\} - \frac{x \, a^2 \cos\{2\pi x/a\}}{4\pi^2} \right]_0^a$$

$$= \frac{2}{a} \left\{ \left[\frac{a^3}{6} - 0 - \frac{a^3}{4\pi^2} \right] - [0 - 0 - 0] \right\} = \frac{a^2}{3} - \frac{a^2}{2\pi^2} = a^2 \left[\frac{1}{3} - \frac{1}{2\pi^2} \right]. \tag{12.45}$$

The square uncertainty present in the position distribution for the ground state is

$$\langle \Delta x^2 \rangle = \langle x^2 \rangle - \langle x \rangle^2 = a^2 \left[\frac{1}{3} - \frac{1}{2\pi^2} \right] - \frac{a^2}{4} = a^2 \left[\frac{1}{12} - \frac{1}{2\pi^2} \right]. \tag{12.46}$$

Equation 12.46 shows that the uncertainty is proportional to the width of the well. This result is reasonable, since we know that, as the number of possible positions available to the particle increases, the more uncertain we will be in guessing the result of a measurement of the particle's position.

EXAMPLE 12.5

Determine the average momentum and the uncertainty in the momentum distribution for a quantum particle in its ground state in a one-dimensional well of width a.

Solution

The momentum operator is $(\hbar/i) \, (d/dx)$. The fourth postulate tells us that

$$\langle p_x \rangle = \frac{\langle \psi_1(x) | p_x | \psi_1(x) \rangle}{\langle \psi_1(x) | \psi_1(x) \rangle} = \frac{\displaystyle\int_0^a \psi_1^*(x) \frac{\hbar}{i} \frac{d}{dx} \psi_1(x) \, dx}{\displaystyle\int_0^a \psi_1^*(x) \psi_1(x) \, dx} = \int_0^a \psi_1^*(x) \frac{\hbar}{i} \frac{d}{dx} \psi_1(x) \, dx, \tag{1}$$

since the integral in the denominator evaluates to unity. Inserting the ground-state eigenfunction gives

$$\langle p_x \rangle = \frac{2\hbar}{ia} \int_0^a \sin\left[\frac{\pi x}{a} \right] \frac{d}{dx} \sin\left[\frac{\pi x}{a} \right] dx = \frac{2\hbar\pi}{ia^2} \int_0^a \sin\left[\frac{\pi x}{a} \right] \cos\left[\frac{\pi x}{a} \right] dx$$

$$= \frac{\hbar}{ia} \left[\sin^2\left[\frac{\pi x}{a} \right] \right]_0^a = 0. \tag{2}$$

Physically, we expect this zero result, because the particle is equally likely to be moving in the negative x direction as the positive direction. Consequently, the plus and minus contributions to $\langle p_x \rangle$ cancel, and the average is zero. Mathematically, we also expect $\langle p_x \rangle$ to be zero as soon as we see that the result will be imaginary if any other value of the integral is obtained. Since an imaginary momentum is physically meaningless, we know before we even perform the integration in Eq. 2 that the result must be zero.

To obtain the uncertainty in the momentum distribution, we must also compute $\langle p_x^2 \rangle$. The quantum operator for p_x^2 was shown in Chapter 11 to be $-\hbar^2(d^2/dx^2)$. Therefore,

$$\langle p_x^2 \rangle = \frac{\langle \psi_1(x)|p_x^2|\psi_1(x)\rangle}{\langle \psi_1(x)|\psi_1(x)\rangle} = -\frac{\int_0^a \psi_1^*(x)\,\hbar^2\frac{d^2}{dx^2}\,\psi_1(x)\,dx}{\int_0^a \psi_1^*(x)\psi_1(x)\,dx} = -\int_0^a \psi_1^*\,\hbar^2\frac{d^2}{dx^2}\,\psi_1(x)\,dx. \quad (3)$$

Inserting the eigenfunctions, we obtain

$$\langle p_x^2 \rangle = -\frac{2\hbar^2}{a}\int_0^a \sin\left[\frac{\pi x}{a}\right]\frac{d^2}{dx^2}\sin\left[\frac{\pi x}{a}\right]dx = \frac{\pi^2\hbar^2}{a^2}\left\{\frac{2}{a}\int_0^a \sin\left[\frac{\pi x}{a}\right]\sin\left[\frac{\pi x}{a}\right]dx\right\}$$

$$= \frac{\pi^2\hbar^2}{a^2}\langle \psi_1(x)|\psi_1(x)\rangle = \frac{\pi^2\hbar^2}{a^2}, \quad (4)$$

because the eigenfunction is normalized, so that $\langle \psi_1(x)|\psi_1(x)\rangle = 1$. The uncertainty in the momentum distribution is given by

$$\langle \Delta p_x^2 \rangle = \langle p_x^2 \rangle - \langle p_x \rangle^2 = \frac{\pi^2\hbar^2}{a^2} - 0^2 = \frac{\pi^2\hbar^2}{a^2}. \quad (5)$$

A principal component of the uncertainty in the momentum is associated with the fact that p_x can be either positive or negative. Therefore, we never know if a measurement of p_x will yield a positive or negative value.

For related exercises, see Problems 12.10 and 12.11.

EXAMPLE 12.6

A particle in a one-dimensional infinite potential well is in its quantum mechanical ground state. If we perform a measurement to locate this particle, what is the probability that we will find it in the range $0.5a \le x \le 0.75a$, where a is the width of the well?

Solution

The normalized probability distribution is

$$P(x)\,dx = |\psi_1(x)|^2\,dx = \frac{2}{a}\sin^2\left[\frac{\pi x}{a}\right]dx. \quad (1)$$

The total probability of finding the particle somewhere within the stipulated range is

$$P(0.5a \le x \le 0.75a) = \frac{2}{a}\int_{0.5a}^{0.75a}\sin^2\left[\frac{\pi x}{a}\right]dx = \frac{2}{a}\int_{0.5a}^{0.75a}\frac{[1 - \cos\{2\pi x/a\}]}{2}\,dx$$

$$= \frac{2}{a}\left[\frac{x}{2} - \frac{a}{4\pi}\sin\left[\frac{2\pi x}{a}\right]\right]_{0.50a}^{0.75a}$$

$$= \frac{2}{a}\left[\frac{0.75a}{2} - \frac{a}{4\pi}\sin[1.5\pi] - \frac{0.50a}{2} + \frac{a}{4\pi}\sin[\pi]\right]$$

$$= 0.75 - \frac{1}{2\pi}(-1) - 0.50 + 0 = 0.40915\dots. \quad (2)$$

For a related exercise, see Problem 12.11.

There are marked similarities between the quantum free particle and the corresponding classical system. (See Problem 12.7.) The same is true for a quantum versus a classical particle in an infinite potential well, although the similarities are less obvious. The most troublesome point concerns the quantization of the translational energy levels. Our everyday experience runs contrary to this result. For example, we know that we can drive our car at any speed we wish. (The Highway Patrol, however, may have other ideas.) Our experience suggests that the translational energy levels are continuous rather than quantized. The fact that the quantum particle cannot be stationary seems particularly absurd. These observations are correct and they are also compatible with the predictions of quantum mechanics. However, to see that this is so requires that we do some arithmetic.

Let us first examine the minimum energy of a 0.1000-kg ball in a one-dimensional box of width 0.5 m. Equation 12.39 tells us that, in the ground state, the translational energy of the ball is $E_x = h^2/(8ma^2)$. Inserting the data shows this minimum energy to be

$$E_x = \frac{(6.62608 \times 10^{-34}\,\text{J s})^2}{8(0.1000\,\text{kg})(0.5\,\text{m})^2} = 2.195 \times 10^{-66}\,\text{J}.$$

With that energy, the speed of the ball will be

$$v_x = \left[\frac{2E_x}{m}\right]^{1/2} = \left[\frac{2(2.195 \times 10^{-66})\,\text{J}}{0.1000\,\text{kg}}\right]^{1/2} = 6.626 \times 10^{-33}\,\text{m s}^{-1},$$

since $E_x = 0.5\,mv_x^2$. At this speed, it will take the ball 7.55×10^{31} seconds, or 2.39×10^{24} years, to move the 0.50 m to cross the box. Since this is about 10^{14} times the estimated age of the universe, we will not be able to detect any movement of the ball. It will appear to our senses that the ball is stationary.

An equally important consideration is the spacing between the translational energy levels. Our senses tell us that this spacing is zero and that the energy levels are continuous. Suppose we have helium atoms in a 0.5-m, one-dimensional infinite well at a temperature of 300 K. We have learned that the average translational energy of such a system is 1.5RT, which is 3,741 J mol^{-1}. Using Eq. 12.39 we may compute the average translational quantum number for these helium atoms. The energy per atom is

$$E_x = \frac{n^2 h^2}{8ma^2} = \frac{n_x^2(6.626 \times 10^{-34}\,\text{J s})^2}{8(6.647 \times 10^{-27}\,\text{kg})(0.5\,\text{m})^2} = 3.303 n_x^2 \times 10^{-41}\,\text{J},$$

since the mass of the helium atom is 6.647×10^{-27} kg. Therefore, a mole of helium atoms has a total average energy of

$$NE_x = (6.022 \times 10^{23}\,\text{mol}^{-1})(3.303\,n_x^2 \times 10^{-41}\,\text{J}) = 1.989 \times 10^{-17} n_x^2\,\text{J mol}^{-1},$$

where N is Avogadro's constant. The average translational quantum state of the helium atoms can be estimated by equating this energy to 3,741 J. The result is

$$\langle n_x \rangle = \left[\frac{3{,}741\,\text{J}}{1.989 \times 10^{-17}\,\text{J}}\right]^{1/2} \approx 1.37 \times 10^{10}.$$

We now ask, "What is the helium-atom translational energy spacing between the $n_x = 1.37 \times 10^{10}$ and $n_x = 1.37 \times 10^{10} + 1$ states?" This spacing is given by

$$\Delta E_x = E_x(n = 1.37 \times 10^{10} + 1) - E_x(n_x = 1.37 \times 10^{10})$$

$$= \frac{h^2}{8ma^2}[(1.37 \times 10^{10} + 1)^2 - (1.37 \times 10^{10})^2]$$

$$\approx \frac{h^2}{8ma^2}[2(1.37 \times 10^{10})] = 9.05 \times 10^{-31} \text{ J}.$$

This spacing corresponds to an energy of 5.45×10^{-10} kJ mol^{-1}. It would be very difficult to detect the presence of such a small energy spacing experimentally. Consequently, the translational energy levels appear to be continuous even though they are not. Problem 12.12 views this situation from a different perspective.

Sometimes, laboratory equipment is three dimensional rather than one dimensional. Let us, therefore, examine the properties of a quantum particle of mass m in a three-dimensional infinite well whose shape is a parallelepiped with rectangular faces of dimensions a, b, and c. Figure 12.7 illustrates such an infinite well.

Our previous treatment of the one-dimensional system demonstrates that outside the well where the potential is infinite, all eigenfunctions $\psi(x, y, z)$ will be zero. There is no chance that our particle can penetrate an infinitely high potential barrier. Inside the well, the three-dimensional Schrödinger equation is

$$-\frac{\hbar^2}{2m}\left[\frac{\partial^2}{\partial x^2} + \frac{\partial^2}{\partial y^2} + \frac{\partial^2}{\partial z^2}\right]\psi(x, y, z) - E\,\psi(x, y, z) = 0, \qquad \textbf{(12.47)}$$

which is identical to Eq. 12.15 for the three-dimensional free particle. The solutions, however, will not be the same, because $\psi(x, y, z)$ must satisfy the boundary conditions. The eigenfunctions can be obtained by separation of variables, via the same procedure employed for the three-dimensional free particle. Problem 12.13 requests the solution of Eq. 12.47 for the three-dimensional particle in an infinite rectangular parallelepiped well. When the equation is solved, it is seen that the eigenfunctions are

$$\psi(x, y, z) = \left[\frac{8}{abc}\right]^{1/2} \sin\left[\frac{n_x \pi x}{a}\right] \sin\left[\frac{n_y \pi y}{b}\right] \sin\left[\frac{n_z \pi z}{c}\right] \qquad \textbf{(12.48)}$$

for $0 \le x \le a$, $0 \le y \le b$, and $0 \le z \le c$

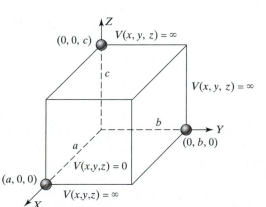

▶ **FIGURE 12.7**
An infinite rectangular parallelpiped potential well. When the particle has coordinates in the range $0 \le x \le a$, $0 \le y \le b$, and $0 \le z \le c$, the potential $V(x, y, z)$ is zero. For all other sets of coordinates, the potential is infinite.

and

$$\boxed{\psi(x, y, z) = 0}\quad \text{otherwise.}$$

The associated energy eigenvalues are

$$\boxed{E(n_x, n_y, n_z) = \frac{h^2}{8m}\left[\frac{n_x^2}{a^2} + \frac{n_y^2}{b^2} + \frac{n_z^2}{c^2}\right]}\quad (n_x, n_y, n_z = 1, 2, 3, \ldots). \quad \textbf{(12.49)}$$

Equations 12.48 and 12.49 illustrate a general principle of quantum theory: When a single bound particle moves in one-dimensional space, the eigenfunctions and eigenvalues depend upon a single quantum number; when the particle moves in three-dimensional space, three quantum numbers are required to specify the eigenfunctions and eigenvalues. For the system under consideration, these quantum numbers are n_x, n_y, and n_z, all of which must be integers in the range from unity to infinity. In general, if N particles move in n-dimensional space, Nn quantum numbers must be specified to determine the eigenfunctions and corresponding eigenvalues.

Let us again turn our attention to the question of degeneracy, which we first discussed in Chapter 7 in relation to the development of the Debye heat capacity equations for a solid. We addressed the topic again in Chapters 10 and 11. Degenerate eigenfunctions have equal eigenvalues. *It is impossible for one-dimensional, bound systems to have degenerate eigenfunctions.* However, multidimensional systems can, and frequently do, have degeneracies. When two or more eigenfunctions are degenerate, they do not have to be orthogonal since their eigenvalues are equal, but they frequently are. If they are not orthogonal, linear combinations of the nonorthogonal eigenfunctions can always be developed that are orthogonal. (See Problem 12.15.) As an example of degeneracy, let us consider the three-dimensional particle in an infinite potential well for which $a = b = c$. Under these conditions, the eigenfunctions are given by

$$\psi(x, y, z) = \left[\frac{8}{a^3}\right]^{1/2} \sin\left[\frac{n_x \pi x}{a}\right] \sin\left[\frac{n_y \pi y}{a}\right] \sin\left[\frac{n_z \pi z}{a}\right], \quad \textbf{(12.50)}$$

and the energy eigenvalues are

$$E(n_x, n_y, n_z) = \frac{h^2}{8ma^2}[n_x^2 + n_y^2 + n_z^2]. \quad \textbf{(12.51)}$$

Equation 12.51 shows that all eigenstates for which the sum of the squares of the three quantum numbers are identical will be degenerate. For example, $E(1, 1, 2) = E(1, 2, 1) = E(2, 1, 1) = 6h^2/8ma^2$. Therefore, we say that the energy eigenvalue $6h^2/(8ma^2)$ is *threefold degenerate* ($g = 3$). The three degenerate eigenfunctions are

$$\psi(x, y, z) = \left[\frac{8}{a^3}\right]^{1/2} \sin\left[\frac{\pi x}{a}\right] \sin\left[\frac{\pi y}{a}\right] \sin\left[\frac{2\pi z}{a}\right] = \psi_{112},$$

$$\psi(x, y, z) = \left[\frac{8}{a^3}\right]^{1/2} \sin\left[\frac{\pi x}{a}\right] \sin\left[\frac{2\pi y}{a}\right] \sin\left[\frac{\pi z}{a}\right] = \psi_{121},$$

and

$$\psi(x, y, z) = \left[\frac{8}{a^3}\right]^{1/2} \sin\left[\frac{2\pi x}{a}\right] \sin\left[\frac{\pi y}{a}\right] \sin\left[\frac{\pi z}{a}\right] = \psi_{211},$$

where the subscripts denote the values of the n_x, n_y, and n_z translational quantum numbers, respectively. ψ_{112} has a single node along the z-coordinate; ψ_{121} has one node along y, and ψ_{211} has one node along x.

EXAMPLE 12.7

What is the degeneracy of the energy eigenvalue $14h^2/(8ma^2)$ for a three-dimensional particle in an infinite potential well with $a = b = c$? How many nodes are present in each coordinate for each eigenfunction with that energy?

Solution

We need to know what values of n_x, n_y, and n_2 make the sum $n_x^2 + n_y^2 + n_z^2$ equal to 14. A little thought shows that the answer is some combination of 1, 2, and 3. Six different combinations have this set of three quantum numbers:

ψ_{123}, for which there are no nodes along x, one node along y, and two nodes along z;

ψ_{132}, for which there are no nodes along x, two nodes along y, and one node along z;

ψ_{213}, for which there is one node along x, no nodes along y, and two nodes along z;

ψ_{231}, for which there is one node along x, two nodes along y, and no nodes along z;

ψ_{312}, for which there are two nodes along x, no nodes along y, and one node along z;

and

ψ_{321}, for which there are two nodes along x, one node along y, and no nodes along z.

Therefore, the total number of nodes is always three.
    For a related exercise, see Problem 12.14.

12.1.3   Superposition of States

This section relies heavily upon the concepts introduced in Chapter 10. The discussion in Section 10.1.4 is particularly relevant and should be reviewed before the reader proceeds with the current section.
    Let us again consider the example of choosing a card at random from a deck of 52 cards and placing the chosen card face down on a table. Since each card is equally likely to be selected, the f_i probability factors that are proportional to the probability that card i lies on the table are all equal. The analysis presented in Chapter 10 shows that the probability that card k lies on the table is

$$P_k = Nf_k,$$

where

$$N = \left[\sum_{i=1}^{52} f_i \right]^{-1} = [f_1 + f_2 + f_3 + \cdots + f_{52}]^{-1}$$

is the normalization constant. Note that N^{-1} is the sum of all the quantities to which the individual probabilities are proportional. In this case, all the f_i are equal, so that

$$P_k = \frac{f_k}{\displaystyle\sum_{i=1}^{52} f_i} = \frac{1}{52}.$$

If we perform a measurement by looking at the card lying on the table and find it to be the five of diamonds, all the probability factors become zero save the one for the diamond five. After the measurement, the probabilities are

$$P_i = 0 \quad \text{unless } i \text{ corresponds to the five of diamonds}$$

and

$$P_{5 \text{ of diamonds}} = \frac{f_{5 \text{ of diamonds}}}{f_{5 \text{ of diamonds}}} = 1.$$

Since the absolute square of the wave function, $|\psi|^2$, for a quantum mechanical system represents the probability distribution, a similar situation should exist for wave functions.

Let us consider a few examples to see how this works. The solution of the stationary-state Schrödinger equation gives the set of eigenfunctions and associated energy eigenvalues that the system can have. This is similar to a listing of all the possible cards in a deck: deuce of spades, trey of spades, four of spades, ... , and so on. Assume we have solved the stationary-state Schrödinger equation to obtain the possible eigenfunctions for a system, which we denote by $\phi_1, \phi_2, \phi_3, \dots, \phi_n, \dots$, with associated eigenvalues $E_1, E_2, E_3, \dots, E_n, \dots$. Assume further that each eigenfunction has been normalized, so that $\langle \phi_i | \phi_i \rangle = 1$ for all values of i.

Suppose we know from some measurement that our quantum mechanical system is equally likely to be in any quantum state in the range $1 \leq n \leq 52$. We might represent this state of affairs by writing the wave function as

$$\psi = N[\phi_1 + \phi_2 + \phi_3 + \cdots + \phi_{52}], \tag{12.52}$$

where N is a normalization constant. Such a wave function is said to be a *superposition of states*, in that it is a linear sum of possible eigenfunctions of the system. Since $\psi^*\psi \, d\tau$ is the probability distribution for the system, we must have

$$\langle \psi | \psi \rangle = |N|^2 \langle \phi_1 + \phi_2 + \phi_3 + \cdots + \phi_{52} | \phi_1 + \phi_2 + \phi_3 + \cdots + \phi_{52} \rangle = 1. \tag{12.53}$$

If $E_i \neq E_j$, then ϕ_i must be orthogonal to ϕ_j, so that $\langle \phi_i | \phi_j \rangle = 0$. Under these conditions, the only integrals that will be nonzero in Eq. 12.53 are those of the form $\langle \phi_k | \phi_k \rangle$, which will be unity, because the eigenfunctions have been normalized. Therefore, that equation becomes

$$\langle \psi | \psi \rangle = |N|^2 [\langle \phi_1 | \phi_1 \rangle + \langle \phi_2 | \phi_2 \rangle + \langle \phi_3 | \phi_3 \rangle + \cdots + \langle \phi_{52} | \phi_{52} \rangle]$$
$$= N^2 [1 + 1 + 1 + \cdots + 1] = 52N^2 = 1. \tag{12.54}$$

Hence, the normalization constant is $[1/52]^{1/2}$, and Eq. 12.54 can be written as

$$\langle \psi | \psi \rangle = \left[\frac{1}{52} + \frac{1}{52} + \frac{1}{52} + \cdots + \frac{1}{52} \right]. \tag{12.55}$$

Since $\langle\psi|\psi\rangle$ is the sum of all the probabilities, Eq. 12.55 indicates that, like the card on the table, the system has an equal chance of being in any of the first 52 eigenstates. Consequently, the probability of it being in any one particular eigenstate is 1/52.

As a second example, consider a modified deck of cards to which we have added an extra ace of spades. The deck now contains 53 cards, and the probability factors are

$$f_i = 1 \quad \text{for } i \text{ being any card other than the ace of spades}$$

and

$$f_{\text{spade ace}} = 2, \quad \text{since there are two spade aces in the deck.}$$

The normalization constant for the distribution giving the probabilities of choosing cards from this modified deck is

$$N = \left[\sum_{i=1}^{52} f_i \right]^{-1} = [1 + 1 + 1 + \cdots + 2]^{-1} = \frac{1}{53},$$

where the summation index runs over the 52 different types of cards in the deck. Consequently, the probability that any given card will be chosen is 1/53, except for the spade ace, whose probability is 2/53.

The analogous quantum mechanical situation is one in which we know that the system has an equal probability of being in quantum states $n = 1, 2, 3, \ldots, 51$, but the probability of finding the system in the quantum state with $n = 52$ is twice that for any of the other eigenstates. This situation can be represented by writing the wave function as

$$\psi = N[\phi_1 + \phi_2 + \phi_3 + \cdots + \sqrt{2}\,\phi_{52}]. \tag{12.56}$$

The normalization constant is now given by

$$\langle\psi|\psi\rangle = |N|^2[\langle\phi_1|\phi_1\rangle + \langle\phi_2|\phi_2\rangle + \langle\phi_3|\phi_3\rangle + \cdots + 2\langle\phi_{52}|\phi_{52}\rangle]$$

$$= |N|^2[1 + 1 + 1 + \cdots + 2] = 53|N|^2 = 1, \tag{12.57}$$

so that we have $N = [1/53]^{1/2}$. The probabilities of finding the quantum system in different eigenstates are 1/53 for every state, except $n = 52$, for which the probability is 2/53. The analogy between the deck of cards and the quantum probabilities is clear: The probability of choosing card k is Nf_k, whereas the probability of finding the quantum system in eigenstate k is $|N|^2|C_k|^2$, where C_k is the coefficient of the eigenfunction ϕ_k in the wave function.

The foregoing discussion demonstrates that if the wave function for a quantum system is

$$\psi = N[C_1\phi_1 + C_2\phi_2 + C_3\phi_3 + \cdots + C_k\phi_k + \cdots], \tag{12.58}$$

then the probability that a measurement will find the system in eigenstate ϕ_k is $N^2|C_k|^2$, where the normalization constant is given by

$$\langle\psi|\psi\rangle = |N|^2[|C_1|^2\langle\phi_1|\phi_1\rangle + |C_2|^2\langle\phi_2|\phi_2\rangle + |C_3|^2\langle\phi_3|\phi_3\rangle + \cdots + |C_k|^2\langle\phi_k|\phi k\rangle$$

$$+ \cdots] = |N|^2[|C_1|^2 + |C_2|^2 + |C_3|^3 + \cdots + |C_k|^2 + \cdots]$$

$$= |N|^2 \sum_{m=1}^{\infty} |C_m|^2 = 1,$$

so that

$$N = \left[\sum_{m=1}^{\infty} |C_m|^2 \right]^{-1/2} \tag{12.59}$$

and

$$P_k = \left[\sum_{m=1}^{\infty} |C_m|^2 \right]^{-1} |C_k|^2. \tag{12.60}$$

This result is general for any quantum mechanical system. *If ψ is written as a superposition of orthogonal eigenfunctions that are individually normalized, the probability of finding the system in eigenstate k is given by Eq. 12.60.*

Now, suppose that we perform a measurement on the quantum system whose wave function is given by Eq. 12.58 and find that system is in eigenstate ϕ_3. This measurement changes the probabilities in the same way that looking at a randomly drawn playing card changes the probability that the card is the ace of spades. If we find the system to be in state ϕ_3, all the expansion coefficients in Eq. 12.58 go to zero, save for C_3, which becomes unity. The wave function for the system, therefore, becomes $\psi = \phi_3$.

Whenever the wave function is a superposition of states, the relationship between the average energy and the number of nodes present in the wave function is lost. Such relationships obtain only when the wave function is an eigenfunction of the quantum mechanical operators for the system. When a superposition of states is present, additional nodes are often introduced.

EXAMPLE 12.8

Consider a particle in a one-dimensional infinite potential well of width a. Suppose that we know that the particle is in either its quantum mechanical ground state or its first excited state and that the probabilities of finding the system in either of these two states are equal. **(A)** Write down an appropriate expression for the system's wave function. Normalize the wave function. **(B)** If the energy of the system is measured, what are the possible results of this measurement? **(C)** If the measurement reveals the energy of the system to be $h^2/(2ma^2)$, what is the wave function of the system subsequent to the measurement?

Solution

(A) Since the two states have equal probabilities, we wish to use a linear expansion with equal coefficients. That is,

$$\psi(x) = N [\phi_1 + \phi_2], \tag{1}$$

where N is the normalization constant and ϕ_1 and ϕ_2 are the ground-state and first excited-state eigenfunctions for a particle in an infinite potential well, respectively. Using Eq. 12.38, we have

$$\phi_1 = \left[\frac{2}{a} \right]^{1/2} \sin\left[\frac{\pi x}{a} \right] \tag{2}$$

and

$$\phi_2 = \left[\frac{2}{a} \right]^{1/2} \sin\left[\frac{2\pi x}{a} \right]. \tag{3}$$

Normalization requires that

$$\langle \psi(x)|\psi(x) \rangle = 1 = N^2 \langle \phi_1 + \phi_2|\phi_1 + \phi_2 \rangle. \tag{4}$$

Expanding Eq. 4, we obtain

$$N^2[\langle\phi_1|\phi_1\rangle + 2\langle\phi_1|\phi_2\rangle + \langle\phi_2|\phi_2\rangle] = 2N^2, \qquad (5)$$

since ϕ_1 and ϕ_2 are normalized, so that $\langle\phi_1|\phi_1\rangle = \langle\phi_2|\phi_2\rangle = 1$ and $\langle\phi_1|\phi_2\rangle = 0$ (because ϕ_1 and ϕ_2 must be orthogonal). Therefore,

$$N = (2)^{-1/2}. \qquad (6)$$

(B) The only two possible eigenstates are ϕ_1 and ϕ_2. These states have energies given by Eq. 12.39, viz., $E_1 = h^2/(8ma^2)$ and $E_2 = 2^2h^2/(8ma^2) = h^2/(2ma^2)$. These are the only two values that we could obtain in a measurement of the system's energy.

(C) If the measurement shows that $E = E_2 = h^2/(2ma^2)$, then the system must be in the first excited state, so that we will have $\psi(x) = \phi_2 = [2/a]^{1/2} \sin[2\pi x/a]$.

For a related exercise, see Problem 12.17.

EXAMPLE 12.9

(A) Suppose we have a particle in an infinite, one-dimensional potential well and we know that the particle is in its $n_x = 1$ or $n_x = 2$ eigenstate. Suppose further that the particle is twice as likely to be in the $n_x = 2$ eigenstate than the $n_x = 1$ state. Write down an appropriate wave function for this system. **(B)** Normalize the wave function obtained in (A). **(C)** If the energy of 10^6 such systems were measured and the average energy computed from the data, what would the average energy be?

Solution

(A) The $n_x = 1$ and $n_x = 2$ eigenfunctions are given by Eq. 12.38. They are

$$\phi_1(x) = \left[\frac{2}{a}\right]^{1/2} \sin\left[\frac{\pi x}{a}\right] \qquad (1)$$

and

$$\phi_2(x) = \left[\frac{2}{a}\right]^{1/2} \sin\left[\frac{2\pi x}{a}\right]. \qquad (2)$$

If the system is in one of these two eigenstates, the wave function must be a superposition of ϕ_1 and ϕ_2:

$$\psi(x) = N[C_1\phi_1(x) + C_2\phi_2(x)]. \qquad (3)$$

If it is twice as likely that the system is in state $n_x = 2$ than $n_x = 1$, we must have $|C_2|^2 = 2|C_1|^2$. We can represent this state of affairs by taking $C_1 = 1$ and $C_2 = \sqrt{2}$. Therefore, the desired superposition is

$$\psi(x) = N[\phi_1(x) + \sqrt{2}\,\phi_2(x)] = N\left\{\left[\frac{2}{a}\right]^{1/2}\sin\left[\frac{\pi x}{a}\right] + \sqrt{2}\left[\frac{2}{a}\right]^{1/2}\sin\left[\frac{2\pi x}{a}\right]\right\}. \qquad (4)$$

(B) Normalization requires that we have

$$\langle\psi(x)|\psi(x)\rangle = |N|^2\,[\langle\phi_1(x)|\phi_1(x)\rangle + 2\langle\phi_2(x)|\phi_2(x)\rangle + 2\sqrt{2}\,\langle\phi_1(x)|\phi_2(x)\rangle] = 1. \qquad (5)$$

Because the eigenfunctions are normalized, $\langle\phi_1(x)|\phi_1(x)\rangle = \langle\phi_2(x)|\phi_2(x)\rangle = 1$. Since the energy eigenvalues of the two eigenfunctions are not the same, ϕ_1 must be orthogonal to ϕ_2. Therefore, $\langle\phi_1(x)|\phi_2(x)\rangle = 0$. As a result, Eq. 5 can be written as

$$\langle\psi(x)|\psi(x)\rangle = |N|^2[1 + 2 + 0] = 1, \qquad (6)$$

which gives

$$N = [1/3]^{1/2}. \qquad (7)$$

(C) The probability that the system is in the eigenstate with $n_x = 1$ is given by Eq. 12.60,

$$P(n_x = 1) = \frac{|C_1|^2}{|C_1|^2 + |C_2|^2} = \frac{1^2}{1^2 + \sqrt{2}^2} = \frac{1}{3}. \tag{8}$$

The probability that the system is in the eigenstate with $n_x = 2$ is

$$P(n_x = 2) = \frac{|C_2|^2}{|C_1|^2 + |C_2|^2} = \frac{\sqrt{2}^2}{1^2 + \sqrt{2}^2} = \frac{2}{3}. \tag{9}$$

Therefore, one-third of the time, we will find the system in the $n_x = 1$ eigenstate, and two-thirds of the time it will be in the $n_x = 2$ state. Hence, 333,333 measurements will find the system in the $n_x = 1$ state, and 666,667 measurements will find it in the $n_x = 2$ state. When the system is in the eigenstate with $n_x = 1$, the measured energy will be $h^2/(8ma^2)$. When it is in the state with $n_x = 2$, the measured energy will be $4h^2/(8ma^2)$. Therefore, the average energy will be

$$\langle E \rangle = \frac{(333{,}333)\dfrac{h^2}{8ma^2} + (666{,}667)\dfrac{4h^2}{8ma^2}}{1{,}000{,}000} = \frac{9}{3}\frac{h^2}{8ma^2} = \frac{3h^2}{8ma^2}. \tag{10}$$

For related exercises, see Problems 12.17, 12.18, and 12.19.

12.1.4 The Uncertainty Principle

When a particle in a one-dimensional, infinite potential well is in its quantum mechanical ground state, there is an uncertainty in the particle's position and a corresponding uncertainty in its momentum. The uncertainty in position is shown by Eq. 12.46 to be

$$\langle \Delta x^2 \rangle^{1/2} = a \left[\frac{1}{12} - \frac{1}{2\pi^2} \right]^{1/2}. \tag{12.61}$$

In Example 12.5, we found that the uncertainty in momentum is $\langle \Delta p_x^2 \rangle^{1/2} = \pi\hbar/a$. The uncertainty product of these two quantities is, therefore,

$$\langle \Delta x^2 \rangle^{1/2} \langle \Delta p_x^2 \rangle^{1/2} = a \left[\frac{1}{12} - \frac{1}{2\pi^2} \right]^{1/2} \frac{\pi\hbar}{a} = \left[\frac{\pi^2}{12} - \frac{1}{2} \right]^{1/2} \hbar$$

$$= 0.56786\ldots\hbar. \tag{12.62}$$

In Problem 12.10, it is shown that the uncertainty product for the n_x quantum state is

$$\langle \Delta x^2 \rangle^{1/2} \langle \Delta p_x^2 \rangle^{1/2} = \hbar \left[\frac{n_x^2 \pi^2}{12} - \frac{1}{2} \right]^{1/2}. \tag{12.63}$$

Consequently, this product has its minimum value when $n_x = 1$. That is, the combined uncertainty product of position and momentum is never less than $0.56786\ldots\hbar$ for any eigenfunction of a particle in a one-dimensional, infinite potential well.

We now raise the question, "What happens if the wave function is a superposition of states rather than a single eigenfunction?" Suppose, for example, the wave function represents a system in which it is equally likely that the particle is in either the $n_x = 1$ or the $n_x = 2$ eigenstate. Then, we have

$$\psi(x) = (2)^{-1/2}[\phi_1 + \phi_2] = (2)^{-1/2}\left\{ \left[\frac{2}{a} \right]^{1/2} \sin\left[\frac{\pi x}{a} \right] + \left[\frac{2}{a} \right]^{1/2} \sin\left[\frac{2\pi x}{a} \right] \right\}, \tag{12.64}$$

where ϕ_1 and ϕ_2 are the eigenfunctions for the ground and first excited states, respectively. For this superposition, the average momentum and average square momentum are respectively given by

$$\langle p_x \rangle = \frac{1}{2}\left\langle \phi_1 + \phi_2 \left| \frac{\hbar}{i} \frac{\partial}{\partial x} \right| \phi_1 + \phi_2 \right\rangle$$

and

$$\langle p_x^2 \rangle = \frac{1}{2}\left\langle \phi_1 + \phi_2 \left| -\hbar^2 \frac{\partial^2}{\partial x^2} \right| \phi_1 + \phi_2 \right\rangle.$$

It can be shown that $\langle p_x \rangle = 0$ and $\langle p_x^2 \rangle = 24.674\ldots \hbar^2/a^2$. (See Problem 12.20.) The corresponding uncertainty in momentum for this superposition is, therefore, $\langle \Delta p_x^2 \rangle^{1/2} = [\langle p_x^2 \rangle - \langle p_x \rangle^2]^{1/2} = 4.96\ldots \hbar/a$. This uncertainty is significantly greater than that for a system known to be in the ground state, in which case $\langle \Delta p_x^2 \rangle^{1/2} = \pi\hbar/a$. Consequently, the uncertainty in momentum for the superposition has increased by about 50% compared with that of the ground state. This finding is not surprising when we consider the fact that the translational energy of the first excited state is four times as great as that of the ground state. As a result, we would expect measurements of the momentum for a system whose wave function is described by Eq. 12.64 to span a much greater range than would be the case for a system known to be in the ground state. The broader momentum distribution will be reflected by a greater uncertainty in the momentum.

What about the uncertainty in the *position* of a quantum particle whose wave function is given by Eq. 12.64? The qualitative answer may be seen by examining Figure 12.8, which shows $\psi(x)$ as a function of x for the case where $a = 1$ Å. Most of the wave amplitude is now localized between $0 \le x \le 0.6$ Å rather than being spread over the entire well. In particular, $\langle \Delta x^2 \rangle^{1/2}$ is now equal to $0.1386\ldots$ Å rather than its value of $0.180\ldots$ Å when the particle is in a 1Å well in the $n_x = 1$ eigenstate. (See Problem 12.20.) Consequently, the particle is more localized, with a smaller uncertainty in position. However, the uncertainty in momentum has increased from $\pi\hbar$ to $4.96729\ldots \hbar$. Basically, we can reduce the uncertainty in the particle's position only by increasing the uncertainty in its momentum.

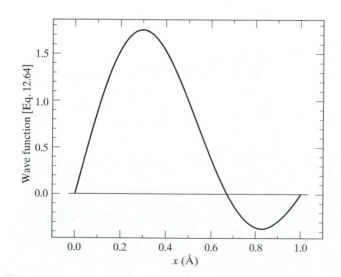

◀ **FIGURE 12.8**
Wave function for a particle in an infinite potential well with a width of 1 Å for the case of the two-state superposition given by Eq. 12.64 in which the system is equally likely to be found in either the ground or first-excited eigenstate.

Figure 12.9 shows what happens as we continue this process of adding more and more eigenfunctions into the wave function superposition. Every time we add another eigenfunction, the momentum distribution broadens and the corresponding momentum uncertainty increases. However, as the figure shows, the particle becomes more and more localized, with smaller and smaller uncertainty in its position. We can reduce the uncertainty in position only at the expense of increasing the uncertainty in the momentum distribution. In every case, however, the uncertainty product, $\langle\Delta x^2\rangle^{1/2}\langle\Delta p_x^2\rangle^{1/2}$, increases. For a particle in a one-dimensional infinite well, it is never smaller than that for the $n_x = 1$ eigenstate: $0.56786\ldots\hbar$.

The preceding analysis suggests that we can never simultaneously reduce both the position and momentum uncertainties to zero. That is, we can never precisely know both the position and momentum of a quantum mechanical particle. With somewhat more quantum theory than we can cover in this text, it can be shown that for any two variables u and w, we always have

$$\langle\Delta u^2\rangle\langle\Delta w^2\rangle \geq \frac{1}{4}|[u, w]|^2, \tag{12.65A}$$

where $[u, w]$ is the commutator of the variables u and w. When u and w are conjugate variables, such as x and p_x, y and p_y, z and p_z, or E and t (energy and time are also conjugate variables), the second postulate tells us that $[x, p_x] = [y, p_y] = [z, p_z] = i\hbar$. Therefore, we have

$$\langle\Delta q^2\rangle\langle\Delta p_q^2\rangle \geq \frac{\hbar^2}{4} \tag{12.65B}$$

and

$$\langle\Delta q^2\rangle^{1/2}\langle\Delta p_q^2\rangle^{1/2} \geq \frac{\hbar}{2} \tag{12.65C}$$

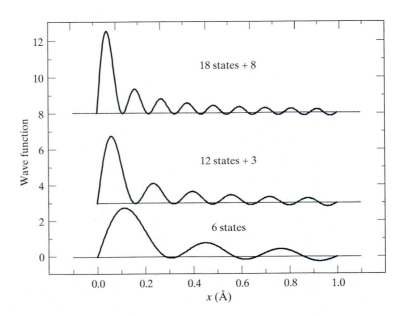

▶ **FIGURE 12.9**
Wave functions for a particle in an infinite potential well with a width of 1 Å for superpositions involving the lowest energy N eigenstates, with $N = 6$, $N = 12$, and $N = 18$. In each superposition, the system is assumed to have equal probability of being found in any of the eigenstates included in the superposition. The three wave functions are displaced upwards in the plot by the amount indicated in order to increase the visual clarity of the graph.

for $q = x, y,$ or z and $p_q = p_x, p_y,$ or p_z. Equations 12.65A through 12.65C are quantitative statements of the *Heisenberg uncertainty principle*, whose existence fully resolves the dilemma posed by the deterministic nature of classical mechanics: Since we can never simultaneously know both the position and momentum coordinates of any system, there is no way to predict the entire future and past behavior of the system.

The uncertainty principle is closely related to the fact that the lowest energy for a particle in an infinite one- or three-dimensional potential well is greater than zero. Because the size of the well is finite, the uncertainty in position must also be finite. If it were possible for the system to exhibit a zero translational energy in some eigenstate, we would necessarily have $\langle p_q \rangle = \langle p_q^2 \rangle = 0$, so that the uncertainty in momentum, $\langle \Delta p_q^2 \rangle = \langle p_q^2 \rangle - \langle p_q \rangle^2$, would also be zero. As a result, the uncertainty product $\langle \Delta q^2 \rangle \langle \Delta p_q^2 \rangle$ would be zero. Since this would violate the uncertainty principle, we may be certain that there exists no eigenfunction for a particle in an infinite potential well for which the translational-energy eigenvalue is zero.

It is possible to use the uncertainty principle to estimate a lower limit for the momentum or energy of a system. For example, consider a one-dimensional system. If we maximize the uncertainty in position and assume that the uncertainty product has its minimum possible value, a lower limit for the uncertainty in momentum can be computed. We can then assert that we can never know that the momentum is lower than $\langle \Delta p^2 \rangle_{min}^{1/2}$, because this is the minimum measurement uncertainty. Consequently, $\langle \Delta p^2 \rangle_{min}^{1/2}$ is an approximate lower limit for p. Example 12.10 illustrates the procedure.

EXAMPLE 12.10

An electron is contained in a one-dimensional, infinite potential well whose width is 1×10^{-10} m (1 Å). **(A)** What is the maximum possible value for the uncertainty in the electron's position? **(B)** What is the minimum possible uncertainty in the momentum of the electron? **(C)** Estimate the minimum kinetic energy the electron can have. Estimate its minimum kinetic energy in units of kJ mol^{-1}.

Solution

(A) The uncertainty in position for a particle in a one-dimensional, infinite potential in eigenstate n is

$$\langle \Delta x^2 \rangle^{1/2} = a \left[\frac{1}{12} - \frac{1}{2n^2\pi^2} \right]^{1/2}. \tag{1}$$

(See Problem 12.10.) The maximum value for $\langle \Delta x^2 \rangle^{1/2}$ occurs when n becomes large. In this limit, we have $[\langle \Delta x^2 \rangle^{1/2}]_{max} = a(12)^{-1/2}$. With $a = 1 \times 10^{-10}$ m, the maximum value of $\langle \Delta x^2 \rangle^{1/2}$ is

$$[\langle \Delta x^2 \rangle^{1/2}]_{max} = \frac{(1 \times 10^{-10} \text{ m})}{(12)^{1/2}} = 2.89 \times 10^{-11} \text{ m}. \tag{2}$$

The minimum uncertainty in p_x is, therefore,

$$[\langle \Delta p_x^2 \rangle^{1/2}]_{min} = \frac{\hbar}{2[\langle \Delta x^2 \rangle^{1/2}]_{max}} = \frac{6.626 \times 10^{-34} \text{ J s}}{4(3.14159)(2.89 \times 10^{-11} \text{ m})}$$

$$= 1.82 \times 10^{-24} \text{ kg m s}^{-1}. \tag{3}$$

The minimum magnitude of the momentum should also be about this value. At least, no measurement could ever show it to be less, since the uncertainty in the measurement could not be less than 1.82×10^{-24} kg m s^{-1}. The corresponding minimum kinetic energy is

$$T_{min} \approx \frac{p_x^2}{2m} = \frac{(1.82 \times 10^{-24} \text{ kg m s}^{-1})^2}{2(9.11 \times 10^{-31} \text{ kg})} = 1.82 \times 10^{-18} \text{ J per electron.} \tag{4}$$

Multiplying by Avogadro's constant gives the energy per mole:

$$T_{min} \text{ per mole} = T_{min} \, N$$

$$= 1.82 \times 10^{-18} \text{ J electron}^{-1} \times (6.022 \times 10^{23} \text{ electrons mol}^{-1})$$

$$= 1.09 \times 10^6 \text{ J mol}^{-1} = 1.09 \times 10^3 \text{ kJ mol}^{-1} \tag{5}$$

For a related exercise, see Problem 12.21.

The preceding discussion deals with a system whose energy states are discrete. The same principles hold for the continuum states of a quantum mechanical free particle. If we know that the system has a momentum $+k\hbar$, then its wave function is Ae^{ikx}, and the uncertainty in the particle's position is infinite. As our knowledge of the system's momentum decreases, more momenta will be possible, and the system wave function will become a superposition of free-particle eigenfunctions, such as $A_1 e^{ik_1 x} + A_2 e^{ik_2 x} + A_3 e^{ik_3 x} + \dots$. In that case, the uncertainty in momentum increases while the uncertainty in position decreases. This is very similar to the case of a particle in a one-dimensional potential well considered earlier. The only significant difference resides in the fact that the superposition of free-particle eigenstates can be a continuous sum over the wave number k, since all values of k are possible. Such a continuous summation (integration) of free-particle eigenfunctions is called a *wave packet*. Example 12.11 illustrates this concept.

EXAMPLE 12.11

A one-dimensional free particle is known to have positive momentum in the range $k_1 \hbar$ to $k_2 \hbar$, with all momenta in the range equally probable. **(A)** Find an appropriate form for the wave function that represents this state of knowledge of the system. **(B)** Obtain an expression for the spatial probability distribution of the system. Plot the result for the specific case where $k_2 = 6.00$ Å^{-1} and $k_1 = 4.00$ Å^{-1}. Show the corresponding free-particle wave number probability distribution in a separate plot. **(C)** Comment on the spatial localization of the particle. Why is the particle more localized in space than a simple free particle with momentum $k_1 \hbar$?

Solution

(A) Since the system is moving with positive momentum, Eq. 12.12 and the subsequent discussion show that the wave function will be $\psi(x) = Ae^{ikx}$ if we know that the momentum is $k \hbar$. If k can have any value between k_1 and k_2, and if all momentum states are equally probable, the superposition of eigenstates that represents our state of knowledge is

$$\Phi(x) = \int_{k=k_1}^{k=k_2} Ae^{ikx} \, dk = \frac{A}{ix} [e^{ikx}]_{k=k_1}^{k=k_2} = \frac{A}{ix} [e^{ik_2 x} - e^{ik_1 x}]. \tag{1}$$

If some momenta are more probable than others, we would represent the wave packet by

$$\Phi(x) = \int_{k=k_1}^{k=k_2} A(k)\, e^{ikx}\, dk, \tag{2}$$

where $A(k)$ is proportional to the square root of the probability of observing momentum $k\,\hbar$.

(B) The spatial probability distribution for the system is given by $\Phi^*(x)\Phi(x)\,dx$. Using Eq. 1, we obtain

$$\Phi^*(x)\Phi(x)\,dx = -\frac{A^*}{ix}[e^{-ik_2x} - e^{-ik_1x}]\frac{A}{ix}[e^{ik_2x} - e^{ik_1x}]\,dx$$

$$= \frac{|A|^2}{x^2}[1 + 1 - \exp\{i(k_2 - k_1)x\} - \exp\{-i(k_2 - k_1)x\}]\,dx. \tag{3}$$

Using the fact that $e^{iz} + e^{-iz} = [\cos z + i \sin z] + [\cos z - i \sin z] = 2 \cos z$, we can write Eq. 3 in the form

$$\Phi^*(x)\Phi(x)\,dx = \frac{|A|^2}{x^2}[2 - 2\cos\{(k_2 - k_1)x\}]\,dx = \frac{2|A|^2}{x^2}[1 - \cos\{(k_2 - k_1)x\}]\,dx, \tag{4}$$

which is the desired expression. The plot of $\dfrac{\Phi^*(x)\Phi(x)}{|A|^2}$ versus x for the case where $k_2 = 6.00\ \text{Å}^{-1}$ and $k_1 = 4.00\ \text{Å}^{-1}$ is shown in Figure 12.10. The corresponding probability distribution of free-particle wave numbers is shown in Figure 12.11.

(C) An examination of Figure 12.10 shows that the wave packet is essentially localized in the range $-3\ \text{Å} \le x \le 3\ \text{Å}$. This is much more localization than is the case

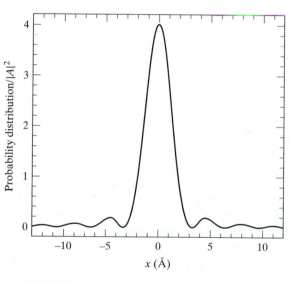

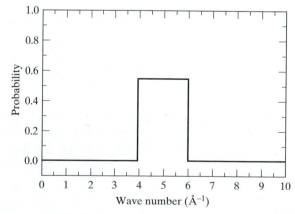

▲ FIGURE 12.10
The spatial probability distribution for a quantum mechanical wave packet whose momentum wave number probability distribution is that shown in Figure 12.11.

▲ FIGURE 12.11
The probability distribution of free-particle wave numbers that produces the quantum mechanical wave packet shown in Figure 12.10.

for a free particle with momentum $k_1\,\hbar$. The increase in uncertainty in the momentum probability distribution reduces the uncertainty in the position distribution.

For related exercises, see Problems 12.24, 12.25 and 12.26.

12.1.5 Tunneling

When a classical particle strikes a potential barrier greater than the energy possessed by the particle, it is reflected by the barrier without penetrating it. This is not the case for a quantum particle. Figure 12.12 shows a particle with momentum $k\hbar$ and corresponding energy E moving in the positive x direction in Region I. At the point $x = 0$, the particle strikes a potential barrier whose magnitude is $V_o > E$. As a result, some of the quantum mechanical wave is reflected with the reflected wave moving back through Region I with momentum $-k\hbar$. If the particle were classical, this is the only thing that would happen. Quantum mechanically, however, some of the matter wave penetrates into Region II, where the wave is now denoted as $\psi_{II}(x)$. When this wave reaches the end of the barrier at the point $x = a$, some of *it* escapes into Region III, where it continues to move in the positive x direction with momentum $k\hbar$ and wave function $\psi_{III}(x)$. The quantum particle can traverse a barrier with an energy lower than the barrier crest. It is as if the particle had dug a tunnel through the barrier to permit its passage. The phenomenon is, therefore, called *tunneling*.

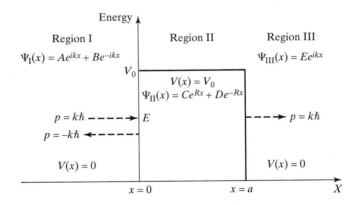

▶ **FIGURE 12.12**
One-dimensional barrier penetration and tunneling through a square barrier of height $V_o > E$.

To determine the probability that a quantum particle will tunnel through the barrier shown in the figure, we first need to obtain the wave function in each region. In Region I, the potential is zero. Therefore, the solutions of the Schrödinger equation are the free-particle solutions given by Eq. 12.9, namely,

$$\psi_I(x) = Ae^{ikx} + Be^{-ikx}, \tag{12.66}$$

where the first term represents the particles moving toward the barrier with momentum $k\hbar$ and the second term represents that fraction of the particles which strike the barrier and are reflected backward with momentum $-k\hbar$.

In Region II, the stationary-state Schrödinger equation is

$$-\frac{\hbar^2}{2m}\frac{\partial^2}{\partial x^2}\psi_{II}(x) + V_o\psi_{II}(x) = E_x\psi_{II}(x). \tag{12.67}$$

Rearrangement of Eq. 12.67 produces

$$\frac{\partial^2}{\partial x^2}\psi_{II}(x) - \frac{2m}{\hbar^2}[V_o - E]\psi_{II}(x) = \frac{\partial^2}{\partial x^2}\psi_{II}(x) - R^2\psi_{II}(x) = 0, \tag{12.68}$$

where

$$R^2 = \frac{2m}{\hbar^2}[V_o - E]. \tag{12.69}$$

The form of Eq. 12.68 is identical to that of Eq. 12.6, except that the sign of the second term is negative. Hence, the solution of Eq. 12.69 is (see Problem 12.22)

$$\psi_{\text{II}}(x) = Ce^{Rx} + De^{-Rx}. \tag{12.70}$$

In Region III, the solution of the Schrödinger equation is once again the free-particle wave function, since we have $V(x) = 0$ in this region:

$$\psi_{\text{III}}(x) = Ee^{ikx}. \tag{12.71}$$

Only the term with positive momentum appears in $\psi_{\text{III}}(x)$, because there are no particles moving in the negative x direction.

The fraction of particles that undergo tunneling, τ, is the number of particles found in Region III, divided by the number of particles in Region I moving toward the barrier with momentum $k\,\hbar$. The latter quantity is the total number of particles, N, times the probability that the particles are in Region I with momentum $k\,\hbar$. This probability is given by $[Ae^{ikx}]^*[Ae^{ikx}]\,dx$. In the same manner, the number of particles found in Region III is $N[\psi_{\text{III}}(x)]^*[\psi_{\text{III}}(x)]\,dx$. Consequently, the tunneling fraction is

$$\tau = \frac{N[\psi_{\text{III}}(x)]^*[\psi_{\text{III}}(x)]\,dx}{N[Ae^{ikx}]^*[Ae^{ikx}]\,dx} = \frac{N[Ee^{ikx}]^*[Ee^{ikx}]\,dx}{N[Ae^{ikx}]^*[Ae^{ikx}]\,dx} = \frac{|E|^2}{|A|^2}, \tag{12.72}$$

since $[e^{ikx}]^*[e^{ikx}] = e^{-ikx}\,e^{ikx} = e^0 = 1$. Equation 12.72 shows that the tunneling fraction can be found if we can obtain the ratio of the absolute squares of E and A.

The coefficients $A, B, C, D,$ and E are obtained by requiring that the wave function and its derivatives be continuous at the boundaries of the region at $x = 0$ and $x = a$, as demanded by the first postulate. The condition that the wave function itself be continuous at the boundaries yields two equations: $\psi_{\text{I}}(x = 0) = \psi_{\text{II}}(x = 0)$ and $\psi_{\text{II}}(x = a) = \psi_{\text{III}}(x = a)$. Substituting Eqs. 12.66, 12.70, and 12.71 gives

$$A + B = C + D \tag{12.73}$$

and

$$Ce^{Ra} + De^{-Ra} = Ee^{ika}. \tag{12.74}$$

The derivatives of the wave functions in each region are, respectively,

$$\frac{d\psi_{\text{I}}(x)}{dx} = ikAe^{ikx} - ikBe^{-ikx}, \tag{12.75}$$

$$\frac{d\psi_{\text{II}}(x)}{dx} = RCe^{Rx} - RDe^{-Rx}, \tag{12.76}$$

and

$$\frac{d\psi_{\text{III}}(x)}{dx} = ikEe^{ikx}. \tag{12.77}$$

The requirement that these derivatives be continuous at the boundaries yields two more equations:

$$ikA - ikB = RC - RD \tag{12.78A}$$

and

$$RCe^{Ra} - RDe^{-Ra} = ikEe^{ika}. \tag{12.78B}$$

Equations 12.73, 12.74, and 12.78A and, 12.78B permit the ratio $|E|^2/|A|^2$ to be obtained in terms of E, V_o, and the mass of the particle. The details are left to the reader in Problem 12.23; the various parts of the problem serve as a guide to the solution. The resulting tunneling fraction is

$$\tau = \frac{|E|^2}{|A|^2} = \frac{16k^2R^2}{\left[\dfrac{(R^2 + k^2)^2}{e^{2Ra}} + \dfrac{(R^2 + k^2)^2}{e^{-2Ra}} + 2(R^2 - k^2)^2 - 8k^2R^2\right]}. \qquad \textbf{(12.79)}$$

If V_o, the energy E, and the mass m of the particle are known, Eq. 12.79 may be used to compute the exact tunneling fraction through the one-dimensional square barrier. In many cases, however, the energy will be substantially less than the barrier height, and the barrier width, a, will be sufficiently large to permit a very convenient limiting form for τ to be obtained. The exponential quantities in Eq. 12.79 are

$$e^{2Ra} = \exp\left[2\left\{\frac{2m}{\hbar^2}[V_o - E]\right\}^{1/2} a\right]$$

and

$$e^{-2Ra} = \exp\left[-2\left\{\frac{2m}{\hbar^2}[V_o - E]\right\}^{1/2} a\right].$$

If $V_o - E$, m, and a are sufficiently large, we will have $e^{-2Ra} \ll e^{2Ra}$, in which case the second term in the denominator of Eq. 12.79 will be much larger than the first term. That is,

$$\frac{(R^2 + k^2)^2}{e^{-2Ra}} \gg \frac{(R^2 + k^2)^2}{e^{2Ra}}.$$

Since the exponential function e^{2Ra} increases much more rapidly than $[2(R^2 - k^2)^2 - 8k^2R^2]$, the second term in the denominator will also be much larger than $[2(R^2 - k^2)^2 - 8k^2R^2]$. Therefore, the tunneling fraction will be approximated by the expression

$$\tau = \frac{16k^2R^2}{\dfrac{(R^2 + k^2)^2}{e^{-2Ra}}} = \frac{16k^2R^2}{(R^2 + k^2)^2}e^{-2Ra}. \qquad \textbf{(12.80)}$$

Substituting the definitions of k^2 and R^2 from Eqs. 12.7 and 12.69, respectively, gives

$$16k^2R^2 = 16\frac{2mE}{\hbar^2}\frac{2m(V_o - E)}{\hbar^2} = 16\left[\frac{2m}{\hbar^2}\right]^2 E(V_o - E)$$

and

$$(R^2 + k^2)^2 = \left[\frac{2m}{\hbar^2}[V_o - E] + \frac{2m}{\hbar^2}E\right]^2 = \left[\frac{2m}{\hbar^2}\right]^2 V_o^2.$$

Combining these expressions with Eq. 12.80 produces

$$\boxed{\tau = \frac{16E(V_o - E)}{V_o^2}\exp\left[-2\left\{\frac{2m}{\hbar^2}[V_o - E]\right\}^{1/2} a\right].} \qquad \textbf{(12.81)}$$

This equation is valid for one-dimensional square barriers when $(V_o - E)$, m, and a are large.

Equation 12.81 demonstrates that the tunneling fraction decreases exponentially with the width of the barrier and with the square root of the product of the mass and $(V_o - E)$. Consequently, we expect tunneling processes to become important for very light particles, such as electrons and protons, encountering narrow barriers with an energy that is not too far below the barrier crest. As V_o or a become infinitely large, τ approaches zero. (Even quantum mechanical particles cannot tunnel through infinitely high or infinitely wide barriers.)

EXAMPLE 12.12

The double-helix structure of DNA is stabilized by hydrogen bonds between base pairs of the DNA sequence, some of which have the form $N_{(1)}-H \cdots\cdots N_{(2)}$. Sometimes it is possible for the hydrogen atom to move from one nitrogen atom to the next by tunneling through the potential barrier. Let us assume that the barrier is one dimensional and square, with a height and width of 1.12×10^{-19} J and 0.3×10^{-10} m, respectively. If the energy of the hydrogen atom is 2.18×10^{-20} J, what is the probability that the atom will tunnel through the barrier when it is incident upon it? If the hydrogen atom were a classical particle, what would be the probability of tunneling?

Solution

We need to compute the value of R. Using Eq. 12.67, we have

$$R^2 = \frac{2m}{\hbar^2}[V_o - E] = \frac{(2)(1.67 \times 10^{-27} \text{ kg})(1.12 \times 10^{-19} - 2.18 \times 10^{-20}) \text{ J}}{(1.055 \times 10^{-34} \text{ J s})^2}$$

$$= 2.17 \times 10^{22} \text{ m}^{-2}. \tag{1}$$

Taking the square roots of both sides of Eq. 1, we obtain $R = 1.65 \times 10^{11}$ m^{-1} and

$$2Ra = 2(1.65 \times 10^{11} \text{ m}^{-1})(0.3 \times 10^{-10} \text{ m}) = 9.90. \tag{2}$$

The exponential factor in Eq. 12.81 is thus

$$e^{-2Ra} = \exp\left[-2\left\{\frac{2m}{\hbar^2}[V_o - E]\right\}^{1/2} a\right] = \exp(-9.90) = 5.02 \times 10^{-5}. \tag{3}$$

The tunneling fraction or probability given by Eq. 12.81 is then

$$\tau = \frac{16E(V_o - E)}{V_o^2} \exp\left[-2\left\{\frac{2m}{\hbar^2}[V_o - E]\right\}^{1/2} a\right]$$

$$= \frac{16(2.18 \times 10^{-20} \text{ J})(1.12 \times 10^{-19} - 2.18 \times 10^{-20} \text{ J})}{(1.12 \times 10^{-19} \text{ J})^2}(5.02 \times 10^{-5})$$

$$= 1.26 \times 10^{-4}. \tag{4}$$

The result obtained in Eq. 4 shows that the tunneling probability is very small. If the hydrogen atom were a classical particle, the tunneling probability would be exactly zero, since classical particles cannot tunnel.

For a related exercise, see Problem 12.27.

The ability of electrons to tunnel through potential barriers is the basis of the *scanning tunneling microscope* (STM) invented by Gerd Binnig and Heinrich Rohrer. This instrument permits scientists to image individual atoms and molecules of a surface, as well as surface absorbates. A scanning tunneling microscope contains a platinum-rhodium needle which acts as a probe that is attached to cylinder of piezoelectric ceramic which expands or contracts in response to an electric current that is passed through it. When the needle is brought into close proximity to a surface, a potential difference is established between the surface and the probe. In response, electrons tunnel through this potential barrier to create an electric current that modifies the potential difference. The amount of modification is measured by sensitive electronics that are designed to maintain a constant potential difference between probe and surface. The latter is achieved by controlling the current passing through the ceramic cylinder by contracting or extending the probe, thereby changing the tunneling distance between surface and probe. As the probe is systematically scanned across the surface, the STM prints out a map of the distance of the probe above the surface at each scanning point. Because the tunneling probability is a highly sensitive function of the tunneling distance [a in Eq. 12.81], the STM can detect tiny, atomic-scale variations in the height of the surface and thereby image surface atoms and absorbed molecules. In 1986, Binnig and Rohrer received the Nobel prize in physics for their contribution to science.

Figure 12.13 shows a schematic diagram of the essential components of an STM instrument. Figure 12.14 illustrates the type of data that can be obtained using an STM. This figure is an STM image of cyclopentene absorbed on a Ag(111) surface. The small, round balls seen in the image are

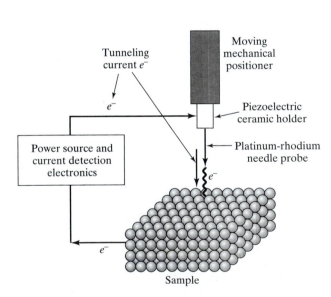

▲ FIGURE 12.13

Schematic diagram of the essential components of a scanning tunneling microscope (STM). Electrons tunnel from the probe needle to the surface. The tunneling current is measured by sensitive electronics. This measurement is used to control the current being passed through the ceramic probe holder, which serves to contract or extend the holder so as to vary the probe distance from the sample surface and thereby maintain a constant tunneling current.

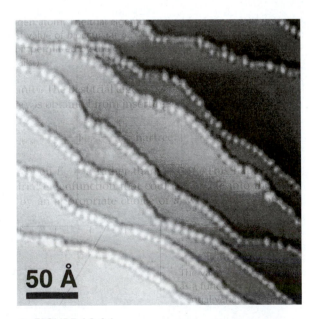

50 Å

▲ FIGURE 12.14

Scanning tunneling microscope image of cyclopentene molecules absorbed on a Ag(111) surface, which consists of large flat planes separated by steps one atomic layer high. Due to the lower coordination of atoms at step edges, these atoms are particularly reactive, so molecules preferentially bond and react at those locations. In this image, each round spot is an individual cyclopentene molecule bonded at a step edge. [Copyright R. J. Hamers and X. Chen, reproduced with permission.]

individual cyclopentene molecules. The data show that the chemisorbed molecules reside along well-defined, specific lines that correspond to the location of steps on the underlying Ag(111) surface. [The author is indebted to Drs. R. J. Hamers and X. Chen for permission to use this image and for providing the negative for reproduction.]

12.2   Rotational Energy

12.2.1   Center of Mass and Relative Coordinates

To exhibit rotational motion, a molecule must contain at least two atoms. A two-atom system requires six coordinates to describe the positions of the atoms. Therefore, we expect that six quantum numbers will be needed to specify the quantum mechanical energy state of the system. If we employ a center-of-mass, relative coordinate system, it becomes clear that three of these quantum numbers determine the translational energy of the center of mass of the molecule, two determine the rotational state, and the sixth quantum number determines the vibrational energy. The discussion that follows shows why this is true.

If the Cartesian coordinates of atoms A and B are denoted by (X_A, Y_A, Z_A) and (X_B, Y_B, Z_B), respectively, the kinetic energy of the system may be expressed as

$$T = \frac{m_A}{2}[v_{XA}^2 + v_{YA}^2 + v_{ZA}^2] + \frac{m_B}{2}[v_{XB}^2 + v_{YB}^2 + v_{ZB}^2], \quad \text{(12.82)}$$

where the v_i are the Cartesian velocities. In this form, the translational, rotational, and vibrational energies cannot be easily identified. The motion of atom A along the X coordinate, for example, contains components of each type of energy. If we transform to a coordinate system in which we use three coordinates to explicitly label the location of the center of mass and employ the remaining coordinates to express the relative positions of the atoms, it becomes much easier to separate the translational from rotational and vibrational energy. For the two-atom system, the X, Y, Z-coordinates of the center of mass are, respectively,

$$X_C = \frac{m_A X_A + m_B X_B}{M}, \quad \text{(12.83A)}$$

$$Y_C = \frac{m_A Y_A + m_B Y_B}{M}, \quad \text{(12.83B)}$$

and

$$Z_C = \frac{m_A Z_A + m_B Z_B}{M}, \quad \text{(12.83C)}$$

where $M = m_A + m_B$. The relative $X, Y,$ and Z positions are

$$X_R = X_A - X_B, \quad \text{(12.84A)}$$

$$Y_R = Y_A - Y_B, \quad \text{(12.84B)}$$

and

$$Z_R = Z_A - Z_B. \quad \text{(12.84C)}$$

These coordinates are illustrated in Figure 12.15. We could equally well define X_R as $X_B - X_A$, so long as we maintain consistency throughout the analysis.

$$\left. \begin{array}{l} X_R = X_A - X_B \\ Y_R = Y_A - Y_B \\ Z_R = Z_A - Z_B \end{array} \right\} \text{Relative coordinates}$$

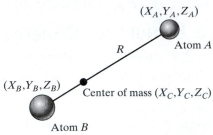

▶ **FIGURE 12.15**
Center-of-mass, relative coordinate system for a two-atom system. The rectangular Cartesian coordinates of atoms A and B are denoted by subscripts A and B, respectively. The center-of-mass and relative coordinates are denoted by subscripts, C and R, respectively.

We now need to invert Eqs. 12.83 and 12.84 to express the Cartesian coordinates for atoms A and B in terms of the center-of-mass and relative coordinates. This may easily be done by combining the corresponding A, B and C equations in 12.83 and 12.84. (See Problem 12.28.) The results are

$$X_A = X_C + \frac{m_B}{M}X_R, \quad Y_A = Y_C + \frac{m_B}{M}Y_R, \quad Z_A = Z_C + \frac{m_B}{M}Z_R, \quad \textbf{(12.85)}$$

and

$$X_B = X_C - \frac{m_A}{M}X_R, \quad Y_B = Y_C - \frac{m_A}{M}Y_R, \quad Z_B = Z_C - \frac{m_A}{M}Z_R. \quad \textbf{(12.86)}$$

Taking the derivatives of both sides of Eqs. 12.85 and 12.86 with respect to time gives us the Cartesian velocities of atoms A and B in terms of the Cartesian velocities of the center of mass and those for the relative motion. These results are

$$v_{XA} = v_{XC} + \frac{m_B}{M}v_{XR}, \quad v_{YA} = v_{YC} + \frac{m_B}{M}v_{YR}, \quad v_{ZA} = v_{ZC} + \frac{m_B}{M}v_{ZR}, \quad \textbf{(12.87)}$$

and

$$v_{XB} = v_{XC} - \frac{m_A}{M}v_{XR}, \quad v_{YB} = v_{YC} - \frac{m_A}{M}v_{YR}, \quad v_{ZB} = v_{ZC} - \frac{m_A}{M}v_{ZR}. \quad \textbf{(12.88)}$$

The final step is to square each of the Cartesian velocities for atoms A and B and substitute the results into Eq. 12.82 to obtain the kinetic energy in terms of the velocities of the center of mass and relative motion. The result (see Problem 12.28) is

$$T = \frac{M}{2}[v_{XC}^2 + v_{YC}^2 + v_{ZC}^2] + \frac{\mu_{AB}}{2}[v_{XR}^2 + v_{YR}^2 + v_{ZR}^2], \quad \textbf{(12.89)}$$

where

$$\mu_{AB} = \frac{m_A m_B}{M}. \quad \textbf{(12.90)}$$

μ_{AB} is called the A–B *reduced mass*. In the form of Eq. 12.89, it is simple to identify the terms associated with the center-of-mass translational energy and those representing the contributions from molecular rotation and vibration. The first three terms, each of which is multiplied by the mass of the total system, make up the translational energy of the center of mass. Accordingly, the solutions of the corresponding quantum mechanical problem for this part of the energy are given by Eqs. 12.48 and 12.49, provided that the diatomic molecule is bound within an infinite rectangular parallelepiped potential well. The three quantum numbers describing the translational energy along the X_C, Y_C, and Z_C coordinates are n_x, n_y, and n_z in Eq. 12.49. The internal molecular rotational and vibrational energies are represented by the terms multiplied by the reduced mass. As we shall see, two of the remaining three quantum numbers describe the rotational eigenstate, while the third quantum number describes the vibrational quantum state.

The preceding result may be generalized to apply to any system, regardless of the number of atoms present. If we have an N-atom system, $3N$ coordinates will be required to stipulate the atomic positions. If three of these are taken to be the Cartesian coordinates of the system's center of mass and the remaining $3N - 3$ coordinates are used to express the Cartesian components of $N - 1$ independent vectors giving the relative positions of the atoms, the kinetic energy will always have the form

$$T = \frac{M}{2}[v_{XC}^2 + v_{YC}^2 + v_{ZC}^2] + \sum_{i=1}^{N-1} \frac{\mu_i}{2}[v_{Xi}^2 + v_{Yi}^2 + v_{Zi}^2], \qquad (12.91)$$

where μ_i is the reduced mass for the two "particles" connected by vector i and v_{xi}, v_{yi}, and v_{zi} are the Cartesian velocity components for that vector. Example 12.13 illustrates this concept.

EXAMPLE 12.13

Consider a three-atom system requiring nine coordinates to stipulate the atomic positions. Suppose we take three of these coordinates to be Cartesian coordinates of the center of mass, a second set of three to be the X, Y, and Z components of the vector connecting atom B to atom A, and the last three to be the X, Y, and Z components of the vector connecting atom C to the A–B center of mass. Express the kinetic energy of the system in terms of the derivatives of these coordinates with respect to time.

Solution

This center-of-mass, relative coordinate system is illustrated in Figure 12.16. The two relative position vectors and the center-of-mass coordinates are independent, in that a change in the **AB** vector can be made without affecting the **C–AB** vector or the coordinates of the center of mass. Therefore, the kinetic energy must have the form

$$T = \frac{M}{2}[v_{XC}^2 + v_{YC}^2 + v_{ZC}^2] + \frac{\mu_{AB}}{2}[v_{X1}^2 + v_{Y1}^2 + v_{Z1}^2] + \frac{\mu_{AB-C}}{2}[v_{X2}^2 + v_{Y2}^2 + v_{Z2}^2], \quad (1)$$

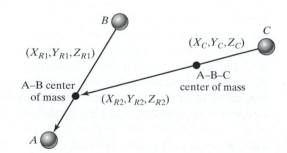

▶ FIGURE 12.16
Center-of-mass, relative coordinate system for a three-atom system.

where

$$M = m_A + m_B + m_C, \tag{2}$$

$$\mu_{AB} = \frac{m_A m_B}{m_A + m_B}, \tag{3}$$

and

$$\mu_{AB-C} = \frac{(m_A + m_B)m_C}{M}. \tag{4}$$

Notice that the reduced-mass terms are always of the form $m_i m_j/(m_i + m_j)$, where m_i and m_j are the masses of the two points connected by the relative position vector.

If we had chosen to represent the system using the Cartesian coordinates of the center of mass and the X, Y, and Z components of the vectors connecting atom B to atom A and atom B to atom C, as shown in Figure 12.17, the kinetic energy would not have the form given in Eq. 12.91, because these coordinates are not all independent. For example, there is no way to alter the **BC** vector without changing either the center-of-mass coordinates, or the **AB** vector, or both. Therefore, the kinetic energy will contain cross terms of the form $v_{X1} v_{X2}$, $v_{Y1} v_{Y2}$, and $v_{Z1} v_{Z2}$, rather than having the quadratic form given in Eq. 12.91.

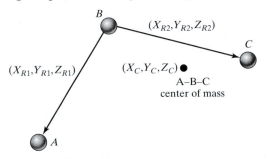

▲ FIGURE 12.17

A possible set of coordinates for a three-body system that are not mutually independent. Therefore, the kinetic-energy expression will contain cross terms of the Cartesian velocity components.

For related exercises, see Problems 12.29 and 12.30.

12.2.2 The Two-Atom Rigid Rotor

The simplest molecular system that contains rotational energy is a diatomic molecule. The classical Hamiltonian for this system has the form

$$\mathbf{H} = T + V = T_C + T_R + V(q_R). \tag{12.92}$$

In Eq. 12.92, T_C and T_R are the kinetic energies associated with motion of the center of mass and relative motion, respectively. $V(q_R)$ is the intramolecular potential energy, which depends only upon the relative coordinates q_R. If this were not true, molecular properties such as the bond energy, equilibrium distance, polarizability, dipole moment, etc., would have one set of values when the diatomic molecule is in Chicago, another set when it is London, and still another when it visits Kathmandu. Since we have already explored the quantum mechanical properties associated with translational motion of the center of mass, only the quantum solutions for the relative motion are of concern to us. The classical Hamiltonian for the relative motion is

$$H_R = T_R + V(q_R) = \frac{\mu_{AB}}{2}[v_{XR}^2 + v_{YR}^2 + v_{ZR}^2] + V(X_R, Y_R, Z_R), \tag{12.93}$$

where the relative coordinates are those defined in Figure 12.15. The corresponding quantum mechanical Hamiltonian is

$$\mathcal{H} = -\frac{\hbar^2}{2\mu_{AB}}\nabla^2 + V(X_R, Y_R, Z_R). \tag{12.94}$$

Since we wish to examine molecular rotation, the arguments presented in Chapters 9 and 11 suggest that we should use a spherical polar coordinate system centered at the center of mass of the diatomic system. (See Figure 9-3.) In this coordinate system, the ∇^2 operator is given by Eq. 9.35:

$$\nabla^2 = \frac{1}{r^2}\frac{\partial}{\partial r}\left[r^2\frac{\partial}{\partial r}\right] + \frac{1}{r^2 \sin\theta}\frac{\partial}{\partial \theta}\left[\sin\theta\frac{\partial}{\partial \theta}\right] + \frac{1}{r^2 \sin^2\theta}\frac{\partial^2}{\partial\phi^2}. \tag{12.95}$$

Combination of Eqs. 12.94 and 12.95 gives the stationary-state Schrödinger equation for the relative motion:

$$\mathcal{H}\psi(r, \theta, \phi) = \left\{-\frac{\hbar^2}{2\mu_{AB}}\left(\frac{1}{r^2}\frac{\partial}{\partial r}\left[r^2\frac{\partial}{\partial r}\right] + \frac{1}{r^2 \sin\theta}\frac{\partial}{\partial \theta}\left[\sin\theta\frac{\partial}{\partial \theta}\right]\right.\right.$$
$$\left.\left. + \frac{1}{r^2 \sin^2\theta}\frac{\partial^2}{\partial\phi^2}\right) + V(r, \theta, \phi)\right\}\psi(r, \theta, \phi) = E\psi(r, \theta, \phi). \tag{12.96}$$

The energy E appearing in Eq. 12.96 is the combined rotational and vibrational energy of the diatomic system.

Later in the chapter, we will examine the solution of this equation for a physically realistic intramolecular potential. However, we will first consider rotation and vibration separately. Although not rigorously correct, the solutions of the separated equations are good first approximations, and they are much easier to obtain than the solution of Eq. 12.96 itself. In addition, they will introduce the reader to some of the most important concepts in quantum chemistry and provide the basis for the treatment of atomic structure in the next chapter.

Let us assume that the interatomic distance R between the two atoms in Figure 12.15 is fixed and that the rigid diatomic system is rotating in field-free space, so that $V(r, \theta, \phi) = 0$ for all values of θ and ϕ. Such a system is called a *rigid rotor*. Under these conditions, Eq. 12.96 becomes

$$\mathcal{H}\psi(\theta, \phi) = -\frac{\hbar^2}{2\mu_{AB}R^2}\left[\frac{1}{\sin\theta}\frac{\partial}{\partial \theta}\left(\sin\theta\frac{\partial}{\partial \theta}\right) + \frac{1}{\sin^2\theta}\frac{\partial^2}{\partial\phi^2}\right]\psi(\theta, \phi) = E_R\psi(\theta, \phi). \tag{12.97}$$

Since r is fixed at the value R, the eigenfunctions, $\psi(\theta, \phi)$, depend only upon θ and ϕ. As a result, the derivatives of $\psi(\theta, \phi)$ with respect to r vanish, and the energy E_R is now pure rotational energy.

The quantity $\mu_{AB}R^2$ appearing in Eq. 12.97 is called the *moment of inertia* of the diatomic system. In general, the moment of inertia, I_k, for rotation of an N-atom rigid body about its principal axis k is defined to be

$$\boxed{I_k = \sum_{i=1}^{N} m_i r_{ik}^2}. \tag{12.98}$$

In Eq. 12.98, the summation runs over all N atoms in the molecule; m_i is the mass of atom i, and r_{ik} is the perpendicular distance from atom i to the kth principal axis of rotation. We shall defer further discussion of the definition and determination of the principal axes of rotation to Section 12.2.5. Suffice it here to say that, for a diatomic molecule, the principal axes correspond to the bond axis and any two mutually perpendicular axes through the center of mass and perpendicular to the bond axis. Using this fact, Example 12.14 shows that, for a diatomic rigid rotor, $I_k = \mu_{AB}R^2$.

EXAMPLE 12.14

Show that, for a diatomic rigid rotor, the moment of inertia for rotation about any axis through the center of mass and perpendicular to the bond axis is given by $\mu_{AB}R^2$.

Solution

Figure 12.18 defines the variables of the problem. By definition, the moment of inertia about the principal axis shown in the figure is

$$I = m_A r_A^2 + m_B r_B^2. \tag{1}$$

Since the principal axis passes through the center of mass, we must have

$$m_A r_A = m_B r_B. \tag{2}$$

Solving Eq. 2 for r_A in terms of the masses and r_B and then substituting into Eq. 1 produces

$$I = m_A \left[\frac{m_B r_B}{m_A}\right]^2 + m_B r_B^2 = \left[\frac{m_B^2}{m_A} + m_B\right] r_B^2. \tag{3}$$

Examining the diagram, we see that we also have

$$R = r_A + r_B, \tag{4}$$

so that

$$r_B = R - r_A = R - \frac{m_B r_B}{m_A}. \tag{5}$$

We may now use Eq. 5 to obtain r_B in terms of R:

$$r_B \left[1 + \frac{m_B}{m_A}\right] = r_B \left[\frac{M}{m_A}\right] = R. \tag{6}$$

Therefore,

$$r_B = \frac{m_A R}{M}. \tag{7}$$

Substitution of Eq. 7 into Eq. 3 produces

$$I = \left[\frac{m_B^2}{m_A} + m_B\right]\left[\frac{m_A}{M}\right]^2 R^2 = \left[\frac{m_B^2 m_A + m_B m_A^2}{M^2}\right]R^2 = \frac{m_A m_B}{M}\left[\frac{m_A + m_B}{M}\right]R^2$$

$$= \frac{m_A m_B}{M}R^2 = \mu_{AB}R^2. \tag{8}$$

For a related exercise, see Problem 12.31.

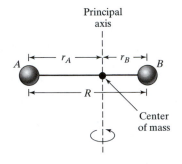

Principal axis

A $\xleftarrow{r_A}\xrightarrow{}\xleftarrow{r_B}\xrightarrow{}$ B

$\xleftarrow{\quad R \quad}$

Center of mass

▲ FIGURE 12.18
Variables related to the determination of the moment of inertia for a diatomic rigid rotor.

To obtain the eigenfunctions for the rigid rotor, we follow our usual procedure of assuming that the solution may be written in the separable form

$$\psi(\theta, \phi) = \Theta(\theta)\, \Phi(\phi). \tag{12.99}$$

To verify that a solution of this form does indeed exist, we need to demonstrate that substituting Eq. 12.99 into Eq. 12.97 and then dividing by $\Theta(\theta)\Phi(\phi)$ results in a separation of variables in the differential equation. Before substituting Eq. 12.99, we clear the $\sin\theta$ terms from the denominator by multiplying both sides by $-2I\sin^2\theta/\hbar^2$, where $I = \mu_{AB}R^2$. This operation yields

$$\left[\sin\theta\frac{\partial}{\partial\theta}\left[\sin\theta\frac{\partial}{\partial\theta}\right] + \frac{\partial^2}{\partial\phi^2}\right]\psi(\theta,\phi) + \frac{2IE_R\sin^2\theta}{\hbar^2}\psi(\theta,\phi) = 0. \quad \textbf{(12.100)}$$

Direct substitution of Eq. 12.99 into 12.100 produces

$$\Phi(\phi)\left\{\sin\theta\frac{\partial}{\partial\theta}\left[\sin\theta\frac{\partial}{\partial\theta}\right]\right\}\Theta(\theta) + \frac{2IE_R\sin^2\theta}{\hbar^2}\Theta(\theta)\Phi(\phi) + \Theta(\theta)\frac{\partial^2}{\partial\phi^2}\Phi(\phi) = 0.$$

Dividing this result by $\Theta(\theta)\Phi(\phi)$ puts the equation into the form

$$\frac{\left\{\sin\theta\frac{\partial}{\partial\theta}\left[\sin\theta\frac{\partial}{\partial\theta}\right]\right\}\Theta(\theta)}{\Theta(\theta)} + \frac{2IE_R\sin^2\theta}{\hbar^2} + \frac{\frac{\partial^2}{\partial\phi^2}\Phi(\phi)}{\Phi(\phi)} = 0. \quad \textbf{(12.101)}$$

Inspection of Eq. 12.101 shows that the variables are separated: The first two terms depend only upon θ, while the third term is a function solely of ϕ. Consequently, we now know that a separable solution for the eigenfunctions of the rigid rotor exists.

The only way Eq. 12.101 can hold for all values of θ and ϕ is for the terms that depend only upon θ to be equal to a constant, say, M^2, while those dependent only upon ϕ are equal to the negative of the same constant. For this reason, the single equation separates into two equations, each dependent only upon one of the two variables. The equation involving θ is

$$\frac{\left\{\sin\theta\frac{\partial}{\partial\theta}\left[\sin\theta\frac{\partial}{\partial\theta}\right]\right\}\Theta(\theta)}{\Theta(\theta)} + \frac{2IE_R\sin^2\theta}{\hbar^2} = M^2,$$

while the one involving ϕ has the form

$$\frac{\frac{\partial^2}{\partial\phi^2}\Phi(\phi)}{\Phi(\phi)} = -M^2.$$

If we multiply the θ-equation by $\Theta(\theta)$ and the ϕ-equation by $\Phi(\phi)$, we obtain the forms seen in most textbooks, viz.,

$$\left\{\sin\theta\frac{\partial}{\partial\theta}\left[\sin\theta\frac{\partial}{\partial\theta}\right]\right\}\Theta(\theta) + \frac{2IE_R\sin^2\theta}{\hbar^2}\Theta(\theta) - M^2\Theta(\theta) = 0 \quad \textbf{(12.102)}$$

and

$$\frac{\partial^2}{\partial\phi^2}\Phi(\phi) + M^2\Phi(\phi) = 0. \quad \textbf{(12.103)}$$

The ϕ-equation has the same form we have previously seen in obtaining the Debye–Hückel expression for the mean ionic activity coefficient and for the particle in an infinite potential well. Since M^2 is positive, the solutions for $\Phi(\phi)$ have a sine–cosine, or complex exponential, form. That is,

$$\Phi(\phi) = Ce^{iM\phi}, \quad \textbf{(12.104)}$$

where C is a constant that can be used to normalize the eigenfunction. The normalization condition is

$$\langle \Phi(\phi) | \Phi(\phi) \rangle = \int_0^{2\pi} C^* e^{-iM\phi} C e^{iM\phi} \, d\phi = |C|^2 \int_0^{2\pi} d\phi = 2\pi |C|^2 = 1.$$

This requires that we have $C = (2\pi)^{-1/2}$, so that the eigenfunctions for the rigid rotor are

$$\boxed{\psi(\theta, \phi) = \Theta(\theta)(2\pi)^{-1/2} e^{iM\phi}}. \qquad \textbf{(12.105)}$$

All that remains is to find $\Theta(\theta)$ by solving Eq. 12.102.

Before proceeding to the problem of finding $\Theta(\theta)$, we may employ the first postulate to learn something very important about the values the constant M in Eqs. 12.102 through 12.105 may take. The first postulate requires that quantum mechanical eigenfunctions be single valued. Since the angles ϕ and $\phi + 2\pi$ are physically the same, the single-valued requirement means we must have $\Phi(\phi) = \Phi(\phi + 2\pi)$; otherwise $\Phi(\phi)$ will have two different values at the same physical angle. Figure 11.18 shows an example of such a situation. Substituting Eq. 12.104 for $\Phi(\phi)$ gives the requirement that

$$(2\pi)^{-1/2} e^{iM\phi} = (2\pi)^{-1/2} e^{iM(\phi + 2\pi)} = (2\pi)^{-1/2} e^{iM\phi} e^{iM2\pi}.$$

Division of both sides by $(2\pi)^{-1/2} e^{iM\phi}$ shows that M must be such that

$$e^{iM2\pi} = \cos(2M\pi) + i \sin(2M\pi) = 1.$$

This equation will be satisfied only if M is an integer, either positive, negative, or zero. Therefore, M is a discrete variable. For reasons that will become clear when we discuss magnetic spectroscopy, M is called the *magnetic quantum number*.

The equation involving θ is somewhat difficult to solve, so we will simply give the solutions for $\Theta(\theta)$ and E_R without proof and then demonstrate how these solutions may be verified. The 16 eigenfunctions with the lowest eigenvalues for E_R are given in Table 12.1. The rigid-rotor eigenfunctions are the product of $\Theta(\theta)$ and $\Phi(\phi)$. The form of these eigenfunctions was well known to mathematicians long before the advent of quantum mechanics. They called the product $\Theta(\theta)\Phi(\phi)$ a *spherical harmonic* and used the notation $Y(\theta, \phi)$ to designate it. Written in terms of this nomenclature, the eigenfunctions are

$$\psi(\theta, \phi) = \Theta_J^M(\theta) \Phi_M(\phi) = Y_J^M(\theta, \phi). \qquad \textbf{(12.106)}$$

The rotational energies that a rigid rotor can have are listed in the third column of the table. An examination of these values shows that they may be represented by the expression

$$\boxed{E_R = E_J = \frac{J(J + 1)\hbar^2}{2I}} \qquad \text{for } J = 0, 1, 2, 3, \ldots. \qquad \textbf{(12.107)}$$

When the θ-equation is solved exactly, Eq. 12.107 can be explicitly derived without recourse to the preceding fitting argument. The discrete variable J is called the *azimuthal quantum number*, because of its relationship to the azimuthal angle θ. We now see that the solution of the rigid-rotor equation, which involves two variables, θ and ϕ, results in two quantum numbers, as expected: J and M. The solutions of the θ-equation depend upon both of these quantum numbers, while $\Phi(\phi)$ is dependent only upon M. It is customary to indicate this dependence when writing the notation for the eigenfunctions. The method commonly used to do this is indicated in Eq. 12.106 and in Table 12.1.

Before Schrödinger and Heisenberg formulated quantum theory, spectroscopists had observed atomic spectral lines associated with the $J = 0, 1, 2,$ and 3 rotational eigenfunctions for the hydrogen atom. They labeled these lines "sharp," "principal," "diffuse," and "fundamental." Because of this historical precedent, these eigenstates are now frequently called s, p, d, and f states, respectively, where the notation derives from the first letter of the designation of the corresponding spectral lines. When we turn our attention to atomic structure and the hydrogen atom in Chapter 13, we will see that the two quantum particles involved in the rotation are the proton nucleus and the electron. For those particles, the rotational eigenfunctions are called

Table 12.1 Eigenfunctions and eigenvalues for the rigid rotor

$\Theta_J^M(\theta)$	$\Phi_M(\phi)$	E_R	J	M	Designation
$\left[\dfrac{1}{2}\right]^{1/2}$	$\left[\dfrac{1}{2\pi}\right]^{1/2}$	0	0	0	s
$\left[\dfrac{3}{2}\right]^{1/2}\cos\theta$	$\left[\dfrac{1}{2\pi}\right]^{1/2}$	$\dfrac{2\hbar^2}{2I}$	1	0	p
$\left[\dfrac{3}{4}\right]^{1/2}\sin\theta$	$\left[\dfrac{1}{2\pi}\right]^{1/2}e^{i\phi}$	$\dfrac{2\hbar^2}{2I}$	1	1	p
$\left[\dfrac{3}{4}\right]^{1/2}\sin\theta$	$\left[\dfrac{1}{2\pi}\right]^{1/2}e^{-i\phi}$	$\dfrac{2\hbar^2}{2I}$	1	-1	p
$\left[\dfrac{5}{8}\right]^{1/2}[3\cos^2\theta-1]$	$\left[\dfrac{1}{2\pi}\right]^{1/2}$	$\dfrac{6\hbar^2}{2I}$	2	0	d
$\left[\dfrac{15}{4}\right]^{1/2}\cos\theta\sin\theta$	$\left[\dfrac{1}{2\pi}\right]^{1/2}e^{i\phi}$	$\dfrac{6\hbar^2}{2I}$	2	1	d
$\left[\dfrac{15}{4}\right]^{1/2}\cos\theta\sin\theta$	$\left[\dfrac{1}{2\pi}\right]^{1/2}e^{-i\phi}$	$\dfrac{6\hbar^2}{2I}$	2	-1	d
$\left[\dfrac{15}{16}\right]^{1/2}\sin^2\theta$	$\left[\dfrac{1}{2\pi}\right]^{1/2}e^{2i\phi}$	$\dfrac{6\hbar^2}{2I}$	2	2	d
$\left[\dfrac{15}{16}\right]^{1/2}\sin^2\theta$	$\left[\dfrac{1}{2\pi}\right]^{1/2}e^{-2i\phi}$	$\dfrac{6\hbar^2}{2I}$	2	-2	d
$\left[\dfrac{7}{8}\right]^{1/2}[5\cos^3\theta-3\cos\theta]$	$\left[\dfrac{1}{2\pi}\right]^{1/2}$	$\dfrac{12\hbar^2}{2I}$	3	0	f
$\left[\dfrac{21}{32}\right]^{1/2}[5\cos^2\theta-1]\sin\theta$	$\left[\dfrac{1}{2\pi}\right]^{1/2}e^{i\phi}$	$\dfrac{12\hbar^2}{2I}$	3	1	f
$\left[\dfrac{21}{32}\right]^{1/2}[5\cos^2\theta-1]\sin\theta$	$\left[\dfrac{1}{2\pi}\right]^{1/2}e^{-i\phi}$	$\dfrac{12\hbar^2}{2I}$	3	-1	f
$\left[\dfrac{105}{16}\right]^{1/2}\sin^2\theta\cos\theta$	$\left[\dfrac{1}{2\pi}\right]^{1/2}e^{2i\phi}$	$\dfrac{12\hbar^2}{2I}$	3	2	f
$\left[\dfrac{105}{16}\right]^{1/2}\sin^2\theta\cos\theta$	$\left[\dfrac{1}{2\pi}\right]^{1/2}e^{-2i\phi}$	$\dfrac{12\hbar^2}{2I}$	3	-2	f
$\left[\dfrac{35}{32}\right]^{1/2}\sin^3\theta$	$\left[\dfrac{1}{2\pi}\right]^{1/2}e^{3i\phi}$	$\dfrac{12\hbar^2}{2I}$	3	3	f
$\left[\dfrac{35}{32}\right]^{1/2}\sin^3\theta$	$\left[\dfrac{1}{2\pi}\right]^{1/2}e^{-3i\phi}$	$\dfrac{12\hbar^2}{2I}$	3	-3	f

orbitals, and the quantum number designations are changed from J and M to l and m, respectively. The student reader will probably remember frequent occasions in introductory and organic chemistry when s, p, d, and f orbitals were discussed. The quantum mechanical solutions to the rigid rotor demonstrate the origin of these orbitals or eigenfunctions. The discussion that follows explores their characteristics.

Although we have not solved Eq. 12.102, it is a straightforward procedure to verify that the eigenfunctions listed in Table 12.1 are indeed solutions of that equation, provided that E_R and M have the values listed in the table. Example 12.15 illustrates the procedure for the $\Theta_1^0(\theta)$ eigenfunction.

EXAMPLE 12.15

Show that the function $\Theta(\theta) = [3/2]^{1/2} \cos\theta$ is a solution of Eq. 12.102, provided that $M = 0$ and $E_R = 2\hbar^2/(2I)$.

Solution

Equation 12.102 is

$$\left\{ \sin\theta \frac{\partial}{\partial\theta} \left[\sin\theta \frac{\partial}{\partial\theta} \right] \right\} \Theta(\theta) + \frac{2IE_R \sin^2\theta}{\hbar^2} \Theta(\theta) - M^2 \Theta(\theta) = 0. \tag{1}$$

Substitution of $\Theta(\theta) = [3/2]^{1/2} \cos\theta$ into Eq. 1 gives

$$\left\{ \sin\theta \frac{\partial}{\partial\theta} \left[\sin\theta \frac{\partial}{\partial\theta} \right] \right\} \cos\theta + \frac{2IE_R \sin^2\theta}{\hbar^2} \cos\theta - M^2 \cos\theta$$

$$= -\sin\theta \frac{\partial}{\partial\theta} \sin^2\theta + \frac{2IE_R \sin^2\theta}{\hbar^2} \cos\theta - M^2 \cos\theta$$

$$= -2\sin^2\theta \cos\theta + \frac{2IE_R \sin^2\theta}{\hbar^2} \cos\theta - M^2 \cos\theta = 0, \tag{2}$$

after dividing out the normalization constant $[3/2]^{1/2}$. Collecting terms in $\sin^2\theta \cos\theta$ and $\cos\theta$, we obtain

$$\sin^2\theta \cos\theta \left[-2 + \frac{2IE_R}{\hbar^2} \right] - M^2 \cos\theta = 0. \tag{3}$$

For the function $\Theta(\theta) = [3/2]^{1/2} \cos\theta$ to be a solution, the left-hand side of Eq. 3 must be zero for all values of the variable θ. The only way this can be achieved is for the coefficients of $\sin^2\theta \cos\theta$ and $\cos\theta$ to be zero. Therefore, $\Theta(\theta) = [3/2]^{1/2} \cos\theta$ will be a solution if and only if

$$M^2 = 0, \tag{4}$$

which requires that the magnetic quantum number M be zero, and

$$-2 + \frac{2IE_R}{\hbar^2} = 0. \tag{5}$$

Equation (5) requires the rotational energy eigenvalue associated with the solution $\Theta(\theta) = [3/2]^{1/2} \cos\theta$ to be

$$E_R = \frac{2\hbar^2}{2I}. \tag{6}$$

Thus, the solution $\Theta(\theta) = [3/2]^{1/2} \cos\theta$ with $M = 0$ and $E_R = 2\hbar^2/(2I)$ has been verified.

For related exercises, see Problems 12.32 and 12.33.

Now that we have the rigid-rotor eigenfunctions, we may compute the average values of any function that depends upon the angles θ and ϕ, using the fourth postulate. In general, the average value of a function $F(\theta,\phi)$ is given by

$$\langle F(\theta,\phi) \rangle = \int_{\text{all }\phi} \int_{\text{all }\theta} [Y_J^M(\theta,\phi)]^* F(\theta,\phi) Y_J^M(\theta,\phi) d\tau,$$

where $d\tau$ is the spherical polar volume element for the angles θ and ϕ. To compute $\langle F(\theta, \phi)\rangle$, we need to know the correct form for $d\tau$ and the limits on the integrals required to sum over all of angular space exactly one time.

The form of the spherical polar volume element can be deduced by examining Figure 12.19. Let the coordinates of the point represented by the black dot in the figure be (r, θ, ϕ). If we now displace r by an amount dr and also increment θ by $d\theta$, these displacements will trace out the shaded area in the figure. If the displacements are infinitesimal, the shaded area will have the form of a rectangle with side lengths dr and $r\,d\theta$ since the arc length is equal to the angle times the length of the radial vector. The product of the lengths of the sides, $r\,d\theta\,dr$, gives the area traced out by the coordinate displacements. If we now rotate the shaded area about the Z-axis by an amount $d\phi$, a volume $d\tau$ in the shape of a rectangular parallelepiped will be formed. The figure shows that the length of the projection of r into the X–Y plane is r $\sin\theta$. Therefore, the length of the third side of the parallelepiped is $r \sin\theta\,d\phi$. The magnitude of the volume element is given by

$$d\tau = (\text{area})(\text{third side length}) = (r\,d\theta\,dr)(r\sin\theta\,d\phi) = r^2 dr \sin\theta\,d\theta\,d\phi.$$

(12.108)

The limits on the angular variables required to trace all of angular space exactly once are

$$0 \leq \theta \leq \pi$$

and

$$0 \leq \phi \leq 2\pi.$$

(12.109)

▶ FIGURE 12.19
Spherical polar volume element.

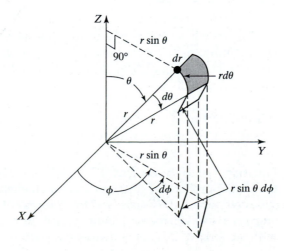

If we start with the radial vector directed along the Z-axis and rotate it through an angle θ of π radians, the tip of the vector will trace out a semicircle in the Z–Y plane. If this surface is now rotated about the Z-axis through an angle ϕ of 2π radians, the semicircle will trace out the surface of a sphere exactly one time. Note that if we allowed the radial vector to rotate through an angle θ of 2π radians, so that a circle is formed, rotation of this circle through an angle ϕ of 2π radians would sweep angular space *twice*. These considerations show that the limits given in Eq. 12.109 properly sum over all angular space exactly once.

EXAMPLE 12.16

Show that the normalization constant for the $\Theta_2^0(\theta)$ eigenfunction is $[5/8]^{1/2}$, as given in Table 12.1.

Solution

The normalization requirement is

$$\langle \Theta_2^0(\theta) | \Theta_2^0(\theta) \rangle = 1. \tag{1}$$

The table shows that $\Theta_2^0(\theta)$ is given by $N[3\cos^2\theta - 1]$. Substitution into Eq. 1 gives

$$\int_0^\pi N^*[3\cos^2\theta - 1]^* N[3\cos^2\theta - 1]\sin\theta\,d\theta = 1, \tag{2}$$

since the θ part of the spherical polar volume element is $\sin\theta\,d\theta$ and the limits on θ are 0 to π. Expanding Eq. 2, we obtain

$$N^2\int_0^\pi [9\cos^4\theta - 6\cos^2\theta + 1]\sin\theta\,d\theta = 1. \tag{3}$$

Integration produces

$$N^2\left[-\frac{9\cos^5\theta}{5} + \frac{6\cos^3\theta}{3} - \cos\theta\right]_0^\pi = N^2\left[\frac{9}{5} - 2 + 1 - \left\{-\frac{9}{5} + 2 - 1\right\}\right]$$

$$= \frac{8N^2}{5} = 1. \tag{4}$$

Solving for N, we obtain $N = [5/8]^{1/2}$, as in Table 12.1.
For related exercises, see Problems 12.32 and 12.34.

The fourth postulate tells us that the spatial probability distribution for our diatomic rigid rotor is given by $[Y_J^M(\theta,\phi)]^*[Y_J^M(\theta,\phi)]d\tau$, which is the same as $[\Theta_J^M(\theta)\Phi_M(\phi)]^*[\Theta_J^M(\theta)\Phi_M(\phi)]d\tau$. Substituting Eq. 12.104 for $\Phi_M(\phi)$ gives

$$[\Theta_J^M(\theta)\Phi_M(\phi)]^*[\Theta_J^M(\theta)\Phi_M(\phi)] = \frac{1}{2\pi}e^{-iM\phi}e^{iM\phi}|\Theta_J^M(\theta)|^2 = \frac{1}{2\pi}|\Theta_J^M(\theta)|^2. \tag{12.110}$$

Equation 12.110 shows that $|Y_J^M(\theta,\phi)|^2$ is independent of ϕ. Consequently, the spatial probability distributions for all eigenfunctions exhibit cylindrical symmetry about the Z-axis, and the rotor is equally likely to be found at any angle ϕ. However, the probability distribution is strongly dependent upon θ, with the form of this dependence being determined by $\Theta_J^M(\theta)$.

We can confidently predict some of the characteristics of the eigenfunctions without any examination of their specific mathematical forms. Equation 12.107 and Table 12.1 show us that the eigenfunction for the ground state of the rotor is the $Y_0^0(\theta, \phi)$ spherical harmonic. The first excited state is three fold degenerate with eigenfunctions $Y_1^0(\theta, \phi)$ and $Y_1^{\pm 1}(\theta, \phi)$. The eigenfunctions $Y_2^M(\theta, \phi)$ and $Y_3^M(\theta, \phi)$ correspond to the five- and sevenfold-degenerate second and third excited states, respectively. In general, we expect the ground state to be nodeless, the first excited state to have one node, the second excited state to have two nodes, and so on. It is, therefore, clear that the number of nodes possessed by the different eigenfunctions is equal to the azimuthal quantum number J. This behavior is similar to that seen for a particle in an infinite one-dimensional potential well, wherein we found the number of nodes to be equal to one less than the translational quantum number $n_x - 1$. The difference between the two cases resides in the fact that the lowest possible value of the translational quantum number is unity, whereas the lowest value of J is zero. Since no value of the angle ϕ makes $\Phi_M(\phi) = (2\pi)^{-1/2} e^{iM\phi}$ equal to zero, we know that all the angular nodes are in the θ coordinate and the $\Theta_J^M(\theta)$ function.

The angular dependence of the $\Theta_J^M(\theta)$ functions is most conveniently viewed with the use of polar plots. In this type of representation, the absolute value of $\Theta_J^M(\theta)$ is plotted as the length of the radial vector that makes an angle θ with the Z-axis. The sign of $\Theta_J^M(\theta)$ is generally indicated by either shading the positive regions or simply writing the sign of the wave function in different angular regions. Figures 12.20 through 22 show polar plots of $\Theta_J^M(\theta)$ for $J = 0$, 1, 2, and 3. The eigenfunctions with $J = 0$ and $J = 1$ are shown in Figure 12.20. The eigenfunction for the ground state is $Y_0^0(\theta, \phi) = (2\pi)^{-1/2} \Theta_0^0(\theta)$, where $\Theta_0^0(\theta) = [1/2]^{1/2}$. Since $\Theta_0^0(\theta)$ is a constant, its polar plot is a circle whose radius is $[1/2]^{1/2}$. The ground state s eigenfunction is, therefore, spherically symmetric. As expected, the ground state is nodeless.

The $J = 1$ p states all exhibit one angular nodal axis or plane, as the preceding analysis indicates they should. These nodes are evident in Figure 12.20. When $M = 0$, the nodal plane is perpendicular to the Z-axis. For the $M = \pm 1$ eigenstates, the Z-axis is a nodal axis. In each case, one angular portion of the eigenfunction is negative while the other portion is positive. When we discuss chemical bonding, we shall find that this difference in sign in different regions of space plays an important role in the quantum representation of the chemical bond.

The $J = 2$, $M = 0, \pm 1, \pm 2$ d states all have two angular nodes. Figure 12.21 shows polar plots of the $\Theta_2^0(\theta)$ and $\Theta_2^{\pm 1}(\theta)$ eigenfunctions. However, $\Theta_2^{\pm 2}(\theta) = [15/16]^{1/2} \sin^2 \theta$ apparently has only one nodal axis. This is a specific example of the degenerate nodes we discussed in Section 12.1, in which we examined the quantum mechanics for a particle in an infinite potential well. The nodal axes and nodal surfaces are found by equating the eigenfunction to zero and solving for the angle. For the $\Theta_2^{\pm 2}(\theta)$ eigenfunction, this requires that $[\sin \theta \sin \theta] = 0$. This equation has two solutions, which are obtained by setting each multiplicative factor to zero. That is, we have

$$\sin \theta = 0 \qquad \text{or} \qquad \sin \theta = 0.$$

In this case, both equations give the same solution, $\theta = 0$ or π radians, so that the only node lies along the Z-axis. The fact that both equations give the same result means that the Z-axis is a doubly degenerate nodal axis. Again, we see that the eigenfunctions often have different signs in different regions of angular space.

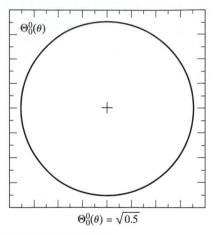

$$\Theta_0^0(\theta) = \sqrt{0.5}$$

▲ FIGURE 12.20A

The spherically symmetric $J = M = 0$, or s, function. This function is positive at all angles relative to the Z-axis and contains no nodes.

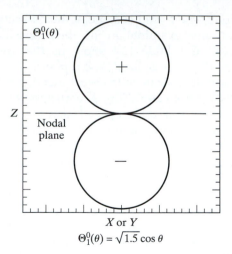

$$X \text{ or } Y$$
$$\Theta_1^0(\theta) = \sqrt{1.5}\cos\theta$$

▲ FIGURE 12.20B

The $\Theta_1^0(\theta)$ rotational p eigenfunction with $J = 1$ and $M = 0$. This function has its greatest magnitude along the Z-axis. It contains one node at $\theta = 90°$ or $270°$, as shown in the figure.

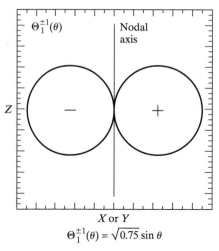

$$X \text{ or } Y$$
$$\Theta_1^{\pm1}(\theta) = \sqrt{0.75}\sin\theta$$

▲ FIGURE 12.20C

The $\Theta_1^{\pm1}(\theta)$ rotational p eigenfunction with $J = 1$ and $M = \pm1$. This function has its greatest magnitude perpendicular to the Z-axis. It contains one node at $\theta = 0°$ or $180°$, as shown in the figure.

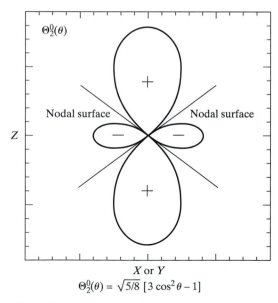

$$X \text{ or } Y$$
$$\Theta_2^0(\theta) = \sqrt{5/8}\,[3\cos^2\theta - 1]$$

▲ FIGURE 12.21A

The $\Theta_2^0(\theta)$ rotational d eigenfunction with $J = 2$, $M = 0$. This function has its greatest magnitude along the Z-axis. It contains two nodes at $\theta = 54.736°$ or $234.736°$ and at $\theta = 125.26°$ or $305.26°$, as shown in the figure.

The f functions with $J = 3$ each have three nodal axes or nodal surfaces. $\Theta_3^0(\theta)$ has three nodal surfaces; $\Theta_3^{\pm1}(\theta)$ exhibits two nodal surfaces and a nodal axis. The $\Theta_3^{\pm2}(\theta)$ and $\Theta_3^{\pm3}(\theta)$ eigenstates each have a degenerate nodal axis. (See Problem 12.35.) Polar plots of these functions are shown in Figure 12.22.

The probability distribution corresponding to a rotational s function exhibits the same spherical symmetry as that shown in Figure 12.20. The probability distributions, $|\Theta_J^M(\theta)|^2$, for the p, d, and f rotational eigenfunctions are shown in Figures 12.23 through 12.25, respectively.

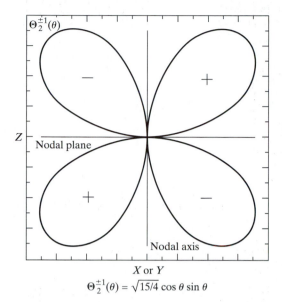

$$\Theta_2^{\pm 1}(\theta) = \sqrt{15/4} \cos \theta \sin \theta$$

▲ FIGURE 12.21B

The $\Theta_2^{\pm 1}(\theta)$ rotational d eigenfunction with $J = 2$ and $M = \pm 1$. This function has its greatest magnitude at 45° and 135° angles with the Z-axis. It contains two nodes at $\theta = 0°$ or 180° and at $\theta = 90°$ or 270°, as shown in the figure.

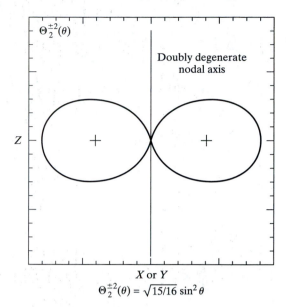

$$\Theta_2^{\pm 2}(\theta) = \sqrt{15/16} \sin^2 \theta$$

▲ FIGURE 12.21C

The $\Theta_2^{\pm 2}(\theta)$ rotational d eigenfunction with $J = 2$ and $M = \pm 2$. This function has its greatest magnitude in a direction perpendicular to the Z-axis. It contains a doubly degenerate node at $\theta = 0°$ or 180°, as shown in the figure.

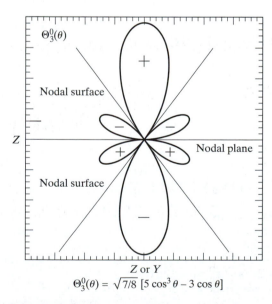

$$\Theta_3^0(\theta) = \sqrt{7/8} \, [5 \cos^3 \theta - 3 \cos \theta]$$

▲ FIGURE 12.22A

The $\Theta_3^0(\theta)$ rotational f eigenfunction with $J = 3$ and $M = 0$. This function has its greatest magnitude along the Z-axis. It contains three nodes at $\theta = 39.232°$ or 219.232°, at $\theta = 93°$ or 270°, and at $\theta = 140.768°$ or 320.769°, as shown in the figure.

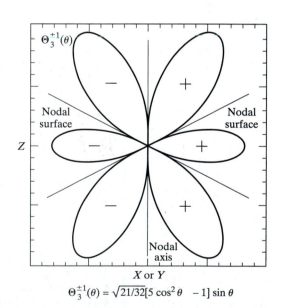

$$\Theta_3^{\pm 1}(\theta) = \sqrt{21/32}[5 \cos^2 \theta - 1] \sin \theta$$

▲ FIGURE 12.22B

The $\Theta_3^{\pm 1}(\theta)$ rotational f eigenfunction with $J = 3$ and $M = \pm 1$. This function has its greatest magnitude at angles of 31.717° or 211.717° and 148.283° or 328.283° relative to the Z-axis. It contains three nodes at $\theta = 0°$ or 180°, at $\theta = 63.435°$ or 243.435°, and at $\theta = 116.565°$ or 269.565°, as shown in the figure.

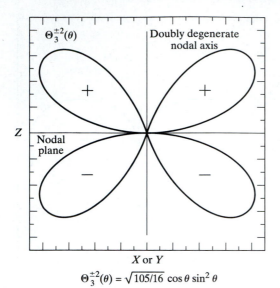

$$\Theta_3^{\pm 2}(\theta) = \sqrt{105/16}\ \cos\theta \sin^2\theta$$

▲ FIGURE 12.22C

The $\Theta_3^{\pm 2}(\theta)$ rotational f eigenfunction with $J = 3$ and $M = \pm 2$. This function has its greatest magnitude at angles of 70.529° or 250.529° and 109.471° or 289.471° relative to the Z-axis. It contains a doubly degenerate node at $\theta = 0°$ or 180° and a third node at $\theta = 90°$ or 270°, as shown in the figure.

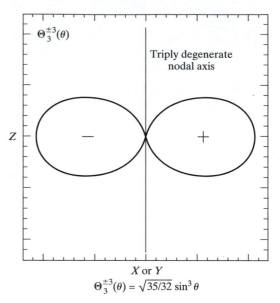

$$\Theta_3^{\pm 3}(\theta) = \sqrt{35/32}\ \sin^3\theta$$

▲ FIGURE 12.22D

The $\Theta_3^{\pm 3}(\theta)$ rotational f eigenfunction with $J = 3$ and $M = \pm 3$. This function has it greatest magnitude at angles of 90° or 270° relative to the Z-axis. It contains a triply degenerate node at $\theta = 0°$ or 180°, as shown in the figure.

EXAMPLE 12.17

At what angles θ does $\Theta_2^0(\theta)$ exhibit nodal surfaces?

Solution

From Table 12.1, we find that $\Theta_2^0(\theta) = [5/8]^{1/2}[3\cos^2\theta - 1]$. Consequently, we will observe a nodal surface or axis when we have

$$[3\cos^2\theta - 1] = 0. \tag{1}$$

Solving for $\cos\theta$, we obtain

$$\cos\theta = \pm\left[\frac{1}{3}\right]^{1/2}. \tag{2}$$

Therefore,

$$\theta = \cos^{-1}\left[\pm\left(\frac{1}{3}\right)^{1/2}\right]. \tag{3}$$

Equation 3 yields $\theta = 54.7356°$ or $234.7356°$, and $\theta = 125.2644°$ or $305.2644°$. Consequently, $\Theta_2^0(\theta)$ will be zero on a surface that forms an angle $\theta = 54.7356°$ or $234.7356°$ with the Z-axis and on a surface that forms an angle $\theta = 125.2644°$ or $\theta = 305.2644°$ with the Z-axis. The intersections of these two surfaces with either the Z–X or Z–Y plane are shown in Figure 12.21A.

For a related exercise, see Problem 12.37.

The plot shown in Figure 12.20A, along with the associated discussion, makes it clear that the ground-state s eigenfunction is spherically symmetric. A common mistake is to assume that this fact means that the rigid rotor is

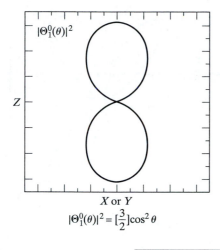

$$|\Theta_1^0(\theta)|^2 = [\tfrac{3}{2}]\cos^2\theta$$

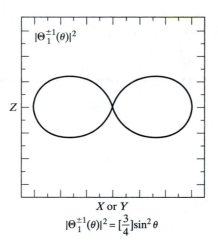

$$|\Theta_1^{\pm1}(\theta)|^2 = [\tfrac{3}{4}]\sin^2\theta$$

◀ FIGURE 12.23
Polar plots of the absolute squares of the rotational p functions.

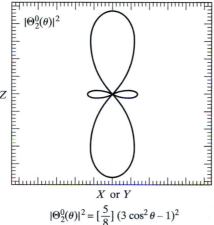

$$|\Theta_2^0(\theta)|^2 = [\tfrac{5}{8}](3\cos^2\theta - 1)^2$$

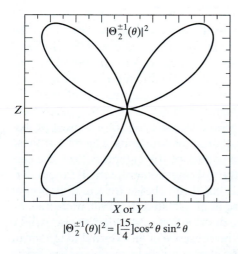

$$|\Theta_2^{\pm1}(\theta)|^2 = [\tfrac{15}{4}]\cos^2\theta\,\sin^2\theta$$

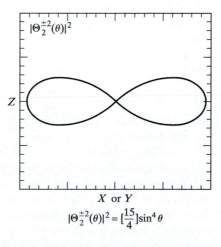

$$|\Theta_2^{\pm2}(\theta)|^2 = [\tfrac{15}{4}]\sin^4\theta$$

◀ FIGURE 12.24
Polar plots of the absolute squares of the rotational d functions

equally likely to be found at any pair of angles (θ, ϕ). However, this statement is true for ϕ, but not for θ. The reason is that the spherical polar volume element weights the θ angles by a $\sin\theta$ factor, but the ϕ weighting is uniform. Hence, the spatial probability distribution for the rigid-rotor ground state is

$$|Y_0^0(\theta, \phi)|^2 d\tau = \left[\frac{1}{4\pi}\right]\sin\theta\,d\theta\,d\phi.$$

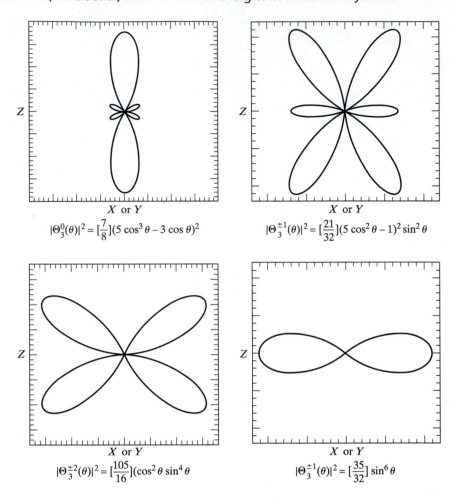

$$|\Theta_3^0(\theta)|^2 = [\tfrac{7}{8}](5\cos^3\theta - 3\cos\theta)^2$$

$$|\Theta_3^{\pm1}(\theta)|^2 = [\tfrac{21}{32}](5\cos^2\theta - 1)^2\sin^2\theta$$

$$|\Theta_3^{\pm2}(\theta)|^2 = [\tfrac{105}{16}](\cos^2\theta\,\sin^4\theta$$

$$|\Theta_3^{\pm1}(\theta)|^2 = [\tfrac{35}{32}]\sin^6\theta$$

▶ **FIGURE 12.25**
Polar plots of the absolute squares of the rotational *f* functions.

This distribution is independent of ϕ, but it varies with $\sin\theta$. Therefore, the most probable angle the rotor makes with the Z-axis is $\theta = \pi/2$. The least probable angle is $\theta = 0$ or π. That is, the rotor is much more likely to be perpendicular to the Z-axis than parallel to it. Physically, the reason for this is that there is only one possible way for the rotor to be aligned with the Z-axis, but an infinitude of orientations that permit the rotor to assume a perpendicular alignment. Consequently, two factors determine the rotor's most probable orientation and the average values of angular functions. The first of these is the weight provided by the absolute square of the eigenfunction; the second is the geometric factor present in the volume element, which favors perpendicular alignment. (See Problem 12.38.) If the choice for the direction of the Z-axis is arbitrary, as it is in the case of an isolated diatomic molecule, all directions are equally probable, and the distribution of rotation angles will be spherically symmetric. If the rotor is in the presence of an external magnetic or electric field, so that it is not isolated, the choice for the direction of the Z-axis is no longer arbitrary, and all directions of the rotor are no longer equally probable, even though the eigenfunction may possess spherical symmetry.

12.2.3   Quantum Mechanical Angular Momentum

Our classical analysis of rotation in Chapter 11 demonstrated that rotational energy and angular momentum are closely related. In Section 11.1.3, it was shown that [see Eq. 11.22] the classical Hamiltonian for the rotation of a single particle is given by

$$\mathbf{H} = T + V = \frac{P_R^2}{2m} + \frac{P_\theta^2}{2mR^2} + V(R, \theta),$$

where P_R and P_θ are the radial and angular momenta, respectively, and $V(R, \theta)$ is the potential energy. This result is directly applicable to the rigid rotor if we replace m with the reduced mass, μ_{AB}, and set both P_R and $V(R, \theta)$ to zero, since there is no radial momentum, because R is fixed and the rotor is moving in field-free space. The classical result for the rigid rotor is, therefore,

$$\mathbf{H} = T + V = E_R = \frac{P_\theta^2}{2\mu_{AB}R^2} = \frac{P_\theta^2}{2I}. \tag{12.111}$$

A comparison of the quantum mechanical energy given in Eq. 12.107 with the corresponding classical result in Eq. 12.111 suggests that the square of the quantum mechanical angular momentum is

$$\boxed{[P_\theta^2]_{\text{quantum}} = L^2 = J(J + 1)\hbar^2} \quad \text{for} \quad J = 0, 1, 2, 3, \dots, \tag{12.112}$$

where we have denoted the quantum mechanical angular momentum with the symbol L, as is customary in most texts. Equation 12.112 is also suggested by the Bohr correspondence principle, which requires that quantum mechanical results transform smoothly into the corresponding classical results as the mass or energy becomes very large. The only way the transition from quantum to classical mechanics can be smooth is for Eq. 12.112 to hold. A detailed mathematical treatment of angular momentum, which requires more quantum theory than will be developed in this text, shows Eq. 12.112 to be correct. Thus, the azimuthal quantum number describes the quantization of the angular momentum as well as the rotational energy.

In Section 11.1.4, we found that the classical Z-component of angular momentum is given by

$$M_z = [xp_y - yp_x].$$

We may obtain the quantum mechanical operator corresponding to M_z by using the second postulate, which directs us to simply substitute the operators listed in Table 11.1 into the classical expression in rectangular Cartesian coordinates. Since the operators for p_x and p_y are $(\hbar/i)(\partial/\partial x)$ and $(\hbar/i)(\partial/\partial y)$, respectively, substitution into M_z yields

$$L_z = \frac{\hbar}{i}\left[x\frac{\partial}{\partial y} - y\frac{\partial}{\partial x}\right],$$

where we again use the L notation to denote a quantum mechanical angular momentum. Since the eigenfunctions for rotation are expressed in spherical polar coordinates, we need to transform L_z into this coordinate system. This transformation is straightforward, but somewhat lengthy. Problem 12.39 provides the essential equations needed to execute the transformation. The result obtained in that problem is

$$\boxed{L_z = \frac{\hbar}{i}\frac{\partial}{\partial \phi}.} \tag{12.113}$$

Equation 12.113 is an extremely important result. A great deal of magnetic spectroscopy that we will study in Chapter 16, as well as atomic structure and chemical bonding, depend upon L_z.

We may now employ Eq. 12.113 along with the rigid-rotor eigenfunctions given in Eqs. 12.105 and 12.106 to obtain the average values of L_z and L_z^2. The fourth postulate tells us that

$$\langle L_z \rangle = \int_{\phi=0}^{2\pi} \int_{\theta=0}^{\pi} [\Theta_J^M(\theta)\,\Phi_M(\phi)]^* L_z [\Theta_J^M(\theta)\,\Phi_M(\phi)]\sin\theta\,d\theta\,d\phi. \quad \textbf{(12.114)}$$

Substituting the eigenfunctions and the L_z operator produces

$$\langle L_z \rangle = \frac{\hbar}{2\pi i} \int_0^{\pi} [\Theta_J^M(\theta)]^* [\Theta_J^M(\theta)]\sin\theta\,d\theta \int_0^{2\pi} e^{-iM\phi}\frac{\partial}{\partial\phi}e^{iM\phi}\,d\phi$$

$$= \frac{\hbar}{2\pi i}iM \int_0^{\pi} [\Theta_J^M(\theta)]^* [\Theta_J^M(\theta)]\sin\theta\,d\theta \int_0^{2\pi} e^{-iM\phi}e^{iM\phi}\,d\phi$$

$$= \frac{M\hbar}{2\pi} \int_0^{\pi} [\Theta_J^M(\theta)]^* [\Theta_J^M(\theta)]\sin\theta\,d\theta \int_0^{2\pi} d\phi, \quad \textbf{(12.115)}$$

since $e^{-iM\phi}e^{iM\phi} = e^0 = 1$. The integral over θ in Eq. 12.115 is unity, because $\Theta_J^M(\theta)$ is normalized. The integral over ϕ is 2π, so that the final result is

$$\boxed{\langle L_z \rangle = M\hbar} \quad \textbf{(12.116)}$$

A similar procedure (see Problem 12.40) using $L_z^2 = [(\hbar/i)\partial/\partial\phi][(\hbar/i)\partial/\partial\phi] = -\hbar^2(\partial^2/\partial\phi^2)$ shows that

$$\langle L_z^2 \rangle = M^2\hbar^2. \quad \textbf{(12.117)}$$

Using Eqs. 12.116 and 12.117, we may now compute the uncertainty in a measurement of L_z. The square uncertainty in L_z is

$$\langle \Delta L_z^2 \rangle = \langle L_z^2 \rangle - \langle L_z \rangle^2 = M^2\hbar^2 - (M\hbar)^2 = 0. \quad \textbf{(12.118)}$$

Equation 12.118 tells us that the distribution of L_z values that we might obtain in a series of measurements on a large group of rigid rotors, each of which is in an eigenstate whose eigenfunction is $Y_J^M(\theta, \phi)$, is a delta function. That is, every time we measure L_z, we always obtain the same result: $M\hbar$. Therefore, we know with certainty the values of both L^2 and L_z.

Our simultaneous knowledge of L^2 and L_z allows us to obtain some information concerning the direction of the angular momentum vector. This knowledge also permits us to deduce an important connection between the azimuthal and magnetic quantum numbers. The total square of any vector is the sum of the squares of its Cartesian components. Therefore,

$$L^2 = L_x^2 + L_y^2 + L_z^2. \quad \textbf{(12.119)}$$

Substituting Eqs. 12.112 and 12.116 into Eq. 12.119 produces

$$J(J+1)\hbar^2 = L_x^2 + L_y^2 + M^2\hbar^2.$$

Since both L_x^2 and L_y^2 must be positive, we must have

$$J(J+1)\hbar^2 \geq M^2\hbar^2,$$

and hence,

$$J(J + 1) \geq M^2. \qquad \textbf{(12.120)}$$

We know that both J and M are integers. Consequently, we cannot have $|M| > J$. For example, if we were to have $M = J + 1$ or $M = -J - 1$, then $M^2 = (J + 1)^2$, which exceeds $J(J + 1)$, so that the inequality in Eq. 12.120 would be violated. Physically, such a situation would mean that we have a vector **L** whose magnitude is less than that of one of its components, L_z. We know that this is impossible; therefore, the magnetic quantum number must lie in the range

$$\boxed{-J \leq M \leq J}. \qquad \textbf{(12.121)}$$

There are exactly $(2J + 1)$ integer values of M that lie in this range. Since Eq. 12.107 shows that the energy is dependent only upon J, and we now know that there are $(2J + 1)$ values of M for each possible azimuthal quantum number, we may conclude that the Jth rotational-energy state will be $(2J + 1)$-fold degenerate. The eigenfunctions listed in Table 12.1 illustrate this fact; all values of M satisfy Eq. 12.121, and the degeneracies of the $J = 0$, 1, 2, and 3 quantum states are 1, 3, 5, and 7, respectively.

12.2.4   Angular Momentum Uncertainty

Let us now examine what we know about the orientation of **L** when the rotor is in the $Y_J^M(\theta, \phi)$ eigenstate. Equation 12.112 shows us that the magnitude of **L** is $[J(J + 1)]^{1/2}\hbar$, and Eqs. 12.114 and 12.116 show that $L_z = M\hbar$. Consequently, the orientation of **L** must be that illustrated in Figure 12.26. Inspection of the figure shows that the azimuthal angle θ is uniquely determined by L and L_z. That is, we must have

$$\cos \theta = \frac{L_z}{L} = \frac{M\hbar}{[J(J + 1)]^{1/2}\hbar} = \frac{M}{[J(J + 1)]^{1/2}}$$

so that

$$\theta = \cos^{-1}\left[\frac{M}{[J(J + 1)]^{1/2}}\right]. \qquad \textbf{(12.122)}$$

Since we can simultaneously measure both L^2 and L_z, we can always determine θ.

The question now arises, "Can we also determine the angle ϕ made by the projection of **L** onto the X–Y plane?" An examination of Figure 12.26 shows that, to determine ϕ, we must be able to measure both L_x and L_y, the x and y components of L. Actually, we need only measure either L_x or L_y, along with L^2 and L_z, to be able to determine the magnitude of all the angular momentum components, because they are related by Eq. 12.119. Therefore, the question concerning the determination of ϕ depends upon the answer to the question, "Can we simultaneously measure L_x, L_y, and L_z exactly?" The answer to this question is "Sometimes we can, and sometimes we can't." We will be able to simultaneously measure both L_x and L_y only when their uncertainty product is zero.

The value of the uncertainty product for L_x and L_y is given by Eq. 12.65A, viz.,

$$\langle \Delta L_x^2 \rangle \langle \Delta L_y^2 \rangle \geq \frac{1}{4}|[L_x, L_y]|^2, \qquad \textbf{(12.123)}$$

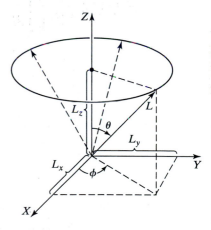

▲ FIGURE 12.26
Orientation of the angular momentum vector **L** for a rigid rotor.

where $[L_x, L_y]$ is the commutator between the quantum mechanical operators corresponding to L_x and L_y. The evaluation of this commutator requires more quantum theory than we can cover in this text. Therefore, we present the result without proof. The commutator of L_x and L_y is $i\hbar L_z$. Using this result in Eq. 12.123, we obtain

$$\langle \Delta L_x^2 \rangle \langle \Delta L_y^2 \rangle \geq \frac{1}{4}|i\hbar L_z|^2 = \frac{M^2\hbar^4}{4}, \qquad (12.124)$$

since $L_z = M\hbar$. We conclude that the uncertainty product of L_x and L_y will be zero only if we are guaranteed that the magnetic quantum number is zero. This is true only if the rotational eigenfunction is a spherically symmetric s function for which $J = 0$. In that case, we can simultaneously determine all the components of the angular momentum. If $J = 0$, we must also have $M = 0$, so that the result is

$$L^2 = L = L_z = L_x = L_y = 0 \qquad \text{for } J = M = 0. \qquad (12.125)$$

Equation 12.125 explains why we can have a rotational ground state with an energy of zero, but not a translational ground state with $E_{\text{translation}} = 0$ for a particle in an infinite potential well. If $E_{\text{translation}} = 0$, we would know the linear momentum exactly, with a finite uncertainty in the coordinate. Thus, the uncertainty principle, Eq. 12.65B, would be violated, because the $[x, p_x]$ commutator is never zero. In fact, it is always $i\hbar$. When $E_{\text{rotation}} = 0$, we can determine both L_x and L_y, but this does not violate the uncertainty principle, since $[L_x, L_y]$ is zero when $J = M = 0$.

If the rotational state has $J > 0$, so that M does not have to be zero, we have $\langle \Delta L_x^2 \rangle \langle \Delta L_y^2 \rangle \geq 0$, and we cannot measure both L_x and L_y simultaneously. Therefore, we cannot determine the angle ϕ in Figure 12.26. In the case where $J = M = 0$, so that Eq. 12.125 holds, there is no angular momentum vector, because $L = 0$. Hence, ϕ is undefined and, consequently, cannot be measured. The final conclusion is that, although we can determine all the angular momenta when the rotor is in an s eigenstate, the angle ϕ can *never* be measured. All we can know is that the tip of the angular momentum vector lies somewhere around the cone illustrated in Fig. 12.26. We could have reached this conclusion by examining the uncertainty product for ϕ and L_z. This product is not zero; therefore, we can never simultaneously know both the Z-component of the angular momentum and the rotation angle about Z. (See Problem 12.45.)

The foregoing discussion shows that L_z may be exactly measured, so that the square uncertainty in the L_z distribution, $\langle \Delta L_z^2 \rangle$, is zero. *When this is the case for a quantum mechanical variable, the system wave function is always an eigenfunction of the operator corresponding to that variable.* Example 12.18 illustrates this point for L_z.

EXAMPLE 12.18

A rigid rotor is in an eigenstate whose eigenfunction is $Y_J^M(\theta, \phi) = \Theta_J^M(\theta)\Phi_M(\phi)$. We have already found that the uncertainty associated with a measurement of L_z in this state is zero. Show that the system wave function is an eigenfunction of L_z.

Solution

The operator corresponding to L_z is given by Eq. 12.113. If $Y_J^M(\theta, \phi)$ is an eigenfunction of L_z, we must have

$$L_z Y_J^M(\theta, \phi) = (\text{constant}) Y_J^M(\theta, \phi). \qquad (1)$$

Inserting the operator for L_z, we obtain

$$\frac{\hbar}{i}\frac{\partial}{\partial\phi}Y_J^M(\theta,\phi) = \frac{\hbar}{i}\frac{\partial}{\partial\phi}\Theta_J^M(\theta)\Phi_M(\phi) = \frac{\hbar}{i}\Theta_J^M(\theta)\frac{\partial}{\partial\phi}\Phi_M(\phi), \qquad (2)$$

since $\Theta_J^M(\theta)$ is independent of ϕ and, therefore, may be factored to the left of the operator. The function $\Phi_M(\phi)$ is given by $\Phi_M(\phi) = (2\pi)^{-1/2}e^{iM\phi}$, so that Eq. 2 becomes

$$\frac{\hbar}{i}\frac{\partial}{\partial\phi}Y_J^M(\theta,\phi) = \frac{\hbar}{i}\Theta_J^M(\theta)\frac{\partial}{\partial\phi}(2\pi)^{-1/2}e^{iM\phi} = M\hbar\Theta_J^M(\theta)(2\pi)^{-1/2}e^{iM\phi}$$

$$= M\hbar Y_J^M(\theta,\phi). \qquad (3)$$

Thus, Eq. 1 is satisfied, and $Y_J^M(\theta,\phi)$ is an eigenfunction of L_z with an eigenvalue equal to $M\hbar$, which is exactly the value always obtained in a measurement of L_z.

For related exercises, see Problems 12.61 and 12.62.

EXAMPLE 12.19

A rigid rotor is in the $J = 2$ eigenstate. What are the possible angles that **L** can make with the Z-axis?

Solution

Equation 12.122 gives us the possible θ angles:

$$\theta = \cos^{-1}\left[\frac{M}{[J(J+1)]^{1/2}}\right]. \qquad (1)$$

In this case, we have $J = 2$, so that the possible values of M are $-2, -1, 0, 1$, and 2. The possible angles between **L** and the Z-axis are

$$\theta = \cos^{-1}\left[\frac{-2}{(6)^{1/2}}\right] = \cos^{-1}(-0.816496\ldots) = 144.74°, \qquad (2)$$

$$\theta = \cos^{-1}\left[\frac{-1}{(6)^{1/2}}\right] = \cos^{-1}(-0.408248\ldots) = 114.09°, \qquad (3)$$

$$\theta = \cos^{-1}\left[\frac{0}{(6)^{1/2}}\right] = \cos^{-1}(0) = 90.000°, \qquad (4)$$

$$\theta = \cos^{-1}\left[\frac{1}{(6)^{1/2}}\right] = \cos^{-1}(0.408248\ldots) = 65.91°, \qquad (5)$$

and

$$\theta = \cos^{-1}\left[\frac{2}{(6)^{1/2}}\right] = \cos^{-1}(0.816496\ldots) = 35.26°. \qquad (6)$$

While it is always possible to measure L_z, this possibility does not mean we have to measure its value. Let us suppose that we have a rigid rotor whose angular momentum has been measured and determined to be $L = \sqrt{2}\,\hbar$, so that we know that $J = 1$. Let us further suppose that we know that L_z is either $\hbar$ or $-\hbar$ (i.e. $M = 1$ or -1) and that these states are equally probable. In this case, the wave function will be a superposition of the $M = 1$ and $M = -1$ eigenfunctions with equal weight:

$$\psi(\theta,\phi) = C[Y_1^1(\theta,\phi) \pm Y_1^{-1}(\theta,\phi)]. \qquad (12.126)$$

Substitution of these eigenfunctions from Table 12.1 produces

$$\psi(\theta, \phi) = C\left[\frac{3}{8\pi}\right]^{1/2} \sin\theta\, [e^{i\phi} \pm e^{-i\phi}] = N \sin\theta\, [e^{i\phi} \pm e^{-i\phi}].$$

If we use the plus sign, we obtain $e^{i\phi} + e^{-i\phi} = [\cos\phi + i\sin\phi] + [\cos\phi - i\sin\phi] = 2\cos\phi$. After normalization, the wave function becomes

$$\psi(\theta, \phi) = \left[\frac{3}{4\pi}\right]^{1/2} \sin\theta \cos\phi = p_x. \qquad \textbf{(12.127A)}$$

If we use the minus sign in Eq. 12.126, the result is

$$\psi(\theta, \phi) = \left[\frac{3}{4\pi}\right]^{1/2} \sin\theta \sin\phi = p_y. \qquad \textbf{(12.127B)}$$

Unlike the eigenfunctions listed in Table 12.1, the two wave functions given in Eqs. 12.127A and 12.127B are not cylindrically symmetric about the Z-axis. This is shown in Figures 12.27 and 12.28, which are polar plots of these functions and their absolute squares in the X–Y plane. The function in Eq. 12.127A has a large probability of lying along the X-axis with a Y–Z nodal plane. It is, therefore, called the p_x function. The function in Eq. 12.127B has a maximum probability along the Y-axis with an X–Z nodal plane. Consequently, it is called the p_y function. Basically, we have reduced the uncertainty in the ϕ coordinate by giving up some knowledge about the conjugate Z-component of angular momentum. Wave functions such as these are the appropriate form when we do not know the value of L_z. Problem 12.62 explores the effect of this absence of knowledge of L_z upon the uncertainty in the L_z distribution.

It was noted in Section 12.1.3 that when the wave function is a superposition of states, additional nodes are often introduced. The superposition in Eq. 12.126 that leads to the p_x and p_y orbitals is a case in point. When the wave function for the rotor is the eigenfunction $Y_1^1(\theta, \phi)$ or $Y_1^{-1}(\theta, \phi)$, we have $J = 1$, and there is a single node at $\theta = 0$ or π, which are equivalent

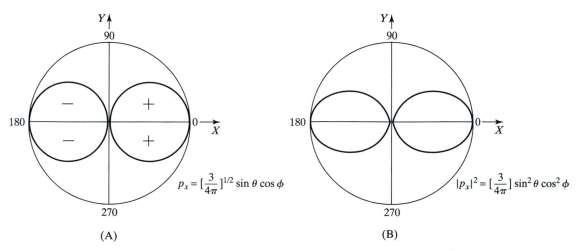

▲ FIGURE 12.27
(A) A polar plot of the p_x superposition in the X–Y plane. (B) A polar plot of $|p_x|^2$ in the X–Y plane.

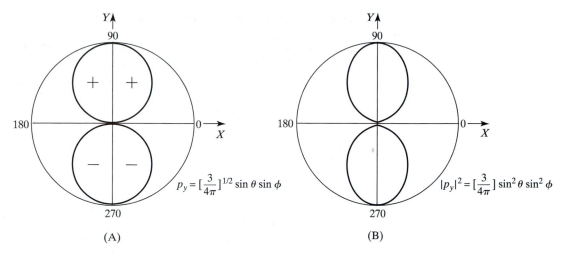

▲ FIGURE 12.28
(A) A polar plot of the p_y superposition in the X–Y plane. (B) A polar plot of $|p_y|^2$ in the X–Y plane.

angles for an isolated rigid rotor. The superpositions p_x and p_y have two nodes. The node at $\theta = 0$ or π is still present, but we now have an additional node at the equivalent angles $\phi = \pi/2$ or $3\pi/2$ for p_x and at $\phi = 0$ or π for p_y.

12.2.5   Rotation of Polyatomic Molecules

The solution of the Schrödinger equation for the rotation of a general many-atom molecule is extremely difficult and often impossible. We shall, therefore, restrict our attention to some simple cases in which the symmetry of the molecule permits us to obtain the expressions for the quantized rotational energy levels.

The moments of inertia for a polyatomic molecule are still defined by Eq. 12.98. However, for a nonlinear molecule, there are now three possible moments of inertia corresponding to rotation about three principal axes. If we denote these three axes as a, b, and c, the corresponding principal moments of inertia have the form

$$I_u = \sum_{i=1}^{N} m_i r_{\mu,i}^2 . \qquad \textbf{(12.128)}$$

In Eq. 12.128, the summation runs over all N atoms in the molecule, the m_i are the atomic masses, and $r_{u,i}$ is the perpendicular distance from atom i to the uth principal axis, with $u = a, b,$ or c. If we can locate the principal axes of a polyatomic molecule, it will be straightforward to calculate the principal moments of inertia. (See Problem 12.46.)

The principal axes can be located by examining the *products of inertia, I_{xy}, I_{xz},* and I_{yz}, where

$$I_{xy} = -\sum_{i=1}^{N} m_i x_i y_i , \qquad \textbf{(12.129)}$$

with analogous definitions for I_{xz} and I_{yz}. In Eq. 12.129, x_i and y_i are the Cartesian coordinates for atom i. It is possible to show that, for every molecule, there exist three principal axes passing through the molecule's center of mass such that $I_{xy} = I_{xz} = I_{yz} = 0$. The principal axes may not be unique;

that is, in some cases, there is more than one set of three axes that satisfy the condition that $I_{xy} = I_{xz} = I_{yz} = 0$. In general, any set will suffice.

If the molecule possesses symmetry, the problem of finding the principal axes is greatly simplified. *It is not hard to demonstrate that a molecular symmetry axis coincides with one of the principal axes and that a symmetry plane contains two principal axes and is perpendicular to the third.* Example 12.20 illustrates this general principle.

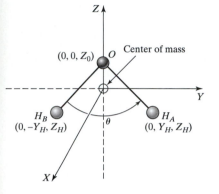

▲ FIGURE 12.29
Principal axes of H_2O.

EXAMPLE 12.20

Consider the water molecule. Let the origin of the Cartesian coordinate system be the molecular center of mass with the axes oriented so that the Z-axis bisects the H–O–H angle, with the molecular plane being the Y–Z plane. With these choices, the atomic coordinates are shown in Figure 12.29. Show that the principal axes of this molecule coincide with the X-, Y-, and Z-axes of the coordinate system.

Solution

First we note that H_2O has two symmetry planes: the X–Z plane, which bisects the H–O–H angle, and the plane of the molecule itself, the Y–Z plane. The general principle stated in the paragraph preceding this example tells us that each of these planes contains two symmetry axes that are perpendicular to the third. The Z-axis is a rotational symmetry axis for the molecule. This axis is called a C_2 axis, because, as we rotate H_2O about it through 2π radians, the original structure of H_2O is reproduced twice, once after a rotation of π radians and again after a rotation of 2π radians. Thus, the Z-axis must be a principal axis. Since the X–Z plane is perpendicular to the third principal axis, and we know that Z is a principal axis, the third principal axis must be the Y-axis. Finally, we know that the Y–Z plane is perpendicular to the other principal axis, so that must be the X-axis. Consequently, in this case, it is a simple matter to use the general theorem to determine the principal axes of H_2O. However, let us show that we also have $I_{xy} = I_{xz} = I_{yz} = 0$ for these axes.

Using Eq. 12.129, we have

$$I_{xy} = -\sum_{i=1}^{N} m_i x_i y_i \quad \text{and} \quad I_{xz} = -\sum_{i=1}^{N} m_i x_i z_i. \tag{1}$$

Since all atoms lie in the Y–Z plane, $x_i = 0$ for all i. Therefore we have $I_{xy} = I_{xz} = 0$. The remaining product of inertia is

$$I_{yz} = -\sum_{i=1}^{N} m_i y_i z_i = -[m_O(0)(Z_O) + m_H(Y_H)(Z_H) + m_H(-Y_H)(Z_H)] = 0. \tag{2}$$

We see, then, that all three products of inertia are zero, and we have located the principal axes of the H_2O molecule.

Equation 12.111 shows that if we have rotation about a single axis, we expect the classical energy of rotation to be given by

$$\mathbf{H} = E_R = \frac{P_\theta^2}{2I},$$

where I refers to the moment of inertia about the axis of rotation. When we have a polyatomic rigid body rotating about three principal axes, the classical energy of rotation is

$$\mathbf{H} = E_R = \frac{P_a^2}{2I_a} + \frac{P_b^2}{2I_b} + \frac{P_c^2}{2I_c} \quad \text{(for nonlinear molecules)}, \tag{12.130}$$

where $I_a \leq I_b \leq I_c$ are the principal moments of inertia and P_a, P_b, and P_c are the classical rotational angular momenta about these axes. If all principal moments of inertia are equal, the rigid body is called a *spherical top*. If we have $I_a = I_b \neq I_c$ or $I_b = I_c \neq I_a$, the rigid body is termed a *symmetric top*. If the three moments are different, we call the rotor an *asymmetric top*. A linear molecule is a special case in which the moment of inertia about the molecular axis is zero, because all the r_i in Eq. 12.128 are zero. There are, therefore, only two terms in the classical energy expression. Furthermore, the remaining two principal moments of inertia will be equal, since all axes perpendicular to the molecular axis are equivalent. Using these definitions, we conclude that a linear molecule is a symmetric top.

The Schrödinger equation for a complex polyatomic rigid body is too difficult to treat in this text. However, we may obtain useful results for the rotational energies of these systems by analogy with the systems we have already considered. So far, we have found that the energies for systems with one degree of freedom, such as the one-dimensional free particle and the particle in an infinite potential well, are dependent upon one quantum number. The harmonic oscillator treated in the next section is another example. Systems that have two degrees of freedom, such as the two-particle rigid rotor, require two quantum numbers to specify their energy, and systems with three degrees of freedom have three quantum numbers. A polyatomic rigid body independently rotating about three principal axes has three degrees of rotational freedom. Therefore, we expect the energy states to depend upon three quantum numbers. This is indeed the case; we denote the three quantum numbers as J, K, and M.

A spherical top, such as CCl_4 or CH_4, has $I_a = I_b = I_c$. Under these conditions, Eq. 12.130 becomes

$$E_R = \frac{P_a^2 + P_b^2 + P_c^2}{2I}, \tag{12.131}$$

where $I = I_a = I_b = I_c$. The square of the total angular momentum is expected to be the sum of the squares of the individual components. Consequently, it appears that Eq. 12.131 can be written in the form

$$E_R = \frac{L^2}{2I},$$

which is the same form as that obtained for the two-particle rigid rotor. We might, therefore, expect that Eq. 12.111 will still describe the square of the total angular momentum. If this is true, the quantum mechanical expression for the rotational energy levels of a spherical top will be

$$\boxed{E_J = \frac{J(J + 1)\hbar^2}{2I}} \qquad \text{for } J = 0, 1, 2, 3, \dots . \tag{12.132}$$

For the two-particle rigid rotor, we found that we had to have $-J \leq M \leq J$ in order to prevent one component of the angular momentum from being larger than the angular momentum itself. This led to an energy-state degeneracy of $2J + 1$. To carry our analogy one step further, we might anticipate that the other two quantum numbers, K and M, have to lie in the ranges $-J \leq K \leq J$ and $-J \leq M \leq J$ for the same reasons. If this is true, then there will be $2J + 1$ values of K and $2J + 1$ values of M for each value of J. Consequently, the energy states given by Eq. 12.132 will have a degeneracy of $(2J + 1)^2$.

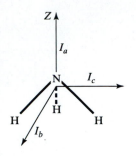

▲ FIGURE 12.30
The NH_3 molecule is a symmetric top rotor with one principal axis being the C_3 molecular axis, shown as the Z-axis in the figure. The other two principal axes are in the X–Y plane. The $Z–I_b$ and $Z–I_c$ planes each bisect one of the H–N–H angles. Since $I_b = I_c \neq I_a$, NH_3 is a symmetric top.

We may obtain similar expressions for symmetric-top rotors by extending the type of reasoning used for the spherical top. Let us consider NH_3, which is a symmetric top. The molecular axis has three fold rotational symmetry. (A 360° rotation about this axis produces the same configuration three times.) Because of this, it is called a C_3 axis. Since this axis is a molecular symmetry axis, it must be one of the principal axes of the molecule. The remaining two principal axes are known to be perpendicular to Z. So they lie in the X–Y plane. The structure of NH_3 is shown in Figure 12.30.

Since we now have $I_b = I_c$, we can write Eq. 12.131 in the form

$$E_R = \frac{P_b^2 + P_c^2}{2I_b} + \frac{P_z^2}{2I_a}. \tag{12.133}$$

Using the same reasoning as that employed for the spherical top, we might expect to have $L^2 = P_b^2 + P_c^2 + P_a^2$ and $L_z^2 = P_z^2 = P_a^2$, so that $L^2 - L_z^2 = P_b^2 + P_c^2$. With these substitutions, Eq. 12.133 becomes

$$E_R = \frac{L^2 - L_z^2}{2I_b} + \frac{L_z^2}{2I_a}.$$

If Eq. 12.112 still gives L^2 and Eq. 12.116 L_z, the quantum mechanical expression for the energy levels of a symmetric top will be

$$E_{JK} = \frac{J(J+1)\hbar^2 - K^2\hbar^2}{2I_b} + \frac{K^2\hbar^2}{2I_a}.$$

Collecting terms, we obtain

$$E_{JK} = \frac{J(J+1)\hbar^2}{2I_b} + \frac{K^2\hbar^2}{2}\left[\frac{1}{I_a} - \frac{1}{I_b}\right] \tag{12.134}$$

for $J = 0, 1, 2, 3, \ldots$ and $-J \leq K \leq J$,

where the component of angular momentum along the molecular axis, L_z, is now $K\hbar$ rather than $M\hbar$, as was the case for the two-atom rigid rotor. Figure 12.31 shows the lower rotational energy levels for the symmetric-top molecule NH_3.

As previously noted, because both principal moments of inertia are always equal, a linear polyatomic rotor has only two terms in the classical expression for the energy:

$$E_R = \frac{P_x^2 + P_y^2}{2I}.$$

The square of the total angular momentum is now just $P_x^2 + P_y^2$, so that the quantum mechanical expression for the energy levels is

$$E_J = \frac{J(J+1)\hbar^2}{2I} \quad \text{for } J = 0, 1, 2, 3, \ldots \text{ and linear molecules.} \tag{12.135}$$

Reasoning by analogy, as we have done in the development of the preceding equations, is very useful in science and engineering. It often permits us to deduce important information without actual derivating or measuring anything. However, it can sometimes lead us astray because of effects that are present in the system under consideration that are missing or inconse-

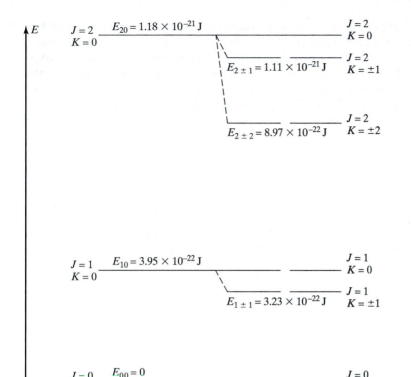

The rotational energy levels of NH_3. The molecule is a symmetric top. The levels on the left of the diagram have $K = 0$. The level splitting for nonzero values of K are shown on the right.

quential in the systems upon which we are basing our logic. Consequently, due caution must be exercised when we proceed in this fashion. Occasionally, our deductions will be incorrect. For the rotation of rigid polyatomic molecules, the results turn out to be right.

The expression for the quantum mechanical energy levels of asymmetric-top rigid bodies are highly complex and follow no simple pattern. Therefore, we will not address them in this text.

12.3   Vibrational Energy

12.3.1   The Classical Harmonic Oscillator

Up to now in this chapter, we have seen that, for the diatomic system, three degrees of freedom and three quantum numbers are associated with the translational motion of the center of mass. An additional two degrees of freedom and two quantum numbers describe the rotation of the molecule. The final degree of freedom describes the vibrational motion. Vibration is, therefore, a one-dimensional problem for a diatomic molecule if we follow the procedure of separating the rotational and vibrational motions, as presented in the last section.

The nature of the vibrational eigenfunctions and the corresponding vibrational-energy eigenvalues depend upon the interatomic bond potential produced by the electron distribution about the two vibrating nuclei. This problem will be examined in detail in Chapter 14. For the present, we shall consider two useful approximations to the potential. The first of these is to assume that the forces binding the diatomic molecule are similar to those acting on a spring that obeys Hooke's law. In accordance with this analogy, the force acting between the two particles is proportional to the displacement from the equilibrium position, or

$$F = -k[r - r_{eq}] = -kx, \qquad \text{(12.136)}$$

where r is the interatomic distance between the two particles and r_{eq} is their equilibrium separation. The minus sign in Eq. 12.136 means that the force acts so as to resist displacement from equilibrium; that is, it always acts in the direction opposite to the displacement. The proportionality constant k is called the *vibrational force constant*.

In Eq. 1.46, we defined the potential in terms of the force by

$$F = -\frac{\partial V}{\partial r}.$$

If F is given by Eq. 12.136, the potential is

$$V(r) = \int dV = -\int F\,dr = k\int (r - r_{eq})dr = \frac{k}{2}(r - r_{eq})^2 + \text{constant.} \quad \textbf{(12.137)}$$

If we take the reference point for vibrational energy to be zero when $r = r_{eq}$, then the constant in Eq. 12.137 is zero, and we have

$$\boxed{V(x) = \frac{k}{2}(r - r_{eq})^2 = \frac{kx^2}{2}.} \quad \textbf{(12.138)}$$

The bond potential is, therefore, parabolic in form, with a curvature given by $d^2V/dx^2 = k$. A two-body system vibrating in such a potential field is called a *harmonic oscillator*. Figure 12.32 illustrates the potential field for such an oscillator whose vibrational frequency matches that of LiH.

The classical motion of an oscillator in the potential field described by Eq. 12.138 is given by the solution of the one-dimensional Newtonian equation of motion [Eq. 11.6A]:

$$\mu\frac{d^2r}{dt^2} + \frac{dV(r)}{dr} = \mu\frac{d^2r}{dt^2} + k(r - r_{eq}) = 0. \quad \textbf{(12.139)}$$

Here, we have replaced the mass m in Eq. 11.6A with the reduced mass, since we are now concerned with the relative vibrational motion of the two particles. (See Problem 11.6.) Because $x = r - r_{eq}$, we have $dx = dr$, and with this substitution, Equation 12.139 becomes

$$\mu\frac{d^2x}{dt^2} + kx = 0.$$

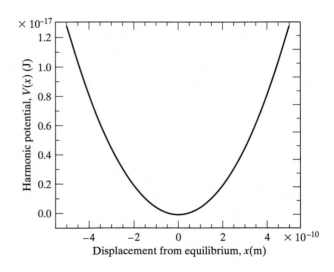

▶ FIGURE 12.32
Harmonic bond potential whose force constant is fitted to the measured vibrational frequency of LiH using Eq. 12.145.

Rearranging terms produces

$$\frac{d^2x}{dt^2} + \frac{kx}{\mu} = 0. \tag{12.140}$$

Equation 12.140 has the same form as that encountered in our solution of the particle in an infinite one-dimensional potential well and as the equation for $\Phi(\phi)$ in the rigid-rotor problem. The solution is

$$x = A \sin\left[\left(\frac{k}{\mu}\right)^{1/2} t\right] + B \cos\left[\left(\frac{k}{\mu}\right)^{1/2} t\right]. \tag{12.141}$$

If we take our initial time to be a point at which $x = 0$, then at $t = 0$, Eq. 12.141 becomes

$$0 = A \sin\left[\left(\frac{k}{\mu}\right)^{1/2} (0)\right] + B \cos\left[\left(\frac{k}{\mu}\right)^{1/2} (0)\right] = 0 + B,$$

so that $B = 0$ and the result is

$$x = A \sin\left[\left(\frac{k}{\mu}\right)^{1/2} t\right]. \tag{12.142}$$

Thus, a classical harmonic oscillator exhibits simple sinusoidal motion about the equilibrium position.

The total vibrational energy of the classical oscillator is the sum of the kinetic and potential energies:

$$E_{vib} = \frac{\mu}{2}\left(\frac{dx}{dt}\right)^2 + \frac{kx^2}{2}.$$

It is easy to show (see Problem 12.49) that $E_{vib} = kA^2/2$. We conclude that the classical energy is continuous and proportional to the square of the vibrational amplitude. Equation 12.142 shows us that the maximum and minimum displacements a classical oscillator can have are A and $-A$, respectively, since the sine function varies between 1 and -1. These limiting displacements are called the *classical turning points*, because, when x reaches a value of A or $-A$, the oscillator "turns around" and moves in the opposite direction. In terms of the total vibrational energy, the classical turning points are given by

$$x_{turn} = \pm\left[\frac{2E_{vib}}{k}\right]^{1/2}. \tag{12.143}$$

The classical period τ is the time required for the displacement to complete one sinusoidal cycle. Equation 12.142 shows that this is

$$\boxed{\tau = 2\pi\left(\frac{\mu}{k}\right)^{1/2}.} \tag{12.144}$$

The vibrational frequency is the reciprocal of the period. Therefore, the classical vibrational frequency is

$$\boxed{\nu_o = \frac{1}{\tau} = \frac{1}{2\pi}\left(\frac{k}{\mu}\right)^{1/2}.} \tag{12.145}$$

EXAMPLE 12.21

The harmonic force constant for HF is 965.6 J m^{-2}. If we assume that the vibrational motion of HF is that of a classical harmonic oscillator, what is the vibrational frequency of HF? If the oscillator has a total vibrational energy of 1.233×10^{-19} J, plot the variation of the displacement with time, and determine the classical turning points.

Solution

The reduced mass of HF is

$$\mu = \frac{m_H m_F}{m_H + m_F} = \frac{(0.001008)(0.0190)}{(0.001008 + 0.0190)(6.022 \times 10^{23})} \text{ kg} = 1.590 \times 10^{-27} \text{ kg}. \quad (1)$$

The vibrational frequency is given by Eq. 12.145:

$$\nu_o = \frac{1}{2\pi}\left(\frac{k}{\mu}\right)^{1/2} = \frac{1}{2(3.14159)}\left[\frac{965.6 \text{ J m}^{-2}}{1.590 \times 10^{-27} \text{ kg}}\right]^{1/2} = 1.240 \times 10^{14} \text{ s}^{-1}. \quad (2)$$

The amplitude or classical turning points of the oscillator is determined by k and the total energy:

$$A^2 = \frac{2E_{vib}}{k} = \frac{2(1.233 \times 10^{-19} \text{ J})}{965.6 \text{ J m}^{-2}} = 2.554 \times 10^{-22} \text{ m}^2. \quad (3)$$

Thus,

$$A = 1.60 \times 10^{-11} \text{ m}. \quad (4)$$

The time dependence of x is now given by Eq. 12.142:

$$x = A \sin\left[\left(\frac{k}{\mu}\right)^{1/2} t\right] = 1.60 \times 10^{-11} \sin\left[7.793 \times 10^{14} t\right] \text{ m}, \quad (5)$$

since $(k/\mu)^{1/2} = [965.6 \text{ J m}^{-2}/1.590 \times 10^{-27} \text{ kg}]^{1/2} = 7.793 \times 10^{14} \text{ s}^{-1}$. To use Eq. 5, the time must be expressed in seconds. The required plot is shown in Figure 12.33.

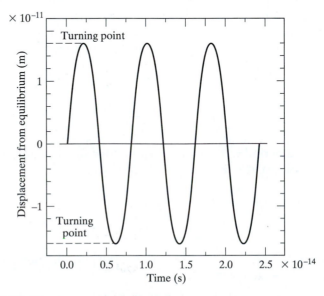

▲ FIGURE 12.33
The classical time dependence of the oscillator bond length for a harmonic oscillator with an energy of 1.233×10^{-19} J whose force constant is equal to that of HF.

12.3.2   Quantum Mechanical Harmonic Oscillator

The quantum mechanical Hamiltonian for the relative motion of a harmonic oscillator is

$$\mathcal{H} = -\frac{\hbar^2}{2\mu}\nabla^2 + V(x) = -\frac{\hbar^2}{2\mu}\frac{d^2}{dx^2} + \frac{kx^2}{2}. \qquad (12.146)$$

The corresponding stationary-state Schrödinger equation is

$$\left[-\frac{\hbar^2}{2\mu}\frac{d^2}{dx^2} + \frac{kx^2}{2}\right]\psi_v(x) = E_v\psi_v(x). \qquad (12.147)$$

The eigenfunctions, $\psi_v(x)$, give the probability distributions for the oscillator while the corresponding eigenvalues, E_v, are the permitted vibrational energies for the system. We know that this one-dimensional system will require exactly one quantum number to specify its state. We have denoted this vibrational quantum number as v and so labeled both the eigenfunctions and eigenvalues in Eq. 12.147.

Equation 12.147 can be solved by series or operator methods. The results, which we can verify by direct substitution, all have the form

$$\psi_v(x) = N_v H_v(y) \exp\left[-\frac{y^2}{2}\right], \qquad (12.148)$$

where $y = \alpha^{1/2}x$, with $\alpha = (\mu k)^{1/2}/\hbar$. The normalization constant is given by

$$N_v = \left[\left(\frac{\alpha}{\pi}\right)^{1/2}\frac{1}{2^v v!}\right]^{1/2}. \qquad (12.149)$$

→ wrong

$$N_v = \left[\left(\frac{\alpha}{\pi}\right)^{\frac{1}{2}}\frac{1}{2^v v!}\right]^{\frac{1}{2}}$$

In Eq. 12.148, $H_v(y)$ is a polynomial in y whose leading power is y^v. $H_v(y)$ is called a *Hermite polynominal*. Hermite polynomials and their associated vibrational-energy eigenvalues are listed in Table 12.2.

Table 12.2   Energy eigenvalues and Hermite polynomials for the harmonic oscillator

v	E_v	$H_v(y)$
0	$\frac{1}{2}h\nu_o$	1
1	$\frac{3}{2}h\nu_o$	$2y$
2	$\frac{5}{2}h\nu_o$	$4y^2 - 2$
3	$\frac{7}{2}h\nu_o$	$8y^3 - 12y$
4	$\frac{9}{2}h\nu_o$	$16y^4 - 48y^2 + 12$
5	$\frac{11}{2}h\nu_o$	$32y^5 - 160y^3 + 120y$
6	$\frac{13}{2}h\nu_o$	$64y^6 - 480y^4 + 720y^2 - 120$

Note: ν_o = classical vibration frequency = $(1/2\pi)(k/\mu)^{1/2}$.

The energy eigenvalues for a quantum harmonic oscillator given in the table indicate that they are related to the vibrational quantum number by

$$E_v = (v + 0.5)h\nu_o \qquad (12.150)$$

Therefore, the ground-state energy is $E_o = h\nu_o/2$. This quantity is called the *zero-point energy*, because, even at absolute zero, a quantum mechanical oscillator will still have that amount of vibrational energy. The fact that the oscillator cannot have a zero energy in the ground state is not surprising, since we now know that such a result would violate the uncertainty principle. It would permit us to know the momentum p_x exactly when the uncertainty in the oscillator's position is finite because it is a bound system. The spacing between the permitted energy states is a constant equal to $h\nu_o$. For molecular systems, this quantity of energy is very large, so we can never accurately consider molecular vibrational energy levels to be continuous. We saw the effect of doing so in Chapter 7, where we found that the assumption of continuous classical vibrational energies led to poor results for the heat capacities of materials at temperatures below 200 K. However, as the mass increases, ν_o becomes very small, and both the energy-level spacing and the zero-point energy approach zero for massive particles. This is another example of the Bohr correspondence principle, which requires quantum mechanical results to approach the predictions of classical mechanics in the limit of large mass or energy. (See Problem 12.50.)

Using Eqs. 12.148 and 12.149 along with the data given in Table 12.2, we can easily obtain the eigenfunctions for the various quantum states of a harmonic oscillator. The first three of these are

$$\psi_0(x) = \left[\frac{\alpha}{\pi}\right]^{1/4} \exp\left[-\frac{\alpha x^2}{2}\right],$$

$$\psi_1(x) = \left[\frac{4\alpha^3}{\pi}\right]^{1/4} x \exp\left[-\frac{\alpha x^2}{2}\right],$$

and

$$\psi_2(x) = \left[\frac{\alpha}{4\pi}\right]^{1/4} [1 - 2\alpha x^2] \exp\left[-\frac{\alpha x^2}{2}\right].$$

It is useful to examine the form of these eigenfunctions. Let us do this for the specific case of the HF molecule, for which $\mu = 1.590 \times 10^{-27}$ kg and $k = 965.6$ J m^{-2}, so that $\alpha = (\mu k)^{1/2}/\hbar = 1.175 \times 10^{22}$ m^{-2}. Figure 12.34 shows the $v = 0, 1,$ and 2 vibrational-state eigenfunctions. As expected, we see that the ground state is nodeless. The first excited state has one node, the second excited state two. In general, the vth eigenstate has v nodes. Figure 12.35 shows these eigenfunctions and eigenvalues superimposed upon the harmonic potential-energy function that has been fitted to the data for HF.

We see from Figure 12.34 that the quantum oscillator can penetrate into classically forbidden regions beyond the classical turning points. These regions are the shaded portions in the figure. This is another example of quantum mechanical tunneling. A careful examination of the relative percentages of the wave function in the classical forbidden regions for each vibrational state suggests that the tunneling probability is decreasing as the total vibrational energy increases. This qualitative observation is confirmed quantitatively in Problem 12.51. It appears to contradict our treatment of tunneling in the case of a one-dimensional square barrier, in which we found that the tunneling probability increases as E increases. Therefore, we need to consider the phenomenon in more detail.

One of the most striking differences between classical and quantum oscillators is their respective probability distributions. Figures 12.34 and 12.35 show that the quantum oscillator in its ground state is most likely to be found in the neighborhood of the equilibrium separation at $x = 0$. In contrast, a classical oscillator is most likely to be found at or near the classical turning points, where the oscillator has come to rest. Classically, the oscillator spends more time in regions where it is stationary or moving very slowly than in regions where its speed is large, such as at the point $x = 0$. In Problem 12.55, it is shown that the classical normalized probability distribution is

$$P(x)dx = \begin{cases} \dfrac{dx}{\pi[A^2 - x^2]^{1/2}} & \text{for } -A \leq x \leq A \\ 0 & \text{for } x < -A \quad \text{or} \quad x > A, \end{cases} \qquad \textbf{(12.151)}$$

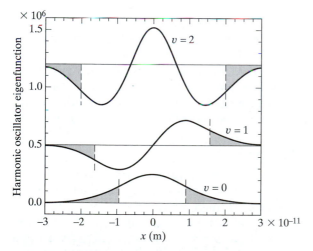

▲ FIGURE 12.34
The $v = 0$, 1, and 2 vibrational-state eigenfunctions for a harmonic oscillator whose reduced mass and vibrational force are chosen to match those of HF. The $v = 1$ and $v = 2$ eigenfunctions are displaced upwards by 5×10^5 and 1.2×10^6, respectively, to enhance the visual clarity of the plot. The vertical dashed lines indicate the classical turning points for energies of $hv_o/2$, $3hv_o/2$, and $5hv_o/2$, which are the eigenvalues for the $v = 0$, 1, and 2 states, respectively. The shaded areas correspond to classically forbidden regions that are penetrated by the quantum oscillator. As expected, the ground state is nodeless. The first excited state has one node, the second excited state two nodes. In general, the vth eigenstate has v nodes.

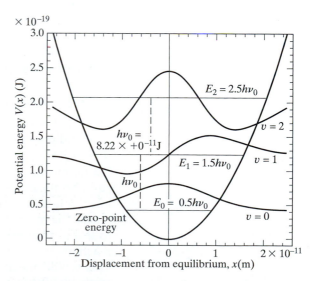

▲ FIGURE 12.35
The energy eigenvalues for a quantum mechanical harmonic oscillator whose reduced mass and vibrational force are chosen to match those of HF. The first three energy eigenvalues are shown superimposed over the harmonic potential for HF. The qualitative form of the associated eigenfunctions is also shown for each state. Note that the spacing between any two adjacent eigenvalues is always hv_o.

where $\pm A$ are the classical turning points. Equation 12.143 tells us that when the oscillator has an energy $(v + 0.5)h\nu_o$,

$$A = \left[\frac{(2v + 1)h\nu_o}{k}\right]^{1/2} = \left[\frac{(2v + 1)h(k/\mu)^{1/2}}{2\pi k}\right]^{1/2} = \left[\frac{(2v + 1)\hbar}{(\mu k)^{1/2}}\right]^{1/2}$$

$$= \left[\frac{(2v + 1)}{\alpha}\right]^{1/2}.$$

Therefore, in the $v = 0$ state, Eq. 12.151 can be written in the form

$$P(x) \, dx = \frac{dx}{\pi[\alpha^{-1} - x^2]^{1/2}} \qquad \text{for } -A \le x \le A. \qquad \textbf{(12.152)}$$

The corresponding normalized probability distribution for a quantum harmonic oscillator in its ground state is

$$|\psi(x)|^2 dx = \left[\frac{\alpha}{\pi}\right]^{1/2} \exp[-\alpha x^2] dx. \qquad \textbf{(12.153)}$$

Let us examine these two distributions for the case of an H_2 molecule with a vibration frequency of $1.350 \times 10^{14} \, s^{-1}$. Using this value and the H_2 reduced mass, we find that $\alpha = 6.730 \times 10^{21} \, m^{-2}$ and $\alpha^{-1} = 1.486 \times 10^{-22} \, m^2$. Figure 12.36 compares the classical and quantum mechanical distributions. As expected, the classical oscillator tends to be in the vicinity of the turning points, whereas the quantum oscillator leans toward the equilibrium distance. Molecules, of course, have distributions similar to the quantum result.

Now let us make the same comparison for the $v = 6$ quantum state for an H_2 molecule. This state is shown in Figure 12.37. Notice how the quantum mechanical distribution is beginning to resemble the classical result, with the most probable positions moving toward the classical turning points as v and the vibrational energy increase. This is another illustration of the Bohr correspondence principle. The probability of tunneling into the classical forbidden regions decreases as the system becomes more and more like a classical oscillator.

12.3.3   Anharmonic Effects

The parabolic harmonic potential cannot accurately represent the actual potential present in a real molecule for all values of the internuclear separation. If it did, chemical bonds, once formed, would never dissociate. Since the harmonic potential increases without bound as the displacement from equilibrium increases, no matter how much vibrational energy we might insert into a H–H molecule, for example, we could never break the chemical bond. The molecule would simply oscillate with greater and greater amplitude, but the two atoms would never dissociate. As a result, most chemical reactions would never occur, and chemistry as a science would not exist. This would mean that you would not be reading this physical chemistry textbook. Of course, it would also mean that life as we know it would not exist, so we can all agree that it is indeed fortunate that molecules do not have harmonic interatomic potentials.

The experimental fact is that *all* chemical bonds dissociate if a sufficient quantity of energy is inserted. This means that the interatomic potential must exhibit a minimum at the measured equilibrium separation and then increase as we move away from equilibrium. When the bond distance is compressed, the potential will continue to rise to very large values because

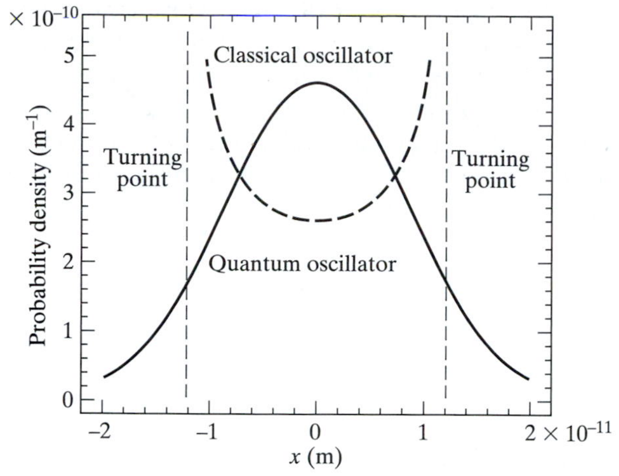

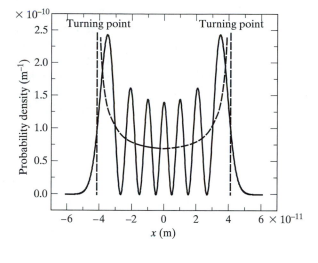

▲ **FIGURE 12.36**
The quantum mechanical and classical probability distributions for the case of an H_2 molecule with a vibration frequency of 1.350×10^{14} s^{-1} in its ground state with $v = 0$ and an energy of $h\nu_o/2$. The figure shows that the classical oscillator (dashed curve) has a probability distribution that is zero outside the classical turning points and exhibits maxima in the neighborhood of those points. In contrast, the quantum mechanical oscillator (solid curve) has a probability distribution which shows that the oscillator can tunnel into classical forbidden regions. The maximum occurs at the equilibrium separation $x = 0$.

▲ **FIGURE 12.37**
The quantum mechanical and classical probability distributions for the case of an H_2 molecule with a vibration frequency of 1.350×10^{14} s^{-1} in its sixth excited state with $v = 6$ and an energy of $13h\nu_o/2$. The figure shows that the classical oscillator (dashed curve) has a probability distribution that is zero outside the classical turning points and exhibits maxima in the neighborhood of those points. The quantum mechanical oscillator (solid curve) has a probability distribution which shows that the oscillator can tunnel into classical forbidden regions. However, the general shape of the quantum probability distribution now resembles the classical distribution much more closely than for the $v = 0$ case shown in Figure 12.36. Note how the probability maxima are approaching the turning points and the distribution is seen to oscillate around the classical curve. This is an example of the Bohr correspondence principle.

the nuclei will not undergo fusion unless we impart a huge compression energy, such as that which can be realized in a nuclear accelerator. When the bond is extended, however, the potential must eventually reach a limiting value, after which there is no further increase in the potential with increasing interatomic distance. The difference between the potential at the equilibrium distance and this asymptotic limiting value is called the *well depth*. The difference between the well depth and the zero-point vibrational energy is the measured dissociation energy of the bond. In general, the quantum mechanical vibrational energy states for this type of potential will be different from those predicted by a harmonic model. These differences are referred to as *anharmonic effects*.

We will address quantum mechanical methods for computing the actual interatomic bond potential in Chapter 14. For now, we can determine the nature of anharmonic effects by employing a simple model potential that has the characteristics we have just described. One of the most commonly used such models is the *Morse function*, which has the form

$$V_M(x) = D[1 - \exp\{-\beta x\}]^2 ,$$

(12.154)

where x is, again, the displacement from equilibrium and D and β are parameters that can be adjusted to describe different types of chemical bonds. At equilibrium, $V_M(x = 0) = 0$; at large displacements, $V_M(x \to \infty) = D$. Therefore, the value of D gives the well depth for the potential. The value of β can be obtained from the measured fundamental vibrational frequency and D. The vibrational force constant k is the curvature at the equilibrium point. For a Morse potential, the curvature is

$$\frac{d^2 V_M}{dx^2} = 2D\beta^2[2e^{-2\beta x} - e^{-\beta x}].$$

At the equilibrium point, $x = 0$, and we have $d^2 V_M/dx^2 = 2D\beta^2 = k = 4\pi^2 \mu v_o^2$. Therefore,

$$\beta = \left[\frac{2\pi^2 \mu v_o^2}{D}\right]^{1/2}. \tag{12.155}$$

Example 12.22 illustrates the fitting procedure and shows a plot of a Morse potential. Clearly, it has all the characteristics of a bond potential. (See Problem 1.22 for an additional example.)

EXAMPLE 12.22

The H_2 molecule has a measured vibrational frequency of $1.317 \times 10^{14}\ \text{s}^{-1}$. Its well depth is $7.607 \times 10^{-19}\ \text{J}$. Find values of the Morse parameters that provide a fit of the potential to these data. Plot the potential as a function of x. On the same graph, plot the corresponding harmonic potential.

Solution

To obtain the correct well depth, we take $D = 7.607 \times 10^{-19}\ \text{J}$. The reduced mass of H_2 is

$$\mu = \frac{m_H m_H}{m_H + m_H} = \frac{m_H}{2} = \frac{0.001008}{2(6.022 \times 10^{23})}\ \text{kg} = 8.369 \times 10^{-28}\ \text{kg}. \tag{1}$$

The best value of β is given by Eq. 12.155:

$$\beta = \left[\frac{2\pi^2 \mu v_o^2}{D}\right]^{1/2} = \left[\frac{2(3.14159)^2(8.369 \times 10^{-28}\ \text{kg})(1.317 \times 10^{14}\ \text{s}^{-1})^2}{7.607 \times 10^{-19}\ \text{J}}\right]^{1/2}$$

$$= 1.941 \times 10^{10}\ \text{m}^{-1}. \tag{2}$$

The Morse fit is

$$V_M(x) = 7.607 \times 10^{-19}\ [1 - \exp\{-1.941 \times 10^{10} x\}]^2\ \text{J}. \tag{3}$$

The harmonic force constant is

$$k = 4\pi^2 \mu v_o^2 = 4(3.14159)^2(8.369 \times 10^{-28}\ \text{kg})(1.317 \times 10^{14}\ \text{s}^{-1})^2$$

$$= 573.1\ \text{J m}^{-2}. \tag{4}$$

The corresponding harmonic potential is

$$V_H(x) = \frac{kx^2}{2} = \frac{573.1 x^2}{2} = 286.5 x^2\ \text{J}. \tag{5}$$

Figure 12.38 shows a comparison of the Morse and harmonic potentials in Eqs. 4 and 5. Notice that the curvatures at $x = 0$ are the same.

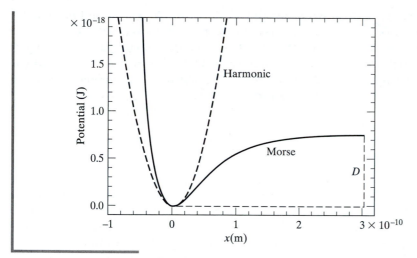

◀ FIGURE 12.38
Harmonic (dashed curve) and Morse (solid curve) potentials fitted to data for the H_2 molecule with a measured vibrational frequency of 1.317×10^{14} s^{-1} and a well depth D of 7.607×10^{-19} J. Note that the curvatures of the two potentials are the same at $x = 0$.

We can anticipate qualitatively the nature of the differences between the quantum mechanical energy levels for a Morse oscillator and those for the harmonic potential. The harmonic energies and the spacings between energy levels are both directly proportional to the vibrational frequency ν_o. Equation 12.145 shows that ν_o is proportional to the square root of the curvature k. An examination of Figure 12.38 reveals that the curvature of the Morse potential decreases as x increases. As the energy rises, the amplitude of vibration will increase, thereby causing the oscillator to sample regions of smaller curvature. As a result, the vibrational frequency will decrease as the vibrational quantum number increases. The Morse vibrational levels are, therefore, expected to lie below the harmonic levels for the same quantum number, and the spacings between adjacent energy states will decrease at higher vibrational energies. When the vibrational energy exceeds D, the diatomic molecule is no longer bound. It dissociates into two atoms. Consequently, it becomes an unbound system that exhibits a continuous range of energies. As a result, there will be a finite number of quantized vibrational energy levels for the Morse oscillator, whereas a harmonic oscillator exhibits an infinite number. Figure 12.39 illustrates this point.

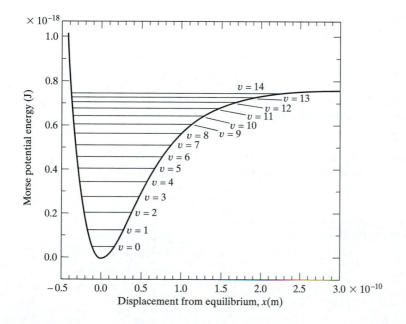

◀ FIGURE 12.39
The Morse potential obtained in Example 12.22 and shown in Figure 12.38 for H_2. The positions of the first 14 vibrational energy levels for H_2 are indicated. The decreasing vibrational-state spacing with increasing vibrational quantum number is apparent.

To obtain the vibrational energy levels for the Morse potential, we must substitute Eq. 12.154 into the one-dimensional vibrational Schrödinger equation and solve for the eigenfunctions and energy eigenvalues. If we again assume that we may separate rotational and vibrational motion, as we did to obtain the results for the rigid rotor and harmonic oscillator, the Morse vibrational energy eigenvalues turn out to be

$$E_v = (v + 0.5)h\nu_o - \frac{h^2\nu_o^2}{4D}(v + 0.5)^2.$$ (12.156)

The first term in Eq. 12.156 is the same as that for a harmonic oscillator. The second term is called the *anharmonicity*. As expected, this term is negative, so we always have $[E_v]_{Morse} < [E_v]_{harmonic}$. Since the magnitude of the second term depends upon v^2, it will become increasingly important as the vibrational energy increases. When v is small, however, the anharmonicity correction can often be ignored. In the case of H_2, the anharmonicity changes the $v = 0, 1$, and 2 eigenvalues of the harmonic oscillator by 1.44%, 4.31%, and 7.18%, respectively. Therefore, a harmonic approximation gives reasonably good answers for the low-lying vibrational states. (See Problem 12.59.) Figure 12.39 shows the positions of the successive vibrational levels for H_2 when its bond potential is described by a Morse function that uses the parameters obtained in Example 12.22. The decreasing vibrational-state spacing with increasing vibrational quantum number is clearly shown in the figure.

Equation 12.156 gives the vibrational energy levels if the interatomic potential is accurately described by a Morse potential. The actual bond potential will, of course, deviate to some extent from a Morse function. The common experimental practice is, therefore, to fit the measured vibrational energy levels to an expression of the form

$$E_v = (v + 0.5)h\nu_o - \omega_\varepsilon\chi_e(v + 0.5)^2,$$ (12.157)

where $\omega_\varepsilon\chi_e$ is used as a least-squares fitting parameter. If the interatomic potential is close to a Morse function, we should find that $h^2\nu_o^2/(4D) = \omega_\varepsilon\chi_e$. The extent to which this equality fails to hold is a measure of the deviation of the actual bond potential from a Morse function. Example 12.23 illustrates this idea.

EXAMPLE 12.23

When the measured vibrational energy levels of H_2 are fitted to Eq. 12.157, it is found that $\omega_\varepsilon\chi_e = 2.343 \times 10^{-21}$ J. The vibrational frequency of H_2 is 1.317×10^{14} s^{-1}, and the experimental well depth is 7.607×10^{-19} J. Use Eq. 12.156 and the measured value of $\omega_\varepsilon\chi_e$ to compute the value of the Morse parameter D. What is the percent error in the predicted well depth?

Solution

If the bond potential is accurately described by a Morse function, the fitted coefficient of the anharmonicity term should be given by

$$\omega_\varepsilon\chi_e = \frac{h^2\nu_o^2}{4D} = 2.343 \times 10^{-21} \text{ J.}$$ (1)

Solving for D, we obtain

$$D = \frac{h^2\nu_o^2}{4(2.343 \times 10^{-21} \text{ J})} = \frac{(6.626 \times 10^{-34} \text{ J s})^2(1.317 \times 10^{14} \text{ s}^{-1})^2}{4(2.343 \times 10^{-21} \text{ J})}$$

$$= 8.125 \times 10^{-19} \text{ J.}$$ (2)

If the potential is a Morse function, we would predict a well depth of 8.125×10^{-19} J. The percent error in this prediction is

$$\% \text{ error} = \frac{100(8.125 - 7.607) \times 10^{-19} \text{ J}}{7.607 \times 10^{-19} \text{ J}} = 6.8\% \tag{3}$$

For related exercises, see Problems 12.56 and 12.57.

We shall defer a discussion of polyatomic vibrational energy levels until Chapter 15.

12.4   Coupled Vibrational–Rotational Energy

If the rigid-rotor–harmonic-oscillator approximations were accurate descriptions of the vibrational–rotational motion of a diatomic molecule, we would expect the total vibrational–rotational energy to be simply the sum of those two energies:

$$E_{vJ} = (v + 0.5)h\nu_o + J(J + 1)\frac{\hbar^2}{2I}. \tag{12.158}$$

However, the approximations made in Sections 12.2 and 12.3 are not consistent. The rotational energy levels for the two-particle rigid rotor were obtained using the assumption that the interparticle distance is fixed. In contrast, the vibrational levels for the harmonic and Morse oscillators were obtained assuming that the problem is one dimensional, in that rotation does not occur. If both rotation and vibration occur simultaneously, the problem becomes three dimensional and much more difficult to solve. We can, however, qualitatively assess the effect on the combined vibrational–rotational energy levels.

If the interatomic distance can extend during rotation, it will do so because of the centrifugal force on the atoms caused by the rotational angular momentum. This effect is analogous to that observed when swinging a ball attached to a rubber band around your head. The centrifugal force causes the rubber band to stretch, with the amount of stretch increasing as the rotational energy increases. As the bond distance increases due to the centrifugal force, the moment of inertia of the rotor will increase. Since the rotational energies are inversely proportional to the moment of inertia [see Eq. 12.107], the centrifugal effect will decrease the rotational-energy eigenvalues.

The combination of anharmonicity and centrifugal effects causes the rotational and vibrational motions to become coupled. An increase in rotational energy increases the centrifugal effect, which stretches the bond into regions where the vibrational potential curvature is less. This results in a decrease in the vibrational-energy eigenvalues. At the same time, an increase in vibrational energy stretches the bond and thereby increases the moment of inertia, which produces a decrease in the rotational-energy eigenvalues.

Let us suppose we were to solve the Schrödinger equation for the Morse potential under conditions in which rotation and vibration occur simultaneously. What type of modifications to Eq. 12.158 might we expect to find? The discussion in the previous section shows that we would certainly expect to see a term of the form $-[h^2\nu_o^2/(4D)](v + 0.5)^2$, in order to account for the effect of the anharmonicity in the bond potential. The above discussion suggests that centrifugal effects would produce a negative term which decreases the rotational-energy eigenvalues. We would expect this term to become larger as J increases, since the centrifugal force increases with increasing J. Finally, the coupling effect should produce a negative term that depends upon both v and J that decreases the vibrational–rotational energies as either v or J increases.

Qualitatively, this is exactly the type of result we obtain when the Schrödinger equation for a rotating Morse oscillator is solved. A solution that is reasonably accurate provided that the total rotational–vibrational energy is not too large is

$$E_{vJ} = (v + 0.5)hv_o - \frac{h^2 v_o^2}{4D}(v + 0.5)^2 + J(J + 1)\frac{\hbar^2}{2I} - \frac{\hbar^4}{4\mu^2\beta^2 R_e^6 D}[J(J + 1)]^2$$

$$- \frac{3[1 - (1/\beta R_e)]\hbar^2 hv_o}{4\mu\beta R_e^3 D}(v + 0.5)J(J + 1). \tag{12.159}$$

The first and third terms in Eq. 12.159 are the harmonic-oscillator–rigid-rotor energy eigenvalue expressions in Eq. 12.158. The second term is the anharmonicity term discussed in the previous section. The negative fourth term is due to centrifugal distortion. Its magnitude becomes much larger as J increases, as we anticipated. The final term is the result of vibration–rotation coupling. It is negative and depends upon both v and J, as it should. As either v or J increases, its presence produces a decrease in the vibrational-rotational eigenvalue E_{vJ}.

Equation 12.159 is the result of an approximate solution of the Schrödinger equation for a rotating Morse oscillator. A real molecule exhibits rotational–vibrational energy levels that deviate somewhat from the predictions of that equation. Consequently, it is common practice to use a form such as

$$E_{vJ} = (v + 0.5)hv_o + J(J + 1)\frac{\hbar^2}{2I} - \omega_e\chi_e[v + 0.5]^2 - A[J(J + 1)]^2$$

$$- B(v + 0.5)J(J + 1) \tag{12.160}$$

to fit measured vibrational–rotational energy levels of real molecules. We shall return to these forms in Chapter 15.

Summary: Key Concepts and Equations

1. When a system has an energy greater than the potential, it is called an unbound system. The free particle is an example of such a system. The energies of unbound systems are continuous variables, whereas bound systems exhibit quantized energy levels. Therefore, the free particle has continuous energy levels and momenta. The eigenfunctions of the Hamiltonian for the free particle are

$$\psi(x) = Ae^{ikx} + Be^{-ikx},$$

where k is called the wave number. Its relationship to the energy is $k^2 = 2mE_x/\hbar^2$, so that the energy eigenvalues for a one-dimensional free particle moving in the x direction are

$$E_x = \frac{k^2\hbar^2}{2m},$$

with k allowed to assume any value. The eigenfunction $\psi(x) = Ae^{ikx}$ represents a particle moving in the positive x direction with momentum $k\hbar$. If the eigenfunction is $\psi(x) = Be^{-ikx}$, the particle is moving in the negative x direction with momentum $-k\hbar$. In both cases, we have no knowledge of the particle's location.

2. A free particle moving in three-dimensional space has eigenfunctions that are the products of the corresponding one-dimensional eigenfunctions and eigenvalues

that are the sum of the one-dimensional eigenvalues. Thus, the three-dimensional free particle has eigenfunctions

$$\psi(x, y, z) = F(x)G(y)H(z)$$
$$= [A_x e^{ik_x x} + B_x e^{-ik_x x}][A_y e^{ik_y y} + B_y e^{-ik_y y}][A_z e^{ik_z z} + B_z e^{-ik_z z}]$$

and eigenvalues

$$E = E_x + E_y + E_z = \frac{k_x^2 \hbar^2}{2m} + \frac{k_y^2 \hbar^2}{2m} + \frac{k_z^2 \hbar^2}{2m}.$$

Because the free particle is unbound, the energies and momenta are continuous variables.

3. The simplest example of a bound system is a particle in an infinite one-dimensional potential well. Since the potential is infinite at the boundaries, all translational energy states are quantized. The particle cannot penetrate the infinitely high potential barriers, so its eigenfunctions in this region are zero. Within the well, application of the continuity requirements of the first postulate at the boundaries produces quantization of the energy. This is a general result: Quantization, when it occurs, is always the result of the requirements imposed upon the eigenfunctions by the first postulate. The allowed eigenfunctions within the well are

$$\psi(x) = \left[\frac{2}{a}\right]^{1/2} \sin\left[\frac{n_x \pi x}{a}\right] \quad \text{for } 0 \le x \le a \quad \text{with } n_x = 1, 2, 3, \ldots .$$

The corresponding translational eigenvalues are

$$E_x = \frac{n_x^2 \hbar^2 \pi^2}{2ma^2} = \frac{n_x^2 h^2}{8ma^2}.$$

Inserting numerical data demonstrates that the spacing between adjacent translational energy states is too small to be detected. Therefore, translational energy states appear to be continuous. Like the free particle, a particle in a three-dimensional well in the shape of a parallelepiped with rectangular faces has eigenfunctions that are the products of the corresponding one-dimensional eigenfunctions and eigenvalues that are the sum of the one-dimensional eigenvalues.

4. Examination of the eigenfunctions for a particle in a one-dimensional infinite potential well shows that the n_x eigenfunction possesses $n_x - 1$ nodes. The ground-state eigenfunction is nodeless. This is always the case for any quantum mechanical system. Also, the mth excited state has m nodes. In a one-dimensional system, all the nodes lie along the sole coordinate of the system. For a multidimensional system, the nodes are distributed among the coordinates in a manner that depends upon the system. Since it is possible for two nodes to appear at the same point when the system is multidimensional, we cannot always determine the qualitative ordering of energy states by a simple node count. The situation is also complicated by the fact that nodes along some coordinates may serve to increase the energy more than nodes along other coordinates.

5. One-dimensional systems cannot have degenerate eigenfunctions. However, multidimensional systems often exhibit such degeneracies. When two eigenfunctions have degenerate eigenvalues, they do not have to be orthogonal, but in most cases they are. When they are not, appropriate linear combinations of the degenerate eigenfunctions that are orthogonal can always be developed. The eigenfunctions for every system examined in this chapter are orthogonal even when they are degenerate.

6. The wave function for a system can be any combination of the allowed eigenfunctions of the system. If we know that the system is in a particular eigenstate, the wave function will be the eigenfunction corresponding to that eigenstate. If, however, we know only that a system is in either state 1, 2, 3, ..., or M, the wave

function will be a linear combination (superposition) of the eigenfunctions corresponding to those states. That is, we will have

$$\psi = N[C_1\phi_1 + C_2\phi_2 + C_3\phi_3 + \cdots + C_M\phi_M].$$

If the eigenfunctions $\phi_1, \phi_2, \ldots, \phi_M$ are orthogonal and individually normalized, the probability of finding the system in eigenstate k is given by

$$P(k) = \left[\sum_{m=1}^{\infty} |C_m|^2\right]^{-1} |C_k|^2.$$

The factor in brackets is the normalization constant for the superposition.

7. When a physical system is described by using matter waves and quantum theory, we find that it is impossible to simultaneously know both the momentum and position of a particle. The more exactly the particle's position is determined, the less knowledge we have of the momentum, and vice versa. In general, quantum mechanics predicts that the product of the square uncertainties for any two variables u and w always obeys the inequality

$$\langle \Delta u^2 \rangle \langle \Delta w^2 \rangle \geq \frac{1}{4}|[u, w]|^2,$$

where $[u, w]$ is the commutator of the variables u and w. If u and w are conjugate variables, such as position and momentum, the commutator is $\pm i\hbar$, so that, for x and p_x, we have

$$\langle \Delta x^2 \rangle \langle \Delta p_x^2 \rangle \geq \frac{\hbar^2}{4}.$$

This relationship is called the uncertainty principle. In view of that principle, it is not surprising that the particle in a one-dimensional well and the harmonic oscillator have nonzero ground-state energies. If the energy were zero, we would know the momentum precisely (it would be zero) while the uncertainty in position is finite. As a result, we would violate the uncertainty principle.

8. Classical systems cannot enter into a region of space where the potential energy exceeds the total energy of the system. However, the wave character of quantum theory permits this to occur. The phenomenon is called tunneling. For the simple case of a one-dimensional particle with energy E impinging upon a square potential barrier whose height is V_o, the tunneling probability depends upon E, V_o, the mass m of the particle, and the width a of the barrier. When $(V_o - E)$, m, and a are all large, this probability is given approximately by

$$\tau = \frac{16E(V_o - E)}{V_o^2} \exp\left[-2\left\{\frac{2m}{\hbar^2}[V_o - E]\right\}^{1/2} a\right].$$

Consequently, tunneling is favored by light particles having energy nearly equal to a potential barrier that is narrow. Numerical calculations show that we expect the phenomenon to be important for electrons and, perhaps, hydrogen atoms.

9. An N-atom system requires $3N$ coordinates to stipulate the particle positions. If three of these coordinates are taken to be the Cartesian coordinates of the system's center of mass and the remaining $3N - 3$ coordinates are used to express the Cartesian components of $N - 1$ independent vectors giving the relative positions of the atoms, the kinetic energy will always have the form

$$T = \frac{M}{2}[v_{XC}^2 + v_{YC}^2 + v_{ZC}^2] + \sum_{i=1}^{N-1} \frac{\mu_i}{2}[v_{Xi}^2 + v_{Yi}^2 + v_{Zi}^2],$$

where μ_i is the reduced mass for the two particles connected by vector i and v_{Xi}, v_{Yi}, and v_{Zi} are the Cartesian velocity components of that vector.

10. The eigenfunctions $\psi(\theta, \phi)$ for a two-atom rigid rotor depend upon two angular variables θ and ϕ, which are the azimuthal angle the rotor makes with the Z-axis and the rotation angle about Z of the projection of the rotor into the X–Y plane, respectively. Consequently, we expect that the eigenfunctions will be dependent upon two quantum numbers. The solutions have the form

$$\psi(\theta, \phi) = \Theta_J^M(\theta)\,\Phi_M(\phi) = Y_J^M(\theta, \phi),$$

where $Y_J^M(\theta, \phi)$ is called a spherical harmonic. The portion of the solution dependent upon ϕ is

$$\Phi_M(\phi) = (2\pi)^{-1/2}e^{-iM\phi}.$$

The requirement of the first postulate that the wave function be single valued restricts M to integer values between $-\infty$ and $+\infty$. The $\Theta_J^M(\theta)$ solutions are more difficult to write in closed form. The 16 eigenfunctions with the lowest rotational energies are given in Table 12.1. The two quantum numbers are J and M. These are called the azimuthal and magnetic quantum numbers, respectively. $\Theta_J^M(\theta)$ depends upon both J and M, whereas $\Phi_M(\phi)$ is dependent only upon M. The value of J determines the rotational energy and angular momentum via the equations

$$E_{\text{rot}} = J(J + 1)\frac{\hbar^2}{2I}$$

and

$$L^2 = J(J + 1)\hbar^2.$$

The Z-component of angular momentum is determined by M according to the equation

$$L_z = M\hbar.$$

The possible values of J are $0, 1, 2, 3, \ldots, \infty$. The eigenfunctions with $J = 0, 1, 2$, and 3 are often called s, p, d, and f functions, respectively. To ensure that we do not have $L_z^2 > L^2$, the magnetic quantum must lie in the range $-J \leq M \leq J$.

11. The rigid-rotor eigenfunctions possess J nodal surfaces or axes. Since $\Phi_M(\phi)$ contains no nodes, all of the nodes are in the θ coordinate and the $\Theta_J^M(\theta)$ part of the eigenfunction. In some cases, nodal surfaces or axes are superimposed so that they appear as a single node. Consequently, a simple node count is not always sufficient to determine the energy state of the rotor. The spatial forms of the s, p, d, and f functions are shown in Figures 12.20 through 12.22. As we shall see in Chapter 14, the positive and negative regions of these orbitals play an important role in chemical bonding. The absolute squares of the rigid-rotor eigenfunctions are shown in Figures 12.23 through 12.25.

12. Because the commutator of L_x and L_y is $i\hbar L_z$, the square uncertainty product between L_x and L_y is

$$\langle \Delta L_x^2 \rangle \langle \Delta L_y^2 \rangle \geq \frac{1}{4}|i\hbar L_z|^2 = \frac{M^2\hbar^4}{4}.$$

Therefore, we cannot know both L_x and L_y unless the system is in a spherically symmetric $J = 0\,s$ orbital where M must be zero. If this is not the case, the most we can ever know about the direction of the angular momentum vector is the angle θ it makes with the Z-axis. Its rotation angle ϕ about Z cannot be determined.

13. Polyatomic rigid bodies possess three principal moments of inertia: I_a, I_b, and I_c. Each is given by

$$I_u = \sum_{i=1}^{N} m_i\, r_{u,i}^2$$

where the summation runs over all N atoms in the molecule, m_i are the atomic masses, and $r_{u,i}$ is the perpendicular distance from atom i to the uth principal axis, with $u = a$, b, or c. The principal axes can be located by examining the products of inertia I_{xy}, I_{xz}, and I_{yz}, where

$$I_{xy} = -\sum_{i=1}^{N} m_i x_i y_i,$$

with analogous definitions for I_{xz} and I_{yz}. In this equation, x_i and y_i are the Cartesian coordinates of atom i. If the principal axes are correctly identified, we always have $I_{xy} = I_{xz} = I_{yz} = 0$. A molecular symmetry axis always coincides with one of the principal axes, and a symmetry plane always contains two principal axes that are perpendicular to the third.

14. If all principal moments of inertia of a rigid body are equal, the body is called a spherical top. If $I_a = I_b \neq I_c$ or $I_b = I_c \neq I_a$, the rigid body is termed a symmetric top. If the three moments are different, we call the rotor an asymmetric top. Using a classical argument, we deduced that the quantum mechanical expressions for the rotational energy levels of spherical and symmetric tops are

$$E_J = \frac{J(J+1)\hbar^2}{2I} \qquad \text{for } J = 0, 1, 2, 3, \ldots$$

and

$$E_{J,K} = \frac{J(J+1)\hbar^2}{2I_b} + \frac{K^2\hbar^2}{2}\left[\frac{1}{I_a} - \frac{1}{I_b}\right] \qquad \text{for } J = 0, 1, 2, 3, \ldots \text{ and } -J \leq K \leq J,$$

respectively.

15. A classical harmonic oscillator has a potential $V(x) = kx^2/2$, where x is the displacement from the equilibrium position. By solving the Newtonian equations of motion, we found that the displacement is the simple sinusoidal function

$$x = A \sin\left[\left(\frac{k}{\mu}\right)^{1/2} t\right].$$

The vibrational period, frequency, and classical turning points are

$$\tau = 2\pi\left(\frac{\mu}{k}\right)^{1/2}, \quad \nu_o = \frac{1}{\tau} = \frac{1}{2\pi}\left(\frac{k}{\mu}\right)^{1/2}, \quad \text{and } x_{\text{turn}} = \pm\left[\frac{2E_{\text{vib}}}{k}\right]^{1/2},$$

respectively.

16. The quantum mechanical eigenfunctions for the harmonic oscillator can be obtained by solving the one-dimensional Schrödinger equation via either series techniques or operator methods. The solutions have the form

$$\psi_v(x) = N_v H_v(y) \exp\left[-\frac{y^2}{2}\right],$$

where $y = \alpha^{1/2}x$, with $\alpha = (\mu k)^{1/2}/\hbar$, and $N_v = [(\alpha/\pi)^{1/2}/2^v v!]^{1/2}$. $H_v(y)$ is a polynomial in y whose leading power is y^v. It is called a Hermite polynomial. Some of these polynomials and the associated vibrational energy eigenvalues are listed in Table 12.2. The energy eigenvalues may be written in the form

$$E_v = (v + 0.5)h\nu_o.$$

The ground-state energy eigenvalue is, therefore, $h\nu_o/2$. This quantity is called the zero-point energy. The vth harmonic-oscillator eigenfunction has v nodes.

17. Interatomic potentials for real molecules cannot be harmonic. If they were, chemical bonds could never be broken. The experimental potential must increase and eventually approach an asymptotic limiting value as the displacement from equi-

librium increases. The Morse potential is a reasonable approximation to an experimental interatomic potential. Its form is

$$V_M = D[1 - \exp\{-\beta x\}]^2.$$

If vibrational and rotational motion are assumed to be separable, the solution of the Schrödinger equation with this potential gives vibrational eigenvalues

$$E_v = (v + 0.5)h\nu_o - \frac{h^2\nu_o^2}{4D}(v + 0.5)^2.$$

The first term is the same as that for a harmonic oscillator. The second term is called the anharmonicity. As expected, this term is negative, so that we always have $[E_v]_{\text{Morse}} < [E_v]_{\text{harmonic}}$. Since the magnitude of the second term depends upon v^2, it becomes increasingly important as the vibrational energy increases. When E_v exceeds D, the vibrational energy is continuous because the system becomes unbound for these energies.

18. Vibrational and rotational motions are actually coupled. Therefore, the vibrational eigenvalues depend upon the rotational energy, and vice versa. When the Schrödinger equation for a diatomic system with a Morse potential is solved approximately, the coupled energy eigenvalues are

$$E_{vJ} = (v + 0.5)h\nu_o - \frac{h^2\nu_o^2}{4D}(v + 0.5)^2 + J(J + 1)\frac{\hbar^2}{2I} - \frac{\hbar^4}{4\mu^2\beta^2R_e^6D}[J(J + 1)]^2$$

$$- \frac{3[1 - (1/\beta R_e)]\hbar^2h\nu_o}{4\mu\beta R_e^3D}(v + 0.5)J(J + 1).$$

The first and third terms are the harmonic-oscillator–rigid-rotor energy eigenvalue expressions. The second term is the anharmonicity. The negative fourth term is due to centrifugal distortion. Its magnitude becomes much larger as J increases. The final term is the result of vibrational–rotational coupling.

Problems

Problems that require the use of some type of computational device are marked with an asterisk (*). Problems that require some type of plotting routine are indicated with a pound sign (#).

12.1 Verify that Eq. 12.9 is a solution of the stationary-state Schrödinger equation for the one-dimensional free particle.

12.2 Show that the average square momentum for a free particle whose eigenfunction is $\psi(x) = Ae^{ikx}$ is $k^2\hbar^2$.

12.3 Show that a quantum free particle whose wave function is $\psi(x) = Be^{-ikx}$ has an average momentum $-k\hbar$ and an average square momentum $k^2\hbar^2$, so that the square uncertainty in the momentum is exactly zero.

12.4 A quantum free particle of mass m is moving in three-dimensional space. The translational energy for motion in the x direction is twice that for motion in the y direction and one-half that for motion in the z direction. It is known that the free particle is moving in the negative z and positive x and y directions. Is this sufficient information to determine the angle

the total momentum vector makes with the z-axis? If not, state what additional information is needed to determine this angle. If it is sufficient, compute the value of the angle.

12.5 The functions $\psi_2(x) = [2/a]^{1/2}\sin[2\pi x/a]$ and $\psi_3(x) = [2/a]^{1/2}\sin[3\pi x/a]$ are eigenfunctions for a particle in an infinite one-dimensional well. Show that these eigenfunctions are orthogonal. Could their orthogonality have been deduced without integrating? How? [Hint: $\sin(ax)$ can be written in the form $(e^{iax} - e^{-iax})/(2i)$.]

12.6 The functions $\psi_n(x) = [2/a]^{1/2}\sin[n\pi x/a]$ and $\psi_m(x) = [2/a]^{1/2}\sin[m\pi x/a]$ are eigenfunctions for a particle in an infinite one-dimensional well. Show that, if $n \neq m$, these two eigenfunctions are orthogonal. Could their orthogonality have been deduced without integrating? How? [Hint: $\sin(ax)$ can be written in the form $(e^{iax} - e^{-iax})/(2i)$.]

12.7 The classical kinetic energy of a one-dimensional free particle moving in the x direction is $p_x^2/(2m)$, where p_x is a continuous variable. Is there any similarity between the classical system and the quantum free particle? Discuss the similarities and differences.

12.8 Show that letting the translational quantum number for a one-dimensional particle in an infinite well be zero reduces the wave function for the system to the trivial solution.

12.9 A particle whose mass is 1.67×10^{-27} kg is in a one-dimensional infinite potential well of width 3 Å. The eigenfunction for the quantum state of this particle has a wavelength of 0.4 Å.

(A) Compute the particle's translational energy.

(B) Calculate the magnitude of the momentum of the particle.

12.10 Consider a particle of mass m confined in an infinite one-dimensional potential well of length a. Suppose the system is in quantum state n.

(A) Determine the uncertainty in position of the particle, $\Delta x = [\Delta x^2]^{1/2}$, as a function of n, m, a, and fundamental constants.

(B) Determine the uncertainty in the momentum of the particle, Δp_x, as a function of n, m, a, and fundamental constants.

(C) In what quantum state is the product $\Delta x \, \Delta p_x$ a minimum? What is its value in this state in terms of $\hbar$?

12.11[#] A Las Vegas casino sets up a gambling game using a "particle" in a one-dimensional infinite potential well as the device. Players place bets that, upon measurement, the "particle" will be found within a distance b of the center of the box, located at the point $x = a/2$.

(A) Plot the probability that the "particle" will be found within a distance b of the center as a function of the ratio b/a for the $n_x = 1$ ground state of the system. On the same graph, plot the result if the system is in the first excited state, $n_x = 2$.

(B) The casino desires to fix the value of b such that the player has a 48% chance of winning an even-money bet. If the casino announces that the "particle" will be in its ground state during the game, what value of b should the casino use as the cutoff. That is, if the "particle" is found within distance b of the center, the player wins. If the "particle" is beyond this distance, the casino wins.

(C) The casino, wishing to increase its advantage, has a hidden device that permits the "particle" to be excited into the $n_x = 2$, first excited state without the player's knowledge. If the particle is so excited, what are the player's chances of winning the bet?

12.12 Suppose we have a device that can detect translational energies as small as 10^{-28} J atom^{-1}. Suppose also that we have 1 mole of helium atoms in a one-dimensional container of width 0.1000 m. Estimate the temperature these atoms would need to have before the spacing between the average translational quantum state and the next higher state

could be detected. The mass of the helium atom is 6.647×10^{-27} kg.

12.13 Solve Eq. 12.47 to obtain the eigenfunctions for a three-dimensional particle in an infinite rectangular parallelepiped well. Show that the total translational energy is given by

$$ E = \frac{h^2}{8m} \left[\frac{n_x^2}{a^2} + \frac{n_y^2}{b^2} + \frac{n_z^2}{c^2} \right]. $$

12.14 Determine the degeneracy g of the lowest 12 energy states for a particle in an infinite three-dimensional potential well whose dimensions are such that $a = b = c$. For each eigenstate, specify the number of nodes present and how many are in each coordinate. This problem illustrates the difficulty encountered in trying to determine the energy of a multidimensional system using a simple node count.

12.15 It is stated in the text that degenerate eigenfunctions of an operator need not be orthogonal, but in the event that they are not, linear combinations of the nonorthogonal eigenfunctions can always be developed that are orthogonal. In this problem, you will explore how this can be done. The first step is to repeat Problem 11.23.

(A) Let ϕ_1 and ϕ_2 be degenerate, normalized eigenfunctions of the operator G. That is, the eigenvalues of G associated with ϕ_1 and ϕ_2 are both the same. Show that any linear combination of ϕ_1 and ϕ_2 is also an eigenfunction of G with the same eigenvalue.

(B) Assume that ϕ_1 is not orthogonal to ϕ_2. This means that the integral $\langle \phi_1 | \phi_2 \rangle$ will be equal to S, where S is not zero. In (A), we found that any arbitrary linear combination of ϕ_1 and ϕ_2 is also an eigenfunction of G. Therefore, instead of using the eigenfunction ϕ_2, we are free to substitute an eigenfunction $\phi_3 = c_1 \phi_1 + c_2 \phi_2$, with c_1 and c_2 arbitrary constants. Find the values of c_1 and c_2 in terms of S that make ϕ_3 orthogonal to ϕ_1 and, at the same time, normalize ϕ_3.

12.16 Sam struggled throughout the entire first-semester course in physical chemistry. At times, his grade was so low that there appeared to be no hope. However, as the semester progressed, his creativity caught the professor's eye. He also began to understand the subject. When the semester ended, Sam had a high D, which his professor raised to a C as a reward for significant improvement and for creative thought about the subject. Sam continues to improve with each passing day. After studying his professor's lecture notes on the quantum particle in an infinite, one-dimensional potential well, he posed the following problem to his professor and the class:

The solutions to the stationary-state Schrödinger equation for the particle in an infinite well should be independent of the coordinate system chosen. In class and in the text, the chosen coordinate system placed the left boundary of the well at the origin, $x = 0$, and the right boundary at $x = a$. Let us consider choosing a coordinate system such that the well is symmetrically placed with respect to the origin. That is, we take the center of well to be at the point $x = 0$ and the left and right boundaries at $x = -a/2$ and $a/2$, respectively. This coordinate system is illustrated in the following diagram:

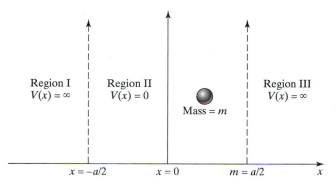

If we now solve the equation by using the procedure presented in Sam's lecture and in the text, we will have $\psi_I(x) = \psi_{III}(x) = 0$, due to the infinite potential. In Region II, the general solution will still be given by Eq. 12.26, viz.,

$$\psi_{II}(x) = C_1 \sin(kx) + C_2 \cos(kx), \quad \text{with } k^2 = \frac{2mE}{\hbar^2}.$$

If we now take $C_2 = 0$, as was done in class, we must have $C_1 \sin(kx) = 0$ at the well boundaries, which are now $x = -a/2$ and $x = a/2$. That is, we require that

$$\sin\left(\frac{ka}{2}\right) = \sin\left(-\frac{ka}{2}\right) = -\sin\left(\frac{ka}{2}\right) = 0.$$

This can be true only if $ka/2$ is an integral multiple of π radians; that is, we must have

$$\frac{ka}{2} = n_x\pi, \quad \text{with } n_x = 1, 2, 3, \dots.$$

Squaring both sides and substituting $2mE/\hbar^2$ for k^2, we find that the energy expression becomes

$$E = \frac{4n_x^2\pi^2\hbar^2}{2ma^2} = \frac{4n_x^2h^2}{8ma^2}.$$

The corresponding wave functions are

$$\psi_{II}(x) = \left[\frac{2}{a}\right]^{1/2} \sin\left[\frac{2n_x\pi x}{a}\right].$$

Obviously, these solutions are not the ones derived in class and in the text. [See Eqs. 12.38 and 12.39.] What is wrong? Do the solutions depend upon the coordinate system chosen? If so, explain why. If not,

show that the solutions in the symmetric coordinate system are identical to those derived in class. [What do you think of Sam's performance in physical chemistry now?]

12.17 A particle in an infinite potential well is known to be in either the $n_x = 2$ or $n_x = 3$ eigenstates. The eigenfunctions of these states are $\phi_2(x) = [2/a]^{1/2} \sin[2\pi x/a]$ and $\phi_3(x) = [2/a]^{1/2} \sin[3\pi x/a]$, respectively.

(A) Write an appropriate wave function for the system that reflects our knowledge of the state of the system.

(B) What energies might be obtained if the energy of the particle is measured? What is the probability of obtaining each of these values?

12.18 The wave function for a particle in an infinite, one-dimensional potential well of width a is

$$\psi(x) = N\left\{\left[\frac{2}{a}\right]^{1/2} \sin\left[\frac{\pi x}{a}\right] + \sqrt{2}\left[\frac{2}{a}\right]^{1/2} \sin\left[\frac{2\pi x}{a}\right]\right.$$
$$\left. + \sqrt{3}\left[\frac{2}{a}\right]^{1/2} \sin\left[\frac{3\pi x}{a}\right]\right\}.$$

(A) If we have 10^6 such particles and we measure the energy of the particle in each of the 10^6 systems, on average, how many times will we obtain the result $E = 4h^2/(8ma^2)$? How many times will we obtain the result $E = 16h^2/(8ma^2)$?

(B) If the energy is measured and found to be $E = 9h^2/(8ma^2)$, what is the wave function of the system after this measurement is made?

12.19* A particle in an infinite, one-dimensional potential well is equally likely to be in any of the six quantum states of lowest energy.

(A) Write down an appropriate form for the normalized wave function of this system.

(B) Plot the wave function over the range $0 \le x \le a$ if $a = 1$ Å.

(C) Are there other possibilities for the wave function that reflect our state of knowledge of the system? If not, why not? If so, illustrate two of them with appropriate equations and plots.

12.20* The eigenfunctions for a particle in a one-dimensional infinite potential well of width a are

$$\psi(x) = \left[\frac{2}{a}\right]^{1/2} \sin\left[\frac{n_x\pi x}{a}\right], \quad \text{for } n_x = 1, 2, 3, 4, 5, 6, \dots.$$

The energies of these eigenstates are

$$E_n = \left[\frac{n_x^2 h^2}{8ma^2}\right], \quad \text{for } n_x = 1, 2, 3, 4, 5, 6, \dots.$$

Suppose the only information we have about the particle is that it is either in the $n_x = 1$ or in the $n_x = 2$ eigenstate, but we do not know which one.

The wave function will, therefore, be a superposition of the $n_x = 1$ and $n_x = 2$ eigenstates. That is,

$$\psi(x) = C\left\{ \left[\frac{2}{a}\right]^{1/2} \sin\left[\frac{\pi x}{a}\right] + \left[\frac{2}{a}\right]^{1/2} \sin\left[\frac{2\pi x}{a}\right] \right\}$$

$$= C\left[\frac{2}{a}\right]^{1/2}\left\{ \sin\left[\frac{\pi x}{a}\right] + \sin\left[\frac{2\pi x}{a}\right] \right\}.$$

(A) Normalize this wave function. That is, adjust C such that

$$\int_0^a \psi^*(x)\psi(x)\,dx = \langle \psi(x)|\psi(x)\rangle = 1.$$

[*Hint:* Remember, eigenfunctions that have different eigenvalues are always orthogonal.]

(B) Show that the average position for the wave function is given by

$$\langle x \rangle = a\left[\frac{1}{2} - \frac{16}{9\pi^2}\right].$$

In doing this part of the problem, you may find it helpful to remember the trigonometric identities $\sin(2ax) = 2\sin(ax)\cos(ax)$ and $\cos(2ax) = \cos^2(ax) - \sin^2(ax)$.

(C) Using any numerical integration method with $a = 1$ Å, obtain the average square position of the particle. That is, obtain the value of $\langle x^2 \rangle$ with $a = 1$ Å.

(D) Obtain an analytic expression for $\langle p_x \rangle$ for the wave function. [See helpful note in (B).]

(E) Obtain an analytic expression for $\langle p_x^2 \rangle$ for the wave function. Recall the orthogonality of eigenfunctions with different eigenvalues.

(F) Evaluate the uncertainty product $\langle \Delta x^2 \rangle^{1/2}\langle \Delta p_x^2 \rangle^{1/2}$ for the wave function when $a = 1$ Å. Show that the result satisfies the requirements of the uncertainty principle. Compare the size of the uncertainty product with that obtained when the system is in the $n_x = 1$ eigenstate. Is it smaller or larger? Give a physical reason for this result.

12.21 An electron is contained in a three-dimensional infinite potential well whose shape is that of a rectangular parallelepiped with dimensions $a = b = c = 1.0 \times 10^{-10}$ m. Use the uncertainty principle to estimate the minimum kinetic energy this electron can have.

12.22 Show that the function $\psi_{II}(x) = Ce^{Rx} + De^{-Rx}$ is a solution of the Schrödinger equation

$$\frac{\partial^2}{\partial x^2}\psi_{II}(x) - R^2\psi_{II}(x) = 0.$$

12.23 The boundary conditions at $x = 0$ and $x = a$ in Figure 12.12 lead to four equations involving the coefficients in the wave function describing tunneling through a square barrier. These equations, which are derived in the text, are

1. $A + B = C + D$,
2. $Ce^{Ra} + De^{-Ra} = Ee^{ika}$,
3. $ikA - ikB = RC - RD$,

and

4. $RCe^{Ra} - RDe^{-Ra} = ikEe^{ika}$.

(A) Combine Eqs. 1 and 3 to show that

$$A = \frac{C(R + ik) - D(R - ik)}{2ik}.$$

(B) Combine Eqs. 2 and 4 to show that

$$D = \frac{E(R - ik)e^{ika}}{2Re^{-Ra}}.$$

(C) Use Eq. 2 with the result from (B) to show that

$$C = \frac{Ee^{ika}}{e^{Ra}}\left[\frac{R + ik}{2R}\right].$$

(D) Use the results obtained from (A), (B), and (C) to show that

$$2ikA = \frac{Ee^{ika}}{2R}\left[\frac{(R + ik)^2}{e^{Ra}} + \frac{(R - ik)^2}{e^{-Ra}}\right].$$

(E) Let

$$F = \left[\frac{(R + ik)^2}{e^{Ra}} + \frac{(R - ik)^2}{e^{-Ra}}\right].$$

Show that

$$|F|^2 = F^*F = \frac{(R^2 + k^2)^2}{e^{2Ra}} + \frac{(R^2 + k^2)^2}{e^{-2Ra}}$$

$$+ 2(R^2 - k^2)^2 - 8k^2R^2.$$

(F) By taking the absolute squares of both sides of the expression derived in (D), show that

$$|A|^2 = \frac{|E|^2}{16k^2R^2}\left[\frac{(R^2 + k^2)^2}{e^{2Ra}} + \frac{(R^2 + k^2)^2}{e^{-2Ra}}\right.$$

$$\left. + 2(R^2 - k^2)^2 - 8k^2R^2\right].$$

(G) Show that the fraction of particles transmitted through the square barrier shown in Figure 12.12 is

$$\tau = \frac{16k^2R^2}{\left[\dfrac{(R^2 + k^2)^2}{e^{2Ra}} + \dfrac{(R^2 + k^2)^2}{e^{-2Ra}} + 2(R^2 - k^2)^2 - 8k^2R^2\right]}.$$

The next three problems explore the concept of wave packets in greater detail. The problems are constructed so that they may be assigned as a single unit or separately.

12.24[#] Suppose we have a quantum mechanical free particle whose eigenfunction is given by Eq. 12.8 with $C_1 = 0$. That is, we have $\psi(x) = C_2 \cos(kx)$.

(A) By writing the wave function in complex exponential form, show that there are exactly two possible momenta for the system: $+k\hbar$ and $-k\hbar$.

(B) Keeping in mind that k is a continuous variable for the free particle, let us assume that we have a system for which we know that the magnitude of the momentum lies in the range $(k_o - \Delta k)\hbar \le p_x \le (k_o + \Delta k)\hbar$, with all values of p_x equally probable. By summing the eigenfunctions $C_2 \cos(kx)$ over this range with equal weights for each value of k, obtain the unnormalized form of the superposition of states that represents this state of affairs. Show that the result may be put into the form

$$\Phi(x) = \frac{2C_2 \cos(k_o x) \sin(\Delta k x)}{x}.$$

The result is called a quantum mechanical wave packet. [*Hint:* Remember, you sum a continuous distribution by integration.]

(C) For the case $k_o = 5.09 \text{ Å}^{-1}$ and $\Delta k = 1.00 \text{ Å}^{-1}$, plot the unnormalized wave packet obtained in (B) over the range $-4\pi \text{ Å} \le x \le 4\pi \text{ Å}$. Is the particle represented by the wave packet more localized or less localized than that for the eigenfunction $\psi(x) = C_2 \cos(k_o x)$? Explain.

12.25[*] It is known that the summation of all free-particle eigenfunctions of the form $\psi(x) = C_2 \cos(kx)$ over the range $k_o - \Delta k \le k \le k_o + \Delta k$ with equal weight has the form

$$\Phi(x) = \frac{2C_2 \cos(k_o x) \sin(\Delta k x)}{x}.$$

(See Problem 12.23.)

(A) For the case $k_o = 5.09 \text{ Å}^{-1}$ and $\Delta k = 1.00 \text{ Å}^{-1}$, plot the unnormalized wave packet $\psi(x)$ over the range $-4\pi \text{ Å} \le x \le 4\pi \text{ Å}$. [If Problem 12.24 has been done, this plot may be omitted, as it is the same as that requested in (C) of that problem.]

(B) For the specific case given in (A), obtain the probability distribution function for observing momenta in the range from p_x to $p_x + dp_x$. Normalize this distribution. [*Hint:* Remember, all magnitudes of momentum in the range $(k_o - \Delta k)\hbar \le p_x \le (k_o + \Delta k)\hbar$ are equally probable.]

(C) Determine the average momentum $\langle p_x \rangle$ and the average square momentum, $\langle p_x^2 \rangle$, in terms of $\hbar$. Calculate the uncertainty in the momentum distribution in terms of $\hbar$.

12.26[*] It is known that the summation of all free-particle eigenfunctions of the form $\psi(x) = C_2 \cos(kx)$ over

the range $k_o - \Delta k \le k \le k_o + \Delta k$ with equal weight has the form

$$\Phi(x) = \frac{2C_2 \cos(k_o x) \sin(\Delta k x)}{x}.$$

(See Problem 12.24.)

(A) For the case $k_o = 5.09 \text{ Å}^{-1}$ and $\Delta k = 1.00 \text{ Å}^{-1}$, normalize this wave function and obtain the value of C_2. Assume that $\Phi(x) = 0$ for x values greater than 4π or less than -4π. [*Hint:* The integration is best done numerically.]

(B) For the case $k_o = 5.09 \text{ Å}^{-1}$ and $\Delta k = 1.00 \text{ Å}^{-1}$, compute the average position $\langle x \rangle$ and average square position $\langle x^2 \rangle$. [*Hint:* The integration for $\langle x^2 \rangle$ is best done numerically.]

(C) The uncertainty in the momentum distribution for the system described by $\Phi(x)$ is $\langle \Delta p_x^2 \rangle^{1/2} = 5.122 \times 10^{10} \hbar$. (See Problem 12.25.) Show that the uncertainty principle is satisfied for this wave packet.

12.27 Let us assume that the potential barrier to electron tunneling in an STM is a one-dimensional square barrier whose height is 2.00×10^{-19} J. If the tunneling electron has an energy of 1.00×10^{-20} J, compute the value of the tunneling probability if the tunneling distance a is 0.500×10^{-10} m. What is the tunneling probability if the tunneling distance is 1.500×10^{-10} m? [*Hint:* Can you use the limiting form for the tunneling probability given by Eq. 12.81? Be certain to investigate this point.]

12.28 (A) Show that inversion of Eqs. 12.83 and 12.84 leads to Eqs. 12.85 and 12.86.

(B) Show that substitution of Eqs. 12.87 and 12.88 into Eq. 12.82 gives a kinetic-energy expression that has the form of Eq. 12.89 with a reduced mass as defined by Eq. 12.90.

12.29 An investigator wishes to represent a four-atom system by using an independent center-of-mass, relative coordinate system. She labels the Cartesian coordinates of the center of mass as (X_C, Y_C, Z_C). The Cartesian coordinates of the vector from atom B to atom A are denoted as (X_{R1}, Y_{R1}, Z_{R1}). The Cartesian coordinates of the vector CD are represented by (X_{R2}, Y_{R2}, Z_{R2}). Finally, the investigator denotes the Cartesian coordinates of the vector from the center of mass of AB to the center of mass of CD as (X_{R3}, Y_{R3}, Z_{R3}).

(A) Are these coordinates independent? If not, state why not.

(B) What is the form of the kinetic energy of the four-atom system, expressed in terms of the derivatives of these coordinates with respect to time (i.e., velocities)?

12.30 An investigator wishes to represent a five-atom system by using an independent center-of-mass, relative coordinate system. Suggest a possible choice of coordinates that will accomplish this objective.

12.31 Consider a linear triatomic molecule ABC with masses m_1, m_2, and m_3 and bond lengths R_1 and R_2, as shown in the following diagram:

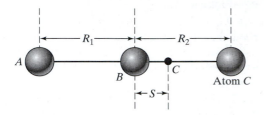

Let point C be the system center of mass, which lies at a distance S from atom B.

(A) Show that

$$S = \frac{(m_3 R_2 - m_1 R_1)}{M},$$

where $M = m_1 + m_2 + m_3$.

(B) The moment of inertia is given by

$$I = \sum_{i=1}^{3} m_i r_i^2$$

where r_i is the perpendicular distance from atom i to the principal axis, which will be through the center of mass C and perpendicular to the bond axis. From the figure, it may be seen that

$$I = m_1(R_1 + S)^2 + m_2 S^2 + m_3(R_2 - S)^2.$$

Show that I may also be expressed as

$$I = m_1 R_1^2 + m_3 R_2^2 - M^{-1}(m_1 R_1 - m_3 R_2)^2.$$

12.32 (A) Show that the function $\Theta(\theta) = N \sin^2 \theta$ is a solution of the θ-equation for the rigid rotor, provided that $E_R = 3\hbar^2/I$ and $m = 2$ or $m = -2$, where I is the moment of inertia of the rotor.

(B) Normalize the wave function in (A).

12.33 A member of the class suggests that the function $\Theta(\theta) = C\theta$, where C is a constant, might be a solution of the θ-equation for the diatomic rigid rotor. On the surface, this is a very reasonable suggestion. Show, however, that that function cannot be made to satisfy the θ-equation, regardless of the values used for E_R and m.

12.34 Normalize the $\Theta_3^1(\theta) = N[5 \cos^2 \theta - 1]\sin \theta$ eigenfunction. Check your answer against the result shown in Table 12.1.

12.35 Consider the H_2 molecule. Assume that the molecule's rotational motion may be treated as that of a quantum mechanical rigid rotor. (The H_2 bond length is 0.740 Å.)

(A) Derive a formula giving the energy spacing between the Jth and the $(J + 1)$st rotational energy levels in terms of J and the moment of inertia of H_2.

(B) Evaluate the spacing found in (A) for $J = 12$. Is this energy sufficiently large to permit its measurement? Is the system essentially classical or not?

12.36 The $\Theta_3^{\pm2}(\theta)$ and $\Theta_3^{\pm3}(\theta)$ functions appear to exhibit two and one nodal axes or planes, respectively. Since $J = 3$, we would expect to observe three angular nodes. Have we made an error? Explain what is happening.

12.37 Determine the location of the nodal angular surfaces or axes for the $\Theta_3^0(\theta)$ function.

12.38 A rigid rotor is in the $J = 1$, $M = 0$ eigenstate.

(A) What is the rotational energy of the rotor in terms of I and $\hbar$?

(B) At what set of angles (θ, ϕ) are we most likely to find this rotor?

12.39 In this problem, we explore the concept of angular momentum in greater detail. We seek to show that the quantum mechanical operator corresponding to the Z component of angular momentum, when expressed in spherical polar coordinates, is $L_z = (\hbar/i)(\partial/\partial\phi)$. We begin with L_z in rectangular Cartesian coordinates. Using the second postulate, the text shows that

$$L_z = \frac{\hbar}{i}\left[x\frac{\partial}{\partial y} - y\frac{\partial}{\partial x}\right].$$

We now need to transform L_z into the corresponding equation in spherical polar coordinates. The transformation equations are

$$x = R \sin \theta \cos \phi,$$
$$y = R \sin \theta \sin \phi,$$
$$z = R \cos \theta,$$

and

$$R^2 = x^2 + y^2 + z^2.$$

The chain rule for the conversion of derivatives gives the result

$$\frac{\partial}{\partial y} = \frac{\partial}{\partial R}\frac{\partial R}{\partial y} + \frac{\partial}{\partial\theta}\frac{\partial\theta}{\partial y} + \frac{\partial}{\partial\phi}\frac{\partial\phi}{\partial y}$$

and

$$\frac{\partial}{\partial x} = \frac{\partial}{\partial R}\frac{\partial R}{\partial x} + \frac{\partial}{\partial\theta}\frac{\partial\theta}{\partial x} + \frac{\partial}{\partial\phi}\frac{\partial\phi}{\partial x}.$$

Use the chain rule in conjunction with the transformation equations to show that, in spherical polar coordinates,

$$L_z = \frac{\hbar}{i}\frac{\partial}{\partial\phi}.$$

12.40 Show that $\langle L_z^2 \rangle = M^2\hbar^2$ when the rigid-rotor eigenfunction is $Y_J^M(\theta, \phi)$.

12.41 A rigid rotor is known to be in a state whose eigenfunction is $Y_4^3(\theta, \phi)$.

(A) What is the rotational energy of the rotor in terms of I and $\hbar$?

(B) What is the magnitude of the angular momentum of the rotor?

(C) Determine the z component of the angular momentum.

(D) What angle does the angular momentum vector make with the Z-axis?

(E) Can we compute the angle ϕ associated with the angular momentum vector in this state? If so, determine the value of ϕ. If not, state why not.

12.42 What do we know about the $Y_4^5(\theta, \phi)$ eigenfunction for the rigid rotor?

12.43. A rigid rotor is in a $J = 6$ rotational state. What are the possible angles the angular momentum vector might make with the Z-axis? Can we compute the angle ϕ?

12.44 Suppose we have a rigid rotor, the magnitude of whose angular momentum is known to be $L = \sqrt{6}\hbar$. Let us further assume that we know that L_z is either $\hbar$ or $-\hbar$, with equal probability of each state, but we do not know which one we have. Obtain two different wave functions that properly express this situation. (You do not need to normalize the wave functions.) Are these wave functions cylindrically symmetric about the Z-axis? One of the wave functions is generally called the d_{xz} orbital; the other is called the d_{yz} orbital. Which of your two wave functions is the d_{xz} orbital, and which is the d_{yz} orbital?

12.45 Use Eqs. 12.65A and 12.113 to show that the uncertainty product $\langle \Delta \phi^2 \rangle \langle \Delta L_z^2 \rangle$ is not zero, so that it is impossible to simultaneously know both the Z component of the angular momentum and the rotation angle about the Z-axis.

12.46 The H_2O bond angle is 104.45°. The O–H distance is 0.958×10^{-10} m. Use these data and the results obtained in Example 12.20 to compute the three principal moments of inertia of H_2O.

12.47 The geometry of the CH_4 molecule is tetrahedral. This configuration may be conveniently pictured by drawing a cube with the carbon atom in the center of the cube and the hydrogen atoms along the opposite diagonals of the top and bottom faces, as illustrated in the following figure:

(A) Show that X-, Y-, and Z-axes with an origin at the center of the cube and drawn perpendicular to the faces of the cube correspond to the principal axes of CH_4.

(B) Show that the moment of inertia about any principal axis is $I = 8m_H R^2/3$, where R is the C–H bond length and m_H is the mass of the hydrogen atom. Obtain an expression for the spacing between the

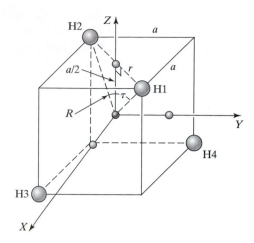

$J = 0$ and the $J = 1$ rotational states of CH_4 in terms of m_H and R if the rotational states are accurately described by the rotation of a rigid body.

12.48 The moments of inertia for NH_3 are $I_b = I_c = 2.816 \times 10^{-47}$ kg m^2 and $I_a = 4.43 \times 10^{-47}$ kg m^2. Compute the rotational energy states for $J = 0$, 1, and 2 of NH_3 with all allowed values of K.

12.49 Show that the total vibrational energy of a classical one-dimensional harmonic oscillator is $E_{vib} = kA^2/2$.

12.50 The classical vibration frequency of HF is 1.240×10^{14} s^{-1}.

(A) Compute the spacing between the HF vibrational levels, assuming that the vibrational eigenstates are accurately described by a harmonic oscillator approximation. What is the energy in units of kJ mol^{-1}. Is this large or small relative to the energy of a C–H bond?

(B) If the masses of hydrogen and fluorine were 1 g and 19 g, respectively, what would the vibrational energy spacing be in units of kJ mol^{-1}? Would this system behave classically?

12.51 (A) Compute the probability that a quantum harmonic oscillator in its ground state with a fundamental vibration frequency equal to 1.350×10^{14} s^{-1} and a reduced mass equal to that of H_2 will be found in the classically forbidden region of space.

(B) Compute the probability that a quantum harmonic oscillator in its first excited state with a fundamental vibration frequency equal to 1.350×10^{14} s^{-1} and a reduced mass equal to that of H_2 will be found in the classically forbidden region of space.

12.52 (A) Show that the wave function $\psi(x) = Nx \exp[-ax^2/2]$ is a solution of the quantum mechanical harmonic oscillator, provided that $a = (k\mu)^{1/2}/\hbar$ and $E = 1.5h\nu_o$.

(B) Normalize the wave function.

(C) Compute the uncertainty in position for an oscillator in this quantum state.

12.53 (A) Assume that the total vibrational and rotational energy of H_2 may be obtained simply by adding together the rotational energy of a rigid rotor with R equal to the equilibrium bond distance and the vibrational energy of a harmonic oscillator. Under these conditions, compute the total vibrational and rotational energy for H_2 in the following quantum states (the equilibrium bond distance for H_2 is 0.74000×10^{-10} m, and the fundamental vibration frequency is $1.3500 \times 10^{14}\ s^{-1}$):

$$v = 0, \quad J = 0; \quad v = 1, \quad J = 0.$$

$$v = 0, \quad J = 1; \quad v = 1, \quad J = 1.$$

$$v = 0, \quad J = 2; \quad v = 1, \quad J = 2.$$

$$v = 0, \quad J = 3; \quad v = 1, \quad J = 3.$$

$$v = 0, \quad J = 4; \quad v = 1, \quad J = 4.$$

$$v = 0, \quad J = 5; \quad v = 1, \quad J = 5.$$

(B) Compute the energy spacing between the following pairs of states:

$$(v = 0, J = 0) \quad \text{and} \quad (v = 1, J = 1).$$

$$(v = 0, J = 1) \quad \text{and} \quad (v = 1, J = 2).$$

$$(v = 0, J = 2) \quad \text{and} \quad (v = 1, J = 3).$$

$$(v = 0, J = 3) \quad \text{and} \quad (v = 1, J = 4).$$

$$(v = 0, J = 4) \quad \text{and} \quad (v = 1, J = 5).$$

What type of regular progression do you notice in the values obtained?

12.54 The fundamental vibration frequency of H_2 is $1.350 \times 10^{14}\ s^{-1}$. Assume that H_2 vibrates like a harmonic oscillator. Compute the H_2 vibrational force constant and the expected fundamental vibrational frequencies for HD and D_2. The vibrational force constant is the same for HD and D_2 as for H_2.

12.55. In this problem, we will compare the difference between the classical probability distribution for the position of a harmonic oscillator and that for a quantum mechanical harmonic oscillator in the ground state. It is shown in the text that the dependence of the displacement on time for a classical oscillator is $x(t) = A \sin[(k/\mu)^{1/2} t]$ for an oscillator with $x(t = 0) = 0$. The probability of finding the classical oscillator in the range from x to $x + dx$ is proportional to the time the oscillator spends in going from one end of the range to the other. This residence time is $t_r = dx(dx/dt)^{-1}$; that is, the time required to traverse the distance dx is dx divided by the velocity, which is (dx/dt). Use this relationship to obtain the normalized probability distribution function for the position of the classical harmonic oscillator. {Hint:

Write $\cos(z)$ as $(1 - \sin^2 z)^{1/2}$, and then use the fact that $x = A \sin[(k/u)^{1/2}t]$.}

12.56 When the measured vibrational energy levels of $H^{35}Cl$ (Cl^{35} is a chlorine isotope of mass 35 amu) are fitted to Eq. 12.157, it is found that $\omega_e \chi_e = 1.034 \times 10^{-21}$ J. The vibrational frequency of $H^{35}Cl$ is $8.960 \times 10^{13}\ s^{-1}$, and the experimental well depth is 7.394×10^{-19} J. Use Eq. 12.156 and the measured value of $\omega_e \chi_e$ to compute the value of the Morse parameter D. What is the percent error in the predicted well depth?

12.57 The experimental well depth for the H_2 molecule is 7.607×10^{-19} J, and the vibrational frequency is $1.317 \times 10^{14}\ s^{-1}$. Use Eq. 12.156 to show that the $v = 17$ vibrational state is just about at the dissociation point.

12.58 In this problem, we will examine a very simple model for the vibrational bond potential. Let

$$V(x) = \infty \qquad \text{for } x < 0,$$

$$V(x) = 0 \qquad \text{for } 0 \le x \le a,$$

and

$$V(x) = V_o \quad \text{for } x > a.$$

That is, the vibrational potential has the form shown in the following diagram:

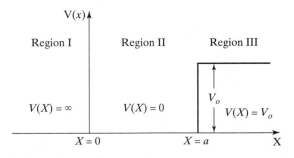

In Region I, we have $\psi_I = 0$ for $x < 0$. For the case where $E < V_o$,

(A) solve the Schrödinger equation to obtain $\psi_{II}(x)$.

(B) Solve the Schrödinger equation to obtain $\psi_{III}(x)$.

(C) By application of the boundary conditions $\psi_I(x = 0) = \psi_{II}(x = 0)$, $\psi_{II}(x = a) = \psi_{III}(x = a)$, $\psi_{III}(x \to \infty) = a$ finite number and $d\psi_{III}(x)/dx |_{x=a} = d\psi_{II}(x)/dx |_{x=a}$, show that the energy must be such as to satisfy the equation

$$\tan\left[\left\{ \frac{8\pi^2 mE}{h^2} \right\}^{1/2} a \right] = -\left[\frac{E}{V_o - E} \right]^{1/2}.$$

(D) Let us now make the model fit H_2 as closely as possible. The equilibrium bond distance for H_2 is

0.74×10^{-10} m. Let us take a to be equal to this value. The H_2 well depth is 7.607×10^{-19} J. Therefore, we will set V_o equal to that value. Finally, let us replace the mass m in the equation obtained in (C) with the reduced mass of H_2. Numerically determine the vibrational energy states for this system. How many are present?

12.59 Let us represent the interatomic potential for $H^{35}Cl$ by a Morse potential. The measured $H^{35}Cl$ vibration frequency and well depth are 8.960×10^{13} s^{-1} and 7.394×10^{-19} J, respectively. Compute the ratio of the anharmonicity term to the harmonic term for the $v = 0, 1, 2,$ and 3 vibrational states of the oscillator.

12.60 Compute the uncertainty product for x and p_x for a harmonic oscillator in its ground state, and show that it obeys the uncertainty principle.

12.61 Show that $\psi(x) = Ae^{ikx}$ is an eigenfunction of p_x and also of p_x^2. What are the associated eigenvalues?

12.62 Show that when the wave function for a rigid rotor is either the p_x function given by Eq. 12.127A or the p_y function given by Eq. 12.127B, the uncertainty associated with a measurement of L_z is no longer zero. Consequently, show that the p_x and p_y functions are not eigenfunctions of L_z.

12.63 I was ready to quit and go to the next chapter on the electronic structure of atoms, but Sam wants to present one more problem to the class. He notes that the quantum solution for the ground state of a particle in a one-dimensional, infinite potential well of width 1 Å has the form shown in the following plot: Sam notes that at the points $x = 0.0$ and $x = 1.0$ Å, $\psi(x)$

has discontinuities (cusps), such as the one shown in Figure 11.16. But the first postulate forbids wave functions to have such discontinuities! Sam wishes to know if a mistake has been made in solving the Schrödinger equation. If so, he would like you to correct this mistake and present the right solution. If not, he would like you to explain why the first postulate is apparently violated. Has quantum theory fallen apart? Should we prepare a letter to the Nobel Committee in Sweden suggesting that the Nobel prizes awarded to Schrödinger and Heisenberg be posthumously rescinded?

CHAPTER 3

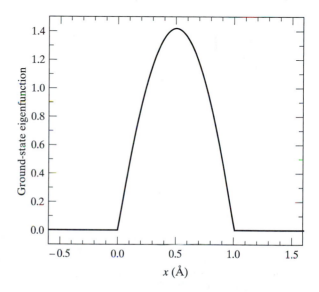

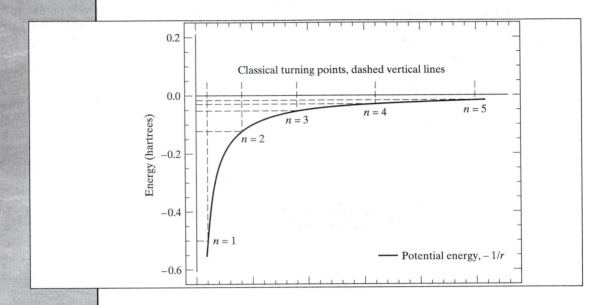

The Electronic Structure of Atoms

The electronic structure of atoms and molecules can be understood and quantitatively predicted only if we consider both quantum theory and special relativity theory. The fact that these theories, seemingly so different and separated in time by over 20 years, provide a quantitative explanation of atomic and molecular structure is perhaps the most compelling evidence that they each contain essential elements of the truth concerning the nature of our universe. At some future date, we may see these two great theories replaced by others that encompass even more of the truth. At the present time, however, they represent the best that mankind has to offer.

In this chapter, we shall first investigate the hydrogen atom and hydrogen-like ions (ions containing a single electron). These systems provide the basis for the treatment of more complex, many-electron atoms and molecules. A study of these quickly reveals the fact that the Schrödinger equation cannot be solved exactly when more than one electron is present. Consequently, it will be necessary to introduce some methods by which approximate solutions can be obtained. At that point, we will find ourselves at an impasse until some results from special relativity are introduced. Due to limitations of space, time, and background, we shall only briefly touch upon these results. However, the brevity of the treatment should not be taken as an indicator of the importance of the theory. With this background, a consideration of exchange symmetry leads directly to the Pauli exclusion principle and the foundation of the periodic table. At the end of the chapter, we shall be well prepared to address the question of molecular structure and bonding.

13.1 One-Electron Atoms and Ions

One-electron atoms and ions comprise the hydrogen, deuterium, and tritium atoms and ions such as He^+, Li^{2+}, Be^{3+}, etc. The quantum mechanical eigenfunctions for these systems have the same mathematical form. They differ only because the nuclear charges on the ions are larger than that for hydrogen and the hydrogen isotopes and because of very slight differences in the reduced mass. It is, therefore, convenient to examine such systems as one general class of particles.

13.1.1 The Coulombic Potential and Atomic Units

Because electrons and nuclei are charged, the primary contribution to the potential energy in atoms and molecules is electrostatic in nature. There are also contributions to the total potential from magnetic interactions. In most cases, however, the magnetic terms are small. We shall, therefore, ignore them in our initial investigations of atomic and molecular structure.

The coulombic potential between two particles with charges Q_1 and Q_2 was introduced in Eq. 1.49. Its form is

$$V(r) = \frac{Q_1 Q_2}{4\pi\varepsilon_o r}, \tag{13.1}$$

where Q_1 and Q_2 are expressed in coulombs and ε_o is the permittivity of the vacuum, 8.85419×10^{-12} J^{-1} C^2 m^{-1}. The corresponding electrostatic force is the negative derivative of $V(r)$:

$$F(r) = \text{force} = -\frac{\partial V(r)}{\partial r} = \frac{Q_1 Q_2}{4\pi\varepsilon_o r^2}. \tag{13.2}$$

Up to this point, we have worked almost exclusively in the SI system of units, so that $V(r)$ is expressed in joules. However, this is not a satisfactory procedure for atomic- and molecular-level calculations. The nature of the difficulty can be seen by inspecting the coefficient of the centrifugal distortion term in Eq. 12.159, which is $\hbar^4/(4\mu^2\beta^2 R_e^6 D)$. If we employ SI units, the value of $\hbar^4$ is 1.9276×10^{-133} J^4 s^4. The magnitude of this quantity is so small that many computers will incur an underflow error that causes $\hbar^4$ to be set to zero. (Try it on your own calculator.) Of course, we can always avoid this difficulty by first dividing by R_e^6 before computing $\hbar^4$, but when we are executing computationally intensive procedures on a computer, the problem is too complex to allow us to foresee all the overflow and underflow danger points. Therefore, we need to employ a system of units in which the quantities of interest are much closer to unity. Such units give us the maximum possible protection against both underflows and overflows. This type of problem was discussed in Chapter 1.

Many possible procedures might be used to circumvent the difficulties caused by the exceptionally small size of most molecular quantities. We shall describe the one that is most commonly employed. We first convert from SI units to the cgs (cm gram s) system. In this system, the electrostatic force between two charges is written in the form

$$F(r) = \frac{Q_{s1} Q_{s2}}{r^2}, \tag{13.3}$$

where the charges are now expressed in statcoulombs (statC) and the subscript s denotes electrostatic units of charge. One statcoulomb is defined to be the quantity of charge needed on each particle to produce a force of 1 dyne (1 g cm^2 s^{-2}) when the two charges are separated by a distance of 1 cm. We may combine Eqs. 13.2 and 13.3 to obtain the relationship between the coulomb and the statcoulomb. Suppose we have two particles, each with a charge of 1 statC, separated by a distance of 1 cm. We would then have

$$F = 1 \text{ dyne} = \frac{Q^2}{4\pi\varepsilon_o(0.01 \text{ m})^2} = 10^{-5} \text{ newton (N)},$$

since 1 dyne is equivalent to 10^{-5} N. Solving for Q, we obtain

$$Q = [(10^{-5} \text{ kg m s}^{-2})(4)(3.1415927)(8.85419$$
$$\times 10^{-12} \text{ kg}^{-1} \text{ m}^{-2} \text{ s}^2 \text{ C}^2 \text{ m}^{-1})10^{-4} \text{ m}^2]^{1/2} = 3.335641 \times 10^{-10} \text{ C}.$$

Therefore, 1 statC is equivalent to 3.335641×10^{-10} C, or 1 C is equivalent to 2.99792×10^9 statC. Since the statcoulomb is a much smaller unit of charge than the coulomb, we are getting closer to a set of units that will be convenient for atomic and molecular calculations.

Electrons and protons have equal charge magnitudes that are opposite in sign. This magnitude is 1.602177×10^{-19} C, or 4.80321×10^{-10} statC, which we shall denote with the symbol e. In this notation, the charges on the electron and proton are $-e$ and $+e$, respectively. Suppose we have a proton and an electron separated by a distance equal to the radius of the first Bohr orbit, $a_o = 0.529177 \times 10^{-8}$ cm. Then the potential energy of this system in SI units is given by Eq. 13.1:

$$V(r = a_o) = -\frac{e^2}{4\pi\varepsilon_o r} = -\frac{(1.602177 \times 10^{-19})^2}{4\pi\varepsilon_o a_o}$$

$$= -\frac{(1.602177 \times 10^{-19})^2}{4(3.1415927)(8.85419 \times 10^{-12})(0.529177 \times 10^{-10})} \text{ J}$$

$$= -4.35975 \times 10^{-18} \text{ J}.$$

If we use electrostatic units, the magnitude is closer to unity, but is still very small:

$$V(r = a_o) = -\frac{Q_s^2}{r} = -\frac{(4.80321 \times 10^{-10})^2}{0.529177 \times 10^{-8}} = -4.35975 \times 10^{-11} \text{ ergs}.$$

If we work with energies of this magnitude in atomic and molecular calculations, we are in danger of underflows and overflows. Clearly, we need an additional modification of the units.

In defining new units of charge, distance, mass, time, and energy, we must be careful to satisfy the basic relationship between units. For example, the energy must be related to mass, distance, and time by

$$E = \frac{1}{2}mv^2 = \frac{1}{2}m\left[\frac{d}{t}\right]^2. \qquad (13.4)$$

That is, we can define any three of the four variables energy, distance, mass, and time, but the fourth will be fixed by Eq. 13.4.

The first step is to define $e = 1$ as the atomic charge unit. Next, we take the unit of distance to be the radius of the first Bohr orbit, which we call a *bohr*. That is, we define

a_o = Bohr radius = 0.529177×10^{-10} m = 1 atomic unit of distance (1 bohr).

These two choices define a new energy unit, since we now have

$$V(r = a_o) = -\frac{e^2}{a_o} = -\frac{(1)^2}{1} = -1 \text{ energy unit } (-1 \text{ hartree}).$$

But we know from our earlier analysis that $V(r = a_o) = -4.35975 \times 10^{-11}$ erg. Therefore, our new energy unit, the *hartree,* is equivalent to 4.35975×10^{-11} erg or 4.35975×10^{-18} J. Having defined charge, distance, and energy, we are free to choose one additional unit. We will take our unit of mass to be the electronic mass:

m_e = electronic mass = 1 atomic mass unit = 9.10939×10^{-31} kg.

Our time unit will now be fixed by Eq. 13.4. When we have 1 mass unit, moving 1 distance unit in a time equal to 1 time unit, Eq. 13.4 shows that $E = 0.5$ energy unit. Hence, we must have

$$0.5 \text{ energy unit} = 0.5 \text{ hartree} = 0.5 \,(1 \text{ mass unit})\left[\frac{1 \text{ distance unit}}{1 \text{ time unit}}\right]^2.$$

Solving for the time unit, we obtain

$$1 \text{ time unit} = \left[\frac{1 \text{ mass unit}}{1 \text{ energy unit}}\right]^{1/2} (1 \text{ distance unit})$$

$$= \frac{(9.10939 \times 10^{-31} \text{ kg})^{1/2}(0.529177 \times 10^{-10} \text{ m})}{(4.35975 \times 10^{-18} \text{ kg m}^2 \text{ s}^{-2})^{1/2}} = 2.41888 \times 10^{-17} \text{ s}.$$

These results completely define the system of units we call *atomic units* (au). Let us now compute the value of $\hbar$ in atomic units. We have

$$\hbar = \frac{h}{2\pi} = \frac{6.62608 \times 10^{-34} \text{ J s}}{2(3.141927)} \times \frac{1 \text{ hartree}}{4.35975 \times 10^{-18} \text{ J}} \times \frac{1 \text{ au of time}}{2.41888 \times 10^{-17} \text{ s}}$$

$$= 1.00000 \text{ au}.$$

Notice that these results are all internally consistent. In Eq. 11.37, we found that the Bohr radius is given by

$$a_o = \left[\frac{4\pi\varepsilon_o\hbar^2}{m_e e^2}\right]$$

if we employ SI units. If we use electrostatic units of charge, the equation becomes

$$a_o = \left[\frac{\hbar^2}{m_e e^2}\right].$$

Using atomic units, we obtain $a_o = (1)^2/[(1)(1)^2] = 1$ bohr, which is in accord with our definition.

The advantage of atomic units is that all the quantities are now in the neighborhood of unity. Atomic and molecular distances are generally on the order of angstroms ($1\text{ Å} = 10^{-10}$ m), so with our distance unit being the Bohr radius, distances will generally lie in the range 0.4 to 20 bohr. The charges that appear in the electrostatic potential will be -1 for the electron and between $+1$ and $+106$ for the elements in the periodic table. The electronic mass and $\hbar$, which appear in the kinetic-energy operator for the electron, will be unity. Finally, the magnitude of the coulombic potentials, as shown above, will be on the order of unity as well. Figure 13.1 illustrates the basic definitions of the atomic unit system.

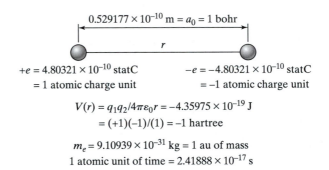

0.529177×10^{-10} m $= a_0 = 1$ bohr

r

$+e = 4.80321 \times 10^{-10}$ statC $\qquad -e = -4.80321 \times 10^{-10}$ statC
$= 1$ atomic charge unit $\qquad\qquad = -1$ atomic charge unit

$V(r) = q_1 q_2/4\pi\varepsilon_0 r = -4.35975 \times 10^{-19}$ J
$= (+1)(-1)/(1) = -1$ hartree

$m_e = 9.10939 \times 10^{-31}$ kg $= 1$ au of mass
1 atomic unit of time $= 2.41888 \times 10^{-17}$ s

▶ FIGURE 13.1
Definitions of atomic units. The choice of distance, charge, energy, and mass units fixes the atomic unit of time, as explained in the text.

The atomic unit system is sufficient for all of our needs. However, it has become common practice to report atomic and molecular energies in units called *electron volts* (eV). This unit is defined to be the energy acquired when a particle with charge e is accelerated through a potential field of 1 volt. Since a volt-coulomb is a joule, 1 eV is equivalent to $(1.602177 \times 10^{-19}$ C) (1 volt) $= 1.602177 \times 10^{-19}$ J. In terms of atomic units, 1 eV is equivalent to 1.602177×10^{-19} J $\times 1$ hartree/$(4.35975 \times 10^{-18}$ J) $= 0.0367493$ hartree. Conversely, we can say that 1 hartree is equivalent to 27.2114 eV. Scientists also often express energy in terms of the equivalent wave number, usually in cm^{-1}. In these terms, we have 1 hartree $= hc\bar{\nu}$, so that the equivalent wave number is $\bar{\nu} = 4.35975 \times 10^{-18}$ J$/[(6.62608 \times 10^{-34}$ J s$)$ $(2.99792 \times 10^{10}$ cm s$^{-1})] = 219,475$ cm^{-1}. Using the fact that 1 hartree is equal to 27.2114 eV, we find that 1 eV is the equivalent of

$$\frac{219,475 \text{ cm}^{-1}}{\text{hartree}} \times \frac{1 \text{ hartree}}{27.2114 \text{ eV}} = 8,065.6 \text{ cm}^{-1}.$$

All of the units and conversion factors we have discussed are summarized in Table 13.1. Fundamental constants expressed in different units and other conversion factors are given in Table 13.2. More detailed conversion factors are given in the tables in Appendix C.

Table 13.1   Conversion factors

Atomic Unit	SI Unit	Electrostatic Unit
1 charge unit	1.602177×10^{-19} C	4.80321×10^{-10} statC
1 mass unit	9.10939×10^{-31} kg	9.10939×10^{-28} g
1 distance unit (1 bohr)	0.529177×10^{-10} m	0.529177×10^{-8} cm
1 time unit	2.41888×10^{-17} s	2.41888×10^{-17} s
1 energy unit (1 hartree)	4.35975×10^{-18} J	4.35975×10^{-11} ergs

Table 13.2   Fundamental constants and other conversion factors

Constant	Atomic Units	SI Value	Electrostatic Value
e	1	1.602177×10^{-19} C	4.80321×10^{-10} statC
$\hbar$	1	1.05457×10^{-34} J s	1.05457×10^{-27} erg s
h	2π	6.62608×10^{-34} J s	6.62608×10^{-27} erg
m_e	1	9.10939×10^{-31} kg	9.10939×10^{-28} g
m_p	1,836.1	1.67262×10^{-27} kg	1.67262×10^{-24} g
(mass of proton)			
a_o	1	0.529177×10^{-10} m	0.529177×10^{-8} cm
c	137.036	2.997924×10^{8} m s^{-1}	2.997924×10^{10} cm s^{-1}
(speed of light in a vacuum)			

1 hartree = 27.2114 eV = 2,625.5 kJ mol^{-1} = 627.51 kcal mol^{-1}.
1 hartree is equivalent to 219,475 cm^{-1}.
1 eV is equivalent to 8,065.6 cm^{-1}.

EXAMPLE 13.1

The coefficient for the centrifugal distortion term in Eq. 12.159 is $\hbar^4/4\mu^2\beta^2R_e^6D$, where μ is the reduced mass of the diatomic system and β, R_e, and D are the parameters in the Morse potential. In Chapter 12, we fitted a Morse function to the vibration frequency, well depth, and equilibrium distance for the H_2 molecule. The results were $D = 7.607 \times 10^{-19}$ J, $\beta = 1.941 \times 10^{10}$ m^{-1} and $R_e = 0.740 \times 10^{-10}$ m. **(A)** Compute the value of the coefficient of the distortion term in SI units. **(B)** Use atomic units to compute the value of the coefficient. Comment on the two calculations.

Solution

(A) The reduced mass of H_2 is

$$\mu = \frac{m_H m_H}{m_H + m_H} = \frac{m_H}{2} = \frac{1.0079 \text{ g mol}^{-1}}{2} \times \frac{1 \text{ kg}}{1{,}000 \text{ g}} \times \frac{1}{6.022 \times 10^{23} \text{ mol}^{-1}}$$
$$= 8.368 \times 10^{-28} \text{ kg.} \tag{1}$$

Straightforward substitution gives

$$\frac{\hbar^4}{4\mu^2\beta^2R_e^6D} = \frac{(1.055 \times 10^{-34})^4}{4(8.368 \times 10^{-28})^2(1.941 \times 10^{10})^2(0.740 \times 10^{-10})^6(7.607 \times 10^{-19})} \text{J}$$
$$= 9.398 \times 10^{-25} \text{ J.} \tag{2}$$

(B) We first convert all quantities to atomic units:

$$\mu = 8.368 \times 10^{-28} \text{ kg} \times \frac{1 \text{mass unit}}{9.10939 \times 10^{-31} \text{ kg}} = 918.6; \quad \text{(3)}$$

$$\beta = 1.941 \times 10^{10} \text{ m}^{-1} \times \frac{0.5292 \times 10^{-10} \text{ m}}{\text{bohr}} = 1.0272 \text{ bohr}^{-1}; \quad \text{(4)}$$

$$R_e = 0.740 \times 10^{-10} \text{ m} \times \frac{1 \text{ bohr}}{0.5292 \times 10^{-10} \text{ m}} = 1.398 \text{ bohr}; \quad \text{(5)}$$

$$D = 7.607 \times 10^{-19} \text{ J} \times \frac{1 \text{ hartree}}{4.35975 \times 10^{-18} \text{ J}} = 0.17448 \text{ hartree}. \quad \text{(6)}$$

Also, we know that $\hbar = 1$; thus,

$$\frac{\hbar^4}{4\mu^2\beta^2 R_e^6 D} = \frac{(1)^4}{4(918.6)^2(1.0272)^2(1.398)^6(0.17448)} = 2.156 \times 10^{-7} \text{ hartree}. \quad \text{(7)}$$

The conversion factors given in Table 13.1 show that the results obtained in Eqs. 2 and 7 are equivalent. Note, however, that there was no difficulty at all in calculating the result of Eq. 7, whereas care had to be used in performing the arithmetic in Eq. 2 to avoid an underflow. Note also how close to unity the numbers are when expressed in atomic units.

For related exercises, see Problems 13.1, 13.2, and 13.3.

13.1.2 Eigenfunctions and Eigenvalues

The hydrogen-like system with which we are concerned is shown in Figure 13.2. One electron with charge $-e$ interacts with a nucleus whose atomic number is Z, so that it possesses a charge $+Ze$. The instantaneous separation of the particles is denoted by r, which, in rectangular Cartesian coordinates, is given by $[x^2 + y^2 + z^2]^{1/2}$. This system has six degrees of freedom. Three are associated with motion of the center of mass whose translational eigenfunctions and eigenvalues were examined in Chapter 12, and the remaining three are associated with the relative motion of the electron and nucleus. In Chapter 12, we found that, by transforming to a center of mass, relative coordinate system, the kinetic energy can be separated into two sets of terms, the first of which describes the translational motion of the center of mass and the second the relative motion.

The classical relative kinetic energy T is given by $(\mu/2)[v_{xr}^2 + v_{yr}^2 + v_{zr}^2]$. [See Eq. 12.89.] If we express this relationship in terms of the relative Cartesian momenta $P_{(x, y, z)r} = \mu v_{(x, y, z)r}$, the result is

$$T = \frac{1}{2\mu}[P_{xr}^2 + P_{yr}^2 + P_{zr}^2],$$

where μ is the reduced mass of the hydrogen-like particle and P_{xr}, P_{yr}, and P_{zr} are, respectively, the x, y, and z components of the relative momentum vector. The interparticle potential is coulombic, so that the classical Hamiltonian is

$$\mathbf{H} = T + V = \frac{1}{2\mu}[P_{xr}^2 + P_{yr}^2 + P_{zr}^2] + \frac{(+Ze)(-e)}{4\pi\varepsilon_0[x^2 + y^2 + z^2]^{1/2}}, \quad \text{(13.5A)}$$

where we are using SI units for the potential. This is the same Hamiltonian we would have for a single particle of mass μ and charge $-e$ interacting with a stationary charge $+Ze$ at the origin. This is the picture represented by Figure 13.2. We can make the equations easier to write by switching to

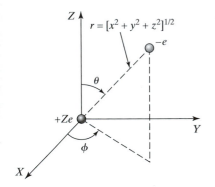

▲ FIGURE 13.2
A one-electron hydrogen-like system. The charge on the nucleus is $+Ze$. The relative motion of the nucleus and electron can be viewed as that of a single particle with a mass equal to the reduced mass of the system moving in three-dimensional space. The motion of the system is described by a spherical polar coordinate system.

cgs–electrostatic units. This change removes the $4\pi\varepsilon_o$ factor in the denominator of the potential, so that Eq. 13.5A becomes

$$\mathbf{H} = T + V = \frac{1}{2\mu}[P_{xr}^2 + P_{yr}^2 + P_{zr}^2] + \frac{(+Ze)(-e)}{[x^2 + y^2 + z^2]^{1/2}}. \qquad \textbf{(13.5B)}$$

The corresponding quantum mechanical Hamiltonian is obtained by direct operator substitution into Eq. 13.5B. The result was shown in Chapter 11 to be

$$\mathcal{H} = -\frac{\hbar^2}{2\mu}\nabla^2 + V(x, y, z) = -\frac{\hbar^2}{2\mu}\nabla^2 - \frac{Ze^2}{(x^2 + y^2 + z^2)^{1/2}}. \qquad \textbf{(13.6)}$$

We now seek solutions to the stationary-state Schrödinger equation, $\mathcal{H}\psi(x, y, z) = E\psi(x, y, z)$, for the eigenfunctions and eigenvalues. Because the Cartesian variables appear beneath the radical in Eq. 13.6, they cannot be separated in this coordinate system. To obtain the solutions, we must transform to spherical coordinates, so that Eq. 13.6 becomes

$$\mathcal{H} = -\frac{\hbar^2}{2\mu}\nabla^2 - \frac{Ze^2}{r}, \qquad \textbf{(13.7)}$$

with the ∇^2 operator given by Eq. 12.95 and also by Eq. 9.35. The Schrödinger equation to be solved is, therefore,

$$-\frac{\hbar^2}{2\mu}\left\{\frac{1}{r^2}\frac{\partial}{\partial r}\left[r^2\frac{\partial}{\partial r}\right] + \frac{1}{r^2\sin\theta}\frac{\partial}{\partial\theta}\left[\sin\theta\frac{\partial}{\partial\theta}\right] + \frac{1}{r^2\sin^2\theta}\frac{\partial^2}{\partial\phi^2}\right\}\psi(r, \theta, \phi)$$

$$-\frac{Ze^2}{r}\psi(r, \theta, \phi) = E\psi(r, \theta, \phi). \qquad \textbf{(13.8)}$$

As usual, we seek a separable solution whose form is

$$\psi(r, \theta, \phi) = R(r)G(\theta, \phi). \qquad \textbf{(13.9)}$$

Before substituting Eq. 13.9 into Eq. 13.8 to see if the variables separate, we first rearrange Eq. 13.8 into a more convenient form. Let us multiply both sides by r^2 and move the term on the right-hand side to the left. This produces

$$-\frac{\hbar^2}{2\mu}\left\{\frac{\partial}{\partial r}\left[r^2\frac{\partial}{\partial r}\right] + \frac{1}{\sin\theta}\frac{\partial}{\partial\theta}\left[\sin\theta\frac{\partial}{\partial\theta}\right] + \frac{1}{\sin^2\theta}\frac{\partial^2}{\partial\phi^2}\right\}\psi(r, \theta, \phi)$$

$$-r^2\left[E + \frac{Ze^2}{r}\right]\psi(r, \theta, \phi) = 0. \qquad \textbf{(13.10)}$$

Substitution of Eq. 13.9 into 13.10 and then factoring those functions not affected by the differential operators to the left yields

$$G(\theta, \phi)\left\{\left(-\frac{\hbar^2}{2\mu}\right)\frac{\partial}{\partial r}\left[r^2\frac{\partial}{\partial r}\right] - r^2\left[E + \frac{Ze^2}{r}\right]\right\}R(r)$$

$$+ R(r)\left\{-\frac{\hbar^2}{2\mu}\left[\frac{1}{\sin\theta}\frac{\partial}{\partial\theta}\left[\sin\theta\frac{\partial}{\partial\theta}\right] + \frac{1}{\sin^2\theta}\frac{\partial^2}{\partial\phi^2}\right]\right\}G(\theta, \phi) = 0.$$

We now divide both sides by $R(r)G(\theta, \phi)$, producing

$$\frac{\left\{\left(-\frac{\hbar^2}{2\mu}\right)\frac{\partial}{\partial r}\left[r^2\frac{\partial}{\partial r}\right] - r^2\left[E + \frac{Ze^2}{r}\right]\right\}R(r)}{R(r)}$$

$$+ \frac{\left\{-\frac{\hbar^2}{2\mu}\left[\frac{1}{\sin\theta}\frac{\partial}{\partial\theta}\left[\sin\theta\frac{\partial}{\partial\theta}\right] + \frac{1}{\sin^2\theta}\frac{\partial^2}{\partial\phi^2}\right]\right\}G(\theta, \phi)}{G(\theta, \phi)} = 0. \qquad \textbf{(13.11)}$$

The first term in Eq. 13.11 depends only upon the radial coordinate r, while the second term is a function only of the angular coordinates θ and ϕ. Thus, the variables have been separated, and a solution of the form of Eq. 13.9 exists.

The only way Eq. 13.11 can hold for all values of r, θ, and ϕ is for the first term to be equal to a constant, say, $-K$, while the second term is equal to $+K$, so that the sum is always zero regardless of the values of the variables. This requirement produces two differential equations. The first depends only upon r and is, therefore, called the radial equation. It is

$$\left\{ \left(-\frac{\hbar^2}{2\mu} \right) \frac{\partial}{\partial r} \left[r^2 \frac{\partial}{\partial r} \right] - r^2 \left[E + \frac{Ze^2}{r} \right] \right\} R(r) = -K\, R(r). \tag{13.12}$$

The second equation involves only the angular variables:

$$\left\{ -\frac{\hbar^2}{2\mu} \left[\frac{1}{\sin\theta} \frac{\partial}{\partial\theta} \left[\sin\theta \frac{\partial}{\partial\theta} \right] + \frac{1}{\sin^2\theta} \frac{\partial^2}{\partial\phi^2} \right] \right\} G(\theta,\phi) = K\, G(\theta,\phi). \tag{13.13}$$

These two equations must now be solved to obtain $R(r)$, $G(\theta,\phi)$, the constant K, and the energy eigenvalue E.

Actually, we have already obtained the solutions of the angular equation when we examined the rigid rotor. Equation 12.97 for the rotor is

$$-\frac{\hbar^2}{2\mu R^2} \left\{ \frac{1}{\sin\theta} \frac{\partial}{\partial\theta} \left[\sin\theta \frac{\partial}{\partial\theta} \right] + \frac{1}{\sin^2\theta} \frac{\partial^2}{\partial\phi^2} \right\} Y_J^M(\theta,\phi)$$

$$= E_R \psi(\theta,\phi) = \frac{J(J+1)\hbar^2}{2\mu R^2} Y_J^M(\theta,\phi),$$

since the rotational eigenvalues were found to be $J(J+1)\hbar^2/(2\mu R^2)$ [see Eq. 12.107] and the eigenfunctions are the spherical harmonics $Y_J^M(\theta,\phi)$. If we multiply this equation on both sides by R^2, the result is

$$-\frac{\hbar^2}{2\mu} \left\{ \frac{1}{\sin\theta} \frac{\partial}{\partial\theta} \left[\sin\theta \frac{\partial}{\partial\theta} \right] + \frac{1}{\sin^2\theta} \frac{\partial^2}{\partial\phi^2} \right\} Y_J^M(\theta,\phi) = \frac{J(J+1)\hbar^2}{2\mu} Y_J^M(\theta,\phi). \tag{13.14}$$

A comparison of Eqs. 13.14 and 13.13 shows that they are identical differential equations. Therefore, the unknown eigenfunction $G(\theta,\phi)$ must be the spherical harmonic $Y_J^M(\theta,\phi)$, and the constant must be given by $K = J(J+1)\hbar^2/(2\mu)$. Since the nucleus and the electron constitute a two-particle system, it is not surprising that the angular part of the eigenfunction is identical to that of a two-particle diatomic system. When the two particles are a nucleus and an electron, it is customary to change the quantum number notation from J and M to ℓ and m, respectively. We also call the rotational eigenfunctions *orbitals*. In these terms, the rotational orbitals for hydrogen-like atoms and ions are the spherical harmonics $Y_\ell^m(\theta,\phi)$, and the constant

$$K = \frac{\ell(\ell+1)\hbar^2}{2\mu}. \tag{13.15}$$

Therefore, we have half of the solution and two of the three quantum numbers we knew we would obtain. We also know that the total square angular momentum of the electron as it moves about the nucleus is given by

$$L^2 = \ell(\ell+1)\hbar^2, \quad \text{with } \ell = 0, 1, 2, 3, \ldots, \tag{13.16}$$

and the Z-component of this angular momentum is

$$L_z = m\hbar \quad \text{for } -\ell \le m \le \ell. \tag{13.17}$$

In fact, all of the characteristics of the solutions of the rigid rotor are also characteristics of the eigenfunctions for hydrogen-like atoms. The functional forms for the s, p, d, and f orbitals are exactly those shown in Figures 12.20 through 12.22. The corresponding absolute squares of these orbitals are shown in Figures 12.23 through 12.25.

All that remains is to solve the radial equation to obtain $R(r)$ and the energy eigenvalue E, which we know will depend upon a third quantum number. Substitution of Eq. 13.15 into Eq. 13.12 followed by multiplication of both sides by $-2\mu/\hbar^2$ and rearrangement gives

$$\frac{\partial}{\partial r}\left[r^2\frac{\partial}{\partial r}\right]R(r) + \frac{2\mu r^2}{\hbar^2}\left[E + \frac{Ze^2}{r}\right]R(r) - \ell(\ell+1)R(r) = 0. \quad \textbf{(13.18)}$$

This equation can be solved by series methods; however, we shall forgo the technique here. Instead, we simply present the solutions and then demonstrate how their validity may be verified. The solutions for the four eigenstates of lowest energy are given in Table 13.3.

Table 13.3   Normalized radial eigenfunctions and eigenvalues for hydrogen-like atoms

n	ℓ	$R_n^\ell(r)$	E_n	Designation
1	0	$2\left[\dfrac{Z}{a_o}\right]^{3/2}\exp\left[-\dfrac{\rho}{2}\right]$	$-\dfrac{Z^2\mu e^4}{2\hbar^2}$	$1s$
2	0	$\left[\dfrac{Z}{2a_o}\right]^{3/2}[2-\rho]\exp\left[-\dfrac{\rho}{2}\right]$	$-\dfrac{Z^2\mu e^4}{8\hbar^2}$	$2s$
2	1	$\dfrac{1}{3^{1/2}}\left[\dfrac{Z}{2a_o}\right]^{3/2}\rho\exp\left[-\dfrac{\rho}{2}\right]$	$-\dfrac{Z^2\mu e^4}{8\hbar^2}$	$2p$
3	0	$\dfrac{1}{9(3)^{1/2}}\left[\dfrac{Z}{a_o}\right]^{3/2}[6-6\rho+\rho^2]\exp\left[-\dfrac{\rho}{2}\right]$	$-\dfrac{Z^2\mu e^4}{18\hbar^2}$	$3s$
3	1	$\dfrac{1}{9(6)^{1/2}}\left[\dfrac{Z}{a_o}\right]^{3/2}[4\rho-\rho^2]\exp\left[-\dfrac{\rho}{2}\right]$	$-\dfrac{Z^2\mu e^4}{18\hbar^2}$	$3p$
3	2	$\dfrac{1}{9(30)^{1/2}}\left[\dfrac{Z}{a_o}\right]^{3/2}\rho^2\exp\left[-\dfrac{\rho}{2}\right]$	$-\dfrac{Z^2\mu e^4}{18\hbar^2}$	$3d$
4	0	$\dfrac{1}{96}\left[\dfrac{Z}{a_o}\right]^{3/2}[24-36\rho+12\rho^2-\rho^3]\exp\left[-\dfrac{\rho}{2}\right]$	$-\dfrac{Z^2\mu e^4}{32\hbar^2}$	$4s$
4	1	$\dfrac{1}{32(15)^{1/2}}\left[\dfrac{Z}{a_o}\right]^{3/2}[20-10\rho+\rho^2]\rho\exp\left[-\dfrac{\rho}{2}\right]$	$-\dfrac{Z^2\mu e^4}{32\hbar^2}$	$4p$
4	2	$\dfrac{1}{96(5)^{1/2}}\left[\dfrac{Z}{a_o}\right]^{3/2}[6-\rho]\rho^2\exp\left[-\dfrac{\rho}{2}\right]$	$-\dfrac{Z^2\mu e^4}{32\hbar^2}$	$4d$
4	3	$\dfrac{1}{96(35)^{1/2}}\left[\dfrac{Z}{a_o}\right]^{3/2}\rho^3\exp\left[-\dfrac{\rho}{2}\right]$	$-\dfrac{Z^2\mu e^4}{32\hbar^2}$	$4f$

Note: $\rho = 2Zr/(na_o)$ with $a_o = \hbar^2/(\mu e^2) \approx$ the Bohr radius.
The results can be easily converted to SI units by replacing e^2 with $e^2/(4\pi\varepsilon_o)$.
If we use atomic units, $a_o = 1$ and $\rho = 2Zr/n$.

EXAMPLE 13.2

Verify that the radial function $R(r) = N \exp[-Zr/a_o]$ is a solution of the radial equation for hydrogen-like particles, provided that $\ell = 0$ and $E = -Z^2e^2/(2a_o)$. Assume that we may set $\mu \approx m_e$, so that $a_o = \hbar^2/(\mu e^2)$. (See the discussion following this example.)

Solution

The radial equation 13.18 is

$$\frac{\partial}{\partial r}\left[r^2 \frac{\partial}{\partial r}\right] R(r) + \frac{2\mu r^2}{\hbar^2}\left[E + \frac{Ze^2}{r}\right] R(r) - \ell(\ell + 1)R(r) = 0. \tag{1}$$

Equation 1 must hold for all values of the radial coordinate. Substituting $R(r) = N \exp[-Zr/a_o]$ and dividing by the normalization constant N gives

$$\frac{\partial}{\partial r}\left[r^2 \frac{\partial}{\partial r}\right]\exp\left[-\frac{Zr}{a_o}\right] + \frac{2\mu r^2}{\hbar^2}\left[E + \frac{Ze^2}{r}\right]\exp\left[-\frac{Zr}{a_o}\right] - \ell(\ell + 1)\exp\left[-\frac{Zr}{a_o}\right] = 0. \tag{2}$$

Expansion of the first term in Eq. 2 produces

$$\frac{\partial}{\partial r}\left[r^2 \frac{\partial}{\partial r}\right]\exp\left[-\frac{Zr}{a_o}\right] = -\frac{Z}{a_o}\frac{\partial}{\partial r}r^2 \exp\left[-\frac{Zr}{a_o}\right] = -\frac{Z}{a_o}\left[2r - \frac{Zr^2}{a_o}\right]\exp\left[-\frac{Zr}{a_o}\right]$$

$$= \left[\frac{Z^2r^2}{a_o^2} - \frac{2Zr}{a_o}\right]\exp\left[-\frac{Zr}{a_o}\right]. \tag{3}$$

Combining Eqs. 2 and 3, we obtain

$$\left[\frac{Z^2r^2}{a_o^2} - \frac{2Zr}{a_o}\right]\exp\left[-\frac{Zr}{a_o}\right] + \frac{2\mu Er^2}{\hbar^2}\exp\left[-\frac{Zr}{a_o}\right] + \frac{2\mu Ze^2r}{\hbar^2}\exp\left[-\frac{Zr}{a_o}\right]$$

$$- \ell(\ell + 1)\exp\left[-\frac{Zr}{a_o}\right] = 0. \tag{4}$$

Dividing out the exponential term on both sides and then collecting terms with factors of r^0, r, and r^2 gives

$$r^2\left[\frac{Z^2}{a_o^2} + \frac{2\mu E}{\hbar^2}\right] + r\left[-\frac{2Z}{a_o} + \frac{2\mu Ze^2}{\hbar^2}\right] - r^0\ell(\ell + 1) = 0. \tag{5}$$

For $R(r) = N \exp[-Zr/a_o]$ to be a solution, the left-hand side of Eq. 5 must be zero for all values of r. This can be so only if all terms are individually zero. Consequently, we must have $\ell = 0$ to make the third term zero. To make the first term vanish, we must have

$$\left[\frac{Z^2}{a_o^2} + \frac{2\mu E}{\hbar^2}\right] = 0. \tag{6}$$

This requires that $E = -Z^2\hbar^2/(2\mu a_o^2)$. Since $a_o = \hbar^2/(\mu e^2)$, the expression for E may also be written in the form

$$E = -\frac{Z^2\hbar^2}{2\mu a_o}\frac{\mu e^2}{\hbar^2} = -\frac{Z^2e^2}{2a_o} = -\frac{Z^2\mu e^4}{2\hbar^2} \tag{7}$$

as is required by the problem. Finally, the coefficient of r in Eq. 5 must be zero. Because we have no additional parameters to adjust to ensure that this is true, it must already be zero if $R(r) = N \exp[-Zr/a_o]$ is truly a solution. Therefore, we check to see if it is. Using the fact that $a_o = \hbar^2/(\mu e^2)$, we have

$$\left[-\frac{2Z}{a_o} + \frac{2\mu Ze^2}{\hbar^2}\right] = \left[-\frac{2Z\mu e^2}{\hbar^2} + \frac{2\mu Ze^2}{\hbar^2}\right] = 0, \tag{8}$$

as required. Therefore, if we have $E = -Z^2e^2/(2a_o)$ and $\ell = 0$, $R(r) = N \exp[-Zr/a_o]$ will be a solution of the radial equation.

For related exercises, see Problems 13.4 and 13.5.

As can be seen from Table 13.3, the energy eigenvalues have the form $[-(\text{constant})Z^2\mu e^4/(2\hbar^2)]$, where the constants for the four lowest-energy states are $1, \frac{1}{4}, \frac{1}{9}$, and $\frac{1}{16}$. The constants for the next three states turn out to be $\frac{1}{25}, \frac{1}{36}$, and $\frac{1}{49}$. A little thought suggests that the eigenvalues are given by

$$E_n = -\frac{Z^2\mu e^4}{2n^2\hbar^2} \qquad \text{for } n = 1, 2, 3, 4, \ldots. \qquad \textbf{(13.19A)}$$

when electrostatic–cgs units are employed.

If we use SI units, e will be expressed in coulombs and e^4 will be replaced by the expression $e^4/(4\pi\varepsilon_o)^2$ in Eq. 13.19A. The resulting energy eigenvalues in these units are

$$E_n = -\frac{Z^2\mu e^4}{2(4\pi\varepsilon_o)^2n^2\hbar^2} \qquad \text{for } n = 1, 2, 3, 4, \ldots \text{ and with } e \text{ in coulombs.}$$

$$\textbf{(13.19B)}$$

This deductive inference is shown to be rigorously correct by solving Eq. 13.18 exactly. The constant n appearing in the eigenvalue expression is called the *principal quantum number*. The fact that the energy eigenvalues do not depend upon the azimuthal quantum number ℓ is surprising, since ℓ appears explicitly in the radial differential equation.

The factors $\mu e^2/\hbar^2$ in Eq. 13.19A and $\mu e^2/(4\pi\varepsilon_o\hbar^2)$ in Eq. 13.19B are almost equal to the reciprocal of the Bohr radius a_o. They differ only in the substitution of μ for the electron mass m_e. These two quantities are nearly the same. For the hydrogen atom, $\mu = m_e m_p/(m_e + m_p)$, where m_p is the mass of the proton. In atomic units, $m_e = 1$, $m_p = 1{,}836.1$, and $\mu = (1)(1{,}836.1)/1{,}837.1 = 0.99946$. The difference lies in the fourth significant digit. For heavier nuclei, the difference is even less. (See Problems 13.6 and 13.7.) It is, therefore, customary to write the hydrogen-like atom or ion eigenvalues as

$$E_n = -\frac{Z^2e^2}{2n^2a_o} \qquad \textbf{(13.20A)}$$

for electrostatic units and

$$E_n = -\frac{Z^2e^2}{(4\pi\varepsilon_o)2n^2a_o} \qquad \textbf{(13.20B)}$$

if we employ SI units. In atomic units, the result is

$$E_n = -\frac{Z^2}{2n^2} \text{ hartrees} . \qquad \textbf{(13.20C)}$$

Equations 13.20A through 13.20C are of course, not exact, since we have written $a_o = \hbar^2/(\mu e^2)$ instead of $\hbar^2/(m_e e^2)$. However, if we are satisfied with four digits of accuracy, these expressions will be satisfactory.

Examination of Table 13.3 shows that the radial eigenfunctions depend upon both n and ℓ, as we might have anticipated, given that ℓ appears explicitly in the radial equation and we knew that the energy eigenvalues had to depend upon a third quantum number. Consequently, those eigenfunctions are generally written as $R_n^\ell(r)$. All radial eigenfunctions have the form

$$R_n^\ell(r) = N[\text{polynomial in } r]\exp\left[-\frac{Zr}{na_o}\right],$$

provided we assume that $a_o = \hbar^2/(\mu e^2)$. The leading power of r in the polynomial is always r^{n-1}.

The eigenvalue of lowest energy for hydrogen-like atoms and ions is achieved when $n = 1$. Therefore, this is the ground state. From Tables 12.1 and 13.3, we find that the ground-state eigenfunction is

$$\psi_{100}(r, \theta, \phi) = R_1^0(r)Y_0^0(\theta, \phi) = \left[\frac{1}{4\pi}\right]^{1/2} 2\left[\frac{Z}{a_o}\right]^{3/2} \exp\left[-\frac{Zr}{a_o}\right]$$

$$= \left[\frac{Z^3}{\pi a_o^3}\right]^{1/2} \exp\left[-\frac{Zr}{a_o}\right], \tag{13.21}$$

since $m = 0$ if $\ell = 0$ and $Y_0^0(\theta, \phi) = [1/4\pi]^{1/2}$. The subscripts on ψ denote the quantum numbers n, ℓ, and m. Examination of Eq. 13.21 shows that $\psi_{100}(r, \theta, \phi)$ is never zero for finite values of r. The ground state is, therefore, nodeless, as we expect it to be. The first excited state corresponds to $n = 2$, the second excited state to $n = 3$, and so on. Since the energy eigenvalues given by Eqs. 13.19 and 13.20 depend only upon a single quantum number, we expect there to be a direct correlation between the number of nodes exhibited by the eigenfunctions and the energy of the system. We saw similar behavior in one-dimensional systems, where the energy was dependent upon a single quantum number. Therefore, we expect that the first excited state will have one node along one of the three coordinates, the second excited state will have two nodes, and, in general, the mth excited state will possess m nodes. The total number of nodes will be $n - 1$. Some of these nodes will appear in the θ coordinate, while others are in the radial coordinate. Since the dependence of the eigenfunctions upon ϕ is given by $\Phi(\phi) = (2\pi)^{-1/2}e^{im\phi}$, there are no nodes in the ϕ coordinate. In our treatment of the rigid rotor, we found that the total number of angular nodes in the spherical harmonic $Y_\ell^m(\theta, \phi)$ is equal to ℓ. Therefore,

$$\text{(Number of radial nodes)}$$

$$= \text{(total number of nodes)} - \text{(Number of angular nodes)}$$

$$= (n - 1) - \ell = n - 1 - \ell. \tag{13.22}$$

Figures 13.3 through 13.5 show the radial eigenfunctions for s, p, and d orbitals, respectively. As expected, the 1s orbital has no radial nodes. The 2s and 3s radial orbitals exhibit one and two radial nodes, respectively, as predicted by Eq. 13.22, since $\ell = 0$ for these states. The p orbitals shown in Figure 13.4 have $\ell = 1$. Consequently, we expect the 2p and 3p eigenfunctions to have zero and one radial node, respectively. This is indeed seen to be the case. The d orbitals shown in Figure 13.5 have $\ell = 2$. With this value of ℓ, Eq. 13.22 tells us that the 3d, 4d, and 5d radial orbitals should exhibit zero, one, and two radial nodes, respectively. The results shown in Figure 13.5 confirm this prediction. The cusps at $r = 0$ seen in Figure 13.3 for the s orbitals are a conse-

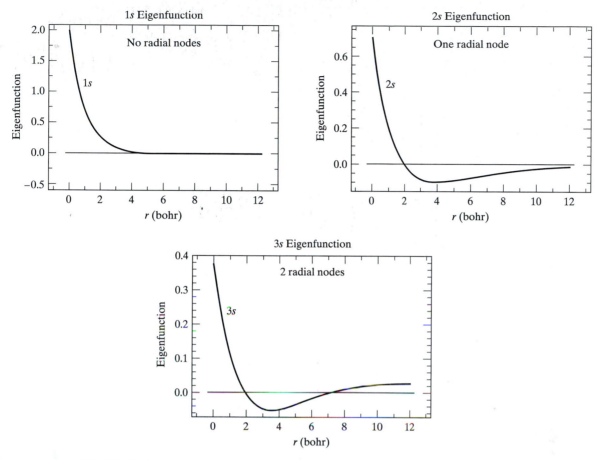

▲ FIGURE 13.3

The radial eigenfunctions of s-type orbitals for hydrogen. The cusp at $r = 0$ is due to the aphysical infinity in the model coulombic potential at this point. Since there are no angular nodes, the number of radial nodes is $n - 1$, as described in the text.

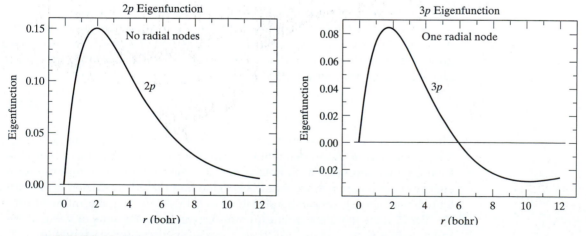

▲ FIGURE 13.4

The radial eigenfunctions of p-type orbitals for hydrogen. The number of radial nodes is $n - 1 - \ell$. For p orbitals, $\ell = 1$; therefore, the number of radial nodes is $n - 2$.

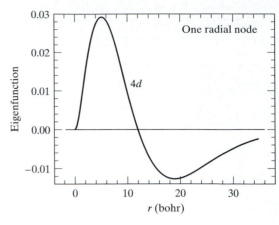

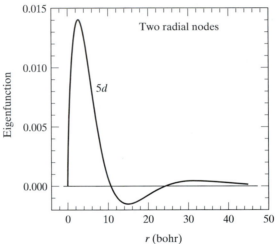

▲ FIGURE 13.5

The radial eigenfunctions of *d*-type orbitals for hydrogen. The number of radial nodes is $n - 1 - \ell$. For *d* orbitals, $\ell = 2$; therefore, the number of radial nodes is $n - 3$.

quence of the aphysical infinity present in our model coulombic potential at $r = 0$. This point has been discussed in detail in Chapters 11 and 12.

Equation 13.22 permits us to reach an important conclusion about the possible values of ℓ for a given value of the principal quantum number. We know that the number of radial nodes must be zero or positive. Therefore, Eq. 13.22 shows that we must always have

$$\boxed{\ell \leq n - 1}. \tag{13.23}$$

This requirement is illustrated in Table 13.3, in which all the eigenfunctions have principal and azimuthal quantum numbers that obey Eq. 13.23. Since the total angular momentum is determined by the value of ℓ, while the total energy depends upon n, we should expect to find that ℓ is restricted by n. With a finite total energy, we cannot possibly have an infinite angular momentum.

Because the magnetic quantum number can have any integer value in the range $-\ell \leq m \leq \ell$, there are $2\ell + 1$ eigenfunctions for each value of ℓ. Equations 13.19A and 13.19B show that all these eigenstates are degenerate in that they have the same energy. We also know that all the eigenfunctions with different ℓ, but the same n, are degenerate. This degeneracy was not expected, since ℓ appears in the radial equation. For that reason, it is often called an *accidental degeneracy*. The total number of eigenstates with the same energy for hydrogen-like atoms and ions is, therefore, the sum of $(2\ell + 1)$ over all possible values of ℓ for a given n; that is,

$$g_n = \sum_{\ell=0}^{n-1} (2\ell + 1) = 1 + 3 + 5 + \cdots + [2(n-1) + 1] = n^2. \quad \textbf{(13.24)}$$

If the potential were anything other than a coulombic potential dependent upon $1/r$, the accidental degeneracy between n and ℓ would not be present. States with the same value of n, but different values of ℓ, would not, in general, have the same energy. This point becomes very important when more than one electron is present in an atom or ion. In such a case, the potential will not have a simple dependence on $1/r$. Consequently, s, p, d, and f orbitals, which have different values of ℓ, will no longer be degenerate.

13.1.3   Atomic Spectra

In Chapter 11, we learned that the observation of discrete emission and absorption spectra created great difficulties for classical theory. Experimentally, it was found that hydrogen-like atoms and ions emitted and absorbed electromagnetic radiation only at certain wavelengths. Empirical fitting of the observed spectra demonstrated that the only wavelengths emitted or absorbed were those given by

$$\lambda^{-1} = 109{,}677 \left[\frac{1}{n^2} - \frac{1}{m^2} \right] cm^{-1}$$

for $n = 1, 2, 3, 4, \ldots$ and $m = n + 1, n + 2, n + 3, \ldots,$

regardless of the intensity of the excitation energy employed in the experiment. The set of emission or absorption frequencies computed by setting $n = 1$ is called the *Lyman series*. The set obtained by using $n = 2$ is termed the *Balmer series*. The $n = 3$ frequencies constitute the *Paschen series*. The constant $109{,}677$ cm^{-1} is called the *Rydberg constant*. It is usually given the symbol R_H.

Quantum theory completely explains these observations and quantitatively predicts the Rydberg constant. Figure 13.6 shows the coulombic potential expressed in hartrees for the hydrogen atom and the location of the energy eigenvalues for the system. In atomic units, the hydrogen-atom

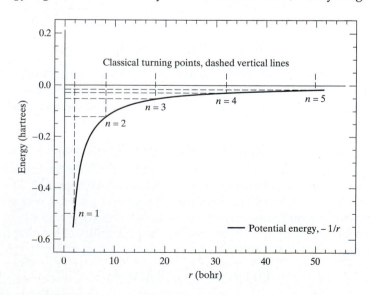

◄ **FIGURE 13.6**
The coulombic potential in atomic units as a function of radial distance for a hydrogen atom. The hydrogen-atom energy eigenvalues are shown as horizontal dashed lines. The point at which these horizontal lines intersect the potential are the classical turning points, shown by the vertical dashed lines.

potential is $V(r) = -1/r$, and the energy eigenvalues given by Eq. 13.20C are $E_n = -1/(2n^2)$. As can be seen, the ground state, $n = 1$, lies far below the excited states. Transitions upward from this state give rise to the Lyman series. The first excited state with $n = 2$ is at -0.125 hartree. Transitions from this state produce the Balmer absorption bands. The figure makes it clear that the spacings between adjacent energy states become smaller as n increases. At sufficiently large n, both the energy eigenvalues and the spacings between adjacent states approach a limiting value of zero. Consequently, at large n, the system approaches classical behavior. When the energy exceeds the limiting value of zero, the electron is unbound, in that the energy exceeds the potential. The discussion in Chapter 12 shows that, in such a case, the energy of the system will be a continuous variable. The variation of the classical turning points with n illustrated in the figure also reflects the fact that the system becomes unbound at large n. As n increases, we see that the turning points tend toward infinity.

Using Eq. 13.20C, we can easily compute the energy spacing between eigenstates with principal quantum numbers n and m. The result is

$$\Delta E = E_m - E_n = -\frac{Z^2}{2m^2} - \left[-\frac{Z^2}{2n^2} \right]$$

$$= \frac{Z^2}{2} \left[\frac{1}{n^2} - \frac{1}{m^2} \right] \text{ hartrees.} \tag{13.25}$$

If we express the energy spacing in terms of the wavelength of incident electromagnetic radiation whose energy is equal to ΔE, we obtain $\Delta E = hc/\lambda$, so that Eq. 13.25 becomes

$$\lambda^{-1} = \frac{Z^2}{2hc} \left[\frac{1}{n^2} - \frac{1}{m^2} \right] \text{bohr}^{-1}. \tag{13.26}$$

The constant appearing in Eq. 13.26 is the Rydberg constant. For hydrogen, its value is

$$R_H = \frac{Z^2}{2hc} = \frac{(1)^2}{2(2)(3.1415927)(137.036)} \text{ bohr}^{-1}$$

$$= (0.000580705 \text{ bohr}^{-1}) \left[\frac{1 \text{ bohr}}{0.529177 \times 10^{-10} \text{m}} \right]$$

$$= 1.09737 \times 10^7 \text{m}^{-1} = 1.09737 \times 10^5 \text{ cm}^{-1}.$$

If the reduced mass correction discussed on page 695 is incorporated, the result becomes $R_H = 1.0968 \times 10^5 \text{ cm}^{-1}$. Quantum theory predicts the observed value of the Rydberg constant to the accuracy of the constants used in this calculation.

13.1.4   Average Values and the Virial Theorem

Average values for atomic systems are computed in exactly the same manner as those for the more elementary systems considered in Chapter 12. The fourth postulate tells us the probability distribution is given by $\psi^*(r, \theta, \phi)\psi(r, \theta, \phi)d\tau$. Therefore, the normalization requirement is $\langle \psi^*(r, \theta, \phi) | \psi(r, \theta, \phi) \rangle = 1$. The radial probability distribution that gives us the probability of locating the electron in a region between r and $r + dr$ is

$$P(r)dr = [R_n^\ell(r)]^* R_n^\ell(r) r^2 dr. \tag{13.27}$$

Consequently, the average value of any function F of r is

$$\langle F(r) \rangle = \langle R_n^\ell(r) | F(r) | R_n^\ell(r) \rangle = \int_{r=0}^{\infty} [R_n^\ell(r)]^* F(r) R_n^\ell(r) r^2 \, dr. \quad \textbf{(13.28)}$$

If we wish to obtain the average value of a function that depends upon r, θ, and ϕ, the required equation is

$$\langle G(r, \theta, \phi) \rangle = \langle \psi(r, \theta, \phi) | G(r, \theta, \phi) | \psi(r, \theta, \phi) \rangle$$

$$= \int_{\phi=0}^{2\pi} \int_{\theta=0}^{\pi} \int_{r=0}^{\infty} \psi^*(r, \theta, \phi) G(r, \theta, \phi) \psi(r, \theta, \phi) d\tau$$

$$= \int_{\phi=0}^{2\pi} \int_{\theta=0}^{\pi} \int_{r=0}^{\infty} \psi^*(r, \theta, \phi) G(r, \theta, \phi) \psi(r, \theta, \phi) r^2 dr \sin \theta \, d\theta \, d\phi. \quad \textbf{(13.29)}$$

The radial probability distribution for the s, p, and d orbitals given by Eq. 13.27 are shown in Figures 13.7 through 13.9, respectively. In each case, the total area beneath these curves is unity, since the radial eigenfunctions are normalized.

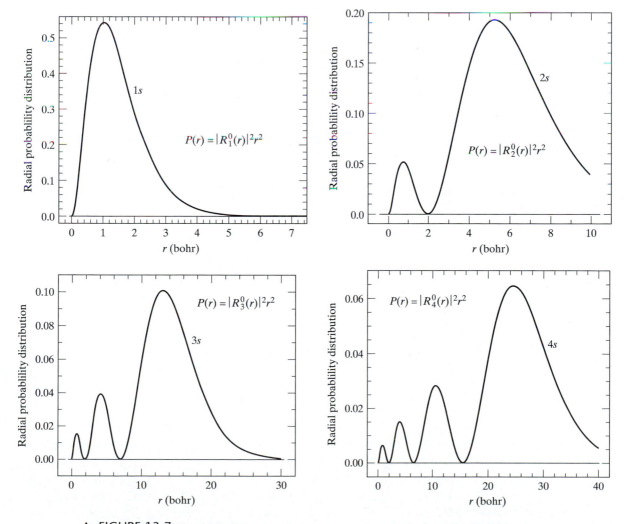

▲ **FIGURE 13.7**
The radial probability distributions $P(r)$ for $1s$, $2s$, $3s$, and $4s$ orbitals for the hydrogen atom. Radial distances are expressed in bohr. Each distribution is obtained from Eq. 13.27.

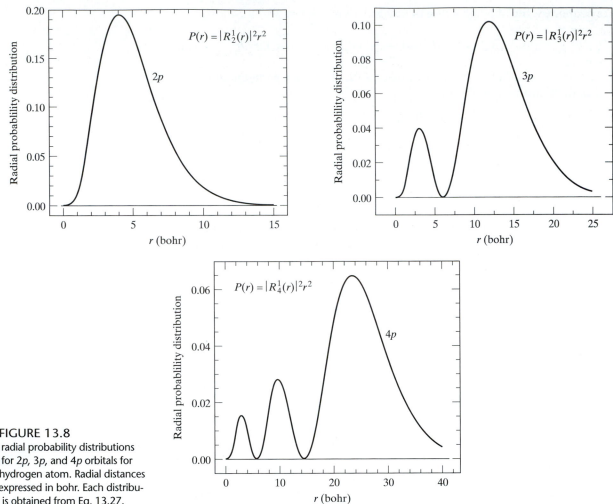

▶ FIGURE 13.8
The radial probability distributions $P(r)$ for 2p, 3p, and 4p orbitals for the hydrogen atom. Radial distances are expressed in bohr. Each distribution is obtained from Eq. 13.27.

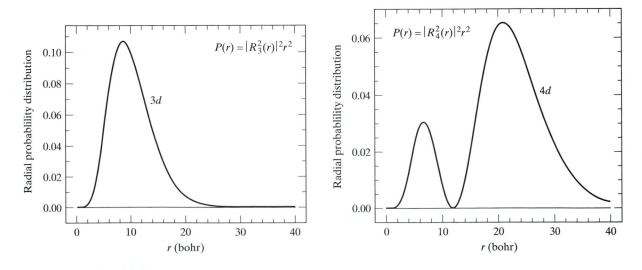

▲ FIGURE 13.9
The radial probability distributions, $P(r)$ for 3d and 4d orbitals for the hydrogen atom. Radial distances are expressed in bohr. Each distribution is obtained from Eq. 13.27.

EXAMPLE 13.3

(A) Determine the average value of r for an electron in the ground state of a hydrogen-like atom or ion. **(B)** Determine the most probable value of r for this eigenstate.

Solution

(A) From Eq. 13.28,

$$\langle r \rangle = \langle R_1^0(r)|r|R_1^0 \rangle. \tag{1}$$

The ground-state radial wave function, given in Table 13.3, is $R_1^0(r) = 2[Z/a_o]^{3/2} \exp[-Zr/a_o]$. In atomic units, this wave function is $R_1^0 = 2Z^{3/2}e^{-Zr}$. Substitution into Eq. 1 produces

$$\langle r \rangle = 4Z^3 \int_{r=0}^{\infty} re^{-2Zr}r^2\,dr = 4Z^3 \int_{r=0}^{\infty} r^3 e^{-2Zr}\,dr. \tag{2}$$

Integrals of the form that appear in Eq. 2 frequently arise in dealing with problems in atomic structure. They have a simple, closed-form solution. In general,

$$\int_0^{\infty} x^n e^{-ax}\,dx = \frac{n!}{a^{n+1}} \quad \text{if } a > 0 \text{ and } n \text{ is an integer.} \tag{3}$$

Clearly, the integral in Eq. 2 has this form. Therefore, we obtain

$$\langle r \rangle = 4Z^3 \frac{3!}{(2Z)^4} = \frac{24Z^3}{16Z^4} = \frac{3}{2Z} \text{ bohr} = \frac{0.7938 \times 10^{-10} \text{ m}}{Z}. \tag{4}$$

Note that the average distance of the electron from the positively charged nucleus decreases as Z increases. Since an increase in Z produces an increase in the attractive forces between the positive nucleus and the negatively charged electron, this type of behavior is expected.

(B) The most probable value of r is the value that maximizes the radial probability distribution. For the ground state, this distribution is given by Eq. 13.27:

$$P(r) = |R_1^0(r)|^2 r^2 = 4Z^3 r^2 \exp[-2Zr] \tag{5}$$

in atomic units. For a maximum value of $P(r)$, we require that $dP(r)/dr = 0$. Taking the derivative of $P(r)$ with respect to r yields

$$\frac{dP(r)}{dr} = 4Z^3[2r \exp(-2Zr) - 2Zr^2 \exp(-2Zr)] = 8Z^3 \exp(-2Zr)[r - Zr^2]. \tag{6}$$

The derivative will be zero when we have

$$r - Zr^2 = 0. \tag{7}$$

Therefore,

$$\boxed{r_{max} = \frac{1}{Z} \text{ bohr}}. \tag{8}$$

For related exercises, see Problems 13.9, 13.10, and 13.11.

In general, the most probable radial position for the electron in a hydrogen-like atom or ion, r_{mp}, and the average position, $\langle r \rangle$, both increase

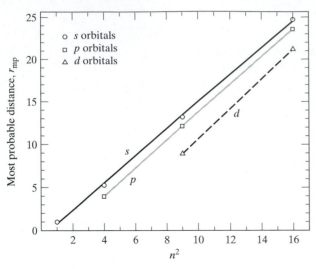

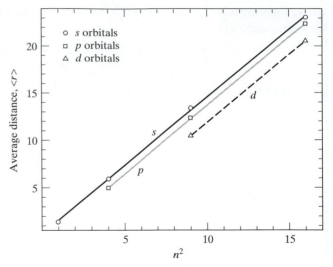

▲ FIGURE 13.10
The most probable radial distance, expressed in bohr, as a function of the square of the principal quantum number for the s, p, and d hydrogen orbitals. The lines are least-squares fits to the computed points. In each case, there is a near-linear variation with n^2.

▲ FIGURE 13.11
The average radial distance, expressed in bohr, as a function of the square of the principal quantum number for the s, p, and d hydrogen orbitals. The lines are least-squares fits to the computed points. In each case, there is a near-linear variation with n^2.

as the principal quantum number increases. The Bohr model predicts that the radii of the circular orbits are proportional to n^2. [See Eq. 11.36.] Figures 13.10 and 13.11 show the variation of r_{mp} and $\langle r \rangle$, respectively, as a function of n^2 for the s, p, and d orbitals of hydrogen. As can be seen, the dependence is very nearly linear, and the values of r_{mp} are similar to those predicted using Eq. 11.36. Quantum mechanically, however, the electron samples all regions of radial space, whereas in the Bohr model, the electron is confined to a circular orbit with a fixed radius.

We may obtain an example of a very important principle for systems whose potentials are electrostatic by computing the average potential and kinetic energies of a hydrogen-like atom or ion in its ground state. The average potential is given by

$$\langle V \rangle = \left\langle -\frac{Ze^2}{r} \right\rangle = -Z\left\langle \frac{1}{r} \right\rangle \tag{13.30}$$

if we use atomic units. Employing the ground-state radial eigenfunction in Table 13.3, expressed in atomic units, we obtain

$$\langle V \rangle = -Z\left\langle R_1^0(r) \left| \frac{1}{r} \right| R_1^0(r) \right\rangle = -4Z^4 \int_{r=0}^{\infty} e^{-Zr} \frac{1}{r} e^{-Zr} r^2 \, dr = -4Z^4 \int_{r=0}^{\infty} r e^{-2Zr} \, dr$$

$$= -4Z^4 \left[\frac{1}{(2Z)^2} \right] = -Z^2 \text{ hartrees,}$$

since $\int_0^{\infty} x^n e^{-ax} \, dx = n!/a^{n+1}$ if a is positive and n is an integer. The average kinetic energy may be obtained from the fact that we must have $\langle E \rangle = E = \langle T + V \rangle = \langle T \rangle + \langle V \rangle$; therefore,

$$\langle T \rangle = E - \langle V \rangle. \tag{13.31}$$

The energy given by Eq. 13.20C for the ground state is $E_1 = -Z^2/2$ hartrees. From Eq. 13.31, we obtain $\langle T \rangle = -Z^2/2 - (-Z^2) = Z^2/2$ hartrees. Comparison of $\langle V \rangle$ and $\langle T \rangle$ shows that, for this state,

$$\boxed{\langle V \rangle = -2 \langle T \rangle}.$$

(13.32)

Equation 13.32 is an example of the *virial theorem,* which holds for any system whose potential is electrostatic. Problems 13.12 and 13.13 provide additional examples.

13.2   Electron Spin

The preceding section solves the Schrödinger equation exactly for hydrogen-like atoms and ions. The result is a set of eigenfunctions and energy eigenvalues that are completely specified by the choice of three quantum numbers, n, ℓ, and m. These quantum numbers also determine the magnitude of the angular momentum of the two-body system and the Z component of **L**. Quantum mechanics has nothing else to say about the system. Yet, experimental observation shows that something is still missing from our description of the system. To understand how experiment reveals this deficiency, we first need to present a few basic laws from electricity and magnetism.

13.2.1   Magnetic Field Interaction with Charged Particles

We expect a molecule with an electric dipole moment to interact with an electric field. When we played with magnets as children, we expected metals to be attracted in the magnetic field produced by our permanent magnet. Most adults have seen movies or pictures of large metallic objects, such as automobiles, being lifted by electromagnets produced by a rotating charge about a metallic core. This type of observation suggests that rotating charges which possess angular momentum are capable of producing magnetic fields and magnetic dipole moments.

According to the laws of electricity and magnetism, a particle with charge Q that possesses angular momentum **L** also has a magnetic dipole moment μ_m given by

$$\mu_m = \gamma \mathbf{L}.$$

(13.33)

The proportionality constant γ, called the *magnetogyric ratio,* is related to the charge and mass of the particle by

$$\gamma = \frac{Q}{2mc},$$

(13.34)

where m is the mass of the particle and c is the speed of light in a vacuum. Note that the vectors μ_m and **L** are collinear.

Let us now suppose that we place this charged particle in a magnetic field. In this situation the magnetic moment of the particle is going to interact with the field to produce a change in energy of the system. The laws of magnetism show that this energy change is given by the product of the magnetic induction $\mathcal{B}$ and the component of μ_m in the direction of the **B**-field. If we take the direction of **B** to be the Z direction, we have

$$\Delta E_{mag} = -[\text{Comp}_B \, \mu_m]\mathcal{B} = -[\mu_m]_z \mathcal{B}.$$

(13.35)

Combining Eqs. 13.33 through 13.35, we obtain

$$\Delta E_{mag} = -\gamma L_z \mathcal{B} = -\frac{Q}{2mc} L_z \mathcal{B}. \tag{13.36}$$

If the charged particle in question is the electron in a hydrogen-like atom, we have $Q = -e$, $m = m_e$, and $L_z = m\hbar$, so that the energy change produced by the magnetic field is

$$\boxed{\Delta E_{mag} = \frac{e\hbar}{2m_e c} m \mathcal{B}}. \tag{13.37}$$

Notice that ΔE_{mag} depends upon the value of m. This is the reason m is called the *magnetic quantum number*.

The constant $e\hbar/(2m_e c)$ appearing in Eq. 13.37 is called the *Bohr magneton*. Its value in mixed SI–Gaussian units or atomic–Gaussian units is

$$\mu_B = \frac{e\hbar}{2m_e c} = 9.27402 \times 10^{-28} \text{ J gauss}^{-1} = 2.1272 \times 10^{-10} \text{ hartrees gauss}^{-1}.$$

Example 13.4 demonstrates how an applied external magnetic field can affect the energy levels in the hydrogen atom.

EXAMPLE 13.4

A hydrogen atom is in an excited $2p$ eigenstate. **(A)** What is the energy of this eigenstate in hartrees? **(B)** What are the possible magnitudes of the Z component of the angular momentum of the electron? **(C)** If this hydrogen atom were subjected to an applied external magnetic field of 10^5 gauss, what would happen to the $2p$ energy states of the atom if we assume that the only angular momentum present is that due to the orbital motion of the electron?

Solution

(A) Equation 13.20C gives the energy eigenvalues for the hydrogen atom with $Z = 1$:

$$E_2 = -\frac{1}{2n^2} = -\frac{1}{8} \text{ hartree.} \tag{1}$$

(B) The Z component of **L** depends upon the magnetic quantum number. For a $2p$ state with $\ell = 1$, there are three possible values of m: -1, 0, and $+1$. Therefore, there are three possible directions for **L** and three possible Z components of **L**, namely,

$$L_z = m\hbar = -\hbar, 0, \text{ or } +\hbar. \tag{2}$$

(C) The change in energy produced by the magnetic field is given by Eq. 13.37. As a result of that change, the three $2p$ energy states will no longer be degenerate with an energy $-\frac{1}{8}$ hartree. The energy will change as follows:

$$[\Delta E_{mag}]_{m=1} = \mu_B \mathcal{B}, \tag{3}$$

$$[\Delta E_{mag}]_{m=0} = 0, \tag{4}$$

and

$$[\Delta E_{mag}]_{m=-1} = -\mu_B \mathcal{B}. \tag{5}$$

Therefore, the new energies are $E(2p_1) = -\frac{1}{8} + \mu_B \mathcal{B}$ hartrees, $E(2p_0) = -\frac{1}{8}$ hartree, and $E(2p_{-1}) = -\frac{1}{8} - \mu_B \mathcal{B}$ hartrees. If $\mathcal{B} = 10^5$ gauss, we have

$$\mu_B \mathcal{B} = 2.1272 \times 10^{-10} \text{ hartree G}^{-1} \times 10^5 \text{ G} = 2.1272 \times 10^{-5} \text{ hartree.} \qquad \textbf{(6)}$$

We see that the changes in energy produced by magnetic fields are very small, but they are not zero. The splitting of energy states by magnetic fields is called the *Zeeman effect*.

For a related exercise, see Problem 13.22.

13.2.2   The Stern–Gerlach Experiment

With the background presented in the previous section, we are now in position to understand the significance of the Stern–Gerlach experiment, first performed in 1926. Stern and Gerlach constructed a magnet that produced an inhomogeneous field. This is done by having one pole of the magnet terminate in a sharp edge while the second pole is a flat plate. Such a design forces the magnetic lines of force together in the region of the edge and thereby increases the field strength. A schematic diagram of such an apparatus is shown in Figure 13.12. The experiment consisted of directing a beam of atoms through the inhomogeneous field along the line indicated by the dashed arrow in the figure and observing where the atomic beam strikes a detector at the right. In the actual experiment, Stern and Gerlach used a beam of Na atoms, because such a beam was relatively easy to produce and detect in 1926. However, the result would have been qualitatively the same had they employed a beam of hydrogen atoms. We shall, therefore, discuss the analogous experiment conducted with a hydrogen-atom beam.

With a magnetic field produced by the apparatus shown in Figure 13.12, the magnetic induction depends upon the z-coordinate. The closer to the sharp edge we are, the higher the field strength. As a first approximation, we may assume that B varies linearly with z; that is,

$$\mathcal{B} = \mathcal{B}_o + az,$$

where a is a constant. Under these conditions, the change in energy produced by the magnetic field is obtained with the use of Eq. 13.37, with $\mathcal{B}$ replaced by the preceding expression:

$$\Delta E_{\text{mag}} = \frac{e\hbar}{2m_e c} m(\mathcal{B}_o + az).$$

The force on the atoms in the field is given by the negative of the derivative of this expression. Thus,

$$\mathbf{F} = F_z = -\frac{\partial \Delta E_{\text{mag}}}{\partial z} = -\frac{ae\hbar}{2m_e c} m, \qquad \textbf{(13.38)}$$

since there are no forces in either the y or the x direction.

Now, imagine sending a beam of hydrogen atoms through this magnetic field along the line indicated by the dashed arrow. At room temperature, the Boltzmann distribution assures us that virtually all of the hydrogen atoms will be in the ground state, with $n = 1$ and $\ell = m = 0$. With $m = 0$, Eq. 13.38 shows that there should be no forces acting on the atoms and that they should, therefore, move through the field without

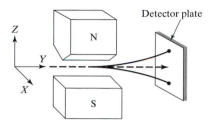

▲ FIGURE 13.12
Stern–Gerlach experiment. The north–south poles of the magnet produce an inhomogeneous field whose field strength increases in the $+Z$ direction. The atomic beam is initially directed along the y-axis. With no deflection by the magnetic field, the beam would strike at the point indicated by the arrow. In the actual experiment, the beam splits into two components of equal intensity, which strike the detector plate at the points indicated by the black dots.

deflection and strike the detector at the point indicated by the arrow. *This is not what occurs!* Stern and Gerlach found that the beam splits into two components of equal intensity. One component deflects in the positive z direction, the second in the negative direction. Furthermore, the magnitude of the upward deflection is exactly equal to that of the downward deflection. This result is sometimes called the *anomalous Zeeman effect*, because its origin was not understood in 1926.

The results of the experiment permitted Stern and Gerlach to reach several important conclusions concerning the system. First, the fact that the beam splits into two components means that there must be two possible values of the z component of angular momentum present in hydrogen atoms. This angular momentum cannot come from the orbital motion of the electron about the nucleus, since m must equal zero in a 1s ground-state hydrogen atom. Consequently, there must exist some other source of angular momentum not described by quantum theory. This type of angular momentum is termed *spin*.

If we presume that spin angular momentum must follow the same quantization rules as L^2 and L_z, the square of the spin angular momentum, S^2, and its z component must be given by expressions whose forms are

$$S^2 = s(s + 1)\hbar^2$$

and

$$S_z = m_s\hbar, \tag{13.39}$$

respectively, with $m_s = -s, -s + 1, -s + 2, \ldots, s - 1, s$. Since the Stern–Gerlach experiment demonstrates that exactly two possible values of the z component of the spin angular momentum are present, it immediately follows that s for an electron must be $\frac{1}{2}$. This gives two possible values for S_z, as m_s can have only the values $-\frac{1}{2}$ and $+\frac{1}{2}$.

If we now go back to Eq. 13.36 and write the corresponding expression for spin angular momentum, we obtain

$$\Delta E_{\text{mag}} = -\gamma_s S_z B = -\gamma_s \hbar m_s (B_o + az), \tag{13.40}$$

where γ_s is the magnetogyric ratio for spin and S_z is the z component of the spin angular momentum. The force acting on the hydrogen atoms in the inhomogeneous field should, therefore, be

$$F_z = \gamma_s a \hbar m_s.$$

From a knowledge of the magnitude of the splitting and the magnetic field strength, Stern and Gerlach were able to deduce the magnetogyric ratio for the electron. It turns out to be a factor of two larger than the corresponding ratio for the angular momentum produced by the orbital motion of the electron, so that

$$\gamma_s = -\frac{e}{m_e c}. \tag{13.41}$$

The Stern–Gerlach experiment established that the electron has two sources of angular momentum. One is its orbital motion, for which $L^2 = \ell(\ell + 1)\hbar^2, L_z = m\hbar$, with $-\ell \leq m \leq \ell$, and $\gamma = -e/(2m_e c)$. The second is electron spin, for which $S^2 = s(s + 1)\hbar^2, S_z = m_s\hbar$, with $-s \leq m_s \leq +s$, and $\gamma_s = -e/(m_e c)$. However, the experiment gave no clue as to why any of this was true. The answer to that question was provided by P. A. M. Dirac in 1928.

13.2.3   Relativistic Quantum Mechanics

In 1928, P. A. M. Dirac demonstrated that the intrinsic spin angular momentum of the electron is relativistic in origin. Dirac realized that a proper treatment of the hydrogen atom required that the postulates of quantum theory be combined with those of special relativity theory in order to obtain a complete description of the system. The mathematics by which Dirac accomplished this are beyond the scope of the text. We shall, therefore, have to be content with an examination of the principal results he obtained.

Dirac was able to show that, when the postulates of quantum theory are combined with the requirements of special relativity, the wave function for the system becomes a (4×1) column vector rather than a single eigenfunction. That is, he found that the solution of the equation for a particle in an infinite potential well, when treated relativistically, has the form

$$\psi(x) = \begin{bmatrix} a(x) \\ b(x) \\ c(x) \\ d(x) \end{bmatrix}.$$

Consequently, four elements make up $\psi(x)$, rather than just one, as we have when the problem is treated without considering relativity requirements. With $\psi(x)$ being a (4×1) matrix, we use standard matrix algebra to execute the operations of quantum theory. For example, the absolute square of $\psi(x)$, $\psi^*(x)\psi(x)$, is obtained by taking the complex transpose of the (4×1) column vector to obtain a (1×4) row vector and multiplying it by the (4×1) column vector. That is,

$$|\psi(x)|^2 = \psi^*(x)\psi(x) = [a^*(x)\, b^*(x)\, c^*(x)\, d^*(x)] \begin{bmatrix} a(x) \\ b(x) \\ c(x) \\ d(x) \end{bmatrix}$$

$$= a^*(x)\,a(x) + b^*(x)\,b(x) + c^*(x)\,c(x) + d^*(x)\,d(x)$$

$$= |a(x)|^2 + |b(x)|^2 + |c(x)|^2 + |d(x)|^2.$$

The absolute square of the wave function is, therefore, the sum of the absolute squares of each of the elements of the (4×1) column vector.

Dirac was also able to show that the nature of the relativistic Hamiltonian produces a fourth quantum number, m_s, that is associated with an intrinsic angular momentum. The three quantum numbers n, ℓ, and m obtained from the nonrelativistic Schrödinger equation are still present. The theory, therefore, provides the theoretical foundation for the empirical results obtained from the Stern–Gerlach experiment.

If chemists had to work with the full relativistic form of the wave function, chemistry would become much more difficult. Fortunately, we can simplify the situation by taking into account the fact that the velocity of molecules and electrons is generally much less than the speed of light in a vacuum. As we approach the limit where v/c is very small, two of the four elements in the (4×1) column vector become negligibly small. If we ignore these small components, the (4×1) column vector becomes a (2×1) vector whose form is either

$$\psi(x) = \begin{bmatrix} \phi(x) \\ 0 \end{bmatrix} = \phi(x) \begin{bmatrix} 1 \\ 0 \end{bmatrix}$$

or

$$\psi(x) = \begin{bmatrix} 0 \\ \phi(x) \end{bmatrix} = \phi(x) \begin{bmatrix} 0 \\ 1 \end{bmatrix}. \tag{13.42}$$

It also turns out that the function $\phi(x)$ in Eq. 13.42 is identical to the solution of the nonrelativistic Schrödinger equation. Therefore, as long as v/c is small, we can write

$$\psi_{\text{rel}}(x) = \psi(x) \begin{bmatrix} 1 \\ 0 \end{bmatrix}$$

or

$$\psi_{\text{rel}}(x) = \psi(x) \begin{bmatrix} 0 \\ 1 \end{bmatrix}, \tag{13.43}$$

where $\psi(x)$ is the nonrelativistic wave function. Consequently, all we have to do to take into account the requirements of relativity theory is multiply our nonrelativistic results by a (2×1) column vector, either $\begin{bmatrix} 1 \\ 0 \end{bmatrix}$ or $\begin{bmatrix} 0 \\ 1 \end{bmatrix}$. These column vectors are called *spin functions*. The first turns out to be associated with $m_s = -\frac{1}{2}$ and $S_z = -\hbar/2$. We generally refer to this (2×1) column vector as the β spin function or the *spin-down* function. The second (2×1) vector is associated with $m_s = +\frac{1}{2}$ and $S_z = \hbar/2$. It is customary to denote this vector as the α, or *spin-up*, spin function.

Since the α and β spin functions are associated with different m_s quantum numbers and different eigenvalues for S_z, we expect them to be orthogonal. That is indeed the case. Consider the matrix multiplication of α^* and β:

$$\alpha^* \times \beta = [0 \; 1] \begin{bmatrix} 1 \\ 0 \end{bmatrix} = (0)(1) + (1)(0) = 0.$$

Hence, α is orthogonal to β. It is also important to note that the spin functions are normalized. The matrix products of $\alpha^* \times \alpha$ and $\beta^* \times \beta$ are, respectively,

$$\alpha^* \times \alpha = [0 \; 1] \begin{bmatrix} 0 \\ 1 \end{bmatrix} = (0)(0) + (1)(1) = 1$$

and

$$\beta^* \times \beta = [1 \; 0] \begin{bmatrix} 1 \\ 0 \end{bmatrix} = (1)(1) + (0)(0) = 1.$$

The foregoing discussion indicates that the ground-state wave function for the hydrogen atom should be written as either $\psi_{1s}(r, \theta, \phi)\,\alpha$ or $\psi_{1s}(r, \theta, \phi)\,\beta$. In the former case, the quantum number assignments are $n = 1$, $\ell = m = 0$, and $m_s = \frac{1}{2}$. In the latter, they are $n = 1$, $\ell = m = 0$, and $m_s = -\frac{1}{2}$. Orbitals of this type that include both a spatial and spin part are called *spin orbitals*. At this point, it is appropriate to note that the classical picture of spin being similar to an electron spinning on its axis in either a clockwise or counterclockwise direction is inappropriate: Spin is a relativistic effect that has no simple classical analogue.

The fact that we must combine the postulates of both quantum theory and special relativity to obtain an accurate theoretical description of atomic structure is compelling evidence that each of these theories contains essential elements of the truth concerning the nature of our universe.

13.3   Addition of Angular Momentum

...a rose by any other name would smell as sweet.
   —Romeo and Juliette, Act II, sc. 2, William Shakespeare (1564–1616)

Civilization began with a rose. A rose is a rose is a rose is a rose.
   —Gertrude Stein (1874–1946)

The basic principle contained in the preceding well-known quotations finds application in science as well as human behavior. We have seen that the hydrogen-atom electron has angular momentum from two sources: its orbital motion and an intrinsic angular momentum due to relativistic effects. The amount of angular momentum present from orbital motion depends upon ℓ and m, as described by Eqs. 13.16 and 13.17. The relativistic spin angular momentum is dependent upon m_s. Equation 13.39 with $s = \frac{1}{2}$ gives the magnitudes of S^2 and S_z. Although we assign different symbols and names to the angular momentum produced by orbital motion and that resulting from relativistic effects, the angular momentum from one source is identical to that from the other. We do not have green angular momentum and purple angular momentum: "*A rose is a rose is a rose is a rose*," and angular momentum is angular momentum is angular momentum.

In view of this consideration, we expect the two components of angular momentum to add vectorially, subject to the quantum mechanical quantization constraints suggested by our examination of the rigid rotor. That is, we expect the total angular momentum to be given by the vector sum

$$\mathbf{J} = \mathbf{L} + \mathbf{S}, \tag{13.44}$$

with the magnitude of $\mathbf{J}$ quantized in the same manner as $\mathbf{L}$ and $\mathbf{S}$:

$$J^2 = j(j + 1)\hbar^2 \tag{13.45}$$

and

$$J_z = m_j\hbar, \qquad \text{with } -j \le m_j \le j. \tag{13.46}$$

Let us take the hydrogen atom as a simple example. In the ground state, $\ell = m = 0$, so there is no orbital angular momentum. As a result, $\mathbf{L} = 0$ and $\mathbf{J} = \mathbf{S}$. Therefore, $j = s = \frac{1}{2}$, and both m_j and m_s can take values $\pm\frac{1}{2}$. Now let us examine a hydrogen atom in the $2p$ excited state. In this eigenstate, we have $\ell = 1$, so that m can assume values of -1, 0, and 1. The spin contribution is still described by $s = \frac{1}{2}$, and $m_s = \pm\frac{1}{2}$. Since the z components of $\mathbf{L}$ and $\mathbf{S}$ are collinear, they add as scalar quantities to produce the z component of the total angular momentum:

$$J_z = L_z + S_z. \tag{13.47}$$

Thus, $m_j\hbar = m\hbar + m_s\hbar$, and therefore, $m_j = m + m_s$. The possible combinations are summarized in Table 13.4. This table shows that when the z components of $\mathbf{L}$ and $\mathbf{S}$ point in the same direction, we have $m_j = \pm\frac{3}{2}$. We

Table 13.4   Addition of angular momentum for a hydrogen atom in the $2p$ eigenstate

m	m_s	$m_j = m + m_s$
1	$\frac{1}{2}$	$\frac{3}{2}$
1	$-\frac{1}{2}$	$\frac{1}{2}$
0	$\frac{1}{2}$	$\frac{1}{2}$
0	$-\frac{1}{2}$	$-\frac{1}{2}$
-1	$\frac{1}{2}$	$-\frac{1}{2}$
-1	$-\frac{1}{2}$	$-\frac{3}{2}$

know that m_j is bounded by j in that it must lie in the range $-j \leq m_j \leq j$. Therefore, if m_j can equal $\pm\frac{3}{2}$, we must have a total angular momentum state with $j = \frac{3}{2}$. With $j = \frac{3}{2}$, m_j can have the values $-\frac{3}{2}, -\frac{1}{2}, \frac{1}{2}$, and $\frac{3}{2}$. This accounts for four of the six possible angular momentum states listed in the table. We still have two remaining m_j states with values of $\frac{1}{2}$ and $-\frac{1}{2}$. Consequently, there must be a second total angular momentum state with $j = \frac{1}{2}$.

We can summarize the results obtained in the preceding analysis as follows: A hydrogen atom in an excited $2p$ state has $s = \frac{1}{2}$ and $\ell = 1$. The spin and orbital angular momenta add vectorially to produce two total angular momentum states, one with $j = \frac{3}{2}$ and a second with $j = \frac{1}{2}$. Consequently, a hydrogen atom in a $2p$ state exhibits two possible total angular momenta, given by Eq. 13.45: $J = [\frac{3}{2}(\frac{3}{2} + 1)\hbar^2]^{1/2} = (\sqrt{15}/2)\hbar$ and $J = [\frac{1}{2}(\frac{1}{2} + 1)\hbar^2]^{1/2} = (\sqrt{3}/2)\hbar$. When the atom is in the $j = \frac{3}{2}$ state, it exhibits z components of angular momentum of $\frac{3}{2}\hbar, \frac{1}{2}\hbar, -\frac{1}{2}\hbar$, and $-\frac{3}{2}\hbar$, as indicated by Eq. 13.46. In the $j = \frac{1}{2}$ angular momentum state, the z component can be only $\frac{1}{2}\hbar$ or $-\frac{1}{2}\hbar$. Problems 13.24 and 13.25 provide additional examples for a hydrogen atom in excited $3d$ and $4f$ eigenstates, respectively.

13.3.1 Term Symbols

The foregoing summary is rather lengthy. To provide a convenient notation that permits all of the information to be easily conveyed, scientists employ atomic *term symbols*, of the form

$$n\,{}^{2S+1}L_J,$$

where n is the principal quantum number for the eigenstate being described. The capital letters $S, P, D, F, G, \ldots$ are used to denote the total orbital angular momentum quantum number $L = 0, 1, 2, 3, 4, \ldots$, respectively. For a hydrogen atom with only one electron, L and ℓ are identical, so that the capital letters $S, P, D, F, G, \ldots$ correspond exactly to the orbital designations $s, p, d, f, g, \ldots$. The left superscript on the L symbol is called the *multiplicity*, in that it is related to the spin degeneracy of the system. Its value is twice the total spin, plus one. The quantum number associated with the total angular momentum is given as a right-hand subscript of L. For a hydrogen atom with only one electron, J and j are identical.

For the hydrogen-atom ground state, we have $\ell = L = 0$, so that the L symbol will be a capital S. The total spin is $\frac{1}{2}$, since we have only one electron. With this value of S, the multiplicity is $2(\frac{1}{2}) + 1 = 2$, which we call a *doublet*. The total angular momentum quantum number is $J = j = s = \frac{1}{2}$. Therefore, the ground state of the hydrogen atom is described by the term $1\,{}^2S_{1/2}$, which we read as "one, doublet S, one-half." The two possible $2p$ angular momentum states have $\ell = L = 1$ with $S = \frac{1}{2}$ and $J = j = \frac{3}{2}$ or $\frac{1}{2}$. Therefore, the two states have the term symbols $2\,{}^2P_{3/2}$ and $2\,{}^2P_{1/2}$, respectively.

13.3.2 Spin–Orbit Coupling

The total angular momentum states specified by the values of J do not have the same energy: There is a small splitting between these states, due to the interaction of the orbital and spin angular momenta. When the electron has orbital angular momentum, a small magnetic moment given by Eqs. 13.33 and 13.34 and an equally small magnetic field are produced. This magnetic field interacts with the spin magnetic moment to produce an energy change whose magnitude depends upon the magnitude of the orbital, spin, and total angular momentum present. That is, for the hydrogen atom, it depends upon ℓ, s, and j. For atoms with more than one electron, it will depend upon the total orbital, spin, and total angular momentum quantum numbers, L, S, and J. This type of interaction is called *L–S coupling* or *spin–orbit coupling*. As a result of such coupling, energy states with different total angular momenta will be split, so that they are no longer degenerate. The observation of such split states in the spectrum of atomic hydrogen led Goudsmit and Uhlenbeck to propose the existence of spin angular momentum in 1925. The Stern–Gerlach experiment later confirmed their hypothesis.

A more detailed treatment of the problem shows that the magnitude of the spin–orbit splitting is proportional to the dot product of **L** and **S**. Squaring Eq. 13.44, we obtain

$$J^2 = \mathbf{J} \cdot \mathbf{J} = [\mathbf{L} + \mathbf{S}] \cdot [\mathbf{L} + \mathbf{S}] = L^2 + S^2 + 2\,\mathbf{L} \cdot \mathbf{S}.$$

Solving for $\mathbf{L} \cdot \mathbf{S}$ yields

$$\mathbf{L} \cdot \mathbf{S} = \frac{1}{2}[J^2 - L^2 - S^2]. \tag{13.48}$$

Using Eqs. 13.39 and 13.45, along with the result for L^2 obtained in Chapter 12, we see that Eq. 13.48 becomes

$$\mathbf{L} \cdot \mathbf{S} = [j(j + 1) - \ell(\ell + 1) - s(s + 1)]\frac{\hbar^2}{2}. \tag{13.49}$$

The change in energy associated with spin–orbit splitting is, therefore,

$$\boxed{\Delta E_{\text{spin–orbit}} = C[j(j + 1) - \ell(\ell + 1) - s(s + 1)]\frac{\hbar^2}{2}.} \tag{13.50}$$

The spin–orbit splitting constant in Eq. 13.50 is approximately $C = 1.67 \times 10^{-6}\, Z^4$ hartrees for hydrogen-like atoms or ions with nuclear charge Z. Since C increases rapidly with increasing atomic number, spin–orbit interactions become much more important as Z increases. When $\ell = 0$, we have $\mathbf{J} = \mathbf{S}$, so that Eq. 13.50 predicts that $\Delta E_{\text{spin–orbit}}$ will be zero. This is the expected result, since with no orbital angular momentum, there will be no magnetic field with which the spin magnetic moment can interact.

EXAMPLE 13.5

Compute the expected spin–orbit splitting between the $2\,^2P_{1/2}$ and $2\,^2P_{3/2}$ states of the hydrogen atom.

Solution

For both angular momentum states, $s = 1/2$ and $\ell = 1$. Using Eq. 13.50 for the $2\,^2P_{3/2}$ state, we obtain

$$\{\Delta E_{\text{spin–orbit}}\}_{j=3/2} = C[j(j + 1) - \ell(\ell + 1) - s(s + 1)]\frac{\hbar^2}{2}$$

$$= C[1.5(1.5 + 1) - 1(1 + 1) - 0.5(0.5 + 1)]\frac{\hbar^2}{2}$$

$$= \frac{C\hbar^2}{2}. \tag{1}$$

For the $2\,^2P_{1/2}$ state, the result is

$$\{\Delta E_{\text{spin–orbit}}\}_{j=1/2} = C[0.5(0.5 + 1) - 1(1 + 1) - 0.5(0.5 + 1)]\frac{\hbar^2}{2} = -C\hbar^2. \tag{2}$$

Therefore, the energy spacing (splitting) between the two states is

$$\text{Splitting} = \{\Delta E_{\text{spin-orbit}}\}_{j=3/2} - \{\Delta E_{\text{spin-orbit}}\}_{j=1/2} = \frac{C\hbar^2}{2} - (-C\hbar^2) = \frac{3C\hbar^2}{2}. \quad \textbf{(3)}$$

For hydrogen with $Z = 1$, $C = 1.67 \times 10^{-6}$ hartree and $\hbar = 1$, so that

$$\text{Splitting} = \frac{3}{2}(1.67 \times 10^{-6} \text{ hartree}) = 2.51 \times 10^{-6} \text{ hartree} = 0.551 \text{ cm}^{-1}. \quad \textbf{(4)}$$

Equation 13.20C shows us that the energy of the $2p$ state is $-\frac{1}{8}$ hartree; we see, then, that spin–orbit interactions are very small in hydrogen relative to the total energy. This will *not* be the case for atoms in which Z is much larger.

For an additional exercise, see Problem 13.26.

Figure 13.13 shows the nature of the spin–orbit splittings for the lowest three principal quantum states of the hydrogen atom. There is no splitting of the $1\,^2S_{1/2}$, $2\,^2S_{1/2}$, and $3\,^2S_{1/2}$ states, since we have no orbital angular momentum to couple with the spin angular momentum. In the absence of L–S coupling, the $2\,^2P_{3/2}$ and $2\,^2P_{1/2}$ states are degenerate. The same is true for the $3\,^2P_{3/2}$ and $3\,^2P_{1/2}$ states and the $3\,^2D_{5/2}$ and $3\,^2D_{3/2}$ states. In the actual system, spin–orbit interactions split these states as shown.

13.4 Many-electron Atoms

13.4.1 General Form of the Hamiltonian

All the systems we have examined up to this point are sufficiently simple that we have been able to obtain rigorous analytical solutions for the eigenfunctions and eigenvalues of the Schrödinger equation describing them. This happy state of affairs comes to an abrupt halt when we begin to investigate atoms and molecules containing more than one electron. From this point onward, we shall have to be content with approximate solutions. However, such solutions can often be as accurate as experimental measurement if enough effort is put into the calculation. The fact that things are about to become much more difficult

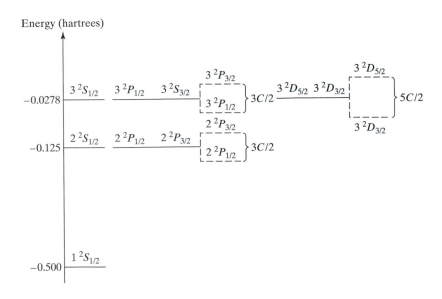

▶ FIGURE 13.13
Spin–orbit splitting for the $n = 1$, 2, and 3 hydrogen-atom energy levels. Solid lines indicate the energy levels with no spin–orbit splitting. Dashed lines show the qualitative nature of the spin–orbit splitting. The actual splittings are much smaller than shown. The figure is not drawn to scale. C is the spin–orbit coupling constant.

should not be a cause for concern. In fact, it is a reason to rejoice. After all, who would pay us to do science or engineering if it were easy?

Consider an atom that contains a nucleus of mass m_n with a positive charge of $+Ze$ and N electrons. For this system, we have a total of $N + 1$ particles, each of which can have kinetic energy. Consequently, we expect the quantum mechanical Hamiltonian to contain $N + 1$ kinetic-energy terms of the form $-\hbar^2/(2m)\nabla^2$. Each of the electrons will be attracted to the positively charged nucleus. The attractive coulombic interactions will lead to N terms in the potential of the general form $-Ze^2/r_i$, where r_i is the distance of electron i from the nucleus. In writing the coulombic potential in this form, we are using electrostatic units to avoid having to put the factor $4\pi\varepsilon_o$ in the denominator. This is the same type of potential term we have for a hydrogen-like atom or ion. Finally, each pair of electrons with the same charge will experience a repulsive interaction that produces a potential term whose form is e^2/r_{ij}, where r_{ij} is the distance between electrons i and j. The number of such pairwise repulsive interactions is the number of combinations of N objects (electrons) taken two at a time:

$$C(N, 2) = \frac{N!}{(N - 2)!\ 2!} = \frac{N(N - 1)}{2}.$$

Thus, if we are studying the lithium atom, which has three electrons, we find that there are $3(2)/2 = 3$ electron–electron repulsion terms in the potential. For sodium, with 11 electrons, the number is $11(10)/2 = 55$ terms. Consequently, the Hamiltonian for the sodium atom will contain 12 kinetic-energy terms, 11 nuclear–electron attraction terms, and 55 electron–electron repulsion terms.

With the preceding considerations in mind, we see that the Hamiltonian for an N-electron atom is

$$\mathcal{H} = -\frac{\hbar^2}{2m_n}\nabla_n^2 - \frac{\hbar^2}{2m_e}\sum_{i=1}^{N}\nabla_i^2 - \sum_{i=1}^{N}\frac{Ze^2}{r_i} + \sum_{i=1}^{N}\sum_{j=i+1}^{N}\frac{e^2}{r_{ij}}. \qquad \textbf{(13.51)}$$

Before proceeding, we need to examine the way the electron–electron repulsions are included in Eq. 13.51. Let us again consider the case of a lithium atom, with three electrons that we denote as 1, 2, and 3. This atom contains three pairwise electron–electron repulsions e^2/r_{12}, e^2/r_{13}, and e^2/r_{23}, where the three terms represent the repulsion between electrons 1 and 2, 1 and 3, and 2 and 3, respectively. If we were to try to include these three terms by using a double sum in which both summation indices run over all values from one to three, we would count every electron–electron repulsion twice, and we would include terms whose form is e^2/r_{ii} that make no physical sense, since r_{ii} is zero. The result would be

$$\sum_{i=1}^{3}\sum_{j=1}^{3}\frac{e^2}{r_{ij}} = \frac{e^2}{r_{11}} + \frac{e^2}{r_{12}} + \frac{e^2}{r_{13}} + \frac{e^2}{r_{21}} + \frac{e^2}{r_{22}} + \frac{e^2}{r_{23}} + \frac{e^2}{r_{31}} + \frac{e^2}{r_{32}} + \frac{e^2}{r_{33}}.$$

On the other hand, if we use a double sum in which the j index must be larger than i, we obtain

$$\sum_{i=1}^{3}\sum_{j=i+1}^{3}\frac{e^2}{r_{ij}} = \frac{e^2}{r_{12}} + \frac{e^2}{r_{13}} + \frac{e^2}{r_{23}},$$

which is exactly the result we want.

The various terms that appear in the Hamiltonian differ in the number of electrons upon whose coordinates each term is dependent. The first term in Eq. 13.51, $-\hbar^2/(2m_n)\,\nabla_n^2$, represents the kinetic energy of the nucleus. The ∇_n^2 operator has the mathematical form given in Eqs. 12.95 and 9.35. It contains the coordinates of the nucleus and derivatives with respect to the coordinates of the nucleus, but it does not contain the coordinates of any of the N electrons or derivatives with respect to these coordinates. Therefore, we call this term a *zero-electron* term. By contrast, terms of the form $-\hbar^2/(2m_e)\,\nabla_e^2$ and $-Ze^2/r_i$ each contain the coordinates of precisely one electron. Consequently, we call them *one-electron* terms. Finally, the electron–electron repulsion terms each depend upon the distance between two electrons. To compute this distance, we need the coordinates of both electrons. Therefore, the electron–electron repulsions are *two-electron* terms.

We can simplify the notation by defining a *one-electron* operator h_i which includes all the terms that depend only upon the coordinates of electron i. Since these are the kinetic energy and the nuclear–electron attraction of electron i, we have

$$h_i = -\frac{\hbar^2}{2m_e}\nabla_i^2 - \frac{Ze^2}{r_i}. \tag{13.52}$$

With this notation, Eq. 13.51 becomes

$$\mathcal{H} = -\frac{\hbar^2}{2m_n}\nabla_n^2 + \sum_{i=1}^{N} h_i + \sum_{i=1}^{N}\sum_{j=i+1}^{N}\frac{e^2}{r_{ij}}. \tag{13.53}$$

The prospect of trying to solve the stationary-state Schrödinger equation, $\mathcal{H}\psi = E\psi$, with $\mathcal{H}$ given by Eq. 13.53 is daunting, to say the least. However, with a few clever approximations, it is not nearly as difficult as it might seem. The first of these is to note that the mass term that appears in the denominator of the nuclear kinetic-energy term is much larger than the mass of the electron. This large mass means that the nuclei are moving very slowly relative to the lighter electrons. We can, therefore, assume that we can omit the kinetic-energy terms for the nuclei from the Hamiltonian. This assumption amounts to saying that we can solve for the electronic eigenfunctions in the field of stationary nuclei. In effect, we have separated the problems of nuclear and electronic motions. This approximation is called the *Born–Oppenheimer separation* or the *Born–Oppenheimer approximation*. It reduces Eq. 13.53 to

$$\mathcal{H} = \sum_{i=1}^{N} h_i + \sum_{i=1}^{N}\sum_{j=i+1}^{N}\frac{e^2}{r_{ij}}. \tag{13.54}$$

13.4.2   An Example: The Helium Atom

To see the nature of our problem when there is more than one electron, let us consider the helium atom, for which $N = 2$. Equation 13.54 for this atom is

$$\mathcal{H} = h_1 + h_2 + \frac{e^2}{r_{12}}, \tag{13.55}$$

where

$$h_1 = -\frac{\hbar^2}{2m_e}\nabla_1^2 - \frac{Ze^2}{r_1}$$

and

$$h_2 = -\frac{\hbar^2}{2m_e}\nabla_2^2 - \frac{Ze^2}{r_2}.$$

We see that h_1 depends only upon the coordinates of Electron 1 and h_2 only upon the coordinates of Electron 2. We now seek the eigenfunctions and eigenvalues for the stationary-state Schrödinger equation, $\mathcal{H}\psi = E\psi$, where $\mathcal{H}$ is given by Eq. 13.55 and ψ is a function of the six coordinates of the two electrons. We can attempt to solve this problem by trying to separate the variables. That is, we substitute a solution of the form

$$\psi(r_1, \theta_1, \phi_1, r_2, \theta_2, \phi_2) = \chi_1(r_1, \theta_1, \phi_1)\,\chi_2(r_2, \theta_2, \phi_2), \qquad \textbf{(13.56)}$$

where (r_1, θ_1, ϕ_1) and (r_2, θ_2, ϕ_2) are the coordinates of Electrons 1 and 2, respectively. The unknown functions $\chi_1(r_1, \theta_1, \phi_1)$ and $\chi_2(r_2, \theta_2, \phi_2)$ are dependent only upon the coordinates of Electrons 1 and 2, respectively.

To see if a solution of the form of Eq. 13.56 exists, we follow our usual procedure. The first step is to substitute the separated form into the differential equation, factor as much to the left of each operator as possible, and then divide both sides by the separated function. If this produces a separation of the variables, a separable solution exists. If it doesn't, it's back to the drawing board. Substitution into the Schrödinger equation produces

$$\mathcal{H}\psi = \left[h_1 + h_2 + \frac{e^2}{r_{12}}\right]\chi_1(r_1, \theta_1, \phi_1)\,\chi_2(r_2, \theta_2, \phi_2) = E\psi$$

$$= E\,\chi_1(r_1, \theta_1, \phi_1)\,\chi_2(r_2, \theta_2, \phi_2),$$

which may be written as

$$\chi_2(r_2, \theta_2, \phi_2)\,h_1\chi_1(r_1, \theta_1, \phi_1) + \chi_1(r_1, \theta_1, \phi_1)\,h_2\,\chi_2(r_2, \theta_2, \phi_2)$$

$$+ \frac{e^2}{r_{12}}\chi_1(r_1, \theta_1, \phi_1)\,\chi_2(r_2, \theta_2, \phi_2)$$

$$= E\,\chi_1(r_1, \theta_1, \phi_1)\,\chi_2(r_2, \theta_2, \phi_2).$$

Dividing by $\chi_1(r_1, \theta_1, \phi_1)\,\chi_2(r_2, \theta_2, \phi_2)$, we obtain

$$\frac{h_1\chi_1(r_1, \theta_1, \phi_1)}{\chi_1(r_1, \theta_1, \phi_1)} + \frac{h_2\chi_2(r_2, \theta_2, \phi_2)}{\chi_2(r_2, \theta_2, \phi_2)} + \frac{e^2}{r_{12}} - E = 0. \qquad \textbf{(13.57)}$$

The first term in Eq. 13.57 depends only upon the coordinates of Electron 1 and the second term only upon the coordinates of Electron 2. E is a constant that is independent of the electron coordinates. However, the electron–electron repulsion term ends our hopes for a separable solution: r_{12} is a two-electron term. Furthermore, the coordinates of the electrons appear beneath a radical; therefore, they can never be separated. Separable solutions for many-electron systems do not exist. As a result, we can never obtain analytically exact eigenfunctions or eigenvalues for any many-electron system. Some approximations are, therefore, needed.

13.4.3   The One-electron Approximation

The term in Eq. 13.57 that is preventing us from obtaining a separable solution of the He-atom equation is the electron–electron repulsion term. The same situation will occur if we try to obtain a separable solution for any other atom in the periodic table. It is always the two-electron terms that thwart our efforts. A simple way to solve this difficulty is to assume that these terms are sufficiently small that we may omit them from the Hamiltonian without incurring significant error. This procedure is called the *one-electron approximation.* If the electron–electron repulsion terms are omitted, the electrons will act as if they are independent particles whose classical motion and quantum mechanical eigenfunctions and eigenvalues are unaffected by the presence of other electrons. For this reason, the term *independent-electron approximation* is also employed to describe the procedure.

If we invoke the one-electron approximation, we can easily obtain the eigenfunctions and eigenvalues for every atom in the periodic table. Let us consider the helium atom as an example. With the electron–electron repulsion term omitted from Eq. 13.57, we have

$$\frac{h_1\chi_1(r_1, \theta_1, \phi_1)}{\chi_1(r_1, \theta_1, \phi_1)} + \frac{h_2\chi_2(r_2, \theta_2, \phi_2)}{\chi_2(r_2, \theta_2, \phi_2)} - E = 0. \qquad \textbf{(13.58)}$$

The variables are now separated, with the first two terms dependent only upon the coordinates of Electrons 1 and 2, respectively. The left side of the equation must be identically zero for all values of the coordinates. This can happen only if each of the first two terms equals a constant such that the sum of the two constants minus E is zero. That is, we must have

$$\frac{h_1\chi_1(r_1, \theta_1, \phi_1)}{\chi_1(r_1, \theta_1, \phi_1)} = E_1$$

and

$$\frac{h_2\chi_2(r_2, \theta_2, \phi_2)}{\chi_2(r_2, \theta_2, \phi_2)} = E_2,$$

with

$$E_1 + E_2 - E = 0. \qquad \textbf{(13.59)}$$

Rearranging the two differential equations gives

$$h_1\chi_1(r_1, \theta_1, \phi_1) = E_1\chi_1(r_1, \theta_1, \phi_1) \qquad \textbf{(13.60)}$$

and

$$h_2\chi_2(r_2, \theta_2, \phi_2) = E_2\chi_2(r_2, \theta_2, \phi_2). \qquad \textbf{(13.61)}$$

Solution of 13.60 and 13.61 yields $\chi_1(r_1, \theta_1, \phi_1)$, $\chi_2(r_2, \theta_2, \phi_2)$, E_1, and E_2. Since the eigenfunction for He is the product of χ_1 and χ_2 and the eigenvalue is the sum of E_1 and E_2, we will have our problem solved once the solutions of Eqs. 13.60 and 13.61 are obtained.

Not only does the one-electron approximation permit us to separate the many-electron Schrödinger equation, but it also makes the solution of the resulting separated equations trivial. We have already solved both Eqs. 13.60 and 13.61. The quantities h_1 and h_2 are just hydrogen-like Hamiltonians for a particle with nuclear charge Z. Therefore, the eigenfunctions are the hydrogen-like eigenfunctions, and the eigenvalues are those for hydrogen-like atoms or ions. That is,

$$\chi_1(r_1, \theta_1, \phi_1) = R_{n_1}^{\ell_1}(r_1)\, Y_{\ell_1}^{m_1}(\theta_1, \phi_1)[\alpha(1) \text{ or } \beta(1)] \qquad \textbf{(13.62)}$$

and

$$\chi_2(r_2, \theta_2, \phi_2) = R_{n_2}^{\ell_2}(r_2)\, Y_{\ell_2}^{m_2}(\theta_2, \phi_2)[\alpha(2) \text{ or } \beta(2)], \qquad \textbf{(13.63)}$$

with

$$E_1 = \frac{Z^2 \mu e^4}{2 n_1^2 \hbar^2} = -\frac{Z^2}{2 n_1^2} \text{ hartrees}$$

and

$$E_2 = -\frac{Z^2 \mu e^4}{2 n_2^2 \hbar^2} = -\frac{Z^2}{2 n_2^2} \text{ hartrees for } n_1 \text{ and } n_2 = 1, 2, 3, 4, \dots. \qquad \textbf{(13.64)}$$

Note that the specification of the eigenfunction now requires eight quantum numbers. Six of these—n_1, ℓ_1, m_1, n_2, ℓ_2, and m_2—are due to the presence of two electrons moving in three-dimensional space. The remaining two are the relativistic spin quantum numbers m_s for each electron.

To simplify the notation, we shall write $\chi_{n\ell m}\alpha(1)$ and $\chi_{n\ell m}\beta(2)$ for $R_n^{\ell}(r_1)$ $Y_{\ell}^m(\theta_1, \phi_1)\alpha(1)$ and $R_n^{\ell}(r_2)Y_{\ell}^m(\theta_2, \phi_2)\beta(2)$, respectively. Thus, when Electron 1 is in the ground-state hydrogen-like orbital with alpha spin, we will write $\chi_{1s}\alpha(1)$. If the second electron is in an excited $2p$ orbital with $m = 1$ and beta spin, we will write $\chi_{2p_1}\beta(2)$. In this notation, it would appear from what has been said to this point that the one-electron eigenfunction for the ground state of the helium atom is either

$$\psi = \chi_{1s}\alpha(1)\, \chi_{1s}\alpha(2),$$

$$\psi = \chi_{1s}\alpha(1)\, \chi_{1s}\beta(2),$$

$$\psi = \chi_{1s}\beta(1)\, \chi_{1s}\alpha(2),$$

or

$$\psi = \chi_{1s}\beta(1)\, \chi_{1s}\beta(2), \qquad \textbf{(13.65)}$$

and the energy eigenvalue, from Eq. 13.64, is

$$E = E_1 + E_2 = -\frac{2^2}{2(1)^2} - \frac{2^2}{2(1)^2} = -4 \text{ hartrees.} \qquad \textbf{(13.66)}$$

If the one-electron approximation were sufficiently accurate, we would have good solutions for every atom in the periodic table and could proceed at this point to investigate molecular structure and bonding. However, Mother Nature is not going to be that kind to us. The fact is that the simple product-type eigenfunctions in Eq. 13.65 do not give the correct electron distribution, nor do they satisfy important symmetry requirements that we will discuss later. The eigenvalues in Eq. 13.66 contain very large errors, which is not really surprising, since the electron–electron repulsion term we threw away in order to obtain that equation and Eq. 13.65 is actually very large. Before we can proceed, we must take some corrective actions that make the one-electron eigenfunctions and eigenvalues more accurate.

13.4.4   Electron Shielding

Let us first determine how much error is present in the one-electron eigenvalues for the ground state of the helium atom. Equation 13.66 shows the total ground-state energy to be -4 hartrees, or -108.84 eV. Since the electron–electron repulsion that we omitted is positive, we can be certain that this energy is too low. That is, the helium atom is predicted to be more stable (i.e., to have lower energy) than it actually is. The first ionization potential for helium, I_1, is defined to be the energy required to remove an electron from the neutral atom: $He \longrightarrow He^+ + e^-$. This energy is the energy difference between He^+ and He:

$$I_1 = E(He^+) - E(He).$$

Since He^+ is a one-electron ion, we know that its ground-state energy is $-Z^2/(2n^2) = -(2)^2/[2(1)^2] = -2$ hartrees. Therefore, the one-electron approximation for He gives a first ionization potential $I_1 = -2 - (-4) = 2$ hartrees $= 54.42$ eV. The experimentally observed result is 24.59 eV. Thus, the one-electron result contains an error of 121.3%! That is, our overestimate of the helium atom stability makes it require 29.83 eV (2,878.2 kJ mol^{-1}) more energy to remove an electron than is actually the case. Clearly, we must do something about this huge error before the one-electron energies have any hope of being useful approximations.

It is important to note that the one-electron approximation always gives the result that the first ionization potential is exactly equal to the negative of the orbital energy (ε) of the electron being removed. For the helium atom, the orbital energy is $\varepsilon = -Z^2/(2n^2) = -2^2/[2(1)^2] = -2$ hartrees, and we have $I_1 = -\varepsilon = -(-2 \text{ hartrees})$. After we improve the one-electron approximation, we will find that the approximation $I_1 \approx -\varepsilon$ is reasonably good. We generally refer to this approximation as *Koopman's theorem*.

We can improve the accuracy of the one-electron approximation dramatically by incorporating the concept of electron shielding. Consider the case of the helium atom, for which Electron 1 is in its ground-state 1s orbital while the second electron is 10 cm from the nucleus, so that it is essentially ionized. This situation is illustrated in Figure 13.14A. Imagine that you are standing on the second electron looking in at the He^+ ion. Would you expect to experience a net attractive force of $-2e^2/(10)^2$ dynes or $-e^2/(10)^2$ dynes? That is, would you expect to feel the attraction of one or two positive charges? At this distance, it is clear that the He^+ ion would appear to be a single particle with a net charge of $+e$, and the attractive force would be $-e^2/(10)^2$ dynes, not twice this quantity. Basically, the one remaining electron on the He^+ ion has canceled out one positive charge. In the nomenclature of this section, we say that one positive charge is *completely shielded* by the one remaining electron. If, on the other hand, the helium atom is in its ground state, with both electrons near the nucleus, as shown in Figure 13.14B, the net positive charge seen by Electron 2 will be much closer to $+2e$, since not all of the electron density of Electron 1 will lie between Electron 2 and the nucleus. Therefore, Electron 1 will not completely shield one proton from Electron 2. If we let the effective charge, Z_e, be the net positive charge seen by Electron 2, we expect Z_e to lie

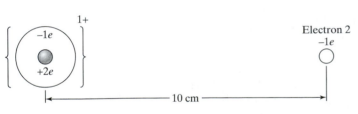

(A) Helium-atom electron far from nucleus, $Z_e = 1$

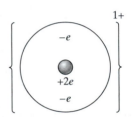

(B) Normal ground-state helium atom, $-1 \leq Z_e \leq 2$

▲ FIGURE 13.14
Variation of effective nuclear charge with distance for He.

between 1 and 2. At large distances, Z_e will approach unity. As Electron 2 approaches the nucleus, we expect to see Z_e approach its limiting value of 2.

If we use first-order perturbation theory, which we will discuss in Chapter 14, it is possible to obtain an analytic expression for the effective charge seen by Electron 2 as a function of its distance from the helium nucleus, r_2. This result is

$$Z_e = 1 + \{2r_2 + 1\}\exp[-4r_2]. \qquad (13.67)$$

Examination of Eq. 13.67 shows that it has the right limiting behavior. When $r_2 \to \infty$, Z_e is unity, as our qualitative analysis suggested that it should be. When $r_2 \to 0$, Z_e approaches 2, the actual nuclear charge on helium. Figure 13.15 presents a plot of the electron–nuclear attraction for Electron 2, which we write as $V(r_2) = -Z_e e^2/r_2$, where Z_e is given by Eq. 13.67, compared with the completely unshielded case, where $V(r_2) = -2e^2/r_2$, and the totally shielded case with $V(r_2) = -e^2/r_2$. We see that Z_e varies continuously as r_2 changes.

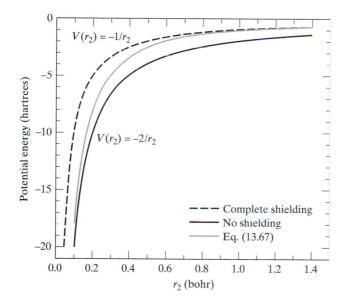

◀ **FIGURE 13.15**
The effect of shielding in the helium atom. The potential experienced by the second $1s$ electron is shown as a function of the distance of the electron from the nucleus. All quantities are expressed in atomic units. The dark solid curve is the result if the second electron is totally unshielded by the first electron. The dashed curve is the result if the first electron completely shields one nuclear positive charge, regardless of the radial distance of the second electron. The light-shaded solid curve is the actual effective potential predicted by a perturbation calculation, Eq. 13.67.

When we omit all the electron–electron repulsion terms and leave the electron–nuclear attraction terms unaltered, we produce a situation that is far from reality. For example, consider the electrons in a Co atom, with $Z = 27$. Every one of the electron–nuclear attraction terms is being written as $-27e^2/r_i$. This is absolutely correct if we also include the 351 electron– electron repulsions, since these positive contributions will cancel an appropriate amount of the attractive interaction and produce the correct answer. If, however, we omit the repulsions and leave the attractive terms unaltered, we are doomed to obtain a result that is far too stable (too negative). Because we cannot easily insert the repulsion terms without destroying the one-electron approximation, we must reduce the magnitude of the attractive interactions to a level that is more reasonable. The discussion in the previous paragraph suggests that the way to do this is to replace Z with an effective charge Z_e in all equations. That is, the energy eigenvalues given by Eq. 13.64 become

$$\boxed{E_n = -\frac{Z_e^2}{2n^2} \text{ hartrees}}. \qquad (13.68)$$

The hydrogen-like eigenfunctions are similarly modified. In addition, the radial part of these eigenfunctions are often modified so that only the highest power in r is retained in the polynomial multiplying the exponential factors in Table 13.3. Little is lost in doing this, since these functions are not the correct solutions of the many-electron-atom equation in any event. With such modifications, the approximate radial solutions become

$$R_n^{\ell}(r) = Nr^{n-1} \exp\left[-\frac{Z_e r}{n}\right] = Nr^{n-1} \exp[-\zeta r], \tag{13.69}$$

where we have written $\zeta = Z_e/n$ and expressed the function in atomic units. Radial orbitals of this type were first introduced by Slater. They are, therefore, called *Slater-type orbitals*, or STOs. This shielding method will significantly improve the one-electron results if we can obtain reasonable values for the effective charges seen by electrons in different orbitals with different average distances from the nucleus.

If we were concerned only with helium, we could use the fact that $I_1 = E(\text{He}^+) - E(\text{He})$, along with Eq. 13.68 and the measured first ionization potential, to obtain Z_e. Since we have two electrons in $1s$ orbitals for He, the ground-state helium-atom energy will be

$$E(\text{He}) = 2\left[-\frac{Z_e^2}{2(1)^2}\right] = -Z_e^2 \text{ hartrees.}$$

The ground-state energy of He^+ is known exactly from our solution of the hydrogen-like atom. It is $E(\text{He}^+) = -2^2/[2(1)^2] = -2$ hartrees. The measured first ionization potential is 24.59 eV, or 0.9037 hartree. Therefore, we have $-2 - (-Z_e^2) = 0.9037$. Solving for Z_e, we obtain

$$Z_e = (2.9037)^{1/2} = 1.704.$$

It is customary to write the effective charge in terms of a shielding factor S that is specific for a given orbital. That is, we define the effective charge seen by electrons in orbital a in terms of the actual nuclear charge Z and a shielding factor for that orbital:

$$[Z_e]_a = Z - S_a. \tag{13.70}$$

In these terms, if we use the preceding estimate obtained for Z_e, we might write

$$[Z_e]_{1s} = Z - S_{1s} = Z - 0.296,$$

so that, for a helium atom, we would have

$$[Z_e]_{1s} \text{ for He} = 2 - 0.296 = 1.704.$$

This method is not unreasonable. However, when we consider bonding and molecular structure, we will be more concerned with properly representing the electron distribution and the energy in the ground state than with ionization potentials. Therefore, effective charges are generally obtained by requiring that they produce the most accurate possible result in a variational calculation of the ground-state energy, using Slater-type orbitals. Such calculations will be discussed in greater detail in Chapter 14. Clementi and Raimondi have reported the results of variational calculations for all atoms with atomic numbers in the range $2 \leq Z \leq 86$. The values for He through Kr are given in Table 13.5. Note that this variational method produces an effective charge for the $1s$ helium electron of 1.6875, which is close to the 1.704 figure obtained from our simple ionization-potential analysis.

Table 13.5   Effective charges in terms of $\zeta = \dfrac{Z_e}{n}$ for ground-state neutral atoms

Atom	Z	1s	2s	2p	3s	3p	4s	3d	4p
He	2	1.6875							
Li	3	2.6906	0.6396						
Be	4	3.6848	0.9560						
B	5	4.6795	1.2881	1.2107					
C	6	5.6727	1.6083	1.5679					
N	7	6.6651	1.9237	1.9170					
O	8	7.6579	2.2458	2.2266					
F	9	8.6501	2.5638	2.5500					
Ne	10	9.6421	2.8792	2.8792					
Na	11	10.6259	3.2857	3.4009	0.8358				
Mg	12	11.6089	3.6960	3.9129	1.1025				
Al	13	12.5910	4.1068	4.4817	1.3724	1.3552			
Si	14	13.5745	4.5100	4.9725	1.6344	1.4284			
P	15	14.5578	4.9125	5.4806	1.8806	1.6288			
S	16	15.5409	5.3144	5.9885	2.1223	1.8273			
Cl	17	16.5239	5.7152	6.4966	2.3561	2.0387			
Ar	18	17.5075	6.1152	7.0041	2.5856	2.2547			
K	19	18.4895	6.5031	7.5136	2.8933	2.5752	0.8738		
Ca	20	19.4730	6.8882	8.0207	3.2005	2.8861	1.0995		
Sc	21	20.4566	7.2868	8.5273	3.4466	3.1354	1.1581	2.3733	
Ti	22	21.4409	7.6883	9.0324	3.6777	3.3679	1.2042	2.7138	
V	23	22.4256	8.0907	9.5364	3.9031	3.5950	1.2453	2.9943	
Cr	24	23.4138	8.4919	10.0376	4.1226	3.8220	1.2833	3.2522	
Mn	25	24.3957	8.8969	10.5420	4.3393	4.0364	1.3208	3.5094	
Fe	26	25.3810	9.2995	11.0444	4.5587	4.2593	1.3585	3.7266	
Co	27	26.3668	9.7025	11.5462	4.7741	4.4782	1.3941	3.9518	
Ni	28	27.3526	10.1063	12.0476	4.9870	4.6950	1.4277	4.1765	
Cu	29	28.3386	10.5099	12.5485	5.1981	4.9102	1.4606	4.4002	
Zn	30	29.3245	10.9140	13.0490	5.4064	5.1231	1.4913	4.6261	
Ga	31	30.3094	11.2995	13.5454	5.6654	5.4012	1.7667	5.0311	1.5554
Ge	32	31.2937	11.6824	14.0411	5.9299	5.6712	2.0109	5.4171	1.6951
As	33	32.2783	12.0635	14.5368	6.1985	5.9499	2.2360	5.7928	1.8623
Se	34	33.2622	12.4442	15.0326	6.4678	6.2350	2.4394	6.1590	2.0718
Br	35	34.2471	12.8217	15.5282	6.7395	6.5236	2.6382	6.5187	2.2570
Kr	36	35.2316	13.1990	16.0235	7.0109	6.8114	2.8289	6.8753	2.4423

EXAMPLE 13.6

(A) Use Eq. 13.68, the data in Table 13.5, and Koopman's approximation to estimate the first ionization potential of Li. For reasons that will be discussed later in the chapter, the electron removed in the first ionization is in a $2s$ orbital. **(B)** The first and second ionization potentials of Li are 5.32 eV and 75.64 eV, respectively. The second ionization potential is the energy required to remove an electron from the singly charged ion, Li^+. For Li^+, this electron will be in a $1s$ orbital. Use these data to compute the value of $\zeta = Z_e/n$ for the $2s$ and $1s$ electrons in Li needed to give the exact values of the first and second ionization potentials. Compare your result with the variational ones listed in the table.

Solution

(A) The energy of the $2s$ electron in Li is given by Eq. 13.68:

$$E_{2s} = -\frac{[Z_e]_{2s}^2}{2n^2}. \tag{1}$$

Using the data in Table 13.5 on page 723, we have ζ_{2s} for Li $= 0.6396 = Z_e/n = Z_e/2$. Thus, $[Z_e]_{2s} = 1.2792$, which gives

$$E_{2s} = -\frac{(1.2792)^2}{2(2)^2} = -0.2045 \text{ hartree} = -5.566 \text{ eV}. \tag{2}$$

Koopman's approximation for the first ionization potential produces

$$I_1 \approx -E_{2s} = 5.57 \text{ eV}. \tag{3}$$

This result is in error by 4.7%, which is a huge improvement over the uncorrected one-electron results.

(B) We could estimate effective charges by requiring that Eq. 13.68 and Koopman's approximation give the exact answers for the ionization potentials. That is, we would require that

$$I_1 = (5.32 \text{ eV}) \times \frac{1 \text{ hartree}}{27.211 \text{ eV}} = -E_{2s} = \frac{[Z_e]_{2s}^2}{2n^2} = \frac{[Z_e]_{2s}^2}{8}. \tag{4}$$

This gives

$$[Z_e]_{2s} = \left[\frac{8(5.32)}{27.211}\right]^{1/2} = 1.251. \tag{5}$$

Thus, $\zeta_{2s} = 1.251/2 = 0.6255$, which is very close to the results obtained from the variational calculations in Table 13.5.

For the second ionization potential, we have

$$I_2 \approx -E_{1s} = \frac{[Z_e]_{1s}^2}{2n^2} = \frac{[Z_e]_{1s}^2}{2(1)^2} = \frac{[Z_e]_{1s}^2}{2} = (75.64 \text{ eV}) \times \frac{1 \text{ hartree}}{27.211 \text{ eV}}. \tag{6}$$

Therefore,

$$[Z_e]_{1s} = \zeta_{1s} = \left[\frac{2(75.64)}{27.211}\right]^{1/2} = 2.358. \tag{7}$$

If we do not use Koopman's approximation, the second ionization potential is given by

$$I_2 = E(Li^{2+}) - E(Li^+). \tag{8}$$

The energy of Li^{2+} is known exactly, because Li^{2+} is a one-electron ion. Therefore, we have

$$I_2 = (75.64 \text{ eV}) \times \frac{1 \text{ hartree}}{27.211 \text{ eV}} = 2.780 = -\frac{3^2}{2} - 2\left[-\frac{[Z_e]_{1s}^2}{2}\right], \qquad (9)$$

since we have two $1s$ electrons in Li^+. Solving for $[Z_e]_{1s}$, we obtain

$$[Z_e]_{1s} = \zeta_{1s} = (4.5000 + 2.780)^{1/2} = 2.698. \qquad (10)$$

The result listed in Table 13.5 is 2.6906. Therefore, Eq. 10 yields a good approximation to the variational results. The value obtained using Koopman's approximation deviates by -12.6%.

The data in Table 13.5 can be put in more succinct form by computing the shielding factors for each orbital as a function of the number of electrons in different orbitals and then using a least-squares method to fit the results. The fitted results for the orbital shielding factors obtained by Clementi and Raimondi are

$$S(1s) = 0.3[N(1s) - 1] + 0.0072[N(2s) + N(2p)]$$
$$+ 0.0158[N(3s) + N(3p) + N(4s) + N(3d) + N(4p)],$$
$$S(2s) = 1.7208 + 0.3601[N(2s) + N(2p) - 1]$$
$$+ 0.2062[N(3s) + N(3p) + N(4s) + N(3d) + N(4p)],$$
$$S(2p) = 2.5787 + 0.3326[N(2p) - 1] - 0.0773N(3s)$$
$$- 0.0161[N(3p) + N(4s)] - 0.0048N(3d) + 0.0085N(4p),$$
$$S(3s) = 8.4927 + 0.2501[N(3s) + N(3p) - 1] + 0.0778N(4s)$$
$$+ 0.3382N(3d) + 0.1978N(4p),$$
$$S(3p) = 9.3345 + 0.3803[N(3p) - 1] + 0.0526N(4s)$$
$$+ 0.3289N(3d) + 0.1558N(4p),$$
$$S(4s) = 15.505 + 0.0971[N(4s) - 1] + 0.8433N(3d) + 0.0687N(4p),$$
$$S(3d) = 13.5894 + 0.2693[N(3d) - 1] - 0.1065N(4p),$$

and

$$S(4p) = 24.7782 + 0.2905[N(4p) - 1], \qquad \textbf{(13.71)}$$

where $N(n\ell)$ is the number of electrons in the orbital with quantum numbers n and ℓ.

EXAMPLE 13.7

(A) Use the data in Table 13.5 to obtain the form of a normalized Slater $3s$ orbital for the Na atom. **(B)** Using the data in Table 13.3, write down the form of a hydrogen-like $3s$ orbital with an effective charge equal to that of the sodium $3s$ electron. On the same graph, plot the Slater $3s$ orbital and hydrogen-like $3s$ orbital when both have the Na $3s$ effective charge for Z. On a separate graph, plot the radial probability distribution functions for both the Slater and hydrogen-like $3s$ orbitals.

Solution

(A) In atomic units, a Slater orbital has the form

$$\psi_s(r, \theta, \phi) = N_s r^{n-1} \exp\left[-\frac{Z_e r}{n}\right] Y_\ell^m(\theta, \phi), \tag{1}$$

where the subscript s denotes a Slater-type orbital. For a $3s$ orbital, we have $\ell = m = 0$, so that

$$Y_0^0(\theta, \phi) = [4\pi]^{-1/2} \tag{2}$$

and

$$R_3^0(r) = N_s r^2 \exp\left[-\frac{Z_e r}{3}\right]. \tag{3}$$

The normalization constant is given by requiring that we have

$$\langle [R_3^0(r)]^*|R_3^0(r)\rangle = 1 = N_s^2 \int_{r=0}^{\infty} r^2 \exp\left[-\frac{Z_e r}{3}\right] r^2 \exp\left[-\frac{Z_e r}{3}\right] r^2 dr$$

$$= N_s^2 \int_0^\infty r^6 \exp\left[-\frac{2Z_e r}{3}\right] dr = N_s^2 \frac{6! 3^7}{2^7 Z_e^7} = N_s^2 \frac{98,415}{8Z_e^7}. \tag{4}$$

Therefore, the normalization constant is

$$N_s = \left[\frac{8Z_e^7}{98,415}\right]^{1/2}. \tag{5}$$

Table 13.5 gives

$$\zeta(3s)_{\text{Na}} = \frac{Z_e}{n} = \frac{Z_e}{3} = 0.8358. \tag{6}$$

Therefore, $Z_e = 2.5074$. The normalized Slater function is

$$\psi_s(r, \theta, \phi) = \left[\frac{8Z_e^7}{98,415}\right]^{1/2} [4\pi]^{-1/2} r^2 \exp\left[-\frac{Z_e r}{3}\right]. \tag{7}$$

Insertion of Z_e produces

$$\psi_s(r, \theta, \phi) = 0.06349 r^2 \exp[-0.8358r]. \tag{8}$$

(B) The $3s$ hydrogen-like orbital for a nucleus with an effective charge $Z_e = 2.5074$ is

$$\psi_H(r, \theta, \phi) = \frac{1}{9(3)^{1/2}} Z_e^{3/2} [4\pi]^{-1/2} [6 - 6\rho + \rho^2] \exp\left[-\frac{\rho}{2}\right], \tag{9}$$

where $\rho = 2Z_e r/n$ in atomic units. Inserting Z_e and $n = 3$ gives $\rho = 1.6716r$. Thus,

$$\psi_H(r, \theta, \phi) = \frac{1}{9(3)^{1/2}} Z_e^{3/2} [4\pi]^{-1/2} [6 - 10.0296r + 2.7942r^2] \exp[-0.8358r]$$

$$= 0.07185 [6 - 10.0296r + 2.7942r^2] \exp[-0.8358r]. \tag{10}$$

The radial probability distribution functions for the Slater and Hydrogen-like orbitals are given by the absolute squares of the wave functions in Eqs. 8 and 10, respectively.

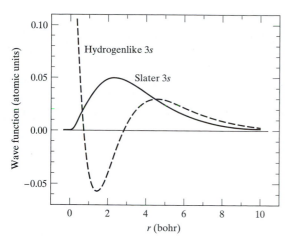

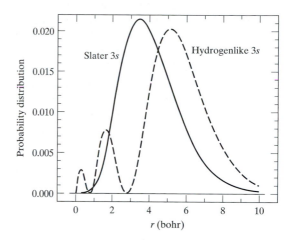

▲ **FIGURE 13.16**

Slater and hydrogen-like 3s radial orbitals for a Na atom, obtained with the use of an effective charge $Z_e = 2.5074$ for both orbitals. At distances greater than 4 bohr, the two orbitals are similar. However, the nodal structure of the hydrogen-like orbital is lost when Slater orbitals are used.

▲ **FIGURE 13.17**

The radial probability density from Eq. 13.27 for Slater and hydrogen-like 3s radial orbitals for a Na atom, obtained with the use of an effective charge $Z_e = 2.5074$ for both orbitals.

Figure 13.16 shows a comparison of the Slater and hydrogen-like 3s orbitals for a Na atom, using an effective charge $Z_e = 2.5074$ for both orbitals. Figure 13.17 compares the radial probability distributions for these two orbitals. Examination of the plots shows that the nodal structure in the radial function is lost when we use a Slater-type orbital. The probability distribution for the hydrogen-like orbital extends to larger values than does the corresponding Slater function. This causes us no concern, because both functions are only second approximations to the correct eigenfunction. More accurate methods for solving the Schrödinger equation will be discussed in Chapter 14.

13.4.5 Electron Exchange Symmetry

Shielding modifications improve one-electron energies significantly and provide a good starting point from which much more accurate results can be obtained by using procedures that will be discussed in Chapter 14. However, the level of accuracy is still low, and the eigenfunctions are still seriously deficient. We shall use the helium atom to illustrate the nature of the problem.

After replacing the actual nuclear charge with an effective charge and after substituting Slater-type orbitals for the radial part of the hydrogen-like eigenfunctions, we can apparently write the ground-state helium-atom wave function in any one of the four ways given in Eq. 13.65; that is,

$$\psi(1,2) = \chi_{1s}\alpha(1)\,\chi_{1s}\alpha(2),$$

$$\psi(1,2) = \chi_{1s}\alpha(1)\,\chi_{1s}\beta(2),$$

$$\psi(1,2) = \chi_{1s}\beta(1)\,\chi_{1s}\alpha(2),$$

or

$$\psi(1,2) = \chi_{1s}\beta(1)\,\chi_{1s}\beta(2),$$

where $\chi_{1s}(1) = N\exp[-Z_e r_1]\,Y_0^0(\theta_1, \phi_1)$, with a similar definition for $\chi_{1s}(2)$. The notation $\psi(1,2)$ represents the fact that the wave function depends upon the coordinates of both Electrons 1 and Electron 2. It is simply shorthand for the more elaborate notation $\psi(r_1, \theta_1, \phi_1, r_2, \theta_2, \phi_2)$. Each of the foregoing wave functions is the product of two spin orbitals, one for each electron. As we shall see, all these possibilities fail to satisfy an important symmetry requirement on the wave function.

The fourth postulate requires that the electron distribution about the helium atom be given by $|\psi(1, 2)|^2\, d\tau$. Although we have labeled the two electrons as 1 and 2, that notation is just for our convenience in writing the equations. All electrons are identical; one does not carry a label "1" and the other "2." To paraphrase Gertrude Stein, "An electron is an electron is an electron is an electron." In view of this, we know that the calculated electron distribution cannot depend upon which electron we label as 1 and which as 2. If we were to exchange the labels in our equations, the fourth postulate would require that we have

$$|\psi(1, 2)|^2 = |\psi(2, 1)|^2. \tag{13.72}$$

Interchanging labels cannot alter the value of the absolute square of $\psi(1, 2)$.

Let us now extract the square root of both sides of Eq. 13.72, keeping in mind that we are dealing with absolute squares. The result is

$$\psi(1, 2) = e^{i\delta}\psi(2, 1). \tag{13.73}$$

It is easy to verify that Eq. 13.73 is correct by taking the absolute square of both sides. This operation gives

$$\psi^*(1, 2)\psi(1, 2) = |\psi(1, 2)|^2 = [e^{i\delta}\psi(2, 1)]^*[e^{i\delta}\psi(2, 1)]$$
$$= e^{-i\delta}\psi^*(2, 1)e^{i\delta}\psi(2, 1) = |\psi(2, 1)|^2,$$

as required. The factor $e^{i\delta}$ is called the *phase factor*. Therefore, the indistinguishability of electrons requires that exchanging electron labels alter the wave function by no more than a phase factor.

We may deduce the nature of the phase angle δ by considering what must occur as we execute multiple exchanges of electron labels. Figure 13.18 illustrates a series of such multiple exchanges for a three-electron system. In the first step, Electrons 1 and 2 are exchanged. The new wave function is the same as the original function, except for the exchange of labels and multiplication by the factor $e^{i\delta}$. In the second step, Electrons 1 and 3 are switched, thereby producing the same wave function with switched labels and an additional factor of $e^{i\delta}$. In the next three steps, Electrons 1 and 2, 1 and 3, and 1 and 2 are successively exchanged. After five exchanges, we have the same wave function with altered labels, but it is now multiplied by $e^{5i\delta}$. In the system shown in the figure, the sixth exchange reproduces the same labels that were present in the original wave function, but $\psi(1, 2, 3)$ is now multiplied by $e^{6i\delta}$. Since the net effect of these six electron exchanges is the same as having no exchanges, the final result must be the same as our initial wave function. That is, we must have $\psi(1, 2, 3) = \psi(1, 2, 3)e^{6i\delta}$. This can be true only if $e^{6i\delta}$ is unity.

We may easily generalize the type of result illustrated in Figure 13.18. Suppose we have an N-electron system on which we execute K total exchanges such that the result of the final exchange is to reproduce the electron assignments in the original wave function. That is, we have

$$\psi(1, 2\ 3, 4, \ldots, N) \xrightarrow{\;K \text{ Exchanges}\;} \psi(1, 2, 3, 4, \ldots, N)e^{ik\delta}.$$

The only way this can happen is for K to be an even integer. An odd number of pairwise exchanges can never reproduce the original wave function labels, regardless of the value of N. Therefore, we must have $K = 2k$, where k is an integer. Since the net effect of the K exchanges is the same as having no

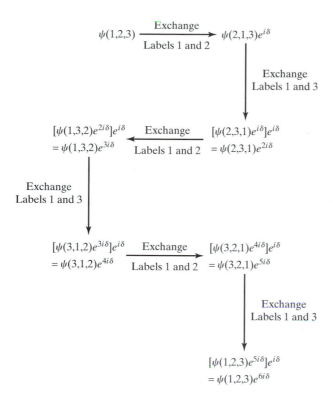

A series of six electron exchanges in a three-electron system. Each exchange of electron labels introduces an additional factor of $e^{i\delta}$ into the wave function. The six exchanges illustrated reproduce the original wave function. Therefore, we must have $\psi(1, 2, 3) = \psi(1, 2, 3,)e^{6i\delta}$.

exchanges, it follows that $e^{iK\delta} = e^{2ik\delta} = 1$. This allows us to determine the phase angle δ. We know that

$$e^{2ik\delta} = \cos(2k\delta) + i\sin(2k\delta) = 1. \tag{13.74}$$

Since k is an integer, Eq. 13.74 is satisfied only if we have $\delta = 0$ or π. We could, of course, also have δ equal to any integral multiple of π, but that would not change the conclusions reached in what follows.

The preceding analysis suggests that the fourth postulate of quantum mechanics predicts that there could be two classes of particles in the universe: those which have $\delta = 0$ and those which have $\delta = \pi$. The results of a variety of experiments demonstrate that this is precisely the case. It is found that particles which have half-integral spin quantum numbers $\frac{1}{2}, \frac{3}{2}, \frac{5}{2}, \ldots$ all have $\delta = \pi$, whereas those particles which have integral spin quantum numbers $0, 1, 2, \ldots$ have $\delta = 0$. The former are called *fermions*, the latter *bosons*. Thus, fermions, with $\delta = \pi$, have wave functions that are *antisymmetric* with respect to particle exchange; that is,

$$\psi_{\text{fermion}}(1, 2) = e^{i\pi}\psi_{\text{fermion}}(2, 1) = [\cos\pi + i\sin\pi]\,\psi_{\text{fermion}}(2, 1)$$

$$= -\psi_{\text{fermion}}(2, 1).$$

The wave functions for fermions change sign when particles are exchanged. By contrast, for bosons, we have

$$\psi_{\text{boson}}(1, 2) = e^{i0}\psi_{\text{boson}}(2, 1) = [\cos 0 + i\sin 0]\,\psi_{\text{boson}}(2, 1) - \psi_{\text{boson}}(2, 1).$$

The wave functions for bosons are unaltered by particle exchange; therefore, we say that they are *symmetric* with respect to exchange.

Since electrons have $s = \frac{1}{2}$, the electron is a fermion whose wave function must always be antisymmetric with respect to electron exchange. For this reason, none of the one-electron wave functions written in Eq. 13.65 are acceptable. The functions $\psi(1, 2) = \chi_{1s}\alpha(1)\,\chi_{1s}\alpha(2)$ and $\psi(1, 2) = \chi_{1s}\beta(1)\,\chi_{1s}\beta(2)$ are both symmetric to electron exchange. This would be satisfactory for a boson, but not for a fermion. The other two wave functions, $\psi(1, 2) = \chi_{1s}\alpha(1)\,\chi_{1s}\beta(2)$ and $\psi(1, 2) = \chi_{1s}\beta(1)\,\chi_{1s}\alpha(2)$, are neither symmetric nor antisymmetric to electron exchange; therefore, they could not properly represent the wave function of any particle.

The foregoing discussion indicates that we need a general method which permits us to write wave functions that have the required antisymmetry to exchange. One of the properties of determinants that we encountered in college algebra provides the answer. When we exchange two rows or two columns of a determinant, the magnitude remains unaltered, but the sign changes. The following 2×2 determinant is a simple example:

$$D = \begin{vmatrix} 5 & 9 \\ 3 & 4 \end{vmatrix} = 20 - 27 = -7.$$

If we exchange the two columns, the result is

$$D_{\text{column exchange}} = \begin{vmatrix} 9 & 5 \\ 4 & 3 \end{vmatrix} = 27 - 20 = 7 = -D.$$

Notice that the magnitude remains unaltered, but the sign changes. If, instead, we exchange the two rows, we obtain the same result:

$$D_{\text{row exchange}} = \begin{vmatrix} 3 & 4 \\ 5 & 9 \end{vmatrix} = 27 - 20 = 7 = -D.$$

Clearly, a determinant is antisymmetric to row or column exchange.

The preceding exercise demonstrates that all we need do to ensure that our wave functions have the required antisymmetry with respect to electron exchange is to write them as a determinant. Thus, while we cannot use the wave functions in Eq. 13.65 for the helium atom, we can use

$$\psi(1, 2) = N\begin{vmatrix} \chi_{1s}\alpha(1) & \chi_{1s}\alpha(2) \\ \chi_{1s}\beta(1) & \chi_{1s}\beta(2) \end{vmatrix}, \tag{13.75}$$

where N is the normalization constant. Notice how the determinant is written. The different one-electron functions are placed in different rows. The assignment of electrons to these one-electron spin orbitals differ from column to column. Electron 1 appears only in Column 1, Electron 2 only in Column 2. As a result, exchanging electrons is equivalent to interchanging Columns 1 and 2 in the determinant, which produces a change in sign. This assures us that our wave function will have the correct antisymmetry with respect to electron exchange. Determinants with this property are called *Slater determinants*. In general, if the spatial functions of the spin orbitals are individually normalized, the normalization constant for an $N \times N$ Slater determinant will be $[1/N!]^{1/2}$. Therefore, N in Eq. 13.75 is $[1/2!]^{1/2} = [1/2]^{1/2}$. (See Example 13.8.)

The two-electron helium atom requires two spin orbitals and a 2×2 determinant. The Li atom, with three electrons, requires three spin orbitals and a 3×3 Slater determinant. An 11×11 Slater determinant is needed to properly represent the wave function for a sodium atom.

The two-electron wave function in Eq. 13.75 has a special property not present in systems with more than two electrons: It can be written as the product of spatial orbitals and a spin function. This is easily seen by expanding the Slater determinant for helium in that equation. The result is

$$\psi(1, 2) = N[\chi_{1s}\alpha(1)\,\chi_{1s}\beta(2) - \chi_{1s}\beta(1)\,\chi_{1s}\alpha(2)]$$

$$= N\chi_{1s}(1)\chi_{1s}(2)\,[\alpha(1)\,\beta(2) - \beta(1)\,\alpha(2)]. \tag{13.76}$$

We see that this one-electron ground-state wave function has a symmetric spatial part, $\chi_{1s}(1)\chi_{1s}(2)$, and an antisymmetric spin function, $[\alpha(1)\beta(2) - \beta(1)\alpha(2)]$. Therefore, the total wave function is antisymmetric, as required.

Examples 13.8 and 13.9 illustrate how we work with such antisymmetric, many-electron wave functions to compute normalization constants and average values.

EXAMPLE 13.8

(A) Let χ_{1s} be a Slater-type orbital with the effective charge for helium given in Table 13.5. Write down the form of this orbital and normalize it. (B) Obtain the normalization constant for $\psi(1, 2)$ in Eq. 13.76.

Solution

(A) The Slater $1s$ orbital for helium is

$$\chi_{1s} = N\exp[-1.6875r]\,Y_0^0(\theta, \phi) = N[4\pi]^{-1/2}\exp[-1.6875r], \qquad (1)$$

where the effective charge, $Z_e = 1.6875$, is obtained directly from the table. The normalization requirement is

$$N^2[4\pi]^{-1}\int_{\phi=0}^{2\pi}\int_{\theta=0}^{\pi}\int_{r=0}^{\infty}\chi_{1s}^*\chi_{1s}r^2\,dr\,\sin\theta\,d\theta\,d\phi = 1. \qquad (2)$$

Inserting the χ_{1s} function and integrating over θ and ϕ gives

$$N^2[4\pi]^{-1}\,4\pi\int_0^{\infty}\exp[-(2)(1.6875)r]r^2\,dr = N^2\left[\frac{2!}{[(2)(1.6875)]^3}\right] = 1. \qquad (3)$$

Solving for N^2, we obtain

$$N^2 = [2^2\,(1.6875)^3] = 19.2217. \qquad (4)$$

Taking square roots of both sides of Eq. 4 yields the value of N:

$$N = 4.384. \qquad (5)$$

(B) As usual, the normalization requirement is

$$\langle\psi(1, 2)|\psi(1, 2)\rangle = N^2\langle\chi_{1s}(1)\chi_{1s}(2)\,[\alpha(1)\,\beta(2)$$
$$-\,\beta(1)\,\alpha(2)]|\chi_{1s}(1)\chi_{1s}(2)\,[\alpha(1)\,\beta(2) - \beta(1)\,\alpha(2)]\rangle = 1. \qquad (6)$$

Let us first examine the multiplication of the spin functions in Eq. 6. Multiplying out the spin terms, we obtain

$$[\alpha^*(1)\,\beta^*(2) - \beta^*(1)\,\alpha^*(2)][\alpha(1)\,\beta(2) - \beta(1)\,\alpha(2)] = \alpha^*(1)\,\alpha(1)\,\beta^*(2)\,\beta(2)$$
$$+\,\beta^*(1)\,\beta(1)\,\alpha^*(2)\,\alpha(2) - \alpha^*(1)\,\beta(1)\,\beta^*(2)\,\alpha(2)$$
$$-\beta^*(1)\,\alpha(1)\,\alpha^*(2)\,\beta(2). \qquad (7)$$

Recalling that $\alpha^*(1)\,\alpha(1) = [0\ 1]\begin{bmatrix}0\\1\end{bmatrix} = 1$, that $\beta^*(2)\beta(2) = [1\ 0]\begin{bmatrix}1\\0\end{bmatrix} = 1$, and that $\alpha^*(1)\beta(1) = [0\ 1]\begin{bmatrix}1\\0\end{bmatrix} = 0$ because α and β are orthogonal, we see that the spin function multiplication yields

$$[\alpha^*(1)\,\beta^*(2) - \beta^*(1)\,\alpha^*(2)][\alpha(1)\,\beta(2) - \beta(1)\,\alpha(2)] = 1 + 1 - 0 - 0 = 2. \qquad (8)$$

Integration over the spatial coordinates of Electrons 1 and 2 gives

$$\langle\chi_{1s}(1)\chi_{1s}(2)|\chi_{1s}(1)\chi_{1s}(2)\rangle = \langle\chi_{1s}(1)|\chi_{1s}(1)\rangle\,\langle\chi_{1s}(2)|\chi_{1s}(2)\rangle = (1)(1) = 1, \qquad (9)$$

since we ensured that χ_{1s} was normalized to unity in (A). Therefore, we obtain

$$\langle\psi(1, 2)|\psi(1, 2)\rangle = N^2[(1)(1)(2)] = 2N^2 = 1, \tag{10}$$

so that the normalization constant for the Slater determinant is

$$N = [2]^{-1/2}. \tag{11}$$

In general, if the spatial parts of the spin orbitals are individually normalized, the normalization constant for an $N \times N$ Slater determinant will be $[1/N!]^{1/2}$.
For a related exercise, see Problem 13.31.

EXAMPLE 13.9

Compute the average distance of a $1s$ electron in helium from the nucleus, using the effective charges given in Table 13.5 and the normalized Slater determinant obtained in Example 13.8.

Solution

The normalized Slater determinant obtained in Example 13.8 is

$$\psi(1, 2) = \left[\frac{1}{2}\right]^{1/2} \chi_{1s}(1)\chi_{1s}(2) [\alpha(1)\beta(2) - \beta(1)\alpha(2)], \tag{1}$$

where the χ_{1s} functions are individually normalized and given by

$$\chi_{1s} = 4.384 \exp[-1.6875r] Y_0^0(\theta, \phi) \tag{2}$$

if the effective charge is taken from Table 13.5. The average distance of Electron 1 from the nucleus is obtained from

$$\langle r_1 \rangle = \frac{1}{2} \langle [\chi_{1s}(1)\chi_{1s}(2)[\alpha(1)\beta(2) - \beta(1)\alpha(2)]]|r_1|$$

$$[\chi_{1s}(1)\chi_{1s}(2) [\alpha(1)\beta(2) - \beta(1)\alpha(2)]]\rangle. \tag{3}$$

Multiplying the spin functions gives the same result as that obtained in Example 13.8; that is,

$$[\alpha^*(1)\beta^*(2) - \beta^*(1)\alpha^*(2)][\alpha(1)\beta(2) - \beta(1)\alpha(2)] = 1 + 1 - 0 - 0 = 2. \tag{4}$$

This factor of two, multiplied by the square of the normalization constant gives unity, so that we have

$$\langle r_1 \rangle = \langle \chi_{1s}(1)|r_1|\chi_{1s}(1)\rangle\langle\chi_{1s}(2)|\chi_{1s}(2)\rangle, \tag{5}$$

where the first integration is performed over the coordinates of Electron 1 and the second over the coordinates of Electron 2. Since we have already normalized $\chi_{1s}(2)$, we know that $\langle\chi_{1s}(2)|\chi_{1s}(2)\rangle = 1$. Thus,

$$\langle r_1 \rangle = \langle\chi_{1s}(1)|r_1|\chi_{1s}(1)\rangle = (4.384)^2[4\pi]^{-1} \int_{\phi=0}^{2\pi} \int_{\theta=0}^{\pi} \int_{r=0}^{\infty} \chi_{1s}{}^*r_1 \chi_{1s}r_1^2 dr_1 \sin\theta \, d\theta \, d\phi$$

$$= (4.384)^2 \int_{r_1=0}^{\infty} r_1^3 \exp[-3.375r_1]dr_1 = 19.219\left[\frac{3!}{(3.375)^4}\right] = 0.889 \text{ bohr.} \tag{6}$$

Notice that the products of the spin terms and the square of the normalization constant equal unity. Thus, all integrals except the one for which we seek the average value are unity. Therefore, the problem is worked essentially as we would work a one-electron type of problem, except for the replacement of the actual nuclear charge with an effective charge. This is the nature of the one-electron approximation.

13.4.6   The Pauli Exclusion Principle

The one-electron spin orbital $\chi_{1s}\alpha(1)$ is shorthand notation for the product of a 1s spatial orbital and an alpha spin function. Written out in more detail, it is

$$\chi_{1s}\alpha(1) = [4Z_e^3]^{1/2}\exp[-Z_e r_1]\, Y_0^0(\theta_1, \phi_1)\begin{bmatrix}0\\1\end{bmatrix}.$$

The four quantum numbers that specify this orbital are $n = 1$, $\ell = m = 0$, and $m_s = \frac{1}{2}$. Suppose we wish to assign the two helium electrons to two spin orbitals, each identical to $\chi_{1s}\alpha$. Since the wave function must be antisymmetric to electron exchange, we must write it in determinant form. That is, we would have

$$\psi(1, 2) = \left[\frac{1}{2!}\right]^{1/2}\begin{vmatrix}\chi_{1s}\alpha(1) & \chi_{1s}\alpha(2)\\ \chi_{1s}\alpha(1) & \chi_{1s}\alpha(2)\end{vmatrix}. \tag{13.77}$$

Notice that the Slater determinant in Eq. 13.77 has two rows that are identical. *When any two rows or columns of a determinant are identical, the determinant is always zero.* Thus, our attempt to put both electrons into spin orbitals that have all four quantum numbers the same produces $\psi(1, 2) = 0$, which is the trivial solution of the Schrödinger equation that violates the fourth postulate of quantum mechanics.

A little thought shows that anytime we attempt to place two or more electrons into spin orbitals that have all quantum numbers the same, we will produce a Slater determinant with two or more rows identical and thereby obtain the mathematically correct, but scientifically unacceptable, trivial solution for the wave function. If we wish to avoid the trivial solution, we must avoid using identical spin orbitals for two or more electrons. This statement is known as the *Pauli exclusion principle*. Simply stated, it says,

> *The quantum number assignments cannot be the same for the spin orbitals of any two electrons if we wish to avoid the trivial solution of the Schrödinger equation.*

It is important to realize that the Pauli principle exists only because of the antisymmetry requirement on fermion wave functions. *There is no Pauli principle for bosons.* The consequences of this simple statement are profound, as will become apparent when we consider the periodic table and electron configurations in the next section.

13.4.7   Electronic Configurations and the Periodic Table

In 1869, Dmitri Mendeleev (1834–1907) advanced one of the most startling and profound theories ever found in the history of science. Mendeleev, a professor of chemistry at the University of Saint Petersberg, discovered that certain physical and chemical properties of the elements exhibited periodic behavior if the elements were arranged in order of increasing atomic mass. He phrased this discovery as the *periodic law:*

> *The properties of simple bodies [elements], the composition of their compounds, as well as the properties of these last, are a periodic function of the atomic weights of the elements.*

Mendeleev's discovery eventually led to the modern periodic table.

Prior to Mendeleev, other scientists had attempted to organize the elements in some systematic fashion, without notable success. Mendeleev succeeded where they had failed because he was brilliant enough to leave blanks in his table for elements that had not yet been discovered. He also corrected some of the atomic mass values so that the elements fell into their proper positions in the table. The missing elements were scandium (Sc), gallium (Ga), germanium (Ge), and technetium (Tc). Mendeleev boldly predicted that the elements which would eventually fill the blanks in his table were yet to be discovered! To bring indium (In) and uranium (U) into the proper positions in the table, he corrected their mass values.

The prediction that there were missing elements which would eventually be discovered was startling by itself. Mendeleev, however, went much further: He predicted the chemical and physical properties of some of the missing elements. The stunning accuracy of his predictions was, in large measure, responsible for the widespread acceptance of the periodic table. Table 13.6 shows the most striking example of his predictions. In 1869, none of the rare gases had been discovered. Since those elements come at the end of the period, Mendeleev had no way to predict their existence. However, if just one of the rare-gas elements had been known at the time, there can be little doubt that he would have been able to successfully predict the existence of the others.

As a result of his discovery, Mendeleev became one of the most influential scientists of his day. His book went through eight editions during his lifetime and five more after his death in 1907. A measure of his stature can be seen by the czar's reaction to Mendeleev's remarriage after his divorce in 1876. At the age of 42, Mendeleev fell in love with a 17-year-old art student. (Men never change!) After arranging a divorce, he learned that both civil and ecclesiastic Russian law forbade the remarriage of a divorced person until seven years had elapsed. Not wishing to wait seven years, Mendeleev found a priest who was willing to perform the ceremony for a reported 10,000 rubles. When word of his remarriage spread, the civil authorities took no action against this world-renowned chemist even though he was officially a bigamist. A divorced nobleman who also wanted to remarry prior to the seven-year waiting period appealed to the czar for special dispensation, citing Mendeleev's case in support of his petition. The czar is reputed to have replied; "I admit that Mendeleev has two wives, but I have only one Mendeleev."

Table 13.6	Comparison of the observed properties of germanium (called eka-silicon by Mendeleev) with those predicted by Mendeleev	
Property	Eka-silicon (Prediction Made in 1871)	Observed for Germanium, Discovered in 1886
Atomic mass	72	72.59
Density (g cm^{-3})	5.5	5.35
Specific heat (J g^{-1} K^{-1})	0.305	0.309
Melting point (°C)	High	947
Color	Dark gray	Grayish white
Formula of oxide	XO_2	GeO_2
Density of oxide (g cm^{-3})	4.7	4.70
Formula of chloride	XCl_4	$GeCl_4$
Boiling point of chloride	A little under 100° C	84° C
Density of chloride (g cm^{-3})	1.9	1.887

With quantum theory, it is relatively easy to predict the structure of the periodic table. The brilliance of Mendeleev's discovery derives from the fact that he made it 58 years before Schrödinger presented his first paper on quantum mechanics. Figure 13.19 shows a modern periodic table with the various rows labeled with the one-electron orbitals that are filling as elements are added.

Since there are two possible values of the spin magnetic quantum number m_s, a maximum of two electrons can be assigned the same set of quantum numbers n, ℓ, and m without producing a violation of the Pauli exclusion principle. In elementary chemistry courses, this fact is usually stated in the form "Orbitals contain a maximum of two electrons with paired spins." We now see that the phrase "paired spins" means one electron with spin function α and one with spin function β.

The restriction on the number of electrons that may be placed in a spatial orbital leads to a gradual filling of higher energy orbitals as the number of electrons increases with the atomic number. As usual, nature endeavors to place the successive electrons into the most stable (lowest energy) orbitals possible. Therefore, we need to determine the energy ordering of the available orbitals in order to predict the ground-state wave function for many-electron atoms. For elements with $Z \leq 18$, this can be done by using the shielding rules developed by Clementi and Raimondi. More accurate calculations are needed for heavier elements.

For helium, the answer is obvious. The one-electron orbital with the lowest possible energy is the $1s$ orbital, with $n = 1$ and $\ell = m = 0$. Therefore, we expect the best one-electron wave function for helium to be that given by the Slater determinant in Eq. 13.75. It is common practice to denote this wave function with the notation $1s^2$, which means that the electronic configuration of helium is described by a wave function in which both electrons are in $1s$ spatial orbitals with appropriate effective charges, one with spin function α, the other with spin function β. The Pauli principle prevents us from assigning additional electrons to the $1s$ orbital. Therefore, the first row of the periodic table contains two elements: $H(1s^1)$ and $He(1s^2)$. We conclude that the first row elements are those for which the electrons are filling orbitals with $n = 1$.

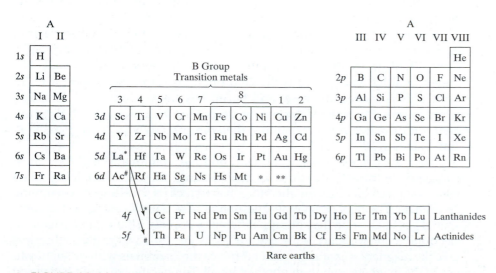

▲ **FIGURE 13.19**
The periodic table. The elements labeled with * and ** have been tentatively assigned the symbols Uun and Uuu. Their existence is as yet unconfirmed.

In the case of Li, with $Z = 3$, it is clear that the first two electrons will fill the $1s$ orbital with paired spins. What about the third electron? It might go into a $2s$ orbital or into one of the three $2p$ orbitals. For one-electron hydrogen-like atoms and ions, the $2s$ and $2p$ orbitals are degenerate. It is, therefore, not obvious which will be lower in energy for Li. If the electron goes into a $2s$ orbital, the electronic configuration of Li will be $1s^2 2s^1$; if the electron enters a $2p$ orbital, we will have $1s^2 2p^1$. The essential question is, therefore, which configuration has the lower energy? Using the shielding rules in Eq. 13.71 enables us to answer this question. The shielding factor for the $1s$ orbital, $S(1s)$, is the same for both the $1s^2 2s^1$ and the $1s^2 2p^1$ configuration. The $2s$ shielding factor for the $1s^2 2s^1$ configuration is $S(2s) = 1.7208$. The $2p$ shielding factor for the $1s^2 2p^1$ electron arrangement is 2.5787. Therefore, the effective charges for $2s$ and $2p$ orbitals are $Z_e(2s) = 3 - 1.7208 = 1.2792$ and $Z_e(2p) = 3 - 2.5787 = 0.4213$, respectively. The corresponding orbital energies are

$$E(2s) = -\frac{Z_e^2}{2n^2} = -\frac{(1.2792)^2}{2(2)^2} \text{ hartree} = -0.2045 \text{ hartree}$$

and

$$E(2p) = -\frac{(0.4213)^2}{2(2)^2} \text{ hartree} = -0.02219 \text{ hartree}.$$

We see that the $2s$ orbital is considerably more stable than the $2p$ orbital. Hence, the electronic configuration of Li is $1s^2 2s^1$, not $1s^2 2p^1$. Once there is more than one electron present, the accidental degeneracy between the s, p, d, and f orbitals is removed.

Similar calculations demonstrate that for beryllium, the configuration $1s^2 2s^2$ is more stable than either $1s^2 2s^1 2p^1$ or $1s^2 2p^2$. Consequently, the beryllium configuration has completely filled $1s$ and $2s$ orbitals. The groups containing Li and Be are usually labeled IA and IIA, respectively. Elements in these two groups are in the process of filling s orbitals. All elements in Group IA have an outer s^1 electronic configuration. Since the chemical properties of an element are dependent primarily upon the outer electronic structure, the elements in a given group exhibit similar chemical behavior.

When we come to boron, with $Z = 5$, we cannot place another electron in either the $1s$ or the $2s$ spatial orbital without violating the Pauli principle. Consequently, the fifth electron must enter a different spatial orbital. The only orbital left with $n = 2$ is the $2p$ orbital. Therefore, we expect it to fill next. This is indeed the case, and boron has the configuration $1s^2 2s^2 2p^1$, which means the Slater determinant describing boron is a 5×5 determinant with the five rows containing the spin orbitals $\chi_{1s}\alpha$, $\chi_{1s}\beta$, $\chi_{2s}\alpha$, $\chi_{2s}\beta$, and $\chi_{2p}\alpha$ or $\chi_{2p}\beta$.

The next five elements, C, N, O, F, and Ne, continue to add their most energetic electrons to the three $2p$ orbitals until they are completely filled when we reach Ne. Carbon, therefore, has two of its six electrons occupying $2p$ orbitals. The only remaining question is, Are both electrons in the same spatial p orbital, or do they occupy different p orbitals. That is, do we have an electronic configuration $1s^2 2s^2 2p_{-1}^2$ or $1s^2 2s^2 2p_{-1}^1 2p_0^1$, where the subscripts give the magnetic quantum numbers? (Since the $2p$ orbitals with different values of the magnetic quantum number are all degenerate, we could have used any of three orbitals. All are equivalent.) The answer is that the electrons tend to occupy different degenerate spatial orbitals with the same spin function if permitted by the Pauli principle. The reason is that this configuration minimizes the destabilizing repulsive interaction between negatively charged electrons. If the electrons occupy different orbitals, the average distance between them will be greater and their mutual repulsion minimized. Since the one-electron approximation ignores the electron–electron repulsions, there is no way for us to predict this result with our present level of theory. It can be predicted, however, using methods we will introduce in Chapter 14. For the moment, we shall simply state the result as a qualitative rule:

> *In the ground state, electrons always occupy different degenerate orbitals with the same spin, either α or β, if permitted by the Pauli principle.*

Application of this principle tells us that the electronic configurations of the above elements are $C[1s^2 2s^2 2p_{-1}^1 2p_0^1]$, $N[1s^2 2s^2 2p_{-1}^1 2p_0^1 2p_1^1]$, $O[1s^2 2s^2 2p_{-1}^2 2p_0^1 2p_1^1]$, $F[1s^2 2s^2 2p_{-1}^2 2p_0^2 2p_1^1]$, and $Ne[1s^2 2s^2 2p_{-1}^2 2p_0^2 2p_1^2]$. The groups containing these elements are labeled IIIA–VIIIA in Figure 13.19. Elements in these groups are in the process of filling p orbitals, and all elements in a given group have the same outer electronic configuration. For example, the outer configuration of all elements in Group VIA is either $p_{-1}^2 p_0^1 p_1^1$, $p_{-1}^1 p_0^2 p_1^1$, or $p_{-1}^1 p_0^1 p_1^2$. All of these configurations are energetically equivalent. Figure 13.20 qualitatively illustrates the preceding rule.

When $Z = 11$, for Na, the $1s$, $2s$, and $2p$ orbitals are filled. Calculations similar to those previously executed show that the orbital with the next lowest energy is the $3s$ orbital. Consequently, the most energetic electrons in Na and Mg fill this orbital, and their electronic configurations are $[Ne]3s^1$ and $[Ne]3s^2$, respectively, where the notation $[Ne]$ denotes the electronic configuration of neon: $1s^2 2s^2 2p_{-1}^2 2p_0^2 2p_1^2$. Since the most energetic Na and Mg electrons fill an s orbital, they are listed in groups IA and IIA, respectively. Elements 13–18 (Al–Ar) fill the $3p$ orbitals according to the

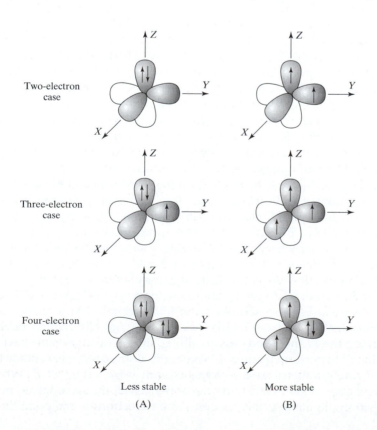

Two-electron case

Three-electron case

Four-electron case

Less stable

(A)

More stable

(B)

◀ FIGURE 13.20
Qualitative illustration of the additional stability produced by placing electrons in different degenerate orbitals when permitted by the Pauli principle. In (A), for the two-, three-, and four-electron cases, one or more spatial p-orbital contains two electrons. The electrons are, therefore, in close proximity, with the result that their mutual repulsion is large. In (B), the electrons are placed in different degenerate spatial orbitals, thereby minimizing their mutual repulsion, which results in a more stable arrangement. The shaded lobe of the orbitals point in the positive direction. Arrows indicate electrons, with the direction of the arrow specifying the spin orientation.

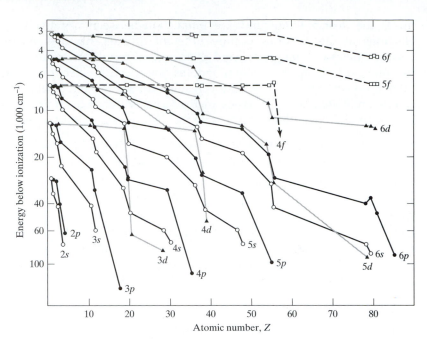

▶ **FIGURE 13.21**
Outer electron orbital energies as a
function of atomic number. [After
W. Kauzmann, *Quantum Chemistry*
(New York: Academic Press, 1957),
p. 326.]

aforementioned qualitative rule. Therefore, phosphorus is $[\text{Ne}]3s^2 3p^1_{-1} 3p^1_0 3p^1_1$,
not $[\text{Ne}]3s^2 3p^2_1 3p^1_0$.

At element number 19, the simple shielding rules in Eq. 13.71 are no
longer sufficiently accurate to predict the correct energy ordering of the
orbitals. Figure 13.21 shows that the lowest unfilled orbital when $Z = 19$ is the
$4s$ orbital. Consequently, K and Ca are I and IIA elements with electronic configurations $[\text{Ar}]4s^1$ and $[\text{Ar}]4s^2$, respectively. At element 21, Figure 13.21
shows the $3d$ orbital to be lowest energy unfilled orbital. Therefore, the five
$3d$ orbitals with $m = -2, -1, 0, 1,$ and 2 fill as we move from Sc($Z = 21$) to
Zn($Z = 30$). Elements in which the outer d orbitals are filling are called *transition metals.* Their groups in the periodic table are generally labeled with the letter B and given the numbers shown in Figure 13.19. After the $3d$ orbital fills at
element 30, the $4p, 5s, 4d, 5p,$ and $6s$ orbitals fill, approximately in that order. At
this point, the next electron enters the $5d$ orbital, so that lanthanum has the
structure $[\text{Xe}]6s^2 5d^1$. The next 14 elements then begin to fill the $4f$ orbital. This
series of 14 elements is called the lanthanide rare-earth series. The same thing
happens at element number 89, actinium, whose structure is $[\text{Rn}]7s^2 6d^1$. The
next 14 elements fill the $5f$ orbital to form the actinide rare-earth series.

As the number of electrons in the atom increases, it becomes more difficult to predict the precise electronic configuration. The energies of the possible configurations become nearly the same, and much more accurate
solutions of the stationary-state Schrödinger equation are required to make
the correct assignments. For example, in the $3d$ transition series, chromium
has the configuration $[\text{Ar}]4s^1 3d^5$ instead of the expected $[\text{Ar}]4s^2 3d^4$, and copper's configuration is $[\text{Ar}]4s^1 3d^{10}$ rather than $[\text{Ar}]4s^2 3d^9$. Qualitatively, these
deviations occur because half-filled and filled orbitals add an extra measure
of stability that distorts the regular filling of the orbitals. Similar deviations
occur when the $4d$ and $5d$ orbitals are filling. In general, it becomes exceptionally difficult to perform quantum calculations with sufficient accuracy to predict the configurations for the elements with large values of Z. When the
nuclear charge becomes large, the electron–nuclear attractive forces are sufficiently strong to increase the velocities of the electrons to the point that they

are approaching the speed of light. Under these conditions, the calculations must be done relativistically. Table 13.7 on page 740 lists the experimentally determined electronic configurations of all atoms.

EXAMPLE 13.10

By examination of Figure 13.19 alone, write down the electronic configuration of the following elements, and state how the result is determined: **(A)** Si; **(B)** Sb; **(C)** Re; **(D)** Sr.

Solution

(A) Silicon is in Group IVA. Figure 13.19 shows that the silicon period is filling the $3p$ shell, so the $3s$ shell is already filled. Therefore, the electronic configuration must be $[\text{Ne}]3s^2 3p^2$, with the two $3p$ electrons in separate p orbitals, such as $[\text{Ne}]3s^2 3p_0^1 3p_1^1$, with the same spin function.

(B) Antimony is in Group VA, so the outer structure must be p^3. Figure 13.19 shows that the $5p$ orbital is filling, with the $5s$ and $4d$ orbitals already filled. Therefore, the configuration is $[\text{Kr}]5s^2 4d^{10} 5p^3$, with the three $5p$ electrons all in different $5p$ orbitals with the same spin function.

(C) Rhenium is a transition metal in Group 7B. Figure 13.19 shows that the $5d$ orbital is filling, with $6s$ already filled. We must be careful about predicting the electronic structure of transition elements, particularly when $Z = 75$, as it is for rhenium. However, here we are safe, because rhenium has exactly five d electrons without taking one from the $6s$ orbital to complete a half-filled orbital. Therefore, we expect to see the configuration $[\text{Xe}]6s^2 5d^5$, with all five of the $5d$ electrons in different $5d$ orbitals with the same spin function.

(D) Strontium is in Group IIA. Figure 13.19 shows that the $5s$ orbital is filling. Therefore, the electronic configuration is $[\text{Kr}]5s^2$, with no unpaired electrons.

For related exercises, see Problems 13.34 and 13.35.

13.4.8   L–S Coupling and Many-electron Term Symbols

The electronic configurations discussed in the previous section are determined either by spectroscopic experiments or by quantum mechanical computation of the orbital energies. The results provide information concerning the number of electrons in each of the one-electron types of orbitals. If all orbitals are doubly occupied, we will have complete information concerning the number of electrons, with m_s equal to $+\frac{1}{2}$ and $-\frac{1}{2}$. We will also know the number of electrons with different magnetic quantum numbers. If, however, the orbitals are only partially filled, this information is missing. Also missing are the total angular momentum states that determine large coulombic interactions and the spin–orbit splittings. To provide this information, we must stipulate the atomic term symbol.

When we have only one electron, the total orbital and spin quantum numbers are the same, respectively, as ℓ and s. If k electrons are present, however, each electron has orbital angular momentum $\mathbf{l}_i$ and spin angular momentum $\mathbf{s}_i$, where the subscript denotes the particular electron in question. If we wish, we can couple the orbital angular momenta for all electrons to obtain a total orbital angular momentum $\mathbf{L}$ and then couple the individual spin angular momenta to obtain the total spin angular momentum $\mathbf{S}$. The total angular momentum $\mathbf{J}$ is then obtained by vector addition of $\mathbf{L}$ and $\mathbf{S}$ according to the quantization requirements imposed by quantum theory. We could also first

Table 13.7   Experimentally determined electronic configurations of the elements

Z	Atom	Configuration	Z	Atom	Configuration	Z	Atom	Configuration
1	H	$1s^1$	36	Kr	$[Ar]4s^23d^{10}4p^6$	71	Lu	$[Xe]6s^24f^{14}5d^1$
2	He	$1s^2$	37	Rb	$[Kr]5s^1$	72	Hf	$[Xe]6s^24f^{14}5d^2$
3	Li	$[He]2s^1$	38	Sr	$[Kr]5s^2$	73	Ta	$[Xe]6s^24f^{14}5d^3$
4	Be	$[He]2s^2$	39	Y	$[Kr]5s^24d^1$	74	W	$[Xe]6s^24f^{14}5d^4$
5	B	$[He]2s^22p^1$	40	Zr	$[Kr]5s^24d^2$	75	Re	$[Xe]6s^24f^{14}5d^5$
6	C	$[He]2s^22p^2$	41	Nb	$[Kr]5s^14d^4$	76	Os	$[Xe]6s^24f^{14}5d^6$
7	N	$[He]2s^22p^3$	42	Mo	$[Kr]5s^14d^5$	77	Ir	$[Xe]6s^24f^{14}5d^7$
8	O	$[He]2s^22p^4$	43	Tc	$[Kr]5s^24d^5$	78	Pt	$[Xe]6s^14f^{14}5d^9$
9	F	$[He]2s^22p^5$	44	Ru	$[Kr]5s^14d^7$	79	Au	$[Xe]6s^14f^{14}5d^{10}$
10	Ne	$[He]2s^22p^6$	45	Rh	$[Kr]5s^14d^8$	80	Hg	$[Xe]6s^24f^{14}5d^{10}$
11	Na	$[Ne]3s^1$	46	Pd	$[Kr]4d^{10}$	81	Tl	$[Xe]6s^24f^{14}5d^{10}6p^1$
12	Mg	$[Ne]3s^2$	47	Ag	$[Kr]5s^14d^{10}$	82	Pb	$[Xe]6s^24f^{14}5d^{10}6p^2$
13	Al	$[Ne]3s^23p^1$	48	Cd	$[Kr]5s^24d^{10}$	83	Bi	$[Xe]6s^24f^{14}5d^{10}6p^3$
14	Si	$[Ne]3s^23p^2$	49	In	$[Kr]5s^24d^{10}5p^1$	84	Po	$[Xe]6s^24f^{14}5d^{10}6p^4$
15	P	$[Ne]3s^23p^3$	50	Sn	$[Kr]5s^24d^{10}5p^2$	85	At	$[Xe]6s^24f^{14}5d^{10}6p^5$
16	S	$[Ne]3s^23p^4$	51	Sb	$[Kr]5s^24d^{10}5p^3$	86	Rn	$[Xe]6s^24f^{14}5d^{10}6p^6$
17	Cl	$[Ne]3s^23p^5$	52	Te	$[Kr]5s^24d^{10}5p^4$	87	Fr	$[Rn]7s^1$
18	Ar	$[Ne]3s^23p^6$	53	I	$[Kr]5s^24d^{10}5p^5$	88	Ra	$[Rn]7s^2$
19	K	$[Ar]4s^1$	54	Xe	$[Kr]5s^24d^{10}5p^6$	89	Ac	$[Rn]7s^26d^1$
20	Ca	$[Ar]4s^2$	55	Cs	$[Xe]6s^1$	90	Th	$[Rn]7s^26d^2$
21	Sc	$[Ar]4s^23d^1$	56	Ba	$[Xe]6s^2$	91	Pa	$[Rn]7s^25f^26d^1$
22	Ti	$[Ar]4s^23d^2$	57	La	$[Xe]6s^25d^1$	92	U	$[Rn]7s^25f^36d^1$
23	V	$[Ar]4s^23d^3$	58	Ce	$[Xe]6s^24f^15d^1$	93	Np	$[Rn]7s^25f^46d^1$
24	Cr	$[Ar]4s^13d^5$	59	Pr	$[Xe]6s^24f^3$	94	Pu	$[Rn]7s^25f^6$
25	Mn	$[Ar]4s^23d^5$	60	Nd	$[Xe]6s^24f^4$	95	Am	$[Rn]7s^25f^7$
26	Fe	$[Ar]4s^23d^6$	61	Pm	$[Xe]6s^24f^5$	96	Cm	$[Rn]7s^25f^76d^1$
27	Co	$[Ar]4s^23d^7$	62	Sm	$[Xe]6s^24f^6$	97	Bk	$[Rn]7s^25f^9$
28	Ni	$[Ar]4s^23d^8$	63	Eu	$[Xe]6s^24f^7$	98	Cf	$[Rn]7s^25f^{10}$
29	Cu	$[Ar]4s^13d^{10}$	64	Gd	$[Xe]6s^24f^75d^1$	99	Es	$[Rn]7s^25f^{11}$
30	Zn	$[Ar]4s^23d^{10}$	65	Tb	$[Xe]6s^24f^9$	100	Fm	$[Rn]7s^25f^{12}$
31	Ga	$[Ar]4s^23d^{10}4p^1$	66	Dy	$[Xe]6s^24f^{10}$	101	Md	$[Rn]7s^25f^{13}$
32	Ge	$[Ar]4s^23d^{10}4p^2$	67	Ho	$[Xe]6s^24f^{11}$	102	No	$[Rn]7s^25f^{14}$
33	As	$[Ar]4s^23d^{10}4p^3$	68	Er	$[Xe]6s^24f^{12}$	103	Lr	$[Rn]7s^25f^{14}6d^1$
34	Se	$[Ar]4s^23d^{10}4p^4$	69	Tm	$[Xe]6s^24f^{13}$			
35	Br	$[Ar]4s^23d^{10}4p^5$	70	Yb	$[Xe]6s^24f^{14}$			

Data taken from CRC *Handbook of Chemistry and Physics,* 78th ed., 1997–98.

couple $\mathbf{l}_i$ and $\mathbf{s}_i$ to obtain $\mathbf{j}_i$ for electron i and then couple all the individual $\mathbf{j}_i$ to obtain $\mathbf{J}$. The former procedure is called *L–S,* or *Russell–Saunders, coupling;* the latter scheme is termed *j–j coupling.* When magnetic interactions and spin–orbit coupling are large, j–j coupling gives the better description of the atomic state. When electrostatic interactions are dominant, L–S coupling is preferred. Since the spin–orbit coupling constant is proportional to Z^4, magnetic interactions and j–j coupling become much more important for the heavier elements. For the lighter elements, the spin–orbit interactions are very small, and L–S coupling is the more appropriate method for representing the angular momentum states.

A large majority of the molecules that are of interest to chemists involve the lighter elements with atomic numbers less than 18. L–S coupling is, therefore, the scheme with which we are primarily concerned. If we focus our attention on the individual orbital and spin angular momenta, $\mathbf{l}$ and $\mathbf{s}$, it is not obvious how we should execute the vector addition of these quantities subject to the quantization rules of quantum theory. However, the Z-components of $\mathbf{l}$ and $\mathbf{s}$ are collinear, so that they add as scalars. This type of addition is much easier to accomplish. Since $L_z = m\hbar$ and $S_z = m_s\hbar$, we have, for a k-electron system,

$$[\mathbf{L}]_z = \text{Z-component of total orbital angular momentum}$$
$$= m_1\hbar + m_2\hbar + m_3\hbar + \cdots + m_k\hbar = \hbar \sum_{i=1}^{k} m_i \qquad (13.78)$$

and

$$[\mathbf{S}]_z = \text{Z-component of total spin angular momentum}$$
$$= m_{s1}\hbar + m_{s2}\hbar + m_{s3}\hbar + \cdots + m_{sk}\hbar = \hbar \sum_{i=1}^{k} m_{si}. \qquad (13.79)$$

Equations 13.78 and 13.79 show that if we define M and M_s as the total magnetic and total spin magnetic quantum numbers such that $[\mathbf{L}]_z = M\hbar$ and $[\mathbf{S}]_z = M_s\hbar$, then

$$M = \sum_{i=1}^{k} m_i \qquad (13.80)$$

and

$$M_s = \sum_{i=1}^{k} m_{si}. \qquad (13.81)$$

If we know the individual Z-components of orbital and spin angular momentum, simple addition via Eqs. 13.80 and 13.81 provides the values of M and M_s for any assignment of electrons to particular one-electron orbitals.

To find the total orbital and spin angular momentum states, $\mathbf{L}$ and $\mathbf{S}$, permitted by the vector addition of the $\mathbf{l}_i$ and $\mathbf{s}_i$, subject to quantization rules, we write down all possible orbital assignments for the k-electron system, compute M and M_s for each assignment, and then use the fact that we must have $-L \le M \le L$ and $-S \le M_s \le S$ to determine the possible values of L and S quantum numbers that produce the squares of the total angular momenta, $|\mathbf{L}|^2 = L(L+1)\hbar^2$ and $|\mathbf{S}|^2 = S(S+1)\hbar^2$. To obtain the various possible total angular momentum states, we can add the possible values of M and M_s as scalars, since the Z-components of the total orbital and spin angular momenta are collinear, and thereby obtain the possible values of M_J, the total magnetic quantum number. Using the fact that we must have $-J \le M_J \le J$, the ensemble of possible M_J values will immediately allow us to determine the possible values of the total angular momentum quantum number J that produces a total angular momentum squared of $|\mathbf{J}|^2 = J(J+1)\hbar^2$. In executing this coupling scheme, we can simplify our task by noting that those electrons in configurations that make up completely filled s, p, d, or f subshells need not be considered at all, because it is always true that $\sum_{i=1}^{k} m_i = \sum_{i=1}^{k} m_{si} = 0$ for these electrons. Therefore, they make no contribution to either M or M_s. In this context, "subshell" refers to the complete set of s, p, d, or f orbitals for a given value of the principal quantum number.

Let us first consider an example for which we do not have to take into account the Pauli exclusion principle in deciding which electron assignments are possible. In its ground state, carbon has the electronic configuration $1s^2 2s^2 2p^2$. Suppose we excite the ground-state atom so that the new configuration is $1s^2 2s^2 2p^1 3p^1$ and then ask what are the possible angular momentum states for this excited carbon atom. In solving this problem, we first note that we do not need to be concerned with the $1s^2$ and $2s^2$ electrons, since they are in filled s subshells. The angular momentum states will be completely determined by the two electrons in the $2p$ and $3p$ orbitals. The second factor to be noted is that we have already satisfied the exclusion principle for these two electrons, as the first has $n = 2$ and the second $n = 3$. Therefore, the remaining quantum number assignments can be arbitrary without making all four quantum numbers the same for both electrons. The possible values of m for an electron in a p orbital are 1, 0, and −1. The possible m_s values are $+\frac{1}{2}$ and $-\frac{1}{2}$.

Each electron has three possible m quantum numbers and two possible values of m_s. Thus, six combinations are possible for each electron. These two sets of six can be combined in 6×6, or 36, possible ways. The 36 possibilities are listed in Table 13.8. All these combinations are permitted, since the Pauli principle is already satisfied.

We now wish to examine the 36 possible sets of M and M_s values and determine the possible total L and S quantum numbers that must be present to yield these 36 combinations of values. The best way to do this is to identify the largest values of M and M_s that appear in the table. In the current example, this is $M = 2$ and $M_s = 1$. For these values of the magnetic quantum numbers to exist, we must have an angular momentum state with $L = 2$ and $S = 1$. That is, we must have a state in which the total orbital angular momentum squared is $2(2 + 1)\hbar^2 = 6\hbar^2$ and the total spin angular momentum squared is $1(1 + 1)\hbar^2 = 2\hbar^2$. When such a state is present, there are five possible values of $M(-2, -1, 0, 1,$ and $2)$ and three possible values of $M_s(-1, 0,$ and $1)$. There are, therefore, 15 combinations of M and M_s that must be present for an $L = 2$, $S = 1$ angular momentum state to exist. These 15 combinations are marked with a pound sign (#) in Table 13.8.

After removing the 15 combinations present in the $L = 2$, $S = 1$ state, we have $36 - 15 = 21$ combinations left. The largest value of M of the remaining combinations is $M = 2$. The value of M_s that occurs with this $M = 2$ state is $M_s = 0$. Thus, we must also have an angular momentum state with $L = 2$, $S = 0$ present. For this state, there are five possible values of M: $-2, -1, 0, 1,$ and 2; but only $M_s = 0$ is possible. Therefore, the $L = 2$, $S = 0$ angular momentum state must produce five combinations of M and M_s values. These combinations are marked with an asterisk (*) in Table 13.8.

We have now found two angular momentum states: $L = 2$, $S = 1$ and $L = 2$, $S = 0$. These two states account for $15 + 5$, or 20, of the 36 combinations listed in the table. When the 20 combinations are removed, the largest value of M in the remaining 16 combinations is $M = 1$. The largest value of M_s associated with an $M = 1$ state is also 1. Consequently, we must have an $L = 1$, $S = 1$ angular momentum state present. Such a state permits M values of $-1, 0,$ and 1 and M_s values of $-1, 0,$ and 1. These possibilities combine to form nine combinations that are indicated by the dollar sign ($) in Table 13.8.

Twenty-nine of the 36 combinations have now been assigned to different angular momentum states. Seven remain. The largest M value of the remain-

Table 13.8 Possible angular momentum assignments for two p electrons

2p Electron		3p Electron		Total		Possible L-S States					
m	m_s	m	m_s	M	M_s	2–1	2–0	1–1	1–0	0–1	0–0
1	1/2	1	1/2	2	1	#					
		1	−1/2	2	0	#					
		0	1/2	1	1	#					
		0	−1/2	1	0	#					
		−1	1/2	0	1	#					
		−1	−1/2	0	0	#					
1	−1/2	1	1/2	2	0		*				
		1	−1/2	2	−1	#					
		0	1/2	1	0		*				
		0	−1/2	1	−1	#					
		−1	1/2	0	0		*				
		−1	−1/2	0	−1	#					
0	1/2	1	1/2	1	1			$			
		1	−1/2	1	0			$			
		0	1/2	0	1			$			
		0	−1/2	0	0			$			
		−1	1/2	−1	1	#					
		−1	−1/2	−1	0	#					
0	−1/2	1	1/2	1	0				@		
		1	−1/2	1	−1			$			
		0	1/2	0	0				@		
		0	−1/2	0	−1			$			
		−1	1/2	−1	0		*				
		−1	−1/2	−1	−1	#					
−1	1/2	1	1/2	0	1					&	
		1	−1/2	0	0					&	
		0	1/2	−1	1			$			
		0	−1/2	−1	0			$			
		−1	1/2	−2	1	#					
		−1	−1/2	−2	0	#					
−1	−1/2	1	1/2	0	0						%
		1	−1/2	0	−1					&	
		0	1/2	−1	0				@		
		0	−1/2	−1	−1			$			
		−1	1/2	−2	0		*				
		−1	−1/2	−2	−1	#					

ing seven is $M = 1$. The largest M_s values associated with an $M = 1$ magnetic quantum number is zero. Hence, we know that there is an $L = 1$, $S = 0$ state present, which generates three combinations in the table: ($M = 1$, $M_s = 0$), ($M = 0$, $M_s = 0$), and ($M = -1$, $M_s = 0$). These three combinations are marked with an "at" sign (@).

Only four combinations remain. All have $M = 0$, so only $L = 0$ states remain. One set of three states has M_s values of 1, 0, and -1, so we must have an $L = 0$, $S = 1$ state. These combinations are labeled with an ampersand (&) in the table. The only remaining combination has $M = M_s = 0$, so that an $L = 0$, $S = 0$ state is present, which we denote with the percent sign (%). We have now identified all the total orbital and spin angular momentum states present for our $1s^2 2s^2 2p^1 3p^1$ carbon atom, viz.,

$$L = 2, \quad S = 1;$$
$$L = 2, \quad S = 0;$$
$$L = 1, \quad S = 1;$$
$$L = 1, \quad S = 0;$$
$$L = 0, \quad S = 1;$$

and

$$L = 0, \quad S = 0.$$

The term symbols corresponding to these states are ^3D, ^1D, ^3P, ^1P, ^3S, and ^1S, respectively.

All that remains is to combine the Z-components of the total orbital and spin angular momenta to obtain the possible Z-components of the total angular momentum. We shall do this for the ^3D state as an example. When $L = 2$, we can have $M = -2, -1, 0, 1$, or 2. With $S = 1$, the possible values of M_s are $-1, 0$, and 1. There are, therefore, $3 \times 5 = 15$ combinations that yield 15 different values of M_J:

M	M_s	M_J	$J = 3$	$J = 2$	$J = 1$
2	1	3	#		
2	0	2	#		
2	-1	1	#		
1	1	2		*	
1	0	1		*	
1	-1	0	#		
0	1	1			$
0	0	0		*	
0	-1	-1	#		
-1	1	0			$
-1	0	-1		*	
-1	-1	-2	#		
-2	1	-1			$
-2	0	-2		*	
-2	-1	-3	#		

Since we can have $M_J = 3$, there must be a total angular momentum state with $J = 3$ for which we can have $M_J = -3, -2, -1, 0, 1, 2,$ or 3. This accounts for 7 of the 15 possible M_J values, which we mark with a pound sign in the preceding table. The largest remaining M_J value is 2, so we must also have a $J = 2$ state with M_J values of 2, 1, 0, -1, and -2, indicated with an asterisk in the table. After eliminating the five additional M_J states, we find that only three remain. The largest M_J among these is $M_J = 1$. Thus, the last total angular momentum state is $J = 1$, which accounts for the last three M_J values. The complete specification of the ^3D state is, therefore, $^3D_{3,2,1}$, which is shorthand notation for three different states: $^3D_3$, $^3D_2$, and $^3D_1$. A similar treatment of the other states shows that the complete array of possible term states is $^3D_3$, $^3D_2$, $^3D_1$, $^1D_2$, $^3P_2$, $^3P_1$, $^3P_0$, $^1P_1$, $^3S_1$, and $^1S_0$.

When the Pauli exclusion principle does not have to be considered, there are a few simple rules that permit you to write all the term states without recourse to the foregoing procedure of examining all the M and M_s combinations. The maximum spin state is obtained when all the Z-components of the spin are aligned in the same direction. That is,

$$S_{max} = \frac{k}{2},\tag{13.82}$$

where k is the number of electrons being considered. (Remember, we omit electrons in filled subshells.) The minimum spin is zero if k is even and $\frac{1}{2}$ if k is odd. The possible total spin quantum numbers are the values from S_{min} to S_{max}, by integer steps. Thus, if we were considering an excited state of nitrogen with the configuration $1s^2 2s^1 2p^1 3p^1 4p^1 5p^1$, we would have $k = 5$. S_{max} would be $\frac{5}{2}$, S_{min} would be $\frac{1}{2}$, and the possible total spin quantum numbers would be $\frac{1}{2}, \frac{3}{2},$ and $\frac{5}{2}$.

The maximum orbital angular momentum state is obtained from

$$L_{max} = \ell_1 + \ell_2 + \ell_3 + \cdots + \ell_k.\tag{13.83}$$

The minimum L is zero if all the ℓ_i are equal. If one ℓ_i is larger than all the others, the minimum L is given by

$$L_{min} = \ell_{max} - \text{sum of all other } \ell_i,\tag{13.84}$$

subject to the constraint that $L_{min} \geq 0$. Thus, the excited nitrogen configuration $1s^2 2s^2 2p^1 3p^1 4f^1$ has $L_{max} = 1 + 1 + 3 = 5$ and $L_{min} = 3 - 1 - 1 = 1$. The possible total orbital angular momentum quantum numbers are the values from L_{min} to L_{max}, by integer steps.

For given values of the L and S quantum numbers, the possible total angular momentum quantum numbers J are the values from $L + S$ to $|L - S|$, by integer steps. (Note the absolute-value bars about $L - S$; J is never negative.)

Using these simple rules, we can easily obtain the term states for the excited $1s^2 2s^2 2p^1 3p^1$ carbon atom. From Eq. 13.82 with $k = 2$, we have $S_{max} = 1$ and $S_{min} = 0$. Next, using Eq. 13.83, we obtain $L_{max} = 1 + 1 = 2$, and since all the electrons are in p orbitals, $L_{min} = 0$. Therefore, the possible values of L are 0, 1, and 2. This immediately gives the terms ^3D, ^1D, ^3P, ^1P, ^3S, and ^1S. Applying of the rule for J states then produces the possible values of J. For example, when $L = 2$ and $S = 1$, we have $J_{max} = 2 + 1 = 3$ and $J_{min} = |2 - 1| = 1$. Therefore, the possible J states are $J = 3, 2,$ and 1, and we have the terms $^3D_3$, $^3D_2$, and $^3D_1$.

When we must take the Pauli exclusion principle into consideration in making the electron assignments to various m and m_s states, the preceding simplified rules cannot be used. Instead, the actual combinations must be written down and examined. As an example, let us consider the possible angular momentum states for a ground-state carbon atom whose electronic

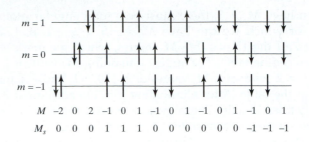

▶ FIGURE 13.22
Pauli-allowed electron combinations for a $2p^2$ configuration such as that present in a ground-state carbon atom with the electronic configuration $1s^2 2s^2 2p^2$.

configuration is $1s^2 2s^2 2p^2$. We do not have to worry about the four electrons in $1s$ and $2s$ orbitals, since these subshells are filled. Only the two $2p$ electrons need be considered. However, the Pauli principle is now not automatically satisfied. Both electrons have $n = 2$ and $\ell = 1$. Therefore, we must be careful not to make both m and m_s the same for those electrons. We can list the allowed combinations in a table such as Table 13.8, or we can construct a diagram like that shown in Figure 13.22, in which we indicate m_s values of $+\frac{1}{2}$ and $-\frac{1}{2}$ with up and down arrows, respectively, and the m values by the line on which we place the arrows. The Pauli principle can be satisfied by never permitting both arrows to point in the same direction on the same line.

Notice how the Pauli principle operates to reduce the number of possible combinations. We now have only 15 instead of the 36 for the $2p^1 3p^1$ configuration listed in Table 13.8. As before, we identify the largest value of M. In this case, it is $M = 2$. The value of M_s associated with $M = 2$ is $M_s = 0$. Therefore, we have a Pauli-allowed term state $L = 2$ and $S = 0$. This state accounts for 5 of the 15 combinations shown in Figure 13.22. The largest value of M in the remaining 10 states is $M = 1$, with an associated M_s value of 1. Therefore, we have an $L = 1$, $S = 1$ term state. This state permits M and M_s values of -1, 0, and 1, so that 9 more combinations are eliminated from the 15. Thus, only one combination, $M = M_s = 0$ is left, so that an $L = 0$, $S = 0$ term state is present. The allowed term symbols are, therefore, ^1D, ^3P, and ^1S. The rule for obtaining the possible J states shows the complete set of term states to be $^1D_2$, $^3P_2$, $^3P_1$, $^3P_0$, and $^1S_0$.

Once the allowed angular momentum states are determined, it is a simple matter to compute the alignment of $\mathbf{L}$ and $\mathbf{S}$ that produces the total angular momentum vector $\mathbf{J}$ for a particular state. As an example, consider the $^3P_2$ state of a ground-state carbon atom. In this state, we have

$$|\mathbf{L}| = [L(L+1)]^{1/2}\hbar = [1(1+1)]^{1/2}\hbar = \sqrt{2}\hbar,$$
$$|\mathbf{S}| = [S(S+1)]^{1/2}\hbar = [1(1+1)]^{1/2}\hbar = \sqrt{2}\hbar,$$

and

$$|\mathbf{J}| = [J(J+1)]^{1/2}\hbar = [2(2+1)]^{1/2}\hbar = \sqrt{6}\hbar.$$

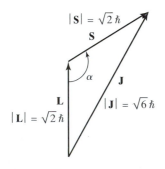

▲ FIGURE 13.23
The vector addition of $\mathbf{L}$ and $\mathbf{S}$ in a $^3P_2$ state to produce the total angular momentum vector $\mathbf{J}$. It is shown in the text that the angle α between $\mathbf{L}$ and $\mathbf{S}$ in this state must be 120°.

Therefore, the vector addition of $\mathbf{L}$ and $\mathbf{S}$ to produce $\mathbf{J}$ must be that shown in Figure 13.23. Using the law of cosines, we obtain $J^2 = L^2 + S^2 - 2LS\cos\alpha$. Thus, the angle between $\mathbf{L}$ and $\mathbf{S}$ is

$$\alpha = \cos^{-1}\left[\frac{L^2 + S^2 - J^2}{2LS}\right] = \cos^{-1}\left[\frac{2\hbar^2 + 2\hbar^2 - 6\hbar^2}{4\hbar^2}\right] = \cos^{-1}\left[-\frac{1}{2}\right] = 120°.$$

Figure 13.24 shows the nature of the vector addition of L and S for all the angular momentum states of an excited-state carbon atom with the configuration $1s^2 2s^2 2p^1 3p^1$.

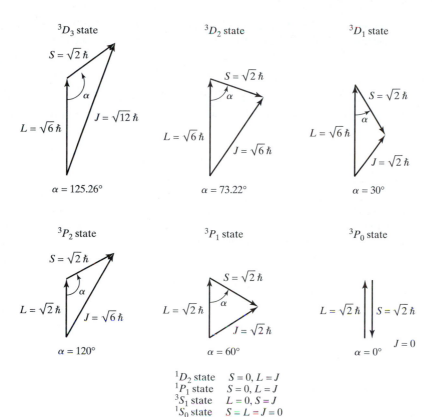

$^3D_3$ state

$^3D_2$ state

$^3D_1$ state

$\alpha = 125.26°$

$\alpha = 73.22°$

$\alpha = 30°$

$^3P_2$ state

$^3P_1$ state

$^3P_0$ state

$\alpha = 120°$

$\alpha = 60°$

$\alpha = 0°$

$^1D_2$ state $S = 0, L = J$
$^1P_1$ state $S = 0, L = J$
$^3S_1$ state $L = 0, S = J$
$^1S_0$ state $S = L = J = 0$

◀ **FIGURE 13.24**
Vector addition of L and S to produce J in the angular momentum states of an excited carbon atom with the configuration $1s^2 2s^2 2p^1 3p^1$.

EXAMPLE 13.11

Determine the possible term states for an excited nitrogen atom whose electronic configuration is $1s^2 2s^2 2p^2 3p^1$.

Solution

The electrons in unfilled subshells are the $2p^2$ and $3p^1$ electrons. All others make no contribution to the angular momentum term states. This problem is a combination of the two cases we have just considered. When writing down the possible m and m_s combinations of the $2p^2$ electrons, we must be careful to satisfy the Pauli principle, since we cannot have both m and m_s the same for these two electrons. This leads to the same result as that obtained for ground-state carbon, whose unfilled configuration is also $2p^2$. The 15 Pauli-allowed combinations of m and m_s for the two electrons are those shown in Figure 13.22.

In contrast, all possible values of m and m_s for the $3p^1$ electron are permitted, since the Pauli principle is already satisfied. In a p orbital, m can take the values $-1, 0,$ and $1,$ and m_s can be $+\frac{1}{2}$ or $-\frac{1}{2}$. There are, therefore, 6 possible combinations of m and m_s for the $3p^1$ electron. Each of these 6 combinations can be combined with the 15 Pauli-allowed combinations of the $2p^2$ electrons given in Figure 13.22, to produce a total of $6 \times 15 = 90$ possible combinations of the m and m_s values of the three electrons. We can then execute the same procedure used to determine the possible term states of the $2p^1 3p^1$ configuration considered in the previous discussion. This method will work very well; however, it is somewhat laborious. We shall, therefore, proceed in a slightly different fashion for illustrative purposes.

The $2p^2$ configuration can be arranged in the 15 combinations shown in Figure 13.22. These combinations produce three orbital and spin angular momentum states: $[L = 2, S = 0](^1D), [L = 1, S = 1](^3P),$ and $[L = 0, S = 0](^1S).$ For the $3p^1$ electron, we have $\ell = 1,$ and $s = \frac{1}{2}.$ We can now couple the orbital and spin angular momenta of each of the states arising from the $2p^2$ configuration with the

ℓ and s values of the $3p^1$ electron. We first couple the $[L = 2, S = 0]$ ^1D state with the $3p^1$ electron. When we couple the orbital angular momenta, we obtain

$$L_{\max} = 2 + 1 = 3 \tag{1}$$

and

$$L_{\min} = 2 - 1 = 1. \tag{2}$$

Therefore, the possible L values are 3, 2, and 1. Coupling the spin angular momenta gives

$$S = 0 + \frac{1}{2} = \frac{1}{2}. \tag{3}$$

Therefore, combining the $^1D_2$ term from the $2p^2$ electrons with the $3p^1$ electron produces the possible term states ^2F, ^2D, and ^2P. Using the fact that $|L - S| \le J \le L + S$, we find that the possible J states for these terms are $^2F_{7/2}$, $^2F_{5/2}$, $^2D_{5/2}$, $^2D_{3/2}$, $^2P_{3/2}$, and $^2P_{1/2}$.

We now couple the $[L = 1, S = 1]$ ^3P state from the $2p^2$ configuration with the $3p^1$ electron. For the orbital angular momentum coupling, we obtain

$$L_{\max} = 1 + 1 = 2 \quad \text{and } L_{\min} = 1 - 1 = 0. \tag{4}$$

Thus, the possible L values are 2, 1, and 0. The spin coupling produces

$$S_{\max} = 1 + \frac{1}{2} = \frac{3}{2} \quad \text{and } S_{\min} = 1 - \frac{1}{2} = \frac{1}{2}. \tag{5}$$

Therefore, we have possible S values of $\frac{3}{2}$ and $\frac{1}{2}$. As a result, the terms are $^4D_{7/2,\,5/2,\,3/2,\,1/2}$, $^4P_{5/2,\,3/2,\,1/2}$, $^4S_{3/2}$, $^2D_{5/2,\,3/2}$, $^2P_{3/2,\,1/2}$, and $^2S_{1/2}$.

When we couple the $[L = 0, S = 0]$ ^1S state from the $2p^2$ configuration with the $3p^1$ electron, we obtain

$$L_{\max} = L_{\min} = 0 + 1 = 1 \quad \text{and } S_{\max} = S_{\min} = 0 + \frac{1}{2} = \frac{1}{2}. \tag{6}$$

Therefore, we have only the terms $^2P_{3/2,\,1/2}$.

In summary, the possible term states are $^2F_{7/2}$, $^2F_{5/2}$, $^2D_{5/2}$, $^2D_{3/2}$, $^2P_{3/2}$, $^2P_{1/2}$, $^4D_{7/2}$, $^4D_{5/2}$, $^4D_{3/2}$, $^4D_{1/2}$, $^4P_{5/2}$, $^4P_{3/2}$, $^4P_{1/2}$, $^4S_{3/2}$, and $^2S_{1/2}$. We also find that the ^2D term occurs twice and the ^2P term three times.

As a check on the results, we can count the number of m–m_s combinations that must be present to produce each of the foregoing terms. For example, the ^2F state requires seven values of m (-3, -2, -1, 0, 1, 2, and 3) and two values of m_s ($\frac{1}{2}$ and $-\frac{1}{2}$). Thus, we need $2 \times 7 = 14$ combinations of m and m_s values. The following table summarizes the number of combinations required for each state and the number of times the state appears:

Term State	No. m	No. m_s	Combinations	No. of such Terms	Total Combinations
^2F	7	2	$7 \times 2 = 14$	1	14
^2D	5	2	$5 \times 2 = 10$	2	20
^2P	3	2	$3 \times 2 = 6$	3	18
^4D	5	4	$5 \times 4 = 20$	1	20
^4P	3	4	$3 \times 4 = 12$	1	12
^4S	1	4	$1 \times 4 = 4$	1	4
^2S	1	2	$1 \times 2 = 2$	1	2
				Sum of Combinations $= 90$	

As we noted earlier, there are 90 possible combinations of m and m_s for the three electrons in the $2p^2$ and $3p^1$ orbitals. Therefore, it appears that we have correctly identified all of the possible term states.

For related exercises, see Problems 13.39 and 13.40.

13.4.9   Hund's Rules

The atomic term states obtained from L–S coupling differ in energy. States with different spin and orbital angular momentum exhibit different energies because the total electrostatic repulsion energy between electrons is different. States with different total angular momentum also have different energies, because the different magnetic interactions produce different spin–orbit couplings. If the atomic number is small ($Z \le 36$), the magnetic interactions are small relative to the electrostatic terms. (See Example 13.5 and Problem 13.26.) Therefore, for light atoms, the spin and orbital angular momenta are more important in determining the energy than the total angular momentum. This situation can be expressed in a set of generalizations known as *Hund's rules:*

1. The term state with maximum multiplicity and hence maximum spin quantum number is the lowest in energy. States with lower multiplicities follow in order of decreasing multiplicity.

2. For term states with equal multiplicities, the state with the maximum value of L is the lowest in energy.

3. Term states that have equal S and L values differ in energy because of spin–orbit coupling. If there is a single subshell (given values of n and ℓ) that is less than half filled, the term state with minimum J is lowest in energy. If the subshell is more than half filled, the state with maximum J is lowest in energy.

The qualitative logic that leads to the first two generalizations derives from the fact that the electrostatic interaction between like-charged electrons is repulsive and destabilizing. Anything that tends to reduce this repulsive interaction lowers the energy of the system and makes it more stable. Maximum multiplicity is achieved when the m_s quantum numbers of the electrons are the same. With the same n, ℓ, and m_s quantum numbers, the electrons must reside in different spatial orbitals so that they will have different magnetic quantum numbers and will satisfy the Pauli principle. When the spatial orbital occupancy is different, the average distance between the electrons is maximized and their mutual repulsions are minimized. Thus, the energy is made as low as possible. (See Figure 13.20.) This is the same qualitative rule we introduced in our discussion of the periodic table and electronic configurations.

When the multiplicities of two term states are the same, the state with maximum L will be lowest, because larger L states require the electrons to be spread over a wider range of m values than states with smaller L values. For example, consider the $^1D_2$ and $^1S_0$ states that arise from the $1s^2 2s^2 2p^2$ configuration we considered in the previous section. To obtain the $^1D_2$ term, we had to include five combinations, such as the following:

Electron 1		Electron 2		Total	
m	m_s	m	m_s	M	M_s
1	$\frac{1}{2}$	1	$-\frac{1}{2}$	2	0
1	$\frac{1}{2}$	0	$-\frac{1}{2}$	1	0
1	$\frac{1}{2}$	−1	$-\frac{1}{2}$	0	0
0	$\frac{1}{2}$	−1	$-\frac{1}{2}$	−1	0
−1	$\frac{1}{2}$	−1	$-\frac{1}{2}$	−2	0

Electron 1		Electron 2		Total	
m	m_s	m	m_s	M	M_s
0	$\frac{1}{2}$	0	$-\frac{1}{2}$	0	0

Of course, other sets of combinations could have been chosen that lead to the $^1D_2$ term, but all of them have the characteristic that three of the combinations must have the electrons in different spatial orbitals (with different m values) and two of them in the same spatial orbital (with the same m values). In contrast, the $^1S_0$ term arises from the single combination shown in the margin. Here, the electrons are always in the same spatial orbital. Therefore, the average electron–electron repulsion in the $^1D_2$ state will be less than that for the $^1S_0$ state. Maximum L is lowest in energy if the multiplicities are the same.

When L and S are identical for two states, the energy differences are due to spin–orbit coupling, which is small for the light atoms. The approximate energy change due to spin–orbit coupling for a one-electron system is given by Eq. 13.50:

$$\Delta E_{spin-orbit} = C[j(j+1) - \ell(\ell+1) - s(s+1)]\frac{\hbar^2}{2}.$$

For a many-electron system, the analogous equation is

$$\Delta E_{spin-orbit} = C'[J(J+1) - L(L+1) - S(S+1)]\frac{\hbar^2}{2}, \qquad \textbf{(13.85)}$$

where the many-electron spin–orbit coupling constant C' is not the same as C. If $C' > 0$, Eq. 13.85 shows that we minimize $\Delta E_{spin-orbit}$ if J is minimum. If, however, $C' < 0$, the opposite is true: The *maximum* value of J minimizes $\Delta E_{spin-orbit}$. The spin–orbit coupling constant reflects the nature of the magnetic field produced by the orbital angular momentum of the electrons that interacts with the spin magnetic moment of the electrons. Our previous discussion of magnetic moments indicates that the magnetic field produced by the orbital motion is dependent upon the magnitude and sign of the charge that has the orbital angular momentum.

To understand the third generalization contained in Hund's rules, we first imagine a subshell that is less than half filled which contains exactly one electron. In this case, the magnetic field is produced by the orbital angular momentum of this single particle whose charge is $-e$. Now consider the identical subshell when it is completely filled. We know that these electrons make no contribution to the angular momentum states of the atom, since the Z-components of the electrons add exactly to zero, so that we have $L = 0$. Consequently, the magnetic field produced by such a filled subshell will be zero, and there is no spin–orbit splitting. Suppose we now insert a positron (a particle with the mass of an electron, but with a positive charge of $+e$) on top of one of the electrons. The $-e$ charge on the electron and the $+e$ charge on the positron will cancel, and we will be left with a subshell that lacks one electronic charge being filled. Therefore, we will produce a magnetic field, and spin–orbit splitting will occur. Since the filled subshell of electrons produced no magnetic field, but the filled subshell plus one positron produced a magnetic field, we can safely assert that the magnetic field is that produced by the single positron in this subshell. That particle has a charge of the same magnitude as that of a single electron, but the sign is opposite. Therefore, we can expect the sign on the spin–orbit coupling constants for subshells that are less than half filled to be opposite that for subshells that are more than half filled. Our treatment of spin–orbit coupling for hydrogen suggests that C' will be positive for subshells less than half filled and, therefore, negative for more than half-filled subshells. As a result, we obtain the third of Hund's rules.

EXAMPLE 13.12

Order the term states for ground-state carbon, and qualitatively predict the relative spacings between the states.

Solution

The five term states are $^3P_2$, $^3P_1$, $^3P_0$, $^1D_2$, and $^1S_0$. Using Hund's rules, we know that the state with maximum multiplicity, ^3P, is the ground state. Since the carbon $2p$ shell is less than half filled, we expect the minimum J state to have the lowest energy. Therefore, the relative energy ordering is $E(^3P_0) < E(^3P_1) < E(^3P_2)$.

The $^1D_2$ and $^1S_0$ terms have the same multiplicity. Therefore, the second of Hund's rules tells us that the largest L state has the lower energy. This is the $^1D_2$ term with $L = 2$. The resulting total energy ordering is $E(^3P_0) < E(^3P_1) < E(^3P_2) < E(^1D_2) < E(^1S_0)$.

Because the energy differences in S and L are due to electrostatic coulombic repulsions, we expect these differences to be on the order of 1 eV or so. On the other hand, the spin–orbit splittings are very small, typically on the order of 10^{-2} to 10^{-4} eV, or, in terms of cm^{-1}, on the order of 1 to 100 cm^{-1}.

The measured experimental spacings are shown in the following diagram:

$$\underline{\qquad\qquad}\ ^1S_0\ E = 2.684\ \text{eV}$$

$$\underline{\qquad\qquad}\ ^1D_2\ E = 1.264\ \text{eV}$$

$$\underline{\qquad\qquad}\ ^3P_2\ E = 0.00539\ \text{eV} = 43.5\ \text{cm}^{-1}$$
$$\underline{\qquad\qquad}\ ^3P_1\ E = 0.00203\ \text{eV} = 16.4\ \text{cm}^{-1}$$
$$\underline{\qquad\qquad}\ ^3P_0\ E = 0$$

The diagram is not drawn to scale; the spin–orbit splittings are too small to permit a scale drawing. Figure 13.25 shows a similar energy-level diagram for the angular momentum states of an excited carbon atom with the configuration $1s^2 2s^2 2p^1 3p^1$.

For related exercises, see Problems 13.36, 13.37, 13.38, 13.39, and 13.45.

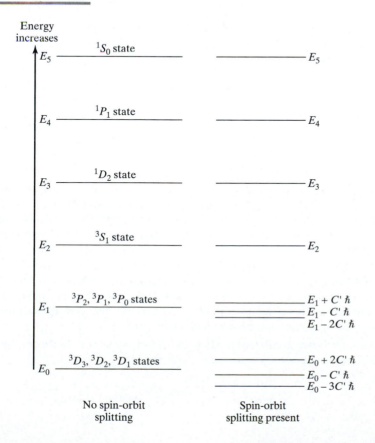

◀ FIGURE 13.25
Qualitative ordering of energy states for an excited carbon atom with the electronic configuration $1s^2 2s^2 2p^1 3p^1$. The left-hand column shows the ordering in the hypothetical case where there is no spin–orbit coupling. The right-hand column shows the ordering with spin–orbit coupling present. The spin–orbit splittings are computed using Eq. 13.85.

Summary: Key Concepts and Equations

1. In executing complex computer calculations on atomic or molecular systems, it is important to use a set of units that make the input quantities as close to unity as possible, in order to provide the maximum protection against underflow and overflow errors. The units most commonly used to achieve this are called atomic units. In this system, the charge on the electron in statcoulombs, the Bohr radius, the mass of the electron, and the electrostatic potential energy of two electrons separated by the Bohr radius are all taken to be unity. These definitions give $\hbar$ the value 1.00000. Conversion factors and the values of atomic constants are given in Tables 13.1 and 13.2.

2. We are always free to define any system of units that may be convenient for solving a particular problem. However, in defining new units of charge, distance, mass, time, and energy, we must be careful to satisfy the basic relationship between units. The energy must be related to mass, distance, and time by

$$E = \frac{1}{2}mv^2 = \frac{1}{2}m\left[\frac{d}{t}\right]^2.$$

That is, we can define any three of the four variables, energy, distance, mass, and time, but the fourth will then be fixed by the preceding relationship.

3. The Schrödinger equation for hydrogen-like atoms and ions separates into a radial part and an angular part. The angular eigenfunctions are identical to those for the rigid rotor. That is, they are the spherical harmonics $Y_\ell^m(\theta, \phi)$ that define the s, p, d, and f orbitals for azimuthal quantum numbers $\ell = 0, 1, 2$, and 3, respectively. The radial eigenfunctions are dependent upon ℓ and the principal quantum number n. They have the form

$$R_n^\ell(r) = N[\text{polynomial in } r \text{ with a leading power of } r^{n-1}]\exp\left[-\frac{Zr}{na_o}\right].$$

4. The energy eigenvalues for hydrogen-like atoms and ions depend only upon the principal quantum number. This is unexpected, since ℓ appears in the radial equation. The result is

$$E_n = -\frac{Z^2\mu e^4}{(4\pi\varepsilon_o)^2 2n^2\hbar^2} \qquad \text{for } n = 1, 2, 3, 4, \ldots \text{ in SI units,}$$

$$E_n = -\frac{Z^2\mu e^4}{2n^2\hbar^2} \qquad \text{for } n = 1, 2, 3, 4, \ldots \text{ in electrostatic units, and}$$

$$E_n = -\frac{Z^2}{2n^2} \qquad \text{for } n = 1, 2, 3, 4, \ldots \text{ in atomic units.}$$

This means that an eigenfunction with quantum numbers $(n\ell m)$ has the same energy eigenvalue as an eigenfunction whose quantum numbers are $(n\ell'm')$. This degeneracy would not occur if the potential energy had any other form than C/r, where C is a constant. In most applications, we can take μ to be equal to the electronic mass.

5. The number of nodes in the hydrogen-like eigenfunctions is equal to $n - 1$. The number of angular nodes in the spherical harmonic $Y_\ell^m(\theta, \phi)$ is ℓ. Therefore, the number of radial nodes must be $n - 1 - \ell$. Since this number must be zero or positive, we obtain an important restriction on the possible values of the azimuthal quantum number: We must have $\ell \leq n - 1$.

6. Quantum theory accurately predicts the atomic spectra of hydrogen-like atoms and ions. The spectral transitions are described by

$$\lambda^{-1} = \frac{Z^2\mu e^4}{2\hbar^2 hc}\left[\frac{1}{n^2} - \frac{1}{m^2}\right] \text{cm}^{-1} = \frac{Z^2}{2hc}\left[\frac{1}{n^2} - \frac{1}{m^2}\right] \text{bohr}^{-1},$$

where the constant $(Z^2 \mu e^4)/(2\hbar^2 hc)$ is the Rydberg constant in electrostatic units. Its value is 109,677 cm^{-1}.

7. For systems with electrostatic potentials, we generally find that $\langle V \rangle = -2\langle T \rangle$. That is, the average potential energy is the negative of twice the average kinetic energy. This relationship is known as the *virial theorem*.

8. The Stern–Gerlach experiment proved that the electron has an intrinsic angular momentum that we call spin, in addition to the angular momentum it may possess because of its orbital motion in an eigenstate with $\ell > 0$. The experiment shows that this intrinsic angular momentum, which we denote with the symbol S, has a magnitude $(\sqrt{3}/2)\hbar$, so that $S^2 = (3/4)\hbar^2$. If we use the usual angular momentum quantization rule, we have $S^2 = s(s + 1)\hbar^2$. Therefore, we must have $s = \frac{1}{2}$. The Z component of spin angular momentum is given by $S_z = m_s \hbar$, where m_s is restricted to lie in the range $-s \leq m_s \leq s$. With $s = \frac{1}{2}$, m_s has only two possible values: $-\frac{1}{2}$ and $+\frac{1}{2}$. This is confirmed by the Stern–Gerlach experiment, which showed that a beam of atoms in a ^2S term state is split into two beams because of the interaction of the inhomogeneous magnetic field with the two possible Z components of the spin angular momentum.

9. P. A. M. Dirac demonstrated that spin angular momentum is a relativistic effect. He also found that the eigenfunctions of a system actually have four components in the form of a 4×1 vector. When the velocity of the electrons is small relative to the speed of light in a vacuum, two of these four components become negligibly small and may be ignored. The remaining 2×1 vector factors into a product of the nonrelativistic eigenfunction times a simple 2×1 vector. That is, the relativistic wave function in the limit $v/c \to 0$ has the form

$$\psi(x) = \begin{bmatrix} \phi(x) \\ 0 \end{bmatrix} = \phi(x) \begin{bmatrix} 1 \\ 0 \end{bmatrix} \quad \text{or} \quad \psi(x) = \begin{bmatrix} 0 \\ \phi(x) \end{bmatrix} = \phi(x) \begin{bmatrix} 0 \\ 1 \end{bmatrix},$$

where $\phi(x)$ is the nonrelativistic eigenfunction and the two vectors $\begin{bmatrix} 1 \\ 0 \end{bmatrix}$ and $\begin{bmatrix} 0 \\ 1 \end{bmatrix}$ are called spin functions. They are assigned the notation β and α, respectively. The combination of the nonrelativistic eigenfunction times a spin function is called a spin orbital. The β spin function turns out to be associated with $m_s = -\frac{1}{2}$ and $S_z = -\hbar/2$, while the α spin function is associated with $m_s = +\frac{1}{2}$ and $S_z = \hbar/2$. For this reason, they are sometimes referred to as the spin-down and spin-up functions, respectively.

10. The orbital and spin angular momenta add vectorially, subject to the quantization rules, to produce a total angular momentum **J**. The magnitude of **J** must be quantized in the same way as **L** and **S**. Therefore, we have

$$J^2 = j(j + 1)\hbar^2 \quad \text{and} \quad J_z = m_j \hbar,$$

with m_j restricted to lie in the range $-j \leq m_j \leq j$. The most straightforward method for finding the possible values of j is to take advantage of the fact that the Z-components of orbital, spin, and total angular momentum are collinear and, therefore, add as scalars. This consideration leads to the result $m_j = m + m_s$. By combining m and m_s in all ways permitted by the Pauli principle, the possible values of j may easily be deduced. The text also gives some simplified methods for finding the allowed values of j.

11. Angular momentum states with the same values of L and S, but different total angular momentum quantum numbers, have different energies because of spin–orbit coupling in which the magnetic moment due to spin interacts with the magnetic field produced by the orbital angular momentum of the electrons. The magnitude of the spin–orbit interaction is proportional to the dot product of **L** and **S**. It is shown in the text that this leads to the result

$$\Delta E_{\text{spin–orbit}} = C[j(j + 1) - \ell(\ell + 1) - s(s + 1)]\frac{\hbar^2}{2}.$$

The spin–orbit splitting constant C is approximately $1.67 \times 10^{-6} Z^4$ hartrees for hydrogen-like atoms or ions with nuclear charge Z. In view of this, we expect that the importance of spin–orbit splitting will become very large as Z increases. Indeed, for heavier elements, it becomes as important as the coulombic interaction terms in the potential.

12. The Schrödinger equation for systems of many-electron atoms and molecules cannot be separated. Therefore, no analytically exact solutions for these systems exist. The problem resides in the electron–electron repulsion terms that depend simultaneously upon the coordinates of two electrons. In the one-electron approximation, these terms are omitted, so that separation of the Schrödinger equation is possible. Under these conditions, the solutions are simple products of hydrogen-like eigenfunctions. The energy eigenvalues are sums of the corresponding hydrogen-like eigenvalues.

13. The electron–electron repulsion terms that are omitted to obtain the one-electron approximations are actually very large. Consequently, the resulting energy eigenvalues are much too stable, and the electron distributions predicted by the simple hydrogen-like product eigenfunctions are far too contracted. If the concept of shielding is introduced, the one-electron predictions can be significantly improved. The concept is to replace the actual nuclear charge that appears in the eigenfunctions and eigenvalues with an effective charge

$$Z_e(a) = Z - S_a,$$

where $Z_e(a)$ is the effective charge seen by an electron in orbital a and S_a is the shielding factor for electrons in that orbital. In most cases, effective charges are computed using variational methods that will be described in Chapter 14. Table 13.5 lists values obtained in this manner for the first 36 elements in the periodic table. With the use of effective charges, the energy eigenvalues become

$$E_n = -\frac{Z_e^2}{2n^2} \text{ hartrees.}$$

The one-electron hydrogen-like orbitals are usually replaced with Slater-type orbitals whose form, in atomic units, is

$$\psi_s(r, \theta, \phi) = N_s r^{n-1} \exp\left[-\frac{Z_e r}{n}\right] Y_\ell^m(\theta, \phi).$$

14. The indistinguishability of the electrons requires that the electron distribution be independent of the electron labels. This means we must have $|\psi(1, 2)|^2 = |\psi(2, 1)|^2$. The relationship of $\psi(1, 2)$ to $\psi(2, 1)$ is, therefore,

$$\psi(1, 2) = e^{i\delta}\psi(2, 1),$$

where $e^{i\delta}$ is called the phase factor. Since an even number of electron exchanges is required to reproduce the original electron labels, the phase factor $e^{2ik\delta}$ must be unity. Because k is an integer, we must have $\delta = 0$ or π. Particles with half-integral spin have $\delta = \pi$ and are called fermions; Particles with integral spins have $\delta = 0$ and are called bosons. Since $e^{i\pi} = -1$, all fermion wave functions must be antisymmetric with respect to electron exchange. Therefore, simple product functions are not acceptable wave functions for electrons. The easiest way to ensure that $\psi(1, 2)$ is antisymmetric is to write it in the form of a Slater determinant in which the different spin orbitals make up the different rows of the determinant and the electron assignments to these spin orbitals differ from column to column. For the helium atom, an acceptable antisymmetric wave function is

$$\psi(1, 2) = N \begin{vmatrix} \chi_{1s}\alpha(1) & \chi_{1s}\alpha(2) \\ \chi_{1s}\beta(1) & \chi_{1s}\beta(2) \end{vmatrix}.$$

Since interchanging Electrons 1 and 2 is equivalent to interchanging Columns 1 and 2 of the Slater determinant, it follows that $\psi(1, 2) = -\psi(2, 1)$, as required.

15. The requirement that the electronic wave function be antisymmetric with respect to electron exchange leads directly to the result that no two spin orbitals can be identical without producing the trivial solution $\psi = 0$ for the wave function. This result is usually expressed as the Pauli exclusion principle: *The quantum number assignments cannot be the same for the spin orbitals of any two electrons if we wish to avoid the trivial solution of the Schrödinger equation.* There is no Pauli principle for bosons.

16. The combination of the Pauli exclusion principle and the requirement that electrons occupy the available orbitals with the lowest energy leads directly to the periodic table. As the number of electrons increases, it becomes more difficult to accurately compute the energies of the possible electron configurations. Therefore, the electronic configurations for elements with large values of Z must usually be determined experimentally. These experimentally determined results are given in Table 13.7.

17. The angular momentum states of atoms can be determined using either L–S or j–j coupling methods. When $Z \leq 36$, L – S coupling is the better procedure. In this method, we first couple the orbital angular momenta of all electrons to obtain a total L quantum number. The spin angular momenta are then coupled to obtain a total spin quantum number S. In general, this can always be accomplished by considering all Pauli-allowed combinations of m and m_s for all electrons. In executing the procedure, we need to consider only those electrons in unfilled subshells. Once the L–S term states are determined, the possible total angular momentum quantum numbers can be obtained using the fact that we always have

$$|L - S| \leq J \leq L + S.$$

18. The angular momentum term states differ in energy. States with different L and S have different energies because the magnitude of the electron–electron repulsion differs between the states. States with the same L and S, but different J, differ because of spin–orbit interactions. The relative energy ordering of the possible term states is given by a set of generalizations known as Hund's rules:

1. The term state with maximum multiplicity and hence maximum spin quantum number is the lowest in energy. States with lower multiplicities follow in order of decreasing multiplicity.

2. For term states with equal multiplicities, the state with the maximum value of L is the lowest in energy.

3. Term states that have equal S and L values differ in energy because of spin–orbit coupling. If there is a single subshell (given n and ℓ values) that is less than half filled, the term state with minimum J is lowest in energy. If the subshell is more than half filled, the state with maximum J is lowest in energy.

Problems

Problems that require the use of some type of computational device are marked with an asterisk (*). Problems that require some type of plotting routine are indicated with a pound sign (#).

13.1 The Bohr model of the hydrogen atom [see Eq. 11.41] predicts the hydrogen atom ground-state energy to be $E = -e^2/(8\pi\varepsilon_o a_o)$, where e is in coulombs and a_o is the Bohr radius.

(A) What system of units is being used in this equation?

(B) What is the equivalent expression in electrostatic units?

(C) What is the value of the ground-state energy of the Bohr hydrogen atom in hartrees? in eV?

13.2 In Chapter 12, the energy of a rigid rotor was found to be $E_{rot} = J(J + 1)\hbar^2/(2I)$, where I is the moment of inertia of the rotor. Use the data in Example 13.1 to compute the value of $\hbar^2/(2I)$ for the Bohr model of the H_2 molecule in

(A) SI units and

(B) atomic units.

13.3 Molecular dynamics is the study of the motion of atoms and molecules in chemical reactions. It turns out that the atomic unit of energy is too large and the atomic mass unit is too small to be convenient for molecular dynamics calculations. A more commonly used set of units in such calculations defines the mass of the hydrogen atom to be equal to its atomic mass expressed in g mol⁻¹, 1.0079. With this choice, the masses of the other atoms are their respective atomic masses. The unit of distance is taken to be the angstrom (10^{-10} m), and the energy unit is the electron volt. These units are sometimes called molecular units.

(A) Determine the value of one molecular time unit in seconds.

(B) The H_2 vibration frequency is 1.350×10^{14} s⁻¹. What is its vibrational frequency in molecular time units?

(C) What is the H_2 vibrational period in molecular time units?

13.4 Show that the radial function

$$R(r) = N\left[1 - \frac{br}{2}\right]\exp\left[-\frac{br}{2}\right]$$

is a solution of the radial equation for hydrogen-like atoms, provided that we have

$$b = \frac{Z}{a_o}, \ell = 0, \text{ and } E = -\frac{Z^2 e^2}{8a_o} \text{ in electrostatic units.}$$

Assume that we may set $\mu \approx m_e$ so that $a_o = \hbar^2/(\mu e^2)$ in electrostatic units.

13.5 Show that the radial function $R(r) = Nr \exp[-br/2]$ is a solution of the radial equation for hydrogen-like atoms, provided that we have $b = Z/a_o$, $\ell = 1$, and $E = -Z^2 e^2/(8a_o)$, where $a_o = \hbar^2/(\mu e^2)$, which is the Bohr radius to four significant digits.

13.6 Calculate the percent error associated with the approximation $a_o = \hbar^2/(m_e e^2) \approx \hbar^2/(\mu e^2)$, where μ is the reduced mass of the hydrogen-like particle, for

(A) a hydrogen atom,

(B) a deuterium atom, and

(C) a tritium atom.

13.7 With the approximation $a_o = \hbar^2/(\mu e^2)$, the ground-state energies of H, D, and T are all equal to $-\frac{1}{2}$ hartree. Compute the actual energies of these atoms and the percent error involved in the approximation.

13.8 Consider the $n = 5$, $\ell = 3$, and $m = 2$ hydrogen-atom eigenfunction.

(A) What is the orbital angular momentum of the electron in this state?

(B) What is the Z component of orbital angular momentum?

(C) How many radial nodes are present? How many angular nodes? Which angular coordinates exhibit the nodes?

(D) What is the energy of this eigenstate in hartrees?

(E) What is the degeneracy of this energy state?

13.9 Compute the average radial distance of the electron from the nucleus of a hydrogen-like atom or ion excited to the $2p$ eigenstate as a function of Z.

13.10 Determine the average value of z^2 for the electron in a hydrogen-like atom or ion in its ground state as a function of Z, where z is the z-axis coordinate of the electron in a rectangular Cartesian coordinate system and Z is the atomic number of the atom or ion.

13.11 Compute the most probable distance of the electron from the nucleus for the ground state of a hydrogen-like atom or ion as a function of Z, the atomic number of the atom or ion.

13.12 Show that the virial theorem is satisfied for a hydrogen-like system in the $2p$ excited state. Use atomic units.

13.13 Without performing any integrations, compute the average kinetic energy of a hydrogen-like system when it is in the $n = 5$, $\ell = 2$, and $m = -1$ eigenstate.

13.14 Spectral lines arising from transitions from the $n = 1$ hydrogen-atom energy level to a higher level $m > 1$ are called Lyman transition energies or Lyman lines. Spectral lines arising from transitions from the $n = 2$ hydrogen-atom energy level to a higher level $m > 2$ are called Balmer transitions or Balmer lines. Calculate the frequency, wavelength, and wave number of the Lyman spectral line of lowest energy for the hydrogen atom. Compute the energy of this transition, in kJ mol⁻¹. Is the spacing between energy levels large or small?

The next three problems deal with a particle in an infinite spherical well. The problems are constructed so that they can be assigned as a unit or individually.

13.15[#] In this problem, we shall investigate a problem similar to that of hydrogen-like atoms and ions. Consider a single particle of mass m in a spherical potential well whose radius is R_o. Let the potential within the well be zero and infinite outside the well. That is,

$$V(r) = 0 \text{ for } r \le R_o \text{ and } V(r) = \infty \text{ for } r > R_o.$$

Since the potential for $r > R_o$ is infinite, we know that we must have $\psi(r, \theta, \phi) = 0$ for $r > R_o$.

(A) Set up the Schrödinger equation for this system when $r \le R_o$ (inside the potential well), and show that the variables can be separated.

(B) Obtain the angular eigenfunctions for this system and the resulting form of the radial differential equation $R(r)$ that must be solved to obtain the radial eigenfunctions.

(C) Using the substitution $R(r) = F(r)/r$, make use of the first postulate to obtain the eigenfunctions and energy eigenvalues for the spherically symmetric case ($\ell = 0$). [*Hint:* At $r = 0$, we must have $F(r) = 0$ to avoid having an infinity in the wave function.]

(D) Plot $R(r)/A$ (A is the normalization constant) versus r if $R_o = 10$ bohr for the four lowest energy eigenstates.

(E) How is the number of radial nodes related to the principal quantum number for this system? (For a very similar example, see Problem 9.7.)

13.16 The spherically symmetric eigenfunctions of a particle in an infinite spherical well whose radius is R_o are

$$\psi(r, \theta, \phi) = R_n^0(r)\, Y_0^0(\theta, \phi) \quad \text{for } r \leq R_o$$

$$\text{and } \psi(r, \theta, \phi) = 0 \quad \text{for } r > R_o.$$

(See Problem 13.15.) $Y_0^0(\theta, \phi)$ is a spherical harmonic equal to $(4\pi)^{-1/2}$, and $R_n^0(r) = (A\,\sin[n\pi r/R_o])/r$, with A serving as the normalization constant. Normalize $R_n^0(r)$, and determine the value of A in terms of R_o.

13.17 The normalized eigenfunctions for a particle in an infinite spherical well whose radius is R_o are

$$\psi(r, \theta, \phi) = R_n^0(r)\, Y_0^0(\theta, \phi) \text{ for } r \leq R_o$$

$$\text{and } \psi(r, \theta, \phi) = 0 \text{ for } r > R_o.$$

See Problems 13.15 and 13.16. $Y_0^0(\theta, \phi)$ is a spherical harmonic equal to $(4\pi)^{-1/2}$, and $R_n^0(r) = [2/R_o]^{1/2}$ $(\sin[n\pi r/R_o])/r$. Compute the average radial position of the particle, $\langle r \rangle$, as a function of the quantum number n.

13.18 (A) How many radial nodes are present in the $R_3^0(r)$ radial eigenfunction for hydrogen-like systems?

(B) For the hydrogen atom, at what values of the radial coordinate will the radial nodes appear?

13.19 Just as a harmonic oscillator can tunnel into classically forbidden regions where $V > E$, the electron can do the same in hydrogen-like systems.

(A) Determine the classical turning point for the hydrogen-atom electron in its ground state.

(B) Compute the probability that the electron will be at a value of r equal to or greater than its classical turning point.

13.20 Determine the uncertainty in position for the electron in a hydrogen-like atom in the ground state. Explain the variation of the uncertainty with atomic number.

13.21 In Table 13.3, the $R_2^1(r)$ radial eigenfunction is shown to have the form

$$R_2^1(r) = N\frac{Zr}{a_o}\, e^{-Zr/a_o}$$

Show that the normalization constant for this eigenfunction is that given in the table.

13.22 A hydrogen atom is in an excited $3d$ eigenstate.

(A) What is the energy of this eigenstate, in hartrees?

(B) What are the possible magnitudes of the Z component of the angular momentum of the electron?

(C) If this hydrogen atom were subjected to an applied external magnetic field of 10^5 gauss, what would happen to the $3d$ energy states of the atom if we assume that the only angular momentum present is that due to the orbital motion of the electron? What is the percent change in the energy of each state?

13.23 A hydrogen atom is in the ground state with energy $-\frac{1}{2}$ hartree. If this hydrogen atom were subjected to an applied external magnetic field of 10^5 gauss, by how much would the $1s$ energy state of the atom change?

13.24 A hydrogen atom is in an excited $3d$ eigenstate.

(A) What are the possible values of ℓ and m in this state?

(B) Construct a table similar to Table 13.4 showing the possible values of m_j that can be obtained by the vector addition of **L** and **S**. What values of j characterize the total possible angular momentum states of a hydrogen atom in this eigenstate?

(C) What are the possible magnitudes of the total angular momentum vector J for a hydrogen atom in that eigenstate?

(D) Give the term symbols describing the possible angular momentum states of a hydrogen atom in that eigenstate.

(E) Determine the angle **J** makes with the Z-axis in each possible m_j state.

13.25 Repeat Problem 13.24 for a hydrogen atom in an excited $4f$ eigenstate.

13.26 (A) The angular momentum states for a hydrogen atom in the $3d$ eigenstate are $3\,^2D_{5/2}$ and $3\,^2D_{3/2}$. (See Problem 13.24.) Compute the spin–orbit splitting between the angular momentum states of a hydrogen atom excited to the $3d$ eigenstate. Compute the splitting for these states for the He$^+$ ion.

(B) The angular momentum states for a hydrogen atom in the $4f$ eigenstate are $4\,^2F_{7/2}$ and $4\,^2F_{5/2}$. (See Problem 13.25.) Compute the spin–orbit splitting between these angular momentum states.

13.27 Write down the Hamiltonian for the lithium atom. Identify which term or terms are omitted when we make the Born–Oppenheimer approximation. Which terms are one-electron terms? How many two-electron terms are present? Which terms are omitted by the one-electron approximation?

13.28 Use appropriate summation notation to write the quantum mechanical Hamiltonian for a cobalt atom $(Z = 27)$, using the Born–Oppenheimer approximation. How many electron–nuclear attraction terms are present? How many electron–electron repulsion terms are present?

13.29 Suppose we are in a universe in which the spin quantum number for the electron is $s = \frac{5}{2}$ instead of $s = \frac{1}{2}$, as it is in our universe.

(A) Describe in qualitative terms what the results of a Stern–Gerlach type of experiment on the hydrogen atom would be in the hypothetical universe.

(B) What are the possible values of S_z for the electron in that universe?

(C) What is the value of S^2 for the electron in that universe?

(D) What are the possible angles the spin angular momentum vector can make with the z-axis in the assumed universe?

(E) With $s = \frac{1}{2}$, there are 10 elements contained in each transition metal series in the periodic table. How many would there be if s for the electron were $\frac{5}{2}$?

13.30 Use the Clementi–Raimondi shielding rules to compute the expected ionization potential of the boron atom. Look up the experimental value and determine the percent error in your computed value. What ionization potential would be predicted if we use Koopman's theorem with the Clementi–Raimondi rules?

13.31 Write down the Slater orbital representing a $3p$ orbital with $L_z = \hbar$. Normalize the orbital.

13.32 Write down the quantum mechanical Hamiltonian for the boron atom. Indicate the physical significance of each term in the Hamiltonian. Which term or terms are omitted by the Born–Oppenheimer approximation? Which term or terms are omitted by the one-electron approximation? Calculate the uncorrected, one-electron energy for this atom.

13.33 The one-electron electronic configuration of the Be atom $(Z = 4)$ is usually written in the form $1s^2 2s^2$. Use a Slater determinant to write a one-electron type of wave function for the Be atom that has the proper symmetry with respect to electron exchange. What is the value of the normalization constant for the Slater determinant if the individual $1s$ and $2s$ Slater-type orbitals are normalized?

13.34 Using Figure 13.19, predict the electronic configurations of the following elements, and then check your predictions against the spectroscopically determined results given in Table 13.7.

(A) Cd;

(B) Br;

(C) W;

(D) Nd.

In each case, state how many unpaired electrons are present. Of the four predictions you made, which is the most likely to be in error? Why?

13.35 Consider a hypothetical situation in which the electron is a boson with a spin quantum number of 1 instead of $\frac{1}{2}$. Would Mendeleev have noticed a periodic behavior for the elements if this were the case? What would the "periodic table" look like if the electron were a boson?

13.36 Order, with respect to energy, the term states obtained in the text for an excited carbon atom with the electronic configuration $1s^2 2s^2 2p^1 3p^1$.

13.37 Determine the term symbol in the ground state for the elements from Li $(Z = 3)$ to Ar $(Z = 18)$.

13.38 Determine all the possible term states for an excited carbon atom with the electronic configuration $1s^2 2s^2 2p^1 3d^1$. Order these term states with respect to their energies.

13.39 When carbon forms four bonds, its electronic configuration becomes $1s^2 2s^1 2p^3$. This configuration is called the *valence state* of carbon. Determine all the possible term states for the configuration.

13.40 Determine all the possible term states for an excited nitrogen atom whose electronic configuration is $1s^2 2s^2 2p^2 3d^1$.

13.41 Example 13.12 gives the spin–orbit splitting between the $^3P_0$, $^3P_1$, and $^3P_2$ states.

(A) Compute the magnitude of these splittings, in hartrees.

(B) Use Eq. 13.85 and the measured splitting between the $^3P_0$ and $^3P_1$ states to determine the spin–orbit splitting constant for carbon.

(C) Use the results obtained in (B) to compute the splitting between the $^3P_1$ and $^3P_2$ states in carbon. Calculate the percent error in your result.

13.42 (A) Using the one-electron approximation and hydrogen-like orbitals without shielding, calculate the most probable radial position for a $2s$ electron in the lithium atom ground state.

(B) Repeat (A), inserting the effects of shielding with the use of the Clementi–Raimondi shielding rules. Explain qualitatively the reason for the difference in the results obtained in (A) and (B).

13.43 Muonium is a transient atom with a proton nucleus and a negative muon. The muon is an elementary particle with a charge of $-e$ and a mass 206.77 times greater than that of the electron. Compute the ground-state energy of muonium in atomic units and in eV. (The author is indebted to Fredrick L. Minn, M.D., Ph.D, for providing this problem and the associated solution.)

13.44 Let us assume that we are in a universe in which the gravitational attraction of two electrons has exactly the same magnitude (but, of course, different signs) as their electrostatic repulsion for all positions of the electrons. The gravitational attraction has the form

$$V_G(r_{12}) = -\frac{Gm_1 m_2}{r_{12}},$$

where G is the gravitation constant. The assumptions of the problem make it clear that G must be very different in this universe than in ours.

(A) Using the Born–Oppenheimer approximation, write down the Schrödinger equation for the lithium atom in the hypothetical universe.

(B) Do the variables separate? If so, write down the form of the separated parts. If not, state which term or terms prevent the separation.

(C) Describe how the periodic table would change in this new universe. Remember, the Pauli principle would still operate. (The author is indebted to Fredrick L. Minn, M.D., Ph.D, for providing this problem and the associated solution.)

13.45 L–S coupling is most appropriate for lighter atoms with $Z \leq 36$. One member of the class has determined that one of the elements with $Z \leq 36$ has a ground-state term state $^7S_3$. Which element is it? Is there more than one possibility? If so, what are they? If there is only one, how do we know that?

13.46 Melrose and Scerri [*J. Chem. Educ.* 1996, **73**, 498] and Vanquickenborne, Pierloot, and Devoghel [*J. Chem. Educ.* 1994, **71**, 469] have presented explanations of the electronic configurations of the $3d$ transition metal elements and their cations. To do this, both groups used the weighted-average energy of all term states for each configuration to represent the energy of the system. Both groups ignored spin–orbit splittings. For example, the electronic configuration of Ti is $[Ar]4s^23d^2$. If we ignore the J states involved in spin–orbit splitting, the Pauli-allowed term states for this configuration are ^3F, ^1D, ^3P, ^1G, and ^1S. The two groups of researchers used a weighted-average energy given by

$$\langle E \rangle_{4s^23d^2} = \frac{21E(^3F) + 5E(^1D) + 9E(^3P) + 9E(^1G) + E(^1S)}{45},$$

where $E(^3F)$ is the energy of the ^3F term state, with similar definitions for the other terms. Explain why the coefficients of the term-state energies are 21, 5, 9, 9, and 1, respectively, and why the number in the denominator is 45.

13.47 I was about to leave this chapter and move to molecular electronic structure and bonding, but I see that Sam has his hand up again. "What is it, Sam?"

"Well, Sir, when we discussed thermodynamics, you never said that there was a limit to the amount of energy we could put into a system, but just look at the expression for the hydrogen-atom energy level. When the hydrogen atom is in its ground state, the energy is -0.5000 hartree. As we put in more and more energy, the atom becomes more and more excited, and the principal quantum number gets bigger and bigger. But n can never become infinite, and when we have $n = \infty$, the energy is zero. So the amount of energy we can add to the hydrogen atom seems to have an upper limit equal to the difference between zero and -0.5000 hartree, which is $+0.5000$ hartree. Does this mean that we can never add more than half a hartree to a hydrogen atom, no matter what kind of engineering design we might use? If this isn't right, what am I doing wrong? If it is right, why can't we add more energy than that?"

Can you respond to Sam's question?

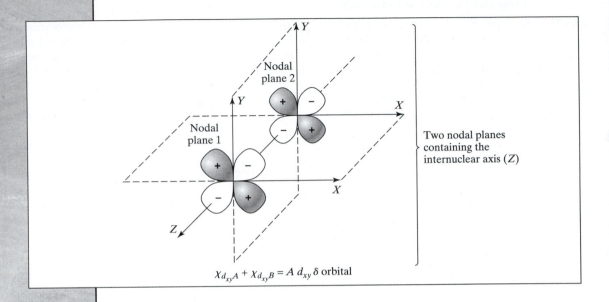

$$\chi_{d_{xy}A} + \chi_{d_{xy}B} = A\ d_{xy}\ \delta \text{ orbital}$$

Molecular Structure and Bonding

The chemical bond and molecular structure are results of nature's effort to minimize the sum of the kinetic and potential energies of any given system. The H_2 equilibrium bond length is 0.74×10^{-10} m because, at this internuclear separation, the sum of the electronic kinetic energy and all coulombic potential terms has its minimum value. The bond angles in methane are tetrahedral because this structure produces the lowest value of the total energy of CH_4. Since the solution of the Schrödinger equation gives us a system's energy, we can, in principle, use quantum mechanics to determine the H_2 equilibrium bond length, the tetrahedral structure of CH_4, or any other molecular structure that may be of interest.

While the preceding statements are correct, there is a great deal of difficulty associated with the task of obtaining sufficiently accurate solutions of the molecular Schrödinger equation. In this chapter, we shall examine the fundamental principles associated with the study of molecular structure and bonding and describe some of the computational methods most frequently used in such studies. Although in most cases the calculations are difficult, computer programs now exist that will do them for us. We just need to gain an understanding of what these programs do, learn how to interpret their output, and realize the limits of their accuracy. With these objectives, we shall introduce perturbation and variational methods. The use of the latter technique in molecular orbital calculations will then be described. The discussion concludes with a description of some more approximate semiempirical methods.

I see that Sam is leaning forward in his chair, pencil ready, so let's begin.

14.1 The Nature of the Chemical Bond

As noted in the first sentence of this chapter, if a chemical bond forms, it does so because that structure produces the lowest possible value of the sum of the kinetic and potential energies of the system. To see how this concept operates, let us examine a very simple model of a hydrogen atom.

In the Bohr model of the ground state of the hydrogen atom, the electron moves in a circular orbit of radius a_o about the nucleus. In the quantum mechanical description, the electron has a probability distribution over all space, but its most probable radial position in the ground state is the Bohr radius a_o. Why is this true? Since the potential energy has the form $V(r) = -1/r$ hartrees, $V(r)$ attains a minimum when r approaches zero. Why doesn't the electron fall into the nucleus so as to make $V(r) = -\infty$ hartrees? Why is the most stable structure one in which the most probable electronic position is at $r = a_o$? The answer to these questions provides the fundamental explanation of why chemical bonds form.

The Hamiltonian for the hydrogen atom is

$$\mathcal{H} = T + V = T - \frac{1}{r}, \tag{14.1}$$

in atomic units. We know that the kinetic-energy term has the form $-(\hbar^2/2\mu)\nabla^2$. In Chapter 13, we used this operator and obtained the exact quantum mechanical eigenvalues and eigenfunctions. Now, we shall simplify the problem in such a fashion that the answers to the questions raised in the previous paragraph become clear. Let the kinetic energy of the electron in the spherically symmetric ground state be that for a particle of mass m_e in an infinite spherical potential well of radius R_o, where the potential inside the well ($r \leq R_o$) is zero and outside the well is infinite. The energy eigenvalues and eigenfunctions for this system were requested in Problems 13.15 and 13.16. The results for the ground state are

$$T = \frac{\hbar^2 \pi^2}{2m_e R_o^2} \quad \text{and} \quad \psi(r) = \left(\frac{2}{R_o}\right)^{1/2} \frac{\sin\left[\frac{\pi r}{R_o}\right]}{r}. \tag{14.2}$$

Let us assume that we can replace $1/r$ in Eq. 14.1 with $\langle 1/r \rangle$, the average of the reciprocal of the radial position for the electron in the infinite spherical well. This average is given by

$$\left\langle \frac{1}{r} \right\rangle = \int_0^{R_o} \psi^*(r) \frac{1}{r} \psi(r) r^2 dr = \frac{2}{R_o} \int_0^{R_o} \frac{\sin^2\left[\frac{\pi r}{R_o}\right]}{r} dr = \frac{2.438}{R_o}. \quad (14.3)$$

where we have evaluated the integral numerically. With these assumptions, we obtain

$$E = T + V = \frac{\hbar^2 \pi^2}{2m_e R_o^2} - \left\langle \frac{1}{r} \right\rangle. \quad (14.4)$$

Combination of Eqs. 14.3 and 14.4 produces

$$E = T + V = \frac{\hbar^2 \pi^2}{2m_e R_o^2} - \frac{2.438}{R_o} = \frac{\pi^2}{2R_o^2} - \frac{2.438}{R_o} \text{ hartrees}, \quad (14.5)$$

since $\hbar = m_e = 1$ in atomic units. Figure 14.1 shows the variation of E with R_o.

Equation 14.5 makes it very clear why the electron is not going to fall into the nucleus: That would require that R_o go to zero. As R_o decreases, the potential decreases and eventually approaches $-\infty$ as R_o approaches zero. However, at the same time, the kinetic energy is increasing even faster. Therefore, the electron position that minimizes the *total* energy is a compromise between a potential-energy term that decreases as R_o decreases and a kinetic-energy term that increases as R_o decreases. We can easily determine the value of R_o that minimizes the total energy by taking the derivative of E with respect to R_o in Eq. 14.5 and setting it to zero. We get

$$\frac{dE}{dR_o} = -\frac{\pi^2}{R_o^3} + \frac{2.438}{R_o^2} = 0.$$

Solving for R_o, we obtain $[R_o]_{min} = \pi^2/2.438 = 4.049$ bohr. Substituting this value of R_o into Eq. 14.5 gives a minimum energy $E_{min} = -0.301$ hartree. The true ground-state energy of the hydrogen atom is -0.5000 hartree, so our crude model yields a result that is 40% too large.

The situation is much the same when we consider the formation of a chemical bond. Figure 14.2 shows the approximate changes in the total energy, the average kinetic energy, and the average potential energy for two hydrogen

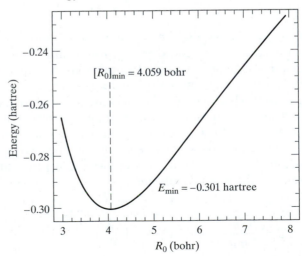

▶ FIGURE 14.1
Variation of the total energy as a function of the average distance of the electron from the nucleus in the model of the hydrogen atom described in the text. The minimum at $R_o = 4.059$ bohr is the result of a kinetic energy that is increasing as R_o decreases and a potential energy that is decreasing as R_o decreases.

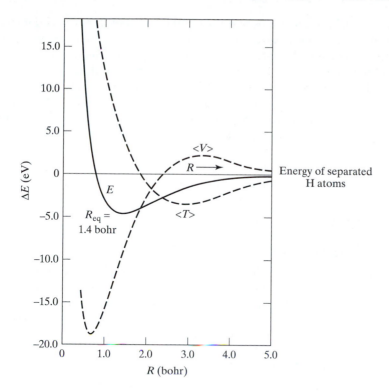

◀ FIGURE 14.2

Approximate total energy, average kinetic energy, and average potential energy of H_2 as a function of the internuclear distance R. [W. Kolos and L. Wolneiwicz, *J. Chem. Phys.* **43**, 2429 (1965)]

atoms as a function of the distance R between the nuclei relative to their values at $R = \infty$ [W. Kolos and L. Wolniewicz, *J. Chem. Phys.* **43**, 2429 (1965)].

When we have two hydrogen atoms, each in the ground state, separated by a large distance, the average kinetic energy of each electron is $\frac{1}{2}$ hartree. The average potential is -1 hartree per electron. The total energy is the sum, which is -1 hartree. As we begin to bring the two atoms into closer proximity so that the system begins to resemble an H_2 molecule, the electron density between the nuclei begins to increase. This increases the effective volume of space available to the electrons, which in turn decreases their kinetic energy, as Eq. 14.2 suggests. At the same time, the average potential energy is seen to increase, because we are moving electron density away from the positively charged nuclei into the space between them. However, the kinetic energy decreases faster than the potential energy increases, so the total-energy curve is seen to decrease. The system is becoming more stable.

As the distance between the nuclei continues to decrease, the average kinetic energy attains a minimum around $R \approx 2.8$ bohr. Further reduction of the internuclear distance now decreases the effective volume of space available to the electrons. Hence, the average kinetic energy begins to increase with decreasing R. Note, however, that although the average kinetic energy is increasing, the total energy continues to drop because the average potential is decreasing rapidly. At smaller internuclear distances, the electrons are strongly attracted by both positively charged nuclei. This attraction becomes even stronger as R decreases.

Since $\langle T \rangle$ tends to increase with $1/R^2$ [see Eq. 14.2], whereas $\langle V \rangle$ decreases approximately as $-C/R$, where C is a constant, we eventually reach a value of R where the rapid increase in $\langle T \rangle$ causes the total energy to reach its minimum value. Further decreases in R result in an increase of E. For H_2, this point occurs around $R \approx 1.4$ bohr, which is the equilibrium bond distance for the H_2 molecule.

Figure 14.2 demonstrates that the equilibrium bond length is determined by the competition between an increasing average kinetic energy, which is rising with decreasing R because of the decrease in the effective volume available to the electrons, and a decreasing average potential, which is dropping because the electrons are being pushed closer to the positively charged nuclei as R decreases. At a very small separation (around $R \approx 0.75$ bohr in the figure), the average potential energy begins to rise because of the mutual repulsion of the nuclei. It is important to note that the nuclear–nuclear repulsion becomes significant only at separations much less than the equilibrium bond distance. It is the competition between kinetic and potential energies that is primarily responsible for determining the equilibrium distance, not the competition between the nuclear–electron attraction and nuclear–nuclear repulsion potential terms.

14.2   Approximation Methods

If we hope to conduct quantum mechanical studies of atomic and molecular structure, bonding, and spectroscopy with reasonable accuracy, we must have better methods than the crude shielding approximations described in Chapter 13. Since the advent of high-speed computers, a large number of such methods have been developed. We shall discuss only two in this text.

14.2.1   Perturbation Theory

In this method, the Hamiltonian is partitioned into two parts, $\mathcal{H}_o$ and $\mathcal{H}'$, such that

$$\mathcal{H} = \mathcal{H}_o + \mathcal{H}'. \tag{14.6}$$

$\mathcal{H}_o$ is called the *zeroth-order Hamiltonian*, and $\mathcal{H}'$ is labeled the *perturbation*. The idea is to choose $\mathcal{H}_o$ such that the Schrödinger equation $\mathcal{H}_o \psi_o = E_o \psi_o$ can be solved exactly for the zeroth-order eigenfunctions ψ_o and the corresponding zeroth-order energy eigenvalues E_o. It is then assumed that the perturbation $\mathcal{H}'$ is sufficiently small that its presence does not appreciably alter the eigenfunctions for the system. If this is true, the average effect of $\mathcal{H}'$ on the energy can be obtained simply by computing the average value of $\mathcal{H}'$ with the use of the zeroth-order eigenfunctions and adding the result to the zeroth-order energy.

If the eigenfunctions are unaltered by the presence of $\mathcal{H}'$, the average value of the perturbation is given by the fourth postulate, viz.,

$$\langle \mathcal{H}' \rangle = \langle \psi_o | \mathcal{H}' | \psi_o \rangle = \int_{\text{all space}} \psi_o^* \mathcal{H}' \psi_o \, d\tau, \tag{14.7}$$

provided that ψ_o is normalized so that $\langle \psi_o | \psi_o \rangle = 1$. The corrected energy is then

$$E \approx E_o + \langle \mathcal{H}' \rangle. \tag{14.8}$$

In this form, the method is called *first-order perturbation theory*, and E in Eq. 14.8 is the *first-order energy*.

We have already encountered several examples of first-order perturbation theory in Chapter 13. The one-electron approximation is clearly such an example. In that case, we let $\mathcal{H}_o$ be the sum of all electronic kinetic-energy and nuclear–electron attraction terms. In effect, we took $\mathcal{H}'$ to be the sum of all electron–electron repulsion terms. The one-electron eigenfunctions this procedure produces are the zeroth-order eigenfunctions, and the uncorrected one-electron energy eigenvalues are the zeroth-order energies. We can now finish this perturbation theory calculation by computing the first-order correction to our zeroth-order energies. Equation 14.7 shows that we simply have to compute the average value of the electron-electron repulsion terms and add that to the zeroth-order energies. Example 14.1 illustrates the method for the helium atom.

EXAMPLE 14.1

Use first-order perturbation theory to compute the ground-state energy of a helium atom.

Solution

The Hamiltonian for the helium atom, in atomic units, is

$$\mathcal{H} = -\frac{1}{2}\nabla_1^2 - \frac{1}{2}\nabla_2^2 - \frac{2}{r_1} - \frac{2}{r_2} + \frac{1}{r_{12}}, \tag{1}$$

provided that we employ the Born–Oppenheimer approximation and omit the nuclear kinetic-energy term. The zeroth-order Hamiltonian is taken to be

$$\mathcal{H}_o = -\frac{1}{2}\nabla_1^2 - \frac{1}{2}\nabla_2^2 - \frac{2}{r_1} - \frac{2}{r_2}, \tag{2}$$

and the perturbation is

$$\mathcal{H}' = \frac{1}{r_{12}}. \tag{3}$$

We solved the zeroth-order problem in Chapter 13. The normalized, zeroth-order eigenfunction for the ground state, in atomic units, is

$$\psi_o(1, 2) = \chi_{1s}(1)\chi_{1s}(2) = \left[\frac{8}{\pi}\right]^{1/2} \exp[-2r_1]\left[\frac{8}{\pi}\right]^{1/2}\exp[-2r_2]$$

$$= \left[\frac{8}{\pi}\right]\exp[-2r_1]\exp[-2r_2], \tag{4}$$

and the zeroth-order ground-state energy is

$$E_o = -\frac{Z^2}{2n_1^2} - \frac{Z^2}{2n_2^2} = -\frac{2^2}{2(1)^2} - \frac{2^2}{2(1)^2} = -4 \quad \text{hartree.} \tag{5}$$

To obtain the first-order correction to the energy, we compute the average of $\mathcal{H}'$:

$$\langle\mathcal{H}'\rangle = \langle\psi_o(1, 2)|\mathcal{H}'|\psi_o(1, 2)\rangle$$

$$= \frac{64}{\pi^2}\int_{\phi_1=0}^{2\pi}\int_{\theta_1=0}^{\pi}\int_{r_1=0}^{\infty}\int_{\phi_2=0}^{2\pi}\int_{\theta_2=0}^{\pi}\int_{r_2=0}^{\infty}\frac{\exp[-4r_1]\exp[-4r_2]}{r_{12}}r_1^2 dr_1 \sin\theta_1 d\theta_1 d\phi_1 \times r_2^2 dr_2\sin\theta_2 d\theta_2 d\phi_2 = \frac{5}{4}\text{ hartrees.} \tag{6}$$

The integral in Eq. 6 can be obtained in closed analytic form, but the procedure is very lengthy and somewhat laborious. Therefore, we simply give the result without actually integrating. The first-order energy is

$$E = E_o + \langle\mathcal{H}'\rangle = -4 + \frac{5}{4} = -\frac{11}{4} \quad \text{hartrees} = -74.83 \text{ eV.} \tag{7}$$

The measured energy of the helium atom is -79.01 eV. The first-order perturbation result is, therefore, in error by 5.28%. The problem here is that the electron–electron repulsion is an extremely large perturbation. Consequently, the assumption that it does not alter the zeroth-order eigenfunctions is poor. As a result, the computed first-order energy is not very accurate.

As noted in Example 14.1, the application of first-order perturbation theory using the electron–electron repulsions as the perturbation is not well advised because these terms are so large. In contrast, perturbation theory is a very good method for computing the magnitude of the spin–orbit coupling. The spin–orbit L–S coupling is very small relative to the rest of the Hamiltonian, as long as the atomic number is small. Consequently, we expect L–S coupling to have little effect on the zeroth-order wave functions.

In Chapter 13, we learned that the energy change due to spin–orbit coupling is proportional to the dot product of $\mathbf{L}$ and $\mathbf{S}$. Consequently, with spin–orbit interaction included, the Hamiltonian for the hydrogen atom, in atomic units, is

$$\mathcal{H} = -\frac{1}{2}\nabla^2 - \frac{1}{r} + C\,\mathbf{L}\cdot\mathbf{S}.$$

Although it was not explicitly discussed in Chapter 13, our procedure was to disregard the effect of the spin–orbit term on the hydrogen-atom eigenfunctions and write the spin–orbit term as a perturbation. The effect of this perturbation on the energy eigenvalues is given by Eqs. 14.7 and 14.8:

$$\langle C\,\mathbf{L}\cdot\mathbf{S}\rangle = C\langle\psi_o|\mathbf{L}\cdot\mathbf{S}\,|\psi_o\rangle$$

Here, ψ_o is the zeroth-order hydrogen-atom eigenfunction. In Eq. 13.48, we found that $\mathbf{L}\cdot\mathbf{S}$ could be written in the form $\frac{1}{2}[J^2 - L^2 - S^2]$. As long as the eigenfunctions are still the zeroth-order results, we can replace J^2, L^2, and S^2 with $j(j+1)\hbar^2$, $\ell(\ell+1)\hbar^2$, and $s(s+1)\hbar^2$, respectively. However, this is true only within the perturbation theory approximation. For these reasons, our calculations of spin–orbit splitting in Chapter 13 were only approximations, but we expect them to be accurate because the $C\,\mathbf{L}\cdot\mathbf{S}$ term in the Hamiltonian is very small. Example 14.2 provides a quantitative illustration of how the error in a perturbation calculation depends upon the relative magnitude of the perturbation.

EXAMPLE 14.2

To illustrate the point that the error incurred in a first-order perturbation calculation depends upon the relative size of the perturbation, let us consider the problem of computing the ground-state energy of the hydrogen atom via a perturbation method. In atomic units, the hydrogen-atom Hamiltonian is $\mathcal{H} = -\frac{1}{2}\nabla^2 - 1/r$. Let us write this in the form

$$\mathcal{H} = -\frac{1}{2}\nabla^2 - \frac{f}{r} + \frac{(f-1)}{r},$$

where f is a positive constant. By letting

$$\mathcal{H}_o = -\frac{1}{2}\nabla^2 - \frac{f}{r}$$

and

$$\mathcal{H}' = \frac{(f-1)}{r},$$

compute the ground-state energy of the hydrogen atom as a function of f, using first-order perturbation theory. Plot the percent error in the calculation for the range $0 \le f \le 2$. Discuss the significance of your results.

Solution

The first order of business is to obtain the zeroth-order eigenfunctions and eigenvalues. We have already done this for this problem. The Hamiltonian for a hydrogen-like atom or ion, in atomic units, is $\mathcal{H} = -\frac{1}{2}\nabla^2 - Z/r$. In Chapter 13, we found that the ground-state eigenfunction and energy, in atomic units, are, respectively,

$$\psi(r, \theta, \phi) = \left[\frac{Z^3}{\pi}\right]^{1/2} e^{-Zr} \quad \text{and} \quad E = -\frac{Z^2}{2} \text{ hartrees.} \tag{1}$$

Examination of the zeroth-order Hamiltonian in this problem shows that $\mathcal{H}_o$ is identical to the Hamiltonian for the true hydrogen-like atom or ion, except for the substitution of f for Z. Therefore, the zeroth-order results are

$$\psi_o(r, \theta, \phi) = \left[\frac{f^3}{\pi}\right]^{1/2} e^{-fr} \quad \text{and} \quad E_o = -\frac{f^2}{2} \text{ hartrees.} \tag{2}$$

We now need to average the perturbation $\mathcal{H}' = (f - 1)/r$ over the zeroth-order eigenfunction. This average is

$$\langle \mathcal{H}' \rangle = \int_{\phi=0}^{2\pi} \int_{\theta=0}^{\pi} \int_{r=0}^{\infty} \psi_o^*(r, \theta, \phi) \frac{(f - 1)}{r} \psi_o(r, \theta, \phi) r^2 \, dr \sin\theta \, d\theta \, d\phi$$

$$= \left[\frac{f^3}{\pi}\right](f - 1) \int_{\phi=0}^{2\pi} \int_{\theta=0}^{\pi} \int_{r=0}^{\infty} \frac{e^{-2fr}}{r} r^2 \, dr \sin\theta \, d\theta \, d\phi$$

$$= 4f^3(f - 1) \int_{r=0}^{\infty} e^{-2fr} r \, dr = 4f^3(f - 1)\frac{1}{4f^2} = f(f - 1). \tag{3}$$

The first-order energy is, therefore,

$$E = E_o + \langle \mathcal{H}' \rangle = -\frac{f^2}{2} + f(f - 1) = \frac{f^2}{2} - f \text{ hartrees.} \tag{4}$$

The correct answer for the hydrogen atom with $Z = 1$ is $-\frac{1}{2}$ hartree. The percent error in the first-order perturbation result is

$$\% \text{ error} = 100 \times \frac{\dfrac{f^2}{2} - f - (-1/2)}{(1/2)} = 100[f^2 - 2f + 1]. \tag{5}$$

A plot of the percent error as a function of f is shown in Figure 14.3. When $f = 1$, the perturbation is zero and the answer is exact. As f deviates from unity, the size of the perturbation increases, and so does the percent error in the first-order estimate of the ground-state energy of the hydrogen atom.

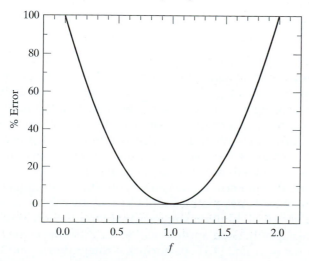

◀ FIGURE 14.3
The percent error in a perturbation calculation on a hydrogen-like atom as a function of the relative size of the perturbation term. In this example, $\mathcal{H}_o = -\frac{1}{2}\nabla^2 - f/r$, while the perturbation is $\mathcal{H} = (f - 1)/r$, where f is a constant. (See Example 14.2 for details.)

For related exercises, see Problems 14.1, 14.2, 14.3, and 14.27.

14.2.2 Variational Methods

First-order perturbation theory is accurate only in cases where the perturbation is very small. When we are concerned with the study of the electronic structure of many-electron atoms and molecules, the perturbation is the sum of all electron–electron interaction terms. This sum is very large, so that first-order perturbation calculations are not usually useful. A more powerful approximation method is needed.

Since the molecular equilibrium structure is the one that minimizes the total energy, the energy eigenvalue in the Schrödinger equation is the quantity of primary concern. As we have seen in Chapter 13, the presence of electron–electron repulsions prevents the separation of the variables, so that exact, closed-form analytic solutions cannot be obtained. However, integration is much easier than solving multidimensional, partial differential equations subject to boundary conditions. If we can recast the problem so that the required mathematical operations involve integrations rather than solution of the Schrödinger equation, perhaps we can make some headway in the study of molecular structure.

14.2.2.1 Rayleigh–Ritz Variational Principle

Suppose we have an N-electron system whose eigenfunction is $\psi(1, 2, \ldots, N)$. Our problem is to solve the equation $\mathcal{H}\psi(1, 2, \ldots, N) = E\psi(1, 2, \ldots, N)$ for the energy eigenvalue. Let us multiply both sides of this equation on the left by $\psi^*(1, 2, \ldots, N)$. This operation produces

$$\psi^*(1, 2, \ldots, N)\mathcal{H}\psi(1, 2, \ldots, N) = \psi^*(1, 2, \ldots, N)E\psi(1, 2, \ldots, N). \quad \textbf{(14.9)}$$

We now integrate both sides of Eq. 14.9 over the space of all N electrons to obtain

$$\langle\psi(1, 2, \ldots, N)|\mathcal{H}|\psi(1, 2, \ldots, N)\rangle = \langle\psi(1, 2, \ldots, N)|E|\psi(1, 2, \ldots, N)\rangle$$

$$= E\langle\psi(1, 2, \ldots, N)|\psi(1, 2, \ldots, N)\rangle. \quad \textbf{(14.10)}$$

Note that the integrals which appear in this equation are $3N$-dimensional integrals over the coordinates of all N electrons. Equation 14.10 is a simple linear equation for E that we can easily solve. The solution is

$$E = \frac{\langle\psi(1, 2, \ldots, N)|\mathcal{H}|\psi(1, 2, \ldots, N)\rangle}{\langle\psi(1, 2, \ldots, N)|\psi(1, 2, \ldots, N)\rangle} = \frac{\langle\psi|\mathcal{H}|\psi\rangle}{\langle\psi|\psi\rangle}, \quad \textbf{(14.11)}$$

where we have replaced the notation $\psi(1, 2, \ldots, N)$ with ψ for brevity.

On the surface, it would appear that Eq. 14.11 solves our problem: We simply compute two integrals, $\langle\psi|\mathcal{H}|\psi\rangle$ and $\langle\psi|\psi\rangle$, take their ratio, and obtain the energy eigenvalue. It may not be a walk in the park to compute these integrals, but it is far easier than solving the molecular Schrödinger equation. Unfortunately, there is a huge roadblock that prevents us from computing the integrals: We do not know the eigenfunction ψ for the system, since we have not solved the Schrödinger equation.

Because we do not know ψ and cannot solve the Schrödinger equation to obtain it, perhaps we can proceed simply by making a guess as to what ψ is. Suppose our chemical intuition suggests that the function λ, which we just invent, is a reasonably good approximation for ψ. If this is the case, we can substitute λ for ψ in Eq. 14.11, perform the integrations, and compute an energy E_v that we might hope will be close to the true energy eigenvalue E. That is, we compute

$$E_v = \frac{\langle \lambda | \mathcal{H} | \lambda \rangle}{\langle \lambda | \lambda \rangle} \qquad (14.12)$$

and hope that we find $E_v \approx E$.

The above procedure would be ridiculous if it were not for one critically important fact: We know an important relationship between E_v and E. Since nature will adjust the electron distribution about the nuclei so as to minimize the total energy, the electron distribution predicted by $\psi^*\psi \, d\tau$ yields the lowest possible energy of the system. Any other distribution (e.g., the distribution predicted by $\lambda^*\lambda \, d\tau$) will be less stable and, therefore, produce a higher energy. For this reason, *we are guaranteed to have* $E_v \geq E$. This simple inequality is called the *Rayleigh–Ritz variational theorem*.

Since E_v must be an upper limit to E, we have a way to make our approximate answer better and better. We simply insert parameters into our trial eigenfunction λ and then adjust those parameters so as to minimize E_v in the certain knowledge that, as we make E_v smaller and smaller, we must be approaching the exact energy E. If, for example, λ is functionally dependent upon the parameters A, B, and C, E_v will also depend upon the values assigned to A, B, and C. We can then require, either analytically or numerically, that the values of A, B, and C be those which give $\partial E_v / \partial A = \partial E_v / \partial B = \partial E_v / \partial C = 0$ so as to produce a minimum result for E_v, which the Rayleigh–Ritz variational theorem tells us will be the best answer we can obtain using the functional form chosen for λ. This procedure is called a *variational calculation*, because we are varying the parameters A, B, and C so as to produce a minimum value for E_v. The computed approximate energy, E_v, is called the *variational energy*. Such variational calculations are the primary means by which molecular quantum mechanics is actually executed.

EXAMPLE 14.3

Using the trial eigenfunction $\lambda = Nx(a - x)$ for $0 \leq x \leq a$ and $\lambda = 0$ otherwise, compute the variational energy for a particle of mass m in an infinite potential well of with a. Show that the Rayleigh–Ritz variational principle is satisfied by comparing E_v with the exact ground-state energy of the system. Plot the normalized trial wave function and the true ground-state eigenfunction on the same graph.

Solution

Let us first normalize the trial ground-state eigenfunction. Normalization requires that we have

$$\langle \lambda | \lambda \rangle = 1 = N^2 \int_0^a x^2(a - x)^2 dx = N^2 \int_0^a [x^4 - 2ax^3 + a^2x^2] dx$$

$$= N^2 \left[\frac{x^5}{5} - \frac{2ax^4}{4} + \frac{a^2x^3}{3} \right]_0^a = N^2 a^5 \left[\frac{1}{5} - \frac{1}{2} + \frac{1}{3} \right] = \frac{N^2 a^5}{30}. \qquad (1)$$

Therefore, the normalization constant is $N = [30/a^5]^{1/2}$. The variational energy is given by Eq. 14.12:

$$E_v = \frac{\langle \lambda | \mathcal{H} | \lambda \rangle}{\langle \lambda | \lambda \rangle} = \langle \lambda | \mathcal{H} | \lambda \rangle, \qquad (2)$$

since the denominator is now unity because we have normalized λ. The Hamiltonian for a particle of mass m in a one-dimensional infinite potential well is $\mathcal{H} = -(\hbar^2/2m)(d^2/dx^2)$. When $\mathcal{H}$ operates on λ, we obtain

$$\mathcal{H}\lambda = -\frac{\hbar^2}{2m}\frac{d^2}{dx^2}N(ax - x^2) = -\frac{\hbar^2 N}{2m}\frac{d}{dx}(a - 2x) = \frac{N\hbar^2}{m}. \tag{3}$$

Hence, the variational energy is

$$E_v = \langle\lambda|\mathcal{H}|\lambda\rangle = \frac{N^2\hbar^2}{m}\int_0^a (ax - x^2)dx = \frac{N^2\hbar^2}{m}\left[\frac{ax^2}{2} - \frac{x^3}{3}\right]_0^a$$

$$= \frac{N^2\hbar^2 a^3}{m}\left[\frac{1}{2} - \frac{1}{3}\right] = \frac{N^2\hbar^2 a^3}{6m}. \tag{4}$$

Insertion of the normalization constant gives

$$E_v = \frac{\hbar^2 a^3}{6m}\frac{30}{a^5} = \frac{5\hbar^2}{ma^2} = \frac{5h^2}{4\pi^2 ma^2} = \frac{0.12665\ldots h^2}{ma^2}. \tag{5}$$

The exact ground-state energy eigenvalue is

$$E = \frac{h^2}{8ma^2} = \frac{0.125h^2}{ma^2}.$$

A comparison of E with E_v shows that $E_v > E$, and the Rayleigh–Ritz variational principle is satisfied. A plot of λ and the correct ground-state eigenfunction, $\psi(x) = [2/a]^{1/2}\sin[\pi x/a]$, is shown in Figure 14.4. The two functions are nearly the same. This is the reason E_v is a good approximation to the true ground-state energy.

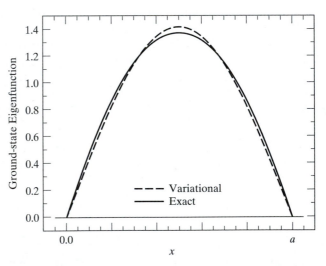

▲ **FIGURE 14.4**
Comparison of the exact and variational eigenfunctions for the ground state of a quantum particle in a one-dimensional, infinite potential well. The variational eigenfunction has the form $\lambda = Nx(a - x)$ for $0 \leq x \leq a$ and $\lambda = 0$ otherwise. (See Example 14.3 for details.)

For a related exercise, see Problem 14.4.

Example 14.3 illustrates the Rayleigh–Ritz variational principle, but it does not demonstrate how adjustable parameters function to lower the variational energy. In fact, if it is possible to adjust the parameters in a trial eigenfunction so as to produce the exact answer, variational calculations will do this. If, for example, we were to conduct a variational calculation on the hydrogen atom using the trial eigenfunction $\lambda = N \exp[-br]$, with b being an adjustable parameter, minimization of E_v with respect to b would produce the exact ground-state eigenfunction and the exact energy eigenvalue. This calculation is carried out in Example 14.4.

EXAMPLE 14.4

Using the trial eigenfunction $\lambda = N \exp[-br]$, obtain the variational energy for a ground-state hydrogen atom. Minimize the result with respect to b.

Solution

Let us first normalize the trial eigenfunction. This is not required, but we will need to compute the normalization integral anyway when we execute the variational calculation. We require that

$$\langle \lambda | \lambda \rangle = 1 = N^2 \int_{\phi=0}^{2\pi} \int_{\theta=0}^{\pi} \int_{r=0}^{\infty} \exp[-2br] r^2 \, dr \, \sin\theta \, d\theta \, d\phi$$

$$= 4\pi N^2 \int_0^\infty \exp[-2br] r^2 \, dr = 4\pi N^2 \left[\frac{2!}{(2b)^3} \right] = \frac{\pi N^2}{b^3}. \qquad (1)$$

Therefore, the normalization constant is $N^2 = b^3/\pi$. The hydrogen-atom Hamiltonian, in atomic units, is

$$\mathcal{H} = -\frac{1}{2} \nabla^2 - \frac{1}{r}. \qquad (2)$$

The variational calculation requires that we evaluate the integral $\langle \lambda | \mathcal{H} | \lambda \rangle$. Therefore, we need to first find the result of operating on λ with $\mathcal{H}$. We have

$$\mathcal{H}\lambda = \left\{ -\frac{1}{2} \left[\frac{1}{r^2} \frac{\partial}{\partial r} \left\{ r^2 \frac{\partial}{\partial r} \right\} + \frac{1}{r^2 \sin\theta} \frac{\partial}{\partial \theta} \left\{ \sin\theta \frac{\partial}{\partial \theta} \right\} + \frac{1}{r^2 \sin^2\theta} \frac{\partial^2}{\partial \phi^2} \right] - \frac{1}{r} \right\} N \exp(-br)$$

$$= -\frac{N}{2r^2} \frac{\partial}{\partial r} r^2 (-b) \exp(-br) - \frac{N \exp(-br)}{r} = \frac{Nb}{2r^2} [2r \exp(-br)$$

$$- br^2 \exp(-br)] - \frac{N \exp(-br)}{r} = N \exp(-br) \left[\frac{(b-1)}{r} - \frac{b^2}{2} \right], \qquad (3)$$

since the angular derivatives produce zero when operating on a function that depends only upon the radial coordinate r. The required integral is, therefore,

$$E_v = \langle \lambda | \mathcal{H} | \lambda \rangle = N^2 \int_{\phi=0}^{2\pi} \int_{\theta=0}^{\pi} \int_{r=0}^{\infty} \exp[-2br] \left[\frac{(b-1)}{r} - \frac{b^2}{2} \right] r^2 \, dr \, \sin\theta \, d\theta \, d\phi$$

$$= 4\pi N^2 \left[\frac{(b-1)}{(2b)^2} - \frac{b^2}{2} \frac{2!}{(2b)^3} \right]. \qquad (4)$$

Inserting the value of N^2 produces

$$E_v = 4\pi \frac{b^3}{\pi} \left[\frac{(b-1)}{(2b)^2} - \frac{b^2}{8b^3} \right] = b(b-1) - \frac{b^2}{2} = \frac{b^2}{2} - b. \qquad (5)$$

Equation 5 shows that the variational energy is dependent upon the value we choose for the parameter b. The Rayleigh–Ritz variational theorem tells us that we wish to choose b so as to minimize E_v. Figure 14.5 shows the dependence of the variational energy upon b. In this case, the minimization task is easy. We simply require that

$$\frac{dE_v}{db} = b - 1 = 0, \tag{6}$$

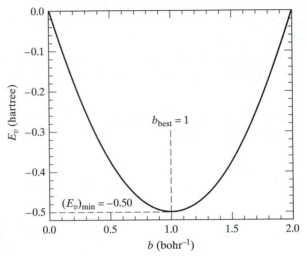

▲ FIGURE 14.5
The variational energy E_v as a function of the exponential variational parameter b for the ground state of the hydrogen atom. The trial eigenfunction is $\lambda = N \exp(-br)$. The lowest and, therefore, best value of b occurs at $b = 1$. This produces the exact result, the possibility of which is permitted by the choice of the trial eigenfunction. (See Example 14.4 for the details.)

so that the best value of b is unity. The best trial eigenfunction is $\lambda = \pi^{-1/2} \exp[-r]$, and the best variational energy is obtained from inserting $b = 1$ into Eq. 5 to obtain

$$[E_v]_{best} = \frac{1}{2} - 1 = -\frac{1}{2} \quad \text{hartree.} \tag{7}$$

These results are exact; we obtain $E_v = E$ rather than $E_v > E$. This happy result occurs because we chose a trial eigenfunction that could be made into the exact ground-state eigenfunction by an appropriate choice of b, which the variational

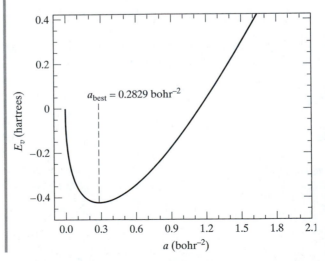

◀ FIGURE 14.6
The variational energy E_v as a function of the exponential variational parameter a for the ground state of the hydrogen atom. The trial eigenfunction is $\lambda = N \exp(-ar^2)$. The lowest and, therefore, best value of a occurs at $a = 8/(9\pi) = 0.2829$ bohr^{-2}. In this case, we see that $E_v > E$, as required by the variational principle. (See Problem 14.5 for details.)

procedure produces. Had we chosen a Gaussian function rather than an exponential, the variational method would still have given us the best answer possible for our trial eigenfunction, but it would not be exact. In solving Problem 14.5, you will show that if we choose $\lambda = N \exp[-ar^2]$ for our trial eigenfunction, we obtain a variational energy given by $E_v = 3a/2 - 2[2a/\pi]^{1/2}$. Figure 14.6 shows the dependence of E_v upon a in this case. As can be seen, E_v attains a minimum value of $-0.424\ldots$ hartree, so that we have $E_v > E$.

For a related exercise, see Problem 14.6.

14.2.2.2   Linear Variational Theory

In Example 14.4 and in Problems 14.5 and 14.6, we carried out variational calculations in which the variation parameter is nonlinear. This means that the parameter does not appear as a multiplicative coefficient to the first power. Instead, in the example, it appears in the exponent. Because the system being examined is so simple, we were able to execute the calculation with little difficulty. However, when we employ variational methods to execute similar calculations for complex molecular systems, the presence of a nonlinear variational parameter introduces a great deal of additional difficulty. As we shall see, if we restrict the variational parameters to linear coefficients, the equations that must be solved to obtain their values and the corresponding best variational energy are linear algebraic equations, which are much easier to handle. Consequently, scientists often conduct variational calculations using only linear parameters. When this is done, the method is called *linear variational theory.*

If the variational parameters are all linear, we can easily obtain the general equations that must be solved to find the parameter values that minimize E_v. Suppose we take as our trial eigenfunction

$$\lambda = \sum_{i=1}^{M} a_i \chi_i, \tag{14.13}$$

where the χ_i are arbitrary functions of our choice and the a_i are linear variational parameters. The χ_i can be anything we wish, but they cannot contain variational parameters. In this form, we have M parameters with respect to which we wish to minimize E_v. Therefore, we intend to require that $\partial E_v/\partial a_1 = \partial E_v/\partial a_2 = \partial E_v/\partial a_3 = \cdots = \partial E_v/\partial a_M = 0$. This will give us M equations, which should permit us to find the values of the M linear variational parameters.

The variational energy that we wish to minimize is given by Eq. 14.12:

$$E_v = \frac{\langle \lambda | \mathcal{H} | \lambda \rangle}{\langle \lambda | \lambda \rangle} = \frac{N}{D},$$

where N and D represent the numerator and denominator, respectively. The derivative of E_v with respect to the kth parameter is

$$\frac{\partial E_v}{\partial a_k} = \frac{1}{D^2}\left[D\frac{\partial N}{\partial a_k} - N\frac{\partial D}{\partial a_k}\right] = \frac{1}{D}\left[\frac{\partial N}{\partial a_k} - \frac{N}{D}\frac{\partial D}{\partial a_k}\right]. \tag{14.14}$$

The ratio N/D that appears in Eq. 14.14 is the value of the variational energy E_v. Consequently, the equations that must hold to produce a minimum value of E_v are

$$\frac{\partial E_v}{\partial a_k} = \frac{1}{D}\left[\frac{\partial N}{\partial a_k} - E_v\frac{\partial D}{\partial a_k}\right] = 0 \quad \text{for } k = 1, 2, 3, \ldots, M. \tag{14.15}$$

Since D is not infinite, we must have

$$\left[\frac{\partial N}{\partial a_k} - E_v \frac{\partial D}{\partial a_k}\right] = 0 \quad \text{for } k = 1, 2, 3, \ldots, M. \tag{14.16}$$

Equation 14.16 is a set of M equations whose solution will give us the values of the variational parameters that produce the lowest possible variational energy for the choices we have made for the χ_i functions.

Since we know the functional form of the numerator and the denominator in Eq. 14.12, we can determine the form of the derivatives that appear in Eq. 14.16 and thereby obtain a set of equations that permits us to execute linear variational calculations with relative ease. To simplify the notation and make the derivation easier to follow, we shall obtain these equations for the case where $M = 2$ and then generalize the result.

If our trial eigenfunction contains only two terms with linear coefficients, we have

$$\lambda = \sum_{i=1}^{2} a_i \chi_i = a_1 \chi_1 + a_2 \chi_2. \tag{14.17}$$

The denominator of Eq. 14.12 is

$$\langle \lambda | \lambda \rangle = \langle a_1 \chi_1 + a_2 \chi_2 | a_1 \chi_1 + a_2 \chi_2 \rangle. \tag{14.18}$$

In the interests of simplification, let us assume that we have chosen the χ_i to be real functions so that the variational parameters will also be real. In that case, Eq. 14.18 becomes

$$\langle \lambda | \lambda \rangle = a_1^2 \langle \chi_1 | \chi_1 \rangle + a_2^2 \langle \chi_2 | \chi_2 \rangle + 2a_1 a_2 \langle \chi_1 | \chi_2 \rangle, \tag{14.19}$$

where we have made use of the fact that $\langle \chi_1 | \chi_2 \rangle = \langle \chi_2 | \chi_1 \rangle$. Let us now take the derivative of Eq. 14.19 with respect to a_1. The result is

$$\frac{\partial \langle \lambda | \lambda \rangle}{\partial a_1} = \frac{\partial D}{\partial a_1} = 2a_1 \langle \chi_1 | \chi_1 \rangle + 2a_2 \langle \chi_1 | \chi_2 \rangle. \tag{14.20}$$

The integrals $\langle \chi_1 | \chi_1 \rangle$ and $\langle \chi_1 | \chi_2 \rangle$ are just numbers that we can compute once χ_1 and χ_2 are chosen. These integrals are commonly called *overlap integrals* and are given the symbol S_{ij} for the integral $\langle \chi_i | \chi_j \rangle$. In this notation, Eq. 14.20 becomes

$$\frac{\partial D}{\partial a_1} = 2a_1 S_{11} + 2a_2 S_{12}. \tag{14.21}$$

We obtain a similar form for the numerator of Eq. 14.12, except the integrals now contain the Hamiltonian for the system. Inserting the trial eigenfunction into the numerator of Eq. 14.12 and then expanding gives

$$N = \langle \lambda | \mathcal{H} | \lambda \rangle = \langle a_1 \chi_1 + a_2 \chi_2 | \mathcal{H} | a_1 \chi_1 + a_2 \chi_2 \rangle$$

$$= a_1^2 \langle \chi_1 | \mathcal{H} | \chi_1 \rangle + a_2^2 \langle \chi_2 | \mathcal{H} | \chi_2 \rangle + a_1 a_2 \langle \chi_1 | \mathcal{H} | \chi_2 \rangle + a_2 a_1 \langle \chi_2 | \mathcal{H} | \chi_1 \rangle. \tag{14.22}$$

The integrals in this equation are called *Hamiltonian matrix elements*. They are generally assigned the notation H_{ij} for the integral $\langle \chi_i | \mathcal{H} | \chi_j \rangle$. In this notation, Eq. 14.22 becomes

$$N = \langle \lambda | \mathcal{H} | \lambda \rangle = a_1^2 H_{11} + a_2^2 H_{22} + a_1 a_2 H_{12} + a_2 a_1 H_{21}. \tag{14.23}$$

To obtain a minimum in E_v, we need the derivative of N with respect to both a_1 and a_2. These are easily obtained. The result when we differentiate with respect to a_1 is

$$\frac{\partial N}{\partial a_1} = 2a_1 H_{11} + a_2 H_{12} + a_2 H_{21}. \tag{14.24}$$

The second postulate requires that $\mathcal{H}$ be Hermitian. Therefore, we have

$$H_{12} = \langle \chi_1 | \mathcal{H} | \chi_2 \rangle = \langle \mathcal{H}\chi_1 | \chi_2 \rangle = \langle \chi_2 | \mathcal{H} | \chi_1 \rangle = H_{21},$$

since we have chosen the χ_i to be real. The equality between H_{12} and H_{21} permits Eq. 14.24 to be written in the form

$$\frac{\partial N}{\partial a_1} = 2a_1 H_{11} + 2a_2 H_{12}. \tag{14.25}$$

If we insert Eqs. 14.21 and 14.25 into 14.16, the result is

$$a_1[H_{11} - E_v S_{11}] + a_2[H_{12} - E_v S_{12}] = 0, \tag{14.26}$$

where we have divided out the factor of two. If we now repeat the derivation, but take the derivatives with respect to a_2 instead of a_1, we obtain

$$a_1[H_{21} - E_v S_{21}] + a_2[H_{22} - E_v S_{22}] = 0. \tag{14.27}$$

Equations 14.26 and 14.27 are linear homogeneous equations with unknowns a_1 and a_2. The solution of these equations gives the values of a_1 and a_2 that minimize the variational energy.

Using Cramer's rule, which we first discussed in Chapter 1 in connection with linear least-squares analysis, we know that the solution of simultaneous linear equations can be written as the ratio of two determinants. The determinant of the coefficients of the unknowns is in the denominator, while the numerator is a determinant identical to that in the denominator, except that the coefficients of the unknown whose solution is being computed are replaced with the constants on the right-hand sides of the equations. Thus, the solution for a_1 is

$$a_1 = \frac{\begin{vmatrix} 0 & H_{12} - E_v S_{12} \\ 0 & H_{22} - E_v S_{22} \end{vmatrix}}{\begin{vmatrix} H_{11} - E_v S_{11} & H_{12} - E_v S_{12} \\ H_{21} - E_v S_{21} & H_{22} - E_v S_{22} \end{vmatrix}}. \tag{14.28}$$

With an entire column having nothing but zero elements, the determinant in the numerator is zero. We will obtain a similar result for a_2. The solution $a_1 = a_2 = 0$ gives the result $\lambda = 0$, which is the trivial solution of the Schrödinger equation. If we wish to avoid the trivial solution, the determinant in the denominator of Eq. 14.28 must also be zero, in which case the right-hand side has the indeterminate form $0/0$. That is, we must have

$$\boxed{\begin{vmatrix} H_{11} - E_v S_{11} & H_{12} - E_v S_{12} \\ H_{21} - E_v S_{21} & H_{22} - E_v S_{22} \end{vmatrix} = 0.} \tag{14.29}$$

Equation 14.29 is called the *secular equation*. It is the actual starting point for linear variational calculations.

If we expand the determinant in Eq. 14.29, we obtain a quadratic equation in E_v. This quadratic has two roots, both of which will be real, since the Hamiltonian is a Hermitian operator and all Hermitian operators have real eigenvalues. One of these roots will be less than the other. The lower energy root for E_v is the best approximation to the ground-state energy of the system that our two-term function with linear variational parameters can produce.

We can now take that root, substitute it into Eqs. 14.26 and 14.27, and solve for the ratio a_1/a_2. The normalization condition on the eigenfunction will then give us the values of each variational parameter and the form of our approximate ground-state eigenfunction. We then repeat the procedure with the higher energy root for E_v to obtain an approximate eigenfunction for the first excited state.

Notice that when we chose to approximate the ground-state eigenfunction with a linear combination of two functions, we obtained two linear, homogeneous equations for the two variational parameters that led to a 2×2 secular equation. The lower of the two roots of the secular equation for E_v produced an upper limit for the ground-state energy of the system and the higher root a similar upper limit for the first excited state. In general, if we write our trial eigenfunction as a linear combination of M functions, we will obtain M homogeneous, linear equations whose forms are

$$a_1[H_{11} - E_v S_{11}] + a_2[H_{12} - E_v S_{12}] + \cdots + a_M[H_{1M} - E_v S_{1M}] = 0,$$
$$a_1[H_{21} - E_v S_{21}] + a_2[H_{22} - E_v S_{22}] + \cdots + a_M[H_{2M} - E_v S_{2M}] = 0,$$
$$a_1[H_{31} - E_v S_{31}] + a_2[H_{32} - E_v S_{32}] + \cdots + a_M[H_{3M} - E_v S_{3M}] = 0$$
$$\vdots$$
$$a_1[H_{M1} - E_v S_{M1}] + a_2[H_{M2} - E_v S_{M2}] + \cdots + a_M[H_{MM} - E_v S_{MM}] = 0.$$

$$(14.30)$$

To avoid the trivial solution, $a_i = 0$ for all i, of these equations, the secular determinant must equal zero. That is, we must have

$$\begin{vmatrix} H_{11} - E_v S_{11} & H_{12} - E_v S_{12} & H_{13} - E_v S_{13} & \ldots & H_{1M} - E_v S_{1M} \\ H_{21} - E_v S_{21} & H_{22} - E_v S_{22} & H_{23} - E_v S_{23} & \ldots & H_{2M} - E_v S_{2M} \\ & & \vdots & & \\ H_{M1} - E_v S_{M1} & H_{M2} - E_v S_{M2} & H_{M3} - E_v S_{M3} & \ldots & H_{MM} - E_v S_{MM} \end{vmatrix} = 0.$$

$$(14.31)$$

When expanded, this $M \times M$ determinant yields an Mth-order polynomial in E_v that has M roots. The lowest root is the best approximation to the ground-state energy, the next-lowest root is an upper limit for the first excited-state energy, and so on. The approximate eigenfunction corresponding to each of these energy states can be obtained by substituting the corresponding root for E_v into Eq. 14.30 and solving for all the variational parameters in terms of any one of them. Normalization will then determine this last parameter, and the problem will be completely solved.

The best part of this procedure is that the solution of large sets of linear, homogeneous equations is a simple problem to program for solution by a computer. Therefore, in actual practice, we do not have to execute any of the foregoing operations; all we need do is input our choices for the functions $\chi_1, \chi_2, \chi_3, \ldots, \chi_M$, and the computer will execute all the integrals, set up and solve the secular equation for the roots of E_v, substitute the results back into the linear equations, compute the values of the variational parameters, and, finally, print out the results. While the computer is working, we can drink coffee, go to a movie, watch a ball game, or even study. Example 14.5 illustrates the essential components of a linear variational calculation.

EXAMPLE 14.5

Carry out a linear variational calculation for a particle of mass m in a one-dimensional infinite potential well of width a. Use two parameters, and do not choose any trigonometric functions for either χ_1 or χ_2. Compare the result with the exact ground-state energy.

Solution

The first step is to select the expansion functions we want to use. We are free to use anything we wish (except trigonometric functions in this case). However, if we want the variational energy to be a good approximation to the exact ground-state energy, we should be careful to choose functions that satisfy the requirements of the first postulate and that have characteristics similar to those we expect the exact eigenfunction to have. In this case, we want both χ_1 and χ_2 to go to zero at $x = 0$ and $x = a$. We also want functions that do not have nodes, since we know that the ground state will be nodeless. Suppose we select $\chi_1 = a^2x - x^3$ and $\chi_2 = a^2x^5 - x^7$. These choices are not unique, but they do have the desired characteristics. In most cases, the success or failure of a variational calculation depends upon the care with which the χ's are selected.

With these choices, our trial eigenfunction is

$$\lambda = c_1[a^2x - x^3] + c_2[a^2x^5 - x^7], \tag{1}$$

where we have written the variational parameters as c_1 and c_2 to avoid confusion with the width a of the well. The secular equation that minimizes E_v, is

$$\begin{vmatrix} H_{11} - E_vS_{11} & H_{12} - E_vS_{12} \\ H_{21} - E_vS_{21} & H_{22} - E_vS_{22} \end{vmatrix} = 0 = \begin{vmatrix} H_{11} - E_vS_{11} & H_{12} - E_vS_{12} \\ H_{12} - E_vS_{12} & H_{22} - E_vS_{22} \end{vmatrix}. \tag{2}$$

Expanding the determinant in Eq. 2 and collecting terms in the different powers of E_v, we obtain

$$[S_{11}S_{22} - S_{12}^2]E_v^2 - E_v[S_{11}H_{22} + S_{22}H_{11} - 2S_{12}H_{12}] + [H_{11}H_{22} - H_{12}^2] = 0. \tag{3}$$

Therefore, we need to compute the values of S_{11}, S_{22}, S_{12}, H_{11}, H_{22}, and H_{12}, which is the same as H_{21}, since $\mathcal{H}$ is a Hermitian operator. The overlap integrals are

$$S_{11} = \langle\chi_1|\chi_1\rangle = \int_0^a (a^2x - x^3)^2 dx = \int_0^a [a^4x^2 - 2a^2x^4 + x^6]dx$$

$$= \left[\frac{a^4x^3}{3} - \frac{2a^2x^5}{5} + \frac{x^7}{7}\right]_0^a = a^7\left[\frac{1}{3} - \frac{2}{5} + \frac{1}{7}\right] = \frac{8a^7}{105} \tag{4}$$

$$S_{22} = \langle\chi_2|\chi_2\rangle = \int_0^a (a^2x^5 - x^7)^2 dx = \int_0^a [a^4x^{10} - 2a^2x^{12} + x^{14}]dx$$

$$= \left[\frac{a^4x^{11}}{11} - \frac{2a^2x^{13}}{13} + \frac{x^{15}}{15}\right]_0^a = a^{15}\left[\frac{1}{11} - \frac{2}{13} + \frac{1}{15}\right] = \frac{8a^{15}}{2,145} \tag{5}$$

and

$$S_{12} = S_{21} = \langle\chi_1|\chi_2\rangle = \int_0^a (a^2x - x^3)(a^2x^5 - x^7)dx = \int_0^a [a^4x^6 - 2a^2x^8 + x^{10}]dx$$

$$= \left[\frac{a^4x^7}{7} - \frac{2a^2x^9}{9} + \frac{x^{11}}{11}\right]_0^a = a^{11}\left[\frac{1}{7} - \frac{2}{9} + \frac{1}{11}\right] = \frac{8a^{11}}{693}. \tag{6}$$

The values of the Hamiltonian matrix elements are

$$H_{11} = \langle\chi_1|\mathcal{H}|\chi_1\rangle = \int_0^a (a^2x - x^3)\frac{-\hbar^2}{2m}\frac{d^2}{dx^2}(a^2x - x^3)dx = \frac{3\hbar^2}{m}\int_0^a (a^2x^2 - x^4)dx$$

$$= \frac{3\hbar^2}{m}\left[\frac{a^5}{3} - \frac{a^5}{5}\right] = \frac{2a^5\hbar^2}{5m}, \tag{7}$$

$$H_{22} = \langle \chi_2|\mathcal{H}|\chi_2 \rangle = \int_0^a (a^2x^5 - x^7)\frac{-\hbar^2}{2m}\frac{d^2}{dx^2}(a^2x^5 - x^7)dx$$

$$= \frac{-\hbar^2}{2m}\int_0^a (a^2x^5 - x^7)(20a^2x^3 - 42x^5)dx = \frac{-\hbar^2}{2m}\int_0^a (20a^4x^8 - 62a^2x^{10} + 42x^{12})dx$$

$$= \frac{-\hbar^2}{2m}\left[\frac{20a^{13}}{9} - \frac{62a^{13}}{11} + \frac{42a^{13}}{13}\right] = \frac{118a^{13}\hbar^2}{1{,}287\,m}, \tag{8}$$

and

$$H_{21} = H_{12} = \langle \chi_2|\mathcal{H}|\chi_1 \rangle = \int_0^a (a^2x^5 - x^7)\frac{-\hbar^2}{2m}\frac{d^2}{dx^2}(a^2x - x^3)dx$$

$$= \frac{3\hbar^2}{m}\int_0^a (a^2x^5 - x^7)x\,dx = \frac{3\hbar^2}{m}\left[\frac{a^9}{7} - \frac{a^9}{9}\right] = \frac{6a^9\hbar^2}{63m}. \tag{9}$$

We now need to compute the coefficients of E_v^2, E_v, and the constant term in Eq. 3. These are

$$S_{11}S_{22} - S_{12}^2 = \frac{8a^7}{105}\frac{8a^{15}}{2{,}145} - \frac{8a^{11}}{693}\frac{8a^{11}}{693} = 0.000150896a^{22}, \tag{10}$$

$$[S_{11}H_{22} + S_{22}H_{11} - 2S_{12}H_{12}] = \frac{(8)(118)a^{20}\hbar^2}{(105)(1{,}287)m} + \frac{8(2)a^{20}\hbar^2}{(2{,}145)(5)m} - \frac{(2)(8)(6)a^{20}\hbar^2}{(693)(63)m}$$

$$= 0.006278589\frac{a^{20}\hbar^2}{m}, \tag{11}$$

and

$$H_{11}H_{22} - H_{12}^2 = \frac{(2)(118)a^{18}\hbar^4}{(5)(1{,}287)m^2} - \frac{6a^9\hbar^2}{63m}\frac{6a^9\hbar^2}{63m} = 0.027604142\frac{a^{18}\hbar^4}{m^2}, \tag{12}$$

respectively. The lowest root of the quadratic equation in Eq. 3 is

$$E_v = \frac{0.006278589a^{20}\hbar^2/m - [(0.006278589)^2a^{40}\hbar^4/m^2 - 0.000016661358a^{40}\hbar^4/m^2]^{1/2}}{2(0.000150896)a^{22}}$$

$$= \frac{0.006278589 - 0.004770667}{2(0.000150896)}\frac{\hbar^2}{ma^2} = 4.99656\frac{\hbar^2}{ma^2} = 0.12656\frac{\hbar^2}{ma^2}. \tag{13}$$

The exact ground-state energy is $h^2/(8ma^2) = 0.12500h^2/(ma^2)$. Clearly, the variational principle is satisfied.

By substituting the variational energy computed in Example 14.5 into either Eq. 14.26 or 14.27, we may compute the ratio c_2/c_1. This ratio is given by

$$\frac{c_2}{c_1} = \frac{H_{11} - E_vS_{11}}{E_vS_{12} - H_{12}}. \tag{14.32}$$

After normalization, the result is (see Problem 14.8)

$$\lambda_- = 3.91312a^{-7/2}\left[(a^2x - x^3) - \frac{0.514134}{a^4}(a^2x^5 - x^7)\right],$$

where we have written λ_- to denote that this is the eigenfunction associated with the lower of the two variational roots. Figure 14.7 shows a comparison between the exact eigenfunction and λ_- for the case where $a = 1$ Å. In this case, we see that the variational result is excellent. As expected, both eigenfunctions are nodeless.

The larger energy root in Example 14.5 is an upper limit to the first excited state. (See Problem 14.9.) Although we have not proven this, it can be shown that if we use M expansion functions in Eq. 14.13, the energy-ordered M roots for the variational energy are upper limits to the ground state and the next $M - 1$ excited states, in turn, and the corresponding eigenfunctions obtained by substituting the variational energies into Eq. 14.30 are approximations to the ground state and successive excited states. The normalized, approximate eigenfunction corresponding to the first excited state in Example 14.5 is

$$\lambda_+ = 3.06657a^{-7/2}\left[(a^2x - x^3) - \frac{7.2981122}{a^4}(a^2x^5 - x^7)\right].$$

(See Problem 14.9.) Figure 14.8 shows a comparison of λ_+ with the exact eigenfunction for the first excited state when $a = 1$ Å. As expected, both of these functions have one node. However, the approximate solution is not nearly so good as that for the ground state shown in Figure 14.7. The computed excited-state energy is too high by 85%, whereas the ground-state energy obtained in Example 14.5 is too high by only 1.2%. This type of result is characteristic of variational calculations; the ground-state energy and eigenfunction are much better approximations to the exact results than are those of the excited states.

Since λ_- and λ_+ have different energy eigenvalues, we expect these eigenfunctions to be orthogonal. That is, we expect to have $\langle\lambda_-|\lambda_+\rangle = 0$. In Problem 14.10, this is shown to be the case.

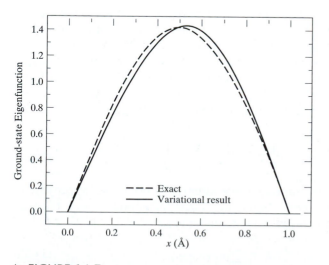

▲ FIGURE 14.7
Comparison of the variational eigenfunction for the ground state of a quantum particle in an infinite potential well with the exact result when $a = 1$ Å. The variational trial function in this case has the form $\lambda = c_1[a^2x - x^3] + c_2[a^2x^5 - x^7]$ for $0 \le x \le a$ and $\lambda = 0$ otherwise. (See Example 14.5 and Problem 14.8 for details.)

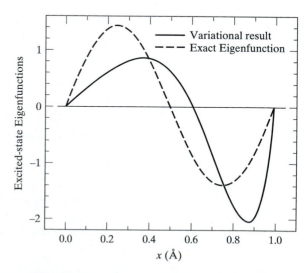

▲ FIGURE 14.8
Comparison of the variational eigenfunction for the first excited state of a quantum particle in an infinite potential well with the exact result when $a = 1$ Å. The variational trial function in this case has the form $\lambda = c_1[a^2x - x^3] + c_2[a^2x^5 - x^7]$ for $0 \le x \le a$ and $\lambda = 0$ otherwise. (See Example 14.5 and Problem 14.9 for details.)

14.3   Molecular Orbital Methods

Molecular orbital methods are the most frequently used procedures for the investigation of molecular electronic energies and structures. These techniques, which are based upon the variational principle, can produce energies and structural data with an accuracy approaching that available from the best experimental measurements if enough effort is devoted to the calculation. We shall introduce the concepts underlying molecular orbital methods by applying them to the one-electron hydrogen-molecule ion, H_2^+, for which we do not need to be concerned with the Pauli principle and electron-exchange antisymmetry requirements. Many-electron molecules will then be examined in a more qualitative fashion.

14.3.1   The Hydrogen-Molecule Ion

The H_2^+ ion is the molecular analogue of the hydrogen atom for atomic systems, in that only one electron is present. There are no electron–electron repulsion terms in the Hamiltonian, and antisymmetry requirements with respect to electron exchange are moot. Because of this, we can obtain numerically exact solutions of the Schrödinger equation for this molecule by transforming to a confocal elliptical coordinate system. However, such exact solutions cannot be obtained for any molecule that has more than one electron. We shall, therefore, ignore the exact solutions and use a molecular orbital approach.

Figure 14.9 defines the coordinate system we shall use. The internuclear axis is chosen to lie along the Z-axis. The internuclear separation is given by R, and the distances of the electron from nuclei A and B are labeled r_a and r_b, respectively. If we ignore spin–orbit terms, the Hamiltonian for the system, in atomic units, is

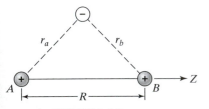

▲ FIGURE 14.9
The hydrogen-molecule ion.

$$\mathcal{H} = -\frac{1}{2m_H}\nabla_A^2 - \frac{1}{2m_H}\nabla_B^2 - \frac{1}{2}\nabla^2 - \frac{1}{r_a} - \frac{1}{r_b} + \frac{1}{R}. \qquad (14.33)$$

The first two terms in Eq. 14.33 are the kinetic-energy terms for the hydrogen nuclei, where m_H is the proton mass in atomic units. The third term is the kinetic energy of the electron, and the next two terms are the electron–nuclear attractions. The last term is the mutual repulsion between the positively charged nuclei. If we make the Born–Oppenheimer approximation (see Section 13.4.1), the nuclear kinetic-energy terms drop out, and the $1/R$ term becomes a constant, since the nuclear positions are now fixed. With this approximation, the Hamiltonian becomes

$$\mathcal{H} = -\frac{1}{2}\nabla^2 - \frac{1}{r_a} - \frac{1}{r_b} + \frac{1}{R}. \qquad (14.34)$$

In a molecular orbital approach, we attempt to obtain an approximate solution of the Schrödinger equation using linear variational theory. The trial eigenfunction is generally written as the linear expansion

$$\lambda = \sum_{i=1}^{M} a_i \chi_i \qquad (14.35)$$

where the a_i are the variational parameters. As usual, we are free to choose the expansion functions χ_i in any manner we wish. However, as noted in Example 14.5, if we want to obtain an accurate result, we need to pick functions that obey the first postulate and that we have some reason to believe are similar to the exact eigenfunctions. The set of functions χ_i is called the *basis set*.

We can obtain clues as to the type of expansion functions we should use by examining what happens to the system as R becomes either very large or very small. In the limit of large R, the H_2^+ molecule begins to resemble either $H_A + H_B^+$ or $H_A^+ + H_B$; that is, we have a hydrogen ion and a hydrogen atom. The eigenfunctions in this limit are just the hydrogen-atom eigenfunctions centered either on nucleus A or nucleus B, depending upon which proton gains the electron as R increases. With this point in mind, it seems reasonable to use atomic orbitals [either hydrogen-atom orbitals or Slater-type orbitals (STOs) with $Z = 1$ as our expansion functions. If we make this choice, we know that our result will be exact in the limit of large R. In the limit where $R \rightarrow 0$, the two protons form a single nucleus with a charge of $+2$. The system, therefore, becomes effectively a He^+ ion for which hydrogen-like eigenfunctions with $Z = 2$ are the exact solutions. These observations argue that an expansion of hydrogen-like orbitals or STOs with an effective charge Z_e intermediate between 1 and 2 will be reasonably accurate when the value of R is in the neighborhood of the H_2^+ equilibrium distance.

Using the preceding logic, we write Eq. 14.35 in the form

$$\lambda = a_1 \chi_{1sA} + a_2 \chi_{1sB} + a_3 \chi_{2sA} + a_4 \chi_{2sB} + a_5 \chi_{2pzA} + a_6 \chi_{2pzB} + a_7 \chi_{2pxA}$$
$$+ a_8 \chi_{2pxB} + a_9 \chi_{2pyA} + a_{10} \chi_{2pyB}. \tag{14.36}$$

The notation used is the same as that introduced in Chapter 13. χ_{1sA} is a $1s$ STO centered on nucleus A and χ_{1sB} is a $1s$ STO centered on nucleus B. Their normalized forms in atomic units are

$$\chi_{1sA} = \left[\frac{Z_e^3}{\pi} \right]^{1/2} \exp[-Z_e r_a]$$

and

$$\chi_{1sB} = \left[\frac{Z_e^3}{\pi} \right]^{1/2} \exp[-Z_e r_b], \tag{14.37}$$

respectively, where Z_e is an effective charge that is stipulated prior to the calculation. The χ_{2px} and χ_{2py} expansion functions are the p_x and p_y orbitals defined in Eqs. 12.127A and 12.127B. The mathematical form of these orbitals ensures that $|p_x|^2$ and $|p_y|^2$ have their maximum values along the x- and y-axes, respectively. Equation 14.36 is clearly a linear combination of atomic orbitals (LCAO). The trial eigenfunction λ is called a *molecular orbital* (MO), and the entire procedure is referred to as an LCAO–MO calculation. This approach is the one most frequently employed in molecular orbital calculations.

The use of Eq. 14.36 will produce a 10×10 secular equation whose solution will yield 10 variational energies that we now call *orbital energies*, as well as 10 associated molecular orbitals. While there would be no difficulty carrying out this calculation with a computer, it is instructive to use a smaller expansion so that we can actually execute the calculation here and, in the process, gain an intuitive feel for what is happening in a molecular orbital calculation.

Because the $n = 1$ atomic orbitals are lower in energy than any of the others in our molecular orbital expansion, they will be the major contributors to the H_2^+ ground state. For illustrative purposes, let us take $M = 2$ in Eq. 14.35 and write

$$\lambda = a_1 \chi_{1sA} + a_2 \chi_{1sB}. \tag{14.38}$$

This choice leads to a 2×2 secular equation whose form is

$$\begin{vmatrix} H_{11} - E_v S_{11} & H_{12} - E_v S_{12} \\ H_{21} - E_v S_{21} & H_{22} - E_v S_{22} \end{vmatrix} = 0. \tag{14.39}$$

In this case, the overlap integrals S_{11} and S_{22} are both unity, since χ_1 and χ_2 are individually normalized. Also, we know that $S_{12} = S_{21}$ and that $H_{12} = H_{21}$ because $\mathcal{H}$ is Hermitian and χ_1 and χ_2 are real. Finally, the symmetry of the problem ensures that we will have

$$H_{11} = \langle \chi_1 | \mathcal{H} | \chi_1 \rangle = H_{22} = \langle \chi_2 | \mathcal{H} | \chi_2 \rangle.$$

χ_1 is a $1s$ orbital centered on proton A, whereas χ_2 is a $1s$ orbital centered on proton B. Since the two protons in H_2^+ are chemically equivalent, there cannot be any energetic difference between a $1s$ orbital centered on A and one centered on B. Therefore, we must have $H_{11} = H_{22}$ for this molecule.

If we insert these equalities in the secular equation, we obtain

$$\begin{vmatrix} H_{11} - E_v & H_{12} - E_v S_{12} \\ H_{12} - E_v S_{12} & H_{11} - E_v \end{vmatrix} = 0. \tag{14.40}$$

Expansion of this determinant produces

$$[H_{11} - E_v]^2 = [H_{12} - E_v S_{12}]^2.$$

Taking square roots of both sides, we obtain the two roots for E_v:

$$H_{11} - E_v = \pm [H_{12} - E_v S_{12}]. \tag{14.41}$$

If we use the plus sign on the right-hand side, Eq. 14.41 becomes $H_{11} - H_{12} = E_v[1 - S_{12}]$, so that

$$E_v^- = \frac{H_{11} - H_{12}}{1 - S_{12}}. \tag{14.42}$$

The second root is obtained using the minus sign in Eq. 14.41. This gives

$$E_v^+ = \frac{H_{11} + H_{12}}{1 + S_{12}}. \tag{14.43}$$

The molecular orbitals corresponding to each of these energies are easily obtained by substituting E_v into one of the linear homogeneous equations that produced the secular equation and solving for the ratio a_2/a_1. The result is given by Eq. 14.32:

$$\frac{a_2}{a_1} = \frac{H_{11} - E_v S_{11}}{E_v S_{12} - H_{12}} = \frac{H_{11} - E_v}{E_v S_{12} - H_{12}}. \tag{14.44}$$

Substitution of E_v^+ for E_v produces

$$\left[\frac{a_2}{a_1}\right]_+ = \frac{H_{11} - \dfrac{H_{11} + H_{12}}{1 + S_{12}}}{\dfrac{H_{11} + H_{12}}{1 + S_{12}} S_{12} - H_{12}} = \frac{\dfrac{H_{11} + H_{11}S_{12} - H_{11} - H_{12}}{1 + S_{12}}}{\dfrac{H_{11}S_{12} + H_{12}S_{12} - H_{12} - H_{12}S_{12}}{1 + S_{12}}}$$

$$= \frac{H_{11}S_{12} - H_{12}}{H_{11}S_{12} - H_{12}} = 1.$$

That is, we must have $a_1 = a_2$ when the variational energy is E_v^+ This means that the associated molecular orbital is

$$\lambda_+ = a_1[\chi_1 + \chi_2]. \tag{14.45}$$

The normalization requirement is

$$\langle \lambda_+ | \lambda_+ \rangle = 1 = a_1^2 \langle \chi_1 + \chi_2 | \chi_1 + \chi_2 \rangle = a_1^2 [\langle \chi_1 | \chi_1 \rangle + \langle \chi_2 | \chi_2 \rangle + 2 \langle \chi_1 | \chi_2 \rangle]$$

$$a_1^2 [S_{11} + S_{22} + 2S_{12}] = a_1^2 [1 + 1 + 2S_{12}].$$

Therefore, the normalization constant is $a_1 = [2 + 2S_{12}]^{-1/2}$, and the normalized molecular orbital is

$$\lambda_+ = [2 + 2S_{12}]^{-1/2}[\chi_{1sA} + \chi_{1sB}]$$

$$= [2 + 2S_{12}]^{-1/2}\left[\left(\frac{Z_e^3}{\pi}\right)^{1/2} \exp[-Z_e r_a] + \left(\frac{Z_e^3}{\pi}\right)^{1/2} \exp[-Z_e r_b]\right]. \qquad \textbf{(14.46)}$$

If we execute a similar analysis using the variational energy E_v^-, we find (see Problem 14.14) that the associated normalized molecular orbital is

$$\lambda_- = [2 - 2S_{12}]^{-1/2}[\chi_{1sA} - \chi_{1sB}]$$

$$= [2 - 2S_{12}]^{-1/2}\left[\left(\frac{Z_e^3}{\pi}\right)^{1/2} \exp[-Z_e r_a] - \left(\frac{Z_e^3}{\pi}\right)^{1/2} \exp[-Z_e r_b]\right]. \qquad \textbf{(14.47)}$$

To obtain the actual variational energies for the ground and excited states of H_2^+, we need to execute the H_{11}, H_{12}, and S_{12} integrals as a function of the internuclear distance R and the parameter Z_e in the χ_{1sA} and χ_{1sB} expansion functions. However, we can determine which of our two orbitals, λ_+ or λ_-, corresponds to the ground state and which to the first excited state by simply plotting the values of these orbitals at various points along the Z-axis for some choice of the parameter Z_e and R. Figure 14.10 shows such a plot when $Z_e = 1$ and R is taken to be the H_2^+ equilibrium distance, 2.00 bohr. The λ_+ MO has no nodes and a significant electron density between the positively charged nuclei. In contrast, λ_- has a node at the point $Z = 1.00$ bohr and, consequently, a zero electron density at that point. Therefore, we expect λ_+ to be the ground-state orbital with energy E_v^+ and the excited state to have energy E_v^- with its corresponding MO given by λ_-. Note that both λ_+ and λ_- exhibit cusps at the nuclei where either r_a or r_b is zero. This behavior is the result of the aphysical infinity in the coulombic potential at these points.

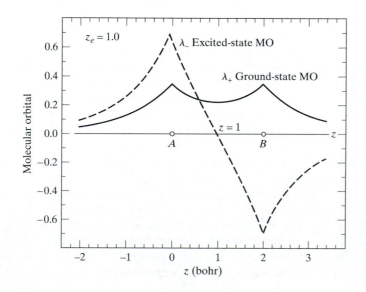

◀ **FIGURE 14.10**
The H_2^+ molecular orbitals for the ground and first excited states obtained from a linear variational calculation using the two-term expansion given in Eq. 14.38 with $Z_e = 1$ and $R = 2$ bohr. The figure shows the values of λ_+ and λ_- at points along the bond axis (Z-axis), whose origin is taken to be at proton A. As can be seen, λ_+ is nodeless and is, therefore, the ground state MO. λ_- has a node midway between the nuclei, at $Z = 1$. Consequently, it represents an approximation of the eigenfunction for the first excited state.

The three integrals H_{11}, H_{12}, and S_{12} can be evaluated in closed analytic form by converting to confocal elliptical coordinates. Since we shall have no further need of this procedure beyond obtaining the solutions to the H_2^+ problem, we present the results without discussing this coordinate system or giving the details of the integration procedure. The results, in atomic units, are

$$H_{11} = \frac{Z_e^2}{2} - Z_e + \frac{(Z_e R + 1)\exp[-2Z_e R]}{R},$$ **(14.48)**

$$H_{12} = -\frac{S_{12}Z_e^2}{2} + (Z_e - 2)(Z_e^2 R + Z_e)\exp[-Z_e R] + \frac{S_{12}}{R},$$ **(14.49)**

and

$$S_{12} = \left[Z_e R + \frac{Z_e^2 R^2}{3} + 1 \right]\exp[-Z_e R].$$ **(14.50)**

Substituting these expressions into Eqs. 14.42 and 14.43 yields E_v^- and E_v^+ as a function of R for any given value of the exponential parameter Z_e.

When R becomes large and the system begins to resemble a proton and a hydrogen atom, we know that the exact eigenfunction has $Z_e = 1$. At other values of R, the best value of Z_e will differ from unity. Figure 14.11 shows the ground- and excited-state energies as a function of R if we take $Z_e = 1$ at all values of R. As we expect, the ground state has energy E_v^+ given by Eq. 14.43, while the excited state has energy E_v^-. The calculations clearly predict that the H_2^+ ion has a chemical bond in the ground state, but not in the first excited state, since there is a minimum at $R = 2.49$ bohr for E_v^+, but no minimum for E_v^-. Therefore, we expect that if we were to excite a ground-state H_2^+ ion to its first excited state, the molecule would spontaneously dissociate. This is precisely what is observed to occur experimentally. Quantitatively, the predicted H_2^+ equilibrium distance of 2.49 bohr is too large by 24.5%. The observed equilibrium bond length is 2.00 bohr. The predicted well depth is the difference between the ground-state energy at $R = 2.49$ bohr ($E_v^+ = -0.56483$ hartree) and that at $R = \infty$ ($E_v^+ = -0.5000$ hartree). This difference of 0.06483 hartree, or 1.76 eV, is 36.8% less than the observed well depth of 2.79 eV. Before we executed the calculations, we knew that the predicted well depth of the chemical bond would be too small. Since the energy at infinite separation is exact and that at $R = 2.49$ bohr must lie above the exact answer because of the variational principle, it follows that the well will be too shallow. However, even this very simple calculation with a two-term expansion for the molecular orbital gives a fair result that correctly predicts the bound nature of the ground state and the unbound character of the first excited state.

If we use Z_e as an adjustable variational parameter, the molecular orbital results are significantly improved. Since Z_e is not a linear parameter, minimization requires a one-dimensional grid search at each value of R. The results of such a series of grid searches at 19 different values of R in the range 0.8 bohr $\leq R \leq 5.5$ bohr are shown as open circles in Figure 14.11. These results lie below those obtained with $Z_e = 1$. The energy always decreases when we use more variational parameters. The potential minimum in the ground state now lies at $R = 2.00$ bohr, which is in near-exact agreement with experiment. This is a general characteristic of molecular orbital calculations; equilibrium distances and angles are usually predicted with excellent accuracy. The new well depth is 0.08650 hartree, or 2.35 eV. This is still too shallow by 0.44 eV, but the percent error has been reduced from 36.8% with

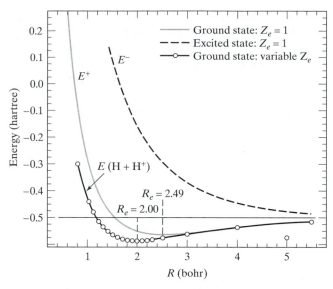

▲ FIGURE 14.11
The ground- and first excited-state energies for H_2^+ as a function of the internuclear distance R with $Z_e = 1$ obtained from a linear variational calculation using the two-term expansion given in Eq. 14.38. The curve connecting the open circles is obtained by treating the effective charge Z_e as a variational parameter and minimizing the variational energy with respect to Z_e at each internuclear distance at which an open circle appears. The values of Z_e obtained in these calculations are shown in Figure 14.12.

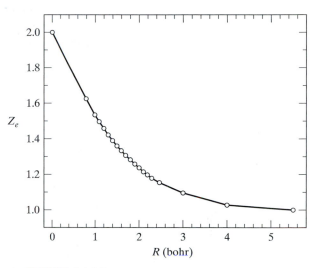

▲ FIGURE 14.12
The value of the effective charge Z_e that minimizes the variational energy for H_2^+, obtained using the trial eigenfunction given in Eq. 14.38. As can be seen, the optimum value of Z_e approaches unity at large values of R, where the system becomes a proton and a hydrogen atom. As R approaches zero, the system begins to resemble a He^+ ion. Therefore, the best value of Z_e approaches 2, the nuclear charge for He^+.

$Z_e = 1$ to 15.8%. It is much more difficult to obtain accurate energies than it is equilibrium distances and angles. Problems 14.17, 14.18, and 14.19 explore other aspects of the simple MO solutions for the H_2^+ ion.

Figure 14.12 shows the variation of Z_{best} as a function of R. As $R \to \infty$, Z_{best} approaches unity, as we knew it must, since the system becomes a hydrogen atom at large R. As R decreases, the value of Z that minimizes the variational energy monotonically increases. When $R = 0$, $Z_{best} = 2$. We were able to predict this result without the need to do variational calculations because, when $R = 0$, the H_2^+ ion becomes a He^+ ion with a nuclear charge $Z = 2$. When the situation is viewed in this context, it is easy to understand why the best effective charge changes as R varies.

Chemists have adopted the practice of designating molecular orbitals formed primarily from pairs of atomic orbitals with the use of a set of nomenclature rules that specify the atomic orbitals involved in the formation of the MO, the number of nodal planes containing the internuclear axis, which is always taken to be the Z-axis, and whether the MO produces a chemical bond. The s and p_z orbitals have $m = 0$, and there is no value of the angle ϕ about the Z-axis for which the MO is always zero. Thus, there are no nodal planes that contain the Z-axis. On the other hand, the p_x and p_y orbitals constructed from linear combinations of the $m = 1$ and $m = -1$ eigenfunctions of the hydrogen atom each have one nodal plane containing the Z-axis. This may be easily seen by examining Eqs. 12.127A and 12.127B, which define the p_x and p_y orbitals. These equations are $p_x = [3/4\pi]^{1/2} \sin\theta \cos\phi$ and $p_y = [3/4\pi]^{1/2} \sin\theta \sin\phi$. When $\phi = \pi/2$ or $3\pi/2$, $\cos\phi = 0$, and the Y–Z plane is a nodal plane that contains the Z-axis. When $\phi = 0$ or π, $\sin\phi = 0$, and the X–Z plane is a nodal plane that contains the Z-axis. If we construct a

MO from atomic orbitals with $m = \pm 2$, there will be two nodal planes containing the Z-axis. Orbitals that have no nodal plane containing the internuclear axis are labeled sigma (σ), those with one nodal plane are designated as pi (π), and those with two are denoted as delta (δ). Two examples are the $d_{x^2-y^2}$ and d_{xy} orbitals, whose ϕ dependencies are $\cos(2\phi)$ and $\sin(2\phi)$, respectively. Figures 14.13 and 14.14 illustrate σ, π, and δ molecular orbitals. Orbitals that have a minimum in the curve of energy versus distance are called *bonding* orbitals. Those with no minimum are designated as *antibonding* and labeled with a superscript asterisk. Table 14.1 summarizes these nomenclature rules. With the use of these rules, the MO we labeled as λ_+ for the H_2^+ ion is more conventionally called a $1s\sigma$ bonding orbital. The λ_- orbital that contains no energy minimum is the $1s\sigma^*$ antibonding orbital. Since the single electron in H_2^+ will go into the orbital with the lowest energy, we expect the electronic configuration of H_2^+ to be $(1s\sigma)^1$: one electron in a $1s$ sigma bonding MO.

Table 14.1	Molecular orbital nomenclature		
Atomic Orbitals	m	**Nodal Planes**	**Designation**
s	0	0	σ
p_z	0	0	σ
p_x or p_y	± 1	1	π
d_{zx} or d_{zy}	± 1	1	π
$d_{x^2-y^2}$	± 2	2	δ
d_{xy}	± 2	2	δ

14.3.2   Qualitative Treatment of Many-electron Diatomic Molecules

Suppose we were to carry out a molecular orbital calculation on the H_2^+ ion using the 10-term expansion given in Eq. 14.36. Without actually integrating or solving the 10×10 secular equation, we can deduce the qualitative nature of the solutions by taking into account a few general principles:

1. The lower energy atomic orbitals will tend to produce the lower energy molecular orbitals.

2. High-energy atomic orbitals do not usually mix significantly with much lower energy orbitals to form molecular orbitals. The atomic orbitals that make the major contributions to a molecular orbital tend to have similar energies.

3. A molecular orbital constructed from a particular pair of atomic orbitals that contains nodes will have a higher energy than one without nodes.

4. σ, π, and δ atomic orbitals do not mix to form diatomic molecular orbitals.

With these points in mind, we expect the lowest energy MO from our 10-term expansion to be primarily a bonding combination of the two $1s$ orbitals, since these are the lowest energy atomic orbitals in the expansion. Our treat-

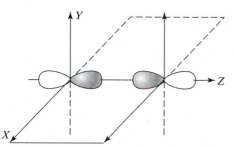

$\chi_{1sA} + \chi_{1sB} = A \ 1s\sigma$ orbital

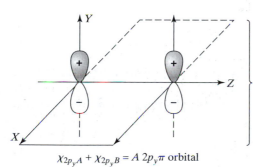

$\chi_{2p_zA} - \chi_{2p_zB} = A \ 2p_z\sigma$ orbital

No nodal planes containing the internuclear axis (Z)

$\chi_{2p_yA} + \chi_{2p_yB} = A \ 2p_y\pi$ orbital

One nodal planes containing the internuclear axis (Z)

◀ FIGURE 14.13
Reflection symmetry of sigma and pi molecular orbitals. A sigma orbital has no nodal planets containing the internuclear axis. A pi MO has one such nodal plane.

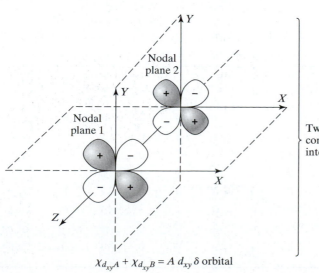

Nodal plane 2

Nodal plane 1

$\chi_{d_{xy}A} + \chi_{d_{xy}B} = A \ d_{xy} \ \delta$ orbital

Two nodal planes containing the internuclear axis (Z)

◀ FIGURE 14.14
Reflection symmetry of delta molecular orbitals. A delta MO has two nodal planes of reflection symmetry containing the internuclear axis.

ment of the H_2^+ ion further indicates that a bonding MO is produced when the variational coefficients a_1 and a_2 multiplying the χ_{1sA} and χ_{1sB} orbitals are equal. The remaining coefficients will be very small in magnitude, because the energies of their corresponding atomic orbitals are significantly larger than that for the 1s orbitals. In fact, the nomenclature rules summarized in Table 14.1 imply that some of these other coefficients must be zero. If it were possible for some of the atomic orbitals in a molecular orbital to have no nodal planes containing the internuclear axis while others have one or two, how could we label the orbital? The fact that these nomenclature rules have been adopted suggests that such mixing does not occur. That is indeed the case: Sigma, pi, and delta orbitals do not mix to form diatomic molecular orbitals. If the coefficients multiplying the atomic orbitals that form σ molecular orbitals are nonzero, those that would form π orbitals must have zero coefficients. For this reason, the coefficients a_7, a_8, a_9, and a_{10} that multiply p_x and p_y orbitals on nuclei A and B must be zero in a molecular orbital that has a_1 and a_2 nonzero. Since the 2s and $2p_z$ orbitals also combine to form σ orbitals, a_3, a_4, a_5, and a_6 can have nonzero values in an orbital in which a_1 and a_2 are both large. These expectations are reflected in Table 14.2.

Our examination of the H_2^+ ion shows that we can form one more orbital from the 1s atomic orbitals by letting $a_1 = -a_2$. As Figure 14.10 shows, this combination produces a node halfway between the two nuclei. Figure 14.11 makes it clear that this orbital has antibonding character and a higher energy than the $1s\sigma$ MO. Therefore, it is appropriate to label it $1s\sigma^*$. Again, the coefficients a_3, a_4, a_5, and a_6 can have nonzero values, but a_7, a_8, a_9, and a_{10} must be zero.

There are no other combinations of the 1s atomic orbitals that we can form to produce another MO. Therefore, the MO with the next lowest energy must make use of the 2s orbitals, which we found in Chapter 13 to have a lower energy than the 2p orbitals when shielding effects are taken into account. The 2s orbitals can be combined in the same manner as the 1s orbitals to produce a lower energy $2s\sigma$ bonding orbital and a higher energy $2s\sigma^*$ antibonding MO.

Table 14.2   Qualitative form of the molecular orbitals for H_2^+ obtained from the 10-term expansion given in Eq. 14.36. The notations + and − denote variational coefficients of equal and large magnitude, but different sign. The notation s indicates coefficients whose magnitudes are small relative to the ones designated as + or −. A zero entry means that the variational coefficient is zero.

a_1	a_2	a_3	a_4	a_5	a_6	a_7	a_8	a_9	a_{10}	Orbital
+	+	s	s	s	s	0	0	0	0	$1s\sigma$
+	−	s	s	s	s	0	0	0	0	$1s\sigma^*$
s	s	+	+	s	s	0	0	0	0	$2s\sigma$
s	s	+	−	s	s	0	0	0	0	$2s\sigma^*$
s	s	s	s	+	−	0	0	0	0	$2p_z\sigma$
0	0	0	0	0	0	+	+	0	0	$2p_x\pi$
0	0	0	0	0	0	0	0	+	+	$2p_y\pi$
0	0	0	0	0	0	+	−	0	0	$2p_x\pi^*$
0	0	0	0	0	0	0	0	+	−	$2p_y\pi^*$
s	s	s	s	+	+	0	0	0	0	$2p_z\sigma^*$

Again, we know that a_7, a_8, a_9, and a_{10} must be zero and that a_1, a_2, a_5, and a_6 have small magnitudes, since the energies of the $1s$ and $2p_z$ orbitals are significantly different from that for the $2s$ orbitals.

In the atom, the p_z, p_y, and p_x orbitals are degenerate. In a molecule, however, this is not the case. The p_z orbital, whose electron density resides primarily between the positively charged nuclei, produces a more stable situation than either a p_x or p_y orbital, each of which has a nodal plane containing the bond axis. Consequently, these orbitals place no electron density along the bond axis. Therefore, we expect the atomic orbitals primarily involved in producing those MOs whose energies lie just above that for the $2s\sigma^*$ orbital to be χ_{2pzA} and χ_{2pzB}. To produce a combination of these atomic orbitals that contains no node along the bond axis, we must form the difference of χ_{2pzA} and χ_{2pzB}. The reason for this can be seen by reference to Figure 14.15. If χ_{2pzA} and χ_{2pzB} are added, the plus and minus lobes of the two orbitals combine to produce a node halfway between nuclei A and B. If they are subtracted, this does not occur. Hence, the $2p_z\sigma$ bonding MO has $a_5 = -a_6$, as indicated in Table 14.2.

At this point, our qualitative analysis fails us. We might expect to find that the antibonding $2p_z\sigma^*$ MO is lower in energy than the $2p\pi$ orbitals, but this is not the case. Determination of this fact requires that we employ more powerful, quantitative methods discussed in the next section. Calculations of this type show that the pi orbitals formed by combining p_x and p_y orbitals are lower in energy than the $2p_z\sigma^*$ MO. Bonding $2p_x\pi$ orbitals are formed by adding the atomic χ_{2pxA} and χ_{2pxB} atomic orbitals, while antibonding $2p_x\pi$ orbitals are formed by subtracting those same atomic orbitals, as illustrated in Figure 14.16. Since the p_x and p_y atomic orbitals are degenerate for diatomic molecules, the corresponding $2p_x\pi$ and $2p_y\pi$ bonding MOs are also degenerate, as are the $2p_x\pi^*$ and $2p_y\pi^*$ antibonding MOs. Because the atomic orbitals that form σ and π MOs do not mix, we expect the coefficients a_1 through a_6 to be zero for π molecular orbitals.

For the H_2^+ ion, the energy ordering of the molecular orbitals listed in Table 14.2 and for those of higher energy is $E(1s\sigma) < E(1s\sigma^*) < E(2s\sigma) < E(2s\sigma^*) < E(2p_z\sigma) < E(2p_x\pi) = E(2p_y\pi) < E(2p_x\pi^*) = E(2p_y\pi^*) < E(2p_z\sigma^*) < E(3s\sigma) < E(3s\sigma^*) < E(3p_z\sigma) < E(3p_x\pi) = E(3p_y\pi) < E(3p_x\pi^*) = E(3p_y\pi^*) < E(3p_z\sigma^*)$. When more than one electron is present, this ordering is sometimes altered. The situation is similar to the one we observed in our examination of atomic electronic configurations, where we often found exceptions to the expected order of filling of the available orbitals.

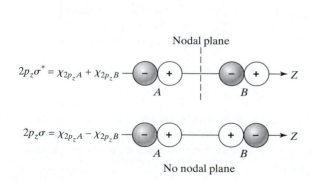

▲ FIGURE 14.15
Bonding and antibonding combinations of $2p_z$ orbitals.

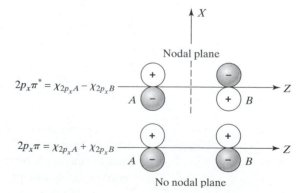

▲ FIGURE 14.16
Bonding and antibonding combinations of $2p_x$ orbitals.

We can now qualitatively predict molecular electronic configurations for homonuclear diatomics in a manner analogous to that employed to obtain atomic configurations. Each MO can be converted to a molecular spin orbital by multiplication by a spin function, either α or β. The total molecular wave function must be antisymmetric to electron exchange, so we cannot have the same molecular spin orbital for two electrons without obtaining the trivial solution of the Schrödinger equation. As a result, the Pauli principle applies, and each MO can contain a maximum of two electrons. Nature will fill the MOs in order of increasing energy, so that molecular electronic configurations can be qualitatively predicted, at least to the extent that the preceding energy ordering holds.

EXAMPLE 14.6

Predict ground-state molecular electronic configurations for the following diatomic molecules: **(A)** H_2; **(B)** He_2; **(C)** O_2; **(D)** F_2. **(E)** Describe, in semiquantitative terms, the ground-state wave function for He_2.

Solution

(A) H_2 has two electrons. Consequently, they will both enter the $1s\sigma$ bonding MO, one with spin α, the other with spin β, to give the configuration $1s\sigma\alpha(1)\,1s\sigma\beta(2)$, or, for brevity, $(1s\sigma)^2$.

(B) He_2 has four electrons. The first two will be in the $1s\sigma$ bonding MO with paired spins. The second two must go into the $1s\sigma^*$ MO with paired spins. The configuration is, therefore, $(1s\sigma)^2(1s\sigma^*)^2$.

(C) O_2 has 16 electrons, which will require at least 8 MOs to accommodate. Filling the orbitals in order of increasing energy, we obtain $(1s\sigma)^2(1s\sigma^*)^2(2s\sigma)^2(2s\sigma^*)^2$ $(2p_z\sigma)^2(2p_x\pi)^2(2p_y\pi)^2(2p_x\pi^*)^1(2p_y\pi^*)^1$ as the electronic configuration. Note that Hund's rules still apply. The electrons occupy different degenerate orbitals if permitted by the Pauli principle, so as to minimize the mutual repulsion of electrons.

(D) F_2 has 18 electrons. Therefore, the nine MOs of lowest energy will be filled. These MOs are $(1s\sigma)^2(1s\sigma^*)^2(2s\sigma)^2(2s\sigma^*)^2(2p_z\sigma)^2(2p_x\pi)^2(2p_y\pi)^2(2p_x\pi^*)^2(2p_y\pi^*)^2$.

(E) The electronic configuration for He_2 obtained in (B) is $(1s\sigma)^2(1s\sigma^*)^2$. The four spin molecular orbitals are $1s\sigma\alpha$, $1s\sigma\beta$, $1s\sigma^*\alpha$, and $1s\sigma^*\beta$. The requirement that the overall wave function be antisymmetric with respect to electron exchange means that the ground-state wave function must be written as a 4×4 Slater determinant whose form is

$$\psi(1, 2, 3, 4) = (1/4!)^{1/2} \begin{vmatrix} 1s\sigma\alpha(1) & 1s\sigma\alpha(2) & 1s\sigma\alpha(3) & 1s\sigma\alpha(4) \\ 1s\sigma\beta(1) & 1s\sigma\beta(2) & 1s\sigma\beta(3) & 1s\sigma\beta(4) \\ 1s\sigma^*\alpha(1) & 1s\sigma^*\alpha(2) & 1s\sigma^*\alpha(3) & 1s\sigma^*\alpha(4) \\ 1s\alpha^*\beta(1) & 1s\alpha^*\beta(2) & 1s\alpha^*\beta(3) & 1s\alpha^*\beta(4) \end{vmatrix}.$$

For related exercises, see Problems 14.20 and 14.21.

Our treatment of H_2^+ indicates that electrons in bonding molecular orbitals tend to enhance the formation of a chemical bond, whereas electrons in antibonding orbitals tend to retard bond formation. It is, therefore, convenient to define the bond order as

$$\rho = \frac{(\text{no. bonding electrons}) - (\text{no. antibonding electrons})}{2}. \qquad \textbf{(14.51)}$$

Qualitatively, the larger the bond order, the stronger we expect the chemical bond to be.

EXAMPLE 14.7

Determine the bond order for the four diatomic molecules in Example 14.6.

Solution

The electronic structure of H_2 is $(1s\sigma)^2$. Two of the electrons are in bonding orbitals and none are in antibonding orbitals. This gives a bond order $\rho = (2 - 0)/2 = 1$, and we say that H_2 exhibits a single bond.

The electronic structure of He_2 is $(1s\sigma)^2(1s\sigma^*)^2$. Two of the electrons are in bonding orbitals and two are in antibonding orbitals. The bond order is $\rho = (2 - 2)/2 = 0$, and we say that He_2 is not stable. Helium is a monatomic gas.

The electronic configuration of O_2 is $(1s\sigma)^2(1s\sigma^*)^2(2s\sigma)^2(2s\sigma^*)^2(2p_z\sigma)^2(2p_x\pi)^2$ $(2p_y\pi)^2(2p_x\pi^*)^1(2p_y\pi^*)^1$. There are 10 electrons in bonding orbitals and 6 in anti-bonding MOs. Therefore, $\rho = (10 - 6)/2 = 2$, and we say that O_2 has a double bond.

The electronic configuration of F_2 is $(1s\sigma)^2(1s\sigma^*)^2(2s\sigma)^2(2s\sigma^*)^2(2p_z\sigma)^2(2p_x\pi)^2$ $(2p_y\pi)^2(2p_x\pi^*)^2(2p_y\pi^*)^2$. There are 10 electrons in bonding orbitals and 8 in anti-bonding MOs. This gives $\rho = (10 - 8)/2 = 1$, and we say that F_2 has a single chemical bond.

The same type of qualitative analysis can be carried out for heteronuclear diatomics. In this case, however, qualitative principle number 2 becomes very important. Since the atoms are now different, the corresponding orbitals on the two centers may have very different energies and, therefore, may tend not to mix in the formation of molecular orbitals. Consider, for example, the HF molecule. The higher nuclear charge on a fluorine atom makes the $1s$ orbital in fluorine much lower in energy than the $1s$ orbital in hydrogen. As a result, we do not expect to see a molecular orbital in which the coefficients multiplying these two $1s$ orbitals both have large magnitudes. The energy of the hydrogen $1s$ orbital is much closer to that of the $2s$ and $2p_z$ fluorine orbitals. Consequently, we expect a MO to form with major contributions from these three orbitals. The qualitative correspondence between the atomic orbital energies and the energies of the molecular orbitals for HF is shown in Figure 14.17. Such a presentation is called a *correlation diagram*.

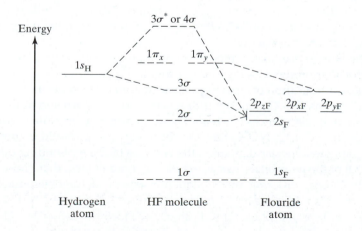

◀ FIGURE 14.17
Qualitative correlation diagram for HF.

Using the effective charges given in Table 13.5, we may easily determine that the energy of the 1s orbital in flourine ($1s_F$) lies far below that for the hydrogen 1s orbital ($1s_H$). The $2p_F$ orbitals are degenerate, and their energy and that for $2s_F$ are also significantly below that for the $1s_H$ orbital. However, the energy match for $2s_F$, $2p_F$, and $1s_H$ is the best available for HF. If we were to carry out a molecular orbital calculation for HF using the six-term LCAO–MO expansion

$$\lambda = a_1\, \chi_{1sF} + a_2\, \chi_{2sF} + a_3\, \chi_{2pzF} + a_4\, \chi_{2pxF} + a_5\, \chi_{2pyF} + a_6\, \chi_{1sH},$$

we would expect six molecular orbitals to be produced. The lowest of these, 1σ, would be essentially a $1s_F$ orbital with $|a_1|$ large, $|a_2|$, $|a_3|$, and $|a_6|$ very small, and $a_4 = a_5 = 0$. The MO of next higher energy would be primarily a combination of the $2s_F$ and $2p_{zF}$ orbitals. $|a_2|$ and $|a_3|$ would be large, $|a_1|$ and $|a_6|$ would be very small, and $a_4 = a_5 = 0$. This mixture of the $2s_F$ and $2p_{zF}$ atomic orbitals is often called an *sp hybrid orbital*. Figure 14.18 illustrates how the 2s and $2p_z$ orbitals combine to form the *sp* hybrid.

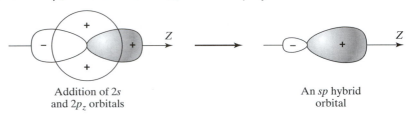

Addition of 2s
and $2p_z$ orbitals

An *sp* hybrid
orbital

The next higher energy MO would be a bonding combination of an *sp* hybrid on fluorine and χ_{1sH} to produce the 3σ MO in which $|a_2|$, $|a_3|$, and $|a_6|$ are large, $|a_1|$ is small, and $a_4 = a_5 = 0$ because π and σ orbitals do not mix. χ_{2pxF} and χ_{2pyF} would form two degenerate π molecular orbitals, in each of which $a_1 = a_2 = a_3 = a_6 = 0$. The $1\pi_x$ MO would be identical to χ_{2pxF}, with $|a_4|$ large and $|a_5|$ small. $1\pi_y$ would have variational coefficients identical to those of $1\pi_x$, except that a_4 and a_5 would be reversed. The MO with the highest energy would be the antibonding combination of a $\chi_{2sF}-\chi_{2pzF}$ hybrid and χ_{1sH} that has a node along the H–F bond axis. It is not obvious that 4σ has a higher energy than the 1π molecular orbitals. To be certain of this fact, we would have to conduct calculations of the type described in the next section.

The foregoing description indicates that the electronic configuration for HF is $(1\sigma)^2(2\sigma)^2(3\sigma)^2(1\pi_x)^2(1\pi_y)^2$, which is often written as $(1\sigma)^2(2\sigma)^2(3\sigma)^2(1\pi)^4$, with the two degenerate π orbitals written as one. It is also customary to omit the asterisk denoting antibonding from orbitals in dealing with heteronuclear diatomics or polyatomic molecules. Thus, you will usually see the notation 4σ instead of $3\sigma^*$; we simply number the sigma and pi orbitals in order of increasing energy.

If the atomic orbital energies of the two atoms are similar, the homonuclear molecular orbitals can be used to provide a qualitative description of the electronic configuration of a heteronuclear diatomic molecule. In this context, "similar" means that the energy differences between corresponding orbitals are less than the differences between the orbitals on a given atom or ion. Some examples are NO, CO, NO^+, and CO^-. In doing this, it should be noted that, usually, for heteronuclear diatomics, the energies of the $p\pi$ bonding orbitals lie below that of the $p\sigma$ orbitals, just as the $p\pi^*$ energies are less than those for $p\sigma^*$. Thus, the energy ordering most frequently observed for heteronuclear diatomics is $E(1s\sigma) < E(1s\sigma^*) < E(2s\sigma) < E(2s\sigma^*) < E(2p_x\pi) = E(2p_y\pi) < E(2p_z\sigma) < E(2p_x\pi^*) = E(2p_y\pi^*) < E(2p_z\sigma^*) < E(3s\sigma) < E(3s\sigma^*) < E(3p_x\pi) =$

Slater ex. pg 281

1986

passé
evry

$E(3p_y\pi) < E(3p_z\sigma) < E(3p_x\pi^*) = E(3p_y\pi^*) < E(3p_z\sigma^*)$. Example 14.8 applies this procedure to NO.

EXAMPLE 14.8

Using the fact that the atomic orbital energies for N and O are similar, develop a qualitative description of the molecular electronic structure of the NO molecule.

Solution

NO contains 15 electrons. Therefore, we require at least eight molecular orbitals for the ground state. If the energies of the atomic orbitals for N and O are similar, we can employ the homonuclear MOs with the qualitative energy ordering for heteronuclear diatomics. This produces the configuration

$$(1s\sigma)^2(1s\sigma^*)^2(2s\sigma)^2(2s\sigma^*)^2(2p_x\pi)^2(2p_y\pi)^2(2p_z\sigma)^2(2p_x\pi^*)^1. \qquad \textbf{(1)}$$

The NO bond order is $\rho = (10 - 5)/2 = 2.5$. Because the diatomic is heteronuclear, this configuration is usually written in the form

$$(1\sigma)^2(2\sigma)^2(3\sigma)^2(4\sigma)^2(1\pi)^4(5\sigma)^2(2\pi)^1. \qquad \textbf{(2)}$$

For additional exercises, see Problems 14.22 and 14.23.

14.3.3   Molecular Term Symbols for Diatomics

The energy for a given molecular electronic configuration of a diatomic molecule is dependent upon the angular momentum state of the system in much the same way as is the case for atoms. The ground-state electronic configuration for an oxygen atom, $1s^2 2s^2 2p^4$, gives rise to three Pauli-allowed angular momentum states, ^3P, ^1D, and ^1S. These states differ significantly in energy because of the effect of electron–electron repulsion terms in the Hamiltonian. The same is true for diatomic molecules.

We characterize the angular momentum state of a diatomic molecule by examining the quantum number associated with the total Z component of angular momentum. Since the Z components of angular momentum of the individual electrons add as scalars, we have

$$[L_Z]_{\text{total}} = M\hbar = m_1\hbar + m_2\hbar + m_3\hbar + \ldots,$$

so that

$$M = m_1 + m_2 + m_3 + \ldots, \qquad \textbf{(14.52)}$$

where M is the total magnetic quantum number and the m_i are the magnetic quantum numbers for the occupied molecular orbitals. The total spin angular momentum state is represented by the multiplicity, $2S + 1$, where S is the total spin quantum number. The values of $|M|$ and the multiplicity are given by a molecular term symbol whose form is similar to that used for atoms, except for the substitution of capital Greek letters for the S, P, D, F, etc., notation. Thus, a molecular term symbol has the form $^{2S+1}|M|$, where $|M|$ is represented by a Greek letter given in Table 14.3.

Because the total spin and magnetic quantum numbers for filled shells are always zero, we need only consider the contributions from unfilled orbitals in the determination of $|M|$ and S. As with atoms, Hund's rules still apply. The angular momentum states with maximum S will be the lowest in energy. For

Table 14.3   Angular momentum designations

$|M|$	Greek Letter
0	Σ
1	Π
2	Δ
3	Φ

states with equal S, the state with the maximum value of $|M|$ will have the lowest energy. Thus, H_2^+, whose configuration is $(1s\sigma)^1$, has $M = 0$ (since this value of M characterizes a sigma orbital) and $S = 1/2$. The term symbol is, therefore, $^2\Sigma$. An oxygen molecule with the ground-state configuration determined in Example 14.6, viz.,

$$(1s\sigma)^2(1s\sigma^*)^2(2s\sigma)^2(2s\sigma^*)^2(2p_z\sigma)^2(2p_x\pi)^2(2p_y\pi)^2(2p_x\pi^*)^1(2p_y\pi^*)^1,$$

can have $|M| = 2$ if both electrons in the $2p_x\pi^*$ and $2p_y\pi^*$ MOs have $m = 1$ or $m = -1$. In this case, the spins of the two electrons would have to be paired to ensure that their quantum numbers are not identical. Thus, if $|M| = 2$, we have $S = 0$. On the other hand, if one of these electrons has $m = 1$ while the other has $m = -1$, so that $|M| = 0$, we can have either $S = 0$ or $S = 1$. Therefore, the possible angular momentum states for O_2 are $^1\Delta$, $^1\Sigma$, and $^3\Sigma$. Using Hund's rules, we know that the relative energies of these states are $E(^3\Sigma) < E(^1\Delta) < E(^1\Sigma)$, so that the ground state is $^3\Sigma$.

Molecular term symbols are also used to convey information concerning the symmetry of the molecular orbitals involved in the electronic configuration. As we shall see in Chapter 15, such symmetry information plays an important role in determining the spectroscopy of the molecule. For homonuclear diatomics, two types of symmetry information are included. The first of these is symmetry with respect to inversion of the orbitals through the origin. Such inversion converts the Cartesian coordinates (x, y, z) into $(-x, -y, -z)$. The point of interest is what happens to the molecular orbitals when this replacement is made. Figure 14.19 shows the inversion operation for sigma and pi orbitals for homonuclear diatomics. In each case, the operation converts point P to point P'. For a $1s\sigma$ bonding orbital, a positive point P is transformed into a positive point P', so the molecular orbital does not change sign. It is, therefore, symmetric with respect to inversion through the origin. Such symmetric orbitals are denoted with a subscript g, which stands for the German *gerade* ("even"), and the orbital notation becomes $1s\sigma_g$. In contrast, inversion for the $1s\sigma^*$ antibonding MO transforms a negative point P into a positive point P'. The orbital is, therefore, antisymmetric (*ungerade*, or "odd") with respect to inversion. This property is denoted by a subscript u, so that the notation is $1s\sigma_u^*$. The *gerade–ungerade* symmetries of the common molecular orbitals are summarized in Table 14.4.

If the molecular electronic configuration contains an odd number of electrons in MOs with u symmetry, the overall wave function will have u symmetry. If the number of electrons in *ungerade* orbitals is even, the total symmetry will be g. This is a consequence of the fact that when we invert the wave function through the origin, each electron in an *ungerade* orbital produces a sign change. An even number of such sign changes results in no sign change on the total wave function; an odd number will change the sign on the total wave function.

Table 14.4	*Gerade–ungerade* symmetries of simple homonuclear diatomic MOs		
MO	**Symmetry**	**MO**	**Symmetry**
$s\sigma$	g	$p_z\sigma^*$	u
$s\sigma^*$	u	$p_x\pi$ or $p_y\pi$	u
$p_z\sigma$	g	$p_x\pi^*$ or $p_y\pi^*$	g

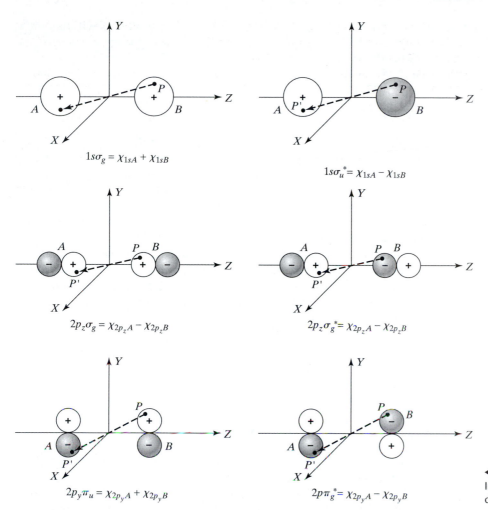

$$1s\sigma_g = \chi_{1sA} + \chi_{1sB}$$

$$1s\sigma_u^* = \chi_{1sA} - \chi_{1sB}$$

$$2p_z\sigma_g = \chi_{2p_zA} - \chi_{2p_zB}$$

$$2p_z\sigma_g^* = \chi_{2p_zA} - \chi_{2p_zB}$$

$$2p_y\pi_u = \chi_{2p_yA} + \chi_{2p_yB}$$

$$2p\pi_g^* = \chi_{2p_yA} - \chi_{2p_yB}$$

◀ FIGURE 14.19
Inversion symmetry of simple diatomic molecular orbitals.

The second type of symmetry conveyed by the term symbol involves whether the total wave function changes sign $(-)$ or does not change sign $(+)$ upon reflection in a plane that contains both nuclei. Such a reflection causes the angle ϕ to be replaced with $-\phi$, as shown in Figure 14.20. The part of the atomic orbitals involving the variable ϕ is $\Phi(\phi) = (2\pi)^{-1/2} e^{-im\phi}$. Therefore, replacing ϕ with $-\phi$ is equivalent to replacing m with $-m$. Because of this property, sigma orbitals, which are constructed from atomic orbitals with $m = 0$, undergo no change upon reflection. This can be seen in Figure 14.13, which shows the symmetry of sigma molecular orbitals with respect to reflection in a plane containing both nuclei. However, π_{+1} and π_{-1} orbitals, which comprise atomic orbitals with $m = 1$ or $m = -1$, respectively, will undergo the transformation $\pi_{+1} \longleftrightarrow \pi_{-1}$. That is, π_{+1} will be converted into π_{-1} and vice versa. Consequently, if the molecular electronic configuration is $(\pi_{+1})^1(\pi_{-1})^1$, reflection through a plane containing the Z-axis is equivalent to exchanging electrons in these two π orbitals. We know that the total wave function must be antisymmetric with respect to electron exchange. Therefore, if the antisymmetry is due to an antisymmetric spatial part, exchanging the electrons in the $(\pi_{+1})^1$ and $(\pi_{-1})^1$ orbitals will produce an overall sign change on the wave function. If, however, the antisymmetry is produced by the spin part of the wave function, no change will occur upon reflection. In Eq. 13.76, we found that singlet spin functions are antisymmetric with respect to electron exchange, while the spatial

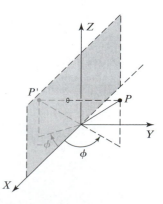

▲ FIGURE 14.20
Reflection through a plane containing the Z-axis. This operation converts the angle ϕ into $-\phi$.

Table 14.5   Configuration symmetries with respect to reflection through a plane containing the Z-axis

Configuration	Spin State	Symmetry	Configuration	Spin State	Symmetry
σ^1	any	+	π^2	triplet	−
σ^2	any	+	π^4	any	+
π^2	singlet	+	π^3	any	none

part of the wave function is symmetric. For triplet states, the opposite is true. We can, therefore, conclude that the configuration $(\pi_{+1})^1(\pi_{-1})^1$ will produce a sign change of the wave function upon reflection through a plane containing the Z-axis if the spin state is triplet, but not if it is singlet. Configurations containing two electrons in each π orbital, $(\pi_{+1})^2(\pi_{-1})^2$, will produce either two sign changes upon reflection or none. In either case, the wave function remains unaltered. If the configuration does not produce a sigma state, there is no symmetry with respect to reflection through a plane containing the Z-axis. For example, the $^2\Pi$ state of O_2^+ has the configuration

$$(1s\sigma_g)^2(1s\sigma_u{}^*)^2(2s\sigma_g)^2(2s\sigma_u{}^*)^2(2p_z\sigma_g)^2(2p_{+1}\pi_u)^2(2p_{-1}\pi_u)^2(2p_{+1}\pi_g{}^*)^1.$$

Upon reflection, this configuration becomes

$$(1s\sigma_g)^2(1s\sigma_u{}^*)^2(2s\sigma_g)^2(2s\sigma_u{}^*)^2(2p_z\sigma_g)^2(2p_{+1}\pi_u)^2(2p_{-1}\pi_u)^2(2p_{-1}\pi_g{}^*)^1,$$

which is neither symmetric nor antisymmetric. Consequently, the question of (+) or (−) symmetry is moot. Table 14.5 summarizes the rules for reflection symmetry through a plane containing the Z-axis.

The symmetry with respect to inversion through the origin is specified by a g or u right-hand subscript on the molecular term symbol. The symmetry with respect to reflection through a plane containing the Z-axis is given by a (+) or (−) right-hand superscript. If there is no symmetry, the corresponding subscript or superscript is omitted. Thus, Π and Δ states have no reflection symmetry, and the (+) or (−) superscript is omitted. Heteronuclear diatomics have no inversion symmetry, and the g or u subscript is omitted.

EXAMPLE 14.9

Determine the molecular term symbol for the ground state of each of the following diatomic molecules: **(A)** H_2; **(B)** O_2; **(C)** NO. Stipulate the appropriate symmetry elements on the term symbol.

Solution

(A) The H_2 electronic configuration is $(1s\sigma_g)^2$. Both electrons have $m = 0$, so that $M = 0$, and the orbital angular momentum state is Σ. The spins are paired to satisfy the Pauli principle. Hence, we have a singlet state. Figure 14.19 shows that the $1s\sigma$ MO has g symmetry. There are an even number (0) of electrons in *ungerade* MOs. Therefore, the total symmetry is g. Table 14.5, along with the associated analysis, shows that sigma orbitals always have (+) symmetry. Therefore, we obtain $^1\Sigma_g^+$ for the H_2 ground state.

(B) The O_2 ground-state electronic configuration is

$$(1s\sigma_g)^2(1s\sigma_u{}^*)^2(2s\sigma_g)^2(2s\sigma_u{}^*)^2(2p_z\sigma_g)^2(2p_x\pi_u)^2(2p_y\pi_u)^2(2p_x\pi_g{}^*)^1(2p_y\pi_g{}^*)^1.$$

To get the angular momentum state with the lowest energy, we need maximum multiplicity. That is produced when we have the two electrons in the $2p_x\pi_g{}^*$ and $2p_y\pi_g{}^*$ antibonding orbitals with the same spin function, giving a total spin quantum number of 1 and a multiplicity of 3. This requires that the electrons have m values of $+1$ and -1, so that $|M| = 0$, which gives us a Σ state. The MOs with *ungerade* symmetry are the $1s\sigma_u{}^*$, $2s\sigma_u{}^*$, $2p_x\pi_u$, and $2p_y\pi_u$ orbitals. These molecular orbitals contain a total of eight electrons, an even number, so the total symmetry with respect to inversion is g. Reflection through a plane containing the Z-axis effectively exchanges the electrons in the $p_x\pi$ and $p_y\pi$ orbitals. Since the spin state is triplet, the spin function is symmetric with respect to electron exchange, forcing the spatial part of the wave function to be antisymmetric. As a result, each $p_x\pi \longleftrightarrow p_y\pi$ exchange produces a sign change. In the stated configuration, three such exchanges will occur, so that the total reflection symmetry will be $(-)$. The ground-state O_2 term symbol is, therefore, $^3\Sigma_g^-$.

(C) The NO ground-state electronic configuration is

$$(1s\sigma)^2(1s\sigma^*)^2(2s\sigma)^2(2s\sigma^*)^2(2p_x\pi)^2(2p_y\pi)^2(2p_z\sigma)^2(2p_x\pi^*)^1.$$

We might also write this configuration in the form $(1\sigma)^2(2\sigma)^2(3\sigma)^2$ $(4\sigma)^2(1\pi)^4(5\sigma)^2(2\pi)^1$. The filled orbitals make no contribution to the angular momentum state. Therefore, the one electron in the unfilled $2p_x\pi^*$ or 2π orbital will produce $|M| = 1$ and $S = 1/2$, giving us a $^2\Pi$ state. There is no reflection or inversion symmetry present, so this is the total term symbol.

For related exercises, see Problems 14.24 and 14.25.

14.3.4   Self-consistent-field LCAO–MO Calculations

Accurate molecular orbital calculations for large many-electron molecules are complex and difficult. However, two factors work in our favor: The molecular Hamiltonian, although complex, always has a well-known form, and the mathematical operations that must be conducted are always the same. For these reasons, the task can be programmed for execution on modern, high-speed computers. All that the scientist or engineer must do is provide the computer with certain critical information in the form of a data file and then know how to interpret the output provided by the program. The most commonly used programs to execute such calculations are various versions of GAUSSIAN and GAMES. Most universities and research installations have these software packages installed on their mainframe computers. In this section, we will investigate the theory underlying these computer codes and discuss the interpretation of their output by means of an actual example.

Let us first examine the form of the molecular Hamiltonian for a molecule containing M nuclei and N electrons. We always make the Born–Oppenheimer approximation and omit the kinetic-energy terms for the nuclei. In addition, it is assumed that the magnetic spin–orbit coupling terms are very small and may be omitted. This assumption means that we cannot accurately study the electronic structure of heavy elements $(Z \geq 40)$ unless we include the spin–orbit terms and do the calculations in a manner that takes relativistic effects into account. Fortunately, most of the molecules of interest to chemists do not contain these heavier elements in which the spin–orbit interactions are large. The Hamiltonian will contain N, ∇^2 terms for the kinetic energies of the N electrons. There will be $M \times N$ nuclear–electron attraction terms present and $N(N-1)/2$ electron–electron repulsion terms. Finally, we will have $M(M-1)/2$ nuclear–nuclear repulsion terms.

EXAMPLE 14.10

Write down the Hamiltonian for the H_2O molecule. Which are the one-electron, two-electron, and zero-electron terms? How many of each type are there?

Solution

There are 10 electrons and three nuclei present in H_2O. The Hamiltonian, in atomic units, is

$$\mathcal{H} = -\frac{1}{2} \sum_{i=1}^{10} \nabla_i^2 - \sum_{\alpha=1}^{3} \sum_{i=1}^{10} \frac{Z_\alpha}{r_{i\alpha}} + \sum_{i=1}^{9} \sum_{j=i+1}^{10} \frac{1}{r_{ij}} + \sum_{\alpha=1}^{2} \sum_{\beta=\alpha+1}^{3} \frac{Z_\alpha Z_\beta}{R_{\alpha\beta}}. \tag{1}$$

In Eq. 1, Greek summation indices are used for the nuclei and lowercase English letters for the electrons, r_{ij} is the distance between electrons i and j, $R_{\alpha\beta}$ is the distance between nuclei α and β, and Z_α and Z_β are the charges on nuclei α and β, respectively. If we label the hydrogen nuclei in H_2O as 2 and 3 and oxygen as 1, we have $Z_2 = Z_3 = 1$ and $Z_1 = 8$.

There are 10 kinetic energy terms for the 10 electrons and 30 electron–nuclear attraction terms, as each of the 10 electrons is attracted to each of the three nuclei. These 40 terms are contained in the first two summations. The $(10)(9)/2 = 45$ electron–electron repulsion terms are contained in the second double summation. These are the two-electron terms. As explained in Chapter 13, the use of this form of the double sum prevents counting the repulsions twice. The $(3)(2)/2 = 3$ nuclear–nuclear repulsion terms are in the third double summation. These three terms, which do not depend upon the coordinates of any electrons, are the zero-electron terms. In all, the Hamiltonian contains 88 terms.

For a related exercise, see Problem 14.26.

The essential steps in the variational solution of the molecular Schrödinger equation are qualitatively the same as those we used to treat the hydrogen-molecular ion. First, we select a linear variational expansion for the trial wave function. This expansion is called the *basis set*. Second, we compute all of the integrals in the H_{ij} and S_{ij} matrix elements that go into the secular equation. Third, the secular equation is solved to obtain all the variational or orbital energies. Fourth, the orbital energies are substituted back into the linear equations that produced the secular equation. The linear equations are then solved, and the molecular orbitals corresponding to each of the orbital energies are obtained. Finally, the total ground-state energy of the system is calculated with the use of the orbital energies and the integral values computed in the second step.

The major difficulty associated with executing the above procedure occurs in the second step: The integrals involving the electron–electron repulsion terms cannot be integrated analytically if we use exponential atomic orbitals in the basis set, yet such exponential functions are precisely the type we wish to employ. The solution to this problem is to approximate exponential Slater-type orbitals with linear expansions of Gaussian functions that are fitted to the desired STOs by least–squares methods. (Gaussian functions contain the square of the variable in the exponential.) These fits need be done only once and the results included in the computer programs. When the exponential STOs are replaced with Gaussian expansions, all integrals can be done in closed analytic form. This brings the computational time required into an acceptable range for many molecules of interest.

To illustrate the fitting procedure, let us suppose that we wish to use a $1s$ STO in our trial eigenfunction. In atomic units, this orbital has the form

$\chi_{1s} = \pi^{-1/2}\exp(-r)$. The use of χ_{1s} in our trial eigenfunction will result in our having to compute the required H_{ij} integrals numerically. To avoid this, we write χ_{1s} as an expansion of four Gaussian functions:

$$\chi_{1s} = \sum_{i=1}^{4} a_i \exp[-b_i r^2].$$

Here, the a_i and b_i ($i = 1, 2, 3, 4$) are fitting coefficients whose values are obtained by least-squares methods. In practice, as many as six Gaussian functions are used in the fitting expansion for the STOs. Figure 14.21 shows the quality of the fit that can be obtained with these methods. In this case, the fit is so good that plots of χ_{1s} and the expansion are essentially indistinguishable. Therefore, we show the expansion as a series of points, with the line being the χ_{1s} STO. The sum of the square deviations of the fit from the χ_{1s} STO over the range $0 \leq r \leq 5$ bohr is 2.4×10^{-4} for 1,000 equally-spaced points.

There is a minor problem in the third step. For reasons that are not obvious unless we take the time to develop all of the equations, it turns out that the H_{ij} matrix elements contain the variational coefficients whose values we seek. Since we do not know these coefficients in advance, we cannot calculate H_{ij}. The solution is conceptually the same as that which we employed in Chapter 9 to solve ionic equilibrium problems. There, we needed the activity coefficients to compute the equilibrium concentrations, but without the concentrations in advance, we could not compute the activity coefficients. Our solution to this problem was to guess the values of the activity coefficients, solve for the equilibrium concentrations, compute new activity coefficients, and then iterate until we converged to a consistent solution. We adopt the same strategy in molecular orbital calculations: We guess the values of the variational coefficients, compute the H_{ij}, solve the secular equation to obtain new values for the coefficients, and then iterate with the new values until a consistent solution is obtained. For this reason, the procedure is called a *self-consistent-field LCAO–MO* method or, for brevity, *SCF–LCAO–MO*. Sometimes, it is simply termed an SCF calculation.

The information that must be supplied to a molecular orbital program consists of (1) a coded specification of the basis set to be used, (2) the multiplicity needed to specify the angular momentum state, (3) the charge, if any, on the molecule, (4) the atoms that are present in the molecule, and (5) the

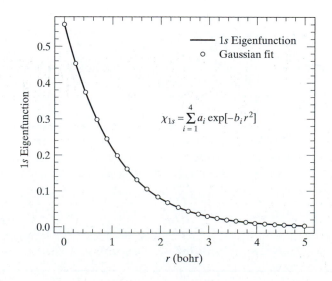

◀ FIGURE 14.21

Comparison of a $1s$ Slater-type orbital (STO) with a four-term Gaussian expansion fitted to the STO using least-square methods. The Gaussian expansion is $\chi_{1s} = \sum_{i=1}^{4} a_i \exp[-b_i r^2]$. A nonlinear least-squares fitting procedure yields the following values for the fitting coefficients: $a_1 = 0.118497$, $a_2 = 0.0694646$, $a_3 = 0.14884$, and $a_4 = 0.221668$. The exponential parameters are $b_1 = 0.156435$, $b_2 = 77.8347$, $b_3 = 4.88369$, and $b_4 = 0.735092$. Plots of χ_{1s} and $\sum_{i=1}^{4} a_i \exp[-b_i r^2]$ cannot be distinguished; therefore, the Gaussian fit is shown as a series of points. The solid line is χ_{1s}.

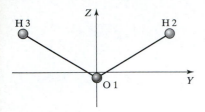

▲ FIGURE 14.22
H_2O orientation and atom labeling. The center of mass lies at the origin, with the molecule in the Y–Z plane.

coordinates of the nuclei of the atoms. Let us now examine the actual results of an SCF–LCAO–MO calculation on the H_2O molecule whose Hamiltonian was developed in Example 14.10.

Figure 14.22 shows the orientation we specified for the molecule in our input to the program. H_2O was placed in the Y–Z plane, with the oxygen atom labeled as Atom 1. The hydrogen atom with positive Y- and Z-coordinates was specified as Atom 2 and the second hydrogen as Atom 3. The atoms were arranged symmetrically about the Z-axis as shown, with the center of mass at the origin. The two O–H bond lengths were 0.96 Å, and the H–O–H bond angle was 104.5°. In our first calculation, we used a basis set consisting of only those orbitals occupied in the atomic ground states of oxygen and hydrogen. Such a basis set is often called a *minimal basis set*. The general form for our molecular orbitals was, therefore,

$$\lambda = a_1\chi_{1sO} + a_2\chi_{2sO} + a_3\chi_{2pxO} + a_4\chi_{2pyO} + a_5\chi_{2pzO} + a_6\chi_{1sH2} + a_7\chi_{1sH3}.$$
(14.53)

With seven expansion functions in the LCAO, we will obtain seven orbital energies and seven molecular orbitals. These results are given in Table 14.6, which shows that the MO with the lowest energy is λ_1, with an orbital energy $\varepsilon_1 = -20.2421$ hartrees. Only the coefficient a_1 multiplying χ_{1sO} makes a significant contribution to this molecular orbital. (Remember, it is the square of the expansion coefficient that determines the relative importance of the expansion function.) λ_1 is essentially just a $1s$ orbital centered on the oxygen atom. We anticipated that this would be the case because the energy of the $1s$ oxygen orbital is much lower than that of the other orbitals in the calculation.

The next-lowest-energy orbital is λ_2, which the results show to be essentially an oxygen $2s$ orbital. The relative absolute squares of the expansion coefficients are 0.054238, 0.69642, 0.01655, and 0.02513 for the χ_{1sO}, χ_{2sO}, χ_{2pzO}, and hydrogen orbitals, respectively. $|a_2|^2$ for χ_{2sO} is an order of magnitude larger than the other absolute squares.

The molecular orbitals λ_3 and λ_4 are primarily responsible for forming the two OH bonds in the molecule. λ_3 combines the oxygen $2p_y$ orbital with the two $1s$ hydrogen orbitals in such a way that there are no nodes along the OH bond axes. The X–Z plane is a nodal plane, but this plane does not intersect the bond axes. Consequently, we expect this MO to function as a bonding orbital. Figure 14.23 illustrates the point. Notice that the negative lobe of χ_{2pyO} points

Table 14.6   Molecular orbitals and orbital energies for H_2O obtained from SCF–LCAO–MO calculations using the minimal basis set given in Eq. 14.53

Orbital	Molecular Orbitals						
Energies (ε), hartrees	-20.24210	-1.26700	-0.61642	-0.45270	-0.39107	0.60292	0.73902
AO	λ_1	λ_2	λ_3	λ_4	λ_5	λ_6	λ_7
$a_1\,\chi_{1sO}$	0.99414	-0.23289	0.00000	-0.10304	0.00000	-0.13199	0.00000
$a_2\,\chi_{2sO}$	0.02650	0.83452	0.00000	0.53601	0.00000	0.87982	0.00000
$a_3\,\chi_{2pxO}$	0.00000	0.00000	0.00000	0.00000	1.00000	0.00000	0.00000
$a_4\,\chi_{2pyO}$	0.00000	0.00000	-0.60636	0.00000	0.00000	0.00000	0.98893
$a_5\,\chi_{2pzO}$	0.00432	0.12864	0.00000	-0.77606	0.00000	0.74225	0.00000
$a_6\,\chi_{1sH2}$	-0.00594	0.15853	-0.44520	-0.27897	0.00000	-0.79510	-0.83615
$a_7\,\chi_{1sH3}$	-0.00594	0.15853	0.44520	-0.27897	0.00000	-0.79510	0.83615

λ_3 $\varepsilon_3 = -0.61642$ hartree

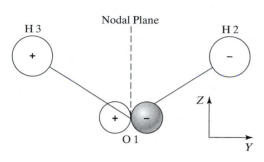

λ_4 $\varepsilon_4 = -0.45270$ hartree

$\lambda_3 = -0.60636\,\chi_{2p_yO} + 0.44520[-\chi_{1sH2} + \chi_{1sH3}]$

$\lambda_4 = -0.10304\,\chi_{1sO} + 0.53601\,\chi_{2sO} - 0.77606\,\chi_{2p_zO} - 0.27897[\chi_{1sH2} + \chi_{1sH3}]$

▲ **FIGURE 14.23**
Molecular orbitals λ_3 and λ_4 for H_2O, calculated with the use of a minimal basis set.

toward the hydrogen $1s$ orbital with the negative expansion coefficient, while the positive lobe points toward the hydrogen orbital with the positive coefficient. As a result, there are no nodes along the two O–H bond axes.

λ_4 functions as a second bonding MO. This MO combines an sp hybrid on the oxygen atom with the two $1s$ hydrogen orbitals in such a manner that no node exists along either of the O–H bond axes. The sp hybrid orbital produced by the combination $0.53601\chi_{2sO} - 0.77606\chi_{2p_zO}$ results in an orbital whose qualitative form is illustrated in the figure. The overlap of the negative lobe of the $2p_z$ orbital with the positive $2sO$ orbital produces a small negative lobe in the positive Z direction. The overlap of the positive lobe with the $2sO$ orbital yields a much larger positive lobe in the negative Z direction. (See Figure 14.18.) The coefficients multiplying both $1s$ hydrogen orbitals are negative. As a result, the negative portions of the orbitals on oxygen and hydrogen point toward each other, so that no nodes along the bond axes are produced.

λ_5 is a π orbital perpendicular to the molecular plane. It involves only the $2p_{xO}$ orbital. Consequently, this orbital holds a lone electron pair on the oxygen atom and does not contribute appreciably to the formation of stable O–H bonds. The remaining two orbitals, λ_6 and λ_7, are much higher in energy. An examination of the properties of these orbitals in a manner analogous to that shown in Figure 14.23 clearly demonstrates why they are high-energy antibonding orbitals. (See Problems 14.28 and 14.29.)

We no longer need an approximate set of rules for the energy ordering of the orbitals: The calculation gives us quantitative orbital energies, so it is obvious which orbitals will be filled by the electrons. For H_2O, the 10 electrons will occupy MOs λ_1 through λ_5 with paired spins, so the total spin will be zero and the multiplicity unity. This is expressed by writing the electronic configuration as $(\lambda_1)^2(\lambda_2)^2(\lambda_3)^2(\lambda_4)^2(\lambda_5)^2$. Molecular orbitals λ_6 and λ_7 are present, but empty. Such empty orbitals are called *virtual orbitals.* The antisymmetrized wave function is therefore a 10×10 Slater determinant in which the 10 spin orbitals are λ_1 through λ_5, each combined with spin functions α and β. This Slater determinant for the ground state is called the *Hartree–Fock configuration,* or the HF configuration for brevity.

The total energy of the system must be computed with care. The electronic contribution to the total energy can be obtained by summing the orbital energies for each electron, provided that we then subtract the total electron–electron repulsion energy that has already been computed during the calculations. This subtraction is necessary because the orbital energies each contain the repulsion terms for the electron in that orbital with respect to all the remaining electrons in the system. The summation of orbital energies, therefore, counts the electron–electron repulsion terms twice. This is easily seen by considering the case of H_2. With two electrons, its configuration would be $(\lambda_1)^2$. The orbital energy for λ_1 counts the kinetic energy of the electron in that orbital, its attractive interaction with the two nuclei, and the repulsion term with the second electron. When we sum the two orbital energies $\varepsilon_1 + \varepsilon_1$, we have properly counted two kinetic-energy terms and two sets of electron–nuclear attraction terms, but we have included the single electron–electron repulsion term twice. For this reason, the total electronic energy is given by

$$E_{\text{electronic}} = \sum_{i=1}^{N} \varepsilon_i - \text{total electron–electron repulsion energy.} \quad \textbf{(14.54)}$$

The total energy is then obtained by adding the nuclear–nuclear repulsion energy to $E_{\text{electronic}}$:

$$E_{\text{total}} = E_{\text{electronic}} + E_{\text{nuclear repulsion}}. \quad \textbf{(14.55)}$$

For H_2O, using the minimal basis set described by Eq. 14.53, we find that the results given by the GAUSSIAN program are $E_{\text{electronic}} = -84.1315097983$ hartrees and $E_{\text{nuclear repulsion}} = 9.1681906860$ hartrees. These data permit us to compute the total electron–electron repulsion energy if we wish. Using Eq. 14.54, we see that this repulsion energy is the difference between the sum of the orbital energies for the 10 electrons and the total electronic energy. Using the orbital energies given in Table 14.6 and taking into account the fact that each orbital contains two electrons, we compute

$$E_{\text{electron–electron repulsion}} =$$
$$2[-20.24210 - 1.26700 - 0.61642 - 0.45270 - 0.39107]$$
$$- (-84.1315097983) = 38.19293 \text{ hartrees.}$$

The ionization potential for H_2O can be computed either by performing a second SCF–LCAO–MO calculation on H_2O^+ and subtracting the two total energies or by using Koopman's theorem. This approximation says that $I_1 \approx -\varepsilon_{\text{homo}}$, where $\varepsilon_{\text{homo}}$ is the orbital energy of the highest-energy occupied MO. In the present case, this gives $I_1 \approx 0.39107$ hartree $= 10.64$ eV. The measured first ionization potential for H_2O is 12.62 eV. Thus, the result obtained from the minimal basis set calculation is low by 15.69%.

These results can be significantly improved by using a larger basis set in the calculations. For example, if we write λ as a 13-term expansion that includes the $1s$, $2s$, $2p_x$, $2p_y$, $2p_z$, $3s$, $3p_x$, $3p_y$, and $3p_z$ orbitals on oxygen and the $1s$ and $2s$ orbitals on each hydrogen atom, we obtain a total H_2O energy of -85.1521309710 hartrees. Since this energy is lower than that obtained with the minimal basis set, the variational principle guarantees that it is closer to the exact answer. With this basis set, the orbital energy for λ_5 is -0.50124 hartree, so that the ionization potential predicted using Koopman's approximation is now 13.64 eV. The error in this result is 8.08%, which is about half that for the minimal basis set.

It is a relatively simple matter to obtain equilibrium geometries using SCF methods. For example, if we calculate the H_2O energy at different H–O–H angles, using a 25-term basis set with the O–H bond distance fixed at 0.96 Å, the results shown in Figure 14.24 are obtained. The minimum-energy structure corresponds to an angle of 105.35°. The experimentally measured H–O–H angle reported in the 78*th* edition of the *Handbook of Chemistry and*

Physics is 104.51°. We see that the agreement with experiment is very good: The difference between the calculated and measured results is only 0.84°. As we found in our calculations on H_2^+, equilibrium bond lengths and angles are accurately predicted by molecular orbital calculations. Accurate energies and ionization potentials are more difficult to obtain.

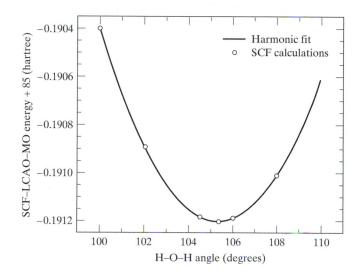

◀ **FIGURE 14.24**
SCF calculation of the equilibrium angle for H_2O. In this calculation, a 25-term basis set was used with an O–H bond length of 0.96 Å. Each point of the graph represents the results obtained in a single SCF–LCAO–MO calculation at the H–O–H bond angle given on the abscissa. The ordinate gives the resulting total SCF energy + 85 hartrees. The curve is a least-squares fit of a harmonic potential of the form $V(\theta) = k_b(\theta - \theta_0)^2$, where θ is the H–O–H angle. The predicted equilibrium angle, $\theta°$, is 105.35°.

14.3.5   Configuration Interaction Calculations

In an SCF calculation, the final ground-state wave function is written in terms of a Slater determinant in which the electrons occupy the lowest energy molecular orbitals obtained in the calculation, subject to the constraints imposed by the Pauli principle. This arrangement of electrons is called the *Hartree–Fock (HF) configuration*. In many cases, the highest–energy occupied molecular orbital (HOMO) has an energy just slightly below that for the lowest–energy unoccupied molecular orbital (LUMO). In such a situation, we would expect electronic configurations in which electrons are excited to the LUMO or other MOs of similar energy to make important contributions to the ground-state wave function. This concept is simply an extension of the general rule that atomic orbitals of similar energies tend to mix to form molecular orbitals. Here, entire electronic configurations tend to mix to form the ground-state wave function. Each configuration is represented by an appropriate Slater determinant so that we can write the total wave function in the form

$$\psi(1, 2, 3, \ldots, N) = \sum_{i=0}^{K} C_i D_i \qquad (14.56)$$

where the D_i are different $N \times N$ Slater determinants representing K + 1 different electronic configurations (with D_0 being the HF configuration) and the C_i are linear variational coefficients. When the final wave functions and corresponding energy eigenvalues are obtained from a variational calculation using Eq. 14.56 as the trial eigenfunction, the procedure is called a *configuration interaction (CI)* calculation.

A CI calculation is usually conducted using the following procedure: We first execute an SCF–LCAO–MO calculation using an M-term basis set. This yields M molecular orbitals and $2M$ spin orbitals. The HF configuration is obtained by assigning the N electrons to the lowest energy N spin orbitals, which leaves $2M - N$ virtual spin orbitals. Additional electronic configurations are generated by promoting one of the electrons in the HF configuration to one of the virtual spin orbitals, being careful to keep the multiplicity of the new configuration the same as that of the HF configuration. Since we have excited one electron in this process, the new configurations generated are called *single excitations* or, simply, *singles*. Let us use our minimal-basis-set SCF calculation on H_2O as an example. The HF configuration is $(\lambda_1)^2(\lambda_2)^2(\lambda_3)^2(\lambda_4)^2(\lambda_5)^2$, with λ_6 and λ_7 empty virtual orbitals. We could generate two additional configurations by promoting one of the electrons in λ_5 to either λ_6 or λ_7 to form $D_1 = (\lambda_1)^2(\lambda_2)^2(\lambda_3)^2(\lambda_4)^2(\lambda_5)^1(\lambda_6)^1$ and $D_2 = (\lambda_1)^2(\lambda_2)^2(\lambda_3)^2(\lambda_4)^2(\lambda_5)^1(\lambda_7)^1$. Both of these configurations are single excitations into the virtual orbitals. If we form new configurations by exciting two electrons from the HF configuration, these new configurations are called *double excitations* or *doubles*. Additional excitations are labeled *triples, quadruples*, and so on. If we execute the calculation using all possible configurations of the same multiplicity that can be formed from the $2M$ spin orbitals, the procedure is called a *full CI calculation*. When a sufficiently large basis set is used in the LCAO expansion, the accuracy of the results from a full CI calculation will be comparable to that which can be obtained experimentally.

Equation 14.56 contains a $(K + 1)$-term expansion of the wave function. This expansion leads to a $(K + 1, K + 1)$ secular equation that produces $K + 1$ energy eigenvalues and $K + 1$ eigenfunctions. The lowest of these eigenvalues is the best approximation for the lowest energy state of the multiplicity used in the calculations. As expected, it is an upper limit to the exact energy eigenvalue. The remaining K variational energies are upper limits to the successive excited states of that multiplicity.

14.3.6 Møller–Plesset Perturbation Theory

SCF calculations yield accurate equilibrium structures, but only fair results for bond energies, ionization potentials, and potential barriers to chemical reactions that we shall examine in more detail in Chapters 19 and 20. Configuration interaction methods, in contrast, produce very accurate energies. Unfortunately, CI methods require extremely large amounts of computational time, even on the most powerful computers, when the number of electrons in the system becomes large (≥ 50). For this reason, it has become common practice to revert to perturbation methods to improve the SCF results.

The most commonly used perturbation method is Møller–Plesset theory. In this approach, we first carry out an SCF–LCAO–MO calculation using an extended basis set. The molecular orbitals resulting from these calculations are regarded as the zeroth-order eigenfunctions and the HF energy as the zeroth-order energy. Second-, third-, fourth-, and fifth-order perturbation corrections to this energy are then computed by executing integrals that involve the SCF molecular orbitals, including the virtual orbitals. Since the SCF molecular orbitals are linear combinations of the basis set functions, the actual integrals involve the Gaussian functions that have been employed to represent the STOs in the SCF calculation. The final ground-state energy is

$$E_{\text{ground state}} = E_{\text{HF}} + E^{(2)} + E^{(3)} + E^{(4)} + E^{(5)} + \ldots, \qquad \textbf{(14.57)}$$

where E_{HF} is the Hartree–Fock energy and the $E^{(i)}$ are the successive Møller–Plesset perturbation corrections. In principle, an infinite number of such corrections could be added, but in practice we usually add only $E^{(2)}$, $E^{(2)} + E^{(3)}$, $E^{(2)} + E^{(3)} + E^{(4)}$, or $E^{(2)} + E^{(3)} + E^{(4)} + E^{(5)}$. Such calculations are called MP2, MP3, MP4, and MP5 methods, respectively.

14.4   Valence-bond Methods

Although the large majority of molecular electronic structure calculations are done using some form of molecular orbital theory, most of our qualitative thinking about chemical bonding employs a valence-bond (VB) model in which a chemical bond is viewed as a pairing of electrons between atomic orbitals located on the bonded atoms. The pairing can be produced by unpaired electrons on each atom combining to form an electron pair with opposing spin functions (a covalent bond) or by the donation of an electron pair on one atom into an empty orbital on a second atom (a coordinate covalent bond). In either case, the electrons are viewed as being localized between the bonded atoms, rather than being spread over the entire molecule in a molecular orbital. Thus, when a chemist says that the bonding in benzene is one of the forms shown in Figure 14.25, he or she is using a valence-bond model that has the electrons localized between bonded centers. Since VB models are rarely employed in quantitative calculations, we shall give only a qualitative description of how this electron-pairing model translates into a quantum mechanical representation of the electronic structure.

▶ FIGURE 14.25

Valence-bond structures of benzene. Structures I and II are the Kekule structures, and III, IV, and V are the Dewar structures.

I   II   III   IV   V

The basis of the VB method is the *perfect-pairing* assumption. This approximation assumes that we can generate a good representation of the molecular eigenfunction by executing the following four-step procedure:

1. Form a set of atomic orbitals (AOs) on each atom into which the bonding pair of electrons will be placed. This set may be a single orbital or an LCAO.

2. Form spin orbitals from the sets of AOs in such a manner that the bonded atoms have paired spins.

3. Form all possible combinations of these spin orbitals using Slater determinants.

4. Combine the Slater determinants that have the required spin assignments in an appropriate linear combination that represents the desired bonding scheme.

This procedure is best understood by applying it to actual examples. Let us first consider the H_2 molecule, with the two atoms labeled A and B.

The first step is to choose a set of atomic orbitals on each atom that will provide the orbital into which the bonded pair of electrons is placed. For two hydrogen atoms, the obvious choice is $1s$ AOs on both atoms, but we could choose a more complex LCAO. Let a and b represent the set chosen for atoms A and B, respectively. In Step 2, we form spin orbitals such that the bonded atoms have paired spins. The spin-pairing requirement gives us two possible choices: $\lambda_1 = a\alpha$ and $\lambda_2 = b\beta$, or $\lambda_1 = a\beta$ and $\lambda_2 = b\alpha$. We now write Slater determinants for each of these choices. The two possibilities are

$$D_1 = (2)^{-1/2} \begin{vmatrix} a\alpha(1) & a\alpha(2) \\ b\beta(1) & b\beta(2) \end{vmatrix} = |a\alpha b\beta|$$

and

$$D_2 = (2)^{-1/2} \begin{vmatrix} a\beta(1) & a\beta(2) \\ b\alpha(1) & b\alpha(2) \end{vmatrix} = |a\beta b\alpha|, \tag{14.58}$$

where we have introduced the notation $|a\alpha b\beta|$ and $|a\beta b\alpha|$ for the Slater determinants to simplify the notation. In the final step, we combine D_1 and D_2 in a way that represents the desired chemical bonding scheme. With two determinants, we have two possibilities:

$$\psi_1^{VB} = N[D_1 - D_2] \tag{14.59}$$

or

$$\psi_2^{VB} = N[D_1 + D_2]. \tag{14.60}$$

We now need to examine the properties of ψ_1^{VB} and ψ_2^{VB} to determine which we wish to use as our approximate VB eigenfunction. If we expand D_1 and D_2, Eq. 14.59 becomes

$$\psi_1^{VB} = \frac{N}{(2)^{1/2}}[a(1)\alpha(1)b(2)\beta(2) - b(1)\beta(1)a(2)\alpha(2) -$$

$$a(1)\beta(1)b(2)\alpha(2) + b(1)\alpha(1)a(2)\beta(2)]$$

$$= \frac{N}{(2)^{1/2}}[a(1)b(2) + b(1)a(2)][\alpha(1)\beta(2) - \beta(1)\alpha(2)], \tag{14.61}$$

and Eq. 14.60 becomes

$$\psi_2^{VB} = \frac{N}{(2)^{1/2}}[a(1)b(2) - b(1)a(2)][\alpha(1)\beta(2) + \beta(1)\alpha(2)]. \tag{14.62}$$

Regardless of what we have chosen for a and b, we know that the symmetry of the problem requires that a and b be identical except for the fact that the orbitals are on different atoms. Therefore, the difference $a(1)b(2) - b(1)a(2)$ appearing in Eq. 14.62 will have a node halfway between the bonded atoms. That orbital will, therefore, be an antibonding orbital or, in VB nomenclature, a *no bond* orbital. The sum $a(1)b(2) + b(1)a(2)$ in Eq. 14.61 will be nodeless and, therefore, produce a *valence bond.* This result is general. If we wish to have a bonded VB structure between two atoms, we take the difference of the Slater determinants with spins exchanged. A no-bond structure is produced by the sum. When there is more than one bond, we multiply the Slater determinant by $(-1)^n$, where n is the number of spin exchanges between the Slater determinants. A no-bond structure is always produced by the sum of Slater determinants.

The relationship between a simple MO description of the bonding in H_2 and a VB model may be seen by comparing the MO and VB eigenfunctions. The MO electronic configuration is $(1s\sigma)^2$, where $1s\sigma\alpha = N[a + b]\alpha$ and $1s\sigma\beta = N[a + b]\beta$. The approximate MO eigenfunction is

$$\psi_{MO} = N(2)^{-1/2} \begin{vmatrix} [a + b]\alpha(1) & [a + b]\alpha(2) \\ [a + b]\beta(1) & [a + b]\beta(2) \end{vmatrix}$$

$$= \frac{N}{2^{1/2}}\{[a(1) + b(1)][a(2) + b(2)]\}\{\alpha(1)\beta(2) - \beta(1)\alpha(2)\}$$

$$= \frac{N}{2^{1/2}}[\{a(1)b(2) + b(1)a(2)\} + \{a(1)a(2) + b(1)b(2)\}][\alpha(1)\beta(2) - \beta(1)\alpha(2)]. \tag{14.63}$$

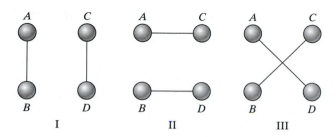

Possible VB structures for a four-atom, four-electron system

The underlined term in Eq. 14.63 corresponds to the approximate VB eigenfunction. However, the MO description also includes a second term in the spatial part of the wave function that represents an ionic state in which both electrons are either on atom A or on atom B. If ionic contributions are important, we would expect the simple MO description to provide a better approximation to the exact energy and eigenfunction. If such contributions are not important, the simple VB model will be superior. For H_2, in which ionic contributions are negligible, the VB model is better. We generally do not use the VB model, however, in quantitative calculations, because the nonorthogonality of the a and b orbitals makes such calculations computationally very time consuming.

If more than one bond is involved in the VB structure, the process becomes more complicated, but the overall procedure remains the same. Consider the case of four atoms, A, B, C, and D, each containing a single unpaired electron whose bonding we wish to describe using a VB model. There are three possible bonding schemes, as shown in Figure 14.26. If a, b, c, and d are the LCAO sets on atoms A, B, C, and D, respectively, the perfect-pairing procedure permits us to write down six possible pairings of the electrons. Notice that the number of possible pairings is just the number of combinations of four things taken two at a time: $C(4, 2) = 4!/(2!\,2!) = 6$. The six Slater determinants for these pairings are

$$D_1 = |a\alpha\,b\beta\,c\alpha\,d\beta|,$$

$$D_2 = |a\beta\,b\alpha\,c\alpha\,d\beta|,$$

$$D_3 = |a\alpha\,b\alpha\,c\beta\,d\beta|,$$

$$D_4 = |a\beta\,b\beta\,c\alpha\,d\alpha|,$$

$$D_5 = |a\beta\,b\alpha\,c\beta\,d\alpha|,$$

and

$$D_6 = |a\alpha\,b\beta\,c\beta\,d\alpha|.$$

To represent bonding structure I, we must have A–B spins and C–D spins paired. This is the case for Slater determinants D_1, D_2, D_5, and D_6. If we take the sign on D_1 to be positive, the sign on D_2 must be negative, since we have interchanged the A–B spins. Therefore, with one pair exchanged, we have $n = 1$ and $(-1)^n = -1$. The sign on D_5 will be positive, because both A–B and C–D spins have been exchanged and $n = 2$, so that $(-1)^2 = 1$. Finally, the sign on D_6 will be negative, since only the C–D spins have been exchanged. The VB wave function for structure I is, therefore,

$$\psi_I^{VB} = [D_1 - D_2 + D_5 - D_6].$$

For bonding structure II, the A–C and B–D spins must be paired. The result is

$$\psi_{II}^{VB} = [D_2 - D_3 - D_4 + D_6].$$

For structure III, in which the A–D and B–C spins are paired,

$$\psi_{III}^{VB} = [D_1 - D_3 - D_4 + D_5].$$

In general, the number of Slater determinants required to represent a VB structure is 2^m, where m is the number of electron pair bonds or no-bonds involved in the structure. For the current four-atom, four-electron system, there are two bonds. Therefore, we need $2^2 = 4$ Slater determinants to represent the VB eigenfunction for each possible structure.

The total VB wave function is now written as a linear combination of each of these structures, and a variational calculation is carried out to obtain the minimum possible energy. That is, we write

$$\psi_{total}^{VB} = C_I \psi_I^{VB} + C_{II} \psi_{II}^{VB} + C_{III} \psi_{III}^{VB}. \tag{14.64}$$

The calculation is actually easier than this analysis suggests. We can write the wave function for Structure III as the sum of the wave functions for Structures I and II; that is,

$$\psi_{III}^{VB} = \psi_I^{VB} + \psi_{II}^{VB} = [D_1 - D_2 + D_5 - D_6] + [D_2 - D_3 - D_4 + D_6]$$

$$= [D_1 - D_3 - D_4 + D_5]. \tag{14.65}$$

Combining Eqs. 14.64 and 14.65 produces

$$\psi_{total}^{VB} = C_I \psi_I^{VB} + C_{II} \psi_{II}^{VB} + C_{III}[\psi_I^{VB} + \psi_{II}^{VB}]$$

$$= [C_I + C_{III}]\psi_I^{VB} + [C_{II} + C_{III}]\psi_{II}^{VB} = B_I \psi_I^{VB} + B_{II} \psi_{II}^{VB}, \tag{14.66}$$

where B_I and B_{II} now serve as the variational coefficients. The VB structure with crossed bonds is already included when one considers a linear combination of all VB structures that do not have crossed bonds. This is sometimes called the *VB no-crossing rule*. Have you ever wondered why organic chemists consider the five benzene structures shown in Figure 14.26, but not the structures shown in the margin? Now you know: The no-crossing rule ensures us that these bonding structures have already been included when we consider just the Kekule and Dewar structures shown in Figure 14.26.

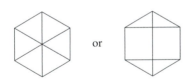

If we completed the variational calculation using Eq. 14.66 as a trial eigenfunction, the magnitudes of B_I and B_{II} would give us the relative importance of structures I and II in the overall bonding in the molecule. If A, B, C, and D form a square, we will obtain $B_I = B_{II}$. In this case, we say that Structures I and II are *resonance forms*. The difference between the energy obtained using Eq. 14.66 and that when ψ is either ψ_I^{VB} or ψ_{II}^{VB} alone is called the resonance stabilization energy. If $B_I > B_{II}$, we say that Structure II contributes to the $ABCD$ bonding, but not to the same extent as structure I; it is not a resonance structure with I.

14.5   Semiempirical Methods

All of the theories and techniques discussed in the previous two sections are classified as *ab initio* methods. This designation means that, for the Hamiltonian employed in the calculation, no simplifying assumptions are present other than those inherent in any variational procedure. The integrals and solutions of the secular equation are executed to the accuracy permitted by the computer being used.

There is a second class of quantum mechanical procedures that we label semiempirical methods. In these methods, the general form of a molecular orbital, variational calculation is retained, but many of the integrals are arbitrarily set to zero. Others are replaced by parameters that are adjusted to force the results to fit measured data of various types. This procedure makes the choice of the LCAO expansion functions irrelevant, since the integrals over the basis set functions are never actually executed. The advantages of these procedures include a substantial reduction in the required computation time and, consequently, an increase in the ability to execute quantum calculations on much larger molecules. In addition, if the calibration of the parameters that are substituted for the integrals is carefully done, the results can be surprisingly accurate. The major disadvantages are the loss of the variational principle (because the integrals are not computed) and the danger of completely spurious results. For example, some semiempirical methods predict CH_5 to be a stable molecule. If care is taken not to employ these methods for molecular structures significantly different from those used in the calibration of the parameters, this danger is almost eliminated.

The various semiempirical procedures differ in what terms are eliminated from the Hamiltonian, in the manner in which the integrals are treated, and in the calibration procedures employed to adjust the parameters. Most of these methods are designated by acronyms that sometimes indicate the nature of the approximations being made. Some examples are CNDO (complete neglect of differential overlap); INDO (intermediate neglect of differential overlap); MINDO (modified neglect of differential overlap); MNDO (modified neglect of diatomic overlap); MINDO/1, MINDO/2, MINDO/2', and MINDO/3, which are different versions of MINDO; PM3 (parametric method 3); AM1; and SAM1. The more sophisticated of these techniques yield equilibrium bond lengths with average absolute errors of about 0.04 to 0.05 Å and equilibrium bond angles with errors of a few degrees. Calculated dipole moments contain errors in the range of 0.3 to 0.4 debye. Errors in the predicted ionization energies lie in the range 0.2 to 1.0 eV. Bond energies are generally poor.

Summary: Key Concepts and Equations

1. Chemical bonds are the result of nature's tendency to minimize the sum of the kinetic and potential energy of a system. When a bond forms, it exhibits a characteristic equilibrium distance that is the result of a competition between kinetic energy and potential energy. The kinetic energy decreases when the internuclear distance increases, thereby providing more space for delocalization of the electrons. In contrast, the potential energy decreases if the internuclear distance decreases, thereby bringing the electrons into closer proximity to the positively charged nuclei. The equilibrium separation is that distance which minimizes the sum of the two energies.

2. A perturbation calculation proceeds by partitioning the Hamiltonian into two terms:

$$\mathcal{H} = \mathcal{H}_o + \mathcal{H}'.$$

$\mathcal{H}_o$ is called the zeroth-order Hamiltonian and $\mathcal{H}'$ is termed the perturbation. The idea is to first solve the Schrödinger equation by using $\mathcal{H}_o$ as the Hamiltonian to obtain the corresponding zeroth-order ground-state eigenfunction ψ_o and zeroth-order energy eigenvalue E_o. It is then assumed that $\mathcal{H}'$ is sufficiently small that the exact ground-state eigenfunction is essentially the same as ψ_o. The effect of the perturbation is included by computing the average value of $\mathcal{H}'$ over the zeroth-order ground-state eigenfunction. The first-order corrected energy is given by

$$E = E_o + \langle \psi_o | \mathcal{H}' | \psi_o \rangle.$$

If $\mathcal{H}'$ is large, first-order perturbation theory is highly inaccurate.

3. The variational energy is defined by the integral ratio

$$E_v = \frac{\langle \lambda | \mathcal{H} | \lambda \rangle}{\langle \lambda | \lambda \rangle},$$

where λ is an arbitrary approximation to the exact eigenfunction for the Hamiltonian $\mathcal{H}$. Since the electron distribution produced by the exact eigenfunction is the distribution that yields the minimum possible ground-state energy, any other distribution, such as that given by $|\lambda|^2 d\tau$, must produce a higher energy. This means that we must always have $E_v \geq E_{\text{exact}}$. This statement, called the Rayleigh–Ritz variational principle, implies that E_v is always an upper limit to E_{exact}. This permits us to insert parameters into our trial eigenfunction λ and minimize E_v with respect to these parameters with the certain knowledge that the lower the value of E_v, the closer it must be to E_{exact}. This concept is the foundation of most quantum mechanical calculations involving the electronic structure of atoms and molecules.

4. When the variational parameters in the trial eigenfunction are linear, the equations that must be solved to minimize E_v are linear homogeneous equations whose unknowns are the linear variational coefficients. If

$$\lambda = \sum_{i=1}^{M} a_i \chi_i,$$

where the χ_i are known expansion functions and the a_i are the linear variational coefficients, the minimum variational energies for the ground state and the $M - 1$ lowest excited states are the roots of the secular equation given by

$$\begin{vmatrix} H_{11} - E_v S_{11} & H_{12} - E_v S_{12} & H_{13} - E_v S_{13} & \cdots & H_{1M} - E_v S_{1M} \\ H_{21} - E_v S_{21} & H_{22} - E_v S_{22} & H_{23} - E_v S_{23} & \cdots & H_{2M} - E_v S_{2M} \\ \vdots & \vdots & \vdots & & \vdots \\ H_{M1} - E_v S_{M1} & H_{M2} - E_v S_{M2} & H_{M3} - E_v S_{M3} & \cdots & H_{MM} - E_v S_{MM} \end{vmatrix} = 0,$$

where $H_{ij} = \langle \chi_i | \mathcal{H} | \chi_j \rangle$ and the overlap integral $S_{ij} = \langle \chi_i | \chi_j \rangle$. E_v has M real roots, the lowest of which is an upper limit to the exact ground-state energy. The higher energy roots are upper limits to the successive excited states. Each root yields a corresponding approximate eigenfunction, so that M eigenfunctions will be obtained. These eigenfunctions are necessarily orthogonal if their corresponding eigenvalues are different, since $\mathcal{H}$ is a Hermitian operator.

5. When linear variational theory is applied to molecular systems, the χ_i expansion functions are usually taken to be atomic orbitals centered on the various atoms in the molecule. In the majority of cases, Slater-type orbitals (STOs) are used. The trial eigenfunction λ formed by the linear combination of atomic orbitals (LCAO) is called a molecular orbital (MO).

Application of this procedure to the hydrogen-molecule ion using a two-term expansion with $1s$ STOs on each proton produces two variational energies and two corresponding eigenfunctions. The lower energy ground-state eigenfunction is nodeless, whereas the higher energy excited-state eigenfunction has a nodal plane at the midpoint between the two protons. Calculation of these two energies as a function of the internuclear distance R shows that the ground state contains a minimum that lies below the energy of the separated $H + H^+$ system. This minimum denotes the formation of a chemical bond. Consequently, the MO is called a bonding orbital. In contrast, the excited-state energy plotted as a function of R exhibits no minimum. It is, therefore, called an antibonding orbital, in that excitation to it results in dissociation of H_2^+.

6. The nomenclature rules for diatomic molecular orbitals are based upon the number of nodal planes that contain the internuclear axis, the bonding or antibonding character of the orbital, the atomic orbitals that make the major contribution to the MO, and, in the case of homonuclear diatomics, the symmetry of the orbital with respect to inversion through the origin. The orbital is labeled σ, π, or δ, depending upon

whether there are zero, one, or two nodal planes, respectively, that contain the internuclear axis. Antibonding orbitals in homonuclear molecules are designated by a superscript asterisk. The symmetry with respect to inversion is given by a right subscript–g or u, for symmetric and antisymmetric orbitals, respectively. The atomic orbitals involved in forming the MO are listed before the σ, π, or δ notation.

7. There are four useful approximations or rules concerning the formation of molecular orbitals that lead to a significant degree of understanding and predictive power without the necessity of doing actual calculations. These are: (1) Lower energy atomic orbitals will tend to produce lower energy molecular orbitals. (2) High-energy atomic orbitals do not usually mix significantly with much lower energy orbitals to form molecular orbitals. The atomic orbitals that make the major contributions to a molecular orbital tend to have similar energies. (3) A molecular orbital constructed from a particular pair of atomic orbitals that contains nodes will have a higher energy than a pair without nodes. (4) σ, π, and δ atomic orbitals do not mix to form diatomic molecular orbitals. Points (1) and (2) are useful approximations. Points (3) and (4) are rigorously correct rules.

8. The approximate ordering of molecular orbital energies for homonuclear diatomics is $E(1s\sigma) < E(1s\sigma^*) < E(2s\sigma) < E(2s\sigma^*) < E(2p_z\sigma) < E(2p_x\pi) = E(2p_y\pi) < E(2p_x\pi^*) = E(2p_y\pi^*) < E(2p_z\sigma^*) < E(3s\sigma) < E(3s\sigma^*) < E(3p_z\sigma) < E(3p_x\pi) = E(3p_y\pi) < E(3p_x\pi^*) = E(3p_y\pi^*) < E(3p_z\sigma^*)$. For heteronuclear diatomics, the energy ordering often must be determined by actual calculation. If, however, the atomic orbital energies of the two atoms are similar, homonuclear molecular orbitals can be used to provide a qualitative description of the electronic configuration of a heteronuclear diatomic, provided that the energy ordering is altered to place the $p\pi$ bonding orbital energies below that of the $p\sigma$ orbitals. This altered ordering is $E(1s\sigma) < E(1s\sigma^*) < E(2s\sigma) < E(2s\sigma^*) < E(2p_x\pi) = E(2p_y\pi) < E(2p_z\sigma) < E(2p_x\pi^*) = E(2p_y\pi^*) < E(2p_z\sigma^*) < E(3s\sigma) < E(3s\sigma^*) < E(3p_x\pi) = E(3p_y\pi) < E(3p_z\sigma) < E(3p_x\pi^*) = E(3p_y\pi^*) < E(3p_z\sigma^*)$.

9. Angular momentum states for diatomics are characterized using a molecular term symbol similar in form to the atomic term symbols introduced in Chapter 13. The multiplicity is again given by a left superscript. The orbital angular momentum state is determined by summing the Z components of orbital angular momentum for all electrons to obtain the total magnetic quantum number M. Since these components add as scalars, we have

$$M = m_1 + m_2 + m_3 + \ldots,$$

where the m_i are the magnetic quantum numbers for the occupied molecular orbitals. The resulting angular momentum state is labeled with capital Greek letters Σ, Π, Δ, and Φ for $|M| = 0, 1, 2,$ or 3, respectively. The energy ordering of the term states follows Hund's rules introduced in Chapter 13. For homonuclear diatomics, the total inversion symmetry is indicated by a right subscript, either g or u. If there is an even number of electrons in MOs with u symmetry, the total symmetry is g; otherwise, it is u. The symmetry with respect to reflection through a plane containing both nuclei is specified by a right superscript $(+)$ if the total wave function is unaltered by such reflection. If the wave function changes sign, the superscript $(-)$ designation is used. Only diatomic molecules in Σ states have this type of reflection symmetry.

10. If we make the Born–Oppenheimer approximation and omit the spin–orbit coupling terms, the total molecular Hamiltonian for a molecule containing N electrons and M nuclei has N kinetic-energy terms for the electrons, $M \times N$ nuclear–electron attraction terms, $N(N-1)/2$ electron–electron repulsion terms, and $M(M-1)/2$ nuclear–nuclear repulsion terms.

11. When LCAO–MO methods are applied to many-electron molecules, two major difficulties arise. First, the integrals involving the electron–electron repulsion terms have no analytic solution if we use exponential STOs in our basis set. This problem is solved by representing the STOs with Gaussian expansions fitted to them by

least–squares methods. The actual integrals, therefore, involve Gaussian, instead of exponential, functions. When the functions are Gaussians, closed-form analytic solutions are available. The second problem involves the computation of the H_{ij} matrix elements in the secular equation. For reasons that are not obvious, it turns out that these matrix elements contain the variational coefficients. Since we do not know the values of these coefficients in advance, we must perform an iterative solution until convergence is attained. For this reason, the method is called a *self-consistent-field* (SCF) calculation. The entire procedure is labeled an SCF–LCAO–MO method.

12. If M STOs are used in an SCF–LCAO–MO calculation, we obtain M orbital energies and $2M$ spin orbitals. The ground state is represented by assigning the N electrons to the N spin orbitals of lowest energy, which leaves $2M - N$ spin orbitals empty. The empty orbitals are called virtual spin orbitals. The electronic configuration thus produced is called the Hartree–Fock (HF) configuration, whose electronic energy is given by

$$E_{\text{electronic}} = \sum_{i=1}^{N} \varepsilon_i - \text{total electron–electron repulsion energy,}$$

where the ε_i are the orbital energies obtained in the SCF calculation. The total electron–electron repulsion energy must be subtracted to avoid counting the repulsions twice. The total energy is then obtained by adding the nuclear–nuclear repulsion energy to $E_{\text{electronic}}$:

$$E_{\text{total}} = E_{\text{electronic}} + E_{\text{nuclear repulsion}}.$$

13. We can improve the results of an SCF calculation and approach experimental accuracy if we conduct configuration interaction (CI) calculations. In this procedure, additional electronic configurations are generated by promoting electrons from the HF configuration of an SCF calculation into the virtual spin orbitals while maintaining the same overall multiplicity. If the new configurations are generated by promoting one electron, they are called single excitations, or just singles. When two, three, or four electrons are promoted, the new configurations are labeled doubles, triples, and quadruples, respectively. The total wave function is then written as a linear combination of all these configurations:

$$\psi(1, 2, 3, \ldots, N) = \sum_{i=0}^{K} C_i D_i.$$

Another linear variational calculation is then executed to minimize the energy and obtain the expansion coefficients C_i that determine the importance of the various configurations in the overall wave function.

14. Most of the chemist's qualitative thinking about chemical bonding is done using a valence-bond model in which chemical bonds are viewed as a pair of electrons having opposite spin functions localized between the bonded atoms. When we draw lines between atoms to represent a chemical bond, we are using a valence-bond model. We can translate this qualitative picture into a quantum mechanical representation of the wave function with the use of a four-step procedure called the perfect-pairing method: (1) Form a set of atomic orbitals (AOs) on each atom into which the bonding pair of electrons will be placed. This set may be a single orbital or an LCAO. (2) Form spin orbitals from the sets of AOs in such a manner that the bonded atoms have paired spins. (3) Using Slater determinants, form all possible antisymmetrized product functions from the spin orbitals obtained in (2). (4) Combine the Slater determinants that have the required spin assignments in an appropriate linear combination that represents the desired bonding scheme. The "appropriate linear combination" for a valence bond between two atoms is obtained if each Slater determinant is multiplied by $(-1)^n$, where n is the number of spin exchanges between the determinant and any arbitrary reference determinant. The "appropriate linear combination" is always the sum of the Slater determinants with the required spin assignments if we wish to obtain a no-bond or antibonding result.

15. The total valence bond wave function is written as a linear combination of the wave functions for the various valence bond structures that the chemist deems important. For example, if there are two such structures, the total wave function is given by

$$\psi_{total}^{VB} = C_I \psi_I^{VB} + C_{II} \psi_{II}^{VB}.$$

A variational calculation is then performed to minimize the energy with respect to C_I and C_{II}. Structures that have equal variational coefficients are called resonance forms. In forming the linear combination, it is not necessary to include valence bond structures containing crossed bonds, since these are always linear combinations of structures without crossed bonds. This is the VB no-crossing rule.

Problems

Problems that require the use of some type of computational device are marked with an asterisk (*). Problems that require some type of plotting routine are indicated with a pound sign (#).

14.1 Consider a particle of mass m in a one-dimensional infinite potential well of width a for which the potential is $V(x) = 0$ for $0 \leq x \leq a$ and $V(x) = \infty$ for $x < 0$ or $x > a$. Suppose we alter the potential such that we now have

$$V(x) = \infty \quad \text{for } x < 0 \text{ or } x > a,$$

$$V(x) = 0 \quad \text{for } 0 \leq x < 0.8a,$$

$$V(x) = -V_o \quad \text{for } 0.8a \leq x \leq 0.9a$$

with

V_o a positive constant,

and

$$V(x) = 0 \quad \text{for } 0.9a < x \leq a.$$

That is, the potential now has the form shown in the following figure:

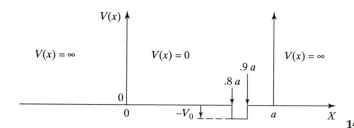

Use first-order perturbation theory to obtain an expression for the first-order ground-state energy for this system in terms of m, a, V_o, and h.

14.2# In Chapter 12, we solved the problem of the one-dimensional harmonic oscillator. The Hamiltonian for this system is

$$\mathcal{H} = -\frac{\hbar^2}{2\mu}\frac{d^2}{dx^2} + \frac{kx^2}{2}.$$

The resulting ground-state eigenfunction is

$$\psi(x) = \left[\frac{\alpha}{\pi}\right]^{1/4} \exp[-\alpha x^2/2],$$

with $\alpha = (\mu k)^{1/2}/\hbar$. The corresponding ground-state energy is

$$E = \frac{h}{4\pi}[k/\mu]^{1/2} = \frac{h\nu_o}{2}.$$

Let us now partition the potential term into a new potential term and a perturbation term. That is, we write the Hamiltonian in the form

$$\mathcal{H} = -\frac{\hbar^2}{2\mu}\frac{d^2}{dx^2} + \frac{fkx^2}{2} + \frac{(1-f)kx^2}{2}$$

$$= \mathcal{H}_o + \frac{(1-f)kx^2}{2},$$

where we take the first two terms to be $\mathcal{H}_o$ and the term $(1-f)kx^2/2$ to be $\mathcal{H}'$.

(A) Obtain the zeroth-order ground-state eigenfunction and eigenvalue with the above choice for $\mathcal{H}_o$.

(B) Obtain an expression for the first-order ground-state energy of the system as a function of f, h, and the classical frequency of the oscillator, ν_o.

(C) Obtain an expression for the percent error in the perturbation result in terms of f alone.

(D) Plot the percent error as a function of f over the range $0.1 \leq f \leq 2$.

14.3 A phenomenon seen in many organizations, particularly among the faculty of a chemistry department, is the constant battle between the younger members of the organization (the Young Turks) and the more senior members (the Old Guard). Whenever a new problem arises, the Young Turks usually wish to dismantle the present system, start fresh, and obtain new, innovative solutions to the problem. The Old Guard is much more conservative. Its members usually wish to employ the old system and simply use it to generate acceptable solutions to the new problem.

The Young Turks accuse the Old Guard of being mired in the past—of having lost the capacity for creative thought. The Old Guard looks upon the Young Turks as one might view a three-year-old child throwing a temper tantrum. Relate this situation to first-order perturbation theory. How are they similar? [*Author's advice:* Since "young" refers to your mental state, not to your chronological age, be a Young Turk all your life. First-order perturbation theory is not particularly accurate.]

14.4[#] Compute the variational energy for a particle of mass m in an infinite potential well of width a, using the trial wave function $\lambda = N \sin^2[\pi x/a]$ for $0 \leq x \leq a$ and $\lambda = 0$ otherwise. Show that the Rayleigh–Ritz variational principle is satisfied by comparing E_v with the true ground-state energy of the system. Plot the normalized trial wave function and the true ground-state eigenfunction on the same graph for the case where $a = 1$ Å.

14.5 Use the variational wave function $\lambda = N \exp[-ar^2]$, where N and a are constants, in a variational calculation of the energy of the hydrogen atom, and determine the value of a that minimizes the energy. Compare this minimum value with the true hydrogen-atom ground-state energy. Use atomic units. Is the variational energy too high or too low?

14.6 Use the variational wave function

$$\lambda = N \exp[-bx^2] \qquad \text{for all } x,$$

where b is a variational parameter, to obtain the best possible energy for the ground state of a harmonic oscillator.

(A) Obtain the variational energy E_v as a function of b.

(B) Minimize E_v with respect to b, and obtain $(E_v)_{min}$.

(C) Show that $(E_v)_{min} = h\nu_o/2$, where ν_o is the classical vibrational frequency of the oscillator.

(D) The result of (C) is exact, whereas the one obtained in Problem 14.7 (see next) is not. Why?

14.7 The ground-state eigenfunction for a harmonic oscillator is $\psi(x) = N \exp[-\alpha x^2/2]$, where $\alpha = (\mu k)^{1/2}/h$, in which μ and k are the reduced mass and vibrational force constant, respectively. The ground-state energy eigenvalue is $E = h\nu_o/2$, with the vibrational frequency given by $\nu_o = [1/(2\pi)][k/\mu]^{1/2}$. An investigator wishes to conduct a variational calculation on a harmonic oscillator using the trial eigenfunction

$$\lambda = N(A^2 - x^2) \quad \text{for } -A \leq x \leq A$$

and

$$\lambda = 0 \text{ for } x > A \text{ or } x < -A.$$

Using A as a variational parameter, obtain the best possible variational energy for this trial eigenfunction. Compare your result with the exact ground-state energy. Is the variational principle obeyed?

14.8 Obtain the approximate normalized eigenfunction for the ground state of the system examined in Example 14.5. [*Hint:* The variational eigenfunction can be written in the form $\lambda = c_1[\chi_1 + (c_2/c_1)\chi_2]$ so that c_1 acts as a normalization constant. Remember that the overlap integrals have already been computed in Example 14.5.]

14.9 Show that the higher energy root for E_v in Example 14.5 is an upper limit to the energy of the first excited state for a particle of mass m in a one-dimensional infinite potential well of width a. Obtain the approximate normalized eigenfunction for this excited state of the system. [See hint in Problem 14.8.]

14.10 Show that the variational eigenfunction for the first excited state obtained in Example 14.5 is orthogonal to that for the ground state in that example. These variational eigenfunctions are given in the discussion following Example 14.5.

14.11 Consider a one-dimensional particle in an infinite potential well of width a. The correct ground-state wave function for such a system is

$$\psi = \left[\frac{2}{a}\right]^{1/2} \sin\left[\frac{\pi x}{a}\right] \quad \text{for } 0 \leq x \leq a$$

and

$$\psi = 0 \text{ for } x < 0 \text{ or } x > a.$$

Show that the trial eigenfunction

$$\lambda = N[x^3 - a^2 x] \quad \text{for } 0 \leq x \leq a$$

and

$$\lambda = 0 \text{ for } x < 0 \text{ or } x > a$$

gives a variational energy greater than that for the exact ground state.

14.12 Consider a one-dimensional particle in an infinite potential well of width a. The correct ground-state wave function for such a system is

$$\psi = \left[\frac{2}{a}\right]^{1/2} \sin\left[\frac{\pi x}{a}\right] \quad \text{for } 0 \leq x \leq a$$

and

$$\psi = 0 \text{ for } x < 0 \text{ or } x > a.$$

Use linear variational theory with the trial eigenfunction

$$\lambda = c_1[ax - x^2] + c_2[ax^3 - x^4] \quad \text{for } 0 \leq x \leq a$$

and

$$\lambda = 0 \qquad \text{for } x < 0 \text{ or } x > a,$$

where c_1 and c_2 are variational parameters, to obtain the lowest variational energy for this approximate eigenfunction. Compare the result with the true ground-state energy, and show that the variational principle is satisfied.

14.13 A linear variational calculation is carried out on a molecule using the expansion $\lambda = \sum_{i=1}^{15} a_i \chi_i$.

(A) How many variational energies will be obtained in the calculation? How many of these will be either imaginary or complex?

(B) How many molecular orbitals will result from the calculation?

(C) How many nodes will the MO corresponding to the next-lowest variational energy have?

(D) What is the value of the integral $\langle \lambda_i | \lambda_j \rangle$ if λ_i and λ_j are molecular orbitals corresponding to different variational energies? Explain.

14.14 Show that the use of the variational energy E_v given in Eq. 14.42 leads to the associated molecular orbital λ_- in Eq. 14.47.

14.15 Show that the two molecular orbitals for the hydrogen-molecule ion derived in the text, λ_+ and λ_-, are orthogonal. Could this have been predicted without integrating? How?

The next four problems examine different aspects of the simple molecular orbital solution of the H_2^+ ion developed in the text. The problems are constructed so as to permit them to be assigned individually or as a unit.

14.16* In the text, it is shown that if the variational wave function $\lambda = a_1 \chi_{1sA} + a_2 \chi_{1sB}$, where

$$\chi_{1sA} = \left[\frac{Z_e^3}{\pi}\right]^{1/2} \exp[-Z_e r_a]$$

and

$$\chi_{1sB} = \left[\frac{Z_e^3}{\pi}\right]^{1/2} \exp[-Z_e r_b],$$

is used for the H_2^+ molecule, the two resulting MO energies are

$$E_v^+ = \frac{[H_{11} + H_{12}]}{[1 + S_{12}]}$$

and

$$E_v^- = \frac{[H_{11} - H_{12}]}{[1 - S_{12}]}.$$

(A) Using an effective charge $Z_e = 1$, compute the two MO energies in atomic units as a function of R over the range 1 bohr $< R <$ 5.5 bohr. Make a careful plot of your results. Compare your plot with that shown in Figure 14.11.

(B) Using the results obtained in (A), determine the computed equilibrium internuclear separation and the predicted well depth for the hydrogen-molecule ion. Compute the percent error between these results and the experimentally determined values of 2.004 bohr and 0.1025 hartree, respectively.

14.17* In the text, it is shown that if the variational wave function $\lambda = a_1 \chi_{1sA} + a_2 \chi_{1sB}$, where

$$\chi_{1sA} = \left[\frac{Z_e^3}{\pi}\right]^{1/2} \exp[-Z_e r_a]$$

and

$$\chi_{1sB} = \left[\frac{Z_e^3}{\pi}\right]^{1/2} \exp[-Z_e r_b],$$

is used for the H_2^+ molecule, the two resulting MO energies are

$$E_v^+ = \frac{[H_{11} + H_{12}]}{[1 + S_{12}]}$$

and

$$E_v^- = \frac{[H_{11} - H_{12}]}{[1 - S_{12}]}.$$

Using Z_e as a variational parameter, determine the value of Z_e (to three significant digits) that yields the best result for the well depth of the hydrogen-molecule ion. (Numerical methods will need to be used for this part of the problem.) Compare your result with the experimental result given in Problem 14.16. What is the percent error in your result?

14.18* It is shown in the text that if the variational wave function $\lambda = a_1 \chi_{1sA} + a_2 \chi_{1sB}$, where

$$\chi_{1sA} = \left[\frac{Z_e^3}{\pi}\right]^{1/2} \exp[-Z_e r_a]$$

and

$$\chi_{1sB} = \left[\frac{Z_e^3}{\pi}\right]^{1/2} \exp[-Z_e r_b],$$

is used for the H_2^+ molecule, the two resulting MO energies are

$$E_v^+ = \frac{[H_{11} + H_{12}]}{[1 + S_{12}]}$$

and

$$E_v^- = \frac{[H_{11} - H_{12}]}{[1 - S_{12}]}.$$

The best result for the well depth of the H_2^+ ion is obtained with $Z = 1.24$. Using this value, determine the predicted equilibrium bond distance for the hydrogen-molecule ion correct to three significant digits. The measured equilibrium distance is 2.004 bohr. What is the precent error in your result?

14.19* It is shown in the text that if the variational wave function $\lambda = a_1 \chi_{1sA} + a_2 \chi_{1sB}$, where

$$\chi_{1sA} = \left[\frac{Z_e^3}{\pi}\right]^{1/2} \exp[-Z_e r_a]$$

and

$$\chi_{1sB} = \left[\frac{Z_e^3}{\pi}\right]^{1/2} \exp[-Z_e r_b],$$

is used for the H_2^+ molecule, the two resulting MO energies are

$$E_v^+ = \frac{[H_{11} + H_{12}]}{[1 + S_{12}]}$$

and

$$E_v^- = \frac{[H_{11} - H_{12}]}{[1 - S_{12}]}.$$

The best result for the well depth of H_2^+ ion is obtained with $Z_e = 1.24$. With this effective charge, the computed equilibrium distance for H_2^+ is 2.001 bohr.

(A) Calculate the H_2^+ ground-state energy for R values of $R_{eq} + dx$, $R_{eq} + 2dx$, $R_{eq} - dx$, and $R_{eq} - 2dx$, where $dx = 0.01$ bohr. Assuming that the ground-state energy near the bottom of the well may be fitted accurately by a parabolic (harmonic) function, use least–squares methods to fit the function $E_v^+ = k(R - R_{eq})^2/2 + D$ to the results at $R = R_{eq}$ and the four values calculated in (A). That is, determine by least–squares methods the values of k and D that yield the best fit to your data.

(B) Assuming that the vibrational motion of H_2^+ in the ground state may be represented by a harmonic oscillator, use the results obtained in (A) to compute the fundamental vibration frequency for the hydrogen-molecule ion. Express your result in cm^{-1}. The measured vibrational frequency is 2,297 cm^{-1}. Calculate the percent error in your answer.

14.20 (A) Write down simple MO descriptions of the electronic configuration for each of the following diatomic molecules: N_2, N_2^+, N_2^-, O_2, O_2^+, and O_2^-.

(B) Which of the following has the greatest bond strength: N_2, N_2^-, or N_2^+? O_2, O_2^-, or O_2^+? Justify your answers.

14.21 Write down an appropriate form for the ground-state wave function of the He_2^+ molecule ion. Is He_2 a bound molecule? Is He_2^+ a bound molecule? If we were to pass an electrical discharge through a container of $He(g)$ so that we ionized a significant fraction of the atoms, discuss qualitatively what you might expect to find as the principal components inside the container a few moments after the discharge. At a much later time, what would you expect to find?

14.22 Develop a qualitative description of the molecular electronic configuration for each of the following diatomic molecules:

(A) CO;

(B) CN;

(C) OH;

(D) BeF. Write the molecular term symbol for each of these ground-state configurations. Stipulate the appropriate symmetry elements (if any) on the term symbol.

14.23 (A) Describe a simple LCAO–MO expansion that might be employed to carry out a molecular orbital calculation for the NH molecule.

(B) Discuss qualitatively the expected nature of the molecular orbitals that would be obtained in the calculation. What would you expect the energy ordering of these orbitals to be? Write down an appropriate representation of the MO electronic configuration. What is the molecular term symbol for the ground state? What are the term symbols for the excited angular momentum states? Arrange all angular momentum states in order of increasing energy.

14.24 The homonuclear diatomic molecules Li_2, N_2, F_2, and F_2^+ have molecular orbitals whose relative energies obey the simple energy-ordering scheme given in the text. Write down appropriate electronic configurations for each of these molecules. Make certain to stipulate the g or u symmetries of the MOs. Determine the complete term symbols for all possible angular momentum states of the various configurations, and arrange the term symbols in order of increasing energy if there is more than one.

14.25 The simple energy ordering given in the text for homonuclear diatomics predicts that the electronic configuration of B_2 should be $(1s\sigma_g)^2(1s\sigma_u^*)^2(2s\sigma_g)^2(2s\sigma_u)^2(2p_z\sigma_g)^2$. If this is the case, what would the full ground-state term symbol for B_2 be? Experimentally, it is found that the ground-state term symbol is actually $^3\Sigma_g^-$. What has occurred? Suggest a reasonable possibility for the actual electronic configuration for the ground state of B_2.

14.26 Using appropriate summation notation, write down the Hamiltonian for the benzene molecule. How many one-electron, two-electron, and zero-electron terms are present?

14.27[#]Consider a perturbed hydrogen atom whose Hamiltonian, in atomic units, is

$$\mathcal{H} = -\frac{1}{2}\nabla^2 - \frac{1}{r} + \frac{b}{r^2},$$

where b is a positive constant. The Schrödinger equation for this Hamiltonian can be solved exactly for the energy eigenvalues. The result for the ground state is

$$E = -\frac{1}{2\beta^2},$$

where

$$\beta = \frac{1 + \{1 + 8b\}^{1/2}}{2}.$$

Use first-order perturbation theory in which the perturbation is $\mathcal{H}' = b/r^2$ to compute the ground-state energy of the perturbed system for $b = 0.01$, 0.05,

0.10, 0.50, 1.0, and 10.0. In each case, compute the percent error in the perturbation calculation, and plot the percent error as a function of b.

14.28 Prepare a diagram similar to that shown in Figure 14.23 for the λ_6 MO obtained in the minimal-basis-set SCF calculation for H_2O described in the text. The resulting variational coefficients are given in Table 14.6. Discuss why this orbital has antibonding characteristics that make it a high-energy orbital.

14.29 Prepare a diagram similar to that shown in Figure 14.23 for the λ_7 MO obtained in the minimal-basis-set SCF calculation for H_2O described in the text. The resulting variational coefficients are given in Table 14.6. Discuss why this orbital has antibonding characteristics that make it a high-energy orbital.

14.30 The SCF–LCAO–MO calculation for H_2O described in the text placed the nuclei in positions such that the O–H distances were both 0.96 Å and the H–O–H angle was 104.5°. The calculation reported that the total nuclear–nuclear repulsion energy was 9.16819 hartrees in that configuration. Verify that this result is correct.

14.31 How many total molecular orbitals were obtained in the LCAO–MO calculation discussed in the text in which 13 expansion functions were used for the H_2O molecule? How many of these orbitals are virtual orbitals?

14.32 Leigh carries out an SCF–LCAO–MO calculation on ethyl alcohol (C_2H_6O), using a minimal-basis-set expansion for her LCAO. How many molecular orbitals will she obtain from her calculation? How many of these orbitals will be filled and how many will be virtual orbitals?

14.33 The minimal-basis-set SCF calculation for H_2O described in the text produced seven molecular orbitals. Suppose an investigator carries out a CI calculation for H_2O employing this minimal-basis-set SCF calculation as a starting point. He then uses all possible single excitations with a multiplicity of unity in his CI expansion.

(A) How many total configurations are present in his CI expansion?

(B) What are these configurations?

(C) How many singlet-state energies would be obtained in the CI calculations?

(D) Which variational coefficient do you expect to have the largest magnitude for the eigenfunction corresponding to the lowest variational energy? Explain.

14.34 Sometimes biradical structures make an important contribution in organic reactions. Suppose we have the four-atom system considered in the text and wish to include in our valence-bond wave function the contribution of the following biradical structure. In terms of the Slater determinants D_1, D_2, D_3, D_4, D_5, and D_6 defined in the text, what is the appropri-

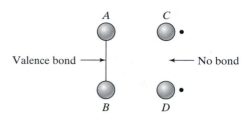

ate combination of the D_i to represent this structure that includes one valence bond and a no-bond structure between atoms C and D?

The next five problems examine the VB wave functions for Kekule and Dewar structures for benzene. The problems are constructed so that they can be assigned separately or as a unit.

14.35 Let us consider only the six π electrons in benzene. These six electrons are on six atoms arranged in a regular hexagon as shown in the following diagram:

The black dots represent the six π electrons. In the perfect-pairing approximation used in VB theory, three of these electrons must have α spin functions and three β spin functions.

(A) How many Slater determinants can be constructed that satisfy the perfect-pairing approximation?

(B) Using the notation $(\alpha\alpha\alpha\beta\beta\beta)$ to represent the Slater determinant $|a\alpha\, b\alpha\, c\alpha\, d\beta\, e\beta\, f\beta|$, write down all of the possible VB determinants for the preceding six-electron system. How many of these will be needed to represent each Kekule or Dewar structure?

14.36 Let us consider only the six π electrons in benzene. These six electrons are on six atoms arranged in a regular hexagon as shown in the following diagram:

The black dots represent the six π electrons. In the perfect-pairing approximation used in VB theory, three of these electrons must have α spin functions and three β spin functions. This yields the following 20 Slater determinants that satisfy the perfect-pairing requirement:

$$D_1 = (\alpha\alpha\alpha\beta\beta\beta); \quad D_2 = (\alpha\alpha\beta\alpha\beta\beta);$$
$$D_3 = (\alpha\alpha\beta\beta\alpha\beta); \quad D_4 = (\alpha\alpha\beta\beta\beta\alpha);$$

$D_5 = (\alpha\beta\alpha\alpha\beta\beta); \quad D_6 = (\alpha\beta\alpha\beta\alpha\beta);$

$D_7 = (\alpha\beta\alpha\beta\beta\alpha); \quad D_8 = (\beta\alpha\alpha\alpha\beta\beta);$

$D_9 = (\beta\alpha\alpha\beta\alpha\beta); \quad D_{10} = (\beta\alpha\alpha\beta\beta\alpha);$

$D_{11} = (\alpha\beta\beta\alpha\alpha\beta); \quad D_{12} = (\alpha\beta\beta\alpha\beta\alpha);$

$D_{13} = (\alpha\beta\beta\beta\alpha\alpha); \quad D_{14} = (\beta\alpha\beta\alpha\alpha\beta);$

$D_{15} = (\beta\alpha\beta\alpha\beta\alpha); \quad D_{16} = (\beta\alpha\beta\beta\alpha\alpha);$

$D_{17} = (\beta\beta\alpha\alpha\alpha\beta); \quad D_{18} = (\beta\beta\alpha\alpha\beta\alpha);$
$D_{19} = (\beta\beta\alpha\beta\alpha\alpha); \quad D_{20} = (\beta\beta\beta\alpha\alpha\alpha).$

The notation is that described in the text for the four-electron, four-atom system. In terms of these determinants, construct VB wave functions for the two Kekule structures of benzene shown in Figure 14.25.

14.37 Let us consider only the six π electrons in benzene. These six electrons are on six atoms arranged in a regular hexagon as shown in the following diagram:

The black dots represent the six π electrons. In the perfect-pairing approximation used in VB theory, three of these electrons must have α spin functions and three β spin functions. This yields a total of 20 Slater determinants that satisfy the perfect-pairing requirement. These determinants are given in Problem 14.36. In terms of these determinants, construct VB wave functions for the three Dewar structures of benzene shown in Figure 14.25.

14.38 Let us consider only the six π electrons in benzene. These six electrons are on six atoms arranged in a regular hexagon as shown in the following diagram:

The black dots represent the six π electrons. In the perfect-pairing approximation used in VB theory, three of these electrons must have α spin functions and three β spin functions. This yields a total of 20 Slater determinants that satisfy the perfect-pairing requirement.

(A) Use these determinants, listed in Problem 14.36, to construct the wave function for the VB structure

The VB wave functions obtained in Problems 14.36 and 14.37 for the Kekule and Dewar structures shown in Figure 14.25 are

$\psi_I^{VB} = [D_2 - D_3 - D_5 - D_{15} + D_6 + D_{16} + D_{18} - D_{19}],$

$\psi_{II}^{VB} = [D_6 - D_7 - D_9 - D_{11} + D_{10} + D_{12} + D_{14} - D_{15}],$

$\psi_{III}^{VB} = [D_3 - D_4 - D_6 - D_{14} + D_7 + D_{15} + D_{17} - D_{18}],$

$\psi_{IV}^{VB} = [D_1 - D_2 - D_6 - D_{10} + D_{11} + D_{15} + D_{19} - D_{20}],$

and

$\psi_V^{VB} = [D_5 - D_6 - D_8 - D_{12} + D_9 + D_{13} + D_{15} - D_{16}].$

(B) Show that the VB wave function for the foregoing structure can be written as a linear combination of the VB wave functions for structures I–V and that the VB no-crossing rule holds.

14.39 A variational calculation on benzene is carried out using the trial wave function

$$\psi_{total}^{VB} = C_I \psi_I^{VB} + C_{II} \psi_{II}^{VB} + C_{III} \psi_{III}^{VB} + C_{IV} \psi_{IV}^{VB} + C_V \psi_V^{VB},$$

where ψ_I^{VB} and ψ_{II}^{VB} are the VB wave functions for the two Kekule structure of benzene and ψ_{III}^{VB}, ψ_{IV}^{VB}, and ψ_V^{VB} are the wave functions for the three Dewar structures. After completion of the calculation, what relationship would be observed between C_I, C_{II}, C_{III}, C_{IV}, and C_V? Are any structures in resonance?

14.40 Sam and Leigh have a demonstration and two questions for the class. I'll just turn it over to them.

"Leigh and I each have a dewar. One is filled with liquid N_2, the other with liquid O_2. As you see, we also have two large, powerful permanent horseshoe magnets on the table in front of us. I'm going to pour the contents of my dewar over this magnet while Leigh pours the contents of her dewar over the magnet in front of her. Watch what happens!"

The contents of Sam's dewar pour over his magnet without sticking to it. The liquid hits the tabletop and runs over the surface, where it rapidly vaporizes. In contrast, most of the contents of Leigh's dewar stick to the magnet, completely filling the space between the magnet's poles with liquid that is held suspended in the air. Only a small amount of the liquid reaches the tabletop.

"Isn't that amazing! Our questions are

(A) Which dewar held the liquid oxygen, Leigh's or mine? and

(B) How can we use simple molecular orbital theory to explain the results of the experiment we have just witnessed?"

Anyone want to respond before we turn our attention to spectroscopy?

[*Author's note:* This experiment really works, provided that the magnetic fields produced by the magnets are strong enough.]

SECTION 3

Spectroscopy

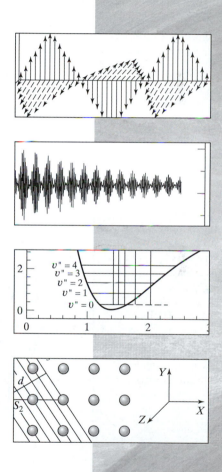

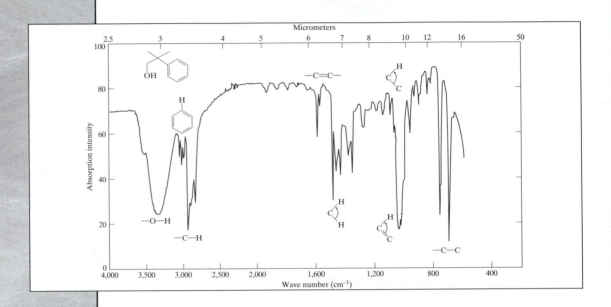

CHAPTER 15

Rotational, Vibrational, and Electronic Spectra

Spectroscopy is the most powerful tool available to the scientist for probing the microscopic world of atoms and molecules. Using this tool, we can determine molecular structure in great detail with an extraordinary level of accuracy. Molecular events that occur on widely differing time scales, ranging from days to picoseconds and, most recently, even femtoseconds, can be monitored and studied in remarkable depth. Modern spectroscopy permits the transition states of chemical reactions to be probed and examined. Taken together, quantum mechanics and spectroscopy provide a powerful weapon—a skeleton key—that allows us to unlock the innermost secrets of molecular dynamics and structure.

In this chapter, we introduce the basic concepts upon which spectroscopic methods rest. Then we turn our attention to molecular spectroscopy and electronic spectroscopy, which probe the internal rotational and vibrational motions of molecules and their electronic structure. Spectroscopic analysis of these motions leads us to information concerning equilibrium bond lengths and bond angles, rotational moments of inertia, force constants, and vibrational frequencies in the ground and electronically excited states. When combined with statistical methods to be introduced in Section 4, this information permits us to compute many of the thermodynamic quantities that occupied our attention during the first portion of the text. As we proceed, it will become evident that spectroscopic measurements also provide the analytical chemist with the means to gather information on chemical composition that would be impossible to obtain in the absence of such measurements.

15.1 Types of Spectroscopy

The opening sentence of Chapter 2 lays down the first important general principle of spectroscopy: *All measurements in science are difference measurements.* So it is with spectroscopic measurements, which determine the spacings between discrete energy states, not the absolute values of the energies of the states themselves. Because of this feature, it is common practice to characterize different types of spectroscopy in terms of the magnitudes of the energy-level spacings that are generally observed. The spacings in turn are usually given in terms of the wavelength λ or wave number ν, using the relationship

$$E = h\nu = \frac{hc}{\lambda} = hc\bar{\nu}, \tag{15.1}$$

where c is the speed of light in a vacuum. Table 15.1 lists the approximate wavelengths and energies often associated with different types of spectroscopy.

The two general types of spectroscopy are also classified by whether the experimentalist is measuring the energy required to excite the system from a low-lying energy state to a state higher in energy or whether he or she is measuring the energy emitted when a system in an excited state undergoes a

Table 15.1 Types of spectroscopy		
Type	**Approximate Energy Spacing (kJ mol^{-1})**	**Approximate Wavelength (cm)**
Nuclear magnetic resonance (NMR)	10^{-6}	10^4
Electron spin resonance (ESR)	10^{-3}	10
Rotational	10^{-1}	0.1 (microwave)
Vibrational	10	10^{-3} (infrared)
Electronic	10^3–10^4	10^{-5}–10^{-6} (ultraviolet–visible)
Nuclear	10^8	10^{-10}

transition to a state with lower energy. These two types are called *absorption* and *emission* spectroscopy, respectively. In this chapter, we shall be primarily concerned with absorption measurements.

15.2 Transition Probabilities, Selection Rules, and Bandwidths

Spectroscopic absorption measurements are usually executed by placing an ensemble of molecules in the path of a beam of electromagnetic radiation. Such radiation is characterized by a sinusoidal oscillation of electric and magnetic field vectors perpendicular to the direction of propagation, as shown in Figure 15.1. The idea is to couple the electric field of the radiation to the molecules and thereby induce a transfer of energy, causing the molecules to move from the ground state to an excited state or vice versa. An absorption measurement consists of determining the number of photons absorbed by the system at each wavelength investigated in the experiment. If none of the molecules absorb any of the radiation, the spectroscopist is going to obtain very little, if any, information; the time taken to do the experiment would have been better spent watching grass grow. For this reason, we need to know the conditions under which we can expect transitions between energy states to occur when the excitation energy is being supplied by the electric field of electromagnetic radiation.

▶ **FIGURE 15.1**
Three cycles of an electromagnetic wave propagating in the Y direction. In this case, the sinusoidal oscillation of the electric field vectors $\mathcal{E}$ all lie in the Y–Z plane. The corresponding sinusoidal oscillation of the magnetic field vectors **B** all lie in the X–Y plane. Such an electromagnetic wave is said to be *plane polarized*. In *unpolarized* radiation, the vectors $\mathcal{E}$ and **B** are randomly oriented about the direction of propagation of the wave, which in this example is the Y direction.

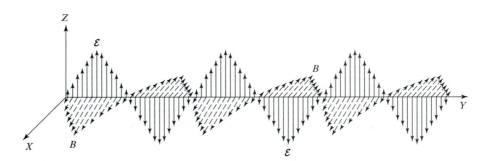

For simplicity, let us consider a hypothetical system that has only two energy states, with eigenvalues E_o and E_1, respectively. If the potential is independent of time, as we expect it to be, the discussion in Chapter 11 and Eq. 11.76 show us that the eigenfunctions in the ground and excited state are $\Psi_o = \psi_o \exp[-iE_ot/\hbar]$ and $\Psi_1 = \psi_1 \exp[-iE_1t/\hbar]$, respectively, where ψ_o and ψ_1 are solutions of the stationary-state Schrödinger equation. From our general treatment of probability in Chapter 10, we also know that, in the absence of any information concerning the state of the system, the wave function will be a linear combination of Ψ_o and Ψ_1, with the squares of the expansion coefficients giving the probability of finding a randomly chosen molecule in the ground or excited state. That is, we will have

$$\Psi = a_o\Psi_o + a_1\Psi_1 = a_o\psi_o \exp\left[-\frac{iE_ot}{\hbar}\right] + a_1\psi_1 \exp\left[-\frac{iE_1t}{\hbar}\right]. \quad (15.2)$$

If, however, we know that all the molecules are in the ground state, a_o will be unity and a_1 will be zero, so that $\Psi = \Psi_o$. Furthermore, in the absence of any change in the system potential, Ψ will remain equal to Ψ_o indefinitely.

Suppose now that at time $t = 0$, we subject the system to an electromagnetic radiation field. The presence of this field may produce a change in the potential energy of the molecules. We know that when a particle with charge q is placed in an electric field $\mathcal{E}$, it is acted on by a force $\mathbf{F} = q\mathcal{E}$. Since the x component of force is given by the negative of the rate of change of potential energy with x, we have

$$-\frac{\partial V}{\partial x} = F_x = q\mathcal{E}_x.$$

Separating the variables and then integrating both sides produces

$$\Delta V = -\int_0^x q\mathcal{E}_x dx = -q\mathcal{E}_x x, \tag{15.3}$$

provided that the electric-field strength is constant over the range from zero to x. For an arbitrary displacement $d\mathbf{s}$ in the presence of the field, the change in potential is the negative of the work done during the displacement which is given by Eq. 2.24:

$$dV = -dw = -\mathbf{F} \cdot d\mathbf{s} = -q\mathcal{E} \cdot d\mathbf{s} = -q[\mathcal{E}_x \mathbf{i} + \mathcal{E}_y \mathbf{j} + \mathcal{E}_z \mathbf{k}]$$
$$\cdot [dx\,\mathbf{i} + dy\,\mathbf{j} + dz\,\mathbf{k}] = -q[\mathcal{E}_x dx + \mathcal{E}_y dy + \mathcal{E}_z dz]. \tag{15.4}$$

Let us assume that the electric field is uniform, so that $\mathcal{E}_x = \mathcal{E}_y = \mathcal{E}_z$, and that it is constant over the dimensions of the molecule. Under these conditions, integrating both sides of Eq. 15.4 gives

$$\Delta V = -q\mathcal{E}\int_0^x \int_0^y \int_0^z [dx + dy + dz] = -q\mathcal{E}[x + y + z]. \tag{15.5}$$

Clearly, Eq. 15.5 reduces to Eq. 15.3 when the field is plane polarized (see Figure 15.1) in the x direction and $\mathcal{E}_y = \mathcal{E}_z = 0$.

When the electromagnetic radiation field is turned on at time $t = 0$, the molecules will be acted on by a new potential, which will be the sum of the old potential produced by the electron–nuclear attraction terms, the electron–electron repulsions, the nuclear–nuclear repulsions, and the change in potential ΔV produced by the electric field. Therefore, the new Hamiltonian will be

$$\mathcal{H} = \mathcal{H}_o + \Delta V = \mathcal{H}_o - q\mathcal{E}[x + y + z] = \mathcal{H}_o + \mathcal{H}'. \tag{15.6}$$

Because of the presence of the perturbing electric field, some of the ground-state molecules may undergo a transition to the excited state, which will cause a_1 to become nonzero and a_o to fall below unity. To determine the probability that such a transition will occur, we need to compute the value of a_1 at some time t^* after the electromagnetic field has been turned on at time $t = 0$. If we can execute such a calculation, the transition probability at time t^* will be given by $|a_1|^2$.

To determine a_1 at time t^*, we need to know how the electric field varies with time. In many cases, the electric field produced by electromagnetic radiation will vary sinusoidally with time, as shown in Figure 15.1. If this is the case, we can represent $\mathcal{E}$ by

$$\mathcal{E} = \mathcal{E}_o \sin[2\pi\nu t], \tag{15.7}$$

where ν is the frequency of the electromagnetic radiation. With $\mathcal{E}$ given by Eq. 15.7, it is clear that $\mathcal{H}$ is no longer independent of time. Therefore, we must use the time-dependent form of the Schrödinger equation, Eq. 11.68, to determine how Ψ varies with time. Using this equation, we have

$$\mathcal{H}\Psi = i\hbar \frac{\partial \Psi}{\partial t}. \tag{15.8}$$

Combining Eqs. 15.2 and 15.6 with 15.8, we obtain

$$[\mathcal{H}_o + \mathcal{H}'][a_o\Psi_o + a_1\Psi_1] = [\mathcal{H}_o + \mathcal{H}']\left\{ a_o\psi_o \exp\left[-\frac{iE_o t}{\hbar}\right] + a_1\Psi_1 \exp \right.$$

$$\left. \left[-\frac{iE_1 t}{\hbar}\right] \right\} = i\hbar \frac{\partial}{\partial t}\left\{ a_o\psi_o \exp\left[-\frac{iE_o t}{\hbar}\right] + a_1\psi_1 \exp\left[-\frac{iE_1 t}{\hbar}\right] \right\}. \tag{15.9}$$

Equation 15.9 is a differential equation that tell us how Ψ varies with time. Since we know that $\Psi = \Psi_o$ at $t = 0$, we should be able to determine Ψ at time t^* by solving the differential equation. We first need to simplify the equation by employing the facts that $\mathcal{H}_o\Psi_o = E_o\Psi_o$ and $\mathcal{H}_o\Psi_1 = E_1\Psi_1$. Using these two equations, we may write the left-hand side (LHS) of Eq. 15.9 in the form

$$\text{LHS} = \left\{ a_o E_o \psi_o \exp\left[-\frac{iE_o t}{\hbar}\right] + a_1 E_1 \psi_1 \exp\left[-\frac{iE_1 t}{\hbar}\right] \right\}$$

$$+ \left\{ a_o \mathcal{H}' \psi_o \exp\left[-\frac{iE_o t}{\hbar}\right] + a_1 \mathcal{H}' \psi_1 \exp\left[-\frac{iE_1 t}{\hbar}\right] \right\}. \tag{15.10}$$

Because ψ_o and ψ_1 are independent of time, the right-hand side (RHS) of Eq. 15.9 is

$$\text{RHS} = i\hbar \left\{ \frac{\partial a_o}{\partial t} \psi_o \exp\left[-\frac{iE_o t}{\hbar}\right] + \frac{\partial a_1}{\partial t} \psi_1 \exp\left[-\frac{iE_1 t}{\hbar}\right] \right\}$$

$$+ i\hbar \left\{ a_o\left(-\frac{iE_o}{\hbar}\right)\psi_o \exp\left[-\frac{iE_o t}{\hbar}\right] + a_1\left(-\frac{iE_1}{\hbar}\right)\psi_1 \exp\left[-\frac{iE_1 t}{\hbar}\right] \right\}$$

$$= i\hbar \left\{ \frac{\partial a_o}{\partial t} \psi_o \exp\left[-\frac{iE_o t}{\hbar}\right] + \frac{\partial a_1}{\partial t} \psi_1 \exp\left[-\frac{iE_1 t}{\hbar}\right] \right\}$$

$$+ \left\{ a_o E_o \psi_o \exp\left[-\frac{iE_o t}{\hbar}\right] + a_1 E_1 \psi_1 \exp\left[-\frac{iE_1 t}{\hbar}\right] \right\}. \tag{15.11}$$

A direct comparison of Eq. 15.11 with Eq. 15.10 shows that the second set of terms in the former is identical to the first set of terms in the latter. These terms, therefore, cancel, and Eq. 15.9 becomes

$$\left\{ a_o \mathcal{H}' \psi_o \exp\left[-\frac{iE_o t}{\hbar}\right] + a_1 \mathcal{H}' \psi_1 \exp\left[-\frac{iE_1 t}{\hbar}\right] \right\}$$

$$= i\hbar \left\{ \frac{\partial a_o}{\partial t} \psi_o \exp\left[-\frac{iE_o t}{\hbar}\right] + \frac{\partial a_1}{\partial t} \psi_1 \exp\left[-\frac{iE_1 t}{\hbar}\right] \right\}. \tag{15.12}$$

We can now put Eq. 15.12 into a more convenient form by taking advantage of the fact that ψ_o and ψ_1 are both normalized, so that we have $\langle \psi_o | \psi_o \rangle = \langle \psi_1 | \psi_1 \rangle = 1$. We also know that $\langle \psi_o | \psi_1 \rangle = 0$, since $E_o \neq E_1$, which requires that ψ_o be orthogonal to ψ_1. To take advantage of these relationships, we multiply both sides of Eq. 15.12 on the left by ψ_1^* and integrate over all spatial coordinates. On the left side, we get

$$\text{LHS} = a_o\langle\psi_1|\mathcal{H}'|\psi_o\rangle \exp\left[-\frac{iE_ot}{\hbar}\right] + a_1\langle\psi_1|\mathcal{H}'|\psi_1\rangle \exp\left[-\frac{iE_1t}{\hbar}\right].$$

On the right side, the result is

$$\text{RHS} = i\hbar\left\{\frac{\partial a_o}{\partial t}\langle\psi_1|\psi_o\rangle \exp\left[-\frac{iE_ot}{\hbar}\right] + \frac{\partial a_1}{\partial t}\langle\psi_1|\psi_1\rangle \exp\left[-\frac{iE_1t}{\hbar}\right]\right\}$$

$$= i\hbar\left\{\frac{\partial a_1}{\partial t}\exp\left[-\frac{iE_1t}{\hbar}\right]\right\}.$$

Equating LHS and RHS, we obtain

$$i\hbar\left\{\frac{\partial a_1}{\partial t}\exp\left[-\frac{iE_1t}{\hbar}\right]\right\}$$

$$= a_o\langle\psi_1|\mathcal{H}'|\psi_o\rangle \exp\left[-\frac{iE_ot}{\hbar}\right] + a_1\langle\psi_1|\mathcal{H}'|\psi_1\rangle \exp\left[-\frac{iE_1t}{\hbar}\right]. \quad \textbf{(15.13)}$$

Division of Eq. 15.13 by $i\hbar \exp[-(iE_1t)/\hbar]$ produces

$$\frac{\partial a_1}{\partial t} = (i\hbar)^{-1}a_o\langle\psi_1|\mathcal{H}'|\psi_o\rangle \exp\left[-\frac{i(E_o - E_1)t}{\hbar}\right] + (i\hbar)^{-1}a_1\langle\psi_1|\mathcal{H}'|\psi_1\rangle. \quad \textbf{(15.14)}$$

With the substitution of Eq. 15.6 for $\mathcal{H}'$ and Eq. 15.7 for $\mathcal{E}$, the integrals in Eq. 15.14 become

$$\langle\psi_1|\mathcal{H}'|\psi_o\rangle = -\langle\psi_1|q\mathcal{E}[x + y + z]|\psi_o\rangle = -q\mathcal{E}_o \sin(2\pi\nu t)\langle\psi_1|x + y + z|\psi_o\rangle$$

$$= -q\mathcal{E}_o \sin(2\pi\nu t)H_{10} \quad \textbf{(15.15)}$$

and

$$\langle\psi_1|\mathcal{H}'|\psi_1\rangle = -\langle\psi_1|q\mathcal{E}[x + y + z]|\psi_1\rangle = -q\mathcal{E}_o \sin(2\pi\nu t)\langle\psi_1|x + y + z|\psi_1\rangle$$

$$= -q\mathcal{E}_o\sin(2\pi\nu t) H_{11}, \quad \textbf{(15.16)}$$

where we have written H_{10} and H_{11} for the integrals $\langle\psi_1|x + y + z |\psi_o\rangle$ and $\langle\psi_1|x + y + z|\psi_1\rangle$, respectively. These integrals are called *transition matrix elements*. It is common to see H_{10} and H_{11} written in terms of the dipole moment μ. In Chapter 1, we found that two charges q and $-q$ separated by a distance r possess a dipole moment vector given by $\mu = q\mathbf{r}$. If we write the vector $\mathbf{r}$ in terms of its Cartesian components, the dipole moment has the form

$$\mu = q[i x + j y + k z] = \mu_x\mathbf{i} + \mu_y\mathbf{j} + \mu_z\mathbf{k}.$$

Therefore, the expression qH_{10} that appears in Eq. 15.15 can be written in the form

$$qH_{10} = q\langle\psi_1|x + y + z|\psi_o\rangle = \langle\psi_1|\mu_x + \mu_y + \mu_z|\psi_o\rangle.$$

Combination of Eqs. 15.14, 15.15 and 15.16 produces

$$\frac{\partial a_1}{\partial t} = -\frac{a_oq\mathcal{E}_oH_{10}}{i\hbar} \sin(2\pi\nu t) \exp\left[-\frac{i(E_o - E_1)t}{\hbar}\right] - \frac{a_1q\mathcal{E}_oH_{11}}{i\hbar} \sin(2\pi\nu t).$$

$$\textbf{(15.17)}$$

We can now simplify Eq. 15.17 by invoking what is, in effect, a perturbation approximation. At time $t = 0, a_o = 1$ whereas $a_1 = 0$. If the effect of the electromagnetic field is small, a_1 will remain small relative to a_o, which will continue to be about unity. We shall, therefore, ignore the second term in Eq. 15.17, since a_1 is small, and set a_o to unity in the first term. These approximations produce

$$\frac{\partial a_1}{\partial t} = -\frac{q\mathcal{E}_o H_{10}}{i\hbar} \sin(2\pi\nu t) \exp\left[-\frac{i(E_o - E_1)t}{\hbar}\right], \tag{15.18}$$

which can be easily integrated to obtain a_1 at time t^*.

Separating the variables and integrating both sides between corresponding limits gives

$$\int_{a_1=0}^{a_1(t^*)} da_1 = a_1(t^*) = -\frac{q\mathcal{E}_o H_{10}}{i\hbar} \int_0^{t^*} \sin(2\pi\nu t) \exp\left[-\frac{i(E_o - E_1)t}{\hbar}\right] dt. \tag{15.19}$$

Since $\sin(ax) = (e^{iax} - e^{-iax})/(2i)$, Eq. 15.19 may be written in the form

$$a_1(t^*) = \frac{q\mathcal{E}_o H_{10}}{2\hbar} \int_0^{t^*} \exp\left[-\frac{it}{\hbar}(\hbar 2\pi\nu - \Delta E)\right] dt$$

$$-\frac{q\mathcal{E}_o H_{10}}{2\hbar} \int_0^{t^*} \exp\left[\frac{it}{\hbar}(\hbar 2\pi\nu + \Delta E)\right] dt, \tag{15.20}$$

where $\Delta E = E_1 - E_o$. Noting that $\hbar 2\pi\nu = h\nu$, we find that the result of the integrations in Eq. 15.20 is

$$a_1(t^*) = \frac{iq\mathcal{E}_o H_{10}}{2}\left[\frac{\exp\left[-\frac{it^*}{\hbar}(h\nu - \Delta E)\right] - 1}{h\nu - \Delta E} + \frac{\exp\left[\frac{it^*}{\hbar}(h\nu + \Delta E)\right] - 1}{h\nu + \Delta E}\right]. \tag{15.21}$$

The transition probability at $t = t^*$ is proportional to $|a_1(t^*)|^2$. Before we examine this quantity, however, we can use Eq. 5.21 to deduce the four critical points upon which spectroscopy is based:

1. In order for the electromagnetic radiation field to produce a transition that makes a_1 larger than zero, q must not be zero. That is, there must be a charge separation so that the molecule possesses a permanent or induced dipole moment such as that discussed in Chapter 1. (See Figure 1.11 and the associated discussion.)

2. To have a nonzero transition probability, the transition matrix element connecting the initial and final states must not be zero. If $H_{10} = 0$ in the foregoing example, a_1 will remain zero, and the $0 \rightarrow 1$ transition will not occur. In this event, the $0 \rightarrow 1$ transition is said to be *forbidden*. The conditions that must be met to make the transition matrix element nonzero are called *selection rules*. These selection rules are only approximations, because of the assumptions we made in the derivation of Eq. 15.21.

3. When the energy of the incident radiation, $h\nu$, approaches the energy spacing between the states, $h\nu - \Delta E$ approaches zero and the transition probability increases rapidly. Therefore, we expect a maximum transition probability when the radiation energy and energy spacing are in *resonance* (i.e., are equal).

4. The preceding analysis is unchanged if we designate Ψ_o as the excited state and Ψ_1 as the ground state. The only difference is that we are

now deriving the probability of emission as the system undergoes a transition from an excited state to a state of lower energy. Since this emission is produced by the incident electromagnetic radiation, it is called *stimulated emission*. In this case, $\Delta E = E_1 - E_o$ is negative. Therefore, the sum $h\nu + \Delta E$ appearing in the denominator of the second term in Eq. 15.21 approaches zero when resonance is achieved between the photon energy and the energy spacing. As a result, the probability of stimulated emission attains a maximum at the resonance point, as does the probability of stimulated absorption. Finally, we note that the symmetry of Eq. 15.21 shows that the probabilities of stimulated absorption and stimulated emission are equal.

A qualitative appreciation of each of the four principles may be obtained by considering a few crude analogies. Suppose I bring a steel box weighing 50 kg into the class. The sides of the box are smooth, and, in addition, an excellent lubricating oil has been poured over the surface of the box. I ask a member of the class, who is also a starting linebacker on the varsity football team, to please lift the box and put it on the table. When he attempts to do this, he fails: The box does not move an inch off the floor. "Have you been sick lately?" I inquire. The student responds, "I can't get a hold on the thing." His point is well taken: It is not enough to have sufficient strength and energy to lift the box; there must also be a means to couple the student's strength to the box so that his energy can be transferred and the box lifted. If a coupling mechanism is provided, such as a pair of handles on the sides of the box, there will be no further difficulty.

The same principle applies to spectroscopy: It is not enough for the electromagnetic field to have sufficient energy to excite the molecules; there must be a means of coupling the electric field to the molecules so that the energy of the field can be transferred and the molecule excited. If a coupling mechanism is present, excitation will occur, but not otherwise. The coupling mechanism in the analogy in the previous paragraph is a pair of handles on the box. For the electric field, the coupling mechanism is a molecular charge separation or dipole moment with which the field can interact. When there is no dipole moment, the electric field cannot couple with the molecule, and no excitations are observed.

The requirement that the transition matrix element between the initial and final states, $\langle \psi_i | x + y + z | \psi_f \rangle$, not be zero is analogous to the need to have a road connecting town A and town B before you can drive from one to the other. You may own a Ferrari capable of hitting 250 mph, but with no road between points A and B, you will not be making the drive.

Let us now examine the transition probability in more quantitative detail. We first note that, although it might appear that Eq. 15.21 predicts that $a_1(t^*)$ will become infinite when $h\nu = \Delta E$ or $h\nu = -\Delta E$, this is not the case. The limit of each term in Eq. 15.21 at this point has the indeterminate form $0/0$. Consider, for example, the first term. In the limit where $h\nu \to \Delta E$, we may use L'hôpital's rule to obtain

$$\lim_{h\nu \to \Delta E} \left[\frac{\exp\left[-\dfrac{it^*}{\hbar}(h\nu - \Delta E) \right] - 1}{(h\nu - \Delta E)} \right] = -\frac{it^*}{\hbar} \lim_{h\nu \to \Delta E} \left\{ \exp\left[-\frac{it^*}{\hbar}(h\nu - \Delta E) \right] \right\}$$

$$= -\frac{it^*}{\hbar}.$$

Therefore, we expect both $a_1(t^*)$ and $|a_1(t^*)|^2$ to be finite at all values of $h\nu$. With a considerable amount of algebra and using the fact that $e^{iax} + e^{-iax} = 2\cos(ax)$, we find that

$$|a_1(t^*)|^2 = 2K^2 \left\{ \frac{1 - \cos\left(\frac{t^*(h\nu - \Delta E)}{\hbar}\right)}{(h\nu - \Delta E)^2} + \frac{1 - \cos\left(\frac{t^*(h\nu + \Delta E)}{\hbar}\right)}{(h\nu + \Delta E)^2} \right.$$
$$\left. + \frac{1 - \cos\left(\frac{t^*(h\nu - \Delta E)}{\hbar}\right) - \cos\left(\frac{t^*(h\nu + \Delta E)}{\hbar}\right) + \cos\left(\frac{2t^*h\nu}{\hbar}\right)}{(h\nu)^2 - (\Delta E)^2} \right\},$$

where $K = q\mathcal{E}_o H_{10}/2$. It is instructive to examine the manner in which the transition probability $|a_1(t^*)|^2$ varies as a function of the electromagnetic energy $h\nu$ for various values of t^*. Let us assume that the energy spacing is $\Delta E = 6.626 \times 10^{-21}$ J, so that the electromagnetic radiation will be in resonance when $\nu = 10^{13}$ s^{-1}. Figures 15.2 and 15.3 show plots of $|a_1(t^*)|^2/(2K_2)$ as a function of the electromagnetic energy $h\nu$ when $t^* = 0.13$ ps (1.3×10^{-13} s). Physically, this means that we have been collecting spectral data for 0.13 ps with the electromagnetic radiation field incident on the sample. Figure 15.2 shows the calculated spectrum over the range $-3 \times 10^{-20} \leq h\nu \leq 3 \times 10^{-20}$ J, where negative values denote emission. Figure 15.3 is the same spectrum, but over the range $-1 \times 10^{-20} \leq h\nu \leq 1 \times 10^{-20}$ J. Several important features should be noted. First, as we anticipated, the transition probabilities are finite at all radiation energies. Second, the transition probability for stimulated emission is the same as that for stimulated absorption. Third, the band maximum corresponds to the resonance point, at which we have $h\nu = 6.626 \times 10^{-21}$ J $= \Delta E$ in this example.

The full-width at half maximum (FWHM) of the absorption or emission band is of particular interest. The smaller the FWHM is, the more well defined the energy spacing ΔE becomes. The absorption bands seen in Figures 15.2 and 15.3 are very broad. The FWHM corresponds to an energy

▶ **FIGURE 15.2**
Hypothetical stimulated absorption–emission spectrum for the two-state system leading to Eq. 15.21. The figure shows a plot of $|a_1(t^*)|^2/(2K^2)$ versus the energy of the radiation field, $h\nu$, when the energy-state spacing ΔE is 6.626×10^{-21} J and the data collection time t^* is 0.13 ps (1.3×10^{-13} s). Negative radiation energies refer to emission, positive energies to absorption. The figure shows the spectrum over the energy range from -3×10^{-20} J $\leq h\nu \leq 3 \times 10^{-21}$ J.

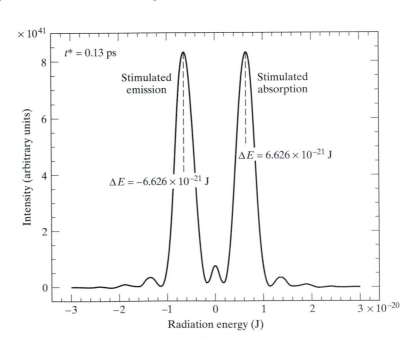

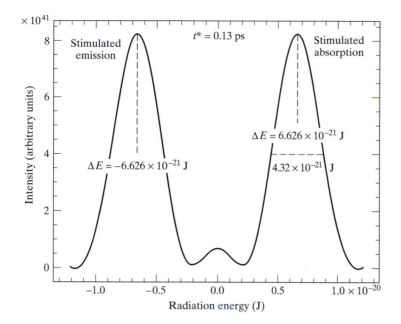

The same spectrum as shown in Figure 15.2, but for the energy range -1×10^{-21} J $\leq h\nu \leq 1 \times 10^{-21}$ J. The full width at half maximum (FWHM) for both absorption and emission bands is 4.32×10^{-21} J.

spread of 4.32×10^{-21} J. A considerable amount of absorption or emission is obtained even when the radiation is far from being in resonance. When we have only a single band in the spectrum, this added breadth creates no real problem; the position of the maximum is still very clear. However, if we have several absorption bands present, such a large FWHM might cause substantial band overlap and thereby make it very difficult to determine the resonance position or, in some cases, even determine how many bands are present. The problem is that we have not collected sufficient data to define the resonance position of the band accurately. The electromagnetic field has been in operation for only 0.13 ps, and therefore, our data are limited.

Figures 15.4 and 15.5 show the same spectrum when t^* is 5.0 and 10.0 ps, respectively. The FWHMs for these two cases are 0.117×10^{-21} J and 0.056×10^{-21} J, respectively. We see that, as the data collection time increases, the resonance position becomes more well defined. When t^* is sufficiently long, we obtain a very narrow band.

The preceding observations suggest that the FWHM of the adsorption band is nearly inversely proportional to the data-sampling time. If this is correct, the product of t^* and the FWHM should be a constant. For the three cases shown in Figures 15.3, 15.4, and 15.5, this product has the value 5.6×10^{-34} J s, 5.8×10^{-34} J s, and 5.6×10^{-34} J s, respectively. The product has the units of Planck's constant and is very close to that value. Therefore,

$$t^* \approx \frac{h}{\text{FWHM}}.$$

This feature of absorption spectroscopy provides scientists with a very important tool for the investigation of the lifetimes of short-lived, transient species. If the molecule whose spectrum is being taken has a lifetime Δt, then the data collection time cannot exceed Δt, and we therefore expect the FWHM of the absorption band to be approximately $h/\Delta t$. Measurement of

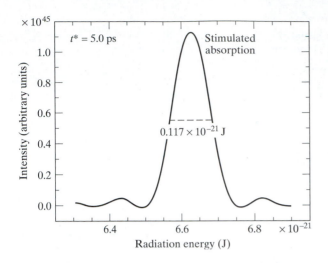

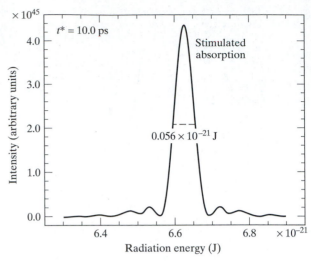

▲ **FIGURE 15.4**
Hypothetical stimulated absorption spectrum for the two-state system leading to Eq. 15.21 in the text. The figure shows a plot of $|a_1(t^*)|^2/(2K^2)$ versus the energy of the radiation field, $h\nu$, when the energy-state spacing ΔE is 6.626×10^{-21} J and the data collection time t^* is 5.0 ps (5.0×10^{-12} s). The full width at half maximum (FWHM) is 0.117×10^{-21} J. This is a factor of 37 less than that shown in Figure 15.3 for the same system. The decrease is due to the additional data collection time, which serves to define the resonance position more accurately.

▲ **FIGURE 15.5**
Hypothetical stimulated absorption spectrum for the two-state system leading to Eq. 15.21 in the text. The figure shows a plot of $|a_1(t^*)|^2/(2K^2)$ versus the energy of the radiation field, $h\nu$, when the energy-state spacing ΔE is 6.626×10^{-21} J and the data collection time t^* is 10.0 ps (1.0×10^{-11} s). The full width at half maximum (FWHM) is 0.056×10^{-21} J. This is a factor of 77 less than that shown in Figure 15.3 for the same system. The decrease is due to the additional data collection time, which serves to define the resonance position more accurately.

▶ **FIGURE 15.6**
(A) A hypothetical absorption spectrum consisting of 11 well-resolved bands. In this case, the spectrometer has sufficient resolving power to distinguish the closely spaced frequencies at which the absorptions occur. (B) The same spectrum taken with a spectrometer whose resolving power is limited.

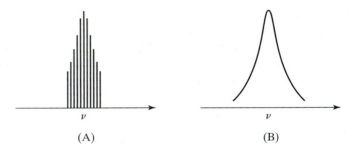

the bandwidth at half maximum thereby provides a measure of the lifetime of the transient species.

Factors other than sampling time also affect the bandwidth. The inherent resolution of the spectrometer is an important factor. If the spectrometer cannot distinguish a radiation frequency ν from another frequency $\nu + \Delta\nu$, two bands, one with resonance frequency ν and the second with resonance frequency $\nu + \Delta\nu$, may appear as a single broad band. If there are many such closely spaced absorption bands, the appearance of the spectrum may be a single very broad band, as illustrated in Figure 15.6.

15.3 Rotational Spectroscopy

15.3.1 Diatomic Molecules

To the extent that we may treat a diatomic molecule as a linear rigid rotor, the rotational-energy states of the molecule are given by Eq. 12.107, viz.,

$$E_J = \frac{J(J+1)\hbar^2}{2I} \quad \text{for } J = 0, 1, 2, 3, \ldots, \tag{15.22}$$

where I is the molecule's moment of inertia. Since there are $2J + 1$ different values of the magnetic quantum number M for a given J, each rotational-energy state has a degeneracy of $2J + 1$. Equation 15.22 shows that the rotational-state energies increase quadratically with J. The energy spacing between adjacent states J and $J + 1$ is

$$\Delta E_{J,J+1} = \frac{(J + 1)(J + 2)\hbar^2}{2I} - \frac{J(J + 1)\hbar^2}{2I} = \frac{\hbar^2}{2I}[J^2 + 3J + 2 - J^2 - J]$$

$$= \frac{(2J + 2)\hbar^2}{2I}. \tag{15.23}$$

The adjacent-state energy spacing, therefore, increases linearly with J.

It is common practice to express the preceding results in terms of the rotational constant of the molecule, namely,

$$B = \frac{\hbar^2}{2Ih} = \frac{h}{8\pi^2 I}. \tag{15.24}$$

In these terms, Eqs. 15.22 and 15.23 become

$$\boxed{E_J = J(J + 1)Bh} \tag{15.25}$$

and

$$\Delta E_{J,J+1} = 2(J + 1)Bh. \tag{15.26}$$

Equation 15.25 makes it clear that the factor Bh plays the same role with respect to rotational-energy that $h\nu_o$ plays with respect to the vibrational energy of a harmonic oscillator. Like the vibrational frequency ν_o, B must have units of $(\text{time})^{-1}$. Classically, ν_o represents the number of times per second the oscillator vibrates. Similarly, B is directly related to the classical rotation frequency, or the number of times per second the rigid rotor undergoes a complete revolution. (See Problem 15.42.)

Our examination of transition probabilities in the previous section shows that if we wish to induce a transition in molecule X from rotational state J to rotational state J', X must possess a permanent dipole moment so that the electric field can couple with the rotational motion of the molecule. Thus, if X is HCl, OH, or NaH, the electric field will couple, and we will observe spectroscopic transitions between rotational states if the frequencies are in resonance with the energy spacings. On the other hand, molecules such as H_2, O_2, and Cl_2 have a zero dipole moment, and no rotational transitions will be observed. Figure 15.7 illustrates this point.

If the molecule has a permanent dipole moment, a transition between quantum states (J, M) and (J', M') is possible when the radiation energy is in resonance with the energy spacing, provided that the transition matrix element connecting the states is not zero. That is, we must have

$$H_{JM,J'M'} = \langle Y_J^M(\theta, \phi)|x + y + z|Y_{J'}^{M'}(\theta, \phi)\rangle \neq 0. \tag{15.27}$$

In Eq. 15.27, $Y_J^M(\theta, \phi)$ and $Y_{J'}^{M'}(\theta, \phi)$ are the spherical harmonic rigid-rotor eigenfunctions for rotational states (J, M) and (J', M'), respectively. In general, $H_{JM,J'M'}$ will be nonzero if

$$\boxed{J' - J = \Delta J = \pm 1}. \tag{15.28}$$

Equation 15.28 is the selection rule for pure rotational spectra of linear rigid rotors.

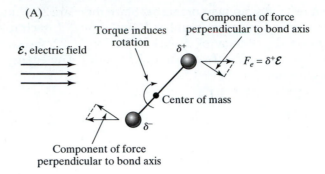

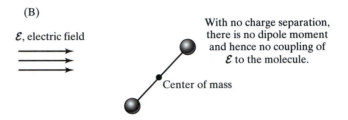

▶ FIGURE 15.7
(A) An illustration of how the presence of a dipole moment permits the electric field to couple to the molecule and thereby produce a torque that induces a change in the rotational quantum state of the molecule. (B) With no dipole moment, there is no coupling and, therefore, no rotational transitions.

EXAMPLE 15.1

Show that the transition matrix element for the rotational transition $(J = 0, M = 0) \rightarrow (J = 1, M = 0)$ is not zero.

Solution

Using Table 12.1, we find that the eigenfunctions for the $(J = M = 0)$ and the $(J = 1, M = 0)$ rotational states are

$$Y_0^0(\theta, \phi) = (4\pi)^{-1/2} \tag{1}$$

and

$$Y_1^0(\theta, \phi) = \left[\frac{3}{4\pi}\right]^{1/2} \cos \theta. \tag{2}$$

In spherical polar coordinates,

$$x = R \sin \theta \cos \phi, \tag{3}$$

$$y = R \sin \theta \sin \phi, \tag{4}$$

and

$$z = R \cos \theta. \tag{5}$$

The transition matrix element is, therefore,

$$H_{00,10} = \langle Y_0^0(\theta, \phi) | x + y + z | Y_1^0(\theta, \phi) \rangle$$

$$= \frac{3^{1/2}R}{4\pi} \int_{\phi=0}^{2\pi} \int_{\theta=0}^{\pi} \cos \theta \left[\sin \theta \cos \phi + \sin \theta \sin \phi + \cos \theta\right] \sin \theta \, d\theta \, d\phi. \tag{6}$$

The first and second integrals are zero because $\int_{\phi=0}^{2\pi} \cos \phi \, d\phi = \int_{\phi=0}^{2\pi} \sin \phi \, d\phi = 0$. However, the third integral is

$$H_{00,10} = \frac{3^{1/2}R}{4\pi} \int_0^{2\pi} d\phi \int_{\theta=0}^{\pi} \cos^2 \theta \sin \theta \, d\theta = -\frac{2\pi R}{3^{1/2} 4\pi} \cos^3 \theta \Big|_0^{\pi} = \frac{R}{3^{1/2}}. \tag{7}$$

Since $H_{00,10} \neq 0$, the transition is allowed.
For related exercises, see Problems 15.3 and 15.4.

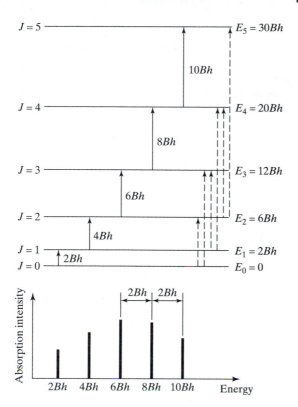

◀ FIGURE 15.8
Allowed transitions with $\Delta J = +1$ and the corresponding rotational spectrum for a linear rigid rotor. Forbidden transitions with $\Delta J = +2$ and $+3$ are shown by the dashed arrows. Absorptions intensities are arbitrarily drawn and are for illustrative purposes only. Note that the line spacings between adjacent absorption lines are $2Bh$ in every case.

Equation 15.28 tells us that we can observe pure rotational transitions only between adjacent states. This situation is illustrated in Figure 15.8, in which the allowed transitions are shown as solid upward arrows. ΔE_{01} for the $J = 0$ to $J = 1$ transition is $2Bh$, in accordance with Eq. 15.26. When the energy of the electromagnetic radiation is equal to $2Bh$, a resonance condition exists, and we will observe the absorption of radiation as molecules are rotationally excited from the ground state to the first excited state. An absorption band will, therefore, appear in the spectrum at the frequency or wavelength corresponding to this energy. The $J = 1$ to $J = 2$ transition is associated with an energy spacing $\Delta E_{12} = 6Bh - 2Bh = 4Bh$. Consequently, a second absorption band will appear in the rotational spectrum at this energy. Analysis of the energy spacings shows that the full rotational absorption spectrum consists of a series of equally spaced absorption bands separated by an energy $2Bh$.

Measurement of the rotational band spacing allows us to compute the moment of inertia and the bond length of the molecule. If the band spacing is denoted by $\Delta \varepsilon$, the previous analysis shows that

$$\Delta \varepsilon = 2Bh = \frac{h^2}{4\pi^2 I}. \tag{15.29}$$

Solving Eq. 15.29 for the moment of inertia, we obtain

$$I = \frac{h^2}{4\pi^2 \Delta \varepsilon}. \tag{15.30}$$

Since the diatomic rigid-rotor moment of inertia is μR^2, the length of the rotor is

$$R = \left[\frac{h^2}{4\pi^2 \Delta \varepsilon \mu} \right]^{1/2} = \hbar [\mu \Delta \varepsilon]^{-1/2}. \tag{15.31}$$

EXAMPLE 15.2

A microwave spectrum of gaseous BeH shows a series of approximately equally spaced absorption bands with a band separation of 20.62 cm^{-1}. Compute the equilibrium bond length of BeH and its moment of inertia.

Solution

The calculation is best done using atomic units. The reduced mass of BeH is

$$\mu = \frac{m_{Be} m_H}{m_{Be} + m_H} = \left[\frac{(9.012)(1.0079)}{(9.012 + 1.0079)} \right] g \ mol^{-1} \times \frac{1}{6.02214 \times 10^{23} \ mol^{-1}} \times \frac{1 \ kg}{1,000 \ g}$$

$$= 1.505 \times 10^{-27} \ kg. \tag{1}$$

The conversion factors in Table 13.1 can now be used to convert to atomic mass units:

$$\mu = 1.505 \times 10^{-27} \ kg \times \frac{1}{9.10939 \times 10^{-31} \ kg} = 1,652 \ \text{atomic mass units.} \tag{2}$$

The band spacing in hartrees is obtained by using the conversion factor in Table 13.2:

$$\Delta\varepsilon = 20.62 \ cm^{-1} \times \frac{1 \ hartree}{219475 \ cm^{-1}} = 9.395 \times 10^{-5} \ hartree. \tag{3}$$

The data obtained may now be inserted directly into Eq. 15.31. Since $\hbar$ is unity in atomic units, we get

$$R = (1) \left[(1,652)(9.395 \times 10^{-5}) \right]^{-1/2} = 2.538 \ bohr = 1.343 \times 10^{-10} \ m. \tag{4}$$

The moment of inertia of BeH is

$$I = \mu R^2 = (1.505 \times 10^{-27} \ kg)(1.343 \times 10^{-10} \ m)^2 = 2.715 \times 10^{-47} \ kg \ m^2, \tag{5}$$

$$I = (1,652)(2.538)^2 = 1.064 \times 10^4 \quad \text{atomic units.} \tag{6}$$

Notice how much closer the result is to unity when it is expressed in atomic units. For related exercises, see Problems 15.5, 15.6, 15.7, and 15.8.

15.3.2 Absorption Band Intensities

The pure rotational absorption spectrum illustrated in Figure 15.8 indicates that the absorption intensities associated with different transitions are different. The observed intensities are due, in part, to the intrinsic characteristics of the molecule and, in part, to the design of the experimental apparatus.

When a beam of electromagnetic radiation passes through a cell containing a molecule whose spectrum is to be taken, the amount of radiation absorbed is directly proportional to the intensity of the radiation (number of photons incident per unit time), the concentration of the molecules in the cell, and the distance traveled by the electromagnetic beam. Thus, if the radiation intensity is denoted by I, the concentration by $[C]$, and the differential change in the path length through the cell by $d\ell$, the change in radiation intensity due to absorption will be

$$dI \ \alpha - I[C]d\ell. \tag{15.32}$$

Converting Eq. 15.32 to an equality by multiplication on the right by a constant, we obtain

$$dI = -\varepsilon I[C]d\ell, \tag{15.33}$$

where ε is called the *extinction coefficient* or the *molar absorption coefficient* of the molecule. The radiation intensity, concentration, and path length are dependent upon the design of the apparatus and experimental conditions. In contrast, ε is characteristic of the molecule whose spectrum is being recorded.

Separating variables in Eq. 15.33, we may easily integrate between corresponding limits to obtain the transmitted radiation intensity as a function of ε, $[C]$, and ℓ. When the path length is zero, the transmitted intensity will be exactly the incident intensity I_o. Therefore,

$$\int_{I=I_o}^{I} \frac{dI}{I} = -\varepsilon[C] \int_{\ell=0}^{\ell} d\ell,$$

which gives

$$\ln\left[\frac{I}{I_o}\right] = -\varepsilon[C]\ell. \tag{15.34}$$

Equation 15.34 is called the *Beer–Lambert law*. It shows that the ratio of the transmitted intensity to the incident intensity varies exponentially with the path length, the concentration of the molecules, and the extinction coefficient.

EXAMPLE 15.3

The pure rotational spectrum of HF is taken in a certain microwave spectrometer. The transmitted radiation intensity I' at a particular wavelength is found to be 90% that of the incident intensity. If the path length of the radiation beam is increased by 40% and the concentration of HF in the sample cell is doubled, while the incident radiation intensity remains unchanged, what will be the ratio of the new transmitted intensity I'' to I' at the same wavelength?

Solution

Let the concentration and path length be $[C]_o$ and ℓ_o, respectively. From Eq. 15.34, we have

$$\ln\left[\frac{I'}{I_o}\right] = -\varepsilon[C]_o\ell_o = \ln(0.90). \tag{1}$$

In the second experiment, we have $\ell = 1.4\ell_o$ and $[C] = 2[C]_o$. The new ratio of I to I_o will then be

$$\ln\left[\frac{I''}{I_o}\right] = -\varepsilon\, 2[C]_o\, (1.4\ell_o) = 2.8(-\varepsilon[C]_o\ell_o). \tag{2}$$

Combining Eqs. 1 and 2 gives

$$\ln\left[\frac{I''}{I_o}\right] = 2.8\ln(0.90) = \ln(0.90)^{2.8}. \tag{3}$$

Performing exponentiation on both sides produces

$$\frac{I''}{I_o} = (0.90)^{2.8} = 0.745. \tag{4}$$

Consequently, the ratio of I'' to I' is

$$\frac{I''}{I'} = \frac{0.745 I_o}{0.90 I_o} = 0.827. \tag{5}$$

For related exercises, see Problems 15.9 and 15.10.

Equation 15.21 makes it clear that the transition probability for a given system, and hence the molar absorption coefficient, is a function of the frequency of the radiation. Consequently, we generally write the coefficient as $\varepsilon(\nu)$ to denote this dependence. As we have already noted, the transition probability and $\varepsilon(\nu)$ attain a maximum when $h\nu$ equals the energy spacing and the resonance condition is satisfied. The total, or *integrated molar absorption coefficient*, sums (integrates) $\varepsilon(\nu)$ over all frequencies for each transition.

The relative intensities of spectral lines are generally observed to vary from line to line, as illustrated qualitatively in Figure 15.8. These relative intensities depend upon the transition rates for absorption and emission and upon the design of the apparatus. Consider the pure rotational transition $J \to J + 1$. Einstein wrote the transition rate per molecule, w, from state J to state $J + 1$ as

$$w_J = B_J \rho, \tag{15.35}$$

where B_J is the *Einstein coefficient of stimulated absorption* for state J and ρ is the energy density of the radiation at the frequency of the transition. The total transition rate is just the rate per molecule times the number N_J of molecules in state J:

$$\boxed{W_J = N_J w_J = N_J B_J \rho}. \tag{15.36}$$

If the system is in thermal equilibrium at temperature T, N_J is given by the Boltzmann distribution, which we shall derive in Chapter 17:

$$\boxed{N_J = C g_J \exp\left[-\frac{E_J}{kT} \right]}. \tag{15.37}$$

In Eq. 15.37, g_J is the degeneracy of state J, E_J is the state energy, k is the Boltzmann constant, and C is, in effect, a normalization constant.

If an experiment is designed so that only the quantity of radiation absorbed is measured, Eqs. 15.36 and 15.37 will determine the relative band intensities. However, Eq. 15.21 shows that the probability of stimulated emission at a given transition frequency is equal to that for stimulated absorption. As a result, if the experiment is designed so as to measure the *total* radiation change, the relative band intensities will depend upon the absorption minus the emission. Since we know that the Einstein coefficients for stimulated absorption and emission are the same, the rate of stimulated emission per molecule from state $J + 1$ to state J is

$$w_J = B_J \rho. \tag{15.38}$$

In addition, even in the absence of a radiation field, nature will tend to produce emission spontaneously, so as to move the system to the lowest possible energy state. Einstein demonstrated that the coefficient of spontaneous emission, A, is related to B_J and to the transition frequency by

$$A_J = \left(\frac{8\pi h\nu^3}{c^3} \right) B_J, \tag{15.39}$$

where c is the speed of light in a vacuum. The overall rate of emission per molecule is, therefore,

$$[w_J']_{\text{total}} = \left\{ B_J \rho + \left(\frac{8\pi h\nu^3}{c^3} \right) B_J \right\}. \tag{15.40}$$

If the emission rate per molecule is multiplied by the number of molecules in the $J + 1$ state, we obtain the total emission rate:

$$[W_J']_{\text{total}} = N_{J+1}[w_J']_{\text{total}} = B_J N_{J+1}\left\{\rho + \left(\frac{8\pi h\nu^3}{c^3}\right)\right\}. \qquad (15.41)$$

If we collect all radiation in the spectroscopic experiment, the net absorption will be

$$W_{\text{net}} = W_J - W_J' = N_J B_J \rho - B_J N_{J+1}\left\{\rho + \left(\frac{8\pi h\nu^3}{c^3}\right)\right\}. \qquad (15.42)$$

For pure rotational transitions, the transition frequencies are sufficiently low that spontaneous emission may be ignored. (See Example 15.4 and Problem 15.11.) Under these conditions, Eq. 15.42 becomes

$$\boxed{W_{\text{net}} = [N_J - N_{J+1}]B_J\rho}. \qquad (15.43)$$

The relative band intensities will depend upon the difference between the number of molecules in the lower and upper energy states involved in the transition.

EXAMPLE 15.4

The $J = 0 \rightarrow J = 1$ pure rotational transition of BeH is seen at a wave number of 20.62 cm^{-1}. Assuming that the energy density per unit frequency of the radiation field at the location of the molecule is that from a blackbody at 1,000 K [see Eq. 11.29], do you expect spontaneous emission to be important in this case?

Solution

Equation 11.29 gives us Planck's expression for the energy density per unit wavelength:

$$u(T, \lambda)d\lambda = \frac{8\pi hc}{\lambda^5}\left\{\exp\left[\frac{hc}{\lambda kT}\right] - 1\right\}^{-1}d\lambda. \qquad (1)$$

This equation may be easily converted to an energy density per unit frequency by using the facts that

$$\nu = \frac{c}{\lambda} \qquad (2)$$

and

$$\left|\frac{d\nu}{d\lambda}\right| = \frac{c}{\lambda^2}. \qquad (3)$$

Substituting of Eqs. 2 and 3 into Eq. 1 produces

$$\mu(T, \nu)d\nu = \frac{8\pi h\nu^3}{c^3}\left\{\exp\left[\frac{h\nu}{kT}\right] - 1\right\}^{-1}d\nu = \rho d\nu. \qquad (4)$$

Therefore,

$$\rho = \frac{8\pi h\nu^3}{c^3}\left\{\exp\left[\frac{h\nu}{kT}\right] - 1\right\}^{-1}. \qquad (5)$$

The transition frequency for the $J = 0 \rightarrow J = 1$ rotational transition of BeH is related to the transition wave number $\bar{\nu}$ by

$$\nu = c\bar{\nu} = (2.9979 \times 10^{10} \text{ cm s}^{-1})(20.62 \text{ cm}^{-1}) = 6.182 \times 10^{11} \text{ s}^{-1}. \qquad (6)$$

Equation 15.42 shows that the importance of spontaneous emission relative to stimulated emission depends upon the ratio

$$\frac{\text{spontaneous}}{\text{stimulated}} = \frac{\dfrac{8\pi h\nu^3}{c^3}}{\rho}. \tag{7}$$

Inserting Eq. 5 into Eq. 6 gives

$$\frac{\text{spontaneous}}{\text{stimulated}} = \left\{ \exp\left[\frac{h\nu}{kT}\right] - 1 \right\} = \left\{ \exp\left[\frac{(6.6261 \times 10^{-34})(6.182 \times 10^{11})}{1.381 \times 10^{-23}(1000)}\right] - 1 \right\}$$

$$= 0.0301. \tag{8}$$

Thus, spontaneous emission is responsible for only about 3% of the total emission. For a related exercise, see Problem 15.11.

In a microwave measurement of a pure rotational spectrum, monochromatic (single-wavelength) radiation is usually generated using an electronic device called a *klystron*. This radiation source is capable of producing a sufficiently broad range of wavelengths, so that the absorption intensity as a function of radiation frequency can be obtained. The radiation beam is directed into the sample cell along a hollow metal pipe called a *waveguide*. If the molecule possesses a permanent or induced dipole moment, stimulated absorption from the radiation beam will occur at frequencies that are in resonance with the rotational-energy spacings. The intensity of radiation in the transmitted beam at frequency ν, $I'(\nu)$, is measured by a suitable detector, which, for microwave spectroscopy, is usually a crystal diode in electrical contact with a semiconductor. The difference in intensity between the incident and transmitted beams, $I(\nu) - I'(\nu)$, gives the magnitude of the absorption at frequency ν. A plot of this difference versus ν gives the rotational spectrum. Figure 15.9 shows a schematic of such a microwave spectrometer.

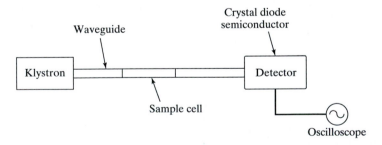

▶ **FIGURE 15.9**
Schematic diagram of the essential components of a rotational microwave spectrometer. The sample is a gas maintained at pressures in the range from 0.01 to 0.10 torr. Higher pressures induce broadening of the spectral bands.

The preceding discussion tells us that we will produce not only stimulated absorption from the microwave beam, but also stimulated emission. Some spontaneous emission will also occur. However, the direction of the emitted radiation will be randomly distributed over all possible spherical polar angles θ and ϕ about the emitting center. As a result, very little of the emitted radiation will reach the crystal diode detector. Therefore, the measurement, in this case, gives us only the extent of stimulated absorption, and because of that, the rotational band intensities will be accurately given by Eqs. 15.36 and 15.37, not Eq. 15.43.

Consider two rotational absorption bands in the foregoing experiment. Let the first result from the transition $J \to J + 1$ and the second from $J' \to J' + 1$. Equations 15.36 and 15.37 tell us that the ratio of the absorption intensities for these two transitions will be

$$\frac{I_J}{I_J} = \frac{N_J B_J}{N_{J'} B_{J'}} = \frac{B_J}{B_{J'}} \frac{g_J}{g_{J'}} \exp\left[-\frac{(E_J - E_{J'})}{kT}\right]. \tag{15.44}$$

In general, the Einstein coefficients B_J and $B_{J'}$ are not the same, because the transition matrix elements for the two transitions are not identical. However, their ratio is usually not too far removed from unity, and the intensity ratio is affected far more by the exponential Boltzmann factor than by the Einstein coefficient ratio. As a result, we expect the relative intensities of the absorption bands to reflect primarily the Boltzmann factors.

EXAMPLE 15.5

The $J = 0 \to J = 1$ pure rotational transition of BeH is seen at a wave number of 20.62 cm^{-1}. Assuming that the Einstein absorption coefficients are independent of J, compute the expected absorption intensity ratio between the $J = 0 \to J = 1$ and the $J = 3 \to J = 4$ microwave absorption bands for BeH at 300 K.

Solution

Equation 15.26 shows that the $J = 0 \to J = 1$ transition occurs at $\Delta E = 2Bh$. Therefore, we have, from Table 13.2,

$$Bh = \frac{20.62 \text{ cm}^{-1}}{2} \times \frac{2625.5 \text{ kJ mol}^{-1}}{219{,}475 \text{ cm}^{-1}} \times \frac{1000 \text{ J}}{\text{kJ}} \times \frac{1}{6.0221 \times 10^{23} \text{ mol}^{-1}}$$

$$= 2.048 \times 10^{-22} \text{ J}. \tag{1}$$

The energy of the initial states for the two transitions are obtained with the use of Eq. 15.25:

$$E_0 = J(J + 1)Bh = 0; \tag{2}$$

$$E_3 = J(J + 1)Bh = 3(4)(2.048 \times 10^{-22}) \text{ J} = 2.458 \times 10^{-21} \text{ J}. \tag{3}$$

Since the degeneracy $g_J = 2J + 1$, we obtain

$$\frac{I_0}{I_3} = \frac{g_0}{g_3} \exp\left[-\frac{(E_0 - E_3)}{kT}\right] = \frac{1}{7} \exp\left[\frac{2.458 \times 10^{-21} \text{ J}}{(1.381 \times 10^{-23} \text{ J K}^{-1})(300 \text{ K})}\right] = 0.259. \tag{4}$$

For related exercises, see Problems 15.12, 15.14, 15.15, 15.16, and 15.17.

15.3.3   The Stark Effect

In Chapter 13, we found that the magnetic moment of a molecule or an atom will interact with an imposed magnetic field $\mathcal{B}$ to produce an energy change given by [See Eq. 13.35]

$$\Delta E_{\text{mag}} = -[\mu_m]_z \mathcal{B},$$

where the Z-axis is taken to be the direction of the magnetic field. The same type of effect occurs when we place a molecule that possesses an electric dipole moment μ_e in an electric field $\mathcal{E}$: The interaction of the electric field with the electric dipole produces the energy change. If the Z-axis is taken to be the direction of the electric field, the energy change is

$$\Delta E_{\text{electric}} = \Delta E_{\text{Stark}} = -[\mu_e]_z E = \mathcal{E} \mu_e \cos \theta, \tag{15.45}$$

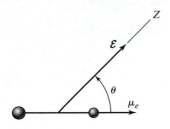

▲ FIGURE 15.10
The rotational Stark effect. θ is the angle between the electric dipole vector μ_e and the electric field $\mathcal{E}$ which is taken to lie along the Z-axis.

where θ is the angle between μ_e and the Z-axis, as shown in Figure 15.10. The energy change ΔE_{Stark} induced by the electric field is called the *Stark effect*.

Now consider a diatomic rigid rotor whose Hamiltonian $\mathcal{H}$ is given by Eq. 12.97. The rotational eigenfunctions are the spherical harmonics $Y_J^M(\theta, \phi)$, and the energy eigenvalues, given by Eq. 15.25, are $E_J = J(J + 1)Bh$. Suppose we now place this rigid rotor in a uniform electric field $\mathcal{E}$. Since the presence of this field will change the energy in the manner described by Eq. 15.45, we expect the Hamiltonian of the system to undergo a similar change, so that the new Hamiltonian is

$$\mathcal{H} = \mathcal{H}_o + \mu_e \mathcal{E} \cos \theta, \tag{15.46}$$

where $\mathcal{H}_o$ is the zeroth-order Hamiltonian when the electric field is zero. Because the change in energy due to the field is very small, perturbation theory will provide an accurate measure of that change.

The first-order perturbation correction involves simply averaging the perturbation $\mathcal{H}' = \mu_e \mathcal{E} \cos \theta$ over the zeroth-order eigenfunctions and adding the result to the zeroth-order energies. [See Eqs. 14.7 and 14.8.] Therefore, the first-order energy eigenvalues for quantum state J are

$$E_J = E_{Jo} + \langle \mathcal{H}' \rangle$$

$$= J(J + 1)Bh + \mu_e \mathcal{E} \int_{\phi=0}^{2\pi} \int_{\theta=0}^{\pi} [Y_J^M(\theta, \phi)]^* \cos \theta \, Y_J^M(\theta, \phi) \sin \theta \, d\theta \, d\phi.$$

Unfortunately, first-order perturbation theory does not give us an improved result, because the integral $\langle \mathcal{H}' \rangle$ is zero for all values of J and M. (See Problem 15.18.) It is, therefore, necessary to use second-order perturbation theory to obtain the energy change due to the Stark effect. Since this is the only point at which second-order theory will be needed in the text, we simply give the result without derivation. For quantum state (J, M) with $J \neq 0$,

$$E_{JM} = J(J + 1)Bh - \frac{I\mu_e^2 \mathcal{E}^2}{\hbar^2}\left[\frac{3M^2 - J(J + 1)}{J(J + 1)(2J - 1)(2J + 3)}\right]. \tag{15.47}$$

For the $J = 0$ ground state, we must have $M = 0$, so the term in brackets in Eq. 15.47 becomes

$$\left[\frac{-J(J + 1)}{J(J + 1)(2J - 1)(2J + 3)}\right] = \left[\frac{-1}{(2J - 1)(2J + 3)}\right] = \frac{1}{3}.$$

This gives a ground-state energy

$$E_{00} = -\frac{I\mu_e^2 \mathcal{E}^2}{3\hbar^2}. \tag{15.48}$$

Note that the presence of the electric field partially removes the $2J + 1$ degeneracy of the rotational states. States with $M = 0$ are nondegenerate, while states with $M \neq 0$ are twofold degenerate. Measurement of the spacings between the Stark-perturbed rotational lines allows us to obtain highly accurate values for the permanent electric dipole moment of the molecule. (See Problem 15.19.) The Stark effect is also useful for enhancing signal detection in spectroscopic measurements. It is easier to amplify an alternating signal than a steady one. Consequently, microwave spectrometers often impose an alternating electric field whose strength is on the order of 10^5 V m^{-1}. The presence of this oscillating field causes the resonant frequency of the rotational spectral lines to exhibit a corresponding oscillation, thereby making the detected signals easier to amplify.

15.3.4 Centrifugal Distortion

Microwave rotational spectra of real diatomic molecules differ slightly from that predicted by the equations developed in the foregoing sections. Real molecules are not rigid; they exhibit vibrational motion and, if sufficient energy is inserted into the molecule, will undergo dissociation to their constituent atoms. For this reason, the energy eigenvalues of the rigid rotor are not precisely equal to those of the real molecule. As a result, the rotational bands, which are predicted to lie at energies of $2Bh$, $4Bh$, $6Bh$, etc., are actually displaced slightly from these values.

In Chapter 12, we noted that the effect of rotational motion is to create a centrifugal force that acts to increase the bond length of the diatomic molecule. The larger average bond length results in an increase of the effective moment of inertia of the molecule. Since the rotational constant is given by $B = \hbar^2/(2Ih)$, the increase in the moment of inertia produces a smaller rotational constant. The energy states are, therefore, displaced toward lower values, which reduces the adjacent-level spacing and moves the rotational bands toward lower energies. The spectroscopist refers to such a displacement as a *redshift* of the band; when the displacement is toward higher energies, the term is *blueshift*.

If the interatomic potential is accurately described by a Morse potential, we can use Eq. 12.159 to provide a quantitative estimate of the magnitude of the centrifugal distortion effect. Replacing ℓ with J in this equation and focusing only upon the rotational terms, we obtain

$$E_J = J(J + 1)Bh - \frac{\hbar^4}{4\mu^2\beta^2 R_e^6 D}[J(J + 1)]^2 , \qquad (15.49)$$

where β, R_e, and D are the Morse parameters that are related to the vibrational frequency, the equilibrium bond length, and the dissociation energy, respectively. Equation 12.155 relates β to the reduced mass of the diatomic molecule, the vibration frequency ν_o, and D:

$$\beta^2 = \frac{2\pi^2\mu\nu_o^2}{D}.$$

If we use this result along with the definition $B = \hbar^2/(2\mu R_e^2 h)$, the distortion constant D_J can be expressed in a more compact form as

$$D_J = \frac{\hbar^4}{4\mu^2\beta^2 R_e^6 D} = \frac{\hbar^4}{4\mu^2 R_e^4} \times \frac{1}{\beta^2 R_e^2 D} = B^2 h^2 \times \frac{D}{2\pi^2\mu\nu_o^2 R_e^2 D}$$

$$= \frac{B^2}{\nu_o^2} \times \frac{2h^2}{4\pi^2\mu R_e^2} = \frac{B^2}{\nu_o^2} \times \frac{4\hbar^2}{2\mu R_e^2} = \frac{4B^3 h}{\nu_o^2}. \qquad (15.50)$$

Consequently, if the intermolecular potential is accurately described by a Morse potential, the rotational-energy eigenvalues will be reasonably well represented by

$$E_J = J(J + 1)Bh - D_J J^2(J + 1)^2 , \qquad (15.51)$$

with D_J given by Eq. 15.50.

As expected, the centrifugal distortion term lowers the rotational-energy eigenvalues. As the vibrational force constant increases, so does the vibrational frequency, and the bond becomes more difficult to stretch. We would expect this to lower the extent of centrifugal distortion. Equation 15.50 shows this to be the case: D_J decreases rapidly as ν_o increases. In actual applications, D_J is treated as an adjustable parameter to be fitted to the measured rotational spectra.

15.3.5   Polyatomic Molecules

In Chapter 12, we obtained approximate expressions for the rotational-energy states of spherical- and symmetric-top molecules. In this section, we will examine the expected appearance of the microwave spectra of such molecules.

Spherical-top molecules have a high degree of symmetry, so that all three principal moments of inertia are equal. Some examples are CH_4, CCl_4, SF_6, and SiH_4. Such highly symmetric molecules have no permanent dipole moment; therefore, we do not expect to observe their pure rotational spectra. However, centrifugal distortion in high rotational states can produce deviations from perfect spherical top symmetry, so that the molecule momentarily possesses a dipole moment. If the concentration of such distorted molecules is sufficiently high, weak microwave absorptions corresponding to rotational transitions between high J-states can sometimes be observed. SiH_4 is an example: Its pure rotational spectrum has been detected through the use of long path lengths and high pressures. Note that this kind of effect cannot occur in a homonuclear diatomic molecule. No matter what the extent of centrifugal distortion, homonuclear diatomic molecules will never exhibit a dipole moment.

Symmetric-top molecules have two of the three principal moments of inertia equal. Examples are NH_3, XeF_4, NF_3, and all diatomic molecules. In Eq. 12.134, we showed that the energy states for such a rigid rotor can be represented with reasonable accuracy by the expression

$$E_{J,K} = \frac{J(J + 1)\hbar^2}{2I_b} + \frac{K^2\hbar^2}{2}\left[\frac{1}{I_a} - \frac{1}{I_b}\right]$$

$$\text{for } J = 0, 1, 2, 3, \ldots \quad \text{and} -J \leq K \leq J, \tag{15.52}$$

where I_a is the unique moment of inertia and I_b is the moment that is common to two of the principal axes. The second term in Eq. 15.52 can be either positive or negative, depending upon the relative magnitudes of I_a and I_b. If we have $I_a < I_b$, so that the second term is positive, the molecule is said to be a *prolate top*. If we have $I_a > I_b$, the rotor is called an *oblate top*. The prolate and oblate tops are sometimes described as being "saucer-like" and "cigar-like," respectively.

The electric dipole selection rules for pure rotational transitions in symmetric-top molecules are

$$\boxed{\Delta J = 0 \quad \text{or} \pm 1 \quad \text{and } \Delta K = 0}. \tag{15.53}$$

Some texts give the ΔJ selection rule as $\Delta J = \pm 1$. Since $\Delta K = 0$, the selection rule $\Delta J = 0$ for a pure rotational transition means that there is no transition at all. Therefore, the two statements for the ΔJ selection rule are equivalent in this case. However, for vibrational–rotational spectra, which we will discuss in the next section, $\Delta J = 0$ transitions are possible. Therefore, the best statement of the selection rule is the one given in Eq. 15.53. A derivation of these selection rules requires the development of appropriate eigenfunctions for symmetric-top rotors. Since this is beyond the scope of this text, Eq. 15.53 is presented without proof.

EXAMPLE 15.6

The moments of inertia of NH_3 are $I_b = I_c = 2.816 \times 10^{-47}$ kg m^2 and $I_a = 4.43 \times 10^{-47}$ kg m^2. **(A)** Is NH_3 a prolate or an oblate symmetric top? **(B)** Calculate the energies for all NH_3 rotational states with $J = 0, 1, 2$. Plot these states on an energy-level diagram, and indicate the allowed transitions with appropriate arrows. Indicate the degeneracy of the levels. **(C)** Describe the expected appearance of a microwave rotational spectrum of NH_3 that results from the $J = 0, 1$, and 2 rotational states.

Solution

(A) Since $I_a > I_b$, NH_3 is an oblate symmetric top.

(B) The rotational-energy levels for a symmetric top are given by Eq. 15.52:

$$E_{J,K} = \frac{J(J+1)\hbar^2}{2I_b} + \frac{K^2\hbar^2}{2}\left[\frac{1}{I_a} - \frac{1}{I_b}\right]. \tag{1}$$

Substituting the values of the moments of inertia gives

$$E_{J,K} = \frac{(6.626 \times 10^{-34}\,\text{J s})^2}{4(3.14159)^2(2)(2.816 \times 10^{-47}\,\text{kg m}^2)}J(J+1)$$

$$+ K^2\frac{(6.626 \times 10^{-34}\,\text{J s})^2}{4(3.14159)^2(2)}\left[\frac{1}{4.43 \times 10^{-47}} - \frac{1}{2.816 \times 10^{-47}}\right]\text{kg}^{-1}\,\text{m}^{-2}, \tag{2}$$

so that

$$E_{J,K} = 1.975 \times 10^{-22}J(J+1) - 7.194 \times 10^{-23}K^2\,\text{J}. \tag{3}$$

For any given J, K must lie in the range $-J \le K \le J$, so the possible energy states are those shown in the following table:

J	K	$E_{J,K}$ (J)	J	K	$E_{J,K}$ (J)
0	0	0	2	−1	1.11×10^{-21}
1	−1	3.23×10^{-22}	2	0	1.18×10^{-21}
1	0	3.95×10^{-22}	2	1	1.11×10^{-21}
1	1	3.23×10^{-22}	2	2	8.97×10^{-22}
2	−2	8.97×10^{-22}			

(C) The only allowed transitions are $(J = 0, K = 0) \rightarrow (J = 1, K = 0)$, $(J = 1, K = 0) \rightarrow (J = 2, K = 0)$, $(J = 1, K = 1) \rightarrow (J = 2, K = 1)$, and $(J = 1, K = -1) \rightarrow (J = 2, K = -1)$. The energies and wave numbers for these transitions are as follows:

Transition	ΔE (J)	Wave Number (cm^{-1})
$(J = 0, K = 0) \longrightarrow (J = 1, K = 0)$	3.95×10^{-22}	19.88
$(J = 1, K = 0) \longrightarrow (J = 2, K = 0)$	7.86×10^{-22}	39.57
$(J = 1, K = 1) \longrightarrow (J = 2, K = 1)$	7.86×10^{-22}	39.57
$(J = 1, K = -1) \longrightarrow (J = 2, K = -1)$	7.86×10^{-22}	39.57

Therefore, only two rotational bands will appear from the $J = 0, 1$, or 2 states. These occur at 19.88 and 39.57 cm^{-1}. The results are illustrated in Figure 15.11.

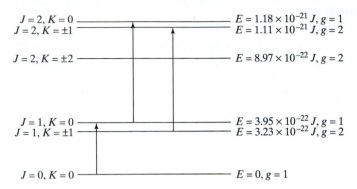

$J = 2, K = 0$ —— $E = 1.18 \times 10^{-21}$ J, $g = 1$
$J = 2, K = \pm 1$ —— $E = 1.11 \times 10^{-21}$ J, $g = 2$

$J = 2, K = \pm 2$ —— $E = 8.97 \times 10^{-22}$ J, $g = 2$

$J = 1, K = 0$ —— $E = 3.95 \times 10^{-22}$ J, $g = 1$
$J = 1, K = \pm 1$ —— $E = 3.23 \times 10^{-22}$ J, $g = 2$

$J = 0, K = 0$ —— $E = 0$, $g = 1$

▲ FIGURE 15.11
Rotational-energy levels for NH_3. Allowed transitions are indicated by the arrows. The energies and state degeneracies are given to the right, the quantum number assignments to the left.

The rotational-energy levels of linear polyatomic molecules are given by Eq. 12.135:

$$E_J = \frac{J(J + 1)\hbar^2}{2I} = J(J + 1)Bh \qquad \text{for } J = 0, 1, 2, \ldots. \tag{15.54}$$

This equation is identical to that for a diatomic molecule. The selection rule is also the same: $\Delta J = \pm 1$. Therefore, the analysis of the rotational spectrum of a linear polyatomic molecule is the same as that for a diatomic molecule.

The rotational spectra for asymmetric-top molecules are quite complex. It is possible to unravel the details of the rotational spectra of such molecules by using computer codes that guess the molecular structure, compute the moments of inertia, predict the location of the spectral lines, and then iteratively refine the assumed structure until agreement is obtained between measurement and calculation. These methods will not be covered in this text.

Microwave spectra of polyatomic molecules are sometimes very useful in that they provide a means of determining the structure of the molecule. In the case of diatomic molecules, we have already seen how a measurement of the rotational band spacings allows the equilibrium distance to be determined. This distance is, of course, an average over the vibrational motion of the molecule, which is not actually a rigid rotor.

To determine the structure of a polyatomic molecule, we must measure many more variables than a single bond distance. In general, if the molecule contains N atoms, we will need about $N(N - 1)/2$ variables to specify its structure. Symmetry considerations can reduce this number, but we will generally require more than a single rotational spectrum to do the job. The additional data can often be obtained by measuring the rotational spectra of various isotopically labeled molecules. If the Born–Oppenheimer approximation is accurate (and it usually is at the energies characteristic of chemical processes), the molecular structure will be independent of isotopic mass. Such measurements can, therefore, provide enough data to uniquely determine that structure. Problem 15.24 illustrates the method for the case of HCN.

15.4   Vibrational Spectroscopy

When the wave number of the incident electromagnetic radiation increases into the range from 0.5×10^3 cm^{-1} to 3×10^3 cm^{-1}, the photons have sufficient energy to induce vibrational as well as rotational transitions. Therefore, we expect to see absorptions corresponding to transitions in which both the vibrational and rotational states of the molecule are changed, provided that the molecule possesses either a permanent or an induced dipole moment and provided that the appropriate selection rules are satisfied. The measurement of these transition energies provides a very useful "fingerprint" of the molecule that finds numerous applications in all fields of chemistry and chemical engineering.

15.4.1   Experimental Methods

Many spectrometers contain three essential elements: the radiation source, the dispersing element, and the detector. The radiation source can be any device that is capable of producing sufficiently intense radiation whose wavelengths span the region for which absorptions are expected to occur. For radiation in the far infrared (IR), a mercury arc inside a quartz envelope can be used. The energy from the mercury arc heats the quartz jacket, causing it to emit radiation as an intense blackbody. The emitted radiation falls in the far IR region of the spectrum. Heated ceramic filaments containing rare-earth oxides emit radiation in the near IR. A tungsten–iodine lamp produces radiation in the visible region of the spectrum, while an electric discharge through deuterium or xenon produces radiation in the near ultraviolet (UV). Lasers generate monochromatic radiation that can often be tuned over a range of frequencies.

In a simple absorption measurement, the radiation produced by the source must be dispersed into its component wavelengths to permit the absorption at each wavelength to be measured. In the simplest case, this can be done using a prism that spatially separates the different wavelengths in the same way that raindrops separate the wavelengths from the sun to produce a rainbow. In more sophisticated instruments, the dispersion is accomplished using a diffraction grating, which consists of a plate in which fine grooves have been cut about 1,000 nm apart. The grating is coated with a reflective substance. As the waves reflect from the surface, interference patterns develop between waves of different wavelengths. Constructive interference occurs at different angles for different wavelengths, thus permitting a separation of the wavelength components of the radiation. Figure 15.12 shows a schematic of such a diffraction grating. The phenomenon of constructive interference is discussed next and again in Chapter 16.

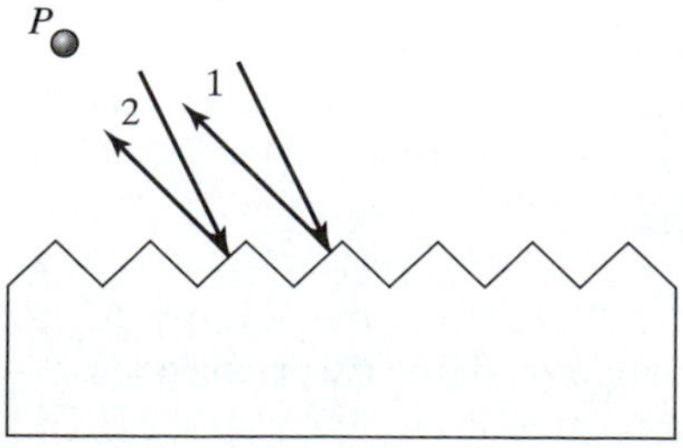

◀ FIGURE 15.12

Schematic diagram of a diffraction grating. Grooves in the grating whose spacing is characteristic of the wavelength of the radiation being dispersed by the element cause the waves to be reflected from the surface as shown. The path length traveled by Waves 1 and 2 to reach point *P* is not the same: Wave 1 must travel a larger distance to reach that point. As a result, interference patterns develop such that constructive interference occurs at specific angles, depending upon the wavelength. This interference permits the wavelengths present in the radiation to be dispersed, or separated.

The detector can be any device that permits radiation to be converted into an electrical signal whose intensity can be measured. Semiconductor devices are often used. In the optical and UV regions of the spectrum, photomultipliers are employed. These devices convert each photon into an ejected electron, which is accelerated through a potential field. When the high-speed electron strikes a suitably coated screen, a large number of additional electrons are ejected and then converted to an electrical current in an external circuit for measurement. As previously mentioned, crystal diodes are used to detect microwave radiation.

Most commercially available modern spectrometers operating in the near and mid-IR spectral region use Fourier transform techniques so that the dispersion step is unnecessary. We shall discuss the principle of the Fourier transform infrared (FTIR) spectrometer in some detail, since this principle also forms the foundation for diffraction and nuclear magnetic resonance (NMR) spectroscopy, which will be examined in Chapter 16.

An FTIR instrument uses an interferometer—typically a Michelson interferometer—whose basic setup is shown in Figure 15.13. A suitable source produces a continuum of frequencies spanning the range required for the particular type of spectrum being recorded. Radiation with such a continuum of frequencies is sometimes called *white light* by analogy with radiation containing a continuum of frequencies in the visible region of the electromagnetic spectrum. The radiation strikes a beam splitter that separates the radiation into two components, labeled Ψ_1 and Ψ_2 in the figure. Ψ_1 continues in the same direction until it strikes a fixed mirror, reflects backward, and again strikes the splitter on the opposite side, this time being partially reflected downward. At the same time, Ψ_2 moves upward toward a mirror that is moving with a fixed velocity. It strikes this mirror, reflects downward, and strikes the splitter a second time. Some of Ψ_2 is transmitted to combine with the portion of Ψ_1 that was reflected downward. The recombined portions of Ψ_1 and Ψ_2 continue into the sample cell, where they induce transitions that result in absorption of some of the radiation at the fre-

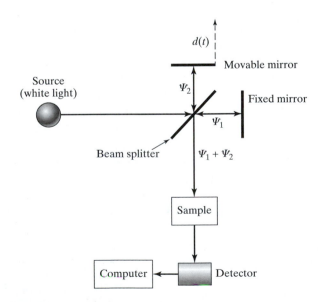

▲ FIGURE 15.13
Schematic diagram of an FTIR spectrometer using a Michelson interferometer.

quencies that are in resonance with the transitions. The transmitted portion of $\Psi_1 + \Psi_2$ is then sent to an appropriate detector and from there to a computer, where it is analyzed to obtain the absorption spectrum of the sample.

We now need to analyze what will be seen at the detector in some simple cases. Let us first assume that only one wavelength λ is produced by the source and that the sample cell is empty. Under these conditions, we can represent Ψ_1 by a simple sine function of the form

$$\Psi_1 = A \sin\left[\frac{2\pi x}{\lambda}\right], \tag{15.55}$$

where x is the distance traveled by the wave after it has left the source. Ψ_2, however, has traveled a greater distance than Ψ_1, since the mirror from which Ψ_2 is reflected is moving upward at a constant speed v_o. The additional distance traveled by Ψ_2 is, therefore, $d(t) = v_o t$. Consequently, when Ψ_2 rejoins Ψ_1 at the beam splitter, it will be described by

$$\Psi_2 = A \sin\left[\frac{2\pi[x + d(t)]}{\lambda}\right]. \tag{15.56}$$

Because the sample cell is empty, no emissions or absorptions will occur, and the radiation beam reaching the detector will be the sum

$$S(t) = \Psi_1 + \Psi_2 = A\left\{ \sin\left[\frac{2\pi x}{\lambda}\right] + \sin\left[\frac{2\pi[x + d(t)]}{\lambda}\right] \right\}. \tag{15.57}$$

Equation 15.57 shows that the resulting intensity will depend upon the time at which beam Ψ_2 struck the moving mirror. If, at the moment of impact, $d(t) = n\lambda$, where n is an integer, we will have

$$S(t) = A\left\{ \sin\left[\frac{2\pi x}{\lambda}\right] + \sin\left[\frac{2\pi[x + n\lambda]}{\lambda}\right] \right\} = 2A \sin\left[\frac{2\pi x}{\lambda}\right], \tag{15.58}$$

since

$$\sin\left[\frac{2\pi[x + n\lambda]}{\lambda}\right] = \sin\left[\frac{2\pi x}{\lambda} + 2n\pi\right] = \sin\left[\frac{2\pi x}{\lambda}\right].$$

In other words, the beam reaching the detector will have twice the amplitude and four times the intensity (which is proportional to the square of the amplitude) as Ψ_1 alone if the difference in path length is an integral multiple of the wavelength. This condition is called *constructive interference* (CI). If, on the other hand, we have $d(t) = (n + \frac{1}{2})\lambda$ at the moment of impact, the result will be $S(t) = 0$. This result, which occurs when the difference in path length is a half-integral multiple of the wavelength, is called *destructive interference* (DI). When $d(t)$ falls in between these extremes, the intensity detected will be an intermediate value. (See Problem 15.27.)

To illustrate the preceding concepts, suppose that the speed of the mirror in our FTIR spectrometer is $v_o = 10^{-2}$ cm s^{-1}, that $A = 1$ arbitrary unit and that the distance from the source to the detector for the path followed by Ψ_1 is $10^4 \lambda$. In this case, Eq. 15.57 becomes

$$S(t) = \sin[2 \times 10^4 \pi] + \sin\left[2 \times 10^4 \pi + \frac{0.02\pi t}{\lambda}\right] = \sin\left[\frac{0.02\pi t}{\lambda}\right].$$

In this case, we expect a sinusoidal signal at the detector. Let us assume that $S(t)$ is found to have the form shown in Figure 15.14. Inspection shows that the

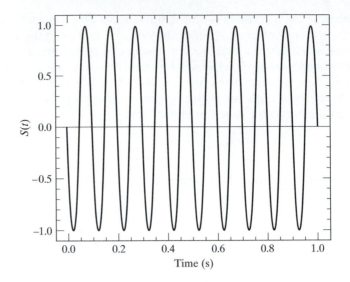

▶ FIGURE 15.14
Output from a Michelson interferometer when the incident radiation contains a single wavelength equal to 10^{-3} cm and the moving mirror has a speed of 10^{-2} cm s^{-1}. It is assumed that the distance from the source to the detector for waves not reflecting from the moving mirror is $10^4\,\lambda$. The signal at the detector has a modulation frequency of 10 s^{-1}. With this information, the wavelength of the incident radiation can be easily calculated. (See text for details.)

sine function has gone through 10 complete oscillations in 1 second. Therefore, the detected signal has a modulation frequency $\nu_d = 10$ s^{-1}. As usual, the frequency is the speed divided by the wavelength ($\nu_d = v_o/\lambda$), so that the single wavelength present in the radiation must be

$$\frac{v_o}{\nu_d} = \frac{10^{-2}\text{ cm s}^{-1}}{10\text{ s}^{-1}} = 10^{-3}\text{ cm}.$$

The corresponding frequency of the radiation is, therefore,

$$\nu = \frac{c}{\lambda} = \frac{2.9979 \times 10^{10}\text{ cm s}^{-1}}{10^{-3}\text{ cm}} = 2.9979 \times 10^{13}\text{ s}^{-1}.$$

In general, we expect to have

$$\nu = \frac{c}{\lambda} = \frac{c\nu_d}{v_o}. \tag{15.59}$$

Consequently, a determination of the radiation frequency is directly related to a determination of the signal modulation frequency and the speed of the moving mirror.

In the preceding case, it is very easy to determine ν_d from a simple inspection of Figure 15.14. However, we need not rely on inspection to determine the frequencies present in $S(t)$. Instead, the frequencies present in any signal $S(t)$ can always be determined by taking what is known as the *Fourier transform* (FT) of $S(t)$, defined as

$$F(\nu) = (2\pi)^{-1/2} \int_{t=0}^{\infty} S(t) e^{i2\pi\nu}\, dt. \tag{15.60}$$

The frequencies present in $S(t)$ are then obtained by plotting $|F(\nu)|^2$ versus ν. Such a plot is called a *power spectrum*. Figure 15.15 shows the power spectrum of $S(t)$ given in Figure 15.14. The power spectrum clearly shows that $\nu_d = 10$ s^{-1}. The fact that Eq. 15.60 yields all the frequencies present in $S(t)$ is by no means obvious. The proof that it does requires that we undertake a study of Fourier series expansions and their properties. While this is not difficult, it is time consuming. Therefore, we present the statement without proof. The computer connected to the FTIR spectrometer takes the Fourier transform

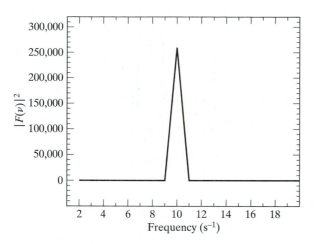

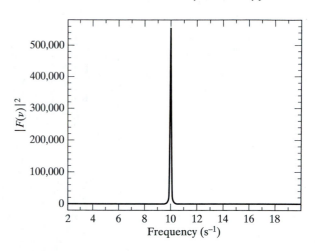

▲ **FIGURE 15.15**
The power spectrum of the Fourier transform (FT) of the data shown in Figure 15.14. The FT clearly identifies the single modulation frequency as that corresponding to $\nu_d = 10 \text{ s}^{-1}$. The resolution of the FT, however, is limited because of the very short duration over which data were collected (1 second).

▲ **FIGURE 15.16**
The power spectrum of the Fourier transform of the data for the system shown in Figure 15.14, but with the data collection time extended to 15.35 s. A comparison of this figure with the power spectrum depicted in Figure 15.15 illustrates the increased signal resolution achieved by increasing the sampling time. Similar behavior is seen in Figures 15.3, 15.4, and 15.5.

of $S(t)$ and then deduces the intensity of each frequency present in the signal. In the actual experiment, $S(t)$ is not measured at infinite time. Therefore, the infinite upper limit in Eq. 15.60 is replaced with t_{max}. If t_{max} is sufficiently large, the error introduced will be very small. This point is illustrated in Fig. 15.16, which shows the resolution of the modulation frequency for the system of Figure 15.14 when t_{max} is increased to 15.35 seconds. A comparison with Figure 15.15 shows the effect of increasing the signal sampling time.

The advantage of Eq. 15.60 may be seen by considering the case when two wavelengths, λ_1 and λ_2, are present in the signal. Let us again assume that $v_o = 10^{-2} \text{ cm s}^{-1}$, $A = 1$ arbitrary unit for each wavelength, and $x = 10^4 \lambda_1$. Under these conditions, the signal reaching the detector is

$$S(t) = \sin[2 \times 10^4 \pi] + \sin\left[2 \times 10^4 \pi + \frac{0.02\pi}{\lambda_1}\right] + \sin\left[\frac{2 \times 10^4 \pi \lambda_1}{\lambda_2}\right]$$

$$+ \sin\left[\frac{2 \times 10^4 \pi \lambda_1}{\lambda_2} + \frac{0.02\pi t}{\lambda_2}\right] = \sin\left[\frac{0.02\pi t}{\lambda_1}\right] + \sin\left[\frac{2 \times 10^4 \pi \lambda_1}{\lambda_2}\right]$$

$$+ \sin\left[\frac{2 \times 10^4 \pi \lambda_1}{\lambda_2} + \frac{0.02\pi t}{\lambda_2}\right].$$

Figure 15.17 shows the resulting signal reaching the detector for the case where $\lambda_1 = 10^{-3}$ cm and $\lambda_2 = 2.17 \times 10^{-3}$ cm. It is now not so simple to determine the modulation frequencies that are present. However, the power spectrum of $S(t)$ given in Figure 15.18 shows that these frequencies are approximately 5.0 s^{-1} and 10 s^{-1}. If we take spectral data for 21.19 seconds and compute the power spectrum of this signal, the result shown in Figure 15.19 is obtained. The two modulation frequencies are now more accurately defined as 4.62 and 10.00 s^{-1}. The two wavelengths are, therefore, 2.16×10^{-3} cm and 1×10^{-3} cm. Using these values in Eq. 15.59 yields frequencies of $1.39 \times 10^{13} \text{ s}^{-1}$ and $2.998 \times 10^{13} \text{ s}^{-1}$, respectively. When a continuous range of

▶ FIGURE 15.17
Output from a Michelson interfer-ometer when the incident radiation contains two wavelengths equal to 10^{-3} cm and 2.17×10^{-3} cm. The moving mirror has a speed of 10^{-2} cm s^{-1}. It is assumed that the distance from the source to the detector for waves not reflecting from the moving mirror is 10 cm. Extraction of the frequencies pre-sent in these data is more difficult than for the case of a single fre-quency, shown in Figure 15.14.

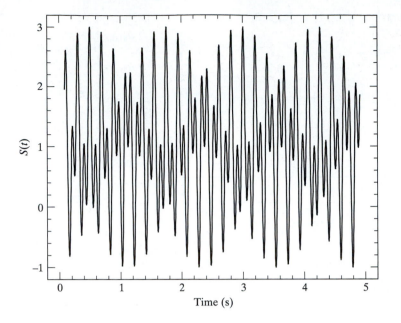

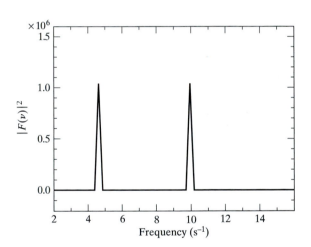

▲ FIGURE 15.18
The power spectrum of the Fourier transform (FT) of the data shown in Figure 15.17. This power spectrum indicates that the two modulation frequencies are approximately 5.0 s^{-1} and 10 s^{-1}. The resolution of the power spectrum can be increased by taking spectral data for a longer period, a situation illustrated in Figure 15.19.

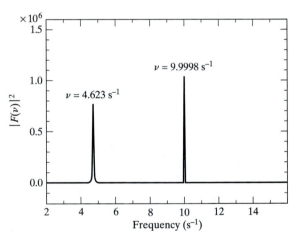

▲ FIGURE 15.19
The power spectrum of the Fourier transform of the data for the system shown in Figure 15.17, but with the data collection time extended to 21.19 s. A comparison of this figure with the power spectrum of Figure 15.18 illustrates the increased resolution achieved by increasing the signal sampling time. Similar behavior is seen in Figures 15.3, 15.4, and 15.5. With the increased resolution, the two modulation frequencies shown in the figure are obtained.

wavelengths is present, the value of the Fourier transform and the associated power spectrum becomes even greater. There would be no way to determine the frequencies present without the help of such mathematical analysis. Problem 15.28 provides an even more complex example illustrating this point.

The FTIR absorption spectrum is taken by measuring two frequency spectra, one with the sample cell empty, the second with the cell containing the material whose spectrum is to be measured. The difference in the power spectra yields the desired absorption spectrum. A particular advantage of an FTIR spectrometer is that the moving mirror's position is determined by using a laser fringing technique that provides such precision on the computed spectra that spectral subtraction of power spectrum band intensities can be performed with great confidence. Since radiation at every frequency reaches the detector and all this radiation is used to compute the power spectrum, an FTIR spectrometer gives a better signal-to-noise ratio than a simple dispersion instrument. In addition, the signal-to-noise ratio can be improved by making repeated spectral scans and having the computer average the results. Such averaging tends to cancel out the random noise present in the measurements. Note that in the UV–visible region of the spectrum, the radiation sources are brighter (more intense) and dispersion instruments compete favorably with FT spectrometers.

15.4.2   Diatomic Molecules

The vibrational–rotational energy levels of a diatomic molecule are described with reasonable accuracy by assuming that the interatomic interaction is given by a Morse potential fitted to the measured dissociation energy, vibration frequency, and equilibrium bond length. The approximate energy levels for such a potential are given by Eq. 12.159, viz.,

$$E_{vJ} = (v + 0.5)h\nu_o - \frac{h^2\nu_o^2}{4D}(v + 0.5)^2 + J(J + 1)Bh - \frac{\hbar^4}{4\mu^2\beta^2R_e^6D}[J(J + 1)]^2$$

$$- \frac{3[1 - (1/\beta R_e)]\hbar^2h\nu_o}{4\mu\beta R_e^3D}(v + 0.5)J(J + 1), \qquad \textbf{(15.61)}$$

where we have again replaced ℓ with J and $\hbar^2/(2I)$ with Bh and where D, β, and R_e are the Morse potential parameters described previously. If the coefficients of the second, fourth, and fifth terms are replaced with adjustable parameters to be fitted to the measured spectra, the results are quite accurate.

For simplicity, let us investigate the expected nature of vibrational–rotational spectra, using only the harmonic oscillator-rigid rotor terms in Eq. 15.61; that is,

$$\boxed{E_{vJ} = (v + 0.5)h\nu_o + J(J + 1)Bh}. \qquad \textbf{(15.62)}$$

For such a system, the rotational selection rule is that which we obtained for rotational spectroscopy, $\Delta J = \pm 1$. The selection rule for vibrational transitions may be estimated by determining the conditions under which the transition matrix element for a harmonic oscillator is nonzero. That is, we wish to know when $H_{ij} = \langle \Psi_i(x)|x|\Psi_j(x)\rangle$ is nonzero, where $\psi_i(x)$ and $\psi_j(x)$ are the harmonic oscillator eigenfunctions. The answer is that H_{ij} will be nonzero only if $i = j \pm 1$. That is, the change in the vibrational quantum number Δv, must be ± 1. Only transitions between adjacent vibrational states are allowed. (See Example 15.7 and Problem 15.29).

In this case, the first-order perturbation approximation leading to Eq. 15.21 is less accurate than for pure rotation of a diatomic. The reason is our assumption that the charge separation q is a constant, an assumption that we made when q was factored out of the integral over the spatial coordinates. For rotation, the dipole moment and charge separation are independent of the rotational angles, but they are not independent of the vibrational amplitude. Hence, the vibrational selection rules are not as accurate as those for rotation.

EXAMPLE 15.7

Show that the electric dipole transition from $v = 0$ to $v = 1$ for a harmonic oscillator is allowed.

Solution

To be an allowed transition, the transition matrix element between the $v = 0$ and $v = 1$ harmonic oscillator states must be nonzero. That is, we must have

$$\langle \psi_o(x)|x|\psi_1(x)\rangle \neq 0, \tag{1}$$

where the eigenfunctions are

$$\psi_o(x) = \left[\frac{\alpha}{\pi}\right]^{1/4} \exp\left[-\frac{\alpha x^2}{2}\right] \tag{2}$$

and

$$\psi_1(x) = \left[\frac{4\alpha^3}{\pi}\right]^{1/4} x \exp\left[-\frac{\alpha x^2}{2}\right]. \tag{3}$$

The transition matrix element is, therefore,

$$H_{01} = \left[\frac{4\alpha^4}{\pi^2}\right]^{1/4} \int_{x=-\infty}^{\infty} \exp\left[-\frac{\alpha x^2}{2}\right] x^2 \exp\left[-\frac{\alpha x^2}{2}\right] dx$$

$$= \left[\frac{4\alpha^4}{\pi^2}\right]^{1/4} \int_{x=-\infty}^{\infty} x^2 \exp[-ax^2]\, dx. \tag{4}$$

Since the integrand is always positive for all values of x in the range $-\infty \leq x \leq \infty$, it follows that $H_{01} \neq 0$, and the transition is allowed.

For a related exercise, see Problem 15.29.

Equation 15.62 and the associated selection rules allow us to predict the energies of all allowed transitions for a system that obeys the harmonic oscillator–rigid rotor approximations. These transitions are summarized in Table 15.2 for vibrational quantum numbers 0 and 1 and for rotational states in the range $0 \leq J \leq 7$. Some of the transitions are illustrated in Figure 15.20.

To the extent that the vibrational–rotational energy states are described by Eq. 15.62, the FTIR spectrum of a polar diatomic molecule will consist of a series of equally spaced bands exactly like those for pure rotational transitions of a diatomic rigid rotor. The spacings between these bands will be $2Bh$. We can now observe two sets of rotational transitions: those for which $\Delta J = +1$ and those for which $\Delta J = -1$. Transitions in the former group are called the *R branch* of the spectrum; those in the latter group are labeled the *P branch*. As in the case of pure rotational spectra, the band intensities for these transitions depend primarily upon the population of the initial state. Since a large majority of the transitions originate from the ground vibrational state, differences in the rotational-state energies and degeneracies determine the relative band intensities.

As an example, let us predict the appearance of the FTIR spectrum for the cation $H^{35}Cl$ at 298 K. The measured equilibrium H–Cl bond length is 1.2746×10^{-10} m. The vibration frequency is 8.9630×10^{13} s^{-1}. Since most spectroscopic data in the infrared region are reported in terms of cm^{-1}, we shall work in that unit. The $H^{35}Cl$ rotational constant is given by Eq. 15.24:

$$B = \frac{h}{8\pi^2 I} = \frac{h}{8\pi^2 \mu R^2}.$$

Table 15.2 Allowed transitions for a harmonic oscillator–rigid rotor system whose selection rules are $\Delta J = \pm 1$ and $\Delta v = \pm 1$

Initial State v	J	Final State v	J	Initial Energy	Final Energy	ΔE
				$\Delta v = +1, \quad \Delta J = +1, \quad R$ Branch		
0	0	1	1	$0.5 h v_o$	$1.5 h v_o + 2Bh$	$h v_o + 2Bh$
0	1	1	2	$0.5 h v_o + 2Bh$	$1.5 h v_o + 6Bh$	$h v_o + 4Bh$
0	2	1	3	$0.5 h v_o + 6Bh$	$1.5 h v_o + 12Bh$	$h v_o + 6Bh$
0	3	1	4	$0.5 h v_o + 12Bh$	$1.5 h v_o + 20Bh$	$h v_o + 8Bh$
0	4	1	5	$0.5 h v_o + 20Bh$	$1.5 h v_o + 30Bh$	$h v_o + 10Bh$
0	5	1	6	$0.5 h v_o + 30Bh$	$1.5 h v_o + 42Bh$	$h v_o + 12Bh$
0	6	1	7	$0.5 h v_o + 42Bh$	$1.5 h v_o + 56Bh$	$h v_o + 14Bh$
				$\Delta v = +1, \quad \Delta J = -1, \quad P$ Branch		
0	1	1	0	$0.5 h v_o + 2Bh$	$1.5 h v_o$	$h v_o - 2Bh$
0	2	1	1	$0.5 h v_o + 6Bh$	$1.5 h v_o + 2Bh$	$h v_o - 4Bh$
0	3	1	2	$0.5 h v_o + 12Bh$	$1.5 h v_o + 6Bh$	$h v_o - 6Bh$
0	4	1	3	$0.5 h v_o + 20Bh$	$1.5 h v_o + 12Bh$	$h v_o - 8Bh$
0	5	1	4	$0.5 h v_o + 30Bh$	$1.5 h v_o + 20Bh$	$h v_o - 10Bh$
0	6	1	5	$0.5 h v_o + 42Bh$	$1.5 h v_o + 30Bh$	$h v_o - 12Bh$
0	7	1	6	$0.5 h v_o + 56Bh$	$1.5 h v_o + 42Bh$	$h v_o - 14Bh$

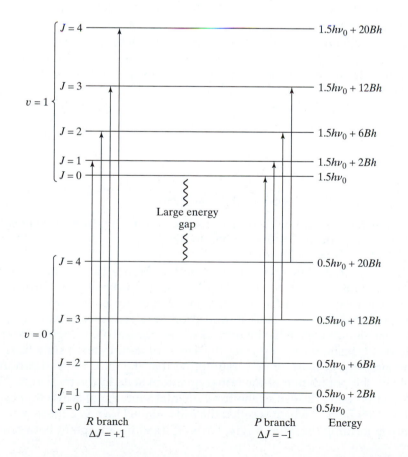

◀ **FIGURE 15.20**
Allowed transitions for a diatomic molecule described by a harmonic oscillator–rigid rotor quantization. Allowed transitions are shown by upward arrows with the R branch on the left and the P branch on the right.

The reduced mass can be computed from the isotopic masses—that is, $^1H = 1.0078$ and $^{35}Cl = 34.969$. We have

$$\mu = \frac{m_H m_{Cl}}{m_H + m_{Cl}} = \frac{(1.0078)(34.969)}{1.0078 + 36.696} \text{ g mol}^{-1} \times \frac{1 \text{ kg}}{1,000 \text{ g}}$$

$$\times \frac{1}{6.02214 \times 10^{23} \text{ mol}^{-1}} = 1.6266 \times 10^{-27} \text{ kg}.$$

Therefore, the rotational constant is

$$B = \frac{6.62608 \times 10^{-34} \text{ J s}}{8(3.1415927)^2(1.6266 \times 10^{-27} \text{ kg})(1.2746 \times 10^{-10} \text{ m})^2}$$

$$= 3.1756 \times 10^{11} \text{ s}^{-1}.$$

If we equate Bh to $hc\bar{\nu}$, we obtain $\bar{\nu} = B/c$, so that

$$\bar{\nu} = \frac{3.1756 \times 10^{11} \text{ s}^{-1}}{2.9979 \times 10^{10} \text{ cm s}^{-1}} = 10.593 \text{ cm}^{-1}.$$

$2Bh$ is, therefore, equivalent to 21.186 cm^{-1}. In terms of cm^{-1}, the vibration frequency is

$$\frac{8.9630 \times 10^{13} \text{ s}^{-1}}{2.9979 \times 10^{10} \text{ cm s}^{-1}} = 2989.7 \text{ cm}^{-1}.$$

Using the results given in Table 15.2, we can easily predict that the first eight R-branch transitions will occur at 3010.9, 3032.1, 3053.3, 3074.4, 3095.6, 3116.9, 3138.0 and 3159.2 cm^{-1}. The first eight P-branch transitions will appear at 2968.5, 2947.3, 2926.1, 2905.0, 2883.8, 2862.6, 2841.4 and 2820.2 cm^{-1}.

We expect the relative band intensities to be determined primarily by the relative populations of the initial quantum states. These are given by Eq. 15.37: $N_J = Cg_J \exp[-E_J/(kT)]$. Thus, the ratio of the peak heights for a transition originating from state J to one originating from the $J = 0$ state will be proportional to

$$\frac{N_J}{N_0} = \frac{g_J}{g_0} \exp\left[-\frac{E_J - E_0}{kT}\right] = (2J + 1) \exp\left[-\frac{J(J + 1)Bh}{kT}\right].$$

At $T = 298 K$, the constant

$$\frac{Bh}{kT} = \frac{(3.1756 \times 10^{11} \text{ s}^{-1})(6.62608 \times 10^{-34} \text{ J s})}{1.38066 \times 10^{-23} \text{ J K}^{-1}(298 K)} = 0.051142.$$

Therefore, the band intensity for a transition originating from state J relative to one arising from $J = 0$ is equal to $(2J + 1) \exp[-J(J + 1)(0.051142)]$. Table 15.3 summarizes the results.

The expected FTIR spectrum is shown in Figure 15.21, where we plot the relative intensity of the band in the downward direction to correspond to the usual experimental practice. Vertical lines are used to represent the absorption bands. The rotational transition involved in each band is given by the small numbers at the band peaks, with the initial state given first. Note that transitions arising from the same initial rotational state have the same intensity, because of our assumption that the intensity is determined solely by the population of the initial quantum state in the transition. Note also that no band is seen at the fundamental vibrational wave number of $H^{35}Cl$, 2,989.7 cm^{-1}, which is called the *band origin*. However, there is no difficulty in locating the band origin, since it lies at the midpoint between the

Table 15.3 Relative band FTIR band intensities for $[H^{35}Cl]^{+}$ at 298 K	
Initial Rotational State	**Relative Intensity**
0	1.0000
1	2.7083
2	3.6788
3	3.7894
4	3.2362
5	2.3718
6	1.5174
7	0.8556
8	0.4278
9	0.1905

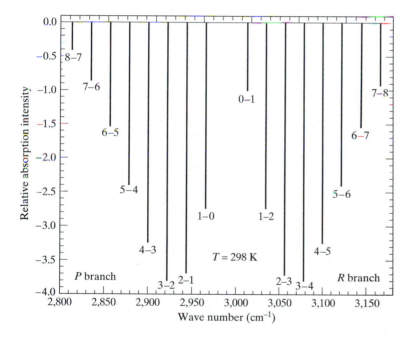

◄ **FIGURE 15.21**
Predicted $H^{35}Cl$ FTIR absorption spectrum at 298 K. The absorption intensities relative to the ($v = 0$, $J = 0) \rightarrow (v = 1, J = 1$) transition (which is taken to be unity) are plotted in the downward direction. The band positions are computed by means of Eq. 15.62, along with the experimentally measured equilibrium bond length and fundamental vibrational frequency. The numbers at the end of each absorption band give the initial and final rotational states of the molecule.

R and P branches, which are easily identified because they are separated by a gap of $4Bh$, twice that for all other band spacings in the spectrum. (For related examples, see Problems 15.30, 15.31 and 15.32.)

   In the foregoing discussion, we have assumed that all vibrational-rotational transitions originate from the $v = 0$ state. For most diatomic molecules at temperatures around 300 K, this is an excellent approximation. The ratio of the population of the $v = 1$ state to the $v = 0$ state for a harmonic quantization is given by

$$\frac{N_1}{N_o} = \frac{\exp\left[\dfrac{1.5h\nu_o}{kT}\right]}{\exp\left[\dfrac{0.5h\nu_o}{kT}\right]} = \exp\left[\dfrac{h\nu_o}{kT}\right].$$

Most diatomic molecules have vibrational frequencies in the range from $3 \times 10^{13} \text{ s}^{-1}$ to 10^{14} s^{-1}. Taking the lower limit of this range, we obtain $N_1/N_o = 8.26 \times 10^{-3}$ at 300 K. For a frequency at the upper limit, the result is $N_1/N_o = 1.14 \times 10^{-7}$. We see that most of the molecules reside in the $v = 0$ ground vibrational state. Nevertheless, the populations of the excited vibrational states are not zero. At elevated temperatures, these populations can become sufficient that transitions from higher v states can be observed. Such transitions are called *hot bands*.

In addition to the hot bands, it is possible to observe weak "forbidden" vibrational transitions with $\Delta v > 1$. Such transitions can be observed because the selection rule $\Delta v = \pm 1$ depends upon the accuracy of the first-order time-dependent perturbation treatment leading to Eq. 15.21. Deviations are also seen because the charge separation is not a constant, as assumed; instead, it actually depends upon the vibrational motion of the molecule. Most importantly, we observe deviations from the simple selection rule because molecules do not behave as perfect harmonic oscillators. Recall that, to obtain the result $\Delta v = \pm 1$, we made use of the harmonic oscillator eigenfunctions to determine the conditions under which the transition matrix element is zero. When the interatomic potential is sufficiently anharmonic, the true eigenfunctions can exhibit significant deviations from those for a harmonic oscillator. As a result, the simple selection rule breaks down, and transitions with $\Delta v > 1$ are observed. Such transitions are called *overtones* or *harmonics*. For example, a transition from $v = 0$ to $v = 2$ is called the first overtone or the second harmonic, while the $v = 0$ to $v = 3$ transition is the second overtone or the third harmonic.

The ability to spectroscopically observe hot bands and overtones allows us to obtain a measure of the anharmonicity terms in the expression for the energy level. It is customary to write the vibrational energy of a diatomic molecule in the form

$$E_v^e = \frac{E_v}{hc} = (v + 0.5)\omega_e - x_e\omega_e(v + 0.5)^2 + y_e\omega_e(v + 0.5)^3 + \dots \quad \textbf{(15.63)}$$

where ω_e is the vibrational wave number, which is equal to v_o/c, and the constants x_e and y_e are anharmonicity parameters that are fitted to the measured spectral data. If the interatomic potential were a Morse potential, we would expect to have $hcx_e\omega_e = h^2v_o^2/(4D)$ and $y_e = 0$.

EXAMPLE 15.8

The band origin for a particular diatomic molecule is found at $4{,}161.14 \text{ cm}^{-1}$. The first and second overtones are seen at $8{,}087.11 \text{ cm}^{-1}$ and $11{,}782.35 \text{ cm}^{-1}$, respectively. Use these data to obtain ω_e, $x_e\omega_e$, and $y_e\omega_e$ for this molecule. Use the result to predict the wave number for the third overtone?

Solution

The wave number associated with the $v = 0$ to $v = 1$ transition can be obtained directly from Eq. 15.63:

$$\Delta E_{01}^e = E_1^e - E_o^e = 1.5\omega_e - x_e\omega_e(1.5)^2 + y_e\omega_e(1.5)^3 - [0.5\omega_e - x_e\omega_e(0.5)^2$$
$$+ y_e\omega_e(0.5)^3] = \omega_e - 2.0x_e\omega_e + 3.25y_e\omega_e = 4{,}161.14 \text{ cm}^{-1}. \quad \textbf{(1)}$$

The energy change for the first overtone is

$$\Delta E_{02}^e = E_2^e - E_0^e = 2.5\omega_e - x_e\omega_e(2.5)^2 + y_e\omega_e(2.5)^3 - [0.5\omega_e - x_e\omega_e(0.5)^2$$
$$+ y_e\omega_e(0.5)^3] = 2\omega_e - 6.0x_e\omega_e + 15.5y_e\omega_e = 8{,}087.11 \text{ cm}^{-1}. \quad (2)$$

For the second overtone, the result is

$$\Delta E_{03}^e = E_3^e - E_0^e = 3.5\omega_e - x_e\omega_e(3.5)^2 + y_e\omega_e(3.5)^3 - [0.5\omega_e - x_e\omega_e(0.5)^2$$
$$+ y_e\omega_e(0.5)^3] = 3\omega_e - 12.0x_e\omega_e + 42.75y_e\omega_e = 11{,}782.35 \text{ cm}^{-1}. \quad (3)$$

Equations 1, 2, and 3 are three linear equations with three unknowns, so we may solve for ω_e, $x_e\omega_e$, and $y_e\omega_e$. Using Cramer's rule to write the solutions as ratios of determinants, we obtain $\omega_e = 4{,}400.6 \text{ cm}^{-1}$, $x_e\omega_e = 120.92 \text{ cm}^{-1}$, and $y_e\omega_e = 0.74 \text{ cm}^{-1}$. With these results, we can predict that the third overtone will be observed at

$$\Delta E_{04}^e = E_4^e - E_0^e = 4.5\omega_e - x_e\omega_e(4.5)^2 + y_e\omega_e(4.5)^3 - [0.5\omega_e - x_e\omega_e(0.5)^2$$
$$+ y_e\omega_e(0.5)^3] = 4\omega_e - 20.0x_e\omega_e + 91.0y_e\omega_e$$
$$= 4(4{,}400.6) - 20(120.92) + 91(0.74) = 15{,}251 \text{ cm}^{-1}. \quad (4)$$

The best procedure is to fit ω_e, $x_e\omega_e$ and $y_e\omega_e$, to a series of overtone data by least-squares methods, rather than fit three spectral bands exactly, as we have done in this example.

For related examples, see Problems 15.35, 15.36, and 15.37.

Table 15.4 lists vibrational wave numbers, anharmonicity factors, rotational constants, equilibrium distances, and dissociation energies (D_e) for selected diatomic molecules in their electronic ground state.

Table 15.4 Spectroscopic and structural properties of diatomic molecules in their electronic ground state[a]

Molecule	ω_e (cm^{-1})	$x_e\omega_e$ (cm^{-1})	$y_e\omega_e$ (cm^{-1})	B/c(cm^{-1})	R_e(Å)	D_e(eV)
^{12}CH	2,861.6	64.3	—	14.457	1.1198	3.47
^{35}Cl^{35}Cl	564.9	4.0	—	0.2438	1.988	2.475
^{35}Cl^{19}F	793.2	9.9	—	0.51651	1.628	2.616
^{12}C^{16}O	2,170.21	13.461	0.0308	1.9313	1.1281	11.108
H–H	4,395.2	117.99	0.29	60.809	0.7416	4.476
H^{35}Cl	2,989.74	52.05	0.056	10.5909	1.2746	4.430
H^{19}F	4,138.52	90.069	0.980	20.939	0.9171	≤6.40
^{127}I^{127}I	214.57	0.6127	−0.000895	0.03735	2.666	1.5417
^{7}Li^{7}Li	351.43	2.592	−0.0058	0.6727	2.672	1.03
^{7}LiH	1,405.1	13.228	0.1633	7.5131	1.5953	2.5
^{14}N^{14}N	2,359.61	14.456	0.00751	2.010	1.094	9.756
^{14}N^{16}O	1,904.03	13.97	−0.0012	1.7046	1.1508	6.49
^{16}O^{16}O	1,580.36	12.073	0.0546	1.44566	1.20739	5.080
^{16}OH	3,735.21	82.81	—	18.871	0.9706	4.35

[a]Source of Data: G. Herzberg, *Spectra of Diatomic Molecules* (Princeton, NJ: Van Nostrand, 1950). Dash indicates data not available.

It is possible to use Eq. 15.63 along with fitted anharmonicity factors to obtain an estimate of the dissociation energy for the molecule. The real inter-atomic potential for a diatomic molecule has a form similar to that for a Morse oscillator. In Chapter 12, we found that the spacings between adjacent energy eigenvalues for such a potential decrease as the vibrational quantum number v increases. (See Figure 12.39.) At the dissociation limit, the system is unbound and the energies become continuous. Therefore, we expect the spacing between adjacent energy states to approach zero as v approaches some maximum value at the dissociation limit.

Using Eq. 15.63, we can easily obtain an expression for the adjacent energy-state spacing. In wave numbers, this spacing is given by

$$\Delta\omega_v = E_{v+1}^e - E_v^e = (v + 1.5)\omega_e - x_e\omega_e(v + 1.5)^2 + y_e\omega_e(v + 1.5)^3$$
$$- [(v + 0.5)\omega_e - x_e\omega_e(v + 0.5)^2 + y_e\omega_e(v + 0.5)^3].$$

Expanding and collecting terms produces

$$\Delta\omega_v = \omega_e - x_e\omega_e[v^2 + 3v + 2.25 - v^2 - v - 0.25]$$
$$+ y_e\omega_e[v^3 + 4.5v^2 + 6.75v + 3.375 - v^3 - 1.5v^2 - 0.75v - 0.125],$$

which gives

$$\Delta\omega_v = \omega_e - 2x_e\omega_e[v + 1] + y_e\omega_e[3v^2 + 6v + 3.25]. \qquad \textbf{(15.64)}$$

If we ignore the $y_e\omega_e$ term because of its small size, we obtain

$$\Delta\omega_v = \omega_e - 2x_e\omega_e[v + 1]. \qquad \textbf{(15.65)}$$

These approximations suggest that the adjacent-state energy spacings decrease linearly with the vibrational quantum number. The maximum vibrational quantum number may be determined by setting $\Delta\omega_v$ to zero and solving for v. This gives

$$v_{max} = v_m = \frac{\omega_e}{2x_e\omega_e} - 1. \qquad \textbf{(15.66)}$$

In general, Eq. 15.66 will produce a noninteger result for v_{max}. Since we know that v takes on only integer values, the maximum vibrational quantum number will be

$$v_m = \text{trunc}\left[\frac{\omega_e}{2x_e\omega_e} - 1\right], \qquad \textbf{(15.67)}$$

where the trunc operator truncates the expression to yield the integer portion only. Because we expect $E_{v_m}^e$ to be very near the dissociation limit, the dissociation energy will be approximately that amount of energy, minus the energy of the $v = 0$ state, or

$$D_e \approx E_{v_m}^e - E_o^e. \qquad \textbf{(15.68)}$$

EXAMPLE 15.9

Use the data in Table 15.4 to estimate the dissociation energy for H_2.

Solution

The maximum vibrational quantum number for H_2 can be estimated using Eq. 15.67:

$$v_m = \text{trunc}\left[\frac{\omega_e}{2x_e\omega_e} - 1\right] = \text{trunc}\left[\frac{4{,}395.2}{2(117.99)} - 1\right] = \text{trunc}[17.614\ldots] = 17. \qquad \textbf{(1)}$$

The energy of vibrational state v, expressed in wave numbers, is given by

$$E_v^e = (v + 0.5)\omega_e - x_e\omega_e(v + 0.5)^2, \qquad (2)$$

where we have dropped the term containing $y_e\omega_e$ to be consistent with Eq. 15.67. Inserting $v = v_m = 17$ in Eq. 2 produces

$$E_{17}^e = 17.5\omega_e - x_e\omega_e(17.5)^2 = 17.5(4{,}395.2) - 117.99(17.5)^2 \text{ cm}^{-1} = 40{,}781.6 \text{ cm}^{-1}. \qquad (3)$$

The wave number for the ground state is

$$E_0^e = 0.5\omega_e - x_e\omega_e(0.5)^2 = 0.5(4395.2) - 117.99(0.25) \text{ cm}^{-1} = 2{,}168.1 \text{ cm}^{-1}. \qquad (4)$$

The dissociation energy can now be estimated using Eq. 15.68:

$$D_e \approx 40{,}781.6 - 2{,}168.1 \text{ cm}^{-1} = 38{,}614 \text{ cm}^{-1}. \qquad (5)$$

Using the conversion factor $1 \text{ eV} = 8{,}065.7 \text{ cm}^{-1}$, we obtain

$$D_e \approx 38{,}614 \text{ cm}^{-1} \times \frac{1 \text{ eV}}{8{,}065.7 \text{ cm}^{-1}} = 4.79 \text{ eV}. \qquad (6)$$

The correct answer is 4.476 eV, so our estimate is too high by 7.0%. This is due primarily to the fact that Eq. 15.63 is only an approximate least-squares fit to the measured overtone spectra.

For a related exercise, see Problem 15.38.

If we have sufficient information about overtones and hot bands so that the actual spacings between the vibrational states are known, a much more accurate value for the dissociation energy can be obtained. Inspection of Figure 12.39 indicates that the dissociation energy is the sum of all the vibrational state spacings from $v = 0$ up to $v = v_m$, plus a small remaining energy ε between the last bound vibrational state and the dissociation limit. That is,

$$D_e = \Delta\omega_0 + \Delta\omega_1 + \Delta\omega_2 + \cdots + \Delta\omega_{v_m} + \varepsilon = \sum_{v=0}^{v=v_m} \Delta\omega_v + \varepsilon. \qquad (15.69)$$

If we plot $\Delta\omega_v$ versus v, the area under the histogram will equal the summation in Eq. 15.69, since

$$\text{Histogram area} = \sum_{v=0}^{v=v_m} \Delta\omega_v \Delta v = \sum_{v=0}^{v=v_m} \Delta\omega_v, \qquad (15.70)$$

because $\Delta v = 1$ for all v. In practice, we may not have $\Delta\omega_v$ values all the way up to the dissociation point, in which case we can use the fact that Eq. 15.65 suggests that $\Delta\omega_v$ varies linearly with v to extrapolate the data and thereby determine the missing values of $\Delta\omega_v$ and also estimate the value of ε in Eq. 15.69. This procedure, which is illustrated in the next example, was originated by Birge and Spooner in 1926. It is, therefore, called the *Birge–Spooner extrapolation*.

EXAMPLE 15.10

The overtone spectra of H_2 give the following data:

v	$\Delta\omega_v$ (cm^{-1})	v	$\Delta\omega_v$ (cm^{-1})
0	4,161.14	7	2,543.14
1	3,925.97	8	2,292.26
2	3,695.24	9	2,026.26
3	3,468.01	10	1,736.60
4	3,241.56	11	1,414.98
5	3,013.73	12	1,049.18
6	2,782.18	13	621.96

Use the Birge–Spooner extrapolation method to obtain the dissociation energy of H_2. Compute the percent error in the result.

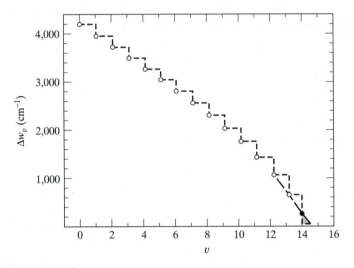

▲ FIGURE 15.22

The Birge–Spooner extrapolation to obtain the dissociation energy for H_2 from overtone spectral data. The dark circle for $v = 14$ is obtained by a linear extrapolation, as described in the text. The energy separating the last bound state from the dissociation point is estimated from the near-triangular area that is shaded in the figure.

Solution

Over the range where data are available, we may sum Eq. 15.69 in a straightforward fashion. This partial sum is

$$X = \sum_{v=0}^{13} \Delta\omega_v = 4{,}161.14 + 3{,}925.97 + 3{,}695.24 + 3{,}468.01 + 3{,}241.56 + 3{,}013.73$$

$$+ 2{,}782.18 + 2{,}543.14 + 2{,}292.26 + 2{,}026.26 + 1{,}736.6 + 1{,}414.98$$

$$+ 1{,}049.18 + 621.96 \text{ cm}^{-1} = 35{,}972.21 \text{ cm}^{-1}. \tag{1}$$

We now use the last two points, at $v = 12$ and $v = 13$, to fit a straight line to permit extrapolation to higher values of v. The linear fit to the points at $v = 12$ and 13 is

$$\Delta\omega_v = 6{,}175.82 - 427.22v \quad \text{for} \quad v \geq 12. \tag{2}$$

The linear extrapolation is shown as the dashed line in Figure 15.22. When $v = 14$, Eq. 2 gives $\Delta\omega_{14} = 6{,}175.82 - 427.22(14) = 195.02 \text{ cm}^{-1}$. This point is shown as the dark circle in the figure. Including it in the sum obtained in Eq. 1 gives

$$X' = X + 195.02 \text{ cm}^{-1} = 35{,}972.21 + 195.02 \text{ cm}^{-1} = 36{,}167.23 \text{ cm}^{-1}. \tag{3}$$

The last bound vibrational state, $v = 15$, is not exactly at the dissociation limit. If we solve Eq. 2 for the value of v for which $\Delta\omega_v = 0$, the result is

$$v_{\max} = v_m = \frac{6{,}175.82}{427.22} = 14.456. \tag{4}$$

A linear extrapolation from $v = 14$ to $v = 14.456$ produces the small shaded triangular area shown in the figure. The area of this triangle provides an estimate of ε, the energy gap between the $v = 15$ state and the dissociation limit. The area is

$$\text{Area} \approx \frac{(\text{height})(\text{base})}{2} = \frac{(195.02 \text{ cm}^{-1})(0.456)}{2} = 44.46 \text{ cm}^{-1}. \tag{5}$$

Adding this quantity to X' in Eq. 3 gives

$$D_e = 36{,}167.23 + 44.46 \text{ cm}^{-1} = 36{,}211.69 \text{ cm}^{-1}. \tag{6}$$

In eV, this result is equivalent to $D_e = 4.490$ eV. The reported result in Table 15.4 is 4.476 eV, so the extrapolation procedure is accurate to within an error of 0.31%.
For related exercises, see Problems 15.39 and 15.40.

15.4.3   Raman Spectroscopy

In 1928, C. V. Raman was investigating the scattering of light by liquids. In the course of his studies, he discovered that when a substance is irradiated with light of a given frequency ν, the scattered light contains not only radiation with frequency ν, but also components with both higher and lower frequencies. The red-shifted lines with frequencies lower than ν are called *Stokes lines*, while the blue-shifted lines whose frequencies are greater than ν are referred to as *anti-Stokes lines*. This finding has since proved to be of great importance in the investigation of molecular structure. Two years after his discovery, Raman was awarded the Nobel prize.

The Raman effect can be qualitatively understood by analogy with Compton scattering of photons. (See Section 11.2.6.) In Compton scattering, a high-energy X-ray photon is scattered by an electron, which is ejected in the process. As a result of the interaction, the ejected electron gains kinetic energy, while the frequency of the photon decreases in the amount required to conserve energy and linear momentum. In Raman scattering, a photon is scattered after colliding with a molecule. In this case, however, the frequency of the scattered photon is almost always unchanged, because of the huge difference in mass between the molecule and the photon. This type of elastic scattering is called *Rayleigh scattering*. On rare occasions, the collision process can cause a change in the vibrational–rotational quantum state of the molecule, which results in *Raman scattering*. If E_{vJ}^i and E_{vJ}^f are the initial and final vibrational–rotational energies of the molecule, conservation of energy requires that we have

$$\boxed{h\nu' = h\nu - (E_{vJ}^f - E_{vJ}^i)}, \tag{15.71}$$

where ν' and ν are the final and initial photon frequencies, respectively. Rearrangement of Eq. 15.71 produces

$$\boxed{h(\nu' - \nu) = h\,\Delta\nu = E_{vJ}^i - E_{vJ}^f}, \tag{15.72}$$

▶ **FIGURE 15.23**
Raman scattering. Case A results in molecular excitation and the observation of a Stokes line. Case B produces molecular de-excitation and an anti-Stokes line.

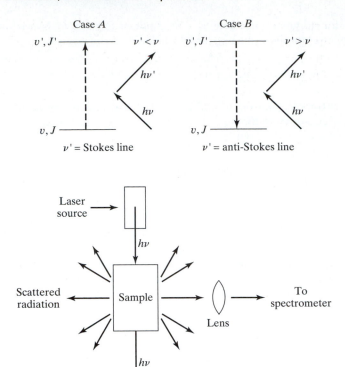

▶ **FIGURE 15.24**
Schematic illustration of a Raman spectrometer. The source is a high-intensity laser. The scattered radiation is collected at right angles to the incident beam and directed into the electronics of the spectrometer for analysis.

$\Delta \nu$ is called the *Raman shift*. If we have $E_{vJ}^f > E_{vJ}^i$, $\Delta \nu$ will be negative, so that the scattered final photon frequency will be less than its initial frequency (Stokes scattering). The nomenclature derives from the analogous red-shifted frequencies commonly observed in electronic emission spectroscopy that were first explained by Sir George Stokes in 1852. (See Section 15.5.4.) If $E_{vJ}^f < E_{vJ}^i$, $\Delta \nu$ is positive, and anti-Stokes scattering will be observed. These two situations are illustrated in Figure 15.23.

Since we are far removed from the resonance point, Eq. 15.21 suggests that the probability of Raman scattering will be very small. This is certainly the case. Measurements show that only about one photon in 10^7 undergoes Raman scattering (about your chances of winning the lottery). For this reason, Raman spectrometers generally employ very high-intensity laser sources, so as to bring a large number of photons to bear on the sample and thereby enhance the Raman signal. Figure 15.24 illustrates the configuration of a typical Raman spectrometer. The radiation is collected at right angles to the path of the beam in order to avoid, as much as possible, background radiation due to the high-intensity incident beam itself. In addition, the solid-angle element, $\sin \theta \, d\theta \, d\phi$, is maximal for a scattering angle of 90°, so that measuring the scattered radiation at this angle leads to maximum efficiency in signal collection.

Raman spectroscopy offers several advantages over IR absorption measurements. For one, it is much easier to measure spacings below 100 cm^{-1} using Raman techniques, since these frequencies appear as differences between the Rayleigh and Raman lines. This allows us to avoid absorption measurements at these low wave numbers. Second, for measurements in solution, interference from the solvent is often much less in Raman spectra than in IR absorption measurements. An important example is aqueous solutions in which there are strong absorptions from the water in the range from 1,500 to 3,000 cm^{-1}, but only very weak vibrational Raman scattering.

An important advantage of Raman spectroscopy lies in the fact that it permits us to determine vibrational–rotational state spacings that cannot be obtained using FTIR methods because the molecule lacks a permanent or induced dipole moment. We can see why this is so by considering a classical model of Raman scattering. Recall from Eq. 7.1 that when electromagnetic radiation of frequency ν is imposed on a molecule, the electric field will produce an induced molecular dipole moment whose magnitude is

$$\mu^* = \alpha\mathcal{E}, \tag{15.73}$$

where α is the polarizability of the molecule. This induced dipole moment permits the electric field to couple to the molecular motion. In a classical model, the scattered radiation is always at the frequencies with which the induced dipole moment oscillates. The imposed electric field, given by $\mathcal{E} = \mathcal{E}_o \sin[2\pi\nu t]$, will oscillate at the same frequency as the radiation. Just as the permanent dipole moment of a molecule varies with the vibrational motion, the polarizability will also depend upon the vibrational coordinate, which we shall denote by Q. In general, we can write the dependence of α upon Q in a Taylor series expansion about the equilibrium position Q_e:

$$\alpha = \alpha_e + \sum_{n=1}^{\infty} \frac{d^n\alpha}{dQ^n}\big|_{Q=Q_e}(Q - Q_e)^n. \tag{15.74}$$

For simplicity, let us assume that we may truncate the series expansion in Eq. 15.74 after the linear term. This assumption produces

$$\alpha = \alpha_e + \left(\frac{d\alpha}{dQ}\right)_e (Q - Q_e). \tag{15.75}$$

Let us further assume that the vibrational coordinate exhibits classical harmonic motion at a frequency ν_o, so that

$$Q - Q_e = A \sin(2\pi\nu_o t). \tag{15.76}$$

It is now simple to determine the frequencies with which the dipole moment will oscillate. Combining the foregoing results, we obtain

$$\mu^* = \left[\alpha_e + \left(\frac{d\alpha}{dQ}\right)_e A \sin(2\pi\nu_o t)\right]\mathcal{E}_o \sin[2\pi\nu t]$$

$$= \alpha_e\mathcal{E}_o \sin[2\pi\nu t] + \left(\frac{d\alpha}{dQ}\right)_e \mathcal{E}_o A \sin(2\pi\nu_o t) \sin[2\pi\nu t]. \tag{15.77}$$

Using the trigonometric relationships

$$\cos(C + D) = \cos C \cos D - \sin C \sin D$$

and

$$\cos(C - D) = \cos C \cos D + \sin C \sin D,$$

we have

$$\sin C \sin D = 0.5[\cos(C - D) - \cos(C + D)]. \tag{15.78}$$

Equation 15.78 permits us to write Eq. 15.77 in the form

$$\mu^* = \alpha_e\mathcal{E}_o\sin[2\pi\nu t] + 0.5\left(\frac{d\alpha}{dQ}\right)_e \mathcal{E}_o A[\cos\{2\pi(\nu - \nu_o)t\} - \cos\{2\pi(\nu + \nu_o)t\}]. \tag{15.79}$$

Equation 15.79 shows us that the scattered frequencies are going to occur at ν, which is the Rayleigh line, at $\nu - \nu_o$, which is the Stokes line, and at $\nu + \nu_o$, which is the anti-Stokes line. However, we will observe the Stokes and anti-Stokes lines only if $(d\alpha/dQ)_e \neq 0$. That is, in order for us to observe a Raman spectrum, the polarizability of the molecule must change with the vibrational coordinate Q at the equilibrium point. Vibrational modes that do this are said to be *Raman active*.

In general, $(d\alpha/dQ)_e$ will be nonzero if the vibrational motion produces an effective change in the volume of the molecule. For a diatomic molecule, the expansion and contraction of the bond produces such a change. Consequently, the vibrational mode of a diatomic molecule is always Raman active. The situation in polyatomic molecules is more complex. As we shall see in the next section, we can have vibrational modes that are both IR and Raman active, IR active and Raman inactive, IR inactive and Raman active, or IR and Raman inactive . If a molecule possesses a center of symmetry, the antisymmetric vibrational modes will be IR active, while the symmetric modes will be Raman active. The best way to predict the IR and Raman activities of a particular vibrational mode involves the use of group theory, which we shall not cover in this text.

The selection rules for Raman spectra are different from those for IR absorption spectra, since the former is associated with a scattering process whereas the latter involves the resonant absorption of a photon. For pure rotational spectra, the selection rules are

$$\Delta J = 0, \pm 2 \text{ for linear molecules}$$

and

$$\Delta J = 0, \pm 1, \text{ or } \pm 2 \text{ and } \Delta K = 0 \text{ for symmetric rotors.} \qquad \textbf{(15.80)}$$

For vibrational–rotational transitions, we must have

$$\Delta v = \pm 1 \qquad \text{and} \qquad \Delta J = 0, \pm 2. \qquad \textbf{(15.81)}$$

The vibrational Stokes lines have $\Delta v = +1$, while the anti-Stokes lines have $\Delta v = -1$. The anti-Stokes lines are much less intense, because of the low population of the excited vibrational states. In Raman spectroscopy, transitions with $\Delta J = +2$ are called the *S branch*. Those with $\Delta J = -2$ are labeled the *O branch*, and those for which $\Delta J = 0$ are said to form the *Q branch*. Since we cannot observe $\Delta J = 0$ transitions in IR absorption spectroscopy, an FTIR spectrum exhibits no set of lines corresponding to the Raman *Q* branch. With the preceding selection rule for linear molecules, it is easy to show (see Problem 15.43) that the line spacings in the *S* and *O* branches are $4Bh$ and that the *Q* branch is separated from both the *S* and *O* branches by $6Bh$. Example 15.11 illustrates how we can predict the appearance of a Raman spectrum.

EXAMPLE 15.11

Using the data given in Table 15.4, predict the form of the vibrational–rotational Raman spectrum for $H^{19}F$ at a temperature of 298 K when the incident radiation has a wave number of 23,004 cm^{-1}. Assume that the vibrational–rotational energy states are given by Eq. 15.62 and that band intensities are determined solely by the population of the initial state in the transition leading to the spectral band. At what wavelength would Rayleigh scattering be observed?

Transition	ΔE^e (cm^{-1})	Spectral Line (cm^{-1})
	S Branch ($\Delta J = +2$)	
$(v = 0, J = 0 \longrightarrow v = 1, J = 2)$	$6{,}333.41 - 2{,}069.26 = 4{,}264.15$	18,739.85
$(v = 0, J = 1 \longrightarrow v = 1, J = 3)$	$6{,}459.05 - 2{,}111.14 = 4{,}347.91$	18,656.09
$(v = 0, J = 2 \longrightarrow v = 1, J = 4)$	$6{,}626.56 - 2{,}194.89 = 4{,}431.67$	18,572.33
$(v = 0, J = 3 \longrightarrow v = 1, J = 5)$	$6{,}835.95 - 2{,}320.53 = 4{,}515.42$	18,488.58
$(v = 0, J = 4 \longrightarrow v = 1, J = 6)$	$7{,}087.22 - 2{,}488.04 = 4{,}599.18$	18,404.82
$(v = 0, J = 5 \longrightarrow v = 1, J = 7)$	$7{,}380.36 - 2{,}697.43 = 4{,}682.93$	18,321.07
	O Branch ($\Delta J = -2$)	
$(v = 0, J = 2 \longrightarrow v = 1, J = 0)$	$6{,}207.78 - 2{,}194.89 = 4{,}012.89$	18,991.11
$(v = 0, J = 3 \longrightarrow v = 1, J = 1)$	$6{,}249.66 - 2{,}320.53 = 3{,}929.13$	19,074.87
$(v = 0, J = 4 \longrightarrow v = 1, J = 2)$	$6{,}333.41 - 2{,}488.04 = 3{,}845.37$	19,158.63
$(v = 0, J = 5 \longrightarrow v = 1, J = 3)$	$6{,}459.05 - 2{,}697.43 = 3{,}761.62$	19,242.38
$(v = 0, J = 6 \longrightarrow v = 1, J = 4)$	$6{,}626.56 - 2{,}948.70 = 3{,}677.86$	19,326.14
$(v = 0, J = 7 \longrightarrow v = 1, J = 5)$	$6{,}835.95 - 3{,}241.84 = 3{,}594.11$	19,409.89
	Q Branch ($\Delta J = 0$)	
$(v = 0, J \longrightarrow v = 1, J)$	$6{,}207.78 - 2{,}069.26 = 4{,}138.52$	18,865.48

Solution

We know that HF will be Raman active, because all diatomics have Raman-active vibrational modes. If the energies of the vibrational–rotational states are given by Eq. 15.62 with $B/c = 20.939$ cm^{-1} and $\omega_e = 4{,}138.52$ cm^{-1}, then

$$E_{vJ}^e = [4{,}138.52\,(v + 0.5) + 20.939J(J + 1)]\,\text{cm}^{-1}. \qquad (1)$$

The Stokes lines will come from $v = 0$ to $v = 1$ transitions. The anti-Stokes lines arising from the $v = 1$ to $v = 0$ transition will be too faint to detect, due to the very low population of the $v = 1$ vibrational state at 298 K. Some of the allowed transitions are given in the table above.

The S branch contains those transitions for which $\Delta J = +2$. From Eq. 15.71, we see that the lines associated with these transitions appear at wave numbers that are the difference between the incident radiation and the spacing between the initial and final quantum states—in this case, 23,004 cm^{-1} − ΔE^e. The O branch contains transitions with $\Delta J = -2$ and the Q branch the line with $\Delta J = 0$.

The population of the initial rotational state at 298 K is given by Eq.15.37:

$$N_J = C(2J + 1)\exp\left[-\frac{20.939hcJ(J + 1)}{kT}\right]$$

$$= C(2J + 1)\exp\left[-\frac{20.939\ \text{cm}^{-1}(6.62608 \times 10^{-34}\ \text{J s})(2.9979 \times 10^{10}\ \text{cm s}^{-1})J(J + 1)}{(1.38066 \times 10^{-23}\ \text{J K}^{-1})(298\ \text{K})}\right]$$

$$= C(2J + 1)\exp[-0.10109\,J(J + 1)]. \qquad (2)$$

The relative populations of the various rotational states are, therefore, as given in the table below.

The sum of all these populations is 10.22C. Since the transition in the Q branch can originate from any initial J state, its intensity will be proportional to 10.22 C. Figure 15.25 shows the predicted spectrum. The notation at the top of each line indicates the initial and final vJ states producing the transition. Note that the

J	N_J	J	N_J
0	C	4	1.192C
1	2.451C	5	0.530C
2	2.726C	6	0.186C
3	2.081C	7	0.052C

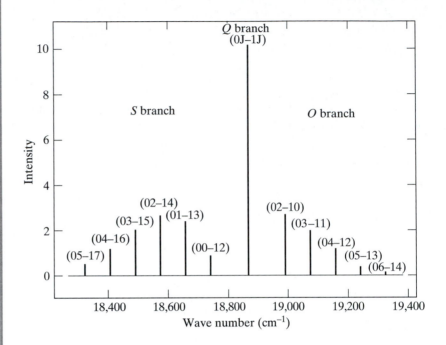

▲ FIGURE 15.25

Predicted Raman spectrum of HF at 298 K in a spectrometer in which the incident radiation frequency is 23,004 cm^{-1}. The relative band intensities are computed under the assumption that they depend only upon the population of the initial states. Note that the band spacings are different from those observed in an FTIR absorption spectrum. This difference arises because of the difference in the selection rules.

spacings between the Raman lines are different from those seen in IR absorption measurements. The differences arise from the selection rules. (See Problem 15.43.) The Rayleigh line appears at the same wave number as the incident radiation, which in this case is 23,004 cm^{-1}. Therefore, it is not seen on the scale shown in the figure.

For related exercises, see Problems 15.43, 15.44, and 15.45.

15.4.4 Polyatomic Molecules

a. Spectral Characteristics

The IR and Raman spectra of polyatomic molecules are considerably more complex than those of diatomic molecules because of the increased number of vibrational modes that are present. For a molecule containing N atoms, there are $3N$ modes of motion, or *degrees of freedom*. Three of these motions are translations of the center of mass in the X, Y, and Z directions. For a nonlinear molecule, three additional motions are associated with rotation about the three principal axes.

When the molecule is linear, rotation occurs only about the two principal axes perpendicular to the bond axis. The remaining degrees of freedom are various vibrational modes, each having its own characteristic frequency. There are, therefore, $3N - 6$ and $3N - 5$ vibrational modes for nonlinear and linear molecules, respectively. When N becomes large, the vibrational spectrum can become very complex, with many absorption bands that often overlap.

The situation, however, is not quite as complicated as the preceding analysis might suggest. Some of the vibrational modes may not be IR or Raman active, so that absorptions or Raman shifts corresponding to the frequencies of these modes will not appear in the spectrum. In addition, degenerate modes with the same vibration frequency may be present. Such modes will appear as a single band in the spectrum. There also exists the possibility of accidental degeneracies that cause multiple bands to appear as one. Nevertheless, the spectrum will usually be complex when N is large.

The following qualitative generalizations help us interpret and understand polyatomic vibrational spectra:

1. If a molecule possesses a permanent dipole moment, its vibrational modes will generally be IR active. If it does not, some of the modes may still be IR active if the vibrational motion is such that the molecule gains a momentary dipole moment because the motion distorts the symmetry of the molecule.

2. Polyatomic molecules have large masses and, therefore, large moments of inertia. Since the rotational spacings are inversely proportional to the moments of inertia, these spacings are often very small in polyatomic molecules. As a result, the individual rotational lines usually cannot be resolved, and even the P–R branches may appear as a single broad band if the molecule is sufficiently large. This type of situation is illustrated in Figure 15.6.

3. When the spectrum is taken in solution, in a pure liquid or a crystalline solid, the rotational motion is frozen out due to the steric hindrance of the solvent or the intermolecular forces present in the condensed phase. Rotational structure is, therefore, not observed.

4. Vibrational bands in solution or in pure liquids are generally much broader than in the gas phase or in a crystalline solid. The reason is that the molecules whose spectrum is being taken appear in a range of different liquid environments, and the various environments produce small shifts in the vibration frequencies, so that the overall result is a broad band centered at the mode frequency rather than a sharp line at this frequency. This broadening effect can produce overlap between neighboring absorption peaks that complicates the spectrum.

5. The vibrational modes that are due predominantly to the stretching or bending of chemical bonds present in a given functional group, such as $-C-O-C-$, $-C-H$, $-C=O$, and $-C\equiv N$, usually appear in a well-defined range of frequencies. Table 15.5 lists the characteristic frequencies of different modes for particular functional groups.

6. Vibrational modes that primarily involve angle-bending, rocking, twisting, or deformational motions usually have much lower frequencies than those associated with bond stretching.

7. The coordinates describing the vibrational motion of polyatomics, which we call *normal-mode coordinates*, must be such that they produce no rotational torque and no motion of the center of mass of the molecule.

Table 15.5 Characteristic frequency ranges for various groups

Group	Bond-stretching Wave Number (cm^{-1})	Bond-bending Wave Number (cm^{-1})
$\equiv$C—H	$\approx$3,300	$\approx$700 (C$\equiv$C—H)
$=$C—H	$\approx$3,020	$\approx$1,100 (C$=$C—H)
—C—H	2,850–2,960	$\approx$1,000 (C—C—H), $\approx$ 1,450 (H—C—H)
—O—H	3,200–3,680	—
—S—H	$\approx$2,670	—
—N—H	3,100–3,500	—
—C$=$O	1,650–1,850	—
—C$\equiv$N	$\approx$2,100	—
—C$\equiv$C—	2,050–2,250	$\approx$300 (C$\equiv$C—C)
—C$=$C—	1,620–1,680	—
—C—C—	700–1,250	—
—C—F	$\approx$1,100	—
—C—Cl	$\approx$650	—
—C—Br	$\approx$560	—

Source: G. Herzberg, *Infrared and Raman Spectra* (Princeton, NJ: D. van Nostrand Co., Inc., 1945) p. 195.

Figure 15.26 shows the FTIR spectrum of liquid 2-methyl-2-phenyl-propanol. This nonlinear molecule contains 25 atoms. Therefore, it has 69 vibrational modes. Since the molecule has a permanent dipole moment, we expect to see absorptions in the IR due to these modes. And indeed, inspection of the figure reveals a wealth of vibrational structure in the spectrum of the compound. Also, the rotational fine structure of the vibrational modes does not appear, since molecular rotation is frozen out by steric hindrance in the liquid phase. The vibrational bands are broad due to the presence of different liquid environments. That is, some molecules may be surrounded by n nearest neighbors, while others have $n - 1$ or $n + 1$ molecules in their near vicinity.

Using Table 15.5, we may identify the major features of the spectrum of the compound without difficulty. The intense absorption around 3,350 cm^{-1} is clearly due to the —O—H stretching motion. The large dipole moment associated with this mode contributes significantly to the intensity of the absorption peak. The series of smaller absorptions between 3,000 and 3,100 cm^{-1} are the result of $=$C—H stretching motions from the hydrogens in the phenyl ring. The ortho, meta, and para hydrogens in the ring give rise to the multiplet structure in this frequency range. The series of absorptions between 2,850 and 2,960 cm^{-1} are due to C—H stretching motions in the —CH_3 and —CH_2OH

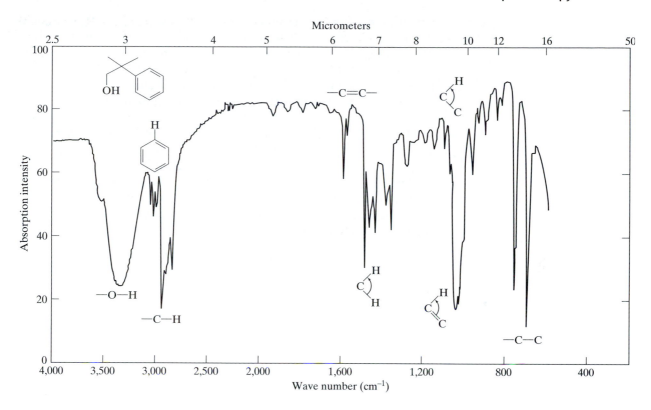

▲ FIGURE 15.26

FTIR spectrum of 2-methyl-2-phenylpropanol. Wave numbers are given on the bottom horizontal axis and wavelengths on the top horizontal axis. The author thanks Professor K. D. Berlin and Dr. M. M. Madler for providing this spectrum.

groups. The fact that there are eight such C—H bonds is, in part, responsible for the greater absorption intensity of these stretching motions relative to the ring hydrogens. The next major feature in the spectrum is seen at 1,600 cm^{-1}, which is the characteristic frequency for the —C≡C— double-bond stretching motion of the carbon atoms in the phenyl ring. The complex multiplet between 1,350 and 1,500 cm^{-1} is due to the various H—C—H bending vibrations in the methyl and CH$_2$OH groups. The intense broad peak between 1,000 and 1,100 cm^{-1} is the result of a combination of —C≡C—H bends in the phenyl ring and —C—C—H bends in the aliphatic group attached to the ring. The two absorption peaks at 700 and 760 cm^{-1} are —C—C— single-bond stretching motions. The minor features in the spectrum at frequencies below 1,500 cm^{-1} are the result of various wagging, twisting, or deformational motions that are very difficult to identify without a much more sophisticated analysis of the spectrum. (For additional exercises, see Problems 15.48 and 15.49.)

b. Normal-mode Coordinates

Our study of the vibrational motion of diatomics makes it clear that the problem is much easier to resolve if we can assume the vibrational motion to be harmonic. The energy eigenvalues have the very simple form $E_v = (v + 0.5)h\nu_o$, and the vibrational eigenfunctions are just polynomials in the vibrational coordinate, multiplied by a Gaussian function. The same is true for polyatomic molecules: If all $3N - 6$ vibrational modes are harmonic, we will have simple expressions for the energy eigenvalues of each mode and correspondingly simple eigenfunctions. It is, therefore, useful to examine the conditions under which such an approximation might be acceptably accurate.

We first need to identify the conditions that must be present for harmonic vibrational motion to occur. In Chapter 12, we obtained the classical solutions of the one-dimensional harmonic oscillator. (See Section 12.3.) The essential characteristics of the system that lead to harmonic motion are a kinetic energy whose form is $T = (\mu/2)(dx/dt)^2 = (\mu/2)v_x^2$ and a potential energy given by $V(x) = kx^2/2$, where k is a constant and x is the vibrational coordinate. When these conditions are met, the solution of the Newtonian equation of motion shows that the vibrational coordinate undergoes harmonic sinusoidal motion of the form $x = A \sin[(k/\mu)^{1/2}t]$. The corresponding quantum mechanical solutions yield harmonic oscillator eigenfunctions and eigenvalues.

Now suppose we have a three-dimensional oscillator whose kinetic and potential energies have the forms

$$T = \left[\frac{\mu_1 v_{q_1}^2}{2} + \frac{\mu_2 v_{q_2}^2}{2} + \frac{\mu_3 v_{q_3}^2}{2} \right]$$

and

$$V(q_1, q_2, q_3) = \left[\frac{aq_1^2}{2} + \frac{bq_2^2}{2} + \frac{cq_3^2}{2} \right],$$

respectively, where a, b, and c are constants and q_1, q_2, and q_3 are the three vibrational coordinates. With kinetic and potential energies of this form, the three vibrational motions will be completely independent, and each will exhibit harmonic motion with solutions of the form

$$q_1 = A_1 \sin\left[\left(\frac{a}{\mu_1} \right)^{1/2} t \right], q_2 = A_2 \sin\left[\left(\frac{b}{\mu_2} \right)^{1/2} t \right], \text{ and } q_3 = A_3 \sin\left[\left(\frac{c}{\mu^3} \right)^{1/2} t \right].$$

The Schrödinger equation for this system will also separate into three independent harmonic oscillator-type equations. (See Problem 15.50.)

We can now extend the analysis to the case where we have $3N - 6$ vibrational modes. If they are to be independent and if each is to exhibit harmonic motion, we must have a set of vibrational coordinates $Q_1, Q_2, Q_3, \ldots, Q_{3N-6}$ such that the potential and kinetic energies can be written in the form

$$V(Q_1, Q_2, \ldots, Q_{3N-6}) = \sum_{i=1}^{3N-6} a_i Q_i^2 \qquad \textbf{(15.82)}$$

and

$$T = \sum_{i=1}^{3N-6} b_i v_i^2, \qquad \textbf{(15.83)}$$

respectively, where $v_i = \partial Q_i / \partial t$ and the a_i and b_i are constants. The vibrational coordinates $Q_1, Q_2, \ldots, Q_{3N-6}$ that achieve this objective are called *normal-mode coordinates*, and the vibrational modes they describe are referred to as *normal modes*.

We have already seen that a harmonic potential accurately describes a real diatomic molecule only when the vibrational amplitude and energy are small. At higher energies, anharmonic terms become increasingly important. The same is true for polyatomics: If the vibrational energies are low and the amplitudes small, Eqs. 15.82 and 15.83 can adequately represent the vibrational potential and kinetic energies, provided that care is exercised in choosing the coordinates. At higher energies, the normal-mode description will break down, just as the harmonic oscillator description of diatomic vibration does. Nevertheless, the normal-mode description is useful, particularly when the molecules are in or near their vibrational ground states.

Numerous computer codes are available that will compute the appropriate normal mode coordinates and the corresponding mode frequencies, given the potential-energy function for the molecule and given the molecular structure. The vibrational spectroscopist often approaches this problem in reverse: The mode frequencies are obtained from a combination of Raman and FTIR measurements, and then a molecular potential that reproduces the measured frequencies is developed by means of iterative methods. While such codes are highly useful for conducting research, they are not very instructive. A better understanding of the nature of normal-mode coordinates can be obtained by examining a specific example. We shall use a water molecule for this purpose.

With three atoms in a nonlinear arrangement, H_2O has three vibrational modes. Since the molecule has a permanent dipole moment, all modes are IR active. Their measured frequencies, expressed in wave numbers, are 3,756, 3,652, and 1,595 cm^{-1}. Table 15.5 indicates that the first two modes should be associated with —O—H stretching motions, while the low-frequency mode is a H—O—H bend. With three vibrational modes, we will need three normal-mode coordinates Q_1, Q_2, and Q_3 that must satisfy Eqs. 15.82 and 15.83 if we are to observe harmonic motion. Since the molecule is planar and all motions out of the plane are either translations or rotations, we know that the Q_i will depend only upon the coordinates defining the molecular plane. Let us take this to be the X–Y plane.

H_2O has a plane of symmetry perpendicular to the molecular plane that bisects the H—O—H angle. This H—O—H bisector is a twofold symmetry axis for the molecule, in that rotation about that axis reproduces the H_2O structure exactly twice. A reasonable starting point in our attempt to obtain an appropriate set of normal-mode coordinates is to construct three coordinates that are either symmetric or antisymmetric with respect to rotation about this symmetry axis. We choose these coordinates such that two of them essentially represent a stretching of the O—H bonds while the third bends the molecule. In addition, the vibrational coordinates must produce neither rotational torque on the molecule nor motion of its center of mass. Figure 15.27 suggests three reasonable choices.

In Mode 1, the hydrogen atoms are each displaced in the negative Y direction by an amount S_1 that, for the moment, we will assume is our normal-mode vibrational coordinate. To maintain a fixed center of mass, the oxygen atom must be displaced upward by an amount Δy_o during this phase of the vibrational motion. The requirement that the center of mass remain fixed means that we must have

$$-m_H S_1 - m_H S_1 + m_o \Delta y_o = 0, \tag{15.84}$$

where m_H and m_o are the masses of hydrogen and oxygen, respectively. This gives an oxygen-atom displacement in Mode 1 of

$$\Delta y_o = \frac{2m_H}{m_o} S_1. \tag{15.85}$$

All other displacements are zero in this mode. These displacements are symmetric with respect to a 180° rotation about the Y-axis in that such a rotation does not change the magnitude or the sign of the displacements. In addition, we see that the vibrational motion in this mode does not change the overall symmetry of the H_2O molecule. Also, it is clear that this type of displacement

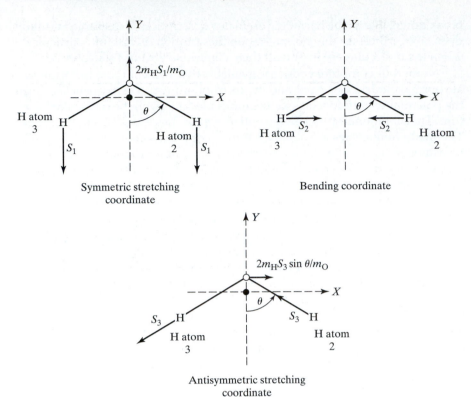

▶ **FIGURE 15.27**
Vibrational coordinates for H_2O.
The black dot at the origin is the
center of mass of the molecule.

stretches the —O—H bonds. Consequently, we might call S_1 a symmetric stretching coordinate.

In Mode 2, the oxygen atom remains stationary while hydrogen atoms number 2 and 3 are displaced along the X-axis by amounts $\Delta x_2 = -S_2$ and $\Delta x_3 = S_2$, respectively. All other displacements in this mode are zero. Because the predominant action of the vibrational coordinate is to bend the molecule, S_2 is called a bending coordinate.

In Mode 3, the displacements are $\Delta x_2 = -S_3 \sin \theta$, $\Delta y_2 = S_3 \cos \theta$, $\Delta x_3 = -S_3 \sin \theta$, and $\Delta y_3 = -S_3 \cos \theta$, where θ is one-half of the H—O—H bond angle, as shown in Figure 15.27. To maintain a fixed center of mass, we must have

$$m_H \Delta x_2 + m_H \Delta x_3 + m_o \Delta x_o = -2S_3 m_H \sin \theta + m_o \Delta x_o = 0. \quad \textbf{(15.86)}$$

Therefore, the oxygen-atom displacement must be $\Delta x_o = (2m_H/m_o)S_3 \sin \theta$. All other displacements are zero. If we rotate the molecule 180° about the Y-axis, all the displacements change directions. Therefore, this vibrational coordinate is antisymmetric. Also, it either stretches or compresses the O—H bonds. We can, therefore, reasonably refer to S_3 as an antisymmetric stretching coordinate. Table 15.6 summarizes all the displacements for the three modes.

We must now determine whether the three vibrational coordinates S_1, S_2, and S_3 satisfy Eqs. 15.82 and 15.83 so that their motion will be harmonic. The total displacement for all of the atoms may be obtained by summing the displacements for the three modes given in the table. The result is

$$\Delta x_o = \frac{2m_H}{m_o} S_3 \sin \theta, \ \Delta y_o = \frac{2m_H}{m_o} S_1, \ \Delta z_o = 0, \quad \textbf{(15.87A)}$$

$$\Delta x_{H2} = -S_2 - S_3 \sin \theta, \ \Delta y_{H2} = -S_1 + S_3 \cos \theta, \ \Delta z_{H2} = 0, \quad \textbf{(15.87B)}$$

Table 15.6 Vibrational displacements for H_2O

Mode	Atom	x Displacement	y Displacement	z Displacement
1	O	0	$\dfrac{2m_H}{m_o}S_1$	0
	H2	0	$-S_1$	0
	H3	0	$-S_1$	0
2	O	0	0	0
	H2	$-S_2$	0	0
	H3	S_2	0	0
3	O	$\dfrac{2m_H}{m_o}S_3 \sin\theta$	0	0
	H2	$-S_3 \sin\theta$	$S_3 \cos\theta$	0
	H3	$-S_3 \sin\theta$	$-S_3 \cos\theta$	0

and

$$\Delta x_{H3} = S_2 - S_3 \sin\theta, \ \Delta y_{H3} = -S_1 - S_3 \cos\theta, \ \Delta z_{H3} = 0. \quad \text{(15.87C)}$$

In Cartesian coordinates, the kinetic energy is given by

$$T = \sum_{i=1}^{3} 0.5m_i\left[\left(\frac{\partial \Delta x_i}{\partial t}\right)^2 + \left(\frac{\partial \Delta y_i}{\partial t}\right)^2 + \left(\frac{\partial \Delta z_i}{\partial t}\right)^2\right]. \quad \text{(15.88)}$$

If we now take the derivatives of Eqs. 15.87 A–C with respect to time, making certain to remember that θ is fixed at the equilibrium angle for H_2O, square each result, and sum the derivatives for the three atoms multiplied by the appropriate mass factors, the result is (See Problem 15.47)

$$T = \left[m_H + \frac{2m_H^2}{m_o}\right]\left(\frac{\partial S_1}{dt}\right)^2 + m_H\left(\frac{\partial S_2}{dt}\right)^2 + \left[m_H + \frac{2m_H^2}{m_o}\sin^2\theta\right]\left(\frac{\partial S_3}{dt}\right)^2$$

$$= \left[m_H + \frac{2m_H^2}{m_o}\right]v_{S_1}^2 + m_H v_{S_2}^2 + \left[m_H + \frac{2m_H^2}{m_o}\sin^2\theta\right]v_{S_3}^2. \quad \text{(15.89)}$$

Equation 15.89 shows that the S_1, S_2, and S_3 vibrational coordinates satisfy Eq. 15.83. The kinetic-energy expression contains only squares of the S_1, S_2, and S_3 velocities, multiplied by constants. If the molecular potential energy for small vibrational amplitudes has the quadratic form of Eq. 15.82, viz.,

$$V = a_1 S_1^2 + a_2 S_2^2 + a_3 S_3^2,$$

then the motion will be harmonic and the vibrational coordinates S_1, S_2, and S_3 will be the normal-mode coordinates Q_1, Q_2, and Q_3 that we seek. In this case, the total vibrational energy will be

$$E_{\text{vib}} = (v_1 + 0.5)h\nu_{o1} + (v_2 + 0.5)h\nu_{o2} + (v_3 + 0.5)h\nu_{o3}, \quad \text{(15.90)}$$

where ν_{o1}, ν_{o2}, and ν_{o3} are the three measured H_2O vibrational frequencies and v_1, v_2, and v_3 are the three vibrational quantum numbers.

If the potential energy is not quadratic when expressed in terms of S_1, S_2, and S_3, even for small vibrational amplitudes, then these coordinates are not proper normal-mode coordinates. In this event, it may be necessary to devise totally different vibrational coordinates. However, it may be possible to take linear combinations of S_1, S_2, and S_3 such that Eqs. 15.82 and 15.83 are satisfied. If this can be done, the linear combinations producing that result will be the normal-mode coordinates. This procedure, however, becomes very laborious and difficult when the number of atoms in the molecule is large. For this reason, normal-mode analysis is generally done using appropriate computer codes.

A symmetric molecule does not possess a permanent dipole moment. However, this does not mean that all of its vibrational modes will be IR inactive. If the vibrational motion is such that it causes the molecule to gain a momentary dipole moment, the mode will still be IR active. If we know the form of the normal-mode coordinates, we can usually predict which modes will be active in such symmetric molecules. Example 15.12 illustrates this point for CO_2.

EXAMPLE 15.12

The normal-mode coordinates for a linear CO_2 molecule are depicted in the following diagram, where the + and − notation denotes motion out and into the plane of the paper, respectively:

Mode A Mode B Mode C Mode D

How many vibrational modes does CO_2 have? Which of the normal modes are degenerate? Which modes will be IR active? Why? How could we determine the frequency of the IR inactive mode? Which of the modes will have the lowest frequency?

Solution

Since CO_2 is linear, the number of vibrational modes is

$$\text{no. of modes} = 3N - 5 = 3(3) - 5 = 4. \tag{1}$$

We know that the bending frequency cannot depend upon whether the bending occurs in the plane of the paper or perpendicular to it. Therefore, Modes C and D must be degenerate bending modes.

CO_2 is symmetric and, therefore, has no dipole moment. However, when the molecule vibrates in Mode B, C, or D, the symmetry is broken and the molecule gains a temporary dipole moment that permits it to couple to the electric vector of the incident electromagnetic radiation. As a result, Modes B, C, and D are IR active. Mode A, on the other hand, is always symmetric during every phase of the molecule's vibrational motion. Hence, the molecule never gains a dipole moment. This mode is, therefore, IR inactive. Its frequency could be determined by taking a Raman spectrum of CO_2, since vibrational motion in Mode A changes the effective molecular volume. Consequently, the polarizability of the molecule will vary with the vibrational motion, $(\partial \alpha / \partial Q_A)_e$ is not zero, and Mode A is Raman active.

The bending modes usually have lower frequencies than the stretching modes. Accordingly, we would expect Modes C and D to have a common frequency which is lower than that of either Mode A or Mode B, each of which is a stretching mode.

For additional exercises, see Problems 15.51 and 15.52.

15.5   Electronic Spectroscopy

In electronic spectroscopy, we monitor transition energies between different electronic states of atoms or molecules. As noted in Table 15.1, the energy spacings from such transitions are generally in the range from 10^3 to 10^4 kJ mol^{-1}, so that the wavelengths of resonant electromagnetic radiation are in the UV-to-visible range (10^{-5} to 10^{-6} cm). The corresponding wave numbers for electronic transitions are 10^5 to 10^6 cm^{-1}.

In most cases, electronic spectra are far more difficult to analyze than rotational, IR, or Raman spectra. They are also more difficult to handle than the magnetic spectra we shall discuss in Chapter 16. Generally, a multitude of different electronic states are present, each of which contains a large number of vibrational–rotational states between which transitions can occur. The selection rules for electronic transitions are more complex than those for microwave, IR, or Raman spectroscopy. Moreover, since electrons carry a negative charge, it is always possible for the electric field to couple to the electronic states. This factor increases the number of spectral lines and thereby adds to the level of difficulty associated with the analysis of the spectra. Nevertheless, electronic spectroscopy contributes significantly to our knowledge of equilibrium structure, dissociation energies, moments of inertia, and vibrational frequencies in both the ground state and excited electronic states. Electronic spectroscopy also serves as a very useful analytical tool. We shall, therefore, discuss some of the less difficult aspects of this type of spectroscopy.

15.5.1   Electronic State Designations and Selection Rules

For diatomic molecules, the electronic states are partially represented by their term symbols. This, however, is not sufficient: Different electronic configurations can give rise to the same term symbol. Therefore, we precede the symbol with letters. The ground state is assigned the letter X. Other states with the same multiplicity as the ground state are denoted by capital letters A, B, C, etc., in order of increasing energy or in order of discovery. Electronic states whose multiplicities differ from that of the ground state are assigned lowercase letters a, b, c, etc.

Let us consider H_2 as a simple example. The ground-state electronic configuration is $(1s\sigma_g)^2$. This configuration produces only one possible angular momentum state: $^1\Sigma_g^+$. This state is, therefore, given the designation $X\,^1\Sigma_g^+$. Suppose we now excite one of the electrons in a $1s\sigma_g$ orbital to the $1s\sigma_u^*$ antibonding MO. There are two angular momentum states produced by this $(1s\sigma_g)^1(1s\sigma_u^*)^1$ configuration: $^1\Sigma_u^+$ and $^3\Sigma_u^+$. The first term state has the same multiplicity as the ground state. It is, therefore, denoted with a capital letter, while the triplet sigma state is assigned a lowercase letter. The actual designations are $B\,^1\Sigma_u^+$ and $b\,^3\Sigma_u^+$.

EXAMPLE 15.13

What are the angular momentum states for a ground-state O_2 molecule? Suggest reasonable spectroscopic designations for these states.

Solution

With 16 electrons, the simple MO description of the O_2 ground state is

$$(1s\sigma_g)^2(1s\sigma_u^*)^2(2s\sigma_g)^2(2s\sigma_u^*)^2(2p_0\sigma_g)^2(2p_{+1}\pi_u)^2(2p_{-1}\pi_u)^2(2p\pi_g^*)^2.$$

Maximum multiplicity, and hence lowest energy, is achieved by having the spins of the two electrons in the π_g^* antibonding orbital be the same. This assignment requires that their magnetic quantum numbers be different ($+1$ and -1), so that the total magnetic quantum number is zero and we have a sigma state. The ground-state term is, therefore, $^3\Sigma$. There is an even number of electrons in *ungerade* molecular orbitals, so the inversion symmetry is g. The reflection symmetry through a plane containing the bond axis is $(-)$, since the configuration $(2p_{-1}\pi_g^*)^1(2p_{+1}\pi_g^*)^1$ changes sign upon reflection from a triplet state. (See Chapter 14.) Consequently, the complete term symbol is $^3\Sigma_g^-$. We could also have this configuration with the spins paired to produce a singlet state. The spatial wave function for the singlet state is symmetric with respect to electron exchange, so the reflection symmetry in this state is $(+)$. Thus, the term symbol is $^1\Sigma_g^+$. If both magnetic quantum numbers are the same, $+1$ or -1, we have $|M| = 2$, so that a Δ state is produced. The spins must be paired to avoid a violation of the Pauli exclusion principle, so the term symbol for the state is $^1\Delta_g$. There is no reflection symmetry in a Δ state.

The spectroscopic designations for the ground state will be $X^3\Sigma_g^-$. Since the multiplicities of the two other states are different from that of the ground state, we use lowercase letters and denote these states as $a^1\Delta_g$ and $b^1\Sigma_g^+$, in order of increasing energy.

For a related exercise, see Problem 15.53.

The selection rules governing transitions between electronic states are complex. The most fundamental of these rules is that the change in the total spin quantum number, ΔS, must be zero. That is, singlet $\longrightarrow$ triplet, triplet $\longrightarrow$ singlet, etc., transitions are forbidden. This follows immediately from the fact that the electric dipole transition matrix element between two states will be zero if the spin function of the final state is different from that of the initial state. However, that result is obtained only if we ignore the spin–orbit coupling terms in the Hamiltonian. When spin–orbit coupling becomes important, the $\Delta S = 0$ selection rule breaks down. Such a breakdown is frequently observed even in molecules containing no heavy atoms, for which we know that spin–orbit coupling will be large. The resulting transitions are *weakly allowed*. The extinction coefficients for such transitions are small, but observable.

Each bound electronic state contains vibrational levels and, within these, rotational states. Since the form of the interatomic potential is different for different electronic states, the vibrational wave functions for one electronic state need not be orthogonal to those of another. As a result, the IR and Raman vibrational selection rule $\Delta v = \pm 1$ no longer holds. However, the rotational eigenfunctions are not affected by the form of the interatomic potential. Consequently, we still have the same rotational selection rules as those for microwave pure rotational spectra. For diatomics, for example, $\Delta J = \pm 1$, with $\Delta J = +1$ corresponding to the R branch of the spectral band and $\Delta J = -1$ to the P branch.

15.5.2 Potential Curves for Electronic States of Diatomics

The potential curves for the bound electronic states of a diatomic molecule can be computed by solving the molecular Schrödinger equation using methods described in Chapter 14. If a sufficiently large number of vibrational eigenvalues have been measured, a parametrized potential can be iteratively adjusted until the solutions of the vibrational Schrödinger equation match the measured results. An easy, but approximate, method for obtaining electronic-potential curves is to fit the parameters of a Morse potential to the measured fundamental vibrational frequency, bond dissociation energy, and equilibrium distance by using Eq. 12.155 and the procedure illustrated in Example 12.22.

The results of this fitting procedure are illustrated in Figure 15.28 for the $X^1\Sigma_g^+$ ground state and the $B^1\Sigma_u^+$ excited state of H_2. The bond dissociation energies, vibrational wave numbers, and equilibrium distances in these two states are given in Table 15.7.

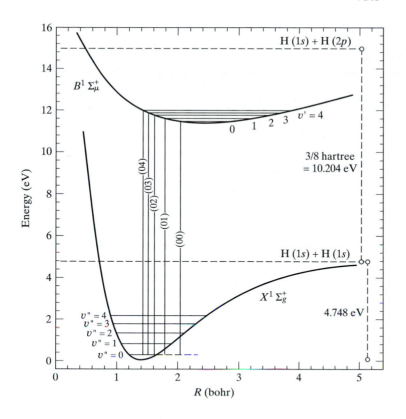

◀ FIGURE 15.28
Morse potential fits for the $X^1\Sigma_g^+$ and the $B^1\Sigma_u^+$ states of H_2. The potential minimum for the ground $X^1\Sigma_g^+$ state is taken as the energy zero. The dissociation products and the first five vibrational levels of each state are shown in the figure. The $(0v')$ bands for $v' = 0, 1, 2, 3,$ and 4 are also illustrated.

Table 15.7	Molecular constants for H_2[a]		
State	**Dissociation Energy (eV)**	ω_e **(cm^{-1})**	R_e **(bohr)**
$X^1\Sigma_g^+$	4.476	4,395.2	1.4014
$B^1\Sigma_u^+$	?	1,356.9	2.4428

[a]Herzberg, *Spectra of Diatomic Molecules* (New York: D. Van Nostrand Co., Inc., 1950).

The data listed in Herzberg do not include the dissociation energy from the $B^1\Sigma_u^+$ state. However, we are given the dissociation products and the energy gap between the minimum of the $B^1\Sigma_u^+$ and $X^1\Sigma_g^+$ potentials. The $B^1\Sigma_u^+$ state dissociates to a hydrogen atom in the $1s$ ground state and a second hydrogen atom in an excited $2p$ state. The energy of the dissociated atoms is, therefore,

$$-\frac{1}{2(1)^2} - \frac{1}{2(2)^2} = -\frac{5}{8} \quad \text{hartree.}$$

The $X^1\Sigma_g^+$ state dissociates to two ground-state hydrogen atoms whose combined energy is -1 hartree. Hence, the energy spacing between the limits of the dissociated atoms is $-\frac{5}{8} - (-1) = \frac{3}{8}$ hartree, which is 10.204 eV. The energy spacing between the two potential minima is 11.37 eV.

Using the foregoing data and the fitting procedure described in Example 12.22, we obtain the following two Morse potentials for these states (see Problem 15.54):

$$E(X^1\Sigma_g^+) = 4.748[\exp\{-1.0274(R - 1.4014)\} - 1]^2 \text{ eV} \quad \textbf{(15.91)}$$

and

$$E(B^1\Sigma_u^+) = 3.582[\exp\{-0.3652(R - 2.4428)\} - 1]^2 \text{ eV}. \quad \textbf{(15.92)}$$

These data permit us to construct the energy-state diagram shown in Figure 15.28. The two curves are plots of Eqs. 15.91 and 15.92, where the zero of energy is taken to be the minimum of the H_2 ground-state potential curve. The H_2 dissociation energy in the $B^1\Sigma_u^+$ state can now be obtained as the simple difference $(10.204 + 4.748) - 11.37 \text{ eV} = 3.582 \text{ eV}$.

15.5.3   Band Structure in Molecular Electronic Transitions

Electronic absorption spectra consist of a series of bands. Each band results from a transition from a particular vibrational level in an electronic state of lower energy to a vibrational level in a higher-energy electronic state. The vibrational quantum numbers and spectroscopic constants for the lower energy state are generally labeled with a double prime ("), while those for the higher energy state have a single prime. The $X^1\Sigma_g^+, v'' = 0 \rightarrow B^1\Sigma_u^+, v' = 0, 1, 2, 3, 4$ transitions for H_2 are shown in Figure 15.28. The wave number associated with a $v'' = i \rightarrow v' = j$ transition is generally labeled $\Delta\omega_{ij}$, and the resulting band is called the (ij) band. Using this nomenclature, we see that Figure 15.28 illustrates the (00), (01), (02), (03), and (04) transitions.

Electronic transitions occur on a time scale that is very short relative to the time required for the heavy nuclei to move. This means that the positions of the nuclei in the excited electronic state will be the same as they were at the moment of transition from the lower state. Consequently, all of the transitions seen in Figure 15.28 are "vertical," because the H_2 internuclear distance must be the same before and after the transition. The requirement that transitions must be vertical is called the *Franck–Condon principle*. This principle plays a major role in determining the intensity of the various electronic absorption bands. We know that the most probable position for a quantum mechanical harmonic oscillator in its ground state is at the equilibrium position. (See Figure 12.36.) As the vibrational quantum number increases, the most probable position shifts toward the classical turning points. (See Figure 12.37.) All the transitions shown in Figure 15.28 originate from the ground vibrational level of the $X^1\Sigma_g^+$ electronic state. Therefore, most of the H_2 molecules reside in a region around $R = R_e$. This leads us to the conclusion that the (04) band will be significantly more intense than the (00) band. The reason is that the former transition can originate from molecules near the equilibrium position as well as molecules at larger distances. In contrast, the (00) transition requires that the H_2 molecules in the ground state have very large internuclear separations in order to access the $v' = 0$ vibrational level of the $B^1\Sigma_u^+$ state. Very few molecules have such extended bonds; consequently, we expect the (00) band to be very weak relative to the (03) and (04) bands.

An expression for the wave number associated with the $(v''v')$ band can be obtained using Eq. 15.63, truncated after the first anharmonicity correction. If we let ω_e be the wave number corresponding to the energy difference between the potential minima of the two electronic states, the transition wave number for the $(v''v')$ band will be

$$\Delta\omega_e = \omega_e + (v' + 0.5)\omega_e' - (x_e\omega_e)'(v' + 0.5)^2 - (v'' + 0.5)\omega_e''$$

$$+ (x_e\omega_e)''(v'' + 0.5)^2. \quad \textbf{(15.93)}$$

If the frequency and anharmonicity factor are known for the initial state, the measurement of the band frequencies for a few transitions will permit these quantities to be obtained for the excited state. (See Problem 15.55.)

When electronic spectra are obtained for liquids or for molecules in solution, the variation of the environment from one molecule to the next produces a broadening of the band structure, just as it does in IR and Raman spectroscopy. In some cases, this broadening effect is sufficient to cause the individual bands to merge into a single broad peak for each electronic transition. When that occurs, we lose a great deal of information related to the structure and nature of the individual electronic states.

If high-resolution electronic spectra are taken in the gas phase at very low pressures at which intermolecular collisions are infrequent, it is often possible to observe rotational structure in each of the vibrational bands. When the molecule becomes very large, however, the increased moments of inertia make the rotational band spacings so small that it becomes impossible to resolve individual rotational lines. (See Figure 15.6.) For diatomic molecules, we may use a rigid-rotor approximation for the rotational levels. With rotational energy included, Eq. 15.93 becomes

$$\Delta\omega_e^r = \omega_e + (v' + 0.5)\omega_e' - (x_e\omega_e)'(v' + 0.5)^2 + B_e'J'(J' + 1)$$
$$- (v'' + 0.5)\omega_e'' + (x_e\omega_e)''(v'' + 0.5)^2 - B_e''J''(J'' + 1)$$
$$= \Delta\omega_e + B_e'J'(J' + 1) - B_e''J''(J'' + 1), \tag{15.94}$$

where $B_e = B/c$. As noted previously, the same selection rules hold for rotation in electronic and microwave spectroscopy. Therefore, we will have a P branch wherein $J' = J'' - 1$ and an R branch in which $J' = J'' + 1$. It is easy to show (see Problem 15.56) that the transition wave numbers for the R and P branches are

$$\Delta\omega_e^r = \Delta\omega_e + [B_e' - B_e''](J'')^2 + [3B_e' - B_e'']J'' + 2B_e' \quad (R \text{ branch}) \tag{15.95}$$

and

$$\Delta\omega_e^r = \Delta\omega_e + [B_e' - B_e''](J'')^2 - [B_e' - B_e'']J'' \quad (P \text{ branch}). \tag{15.96}$$

Because equilibrium distances vary significantly from one electronic state to another, there is usually a large difference between B_e' and B_e''. Let us suppose that for a given pair of electronic states, we have $B_e' > B_e''$. As a hypothetical example, let us also assume that $B_e' = 5.4 \text{ cm}^{-1}$ while $B_e'' = 5 \text{ cm}^{-1}$. In that case, the wave numbers for the P branch lines will be given by

$$\Delta\omega_e^r = \Delta\omega_e + 0.4(J'')^2 - 10.4J''.$$

Figure 15.29 shows a plot of $(\Delta\omega_e^r - \Delta\omega_e)$ versus J''. In this case, we see that the P branch lines exhibit a maximum redshift of 67.6 cm^{-1} from $\Delta\omega_e$ at $J'' = 13$. We can also see that the line positions get closer and closer together as we approach the position of maximum shift at $J'' = 13$. Since more rotational lines congregate at wave numbers around this point, the line intensity reaches a maximum. As a result, the position of this maximum is obvious upon examination of the spectrum. The point of maximum shift is called the *band head*, and the parabola seen in the figure is called a *Fortrat parabola*. If we have $B_e' < B_e''$, we can demonstrate (see Problem 15.57) that the R branch of the spectrum will usually exhibit a band head, but in this case blue-shifted from $\Delta\omega_e$. Consequently, a simple inspection of the spectrum will tell us whether we have $B_e' > B_e''$ or $B_e' < B_e''$. In the former case, we know that

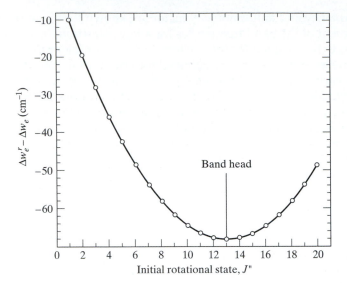

▶ **FIGURE 15.29**
A Fortrat parabola for the *P* branch transitions for the hypothetical case in which $B_e' = 5.4$ cm^{-1} while $B_e'' = 5.0$ cm^{-1}. In this case, the band head is seen at $J'' = 13$, with a maximum redshift of 67.6 cm^{-1}. (See text for details.)

the equilibrium distance in the higher energy state will be less than that in the lower energy state. In the latter case, the reverse will be true.

15.5.4 Emission Processes and Lasers

Lasers make use of emission spectroscopy to produce a highly intense, monochromatic beam of radiation. In addition, the laser output is highly directional and both spatially and temporally coherent. Coherent radiation consists of in-phase waves whose phase angle varies in a smooth fashion along the beam axis. The availability of this type of radiation source has revolutionized spectroscopy. For example, it has made Raman spectroscopy a useful tool in spite of the fact that only one photon in 10^7 is Raman scattered. The intensity of the laser source is so high that, even with this low yield, there is sufficient Raman signal to analyze.

Before discussing the general principle of laser operation, we shall qualitatively examine various aspects of emission spectroscopy. Suppose we excite a molecule by using a source that produces a wide range of wavelengths. Because of the $\Delta S = 0$ selection rule, the major result will be absorption into an array of different vibrational levels of several electronic states with the same multiplicity as the lower energy state, as illustrated in Figure 15.30. In this figure, electronic states A and B have the same multiplicity as the ground state X, while the multiplicities of states a and b differ from that of the ground state. After excitation, several processes can occur. The excited vibrational levels can relax to the ground vibrational state within each electronic state. Alternatively, "internal conversion" between vibrational levels in different electronic states with the same multiplicity may take place. This is usually followed by vibrational relaxation to the $v' = 0$ state. Much less frequently because of the $\Delta S = 0$ selection rule, "intersystem crossing" between vibrational levels in electronic states with different multiplicities can occur. We then observe emission processes between the ground vibrational level of each electronic state and various vibrational levels of the ground state. When the emission occurs between states having the same multiplicity, the process is allowed, so that the lifetime of the excited state is short. Such emissions are called *fluorescence*. If the emission involves states of differing multiplicities, the process is forbidden. The excited state is then *metastable*, in that its lifetime is relatively long. Emissions of this type are termed *phosphorescence*.

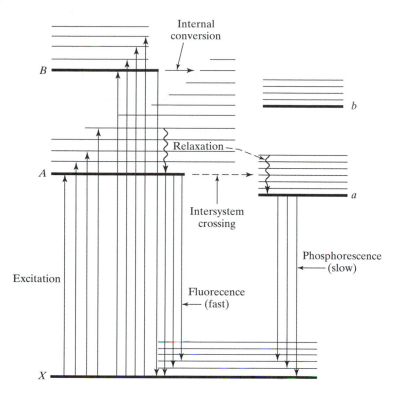

◀ FIGURE 15.30
Absorption and emission processes. Electronic states *A* and *B* have the same multiplicity as the ground state *X*. The multiplicities of states *a* and *b* differ from that of the ground state. The figure illustrates the result of broadband absorptions into various vibrational levels of states *A* and *B*. Internal conversion and intersystem crossing processes can then occur, followed by emission between states of the same multiplicity (fluorescence) or between states of different multiplicities (phosphorescence).

If a monochromatic radiation source is employed, the initial excitation will be to a single excited vibrational level in either the *A* or *B* electronic state shown in Figure 15.30. Subsequently, internal conversion and vibrational relaxation to the $v' = 0$ state will occur. At this point, fluorescence to different vibrational levels of the electronic ground state will produce a variety of emission bands. The figure shows that most of these bands will be red-shifted relative to the wavelength of the radiation source. This redshift was initially explained by Sir George Stokes. His first paper on the subject, "On the Change of Refrangibility [color] of Light," was published in 1852. Stokes had observed that the spectrum of the fluorescent light covered a wide range and that the emitted wavelengths were always longer than that of the monochromatic radiation source. He was of the opinion that this would always be the case. His view became known as *Stoke's law,* and the red-shifted emission lines were called *Stokes lines.* E. Lommel was the first to question the general validity of Stokes's law. Subsequently, E. Nichols and E. Merritt showed that when the initial excitation is from an excited electronic state, such as the *A* or *B* state shown in the figure, blue-shifted fluorescent emission lines could be observed. These lines were labeled *anti-Stokes lines.*

The analogy between the red- and blue-shifted lines seen in Raman spectra and those observed in fluorescence led directly to the nomenclature in which the red-shifted Raman bands are called Stokes lines while the blue-shifted bands are labeled anti-Stokes lines. It should be realized, however, that although similar terminology is used, the two phenomena are very different. The Raman effect is a scattering processes similar to Compton scattering; fluorescence is the result of a resonant photon absorption followed by spontaneous emission.

Now suppose we have a system that we excite or "pump" into a higher energy state, as illustrated in Figure 15.31. Although we are discussing the process with reference to electronic states, lasers can operate in any spectro-

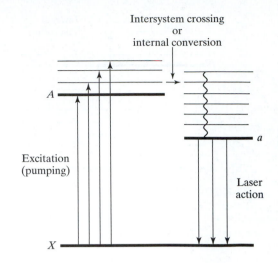

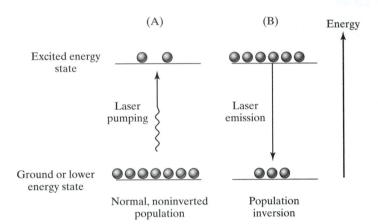

▶ FIGURE 15.31
Principle of laser action.

▶ FIGURE 15.32
Population inversion leading to laser action. (A) Thermal distribution of molecules in lower and excited energy states in which the lower energy state contains the majority of the molecules. (B) A state of population inversion caused by pumping. In this state, the population density of the excited state exceeds that of the lower energy state.

scopic range. After excitation to the upper state (State A in the figure), the incident radiation may stimulate direct emission back to the ground state, but let us assume that some of the time either internal conversion or intersystem crossing to State a occurs. Let us further assume that State a is a "bottleneck," in that the probability of emission back to the ground state from State a is very low. Under these conditions, it is clear that as pumping continues, the population of State a will continue to build until it reaches a point where the number of molecules in State a exceeds that in the ground state. This situation, which is illustrated in Figure 15.32, is called a *population inversion*. Such inversion is a necessary condition for laser action.

Sometime after the population inversion between states a and X is established, a few molecules in State a undergo emission to the ground state, thereby emitting photons whose energy is in exact resonance with the a–X energy spacing. If the photons are emitted at an angle to the axis of the laser cavity, they exit from the system and play no further role in the laser action. If, however, they are emitted along the cavity axis, mirrors at the ends of the cavity cause the photons to be reflected back and forth, thereby establishing resonant standing waves. (See Figure 15.33.) Those photons having wavelengths that are integral or half-integral multiples of the cavity length L undergo constructive interference and are amplified in intensity. Photons

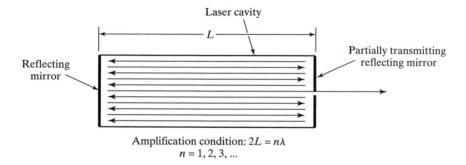

Laser cavity

L

Reflecting mirror

Partially transmitting reflecting mirror

Amplification condition: $2L = n\lambda$
$n = 1, 2, 3, \ldots$

▲ FIGURE 15.33
The principle of laser action: Photons emitted from the excited state reflect back and forth within the laser cavity. This action stimulates even more emission and absorption between the states shown in Figure 15.32. However, because of population inversion, emission dominates. As the waves reflect within the cavity, those with wavelengths $2L/n$, for integer n, undergo constructive interference. As a result, their intensity builds. Other wavelengths suffer destructive interference. This produces a situation in which the photons emerging through the partially transmitting mirror on the right are nearly monochromatic and, therefore, both spatially and temporally coherent.

with other wavelengths suffer destructive interference, which diminishes their intensity. The amplification condition is, therefore,

$$\lambda = \frac{2L}{n} \quad (n = 1, 2, 3, \ldots). \tag{15.97}$$

The amplified waves stimulate even more emission from State a, and the intensity of the standing wave builds. The waves also stimulate absorption from the ground state, reducing their intensity, but because there are many more molecules in State a than in the ground state, due to the population inversion, emission dominates over absorption. As a result, highly monochromatic, highly directional, very intense and coherent radiation builds within the cavity. One end mirror is constructed so that it is partially transmitting, thus allowing some of the radiation to leave the cavity as a laser beam. The output may be a very brief pulse, or it may be a continuous wave (CW).

Because the laser beam is highly directional, it cannot be seen from the side unless there are macroscopic particles present that scatter the beam's photons. This type of scattering is called the *Tyndall effect*. The magnitude of the scattered radiation depends upon the particle density and size compared with the wavelength of the photon. Consequently, the experimental measurement of scattered radiation intensity can be used to determine particle density and size.

Summary: Key Concepts and Equations

1. Spectroscopic experiments measure the energy spacings between different quantum states. These energy spacings vary from 10^{-6} kJ mol^{-1} for nuclear magnetic resonance to 10^4 kJ mol^{-1} for electronic energy states. Consequently, the wavelengths of electromagnetic radiation required to be in resonance with such spacings varies from 10^{-6} to 10^4 cm.

2. In order for the electric vector of incident electromagnetic radiation to couple to a molecule, transfer energy, and produce a transition, the molecule must have a

permanent or induced dipole moment. In addition, the transition matrix element connecting the initial and final states cannot be zero. This element is given by

$$H_{ij} = \langle \psi_i | x + y + z | \psi_j \rangle = q^{-1} \langle \psi_i | \mu_x + \mu_y + \mu_z | \psi_j \rangle,$$

where q represents the charge separation in the molecule and μ_x, μ_y, and μ_z are the x, y, and z components of the dipole moment vector. The conditions that ensure a nonzero value for H_{ij} give us the selection rules for various types of spectroscopy. The maximum probability for absorption and stimulated emission occur when the energy of the incident radiation equals the energy spacing between the quantum states. The precision with which this resonant frequency can be determined increases as the data collection time increases.

3. If the rotational energy states of a diatomic molecule are represented by those of a rigid rotor, the spacing between quantum states J and $J + 1$ is given by

$$\Delta E_{J, J+1} = 2(J + 1)Bh.$$

Since the selection rule for pure rotational transitions is $\Delta J = \pm 1$, rotational spectra of diatomic molecules comprise a set of absorption bands whose spacing, $\Delta \varepsilon$, is uniform and equal to $2Bh$, where

$$B = \frac{h}{8\pi^2 I} = \frac{h}{8\pi^2 \mu R^2}$$

is the rotational constant. Consequently, a measurement of the rotational band spacing permits the bond length to be computed using

$$R = \hbar [\mu \Delta \varepsilon]^{-1/2}.$$

4. The intensity of absorption lines depends upon the concentration of molecules in the sample cell ($[C]$), the path length ℓ of the incident radiation beam through the cell, and the molar absorption coefficient ε. The first two of these parameters are dependent upon the design of the spectrometer. In contrast, ε is characteristic of the molecule whose spectrum is being taken. The ratio of the transmitted to the incident intensity is given by the Beer–Lambert law:

$$\ln \left[\frac{I}{I_o} \right] = -\varepsilon [C] \ell.$$

5. If an experiment measures only the quantity of radiation absorbed, the absorption intensity is dependent upon the total transition rate out of state J, given by

$$W_J = N_J B_J \rho,$$

where N_J is the number of molecules in state J, B_J is the Einstein coefficient of stimulated absorption for state J, and ρ is the energy density of the radiation at the frequency of the transition. If the system is in thermal equilibrium at temperature T, then

$$N_J = C g_J \exp \left[-\frac{E_J}{kT} \right],$$

where C is a constant, g_J and E_J are the degeneracy and energy respectively, of state J, and k is the Boltzmann constant. In many cases, the variation of B_J with J is negligible compared with the variation of N_J with T. Consequently, we can usually assume that the relative intensity of two different absorption bands depends almost entirely upon the populations of the initial states for the transitions giving rise to the bands.

If an experiment measures all the radiation, absorbed and emitted, the net absorption is given approximately by

$$W_{net} = [N_J - N_{J+1}]B_J\rho.$$

In rotational, vibrational, and electronic spectroscopy, we generally measure only the radiation absorbed. Therefore, the previous equation is the one of interest. However, in some types of spectroscopy that will be discussed in Chapter 16, all of the radiation is collected. In these cases, the relative band intensities will depend upon the difference in population in the upper and lower states involved in the transition.

6. If an electric field is imposed upon a molecule possessing an electric dipole moment, an energy change called the Stark effect is produced. In general, the total change is very small. Its magnitude depends upon the square of the electric dipole moment, the square of the magnitude of the electric field, the molecular moment of inertia, and the M and J quantum numbers of the rotational state of the molecule. Since these small shifts in energy can be measured, the Stark effect provides a method for obtaining the electric dipole moments of molecules.

7. Since molecules are not rigid rotors, their actual rotational-energy eigenvalues will be slightly different from those predicted with a rigid-rotor approximation. The rotational motion produces a centrifugal force that elongates the molecular bond, thereby increasing the moment of inertia and decreasing the rotational energy eigenvalues. The magnitude of this centrifugal distortion effect is usually represented by the addition of a negative term to the expression for the rigid rotor. With this addition, the rotational eigenvalues are given by

$$E_J = J(J + 1)Bh - D_J J^2(J + 1)^2,$$

where D_J is the centrifugal distortion constant. If the interatomic potential can be represented by a Morse function, then

$$D_J = \frac{4B^3h}{\nu_o^2}.$$

8. Spherical-top molecules have all three moments of inertia equal. Their symmetry means that they have no permanent electric dipole moment. Consequently, we usually do not observe a pure rotational spectrum for such molecules. In a few cases, centrifugal distortion produces a sufficiently large momentary dipole moment to permit very weak rotational bands to be observed for transitions from very high J states.

   Symmetric-top molecules have two of their three principal moments of inertia equal. The rotational-energy eigenvalues for such molecules can be represented with reasonable accuracy by the expression

$$E_{J,K} = \frac{J(J + 1)\hbar^2}{2I_b} + \frac{K^2\hbar^2}{2}\left[\frac{1}{I_a} - \frac{1}{I_b}\right] \quad \text{for} \quad J = 0, 1, 2, 3, \dots. \quad \text{and} \quad -J \le K \le J,$$

where I_a is the unique moment of inertia and I_b is the moment that is common to two of the principal axes. The electric dipole selection rules for symmetric-top molecules are $\Delta J = 0$ or ± 1 and $\Delta K = 0$.

   The rotational-energy levels of linear polyatomic molecules are given by

$$E_J = \frac{J(J + 1)\hbar^2}{2I} = J(J + 1)Bh \quad \text{for} \quad J = 0, 1, 2, \dots.$$

This is identical to that for a diatomic molecule. The selection rule is also the same: $\Delta J = \pm 1$.

9. If we represent the combined rotational–vibrational energy eigenvalues by using the simple harmonic oscillator–rigid-rotor expression

$$E_{vJ} = (v + 0.5)h\nu_o + J(J + 1)Bh,$$

the rotational–vibrational selection rules are $\Delta J = \pm 1$ and $\Delta v = \pm 1$. Only transitions between adjacent vibrational states are allowed. With this approximation, the FTIR spectrum of a polar diatomic molecule will consist of a series of equally spaced bands exactly like those for pure rotational transitions of a diatomic rigid rotor. The spacings between these bands will be $2Bh$. Two sets of rotational transitions will be observed: those for which $\Delta J = +1$ and those for which $\Delta J = -1$. Transitions in the former group are called the R branch of the spectrum; those in the latter group are labeled the P branch. The band intensities for these transitions depend primarily upon the population of the initial state. Since a large majority of the transitions originate from the ground vibrational state, differences in the rotational-state energies and degeneracies determine the relative band intensities.

10. When the interatomic potential deviates from harmonic form, the eigenfunctions exhibit significant deviations from those for a harmonic oscillator. As a result, the $\Delta v = \pm 1$ selection rule breaks down, and transitions with $\Delta v > 1$ are observed. Such transitions are called overtones or harmonics. In general, we write the vibrational energy of a diatomic in the form

$$E_v^e = \frac{E_v}{hc} = (v + 0.5)\omega_e - x_e\omega_e(v + 0.5)^2 + y_e\omega_e(v + 0.5)^3 + \dots,$$

where ω_e is the vibrational wave number, which is equal to ν_o/c, and x_e and y_e are anharmonicity parameters that are fitted to the measured spectral data. If the interatomic potential is a Morse function, we have $hcx_e\omega_e = h^2\nu_o^2/(4D)$ and $y_e = 0$.

11. The bond dissociation energy D_e is the sum of all the vibrational state spacings from $v = 0$ up to a maximum value of $v = v_m$, plus a small remaining energy ε between the last bound vibrational state and the dissociation limit. That is,

$$D_e = \Delta\omega_o + \Delta\omega_1 + \Delta\omega_2 + \dots + \Delta\omega_{v_m} + \varepsilon = \sum_{v=0}^{v=v_m}\Delta\omega_v + \varepsilon.$$

We usually do not have $\Delta\omega_v$ values all the way up to the dissociation point. If we do not, we can execute a linear extrapolation of the data to determine the missing values of $\Delta\omega_v$ and ε. This procedure is called the Birge–Spooner extrapolation.

12. When a photon is scattered after colliding with a molecule, the frequency of the scattered photon is almost always unchanged, because of the huge difference in mass between the molecule and the photon. This type of elastic scattering is called *Rayleigh scattering*. On rare occasions, the collision process can cause a change in the vibrational–rotational quantum state of the molecule, which results in Raman scattering. If E_{vJ}^i and E_{vJ}^f are, respectively, the initial and final vibrational–rotational energies of the molecule, we must observe

$$h(\nu' - \nu) = h\,\Delta\nu = E_{vJ}^i - E_{vJ}^f,$$

where ν' and ν are the final and initial photon frequencies, respectively. $\Delta\nu$ is called the Raman shift. If $E_{vJ}^f > E_{vJ}^i$, $\Delta\nu$ will be negative, so that the scattered photon frequency will be less than the photon's initial frequency (Stokes scattering). If $E_{vJ}^f < E_{vJ}^i$, then $\Delta\nu$ is positive and anti-Stokes scattering will be observed.

   In order for a vibrational mode to be Raman active, the rate of change of the polarizability of the molecule with the vibrational coordinate evaluated at the equilibrium point must be nonzero. This will generally be the case if the vibrational motion produces an effective change in the volume of the molecule. For a diatomic molecule, the expansion and contraction of the bond produce such a change. Consequently, the vibrational mode of a diatomic molecule is always Raman active.

The selection rules for Raman spectra are different from those for IR absorption spectroscopy. For pure rotational spectra, the selection rules are

$$\Delta J = 0, \pm 2 \text{ for linear molecules}$$

and

$$\Delta J = 0, \pm 1, \text{ or } \pm 2 \text{ and } \Delta K = 0 \text{ for symmetric rotors.}$$

For vibrational–rotational transitions, we must have

$$\Delta v = \pm 1 \quad \text{and} \quad \Delta J = 0, \pm 2.$$

Transitions with $\Delta J = +2$ are called the S branch, those with $\Delta J = -2$ are labeled the O branch, and those for which $\Delta J = 0$ are said to form the Q branch. The line spacings in the S and O branches are $4Bh$, and the Q branch is separated from both the S and the O branch by $6Bh$.

13. Nonlinear polyatomic molecules with N atoms possess $3N - 6$ vibrational modes. Linear polyatomics have $3N - 5$ vibrational modes. If a molecule possesses a permanent dipole moment, its vibrational modes will generally be IR active. If it does not, some of the modes may still be IR active if the vibrational motion is such that the molecule gains a momentary dipole moment due to distortion of its molecular symmetry. Rotational lines usually cannot be resolved for large polyatomics, because of the small spacings between the lines. When the spectrum is taken in solution, for a neat liquid or in a crystalline solid, the rotational motion is frozen out due to the steric hindrance. Rotational structure is, therefore, not observed. Vibrational bands in solution or in pure liquids are generally much broader than in the gas phase or in a crystalline solid, because of the range of different liquid environments that are present. Vibrational modes that are due predominantly to the stretching or bending of chemical bonds in a given functional group, such as $-C-O-C-$, $-C-H$, $-C=O$, and $-C\equiv N$, usually appear in a well-defined frequency range. Modes that involve primarily angle-bending, rocking, twisting, or deformational motions usually have much lower frequencies than those associated with bond stretching.

14. The coordinates describing the vibrational motion of polyatomics are called normal-mode coordinates. They must be such that the motion produces neither rotational torque nor translation of the center of mass of the molecule. In order for the vibrational modes to be separable and have harmonic form, the kinetic energy must depend only upon a linear combination of squares of the normal-mode velocities and the potential energy must be a linear combination of the squares of the normal-mode coordinates themselves. The second of these requirements is usually possible only when the vibrational energy is small.

15. For diatomic molecules, the electronic states are represented by their term symbols, preceded by letters. The ground state is assigned the letter X. Other states with the same multiplicity are denoted by the capital letters A, B, C, etc., in order of increasing energy or in order of discovery. Electronic states whose multiplicities differ from that of the ground state are assigned the lowercase letters a, b, c, etc.

16. The selection rules for electronic transitions are

$$\Delta S = 0 \text{ (no change in total spin quantum number)}$$

and

$$\Delta J = \pm 1 \text{ (for diatomics).}$$

There is no selection rule for changes in vibrational quantum number between electronic states; all values of Δv are allowed.

17. Electronic transitions occur on a time scale that is very short relative to the time required for the heavy nuclei to move. Consequently, the positions of the nuclei in the excited electronic state will be the same as they were at the moment of transition from the lower state. The transitions will, therefore, be vertical in any plot of energy versus nuclear structure. The requirement that transitions be vertical is called the Franck–Condon principle.

18. The wave number associated with the $(v''v')$ band can be expressed in terms of the harmonic energies of each state and the associated anharmonicity corrections as

$$\Delta\omega_e = \omega_e + (v' + 0.5)\omega_e' - (x_e\omega_e)'(v' + 0.5)^2 - (v'' + 0.5)\omega_e'' + (x_e\omega_e)''(v'' + 0.5)^2,$$

where ω_e is the wave number corresponding to the energy difference between the potential minima of the two electronic states. With rotational energy included, this equation becomes

$$\Delta\omega_e^r = \omega_e + (v' + 0.5)\omega_e' - (x_e\omega_e)'(v' + 0.5)^2 + B_e'J'(J' + 1) - (v'' + 0.5)\omega_e''$$

$$+ (x_e\omega_e)''(v'' + 0.5)^2 - B_e''J''(J'' + 1) = \Delta\omega_e + B_e'J'(J' + 1) - B_e''J''(J'' + 1),$$

where $B_e = B/c$. The transition wave numbers for the R and P branches are

$$\Delta\omega_e^r = \Delta\omega_e + [B_e' - B_e''](J'')^2 + [3B_e' - B_e'']J'' + 2B_e' \quad (R \text{ branch})$$

and

$$\Delta\omega_e^r = \Delta\omega_e + [B_e' - B_e''](J'')^2 - [B_e' + B_e'']J'' \quad (P \text{ branch}).$$

If $B_e' > B_e''$, the P branch will exhibit a band head that is red-shifted from $\Delta\omega_e$. If $B_e' < B_e''$, a blue-shifted band head will be seen in the R branch.

19. Emission processes frequently involve internal conversion between vibrational levels in different electronic states with the same multiplicity. When this process occurs between electronic states with different multiplicities, it is called intersystem crossing. The latter process is much less frequent, because of the $\Delta S = 0$ selection rule. Such internal conversion or intersystem crossing is generally followed by rapid vibrational relaxation to the $v' = 0$ state. Emission between electronic states of the same multiplicity is called fluorescence. If the multiplicities differ, the phenomenon is termed *phosphorescence*. The ΔS selection rule usually makes the lifetimes of excited states undergoing fluorescence much shorter than those involved in phosphorescence.

20. Lasers operate by producing a population inversion between an excited state and the ground state. The production of such an inversion requires that the probability of emission between the excited and ground states be very small. When emission between these states does occur, the photon that is emitted is trapped by mirrors within an optical cavity. Photons whose wavelength is an integral or half-integral multiple of the length of the cavity undergo constructive interference, which establishes resonant standing waves that stimulate even more emission, and the intensity of the standing wave builds. The waves also stimulate absorption from the ground state that reduces their intensity, but because there are many more molecules in the excited state than in the ground state due to the population inversion, emission dominates over absorption. As a result, highly monochromatic, highly directional, very intense, and coherent radiation builds within the cavity. One end mirror is constructed so that it is partially transmitting, thus allowing some of the radiation to leave the cavity as a laser beam. The output may be a very brief pulse, or it may be a continuous wave (CW).

Problems

Problems that require the use of some type of computational device are marked with an asterisk (*). Problems that require some type of plotting routine are indicated with a pound sign (#).

15.1 The spacing between two energy levels is 0.2380 kJ mol^{-1}. What are the wavelength and wave number of the electromagnetic radiation needed to produce the resonance condition? Refer to Table 15.1 and state what type of energy levels we might reasonably expect to be associated with that wavelength and wave number?

15.2 A spectroscopist wishes to measure the Lyman series absorption lines for the hydrogen atom. What range of wavelengths must the electromagnetic radiation produced by the spectrometer span to permit such measurements to be made? Are these wavelengths in the microwave, IR, or UV–visible range of the spectrum?

15.3 The selection rule for microwave rotational spectra is $\Delta J = \pm 1$. In this problem, you will show that this rule holds in two specific cases. The normalized eigenfunctions for the $(J = 1, M = 0)$, $(J = 2, M = 0)$, and $(J = 3, M = 0)$ rotational states are

$$Y_1^0(\theta, \phi) = [3/4\pi]^{1/2} \cos\theta,$$

$$Y_2^0(\theta, \phi) = [5/16\pi]^{1/2}[3\cos^2\theta - 1],$$

and

$$Y_3^0(\theta, \phi) = [7/16\pi]^{1/2}[2\cos\theta - 5\cos\theta\sin^2\theta],$$

respectively. Using these data,

(A) show that the $(J = 1, M = 0) \longrightarrow (J = 2, M = 0)$ rotational transition is allowed, and

(B) show that the $(J = 1, M = 0) \longrightarrow (J = 3, M = 0)$ rotational transition is forbidden.

15.4 The pure rotational transition $(J = M = 1) \longrightarrow (J = 1, M = 0)$ has $\Delta J = 0$ and hence violates the rotational selection rule. Show that the transition matrix element $H_{11, 10}$ is zero in this case.

15.5 To the extent that the Born–Oppenheimer approximation holds, the Hamiltonian is independent of nuclear mass. This means that molecular structures, bond lengths, and bond angles will not depend upon the isotopic mass of the atoms that make up the molecule. Suppose we mix BeH and BeD (Be^{2}H) and take the microwave rotational spectrum of the mixture. Using the data obtained in Example 15.2, determine the wave numbers for the first three rotational absorption bands for both BeH and BeD.

15.6 The equilibrium bond length for BH is 1.2325 $\times$ 10^{-10} m.

(A) If the rotational spectrum can be described by that for a rigid rotor, compute the expected energy spacing between the rotational absorption bands.

(B) At what energy would we expect to observe the absorption band corresponding to a transition from the $J = 4$ to the $J = 5$ rotational state?

(C) At what wave number would the transition in (B) be observed?

15.7 Since chlorine has isotopes of nominal mass 35 and 37, HCl is a mixture of H^{35}Cl and H^{37}Cl. How much spectral resolution in terms of cm^{-1} must our spectrometer have to permit us to separate the lowest energy rotational absorption band of H^{35}Cl from that of H^{37}Cl? The atomic masses of ^{35}Cl and ^{37}Cl are 34.9688 and 36.9659, respectively, and the HCl equilibrium bond length is 1.3152 $\times$ 10^{-10} m. To the extent that the Born–Oppenheimer approximation holds, the equilibrium bond lengths of H^{35}Cl and H^{37}Cl are the same. What is the separation between the H^{35}Cl and H^{37}Cl absorption bands for the $J = 7$ to $J = 8$ transition?

15.8 Be^{2}H (BeD), ^{11}BH, and H^{35}Cl are mixed in equal proportions in a container, and the pure rotational microwave spectrum of the mixture is taken. The reduced masses and equilibrium bond lengths of these molecules are given in the following table:

Molecule	μ (atomic mass units)	R (bohr)
BeD	2,996.2	2.538
^{11}BH	1,682.9	2.329
H^{35}Cl	1,785.8	2.4854

The rotational absorption spectrum shows bands at the following wave numbers: 11.37, 19.90, 22.74, 24.04, 34.11, 39.80, 45.48, 48.08, 59.70, 72.12, 79.60, and 96.16 cm^{-1}. The 11.37 cm^{-1} absorption band is the lowest-energy band in the spectrum. *Using only these data and without doing any actual calculations,* identify each of the absorption bands in this combined spectrum.

15.9 An investigator finds that if he increases the path length of the electromagnetic radiation beam through the sample cell in an absorption measurement by 50%, the ratio of the transmitted radiation intensity to the incident intensity drops by a factor of two, provided that the measurements are made at the same wavelength and concentration. Is this sufficient information to determine the fraction of the radiation transmitted in the original experiment? If so, compute that fraction. If not, state what additional information is required to do the calculation.

15.10 Suppose we take an absorption spectrum by using a sample cell with reflecting mirrors on both sides such that the electromagnetic beam makes multiple passes through the cell before exiting to the detector. Such a cell is illustrated in the following diagram:

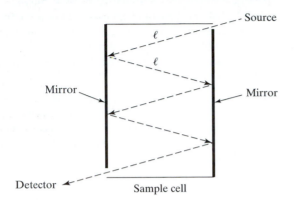

In the figure, the electromagnetic beam traverses the cell five times before exiting to the detector.

(A) Consider a similar cell in which the beam traverses the cell n times before exiting to the detector. Derive a general equation relating the fraction of the incident intensity transmitted in such a cell to the fraction transmitted for a cell using only a single pass of path length ℓ.

(B) If the fraction of the incident intensity transmitted in a cell using a single pass is f, derive an equation, in terms of f, that gives the value n must be to ensure the absorption of at least 80% of the incident radiation.

(C) If $f = 0.9$, determine n for the conditions given in (B).

15.11 (A) Compute the ratio of spontaneous emission to stimulated emission for a rotational transition whose transition frequency is 2×10^{11} s^{-1}, assuming that the incident radiation density per unit frequency is that for a blackbody radiator at a temperature of 500 K.

(B) At what temperature would spontaneous emission be 20% that of stimulated emission?

15.12 A scientist desires to use pure rotational band intensities as a measure of a system's temperature. Let the rotational system be H^{35}Cl, for which $2Bh = hc\bar{\nu} = (6.6261 \times 10^{-34}$ J s$)(2.9979 \times 10^{10}$ cm s$^{-1})$ $(20.70$ cm$^{-1})$ $= 4.112 \times 10^{-22}$ J. Suppose the scientist wishes that to use the ratio of the band intensity at $6Bh$ to that at $4Bh$ as a measure of the temperature of the system. Obtain a calibration curve of this ratio as a function of temperature, assuming that the absorption intensity is determined solely by the population of the initial state and that HCl may be treated as a rigid rotor. At $T = 50$ K, how accurately must this ratio be measured to obtain three significant digits of accuracy for the system's temperature?

15.13 Consider a hypothetical situation in which the pure rotational selection rule is

$$\Delta J = \pm 2 \quad \text{instead of} \quad \Delta J = \pm 1.$$

(A) Derive a general formula giving the observed absorption frequencies for the allowed transitions from rotational state J for the rigid rotor.

(B) Describe, by means of an appropriate sketch, the expected appearance of a rotational spectrum of a rigid rotor for this selection rule.

15.14 Consider a hypothetical situation in which the pure rotational selection rule is

$$\Delta J = \pm 2 \quad \text{or} \quad \pm 1 \quad \text{instead of just} \quad \Delta J = \pm 1.$$

(A) Obtain expressions for the energies at which allowed rotational transitions of a diatomic rigid rotor will be observed for this selection rule.

(B) For the case of a diatomic molecule with the preceding selection rule for which $2Bh$ corresponds to a wave number of 20.7 cm^{-1}, calculate the expected intensity ratio of the absorption band at $10Bh$ to that at $6Bh$ at a temperature of 400 K. (You may assume that the absorption intensity is determined solely by the population of the initial state.)

(C) Repeat (B) for the actual rotational selection rule $\Delta J = \pm 1$.

15.15$^\#$ The absorption band for the pure rotational transition from $J = 0$ to $J = 1$ in a diatomic molecule possessing a permanent electric dipole moment is observed at an energy corresponding to 18.00 cm^{-1}.

(A) If we arbitrarily assign a value of 1.000 to the intensity of this band at 298 K, compute and plot the relative intensities of the 18 absorption bands with lowest energy for this molecule at 298 K.

(B) Use the same procedure to compute and plot the relative intensities of the same 18 absorption bands at 150 K. (Assume that the absorption intensity is determined solely by the population of the initial state and that the spectrometer is capable of spanning the necessary range of frequencies.) Make separate plots of the results of (A) and (B) by plotting the relative intensities versus the wave number at which the transition occurs. What happens to the intensities as the temperature drops?

15.16* Rotational transitions in a particular sample of the molecule described in Problem 15.15 can be detected if their relative intensities are greater than or equal to 1% that for the $J = 0$ to $J = 1$ transition. What is the highest energy transition that can be detected at a temperature of 298 K? What is its intensity relative to the $J = 0$ to $J = 1$ transition at that temperature? (Assume that the absorption intensity is determined solely by the population of the initial state and that the spectrometer is capable of spanning the necessary range of frequencies.)

15.17 An investigator wishes to use the intensity of one band in the microwave rotational spectrum of the molecule described in Problem 15.15 as a means of detecting changes in temperature. The investigator states that the temperatures will range from a low of 298 K to a high of 400 K. The most intense spectral band at 298 K is the band at 72.00 cm^{-1} that corresponds to the $J = 3$ to $J = 4$ transition, while the least intense observable band is the one corresponding to the $J = 12$ to $J = 13$ transition at 234.00 cm^{-1}. (See Problem 15.15.) The investigator would like to have the advice of the physical chemistry class as to which rotational band should be selected so as to provide the maximum sensitivity to temperature changes. One member of the class suggests that the band at 72.00 cm^{-1} be used. Sam, on the other hand, believes that the 234.00 cm^{-1} band is the best choice. What do you think? Why? (Assume that the absorption intensity is determined solely by the population of the initial state and that the spectrometer is capable of spanning the necessary range of frequencies.)

15.18 It is stated in the text that the first-order perturbation correction to the zeroth-order rigid-rotor energy is zero for the Stark effect. In this problem, you will prove that this assertion is correct. We seek to show that

$$\langle \mathcal{H}' \rangle = \mu_e \mathcal{E} \int_{\phi=0}^{2\pi} \int_{\theta=0}^{\pi} [Y_J^M(\theta, \phi)]^* \cos\theta\, Y_J^M(\theta, \phi)$$
$$\sin\theta\, d\theta\, d\phi = 0.$$

Hint: (1) Recall that the spherical harmonic is separable into a θ part and a ϕ part and may be written as $Y_J^M(\theta, \phi) = \Theta_J^M(\theta)\Phi_M(\phi) = (2\pi)^{-1/2} e^{iM\phi} \Theta_J^M(\theta)$. (2) The $\Theta_J^M(\theta)$ function depends upon either sines or cosines of θ. Hence, it is either symmetric or antisymmetric about $\theta = \pi/2$. That is, we always have $\Theta_J^M(\theta) = \pm\Theta_J^M(\pi - \theta)$.

15.19 Determine the transition energies for all of the $J = 0$ to $J = 1$ and $J = 1$ to $J = 2$ transitions, with $\Delta J = \pm 1$ and $\Delta M = 0, \pm 1$ for a polar rigid rotor in an electric field $\mathcal{E}$ in terms of the Bh and the constant $C = I\mu_e^2\mathcal{E}^2/\hbar^2$. Obtain an expression for the electric dipole moment in terms of $\mathcal{E}$, I, $\hbar$, and B and the largest energy spacing for the stated transitions.

15.20 (A) Assuming that J may be treated as a continuous variable, use Eq. 15.37 and the approximation that the band intensities for pure rotational transitions are determined solely by the population of the initial state to obtain a formula in terms of B, h, and T giving the most intense band in a rotational microwave spectrum.

(B) The most intense pure rotational band at 298 K for the molecule described in Problem 15.15 is the

$J = 3$ to $J = 4$ transition. Compare this result with the prediction of (A).

15.21 Derive an expression in terms of B, D_J, and J giving the spacing between adjacent rotational states for a diatomic rotor whose rotational energy levels are given by Eq. 15.51. Does the deviation from the rigid-rotor result increase or decrease as J increases? Explain.

15.22 The equilibrium bond length of $^{12}C^{16}O$ is 1.1150×10^{-8} m. The atomic weight of ^{16}O is 15.9949.

(A) Compute the rotational constant for $^{12}C^{16}O$ in units of s^{-1}.

(B) Compute the value of $2Bh$ for this molecule. What is the equivalent value of $2Bh$ in wave numbers?

(C) Assuming that the rotational spectrum of $^{12}C^{16}O$ is described by a rigid-rotor quantization, predict the position in cm^{-1} of the five lowest energy pure rotational bands.

15.23 The measured pure rotational spectrum of $^{12}C^{16}O$ shows that the five bands with the lowest energy are at 3.86337 cm^{-1}, 7.72659 cm^{-1}, 11.5895 cm^{-1}, 15.4520 cm^{-1}, and 19.3138 cm^{-1}.

(A) Demonstrate that this spectrum cannot be described as that of a rigid-rotor.

(B) If $^{12}C^{16}O$ is treated as a rigid rotor, the five lowest energy rotational bands are at 3.8634 cm^{-1}, 7.7268 cm^{-1}, 11.5902 cm^{-1}, 15.4536 cm^{-1}, and 19.317 cm^{-1}. (See Problem 15.22.) Plot the magnitude of the difference between the observed transition energies and those predicted by a rigid-rotor approximation as a function of the J state from which the transition originates.

(C) Use the data given in (B) to obtain the $^{12}C^{16}O$ rotational constant in units of s^{-1}.

(D) The measured fundamental vibrational frequency of $^{12}C^{16}O$ is 6.6381×10^{13} s^{-1}. Compute the value of D_J that would result if the interatomic potential for $^{12}C^{16}O$ were a Morse potential. Express the result in both J and cm^{-1}.

(E) The magnitude of the difference between the observed rotational bands and those predicted by a rigid-rotor approximation is given by

$$\text{Difference Magnitude} = 4D_J[J + 1]^3.$$

(See Problem 15.21.) Use the value of D_J obtained in (D) to compute the expected differences, and compare them with the experimentally measured results.

15.24* In this problem, we will examine how rotational spectra can be employed to determine the molecular structure of a polyatomic molecule. We use HCN as an example. The various parts of the problem serve as a procedure guide to the method.

HCN is a linear molecule whose structure is illustrated in the following diagram, in which A, B, and C are H, C, and N, respectively, with corresponding masses m_1, m_2, and m_3.

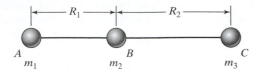

The moment of inertia of the molecule is given by

$$I = m_1 R_1^2 + m_3 R_2^2 - \frac{1}{M}(m_1 R_1 - m_3 R_2)^2,$$

where $M = m_1 + m_2 + m_3$. (See Problem 12.31.)

(A) A microwave rotational spectrum of HCN is taken. A series of absorption bands equally spaced by 2.95872 cm^{-1} is observed. Compute the moment of inertia of HCN. Express your result in units of kg m^2 and also in (atomic weight units) Å^2.

(B) A second microwave rotational spectrum of DCN is taken. A series of absorption bands equally spaced by 2.41642 cm^{-1} is observed. Compute the moment of inertia of DCN. Express your result in units of kg m^2 and also in (atomic weight units) Å^2.

(C) The atomic weights of H, D, C, and N are 1.0078, 2.0141, 12.0000, and 14.0032, respectively. Using the data obtained in (A) and (B), along with the expression for the moment of inertia, determine the bond lengths R_1 and R_2 for HCN and DCN. (*Hint:* Express the masses and distances in atomic weight units and angstroms, respectively. Then solve the moment-of-inertia equation for HCN for R_1 in terms of R_2. Finally, substitute this result into the moment-of-inertia equation for DCN, and solve the resulting equation numerically for R_2 to four significant digits.)

15.25 Which of the following molecules do we expect to exhibit pure rotational absorption spectra: N_2, H_2O, NF_3, C_6H_6, $H_2C{=}CH_2$, CO_2, H_2, O_2, CH_4, NO, HCN?

15.26 Describe qualitatively how you might determine the equilibrium structure of NH_3 using rotational microwave spectroscopy. Specify all measurements that would be required. Should one of the measurements be the rotational level spacings for NH_2D? Explain.

15.27 Prove that if the ratio of the path length difference Δp to the wavelength between two waves with the same wavelength and amplitude equals k, where k is an integer, constructive interference will be observed. Also, demonstrate that if $\Delta p/\lambda = k + \frac{1}{2}$, destructive interference will result.

15.28* In this problem, we will examine a more complex case involving incident radiation in an FTIR spectrometer. Suppose the incident radiation contains

four wavelengths: $\lambda_1 = 10^{-3}$ cm, $\lambda_2 = 1.7 \times 10^{-3}$ cm, $\lambda_3 = 2.1 \times 10^{-3}$ cm, and $\lambda_4 = 2.9 \times 10^{-3}$ cm. Suppose further that the amplitudes of these four waves are $A_1 = 1$ arbitrary unit, $A_2 = 1.5$ arbitrary units, $A_3 = 2.0$ arbitrary units, and $A_4 = 2.0$ arbitrary units, respectively.

(A) If the speed of the mirror is 10^{-2} cm s^{-1} and the path length from the source to the detector via the fixed mirror is 10 cm, obtain an equation giving the sum of the four waves that impinge upon the detector. Plot this sum over the range $t = 0$ to $t = 20$ s.

(B) Compute the modulation frequencies that are expected at the detector. Sketch the form of the power spectrum you would expect to see if you were to take the Fourier transform of the result obtained in (A).

15.29 Show that the electric dipole transition from $v = 0$ to $v = 2$ for a harmonic oscillator is forbidden.

15.30\# The equilibrium bond length and vibrational frequency for HF are 0.9171×10^{-10} m and 1.24069×10^{14} s^{-1}, respectively. The isotopic masses are ^{1}H = 1.0079 and ^{19}F = 18.9984. Assuming that the vibrational–rotational states are described by Eq. 15.62 with the selection rules $\Delta J = \pm 1$ and $\Delta v = \pm 1$ and that the relative intensities are determined solely by the initial-state populations, predict the appearance of the FTIR vibrational–rotational spectrum of HF at 298 K. Report your results in the form of a graph similar to Figure 15.21 for the first seven absorption bands in both the R and P branches.

15.31 The FTIR spectrum of gaseous $H^{127}I$ exhibits absorption bands at the following wave numbers: 2,257.1, 2,270.2, 2,283.3, 2,296.4, 2,322.6, 2,335.7, and 2,348.8 cm^{-1}.

(A) What is the vibrational frequency of HI, expressed in cm^{-1}? Compute the harmonic vibrational force constant for $H^{127}I$.

(B) Compute the moment of inertia of $H^{127}I$ from the given data.

(C) Calculate the equilibrium H–I bond distance using the given data. Assume that Eq. 15.62 describes the vibrational–rotational quantum states of $H^{127}I$ with sufficient accuracy. The atomic mass of ^{127}I is 126.9045.

15.32 The FTIR spectrum of an unknown diatomic molecule is taken. The result shows a series of absorption bands at 2,932.229, 2,954.236, 2,976.243, 3,020.257, 3,042.264, 3,064.271, and 3,086.278 cm^{-1}. It is also observed that the intensity of the band at 3,086.278 cm^{-1} is a factor of 1.9752 greater than the intensity of the band at 3,042.264 cm^{-1}.

(A) What is the wave number of the band origin?

(B) What transition produces the band at 3,042.264 cm^{-1}?

(C) At what temperature was the spectrum taken? Assume that Eq. 15.62 describes the vibrational–rotational quantum states of the molecule with sufficient accuracy and that the relative band intensities depend only upon the initial-state population.

15.33 The equilibrium bond length for $H^{79}Br$ is 1.413 Å. Assume that the atomic weights are H = 1.0078 and ^{79}Br = 78.9183.

(A) Calculate the value of the rotational constant for HBr. Express the answer in (a) s^{-1} and (b) cm^{-1}.

(B) A microwave spectrum of HBr is taken. At what wavelength will the $J = 2 \rightarrow J = 3$ rotational transition be observed? (Assume that HBr behaves as a rigid rotor.)

(C) Determine the most populated rotational state of HBr at a temperature of 800 K.

(D) The band origin for HBr is at 2,649.67 cm^{-1}. If the vibrational–rotational energy levels of HBr are given by a harmonic oscillator–rigid-rotor expression, calculate the energy in joules for the $(v = 0, J = 2)$ vibrational–rotational state. At what wave number will the absorption band for the $(v = 0, J = 2) \rightarrow (v = 1, J = 3)$ transition be observed?

15.34* Assuming that the energy states for a rotating Morse oscillator are given by Eq. 15.61, compute the energies in eV for the following absorption bands for an HI molecule described by a Morse potential whose parameter values are D = 3.194 eV, α = 0.9468 $bohr^{-1}$, and R_e = 3.032 bohr.

Initial State		Final State	
v	J	v	J
0	0 →	0	1
0	1 →	0	2
0	2 →	0	3
0	3 →	0	4
0	4 →	0	5
0	5 →	0	6
0	0 →	1	1
1	0 →	2	1
4	0 →	5	1

Compute the corresponding results if the system is treated as a rigid-rotor–harmonic oscillator having a fundamental vibrational frequency corresponding to 2,309.5 cm^{-1}.

15.35 Using the data in Table 15.4, predict the wave numbers for the first, second, and third vibrational overtones for $^{14}N^{14}N$.

15.36 Using the data in Table 15.4, predict the wave numbers for the first two vibrational hot bands arising from the $v = 1$ vibration state of $^{16}O^{16}O$.

15.37 (A) Assuming that we can ignore centrifugal distortion effects on rotational-energy levels, use the data in Table 15.4 to predict the wave number at which the hot band for the transition $(v = 1, J = 2) \rightarrow (v = 2, J = 3)$ would be observed in ^{7}LiH.

(B) Compute the ratio of the intensity of the transition in (A) to that for the transition $(v = 0, J = 2) \rightarrow (v = 1, J = 3)$ at 300 K. What is the intensity ratio at 600 K? Assume that the intensities are determined solely for the population of the initial states involved in the transitions.

15.38 Use the data in Table 15.4 and Eq. 15.65 to estimate the dissociation energy of $^{14}N^{14}N$.

15.39 The overtone spectra of H_2^+ give the following data:

v	$\Delta\omega_v (cm^{-1})$	v	$\Delta\omega_v (cm^{-1})$
0	2,191	8	1,257
1	2,064	9	1,145
2	1,941	10	1,033
3	1,821	11	918
4	1,705	12	800
5	1,591	13	677
6	1,479	14	548
7	1,368		

Use the Birge–Spooner extrapolation method to obtain the dissociation energy of H_2^+. The measured dissociation energy is 2.648 eV. Compute the percent error in your result.

15.40 The band origin and overtone spectra for HgH show that the successive vibrational spacings are 1,203.7, 965.6, 632.4, and 172 cm^{-1}. Determine the dissociation energy as accurately as possible from these data. The measured dissociation energy is 0.376 eV. Compute the percent error in your result.

15.41* (A) Use Eq. 15.64 to write $\Delta\omega_v$ as a quadratic function of v.

(B) Using the data given in Example 15.10 for H_2, execute a linear least-squares fitting of the function

$$\Delta\omega_v = a_o + a_1 v + a_2 v^2$$

to the data and determine the best values for a_o, a_1, and a_2. Plot the results of the fit versus v, and compare them with the data.

(C) By combining the results of (A) and (B), determine the values of ω_e, $x_e\omega_e$, and $y_e\omega_e$ for H_2.

15.42 The classical angular momentum of a diatomic rigid rotor is $I\omega$, where I is the moment of inertia and ω is the angular velocity, which is related to the angular frequency by $\omega = 2\pi\nu$. Consider a diatomic rigid rotor in the $J = 1$ rotational state.

(A) Show that the classical angular frequency is related to the rotational constant by $\nu = 2[J(J + 1)]^{1/2}B$ if we assume that the angular momentum is quantized.

(B) Using the data given in Table 15.4, determine how many times a classical $H^{35}Cl$ molecule will vibrate in the time required for it to undergo one complete rotation when the molecule is in the $J = 1$ rotational state.

15.43 (A) Show that the minimum Raman shift for a line in the S branch of a Raman spectrum for a diatomic molecule is $h\nu_o + 6Bh$, provided that Eq. 15.62 accurately describes the vibrational–rotational energies.

(B) Show that the maximum Raman shift for a line in the O branch is $h\nu_o - 6Bh$.

(C) Show that the spacings between the lines in the S and O branches are both $4Bh$.

(D) Show that the spacings between the Q branch Raman line and the closest S and O branch lines are both $6Bh$.

15.44# Using the data given in Table 15.4, predict the form of the vibrational–rotational Raman spectrum for $^{12}C^{16}O$ at a temperature of 200 K when the incident radiation has a wave number of 23,004 cm^{-1}. Assume that the vibrational–rotational energy states are given by Eq. 15.62 and that band intensities are determined solely by the population of the initial state in the transition leading to the spectral band. Report your results in the form of a plot similar to Figure 15.25, which shows the first eight S branch lines, the first seven O branch lines, and the Q branch line. Scale the Q branch line to twice the intensity of the most intense S branch line to improve the visual clarity of the line, but compute its actual expected intensity.

15.45 A Raman sample cell contains a gaseous mixture of two diatomic molecules that are listed in Table 15.4. A Raman spectrum of the mixture shows a series of vibrational–rotational lines whose Raman shifts are 3,300.64, 3,543.88, 3,761.62, 3,787.11, 3,845.38, 3,929.13, 4,012.89, 4,030.35, 4,138.52, 4,264.15, 4,347.91, 4,395.20, 4,431.67, 4,515.42, 4,760.05, 5,003.29, 5,246.53, and 5,489.76 cm^{-1}.

(A) Identify the two molecules that are in the sample cell.

(B) Which Raman lines correspond to the Q branch for each molecule?

(C) Identify the S and O branch lines for each molecule.

15.46 Using the data given in Table 15.4, predict the Raman shift wave number for the $(v = 0, J = 1 \rightarrow v = 1, J = 3)$ transition that would be observed in the S branch of the Raman spectrum for $H^{35}Cl$, assuming

that Eq. 15.62 describes the vibrational–rotational energy levels of $H^{35}Cl$.

15.47 Show that the combination of Eqs. 15.87A, 15.87B, 15.87C, and 15.88 leads to Eq. 15.89, as long as θ is kept fixed at the equilibrium H_2O angle.

15.48 The figure at the top of the next page shows the FTIR spectrum of 2-methyl-2-(4-methylphenyl) propanonitrile.

(A) Identify the bands that are likely due to C—H stretching modes in the phenyl ring.

(B) Which bands are due to the C—H stretches in the —CH$_3$ groups?

(C) Which band identifies the presence of the —C≡N group?

(D) Identify the band or bands produced by stretching modes of the —C=C— bonds in the phenyl ring.

(E) Which bands are most likely due to the —C=C—H bending modes?

(F) Which bands are most likely the result of —C—C—H bending modes?

(G) What would be a reasonable assignment for the strong absorption seen around 800 cm^{-1}?

15.49 The figure at the bottom of the next page shows the FTIR spectrum of 4-bromo-N-isopropylaniline.

(A) Identify the absorption band due to the —N—H stretching mode.

(B) Which bands are due to the C—H stretching modes in the phenyl ring?

(C) What absorption bands are the result of C—H stretching motions in the —CH$_3$ groups?

(D) What strong absorption is the likely result of the —C=C— stretching modes in the phenyl ring?

(E) Which bands are associated with the —C=C—H and —C—C—H bending modes?

(F) There is a strong absorption around 1,450 cm^{-1}. What modes produce this absorption?

(G) Which absorption band is due to the —C—Br stretching mode?

15.50 Consider a three-dimensional oscillator whose potential energy is given by

$$V(x, y, z) = \frac{k_1 x^2}{2} + \frac{k_2 y^2}{2} + \frac{k_3 z^2}{2}.$$

Let μ be the reduced mass of the oscillator.

(A) Write down the Schrödinger equation for this system in rectangular Cartesian coordinates.

(B) Let $\psi(x, y, z) = F(x)G(y)H(z)$, and show that the variables separate in the Schrödinger equation.

(C) Show that the separable equations all have the form of Eq. 12.147 so that the solutions are just harmonic oscillator eigenfunctions with eigenvalues $E_x = (v_x + 0.5)h\nu_x$, $E_y = (v_y + 0.5)h\nu_y$,

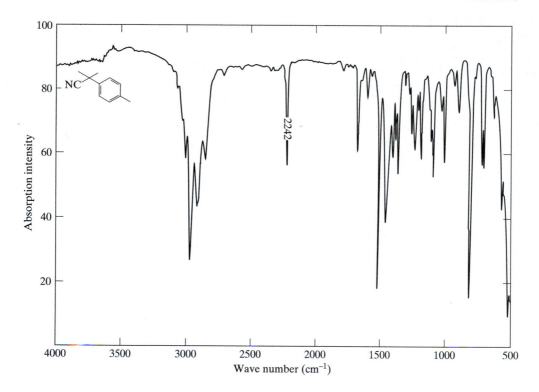

FTIR spectrum of 2-methyl-2-(4-methylphenyl) propanonitrile. The author thanks Professor K. D. Berlin and Dr. M. M. Madler for providing this spectrum.

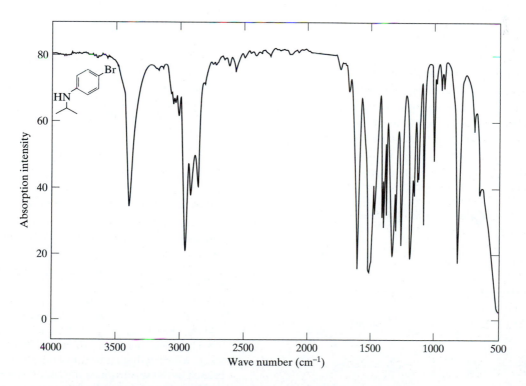

FTIR spectrum of 4-bromo-*N*-isopropylaniline. The author thanks Professor K. D. Berlin and Dr. M. M. Madler for providing this spectrum.

and $E_z = (v_z + 0.5)h\nu_z$. Therefore, the coordinates x, y, and z function as normal-mode coordinates.

15.51 Carbon suboxide is a linear symmetric molecule whose structure is $O=C=C=C=O$. Eight molecular motions of this molecule are shown in the following diagram:

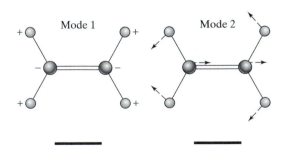

(A) How many vibrational modes does carbon suboxide have?

(B) Which of the molecular motions shown does not correspond to one of the vibrational modes of carbon suboxide?

(C) Which of the stretching modes shown above will be IR active? Which will be Raman active?

(D) Which of the molecular motions will be doubly degenerate vibrational modes of carbon suboxide? (The arrows show the normal-mode displacements in each mode.)

15.52 Consider the ethylene molecule, (C_2H_4).

(A) How many vibrational modes does C_2H_4 have?

(B) The following diagram shows 6 molecular motions of the planar C_2H_4 molecule, each of which may be a translation, a rotation, or a vibration:

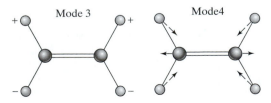

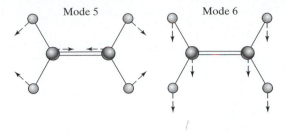

For each motion, indicate whether it is a translation, rotation, or vibration. In the case of the vibrations, the arrows represent the normal-mode displacement coordinates.

(C) Which of the vibrational motions shown would be expected to be IR active? The atoms labeled with the $+$ sign are moving out of the plane of the paper, while those labeled with the $-$ sign are moving into the plane of the paper.

15.53 Consider the ground state of the Li_2 molecule.

(A) What is the appropriate spectroscopic designation for the ground electronic state of this molecule?

(B) Suppose we excite Li_2 into the configuration $(1s\sigma_g)^2(1s\sigma_u^*)^2(2s\sigma_g)^1(2s\sigma_u^*)^1$. What are appropriate spectroscopic designations for the term states arising from this configuration?

15.54 Using the data given in Table 15.7, and referring to the discussion immediately following this table along with the procedure illustrated in Example 12.22, obtain Morse potential fits for the $X^1\Sigma_g^+$ and the $B^1\Sigma_u^+$ states of H_2 and thereby derive Eqs. 15.91 and 15.92.

15.55 A spectroscopist measures the vibrational bands between the $X^1\Sigma_g^+$ and the $B^1\Sigma_u^+$ states of H_2. The (02), (03), and (04) bands are found at 11.5043 eV, 11.6577 eV, and 11.8062 eV, respectively. The vibrational wave number and anharmonicity factor for the $X^1\Sigma_g^+$ ground state are known to be 4,395.2 cm^{-1} and 117.99 cm^{-1}, respectively. Use these data to determine the vibrational wave number and anharmonicity factor for the $B^1\Sigma_u^+$ state and the energy spacing between the potential minima of the two states.

15.56 Using the rigid-rotor selection rule, derive Eqs. 15.95 and 15.96 for the R and P branches of the electronic band structure.

15.57 The $A^3\Sigma_u^+$ state of O_2 has a rotational constant $B/c = B_e' = 1.05$ cm^{-1}. The O_2 $X^3\Sigma_g^-$ ground state has $B_e'' = 1.446$ cm^{-1}. Thus, in this case, we have $B_e' < B_e''$. At what rotational quantum number J'' will the band head in the $X^3\Sigma_g^- \rightarrow A^3\Sigma_u^+$ vibrational bands appear? Plot the difference $(\Delta\omega_e^r - \Delta\omega_e)$ versus J'' to show that a Fortrat parabola exists in the R branch of the spectrum. Is the band head red-shifted or blue-shifted from $\Delta\omega_e$? What does this tell us about the equilibrium O_2 bond length in the two electronic states? The $X^3\Sigma_g^- \rightarrow A^3\Sigma_u^+$ vibrational bands are called Herzberg bands.

15.58 The R branch of the electronic band structure usually exhibits a maximum blueshift of $\Delta\omega_e^r - \Delta\omega_e$ with respect to J'', the rotational quantum number of the initial electronic state, if $B_e' < B_e''$. The value of J'' that produces the maximum shift is called the band head. Show that if the rotational constants are such that $B_e'' > 3B_e'$, no band head will be observed in the R branch.

15.59 I'm ready to move onto nuclear magnetic resonance, but as usual, Sam has his hand up. It seems that he has a few questions about the James Bond (Agent 007) movie *Goldfinger*. For those members of the class who have not seen this movie, the scene is as follows: The archvillian Goldfinger has secret agent James Bond strapped to a platform. A large laser is moved into position with the beam directed at the foot of the platform between Bond's feet. When Goldfinger activates the laser, we see a brilliant golden-colored beam emerge and strike the platform. Immediately, the platform begins to melt under the action of the laser beam, which moves upward, threatening to cut 007 in half. Goldfinger announces that he expects Bond to die. Sam has the following questions:

(A) What is the approximate frequency of the electromagnetic radiation whose energy is in resonance with the transition responsible for Goldfinger's laser?

(B) What, if anything, do we know about the composition and color of the platform to which Bond is tied?

(C) What do we know about vibrational relaxation processes occurring in the platform under the action of the laser?

(D) What does this scene tell us, if anything, about the air quality in Goldfinger's laboratory? Explain.

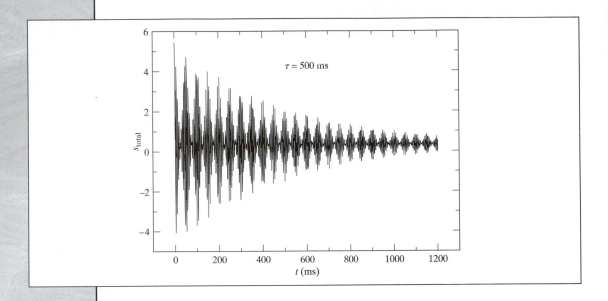

Magnetic and Diffraction Spectroscopy

Nuclear magnetic resonance (NMR) and X-ray diffraction spectroscopy are the two most powerful weapons the scientist has to investigate molecular structure. These spectroscopic methods are so important that it is not an overstatement to say that no modern research facility could exist without excellent NMR and X-ray equipment. Chemists in all areas rely on such methods to probe the structure and, in some cases, the dynamics of large, complex molecules. If a chemistry department were allowed to keep only one spectroscopic instrument, there is little doubt that the instrument of choice would be the NMR spectrometer. Today, NMR spectroscopy finds application not only in the chemical laboratories of the world, but also in medicine, where it is used as a noninvasive imaging technique.

In this chapter, we shall explore the principles and practice of NMR spectroscopy. This study will lead us naturally to electron spin resonance (ESR) spectroscopy. In the last portion of the chapter, some aspects of X-ray diffraction will be presented.

16.1 Nuclear Magnetic Resonance

Nuclear magnetic resonance was originally developed to measure the magnetic moments of various nuclei. Since there are a very limited number of different nuclei, and their magnetic moments need be measured only once, NMR was not initially regarded as a particularly important experimental technique. This perception changed dramatically when chemists discovered the chemical shift effect. With that discovery, NMR was no longer a curious phenomenon of interest only to physicists. Instead, it rapidly became a spectroscopic method of awesome power and importance.

16.1.1 Magnetic Field Interaction with a Circular Conducting Loop

In NMR spectroscopy, we measure transition energies between nuclear spin states in the presence of a superimposed magnetic field. We may gain an understanding of the basis of NMR by first considering the simple classical case of a charge of magnitude q rotating in a circular conducting loop of radius r with a rotation frequency ν. Figure 16.1 illustrates such a system.

In Chapter 13, we noted that a rotating charge possesses angular momentum and a magnetic moment. The magnitude of the magnetic moment is equal to the current I times the area A traced out by the loop. Mathematically,

$$|\mu| = IA, \tag{16.1}$$

provided that we express all quantities in SI units. If the current is produced by a charge q rotating with frequency ν, we have

$$I = q\nu. \tag{16.2}$$

Inserting this result and the fact that, for a circular loop, $A = \pi r^2$ into Eq. 16.1, we obtain

$$|\mu| = q\nu\pi r^2. \tag{16.3}$$

Multiplying numerator and denominator in Eq. 16.3 by $2m$, where m is the mass of the charge carrier, produces

$$|\mu| = \frac{2\pi\nu qmr^2}{2m}.$$

Since the angular velocity $v_\theta = 2\pi\nu$, we may write this equation in the form

$$|\mu| = \frac{qv_\theta mr^2}{2m}. \tag{16.4}$$

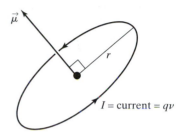

▲ FIGURE 16.1
A circular conducting loop of radius r with a charge q rotating with frequency ν to produce a magnetic dipole moment μ.

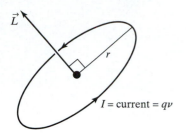

▲ FIGURE 16.2

The angular momentum **L** associated with the rotation of a charged particle of mass m and charge q in a circular loop of radius r. A comparison with Figure 16.1 illustrates the important fact that **L** and $\boldsymbol{\mu}$ are collinear and perpendicular to the loop.

Inspection of Eq. 11.21 shows that the quantity $v_\theta m r^2$ in Eq. 16.4 is the angular momentum P_θ. Therefore, the magnetic moment of a classical circular conducting loop is

$$|\mu| = \frac{q}{2m} P_\theta. \tag{16.5}$$

Quantum mechanically, we know that the angular momentum is quantized. We incorporate this concept into Eq. 16.5 by writing it in the form

$$\boldsymbol{\mu} = \frac{q}{2m}\mathbf{L}, \tag{16.6}$$

where **L** represents the quantized vector orbital angular momentum of the system. Equation 16.6 shows that $\boldsymbol{\mu}$ and **L** are collinear. Since **L** is perpendicular to the current loop, so is $\boldsymbol{\mu}$. This situation is illustrated in Figures 16.1 and 16.2. Equation 16.6 becomes identical to Eqs. 13.33 and 13.34 if we identify the quantity $q/(2m)$ as the magnetogyric ratio γ. The factor c (the speed of light) that appears in Eqs. 13.34 through 13.37 is due to our use of cgs units in those equations as opposed to the SI units we are using here.

The discussion in Chapter 13 pointed out that when a system possessing a magnetic moment is placed in a magnetic field **B**, a change in energy is produced. The reason for this change is illustrated in Figure 16.3. The magnetic moment produced by the current loop effectively acts like a compass, in that $\boldsymbol{\mu}$ will attempt to align itself with **B**. The force on the loop produced by the magnetic field that tends to align $\boldsymbol{\mu}$ with **B** causes the energy change, whose magnitude is given by Eqs. 13.35 through 13.37. The force depends upon the component of $\boldsymbol{\mu}$ in the direction of **B**, which we will take to be the Z-axis. This will in no way bias any measurements we make, since we are free to choose the coordinate system as we wish. For convenience, we repeat the equations obtained in Chapter 13 here. The energy change is

$$\Delta E_{\text{mag}} = -[\text{comp}_z\, \boldsymbol{\mu}]B = -\mu_z B. \tag{16.7A}$$

Using Eq. 16.6, we see that this takes the form

$$\Delta E_{\text{mag}} = -\frac{q}{2m}L_z B. \tag{16.7B}$$

Since the Z component of the orbital angular momentum is given by $m'\hbar$, Eq. 16.7B becomes

$$\Delta E_{\text{mag}} = -\frac{q\hbar}{2m}m'B. \tag{16.8}$$

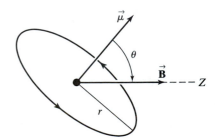

▶ FIGURE 16.3

The interaction of a magnetic moment of a circular conducting loop with a magnetic field in the Z direction.

Note that we are temporarily writing the magnetic quantum number as m' rather than m to avoid confusion with the mass of the charge carrier.

If the current in our loop is being carried by a proton, we will have $q = e = 1.602177 \times 10^{-19}$ C and $m = m_p = 1.67262 \times 10^{-27}$ kg. Under these conditions, the constant in Eq. 16.8 is

$$\frac{q\hbar}{2m} = \frac{e\hbar}{2m_p} = \frac{(1.602177 \times 10^{-19}\,\text{C})(1.05457 \times 10^{-34}\,\text{J s})}{2(1.67262 \times 10^{-27}\,\text{kg})}$$

$$= 5.0508 \times 10^{-27}\,\text{C m}^2\,\text{s}^{-1}.$$

Since a joule is a volt-coulomb, we can write this result as

$$\frac{e\hbar}{2m_p} = 5.0508 \times 10^{-27}\,\text{J m}^2\,\text{s}^{-1}\,\text{V}^{-1}.$$

This quantity, called the *nuclear magneton*, is given the symbol μ_N. The SI unit for magnetic flux density is the tesla (T), which is defined to be one volt s m^{-2}. We can, therefore, express the nuclear magneton in units of J T^{-1} and write $\mu_N = 5.0508 \times 10^{-27}$ J T^{-1}. In NMR experiments, it is common practice to express the magnetic flux in terms of gauss (G), where 1 gauss is 10^{-4} T. In these units, the nuclear magneton is

$$\boxed{\mu_N = 5.0508 \times 10^{-27}\,\text{J T}^{-1} \times \frac{1\,\text{T}}{10^4\,\text{G}} = 5.0508 \times 10^{-31}\,\text{J G}^{-1}.} \qquad \textbf{(16.9)}$$

The energy change of a circular conducting loop produced by a magnetic field **B** is, therefore,

$$\boxed{\Delta E_{\text{mag}} = -\mu_N m B} \qquad \textbf{(16.10)}$$

when the charge carrier is a proton. In Eq. 16.10, we revert to m as the notation for the magnetic quantum number since there is no longer any confusion with the mass of the charge carrier.

16.1.2   Nuclear Magnetic Moments

We learned in Chapter 13 that atomic particles often possess an intrinsic spin angular momentum due to relativistic effects. We denoted this angular momentum by **S**. We also learned that the same quantization rules hold for spin and orbital angular momentum. For the electron, these facts gave us

$$S^2 = s(s + 1)\hbar^2 \qquad \text{with } s = \frac{1}{2}$$

and

$$S_z = m_s\hbar \qquad \text{with } -s \leq m_s \leq s.$$

The same type of result is obtained for *all* atomic particles. They all possess an intrinsic relativistic spin angular momentum (**S**) that is quantized in the same manner as orbital angular momentum. That is, we have

$$\boxed{S^2 = I(I + 1)\hbar^2}, \qquad \textbf{(16.11)}$$

where I is the spin quantum number associated with a particular type of atomic particle. The Z component of spin angular momentum has its usual form

$$S_z = m_I \hbar \quad \text{with} \ -I \le m_I \le I \ , \tag{16.12}$$

where m_I is the magnetic spin quantum number, restricted to lie in the range $-I \le m_I \le I$. Equations 16.11 and 16.12 show that the relativistic spin angular momentum for nuclear particles is analogous to that for electron spin. The notational differences include a change from s to I for the spin quantum number and from m_s to m_I for the magnetic spin quantum number. The real differences are that some nuclear particles have values of I other than $\frac{1}{2}$ and magnetic moments that are different from the electron spin magnetic moment. Figure 16.4 illustrates this point for a ^{11}B nucleus with $I = \frac{3}{2}$ and possible m_I values of $\frac{3}{2}, \frac{1}{2}, -\frac{1}{2},$ and $-\frac{3}{2}$.

▶ FIGURE 16.4
Possible orientations of the nuclear spin angular momentum vector for a ^{11}B nucleus with $I = \frac{3}{2}$. In this case, m_I can take values of $\frac{3}{2}, \frac{1}{2}, -\frac{1}{2}, -\frac{3}{2}$. Since $\theta = \cos^{-1}(S_Z/S)$, the possible angles that **S** can make with the Z-axis are 39.23°, 75.04°, 105.0° and 140.8°, as shown in the figure.

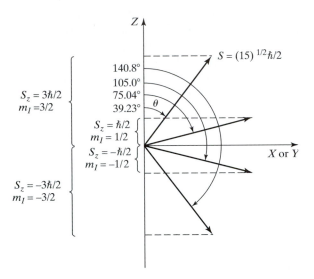

If nuclear particles behaved like circular conducting loops, we could predict their magnetic moments by combining Eq. 16.11 with 16.6 to obtain

$$\mu = \frac{q}{2m}S = \frac{q\hbar}{2m}[I(I+1)]^{1/2}. \tag{16.13}$$

Under these conditions, the Z component of the magnetic moment would be

$$\mu_z = \frac{q}{2m}S_z = \frac{q\hbar}{2m}m_I. \tag{16.14}$$

However, we cannot expect the simplistic model of a circular conducting loop to accurately describe a relativistic angular momentum effect for a nuclear particle. In fact, at present, there is no way to compute nuclear magnetic moments and spin quantum numbers; they must be determined by experimental measurement. The results of such experiments are usually expressed in the form

$$\mu = g_N \frac{e}{2m_p}S = g_N \frac{e\hbar}{2m_p}[I(I+1)]^{1/2} = g_N \mu_N [I(I+1)]^{1/2} \ , \tag{16.15}$$

Table 16.1 Nuclear g-factors, spin quantum numbers, relative abundance, and resonant frequencies[a]

Isotope	I (spin)	Relative Abundance	ν(MHz)[b]	g_N
^{1}H	1/2	99.985	42.5764	5.5856948
^{2}H	1	0.015	6.53573	0.8574382
^{11}B	3/2	80.1	13.6626	1.7924327
^{13}C	1/2	1.10	10.7081	1.4048236
^{15}N	1/2	0.366	4.3172	−0.5663776
^{14}N	1	99.634	3.0776	0.4037610
^{19}F	1/2	100.0	40.0765	5.257736
^{29}Si	1/2	4.67	8.4653	−1.11058
^{31}P	1/2	100.0	17.2510	2.26320
^{35}Cl	3/2	75.77	4.1764	0.5479162
^{79}Br	3/2	50.69	10.7039	2.106400
^{77}Se	1/2	7.63	8.1566	1.070084
^{89}Y	1/2	100.0	2.0949	−0.2748308
^{103}Rh	1/2	100.0	1.3476	−0.17680
^{107}Ag	1/2	51.839	1.7330	−0.2273592
^{109}Ag	1/2	48.161	1.9924	−0.2613812
^{111}Cd	1/2	12.80	9.0689	−1.1897722
^{113}Cd	1/2	12.22	9.4868	−1.2446018
^{117}Sn	1/2	7.68	15.2606	−2.00208
^{129}Xe	1/2	26.4	11.8601	−1.5559526

[a]Source of data: *Handbook of Chemistry and Physics*, 78th ed. (CRC Press, Boca Raton, FL, 1998).
[b]Data are for an applied magnetic field of 10^4 gauss = 1 tesla.

where g_N is called the *nuclear g-factor*. The corresponding Z component of the measured magnetic moment is

$$\mu_z = g_N \frac{e}{2m_p} S_z = g_N \frac{e\hbar}{2m_p} m_I = g_N \mu_N m_I . \tag{16.16}$$

The ratio μ/S is called the *magnetogyric ratio* γ. Using Eq. 16.15, we see that

$$\gamma = g_N \frac{e}{2m_p} = \frac{g_N \mu_N}{\hbar} . \tag{16.17}$$

Some measured nuclear g-factors and spin quantum numbers for selected nuclei are given in Table 16.1.

Combination of Eqs. 16.7A and 16.16 shows that the energy change produced by a magnetic field B is

$$\Delta E_{mag} = -g_N \mu_N B m_I . \tag{16.18}$$

EXAMPLE 16.1

In terms of the spin magnetic quantum number m_I, compute the Z component of the spin magnetic moment for a bare ^{13}C nucleus if the system behaved as a circular conducting loop. Using Table 16.1, calculate the measured value of μ_z and compare the two results. Does the conducting loop model give a reasonable answer?

Solution

If the magnetic moment of the ^{13}C nucleus could be computed with the use of the circular conducting loop model, we could employ Eq. 16.14 to obtain μ_z:

$$\mu_z = \frac{q}{2m} S_z = \frac{q\hbar}{2m} m_I. \tag{1}$$

The charge on a bare ^{13}C nucleus is $6 \times 1.602177 \times 10^{-19}$ C $= 9.61306 \times 10^{-19}$ C. The mass is

$$m = 13.00335 \times \frac{1.67262 \times 10^{-27} \text{ kg}}{1.0078} = 2.1581 \times 10^{-26} \text{ kg}. \tag{2}$$

Substituting these values into Eq. 1 and replacing $\hbar$ with $h/(2\pi)$ produces

$$\mu_z = \frac{(9.61306 \times 10^{-19} \text{ C})(6.62608 \times 10^{-34} \text{ J s})}{4(3.1415927)(2.1581 \times 10^{-26} \text{ kg})} m_I = 2.3488 \times 10^{-27} m_I \text{ J T}^{-1}. \tag{3}$$

The measured value of μ_z is given by Eq. 16.16:

$$\mu_z = g_N \frac{e}{2m_p} S_z = g_N \frac{e\hbar}{2m_p} m_I = g_N \mu_N m_I. \tag{4}$$

Substituting the values of g_N from Table 16.1 and the nuclear magneton gives

$$\mu_z = 1.4048236(5.0508 \times 10^{-27} \text{ J T}^{-1})m_I = 7.0955 \times 10^{-27} m_I \text{ J T}^{-1}. \tag{5}$$

The ratio of the two results is

$$\frac{[\mu_z]_{\text{loop}}}{[\mu_z]_{\text{measured}}} = \frac{2.3488 \times 10^{-27} m_I \text{ J T}^{-1}}{7.0955 \times 10^{-27} m_I \text{ J T}^{-1}} = 0.331. \tag{6}$$

The circular conducting loop model produces an answer that is one-third that of the measured Z component of the nuclear spin magnetic moment. Although the model is not very accurate, it does, in this case, give the right order of magnitude.

For related exercises, see Problems 16.1 and 16.2.

16.1.3   Nuclear Spin Energy Levels in a Magnetic Field

Consider a bare ^{1}H nucleus in the absence of a magnetic field. With a nuclear spin quantum number of $I = \frac{1}{2}$, the nucleus may be in either of two nuclear spin states, one with $I = \frac{1}{2}$ and $m_I = \frac{1}{2}$ and a second state with $I = \frac{1}{2}$ and $m_I = -\frac{1}{2}$. In the absence of a magnetic field, these nuclear states are degenerate. The situation is illustrated in Figure 16.5.

When a magnetic field is imposed, however, the degeneracy between the two states is lifted. Both states experience an energy change described by Eq. 16.18. The state with $m_I = \frac{1}{2}$ will decrease in energy by an amount $\Delta E_{\text{mag}} = -g_N \mu_N B/2$, while the state with $m_I = -\frac{1}{2}$ will increase by a like amount, as shown in the figure. This change will produce an energy spacing between the two states equal to $g_N\mu_N B$. *Nuclear magnetic resonance spectroscopy is concerned with the measurement of the frequencies of electromagnetic radiation required to induce transitions between these split energy states.*

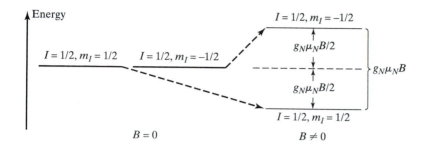

◀ **FIGURE 16.5**
Splitting of nuclear energy states for particles with $I = \frac{1}{2}$ in the presence of a magnetic field.

Equation 16.18 shows that the energy changes produced by an imposed magnetic field are directly proportional to the magnitude of the field. This situation is illustrated in Figure 16.6. The energy of each state varies linearly with B, with a slope whose magnitude is $g_N\mu_N/2$, as shown in the figure. In order to induce absorption or stimulated emission between these nuclear states with a high probability, the energy of the electromagnetic radiation must be equal to the energy spacing. This resonance condition is

$$\boxed{h\nu = g_N\mu_N B}.\qquad (16.19)$$

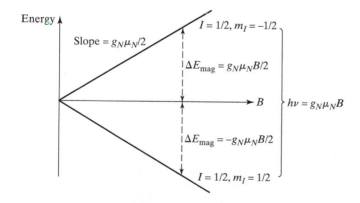

◀ **FIGURE 16.6**
Nuclear spin state splitting for nuclei with $I = \frac{1}{2}$. The resonance condition for absorption and stimulated emission is $h\nu = g_N\mu_N B$.

Since we are able to vary the energy spacing between nuclear spin states by varying the imposed magnetic field, a nuclear magnetic resonance spectrum can be taken by either of two methods. We can fix the magnetic field and vary the frequency until absorption occurs, as is done in rotational, vibrational, and electronic spectroscopy. Or we can fix the frequency of the electromagnetic radiation and vary the **B** field until the resonance condition is satisfied. Example 16.2 illustrates both alternatives.

EXAMPLE 16.2

(A) At what frequency will we observe spectroscopic transitions between the nuclear spin states of a bare proton when the imposed magnetic field is 23,486 G? To what wavelength does this frequency correspond? **(B)** If we fix the frequency at $1.75 \times 10^8 \text{ s}^{-1}$ (175 MHz), what magnetic field will be required to satisfy the resonance condition for spectroscopic transitions between nuclear spin states of a proton?

Solution

(A) Using the resonance condition given by Eq. 16.19, we find that the required frequency is

$$\nu = \frac{g_N \mu_N B}{h} = \frac{(5.58569)(5.0508 \times 10^{-31} \text{ J G}^{-1})(23,486 \text{ G})}{6.62608 \times 10^{-34} \text{ J s}} = 1.000 \times 10^8 \text{ s}^{-1}$$

$$= 100.0 \text{ MHz.} \tag{1}$$

The corresponding wavelength is

$$\lambda = \frac{c}{\nu} = \frac{2.9979 \times 10^{10} \text{ cm s}^{-1}}{1.000 \times 10^8 \text{ s}^{-1}} = 2.9979 \times 10^2 \text{ cm} = 299.79 \text{ cm} = 2.9979 \text{ m.} \tag{2}$$

(B) When the frequency is fixed, the magnetic field required to achieve resonance between the nuclear spin states is

$$B = \frac{h\nu}{g_N \mu_N} = \frac{(6.62608 \times 10^{-34} \text{ J s})(1.75 \times 10^8 \text{ s}^{-1})}{5.58569(5.0508 \times 10^{-31} \text{ J G}^{-1})} = 41,102 \text{ G.} \tag{2}$$

In modern, high-field NMR spectrometers, the magnetic field is always fixed and the frequency is varied to achieve resonance. In older instruments, the opposite method was used.

For related exercises, see Problems 16.3, 16.4, and 16.5.

Let us now examine the more complex case of a nucleus whose spin quantum number is greater than $\frac{1}{2}$. For example, consider ^{11}B, with $I = \frac{3}{2}$ and whose possible nuclear spin–angular momentum orientations are shown in Figure 16.4. (See also Figures 16.7 and 16.8) When there is no applied magnetic field, four degenerate nuclear states exist, each with $I = \frac{3}{2}$, but with different spin magnetic quantum numbers $m_I = -\frac{3}{2}, -\frac{1}{2}, \frac{1}{2}$, and $\frac{3}{2}$. This degeneracy is lifted when a magnetic field is applied. If the nuclear g-factor is positive, the states with negative values of m_I increase in energy by an amount given by Eq. 16.18, while the energies of those states with positive values of m_I decrease. In this case, the energy spacing between adjacent states is given by

$$\Delta E_{\text{adjacent states}} = \Delta E_{\text{mag}}(m_I + 1) - \Delta E_{\text{mag}}(m_I) = g_N \mu_N B(m_I + 1) - g_N \mu_N B m_I$$

$$= g_N \mu_N B. \tag{16.20}$$

This is the same adjacent state spacing obtained for nuclei with $I = \frac{1}{2}$. The form of Eq. 16.18 ensures that that will always be true. That is, we have $\Delta E_{\text{adjacent states}} = g_N \mu_N B$, regardless of the value of I.

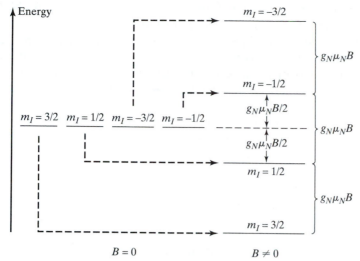

◀ FIGURE 16.7
Splitting of nuclear energy states for particles with $I = \frac{3}{2}$ in the presence of a magnetic field.

The selection rule for transitions between nuclear spin states is

$$\boxed{\Delta m_I = \pm 1}. \qquad\qquad (16.21)$$

Since the spacing between adjacent states is the same regardless of I, Eq. 16.19 expresses the resonance condition for all nuclei.

The nuclear spin state spacing is again a linear function of the applied magnetic field. This is seen in Figure 16.8, in which the energy changes of each nuclear state are plotted as a function of the applied field.

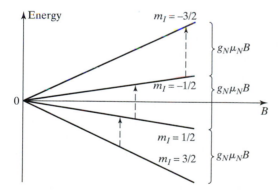

◀ FIGURE 16.8
Nuclear state splitting for $I = \frac{3}{2}$ nuclei. Dashed arrows indicate allowed transitions.

Although the resonance condition is the same for all nuclei, it is more difficult to interpret NMR spectra of nuclei with spin quantum numbers greater than $\frac{1}{2}$ compared with those for which $I = \frac{1}{2}$. The difficulty arises because nuclei with $I > \frac{1}{2}$ have a nuclear magnetic quadrupole moment as well as a magnetic dipole moment. The presence of a quadrupole moment causes the spectral lines to broaden considerably, which, as we have already seen in the case of vibrational spectra, may cause the bands to overlap. Such overlap makes the interpretation of the spectrum more difficult.

16.1.4 The Chemical Shift

As mentioned earlier, NMR spectroscopy permits the measurement of nuclear spin magnetic moments with great accuracy. It also serves as an experimental demonstration of the existence of a relativistic spin angular momentum for some particles. If this were the complete story for NMR spectroscopy, it would be of little interest or utility to the chemist.

The situation changed dramatically when chemists discovered the effect we now call the *chemical shift*. The electronic ground state of most molecules has total spin and magnetic quantum numbers equal to zero. When this is the case, there is no net spin or orbital angular momentum and, hence, no magnetic field. However, when a large external magnetic field B_o is imposed, the electronic wave function is perturbed, with the result that a small internal magnetic field develops because of the electron structure around the various nuclei. Using Le Chatelier's principle, we can predict that the direction of this internal field will be opposite from that of the applied external field B_o. This follows because the application of a perturbation or stress to an equilibrium system always shifts the equilibrium in a direction that tends to remove the perturbation or stress. Hence, when we perturb the equilibrium electronic structure with an applied magnetic field, the electronic wave function shifts in such a manner that an opposing magnetic field is established which tends to eliminate the perturbation.

The magnitude of the internal field about nucleus i is proportional to the magnitude of the perturbation B_o, so that

$$\text{Internal magnetic field about nucleus } i = B_i^{int} \propto B_o.$$

Writing this proportionality as an equality, we have

$$B_i^{int} = -\sigma_i B_o, \tag{16.22}$$

where the minus sign reflects the fact that the internal field is opposite to B_o. The total field acting on nucleus i will, therefore, be

$$B_i = B_o + B_i^{int} = B_o - \sigma_i B_o = B_o(1 - \sigma_i). \tag{16.23}$$

The proportionality constant σ_i is called the *shielding constant* for nucleus i. Each nucleus in the molecule will have its own characteristic shielding constant. If the electronic environments about two nuclei are identical, the shielding constants for these nuclei will be the same. An example is the four hydrogen nuclei in CH_4. All hydrogen atoms are chemically equivalent; therefore, their shielding constants are identical. In contrast, the protons in ethanal, H_3C-CHO, are not equivalent: The electronic environment about the aldehyde proton is not the same as that about the methyl hydrogens. Consequently, the aldehyde and CH_3 protons have different shielding constants. If the temperature is so low that the CH_3 group cannot rotate freely about the C–C single bond, the methyl protons may become nonequivalent. In this event, we could see different shielding constants for CH_3 protons in and out of the plane defined by the –CHO group.

The existence of the internal magnetic field does not alter the resonance condition described by Eq. 16.19. In order to observe adjacent state transitions between nuclear spin states, we still must have $h\nu = g_N \mu_N B$. However, the B field at the nucleus is now that given by Eq. 16.23. Therefore, the resonance condition for nucleus i is

$$\boxed{h\nu_i = g_N \mu_N B_o(1 - \sigma_i)}. \tag{16.24}$$

Nuclei with different shielding constants exhibit different NMR frequencies. If we fix the frequency and scan over the applied field to achieve resonance, the required resonance field for nucleus i is

$$B_{oi} = \frac{h\nu}{g_N \mu_N (1 - \sigma_i)}. \qquad (16.25)$$

The shift in resonance frequency or field produced by the electronic environment around the nucleus is called the *chemical shift*. Formally, the chemical shift for nucleus i (δ_i) is defined to be the difference in the shielding constant for some convenient reference nucleus and that for nucleus i times 10^6; that is,

$$\boxed{\delta_i = (\sigma_{\text{ref}} - \sigma_i)10^6}. \qquad (16.26)$$

For proton NMR, the reference is usually chosen to be the shielding constant for the 12 equivalent hydrogen atoms in tetramethylsilane, $(CH_3)_4Si$. The reference for ^{13}C NMR is the shielding constant for the four equivalent carbon atoms in $(CH_3)_4Si$.

By combining Eqs. 16.24 and 16.26, we can obtain an expression for the resonant frequency of nucleus i in terms of its chemical shift, the applied magnetic field, and the shielding factor for the reference compound. We first use Eq. 16.26 to obtain σ_i in terms of δ_i and σ_{ref}:

$$\sigma_i = \sigma_{\text{ref}} - \frac{\delta_i}{10^6}. \qquad (16.27)$$

Substitution of this result into Eq. 16.24 produces

$$\boxed{\nu_i = \frac{g_N \mu_N}{h} B_0 \left[1 - \sigma_{\text{ref}} + \frac{\delta_i}{10^6} \right]}. \qquad (16.28)$$

Let us now consider the difference between the resonant frequencies for two nuclei of the same type, but with different chemical shifts, which we denote as nucleus A and nucleus B. The difference in their resonant frequencies can be obtained directly from Eq. 16.28:

$$\nu_A - \nu_B = \frac{g_N \mu_N}{h} B_0 \left[1 - \sigma_{\text{ref}} + \frac{\delta_A}{10^6} \right] - \frac{g_N \mu_N}{h} B_0 \left[1 - \sigma_{\text{ref}} + \frac{\delta_B}{10^6} \right]. \qquad (16.29)$$

This yields

$$\boxed{\nu_A - \nu_B = \frac{g_N \mu_N}{h} B_0 [\delta_A - \delta_B] 10^{-6}}. \qquad (16.30)$$

Equation 16.30 reveals an extremely important fact about NMR spectra: The difference in resonant frequencies between two similar nuclei is not only proportional to the difference in their δ values, but also to the applied external magnetic field. Therefore, the separation of the resonant frequencies for two ^1H or two ^{13}C nuclei will be twice as great in a spectrometer employing a fixed external field of 1.4×10^5 G than in a spectrometer using a field of 0.7×10^5 G.

Experiment shows that protons and carbon nuclei in different functional groups tend to have chemical shifts in the same general range from compound to compound, in much the same way that stretching and bending vibrational frequencies for given functional groups tend to fall in a narrow frequency range. This result means that an NMR measurement of all the resonant frequencies or magnetic fields for a molecule has the potential to tell us the different types of nuclei that are present in the molecule. All that we need to

Table 16.2 Proton NMR chemical shifts relative to $(CH_3)_4Si$ as reference[a]

Proton Group	δ	Proton Group	δ
$(CH_3)_4Si$	0	$HC\equiv C-$	2.8–3.0
$(CH_3)_4C$	1.0–1.1	CH_3-X	3.4–4.0
CH_3-CH_2-	1.2–1.3	$-CH_2-X$	3.5–4.5
$R-SH$	1.2–1.7	$-CH-X$	3.8–5.1
$-CH_2-$ in a ring	1.4–1.6	X = OH, OR, OPh,	
$(CH_3)_3CH$	1.6	OCOR, OCOPh	
CH_3-X	2.0–2.8	CH_3-X	2.2–4.3
$-CH_2-X$	2.1–2.5	$-CH_2-X$	3.4–4.4
$CH-X$	2.5–2.8	$-CH-X$	4.0–4.3
X = CHO, COR,		X = F, Cl, Br, I	
COPh, COOH,		PhSH	3.7
COOR, $CONH_2$		$CH_3NO_2, -CH_2NO_2$	4.3–4.6
$-CH_2-$ ring ketones	2.0–2.5	$PhNH_2$	3.6–4.7
$(CH_3CO)_2O$	2.2–2.3	$-CH=CH-$ conj.	4.4–8.0
CH_3-X	2.3–3.4	$-CH=CH-$ nonconj	4.6–5.7
$-CH_2-X$	2.5–3.5	$CH_2=C$ (terminal)	4.7–5.0
$CH-X$	2.9–3.0	$CH_2=C=CH_2$	4.9–5.0
X = N, $NHCOCH_3$,		$CH_2=C(CH_3)$	4.8
$NHSO_2Ph$,		$(CH_3)_2C=CHCH_3$	5.1
Quart. salt		$H-Ph-X$	7.2–8.1
CH_3CN	2.1	$Ph-H$	7.2–7.3
$CH_2=C(CH_3)_2$	1.7–1.8	RCHO, PhCHO	9.7–10.0
$-CH_2-NH_2$	2.7–2.8	RCOOH, PhCOOH	10.2–11.4
CH_3Ph,	2.3	RSO_3H, $PhSO_3H$	11.0–11.4
CH_3CH_2Ph,			
$PhCH_2CH_2Ph$			
$(CH_3)_2CHPh$	2.6–2.9		

[a]Source of data: *Handbook of Chemistry and Physics*, 78th ed. (CRC Press, Boca Raton, FL, 1998).

accomplish this is a table listing the expected chemical shifts for different functional groups. These shifts are given in Tables 16.2 and 16.3. Notice that they are much larger for ^{13}C than for ^1H. This is not unexpected, since there are many more electrons about a carbon nucleus than about a proton. These additional electrons are perturbed by the applied magnetic field and thereby produce small internal magnetic fields that give rise to the larger chemical shifts.

EXAMPLE 16.3

Using the data in Tables 16.1 and 16.2, estimate the difference in the frequencies required to produce proton resonance for the aldehyde and methyl group protons in ethanal if the spectrum is taken with a 300-MHz NMR spectrometer that uses a

Table 16.3	^{13}C NMR chemical shifts relative to $(CH_3)_4Si$ as reference[a]

Carbon Group	δ
—C=O ketones	206–212
—C=O aldehydes	199–200
—C=O α, β unsaturated carbonyls	
$CH_3CH=CHCHO$	192.4
$CH_2=CHCOCH_3$	169.9
—C=O carboxylic acids	166.0–178.1
—C=O amides	165–172.7
—C=O esters	165.5–170.3
C=C aromatic	128.5
C=C alkenes	123.2–136.2
—C≡N nitriles	117.7
—C≡C- alkynes	71.9–73.9
—C—O esters	57.6–67.9
—C—O alcohols	49.0–57.0
—C—NH_2 amines	26.9–35.9
—S—CH_3 sulfides	15.6
—C—H alkanes, cycloalkanes	−2.3–25.2

[a]Source of data: *Handbook of Chemistry and Physics*, 78th ed. (CRC Press, Boca Raton, FL, 1998).

fixed external field of 70,460 G. What is the difference in the resonant frequencies if the spectrum is taken with a 600-MHz spectrometer that employs an external field of 140,920 G?

Solution

The first step is to obtain a reasonable estimate for the chemical shifts of the protons in the –CHO and –CH_3 groups of the CH_3–CHO molecule. Table 16.2 tells us that CH_3– protons have chemical shifts in the range 2.0–2.8 when they are bonded to an aldehyde group. The aldehyde chemical shifts lie in the range from 9.7 to 10.0. Let us take intermediate values as estimates. This gives

$$\delta_{CH_3} \approx 2.4 \tag{1}$$

and

$$\delta_{CHO} \approx 9.85. \tag{2}$$

We can now use Eq. 16.30 to compute the difference in resonant frequencies:

$$\nu_{CHO} - \nu_{CH3} = \frac{g_N \mu_N}{h} B_o [\delta_{CHO} - \delta_{CH3}] 10^{-6}. \tag{3}$$

Substituting the data gives

$$\nu_{CHO} - \nu_{CH3} = \frac{(5.58569)(5.0508 \times 10^{-31} \text{ J G}^{-1})}{6.62608 \times 10^{-34} \text{ J s}} (70{,}460 \text{ G})(9.85 - 2.4)10^{-6}$$

$$= 2.24 \times 10^3 \text{ s}^{-1}. \tag{4}$$

Since there are 10^6 s^{-1} in 1 MHz and 1 hertz is the same as 1 s^{-1}, the separation in resonant frequencies is

$$\nu_{CHO} - \nu_{CH3} = 0.00224 \text{ MHz.} \tag{5}$$

The frequency required for resonance for the aldehyde groups is 0.00224 MHz greater than that for the methyl protons. If we employed a 600-MHz spectrometer with an external magnetic field of 140,920 G, the resonance positions of the methyl and aldehyde protons would be separated by twice this amount, or 0.00448 MHz.

For related exercises, see Problems 16.6 and 16.7.

16.1.5   Spin–Spin Coupling

Example 16.3 suggests that an NMR proton spectrum of ethanal (CH_3–CHO) should consist of two resonant absorption bands: one for the aldehyde proton and a second for the three equivalent protons in the –CH_3 group. Also, the calculations indicate that these two resonance peaks should be separated by approximately 0.00224 MHz if the spectrum is taken with a 300-MHz instrument and by 0.00448 MHz if a 600-MHz spectrometer is used. Since there are three methyl protons and only one aldehyde proton, we might also infer that the absorption from the methyl group will be three times as intense as that from the aldehyde proton.

When an NMR spectrum of ethanal is taken with a 300-MHz spectrometer, we find that the foregoing expectations are fulfilled in part. There are two groups of absorption bands separated by approximately 0.00224 MHz, with the aldehyde group appearing at a higher frequency than the –CH_3 bands. The total absorption intensity for the protons in the –CH_3 group is about three times that for the –CHO proton. However, the absorption due to the aldehyde proton actually appears as four equally spaced peaks, with the two center peaks having about three times the intensity of the two outer peaks. This grouping is called a *quartet*. The –CH_3 absorption appears as a *doublet*, with both peaks having about the same intensity.

The interaction that causes the splitting of single NMR absorption bands into various multiplets arises from the interaction of the nuclear spin magnetic moments of nuclei in close proximity. For this reason, the effect is called *spin–spin coupling*. The existence of such an interaction is not surprising. If we had two small circular conducting loops, each would act like a small solenoid, producing a magnetic moment and a magnetic field. If the loops were near each other, we would certainly expect an interaction between the magnetic field produced by one loop and the spin magnetic moment of the second loop, as illustrated in Figure 16.9. However, in the gas or liquid phase, rapid molecular rotation causes these direct interactions to average to zero. The figure shows two orientations of Loop B whose interactions with the magnetic field produced by Loop A would cancel and average to zero. The only time such direct spin–spin interactions can be observed is in a well-ordered solid at low temperatures.

There is, however, an indirect mechanism by means of which the spin of one nucleus can affect the spin energy levels of a second nucleus. In this mechanism, the nuclear magnetic moment of Spin A interacts with the electrons on Nucleus A to affect a small change in their energy state. The energy change of the electrons on Nucleus A in turn produces a similar small change in the energy of the bonding electrons on Nucleus B. These electrons interact with the spin of Nucleus B to induce a small alteration of its nuclear spin energy levels. This type of indirect coupling is unaffected by molecular rotation, so its effects are observed in both gases and liquids. However, because the interaction must be transmitted through chemical bonds, its

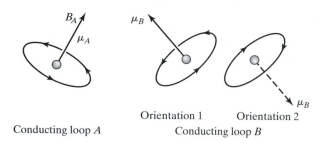

◀ FIGURE 16.9
Direct interaction between the magnetic field produced by conducting Loop *A* and the magnetic moment of conducting Loop *B*. Rapid molecular rotation in gases and liquids produces different orientations such that these direct interactions average to zero.

magnitude falls off rapidly as the number of chemical bonds separating the two spins increases. If four or more bonds are involved, the spin–spin interaction is essentially zero, as illustrated in Figure 16.10.

Logically, we would expect the magnitude of spin–spin coupling on the energy levels to be proportional to the product of the components of the two magnetic moments in the direction of the applied field. Equation 16.16 shows that these components are each proportional to S_z, which is given by $m_I\hbar$. Therefore, the spin–spin coupling term produces an energy change whose form is $\Delta E_{\text{spin–spin}} \propto m_{IA}m_{IB}$. Writing this proportionality in form of an equality, we have

$$\Delta E_{\text{spin–spin}} = J_{AB}\hbar m_{IA}m_{IB},\qquad (16.31)$$

where J_{AB} is called the *A–B spin–spin coupling constant*. With the insertion of Planck's constant as part of the proportionality constant in Eq. 16.31, J_{AB} has

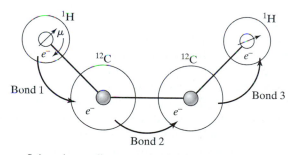

Spin–spin coupling transmitted through three bonds

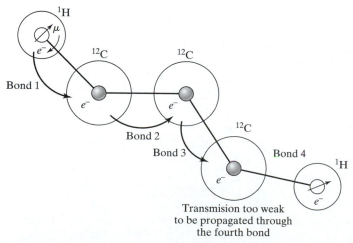

Transmission too weak to be propagated through the fourth bond

Spin–spin coupling is too weak to be transmitted through four bonds

◀ FIGURE 16.10
Spin–spin coupling through chemical bonds.

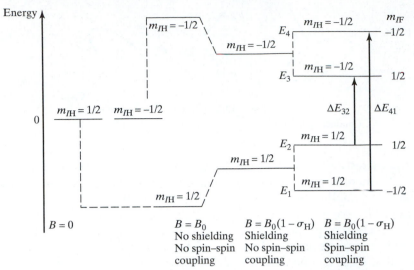

▶ **FIGURE 16.11**
Schematic diagram of hydrogen spin energy levels in HF with no external field; with an external field, but no shielding or spin–spin interaction; with an external field and shielding, but no spin–spin interaction; and with all interactions present. The energy splittings are not shown to scale. Note that transitions between states associated with different values of the fluorine magnetic quantum number, m_{IF}, do not occur, because the spectrometer is not tuned to the frequency required to induce transitions between the nuclear spin energy levels of fluorine.

units of frequency. The total change in energy for Nucleus A produced by the imposition of an external magnetic field when Nucleus B is nearby is obtained by combining Eqs. 16.18, 16.23, and 16.31, giving

$$\Delta E_A = -g_N \mu_N B_o (1 - \sigma_A) m_{IA} + J_{AB} h m_{IA} m_{IB}. \qquad (16.32)$$

To see how spin–spin coupling affects the observed NMR spectra, let us consider the simple case of HF. Both ^{1}H and ^{19}F have $I = \frac{1}{2}$, so that $m_{IH} = \pm\frac{1}{2}$ and $m_{IF} = \pm\frac{1}{2}$. Figure 16.11 illustrates the proton energy levels when there is no external magnetic field; when an external field is applied, but there is no shielding or spin–spin coupling present; when the chemical shift is present, but spin–spin coupling is not; and, finally, when all interactions are active. When there is no shielding or spin–spin coupling, the two proton spin states are split, with the $m_{IH} = -\frac{1}{2}$ state increasing in energy while the $+\frac{1}{2}$ state decreases by an amount given by Eq. 16.18. When shielding is taken into account, the splitting between the two states decreases because of the reduction in the effective magnetic field, as described by Eq. 16.23. The presence of spin–spin coupling between the ^{1}H and ^{19}F nuclei causes a further change in the energy of each state that depends upon the spin magnetic quantum number of the fluorine nucleus, as described by Eq. 16.32. Because m_{IF} for ^{19}F can take values of $+\frac{1}{2}$ and $-\frac{1}{2}$, the m_{IH} $+\frac{1}{2}$ and $-\frac{1}{2}$ spin states are each further split into two levels.

Using Eq. 16.32, we can easily write down the energies of the four states shown in Figure 16.11:

$$E_1 = -\frac{g_H \mu_N B_o (1 - \sigma_H)}{2} - \frac{J_{HF} h}{4}; \qquad (16.33)$$

$$E_2 = -\frac{g_H \mu_N B_o (1 - \sigma_H)}{2} + \frac{J_{HF} h}{4}; \qquad (16.34)$$

$$E_3 = \frac{g_H \mu_N B_o (1 - \sigma_H)}{2} - \frac{J_{HF} h}{4}; \qquad (16.35)$$

$$E_4 = \frac{g_H \mu_N B_o (1 - \sigma_H)}{2} + \frac{J_{HF} h}{4}. \qquad (16.36)$$

The selection rule for NMR transitions is $\Delta m_{I\mathrm{H}} = \pm 1$. Therefore, we can observe two transitions, as shown in the figure. The energies of the allowed transitions are

$$\Delta E_{41} = E_4 - E_1 = g_\mathrm{H}\mu_N B_o(1 - \sigma_\mathrm{H}) + \frac{J_\mathrm{HF} h}{2} \qquad \text{(16.37A)}$$

and

$$\Delta E_{32} = E_3 - E_2 = g_\mathrm{H}\mu_N B_o(1 - \sigma_\mathrm{H}) - \frac{J_\mathrm{HF} h}{2}. \qquad \text{(16.37B)}$$

Consequently, the ^{1}H resonance will appear as a doublet in the NMR spectrum at the energies given by Eqs. 16.37A and 16.37B. If we are using a modern NMR spectrometer that fixes the applied magnetic field and scans over frequencies to obtain the spectrum, the resonant frequencies (ν' and ν'') are those for which $h\nu' = \Delta E_{41}$ and $h\nu'' = \Delta E_{32}$. These frequencies are

$$\nu' = \frac{g_\mathrm{H}\mu_N B_o(1 - \sigma_\mathrm{H})}{h} + \frac{J_\mathrm{HF}}{2} = \frac{\gamma_\mathrm{H} B_o(1 - \sigma_\mathrm{H})}{2\pi} + \frac{J_\mathrm{HF}}{2} \qquad \text{(16.38)}$$

and

$$\nu'' = \frac{g_\mathrm{H}\mu_N B_o(1 - \sigma_\mathrm{H})}{h} - \frac{J_\mathrm{HF}}{2} = \frac{\gamma_\mathrm{H} B_o(1 - \sigma_\mathrm{H})}{2\pi} - \frac{J_\mathrm{HF}}{2}. \qquad \text{(16.39)}$$

Hence, the frequency spacing between the doublet peaks produced by the spin–spin coupling is

$$\boxed{\Delta \nu = \nu' - \nu'' = J_\mathrm{HF}}. \qquad \text{(16.40)}$$

Equation 16.40 is a general result of NMR spectra. The frequency spacing between multiplet peaks produced by spin–spin coupling is always equal to the spin–spin coupling constant between the nuclei.

It is important to note that the spin–spin peak spacing is independent of the magnitude of the applied magnetic field. This is in direct contrast to the chemical shift, which is proportional to the applied field. As we shall see in a subsequent section, the independence of the spin–spin peak spacing plays a very important role in the interpretation of NMR spectra. Some representative values of proton–proton spin–spin coupling constants are given in Table 16.4.

Table 16.4	Selected proton–proton spin–spin coupling constants
Coupling Protons	J_HH **(Hz)**
HC–CH	+5 to +9
C=CH$_2$	−3 to +3
C=C (with H, H)	+5 to +12
H\C=C\H	+12 to 19
HCOH	+5 to +10
CH–CHO	+1 to +3

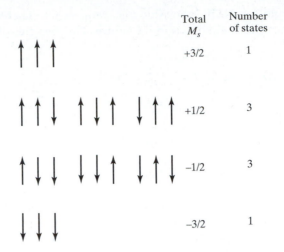

Total M_s	Number of states
+3/2	1
+1/2	3
−1/2	3
−3/2	1

▶ FIGURE 16.12
Possible spin alignments for three equivalent protons and the number of ways each total M_s state can be achieved.

We are now in position to understand and predict the spin–spin splitting pattern seen for ethanal that we described at the beginning of this section. The $-CH_3$ resonance peak is observed to be split into a doublet, with the lines having equal intensity. This result is completely analogous to the example of HF, in which the proton resonance peak is split into a doublet by spin–spin coupling with the ^{19}F nucleus, which has $I = \frac{1}{2}$. In a similar manner, the aldehyde proton has spin $I = \frac{1}{2}$, so it can have $m_I = \pm\frac{1}{2}$. These two possible states split the $+\frac{1}{2}$ and $-\frac{1}{2}$ spin states of the $-CH_3$ protons into two further states. The $\Delta m_I = \pm 1$ selection rule gives two allowed transitions and two NMR peaks with a spacing equal to the spin–spin coupling constant between the $-CHO$ and $-CH_3$ protons. Table 16.4 indicates that J lies in the range from 1 to 3 Hz for this coupling. Notice that we are ignoring the spin–spin coupling between the equivalent protons of the $-CH_3$ in predicting a doublet for this group. It can be shown quantum mechanically that transitions between spin states split by the interaction of equivalent nuclei are forbidden. Therefore, *spin–spin couplings between equivalent nuclei do not affect the spectrum and, consequently, may be ignored.*

The $-CHO$ proton appears as a quartet in the NMR ethanal spectrum, because of the four possible combinations of the three equivalent proton spins in the methyl group. All spins can be aligned with $m_I = \frac{1}{2}$ for each spin to give a total spin magnetic quantum number $M_I = +\frac{3}{2}$. Alternatively, we can have all spins aligned with $m_I = -\frac{1}{2}$ to produce $M_I = -\frac{3}{2}$. Two spins can have $m_I = \frac{1}{2}$ and the third $m_I = -\frac{1}{2}$. Or the reverse may hold. This gives four possible total spin states, as shown in Figure 16.12. The four possible total spin states split both the $\pm\frac{1}{2}$ spin states of the aldehyde proton into four different states, as shown in Figure 16.13. The $\Delta m_I = \pm 1$ selection rule gives us four different allowed transitions between these states, shown also in the figure. The same analysis as that used for the case of HF shows that the resonant frequencies for these four transitions are (see Problem 16.8)

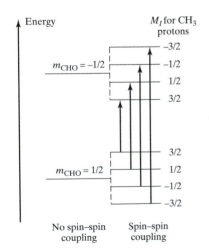

▲ FIGURE 16.13
Spin–spin coupling for a three-spin system. The relative energies of the spin states are not shown to scale.

$$\nu_1 = \frac{g_H \mu_N B_o (1 - \sigma_{CHO})}{h} + \frac{3J_{CHO-CH}}{2}, \quad \textbf{(16.41A)}$$

$$\nu_2 = \frac{g_H \mu_N B_o (1 - \sigma_{CHO})}{h} + \frac{J_{CHO-CH}}{2}, \quad \textbf{(16.41B)}$$

$$\nu_3 = \frac{g_H \mu_N B_o (1 - \sigma_{CHO})}{h} - \frac{J_{CHO-CH}}{2}, \quad \textbf{(16.41C)}$$

and

$$\nu_4 = \frac{g_H \mu_N B_o (1 - \sigma_{CHO})}{h} - \frac{3J_{CHO-CH}}{2}. \qquad \textbf{(16.41D)}$$

The spacings between these bands is again equal to the spin–spin coupling constant, J_{CHO-CH}.

The peak intensities of the two middle peaks whose frequencies are given by Eqs. 16.41B and 16.41C are about three times that of the two outer peaks, because three different spin states give rise to $M_I = \pm\frac{1}{2}$, whereas only one gives either $M_I = \frac{3}{2}$ or $-\frac{3}{2}$. All of these states are shown in Figure 16.12.

In general, N equivalent spins can be arranged $N + 1$ different ways, with N, $N - 1$, $N - 2$, ... , 2, 1, 0 spins with $m_I = +\frac{1}{2}$. Therefore, a proton adjacent to N equivalent nuclei, each with $I = \frac{1}{2}$, will be split into $N + 1$ distinct peaks. The relative intensities of these $N + 1$ peaks will be proportional to the number of different spin assignments that give rise to the same total M_I state. If a particular M_I state has n nuclei with $m_I = +\frac{1}{2}$, this will be the number of combinations of N spins taken n at a time, which is $N!/[n!(N - n)!]$. When $N = 3$, as it does for the $-CH_3$ group protons, the number of ways to have two spins with $m_I = \frac{1}{2}$ is $3!/[2!(3 - 2)!] = 3$. The actual three combinations are shown in Figure 16.12.

EXAMPLE 16.4

Consider propanoic acid (H_3C-CH_2-COOH). **(A)** Qualitatively predict the nature of the spin–spin splitting of the $-CH_3$, $-CH_2-$, and acidic protons that we would expect to observe in an ^1H NMR spectrum taken with a 400-MHz spectrometer. **(B)** What would be the approximate ratio of the total absorption intensities for each of these groups of protons? Explain. **(C)** What are the approximate intensity ratios for each of the spin–spin bands within each individual proton group?

Solution

(A and C) The two equivalent $-CH_2-$ protons are split by the three $-CH_3$ protons, but not by the acidic proton, because the $-COOH$ proton is four chemical bonds (H–C, C–C, C–O, and O–H) distant from the $-CH_2-$ protons. (See Figure 16.10.) The three $-CH_3$ protons can have four different total spin arrangements that produce $M_I = \frac{3}{2}, \frac{1}{2}, -\frac{1}{2}$, and $-\frac{3}{2}$, as illustrated in Figure 16.12. Thus, the $-CH_2-$ proton resonance is split into four different bands whose intensity ratios are 1:3:3:1, since the $M_I = \frac{3}{2}$ and $-\frac{3}{2}$ spin states can be obtained only in one way $(3!/[3!(3 - 3)!] = 1)$, while the $M_I = \frac{1}{2}$ and $-\frac{1}{2}$ states can be obtained in three ways $(3!/[1!(3 - 1)!] = 3.)$ Therefore, the qualitative form of the $-CH_2-$ NMR bands is as shown in the following diagram:

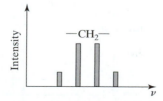

The three equivalent $-CH_3$ protons are split by the two $-CH_2-$ protons, but not by the acidic proton, because the $-COOH$ proton is five chemical bonds distant from the $-CH_3$ protons. The two $-CH_2-$ protons can have three different total spin arrangements that produce $M_I = 1, 0,$ and -1. Thus, the $-CH_3$ proton resonance is split into three different bands. The intensity ratios of these three bands are 1:2:1, since the $M_I = 1$ and -1 spin states can be obtained only in one way $(2!/[2!(2-2)!] = 1)$, while the $M_I = 0$ state can be obtained in two ways $(2!/[1!(2-1)!] = 2.)$ Therefore, the qualitative form of the $-CH_3$ NMR bands is as shown in the following diagram:

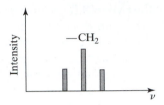

The acidic proton appears as a single absorption, since there are no other nuclei with a nuclear spin quantum number greater than zero within a distance of three or fewer chemical bonds.

(B) The total absorption intensity of the $-CH_3$ protons will be about 1.5 times that of the $-CH_2-$ protons, since there are three equivalent hydrogen nuclei undergoing transitions, whereas the $-CH_2-$ group has only two protons. For the same reason, the $-CH_3$ and $-CH_2-$ protons will have total absorption intensities about three and two times that of the acidic proton, respectively.

For related exercises, see Problems 16.9 and 16.10.

Spin–spin splitting due to nuclei with $I > \frac{1}{2}$ requires special attention. Let us first consider the expected form for the ^1H NMR spectrum of HD ($^1H^2H$). Table 16.1 shows that I for ^1H and ^2H are $\frac{1}{2}$ and 1, respectively. Suppose we are at resonance for ^1H in a 400-MHz NMR spectrometer. Then the external field B_o will cause the $m_I = \frac{1}{2}$ and $-\frac{1}{2}$ nuclear spin states to split in the manner described by Figure 16.5. Each of these states will be further split by spin–spin coupling to the deuterium nucleus. Since $I = 1$ for ^2H, we may have $m_{ID} = 1, 0,$ or -1. Consequently, we might expect to see each of the ^1H spin states split into three states, as shown in Figure 16.14. There would be three allowed ^1H transitions, so the ^1H spectrum would exhibit three bands whose spacing is the H–D spin–spin coupling constant, which is found to be 42.8 Hz. These expectations are confirmed by experiment.

Now consider the proton NMR spectrum of $^1H^{79}Br$. In this case, the spin quantum number for ^{79}Br is $I = \frac{3}{2}$. Following the same logic as that used for

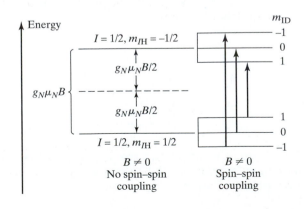

▶ FIGURE 16.14
Spin–spin splitting for HD. The allowed NMR ^1H transitions are shown by the solid upward arrows.

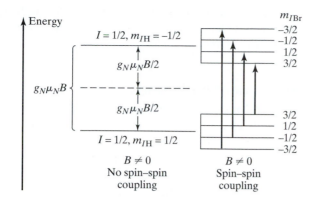

◀ **FIGURE 16.15**
The expected spin–spin splitting for HBr if there were no quadrupole moment on the bromine atom. The allowed NMR ^1H transitions shown by the solid upward arrows are not observed in the HBr NMR spectrum. Instead, a single broad band appears.

HD, we would expect the proton spectrum to be split into four bands, because ^{79}Br can exhibit spin magnetic quantum numbers of $\frac{3}{2}, \frac{1}{2}, -\frac{1}{2}$, and $-\frac{3}{2}$. Our expectation is illustrated in Figure 16.15. However, the experimental result shows that the proton spectrum appears as a single band. There is no spin–spin splitting!

The theory behind this surprising result goes beyond the scope of the text. We must, therefore, be content with a qualitative explanation. When the spin quantum number is greater than $\frac{1}{2}$, the nucleus possesses a magnetic quadrupole moment in addition to its magnetic dipole moment. The presence of this quadrupole moment causes the NMR absorption bands to broaden. It also causes the spin orientation to undergo rapid transitions among the possible nuclear spin states. The larger the quadrupole moment, the more rapid these transitions become. When $I \geq \frac{3}{2}$, the transitions are so fast that adjacent nuclei experience only an average magnetic dipole moment rather than the individual spin states. As a result, spin–spin splitting is not seen. This situation is illustrated in Figure 16.16. For ^2H and ^{14}N,

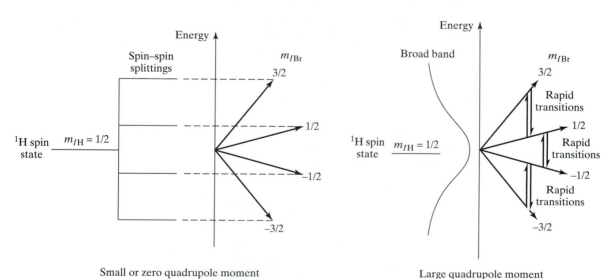

▲ **FIGURE 16.16**
Effect of a large ($I \geq \frac{3}{2}$) quadrupole moment for a ^{79}Br nucleus. The large quadrupole moment causes rapid spin transitions in the nucleus. As a result, the spin–spin splitting peaks are broadened and appear as a single broad NMR band. Splittings and the width of the NMR band are not shown to scale.

for which $I = 1$, the quadrupole moment is relatively small, so that the transition rate between nuclear spin states is slow relative to the time scale for NMR measurements. Therefore, we see spin–spin splitting for $^1H^2H$, but not for $^1H^{79}Br$.

16.1.6   Experimental Methods

The nature of NMR spectra and their interpretation are, in large measure, dependent upon the characteristics of the spectrometer employed. Questions related to sensitivity, signal-to-noise ratio, and resolution depend upon the type of experiment being conducted and the design of the spectrometer. For this reason, we must know something about the experiment and the instrument before turning to actual examples of NMR spectra.

Equation 16.19 and the data in Table 16.1 show that a magnetic field of 1.4092×10^5 G is required to produce ^1H NMR resonance at a frequency of 600 MHz. Even higher fields are needed for 750- and 1,000-MHz spectrometers to produce ^1H resonance at these frequencies. The only practical means of obtaining such high, uniform magnetic fields are electromagnets. If these devices are operated around room temperature, there is a huge heat dissipation problem due to the large currents needed. To circumvent this difficulty, the electromagnets are operated at liquid helium temperatures (≈ 4–12 K), so that the circuits are superconducting and there is no electrical resistance and, hence, no heat dissipation. With no electrical resistance, the current, once started, will continue indefinitely, as long as no work is done by the magnetic field.

Before the development of Fourier transform spectroscopic methods, NMR spectra were typically obtained by slowly scanning over magnetic field to locate the resonance lines at a fixed spectrometer frequency. Figure 16.17 shows the principal components of such a spectrometer. A large electromagnet provides a magnetic field near the resonance position for the fixed frequency being employed by the spectrometer. A smaller electromagnet permits this field to be varied continuously over the range of a few gauss. Such an instrument is, therefore, called a *continuous-wave*, or CW, spectrometer. The sample is placed in the magnetic field and irradiated with a radiofrequency (rf) signal whose frequency is fixed. As the magnetic field is scanned, the intensity of the rf field is measured by a suitable detector, as shown in the figure. When the magnetic field is such that the resonance condition described by Eq. 16.19 is satisfied, absorption from the lower spin state and stimulated emission from the upper spin state both take place. Since there are more molecules in the lower energy spin state, absorption dominates over emission, and there is a decrease in the intensity of the rf signal. A plot of this decrease versus the magnetic field provides the NMR spectrum.

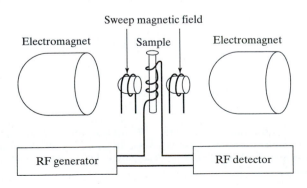

▶ **FIGURE 16.17**
Schematic diagram of an NMR apparatus using a fixed frequency with a magnetic field sweep to find the absorption bands. This type of NMR spectrometer is called a continuous-wave, or CW, spectrometer.

The preceding discussion points out an important difference between microwave, IR, and UV spectrometers, on the one hand, and the instruments used to obtain NMR spectra, on the other. In the former, only the absorbed radiation from the incident beam is measured; the radiation produced by stimulated emission is scattered at all angles, so that very little reaches the detector. Consequently, the band intensities depend only upon the population of the lower energy state, and Eqs. 15.36 and 15.37 can be used to compute these quantities. In contrast, the rf coil in an NMR spectrometer measures the difference between the absorbed and emitted radiation. Therefore, the band intensities depend upon the difference in population between the upper and lower spin states, as described by Eq. 15.43. Since the splitting between spin states is very small, this difference is correspondingly small. Consequently, NMR band intensities are very low, and sensitivity is a major problem.

Let us assume that the energy of the lower spin state is E_o. In this case, Eq. 16.19 shows that the energy of the upper state will be $E_o + g_N \mu_N B$, which, at resonance, will be equal to $E_o + h\nu$, where ν is the frequency of the spectrometer. Equation 15.43 tells us that the intensity of the NMR band will be proportional to the difference in the population of molecules in these two states. Using Eq. 15.37, we have

$$\text{Intensity} \propto [N_{\text{lower}} - N_{\text{upper}}] = C\left[\exp\left\{-\frac{E_o}{kT}\right\} - \exp\left\{-\frac{E_o + h\nu}{kT}\right\}\right]. \quad \textbf{(16.42)}$$

This result can be written in the form

$$\text{Intensity} = C \exp\left\{-\frac{E_o}{kT}\right\}\left[1 - \exp\left\{-\frac{h\nu}{kT}\right\}\right] = C'\left[1 - \exp\left\{-\frac{h\nu}{kT}\right\}\right], \quad \textbf{(16.43)}$$

since E_o is a constant for a given type of nucleus. The quantity within the brackets in Eq. 16.43 is usually very small. For this reason, it is very difficult to obtain NMR spectra on small samples. For example, consider the ^{1}H NMR spectra of HBr. Since the spin quantum number of Br is $\frac{3}{2}$, there is no spin–spin splitting, and the proton spectrum will be a single band. (See Figure 16.16.) If the frequency of the spectrometer is 100 MHz, the intensity of this band at 298 K will be

$$\text{Intensity} = C'\left[1 - \exp\left\{-\frac{(6.626 \times 10^{-34}\,\text{J s})(100 \times 10^6\,\text{s}^{-1})}{(1.381 \times 10^{-23}\,\text{J K}^{-1})(298)}\right\}\right]$$

$$= C'[1 - 0.9999839] = 0.0000161C'.$$

Notice that the difference of the two terms within the brackets is only 16.1 parts per million (ppm). These low intensities mean that if we have only a small amount of sample, it will generally be necessary to take repeated spectral scans to collect enough signal to obtain the NMR spectrum.

Inspection of Eqs. 16.19 and 16.43 shows us that if we operate our spectrometer at higher fixed magnetic fields, the resonant frequency for NMR transitions will be larger. The larger frequency in Eq. 16.43 will produce a more intense signal, and consequently, the spectrometer will have greater sensitivity. This gain in sensitivity is particularly important in biological applications of NMR, where the quantity of sample in many cases is in the nanogram range or even smaller. Problem 16.11 illustrates the importance of this factor.

The difficulty presented by the low intensity of NMR spectral bands can sometimes be overcome by using high-field spectrometers. However, when the sample size is small or when NMR spectra for isotopes with low relative abundance are desired, this will not solve the problem. Typical examples include NMR measurements on biological materials for which very little sample is available and ^{13}C measurements where the relative abundance is only 1.10%. In such cases, it is necessary to take a large number of spectral scans of the sample and feed the resulting data into a computer that sums the results, thereby enhancing the signal-to-noise ratio by canceling the random noise present in the spectrometer. If the noise is random, the signal-to-noise ratio (S/N) is proportional to $M^{-1/2}$, where M is the number of spectral scans taken. Since $M^{-1/2}$ decreases very slowly with increasing M, a considerable number of scans is required to obtain large S/N ratios. If a CW spectrometer is used for such measurements, each scan can take up to 15 minutes. The reason the scan time is so protracted is that the spectral resolution is dependent upon the time the instrument spends monitoring the signal at each magnetic field. Rapid scanning will permit more scans to be taken in a given amount of time, but the price will be significantly decreased resolution. This point was discussed in quantitative detail in Chapter 15. The effect of limited sampling time can be seen by comparing Figures 15.3, 15.4, and 15.5. Figures 15.15 and 15.16, along with Figures 15.18 and 15.19, provide two additional examples of this effect. The need to have a large number of scans, combined with long scan times, produces unrealistic experimental times. For example, if 10,000 scans are required, each of which takes 15 minutes, a single spectrum would require about 100 days of spectrometer time.

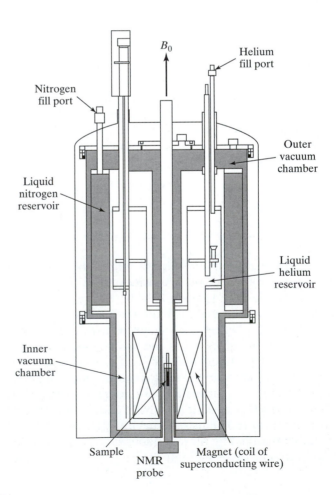

▶ FIGURE 16.18
Schematic cross section of a Fourier transform, superconducting NMR spectrometer.

Because chemists occasionally like to do other things than sit in the laboratory taking data, a different technique is needed.

The solution to the problem is to utilize Fourier transform (FT) methods, which permit the entire spectrum to be obtained with a single rf pulse in approximately 2 seconds. This effectively eliminates the 15-minute scan time and makes the investigation of small samples practical. Figure 16.18 shows a schematic of the essential components of a modern FT NMR spectrometer. The superconducting magnet is maintained at liquid-helium temperatures by immersing the current-carrying coils in a reservoir of liquid helium surrounded by liquid nitrogen to reduce the evaporation rate. The entire assembly is contained within a vacuum chamber, so that the housing of the spectrometer is, in reality, a large thermos bottle, albeit a very expensive one. (A modern 600 FT NMR spectrometer with the required computers and other accessories costs about $1.5 million.) The sample is again placed in the magnetic field within an NMR probe that contains the electronics required to execute the procedure to be described shortly. Figures 16.19 and 16.20 are photographs of 600-, 400-, and 300-MHz FT NMR spectrometers.

Before discussing FT NMR techniques, we need to examine what happens when nuclei are excited from the lower to the upper nuclear spin state. Let us take the case of an isotope with $I = \frac{1}{2}$, so that there are two spin states. Before excitation, but after application of the external magnetic field, the nuclei are either in the lower energy state with S_z aligned with the field or in the higher energy state with S_z opposed to the field. At equilibrium, slightly more nuclei will be in the lower state, so that there will be a net polarization of the spin angular momentum in the direction of the field (the Z-axis). Since the angular momentum and magnetic dipole vectors are collinear, the net

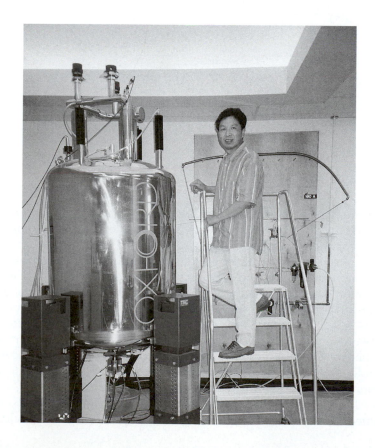

◀ **FIGURE 16.19**
The superconducting magnet-and-probe assembly for a 600-MHz FT NMR spectrometer. Liquid nitrogen and helium for the required cooling are inserted into the dewars from the top of the assembly. The NMR probe containing the electronics required to acquire the spectrum is inserted from the bottom. Dr. Feng Qiu, manager, Statewide Shared NMR Facility at Oklahoma State University stands on the ladder to provide a reference point for judging the size of the apparatus. The entire building is guarded by a safety fence to protect individuals wearing pacemakers from the magnetic field. The author thanks Dr. Qiu for his assistance with this illustration and the ones that follow.

(A)

(B)

▲ FIGURE 16.20

Photographs of (A) a 300-MHz FT NMR spectrometer and (B) a 400-MHz FT NMR spectrometer. Dr. Feng Qiu, manager, Statewide Shared NMR Facility at Oklahoma State University stands beside the spectrometers to provide a reference point for judging the size of the instruments.

▶ FIGURE 16.21

Net magnetization of nuclear spin angular momentum due to an externally applied magnetic field B_o before and after excitation. The arrows show the classical precession of the spin angular momentum of the excited nuclear spins about the Z-axis.

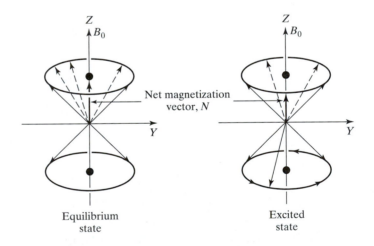

magnetization vector **N** will also lie along the Z-axis. This situation is illustrated in Figure 16.21, in which we have exaggerated the population difference for illustrative purposes. When some of the nuclei are excited by rf radiation with frequency ν, there is a decrease in the net magnetization vector, since more nuclei now have their spins opposed to the field. Classically, the excitation energy causes the angular momentum vectors of the excited nuclei to rotate about the Z-axis with frequency ν. This type of rotational motion is called *precession*. After excitation, the system will begin to move back toward equilibrium as the excited spins undergo transitions to the lower energy state in a process called *relaxation*. As relaxation occurs, the magnitude of **N** will gradually increase toward its equilibrium value. Relaxation is discussed in greater detail later in this section.

An FT NMR spectrometer operates at a fixed magnetic field B_o that is maintained constant with the use of an electronic frequency/field lock.

However, instead of slowly scanning over all frequencies to obtain the spectrum, the sample is irradiated for a few microseconds with a powerful rf signal whose magnetic vector is oriented perpendicular to the Z-axis. The frequency of the rf pulse is that given by Eq. 16.19 for the particular nucleus whose spectrum is desired. Since the nuclear g-factors vary significantly for different nuclei, a different NMR probe is usually required for each isotope studied. The operation of the spectrometer depends critically upon the effect of this rf signal, as described next.

In Chapter 12, we learned that the uncertainty product between any two variables u and v is related to the commutator between these variables. The quantitative relationship was presented without proof in Eq. 12.65A, viz.,

$$\langle \Delta u^2 \rangle \langle \Delta v^2 \rangle \geq \frac{1}{4} |[u, v]|^2,$$

where $[u, v]$ is the u–v commutator. For the rf pulse in an FT NMR spectrometer, the two variables of interest are the duration or time of the pulse, t, and the energy of the pulse, E. The operator corresponding to the energy is $i\hbar(\partial/\partial t)$. (See Table 11.1.) It is easy to show (see Problem 16.13) that the commutator $[t, i\hbar (\partial/\partial t)] = -i\hbar$. Therefore, the product of the square uncertainties in pulse time and energy of the rf signal is

$$\langle \Delta t^2 \rangle \langle \Delta E^2 \rangle \geq \frac{1}{4} |-i\hbar|^2 = \frac{\hbar^2}{4}. \tag{16.44}$$

Since the energy of the rf radiation is $h\nu$, Eq. 16.44 may be written in the form

$$\langle \Delta t^2 \rangle \langle h^2 \Delta \nu^2 \rangle = h^2 \langle \Delta t^2 \rangle \langle \Delta \nu^2 \rangle \geq \frac{\hbar^2}{4}.$$

Using the fact that $\hbar = h/(2\pi)$, we find that this result is equivalent to

$$\langle \Delta t^2 \rangle \langle \Delta \nu^2 \rangle \geq \frac{1}{16\pi^2}. \tag{16.45}$$

Equation 16.45 shows us that we can never know both the duration and frequency of the rf pulse exactly.

We can obtain an estimate of the minimum uncertainty in the rf frequency by maximizing the uncertainty in the pulse duration. Let us take this maximum uncertainty to be the pulse duration itself, t_d. In effect, we are saying that if the pulse duration is about 10 μs, then we surely know that the duration lies somewhere in the range 10 μs $\pm$ 10 μs. The actual uncertainty is probably much less than t_d, so we are safe in saying that t_d is an upper limit for Δt. Inserting this estimate into Eq. 16.45 produces

$$\langle \Delta \nu^2 \rangle \geq \frac{1}{16 t_d^2 \pi^2}.$$

Therefore, the minimum frequency spread of the rf pulse is about

$$\langle \Delta \nu^2 \rangle_{\text{min}} \approx \frac{1}{16 t_d^2 \pi^2},$$

so that

$$\Delta \nu_{\text{min}} \approx \frac{1}{4\pi t_d}. \tag{16.46}$$

This minimum frequency spread of the pulse is called the *natural spectral width*.

The preceding analysis shows that when the sample is irradiated with an rf pulse of short duration, it is actually being subjected to a range of frequencies centered around the oscillator frequency given by Eq. 16.19. If this range is sufficiently broad to cover the resonant frequencies for all the NMR spectral transitions, we can excite all of those transitions with a single pulse and avoid the necessity to scan over frequency. Example 16.5 illustrates how the rf pulse duration may be adjusted to ensure this possibility.

EXAMPLE 16.5

What is the maximum rf pulse duration that can be used in a 400-MHz FT NMR spectrometer and still have the pulse contain frequencies components for all expected ^{1}H nuclear spin transitions?

Solution

The resonance frequency for proton i is given by Eq. 16.19:

$$h\nu_i = g_N\mu_N B_{oi}. \tag{1}$$

The difference in resonance frequencies between two proton groups A and B is given by Eq. 16.30:

$$\nu_A - \nu_B = \frac{g_N\mu_N}{h}B_o[\delta_A - \delta_B]10^{-6}. \tag{2}$$

In a 400-MHz spectrometer,

$$B_o = \frac{(400 \times 10^6)h}{g_N\mu_N}. \tag{3}$$

Combining Eqs. 2 and 3, we obtain

$$\nu_A - \nu_B = \frac{g_N\mu_N}{h}\frac{(400 \times 10^6)h}{g_N\mu_N}[\delta_A - \delta_B]10^{-6} = 400[\delta_A - \delta_B]. \tag{4}$$

Table 16.2 shows that the maximum difference $\delta_B - \delta_A$ is about 12 for ^{1}H NMR spectra. Therefore, we need a spread of about

$$\Delta\nu \approx 400(12) \text{ s}^{-1} = 4{,}800 \text{ s}^{-1} = 4{,}800 \text{ Hz} \tag{5}$$

to cover all expected proton transition frequencies. The pulse duration required to produce a minimum spread equal to this amount is given by Eq. 16.46:

$$t_d \approx \frac{1}{4\pi\Delta\nu_{min}} = \frac{1}{4(3.1416)(4800 \text{ s}^{-1})} = 16.6 \times 10^{-6} \text{ s} = 16.6 \text{ }\mu\text{s}. \tag{6}$$

If the pulse duration is shorter than 16.6 μs, the frequency spread will be even larger. In actual practice, pulses between 6 and 10 μs are used. With such short pulses, we can easily cover the range of frequencies required to excite all ^{1}H transitions.

For a related exercise, see Problem 16.14.

Once excitation at every resonance frequency is produced with the rf pulse, the excited spins begin to decay toward equilibrium. In Chapter 19, we shall learn that the rate of this relaxation process is such that the number of excited spins decreases exponentially with time. The final task the spectrometer must perform is to measure the frequencies that have been absorbed from the rf pulse in the excitation process. This is accomplished by placing detectors that are sensitive to the net magnetization vector in the X–Y plane perpendicular to $\mathbf{B}_o$. If $\mathbf{N}$ is still aligned along the Z-axis, as shown in Figure 16.21, the detectors will measure no signal, since the component of **N** in the X–Y plane will be zero. However, the rf pulse has its magnetic vector aligned perpendicular to the Z-axis. (The NMR spectroscopist refers to this as a 90° pulse.) As a result, all components making up the net magnetization vector are momentarily rotated 90° into the X–Y plane, where they now precess about the Z-axis with frequencies equal to their individual excitation frequencies. Figure 16.22 shows one of these components, $\mathbf{n}_i$, and the arrangement of detectors. The presence of the second detector at right angles to the first permits the direction of precession to be determined.

The electronics in the NMR probe are designed so as to make it appear that the components making up the net magnetization vector are precessing at frequencies equal to their actual frequency, minus some reference frequency ν_{ref}. It is as if the X–Y plane and detectors shown in Figure 16.22 are rotating with the components making up **N** at a frequency ν_{ref}. The reference frequency might, for example, be the resonance frequency for tetra-methylsilane (TMS). Since the detector on the Y-axis measures only the component of $\mathbf{n}_i$ along Y, the signal will vary periodically with a frequency $\nu - \nu_{\text{ref}}$. If the system did not relax back to the equilibrium state the signal due to $\mathbf{n}_i$ would be

$$s_i = n_i \cos[2\pi(\nu_i - \nu_{\text{ref}})t] \qquad (16.47)$$

at all values of t. However, exponential relaxation does occur with some time constant τ.

Relaxation occurs for two reasons. First, the fluctuating local magnetic fields arising from molecular motion stimulate emission from the upper to lower spin states. The time constant for this effect is called the *spin–lattice relaxation time*. It is generally given the symbol T_1. The effect is illustrated in Figure 16.23. The second effect involves the redistribution of the individual spin magnetic moments as they precess about the Z-axis shown in Figure 16.22. Since each spin precesses at a different frequency, there will be a natural randomization of the directions of the magnetic dipole vectors in the X–Y plane. This randomization decreases the net magnetization along the Y-axis. The time constant for this effect is represented by T_2 and is called the *spin–spin*

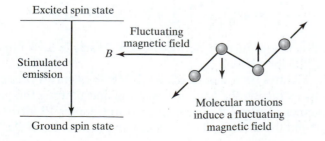

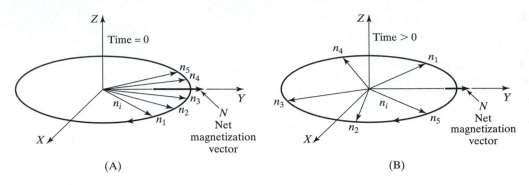

▲ FIGURE 16.24
Spin–spin relaxation. The difference in frequencies of the components of the net magnetization vector causes the components to precess about the Z–axis at different frequencies. As a result, the individual magnetization vectors tend to randomize with time. This randomization process causes a decrease in the magnitude of the net magnetization vector N as shown. The randomization is illustrated for a hypothetical case in which five signals with different frequencies are responsible for N. The time constant T_2 for this process is called the spin–spin relaxation time. (A) Time zero is immediately subsequent to the 90° rf pulse. (B) The orientation of the spin magnetic dipoles at some later time.

relaxation time. Figure 16.24 shows how the effect serves to reduce the magnitude of the net magnetization vector.

Both effects produce an exponential decrease in the NMR signal with time, so that the number of excited spins at time t is given by

$$n(t) = n_o \exp\left[-\frac{t}{T_1}\right]\exp\left[-\frac{t}{T_2}\right] = n_o \exp\left[-t\left\{\frac{1}{T_1} + \frac{1}{T_2}\right\}\right] = n_o \exp\left(-\frac{t}{\tau}\right),$$

(16.48A)

where τ, the total relaxation time, is given by

$$\boxed{\frac{1}{\tau} = \frac{1}{T_1} + \frac{1}{T_2}}.$$

(16.48B)

Equations 16.48A and 16.48B indicate that the amplitude of the signal s_i due to the spin magnetic moment $\mathbf{n}_i$ will decrease exponentially with time. Therefore, we expect that s_i will have the form

$$s_i = n_i \exp\left(-\frac{t}{\tau}\right)\cos[2\pi(\nu_i - \nu_{\text{ref}})t].$$

(16.49A)

The total signal recorded by the detector is the sum over all excitation frequencies:

$$s_{\text{total}} = \sum_i n_i \exp\left(-\frac{t}{\tau}\right)\cos[2\pi(\nu_i - \nu_{\text{ref}})t]$$

$$= \exp\left(-\frac{t}{\tau}\right)\sum_i n_i \cos[2\pi(\nu_i - \nu_{\text{ref}})t].$$

(16.49B)

s_{total} is called a free-induction decay, or FID. Example 16.6 illustrates the nature of the FID for ethanal.

Once the FID has been measured, the NMR frequencies present in it are determined by extracting the Fourier transform and plotting the power spectrum in a manner identical to that described in Chapter 15 for FTIR spectra (see Eq. 15.60 and the associated discussion). In practice, the FID is measured for a few seconds, and the Fourier transform is computed by the electronic circuitry. Thus, the 10- to 15-minute scan time required by a CW spectrometer is replaced by a FID measurement requiring a few seconds in an FT instrument. This permits many more spectral scans to be executed in a given time and thereby makes the investigation of very small samples feasible. Example 16.6 also shows the power spectrum of the FID computed for ethanal.

EXAMPLE 16.6

Use the data in Table 16.2 to predict the form of the FID that would be obtained for a ^{1}H NMR spectrum of ethanal by means of a 60-MHz FT NMR spectrometer. For simplicity, assume that the amplitudes of all frequency components in the rf pulse are the same and that the spin relaxation time constant τ is 500 ms for all ^{1}H spins. To reduce resolution problems, use a spin–spin coupling constant J_{CH-CHO} of 20 Hz instead of the expected value of about 2 Hz listed in Table 16.4. Let the reference frequency ν_{ref} be that for TMS.

Solution

We first need to estimate the frequencies at which ^{1}H resonance will be obtained for CH_3–CHO. The previous discussion on chemical shift and spin–spin coupling indicates that the ^{1}H spectrum of ethanal will consist of a doublet with equally intense components for the $-CH_3$ resonance due to spin–spin splitting from the aldehyde proton and a quartet for the $-CHO$ proton in which the resonance for the two intermediate bands will have about three times the intensity of that of the outer bands. The total intensity of the $-CH_3$ resonance will be about three times that for $-CHO$, because the $-CH_3$ group possesses three ^{1}H spins.

The resonant frequencies for the methyl protons are given by Eqs. 16.38 and 16.39 if we replace J_{HF} with the CH–CHO spin–spin coupling constant and $B_o(1 - \sigma_H)$ with the appropriate resonant magnetic field for $-CH_3$ protons given by Eq. 16.27. We have

$$\nu_1^{CH_3} = \frac{g_H \mu_N B_o(1 - \sigma_{ref} + 10^{-6}\delta_{CH_3})}{h} + \frac{J_{CH-CHO}}{2} \qquad (1)$$

and

$$\nu_2^{CH_3} = \frac{g_H \mu_N B_o(1 - \sigma_{ref} + 10^{-6}\delta_{CH_3})}{h} - \frac{J_{CH-CHO}}{2}. \qquad (2)$$

In a 60-MHz spectrometer,

$$B_o = \frac{(60 \times 10^6 \text{ s}^{-1})h}{g_H \mu_N}. \qquad (3)$$

Substituting this result into Eqs. 1 and 2 produces

$$\nu_1^{CH_3} = 60 \times 10^6 - 60 \times 10^6 \sigma_{ref} + 60\delta_{CH_3} + \frac{J_{CH-CHO}}{2} \qquad (4)$$

and

$$\nu_2^{CH_3} = 60 \times 10^6 - 60 \times 10^6 \sigma_{ref} + 60\delta_{CH_3} - \frac{J_{CH-CHO}}{2}. \qquad (5)$$

The frequencies for the $-CHO$ quartet are given by Eqs. 16.41 A–D. Using Eq. 3, we find that

$$\nu_1^{CHO} = 60 \times 10^6 - 60 \times 10^6 \sigma_{ref} + 60\delta_{CHO} + \frac{3J_{CH-CHO}}{2}, \qquad (6)$$

$$\nu_2^{CHO} = 60 \times 10^6 - 60 \times 10^6 \sigma_{ref} + 60\delta_{CHO} + \frac{J_{CH-CHO}}{2}, \qquad (7)$$

$$\nu_3^{CHO} = 60 \times 10^6 - 60 \times 10^6 \sigma_{ref} + 60\delta_{CHO} - \frac{J_{CH-CHO}}{2}, \tag{8}$$

and

$$\nu_4^{CHO} = 60 \times 10^6 - 60 \times 10^6 \sigma_{ref} + 60\delta_{CHO} - \frac{3J_{CH-CHO}}{2}. \tag{9}$$

The resonant frequency for the reference TMS is

$$\nu_{TMS} = \nu_{ref} = \frac{g_H \mu_N B_o (1 - \sigma_{ref})}{h} = 60 \times 10^6 - 60 \times 10^6 \sigma_{ref}, \tag{10}$$

since $\delta_{TMS} = 0$. The detector measures a difference in frequency between the ethanal resonance lines and that for the reference. This difference is

$$\nu_1^{CH_3} - \nu_{ref} = 60 \times 10^6 - 60 \times 10^6 \sigma_{ref} + 60\delta_{CH_3} + \frac{J_{CH-CHO}}{2}$$

$$- [60 \times 10^6 - 60 \times 10^6 \sigma_{ref}] = 60\delta_{CH_3} + \frac{J_{CH-CHO}}{2}. \tag{11}$$

Similar results are obtained for each of the remaining five frequency components:

$$\nu_2^{CH_3} - \nu_{ref} = 60\delta_{CH_3} - \frac{J_{CH-CHO}}{2}; \tag{12}$$

$$\nu_1^{CHO} - \nu_{ref} = 60\delta_{CHO} + \frac{3J_{CH-CHO}}{2}; \tag{13}$$

$$\nu_2^{CHO} - \nu_{ref} = 60\delta_{CHO} + \frac{J_{CH-CHO}}{2}; \tag{14}$$

$$\nu_3^{CHO} - \nu_{ref} = 60\delta_{CHO} - \frac{J_{CH-CHO}}{2}; \tag{15}$$

$$\nu_4^{CHO} - \nu_{ref} = 60\delta_{CHO} - \frac{3J_{CH-CHO}}{2}. \tag{16}$$

Using the data in Tables 16.2 and 16.4, we estimate that $\delta_{CH_3} \approx 2.65$ and $\delta_{CHO} \approx 9.85$. We also take $J_{CH-CHO} \approx 20$ Hz, as requested in the problem. Inserting these values into Eqs. 11–16 yields the following results:

$$\nu_1^{CH_3} - \nu_{ref} = 60\delta_{CH_3} + \frac{J_{CH-CHO}}{2} = 60(2.65) + \frac{20}{2} = 169 \text{ Hz}; \tag{17}$$

$$\nu_2^{CH_3} - \nu_{ref} = 60\delta_{CH_3} - \frac{J_{CH-CHO}}{2} = 60(2.65) - \frac{20}{2} = 149 \text{ Hz}; \tag{18}$$

$$\nu_1^{CHO} - \nu_{ref} = 60\delta_{CHO} + \frac{3J_{CH-CHO}}{2} = 60(9.85) + \frac{3(20)}{2} = 621 \text{ Hz}; \tag{19}$$

$$\nu_2^{CHO} - \nu_{ref} = 60\delta_{CHO} + \frac{J_{CH-CHO}}{2} = 60(9.85) + \frac{(20)}{2} = 601 \text{ Hz}; \tag{20}$$

$$\nu_3^{CHO} - \nu_{ref} = 60\delta_{CHO} - \frac{J_{CH-CHO}}{2} = 60(9.85) - \frac{(20)}{2} = 581 \text{ Hz}; \tag{21}$$

$$\nu_4^{CHO} - \nu_{ref} = 60\delta_{CHO} - \frac{3J_{CH-CHO}}{2} = 60(9.85) - \frac{3(20)}{2} = 561 \text{ Hz}. \tag{22}$$

The signal for each of these excitations has the form given by Eq. 16.49A:

$$s_i = n_i \exp\left(-\frac{t}{\tau}\right) \cos[2\pi(\nu_i - \nu_{ref})t]. \tag{23}$$

Thus, the signal due to the band at frequency $\nu_1^{CH_3}$ is

$$s_1^{CH_3} = n_1^{CH_3} \exp\left[-\frac{t}{500}\right] \cos[2\pi(0.169)t], \tag{24}$$

provided that we measure t in ms. Similar forms give the other five components of the total signal.

We can obtain the expected intensity ratios by adjusting the n_i appropriately. In doing this, however, we must remember that the intensity of a wave is proportional to the absolute square of its amplitude. The $-CH_3$ resonance will be split into a doublet, with each line having one-half of the total intensity. Therefore, we take $n_1^{CH_3} = n_2^{CH3} = (\frac{1}{2})^{1/2}$. The $-CHO$ resonance will be a quartet, with ν_2 and ν_3 having $\frac{3}{8}$ of the total intensity and ν_1 and ν_4 having $\frac{1}{8}$. We can obtain this result by taking $n_1^{CHO} = n_4^{CHO} = (\frac{1}{8})^{1/2}$ and $n_2^{CHO} = n_3^{CHO} = [\frac{3}{8}]^{1/2}$. Finally, the total CH_3 intensity will be three times that of the CHO resonance. This can be achieved by multiplying the sum of the two CH_3 bands by $(3)^{1/2}$. With these points in mind, we find that the total FID signal is

$$s_{\text{total}} = (3)^{1/2}[s_1^{CH_3} + s_2^{CH3}] + \left[\left(\frac{1}{8}\right)^{1/2}\{s_1^{CHO} + s_4^{CHO}\} + \left(\frac{3}{8}\right)^{1/2}\{s_2^{CHO} + s_3^{CHO}\}\right]$$

$$= (3)^{1/2} \exp\left[-\frac{t}{500}\right][\cos[2\pi(0.169)t] + \cos[2\pi(0.149)t]]$$

$$+ \exp\left[-\frac{t}{500}\right]\left[\left(\frac{1}{8}\right)^{1/2}\{\cos[2\pi(0.621)t] + \cos[2\pi(0.561)t]\}\right]$$

$$+ \exp\left[-\frac{t}{500}\right]\left[\left(\frac{3}{8}\right)^{1/2}\{\cos[2\pi(0.601)t] + \cos[2\pi(0.581)t]\}\right]. \tag{25}$$

Figure 16.25 shows a plot of this FID over the range from 0 to 1,200 ms. The Fourier transform power spectrum of this FID is shown in Figure 16.26.

Note that, to obtain these results, we have assumed (1) that all frequencies in the rf pulse have equal intensities, (2) an exaggerated value for $J_{\text{CH–CHO}}$, in order to

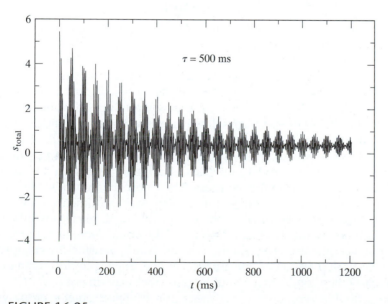

▲ **FIGURE 16.25**
FID signal for ethanal, assuming a CHO–CH$_3$ spin–spin splitting constant of 20 Hz, a total relaxation time constant of 500 ms, and that all frequencies in the rf pulse have equal intensities. (See Example 16.6 for details.)

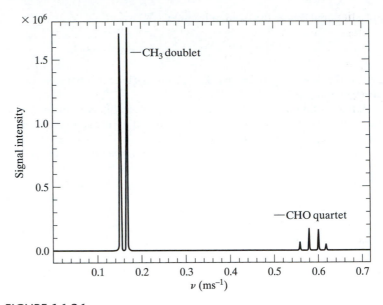

▲ FIGURE 16.26
The Fourier transform of the ethanal FID shown in Figure 16.25. (See Example 16.6 for details.)

make the spin–spin splitting easier to resolve in the plot, and (3) an arbitrary time constant $\tau = 500$ ms for relaxation for all ^{1}H spins. These assumptions cause the results shown in Figure 16.25 to deviate from the FID that would be obtained in an actual experiment. Nevertheless, the calculations contain all the essential ingredients of a more exact treatment. Consequently, Figure 16.25 is quite similar to the experimental FID.

For a related exercise, see Problem 16.15.

16.1.7 Spin-decoupling Techniques

In some cases, it is desirable to suppress the spin–spin coupling between nuclei. When this is done, the number of spectral lines is decreased dramatically, thereby making interpretation of the spectrum much easier.

Consider the case of a ^{13}C NMR spectrum. The natural abundance of ^{13}C is only 1.1%. Because of this, ^{13}C nuclei will rarely be adjacent in the molecule, and spin–spin coupling between ^{13}C nuclei will produce only minor perturbations on the ^{13}C NMR spectrum. However, almost every ^{13}C nucleus will be bonded to one or more protons. In addition, adjacent carbons will usually have several protons attached. The resulting ^{1}H–^{13}C spin–spin couplings will generally produce doublets, triplets, quartets, and higher multiplets for each nonequivalent ^{13}C nucleus in the molecule. These absorption bands will, in many cases, overlap to produce complex multiplets that are difficult to interpret. If this coupling were suppressed, each nonequivalent ^{13}C nucleus would appear as a single line in the spectrum. This not only would remove the problem of band overlap, but also, it would increase the band intensities by collapsing all spin–split multiplets under one absorption line. Such spin decoupling can be accomplished by conducting a double- resonance experiment.

In a double resonance experiment, the FT spectrometer employs two simultaneous strong rf pulses. One of these is centered in the frequency range for the spin transitions being observed in the spectrum. For example, the magnetic field employed on a 400-MHz spectrometer is that required to produce ^{1}H resonance at a frequency of 400 MHz. Using Eq. 16.19, we find that

$$B_o = \frac{h\nu}{g_H\mu_H} = \frac{(6.62608 \times 10^{-34}\,\text{J s})(400 \times 10^6\,\text{s}^{-1})}{(5.58569)(5.0508 \times 10^{-31}\,\text{J G}^{-1})} = 93{,}946\,\text{G}.$$

If we are taking a ^{13}C spectrum with this spectrometer, the ^{13}C resonant frequency will be about

$$\nu_C = \frac{g_C\mu_N B_o}{h} = \frac{(1.4048)(5.0508 \times 10^{-31}\,\text{J G}^{-1})(93{,}946\,\text{G})}{6.62608 \times 10^{-34}\,\text{J s}}$$

$$= 100.60 \times 10^6\,\text{Hz} = 100.60\,\text{MHz}.$$

Therefore, we expect that all the ^{13}C absorptions will occur at frequencies that span a narrow range around 100.60 MHz. The first rf pulse will cover this range. The second rf pulse will be centered at 400 MHz, with a spread sufficient to cover all of the expected proton resonances. The effect of this second pulse is to induce rapid proton transitions between the two nuclear spin states, thus producing the same effect as a large quadruple moment. The adjacent ^{13}C nuclei now sense an *average* spin–spin coupling, rather than two distinct couplings due to protons in the two spin states. As a result, there is no splitting of the ^{13}C nuclear spin energy states and hence no ^{1}H–^{13}C spin–spin splitting. The NMR spectroscopist calls this effect *spin decoupling.* Such decoupling techniques can be employed to remove spin–spin splitting between any two types of nuclei. All that is required is to irradiate the sample with a strong rf pulse that covers the range of transition frequencies for the nuclei whose spins you wish to decouple.

16.1.8   Interpretation of NMR Spectra

The basic principles described in this section are sufficient to permit the interpretation of most NMR spectra, provided that the absorption bands do not overlap to form complex patterns. If that occurs, it is still possible to analyze the spectrum by computer or by more sophisticated experiments, such as two-dimensional NMR, to determine the molecular structure. We can avoid the need for computer analysis if the spectra exhibit three characteristics:

1. The resonant frequency separation between magnetically nonequivalent nuclei is large relative to the spin–spin coupling constants between them.

2. The absorption bands are narrow. The broad bands which result in the NMR spectra of nuclei with $I > \frac{1}{2}$ that have nuclear quadrupole moments make absorption band overlap more likely to occur.

3. There is a unique spin–spin coupling constant between any two sets of chemically equivalent $I = \frac{1}{2}$ nuclei. We have explicitly invoked this assumption in all the preceding examples.

Spectra that satisfy these requirements are said to be *first order.*

The use of modern, high-field NMR spectrometers makes it relatively easy to satisfy the first requirement for small molecules. Equation 16.30 shows that the resonant frequency spacing between two nonequivalent nuclei is directly proportional to the magnitude of the externally applied magnetic field B_o. In contrast, the spin–spin splitting is independent of B_o. We can, therefore, increase the frequency separation between two absorption lines by increasing

B_o while leaving the spin–spin splitting unchanged. If B_o is made large enough, the first requirement will be satisfied. This advantage, combined with the increased sensitivity previously discussed, are the reasons scientists are willing to expend large sums of money to acquire higher field NMR spectrometers. When acquiring spectra for nuclei other than ^1H, it is also possible to employ double-resonance spin-decoupling techniques to avoid band overlap.

If we have first-order spectra and sufficient data on chemical shifts, band assignments are relatively easy to make. Occasionally, electronic transients produce spurious signals that can cause concern; however, certain experimental methods permit the identification of such transients.

EXAMPLE 16.7

Figure 16.27 shows the ^1H NMR spectrum of ethyl 3-methylbenzoate. The spectrum was recorded on a 300-MHz FT NMR spectrometer with TMS as the reference. The positions of the bands are given in terms of δ (parts per million, or ppm). The conversion between δ and the frequency required for resonance is given by Eq. 16.28. (See Shankar Subramanian, *Modified Heteroarotinoids: Potential Anticancer Agents*, Ph.D. dissertation, Oklahoma State University, 1993.) We see that the resonant frequency separations are much larger than the spin–spin splittings for many of the bands, but not for all. Identify the proton groups giving rise to each of the spectral lines. Are any electronic transients present?

Solution

The first step is to identify the spin–spin splitting expected. The $-CH_3$ protons in the ethyl group will appear as a triplet, since they are adjacent to the $-CH_2-$ group that has three possible orientations of the two equivalent proton spins. This triplet appears at $\delta = 1.39$, which is near the expected chemical shift for the group, according to Table 16.2, which gives the chemical shift range as 1.2 to 1.3. Note that the line intensities in the triplet are about 1:2:1, as expected.

The $-CH_2-$ protons are adjacent to an ester group. The table suggests that δ should be in the range 3.8–5.1. We expect the $-CH_2-$ resonance to appear as a quartet with an intensity ratio 1:3:3:1, because of the analysis presented in Figures 16.12

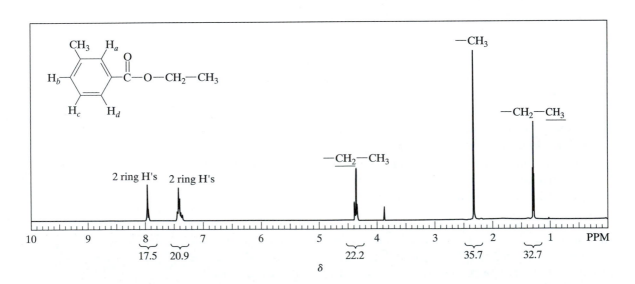

▲ FIGURE 16.27
300-MHz ^1H NMR spectrum of ethyl 3-methylbenzoate. The total integrated intensities for each gruop of NMR bands are shown beneath the abscissa scale.(Shankar Subramanian, *Modified Heteroarotinoids: Potential Anticancer Agents*, Ph.D. Dissertation, Oklahoma State University, 1993.)

and 16.13. This quartet appears at $\delta = 4.35$. Note that its integrated total intensity of 22.2 is about two-thirds that for the CH_3 groups, as expected.

The ring $-CH_3$ group should appear as a singlet, since there are no adjacent spins within a distance of three or fewer chemical bonds to split the resonance line. The table indicates a chemical shift of 2.3 for the CH_3Ph grouping. In this case, the table is right on target: The resonance line for $-CH_3$ appears at $\delta = 2.37$.

The chemical shifts for all ring protons should lie near the range from 7.2 to 7.3, according to Table 16.2. The bands between $\delta = 7$ and $\delta = 8$ must, therefore, be due to resonances from these protons. We would expect the resonance lines from H_b and H_d to each appear as doublets, since each line will be split by the single proton H_c. On the other hand, H_c resonance should appear as a triplet if we regard H_b and H_d as two equivalent protons or as a quartet if they are sufficiently different in their chemical environment. The resonance line for H_a should be a singlet, because there are no other spins within three chemical bonds. Therefore, if the chemical shifts of these protons were sufficiently large, we would observe two doublets from H_b and H_d resonance, a quartet from H_c, and a singlet from H_a. The spectrum makes it clear, however, that this is not the case; the chemical shifts of the ring protons are sufficiently close that these bands overlap to produce two multiplets. Since the two multiplets have about equal integrated total intensities, we conclude that two proton resonances are involved in each. The multiplet around $\delta = 7.3$ appears to contain four or five peaks, whereas the one at $\delta = 7.85$ seems to contain three. Consequently, it appears likely that a triplet from H_c and the doublet from H_b make up the multiplet at $\delta = 7.3$. The singlet from H_a and the doublet from H_d probably form the multiplet at $\delta = 7.85$. This type of result shows why spectral interpretation is easier when the chemical shifts are large.

Finally, we see a small singlet at $\delta = 3.9$. As there are no other protons left to account for this band, it must be a small electronic transient or an impurity.

For related exercises, see Problems 16.16, 16.19, and 16.20.

EXAMPLE 16.8

Figure 16.28 shows a spin-decoupled ^{13}C NMR spectrum of ethyl 3-methylbenzoate in $DCCl_3$ solvent. The spectrum was recorded on a 300-MHz FT NMR spectrometer with TMS as the reference. The positions of the bands are given in terms

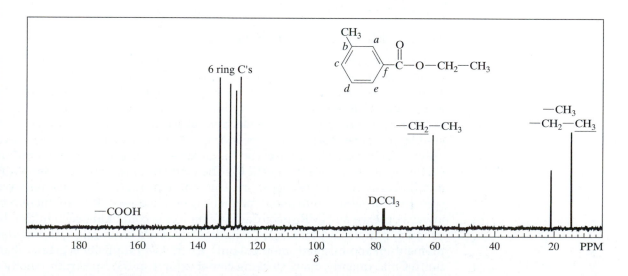

▲ FIGURE 16.28
300-MHz spin-decoupled ^{13}C NMR spectrum of ethyl 3-methylbenzoate. (Shankar Subramanian, *Modified Heteroarotinoids: Potential Anticancer Agents*, Ph.D. Dissertation, Oklahoma State University, 1993.)

of δ, as described in Example 16.7. (See Shankar Subramanian, *Modified Heteroarotinoids: Potential Anticancer Agents*, Ph.D. Dissertation, Oklahoma State University, 1993.) Identify the ^{13}C atoms giving rise to each of the spectral lines.

Solution

We can see that the spectrum is spin decoupled, since none of the ^{13}C resonance lines are split by the attached protons. Table 16.3 lists alkane ^{13}C atoms as having chemical shifts in the range from -2.3 to 25.2. We can, therefore, conclude that the two resonance lines at $\delta = 14.35$ and 21.26 are due to the ring $-CH_3$ and the ethyl $-CH_3$ groups. It turns out that the $\delta = 14.35$ line is due to the ethyl $-CH_3$ ^{13}C atom, but there is no way to predict this without some additional experiments or experience dealing with NMR spectral interpretation.

The table gives the $-C-O$ chemical shift for ^{13}C esters as 57.6 to 67.9. This means we can identify the band at $\delta = 60.87$ as being due to the $-CH_2-$ ^{13}C atom. Aromatic $C=C$ carbons have chemical shifts around $\delta = 128.5$. The spectrum shows six resonance lines between 126.69 and 136.08, due to ^{13}C resonance of carbon atoms $a, b, c, d, e,$ and f shown in the figure. Additional experiments involving isotopic enrichment at known positions, two-dimensional NMR spectroscopy, or more experience is required to identify which line is associated with which carbon nucleus.

^{13}C resonance for a $C=O$ ester carbon occurs around 165.5 to 170.3, according to Table 16.3. Therefore, the small band at $\delta = 166.80$ results from this carbon atom.

We have now accounted for all spectral bands except the small triplet around $\delta = 77$. By a process of elimination, we know that this band must be due to the $DCCl_3$ solvent. Its structure also identifies it. The chlorine nuclei, ^{37}Cl and ^{35}Cl, both have spin quantum numbers $I = \frac{3}{2}$. Therefore, they possess nuclear quadrupole moments which cause rapid spin transitions, and it follows that spin–spin splitting is not produced by these nuclei. However, the deuterium nucleus has $I = 1$. Consequently, its quadrupole moment is very small, and spin–spin splitting will still be observed. Note that the second rf field is at the resonance frequency for ^1H, not ^2H. The deuterium spins are not decoupled. With $I = 1$, m_{ID} can have values of 1, 0, and -1. Hence, we expect the ^{13}C resonance line from the $DCCl_3$ solvent to appear as a triplet with all lines of equal intensity. This is exactly what is seen in the spectrum. We conclude that the triplet at $\delta = 77$ is due to the carbon in the $DCCl_3$ solvent.

For related exercises, see Problems 16.22 and 16.23.

16.2   Electron Spin Resonance

Nuclear magnetic resonance spectroscopy is based on the fact that the nuclear spin energy states can be split by imposing a magnetic field. Transitions can then be observed between these states. Since the electron has a spin quantum number of $\frac{1}{2}$, everything that we have said about ^1H NMR spectroscopy also applies to the electron. In particular, its spin states can be split by a magnetic field and spectroscopic transitions between these states observed. This type of spectroscopy is called *electron spin resonance* (ESR) or *electron paramagnetic resonance* (EPR). The primary difference between NMR and ESR spectroscopy resides in the fact that nuclei with $I > 0$ always have spin angular momentum and a magnetic dipole moment. We can, therefore, always observe an NMR spectrum for a molecule containing such nuclei. In contrast, a molecule, an ion, or a radical has a net electronic spin angular momentum and magnetic moment only when its multiplicity is greater than unity. Consequently, many systems do not exhibit an ESR spectrum. Another difference involves the electronic mass, which is significantly smaller than that for nuclear isotopes. As a result, transition energies in ESR experiments generally lie in the microwave region of the spectrum, rather than the radio frequency range that characterizes NMR transitions. Finally, the electronic charge is negative, so that the energies of the split spin states are reversed. That is, the states with $m_s < 0$ are now the lower energy states.

16.2.1   Energies of Electron Spin States in a Magnetic Field

Let us examine how the classical conducting loop model described in Section 16.1 would be altered if the charge carrier were an electron rather than a proton. Equation 16.8 would still be obtained, but the charge q would now be $q = -e = -1.602177 \times 10^{-19}$ C, and the mass would be that for an electron, namely, $m = m_e = 9.10939 \times 10^{-31}$ kg. Under these conditions, the constant in Eq. 16.8 would become

$$\frac{q\hbar}{2m} = -\frac{e\hbar}{2m_e} = -\frac{(1.6022177 \times 10^{-19} \text{ C})(1.05457 \times 10^{-34} \text{ J s})}{2(9.10939 \times 10^{-31} \text{ kg})}$$

$$= -9.2742 \times 10^{-24} \text{ C m}^2 \text{ s}^{-1} = -9.2742 \times 10^{-24} \text{ J T}^{-1}$$

$$= -9.2742 \times 10^{-28} \text{ J G}^{-1}.$$

The negative of this quantity was first introduced in Chapter 13. It is called the *Bohr magneton* and given the symbol μ_B. Substituting this result into Eq. 16.8 and replacing m with m_s, the electron spin magnetic quantum number, gives an energy change of

$$\Delta E_{mag} = \mu_B m_s B. \tag{16.50}$$

for the circular conducting loop.

It was pointed out in Section 16.1.2 that the simple conducting loop model gives poor results when it is used to describe a relativistic angular momentum effect. In general, magnetic moments must be determined by experiment. The result for electrons is expressed in the same manner as that used for nuclei, viz.,

$$\boldsymbol{\mu} = -g_e \frac{e}{2m_e} \mathbf{S}, \tag{16.51}$$

where g_e is called the *electronic g-factor*. The Z component of $\boldsymbol{\mu}$,

$$\mu_z = -g_e \frac{e}{2m_e} S_z = -g_e \frac{e\hbar}{2m_e} m_s = g_e \mu_B m_s, \tag{16.52}$$

determines the energy change in a magnetic field. The energy change produced by an applied magnetic field of magnitude B is, therefore,

$$\boxed{\Delta E_{mag} = g_e \mu_B m_s B}. \tag{16.53}$$

Equation 16.53 shows that the change in energy will be negative for spin states with $m_s < 0$.

For pure electron spin, the value of g_e is found to be 2.0023. This value may generally be used for most organic molecules, in which no net orbital angular momentum is present. However, in transition metal complexes, in which unpaired electrons in d orbitals possess orbital angular momentum ($\ell > 0$), the coupling of spin and orbital angular momentum often produces g_e values significantly different from the free-electron value.

Most commercial ESR spectrometers obtain the spectrum by fixing the frequency and scanning over the magnetic field. Usually, frequencies in the range from 9 to 10 GHz are used. The apparatus is similar to that shown in

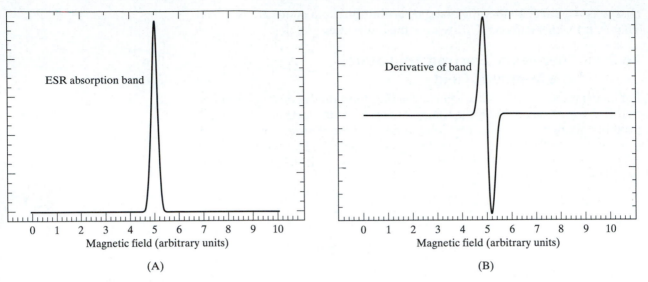

▲ **FIGURE 16.29**
(A) Typical line shape for an ESR absorption band. (B) Derivative of the line with respect to the magnetic field. ESR signals are commonly presented as the derivative of the band rather than as the band itself.

Figure 16.17, except that the rf source is replaced with a klystron that generates radiation in the microwave region of the spectrum. The rf detector is replaced with one that is sensitive to microwave radiation. The method of detection usually employed measures the first derivative of the absorption band with respect to the magnetic field. This markedly changes the appearance of the spectrum. To visualize the change, consider an absorption band which is Gaussian in shape, such as that shown in Figure 16.29A. The derivative of this band with respect to the magnetic field is shown in Figure 16.29B. In general, ESR absorption bands have this form.

With these changes, all of the results derived for ^{1}H NMR spectroscopy hold for ESR measurements. Since the selection rule for transitions is $\Delta m_s = \pm 1$, the condition for ESR resonance is the analogue of Eq. 16.19:

$$h\nu = g_e \mu_B B \,. \tag{16.54}$$

EXAMPLE 16.9

An ESR spectrum of an organic radical exhibits an ESR absorption band at a magnetic field of 0.3294 T. What frequency is being used in the spectrometer?

Solution

The answer can be obtained directly from Eq. 16.54. Solving this equation for the frequency, we obtain

$$\nu = \frac{g_e \mu_B B}{h}. \tag{1}$$

For most organic radicals, $g_e = 2.0023$. Thus,

$$\nu = \frac{(2.0023)(9.2742 \times 10^{-24}\,\text{J T}^{-1})(0.3294\,\text{T})}{6.62608 \times 10^{-34}\,\text{J s}} = 9.231 \times 10^9\,\text{s}^{-1}$$

$$= 9.231 \times 10^9\,\text{Hz} = 9.231\,\text{GHz}. \tag{2}$$

16.2.2 Hyperfine Splitting

In Section 16.1, we learned that the spin magnetic moments of nuclei in close proximity couple to produce spin–spin splitting of the NMR absorption bands. The same effect occurs in ESR spectra: Nuclear magnetic moments couple to the electronic spin magnetic moment to produce spin–spin splitting of the ESR bands. In this case, the effect is called *hyperfine splitting*.

Consider first the case of a bare hydrogen atom in an ^2S state. Since there is no electronic orbital angular momentum, the magnetic moment is due solely to the spin. This moment is given by Eq. 16.51. The interaction of the spin magnetic moment with the applied field produces an energy change described by Eq. 16.53. The spin angular momentum of the proton also produces a small magnetic field that couples to the electronic spin magnetic moment and thereby affects the electron–spin energy state. The magnitude of this effect is given by an expression analogous to Eq. 16.31, viz.,

$$\Delta E_{\text{spin–spin}} = ahm_s m_{IH}, \qquad (16.55)$$

where m_s and m_{IH} are the spin magnetic quantum numbers of the electron and proton, respectively, and a is a constant called the *hyperfine splitting constant*. The total change in energy of the electron's spin state is, therefore, given by combining Eqs. 16.53 and 16.55:

$$\boxed{\Delta E_{\text{mag}} = g_e \mu_B m_s B + ahm_s m_{IH}}. \qquad (16.56)$$

Since m_s and m_{IH} can each take values of $\frac{1}{2}$ and $-\frac{1}{2}$, the magnetic interactions split the degenerate electronic spin states into four levels given by

$$\Delta E_{\text{mag1}} = -\frac{g_e \mu_B B}{2} + \frac{ah}{4} \qquad \left(m_s = m_{IH} = -\frac{1}{2} \right),$$

$$\Delta E_{\text{mag2}} = -\frac{g_e \mu_B B}{2} - \frac{ah}{4} \qquad \left(m_s = -\frac{1}{2}, m_{IH} = \frac{1}{2} \right),$$

$$\Delta E_{\text{mag3}} = \frac{g_e \mu_B B}{2} + \frac{ah}{4} \qquad \left(m_s = m_{IH} = \frac{1}{2} \right),$$

and

$$\Delta E_{\text{mag4}} = \frac{g_e \mu_B B}{2} - \frac{ah}{4} \qquad \left(m_s = \frac{1}{2}, m_{IH} = -\frac{1}{2} \right). \qquad (16.57)$$

This splitting is shown in Figure 16.30.

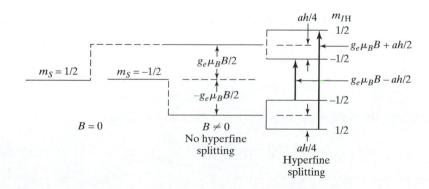

◀ FIGURE 16.30
Hyperfine splitting in a bare hydrogen atom.

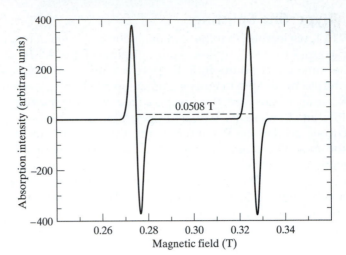

▶ FIGURE 16.31
ESR spectrum for a single hydrogen atom. The two peaks are due to the hyperfine splitting produced by the proton nucleus with $I = \frac{1}{2}$ and $m_{IH} = \frac{1}{2}$ and $-\frac{1}{2}$. The magnitude of the splitting, 0.0508 T, is the hyperfine splitting constant.

ESR transitions occur for $\Delta m_s = \pm 1$, so there are two allowed transitions, one with $m_{IH} = \frac{1}{2}$ and one for which $m_{IH} = -\frac{1}{2}$. These two transitions are indicated by the solid arrows in Figure 16.30. Their energies are $g_e\mu_B B + ah/2$ and $g_e\mu_B B - ah/2$, so that the transition frequencies are

$$\nu_1 = \frac{g_e\mu_B B}{h} + \frac{a}{2} \qquad \textbf{(16.58A)}$$

and

$$\nu_2 = \frac{g_e\mu_B B}{h} - \frac{a}{2}. \qquad \textbf{(16.58B)}$$

The ESR band separation, $\nu_1 - \nu_2$, is, therefore, equal to the hyperfine splitting constant. The similarity to NMR spectra, in which the spin–spin splitting is equal to J, is obvious. Figure 16.31 shows the form of an ESR spectrum of ^{1}H for a spectrometer using a fixed frequency of 8.408 GHz. At this frequency, the ^{1}H resonance occurs at 0.300 T. The hyperfine splitting, however, produces two absorption bands, at 0.2746 and 0.3254 T, with a separation of 0.0508 T. Consequently, the ^{1}H hyperfine splitting constant is 0.0508 T, which is equivalent to 1.433 GHz.

When a is defined by Eq. 16.56, it has units of frequency. If we fix the frequency and scan over magnetic field to find the resonance lines, the two absorptions will be seen at fields given by solving Eqs. 16.58A and 16.58B. The result is

$$B_1 = \frac{h\nu}{g_e\mu_B} - \frac{ah}{2g_e\mu_B}$$

and

$$B_2 = \frac{h\nu}{g_e\mu_B} + \frac{ah}{2g_e\mu_B}.$$

Therefore, the field separation between the ESR bands is

$$\Delta B = B_2 - B_1 = \frac{ah}{g_e\mu_B}. \qquad \textbf{(16.59)}$$

Table 16.5   Hyperfine coupling constants for atoms

Nucleus	a (GHz)	a (T)	Nucleus	a (GHz)	a (T)
^{1}H	1.433	0.0508	^{19}F	48.203	1.720
^{2}H	0.219	0.0078	^{31}P	10.20	0.364
^{13}C	3.167	0.1130	^{35}Cl	4.708	0.168
^{14}N	1.547	0.0552	^{37}Cl	3.924	0.140

This gives us two methods for expressing the hyperfine coupling constant: We can write either

$$a = \Delta\nu \qquad (16.60)$$

or

$$a = \frac{g_e\mu_B}{h}\Delta B. \qquad (16.61)$$

Both methods are used. Table 16.5 lists some hyperfine coupling constants for different nuclei in both units.

Since hyperfine splitting is the result of coupling between the magnetic moments of the nuclei and the electron, we expect the hyperfine constant to be proportional to the product of $g_e\mu_B$ and $g_N\mu_N$ that determine the magnitudes of these moments. We might also expect it to be proportional to the proximity of the nucleus to the electron. This quantity is measured by the electron density at the nucleus, which is proportional to the absolute square of the electronic wave function evaluated at the nucleus, ($|\psi(0)|^2$). Therefore, we expect to have

$$a \propto g_e\mu_B g_N\mu_N|\psi(0)|^2.$$

A more exact analysis shows the proportionality constant to be $8\pi/(3h)$ if we express a in terms of Hz. The hyperfine coupling constant can, therefore, be computed as

$$a = \frac{8\pi}{3h}g_e\mu_B g_N\mu_N|\psi(0)|^2. \qquad (16.62)$$

Equation 16.62 is called the *Fermi contact interaction*. The fact that the hyperfine splitting constant depends upon the magnitude of the electronic wave function at the nucleus permits the electron spin density at various nuclei to be inferred from an ESR spectrum. Example 16.10 illustrates this point.

Let us now consider the more complex case of the ESR spectrum of CH_4^-. Here, the resonance band will be split by the four equivalent proton spins on the carbon. ^{12}C has $I = 0$, so it will not contribute to the hyperfine splitting. The four ^{1}H spins can be aligned in five different ways to produce $M_{IH} = 2, 1, 0, -1,$ and -2. The ESR band will, therefore, be split into five bands. The intensity ratio of these bands is given by the number of ways each of the M_{IH} spin states can be obtained. To obtain $M_{IH} = 2$, all ^{1}H spins must have $m_{IH} = \frac{1}{2}$. This can occur in $4!/[4!(4-4)!] = 1$ way. $M_{IH} = 1$ results whenever we have three spins with $m_{IH} = \frac{1}{2}$ and one with $m_{IH} = -\frac{1}{2}$. This can happen in $4!/[3!(4-3)!] = 4$ ways. Finally, $M_{IH} = 0$ results if we have two spins

▶ **FIGURE 16.32**
ESR spectrum for the methane negative ion, CH_4^-. The hyperfine splitting produced by the four equivalent protons produces five ESR bands, because the four nuclear spins can be aligned so as to yield M_s = 2, 1, 0, −1, or −2. The intensity ratio of 1:4:6:4:1 is directly related to the number of ways each of these M_s's can be achieved. (See text for details.) The splitting between peaks is the proton hyperfine splitting constant, 0.0508 T.

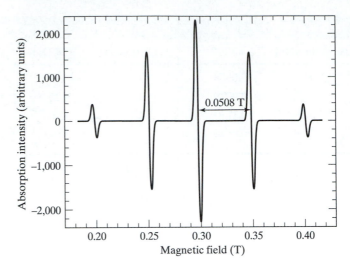

with $m_{IH} = \frac{1}{2}$ and two with $m_{IH} = -\frac{1}{2}$. The number of ways this can occur is $4!/[2!(4-2)!] = 6$. Therefore, the intensity ratios are 1:4:6:4:1 for the five bands. The splitting between bands is equal to the ^1H hyperfine splitting constant for CH_4^-. Figure 16.32 shows the expected spectrum.

EXAMPLE 16.10

(A) An ESR spectrum of the benzene negative ion $(C_6H_6^-)$ is taken. Describe the general appearance of this spectrum. **(B)** The measured hyperfine splitting constant for $C_6H_6^-$ is 0.000375 T, or 3.75 G. Diagram the ESR spectrum for this ion. Why is the hyperfine splitting constant so small relative to the values listed in Table 16.5?

Solution

(A) The symmetry of the molecule ensures that the hyperfine constant will be the same for all ^1H spins. Therefore, the ESR absorption band will be split by six equivalent protons with equal splitting constants. The six ^1H spins can all be aligned to produce $M_{IH} = 3$ or −3. If five spins are aligned and one is opposed, we have $M_{IH} = 2$ or −2. When four are aligned with two opposed, the result is $M_{IH} = 1$ or −1. With three aligned in one direction and the other three in the opposite direction, $M_{IH} = 0$. Therefore, we expect seven ESR bands. The relative intensities of these bands depend upon the number of ways we can achieve each of the aforesaid proton spin alignments. For $M_{IH} = 3$ or −3, the result is $6!/[6!(6-6)!] = 1$ way. With $M_{IH} = 2$ or −2, we have $6!/[5!(6-5)!] = 6$ ways. When $M_{IH} = 1$ or −1, we get $6!/[4!(6-4)!] = 15$ ways. Finally, three spins with $m_I = \frac{1}{2}$ and three with $m_I = -\frac{1}{2}$ can be obtained in $6!/[3!(6-3)!] = 20$ ways. Therefore, the seven ESR bands will have intensity ratios 1:6:15:20:15:6:1. The splitting between the bands will equal the hyperfine splitting constant.

(B) Figure 16.33 shows the ESR spectrum of $C_6H_6^-$ with the expected splittings and band intensities. The hyperfine splitting constant is small because of its dependence upon the electron density of the unpaired electron at the ^1H nuclei, as described by Eq. 16.62. The molecular orbital containing the extra electron is a virtual π orbital made up primarily of carbon p-type orbitals. These atomic orbitals have a node in the plane of the molecule. Consequently, if they were the only orbitals present in the actual wave function, we would have $\psi(0) = 0$, and the

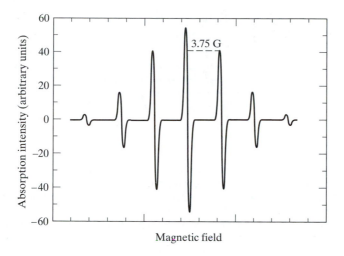

ESR spectrum for the benzene negative ion ($C_6H_6^-$). The hyperfine splitting produced by the six equivalent protons produces seven ESR bands, because the six nuclear spins can be aligned so as to yield $M_s = 3, 2, 1, 0, -1, -2,$ or -3. The intensity ratio of 1:6:15:20:15:6:1 is directly related to the number of ways each of these M_s's can be achieved. (See text for details.) The splitting between peaks is the proton hyperfine splitting constant, 0.000375 T. Its small value is due to the very low electron density of the orbital containing the unpaired electron in the benzene plane.

hyperfine splitting constant would be zero. However, the true wave function has contributions from orbitals that have some density in the molecular plane, so that a is larger than zero. We know that the contribution from such orbitals is small; otherwise we would observe a much larger splitting constant.

For related exercises, see Problems 16.26, 16.27, 16.28, 16.29, and 16.30.

The band splitting seen in Figure 16.33 shows that the hyperfine splitting constant for $C_6H_6^-$ is 3.75 G. In this case, the molecular symmetry ensures that one-sixth of the electron spin density will be located near each ^1H nucleus. Since the situation leads to a hyperfine splitting constant of 3.75 G, we may infer from Eq. 16.62 that if all of the spin density were located around a single ^1H nucleus, we would observe a splitting constant six times as large. That is, we would have $a = 6 \times 3.75\,G = 22.5\,G$. If we assume that all conjugated aromatic radicals have electronic wave functions for the unpaired electron that are essentially the same as that for $C_6H_6^-$, we can infer that the hyperfine splitting constant for carbon atom i for all such organic radicals will be given by

$$a_i^H = (22.5\,G)\rho_i \qquad (16.63)$$

where ρ_i is the spin density of the unpaired electron on carbon atom i and a_i^H is the hyperfine constant for the hydrogen atom attached to carbon i. Equation 16.63 is called the *McConnell equation*. Although an approximation, it is nevertheless useful in that it permits unpaired electron densities at various carbon sites to be estimated from the results of ESR spectral measurements. Problem 16.28 illustrates the use of this equation.

In conjugated aromatic radicals and ions, the electron spin density is spread over the entire conjugated system, so that all attached ^1H spins contribute to the hyperfine splitting. The benzene negative ion just examined is an example. The naphthalene negative ion is another. (See Problem 16.28.) In nonconjugated aliphatic radicals, however, this is not the case; the electron spin density tends to be localized around the site of the radical, so that only nearby ^1H spins contribute to the hyperfine splitting. The ESR spectrum of the ethyl radical (see Problem 16.27) shows that the ^1H spins in the alpha and beta positions both contribute about equally to the hyperfine splitting. The

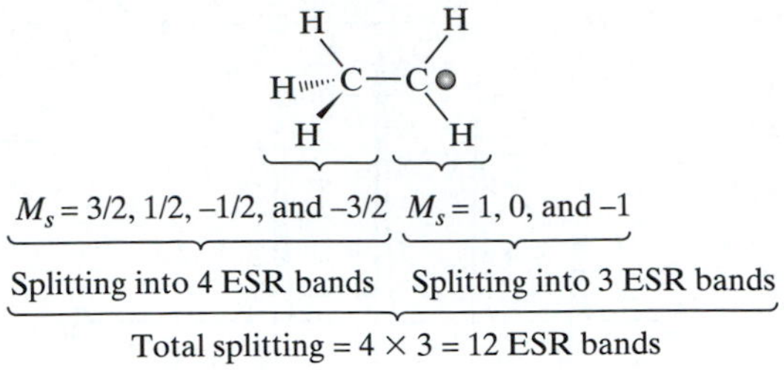

$M_s = 3/2, 1/2, -1/2, \text{ and } -3/2$ $M_s = 1, 0, \text{ and } -1$

Splitting into 4 ESR bands   Splitting into 3 ESR bands

Total splitting = 4 × 3 = 12 ESR bands

▲ **FIGURE 16.34A**
Hyperfine splitting in the ethyl radical. The two protons on the carbon atom containing the unpaired electron have the three possible spin states shown. Therefore, three ESR bands are produced. The three protons on the beta carbon have four possible spin states. As a result, each of the three bands produced by the hyperfine splitting generated by the alpha protons are further split into four bands. The resulting ESR spectrum, therefore, contains a total of 4 × 3 = 12 bands. If there were a gamma carbon with protons, they would further split the ESR bands. However, protons on carbons further removed from the unpaired electron would not produce hyperfine splitting.

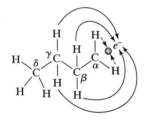

▲ **FIGURE 16.34B**
Hyperfine coupling in organic radicals. The figure illustrates the expected hyperfine coupling between protons in the butyl radical with the unpaired electron on the alpha carbon. Protons on the alpha, beta, and gamma carbons couple with the spin magnetic moment of the unpaired electron to produce hyperfine splitting. These couplings are indicated by arrows. The coupling of protons on the delta carbon and those even further removed from the unpaired electron are too weak to produce observable hyperfine splitting.

[1]H spins in the gamma position exert a much smaller effect, and those beyond the gamma position have a negligible effect. The ethyl radical, therefore, exhibits 12 ESR bands, as shown in Figure 16.34A. Problem 16.30 examines the expected result for the *n*-butyl radical. Figure 16.34B illustrates the hyperfine coupling expected in this radical.

16.3   X-ray Diffraction Spectroscopy

The most accurate means for determining the molecular structure of solid crystals is X-ray diffraction spectroscopy. The basis of this method is the interference of coherent electromagnetic waves that have traveled different distances through space. This is also the basis of operation of the Fourier transform IR spectrometers that we examined in Section 15.4.1. The two techniques, however, differ in their complexity. The results from an FTIR spectrometer can be analyzed simply by extracting the Fourier transform of the transmitted signal. By contrast, the Fourier analysis of a complex X-ray diffraction pattern requires iterative computer fitting of the measured intensity patterns. It is, therefore, much more computationally intense. For this reason, we shall restrict the discussion in this section to the use of X-ray diffraction to determine the structure of simple crystalline solids. In this connection, a review of the discussion of crystalline forms, unit cells, and primitive translations in Section 7.2.2 will be useful.

16.3.1   Crystal Indices

The objectives of an X-ray diffraction measurement are to determine the dimensions and shape of the unit cell and to identify the detailed structure of the molecules contained within the unit cell. To achieve these objectives, we must be able to mathematically express the nature of the measured interference patterns in terms of the positions of the various atoms within the crystal. Since these atoms generally occupy well-defined planes, it follows that we must develop a means by which specific crystalline planes can be designated. This is done by assigning each type of plane a set of *crystal indices*.

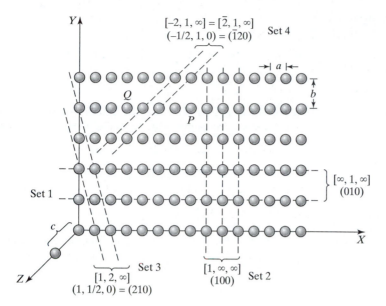

The Weiss and Miller indices of planes parallel to the *Z*-axis of an orthorhombic crystal. The lines represent the edge view of the planes, which are perpendicular to the plane of the figure.

Because all crystals are three dimensional, every indexing method must assign three indices to specify a particular plane. Two common indices that are used are termed *Weiss* and *Miller* indices.

The nature of the two indices is best understood by examining some specific examples. Consider an orthorhombic crystal of atoms or ions in which the unit cell dimensions are all different and the angles between the primitive translations are all 90°. Figure 16.35 shows a two-dimensional slice through such a crystal. The lines indicate the intersections of various perpendicular crystal planes with this slice. To designate any set of planes, we first choose as the origin of our coordinate system any convenient point not in the set of planes to be described. The *X*-, *Y*-, and *Z*-axes are taken to be coincident with the primitive translational vectors of the crystal. For an orthorhombic crystal, the axes will, therefore, be mutually perpendicular. To obtain the Weiss indices for a particular set of planes, we simply count the number of unit cell distances in the *X*, *Y*, and *Z* directions from our origin to the intersections of the axes with the plane. For the planes labeled "Set 1" in the figure, these distances are infinite for the *X*- and *Z*-axes and unity for the *Y*-axis. (The reason for the former is that parallel planes intersect only at an infinite distance.) The Weiss indices for these planes are, therefore, $[\infty, 1, \infty]$, where we list them in the order *X*, *Y*, and *Z* and enclose the three values in brackets to denote the fact that they are Weiss indices. The crystal planes contained in Set 2 can be denoted by the Weiss indices $[8, \infty, \infty]$ since there are eight unit cell distances separating the origin from the intersection with the first member of the planes of the set. However, we could have chosen an origin just to the left of the first plane. Had that been done, the result would be $[1, \infty, \infty]$. This fact shows that crystal indices provide information only about the *relative* distances to the intersection points. Mathematically, this is equivalent to saying that we can always divide the Weiss indices by any convenient number. Therefore, the designations $[8, \infty, \infty]$ and $[1, \infty, \infty]$ are equivalent, since dividing the first by 8 produces the second. When a negative index is present, it is written with a bar over it.

EXAMPLE 16.11

What are appropriate Weiss indices for the planes contained in Sets 3 and 4 in Figure 16.35?

Solution

Using the origin of the coordinate system shown in the figure, we calculate that the number of unit cells to the intersection points with the planes in Set 3 are as follows:

$$X: 1 \text{ unit cell distance, } a; \quad Y: 2 \text{ unit cell distances, } 2b;$$

$$Z: \text{ an infinite number of unit cell distances, because the planes are parallel to the } Z\text{-axis.}$$

The Weiss indices for these planes are, therefore, $[1, 2, \infty]$.

For the planes in Set 4, it is most convenient to change the origin to the point labeled P in the figure. With this choice, the intersection distances are as follows:

$$X: 2 \text{ unit cells distance in the negative } X \text{ direction;}$$

$$Y: 1 \text{ unit cell distance;}$$

$$Z: \text{ an infinite distance.}$$

The appropriate Weiss indices for the planes in Set 4 are thus $[\bar{2}, 1, \infty]$. Note that the bar over the 2 denotes that the intersection along the X-axis is in the negative direction. If we choose point Q as our origin, the resulting indices would be $[1, \frac{\bar{1}}{2}, \infty]$. Since we can divide the Weiss indices by any convenient number, it is clear that division of this result by $-\frac{1}{2}$ produces $[\bar{2}, 1, -\infty]$, which is the same as $[\bar{2}, 1, \infty]$. Consequently, the two designations are equivalent.

For related exercises, see Problems 16.31 and 16.32.

The determination of the distance between crystal planes is a matter of prime concern in the analysis of X-ray diffraction data. As we shall see in the next section, this distance is more conveniently expressed in terms of the *reciprocals* of the Weiss indices than the Weiss indices themselves. For this reason, it is useful to define a second set of crystal indices, called *Miller indices*, that are these reciprocals with fractions cleared by multiplication by an appropriate integer. For example, the $[\infty, \frac{1}{2}, \infty]$ Weiss plane becomes the $(1/\infty, 1/1/2, 1/\infty)$, or the (020), Miller plane. The $[1, 2, \infty]$ Weiss plane is the same as the $(1/1, 1/2, 1/\infty)$, or $(1\frac{1}{2}0)$, Miller plane. Clearing the fraction by multiplying by two, we obtain the usual Miller notation for this set of planes: (210). Negative indices are still denoted by an overbar. Therefore, the $[\bar{2}, 1, \infty]$ Weiss plane is the same as the $(\frac{\bar{1}}{2}10)$ Miller plane. Clearing the fractions, we have $(\bar{1}20)$ as the usual Miller designation for this set of planes. Figure 16.36 shows perspectival views of the $(20\bar{2})$ and the $(11\bar{1})$ Miller planes for an orthorhombic crystal.

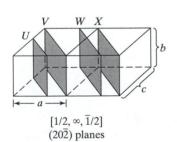

$[1/2, \infty, \bar{1}/2]$
$(20\bar{2})$ planes

(A)

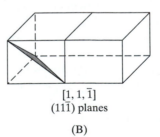

$[1, 1, \bar{1}]$
$(11\bar{1})$ planes

(B)

◀ **FIGURE 16.36**
Perspective view of the $(20\bar{2})$ and $(11\bar{1})$ Miller planes of an orthorhombic crystal. The Weiss indices for the planes are also given.

16.3.2   Separation of Crystal Planes

To analyze X-ray diffraction data, we need a general expression for the distance separating crystal planes whose Miller indices are $(hk\ell)$. We will derive this expression for planes parallel to the Z-axis and then generalize the result so that it is applicable to an arbitrary plane. We begin by considering a simple relationship between the sides of the right triangle shown in Figure 16.37. In the figure, S_1 and S_2 are the lengths of the sides of the right triangle, and S_3 is the length of the hypotenuse. The perpendicular distance from the vertex of the right angle to the hypotenuse is d. We have, then,

$$d = S_1 \sin \alpha, \tag{16.64}$$

where α is the angle formed by S_1 and S_3. Using the large triangle, we may express $\sin \alpha$ in terms of S_2 and S_3 as

$$\sin \alpha = \frac{S_2}{S_3}. \tag{16.65}$$

Combining Eqs. 16.64 and 16.65 yields the desired result:

$$d = \frac{S_1 S_2}{S_3}. \tag{16.66}$$

Equation 16.66 permits us to obtain the distance between planes in terms of the Weiss indices. Consider first the $[1, 2, \infty]$ planes of an orthorhombic crystal, shown in Figure 16.38. The distance we seek between the $[1, 2, \infty]$ planes is shown as Δ in the figure. Since the $[1, 2, \infty]$ plane labeled 2 bisects S_1 and is parallel to S_3, it must also bisect the distance d. Therefore,

$$\Delta = \frac{d}{(1)(2)} = \frac{d}{2}. \tag{16.67}$$

Notice that if we write the Weiss indices for the plane in the form $[hk\,\infty]$, Eq. 16.67 becomes

$$\Delta = \frac{d}{hk}. \tag{16.68}$$

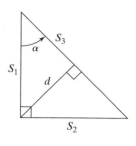

▲ FIGURE 16.37
Relationship between the side lengths of a right triangle and the perpendicular distance from the vertex of the right angle to the opposite side, d.

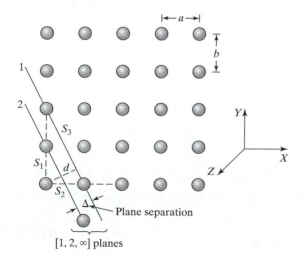

[1, 2, ∞] planes

◄ FIGURE 16.38
The separation of the Weiss $[1, 2, \infty]$ planes of an orthorhombic crystal.

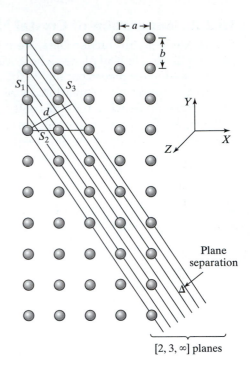

The separation of the Weiss $[2, 3, \infty]$ planes of an orthorhombic crystal.

Now consider the separation between the Weiss $[2, 3, \infty]$ planes shown in Figure 16.39. In this case, we see that the plane separation is $\Delta = d/6$. This result may again be written in the form

$$\Delta = \frac{d}{hk} = \frac{d}{(2)(3)} = \frac{d}{6}. \tag{16.69}$$

Equations 16.68 and 16.69, which we have shown to hold for the $[1, 2, \infty]$ and $[2, 3, \infty]$ Weiss planes, respectively, are true in general.

Using Eq. 16.66, we may now write the plane separation in terms of the side lengths shown in Figures 16.38 and 16.39. This result is

$$\Delta = \frac{S_1 S_2}{hk S_3}. \tag{16.70}$$

The lengths of the sides can be expressed in terms of the Weiss indices and the unit cell spacings. For an orthorhombic crystal, we have

$$S_1 = kb,$$

$$S_2 = ha,$$

and

$$S_3 = [S_2^2 + S_1^2]^{1/2} = [h^2 a^2 + k^2 b^2]^{1/2}. \tag{16.71}$$

Combination of Eqs. 16.70 and 16.71 produces

$$\Delta = \frac{kbha}{hk[h^2 a^2 + k^2 b^2]^{1/2}} = \frac{ab}{[h^2 a^2 + k^2 b^2]^{1/2}}. \tag{16.72}$$

Squaring both sides of Eq. 16.72 and taking the reciprocal gives

$$\frac{1}{\Delta^2} = \frac{[h^2 a^2 + k^2 b^2]}{a^2 b^2} = \frac{h^2}{b^2} + \frac{k^2}{a^2} \quad \text{(for planes parallel to } Z\text{)}. \tag{16.73}$$

Equation 16.73 gives the plane separation for crystals whose primitive translational vectors intersect at 90° angles, in terms of Weiss indices for planes parallel to the Z-axis. Notice the form of this equation. The Weiss indices h and k tell us that the distance from the origin to the intersection of the X-axis with the plane is ha and that for Y is kb. Observe, however, that the Weiss index h is associated with b, whereas the k index is in the term containing a. While this result creates no real problem, it must be kept in mind if we use Eq. 16.73 and Weiss indices to compute plane separations. A more intuitive result is obtained if we convert to Miller indices. The Miller indices corresponding to $[h, k, \infty]$ are $(1/h \ 1/k \ 1/\infty)$. If we clear the fractions by multiplying by hk, we obtain $(kh0)$ as the corresponding Miller indices. Thus, if the h and k indices in Eq. 16.73 are regarded as Miller indices, each index is associated with its corresponding unit cell distance. Example 16.12 illustrates the significance of this point.

EXAMPLE 16.12

An orthorhombic crystal has unit cell distances of 2.7 Å, 2.9 Å, and 3.2 Å along the X, Y, and Z primitive translations, respectively. **(A)** Compute the spacing between the $[4, 5, \infty]$ planes. **(B)** Compute the spacing between the (340) planes.

Solution

(A) The $[4, 5, \infty]$ plane is expressed with Weiss indices. Therefore, the reciprocal of the square of the distance is given by

$$\frac{1}{\Delta^2} = \frac{h^2}{b^2} + \frac{k^2}{a^2} = \frac{4^2}{2.9^2} + \frac{5^2}{2.7^2} = 5.332 \ \text{Å}^{-2}. \tag{1}$$

Solving for Δ, we obtain

$$\Delta = 0.433 \ \text{Å}. \tag{2}$$

(B) The (340) plane is designated with the use of Miller indices. Therefore, each index is associated with its corresponding unit cell spacing. This produces

$$\frac{1}{\Delta^2} = \frac{3^2}{2.7^2} + \frac{4^2}{2.9^2} = 3.137 \ \text{Å}^{-2}, \tag{3}$$

and it follows that

$$\Delta = 0.565 \ \text{Å}. \tag{4}$$

For related exercises, see Problems 16.33 and 16.34.

When Eq. 16.73 is generalized to arbitrary planes, the result for the Miller $(hk\ell)$ planes of any crystal in which the angles between the primitive translations are all 90° is

$$\boxed{\frac{1}{\Delta_{hkl}^2} = \frac{h^2}{a^2} + \frac{k^2}{b^2} + \frac{\ell^2}{c^2},} \tag{16.74}$$

where a, b, and c are the unit cell spacings along X, Y, and Z, respectively. The form of this equation ensures that the spacing for the $(nh \, nk \, n\ell)$ Miller planes is n^{-1} times that for the $(hk\ell)$ planes. This follows immediately from Eq. 16.74. For the $(nh\,nk\,n\ell)$ planes, we have

$$\frac{1}{\Delta^2_{nh\,nk\,nl}} = \frac{n^2 h^2}{a^2} + \frac{n^2 k^2}{b^2} + \frac{n^2 \ell^2}{c^2} = n^2 \left[\frac{h^2}{a^2} + \frac{k^2}{b^2} + \frac{\ell^2}{c^2}\right] = \frac{n^2}{\Delta^2_{hkl}}.$$

Taking reciprocals of both sides and then extracting the square roots gives

$$\Delta_{nh\,nk\,nl} = \frac{\Delta_{hkl}}{n}. \tag{16.75}$$

16.3.3 The Bragg Law

The basis of X-ray diffraction spectroscopy is similar to the principle underlying the operation of an FTIR spectrometer. Consequently, a review of the discussion in Section 15.4.1 is appropriate at this point. In an FTIR instrument, coherent photons with a continuous range of frequencies experience different path lengths to the detector, thereby creating an interference pattern whose frequency components can be determined by extracting the Fourier transform of the pattern. The signal from an X-ray spectrometer is similar in that coherent photons with a single wavelength also form an interference pattern due to differing path lengths. However, the analysis of the X-ray data involves determining the actual path lengths or phase angle differences that lead to the interference pattern, rather than just the frequency spectrum. This is a considerably more difficult task.

In an X-ray diffraction experiment, a monochromatic beam of coherent electromagnetic radiation whose wavelength is about 10^{-10} m (1 Å) is directed toward a solid sample whose structure is to be determined. The high-energy X rays strike the crystal and penetrate deeply into its bulk. Occasionally, a photon encounters one of the atoms of the crystal. When this happens, the photon is reflected, with the angle of reflection generally being equal or nearly equal to the angle of incidence, as illustrated in Figure 16.40. Such scattering is called *specular reflection*. Since the X-ray photons strike different atoms, the electromagnetic waves in the reflected radiation are no longer coherent. Instead, the waves are now out of phase, because they have

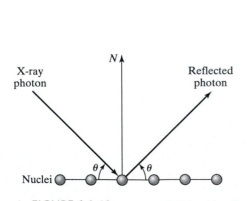

▲ FIGURE 16.40

The specular reflection of X rays after impact with an atomic nucleus in the crystal.

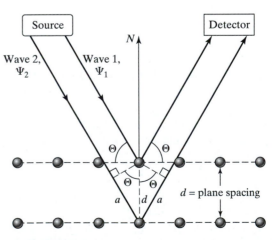

▲ FIGURE 16.41

The Bragg reflection law.

traversed different distances. The determination of the crystal structure rests upon the analysis of the interference patterns that develop between these waves.

Figure 16.41 illustrates two coherent electromagnetic waves labeled Ψ_1 and Ψ_2, incident at an angle Θ upon a set of crystal planes whose spacing is d. Ψ_1 and Ψ_2 reflect specularly from the first and second planes, respectively. The figure shows that the distance traversed by the two waves from source to detector is different. If Ψ_1 travels a distance X_o, the path length for Ψ_2 is $X_o + 2a$. The geometry of the figure shows that

$$a = d \sin \Theta,$$

so that the difference in path length between the two waves is

$$\Delta X = 2a = 2d \sin \Theta. \tag{16.76}$$

Therefore, the sum of Ψ_1 and Ψ_2 at the detector is given by

$$\Psi_{total} = \Psi_1 + \Psi_2 = A \sin\left[\frac{2\pi X_o}{\lambda}\right] + A \sin\left[\frac{2\pi(X_o + \Delta X)}{\lambda}\right], \tag{16.77}$$

where A and λ are the amplitude and wavelength of the coherent radiation. Using the trigonometric identity $\sin(\alpha + \beta) = \sin \alpha \cos \beta + \sin \beta \cos \alpha$, we see that Eq. 16.77 may be written in the form

$$\Psi_{total} = A \sin\left[\frac{2\pi X_o}{\lambda}\right] + A \sin\left[\frac{2\pi X_o}{\lambda}\right] \cos\left[\frac{2\pi \Delta X}{\lambda}\right]$$

$$+ A \cos\left[\frac{2\pi X_o}{\lambda}\right] \sin\left[\frac{2\pi \Delta X}{\lambda}\right]. \tag{16.78}$$

In order for Ψ_1 and Ψ_2 to add constructively, the third term in Eq. 16.78 must vanish and the second term must be equal to $A \sin[2\pi X_o/\lambda]$. For this to occur, we must have $[2\pi \Delta X/\lambda]$ equal to an integral multiple of 2π radians—that is,

$$\left[\frac{2\pi \Delta X}{\lambda}\right] = 2\pi n \text{ for integer } n.$$

This stipulation requires that

$$\Delta X = n\lambda \tag{16.79}$$

as the condition for constructive interference of the waves that leads to an intensity maximum at the detector. Combining Eqs. 16.76 and 16.79 produces

$$\boxed{2d \sin \Theta = n\lambda}. \tag{16.80}$$

Equation 16.80 is called the *Bragg equation* or the *Bragg law*. It is the fundamental relationship upon which the analysis of X-ray interference patterns is based. It tells us that if we scan over all possible angles of incidence, angles Θ, in Figure 16.41 and measure the reflected radiation intensity as a function of Θ, intensity maxima will be obtained whenever $\Theta = \sin^{-1}[n\lambda/(2d)]$ for integer n. Since the wavelength of the radiation is known, it follows that if we measure the angles at which these intensity maxima occur, the plane spacing can be determined. Problem 16.35 examines the conditions for complete destructive interference of the reflected X rays.

EXAMPLE 16.13

X rays with $\lambda = 1.54$ Å are reflected from the (100) planes of a crystal and are observed to produce intensity maxima at angles of incidence of $10.82°$, $22.06°$, $34.30°$, $48.70°$, and $69.89°$, measured from the surface plane. What is the spacing between the (100) planes? What is the value of n in the Bragg equation for each of the observed maxima?

Solution

Solving the Bragg equation for d, we obtain

$$d = \frac{n\lambda}{2\sin\Theta}. \tag{1}$$

At the first intensity maximum $\Theta = 10.82°$, and we have

$$d = \frac{n(1.54\text{ Å})}{2\sin(10.82°)} = 4.10n_1, \tag{2}$$

where n_1 represents n at the first maximum. At the second maximum, the result is

$$d = \frac{n(1.54\text{ Å})}{2\sin(22.06°)} = 2.05n_2. \tag{3}$$

The results at the other maxima are

$$d = 1.37n_3 = 1.02n_4 = 0.820n_5. \tag{4}$$

Since the plane spacing is constant, all of these results must yield the same value of d. Thus, we have

$$4.10n_1 = 2.05n_2 = 1.37n_3 = 1.02n_4 = 0.820n_5, \tag{5}$$

which requires that

$$\frac{n_2}{n_1} = \frac{4.10}{2.05} = 2, \frac{n_3}{n_1} = \frac{4.10}{1.37} = 3, \frac{n_4}{n_1} = \frac{4.10}{1.02} = 4, \text{ and } \frac{n_5}{n_1} = \frac{4.10}{0.820} = 5. \tag{6}$$

Therefore, if $n_1 = 1$, the other integers must be $n_2 = 2$, $n_3 = 3$, $n_4 = 4$, and $n_5 = 5$. Equation 2 then gives a plane spacing of 4.10 Å.

The successive maxima for increasing values of n are referred to as the *first-*, *second-*, *third-*, *fourth-*, and *fifth-order reflections* from the (100) planes. We can also utilize Eq. 16.75 to label these intensity maxima in a different way. Equation 16.80 shows that the nth-order reflection occurs at an angle of incidence of

$$\Theta_n = \sin^{-1}\left[\frac{n\lambda}{2d}\right] = \sin^{-1}\left[\frac{\lambda}{2(d/n)}\right]. \tag{7}$$

Therefore, the nth-order reflection from the (100) planes with spacing d is equivalent to the first-order reflection from a set of planes whose spacing is d/n. Equation 16.75 shows that the (n00) planes are such a set of planes. Therefore, the successive reflections in this example can be labeled as the (100), (200), (300), (400), and (500) reflections.

For related exercises, see Problems 16.36 and 16.37.

16.3.4 Powder X-ray Diffraction

The identification of complex crystalline structures requires that a single crystal be mounted in an X-ray diffractometer and the interference patterns measured for many different crystalline orientations relative to the incident X-ray beam. The deconvolution of the diffraction data then requires iterative Fourier analysis on a computer. If, however, we are deal-

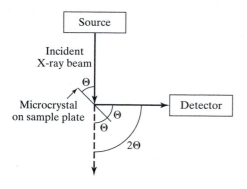

A modern X-ray powder diffractometer in which the reflected radiation is detected electronically.

ing with a substance with a simple crystal structure, we can significantly reduce the time required to accumulate the data by employing the powder method.

In a powder X-ray diffraction experiment, the monochromatic, coherent X-ray beam is directed into a sample holder that contains powdered microcrystals of the material whose structure is to be identified. The large number of randomly arranged tiny crystals ensures that some of the crystals will be oriented at the value of Θ required to satisfy the Bragg equation for each set of Miller planes. Consequently, the data for all planes are obtained in a single experiment.

In a modern powder diffractometer, the sample is spread over a sample plate and the reflected radiation is measured electronically. Figure 16.42 shows a schematic of the reflection of the X-ray beam from one such randomly oriented microcrystal on the sample plate. The reflection angles corresponding to the intensity maxima are easily determined, since the angle of incidence measured from the surface of the crystal is always one-half of the angle between the direction of the beam and the detector. Once the angles for the interference maxima are determined, all that remains is to assign these maxima to reflections from specific crystal planes. This process of assigning intensity maxima to specific reflections is called *indexing*.

The powder diffraction pattern for a primitive cubic crystal provides a particularly simple example. Equation 16.74 tells us that the plane spacing in such a crystal is

$$\frac{1}{\Delta^2} = \frac{h^2 + k^2 + \ell^2}{a^2},$$

so that

$$\Delta = \frac{a}{(h^2 + k^2 + \ell^2)^{1/2}}.$$

The Bragg equation will be satisfied when we have

$$2\,\Delta \sin \Theta = \lambda.$$

The reflection intensity maxima will, therefore, occur whenever

$$\sin \Theta = \frac{\lambda}{2\Delta} = \frac{\lambda}{2a}(h^2 + k^2 + \ell^2)^{1/2}. \tag{16.81}$$

Suppose we conduct a powder diffraction measurement with $\lambda = 1.54$ Å and find that the angles for the intensity maxima are 7.02°, 9.95°, 12.22°, 14.15°, 15.86°, 17.42°, 20.22°, 21.51°, 22.73°, 23.91°, 25.04, and 26.14°. Equation 16.81

shows that the minimum angle for constructive interference is obtained when $(h^2 + k^2 + \ell^2)^{1/2}$ has its minimum value. This occurs for the Miller (100), (010), and (001) planes, which are all equivalent in the primitive cubic lattice. For these planes, we have

$$\sin\Theta = \sin(7.02°) = \frac{\lambda}{2a}(1^2 + 0^2 + 0^2)^{1/2} = \frac{\lambda}{2a} = 0.1222.$$

The unit cell spacing is, therefore,

$$a = \frac{\lambda}{2(0.1222)} = \frac{1.54 \text{ Å}}{2(0.1222)} = 6.301 \text{ Å}.$$

Inserting this result into Eq. 16.81 produces

$$h^2 + k^2 + \ell^2 = \left[\frac{2a}{\lambda}\right]^2 \sin^2\Theta = 66.97 \sin^2\Theta. \tag{16.82}$$

The indexing of the intensity maxima follows immediately from Eq. 16.82, as shown in Table 16.6. Notice that the sum $h^2 + k^2 + \ell^2$ can have only the values 1, 2, 3, 4, 5, 6, 8, 9, 10, 11, 12 ... for a primitive cubic crystal; the value $h^2 + k^2 + \ell^2 = 7$ is missing. This omission is characteristic of the primitive cubic crystal.

The general appearance of the powder diffraction pattern is often sufficient to identify a crystal structure. Consider, for example, a face-centered cubic lattice such as that illustrated in Figure 16.43. Suppose the detector of our powder diffractometer is positioned so as detect the intensity maximum for first-order Bragg reflection from the (100) Miller planes labeled "1" and "2" in the figure. Then Eq. 16.80 shows that we have

$$2a \sin\Theta = \lambda,$$

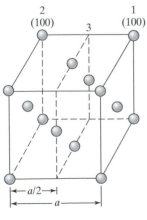

▲ FIGURE 16.43
X-ray diffraction from a face-centered cubic crystal showing systematic absences.

Table 16.6	Indexing the intensity maxima for a primitive cubic crystal	
Θ_{max} (deg)	$66.97 \sin^2\Theta_{max} = h^2 + k^2 + \ell^2$	**Miller Planes**
7.02	1.00	(100), (010), (001)
9.95	2.00	(110), (101), (011)
12.22	3.00	(111)
14.15	4.00	(200), (020), (002)
15.86	5.00	(210), (201), (021), (012), (120), (102)
17.42	6.00	(211), (121), (112)
20.22	8.00	(220), (202), (022)
21.51	9.00	(221), (212), (122), (300), (030), (003)
22.73	10.00	(310), (301), (130), (103), (013), (031)
23.91	11.00	(311), (131), (113)
25.04	12.00	(222)
26.14	13.00	(320), (302), (032), (023), (230), (203)

so that

$$\sin \Theta = \frac{\lambda}{2a}.$$

Simultaneously, the plane of atoms labeled "3" will also reflect the X-ray beam. The difference in path length for this plane of atoms can be computed with the use of Eq. 16.76:

$$\Delta X = 2\left[\frac{a}{2}\right]\sin \Theta = 2\left[\frac{a}{2}\right]\frac{\lambda}{2a} = \frac{\lambda}{2}.$$

The solution of Problem 16.35 reveals that the condition for the destructive interference of two waves is that their path length difference satisfy the equation $\Delta X = [n + \frac{1}{2}]\lambda$, where n is an integer. Therefore, the reflections from plane 3 tend to cancel those from the (100) planes 1 and 2.

  The intensity of the reflections in a crystal is dependent upon the number of atoms in the plane per unit cell and upon the electron density about each atom. The ability of an atom to reflect the X-ray beam is measured for each type of atom by its *scattering factor*. If all atoms in the crystal are identical, the relative reflected intensity will depend only upon the number of atoms in the plane per unit cell. For a face-centered cubic crystal, the atoms at the vertices of the cell are shared by eight unit cells (four above and four below), while the face-centered atoms are shared by the two adjacent unit cells. Therefore, the effective number of atoms in the (100) planes 1 and 2 are

$$\frac{4 \text{ vertex atoms}}{8} + \frac{1 \text{ face-centered atom}}{2} = 1.$$

The effective number of scattering centers in plane 3 is

$$\frac{4 \text{ face-centered atoms}}{2} = 2.$$

As a result, the scattering from plane 3 will be twice as intense as that from either plane 1 or plane 2. In fact, it will be exactly equal to the sum of the reflected radiation from both (100) planes. There will, therefore, be complete cancellation of the three reflections, and no intensity maximum for the (100) planes will be observed. Such a cancellation of an expected intensity maximum is called a *systematic absence*.

  Although we will not observe an intensity maximum for the (100) planes in a face-centered cubic crystal in which the atoms or ions have equal scattering factors, the (200) reflections will still be present. This is easily seen by noting that first-order scattering from the (200) planes is equivalent to second-order scattering from the (100) planes. For this case, $n = 2$ in Eq. 16.80, and the result is $2a \sin \Theta = 2\lambda$, so that $\sin \Theta = \lambda/a$. The path length difference for plane 3 now becomes $\Delta X = 2[a/2]\sin \Theta = 2[a/2]\lambda/a = \lambda$, so the waves now add constructively, and an intensity maximum is observed for the (200) planes. It is easy to show (see Problem 16.38) that only the (200), (400), (600), ... , (2n00) reflections give intensity maxima for a face-centered cubic crystal when the atoms have equal scattering factors. A specific example is KCl. The number of electrons about K^+ and Cl^- are the same, so the two ions exhibit nearly identical scattering factors. Systematic absences are, therefore, observed in the powder diffraction spectrum. In contrast, NaCl, with a similar structure, does not exhibit such absences, because the scattering factor for the Na^+ ions is significantly less than that for the Cl^- ions.

A detailed treatment demonstrates that, for a primitive lattice, there are no systematic absences. A face-centered lattice with equal scattering factors shows intensity maxima only for the Miller planes whose indices are either all even or all odd. For a body-centered crystal with equal scattering factors, we observe intensity maxima only for the Miller planes for which the sum of the Miller indices is even.

Summary: Key Concepts and Equations

1. When a nucleus possesses spin angular momentum $\mathbf{S}$, it also has a magnetic dipole moment given by

$$\boldsymbol{\mu} = g_N \frac{e}{2m_p} \mathbf{S},$$

where e and m_p are the electronic charge and proton mass, respectively, and g_N is a constant characteristic of the nucleus called the nuclear g-factor. g-factors must be determined by experimental measurement. When the nucleus is placed in an externally applied magnetic field $\mathbf{B}$, there is a coupling between the field and the nuclear magnetic moment. This coupling produces a change in the energy of the system that is proportional to the magnitude of the field and the component of the magnetic moment in the direction of the field. If the direction of the field is taken to be along the Z-axis, the energy change is

$$\Delta E_{\text{mag}} = -\mu_z B = -g_N \frac{e}{2m_p} S_z B = -g_N \frac{e\hbar}{2m_p} m_I B = -g_N \mu_N m_I B,$$

where $\mu_N = e\hbar/(2m_p)$ is called the nuclear magneton.

2. If the spin of the nucleus is I, the spin magnetic quantum number m_I can take values that advance by integer steps over the range $-I \leq m_I \leq I$. There are, therefore, $2I + 1$ different possible values of m_I. These states are degenerate in the absence of a magnetic field, but are split when a field is applied. The positive values of m_I have their spin magnetic moments aligned with the field. This produces additional stability and lowers the energy. The spin states with negative values of m_I are aligned opposite to the field and, therefore, increase in energy. Nuclear magnetic resonance spectroscopy is concerned with the measurement of the frequencies of electromagnetic radiation required to induce transitions between these split energy states. The selection rule for spin transitions is $\Delta m_I = \pm 1$. Since the splitting between adjacent spin states is given by $g_N \mu_N B$, the condition for resonance is

$$h\nu = g_N \mu_N B.$$

3. Since both the frequency and the magnetic field can be varied in an NMR experiment, we can either fix the field and vary the frequency to find the resonance absorption bands, or we can fix the frequency and scan over magnetic field. Older NMR spectrometers fixed the frequency and scanned the magnetic field. Modern instruments do the opposite.

4. The utility of NMR spectroscopy rests upon the fact that the resonance frequency for a particular nucleus depends upon its chemical environment. When a large external magnetic field is imposed, the electronic wave function is perturbed, with the result that a small internal magnetic field develops because of the electron structure around the various nuclei. Using Le Chatelier's principle, we can predict that the direction of this internal field will be opposite from that of the applied external field B_o. These small fields vary as the electronic environment around the nucleus changes. Therefore, not all nuclei of a given type will exhibit

resonant absorption at the same frequency for a fixed magnetic field. The required resonance frequency for nucleus i is

$$\nu_i = \frac{g_N \mu_N B_o}{h}(1 - \sigma_i),$$

where σ_i is the diamagnetic shielding constant.

5. The chemical shift for nucleus i (δ_i) is defined to be the difference of the shielding constant for some convenient reference nucleus and that for nucleus i, multiplied by 10^6; that is,

$$\delta_i = (\sigma_{\text{ref}} - \sigma_i)\, 10^6.$$

With this definition, we can show that the difference in resonant frequencies for nuclei A and B at a fixed magnetic field is

$$\nu_A - \nu_B = \frac{g_N \mu_N}{h} B_o [\delta_A - \delta_B] 10^{-6}.$$

Consequently, the difference in resonant frequencies between two nuclei of the same type is proportional to the difference in their δ values and to the applied external magnetic field.

6. The interaction of the magnetic field produced by one nuclear spin can couple to the magnetic dipole moment of a nearby spin to produce an energy change. This type of interaction, called spin–spin coupling, is seen only between nonequivalent nuclei within a distance of three chemical bonds from one another. If nuclear spin A is coupled to N equivalent B spins with $I = \frac{1}{2}$, the NMR absorption band will be split into $N + 1$ lines, since there are $N + 1$ different possible alignments of N equivalent spin states. The magnitude of the splitting is equal to the spin–spin splitting constant. The intensities of the split lines are proportional to the number of different ways each possible alignment of the N equivalent spins can be achieved. If the nuclei have $I = \frac{1}{2}$, so that there are two possible m_I values, then the number of ways to align N spins with m spins in the α state and $N - m$ spins in the β state is

$$\text{no. of ways} = \frac{N!}{m!(N - m)!}.$$

7. Nuclei with $I > \frac{1}{2}$ possess nuclear quadrupole moments. The presence of such a moment induces rapid transitions between the various spin states. These transitions tend to average out the spin–spin coupling and produce broadened NMR absorption bands. If $I \geq \frac{3}{2}$, we generally do not observe spin–spin splittings.

8. Because the rf coil in an NMR spectrometer measures the sum of the absorption and the stimulated emission, NMR band intensities depend upon the population difference between the upper and lower spin states. These differences are very small, generally in the parts-per-million range. As a result, the NMR band intensities are low, and sensitivity is a problem. Since the population difference is almost linearly dependent upon the frequency of the spectrometer, sensitivity can be increased by using instruments that employ higher magnetic fields and hence higher frequencies.

9. Older, continuous-wave (CW) NMR spectrometers operate by fixing the frequency and slowly scanning over the magnetic field to find the fields at which resonance occurs. The sweep must be done slowly to permit the accumulation of sufficient signal at each field to produce the line resolution usually required. Typically, a complete scan will take about 15 minutes. Whenever sensitivity is a problem, repeated scans must be executed to allow random noise to be averaged out. The combination of a 15-minute sweep time and the need for repeated scans leads to unacceptably long experimental times. This problem is partially removed by the use of a Fourier transform (FT) NMR spectrometer.

10. An FT NMR spectrometer operates at a fixed magnetic field B_o that is maintained constant using an electronic frequency/field lock. The sample is placed in an NMR probe that contains the electronics needed to produce a powerful rf signal for a few microseconds whose magnetic vector is oriented perpendicular to the B_o field. The frequency of the rf pulse is centered in the range of resonant frequencies exhibited by the particular nucleus whose spectrum is desired. Because of the short duration of the pulse, the uncertainty principle spreads the frequencies over a range sufficient to cover all resonance lines. Since the nuclear g-factors vary significantly for different nuclei, a different NMR probe is usually required for each nucleus studied. After excitation, the net magnetization vector is rotated 90° by the powerful rf pulse into the X–Y plane, where the magnetic moments of the excited spins precess about the Z-axis. Detectors located along the X- and Y-axes sense the component of magnetic moment in their direction. Since the excited spins are precessing about Z with different frequencies, the result is a complex oscillatory pattern that decays exponentially with time, due to spin relaxation. This pattern is called a free-induction decay (FID). The power spectrum obtained from the Fourier transform of the FID provides the NMR spectrum.

11. Spin–spin splitting between different types of nuclei can be suppressed by conducting a double-resonance spin-decoupling experiment. In a double-resonance experiment, the FT spectrometer employs two simultaneous strong rf pulses. One of these is centered in the frequency range of the nuclear spin transitions being observed in the spectrum. The second pulse is centered at the frequency for the spins to be decoupled, with a spread sufficient to cover all of the expected resonances for that type of nucleus. The effect of this second pulse is to induce rapid transitions between the nuclear spin states. This has the same effect as a large quadrupole moment. The nuclei whose spectrum is being recorded now "sense" an average spin–spin coupling, rather than distinct couplings, due to nearby nuclei in specific spin states. As a result, there is no spin–spin splitting.

12. The interpretation of NMR spectra is straightforward, provided that the spectra satisfy the following first-order conditions: (1) The resonant frequency separation between magnetically nonequivalent nuclei must be large relative to the spin–spin coupling constants between them; (2) the absorption bands must be narrow; and (3) there must be a unique spin–spin coupling constant between any two sets of chemically equivalent $I = \frac{1}{2}$ nuclei.

13. Nuclear magnetic resonance spectroscopy is based on the fact that the nuclear spin energy states can be split by imposing a magnetic field. Transitions can then be observed between these states. Since the electron has a spin quantum number of $\frac{1}{2}$, everything that has been said about ^{1}H NMR spectroscopy also applies to electron spin resonance (ESR) spectroscopy. The change in the electron spin energy states produced by an applied magnetic field is given by

$$\Delta E_{\text{mag}} = g_e \mu_B m_s B.$$

This equation is similar to that for nuclear spins. The differences are (1) a change of sign, since the electron is negatively charged, whereas nuclei have positive charges; (2) replacement of the nuclear magneton μ_N with the Bohr magneton μ_B, necessitated by the difference in the nuclear versus the electronic mass; and (3) replacement of the nuclear g-factor with g_e. For the free electron and many organic radicals and ions, $g_e = 2.0023$. In addition, ESR spectro-meters normally record the derivative of the band rather than the absorption band itself. Therefore, ESR band shapes are different from those seen in NMR spectra.

14. In NMR spectroscopy, band splitting occurs because of spin–spin coupling. The same interaction is present in ESR spectroscopy, where the effect is called hyperfine splitting. The splitting between the ESR bands equals the hyperfine splitting constant.

15. Distinct crystal planes are specified by the assignment of three indices. Weiss indices give the relative distances, in terms of unit cell spacings along the primitive

translational vectors, from an arbitrary origin to the intersection with the plane to be specified. Negative values are indicated by an overbar. Miller indices are the reciprocals of the Weiss indices, with fractions cleared by multiplication by an appropriate integer.

16. The distance Δ between the $(hk\ell)$ Miller planes for a crystal with 90° angles between the primitive translational vectors is given by

$$\frac{1}{\Delta^2} = \frac{h^2}{a^2} + \frac{k^2}{b^2} + \frac{\ell^2}{c^2},$$

where a, b, and c are the unit cell spacings along X, Y, and Z, respectively. The form of this equation ensures that the spacing for the $(nh\,nk\,n\ell)$ Miller planes is n^{-1} times that for the $(hk\ell)$ planes.

17. In an X-ray diffraction experiment, a monochromatic beam of coherent electromagnetic radiation whose wavelength is about 10^{-10} m (1 Å) is directed toward a solid sample whose structure is to be determined. The high-energy X rays strike the crystal and penetrate deeply into its bulk. Occasionally, a photon encounters one of the atoms of the crystal. When this happens, the photon is specularly reflected. Since the X-ray photons will strike different atoms, the electromagnetic waves in the reflected radiation are no longer coherent. Instead, the waves are now out of phase because they have traversed different distances. The identification of the crystal structure rests upon the analysis of the interference patterns that develop between these waves.

The path length difference between two waves scattered from crystal planes whose spacing is d is $\Delta X = 2d \sin \Theta$. This result leads directly to the fact that, for constructive interference between waves and an intensity maximum at the detector, we must have

$$2d \sin \Theta = n\lambda,$$

where Θ is the angle of incidence of the X rays, measured from the surface plane, λ is the wavelength of the radiation, and n is an integer. This relationship is called the Bragg equation.

18. In a powder X-ray diffraction experiment, the monochromatic, coherent X-ray beam is directed into a sample holder that contains powdered microcrystals of the material whose structure is to be identified. The large number of randomly arranged tiny crystals ensures that some of the crystals will be oriented at the value of Θ required to satisfy the Bragg equation for each set of Miller planes. Consequently, the data for all planes are obtained in a single experiment. The assignment of each observed intensity maximum to a given set of crystal planes is called indexing. For simple crystals, indexing is a straightforward process. Once it is accomplished, the crystal form can be identified and the unit cell distances determined. When a powder diffraction pattern is indexed, systematic absences due to the crystal structure are often observed. A detailed treatment demonstrates that, for a primitive lattice, there are no systematic absences. A face-centered lattice with equal scattering factors shows intensity maxima only for the Miller planes whose indices are either all even or all odd. For a body-centered crystal with equal scattering factors, we observe intensity maxima only for the Miller planes for which the sum of the Miller indices is even.

Problems

Problems that require the use of some type of computational device are marked with an asterisk (*). Problems that require some type of plotting routine are indicated with a pound sign (#).

16.1 (A) Compute the value the Z component of the spin magnetic moment of ^{19}F would have if the bare nucleus behaved as a circular conducting loop. The mass of ^{19}F is 3.1531×10^{-26} kg.

(B) Use the data in Table 16.1 to obtain the measured value for μ_z. You may express your answer in terms of m_I. What is the ratio of the results obtained in (A) and (B)?

16.2 Use the data given in Table 16.1 to compute the magnitude of the measured spin magnetic moment and magnetogyric ratio for

(A) ^{1}H,

(B) ^{13}C, and

(C) ^{11}B.

16.3 Many NMR spectrometers use fixed magnetic fields such that ^{1}H resonance occurs at a frequency of 300 MHz, 400 MHz, 500 MHz, 600 MHz, or 750 MHz. Instruments using even higher frequencies will certainly be developed in the near future.

(A) Determine the magnetic fields required to produce proton resonance at the given frequencies.

(B) Determine the frequencies at which we would observe resonance for ^{13}C and ^{19}F bare nuclei in each of the indicated spectrometers.

16.4 Table 16.1 states that the resonance frequency for a bare ^{31}P nucleus under an applied magnetic field of 1 T is 17.2510 MHz. Verify the accuracy of this statement.

16.5 An unknown nuclear particle is placed in an NMR spectrometer. When the applied field is 23,486 G, the particle is found to undergo nuclear spin transitions when the electromagnetic radiation has a frequency of 9.809 MHz. What is the nuclear particle whose spin transitions are being observed in the spectrometer?

16.6 Consider a molecule of *p*-methylbenzoic acid:

An NMR ^{1}H spectrum of this molecule is taken in a 400-MHz spectrometer.

(A) If there were no chemical shift between protons, what magnetic field would be required to satisfy the NMR resonance condition?

(B) Estimate the frequency separations between the resonance fields for the CH_3, ring, and acidic protons.

16.7 An NMR ^{13}C spectrum of *p*-methylbenzoic acid (see Problem 16.6) is taken with a 400-MHz spectrometer.

(A) If there were no chemical shift between carbon nuclei, what frequency would be required to satisfy the NMR resonance condition?

(B) Estimate the frequency differences for the CH_3, ring, and acidic ^{13}C nuclei when the chemical shift is present.

16.8 Verify the validity of Eqs. 16.41A, 16.41B, 16.41C, and 16.41D.

16.9 The propane molecule is $H_3C–CH_2–CH_3$.

(A) Qualitatively speaking, what is the nature of the spin–spin splitting of the $–CH_3$ and $–CH_2–$ protons that we would expect to observe in a ^{1}H NMR spectrum taken with a 400-MHz spectrometer.

(B) What would be the approximate ratio of the absorption intensities for each of these groups of protons? Explain.

(C) What are the approximate intensity ratios for each of the spin–spin bands within each individual proton group?

16.10 The propanal molecule is $H_3C–CH_2–CHO$.

(A) Qualitatively speaking, what is the nature of the spin–spin splitting of the $–CH_3$, $–CH_2–$, and $–CHO$ protons that we would expect to observe in a ^{1}H NMR spectrum taken with a 400-MHz spectrometer.

(B) What would be the approximate ratio of the absorption intensities for each of these groups of protons? Explain.

(C) What are the approximate intensity ratios for each of the spin–spin bands within each individual proton group?

(D) Use Table 16.4 to estimate the spacing between the NMR bands for the $–CH_2–$ protons.

16.11 Use Eq. 16.43 to obtain the sensitivity of 200-, 300-, 400-, 500-, 600-, and 750-MHz NMR spectrometers relative to that of a spectrometer operating at 100 MHz at a temperature of 298 K. How does the sensitivity vary with frequency?

16.12 Prove that in the limit of small energy spacing $h\nu$ between the nuclear spin states, we expect the sensitivity of an NMR spectrometer operating at a frequency ν relative to that of a spectrometer operating at 100 MHz to be a linear function of ν. (*Hint*: Expand the exponential in a power series.)

16.13 Prove that the commutator $[t, i\hbar\,(\partial/\partial t)] = -i\hbar$.

16.14 What is the maximum rf pulse duration that can be used in a 400-MHz FT NMR spectrometer and still have the pulse contain frequencies components for all expected ^{13}C nuclear spin transitions?

16.15* The propanoic acid molecule is $CH_3–CH_2–COOH$.

(A) Qualitatively describe the expected appearance of the ^{1}H NMR spectrum of this molecule if the spectrum is first order. Give the expected line intensities.

(B) Using Tables 16.2 and 16.4, estimate the ^{1}H chemical shifts and spin–spin coupling constants.

(C) Use intermediate values of the data estimated in (B) to compute the expected resonance frequencies for all lines in the NMR spectrum if a 60-MHz spectrometer is used.

(D) Assuming a spin relaxation time constant of 500 ms, obtain a mathematical function describing the FID that would be expected if the NMR spectrum of propanoic acid were taken on a 60-MHz NMR spectrometer using the resonant frequency for $Si(CH_3)_4$ as ν_{ref}. Plot this function for times from zero to 1,200 ms.

16.16 How could you use ^1H NMR spectroscopy to decide between the following three structures for a molecule with the molecular formula $C_{10}H_{11}Br$?

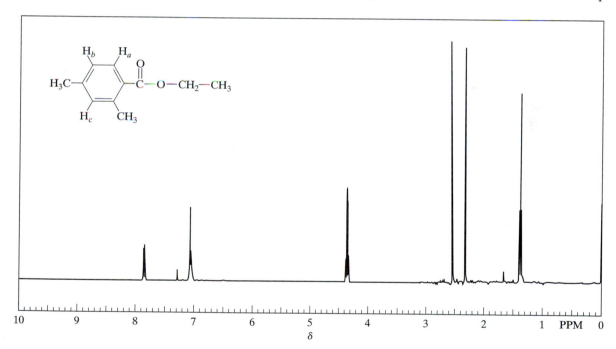

Compound A

Compound B

Compound C

16.17 An investigator needs a ^1H NMR spectrum of a biological sample. The amount of material available is so small that it is estimated that 15,000 spectral scans will be required to obtain an accurate spectrum with sufficient resolution. Estimate the amount of spectrometer time required to obtain the data if a CW spectrometer is used. Estimate the time required for an FT NMR spectrometer. Which would you use?

16.18 Qualitatively describe the appearance of the ^1H NMR spectrum expected for each of the following molecules:
(A) $CH_3–CF_3$;

(B) $CH_3–CBr_3$. Explain your reasoning.

16.19 The following figure shows the ^1H NMR spectrum of ethyl 2,4-dimethylbenzoate:
The spectrum was recorded on a 300-MHz FT NMR spectrometer with TMS as the reference. The positions of the bands are given in terms of δ (parts per million, or ppm). The conversion between δ and the frequency required for resonance is given by Eq. 16.28. (See Shankar Subramanian, *Modified Heteroarotinoids: Potential Anticancer Agents, Ph.D. dissertation,* Oklahoma State University, 1993.) The spectrum shows that the chemical shifts are much larger than the spin–spin splittings for many, but not all, of the bands. Identify the proton groups giving rise to each of the spectral lines.

16.20 The figure at the top of page 962 shows the ^1H NMR spectrum of 6-amino-2,3-dihydro-1,4-benzodioxan:
The spectrum was recorded on a 300-MHz FT NMR spectrometer with TMS as the reference. The positions of the bands are given in terms of δ (parts per million, or ppm). The conversion between δ and the frequency required for resonance is given by Eq. 16.28. (See Shankar Subramanian, *Modified Heteroarotinoids: Potential Anticancer Agents,* Ph.D. dissertation, Oklahoma State University, 1993.) Identify the proton groups giving rise to each of the spectral lines.

16.21 (A) A spin-decoupled, natural-abundance ^{13}C NMR spectrum of propanone (acetone) $[CH_3–CO–CH_3]$ is taken on a 400-MHz spectrometer. Describe qualitatively the expected appearance of the spectrum.

(B) A spin-decoupled ^{13}C NMR spectrum of ^{13}C-enriched propanone is taken on a 400-MHz spectro-

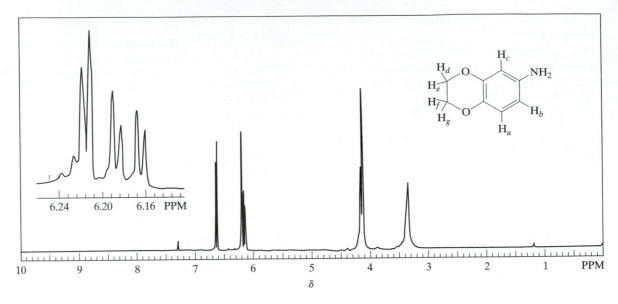

meter. Describe qualitatively the expected appearance of the spectrum. Assume that the ^{13}C enrichment is in excess of 70%.

(C) A natural-abundance ^{13}C NMR spectrum of propanone is taken on a 400-MHz spectrometer without spin decoupling. Describe qualitatively the expected appearance of the spectrum.

(D) A ^{13}C NMR spectrum of ^{13}C-enriched propanone is taken on a 400-MHz spectrometer without spin decoupling. How many absorption bands might be seen? Explain.

16.22 The figure at the bottom of this page shows a spin-decoupled ^{13}C NMR spectrum of ethyl p-toluate in $DCCl_3$ solvent:

The spectrum was recorded on a 300-MHz FT NMR spectrometer with TMS as the reference and $DCCl_3$ as the solvent. The positions of the bands are given in terms of δ, as described in Example 16.7. (See Githarangi Mahika Weerasekare, *Chiral α-Substituted Sulfoxides: Liquid Crystals and Potential Anticancer Agents*, Ph.D. dissertation, Oklahoma State University, 1994.) Identify the ^{13}C atoms giving rise to each of the spectral lines.

16.23 The figure at the top of the next page shows a spin-decoupled ^{13}C NMR spectrum of thiochroman-6-ethanol in $DCCl_3$ solvent.

The spectrum was recorded on a 300-MHz FT NMR spectrometer with TMS as the reference and $DCCl_3$ as the solvent. The positions of the bands are given in

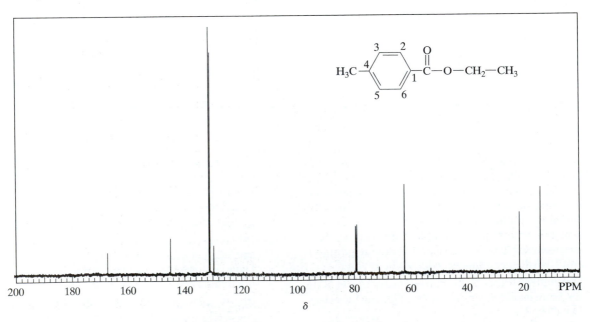

▲ Spin-decoupled ^{13}C NMR spectrum of ethyl p-toluate.

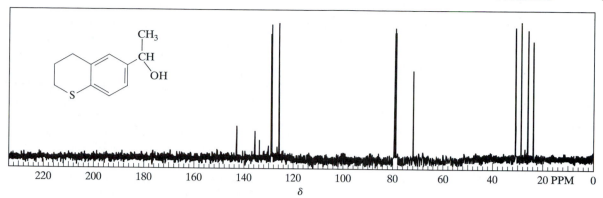

▲ Spin-decoupled ^{13}C NMR spectrum of thiochroman-6-ethanol

terms of δ, as described in Example 16.7. (See Githarangi Mahika Weerasekare, *Chiral α-Substituted Sulfoxides: Liquid Crystals and Potential Anticancer Agents*, Ph.D. dissertation, Oklahoma State University, 1994.) Identify the ^{13}C atoms giving rise to each of the spectral lines.

16.24 Consider the ^{13}C NMR spectrum of thiochroman-6-ethanol in DCCl$_3$ solvent shown at the top of the page. The spectrum has four resonance lines between $\delta = 20$ and $\delta = 30$. These bands are due to the ^{13}C spins contained in the $-CH_2-$ groups in the heterocyclic ring and the $-CH_3$ group. Suggest four experimental methods that might be used to determine which of these four bands results from the ^{13}C spin in the $-CH_3$ group. Discuss any problems that might arise in the execution of your suggestions.

16.25 Two ESR sample tubes are resting on the laboratory bench. One contains N$_2$, the other O$_2$. Explain how we might use ESR spectral measurements to determine which sample tube holds the O$_2$.

16.26 Describe the expected appearance of an ESR spectrum of the tertiary butyl radical, whose molecular structure is

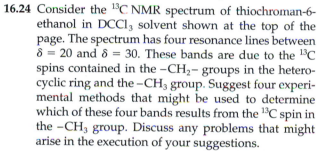

16.27[#] Consider the ethyl radical (CH$_3$–CH$_2$·). The measured hyperfine splitting constant for the ^{1}H nuclei on the $-CH_2$ group is -22.4 G. That for the ^{1}H nuclei on the $-CH_3$ group is 26.9 G.

(A) How many bands will be seen in the ESR spectrum of this radical?

(B) Assume that in the absence of hyperfine splitting, the ESR resonance line for the ethyl radical is seen at a field of G_o gauss. Determine the positions of all lines, in terms of G_o, when hyperfine splitting is present.

(C) What are the expected relative intensities of the bands? Prepare a plot of the spectrum, showing the positions and intensities of the bands.

16.28[#] Consider the naphthalene negative ion, shown above:

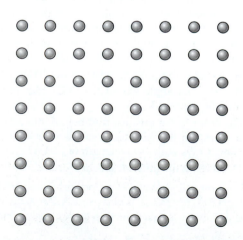

The hyperfine coupling constant for ^{1}H nuclei attached to the alpha carbons is 4.90 G. The coupling constant for hydrogen atoms attached to the beta carbons is 1.83 G.

(A) Estimate the electron spin density at the alpha and beta carbons in this ion. How does the result illustrate the approximate nature of Eq. 16.63?

(B) How many bands would be seen in an ESR spectrum of the ion?

(C) Plot the spectrum, showing the positions and intensities of all bands.

16.29 (A) Qualitatively describe the expected appearance of the ESR spectrum of a ^{14}N atom.

(B) What spacing, expressed in gauss, would be observed between the ESR bands?

16.30 Predict the number of bands that we would see in an ESR spectrum of the *n*-butyl radical (CH$_3$–CH$_2$–CH$_2$–CH$_2$·). Explain the reasoning behind your prediction.

16.31 The following diagram shows the *X–Y* plane of a cubic crystal:

Use this diagram to illustrate sets of crystal planes that are described by the following Miller indices:

(A) (330);

(B) (110);

(C) (230).

16.32 The following diagram shows two sets of crystal planes in the X–Y plane of a cubic crystal:

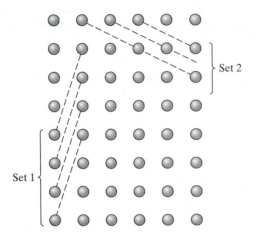

Determine the Weiss and Miller indices that describe these two sets of planes.

16.33 An orthorhombic crystal has unit cell distances $a = 5.5$ Å, $b = 6.7$ Å, and $c = 7.1$ Å. What is the spacing between the following planes?

(A) $[1, 4, \infty]$;

(B) (321).

16.34 An orthorhombic crystal has unit cell distances $a = 5.5$ Å, $b = 6.7$ Å, and $c = 7.1$ Å. The spacing between the Miller (230) planes is found to be 1.603 times that for another set of Miller planes whose indices are $(hk0)$. Determine reasonable values for h and k.

16.35 Determine the condition for complete destructive interference (complete cancellation of waves) for coherent X rays of wavelength λ scattering from planes with a separation d.

16.36 An X-ray diffraction pattern using radiation with a wavelength 1.54 Å is obtained from the (111) planes of a cubic crystal. The pattern exhibits a maximum intensity for an angle of incidence of 13.92°, measured from the surface of the crystal.

(A) What is the spacing between the (111) planes?

(B) At what angle would the (333) intensity maximum be observed?

(C) What is the greatest number of intensity maxima that can be observed from these crystal planes, using X rays with a wavelength of 1.54 Å?

(D) What is the unit cell spacing for the cubic crystal?

16.37 The unit cell spacings of an orthorhombic crystal are $a = 4.7$ Å, $b = 5.2$ Å, and $c = 6.1$ Å. An X-ray diffraction pattern is taken of this crystal, using radiation whose wavelength is 1.54 Å. Reflections from the $(hk\ell)$ Miller planes exhibits a first-order maximum at 30.53°. From what Miller planes did the reflections occur?

16.38 Show that intensity maxima for reflections from the $(n00)$ Miller planes of a face-centered cubic lattice in which the atoms have equal scattering factors will be observed only if n is an even integer.

16.39 A powder diffraction spectrum of a cubic lattice of a crystal having only one type of atom exhibits intensity maxima at 14.15°, 20.22°, 25.04°, 29.26°, 33.13°, 36.77°, 43.73°, 47.16°, 50.61°, 54.15°, 57.85°, 61.79°, and 66.13°. No other maxima are observed.

(A) If the crystal structure is known to be either primitive or face-centered cubic, which is it? Prove that your answer is correct.

(B) What is the unit cell spacing of the crystal if the wavelength of the X ray is 1.54 Å?

16.40 A powder X-ray diffraction spectrum using X rays with $\lambda = 1.54$ Å is taken for a primitive tetragonal crystal with $a = b \neq c$ and $\alpha = \beta = \gamma = 90°$. The first five intensity maxima are observed at angles of 14.382°, 17.939°, 23.308°, 30.094°, and 41.620°. Index these reflections and determine the unit cell spacings for the crystal. Prove that your answer is correct.

16.41 Powder X-ray diffraction patterns are obtained for three samples, each containing a single type of atom. The first sample contains microcrystals of a primitive cubic crystal, the second sample is an ensemble of microcrystals for a face-centered cubic crystal, and the third sample holds body-centered cubic crystals. All samples have reflections from the following Miller planes: (100), (110), (111), (200), (210), (211), (220), (221), (300), (310), (311), (222), (320), (321), (400), (420), and (422). Which of these reflections might produce intensity maxima in the diffraction patterns for each sample? What condition on the spacing between the planes must be satisfied to observe intensity maxima.

16.42 It looks like Sam has another question.

"What is it, Sam?"

"Well, sir, your description of the appearance of the free-induction decay [FID] observed for an FT NMR spectrum such as the one you show in Figure 16.25 seems incomplete to me."

"Why is that, Sam?"

"Well, maybe I'm not looking at this correctly, but it seems to me that it is possible to have an NMR spectrum in which the FID does not oscillate at all."

"You think that Figure 16.25 is wrong?"

"No, sir. I just think that it doesn't have to be characteristic of all possible FIDs that we might observe. In particular, I think we might see a simple exponential decay with no oscillations at all."

Do you agree with Sam? If not, what is wrong with his reasoning? If you do agree, when might we observe an FID that is a simple exponential decay with no oscillations? Be certain to justify your answer.

Statistical Mechanics

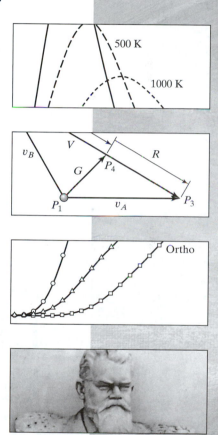

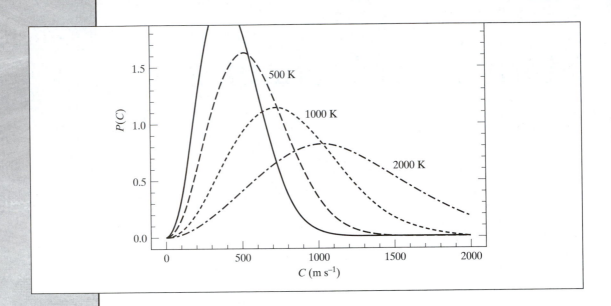

CHAPTER 17

Molecular-Energy Distributions: Kinetic Theory of Gases

If the molecular-energy states and the distribution of molecules among these energy states are known, all the thermodynamic quantities discussed in Chapters 2–9 can be computed. The energies of the available eigenstates must be obtained by solving the molecular Schrödinger equation. The distribution of molecules among these states is obtained from statistical mechanics. The combination of the two produces thermodynamics.

We have already seen several examples of the use of statistical mechanics that were presented without derivation. In Chapter 7, we employed the average vibrational energy of an ensemble of harmonic oscillators to obtain the Einstein and Debye heat capacity expressions for solids. The Boltzmann distribution was used to obtain the Debye–Hückel expression for ionic activity coefficients in Chapter 9. It was also employed to compute the intensities of spectral absorption bands in Chapters 15 and 16. In this chapter and in Chapter 18, those results will be derived from first principles.

Once the fundamental equations have been obtained, we shall use them to examine the translational motion of gases. In particular, expressions for the distribution of molecular speeds, the collision frequency, and the mean free path will be derived. These expressions will be of central importance in Chapters 19 and 20, in which we examine chemical reaction kinetics and dynamics.

17.1 The Boltzmann Distribution

The Boltzmann distribution lies at the heart of statistical thermodynamics, statistical theories of chemical reaction rates, band intensities, the Debye–Hückel theory of ionic activity coefficients, solid-state heat capacities, and many other phenomena that we will not cover in this text. Throughout the first 16 chapters, we have repeatedly employed the distribution without derivation or explanation. That omission is corrected in this section.

17.1.1 The Basic Problem

In Chapters 12–14, we found that the translational, rotational, vibrational, and electronic energy states of atoms and molecules are quantized. The solution of the Schrödinger equation provides a listing of the possible eigenfunctions and energy eigenvalues a molecule may have. In effect, this listing informs us of the possible energy states each molecule may occupy. It does not, however, tell us in which of these possible states a particular molecule may reside at any given moment of time, nor does it tell us the overall distribution of molecules among the various energy states. In general, we found that the translational energy states are very closely spaced and independent of the quantum numbers defining the internal rotational, vibrational, and electronic energy states. We also determined that the internal energy states are all coupled: The rotational states depend upon the nature of the vibrational eigenfunctions, which, in turn, depend upon the electronic states. Figure 15.28 illustrates this dependence. When the vibrational and rotational energies are small, we can usually assume them to be independent without incurring significant error, but this approximation must be made with care.

The available molecular-energy states are illustrated in Figure 17.1. (The actual energy spacings are not reproduced to scale.) If we have one mole of a compound, each of the 6.022×10^{23} molecules will reside in a particular translational state and a particular rotational–vibrational–electronic energy state. If we knew how many molecules were in each possible energy state, we could compute the molar internal energy of the system, U, simply by summing the products of the energy of each quantum state and the number of molecules in that quantum state. Since the heat capacity is related to U by $C_v = (\partial U / \partial T)_V$, we could obtain C_v by computing U at different temperatures at a fixed volume and numerically extracting the derivative. This type of

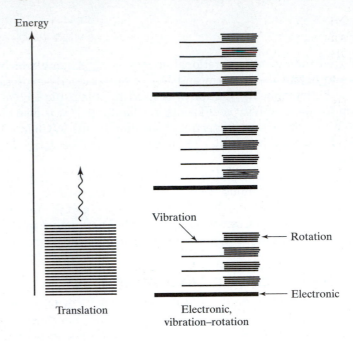

▲ FIGURE 17.1

Manifold of molecular-energy states over which molecules are distributed. The relative spacings of the energy states are not drawn to scale.

analysis demonstrates that we are on the verge of being able to combine quantum mechanics with statistical mechanics to produce thermodynamics.

To compute thermodynamic quantities, we must first solve the Schrödinger equation with sufficient accuracy that the energies of the various states shown in Figure 17.1 are known. We must then determine how many molecules reside in each energy state when the system attains equilibrium. It may be that we will need to consider the possibility that a different number of molecules may be in each particular energy state and then weigh each possible number by the appropriate probability.

Let us state the second problem in more general terms. We seek to find the probability that nature will distribute N objects (molecules) into s different aspects (energy states) in a particular manner that we denote as distribution d_i, subject to the constraints that both mass and total energy be conserved. The discussion in Chapter 10 shows that the probability of obtaining distribution d_i is

$$P(d_i) = \frac{\text{(number of ways to obtain distribution } d_i)}{\text{total number of ways to obtain all possible arrangements}}. \quad \textbf{(17.1)}$$

For example, distribution d_i might represent the case where one-third of the molecules are in the ground state, one-third in the first excited translational state and the ground internal states, and the remaining one-third in the second excited translational state and the ground rotational, vibrational, and electronic states. Equation 17.1 tells us that the probability that we will observe this distribution is equal to a number of different ways it can be generated, divided by the total number of ways to distribute the 6.022×10^{23} molecules among all the energy states. Our basic problem is to obtain appropriate expressions for the numerator and denominator of that equation.

17.1.2   The Thermodynamic Probability

In most cases, there is a near infinitude of different aspects (energy states) over which we can distribute the molecules. In addition, the number of objects (molecules) to be distributed for one mole is extremely large. These two factors make the general solution to our problem difficult to envision. It is much easier if we first examine less complex systems. We can then gradually increase the complexity of the system until the general solution becomes evident.

Let us consider the case where the number of objects, N, is three and the number of aspects, s, is two. A specific example of such a system is the random toss of three coins. Each coin can exhibit two aspects upon landing: a head (H) or a tail (T). This system is so simple that we have the luxury of actually writing down all possible distributions. (See Table 17.1.) Notice that there are eight possible arrangements, which is exactly 2^3, or s^N. The number of ways to obtain three heads is precisely one, which is the distribution count for that result shown in the table. Using Eq. 17.1, we can find the probability of obtaining three heads (H^3) on the random toss of three coins:

Table 17.1   Possible arrangements and distributions for the random toss of three coins

Arrangement				
Coin 1	Coin 2	Coin 3	Distribution	Distribution Count
H	H	H	3 Heads	1
H	H	T	2 Heads, 1 Tail	
H	T	H	2 Heads, 1 Tail	3
T	H	H	2 Heads, 1 Tail	
H	T	T	2 Tails, 1 Head	
T	H	T	2 Tails, 1 Head	3
T	T	H	2 Tails, 1 Head	
T	T	T	3 Tails	1

$$P(H^3) = \frac{\text{number of ways to obtain three heads}}{\text{total number of possible arrangements}} = \frac{1}{8}.$$

The corresponding result for two heads and one tail (H^2T) or two tails and one head (HT^2) is $\frac{3}{8}$, while the probability that three tails will appear is $\frac{1}{8}$.

The important part of this example is to note that the distribution counts are simply the expansion coefficients of $(H + T)^3$. That is,

$$(H + T)^3 = 1H^3 + 3H^2T + 3HT^2 + 1T^3,$$

and the coefficients of the four terms on the right are exactly the corresponding distribution counts in Table 17.1. The reason for this may be seen by recalling our beginning algebra course. When we were asked to expand $(H + T)^3$, we first computed $(H + T)^2$ and obtained

$$(H + T)^2 = HH + HT + TH + TT.$$

We then multiplied this result by $H + T$ to produce

$$(H + T)(H + T)^2 = (H + T)(HH + HT + TH + TT)$$

$$= HHH + HHT + HTH + HTT + THH + THT + TTH + TTT.$$

If we left the result in this form, our algebra teacher awarded us only partial credit because we did not "collect terms," which is a mathematician's way of saying that we need to execute the distribution count and thereby determine how many of the arrangements have three heads, how many have two heads and a tail, how many two tails and a head, and how many three tails. The results are precisely the distribution counts given in the table.

The foregoing analysis tells us that the number of ways a given distribution can be achieved when there are two possible aspects is the same as the binomial expansion coefficient for that distribution, which, for three coins, can be written in the form

$$C(3, n_1, n_2) = \frac{3!}{n_1! n_2!}, \tag{17.2}$$

where n_1 and n_2 are the number of coins that show heads and tails, respectively. The accuracy of Eq. 17.2 can be easily verified by computing the four possible expansion coefficients $C(3, 3, 0)$, $C(3, 2, 1)$, $C(3, 1, 2)$, and $C(3, 0, 3)$.

Now suppose we execute the random toss of N coins where N is unlimited. When N is large, we can no longer write down all the possible distributions and thereby obtain the distribution counts. However, we have no need to do so: We already know that the distribution counts are simply the binomial expansion coefficients for the expression $(H + T)^N$. These coefficients are

$$\boxed{C(N, n_1, n_2) = \frac{N!}{n_1! n_2!}.} \tag{17.3}$$

We also know that the total number of possible arrangements is 2^N, since each coin can have two possible arrangements and there are N coins. Equations 17.1 and 17.3 permit us to compute the probability of all possible distributions. Example 17.1 illustrates the method.

EXAMPLE 17.1

If we randomly toss 10 coins onto a table, what is the probability that we will obtain exactly five heads and five tails?

Solution

The answer is provided by Eq. 17.1:

$$P(H^5 T^5) = \frac{\text{number of ways to obtain five heads and five tails}}{\text{total number of possible arrangements}}. \tag{1}$$

The total number of possible arrangements is $2^{10} = 1{,}024$. The number of ways to obtain five heads and five tails is given by Eq. 17.3:

$$\text{Number of ways to obtain five heads and five tails} = C(10, 5, 5) = \frac{10!}{5!5!}$$

$$= \frac{(10)(9)(8)(7)(6)}{5!} = \frac{30{,}240}{120} = 252 \text{ ways.} \qquad (2)$$

Therefore, the result we seek is

$$P(H^5T^5) = \frac{252}{1{,}024} = 0.24609\ldots. \qquad (3)$$

For related exercises, see Problems 17.1, 17.2, 17.3, and 17.9.

Instead of using coins as our working example, we could have equally well used N nuclear spins, each with two aspects, say, α and β spin functions. In Chapter 16, we frequently used the result that the number ways to obtain $n_1 \alpha$ spins and $n_2 \beta$ spins out of a total of N spins is $N!/(n_1!n_2!)$. Equation 17.3 and the preceding discussion make it clear why this result obtains.

Let us now increase the complexity of our example by considering a case in which there are more than two aspects. A simple example is the dice used in many board games. Each die has six faces. One face has a single dot, a second face two dots, and so on, with the last face containing six dots. We can, therefore, observe any number from 1 to 6 on the up side of a die on a single throw. Suppose we throw two dice and ask what are the possible arrangements, distributions, and distribution counts for this situation wherein $N = 2$ and $s = 6$. All these are listed in Table 17.2, with f_i representing the situation in which i dots are showing on the up face of the die. Inspection of the results shows that we have 21 possible distributions. The distribution counts for the distributions that have the form f_i^2 are all one, while those for the distributions $f_if_j(i \neq j)$ are two. When we turn our attention to molecules distributed over energy states, such distributions are called *microstates*. Since each die can exhibit six possible faces, the total number of arrangements is $6^2 = 36$, which again has the form s^N.

It is easy to see that the distribution counts are once more just the expansion coefficients for the hexanomial $(f_1 + f_2 + f_3 + f_4 + f_5 + f_6)^2$. These hexanomial expansion coefficients have a form analogous to Eq. 17.2; that is,

$$C(2, n_1, n_2, n_3, n_4, n_5, n_6) = \frac{2!}{n_1!n_2!n_3!n_4!n_5!n_6!} \qquad (17.4)$$

where n_i is the number of dice that show i dots on the up face. For the case of two dice, we will have $n_1 + n_2 + n_3 + n_4 + n_5 + n_6 = 2$. If we increase the complexity of the example by throwing N dice, the distribution counts are still the hexanomial expansion coefficients for $(f_1 + f_2 + f_3 + f_4 + f_5 + f_6)^N$, which are

$$C(N, n_1, n_2, n_3, n_4, n_5, n_6) = \frac{N!}{n_1!n_2!n_3!n_4!n_5!n_6!}. \qquad (17.5)$$

The total number of arrangements of the N dice is $s^N = 6^N$, where $N = n_1 + n_2 + n_3 + n_4 + n_5 + n_6$.

Table 17.2 Possible arrangements and distributions for the random toss of two dice

Arrangements		Distribution	Distribution Count	Sum of dice
Die 1	Die 2	(microstate)		
f_1	f_1	Two ones	1	2
f_1	f_2	One 1, One 2		3
f_2	f_1	One 1, One 2	2	3
f_1	f_3	One 1, One 3		4
f_3	f_1	One 1, One 3	2	4
f_1	f_4	One 1, One 4		5
f_4	f_1	One 1, One 4	2	5
f_1	f_5	One 1, One 5		6
f_5	f_1	One 1, One 5	2	6
f_1	f_6	One 1, One 6		7
f_6	f_1	One 1, One 6	2	7
f_2	f_2	Two twos	1	4
f_2	f_3	One 2, One 3		5
f_3	f_2	One 2, One 3	2	5
f_2	f_4	One 2, One 4		6
f_4	f_2	One 2, One 4	2	6
f_2	f_5	One 2, One 5		7
f_5	f_2	One 2, One 5	2	7
f_2	f_6	One 2, One 6		8
f_6	f_2	One 2, One 6	2	8
f_3	f_3	Two threes	1	6
f_3	f_4	One 3, One 4		7
f_4	f_3	One 3, One 4	2	7
f_3	f_5	One 3, One 5		8
f_5	f_3	One 3, One 5	2	8
f_3	f_6	One 3, One 6		9
f_6	f_3	One 3, One 6	2	9
f_4	f_4	Two fours	1	8
f_4	f_5	One 4, One 5		9
f_5	f_4	One 4, One 5	2	9
f_4	f_6	One 4, One 6		10
f_6	f_4	One 4, One 6	2	10
f_5	f_5	Two fives	1	10
f_5	f_6	One 5, One 6		11
f_6	f_5	One 5, One 6	2	11
f_6	f_6	Two sixes	1	12

EXAMPLE 17.2

Six dice are randomly thrown. What is the probability that the sum of the up faces of the dice will add to eight?

Solution

There are only two distributions that give a total of eight: $f_1^5 f_3$ and $f_1^4 f_2^2$. The number of ways to obtain five ones and one three is

$$C(6, 5, 0, 1, 0, 0, 0) = \frac{6!}{5!0!1!0!0!0!} = 6. \tag{1}$$

The number of ways to obtain four ones and two twos is

$$C(6, 4, 2, 0, 0, 0, 0) = \frac{6!}{4!2!0!0!0!0!} = 15. \tag{2}$$

There are, therefore, $6 + 15$ ways to obtain a total of eight dots on the up faces of the six dice. The total number of arrangements is

$$\text{Total arrangements} = s^N = 6^6 = 46{,}656. \tag{3}$$

Consequently, the probability we seek is

$$\text{Probability of a total of eight} = \frac{\text{number of ways to get a total of eight}}{\text{total arrangements}} = \frac{6 + 15}{46{,}656}$$

$$= 0.0004501 \dots . \tag{4}$$

For a related exercise, see Problem 17.9.

We can now generalize the preceding development. If each object (molecule) can exhibit s different aspects (can be in s different energy states), then

$$\text{Total arrangements} = s^N. \tag{17.6}$$

If N objects are distributed among these s aspects, the number of ways to obtain a particular distribution will be the s-nomial expansion coefficient for that particular distribution in the expression $(f_1 + f_2 + f_3 + \cdots + f_{s-1} + f_s)^N$. This expansion coefficient has the same form as that given in Eqs. 17.3 and 17.5, viz.,

$$C(N, n_1, n_2, n_3, \dots, n_{s-1}, n_s) = \frac{N!}{n_1! n_2! n_3! \dots n_{s-1}! n_s!},$$

where the n_i ($i = 1, 2, 3, \dots, s$) are the number of objects exhibiting the ith aspect. This equation can be written in more compact form by using product notation in the denominator:

$$\boxed{C(N, n_1, n_2, n_3, \dots, n_{s-1}, n_s) = \frac{N!}{\displaystyle\prod_{i=1}^{s} n_i!} = W}. \tag{17.7}$$

Equation 17.7, which gives the number of ways a particular distribution can be obtained, is called the *thermodynamic probability* W. In these terms, the probability of observing any given distribution d_i is

$$\boxed{P(d_i) = \frac{W(d_i)}{s^N}}. \tag{17.8}$$

In Eq. 17.8, $W(d_i)$ represents the thermodynamic probability for distribution d_i.

17.1.3   Statistical Evaluation of Average Properties of a System

To understand the problem we face in attempting to obtain thermodynamic quantities using statistical methods, let us address a much easier problem. Suppose we wish to compute the average number of spots on the up faces when two dice are thrown at random. The principles deduced in Chapter 10 tell us that this average value is given by

$$\langle X \rangle = \sum_{\alpha=1}^{21} P_\alpha X_\alpha,$$

where X_α is the sum of the up faces for microstate or distribution α, whose probability is P_α. The upper limit of the summation is 21, since we can see from Table 17.2 and Figure 17.2 that there are 21 different distributions. Using the sums given in the final column of the table and the fact that $P_\alpha =$ distribution count/36, we can compute the average sum as follows:

$$\langle X \rangle = \frac{1}{36}(2) + \frac{2}{36}(3) + \frac{2}{36}(4) + \frac{2}{36}(5) + \frac{2}{36}(6) + \frac{2}{36}(7) + \frac{1}{36}(4) + \frac{2}{36}(5)$$

$$+ \frac{2}{36}(6) + \frac{2}{36}(7) + \frac{2}{36}(8) + \frac{1}{36}(6) + \frac{2}{36}(7) + \frac{2}{36}(8) + \frac{2}{36}(9) + \frac{1}{36}(8)$$

$$+ \frac{2}{36}(9) + \frac{2}{36}(10) + \frac{1}{36}(10) + \frac{2}{36}(11) + \frac{1}{36}(12) = 7.$$

This procedure can be applied to the computation of the average value of *any* property of a system. For example, if we are dealing with N molecules distributed over all possible quantum states, the average thermodynamic energy of the system is

$$U = \langle E \rangle = \sum_{\alpha=1}^{n^D} P_\alpha E_\alpha, \tag{17.9}$$

where E_α is the energy of the αth distribution or microstate and P_α is the probability that such a distribution will be observed. The upper limit of the summation is the total number of distributions, which we denote by n^D. Throughout the remainder of this text, we will adopt the convention of using a lowercase Greek letter as the summation index when the sum is taken over the microstates of a system. Equation 17.9 is the basic equation underlying statistical thermodynamics.

When we have relatively few distributions, the summation in Eq. 17.9 is easy to compute in a straightforward fashion. However, when 6.022×10^{23} molecules are distributed over a large number of energy states, the number of microstates is so large that it is effectively impossible to individually compute the energy for each one and then calculate the sum. To achieve this objective, we are going to have to find a more efficient procedure than the brute-force method we used to compute the average sum for the throw of two dice. A discussion of the general approach taken in statistical thermodynamics will be deferred until Chapter 18.

There is one special case where the summation contained in Eq. 17.9 can be executed with ease: *If the number of objects is very large compared with the number of aspects, the summation effectively reduces to a single term.* To see why this is so, let us return to our first example involving the distribution

Microstates with distribution counts of 1

| Die #1 = Die #2 = 1 |
| Sum = 2 |

| Die #1 = Die #2 = 2 |
| Sum = 4 |

| Die #1 = Die #2 = 3 |
| Sum = 6 |

| Die #1 = Die #2 = 4 |
| Sum = 8 |

| Die #1 = Die #2 = 5 |
| Sum = 10 |

| Die #1 = Die #2 = 6 |
| Sum = 12 |

Microstates with distribution counts of 2

Die #1	Die #2
1	2
2	1
Sum = 3	

Die #1	Die #2
1	3
3	1
Sum = 4	

Die #1	Die #2
1	4
4	1
Sum = 5	

Die #1	Die #2
1	5
5	1
Sum = 6	

Die #1	Die #2
1	6
6	1
Sum = 7	

Die #1	Die #2
2	3
3	2
Sum = 5	

Die #1	Die #2
2	4
4	2
Sum = 6	

Die #1	Die #2
2	5
5	2
Sum = 7	

Die #1	Die #2
2	6
6	2
Sum = 8	

Die #1	Die #2
3	4
4	3
Sum = 7	

Die #1	Die #2
3	5
5	3
Sum = 8	

Die #1	Die #2
3	6
6	3
Sum = 9	

Die #1	Die #2
4	5
5	4
Sum = 9	

Die #1	Die #2
4	6
6	4
Sum = 10	

Die #1	Die #2
5	6
6	5
Sum = 11	

◀ FIGURE 17.2
The 21 microstates for the random toss of two six-sided dice. There are 6 microstates with distribution counts of 1 and 15 with distribution counts of 2, as shown in the figure. The average of any quantity associated with the random toss of two such dice is the average over these 21 microstates, as described by Eq. 17.9.

of N coins between two aspects, heads and tails. The probability that we will observe m heads and $N - m$ tails is

$$P(H^m T^{N-m}) = \frac{N!}{m!\,(N - m)!\,2^N}.$$

Now, the most probable distribution of six coins is the one with an equal number of heads and tails. (See Problem 17.2.) In Example 17.3, we shall prove that this is always true, regardless of the value of N. Therefore, if we have 200 coins, the most probable distribution will be the one with 100 heads and 100 tails. The probability that this distribution will be observed is

$$P(H^{100} T^{100}) = \frac{200!}{100!\,100!\,2^{200}}.$$

The central point we wish to investigate involves the relative probabilities of distributions other than the most probable one. For example, we might inquire about the probability that a 1% deviation from the most probable distribution for the system of 200 coins will be observed. With 100 heads and 100 tails, a 1% deviation corresponds to changing one tail to a head or vice versa. This change gives us 101 heads and 99 tails. The corresponding probability is

$$P(H^{101} T^{99}) = \frac{200!}{101!\,99!\,2^{200}}.$$

The ratio of these probabilities is, therefore,

$$\frac{P(H^{101}T^{99})}{P(H^{100}T^{100})} = \frac{\dfrac{200!}{101!\ 99!\ 2^{200}}}{\dfrac{200!}{100!\ 100!\ 2^{200}}} = \frac{100!\ 100!}{101!\ 99!} = \frac{100}{101} = 0.99009901\ldots.$$

Consequently, 1% deviations from the most probable distribution with 200 coins are very likely, and many microstates will contribute to the summation in Eq. 17.9.

Now suppose we have 2,000 coins. A 1% deviation from the most probable distribution corresponds to the displacement of 10 coins from tails to heads or vice versa. The probability ratio for the 1% deviation distribution to the most probable distribution is now

$$\frac{P(H^{1,010}T^{990})}{P(H^{1,000}T^{1,000})} = \frac{\dfrac{2,000!}{1,010!\ 990!\ 2^{2,000}}}{\dfrac{2,000!}{1,000!\ 1,000!\ 2^{2,000}}} = \frac{1,000!\ 1,000!}{1,010!\ 990!} = 0.90488114\ldots.$$

The relative probability of observing a 1% deviation from the most probable distribution is still large, but it is significantly less than for 200 coins. When we have 20,000 coins, the result is

$$\frac{P(H^{10,100}T^{9,900})}{P(H^{10,000}T^{10,000})} = 0.3678917\ldots.$$

If 200,000 coins are present, the probability ratio decreases to 4.5395×10^{-5}, and with 2×10^6 coins, the ratio of the probability of a 1% deviation to that of the most probable distribution is 3.7141×10^{-44}. When we reach 2×10^7 coins, the ratio is an infinitesimal 5.99×10^{-435}.

The same type of result is obtained when we consider a deviation of 0.1% from the most probable distribution. When 2×10^9 coins are present, the ratio of the probability of a 0.1% deviation to that of the most probable distribution is again about 6×10^{-435}. Even when we consider very small deviations of only 0.0001% from the most probable distribution, the probability ratio is on the order of 10^{-435} when 2×10^{15} coins are present. Figure 17.3 illustrates how rapidly the ratio drops for 1% and 0.1% deviations. Problems 17.4 and 17.5 explore these points in more detail.

The conclusion we reach from the foregoing calculations is that as N becomes large with $N \gg s$, the relative probability of observing small deviations from the most probable distribution becomes extremely small. When N reaches 6.022×10^{23}, this ratio is essentially zero if the number of aspects remains small. (See Problem 17.5.) *Therefore, for large N and small s, the only distribution that must be considered in the computation of average values is the most probable distribution.*

In the remainder of this section, we shall examine the special case of large N and small s. Although this special case does not describe the situation in which we are distributing Avogadro's number of molecules among the available quantum states, it does lead to some interesting and useful results that will be valuable when we turn our attention to statistical thermodynamics in Chapter 18.

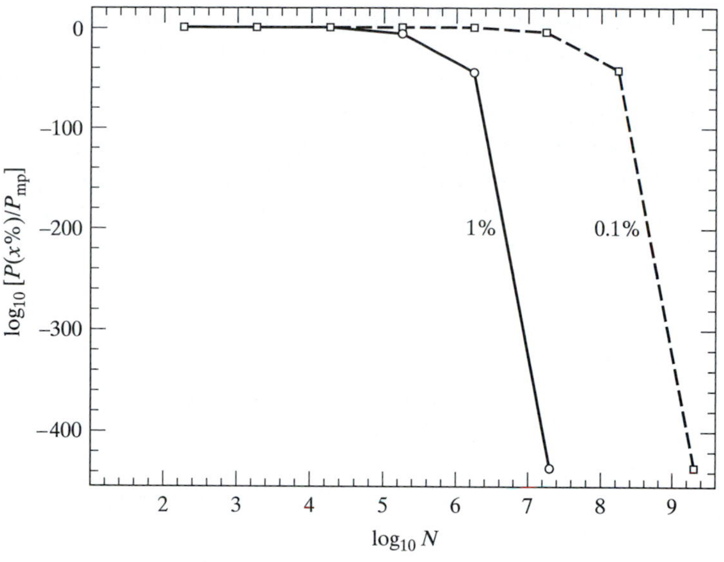

◀ FIGURE 17.3
Ratio of the probability of an $x\%$ deviation from the probability of the most probable distribution, P_{mp}, to that for the most probable distribution when N objects are distributed among s aspects as a function of N for the case $N \gg s$. The figure shows the results for $x = 1\%$ and 0.1%.

It is often stated that "statistical theory breaks down when the number of molecules becomes small." Such a statement is somewhat misleading. Statistical theory can be applied regardless of the number of molecules present. Our statistical treatment of three coins and two dice are examples. However, equations developed under the assumption that only the most probable distribution needs to be considered will break down when N becomes small or s becomes much larger than N.

17.1.4   Mathematical Methods

The analysis presented in the previous section demonstrates that when we are dealing with a system for which N is large and s is small, the most probable distribution determines the properties of the system. Since s and N are fixed for a given system, Eq. 17.8 shows that the most probable distribution is the one that maximizes W, subject to the constraint that both mass and energy be conserved. We are, therefore, faced with the pure mathematical problem of finding the absolute maximum of a function dependent upon s variables with two constraining conditions.

To understand the importance of the constraints on the maximization, let us examine a much easier problem. Consider a function of x and y defined by

$$Z = f(x, y) = xy. \tag{17.10}$$

Suppose we were to ask, What is the maximum value of Z? In this form, the question has little significance, since we obviously obtain a maximum value of Z if we set $x = y = \infty$. The same situation exists for the thermodynamic probability. Using Eq. 17.8, we have

$$W = \frac{N!}{\displaystyle\prod_{i=1}^{s} n_i!},$$

so that W has a maximum value of infinity if we simply set $N = \infty$. However, in the actual physical situation, we are not free to set N to any value that strikes our fancy; N must equal the total number of molecules in the system. That is to say, mass must be conserved in the adjustment of the values of the n_i. Maximization problems of this type without constraints on the variables have no physical significance.

Let us now alter the problem to include a simple constraint. Suppose we inquire what the maximum value of Z is, subject to the constraint that the sum of x and y be fixed at the value 50. This constraint can be expressed mathematically by writing

$$x + y = 50. \tag{17.11}$$

With this constraint, Z becomes

$$Z = xy = x(50 - x) = 50x - x^2$$

after substituting the constraint for y. The maximum value of Z can now be obtained by the usual procedure of requiring the derivative to vanish at any extremum. This method gives

$$\frac{dZ}{dx} = 50 - 2x = 0,$$

so that the value of x which produces an extremum in Z is $x = 25$. Since d^2Z/dx^2 is negative, (-2), we know that the extremum is a maximum. Equation 17.11 then requires that we have $y = 50 - x = 25$. This leads to $Z_{max} = (25)^2 = 625$.

Notice how imposing a constraint on the variables makes the maximization problem meaningful. With the constraint, Z is maximized for one particular set of values of the variables, namely, $x = y = 25$. Without the constraint, the maximization problem has no significance. The same is true for the thermodynamic probability.

When the function to be maximized or minimized is simple, such as that in the example just presented, the usual substitution method works well. However, if we are attempting to find an extremum for a complex function of many variables with several constraints, it is often useful to employ an alternative procedure called the *method of Lagrangian multipliers*. The principle underlying this procedure may be seen by applying the technique to the preceding problem in which we wish to maximize Z defined by Eq. 17.10, subject to the constraint expressed by Eq. 17.11.

To apply the technique, we first multiply the constraint condition by a constant α, which is called a *Lagrangian multiplier*. This operation gives

$$50\alpha = \alpha(x + y). \tag{17.12}$$

Next, we add Eqs. 17.10 and 17.12 to obtain a function $F(x, y; \alpha)$ that contains both the constraint and the function to be maximized:

$$F(x, y; \alpha) = Z + 50\alpha = xy + \alpha(x + y). \tag{17.13}$$

At any maximum or minimum, the differential change in Z must be zero. That is, at an extremum, we have

$$dF(x, y; \alpha) = d[Z + 50\alpha] = dZ = 0, \tag{17.14}$$

since 50α is a constant whose differential is zero. Substituting Eq. 17.13 for $F(x, y; \alpha)$ in Eq. 17.14 yields

$$d[xy + \alpha(x + y)] = x\,dy + y\,dx + \alpha\,dx + \alpha\,dy$$

$$= (x + \alpha)dy + (y + \alpha)dx = 0. \qquad \textbf{(17.15)}$$

Equation 17.15 contains three variables: dx, dy, and α. There is one constraint. Consequently, we may take any two of the variables to be independent and allow the constraint to be satisfied by the third variable. If dx and dy are taken as the independent variables, Eq. 17.15 requires that the coefficients of both of those quantities vanish, so that the left-hand side will always add to zero regardless of the values assigned to dx and dy. This gives

$$y + \alpha = x + \alpha = 0,$$

which requires that we have $y = x = -\alpha$. Substituting this result into Eq. 17.11 yields

$$-\alpha - \alpha = -2\alpha = 50,$$

so that

$$-\alpha = x = y = 25.$$

This is the same result as that obtained with the substitution method normally employed in beginning calculus courses.

The general method of Lagrangian multipliers may be summarized as follows: If we wish to find an extremum of a function Z that depends upon n variables subject to K constraints, then we multiply each constraint equation on both sides by a Lagrangian multiplier. All equations are then added or subtracted to produce a function F that contains $n + K$ variables consisting of the n original variables and the K Lagrangian multipliers. At an extremum, we must have $dF = 0$. This requirement yields an equation involving the differentials of the n original variables and the K Lagrangian multipliers. Since there are K constraints, n of these variables may be regarded as being independent. These are chosen to be the differentials of the original variables, which then permits the coefficients of those variables to be equated to zero. The solutions of these equations are substituted into the K constraint equations, and this set of equations is solved for the Lagrangian multipliers.

One problem remains that must be addressed before we can apply the method of Lagrangian multipliers to the maximization of the thermodynamic probability: Because W contains factorials that cannot be differentiated, we will be unable to take the differential of W unless a suitable approximation for these expressions can be found. It turns out that it is easier to handle the problem of differentiating $\ln(n!)$ than differentiating $n!$ itself. Therefore, we focus our attention on $\ln(n!)$.

Since $n! = (1)(2)(3)(4)\dots(n - 1)(n)$, we can write $\ln(n!)$ in the form

$$\ln(n!) = \ln(1) + \ln(2) + \ln(3) + \dots + \ln(n - 1) + \ln(n) = \sum_{k=1}^{n} \ln(k).$$

$$\textbf{(17.16)}$$

Equation 17.16 can also be expressed as

$$\ln(n!) = \sum_{k=1}^{n} \ln(k)\,\Delta k \qquad (17.17)$$

if we take $\Delta k = 1$. In this form, it is clear that $\ln(n!)$ is the area under the histogram shown in Figure 17.4.

If Δk in Eq. 17.17 were permitted to take on continuous values that could approach zero, we could obtain an alternative expression for $\ln(n!)$. The fundamental theorem of integral calculus tells us that

$$\lim_{\Delta k \longrightarrow 0} \sum_{k=1}^{n} \ln(k)\,\Delta k = \int_{k=1}^{n} \ln(k)\,dk. \qquad (17.18)$$

The integral on the right-hand side of Eq. 17.18 is the area under a plot of $\ln(k)$ versus k from $k = 1$ to $k = n$. Figure 17.4 shows this area for the case where $n = 20$. An inspection of this plot for the case where $n = 2$ makes it clear that the area under the continuous curve (shown by the darker shaded area alone) is always less than the area beneath the histogram (shown light shaded). Therefore,

$$\sum_{k=1}^{n} \ln(k)\,\Delta k > \int_{k=1}^{n} \ln(k)\,dk \qquad (17.19)$$

for all n. The figure also shows that as n becomes large, the discrete sum becomes a very tight upper bound on the area beneath the continuous curve. Therefore, for large n,

$$\sum_{k=1}^{n} \ln(k)\,\Delta k \approx \int_{k=1}^{n} \ln(k)\,dk. \qquad (17.20)$$

▶ FIGURE 17.4
Stirling's approximation. In this approximation, the area beneath the histogram, which is given by $\sum_{k=1}^{n}\ln(k)\,\Delta k$ with $\Delta k = 1$, is replaced by the area beneath $\int_{k=1}^{n}\ln(k)\,dk$, which is the area beneath the continuous curve. When $n = 2$, the area under the histogram is given by the light shaded area in the figure. The area beneath the continuous curve between $k = 1$ and $k = 2$ is shown by the darker shaded area alone. A comparison of these areas shows that $\sum_{k=1}^{n}\ln(k)\,\Delta k > \int_{k=1}^{n}\ln(k)\,dk$. The figure shows that these areas become nearly equal for large n.

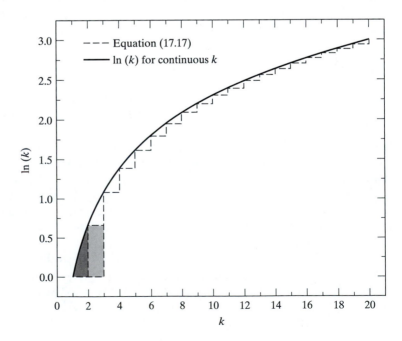

We can put Eq. 17.20 in a more convenient form by using Eq. 17.17 and performing integration by parts. This produces

$$\int_{k=1}^{n} \ln(k)\,dk = [k \ln(k) - k]_{1}^{n} = n \ln(n) - n + 1.$$

For large n, this equation becomes

$$\boxed{\ln(n!) \approx n \ln(n) - n + 1 \approx n \ln(n) - n}, \qquad \textbf{(17.21)}$$

since, for large n, $n + 1 \approx n$. Equation 17.21 is known as *Stirling's approximation*. Table 17.3 illustrates the accuracy of the approximation. When n becomes as large as 6.022×10^{23}, the percent error incurred using Stirling's approximation will be essentially zero.

Table 17.3	The accuracy of Stirling's approximation		
n	$\ln(n!)$	$n \ln(n) - n$	**% Error**
10	15.1044	13.0259	-13.76
20	42.3356	39.9146	-5.72
30	74.6582	72.0359	-3.15
40	110.3206	107.5552	-2.51
50	148.4778	145.6012	-1.94
60	188.6282	185.6607	-1.57

In the next section, we shall use these methods to obtain the distribution of molecules across the molecular-energy states that produces the maximum value of the thermodynamic probability.

EXAMPLE 17.3

A large number of marbles, N, is to be distributed among s different boxes. Find the distribution of marbles among the boxes that maximizes the thermodynamic probability. Note that if $s = 2$, this is the same problem as the distribution of N coins between heads and tails.

Solution

Since $\ln W$ will be a maximum whenever W is maximum, we can maximize W by maximizing $\ln W$. This procedure permits us to avoid the problem presented by the factorials in W. Using Eq. 17.8, we have

$$\ln W = \ln N! - \ln \prod_{i=1}^{s} n_i! = \ln N! - \sum_{i=1}^{s} \ln n_i! \qquad \textbf{(1)}$$

since the logarithm of a product is the sum of the logarithms of the individual factors. Using Stirling's approximation for the factorials, we find that Eq. 1 becomes

$$\ln W = N \ln N - N - \sum_{i=1}^{s} (n_i \ln n_i - n_i). \qquad \textbf{(2)}$$

It is important to note that the instant we use Stirling's approximation, we are implicitly requiring that all the n_i be very large. This will be the case only if we have $N \gg s$. We now wish to maximize $\ln W$ subject to the constraint that mass be conserved. This conservation requirement means that the sum of the marbles in each box must be N:

$$\sum_{i=1}^{s} n_i = N. \tag{3}$$

The constraint equation 3 is now multiplied by the Lagrangian multiplier α to produce

$$\alpha N = \alpha \sum_{i=1}^{s} n_i. \tag{4}$$

Adding Eqs. 2 and 4 yields

$$F(n_i; \alpha) = \ln W + \alpha N = N \ln N - N - \sum_{i=1}^{s} (n_i \ln n_i - n_i) + \alpha \sum_{i=1}^{s} n_i. \tag{5}$$

When W attains a maximum value, we have

$$dF(n_i; \alpha) = d \ln W = 0, \tag{6}$$

since αN is a constant whose differential is zero. Substituting Eq. 5 for $F(n_i; \alpha)$, into Eq. 6 produces

$$-\sum_{i=1}^{s} \left[\ln n_i dn_i + \frac{n_i dn_i}{n_i} - dn_i \right] + \sum_{i=1}^{s} \alpha dn_i = 0. \tag{7}$$

In obtaining Eq. 7, we have made use of the fact that the differential of $N \ln N - N$ is zero because N is a fixed constant. Collecting terms in Eq. 7 gives

$$\sum_{i=1}^{s} [\alpha - \ln n_i] dn_i = 0. \tag{8}$$

We now have $s + 1$ variables with one constraint. Therefore, there are s independent variables, which we take to be $dn_1, dn_2, dn_3, \dots, dn_s$. With these variables all independent, we must have their coefficients in Eq. 8 equal to zero to ensure that the left side will always be zero, no matter what values are assigned to the differentials in the number of marbles in each box. This gives us

$$\alpha - \ln n_i = 0 \text{ for all } n_i. \tag{9}$$

Therefore,

$$\ln n_i = \alpha \tag{10}$$

and

$$n_i = e^{\alpha}. \tag{11}$$

Equation 11 tells us that the most probable distribution is the one with an equal number of marbles in each box, as illustrated in Figure 17.5. By combining Eqs. 3 and 11, we can determine the actual number:

$$\sum_{i=1}^{s} n_i = \sum_{i=1}^{s} e^{\alpha} = s e^{\alpha} = N. \tag{12}$$

Combining Eqs. 11 and 12 produces

$$n_i = e^{\alpha} = \frac{N}{s} \tag{13}$$

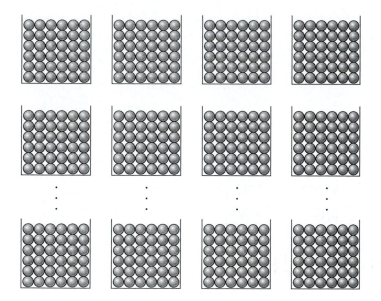

in every box. Thus, if we are distributing 10^{23} marbles into 10^{10} boxes, the most probable distribution is the one that has $10^{23}/10^{10} = 10^{13}$ marbles in each box.

For related exercises, see Problems 17.6, 17.7, and 17.8.

17.1.5   Maximization of *W* for Molecular-energy Distributions

We are now prepared to obtain the most probable distribution of N mole-cules among s energy states, which alone determines the thermodynamic properties of a system, so long as N is large and s is small. We seek to maxi-mize W or $\ln W$, subject to the constraints that mass and energy be con-served. $\ln W$ is obtained directly from Eq. 17.7 and Stirling's approximation:

$$\ln W = \ln N! - \ln \prod_{i=1}^{s} n_i! = \ln N! - \sum_{i=1}^{s} \ln n_i!$$

$$= N \ln N - N - \sum_{i=1}^{s} (n_i \ln n_i - n_i). \qquad (17.22)$$

Conservation of mass requires that the sum of the number of molecules in all energy states be equal to N:

$$\sum_{i=1}^{s} n_i = N. \qquad (17.23)$$

The sum of the energies of the molecules in each state must likewise be equal to the total energy E of the system; that is,

$$\sum_{i=1}^{s} n_i \varepsilon_i = E, \qquad (17.24)$$

where ε_i is the energy the molecule possesses in energy state i.

We now employ the method of Lagrangian multipliers to find the distribution that maximizes ln W and hence W. Multiplying the constraint equations by the Lagrangian multipliers α and β produces

$$\alpha N = \alpha \sum_{i=1}^{s} n_i \qquad (17.25\text{A})$$

and

$$\beta E = \beta \sum_{i=1}^{s} n_i \varepsilon_i. \qquad (17.25\text{B})$$

Adding Eqs. 17.25A and 17.25B and subtracting Eq. 17.22 gives

$$F(n_i; \alpha, \beta) = \alpha N + \beta E - \ln W$$

$$= \alpha \sum_{i=1}^{s} n_i + \beta \sum_{i=1}^{s} n_i \varepsilon_i - N \ln N + N + \sum_{i=1}^{s} (n_i \ln n_i - n_i).$$

$$(17.26)$$

When W attains a maximum, we must have $dF(n_i; \alpha, \beta) = d \ln W = 0$. Applied to Eq. 17.26, this requirement gives

$$\sum_{i=1}^{s} \alpha \, dn_i + \sum_{i=1}^{s} \beta \varepsilon_i \, dn_i + \sum_{i=1}^{s} \left[\ln n_i \, dn_i + \frac{n_i \, dn_i}{n_i} - dn_i \right] = 0. \quad (17.27)$$

Collecting terms in Eq. 17.27, we obtain

$$\sum_{i=1}^{s} [\alpha + \beta \varepsilon_i + \ln n_i] dn_i = 0. \qquad (17.28)$$

Equation 17.28 contains $s + 2$ variables with two constraints. Consequently, we have s independent variables, which we take to be the dn_i ($i = 1, 2, 3, \ldots, s$). This requires that the coefficients of the dn_i all be zero. Therefore,

$$[\alpha + \beta \varepsilon_i + \ln n_i] = 0,$$

which gives

$$\ln n_i = -\alpha - \beta \varepsilon_i,$$

or

$$n_i = \exp[-\alpha - \beta \varepsilon_i] = \exp(-\alpha) \exp(-\beta \varepsilon_i). \qquad (17.29)$$

Equation 17.29 gives us the distribution that maximizes the thermodynamic probability. All that remains is to use the constraint equations to determine the Lagrangian multipliers.

17.1.6   Identification of α

The Lagrangian multipliers are obtained by substituting Eq. 17.29 into the two constraint equations. We first use Eq. 17.23 to obtain α in terms of β. Combination of Eqs. 17.23 and 17.29 produces

$$\sum_{i=1}^{s} n_i = \sum_{i=1}^{s} \exp(-\alpha) \exp(-\beta \varepsilon_i) = \exp(-\alpha) \sum_{i=1}^{s} \exp(-\beta \varepsilon_i) = N. \quad (17.30)$$

Therefore,

$$\exp(-\alpha) = \frac{N}{\sum_{i=1}^{s} \exp(-\beta\varepsilon_i)}. \tag{17.31}$$

The quantity in the denominator of Eq. 17.31 plays a critical role in statistical thermodynamics; it is called the *molecular partition function*. In some texts, it is given the symbol Q, in others z. We shall use the latter notation, so that

$$\exp(-\alpha) = \frac{N}{z}. \tag{17.32}$$

Combination of Eqs. 17.29 and 17.32 yields

$$\boxed{n_i = \frac{N}{z} \exp(-\beta\varepsilon_i)}. \tag{17.33}$$

All that remains is to determine the second Lagrangian multiplier, β.

Before proceeding, it is worthwhile to examine the molecular partition function in greater detail. By definition,

$$z = \sum_{i=1}^{s} \exp(-\beta\varepsilon_i)$$
$$= \exp(-\beta\varepsilon_1) + \exp(-\beta\varepsilon_2) + \exp(-\beta\varepsilon_3) + \cdots + \exp(-\beta\varepsilon_s). \tag{17.34}$$

It is important to remember that the summation in Eq. 17.34 runs over the individual molecular quantum states, whereas the summation in Eq. 17.9 runs over the distributions or microstates for N molecules distributed among s aspects. In the former case we will employ a lowercase English letter, in the latter case a lowercase Greek letter. Notice what happens if energy states 2 and 3 are degenerate with a common energy ε_o. Under this condition, Eq. 17.34 becomes

$$z = \exp(-\beta\varepsilon_1) + \exp(-\beta\varepsilon_o) + \exp(-\beta\varepsilon_o) + \exp(-\beta\varepsilon_4) + \cdots + \exp(-\beta\varepsilon_s)$$
$$= \exp(-\beta\varepsilon_1) + 2\exp(-\beta\varepsilon_o) + \exp(-\beta\varepsilon_4) + \cdots + \exp(-\beta\varepsilon_s).$$

Since the terms $\exp(-\beta\varepsilon_2)$ and $\exp(-\beta\varepsilon_3)$ are both equal to $\exp(-\beta\varepsilon_o)$, their sum can be written as $2\exp(-\beta\varepsilon_o)$, where the degeneracy of the state appears as a coefficient in the partition function. This means that we can write z in the form

$$\boxed{z = \sum_{k=1}^{s} g_k \exp(-\beta\varepsilon_k)}, \tag{17.35}$$

where g_k is the degeneracy of energy state k. If z is written without the degeneracy factor, the summation runs over all energy states. If z is written with the degeneracy factor, the sum includes only those states with different energies. Either method can be used.

17.1.7 Determination of β: The Molecular One-Dimensional Translational Partition Function

To determine β, we use the second constraint equation (17.24), together with Eq. 17.33:

$$\sum_{i=1}^{s} n_i \varepsilon_i = \sum_{i=1}^{s} \frac{N}{z} \exp(-\beta \varepsilon_i) \varepsilon_i = \frac{N}{z} \sum_{i=1}^{s} \exp(-\beta \varepsilon_i) \varepsilon_i = E. \qquad \textbf{(17.36)}$$

The form of Eq. 17.36 permits us to make some qualitative observations about the nature of β. Since the exponent must be unitless, β must have the units $1/(\text{energy molecule}^{-1})$. It is, therefore, an intensive quantity independent of the volume or the number of molecules present in the system.

To proceed further at this point in our development, we can choose a system whose energy states are known, so that β can be evaluated. This procedure will provide us with the value of β for the system of choice. For simplicity, let us choose the particle in a one-dimensional infinite potential well that has only translational energy. The energy eigenvalues for this system were obtained in Chapter 12. They are

$$\varepsilon_i = \frac{i^2 h^2}{8ma^2} \quad (i = 1, 2, 3, 4, \ldots, \infty), \qquad \textbf{(17.37)}$$

where a is the width of the potential well. Substituting this expression into Eq. 17.36 produces

$$E = \frac{N}{z} \sum_{i=1}^{\infty} \frac{i^2 h^2}{8ma^2} \exp\left[-\frac{\beta i^2 h^2}{8ma^2}\right] = \frac{Nh^2}{8ma^2 z} \sum_{i=1}^{\infty} i^2 \exp\left[-\frac{\beta i^2 h^2}{8ma^2}\right]. \qquad \textbf{(17.38)}$$

For this system, the molecular partition function is

$$z = \sum_{i=1}^{\infty} \exp(-\beta \varepsilon_i) = \sum_{i=1}^{\infty} \exp\left[-\frac{\beta i^2 h^2}{8ma^2}\right]. \qquad \textbf{(17.39)}$$

In Chapter 10, we demonstrated that when the successive terms of a summation are very closely spaced, they can be treated as being continuous, so that the summation can be replaced with the corresponding integral with no significant loss of accuracy. In Chapter 12, we found that the translational energy eigenvalues form such a closely spaced set. Therefore, the summations contained in Eqs. 17.38 and 17.39 can be replaced with the corresponding integrals. We may also replace the lower limit $i = 1$ with $i = 0$, since the near-zero spacing between these states ensures that such a replacement will make no measurable difference. With these approximations, the molecular partition function becomes

$$z = \int_{0}^{\infty} \exp\left[-\frac{\beta i^2 h^2}{8ma^2}\right] di. \qquad \textbf{(17.40)}$$

This integral is a standard form that can be found in any table of integrals. The result of integrating it is

$$z = \frac{1}{2}\left[\frac{8\pi ma^2}{\beta h^2}\right]^{1/2}. \qquad \textbf{(17.41)}$$

A similar replacement of the summation in Eq. 17.38 yields the average energy as a function of β and the other constants of the problem:

$$E = \frac{Nh^2}{8ma^2 z} \int_{0}^{\infty} i^2 \exp\left[-\frac{\beta i^2 h^2}{8ma^2}\right] di. \qquad \textbf{(17.42)}$$

The integral in Eq. 17.42 is also a standard form that we have encountered on several previous occasions. Integrating that equation produces

$$E = \frac{Nh^2}{8ma^2z} \frac{2ma^2}{\beta h^2} \left[\frac{8\pi ma^2}{\beta h^2} \right]^{1/2} = \frac{N}{4\beta z} \left[\frac{8\pi ma^2}{\beta h^2} \right]^{1/2}. \tag{17.43}$$

Inserting the expression for z from Eq. 17.41 gives

$$E = \frac{N}{2\beta}. \tag{17.44}$$

Equation 17.44 makes it clear that if we knew the average energy of one mole of classical particles moving back and forth in a one-dimensional potential well of width a, we could determine the value of the Lagrangian multiplier β. This determination has already been made and used many times throughout our discussion of thermodynamics in Chapters 1–9. As early as Example 1.1, it was shown that the classical energy for such a system is

$$E_{classical} = \frac{PV}{2}. \tag{17.45}$$

Since the potential energy inside the one-dimensional well is zero, the system will behave as an ideal gas, so that for one mole of gas, we will have $PV = RT$. We have already argued that the closely spaced energy eigenstates permit us to treat our system as classical. Therefore,

$$E = E_{classical} = \frac{PV}{2} = \frac{RT}{2} = \frac{N}{2\beta}. \tag{17.46}$$

Solving for β, we obtain

$$\beta = \frac{N}{RT} = \frac{1}{(R/N)T} = \frac{1}{kT}, \tag{17.47}$$

where $k = R/N$ is the Boltzmann constant. The distribution of molecules across the various translational-energy states is, therefore,

$$\boxed{n_i = \frac{N}{z} \exp\left[-\frac{\varepsilon_i}{kT} \right].} \tag{17.48}$$

If we multiply numerator and denominator of the exponent by Avogadro's constant, the ratio can be written in terms of energy per mole. This gives

$$\boxed{n_i = \frac{N}{z} \exp\left[-\frac{E_i}{RT} \right].} \tag{17.49}$$

Equations 17.48 and 17.49 are both expressions for the familiar Boltzmann distribution that we have employed repeatedly in this text. We now see that that distribution is the one which maximizes the thermodynamic probability, subject to the constraints of mass and energy conservation for systems in which $N \gg s$. We still need to prove that $\beta = (kT)^{-1}$ regardless of the system being examined. We get a strong indication that this is the case from the fact that no matter what specific system we choose to examine, the average energy is always given correctly when we take $\beta = (kT)^{-1}$. In Chapter 18, we shall complete the proof. In the meantime, we shall assume that $\beta = (kT)^{-1}$ for all systems.

17.1.8 Physical Interpretation of the Molecular Partition Function: Diatomic Rigid-Rotor Molecular Partition Function

Now that we suspect that $\beta = (kT)^{-1}$, we can obtain an understanding of the physical significance of the molecular partition function. This will be crucial to the development of statistical thermodynamics in Chapter 18. *The molecular partition function represents the effective number of energy states available to the molecules.* If we write the effective number of occupied states as s_{eff}, the foregoing statement means that

$$z = s_{eff} = \sum_{i=1}^{s} \exp\left[-\frac{\varepsilon_i}{kT}\right].$$

We can easily see that z and s_{eff} are identical by considering a few simple cases. Suppose that the energies of all states are zero. The case of N marbles distributed among s identical boxes with $\varepsilon_i = 0$ for all boxes is an example of such a system. For this case, we have $\exp[-\varepsilon_i/kT] = \exp(0) = 1$, which, in turn, gives $z = s_{eff} = \sum_{i=1}^{s} e^o = se^o = s$, and the effective number of boxes available to the marbles equals the total number of boxes.

Now consider the rotational-energy states of a rigid rotor whose rotational constant is that for ^{16}OH. (See Table 15.4.) For this system, $\varepsilon_J = J(J + 1)Bh$ and $g_J = 2J + 1$ for $J = 0, 1, 2, 3, \ldots, \infty$, so that

$$z = s_{eff} = \sum_{J=0}^{\infty} (2J + 1) \exp\left[-\frac{J(J + 1)Bh}{kT}\right].$$

As the temperature of the system approaches 0 K, the exponentials will approach $e^{-\infty} = 0$, unless $J = 0$, in which case the exponential is unity. Thus, as $T \rightarrow 0$ K, we have

$$z = s_{eff} = 1 + e^{-\infty} + e^{-\infty} + e^{-\infty} + \cdots = 1.$$

Table 17.4	Rotational molecular partition function and effective number of states as a function of temperature for a rigid rotor with a rotational constant equal to that for OH
T (K)	$z = s_{eff}$
10	1.0131
50	2.2155
100	4.0355
200	7.7088
300	11.389
400	15.070
500	18.753
750	27.959
1,000	37.166
2,000	73.996
4,000	147.66

The effective number of states is unity, which is exactly the result we expect. At temperatures near 0 K, all molecules will be in the ground state ($J = 0$), and the system will be reduced to effectively one state. As T increases, the system will have sufficient energy to permit more states to be occupied, and s_{eff} will increase. Table 17.4 gives the values of $z = s_{eff}$ for the above system at several different temperatures. Figure 17.6 shows the variation of z with temperature. As expected, the effective number of rotational states available to ^{16}OH increases as the temperature rises.

The concept that z represents the effective number of states available to the system will be of central importance in Chapter 18, when we turn our attention to the development of statistical thermodynamics.

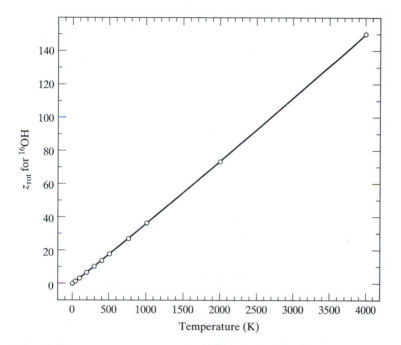

◀ **FIGURE 17.6**
The molecular rotational partition function for ^{16}OH as a function of temperature. The molecular partition function physically represents the effective number of quantum states available to the system at temperature T. We see that z_{rot} for ^{16}OH approaches unity at $T \rightarrow 0$ K, which is the anticipated result, since we expect all ^{16}OH molecules to be in the $J = 0$ quantum state at low temperatures. As T rises, the additional thermal energy permits ^{16}OH molecules to occupy higher rotational states, and as a result, z and s_{eff} increase.

EXAMPLE 17.4

The bond length of the HF molecule is 0.917 angstrom = 9.17×10^{-11} m. Assume that HF behaves as a diatomic rigid rotor. **(A)** Compute the moment of inertia of HF. **(B)** Compute the rotational molecular partition function to four significant digits for HF at 350 K. **(C)** Assume that the HF rotational quantum number J is continuous, and compute the most populated J state at 350 K. Compare your result with the quantum result obtained in (B). **(D)** What is the effective number of rotational states occupied by HF at 350 K?

Solution

(A) The reduced mass of HF is

$$\mu = \frac{(1.008)(19.00)}{20.008} \text{ g mol}^{-1} \times \frac{1 \text{ kg}}{1000 \text{ g}} \times \frac{1}{6.022 \times 10^{23} \text{ mol}^{-1}}$$

$$= 1.590 \times 10^{-27} \text{ kg.} \tag{1}$$

The moment of inertia is, therefore,

$$I = \mu R_{eq}^2 = (1.590 \times 10^{-27} \text{ kg})(9.17 \times 10^{-11} \text{ m})^2 = 1.337 \times 10^{-47} \text{ kg m}^2. \quad (2)$$

(B) The rotational-energy states for a rigid rotor are

$$E_J = J(J + 1)Bh, \text{ with } J = 0, 1, 2, 3, 4\ldots, \infty. \quad (3)$$

Since the magnetic quantum number has $2J + 1$ values for each value of J, the degeneracy of the rotational state J is

$$g_J = 2J + 1. \quad (4)$$

The rotational molecular partition function is given by

$$z = \sum_{J=0}^{\infty} g_J \exp\left[-\frac{E_J}{kT}\right] = \sum_{J=0}^{\infty} (2J + 1) \exp\left[-J(J + 1)\frac{Bh}{kT}\right]. \quad (5)$$

The quantity Bh/kT needed in Eq. 5 is

$$\frac{Bh}{kT} = \frac{h^2}{8\pi^2 IkT} = \frac{(6.626 \times 10^{-34} \text{ J s})^2}{8(3.14159)^2(1.337 \times 10^{-47} \text{ kg m}^2)(350 \text{ K})(1.3807 \times 10^{-23} \text{ J K}^{-1})}$$

$$= 0.0861. \quad (6)$$

Therefore, the partition function is

$$z = \sum_{J=0}^{\infty} (2J + 1) \exp[-0.0861J(J + 1)]. \quad (7)$$

The summation can be carried out with the use of a Microsoft Excel file. The result so obtained is shown in the table on the following page.

Therefore, at $T = 350$ K, we have $z = 11.95$.

(C) If J is continuous, the most populated level is the one for which the derivative of $(2J + 1) \exp[-J(J + 1)0.0861]$ is zero. Thus,

$$\frac{d}{dJ}\{(2J + 1) \exp[-J(J + 1)0.0861]\}$$

$$= 2 \exp[-J(J + 1)0.0861] - 0.0861(2J + 1)^2 \exp[-J(J + 1)0.0861] = 0. \quad (8)$$

Factoring out the exponential, we obtain

$$\exp[-J(J + 1)0.0861]\{2 - 0.0861(2J + 1)^2\} = 0. \quad (9)$$

Therefore, we must have

$$(2J + 1)^2 = \frac{2}{0.0861} = 23.229. \quad (10)$$

Hence,

$$J_{mp} = \frac{((23.229)^{1/2} - 1)}{2} = 1.91. \quad (11)$$

Rounding, we obtain $J_{mp} = 2$.

Equations 17.33 and 17.35 show that the number of molecules in energy state E_J is proportional to $(2J + 1) \exp[-E_J/(kT)]$, which, in the example, is given by $(2J + 1) \exp[-J(J + 1)Bh/(kT)]$. The third column in the Excel file lists the values of this quantity for J states between 0 and 20. Inspection of these values shows that the maximum occurs for $J = 2$. Therefore, the result obtained, assuming a continuous distribution of J values, is correct.

J	$2J + 1$	$(2J + 1) \exp\left[-\dfrac{J(J + 1)Bh}{kT}\right]$
0	1	1
1	3	2.52543238
2	5	2.98272679
3	7	2.49106455
4	9	1.60837536
5	11	0.83101725
6	13	0.34950038
7	15	0.12080836
8	17	0.03452793
9	19	0.0081923
10	21	0.00161814
11	23	0.00026661
12	25	3.67E-05
13	27	4.2255E-06
14	29	4.073E-07
15	31	3.2893E-08
16	33	2.2268E-09
17	35	1.2644E-10
18	37	6.0239E-12
19	39	2.4089E-13
20	41	8.0877E-15

$$\text{sum} = 11.9535714$$

(D) The effective number of rotational states occupied by HF at 350 K is given by the value of the rotational molecular partition function at 350 K:

$$s_{\text{eff}} = z = 11.95. \tag{12}$$

The Excel file shows that virtually all the molecules are in J states between $J = 0$ and $J = 11$. Consequently, $s_{\text{eff}} \approx 12$ is the result we expect.

For related exercises, see Problems 17.10, 17.11, and 17.12.

17.2   The Kinetic Theory of Gases

The final topic to be covered in this text, in Chapters 19 and 20, involves the measurement and theoretical treatment of the kinetics and dynamics of chemical reactions. The translational motion of gas-phase systems is of central importance in such studies. In particular, we will require a description of the distribution of molecular speeds. Since reaction rates depend upon the frequency with which molecules collide, we need to develop appropriate expressions that give these frequencies. In the process, we shall also obtain equations for the distribution of free paths and the mean free path.

17.2.1   The Maxwell–Boltzmann Distribution of Molecular Speeds

In Chapter 12, we found that the translational-energy states of molecules are very closely spaced, so much so that the states can be treated as if they form a continuum. Under these conditions, we can treat the translational motion of gases as if it were classical and still have reason to expect our results to be an accurate representation of the quantum mechanical system. Problems 17.14, 17.15, and 17.16 show that we can make similar approximations for rotational motion if the molecules have large moments of inertia or if the temperature is sufficiently high. In contrast, Problems 17.17 and 17.18 demonstrate that a classical approximation will rarely be accurate for vibration.

If the translational motion of a particle of mass m can be accurately represented as being classical, the energy of the particle may be expressed in terms of the rectangular velocity components v_x, v_y, and v_z:

$$E_{trans} = E_t = \frac{m}{2}[v_x^2 + v_y^2 + v_z^2]. \tag{17.50}$$

Since the velocity components are continuous variables, we cannot discuss the probability of observing a particular velocity v_x. Instead, we must discuss the probability that v_x lies in the range v_x to $v_x + dv_x$. This point was covered in detail in Chapter 10. With that modification, Eq. 17.48 gives us the number of molecules that, at temperature T, will have components of velocity in the range v_x to $v_x + dv_x$, v_y to $v_y + dv_y$, and v_z to $v_z + dv_z$. That number is

$$n(v_x, v_y, v_z)dv_x dv_y dv_z = \frac{N}{z} \exp\left[-\frac{m[v_x^2 + v_y^2 + v_z^2]}{2kT}\right] dv_x dv_y dv_z, \tag{17.51}$$

where the partition function now has its classical form with the summations replaced with integrations:

$$z = \sum_{v_x=-\infty}^{\infty} \sum_{v_y=-\infty}^{\infty} \sum_{v_z=-\infty}^{\infty} \exp\left[-\frac{m[v_x^2 + v_y^2 + v_z^2]}{2kT}\right]$$

$$= \int_{v_x=-\infty}^{\infty} \int_{v_y=-\infty}^{\infty} \int_{v_z=-\infty}^{\infty} \exp\left[-\frac{m[v_x^2 + v_y^2 + v_z^2]}{2kT}\right] dv_x dv_y dv_z. \tag{17.52}$$

By writing the exponential in Eq. 17.52 as the product of three exponentials, we may separate the three-dimensional integral into three one-dimensional integrals:

$$z = \int_{v_x=-\infty}^{\infty} \exp\left[-\frac{mv_x^2}{2kT}\right] dv_x \int_{v_y=-\infty}^{\infty} \exp\left[-\frac{mv_y^2}{2kT}\right] dv_y \int_{v_z=-\infty}^{\infty} \exp\left[-\frac{mv_z^2}{2kT}\right] dv_z.$$

$$\tag{17.53}$$

Each integral in Eq. 17.53 has the standard form

$$\int_{-\infty}^{\infty} \exp(-ax^2)dx = \left[\frac{\pi}{a}\right]^{1/2}.$$

Using this form, we find that Eq. 17.53 becomes

$$z = \left[\frac{2\pi kT}{m}\right]^{1/2}\left[\frac{2\pi kT}{m}\right]^{1/2}\left[\frac{2\pi kT}{m}\right]^{1/2} = \left[\frac{2\pi kT}{m}\right]^{3/2}. \tag{17.54}$$

Combining Eqs. 17.51 and 17.54 gives the normalized distribution of classical molecular velocities:

$$n(v_x, v_y, v_z)dv_x dv_y dv_z = N\left[\frac{m}{2\pi kT}\right]^{3/2} \exp\left[-\frac{m[v_x^2 + v_y^2 + v_z^2]}{2kT}\right] dv_x dv_y dv_z .$$

(17.55)

Many chemical processes depend only upon the magnitude of the velocity vector and not upon its direction. It is, therefore, important for us to convert Eq. 17.55 from an expression that gives us the velocity distribution to one that provides just the distribution of molecular speeds. To obtain this result, we need to convert from rectangular Cartesian velocity coordinates to a system that uses one variable to give the magnitude of the velocity and the other two to furnish information about the direction of motion. That is precisely what is done by spherical polar coordinates, which we have employed frequently throughout the text. The volume element for this coordinate system was developed in the discussion related to Figure 12.19. If C is the magnitude of the velocity and θ and ϕ are the spherical polar angles describing the direction of motion, as shown in Figure 17.7, then the transformation equations between rectangular Cartesian velocities and the spherical polar system are

$$v_x = C \sin\theta \cos\phi,$$

$$v_y = C \sin\theta \sin\phi,$$

and

$$v_z = C \cos\theta.$$ (17.56)

Squaring both sides of the three equations in (17.56) and adding, we obtain

$$C^2 = v_x^2 + v_y^2 + v_z^2.$$ (17.57)

The spherical polar volume, developed in Eq. 12.108, has the form

$$d\tau = dv_x dv_y dv_z = C^2 \sin\theta\, dC\, d\theta\, d\phi.$$ (17.58)

Substituting Eqs. 17.57 and 17.58 into Eq. 17.55 yields the corresponding result in spherical polar coordinates:

$$n(C, \theta, \phi)\, C^2 dC \sin\theta\, d\theta\, d\phi = N\left[\frac{m}{2\pi kT}\right]^{3/2} \exp\left[-\frac{mC^2}{2kT}\right] C^2 \sin\theta\, dC\, d\theta\, d\phi.$$

(17.59)

If we have no interest in the direction of motion, we can determine the total number of molecules moving with speeds from C to $C + dC$ in any direction simply by summing (integrating) over all possible directions. This operation produces

$$n(C)dC = \int_{\phi=0}^{2\pi} \int_{\theta=0}^{\pi} n(C, \theta, \phi) C^2 dC \sin\theta\, d\theta\, d\phi$$

$$= N\left[\frac{m}{2\pi kT}\right]^{3/2} \exp\left[-\frac{mC^2}{2kT}\right] C^2 dC \int_{\phi=0}^{2\pi} d\phi \int_{\theta=0}^{\pi} \sin\theta\, d\theta.$$ (17.60)

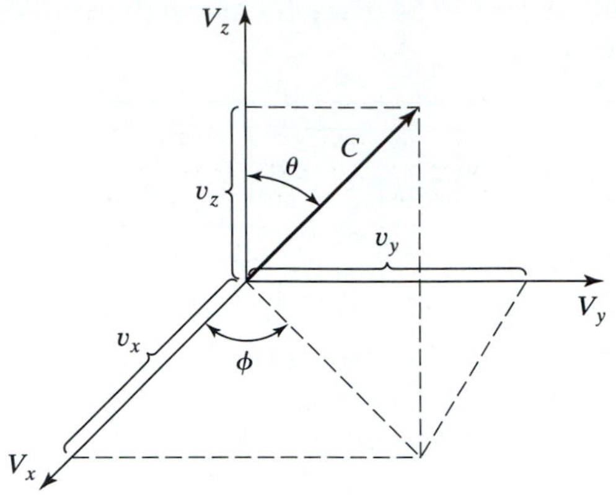

▲ FIGURE 17.7
Spherical polar velocity coordinates.

The angular integrations in Eq. 17.60 give the total solid angle about the origin, 4π. Thus, we obtain

$$n(C)\,dC = 4\pi N\left[\frac{m}{2\pi kT}\right]^{3/2} \exp\left[-\frac{mC^2}{2kT}\right]C^2\,dC\,. \qquad \textbf{(17.61)}$$

Equation 17.61 is called the *Maxwell–Boltzmann distribution* of molecular speeds. It gives us the number of molecules of mass m at temperature T that have speeds in the range from C to $C + dC$. If we divide both sides by N, we obtain the normalized probability distribution for observing a molecule at some speed in that range. Mathematically, this statement is usually written in the form

$$P(C)\,dC = 4\pi\left[\frac{m}{2\pi kT}\right]^{3/2} \exp\left[-\frac{mC^2}{2kT}\right]C^2\,dC\,. \qquad \textbf{(17.62)}$$

Figure 17.8 shows this probability distribution for $O_2(g)$ at various temperatures.

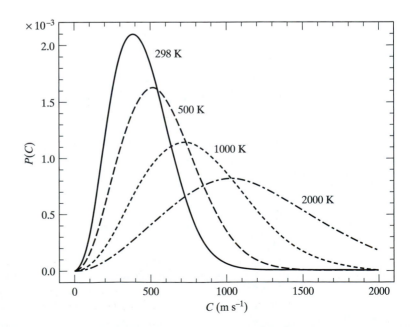

► FIGURE 17.8
Normalized Maxwell–Boltzmann probability distributions of molecular speeds for $O_2(g)$ at temperatures of 298 K, 500 K, 1,000 K, and 2,000 K.

EXAMPLE 17.5

A slotted-disk speed selector consists of a series of slotted rotating disks. When a beam of molecules enters this apparatus, only those molecules with a narrow distribution of speeds will successfully traverse the disks by encountering a slot at each disk. Thus, the distribution of molecules that emerges from the apparatus will have a very narrow distribution of molecular speeds. By varying the angular speed of the disks, the investigator can control the speed of the molecules that emerge from the selector. Suppose a beam of Ar atoms with a Maxwell–Boltzmann speed distribution at 300 K enters the selector with the disks rotating so as to transmit only those atoms whose speeds lie in the range $347 \leq C \leq 353$ m s^{-1}. Compute the fraction of molecules transmitted by the speed selector.

Solution

The probability distribution is given by Eq. 17.62:

$$P(C)\,dC = 4\pi \left[\frac{m}{2\pi kT} \right]^{3/2} \exp\left[-\frac{mC^2}{2kT} \right] C^2\,dC. \qquad (1)$$

Replacing k with R/N, where N is Avogadro's constant, puts Eq. 1 into the form

$$P(C)\,dC = 4\pi \left[\frac{M}{2\pi RT} \right]^{3/2} \exp\left[-\frac{MC^2}{2RT} \right] C^2\,dC, \qquad (2)$$

where M is the molar mass. For argon, $M = 0.039948$ kg mol^{-1}. At 300 K, Eq. 2 becomes

$$P(C)\,dC = 4(3.14159)\left[\frac{(0.039948 \text{ kg mol}^{-1})}{2(3.14159)(8.314 \text{ J mol}^{-1}\text{K}^{-1})(300 \text{ K})} \right]^{3/2}$$

$$\times \exp\left[-\frac{(0.039948 \text{ kg mol}^{-1})C^2}{2(8.314 \text{ J mol}^{-1}\text{K}^{-1})(300 \text{ K})} \right] C^2\,dC, \quad (3)$$

which gives

$$P(C)\,dC = 5.114 \times 10^{-8} \exp[-8.008 \times 10^{-6}\,C^2]C^2\,dC. \qquad (4)$$

The probability that the speed of an argon atom will lie in the range from 347 m s^{-1} to 353 m s^{-1} is just the sum or integral of $P(C)\,dC$ over this range. Thus,

$$P(347 \leq C \leq 353 \text{ m s}^{-1}) = (5.114 \times 10^{-8} \text{ s}^3 \text{ m}^{-3})$$

$$\times \int_{347}^{353} \exp[-8.008 \times 10^{-6}\,C^2]C^2\,dC. \qquad (5)$$

There is no closed-form expression for the integral appearing in Eq. 5. We can, however, easily evaluate this integral numerically. The result is

$$\int_{347}^{353} \exp[-8.008 \times 10^{-6}\,C^2]C^2\,dC \approx 2.756 \times 10^5 \text{ m}^3 \text{ s}^{-3}. \qquad (6)$$

Therefore,

$$P(347 \leq C \leq 353 \text{ m s}^{-1}) = (5.114 \times 10^{-8} \text{ s}^3 \text{ m}^{-3})(2.756 \times 10^5 \text{ m}^3 \text{ s}^{-3})$$

$$= 0.01409. \qquad (7)$$

About 1.41% of the atoms successfully traverse the speed selector. These atoms are the ones with speeds in the shaded area of Figure 17.9.

There is a second way to do this problem. The range from 347 m s^{-1} to 353 m s^{-1} is very small. Over this range, $P(C)$ is nearly constant at the value

$P(C = 350 \text{ m s}^{-1})$. If we make this approximation, the integral over the probability becomes

$$P(347 \le C \le 353 \text{ m sec}^{-1}) = \int_{347}^{353} P(C)dC \approx P(350) \int_{347}^{353} dC$$

$$= (5.114 \times 10^{-8} \text{ s}^3 \text{ m}^{-3}) \exp[-8.008 \times 10^{-6}(350)^2](350)^2[353 - 347] \text{ m}^3 \text{ s}^{-3}$$

$$= 0.01409, \tag{8}$$

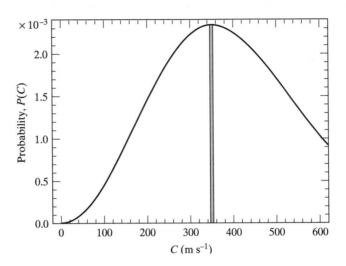

▲ FIGURE 17.9

The normalized Maxwell–Boltzmann probability distribution for Ar atoms at 300 K. The Ar atoms traversing the slotted-disk speed selector in Example 17.5 are those with speeds in the shaded area of the plot. Over this range, $P(C)$ is seen to be nearly constant.

which is correct to four significant digits. Inspection of Figure 17.9 shows why the approximation is so accurate for this problem: $P(C)$ hardly changes over the range $347 \text{ m s}^{-1} \le C \le 353 \text{ m s}^{-1}$. If the range were large or in a different portion of the distribution curve, the approximation would not be as accurate.

For related exercises, see Problems 17.21 and 17.23.

It is a simple matter to convert Eq. 17.61 into a distribution of molecular translational energies. If we define ε_t as the translational energy of a molecule moving with speed C, we have $\varepsilon_t = mC^2/2$. The differential change in ε_t for a differential change in C is given by

$$d\varepsilon_t = mC\,dC = m\left[\frac{2\varepsilon_t}{m}\right]^{1/2} dC,$$

so that

$$dC = [2m\varepsilon_t]^{-1/2}\,d\varepsilon_t. \tag{17.63}$$

A direct transformation of variables in Eq. 17.62 gives us the probability that a molecule will have a translational energy in the range from ε_t to $\varepsilon_t + d\varepsilon_t$. The result is (see Problem 17.22)

$$\boxed{P(\varepsilon_t)d\varepsilon_t = 4(2)^{1/2}\pi\left[\frac{1}{2\pi kT}\right]^{3/2}\exp\left[-\frac{\varepsilon_t}{kT}\right]\varepsilon_t^{1/2}\,d\varepsilon_t}. \tag{17.64}$$

Note that the distribution of translational energies is independent of mass. Replacing k with R/N, where N is Avogadro's constant, converts Eq. 17.64 into an equivalent expression in terms of energy per mole rather than energy per molecule:

$$P(E_t)dE_t = 4(2)^{1/2}\pi\left[\frac{1}{2\pi RT}\right]^{3/2}\exp\left[-\frac{E_t}{RT}\right]E_t^{1/2}dE_t. \qquad (17.65)$$

Here, E_t is the translational energy per mole.

In discussing Eq. 5.89, we noted that the temperature dependence of any variable X that measures the extent or rate of a process for which there is an energy requirement E_o usually has the form

$$X = (\text{constant})\exp\left[-\frac{E_o}{RT}\right], \qquad (17.66)$$

provided that the energy requirement is independent of temperature. Equation 17.65 provides the underlying foundation for this observation. The extent or rate of any process that has an energy requirement E_o is proportional to the probability that the system has an energy equal to or greater than E_o. This probability can be computed by summing Eq. 17.65 from E_o to ∞:

$$P(E_t \geq E_o) = 4(2)^{1/2}\pi\left[\frac{1}{2\pi RT}\right]^{3/2}\int_{E_o}^{\infty}\exp\left[-\frac{E_t}{RT}\right]E_t^{1/2}dE_t. \qquad (17.67)$$

The integral in Eq. 17.67 has no closed-form solution, but we can represent its value with a series expansion obtained by repeated integration by parts. This procedure permits us to write the equation in the form

$$P(E_t \geq E_o) = 4(2)^{1/2}\pi\left[\frac{1}{2\pi RT}\right]^{3/2}RTE_o^{1/2}\exp\left[-\frac{E_o}{RT}\right]$$

$$\times\left\{1 + \left(\frac{RT}{2E_o}\right) - \left(\frac{RT}{2E_o}\right)^2 + \cdots\right\}. \qquad (17.68)$$

If the energy requirement for the process is large compared with RT, we will have $RT/(2E_o) \ll 1$, so that we can truncate the series expansion in Eq. 17.68 after the first term to obtain

$$P(E_t \geq E_o) \approx 4(2)^{1/2}\pi\left[\frac{1}{2\pi RT}\right]^{3/2}RTE_o^{1/2}\exp\left[-\frac{E_o}{RT}\right]$$

$$= KT^{-1/2}\exp\left[-\frac{E_o}{RT}\right], \qquad (17.69)$$

where K is a constant. Since we expect the extent or rate of a process with energy requirement E_o to be proportional to $P(E_t \geq E_o)$, we have

$$X \approx (\text{constant})T^{-1/2}\exp\left[-\frac{E_o}{RT}\right]. \qquad (17.70)$$

Except for the factor $T^{-1/2}$, Eq. 17.70 is identical to Eq. 17.66. If the temperature range being investigated is not very large, as is generally the case,

$T^{-1/2}$ is nearly constant relative to the changes produced by the exponential. Under this condition, Eqs. 17.66 and 17.70 are the same. This behavior lies at the heart of the empirical observation we made in Chapter 5 that led to Eq. 5.89.

17.2.2   Average and Most Probable Speeds

Figure 17.8 shows that the distribution of molecular speeds exhibits a maximum at a value C_{mp} that shifts to increasingly larger values as the temperature increases. C_{mp}, the most probable speed, may be obtained by the usual procedure for finding the extrema of functions: We search for the speed that makes the first derivative of the probability distribution zero. That is, at $C = C_{mp}$, we must have

$$\frac{dP(C)}{dC} = 4\pi \left[\frac{m}{2\pi kT} \right]^{3/2} \left[2C \exp\left[-\frac{mC^2}{2kT} \right] - \frac{mC^3}{kT} \exp\left[-\frac{mC^2}{2kT} \right] \right] = 0. \quad \textbf{(17.71)}$$

Equation 17.71 has two solutions. The first is $C = 0$, which is clearly not the most probable speed that we seek. The other solution is the one that makes

$$2 - \frac{mC^2}{kT} = 0.$$

The solution of this equation is

$$C_{mp} = \left[\frac{2kT}{m} \right]^{1/2} = \left[\frac{2RT}{M} \right]^{1/2}, \quad \textbf{(17.72)}$$

where M is the molar mass.

The average speed can be computed using the methods discussed in Chapter 10, viz.,

$$\langle C \rangle = \frac{\displaystyle\int_0^\infty C P(C)\,dC}{\displaystyle\int_0^\infty P(C)\,dC} = \int_0^\infty C P(C)\,dC, \quad \textbf{(17.73)}$$

provided that the probability distribution is normalized. Inserting Eq. 17.62 into Eq. 17.73 produces

$$\langle C \rangle = 4\pi \left[\frac{m}{2\pi kT} \right]^{3/2} \int_0^\infty \exp\left[-\frac{mC^2}{2kT} \right] C^3\,dC. \quad \textbf{(17.74)}$$

Integrating by parts twice (see Problem 17.24) gives the result

$$\langle C \rangle = \left[\frac{8RT}{\pi M} \right]^{1/2}. \quad \textbf{(17.75)}$$

Note that we always have $\langle C \rangle > C_{mp}$, regardless of the temperature and the mass of the molecule.

The average square speed is obtained using the same methods:

$$\langle C^2 \rangle = \frac{\displaystyle\int_0^\infty C^2 P(C)\,dC}{\displaystyle\int_0^\infty P(C)\,dC} = \int_0^\infty C^2 P(C)\,dC$$

$$= 4\pi \left[\frac{m}{2\pi kT}\right]^{3/2} \int_0^\infty \exp\left[-\frac{mC^2}{2kT}\right] C^4\,dC. \qquad \textbf{(17.76)}$$

The integral in Eq. 17.76 is a standard form whose value is $(3/8)(2kT/m)^2$ $(2\pi kT/m)^{1/2}$. Inserting this result gives

$$\langle C^2 \rangle = 4\pi \left[\frac{m}{2\pi kT}\right]^{3/2} \frac{3}{8}\left(\frac{2kT}{m}\right)^2 \left(\frac{2\pi kT}{m}\right)^{1/2} = \frac{3kT}{m} = \frac{3RT}{M}. \qquad \textbf{(17.77)}$$

With the average square speed given by Eq. 17.77, we can easily obtain the average translational energy, namely,

$$\langle E_t \rangle = \left\langle \frac{MC^2}{2} \right\rangle = \frac{M}{2}\langle C^2 \rangle = \frac{M}{2}\frac{3RT}{M} = \frac{3RT}{2}, \qquad \textbf{(17.78)}$$

which is the result we have employed frequently throughout this text.

The *median* of any distribution is, by definition, that value for which exactly half of the possible values lie below it and half lie above it. Thus, if C_m is the median speed, we must have

$$\int_0^{C_m} P(C)\,dC = \frac{1}{2}. \qquad \textbf{(17.79)}$$

It is relatively easy to show (see Problems 17.25 and 17.26) that

$$C_m = 1.0877\ldots C_{mp} = 1.0877\ldots \left[\frac{2RT}{M}\right]^{1/2}. \qquad \textbf{(17.80)}$$

Figure 17.10 shows the relative positions of the most probable, average, and median speeds for argon atoms at 300 K.

If we express the Maxwell–Boltzmann distribution in terms of a reduced speed, we obtain a single probability distribution function for all gases at all temperatures. In this sense, the result is reminiscent of the *law of corresponding states* that we introduced in Chapter 1. However, that law is only approximate, whereas the Maxwell–Boltzmann distribution law is essentially exact. Let us define a reduced speed as

$$C_R = \frac{C}{C_{mp}} = \left[\frac{M}{2RT}\right]^{1/2} C. \qquad \textbf{(17.81)}$$

Then the differentials are related by

$$dC = \left[\frac{2RT}{M}\right]^{1/2} dC_R. \qquad \textbf{(17.82)}$$

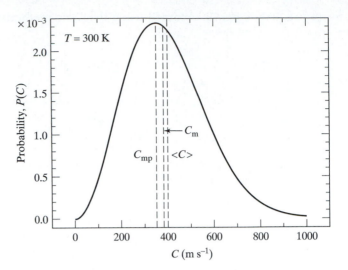

▶ **FIGURE 17.10**
The most probable (C_{mp}), average ($\langle C \rangle$), and median (C_m) speeds for argon atoms with a Maxwell–Boltzmann probability distribution of molecular speeds at 300 K.

If Eqs. 17.81 and 17.82 are substituted into Eq. 17.62, the Maxwell–Boltzmann probability distribution is obtained in terms of the reduced speed. The result is (see Problem 17.25)

$$P(C_R)dC_R = \left(\frac{16}{\pi}\right)^{1/2} \exp[-C_R^2]C_R^2 \, dC_R. \qquad \textbf{(17.83)}$$

Note that $P(C_R) \, dC_R$ is independent of both mass and temperature; therefore, it represents the speed probability distribution for all gases at all temperatures. A plot of $P(C_R)$ versus C_R is shown in Figure 17.11.

EXAMPLE 17.6

Compute C_{mp}, $\langle C \rangle$, and $\langle C^2 \rangle$ for $O_2(g)$ at 298 K. Compute the uncertainty in the speed at this temperature.

Solution

Using Eqs. 17.72, 17.75, and 17.77, we obtain

$$C_{mp} = \left(\frac{2RT}{M}\right)^{1/2} = \left(\frac{2(8.314 \text{ J mol}^{-1} \text{ K}^{-1})(298 \text{ K})}{0.031999 \text{ kg mol}^{-1}}\right)^{1/2} = 393.5 \text{ m s}^{-1}. \qquad \textbf{(1)}$$

Inspection of Figure 17.8 shows this result to be correct. Next,

$$\langle C \rangle = \left(\frac{8RT}{\pi M}\right)^{1/2} = \left(\frac{8(8.314 \text{ J mol}^{-1} \text{ K}^{-1})(298 \text{ K})}{(3.14159)(0.031999 \text{ kg mol}^{-1})}\right)^{1/2} = 444.0 \text{ m s}^{-1}. \qquad \textbf{(2)}$$

As expected, we have $\langle C \rangle > C_{mp}$. Finally,

$$\langle C^2 \rangle = \frac{3RT}{M} = \frac{3(8.314 \text{ J mol}^{-1} \text{ K}^{-1})(298 \text{ K})}{0.031999 \text{ kg mol}^{-1}} = 2.323 \times 10^5 \text{ m}^2 \text{ s}^{-2}. \qquad \textbf{(3)}$$

The uncertainty in the speed is given by

$$\langle \Delta C^2 \rangle^{1/2} = [\langle C^2 \rangle - \langle C \rangle^2]^{1/2} = [2.323 \times 10^5 - (444.0)^2]^{1/2} = 187.5 \text{ m s}^{-1}. \qquad \textbf{(4)}$$

For related exercises, see Problems 17.25, 17.26, 17.27, and 17.28.

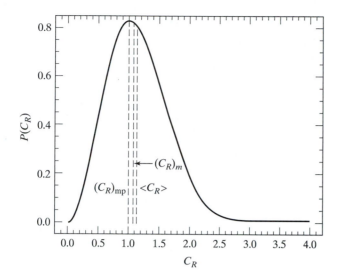

◀ FIGURE 17.11
Normalized probability distribution of reduced speeds. The most probable $(C_R)_{mp}$, median $(C_R)_m$, and average $\langle C_R \rangle$ reduced speeds are shown. This distribution holds for any gas having a Maxwell–Boltzmann distribution of molecular speeds, regardless of the identity of the gas and regardless of the temperature.

The diffusion rate of a gas is proportional to the average speed with which the gas moves. Therefore, if D represents the diffusion rate at a given pressure, we expect to observe $D = b\langle C \rangle$, where b is a constant. This relationship allows us to write a very simple expression for the ratio of diffusion rates for two gases in the same environment at the same temperature:

$$\frac{D_1}{D_2} = \frac{b\langle C \rangle_1}{b\langle C \rangle_2} = \frac{\langle C \rangle_1}{\langle C \rangle_2}. \tag{17.84}$$

Substituting Eq. 17.75 for $\langle C \rangle$ gives

$$\frac{D_1}{D_2} = \frac{\left[\dfrac{8RT}{\pi M_1} \right]^{1/2}}{\left[\dfrac{8RT}{\pi M_2} \right]^{1/2}} = \left[\frac{M_2}{M_1} \right]^{1/2}. \tag{17.85}$$

Equation 17.85 is known as *Graham's law* of diffusion: If the temperature and environment of two gases are the same, the ratio of their diffusion rates is equal to the square root of the reciprocal of their molar mass ratio.

During World War II, Nazi Germany and the United States were locked in a life-and-death struggle to construct an atomic bomb. If Germany had succeeded before the United States did, the Third Reich would have won the war. The consequences of such an event for the world are simply too horrible to imagine. An idea of the ensuing carnage can be visualized in the Nazi death camps and the Holocaust. In order to build the first explosive nuclear device, it was critical to effect a separation of ^{235}U from ^{238}U, since only ^{235}U would undergo the nuclear fission reaction needed to build an atomic bomb. This separation was achieved at the Oak Ridge National Laboratories using Graham's law and the fact that the normal boiling point of UF_6 is 329.4 K, so that that compound can easily be put into the gas phase. (See Problem 17.28.) Equation 17.85, therefore, played a central role in efforts by the United States to rid the world of the Nazi regime.

17.2.3   The Collision Frequency

The collision frequency plays an important role not only in the kinetic behavior of gases, but also in the formulation of collision theories of chemical reaction rates. That molecules cannot react unless they approach one another is self-evident.

If all the molecules in a container were stationary, there would be no collisions and hence no reactions. It seems intuitive that as the velocities of the molecules increase, we will observe an increase in the number of collisions. This expectation is usually correct, but it is not the velocities of the molecules relative to the walls of the container that determine the number of collisions; it is their velocity relative to each other, or their *relative velocity*. Consider an ensemble of automobiles on a crowded highway. Relative to the highway, the cars may all be moving at velocities of perhaps 96.2 km hr^{-1} ($\approx$ 60 miles hr^{-1}) in the direction of the highway, yet no collisions occur! The reason is that the collision frequency depends upon the relative velocity $\mathbf{V}$, not upon the velocities of the cars relative to the stationary highway. The relative velocity between molecules A and B is the vector difference of their velocities with respect to the walls of the container:

$$\mathbf{V} = \mathbf{v}_A - \mathbf{v}_B. \tag{17.86}$$

Since the velocities of the cars on the highway are all nearly the same, we have $\mathbf{V} = 0$, and no collisions occur. If, however, you drive 96 km hr^{-1} in the wrong direction on a one-way street, the magnitude of the relative velocity between your car and the other cars going in the opposite direction at about the same speed is $V = 96$ km hr$^{-1} - (-96$ km hr$^{-1}) = 192$ km hr^{-1}, and we will now observe collisions. (Do not try this experiment in your automobile!)

We now need to define precisely what is meant by the term "collision." Quantum mechanics shows that the electron distribution surrounding the nuclei contained in a molecule extends over all space. Consequently, if molecule A were to pass molecule B at any distance, their respective electron clouds would come into contact, and we would have a "collision." This analysis demonstrates that the term "collision" standing alone is meaningless when we are dealing with molecules. To assign it meaning, we must speak in terms of the frequency with which molecule A passes molecule B at a distance in the range from b to $b + db$. This situation is illustrated in Figure 17.12. Molecule A with laboratory velocity $\mathbf{v}_A$ to $\mathbf{v}_A + d\mathbf{v}_A$ moves toward molecule B, whose laboratory velocity is $\mathbf{v}_B$ to $\mathbf{v}_B + d\mathbf{v}_B$. These laboratory velocities give a relative velocity $\mathbf{V}$ to $\mathbf{V} + d\mathbf{V}$, which we know is the quantity that determines the collision frequency. If we draw the extension of the relative velocity vector as shown in the figure, the perpendicular distance from B to this extension is called the *impact parameter*, which is given the symbol b. A molecular collision is defined to be the passage of molecule A past B at a dis-

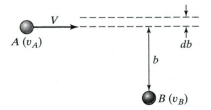

▲ FIGURE 17.12

Collision at impact parameter b to $b + db$ between molecules A and B moving with laboratory velocities $\mathbf{v}_A$ and $\mathbf{v}_B$, respectively, such that their relative velocity is $\mathbf{V}$.

tance in the impact parameter range from b to $b + db$. The frequency with which this event occurs when A and B have laboratory velocities in the ranges from $\mathbf{v}_A$ to $\mathbf{v}_A + d\mathbf{v}_A$ and $\mathbf{v}_B$ to $\mathbf{v}_B + d\mathbf{v}_B$, respectively, and relative velocity $\mathbf{V}$ to $\mathbf{V} + d\mathbf{V}$ is represented by the symbol $Z_{AB}(\mathbf{v}_A, \mathbf{v}_B, \mathbf{V}, b)$. Our task is to find the appropriate expression for Z_{AB}.

To obtain $Z_{AB}(\mathbf{v}_A, \mathbf{v}_B, \mathbf{V}, b)$, we focus our attention on a single A molecule with laboratory velocity $\mathbf{v}_A$ to $\mathbf{v}_A + d\mathbf{v}_A$ in a container holding N_A A molecules m^{-3}. Suppose you are riding on this molecule as it moves through the container for 1 second. Relative to the B molecules inside the container, this journey will move you through a distance of V meters, since the magnitude of the relative velocity is V m s^{-1}. This situation is illustrated in Figure 17.13. Your task is to count the number of times you pass a B molecule whose laboratory velocity is $\mathbf{v}_B$ to $\mathbf{v}_B + d\mathbf{v}_B$ at an impact parameter distance b to $b + db$. If you do the count accurately, you will have determined the number of collisions per second of one A molecule whose velocity is in the range $\mathbf{v}_A$ to $\mathbf{v}_A + d\mathbf{v}_A$ with the B molecules in the container that have velocities $\mathbf{v}_B$ to $\mathbf{v}_B + d\mathbf{v}_B$. If this number is then multiplied by the number of such A molecules in the container, we will have $Z_{AB}(\mathbf{v}_A, \mathbf{v}_B, \mathbf{V}, b)$. Unfortunately (or fortunately, depending upon your point of view), you're too big to ride on the A molecule, and you probably can't count fast enough or high enough, so we need to deduce what your count would have been if you had actually executed this task.

The total number of B molecules that you will pass at impact parameter b to $b + db$ is the number of B molecules in the annular volume between the two right circular cylinders of radii $b + db$ and b, respectively, as shown by the shaded end piece in Figures 17.13 and 17.14. This volume is given by

$$\tau = \pi V(b + db)^2 - \pi V b^2 = 2\pi b V \, db + \pi V (db)^2 = 2\pi b V \, db, \qquad \textbf{(17.87)}$$

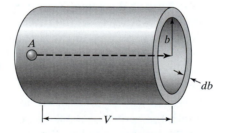

▶ **FIGURE 17.13**
The collision frequency at an impact parameter b to $b + db$ of a single A molecule with laboratory velocity $\mathbf{v}_A$ to $\mathbf{v}_A + d\mathbf{v}_A$, with B molecules having laboratory velocities in the range $\mathbf{v}_B$ to $\mathbf{v}_B + d\mathbf{v}_B$. The magnitude of the relative velocity is V.

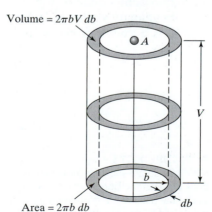

Volume = $2\pi b V \, db$

Area = $2\pi b \, db$

▶ **FIGURE 17.14**
Vertical view of the collision volume at impact parameter b to $b + db$ swept out by molecule A in Figure 17.13 in 1 second. The volume is that between the two right circular cylinders of radii b and $b + db$.

since, in the limit of infinitesimal db, $(db)^2$ is negligible relative to db. If the concentration of B molecules is N_B molecules m^{-3}, the total number of B molecules in volume τ is τN_B. The fraction of this total with velocities $\mathbf{v}_B$ to $\mathbf{v}_B + d\mathbf{v}_B$ is given by Eq. 17.55:

$$f(\mathbf{v}_B)d\mathbf{v}_B = \left[\frac{m_B}{2\pi kT}\right]^{3/2} \exp\left[-\frac{m_B[v_x^2 + v_y^2 + v_z^2]}{2kT}\right]_B dv_{xB}dv_{yB}dv_{zB}.$$

Therefore, the count you would have obtained had you been able to execute the experiment is

$$z_{AB}(\mathbf{v}_A, \mathbf{v}_B, V, b) = \tau N_B f(\mathbf{v}_B)d\mathbf{v}_B = 2\pi bV\,db\,N_B f(\mathbf{v}_B)d\mathbf{v}_B,$$

where $z_{AB}(\mathbf{v}_A, \mathbf{v}_B, V, b)$ is the collision frequency for one A molecule with laboratory velocity in the range $\mathbf{v}_A$ to $\mathbf{v}_A + d\mathbf{v}_A$. If this quantity is multiplied by the total number of A molecules with that velocity, we obtain $Z_{AB}(\mathbf{v}_A, \mathbf{v}_B, V, b)$. The total number of A molecules with laboratory velocities in the range $\mathbf{v}_A$ to $\mathbf{v}_A + d\mathbf{v}_A$ is $N_A f(\mathbf{v}_A)d\mathbf{v}_A$. Therefore,

$$Z_{AB}(\mathbf{v}_A, \mathbf{v}_B, V, b) = N_A f(\mathbf{v}_A)d\mathbf{v}_A z_{AB}(\mathbf{v}_A, \mathbf{v}_B, V, b)$$

$$= N_A N_B V f(\mathbf{v}_A)d\mathbf{v}_A f(\mathbf{v}_B)d\mathbf{v}_B(2\pi b)db. \qquad \textbf{(17.88)}$$

Equation 17.88 shows that the collision frequency is proportional to the concentration of the colliding species. This result is generally called the *law of mass action*. If we are concerned with the collision frequency of A molecules with other A molecules, then Eq. 17.88 must be divided by a factor of two to avoid counting the A–A collisions twice.

Since we know that the collision frequency depends only upon the relative velocity, it should be possible to eliminate the laboratory velocities from Eq. 17.88. This can be done by expressing $\mathbf{v}_A$ and $\mathbf{v}_B$ in terms of $\mathbf{V}$ and the velocity vector of the center of mass, $\mathbf{G}$. Assume that molecules A and B are at the same point in space at time $t = 0$. After 1 second, A has moved a distance v_A in the direction of its laboratory velocity, while B has moved a distance v_B in the direction of $\mathbf{v}_B$. This situation is illustrated in Figure 17.15, which is called a *Newton diagram*. Initially, both molecules are at point P_1. After 1 second, A is at point P_3 while B is at point P_2. Therefore, at $t = 1$ s, the center of mass must lie somewhere along the line connecting P_2 and P_3. We take this point to be P_4. Consequently, in 1 second, the center of mass has moved from P_1 to P_4. A vector connecting these two points must be the velocity vector of the center of mass, $\mathbf{G}$. The relative velocity vector is given by Eq. 17.86. The vector difference of $\mathbf{v}_A$ and $\mathbf{v}_B$ is a vector from P_2 to P_3 as shown in Figure 17.15. Finally, we

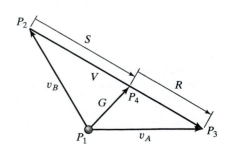

▶ **FIGURE 17.15**
Newton diagram showing conversion equations between the laboratory velocities $\mathbf{v}_A$ and $\mathbf{v}_B$ and the relative and center-of-mass velocities, $\mathbf{V}$ and $\mathbf{G}$, respectively.

assign the distance between points P_2 and P_4 to be S, while that separating P_4 and P_3 is labeled R.

Since the total moment about the center of mass must be zero, we have

$$m_B \mathbf{S} = m_A \mathbf{R}. \tag{17.89}$$

Inspection of Figure 17.15 shows that we also have

$$\mathbf{V} = \mathbf{S} + \mathbf{R}. \tag{17.90}$$

Combination of Eqs. 17.89 and 17.90 gives

$$\mathbf{V} = \mathbf{S} + \frac{m_B}{m_A}\mathbf{S} = \left[1 + \frac{m_B}{m_A}\right]\mathbf{S} = \frac{M}{m_A}\mathbf{S}, \tag{17.91}$$

where $M = m_A + m_B$. Solving Eq. 17.91 for $\mathbf{S}$ and using Eq. 17.89 produces

$$\mathbf{S} = \frac{m_A}{M}\mathbf{V} \tag{17.92A}$$

and

$$\mathbf{R} = \frac{m_B}{M}\mathbf{V}. \tag{17.92B}$$

The transformation equations between the laboratory velocities, on the one hand, and $\mathbf{V}$ and $\mathbf{G}$, on the other, can be obtained directly from Figure 17.15. The two vector equations connecting $\mathbf{V}$ and $\mathbf{G}$ with $\mathbf{v}_A$ and $\mathbf{v}_B$ are

$$\mathbf{v}_A = \mathbf{G} + \mathbf{R} = \mathbf{G} + \frac{m_B}{M}\mathbf{V} \tag{17.93}$$

and

$$\mathbf{v}_B = \mathbf{G} - \mathbf{S} = \mathbf{G} - \frac{m_A}{M}\mathbf{V}, \tag{17.94}$$

where we have used Eqs. 17.92 for $\mathbf{R}$ and $\mathbf{S}$. We may now use the two transformation equations 17.93 and 17.94 to convert $f(\mathbf{v}_A)d\mathbf{v}_A$ and $f(\mathbf{v}_B)d\mathbf{v}_B$ into expressions involving $\mathbf{V}$, and $\mathbf{G}$, and their volume elements. The product of these two distribution functions is

$$f(\mathbf{v}_A)f(\mathbf{v}_B)d\mathbf{v}_A d\mathbf{v}_B = \left[\frac{1}{2\pi kT}\right]^3 [m_A m_B]^{3/2} \exp\left[-\frac{m_B v_B^2 + m_A v_A^2}{2kT}\right]d\mathbf{v}_A d\mathbf{v}_B.$$

Using Eqs. 17.93 and 17.94, we obtain

$$m_B v_B^2 + m_A v_A^2$$

$$= m_B\left[G^2 + \frac{m_A^2}{M^2}V^2 - \frac{2m_A}{M}GV\right] + m_A\left[G^2 + \frac{m_B^2}{M^2}V^2 + \frac{2m_B}{M}GV\right]$$

$$= MG^2 + \left[\frac{m_B m_A^2 + m_A m_B^2}{M^2}\right]V^2 = MG^2 + \frac{m_A m_B}{M}\left[\frac{m_A + m_B}{M}\right]V^2$$

$$= MG^2 + \mu V^2, \tag{17.95}$$

where μ is the $A\text{–}B$ reduced mass. This transformation of variables produces

$$f(\mathbf{v}_A)f(\mathbf{v}_B)d\mathbf{v}_A d\mathbf{v}_B = \left[\frac{(m_A m_B)^{1/2}}{2\pi kT}\right]^3 \exp\left[-\frac{MG^2}{2kT}\right]\exp\left[-\frac{\mu V^2}{2kT}\right]d\mathbf{V}d\mathbf{G}$$

$$= \left[\frac{(m_A m_B)^{1/2}}{2\pi kT}\right]^3 \exp\left[-\frac{MG^2}{2kT}\right]\exp\left[-\frac{\mu V^2}{2kT}\right]dV_x dV_y dV_z dG_x dG_y dG_z. \tag{17.96}$$

Using Eq. 17.96, we may write the collision frequency from Eq. 17.88 as

$$Z_{AB}(\mathbf{G}, \mathbf{V}, b) = N_A N_B V (2\pi b) db \left[\frac{(m_A m_B)^{1/2}}{2\pi kT} \right]^3$$

$$\exp\left[-\frac{MG^2}{2kT} \right] \exp\left[-\frac{\mu V^2}{2kT} \right] dV_x dV_y dV_z dG_x dG_y dG_z. \qquad (17.97)$$

In this form, $Z_{AB}(\mathbf{G}, \mathbf{V}, b)$ gives the frequency of collision between A and B molecules in an impact parameter range from b to $b + db$ with a relative velocity $\mathbf{V}$ to $\mathbf{V} + d\mathbf{V}$ while the velocity of the center of mass lies in the range from $\mathbf{G}$ to $\mathbf{G} + d\mathbf{G}$. However, we know that the collision frequency of molecules within a container is independent of the velocity of the center of mass. This is the same thing as saying that the collision frequency of helium atoms held within a closed flask at 298 K is the same if we hold the flask stationary or if we run across the room while holding the flask in our hand. In effect, we can sum $Z_{AB}(\mathbf{G}, \mathbf{V}, b)$ over all center-of-mass velocities and not affect the total collision frequency. This operation produces

$$Z_{AB}(\mathbf{V}, b) = N_A N_B V (2\pi b) db \left[\frac{(m_A m_B)^{1/2}}{2\pi kT} \right]^3 \exp\left[-\frac{\mu V^2}{2kT} \right] dV_x dV_y dV_z$$

$$\times \int_{G_x=-\infty}^{\infty} \exp\left[-\frac{MG_x^2}{2kT} \right] dG_x \int_{G_y=-\infty}^{\infty} \exp\left[-\frac{MG_y^2}{2kT} \right] dG_y \int_{G_z=-\infty}^{\infty} \exp\left[-\frac{MG_z^2}{2kT} \right] dG_z.$$

$$(17.98)$$

Execution of the three integrals in Eq. 17.98 gives

$$Z_{AB}(\mathbf{V}, b) = N_A N_B V (2\pi b) db \left[\frac{(m_A m_B)^{1/2}}{2\pi kT} \right]^3 \exp\left[-\frac{\mu V^2}{2kT} \right] dV_x dV_y dV_z \left[\frac{2\pi kT}{M} \right]^{3/2}$$

$$= N_A N_B V \left[\frac{\mu_{AB}}{2\pi kT} \right]^{3/2} (2\pi b) db \exp\left[-\frac{\mu V^2}{2kT} \right] dV_x dV_y dV_z. \qquad (17.99)$$

Equation 17.99 almost gives us the final expression; the only remaining problem is that $Z_{AB}(\mathbf{V}, b)$ gives us the collision frequency at impact parameter b to $b + db$ when the relative velocity is $\mathbf{V}$ to $\mathbf{V} + d\mathbf{V}$. Because the relative velocity, rather than its magnitude, is involved, the expression includes information on the direction of the collision relative to the container walls. We really want to know the total collision frequency for all directions, since that is the quantity upon which chemical reaction rates depend. We can add up the collisions for all directions simply by converting the rectangular Cartesian relative velocities in Eq. 17.99 to a spherical coordinate system in which we have

$$V_x = V \sin \theta \cos \phi,$$

$$V_y = V \sin \theta \sin \phi,$$

and

$$V_z = V \cos \theta$$

and then summing (i.e., integrating) over all angles for the relative velocity vector. This gives

$$Z_{AB}(V, b) = N_A N_B V \left[\frac{\mu_{AB}}{2\pi kT} \right]^{3/2} (2\pi b) db \exp\left[-\frac{\mu V^2}{2kT} \right] V^2 dV \int_{\phi=0}^{2\pi} d\phi \int_{\theta=0}^{\pi} \sin \theta \, d\theta.$$

The angular integrals give 4π; thus,

$$Z_{AB}(V, b) = 4\pi N_A N_B \left[\frac{\mu_{AB}}{2\pi kT}\right]^{3/2} (2\pi b)\,db\,\exp\left[-\frac{\mu V}{2kT}\right] V^3\,dV. \quad \textbf{(17.100)}$$

Equation 17.100 is the generalized expression for the gas-phase collision frequency.

EXAMPLE 17.7

An argon atom is moving with a speed of 400 m s^{-1}. A neon atom is moving with a speed of 450 m s^{-1}. The angle between the argon and neon velocity vectors is 115°. **(A)** Compute the magnitude of the relative velocity in m s^{-1}. **(B)** Compute the angle between $\mathbf{V}$ and $\mathbf{v}_{Ar}$ if we define $\mathbf{V} = \mathbf{v}_{Ne} - \mathbf{v}_{Ar}$. **(C)** Compute the speed of the center of mass, in m s^{-1}. **(D)** Compute the total kinetic energy of the two atoms, using $T = (m_{Ar}v_{Ar}^2/2) + (m_{Ne}v_{Ne}^2/2)$. Now compute the total kinetic energy, using the magnitude of the relative velocity and the center-of-mass velocity. What relationship exists between the two calculations of the total kinetic energy? Use the average atomic masses (i.e., atomic weights) for Ne and Ar.

Solution

The Newton diagram for this system is as follows:

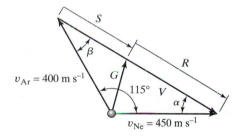

(A) The relative velocity vector can be computed directly by using the law of cosines. This gives

$$V^2 = (400)^2 + (450)^2 - 2(400)(450)\cos 115° = 514{,}643 \text{ m}^2\,\text{s}^{-2}, \quad \textbf{(1)}$$

so that

$$V = 717.39 \text{ m s}^{-1}. \quad \textbf{(2)}$$

(B) The angle between $\mathbf{V}$ and $\mathbf{v}_{Ar}$ is β in the preceding figure. Using the law of cosines again, we have

$$(450)^2 = (400)^2 + (717.39)^2 - 2(400)(717.39)\cos\beta. \quad \textbf{(3)}$$

Solving for $\cos\beta$, we obtain

$$\cos\beta = \frac{(717.39)^2 + (400)^2 - (450)^2}{2(400)(717.39)} = 0.82268, \quad \textbf{(4)}$$

which gives $\beta = 34.646°$.

(C) The distance R in the figure can be computed as follows:

$$V = S + R = 717.39. \quad \textbf{(5)}$$

Since the moment arms about the center of mass must be equal, we have

$$m_{Ar}S = m_{Ne}R \tag{6}$$

which gives

$$S = \frac{m_{Ne}}{m_{Ar}}R = \frac{20.18}{39.945}R = 0.50519R. \tag{7}$$

Substitution of Eq. 7 into Eq. 5 produces

$$S + R = 0.50519R + R = 1.50519R = 717.39. \tag{8}$$

This gives $R = 476.61$ m s^{-1}.
The angle α in the figure is

$$\alpha = 180° - \beta - 115° = 65° - 34.646° = 30.354°. \tag{9}$$

We can now calculate the magnitude of the velocity of the center of mass by using the law of cosines:

$$G^2 = (450)^2 + (476.61)^2 - 2(450)(476.61)\cos(30.354°) = 59,508.57 \text{ m}^2 \text{ s}^{-2}. \tag{10}$$

$$G = 243.94 \text{ m s}^{-1}. \tag{11}$$

(D) To compute the kinetic energies, we need the masses of argon and neon, the total mass, and the reduced mass. These quantities are

$$m_{Ar} = \frac{0.039945 \text{ kg mol}^{-1}}{6.02214 \times 10^{23} \text{ mol}^{-1}} = 6.6330 \times 10^{-26} \text{ kg}, \tag{12}$$

$$m_{Ne} = \frac{0.020180 \text{ kg mol}^{-1}}{6.02214 \times 10^{23} \text{ mol}^{-1}} = 3.3510 \times 10^{-26} \text{ kg}, \tag{13}$$

$$\mu = \frac{m_{Ar}m_{Ne}}{m_{Ar} + m_{Ne}} = \frac{(6.6330)(3.3510) \times 10^{-26}}{(6.633 + 3.351)} = 2.2263 \times 10^{-26} \text{ kg}, \tag{14}$$

and

$$M = m_{Ar} + m_{Ne} = 9.984 \times 10^{-26} \text{ kg}. \tag{15}$$

Using the laboratory speeds to compute the kinetic energy gives

$$T = 0.5m_{Ar}v_{AR}^2 + 0.5m_{Ne}v_{Ne}^2 = 0.5(6.633 \times 10^{-26} \text{ kg})(400)^2 \text{ m}^2 \text{ s}^{-2}$$
$$+ 0.5(3.351 \times 10^{-26} \text{ kg})(450)^2 \text{ m}^2 \text{ s}^{-2} = 8.6993 \times 10^{-21} \text{ J}. \tag{16}$$

Using the relative and center of mass speeds, we obtain

$$T = 0.5MG^2 + 0.5\mu V^2$$
$$= 0.5(9.984 \times 10^{-26})(243.94)^2 + 0.5(2.2263 \times 10^{-26})(717.39)^2 \text{ J}$$
$$= 8.6994 \times 10^{-21} \text{ J}. \tag{17}$$

Both calculations yield the same kinetic energy, as we knew they must because of the analysis leading to Eq. 17.95.

For related exercises, see Problems 17.31 and 17.32.

17.2.4 Hard-sphere Models

Although the electron distribution about a molecule extends over all space, most of the negative charge density is contained within a relatively small volume surrounding the nuclei. We might, therefore, enclose a molecule within a spherical cavity whose radius is sufficiently large to contain perhaps 90% of the total electron density and then say that this is the "size" of the molecule. This type of approximation is often called a *hard-sphere* model.

If we employ a hard-sphere model, it is a relatively simple matter to use Eq. 17.100 to obtain an expression for the collision frequency. Figure 17.16 shows that hard spheres will always collide when the impact parameter is less than or equal to the sum of the radii of the spheres. This sum is called the *hard-sphere collision radius*. In most texts, it is given the symbol σ_{AB}. We may, therefore, obtain the total collision frequency of such hard spheres by summing Eq. 17.100 over all possible relative speeds and impact parameters over the range $0 \le b \le \sigma_{AB}$:

$$
Z_{total} = \int_{V=0}^{\infty} \int_{b=0}^{\sigma_{AB}} Z_{AB}(V, b)
$$

$$
= N_A N_B \int_{V=0}^{\infty} \int_{b=0}^{\sigma_{AB}} 4\pi \left[\frac{\mu_{AB}}{2\pi kT} \right]^{3/2} (2\pi b)\, db\, \exp\left[-\frac{\mu V^2}{2kT} \right] V^3 dV
$$

$$
= N_A N_B 4\pi \left[\frac{\mu_{AB}}{2\pi kT} \right]^{3/2} (\pi\sigma_{AB}^2) \int_{V=0}^{\infty} \exp\left[-\frac{\mu V^2}{2kT} \right] V^3 dV. \qquad (17.101)
$$

The integral over relative speeds can be found by integrating by parts twice. The result is $2k^2 T^2/\mu^2$. Inserting this result into Eq. 17.101 produces

$$
\boxed{ Z_{total} = N_A N_B 4\pi \left[\frac{\mu_{AB}}{2\pi kT} \right]^{3/2} (\pi\sigma_{AB}^2) \frac{2k^2 T^2}{\mu^2} = N_A N_B (\pi\sigma_{AB}^2) \left[\frac{8kT}{\pi\mu} \right]^{1/2}. }
$$

$$(17.102)$$

Equation 17.102 gives the total number of collisions per unit time per unit volume in a systems of hard spheres with concentrations N_A and N_B. The number of collisions per A molecule per unit time is

$$
\boxed{ z_{AB} = \frac{Z_{total}}{N_A} = N_B(\pi\sigma_{AB}^2) \left[\frac{8kT}{\pi\mu} \right]^{1/2}. } \qquad (17.103)
$$

The quantity $\pi\sigma_{AB}^2$ has the units of cross-sectional area. Consequently, it is often called the *hard-sphere collision cross section*. Equation 17.75 shows that the factor under the radical is the average relative speed. It should be remembered that if we are computing the collision frequency between

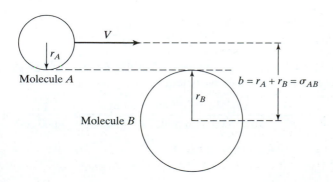

Molecule A

Molecule B

V

r_A

r_B

$b = r_A + r_B = \sigma_{AB}$

◀ FIGURE 17.16
Collision of two hard spheres, showing that the maximum possible impact parameter for collision is $b_{max} = r_A + r_B = \sigma_{AB}$.

identical molecules, we must divide Z_{total} by a factor of two to avoid counting collisions twice.

The units on z_{AB} are reciprocal time, the same as that for a vibrational frequency. Just as the reciprocal of the vibration frequency is the period for a complete vibration, the reciprocal of z_{AB} is the time between collisions suffered by an A molecule. This time is

$$\Delta t = \frac{1}{z_{AB}} = \frac{1}{N_B(\pi\sigma_{AB}^2)\left[\dfrac{8kT}{\pi\mu}\right]^{1/2}}.$$

If we multiply Δt by the average speed of A molecules, we obtain the average distance traveled by a hard-sphere A molecule between collisions. This quantity is called the *mean free path*. It is given by

$$\langle L_{AB}\rangle = \frac{1}{N_B(\pi\sigma_{AB}^2)\left[\dfrac{8kT}{\pi\mu}\right]^{1/2}}\left[\frac{8kT}{\pi m_A}\right]^{1/2} = \frac{(\mu/m_A)^{1/2}}{N_B(\pi\sigma_{AB}^2)}. \qquad (17.104)$$

EXAMPLE 17.8

A container holds helium atoms at 300 K and 1 atm pressure. Assuming that we may treat helium as an ideal gas with a hard-sphere radius of 0.5×10^{-10} m and a mass of 0.0040026 kg mol^{-1}, compute **(A)** the total number of He–He collisions per second per m^3 occurring in the container, **(B)** the average number of collisions per He atom per second, and **(C)** the mean free path of a helium atom in the container. **(D)** What would the mean free path be if the helium pressure inside the container were 10^{-4} torr?

Solution

(A) Z_{total} is given by Eq. 17.102, with the modification that B and A are identical, which requires that we insert a factor of 0.5 to avoid counting He–He collisions twice. These modifications produce

$$Z_{total} = 0.5N_A^2(\pi\sigma_{AA}^2)\left[\frac{8kT}{\pi\mu}\right]^{1/2}, \qquad (1)$$

where $\sigma_{AB} = 2(0.5 \times 10^{-10} \text{ m}) = 1 \times 10^{-10}$ m. For He–He collisions, the reduced mass is

$$\mu = \frac{m_{He}m_{He}}{2m_{He}} = \frac{m_{He}}{2} = \frac{0.0040026 \text{ kg mol}^{-1}}{2(6.02214 \times 10^{23} \text{ mol}^{-1})} = 3.3232 \times 10^{-27} \text{ kg}. \qquad (2)$$

The concentration, in number of atoms m^{-3}, can be obtained from the ideal-gas law; that is,

$$PV = nRT = \frac{n_A RT}{N}, \qquad (3)$$

where N is Avogadro's constant and n_A is the number of He atoms present. Solving for the concentration, we obtain

$$N_A = \frac{n_A}{V} = \frac{NP}{RT} = \frac{(6.02214 \times 10^{23} \text{ mol}^{-1})(1 \text{ atm})}{(0.08206 \text{ L atm mol}^{-1}\text{ K}^{-1})(300 \text{ K})} \times \frac{1 \text{ L}}{1{,}000 \text{ cm}^3} \times \frac{10^6 \text{ cm}^3}{\text{m}^3}$$

$$= 2.446 \times 10^{25} \text{ m}^{-3}. \qquad (4)$$

Insertion of these data into Eq. 1 yields

$$Z_{total} = (0.5)(2.446 \times 10^{25})^2 \, m^{-6}(3.14159)(1.0 \times 10^{-10} \, m)^2$$

$$\times \left[\frac{8(1.3807 \times 10^{-23})(300)}{(3.14159)(3.3232 \times 10^{-27})} \right]^{1/2} m \, s^{-1} = 1.674 \times 10^{34} \, m^{-3} \, s^{-1}. \tag{5}$$

(B) The average number of collisions per helium atom per second is the total number of collisions divided by the number of helium atoms per m^3, or

$$z_{He-He} = \frac{Z_{total}}{N_A} = \frac{1.674 \times 10^{34} \, m^{-3} \, s^{-1}}{2.446 \times 10^{25} \, m^{-3}} = 6.844 \times 10^8 \, s^{-1}. \tag{6}$$

(C) The mean free path is given by Eq. 17.104:

$$\langle L_{He-He} \rangle = \frac{1}{z_{He-He}} \left[\frac{8kT}{\pi m_A} \right]^{1/2}. \tag{7}$$

The mass of He is

$$m_A = m_{He} = \frac{0.0040026 \, kg \, mol^{-1}}{6.02214 \times 10^{23} \, mol^{-1}} = 6.646 \times 10^{-27} \, kg. \tag{8}$$

Substituting the results into Eq. 7 gives

$$\langle L_{He-He} \rangle = \frac{1}{6.844 \times 10^8 \, s^{-1}} \left[\frac{8(1.3807 \times 10^{-23} \, J \, K^{-1})(300 \, K)}{(3.14159)(6.646 \times 10^{-27} \, kg)} \right]^{1/2}$$

$$= 1.841 \times 10^{-6} \, m = 1.841 \times 10^{-4} \, cm. \tag{9}$$

(D) Notice that $\langle L_{He-He} \rangle$ is inversely proportional to N_A, which Eq. 4 shows is directly proportional to the pressure. Therefore, the mean free path is inversely proportional to the pressure. This allows us to write

$$\langle L \rangle = \frac{c}{P}, \tag{10}$$

where c is a constant. The ratio of the mean free paths at pressures P_1 and P_2 is, therefore,

$$\frac{\langle L \rangle_1}{\langle L \rangle_2} = \frac{P_2}{P_1}, \tag{11}$$

which shows that

$$\langle L \rangle_2 = \frac{P_1}{P_2} \langle L \rangle_1. \tag{12}$$

Thus, with $P_1 = 760$ torr and $P_2 = 10^{-4}$ torr, we obtain

$$\langle L_{He-He} \rangle \text{ at } 10^{-4} \text{ torr} = [\langle L_{He-He} \rangle \text{ at } 760 \text{ torr}] \frac{760}{10^{-4}}$$

$$= 1.841 \times 10^{-6} \, m \times \frac{760}{10^{-4}} = 13.99 \, m. \tag{13}$$

Helium has a very small hard-sphere collision cross section and moves very fast because of its small mass. For this reason, the mean free path at a pressure of 10^{-4} torr is large. For many gases, a reasonable rule of thumb is that the mean free path at 10^{-4} torr is about 1 meter.

For related exercises, see Problems 17.33 and 17.34.

▶ **FIGURE 17.17**
The distribution of free paths of molecule A in a container at temperature T and a concentration of B molecules equal to N_B. As illustrated, some free paths are very short, and some are much longer. If the molecules can be treated as hard spheres with collision radius σ_{AB}, Eq. 17.104 gives the average free path.

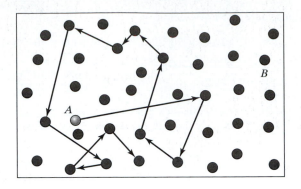

17.2.5 The Distribution of Free Paths

The analysis in the previous section gives us the average, or mean, free path for a system of hard spheres in Maxwell–Boltzmann equilibrium. This average free path is the result of a very broad distribution of possible free paths, as illustrated in Figure 17.17. Although we know the average free path, we do not have the normalized probability distribution for observing a free path in the range from X to $X + dX$. In this section, we obtain that distribution.

Consider the situation illustrated in Figure 17.18. Molecule A moves from point P_1 to P_2 through a distance X_1 without colliding. It then moves from point P_2 to P_3 through an additional distance X_2, again without colliding. Let the probability that the molecule will *not* undergo a collision as it moves through a distance X be given by $f(X)$. With this notation, the probability that the molecule will have been able to move from point P_1 to P_2 without colliding is $f(X_1)$, while that for moving from P_2 to P_3 without colliding is $f(X_2)$. In Chapter 10, we learned that the combined probability that both of these independent events will occur is the product of the individual probabilities. That is,

Probability of no collision in moving from P_1 to $P_3 = f(X_1)f(X_2)$. **(17.105)**

However, the left-hand side of Eq. 17.105 can also be written as $f(X_t) = f(X_1 + X_2)$. Therefore, we must have

$$f(X_1 + X_2) = f(X_1)f(X_2). \quad \textbf{(17.106)}$$

The only function that has the property described by Eq. 17.106 is an exponential function

$$f(X) = \exp(-aX), \quad \textbf{(17.107)}$$

where a is a constant greater than zero. (For a proof of this statement, see Section 18.2.) We must have $a > 0$ because $f(X)$ must approach zero as X becomes very large.

To obtain the normalized probability distribution for free paths, we need an expression that gives us the probability that a molecule will move a distance X without colliding and then undergo a collision as it moves through

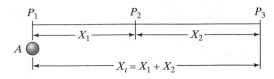

▶ **FIGURE 17.18**
Relationship of free-path probabilities.

an additional distance ΔX. If this combined probability is denoted by $P(X)$ ΔX, we must have

$P(X)\Delta X = [\text{probability of moving through distance } X \text{ without collision}]$

$\times [\text{probability of having a collision while moving through distance } \Delta X]$

$$= f(X)[1 - f(\Delta X)] = f(X) - f(X)f(\Delta X), \qquad (17.108)$$

since the probability of having a collision is one minus the probability that no collision will occur. Using Eq. 17.106, we can write $P(X)\Delta X$ in the form

$$P(X)\Delta X = f(X) - f(X + \Delta X) = -[f(X + \Delta X) - f(X)]. \qquad (17.109)$$

Dividing by ΔX produces

$$P(X) = -\frac{f(X + \Delta X) - f(X)}{\Delta X}. \qquad (17.110)$$

If we now take the limit of Eq. 17.110 as $\Delta X \to 0$, we obtain

$$P(X) = -\lim_{\Delta X \to 0}\left[\frac{f(X + \Delta X) - f(X)}{\Delta X}\right] = -\frac{df(X)}{dX}, \qquad (17.111)$$

because the middle expression is the definition of the derivative. With $f(X) = \exp(-aX)$, Eq. 17.111 gives $P(X) = a\exp(-aX)$. The probability distribution for free paths is, therefore,

$$P(X)dX = a\exp(-aX)dX. \qquad (17.112)$$

It is easy to show that the probability distribution given by Eq. 17.112 is normalized. If we sum $P(X)dX$ over all possible free paths, we obtain

$$\int_{X=0}^{\infty} P(X)dX = a\int_{X=0}^{\infty} \exp(-aX)dX = -\exp(-aX)|_o^\infty = 1,$$

which is the condition for a normalized distribution.

We can find the constant a in Eq. 17.112 by computing the mean free path:

$$\langle L \rangle = \int_{X=0}^{\infty} XP(X)dX = a\int_{X=0}^{\infty} X\exp(-aX)dX = a\left[\frac{1!}{a^2}\right] = \frac{1}{a}. \qquad (17.113)$$

Equation 17.113 demonstrates that $a = 1/\langle L \rangle$, so that the normalized distribution of free paths is

$$\boxed{P(X)dX = \frac{1}{\langle L \rangle}\exp\left[-\frac{X}{\langle L \rangle}\right]dX}. \qquad (17.114)$$

If the molecules are treated as hard spheres, $\langle L \rangle$ is given by Eq. 17.104.

If we define a reduced free path as $X_R = X/\langle L \rangle$, Eq. 17.114 can be used to obtain the normalized probability distribution for reduced free paths. Using the fact that $dX = \langle L \rangle dX_R$, we obtain

$$P(X_R)dX_R = \exp(-X_R)dX_R, \qquad (17.115)$$

which is the same for all gases. Figure 17.19 illustrates the important features of this probability distribution.

EXAMPLE 17.9

The distribution of free paths can be used to obtain a variety of useful equations related to the collision frequency. Suppose an investigator is passing a beam of A molecules into a chamber filled with B molecules. He would like the A molecules

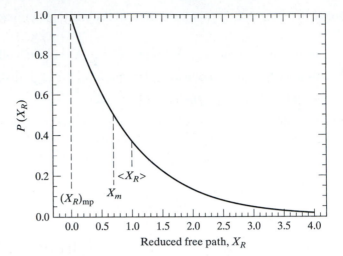

▶ **FIGURE 17.19**
The normalized probability distribution for reduced free paths, Eq. 17.115. The most probable reduced free path $(X_R)_{mp}$, the median reduced free path (X_m), and the average reduced free path $\langle X_R \rangle$, are all shown in the figure.

to undergo exactly one collision before they reach the detector, which is located an average distance of X_o meters away from the entrance slit to the chamber. Obtain an expression in terms of X_o and $\langle L \rangle$ that gives the probability that exactly one collision will occur.

Solution

The investigator desires that the following events occur:

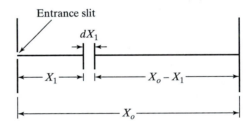

He wants the molecule to move a distance X_1 and then collide between X_1 and $X_1 + dX_1$, with dX_1 infinitesimally small. After the collision, he wants the molecule to move the remaining distance, $X_o - X_1$, without further collision, so that there will be precisely one collision over the path length X_o. Since the investigator does not care where along the path the single collision occurs, we need to obtain an expression giving the probability that the foregoing sequence of events will take place for a particular value of X_1 and then sum that probability over all possible values of X_1 from zero to X_o. The sum will be the desired expression.

The probability of observing a free path length X_1 is

$$P(X_1)dX_1 = \frac{1}{\langle L \rangle} \exp\left[-\frac{X_1}{\langle L \rangle}\right] dX_1. \qquad (1)$$

The probability that there will be no collision over the path length $X_o - X_1$ is given by Eq. 17.107 with $a = 1/\langle L \rangle$:

$$P(\text{no collision over distance } X_o - X_1) = \exp\left[-\frac{X_o - X_1}{\langle L \rangle}\right]. \qquad (2)$$

The combined probability that both of these independent events will take place is the product of the two probabilities:

$P(\text{one collision at } X_1)dX_1 = P(X_1)dX_1 \, P(\text{no collision over distance } X_o - X_1)$

$$= \frac{1}{\langle L \rangle} \exp\left[-\frac{X_1}{\langle L \rangle} \right] dX_1 \exp\left[-\frac{X_o - X_1}{\langle L \rangle} \right]$$

$$= \frac{1}{\langle L \rangle} \exp\left[-\frac{X_o}{\langle L \rangle} \right] dX_1. \tag{3}$$

As noted, the investigator does not care where the single collision occurs. Therefore, we must add up the probabilities that the collision will occur at any point X_1 between 0 and X_o. As usual, we sum a continuous distribution by integration. Therefore,

$$P(\text{one collision}) = \int_{X_1=0}^{X_o} P(\text{one collision at } X_1)dX_1$$

$$= \int_{X_1=0}^{X_o} \frac{1}{\langle L \rangle} \exp\left[-\frac{X_o}{\langle L \rangle} \right] dX_1 = \frac{1}{\langle L \rangle} \exp\left[-\frac{X_o}{\langle L \rangle} \right] \int_{X_1=0}^{X_o} dX_1$$

$$= \frac{X_o}{\langle L \rangle} \exp\left[-\frac{X_o}{\langle L \rangle} \right]. \tag{4}$$

Equation 4 is the desired result.

For related exercises, see Problems 17.37, 17.38, and 17.39.

Summary: Key Concepts and Equations

1. The number of ways N distinguishable objects can be distributed over s different aspects with n_1 objects in aspect 1, n_2 objects in aspect 2, etc., is given by the s-nomial expansion coefficient, which has the form

$$C(N, n_1, n_2, n_3, \ldots, n_{s-1}, n_s) = \frac{N!}{\displaystyle\prod_{i=1}^{s} n_i!} = W.$$

This expression, called the thermodynamic probability, is generally assigned the symbol W. We have also referred to W as the distribution count.

2. The probability that a given distribution d_i of N objects among s aspects will be observed is given by the number of ways to obtain the distribution, divided by the total number of possible arrangements of the N objects in the s aspects. This quantity is equal to s^N. Therefore, the probability that distribution d_i will be observed is

$$P(d_i) = \frac{W(d_i)}{s^N}.$$

3. The average value of some property X of a system is given by

$$\langle X \rangle = \sum_{\alpha=1}^{n^D} X_\alpha P_\alpha,$$

where X_α is the value of X for the αth distribution or microstate of the objects over the available aspects. This is the fundamental equation of statistical mechanics. When N is extremely large and the number of aspects s is small, deviations from the most probable distribution are so unlikely that the foregoing summation reduces to a single term:

$$\langle X \rangle = X_{mp},$$

where X_{mp} is the value of X for the most probable distribution of objects among the available aspects. The most probable distribution is the distribution that maximizes W, subject to the constraints that are appropriate for the system.

4. Using the method of Lagrangian multipliers and Stirling's approximation for $\ln(n!)$, we find that W attains a maximum, subject to the constraints that mass and energy be conserved, whenever the number of molecules in energy state i is given by

$$n_i = \frac{N}{z}\exp\left[-\frac{\varepsilon_i}{kT}\right],$$

where z is the molecular partition function, ε_i is the energy of state i, and k is the Boltzmann constant, which is equal to R/N, in which N is Avogadro's constant. This result is called the Boltzmann distribution. It holds only in cases for which the number of molecules is large relative to the number of available energy states.

5. The molecular partition function is defined by

$$z = \sum_{i=1}^{\infty}\exp\left[-\frac{\varepsilon_i}{kT}\right] = \sum_{n=1}^{\infty}g_n\exp\left[-\frac{\varepsilon_n}{kT}\right].$$

In the first summation, the index i runs over all possible molecular-energy states. If the second summation, containing the degeneracy factor g_n, is used to represent z, the summation index n runs only over those energy states with different energies. Physically, the molecular partition function represents the effective number of energy states available to the molecules.

6. If we assume that translational-energy levels are so closely spaced that they can be treated as being classical, we find that the normalized distribution of molecular velocities in rectangular Cartesian coordinates is

$$n(v_x, v_y, v_z)dv_x dv_y dv_z = N\left[\frac{m}{2\pi kT}\right]^{3/2}\exp\left[-\frac{m[v_x^2 + v_y^2 + v_z^2]}{2kT}\right]dv_x dv_y dv_z.$$

If we convert to a spherical polar coordinate system in which one variable gives the magnitude of the velocity (i.e., the speed) while the other two provide information about the direction of motion, we can obtain the distribution of molecular speeds. This distribution, called the Maxwell–Boltzmann distribution, is

$$n(C)dC = 4\pi N\left[\frac{m}{2\pi kT}\right]^{3/2}\exp\left[-\frac{mC^2}{2kT}\right]C^2 dC.$$

The corresponding normalized probability distribution for observing a molecule with speed in the range from C to $C + dC$ is

$$P(C)dC = 4\pi\left[\frac{m}{2\pi kT}\right]^{3/2}\exp\left[-\frac{mC^2}{2kT}\right]C^2 dC.$$

7. The Maxwell–Boltzmann distribution of speeds shows that the most probable speed, the median speed, the average speed, and the average square speed are

$$C_{mp} = \left[\frac{2RT}{M}\right]^{1/2}, \; C_m = 1.0877\ldots\left[\frac{2RT}{M}\right]^{1/2}, \; \langle C\rangle = \left[\frac{8RT}{\pi M}\right]^{1/2}, \; \text{and } \langle C^2\rangle = \frac{3RT}{M},$$

respectively. In these equations, M is the molar mass. The average square speed leads directly to the conclusion that the average kinetic energy is $3RT/2$. If we use the most probable speed to define a reduced speed as $C_R = C/C_{mp}$, then the Maxwell–Boltzmann distribution for reduced speeds is identical for all gases at all temperatures. Its form is

$$P(C_R)dC_R = \left(\frac{16}{\pi}\right)^{1/2}\exp[-C_R^2]C_R^2\, dC_R.$$

8. To speak meaningfully about a collision frequency, we must speak in terms of the frequency with which two molecules pass within an impact parameter range from b to $b + db$ when their relative velocity lies in the range from V to $V + dV$.

We use the relative velocity because it is this velocity, not the velocities relative to the walls of the container, that determines the collision frequency. The total differential collision frequency per unit volume in the gas phase is given by

$$Z_{AB}(V, b) = 4\pi N_A N_B \left[\frac{\mu_{AB}}{2\pi kT}\right]^{3/2} (2\pi b)\, db \,\exp\left[-\frac{\mu V^2}{2kT}\right] V^3 \, dV.$$

The dependence of the collision frequency upon the product of the concentrations of the colliding species is called the law of mass action.

9. If A and B are assumed to be hard spheres with radii r_A and r_B, respectively, then the total collision frequency per unit volume can be obtained by summing the differential collision frequency over all relative velocities and over impact parameters between zero and $r_A + r_B$, which is called the hard-sphere collision radius σ_{AB}. The result is

$$Z_{total} = N_A N_B (\pi\sigma_{AB}^2)\left[\frac{8kT}{\pi\mu}\right]^{1/2},$$

which gives a collision frequency per A molecule of

$$z_{AB} = \frac{Z_{total}}{N_A} = N_B(\pi\sigma_{AB}^2)\left[\frac{8kT}{\pi\mu}\right]^{1/2}$$

and a hard-sphere mean free path of

$$\langle L_{AB}\rangle = \frac{(\mu/m_A)^{1/2}}{N_B(\pi\sigma_{AB}^2)}.$$

If we are concerned with collisions of A with other A molecules, the right sides of each of these equations must be divided by a factor of two to avoid counting collisions twice.

10. By employing the probability considerations discussed in Chapter 10, we can deduce the normalized probability distribution for free paths in a gas-phase system. The probability that we will observe a free path between X and $X + dX$ is

$$P(X)dX = \frac{1}{\langle L\rangle}\exp\left[-\frac{X}{\langle L\rangle}\right]dX,$$

where $\langle L\rangle$ is the mean free path for the system.

Problems

Problems that require the use of some type of computational device are marked with an asterisk (*). Problems that require some type of plotting routine are indicated with a pound sign (#).

17.1 Twelve coins are randomly tossed onto a table. What is the probability that we will obtain three heads and nine tails?

17.2 Six coins are randomly tossed onto a table.

(A) How many possible distributions of heads and tails can be obtained?

(B) What is the probability of obtaining each of the distributions listed in (A)?

(C) Which distribution is the most probable?

17.3 A particular board game uses eight-sided dice, with the probability of all faces being in the totally concealed position ("down") being equal. A player throws three such dice. What is the probability of his obtaining two fives and a seven in the down positions of the dice?

17.4 Consider N coins randomly tossed onto a table. Compute the ratio of the probability of observing a 2% deviation from the most probable distribution (heads and tails) to the probability of the most probable distribution when we have 500 coins present. (Assume that the most probable distribution corresponds to having 250 heads and 250 tails.)

17.5 Consider N coins randomly tossed onto a table, where N is very large.

(A) Obtain a general equation giving the ratio R of the probability of observing a $p\%$ deviation from the most probable distribution of heads and tails to the probability of the most probable distribution.

(B) Use Stirling's approximation to put the result obtained in (A) into a more convenient form. In particular, show that

$$R = \exp[2n\,\ell n(n) - (n+c)\ell n(n+c) - (n-c)\ell n(n-c)]$$

where

$$n = \frac{N}{2} \text{ and } c = \frac{np}{100}.$$

(C) Compute the ratio for the case $p = 0.01\%$ and $N = 2 \times 10^7$ coins. (Assume that the most probable distribution corresponds to having an equal number of heads and tails.)

(D) Compute the ratio for case $p = 0.01\%$ and $N = 6.0220 \times 10^{23}$.

17.6* A beam of length $3L$ has a mass distribution such that the beam is balanced when a fulcrum is placed a distance L from one end. (See the accompanying figure.)

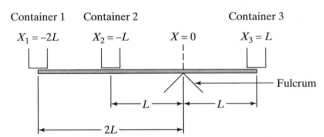

Let the fulcrum be at the origin ($X = 0$). Three weightless containers are placed on the beam at positions $X_1 = -2L$, $X_2 = -L$, and $X_3 = L$. An Avogadro's number N of identical particles of mass m_o each is distributed among the three containers. Use Lagrangian multipliers to determine the distribution of particles among the three containers that maximizes W subject to the constraints that mass be conserved and that the beam remain balanced. That is, determine the fraction of the particles in each container when W is a maximum subject to the said constraints.

17.7* Consider a weightless lever arm of length $3L$ that rotates about an axis perpendicular to the paper on which this problem is written at $X = 0$. (See the accompanying figure.)

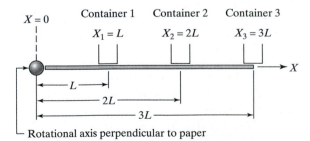

N identical particles, each of mass m_o, are distributed into three containers located at $X_1 = L$, $X_2 = 2L$, and $X_3 = 3L$. Use Lagrangian multipliers to determine

the distribution of particles among the three containers that maximizes W subject to the constraints that mass be conserved and that $I = (7/3)Nm_oL^2$, where I is the moment of inertia about the rotation axis. That is, determine the fraction of the particles in each container that maximizes W subject to the said constraints. Assume that N is large.

17.8 Eight members of the physical chemistry class (four men and four women) go to Las Vegas to enter the annual poker tournament. In this tournament, the players are each given chips and play begins. The player ending up with all the chips is the winner and receives the overall prize provided by the casino. The casino owners decide that the total number of chips initially distributed to the male players will be twice the number of chips initially distributed to the female players. If N chips are to be distributed among the eight players subject to the said constraint, what fraction of the total chips should be received by each player so as to maximize the thermodynamic probability. Assume that N is large.

17.9 Four normal cubic dice are thrown. Compute the probability that the sum of the numbers thrown will add to 15.

17.10* A hypothetical system has four different energy levels with energies and degeneracies as given in the following table:

Level (n)	E(J mol^{-1})	g_n
1	0	1
2	200	2
3	400	3
4	600	1

(A) If N molecules are distributed among these levels at $T = 300$ K, compute the fraction of the molecules that will be in the ground state.

(B) Two molecules are picked out of the system at random. Compute the temperature at which the probability that the sum of the energies of the two molecules will add to 400 J mol^{-1} will be a maximum. What is the value of this maximum probability? Assume that $N \gg 4$.

17.11 A hypothetical system has four different energy levels with energies and degeneracies as given in the following table:

Level (n)	E(J mol^{-1})	g_n
1	0	1
2	200	2
3	400	3
4	600	1

At $T = 300$ K, four molecules are picked out of the system at random. Compute the probability that one of the molecules will be in the $n = 1$ state, one in the $n = 2$ state, one in the $n = 3$ state, and one in the $n = 4$ state.

17.12 The equilibrium bond length for HBr is 1.66 Å.

(A) Calculate the value of the rotational constant for HBr, assuming that a molecule of the compound behaves as a rigid rotor whose length is equal to the HBr equilibrium bond length.

(B) Compute the rotational molecular partition function for HBr at a temperature of 800 K to four significant digits.

(C) Determine the most populated rotational state of HBr at 800 K. Assume that the atomic masses are the average atomic masses, (i.e., atomic weights).

17.13 Obtain the translational molecular partition function for a quantum "particle" of mass m in a three-dimensional potential well whose shape is that of a rectangular parallelepiped whose side lengths are a, b, and c. Assume that the translational quantum states can be treated as being continuous. Express your result in terms of the volume of the well.

17.14 Sometimes, rotational quantum states are very closely spaced. The diatomic rigid-rotor energy eigenvalues are given by $E_J = J(J + 1)Bh$, where $J = 0, 1, 2, 3, \ldots, \infty$. Make the classical approximation that J may be treated as a continuous variable, and derive the form of the rotational molecular partition function for a diatomic rigid rotor.

17.15* The diatomic rigid-rotor rotational molecular partition function has the form $z = kT/(Bh)$ if we assume that the rotational quantum number J may be treated as a classical continuous variable. (See Problem 17.14.) Let us assess the accuracy of this approximation for two systems: $H^{35}Cl$ and $^{35}Cl^{35}Cl$. Use the data given in Table 15.4 to compute the correct rotational molecular partition function for both of these systems at 200 K, assuming that they act as rigid rotors. Compare your results with the value of the classical rotational molecular partition function that is obtained if we assume J to be a continuous variable. Compute the percent error in the classical approximation for each molecule. For which molecule is the approximation more accurate? Explain.

17.16* The diatomic rigid-rotor rotational molecular partition function has the form $z = kT/(Bh)$ if we assume that the rotational quantum number J may be treated as a classical continuous variable. (See Problem 17.14.) The percent error in this classical approximation for $H^{35}Cl$ is -2.52% at $T = 200$ K. (See Problem 17.15.) Let us explore how the accuracy of the classical approximation for rotation varies with temperature.

(A) Compute the rotational molecular partition function for $H^{35}Cl$, assuming a molecule of the compound to behave as a rigid rotor with the rotational

constant given in Table 15.4 at the following temperatures: 10 K, 20 K, 30 K, 40 K, 50 K, 75 K, 100 K, 150 K, 200 K, 250 K, and 300 K.

(B) Plot your results and, on the same graph, show the classical result.

(C) Compute the percent error in the classical approximation at each temperature, and plot your results as a function of T. Why do the results become better at higher temperatures?

17.17 The energy eigenvalues for a one-dimensional harmonic oscillator are $E_v = (v + 0.5)h\nu_o$, where v, the vibrational quantum number, can take the integral values 1, 2, 3, $\ldots$, ∞. Make the classical approximation that v may be treated as a continuous variable, and derive the form of the vibrational molecular partition function for a diatomic harmonic oscillator.

17.18* It has been found that the classical vibrational molecular partition function for a harmonic oscillator that is obtained if we make the assumption that the vibration quantum number v can be treated as a continuous variable is

$$z_c = \frac{kT}{h\nu_o} \exp\left[-\frac{h\nu_o}{2kT}\right],$$

where ν_o is the vibrational frequency. (See Problem 17.17.) For $H^{35}Cl$, ν_o is 8.963×10^{13} s^{-1}. Compute the correct vibrational molecular partition function for $H^{35}Cl$ and z_c at temperatures of 300 K, 1,000 K, 10,000 K, 20,000 K, and 30,000 K. Compute the percent error in the classical approximation at each temperature. Comment on the classical approximation for vibration.

17.19 Show that, in general, the reciprocal of the molecular partition function plays the role of a normalization constant for the probability distribution describing the system.

17.20 Show that the molecular speed probability distribution given by Eq. 17.62 is normalized.

17.21 In Example 17.5, we found that a fraction equal to 0.01409 of a beam of argon atoms would successfully traverse a slotted-disk speed selector if the entering distribution is Maxwell–Boltzmann at 300 K and the selector transmits all atoms with speeds in the range from 347 m s^{-1} to 353 m s^{-1}. If the rotation speed of the disks is changed so that speeds in the range $(450 - \varepsilon)$ m s$^{-1} \le C \le (450 + \varepsilon)$ m s^{-1} are permitted to traverse the selector, what must ε be in order for the fraction of the total atoms transmitted to be the same as that in Example 17.5?

17.22 Starting with the molecular speed probability distribution given by Eq. 17.62, derive Eq. 17.64. That is, show that the probability distribution of molecular energies for a gas whose distribution of speeds is Maxwell–Boltzmann is given by

$$P(\varepsilon_t)d\varepsilon_t = 4(2)^{1/2}\pi\left[\frac{1}{2\pi kT}\right]^{3/2}\exp\left[-\frac{\varepsilon_t}{kT}\right]\varepsilon_t^{1/2}d\varepsilon_t.$$

17.23 Classical particles with translational energy ε_t are incident upon a potential barrier whose height is $V_o = 30$ kJ mol^{-1}.

(A) Compute the fraction F of molecules that will have sufficient energy to overcome this barrier at 300 K.

(B) Repeat (A) at temperatures of 330 K, 360 K, 400 K, and 500 K.

(C) Use your results to plot $\ln(F)$ versus T^{-1}. What type of behavior do you notice? Explain.

17.24 Starting with Eq. 17.74, show that

$$\langle C \rangle = \left[\frac{8RT}{\pi M}\right]^{1/2}.$$

The next two problems cast the Maxwell–Boltzmann probability distribution for molecular speeds into a reduced form and then use the result to obtain the median speed in terms of C_{mp}. The two problems are, therefore, best assigned as a unit, but they are written so that they can be assigned individually.

17.25 Let us define a reduced speed as $C_R = C/C_{mp}$, where C_{mp} is the most probable speed, given by Eq. 17.72. Transform the molecular speed probability distribution for molecular speeds given by Eq. 17.62 into a probability distribution of reduced speeds.

17.26* The Maxwell–Boltzmann probability distribution of reduced speeds is

$$P(C_R)dC_R = \left(\frac{16}{\pi}\right)^{1/2} \exp[-C_R^2]C_R^2 dC_R$$

(see Problem 17.25), where, by definition, $C_R = C/C_{mp}$. Use this distribution to show that the median speed is given by

$$C_m = 1.08777 \ldots C_{mp}.$$

17.27 Obtain an expression in terms of temperature and mass for the uncertainty of speeds when the probability distribution is Maxwell–Boltzmann.

17.28 (A) Compute $\langle C \rangle$ for $^{235}UF_6(g)$ and $^{238}UF_6(g)$ at 298 K if the speed distribution is Maxwell–Boltzmann.

(B) If an ensemble of $^{235}UF_6(g)$ and $^{238}UF_6(g)$ molecules are released into a long diffusion tunnel, what distance would they have to travel before the average distance between the $^{235}UF_6(g)$ and $^{238}UF_6(g)$ molecules is 10 m?

17.29 A physical chemistry professor stands in front of the class and opens a container of $N_2O(g)$, a common anesthetic called "laughing gas" used in dental work. At the same time, the professor's assistant opens a container of benzoyl chloride in the back of the classroom. Benzoyl chloride (C_6H_5COCl) is a lacrimator (a substance that produces tears). If the classroom contains 25 rows of students, in which row do the students begin to laugh and cry at the

same time? [The author is indebted to Fredrick L. Minn, M.D. Ph.D. for providing this problem and the associated solution.]

17.30 The nth moment of the normalized probability distribution function for the variable X is defined to be

$$\langle X^n \rangle = \int_{\text{all } X} X^n P(X) dX.$$

Consequently, the first moment is the average, or mean, value of X, the second moment is the average square, $\langle X^2 \rangle$, etc. Obtain an expression in terms of T and M for the nth moment of the probability distribution for molecular speeds. Verify that your results are compatible with Eqs. 17.75 and 17.77. (Hint: Treat the problem in terms of the reduced speed, and consider odd and even moments as separate cases.)

17.31 One of the most powerful experimental techniques for the study of chemical reaction dynamics is crossed molecular beam measurement. In this method, a well-collimated beam of molecules is directed into a high-vacuum chamber, where it is caused to intersect a second well-collimated molecular beam. The two beams interact, and suitable detectors measure what occurs during the collisions that take place when the beams intersect. Assume that a velocity-selected beam of tritium atoms (^3H) intersects a velocity-selected beam of H_2 molecules at an angle of 90°.

(A) If the tritium atoms each have a speed of 2,500 m s^{-1}, while the H_2 molecules are each moving with a speed of 1,750 m s^{-1}, what fraction of the total kinetic energy is in relative motion and what fraction is in center-of-mass motion?

(B) What fraction of the total kinetic energy would you expect to be effective in promoting chemical reactions between the tritium atoms and H_2 molecules? Explain. The atomic mass of tritium is 3.016029 amu.

17.32 Observer A is riding in an airplane that is proceeding due north at 300 km hr^{-1}. She looks out the window and notices a second plane that is traveling at 400 km hr^{-1} in a southeast direction along a line that makes a 45° angle with due east. Using an apparatus on her plane, Observer A performs a measurement of the speed of the second plane. What speed does she obtain? If there are no other objects in the sky and no other points of reference, can the direction of the second plane be determined? Discuss this point.

17.33 The 78th edition of the CRC *Handbook of Chemistry and Physics* reports that, at an altitude of 10^6 m (621.4 miles), the atmospheric pressure is 7.514×10^{-9} Pa and the number of molecules per m^3 is 5.442×10^{11}. The temperature of the gas at that altitude is estimated to be about 1,000 K. These figures assume the air is dry and that it obeys the ideal-gas equation of state. Let us further assume that the air is pure N_2 gas with a hard-sphere collision radius σ_{AA} of 5×10^{-10} m.

(A) Compute the collision frequency of each N_2 molecule.

(B) Compute the mean free path of N_2 at 10^6 m altitude. You may use the average N_2 molecular mass (i.e., the molecular weight), for the mass of N_2.

(C) If you were an astronaut taking a space walk 10^6 m above the earth's surface, would you expect to be incinerated by the 1000K gas present at that altitude? Explain. (The author thanks Fredrick L. Minn, M.D., Ph.D., for this part of the question.)

17.34 A cubic container, 1 meter on each side, holds one neon atom and one argon atom. Thermal exchange between these atoms and the container walls is such that the distribution of speeds of the two atoms over time is Maxwell–Boltzmann. If $T = 300$ K and the hard-sphere collision radius of the Ne–Ar pair is 4×10^{-10} m, compute the average time between Ar–Ne collisions. Assume that the atomic masses of Ar and Ne are the average atomic masses (i.e., the atomic weights) of those elements.

17.35 In this problem, you will develop the probability distribution of free paths in terms of a reduced free path. Let us define the reduced free path as $X_R = X/\langle L \rangle$, where $\langle L \rangle$ is the mean free path.

(A) Obtain the normalized probability distribution function for observing a reduced free path in the range X_R to $X_R + dX_R$.

(B) Use the expression obtained in (A) to determine the median free path in terms of $\langle L \rangle$.

17.36 (A) Compute the mean square path length in terms of $\langle L \rangle$, the mean free path, for an ensemble of gaseous molecules in Maxwell–Boltzmann equilibrium.

(B) Determine the uncertainty in the free-path distribution in terms of $\langle L \rangle$.

17.37 Consider the system described in Example 17.9. Assume that the total path length to the detector in the experimental apparatus is 0.20 m, that the molecules may be treated as hard spheres with a collision radius of 6×10^{-10} m, and that the gases are ideal. Determine the pressure the investigator should use inside the chamber to maximize the probability of observing exactly one collision of an entering A molecule with a B molecule inside the chamber. Let A be a potassium atom and B a Cl_2 molecule. Assume that the temperature is 298 K and that the masses are the average atomic and molecular masses (i.e., the atomic and molecular weights) of the elements.

17.38 For the system described in Example 17.9, obtain an expression in terms of X_o and $\langle L \rangle$ giving the probability that an entering A molecule will undergo either no collisions or one collision with a B molecule inside the chamber.

17.39 For the system described in Example 17.9, obtain an expression in terms of X_o and $\langle L \rangle$ giving the probability that an entering A molecule will undergo exactly two collisions with a B molecule as it moves through a total path length of X_o inside the chamber. In terms of $\langle L \rangle$, what value of X_o makes this probability a maximum?

17.40 I'm ready to go on to statistical thermodynamics, but Sam is waving both hands in the air. "What is it, Sam?"

"Well, sir, I've studied Example 17.9 and done Problems 17.38 and 17.39, and it seems to me that you're teaching this material about the probability of a certain number of collisions in path length X_o the wrong way."

"Why do you think that, Sam," I ask.

"Well, Example 17.9 and Problems 17.38 and 17.39 show that the probability of zero collisions, one collision, and two collisions in a total path length X_o are

$$P(0) = \exp\left[-\frac{X_o}{\langle L \rangle}\right],$$

$$P(1) = \frac{X_o}{\langle L \rangle} \exp\left[-\frac{X_o}{\langle L \rangle}\right],$$

and

$$P(2) = \frac{X_o^2}{2\langle L \rangle^2} \exp\left[-\frac{X_o}{\langle L \rangle}\right],$$

respectively."

"Those results are correct, Sam," I respond.

"I know they are, sir, but to obtain them, we had to integrate one- and two-dimensional integrals to obtain $P(1)$ and $P(2)$, respectively. Using your method, if we wanted to compute the probability that 75 collisions would occur in a total path length X_o, we would have to perform a 75-dimensional integral. There is a much easier way to get the probability that N collisions will occur in a total path length X_o for any arbitrary N."

"What do you have in mind, Sam?"

"I'll give you a hint. Consider the series expansion of $\exp[X_o/\langle L \rangle]$."

Do you see what method Sam has in mind? What is the appropriate expression for the probability that exactly 75 collisions will occur in a total path length X_o in terms of $\langle L \rangle$ and X_o? What do you think of Sam's suggested method for teaching this point?

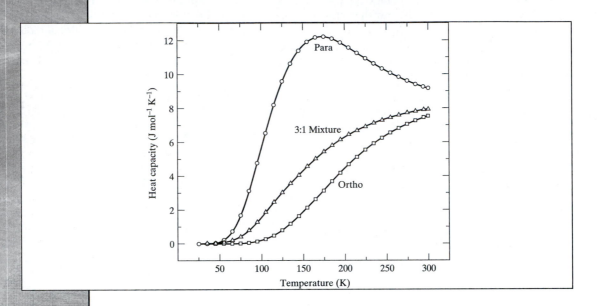

CHAPTER 18

Statistical Thermodynamics

Statistical thermodynamics is the window that connects the microscopic world of quantum mechanics to the macroscopic world of thermodynamics.

In the previous chapter, we introduced the fundamental equation connecting the microscopic world of quantum mechanics to the macroscopic world of thermodynamics. If we deduce all possible distributions of the molecules over the available quantum states and determine the probability of the existence of each of these distributions or microstates, the thermodynamic value of any property of the system, X, is the average value of X over all available microstates. That is,

$$\text{Thermodynamic value of } X = \langle X \rangle = \sum_{\alpha=1}^{n^D} P_\alpha X_\alpha, \tag{18.1}$$

where P_α is the probability of the formation of microstate α, X_α is the value of X for the αth microstate, and n^D is the total number of available microstates. Figure 18.1 illustrates the 15 microstates that would be included in the summation in Eq. 18.1 if the system contained only four molecules and three energy states.

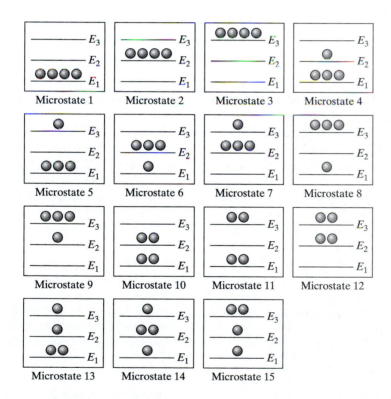

◀ FIGURE 18.1
The 15 microstates for a system consisting of four molecules distributed over three possible energy states. To use Eq. 18.1, we would have to determine the probability of occurrence of each of these microstates before executing the summation.

Our basic problem centers on the fact that n^D is extremely large—almost beyond imagination—so that we cannot execute the summation in Eq. 18.1 by brute force, as we did in Chapter 17 when we computed the average sum on the throw of two dice. We did find in the previous chapter that, for the special case of a large number of molecules and a small number of quantum states (large N, small s), the summation reduces to a single term involving only the most probable distribution. For those cases, we were able to determine the most probable microstate by maximizing the thermodynamic probability. As we shall see, this approach permits us to compute thermodynamic quantities. However, for molecular systems, we usually do not have a situation in which N is large and s is small. In fact, the number of quantum states is usually large compared with N. Therefore, it's back to the drawing board: How can we accurately execute the summation in Eq. 18.1? In this chapter, we shall find that it's not as hard as it might appear.

Once an adequate method is obtained for handling Eq. 18.1, we shall develop equations that permit the important thermodynamic quantities, such as U, H, S, A, G, pressure, and the equilibrium constant, to be computed from the results of quantum theory.

18.1   The Number of Quantum States for Molecular Systems

Suppose we have a system consisting of one mole of some substance, so that there are 6.02214×10^{23} molecules present. Let us first determine why $N = 6.02214 \times 10^{23}$ is insufficient to make the number of molecules much larger than the number of quantum states. If it were, we could reduce Eq. 18.1 to a single term, and our problem would essentially be solved.

Consider a molecule with r atoms. The quantum state of this molecule requires that we assign three translational quantum numbers n_x, n_y, and n_z, three rotational quantum numbers (assuming the molecule to be nonlinear), $3r - 6$ vibrational quantum numbers, and one more quantum number to specify the electronic state. The total number of quantum number assignments is therefore $3r + 1$. Although that is itself a relatively small number, the problem arises when we consider the range of choices we have for each of these assignments. The range of choice available for the rotational quantum numbers is rather small. In Table 17.4 and Example 17.4, we found that the effective number of rotational states available to OH and HF is on the order of 10^2 or less at temperatures characteristic of most chemical processes. The effective number of vibrational states is even less, and that for electronic states will be close to unity because of the much greater energy spacing between quantum levels. The problem arises because of the translational states.

Let us determine the effective number of translational states available to the molecules. In the previous chapter, we found that the effective number of quantum states at a given temperature is determined by the molecular partition function. Suppose we have one molecule of mass m in a three dimensional potential well whose shape is that of a rectangular parallelepiped with side lengths a, b, and c. The energy eigenvalues for this system, obtained in Chapter 12, are given by

$$\varepsilon(n_x, n_y, n_z) = \frac{h^2}{8m}\left[\frac{n_x^2}{a^2} + \frac{n_y^2}{b^2} + \frac{n_z^2}{c^2}\right] \qquad (n_x, n_y, \text{ and } n_z = 1, 2, 3, \ldots, \infty).$$

$$(18.2)$$

Consequently, the translational molecular partition function is

$$
\begin{aligned}
z_{\text{tr}} &= \sum_{n_x=1}^{\infty}\sum_{n_y=1}^{\infty}\sum_{n_z=1}^{\infty}\exp\left\{-\frac{h^2}{8mkT}\left[\frac{n_x^2}{a^2} + \frac{n_y^2}{b^2} + \frac{n_z^2}{c^2}\right]\right\} \\
&= \sum_{n_x=1}^{\infty}\exp\left\{-\frac{h^2 n_x^2}{8ma^2kT}\right\}\sum_{n_y=1}^{\infty}\exp\left\{-\frac{h^2 n_y^2}{8mb^2kT}\right\}\sum_{n_z=1}^{\infty}\exp\left\{-\frac{h^2 n_z^2}{8mc^2kT}\right\}. \quad (18.3)
\end{aligned}
$$

Since the translational energy levels are very closely spaced, we may replace the summations with integrals over the limits from 0 to ∞. This replacement produces

$$z_{tr} = \int_0^\infty \exp\left\{-\frac{h^2 n_x^2}{8ma^2 kT}\right\} dn_x \int_0^\infty \exp\left\{-\frac{h^2 n_y^2}{8mb^2 kT}\right\} dn_y$$

$$\int_0^\infty \exp\left\{-\frac{h^2 n_z^2}{8mc^2 kT}\right\} dn_z.$$

Each of these integrals is a standard form. The result of integrating is

$$z_{tr} = \frac{1}{2}\left[\frac{8\pi ma^2 kT}{h^2}\right]^{1/2} \frac{1}{2}\left[\frac{8\pi mb^2 kT}{h^2}\right]^{1/2} \frac{1}{2}\left[\frac{8\pi mc^2 kT}{h^2}\right]^{1/2}$$

$$= \frac{1}{h^3}[2\pi mkT]^{3/2} abc. \tag{18.4}$$

The volume of a rectangular parallelepiped is the product of the lengths of the sides, so that $V = abc$. This produces

$$\boxed{z_{tr} = s_{eff} = \frac{V}{h^3}[2\pi mkT]^{3/2}} \tag{18.5}$$

for three-dimensional translational motion.

We can now evaluate the effective number of translational states for a typical system. Suppose we have an H_2O molecule at 298 K in a container whose side lengths are $a = b = c = 0.2900$ m, so that the volume of the container is about 24.4 L, which is near the molar volume of an ideal gas at 298 K and 1 atm pressure. Under these conditions, the effective number of translational quantum states is

$$z = s_{eff} = \frac{0.02439 \text{ m}^3}{(6.62608 \times 10^{-34} \text{ J s})^3}$$

$$\times [2(3.14159)(2.992 \times 10^{-26} \text{ kg})(1.3807 \times 10^{-23} \text{ J K}^{-1})(298 \text{ K})]^{3/2}$$

$$= 1.804 \times 10^{30}.$$

Therefore, we have 1.804×10^{30} choices for the set of three translational quantum numbers n_x, n_y, and n_z. This a factor of almost 3×10^6 larger than the number of molecules present in one mole of a compound. As a result, N is not large compared with s, and we must consider all the microstates, not just the most probable one.

The preceding analysis shows that the difficulty arises because of the translational motion. If we consider only rotational, vibrational, or electronic states, the number of available quantum states is drastically reduced, and we will have a system for which N is large compared with s, so that only the most probable distribution or microstate needs to be considered. This is also the situation for nuclear spin states: The total number of such states is very small relative to the number of molecules; consequently, we expect the Boltzmann distribution to describe the population of such states quite accurately.

18.2   The Microstate Probability Distribution

The presence of a huge excess of possible quantum states relative to the number of molecules present means that we are going to have to consider many microstates in the evaluation of Eq. 18.1. To execute the summation, we need the probability that we will observe the αth microstate, P_α. To obtain the requisite probabilities, the fundamental postulate of statistical mechanics is required:

For a thermodynamic system with a fixed volume, composition, and temperature, all distributions or microstates that have the same energy have equal values of P_α.

This postulate means that P_α must depend only upon the energy of the distribution, E_α, and not upon how this energy is distributed among translational, rotational, vibrational, and electronic degrees of freedom.

For the fundamental postulate to hold, P_α must have a specific functional form that we can deduce by considering two microstates, labeled 1 and 2, both at temperature T, but with different compositions, volumes, and energies E_1 and E_2, respectively. These two microstates are illustrated in Figure 18.2. Since the probabilities for these two microstates depend only upon the total energies of the microstates, we must have $P_1 = f(E_1)$ and $P_2 = g(E_2)$, where f and g are, at the moment, unknown functions of the energy. Now consider a third microstate, labeled 3, that is the composite of microstates 1 and 2, so that E_3 is the sum $E_1 + E_2$. The fundamental postulate requires that we have $P_3 = h(E_3) = h(E_1 + E_2)$, where h is an unknown function of the total energy of the microstate. Since the combined probability of simultaneous observation of microstates 1 and 2 is the product of the individual probabilities, we must have

$$h(E_3) = h(E_1 + E_2) = f(E_1)g(E_2). \tag{18.6}$$

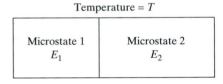

Temperature = T

| Microstate 1 E_1 | Microstate 2 E_2 |

| Microstate 3 $E_3 = E_1 + E_2$ |

▶ **FIGURE 18.2**
Two microstates 1 and 2 with the same temperature, but different compositions and volumes. Microstate 3 is the composite of microstates 1 and 2.

We already know from our examination of the free-path probability distribution in Chapter 17 that the only mathematical function with the property required by Eq. 18.6 is an exponential function. Indeed, it is easy to prove this. We simply take the partial derivative of both sides of that equation with respect to E_1, holding E_2 constant. This operation produces

$$\left(\frac{\partial h(E_3)}{\partial E_1}\right)_{E_2} = \left(\frac{\partial h(E_3)}{\partial E_3}\right)\left(\frac{\partial E_3}{\partial E_1}\right)_{E_2} = \left(\frac{dh(E_3)}{dE_3}\right) = g(E_2)\left(\frac{df(E_1)}{dE_1}\right),$$

since $(\partial E_3/\partial E_1)_{E_2} = 1$. Now we repeat the operation, but this time we take the derivative with respect to E_2 while holding E_1 constant. The result is

$$\left(\frac{\partial h(E_3)}{\partial E_2}\right)_{E_1} = \left(\frac{\partial h(E_3)}{\partial E_3}\right)\left(\frac{\partial E_3}{\partial E_2}\right)_{E_1} = \left(\frac{dh(E_3)}{dE_3}\right) = f(E_1)\left(\frac{dg(E_2)}{dE_2}\right)$$

since $(\partial E_3/\partial E_2)_{E_1} = 1$. The right-hand sides of both partial differential equations are equal to $(dh(E_3)/dE_3)$, so they must be equal to each other. This gives

$$g(E_2)\left(\frac{df(E_1)}{dE_1}\right) = f(E_1)\left(\frac{dg(E_2)}{dE_2}\right).$$

Rearrangement of the factors produces

$$\frac{\left(\dfrac{df(E_1)}{dE_1}\right)}{f(E_1)} = \frac{\left(\dfrac{dg(E_2)}{dE_2}\right)}{g(E_2)} = -\beta. \tag{18.7}$$

Because E_1 and E_2 are independent and the variables are separated in Eq. 18.7, the only possible way for that equation to hold for all values of E_1 and E_2 is for both sides to be equal to a common constant that is independent of both E_1 and E_2. We label this constant $-\beta$.

Equation 18.7 represents two differential equations, which, when solved, give the functions $f(E_1)$ and $g(E_2)$. Separating variables in the equation for f produces

$$\frac{df(E_1)}{f(E_1)} = -\beta \, dE_1.$$

Integration of both sides yields

$$\ln f(E_1) = -\beta E_1 + \text{constant}.$$

The solution for $g(E_2)$ is similar, so that we obtain

$$f(E_1) = c \exp[-\beta E_1]$$

and

$$g(E_2) = b \exp[-\beta E_2]. \tag{18.8}$$

Thus, $f(E)$ and $g(E)$ are the same to within a constant, and both are exponential functions, as we expected before going through the mathematical proof. Equation 18.8 tells us something about the nature of the constant β: Since β has a common value for both microstates in Figure 18.2, it cannot depend upon volume or composition, as these quantities differ for the two microstates; It can, however, be a function of temperature, because both microstates are at the same temperature.

The principal conclusion we reach from our analysis is that the fundamental postulate of statistical mechanics requires that the microstate probabilities have the form

$$P_\alpha = A \exp[-\beta E_\alpha] \qquad (A = \text{constant}). \tag{18.9}$$

We can determine the constant A by the usual normalization procedure for a probability distribution. We must have

$$\sum_{\alpha=1}^{n^D} P_\alpha = A \sum_{\alpha=1}^{n^D} \exp[-\beta E_\alpha] = 1.$$

Therefore,

$$A = \frac{1}{\displaystyle\sum_{\alpha=1}^{n^D} \exp[-\beta E_\alpha]}. \tag{18.10}$$

The quantity in the denominator of Eq. 18.10 has a form similar to that of the molecular partition function. However, in this case, the summation runs over all the distributions or microstates, rather than over the individual molecular-energy levels. This summation is called the *canonical partition function*. We will represent it with an uppercase Z, so that

$$Z = \sum_{\alpha=1}^{n^D} \exp[-\beta E_\alpha].$$ (18.11)

With this notation, the probability that microstate α will occur is

$$P_\alpha = \frac{\exp[-\beta E_\alpha]}{Z}.$$ (18.12)

18.3   Thermodynamic Quantities in Terms of the Canonical Partition Function

Since we have not yet evaluated the energies of each microstate, we do not have the value of E_α for all α. Therefore, at this point, we cannot evaluate the canonical partition function. However, we can utilize Eq. 18.1 to obtain the important thermodynamic quantities in terms of Z. In the process, we will also be able to evaluate the constant β, which, in view of the discussion in Chapter 17, we suspect is $(kT)^{-1}$.

18.3.1   The Internal Energy

The internal system energy averaged over all possible distributions or microstates is given by Eq. 18.1 for X equal to E:

$$U = \langle E \rangle = \sum_{\alpha=1}^{n^D} P_\alpha E_\alpha.$$ (18.13)

Using Eq. 18.12 for P_α, we see that Eq. 18.13 becomes

$$U = \frac{1}{Z} \sum_{\alpha=1}^{n^D} \exp[-\beta E_\alpha] E_\alpha.$$ (18.14)

We now wish to express U in terms of Z. To see how this is done, let us examine the partial derivative of $\ln Z$ with respect to β under the condition that the volume is held constant and the system is closed (there is no change in its composition). We can immediately write

$$\left(\frac{\partial \ln Z}{\partial \beta}\right)_V = \frac{1}{Z}\left(\frac{\partial Z}{\partial \beta}\right)_V.$$ (18.15)

We already know that β is independent of composition and volume. At most, it is a function of temperature. Therefore, the energies of the microstates, E_α, are independent of β, since they depend only upon volume. This dependence of E_α on volume arises because the translational energies depend upon the lengths of the sides of the container holding the molecules. (See Eq. 18.2.) Therefore, the derivative of the canonical partition function with respect to β with V held constant is

$$\left(\frac{\partial Z}{\partial \beta}\right)_V = \frac{\partial}{\partial \beta}\left[\sum_{\alpha=1}^{n^D} \exp[-\beta E_\alpha]\right]_V = -\sum_{\alpha=1}^{n^D} E_\alpha \exp[-\beta E_\alpha].$$ (18.16)

Combination of Eqs. 18.15 and 18.16 produces

$$\left(\frac{\partial \ln Z}{\partial \beta}\right)_V = -\frac{1}{Z} \sum_{\alpha=1}^{n^D} E_\alpha \exp[-\beta E_\alpha].$$ (18.17)

A comparison of Eqs. 18.14 and 18.17 shows that the right-hand side of the latter is the negative of U. Therefore, we have U expressed in terms of Z. The result is

$$U = -\left(\frac{\partial \ln Z}{\partial \beta}\right)_V \qquad (18.18)$$

for a closed system.

18.3.2   The Pressure

If the property of the system that interests us is the pressure, we set X in Eq. 18.1 equal to the pressure p and obtain

$$p = \langle p_\alpha \rangle = \sum_{\alpha=1}^{n^D} P_\alpha p_\alpha, \qquad (18.19)$$

where we now use a lowercase p for pressure to avoid confusion with the probabilities for the microstates. Substituting Eq. 18.12 for P_α gives

$$p = \langle p_\alpha \rangle = \frac{1}{Z} \sum_{\alpha=1}^{n^D} p_\alpha \exp[-\beta E_\alpha]. \qquad (18.20)$$

The quantity p_α is the pressure associated with microstate α. For a particular microstate, the quantum number assignments are all fixed. Consequently, there can be no energy change because of quantum transitions. The only energy change can come from work done on the system by virtue of a change in volume, assuming that we have only p–V work. Under these conditions, the energy change is

$$\left(\frac{\partial E_\alpha}{\partial V}\right)dV = -p_\alpha dV, \qquad (18.21)$$

and we have

$$p_\alpha = -\left(\frac{\partial E_\alpha}{\partial V}\right). \qquad (18.22)$$

Combination of Eqs. 18.20 and 18.22 produces

$$p = \langle p_\alpha \rangle = -\frac{1}{Z} \sum_{\alpha=1}^{n^D} \left(\frac{\partial E_\alpha}{\partial V}\right) \exp[-\beta E_\alpha]. \qquad (18.23)$$

To express p in terms of Z, we need only examine the derivative of $\ln Z$ with respect to volume with temperature held constant for a closed system. This derivative is

$$\left(\frac{\partial \ln Z}{\partial V}\right)_T = \frac{1}{Z}\left(\frac{\partial Z}{\partial V}\right)_T = \frac{1}{Z}\frac{\partial}{\partial V}\left[\sum_{\alpha=1}^{n^D} \exp[-\beta E_\alpha]\right]_T$$
$$-\frac{\beta}{Z} \sum_{\alpha=1}^{n^D} \left[\left(\frac{\partial E_\alpha}{\partial V}\right) \exp[-\beta E_\alpha]\right], \qquad (18.24)$$

since β is independent of volume. A comparison shows that, except for the factor of β, Eq. 18.24 is identical to Eq. 18.23. Therefore, we must have

$$p = \frac{1}{\beta}\left(\frac{\partial \ln Z}{\partial V}\right)_T \qquad (18.25)$$

for any closed system.

18.3.3   Identification of β

We can now determine the value of β in the foregoing equations. For any reversible process in a closed system, the internal energy is given by

$$dU = T\,dS - p\,dV.$$

Therefore, the rate of change of the internal energy with volume for a fixed temperature is

$$\left(\frac{\partial U}{\partial V}\right)_T = T\left(\frac{\partial S}{\partial V}\right)_T - p.$$

However, the Maxwell relationship obtained from dA (see Eq. 5.25) is

$$\left(\frac{\partial S}{\partial V}\right)_T = \left(\frac{\partial p}{\partial T}\right)_V.$$

Combining these equations gives the result we used extensively in the early portions of this text:

$$\left(\frac{\partial U}{\partial V}\right)_T = T\left(\frac{\partial p}{\partial T}\right)_V - p. \tag{18.26}$$

We can evaluate the left-hand side of Eq. 18.26 by substituting Eq. 18.18 for U. This operation produces

$$\left(\frac{\partial U}{\partial V}\right)_T = \frac{\partial}{\partial V}\left[-\left(\frac{\partial \ln Z}{\partial \beta}\right)_V\right]_T = -\frac{\partial}{\partial \beta}\left[\left(\frac{\partial \ln Z}{\partial V}\right)_T\right]_V \tag{18.27}$$

since we can always reverse the order of differentiation. The derivative on the right-hand side of Eq. 18.27 can be seen to be βp from Eq. 18.25. Therefore,

$$\left(\frac{\partial U}{\partial V}\right)_T = -\frac{\partial}{\partial \beta}[\beta p]_V = -p - \beta\left(\frac{\partial p}{\partial \beta}\right)_V. \tag{18.28}$$

We now have two expressions, both equal to $(\partial U/\partial V)_T$ in Eqs. 18.26 and 18.28. The right-hand sides of these equations must, therefore, be equal. Equating these expressions, we obtain

$$T\left(\frac{\partial p}{\partial T}\right)_V - p = -p - \beta\left(\frac{\partial p}{\partial \beta}\right)_V,$$

so that we must have

$$T\left(\frac{\partial p}{\partial T}\right)_V = -\beta\left(\frac{\partial p}{\partial \beta}\right)_V.$$

Dividing both sides by $-\beta(\partial p/\partial T)_V$ produces

$$-\frac{T}{\beta} = \frac{\left(\dfrac{\partial p}{\partial \beta}\right)_V}{\left(\dfrac{\partial p}{\partial T}\right)_V} = \frac{\partial T}{\partial \beta}.$$

Separation of variables gives the equation

$$\frac{\partial \beta}{\beta} = -\frac{\partial T}{T}. \tag{18.29}$$

Integrating both sides of Eq. 18.29, we obtain

$$\ln \beta = -\ln T + \text{constant} = \ln\left[\frac{1}{T}\right] + \text{constant}.$$

Exponentiation of both sides of this equation produces

$$\beta = \frac{1}{(\text{constant})T}. \tag{18.30}$$

If we find the constant for any system, we have found it for all systems, since the analysis is independent of the particular system being examined. We already know from our analysis in Chapter 17 that, for translational motion in one dimension, we have $\beta = 1/(kT)$. Therefore, the constant in Eq. 18.30 is the Boltzmann constant, and we have, in general,

$$\boxed{\beta = \frac{1}{kT}}. \tag{18.31}$$

Equation 18.31 permits us to express U and p in terms of temperature. Using the fact that $d\beta = -dT/(kT^2)$, we see that Eqs. 18.18 and 18.25 become

$$\boxed{U = kT^2\left(\frac{\partial \ln Z}{\partial T}\right)_V} \tag{18.32}$$

and

$$\boxed{p = kT\left(\frac{\partial \ln Z}{\partial V}\right)_T}, \tag{18.33}$$

respectively. Equation 18.32 tells us that if we plot $\ln Z$ versus T, the slope of this plot at any given temperature T_1, multiplied by kT_1^2, yields the internal energy of the system when $T = T_1$. The pressure can be obtained in a similar manner from the slope of a plot of $\ln Z$ versus V at a fixed temperature. Figures 18.3 and 18.4 illustrate these points.

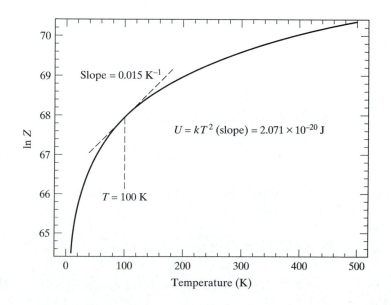

◀ FIGURE 18.3
The internal energy U, computed from the slope of a plot of $\ln Z$ versus temperature. The figure illustrates the computation of U at $T = 100$ K from the slope at that point, which, for this system, is $(\partial \ln Z/\partial T)_V = 0.015$ K^{-1}. The internal energy is obtained from this value by using Eq. 18.32.

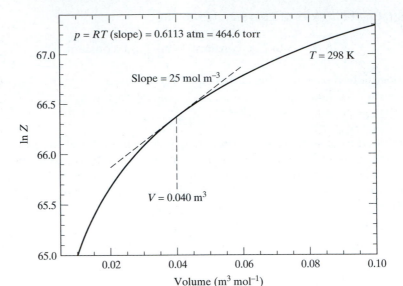

▶ **FIGURE 18.4**
The pressure p, computed from the slope of a plot of ln Z versus volume at a fixed temperature. The figure illustrates the computation of p at $V = 0.040$ m^3 mol^{-1} from the slope at that point with $T = 298$ K. For this system, the slope is $(\partial \ln Z/\partial V)_T = 25$ mol m^{-3}. The pressure is obtained from this value by using Eq. 18.33.

18.3.4 The Entropy

The entropy can now be obtained directly from Eq. 5.22, which gives dU for a closed system:

$$dU = T\,dS - p\,dV,$$

which gives

$$dS = \frac{dU}{T} + \frac{p\,dV}{T}. \tag{18.34}$$

We now use the same mathematical procedure we employed to derive the Gibbs–Helmholtz equations in Chapter 5. We write

$$d\left[\frac{U}{T}\right] = \frac{dU}{T} - \frac{U\,dT}{T^2},$$

so that

$$\frac{dU}{T} = d(T^{-1}U) + \frac{U\,dT}{T^2}. \tag{18.35}$$

Substitution of Eq. 18.35 into Eq. 18.34 produces

$$dS = d(T^{-1}U) + \frac{U\,dT}{T^2} + \frac{p\,dV}{T}. \tag{18.36}$$

Inserting Eqs. 18.32 and 18.33 for U and p, respectively, in the second and third terms of Eq. 18.36 yields

$$dS = d(T^{-1}U) + k\left(\frac{\partial \ln Z}{\partial T}\right)_V dT + k\left(\frac{\partial \ln Z}{\partial V}\right)_T dV$$

$$= d(T^{-1}U) + k\,d(\ln Z) = d[T^{-1}U + k \ln Z]. \tag{18.37}$$

Integrating both sides of Eq. 18.37 gives

$$S = k \ln Z + \frac{U}{T} + \text{constant} = k \ln Z + kT \left(\frac{\partial \ln Z}{\partial T} \right)_V + c. \quad \textbf{(18.38)}$$

The third law requires that the entropy approach a constant for all perfect crystals at absolute zero. If we take this constant to be zero, as suggested by Eq. 4.55, and evaluate the right-hand side of Eq. 18.38 at $T = 0$ K, we find that c must also be zero. Even without the third law, c will add out when we measure differences in entropy. Therefore,

$$S = k \ln Z + \frac{U}{T} = k \ln Z + kT \left(\frac{\partial \ln Z}{\partial T} \right)_V. \quad \textbf{(18.39A)}$$

Using Eq. 18.39A, we may easily compute the entropy for the system shown in Figure 18.3. At $T = 100$ K, the canonical partition function is seen to be 67.94, and the slope $(\partial \ln Z/\partial T)_V$ at this point is 0.015 K^{-1}. Therefore, the entropy per mole at 100 K is

$$S_{100\,K} = R \ln Z + RT \left(\frac{\partial \ln Z}{\partial T} \right)_V$$

$$= (8.314 \, \text{J mol}^{-1} \text{K}^{-1}) \ln(67.94) + (8.314 \, \text{J mol}^{-1} \text{K}^{-1})(100 \, \text{K})(0.015 \, \text{K}^{-1})$$

$$= 47.54 \, \text{J mol}^{-1} \text{K}^{-1}.$$

It is useful and instructive to note that Eq. 18.39A may be written in the form

$$S = k \ln Z + kT \left(\frac{\partial \ln Z}{\partial T} \right)_V = \frac{\partial}{\partial T} [kT \ln Z]_V. \quad \textbf{(18.39B)}$$

We see that in this form the entropy at 100 K can also be obtained from the slope at 100 K of a plot of $kT \ln Z$ versus T.

Table 18.1	Thermodynamic quantities in terms of the canonical partition function	
Quantity	**Equivalent Expression in terms of Z**	
Internal Energy U	$kT^2 \left(\dfrac{\partial \ln Z}{\partial T} \right)_V$	
Pressure p	$kT \left(\dfrac{\partial \ln Z}{\partial V} \right)_T$	
Entropy S	$k \ln Z + \dfrac{U}{T} = k \ln Z + kT \left(\dfrac{\partial \ln Z}{\partial T} \right)_V$	
Helmholtz Free Energy A	$-kT \ln Z$	
Gibbs Free Energy G	$-kT \ln Z + pV = -kT \ln Z + kTV \left(\dfrac{\partial \ln Z}{\partial V} \right)_T$	
Enthalpy H	$kT^2 \left(\dfrac{\partial \ln Z}{\partial T} \right)_V + pV$	
	$= kT^2 \left(\dfrac{\partial \ln Z}{\partial T} \right)_V + kTV \left(\dfrac{\partial \ln Z}{\partial V} \right)_T$	

18.3.5 Helmholtz and Gibbs Free Energies and the Enthalpy

The preceding results allow us to obtain A and G in terms of the canonical partition function by simple substitution. For the Helmholtz free energy, we have

$$A = U - TS = U - kT \ln Z - T\frac{U}{T} = -kT \ln Z. \qquad \textbf{(18.40)}$$

The Gibbs free energy is

$$G = U - TS + PV = A + pV = -kT \ln Z + kTV\left(\frac{\partial \ln Z}{\partial V}\right)_T. \qquad \textbf{(18.41)}$$

The enthalpy of the system is obtained directly from the internal energy:

$$H = U + pV = kT^2\left(\frac{\partial \ln Z}{\partial T}\right)_V + pV = kT^2\left(\frac{\partial \ln Z}{\partial T}\right)_V + kTV\left(\frac{\partial \ln Z}{\partial V}\right)_T.$$

$$\textbf{(18.42)}$$

18.3.6 Summary

Table 18.1 summarizes the relationships of various thermodynamic quantities to the canonical partition function. The easiest way to derive all of the results given in the table is simply to remember the relationship between A and Z and then use the differential dA for closed systems undergoing processes along a reversible path. Example 18.1 illustrates the method.

EXAMPLE 18.1

Starting with the relationship between the Helmholtz free energy and the canonical partition function, derive the remaining expressions in Table 18.1.

Solution

The Helmholtz free energy is given by Eq. 18.40:

$$A = -kT \ln Z. \qquad \textbf{(1)}$$

Equation 5.22 shows that, for any closed system undergoing a reversible process,

$$dA = -S\,dT - p\,dV. \qquad \textbf{(2)}$$

Therefore, the entropy is given by

$$\left(\frac{\partial A}{\partial T}\right)_V = -S = \frac{\partial}{\partial T}[-kT \ln Z]_V = -k \ln Z - kT\left(\frac{\partial \ln Z}{\partial T}\right)_V, \qquad \textbf{(3)}$$

so that

$$S = k \ln Z + kT\left(\frac{\partial \ln Z}{\partial T}\right)_V. \qquad \textbf{(4)}$$

U can now be obtained from $U = A + TS$. This gives

$$U = -kT \ln Z + T\left[k \ln Z + kT\left(\frac{\partial \ln Z}{\partial T}\right)_V\right] = kT^2\left(\frac{\partial \ln Z}{\partial T}\right)_V. \qquad \textbf{(5)}$$

Combining Eqs. 4 and 5 allows S to be written in the form

$$S = k \ln Z + kT \left(\frac{\partial \ln Z}{\partial T} \right)_V = k \ln Z + \frac{U}{T}. \tag{6}$$

Using Eq. 2 again, we obtain, for the pressure,

$$p = -\left(\frac{\partial A}{\partial V} \right)_T = -\frac{\partial}{\partial V} [-kT \ln Z]_T = kT \left(\frac{\partial \ln Z}{\partial V} \right)_T. \tag{7}$$

The Gibbs free energy and the enthalpy are, therefore,

$$G = A + pV = -kT \ln Z + pV = -kT \ln Z + kTV \left(\frac{\partial \ln Z}{\partial V} \right)_T \tag{8}$$

and

$$H = U + pV = kT^2 \left(\frac{\partial \ln Z}{\partial T} \right)_V + pV = kT^2 \left(\frac{\partial \ln Z}{\partial T} \right)_V + kTV \left(\frac{\partial \ln Z}{\partial V} \right)_T, \tag{9}$$

respectively. Just remember that $A = -kT \ln Z$ and $dA = -S\,dT - p\,dV$, and you have the rest!

18.4   Relationship between Z and z for Ideal Systems

Table 18.1 gives us all the thermodynamic quantities if we can obtain the canonical partition function. Since $Z = \sum_{\alpha=1}^{n_D} \exp[-E_\alpha/(kT)]$, it appears that Z can be obtained only if we first compute the energies of all statistically significant distributions or microstates. In principle, this observation is correct; however, we can obtain a very accurate evaluation of Z by taking advantage of its physical significance. We found in Chapter 17 that the molecular partition function z is equal to the effective number of quantum states occupied by the molecules at temperature T. The canonical partition function has exactly the same mathematical form as the molecular partition function. It is, therefore, equal to the effective number of distributions or microstates that can be formed by distributing N molecules among s effective quantum states. Consequently, if we can find a means of obtaining this effective number of microstates, we will have the value of the canonical partition function.

When N is Avogadro's constant and s_{eff} is even larger, it is very difficult to envision the effective number of microstates that can be formed. To find the appropriate expression for this quantity, we shall first consider some simple cases and then gradually make them more complex. By considering how the expression for the effective number of microstates varies as the complexity increases, we will be better able to understand the expression for the general case. (At least, that's the plan.)

A good starting point is the number of distributions or microstates of two indistinguishable cubic dice among six states. All of the possible distributions for this system are written out in Table 17.2 and illustrated in Figure 17.2. There are 6^2, or 36, possible arrangements, which form a total of 21 distinguishable microstates. Let us now increase the complexity of this system by increasing the number of effective states from six to s_{eff}, where s_{eff} is unlimited. We now ask, How many distinguishable microstates can be formed by distributing two objects among the s_{eff} states? First, we can place both objects in the same state. Since there are s_{eff} distinct choices for the state that will hold both objects, there are s_{eff} microstates containing two objects in a single state. Alternatively, we

Table18.2	Number of distinguishable microstates for two indistinguishable objects distributed among s_{eff} states	
Type of Distribution	**Number of Distinguishable Microstates**	
Two objects per state	s_{eff}	
One object per state	$\dfrac{s_{eff}(s_{eff}-1)}{2} = \dfrac{s_{eff}^2 - s_{eff}}{2}$	
Total number of distinguishable distributions or microstates	$s_{eff} + \dfrac{s_{eff}^2 - s_{eff}}{2} = \dfrac{s_{eff}^2 + s_{eff}}{2!}$	

can place the two objects in different states. There are s_{eff} choices for the first object and $s_{eff} - 1$ choices for the second object. The total number of choices is, therefore, $s_{eff}(s_{eff} - 1)$. However, since the objects are indistinguishable (like dice or molecules), this number counts each distribution twice. That is, putting Object 1 in State i and Object 2 in state j gives the same microstate as putting Object 1 in State j and Object 2 in state i. Consequently, the total number of distinguishable microstates with one object per state is $s_{eff}(s_{eff} - 1)/2$. The total number of microstates is

$$s_{eff} + \frac{s_{eff}(s_{eff}-1)}{2} = \frac{s_{eff}^2 + s_{eff}}{2},$$

which we can write as $(s_{eff}^2 + s_{eff})/2!$. This situation is summarized in Table 18.2. When $s_{eff} = 6$, the number of distinguishable microstates is $(6^2 + 6)/2 = 42/2 = 21$. These are the 21 microstates listed in Table 17.2 and Figure 17.2.

Let us now increase the complexity of the system still further. Suppose we wish to distribute *three* indistinguishable objects over s_{eff} states. Now how many microstates can we obtain? There are three possibilities: all three objects in the same state, two objects in one state and the third in a different state, or all three objects in different states. With s_{eff} different states, there are s_{eff} different distributions with all three objects in a single state. When we place a pair of objects in the same state, we have s_{eff} choices for the state that will contain the pair. There are then $s_{eff} - 1$ choices for the remaining object. This gives $s_{eff}(s_{eff} - 1)$ total microstates. When all objects reside in different states, we have s_{eff} choices for the first object, $s_{eff} - 1$ for the second, and $s_{eff} - 2$ for the third. This gives $s_{eff}(s_{eff} - 1)(s_{eff} - 2)$ total choices. However, some of these are indistinguishable; that is, we can distribute Objects 1, 2, and 3 into States i, j, and k in six ways as follows: 1–2–3, 1–3–2, 2–1–3, 2–3–1, 3–1–2, and 3–2–1. Therefore, this total must be divided by $3! = 6$ to prevent counting indistinguishable distributions more than once. Table 18.3 summarizes the number of distinguishable ways each of these events can occur.

Figure 18.5 shows all the possible microstates when $s_{eff} = 3$. The total number of microstates is $s_{eff}^3/3! + s_{eff}^2/2 + s_{eff}/3 = 3^3/3! + 3^2/2 + 3/3 = 10$, which is exactly the number in the figure.

EXAMPLE 18.2

How many distinguishable microstates exist for the distribution of three indistinguishable objects among 12 states?

Solution

The number of distinguishable microstates when three indistinguishable objects are distributed among 12 states is given by

$$\text{no. of microstates} = \frac{s_{\text{eff}}^3}{3!} + \frac{s_{\text{eff}}^2}{2} + \frac{s_{\text{eff}}}{3} = \frac{12^3}{6} + \frac{12^2}{2} + \frac{12}{2} = 366 \text{ microstates.}$$

For related exercises, see Problems 18.2 and 18.3.

Table 18.3   Number of distinguishable microstates for three indistinguishable objects distributed among s_{eff} states

Type of Distribution	Number of Distinguishable Microstates
Three objects per state	s_{eff}
Two objects in one state and the third object in a different state	$s_{\text{eff}}(s_{\text{eff}} - 1) = s_{\text{eff}}^2 - s_{\text{eff}}$
All objects in different states	$\dfrac{s_{\text{eff}}(s_{\text{eff}} - 1)(s_{\text{eff}} - 2)}{3!} = \dfrac{s_{\text{eff}}^3 - 3s_{\text{eff}}^2 + 2s_{\text{eff}}}{3!}$
Total number of distinguishable distributions or microstates	$s_{\text{eff}} + s_{\text{eff}}^2 - s_{\text{eff}} + \dfrac{s_{\text{eff}}^3 - 3s_{\text{eff}}^2 + 2s_{\text{eff}}}{3!}$
	$= \dfrac{s_{\text{eff}}^3}{3!} + \dfrac{s_{\text{eff}}^2}{2} + \dfrac{s_{\text{eff}}}{3}$

◄ **FIGURE 18.5**
The 10 possible microstates for the distribution of three indistinguishable molecules into three possible energy states. The total number is exactly that predicted by the analysis that is summarized in Table 18.3.

Notice that when we had two objects, the number of microstates was a two-term polynomial in s_{eff}, with the leading term being $s_{eff}^2/2!$. When we have three objects, the result is a three-term polynomial, with $s_{eff}^3/3!$ being the leading term.

Let us increase the level of complexity once more by increasing the number of objects to four. Now there are five possibilities: (1) all objects in one state, (2) three objects in a single state, (3) two objects in one state with the remaining two in different states, (4) two objects in each of two states, and (5) all objects in different states. As in the previous two cases, there are s_{eff} microstates with all objects in a single state. If we place three objects in a single state, we have s_{eff} choices for this state and $s_{eff} - 1$ choices for the state that will contain a single object. This gives a total of $s_{eff}(s_{eff} - 1)$ possible microstates. For the third possibility, there are s_{eff} choices for the state that contains two objects. This leaves $s_{eff} - 1$ choices for the third object and $s_{eff} - 2$ choices for the fourth. The total number of choices is $s_{eff}(s_{eff} - 1)(s_{eff} - 2)$. This total must be divided by a factor of two to avoid recounting distributions. That is, the distributions in which objects 1, 2, 3, and 4 are distributed among states i, j, and k as (Objects 1 and 2 in State i, Object 3 in State j, and Object 4 in State k) or (Objects 1 and 2 in State i, Object 4 in State j, and Object 3 in State k) are the same. We wish to count these two distributions only once, so division by two is needed. Figure 18.6 illustrates this situation for the case

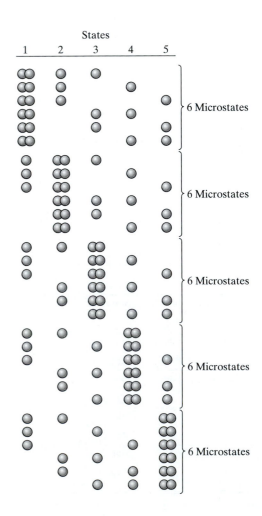

▶ **FIGURE 18.6**
The 30 distinguishable microstates obtained from the distribution of four indistinguishable objects among five states with two objects in one state and the other two in different states.

where $N = 4$ and $s_{eff} = 5$, for which we expect $(5)(4)(3)/2 = 30$ microstates. For the fourth possibility, where two pairs of objects reside in two states, we have s_{eff} choices for the first state and $s_{eff} - 1$ choices for the second state. However this analysis again counts microstates twice. That is, having the 1–2 pair in State i and the 3–4 pair in State j is the same as having the 3–4 pair in State i and the 1–2 pair in State j. Therefore, the number of distinguishable microstates is $s_{eff}(s_{eff} - 1)/2$. Finally, for the fifth possibility, with all objects in different states, we have s_{eff} choices for the first object, $s_{eff} - 1$ for the second, $s_{eff} - 2$ for the third, and $s_{eff} - 4$ for the fourth. The product of these factors must be divided by 4! to avoid counting indistinguishable microstates more than once. This analysis is summarized in Table 18.4. Again, we note that with four objects, the total number of distinguishable microstates is a four-term polynomial in s_{eff} with $s_{eff}^4/4!$ as the leading term.

We are now prepared to extrapolate to the result that is expected when N is Avogadro's constant. In this case, the number of distinguishable microstates should be given by an N-term polynomial in s_{eff} with $s_{eff}^N/N!$ as the leading term. That is,

$$\text{no. of distinguishable microstates} = \frac{s_{eff}^N}{N!} + c_{N-1}s_{eff}^{N-1} + c_{N-2}s_{eff}^{N-2}$$

$$+ \cdots + c_1 s_{eff} \qquad (18.43)$$

where the c_k are constants. It might appear that we have no effective way of evaluating Eq. 18.43, since the number of terms in the sum is 6.02214×10^{23}.

Table 18.4   Number of distinguishable microstates for four indistinguishable objects distributed among s_{eff} states

Type of Distribution	Number of Distinguishable Microstates
Four objects per state	s_{eff}
Three objects in one state	$s_{eff}(s_{eff} - 1) = s_{eff}^2 - s_{eff}$
Two objects in one state and the other two objects in different states	$\dfrac{s_{eff}(s_{eff} - 1)(s_{eff} - 2)}{2} = \dfrac{s_{eff}^3 - 3s_{eff}^2 + 2s_{eff}}{2}$
Two pairs of objects in two states	$\dfrac{s_{eff}(s_{eff} - 1)}{2} = \dfrac{s_{eff}^2 - s_{eff}}{2}$
All objects in different states	$\dfrac{s_{eff}(s_{eff} - 1)(s_{eff} - 2)(s_{eff} - 3)}{4!}$ $= \dfrac{s_{eff}^4}{4!} - \dfrac{s_{eff}^3}{4} + \dfrac{11s_{eff}^2}{4!} - \dfrac{s_{eff}}{4}$
Total number of distinguishable distributions or microstates	$s_{eff} + s_{eff}^2 - s_{eff} + \dfrac{s_{eff}^3 - 3s_{eff}^2 + 2s_{eff}}{2} + \dfrac{s_{eff}^2 - s_{eff}}{2}$ $+ \dfrac{s_{eff}^4}{4!} - \dfrac{s_{eff}^3}{4} + \dfrac{11s_{eff}^2}{4!} - \dfrac{s_{eff}}{4}$ $= \dfrac{s_{eff}^4}{4!} + \dfrac{s_{eff}^3}{4} + \dfrac{11s_{eff}^2}{4!} + \dfrac{s_{eff}}{4}$

However, we know that s_{eff} is much greater than N. In fact, we found in Section 18.1 that the ratio s_{eff}/N is on the order of 10^6. (See also Problem 18.1.) Under these conditions (see Problem 18.3), the number of distinguishable microstates is given by the leading term in Eq. 18.43 because

$$\lim_{s_{eff} \longrightarrow large} \left[\frac{s_{eff}^N}{N!} + c_{N-1} s_{eff}^{N-1} + c_{N-2} s_{eff}^{N-2} + \cdots + c_1 s_{eff} \right] \approx \frac{s_{eff}^N}{N!}, \quad (18.44)$$

provided that the coefficients c_{N-1}, c_{N-2}, etc., do not become extremely large. (See Problem 18.15 for further analysis.) Since the canonical partition function is equal to the number of effective microstates, we have

$$Z = \frac{s_{eff}^N}{N!}. \quad (18.45)$$

The only remaining question concerns the number of effective quantum states present in the system. We found the answer to this question in Chapter 17. The number is given by the molecular partition function z. Therefore,

$$Z = \frac{z^N}{N!}. \quad (18.46)$$

Note that Eq. 18.46 requires that the energies of the various quantum states be independent of the numbers of molecules occupying the states. This, in turn, requires that the molecules be noninteracting, so that they behave as ideal gases.

If we have solved the appropriate Schrödinger equation to find the energy eigenvalues of the various quantum states, we can always evaluate z at any temperature. Inserting the result into Eq. 18.46 produces Z, which, in turn, gives us all of the thermodynamic quantities. (See Problem 18.15 for a further discussion of the accuracy of Eq. 18.46.)

We can now modify Table 18.1 to express the thermodynamic quantities in terms of the molecular partition function. For example, the internal energy is given by

$$U = kT^2 \left(\frac{\partial \ln Z}{\partial T} \right)_V = kT^2 \left(\frac{\partial \ln[z^N/N!]}{\partial T} \right)_V = kT^2 \left[N \left(\frac{\partial \ln z}{\partial T} \right)_V - \left(\frac{\partial \ln N!}{\partial T} \right)_V \right]$$

$$= RT^2 \left(\frac{\partial \ln z}{\partial T} \right)_V, \quad (18.47)$$

since the derivative of $\ln N!$ is zero. Similarly, the expression for the entropy is

$$S = k \ln Z + \frac{U}{T} = k \ln \frac{z^N}{N!} + RT \left(\frac{\partial \ln z}{\partial T} \right)_V = R \ln z + RT \left(\frac{\partial \ln z}{\partial T} \right)_V - k \ln N!$$

$$= R \ln z + RT \left(\frac{\partial \ln z}{\partial T} \right)_V - k[N \ln N - N]$$

$$= R[\ln z - \ln N] + RT \left(\frac{\partial \ln z}{\partial T} \right)_V + R$$

$$= R \ln \left[\frac{z}{N} \right] + RT \left(\frac{\partial \ln z}{\partial T} \right)_V + R = R \ln \left[\frac{z}{N} \right] + \frac{U}{T} + R, \quad (18.48)$$

where we have substituted Sterling's approximation for $\ln(N!)$. All of the pertinent results are summarized in Table 18.5. Problem 18.4 addresses the derivations of the expressions for A, G, H, and p.

Table 18.5   Thermodynamic quantities in terms of the molecular partition function for ideal systems

Quantity	Equivalent Expression in terms of z
Internal Energy U	$RT^2 \left(\dfrac{\partial \ln z}{\partial T} \right)_V$
Pressure p	$RT \left(\dfrac{\partial \ln z}{\partial V} \right)_T$
Entropy S	$R \ln\left[\dfrac{z}{N} \right] + \dfrac{U}{T} + R = R \ln\left[\dfrac{z}{N} \right] + RT \left(\dfrac{\partial \ln z}{\partial T} \right)_V + R$
Helmholtz Free Energy A	$-RT \ln\left[\dfrac{z}{N} \right] - RT$
Gibbs Free Energy G	$-RT \ln\left[\dfrac{z}{N} \right]$
Enthalpy H	$RT^2 \left(\dfrac{\partial \ln z}{\partial T} \right)_V + pV$
	$= RT^2 \left(\dfrac{\partial \ln z}{\partial T} \right)_V + RTV \left(\dfrac{\partial \ln z}{\partial V} \right)_T$

18.5   Computation of Thermodynamic Quantities for Ideal Systems

As noted in the previous section, all we need to obtain the important thermodynamic quantities for ideal systems is appropriate expressions for the molecular partition function, which can always be computed if we have accurate expressions for the translational, rotational, vibrational, and electronic energy states. Since the translational energy of the center of mass is always rigorously separable from the internal rotational–vibrational and electronic energies (see Section 12.2.1), we can express the total molecular energy as

$$\varepsilon = \varepsilon_{tr} + \varepsilon_{int}. \tag{18.49}$$

Therefore, the nth overall energy state is the sum of the translational energy of the ith state and the internal energy of the jth state:

$$\varepsilon_n = (\varepsilon_{tr})_i + (\varepsilon_{int})_j. \tag{18.50}$$

The molecular partition function is given by Eq. 17.35, viz.,

$$z = \sum_k g_k \exp\left[-\frac{\varepsilon_i}{kT} \right] = \sum_n \exp\left[-\frac{\varepsilon_n}{kT} \right], \tag{18.51}$$

where the first summation, with the degeneracy factor, runs over all quantum states with different energies. The second summation, without g, runs over all quantum states. If we substitute Eq. 18.50 into Eq. 18.51, we obtain a very important result:

$$z = \sum_n \exp\left[-\frac{(\varepsilon_{tr} + \varepsilon_{int})_n}{kT} \right] = \sum_i \sum_j \exp\left[-\frac{(\varepsilon_{tr})_i + (\varepsilon_{int})_j}{kT} \right]$$

$$= \sum_i \exp\left[-\frac{(\varepsilon_{tr})_i}{kT} \right] \sum_j \exp\left[-\frac{(\varepsilon_{int})_j}{kT} \right] = z_{tr} z_{int}. \tag{18.52}$$

Equation 18.52 tells us that we can rigorously write the molecular partition function as the product of a translational molecular partition function and an internal molecular partition function. This permits their separate evaluation. It also permits us to evaluate the contributions of translational and internal energy to the various thermodynamic quantities separately. We illustrate this point by examining the internal energy and the entropy.

The internal energy is given by Eq. 18.47:

$$U = RT^2\left(\frac{\partial \ln z}{\partial T}\right)_V.$$

Substituting Eq. 18.52 for z gives

$$U = RT^2\left(\frac{\partial \ln (z_{tr}z_{int})}{\partial T}\right)_V = RT^2\left(\frac{\partial \ln z_{tr}}{\partial T}\right)_V + RT^2\left(\frac{\partial \ln z_{int}}{\partial T}\right)_V = U_{tr} + U_{int}.$$

$$(18.53)$$

For the entropy, we have, from Eq. 18.48, after substituting Sterling's approximation for $\ln(N!)$,

$$S = R \ln z + RT\left(\frac{\partial \ln z}{\partial T}\right)_V - k[N\ln N - N].$$

Substituting Eq. 18.52 for z produces

$$S = R \ln(z_{tr}z_{int}) + RT\left(\frac{\partial \ln(z_{tr}z_{int})}{\partial T}\right)_V - k[N\ln N - N]$$

$$= \left\{R \ln z_{tr} + RT\left(\frac{\partial \ln z_{tr}}{\partial T}\right)_V - k[N\ln N - N]\right\}$$

$$+ \left\{R \ln z_{int} + RT\left(\frac{\partial \ln z_{int}}{\partial T}\right)_V\right\} = S_{tr} + S_{int}. \qquad (18.54)$$

For the entropy and other quantities that involve the entropy, such as A and G, we must be careful. There is only one $\ln(N!)$ term. In Eq. 18.54, we have included it with the terms involving the translational molecular partition function. The reason for this choice will become clear later in the chapter, when we demonstrate that, without the translational quantum states, there would not be a $\ln(N!)$ term. With this choice, we have

$$S_{tr} = R \ln\left[\frac{z_{tr}}{N}\right] + \frac{U_{tr}}{T} + R \qquad (18.55)$$

and

$$S_{int} = R \ln z_{int} + \frac{U_{int}}{T}. \qquad (18.56)$$

The corresponding equations for A_{tr}, A_{int}, G_{tr}, and G_{int} are easily obtained with the use of Eqs. 18.53 through 18.56.

18.5.1   Translational Contributions to Thermodynamics

The translational contributions to the thermodynamic quantities can be obtained directly from z_{tr}, which is given by Eq. 18.5 if we assume that the molecules do not interact:

$$z_{tr} = \frac{V}{h^3} [2\pi m k T]^{3/2}. \tag{18.57}$$

Using Eq. 18.53, we obtain

$$U_{tr} = RT^2 \left(\frac{\partial \ln z_{tr}}{\partial T} \right)_V = RT^2 \frac{\partial}{\partial T} \left[\ln \left\{ \frac{V}{h^3} [2\pi m k T]^{3/2} \right\} \right]_V$$

$$= RT^2 \frac{\partial}{\partial T} \left[\ln(T^{3/2}) + \ln \left\{ \frac{V}{h^3} [2\pi m k]^{3/2} \right\} \right]_V = \frac{3}{2} RT^2 \frac{1}{T} = \frac{3RT}{2}. \tag{18.58}$$

This result comes as no surprise; we have been using $3RT/2$ as the average thermal translational energy since Chapter 1.

The translational contribution to the entropy can be obtained from Eq. 18.55:

$$S_{tr} = R \ln \left[\frac{z_{tr}}{N} \right] + \frac{U_{tr}}{T} + R = R \ln \left\{ \frac{V}{Nh^3} [2\pi m k T]^{3/2} \right\} + \frac{3RT}{2T} + R$$

$$= R \ln \left\{ \frac{V}{Nh^3} [2\pi m k T]^{3/2} \right\} + \frac{5R}{2}. \tag{18.59}$$

Since we have already assumed noninteracting molecules, we have, for one mole, $pV = RT = NkT$, so that $V/N = kT/p$. Substitution of this result into Eq. 18.59 gives

$$\boxed{S_{tr} = R \ln \left\{ \frac{[2\pi m]^{3/2} [kT]^{5/2}}{ph^3} \right\} + \frac{5R}{2}.} \tag{18.60}$$

Equation 18.60 is called the *Sackur–Tetrode equation.*

Separating the logarithmic term in Eq. 18.60 into terms involving mass, temperature, pressure, and constants produces

$$S_{tr} = \frac{3R}{2} \ln(m) + \frac{5R}{2} \ln(T) - R \ln(p) + \left[R \ln \left\{ \frac{[2\pi]^{3/2} k^{5/2}}{h^3} \right\} + \frac{5R}{2} \right]. \tag{18.61}$$

If we use SI units, p must be expressed in newton m^{-2}, mass in kg, and energy in joules. In terms of the molar mass M, expressed in g mol^{-1},

$$m = \frac{M}{1,000N}, \tag{18.62}$$

where N is Avogadro's constant. One newton m^{-2} is 1 Pa, and 1 bar is 10^5 Pa. Therefore, p in Pa $= 10^5 p$ in bar, so that

$$\ln(p) = \ln \left[\frac{10^5 p}{1 \text{ bar}} \right] = \ln \left[\frac{p}{1 \text{ bar}} \right] + \ln(10^5). \tag{18.63}$$

Substituting Eqs. 18.62 and 18.63 into Eq. 18.61 produces

$$S_{tr} = \frac{3R}{2} \ln(M) + \frac{5R}{2} \ln(T) - R \ln(p) + R \left[\ln \left\{ \frac{[2\pi]^{3/2} k^{5/2}}{(1,000N)^{3/2} (10^5) h^3} \right\} + \frac{5}{2} \right],$$

where M is expressed in units of g mol^{-1} and p is in bars. The constant is

$$\left[\ln\left\{ \frac{[2\pi]^{3/2}k^{5/2}}{(1,000N)^{3/2}(10^5)h^3} \right\} + \frac{5}{2} \right]$$

$$= \ln\left\{ \frac{[2(3.14159)]^{3/2}(1.38066 \times 10^{-23})^{5/2}}{[1,000(6.02214 \times 10^{23})]^{3/2}(10^5)(6.62608 \times 10^{-34})^3} \right\} + 2.5$$

$$= \ln(0.025947) + 2.5 = -1.15169.$$

The translational entropy can, therefore, be written in the form

$$\boxed{S_{tr} = \frac{3R}{2}\ln(M) + \frac{5R}{2}\ln(T) - R\ln(p) - 1.15169R}. \qquad \textbf{(18.64)}$$

Equation 18.64 is very convenient for computational purposes.

EXAMPLE 18.3

Compute the entropy of 1 mole of He atoms at 1 bar pressure and 300 K. The value reported in the 78th edition of the CRC *Handbook of Chemistry and Physics* is 126.3 J mol^{-1} K^{-1}. Compute the percent error in the result.

Solution

Since He atoms have only translational energy, we may use Eq. 18.64 to obtain S:

$$S = S_{tr} = \frac{3R}{2}\ln(M) + \frac{5R}{2}\ln(T) - R\ln(p) - 1.15169R$$

$$= R[1.5\ln(4.0026) + 2.5\ln(300) - \ln(1) - 1.15169] = 15.188R$$

$$= 126.27 \text{ J mol}^{-1}\text{ K}^{-1}.$$

The answer is essentially exact; the deviation of helium at this pressure from ideality is very small.

    For related exercises, see Problems 18.5 and 18.6.

    The translational Helmholtz free energy, A_{tr}, can be obtained directly from the preceding results. Since we have $A = U - TS$, the use of Eqs. 18.58 and 18.60 gives

$$A_{tr} = U_{tr} - TS_{tr} = \frac{3RT}{2} - RT\ln\left\{ \frac{[2\pi m]^{3/2}[kT]^{5/2}}{ph^3} \right\} - \frac{5RT}{2}$$

$$= -RT\ln\left\{ \frac{[2\pi m]^{3/2}[kT]^{5/2}}{ph^3} \right\} - RT. \qquad \textbf{(18.65A)}$$

Using Eq. 18.64 for S_{tr}, we can express the translational Helmholtz free energy in a form more convenient for computations:

$$A_{tr} = 1.5RT - T\left[\frac{3R}{2}\ln(M) + \frac{5R}{2}\ln(T) - R\ln(p) - 1.15169R \right]$$

We then have

$$\boxed{A_{tr} = -\frac{3RT}{2}\ln(M) - \frac{5RT}{2}\ln(T) + RT\ln(p) + 2.6517RT}. \qquad \textbf{(18.65B)}$$

The Gibbs free energy is related to A by $G = A + pV$. Therefore, from Eq. 18.65A,

$$G_{tr} = A_{tr} + pV = -RT \ln\left\{\frac{[2\pi m]^{3/2}[kT]^{5/2}}{ph^3}\right\} - RT + pV$$

$$= -RT \ln\left\{\frac{[2\pi m]^{3/2}[kT]^{5/2}}{ph^3}\right\}. \tag{18.66A}$$

Using Eq. 18.65B, we can express G_{tr} in the form

$$G_{tr} = -\frac{3RT}{2}\ln(M) - \frac{5RT}{2}\ln(T) + RT\ln(p) + 2.6517RT + pV,$$

which gives

$$\boxed{G_{tr} = -\frac{3RT}{2}\ln(M) - \frac{5RT}{2}\ln(T) + RT\ln(p) + 3.6517RT}. \tag{18.66B}$$

In Eqs. 18.66A and 18.66B, we have replaced pV with RT, since we have already assumed noninteracting molecules when we evaluated the translational molecular partition function using the eigenvalues for independent particles in a potential well with zero potential energy. (See Problem 18.7.)

18.5.2   The Internal Molecular Partition Function

The internal energy of a molecule comprises rotational, vibrational, and electronic energy. In general, these quantities are all coupled, so that the evaluation of the molecular partition function for internal motion involves the execution of a complex coupled multiple summation. Fortunately, the problem can often be simplified.

In Chapters 13 and 14, we found that the energy spacings between electronic states is usually on the order of several electron-volts per molecule ($1 \text{ eV} = 9.648 \times 10^4 \text{ J mol}^{-1}$). With spacings this large, only the ground electronic state will make any measurable contribution to the molecular partition function. Consider the hydrogen atom as an example. Equation 13.20 tells us that the electronic energy of hydrogen-like atoms and ions is

$$E_n = -\frac{Z^2 e^2}{2n^2 a_o} \quad (n = 1, 2, 3, \ldots),$$

where a_o is the Bohr radius. For hydrogen, $Z = 1$, so that in atomic units,

$$E_n(\text{H}) = -\frac{1}{2n^2} \text{ hartrees} \quad (n = 1, 2, 3, \ldots)$$

when the zero-energy point is the ionized system $\text{H}^+ + 1e^-$. Since we are free to use any convenient reference point, let us take the ground electronic state to be our zero of energy. With this choice, we have

$$E_n(\text{H}) = \frac{1}{2} - \frac{1}{2n^2} = 0.5[1 - n^{-2}] \text{ hartrees} = 1.313 \times 10^6[1 - n^{-2}] \text{ J mol}^{-1},$$

where we have used the conversion factor $1 \text{ hartree} = 27.2105 \text{ eV} = 2.625 \times 10^6 \text{ J mol}^{-1}$.

The molecular electronic partition function for the hydrogen atom written with the degeneracy factor present is

$$z_{el} = \sum_n g_n \exp\left[-\frac{E_n}{RT}\right] = \sum_n n^2 \exp\left[-\frac{1.313 \times 10^6[1 - n^{-2}]}{RT}\right],$$

since the degeneracy of the nth electronic state for hydrogen is n^2. At 298 K,

$$z_{el} = \sum_n n^2 \exp[-530.0(1 - n^{-2})]$$

$$= e^o + 4\exp[-397.5] + 9\exp[-471.1] + \cdots = 1.$$

This is a typical result. The only term in the series of significant magnitude is the ground-state energy term, which is always $g_o \exp[-\varepsilon_o/(RT)]$, where g_o and ε_o are the degeneracy and energy of the ground state, respectively. We can, therefore, write the internal molecular partition function in the form

$$z_{int} = g_o \exp\left[-\frac{\varepsilon_o}{RT}\right] z_{rot-vib}, \tag{18.67}$$

where the rotational–vibrational eigenvalues are those appropriate for the ground electronic state of the molecule.

We can gain an understanding of the problem associated with the evaluation of $z_{rot-vib}$ by treating the case of a diatomic molecule whose ground-state electronic potential-energy curve is described with sufficient accuracy by a Morse potential. In Chapter 12, we found that the vibrational–rotational energy eigenvalues for this potential are given by

$$E_{vJ} = (v + 0.5)h\nu_o - \frac{h^2\nu_o^2}{4D}(v + 0.5)^2 + J(J + 1)\frac{\hbar^2}{2I} - \frac{\hbar^4}{4\mu^2\beta^2 R_e^6 D}[J(J + 1)]^2$$

$$- \frac{3[1 - (1/\beta R_e)]\hbar^2 h\nu_o}{4\mu\beta R_e^3 D}(v + 0.5)J(J + 1), \tag{18.68}$$

provided that E_{vJ} is not too large. The first and third terms are the harmonic oscillator and rigid-rotor energies, respectively. The second term reflects the contribution of vibrational anharmonicity. The fourth term is the result of centrifugal distortion of the rotational-energy eigenvalues, and the last term couples vibration and rotational motion. To reduce the complexity of the notation, let us write this eigenvalue expression in the form

$$E_{vJ} = (v + 0.5)h\nu_o - C_h(v + 0.5)^2 + J(J + 1)\frac{\hbar^2}{2I} - C_d[J(J + 1)]^2$$

$$- C_{vr}(v + 0.5)J(J + 1), \tag{18.69}$$

where C_h, C_d, and C_{vr} are the coefficients of the anharmonicity, centrifugal distortion, and coupling terms, respectively, in Eq. 18.68. The internal molecular partition function for this system is

$$z_{rot-vib} = \sum_{v=0}^{\infty} \sum_{J=0}^{\infty} (2J + 1) \exp\left[-\frac{E_{vJ}}{kT}\right], \tag{18.70}$$

where we have written the molecular partition function with the rotational degeneracy factor $2J + 1$ present. Note that the presence of the coupling term means that we must execute the summation over J for every value of the vibrational quantum number v. Because there are only two quantum numbers over which we must sum, the coupled summation in Eq. 18.70 can be easily executed with a small computer. However, for a large molecule containing three rotational quantum numbers and $3N - 6$ vibrational quantum numbers, the execution of such coupled summations can become difficult even for very powerful workstations.

If the temperature is sufficiently low that higher quantum states make only negligible contributions to $z_{rot-vib}$, we can usually ignore the vibrational–rotational coupling terms. As an example, consider the HI molecule with the interatomic potential treated as a Morse potential with the parameters $D = 3.194$ eV, $\beta = 0.9468$ bohr^{-1}, and $R_e = 3.032$ bohr and a vibrational frequency of 6.924×10^{13} s^{-1}. This system was examined in Problem 15.34. If all quantities are converted to atomic units, a direct calculation using Eq. 18.68 gives the result

$$E_{vJ} = 0.01052(v + 0.5) - 0.0002357(v + 0.5)^2 + 3.006 \times 10^{-5}J(J + 1)$$
$$- 9.341 \times 10^{-10}[J(J + 1)]^2 - 9.173 \times 10^{-7}(v + 0.5)J(J + 1) \text{ hartrees.}$$

The coefficient of the coupling term is a factor of 32.77 smaller than that for the rigid-rotor term and a factor of 1.147×10^4 smaller than the harmonic term. Differences of this magnitude are often the case.

When the coupling terms can be ignored, Eq. 18.70 assumes a very simple form. Substituting E_{vJ} from Eq. 18.69 after omitting the coupling term gives

$$z_{rot-vib} = \sum_{v=0}^{\infty} \exp\left[-\frac{(v + 0.5)h\nu_o - C_h(v + 0.5)^2}{kT} \right] \times$$

$$\sum_{J=0}^{\infty} (2J + 1) \exp\left[-\frac{J(J + 1)\dfrac{\hbar^2}{2I} - C_d[J(J + 1)]^2}{kT} \right]. \quad \textbf{(18.71)}$$

Notice that the two summations are now independent. This means that each sum need be evaluated only once. The first factor in Eq. 18.71 corresponds to a summation over vibrational states, while the second sums over the rotational levels. We can, therefore, express the internal molecular partition function as the product

$$z_{int} = g_o \exp\left[-\frac{\varepsilon_o}{RT} \right] z_{rot} z_{vib}. \quad \textbf{(18.72)}$$

In general, whenever the internal energy can be written as a sum of separable rotational and vibrational energies ($E_{int} = E_{rot} + E_{vib}$), $z_{rot-vib}$ may be expressed as the product of rotational and vibrational molecular partition functions. We shall, therefore, consider rotation and vibration separately, as we have already done for translation.

18.5.3   Rotational-energy Contributions to Thermodynamics

Let us begin with an examination of linear molecules. If we ignore rotational–vibrational coupling terms, the rotational-energy states of a linear molecule can be accurately represented by the eigenvalues for a linear rigid rotor plus a correction term for centrifugal distortion. If the interatomic potential is represented by a Morse function, Eq. 18.68 gives the distortion term. In this case, the rotational molecular partition function is

$$z_{\text{rot}} = \sum_{J=0}^{\infty} (2J + 1) \exp\left[-\frac{J(J + 1)\dfrac{\hbar^2}{2I} - C_d[J(J + 1)]^2}{kT} \right] \tag{18.73}$$

where

$$C_d = \frac{\hbar^4}{4\mu^2\beta^2 R_e^6 D}. \tag{18.74}$$

In very precise work, the coefficient of the distortion term can be determined by fitting C_d to the measured microwave spectrum. However, in most cases, the distortion term makes very little difference. For example, when z_{rot} is computed for HI at 300 K, there is only a 0.194% difference between the results obtained with and without the distortion term. (See Problem 18.8.)

If we omit the distortion term from Eq. 18.73 and write $\hbar^2/(2I)$ in terms of the rotational constant, we obtain

$$z_{\text{rot}} = \sum_{J=0}^{\infty} (2J + 1) \exp\left[-\frac{J(J + 1)Bh}{kT} \right]. \tag{18.75}$$

If B is small or if the temperature is high, the rotational levels will behave as if they form a near continuum. In such a case, the summation in Eq. 18.75 may be replaced with an integral, from which it follows that

$$z_{\text{rot}} = z_{\text{rot}}^c = \int_0^{\infty} (2J + 1) \exp\left[-\frac{J(J + 1)Bh}{kT} \right] dJ.$$

If we let $X = J(J + 1)$, so that $dX = (2J + 1)dJ$, the classical rotational molecular partition function becomes

$$z_{\text{rot}}^c = \int_{X=0}^{\infty} \exp\left[-\frac{BhX}{kT} \right] dX = -\frac{kT}{Bh} \exp\left[-\frac{BhX}{kT} \right]\bigg|_0^{\infty} = \frac{kT}{Bh}. \tag{18.76}$$

Equations 18.75 and 18.76 permit us to compute the rotational contribution to all thermodynamic quantities for heteronuclear diatomic molecules. However, a problem arises with homonuclear diatomics, because we have ignored nuclear spin and the Pauli principle in developing expressions for the molecular partition functions. In Section 13.4.5, we learned that particles with half-integral spin quantum numbers are fermions that must have antisymmetric wave functions with respect to the exchange of identical particles. Particles with integral spin quantum numbers are bosons whose wave functions must be symmetric with respect to identical-particle exchange. When the nuclei are different, this requirement has no effect. However, when we have identical nuclei, the total wave function must be symmetric to exchange if the nuclei are bosons or antisymmetric if they are fermions. Since the vibrational and electronic eigenfunctions for neutral molecules are all symmetric to particle exchange, this leaves the fulfillment of the Pauli principle to the molecular rotational wave function.

The total rotational wave function for a diatomic rigid rotor is the product of a spatial wave function and a Pauli spin function. In Chapter 12, we learned that the spatial part of the wave function is a spherical harmonic, $Y_J^M(\theta, \phi)$.

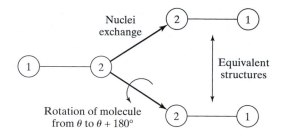

◀ FIGURE 18.7
Nuclear exchange and 180° degree rotation of a diatomic molecule. The two operations are clearly equivalent.

Since nuclear exchange is equivalent to a rotation of the diatomic molecule from θ to $\theta + 180°$ (see Figure 18.7), the spatial wave function will be symmetric to nuclear exchange if we have $Y_J^M(\theta, \phi) = Y_J^M(\theta + \pi, \phi)$; it will be antisymmetric to exchange if $Y_J^M(\theta, \phi) = -Y_J^M(\theta + \pi, \phi)$.

The spin portion of the wave function depends upon the nuclear spin quantum number I and the relative alignment of the spins of the two nuclei. For example, when $I = \frac{1}{2}$, we have two possible spin states α and β, with $m_I = \frac{1}{2}$ and $-\frac{1}{2}$, respectively. There are, therefore, four possible nuclear spin wave functions: $\alpha(1)\alpha(2)$, $\beta(1)\beta(2)$, $[\alpha(1)\beta(2) + \alpha(2)\beta(1)]$, and $[\alpha(1)\beta(2) - \alpha(2)\beta(1)]$. The first three are symmetric with respect to exchange, the fourth antisymmetric. If we represent the spin wave function by $S(1, 2)$, the total rotational wave function is

$$\psi(\theta, \phi) = Y_J^M(\theta, \phi) S(1, 2). \qquad (18.77)$$

If $S(1, 2)$ is antisymmetric to exchange, $Y_J^M(\theta, \phi)$ must be symmetric for fermion nuclei or antisymmetric for boson nuclei. The opposite is true if $S(1, 2)$ is symmetric. Table 18.6 summarizes the various possibilities.

In all cases, we arrive at the conclusion that $Y_J^M(\theta, \phi)$ must be either symmetric or antisymmetric with respect to a 180° rotation of the molecule. This is mathematically equivalent to replacing $Y_J^M(\theta, \phi)$ with $Y_J^M(\theta + \pi, \phi)$. The spherical harmonics are listed in Table 12.1. An examination of these functions shows that when J is even, $Y_J^M(\theta, \phi)$ is symmetric with respect to a rotation of π radians about θ. When J is odd, Y_J^M is antisymmetric. (See Problem 18.13.) Consequently, Table 18.6 shows that we will always eliminate exactly half of the possible rotational eigenstates and eigenvalues because of exchange symmetry requirements. Since we have learned that the molecular partition function counts the effective number of quantum states accessible to the system, we see that our count is too large by a factor of two. Half of the quantum states are eliminated by exchange symmetry requirements when the diatomic molecule is homonuclear. Therefore, we must divide z_{rot} by a factor of

Table 18.6	Exchange symmetry requirements for homonuclear diatomic molecules	
Type of Nuclei	**Required Symmetry**	
	Spatial Wave Function	**Spin Wave Function**
Fermions	Symmetric	Antisymmetric
or	Antisymmetric	Symmetric
Bosons	Symmetric	Symmetric
or	Antisymmetric	Antisymmetric

two. This factor is called the *symmetry number*. It is generally given the symbol σ. Physically, σ represents the number of rotational orientations that are chemically identical. For a heteronuclear molecule, $\sigma = 1$; for a homonuclear molecule, $\sigma = 2$. With this modification, Eqs. 18.75 and 18.76 become, respectively,

$$z_{rot} = \frac{1}{\sigma} \sum_{J=0}^{\infty} (2J + 1) \exp\left[-\frac{J(J + 1)Bh}{kT} \right] \qquad (18.78)$$

and

$$z_{rot}^c = \frac{kT}{\sigma Bh}. \qquad (18.79)$$

These equations may be used to evaluate the rotational contribution to the thermodynamic quantities for both homonuclear and heteronuclear molecules.

If rotation is treated as classical, U_{rot} assumes a very simple form. From Eq. 18.47, $U_{rot} = RT^2(\partial \ln z_{rot}^c/\partial T)_V$. Substituting Eq. 18.79 for z_{rot}^c produces

$$U_{rot} = RT^2 \frac{\partial}{\partial T}\left[\ln T + \ln\left\{ \frac{k}{\sigma Bh} \right\} \right]_V = RT^2 \frac{1}{T} = RT. \qquad (18.80)$$

The rotational contribution to the heat capacity is

$$[C_v]_{rot} = \left(\frac{\partial U_{rot}}{\partial T} \right)_V = R \qquad (18.81)$$

if rotation is treated classically. This is the assumption we made in Chapter 7 when we computed gas-phase heat capacities.

More accurate expressions for U_{rot} and $[C_v]_{rot}$ are obtained by using Eq. 18.78. With this expression, the rotational energy is

$$U_{rot} = RT^2 \left(\frac{\partial \ln z_{rot}}{\partial T} \right)_V = \frac{RT^2}{z_{rot}} \left(\frac{\partial z_{rot}}{\partial T} \right)_V$$

$$= \frac{R}{z_{rot}} \sum_{J=0}^{\infty} (2J + 1)J(J + 1)\frac{Bh}{k} \exp\left[-\frac{J(J + 1)Bh}{kT} \right]. \qquad (18.82)$$

The rotational contribution to the heat capacity is

$$[C_v]_{rot} = \left(\frac{\partial U_{rot}}{\partial T} \right)_V = \frac{2RT}{z_{rot}} \left(\frac{\partial z_{rot}}{\partial T} \right)_V - \frac{RT^2}{z_{rot}^2}\left[\left(\frac{\partial z_{rot}}{\partial T} \right) \right]_V^2 + \frac{RT^2}{z_{rot}} \left(\frac{\partial^2 z_{rot}}{\partial T^2} \right)_V$$

$$= -\frac{U_{rot}^2}{RT^2} + \frac{R}{z_{rot}}\left(\frac{Bh}{kT} \right)^2 \sum_{J=0}^{\infty} (2J + 1)J^2(J + 1)^2 \exp\left[-\frac{J(J + 1)Bh}{kT} \right]. \qquad (18.83)$$

(See Problem 18.9 for a complete derivation of Eq. 18.83.

EXAMPLE 18.4

Use the data in Table 15.4 to compute the rotational energy of HF at 200 K. Assume that the rotational-energy levels are given by a rigid-rotor expression. Compare your result with that for classical rotation.

Solution

The symmetry number for HF is unity. Therefore, the rotational internal energy is

$$U_{rot} = \frac{R}{z_{rot}} \sum_{J=0}^{\infty} (2J + 1)J(J + 1)\frac{Bh}{k} \exp\left[-\frac{J(J + 1)Bh}{kT} \right] \tag{1}$$

where

$$z_{rot} = \sum_{J=0}^{\infty} (2J + 1) \exp\left[-\frac{J(J + 1)Bh}{kT} \right]. \tag{2}$$

The quantities required inside the sums are

$$\frac{Bh}{k} = \frac{(20.939)(2.9979 \times 10^{10})s^{-1}(6.62608 \times 10^{-34}\,J\,s)}{1.38066 \times 10^{-23}\,J\,K^{-1}} = 30.126\ K \tag{3}$$

and

$$\frac{Bh}{kT} = 0.15063 \tag{4}$$

at 200 K. The two summations are, therefore,

$$z_{rot} = \sum_{J=0}^{\infty} (2J + 1) \exp[-0.15063J(J + 1)] \tag{5}$$

and

$$U_{rot} = \frac{8.314\,J\,mol^{-1}}{z_{rot}} \sum_{J=0}^{\infty} 30.126(2J + 1)J(J + 1) \exp[-0.15063J(J + 1)]. \tag{6}$$

These sums can be easily obtained with the use of an Excel spreadsheet. The results are

$$z_{rot}(200\ K) = 6.9825 \tag{7}$$

and

$$\sum_{J=0}^{\infty} 30.126(2J + 1)J(J + 1) \exp[-0.15063J(J + 1)] = 1,325.63\ K. \tag{8}$$

This gives

$$U_{rot} = \frac{8.314\,J\,mol^{-1}\,K^{-1}\,(1,325.63\ K)}{6.9825} = 1,578.4\,J\,mol^{-1}. \tag{9}$$

If rotation were classical, we would have

$$U_{rot} = RT = (8.314\,J\,mol^{-1}\,K^{-1})(200\ K) = 1,662.8\,J\,mol^{-1}. \tag{10}$$

The ratio of these two results is

$$\frac{U_{rot}}{U_{rot}^c} = \frac{1,578.4}{1,662.8} = 0.9492. \tag{11}$$

Thus, the classical approximation is in error by about 5% at 200 K for HF.
     For related exercises, see Problems 18.10, 18.11, and 18.12.

The entropy due to internal degrees of freedom is given by Eq. 18.56:

$$S_{int} = R \ln z_{int} + \frac{U_{int}}{T}.$$

If we write the internal molecular partition function in the product form, we obtain

$$S_{int} = R \ln[z_{el} z_{rot} z_{vib}] + \frac{U_{rot} + U_{vib} + U_{el}}{T} = \left[R \ln z_{rot} + \frac{U_{rot}}{T} \right]$$

$$+ \left[R \ln E_{vib} + \frac{U_{vib}}{T} \right] + \left[R \ln z_{el} + \frac{U_{el}}{T} \right] = S_{el} + S_{rot} + S_{vib}. \quad \textbf{(18.84)}$$

That is, the contribution of internal degrees of freedom to the entropy are separable and additive. If the degeneracy of the ground electronic state is unity and the ground-state electronic energy is taken to be zero, $S_{el} = R \ln(1) = 0$.

The rotational contribution to the entropy for a diatomic rigid rotor is

$$S_{rot} = \left[R \ln z_{rot} + \frac{U_{rot}}{T} \right]. \quad \textbf{(18.85)}$$

With z_{rot} given by Eq. 18.78 and U_{rot} by Eq. 18.82, the result is

$$S_{rot} = R \ln \left\{ \frac{1}{\sigma} \sum_{J=0}^{\infty} (2J + 1) \exp \left[-\frac{J(J+1)Bh}{kT} \right] \right\}$$

$$+ \frac{R}{z_{rot}} \left(\frac{Bh}{kT} \right) \sum_{J=0}^{\infty} (2J + 1)J(J+1) \exp \left[-\frac{J(J+1)Bh}{kT} \right]. \quad \textbf{(18.86)}$$

If the temperature is sufficiently high, we can usually assume that rotation behaves classically. (See Problems 18.10, 18.11, and 18.12.) In this case, we can use Eq. 18.79 for z_{rot}. This produces

$$S_{rot}^c = R \ln \left[\frac{kT}{\sigma Bh} \right] + \frac{RT}{T} = R \ln \left[\frac{kT}{\sigma Bh} \right] + R. \quad \textbf{(18.87)}$$

Notice that the symmetry number now enters the result. Its effect is to reduce the entropy, which is the result we expect, since symmetry constraints reduce the effective number of quantum states accessible to the system. The rotational constant is $B = h/(8\pi^2 I)$. Insertion of this equation into Eq. 18.87 allows the classical entropy to be written in the form

$$S_{rot}^c = R \left[\ln \left\{ \frac{8\pi^2 kTI}{\sigma h^2} \right\} + 1 \right] = R \left[\ln \left\{ \frac{IT}{\sigma} \right\} + \ln \left\{ \frac{8\pi^2 k}{h^2} \right\} + 1 \right] = R \left[\ln \left\{ \frac{IT}{\sigma} \right\} + 105.53 \right],$$

$$\textbf{(18.88)}$$

provided that the moment of inertia, I, is expressed in SI units of kg m^2.

EXAMPLE 18.5

Compute the rotational contribution to the entropy of HF at 50 K and 300 K. Do the computation both quantum mechanically and classically. Compare the two results. Assume that HF is a rigid rotor with a bond distance of 0.9171×10^{-10} m and a rotational constant of 6.277×10^{11} s^{-1}.

Solution

The quantum mechanical computation requires that we evaluate two summations in Eq. 18.86:

$$s_1 = \sum_{J=0}^{\infty} (2J + 1) \exp \left[-\frac{J(J+1)Bh}{kT} \right] \quad \textbf{(1)}$$

and

$$s_2 = \sum_{J=0}^{\infty} (2J + 1)J(J + 1) \exp\left[-\frac{J(J + 1)Bh}{kT}\right]. \tag{2}$$

The quantity

$$\frac{Bh}{k} = \frac{(6.277 \times 10^{11} \text{ s}^{-1})(6.62608 \times 10^{-34} \text{ J s})}{1.38066 \times 10^{-23} \text{ J K}^{-1}} = 30.125 \text{ K}. \tag{3}$$

Using an Excel spreadsheet to compute these two summations at 50 K, we obtain

$$s_1 = 2.0388 \quad \text{and} \quad s_2 = 2.6676. \tag{4}$$

Substituting these data into Eq. 18.86 with $\sigma = 1$ for HF, we obtain

$$S_{\text{rot}}(50 \text{ K}) = R \ln\left\{\sum_{J=0}^{\infty} (2J + 1) \exp\left[-\frac{J(J + 1)Bh}{kT}\right]\right\}$$

$$+ \frac{R}{z_{\text{rot}}}\left(\frac{Bh}{kT}\right) \sum_{J=0}^{\infty} (2J + 1)J(J + 1) \exp\left[-\frac{J(J + 1)Bh}{kT}\right]$$

$$= (8.314 \text{ J mol}^{-1} \text{ K}^{-1}) \ln(2.0388) + \frac{(8.314 \text{ J mol}^{-1} \text{ K}^{-1})}{2.0388} \frac{30.125 \text{ K}}{50 \text{ K}}(2.6676)$$

$$= 5.923 + 6.554 \text{ J mol}^{-1} \text{ K}^{-1} = 12.48 \text{ J mol}^{-1} \text{ K}^{-1}. \tag{5}$$

At 300 K, the results are

$$s_1 = 10.299 \quad \text{and} \quad s_2 = 99.104. \tag{6}$$

Use of these results in Eq. 18.86 with $\sigma = 1$ gives

$$S_{\text{rot}}(300 \text{ K}) = (8.314 \text{ J mol}^{-1} \text{ K}^{-1}) \ln(10.299)$$

$$+ \frac{(8.314 \text{ J mol}^{-1} \text{ K}^{-1})}{10.299} \frac{30.125 \text{ K}}{300 \text{ K}}(99.104)$$

$$= 19.389 + 8.034 \text{ J mol}^{-1} \text{ K}^{-1} = 27.42 \text{ J mol}^{-1} \text{ K}^{-1}. \tag{7}$$

If the rotational energy of HF is assumed to behave classically, the entropy is given by Eq. 18.88:

$$S_{\text{rot}}^c = R\left[\ln\left\{\frac{IT}{\sigma}\right\} + 105.53\right]. \tag{8}$$

The symmetry number is unity. The moment of inertia is

$$I = \mu R^2. \tag{9}$$

The reduced mass, in kg, is

$$\mu = \frac{(1.0078)(19.000)}{(20.0078)(6.02214 \times 10^{23})(1,000)} \text{ kg} = 1.5892 \times 10^{-27} \text{ kg}. \tag{10}$$

Therefore,

$$I = (1.5892 \times 10^{-27} \text{ kg})(0.9171 \times 10^{-10} \text{ m})^2 = 1.3366 \times 10^{-47} \text{ kg m}^2. \tag{11}$$

Substitution into Eq. 8 gives

$$S_{\text{rot}}^c(50 \text{ K}) = 8.314 \text{ J mol}^{-1} \text{ K}^{-1}[\ln[(1.3366 \times 10^{-47})(50)] + 105.53]$$

$$= (8.314)(1.5107) \text{ J mol}^{-1} \text{ K}^{-1} = 12.56 \text{ J mol}^{-1} \text{ K}^{-1}. \tag{12}$$

At 300 K, the result is

$$S_{rot}^c(300\ K) = 8.314\ J\ mol^{-1}\ K^{-1}[\ln[(1.3366 \times 10^{-47})(300)] + 105.53]$$

$$= (8.314)(3.3024)\ J\ mol^{-1}\ K^{-1} = 27.46\ J\ mol^{-1}\ K^{-1}. \qquad \text{(13)}$$

A comparison demonstrates that the classical result is highly accurate, even at temperatures as low as 50 K. The percent errors at the two temperatures are

$$\% \text{ error at 50 K} = 100 \times \frac{12.56 - 12.48}{12.48} = 0.641\% \qquad \text{(14)}$$

and

$$\% \text{ error at 300 K} = 100 \times \frac{27.46 - 27.42}{27.42} = 0.146\%. \qquad \text{(15)}$$

For related exercises, see Problems 18.14 and 18.16.

Now that we have both quantum mechanical and classical expressions for the rotational internal energy and the entropy, it is straightforward to obtain similar equations for the Helmholtz and Gibbs free energies. The rotational contribution to the Helmholtz free energy for a diatomic rigid rotor is

$$\boxed{A_{rot} = U_{rot} - TS_{rot} = U_{rot} - T\left[R \ln z_{rot} + \frac{U_{rot}}{T}\right] = -RT \ln z_{rot}}, $$

$$\text{(18.89)}$$

where the rotational molecular partition function is given by Eq. 18.78.

We must exercise a little care in obtaining G and H for rotation. In general, we have $H = U + pV$ and $G = A + pV$. It might appear that all we need do is add the pV term to A_{rot} to obtain G_{rot}. However, this is not correct. There is only one pV term, and we already included it when we developed the expressions for G_{tr} that led to Eqs. 18.66A and 18.66B. Therefore, for rotation and vibration, we have

$$\boxed{G_{rot} = A_{rot}}$$

and

$$\boxed{G_{vib} = A_{vib}}. \qquad \text{(18.90)}$$

The same considerations lead to the conclusion that $H_{rot} = U_{rot}$ and $H_{vib} = U_{vib}$. (See Problem 18.16.)

For spherical- and symmetric-top molecules, we can obtain the molecular rotational partition function by using the quantum mechanical expressions for their rotational eigenvalues given in Eqs. 12.132 and 12.134, respectively. If this is done, it is important to remember that the rotational degeneracy for a spherical top is $(2J + 1)^2$. (See the discussion related to Eq. 12.132.) There is no general expression for the energy levels of asymmetric-top molecules. The aforesaid procedure will certainly give the correct results, but it is unnecessary unless the temperature is very near absolute zero. We have already found in Example 18.5, and we shall see in several of the problems at the end of the chapter, that rotation behaves classically except at very low temperatures.

This approximation is even more accurate for polyatomic molecules, which have much larger rotational constants due to their larger size and mass. Therefore, the easiest method for computing z_{rot} for polyatomic molecules is to employ the molecular partition function obtained by assuming that the rotational levels may be treated as a continuum. We shall omit the derivation of this result, as it is somewhat complex and requires that we first introduce the concept of classical phase space. We simply give the result:

$$z_{rot}^c = \frac{\sqrt{\pi}}{\sigma} \left(\frac{8\pi^2 I_1 kT}{h^2} \right)^{1/2} \left(\frac{8\pi^2 I_2 kT}{h^2} \right)^{1/2} \left(\frac{8\pi^2 I_3 kT}{h^2} \right)^{1/2} = \frac{\sqrt{\pi}}{\sigma h^3} [8\pi^2 kT]^{3/2} [I_1 I_2 I_3]^{1/2}$$

(18.91)

Here, I_1, I_2, and I_3 are the principal moments of inertia of the molecule and σ is the symmetry number. As in the case of a diatomic molecule, σ is the number of indistinguishable orientations of the molecule that are obtainable from one another by rotations about the principal axes. It is possible to derive a result very close to Eq. 18.91 by assuming that the rotational degrees of freedom about the three principal axes are all independent and that the total energy is simply the sum of three such independent rotations. (See Problem 18.17.)

The internal energy of rotation for nonlinear polyatomic molecules is easily shown to be $3RT/2$. (See Problem 18.18.) The rotational entropy is obtained by inserting Eq. 18.91 into Eq. 18.85. This gives

$$S_{rot} = \left[R \ln z_{rot} + \frac{U_{rot}}{T} \right] = R \ln \left[\frac{\sqrt{\pi}}{\sigma h^3} [8\pi^2 kT]^{3/2} [I_1 I_2 I_3]^{1/2} \right] + \frac{3RT/2}{T}$$

$$= R \ln \left[\frac{\sqrt{\pi}}{\sigma h^3} [8\pi^2 kT]^{3/2} [I_1 I_2 I_3]^{1/2} \right] + 1.5R.$$

(18.92)

By writing the logarithmic factor as the sum of terms, we can express Eq. 18.92 in the form

$$S_{rot} = \frac{3R}{2} \ln T + R \ln \left[\frac{[I_1 I_2 I_3]^{1/2}}{\sigma} \right] + R \ln \left[\frac{\sqrt{\pi}}{h^3} [8\pi^2 k]^{3/2} \right] + 1.5R.$$

Substitution of the constants in SI units gives

$$S_{rot} = \frac{3R}{2} \ln T + R \ln \left[\frac{[I_1 I_2 I_3]^{1/2}}{\sigma} \right] + 158.9R.$$

(18.93)

The contribution of polyatomic rotation to the Helmholtz free energy is

$$A_{rot} = U_{rot} - TS_{rot} = \frac{3RT}{2} - \frac{3RT}{2} \ln T - RT \ln \left[\frac{[I_1 I_2 I_3]^{1/2}}{\sigma} \right] - 158.9RT$$

$$= -\left[\frac{3RT}{2} \ln T + RT \ln \left[\frac{[I_1 I_2 I_3]^{1/2}}{\sigma} \right] + 157.4RT \right].$$

(18.94)

For the reasons discussed earlier, $G_{rot} = A_{rot}$.

18.5.4   Ortho–para Hydrogen

The exchange symmetry requirements on the rotational wave function leads naturally to the existence of two types of H_2. The nuclear spin of hydrogen is $I = \frac{1}{2}$. It is, therefore, a fermion whose total wave function must be antisymmetric with respect to the exchange of identical nuclei. With $I = \frac{1}{2}$, m_I can have two possible values: $+\frac{1}{2}$ and $-\frac{1}{2}$. These two spins states are generally labeled with the notation α and β, respectively. Let us label the two hydrogen nuclei as A and B. There are three ways to combine the nuclear spins of A and B to form a symmetric spin wave function: $\alpha_A \alpha_B$, $\beta_A \beta_B$, and $[\alpha_A \beta_B + \beta_A \alpha_B]$. The interchange of nuclei corresponds to an interchange of the A and B labels. It is easy to see that such an interchange does not alter any of these three spin functions. There is only one way to form an antisymmetric spin function: $[\alpha_A \beta_B - \beta_A \alpha_B]$. If we exchange A–B labels in this function, the result is $[\alpha_B \beta_A - \beta_B \alpha_A]$, which is the negative of $[\alpha_A \beta_B - \beta_A \alpha_B]$. Therefore, the function is antisymmetric with respect to exchange.

When the spin function is symmetric, the spherical harmonic $Y_J^M(\theta, \phi)$ spatial wave function must be antisymmetric to nuclear exchange, which means antisymmetric with respect to rotation about θ by π radians. Only the spherical harmonics with odd values of J satisfy this requirement. (See Problem 18.13.) H_2 molecules with this characteristic are said to be *ortho hydrogen*. If the spin function is antisymmetric, $Y_J^M(\theta, \phi)$ must be symmetric. This means we can have only even values of J. Hydrogen with only even J states is called *para hydrogen*.

The restriction to either even or odd rotational states significantly alters the form of the rotational molecular partition function. For the para and ortho forms of H_2, we have

$$z_{\text{rot}}^{\text{para}} = \frac{1}{4} \sum_{\text{even } J}^{\infty} (2J + 1) \exp\left[-\frac{J(J + 1)Bh}{kT} \right] \tag{18.95}$$

and

$$z_{\text{rot}}^{\text{ortho}} = \frac{3}{4} \sum_{\text{odd } J}^{\infty} (2J + 1) \exp\left[-\frac{J(J + 1)Bh}{kT} \right]. \tag{18.96}$$

Since there are three ways to have a symmetric spin function for H_2, and there is only one way to have an antisymmetric spin function, the ortho partition function is three times more heavily weighted than the para partition function. Notice that the symmetry factors are now omitted, because we have already taken this point into account by summing only over either the even or the odd J rotational states of H_2.

If the two types of hydrogen are in equilibrium, the total rotational molecular partition function will be the sum of Eqs. 18.95 and 18.96. We can produce an equilibrium state by waiting an extremely long period of time or, more conveniently, by keeping the material in the presence of a catalyst that promotes nuclear spin transitions between symmetric and antisymmetric spin states. To promote such transitions, the catalyst must be capable of interacting with the nuclear magnetic moments of the two hydrogen nuclei. Therefore, we need a paramagnetic material that contains unpaired electron spins. Pure carbon or charcoal contains unpaired spins, so it can function effectively as an ortho–para conversion catalyst. In the presence of such a catalyst, we expect the high-temperature ratio of ortho to para hydrogen to be three to one because of the nuclear spin weighting. As the temperature decreases toward absolute zero, the molecules will tend to accumulate in the rotational state with the lowest energy, $J = 0$. Therefore, with a catalyst present, very low-temperature H_2 will tend to approach pure para-H_2.

If we put H_2 in the presence of an ortho–para conversion catalyst and raise the temperature, a three-to-one mixture of ortho to para forms will be

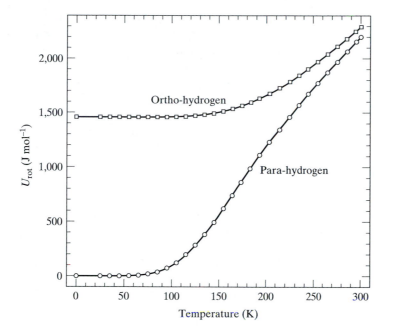

◀ **FIGURE 18.8**
The internal rotational energies of ortho-H_2 and para-H_2 as a function of temperature. Note the presence of zero-point rotational energy for ortho-H_2 and the much greater slope of the U_{para}^{rot} curve. A careful examination of the plot shows that U_{para}^{rot} exhibits an inflection point around 170 K.

produced. If we then remove the catalyst and lower the temperature, the three-to-one mixture will remain, and the heat capacity and internal energy will be a weighted average of these quantities for the pure ortho–para forms.

The internal rotational energies of para- and ortho-hydrogen are shown in Figure 18.8 as a function of temperature. These energies are obtained by using Eq. 18.82, modified so that the summations run only over even or odd values of J for para- and ortho-hydrogen, respectively. The figure reveals a startling difference in the internal energies of the two forms. Notice that ortho-H_2 possesses zero-point internal rotational energy, whereas para-H_2 does not. The zero-point rotational energy for ortho-H_2 is $2NBh$, where B is the rotational constant and N is Avogadro's constant. (See Problem 18.25.) Since $C_v = (\partial U / \partial T)_V$, we can qualitatively predict that $(C_v)_{para}^{rot} > (C_v)_{ortho}^{rot}$, since the slope of the internal-energy curve for para-hydrogen is significantly greater than that for the ortho form. In addition, a careful examination of the (U_{para}^{rot}) curve shows the presence of an inflection point around 170 K. At this point, the slope attains a maximum value. Consequently, we expect $(C_v)_{para}^{rot}$ to exhibit a maximum around 170 K.

The exchange symmetry requirements on the wave function exert a profound influence on the heat capacities and internal energies of the two forms of H_2. Figure 18.9 shows the computed values of $(C_v)_{para}^{rot}$ and $(C_v)_{ortho}^{rot}$ as a function of temperature, using Eq. 18.83 modified as described in the previous paragraph. The figure also shows the heat capacity of the three-to-one mixture. The large differences in heat capacities are evident. As expected from the results seen in Figure 18.8, the para-H_2 rotational heat capacity exceeds that of ortho-H_2. The predicted maximum in $(C_v)_{para}^{rot}$ near 170 K is also evident.

The equilibrium ratio of ortho-H_2 to para-H_2 can be computed with the use of the Boltzmann distribution once the rotational energies of each form have been obtained from Eq. 18.82. The number of H_2 molecules in each state is proportional to $\exp[-U/RT]$, but since the ortho form can exist with three

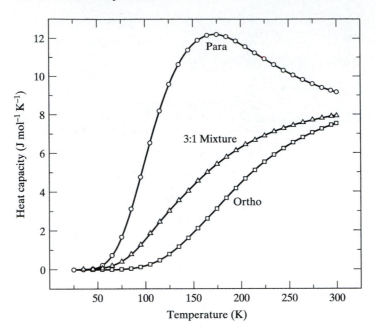

► FIGURE 18.9
The rotational heat capacities of ortho-H_2, para-H_2 and the equilibrium 3:1 mixture as a function of temperature. In the 3:1 mixture, ortho-H_2 constitutes 75% of the mixture. The greater rotational heat capacity of the para form is a reflection of the greater slope of U_{para}^{rot} in Figure 18.8. The maximum in $(C_v)_{para}^{rot}$ around 170 K corresponds to the inflection point in the U_{para}^{rot} curve at this temperature.

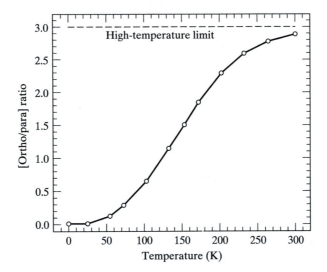

► FIGURE 18.10
The ortho–para-H_2 ratio in the presence of a catalyst as a function of temperature. As the temperature approaches 0 K, all of the H_2 converts to the para form, which has the lower rotational energy. At high temperatures, the equilibrium ratio approaches the 3:1 statistical ratio.

different spin functions, this state is essentially threefold degenerate. Therefore, the ortho–para ratio is

$$\frac{[\text{ortho-H}_2]}{[\text{para-H}_2]} = \frac{3\exp\left[-\dfrac{U_{ortho}^{rot}}{RT}\right]}{\exp\left[-\dfrac{U_{para}^{rot}}{RT}\right]} = 3\exp\left[\frac{(U_{para}^{rot} - U_{ortho}^{rot})}{RT}\right].$$

Figure 18.10 shows this ratio as a function of temperature from 0 to 300 K. As T approaches 0 K, the equilibrium ratio approaches zero, as expected, since all of the molecules seek the lower energy para rotational state of $J = 0$. As the temperature rises, the ortho spin states become increasingly populated. Around $T \approx 130$ K, the two states have equal populations. At higher temperatures, the ortho state dominates, and at 300 K, the ratio is seen to approach the high-temperature limit of 3.

18.5.5   Vibrational-energy Contributions to Thermodynamics

We consider first the case of a diatomic molecule with a single vibrational degree of freedom. In this situation, the vibrational energy can be described by a single quantum number, and the vibrational molecular partition function will involve only one summation. If we write the vibrational energy as the sum of a harmonic term plus a correction term for anharmonicity, we have

$$E_{vib} = (v + 0.5)h\,\nu_o - C_h(v + 0.5)^2. \tag{18.98}$$

If the electronic potential energy can be described accurately using a Morse function, the anharmonic correction coefficient is that given in Eq. 18.68. In more precise work, C_h can be fitted to the measured overtone spectrum of the molecule.

When the vibrational energy is given by Eq. 18.98, the vibrational molecular partition function has the form

$$z_{vib} = \sum_{v=0}^{\infty} \exp\left[-\frac{(v + 0.5)h\nu_o - C_h(v + 0.5)^2}{kT} \right]. \tag{18.99}$$

In general, the vibrational-energy states are so widely spaced that a classical approximation in which the summation in Eq. 18.99 is replaced by an integral is highly inaccurate. (See Problem 18.26.) In fact, at most temperatures, only the first term in the summation makes an appreciable contribution to z_{vib}. That allows us to assess the importance of the anharmonicity correction. If we ignore all terms in Eq. 18.99 except the $v = 0$ ground-state term, we have

$$z_{vib} \approx \exp\left(-\frac{0.5h\nu_o}{kT} \right) \exp\left(0.25\frac{C_h}{kT} \right).$$

If the anharmonicity correction is omitted, the result is

$$z'_{vib} \approx \exp\left(-\frac{0.5h\nu_o}{kT} \right).$$

The percent error in this result is

$$
\begin{aligned}
\% \text{ error} &\approx 100 \times \frac{\exp(-0.5h\nu_o/kT) - \exp(-0.5h\nu_o/kT)\exp(0.25C_h/kT)}{\exp(-0.5h\nu_o/kT)\exp(0.25C_h/kT)} \\
&= 100 \times \frac{1 - \exp(0.25C_h/kT)}{\exp(0.25C_h/kT)} = 100[\exp(-0.25C_h/kT) - 1].
\end{aligned}
\tag{18.100}
$$

Equation 18.100 shows that we expect the error incurred by omitting the anharmonicity correction to decrease as T increases and as C_h decreases. For H_2 with $C_h = 2.344 \times 10^{-21}$ J, the error predicted by our approximate treatment at 300 K is -13.19%. Example 18.6 shows this result to be correct to almost four significant digits. At 1,000 K, we obtain an error of -4.16% for H_2. If we take I_2 as our example, $C_h = 1.217 \times 10^{-23}$ J. The errors predicted by Eq. 18.100 if we ignore the anharmonicity term are -0.073% and -0.022% at 300 K and 1,000 K, respectively. Table 15.4 shows that H_2 has a very large anharmonicity factor; therefore, it represents about the worst case for the error incurred by omitting the anharmonicity correction. I_2 has a very small anharmonicity correction; therefore, it is one of the more favorable cases. Example 18.6 does the calculation correctly.

EXAMPLE 18.6

Use the data given in Table 15.4 to compute z_{vib} for H_2 at 300 K and 1,000 K.

Solution

The vibration frequency for H_2 is $(4,395.2 \text{ cm}^{-1})(2.9979 \times 10^{10} \text{ cm s}^{-1}) = 1.3176 \times 10^{14} \text{ s}^{-1}$. The anharmonicity coefficient is

$$C_h = x_e \omega_e hc = 117.99 \text{ cm}^{-1}(2.9979 \times 10^{10} \text{ cm s}^{-1})(6.62608 \times 10^{-34} \text{ J s})$$

$$= 2.344 \times 10^{-21} \text{ J}. \tag{1}$$

At 300 K, the two constants needed in the summation for z_{vib} are,

$$\frac{h\nu_o}{kT} = \frac{(6.62608 \times 10^{-34} \text{ J s})(1.3176 \times 10^{14} \text{ s}^{-1})}{(1.38066 \times 10^{-23} \text{ J K}^{-1})(300 \text{ K})} = 21.078 \tag{2}$$

and

$$\frac{C_h}{kT} = \frac{2.344 \times 10^{-21} \text{ J}}{(1.38066 \times 10^{-23} \text{ J K}^{-1})(300 \text{ K})} = 0.5659. \tag{3}$$

At $T = 1,000$ K, the results are 6.323 and 0.1698, respectively. Therefore, we have

$$z_{vib} = \sum_{v=0}^{\infty} \exp[-21.078(v + 0.5) + 0.5659(v + 0.5)^2] \quad \text{at } T = 300 \text{ K} \tag{4}$$

and

$$z_{vib} = \sum_{v=0}^{\infty} \exp[-6.323(v + 0.5) + 0.1698(v + 0.5)^2] \quad \text{at } T = 1,000 \text{ K}. \tag{5}$$

A simple calculation shows that only the first term contributes to the summation at 300 K. The result is

$$z_{vib}(300) = 3.051 \times 10^{-5}. \tag{6}$$

At $T = 1,000$ K, we obtain

$$z_{vib}(1,000) = 0.04431. \tag{7}$$

We can assess the importance of the anharmonicity term by setting C_h to zero and repeating the calculation. The results are

$$z_{vib}^{harmonic}(300 \text{ K}) = z_{vib}^h = 2.648 \times 10^{-5} \tag{8}$$

and

$$z_{vib}^h(1,000 \text{ K}) = 0.04244. \tag{9}$$

The percent error incurred by omission of the anharmonicity correction is

$$\% \text{ error at 300 K} = 100 \times \frac{2.648 - 3.051}{3.051} = -13.2\% \tag{10}$$

and

$$\% \text{ error at 1,000 K} = 100 \times \frac{0.04244 - 0.04431}{0.04431} = -4.2\%. \tag{11}$$

For a related exercise, see Problem 18.26.

If the anharmonicity correction is sufficiently small that it may be ignored, the vibrational molecular partition function assumes a very simple form. Without the anharmonicity term, Eq. 18.99 becomes

$$z_{vib} = z_{vib}^h = \sum_{v=0}^{\infty} \exp\left[-\frac{(v + 0.5)h\nu_o}{kT}\right]. \tag{18.101}$$

The exponential can be written as the product of two exponentials. Only one of these depends upon the summation index. The other factor may, therefore, be removed from under the summation to obtain

$$z_{vib}^h = \exp\left[-\frac{0.5h\nu_o}{kT}\right] \sum_{v=0}^{\infty} \exp\left[-\frac{v h\nu_o}{kT}\right].$$

Let us now define $x = \dfrac{h\nu_o}{kT}$. In terms of x, z_{vib}^h is

$$z_{vib}^h = \exp\left[-\frac{x}{2}\right] \sum_{v=0}^{\infty} e^{-vx}$$

$$= e^{-x/2}[1 + e^{-x} + e^{-2x} + e^{-3x} + e^{-4x} + \cdots] = ye^{-x/2}, \tag{18.102}$$

where we have represented the summation with the variable y. It is very easy to perform this summation in closed analytic form as follows:

$$y = [1 + e^{-x} + e^{-2x} + e^{-3x} + e^{-4x} + \cdots]$$

$$= 1 + [e^{-x} + e^{-2x} + e^{-3x} + e^{-4x} + \cdots]$$

$$= 1 + e^{-x}[1 + e^{-x} + e^{-2x} + e^{-3x} + e^{-4x} + \cdots]. \tag{18.103}$$

But the quantity in brackets on the right of Eq. 18.103 is y; therefore,

$$y = 1 + e^{-x}y.$$

Solving for y, we obtain

$$y = 1 + e^{-x} + e^{-2x} + e^{-3x} + e^{-4x} + \cdots = \frac{1}{1 - e^{-x}}. \tag{18.104}$$

Substitution of Eq. 18.104 into Eq. 18.102 gives

$$\boxed{z_{vib}^h = \frac{e^{-x/2}}{1 - e^{-x}}.} \tag{18.105}$$

It should be kept in mind that the preceding summation procedure is valid only if the series being summed converges. (See Problem 18.27.)

EXAMPLE 18.7

Show that Eq. 18.105 gives the same results for the harmonic molecular partition function for H_2 at 300 K and 1,000 K that we obtained in Example 18.6, in which the series was actually summed.

Solution

For H_2 at 300 K and 1,000 K, we have

$$x(300 \text{ K}) = \frac{h\nu_o}{300 \text{ k}} = \frac{(6.62608 \times 10^{-34} \text{ J s})(1.3176 \times 10^{14} \text{ s}^{-1})}{(1.38066 \times 10^{-23} \text{ J K}^{-1})(300 \text{ K})} = 21.078 \tag{1}$$

and

$$x(1{,}000 \text{ K}) = \frac{h\nu_o}{1{,}000 \, k} = \frac{(6.62608 \times 10^{-34} \text{ J s})(1.3176 \times 10^{14} \text{ s}^{-1})}{(1.38066 \times 10^{-23} \text{ J K}^{-1})(1{,}000 \text{ K})} = 6.323. \quad \text{(2)}$$

Thus, at 300 K,

$$\exp[-x(300 \text{ K})] = \exp(-21.078) = 7.0136 \times 10^{-10}, \quad \text{(3)}$$

$$\exp\left[-\frac{x(300 \text{ K})}{2}\right] = \exp\left[-\frac{21.078}{2}\right] = 2.648 \times 10^{-5}, \quad \text{(4)}$$

and

$$z_{\text{vib}}^h(300 \text{ K}) = \frac{2.648 \times 10^{-5}}{1 - 7.0136 \times 10^{-10}} = 2.648 \times 10^{-5}, \quad \text{(5)}$$

which is the same result obtained in Example 18.6. At 1,000 K,

$$\exp[-x(1{,}000 \text{ K})] = \exp(-6.323) = 0.001794, \quad \text{(6)}$$

$$\exp\left[-\frac{x(1{,}000 \text{ K})}{2}\right] = \exp\left[-\frac{6.323}{2}\right] = 0.04236, \quad \text{(7)}$$

and

$$z_{\text{vib}}^h(1{,}000 \text{ K}) = \frac{0.04236}{1 - 0.001794} = 0.04244, \quad \text{(8)}$$

which is identical to the result obtained in Example 18.6.

The simple closed-form expression for z_{vib}^h allows us to obtain very convenient expressions for the vibrational contributions to U, C_v, S, A, H, and G. The internal energy is given by $U_{\text{vib}} = RT^2(\partial \ln z_{\text{vib}}^h/\partial T)_V$ if we assume that vibration may be treated as harmonic. The required derivative is

$$\frac{\partial}{\partial T}\left[\ln\left(\frac{e^{-x/2}}{1 - e^{-x}}\right)\right]_V = \frac{\partial}{\partial T}\left[-\frac{x}{2} - \ln(1 - e^{-x})\right]_V.$$

This derivative is most conveniently obtained by using the chain rule:

$$\frac{\partial}{\partial T}\left[\ln\left(\frac{e^{-x/2}}{1 - e^{-x}}\right)\right]_V$$

$$= \frac{\partial}{\partial T}\left[-\frac{x}{2} - \ln(1 - e^{-x})\right]_V = \frac{\partial}{\partial x}\left[-\frac{x}{2} - \ln(1 - e^{-x})\right]_V \frac{\partial x}{\partial T}$$

$$= \left[-\frac{1}{2} - \frac{e^{-x}}{1 - e^{-x}}\right]\left(-\frac{h\nu_o}{kT^2}\right) = \left[\frac{1}{2} + \frac{e^{-x}}{1 - e^{-x}}\right]\left(\frac{h\nu_o}{kT^2}\right). \quad \text{(18.106)}$$

Substituting Eq. 18.106 into the expression for the internal energy produces

$$U_{\text{vib}} = \frac{Nh\nu_o}{2} + \frac{Nh\nu_o e^{-x}}{1 - e^{-x}} = \frac{Nh\nu_o}{2} + \frac{Nh\nu_o}{e^x - 1}, \quad \text{(18.107)}$$

where we have multiplied the numerator and denominator of the second term by e^x to obtain the final expression. The first term in Eq. 18.107 represents the contribution of the zero-point vibration energy to U_{vib}, which is always present regardless of the temperature. The second term is sometimes called the excitation term, in that it includes the contributions to U_{vib}

from the excitation of molecules into higher vibrational levels at temperatures above 0 K. This equation was first introduced as Eq. 7.18 in Chapter 7, to permit us to examine the heat capacity of solids, which is due solely to contributions from vibrational degrees of freedom. The importance of the excitation term relative to the zero-point energy term is given by the ratio

$$\frac{\dfrac{Nh\nu_o}{e^x - 1}}{\dfrac{Nh\nu_o}{2}} = \frac{2}{e^x - 1}.$$

Figure 18.11 shows this ratio as a function of temperature for H_2, $H^{35}Cl$, and CO. As can be seen, the excitation term is of almost no importance at temperatures below 500 K for these molecules. The excitation term becomes more important as the vibrational frequency decreases, because this reduces the spacing between the harmonic vibrational energy states and thereby makes more of them accessible.

The vibrational contribution to the constant-volume heat capacity is obtained directly from Eq. 18.107:

$$[C_v]_{\text{vib}} = \left(\frac{\partial U_{\text{vib}}}{\partial T}\right)_V = \left(\frac{\partial U_{\text{vib}}}{\partial x}\right)_V \frac{\partial x}{\partial T}. \tag{18.108}$$

Differentiating U_{vib} directly in Eq. 18.107 gives

$$\left(\frac{\partial U_{\text{vib}}}{\partial x}\right)_V = -\frac{Nh\nu_o e^x}{(e^x - 1)^2}.$$

Combining this result with $\partial x/\partial T = -h\nu_o/kT^2$ and Eq. 18.108 gives us

$$[C_v]_{\text{vib}} = \frac{N(h\nu_o)^2 \, e^x}{kT^2(e^x - 1)^2}. \tag{18.109}$$

Multiplying numerator and denominator by k and replacing Nk with R yields

$$\boxed{[C_v]_{\text{vib}} = R\left(\frac{h\nu_o}{kT}\right)^2 \frac{e^x}{(e^x - 1)^2} = \frac{Rx^2 e^x}{(e^x - 1)^2}.} \tag{18.110}$$

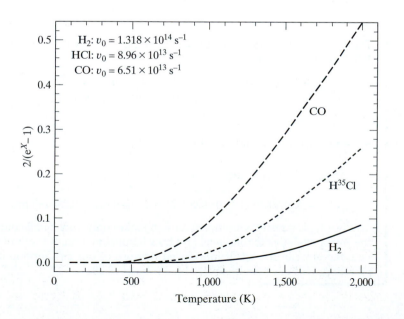

H_2: $\nu_0 = 1.318 \times 10^{14} \text{ s}^{-1}$
HCl: $\nu_0 = 8.96 \times 10^{13} \text{ s}^{-1}$
CO: $\nu_0 = 6.51 \times 10^{13} \text{ s}^{-1}$

◀ **FIGURE 18.11**
The ratio of the harmonic vibrational excitation term to the zero-point energy term. It is shown in the text that this ratio is $2/(e^x - 1)$, where $x = h\nu_o/(kT)$. The figure indicates that the excitation term contributes virtually nothing to the internal energy for H_2 until the temperature reaches 1,400 K. Below 500 K, excitation into higher vibrational energy states rarely occurs for CO, HCl, or H_2. As ν_o decreases, the vibrational energy levels become more closely spaced and, therefore, more accessible. As a result, the excitation term becomes more important.

Equation 18.110 is the expression we utilized in Chapter 7 to derive the Einstein and Debye theories of crystal heat capacities.

The harmonic vibrational contribution to the entropy is obtained from Eq. 18.56 after substituting z_{vib}^h for the molecular partition function:

$$S_{\text{vib}} = R \ln z_{\text{vib}}^h + \frac{U_{\text{vib}}}{T}. \tag{18.111}$$

Substitution of Eqs. 18.105 and 18.107 allows us to express the vibrational entropy in terms of x:

$$S_{\text{vib}} = R \ln\left(\frac{e^{-x/2}}{1 - e^{-x}}\right) + \frac{Nh\nu_o}{2T} + \frac{Nh\nu_o}{T(e^x - 1)}$$

$$= -\frac{Rx}{2} - R \ln(1 - e^{-x}) + \frac{Nh\nu_o}{2T} + \frac{Nh\nu_o}{T(e^x - 1)}. \tag{18.112}$$

Using the facts that

$$\frac{Nh\nu_o}{2T} = \frac{Nkh\nu_o}{2kT} = \frac{Rx}{2}$$

and

$$\frac{Nh\nu_o}{T(e^x - 1)} = \frac{Rx}{e^x - 1},$$

we can express Eq 18.112 in the form

$$\boxed{S_{\text{vib}} = -R \ln(1 - e^{-x}) + \frac{Rx}{e^x - 1}}. \tag{18.113}$$

Notice that the zero-point terms add out of the expression; zero-point energy makes no contribution to the system entropy.

EXAMPLE 18.8

Use the data given in Table 15.4 to compute the vibrational contribution to the entropy of $^{16}O^{16}O$ at 300 K. Assume that the O_2 vibrational motion may be treated as harmonic.

Solution

The vibrational frequency of O_2 given in Table 15.4 and expressed in cm^{-1} is 1,580.36 cm^{-1}. Therefore,

$$x = \frac{h\nu_o}{kT} = \frac{(6.62608 \times 10^{-34}\,\text{J s})(1,580.36\,\text{cm}^{-1})(2.9979 \times 10^{10}\,\text{cm s}^{-1})}{(1.38066 \times 10^{-23}\,\text{J K}^{-1})(300\,\text{K})}$$

$$= 7.5792. \tag{1}$$

Substituting this result into Eq. 18.113, we obtain

$$S_{\text{vib}} = -R \ln(1 - \exp\{-7.5792\}) + \frac{7.5792R}{(\exp(7.5792) - 1)}$$

$$= R[0.0005111 + 0.0038747] = 0.004386R = 0.03646\,\text{J mol}^{-1}\,\text{K}^{-1}. \tag{2}$$

This is a typical result; vibrational motion makes only very small contributions to the entropy relative to rotational and translational motion. It is of interest to evaluate the error in the result introduced because of our assumption of harmonic vibration. This question is explored in Problem 18.28.

The remaining thermodynamic quantities, A, G, and H, can be obtained by a simple combination of the foregoing results. From Eq. 18.107, we have

$$U_{vib} = \frac{Nh\nu_o}{2} + \frac{Nh\nu_o}{e^x - 1} = \frac{NkT}{2}\left(\frac{h\nu_o}{kT}\right) + NkT\left(\frac{h\nu_o}{kT}\right)\frac{1}{e^x - 1} = \frac{RTx}{2} + \frac{RTx}{e^x - 1}.$$

Combining this expression for U_{vib} with Eq. 18.113 allows us to obtain the Helmholtz free energy:

$$A_{vib} = U_{vib} - TS_{vib} = \frac{RTx}{2} + \frac{RTx}{e^x - 1} - T\left[-R\ln(1 - e^{-x}) + \frac{Rx}{e^x - 1}\right].$$

This gives

$$\boxed{A_{vib} = \frac{RTx}{2} + RT\ln(1 - e^{-x})}.$$ **(18.114)**

For reasons discussed previously, $A_{vib} = G_{vib}$ and $H_{vib} = U_{vib}$. We are now in a position to compute thermodynamic quantities for molecular systems.

EXAMPLE 18.9

Compute the entropy of HF at 298.15 K and 1 bar of pressure. Assume that the rotational and vibrational energy states can be described by a rigid-rotor and harmonic-oscillator quantization, respectively. The required data are given in Table 15.4.

Solution

As long as we can separate vibrational and rotational motion, the entropy is given by

$$S = S_{tr} + S_{rot} + S_{vib}.$$ **(1)**

The electronic states make no contribution, since the electronic ground state of HF has a degeneracy of unity and its energy can be taken as our zero of electronic energy. The translational contribution is given by Eq. 18.64:

$$S_{tr} = \frac{3R}{2}\ln(M) + \frac{5R}{2}\ln(T) - R\ln(p) - 1.15169R.$$ **(2)**

The molar mass of HF in grams is 20.008 grams. Thus,

$$S_{tr} = R[1.5\ln(20.008) + 2.5\ln(298.15) - \ln(1) - 1.15169] = 17.586R$$

$$= 146.21 \text{ J mol}^{-1}\text{K}^{-1}.$$ **(3)**

At $T = 298.15$ K, the temperature is sufficiently high that rotation can be treated classically. Under this condition, the rotational entropy is given by

$$S_{rot} = R\ln\left[\frac{kT}{\sigma Bh}\right] + R.$$ **(4)**

Using the data given in Table 15.4, we obtain

$$\frac{Bh}{k} = \frac{(20.939)(2.9979 \times 10^{10}\text{s}^{-1})(6.62608 \times 10^{-34}\text{ J s})}{1.38066 \times 10^{-23}\text{ J K}^{-1}} = 30.126 \text{ K}.$$ **(5)**

The symmetry number of HF is unity. Thus,

$$S_{rot} = R\ln\left[\frac{298.15}{30.126}\right] + R = 3.2922R = 27.371 \text{ J mol}^{-1}\text{K}^{-1}.$$ **(6)**

The vibration frequency of HF, expressed in wave numbers, is 4,138.52 cm^{-1}. The value of x is, therefore,

$$x = \frac{h\nu_o}{kT} = \frac{(6.62608 \times 10^{-34} \, \text{J s})(4,138.52 \, \text{cm}^{-1})(2.9979 \times 10^{10} \, \text{cm s}^{-1})}{(1.38066 \times 10^{-23} \, \text{J K}^{-1})(298.15 \, \text{K})}$$

$$= 19.971. \tag{7}$$

The vibrational contribution to the entropy is given by Eq. 18.113:

$$S_{\text{vib}} = -R \ln(1 - e^{-x}) + \frac{Rx}{e^x - 1} = -R \ln[1 - \exp(-19.971)] + \frac{19.971R}{\exp(19.971) - 1}$$

$$= R[2.12 \times 10^{-9} + 4.24 \times 10^{-8}] = 3.70 \times 10^{-7} \, \text{J mol}^{-1} \, \text{K}^{-1}. \tag{8}$$

The total entropy is, therefore,

$$S = 146.21 + 27.317 + 3.70 \times 10^{-7} \, \text{J mol}^{-1} \, \text{K}^{-1} = 173.53 \, \text{J mol}^{-1} \, \text{K}^{-1}. \tag{9}$$

The value listed for the entropy of HF at 298.15 K and $p = 1$ bar in the 78th edition of the CRC *Handbook of Chemistry and Physics* is 173.8 J mol^{-1} K^{-1}. The percent error in our result is

$$\% \text{ error} = 100 \times \frac{173.53 - 173.8}{173.8} = -0.16\%. \tag{10}$$

This error is due to the deviation of HF from ideal behavior at 298.15 K and 1 bar of pressure and to our assumption that the vibrational and rotational quantum states may be assumed to be uncoupled, with their energies represented by a rigid-rotor–harmonic-oscillator quantization.

We can easily repeat this calculation at any temperature and thereby obtain the entropy of HF as a function of temperature. The results of such calculations are shown in Figure 18.12. The slope of the entropy curve, $(\partial S/\partial T)_V$, is related to the

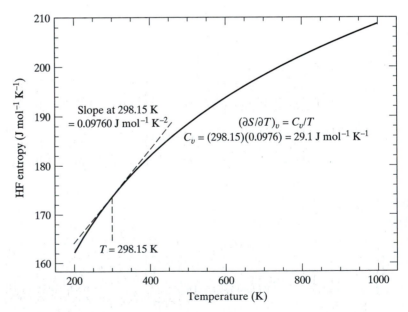

▲ FIGURE 18.12
The entropy of HF(g) as a function of temperature. The figure illustrates the point that, since $(\partial S/\partial T)_V = C_V/T$, we may obtain the heat capacity by determining the slope of the curve at any temperature and multiplying the result by T. When this is done at $T = 298.15$ K, the slope is 0.09760 J mol^{-1} K^{-2}, and the resulting heat capacity is 29.1 J mol^{-1} K^{-1}. The error is −0.10% which is due primarily to the rigid-rotor–harmonic-oscillator quantization approximation we have employed to compute the HF(g) entropy.

constant-volume heat capacity of HF by $C_v = T(\partial S/\partial T)_V$. Therefore, we can obtain the heat capacity by determining the curve's slope at any temperature and multiplying the result by T. This determination is illustrated in the figure at 298.15 K. The result is 29.1 J mol^{-1} K^{-1}. The experimental value at this temperature is 29.13 J mol^{-1} K^{-1}.

For related exercises, see Problems 18.29 and 18.30.

For polyatomic molecules, there are $3N - 6$ or $3N - 5$ vibrational degrees of freedom for nonlinear and linear molecules, respectively. When the vibrational energy is large, the anharmonicities couple all these modes and make the computation of the vibrational molecular partition function extremely difficult. However, if the temperature is not too high, the total vibrational energy will not be much greater than the zero-point vibrational energy. (See Problem 18.31 and Figure 18.11.) Under this condition, we found in Section 15.4.4, that normal-mode coordinates describe the vibrational motion with reasonable accuracy and that these modes are harmonic and independent. This means that the total vibrational energy is the uncoupled sum of the individual mode energies:

$$E_{\text{vib}} = \sum_{i=1}^{3N-6} \varepsilon_i = \sum_{i=1}^{3N-6} (v_i + 0.5)h\nu_{oi}. \tag{18.115}$$

In Eq. 18.115, the ε_i are the normal-mode harmonic energies, each of which is specified by a vibrational quantum number v_i. The vibration frequency of mode i is ν_{oi}. The upper limit of the summation will be $3N - 5$ if the molecule is linear.

Earlier, we found that whenever the total energy can be written as an uncoupled sum of individual mode energies, the molecular partition function is the corresponding product of the individual mode molecular partition functions. That is,

$$z_{\text{vib}} = \sum_{v_1=0}^{\infty} \sum_{v_2=0}^{\infty} \cdots \sum_{v_{3N-6}=0}^{\infty} \exp\left[-\frac{\varepsilon_1 + \varepsilon_2 + \cdots + \varepsilon_{3N-6}}{kT}\right]$$

$$= \sum_{v_1=0}^{\infty} \exp\left[-\frac{\varepsilon_1}{kT}\right] \sum_{v_2=0}^{\infty} \exp\left[-\frac{\varepsilon_2}{kT}\right] \cdots \sum_{v_{3N-6}=0}^{\infty} \exp\left[-\frac{\varepsilon_{3N-6}}{kT}\right]$$

$$= z_1^{\text{vib}} z_2^{\text{vib}} \cdots z_{3N-6}^{\text{vib}}.$$

Since each of the vibrational normal modes are assumed to be harmonic, their molecular partition functions are all given by Eq. 18.105. Therefore, the total vibrational molecular partition function is

$$z_{\text{vib}} = \prod_{i=1}^{3N-6} \left[\frac{e^{-x/2}}{1 - e^{-x}}\right]_i. \tag{18.116}$$

Since the thermodynamic functions all depend upon $\ln(z)$ or derivatives of $\ln(z)$, the contributions of the vibrational modes to U, C_v, S, A, H, and G will all be additive. That is, using the equations developed for diatomics, we can compute the contribution to U for each vibrational mode as if it were the only one present, and then we can simply add the individual contributions for each of the modes. This is the basic assumption we made in Chapter 7 in our development of the Einstein and Debye theories for crystal heat capacities. Example 18.10 illustrates the procedure.

EXAMPLE 18.10

The measured vibration frequencies of NH_3, expressed in wave numbers, are $3,337 \text{ cm}^{-1}$ and 950 cm^{-1} for the symmetric stretch and symmetric deformation, respectively. The two doubly degenerate vibrations are a stretching mode at $3,444 \text{ cm}^{-1}$ and a deformation at $1,627 \text{ cm}^{-1}$. Compute the internal energy in excess of the zero-point energy for NH_3 at 300 K. Which modes make the largest contribution?

Solution

The internal energy in excess of the zero-point vibrational energy is

$$U_{vib} - \frac{Nh\nu_o}{2} = \frac{Nh\nu_o}{e^x - 1} = \frac{RTx}{e^x - 1}. \tag{1}$$

Therefore, we need the value of x for each mode. For the first mode, we have

$$x(3,337) = \frac{h\nu_o}{kT} = \frac{(6.62608 \times 10^{-34} \text{ J s})(3,337 \text{ cm}^{-1})(2.9979 \times 10^{10} \text{ cm s}^{-1})}{(1.38066 \times 10^{-23} \text{ J K}^{-1})(300 \text{ K})}$$

$$= 16.004. \tag{2}$$

Similar calculations for the remaining modes yield

$$x(950) = 4.5561, \quad x(3,444) = 16.517, \quad \text{and } x(1,627) = 7.8029. \tag{3}$$

The total energy in excess of the zero-point energy is just the sum of this quantity for the individual modes:

$$\text{Total } U \text{ in excess of ZPE} = \sum_{i=1}^{6} \left[\frac{RTx}{e^x - 1} \right]_i = RT \sum_{j=1}^{4} g_j \left[\frac{x}{e^x - 1} \right]_j \tag{4}$$

where g_j is the degeneracy of vibrational mode j. Notice that in the second sum, we sum only over those modes with different vibrational frequencies or x_j. Substituting the values of x gives

$$\text{Total } U \text{ in excess of ZPE} = \frac{16.004 \text{ RT}}{e^{16.004} - 1} + \frac{4.5561 \text{ RT}}{e^{4.5561} - 1} + 2\frac{16.517 \text{ RT}}{e^{16.517} - 1} + 2\frac{7.8029 \text{ RT}}{e^{7.8029} - 1}$$

$$= (8.314)(300)[0.00000179 + 0.04836 + 0.00000222 + 0.006378] \text{ J mol}^{-1}$$

$$= 136.54 \text{ J mol}^{-1}. \tag{5}$$

Inspection of the various terms in Eq. 5 makes it clear that the low-frequency vibrational modes make the major contributions to the internal energy in excess of the zero-point energy.

For a related exercise, see Problem 18.32.

18.5.6 The Equilibrium Constant

Consider the general case of a gas-phase equilibrium process

$$aA(g) + bB(g) = cC(g) + dD(g),$$

where the lowercase letters represent the stoichiometric coefficients and the uppercase letters denote the compounds in equilibrium. For this process, the change in chemical potential is

$$\Delta\mu = d\mu_D + c\mu_C - a\mu_A - b\mu_B. \tag{18.117}$$

Under the condition that the molecules do not interact and that the gases are ideal, the partial molar Gibbs free energies (chemical potentials) are equal to the molar free energies of the pure gases at the partial pressure of interest. That is,

$$\Delta\mu = dG_D + cG_C - aG_A - bG_B. \tag{18.118}$$

Table 18.5 shows that these molar Gibbs free energies are given by

$$G = -RT \ln\left[\frac{z}{N}\right]. \tag{18.119}$$

Let us express Eq. 18.119 in a somewhat different form. If we assume that rotational, vibrational, and electronic energies are separable, we have $z = z_{tr}z_{rot}z_{vib}z_{el}$. The translational partition function for noninteracting molecules, divided by Avogadro's constant, is given by $z_{tr}/N = [2\pi m]^{3/2}[kT]^{5/2}/(ph^3)$. (See Eqs. 18.53 through 18.60.) Therefore, the logarithmic factor in Eq. 18.119 can be written

$$\ln\left[\frac{z}{N}\right] = \ln\left[\frac{z_{tr}z_{rot}z_{vib}z_{el}}{N}\right] = \ln[z'_{tr}z_{rot}z_{vib}z_{el}] - \ln(p), \tag{18.120}$$

where

$$z'_{tr} = \frac{[2\pi m]^{3/2}[kT]^{5/2}}{h^3}. \tag{18.121}$$

If we define $z' = z'_{tr}z_{rot}z_{vib}z_{el}$, the molar Gibbs free energy becomes

$$G = -RT \ln z' + RT \ln(p) \tag{18.122}$$

provided that p is in bars.

Substituting Eq. 18.122 into Eq. 18.118 allows us to express the change in the chemical potential as

$$\Delta\mu = -RT[c\{\ln(z'_c) - \ln(p_c)\} + d\{\ln(z'_d) - \ln(p_d)\} - a\{\ln(z'_a) - \ln(p_a)\}$$
$$- b\{\ln(z'_b) - \ln(p_b)\}].$$

Combination of the logarithmic terms produces

$$\Delta\mu = -RT \ln\left[\frac{(z'_c)^c(z'_d)^d}{(z'_a)^a(z'_b)^b}\right] + RT \ln\left[\frac{(p_c)^c(p_d)^d}{(p_a)^a(p_b)^b}\right]. \tag{18.123}$$

In Chapter 5, we learned that at constant temperature and either constant pressure or constant volume, the condition for equilibrium is $\Delta\mu = 0$. Therefore, when the chemical reaction reaches equilibrium, the left-hand side of Eq. 18.123 becomes zero, and we have

$$\ln\left[\frac{(p_c)^c(p_d)^d}{(p_a)^a(p_b)^b}\right]_{eq} = \ln K_p = \ln\left[\frac{(z'_c)^c(z'_d)^d}{(z'_a)^a(z'_b)^b}\right]_{eq},$$

where the subscript eq reminds us that the pressures must be those partial pressures that exist at the equilibrium point. This equilibrium pressure ratio is just the equilibrium constant K_p for an ideal system. (See Eq. 5.58.) Exponentiation of both sides gives us K_p in terms of the molecular partition functions of the reactants and products:

$$K_p = \left[\frac{(z'_c)^c(z'_d)^d}{(z'_a)^a(z'_b)^b}\right]_{eq}. \tag{18.124}$$

Example 18.11 illustrates the computation of equilibrium constants for ideal systems.

EXAMPLE 18.11

Consider the equilibrium process $H_2(g) = H(g) + H(g)$. Use Eq. 18.124 to compute the equilibrium constant for this process at 1,000 K.

Solution

For this reaction, we have, from Eq. 18.124,

$$K_p = \frac{(z_H')^2}{(z_{H2}')}. \tag{1}$$

Since hydrogen atoms have only translational and electronic energy,

$$z_H' = [z_{tr}' z_{el}]_H. \tag{2}$$

The translational molecular partition function is

$$[z_{tr}']_H = \frac{[2\pi m_H]^{3/2} [kT]^{5/2}}{h^3}. \tag{3}$$

The electronic partition function is

$$[z_{el}]_H = \sum_{n=1}^{\infty} g_n \exp\left[-\frac{\varepsilon_n}{kT}\right]_H. \tag{4}$$

As discussed previously, the electronic energy states are so widely spaced that only the ground electronic state contributes to the sum in Eq. 4. Thus,

$$[z_{el}]_H = g_1 \exp\left[-\frac{\varepsilon_{1H}}{kT}\right], \tag{5}$$

where g_1 is the degeneracy of the ground electronic state and ε_{1H} is the ground-state energy. For a hydrogen atom, $g_1 = 2$, since the ground state can be either $\psi_{1s}\alpha$ or $\psi_{1s}\beta$. We could, of course, choose $\varepsilon_{1H} = 0$ as our reference point, but it is more instructive to leave the expression in the form

$$[z_{el}]_H = 2 \exp\left[-\frac{\varepsilon_{1H}}{kT}\right]. \tag{6}$$

For H_2, the molecular partition function is

$$z_{H2}' = [z_{tr}' z_{rot} z_{vib} z_{el}]_{H2}. \tag{7}$$

The translational molecular partition function is

$$[z_{tr}']_{H2} = \frac{[2\pi m_{H2}]^{3/2} [kT]^{5/2}}{h^3}. \tag{8}$$

At 1,000 K, rotation will behave classically, so that

$$[z_{rot}]_{H2} = \frac{kT}{\sigma Bh}. \tag{9}$$

The vibrational molecular partition function is given by Eq. 18.105, namely,

$$[z_{vib}]_{H2} = \frac{e^{-x/2}}{1 - e^{-x}}, \tag{10}$$

where

$$x = \frac{h\nu_o}{kT}. \tag{11}$$

The ground-state electronic degeneracy for H_2 is unity; hence,

$$[z_{el}]_{H2} = \exp\left[-\frac{\varepsilon_{1H2}}{kT}\right]. \tag{12}$$

The ratio

$$\frac{(z'_{tr})_H^2}{(z'_{tr})_{H2}} = \frac{[2\pi]^{3/2}[kT]^{5/2}}{h^3}\left[\frac{m_H^3}{(2m)_H^{3/2}}\right] = \frac{[\pi m_H]^{3/2}[kT]^{5/2}}{h^3}. \tag{13}$$

Substitution of these results into Eq. 1 produces

$$K_p = \frac{[\pi m_H]^{3/2}[kT]^{5/2}}{h^3}\frac{\sigma B h}{kT}\frac{1-e^{-x}}{e^{-x/2}}4\exp\left[-\frac{(2\varepsilon_{1H}-\varepsilon_{1H2})}{kT}\right]. \tag{14}$$

The symmetry number for H_2 is 2, and $B = h/(8\pi^2 I)$. Substituting these quantities into Eq. 14 and then simplifying gives

$$K_p = \frac{[m_H kT]^{3/2}(1-e^{-x})}{\pi^{1/2}hI}\exp\left[-\frac{(2\varepsilon_{1H}-\varepsilon_{1H2})-0.5h\nu_o}{kT}\right]. \tag{15}$$

We now note that the quantity $2\varepsilon_{1H} - \varepsilon_{1H2} - 0.5h\nu_o$ that appears in the numerator of the exponent is just the effective dissociation energy D_o of an $H_2(g)$ molecule. The energy difference between two hydrogen atoms and an H_2 molecule is $2\varepsilon_{1H} - \varepsilon_{1H2}$, but since the molecule already possesses a zero-point vibrational energy, the energy required for dissociation is $2\varepsilon_{1H} - \varepsilon_{1H2} - 0.5h\nu_o$. Thus, we obtain

$$K_p = \frac{[m_H kT]^{3/2}(1-e^{-x})}{\pi^{1/2}hI}\exp\left[-\frac{D_o}{kT}\right]. \tag{16}$$

For a hydrogen molecule,

$$x(H_2) = \frac{h\nu_o}{kT} = \frac{(6.62608\times10^{-34}\text{ J s})(4{,}395.2\text{ cm}^{-1})(2.9979\times10^{10}\text{ cm s}^{-1})}{(1.38066\times10^{-23}\text{ J K}^{-1})(300\text{ K})} = 21.08. \tag{17}$$

Hence, $1-e^{-x} = 1-e^{-21.08} \approx 1$. Using Table 15.4, we find that the H_2 bond length is 0.7416×10^{-10} m. Therefore, the moment of inertia is

$$I_{H2} = \mu R^2 = \frac{m_H}{2}R^2 = \frac{1.6735\times10^{-27}\text{ kg}}{2}(0.7416\times10^{-10}\text{ m})^2$$

$$= 4.6019\times10^{-48}\text{ kg m}^2. \tag{18}$$

At 1,000 K,

$$[m_H kT]^{3/2} = [(1.6735\times10^{-27}\text{ kg})(1.38066\times10^{-23}\text{ J K}^{-1})(1{,}000\text{ K})]^{3/2}$$

$$= 1.1106\times10^{-70}\text{ kg}^3\text{ m}^3\text{ s}^{-3}. \tag{19}$$

The dissociation energy of H_2 is given in Table 15.4 as 4.476 eV, which is equal to 7.1712×10^{-19} J. Therefore,

$$\exp\left[-\frac{D_o}{kT}\right] = \exp\left[-\frac{7.1712\times10^{-19}}{(1.38066\times10^{-23})(1{,}000)}\right]$$

$$= \exp(-51.940) = 2.7706\times10^{-23}. \tag{20}$$

Combining these results, we obtain

$$K_p = \frac{(1.1106\times10^{-70})(1)(2.7706\times10^{-23})}{(3.14159)^{1/2}(6.62608\times10^{-34})(4.6019\times10^{-48})} = 5.693\times10^{-13}. \tag{21}$$

The units are $\text{kg}^3\text{ m}^3\text{ s}^{-3}/(\text{kg}^2\text{ m}^4\text{ s}^{-1}) = \text{kg}/(\text{m s}^2)$, which is equal to $\text{kg m}/(\text{m}^2\text{ s}^2) = \text{newton}/\text{m}^2 = \text{Pa}$. Thus, K_p has the nominal unit of Pa. Since we wish to measure pressure in bars, we need to convert by using the equivalence 1 bar = 10^5 Pa. This gives

$$K_p = 5.693\times10^{-13}\text{ Pa}\times\frac{1\text{ bar}}{10^5\text{ Pa}} = 5.693\times10^{-18}\text{ bar}. \tag{22}$$

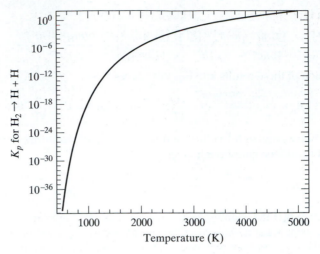

▲ **FIGURE 18.13**
K_p for the reaction $H_2(g) \rightarrow H(g) + H(g)$ as a function of temperature. The computational procedure is described in Example 18.11.

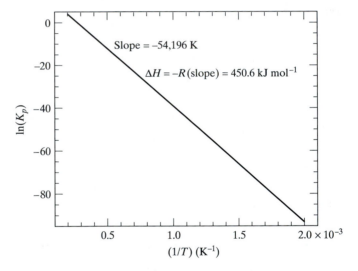

▲ **FIGURE 18.14**
In K_p versus T^{-1}, calculated from the data shown in Figure 18.13. A least-squares fit of a straight line to the data yields a slope of $-54{,}196$ K. The analysis in Chapter 5 leading to Eq. 5.85 shows that, for an ideal system, $\partial \ln K_p / \partial(T^{-1}) = -\Delta H/R$. The value of ΔH obtained from the slope is 450.6 kJ mol^{-1}, which deviates from the experimentally measured value by 4.3%.

If, using this procedure, we compute K_p as a function of temperature, the results shown in Figure 18.13 are obtained. Equation 5.85 shows that the temperature dependence of K_p is described by $(\partial \ln K_p / \partial T) = \Delta \overline{H}/(RT^2)$. Using the fact that $d(T^{-1}) = -T^{-2} dT$, we may write this equation in the form $(\partial \ln K_p / \partial(T^{-1})) = -\Delta \overline{H}/R$. Therefore, we expect a plot of $\ln K_p$ versus T^{-1} to be linear with a slope of $-\Delta \overline{H}/R$ for ideal systems. Figure 18.14 shows such a plot using the data in Figure 18.13. The result is very close to linear. A least-squares fit of a straight line to the data yields a slope of $-54{,}196$ K and an associated value of ΔH of 450.6 kJ mol^{-1}, which is equivalent to 4.67 eV. The deviation of the result from the experimental dissociation enthalpy of H_2 is 4.3%.

For a related exercise, see Problem 18.33.

18.6   Special Case: Systems with a Small Number of Quantum States

The development in Sections 18.2 and 18.3 assumes that the number of quantum states is very large compared to the number of molecules. This will always be the case when we are concerned with systems that possess translational energy. However, when translational energy is not present, the number of molecules will usually exceed the effective number of states by many orders of magnitude. In this situation, the equations obtained in those sections will not be valid. For example, if we are interested in the entropy associated with distributing Avogadro's number of molecules into four different nuclear spin states, we cannot use Eq. 18.48 to compute this quantity.

To obtain appropriate expressions for the thermodynamic quantities in systems in which the number of quantum states is small relative to the number of molecules, we must make use of the result obtained in Chapter 17 that the only microstate of importance is the one which maximizes the thermodynamic probability W. Using Lagrangian multipliers, we determined that the distribution of molecules among the various quantum states that maximizes W is the Boltzmann distribution

$$n_i = \frac{N}{z} \exp\left[-\frac{\varepsilon_i}{kT}\right]. \qquad (18.125)$$

The total internal energy of this microstate is, therefore, just the summation of the product of the number of molecules in quantum state i and the energy of that state; that is,

$$U = \sum_{i=1}^{s} n_i \varepsilon_i = \frac{N}{z} \sum_{i=1}^{s} \varepsilon_i \exp\left[-\frac{\varepsilon_i}{kT}\right], \qquad (18.126)$$

where s is the effective number of quantum states, which we assume to be small. U can easily be expressed in terms of the molecular partition function. The derivative, with respect to temperature, of $\ln(z)$ at constant volume is

$$\left(\frac{\partial \ln(z)}{\partial T}\right)_V = \frac{1}{z}\left(\frac{\partial z}{\partial T}\right)_V = \frac{1}{z}\frac{\partial}{\partial T}\sum_{i=1}^{s} \exp\left[-\frac{\varepsilon_i}{kT}\right] = \frac{1}{zkT^2}\sum_{i=1}^{s} \varepsilon_i \exp\left[-\frac{\varepsilon_i}{kT}\right].$$

Rearrangement of this equation produces

$$NkT^2\left(\frac{\partial \ln(z)}{\partial T}\right)_V = RT^2\left(\frac{\partial \ln(z)}{\partial T}\right)_V = \frac{N}{z}\sum_{i=1}^{s} \varepsilon_i \exp\left[-\frac{\varepsilon_i}{kT}\right] = U. \quad (18.127)$$

This is the same as the result in Table 18.5 that we obtained for the case where the number of quantum states is much larger than the number of molecules. Thus, the internal energy and the enthalpy are independent of the relative number of quantum states and molecules.

We can now obtain the system entropy for a closed system from the formula $dU = T\,dS - p\,dV$. At constant volume, we have $(\partial S/\partial T)_V = (1/T)(\partial U/\partial T)_V$, so that the entropy is given by

$$\int\left(\frac{\partial S}{\partial T}\right)_V dT = S = \int \frac{1}{T}\left(\frac{\partial U}{\partial T}\right)_V dT. \qquad (18.128)$$

This indefinite integral is easily done by parts. Let $u = 1/T$ and $dv = (\partial U/\partial T)_V\,dT = \partial U$. With this choice, $du = -dT/T^2$ and $v = U$. The integral is, therefore,

$$S = \frac{U}{T} + \int \frac{U}{T^2} dT. \tag{18.129}$$

Replacing U in the integral in Eq. 18.129 with the expression derived in Eq. 18.127, we obtain

$$S = \frac{U}{T} + R \int \left(\frac{\partial \ln(z)}{\partial T} \right)_V dT = \frac{U}{T} + R \int \partial \ln(z) = \frac{U}{T} + R \ln(z) + \text{constant.} \tag{18.130}$$

If we take the entropy to be zero for a perfect crystal at $T = 0$ K, where U/T approaches zero and z approaches unity, the constant must be zero. The result when the number of quantum states is small relative to N is, therefore,

$$\boxed{S = R \ln(z) + \frac{U}{T}}. \tag{18.131}$$

The difference between this result and the one in Table 18.5 that holds when the number of states is large compared to N is the $-\ln(N!)$ term, which is present when the number of states is large, but not when s is small relative to N.

The Helmholtz free energy is obtained directly from Eqs. 18.127 and 18.131:

$$\boxed{A = U - TS = U - T[R \ln(z) + U/T] = -RT \ln(z)}. \tag{18.132}$$

Table 18.7 summarizes the relationship between the molecular partition function and various thermodynamic quantities when the effective number of quantum states is small relative to the number of molecules.

Table 18.7 Relationships between various thermodynamic quantities and z when the effective number of quantum states is small relative to N	
Quantity	**Equivalent Expression in terms of z**
Internal Energy U	$RT^2 \left(\dfrac{\partial \ln z}{\partial T} \right)_V$
Pressure p	$RT \left(\dfrac{\partial \ln z}{\partial V} \right)_T$
Entropy S	$R \ln z + \dfrac{U}{T} = R \ln z + RT \left(\dfrac{\partial \ln z}{\partial T} \right)_V$
Helmholtz Free Energy A	$-RT \ln z$

EXAMPLE 18.12

When no magnetic field is present, the three nuclear spin states of deuterium are all degenerate. Let us take the energy of these three spin states to be zero. Compute the entropy for the distribution of Avogadro's number of D atoms into these three states at 300 K.

Solution

Since the energy of all three states is zero, the total energy associated with the nuclear spin states is $U = 0$. Therefore, the entropy is

$$S = R \ln(z) + \frac{U}{T} = R \ln(z). \qquad (1)$$

The molecular partition function for this system is

$$z = \sum_{i=0}^{3} \exp\left[-\frac{\varepsilon_i}{kT}\right] = 1 + 1 + 1 = 3, \qquad (2)$$

since $\varepsilon_i = 0$ for all i.

Equation 2 is the expected result, because the effective number of states here is clearly three. Consequently, the entropy is given by

$$S = R \ln(3) = 9.134 \, \text{J mol}^{-1} \, \text{K}^{-1}. \qquad (3)$$

It is instructive to note what happens if we erroneously use the expression in Table 18.5 for the entropy. We then obtain

$$S = R \ln\left[\frac{z}{N}\right] + \frac{U}{T} + R = R \ln\left[\frac{3}{6.02214 \times 10^{23}}\right] + R = -437.8 \, \text{J mol}^{-1} \, \text{K}^{-1}, \qquad (4)$$

and the entropy would be a negative number, in violation of the third law. This is the reason we included the $-\ln(N!)$ term with the translational contribution to the entropy. The term is present only when translational motion is present or when the number of quantum states is large relative to the number of molecules.

For related exercises, see Problems 18.34, 18.35, 18.36, and 18.37.

18.7   The Boltzmann Expression

In 1896, Ludwig Boltzmann suggested that a system's entropy should be given by

$$S = k \ln W_{max}, \qquad (18.133)$$

where W_{max} is the maximum value for the thermodynamic probability introduced and derived in Chapter 17. Indeed, this expression is even carved on Boltzmann's tombstone in Vienna. (See Figure 18.15.) It specifically relates the

▲ FIGURE 18.15
Ludwig Boltzmann's (1844–1906) tombstone in Vienna, Austria. The tombstone bears the expression $S = k \log W$.

entropy to the number of ways the state of the system can be realized. Since there are more ways to produce chaos than order, Eq. 18.133 explicitly connects the entropy to the extent of chaos that characterizes a system. In this section, we examine the conditions under which that equation is accurate.

In Chapter 17, we demonstrated that if the number of molecules is large relative to the number of quantum states,

$$W_{max} = \frac{N!}{\prod_{i=1}^{s} n_i!},$$

where the number of molecules in quantum state i is given by Eq. 18.125. Taking logarithms of both sides of the expression for W_{max} produces

$$\ln W_{max} = \ln(N!) - \sum_{i=1}^{s} \ln(n_i!). \tag{18.134}$$

Since both N and the n_i are large, we may safely employ Stirling's approximation for the factorial expressions to obtain

$$\ln W_{max} = N \ln N - N - \sum_{i=1}^{s} n_i \ln n_i + \sum_{i=1}^{s} n_i = N \ln N - \sum_{i=1}^{s} n_i \ln n_i. \tag{18.135}$$

Let us examine the form of the last term in Eq. 18.135 when Eq. 18.125 is inserted for the n_i:

$$\sum_{i=1}^{s} n_i \ln n_i = \frac{N}{z} \sum_{i=1}^{s} \exp\left[-\frac{\varepsilon_i}{kT}\right] \left\{\ln N - \ln z - \frac{\varepsilon_i}{kT}\right\}$$

$$= \frac{N \ln N}{z} \sum_{i=1}^{s} \exp\left[-\frac{\varepsilon_i}{kT}\right] - \frac{N \ln z}{z} \sum_{i=1}^{s} \exp\left[-\frac{\varepsilon_i}{kT}\right] - \frac{N}{kTz} \sum_{i=1}^{s} \varepsilon_i \exp\left[-\frac{\varepsilon_i}{kT}\right]. \tag{18.136}$$

But since $z = \sum_{i=1}^{s} \exp[-\varepsilon_i/(kT)]$, and because Eq. 18.126 shows us that $U = (N/z) \sum_{i=1}^{s} \varepsilon_i \exp[-\varepsilon_i/(kT)]$, Eq. 18.136 becomes

$$\sum_{i=1}^{s} n_i \ln n_i = N \ln N - N \ln z - \frac{U}{kT}. \tag{18.137}$$

Combination of Eqs. 18.135 and 18.137 produces

$$\ln W_{max} = N \ln N - \sum_{i=1}^{s} n_i \ln n_i = N \ln N - N \ln N + N \ln z + \frac{U}{kT}$$

$$= N \ln z + \frac{U}{kT}.$$

Therefore, the Boltzmann expression for the entropy is

$$S = k \ln W_{max} = Nk \ln z + \frac{U}{T} = R \ln z + \frac{U}{T}. \tag{18.138}$$

Equation 18.138 is identical to Eq. 18.131, but it is not the same as the expression for the entropy in Table 18.5. Therefore, we conclude that the Boltzmann expression for the entropy is correct when the effective number of quantum states in the system is small relative to N, but the expression does *not* correctly describe the entropy of systems with an effective number of quantum states that is much larger than N.

Summary: Key Concepts and Equations

1. The thermodynamic value of any property of a system, X, is the average value of X over all available microstates. That is,

$$\text{Thermodynamic value of } X = \langle X \rangle = \sum_{\alpha=1}^{n^D} P_\alpha X_\alpha,$$

where P_α is the probability of the formation of microstate α, X_α is the value of X for the αth microstate, and n^D is the total number of available microstates. Therefore, if we wish to compute thermodynamic properties from quantum mechanics and statistical mechanics, we must find a means of executing the foregoing summation.

2. If the effective number of quantum states is small compared to the number of molecules, the only microstate of importance is the one that maximizes the thermodynamic probability that we introduced in Chapter 17. In this microstate, the population of each quantum state is given by the Boltzmann distribution, and the relationships of the molecular partition function to thermodynamic quantities are those listed in Table 18.7. In this situation, the Boltzmann expression for the entropy, $S = k \ln W_{max}$, is valid.

3. When translational motion is present, the effective number of quantum states is very large relative to the number of molecules present. The ratio is on the order of 10^5 to 10^6. In this case, we must concern ourselves with many microstates in performing the summation in Eq. 18.1.

4. The fundamental postulate of statistical mechanics is as follows:

    For a thermodynamic system with a fixed volume, composition, and temperature, all distributions or microstates that have the same energy have equal values of P_α.

    For this postulate to hold, the probability that we will observe microstate α must have the form $P_\alpha = A \exp[-\beta E_\alpha]$, where A is a constant, $\beta = 1/kT$, and E_α is the energy of microstate α. The normalization requirement on the probability distribution gives us the result $A = Z^{-1}$, where

$$Z = \sum_{\alpha=1}^{n^D} \exp[-\beta E_\alpha],$$

    with the summation running over all microstates, is the canonical partition function.

5. It is relatively easy to express the thermodynamic quantities U, H, p, S, A, and G in terms of the canonical partition function. The results are given in Table 18.1. The easiest way to obtain these relationships is from the result that

$$A = -kT \ln Z$$

    and the fact that

$$dA = -S\,dT - p\,dV$$

    for any closed system undergoing a reversible process. The requirement that $(\partial U/\partial V)_T$, expressed in terms of Z, and the same derivative obtained from Maxwell's relationships in Chapter 5, viz., $(\partial U/\partial V)_T = T(\partial p/\partial T)_V - p$, be equal leads to the result that we must have $\beta = 1/(kT)$ for all systems.

6. To obtain the canonical partition function (Z) in terms of the molecular partition function (z), we make use of the fact that z represents the effective number of

quantum states available to the system and Z represents the effective number of microstates that can be formed by distributing N molecules among s_{eff} effective quantum states. Explicit consideration of the cases for which $N = 2, 3$, and 4 with an arbitrary number of effective quantum states such that $s_{eff} \gg N$ shows that the effective number of distributions or microstates that can be formed is given by an N-term polynomial in s_{eff} in which the leading term has the form $s_{eff}^N/N!$. By extrapolating this result to the case for which N is Avogadro's constant and then taking the limit in which the ratio s_{eff}/N is on the order of 10^5 to 10^6, we obtain the result that the effective number of microstates is $s_{eff}^N/N!$. Since $z = s_{eff}$, the canonical partition function is given by $Z = z^N/N!$. This result relates the thermodynamic quantities to z in the manner described in Table 18.5.

7. Since translational motion of the center of mass of a system is rigorously separable from internal motion of the system, the total molecular partition function can always be written as a product of translational and internal molecular partition functions. For noninteracting molecules, the translational molecular partition function is $z_{tr} = V/h^3 [2\pi mkT]^{3/2}$. The use of this result, leads to the Sackur–Tetrode equation for the translational entropy and $3RT/2$ for the translational energy.

8. If electronic, vibrational, and rotational energy are assumed to be separable, the internal molecular partition function becomes $z_{int} = z_{rot}z_{vib}z_{el}$. The electronic energy levels are so widely spaced that only the first term in the electronic partition function makes an appreciable contribution to the sum. Thus, we obtain $z_{el} = g_o \exp[-\varepsilon_o/(kT)]$, where g_o and ε_o are the degeneracy and energy, respectively, of the ground electronic quantum state. If ε_o is taken to be the zero of electronic energy, $z_{el} = g_o$.

9. In precise calculations, the rotational-energy levels are represented by a rigid-rotor expression, with a correction term for centrifugal distortion obtained by fitting the rotational microwave spectrum of the molecule. At low temperatures, the centrifugal correction term is relatively unimportant. If we assume that the rigid-rotor quantized energy levels can be treated as a classical continuum, the diatomic rotational molecular partition function can be obtained in closed form. The result is $z_{rot} = kT/(Bh)$, where B is the rotational constant. Calculations show that the results obtained with the classical expression are highly accurate if the ratio B/T is small. In practice, if we have $T \geq 300$ K, the classical approximation is often sufficiently accurate. For polyatomic molecules, the moments of inertia are usually very large, so that the classical assumption is generally excellent except at temperatures near absolute zero. Thermodynamic quantities for these molecules can, therefore, be computed by using Eq. 18.91.

10. Since the total wave function must be symmetric or antisymmetric to identical particle exchange for bosons and fermions, respectively, the spatial portion of the rotational wave function for homonuclear diatomic molecules must be either symmetric or antisymmetric, depending upon the symmetry of the nuclear spin wave function and whether the identical nuclei are bosons or fermions. The various possibilities are summarized in Table 18.6. Figure 18.7 shows that rotation of π radians about an angle θ is equivalent to nuclear exchange. Therefore, the spherical harmonic $Y_J^M(\theta, \phi)$ that forms the spatial part of the rotational wave function must have only even J quantum numbers if $Y_J^M(\theta, \phi)$ must be symmetric to exchange and only odd J states if it must be antisymmetric. This distinction leads naturally to the existence of two types of molecules. Those with even J states are said to be para, while those with odd values of J are called ortho. At higher temperatures, we can approximate this situation by summing over all rotational states and then dividing the partition function by a symmetry number.

11. In precise calculations, the vibrational-energy levels can be represented by a harmonic quantization with a correction term for anharmonicity. If the correction term is omitted, the molecular harmonic vibrational partition function has the simple closed-form expression $z_{vib} = e^{-x/2}/(1 - e^{-x})$, where $x = h\nu_o/(kT)$. The use of this expression leads to very simple equations for the vibrational contributions

to the various thermodynamic quantities. For polyatomics with $3N - 6$ vibrational degrees of freedom, it is difficult to compute the total vibrational molecular partition function unless the total vibrational energy is sufficiently low that the vibrational motion can be accurately described by uncoupled harmonic normal modes. In this case, z_{vib}^{total} is just the product of $3N - 6$ factors, each of which is that for a single harmonic oscillator. At most temperatures, the only important contribution from vibrational motion comes from the zero-point energy term.

12. For ideal systems, the equilibrium constant can be written as a simple ratio of molecular partition functions. This kind of expression permits reasonably accurate computation of K_p, provided that the conditions are such that we expect the molecules to behave ideally.

Problems

Problems that require the use of some type of computational device are marked with an asterisk (*). Problems that require some type of plotting routine are indicated with a pound sign (#).

18.1 (A) Compute the effective number of available translational states for one argon atom in a three-dimensional potential well with a potential $V(x, y, z) = 0$ whose shape is that of a rectangular parallelepiped, the lengths of whose sides are $a = b = c = 0.200$ m. Assume that the temperature is 298 K.

(B) Let us now assume that we have 1 mole of noninteracting argon atoms in the potential well. Suppose that we can examine each possible translational quantum state to see whether it is occupied by any of the argon atoms. Suppose further that we can examine two such states per second and that we continue our examination 10 hours per day, every day for one year. Approximately how many times will we find that a translational quantum state is occupied by one or more argon atoms?

18.2 (A) How many distinguishable microstates exist when we distribute four indistinguishable objects among 12 different states?

(B) How many distinguishable microstates exist when we distribute four indistinguishable objects among 100 different states? Which term makes the major contribution to the answer?

18.3 (A) Obtain an expression for the ratio of $s_{eff}^4/4!$ to the number of distinguishable microstates obtained when four indistinguishable objects are distributed among s_{eff} different states.

(B) Compute this ratio when s_{eff}/N is 10, 10^2, 10^3, 10^4, 10^5, and 10^6, where N is the number of objects to be distributed into the s_{eff} quantum states.

18.4 Verify the results given in Table 18.5 for A, G, H, and p, using the results derived in the text for U and S.

18.5 Compute the entropy of argon at pressures of 1, 10, and 100 bar for a temperature of 300 K. The experimentally determined argon entropies at 300 K listed in the 78th edition of the CRC *Handbook of Chemistry and Physics* at these pressures at 300 K are 155.0, 135.6, and 114.7 J mol^{-1} K^{-1}, respectively. Compute the percent error in your result as a function of pressure. Why does the error increase as the pressure rises?

18.6 Compute the entropy of argon at a pressure of 10 bar for temperatures of 160 K, 220 K, 300 K, and 380 K. The experimentally determined entropies listed in the 78th edition of the CRC *Handbook of Chemistry and Physics* at these temperatures and the given pressure are 121.8, 128.9, 135.6, and 140.6 J mol^{-1} K^{-1}. Compute the percent error in your result as a function of temperature. Why does the error decrease as the temperature rises?

18.7 Show that the assumption that we can obtain the translational molecular partition function by using the eigenvalues for independent particles in a potential well with zero potential energy leads directly to the ideal-gas equation of state. (*Hint*: Compute the pressure from the translational molecular partition function.)

18.8* Assume that the rotational-energy states of HI are given by those for a rotating Morse oscillator for which we obtain (see Problem 15.34)

$$E_J = +3.006 \times 10^{-5}J(J + 1)$$

$$- 9.341 \times 10^{-10}[J(J + 1)]^2 \text{ hartrees.}$$

(A) Compute z_{rot} for HI at 300 K.

(B) Repeat (A), ignoring the centrifugal distortion term in E_J. What is the percent difference between your results in (A) and (B)?

18.9 Starting with Eq. 18.82, derive Eq. 18.83. Show all details.

18.10* Use the data in Table 15.4 to compute the rotational energy of HF at temperatures of 10 K, 15 K, 25 K, 50 K, 100 K, 150 K, 250 K, 300 K, and 400 K. Compute the ratio $U_{rot}/(RT)$ at each temperature, and plot the results along with that in Example 18.4 as a function of temperature. At approximately what temperature is the error in the classical approximation for U_{rot} for HF 2.5%? Assume that the rotational-energy levels are given by a rigid-rotor expression.

18.11* Use the data in Table 15.4 to compute the rotational heat capacity, $[C_v]_{rot}$, for HF at temperatures between 5 K and 40 K at 5-K intervals and then at 50 K, 100 K, 200 K, and 300 K. Compute the ratio $[C_v]_{rot}/R$ at each temperature, and plot the results as a function of temperature. At approximately what temperature does the error in the classical approximation for $[C_v]_{rot}$ for HF approach 1.0%? Assume that the rotational-energy levels are given by a rigid-rotor expression.

18.12* Compute the rotational energy of hypothetical molecules at 25 K that have rotational constants from $6 \times 10^{10}\,s^{-1}$ to $60 \times 10^{10}\,s^{-1}$ by increments of $6 \times 10^{10}\,s^{-1}$. In each case, compute the ratio $U_{rot}/(RT)$. Plot this ratio as a function of the rotational constant. Arrange the molecules listed in Table 15.4 in order, from those most likely to have their rotational energy described accurately by a classical approximation to those that are least likely.

18.13 Table 12.1 lists the $\Theta_\ell^m(\theta)$ portion of the $Y_\ell^m(\theta, \phi)$ spherical harmonic for $\ell = 0$, 1, and 2. The $\Theta_\ell^m(\theta)$ functions for $\ell = 3$ are as follows,

$$\Theta_3^0(\theta) = \frac{3(14)^{1/2}}{4}\left[\frac{5}{3}\cos^3\theta - \cos\theta\right],$$

$$\Theta_3^{\pm 1}(\theta) = \frac{(42)^{1/2}}{8}\sin\theta[5\cos^2\theta - 1],$$

$$\Theta_3^{\pm 2}(\theta) = \frac{(105)^{1/2}}{4}\sin^2\theta\cos\theta,$$

and

$$\Theta_3^{\pm 3}(\theta) = \frac{(70)^{1/2}}{8}\sin^3\theta.$$

For the spherical harmonics with $\ell \le 3$, show that $\Theta_\ell^m(\theta)$ is symmetric to nuclear exchange in a homonuclear rotor if ℓ is even and antisymmetric if ℓ is odd. Discuss the significance of this result with respect to the rotational molecular partition function.

18.14 Using the data given in Table 15.4 and the information provided by Example 18.5, compute the rotational contribution to the entropy of $^{12}C^{16}O$ at 298.15 K.

18.15 F. Hynne [*Am. J. Phys.*, **49**, 125 (1981)] and H. Kroemer [*Am. J. Phys.* **48**, 962 (1980)] have pointed out that omitting the $c_{N-1}s_{eff}^{N-1}$ term in Eq. 18.44 that includes the contribution of doubly occupied states produces a large error in the value of Z given by Eq. 18.45. Their analysis indicates that the canonical partition function should have the form $Z = fz^N/N!$, where f can be very large.

(A) Show that the presence of the factor f in the expression for Z makes no difference in the computed values of U and C_v.

(B) Show that the only effect of the factor f on the entropy is to alter the value of the constant in Eq. 18.38.

18.16 Compute the rotational contributions to the Helmholtz and Gibbs free energies for $^{12}C^{16}O$ at 298.15 K. Use the data given in Table 15.4 to obtain the rotational constant.

18.17 In this problem, you will derive a result for the classical rotation of an asymmetric polyatomic molecule that is very close to the exact result given in Eq. 18.91. The various parts of the problem serve as a procedure guide.

(A) Suppose we have a particle of mass m rotating in a plane about the origin at a fixed distance R with the potential energy equal to zero. Since R is fixed and the motion is planar, this is a single-variable problem. Let the angle between the radial vector to the particle and the X-axis be ϕ. Figure 11.5 illustrates the model if we replace θ with ϕ. The classical Hamiltonian for this system is given by Eq. 11.22:

$$\mathbf{H} = \frac{P_R^2}{2m} + \frac{P_\phi^2}{2mR^2} + V(R, \phi).$$

In the case under examination, P_R and $V(R, \phi)$ are both zero, so that

$$\mathbf{H} = \frac{P_\phi^2}{2mR^2}.$$

The quantum mechanical operator corresponding to P_ϕ is $(\hbar/i)(\partial/\partial\phi)$. Therefore, the Schrödinger equation for the system is

$$-\frac{\hbar^2}{2mR^2}\frac{\partial^2\psi(\phi)}{\partial\phi^2} = E\psi(\phi).$$

Solve the Schrödinger equation and obtain the eigenfunctions and eigenvalues for this two-dimensional rotational motion. What quantum restrictions are there on the rotational quantum number? Explain.

(B) Write down the quantum mechanical expression for the rotational molecular partition function for the system described in (A). By assuming the rotational levels to be continuous, derive an appropriate expression for the molecular partition function.

(C) Now assume that the Hamiltonian for rotation in three-dimensional space about the three principal axes with moments of inertia I_1, I_2, and I_3 is the sum of Hamiltonians that look exactly like the one in (A). (This is not correct.) Then $E_{rot}^c = E_1 + E_2 + E_3$, where E_i is the energy for rotation about the principal axis whose moment of inertia is I_i. With this assumption, show that total molecular partition function for classical rotation of the polyatomic molecule is the product of molecular partition functions for each of the independent rotations. Determine the form that the total molecular partition function for classical rotation would have. Compare your result with Eq. 18.91.

18.18 Show that if we can treat rotation as being classical, the rotational contribution of a nonlinear polyatomic molecule to U is $1.5RT$.

18.19 The moments of inertia of NH_3 are $I_1 = I_2 = 2.816 \times 10^{-47}$ kg m^2 and $I_3 = 4.43 \times 10^{-47}$ kg m^2. Compute the contribution of the rotational motion of NH_3 to S, A, and G.

18.20 Describe in qualitative terms a method by which we might prepare a sample of pure para-H_2 at 300 K.

18.21 Deuterium (^2H) has a nuclear spin quantum number $I = 1$.

(A) Does ortho–para deuterium exist? Explain.

(B) What is the high-temperature ratio of ortho to para D_2? Explain. (*Hint*: Is deuterium a fermion or a boson? How many spin states can each deuterium atom have? How many symmetric and antisymmetric spin functions can be constructed?)

18.22 (A) Do we have ortho and para forms of HD? Explain.

(B) Do we have ortho and para forms of acetylene? Explain.

18.23 Figure 18.9 indicates that the rotational contribution to the constant-volume heat capacity of ortho-H_2 at 175 K is 3.173 J mol^{-1} K^{-1}, while that for para-H_2 is 12.178 J mol^{-1} K^{-1} at this temperature. Use an appropriately modified form of Eq. 18.83 to verify the correctness of these two results. The rotational constant for H_2 can be found in Table 15.4.

18.24* By computing the equilibrium ortho–para ratio between $T = 125$ K and $T = 135$ K at 1-K intervals, determine, within 1 degree, the temperature at which there are equal numbers of para-H_2 and ortho-H_2 molecules in a system that contains an ortho–para conversion catalyst so that the two forms reach equilibrium rapidly.

18.25 Starting with the fact that $U = RT^2(\partial \ln z/\partial T)_V$, show that ortho-$H_2$ has a rotational zero-point energy equal to $2NBh$, where N is Avogadro's constant and B is the H_2 rotational constant. Evaluate the expected rotational zero-point energy for ortho-H_2. Is the result consistent with Figure 18.8? (*Hint*: Write down the appropriate form for z_{ortho}^{rot}. Expand the summation by writing down the first few terms, factor out anything you can, and then take the natural logarithm of the result. Finally, take the derivative of the natural logarithm of the result with respect to temperature at constant volume, and then examine the limit of the result as $T \to 0$ K.)

18.26 Compute the value of the vibrational molecular partition function for H_2 at 1,000 K, assuming that we can ignore the anharmonicity correction and that v is a continuous variable so that we can replace the summation over v with an integral. Compare your result with the correct answer obtained in Example 18.6.

18.27 Consider the series $y = 1 + 2 + 2^2 + 2^3 + 2^4 + 2^5 + \ldots$. An investigator attempts to sum this series as follows:

$$y = 1 + 2 + 2^2 + 2^3 + 2^4 + 2^5 + \cdots$$
$$= 1 + 2[1 + 2 + 2^2 + 2^3 + 2^4 + 2^5 + \cdots]$$
$$= 1 + 2y.$$

Solving for y, he obtains

$$y - 2y = 1 = -y,$$

so that the sum is -1. The investigator has the nagging feeling that something is wrong. Can you help him out?

18.28* Compute the vibrational contribution to the entropy of $^{16}O^{16}O$ at 300 K when the anharmonicity correction is included in the energy-level expression. Compare the result with the entropy obtained in Example 18.8, in which a harmonic approximation was used. Obtain the required data from Table 15.4.

18.29* (A) Compute the rotational contribution to the entropy of para-H_2 and ortho-H_2 at 298.15 K.

(B) Assuming that the rotation can be treated classically at 298.15 K, use a symmetry number in lieu of explicit consideration of ortho–para forms to compute the rotational contribution to the entropy of 1 mole of H_2 at that temperature. Refer to Table 15.4 for any data you may need.

(C) Use the results of (A) to compute the expected ratio of ortho to para hydrogen at 298.15 K.

(D) Calculate the total entropy of 1 mole of H_2 that has the ratio of ortho to para forms computed in (C). Compare your result with that obtained in (B). How much error is made by using a classical approximation with a symmetry number instead of actually considering the two forms of H_2?

18.30. Using a classical approximation with a symmetry number instead of actually considering the fact that H_2 consists of ortho- and para-H_2 results in only a 0.41% error in the computed rotational entropy at 298.15 K. (See Problem 18.29.) In view of this result, compute the total entropy of H_2 at 298.15 K and 1 bar of pressure. The measured value is 130.7 J mol^{-1} K^{-1}. Compute the percent error in your result.

18.31* If the temperature is not too high, the vibrational energy will be only slightly greater than the zero-point energy. Under these conditions, we can regard the vibrational modes as being uncoupled. In this problem, you will examine how the total vibrational energy of a diatomic molecule whose vibrational mode is treated harmonically varies with the ratio of v_o to T. For a diatomic molecule whose vibrational mode is assumed to be harmonic, compute the ratio of U_{vib} to the vibrational zero-point energy as a function of $x = hv_o/(kT)$ over the range $1.0 \le x \le 6$. Plot the ratio as a function of x. What is the value of the ratio for H_2, O_2, and $^{35}Cl^{35}Cl$ at 300 K?

18.32 Compute the total entropy of NH_3 at 298.15 K and 1 bar of pressure. The required data can be found in Problem 18.19 and Example 18.10. The measured value given in the 78th edition of the CRC *Handbook of Chemistry and Physics* is 192.8 J mol^{-1} K^{-1}. Compute the percent error in your result.

18.33 (A) Obtain an expression for K_p for the equilibrium reaction $H_2(g) + D_2(g) = 2 HD(g)$ in terms of fundamental frequencies, moments of inertia, masses, and temperature.

(B) The atomic mass of deuterium is 2.0141 amu. The vibrational frequencies of H_2, D_2, and HD, expressed in wave numbers, are 4,395.2 cm^{-1}, 3,118.4 cm^{-1}, and 3,817.09 cm^{-1}, respectively. Obtain K_p for the reaction in (A) as a function of temperature alone.

(C) Plot K_p versus T and determine the temperature at which $K_p = 1$.

18.34 In the absence of an external magnetic field, the three nuclear spin states of deuterium (^2H) are all degenerate. Let us take the energy of these unsplit spin states to be our zero for nuclear spin energy. Suppose we place 1 mole of deuterium atoms in a magnetic field of 10^5 gauss, or 10 tesla, at 300 K. Compute the internal spin energy and the entropy of the system. Use Table 16.1 to obtain any data you may need.

18.35 An investigator has 1 mole of $H_2(g)$ molecules at 300 K inside a rectangular parallelepiped, the lengths of whose sides are all 0.2000 m. If the molecules do not interact, the ground-state translational energy is given by $E_{gs} = 3h^2/(8ma^2)$, where m is the mass of H_2 and a is the length of the sides of the cubical container.

(A) Use the Boltzmann distribution to compute the number of $H_2(g)$ molecules in the translational ground state.

(B) Is the calculation meaningful? Discuss this point.

18.36 One mole of marbles is distributed into 28 identical states whose energies are all zero. Compute the entropy of the system.

18.37 One mole of helium atoms at a temperature of 500 K is distributed into four containers at heights of zero, 2,000, 3,000, and 4,000 m above the earth's surface.

(A) If the gravitational energy is assumed to be given by $E_{gravitation} = mgh$ and the helium atoms do not interact, compute the number of helium atoms in each container.

(B) Compute the gravitational contribution to the entropy for the system described in (A).

18.38 In Chapter 7, we introduced the Einstein theory of solid-state heat capacities. In this theory, the $3N - 6$ vibrational degrees of freedom of the crystal are assumed to be harmonic with a common vibrational frequency. Equation 18.110, multiplied by three (since we have approximately $3N$ vibrational degrees of freedom), is then used to compute the crystal heat capacity at constant volume. By treating the common vibrational frequency as an adjustable parameter, we determined that a characteristic temperature $\theta = h\nu_o/k = 230.39$ K gave the best fit to the crystal heat capacity for Mg(s). Use this result to compute the crystal entropy of Mg(s) at 298.15 K. The measured value reported in the 78th edition of the CRC *Handbook of Chemistry and Physics* is 32.7 J mol^{-1} K^{-1}. Compute the percent error in your result.

18.39 "Sir, I have a question."

"What is it, Sam?"

"It concerns the electronic partition function for the hydrogen atom that you discussed."

"Yes, what about it, Sam?"

"Well, sir, I really don't know how to say this, but I don't think your result is correct."

"Why is that, Sam," I ask.

"Well, sir, if we take the ground state of the hydrogen atom as our zero for energy, as you did, the electronic partition function is given by

$$z_{el} = \sum_{n=1}^{\infty} n^2 \exp\left[-\frac{1.313 \times 10^6 [1 - n^{-2}]}{RT}\right].$$

You then expanded this summation at 298 K and wrote

$$z_{el}(T = 298 \text{ K}) = \sum_{n=1}^{\infty} n^2 \exp[-530.0(1 - n^{-2})]$$
$$= e^o + 4 \exp[-397.5]$$
$$+ 9 \exp[-471.1] + \cdots = 1.$$

"Yes. That's correct, Sam. So what's your point?"

"Well, sir, the exponential factor in the series approaches a limiting value for large n that is *not* zero. In fact, its limiting value at $n = \infty$ is $\exp(-530.0) = 6.667 \times 10^{-231}$. Therefore, the exponential factor approaches a constant, while the n^2 factor continues to increase without bound. Consequently, $z_{el}(T = 298.15 \text{ K}) = \infty$, not unity. Not only is the first term not the only important one; it is, in fact, negligible compared to the term with $n = 10^{300}$. The electronic partition function appears to be divergent. I hate to say this, sir, but, I think we've been bamboozled."

(*Author's note*: To bamboozle—to deceive by underhand methods; to dupe, to hoodwink.)

What do you think of Sam's comment? Have *you* been bamboozled? If z_{el} is indeed infinite, what is the significance of such a result? If z_{el} is not infinite, what is wrong with Sam's analysis?

SECTION 5

Kinetics and Dynamics

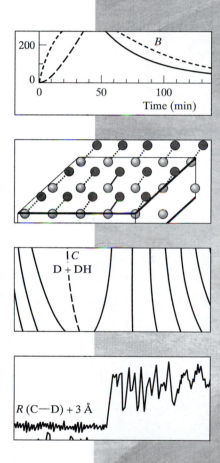

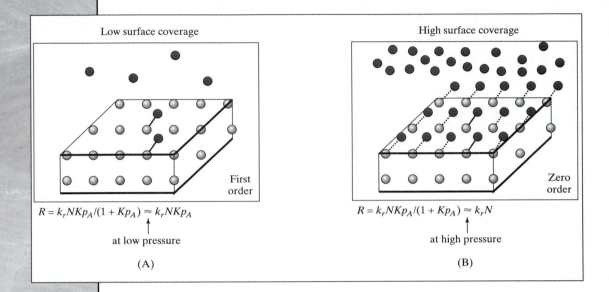

CHAPTER 19

Low surface coverage

First order

$$R = k_r N K p_A / (1 + K p_A) \approx k_r N K p_A$$

↑ at low pressure

(A)

High surface coverage

Zero order

$$R = k_r N K p_A / (1 + K p_A) \approx k_r N$$

↑ at high pressure

(B)

Phenomenological Kinetics

Phenomenological kinetics is the experimental study of rates. For our purposes, the specific rates of interest are those of chemical reactions. In this context, it is important to keep in mind the fact that thermodynamics tells us whether a reaction will be spontaneous under given conditions; it does not, however, give us any information about the rate at which the reaction will occur. In many applications, the reaction rate is the important factor, not the equilibrium properties of the system. An interesting example is the reaction

$$C_{\text{diamond}} \longrightarrow C_{\text{graphite}}.$$

At room temperature and atmospheric pressure, this reaction is spontaneous, in that $\Delta\mu$ is negative. Therefore, the diamond ring you just purchased for 2×10^3 dollars is spontaneously converting into a piece of graphite that you can purchase for less than a penny. Of course, this fact does not keep you awake at night, because the rate at which the process occurs under the given conditions is so slow that the sun will probably exhaust itself before the conversion is complete.

In most cases, experimental and theoretical investigations of kinetics are much more difficult to carry out than similar studies of equilibrium properties, such as activity coefficients, vapor pressures, and equilibrium constants, and also much more difficult to carry out than the spectroscopy of stable systems. The equilibrium system just sits there and awaits your pleasure. You can study it at your leisure. You can take a coffee break or go to a movie or a political rally if you're really at a loss for something to do, and the system will still be there waiting for you when you return. Kinetic systems, on the other hand, are constantly changing. If you wish to conduct a scientific investigation, you must do it on the time scale afforded to you by the reaction. That is, if a reaction starts and finishes in five picoseconds, all your measurements must be completed within that time frame, because that's all the time you have at your disposal. In many cases, this means that experimental studies will involve short-timescale spectroscopic measurements or other fast-response techniques. In addition to these difficulties, we often have situations in which several reactions occur simultaneously, leading to a variety of different products. This concurrent action further complicates the problem. From a theoretical perspective, to investigate kinetics and dynamics, we must have methods that are capable of producing information about the temporal behavior of the system. If we execute the study quantum mechanically, we will often have to work with the time-dependent Schrödinger equation rather than the stationary-state form.

The inherent difficulty of kinetic investigations means that we need every weapon available to us. Statistical mechanics, quantum mechanics, classical mechanics, and spectroscopy must all be brought to bear on the problem if we are to be successful. For this reason, the chapters in this section are the final ones in the text. Their relative position in the book does not reflect a lack of importance; instead, it is an indication of the difficulty associated with this type of investigation.

19.1   Bimolecular Reactions

Reactions involving the collision of two molecules are called *bimolecular reactions*. The molecules involved in the process can be different, so that the reaction is $A + B \longrightarrow$ products, or they can be the same, giving $A + A \longrightarrow$ products. Most reactions involve different reactants, but dimerization processes, such as $NO_2(g) + NO_2(g) \longrightarrow N_2O_4(g)$, are examples of the latter type.

19.1.1   Definition of Terms

Consider a chemical reaction between molecules A and B in which the reaction products are formed in a single collision of the two reactants. Such processes are said to be *concerted*. The rate at which products are formed or the rate at which reactants are consumed depends upon the collision frequency of

A with *B*. In Chapter 17, we derived a general expression for the total differential gas-phase collision frequency. This expression, given as Eq. 17.100, is

$$Z_{AB}(V, b) = 4\pi N_A N_B \left[\frac{\mu_{AB}}{2\pi kT} \right]^{3/2} (2\pi b) db \, \exp\left[-\frac{\mu V^2}{2kT} \right] V^3 dV \; . \quad \text{(19.1)}$$

Equation 19.1 tells us the number of times per second that the center of mass of molecule *A* will pass the center of mass of molecule *B* within the impact parameter range from *b* to *b* + *db*, with the two molecules having a relative speed in the range from *V* to *V* + *dV*. (See Figure 17.12.) Now suppose we have a means by which we can determine or compute the average probability, $\langle P(V, b) \rangle$, that a reaction will occur when a collision in these ranges takes place. If we simply multiply $Z_{AB}(V, b)$ by $\langle P(V, b) \rangle$, the result will be the number of reactions per second occurring from collisions with impact parameter from *b* to *b* + *db* and relative speed from *V* to *V* + *dV*. The total reaction rate is the sum of the reactions occurring at all impact parameters and relative speeds. Formally, this is

$$\text{Reaction Rate} = \mathcal{R} = \sum_{\text{all } V} \sum_{\text{all } b} \langle P(V, b) \rangle \, Z_{AB}(V, b).$$

However, we sum over continuous variables by integration. Therefore,

$$\mathcal{R} = \left[\int_{V=0}^{\infty} \int_{b=0}^{\infty} \langle P(V, b) \rangle \, 4\pi \left[\frac{\mu_{AB}}{2\pi kT} \right]^{3/2} (2\pi b) db \, \exp\left[-\frac{\mu V^2}{2kT} \right] V^3 dV \right] N_A N_B \; .$$

$$\text{(19.2)}$$

Equation 19.2 is a formal expression for the rate of a gas-phase bimolecular reaction. We shall return to this equation in Chapter 20, where we address the theoretical study of reaction rates and dynamics. For our present purposes, it will be sufficient simply to examine the characteristics of the quantity within the outermost set of brackets. Since we are integrating over the impact parameter and relative speed, this quantity is a function of temperature only for a given reaction. If we represent its value by $k(T)$, Eq. 19.2 becomes

$$\mathcal{R} = k(T) N_A N_B \; , \quad \text{(19.3)}$$

where $k(T)$ is called the *specific reaction rate constant* or *rate coefficient*. Physically, $k(T)$ represents a reaction probability averaged over all internal variables, impact parameters, and relative speeds. A form identical to Eq. 19.3 is also found to accurately represent the rate of solution reactions. However, in this case, the rate coefficient cannot be computed by using Eq. 19.2.

We can express the concentration factors in any units we choose, as that will change $k(T)$ by no more than a constant conversion factor. In experimental kinetics measurements, N_A and N_B are usually stated in mol L^{-1}, or molarity. In these units, we have

$$\mathcal{R} = k(T)[A][B], \quad \text{(19.4)}$$

where the brackets denote concentration in mol L^{-1}. Equation 19.4 is a direct result of the law of mass action that we derived in Chapter 17.

To proceed further, we need to express the reaction rate in terms of the derivative of a concentration with respect to time. Consider a chemical reac-

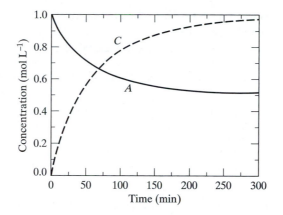

► FIGURE 19.1

The temporal variation of the concentrations of reactant $A(g)$ and product $C(g)$ in a hypothetical reaction whose stoichiometry is $A(g) + B(g) \longrightarrow 2\,C(g)$. As can be seen, this stoichiometry leads to a situation in which twice as many molecules of $C(g)$ are produced as molecules of $A(g)$ consumed.

tion whose balanced equation is $A(g) + B(g) \longrightarrow 2\,C(g)$. Suppose we measure the concentrations of A and C as a function of time by some suitable method and plot the results to obtain Figure 19.1. Intuitively, we would like the rate to be a positive number. Figure 19.1 shows that if we define the rate for this reaction as $d[C]/dt$, then $\mathcal{R}$ will be positive. However, if we define the rate in terms of the rate of change of the concentration of compound A, then $d[A]/dt$ is negative. These two conditions suggest that the rate be defined as the negative of the derivative of the concentration with respect to time for the reactants and as the (positive) derivative for the products. This choice alone, however, will not suffice. If we compute the derivatives $d[C]/dt$ and $-d[A]/dt$ and plot them for the reaction in question, we obtain Figure 19.2. Clearly, the two derivatives are not equal: $d[C]/dt$ exceeds $-d[A]/dt$ by exactly a factor of two, because the stoichiometric coefficient of C is two while that of A is unity. The solution to this difficulty is to include the stoichiometric coefficients in the definition of the rate. For the general reaction

$$aA + bB \longrightarrow cC + dD,$$

where the lowercase letters are the stoichiometric coefficients and the uppercase letters the compounds, we define the rate to be

$$\mathcal{R} = -\frac{1}{a}\frac{d[A]}{dt} = -\frac{1}{b}\frac{d[B]}{dt} = \frac{1}{c}\frac{d[C]}{dt} = \frac{1}{d}\frac{d[D]}{dt}. \tag{19.5}$$

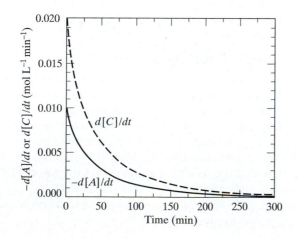

◄ FIGURE 19.2

The temporal variation of the derivatives $-d[A(g)]/dt$ and $d[C(g)]/dt$ for the hypothetical reaction illustrated in Figure 19.1. Because of the stoichiometry, $d[C(g)]/dt$ is twice as large as $-d[A(g)]/dt$ at all times during the reaction.

With this definition, the reaction rate is the same no matter how it is measured or expressed.

Equation 19.3 can now be written as

$$\mathcal{R} = -\frac{d[A]}{dt} = -\frac{d[B]}{dt} = k(T)[A][B]. \tag{19.6A}$$

If we were dealing with the bimolecular concerted reaction $A + A \longrightarrow$ products, the appropriate expression for the reaction rate would be

$$\mathcal{R} = -\frac{1}{2}\frac{d[A]}{dt} = k(T)[A][A] = k(T)[A]^2. \tag{19.6B}$$

When the reaction is $A + B \longrightarrow$ products, the exponents on the concentration factors in Eq. 19.6A are both unity. Such rate processes are said to be first order with respect to A, first order with respect to B, and second order overall ($1 + 1 = 2$). In the case of identical reactants that lead to the rate expression in Eq. 19.6B, we say that the reaction is second order with respect to A and second order overall. A great deal of experimental kinetics is concerned with determining the order of a reaction and measuring the corresponding rate coefficient as a function of temperature.

EXAMPLE 19.1

Suppose we have a concerted reaction involving the simultaneous collision and subsequent reaction of two molecules of A with a single molecule of B, so that the reaction stoichiometry is $2A + B \longrightarrow$ products. **(A)** What is the appropriate expression for the reaction rate, according to the law of mass action? **(B)** What are the orders of the reaction?

Solution

(A) For a concerted reaction, the law of mass action tells us that the collision frequency and the reaction rate are proportional to the concentrations of the colliding species. Therefore, the appropriate rate expression is

$$\mathcal{R} = -\frac{1}{2}\frac{d[A]}{dt} = -\frac{d[B]}{dt} = k(T)[A][A][B] = k(T)[A]^2[B]. \tag{1}$$

(B) The reaction orders are second order with respect to A, first order with respect to B, and third order overall. Reactions involving the simultaneous collision of three molecules are called *termolecular processes*.

For related exercises, see Problems 19.1 and 19.2.

19.1.2   Integrated Rate Expressions

The differential rate equations 19.6A and 19.6B for the two types of concerted bimolecular reactions can be easily integrated to yield an expression that provides a mathematical description of the manner in which the reactant and product concentrations vary with time. When two or more variables are present, the general procedure is to utilize the reaction stoichiometry to reduce the number of variables by one. After that is accomplished, the variables are separated, and the resulting equation is integrated. We illustrate the general procedure for the reaction $A + B \longrightarrow C + D$.

Starting with Eq. 19.6A, the time and concentration variables can be immediately separated to give

$$\frac{d[A]}{[A][B]} = -k(T)dt. \tag{19.7}$$

Since $[B]$ and $[A]$ are coupled variables, we must express either $[B]$ as a function of $[A]$ or both $[A]$ and $[B]$ as a function of a third variable before the left-hand side of Eq. 19.7 can be integrated. Using the latter procedure, we let x be the number of mol L^{-1} of A that have reacted at time t. The one-to-one stoichiometry of the reaction ensures that the number of mol L^{-1} of A that have reacted is equal to the number of mol L^{-1} of B that have reacted. We can express this concept mathematically by writing

$$[A] = [A]_o - x \tag{19.8A}$$

and

$$[B] = [B]_o - x, \tag{19.8B}$$

where $[A]_o$ and $[B]_o$ are the concentrations of A and B at time $t = 0$. Equation 19.8A tells us that $d[A] = -dx$, since $[A]_o$ is a constant. Substituting Eqs. 19.8A and 19.8B into Eq. 19.7 produces

$$\frac{dx}{\{[A]_o - x\}\{[B]_o - x\}} = k(T)dt. \tag{19.9}$$

We can now integrate both sides of Eq. 19.9 between the corresponding limits $t = 0$, at which time $x = 0$, and some arbitrary time t at which the number of mol L^{-1} of A and B that have reacted is x. This operation produces

$$\int_{x=0}^{x} \frac{dx}{\{[A]_o - x\}\{[B]_o - x\}} = \int_{t=0}^{t} k(T)dt = k(T)t. \tag{19.10}$$

The left-hand side of Eq. 19.10 is a standard form that can be found in most tables of integrals. Alternatively, we can utilize the fact that

$$\frac{1}{\{[A]_o - x\}\{[B]_o - x\}} = \frac{1}{[A]_o - [B]_o}\left[\frac{1}{[B]_o - x} - \frac{1}{[A]_o - x}\right] \tag{19.11}$$

to write the left-hand side of Eq. 19.10 as the difference of two integrals. That is,

$$\int_{x=0}^{x} \frac{dx}{\{[A]_o - x\}\{[B]_o - x\}} = \frac{1}{[A]_o - [B]_o}\left[\int_0^x \frac{dx}{[B]_o - x} - \int_0^x \frac{dx}{[A]_o - x}\right]. \tag{19.12}$$

Integrating Eq. 19.12, followed by a little algebra (see Problem 19.3), produces

$$\ln\left[\frac{[A]_o - x}{[B]_o - x}\right] = \ln\left[\frac{[A]_o}{[B]_o}\right] + ([A]_o - [B]_o)k(T)t. \tag{19.13}$$

If we now substitute Eqs. 19.8A and B into Eq. 19.13, the result is

$$\boxed{\ln\left[\frac{[A]}{[B]}\right] = \ln\left[\frac{[A]_o}{[B]_o}\right] + ([A]_o - [B]_o)k(T)t} , \tag{19.14}$$

which is the integrated form of the rate law for a bimolecular concerted reaction with different reactants. Once we have $[A]$ and $[B]$ as functions of time, the concentrations of the reaction products, C and D, can be obtained directly from the stoichiometry and the initial concentrations of C and D.

If the bimolecular reaction involves identical reactants such that $2A \longrightarrow$ products, then Eq. 19.6B provides the differential rate law. From that equation, we obtain

$$\frac{d[A]}{[A]^2} = -2k(T)dt. \tag{19.15}$$

Equation 19.15 can be easily integrated between corresponding limits to yield the integrated form of the rate law:

$$\int_{[A]=[A]_o}^{[A]} \frac{d[A]}{[A]^2} = -2k(T) \int_{t=0}^{t} dt = -2k(T)t.$$

Integrating the left side produces

$$\frac{1}{[A]_o} - \frac{1}{[A]} = -2k(T)t,$$

so that

$$\boxed{\frac{1}{[A]} = \frac{1}{[A]_o} + 2k(T)t}. \tag{19.16}$$

In taking experimental rate data for bimolecular reactions with different reactants, it is important to use initial concentrations such that $[A]_o \neq [B]_o$. If this is not done, the integrated rate expression, Eq. 19.14, reduces to the identity $0 = 0$, which is certainly correct, but not particularly useful. The problem is that if we use identical initial concentrations, the reactions

$$A + B \longrightarrow \text{products} \quad \text{and} \quad A + A \longrightarrow \text{products}$$

cannot be kinetically distinguished. Mathematically, this means that in the limit where $[B]_o \to [A]_o$, Eq. 19.14 must reduce to Eq. 19.16, except for the factor of two on the right-hand side. (See Problem 19.4.)

19.1.3   Determination of $k(T)$ from Rate Data

Equation 19.14 tells us that $\ln([A]/[B])$ will vary linearly with time with a slope that is equal to $([A]_o - [B]_o)k(T)$ and an intercept $\ln([A]_o/[B]_o)$. When the bimolecular reaction involves only a single reactant, Eq. 19.16 shows that $[A]^{-1}$ varies linearly with time with a slope equal to $2k(T)$ and an intercept $[A]_o^{-1}$. These facts provide one straightforward method for testing kinetic data: If a plot of $\ln([A]/[B])$ or $[A]^{-1}$ versus time is linear, we have identified the orders of the reaction, and the slope of the line gives us the value of the specific reaction rate coefficient.

Alternatively, Eqs. 19.14 and 19.16 can be solved for the rate coefficient to obtain

$$k(T) = \frac{\ln\left\{\dfrac{[A][B]_o}{[B][A]_o}\right\}}{\{[A]_o - [B]_o\}t}, \tag{19.17}$$

and

$$k(T) = \frac{[A]^{-1} - [A]_o^{-1}}{2t}, \tag{19.18}$$

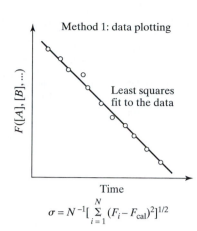

Method 1: data plotting

F([A], [B], ...)

Time

Least squares fit to the data

$$\sigma = N^{-1}[\sum_{i=1}^{N} (F_i - F_{cal})^2]^{1/2}$$

Method 2: Computation of $k(T)$

Step 1: Solve integrated rate expression to obtain $k(T) = g([A], [B], [A]_o, [B]_o, t)$

Step 2: Compute $k(T)$ at each data point

Step 3: Use the results to calculate the average deviation of $k(T)$ from the mean, i.e.,

$$\sigma = N^{-1}[\sum_{i=1}^{N} (k_i(T) - <k(T)>)^2]^{1/2}$$

Step 4: Compare σ with the expected experimental error. If $\sigma <$ expected error, the reaction order is probably correct, and the best value of $k(T)$ is the average.

◀ FIGURE 19.3

The two basic methods of analyzing kinetic data. In the plotting method, a suitable function of the concentrations is plotted against time. The data are fitted to a straight line, and the root-mean-square deviation of the measured points from the line is computed. If σ is less than expected/experienced error, the order of the reaction is probably correct. In the computational method, the integrated rate expression is used to compute $k(T)$ at each data point. The root-mean-square deviation of the results from the mean is used to decide whether the order is correct. (See text for details.)

respectively. The measured rate data can now be substituted into these equations and values of $k(T)$ computed at each time for which concentration data have been obtained. If the correct integrated rate expression is either Eq. 19.14 or Eq. 19.16, the computed values of $k(T)$ will be constant within the experimental error of the data for the equation describing the correct reaction mechanism. If neither of these equations correctly describes the kinetics, $k(T)$ will not be constant for either calculation. Figure 19.3 summarizes the two basic methods of data analysis. Example 19.2 illustrates the two techniques and demonstrates how we work with experimental data that are proportional to the concentration.

EXAMPLE 19.2

Compound A is mixed with compound B in the ratio $[A]_o/[B]_o = 0.370$. The temperature of the mixture is then raised to 350 K, at which point compound A begins to react to form products. The time dependence of the concentration of A is monitored spectroscopically by measuring the integrated absorbance at a wavelength at which only compound A absorbs. The data obtained are given in the table to the right and plotted in Figure 19.4.
(A) Determine from these data whether compound A is reacting with itself to form products or with compound B via the reaction $A + B \longrightarrow$ products. **(B)** Determine

Time (min)	Absorbance (arbitrary units); I_A
0.00	256.0
1.00	230.1
2.00	207.6
3.00	187.9
4.00	170.6
5.00	155.3
7.00	129.4
10.00	99.8
15.00	66.4

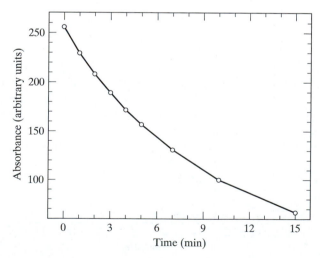

▲ FIGURE 19.4
The measured total absorbance as a function of time for the data in Example 19.2.

the specific rate coefficient for the reaction. Do we have enough data for this determination? If not, what additional data are needed?

Solution

(A) Let us begin by relating the integrated rate expressions to the measured absorbance data. The Beer–Lambert law (see Eq. 15.34) tells us that the total absorbance is proportional to the concentration of A. Therefore,

$$[A] = c\, I_A \tag{1}$$

where c is the constant of proportionality. Thus,

$$c = \frac{[A]_o}{[I_A]_o} = \frac{[A]_o}{256}, \tag{2}$$

and it follows that

$$[A] = \frac{[A]_o I_A}{256}. \tag{3}$$

Substituting Eq. 3 into Eq. 19.16 provides the integrated rate law for a bimolecular reaction involving only a single component in terms of the measured absorbance. The result is

$$\frac{1}{[A]} = \frac{256}{[A]_o I_A} = \frac{1}{[A]_o} + 2k(T)t. \tag{4}$$

Multiplying by $[A]_o/256$ gives

$$\frac{1}{I_A} = \frac{1}{256} + \frac{2[A]_o\, k(T)}{256}t. \tag{5}$$

If this equation describes the time dependence of the measured absorbance, a plot of $[I_A]^{-1}$ versus time should be linear. Figure 19.5 shows that it is not. Therefore, we know that the reaction does not involve compound A reacting with itself.

If the reaction is $A + B \longrightarrow$ products, a plot of $\ln([A]/[B])$ against time should be linear. To test this possibility, we need to express the ratio $[A]/[B]$ in terms of the measured absorbance. The number of mol L^{-1} of A that have reacted is

$$x = [A]_o - [A] = [A]_o - \frac{[A]_o I_A}{256} = [A]_o\left[1 - \frac{I_A}{256}\right]. \tag{6}$$

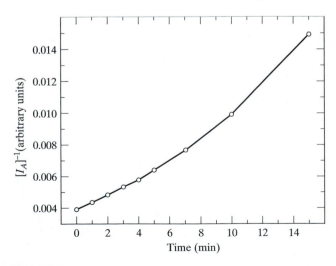

▲ **FIGURE 19.5**
A plot of $1/I_A$ versus time. Equation 5 demonstrates that if the reaction in question involves the reaction of compound A with itself, this plot would have to be linear.

The stoichiometry of the reaction tells us that every time 1 mole of A reacts, 1 mole of B also reacts. Consequently,

$$[B] = [B]_o - x = [B]_o - [A]_o\left[1 - \frac{I_A}{256}\right]. \tag{7}$$

But the ratio of $[A]_o$ to $[B]_o$ is 0.370, so that

$$[B] = \frac{[A]_o}{0.370} - [A]_o\left[1 - \frac{I_A}{256}\right] = [A]_o\left[\frac{1}{0.370} - 1 + \frac{I_A}{256}\right] = [A]_o\left[1.703 + \frac{I_A}{256}\right]. \tag{8}$$

The ratio $[A]/[B]$ can be expressed in terms of the absorbance using Eqs. 3 and 8:

$$\frac{[A]}{[B]} = \frac{\dfrac{[A]_o I_A}{256}}{[A]_o\left[1.703 + \dfrac{I_A}{256}\right]} = \frac{I_A}{256\left[1.703 + \dfrac{I_A}{256}\right]} = \frac{I_A}{436.0 + I_A}. \tag{9}$$

The measured absorbances, therefore, yield the following data:

Time (min)	$[A]/[B]$	$\ln\left[\dfrac{[A]}{[B]}\right]$	Time (min)	$[A]/[B]$	$\ln\left[\dfrac{[A]}{[B]}\right]$
0.00	0.3700	-0.9944	5.00	0.2626	-1.337
1.00	0.3454	-1.063	7.00	0.2289	-1.475
2.00	0.3226	-1.131	10.00	0.1863	-1.681
3.00	0.3012	-1.200	15.00	0.1322	-2.024
4.00	0.2812	-1.269			

Figure 19.6 shows a plot of $\ln([A]/[B])$ versus time. The linearity of the data plotted in this fashion is clearly excellent. The line in the figure is a least-squares fit to the points. The line has a slope of -0.06866 min^{-1}. We now need to relate this slope to $k(T)$.

Substituting Eq. 9 into Eq. 19.14 gives

$$\ln\left[\frac{[A]}{[B]}\right] = \ln\left[\frac{I_A}{436.0 + I_A}\right] = \ln\left[\frac{[A]_o}{[B]_o}\right] + \{[A]_o - [B]_o\}k(T)t$$

$$= \ln(0.370) + [A]_o\left[1 - \frac{[B]_o}{[A]_o}\right]k(T)t = \ln(0.370) - 1.703[A]_ok(T)t. \tag{10}$$

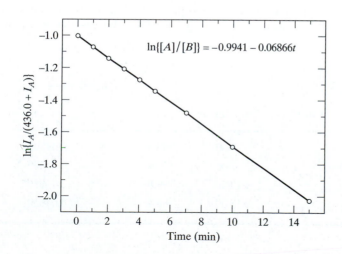

$$\ln\{[A]/[B]\} = -0.9941 - 0.06866t$$

◀ FIGURE 19.6
A plot of $\ln[I_A/(436.0 + I_A)]$ versus time. The text shows that this is equivalent to a plot of $\ln([A]/[B])$ versus time. The linearity of the result is excellent. A least-squares procedure shows that the best straight fit is $\ln[I_A/(436.0 + I_A)] = -0.9941 - (0.06866$ min$^{-1})t$. The linearity of the plot suggests that the reaction involves a second-order concerted process in which compound A reacts with compound B.

Therefore, the slope of the line is $-1.703[A]_o k(350\ K)$. The specific reaction rate coefficient at 350 K is

$$k(350\ K) = -\frac{slope}{1.703[A]_o} = \frac{0.06866\ min^{-1}}{1.703[A]_o} = \frac{0.0403}{[A]_o}\ L\ mol^{-1}\ min^{-1}. \tag{11}$$

Since the initial concentration of A is not given, Eq. 11 is the best we can do. If the initial concentration of A were known, it would be a simple matter to extract the rate coefficient from the measured absorbance data.

We can also use this example to illustrate the second method of analyzing rate data. If we solve Eq. 5 for $[A]_o k(T)$, we obtain

$$[A]_o k(T) = \frac{256[I_A]^{-1} - 1}{2t}. \tag{12}$$

This result permits us to compute the value of the product $[A]_o k(T)$ at each of the eight data points for times greater than zero. The results of this calculation are as follows:

Time (min)	I_A	$\dfrac{256[I_A]^{-1} - 1}{2t}$	Time (min)	I_A	$\dfrac{256[I_A]^{-1} - 1}{2t}$
1.00	230.1	0.05628	5.00	155.3	0.06484
2.00	207.6	0.05829	7.00	129.4	0.06988
3.00	187.9	0.06040	10.00	99.8	0.07826
4.00	170.6	0.06257	15.00	66.4	0.09518

If we have the correct integrated rate expression, the value of $[A]_o k(T)$ should be a constant that is independent of the time. The table, however, shows that this is not the case; the computed value of $[A]_o k(T)$ varies by 69.1% from $t = 1.00$ min to $t = 15.00$ min. This variation exceeds our expected experimental error in the data. Consequently, we can safely conclude that the reaction is not a concerted bimolecular process involving a single reactant.

The same procedure can be used to investigate the possibility that Eq. 19.14 is the correct integrated expression describing how the concentration of compound A varies with time. From Eq. 10, we obtain

$$[A]_o k(T) = \frac{\ln\left[\dfrac{0.370(436 + I_A)}{I_A}\right]}{1.703t}. \tag{13}$$

We now compute the value of the product $[A]_o k(T)$ at each of the eight data points for times greater than zero. The results of this calculation are shown in the following table:

Time (min)	I_A	$\dfrac{\ln\left[\dfrac{0.370(436 + I_A)}{I_A}\right]}{1.703t}$ (min^{-1})	Time (min)	I_A	$\dfrac{\ln\left[\dfrac{0.370(436 + I_A)}{I_A}\right]}{1.703t}$ (min^{-1})
1.00	230.1	0.04033	5.00	155.3	0.04025
2.00	207.6	0.04029	7.00	129.4	0.04030
3.00	187.9	0.04029	10.00	99.8	0.04030
4.00	170.6	0.04027	15.00	66.4	0.04030

The total variation in the product $[A]_o k(T)$ is 8 units in the fourth significant digit. This is probably well within our expected experimental error, so we can safely conclude that Eq. 19.14 accurately describes the temporal variation of $[A]$ and that the reaction is probably a concerted bimolecular reaction involving two reactants.

The best value of $[A]_o k(350\ \text{K})$ is obtained by averaging the computed values at each of the eight data points in the table. This average is 0.0403 min^{-1}, which is the same as the result obtained from the least-squares fit in Figure 19.6. The two methods are equivalent, provided that a least-squares procedure is used to obtain the best straight-line fit to the data.

For related exercises, see Problems 19.5, 19.6, and 19.7.

19.1.4   Pseudo-order Reactions

It sometimes happens that one of the reactants in a bimolecular concerted reaction is present in great excess. For example, if we are investigating the kinetics of a hydrolysis reaction of some compound in aqueous solution, it is clear that the concentration of H_2O greatly exceeds that of the compound being hydrolyzed. In such a situation, the concentration of H_2O will remain effectively constant throughout the reaction. This means that the differential rate expression will be altered. For a hydrolysis reaction of compound A, the bimolecular rate is

$$\mathcal{R} = -\frac{d[A]}{dt} = k(T)[H_2O][A].$$

However, since $[H_2O]$ is effectively a constant, we can write the rate expression in the form

$$\mathcal{R} = -\frac{d[A]}{dt} = k_p(T)[A], \tag{19.19}$$

where $k_p(T) = k(T)[H_2O]$. The reaction now appears to be an overall first-order process. It is, of course, still a concerted second-order reaction, but kinetically, it will behave as if it is first order. Under these conditions, we refer to the reaction as being *pseudo first order*. The prefix "pseudo" means "false." The new rate coefficient, $k_p(T)$, is called the *pseudo first-order reaction rate coefficient*. If the concentration of the reactant that is present in excess is known, $k(T)$ can always be computed from the measured value of $k_p(T)$.

The integrated rate law for a first-order process is simple to obtain from Eq. 19.19. Separating the variables and then integrating between corresponding limits yields

$$\int_{[A]_o}^{[A]} \frac{d[A]}{[A]} = \ln\left\{\frac{[A]}{[A]_o}\right\} = -\int_{t=0}^{t} k_p(T)dt = -k_p(T)t. \tag{19.20}$$

Note that the argument of the ln function is unitless, because the units on $d[A]$ and $[A]$ cancel. Expanding the logarithmic term permits us to write Eq. 19.20 in the form

$$\boxed{\ln[A] = \ln[A]_o - k_p(T)t}, \tag{19.21}$$

which tells us that a plot of $\ln[A]$ versus time for a pseudo first-order process will be linear with a slope equal to $-k_p(T)$. The temporal variation of the concentration of A is exponential with time:

$$\boxed{[A] = [A]_o \exp[-k_p(T)t]}. \qquad (19.22)$$

Usually, it is easier to deal with a first-order reaction than one that is second order. For this reason, experimentalists often choose the initial concentrations so that all but one reactant will be in great excess, thereby forcing the reaction to behave as a pseudo-order process of lower order.

EXAMPLE 19.3

Suppose an investigator is running the reaction examined in Example 19.2 at $T = 350$ K, but chooses her initial concentrations such that $[A]_o/[B]_o = 10^{-4}$ instead of 0.370, as in that example. Therefore, we expect the process to behave as a pseudo first-order reaction. Suppose further that the measured initial absorbance is still 256 arbitrary units. **(A)** Using the results obtained in Example 19.2, find the value of the pseudo first-order reaction rate coefficient. **(B)** Compute the absorbance that will be measured by the investigator as a function of time, and plot the results.

Solution

(A) The pseudo first-order reaction rate coefficient will be

$$k_p(350 \text{ K}) = k(350 \text{ K})[B]_o. \qquad (1)$$

Using the result obtained in Example 19.2, we have

$$k(350 \text{ K}) = \frac{0.04032}{[A]_o} \text{ L mol}^{-1}\text{ min}^{-1} = (0.04032 \text{ min}^{-1})\frac{1}{[A]_o} \qquad (2)$$

if we regard the units on $[A]_o$ to have been cancelled out. Combination of Eqs. 1 and 2 produces

$$k_p(350 \text{ K}) = (0.04032 \text{ min}^{-1})\frac{[B]_o}{[A]_o} = 0.04032 \times 10^4 \text{ min}^{-1} = 403.2 \text{ min}^{-1}. \qquad (3)$$

(B) The integrated rate law for the pseudo first-order reaction is given by Eq. 19.22, viz.,

$$\frac{[A]}{[A]_o} = \exp[-k_p(T)t] = \exp[-403.2t], \qquad (4)$$

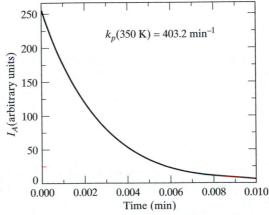

▲ FIGURE 19.7
The calculated absorbance in the pseudo first-order reaction described in Example 19.3. The pseudo first-order rate coefficient is 403.2 min^{-1}.

provided that t is expressed in minutes. Since $[A] = c\,I_A$, we obtain (see Eq. 3 in Example 19.2)

$$\frac{[A]}{[A]_o} = \frac{I_A}{256}. \tag{5}$$

Therefore, the measured absorbance will be

$$I_A = 256\exp[-403.2t]. \tag{6}$$

Figure 19.7 shows the result. Two things are obvious. First, the reaction is much faster, since we have increased the concentration of reactant B substantially. The entire process now takes only about 0.01 minute. As a result, some very fast measurement method will need to be employed. The second point is that the kinetics are much easier to analyze. Compare the ease with which this example was carried out relative to the way Example 19.2 was done.

For related exercises, see Problems 19.8 and 19.9.

19.1.5   Half-life and Relaxation Time

It is often convenient to discuss the kinetics of a reaction in terms of its *half-life* or its *relaxation time*. The former quantity is generally written $t_{1/2}$, while the notation for the latter is τ. Both of these quantities are measures of the reaction rate that are closely related to the specific reaction rate coefficient. The half-life is defined to be the time required for the concentration of a given reactant at time t_o to decrease to one-half of its original value. The relaxation time is similarly defined as the time needed for the original concentration to decrease to $1/e$ of its starting value at $t = t_o$, where e, the base of the natural-logarithm scale, is equal to 2.718281828.... These definitions are illustrated in Figure 19.8.

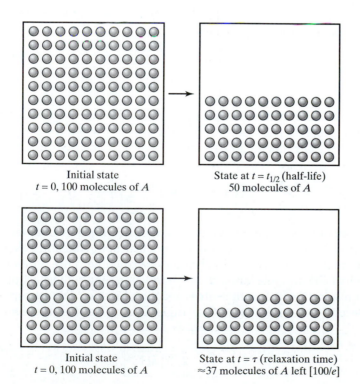

Initial state
$t = 0$, 100 molecules of A

State at $t = t_{1/2}$ (half-life)
50 molecules of A

Initial state
$t = 0$, 100 molecules of A

State at $t = \tau$ (relaxation time)
≈ 37 molecules of A left [100/e]

◀ **FIGURE 19.8**
Cartoon illustration of the half-life ($t_{1/2}$) and the relaxation time (τ) for a reaction involving only one compound, such as $A + A \longrightarrow$ products. When $t = t_{1/2}$, half of the molecules have reacted and half remain (50 in this case). When $t = \tau$, a fraction $1/e = 0.3678\ldots$ of the molecules remain (about 37 in this case). The rest have reacted. Since more molecules have undergone reaction at $t = \tau$ than at $t = t_{1/2}$, we can conclude that we always have $\tau > t_{1/2}$.

The relationship between $t_{1/2}$, τ, and $k(T)$ depends upon the type of reaction involved. If we are dealing with a pseudo first-order process, the integrated rate expression is given by Eq. 19.21:

$$\ln[A] = \ln[A]_o - k_p(T)t.$$

At time $t = 0$, $[A]$ is equal to $[A]_o$. At the point $t = t_{1/2}$, by definition, we will have $[A] = [A]_o/2$. Substituting this requirement into Eq. 19.21 produces

$$\ln\left[\frac{[A]_o}{2}\right] = \ln[A]_o - k_p(T)t_{1/2}.$$

Solving for the half-life, we obtain

$$t_{1/2} = \frac{\ln[A]_o - \ln\left[\dfrac{[A]_o}{2}\right]}{k_p(T)} = \frac{\ln\left[\dfrac{2[A]_o}{[A]_o}\right]}{k_p(T)} = \frac{\ln(2)}{k_p(T)}. \tag{19.23}$$

Equation 19.23 shows us that, for a first-order process, the half-life is inversely proportional to the specific reaction rate coefficient. This is expected behavior, since a reaction that goes at twice the speed should require only half the time to reach any given concentration. In fact, the half-life is always inversely proportional the rate coefficient for any type of reaction. One of the most important features of Eq. 19.23 is that it demonstrates that $t_{1/2}$ is independent of the starting concentration, $[A]_o$, if the process is first order. That is, if it takes 20 minutes for an initial concentration of 0.50 mol L^{-1} to decrease to 0.25 mol L^{-1}, it will take an additional 20 minutes for the 0.25 mol L^{-1} concentration to fall to 0.125 mol L^{-1}, provided that the kinetics are first order.

It is relatively easy (see Problem 19.10) to show that the relaxation time for a first-order reaction is given by

$$\tau = \frac{1}{k_p(T)}. \tag{19.24}$$

Like the half-life, τ is inversely proportional to the rate coefficient and independent of the initial concentration of the reactant when the process is first order. In terms of τ, the integrated rate expression for a first-order reaction is

$$[A] = [A]_o \exp\left[-\frac{t}{\tau}\right]. \tag{19.25}$$

We have already encountered an equation of this form in Chapter 16 when we discussed the relaxation of excited nuclear spin states in NMR experiments. In that case, the relaxation time was written with the symbol T_1 or T_2.

The half-life and relaxation time are independent of the starting concentration only in the case of first-order reaction kinetics. For example, a concerted second-order reaction with a single reactant has (see Problem 19.11)

$$t_{1/2} = \frac{1}{2[A]_o k(T)} \tag{19.26A}$$

and

$$\tau = \frac{e - 1}{2[A]_o k(T)} = \frac{0.8591409\ldots}{[A]_o k(T)}. \qquad \textbf{(19.26B)}$$

As expected, both $t_{1/2}$ and τ are inversely proportional to $k(T)$, but now they also depend inversely upon the starting concentration. Sometimes, we can conveniently determine the order of a reaction by measuring the dependence of the half-life or relaxation time on $[A]_o$.

EXAMPLE 19.4

An investigator measures the concentration of a reactant A as a function of time. Part of his data are as follows:

Time (min)	[A] (mol L⁻¹)
17.5	0.02566
39.6	0.01300
83.8	0.00650

Using the data alone, what can we say about the kinetics of the reaction?

Solution

We know the reaction cannot be a first-order process; this is clear from the fact that it requires about 22.1 minutes for $[A]$ to decrease by half from 0.02566 to 0.01300 mol L^{-1}, whereas to decrease by half from 0.01300 to 0.00650 mol L^{-1} requires 44.2 min. Therefore, $t_{1/2}$ is not independent of the starting concentration, as it must be if we have first-order kinetics.

It appears that the process is a concerted second-order reaction with a single reactant. If this is the case, Eq. 19.26A tells us that we will have

$$2[A]_o t_{1/2} = \frac{1}{k(T)}. \qquad \textbf{(1)}$$

At $t = 17.5$ min, when $[A]_o = 0.02566$ mol L^{-1}, $t_{1/2}$ is about 22.1 min, so that

$$2(0.02566 \text{ mol L}^{-1})(22.1 \text{ min}) = 1.13 \text{ mol min L}^{-1} = \frac{1}{k(T)}, \qquad \textbf{(2)}$$

which implies that $k(T) \approx 0.885$ L mol^{-1} min^{-1}. If this is correct, then the half-life, starting with the concentration $[A]_o = 0.01300$ mol L^{-1} at $t = 39.6$ min, should be

$$t_{1/2} = \frac{1}{2[A]_o k(T)} = \frac{1}{2(0.01300 \text{ mol L}^{-1})(0.885 \text{ L mol}^{-1} \text{min}^{-1})} \approx 43.5 \text{ min.} \qquad \textbf{(3)}$$

The measured half-life starting at $t = 39.6$ min is the difference between 83.8 and 39.6 min, or 44.2 min. This is very close to our estimate of 43.5 min. To be more confident of that estimate, we need more data, but it does appear that $t_{1/2}$ is inversely proportional to $[A]_o$, which would make the kinetics that for a second-order concerted process involving a single reactant.

For related exercises, see Problems 19.10, 19.11, and 19.12.

19.1.6   Temperature Dependence of the Rate Coefficient

Equation 19.1 shows that reaction rate coefficients are dependent on temperature. We have recognized this fact by writing the rate coefficient as $k(T)$ rather than as k, which would suggest that it is a constant. In Chapter 20, we shall explicitly derive the functional form of the temperature dependence. At this point, however, the nature of that dependence will be examined from a more qualitative point of view.

If deuterium atoms are mixed with H_2 molecules at some elevated temperature, the exchange reaction $H_2 + D \longrightarrow HD + H$ will be observed. Since there is no stable H_2D molecule, it is qualitatively clear that before the exchange reaction can occur, we must insert energy into the H_2 molecule to begin the process of H–H bond scission. If the reaction mechanism required complete dissociation of the H_2 bond prior to formation of the HD bond, we would have to insert the H_2 bond dissociation energy, which Table 15.4 shows to be 4.476 eV, or 431.9 kJ mol^{-1}, before any reaction could occur. However, it is not necessary that we completely dissociate the H–H bond before H–D bond formation begins. The formation of the H–D bond releases energy, so that the overall energy requirement for the process is much less than the total H–H dissociation energy.

Figure 19.9 describes semiquantitatively the overall energy changes during the $H_2 + D$ exchange reaction. The ordinate gives the energy of the system in excess of that for ground-state reactants without zero-point vibrational energy. The abscissa is called the *reaction coordinate*. Qualitatively, this coordinate measures the progress of the system as it moves from a configuration that characterizes the reactants to one characteristic of the products. In Chapter 20, we shall explicitly define the reaction coordinate. For the moment, however, we shall simply regard it as a measure of the progress of the reaction, without defining it explicitly. As expected, the figure shows that energy insertion is a prerequisite for the exchange reaction. Initially, the energy of the system rises from zero up to a maximum around 9.6 kcal mol^{-1} ($\approx$40.2 kJ mol^{-1}). Beyond this point, the system energy decreases monotonically until the product configuration is attained. When the nuclear kinetic-energy terms are omitted from the electronic Hamiltonian (via a Born–Oppenheimer approximation), the

▶ **FIGURE 19.9**
The energy variation during the bimolecular isotopic exchange reaction between $H_2(g)$ and $D(g)$ to give $HD(g) + H(g)$. The abscissa, which we call the reaction coordinate, measures the progress of a reactant system as it moves from a configuration characteristic of $H_2 + D$ to one that resembles HD + H. In Chapter 20, this coordinate will be defined in a more rigorous manner. The barrier to reaction for this process is 9.6 kcal mol^{-1}, or 40.2 kJ mol^{-1}.

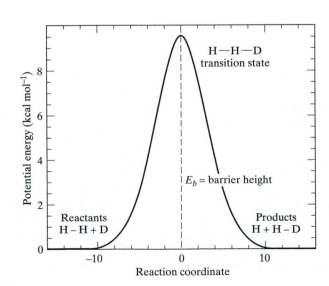

electronic energy is no longer dependent upon isotopic mass. Therefore, the ground-state electronic energy of HD is the same as that of H_2, and the reaction is thermochemically neutral except for differences in zero-point vibrational energy. The nuclear configuration that exists at the maximum of this energy profile is called the *transition state*. Its energy, minus that of the reactants, gives us the *barrier height* for the reaction process. In this case, the barrier is about 40.2 kJ mol^{-1}, 9.31% of that which would be needed if complete H–H dissociation were required for exchange.

Figure 19.10 shows a similar energy profile for a hypothetical exothermic reaction. In this case, the electronic energy of the products, without including the zero-point vibrational energy, is taken as the energy zero. Again, we see that some energy must be inserted to begin reorganizing the nuclear positions required for the reaction. In this case, the energy crests at 35.0 kJ mol^{-1} and then decreases monotonically to zero as the products are formed. The potential-energy barrier to reaction is, therefore, 25.0 kJ mol^{-1}. Since the reaction is exothermic, the product energy lies below that of the reactants, with the difference representing ΔU for the reaction if zero-point vibrational energy is ignored.

With an energy barrier to reaction, we expect that the probability of reaction upon collision will be proportional to the probability that the colliding system will possess sufficient energy to traverse the barrier. Equation 19.1 shows that $k(T)$ is dependent upon the probability of reaction. Therefore, it is reasonable to anticipate that $k(T)$ will be proportional to the probability that the system possesses energy equal to or greater than some minimum amount required to cross the potential barrier. We label this energy requirement E_a and call it the *activation energy* of the reaction.

In Eq. 17.69, we found that the probability that we will observe an energy equal to or greater than E_a for a system in Maxwell–Boltzmann equilibrium is

$$P(E \geq E_a) = KT^{-1/2} \exp\left[-\frac{E_a}{RT}\right], \tag{19.27}$$

provided that $E_a \gg RT$. If the temperature range being examined in the experiments is not very large, the factor $T^{-1/2}$ is nearly constant, so it can be

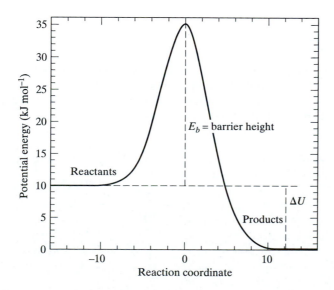

◀ **FIGURE 19.10**
The energy variation during a hypothetical exothermic bimolecular reaction whose potential barrier is 25 kJ mol^{-1} and whose exothermicity is 10 kJ mol^{-1}.

included in K. Therefore, we expect the temperature dependence of the specific reaction rate coefficient to be given approximately by

$$k(T) \approx A \exp\left[-\frac{E_a}{RT}\right], \tag{19.28}$$

where A is a constant called the *frequency factor*. Equation 19.28 is the *Arrhenius equation*. In precise work covering wide temperature ranges, an extended form of the equation is sometimes employed. This form incorporates the preexponential temperature factor by writing

$$k(T) \approx AT^n \exp\left[-\frac{E_a}{RT}\right], \tag{19.29}$$

where the exponent n is usually employed as an adjustable parameter to fit the measured data.

   If the rate coefficient has been measured at several temperatures, Eq. 19.28 shows that a plot of $\ln k(T)$ versus T^{-1} will be linear with a slope equal to $-E_a/R$ and an intercept equal to $\ln A$. Such a representation of the data is called an *Arrhenius plot*.

   The experimentally determined activation energy for a given reaction is usually not the same as the corresponding potential-energy barrier. The reason for the difference is the effect of the average internal and translational energies in the reactants and transition-state complex. If we denote this average as $\langle E_{int} + E_{tr} \rangle$, the activation energy is given approximately by

$$E_a \approx E_b + \langle E_{int} + E_{tr} \rangle_{ts} - \langle E_{int} + E_{tr} \rangle_r \tag{19.30}$$

where E_b is the reaction potential barrier and the subscripts ts and r represent the transition state and reactants, respectively. In most cases, the largest differences between $\langle E_{int} + E_{tr} \rangle_{ts}$ and $\langle E_{int} + E_{tr} \rangle_r$ arise from variations in the vibrational zero-point energies of the two configurations. These differences, although not negligible, are often small relative to E_b, so that it is common practice to think of the measured activation energy as being nearly equal to the reaction barrier.

EXAMPLE 19.5

An investigator measures the half-life of a pseudo first-order reaction at 320 K and finds it to be 111.6 minutes. A second investigator measures the relaxation time of the same pseudo first-order reaction at 340 K and obtains 52.06 minutes. Compute the activation energy for the process.

Solution

Since the process is pseudo first order, the half-life and relaxation time are related to the pseudo first-order reaction rate coefficient by

$$t_{1/2} = \frac{\ln(2)}{k_p(T)} \tag{1}$$

and

$$\tau = \frac{1}{k_p(T)}, \tag{2}$$

respectively. Therefore, the rate coefficients at 320 K and 340 K are

$$k_p(320 \text{ K}) = \frac{\ln(2)}{111.6 \text{ min}} = 0.006211 \text{ min}^{-1} \qquad (3)$$

and

$$k_p(340 \text{ K}) = \frac{1}{52.06 \text{ min}} = 0.01921 \text{ min}^{-1}. \qquad (4)$$

The temperature range is sufficiently narrow that the simple Arrhenius expression should be accurate enough. Therefore, we have

$$k_p(320 \text{ K}) = A \exp\left[-\frac{E_a}{320R}\right] = 0.006211 \text{ min}^{-1} \qquad (5)$$

and

$$k_p(340 \text{ K}) = A \exp\left[-\frac{E_a}{340R}\right] = 0.01921 \text{ min}^{-1}. \qquad (6)$$

Dividing Eq. 6 by Eq. 5 produces

$$\frac{A \exp\left[-\dfrac{E_a}{340R}\right]}{A \exp\left[-\dfrac{E_a}{320R}\right]} = \exp\left[-\frac{E_a}{R}\left\{\frac{1}{340} - \frac{1}{320}\right\}\right] = \frac{0.01921}{0.006211} = 3.093. \qquad (7)$$

Taking logarithms of both sides gives

$$-\frac{E_a}{R}\left\{\frac{1}{340} - \frac{1}{320}\right\} = \ln(3.093). \qquad (8)$$

Solving for E_a, we obtain

$$E_a = -\frac{R \ln(3.093)}{\left\{\dfrac{1}{340} - \dfrac{1}{320}\right\}} = -\frac{8.314 \ln(3.093)}{\left\{\dfrac{1}{340} - \dfrac{1}{320}\right\}} \text{ J mol}^{-1} = 5.107 \times 10^4 \text{ J mol}^{-1}$$

$$= 51.07 \text{ kJ mol}^{-1}. \qquad (9)$$

For related exercises, see Problems 19.13, 19.14, and 19.15.

19.2   Unimolecular Reactions

Many first-order reactions are actually second-order reactions that have been forced to exhibit apparent first-order kinetic behavior by our choice of the initial concentrations. We term these processes *pseudo first-order reactions*. There is, however, a group of reactions involving a single reactant that exhibits true first-order kinetics. These unimolecular reactions generally involve molecules that, for some reason, reside in a metastable state, so that conversion reactions to more stable states can be observed. We have already encountered several examples of such processes. Both fluorescence and phosphorescence emission, discussed in Chapter 15, are unimolecular reactions in which molecules in excited electronic states decay into various vibrational levels of the electronic ground state. The unimolecular relaxation of molecules in excited nuclear spin states are responsible for the exponential decay of the FID signal in an FT NMR spectrum (see Chapter 16). Many nuclei are inherently unstable and consequently undergo unimolecular nuclear reactions of various types. In this section, we examine the kinetics of these reactions.

19.2.1   Random Lifetime Processes

We can gain an understanding of the kinetics describing unimolecular reactions by considering a simple model. Suppose we have an ideal gas confined inside a rectangular parallelepiped container at temperature T and a concentration of N molecules m^{-3}. Random collisions between these molecules will rapidly establish a Maxwell–Boltzmann distribution of molecular speeds. Let us now imagine that we cut a small hole of area A in one side of the container. The hole provides a pathway whereby the molecules inside the box can undergo the "unimolecular reaction" of escaping from the box through the hole. This model is illustrated in Figure 19.11.

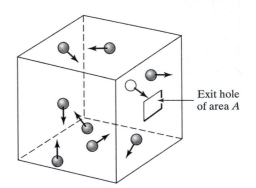

Exit hole of area A

▶ **FIGURE 19.11**
A simple unimolecular "reaction" exhibiting random lifetimes. The box contains molecules of an ideal gas in random motion with a Maxwell–Boltzmann distribution of speeds. One face of the box contains an exit hole of area A.

Our concern here is the rate of escape of molecules from the box. Suppose we were able to focus our attention on one molecule—say, the unshaded molecule in the figure, and follow its motion as it collides with other molecules and the walls of the container. It is clear that, since the motion of the molecule is random, its lifetime inside the box will also be random. It might find the hole and exit the box within the first few nanoseconds after the hole is opened. On the other hand, it might bounce futilely around the box for seconds or minutes without encountering the exit hole. The same is true for all molecules within the container. The result of these random lifetimes is that at a fixed temperature, area of the hole, molecular mass, and molecular concentration, the number of molecules that exit the box in a given time interval Δt is proportional to that interval. That is, twice as many molecules will escape in 10 nanoseconds as will escape in 5 nanoseconds. If ΔN is the change in the number of molecules inside the box during the interval Δt, then $\Delta N \propto -\Delta t$, so that

$$\Delta N = -c\,\Delta t, \tag{19.31}$$

where the negative sign appears because N decreases with time. The proportionality constant, c, will depend upon the area of the hole, the temperature, the mass of the molecules, and the concentration of molecules inside the container. The law of mass action tells us that the frequency with which molecules collide with the exit hole is proportional to the concentration. Therefore, c must be proportional to N. This permits us to write Eq. 19.31 in the form

$$\Delta N = -kN\,\Delta t. \tag{19.32}$$

If we now take the time interval to be infinitesimally small, so that $\Delta t \rightarrow dt$ and $\Delta N \rightarrow dN$, Eq. 19.32 becomes

$$dN = -kN\,dt. \tag{19.33}$$

Equation 19.33 is a differential equation describing a first-order reaction. Integrating both sides between corresponding limits produces

$$\int_{N_o}^{N} \frac{dN}{N} = -k \int_{t=0}^{t} dt,$$

or

$$\boxed{\ln(N) - \ln(N_o) = -kt}, \qquad\qquad \textbf{(19.34A)}$$

or, equivalently,

$$\boxed{N = N_o \exp[-kt]}, \qquad\qquad \textbf{(19.34B)}$$

and we have a true first-order process. So long as the lifetimes of molecules undergoing unimolecular reactions are random, the kinetics will always be described by a first-order rate law.

EXAMPLE 19.6

Carbon-14 (^{14}C) is unstable because the neutron–proton (n/p) ratio in the nucleus exceeds the stable range for this ratio for carbon. The nucleus can be stabilized by beta emission. A beta particle ($_{-1}^{0}\beta$) is an electron ejected from a neutron, with a proton forming in the process. The overall reaction is $_{0}^{1}n \longrightarrow {}_{-1}^{0}\beta + {}_{1}^{1}p$. When this unimolecular reaction occurs for $_{6}^{14}$C, we obtain $_{6}^{14}\text{C} \longrightarrow {}_{-1}^{0}\beta + {}_{7}^{14}\text{N}$. The half-life of $_{6}^{14}$C is 5,715 years. In the upper atmosphere, $_{6}^{14}$C is formed by neutron capture by $_{7}^{14}$N via the reaction $_{7}^{14}\text{N} + {}_{0}^{1}n \longrightarrow {}_{6}^{14}\text{C} + {}_{1}^{1}p$. The result of these two competing processes is a near-constant concentration of $_{6}^{14}$C in the atmosphere. The $_{6}^{14}$C is incorporated into $CO_2(g)$, which is then absorbed by plants and converted into more complex carbon molecules through photosynthesis. These molecules are eaten by animals, which causes their tissues to contain an amount of $_{6}^{14}$C that makes their $_{6}^{14}\text{C}-{}_{6}^{12}\text{C}$ ratio nearly identical to that in the atmosphere. However, once the plant or animal dies, it ceases to replenish the $_{6}^{14}$C in its system. The $_{6}^{14}$C continues to decay, so that the $_{6}^{14}\text{C}-{}_{6}^{12}\text{C}$ ratio decreases after death. This means we can date a once-living plant or animal by measuring its $_{6}^{14}\text{C}-{}_{6}^{12}\text{C}$ ratio.

A wooden art object claimed to have been found in an Egyptian pyramid is offered for sale to an art museum. You are retained to advise the governing board of the museum on this purchase. You find that nondestructive radiocarbon dating of the object reveals a beta emission rate that is 61.5% that of a sample taken from a living tree. What is your advice to the board concerning the purchase? Explain.

Solution

The emission rate is given by Eq. 19.33:

$$\frac{dN}{dt} = -kN. \qquad\qquad \textbf{(1)}$$

Since the emission rate of the wooden art object is 61.5% that of living wood, the ratio of the number of $_{6}^{14}$C atoms in the art object to that in a living sample is

$$\frac{\left(\dfrac{dN}{dt}\right)_{art}}{\left(\dfrac{dN}{dt}\right)_{living}} = \frac{-kN_{art}}{-kN_{living}} = \frac{N_{art}}{N_{living}} = 0.615. \qquad\qquad \textbf{(2)}$$

This means that the wood in the art object has been dead for a time sufficient for the number of $^{14}_{6}C$ atoms to decrease from the amount in the object when it was alive to 0.615 of this amount. Therefore, if t^* is the time elapsed since death of the tree from which the object was made, Eq. 19.34B tells us that we must have

$$N_{art} = N_{living} \exp[-kt^*]. \tag{3}$$

Solving for t^*, we obtain

$$t^* = -\frac{\ln[N_{art}/N_{living}]}{k} = -\frac{\ln(0.615)}{k}. \tag{4}$$

For a first-order reaction, the specific reaction rate coefficient is related to the half-life by

$$k = \frac{\ln(2)}{t_{1/2}}. \tag{5}$$

Combining Eqs. 4 and 5 gives us

$$t^* = -\frac{t_{1/2}\ln(0.615)}{\ln(2)} = -\frac{(5{,}715 \text{ years})\ln(0.615)}{\ln(2)} \approx 4{,}008 \text{ years}. \tag{6}$$

Therefore, it appears that the tree from which the wooden object was made died slightly over 4,000 years ago. The Egyptian pyramids were constructed between the second and the twelfth Dynasties ($\approx$4,800 B.C. to 3,000 B.C.). The oldest is the pyramid at Medum, which was constructed during the reign of Snefenu around 4,750 B.C. In view of these data, you confidently advise the governing board of the museum that the wooden object was not placed in the pyramid at the time of entombment of the reigning pharaoh. It is possible that the object was placed in the pyramid between 1,000 to 2,700 years subsequent to the pharaoh's entombment. Since the object is 4,000 years old, it may very well be a valuable acquisition for the museum. This is a matter you advise them to consider in consultation with competent archaeologists and the financial advisors for the museum.

For related exercises, see Problems 19.16, 19.17, and 19.18.

19.2.2   Distribution of Lifetimes

Let us now examine the distribution of lifetimes for molecules undergoing first-order reactions. The question we wish to answer is "What is the probability that a randomly chosen molecule will exhibit a lifetime in the range t_L to $t_L + dt_L$ if the system is undergoing a first-order reaction?"

In order to exhibit a lifetime t_L, the molecule must not react during the interval from $t = 0$ to $t = t_L$. From Eq. 19.34B, we see that the fraction f of molecules that do not react during this interval is

$$f = \frac{N(t_L)}{N_o} = \exp[-kt_L], \tag{19.35}$$

where $N(t_L)$ is the number of molecules that have not reacted at time t_L. Therefore, the probability that reaction will not occur is f. After having survived to this point, the molecule must now undergo a reaction in the interval between t_L and $t_L + dt_L$. The number of molecules left at time $t_L + dt_L$ is

$$N(t_L + dt_L) = N_o \exp[-k(t_L + dt_L)]. \tag{19.36}$$

The fraction of the molecules remaining at $t = t_L$ that have not reacted at $t_L + dt_L$ is $N(t_L + dt_L)/N(t_L)$. Therefore, the probability that *no* reaction will occur in the time interval dt_L is

$$P(\text{no reaction in time } dt_L) = \frac{N(t_L + dt_L)}{N(t_L)} = \frac{N_o \exp[-k(t_L + dt_L)]}{N_o \exp[-kt_L]}$$

$$= \exp[-k\,dt_L]. \tag{19.37}$$

Since the probability that a reaction will occur is one minus the probability that none will occur, we have

$$P(\text{reaction in time } dt_L) = 1 - P(\text{no reaction in time } dt_L) = 1 - \exp[-k\,dt_L].$$

$$\tag{19.38}$$

The combined probability that a particular molecule will survive up to time t_L and then undergo a reaction in the interval t_L to $t_L + dt_L$ is the product of the individual independent probabilities. This tells us that the probability of observing a lifetime in the range t_L to $t_L + dt_L$ is

$$P(t_L) = f P(\text{reaction in time } dt_L) = f[1 - \exp(-k\,dt_L)].$$

Since the time interval dt_L is infinitesimally small, we can expand the exponential in a power series and truncate it after the linear term. This operation produces

$$P(t_L) = f\left[1 - \left\{1 - k\,dt_L + \frac{(k\,dt_L)^2}{2!} - \frac{(k\,dt_L)^3}{3!} + \cdots\right\}\right] = f k\,dt_L.$$

Using Eq. 19.35, we obtain

$$\boxed{P(t_L) = k \exp[-kt_L]dt_L}. \tag{19.39}$$

Equation 19.39 provides the distribution of lifetimes.

EXAMPLE 19.7

Determine the average lifetime in terms of the specific reaction rate coefficient for molecules undergoing a first-order reaction.

Solution

The average, or mean, lifetime is the first moment of the probability distribution. That is,

$$\langle t_L \rangle = \int_{t_L=0}^{\infty} k t_L \exp[-kt_L]dt_L = k \int_{t_L=0}^{\infty} t_L \exp[-kt_L]dt_L. \tag{1}$$

The integral has the standard form we encountered frequently in Chapter 13:

$$\int_0^{\infty} x^n \exp[-ax]dx = \frac{n!}{a^{n+1}} \qquad \text{for integer values of } n. \tag{2}$$

Thus,

$$\boxed{\langle t_L \rangle = k\,\frac{1!}{k^2} = \frac{1}{k} = \tau}. \tag{3}$$

Equation 3 is the solution to the example. It also demonstrates an important point: The relaxation time τ for a first-order reaction is the average lifetime of the molecules undergoing reaction.

For related exercises, see Problems 19.19, 19.23, 19.24, 19.29, and 19.30.

19.3   Termolecular Reactions

Concerted termolecular reactions involve the simultaneous or near-simultaneous collision of three molecules, followed by the direct generation of products. That termolecular reactions are rare is a consequence of the fact that three-body collisions are very unlikely.

In principle, there are three distinct types of termolecular reactions: The three colliding molecules can all be different, two of them can be the same, or all can be identical. That is,

$$A + B + C \longrightarrow \text{Products}, \quad A + A + B \longrightarrow \text{Products},$$

$$\text{and } A + A + A \longrightarrow \text{Products}$$

are all termolecular reactions. In practice, there are very few processes in which the three molecules are different; in most cases, either two or three are the same.

The most common example of a termolecular reaction is atomic recombination, such as $I + I + M \longrightarrow I_2 + M$. Without the presence of the third body represented by M, the two atoms cannot recombine. The energy released upon bond formation will cause immediate dissociation back to atoms, unless there is a third body to remove a portion of that energy. The third body can be another atom of iodine. In a very real sense, the third body serves as a *catalyst*, which is generally defined to be a substance that enhances the reaction rate without being consumed. When larger radicals recombine, there is usually no need for the presence of a third body, since the large number of internal degrees of freedom can serve as energy sinks that effectively transport the recombination energy away from the re-forming chemical bond and thereby prevent its immediate dissociation. Subsequent collisions of the recombined molecule result in stabilization.

Only a few wholly chemical termolecular reactions have been studied extensively. The reactions

$$2\,NO(g) + O_2(g) \longrightarrow 2\,NO_2(g),$$

$$2\,NO(g) + Cl_2(g) \longrightarrow 2\,NOCl(g),$$

and

$$2\,NO(g) + Br_2(g) \longrightarrow 2\,NOBr(g)$$

all appear to be third-order termolecular reactions. However, it is possible that each involves two bimolecular steps. Although integrated rate expressions for such third-order reactions can be obtained, they are not particularly useful. The analysis of the rate of a termolecular reaction usually involves the use of pseudo-order methods. Problem 19.28 is an example.

19.4   Multiple-Reaction Processes

Many chemical processes involve several reactions that occur either simultaneously, sequentially, or both. The ensemble of reactions that eventually lead to the final products is often called the *reaction mechanism*. Since the kinetic behavior of the combined reactions in a mechanism must be consistent with the measured data, experimental investigations of reaction rates can provide important clues about the reaction mechanism. If we know all the chemical reactions and their associated rate coefficients for a particular chemical process, we can always predict the temporal variations of the concentrations for all species. However, it is not always possible to do the reverse; that is, if we have measured the concentrations as functions of time, it does not follow that we can always determine the reaction mechanism. Nevertheless, this frequently can be accomplished. The possibility provides strong motivation for kinetic studies.

In this section, we consider some multiple-reaction systems whose integrated rate laws can be obtained in closed analytic form. These results will permit us to devise some approximation methods that can be used to treat more complex systems for which we cannot obtain analytic solutions to the differential rate equations.

19.4.1   Opposing Reactions

Opposing reactions that eventually lead to equilibrium frequently occur in multiple-reaction systems. The simplest such system is two first-order opposing reactions

$$A \xrightarrow{k_1} B \quad \text{and} \quad B \xrightarrow{k_2} A$$

Reaction 1   Reaction 2

A and B might, for example, represent the "boat" and "chair" forms of cyclohexane shown in Figure 19.12.

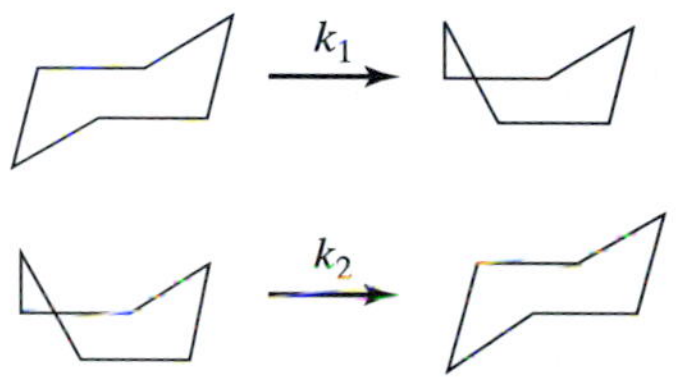

▶ FIGURE 19.12
Simple first-order opposing reactions.

We can obtain the differential equations for the mechanism of two first-order opposing reactions by writing the appropriate rate equations for the individual reactions. The rate expressions for Reaction 1 are

$$\left(\frac{d[A]}{dt}\right)_1 = -k_1[A] \tag{19.40A}$$

and

$$\left(\frac{d[B]}{dt}\right)_1 = k_1[A], \tag{19.40B}$$

since B is formed in the reaction while A is consumed. For Reaction 2, the corresponding rate expressions are

$$\left(\frac{d[A]}{dt}\right)_2 = k_2[B] \tag{19.41A}$$

and

$$\left(\frac{d[B]}{dt}\right)_2 = -k_2[B]. \tag{19.41B}$$

The total rate of change for a given compound is the sum of the rates for the reactions involving that compound. Thus, to obtain $(d[A]/dt)_{total}$, we simply sum $(d[A]/dt)_1$ and $(d[A]/dt)_2$. This gives

$$\left(\frac{d[A]}{dt}\right)_{total} = -k_1[A] + k_2[B]. \tag{19.42A}$$

A similar operation gives us the total rate of change for compound B:

$$\left(\frac{d[B]}{dt}\right)_{\text{total}} = k_1[A] - k_2[B].$$ (19.42B)

Since each compound is involved in two reactions, the expressions for the total rates of change of $[A]$ and $[B]$ both contain two terms. Equations 19.42A and 19.42B are the differential rate equations for the system. If we know the reaction mechanism, such a set of differential equations can always be developed.

It is relatively easy to obtain the integrated rate equations for this system. If we define x as the amount of A consumed and the amount of B produced, we have $[A] = [A]_o - x$, $[B] = [B]_o + x$, and $d[A] = -dx$. Substituting these transformation equations into Eq. 19.42A leads to the result (see Problem 19.21)

$$\ln\{k_1[A] - k_2[B]\} = \ln\{k_1[A]_o - k_2[B]_o\} - (k_1 + k_2)t.$$ (19.43)

Even though Eq. 19.43 is relatively simple, it is much more difficult to obtain the rate coefficients by analyzing kinetic data in multiple-reaction systems. In this case, the integrated rate expression tells us that a plot of $\ln\{k_1[A] - k_2[B]\}$ versus t will be linear with slope $k_1 + k_2$. However, this information is not particularly useful, since there is no way to prepare such a plot without knowing k_1 and k_2 in advance.

Because the opposing reactions will eventually reach equilibrium, we can use the equilibrium constant to assist in the data analysis. At equilibrium, the rates of the two opposing reactions must be equal, so that, from Eqs. 19.42A and 19.42B, we have

$$-k_1[A]_{\text{eq}} + k_2[B]_{\text{eq}} = k_1[A]_{\text{eq}} - k_2[B]_{\text{eq}}.$$ (19.44)

Rearranging this equation gives

$$\frac{[B]_{\text{eq}}}{[A]_{\text{eq}}} = \frac{k_1}{k_2}.$$ (19.45)

In Chapters 5, 8, and 9, we learned that the thermodynamic equilibrium constant is given by

$$K_{\text{eq}} = \frac{(a_B)_{\text{eq}}}{(a_A)_{\text{eq}}},$$ (19.46A)

where $(a_A)_{\text{eq}}$ and $(a_B)_{\text{eq}}$ are the activities of compounds A and B, respectively, at equilibrium. If we take molarity to be our reference function and infinite dilution to be the reference state, Eq. 19.46A can be written in terms of the activity coefficients γ_A and γ_B:

$$K_{\text{eq}} = \frac{(\gamma_B[B])_{\text{eq}}}{(\gamma_A[A])_{\text{eq}}}.$$ (19.46B)

In almost all cases, kinetic measurements are conducted in very dilute solutions. This choice is dictated by the exponential temperature dependence of the rate coefficient. If measurements are made using concentrated solutions, it becomes extremely difficult to maintain constant temperature conditions because of the endo- or exothermicity of the reactions. In effect, the use of concentrated solutions produces a large heat transfer problem that can be avoided simply by employing dilute solutions. When the solutions are dilute, we are approaching the reference state of infinite dilution in which the activity coefficients all approach unity. Under these conditions, Eqs. 19.45 and 19.46B give

$$K_{eq} = \frac{[B]_{eq}}{[A]_{eq}} = \frac{k_1}{k_2}. \qquad (19.47)$$

Since the forward and reverse rates will always be equal at the equilibrium point regardless of the concentrations used in the experiments, the preceding analysis tells us that the rate expressions in Eqs. 19.42A and B should actually be expressed in terms of the activities or fugacities of the reactants rather than their concentrations or pressures. To give the correct expression for the equilibrium constant, the rate equations must have the forms

$$\left(\frac{d[A]}{dt}\right)_{total} = F([A], [B])[-k_1 a_A + k_2 a_B]$$

and

$$\left(\frac{d[B]}{dt}\right)_{total} = F([A], [B])[k_1 a_A - k_2 a_B],$$

where $F([A], [B])$ is an unknown function of the concentrations of A and B. When we equate these expressions at equilibrium, the F functions cancel, and we obtain

$$\frac{k_1}{k_2} = \frac{(a_B)_{eq}}{(a_A)_{eq}} = K_{eq}.$$

If we actually had to employ activities in kinetics studies, the data would be extremely difficult to analyze. Fortunately, as long as we work with dilute solutions, activities can usually be replaced with concentrations. Ionic reactions are sometimes an exception: In Chapter 9, we found that even at very low concentrations, they deviate significantly from ideality; consequently, for such reactions, it may be necessary to work with activities.

Using Eq. 19.47, we can put the integrated rate expression for first-order opposing reactions in a more convenient form. Rearrangement of Eq. 19.43 produces

$$\ln\left\{\frac{k_1[A] - k_2[B]}{k_1[A]_o - k_2[B]_o}\right\} = -(k_1 + k_2)t.$$

Dividing the numerator and denominator of the argument of the logarithm function by k_2 gives

$$\ln\left\{\frac{K_{eq}[A] - [B]}{K_{eq}[A]_o - [B]_o}\right\} = -(k_1 + k_2)t,$$

so that

$$\ln\{K_{eq}[A] - [B]\} = \ln\{K_{eq}[A]_o - [B]_o\} - (k_1 + k_2)t. \qquad (19.48)$$

If the equilibrium constant is known or can be obtained from the data, we can plot $\ln\{K_{eq}[A] - [B]\}$ versus time. The result will be a straight line with slope equal to $-(k_1 + k_2)$. Since the ratio k_1/k_2 is known from the equilibrium

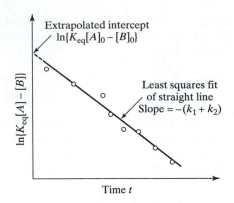

▶ **FIGURE 19.13**

Illustration of the analysis of two opposing first-order reactions, using the measured equilibrium constant $K_{eq} = k_1/k_2$. The figure shows the type of plot to be made and the manner in which the results are utilized to obtain the rate coefficients. (See text for details.)

Data analysis

Since $K_{eq} = k_1/k_2$, we have

$$\text{Slope} = -(k_1 + k_2) = (K_{eq}k_2 + k_2) = -k_2(K_{eq} + 1)$$

so that

$$k_2 = -(\text{slope})/(K_{eq} + 1)$$

constant, the determination of the slope provides values for both rate coefficients, as illustrated in Figure 19.13. Problem 19.22 serves as an example of such an analysis.

Although integrated rate expressions can be obtained for more complex opposing reactions, such as two concerted second-order reactions

$$A + B \xrightarrow{k_1} C + D \qquad \text{and} \qquad C + D \xrightarrow{k_2} A + B,$$

the rate expressions are so complex that it is usually better to analyze these reactions by means of the relaxation methods discussed in Section 19.7.

19.4.2 Concurrent Reactions

When a reactant is undergoing two or more simultaneous reactions, the mechanism is said to involve *concurrent reactions*. If vinyl bromide is photolyzed so that its internal energy is raised to 6.44 eV (621.4 kJ mol^{-1}) in excess of the zero-point energy, the decomposition reactions shown in Figure 19.14 all occur. Each of the reactions is a first-order process, but the presence of five such reactions makes the overall decomposition kinetics more complex.

Let us first consider the simple case of two concurrent first-order relaxation processes

$$A \xrightarrow{k_1} B$$

and

$$A \xrightarrow{k_2} C$$

We encountered an example of such a system when we considered the relaxation of excited nuclear spin states via two different processes. In the first of these, spin–lattice relaxation, the fluctuating local magnetic fields arising from molecular motion induce stimulated emission with a relaxation time T_1. The second process, called spin–spin relaxation, involves the redistribution of individual spin magnetic moments as they precess about the magnetic axis. This relaxation process has a characteristic relaxation time T_2. The differential rate equation for the first process is

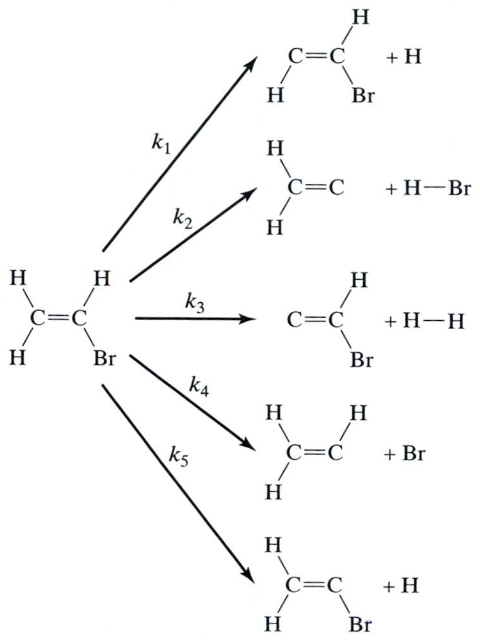

◀ **FIGURE 19.14**
Possible concurrent unimolecular decomposition pathways for vinyl bromide with 6.44 eV of internal energy in excess of the zero-point vibrational energy.

$$\left(\frac{d[A]}{dt}\right)_1 = -k_1[A].$$

For the second reaction, we have

$$\left(\frac{d[A]}{dt}\right)_2 = -k_2[A].$$

Therefore, the total reaction rate is

$$\left(\frac{d[A]}{dt}\right) = -k_1[A] - k_2[A] = -(k_1 + k_2)[A]. \qquad \textbf{(19.49)}$$

Equation 19.49 is the differential equation describing a first-order reaction with a total rate coefficient $k_1 + k_2$. The integrated rate law is

$$\boxed{[A] = [A]_o \exp[-(k_1 + k_2)t]}. \qquad \textbf{(19.50)}$$

Since the relaxation times are the reciprocals of the rate coefficients for first-order reactions and $k_T = k_1 + k_2$, the total relaxation time τ is related to the individual relaxation times τ_1 and τ_2, by

$$\boxed{\frac{1}{\tau} = \frac{1}{\tau_1} + \frac{1}{\tau_2}}. \qquad \textbf{(19.51)}$$

Equation 19.50 shows that a plot of $\ln[A]$ versus time will be linear with a slope equal to $k_1 + k_2$. To obtain the individual rate coefficients, we need information concerning the product yields. If we write the rate for the preceding mechanism in terms of the products B and C, we have

$$\frac{d[B]}{dt} = k_1[A] \qquad \textbf{(19.52A)}$$

and

$$\frac{d[C]}{dt} = k_2[A]. \qquad (19.52B)$$

Dividing Eq. 19.52A by Eq. 19.52B produces

$$\frac{d[B]}{d[C]} = \frac{k_1}{k_2}.$$

If $[B]_o = [C]_o = 0$, which is the usual situation, integration of this differential equation gives

$$\int_0^{[B]} d[B] = \frac{k_1}{k_2} \int_0^{[C]} d[C],$$

or

$$[B] = \frac{k_1[C]}{k_2}. \qquad (19.53)$$

If we have determined the ratio $[B]/[C]$, k_1/k_2 can be computed, and this information, combined with the slope of a plot of $\ln[A]$ versus time, will allow both rate coefficients to be found.

EXAMPLE 19.8

Using theoretical methods that will be discussed in Chapter 20, we can compute the rate of unimolecular decomposition of CH_4 after it has been energized by the insertion of a given amount of internal energy in vibrational–rotational motion. With these methods, the decomposition rates for

$$CH_4 \xrightarrow{k_1} CH_2 + H_2 \qquad \text{and} \qquad CH_4 \xrightarrow{k_2} CH_3 + H$$

Time (tu)[a]	Number N_1 of Reactions Giving $CH_2 + H_2$	Number N_2 of Reactions Giving $CH_3 + H$
15	N.A.	119
20	109	129
25	116	141
30	129	170
35	143	213
40	148	218
45	156	229
50	161	273
55	174	295
60	184	295
65	192	298
70	198	304
75	201	313

[a]Time is given in units of 0.538×10^{-14} s.
N.A. = not available.

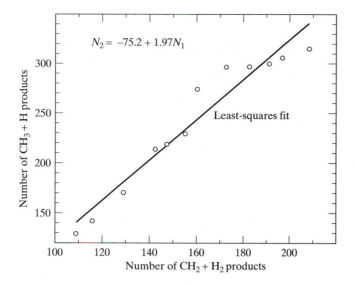

$N_2 = -75.2 + 1.97N_1$

Least-squares fit

Number of $CH_3 + H$ products (y-axis)

Number of $CH_2 + H_2$ products (x-axis)

▲ FIGURE 19.15
Plot of the product yields for the concurrent decomposition reactions of excited CH_4 molecules to yield $CH_3 + H$ and $CH_2 + H_2$. The internal excitation energy present in CH_4 in these calculations is 7.62 eV. The statistical data scatter is the result of the relatively low number (656) of decomposition reactions investigated. The trajectory methods employed to perform the calculations are described in Chapter 20.

have been computed for 7.62 eV of internal excitation. The results at the bottom of page 1114 were obtained.

The total number of CH_4 molecules studied in the calculations was 656. **(A)** Determine k_1 and k_2 as accurately as possible from the data in the table. **(B)** Determine $\tau, \tau_1,$ and τ_2.

Solution

(A) If we plot N_2 versus N_1 and fit the best straight line to the data, Eq. 19.53 tells us that the slope of the line will be the ratio k_2/k_1. The plot is shown in Figure 19.15. The statistical data scatter due to the small number of reactions investigated is evident. The best linear fit to the data gives a slope $k_2/k_1 = 1.97$. We can now use Eq. 19.50 in logarithmic form to obtain the sum $k_1 + k_2$ from the slope of a plot of $\ln(N)$ versus t, where N is the number of CH_4 molecules that have not reacted at time t. From the fact that the total number of CH_4 molecules studied is $N_o = 656$, we can easily obtain the following data from the table:

Time (tu)[a]	N	$\ln(N)$	Time (tu)[a]	N	$\ln(N)$
20	418	6.035	50	222	5.403
25	399	5.989	55	187	5.231
30	357	5.878	60	177	5.176
35	300	5.704	65	166	5.112
40	290	5.670	70	154	5.037
45	271	5.602	75	142	4.956

[a]Time is given in units of 0.538×10^{-14} s.

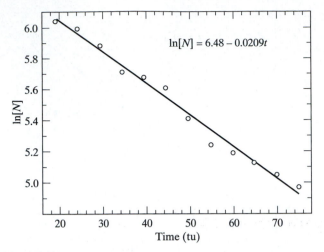

▲ FIGURE 19.16
A plot of ln(N) versus time for the decomposition of excited CH_4 molecules described in Example 19.8. N is the number of CH_4 molecules remaining at time t out of a total of 656 molecules investigated in the calculations. Time is given in molecular time units (tu), for which the conversion factor is 1 tu = 0.538×10^{-14} s. The slope of the least-squares fit of the data yields a total decomposition rate coefficient of 0.0209 tu^{-1}.

Figure 19.16 shows a plot of ln(N) versus time, which Eq. 19.50 tells us should be linear with a slope equal to $-(k_1 + k_2)$. The least-square fit gives the result

$$k_1 + k_2 = -\text{slope} = 0.0209\ tu^{-1} = 0.0209\ tu^{-1} \times \frac{1\ tu}{0.538 \times 10^{-14}\ s}$$

$$= 3.88 \times 10^{12}\ s^{-1}, \tag{1}$$

where 1 tu = 0.538×10^{-14} s. Since the ratio $k_2/k_1 = 1.97$, we have

$$k_1 + 1.97k_1 = 2.97k_1 = 3.88 \times 10^{12}\ s^{-1}. \tag{2}$$

Thus,

$$k_1 = 1.31 \times 10^{12}\ s^{-1} \tag{3}$$

and

$$k_2 = 1.97k_1 = 2.58 \times 10^{12}\ s^{-1}. \tag{4}$$

(B) The individual relaxation times are

$$\tau_1 = \frac{1}{k_1} = \frac{1}{1.31 \times 10^{12}\ s^{-1}} = 7.63 \times 10^{-13}\ s \tag{5}$$

and

$$\tau_2 = \frac{1}{k_2} = \frac{1}{2.58 \times 10^{12}\ s^{-1}} = 3.88 \times 10^{-13}\ s. \tag{6}$$

The overall relaxation time is given by Eq. 19.51:

$$\frac{1}{\tau} = \frac{1}{\tau_1} + \frac{1}{\tau_2} = 1.31 \times 10^{12}\ s^{-1} + 2.58 \times 10^{12}\ s^{-1} = 3.89 \times 10^{12}\ s^{-1}. \tag{7}$$

Thus,

$$\tau = 2.57 \times 10^{-13}\ s. \tag{8}$$

For related exercises, see Problems 19.25, 19.26, 19.27, 19.31, and 19.32.

Second-order concurrent reactions have more complex integrated rate expressions, and the analysis of the associated kinetic data is, therefore, more difficult. For example, when the two concurrent reactions are

$$A \xrightarrow{k_1} B \qquad \text{and} \qquad A + A \xrightarrow{k_2} C,$$

$$\text{Reaction 1} \qquad\qquad \text{Reaction 2}$$

the integrated rate expression is (see Problem 19.25)

$$\ln\left[\frac{[A]_o\{k_1 + 2k_2[A]\}}{[A]\{k_1 + 2k_2[A]_o\}}\right] = k_1 t. \tag{19.54}$$

The transcendental nature of this equation usually means that numerical methods of some type will be required to analyze the kinetic data. If we have some additional information, the data analysis can sometimes be simplified. Problem 19.27 illustrates such simplification.

19.4.3   Consecutive Reactions

Many organic reactions and nuclear decay processes comprise a set of consecutive reactions in which the product of one reaction becomes the reactant for the next. Two common examples are a nuclear decay sequence in which the reactions are all first order and a polymerization process for which the individual reactions are second order.

The simplest possible consecutive sequence of reactions is the two first-order reactions

$$A \xrightarrow{k_1} B \qquad \text{and} \qquad B \xrightarrow{k_2} C.$$

In this sequence, the overall reaction is $A \longrightarrow C$. Compound B, which does not appear in the description of the overall reaction, is called an *intermediate*. Physically, this tells us that if we were to carry out the reaction sequence in the laboratory and make no observations until the reaction was effectively complete, we would be totally unaware of the fact that the intermediate was ever present. Information on intermediates can be obtained only from measurements made during the course of the reaction.

The differential rate equations describing the two reactions are

$$\frac{d[A]}{dt} = -k_1[A], \tag{19.55A}$$

$$\frac{d[B]}{dt} = k_1[A] - k_2[B], \tag{19.55B}$$

and

$$\frac{d[C]}{dt} = k_2[B]. \tag{19.55C}$$

Since the stoichiometry of both reactions is such that the total number of moles of material never changes, we must have $[A] + [B] + [C] = [A]_o + [B]_o + [C]_o$, which permits the time dependence of C to be easily obtained once we have $[A]$ and $[B]$ as functions of time.

In most cases of interest, the initial concentrations of B and C will be zero. Let us, therefore, consider this case. Since the concentration of compound A appears only in Eq. 19.55A, which is that for a first-order reaction, it follows that

$$[A] = [A]_o \exp[-k_1 t]. \tag{19.56}$$

Using this result, we can put Eq. 19.55B for compound B in the form

$$\frac{d[B]}{dt} = k_1[A] - k_2[B] = k_1[A]_o \exp[-k_1 t] - k_2[B]. \tag{19.57}$$

Multiplying this equation by dt and rearranging produces

$$d[B] + k_2[B]dt = k_1[A]_o \exp[-k_1 t]dt. \tag{19.58}$$

Although the variables in Eq. 19.58 are not separated, it is still easy to integrate the equation using an integrating factor that converts the left-hand side to an exact differential. The general principle is the following: If we have a differential equation whose form is

$$dy + f(x)y\, dx = g(x)\, dx, \tag{19.59}$$

where $f(x)$ and $g(x)$ are any arbitrary functions of x alone, the left-hand side of the equation can always be converted to an exact differential by multiplication by an integrating factor

$$I = \exp[\int f(x)dx]. \tag{19.60}$$

That is, the expression $I\{dy + f(x)y\, dx\} = \exp[\int f(x)dx]\{dy + f(x)y\, dx\}$ is always the exact differential of the function $F = y \exp[\int f(x)dx]$. This is easy to show by taking the total differential of F, namely, $dF = (\partial F/\partial y)_x dy + (\partial F/\partial x)_y dx$. If we keep in mind that the derivative of an exponential is the exponential times the derivative of the exponent and that the derivative of an integral is just the integrand, we obtain

$$dF = \left(\frac{\partial F}{\partial y}\right)_x dy + \left(\frac{\partial F}{\partial x}\right)_y dx = \exp[\int f(x)dx]dy + \exp[\int f(x)dx]y f(x)dx$$

$$= \exp[\int f(x)dx][dy + f(x)y\, dx], \tag{19.61}$$

so that dF is exactly $I\{dy + f(x)y\, dx\}$.

The foregoing discussion shows us how to integrate Eq. 19.58. We first compute the integrating factor, in this case,

$$I = \exp[\int k_2 dt] = \exp[k_2 t].$$

Multiplying both sides of Eq. 19.58 by I produces

$$\exp[k_2 t](d[B] + k_2[B]dt) = k_1[A]_o \exp[-k_1 t + k_2 t]dt. \tag{19.62}$$

We now integrate both sides between corresponding limits to obtain

$$\int_{t=0,\,[B]=0}^{t,\,[B]} \exp[k_2 t](d[B] + k_2[B]dt) = k_1[A]_o \int_0^t \exp[-k_1 t + k_2 t]dt.$$

The integral of the left side is $[B]\exp[k_2 t]$. Therefore, the result is

$$[B]\exp[k_2 t]\,\Big|_{0,\,0}^{t,\,[B]} = [B]\exp[k_2 t] = \frac{k_1[A]_o}{k_2 - k_1}\exp[(k_2 - k_1)t]\,\Big|_0^t$$

$$= \frac{k_1[A]_o}{k_2 - k_1}\{\exp[(k_2 - k_1)t] - 1\}. \tag{19.63}$$

Division of Eq. 19.63 by $\exp[k_2 t]$ produces the expression we need for $[B]$:

$$\boxed{[B] = \frac{k_1[A]_o}{k_2 - k_1}[\exp[-k_1 t] - \exp[-k_2 t]]\,.} \tag{19.64}$$

If the initial concentration of B is not zero, the result is (see Problem 19.33)

$$[B] = \frac{k_1[A]_o}{k_2 - k_1} \exp[-k_1 t] + \left[[B]_o - \frac{k_1[A]_o}{k_2 - k_1} \right] \exp[-k_2 t]. \quad \textbf{(19.65)}$$

Figures 19.17 and 19.18 illustrate the temporal behavior of two hypothetical first-order nuclear decay processes in which 1,000 atoms of isotope A are present at $t = 0$. In Figure 19.17, the relaxation times are $\tau_1 = 300$ min and $\tau_2 = 30$ min. In Figure 19.18, they are $\tau_1 = 40$ min and $\tau_2 = 30$ min. In both cases, isotope A decays exponentially, while the final product C rises monotonically to its limiting value of 1,000 atoms. Also in both cases, the number of atoms of the intermediate, isotope B, rises to a maximum and then decays monotonically toward its limiting value of zero. This is characteristic of all intermediates in consecutive reactions of any type. There is, however, a major difference in the behavior of B between the two cases, residing in the nature of the monotonic decay. In Figure 19.17, where we have $\tau_1 \gg \tau_2$, so that $k_1 \ll k_2$, the number of atoms of isotope B remains relatively small throughout the process, and the rate of change of B is near zero

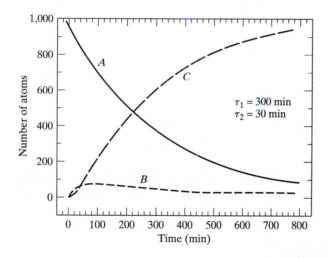

◀ **FIGURE 19.17**
The temporal behavior of the nuclear isotopes A, B, and C undergoing a sequence of two consecutive first-order reactions $A \xrightarrow{k_1} B$ and $B \xrightarrow{k_2} C$ with $A_o = 1,000$ atoms and $B_o = C_o = 0$. In this example, the relaxation times are $\tau_1 = 300$ min and $\tau_2 = 30$ min, so that the corresponding rate coefficients are $k_1 = 0.00333\ldots$ min^{-1} and $k_2 = 0.0333\ldots$ min^{-1}. Under these conditions, with $k_2 \gg k_1$, the number of atoms of the intermediate B remains small, and its time derivative dB/dt is near zero for an extended period.

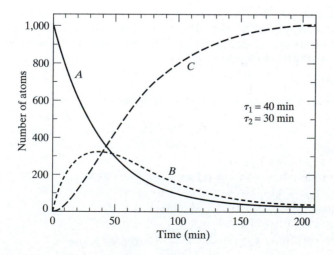

◀ **FIGURE 19.18**
The temporal behavior of the nuclear isotopes A, B, and C for the system described in Figure 19.17. In this case, the relaxation times are $\tau_1 = 40$ min and $\tau_2 = 30$ min, so that the corresponding rate coefficients are $k_1 = 0.025$ min^{-1} and $k_2 = 0.0333\ldots$ min^{-1}. Under these conditions, with $k_2 \approx k_1$, the number of atoms of the intermediate B becomes much larger, and its time derivative dB/dt is near zero only in a narrow range around the maximum and near the end of the reaction.

for times greater than 50 minutes. In Figure 19.18, where τ_1 and τ_2 are comparable in magnitude, the number of intermediate atoms becomes much larger, and the rate at which B changes with time is near zero only in the region near the maximum. This type of behavior will be of critical importance to us in the analysis of complex kinetic systems.

The time at which the intermediate concentration or number of atoms attains a maximum provides information related to the relative values of the rate coefficients. The condition for a maximum in $[B]$ is $d[B]/dt = 0$. From Eq. 19.64, this derivative is

$$\frac{d[B]}{dt} = \frac{k_1[A]_o}{k_2 - k_1}[-k_1 \exp[-k_1 t] + k_2 \exp[-k_2 t]]. \tag{19.66}$$

If the concentration of the intermediate reaches a maximum at $t = t_m$, we must have

$$\frac{d[B]}{dt} = \frac{k_1[A]_o}{k_2 - k_1}[-k_1 \exp[-k_1 t_m] + k_2 \exp[-k_2 t_m]] = 0.$$

This requires that

$$-k_1 \exp[-k_1 t_m] + k_2 \exp[-k_2 t_m] = 0.$$

Rearranging terms produces

$$\exp[(k_2 - k_1)t_m] = \frac{k_2}{k_1}.$$

Solving for t_m, we obtain

$$\boxed{t_m = \frac{\ln(k_2/k_1)}{k_2 - k_1}}. \tag{19.67}$$

Since k_1 can be easily obtained from the first-order decomposition of reactant A, Eq. 19.67 provides a simple method for obtaining k_2. The technique is illustrated in Example 19.9.

EXAMPLE 19.9

A nuclear isotope A decays in a sequence of two reactions. Seventy-five percent of A is found to decay in 20.79 hours. The isotope produced by the decomposition of A attains a maximum amount at $t = 17.77$ hours. What are the rate coefficients and half-lives of the two isotopes?

Solution

Since the decomposition of A is first order, we have

$$A = A_o \exp[-k_1 t]. \tag{1}$$

Therefore,

$$\frac{A}{A_o} = 0.2500 = \exp[-k_1(20.79)]. \tag{2}$$

Taking logarithms of both sides, we obtain

$$k_1 = -\frac{\ln(0.2500)}{20.79 \text{ hours}} = 0.06668 \text{ hours}^{-1}. \tag{3}$$

The half-life of isotope A is, therefore,

$$(t_{1/2})_1 = \frac{\ln(2)}{k_1} = \frac{0.6931}{0.06668 \text{ hours}^{-1}} = 10.39 \text{ hours.} \qquad (4)$$

Equation 19.67 shows that isotope B attains a maximum number of atoms at

$$t_m = \frac{\ln(k_2/k_1)}{k_2 - k_1} = 17.77 \text{ hours.} \qquad (5)$$

Inserting the value of k_1 gives

$$\frac{\ln(k_2/0.06668)}{k_2 - 0.06668} = 17.77 \text{ hours.} \qquad (6)$$

Therefore, we must have

$$F(k_2) = \ln\left(\frac{k_2}{0.06668}\right) - 17.77k_2 + 17.77(0.06668)$$

$$= \ln\left(\frac{k_2}{0.06668}\right) - 17.77k_2 + 1.185 = 0. \qquad (7)$$

This transcendental equation must be solved numerically for k_2. A one-dimensional grid search of $F(k_2)$ over the range $0 \le k_2 \le 1 \text{ hour}^{-1}$ shows that the solution lies in the range $0.046 \le k_2 \le 0.048 \text{ hour}^{-1}$. A second search over this interval yields $k_2 = 0.0470 \text{ hour}^{-1}$. Thus,

$$(t_{1/2})_2 = \frac{\ln(2)}{0.0470 \text{ hour}^{-1}} = 14.75 \text{ hours.} \qquad (8)$$

Figure 19.19 shows a plot of $F(k_2)$ versus k_2 that makes the solution clear.

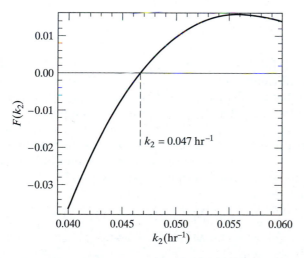

◀ FIGURE 19.19
The results of a one-dimensional grid search to obtain k_2 in Example 19.9. A plot of $F(k_2) = \ln(k_2/0.06668) - 17.77k_2 + 1.185 = 0$ versus k_2 is shown. As can be seen, the root occurs at $k_2 = 0.0470 \text{ hr}^{-1}$.

For related exercises, see Problems 19.33, 19.34, 19.35, and 19.36.

It is relatively easy to extend the analysis to the case of three consecutive first-order reactions in which we have

$$A \xrightarrow{k_1} B, \qquad B \xrightarrow{k_2} C, \qquad \text{and} \qquad C \xrightarrow{k_3} D.$$

The differential rate equations for this mechanism are

$$\frac{d[A]}{dt} = -k_1[A], \qquad (19.68A)$$

$$\frac{d[B]}{dt} = k_1[A] - k_2[B], \qquad (19.68B)$$

and

$$\frac{d[C]}{dt} = k_2[B] - k_3[C].$$ (19.68C)

The differential equations describing the temporal variations of $[A]$ and $[B]$ are identical to Eqs. 19.55A and 19.55B. Therefore, the integrated expressions are the same as Eqs. 19.56 and 19.64 for A and B, respectively, when we have $[B]_o = 0$. As before, the stoichiometry of the mechanism allows us to compute $[D]$ from the results for $[A]$, $[B]$, and $[C]$. Hence, we need concern ourselves only with Eq. 19.68C. Using the result for $[B]$, we can again employ an integrating factor to obtain an expression for $[C]$ as a function of time. For the case in which $[B]_o = [C]_o = 0$, the result is (see Problem 19.37)

$$[C] = \frac{k_1 k_2 [A]_o}{(k_1 - k_2)(k_1 - k_3)} e^{-k_1 t} + \frac{k_1 k_2 [A]_o}{(k_2 - k_1)(k_2 - k_3)} e^{-k_2 t} + \frac{k_1 k_2 [A]_o}{(k_3 - k_1)(k_3 - k_2)} e^{-k_3 t}.$$

(19.69)

An interesting special case occurs whenever all of the rate coefficients for a series of consecutive first-order reactions are equal. If we write the reactions in the form

$$A_1 \xrightarrow{k} A_2, \quad A_2 \xrightarrow{k} A_3, \quad A_3 \xrightarrow{k} A_4, \quad \dots, \quad A_i \xrightarrow{k} A_{i+1}, \quad \dots,$$

then

$$[A_i] = \frac{[A_1]_o (kt)^{i-1} e^{-kt}}{(i-1)!}$$ (19.70)

provided that $[A_2]_o = [A_3]_o = [A_4]_o = \cdots = [A_i]_o = 0$ for all $i \neq 1$ with equal rate coefficients. This result is derived in Problem 19.39, where the various parts of the problem serve as a procedure guide. Problem 19.40 examines some of the characteristics of systems of this type.

It might seem that Eq. 19.70 could be utilized to obtain an approximate kinetic description for polymerization processes, such as the one illustrated in Figure 19.20. If we represent the starting monomer in a polymerization by M_1 and the dimers, trimers, tetramers, etc. by $M_2, M_3, M_4, \dots$, respectively, the overall process is

$$M_1 + M_1 \longrightarrow M_2,$$
$$M_1 + M_2 \longrightarrow M_3,$$
$$M_1 + M_3 \longrightarrow M_4,$$
$$M_1 + M_4 \longrightarrow M_5,$$
$$\vdots$$
$$M_1 + M_i \longrightarrow M_{i+1},$$
$$\vdots$$

In most polymerizations, the monomer is present in great excess, so that each step after the first will behave as a pseudo first-order reaction. In addition, there is probably very little difference between the reaction rates of $M_{20} + M_1$ and $M_{21} + M_1$. Therefore, we can make the approximation that

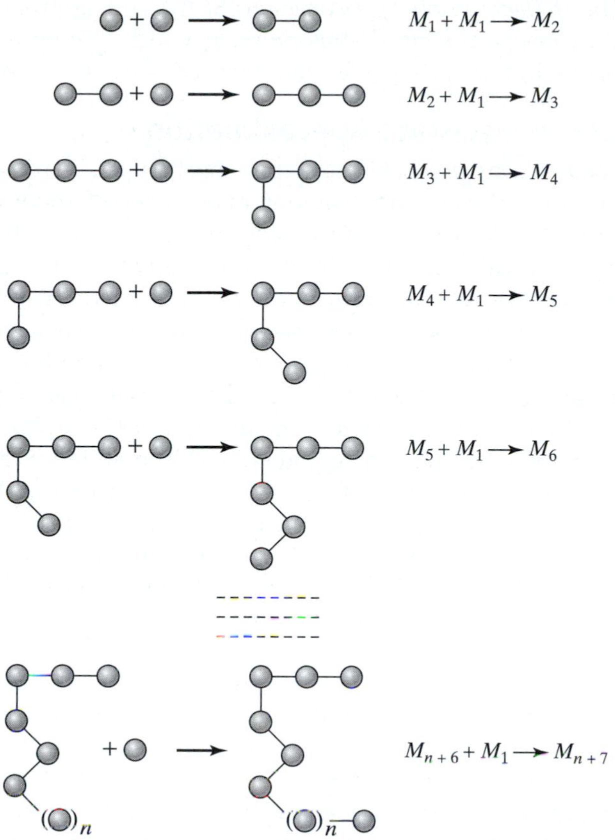

$$M_1 + M_1 \longrightarrow M_2$$

$$M_2 + M_1 \longrightarrow M_3$$

$$M_3 + M_1 \longrightarrow M_4$$

$$M_4 + M_1 \longrightarrow M_5$$

$$M_5 + M_1 \longrightarrow M_6$$

$$M_{n+6} + M_1 \longrightarrow M_{n+7}$$

◀ FIGURE 19.20

A typical polymerization process in which the monomer is represented by ⚫. It is possible to obtain a reasonably good temporal description of such processes by using the kinetics of consecutive reactions. This is done in Problems 19.41 through 19.44.

all of the rate coefficients are equal. However, the first reaction is second order, and although $[M_1]$ may be large, it will not remain constant throughout the process. Therefore, the conditions leading to Eq. 19.70 are not present. Accordingly, to obtain a kinetic description of polymerization, we need to modify the model. This is done in Problems 19.41 through 19.44. The result is a reasonable approximation of the kinetic behavior of polymerization processes.

In many cases, integrated rate expressions for higher order consecutive reactions cannot be obtained in closed analytic form. Even in those cases where the differential equations can be integrated, the result is usually so complex that it is not useful. To handle the kinetics of such complex systems, we need some approximation methods.

19.5   Approximation Methods for Complex Reaction Mechanisms

When several higher order reactions of different types are present in a reaction mechanism, it is often impossible to obtain analytic expressions for the temporal dependence of all the concentrations. Even when this can be done, the expressions are usually so complex that they are not very useful in analyzing experimental kinetic data. Under certain conditions, it is often possible to simplify the kinetic expressions and still obtain a worthwhile description of the behavior of the system. The most useful procedure for this purpose involves the use of the *stationary-state approximation (SSA)*. This approximation leads directly to the concept of a *rate-determining step*, which also permits us to simplify the kinetics. In this section, we examine each of these approximations.

19.5.1   Stationary-state Approximation

When intermediates are present in a reaction mechanism, it often happens that the rate of change of the concentrations or number of molecules of these species is near zero. A case in point is seen in Figure 19.17, where $d[B]/dt$ is near zero at times greater than 60 minutes. Physically, this means that B atoms are being produced by the decomposition of A at nearly the same rate as they are reacting to produce C atoms. B is, therefore, said to be in a *steady state*. We should not assume that this will always be the case when intermediates are present. For example, the same reaction mechanism with different rate coefficients can lead to the kinetic behavior seen in Figure 19.18. In this case, $d[B]/dt$ is near zero only in a very narrow region around its maximum. The important question is "When can we expect $d[B]/dt$ to be near zero throughout much of the reaction, and when will it behave like the system shown in Figure 19.18."

Inspection of Figures 19.17 and 19.18 shows that $d[B]/dt$ remains near zero in those time domains wherein $[B] \ll [A]$ and $[C]$ or $N_B \ll N_A$ and N_C. When the concentration or number of atoms of the intermediate becomes equal to or greater than the reactants or products, we do not observe $d[B]/dt \approx 0$, except in very limited time intervals in the neighborhood of the maximum. This observation is actually a statement of the stationary-state, or steady-state, approximation: *In those time domains in which the concentration or number of atoms of an intermediate is small relative to both reactants and products, the derivative of the concentration of the intermediate with respect to time will be near zero.*

As an example of how the stationary-state approximation works, let us apply it to a mechanism involving two first-order consecutive reactions for which we already have exact solutions. A comparison of the results obtained with the exact answers will permit us to assess the accuracy of the approximation. Our mechanism is

$$A \xrightarrow{k_1} B \qquad \text{and} \qquad B \xrightarrow{k_2} C,$$

so that the overall reaction is $A \longrightarrow C$ and B is an intermediate. Therefore, if $[B] \ll [A]$ and $[C]$, the stationary-state approximation tells us that we can write

$$\frac{d[B]}{dt} = k_1[A] - k_2[B] \approx 0. \tag{19.71}$$

Notice that the approximation converts a difficult differential equation into a simple algebraic equation. That is always the case when we employ this procedure. Consequently, instead of using an integrating factor to solve a differential equation, we can obtain $[B]$ simply by solving Eq. 19.71. The solution is

$$[B] = [B]_{ss} \approx \frac{k_1[A]}{k_2}, \tag{19.72}$$

where we have written $[B]$ as $[B]_{ss}$ to denote the fact that result is obtained by using the stationary-state approximation. Since compound A decays in a simple first-order reaction, we have

$$\boxed{[B]_{ss} = \frac{k_1[A]_o}{k_2} \exp[-k_1 t]}. \tag{19.73}$$

It is apparent that the effort needed to obtain $[B]_{ss}$ is much less than that required to derive the exact solution given by Eq. 19.64.

The requirement that $[B] \ll [A]$ and $[C]$ in order for the stationary-state approximation to be accurate leads to stringent constraints on the relative magnitudes of k_1 and k_2. From Eq. 19.72, we have

$$\frac{[B]_{ss}}{[A]} = \frac{k_1}{k_2}.$$

If $[B] \ll [A]$, we must have $[B]_{ss}/[A] \ll 1$, which means that $k_1 \ll k_2$. In other words, the first reaction producing B must be slow relative to the rate at which B reacts if the inequality between the concentrations is to be maintained. This relationship explains the difference seen in Figures 19.17 and 19.18. In the former case, we have $10k_1 = k_2$, and we see that $[B] \ll [A]$ and $[C]$, so that $d[B]/dt \approx 0$ throughout much of the process. In the latter case, k_1 is nearly equal to k_2. As a result, $[B]$ is not small relative to $[A]$ and $[C]$, and we do not observe $d[B]/dt \approx 0$ except near the maximum.

When $k_1 \ll k_2$, it is easy to show that the exact solution given by Eq. 19.64 reduces to the stationary-state solution of Eq. 19.73. The exact solution is

$$[B] = \frac{k_1[A]_o}{k_2 - k_1}(\exp[-k_1t] - \exp[-k_2t]).$$

If $k_2 \gg k_1$, then

$$\exp[-k_2t] \ll \exp[-k_1t],$$

except at the very beginning of the reaction, which is sometimes called the *induction period*. Therefore, the second exponential is negligible relative to the first, and

$$[B] \approx \frac{k_1[A]_o}{k_2 - k_1}\exp[-k_1t].$$

Since $k_2 \gg k_1$, we also have $k_2 - k_1 \approx k_2$. This gives

$$[B] = [B]_{ss} = \frac{k_1[A]_o}{k_2}\exp[-k_1t],$$

which is identical to Eq. 19.73. The greater the difference between k_1 and k_2, the more accurate the stationary-state approximation will be.

EXAMPLE 19.10

Apply the stationary-state approximation to the system whose exact results are shown in Figure 19.17. Plot the number of atoms of B as a function of time for the exact result and the stationary-state approximation. Where is the result good? Where is it poor? Explain. Repeat the analysis for the system shown in Figure 19.18. Is the stationary-state approximation better or worse? Explain.

Solution

The exact answer is given by Eq. 19.64. The stationary-state approximation is Eq. 19.73. Figure 19.21 shows the two results, using the values of k_1 and k_2 obtained from the relaxation times given in Figure 19.17. For times in the range 60 min $\leq t \leq$ 800 min, the SSA gives a reasonably accurate result. This is not surprising, since, in this time domain, Figure 19.17 clearly shows that we have $[B] \ll [A]$ and $[C]$, which

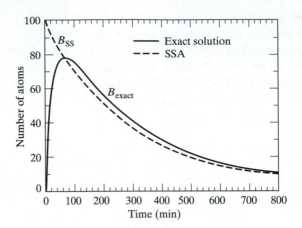

▲ FIGURE 19.21

Comparison of the exact temporal variation of the number of atoms of the intermediate in the sequence of two first-order consecutive reactions shown in Figure 19.17
with the result obtained by using the stationary-state approximation (SSA). As can
be seen, for times greater than 60 minutes, the approximation is reasonably accurate. This relationship holds because in this case we have $k_2 \gg k_1$, which the text
shows is the condition for the SSA to obtain.

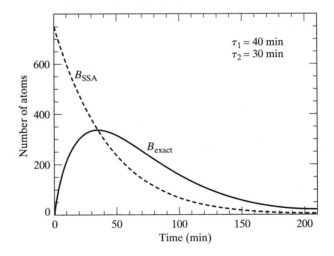

▲ FIGURE 19.22

Comparison of the exact temporal variation of the number of atoms of the intermediate in the sequence of two first-order consecutive reactions shown in Figure 19.18
with the result obtained by using the stationary-state approximation. In this case, the
stationary-state approximation is much less accurate than that seen in Figure 19.21.
The reason is that we now no longer have $k_2 \gg k_1$, as required by the SSA for
accuracy.

is the requirement for accuracy of the stationary-state approximation. At times
less than 60 min, the SSA result is poor. When $t \le 60$ min, we clearly do not have
$[B] \ll [A]$ and $[C]$; in fact, the inequality is reversed, so that $[B] \gg [C]$! We
would, therefore, not expect the SSA to be accurate in that region.

Figure 19.22 shows the same comparison for the values of k_1 and k_2 obtained
from the relaxation times given in Figure 19.18. We no longer have $k_2 \gg k_1$. As a
result, the SSA is not nearly as accurate as was the case for the system in Figure 19.17.

For a related exercise, see Problem 19.38.

EXAMPLE 19.11

Consider the following complex mechanism:

$$A + B \xrightarrow{k_1} C + D;$$

$$C + D \xrightarrow{k_2} A + B;$$

$$C \xrightarrow{k_3} P.$$

(A) Write down differential rate equations for all species. **(B)** Determine the overall reaction. **(C)** Using a stationary-state approximation for the intermediate, show that the rate of production of product P is first order with respect to both A and B and second order overall, provided that $k_3 \gg k_2[D]$.

Solution

(A) Compounds A, B, and D are involved in two reactions. There will, therefore, be two terms on the right side of their differential equations. Compound C is involved in three reactions. Consequently, this differential equation will have three terms on the right side. The differential rate expression for compound P will have only a single term. The results are

$$\frac{d[A]}{dt} = -k_1[A][B] + k_2[C][D] = \frac{d[B]}{dt}, \tag{1}$$

$$\frac{d[D]}{dt} = k_1[A][B] - k_2[C][D], \tag{2}$$

$$\frac{d[C]}{dt} = k_1[A][B] - k_2[C][D] - k_3[C], \tag{3}$$

and

$$\frac{d[P]}{dt} = k_3[C]. \tag{4}$$

(B) In the determination of the overall process, we ignore the equilibrium step and write

$$A + B \longrightarrow C + D$$

and

$$C \longrightarrow P.$$

The overall reaction for these two processes is $A + B \longrightarrow D + P$, with C an intermediate.

(C) If k_3 is large, $[C]$ will remain small and the stationary-state approximation will be accurate. This gives

$$\frac{d[C]}{dt} = k_1[A][B] - k_2[C][D] - k_3[C] \approx 0. \tag{5}$$

Letting $[C] = [C]_{ss}$ and solving for $[C]_{ss}$, we obtain

$$[C]_{ss} = \frac{k_1[A][B]}{k_2[D] + k_3}. \tag{6}$$

Substituting Eq. 6 into Eq. 4 then produces

$$\frac{d[P]}{dt} \approx \frac{k_3 k_1[A][B]}{k_2[D] + k_3}. \tag{7}$$

If $k_3 \gg k_2[D]$, then $k_2[D] + k_3 \approx k_3$, and Eq. 7 becomes

$$\frac{d[P]}{dt} \approx \frac{k_3 k_1[A][B]}{k_3} = k_1[A][B]. \tag{8}$$

It follows that the rate of production of product P is first order with respect to both A and B and second order overall.

For related exercises, see Problems 19.45, 19.46, and 19.47.

19.5.2 Rate-determining Step

The stationary-state approximation leads directly to the concept of a *rate-determining step*. This concept tells us that, in a sequence of consecutive reactions, the rate of production of the final products depends only upon the rate coefficient for the last slow step in the sequence and all rate coefficients preceding it. We may see how the stationary-state approximation leads to this conclusion by considering an example.

Suppose we have the following reaction mechanism:

$$A + B \xrightarrow{k_1} C \quad \text{(second order)};$$

$$C + A \xrightarrow{k_2} D \quad \text{(second order, slow)};$$

$$D \xrightarrow{k_3} F \quad \text{(pseudo first order, fast)};$$

$$F \xrightarrow{k_4} P \quad \text{(pseudo first order, fast)}.$$

The overall reaction is $2A + B \longrightarrow P$, so that compounds C, D, and F are intermediates. If k_3 and k_4 are large relative to k_2, as stated, we are assured that $[D]$ and $[F]$ will remain small throughout the process. Therefore, a stationary-state approximation may be applied to those intermediates. However, since k_2 is small, leading to a slow reaction, the concentration of compound C will tend to build up. This means that we will not have $[C]$ small relative to $[A]$, $[B]$, and $[P]$, and a stationary-state approximation applied to C will not be accurate. Applying the SSA to F gives

$$\frac{d[F]}{dt} = k_3[D] - k_4[F] \approx 0,$$

so that

$$[F]_{ss} = \frac{k_3[D]}{k_4}.$$

The differential rate of production of products P is

$$\frac{d[P]}{dt} = k_4[F] \approx k_4[F]_{ss} = k_3[D].$$

Using the SSA for compound D, we obtain

$$\frac{d[D]}{dt} = k_2[C][A] - k_3[D] \approx 0,$$

which leads to the result

$$[D]_{ss} = \frac{k_2[C][A]}{k_3}.$$

Substitution into the expression for $d[P]/dt$ gives

$$\frac{d[P]}{dt} = k_3[D] \approx k_3[D]_{ss} = k_3\frac{k_2[C][A]}{k_3} = k_2[C][A].$$

We see that the rate of production of P is independent of both k_3 and k_4; it depends only upon the rate coefficient for the last slow step, k_2, and also upon the rate coefficients for all steps preceding the last slow step, k_1 in this case. k_1 is present implicitly, since both $[A]$ and $[C]$ depend upon its value.

19.5.3   Applications

19.5.3.1   Thermal Unimolecular Reactions

Molecules or nuclei that are in metastable states generally exhibit random lifetimes in those states and, consequently, undergo true first-order unimolecular reactions leading to a more stable state. However, we also observe what are apparently thermal, unimolecular first-order gas-phase decomposition reactions of many compounds, including propaldehyde, azomethane, cyclopropane, diethyl and dimethyl ether, and CH_3NC. At elevated temperatures and high pressures, these compounds all undergo reactions whose kinetic data fit a simple first-order rate law. Moreover, experiments show that the measured first-order rate coefficients for these processes are a decreasing function of pressure. Figure 19.23 gives a typical example for CH_3NC at 503 K. In this figure, the ratio of the rate coefficient at pressure p, $k(p)$, to that at high pressure, k_∞, is plotted as a function of pressure expressed in torr. Such a plot is frequently called a *unimolecular falloff curve*.

The explanation for the experimental observations is not immediately obvious. Since propaldehyde, azomethane, cyclopropane, diethyl and dimethyl ether, and CH_3NC are all stable molecules which require substantial energy to break the chemical bonds that are present, why should unimolecular decomposition reactions occur? If we postulate that the molecules are energized via collisions prior to reaction, it would seem that the kinetics should be second order instead of first order. Finally, the dependence of $k(p)$ upon pressure seems inconsistent with Eq. 19.1, which shows that the rate coefficient depends only upon temperature. The answer to all these conundrums was provided by Frederick Lindemann in 1921.

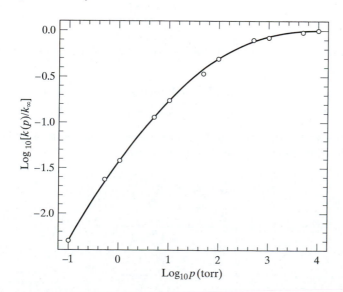

◀ FIGURE 19.23
A unimolecular falloff curve for the decomposition of CH_3NC at 503 K. The ordinate shows the base-10 logarithm of the ratio of the observed first-order rate coefficient at pressure p to that at very high pressure. This quantity is plotted against $\log_{10}p$ on the abscissa. The name "falloff" curve comes from the fact that the apparent rate coefficient "falls off" with decreasing pressure.

The Lindemann mechanism for thermal gas-phase unimolecular reactions assumes that the energy required for the reaction is obtained by collisions with other molecules via the process

$$A + M \xrightarrow{k_a} A^* + M \text{ (activation)},$$

where A is the molecule undergoing unimolecular decomposition, M is any molecule (including A), and A^* represents an A molecule with energy at or above that required for decomposition. This step is called the *activation* step. If internal energy can be gained via collision, it can also be lost in the same manner. Therefore, there is a deactivation step

$$A^* + M \xrightarrow{k_d} A + M \text{ (deactivation)}.$$

At the same time, A^* molecules exhibit random lifetimes in the activated state, so that first-order decomposition reactions

$$A^* \xrightarrow{k_r} P \text{ (products)}$$

occur.

If the reaction mechanism consists of the preceding three processes, the differential rate equations are

$$\frac{d[A]}{dt} = -k_a[A][M] + k_d[A^*][M], \tag{19.74A}$$

$$\frac{d[A^*]}{dt} = k_a[A][M] - k_d[A^*][M] - k_r[A^*], \tag{19.74B}$$

and

$$\frac{d[P]}{dt} = k_r[A^*]. \tag{19.74C}$$

The measured rate at which products are formed is given by Eq. 19.74C. In many cases, the energy required for the unimolecular reaction of A is substantial. Therefore, very few thermally activated molecules will be present, and we expect the stationary-state approximation to be accurate when applied to A^*. Equating Eq. 19.74B to zero gives us the steady-state concentration of A^*:

$$[A^*]_{ss} = \frac{k_a[A][M]}{k_d[M] + k_r}. \tag{19.75}$$

Substituting $[A^*]_{ss}$ into Eq. 19.74C produces the differential reaction rate expression

$$\frac{d[P]}{dt} = \frac{k_r k_a[A][M]}{k_d[M] + k_r}. \tag{19.76}$$

If we represent the observed reaction rate as an apparent first-order process with a rate coefficient $k(p)$, Eq. 19.76 shows that

$$\frac{d[P]}{dt} = k(p)[A] = \left[\frac{k_r k_a[M]}{k_d[M] + k_r} \right][A],$$

so that

$$\boxed{k(p) = \frac{k_r k_a[M]}{k_d[M] + k_r}.} \tag{19.77}$$

Equation 19.77 provides the qualitative explanation for the falloff behavior seen in Figure 19.23. At high pressure, we have $k_d[M] \gg k_r$. In this limit, the observed rate equation reduces to

$$\frac{d[P]}{dt} = \frac{k_r k_a}{k_d}[A],$$

and the reaction will appear to be first order with a fixed value of $k(p)$ that will behave as if it is independent of pressure. We will also observe first-order kinetics if the reaction to form products does not change the total concentration of gas-phase molecules. This makes $[M]$ a constant at a given pressure, so that the kinetics will appear to be first order. However, the value of $k(p)$ will change with pressure.

In the low-pressure limit where $k_d[M] \ll k_r$, the differential rate equation becomes $d[P]/dt = k_a[M][A]$. If the reaction changes the total concentration of gas-phase molecules, the kinetics will be second order. If, however, $[M]$ is constant, we will again observe first-order kinetics, but with an apparent rate coefficient that decreases with decreasing pressure, as seen in Figure 19.23.

Equation 19.77 provides a means of obtaining some information about the rate coefficients for the three steps in the reaction mechanism. Taking the reciprocal of that equation, we obtain

$$\frac{1}{k(p)} = \frac{k_d}{k_r k_a} + \frac{1}{k_a[M]}. \tag{19.78}$$

If we use the ideal-gas equation of state, $[M]$ can be written in the form

$$[M] = \frac{n}{V} = \frac{P}{RT},$$

so that

$$\frac{1}{k(p)} = \frac{k_d}{k_r k_a} + \frac{RT}{k_a P}. \tag{19.79}$$

This suggests that a plot of the reciprocal of the measured apparent first-order rate coefficient against P^{-1} should be linear with a slope RT/k_a and an intercept $k_d/(k_r k_a)$. We could, therefore, obtain k_a from the slope and the ratio k_d/k_r from the intercept. Unfortunately, it is not that simple. The plot is linear over a wide range of pressures, but it deviates from linearity at high pressures. The problem is our assumption that k_r is a constant regardless of the energy present in A^*. In fact, this is not the case. (Problem 19.48 illustrates the point.) Consequently, the Lindemann mechanism provides a means to qualitatively understand thermal gas-phase unimolecular reactions, but more detailed theory is required to actually obtain the rate coefficients for the various steps.

In solution, the concentration of other molecules is always large. We are, therefore, always in the high-pressure limit, where the kinetics are first order. Falloff curves can be observed only in the gas phase.

19.5.3.2   Enzyme Kinetics

Biological reactions in living organisms are almost all catalyzed by materials called *enzymes*. These molecules are generally proteins that are highly specific in their action. A given enzyme is usually effective in catalyzing one and only one type of reaction. This specificity is achieved by the requirement that the structure of the enzyme "fit" an active site in the chemical structure of the biological reactant, generally called the *substrate*. So demanding is this "lock and key" mechanism for catalysis, that often only one stereoisomer of an optically active enzyme can act as an effective catalyst. The high specificity means that living organisms must have millions of enzymes to catalyze the vast number of reactions needed to sustain life.

The effectiveness of catalytic enzymes is remarkable. When they are present, a biological reaction generally proceeds with ease. When they are absent, the reaction usually takes place at a rate too slow to measure. The physiological action of some poisons is to block the active site on the substrate from the enzyme. By doing so, the poison shuts off vital biological reactions necessary for life. Cyanide is an example of such a poison. Sometimes this mechanism of active-site blockage can be used to advantage. Antihistamines generally act in this manner to suppress allergic reactions. The antihistamine blocks the active site of attack used by the agent to which the organism is allergic.

The *Michaelis–Menten mechanism* is the mechanism most commonly employed to describe enzyme catalysis. In this mechanism, the enzyme (E) reacts with the substrate (S) to form an enzyme–substrate complex (ES), which can then either decompose to regenerate the isolated enzyme and substrate or react to form product P and the original enzyme, which can then re-form the ES complex. Thus, the enzyme itself is not consumed in the process. We can represent this mechanism with a set of four chemical equations:

$$E + S \xrightarrow{k_1} ES,$$

$$ES \xrightarrow{k_{-1}} E + S,$$

$$ES \xrightarrow{k_2} E + P,$$

and

$$E + P \xrightarrow{k_{-2}} ES.$$

The rate of reaction of the substrate is

$$\mathcal{R} = -\frac{d[S]}{dt} = k_1[E][S] - k_{-1}[ES]. \tag{19.80}$$

If the ratio $[E]/[S]$ is very small, we are assured that the concentration of the enzyme–substrate complex will be small relative to those of the substrate and product. Under this condition, a stationary-state approximation for ES will be accurate, so that

$$\frac{d[ES]}{dt} = k_1[E][S] - k_{-1}[ES] - k_2[ES] + k_{-2}[E][P] \approx 0. \tag{19.81}$$

Since the total enzyme is conserved throughout the reaction, we must have

$$[E] = [E]_o - [ES], \tag{19.82}$$

where $[E]_o$ is the initial concentration of enzyme in the system. Substituting Eq. 19.82 into Eq. 19.81 gives

$$\frac{d[ES]}{dt} = k_1\{[E]_o - [ES]\}[S] - k_{-1}[ES] - k_2[ES] + k_{-2}\{[E]_o - [ES]\}[P] \approx 0.$$

Solving for the steady-state concentration of the complex, we obtain

$$[ES]_{ss} = \frac{k_1[S] + k_{-2}[P]}{k_1[S] + k_{-2}[P] + (k_{-1} + k_2)}[E]_o. \tag{19.83}$$

Using this steady-state result, we can express the reaction rate in a more useful form. Substitution of Eq. 19.82 into 19.80 produces

$$\mathcal{R} = -\frac{d[S]}{dt} = k_1\{[E]_0 - [ES]\}[S] - k_{-1}[ES]$$

$$= k_1[S][E]_o - \{k_1[S] + k_{-1}\}[ES].$$

Replacing $[ES]$ with its steady-state value from Eq. 19.83 gives

$$\mathcal{R} = k_1[S][E]_o - \frac{\{k_1[S] + k_{-1}\}\{k_1[S] + k_{-2}[P]\}}{k_1[S] + k_{-2}[P] + (k_{-1} + k_2)}[E]_o. \tag{19.84}$$

If we take a common denominator and collect terms in Eq. 19.84, the result is

$$\mathcal{R} = \frac{k_1k_2[S] - k_{-1}k_{-2}[P]}{k_1[S] + k_{-2}[P] + (k_{-1} + k_2)}[E]_o. \tag{19.85}$$

In most kinetic measurements involving enzyme kinetics, the experimental method involves measuring the initial reaction rate near $t = 0$. At this point, we have $[P] = 0$ and $[S] = [S]_o$. Using Eq. 19.85, we can, therefore, write the initial rate as

$$\boxed{\mathcal{R}_o = \frac{k_1k_2[S]_o[E]_o}{k_1[S]_o + (k_{-1} + k_2)}.}$$

Division of numerator and denominator by k_1 produces

$$\boxed{\mathcal{R}_o = \frac{k_2[S]_o[E]_o}{[S]_o + (k_{-1} + k_2)/k_1} = \frac{k_2[S]_o[E]_o}{[S]_o + K_M},} \tag{19.86}$$

where $K_M = (k_{-1} + k_2)/k_1$ is the *Michaelis constant* and the result is called the *Michaelis–Menten equation*. If we take the reciprocal of both sides, we obtain

$$\boxed{\frac{1}{\mathcal{R}_o} = \frac{1}{k_2[E]_o} + \frac{K_M}{k_2[S]_o[E]_o},} \tag{19.87}$$

which is known as the *Lineweaver–Burk equation*. This equation tells us that if we measure the initial reaction rate at a fixed value of $[E]_o$ for different substrate concentrations and plot $1/\mathcal{R}_o$ versus $1/[S]_o$, the result should be linear, provided that the stationary-state condition $[ES]/[S] \ll 1$ holds. In many enzyme reactions, this is indeed the behavior observed. The intercept of such a plot permits us to obtain k_2, after which the Michaelis constant can be extracted from the measured slope. Problems 19.49 and 19.50 are examples.

There are some similarities between thermal gas-phase unimolecular reactions and enzyme kinetics. The Lindemann mechanism predicts that at high pressures decomposition reactions should follow a first-order rate law. The Michaelis–Menten equation shows that in the limit of high substrate concentration with $[S]_o \gg K_M$, $\mathcal{R}_o = k_2[E]_o$, and the initial rate is first order in enzyme concentration. In the other limit, where $[S]_o \ll K_M$, we have $\mathcal{R}_o = (k_2/K_M)[E]_o[S]_o$, and the initial rate of the enzyme-catalyzed reaction is second order. The same situation exists in gas-phase thermal unimolecular reactions, which become second order at low pressures.

19.5.3.3 Heterogeneous Catalysis: Langmuir Isotherms

Most industrial reactions are carried out with the use of solid-state catalysts. Indeed, a great deal of industrial research involves a search for less expensive, longer lasting, and more effective catalysts. Since the reactants are usually in the gas or liquid phase, we have at least two phases present: either gas–solid or liquid–solid. Consequently, such solid-state catalysts are called *heterogeneous*.

Almost any industrial process can be cited as an example of heterogeneous catalysis. Ammonia is commonly made by reacting $H_2(g)$ with $N_2(g)$ in the presence of solid tungsten (W). Without the catalyst, the activation energy for this conversion is about 350 kJ mol^{-1}. With the tungsten catalyst, the activation energy is lowered to 162 kJ mol^{-1}. Solid iron is also employed as a catalyst in the process. The decomposition of HI(g) to $H_2(g)$ and $I_2(g)$ is catalyzed by either Au(s) or Pt(s), which lowers the activation energy from 184 kJ mol^{-1} to 105 and 59 kJ mol^{-1}, respectively. These same solids catalyze the decomposition of $N_2O(g)$ to $N_2(g)$ and $O_2(g)$. A solid bed of SiO_2/Al_2O_3 often serves as the catalyst for the cracking of high-molecular-mass hydrocarbons to gasoline. Sulfuric acid is made by a Pt(s)- or V_2O_5(s)-catalyzed conversion of $SO_2(g)$ to $SO_3(g)$, followed by reaction with H_2O.

The activity, or effectiveness, of a solid catalyst depends in a very complex manner upon the physical and electronic structure of the solid and the relationship of the lattice spacings to the chemical bonds to be made or broken during the reaction. It also depends upon the magnitude of the molar enthalpy of absorption of the reactants, $|\Delta H_{ab}|$. If $|\Delta H_{ab}|$ is very small, absorption of reactants on the surface will be less likely, with the result that the rate of production of products will be small. On the other hand, if $|\Delta H_{ab}|$ is very large, the reactants will readily absorb, but so may the products. In this event, the inability to clear the surface sites for additional reactant absorption will slow the process. It is, therefore, best for $|\Delta H_{ab}|$ to have an intermediate magnitude. Since the effectiveness of the catalyst depends upon surface absorption, anything that increases the effective surface area will enhance the catalytic activity. For this reason, catalysis are sometimes spread in a thin layer over a support material, or *carrier*. *Promoters* are substances that increase the effectiveness of a catalyst. Their mode of action is often to prevent the catalytic particles from coagulating, which would reduce their surface area.

Since the reaction rate in heterogeneous catalysis depends upon surface absorption, we shall begin by considering rate expressions for such a process. The absorption and desorption of a gaseous molecule $A(g)$ on a surface site, which we shall denote by an asterisk (*), can be represented, respectively, by

$$A(g) \xrightarrow{k_a} * - A$$

and

$$* - A \xrightarrow{k_d} * + A(g)$$

Let N be the total number of absorption sites per unit area, p_A be the partial pressure of $A(g)$, and θ be the fraction of the surface sites that are occupied. θ is often called the *surface coverage*. The law of mass action indicates that the rate of absorption will be proportional to the number of $A(g)$ molecules per unit volume (i.e., to p_A) and to the number of available surface sites per unit area. If the lifetime of absorbed molecules is random, we expect the desorption kinetics to be first order, so that the rate is proportional to the number of *−A species per unit area. Thus, the total absorption rate is

$$\left(\frac{d[A_{(g)}]}{dt}\right)_{absorb} = -k_a p_A N(1 - \theta),$$ (19.88)

since $N(1 - \theta)$ is the number of sites per unit area that are vacant. The desorption rate is

$$\left(\frac{d[A_{(g)}]}{dt}\right)_{desorb} = k_d N\theta.$$ (19.89)

At equilibrium, or steady state, the sum of the absorption and desorption rates is zero. Therefore, the equilibrium coverage, θ_e, can be obtained from the fact that $-k_a p_A N(1 - \theta) + k_d N\theta = 0$. This gives

$$k_a p_A N(1 - \theta_e) = k_d N\theta_e.$$ (19.90)

Solving for θ_e, we obtain

$$\theta_e = \frac{k_a p_A}{k_d + k_a p_A}.$$

Division of numerator and denominator by k_d produces

$$\boxed{\theta_e = \frac{K p_A}{1 + K p_A}},$$ (19.91)

where $K = k_a/k_d$. Equation 19.47 shows that K is the equilibrium constant for this system. Equation 19.91 is called the *Langmuir isotherm*.

If the surface coverage has been measured as a function of pressure, we can easily obtain K. The most convenient method for analyzing such data is to plot $1/\theta_e$ versus $1/p_A$. Taking the reciprocal of both sides of Eq. 19.91 shows us that

$$\frac{1}{\theta_e} = 1 + \frac{1}{K p_A}.$$ (19.92)

Therefore, the equilibrium constant can be obtained directly from the slope of the best straight-line fit to the measured data.

EXAMPLE 19.12

The extent of surface coverage for an absorbate $A(g)$ on a particular surface is measured as a function of the pressure of $A(g)$ at 273 K. The data obtained are given in the following table:

p_A (torr)	θ_e	p_A (torr)	θ_e
0.0	0.000	60.0	0.558
10.0	0.174	70.0	0.595
20.0	0.296	80.0	0.627
30.0	0.387	90.0	0.654
40.0	0.457	100.0	0.677
50.0	0.512		

(A) If we assume that these data can be described by a Langmuir isotherm, determine the best value of K. (B) Plot the entire isotherm; that is, plot θ_e versus p_A over the range $0 \leq p_A \leq 150$ torr. (C) At what pressure will the surface coverage be less than 0.1%? At what pressure will the surface be 99% covered?

Solution

(A) Equation 19.92,

$$\frac{1}{\theta_e} = 1 + \frac{1}{Kp_A}, \tag{1}$$

shows that a plot of $1/\theta_e$ versus $1/p_A$ should be linear. This plot is shown in Figure 19.24. The best straight-line fit to the data is

$$\frac{1}{\theta_e} = 1.00 + \frac{(47.5 \text{ torr})}{p_A}. \tag{2}$$

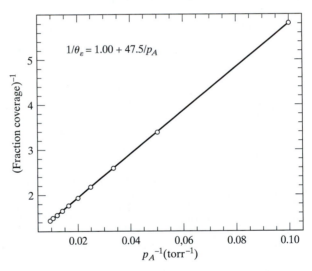

▲ FIGURE 19.24
Plot of the reciprocal of the measured surface coverage versus the reciprocal of the pressure for the data given in Example 19.12. Equation 19.92 in the text shows that the slope of such a plot will be the reciprocal of the equilibrium constant between gas-phase molecules and those absorbed on the surface, provided that the absorption is accurately described by a Langmuir isotherm Eq. 19.91. The best linear fit to the data yields a slope of 47.5 torr, so that $K = 1/47.5$ torr $= 0.0211$ torr^{-1}.

Therefore, K is given by

$$\frac{1}{K} = \text{slope} = 47.5 \text{ torr}, \tag{3}$$

so that

$$K = \frac{1}{47.5 \text{ torr}} = 0.0211 \text{ torr}^{-1}. \tag{4}$$

(B) The plot of the isotherm is shown in Figure 19.25. As the fractional coverage approaches unity, the isotherm will become less accurate because of the formation of multiple layers of coverage.

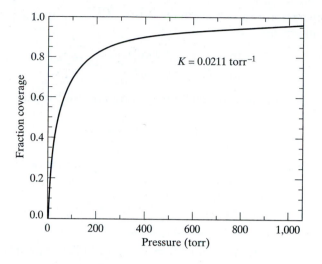

▶ FIGURE 19.25
The Langmuir isotherm, (Eq. 19.91), obtained from the data given in Example 19.12, which yield $K = 0.0211$ torr^{-1}.

(C) Solving Eq. 1 for p_A, we obtain

$$p_A = \frac{\theta_e}{K(1 - \theta_e)}. \tag{5}$$

When $\theta_e = 0.01$, $p_A = 0.479$ torr. When $\theta_e = 0.99$, $p_A = 4.69 \times 10^3$ torr. For related exercises, see Problems 19.51, 19.52, 19.53, and 19.54.

At a given gaseous pressure, it is found that the surface coverage is dependent on the temperature. This is not surprising, since K is the equilibrium constant between $A(g)$ and $*{-}A$ and we found in Eq. 5.85 that the equilibrium constant is temperature dependent. The fact that K varies with temperature means that, to maintain a constant equilibrium surface coverage, we must vary the applied pressure of A with temperature. The pressure required to maintain a constant equilibrium surface coverage can be computed by rearranging Eq. 19.91 to obtain

$$K = \frac{\theta_e}{p_A(1 - \theta_e)}. \tag{19.93}$$

In logarithmic form, Eq. 19.93 is

$$\ln K = \ln\left[\frac{\theta_e}{(1 - \theta_e)}\right] - \ln(p_A).$$

Taking derivatives with respect to temperature at constant equilibrium coverage gives

$$\left(\frac{\partial \ln K}{\partial T}\right)_\theta = -\left(\frac{\partial \ln(p_A)}{\partial T}\right)_\theta. \tag{19.94}$$

The temperature dependence of $\ln(K)$ was developed in the discussion leading to Eq. 5.85. The result is

$$\left(\frac{\partial \ln(K)}{\partial T}\right) = \frac{\overline{\Delta H}}{RT^2}.$$

For the absorption process, we have $\overline{\Delta H} = \overline{\Delta H}_{ab}$, so that Eq. 19.94 can be written in the form

$$\left(\frac{\partial \ln K}{\partial T}\right)_\theta = -\left(\frac{\partial \ln(p_A)}{\partial T}\right)_\theta = \frac{\overline{\Delta H}_{ab}}{RT^2}. \tag{19.95}$$

Separating the variables and then integrating Eq. 19.95 yields

$$\int \partial \ln(p_A) = -\int \frac{\overline{\Delta H}_{ab}}{RT^2} dT.$$

If $\overline{\Delta H}_{ab}$ is independent of temperature, the integration produces

$$\boxed{\ln(p_A) = \left(\frac{\overline{\Delta H}_{ab}}{R}\right)\left(\frac{1}{T}\right) + F(\theta_e)}, \tag{19.96}$$

where $F(\theta_e)$ is an arbitrary function of the equilibrium surface coverage. Equation 19.96 tells us that if we measure the pressure p_A needed to maintain a constant surface coverage, a plot of $\ln(p_A)$ versus T^{-1} will be linear with a slope equal to $\overline{\Delta H}_{ab}/R$. This provides a convenient means of determining the enthalpy of absorption. Problem 19.52 illustrates the point.

The Langmuir isotherm expression in Eq. 19.91 can be expanded to the case in which different compounds are simultaneously absorbing on the same surface. Suppose there are two such compounds, A and B. The general Langmuir mechanism for their absorption–desorption is given by the set of equations

$$A(g) \xrightarrow{k_a} * - A,$$

$$* - A \xrightarrow{k_{da}} * + A(g),$$

$$B(g) + * \xrightarrow{k_b} * - B,$$

and

$$* - B \xrightarrow{k_{db}} * + B(g)$$

The rate of change of $[A(g)]$ is, therefore,

$$\frac{d[A(g)]}{dt} = -k_a p_A N(1 - \theta) + k_{da} N\theta^A,$$

where θ^A and p_A are the fraction of the surface covered by compound A and the partial pressure of $A(g)$, respectively. A similar equation can be written for compound B. When steady-state or equilibrium conditions are attained, we have

$$-k_a p_A N(1 - \theta_e) + k_{da} N\theta_e^A = 0, \tag{19.97}$$

so that

$$\theta_e^A = \frac{k_a p_A(1 - \theta_e)}{k_{da}} = K_A p_A(1 - \theta_e) = K_A p_A(1 - \theta_e^A - \theta_e^B), \tag{19.98}$$

where K_A is the absorption–desorption equilibrium constant for compound A, which is k_a/k_{da}. A similar equation can be written for θ_e^B. When these two equations are combined and solved for the surface fractions covered by A and B, the respective results are (see Problem 19.53)

$$\theta_e^A = \frac{K_A p_A}{1 + K_A p_A + K_B p_B}$$

and

$$\theta_e^B = \frac{K_B p_B}{1 + K_A p_A + K_B p_B}. \qquad (19.99)$$

If we have m different compounds simultaneously absorbing on a single surface, the fraction of the surface covered by compound s is given by

$$\theta_e^s = \frac{K_s p_s}{1 + \sum\limits_{i=1}^{m} K_i p_i}. \qquad (19.100)$$

With the foregoing analysis of absorption–desorption kinetics in hand, we can now turn our attention to heterogeneous surface catalysis. We first consider an example involving the surface absorption of a single reactant, followed by its reaction. The decomposition of $PH_3(g)$ on W at 973 K is an example. The mechanism for such a process consists of the set of reactions

$$A(g) + * \xrightarrow{k_a} * - A,$$

$$* - A \xrightarrow{k_{da}} * + A(g),$$

and

$$* - A \xrightarrow{k_r} \text{products } (F)$$

The rate of production of products is, therefore,

$$\mathcal{R} = \frac{d[F]}{dt} = k_r N \theta_e.$$

At steady state, or equilibrium, θ_e is given by Eq. 19.91. Substitution produces

$$\frac{d[F]}{dt} = \frac{k_r N K p_A}{1 + K p_A}. \qquad (19.101)$$

At low pressure, where $K p_A \ll 1$, Eq. 19.101 becomes

$$\frac{d[F]}{dt} = k_r N K p_A, \qquad (19.102)$$

and the rate is first order with respect to $A(g)$. At high pressure, with $K p_A \gg 1$, the result is

$$\frac{d[F]}{dt} = k_r N, \qquad (19.103)$$

and we have a zero-order rate law. Since $d[F]/dt$ attains a maximum value at high pressure, industrial processes of this type are always run under zero-order conditions. If the second phase is a liquid instead of a gas, we observe zero-order kinetics, since the concentration of reactant near the surface is very large. First-order kinetics in heterogeneous catalytic systems involving a single compound can be observed only when the reactant is a gas at low pressure. Figure 19.26 illustrates these points.

If the reaction of interest is a bimolecular reaction between A and B, then there are two possibilities for the reaction mechanism. Sometimes, one of the reactants can absorb on the surface and then react with the second

Low surface coverage

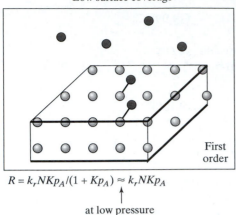

First order

$$R = k_r N K p_A / (1 + K p_A) \approx k_r N K p_A$$

↑
at low pressure

(A)

High surface coverage

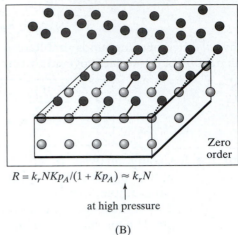

Zero order

$$R = k_r N K p_A / (1 + K p_A) \approx k_r N$$

↑
at high pressure

(B)

▲ **FIGURE 19.26**
Heterogeneous surface catalysis involving a single reactant, ●, chemisorbing on active surface sites, ◯. (A) At low pressure and low surface coverage, the kinetics are first order. (B) At high pressures and high surface coverage, zero-order kinetics are observed in which the rate of production of the products is a constant.

Low surface coverage

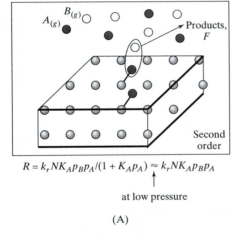

Products, F

Second order

$$R = k_r N K_A p_B p_A / (1 + K_A p_A) \approx k_r N K_A p_B p_A$$

↑
at low pressure

(A)

High surface coverage

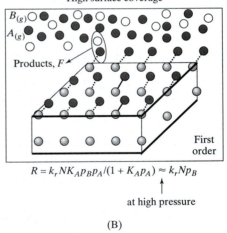

Products, F

First order

$$R = k_r N K_A p_B p_A / (1 + K_A p_A) \approx k_r N p_B$$

↑
at high pressure

(B)

▲ **FIGURE 19.27**
The Rideal–Eley mechanism for heterogeneous surface catalysis. In this mechanism, one reactant, $A(g)$, denoted by ●, chemisorbs on an active surface site, ◯. Subsequently, the chemisorbed molecule reacts with a gas-phase or solution-phase B molecule, ◯, to form products. (A) At low pressure of $A(g)$, the reaction exhibits second-order kinetics. (B) At high pressure and high coverage of A, the reaction will be first order with respect to $B(g)$.

compound in the liquid or gas phase. (See Figure 19.27.) Such a process is said to occur via a *Rideal–Eley* mechanism. We can represent it by the reactions

$$A(g) \xrightarrow{k_a} * - A,$$

$$* - A \xrightarrow{k_{da}} * + A(g)$$

and

$$* - A + B(g) \xrightarrow{k_r} \text{products } (F).$$

The rate of production of products for this mechanism is $d[F]/dt = k_r p_B N \theta_e$. Use of the Langmuir isotherm for θ_e gives

$$\frac{d[F]}{dt} = \frac{k_r K_A N p_B p_A}{1 + K_A p_A}. \qquad (19.104)$$

At high pressure of the absorbed component, a Rideal–Eley reaction becomes first order in the gas- or liquid-phase component and zero order in the surface reactant. At low partial pressure of the absorbed reactant, the reaction is second order.

More commonly, heterogeneous catalysis involves the surface absorption of both reactants, followed by their surface reaction, as shown in Figure 19.28. This type of process is said to occur via a *Langmuir–Hinshelwood* mechanism. The reactions involved are

$$A(g) + * \xrightarrow{k_a} * - A,$$
$$* - A \xrightarrow{k_{da}} * + A(g),$$
$$B(g) + * \xrightarrow{k_b} * - B,$$
$$* - B \xrightarrow{k_{db}} * + B(g),$$

and

$$* - A + * - B(g) \xrightarrow{k_r} \text{products } (F).$$

In this case, we have $d[F]/dt = k_r(N\theta_e^A)(N\theta_E^B)$. Equation 19.99 then gives

$$\frac{d[F]}{dt} = \frac{k_r N^2 K_A K_B p_A p_B}{(1 + K_A p_A + K_B p_B)^2}. \qquad (19.105)$$

Low surface coverage

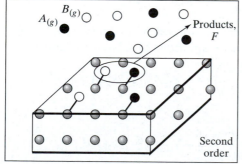

Second order

$$R = k_r N^2 K_B K_A p_B p_A / (1 + K_B p_B + K_A p_A)^2 \approx k_r N^2 K_B K_A p_B p_A$$

at low pressure

(A)

High surface coverage

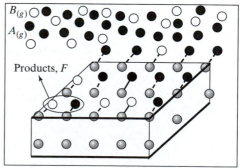

$$R = k_r N^2 K_B K_A p_B p_A / (1 + K_B p_B + K_A p_A)^2$$

(B)

▲ FIGURE 19.28
The Langmuir–Hinshelwood mechanism for heterogeneous surface catalysis. In this mechanism, two reactants, A and B, denoted by ● and ○, respectively, chemisorb on active surface sites, ◉. Subsequent to chemisorption, A and B react as shown in the figure. (A) At low pressures of A and B, the reaction kinetics will appear to be second order. (B) At high pressures, no simple rate law describes the kinetics.

At low partial pressures of both A and B, the kinetics follow a simple second-order rate law. At higher pressures, no simple rate law describes the kinetics. Qualitatively, we can note that if we have $p_A \gg p_B$ or vice versa, the reaction rate may be very slow. For example, in the limit where $p_A \gg p_B$, Eq. 19.105 becomes $d[F]/dt = k_r N^2 K_B p_B/(K_A p_A)$, which may be very small because p_B/p_A is near zero. Physically, the problem is that all the surface sites become occupied by compound A, which decreases the reaction rate to zero. The optimum conditions involve adjusting the partial pressures so that equal amounts of A and B absorb on the surface.

The preceding discussion shows that the kinetics of heterogeneous catalysis depend upon the details of the absorption–desorption reaction mechanism. For this reason, we can often obtain important information about the mechanism by conducting rate measurements in both the high- and low-pressure regions. Problems 19.55 and 19.56 illustrate this point.

19.6   Differential Methods

The problem of analyzing data for complex reaction mechanisms can be simplified if we either measure or compute the rates from the measured concentration data. This procedure enables us to work with the differential equations themselves rather than their integrated forms. Since the differential rate expressions are all linear algebraic equations in the rate coefficients, the process of data analysis is greatly simplified.

As an example, consider the case of concurrent first- and second-order reactions of a single reactant:

$$A \xrightarrow{k_1} B \qquad \text{and} \qquad A + A \xrightarrow{k_2} C.$$

$$\text{Reaction 1} \qquad\qquad \text{Reaction 2}$$

The integrated rate equation for this system is given in Eq. 19.54. This expression is transcendental and, therefore, somewhat difficult to solve. However, the differential equation itself has the simple form

$$\frac{d[A]}{dt} = -k_1[A] - 2k_2[A]^2. \tag{19.106}$$

If we either measure or compute the rates at concentrations $[A]_1$ and $[A]_2$, we will have two simple linear equations to solve for k_1 and k_2:

$$\left(\frac{d[A]}{dt}\right)_1 = -k_1[A]_1 - 2k_2[A]_1^2$$

and

$$\left(\frac{d[A]}{dt}\right)_2 = -k_1[A]_2 - 2k_2[A]_2^2. \tag{19.107}$$

These equations can easily be solved for the two rate coefficients. The correctness of the assumed mechanism can be verified by choosing two different concentrations, $[A]_3$ and $[A]_4$, at which to measure or compute the rates. If the mechanism is correct, the values of k_1 and k_2 will be the same within the experimental error present in determining the rates. By repeating such measurements and calculations several times and averaging the results for the rate coefficients, reasonably accurate values can be determined. In many experimental situations, the concentrations are chosen to be the initial ones. In this case, the procedure is sometimes called the *initial-rate method*.

The procedure of working directly with the rates rather than with time-dependent concentrations makes our task of data analysis easier. However, there is a price to be paid for this advantage. The price comes in the form of a much greater demand for accuracy on the experimental measurements. It is far more difficult to measure the rate of change of the concentration accurately than to measure the concentration itself. If $[A]$ can be measured with an experimental error of 1%, the error in $d[A]/dt$ will generally be significantly greater than 1%. For this reason, the measurements or calculations need to be repeated several times and appropriate averages taken so that random errors will tend to cancel. If we are extracting derivatives numerically from the measured concentrations, an accurate analytical method such as that described in Eqs. 9.173 and 9.174 should be employed. Derivatives should never be obtained by attempting to visually draw tangents to a curve. These points are illustrated in Problems 19.57 and 19.58.

EXAMPLE 19.13

An investigator suspects that a single reactant is simultaneously undergoing first- and second-order reactions via the mechanism

$$A \xrightarrow{k_r} B \quad \text{and} \quad A + A \xrightarrow{k_2} C$$

She measures the initial rates of the reaction when $[A]_o = 0.100$ mol L^{-1} and $[A]_o = 0.200$ mol L^{-1} and obtains -0.000700 mol L^{-1} min^{-1} and -0.00180 mol L^{-1} min^{-1}, respectively. **(A)** What values of k_1 and k_2 do these results suggest? **(B)** How can the investigator obtain evidence that her suspected mechanism is correct?

Solution

The rate of change of $[A]$ is given by Eq. 19.106:

$$\frac{d[A]}{dt} = -k_1[A] - 2k_2[A]^2. \tag{1}$$

Using this result and Eq. 19.107, we find from the data that

$$-0.000700 = -k_1(0.100) - 2k_2(0.100)^2 = -0.100k_1 - 0.0200k_2 \tag{2}$$

and

$$0.00180 = 0.200k_1 + 0.0800k_2. \tag{3}$$

Solving Eq. 2 for k_1 in terms of k_2 gives

$$k_1 = \frac{0.000700 - 0.0200k_2}{0.100} = 0.00700 - 0.200k_2. \tag{4}$$

Substituting Eq. 4 into Eq. 3 produces

$$0.00180 = 0.20[0.00700 - 0.200k_2] + 0.0800k_2 = 0.0014 + 0.040k_2. \tag{5}$$

Thus,

$$k_2 = \frac{0.0018 - 0.0014}{0.04} \text{ L mol}^{-1} \text{ min}^{-1} = 0.0100 \text{ L mol}^{-1} \text{ min}^{-1}, \tag{6}$$

where the units are those for a second-order reaction. Finally, substituting k_2 into Eq. 4 gives

$$k_1 = 0.00700 - 0.200(0.01) = 0.00500 \text{ min}^{-1}, \tag{7}$$

which are the units for a first-order reaction. To obtain evidence that the mechanism does indeed involve concurrent first- and second-order reactions of a single compound, the investigator needs to repeat the rate measurements at different initial concentrations and repeat the calculation. If the mechanism is correct, k_1 and k_2 will be nearly the same, regardless of the values of the initial concentrations used.

For related exercises, see Problems 19.57 and 19.58.

19.7   Relaxation Methods

When we have an equilibrium system with very large forward and reverse rate coefficients, their values can often be most conveniently measured by using some type of experimental relaxation method. These methods involve the introduction of a small perturbation that causes the equilibrium position to shift. The result is a rapid response of the system to move toward the new equilibrium point. The temporal variation of the system response is measured by some suitable detector. Since the response rate depends upon the forward and reverse rate coefficients, the data allow each constant to be determined. To be effective, the rate at which the perturbation is introduced must be large compared with the response rate of the system. Thus, if the system adjusts to the new equilibrium point in 10^{-6} s, the perturbation needs to be introduced in 10^{-8} s or less.

The most common relaxation method is temperature jump, or T-jump. If the molar enthalpy change for the reaction is not zero, there will be a small change in the equilibrium constant if the temperature is altered. This small change shifts the system away from the equilibrium point so that its relaxation to a new equilibrium position can be followed. Very rapid temperature jumps of a few degrees can be achieved through the use of electrical discharges in conducting systems or a powerful laser. If the equilibrium position shifts with pressure, pressure-jump or ultrasonic methods can be employed. In a pressure-jump experiment, the system pressure is rapidly altered by mechanical means. An ultrasonic measurement produces a rapid, small pressure change because of the sinusoidal pressure wave associated with the ultrasonic pulse.

Suppose we have two opposing first-order reactions at equilibrium in which the forward reaction is exothermic. This situation can be represented by the pair of equations

$$A \xrightarrow{k_1} B - \Delta\overline{H}$$

and

$$-\Delta\overline{H} + B \xrightarrow{k_2} A.$$

The rate of change of $[A]$ is

$$\mathcal{R} = -\frac{d[A]}{dt} = k_1[A] - k_2[B]. \tag{19.108}$$

At equilibrium, we have

$$\mathcal{R} = -\frac{d[A]}{dt} = k_1[A]_e - k_2[B]_e = 0, \tag{19.109}$$

where the equilibrium concentrations are represented by $[A]_e$ and $[B]_e$. If the concentrations are small, so that we are near the reference state of zero concentration, we will have

$$K_{eq} = \frac{k_1}{k_2} = \frac{[B]_e}{[A]_e}.$$

If the temperature is now raised a small amount, K_{eq} will decrease, and the reaction will be begin to shift toward compound A, provided that $\Delta \overline{H} < 0$. This means that at the instant the perturbation is introduced, $[A]$ is displaced from its new equilibrium position by an amount Δn_o, and $[B]$ is displaced by a similar amount. If we introduce the variable

$$\Delta n = [A] - [A]_e \qquad (19.110)$$

we will have

$$[A] = [A]_e + \Delta n$$
$$[B] = [B]_e - \Delta n$$

and

$$\frac{d[A]}{dt} = \frac{d\Delta n}{dt}.$$

Substituting these equations into Eq. 19.108 gives

$$-\frac{d\Delta n}{dt} = k_1\{[A]_e + \Delta n\} - k_2\{[B]_e - \Delta n\}$$

$$= \{k_1[A]_e - k_2[B]_e\} + (k_1 + k_2)\Delta n. \qquad (19.111)$$

Equation 19.109 shows that the first term on the right-hand side of Eq. 19.110 vanishes. Thus, the differential rate equation describing the adjustment to the new equilibrium position is

$$\frac{d\Delta n}{dt} = -(k_1 + k_2)\Delta n. \qquad (19.112)$$

Note that, in setting $\{k_1[A]_e - k_2[B]_e\}$ equal to zero, we are assuming that k_1 and k_2 are not affected by the small temperature jump. For this to be true, the temperature change must be very small.

Equation 19.111 describes a first-order reaction whose integrated rate expression is

$$\Delta n = \Delta n_o \exp[-(k_1 + k_2)t]. \qquad (19.113)$$

Thus, if we monitor the change in the number of moles L^{-1} of A, Δn, as a function of time and present the results on an oscilloscope, the trace will be a decaying exponential such as that shown in Figure 19.29. The relaxation

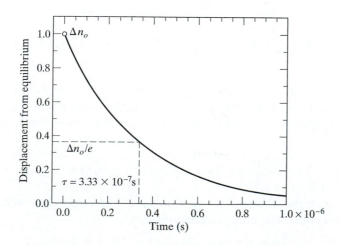

◀ **FIGURE 19.29**
Typical first-order relaxation curve resulting from a rapid temperature jump in an endo- or an exothermic reaction. The time at which the initial displacement from equilibrium, Δn_o, falls to $\Delta n_o/e$ is the relaxation time τ of the system. This relaxation time can be related to the forward and reverse rate coefficients for the equilibrium by using the differential equations describing the kinetics of the process, as shown in the text.

time for the system occurs at the point $\Delta n = \Delta n_o/e$. Since the relaxation time for a first-order process is just the reciprocal of the rate coefficient, we will have

$$\tau = \frac{1}{k_1 + k_2}. \tag{19.114}$$

We could also plot $\ln(\Delta n)$ versus time and obtain $k_1 + k_2$ from the slope of the best linear fit to the data. The ratio of k_1 to k_2 is known from the equilibrium constant, so a measurement of the sum provides both rate coefficients.

If the equilibrium reaction has a second-order process, the same technique can be used, but the expression for the relaxation time will be different. For example, if the equilibrium is

$$A + B \xrightarrow{k_1} C - \Delta\overline{H}$$

and

$$-\Delta\overline{H} + C \xrightarrow{k_2} A + B,$$

then the rate is given by

$$\mathcal{R} = -\frac{d[A]}{dt} = k_1[A][B] - k_2[C]. \tag{19.115}$$

At equilibrium, we have

$$-\frac{d[A]}{dt} = k_1[A]_e[B]_e - k_2[C]_e = 0. \tag{19.116}$$

If we now jump the temperature so that the equilibrium shifts toward A and B, we can write

$$[A] = [A]_e + \Delta n,$$
$$[B] = [B]_e + \Delta n,$$
$$[C] = [C]_e - \Delta n,$$

and

$$\frac{d[A]}{dt} = \frac{d\Delta n}{dt}.$$

Substituting into Eq. 19.115 gives

$$-\frac{d\Delta n}{dt} = k_1\{[A]_e + \Delta n\}\{[B]_e + \Delta n\} - k_2\{[C]_e - \Delta n\}$$

$$= \{k_1[A]_e[B]_e - k_2[C]_e\} + [k_1\{[A]_e + [B]_e\} + k_2]\Delta n + k_1\Delta n^2. \tag{19.117}$$

Equation 19.116 shows us that the $\{k_1[A]_e[B_e] - k_2[C_e]\}$ term in Eq. 19.117 is zero. Furthermore, if the displacement from the original equilibrium position is very small, we can ignore the term in Δn^2 relative to the Δn term. This gives

$$\frac{d\Delta n}{dt} = -[k_1\{[A]_e + [B]_e\} + k_2]\Delta n, \tag{19.118}$$

which is again the differential equation for a first-order approach to a new equilibrium point. The integrated expression is

$$\Delta n = \Delta n_o \exp[-[k_1\{[A]_e + [B]_e\} + k_2]t]. \tag{19.119}$$

The relaxation time for the system will now be

$$\tau = \frac{1}{[k_1\{[A]_e + [B]_e\} + k_2]}.$$ (19.120)

The reciprocal of τ is, therefore,

$$\frac{1}{\tau} = [k_1\{[A]_e + [B]_e\} + k_2].$$ (19.121)

We can now perform a series of measurements using different concentrations of A and B so that $\{[A]_e + [B]_e\}$ varies. Equation 19.121 shows us that if we plot τ^{-1} versus $\{[A]_e + [B]_e\}$, the result will be a straight line with intercept k_2 and slope k_1. Problem 19.59 provides an example of such an analysis. Problem 19.60 extends the preceding analysis to the case where the equilibrium system comprises two bimolecular second-order reactions.

The power of relaxation methods lies in the fact that the kinetics of an equilibrium system's adjustment to a new equilibrium point are always first order if the perturbation is small. The forms for the rate coefficient and relaxation time depend upon the order of the equilibrium reactions.

Summary: Key Concepts and Equations

1. The gas-phase reaction rate for a bimolecular process can be obtained by multiplying the collision frequency by the average reaction probability for collision of the molecules in particular relative speed and impact parameter ranges and then summing over all possible collision speeds and impact parameters. This gives a reaction rate of the form

$$\mathcal{R} = \left[\int_{V=0}^{\infty} \int_{b=0}^{\infty} \langle P(V, b)\rangle 4\pi \left[\frac{\mu_{AB}}{2\pi kT}\right]^{3/2} (2\pi b)db \, \exp\left[-\frac{\mu V^2}{2kT}\right] V^3 dV \right] N_A N_B,$$

where N_A and N_B are the concentrations of A and B, respectively. The quantity inside the brackets depends only upon temperature. It is called the specific reaction rate coefficient. If we write this quantity as $k(T)$, the expression for the reaction rate becomes

$$\mathcal{R} = k(T)N_A N_B.$$

2. If we define the rate for the reaction $aA + bB \longrightarrow cC + dD$ as

$$\mathcal{R} = -\frac{1}{a}\frac{d[A]}{dt} = -\frac{1}{b}\frac{d[B]}{dt} = \frac{1}{c}\frac{d[C]}{dt} = \frac{1}{d}\frac{d[D]}{dt},$$

then

$$\mathcal{R} = -\frac{d[A]}{dt} = -\frac{d[B]}{dt} = k(T)N_A N_B$$

for bimolecular concerted gas-phase reactions. Reactions of this form are said to be second order. If both reactants are the same, the expression becomes

$$\mathcal{R} = -\frac{1}{2}\frac{d[A]}{dt} = k(T)N_A^2.$$

3. The integrated rate expression for a second-order reaction between A and B is

$$\ln\left[\frac{[A]}{[B]}\right] = \ln\left[\frac{[A]_o}{[B]_o}\right] + ([A]_o - [B]_o)k(T)t,$$

where $[A]$ and $[B]$ are the concentrations of A and B, respectively, in mol L^{-1}, and $[A]_o$ and $[B]_o$ are the concentrations at $t = 0$. This result tells us that a plot of $\ln([A]/[B])$ versus time will be linear with a slope equal to $([A]_o - [B]_o)k(T)$.

4. If a bimolecular reaction involves only a single reactant, the integrated rate expression becomes

$$\frac{1}{[A]} = \frac{1}{[A]_o} + 2k(T)t.$$

In this case, a plot of $[A]^{-1}$ versus time will be linear with a slope equal to $2k(T)$.

5. There are two generally used methods for determining the order of a reaction and the specific reaction rate coefficient. In the first method, an appropriate expression involving the concentrations is plotted against time. If the result is linear, the order of the reaction is determined and the rate coefficient can be obtained from the slope. In the second procedure, the integrated rate expression is solved for the rate coefficient. The kinetic data at different times are substituted into this expression and values of $k(T)$ computed. If these values are constant within the bounds of experimental error, the order of the reaction is confirmed, and the best value for $k(T)$ is then obtained by averaging all the computed values for this quantity.

6. Bimolecular second-order reactions in which one of the reactants is present in great excess will appear to be first order because the concentration of the reactant that is in excess will behave as if it is constant throughout the reaction. Under these conditions, the differential rate equation becomes

$$\mathcal{R} = -\frac{d[A]}{dt} = k(T)[A][B] \approx k(T)[B]_o[A] = k_p(T)[A]$$

if $[B]_o \gg [A]_o$. This kind of reaction is called a pseudo first-order reaction, and $k_p(T)$ is the pseudo first-order rate coefficient. The integrated expression for such a process has the exponential form

$$[A] = [A]_o \exp[-k_p(T)t],$$

so a plot of $\ln[A]$ versus time will be linear with a slope equal to $-k_p(T)$.

7. The half-life $t_{1/2}$ and relaxation time τ are defined to be the times required for the concentration $[A]_o$ of a given reactant A at time t_o to decrease to $[A]_o/2$ and $[A]_o/e$, respectively. For a first-order reaction, these times are respectively given by

$$t_{1/2} = \frac{\ln(2)}{k(T)} \quad \text{and } \tau = \frac{1}{k(T)}.$$

As such, $t_{1/2}$ and τ are independent of the initial concentration at $t = t_o$. In general, $t_{1/2}$ and τ are inversely proportional to $k(T)$. For other reaction orders, however, they depend upon $[A]_o$.

8. The temperature dependence of the rate coefficient can usually be accurately described by the exponential form

$$k(T) = AT^n \exp\left[-\frac{E_a}{RT}\right].$$

This is a consequence of the fact that, in most cases, there is an energy requirement for reaction. The required energy, E_a, is called the activation energy. If the temperature range involved in the experiments is not large, n can usually be taken to be zero. The resulting expression is called the Arrhenius equation.

9. When molecules or nuclei reside in metastable states, true unimolecular reactions to more stable states can be observed. If the lifetimes of the molecules in the metastable states are random, the kinetics will be first order. Examples of such true first-order unimolecular reactions include nuclear decay processes, fluorescence and phosphorescence emission from molecules in excited electronic states, and the relaxation of molecules in excited nuclear spin states.

10. All systems undergoing true first-order unimolecular reactions exhibit random lifetimes in their initial states. When the lifetimes are random, the lifetime distribution is

$$P(t_L)dt_L = k \exp[-kt_L]dt_L.$$

$P(t_L)dt_L$ is the probability that a molecule will have a lifetime in the range from t_L to $t_L + dt_L$ when the rate coefficient for the reaction is k. The most probable lifetime is zero. Evaluation of the average lifetime, $\langle t_L \rangle$, shows that it is equal to the relaxation time τ for the system.

11. Concerted termolecular reactions that lead to third-order kinetics are very rare, due to the low probability of three-body collisions. The most common example of such processes are recombination reactions that require a third body to remove the exothermicity of reaction and thereby prevent immediate dissociation. Reactions of $NO(g)$ are the most common examples of chemical reactions that appear to be termolecular processes.

12. Many chemical processes involve several reactions that occur either simultaneously, sequentially, or both. The ensemble of reactions that eventually leads to the final products is often called the reaction mechanism. If we know all the chemical reactions and their associated rate coefficients, we can always predict the temporal variations of the concentrations for all species. However, it is not always possible to do the reverse; that is, if we have measured the concentrations as functions of time, it does not follow that we can always determine the reaction mechanism.

13. The analysis of the kinetics of equilibrium systems demonstrates that rates should be expressed in terms of the activities of the components rather than their concentrations or pressures. However, in almost all cases, kinetic measurements are conducted in very dilute solutions. Therefore, if we take molarity to be the reference function and infinite dilution as the reference state, we will be approaching the reference state, so that the activity coefficients will approach unity and the molarity and activity will become nearly equal.

14. The kinetics of concurrent first-order reactions are described by a first-order rate law in which the total reaction rate coefficient is just the sum of the rate coefficients for each of the reactions. The overall relaxation time is related to the relaxation times for the individual reactions by

$$\frac{1}{\tau} = \frac{1}{\tau_1} + \frac{1}{\tau_2} + \cdots + \frac{1}{\tau_n}.$$

15. An integrated rate expression for consecutive first-order reactions can be obtained using integrating factors to solve the differential equations. If the initial concentrations of all intermediates are zero, each intermediate concentration will rise to a maximum and then decay toward zero an infinite time later. When there are only two consecutive reactions, the time at which the intermediate attains a maximum is given by

$$t_m = \frac{\ln(k_2/k_1)}{k_2 - k_1}.$$

Therefore, the measurement of t_m provides a convenient means of obtaining both reaction rate coefficients, since k_1 can be easily determined from the first-order reaction of the initial reactant.

16. The kinetics of complex systems containing consecutive reactions can often be simplified by employing the stationary-state approximation. This approximation states that in time domains in which the concentration or number of atoms of an intermediate is small relative to the concentrations of both reactants and products, the derivative of the concentration of the intermediate with respect to time will be near zero. In practice, this means that the rate of formation of an intermediate must be slow relative to the rate at which the intermediate reacts. Application of the stationary-state approximation leads directly to the concept of a rate-determining step. This concept tells us that, in a sequence of consecutive reactions, the rate of production of the final products depends only upon the rate coefficient for the last slow step in the sequence and all rate coefficients preceding it. The power of the stationary-state approximation lies in the fact that its use converts differential equations into algebraic equations.

17. The Lindemann mechanism is an attempt to explain the fact that stable gas-phase molecules at elevated temperatures are often observed to undergo apparent first-order unimolecular decomposition reactions. However, the measured rate coefficients decrease with pressure, and at low pressures the kinetics often become second order. In the Lindemann mechanism, molecules are activated above the energy required for reaction via collisions. Subsequent to activation, the molecules may undergo first-order decomposition reactions or collisional deactivation. Use of the stationary-state approximation leads to the result that the rate coefficient for the production of products has the form

$$k(p) = \frac{k_r k_a [M]}{k_d [M] + k_r},$$

where k_a and k_d are, respectively, the activation and collisional deactivation rate coefficients, k_r is the rate coefficient for reaction of activated molecules, and $[M]$ represents the total concentration of gas-phase molecules. At high pressures, we have $k_d[M] \gg k_r$, and therefore, $k(p) = k_r k_a / k_d =$ a constant. However, at lower pressures, $k(p)$ becomes $k_a[M]$, which decreases as the pressure drops. From a qualitative point of view, the Lindemann mechanism explains all the observations. Its quantitative accuracy, however, is limited because of the assumption that k_r is a constant independent of the excitation energy present in the molecule.

18. The Michaelis–Menten mechanism models enzyme kinetics as a sequence of four reactions: (1) the formation of an enzyme–substrate (ES) complex, (2) the decomposition of ES back to enzyme and substrate, (3) the reaction of ES to form the free enzyme and the product, and (4) the reaction of the enzyme with the product to re-form ES. The use of the stationary-state approximation leads to the expression

$$\mathcal{R}_o = \frac{k_2 [S]_o [E]_o}{[S]_o + K_M}$$

for the initial reaction rate, where $K_M = (k_{-1} + k_2)/k_1$ is the Michaelis constant and k_2 is the rate coefficient for the reaction of ES to form the product. Many enzyme-catalyzed reactions are accurately described by this model. If $1/\mathcal{R}_o$ is plotted against $1/[S]_o$ at a fixed enzyme concentration, the values of K_M and k_2 can be determined from the slope and intercept of the curve.

19. Most industrial processes are conducted using a solid-state, heterogeneous catalyst. The mechanisms describing such surface reactions all involve the initial absorption of one or more reactants on the catalytic surface. The extent of surface coverage at equilibrium or steady state, θ_e, is generally described by a Languir isotherm,

$$\theta_e = \frac{Kp_A}{1 + Kp_A},$$

where p_A is the pressure of the compound undergoing surface absorption and K is the equilibrium constant between gas-phase $A(g)$ and absorbed $*-A$ molecules. By measuring the pressure required to maintain a constant surface coverage as a function of temperature, the molar enthalpy of absorption can be determined. If we have m different compounds simultaneously absorbing on a single surface, the fraction of the surface covered by compound s is given by

$$\theta_e^s = \frac{K_s p_s}{1 + \displaystyle\sum_{i=1}^{m} K_i p_i}.$$

20. A surface-catalyzed reaction involving a single reactant has a production rate given by

$$\frac{d[F]}{dt} = \frac{k_r N K p_A}{1 + K p_A},$$

so that the rate will be first order at low pressures and zero order at high pressures. A bimolecular surface-catalyzed reaction in which one reactant absorbs on the surface and then reacts with the second reactant while it is still in the gas or liquid phase is said to occur via a Rideal–Eley mechanism. This type of process has a production rate

$$\frac{d[F]}{dt} = \frac{k_r K_A N p_B p_A}{1 + K_A p_A},$$

where compound A is the one absorbing on the surface. At high pressure, this rate becomes first order; at low pressure, it is second order. When both compounds absorb on the surface and then react, a Langmuir–Hinshelwood reaction mechanism is said to be present. In this case, the production rate is given by

$$\frac{d[F]}{dt} = \frac{k_r N^2 K_A K_B p_A p_B}{(1 + K_A p_A + K_B p_B)^2},$$

which reduces to a simple rate law only in the case where $K_A p_A + K_B p_B \ll 1$. For this special case, the kinetics will appear to be second order.

21. Data analysis can be greatly simplified by measuring or calculating the rate, rather than the temporal dependence of the concentrations. However, the accuracy demands on the experimental measurements are much greater when this procedure is employed.

22. The forward and reverse rate coefficients for equilibrium systems can be conveniently determined by using relaxation methods in which the system is subjected to a small perturbation that shifts it away from the equilibrium point. The rapid response of the system to move toward the new equilibrium point is measured by a suitable detector. Any perturbation that shifts the equilibrium can be employed. The most common examples are temperature and pressure jump. These methods require that both the forward and reverse rate coefficients be large and that the rate at which the perturbation is introduced be large relative to the response rate of the system.

The advantage of relaxation methods lies in the fact that the kinetics of the system's adjustment to a new equilibrium point are always first order if the perturbation is small. The form of the rate coefficient and relaxation time depends upon the order of the equilibrium reactions. If both reactions are first order, the relaxation time for the adjustment to the new equilibrium position is

$$\tau = \frac{1}{k_1 + k_2},$$

where k_1 and k_2 are the forward and reverse rate coefficients, respectively. If the forward reaction is second order and the reverse reaction first order, the result is

$$\tau = \frac{1}{[k_1\{[A]_e + [B]_e\} + k_2]},$$

where $[A]_e$ and $[B]_e$ are the equilibrium concentrations of reactants prior to the introduction of the small perturbation.

Problems

Problems that require the use of some type of computational device are marked with an asterisk (*). Problems that require some type of plotting routine are indicated with a pound sign (#).

19.1 (A) If we measure concentrations in mol L^{-1}, what are the units on the rate coefficient for a concerted bimolecular reaction $A + B \longrightarrow$ products?

(B) What are the units on $k(T)$ for the third-order reaction discussed in Example 19.1?

19.2 An investigator measures the rate coefficient for the reaction $A + B \longrightarrow$ products and finds it to be 0.025 L mol^{-1} min^{-1} at 300 K.

(A) Express $k(T)$ at 300 K in units of m^3 molecule^{-1} s^{-1}.

(B) Express $k(T)$ at 300 K in units of U.S. gallons (slug-mole)$^{-1}$ fortnight^{-1}.

19.3 (A) Show that Eq. 19.11 is correct.

(B) Starting with Eq. 19.12, derive Eqs. 19.13 and 19.14.

19.4 Show that in the limit where $[A]_o \rightarrow [B]_o$, Eq. 19.14 reduces identically to Eq. 19.16, except for the factor of two on the right-hand side.

19.5 For the reaction $2A \longrightarrow B + C$, the following data are obtained for $[A]$ as a function of time at 310 K:

Time (min)	$[A]$ (mol L^{-1})
0	0.800
8	0.659
24	0.487
40	0.387
60	0.302
100	0.218

(A) By suitable means, establish the order of the reaction.

(B) What is the value of the rate coefficient?

(C) Calculate the rate of formation of B at $t = 30$ min.

19.6 The kinetics of a gas-phase dimerization reaction $A(g) + A(g) \longrightarrow A_2(g)$, are investigated by measuring the temporal variation of the pressure at 340 K inside a reaction chamber whose volume is fixed. The data are as follows:

Time (min)	P_{total} (torr)	Time (min)	P_{total} (torr)
0.00	212.0	140.00	155.0
20.00	196.9	160.00	151.5
40.00	185.6	180.00	148.5
60.00	176.7	200.00	145.8
80.00	169.7	250.00	140.4
100.00	163.9	300.00	136.3
120.00	159.1		

Assuming that the pressures are sufficiently low that the gases may be assumed to be ideal, determine the reaction order with respect to compound A and the rate coefficient at 340 K.

19.7 An investigator is measuring kinetic data for a rather slow concerted bimolecular reaction between compounds A and B at $T = 320$ K. The initial concentrations of A and B are $[A]_o = 0.0400$ mol L^{-1} and $[B]_o = 0.0300$ mol L^{-1}. While the experiment is in progress, the investigator leaves the laboratory for a few moments, during which a jealous rival enters, raises the temperature of the reaction vessel to 335 K, and then leaves. The investigator returns and completes the measurements of the temporal variation of $[A]$. The data are as follows:

Time (min)	$[A]$ (mol L^{-1})	Time(min)	$[A]$ (mol L^{-1})
0.00	0.0400	160.00	0.0306
10.00	0.0392	180.00	0.0298
20.00	0.0384	200.00	0.0290
40.00	0.0370	220.00	0.0271
60.00	0.0357	240.00	0.0256
80.00	0.0346	260.00	0.0242
100.00	0.0333	280.00	0.0230
120.00	0.0324	300.00	0.0220
140.00	0.0315		

(A) How will the investigator know that the data have been corrupted by the trick played by the jealous rival?

(B) Can the data be salvaged so that the investigator will be able to obtain the value of the rate coefficient at 320 K?

(C) Is it also possible to obtain the rate coefficient at 335 K? If so, how? If not, what additional information would be needed?

19.8 The reaction discussed in Problem 19.7 is studied again at 320 K. This time, the initial concentrations are $[A]_o = 0.0010$ mol L^{-1} and $[B]_o = 1.000$ mol L^{-1}. The measured concentrations of A are as follows:

Time (min)	$[A]$ (mol L^{-1})
0.00	0.001000
2.00	0.000874
4.00	0.000764
6.00	0.000668
8.00	0.000624
10.00	0.000546
12.00	0.000446
14.00	0.000390
16.00	0.000341
18.00	0.000298
20.00	0.000260
25.00	0.000186

Determine the reaction rate coefficient for the reaction of A with B at 320 K. If you have done Problem 19.7, compare the relative difficulty of the two problems.

19.9 Show that, under the conditions that $[B]$ is constant at a value $[B]_o$ with $[B]_o \gg [A]$, Eq. 19.14 reduces to Eq. 19.22 for a pseudo first-order reaction.

19.10 Derive Eq. 19.24, which gives the relaxation time for a first-order process.

19.11. Derive Eqs. 19.26A and 19.26B, which respectively give the half-life and relaxation time for a second-order concerted reaction involving a single reactant.

19.12 Determine the half-life and relaxation time for the kinetic process described in Example 19.3.

19.13 A compound A undergoes a first-order decomposition reaction at 320 K. An investigator conducts experiments that yield the following concentration data for A as a function of time:

Time (min)	Concentration (mmol L^{-1})
0.00000	10.00000
4.21053	9.66877
8.42105	9.34851
12.63158	9.03885
16.84211	8.73946
21.05263	8.44998
25.26316	8.17009
29.47368	7.89947
33.68421	7.63781
37.89474	7.38482
42.10526	7.14022
46.31579	6.90371
50.52632	6.67504
54.73684	6.45394
58.94737	6.24016
63.15789	6.03347
67.36842	5.83362
71.57895	5.64039
75.78947	5.45356
80.00000	5.27292
84.00000	4.99374
88.00000	4.72933
92.00000	4.47893
96.00000	4.24179
100.00000	4.01720
104.00000	3.80450
108.00000	3.60306
112.00000	3.41229
116.00000	3.23162
120.00000	3.06052

(continued)

Time (min)	Concentration (mmol L^{-1})
124.00000	2.71400
128.00000	2.40672
132.00000	2.13423
136.00000	1.89259
140.00000	1.67831
144.00000	1.48829
148.00000	1.31978
152.00000	1.17035
156.00000	1.03784
160.00000	0.92034
164.00000	0.81614
168.00000	0.72373
172.00000	0.64179
176.00000	0.56913
180.00000	0.50469
184.00000	0.44755
188.00000	0.39687
192.00000	0.35194
196.00000	0.31209
200.00000	0.27676

Without his knowledge, a practical joker alters the system temperature to 340 K partway into the experiment. At a later time, this same practical joker raises the temperature again. The investigator cannot understand his data and concludes that his apparatus must have a serious design flaw. He needs your assistance.

(A) Determine the time at which the practical joker altered the temperature to 340 K and the time at which she raised the temperature a second time.

(B) Determine the specific reaction rate coefficient for the decomposition at 320 K.

(C) Determine the activation energy for the reaction.

(D) Determine the final temperature to which the practical joker raised the system.

19.14 An investigator is conducting kinetic experiments on a first-order reaction $A \longrightarrow P$ for which the rate coefficient at 298 K is $k_o = 0.001$ s^{-1}. The activation energy for the reaction is found to be 40 kJ mol^{-1}. Assume that a simple Arrhenius expression gives the dependence of k upon temperature. The investigator runs the reaction in a container whose temperature is being continuously varied such that the temperature at time t is

$$T = \frac{T_o}{1 - \beta t} \quad \text{for } 0 \le t < \frac{1}{\beta}.$$

where β is a constant whose units are s^{-1} and T_o is the temperature at time $t = 0$.

(A) Obtain an expression in terms of the activation energy E_a, T_o, k_o, β, and t that gives the fraction of compound A that remains at time t, provided that t is in the given range.

(B) What half-life will the investigator obtain in his experiment if β is 0.0001 s^{-1}?

(C) When half of the initial amount of A has been consumed, what is the temperature in the reaction vessel?

(D) After 300 s, what fraction of compound A remains?

19.15 A series of rate measurements is conducted on a pseudo first-order reaction. The data obtained at 300 K, 320 K, 340 K, and 360 K are given in the table at the bottom of the page (time is given in minutes and concentration in moles L^{-1}). Determine the activation energy and frequency factor for the reaction.

19.16 (A) The disintegration rate for a sample containing ^{60}Co as the only radioactive nuclide is found to be 240 atoms min^{-1}. The half-life of ^{60}Co is 5.271 years. Estimate the number of atoms of ^{60}Co in the sample.

(B) How long must this radioactive sample be allowed to react before the disintegration rate falls to 100 atoms min^{-1}?

19.17 Consider two different radioactive isotopes A and B that each decay via $_{-1}^{0}\beta$ emission, but with different half-lives. A scientist prepares a 2-gram sample containing isotope A and a second 2-gram sample containing isotope B. Suppose it is known that each of the samples initially contains the same number of atoms

of radioactive isotope and that the half-life of the first isotope is 49.90 min. The scientist now mixes the two samples together and measures the total $_{-1}^{0}\beta$ emission rate of the combined samples as a function of time. The following data are obtained:

Time (min)	Emissions (counts) per minute
0	2,464
10	2,174
30	1,694
45	1,406
60	1,168
100	713
150	387
200	212

Use the data to determine the half-life of the second radioactive isotope and the total number of radioactive atoms of each type that were present in the original 2-gram mixtures.

19.18 $_{92}^{238}$U decays via a series of nuclear processes to $_{82}^{206}$Pb. One of these decay processes is much slower than the others, so that its rate controls the rate of formation of $_{82}^{206}$Pb. Based on the rate of the slow step, half of the original $_{92}^{238}$U will decay to $_{82}^{206}$Pb in 4.46×10^9 years.

	$T = 300$ K	$T = 320$ K		$T = 340$ K		$T = 360$ K
Time	[A]	[A]	**Time**	[A]	**Time**	[A]
0.00	0.02000	0.02000	0.00	0.02000	0.00	0.02000
200.00	0.01638	0.00777	50.00	0.00789	10.00	0.01065
400.00	0.01341	0.00302	100.00	0.00311	20.00	0.00567
600.00	0.01098	0.00117	150.00	0.00123	30.00	0.00302
800.00	0.00899	0.00046	200.00	0.00048	40.00	0.00161
1000.0	0.00736	0.00018	250.00	0.00019	50.00	0.00086
1200.0	0.00602	0.00007	300.00	0.00008	60.00	0.00046
1400.0	0.00493	0.00003	350.00	0.00003	70.00	0.00024
1600.0	0.00404				80.00	0.00013
1800.0	0.00331				90.00	0.00007
2000.0	0.00271					
2200.0	0.00222					
2400.0	0.00181					
2600.0	0.00148					
2800.0	0.00122					
3000.0	0.00100					

Rocks containing $^{238}_{92}U$ can, therefore, be dated by measuring the ratio of $^{238}_{92}U$ to $^{206}_{82}Pb$ in them. There are two assumptions inherent in this method. First, it is assumed that all of the $^{206}_{82}Pb$ found in a rock is a result of the nuclear decay of $^{238}_{92}U$. This assumption is based on the fact that the most abundant isotope of lead is $^{208}_{82}Pb$. If lead were present from other sources, it should contain large amounts of the 208 isotope. When this is not the case, it appears very likely that the $^{206}_{82}Pb$ isotope is the result of the nuclear decay of $^{238}_{92}U$. The second assumption is that when the rock was formed, it did not contain any $^{206}_{82}Pb$. The ratio of $^{238}_{92}U$ atoms to $^{206}_{82}Pb$ atoms in the oldest rocks found on earth is about 0.627. Determine the approximate age of these rocks. If it took 1×10^9 to 1.5×10^9 years for the earth's crust to cool and become solid, what is the estimated age of the planet, based on the two assumptions?

19.19 What is the probability that a randomly selected molecule in a system undergoing a first-order reaction will survive for at least 2τ, where τ is the relaxation time?

19.20 The measured nuclear spin relaxation time T_1 for ^{13}C NMR in benzene is 29 s. If it is necessary to wait until the free-induction decay (FID) signal (see Chapter 16) has decayed to 80% of its maximum value before pulsing the compound again to take a second scan, how many FT NMR scans can be executed in 10 minutes for this compound. Assume that the spin–spin relaxation time, T_2, is long compared with T_1, so that it may be ignored.

19.21 Starting with Eq. 19.42A, derive Eq. 19.43.

19.22 An investigator obtains kinetic data on two opposing first-order reactions. Her results are as follows:

Time (min)	[A] (mol L^{-1})	[B] (mol L^{-1})
0.0	0.01000	0.00000
10.0	0.00818	0.00182
20.0	0.00689	0.00311
30.0	0.00598	0.00402
40.0	0.00534	0.00466
50.0	0.00488	0.00512
60.0	0.00456	0.00544
70.0	0.00434	0.00566
80.0	0.00418	0.00582
90.0	0.00406	0.00594
100.0	0.00398	0.00602
200.0	0.00380	0.00620
300.0	0.00379	0.00621

Show that these data are described by Eq. 19.43, and in the process, determine the values of k_1 and k_2.

19.23 (A) Determine the median lifetime in terms of the specific reaction rate coefficient for molecules undergoing a first-order reaction.

(B) Compute the probability of observing a lifetime at or between the median and the average lifetime for molecules undergoing a first-order reaction?

19.24 What is the most probable lifetime for molecules undergoing a first-order reaction? Explain qualitatively why this is true.

19.25 Compound A simultaneously undergoes a pseudo first-order reaction and a second-order reaction, so that the mechanism is

$$A \xrightarrow{k_1} B \quad \text{and} \quad A + A \xrightarrow{k_2} C$$

$$\text{Reaction 1} \qquad \qquad \text{Reaction 2}$$

Obtain the integrated rate expression for $[A]$ as a function of time.

19.26 Compound A is undergoing M first-order concurrent reactions. Obtain an integrated rate equation for the overall mechanism, and discuss methods that might be employed to obtain each of the M rate coefficients.

19.27 An investigator carries out a chemical process involving a single reactant A. He obtains the following data:

Time (min)	[A] (mol L^{-1})	1 / [A] (mol L^{-1})
0	1.0	1.0
10	0.82621	1.2103
20	0.69309	1.4428
30	0.58833	1.6997
40	0.50412	1.9836
50	0.43527	2.2974
60	0.37818	2.6442
70	0.33030	3.0275
80	0.28976	3.4511
90	0.25515	3.9192

It is quickly found that these data do not fit any simple reaction rate law. After much effort, the investigator learns an interesting fact: A plot of $\ln[(1/[A]) + 1]$ vs. time is linear. (See accompanying figure.) Being unable to determine the reason behind this behavior, he submits the problem to you for solution.

(A) Show that the reaction mechanism

$$A + A \xrightarrow{k_1} P_1 \quad \text{and} \quad A \xrightarrow{k_2} P_2$$

obeys the integrated rate law

$$\frac{1}{[A]} = a \exp[k_2 t] - b,$$

where

$$a = \frac{2k_1}{\beta k_2} \quad \text{and} \quad b = \frac{2k_1}{k_2},$$

in which

$$\beta = \frac{2k_1[A]_o}{(2k_1[A]_o + k_2)}.$$

(B) In view of the result in (A) and the data given in the problem, determine k_1 and k_2 for the reaction mechanism.

(P.S. If you solved the problem, the investigator extends his profound thanks. In the future, remember to ask him for a consultant's fee.)

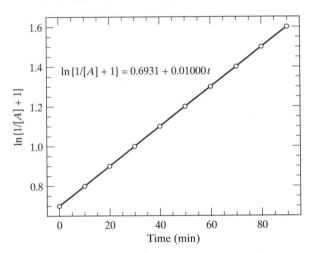

$$\ln\{1/[A] + 1\} = 0.6931 + 0.01000t$$

19.28 The reaction $2\,NO(g) + Cl_2(g) \longrightarrow 2\,NOCl(g)$ is thought to be a third-order concerted termolecular process. The following kinetic data are obtained at 400 K by using the initial concentrations $[NO(g)]_o = 0.000500$ mol L^{-1} and $[Cl_2(g)]_o = 0.800$ mol L^{-1}:

Time (s)	$[NO(g)]$ (mol L^{-1})
0	0.000500
10,000	0.000391
20,000	0.000321
30,000	0.000272
40,000	0.000237
50,000	0.000209
60,000	0.000187
70,000	0.000170
80,000	0.000155
90,000	0.000143
100,000	0.000132

Determine the specific reaction rate coefficient for this process at 400 K.

19.29 Show that the lifetime probability distribution given by Eq. 19.39 is properly normalized.

19.30 An ensemble of atoms is undergoing first-order nuclear decay. The relaxation time for the reaction is 4.52 days. Suppose we focus our attention on a subgroup of these atoms that have been in the reaction vessel for 4.00 days without reacting. What will the average lifetime for atoms in this subgroup be?

19.31 Compound A is forming Compounds B and C by concurrent reactions. The following data are obtained for the reactions:

Time (min)	B (mol L^{-1})	C (mol L^{-1})
5.00	$2.00a$	$4.00a$
10.00	b	$2.00b$
15.00	c	$2.00c$
20.00	d	$2.00d$
25.00	e	$2.00e$

where a, b, c, d, and e represent some set of concentrations. What can be said about the orders and the rate coefficients for the two concurrent reactions? Justify your answer.

19.32 (A) Compound A is forming Compound B in a pseudo first-order reaction. At the same time, Compound A is forming Compound C in a concerted second-order reaction involving two molecules of A. Let the rate coefficients for the two reactions be k_1 and k_2, respectively. If we have $[B]_o = [C]_o = 0.00$, obtain an expression giving $[C]$ as a function of $[B]$, k_1, k_2, and $[A]_o$.

(B) Show that your final result predicts that we will have $[C] = 0.00$ whenever we have $[B] = 0.00$.

19.33 Consider two consecutive first-order nuclear decay reactions with rate coefficients k_1 and k_2 in which

$$A \xrightarrow{k_1} B \quad \text{and} \quad B \xrightarrow{k_2} C.$$

Show that if the initial numbers of atoms of A and B are $[A]_o$ and $[B]_o$, respectively, with $[B]_o \neq 0$, the temporal dependence of the number of atoms of B is given by Eq. 19.65.

19.34 Consider two consecutive first-order nuclear decay reactions with rate coefficients k_1 and k_2 in which

$$A \xrightarrow{k_1} B \quad \text{and} \quad B \xrightarrow{k_2} C.$$

If $k_1 = k_2 = 0.1340$ year^{-1}, determine the time at which B attains a maximum number of atoms in the process.

19.35 Consider two consecutive first-order nuclear decay reactions with rate coefficients k_1 and k_2 in which

$$A \xrightarrow{k_1} B \quad \text{and} \quad B \xrightarrow{k_2} C.$$

In the general case when $[B]_o \neq 0$, does the number of atoms or the concentration of the intermediate B always exhibit a maximum when plotted against time? If so, prove that this is the case. If not, determine the condition that must hold for a maximum to exist.

19.36# If we have $k_2 - k_1 > 0$ for two consecutive first-order nuclear decay reactions with rate coefficients k_1 and k_2 in which

$$A \xrightarrow{k_1} B \quad \text{and} \quad B \xrightarrow{k_2} C$$

then a maximum in the number of atoms or the concentration of B will be obtained if the inequality

$$\frac{k_1[A]_o}{(k_2 - k_1)}\left[1 - \frac{k_1}{k_2}\right] > [B]_o$$

is satisfied, but not otherwise. (See Problem 19.35.)
(A) Let $k_1 = 0.150 \text{ day}^{-1}$ and $k_2 = 0.180 \text{ day}^{-1}$, so that we have $k_2 - k_1 > 0$. For $[A]_o = 0.0100 \text{ mol L}^{-1}$ and $[B]_o = 0.00950 \text{ mol L}^{-1}$, plot $[B]$ versus time, using Eq. 19.65. Do we obtain a maximum? Explain.
(B) Repeat (A) with $[B]_o = 0.00300 \text{ mol L}^{-1}$. Do we obtain a maximum? Is this the expected behavior? Explain.

19.37 For the sequence of three first-order consecutive reactions

$$A \xrightarrow{k_1} B \quad B \xrightarrow{k_2} C \quad \text{and} \quad C \xrightarrow{k_3} D$$

show that the temporal variation of $[C]$ is given by Eq. 19.69, provided that $[B]_o = [C]_o = 0$. You may use the results obtained in the text for the first two reactions in the sequence.

19.38# Consider the sequence of three pseudo first-order consecutive reactions

$$A \xrightarrow{k_1} B \quad B \xrightarrow{k_2} C \quad \text{and} \quad C \xrightarrow{k_3} D$$

If $[A]_o = 10 \text{ mmole L}^{-1}$ and $[B]_o = [C]_o = [D]_o = 0$, with $k_1 = 0.01 \text{ min}^{-1}$, $k_2 = 0.02 \text{ min}^{-1}$, and $k_3 = 0.10 \text{ min}^{-1}$, compute and plot $[C]$ as a function of time. On the same graph, plot the predicted value of $[C]$ as a function of time if we make the stationary-state approximation for it. Plot the percent error from the stationary-state approximation as a function of time. Where is the error largest? Explain.

The next two problems examine a series of first-order consecutive reactions in which all rate coefficients are equal. The problems are constructed so that they can be assigned individually or as a unit.

19.39 Consider the series of consecutive first-order reactions

$$A_1 \xrightarrow{k} A_2, \quad A_2 \xrightarrow{k} A_3, \quad A_3 \xrightarrow{k} A_4, \dots,$$

$$A_i \xrightarrow{k} A_{i+1}, \dots,$$

in which all rate coefficients are equal and all initial concentrations are zero save for $[A_1]_o$.
(A) Obtain $[A_1]$ as a function of time.
(B) Use the result of (A) and an integrating factor as described in the text to obtain $[A_2]$ as a function of time.
(C) Use the result of (B) and an integrating factor as described in the text to obtain $[A_3]$ as a function of time.
(D) Using the results obtained in (A), (B), and (C), infer the form of the equation that gives $[A_i]$ as a function of time. Prove by induction that your inference is correct.

19.40 (A) Using Eq. 19.70, determine the time t_{mi} in terms of k at which compound A_i in the first-order sequence given in Problem 19.39 attains a maximum concentration.
(B) What is the time interval between successive maxima for compounds A_i, A_{i+1}, A_{i+2}, etc.? Does this result suggest a method whereby we might experimentally determine k? Explain.

The next four problems examine a kinetic model of polymerization. The problems are constructed so that they can be assigned individually or as a single unit.

19.41 Monomer M_1 is polymerized inside the reaction vessel shown in the following diagram:

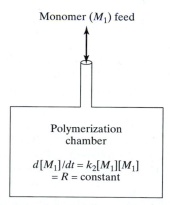

Monomer (M_1) feed

Polymerization chamber

$d[M_1]/dt = k_2[M_1][M_1]$
$= R = $ constant

The overall process is illustrated in Figure 19.20. Monomer is continuously added to the chamber at a rate sufficient to keep the rate of monomer reaction, $d[M_1]/dt$, equal to a constant that we denote by $R \text{ mol L}^{-1} \text{ min}^{-1}$. During the reaction, the temperature is held constant, and the concentration of M_1 is sufficiently large to make all the polymerization steps behave as pseudo first-order reactions with a common pseudo first-order rate coefficient k. That is, we have $k = k_{\text{true}}[M_1] = k'[M_1]$ for all polymerization steps, where $k_{\text{true}} = k'$ is the actual second-order rate coefficient. Subsequent to the addition of monomer, polymerization begins. The first two steps in the process are

$$M_1 + M_1 \xrightarrow{k_2} M_2$$

and

$$M_1 + M_2 \xrightarrow{k=k_{true}[M_1]=k'[M_1]} M_3.$$

(A) Obtain an expression giving $[M_2]$ as a function of time, k, and R. Use an integrating factor to solve the differential equation.

(B) Does $[M_2]$ attain a steady state? If not, why not? If it does, what is the steady-state $[M_2]$, and does the result agree with the prediction of the stationary-state approximation?

19.42 For the polymerization process described in Problem 19.41, the concentration of dimer is given by $[M_2] = R/k - Re^{-kt}/k$.

(A) If the pseudo first-order rate coefficients for all subsequent steps in the polymerization process are equal to k, obtain expressions for $[M_3]$, $[M_4]$, and $[M_5]$ in terms of R, k, and time.

(B) By extrapolating the results obtained in (A), write down the appropriate expression for the concentration of the nth-mer, $[M_n]$.

19.43 The concentration of the nth-mer, $[M_n]$, in the polymerization process described in Problems 19.41 and 19.42 is

$$[M_n] = \frac{R}{k} - \frac{R}{k}\left[1 + kt + \frac{(kt)^2}{2!} + \frac{(kt)^3}{3!} + \cdots + \frac{(kt)^{n-2}}{(n-2)!}\right]e^{-kt}.$$

(A) Does $[M_n]$ attain a steady state? If so, what is its value in the steady state?

(B) Show that at any finite time the concentration of M_∞ is zero.

19.44*(A) Using the results obtained in Problems 19.41 and 19.42 and given in Problem 19.43, find an expression giving the probability of observing any given polymer M_i ($i \geq 2$) in the polymerization chamber at time t.

(B) For the case $k = 0.0100$ min^{-1}, compute and plot the distribution of polymer probabilities $P(i)$ as a function of i at the times $T = 50$ min, 100 min, 200 min, 400 min, 600 min, and 800 min.

19.45 The N_2O_5-catalyzed gas-phase decomposition of O_3 to produce O_2 is presumed to occur via the following mechanism:

$$N_2O_5 \xrightarrow{k_1} NO_2 + NO_3 \qquad \text{(Reaction 1)};$$

$$NO_2 + NO_3 \xrightarrow{k_2} N_2O_5 \qquad \text{(Reaction 2)};$$

$$NO_2 + O_3 \xrightarrow{k_3} NO_3 + O_2 \qquad \text{(Reaction 3)};$$

$$2\,NO_3 \xrightarrow{k_4} 2\,NO_2 + O_2 \qquad \text{(Reaction 4)}.$$

For the first two reactions, the equilibrium constant $K_{12} \ll 1$, with both k_1 and k_2 large relative to k_3 and k_4.

(A) Assuming that Reaction 1 is first order and that all other reactions are second order, write down the differential rate equations for all the species.

(B) By employing the stationary-state approximation, obtain expressions for the steady-state concentrations, and express the rate of reaction of O_3 in terms of $[O_3]$ and $[N_2O_5]$.

19.46 Hydrogen iodide can be synthesized from the elements in a gas-phase reaction. One proposed mechanism for the process is

$$I_2 \xrightarrow{k_1} 2I \qquad \text{(first order)};$$

$$2I \xrightarrow{k_2} I_2 \qquad \text{(second order)};$$

$$H_2 + 2I \xrightarrow{k_3} 2\,HI \qquad \text{(third order)}.$$

Derive a differential rate law for the rate of production of HI if

(A) I_2 and I are in equilibrium throughout the reaction.

(B) Derive the differential rate law if a stationary-state approximation can be applied to I atoms with $k_3[H_2] \gg k_2$. How could you determine experimentally which of these two assumptions, (A) or (B), is correct?

19.47 Compounds A and B are in equilibrium with compound C, which also reacts via a pseudo first-order process to form product P. The reaction mechanism is

$$A + B \xrightarrow{k_1} C;$$

$$C \xrightarrow{k_2} A + B;$$

$$C \xrightarrow{k_3} P.$$

(A) Show that this mechanism will appear to be one for which the rate of change of A is first order with respect to both A and B and second-order overall, provided that a stationary-state approximation is accurate for compound C.

(B) What will be the apparent second-order rate coefficient in terms of k_1, k_2, and k_3?

19.48 The measured apparent first-order rate coefficients for the decomposition of CH_3NC as a function of pressure at 503 K are given in the following table:

p (torr)	$k(p)$ (s^{-1})	p (torr)	$k(p)$ (s^{-1})
10,000	95.1	10.0	17.0
5,010	91.0	5.01	11.0
1,000	80.9	1.00	3.80
501	76.0	0.501	2.30
100	48.0	0.100	0.005
50.1	33.0		

Plot $1/k(p)$ versus $1/p$ and comment on the results. What information, if any, can we obtain from the plot?

19.49 The initial rate of an enzyme-catalyzed reaction is measured as a function of the concentration of the substrate by using an initial enzyme concentration of 1.75×10^{-9} mol L^{-1}. The data obtained are shown in the following table:

$[S]_o$ (mol L^{-1})	Initial Rate $\mathcal{R}_o$ (mol L^{-1} s^{-1})
0.0059	0.000048
0.0108	0.000066
0.0157	0.000077
0.0206	0.000085
0.0255	0.000091
0.0304	0.000095
0.0353	0.000098
0.0402	0.000100
0.0451	0.000103

Use these data to determine k_2 and the Michaelis constant for the reaction at the temperature of the experiment. (For a definition of k_2, see the discussion preceding Eq. 19.80 in the text.)

19.50 Initial rate measurements are conducted on an enzyme-catalyzed reaction for which $k_2 = 9.6 \times 10^4$ s^{-1} and $K_M = 6.17 \times 10^{-3}$ mol L^{-1}, using an initial enzyme concentration of 2.3×10^{-9} mol L^{-1} (See the discussion preceding Eq. 19.80 in the text.)

(A) Use the Michaelis–Menten mechanism to calculate and plot the expected initial rate as a function of $[S]_o$.

(B) Derive the form of the limiting initial rate predicted by the Michaelis–Menten mechanism as $[S]_o$ becomes large. Does this result agree with the plot obtained in (A)?

(C) The maximum number of substrate molecules that can be converted into product by one enzyme molecule in one unit of time is called the *turnover number* for the enzyme. This quantity is just the maximum initial rate divided by $[E]_o$, ($\mathcal{R}_o^{max}/[E]_o$). What is the turnover number for the enzyme in this problem?

(D) Repeat (A) for an initial enzyme concentration of 4.6×10^{-9} mol L^{-1}. Plot the results on the same graph you prepared in (A). Does the limiting rate at large $[S]_o$ change from that obtained in (A)? If not, why not? If it does change, by how much? Does the turnover number for the enzyme in (A) and (B) change?

19.51 The extent of surface coverage for an absorbate $A(g)$ on a particular surface is measured as a function of the pressure of $A(g)$ at 273 K. The data obtained are as follows:

p_A (torr)	θ_e	p_A (torr)	θ_e
100.0	0.2593	600.0	0.6774
200.0	0.4118	700.0	0.7101
300.0	0.5122	800.0	0.7368
400.0	0.5833	900.0	0.7590
500.0	0.6364		

From these data, determine K, assuming that the absorption process is accurately described by a Langmuir isotherm. (For a definition of K, see Eq. 19.91 in the text.)

19.52 To maintain a particular equilibrium surface coverage on a particular surface, the gaseous pressure must be varied in the manner shown in the following table as the temperature is varied:

Temperature (K)	Pressure $A(g)$ (torr)
298	120.0
310	150.5
330	211.7
350	286.3
370	374.8

Determine the molar enthalpy of absorption of $A(g)$ on this surface.

19.53 Starting with Eq. 19.98 and a similar equation for Compound B, derive Eq. 19.99. (*Hint*: Remember that the fractional coverage at steady state or equilibrium, θ_e, can be written as the sum of the fractions covered by each of the absorbing compounds. That is, $\theta_e = \theta_e^A + \theta_e^B + \theta_e^C + \dots$.)

19.54 If N in Eqs. 19.88 and 19.89 is expressed in mol cm^{-2}, what are the units on k_a and k_d in those equations?

19.55[#] An investigator is examining the surface-catalyzed unimolecular decomposition of compound $A(g)$ at 300 K. After cleaning the catalyst, he follows the gas-phase pressure of $A(g)$ as a function of time, starting with an initial pressure of 10.00 torr. His data are as follows:

Time (s)	$\ln(p_A)$ (torr)	Time(s)	$\ln(p_A)$ (torr)
0.0	2.303	600.0	1.664
100.0	2.196	700.0	1.558
200.0	2.090	800.0	1.451
300.0	1.983	900.0	1.345
400.0	1.877	1,000.0	1.239
500.0	1.771		

In a second experiment, the investigator starts with p_A equal to 76,000 torr. He then measures the concentration of decomposition product F as a function of time. His results are shown in the following table:

Time (s)	$[F]$ (mol cm^{-3})
0.0	0.00
1000.0	5.69×10^{-6}
5000.0	2.84×10^{-5}
10,000.0	5.69×10^{-5}
30,000.0	1.71×10^{-4}

(A) Express the low-pressure limiting rate for the surface-catalyzed unimolecular decomposition of compound A(g) at 300 K in terms of the rate of change of the gas-phase pressure of A.

(B) Use the low-pressure data to determine a value for the $k_r NK$ product, where the symbols are as defined in Eq. 19.101.

(C) Use the high-pressure data to obtain the value for the $k_r N$ product. What is the value of K?

(D) Plot the surface-catalyzed rate as a function of p_A at 300 K over the range $0 \le p_A \le 2{,}000$ torr.

19.56 Gaseous reactants A and B undergo heterogeneous surface catalysis to produce products at 300 K. At the initial partial pressures $p_A^o = 1{,}500$ torr and $p_B^o = 50$ torr, the partial pressure of B(g) is measured as a function of time, with the following results:

Time (s)	p_B (torr)	Time(s)	p_B (torr)
0.0	50.00	600.0	23.76
100.0	44.17	700.0	20.99
200.0	39.02	800.0	18.54
300.0	34.47	900.0	16.38
400.0	30.45	1,000.0	14.47
500.0	26.90		

From these data alone, what can be said about the reaction mechanism and the rate coefficients? Justify your answers.

19.57 An investigator suspects that a single reactant is simultaneously undergoing first- and second-order reactions via the mechanism

$$A \xrightarrow{k_1} B \quad \text{and} \quad A + A \xrightarrow{k_2} C.$$

She measures the concentration of A as a function of time during the initial stages of the reaction for two different concentrations. Her data are shown in the following pair of tables:

Data Set 1		Data Set 2	
Time (min)	$[A]$ (mol L^{-1})	Time (min)	$[A]$ (mol L^{-1})
0.00	0.800	0.00	0.400
1.00	0.784	1.00	0.395
2.00	0.768	2.00	0.390
3.00	0.752	3.00	0.385
4.00	0.737	4.00	0.380
5.00	0.723	5.00	0.375
6.00	0.709	6.00	0.371

Using Eqs. 9.173 and 9.174 to obtain $d[A]/dt$ at $t = 3.00$ min for each data set, determine k_1 and k_2.

19.58 To illustrate the accuracy problem that arises when we work with rates instead of concentrations, suppose the investigator who took the data given in Problem 19.57 makes a random $\pm 0.2\%$ error in measuring each concentration, so that her results are as follows:

Data Set 1		Data Set 2	
Time (min)	$[A]$ (mol L^{-1})	Time (min)	$[A]$ (mol L^{-1})
0.00	0.800	0.00	0.400
1.00	0.785	1.00	0.395
2.00	0.769	2.00	0.391
3.00	0.750	3.00	0.384
4.00	0.736	4.00	0.379
5.00	0.723	5.00	0.376
6.00	0.708	6.00	0.371

The results obtained in Problem 19.57 without the random $\pm 0.2\%$ error in the data are $k_1 = 0.00468$ min^{-1} and $k_2 = 0.0107$ L mol^{-1} min^{-1}. Repeat the calculation described in that problem, and determine the percent difference in the results for the rate coefficients obtained from the foregoing two data sets.

19.59 Temperature-jump measurements are taken on an equilibrium system exhibiting reactions of the form

$$A + B \xrightarrow{k_1} C - \Delta \overline{H}$$

and

$$-\Delta \overline{H} + C \xrightarrow{k_2} A + B.$$

The investigator uses different initial concentrations of A and B in three different kinetic measurements. In each case, he determines the relaxation time of the equilibrium system after a small jump in the temperature. The results are given in the following table:

$[A]_e$	$[B]_e$ (mmol L^{-1})	τ (s)
22.36	22.36	2.14×10^{-7}
10.00	10.00	4.55×10^{-7}
7.07	7.07	6.20×10^{-7}

Use these data to determine k_1 and k_2 for the equilibrium reaction.

19.60 Consider an equilibrium system in which there are two second-order bimolecular reactions for the forward and reverse reactions:

$$A + B \xrightarrow{k_1} C + D - \Delta\overline{H}$$

and

$$-\Delta\overline{H} + C + D \xrightarrow{k_2} A + B.$$

Derive the expression connecting the rate coefficients and equilibrium concentrations to the time required for relaxation of the system to a new equilibrium point in a temperature-jump experiment for which the displacement from equilibrium is very small.

19.61 Sam has located some kinetic data that puzzle him. He would like your comments. The data, for the gas-phase decomposition of a single reactant that we shall label $A(g)$, are as follows:

Time (min)	$[A(g)]$ (mol L^{-1})
0.0	0.1000
4.0	0.0961
8.0	0.0923
12.0	0.0887
16.0	0.0852
20.0	0.0819

In an effort to determine the order of the reaction, Sam has plotted $\ln[A]$ versus time to test for first-order kinetics and also $[A]^{-1}$ versus time to test for second-order kinetics. His plots are shown in the diagrams at the bottom of the page, where the lines are least-square fits to the data.

As can be seen, both plots are essentially linear. The plot of $\ln[A]$ versus time has a slope of -0.00998 min^{-1}, which suggests that the reaction is first order with a rate coefficient $k = 0.00998$ min^{-1}. The plot of $[A]^{-1}$ versus time has a slope of 0.110 L mol^{-1} min^{-1}. This suggests that the kinetics are second order with a rate coefficient such that $2k = 0.110$ L mol^{-1} min^{-1}. What is wrong? Can you assist Sam? Here are some suggestions:

(A) The integrated rate expression for a first-order reaction of Compound A is $[A] = [A]_o e^{-kt}$. Write this expression in the form of a power series expansion in time about the point $t = 0$.

(B) The integrated rate law for a second-order reaction with a single component is given by Eq. 19.16. If we combine the factor of two with the rate coefficient, this expression is $[A]^{-1} = [A]_o^{-1} + k_2 t$. Solve the preceding equation for $[A]$, and expand the result in a power series in time about the point $t = 0$.

(C) Let us define the quantity $k_2[A]_o = k$. In terms of k and time, what is the power series expansion for $[A]$ in a second-order reaction with a single component? Compare this expression with the result obtained in (A). Comment on the similarities and the differences.

(D) Compute $[A]$ for both a first-order and a second-order process with $[A]_o = 0.100$ mol L^{-1} and $k = k_2[A]_o = 0.0100$ min^{-1}. Plot $\ln[A]$ versus time for both cases over the interval $0 \leq t \leq 30$ minutes. Comment on the result. Having done (A)–(D), can you help Sam?

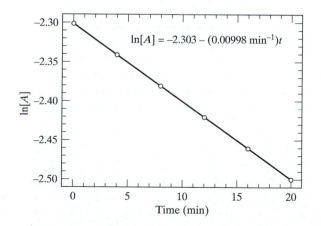

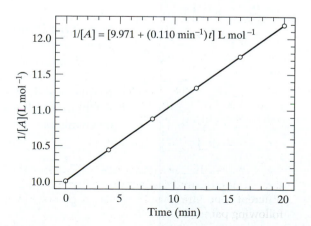

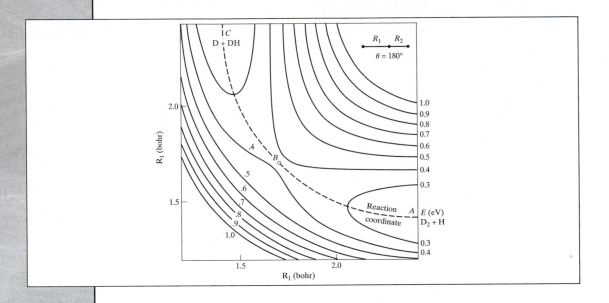

Theoretical Kinetics and Reaction Dynamics

Most of the knowledge scientists have concerning the actual motion of atoms during chemical reactions has been obtained from sophisticated crossed molecular-beam, infrared chemiluminescence and other laser studies, combined with detailed theoretical calculations. In this final chapter, we shall explore some of these methods and examine the extent to which we are currently able to predict reaction rates and mechanisms.

20.1   Collision Theory

Collision theory seeks to predict the rates and mechanisms of chemical reactions by following the details of individual molecular collisions that occur at particular relative velocities, impact parameters, orientations, etc. The results of these calculations are then averaged over all possible values of the collision variables to obtain the measured reaction rate. In principle, the calculations can be done either quantum mechanically or classically. However, if the number of atoms involved in the reaction exceeds four, quantum calculations become intractable for the computers available in 2000.

20.1.1   The General Problem

We can gain an understanding of the general procedures involved in the computation of a reaction rate by considering a "simple" three-body reaction. Suppose we wish to compute the isotopic exchange rate between hydrogen and D_2:

$$D_2 + H \longrightarrow DH + D.$$

If we knew the frequency with which the hydrogen atom collides perpendicularly with D_2 and the probability of reaction when it does so, the product of this frequency and probability would give us the reaction rate for perpendicular collisions. If we also knew the frequency of 45° collisions and the corresponding reaction probability, the product of these two factors would again give us the reaction rate for 45° collisions. The total reaction rate that is observed in the laboratory is the sum of such probabilities. We are, therefore, led to the conclusion that, in order to compute the total reaction rate, (1) we must identify a set of variables that is sufficient to completely describe the details of the collision, (2) we must have a means of computing the probability of reaction and the frequency of collision for any particular set of values of these variables, and (3) we must be able to sum or integrate the differential reaction rates for all possible values of the variables. While these are by no means trivial problems, they are not as difficult as playing the Tchaikovsky D major violin concerto with the New York Philharmonic. Therefore, we should be able to make some progress.

20.1.2   Description of the Collision

Nine variables are needed to specify the details of a collision between H and D_2. See Figure 20.1. First, we need to specify the relative speed between the two centers of mass and the impact parameter for the collision. Once we stipulate that the collision occurs with a relative speed from V to $V + dV$ and an impact parameter from b to $b + db$, Eq. 17.100 gives us the total number of collisions per unit time occurring in these ranges. Next, we need to orient the D_2 molecule relative to the H atom and the initial relative-velocity vector. This can be conveniently done by giving the spherical polar orientation angles, θ and ϕ, for D_2. With the coordinate system employed in the figure, perpendicular collisions are characterized by having $\theta = \pi/2$, while collinear collisions have $\theta = 0$ or π. The D_2 molecule is initially in some vibrational-rotational state whose quantum numbers are v, j, and m, respectively. The initial D_2 bond length is R. Finally, we must orient the D_2 angular momentum vector, which, in effect, selects the D_2 plane of rotation. If we assume a rigid-rotor

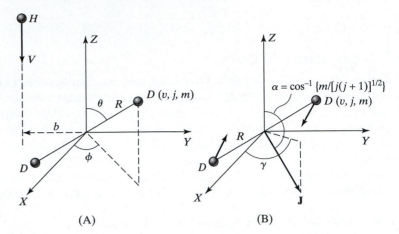

▶ **FIGURE 20.1**
Three-body collision variables for
H + D₂ reaction. **(A)** Center-of-
mass, orientation, and internal variables. **(B)** D₂ rotation plane and
angular momentum variables.

quantization, this vector J has a magnitude $|J| = [j(j + 1)]^{1/2}\hbar$. Its spherical polar orientation angle with the Z-axis, α, is determined by the magnetic quantum number, which can take on $2j + 1$ values running from $-j$ to $+j$ in integer steps. This orientation angle is given by

$$\cos \alpha = \frac{m\hbar}{[j(j + 1)]^{1/2}\hbar} = \frac{m}{[j(j + 1)]^{1/2}}.$$

To complete the orientation of **J**, we must select its rotation angle about the Z-axis. This angle is denoted by γ in the figure. The nine variables that describe the collision are, therefore, $[V, b, \theta, \phi, v, j, m, \gamma, R]$. The variables V and b describe the relative motion of the centers of mass of H and D₂. The remaining seven variables determine the internal details of the collision. In the discussion that follows, we shall denote the set of internal variables as $\{q\}$.

We now need an expression that gives us the frequency of collisions with relative speed from V to $V + dV$, impact parameter from b to $b + db$, orientation angles from θ to $\theta + d\theta$ and from ϕ to $\phi + d\phi$, D₂ bond length from R to $R + dR$, vibrational–rotational state quantum numbers v, j, and m, and angular momentum orientation angle from γ to $\gamma + d\gamma$. The frequency of collision at a relative speed from V to $V + dV$ and an impact parameter from b to $b + db$ is $Z_{AB}(V, b)$, as defined in Eq. 17.100. If $Z_{AB}(V, b)$ is multiplied by the probabilities of having orientation angles from θ to $\theta + d\theta$ and from ϕ to $\phi + d\phi$, bond length from R to $R + dR$, and angular momentum orientation from γ to $\gamma + d\gamma$, while the D₂ molecule has quantum numbers v, j, and m, we will obtain the frequency with which such collisions occur. Denoting this frequency by $Z_{AB}(V, b, \theta, \phi, v, j, m, \gamma, R)$, we have

$$Z_{AB}(V, b, \theta, \phi, v, j, m, \gamma, R) = P_1(\theta, \phi)P_2(R)P_3(\gamma)P_4(v, j, m)Z_{AB}(V, b). \quad \textbf{(20.1)}$$

The probability of observing D₂ oriented at angles from θ to $\theta + d\theta$ and from ϕ to $\phi + d\phi$ is proportional to the total solid angle corresponding to this angular range. Therefore,

$$P_1(\theta, \phi) = C \sin \theta \, d\theta \, d\phi,$$

where C is the proportionality constant. Its value can be determined by requiring that $P_1(\theta, \phi)$ be properly normalized. This requirement gives

$$\int_{\phi=0}^{2\pi}\int_{\theta=0}^{\pi} C \sin\theta \, d\theta \, d\phi = -2\pi C \cos\theta\big|_o^\pi = -2\pi C[-1-1] = 4\pi C = 1.$$

With $C = (4\pi)^{-1}$, the normalized probability distribution that the collision will occur with D_2 having orientation angles in the ranges from θ to $\theta + d\theta$ and from ϕ to $\phi + d\phi$ is

$$P_1(\theta, \phi) = \frac{\sin\theta \, d\theta \, d\phi}{4\pi}.$$

The probability that the D_2 bond length will be in the range from R to $R + dR$ depends upon the nature of the interatomic D_2 potential. If the vibrational quantum number is near zero, it is usually sufficiently accurate to assume that the vibrational motion is harmonic. In that case, the required probability distributions are those obtained in Eqs. 12.152 and 12.153. If we are treating the vibrational motion as classical, $P_2(R)$ is given by Eq. 12.152, viz.,

$$P_2(R) = \frac{dR}{\pi[\alpha^{-1} - (R - R_{eq})^2]^{1/2}} \tag{20.2}$$

where

$$\alpha = \frac{(\mu k)^{1/2}}{\hbar}, \tag{20.3}$$

with R_{eq}, μ, and k being the D_2 equilibrium distance, the reduced mass, and the vibrational force constant, respectively. If we elect to treat the vibrational motion quantum mechanically and D_2 is in its vibrational ground state, Eq. 12.153 provides the distribution we need:

$$P_2(R) = \left[\frac{\alpha}{\pi}\right]^{1/2} \exp[-\alpha(R - R_{eq})^2] dR. \tag{20.4}$$

Both distributions are shown in Figure 12.36.

The rotation angle for the angular momentum vector will be uniformly distributed over the range $0 \le \gamma \le 2\pi$, since the solid-angle element is uniform in the rotation angle about Z. Therefore, $P_3(\gamma) \propto d\gamma$, or $P_3(\gamma) = c\,d\gamma$, where c is the proportionality constant. Normalization produces

$$\int_{\gamma=0}^{2\pi} c\,d\gamma = 1 = 2\pi c,$$

so that the normalized distribution is

$$P_3(\gamma) = \frac{d\gamma}{2\pi}.$$

If the reaction is being carried out thermally at temperature T, the probability of observing the D_2 molecule in the vibrational–rotational quantum state specified by (v, j, m) is just the corresponding Boltzmann factor derived in Chapters 17 and 18, namely,

$$P_4(v, j, m) = \frac{1}{z} \exp\left[-\frac{E_{vj}}{kT}\right], \tag{20.5}$$

where E_{vj} is the vibrational–rotational energy and

$$z = \sum_{v=0}^{\infty} \sum_{j=0}^{\infty} \sum_{m=-j}^{j} \exp\left[-\frac{E_{vj}}{kT}\right] = \sum_{v=0}^{\infty} \sum_{j=0}^{\infty} (2j+1) \exp\left[-\frac{E_{vj}}{kT}\right] \qquad (20.6)$$

is the molecular partition function, provided that E_{vj} is independent of m.

Combining these results with Eq. 20.1 gives us the differential collision frequency. The overall expression is

$$Z_{AB}(V, b, \theta, \phi, v, j, m, \gamma, R) =$$

$$\left[\frac{\sin\theta\, d\theta\, d\phi}{4\pi}\right]\left[\frac{dR}{\pi[\alpha^{-1} - (R - R_{eq})^2]^{1/2}}\right]\left[\frac{d\gamma}{2\pi}\right]\left[\frac{1}{z}\exp\left(-\frac{E_{vj}}{kT}\right)\right]Z_{AB}(V, b).$$

$$(20.7)$$

20.1.3 General Collision Theory Rate

If the number of collisions that occur per unit time for a given set of collision variables is multiplied by the probability that a reaction will take place upon collision, we obtain the reaction rate for collisions of that type. Equation 20.7 gives us the differential collision frequency. If we can now determine the reaction probability, $P_{AB}(V, b, \theta, \phi, \gamma, R, v, j, m)$, for such collisions, the reaction rate for those collisions will be

$$\mathcal{R}(V, b, \theta, \phi, \gamma, R, v, j, m) = P_{AB}(V, b, \theta, \phi, \gamma, R, v, j, m)$$

$$\times Z_{AB}(V, b, \theta, \phi, \gamma, R, v, j, m). \qquad (20.8)$$

The total rate observed in the laboratory is just the sum of this differential rate over all possible collisions; that is,

$$\boxed{\mathcal{R} = \int_{V=0}^{\infty} \int_{b=0}^{\infty} \int_{\theta=0}^{\pi} \int_{\phi=0}^{2\pi} \int_{\gamma=0}^{2\pi} \int_{R=R_{min}}^{R_{max}} \sum_{v=0}^{\infty} \sum_{j=0}^{\infty} \sum_{m=-j}^{j} P_{AB} Z_{AB}}, \qquad (20.9)$$

where, for convenience, we have written Z_{AB} and P_{AB} without the list of arguments. In Eq. 20.9, we are integrating over the continuous variables and summing over the ones that are discrete, as is appropriate.

There remain two major problems. The first involves the actual execution of the integrals and sums in Eq. 20.9. We are faced with the problem of executing a coupled, six-dimensional integral and a triple summation. If we have more than three atoms in the collision, the complexity of the problem will be even greater. Such multidimensional integrals can usually not be successfully attacked using summation procedures such as Simpson's rule. Instead, we will need a new, more powerful method for handling integrals of this nature. The second problem involves the determination of the reaction probability P_{AB} for a particular set of collision variables.

20.1.4 Simple Hard-Sphere Model

Before addressing the problem of multidimensional integration and the determination of the reaction probability, it is useful to develop some simple expressions for the reaction rate. While not quantitatively accurate, they will provide some physical insights into the nature of reaction rates.

To simplify the problem of solving Eq. 20.9, we need to uncouple the integrals and sums and remove the problem of computing the reaction probability. If we assume that the colliding molecules are hard spheres with a collision radius σ_{AB}, we immediately achieve part of the desired result. Under such conditions, the reaction probability will be independent of all the internal variables $\theta, \phi, \gamma, R, v, j,$ and m, since these variables play no role in describing the collision of two hard spheres. This means that we can write Eq. 20.9 in the form

$$\mathcal{R} = \int_{\theta=0}^{\pi} \int_{\phi=0}^{2\pi} \left[\frac{\sin\theta \, d\theta \, d\phi}{4\pi} \right] \int_{\gamma=0}^{2\pi} \frac{d\gamma}{2\pi} \int_{R=R_{min}}^{R_{max}} \left[\frac{dR}{\pi[\alpha^{-1} - (R - R_{eq})^2]^{1/2}} \right]$$

$$\sum_{v=0}^{\infty} \sum_{j=0}^{\infty} \sum_{m=-j}^{j} \left[\frac{1}{z} \exp\left(-\frac{E_{vj}}{kT} \right) \right] \int_{V=0}^{\infty} \int_{b=0}^{\infty} P_{AB}(V, b) Z_{AB}(V, b). \quad (20.10)$$

The probability distributions are all normalized so that the integrals over θ, $\phi, \gamma,$ and R and the summations over $v, j,$ and m are all unity. Thus, we obtain

$$\mathcal{R} = \int_{V=0}^{\infty} \int_{b=0}^{\infty} P_{AB}(V, b) Z_{AB}(V, b). \quad (20.11)$$

We can remove the last remaining difficulty by assuming that $P_{AB}(V, b)$ is a constant if the two hard spheres collide with sufficient energy to overcome the potential-energy barrier to reaction. The hard spheres will collide if the impact parameter lies in the range $0 \le b \le \sigma_{AB}$. If the required energy for reaction is E_a, the system will possess at least this energy if the relative speed of the spheres is in the range $(2E_a/\mu)^{1/2} \le V \le \infty$. Therefore, we take the reaction probability to be

$$P_{AB}(V, b) = p = \text{a constant} \quad \text{for} \quad 0 < b \le \sigma_{AB} \quad \text{and} \quad \left(\frac{2E_a}{\mu} \right)^{1/2} \le V \le \infty$$

and

$$P_{AB}(V, b) = 0, \quad \text{otherwise.} \quad (20.12)$$

Using Eq. 20.12 for the reaction probability and Eq. 17.100 for $Z_{AB}(V, b)$, we obtain

$$\mathcal{R} = 4\pi p N_A N_B \left[\frac{\mu}{2\pi kT} \right]^{3/2} \int_{b=0}^{\sigma_{AB}} 2(\pi b) db \int_{\left(\frac{2E_a}{\mu} \right)^{1/2}}^{\infty} \exp\left[-\frac{\mu V^2}{2kT} \right] V^3 dV. \quad (20.13)$$

Integrating over the impact parameter is straightforward. The result is $\pi\sigma_{AB}^2$, which we recognize as the hard-sphere collision cross section introduced in Chapter 17. The integration over relative velocity can be done by parts. (See Problem 20.1.) The final result is

$$\mathcal{R} = p[\pi\sigma_{AB}^2] \left[\frac{8kT}{\pi\mu} \right]^{1/2} \left[1 + \frac{E_a}{kT} \right] \exp\left[-\frac{E_a}{kT} \right] N_A N_B. \quad (20.14)$$

Since the phenomenological rate expression for a concerted second-order reaction is

$$\mathcal{R} = k(T) N_A N_B,$$

the hard-sphere collision model predicts that the specific reaction rate coefficient is

$$k(T) = p[\pi\sigma_{AB}^2] \left[\frac{8kT}{\pi\mu} \right]^{1/2} \left[1 + \frac{E_a}{kT} \right] \exp\left[-\frac{E_a}{kT} \right]. \quad (20.15)$$

Inspection shows that Eq. 20.15 has a form similar to that of an extended Arrhenius equation in which the rate coefficient is written

$$k(T) = AT^n \exp\left[-\frac{E_a}{(kT)}\right].$$

Anytime the model incorporates the requirement that the system possess an energy equal to or greater than E_a as a prerequisite for reaction, an exponential temperature dependence of this type will be the result.

The constant p is some type of average probability, which means that we must have $p \leq 1$. If we set $p = 1$ and take the activation energy to be zero, so that a reaction occurs on every collision, we will have the following upper limit to the magnitude of the bimolecular rate coefficient:

$$k_{max}(T) = [\pi\sigma_{AB}^2]\left[\frac{8kT}{\pi\mu}\right]^{1/2}. \tag{20.16}$$

Reactions that have rate coefficients at or near this magnitude are said to have *gas-kinetic* rates, because the rate is controlled solely by the gas-kinetic collision frequency. Example 20.1 evaluates this upper limit for a specific case.

EXAMPLE 20.1

Compute the gas-kinetic rate for the $H + D_2$ exchange reaction at 300 K if the collision radius is 1.5×10^{-10} m. Express the result in both $m^3\,s^{-1}$ and $L\,mol^{-1}\,s^{-1}$.

Solution

We need the reduced mass of the $H + D_2$ system. This is given by

$$\mu = \frac{m_H m_{D2}}{m_H + m_{D2}} = \frac{(1.0078)(4.0282)}{1.0078 + 4.0282} g\,mol^{-1} \times \frac{1\,kg}{1{,}000\,g} \times \frac{1}{6.02214 \times 10^{23}\,mol^{-1}}$$

$$= 1.339 \times 10^{-27}\,kg. \tag{1}$$

The factors required to compute $k_{max}(300)$ are

$$\pi\sigma_{AB}^2 = (3.14159)(1.5 \times 10^{-10})^2\,m^2 = 7.069 \times 10^{-20}\,m^2 \tag{2}$$

and

$$\left[\frac{8kT}{\pi\mu}\right]^{1/2} = \left[\frac{8(1.381 \times 10^{-23}\,J\,K^{-1})(300\,K)}{(3.14159)(1.339 \times 10^{-27}\,kg)}\right]^{1/2} = 2{,}807\,m\,s^{-1}. \tag{3}$$

Combining these factors gives

$$k_{max}(300\,K) = [\pi\sigma_{AB}^2]\left[\frac{8kT}{\pi\mu}\right]^{1/2}$$

$$= (7.069 \times 10^{-20}\,m^2)(2807\,m\,s^{-1}) = 1.98 \times 10^{-16}\,m^3\,s^{-1}. \tag{4}$$

Conversion to $L\,mol^{-1}\,s^{-1}$ yields

$$k_{max}(300\,K) = 1.98 \times 10^{-16}\,m^3\,s^{-1} \times \frac{6.022 \times 10^{23}}{mol} \times \left(\frac{100\,cm}{m}\right)^3 \times \frac{1\,L}{1{,}000\,cm^3}$$

$$= 1.19 \times 10^{11}\,L\,mol^{-1}\,s^{-1}. \tag{5}$$

Rate coefficients on the order of $10^{11}\,L\,mol^{-1}\,s^{-1}$ are usually called gas kinetic. For related exercises, see Problems 20.2 and 20.3.

If the experimental rate coefficient and activation energy have been determined, substituting them into Eq. 20.15 permits us to compute the value of the average reaction probability p, which is frequently called the *steric factor*. If p should turn out to be very small, the reaction is regarded to have a large degree of *steric hindrance*. Presumably, the internal structure of the reactants is such that alignment problems make reaction very unlikely. Problems 20.2 and 20.3 serve as examples in this regard.

20.1.5   The Reaction Cross Section

In the derivation of Eq. 20.15, it is the integration over the impact parameter and all internal variables that leads to the cross-sectional term $p[\pi\sigma_{AB}^2]$. For a hard-sphere model, this reaction cross section has the very simple form shown in Figure 20.2. Below the threshold relative speed V_o, $p[\pi\sigma_{AB}^2]$ is zero. For relative speeds in excess of V_o, the reaction cross section is a constant.

In the general case, we again define the reaction cross section $\sigma^2(V)$ as the result of integrating $P_{AB}(2\pi b\,db)P(\{q\})$ over the impact parameter and all internal variables, where $P(\{q\})$ represents the normalized probability distribution functions for the complete set of internal variables $\{q\}$. For a three-body reaction, this definition gives

$$\sigma^2(V) = \int_{b=0}^{\infty} \int_{\theta=0}^{\pi} \int_{\phi=0}^{2\pi} \int_{\gamma=0}^{2\pi} \int_{R=R_{\min}}^{R_{\max}} \sum_{v=0}^{\infty} \sum_{j=0}^{\infty} \sum_{m=-j}^{j} P_{AB}(2\pi b\,db)$$
$$P_1(\theta,\phi)P_2(R)P_3(\gamma)P_4(v,j,m). \tag{20.17}$$

We can also discuss the reaction cross section for a particular vibrational–rotational state. This quantity is

$$\sigma^2(V,v,j) = \int_{b=0}^{\infty} \int_{\theta=0}^{\pi} \int_{\phi=0}^{2\pi} \int_{\gamma=0}^{2\pi} \int_{R=R_{\min}}^{R_{\max}} \sum_{m=-j}^{+j} P_{AB}(2\pi b\,db)P_1(\theta,\phi)P_2(R)P_3(\gamma). \tag{20.18}$$

The total reaction cross section can then be written as a weighted average of reaction cross sections out of particular vibrational–rotational states:

$$\sigma^2(V) = \sum_{v=0}^{\infty} \sum_{j=0}^{\infty} \left[\frac{1}{z}\exp\left(-\frac{E_{vj}}{kT}\right)\right]\sigma_{AB}^2(V,v,j). \tag{20.19}$$

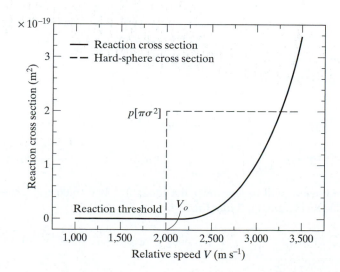

Comparison of a typical reaction cross section as a function of relative speed with a hard-sphere reaction cross section. Above the reaction threshold at $V = V_o$, the hard-sphere model has a constant reaction cross section equal to $p\pi\sigma^2$. An accurately computed or measured reaction cross section will become nonzero at $V = V_o$, and then rise gradually with increasing relative speed, as shown.

Equation 20.19 indicates that the total reaction cross section is, in general, a function of the relative speed. This is certainly not unexpected. Below a certain relative speed V_o, the reactants will have insufficient energy to traverse the potential barrier to reaction so that the reaction cross section will be zero, just as it is in a hard-sphere model. The minimum energy required for reaction is called the *reaction threshold*. When the relative speed exceeds V_o, the reaction cross section becomes positive. As long as V is not in great excess of V_o, σ^2 will generally increase as V increases. This type of behavior is illustrated by the smooth curve in Figure 20.2.

If we combine Eqs. 20.9 and 20.17, the reaction rate coefficient can be expressed as an integral of the reaction cross section over relative speed:

$$k(T) = \int_{V=0}^{\infty} \sigma^2 Z_{AB} = 4\pi \left[\frac{\mu}{2\pi kT}\right]^{3/2} \int_{V=0}^{\infty} \sigma^2 V^3 \exp\left[-\frac{\mu V^2}{2kT}\right] dV. \quad \textbf{(20.20)}$$

Equation 20.20 follows from the fact that the integrals over the impact parameter and all internal variables in Eq. 20.10 have already been done in the computation of the reaction cross section. (Methods by which this might be accomplished will be discussed later in this section.)

Although Eq. 20.20 holds rigorously only at the temperature employed in the computation or measurement of the reaction cross section, it turns out that σ^2 is a rather insensitive function of T. For this reason, we can usually use that equation to compute the rate coefficient at different temperatures and thereby obtain the activation energy. Example 20.2 makes this point.

EXAMPLE 20.2

The total reaction cross section for a bimolecular reaction with a reduced mass of 3.82×10^{-26} kg has been computed at 300 K. If V is expressed in m s^{-1}, the result is

$$\sigma^2 = (3.00 \times 10^{-23} \text{ s}^2)(V - 1{,}600)^2 \quad \text{for} \quad V \geq 1{,}600 \text{ m s}^{-1}$$

$$= 0 \qquad\qquad\qquad\qquad \text{for} \quad V < 1{,}600 \text{ m s}^{-1}.$$

Figure 20.3 shows σ^2 as a function of V. The reaction threshold appears at $V = 1{,}600$ m s^{-1}. **(A)** Compute $k(T)$ at 300 K from the given data. **(B)** Assuming that the reaction cross section can be regarded as being independent of temperature, compute $k(T)$ at 330 K, 360 K, and 400 K. Determine the activation energy for this reaction. Compare the activation energy with the threshold energy. Are they the same? Explain.

Solution

(A) Using Eq. 20.20, we have

$$k(T) = 4\pi \left[\frac{\mu}{2\pi kT}\right]^{3/2} \int_{V=0}^{\infty} \sigma_{AB}^2 V^3 \exp\left[-\frac{\mu V^2}{2kT}\right] dV$$

$$= (1.20 \times 10^{-22})\pi \left[\frac{\mu}{2\pi kT}\right]^{3/2} \int_{V=1{,}600}^{\infty} (V - 1{,}600)^2 V^3 \exp\left[-\frac{\mu V^2}{2kT}\right] dV \text{ m}^3 \text{ s}^{-1}, \quad \textbf{(1)}$$

since the integral will be zero over the range $0 \leq V < 1600$ m s^{-1}. Inserting the data at 300 K converts Eq. 1 into the form

$$k(300) = 6.70 \times 10^{-31} \int_{V=1{,}600}^{\infty} (V - 1{,}600)^2 V^3 \exp[-4.61 \times 10^{-6} V^2] dV \text{ m}^3 \text{ s}^{-1}. \quad \textbf{(2)}$$

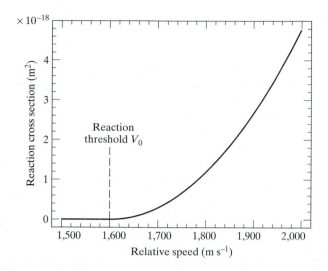

◀ FIGURE 20.3
The reaction cross section in Example 20.2. The reaction threshold is seen at $V = 1,600$ m s^{-1}.

This integral must be integrated numerically because Gaussian integrands of the form $x^{2n} \exp(-ax^2)$ between limits that are not $-\infty$ to $+\infty$ or 0 to $+\infty$ have no closed-form analytic solution. When the integral is evaluated using a trapezoid rule, the result is

$$k(300 \text{ K}) = 1.43 \times 10^{-20} \text{ m}^3 \text{ s}^{-1} = 8.61 \times 10^6 \text{ L mol}^{-1} \text{ s}^{-1}. \tag{3}$$

If it is assumed that the reaction cross section is independent of T, then

$$k(T) = 3.48 \times 10^{-27} T^{-3/2} \int_{V=1,600}^{\infty} (V - 1,600)^2 V^3 \exp\left[-\frac{0.001383 V^2}{T} \right] dV \text{ m}^3 \text{ s}^{-1} \tag{4}$$

after insertion of the constants and the value of μ given in the problem. Evaluation of this integral at 330 K, 360 K, and 400 K yields the data at right. An Arrhenius plot of $\ln[k(t)]$ versus T^{-1} is shown in Figure 20.4. The line is a least-squares fit to the four data points. Taking logarithms of both sides of the Arrhenius equation gives

$$\ln k(T) = \ln A - \frac{E_a}{RT}. \tag{5}$$

T (K)	$k(T)$ m^3 s^{-1}
300	1.43×10^{-20}
330	4.85×10^{-20}
360	1.36×10^{-19}
400	4.32×10^{-19}

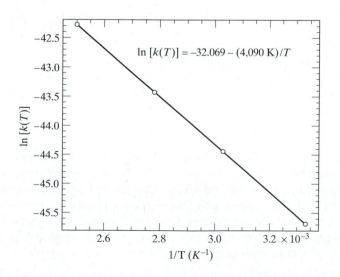

◀ FIGURE 20.4
An Arrhenius plot of the computed rate coefficients in Example 20.2. The slope of the least-squares fit of a straight line is related to the activation energy for the reaction by slope $= -E_a/R$.

Consequently, the slope of the line in the figure is $-E_a/R$. Therefore,

$$E_a = (4{,}089 \text{ K})R = (4{,}090 \text{ K})(8.314 \text{ J mol}^{-1} \text{K}^{-1}) = 3.400 \times 10^4 \text{ J mol}^{-1}. \quad \textbf{(6)}$$

The threshold energy is the energy available when $V = V_o = 1{,}600 \text{ m s}^{-1}$. This is

$$E_o = \text{threshold energy} = 0.5\mu V_o^2 = (0.5)(3.82 \times 10^{-26} \text{ kg})(1{,}600 \text{ m s}^{-1})^2$$

$$= 4.89 \times 10^{-20} \text{ J} = 2.94 \times 10^4 \text{ J mol}^{-1}. \quad \textbf{(7)}$$

Thus, we see that the activation energy exceeds the threshold energy, in this case by about 4,600 J mol^{-1}. This type of inequality is always observed. The threshold energy is the *minimum* energy required for reaction; the activation energy represents the *average* energy required for reaction.

For related exercises, see Problems 20.5, 20.6, and 20.11.

20.1.6   Potential-energy Surfaces

If we wish to obtain more accurate predictions of reaction rates and mechanisms than is possible by using hard-sphere models, we must have a means of accurately computing the reaction probability for a given set of collision variables. Simple qualitative considerations tell us that this probability will depend upon the energy barrier to reaction and upon the interatomic forces that are present during the collision. These quantities are determined by the potential energy of the system, V. The Cartesian components of the force acting on atom i are given by the negative of the derivatives of V with respect to the corresponding coordinates:

$$F_{xi} = -\frac{\partial V}{\partial x_i}, \quad \textbf{(20.21A)}$$

$$F_{yi} = -\frac{\partial V}{\partial y_i}, \quad \textbf{(20.21B)}$$

and

$$F_{zi} = -\frac{\partial V}{\partial z_i}. \quad \textbf{(20.21C)}$$

The set of equations 20.21 was previously introduced in Eq. 11.5. The energy barrier is just the difference between the potential energy of the system in the transition state and that of the equilibrium reactant configuration. Therefore, if we wish to compute the reaction rate, we must first obtain the potential energy of the system.

In principle, there is no fundamental difficulty in computing V; we simply need to solve the stationary-state Schrödinger equation for all statistically significant nuclear configurations of the system. That is, for a system containing M nuclei and N electrons, we need to solve

$$\mathcal{H}\psi(1, 2, 3, \ldots N) = E(\mathbf{r})\psi(1, 2, 3, \ldots N) \quad \textbf{(20.22)}$$

for the energy eigenvalues $E(\mathbf{r})$, where the eigenfunctions depend upon the $3N$ coordinates of the electrons moving in the field of M stationary nuclei whose relative positions are determined by the specification of one coordinate for a two-body system and $3M - 6$ coordinates in the general case. These coordinates are represented by the vector $\mathbf{r}$ in Eq. 20.22. The Hamiltonian for this system in electrostatic units is given by generalizing Eq. 13.51 to a system with M nuclei. Therefore,

$$\mathcal{H} = -\frac{\hbar^2}{2m_e}\sum_{i=1}^{N}\nabla_i^2 - \sum_{\alpha=1}^{M}\sum_{i=1}^{N}\frac{Z_\alpha e^2}{r_{\alpha i}} + \sum_{i=1}^{N}\sum_{j=i+1}^{N}\frac{e^2}{r_{ij}} + \sum_{\alpha=1}^{M}\sum_{\beta=\alpha+1}^{M}\frac{Z_\alpha Z_\beta e^2}{R_{\alpha\beta}},$$

(20.23)

provided that we make the Born–Oppenheimer approximation that electronic motion is so fast relative to nuclear motion, that we may omit the kinetic-energy terms for the nuclei and solve for the electronic eigenvalues $E(\mathbf{r})$ with the nuclei stationary. The set of energy eigenvalues for all significant nuclear configurations is the system's potential energy that we seek.

Although it is easy to state what we need to do to obtain the potential energy for our reacting system, actually performing the calculation is another matter. Let us consider what that would entail. Suppose our system contains just two hydrogen atoms, so that $N = M = 2$. Then the preceding calculation is well within the capabilities of modern computers. With $M = 2$, we need six coordinates to specify the nuclear positions. Three of these, however, determine the position of the center of mass of the system, which has no effect on the solutions of Eq. 20.22. Two additional coordinates can be associated with rotations of H_2 about two axes through the center of mass and perpendicular to the H–H bond axis. Therefore, we need only one coordinate to specify the relative position of the two hydrogen atoms. We can take this coordinate to be the interatomic distance R. We now need to solve Eq. 20.22 at different values of R until we have enough points to plot $E(R)$ versus R to obtain the diatomic potential curve shown in Figure 14.2. This would probably require about 20 values for $E(R)$. To ensure that each calculated eigenvalue is accurate, we would need to use either configuration interaction (see Section 14.3.5) or Møller–Plesset perturbation (see Section 14.3.6) methods.

To compute the $D_2 + H$ exchange reaction rate, we must perform a much more difficult calculation, because we now have $M = N = 3$. We can still solve Eq. 20.22 with reasonable accuracy when three electrons are present; however, the problem lies in the number of times we must solve the equation. With $M = 3$, the specification of the relative positions of the three atoms now requires $3(3) - 6 = 3$ nuclear coordinates, which might be the three interatomic distances or two interatomic distances and the included angle. If 10 energy determinations must be made along each of the coordinates, the number of times we would have to solve Eq. 20.22 is 10^3, or 1,000. The symmetry of the system permits this number to be reduced to 500 or 600. Although very difficult and laborious, such a calculation is still within the capabilities of modern computers. Liu and Siegbahn [B. Liu, *J. Chem. Phys.* **58**, 1925 (1973); P. Siegbahn and B. Liu, *ibid.*, **68**, 2457 (1978); B. Liu, *ibid.*, **80**, 581 (1984)] have conducted such computations by using methods that are accurate to within an estimated error of 0.03 eV (1 eV = 96.48 kJ mol^{-1}). These investigators evaluated the energy eigenvalues for 267 H_3 nuclear configurations. Since this number is significantly less than the estimated 500 times, it is clear that Liu and Siegbahn had to exercise considerable judgment in the selection of nuclear configurations to be examined. Basically, they chose configurations of great importance in the exchange reaction and eliminated those of lesser importance. This is not too difficult to do when the reaction is as simple as the $D_2 + H$ exchange. For more complex reactions, it is usually not obvious which configurations are important and which are not.

Once the potential has been computed at a sufficient number of nuclear configurations for the H_3 or D_2H systems (within the framework of the Born–Oppenheimer approximation, these potentials are the same, since the nuclear masses do not enter into the Hamiltonian), we are now faced with the problem of representing the results. In the case of two hydrogen atoms with only one nuclear coordinate, there is no problem: We simply plot $E(R)$ versus R and connect the computed points with a smooth curve, as shown in Figure 14.2. When we have three nuclear coordinates, a plot of $E(R_1, R_2, \theta)$ versus (R_1, R_2, θ) requires that we have four dimensions available, one for each of the three nuclear coordinates and one for the energy. Such a representation of the system potential is called a *potential-energy hypersurface*. In Rod Serling's "Twilight Zone," they may have had four dimensions, but in our three-dimensional world, we are out of luck. In fact, the problem is even worse than this analysis suggests: Unless we wish to build three-dimensional models, we are confined to the two-dimensional world of the printed page. The best solution available is to fix one of the three nuclear coordinates and plot contour lines of constant energy as a function of the remaining two coordinates. Such a representation is called a *potential-energy surface*. The potential is actually a hypersurface, but the terminology "surface" is frequently used. We shall illustrate such contour plots later in the section.

If we wish to predict the reaction rate of a more complex system, such as $F + H_2 \longrightarrow HF + H$, the computations become more difficult. We still have three nuclear coordinates to consider, but each calculation now requires that we solve the stationary-state Schrödinger equation for an 11-electron system. This is a far more demanding task than that for the three-electron H_3 system. Nevertheless, over a dozen calculations of the potential energy for this system have been reported. [See G. C. Schatz, *Science*, **262**, 1828 (1993).]

When we address a reaction containing four nuclei, the difficulties encountered in obtaining the potential-energy surface by directly solving Eq. 20.22 become severe. The most elementary such reaction is $H_2 + D_2 \longrightarrow 2\,HD$. With four nuclei present, we need $3(4) - 6 = 6$ nuclear coordinates to specify their relative positions. If we must make 10 calculations for each coordinate, it would be necessary to solve Eq. 20.22 a total of 10^6, or 1 million, times. With this requirement, we are beyond the capabilities of even the most powerful modern computers. Even if we attempt to study only "important" configurations, the result of the calculations will not determine the complete hypersurface. When we reach reactions containing five or more atoms, it becomes absurd to consider approaching the problem by a brute-force solution of Eq. 20.22. Also, the problem of representing the potential hypersurface for a four-atom system on a two-dimensional page is more extreme than that for a three-atom reaction: With six nuclear coordinates, we must fix four of these coordinates and be content with illustrating constant-energy contour lines for the remaining two coordinates.

One attractive alternative would appear to be to replace the rigorous solution of Eq. 20.22 by means of configuration interaction or Møller–Plesset methods with much easier semiempirical quantum methods, such as CNDO, INDO, MINDO, MNDO, MINDO/1, MINDO/2, MINDO/2', MINDO/3, PM3, AM1, or SAM1. (See Section 14.5.) These methods are known to yield equilibrium bond lengths with average absolute errors of about 0.04 to 0.05 Å and equilibrium bond angles with errors of a few degrees. In addition, computed endo- and exothermicities are usually reasonably good. Unfortunately, the energies predicted by those methods for configurations far removed from equilibrium are often very poor. Since it is precisely such nonequilibrium configurations that determine the potential barriers to reaction, the semiempirical quantum methods have not proven to be useful in quantitative computations of reaction rates. They are, however, sometimes useful in obtaining qualitative predictions concerning possible reaction mechanisms.

A more successful approach to obtaining reasonably accurate potentials for complex systems involves a partial abandonment of Eq. 20.22. In this empirical approach, chemically and physically reasonable functional forms are selected to represent interatomic two-body potentials, three-body bending interactions, and four-body dihedral angle potentials. For example, we might represent two-body bonding potentials with Morse functions

$$V_M(R) = D[\exp\{-2\beta(R - R_{eq})\} - 2\exp\{-\beta(R - R_{eq})\}]$$

and three-body bending potentials with harmonic-type functions of the form

$$V_b(\theta) = \frac{k(\theta - \theta_o)^2}{2},$$

where k is the bending force constant, θ_o is the equilibrium angle, and θ is the three-body angle. Several more complex forms have been suggested for the dihedral interactions. When this is done, the total potential becomes dependent upon a large number of parameters. In the preceding equations, these parameters are D, β, R_{eq}, k, and θ_o. Their values are adjusted to fit various types of experimental data, such as measured equilibrium structures, reaction energies, measured IR and Raman vibrational frequencies for reactants and products, and measured activation energies. In addition, Eq. 20.22 can be solved for critically important regions of the potential hypersurface around the transition-state region, and the parameters can be adjusted to fit the results of these calculations. For very complex systems, it usually is necessary to make the "parameters" functionally dependent upon the instantaneous nuclear coordinates of the system. At the present time, this empirical approach, which combines chemical intuition, experimental data, and some *ab initio* solutions of Eq. 20.22, has proven to be the most successful for complex systems. Although there is always some uncertainty surrounding the results of classical or quantum mechanical calculations of reaction rates obtained with the use of such empirical surfaces, comparisons with experimental data have shown that the method often yields very useful and fairly accurate predictions of reaction mechanisms and rates.

Let us now consider the contour map method of illustrating potential-energy hypersurfaces once they are obtained by one or some combination of the methods just described. We first consider the D_2H, or H_3, system. For purposes of contour mapping, the three nuclear coordinates are usually taken to be two interatomic distances and the included angle, as shown in Figure 20.5. In most cases, a set of potential contour maps are presented for fixed values of the angle θ. In such maps, constant-energy contours are plotted as a function of R_1 and R_2. A typical result for D_2H with $\theta = 180°$ is shown in

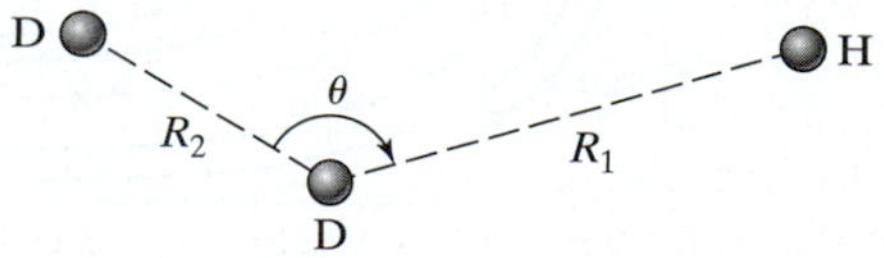

▶ FIGURE 20.5
Nuclear coordinates for the D_2H system involving two interparticle distances and the included angle.

Figure 20.6, which is taken from empirical-type calculations reported by Porter and Karplus [R. N. Porter and M. Karplus, *J. Chem. Phys.* **40**, 1105 (1964)]. In the figure, contour lines are shown for energies of 0.3, 0.5, 1.0, 2.0, 3.0, and 4.0 eV relative to the reactants ($D_2 + H$) and products ($D + DH$), whose potential energy is taken to be zero. In principle, as many contours as desired may be plotted. Of course, the presentation of additional contour lines usually requires that Eq. 20.22 be solved over a finer grid of nuclear configurations if that method is being used to obtain the potential-energy hypersurface. Since each contour line represents the loci of points having the particular energy for that contour, it is clear that no two contours lines can ever cross.

When we are in the region of the potential surface where the H–D distance R_1 is large and the D–D distance R_2 is near the D_2 equilibrium separation, the nuclear configuration resembles the reactants, $D_2 + H$. Point *A* lies in this region of the surface. If we were to stand at Point *A* facing in the direction of decreasing values of R_1, the potential would be near zero, because we have taken our reference point of zero potential energy to be the reactant $D_2 + H$ configuration. While standing at Point *A*, we would see the potential to our left rising sharply to higher values. Looking to our right, we would again see the potential energy rising sharply as R_2 increases. We would, therefore, find ourselves in a potential-energy valley. For this reason, points in the region of Point *A* are said to be in the *reactant valley* or *reactant channel*. In the calculations reported by Porter and Karplus, the D–D and H–D interatomic potentials are represented by Morse functions. Therefore, if we were to examine the form of the D_2H potential surface in a plane through Point *A* and perpendicular to the R_1-axis, we would see a Morse potential similar to the one shown in Figure 12.39.

In regions of the potential surface where the D–D distance R_2 is large and the D–H distance R_1 is near the H–D equilibrium separation, the nuclear configuration resembles the products, DH + H. Point *C* lies in this region of the surface. At that point, we again find ourselves in a potential-energy val-

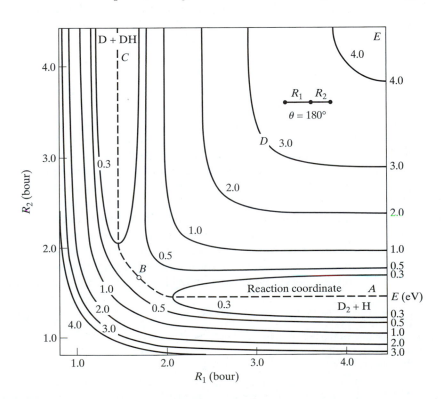

▶ FIGURE 20.6
Contour map for the D_2H system with $\theta = 180°$. Contour-line energies are given in eV, where 1 eV = 96.48 kJ mol^{-1}. [Data taken from R. N. Porter and M. Karplus, *J. Chem. Phys.* **40**, 1105 (1964).]

ley. Because of this, we say that Point *C* lies in the *product valley* or *product channel*. In this region, the potential-energy curve in a plane through Point *C* and perpendicular to the R_2-axis is a Morse function. In regions of the surface for which both R_1 and R_2 are large, such as Point *E*, the nuclear configuration resembles D + D + H. If we move far enough along the diagonal of the figure, where $R_1 = R_2$, we reach a plateau region of the surface whose energy is that for the separated atoms, which in this case is 4.74 eV.

Now suppose you are at Point *A* and wish to "guide" the reactants toward the product valley along the path that requires the minimum expenditure of energy. Along what path would you move the reacting system? This question is analogous to asking a mountain guide how he would direct a group of tourists from one side of a mountain to the other. If we move the reactants from Point *A* to Point *D*, which lies on the 3.0-eV contour line on the diagonal of the figure, the reactant system will need to possess at least 3.0 eV of energy before it can make the trip. Once the system reaches Point *D*, it will be a downhill slide toward Point *C* and the product valley, but before we can enjoy the downhill trip, there is a 3.0-eV "climb" that has to be undertaken. If our mountain guide were to lead his group of tourists on such a "hike," it is a good bet that many would not complete the journey, and the ones that did would probably not tip him.

If, on the other hand, we move in a direction nearly parallel to the R_1-axis, there is a very gradual climb up to the 0.3-eV contour line. When we are in the vicinity of the tip of this contour, we wish to change direction. Continuing along a linear path parallel to R_1 will lead us uphill to the 0.5-eV contour and then on toward the 1.0-eV contour. Instead, we follow the curved path indicated by the dashed line in the figure. This path continues to lead us slowly toward higher potentials, until we attain Point *B* on the diagonal. At *B*, the potential energy along the dashed curve reaches a maximum. If we plot more contour lines in the region of Point *B*, as is done in Figure 20.7, we can see that we attain that point without having to cross the 0.40-eV contour line.

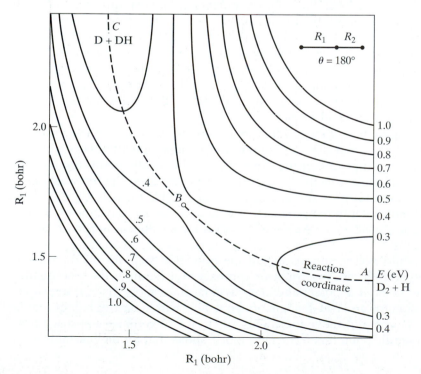

◀ **FIGURE 20.7**
Contour map of the saddle-point region for the D₂H system with $\theta = 180°$. Contour-line energies are given in eV, where 1 eV = 96.48 kJ mol⁻¹. [Data taken from R. N. Porter and M. Karplus, *J. Chem. Phys.* **40**, 1105 (1964).]

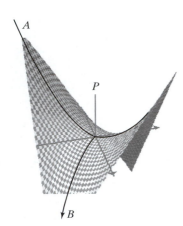

▲ FIGURE 20.8
A perspective image of a saddle point, or *col*. The point labeled *P* is a saddle point in that it is a minimum along curve *A* and a maximum along curve *B*.

Therefore, the maximum potential along the dashed path lies slightly below 0.4 eV. More quantitative calculations show that, on the Porter–Karplus D_2H potential, the maximum lies at 0.396 eV, or 38.2 kJ mol^{-1} above the reactants. If we continue to follow the dashed path, we find that the potential energy decreases continuously until we reach Point *C* in the product valley. Thus, the reactants need only 0.396 eV of energy to succeed in producing D + DH.

An examination of Figure 20.6 shows that no other path with $\theta = 180°$ will transform reactants into products with as little as 0.396 eV of available energy. This minimum-energy path is called the *reaction coordinate*. Inspection of the figure shows that the reaction coordinate is not a rectilinear coordinate; rather, it is a curved path whose exact nature varies with the contour map being plotted. The variation in potential energy along the reaction coordinate is the quantity plotted in Figures 19.9 and 19.10, where the numerical values on the abscissas of the plots are distances along the reaction coordinate, measured from the position of the maximum, which is taken to be zero.

The maximum along the reaction coordinate for a given contour map is a mathematical *saddle point*, or *col*. A saddle point is any point on a surface that is a maximum along one direction and a minimum along a perpendicular direction. This nomenclature is used because of the analogy to the contours of a saddle, which attain a minimum along a direction parallel to the horse's body and a maximum along a direction transverse to the horse's backbone. Figure 20.8 shows a perspective image of a typical saddle point at Point *P*. It is easy to see that Point *B* in Figures 20.6 and 20.7 is a saddle point. We have already seen that this point is a maximum along the reaction coordinate. If we plot the potential energy along the diagonal of the figure, which is perpendicular to the reaction coordinate at *B*, we obtain a curve that rises to infinity at small values of $R_1 = R_2 = R$, goes through a minimum at Point *B*, and then rises to the dissociation plateau at Point *E*. (See Problem 20.7.)

If we take θ to be some angle other than 180°, the shapes and positions of the contour lines change because the potential depends upon the included angle as well as upon the interatomic distances. The contour map for the D_2H system with $\theta = 90°$ shown in Figure 20.9 illustrates this point. Although the figure is qualitatively similar to Figure 20.6, there are important differences, the most important of which is the magnitude of the potential barrier to reaction. In Figure 20.9, the saddle point lies at a much higher energy, since the reaction coordinate now crosses the 1.0-eV contour line on its path toward the saddle point.

An examination of the saddle-point energies on contour maps with various values of θ shows that the minimum possible potential barrier occurs for a collinear approach of the three atoms. The saddle point that has the minimum possible potential barrier to reaction is called the *transition state*. On the Porter–Karplus surface, this barrier height, E_b, is 0.396 eV. More exact, *ab initio* quantum mechanical calculations yield a transition-state barrier height of 0.42 eV, or 40.5 kJ mol^{-1}, also for a collinear approach of the three atoms.

It should not be supposed that the transition state for all three-body reactions corresponds to a collinear approach. For example, the best calculations for the $H_2 + F$ potential-energy hypersurface show that the transition state occurs for an H–H–F angle of about 120° with a barrier height that is about 5.44 kJ mol^{-1} above the energy of $H_2 + F$. A collinear approach produces a saddle point that lies 7.11 kJ mol^{-1} above that for the reactants.

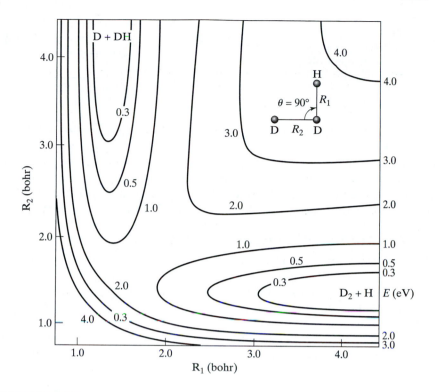

◀ **FIGURE 20.9**
Contour map for the D_2H system
with $\theta = 90°$. Contour-line energies
are given in eV, where
1 eV = 96.48 kJ mol^{-1}. [Data taken
from R. N. Porter and M. Karplus, *J. Chem. Phys.* **40**, 1105 (1964).]

EXAMPLE 20.3

(A) If the reaction $D_2 + H \longrightarrow D + DH$ were to proceed along a path passing through Point D in Figure 20.6 as the point of maximum potential, what would be the configuration of the system when it attained this point of maximum potential energy? **(B)** What is the configuration of the system at the transition state for this reaction?

Solution

(A) All points shown on the contour map in Figure 20.6 are collinear. Point D lies on the diagonal of the figure, along which $R_1 = R_2$. If we read the coordinates of D as accurately as permitted by the figure, we find that $R_1 = R_2 \approx 3.1$ bohr. Therefore, the system configuration at the point of maximum potential along this path would be as shown at right.

(B) For this reaction, the transition state has a collinear arrangement of the three atoms. Therefore, Point B in Figure 20.6 is the transition state. This point lies on the diagonal of the figure, along which $R_1 = R_2$. If we read the coordinates of Point B as accurately as permitted by the figure, we find that $R_1 = R_2 \approx 1.7$ bohr. Therefore, the system configuration at the transition state is as shown at right.

For related exercises, see Problems 20.7, 20.8, 20.9, and 20.10.

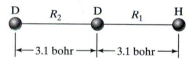

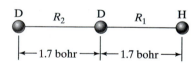

When we are dealing with a four-atom system, it becomes more difficult to illustrate the nature of the potential-energy hypersurface with the use of contour maps. A four-atom system requires the specification of six coordinates to determine the relative nuclear positions. Therefore, the potential is a seven-dimensional hypersurface. To represent these data using two-dimensional contour maps, we must fix four of the six nuclear coordinates. As an example, let us consider the termolecular reaction, $H_2 + 2I \longrightarrow 2HI$. If we require that the four atoms be in a symmetric, collinear arrangement, the six nuclear

▶ **FIGURE 20.10**
Symmetric, collinear arrangement of the H_2I_2 system. Only two nuclear coordinates, R_{II} and R_{HH}, are required to specify the system configuration.

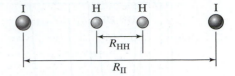

coordinates are reduced to two, as shown in Figure 20.10. We can, therefore, plot a contour map for this configuration. Such a contour map is shown in Figure 20.11, where the energy zero configuration is taken to be that of the separated atoms. Relative to this reference point, the energy of $H_2 + 2I$ and 2 HI are -4.746 eV and -6.388 eV, respectively. Problems 20.9 and 20.10 explore various features of this map and those of other four-body contour maps.

20.1.7   Monte Carlo Integration Methods

If the potential-energy hypersurface of a system can be obtained with sufficient accuracy, we can, in principle, compute the reaction probability for any set of collision variables. Methods for accomplishing this task will be described in subsequent sections. For the moment, let us assume that $P_{AB}(V, b, \{q\})$ can be calculated. If we wish to use this information to compute the specific reaction rate coefficient, there still remains the problem of executing the integrals in Eq. 20.9 for a three-body system and even more extensive integrals for reactions involving more than three atoms.

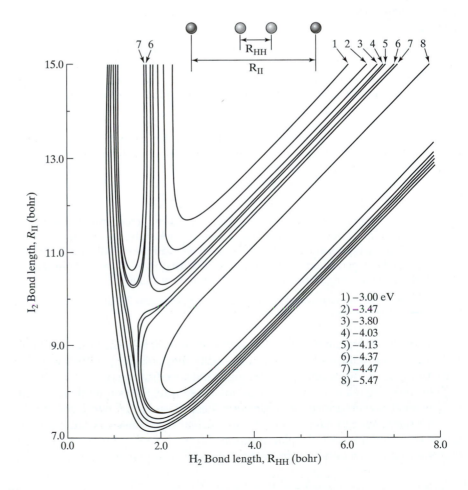

▶ **FIGURE 20.11**
Contour map for the collinear, symmetric I–H–H–I system. Contour-line values are given in eV, where 1 eV = 96.48 kJ mol⁻¹. Data are taken from the semiempirical valence-bond calculations reported by L. M. Raff *et al., J. Chem. Phys.* **52**, 3449 (1970).

In general, the required integrals must be integrated numerically. Most numerical integration methods reduce the integral to a summation in which the integrand is evaluated at closely spaced intervals between the upper and lower limits. These values are then summed according to some algorithm to produce the value of the integral. A typical one-dimensional integral might require that the integrand be evaluated at 100 equally spaced intervals between the upper and lower limits. If the integral is two dimensional, we might need to evaluate the integrand across a 100×100 grid of points in the two dimensions. We would then have to compute the value of the integrand 10,000 times. In general, if the integral has dimension n, we will need to evaluate the integrand about 100^n times. If $n > 4$, this type of summation method becomes so computationally demanding that our computer resources will probably be insufficient. Since the dimensionality of the integrations in the computation of a rate coefficient is always greater than 4, we need an integration method that does not rely on summation techniques. This section describes such a method.

Suppose we have a discrete variable x_i defined for $i = 1, 2, 3, \ldots, K$. In Chapter 10, we found that the first moment, or average value, of x over this range is given simply by summing the values of x_i for all values of i and then dividing the result by the number of x_i present. That is,

$$\langle x \rangle = \frac{\sum\limits_{i=1}^{K} x_i}{K}. \tag{20.24}$$

If x is a continuous variable defined over the range $0 \le x \le K$, we replace the summation in Eq. 20.24 with the corresponding integral to obtain the average:

$$\langle x \rangle = \frac{\displaystyle\int_0^K x \, dx}{\displaystyle\int_0^K dx} = \frac{\displaystyle\int_0^K x \, dx}{K}. \tag{20.25}$$

Thus, $\int_0^K x \, dx = K \langle x \rangle$. The analysis in Chapter 10 has shown that if we wish to obtain the average value of some function of x, $f(x)$, over the range $a \le x \le b$, we get

$$\langle f(x) \rangle = \frac{\displaystyle\int_a^b f(x) \, dx}{\displaystyle\int_a^b dx} = \frac{\displaystyle\int_a^b f(x) \, dx}{b - a},$$

so that

$$\int_a^b f(x) \, dx = (b - a)\langle f(x) \rangle. \tag{20.26}$$

This type of analysis is easily extended to multidimensional integrals. For example, the average value of $f(x, y)$ over the range $a \le x \le b$ and $c \le y \le d$ is given by

$$\langle f(x, y) \rangle = \frac{\displaystyle\int_a^b \int_c^d f(x, y) \, dy \, dx}{\displaystyle\int_a^b \int_c^d dy \, dx} = \frac{\displaystyle\int_a^b \int_c^d f(x, y) \, dy \, dx}{(b - a)(d - c)}.$$

Therefore, the two-dimensional integral may be written in the form

$$\int_a^b \int_c^d = f(x, y)\, dy\, dx = (b - a)(d - c)\langle f(x, y)\rangle. \tag{20.27}$$

The foregoing analysis shows us that we may always express an integral of a function in terms of the average value of that function over the ranges of integration, times the ranges themselves. Therefore, the integral of the n-variable function $f(x_1, x_2, x_3, \ldots, x_n)$ over the ranges $a_1 \le x_1 \le b_1$, $a_2 \le x_2 \le b_2, \ldots, a_n \le x_n \le b_n$ is given by

$$\int_{a_n}^{b_n} \cdots \int_{a_3}^{b_3} \int_{a_2}^{b_2} \int_{a_1}^{b_1} f(x_1, x_2, x_3, \ldots, x_n)\, dx_1\, dx_2\, dx_3 \ldots dx_n$$

$$= \langle f(x_1, x_2, x_3, \ldots, x_n)\rangle \prod_{i=1}^{n} (b_i - a_i). \tag{20.28}$$

For Eq. 20.28 to be an effective method of computating multidimensional integrals, we need a method for finding the average value of the integrand. If this average can be obtained, a few simple multiplications will produce the desired integral. To compute this average, we use the technique of random sampling, which is also called *Monte Carlo* sampling, with reference to the world-famous gambling casino in Monte Carlo. The basic idea is to compute values of the integrand at randomly selected points in the ranges of integration. These values form a discrete set from which the average is computed by using Eq. 20.24. If the number of points at which the integrand is computed is infinite, the result will be exact.

As an example, consider Eq. 20.28. To evaluate this integral using Monte Carlo methods, we first perform a random selection of the variables. Almost all computers are equipped with a random-number generator that, when invoked repeatedly, produces a set of random numbers, $\xi_1, \xi_2, \xi_3, \ldots, \xi_n, \ldots$, whose distribution is uniform over the interval from zero to unity. Since the number of digits that can be represented in the registers of any computer is finite, the set cannot be truly random. It is, therefore, often called a *pseudorandom set*. This is a matter of little concern to the scientist, because the error produced by the very small deviation from true randomness will be negligible relative to the error incurred because of inaccuracies in the potential-energy surface. Using such a random-number generator, we select the variables $x_1, x_2, x_3, \ldots, x_n$ by means of the algorithm

$$x_k = a_k + (b_k - a_k)\xi_k \quad \text{for } k = 1, 2, 3, \ldots, n. \tag{20.29}$$

The use of Eq. 20.29 ensures that the variable x_k will be randomly selected in the appropriate range of integration from a_k to b_k. The selection of all n variables produces point number 1 in the hyperspace of the system. The value of the integrand at this point is $f(x_1^1, x_2^1, x_3^1, \ldots, x_n^1)$, which we shall denote by f_1 for simplicity. The procedure is then repeated to produce point 2 in the system hyperspace, at which the integrand has the value $f_2 = f(x_1^2, x_2^2, x_3^2, \ldots, x_n^2)$. Using Eq. 20.24, we find that the average value of the integrand is

$$\langle f(x_1, x_2, x_3, \ldots, x_n)\rangle = \langle f\rangle = \operatorname*{Lim}_{N \longrightarrow \infty}\left[\frac{1}{N}\sum_{i=1}^{N} f_i\right]. \tag{20.30}$$

Although Eq. 20.30 provides an exact evaluation of the average value of the integrand, it is not really useful, since the calculation of an infinite number of integrand values is even more computationally demanding than summation methods. However, if we employ a finite number of points, the result may still be sufficiently accurate for our purposes. Therefore, we define

$$\langle f \rangle \approx \langle f \rangle' = \frac{1}{N} \sum_{i=1}^{N} f_i \qquad (20.31)$$

and use this approximate average to compute the integral. When this is done, the question of accuracy is of central importance.

The statistical error present in Eq. 20.31 is generally estimated from the square variance, s^2, of $f(\mathbf{x})$ from the approximate mean $\langle f \rangle'$. The square variance is

$$s^2 = \sum_{i=1}^{N} \left[\frac{[f(\mathbf{x}) - \langle f \rangle']^2}{N(N-1)} \right] = \sum_{i=1}^{N} \left[\frac{[f^2(\mathbf{x}) - 2\langle f \rangle' f(\mathbf{x}) + (\langle f \rangle')^2]}{N(N-1)} \right]. \qquad (20.32)$$

Because N is usually large, $N(N-1) \approx N^2$. If we make this replacement in Eq. 20.32 and use Eq. 20.31, we have

$$s^2 = \frac{1}{N} \left[\frac{1}{N} \sum_{i=1}^{N} f_i^2(\mathbf{x}) - 2\langle f \rangle' \frac{1}{N} \sum_{i=1}^{N} f_i(\mathbf{x}) + \left\{ \frac{1}{N} \sum_{i=1}^{N} (\langle f \rangle')^2 \right\} \right]$$

$$= \frac{1}{N} [\langle f^2 \rangle' - 2(\langle f \rangle')^2 + (\langle f \rangle')^2] = \frac{1}{N} [\langle f^2 \rangle' - (\langle f \rangle')^2]. \qquad (20.33)$$

Equation 20.33 follows from the facts that the average square of $f(\mathbf{x})$ is $(1/N) \sum_{i=1}^{N} f_i^2(\mathbf{x})$ and $\sum_{i=1}^{N} (\langle f \rangle')^2 = N(\langle f \rangle')^2$. The variance is related to the square of the uncertainty of the distribution, since the latter quantity is just the average square minus the square of the average. If the sampling is truly random, there is about a 0.68 probability that the absolute error in the average, $\varepsilon = |\langle f \rangle - \langle f \rangle'|$, will be less than s. Therefore, s provides a reasonable estimate of the error. With random sampling, we know that 68% of the time we will have

$$\boxed{s = N^{-1/2} [\langle f^2 \rangle' - (\langle f \rangle')^2]^{1/2} \geq \varepsilon} \qquad (20.34)$$

Thus, the absolute percent error in $\langle f \rangle'$ is such that about 68% of the time,

$$\boxed{|\% \text{ error}| \leq 100 \times \frac{s}{\langle f \rangle} \approx 100 \times \frac{s}{\langle f \rangle'}} \qquad (20.35)$$

Equations 20.34 and 20.35 show that the absolute percent error decreases with $N^{-1/2}$. Furthermore, the convergence of $\langle f \rangle'$ toward $\langle f \rangle$ is independent of the dimensionality of the integral. However, $N^{-1/2}$ decreases very slowly with increasing N. The slow convergence rate of Monte Carlo integration means that it is not competitive with other techniques in many applications. Nevertheless, the method is ideally suited to applications requiring evaluations of multidimensional integrals, where errors on the order of a few percent can be tolerated.

Before proceeding, it is appropriate that we digress for a few moments to discuss the difference between scientific Monte Carlo sampling and the type of sampling seen in public opinion polls. *The analysis just presented depends completely upon the assumption that the sampling of points is totally random, with no systematic bias of any kind.* When this assumption holds, the errors will be randomly positive and negative. Therefore, they tend to add to zero if the sample is large. For this reason, and for this reason alone, reasonably accurate results can be obtained with rather small samples, regardless of the dimensionality of the system. If there is any systematic bias, this is no longer true; systematic error cannot be removed by increasing the sample size.

There is little difficulty in assuring near-total randomness in a scientific calculation. However, this is not the case when public opinion is sampled via polls. This point was discussed previously in Section 10.1. Let us consider some typical examples. The most well-known study of student dishonesty and cheating is based on statistical data obtained from anonymous student questionnaires. The study is forced to treat the responses obtained from these questionnaires as honest. Therefore, the study assumes honest answers from students whose dishonesty is being investigated. The fact that systematic bias will be present in such data is obvious. As a result, the accuracy of the study does not necessarily increase as the size of the sample increases. Politicians and some educators usually ignore this fact. Political surveys are generally replete with systematic bias. In this case, the sampling is far from random. If, for example, we obtain data from telephone calls or an Internet survey, systematic bias is present because not all persons have telephones or computers and not all people will pay the telephone or Internet hookup charges. Therefore, these individuals are automatically excluded from the sample, which produces systematic bias. Other political surveys are based on person-to-person interviews. In this type of sampling, there is not only the difficulty associated with producing a random selection of persons to be interviewed, but frequently, an even more insidious type of systematic bias is present. This bias is associated with the form in which the questions are asked. Suppose a news agency is conducting a poll about the impeachment of the president. If the question asked is "Do you favor the impeachment of Mr. William Clinton?" the statistical data will be different from that obtained if the question asked is "Do you favor the impeachment of the president of the United States?" even though the persons sampled are the same. When a news commentator or newspaper states that the margin of error of a poll is 3 to 6%, he or she is assuming completely random, unbiased sampling, with Eqs. 20.34 and 20.35 used to compute the expected percent error. Since the sampling is *not* random, this claim is bogus. In many cases, the errors are much larger than stated. In other cases, there is deliberate distortion via the sampling methods employed. So, *caveat emptor*! (Let the buyer beware!)

EXAMPLE 20.4

A multidimensional integral is evaluated using Monte Carlo methods. When the integrand is evaluated 5,000 times, the absolute error is about 2.0%. If we wish to reduce the expected absolute error to about 0.2%, how many times must the integrand be evaluated?

Solution

Using Eqs. 20.34 and 20.35, and equating the estimated error to s, we have

$$2.0 \approx \frac{100(5,000)^{-1/2}[\langle f^2 \rangle' - (\langle f \rangle')^2]^{1/2}}{\langle f \rangle}. \tag{1}$$

If we wish to reduce the statistical error to 0.2%, the number of evaluations required is

$$0.2 \approx \frac{100N^{-1/2}[\langle f^2 \rangle' - (\langle f \rangle')^2]^{1/2}}{\langle f \rangle}. \tag{2}$$

Since the averages are reasonably accurate, there will be little change in $\langle f^2 \rangle'$ and $\langle f \rangle'$. Therefore, we expect the ratio $[\langle f^2 \rangle' - (\langle f \rangle')^2]^{1/2}/\langle f \rangle$ to be nearly the same in both calculations. Using this fact, we divide Eq. 1 by Eq. 2 to give

$$\frac{2.0}{0.2} = 10 = \left[\frac{N}{5,000} \right]^{1/2}. \tag{3}$$

Solving for N, we obtain

$$N = 5,000(10)^2 = 500,000. \tag{4}$$

We conclude that a reduction in the expected absolute error by an order of magnitude from 2.0% to 0.20% requires that we execute two orders of magnitude more computations. That is, the integrand must now be evaluated 500,000 times instead of 5,000. Monte Carlo methods can produce answers accurate to 5–10% with little effort, but increasing the accuracy is a computationally demanding task.

For related exercises, see Problems 20.12 and 20.15.

The integrals that appear in the expression for the rate in Eq. 20.9 all have the form $\int_a^b w(x)P(x)\,dx$, where $P(x)$ is the reaction probability and $w(x)$ is a normalized probability distribution function. Using the Monte Carlo methods described, we can express this integral as a Monte Carlo sum:

$$\int_a^b w(x)P(x)\,dx \approx \frac{b-a}{N}\sum_{i=1}^N w(x_i)P(x_i). \qquad (20.36)$$

Here, $w(x_i)$ and $P(x_i)$ are, respectively, the values of the normalized probability distribution for variable x and the reaction probability when we have $x = x_i$ where x_i is a randomly chosen value of x in the range $a \le x \le b$.

We can simplify Eq. 20.36 by defining a new function,

$$G(x) = \int_a^x w(x)\,dx, \qquad (20.37)$$

called the *cumulative distribution function*. Since $w(x)$ is a normalized probability distribution, it follows that

$$G(b) = \int_a^b w(x)\,dx = 1 \qquad (20.38)$$

and

$$G(a) = \int_a^a w(x)\,dx = 0, \qquad (20.39)$$

because definite integrals with equal upper and lower limits are zero. Taking differentials of both sides of Eq. 20.37, we obtain the result

$$dG(x) = w(x)\,dx. \qquad (20.40)$$

Direct substitution of the results in Eqs. 20.37 through 20.40 into the integral in Eq. 20.36 produces

$$\int_a^b w(x)P(x)\,dx = \int_{G(a)}^{G(b)} P(G)\,dG = \int_0^1 P(G)\,dG. \qquad (20.41)$$

If Monte Carlo methods are now used to evaluate the right-hand side of Eq. 20.41, the resulting expression is

$$\int_a^b w(x)P(x)\,dx = \frac{1}{N}\sum_{i=1}^N P(G_i)(1-0) = \frac{1}{N}\sum_{i=1}^N P(G_i). \qquad (20.42)$$

These equations are valid only if $w(x)$ is normalized. If it is not, the probability distribution must be normalized before Eqs. 20.37 through 20.42 can be used.

In the evaluation of the Monte Carlo summation in Eq. 20.42, it is the value of the cumulative distribution function $G(x)$ that is selected randomly in the range of its limiting values, which is always zero to unity. In most

Monte Carlo integrations, we recast the integrals by using appropriate cumulative distribution functions before executing the Monte Carlo sampling procedure. Examples 20.5 and 20.6 illustrate the method.

EXAMPLE 20.5

The integral over the azimuthal orientation angle θ in the computation of the reaction cross section or reaction rate coefficient has the form $\int_{\theta=0}^{\pi} (\sin \theta \, P(\theta) d\theta)/2$, where $P(\theta)$ represents the dependence of the reaction probability upon θ. Show how a cumulative distribution function and Monte Carlo integration can be used to evaluate this integral, assuming that we have a means of computing the reaction probability for any value of θ.

Solution

The normalized distribution function is $w(\theta) = \sin \theta \, d\theta/2$. Therefore, the cumulative distribution function obtained with use of Eq. 20.37 is

$$G(\theta) = \frac{1}{2} \int_0^{\theta} \sin \theta \, d\theta = \frac{1}{2} [-\cos \theta] \Big|_0^{\theta} = \frac{1}{2} [1 - \cos \theta]. \tag{1}$$

The differential of $G(\theta)$ is

$$dG(\theta) = \frac{\sin \theta \, d\theta}{2}, \tag{2}$$

and we have

$$G(\pi) = \frac{1}{2} [1 - \cos(\pi)] = 1 \tag{3}$$

and

$$G(0) = \frac{1}{2} [1 - \cos(0)] = 0. \tag{4}$$

Substituting into the integral over θ produces

$$\int_{\theta=0}^{\pi} \frac{\sin \theta \, P(\theta) d\theta}{2} = \int_{G(0)}^{G(\pi)} P(G) dG = \int_0^1 P(G) dG. \tag{5}$$

A Monte Carlo evaluation of the right-hand side of Eq. 5 yields

$$\int_{\theta=0}^{\pi} \frac{\sin \theta \, P(\theta) d\theta}{2} = \frac{1}{N} \sum_{i=1}^{N} P(G_i). \tag{6}$$

To evaluate the Monte Carlo sum in Eq. 6, we select random values of $G(\theta)$. That is, we take $G(\theta)$ to be a randomly chosen value between its integration limits of 0 and 1. Thus,

$$G_i(\theta) = \xi_i = 0.5[1 - \cos \theta_i], \tag{7}$$

where ξ_i is a random number whose distribution is uniform over the interval from zero to unity. We can now use Eq. 7 to find the value of θ corresponding to this selection of the cumulative distribution function. Solving for $\cos \theta_i$, we obtain

$$\cos \theta_i = 1 - 2\xi_i, \tag{8}$$

so that

$$\theta_i = \cos^{-1}[1 - 2\xi_i]. \tag{9}$$

The procedure is, therefore; (1) select ξ_i; (2) use Eq. 9 to obtain θ_i; (3) compute the value of the reaction probability, $P(\theta_i) = P(G_i)$; (4) repeat steps (1)–(4) N times; and then (5) execute the summation in Eq. 6 to obtain the value of the integral.

The integral over the impact parameter in the computation of the reaction cross section or reaction rate coefficient has the form $\int_0^\infty 2\pi b\, P(b)\, db$, where $P(b)$ represents the dependence of the reaction probability upon the impact parameter. Show how a cumulative distribution function and a Monte Carlo integration can be used to evaluate this integral, assuming that we have a means of computing the reaction probability for any value of b.

Solution

In this example, we must exercise some care. First, the infinite upper limit must be replaced with an appropriate finite limit. We know that beyond a certain impact parameter b_m, the reaction probability must be zero, so that there is no further contribution to the integral for values of $b > b_m$. This is simply saying that if two chemical reactants pass one another at an impact parameter of, say, 10^5 km, then the probability that they will react is zero. Therefore, without error, we can write

$$\int_0^\infty 2\pi b\, P(b)\, db = 2\pi \int_0^{b_m} b\, P(b)\, db. \tag{1}$$

We have not yet normalized the distribution function for the impact parameter, so we must do this before we can use a cumulative distribution function. We require that

$$N \int_0^{b_m} 2\pi b\, db = N\pi b_m^2 = 1. \tag{2}$$

Therefore, the normalization constant is $N = 1/(\pi b_m^2)$. We can produce a normalized probability distribution function by multiplying and dividing by πb_m^2. Thus, we express the integral in the form

$$\int_0^\infty 2\pi b\, P(b)\, db = \pi b_m^2 \int_0^{b_m} \frac{2\pi b\, P(b)\, db}{\pi b_m^2} = \pi b_m^2 \int_0^{b_m} \frac{2b\, P(b)\, db}{b_m^2}. \tag{3}$$

Since the probability distribution function is now normalized, we can compute a cumulative distribution function whose form is

$$G(b) = \int_{b=0}^{b} \frac{2b\, db}{b_m^2} = \left[\frac{b}{b_m} \right]^2. \tag{4}$$

With this definition, we have

$$dG(b) = \frac{2b\, db}{b_m^2}, \tag{5}$$

$$G(0) = \left(\frac{0}{b_m} \right)^2 = 0, \tag{6}$$

and

$$G(b_m) = \left(\frac{b_m}{b_m} \right)^2 = 1. \tag{7}$$

Substitution into Eq. 3 produces

$$\int_0^\infty 2\pi b\, P(b)\, db = \pi b_m^2 \int_{G(0)}^{G(b_m)} P(G)\, dG = \pi b_m^2 \int_0^1 P(G). \tag{8}$$

The Monte Carlo expression for the integral on the right-hand side of Eq. 8 is

$$\int_0^\infty 2\pi b\, P(b)\, db = \pi b_m^2 \frac{1}{N} \sum_{i=1}^N P(G_i). \tag{9}$$

To evaluate the Monte Carlo sum in Eq. 9, we select random values of $G(b)$. That is, we take $G(b)$ to be a randomly chosen value between its integration limits of zero and unity. Thus,

$$G_i(b) = \xi_i = \left[\frac{b_i}{b_m} \right]^2, \tag{10}$$

where ξ_i is a random number whose distribution is uniform over the interval from zero to unity. We can now use Eq. 10 to find the value of b_i corresponding to this selection of the cumulative distribution function. Solving for b_i, we obtain

$$b_i = (\xi_i)^{1/2} b_m. \tag{11}$$

The procedure is, therefore; (1) select ξ_i; (2) use Eq. 11 to obtain b_i; (3) compute the value of the reaction probability $P(b_i) = P(G_i)$; (4) repeat steps (1)–(4) N times, and then (5) execute the summation in Eq. 9 to obtain the value of the integral.

For related exercises, see Problems 20.13 and 20.14.

20.1.8   Trajectory Calculations

In the foregoing sections, we have developed integral expressions for the gas-phase specific reaction rate coefficient. Although the multidimensional integrals are difficult, we found in the previous section that they can be evaluated with relatively little difficulty by using Monte Carlo methods if we have a means of computing the integrand at randomly selected points. We also found that when a cumulative distribution function method is used, the integrand is just the reaction probability P_{AB} multiplied by a few simple constants. Therefore, to complete the calculation, we need a method that permits us to compute P_{AB} for any choice of the initial collision variables. In almost all investigations, P_{AB} is obtained using the method of quasi-classical trajectories.

In a classical trajectory study, we assume that the nuclei are sufficiently massive that they move classically on the potential-energy hypersurface describing the system. In this case, the detailed motion of the nuclei may be determined for any given set of initial conditions by integrating the Hamiltonian equations of motion that were presented in Eqs. 11.12A through 11.12F. For an N-atom system, they comprise a set of $6N$ first-order differential equations involving the derivatives of the coordinates and momenta of each particle with respect to time. In almost all cases, trajectories are computed by using a rectangular Cartesian coordinate system. With this choice, the Hamiltonian equations of motion are

$$\frac{\partial x_i}{\partial t} = \frac{p_{xi}}{m_i}, \tag{20.43A}$$

$$\frac{\partial y_i}{\partial t} = \frac{p_{yi}}{m_i}, \tag{20.43B}$$

$$\frac{\partial z_i}{\partial t} = \frac{p_{zi}}{m_i}, \tag{20.43C}$$

$$\frac{\partial p_{xi}}{\partial t} = -\frac{\partial V}{\partial x_i}, \tag{20.43D}$$

$$\frac{\partial p_{yi}}{\partial t} = -\frac{\partial V}{\partial y_i},$$

(20.43E)

and

$$\frac{\partial p_{zi}}{\partial t} = -\frac{\partial V}{\partial z_i},$$

(20.43F)

where x_i, y_i, and z_i are the Cartesian coordinates of nucleus i, p_{xi}, p_{yi}, and p_{zi} are the corresponding conjugate momenta, and V is the potential-energy hypersurface for the system. Equations 20.43A, 20.43B, and 20.43C are very simple: They are just a statement that in Cartesian coordinates, $p_{xi} = m_i v_{xi}$, where v_{xi} is the x component of velocity of nucleus i. The difficulty lies in evaluating the right-hand sides of Eqs. 20.43D, 20.43E, and 20.43F, which require that derivatives of the potential hypersurface be obtained. This means that if we have solved Eq. 20.22 at a series of points to obtain V, we must fit the computed points to some appropriately chosen functional form so that derivatives can be analytically computed. Numerical computation of the derivatives is both too inaccurate and too slow.

Since V is a complicated function of the coordinates, the equations of motion can generally not be solved analytically. Instead, the $6N$ differential equations must be integrated numerically on a high-speed modern workstation or mainframe computer for each selected set of initial collision variables. The method of choice to effect this numerical integration is usually a Runge–Kutta procedure. We shall not describe the details of the numerical computations here; suffice it to say that the result of such an integration is a detailed numerical listing of all the position and momentum coordinates for each nucleus throughout the time of the collision. This listing is called a *classical trajectory*. If the initial variables are selected from quantized distributions, the method is designated a *quasi-classical trajectory*. However, it should be recognized that even though the initial variables may be selected from properly quantized distribution functions, once the integration of the classical equations of motion begins, all quantization effects will be lost. Thus, the result will still reflect, for the most part, classical, rather than quantum mechanics. Quantum effects such as tunneling will be lost.

Once the trajectory has been computed, we will know the complete details of the collision. Thus, if we are examining a collision of H with D_2, we will know whether the hydrogen atom simply scattered off the D_2 molecule without reacting or whether the D_2 bond broke and a new HD molecule formed. If the former event occurred, an examination of the D–D distance throughout the trajectory will reveal that at the start, this distance was oscillating about the D_2 equilibrium separation. As the hydrogen atom approached, the amplitude of the oscillations probably varied due to energy transfer between H and D_2. At the end of the trajectory, when we find that the H–D distances are again large, the D–D separation will still be oscillating about its equilibrium distance. Therefore, no reaction occurred, and the reaction probability for this collision is zero. If, on the other hand, we find that at the end of the trajectory, one of the H–D distances is small and oscillating about the H–D equilibrium separation while the other H–D distance is very large, we know that the D_2 bond ruptured and a new H–D bond formed. Consequently, the probability of reaction for *this* choice of initial collision variables is unity.

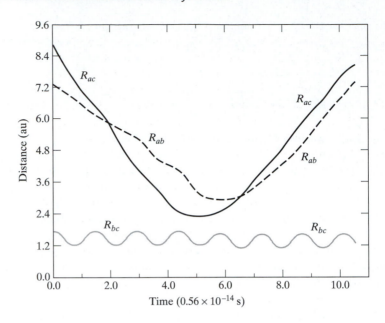

► FIGURE 20.12
A nonreactive collision of H_a with H_b–H_c. The initial collision variables are $j = 5$, $v = 0$, and $V = 1.18 \times 10^4$ m s^{-1}. Distances are given in atomic units, or bohr. Time is expressed in units of 0.0056 ps. Data are taken from M. Karplus, R. N. Porter, and R. D. Sharma, *J. Chem. Phys.* **43**, 3259 (1965).

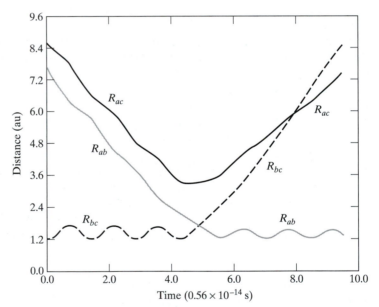

► FIGURE 20.13
A reactive collision of H_a with H_b–H_c to form H_a–H_b. The initial collision variables are $j = 0$, $v = 0$, and $V = 1.32 \times 10^4$ m s^{-1}. Distances are given in atomic units, or bohrs. Time is expressed in units of 0.0056 ps. Data are taken from M. Karplus, R. N. Porter, and R. D. Sharma, *J. Chem. Phys.* **43**, 3259 (1965).

Figures 20.12 and 20.13 illustrate nonreactive and reactive collisions, respectively, for hydrogen exchange reactions between H and H_2. There are two possible reactive channels, viz.,

$$H_a + H_b\text{–}H_c \longrightarrow H_a\text{–}H_b + H_c$$

and

$$H_a + H_b\text{–}H_c \longrightarrow H_a\text{–}H_c + H_b,$$

where we have labeled the hydrogen atoms with subscripts to denote the different possible exchange reactions. Such reactions would be very difficult to detect experimentally, because the products are chemically identical. However, the computer has no difficulty in doing this. As we have already noted, the potential-energy hypersurface for these exchange reactions is identical to that for the $H + D_2$ reaction, since the potential is independent of isotopic mass when we make the Born–Oppenheimer approximation. In the figures, R_{ab}, R_{ac}, and R_{bc} are the internuclear distances between H_a–H_b, H_a–H_c, and H_b–H_c, respectively. The trajectory shown in Figure 20.12 is clearly nonreactive with $P_{AB} = 0$. The R_{bc} distance continues to oscillate about the equilibrium position throughout the trajectory, while both R_{ab} and R_{ac} at first decrease as H_a approaches the H_2 molecule, then reach a minimum, and finally increase as H_a scatters and recedes into the distance. In contrast, the trajectory shown in Figure 20.13 is that of a reactive process in which H_a–H_b is formed during the collision. This is obvious from the fact that the H_b–H_c bond breaks about 0.027 ps into the trajectory. This breakage causes R_{bc} to become large while the H_a–H_b bond forms, and R_{ab} begins an oscillatory motion about the H_2 equilibrium position. Therefore, for this trajectory, $P_{AB} = 1$.

Let us again examine the functional form for the reaction cross section from particular vibrational and rotational states. Equation 20.18 gives us the general form for an atom–diatomic reaction:

$$\sigma^2(V, v, j) = \int_{b=0}^{\infty} \int_{\theta=0}^{\pi} \int_{\phi=0}^{2\pi} \int_{\gamma=0}^{2\pi} \int_{R=R_{min}}^{R_{max}} \sum_{m=-j}^{+j} P_{AB}(2\pi b\, db) P_1(\theta, \phi) P_2(R) P_3(\gamma).$$

The probability distributions $P_1(\theta, \phi), P_2(R)$, and $P_3(\gamma)$ are already normalized. If we now normalize the distribution function for the impact parameter integral as shown in Example 20.6 and also that for the magnetic quantum number (see Problem 20.16) and then use the cumulative distribution method for all integrals, the result is

$$\sigma^2(V, v, j) = \pi b_m^2(2j + 1) \int_{G_b=0}^{1} \int_{G_\theta=0}^{1} \int_{G_\phi=0}^{1} \int_{G_\gamma=0}^{1} \int_{G_R=0}^{1} \int_{G_m=0}^{1}$$

$$P_{AB}(\xi_b, \xi_\theta, \xi_\phi, \xi_\gamma, \xi_R, \xi_m)\, d\xi_b\, d\xi_\theta\, d\xi_\phi\, d\xi_\gamma\, d\xi_R\, d\xi_m,$$

where $\xi_b, \xi_\theta, \xi_\phi, \xi_\gamma, \xi_R$, and ξ_m are the random numbers used to select values of the cumulative distribution functions for b, θ, ϕ, γ, R, and m, respectively. The factor of $2j + 1$ results from normalizing the cumulative distribution function for m. Using a Monte Carlo method to evaluate this multidimensional integral, we obtain

$$\sigma^2(V, v, j) = \pi b_m^2(2j + 1)\frac{1}{N} \sum_{i=1}^{N} P_{AB}(\xi_b, \xi_\theta, \xi_\phi, \xi_\gamma, \xi_R, \xi_m)_i. \qquad \textbf{(20.44)}$$

Suppose we now employ trajectories to compute the reaction probability in Eq. 20.44. Our procedure will be as follows: First, we select the six random numbers $\xi_b, \xi_\theta, \xi_\phi, \xi_\gamma, \xi_R$, and ξ_m. By equating the cumulative distribution functions to these values, we can determine the impact parameter, the orientation angles θ and ϕ, the rotation angle γ for the rotational angular momentum vector, the diatomic internuclear distance R, and the magnetic quantum number m. These variables, along with the relative speed V and the vibrational–rotational state (v, j) at which we are computing the reaction cross section, provide all the initial collision variables that determine the details of the trajectory. Second, we integrate the equations of motion to obtain the trajectory and thereby determine whether we have $P_{AB} = 1$ (and hence a reaction) or $P_{AB} = 0$ (no reaction). This procedure is repeated N times and the reaction cross section is computed from Eq. 20.44. Each Monte Carlo point in the summation requires the computation of a trajectory.

Since P_{AB} is either zero or unity, the summation in Eq. 20.44 will simply be the number of reactive trajectories obtained in the computation of N total trajectories, and the reaction cross section will be given by

$$\sigma(V, v, j) = \frac{N_R(2j + 1)\pi b_m^2}{N},\qquad (20.45)$$

where N_R is the number of reactive trajectories obtained in our sample of N total trajectories. The fact that P_{AB} is classically either zero or unity converts Eqs. 20.34 and 20.35 into particularly simple expressions. The quantity f in these equations is now P_{AB}, so that we have

$$\langle f \rangle' = \langle P_{AB} \rangle' = \frac{1}{N} \sum_{i=1}^{N} [P_{AB}]_i = \frac{N_R}{N}$$

and

$$\langle f^2 \rangle' = \langle (P_{AB})^2 \rangle' = \frac{1}{N} \sum_{i=1}^{N} [P_{AB}]_i^2 = \frac{N_R}{N}.$$

Substituting these results into Eq. 20.34 produces

$$s^2 = \frac{1}{N}[\langle (P_{AB})^2 \rangle' - \{\langle P_{AB} \rangle'\}^2] = \frac{1}{N}\left[\frac{N_R}{N} - \left\{\frac{N_R}{N}\right\}^2\right] = \frac{NN_R - N_R^2}{N^3}.$$

Using this expression in Eq. 20.35 gives us the absolute percent error in a trajectory calculation:

$$|\% \text{ error}| = 100 \times \frac{s}{\langle P_{AB} \rangle'} = 100 \times \frac{N}{N_R} \times \left[\frac{NN_R - N_R^2}{N^3}\right]^{1/2}$$

$$= 100 \times \left[\frac{N - N_R}{NN_R}\right]^{1/2}.\qquad (20.46)$$

EXAMPLE 20.7

The reaction cross section $\sigma^2(V, v, j)$ for $H + D_2 \longrightarrow HD + D$ reactions with the initial D_2 vibrational–rotational states $v = 0$ and $j = 0$ at a relative speed of 1.37×10^4 m s^{-1} is computed with the use of quasi-classical trajectories with a maximum impact parameter $b_m = 2.0 \times 10^{-10}$ m. Two thousand trajectories are computed, among which HD is formed 478 times. **(A)** Compute $\sigma^2(V, v, j)$ for $v = j = 0$ and $V = 1.37 \times 10^4$ m s^{-1} from these data. **(B)** Compute the expected percent error in $\sigma^2(V, v, j)$.

Solution

(A) The reaction cross section is given by Eq. 20.45:

$$\sigma^2(V, v, j) = \frac{N_R(2j + 1)\pi b_m^2}{N} = \frac{(478)(2(0) + 1)(3.14159)(2.0 \times 10^{-10} \text{ m})^2}{2,000}$$

$$= 3.00 \times 10^{-20} \text{ m}^2.\qquad (1)$$

(B) The absolute percent error in the calculation is given by Eq. 20.46:

$$|\% \text{ error}| = 100 \times \left[\frac{N - N_R}{NN_R}\right]^{1/2} = 100 \times \left[\frac{2,000 - 478}{(2,000)(478)}\right]^{1/2}$$

$$= 3.99\%. \tag{2}$$

Notice that it is the number of reactive trajectories obtained in the calculation that is the most important factor in determining the accuracy of the computed reaction cross section and rate coefficient. If we compute the same 2,000 trajectories, but obtain only 5 reactive ones, the absolute percent error, according to Eq. 2 will be 44.7%.

For related exercises, see Problems 20.17 and 20.18.

The preceding discussion demonstrates how quasi-classical trajectories can be used to obtain the reaction probabilities required for the computation of cross sections and rate coefficients. However, it should not be assumed that that is the major reason scientists compute trajectories. Trajectory methods have been utilized to explore the mechanistic details of intra- and intermolecular energy transfer; gas–solid surface scattering; chemisorption; energy transfer and chemical reactions occurring within the confines of a cryogenic matrix at temperatures around 12 K; the molecular events occurring in machining, indentation, polishing, and frictional motion; and the mechanism of complex chemical reactions. It is not an overstatement to say that classical and quasi-classical trajectories are the most powerful and generally useful theoretical tool available to the investigator who wishes to explore the dynamic motions of atoms and molecules in a wide variety of different processes. [Readers interested in more complete details of the computation of trajectories may consult L. M. Raff and D. L. Thompson, "The Classical Trajectory Approach to Reactive Scattering" in *Theory of Chemical Reaction Dynamics*, ed. M. Baer, Vol. 3, Chapter 1, pp. 1–122 (Boca Raton, FL: CRC Press, 1985) and references therein. The presentation in that discussion is directed toward seniors and beginning graduate students.]

We conclude this section on collision theory and trajectories with an illustration that demonstrates the power of these theoretical methods. Consider the gas-phase reaction of *cis*-ethylene-d_2 with F_2. The F_2 bond is very weak (156 kJ mol^{-1}), while the C–F bond is very strong (484 kJ mol^{-1}). Therefore, the reaction can be highly exothermic. (Don't try this in your kitchen!) The first question any chemist asks is "What are the reaction products?" In this case, there are several possibilities. Six of the most likely candidates are shown in Figure 20.14. It is not a simple matter to predict the result even qualitatively, and a quantitative prediction of the relative yields of these products under specified experimental conditions is far more difficult. Yet, such predictions can and have been made using classical trajectory methods. The results are in reasonably good accord with experimental measurements, even though the potential-energy hypersurface employed in the calculations certainly contains inaccuracies.

Since we know the complete details of the atomic motions during the trajectories, it is a simple matter to determine the extent to which each possible set of products is formed. Moreover, the trajectories also provide the detailed mechanisms by which each reaction occurs. The trajectory calculations [L. M. Raff, *J. Chem. Phys.* **95**, 8901 (1991), and *ibid.*, **97**, 7459 (1992)] show that the major product is the fluoroethyl radical. Most trajectories add only one of the two fluorine atoms across the double bond. 1,2-difluoroethane-d_2 does form, but its lifetime is short due to the very large exothermicity of the reaction, which tends to produce unimolecular decomposition. Some of this product— but very little—is seen because of subsequent collisional deactivation. The

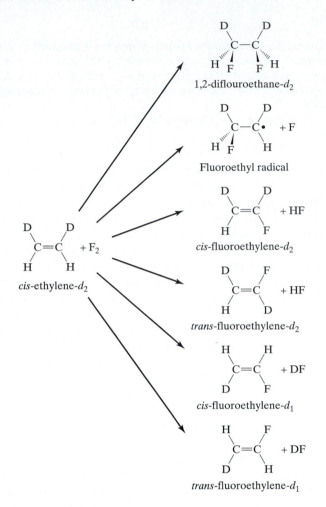

▶ FIGURE 20.14
Possible products in the reaction of
cis-ethylene-d_2 with F_2.

unimolecular decomposition of the 1,2-difluoroethane-d_2 forms both the fluoroethyl radical and substantial amounts of *cis*- and *trans*-fluoroethylene-d_1 and *cis*- and *trans*-fluoroethylene-d_2.

The formation of both the *cis* and *trans* forms of fluoroethylene seems at odds with the usual $\alpha\beta$-addition, *syn*-elimination mechanism generally presented in most courses in organic chemistry. This mechanism is shown in Figure 20.15. In the first step, F_2 undergoes a facial $\alpha\beta$ addition across the $C=C$ double bond to form the eclipsed structure shown in the figure. Before HF or DF elimination can occur, we must have a 120° rotation about the newly formed C–C single bond. Such a rotation produces a different eclipsed structure in which H–F and D–F are in positions such that a one-step, concerted *syn* elimination of either HF or DF can occur. In the former case the product is *trans*-fluoroethylene-d_2, in the latter *trans*-fluoroethylene-d_1. Thus, this reaction mechanism predicts that we will always obtain the *trans*-fluoroethylene product. However, both the experiments and the trajectory calculations predict that we also obtain some *cis* product, although the amount is less than the *trans* product.

Figure 20.16 illustrates the major mechanism predicted by the trajectory calculations for the formation of *cis*-fluoroethylenes. The first two steps in the process are identical to those illustrated in Figure 20.15. However, the mechanistic details now change. Instead of eliminating HF in a single step,

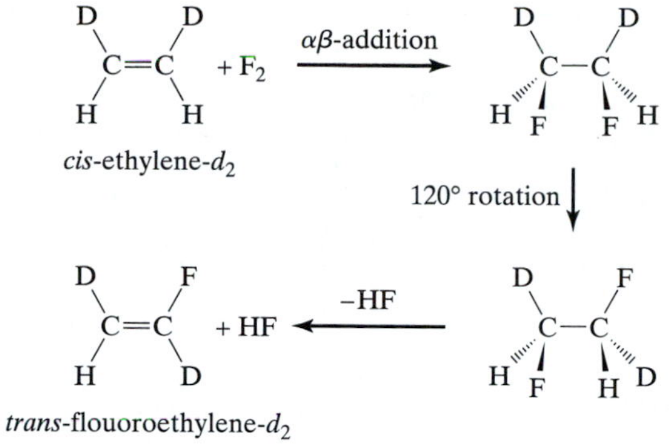

◄ FIGURE 20.15
Mechanism of $\alpha\beta$ addition followed by *syn* elimination.

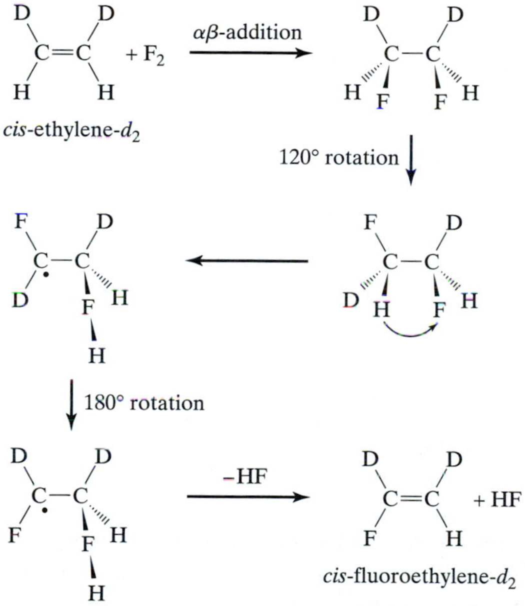

◄ FIGURE 20.16
Reaction mechanism predicted by trajectory calculations for the formation of *cis*-fluoroethylene products in the reaction of *cis*-ethylene-d_2 with F_2.

we observe a preliminary transfer of the mobile hydrogen atom to the adjacent fluorine atom to form the third, intermediate structure. After this transfer, a further 180° rotation of the FDC–group about the C–C single bond occurs to form the fifth structure shown in Figure 20.16. At this point, the C–F bond breaks, forming HF and *cis*-fluoroethylene-d_2.

It should not be thought that the mechanism illustrated in Figure 20.16 is simply a hypothesis being advanced to explain the experimental observation that *cis* products are formed. The mechanism is predicted by the results of the trajectory calculations. Figure 20.17 shows the results of one such trajectory, in which the temporal behavior of the D–F, C–D, and C–F internuclear distances eventually results in the elimination of DF. Distances are given in angstroms and time is in units of 0.01019 ps. The zero of time is taken to be the moment at which 1,2-difluoroethane-d_2 forms in the trajectory. The C–D and C–F distances are displaced upwards by 3Å and 6Å, respectively, to provide better visual clarity. The plot shows that from $t = 0$ to $t \approx 60$ tu (1 tu = 1.019×10^{-14} s), the C–D and C–F distances oscillate about their equilibrium bond lengths, while the D–F distance is

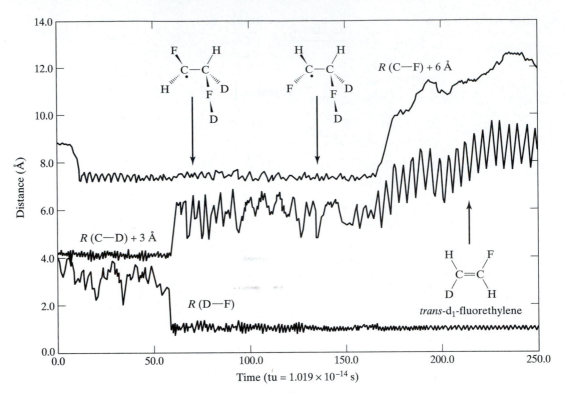

▲ FIGURE 20.17
Details of a trajectory for the reactions of *cis*-ethylene-d_2 with F_2, leading to the formation of *trans*-fluoroethylene-d_1 via the mechanism described in the text and in Figure 20.16. Distances are given in Å and time is in units of 0.01019 ps, with the zero of time taken to be the moment of formation of 1,2-difluoroethane-d_2. Data are taken from L. M. Raff, *J. Chem. Phys.* **95**, 8901 (1991), and *ibid.*, **97**, 7459 (1992).

that characteristic of D and F atoms on adjacent carbon atoms. The structure is that for 1,2-difluoroethane-d_2. Around 60 tu into the process, the C–D distance is seen to increase suddenly by about 3 Å, while simultaneously the D–F distance decreases to that characteristic of the DF molecule. However, the C–F bond does not break; instead, it remains in the vicinity of 1.5 Å from $t \approx 60$ tu to $t \approx 170$ tu. Clearly, a deuterium atom has transferred from one carbon to the adjacent fluorine atom, as shown in the fourth structure in Figure 20.16. During the lifetime of this structure, which is about 110 tu, the HFC–group can rotate freely about the C–C single bond. By computing the H–C–C–H dihedral angle as a function of time from the trajectory data, we can determine the structure of this intermediate at different moments of time. Around $t \approx 70$ tu, the structure is that shown in Figure 20.17. If the C–F bond were to rupture at this moment to eliminate DF, *trans*-fluoroethylene-d_1 would be the product. However, the C–F bond does not break at this point. At $t \approx 130$ tu, a 180° rotation of the HFC–group has occurred, and the structure is such that the elimination of DF at that point would result in the formation of *cis*-fluoroethylene-d_1. When the C–F bond finally does rupture, around 170 tu, the HFC–group has again rotated, so the final product in this trajectory is the *trans* product. In other trajectories, the C–F bond rupture occurs at a time when the structure is such that the *cis* product forms. By computing a large number of trajectories, the *cis*–*trans* product ratio can be predicted.

The fact that trajectory calculations can unravel the intimate details of the reaction mechanism in such complex processes illustrates the power of the method.

20.2   Statistical Theory of Reaction Rates

The calculation of specific reaction rate coefficients using collision theory and quasi-classical trajectories can require very large quantities of computational time, even on the fastest computers, if the system and corresponding potential-energy hypersurface are complex. This is a consequence of our need to compute a sufficiently large number of trajectories to obtain accurate results in the Monte Carlo integrations. However, in return for our efforts, we obtain not only the rate coefficients, but also detailed information concerning the reaction mechanism, energy disposal, and energy transfer occurring in the process. We learn, for example, what happens to the energy that is released in an exothermic process. This energy may appear in the form of relative translational motion of the products or as vibrational–rotational energy of one or more of the products. The trajectory results give us these energy distributions for any set of experimental conditions we may wish to investigate.

In some cases, however, our only interest will be in calculating the reaction rates. When this is the situation, much of our computational effort using trajectory methods will be wasted. It is, therefore, far better to use a different procedure that does not provide the extensive information available from trajectories, but that permits us to obtain rate coefficients with much less computational effort. This can be done using theoretical methods based on statistical mechanics developed in Chapters 17 and 18. In the current section, we shall examine such statistical theories of reaction rates.

20.2.1   The Wigner Variational Theorem

The concepts to be developed in this section hold for any reaction. However, the mathematical notation becomes very complex when the system has a large number of nuclei. We shall, therefore, develop the concepts with reference to atom–diatomic reactions.

Suppose we wish to compute the rate coefficients for the $H + D_2 \longrightarrow HD + D$ reaction and that we have no interest in the reaction mechanism or in energy disposal or energy transfer processes that may occur. Before we can proceed, we still must obtain the potential-energy hypersurface for the system, but as we shall see, we do not need the entire hypersurface; only those regions in the near vicinity of the transition state, along with the reactant valley, are needed. Therefore, we have already significantly reduced the computational effort that we must expend to complete the calculation.

Three nuclear coordinates are required to specify the relative nuclear configuration of the D_2H system. Let us suppose that we take two internuclear distances R_1 and R_2 and the included angle θ as our three coordinates and represent the potential hypersurface by $V(R_1, R_2, \theta)$. We now construct several additional hypersurfaces that are functionally dependent upon the same internuclear coordinates and that contain adjustable parameters $a, b, c, d, \ldots$. These hypersurfaces can, for the moment, be anything that strikes our fancy. We shall denote them by $S_1(R_1, R_2, \theta; a, b, \ldots)$, $S_2(R_1, R_2, \theta; a, b, \ldots)$, $S_3(R_1, R_2, \theta; a, b, \ldots)$, $\ldots$, where the semicolon separates the variables from the parameters defining the surface. All hypersurfaces $V(R_1, R_2, \theta)$, S_1, S_2,

$S_3, \ldots$ span the same space, so it is possible that they may intersect. Figure 20.18 illustrates possible intersections of S_1, S_2, and S_3 with $V(R_1, R_2, \theta)$ in the plane for which $\theta = 180°$. The contour lines are those shown in Figure 20.7 for $V(R_1, R_2, 180°)$, and the curves are hypothetical intersections of S_1, S_2, and S_3 with that surface.

We now divide the hypersurfaces S_1, S_2, $S_3, \ldots$ into two classes: those which completely separate the reactant and product valleys of $V(R_1, R_2, \theta)$ and those which do not effect such a separation. Hypersurface S_1 is member of the first class, since it completely separates the reactant and product valleys of the potential-energy hypersurface. By "completely separates" we mean that if one starts in the reactant valley, it is impossible to reach the product valley along any path without crossing hypersurface S_1 at least once. An examination of the intersection of S_1 with $V(R_1, R_2, \theta)$ in Figure 20.18 shows that S_1 satisfies this condition. Hypersurfaces that permit reactants to reach the product valley without crossing the hypersurface are members of the second class, which do not separate the reactant and product valleys of $V(R_1, R_2, \theta)$. The figure shows that hypersurfaces S_2 and S_3 are members of this class. When we compute reaction rate coefficients using statistical methods, we are concerned only with the first class of hypersurfaces—those which completely separate the reactant and product valleys. Such surfaces are called *dividing surfaces*, since they "divide" the reactant configuration space from the product configuration space on the potential-energy hypersurface.

Suppose we now examine a very large number of reactant systems and follow their trajectories as they attempt to reach the product valley. Some of the reactants will cross the dividing surface and move directly into the product valley. This class of reactions is illustrated by Trajectory 1 in Figure 20.19. Other

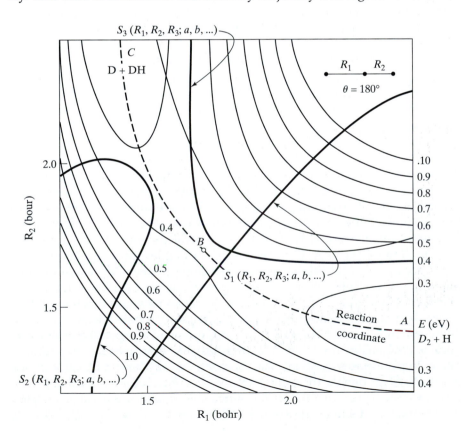

► FIGURE 20.18
Contour map of the saddle-point region for the D_2H system with $\theta = 180°$. Contour-line energies are given in eV, where 1 eV = 96.48 kJ mol. The other lines represent the intersections of three hypersurfaces, $S_1(R_1, R_2, \theta; a, b, \ldots)$, $S_2(R, R, \theta; a, b, \ldots)$ and $S_3(R, R, \theta; a, b, \ldots)$, with the potential-energy hypersurface. [Data for potential-energy hypersurface taken from R. N. Porter and M. Karplus, *J. Chem. Phys.* **40**, 1105 (1964).]

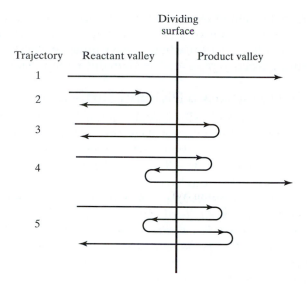

Dividing surface

Trajectory   Reactant valley   Product valley

1

2

3

4

5

Possible trajectories, reactive and nonreactive, that may cross a dividing hypersurface. Trajectory 1 crosses and reacts. Trajectory 2 does not cross and does not react. Trajectory 3 crosses twice and does not react. Trajectory 4 crosses three times and then reacts. Trajectory 5 crosses four times and does not react.

reactant systems will move toward the dividing surface, but fail to reach it and return to the product valley. Systems that do not have sufficient energy to traverse the potential barrier to reaction are members of this class, which is represented by Trajectory 2 in the figure. A third class of reactant systems will cross the dividing surface, but fail to reach the product valley. Instead, they reverse course and return to the reactant valley, as shown by Trajectory 3. Other classes of trajectories cross the dividing surface multiple times. Some of these eventually react, such as Trajectory 4, while others, like Trajectory 5, fail to react.

Figure 20.19 shows that if $\mathcal{F}(T)$ is the total flux of reactant systems crossing a dividing surface per unit volume per unit time at temperature T and $\mathcal{R}(T)$ is the reaction rate at that same temperature, then

$$\boxed{\mathcal{R}(T) \leq \mathcal{F}(T)}. \qquad (20.47)$$

Equation 20.47 is known as the *Wigner variational theorem*. As long as the dividing surface truly separates reactant and product valleys, the validity of this theorem follows from the fact that every reactant system that reacts must cross the dividing surface at least once, but there may be some systems that cross the dividing surface without reacting. The equality in Eq. 20.47 will hold if and only if there are no systems that cross the dividing surface that do not react and no systems that react that cross the dividing surface more than once. That is, the equality will hold only if there are no trajectories like those illustrated by Trajectories 3, 4, and 5 in Figure 20.19.

The Wigner variational theorem permits us to carry out calculations similar in spirit to the variational calculations we executed in obtaining approximate solutions of the Schrödinger equation. In that case, the Rayleigh–Ritz variational theorem assures us that the variational energy λ must be an upper limit to the true ground-state energy of the system. That is, we must have $E \leq \lambda$, regardless of the nature of the approximate wave function employed in the calculations. This fact permitted us to insert parameters in the approximate wave function and then vary those parameters to minimize λ, secure in the knowledge that the lower the value of λ, the closer to the true ground-state energy eigenvalue we must be. A similar idea can now be used with respect to reaction rates: Since we must always have $\mathcal{R}(T) \leq \mathcal{F}(T)$, we can parameterize the definition of the dividing surface and then adjust those parameters so as to minimize $\mathcal{F}(T)$, knowing that the smaller the flux becomes, the closer to the true reaction rate we must be. All statistical methods for computing reaction rates are based on this variational principle.

20.2.2   The Flux across a Dividing Surface

In order to utilize the Wigner variational theorem to compute approximate reaction rates, we must have a means of calculating the flux across a dividing surface. We could, of course, accomplish this by using classical or quasi-classical trajectories, but if we do either, we will be investing almost as much computational effort as if we carried out the entire investigation using trajectories. *Instead, we shall assume that the populations of internal molecular states are always described by a Boltzmann distribution at the temperature of the experiment, regardless of where we are on the potential-energy hypersurface for the system. We shall also assume that the translational energies are described by a Maxwell–Boltzmann distribution.* These assumptions are the foundation upon which all statistical theories of reaction rates rest. The accuracy of the approximations depends upon the relative rates of reaction and the redistribution of energy among the internal molecular states. If the redistribution rate between all internal modes is large compared with the reaction rate, the approximations will be quite accurate. If the opposite is true, the accuracy of statistical theories decreases. A great deal of research, both experimental and theoretical, has been devoted to the investigation of the accuracy of the statistical approximations under a variety of conditions. Throughout the remainder of this section, we shall assume that these approximations are accurate.

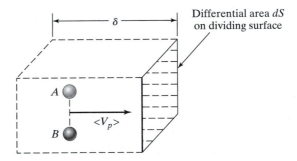

▶ FIGURE 20.20

Number of *A–B* pairs with energy *E* per unit volume crossing a small differential area *dS* on the dividing surface per unit time.

To formulate an appropriate expression for the flux of *A–B* pairs crossing the dividing surface for the reaction $A + B \longrightarrow$ products, let us first determine the number of such *A–B* pairs having a total energy *E* in a unit volume on the reactant side of the dividing surface. Figure 20.20 illustrates this situation. If N_A and N_B are the number of *A* and *B* molecules per unit volume, respectively, then in the unit volume shown in the figure, there will be $N_A N_B$ possible *A–B* pairs. The number of these pairs that have energy *E* is $N_A N_B P_{AB}(E)$, where $P_{AB}(E)$ is the probability that an *A–B* pair will have total energy *E*. If we multiply this number by the average speed of *A–B* pairs perpendicular to the dividing surface, $\langle V_p \rangle$, and the magnitude of the differential surface element, we obtain the number of *A–B* pairs per unit volume with energy *E* crossing the dividing surface through area *dS* per unit time. We may express this quantity in the form

$$dN_{AB}(E) = N_A N_B \langle V_p \rangle P_{AB}(E) dS. \tag{20.48}$$

The number of A–B pairs per unit volume with energy E crossing per unit area per unit time is, therefore, given by

$$\frac{\partial N_{AB}(E)}{\partial S} = N_A N_B \langle V_p \rangle P_{AB}(E). \tag{20.49}$$

The total number of A–B pairs per unit volume with energy E that cross the entire dividing surface per unit time can be obtained simply by summing (integrating) over the contribution for every differential area on the surface. This total is the differential flux at energy E:

$$d\mathcal{F}(E) = \int_S \frac{\partial N_{AB}(E)}{\partial S} dS = \int_S N_A N_B \langle V_p \rangle P_{AB}(E) \, dS. \tag{20.50}$$

The total flux across the dividing surface is obtained by summing $d\mathcal{F}(E)$ over all possible energies:

$$\mathcal{F}_S^{\text{total}} = \sum_{\text{all } E} d\mathcal{F}(E) = N_A N_B \int_S \sum_{\text{all } E} \langle V_p \rangle P_{AB}(E) dS. \tag{20.51}$$

If the system is in Boltzmann equilibrium at temperature T at all points on the potential hypersurface, then

$$P_{AB}(E) = \frac{1}{z_A z_B} \exp\left[-\frac{E}{kT}\right], \tag{20.52}$$

where z_A and z_B are the molecular partition functions for molecules A and B, respectively. Combining Eqs. 20.51 and 20.52 then gives

$$\boxed{\mathcal{F}_S^{\text{total}} = \frac{N_A N_B}{z_A z_B} \int_S \sum_{\text{all } E} \langle V_p \rangle \exp\left[-\frac{E}{kT}\right] dS.} \tag{20.53}$$

The total flux in Eq. 20.53 depends upon the definition of S and the values assigned to the parameters present in the definition of the surface. This is clear because the potential energy at each surface point will vary if we alter the dividing surface. The Wigner variational theorem tells us that we can now vary the dividing surface to obtain a minimum for $\mathcal{F}_S^{\text{total}}$, knowing that by doing so we will be approaching the value of the reaction rate. Computations carried out in this manner are called *variational transition-state theory* (VTST) calculations.

Statistical calculations carried out using the general form of Eq. 20.53 with no restrictions or further assumptions are very difficult and computationally laborious. Since the point of statistical theory is to minimize the effort required to compute the rate coefficients, some simplifications of that expression are required. One possibility is to assume that all molecular-energy states may be treated classically. This allows the sums in the equation to be replaced with integrals that can be evaluated by means of Monte Carlo–type methods. Other possible approximations involve restricting the dividing surface to forms that simplify the calculations. [Readers interested in more complete details related to variational transition-state theory may consult D. G. Truhlar, A. D. Isaacson, and B. C. Garrett, "Generalized Transition State Theory," in *Theory of Chemical Reaction Dynamics*, ed. M. Baer, Vol. IV, Chapter 2, pp. 65–138 (Boca Raton, FL: CRC Press, 1985).] In the current text, we shall present the theory only in its most elementary form.

20.2.3   Simple Transition-state Theory

In this section, we shall reduce Eq. 20.53 to the simple form originally proposed by Henry Eyring [*J. Chem. Phys.* **3**, 107 (1935)]. Although Eyring's derivation differs from that presented here, the final results are identical.

To remove many of the difficulties associated with evaluating the total flux across the dividing surface with the use of Eq. 20.53, we shall introduce several additional approximations. The first of these is to assume that we do not need to examine all possible dividing surfaces in order to minimize $\mathcal{F}_S^{\text{total}}$; we simply state that the optimum dividing surface is the one that passes through the saddle points perpendicularly to the reaction coordinate. The rationale for this approximation is seen in the exponential term, $\exp[-E/(kT)]$, present in the equation. When we choose S in the manner described, the minimum value of the potential energy on the dividing surface will be as large as possible. Thus, $\exp[-E/(kT)]$ will be as small as we can make it. Consequently, we might expect this choice to minimize the total flux. More exact calculations show that that expectation is close to being correct. With this assumption, the intersection of the dividing surface with the potential hypersurfaces shown in Figures 20.6, 20.7, and 20.9 is a straight line along the diagonal of the figure, along which $R_1 = R_2$. This is illustrated in Figure 20.21.

We now insert a second approximation. Since the factor $\exp[-E/(kT)]$ decreases very rapidly as the potential increases, we assume that most of the contribution to the integral over the area of the dividing surface comes from a small differential area dS in the transition-state region, where the potential has its minimum value. If we also assume that the integrand is constant over this small area, it can be factored out of the integral, so that the integral

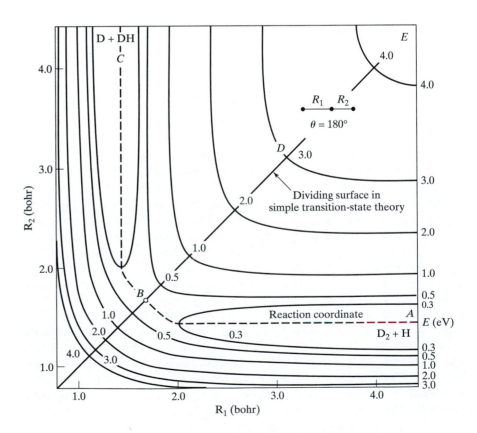

▶ FIGURE 20.21
Contour map for the D_2H system with $\theta = 180°$, as shown in Figure 20.6. The optimum dividing surface in simple transition-state theory is assumed to be a surface through the saddle point and perpendicular to the reaction coordinate. For the D_2H system, the intersection of this surface with the contour surface corresponds to the diagonal in the figure, along which we have $R_1 = R_2$. [Data taken from R. N. Porter and M. Karplus, *J. Chem. Phys.* **40**, 1105 (1964).]

becomes $\int dS$ over the differential area shown in Figure 20.20. We shall represent the value of this integration by σ^2, which has the units of distance squared. Integration over the remainder of the surface gives a near-zero result, since we are assuming that the flux through these regions is essentially zero. With these assumptions, Eq. 20.53 reduces to

$$\mathcal{F}_S^{\text{total}} = \frac{N_A N_B}{z_A z_B} \int_S \sum_{\text{all } E} \langle V_p \rangle \exp\left[-\frac{E}{kT}\right] dS = \frac{N_A N_B}{z_A z_B} \sum_{\text{all } E} \langle V_p^\ddagger \rangle \exp\left[-\frac{E^\ddagger}{kT}\right] \int_S dS$$

$$= \frac{N_A N_B \sigma^2}{z_A z_B} \sum_{\text{all } E} \langle V_p^\ddagger \rangle \exp\left[-\frac{E^\ddagger}{kT}\right], \tag{20.54}$$

where the superscript $\ddagger$ denotes evaluation at the transition state. It should be recognized that, because we are not integrating over the entire dividing surface, the Wigner variational theorem is no longer rigorously correct. In effect, we have neglected some of the flux through the dividing surface, so the inequality $\mathcal{F}_s^{\text{total}} \geq \mathcal{R}$ may be lost.

Our third set of assumptions involves the nature of the molecular-energy states in the transition state and reactant valley and the manner in which we are going to evaluate the summations over these states. We assume that the total energy at all points on the potential hypersurface can be written as the separable sum of translational, rotational, vibrational, and electronic energies. Since we have taken the dividing surface to be perpendicular to the reaction coordinate, the direction of V_p is along that coordinate. As we shall see, motion along the reaction coordinate at the transition state corresponds to one of the $3N - 6$ vibrational modes. However, the potential along the reaction coordinate exhibits a maximum at the transition state. Therefore, motion along this coordinate will behave more like a translation than a vibration. The simple form of transition-state theory assumes that we may treat this motion as an uncoupled classical translation and replace $\langle V_p \rangle$ with the average classical speed along the reaction coordinate, which we will denote by $\langle V_{\text{rc}} \rangle$. Incorporating these approximations into Eq. 20.54 produces

$$\mathcal{F}_S^{\text{total}} = N_A N_B \sigma^2 \frac{\langle V_{\text{rc}} \rangle}{z_{\text{tr}}^A z_{\text{rot}}^A z_{\text{vib}}^A z_{\text{el}}^A z_{\text{tr}}^B z_{\text{rot}}^B z_{\text{vib}}^B z_{\text{el}}^B}$$

$$\sum_{\text{tr}} \sum_{\text{rot}} \sum_{\text{vib}} \sum_{\text{el}} \exp\left[-\frac{E_{\text{tr}}^\ddagger + E_{\text{rot}}^\ddagger + E_{\text{vib}}^\ddagger + E_{\text{el}}^\ddagger}{kT}\right]. \tag{20.55}$$

The summations over translational, rotational, vibrational, and electronic energies at the transition state in Eq. 20.55 all have the form of partition functions. That is,

$$\sum_{\text{tr}} \exp\left[-\frac{E_{\text{tr}}^\ddagger}{kT}\right] = z_{\text{tr}}^\ddagger \tag{20.56A}$$

$$\sum_{\text{rot}} \exp\left[-\frac{E_{\text{rot}}^\ddagger}{kT}\right] = z_{\text{rot}}^\ddagger \tag{20.56B}$$

$$\sum_{\text{vib}} \exp\left[-\frac{E_{\text{vib}}^\ddagger}{kT}\right] = z_{\text{vib}}^\ddagger z_{\text{rc}} \tag{20.56C}$$

and

$$\sum_{\text{el}} \exp\left[-\frac{E_{\text{el}}^\ddagger}{kT}\right] = z_{\text{el}}^\ddagger. \tag{20.56D}$$

In writing the vibrational partition function at the transition state, we have separated out the vibration corresponding to motion along the reaction coordinate, which we will treat as translational motion. Therefore, $z_{vib}^{\ddagger}$ in Eq. 20.56C includes only the remaining $3N - 5$ vibrational modes. With this notation, Eq. 20.55 becomes

$$\mathcal{F}_s^{total} = N_A N_B \sigma^2 \langle V_{rc} \rangle \frac{z_{tr}^{\ddagger} z_{rot}^{\ddagger} z_{vib}^{\ddagger} z_{rc} z_{el}^{\ddagger}}{z_{tr}^A z_{rot}^A z_{vib}^A z_{el}^A z_{tr}^B z_{rot}^B z_{vib}^B z_{el}^B}. \tag{20.57}$$

In this simple form of transition-state theory, we evaluate the molecular translational, rotational, and vibrational partition functions by using the quantum states for a particle in an infinite potential well, classical rigid rotor, and harmonic oscillator, respectively. Therefore, the partition function for translational motion along the reaction coordinate at the transition state is evaluated using the translational energy levels for a reactant A–B pair in the volume shown in Figure 20.20. These energy levels are given by

$$E_n = \frac{n^2 h^2}{8m\delta^2} \quad (n = 1, 2, 3, \ldots, \infty). \tag{20.58}$$

Consequently,

$$z_{rc} = \sum_{n_{rc}=1}^{\infty} \exp\left[-\frac{n_{rc}^2 h^2}{8m_{rc}\delta^2 kT}\right],$$

where m_{rc} is the effective mass of the reactant system as it moves along the reaction coordinate. Previously, we learned that translational quantum states are so closely spaced that this summation may be replaced with an integral with no measurable loss of accuracy. Accordingly, we have

$$z_{rc} = \int_0^{\infty} \exp\left[-\frac{n_{rc}^2 h^2}{8m_{rc}\delta^2 kT}\right] dn = \frac{1}{2}\left[\frac{8\pi m_{rc}\delta^2 kT}{h^2}\right]^{1/2} = \left[\frac{2\pi m_{rc}\delta^2 kT}{h^2}\right]^{1/2}. \tag{20.59}$$

The normalized one-dimensional Maxwell-Boltzmann probability distribution of molecular speeds was obtained in Chapter 17. If we consider only one-dimensional motion, Eqs. 17.52 through 17.55 give

$$P(V) = \left[\frac{m}{2\pi kT}\right]^{1/2} \exp\left[-\frac{mV^2}{2kT}\right].$$

Therefore, the average classical speed along the reaction coordinate is

$$\begin{aligned}
\langle V_{rc} \rangle &= \left[\frac{m_{rc}}{2\pi kT}\right]^{1/2} \int_0^{\infty} V_{rc} \exp\left[-\frac{m_{rc}V_{rc}^2}{2kT}\right] \\
&= \left[\frac{m_{rc}}{2\pi kT}\right]^{1/2} \left[-\frac{kT}{m_{rc}}\exp\left(-\frac{m_{rc}V_{rc}^2}{2kT}\right)\right]\Big|_0^{\infty} \\
&= \left[\frac{m_{rc}}{2\pi kT}\right]^{1/2} \frac{kT}{m_{rc}} = \left[\frac{kT}{2\pi m_{rc}}\right]^{1/2}.
\end{aligned} \tag{20.60}$$

Substituting Eqs. 20.59 and 20.60 into Eq. 20.57 produces

$$\begin{aligned}
\mathcal{F}_S^{total} &= N_A N_B \sigma^2 \left[\frac{kT}{2\pi m_{rc}}\right]^{1/2}\left[\frac{2\pi m_{rc}\delta^2 kT}{h^2}\right]^{1/2} \frac{z_{tr}^{\ddagger} z_{rot}^{\ddagger} z_{vib}^{\ddagger} z_{el}^{\ddagger}}{z_{tr}^A z_{rot}^A z_{vib}^A z_{el}^A z_{tr}^B z_{rot}^B z_{vib}^B z_{el}^B} \\
&= \frac{N_A N_B kT\sigma^2\delta}{h} \frac{z_{tr}^{\ddagger} z_{rot}^{\ddagger} z_{vib}^{\ddagger} z_{el}^{\ddagger}}{z_{tr}^A z_{rot}^A z_{vib}^A z_{el}^A z_{tr}^B z_{rot}^B z_{vib}^B z_{el}^B}.
\end{aligned} \tag{20.61}$$

Inspection of Figure 20.20 shows that the product $\sigma^2 \delta$ is just the volume of the system we are considering. Hence, if we write the volume as V, Eq. 20.61 becomes

$$F_S^{\text{total}} = \frac{N_A N_B kTV}{h} \frac{z_{\text{tr}}^{\ddagger} z_{\text{rot}}^{\ddagger} z_{\text{vib}}^{\ddagger} z_{\text{el}}^{\ddagger}}{z_{\text{tr}}^A z_{\text{rot}}^A z_{\text{vib}}^A z_{\text{el}}^A z_{\text{tr}}^B z_{\text{rot}}^B z_{\text{vib}}^B z_{\text{el}}^B}. \tag{20.62}$$

It is both convenient and instructive to express the ratio of electronic partition functions appearing in Eq. 20.62 in terms of the ground-state electronic energies and degeneracies. In Chapters 17 and 18, we found that all contributions to z_{el} are negligible, save that for the electronic ground state. We can, therefore, write the partition functions in the form

$$z_{\text{el}} = g_{\text{el}} \exp\left[-\frac{\varepsilon_o}{kT}\right], \tag{20.63}$$

where ε_o and g_{el} are the electronic ground-state energy and degeneracy, respectively. Using Eq. 20.63, we find that the ratio of electronic partition functions in Eq. 20.62 becomes

$$\frac{z_{\text{el}}^{\ddagger}}{z_{\text{el}}^A z_{\text{el}}^B} = \frac{g_{\text{el}}^{\ddagger}}{g_{\text{el}}^A g_{\text{el}}^B} \exp\left[-\frac{(\varepsilon_o^{\ddagger} - \varepsilon_o^A - \varepsilon_o^B)}{kT}\right]. \tag{20.64}$$

The energy difference appearing in the exponential is the potential energy at the transition state minus that in the reactant valley. This quantity is the potential-energy barrier to reaction. If we denote this as E_b, Eq. 20.62 can be written in the form

$$F_S^{\text{total}} = \frac{N_A N_B kTV}{h} \frac{g_{\text{el}}^{\ddagger}}{g_{\text{el}}^A g_{\text{el}}^B} \exp\left[-\frac{E_b}{kT}\right] \frac{z_{\text{tr}}^{\ddagger} z_{\text{rot}}^{\ddagger} z_{\text{vib}}^{\ddagger}}{z_{\text{tr}}^A z_{\text{rot}}^A z_{\text{vib}}^A z_{\text{tr}}^B z_{\text{rot}}^B z_{\text{vib}}^B}. \tag{20.65}$$

After any long derivation, it is always advisable to check the units on the result. There are no units on the degeneracies or the exponential or partition functions in Eq. 20.65. The units on the expression are, therefore, those for the factors of $N_A N_B kTV/h$. N_A and N_B are the number of A and B molecules, respectively, per unit volume. V has units of m^3, and the unit on kT/h is s^{-1}. These give an overall unit of (A–B pairs)$(m^{-3})(m^{-3})(m^3)(s^{-1})$, or A–B pairs $m^{-3} s^{-1}$, which are the correct units for a flux or a rate.

We are now in a position to obtain an expression for the specific reaction rate coefficient. The phenomenological rate coefficient is related to the rate by $\mathcal{R} = k(T) N_A N_B$. Assuming that Wigner variational theorem still holds in spite of our approximations, we have $\mathcal{R} \leq F_S^{\text{total}}$. Therefore,

$$k(T) \leq \frac{kTV}{h} \frac{g_{\text{el}}^{\ddagger}}{g_{\text{el}}^A g_{\text{el}}^B} \exp\left[-\frac{E_b}{kT}\right] \frac{z_{\text{tr}}^{\ddagger} z_{\text{rot}}^{\ddagger} z_{\text{vib}}^{\ddagger}}{z_{\text{tr}}^A z_{\text{rot}}^A z_{\text{vib}}^A z_{\text{tr}}^B z_{\text{rot}}^B z_{\text{vib}}^B}. \tag{20.66}$$

Equation 20.66 is the standard transition-state theory expression for the reaction rate coefficient. To use this expression, we must know the barrier height E_b. We must also know the reactant and transition-state structures so that the moments of inertia can be evaluated and the rotational partition functions calculated. All vibrational frequencies for reactants and the transition-state structure must be computed or measured, and the degeneracies of the ground electronic states are required.

As noted at the beginning of this section, statistical methods significantly simplify the task of computing rate coefficients. We no longer require the entire potential-energy hypersurface; all that is needed is the region around the transition state and the reactant valley. If we have these regions with sufficient accuracy, the classical barrier height, moments of inertia, and vibrational frequencies can be obtained. However, there is a price to be paid for this simplification: In reducing Eq. 20.53 to Eq. 20.66, we may have introduced significant error into the latter, even when the potential hypersurface is accurate. The assumption of classical motion along the reaction coordinate means that we have lost all tunneling effects, which makes the computed flux too small. The elimination of contributions to the flux from crossings of the dividing surface in regions other than the transition state also makes the flux smaller than it should be. The assumption of separable motion along the reaction coordinate and separable vibrational–rotational energies introduces errors that may make the flux either too large or too small. In more sophisticated forms, variational transition-state theory eliminates or reduces many of these errors. [For a discussion of such methods, the reader should consult D. G. Truhlar, A. D. Isaacson, and B. C. Garrett, "Generalized Transition State Theory," in *Theory of Chemical Reaction Dynamics*, ed. M. Baer, Vol. 4, Chapter 2, pp. 65–138 (Boca Raton, FL: CRC Press, 1985), and references therein.]

EXAMPLE 20.8

Use simple transition-state theory to compute the ratio of the reaction rates at 300 K for the exchange reactions

$$H + H_2 \longrightarrow H_2 + H$$

and

$$D + H_2 \longrightarrow HD + H$$

where deuterium (D) is ^2H. The computed vibrational frequencies for H_3 and DH_2 at the transition state shown in Figures 20.6 and 20.7 are 1,764 cm^{-1} (symmetric stretch) and 870 cm^{-1} (doubly degenerate bend) for DH_2 and 2,184 cm^{-1} (symmetric stretch) and 979 cm^{-1} (doubly degenerate bend) for H_3.

Solution

Employing Eq. 20.66 for the two rates, along with the assumption that the flux is sufficiently close to the true rate that we can use the equals sign, we have

$$k_{\text{H}}(T) = \frac{kTV}{h} \frac{g_{\text{el}}^{\ddagger}}{g_{\text{el}}^{\text{H}} g_{\text{el}}^{\text{H2}}} \exp\left[-\frac{E_b}{kT}\right] \frac{[z_{\text{tr}}^{\ddagger} z_{\text{rot}}^{\ddagger} z_{\text{vib}}^{\ddagger}]_{\text{H3}}}{z_{\text{tr}}^{\text{H}} z_{\text{rot}}^{\text{H}} z_{\text{vib}}^{\text{H}} z_{\text{tr}}^{\text{H2}} z_{\text{rot}}^{\text{H2}} z_{\text{vib}}^{\text{H2}}} \tag{1}$$

and

$$k_{\text{D}}(T) = \frac{kTV}{h} \frac{g_{\text{el}}^{\ddagger}}{g_{\text{el}}^{\text{D}} g_{\text{el}}^{\text{H2}}} \exp\left[-\frac{E_b}{kT}\right] \frac{[z_{\text{tr}}^{\ddagger} z_{\text{rot}}^{\ddagger} z_{\text{vib}}^{\ddagger}]_{\text{DH2}}}{z_{\text{tr}}^{\text{D}} z_{\text{rot}}^{\text{D}} z_{\text{vib}}^{\text{D}} z_{\text{tr}}^{\text{H2}} z_{\text{rot}}^{\text{H2}} z_{\text{vib}}^{\text{H2}}} . \tag{2}$$

Because the potential-energy hypersurface is independent of isotopic mass, the barrier heights for these reactions are the same. The electronic degeneracies are unity for H_2 and two for H, D, H_3, and DH_2, since each of these is a doublet because of the unpaired electron. Therefore, the degeneracy ratio is the same for both reactions. The molecular partition function for H_2, z_{H2}, is also the same in both reactions. Therefore, the ratio is

$$\frac{k_{\text{H}}(T)}{k_{\text{D}}(T)} = \frac{z_{\text{tr}}^{\text{D}}}{z_{\text{tr}}^{\text{H}}} \left[\frac{\{z_{\text{tr}}^{\ddagger} z_{\text{rot}}^{\ddagger} z_{\text{vib}}^{\ddagger}\}_{\text{H3}}}{\{z_{\text{tr}}^{\ddagger} z_{\text{rot}}^{\ddagger} z_{\text{vib}}^{\ddagger}\}_{\text{DH2}}}\right], \tag{3}$$

since the atomic reactants D and H do not have rotational or vibrational energy. The translational partition function for a particle in an infinite potential well is given by Eq. 18.5. The results for D and H are

$$z_{tr}^D = \frac{V}{h^3}[2\pi m_D kT]^{3/2} \tag{4}$$

and

$$z_{tr}^H = \frac{V}{h^3}[2\pi m_H kT]^{3/2}, \tag{5}$$

respectively. The ratio appearing in Eq. 3 is, therefore,

$$\frac{z_{tr}^D}{z_{tr}^H} = \left[\frac{m_D}{m_H}\right]^{3/2} = \left[\frac{3.34 \times 10^{-27} \text{ kg}}{1.674 \times 10^{-27} \text{ kg}}\right]^{3/2} = 2.818. \tag{6}$$

The ratio of translational partition functions in the transition state has the same form:

$$\frac{(z_{tr}^{\ddagger})_{H3}}{(z_{tr}^{\ddagger})_{DH2}} = \left[\frac{m_{H3}}{m_{DH2}}\right]^{3/2} = \left[\frac{3(1.674)}{3.34 + 2(1.674)}\right]^{3/2} = 0.6507. \tag{7}$$

Figures 20.6 and 20.7 show that the transition-state structure is a symmetric, linear arrangement of the three atoms with an internuclear distance of 0.90×10^{-10} m. For H_3, the center of mass is on the central hydrogen atom. The moment of inertia for a linear molecule is

$$I = \sum_{i=1}^{3} m_i r_i^2, \tag{8}$$

where r_i is the distance from atom i to the center of mass of the molecule. Therefore, for H_3,

$$I_{H3} = (1.674 \times 10^{-27} \text{ kg})(0.90 \times 10^{-10} \text{ m})^2 + (1.674 \times 10^{-10} \text{ kg})(0 \text{ m})^2$$

$$+ (1.674 \times 10^{-10} \text{ kg})(0.90 \times 10^{-10} \text{ m})^2 = 2.71 \times 10^{-47} \text{ kg m}^2, \tag{9}$$

since the mass of hydrogen is 1.674×10^{-27} kg. The deuterium mass is 3.34×10^{-27} kg. To compute I_{DH2}, we need the location of the center of mass for the structure shown at right. Because the total moment about the center of mass must be zero, we have

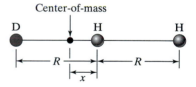

Center-of-mass

$$m_D(-R + x) + m_H x + m_H(R + x) = R(m_H - m_D) + x(2m_H + m_D) = 0. \tag{10}$$

Solving for x, we obtain

$$x = \frac{R(m_D - m_H)}{2m_H + m_D} = R\frac{3.34 - 1.674}{2(1.674) + 3.34} = 0.249R$$

$$= 0.249(0.90 \times 10^{10} \text{ m}) = 0.224 \times 10^{-10} \text{ m}. \tag{11}$$

The moment of inertia about this center of mass for DH_2 in the transition state is, therefore,

$$I_{DH2} = m_D(R - x)^2 + m_H x^2 + m_H(R + x)^2$$

$$= (3.34 \times 10^{-27} \text{ kg})(0.676 \times 10^{-10} \text{ m})^2 + 1.674 \times 10^{-27} \text{ kg}(0.224 \times 10^{-10} \text{ m})^2$$

$$+ 1.674 \times 10^{-27} \text{ m} (1.124 \times 10^{-10} \text{ m})^2 = 3.725 \times 10^{-47} \text{ kg m}^2. \tag{12}$$

In Chapters 17 and 18, we found that, at 300 K, the rotational partition function can be evaluated classically. For a linear system, the result is given by Eq. 18.76:

$$z_{rot}^c = \frac{kT}{\sigma Bh} = \frac{8\pi^2 IkT}{\sigma h^2}. \tag{13}$$

The symmetry number for H_3 is 2, for DH_2, 1. Therefore, the ratio of rotational partition functions in the transition state is

$$\frac{(z^{\ddagger}_{rot})_{H3}}{(z^{\ddagger}_{rot})_{DH2}} = \frac{I_{H3}\sigma_D}{I_{DH2}\sigma_H} = \frac{2.71 \times 10^{-47}\ \text{kg m}^2(1)}{3.725 \times 10^{-47}\ \text{kg m}^2(2)} = 0.364. \tag{14}$$

The vibrational partition function for each mode is given by Eq. 18.105 if we assume a harmonic quantization. That equation is

$$z^h_{vib} = \frac{e^{-x/2}}{1 - e^{-x}}, \tag{15}$$

where $x = h\nu_o/(kT)$. At $T = 300$ K, we obtain, for H_3,

$$x(\text{symmetric stretch}) = \frac{(6.626 \times 10^{-34}\ \text{J s})(2{,}184\ \text{cm}^{-1})(2.998 \times 10^{10}\ \text{cm s}^{-1})}{(1.381 \times 10^{-23}\ \text{J K}^{-1})(300\ \text{K})}$$

$$= 10.471 \tag{16}$$

and

$$x(\text{bending}) = \frac{(6.626 \times 10^{-34}\ \text{J s})(979\ \text{cm}^{-1})(2.998 \times 10^{10}\ \text{cm s}^{-1})}{(1.381 \times 10^{-23}\ \text{J K}^{-1})(300\ \text{K})} = 4.694. \tag{17}$$

For DH_2, the results are

$$x(\text{symmetric stretch}) = \frac{(6.626 \times 10^{-34}\ \text{J s})(1{,}764\ \text{cm}^{-1})(2.998 \times 10^{10}\ \text{cm s}^{-1})}{(1.381 \times 10^{-23}\ \text{J K}^{-1})(300\ \text{K})}$$

$$= 8.458 \tag{18}$$

and

$$x(\text{bending}) = \frac{(6.626 \times 10^{-34}\ \text{J s})(870\ \text{cm}^{-1})(2.998 \times 10^{10}\ \text{cm s}^{-1})}{(1.381 \times 10^{-23}\ \text{J K}^{-1})(300\ \text{K})} = 4.171. \tag{19}$$

The vibrational partition function for H_3 is, therefore,

$$(z^{\ddagger}_{vib})_{H3} = \frac{\exp(-10.471/2)}{1 - \exp(-10.471)} \times 2\frac{\exp(-4.694/2)}{1 - \exp(-4.694)}$$

$$= (0.005324)(2)(0.09654) = 0.001028, \tag{20}$$

where the factor 2 enters because the bending mode is doubly degenerate. For DH_2, the result is

$$(z^{\ddagger}_{vib})_{DH2} = \frac{\exp(-8.458/2)}{1 - \exp(-8.458)} \times 2\frac{\exp(-4.171/2)}{1 - \exp(-4.171)}$$

$$= (0.01457)(2)(0.1262) = 0.003680. \tag{21}$$

The ratio of vibrational partition functions required by Eq. 3 is

$$\frac{(z^{\ddagger}_{vib})_{H3}}{(z^{\ddagger}_{vib})_{DH2}} = \frac{0.001028}{0.003680} = 0.2793. \tag{22}$$

Before proceeding, a comment about the missing asymmetric stretching mode for H_3 and DH_2 in the transition state is appropriate. The normal-mode asymmetric stretching motion for H_3 is as follows:

Phase 1   180° Shift from phase 1

In Phase 1 of the asymmetric stretching motion, we are moving toward the reactant valley, where we have $H_a + H_b - H_c$. When the vibrational phase shifts by 180°, we begin moving toward the product valley where we have $H_a - H_b + H_c$. This vibrational mode is the one corresponding to motion along the reaction coordinate, which we have treated as a translation and have already included in the rate expression. We must not include it a second time; hence, we do not consider the asymmetric stretch in computing $z_{vib}^{\ddagger}$.

Combining all factors, we obtain

$$\frac{k_H(T)}{k_D(T)} = (2.818)(0.6507)(0.364)(0.2793) = 0.186. \tag{23}$$

For related exercises, see Problems 20.23, 20.24, 20.25, and 20.26.

In some applications, Eq. 20.66 is corrected for recrossings by multiplying by a *transmission coefficient*, κ. In this corrected form, the transition-state theory expression for the rate coefficient is

$$k(T) = \frac{\kappa k T V}{h}\, \frac{g_{el}^{\ddagger}}{g_{el}^A g_{el}^B}\, \exp\left[-\frac{E_b}{kT}\right] \frac{z_{tr}^{\ddagger} z_{rot}^{\ddagger} z_{vib}^{\ddagger}}{z_{tr}^A z_{rot}^A z_{vib}^A z_{tr}^B z_{rot}^B z_{vib}^B}. \tag{20.67}$$

In principle, κ accounts for trajectories that recross the dividing surface, such as Trajectories 3, 4, and 5 in Figure 20.19, by multiplying the rate coefficient expression by a number less than unity. While there is nothing wrong with this procedure, it must be recognized that the simple theory provides no method whereby the value of κ can be computed. Therefore, κ essentially assumes the role of a fitting parameter, much like the steric factor in simple collision theory. If full trajectory calculations have been carried out, we can use the results to estimate the value of κ. Problem 20.27 illustrates this type of calculation.

20.2.4 Thermodynamic Expression for the Reaction Rate Coefficient

Simple transition-state theory is frequently used in a semiquantitative fashion to infer something about the nature of the transition state. This is accomplished by noting that Eq. 20.62 looks a great deal like the expression for an equilibrium constant for an ideal gas-phase system.

In Chapter 18, we found that, for the gas-phase reaction

$$aA(g) + bB(g) = cC(g) + dD(g),$$

the equilibrium constant is given by

$$K_{eq} = K_p = \frac{(z_c')^c (z_d')^d}{(z_a')^a (z_b')^b}. \tag{20.68}$$

In Eq. 20.68, z_a, z_b, z_c, and z_d are the molecular partition functions for compounds A, B, C, and D, respectively, and $z_i' = z_i/N$, where N is Avogadro's constant. The similarity of Eq. 20.62 to Eq. 20.68 becomes more obvious if we write Eq. 20.62 in the form

$$\mathcal{F}_S^{total} = \frac{N_A N_B k T V}{h}\, \frac{z_{tr}^{\ddagger} z_{rot}^{\ddagger} z_{vib}^{\ddagger} z_{el}^{\ddagger}}{z_{tr}^A z_{rot}^A z_{vib}^A z_{el}^A z_{tr}^B z_{rot}^B z_{vib}^B z_{el}^B} = N_A N_B \left(\frac{kT}{h}\right) \frac{V z^{\ddagger}}{z_A z_B}. \tag{20.69}$$

Using the equals sign in the Wigner variation theorem and the fact that $\mathcal{R}(T) = k(T)N_A N_B$, we can express the rate coefficient as

$$k(T) = \left(\frac{kT}{h}\right)\frac{Vz^{\ddagger}}{z_A z_B}. \tag{20.70}$$

Multiplying both sides of Eq. 20.70 by Avogadro's constant gives

$$Nk(T) = N\left(\frac{kT}{h}\right)\frac{Vz^{\ddagger}}{z_A z_B} = \left(\frac{kT}{h}\right)\frac{V(z^{\ddagger})'}{z'_A z'_B}. \tag{20.71}$$

This operation has the effect of converting $k(T)$ from units of $m^3\,s^{-1}$ to $m^3\,mol^{-1}\,s^{-1}$. Each translational partition function in Eq. 20.71 contains a factor of V. Therefore, the volume factors cancel. (Problems 20.23 and 20.24 explicitly demonstrate this cancellation.) Consequently, we may take V to be $1\,m^3\,mol^{-1}$ without loss of accuracy. When this is done, we have

$$Nk(T) = \left(\frac{kTV'}{h}\right)\frac{(z^{\ddagger})'}{z'_A z'_B}, \tag{20.72}$$

where $V' = 1\,m^3\,mol^{-1}$.

A comparison of Eqs. 20.68 and 20.72 suggests that the rate coefficient in simple transition-state theory can be written as kTV'/h multiplied by an equilibrium constant between the transition-state complex and the reactants. That is,

$$Nk(T) = \left(\frac{kTV'}{h}\right)K^{\ddagger}. \tag{20.73}$$

Although the reasoning leading to Eq. 20.73 appears to be correct, it falls short in one respect. The expression for the equilibrium constant in Eq. 20.68 requires that each partition function contain the appropriate factors for all degrees of freedom. In Eq. 20.70, however, the partition function for the transition state, $z^{\ddagger}$, is missing the degree of freedom for the vibrational mode corresponding to the reaction coordinate. This factor has been separated and combined with the average speed perpendicular to the dividing surface, to obtain the kT/h factor. Consequently, it is not really correct to equate the ratio $(z^{\ddagger})'/(z'_A z'_B)$ to an equilibrium constant. However, since we do not expect a high level of accuracy from the treatment, Eq. 20.73 is frequently used as the transition-state theory expression for the reaction rate coefficient.

Once we accept Eq. 20.73, it is a simple matter to express the rate coefficient in terms of thermodynamic quantities related to the reaction $A + B \longrightarrow (A - B)^{\ddagger}$. The equilibrium constant can be expressed in terms of the change in the standard chemical potential for the reaction, viz.,

$$\Delta\mu_o^{\ddagger} = -RT \ln K^{\ddagger}, \tag{20.74}$$

where the superscript denotes conversion of the reactants into the transition-state complex. Since we are assuming the system to be ideal, the change in the standard chemical potential is just the change in the standard molar Gibbs free energies for the reactants and the transition state:

$$\Delta\mu_o^{\ddagger} = \Delta\overline{G}_o^{\ddagger} = (\overline{G}_o^{\ddagger})_{A-B} - \overline{G}_o^A - \overline{G}_o^B. \tag{20.75}$$

Combining Eqs. 20.74 and 20.75 produces

$$K^{\ddagger} = \exp\left[-\frac{\Delta G_o^{\ddagger}}{RT}\right]. \tag{20.76}$$

The change in the standard Gibbs free energy can be written in terms of the standard enthalpy and entropy changes for the reaction as

$$\Delta G_o^{\ddagger} = \Delta H_o^{\ddagger} - T\Delta S_o^{\ddagger}. \tag{20.77}$$

Substituting Eqs. 20.76 and 20.77 into 20.73 gives

$$Nk(T) = \left(\frac{kTV'}{h}\right)\exp\left[-\frac{\Delta H_o^{\ddagger} - T\Delta S_o^{\ddagger}}{RT}\right]$$

$$= \left(\frac{kTV'}{h}\right)\exp\left[\frac{\Delta S_o^{\ddagger}}{R}\right]\exp\left[-\frac{\Delta H_o^{\ddagger}}{RT}\right]. \tag{20.78}$$

Using the fact that $\Delta H = \Delta U + \Delta(pV) = \Delta U + RT\Delta n$ for an ideal system, we have $\Delta H_o^{\ddagger} = \Delta U_o^{\ddagger} - RT$, since $\Delta n = -1$ for reactants going to the transition-state complexes. Inserting this expression into Eq. 20.78 gives

$$Nk(T) = \left(\frac{ekTV'}{h}\right)\exp\left[\frac{\Delta S_o^{\ddagger}}{R}\right]\exp\left[-\frac{\Delta U_o^{\ddagger}}{RT}\right], \tag{20.79}$$

where e is the base of the natural-logarithm scale.

The principal use of Eq. 20.79 is in the interpretation of experimental data for the rate coefficient. If we equate the energy change for reactants going to transition-state complexes to the activation energy and then use a simple Arrhenius expression for $Nk(T)$, we have

$$Nk(T) = A\exp\left[-\frac{E_a}{RT}\right] = \left(\frac{ekTV'}{h}\right)\exp\left[\frac{\Delta S_o^{\ddagger}}{R}\right]\exp\left[-\frac{E_a}{RT}\right]. \tag{20.80}$$

Equation 20.80 permits us to extract the *entropy of activation*, $\Delta S_o^{\ddagger}$, from the measured frequency factor. A large negative result for $\Delta S_o^{\ddagger}$ can be interpreted to mean a tightly bound, rigid transition state with a great deal of order. A smaller negative value for $\Delta S_o^{\ddagger}$ suggests a loosely bound, perhaps freely rotating, transition state. Example 20.9 illustrates the procedure.

EXAMPLE 20.9

The frequency factor for the gas-phase reaction of NO with O_3 is experimentally determined to be 6.90×10^9 L mol^{-1} s^{-1} at 300 K. Use simple transition-state theory to compute the entropy of activation for this reaction, and comment on the result.

Solution

Equation 20.80 shows that the frequency factor is given by

$$A = \left(\frac{ekTV'}{h}\right)\exp\left[\frac{\Delta S_o^{\ddagger}}{R}\right]. \tag{1}$$

Before proceeding, we need to make the units compatible. The experimental frequency factor is expressed in units of L mol^{-1} s^{-1}, whereas the right-hand side of Eq. 1 is in m^3 mol^{-1} s^{-1}. Conversion of m^3 to liters on the right gives

$$A = \left(\frac{ekTV'}{h}\right)\exp\left[\frac{\Delta S_o^{\ddagger}}{R}\right] \times \left[\frac{100\text{ cm}}{\text{m}}\right]^3 \times \left[\frac{1\text{L}}{1000\text{ cm}^3}\right] = 10^3\left(\frac{ekTV'}{h}\right)\exp\left[\frac{\Delta S_o^{\ddagger}}{R}\right]\text{L mol}^{-1}\text{ s}^{-1}. \tag{2}$$

Solving Eq. 2 for $\Delta \overline{S}_o^{\ddagger}$, we obtain

$$\Delta \overline{S}_o^{\ddagger} = R \ln \left[\frac{hA}{10^3 (ekTV')} \right]. \tag{3}$$

Substituting the given data into Eq. 3 produces

$$\Delta \overline{S}_o^{\ddagger} = R \ln \left[\frac{(6.626 \times 10^{-34})(6.90 \times 10^9)}{10^3 (2.71828)(1.381 \times 10^{-23})(300)(1)} \right]$$

$$= (8.314 \, \text{J mol}^{-1} \, \text{K}^{-1}) \ln(4.060 \times 10^{-7}) = -122 \, \text{J mol}^{-1} \, \text{K}^{-1}. \tag{4}$$

This rather large negative entropy change suggests that the transition between NO and O_3 is highly restricted and has a substantial degree of order.

For additional exercises, see Problems 20.28 and 20.29.

20.3   Experimental Techniques

It is extremely difficult to obtain detailed experimental information concerning the motion of molecules during a chemical reaction. It is also difficult to conduct measurements that give information related to energy disposal. Such molecular detail can be computed with trajectory methods, but because the potential-energy hypersurface usually cannot be obtained with high accuracy, there is always some uncertainty associated with the reliability of the results.

The two most powerful experimental methods for obtaining detailed information concerning the molecular processes occurring during a chemical reaction are crossed molecular-beam measurements and chemiluminescence techniques. In this section, we shall explore these methods in a qualitative fashion.

20.3.1   Crossed Molecular-beam Measurements

When kinetic measurements are carried out by mixing the reactants and using some suitable means for monitoring the number of product molecules formed as a function of time, a great deal of information concerning the details of the reaction is lost. The results are averages over all relative speeds and internal quantum states of the reactants. Consequently, the dependence of the reaction cross section upon these quantities cannot be obtained from the experiments. It is also impossible to determine the energy disposal during the reaction, as well as the more intricate details of the reaction mechanism. If we wish to investigate the reaction in greater depth, we must have an experimental method that provides data from which this type of information can be extracted.

In a crossed molecular-beam measurement, a beam of A molecules crosses a beam of B molecules within a vacuum chamber whose pressure is maintained in the range from 10^{-7} to 10^{-6} torr. The beams are produced by gaseous expansion from high-pressure chambers, after which the molecules traverse a set of collimating slits and skimmers to produce a narrow, highly directional flow. In some cases, the expansion occurs through a constricted nozzle similar in concept to the nozzle that is often placed on the end of a garden hose to produce a high-velocity stream of water. Expansion through such a nozzle yields a beam of molecules whose rotational temperature is very low. Essentially all the molecules are in the ground rotational state. At this point, it is possible to send the beams through an array of slotted-disk

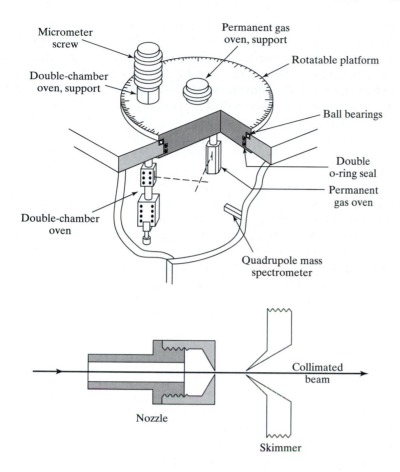

◀ FIGURE 20.22
Schematic diagram of a typical crossed molecular-beam apparatus.

◀ FIGURE 20.23
Schematic diagram of typical nozzle and skimmer.

velocity selectors, an apparatus that removes all molecules in the beam whose speed does not lie within a very narrow range. Lasers can be used to excite the molecules in the resulting beam to higher vibrational states. By combining these techniques, the investigator can produce highly collimated beams whose internal vibrational–rotational states and translational speeds are known within a certain range of values.

Figure 20.22 shows a schematic of a typical crossed molecular-beam apparatus. In most cases, the sources of the beams are suspended from the top of the vacuum chamber, as shown in the figure. The supporting assembly is machined so as to permit rotation of the entire unit in a controlled manner. In the figure, this rotational motion is achieved with micrometer screws. The ovens heat the molecules to the desired temperature prior to expansion. Typical arrangements for the collimating slits, skimmers, and velocity selectors are shown in Figures 20.23 and 20.24. The two beams that are generated are aligned so as to

◀ FIGURE 20.24
A slotted-disk velocity selector. The unselected beam with a Maxwell–Boltzmann distribution of speeds enters from the left as shown. The disks, each containing many slots (only six are shown in this illustration, for simplicity) rotate with an angular velocity that can be controlled by the investigator. Only molecules having a certain narrow range of speeds will strike slots at each disk.

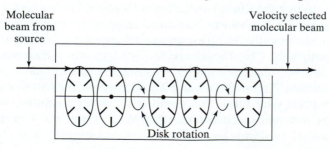

produce a crossing point at which the chemical reaction $A + B \longrightarrow C + D$ can occur. The pressure within the beams is sufficiently low to make the probability of multiple collisions essentially zero. The low background pressure ensures that very few collisions will occur once the reactants and products leave the small intersection volume at the crossing point. Consequently, we know that virtually all of the reactions are the result of single collisions without heterogeneous wall effects. The only remaining step is to detect the products by some suitable means. This is usually done using a quadrupole mass spectrometer. For some reactions, surface ionization detectors can be employed. In most experiments, the detector position is fixed while the beam sources rotate about it. This is the arrangement shown in Figure 20.22. Measurements of the number of product molecules formed as a function of rotation angle permit the distribution of scattering angles for the reaction products to be determined. In principle, the speed distribution for the products can also be measured, by using a velocity selector in front of the mass spectrometer. However, this is usually not done, because of insufficient signal intensity. By performing experiments in which the molecular speeds and internal states of the reactants are varied, the dependence of the reaction cross section upon these variables can be determined.

To illustrate the power of the method, let us consider the reactions

$$K(g) + Br_2(g) \longrightarrow KBr(g) + Br(g)$$

and

$$K(g) + CH_3I(g) \longrightarrow KI(g) + CH_3(g)$$

of potassium with Br_2 and with CH_3I, which have been investigated by D. R. Herschbach and coworkers. In these experiments, a surface ionization detector is used to measure the intensity of product $KBr(g)$ and $KI(g)$ as a function of scattering angle. The laboratory scattering data are converted to the center-of-mass system in which the product scattering angle is given with respect to the initial relative-velocity vector for the reactants. The results for $KBr(g)$ and $KI(g)$, shown in Figure 20.25, tell us several important things about the reactions. First, the total reaction cross section, which is the integral of the data over all scattering angles, is about 10 times larger for reaction with $Br_2(g)$ than with $CH_3I(g)$. Second, we notice a striking difference between the "preferred" direction for scattering in the two reactions. The $KBr(g)$ tends to scatter "forward," in the same direction as the initial relative-velocity vector. That is, the most probable center of mass scattering angle is $\theta = 0°$. In contrast, $KI(g)$ tend to scatter "backwards," in the opposite direction from the initial relative-velocity vector. Here, the most probable scattering angle is $\theta = 180°$.

The difference in scattering patterns for $KI(g)$ and $KBr(g)$ tell us something very important about the reaction mechanism for these processes. When $K(g)$ reacts with $Br_2(g)$, the potassium atom tends to pass at relatively large impact parameters and "pick off" one of the Br atoms as it passes. Once having stripped off the bromine atom, the newly formed $KBr(g)$ molecule continues in the same direction, so that the $KBr(g)$ scattering is "forward" in the center of mass system. This type of mechanism was called *stripping* by Herschbach. In contrast, when $K(g)$ reacts with $CH_3I(g)$, it apparently approaches at a small impact parameter, disrupts the $CH_3–I$ bond, and forms $KI(g)$, which then rebounds "backwards" at large scattering angles. Herschbach called such a mechanism a *rebound* reaction. Figures 20.26A and 20.26B illustrate the two mechanisms.

To explain why these two seemingly similar reactions should exhibit such very different mechanisms, Herschbach proposed a notion he called "harpooning." In simple terms, this hypothesis suggests that the electroneg-

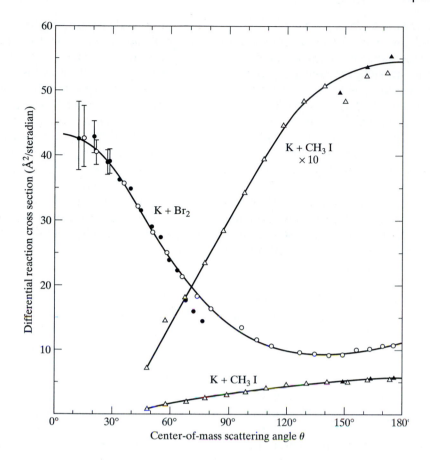

◀ **FIGURE 20.25**
Comparison of center of mass angular distributions for KBr and KI products in the K + Br_2 and K + CH_3I reactions, respectively. Note the multiplication by a factor of 10 to enhance the visual clarity for the KI product. Data are taken from D. R. Herschbach, "Reactive Scattering in Molecular Beams," in *Molecular Beams*, ed. John Ross (New York: John Wiley & Sons, 1966), Chapter 9, pp. 319–393.

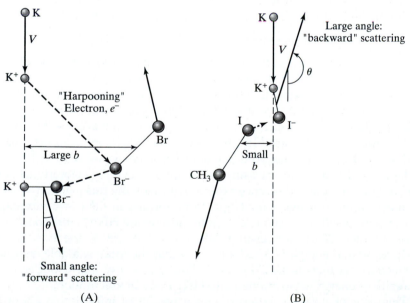

◀ **FIGURE 20.26**
Reaction mechanisms of K reacting with Br_2 and CH_3I **(A)** Stripping mechanism, with Br_2 produced by "harpooning." **(B)** Rebound mechanism with CH_3I.

ativity of $Br_2(g)$ is sufficiently high that the approaching K(g) atom transfers its 4s electron to the bromine atom while still at a large distance. This transferred electron is the "harpoon." With the electron transferred, long-range coulombic forces now operate between $K^+(g)$ and $Br_2^-(g)$, permitting the

$K^+(g)$ ion to "reel in" the "harpooned" $Br^-(g)$ ion while continuing to move in the forward direction. Such a mechanism leads naturally to forward scattering and a very large reaction cross section, since reactions tend to occur at large impact parameters. In contrast, when the reactant is $CH_3I(g)$, the electronegativity is greatly reduced, and the potassium atom cannot transfer its $4s$ electron at large distances. To react, it must approach at small impact parameters, which leads to a much smaller reaction cross section. Once it is close enough, it can disrupt the CH_3–I bond and form $KI(g)$. The close approach, however, means that the forces between $CH_3(g)$ and $KI(g)$ are primarily repulsive, so the $KI(g)$ molecule scatters backwards in the center of mass system.

This type of intricate detail concerning molecular motions during reaction is typical of the information that can be gleaned from crossed molecular-beam experiments. In recognition of their pioneering achievements in the development of these methods, Herschbach and Y. T. Lee were jointly awarded the Nobel prize.

20.3.2   Infrared Chemiluminescence

Chemiluminescence provides a second method that probes the recondite details of exothermic chemical reactions. Accordingly, it is an excellent complement to crossed molecular-beam measurements. In a chemiluminescence measurement, an exothermic gas-phase reaction is carried out at sufficiently low pressures that collisional deactivation of the reaction products does not occur. Instead, the vibrationally and rotationally excited products undergo emission. The intensity of emitted infrared radiation can be measured as a function of wavelength. The results of such measurements tell us which product vibrational–rotational states are populated as a result of the disposal of the reaction exothermicity and reactant translational energy.

For example, consider the reaction of a hydrogen atom with $F_2(g)$:

$$H(g) + F_2(g) \longrightarrow HF(g) + F(g) + 410 \text{ kJ mol}^{-1}.$$

This highly exothermic reaction must dispose of 410 kJ per mole of $F_2(g)$ reacted, plus the relative translational energy of $H(g)$–$F_2(g)$ motion. This energy might appear as relative translational motion between the $HF(g)$ and $F(g)$ products, or it might appear as $HF(g)$ vibrational–rotational energy. Using the data in Table 15.4, we calculate that the $v = 10$ vibrational state of $HF(g)$ lies 390.1 kJ mol^{-1} above the ground vibrational state. Therefore, it is energetically possible for $HF(g)$ to be produced in vibrational states as high as $v = 10$.

Figure 20.27 shows the measured intensity of the $HF(g)$ vibrational emission lines in a chemiluminescence experiment involving this reaction at 300 K. The most probable result is for $HF(g)$ to be formed in the $v = 6$ vibrational state. If these results are used to compute the average $HF(g)$ vibrational energy (see Problem 20.31), the result is 249 kJ mol^{-1}. At 300 K, the average relative translational energy is 3.7 kJ mol^{-1}, so that the total available energy to the reaction products is 413.7 kJ mol^{-1}. Thus, we know that about 60.2% of the available energy is partitioned into $HF(g)$ vibrational motion.

This type of detailed data provides a wealth of information about the energy flow in the reaction. Combined with trajectory calculations, it also provides information concerning the potential-energy hypersurface for the system and the reaction mechanism. For his pioneering research in the development of IR chemiluminescence, John C. Polanyi shared the Nobel prize with Herschbach and Lee.

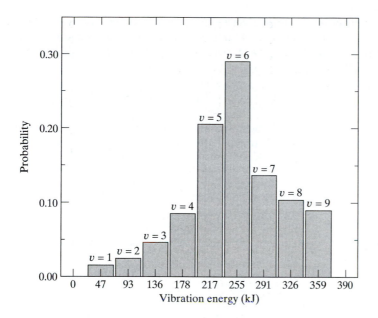

◀ FIGURE 20.27
Chemiluminescence of HF(g) formed in the reaction of H(g) with F_2(g) at 300 K. Based on data of J. C. Polanyi and coworkers.

Summary: Key Concepts and Equations

1. The differential gas-phase reaction rate can be written as the product of the differential collision frequency for reactants with a relative speed from V to $V + dV$, impact parameter from b to $b + db$, and internal collision variables specified by the set $\{q\}$ and the reaction probability $P_{AB}(V, b, \{q\})$ for such collisions. That is,

$$dR(T) = Z_{AB}(V, b, \{q\}) P_{AB}(V, b, \{q\}).$$

If we sum this differential rate over all possible relative speeds, impact parameters, and internal collision variables, we obtain the total $(A + B)$ reaction rate:

$$R(T) = k(T) N_A N_B = \int_{V=0}^{\infty} \int_{b=0}^{\infty} \int_{\{q\}} Z_{AB}(V, b, \{q\}) P_{AB}(V, b, \{q\}).$$

This is the fundamental equation of collision theory. If the reaction is an atom–diatomic process, the expression is

$$R(T) = \int_{V=0}^{\infty} \int_{b=0}^{\infty} \int_{\theta=0}^{\pi} \int_{\phi=0}^{2\pi} \int_{\gamma=0}^{2\pi} \int_{R=R_{min}}^{R_{max}} \sum_{v=0}^{\infty} \sum_{j=0}^{\infty} \sum_{m=-j}^{j} Z_{AB}(V, b, \{q\}) P_{AB}(V, b, \{q\}).$$

2. If we assume that molecules A and B are hard spheres with a collision radius σ_{AB} and that the reaction probability is given by

$$P_{AB}(V, b) = p = \text{a constant for } 0 \leq b \leq \sigma_{AB} \text{ and } \left(\frac{2E_a}{\mu}\right)^{1/2} \leq V \leq \infty$$

and

$$P_{AB}(V, b) = 0 \text{ otherwise,}$$

we obtain a very simple expression for the reaction rate coefficient:

$$k(T) = p[\pi \sigma_{AB}^2] \left[\frac{8kT}{\pi \mu}\right]^{1/2} \left[1 + \frac{E_a}{kT}\right] \exp\left[-\frac{E_a}{kT}\right].$$

The factor p is called the steric factor. When it is evaluated by means of experimental data, it provides a measure of the degree of steric hindrance present in the reaction. In general, such hard-sphere models are highly inaccurate.

3. The reaction cross section, $\sigma^2(V)$, is defined to be the result of integrating the product of the reaction probability and all normalized probability distributions, except that for relative speed, over the impact parameter and all internal variables. The general expression is

$$\sigma^2(V) = \int_{b=0}^{\infty} \int_{\{q\}} P_{AB}(V, b, \{q\})(2\pi b\,db)\,P(\{q\}),$$

where $P(\{q\})$ represents the normalized probability distribution functions for the complete set of internal variables $\{q\}$. For a three-body, atom–diatomic molecule reaction, this definition gives

$$\sigma^2(V) = \int_{b=0}^{\infty} \int_{\theta=0}^{\pi} \int_{\phi=0}^{2\pi} \int_{\gamma=0}^{2\pi} \int_{R=R_{min}}^{R_{max}} \sum_{v=0}^{\infty} \sum_{j=0}^{\infty} \sum_{m=-j}^{j} P_{AB}(2\pi b\,db)$$

$$P_1(\theta, \phi)P_2(R)P_3(\gamma)P_4(v, j, m).$$

The reaction cross section for a particular vibrational–rotational state is

$$\sigma^2(V, v, j) = \int_{b=0}^{\infty} \int_{\theta=0}^{\pi} \int_{\phi=0}^{2\pi} \int_{\gamma=0}^{2\pi} \int_{R=R_{min}}^{R_{max}} \sum_{m=-j}^{+j} P_{AB}(2\pi b\,db)P_1(\theta, \phi)P_2(R)P_3(\gamma).$$

The total reaction cross section can, therefore, be written as a weighted average of reaction cross sections out of particular vibrational–rotational states:

$$\sigma^2(V) = \sum_{v=0}^{\infty} \sum_{j=0}^{\infty} \left[\frac{1}{Z}\exp\left(-\frac{E_{vj}}{kT}\right)\right]\sigma_{AB}^2(V, v, j).$$

4. The potential energy of a system containing M nuclei depends upon their relative positions, which require $3M - 6$ nuclear coordinates to specify. Therefore, the potential energy is a $(3M - 5)$-dimensional function. This function is called the potential-energy hypersurface for the system. In many cases, it is referred to as the potential-energy surface, although it is actually a multidimensional hypersurface. In principle, this hypersurface can be obtained by solving the stationary-state Schrödinger equation for the energy eigenvalues in all important nuclear configurations. Except for a few very simple systems, though, this brute-force method is too computationally demanding to execute.

   The most successful approach to obtaining the potential-energy hypersurface involves the use of chemically and physically reasonable functional forms to represent interatomic two-body potentials, three-body bending interactions, and four-body dihedral angle potentials. The parameters contained in these functions are then fitted to experimental data that include measured equilibrium structures, reaction energies, IR and Raman vibrational frequencies for reactants and products, and activation energies, as well as solutions of the molecular Schrödinger equation at critically important regions of the hypersurface.

5. Potential-energy hypersurfaces are usually represented visually through contour-mapping methods in which all but two of the $3M - 6$ nuclear coordinates are fixed and contours of constant energy are plotted as a function of the remaining two nuclear coordinates. The regions of such contour maps where the nuclear configurations correspond to the reactants and products are called the reactant and product valleys, respectively. The minimum-energy reaction path leading from the reactant valley to the product valley on a given contour map is called the reaction coordinate. The potential energy along the reaction coordinate usually exhibits a maximum, so that there is a potential barrier to reaction. At the point corresponding to the maximum, the potential is a minimum along a direction perpendicular to the reaction coordinate. This point is, therefore, a mathematical saddle point, or col. The saddle point possessing the minimum potential barrier to reaction is called the transition state.

6. Multidimensional integrals generally cannot be numerically evaluated by using methods that rely on the summation of finite segments along each of the integra-

tion variables. The problem is that such methods usually require that the integrand be evaluated on the order of 10^2 times for each of the integration coordinates. Thus, an n-dimensional integral would require 10^{2n} evaluations of the integrand. These methods exceed the computational power of modern computers when $n > 4$.

The most effective solution to the numerical evaluation of multidimensional integrals is the Monte Carlo approach. In this method, the integrand is evaluated N times at randomly selected points in the integration ranges. When this is done, the value of the integral can be approximated by

$$\int_{a_n}^{b_n} \cdots \int_{a_3}^{b_3} \int_{a_2}^{b_2} \int_{a_1}^{b_1} f(x_1, x_2, x_3, \ldots, x_n)\, dx_1\, dx_2\, dx_3 \ldots dx_n = \langle f(x_1, x_2, x_3, \ldots, x_n)\rangle \prod_{i=1}^{n}(b_i - a_i),$$

where

$$\langle f(x_1, x_2, x_3, \ldots, x_n)\rangle = \langle f\rangle' \approx \frac{1}{N} \sum_{i=1}^{N} f(x_1, x_2, x_3 \ldots, x_n)_i$$

is the average value of the integrand. In this equation, $f(x_1, x_2, x_3, \ldots, x_n)_i$ is the value of the integrand at the ith randomly selected point. The absolute percent error in a Monte Carlo integration is about

$$|\% \text{ error}| \approx \frac{100[\langle f^2\rangle' - (\langle f\rangle')^2]^{1/2}}{N^{1/2}\langle f\rangle'},$$

where $\langle f^2\rangle'$ is the average of the square of the integrand obtained by Monte Carlo methods and N is the number of times the integrand is sampled. Note that the absolute percent error decreases with $N^{-1/2}$, so convergence to the correct answer is very slow. However, if 5 to 10% error can be tolerated, the method is excellent. Note also that unless the sampling is absolutely random, the error is generally much larger than the preceding equation indicates. This will be true no matter how many points are sampled, because systematic bias cannot be removed by increasing the sample size.

7. If $w(x)$ is a normalized probability distribution function, then integrals of the form

$$\int_a^b w(x)P(x)\, dx$$

can be put into the form $\int_0^1 P(G)\, dG$ by using a cumulative distribution function, which is defined as

$$G(x) = \int_a^x w(x)\, dx.$$

The transformed integral is then evaluated using Monte Carlo methods. This is the standard procedure for evaluating such integrals.

8. Reaction probabilities for a given set of initial collision variables are generally evaluated with the use of trajectory methods in which it is assumed that the nuclei are sufficiently massive that they move classically on the potential-energy hypersurface. If the initial collision variables are selected randomly from properly quantized distributions, the trajectories are said to be quasi-classical. In a trajectory calculation, the motion of all nuclei are followed as functions of time by numerically solving the Hamiltonian equations of motion, usually with a Runge–Kutta procedure.

The classical reaction probability for a specified set of initial collision variables will be either zero or unity depending upon whether or not a reaction occurs. The results of a trajectory calculation not only provide the probability of reaction, but also yield information related to the scattering of the products, the disposal of the available energy of the reaction, the yields of various possible products, and the reaction mechanism. Because of the wealth of information afforded by trajectory calculations, the method has become the most widely used and the most powerful theoretical weapon available for the investigation of complex chemical reactions. The major difficulty associated with the calculation of trajectories is the development of a sufficiently accurate potential-energy hypersurface.

9. By employing the Wigner variational theorem, statistical theories of reaction rates seek to avoid the intense computational effort required to carry out trajectory calculations. The theorem states that if we have a dividing surface that completely separates the reactant and product valleys of a potential-energy hypersurface such that it is impossible to reach the product valley from the reactant valley without crossing the dividing surface at least once, then we must have

$$\mathcal{R}(T) \leq \mathcal{F}(T),$$

where $\mathcal{F}(T)$ is the total flux across the dividing surface. The inequality follows from the fact that every reacting trajectory must cross the dividing surface, but some trajectories cross the surface without reacting or cross the surface more than once. Wigner's variational theorem allows us to parametrize the definition of the dividing surface and then adjust the parameters so as to minimize $\mathcal{F}(T)$, secure in the knowledge that the lower the value we obtain for the total flux, the closer to the true rate we must be. In this form, the method is called variational transition-state theory.

10. The total flux across a dividing surface could be calculated by using trajectories, but that would defeat the purpose of statistical theory, which is to minimize the computational effort required to obtain the rate coefficient. Instead, the total flux is computed by assuming that

> *the populations of internal molecular states are always described by a Boltzmann distribution at the temperature of the experiment, regardless of where we are on the potential-energy hypersurface for the system, and that the translational energies are always described by a Maxwell–Boltzmann distribution.*

The statement in italics is the foundation upon which all statistical theories of reaction rates rest. The accuracy of these approximations depends upon the relative rates of reaction and the redistribution of energy among the internal molecular states. If the redistribution rate between all internal modes is large compared with the reaction rate, the approximations will be quite accurate. If the two rates are close to each other, the accuracy of statistical theories decreases.

When the foregoing assumptions are made, the total flux across a dividing surface S involves an integral over the entire surface and a sum over all possible system energies. The formal expression is

$$\mathcal{F}_S^{\text{total}} = \frac{N_A N_B}{z_A z_B} \int_S \sum_{\text{all } E} \langle V_p \rangle \exp\left[-\frac{E}{kT}\right] dS,$$

where N_A and N_B are the number of A and B molecules, respectively, per unit volume, $\langle V_p \rangle$ is the average speed of the system perpendicular to the dividing surface, and z_A and z_B are the molecular partition functions for the reactants.

11. In most applications, the general expression for the flux is simplified by invoking the following additional approximations:

    1. The optimum dividing surface is the one that passes through the saddle points perpendicularly to the reaction coordinate.
    2. Integration over the entire dividing surface can be replaced with evaluation over a differential surface element in the region of the transition state. Further, the integrand is constant over this differential element and may be factored out of the integral.
    3. The total energy at all points on the potential-energy hypersurface can be written as the separable sum of translational, rotational, vibrational, and electronic energies.

4. Vibrational motion corresponding to motion along the reaction coordinate may be treated as a classical translational motion.

5. Partition functions for translation, rotation, and vibration can be evaluated by using the energy levels for a particle in an infinite potential well, a rigid rotor, and a harmonic oscillator, respectively.

The preceding approximations lead to the expression

$$\mathcal{F}_S^{\text{total}} = \frac{N_A N_B kTV}{h} \frac{g_{\text{el}}^{\ddagger}}{g_{\text{el}}^A g_{\text{el}}^B} \exp\left[-\frac{E_b}{kT}\right] \frac{z_{\text{tr}}^{\ddagger} z_{\text{rot}}^{\ddagger} z_{\text{vib}}^{\ddagger}}{z_{\text{tr}}^A z_{\text{rot}}^A z_{\text{vib}}^A z_{\text{tr}}^B z_{\text{rot}}^B z_{\text{vib}}^B},$$

where the g_i are the degeneracies of the electronic ground states, E_b is the potential barrier to reaction at the transition state, and the z_i are molecular partition functions. The phenomenological rate coefficient is related to the rate by $\mathcal{R} = k(T)N_A N_B$. If the Wigner variational theorem still holds in spite of the approximations, we have $\mathcal{R} \leq \mathcal{F}_S^{\text{total}}$. Therefore,

$$k(T) \leq \frac{kTV}{h} \frac{g_{\text{el}}^{\ddagger}}{g_{\text{el}}^A g_{\text{el}}^B} \exp\left[-\frac{E_b}{kT}\right] \frac{z_{\text{tr}}^{\ddagger} z_{\text{rot}}^{\ddagger} z_{\text{vib}}^{\ddagger}}{z_{\text{tr}}^A z_{\text{rot}}^A z_{\text{vib}}^A z_{\text{tr}}^B z_{\text{rot}}^B z_{\text{vib}}^B}.$$

To correct for recrossings of the dividing surface, a transmission coefficient κ is sometimes inserted into this expression to give

$$k(T) \leq \frac{\kappa kTV}{h} \frac{g_{\text{el}}^{\ddagger}}{g_{\text{el}}^A g_{\text{el}}^B} \exp\left[-\frac{E_b}{kT}\right] \frac{z_{\text{tr}}^{\ddagger} z_{\text{rot}}^{\ddagger} z_{\text{vib}}^{\ddagger}}{z_{\text{tr}}^A z_{\text{rot}}^A z_{\text{vib}}^A z_{\text{tr}}^B z_{\text{rot}}^B z_{\text{vib}}^B}.$$

12. By invoking one additional approximation, it is possible to write the simple transition-state theory expression for the rate coefficient in thermodynamic form in which the quantities that appear are related to the energy and entropy associated with reactants forming transition-state complexes. The approximation needed is to equate the expression

$$\frac{g_{\text{el}}^{\ddagger}}{g_{\text{el}}^A g_{\text{el}}^B} \exp\left[-\frac{E_b}{kT}\right] \frac{N z_{\text{tr}}^{\ddagger} z_{\text{rot}}^{\ddagger} z_{\text{vib}}^{\ddagger}}{z_{\text{tr}}^A z_{\text{rot}}^A z_{\text{vib}}^A z_{\text{tr}}^B z_{\text{rot}}^B z_{\text{vib}}^B},$$

where N is Avogadro's constant, to an equilibrium constant between transition-state complexes and reactants. Although this replacement appears to be correct, it is not, because of the missing degree of vibrational freedom in $z_{\text{vib}}^{\ddagger}$, corresponding to motion along the reaction coordinate. With the replacement, the transition-state theory rate coefficient becomes

$$k(T) \leq \frac{\kappa kTV'}{h} K^{\ddagger},$$

where $K^{\ddagger}$ is the equilibrium constant and V' is a unit volume of 1 m^3 mol^{-1}.

Replacing $K^{\ddagger}$ with $\exp[-\Delta \overline{G}_o^{\ddagger}/(RT)]$ and then expressing the standard Gibbs free energy of activation in terms of the standard energy and entropy of activation leads to the result

$$Nk(T) = \left(\frac{e\kappa kTV'}{h}\right) \exp\left[\frac{\Delta \overline{S}_o^{\ddagger}}{R}\right] \exp\left[-\frac{\Delta \overline{U}_o^{\ddagger}}{RT}\right].$$

13. A crossed molecular-beam apparatus produces two highly collimated beams whose internal vibrational–rotational states and translational speeds may be known within a certain range of values. The two beams are aligned so as to cross and thereby permit chemical reactions. The pressure within the beams is sufficiently low to make the probability of multiple collisions essentially zero. A low background pressure ensures that very few collisions will occur once the reactants and products leave the small intersection volume at the beam-crossing point. Consequently, all reactions are the result of single collisions without heterogeneous wall effects. The intensity of product molecules scattered in different directions is measured by rotating the beams about a fixed detector, which is usu-

ally a quadrupole mass spectrometer. It is also possible to use velocity selectors to measure the translational energy of the products. This type of data provides direct information concerning the reaction cross section as a function of the initial states of the reactants. The data also frequently permit the investigator to draw accurate conclusions concerning the mechanism of the reaction.

14. In an infrared chemiluminescence measurement, an exothermic gas-phase reaction is carried out at sufficiently low pressures that collisional deactivation of the reaction products does not occur. Instead, the vibrationally and rotationally excited products undergo emission. The intensity of the infrared radiation emitted is measured as a function of wavelength to obtain the distribution of product molecules over the internal vibrational–rotational states of the product molecule. This distribution tells us how the reaction exothermicity and reactant translational energy are distributed among the various degrees of freedom present in the product molecules.

Problems

Problems that require the use of some type of computational device are marked with an asterisk (*). Problems that require some type of plotting routine are indicated with a pound sign (#).

20.1 Starting with Eq. 20.13, obtain Eq. 20.14.

20.2 The rate coefficient for the bimolecular gas-phase reaction $H_2 + I_2 \longrightarrow 2\,HI$ at 300 K has been found to be $2.32 \times 10^{-19}\,L\,mol^{-1}\,s^{-1}$. The measured activation energy for the reaction is $170.3\,kJ\,mol^{-1}$. Using the hard-sphere model developed in the text, along with a hard-sphere collision radius of $4.0 \times 10^{-10}\,m$ (4 Å), determine the steric factor for this reaction. Discuss the reasons the steric factor should be less than unity.

20.3 The rate coefficient for the bimolecular gas-phase reaction $NO_2 + NO_2 \longrightarrow 2\,NO + O_2$ at 300 K has been found to be $1.90 \times 10^{-10}\,L\,mol^{-1}\,s^{-1}$. The measured activation energy for the reaction is $111.3\,kJ\,mol^{-1}$. Using the hard-sphere model developed in the text, along with a hard-sphere collision radius of $8.0 \times 10^{-10}\,m$ (8 Å), determine the steric factor for this reaction. Discuss the reasons the steric factor should be less than unity.

20.4 Show that the units on $k(T)$, when evaluated by using Eq. 20.20, are $m^3\,s^{-1}$. Determine the conversion factor required to convert these units to $L\,mol^{-1}\,s^{-1}$.

20.5* The total reaction cross section for a particular reaction is computed at 300 K. The results obtained at different relative speeds are given in the following table: The reduced mass of the reactants is $6.75 \times 10^{-27}\,kg$.

(A) Using a linear least-squares method, fit the computed total reaction cross section to a cubic equation in V. That is, writing $\sigma^2 = a_o + a_1 V + a_2 V^2 + a_3 V^3$, determine the best values for a_o, a_1, a_2, and a_3.

V (m s^{-1})	σ^2 (m^2)	V (m s^{-1})	σ^2 (m^2)
2,000.0	0.00	2,600.0	2.16×10^{-20}
2,100.0	1.01×10^{-22}	2,700.0	3.43×10^{-20}
2,200.0	8.09×10^{-22}	2,800.0	5.13×10^{-20}
2,300.0	2.71×10^{-21}	2,900.0	7.30×10^{-20}
2,400.0	6.42×10^{-21}	3,000.0	1.00×10^{-19}
2,500.0	1.25×10^{-20}		

(B) Compute the threshold energy.

(C) Using any convenient numerical integration method, determine the value of the specific reaction rate coefficient at 300 K, 330 K, 375 K, and 400 K. Assume that σ^2 is independent of temperature over this range.

(D) Determine the activation energy for the reaction. Compare your result for E_a with the threshold energy. Comment on the result.

20.6 The specific reaction rate coefficients and activation energies for the gas-phase reactions

$$CH_3 + C_4H_8 \text{ (2-butene)} \longrightarrow CH_4 + C_4H_7$$

and

$$CH_3 + CHCl_3 \longrightarrow CH_4 + CCl_3$$

have been determined experimentally. The results are given in the table following Parts (A) and (B).

(A) On the basis of these data alone, what is known about the threshold energies for the two reactions?

(B) Using a hard-sphere model, determine the ratio of reaction cross sections times the corresponding steric factors at 300 K for these two reactions. Discuss

Reaction	k(T) at 300 K (L mol⁻¹ s⁻¹)	E_a (kJ mol⁻¹)
$CH_3 + C_4H_8$	3.23×10^2	32.22
$CH_3 + CHCl_3$	1.30×10^3	24.27

the results in terms of steric hindrance to the hydrogen abstraction reactions.

20.7 Using Figures 20.6, 20.7, and 20.9, plot, as accurately as possible, the potential energy for the D_2H system for the collinear and 90° configurations for structures where $R_1 = R_2$. Show the results on the same graph. Assume that the saddle point for the 90° configuration lies at 1.3 eV and that the dissociated-atom energy is 4.74 eV.

20.8 Figures 20.6, 20.7, and 20.9 demonstrate that the minimum-energy path for the reaction $D_2 + H \longrightarrow D + DH$ corresponds to the collinear approach of the three atoms with $\theta = 180°$. The probability distribution for collisional orientations derived in the equations leading to Eq. 20.7 is

$$P_1(\theta, \phi) = \frac{\sin \theta \, d\theta \, d\phi}{4\pi}.$$

In addition, we know that the saddle-point energies for a collinear ($\theta = 180°$) approach and an approach of $\theta = 112°$ are 38,210 and 77,190 J mol⁻¹, respectively. Using these facts, along with the assumption that the saddle-point energies vary linearly with θ, estimate, as a function of temperature, the angle θ at which most reactions of $H + D_2$ to form $D + DH$ occur. Be certain to provide a logical justification for your answer.

20.9 Figure 20.11 represents a possible potential-energy contour map for a linear, symmetric reaction of two iodine atoms with H_2 to form two HI molecules. Sketch this contour map roughly, and show the following features on your sketch:

(A) the $H_2 + 2I$ reactant valley,

(B) the 2HI product valley,

(C) the saddle point for reaction along a symmetric, linear reaction coordinate, and

(D) the reaction coordinate. What is the approximate barrier height along the reaction coordinate? What is the structure of the system at the saddle point?

20.10 Consider the bimolecular gas-phase reaction $H_2 + I_2 \longrightarrow 2HI$. Suppose we believe that the mechanism for this reaction involves a planar approach of H_2 and I_2, with the H–H and I–I bonds parallel such that at all points during the approach, the H_2I_2 structure is that of a symmetric, isosceles trapezoid. Describe two coordinates which might be

used to plot a contour map that illustrates the potential energy for such a reaction mechanism. If the H_2, I_2, and HI equilibrium distances are 1.40 bohr, 5.04 bohr, and 3.03 bohr, respectively, what values of the two coordinates correspond to the reactant and product valleys?

20.11 Consider a hypothetical three-body reaction $A + BC \longrightarrow$ products. Suppose we have determined that the reaction probability, is given by

$$P(V, b, \theta, \phi, \gamma, R, v, j, m) = K \sin \theta \exp[-\alpha b] F(V, v),$$

where K and α are constants and where

$$F(V, v) = a_v V^3 + b_v V^2 + c_v V + d_v \qquad \text{for } V \geq V_o$$

and

$$F(V, v) = 0 \qquad \text{for } V < V_o.$$

The parameters a_v, b_v, c_v, and d_v are constants whose values depend upon the BC vibrational state. In general,

$$a_v = a_n, b_v = b_n, c_v = c_n, \text{ and } d_v = d_n$$

if the BC molecule is in the nth vibrational state. Therefore, in this model, the reaction probability is independent of ϕ, γ, R, j, and m. Use Eq. 20.18 and other appropriate equations derived in the text to determine the proper form for the reaction cross section for the vibrational–rotational state (v, j). That is, derive $\sigma^2(V, v, j)$ in terms of the constants and parameters of the problem.

20.12 A cubic die is randomly thrown 20 times with the following numbers coming up: 1, 6, 3, 4, 4, 6, 1, 3, 3, 5, 5, 6, 4, 2, 1, 4, 2, 6, 2, 5.

(A) Compute the average of these results.

(B) Compute the square variance of the distribution.

(C) Use the variance to estimate the error in the average. Compare this estimate with the actual percent error.

(D) Approximately how many times would the die need to be thrown to obtain an average within 0.01% of the true average?

20.13 Suppose we have means of computing the total reaction cross section in a certain reaction. The reaction rate coefficient is then given by Eq. 20.20. Using a cumulative distribution function, express this integral in the form of a Monte Carlo summation. (*Hint*: You will first need to normalize the probability distribution function for the relative speed.)

20.14 The total three-body reaction cross section can be written as a weighted average of reaction cross sections out of particular vibrational–rotational states by using a modified form of Eq. 20.19:

$$\sigma^2(V) = \sum_{v=0}^{\infty} \sum_{j=0}^{\infty} \sum_{m=-j}^{+j} \left[\frac{1}{z} \exp\left(-\frac{E_{vj}}{kT} \right) \right] \sigma_{AB}^2(V, v, j).$$

Let us assume that the vibrational–rotational energies can be accurately represented by the sum of separable harmonic-oscillator–rigid-rotor energies. Previously, we found that at room temperature the rotational-energy levels are such that j can be treated as a continuous variable. Under these conditions, the molecular partition function can be written as the product $z = z_{vib} z_{rot}$.

(A) Using these assumptions, obtain an expression for z_{rot} in terms of the rotational constant B, temperature, and other constants.

(B) Assuming that $\sigma_{AB}^2(V, v, j)$ is independent of m, obtain the cumulative probability distribution function $G(j)$, and express the integral over j and the summation over m in Monte Carlo form. Describe how we could determine the value j from a randomly selected value of $G(j)$.

20.15* An engineer wishes to evaluate the two-dimensional integral

$$I = \int_{x=0}^{10} \int_{y=0}^{10} xy^2 \sin\left[\frac{\pi x}{10}\right] dy\, dx.$$

(A) Perform the integration analytically, and obtain the correct answer to eight significant digits.

(B) Write a FORTRAN or C computer code to evaluate the integral via Monte Carlo methods.

(C) Run your computer code using 10^2, 10^3, 10^4, 10^5, and 10^6 points for the evaluation. Compute the percent error in the result for each evaluation.

(D) Plot $\ln[|\% \text{ error}|]$ from (C) versus $\ln[N^{-1/2}]$ where N is the number of points used in the Monte Carlo evaluation of the integral. Comment on the results.

20.16 In this problem, you will explore methods for employing cumulative distribution functions when the variable is discrete. The reaction cross section for the $H + D_2$ reaction at a relative speed V for D_2 vibrational–rotational states v and j depends upon the magnetic quantum number m. Equation 20.18 shows that the dependence upon m is a summation whose functional form is

$$I = \sum_{m=-j}^{+j} P_{AB}(m).$$

Normalize the probability distribution function for m, and obtain the corresponding cumulative distribution function. Describe a method that might be used to effect a random selection of that function. [*Hint*: The unnormalized probability distribution function for m in the preceding summation is $w(m) = 1$ for all m. Since m is a discrete variable, you will need to invoke some type of quantization in selecting the random value for its cumulative distribution function.]

20.17 Karplus, Porter, and Sharma [*J. Chem. Phys.* **43**, 3259 (1965)] computed cross sections for the $H + H_2$ exchange reaction as a function of relative speed for particular vibrational–rotational states. When $v = j = 0$ and $V = 1.76 \times 10^4$ m s^{-1}, they obtained a reaction cross section of 1.15×10^{-20} m^2.

(A) If 400 total trajectories were computed with a maximum impact parameter $b_m = 1.058 \times 10^{-10}$ m, how many trajectories resulted in a reaction?

(B) What is the expected absolute percent error in the result?

20.18 Karplus, Porter, and Sharma [*J. Chem. Phys.* **43**, 3259 (1965)] computed cross sections for the $H + H_2$ exchange reaction as a function of relative speed for particular vibrational–rotational states. When $v = j = 0$ and $V = 9.38 \times 10^3$ m s^{-1}, they obtained a reaction cross section of 3.54×10^{-22} m^2.

(A) If 400 total trajectories were computed with a maximum impact parameter $b_m = 5.03 \times 10^{-11}$ m, how many trajectories resulted in a reaction?

(B) What is the expected absolute percent error in the result?

(C) For $v = j = 0$ and $V = 1.76 \times 10^4$ m s^{-1}, the reaction cross section is 1.15×10^{-20} m^2 for the $H + H_2$ exchange reaction, with an estimated error of 7.16%. (See Problem 20.17.) Using this result and those obtained in (A) and (B), comment on how the accuracy of a trajectory calculation varies with the reaction cross section. The threshold relative speed, V_o, is that speed at which the reaction cross section goes to zero. Why is it so difficult to compute V_o using trajectory methods?

20.19 Figure 20.13 shows the details of a reactive hydrogen exchange trajectory in which $H_a + H_b$–$H_c \longrightarrow H_a$–$H_b + H_c$. Initially, H_b–H_c was in the $j = 0$ rotational state, so there was no molecular rotational angular momentum present. How can we tell from the figure that the newly formed H_a–H_b molecule is not in the $j = 0$ rotational state? Explain your reasoning.

20.20 Figure 20.13 shows the details of a reactive hydrogen exchange trajectory in which $H_a + H_b$–$H_c \longrightarrow H_a$–$H_b + H_c$. The discussion in the text concerning the H_3 potential-energy hypersurface shows that the transition state corresponds to a symmetric, linear arrangement of the three hydrogen atoms, with an H–H distance of about 0.90×10^{-10} m. Using the aforementioned figure, determine the system configuration at the point of reaction around $t \approx 5.0$ time units, where the R_{bc} and R_{ab} distances cross. Did the reaction proceed through a linear transition state? Explain.

20.21 Figure 20.17 indicates that the vibrational motion of the D–F bond is much more erratic between $t \approx 60$ tu and $t \approx 170$ tu than it is beyond 170 tu. Explain qualitatively why this is true. (1 tu $= 1.019 \times 10^{-14}$ s.)

20.22 Figure 20.17 indicates that the C–D vibrational frequency is substantially larger at times less than 60 tu than it is subsequent to that time. Why? (1 tu $= 1.019 \times 10^{-14}$ s.)

20.23 Use simple transition-state theory to compute the reaction rate at 300 K for the hydrogen exchange reaction $H + H_2 \longrightarrow H_2 + H$. Required data can be found in Example 20.8, Figures 20.6 and 20.7, the related discussion in the text, and Table 15.4. You may use a classical expression for the rotational partition functions and a harmonic quantization for the vibrational partition functions. Express the result in units of $m^3\,s^{-1}$ and $cm^3\,mol^{-1}\,s^{-1}$.

20.24 Use simple transition-state theory to compute the reaction rate at 300 K for the hydrogen exchange reaction $D + H_2 \longrightarrow DH + H$. Required data can be found in Example 20.8, Figures 20.6 and 20.7, the related discussion in the text, and Table 15.4. You may use a classical expression for the rotational partition functions and a harmonic quantization for the vibrational partition functions. Express the result in units of $m^3\,s^{-1}$ and $cm^3\,mol^{-1}\,s^{-1}$.

20.25 The ratio of the rate coefficients for the reactions

$$H + H_2 \longrightarrow H_2 + H$$

and

$$D + H_2 \longrightarrow HD + H$$

at 300 K is $k_H(300\text{ K})/k_D(300\text{ K}) = 0.187$. (See Example 20.8.) The deviation of this result from unity in spite of the fact that the potential-energy hypersurfaces for both reactions are identical is called the isotope effect. Is the isotope effect temperature dependent? Explain. If it is, discuss the origin of the temperature dependence within the framework of simple transition-state theory.

20.26 The ratio of the rate coefficients for the reactions

$$H + H_2 \longrightarrow H_2 + H$$

and

$$D + H_2 \longrightarrow HD + H$$

at 300 K is $k_H(300\text{ K})/k_D(300\text{ K}) = 0.187$. (See Example 20.8.) The deviation of this result from unity in spite of the fact that the potential-energy hypersurfaces for both reactions are identical is called the isotope effect. Let us define

$$R(T) = \frac{\dfrac{k_H(T)}{k_D(T)}}{\dfrac{k_H(300\text{ K})}{k_D(300\text{ K})}}.$$

From the data given in Example 20.8 and simple transition-state theory, obtain an expression for $R(T)$ in terms of constants and temperature. Use your result to compute and plot $k_H(T)/k_D(T)$ as a function of T from 300 K to 1,000 K.

20.27 The simple transition-state theory rate coefficient for the $H + H_2$ exchange reaction at 300 K on a Porter–Karplus potential-energy hypersurface is 3.56×10^8 cm^3 mol^{-1} s^{-1}. (See Problem 20.23.) The rate coefficient obtained from trajectory calculations at 300 K is 1.84×10^8 cm^3 mol^{-1} s^{-1} [M. Karplus, R. N. Porter, and R. D. Sharma, *J. Chem. Phys.* **43**, 3259 (1965)]. Using these data, estimate the value of the transmission coefficient in Eq. 20.67.

20.28 The frequency factor for the gas-phase reaction of H_2 with I_2 is experimentally determined to be 1.044×10^{11} L mol^{-1} s^{-1} at 300 K. Use simple transition-state theory to compute the entropy of activation for this reaction, and comment on the result.

20.29 The largest bimolecular gas-phase rate coefficient that we can have is one for a reaction with zero activation energy and a steric factor of unity. Use the simple hard-sphere model to evaluate this rate coefficient at 300 K for a reaction in which the hard-sphere collision radius is 3.0×10^{-10} m and the reduced mass is 2.21×10^{-27} kg. What is the entropy of activation for this process?

20.30 In the text, it is shown that, for the reaction $A + B \longrightarrow$ products, the number of A–B pairs per unit volume is $N_A N_B$, where N_A and N_B are the number of A and B molecules per unit volume, respectively.

(A) If the reaction is $A + A \longrightarrow$ products, derive an expression for the number of A–A pairs per unit volume in terms of N_A.

(B) In the limit of large N_A, show that the expression obtained in (A) becomes $N_A^2/2$.

(C) Write down the simple transition-state theory expression analogous to Eq. 20.66 for the rate coefficient for the reaction $A + A \longrightarrow$ products.

20.31 The data used to plot Figure 20.27 in the text are given in the table on the next page. Since chemiluminescence cannot be observed from the ground vibrational state, the result for $v = 0$ is assumed. Use these data to verify the result given in the text that, on the average, 249 kJ mol^{-1} of vibrational excitation energy is present as HF(g) vibrational energy.

20.32 The populations of various vibrational states of HF(g) formed in the reaction

$$H(g) + F_2(g) \longrightarrow HF(g) + F(g)$$

at 300 K are given in Problem 20.31.

(A) Is this distribution a Boltzmann distribution? If so, compute its temperature. If not, prove that the distribution is not Boltzmann.

HF(g) Vibrational State	Excitation Energy (kJ mol^{-1})	Probability
0	0	0.000
1	47	0.017
2	93	0.026
3	136	0.046
4	178	0.086
5	217	0.205
6	255	0.291
7	291	0.137
8	326	0.103
9	359	0.089
10	390	0.000

(B) What term is used to describe this type of distribution? Could the $H(g) + F_2(g) \longrightarrow HF(g) + F(g)$ reaction be used to make a chemical laser? Discuss this possibility. (*Hint*: Review Section 15.5.4. and examine the listing of HF(g) in Table 15.4.)

20.33 Sam seems to have one last question. "Last chance, Sam. What's the problem?"

"I know everyone is anxious to wrap it up and leave, but I'm curious. Collision theory shows that $k(T)$ for the reaction of two hard spheres A and B with a collision radius σ_{AB} is

$$k(T) = p[\pi\sigma_{AB}^2]\left[\frac{8kT}{\pi\mu}\right]^{1/2}\left[1 + \frac{E_a}{kT}\right]\exp\left[-\frac{E_a}{kT}\right],$$

provided that the reaction probability is given by Eq. 20.12. Suppose we evaluate this rate coefficient using Eq. 20.67, with the assumption that the transition state corresponds to the following structure

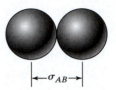

What relationship must exist between the transmission coefficient κ and the steric factor p in order for the two results to be identical?"

"Sam," I reply, "I have the feeling that you already know the answer to your question. Do you?"

With a sly grin, "I think so."

"I think so, too."

As one last problem for the road, can you answer Sam's question?

EPILOGUE

As you probably have guessed, Sam made an "A" in the second-semester course. Sam and Leigh are fictitious students, but each represents a composite of many very real students whom it has been my great pleasure to know and teach during my years on the faculty. Leigh epitomizes that large group of students who are highly skilled in chemistry, physics, and other mathematically oriented subjects. Their presence in the class gladdens the professor's heart. Sam represents that group of students who start slowly and struggle at first, but then gradually develop an understanding of the subject and frequently come to love it for its depth and beauty. In many ways, Sam is every professor's favorite student. Without a doubt, Sam has always been my personal favorite.

It is my sincere hope that those of you who have used this text have found it helpful. I also hope that it has occasionally replaced confusion and uncertainty with insight and knowledge. In addition, I hope that it has brought a smile to your face. Most of all, I hope that it has assisted you in the development of your intellectual powers and your ability to understand complex mathematical concepts in our physical world. Although you may never directly use physical chemistry in your chosen profession, this intellectual capability will stay with you always, assist you in your work, and provide comfort and satisfaction as you grow older. I wish each of you the very best of success and happiness in life.

Sincerely,

Lionel Mischa Raff
Regents Professor
May 21, 1999

BIBLIOGRAPHY

Atkins, P. *Physical Chemistry*, 5th ed. New York: W. H. Freeman and Company, 1994.

Baer, M., ed. *Theory of Chemical Reaction Dynamics*. Vols. 3 and 4. Boca Raton, FL: CRC Press, 1985.

Benson, S. W. *The Foundations of Chemical Kinetics*. New York: McGraw-Hill, 1960.

Brown, T. L., H. E. LeMay, Jr., & B. E. Bursten. *Chemistry: The Central Science*, 6th ed. Englewood Cliffs, NJ: Prentice Hall, 1994.

Davis, J. C., Jr. *Advanced Physical Chemistry*. New York: Ronald Press Co., 1965.

Eyring, H., J. Walter, & G. E. Kimball. *Quantum Chemistry*. New York: Wiley, 1944.

Frank, N. H. *Introduction to Electricity and Optics*. New York: McGraw-Hill, 1950.

Herzberg, G. *Molecular Spectra and Molecular Structure, II: Infrared and Raman Spectra of Polyatomic Molecules*. New York: Van Nostrand, 1960.

Herzberg, G. *Spectra of Diatomic Molecules*. New York: Van Nostrand, 1950.

Hirschfelder, J. O., C. F. Curtiss, & R. B. Bird. *The Molecular Theory of Gases and Liquids*. New York: Wiley, 1954.

Joshi, A. W., ed. *Horizons of Physics*. New Delhi: Wiley Eastern Limited, 1989.

Karplus, M., & R. N. Porter. *Atoms and Molecules*. New York: Benjamin, 1970.

Levine, I. N. *Physical Chemistry*, 1st and 4th eds. New York: McGraw-Hill, 1978, 1995.

MacLaren, J. M., J. B. Pendry, P. J. Rous, D. K. Saldin, G. A. Somorjai, M. A. Van Hove, & D. D. Vvedensky. *Surface Crystallographic Information Service*. Dordrecht, the Netherlands: D. Reidel Publishing Co., 1987.

Margenau, H., & G. M. Murphy. *The Mathematics of Physics and Chemistry*, 2d ed. Princeton, NJ: Van Nostrand, 1956.

Moelwyn-Hughes, E. A. *Physical Chemistry*, 2d ed. New York: Pergamon, 1961.

Pauling, L., & E. B. Wilson. *Introduction to Quantum Mechanics*. New York: McGraw-Hill, 1935.

Petrucci, R. H. *General Chemistry*, 4th ed. New York: Macmillan, 1985.

Ross, John, ed. *Molecular Beams*. Advances in Chemical Physics. Vol. 10. New York: Wiley Interscience, 1966.

Schutte, C. J. H. *The Wave Mechanics of Atoms, Molecules and Ions*. New York: St. Martin's Press, 1968.

Wall, F. T. *Chemical Thermodynamics*, 2d ed. San Francisco: W. H. Freeman and Co., 1965.

Woodbury, G. *Physical Chemistry*. Pacific Grove, CA: Brooks/Cole Publishing Co., 1996.

Appendices

APPENDIX A

Thermodynamic data (all values relate to 298.15 K and 1 bar of pressure)

Compound, Atom, or Ion	$\overline{H}^o$ or ΔH_f^o (kJ mol^{-1})	μ^o or $\overline{G}_f^o$ (kJ mol^{-1})	S^o (J K^{-1} mol^{-1})	C_p^m (J K^{-1} mol^{-1})
Aluminium				
Al(s)	0	0	28.33	24.35
Al(l)	10.56	7.20	39.55	24.21
Al(g)	326.4	285.7	164.54	21.38
Al^{3+}(g)	5,483.17	—	—	—
Al^{3+}(aq)	−531	−485	−321.7	—
Al$_2$O$_3$(s, α)	−1,675.7	−1,582.3	50.92	79.04
AlCl$_3$(s)	−704.2	−628.8	110.67	91.84
Argon				
Ar(g)	0	0	154.84	20.786
Antimony				
Sb(s)	0	0	45.69	25.23
SbH$_3$(g)	145.11	147.75	232.78	41.05
Arsenic				
As(s, α)	0	0	35.1	24.64
As(g)	302.5	261.0	174.21	20.79
As$_4$(g)	299.69	143.9	92.4	
AsH$_3$(g)	66.44	68.93	222.78	38.07
Barium				
Ba(s)	0	0	62.8	28.07
Ba(g)	180	146	170.24	20.79
Ba^{2+}(aq)	−537.64	−560.77	9.6	—
BaO(s)	−553.5	−525.1	70.43	47.78
BaCl$_2$(s)	−858.6	−810.4	123.68	75.14
Beryllium				
Be(s)	0	0	9.50	16.44
Be(g)	324.3	286.6	136.27	20.79
Bismuth				
Bi(s)	0	0	56.74	25.52
Bi(g)	207.1	168.2	187.00	20.79
Bromine				
Br$_2$(l)	0	0	152.23	75.689
Br$_2$(g)	30.907	3.110	245.46	36.02
Br(g)	111.88	82.396	175.02	20.786

Note: Dash indicates value of quantity unknown.

Compound, Atom, or Ion	$\overline{H}^o$ or ΔH_f^o (kJ mol^{-1})	μ^o or $\overline{G}_f^o$ (kJ mol^{-1})	S^o (J K^{-1} mol^{-1})	C_p^m (J K^{-1} mol^{-1})
Br$^-$(g)	−219.07	—	—	—
Br$^-$(aq)	−121.55	−103.96	82.4	−141.8
HBr(g)	−36.40	−53.45	198.70	29.142
Cadium				
Cd(s, γ)	0	0	51.76	25.98
Cd(g)	112.01	77.41	167.75	20.79
Cd^{2+}(aq)	−75.90	−77.612	−73.2	—
CdO(s)	−258.2	−228.4	54.8	43.43
CdCO$_3$(s)	−750.6	−669.4	92.5	—
Cesium				
Cs(s)	0	0	85.23	32.17
Cs(g)	76.06	49.12	175.60	20.79
Cs$^+$(aq)	−258.28	−292.02	133.05	−10.5
Calcium				
Ca(s)	0	0	41.42	25.31
Ca(g)	178.2	144.3	154.88	20.786
Ca^{2+}(aq)	−542.83	−553.58	−53.1	
CaO(s)	−635.09	−604.03	39.75	42.80
CaCO$_3$(s) (calcite)	−1,206.9	−1,128.8	92.9	81.88
CaCO$_3$(s)	−1,207.1	−1,127.8	88.7	81.25
CaF$_2$(s)	−1,219.6	−1,167.3	68.87	67.03
CaCl$_2$(s)	−795.8	−748.1	104.6	72.59
CaBr$_2$(s)	−682.8	−663.6	130	—
Carbon				
C(s) (graphite)	0	0	5.740	8.527
C(s) (diamond)	1.895	2.900	2.377	6.113
C(g)	716.68	671.26	158.10	20.838
C$_2$(g)	831.90	775.89	199.42	43.21
CO(g)	−110.53	−137.17	197.67	29.14
CO$_2$(g)	−393.51	−394.36	213.74	37.11
H$_2$CO$_3$(aq)	−699.65	−623.08	187.4	—
HCO$_3^-$(aq)	−691.99	−586.77	91.2	—
CO$_3^{2-}$(aq)	−677.14	−527.81	−56.9	—
CCl$_4$(l)	−135.44	−65.21	216.40	131.75
CS$_2$(l)	89.70	65.27	151.34	75.7
HCN(g)	135.1	124.7	201.78	35.86
HCN(l)	108.87	124.97	112.84	70.63
CN$^-$(aq)	150.6	172.4	94.1	

Compound, Atom, or Ion	$\overline{H}^o$ or ΔH_f^o (kJ mol^{-1})	μ^o or $\overline{G}_f^o$ (kJ mol^{-1})	S^o (J K^{-1} mol^{-1})	C_p^m (J K^{-1} mol^{-1})
Organic Compounds				
Hydrocarbons				
$CH_4(g)$ methane	−74.81	−50.72	186.26	35.31
$CH_3(g)$ methyl	145.69	147.92	194.2	38.70
$C_2H_2(g)$ ethyne	226.73	209.20	200.94	43.93
$C_2H_4(g)$ ethene	52.26	68.15	219.56	43.56
$C_2H_6(g)$ ethane	−84.68	−32.82	229.60	52.63
$C_3H_6(g)$ propene	20.42	62.78	267.05	63.89
$C_3H_8(g)$ propane	−103.85	−23.49	269.91	73.5
$C_3H_6(g)$ cyclopropane	53.30	104.45	237.55	55.94
$C_4H_8(g)$ 1-butene	−0.13	71.39	305.71	85.65
$C_4H_8(g)$ *cis*-2-butene	−6.99	65.95	300.94	78.91
$C_4H_8(g)$ *trans*-2-butene	−11.17	63.06	296.59	87.82
$C_4H_{10}(g)$ butane	−126.15	−17.03	310.23	97.45
$C_5H_{12}(g)$ pentane	−146.44	−8.20	348.40	120.2
$C_5H_{12}(l)$ pentane	−173.1	—	—	—
$C_6H_6(l)$ benzene	49.0	124.3	173.3	136.1
$C_6H_6(g)$ benzene	82.93	129.72	269.31	81.67
$C_6H_{12}(l)$ cyclohexane	−156	26.8		156.5
$C_6H_{14}(l)$ hexane	−198.7	—	204.3	—
$C_7H_8(g)$ toluene	50.0	122.0	320.7	103.6
$C_7H_{16}(l)$ heptane	−224.4	1.0	328.6	224.3
$C_8H_{18}(l)$ octane	−249.9	6.4	361.1	—
$C_8H_{18}(l)$ isooctane	−255.1	—	—	—
$C_9H_{20}(l)$ nonane	−288.1	—	—	284.4
$C_9H_{20}(g)$ nonane	−228.2	—	—	—
$C_{10}H_8(s)$ napthalene	78.53	—	—	—
$C_{10}H_{22}(l)$ decane	−300.9	—	—	314.4
$C_{10}H_{22}(g)$ decane	−249.5	—	—	—
$C_{11}H_{24}(l)$ undecane	−327.2	—	—	344.9
$C_{11}H_{24}(g)$ undecane	−270.8	—	—	—
$C_{12}H_{26}(l)$ dodecane	−350.9	—	—	375.8
$C_{12}H_{26}(g)$ dodecane	−289.4	—	—	—
$C_{14}H_{10}(s)$ anthracene	129.2	—	207.5	210.5
$C_{60}(s)$ fullerene$_{60}$	2,327.0	2,302.0	426.0	520.0
$C_{60}(g)$ fullerene$_{60}$	2,502.0	2,442.0	544.0	512.0
$C_{70}(s)$ fullerene$_{70}$	2,555.0	2,537.0	464.0	650.0
$C_{70}(g)$ fullerene$_{70}$	2,755.0	2,692.0	614.0	585.0
Alcohols and Phenols				
$CH_3OH(l)$ methanol	−238.66	−166.27	126.8	81.6

Compound, Atom, or Ion	$\overline{H}^o$ or ΔH_f^o (kJ mol^{-1})	μ^o or $\overline{G}_f^o$ (kJ mol^{-1})	S^o (J K^{-1} mol^{-1})	C_p^m (J K^{-1} mol^{-1})
CH$_3$OH(g)	−200.66	−161.96	239.81	43.89
C$_2$H$_5$OH(l) ethanol	−277.69	−174.78	160.7	111.46
C$_2$H$_5$OH(g)	−235.10	−168.49	282.70	65.44
C$_6$H$_5$OH(s) phenol	−165.0	−50.9	146.0	—
C$_3$H$_7$OH(l) 1-propanol	−302.6	—	193.6	143.9
C$_4$H$_9$OH(l) 1-butanol	−327.3	—	225.8	177.2
C$_5$H$_{11}$OH(l) 1-pentanol	−351.6	—	—	208.1
C$_6$H$_{13}$OH(l) 1-hexanol	−377.5	—	287.4	240.4
C$_7$H$_{15}$OH(l) 1-heptanol	−403.3	—	—	272.1
C$_8$H$_{17}$OH(l) 1-octanol	−426.5	—	—	305.2
C$_9$H$_{19}$OH(l) 1-nonanol	−453.4	—	—	—
C$_{10}$H$_{21}$OH(l) 1-decanol	−478.1	—	—	370.6

Carboxylic Acids and Esters

HCO$_2$H(l) formic	−424.72	−361.35	128.95	99.04
CH$_3$CO$_2$H(l) acetic	−484.5	−389.9	159.8	124.3
CH$_3$CO$_2$H(aq)	−485.76	−396.46	178.7	—
CH$_3$CO$_2^-$(aq)	−486.01	−369.31	86.6	−6.3
(CO$_2$H)$_2$(s) oxalic	−827.2	—	—	117
C$_6$H$_5$CO$_2$(s) benzoic	−385.1	−245.3	167.6	146.8
CH$_3$CO$_2$C$_2$H$_5$(l) ethyl acetate	−479.0	−332.7	259.4	170.1

Aldehydes and Ketones

HCHO(g) methanal	−108.57	−102.53	218.77	35.40
CH$_3$CHO(l) ethanal	−192.30	−128.12	160.2	—
CH$_3$CHO(g)	−166.19	−128.86	250.3	57.3
CH$_3$COCH$_3$(l) propanone	−248.1	−155.4	200.4	124.7
C$_2$H$_5$CHO(l) propanal	−215.6	—	—	—
C$_2$H$_5$CHO(g)	−185.6	—	304.5	80.7
C$_3$H$_7$CHO(l) butanal	−239.2	—	246.6	163.7
C$_3$H$_7$CHO(g)	−204.8	—	343.7	103.4
C$_2$H$_5$COCH$_3$(l) butanone	−273.3	—	239.1	158.7
C$_2$H$_5$COCH$_3$(g)	−238.5	—	339.9	101.7

Sugars

C$_6$H$_{12}$O$_6$(s) α–D-glucose	−1,273.3	—	—	—
C$_6$H$_{12}$O$_6$(s) β–D-glucose	−1,286	−910	212	—
C$_6$H$_{12}$O$_6$(s) β–D-fructose	−1,265.6	—	—	—
C$_{12}$H$_{22}$O$_{11}$(s) sucrose	−2,222	−1,543	360.2	—

Compound, Atom, or Ion	$\overline{H}^o$ or ΔH_f^o (kJ mol^{-1})	μ^o or $\overline{G}_f^o$ (kJ mol^{-1})	S^o (J K^{-1} mol^{-1})	C_p^m (J K^{-1} mol^{-1})
Nitrogen Compounds				
CO(NH$_2$)$_2$(s) urea	−333.51	−197.33	104.60	93.14
CH$_3$NH$_2$(g) methyl amine	−22.97	32.16	243.41	53.1
C$_6$H$_5$NH$_2$(l) aniline	31.1	—	—	—
Ethers				
CH$_3$OC(CH$_3$)$_3$(l) methyl *tert*-butyl ether (MTBE)	−313.6	—	265.3	187.5
CH$_3$OC(CH$_3$)$_3$(g)	−283.7	—	—	—
Chlorine				
Cl$_2$(g)	0	0	223.07	33.91
Cl(g)	121.68	105.68	165.20	21.840
Cl$^-$(g)	−233.13	—	—	—
Cl$^-$(aq)	−167.16	−131.23	56.5	−136.4
HCl(g)	−92.31	−95.30	186.91	29.12
HCl(aq)	−167.16	−131.23	56.5	−136.4
Chromium				
Cr(s)	0	0	23.77	23.35
Cr(g)	396.6	351.8	174.50	20.79
CrO$_4^{2-}$(aq)	−881.15	−727.75	50.21	—
Cr$_2$O$_7^{2-}$(aq)	−1,490.3	−1,301.1	261.9	—
Copper				
Cu(s)	0	0	33.150	23.35
Cu(g)	338.32	298.58	166.38	20.79
Cu$^+$(aq)	71.67	49.98	40.6	—
Cu^{2+}(aq)	64.77	65.49	−99.6	—
Cu$_2$O(s)	−168.6	−146.0	93.14	63.64
CuO(s)	−157.3	−129.7	42.63	42.30
CuSO$_4$(s)	−771.36	−661.8	109	100.0
Deuterium				
D$_2$(g)	0	0	144.96	29.20
HD(g)	0.318	−1.464	143.80	29.196
D$_2$O(g)	−249.20	−234.54	198.34	34.27
D$_2$O(l)	−294.60	−243.44	75.94	84.35
HDO(g)	−245.30	−233.11	199.51	33.81
HDO(l)	−289.89	−241.86	79.29	—
Fluorine				
F$_2$(g)	0	0	202.78	31.30
F(g)	78.99	61.91	158.75	22.74

Compound, Atom, or Ion	$\overline{H}^o$ or ΔH_f^o (kJ mol^{-1})	μ^o or $\overline{G}_f^o$ (kJ mol^{-1})	S^o (J K^{-1} mol^{-1})	C_p^m (J K^{-1} mol^{-1})
F$^-$(aq)	−332.63	−278.79	−13.8	−106.7
HF(g)	−271.1	−273.2	173.78	29.13
Helium				
He(g)	0	0	126.15	20.786
Hydrogen				
H$_2$(g)	0	0	130.684	28.824
H(g)	217.97	203.25	114.71	20.784
H$^+$(aq)	0	0	0	0
H$_2$O(l)	−285.83	−237.13	69.91	75.291
H$_2$O(g)	−241.82	−228.57	188.83	33.58
H$_2$O$_2$(l)	−187.78	−120.35	109.6	89.1
Iodine				
I$_2$(s)	0	0	116.135	54.44
I$_2$(g)	62.44	19.33	260.69	36.90
I(g)	106.84	70.25	180.79	20.786
I$^-$(aq)	−55.19	−51.57	111.3	−142.3
HI(g)	26.48	1.70	206.59	29.158
Iron				
Fe(s)	0	0	27.28	25.10
Fe(g)	416.3	370.7	180.49	25.68
Fe^{2+}(aq)	−89.1	−78.90	−137.7	—
Fe^{3+}(aq)	−48.5	−4.7	−315.9	—
Fe$_3$O$_4$(s)	−1,118.4	−1,015.4	146.4	143.43
Fe$_2$O$_3$(s)	−824.2	−742.2	87.40	103.85
FeS(s)	−100.0	−100.4	60.29	50.54
Krypton				
Kr(g)	0	0	164.08	20.786
Lead				
Pb(s)	0	0	64.81	26.44
Pb(g)	195.0	161.9	175.37	20.79
Pb^{2+}(aq)	−1.7	−24.43	10.5	—
PbO(s) yellow	−217.32	−187.89	68.70	45.77
PbO(s) red	−218.99	−188.93	66.5	45.81
PbO$_2$(s)	−277.4	−217.33	68.6	64.64
Lithium				
Li(s)	0	0	29.12	24.77
Li(g)	159.37	126.66	138.77	20.79
Li$^+$(aq)	−248.49	−293.31	13.4	68.6
Magnesium				
Mg(s)	0	0	32.68	24.89
Mg(g)	147.70	113.10	148.65	20.786
Mg^{2+}(aq)	−466.85	−454.8	−138.1	—

Compound, Atom, or Ion	$\overline{H}^o$ or ΔH_f^o (kJ mol^{-1})	μ^o or $\overline{G}_f^o$ (kJ mol^{-1})	S^o (J K^{-1} mol^{-1})	C_p^m (J K^{-1} mol^{-1})
MgO(s)	−601.70	−569.43	26.94	37.15
MgCO$_3$(s)	−1,095.8	−1,012.1	65.7	75.52
MgCl$_2$(s)	−641.32	−591.79	89.62	71.38
Mercury				
Hg(l)	0	0	76.02	27.983
Hg(g)	61.32	31.82	174.96	20.786
Hg^{2+}(aq)	171.1	164.40	−32.2	—
Hg$_2^{2+}$(aq)	172.4	153.52	84.5	—
HgO(s)	−90.83	−58.54	70.29	44.06
Hg$_2$Cl$_2$(s)	−265.22	−210.75	192.5	102
HgCl$_2$(s)	−224.3	−178.6	146.0	—
HgS(s)	−53.6	−47.7	88.3	—
Neon				
Ne(g)	0	0	146.33	20.786
Nitrogen				
N$_2$(g)	0	0	191.61	29.125
N(g)	472.70	455.56	153.30	20.786
NO(g)	90.25	86.55	210.76	29.844
N$_2$O(g)	82.05	104.20	219.85	38.45
NO$_2$(g)	33.18	51.31	240.06	37.20
N$_2$O$_4$(g)	9.16	97.89	304.29	77.28
N$_2$O$_5$(s)	−43.1	113.9	178.2	143.1
N$_2$O$_5$(g)	11.3	115.1	355.7	84.5
HNO$_3$(l)	−174.10	−80.71	155.60	109.87
HNO$_3$(aq)	−207.36	−111.25	146.4	−86.6
NO$_3^-$(aq)	−205.0	−108.74	146.4	−86.6
NH$_3$(g)	−46.11	−16.45	192.45	35.06
NH$_3$(aq)	−80.29	−26.50	111.3	—
NH$_4^+$(aq)	−132.51	−79.31	113.4	79.9
HN$_3$(l)	264.0	327.3	140.6	43.68
HN$_3$(g)	294.1	328.1	238.97	98.87
N$_2$H$_4$(l)	50.63	149.43	121.21	139.3
NH$_4$NO$_3$(s)	−365.56	−183.87	151.08	84.1
NH$_4$Cl(s)	−314.43	−202.87	94.6	—
Oxygen				
O$_2$(g)	0	0	205.138	29.355
O(g)	249.17	231.73	161.06	21.912
O$_3$(g)	142.7	163.2	238.93	39.20
OH$^-$(aq)	−229.99	−157.24	−10.75	−148.5
Phosphorus				
P(s) white	0	0	41.09	23.840

Compound, Atom, or Ion	$\overline{H}^o$ or ΔH_f^o (kJ mol^{-1})	μ^o or $\overline{G}_f^o$ (kJ mol^{-1})	S^o (J K^{-1} mol^{-1})	C_p^m (J K^{-1} mol^{-1})
P(g)	314.64	278.25	163.19	20.786
P$_2$(g)	144.3	103.7	218.13	32.05
P$_4$(g)	58.91	24.44	279.98	67.15
PH$_3$(g)	5.4	13.4	210.23	37.11
PCl$_3$(g)	−287.0	−267.8	311.78	71.84
PCl$_5$(g)	−374.9	−305.0	364.6	112.8
H$_3$PO$_3$(s)	−964.4	—	—	—
H$_3$PO$_3$(aq)	−964.8	—	—	—
H$_3$PO$_4$(s)	−1,279.0	−1,119.1	110.50	106.06
H$_3$PO$_4$(l)	−1,266.9	—	—	—
H$_3$PO$_4$(aq)	−1,277.4	−1,018.7	−222	—
PO$_4^{3-}$(aq)	−2,984.0	−2,697.0	228.86	211.71
Potassium				
K(s)	0	0	64.18	29.58
K(g)	89.24	60.59	160.336	20.786
K$^+$(g)	514.26	—	—	—
K$^+$(aq)	−252.38	−283.27	102.5	21.8
KOH(s)	−424.76	−379.08	78.9	64.9
KF(s)	−576.27	−537.27	66.57	49.04
KCl(s)	−436.75	−409.14	82.59	51.30
KBr(s)	−393.80	−380.66	95.90	52.30
KI(s)	−327.90	−324.89	106.32	52.93
Silicon				
Si(s)	0	0	18.83	20.00
Si(g)	455.6	411.3	167.97	22.25
SiO$_2$(s)	−910.94	−856.64	41.84	44.43
Silver				
Ag(s)	0	0	42.55	25.351
Ag(g)	284.55	245.65	173.00	20.79
Ag$^+$(aq)	105.58	77.11	72.68	21.8
AgBr(s)	−100.37	−96.90	107.1	52.38
AgCl(s)	−127.07	−109.79	96.2	50.79
Ag$_2$O(s)	−31.05	−11.20	121.3	65.86
AgNO$_3$(s)	−124.39	−33.41	140.92	93.05
Sodium				
Na(s)	0	0	51.21	28.24
Na(g)	107.32	76.76	153.71	20.79
Na$^+$(aq)	−240.12	−261.91	59.0	46.4
NaOH(s)	−425.61	−379.49	64.46	59.54
NaCl(s)	−411.15	−384.14	72.13	50.50
NaBr(s)	−361.06	−348.98	86.82	51.38

Compound, Atom, or Ion	$\overline{H}^o$ or ΔH_f^o (kJ mol^{-1})	μ^o or $\overline{G}_f^o$ (kJ mol^{-1})	S^o (J K^{-1} mol^{-1})	C_p^m (J K^{-1} mol^{-1})
NaI(s)	−287.78	−286.06	98.53	52.09
Sulfur				
S(s) rhombic	0	0	31.80	22.64
S(s) monoclinic	0.33	0.1	32.6	23.6
S(g)	278.81	238.25	167.82	23.673
S$_2$(g)	128.37	79.30	228.18	32.47
SO$_2$(g)	−296.83	−300.19	248.22	39.87
SO$_3$(g)	−395.72	−371.06	256.76	50.67
H$_2$SO$_4$(l)	−813.99	−690.00	156.90	138.9
H$_2$SO$_4$(aq)	−909.27	−744.53	20.1	−293
SO$_4^{2-}$(aq)	−909.27	−744.53	20.1	−293
HSO$_4^-$(aq)	−887.34	−755.91	131.8	−84
H$_2$S(g)	−20.63	−33.56	205.79	34.23
H$_2$S(aq)	−39.7	−27.83	121	—
HS$^-$(aq)	−17.6	12.08	62.08	—
SF$_6$(g)	−1,209	−1,105.3	291.82	97.28
Tin				
Sn(s)	0	0	51.55	26.99
Sn(g)	302.1	267.3	168.49	20.26
Sn^{2+}(aq)	−8.8	−27.2	−17	—
SnO(s)	−285.8	−256.9	56.5	44.31
SnO$_2$(s)	−580.7	−519.6	52.3	52.59
Xenon				
Xe(g)	0	0	169.68	20.786
Zinc				
Zn(s)	0	0	41.63	25.40
Zn(g)	130.73	95.14	160.98	20.79
Zn^{2+}(aq)	−153.89	−147.06	112.1	46
ZnO(s)	−348.28	−318.30	43.64	40.25

Sources of data: *Handbook of Chemistry and Physics,* 78th ed., CRC Press, Boca Raton, FL, 1997–1998. P. Atkins, *Physical Chemistry,* 5th ed., New York: W. H. Freeman and Co., 1994.

APPENDIX B

Crystal lattice energies obtained from the Born–Haber cycle. Energies are given in units of kJ mol^{-1}. Data enclosed in parentheses are computed energies. Dash denotes lack of data.

Cation	Anion					
	F	Cl	Br	I	O	S
Li	1,036	853	807	757	(2,799)	2,472
Na	923	786	747	704	(2,481)	2,203
K	821	715	682	649	(2,238)	2,052
Rb	785	689	660	630	(2,163)	1,949
Cs	740	659	631	604	—	1,850
Ag	967	915	904	889	(3,002)	2,677
Be	3,505	3,020	2,914	2,800	4,443	—
Mg	2,957	2,526	2,440	2,327	3,791	3,046*
Ca	2,630	2,258	2,176	2,074	3,401	3,119*
Sr	2,492	2,156		1,963	3,223	2,974*
Ba	2,352	2,056	1,985	1,877	3,054	2,832*

Sources of data: *Handbook of Chemistry and Physics*, 78th ed., CRC Press, Boca Raton, FL, 1997–1998; *D. Cubicciotti, *J. Chem. Phys.* **31**, 1646 (1959).

APPENDIX C

Conversion Factors

Conversion factors are given in scientific notation. An asterisk (*) indicates that the number is exact and all subsequent digits are zero. Other conversion factors have been rounded to the figures given, in accordance with accepted practice. In the conversion tables, the unit in the left-hand column is converted to a new unit by multiplying by the factor appearing in the column for the new unit. Source of data: (*Handbook of Chemistry and Physics,* 78th ed., CRC Press, Boca Raton, FL., 1997–1998.)

C. 1. Energy Conversion Factors

	joule	Btu[†]	calorie[‡]	erg
joule	1	0.00094845	0.23900574	1.0E+7*
Btu[†]	1,054.35	1	251.996	1.05435E+10
calorie[‡]	4.184*	0.0039683	1	4.184E+7*
erg	1.0E−7*	9.48450E−10	2.39006E−08	1

[†]British thermal unit (thermochemical)
[‡]thermochemical calorie

C.2. Energy Equivalences

	Wave Number (cm^{-1})	Frequency MHz	Energy eV	Energy hartrees	Molar Energy kJ / mol	Molar Energy kcal / mol
1 (cm^{-1})	1	2.99793E+04	1.23984E−04	4.55634E−06	1.19627E−02	2.85914E−03
1 MHz	3.33564E−05	1	4.13567E−09	1.151983E−10	3.99031E−07	9.53708E−08
1 eV	8.06554E+03	2.41799E+08	1	3.67493E−02	9.64853E+01	2.30605E+01
1 hartree	2.19475E+05	6.57968E+09	2.72114E+01	1	2.62550E+03	6.27510E+02
1 kJ/mol	8.35935E+01	2.50607E+06	1.03643E−02	3.80880E−04	1	2.39006E−01
1 kcal/mol	3.49755E+02	1.04854E+07	4.33641E−02	1.59360E−03	4.18400E+00	1

C.3 Pressure Conversion Factors

	Pa	kPa	MPa	bar	atmosphere	torr[†]	psi[‡]
Pa	1	1.00000E−03*	1.00000E−06*	1.00000E−05*	9.8692E−06	7.5006E−03	1.45038E−04
kPa	1.00000E+03*	1	1.00000E−03*	1.00000E−02*	9.8692E−03	7.5006E+00	1.45038E−01
MPa	1.00000E+06*	1.00000E+03*	1	1.00000E+01*	9.8692E+00	7.5006E+03	1.45038E+02
bar	1.00000E+05*	1.00000E+02*	1.00000E−01*	1	9.8692E−01	7.5006E+02	1.45038E+01
atmosphere	1.01325E+05	1.01325E+02	1.01325E−01	1.01325E+00	1	7.60000E+02	1.46959E+01
torr[†]	133.322	1.33322E−01	1.33322E−04	1.33322E−03	1.31579E−03	1	1.93367E−02
psi[‡]	6.89476E+03	6.89476E+00	6.89476E−03	6.89476E−02	6.80460E−02	5.17151E+01	1

[†]Torr is essentially identical to mm Hg.
[‡]psi is pounds per square inch.

C.4. Length Conversion Factors

	mm	cm	m	km	inch	foot	mile[‡]
mm	1	1.00000E−01*	1.00000E−03*	1.00000E−06*	3.9370E−02	3.28084E−02	6.2139E−07
cm	1.00000E+01*	1	1.00000E−02*	1.00000E−05*	3.9370E−01	3.28084E−01	6.2139E−06
m	1.00000E+03*	1.00000E+02*	1	1.00000E−03*	3.9370E+01	3.28084E+01	6.2139E−04
km	1.00000E+06*	1.00000E+05*	1.00000E+03*	1	3.9370E+04	3.28084E+03	6.2139E−01
inch	25.40*	2.54*	0.0254*	0.0000254*	1	8.33333E−02	1.57830E−05
foot	304.8*	30.48*	0.3048*	0.0003048*	12*	1	1.89394E−04
mile[‡]	1.6093E+06	1.6093E+05	1.6093E+03	1.6093E+00	63,360*	5,280*	1

[‡]U.S. statute mile

C.5. Volume Conversion Factors

	cubic cm	cubic m	milliliters	liters	cubic inch	cubic foot
cubic cm	1	0.000001*	1.0*	0.001*	6.10236E−02	3.53146E−05
cubic m	1,000,000*	1	1,000,000*	1,000*	6.10236E+04	3.53146E+01
milliliters	1.0*	0.000001*	1	0.001*	6.10236E−02	3.53146E−05
liters	1,000*	0.001*	1,000*	1	6.10236E+01	3.53146E−02
cubic inch	1.63871E+01	1.63871E−05	1.63871E+01	1.63871E−02	1	5.78704E−04
cubic foot	2.83169E+04	2.83169E−02	2.83169E+04	2.83169E+01	1,728*	1

C.6 Mass Conversion Factors

	gram	kg	ounce[‡]	pounde[‡]
gram	1	0.001*	3.52740E−02	2.20462E−03
kg	1,000*	1	3.52740E+01	2.20462E+00
ounce[‡]	2.83495E+01	2.83495E−02	1	0.0625*
pound[‡]	4.53592E+02	4.53592E−01	16*	1

[‡]avoirdupois

C.7 The Gas Constant in Different Units

$R = 8.314510 \text{ Pa m}^3 \text{ mol}^{-1} \text{ K}^{-1}$
$= 8{,}314.510 \text{ Pa L mol}^{-1} \text{ K}^{-1}$
$= 0.08314510 \text{ bar L mol}^{-1} \text{ K}^{-1}$
$= 0.0820578 \text{ L atm mol}^{-1} \text{ K}^{-1}$
$= 62.3641 \text{ L torr mol}^{-1} \text{ K}^{-1}$

C.8. Other Useful Conversion Factors

The conversion factors that follow give the number of the units on the right that are equivalent to one of the left-hand units. Thus, to convert 25 angstroms to meters, we execute the following calculation:

$$(25 \text{ Å}) \times \frac{10^{-10} \text{ m}}{1 \text{ Å}} = 25 \times 10^{-10} \text{ m} = 2.5 \times 10^{-9} \text{ m}.$$

1 angstrom = 1 Å = 10^{-8} cm = 10^{-10} m
1 abampere = 10 amperes (A)
1 abcoulomb = 10 coulombs (C)
1 abohm = 10^{-9} ohm (Ω)
1 abvolt = 10^{-8} volt (V)
1 dyne = 10^{-5} newton (N)
1 $_6^{12}\text{C}$ Faraday = 96,487.0 coulombs (C)
1 U.S. liquid gallon = 3.785412×10^{-3} cubic meter (m^3)
1 Canadian liquid gallon = 4.546090×10^{-3} cubic meter (m^3)
1 light-year = 9.46073×10^{15} meters (m)
1 parsec = 3.085678×10^{16} meters (m)
1 slug = 14.59390 kilograms (kg)
1 statampere = 3.335641×10^{-10} ampere (A)
1 statcoulomb = 3.335641×10^{-10} coulomb (C)
1 statohm = 8.987552×10^{-11} ohm (Ω)
1 statvolt = 2.997925×10^2 volts (V)

APPENDIX D

Ionization constants of acids and bases in aqueous solution at 298.15 K

Acid or Base	Formula	Ionization	K_a or	K_b
Sulfuric	H_2SO_4	1	1×10^2	
		2	1.02×10^{-2}	
Oxalic	$(COOH)_2$	1	5.89×10^{-2}	
		2	6.46×10^{-5}	
Sulfurous	H_2SO_3	1	1.41×10^{-2}	
		2	6.31×10^{-8}	
Phosphoric	H_3PO_4	1	6.92×10^{-3}	
		2	6.17×10^{-8}	
		3	4.79×10^{-13}	
Hydrofluoric	HF		6.31×10^{-4}	
Formic	HCOOH		3.5×10^{-4}	
Lactic	$CH_3CH(OH)COOH$		1.4×10^{-4}	
Analine	$C_6H_5NH_2$			4.31×10^{-10}
Acetic (ethanoic)	CH_3COOH		1.74×10^{-5}	
Butanoic	$CH_3(CH_2)_2COOH$		1.48×10^{-5}	
Propanoic	CH_3CH_2COOH		1.38×10^{-5}	
Pyridine	C_5H_5N			1.80×10^{-9}
Carbonic	H_2CO_3	1	4.47×10^{-7}	
		2	4.68×10^{-11}	
Hydrogen sulfide	H_2S	1	8.91×10^{-8}	
		2	1×10^{-19}	
Hypochlorous	HClO		3.0×10^{-8}	
Hypobromous	HBrO		2.82×10^{-9}	
Boric	H_3BO_3	1	5.37×10^{-10}	
		2	$< 1 \times 10^{-14}$	
Ammonia	NH_3			1.80×10^{-5}
Hydrogen cyanide	HCN		6.17×10^{-10}	
Trimethylamine	$(CH_3)_3N$			6.25×10^{-5}
Phenol	C_6H_5OH		1.29×10^{-10}	
Hypoiodous	HIO		3.16×10^{-11}	
Methylamine	CH_3NH_2			4.31×10^{-4}
Dimethylamine	$(CH_3)_2NH$			4.83×10^{-4}
Triethylamine	$(C_2H_5)_3N$			5.68×10^{-4}
Diethylamine	$(C_2H_5)_2NH$			1.06×10^{-3}

Source of data: *Handbook of Chemistry and Physics*, 78th ed., CRC Press, Boca Raton, FL, 1997–1998.

INDEX

Conversion Factors

Quantity	SI Unit	Conversion Factors
Length	meter (m)	1 m = 100 centimeters (cm) 1 m = 39.3701 inch (in) 1 in = 2.54 cm 1 m = 10^{10} angstrom (Å) 1 m = 6.21388×10^{-4} mile 1 mile = 1.6093 km
Mass	kilogram (kg)	1 kg = 1000 gram (g) 1 kg = 2.205 pounds (lb) 1 lb = 453.6 grams (g) 1 atomic mass unit (amu) $\quad = 1.66054 \times 10^{-27}$ kg
Charge	coulomb (C)	1 C = 2.997925×10^9 statcoulomb (statC)
Temperature	kelvin (K)	0 K = $-273.15°$ Celsius (C) $\quad = -459.67°$ Fahrenheit (F) °F = (9/5) °C + 32 °C = (5/9)[°F − 32] K = °C + 273.15
Volume	cubic meter (m^3)	1 m^3 = 10^6 cubic centimeters (cm^3) $\quad = 1000$ liter (L) 1L = 1.057 quarts (qt) 1 cm^3 = 1 milliliter (ml)
Force	newton (N) = 1 kg m s^{-2}	1 N = 10^5 dynes (dyn) 1 dyn = 1 g cm s^{-2}
Pressure	pascal (Pa) = 1 N m^{-2}	1 atmosphere (atm) = 101325 Pa 1 atm = 760 torr 1 atm = 14.70 lb in^{-2} 1 bar = 10^5 Pa 1 atm = 1.01325 bar
Energy	joule (J) = 1 N m	1 calorie (cal) = 4.184 J 1 J = 10^7 erg 1 L atm = 101.325 J 1 electron volt (eV) = 96.485 kJ mol^{-1} 1 kJ = 1000 J 1 eV = 8065.5 cm^{-1}